This book is due for return on or before the last date shown below.

D1426400

Principles of Regenerative Medicine

Second edition

Anthony Atala
Robert Lanza
James A. Thomson
Robert Nerem

ELSEVIER

AMSTERDAM • BOSTON • HEIDELBERG • LONDON • NEW YORK • OXFORD
PARIS • SAN DIEGO • SAN FRANCISCO • SINGAPORE • SYDNEY • TOKYO

Academic Press is an imprint of Elsevier

Academic Press is an imprint of Elsevier
32 Jamestown Road, London NW1 7BY, UK
30 Corporate Drive, Suite 400, Burlington, MA 01803, USA
525 B Street, Suite 1800, San Diego, CA 92101-4495, USA

First edition 2008
Second edition 2011

Notice
No responsibility is assumed by the publisher for any injury and/or damage to persons or property as a matter of products liability, negligence or otherwise, or from any use or operation of any methods, products, instructions or ideas contained in the material herein. Because of rapid advances in the medical sciences, in particular, independent verification of diagnoses and drug dosages should be made

British Library Cataloguing-in-Publication Data
A catalogue record for this book is available from the British Library

Library of Congress Cataloging-in-Publication Data
A catalog record for this book is available from the Library of Congress

ISBN: 978-0-12-381422-7

For information on all Academic Press publications
visit our website at www.elsevierdirect.com

Typeset by TNQ Books and Journals

Printed and bound in Canada

10 11 12 13   10 9 8 7 6 5 4 3 2 1

Working together to grow
libraries in developing countries

www.elsevier.com | www.bookaid.org | www.sabre.org

ELSEVIER   BOOK AID International   Sabre Foundation

CONTENTS

v

PART 5 • Regulation and Ethics

Tamer Aboushwareb
Department of Urology and Wake Forest Institute for Regenerative Medicine, Wake Forest University School of Medicine, Winston-Salem, NC, USA

Jon D. Ahlstrom
Nephrology, University of Utah and VA Medical Centers, Salt Lake City, UT, USA

Alejandro J. Almarza
Musculoskeletal Research Center, Department of Bioengineering, University of Pittsburgh, Pittsburgh, PA, USA

James M. Anderson
Pathology, Macromolecular Science, and Biomedical Engineering, Case Western Reserve University, Cleveland, OH, USA

Judith Arcidiacono
Center for Biologics Evaluation and Research, FDA, Rockville, MD, USA

Anthony Atala
Wake Forest Institute for Regenerative Medicine, Wake Forest University School of Medicine, Winston-Salem, NC, USA and Department of Urology, Korea University Medical Center, Seoul, Korea

Stephen F. Badylak
McGowan Institute for Regenerative Medicine, University of Pittsburgh, Pittsburgh, PA, USA

Jae Hyun Bae
Wake Forest Institute for Regenerative Medicine, Wake Forest University Health Sciences, Medical Center Boulevard, Winston-Salem, NC, USA; Department of Urology, Korea University Medical Center, Seoul, Korea

Brian G. Ballios
Institute of Medical Science, University of Toronto, Toronto, Ontario, Canada

Ashok Batra
SUNY-Syracuse, Syracuse, NY; US Biotechnology & Pharma Consulting Group, Potomac, MD, USA

M. Douglas Baumann
Department of Chemical Engineering and Applied Chemistry, University of Toronto, Toronto, Ontario, Canada

Ravi V. Bellamkonda
Neurological Biomaterials and Cancer Therapeutics, Coulter Department of Biomedical Engineering, Georgia Institute of Technology/Emory University, Atlanta, GA, USA

Nicole M. Bergmann
Department of Bioengineering, Rice University, Houston, TX, USA

Mickie Bhatia
Stem Cell and Cancer Research Institute, Michael G. DeGroote School of Medicine and Department of Biochemistry and Biomedical studies, McMaster University, Hamilton, Ontario, Canada

Martin A. Birchall

University College London, Center for Stem Cells and Regenerative Medicine and UCL Ear Institute, Royal National Throat Nose and Ear Hospital, London, UK

Helen M. Blau

Baxter Laboratory for Stem Cell Biology, Stanford University School of Medicine, Stanford, CA, USA

Joel D. Boerckel

Woodruff School of Mechanical Engineering, Georgia Institute of Technology

Ali H. Brivanlou

Laboratory of Molecular Embryology, The Rockefeller University, New York, NY, USA

Mara Cananzi

Surgery Unit, UCL Institute of Child Health and Great Ormond Street Hospital, London, UK Department of Paediatrics, University of Padua, Padua, Italy

Arnold I. Caplan

Professor of Biology, Director, Skeletal Research Center, Case Western Reserve University, 10900 Euclid Avenue, Cleveland, OH, USA

Joseph W. Carnwath

Department of Biotechnology, Institute of Farm Animal Genetics, Friedrich-Loeffler-Institut (FLI), Federal Research Institute for Animal Health, Neustadt, Germany

Grant A. Challen

Center for Cell and Gene Therapy, Stem Cell and Regenerative Medicine Center, Department of Pathology and Immunology, Baylor College of Medicine, Houston, TX, USA

George J. Christ

Wake Forest Institute for Regenerative Medicine, Winston-Salem, NC, USA

Hyun Jung Chung

Department of Biological Sciences, Korea Advanced Institute of Science and Technology, Daejeon, Korea

Maegen Colehour

Center for Devices and Radiological Health, FDA, Silver Spring, MD, USA

Michael J. Cooke

Department of Chemical Engineering and Applied Chemistry, University of Toronto, Toronto, Ontario, Canada

V.M. Correlo

3B's Research Group — Biomaterials, Biodegradables and Biomimetics, University of Minho, Headquarters of the European Institute of Excellence on Tissue Engineering and Regenerative Medicine, Taipas, Guimarães, Portugal; IBB — Institute for Biotechnology and Bioengineering, PT Associated Laboratory, Guimarães, Portugal

Benjamin D. Cosgrove

Baxter Laboratory for Stem Cell Biology, Stanford University School of Medicine, Stanford, CA, USA

Stefano Da Sacco

Department of Urology, Childrens Hospital Los Angeles, University of Southern California Keck School of Medicine, Los Angeles, CA, USA

Jiyoung M. Dang

Center for Devices and Radiological Health, FDA, Silver Spring, MD, USA

Richard M. Day
Centre for Gastroenterology & Nutrition, Division of Medicine, University College London, London, UK

Paolo De Coppi
Surgery Unit, UCL Institute of Child Health and Great Ormond Street Hospital, London, UK
Department of Paediatrics, University of Padua, Padua, Italy
Wake Forest Institute for Regenerative Medicine, Winston Salem, NC, USA

Roger E. De Filippo
Department of Urology, Childrens Hospital Los Angeles, University of Southern California Keck School of Medicine, Los Angeles, CA, USA

Mahesh C. Dodla
Neurological Biomaterials and Cancer Therapeutics, Coulter Department of Biomedical Engineering, Georgia Institute of Technology/Emory University, Atlanta, GA, USA

Juan Domínguez-Bendala
Diabetes Research Institute, Cell Transplant Center and Department of Surgery, University of Miami, FL, USA

Ryan P. Dorin
University of Southern California (USC) Institute of Urology, Keck School of Medicine, USC, Los Angeles, CA, USA

Charles N. Durfor
Center for Devices and Radiological Health, FDA, Silver Spring, MD, USA

Rita B. Effros
Department of Pathology and Laboratory Medicine and UCLA AIDS Institute, David Geffen School of Medicine at UCLA, Los Angeles, CA, USA

Jennifer H. Elisseeff
Department of Biomedical Engineering, Johns Hopkins University, Baltimore, MD, USA

Ewa C.S. Ellis
Department of Clinical Science, Intervention and Technology, Division of Transplantation, Liver Cell Lab., Karolinska Institute, Stockholm, Sweden

Juliet A. Emamaullee
Department of Surgery, University of Alberta, Edmonton, Alberta, Canada

Per Fagerholm
Department of Clinical and Experimental Medicine, Division of Ophthalmology, Linköping University, Linköping, Sweden

Qiang Feng
Stem Cell & Regenerative Medicine International, Worcester, MA, USA; Department of Applied Bioscience, Cha University, Seoul, Korea

Donald Fink
Center for Biologics Evaluation and Research, FDA, Rockville, MD, USA

Matthew B. Fisher
Musculoskeletal Research Center, Department of Bioengineering, University of Pittsburgh, Pittsburgh, PA, USA

Andrés J. García
Woodruff School of Mechanical Engineering, Petit Institute for Bioengineering and Bioscience, Georgia Institute of Technology, Atlanta, GA, USA

xi

Svetlana Gavrilov
Department of Genetics and Development, College of Physicians and Surgeons of Columbia University, New York, NY, USA

Dan Gazit
Skeletal Biotechnology Laboratory, Hebrew University—Hadassah Faculty of Dental Medicine, Jerusalem, Israel; Department of Surgery and Cedars-Sinai Regenerative Medicine Institute (CS-RMI), Cedars-Sinai Medical Center, Los Angeles, CA, USA

Zulma Gazit
Skeletal Biotechnology Laboratory, Hebrew University—Hadassah Faculty of Dental Medicine, Jerusalem, Israel; Department of Surgery and Cedars-Sinai Regenerative Medicine Institute (CS-RMI), Cedars-Sinai Medical Center, Los Angeles, CA, USA

Christopher V. Gemmiti
Woodruff School of Mechanical Engineering, Georgia Institute of Technology

Charles A. Gersbach
Department of Biomedical Engineering, Duke University, Durham, NC, USA

Margaret A. Goodell
Center for Cell and Gene Therapy, Stem Cell and Regenerative Medicine Center, Department of Pathology and Immunology, Baylor College of Medicine, Houston, TX, USA

Deborah Lavoie Grayeski
M Squared Associates, Inc., Alexandria, VA, USA

Ronald M. Green
Ethics Institute, Dartmouth College, Hanover, NH, USA

May Griffith
Department of Clinical and Experimental Medicine, Division of Cell Biology, Linköping University, Linköping, Sweden

Robert E. Guldberg
Woodruff School of Mechanical Engineering, Georgia Institute of Technology

Qiongyu Guo
Department of Biomedical Engineering, Johns Hopkins University, Baltimore, MD, USA

M.C. Hacker
Institute of Pharmacy, Pharmaceutical Technology, University of Leipzig, Leipzig, Germany

Joanne Hackett
Department of Clinical and Experimental Medicine, Division of Cell Biology, Linköping University, Linköping, Sweden

Joshua M. Hare
University of Miami, Miller School of Medicine, Interdisciplinary Stem Cell Institute, Miami, Florida, USA

Benjamin S. Harrison
Wake Forest Institute for Regenerative Medicine, Wake Forest University, Medical Center BLVD, Winston-Salem, NC, USA

Konstantinos E. Hatzistergos
University of Miami, Miller School of Medicine, Interdisciplinary Stem Cell Institute, Miami, Florida, USA

Kevin E. Healy
Department of Bioengineering and Department of Materials Science and Engineering, University of California at Berkeley, Berkeley, CA, USA

Stephen L. Hilbert
Children's Mercy Hospital, Kansas City, MO, USA

Jiang Hu
Department of Biologic and Materials Sciences, University of Michigan, Ann Arbor, MI, USA

Alexander Huber
McGowan Institute for Regenerative Medicine, University of Pittsburgh, Pittsburgh, PA, USA

H. David Humes
Department of Internal Medicine, University of Michigan, Ann Arbor, MI, USA

Elizabeth F. Irwin
Department of Bioengineering, Department of Materials Science and Engineering, University of California at Berkeley, Berkeley, CA, USA

Brett C. Isenberg
Department of Biomedical Engineering, Boston University, Boston, MA, USA

Takanori Iwata
Institute of Advanced Biomedical Engineering and Science
Department of Oral and Maxillofacial Surgery, Tokyo Women's Medical University, 8-1 Kawada-cho, Shinjuku-ku, Tokyo, Japan

Sam Janes
Center for Respiratory Research, Rayne Building, University College London, London, UK

Lily Jeng
Tissue Engineering, VA Boston Healthcare System, Boston, MA, USA; Department of Biological Engineering, Massachusetts Institute of Technology, Cambridge, MA, USA

Junfeng Ji
Stem Cell and Cancer Research Institute, Michael G. DeGroote School of Medicine and Department of Biochemistry and Biomedical Studies, McMaster University, Hamilton, Ontario, Canada

Josephine Johnston
The Hastings Center, Garrison, NY, USA

Kimberly A. Johnston
Innovative Biotherapies, Ann Arbor, MI, USA

David L. Kaplan
Department of Biomedical Engineering, Tufts University, Medford, MA, USA

David S. Kaplan
Center for Devices and Radiological Health, FDA, Silver Spring, MD, USA

Sinan Karaoglu
Musculoskeletal Research Center, Department of Bioengineering, University of Pittsburgh, Pittsburgh, PA, USA

Adam J. Katz
Department of Plastic Surgery, Department of Biomedical Engineering, Laboratory of Applied Developmental Plasticity, University of Virginia Health System, Virginia, USA

Jaehyun Kim
Wake Forest Institute for Regenerative Medicine, Wake Forest University Health Sciences, Medical Center Boulevard, Winston-Salem, NC, USA

Erin A. Kimbrel
Stem Cell & Regenerative Medicine International, Worcester, MA, USA and Department of Applied Bioscience, Cha University, Seoul, Korea

Nadav Kimelman
Skeletal Biotechnology Laboratory, Hebrew University—Hadassah Faculty of Dental Medicine, Jerusalem, Israel

Jonathan A. Kluge
McKay Orthopaedic Research Laboratory, University of Pennsylvania, Philadelphia, PA, USA

Chester J. Koh
Division of Pediatric Urology and the Developmental Biology, Regenerative Medicine, and Surgery Program, Children's Hospital Los Angeles, and the University of Southern California (USC) Institute of Urology, Keck School of Medicine, USC, Los Angeles, CA, USA

Yash M. Kolambkar
Woodruff School of Mechanical Engineering, Georgia Institute of Technology

Makoto Komura
Department of Pediatric Surgery, The University of Tokyo Hospital, Tokyo, Japan

Wilfried A. Kues
Department of Biotechnology, Institute of Farm Animal Genetics, Friedrich-Loeffler-Institut (FLI), Federal Research Institute for Animal Health, 31535 Neustadt, Germany

Francois Ng kee Kwong
Department of Histopathology, Cambridge University Hospitals NHS Foundation Trust, Cambridge, UK

Neil Lagali
Department of Clinical and Experimental Medicine, Division of Ophthalmology, Linköping University, Linköping, Sweden

Deepak A. Lamba
Department of Opthalmology, University of Washington, Seattle, WA, USA

Donald W. Landry
Department of Medicine, College of Physicians and Surgeons of Columbia University, New York, NY, USA

Robert Lanza
Stem Cell & Regenerative Medicine International, Worcester, MA, USA and Advanced Cell Technology, Inc., Worcester, MA, USA

Barrett Larson
Hagey Laboratory for Pediatric and Regenerative Medicine, Division of Plastic and Reconstructive Surgery, Department of Surgery, Institute of Stem Cell Biology and Regenerative Medicine, Stanford University School of Medicine, Palo Alto, CA, USA

Malcolm A. Latorre
Department of Biomedical Engineering, Linköping University, Linköping, Sweden

Ellen Lazarus
Center for Biologics Evaluation and Research, FDA, Rockville, MD, USA

Hyukjin Lee
Department of Biological Sciences, Korea Advanced Institute of Science and Technology, Daejeon, Korea

Mark H. Lee
Center for Biologics Evaluation and Research, FDA, Rockville, MD, USA

Sang Jin Lee
Wake Forest Institute for Regenerative Medicine, Wake Forest University Health Sciences, Medical Center Boulevard, Winston-Salem, NC, USA

Gary G. Leisk
Department of Mechanical Engineering, Tufts University, Medford, MA, USA

Feng Li
Stem Cell & Regenerative Medicine International, Worcester, MA, USA and Department of Applied Bioscience, Cha University, Seoul, Korea

Rui Liang
Musculoskeletal Research Center, Department of Bioengineering, University of Pittsburgh, Pittsburgh, PA, USA

Kuanyin K. Lin
Center for Cell and Gene Therapy, Stem Cell and Regenerative Medicine Center, Department of Pathology and Immunology, Baylor College of Medicine, Houston, TX, USA

Xiaohua Liu
Department of Biologic and Materials, University of Michigan, Ann Arbor, MI, USA

Michael T. Longaker
Hagey Laboratory for Pediatric and Regenerative Medicine, Division of Plastic and Reconstructive Surgery, Department of Surgery, Institute of Stem Cell Biology and Regenerative Medicine, Stanford University School of Medicine, Palo Alto, CA, USA

H. Peter Lorenz
Hagey Laboratory for Pediatric and Regenerative Medicine, Division of Plastic and Reconstructive Surgery, Department of Surgery, Institute of Stem Cell Biology and Regenerative Medicine, Stanford University School of Medicine, Palo Alto, CA, USA

Shi-Jiang Lu
Stem Cell & Regenerative Medicine International, Worcester, MA, USA and Department of Applied Bioscience, Cha University, Seoul, Korea

Andrea Lucas-Hahn
Department of Biotechnology, Institute of Farm Animal Genetics, Friedrich-Loeffler-Institut (FLI), Federal Research Institute for Animal Health, Neustadt, Germany

Peter X. Ma
Department of Biologic and Materials, University of Michigan, Ann Arbor, MI, USA

Paolo Macchiarini
Cardiothoracic Surgery, Hospital Careggi, Florence, Italy and University College London, London, UK

Masood A. Machingal
Wake Forest Institute for Regenerative Medicine, Winston-Salem, NC, USA

J.F. Mano
3B's Research Group — Biomaterials, Biodegradables and Biomimetics, University of Minho, Headquarters of the European Institute of Excellence on Tissue Engineering and Regenerative Medicine, Taipas, Guimarães, Portugal and Institute for Biotechnology and Bioengineering, PT Associated Laboratory, Guimarães, Portugal

M. Martins-Green
Department of Cell Biology and Neuroscience, University of California, Riverside, CA, USA

Michael McCall
Department of Surgery, University of Alberta, Edmonton, Alberta, Canada

Richard McFarland

Center for Biologics Evaluation and Research, FDA, Rockville, MD, USA

Melissa K. McHale

Department of Bioengineering, Rice University, Houston, TX, USA

Alexander F. Mericli

Resident, Department of Plastic Surgery, University of Virginia Health System, Virginia, USA

A.G. Mikos

Department of Bioengineering, Rice University, Houston, TX, USA

Vivek J. Mukhatyar

Neurological Biomaterials and Cancer Therapeutics, Coulter Department of Biomedical Engineering, Georgia Institute of Technology/Emory University, Atlanta, GA, USA

Allison Nauta

Hagey Laboratory for Pediatric and Regenerative Medicine, Division of Plastic and Reconstructive Surgery, Department of Surgery, Institute of Stem Cell Biology and Regenerative Medicine, Stanford University School of Medicine, Palo Alto, California, USA Department of Surgery, Georgetown University Hospital, Washington DC, USA

N.M. Neves

3B's Research Group — Biomaterials, Biodegradables and Biomimetics, University of Minho, Headquarters of the European Institute of Excellence on Tissue Engineering and Regenerative Medicine, Taipas, Guimarães, Portugal and Institute for Biotechnology and Bioengineering, PT Associated Laboratory, Guimarães, Portugal

Heiner Niemann

Institute of Farm Animal Genetics, Friedrich-Loeffler-Institut (FLI), Federal Research Institute for Animal Health, Mariensee, Neustadt, Germany

Teruo Okano

Institute of Advanced Biomedical Engineering and Science

Keisuke Okita

Center for iPS Cell Research and Application (CiRA), Institute for Integrated Cell-Material Sciences, Kyoto University, Kyoto, Japan

J.M. Oliveira

3B's Research Group — Biomaterials, Biodegradables and Biomimetics, University of Minho, Headquarters of the European Institute of Excellence on Tissue Engineering and Regenerative Medicine, Taipas, Guimarães, Portugal and IBB — Institute for Biotechnology and Bioengineering, PT Associated Laboratory, Guimarães, Portugal

Virginia E. Papaioannou

Department of Genetics and Development, College of Physicians and Surgeons of Columbia University, New York, NY, USA

Tae Gwan Park

Department of Biological Sciences, Korea Advanced Institute of Science and Technology, Daejeon, Korea

Gadi Pelled

Skeletal Biotechnology Laboratory, Hebrew University—Hadassah Faculty of Dental Medicine, Jerusalem, Israel Department of Surgery and Cedars-Sinai Regenerative Medicine Institute (CS-RMI), Cedars-Sinai Medical Center, Los Angeles, CA, USA

Laura Perin

Department of Urology, Childrens Hospital Los Angeles, University of Southern California Keck School of Medicine, Los Angeles, CA, USA

M. Petreaca
Department of Cell Biology and Neuroscience, University of California, Riverside, CA, USA

Antonello Pileggi
Diabetes Research Institute, Cell Transplant Center, and Department of Surgery, University of Miami, Miami, FL, USA

Jacob F. Pollock
Department of Bioengineering, University of California at Berkeley, Berkeley, CA, USA

Blaise D. Porter
Woodruff School of Mechanical Engineering, Georgia Institute of Technology

Milica Radisic
Institute of Biomaterials and Biomedical Engineering, Department of Chemical Engineering and Applied Chemistry, University of Toronto, Ontario, Canada

Nandini Rao
Department of Biology and Indiana University Center for Regenerative Biology and Medicine, Indiana University-Purdue University, Indianapolis, IN, USA

A.H. Reddi
Lawrence Ellison Center for Tissue Regeneration, University of California, Davis, School of Medicine, Sacramento, CA, USA

Thomas A. Reh
Department of Biological Structure, University of Washington, Seattle, WA, USA

R.L. Reis
3B's Research Group – Biomaterials, Biodegradables and Biomimetics, University of Minho, Headquarters of the European Institute of Excellence on Tissue Engineering and Regenerative Medicine, Taipas, Guimarães, Portugal and Institute for Biotechnology and Bioengineering, PT Associated Laboratory, Guimarães, Portugal

Camillo Ricordi
Diabetes Research Institute, Cell Transplant Center, Departments of Surgery, Medicine, Biomedical Engineering, Microbiology and Immunology, University of Miami, Miami, FL, USA; Wake Forest Institute for Regenerative Medicine, Winston Salem, NC, USA; Karolinska Institutet, Stockholm, Sweden

Philip Roelandt
Interdepartmental Stem Cell Institute Leuven, Catholic University Leuven, Belgium

Caroline Beth Sangan
Centre for Regenerative Medicine, Department of Biology and Biochemistry, University of Bath, Claverton Down, Bath, UK

Justin M. Saul
Wake Forest Institute for Regenerative Medicine, Wake Forest University Health Sciences, Winston-Salem, NC, USA

David V. Schaffer
Department of Chemical and Biomolecular Engineering, Department of Bioengineering, and The Helen Wills Neuroscience Institute, University of California at Berkeley, Berkeley, CA, USA

Gunter Schuch
Institute for Regenerative Medicine, Wake Forest University School of Medicine, Medical Center Blvd, Winston-Salem, NC, USA

Michael V. Sefton
Institute of Biomaterials and Biomedical Engineering, Department of Chemical Engineering and Applied Chemistry, University of Toronto, Ontario, Canada

Sarah Selem
University of Miami, Miller School of Medicine, Interdisciplinary Stem Cell Institute, Miami, FL, USA

A.M. James Shapiro
Department of Surgery, University of Alberta, Edmonton, Alberta, Canada

Heather Sheardown
Department of Chemical Engineering, McMaster University, Hamilton, Ontario, Canada

Dima Sheyn
Skeletal Biotechnology Laboratory, Hebrew University–Hadassah Faculty of Dental Medicine, Jerusalem, Israel

Molly S. Shoichet
Department of Chemical Engineering and Applied Chemistry, Department of Chemistry, Institute of Biomaterials and Biomedical Engineering, University of Toronto, Toronto, Ontario, Canada

Harvir Singh
Laboratory of Molecular Embryology, The Rockefeller University, New York, NY, USA

Sirinrath Sirivisoot
Wake Forest Institute for Regenerative Medicine, Wake Forest University, Medical Center BLVD, Winston-Salem, NC, USA

Daniel Skuk
Research Unit on Human Genetics, CHUL Research Center, Quebec, Canada

Shay Soker
Institute for Regenerative Medicine, Wake Forest University School of Medicine, Medical Center Blvd, Winston-Salem, NC, USA

Myron Spector
Tissue Engineering, VA Boston Healthcare System, Boston, MA, USA; Department of Orthopaedic Surgery, Brigham and Women's Hospital, Harvard Medical School, Boston, MA, USA

David L. Stocum
Department of Biology and Indiana University Center for Regenerative Biology and Medicine, Indiana University-Purdue University, Indianapolis, IN, USA

Stephen C. Strom
Department of Pathology, University of Pittsburgh, PA, USA

James A. Thomson
National Primate Research Center, University of Wisconsin Graduate School, Madison, WI, USA; WiCell Research Institute, Madison, WI, USA; Department of Anatomy, University of Wisconsin Medical School, Madison, WI, USA; Genome Center of Wisconsin, University of Wisconsin-Madison, Madison, WI, USA

David Tosh
Centre for Regenerative Medicine, Department of Biology and Biochemistry, University of Bath, Claverton Down, Bath, UK

Robert T. Tranquillo
Department of Biomedical Engineering, University of Minnesota, Minneapolis, MN, USA

Jacques P. Tremblay
Research Unit on Human Genetics, CHUL Research Center, Quebec, Canada

Catherine M. Verfaillie
Interdepartmental Stem Cell Institute Leuven, Catholic University Leuven, Belgium

Zhan Wang
Institute for Regenerative Medicine, Wake Forest University School of Medicine, Medical Center Blvd, Winston-Salem, NC, USA

Jennifer L. West
Department of Bioengineering, Rice University, Houston, TX, USA

Kevin J. Whittlesey
Office of the Commissioner, FDA, Silver Spring, MD, USA

Chrysanthi Williams
Bose Corporation, ElectroForce Systems Group, Eden Prairie, MN, USA

David F. Williams
Wake Forest Institute for Regenerative Medicine, Wake Forest University Health Sciences, Winston-Salem, NC, USA; Christiaan Barnard Department of Cardiothoracic Surgery, Cape Town, South Africa; University of New South Wales, Graduate School of Biomoedical Engineering, Sydney, Australia; Tsinghua University, Beijing, China, Shanghai Jiao Tong University, China; University of Liverpool, Liverpool, UK

J. Koudy Williams
Institute for Regenerative Medicine, Wake Forest University School of Medicine, Medical Center Blvd, Winston-Salem, NC, USA

Celia Witten
Center for Biologics Evaluation and Research, FDA, Rockville, MD, USA

Savio L-Y. Woo
Musculoskeletal Research Center, Department of Bioengineering, University of Pittsburgh, Pittsburgh, PA, USA

Fiona Wood
Burns service of WA, Burn Injury Research Unit UWA, McComb Research Foundation, Western Australia

Shinya Yamanaka
Center for iPS Cell Research and Application (CiRA), Institute for Integrated Cell-Material Sciences, Kyoto University, Kyoto, Japan
Department of Stem Cell Biology, Institute for Frontier Medical Sciences, Kyoto University, Kyoto, Japan
Yamanaka iPS Cell Special Project, Japan Science and Technology Agency, Kawaguchi, Japan
Gladstone Institute of Cardiovascular Disease, San Francisco, CA, USA

Masayuki Yamato
Institute of Advanced Biomedical Engineering and Science

Saami K. Yazdani
Wake Forest Institute for Regenerative Medicine, Winston-Salem, NC, USA

James J. Yoo
Wake Forest Institute for Regenerative Medicine, Wake Forest University Health Sciences, Medical Center Boulevard, Winston-Salem, NC, USA; Joint Institute for Regenerative Medicine, Kyungpook National University Hospital, Daegu, Korea

Junying Yu
Cellular Dynamics International, Inc., 525 Science Drive, Madison, WI, USA

Bonan Zhong
Stem Cell and Cancer Research Institute, Michael G. DeGroote School of Medicine and Department of Biochemistry and Biomedical Studies, McMaster University, Hamilton, Ontario, Canada

PART 1

Biologic and Molecular Basis for Regenerative Medicine

CHAPTER 1

Molecular Organization of Cells

Jon D. Ahlstrom
Nephrology, University of Utah and VA Medical Centers, Salt Lake City, UT, USA

INTRODUCTION

Multicellular tissues exist in one of two types of cellular arrangements, epithelial or mesenchymal. Epithelial cells adhere tightly to each other at their lateral surfaces and to an organized extracellular matrix (ECM) at their basal domain, thereby producing a sheet of cells resting on a basal lamina with an apical surface. Mesenchymal cells, in contrast, are individual cells with a bipolar morphology that are held together as a tissue within a three-dimensional ECM (see Fig. 1.1). The conversion of epithelial cells into mesenchymal cells, an "epithelial-mesenchymal transition" (EMT), is central to many aspects of embryonic morphogenesis and adult tissue repair, as well as a number of disease states (Hay, 2005; Baum et al., 2008; Thiery et al., 2009). The reverse process whereby mesenchymal cells coalesce into an epithelium is a "mesenchymal-epithelial transition" (MET). Understanding the molecules that regulate this transition between epithelial and mesenchymal states offers important insights into how cells and tissues are organized.

The early embryo is structured as one or more epithelia. An EMT allows the rearrangements of cells to create additional morphological features. Well-studied examples of EMTs during embryonic development include gastrulation in *Drosophila* (Baum et al., 2008), the emigration of primary mesenchyme cells (PMCs) in sea urchin embryos (Shook and Keller, 2003), and gastrulation in amniotes (reptiles, birds, and mammals) at the primitive streak (Hay, 2005). EMTs also occur later in vertebrate development, such as the emigration of neural crest cells from the neural tube (Sauka-Spengler and Bronner-Fraser, 2008), the formation of the sclerotome from epithelial somites, and during palate fusion (Hay, 2005). The reverse process, MET, is likewise crucial to development, and examples include the condensation of mesenchymal cells to form the notochord and somites (Thiery et al., 2009), kidney tubule formation from nephrogenic mesenchyme (Schmidt-Ott, 2006), and the creation of heart valves from cardiac mesenchyme (Nakajima et al., 2000). In the adult organism, EMTs and METs occur during wound healing and tissue remodeling (Kalluri and Weinberg, 2009; Thiery et al., 2009). The conversion of neoplastic epithelial cells into invasive cancer cells has long been considered an EMT process (Thiery, 2002; Thiery et al., 2009). However, there are also examples of tumor cells that have functional cell-cell adhesion junctions, yet are still migratory and invasive as a group (Rørth, 2009). This "collective migration" also occurs during development (Rørth, 2009). Hence, there is debate regarding whether an EMT model accurately describes all epithelial metastatic cancers. Similarly, the fibrosis of cardiac, kidney, lens, and liver epithelial tissue has also long been categorized as an EMT event (Thiery et al., 2009; Iwano et al., 2002). However, recent research in the kidney shows that the myofibroblasts induced following kidney injury *in vivo* are derived from mesenchymal pericytes, rather than the proximal

3

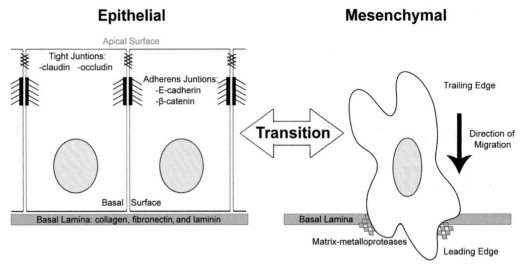

FIGURE 1.1

Epithelial versus mesenchymal. Epithelial cells adhere tightly together by tight junctions and adherens junctions localized near the apical surface. Epithelial cells also have a basal surface that rests on a basal lamina. Mesenchymal cells in contrast do not have well-defined cell-cell adhesion complexes, have front-end/back-end polarity instead of apical/basal polarity, and are characterized by their ability to invade the basal lamina.

epithelial cells (Humphreys et al., 2010). Therefore, the origin of the cells that contribute to fibrotic tissue scarring (epithelial or otherwise) may need to be carefully re-examined.

The focus of this chapter is on the molecules that regulate the organization of cells into epithelium or mesenchyme. We will first look at the cellular changes that occur during an EMT, including changes in cell-cell and cell-ECM adhesions, changes in cell polarity, and the stimulation of invasive cell motility. Then we will consider the molecules and mechanisms that control the EMT or MET, including the structural molecules, transcription factors, and signaling pathways that regulate EMTs.

MOLECULES THAT ORGANIZE CELLS

The conversion of an epithelial sheet into individual migratory cells and back again requires the coordinated changes of many distinct families of molecules.

Changes in cell-cell adhesion

Epithelial cells are held together by specialized cell-cell junctions, including adherens junctions, desmosomes, and tight junctions (Giepmans and van Ijzendoorn, 2009). These junctions are localized in the lateral domain near the apical surface and establish the apical polarity of the epithelium. In order for an epithelial sheet to produce individual mesenchymal cells, cell-cell adhesions must be disrupted. The principal transmembrane proteins that mediate cell-cell adhesions are members of the cadherin superfamily (Stepniak et al., 2009). E-cadherin and N-cadherin are classical cadherins that interact homotypically through their extracellular IgG domains with like-cadherins on adjacent cells. Cadherins are important mediators of cell-cell adhesion. For example, misexpression of E-cadherin is sufficient for promoting cell-cell adhesion and assembly of adherens junctions in fibroblasts (Nagafuchi et al., 1987). In epithelial cancers (carcinomas), E-cadherin acts as a tumor suppressor (Thiery, 2002). In a mouse model for β-cell pancreatic cancer, the loss of E-cadherin is the rate-limiting step for transformed epithelial cells to become invasive (Perl et al., 1998). Although the loss of cadherin-mediated cell-cell adhesion is necessary for an EMT, the loss of cadherins is not always sufficient to generate a complete EMT *in vivo*. For example, neural tube epithelium in mice expresses N-cadherin, but in the N-cadherin knockout mouse an EMT is not induced in the neural tube (Radice et al., 1997). Hence, cadherins are essential for maintaining epithelial integrity, and the loss of cell-cell adhesion due to the reduction of cadherin function is an important step for an EMT.

One characteristic of an EMT is "cadherin switching." Often, epithelia that express E-cadherin will downregulate E-cadherin expression at the time of the EMT, and express different cadherins such as N-cadherin (Christofori, 2003). Cadherin switching may promote motility. For instance, in mammary epithelial cell lines, the misexpression of N-cadherin is sufficient for increased cell motility. Blocking N-cadherin expression results in less motility, but does not alter cellular morphology. Hence, cadherin switching may be necessary for cell motility, but cadherin switching alone is not sufficient to bring about a complete EMT (Maeda et al., 2005).

There are several ways that cadherin expression and function are regulated. Transcription factors that are central to most EMTs, such as Snail-1, Snail-2, Zeb1, Zeb2, Twist, and E2A, all bind to E-boxes on the E-cadherin promoter and repress the transcription of E-cadherin (de Craene, 2005). Post-transcriptionally, the E-cadherin protein is ubiquitinated by the E3-ligase, Hakai, which targets E-cadherin to the proteasome (Fujita et al., 2002). E-cadherin turnover at the membrane is regulated by either caveolae-dependent endocytosis or clathrin-dependent endocytosis (Bryant and Stow, 2004), and p120-catenin prevents endocytosis of E-cadherin at the membrane (Xiao et al., 2007). E-cadherin function can also be disrupted by matrix metalloproteases, which degrade the extracellular domain of E-cadherin (Egeblad and Werb, 2002). Some or all of these mechanisms may occur during an EMT to disrupt cell-cell adhesion.

In summary, cell-cell adhesion is maintained principally by cadherins, and changes in cadherin expression are typical of an EMT.

Changes in cell-ECM adhesion

Altering the way that a cell interacts with the ECM is also important in EMTs. For example, at the time that sea urchin PMCs ingress, the cells have increased adhesiveness for ECM (Shook and Keller, 2003). Cell-ECM adhesion is mediated principally by integrins. Integrins are transmembrane proteins composed of two non-covalently linked subunits, α and β, that bind to ECM components such as fibronectin, laminin, and collagen. The cytoplasmic domain of integrins links to the cytoskeleton and interacts with signaling molecules. Changes in integrin function are required for many EMTs, including neural crest emigration (Delannet and Duband, 1992), mouse primitive streak formation (Hay, 2005), and cancer metastasis (Desgrosellier and Cheresh, 2010). However, the misexpression of integrin subunits is not sufficient to bring about a full EMT *in vitro* (Valles et al., 1996) or *in vivo* (Carroll et al., 1998).

The presence and function of integrins is modulated in several ways. For example, the promoter of the *integrin β6* gene is activated by the transcription factor Ets-1 during colon carcinoma metastasis (Bates, 2005). Most integrins can also cycle between "On" (high affinity) and "Off" (low affinity) states. This "inside-out" regulation of integrin adhesion occurs at the integrin cytoplasmic tail (Hood and Cheresh, 2002). In addition to integrin activation, the "clustering" of integrins on the cell surface also affects the overall strength of integrin-ECM interactions. The increased adhesiveness of integrins due to clustering, known as avidity, can be activated by chemokines, and is dependent on RhoA and phosphatidylinositol 3' kinase (PI3K) activity (Hood and Cheresh, 2002).

In summary, changes in ECM adhesion are required for an EMT. Cell-ECM adhesions are maintained by integrins, and integrins have varying degrees of adhesiveness dependent upon the presence, activity, and avidity of the integrin subunits.

Changes in cell polarity and stimulation of cell motility

Cellular polarity is defined by the distinct arrangement of cytoskeletal elements and organelles in epithelial versus mesenchymal cells. Epithelial polarity is characterized by cell-cell junctions found near the apical-lateral domain (non-adhesive surface), and a basal lamina (adhesive surface) opposite the apical surface. Mesenchymal cells in contrast do not have apical/basal polarity, but rather front-end/back-end polarity, with actin-rich lamellipodia and Golgi localized at the leading edge (Hay, 2005). Molecules that establish cell polarity include Cdc42, PAK1, PI3K, PTEN, Rac, Rho, and the PAR proteins (Moreno-Bueno et al., 2008; McCaffrey and Macara, 2009). Changes in cell polarity help to promote an EMT. In mammary epithelial cells, the activated TGF-β receptor II causes Par6 to activate the E3 ubiquitin ligase Smurf1, and Smurf1 then targets RhoA to the proteasome. The loss of RhoA activity results in the loss of cell-cell adhesion and epithelial cell polarity (Ozdamar et al., 2005).

In order for mesenchymal cells to migrate away from the epithelium, the cells must become motile. Many of the same polarity (Crumbs, PAR, and Scribble complexes), structural (actin, microtubules), and regulatory molecules (Cdc42, Rac1, RhoA) that govern epithelial polarity are also central to cell motility (Nelson, 2009). Cell motility mechanisms also vary depending on whether the environment is two-dimensional or three-dimensional (Friedl and Wolf, 2010). Many mesenchymal cells express the intermediate filament vimentin, and vimentin may be responsible for several aspects of the EMT phenotype (Mendez et al., 2010).

In short, a wide variety of structural, polarity, and regulatory molecules must be reassigned as cells transition between epithelial polarity and mesenchymal migration.

Invasion of the basal lamina

In most EMTs the emerging mesenchymal cells must penetrate a basal lamina that consists of ECM components such as collagen type IV, fibronectin, and laminin. The basal lamina functions to stabilize the epithelium and is a barrier to migratory cells (Erickson, 1987). One mechanism that mesenchymal cells use to breach the basal lamina is to produce enzymes that degrade it. Plasminogen activator is one protease associated with a number of EMTs, including neural crest emigration (Erickson, 1987) and the formation of cardiac cushion cells during heart morphogenesis (McGuire and Alexander, 1993). The type II serine protease, TMPRSS4, also promotes an EMT and metastasis when overexpressed *in vitro* and *in vivo* (Jung et al., 2007). Matrix-metalloproteases (MMPs) are also important for many EMTs. When MMP-2 activity is blocked in the neural crest EMT, neural crest emigration is inhibited, but not neural crest motility (Duong and Erickson, 2004). In mouse mammary cells, MMP-3 overexpression is sufficient to induce an EMT *in vitro* and *in vivo* (Sternlicht et al., 1999). Misexpressing MMP-3 in cultured cells induces an alternatively spliced form of Rac1 (Rac1b), which then causes an increase in reactive oxygen species (ROS) intracellularly, and Snail-1 expression. Either Rac1b activity or ROS are necessary and sufficient to bring about an MMP-3-induced EMT (Radisky et al., 2005). Hence, a number of extracellular proteases are important to bring about an EMT.

While epithelial cells undergoing an EMT will eventually lose cell-cell adhesion, change apical-basal polarity, and gain invasive motility, the EMT program may not necessarily be ordered or linear. For example, in a study where neural crest cells were labeled with cell-adhesion or polarity markers and individual live cells were observed undergoing the EMT in slice culture, neural crest cells changed epithelial polarity either before or after the complete loss of cell-cell adhesion, or lost cell-cell adhesions either before or after cell migration commenced (Ahlstrom and Erickson, 2009). Therefore, while an EMT does consist of several distinct phases, these steps may occur in different orders or combinations, some of which (e.g. the complete loss of cell-cell adhesion) may not always be necessary.

In summary, changes in a wide range of molecules are needed for an EMT as epithelial cells lose cell-cell adhesion, change cellular polarity, and gain invasive cell motility.

THE EMT TRANSCRIPTIONAL PROGRAM

At the foundation of every EMT or MET program are the transcription factors that regulate the gene expression required for these cellular transitions. While many of the transcription factors that regulate EMTs have been identified, the complex regulatory networks are still incomplete. Here are reviewed the transcription factors that are known to promote the various phases of an EMT. Then we will examine how these EMT transcription factors themselves are regulated at the promoter and post-transcriptional levels.

Transcription factors that regulate EMTs

The Snail family of zinc-finger transcription factors, including Snail-1 and Snail-2 (formerly Snail and Slug), are direct regulators of cell-cell adhesion and motility during EMTs (Barrallo-Gimeno and Nieto, 2005; de Craene et al., 2005). The knockout of *Snail-1* in mice is lethal early in gestation, and the presumptive primitive streak cells that normally undergo an EMT still retain apical/basal polarity and adherens junctions, and express E-cadherin mRNA (Carver et al., 2001). Snail-1 misexpression is sufficient for breast cancer recurrence in a mouse model *in vivo*, and high levels of Snail-1 predict the relapse of human breast cancer (Moody et al., 2005). Snail-2 is necessary for the chicken primitive streak and neural crest EMTs (Nieto et al., 1994). One way that Snail-1 or Snail-2 causes a decrease in cell-cell adhesion is by repressing the *E-cadherin* promoter (de Craene et al., 2005). This repression requires the mSin3A co-repressor complex, histone deacetylases, and components of the Polycomb 2 complex (Herranz et al., 2008). Snail-1 is also a transcriptional repressor of the tight junction genes *Claudin* and *Occludin* (de Craene et al., 2005) and the polarity gene *Crumbs3* (Whiteman et al., 2008). The misexpression of Snail-1 and Snail-2 further leads to the transcription of proteins important for cell motility such as fibronectin, vimentin (Cano et al., 2000), and RhoB (del Barrio and Nieto, 2002). Further, Snail-1 promotes invasion across the basal lamina. In Madin-Darby Canine Kidney (MDCK) cells, the misexpression of Snail-1 represses laminin (basement membrane) production (Haraguchi et al., 2008) and indirectly upregulates *mmp-9* transcription (Jorda et al., 2005). Snail and Twist also make cancer cells more resistant to senescence, chemotherapy, and apoptosis, and endow cancer cells with "stem cell" properties (Thiery et al., 2009). Hence, Snail-1 or Snail-2 is necessary and sufficient for bringing about many of the steps of an EMT, including loss of cell-cell adhesion, changes in cell polarity, gain of cell motility, invasion of the basal lamina, and increased proliferation and survival.

Other zinc-finger transcription factors important for EMTs are zinc-finger E-box-binding homeobox 1 (Zeb1, also known as δEF1), and Zeb2 (also known as Smad-interacting protein-1, Sip1). Both Zeb1 and Zeb2 bind to the *E-cadherin* promoter and repress transcription (de Craene et al., 2005). Zeb1 can also bind to and repress the transcription of the polarity proteins Crumbs3, Pals1-associated tight junction proteins (PATJ), and Lethal giant larvae 2 (Lgl2) (Spaderna et al., 2008). Zeb2 is structurally similar to Zeb1, and Zeb2 overexpression is sufficient to downregulate E-cadherin, dissociate adherens junctions, and increase motility in MDCK cells (Comijn et al., 2001).

The lymphoid enhancer-binding factor/T-cell factor (LEF/TCF) transcription factors also play an important role in EMTs. For instance, the misexpression of Lef-1 in cultured colon cancer cells reversibly causes the loss of cell-cell adhesion (Kim et al., 2002). LEF/TCF transcription factors directly activate genes that regulate cell motility, such as the L1 adhesion molecule (Gavert et al., 2005) and the *fibronectin* gene (Gradl et al., 1999). LEF/TCF transcription factors also upregulate genes required for basal lamina invasion, including *mmp-3* and *mmp-7* (Gustavson et al., 2004).

Other transcription factors that have a role in promoting EMTs are the class I bHLH factors E2-2A and E2-2B (Sobrado et al., 2009), the forkhead box transcription factor FOXC2 (Mani et al., 2007), the homeobox protein Goosecoid (Hartwell et al., 2006), and the homeoprotein Six1 (McCoy et al., 2009; Micalizzi et al., 2009).

To summarize, transcription factors that regulate an EMT often do so by directly repressing cell adhesion and epithelial polarity molecules, and by upregulating genes required for cell motility and basal lamina invasion.

Regulation at the promoter level

Given the importance of the Snail, Zeb, and LEF/TCF transcription factors in orchestrating the various phases of an EMT, it is essential to understand the upstream events that regulate these EMT-promoting transcription factors.

The activation of *Snail-1* transcription in *Drosophila* requires the transcription factors Dorsal (NF-κB) and Twist (de Craene et al., 2005). The human *Snail-1* promoter also has functional NF-κB sites (Barbera et al., 2004) and blocking NF-κB reduces *Snail-1* transcription (Strippoli et al., 2008). Additionally, a region of the *Snail-1* promoter is responsive to integrin-linked kinase (ILK) (de Craene et al., 2005), and ILK can activate Snail-1 expression via poly-ADP-ribose polymerase (PARP) (Lee et al., 2006). In mouse mammary epithelial cells, high mobility group protein A2 (HMGA2) and Smads activate *Snail-1* expression, and subsequently *Snail-2*, *Twist*, and *Id2* transcription (Thuault et al., 2008). For *Snail-2* expression, myocardin-related transcription factors (MRTFs) interact with Smads to induce *Snail-2* (Morita et al., 2007) and MRTFs may play a role in metastasis (Medjkane et al., 2009) and fibrosis (Fan et al., 2007). There are also several *Snail-1* transcriptional repressors. In breast cancer cell lines, metastasis-associated protein 3 (MTA3) binds directly to and represses the transcription of *Snail-1* in combination with the Mi-2/NuRD complex (Fujita et al., 2003), as also does lysine-specific demethylase (LSD1) (Wang et al., 2009a). The Ajuba LIM proteins (Ajuba, LIMD1, and WTIP) are additional transcriptional co-repressors of the Snail family (Langer et al., 2008).

The transcription of LEF/TCF genes such as *Lef-1* is activated by Smads (Nawshad and Hay, 2003). The misexpression of Snail-1 results in the transcription of *δEF-1* and *Lef-1* through a yet unknown mechanism (de Craene et al., 2005).

Post-transcriptional regulation of EMT transcription factors

The activity of EMT transcription factors is also regulated at the protein level, including translational control, protein stability (targeting to the proteasome), and nuclear localization. Non-coding RNAs are emerging as important regulators of EMTs. In a breast cancer model, Myc activates the expression of microRNA-9 (miR-9), and miR-9 directly binds to and represses the *E-cadherin* promoter (Ma et al., 2010). Members of the miR-200 family repress the translation of *Zeb1*, and the expression of these miR-200 family members is repressed by Snail-1. Additionally, *Zeb2* transcription can be activated by naturally occurring RNA antisense transcripts (Beltran et al., 2008). It is not yet known whether there are non-coding RNAs that regulate Snail family members. However, the Y-box-binding protein-1 (YB-1) is important for the selective activation of *Snail-1* translation (Evdokimova et al., 2009).

Protein stability is another layer of EMT control. Snail-1 is phosphorylated by GSK-3β and targeted for destruction (Zhou et al., 2004). Therefore, the inhibition of GSK-3β activity by Wnt signaling may have multiple roles in an EMT, leading to the stabilization of both β-catenin and Snail-1. Some proteins that prevent GSK-3β-mediated phosphorylation (and thus promote Snail-1 activation) are lysyl-oxidase-like proteins LOXL2, LOXL3 (Peinado et al., 2007), and ILK (Delcommenne et al., 1998). A Snail-1-specific phosphatase (Snail-1 activator) is C-terminal domain phosphatase (SCP) (Wu et al., 2009). Snail-2 is targeted for degradation by the direct action of p53 and the ubiquitin ligase Mdm2 (Wang et al., 2009b).

In addition to protein translation and stability, the function of Snail-1 also depends upon nuclear localization mediated by Snail-1's nuclear localization sequence. The phosphorylation of human Snail-1 by p21-activated kinase 1 (Pak1) promotes the nuclear localization of Snail-1 (and therefore Snail-1 activation) in breast cancer cells (Yang et al., 2005). In zebrafish,

LIV-1 promotes the translocation of Snail-1 into the nucleus (Yamashita et al., 2004). Snail-1 also contains a nuclear export sequence (NES) that is dependent on the calreticulin (CalR) nuclear export pathway (Dominguez et al., 2003). This NES sequence is activated by the phosphorylation of the same lysine residues targeted by GSK-3β, which suggests a mechanism whereby phosphorylation of Snail-1 by GSK-3β results in the export of Snail-1 from the nucleus and subsequent degradation.

LEF/TCF activity is also regulated by other proteins. β-Catenin is required as a co-factor for LEF/TCF-mediated activation of transcription, and Lef-1 can also associate with co-factor Smads to activate the transcription of additional EMT genes (Labbe et al., 2000). In colon cancer cells, Thymosin β4 stabilizes ILK activity (Huang et al., 2006).

In summary, EMT transcription factors such as Snail-1, Zeb1, and Lef-1 are regulated by a variety of mechanisms, both at the transcriptional level and post-transcriptional level, by non-coding RNA translation control, protein degradation, nuclear localization, and co-factors such as β-catenin.

MOLECULAR CONTROL OF THE EMT

The initiation of an EMT or MET is a tightly regulated event during development and tissue repair because deregulation of cellular organization is disastrous to the organism. A variety of external and internal signaling mechanisms coordinate the complex events of the EMT, and these same signaling pathways are often disrupted or reactivated during disease. EMTs or METs can be induced by either diffusible signaling molecules or ECM components. Below is discussed the role of signaling molecules and ECM in triggering an EMT, and then a summary model for EMT induction is presented.

Ligand-receptor signaling

During development, five main ligand-receptor signaling pathways are employed, namely TGF-β, Wnt, RTK, Notch, and Hedgehog. These pathways, among others, all have a role in triggering EMTs. While the activation of a single signaling pathway can be sufficient for an EMT, in most cases an EMT or MET is initiated by multiple signaling pathways acting in concert.

TGF-β PATHWAY

The transforming growth factor-beta (TGF-β) superfamily includes TGF-β, activin, and the bone morphogenetic protein (BMP) families. These ligands operate through receptor serine/threonine kinases to activate a variety of signaling molecules including Smads, MAPK, PI3K, and ILK. Most of the EMTs studied to date are induced in part, or solely, by TGF-β superfamily members (Zavadil and Bottinger, 2005). During embryonic heart development, TGF-β2 and TGF-β3 have sequential and necessary roles in activating the endocardium to invade the cardiac jelly and form the endocardial cushions (Camenisch et al., 2002a). In the avian neural crest, BMP4 induces *Snail-2* expression (Liem et al., 1995). In the EMT that transforms epithelial tissue into metastatic cancer cells, TGF-β acts as a tumor suppressor during early stages of tumor development, but as a tumor/EMT inducer at later stages (Cui et al., 1996; Zavadil and Bottinger, 2005). TGF-β signaling may combine with other signaling pathways to induce an EMT. For example, in cultured breast cancer cells, activated Ras and TGF-β induce an irreversible EMT (Janda et al., 2002), and, in pig thyroid epithelial cells, TGF-β and epidermal growth factor (EGF) synergistically stimulate the EMT (Grande et al., 2002).

One outcome of TGF-β signaling is to immediately change epithelial cell polarity. In a TGF-β-induced EMT of mammary epithelial cells, TGF-βR II directly phosphorylates the polarity protein, Par6, leading to the dissolution of tight junctions (Ozdamar et al., 2005). TGF-β signaling also regulates gene expression through the phosphorylation and activation of

Smads. Smads are important co-factors in the stimulation of an EMT. For example, Smad3 is necessary for a TGF-β-induced EMT in lens and kidney tissue *in vivo* (Roberts et al., 2006). Smad3/4 also complexes with Snail-1 and co-represses the promoters of cell-cell adhesion molecules (Vincent et al., 2009). Further, TGF-βR I directly binds to and activates PI3K (Yi et al., 2005), which in turn activates ILK and downstream pathways.

ILK is emerging as an important positive regulator of EMTs (Larue and Bellacosa, 2005). ILK interacts directly with growth factor receptors (TGF-β, Wnt, or RTK), integrins, the actin skeleton, PI3K, and focal adhesion complexes. ILK directly phosphorylates Akt and GSK-3β, and results in the subsequent activation of transcription factors such as AP-1, NF-κB, and Lef-1. Overexpression of ILK in cultured cells causes the suppression of GSK-3β activity (Delcommenne et al., 1998), translocation of β-catenin to the nucleus, activation of Lef-1/β-catenin transcription factors, and the downregulation of E-cadherin (Novak et al., 1998). Inhibition of ILK in cultured colon cancer cells leads to the stabilization of GSK-3β activity, decreased nuclear β-catenin localization, the suppression of *Lef-1* and *Snail-1* transcription, and reduced invasive behavior of colon cancer cells (Tan et al., 2001). ILK activity also results in Lef-1-mediated transcriptional upregulation of MMPs (Gustavson et al., 2004). Hence, ILK (inducible by TGF-β signaling) is capable of orchestrating most of the major events in an EMT, including the loss of cell-cell adhesion and invasion across the basal lamina.

WNT PATHWAY

Many EMTs or METs are also regulated by Wnt signaling. Wnts signal through seven-pass transmembrane proteins of the Frizzled family, which activates G-proteins and PI3K, inhibits GSK-3β, and promotes nuclear β-catenin signaling. For example, during zebrafish gastrulation, Wnt11 activates the GTPase Rab5c, which results in the endocytosis of E-cadherin (Ulrich et al., 2005). Wnt6 signaling is sufficient for increased transcription of *Snail-2* in the avian neural crest (Garcia-Castro et al., 2002). Snail-1 expression increases Wnt signaling (Stemmer et al., 2008), which suggests a positive feedback loop.

One of the downstream signaling molecules activated by Wnt signaling is β-catenin. β-Catenin is a structural component of adherens junctions. Nuclear β-catenin is also a limiting factor for the activation of LEF/TCF transcription factors. β-Catenin is pivotal for regulating most EMTs. Interfering with nuclear β-catenin signaling blocks the ingression of sea urchin PMCs (Logan et al., 1999) and, in β-catenin mouse knockouts, the primitive streak EMT does not occur and no mesoderm is formed (Huelsken et al., 2000). β-Catenin is also necessary for the EMT that occurs during cardiac cushion development (Liebner et al., 2004). In breast cancer, β-catenin expression is highly correlated with metastasis and poor survival (Cowin et al., 2005), and blocking β-catenin function in tumor cells inhibits invasion *in vitro* (Wong and Gumbiner, 2003). It is unclear whether β-catenin overexpression alone is sufficient for all EMTs. If β-catenin is misexpressed in cultured cells, it causes apoptosis (Kim et al., 2000). However, the misexpression of a stabilized form of β-catenin in mouse epithelial cells *in vivo* results in metastatic skin tumors (Gat et al., 1998).

SIGNALING BY RTK LIGANDS

The receptor tyrosine kinase (RTK) family of receptors and the growth factors that activate them also regulate EMTs or METs. Ligand binding promotes RTK dimerization and activation of the intracellular kinase domains by auto-phosphorylation of tyrosine residues. These phosphotyrosines act as docking sites for intracellular signaling molecules, which can activate signaling cascades such as Ras/MAPK, PI3K/Akt, JAK/STAT, or ILK. Below we cite a few examples of RTK signaling in EMTs and METs.

Hepatocyte growth factor (HGF, also known as scatter factor) acts through the RTK c-met. HGF is important for the MET in the developing kidney (Woolf et al., 1995). HGF signaling is required for the EMT that produces myoblasts (limb muscle precursors) from somite tissue in the mouse (Thiery, 2002). In epithelial cells, HGF causes an EMT through MAPK and early growth response factor-1 (Egr-1) signaling (Grotegut et al., 2006).

Fibroblast growth factor (FGF) signaling regulates mouse primitive streak formation (Ciruna and Rossant, 2001). FGF signaling also stimulates cell motility and activates MMPs (Suyama et al., 2002; Billottet et al., 2008).

Epidermal growth factor (EGF) promotes the endocytosis of E-cadherin (Lu et al., 2003). EGF can also increase Snail-1 activity via the inactivation of GSK3-β (Lee et al., 2008) and EGF promotes increased *Twist* expression through a JAK/STAT3 pathway (Lo et al., 2007).

Insulin growth factor (IGF) signaling induces an EMT in breast cancer cell lines through the activation of Akt2 and suppression of Akt1 (Irie et al., 2005). In prostate cancer cells, IGF-1 promotes Zeb-1 expression (Graham et al., 2001). In fibroblast cells, constitutively activated IGF-IR increases NF-κB activity and Snail-1 levels (Kim et al., 2007). In several cultured epithelial cell lines, IGFR1 is associated with the complex of E-cadherin and β-catenin, and the ligand IGF-II causes the redistribution of β-catenin from the membrane to the nucleus, activation of the transcription factor TCF-3, and a subsequent EMT (Morali et al., 2001).

Another RTK known for its role in EMTs is the ErbB2/HER-2/Neu receptor, whose ligand is heregulin/neuregulin. Overexpression of HER-2 occurs in 25% of human breast cancers, and the misexpression of HER-2 in mouse mammary tissue *in vivo* is sufficient to cause metastatic breast cancer (Muller et al., 1988). Herceptin® (antibody against the HER-2 receptor) treatment is effective in reducing the recurrence of HER-2-positive metastatic breast cancers. HER-2 signaling activates *Snail-1* expression in breast cancer through an unknown mechanism (Moody et al., 2005). The RTK Axl is also required for breast cancer carcinoma invasiveness (Gjerdrum et al., 2010).

Vascular endothelial growth factor (VEGF) signaling promotes Snail-1 activity by suppression of GSK3-β (Wanami et al., 2008) and results in increased levels of *Snail-1*, *Snail-2*, and *Twist* (Yang et al., 2006). Snail-1 can also activate the expression of VEGF (Peinado et al., 2004). In summary, RTK signaling is important for many EMTs.

NOTCH PATHWAY

The Notch signaling family also regulates EMTs. When the Notch receptor is activated by its ligand Delta, an intracellular portion of the Notch receptor ligand is cleaved and transported to the nucleus where it regulates target genes. Notch1 is required for cardiac endothelial cells to undergo an EMT to make cardiac cushions, and the role of Notch may be to make cells competent to respond to TGF-β2 (Timmerman et al., 2004). In the avian neural crest EMT, Notch signaling is required for the induction and/or maintenance of *BMP4* expression (Endo et al., 2002). Similarly, Notch signaling is required for the TGF-β-induced EMT of epithelial cell lines (Zavadil et al., 2004), and Notch promotes *Snail-2* expression in cardiac cushion cells (Niessen et al., 2008) and cultured cells (Leong et al., 2007).

HEDGEHOG PATHWAY

The hedgehog pathway is also involved in EMTs. Metastatic prostate cancer cells express high levels of hedgehog and *Snail-1*. If prostate cancer cell lines are treated with the hedgehog-pathway inhibitor, cyclopamine, levels of *Snail-1* are decreased. If the hedgehog-activated transcription factor, Gli, is misexpressed, *Snail-1* expression increases (Karhadkar et al., 2004).

Additional signaling pathways

Other signaling pathways that activate EMTs include inflammatory signaling molecules, lipid hormones, ROS species, and hypoxia. Interleukin-6 (Il-6, inflammatory and immune response) can promote Snail-1 expression in breast cancer cells (Sullivan et al., 2009), and Snail-1 in turn can activate Il-6 expression (Lyons et al., 2008), providing a link between inflammation and EMTs (López-Novoa and Nieto, 2009). The lipid hormone prostaglandin E2 (PGE2) induces Zeb1 and Snail activity in lung cancer cells (Dohadwala et al., 2006), and Snail-1 can also induce PGE2 expression (Mann et al., 2006). ROS species can also activate EMTs by PKC and MAPK signaling (Wu, 2006). Hypoxia is important for initiating EMTs during development (Dunwoodie, 2009) and disease (López-Novoa and Nieto, 2009), often through hypoxia-inducible factor-1 (HIF-1), which directly activates *Twist* expression (Yang et al., 2008). Hypoxia also activates lysyl oxidases (LOXs), which stabilize Snail-1 expression (Sahlgren et al., 2008) by inhibiting GSK-3β activity (Peinado et al., 2005).

In addition to diffusible signaling molecules, extracellular matrix molecules also regulate EMTs or METs. This was first dramatically demonstrated when lens or thyroid epithelium was embedded in collagen gels and then promptly underwent an EMT (Hay, 2005). Integrin signaling appears to be important in this process (Zuk and Hay, 1994) and involves ILK-mediated activation of NF-κB, Snail-1, and Lef-1 (Medici and Nawshad, 2010). Other ECM components that regulate EMTs include hyaluronan (Camenisch et al., 2002b), the gamma-2 chain of laminin 5 (Koshikawa et al., 2000), periostin (Ruan et al., 2009), and podoplanin (Martin-Villar et al., 2006; Wicki et al., 2006). In summary, a variety of diffusible signals and ECM components can stimulate EMTs or METs.

A model for EMT induction

Many of the experimental studies on EMT mechanisms focus on individual molecules and, while great progress has been made in discovering EMT pathways, the entire signaling network is still incomplete. Figure 1.2 summarizes many of the various signaling mechanisms, although in actuality only a few of the inductive pathways may be utilized for individual EMTs. From experimental evidence to date, it appears that many of the EMT signaling pathways converge on ILK, the inhibition of GSK-3β, and stimulation of nuclear β-catenin signaling to activate Snail and LEF/TCF transcription factors. Snail, Zeb, and LEF/TCF transcription factors then act on a variety of targets to suppress cell-cell adhesion, induce changes in cell polarity, stimulate cell motility, and promote invasion of the basal lamina.

CONCLUSION

Over the more than 20 years since the term "EMT" was coined (Thiery, 2002), important insights have been made in this rapidly expanding field of research. EMT and MET events occur during development, tissue repair, and disease, and many molecules that regulate the various EMTs or METs have been characterized, thanks in large part to the advent of cell culture models. However, the EMT regulatory network as a whole is still incomplete. Improved

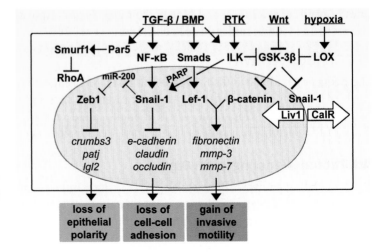

FIGURE 1.2

Induction of an EMT. This figure summarizes some of the important molecular pathways that bring about an EMT. Many of the signaling pathways converge on the activation of Snail-1 and nuclear β-catenin signaling to change gene expression, which results in the loss of epithelial cell polarity, the loss of cell-cell adhesion, and increased invasive cell motility.

understanding of EMT and MET pathways in the future will lead to more effective strategies for tissue engineering and novel therapeutic targets for the treatment of disease.

References

Ahlstrom, J. D., & Erickson, C. A. (2009). The neural crest epithelial-mesenchymal transition in 4D: a "tail" of multiple non-obligatory cellular mechanisms. *Development, 136,* 1801–1812.

Barbera, M. J., Puig, I., Dominguez, D., Julien-Grille, S., Guaita-Esteruelas, S., Peiro, S., et al. (2004). Regulation of Snail transcription during epithelial to mesenchymal transition of tumor cells. *Oncogene, 23,* 7345–7354.

Barrallo-Gimeno, A., & Nieto, M. A. (2005). The Snail genes as inducers of cell movement and survival: implications in development and cancer. *Development, 132,* 3151–3161.

Bates, R. C. (2005). Colorectal cancer progression: integrin alphavbeta6 and the epithelial-mesenchymal transition (EMT). *Cell Cycle, 4,* 1350–1352.

Baum, B., Settleman, J., & Quinlan, M. P. (2008). Transitions between epithelial and mesenchymal states in development and disease. *Semin. Cell Dev. Biol., 19,* 294–308.

Beltran, M., Puig, I., Peña, C., García, J. M., Alvarez, A. B., Peña, R., et al. (2008). A natural antisense transcript regulates Zeb2/Sip1 gene expression during Snail1-induced epithelial-mesenchymal transition. *Genes Dev., 22,* 756–769.

Billottet, C., Tuefferd, M., Gentien, D., Rapinat, A., Thiery, J.-P., Broët, P., et al. (2008). Modulation of several waves of gene expression during FGF-1 induced epithelial-mesenchymal transition of carcinoma cells. *J. Cell Biochem., 104,* 826–839.

Bryant, D. M., & Stow, J. L. (2004). The ins and outs of E-cadherin trafficking. *Trends Cell Biol., 14,* 427–434.

Camenisch, T. D., Molin, D. G. M., Person, A., Runyan, R. B., Gittenberger-de Groot, A. C., McDonald, J. A., et al. (2002a). Temporal and distinct TGFβ ligand requirements during mouse and avian endocardial cushion morphogenesis. *Dev. Biol., 248,* 170–181.

Camenisch, T. D., Schroeder, J. A., Bradley, J., Klewer, S. E., & McDonald, J. A. (2002b). Heart-valve mesenchyme formation is dependent on hyaluronan-augmented activation of ErbB2-ErbB3 receptors. *Nat. Med., 8,* 850–855.

Cano, A., Perez-Moreno, M. A., Rodrigo, I., Locascio, A., Blanco, M. J., del Barrio, M. G., et al. (2000). The transcription factor Snail controls epithelial-mesenchymal transitions by repressing E-cadherin expression. *Nat. Cell Biol., 2,* 76–83.

Carroll, J. M., Luetteke, N. C., Lee, D. C., & Watt, F. M. (1998). Role of integrins in mouse eyelid development: studies in normal embryos and embryos in which there is a failure of eyelid fusion. *Mech. Dev., 78,* 37–45.

Carver, E. A., Jiang, R., Lan, Y., Oram, K. F., & Gridley, T. (2001). The mouse snail gene encodes a key regulator of the epithelial-mesenchymal transition. *Mol. Cell Biol., 21,* 8184–8188.

Christofori, G. (2003). Changing neighbours, changing behaviour: cell adhesion molecule-mediated signalling during tumour progression. *EMBO J., 22,* 2318–2323.

Ciruna, B., & Rossant, J. (2001). FGF signaling regulates mesoderm cell fate specification and morphogenetic movement at the primitive streak. *Dev. Cell, 1,* 37–49.

Comijn, J., Berx, G., Vermassen, P., Verschueren, K., van Grunsven, L., Bruyneel, E., et al. (2001). The two-handed E box binding zinc finger protein SIP1 downregulates E-cadherin and induces invasion. *Mol. Cell, 7,* 1267–1278.

Cowin, P., Rowlands, T. M., & Hatsell, S. J. (2005). Cadherins and catenins in breast cancer. *Curr. Opin. Cell Biol., 17,* 499–508.

Cui, W., Fowlis, D. J., Bryson, S., Duffie, E., Ireland, H., Balmain, A., et al. (1996). TGFβ1 inhibits the formation of benign skin tumors, but enhances progression to invasive spindle carcinomas in transgenic mice. *Cell, 86,* 531–542.

de Craene, B., van Roy, F., & Berx, G. (2005). Unraveling signalling cascades for the Snail family of transcription factors. *Cell. Signal, 17,* 535–547.

del Barrio, M. G., & Nieto, M. A. (2002). Overexpression of Snail family members highlights their ability to promote chick neural crest formation. *Development, 129,* 1583–1593.

Delannet, M., & Duband, J. L. (1992). Transforming growth factor-beta control of cell-substratum adhesion during avian neural crest cell emigration *in vitro. Development, 116,* 275–287.

Delcommenne, M., Tan, C., Gray, V., Rue, L., Woodgett, J., & Dedhar, S. (1998). Phosphoinositide-3-OH kinase-dependent regulation of glycogen synthase kinase 3 and protein kinase B/AKT by the integrin-linked kinase. *Proc. Natl. Acad. Sci. U.S.A., 95,* 11211–11216.

Desgrosellier, J. S., & Cheresh, D. A. (2010). Integrins in cancer: biological implications and therapeutic opportunities. *Nat. Rev. Cancer, 10,* 9–22.

13

Dohadwala, M., Yang, S.-C., Luo, J., Sharma, S., Batra, R. K., Huang, M., et al. (2006). Cyclooxygenase-2-dependent regulation of E-cadherin: prostaglandin E2 induces transcriptional repressors ZEB1 and Snail in non-small cell lung cancer. *Cancer Res., 66,* 5338–5345.

Dominguez, D., Montserrat-Sentis, B., Virgos-Soler, A., Guaita, S., Grueso, J., Porta, M., et al. (2003). Phosphorylation regulates the subcellular location and activity of the Snail transcriptional repressor. *Mol. Cell. Biol., 23,* 5078–5089.

Dunwoodie, S. L. (2009). The role of hypoxia in development of the mammalian embryo. *Dev. Cell, 17,* 755–773.

Duong, T. D., & Erickson, C. A. (2004). MMP-2 plays an essential role in producing epithelial-mesenchymal transformations in the avian embryo. *Dev. Dyn., 229,* 42–53.

Egeblad, M., & Werb, Z. (2002). New functions for the matrix metalloproteinases in cancer progression. *Nat. Rev. Cancer, 2,* 161–174.

Endo, Y., Osumi, N., & Wakamatsu, Y. (2002). Bimodal functions of Notch-mediated signaling are involved in neural crest formation during avian ectoderm development. *Development, 129,* 863–873.

Erickson, C. A. (1987). Behavior of neural crest cells on embryonic basal laminae. *Dev. Biol., 120,* 38–49.

Evdokimova, V., Tognon, C., Ng, T., Ruzanov, P., Melnyk, N., Fink, D., et al. (2009). Translational activation of Snail1 and other developmentally regulated transcription factors by YB-1 promotes an epithelial-mesenchymal transition. *Cancer Cell, 15,* 402–415.

Fan, L., Sebe, A., Peterfi, Z., Masszi, A., Thirone, A. C. P., Rotstein, O. D., et al. (2007). Cell contact-dependent regulation of epithelial-myofibroblast transition via the rho-rho kinase-phospho-myosin pathway. *Mol. Biol. Cell, 18,* 1083–1097.

Friedl, P., & Wolf, K. (2010). Plasticity of cell migration: a multiscale tuning model. *J. Cell Biol., 188,* 11–19.

Fujita, N., Jaye, D. L., Kajita, M., Geigerman, C., Moreno, C. S., & Wade, P. A. (2003). MTA3, a Mi-2/NuRD complex subunit, regulates an invasive growth pathway in breast cancer. *Cell, 113,* 207–219.

Fujita, Y., Krause, G., Scheffner, M., Zechner, D., Leddy, H. E. M., Behrens, J., et al. (2002). Hakai, a c-Cbl-like protein, ubiquitinates and induces endocytosis of the E-cadherin complex. *Nat. Cell Biol., 4,* 222–231.

Garcia-Castro, M. I., Marcelle, C., & Bronner-Fraser, M. (2002). Ectodermal Wnt function as a neural crest inducer. *Science, 297,* 848–851.

Gat, U., DasGupta, R., Degenstein, L., & Fuchs, E. (1998). De novo hair follicle morphogenesis and hair tumors in mice expressing a truncated β-catenin in skin. *Cell, 95,* 605–614.

Gavert, N., Conacci-Sorrell, M., Gast, D., Schneider, A., Altevogt, P., Brabletz, T., et al. (2005). L1, a novel target of β-catenin signaling, transforms cells and is expressed at the invasive front of colon cancers. *J. Cell Biol., 168,* 633–642.

Giepmans, B. N., & van Ijzendoorn, S. C. (2009). Epithelial cell-cell junctions and plasma membrane domains. *Biochim. Biophys. Acta, 1788,* 820–831.

Gjerdrum, C., Tiron, C., Høiby, T., Stefansson, I., Haugen, H., Sandal, T., et al. (2010). Axl is an essential epithelial-to-mesenchymal transition-induced regulator of breast cancer metastasis and patient survival. *Proc. Natl. Acad. Sci. U.S.A., 107,* 1124–1129.

Gradl, D., Kuhl, M., & Wedlich, D. (1999). The Wnt/Wg signal transducer β-catenin controls fibronectin expression. *Mol. Cell. Biol., 19,* 5576–5587.

Graham, T. R., Zhau, H. E., Odero-Marah, V. A., Osunkoya, A. O., Kimbro, K. S., Tighiouart, M., et al. (2008). Insulin-like growth factor-I-dependent up-regulation of ZEB1 drives epithelial-to-mesenchymal transition in human prostate cancer cells. *Cancer Res., 68,* 2479–2488.

Grande, M., Franzen, A., Karlsson, J. O., Ericson, L. E., Heldin, N.-E., & Nilsson, M. (2002). Transforming growth factor-β and epidermal growth factor synergistically stimulate epithelial to mesenchymal transition (EMT) through a MEK-dependent mechanism in primary cultured pig thyrocytes. *J. Cell Sci., 115,* 4227–4236.

Grotegut, S., von Schweinitz, D., Christofori, G., & Lehembre, F. (2006). Hepatocyte growth factor induces cell scattering through MAPK/Egr-1-mediated upregulation of Snail. *EMBO J., 25,* 3534–3545.

Gustavson, M. D., Crawford, H. C., Fingleton, B., & Matrisian, L. M. (2004). Tcf binding sequence and position determines β-catenin and Lef-1 responsiveness of MMP-7 promoters. *Mol. Carcinog., 41,* 125–139.

Haraguchi, M., Okubo, T., Miyashita, Y., Miyamoto, Y., Hayashi, M., Crotti, T. N., et al. (2008). Snail regulates cell-matrix adhesion by regulation of the expression of integrins and basement membrane proteins. *J. Biol. Chem., 283,* 23514–23523.

Hartwell, K. A., Muir, B., Reinhardt, F., Carpenter, A. E., Sgroi, D. C., & Weinberg, R. A. (2006). The Spemann organizer gene, Goosecoid, promotes tumor metastasis. *Proc. Natl. Acad. Sci. U.S.A., 103,* 18969–18974.

Hay, E. D. (2005). The mesenchymal cell, its role in the embryo, and the remarkable signaling mechanisms that create it. *Dev. Dyn., 233,* 706–720.

Herranz, N., Pasini, D., Diaz, V. M., Franci, C., Gutierrez, A., Dave, N., et al. (2008). Polycomb complex 2 is required for E-cadherin repression by the Snail1 transcription factor. *Mol. Cell. Biol., 28,* 4772–4781.

Hood, J. D., & Cheresh, D. A. (2002). Role of integrins in cell invasion and migration. *Nat. Rev. Cancer, 2,* 91–100.

Huang, H. C., Hu, C. H., Tang, M. C., Wang, W. S., Chen, P. M., & Su, Y. (2006). Thymosin B4 triggers an epithelial-mesenchymal transition in colorectal carcinoma by upregulating integrin-linked kinase. *Oncogene, 26,* 2781–2790.

Huelsken, J., Vogel, R., Brinkmann, V., Erdmann, B., Birchmeier, C., & Birchmeier, W. (2000). Requirement for beta-catenin in anterior-posterior axis formation in mice. *J. Cell Biol., 148,* 567–578.

Humphreys, B. D., Lin, S.-L., Kobayashi, A., Hudson, T. E., Nowlin, B. T., Bonventre, J. V., et al. (2010). Fate tracing reveals the pericyte and not epithelial origin of myofibroblasts in kidney fibrosis. *Am. J. Pathol. 176,* 85–97.

Irie, H. Y., Pearline, R. V., Grueneberg, D., Hsia, M., Ravichandran, P., Kothari, N., et al. (2005). Distinct roles of Akt1 and Akt2 in regulating cell migration and epithelial-mesenchymal transition. *J. Cell Biol., 171,* 1023–1034.

Iwano, M., Plieth, D., Danoff, T. M., Xue, C., Okada, H., & Neilson, E. G. (2002). Evidence that fibroblasts derive from epithelium during tissue fibrosis. *J. Clin. Investig., 110,* 341–350.

Janda, E., Lehmann, K., Killisch, I., Jechlinger, M., Herzig, M., Downward, J., et al. (2002). Ras and TGFβ cooperatively regulate epithelial cell plasticity and metastasis: dissection of Ras signaling pathways. *J. Cell Biol., 156,* 299–314.

Jorda, M., Olmeda, D., Vinyals, A., Valero, E., Cubillo, E., Llorens, A., et al. (2005). Upregulation of MMP-9 in MDCK epithelial cell line in response to expression of the Snail transcription factor. *J. Cell Sci., 118,* 3371–3385.

Jung, H., Lee, K. P., Park, S. J., Park, J. H., Jang, Y. S., Choi, S. Y., et al. (2007). TMPRSS4 promotes invasion, migration and metastasis of human tumor cells by facilitating an epithelial-mesenchymal transition. *Oncogene, 27,* 2635–2647.

Kalluri, R., & Weinberg, R. A. (2009). The basics of epithelial-mesenchymal transition. *J. Clin. Investig., 119,* 1420–1428.

Karhadkar, S. S., Steven Bova, G., Abdallah, N., Dhara, S., Gardner, D., Maitra, A., et al. (2004). Hedgehog signalling in prostate regeneration, neoplasia and metastasis. *Nature, 431,* 707–712.

Kim, H.-J., Litzenburger, B. C., Cui, X., Delgado, D. A., Grabiner, B. C., Lin, X., et al. (2007). Constitutively active type I insulin-like growth factor receptor causes transformation and xenograft growth of immortalized mammary epithelial cells and is accompanied by an epithelial-to-mesenchymal transition mediated by NF-κB and Snail. *Mol. Cell Biol., 27,* 3165–3175.

Kim, K., Lu, Z., & Hay, E. D. (2002). Direct evidence for a role of β-Catenin/LEF-1 signalling pathway in induction of EMT. *Cell Biol. Int., 26,* 463–476.

Kim, K., Pang, K. M., Evans, M., & Hay, E. D. (2000). Overexpression of β-catenin induces apoptosis independent of its transactivation function with LEF-1 or the involvement of major G1 cell cycle regulators. *Mol. Biol. Cell, 11,* 3509–3523.

Koshikawa, N., Giannelli, G., Cirulli, V., Miyazaki, K., & Quaranta, V. (2000). Role of cell surface metalloprotease MT1-MMP in epithelial cell migration over laminin-5. *J. Cell Biol., 148,* 615–624.

Labbe, E., Letamendia, A., & Attisano, L. (2000). Association of Smads with lymphoid enhancer binding factor 1/T cell-specific factor mediates cooperative signaling by the transforming growth factor-beta and Wnt pathways. *Proc. Natl. Acad. Sci. U.S.A., 97,* 8358–8363.

Langer, E. M., Feng, Y., Zhaoyuan, H., Rauscher, F. J., III, Kroll, K. L., & Longmore, G. D. (2008). Ajuba LIM proteins are Snail/Slug corepressors required for neural crest development in Xenopus. *Dev. Cell, 14,* 424–436.

Larue, L., & Bellacosa, A. (2005). Epithelial-mesenchymal transition in development and cancer: role of phosphatidylinositol 3′ kinase/AKT pathways. *Oncogene, 24,* 7443–7454.

Lee, J. M., Dedhar, S., Kalluri, R., & Thompson, E. W. (2006). The epithelial-mesenchymal transition: new insights in signaling, development, and disease. *J. Cell Biol., 172,* 973–981.

Lee, M.-Y., Chou, C.-Y., Tang, M.-J., & Shen, M.-R. (2008). Epithelial-mesenchymal transition in cervical cancer: correlation with tumor progression, epidermal growth factor receptor overexpression, and Snail up-regulation. *Clin. Cancer Res., 14,* 4743–4750.

Leong, K. G., Niessen, K., Kulic, I., Raouf, A., Eaves, C., Pollet, I., et al. (2007). Jagged1-mediated Notch activation induces epithelial-to-mesenchymal transition through Slug-induced repression of E-cadherin. *J. Exp. Med., 204,* 2935–2948.

Liebner, S., Cattelino, A., Gallini, R., Rudini, N., Iurlaro, M., Piccolo, S., et al. (2004). β-Catenin is required for endothelial-mesenchymal transformation during heart cushion development in the mouse. *J. Cell Biol., 166,* 359–367.

Liem, J., Karel, F., Tremml, G., Roelink, H., & Jessell, T. M. (1995). Dorsal differentiation of neural plate cells induced by BMP-mediated signals from epidermal ectoderm. *Cell, 82,* 969–979.

Lo, H.-W., Hsu, S.-C., Xia, W., Cao, X., Shih, J.-Y., Wei, Y., et al. (2007). Epidermal growth factor receptor cooperates with signal transducer and activator of transcription 3 to induce epithelial-mesenchymal transition in cancer cells via up-regulation of TWIST gene expression. *Cancer Res., 67,* 9066–9076.

Logan, C., Miller, J., Ferkowicz, M., & McClay, D. (1999). Nuclear beta-catenin is required to specify vegetal cell fates in the sea urchin embryo. *Development, 126,* 345–357.

López-Novoa, J. M., & Nieto, M. A. (2009). Inflammation and EMT: an alliance towards organ fibrosis and cancer progression. *EMBO Mol. Med., 1,* 303–314.

Lu, Z., Ghosh, S., Wang, Z., & Hunter, T. (2003). Downregulation of caveolin-1 function by EGF leads to the loss of E-cadherin, increased transcriptional activity of β-catenin, and enhanced tumor cell invasion. *Cancer Cell, 4,* 499–515.

Lyons, J. G., Patel, V., Roue, N. C., Fok, S. Y., Soon, L. L., Halliday, G. M., et al. (2008). Snail up-regulates proin-flammatory mediators and inhibits differentiation in oral keratinocytes. *Cancer Res., 68,* 4525–4530.

Ma, L., Young, J., Prabhala, H., Pan, E., Mestdagh, P., Muth, D., et al. (2010). miR-9, a MYC/MYCN-activated microRNA, regulates E-cadherin and cancer metastasis. *Nat. Cell Biol., 12,* 247–256.

Maeda, M., Johnson, K. R., & Wheelock, M. J. (2005). Cadherin switching: essential for behavioral but not morphological changes during an epithelium-to-mesenchyme transition. *J. Cell Sci., 118,* 873–887.

Mani, S. A., Yang, J., Brooks, M., Schwaninger, G., Zhou, A., Miura, N., et al. (2007). Mesenchyme Forkhead 1 (FOXC2) plays a key role in metastasis and is associated with aggressive basal-like breast cancers. *Proc. Natl. Acad. Sci. U.S.A., 104,* 10069–10074.

Mann, J. R., Backlund, M. G., Buchanan, F. G., Daikoku, T., Holla, V. R., Rosenberg, D. W., et al. (2006). Repression of prostaglandin dehydrogenase by epidermal growth factor and Snail increases prostaglandin E2 and promotes cancer progression. *Cancer Res., 66,* 6649–6656.

Martin-Villar, E., Megias, D., Castel, S., Yurrita, M. M., Vilaro, S., & Quintanilla, M. (2006). Podoplanin binds ERM proteins to activate RhoA and promote epithelial-mesenchymal transition. *J. Cell Sci., 119,* 4541–4553.

McCaffrey, L. M., & Macara, I. G. (2009). Widely conserved signaling pathways in the establishment of cell polarity. *Cold Spring Harbor Perspect. Biol., 1,* a001370.

McCoy, E. L., Iwanaga, R., Jedlicka, P., Abbey, N.-S., Chodosh, L. A., Heichman, K. A., et al. (2009). Six1 expands the mouse mammary epithelial stem/progenitor cell pool and induces mammary tumors that undergo epithelial-mesenchymal transition. *J. Clin. Invest., 119,* 2663–2677.

McGuire, P. G., & Alexander, S. M. (1993). Inhibition of urokinase synthesis and cell surface binding alters the motile behavior of embryonic endocardial-derived mesenchymal cells *in vitro. Development, 118,* 931–939.

Medici, D., & Nawshad, A. (2010). Type I collagen promotes epithelial-mesenchymal transition through ILK-dependent activation of NF-kB and LEF-1. *Matrix Biol., 29,* 161–165.

Medjkane, S., Perez-Sanchez, C., Gaggioli, C., Sahai, E., & Treisman, R. (2009). Myocardin-related transcription factors and SRF are required for cytoskeletal dynamics and experimental metastasis. *Nat. Cell Biol., 11,* 257–268.

Mendez, M.G., Kojima, S.-I., & Goldman, R.D. (2010). Vimentin induces changes in cell shape, motility, and adhesion during the epithelial to mesenchymal transition. *FASEB J., 24,* 1838–1851.

Micalizzi, D. S., Christensen, K. L., Jedlicka, P., Coletta, R. D., Barón, A. E., Harrell, J. C., et al. (2009). The Six1 homeoprotein induces human mammary carcinoma cells to undergo epithelial-mesenchymal transition and metastasis in mice through increasing TGF-B signaling. *J. Clin. Invest., 119,* 2678–2690.

Moody, S. E., Perez, D., Pan, T.-C., Sarkisian, C. J., Portocarrero, C. P., Sterner, C. J., et al. (2005). The transcriptional repressor Snail promotes mammary tumor recurrence. *Cancer Cell, 8,* 197–209.

Morali, O. G., Delmas, V., Moore, R., Jeanney, C., Thiery, J. P., & Larue, L. (2001). IGF-II induces rapid beta-catenin relocation to the nucleus during epithelium to mesenchyme transition. *Oncogene, 20,* 4942–4950.

Moreno-Bueno, G., Portillo, F., & Cano, A. (2008). Transcriptional regulation of cell polarity in EMT and cancer. *Oncogene, 27,* 6958–6969.

Morita, T., Mayanagi, T., & Sobue, K. (2007). Dual roles of myocardin-related transcription factors in epithelial mesenchymal transition via slug induction and actin remodeling. *J. Cell Biol., 179,* 1027–1042.

Muller, W. J., Sinn, E., Pattengale, P. K., Wallace, R., & Leder, P. (1988). Single-step induction of mammary adenocarcinoma in transgenic mice bearing the activated c-neu oncogene. *Cell, 54,* 105–115.

Nagafuchi, A., Shirayoshi, Y., Okazaki, K., Yasuda, K., & Takeichi, M. (1987). Transformation of cell adhesion properties by exogenously introduced E-cadherin cDNA. *Nature, 329,* 341–343.

Nakajima, Y., Yamagishi, T., Hokari, S., & Nakamura, H. (2000). Mechanisms involved in valvuloseptal endocardial cushion formation in early cardiogenesis: roles of transforming growth factor (TGF)-beta; and bone morpho-genetic protein (BMP). *Anat. Rec., 258,* 119–127.

Nawshad, A., & Hay, E. D. (2003). TGFβ3 signaling activates transcription of the LEF1 gene to induce epithelial mesenchymal transformation during mouse palate development. *J. Cell Biol., 163,* 1291–1301.

Nelson, W. J. (2009). Remodeling epithelial cell organization: transitions between front-rear and apical-basal polarity. *Cold Spring Harbor Perspect. Biol., 1,* a000513.

Niessen, K., Fu, Y., Chang, L., Hoodless, P. A., McFadden, D., & Karsan, A. (2008). Slug is a direct Notch target required for initiation of cardiac cushion cellularization. *J. Cell Biol., 182,* 315–325.

Nieto, M. A., Sargent, M. G., Wilkinson, D. G., & Cooke, J. (1994). Control of cell behavior during vertebrate development by Slug, a zinc finger gene. *Science, 264*, 835–839.

Novak, A., Hsu, S.-C., Leung-Hagesteijn, C., Radeva, G., Papkoff, J., Montesano, R., et al. (1998). Cell adhesion and the integrin-linked kinase regulate the LEF-1 and β-catenin signaling pathways. *Proc. Natl. Acad. Sci. U.S.A., 95*, 4374–4379.

Ozdamar, B., Bose, R., Barrios-Rodiles, M., Wang, H.-R., Zhang, Y., & Wrana, J. L. (2005). Regulation of the polarity protein Par6 by TGFβ peceptors controls epithelial cell plasticity. *Science, 307*, 1603–1609.

Peinado, H., del Carmen Iglesias-de la Cruz, M., Olmeda, D., Csiszar, K., Fong, K. S., Vega, S., et al. (2005). A molecular role for lysyl oxidase-like 2 enzyme in Snail regulation and tumor progression. *EMBO J., 24*, 3446–3458.

Peinado, H., Marin, F., Cubillo, E., Stark, H.-J., Fusenig, N., Nieto, M. A., et al. (2004). Snail and E47 repressors of E-cadherin induce distinct invasive and angiogenic properties *in vivo. J. Cell Sci., 117*, 2827–2839.

Peinado, H., Olmeda, D., & Cano, A. (2007). Snail, Zeb and bHLH factors in tumour progression: an alliance against the epithelial phenotype? *Nat. Rev. Cancer, 7*, 415–428.

Perl, A.-K., Wilgenbus, P., Dahl, U., Semb, H., & Christofori, G. (1998). A causal role for E-cadherin in the transition from adenoma to carcinoma. *Nature, 392*, 190–193.

Radice, G. L., Rayburn, H., Matsunami, H., Knudsen, K. A., Takeichi, M., & Hynes, R. O. (1997). Developmental defects in mouse embryos lacking N-cadherin. *Dev. Biol., 181*, 64–78.

Radisky, D. C., Levy, D. D., Littlepage, L. E., Liu, H., Nelson, C. M., Fata, J. E., et al. (2005). Rac1b and reactive oxygen species mediate MMP-3-induced EMT and genomic instability. *Nature, 436*, 123–127.

Roberts, A. B., Tian, F., Byfield, S. D., Stuelten, C., Ooshima, A., Saika, S., et al. (2006). Smad3 is key to TGF-β-mediated epithelial-to-mesenchymal transition, fibrosis, tumor suppression and metastasis. *Cytokine Growth Factor. Rev., 17*, 19–27.

Rørth, P. (2009). Collective cell migration. *Annu. Rev. Cell Dev. Biol., 25*, 407–429.

Ruan, K., Bao, S., & Ouyang, G. (2009). The multifaceted role of periostin in tumorigenesis. *Cell Mol. Life Sci., 66*, 2219–2230.

Sahlgren, C., Gustafsson, M. V., Jin, S., Poellinger, L., & Lendahl, U. (2008). Notch signaling mediates hypoxia-induced tumor cell migration and invasion. *Proc. Natl. Acad. Sci. U.S.A., 105*, 6392–6397.

Sauka-Spengler, T., & Bronner-Fraser, M. (2008). A gene regulatory network orchestrates neural crest formation. *Nat. Rev. Mol. Cell Biol., 9*, 557–568.

Schmidt-Ott, K. M., Lan, D., Hirsh, B. J., & Barasch, J. (2006). Dissecting stages of mesenchymal-to-epithelial conversion during kidney development. *Nephron. Physiol., 104*, 56–60.

Shook, D., & Keller, R. (2003). Mechanisms, mechanics and function of epithelial-mesenchymal transitions in early development. *Mech. Dev., 120*, 1351–1383.

Sobrado, V. R., Moreno-Bueno, G., Cubillo, E., Holt, L. J., Nieto, M. A., Portillo, F., et al. (2009). The class I bHLH factors E2-2A and E2-2B regulate EMT. *J. Cell Sci., 122*, 1014–1024.

Spaderna, S., Schmalhofer, O., Wahlbuhl, M., Dimmler, A., Bauer, K., Sultan, A., et al. (2008). The transcriptional repressor ZEB1 promotes metastasis and loss of cell polarity in cancer. *Cancer Res., 68*, 537–544.

Stemmer, V., de Craene, B., Berx, G., & Behrens, J. (2008). Snail promotes Wnt target gene expression and interacts with beta-catenin. *Oncogene, 27*, 5075–5080.

Stepniak, E., Radice, G. L., & Vasioukhin, V. (2009). Adhesive and signaling functions of cadherins and catenins in vertebrate development. *Cold Spring Harbor Perspect. Biol., 1*, a002949.

Sternlicht, M. D., Lochter, A., Sympson, C. J., Huey, B., Rougier, J. P., Gray, J. W., et al. (1999). The stromal proteinase MMP3/stromelysin-1 promotes mammary carcinogenesis. *Cell, 98*, 137–146.

Strippoli, R., Benedicto, I., Perez Lozano, M. L., Cerezo, A., Lopez-Cabrera, M., & del Pozo, M. A. (2008). Epithelial-to-mesenchymal transition of peritoneal mesothelial cells is regulated by an ERK/NF-kB/Snail1 pathway. *Dis. Model. Mech., 1*, 264–274.

Sullivan, N. J., Sasser, A. K., Axel, A. E., Vesuna, F., Raman, V., Ramirez, N., et al. (2009). Interleukin-6 induces an epithelial-mesenchymal transition phenotype in human breast cancer cells. *Oncogene, 28*, 2940–2947.

Suyama, K., Shapiro, I., Guttman, M., & Hazan, R. B. (2002). A signaling pathway leading to metastasis is controlled by N-cadherin and the FGF receptor. *Cancer Cell, 2*, 301–314.

Tan, C., Costello, P., Sanghera, J., Dominguez, D., Baulida, J., de Herreros, A. G., et al. (2001). Inhibition of integrin linked kinase (ILK) suppresses beta-catenin-Lef/Tcf-dependent transcription and expression of the E-cadherin repressor, snail, in APC−/− human colon carcinoma cells. *Oncogene, 20*, 133–140.

Thiery, J. P. (2002). Epithelial-mesenchymal transitions in tumour progression. *Nat. Rev. Cancer, 2*, 442–454.

Thiery, J. P., Acloque, H., Huang, R. Y. J., & Nieto, M. A. (2009). Epithelial-mesenchymal transitions in development and disease. *Cell, 139*, 871–890.

17

Thuault, S., Tan, E. J., Peinado, H., Cano, A., Heldin, C.-H., & Moustakas, A. (2008). HMGA2 and Smads co-regulate SNAIL1 expression during induction of epithelial-to mesenchymal transition. *J. Biol. Chem., 283,* 33437–33446.

Timmerman, L. A., Grego-Bessa, J., Raya, A., Bertran, E., Perez-Pomares, J. M., Diez, J., et al. (2004). Notch promotes epithelial-mesenchymal transition during cardiac development and oncogenic transformation. *Genes Dev., 18,* 99–115.

Ulrich, F., Krieg, M., Schotz, E.-M., Link, V., Castanon, I., Schnabel, V., et al. (2005). Wnt11 functions in gastrulation by controlling cell cohesion through Rab5c and E-cadherin. *Dev. Cell, 9,* 555–564.

Valles, A., Boyer, B., Tarone, G., & Thiery, J. (1996). Alpha 2 beta 1 integrin is required for the collagen and FGF-1 induced cell dispersion in a rat bladder carcinoma cell line. *Cell Adhes. Commun., 4,* 187–199.

Vincent, T., Neve, E. P. A., Johnson, J. R., Kukalev, A., Rojo, F., Albanell, J., et al. (2009). A SNAIL1-SMAD3/4 transcriptional repressor complex promotes TGF-beta mediated epithelial-mesenchymal transition. *Nat. Cell Biol., 11,* 943–950.

Wanami, L. S., Chen, H.-Y., Peiró, S., García de Herreros, A., & Bachelder, R. E. (2008). Vascular endothelial growth factor-A stimulates Snail expression in breast tumor cells: implications for tumor progression. *Exp. Cell Res., 314,* 2448–2453.

Wang, S.-P., Wang, W.-L., Chang, Y.-L., Wu, C.-T., Chao, Y.-C., Kao, S.-H., et al. (2009b). p53 controls cancer cell invasion by inducing the MDM2-mediated degradation of Slug. *Nat. Cell Biol., 11,* 694–704.

Wang, Y., Zhang, H., Chen, Y., Sun, Y., Yang, F., Yu, W., et al. (2009a). LSD1 is a subunit of the NuRD complex and targets the metastasis programs in breast cancer. *Cell, 138,* 660–672.

Whiteman, E. L., Liu, C. J., Fearon, E. R., & Margolis, B. (2008). The transcription factor snail represses Crumbs3 expression and disrupts apico-basal polarity complexes. *Oncogene, 27,* 3875–3879.

Wicki, A., Lehembre, F., Wick, N., Hantusch, B., Kerjaschki, D., & Christofori, G. (2006). Tumor invasion in the absence of epithelial-mesenchymal transition: podoplanin-mediated remodeling of the actin cytoskeleton. *Cancer Cell, 9,* 261–272.

Wong, A. S. T., & Gumbiner, B. M. (2003). Adhesion-independent mechanism for suppression of tumor cell invasion by E-cadherin. *J. Cell Biol., 161,* 1191–1203.

Woolf, A. S., Kolatsi-Joannou, M., Hardman, P., Andermarcher, E., Moorby, C., Fine, L. G., et al. (1995). Roles of hepatocyte growth factor/scatter factor and the met receptor in the early development of the metanephros. *J. Cell Biol., 128,* 171–184.

Wu, W.-S. (2006). The signaling mechanism of ROS in tumor progression. *Cancer Metastasis Rev., 25,* 695–705.

Wu, Y., Evers, B. M., & Zhou, B. P. (2009). Small C-terminal domain phosphatase enhances Snail activity through dephosphorylation. *J. Biol. Chem., 284,* 640–648.

Xiao, K., Oas, R. G., Chiasson, C. M., & Kowalczyk, A. P. (2007). Role of p120-catenin in cadherin trafficking. *Biochim. Biophys. Acta, 1773,* 8–16.

Yamashita, S., Miyagi, C., Fukada, T., Kagara, N., Che, Y.-S., & Hirano, T. (2004). Zinc transporter LIVI controls epithelial-mesenchymal transition in zebrafish gastrula organizer. *Nature, 429,* 298–302.

Yang, A. D., Camp, E. R., Fan, F., Shen, L., Gray, M. J., Liu, W., et al. (2006). Vascular endothelial growth factor receptor-1 activation mediates epithelial to mesenchymal transition in human pancreatic carcinoma cells. *Cancer Res., 66,* 46–51.

Yang, M.-H., Wu, M.-Z., Chiou, S.-H., Chen, P.-M., Chang, S.-Y., Liu, C.-J., et al. (2008). Direct regulation of TWIST by HIF-1a promotes metastasis. *Nat. Cell Biol., 10,* 295–305.

Yang, Z., Rayala, S., Nguyen, D., Vadlamudi, R. K., Chen, S., & Kumar, R. (2005). Pak1 phosphorylation of Snail, a master regulator of epithelial-to-mesenchyme transition, modulates Snail's subcellular localization and functions. *Cancer Res., 65,* 3179–3184.

Yi, J. Y., Shin, I., & Arteaga, C. L. (2005). Type I transforming growth factor beta receptor binds to and activates phosphatidylinositol 3-kinase. *J. Biol. Chem., 280,* 10870–10876.

Zavadil, J., & Bottinger, E. P. (2005). TGF-β and epithelial-to-mesenchymal transitions. *Oncogene, 24,* 5764–5774.

Zavadil, J., Cermak, L., Soto-Nieves, N., & Bottinger, E. P. (2004). Integration of TGF-β/Smad and Jagged1/Notch signalling in epithelial-to-mesenchymal transition. *EMBO J., 23,* 1155–1165.

Zhou, B. P., Deng, J., Xia, W., Xu, J., Li, Y. M., Gunduz, M., et al. (2004). Dual regulation of Snail by GSK-3β-mediated phosphorylation in control of epithelial-mesenchymal transition. *Nat. Cell Biol., 6,* 931–940.

Zuk, A., & Hay, E. D. (1994). Expression of β1 integrins changes during transformation of avian lens epithelium to mesenchyme in collagen gels. *Dev. Dyn., 201,* 378–393.

CHAPTER 2

Cell-ECM Interactions in Repair and Regeneration

M. Petreaca, M. Martins-Green
Department of Cell Biology and Neuroscience,
University of California, Riverside, CA, USA

INTRODUCTION

For many years, the extracellular matrix (ECM) was thought to serve only as a structural support for tissues. However, as early as 1966, Hauschka and Konigsberg showed that interstitial collagen promoted the conversion of myoblasts to myotubes, and, shortly thereafter, it was shown that both collagen (Wessells and Cohen, 1968) and glycosaminoglycans (Bernfield et al., 1972) play a crucial role in salivary gland morphogenesis. Based upon these findings as well as other pieces of indirect evidence, Hay (1977) put forth the idea that the ECM is an important component in embryonic inductions, a concept that implicated the presence of binding sites (receptors) for specific matrix molecules on the surface of cells. This led to investigation into detailed mechanisms by which extracellular matrix molecules influence cell behavior. Bissell et al. proposed the model of "dynamic reciprocity," in which ECM molecules interact with receptors on the surface of cells that then transmit signals across the cell membrane to molecules in the cytoplasm; these signals initiate a cascade of events through the cytoskeleton into the nucleus, resulting in the expression of specific genes, whose products, in turn, affect the ECM in various ways (Bissell et al., 1982). It has become clear that this concept is essentially correct (Ingber, 1991; Boudreau et al., 1995); cell-ECM interactions can regulate cell adhesion, migration, growth, differentiation, and programmed cell death (also called apoptosis); modulate cytokine and growth factor activities; and activate intracellular signaling.

Much of our current understanding of the molecular basis of cell-ECM interactions in these events comes from studies involving specific mutations, experimental perturbations *in vivo*, and cell/organ cultures. Below, we will first briefly discuss the composition and diversity of some of the better-known ECM molecules and their receptors, and then discuss selected examples that illustrate the dynamics of cell-ECM interactions during wound healing and regeneration, as well as the potential mechanisms involved in the signal transduction pathways initiated by these interactions. Finally, we will discuss the implications of cell-ECM interactions in regenerative medicine.

COMPOSITION AND DIVERSITY OF THE ECM

The ECM is a molecular complex that consists of collagens and other glycoproteins, hyaluronan, proteoglycans, glycosaminoglycans, and elastins; this complex interacts with molecules such as growth factors, cytokines, and matrix-degrading enzymes and their inhibitors. The distribution and organization of these molecules is not static, but rather varies from tissue to tissue and during development from stage to stage (Ffrench-Constant and Hynes, 1989; Laurie et al., 1989; Sanes et al., 1990; Martins-Green and Bissell, 1995; Tsuda et al., 1998; Werb and

Principles of Regenerative Medicine. DOI: 10.1016/B978-0-12-381422-7.10002-1

Chin, 1998; Zhu et al., 2001; Hynes, 2009). The presence of specific matrix molecules in certain tissues or at particular times during development is critical for tissue function, as shown by targeted mutations in matrix molecules in animals and human diseases resulting from similar mutations (Xu et al., 1998; So et al., 2001; White et al., 2008; Bateman et al., 2009). Mesenchymal cells are immersed in an interstitial matrix that confers specific biomechanical and functional properties to connective tissue (Culav et al., 1999; Suki et al., 2005). In contrast, epithelial and endothelial cells contact a specialized matrix, the basement membrane, via their basal surfaces only, conferring mechanical strength and specific physiological properties to the epithelia (Edwards and Streuli, 1995; Fuchs et al., 1997; Dockery et al., 1998; Breitkreutz et al., 2009). This diversity of composition, organization, and distribution of ECM results not only from differential gene expression of the various molecules in specific tissues, but also from the existence of differential splicing and post-translational modifications of those molecules. For example, alternative splicing may change the binding potential of proteins to other matrix molecules (Ffrench-Constant and Hynes, 1989; Chiquet-Ehrismann et al., 1991; Wallner et al., 1998; Ghert et al., 2001; Mostafavi-Pour et al., 2001;) or to their receptors (Aota et al., 1994; Mould et al., 1994; Akiyama et al., 1995; Cox and Huttenlocher, 1998; White et al., 2008), and variations in glycosylation can lead to changes in cell adhesion (Dean et al., 1990; Anderson et al., 1994; Vlodavsky et al., 1996; Cotman et al., 1999; Zhao et al., 2008). In addition, the presence of divalent cations such as Ca^{2+} (Paulsson, 1988; Ekblom et al., 1994; Wess et al., 1998) can affect matrix organization and influence molecular interactions that are important in the way ECM molecules interact with cells (Sjaastad and Nelson, 1997; Kielty et al., 2002).

Growth factors and cytokines interact with the ECM in a variety of ways that allow them to affect each other (Nathan and Sporn, 1991; Adams and Watt, 1993); they can stimulate cells to alter the production of ECM molecules, their inhibitors, and/or their receptors (Streuli et al., 1993; Schuppan et al., 1998; Gratchev et al., 2005; Gharaee-Kermani et al., 2009). TGFβ, for example, upregulates the expression of matrix molecules and of inhibitors of enzymes that degrade ECM molecules, the combination of which increases ECM levels (Wikner et al., 1990; Bonewald, 1999; Kutz et al., 2001; Gharaee-Kermani et al., 2009). The ECM can also influence the local concentration and biological activity of growth factors and cytokines by serving as a reservoir that binds them and protects them from being degraded, by presenting them more efficiently to their receptors or by affecting their synthesis (Roberts et al., 1988; Flaumenhaft and Rifkin, 1992; Lamszus et al., 1996; Miao et al., 1996; Kagami et al., 1998; Schonherr and Hausser, 2000; Miralem et al., 2001; Rahman et al., 2005; Hynes, 2009). Examples of this include the increased production of TNFα by neutrophils after binding to fibronectin (Nathan and Sporn, 1991), the dependence of HGF (hepatocyte growth factor)-mediated hepatocyte proliferation on heparan sulfate proteoglycans (Sakakura et al., 1999), and the increased ability of VEGF (vascular endothelial growth factor) to induce endothelial cell proliferation and migration when bound to fibronectin (Wijelath et al., 2006). Growth factor binding to ECM molecules may also exert an inhibitory effect; SPARC/osteonectin binds multiple growth factors, preventing receptor binding and/or downstream signaling events (Kupprion et al., 1998; Francki et al., 2003). In some cases, only particular forms of these growth factors and cytokines bind to specific ECM molecules, e.g. PDGF (platelet derived growth factor) (LaRochelle et al., 1991; Pollock and Richardson, 1992), VEGF (Poltorak et al., 1997), and the chemokine cIL-8 (previously called cCAF = -chicken chemotactic and angiogenic factor). cIL-8 is a small cytokine that is overexpressed during wound repair and in the stroma of tumors (Martins-Green and Bissell, 1990; Martins-Green et al., 1992), and is secreted as a 9 kDa protein, although it can be processed by plasmin to yield a 7 kDa protein. Both forms of the protein are found in association with interstitial collagen, but only the smaller form binds to laminin or tenascin, while neither form binds to fibronectin, collagen IV, or heparin (Martins-Green and Bissell, 1995; Martins-Green et al., 1996). Importantly, binding of specific forms of these factors to specific ECM molecules can lead to their localization to particular areas of tissues and affect their biological activities.

A feature of ECM/growth factor interactions that has been more recently characterized involves the ability of specific domains of various ECM molecules, including laminin-5, tenascin-C, and decorin, to bind and activate growth factor receptors (Tran et al., 2005). The EGF-like repeats of laminin and tenascin-C bind and activate the EGFR (Panayotou et al., 1989; Swindle et al., 2001; Schenk et al., 2003; Koshikawa et al., 2005; Iyer et al., 2008). In the case of laminin, the EGF-like repeats can interact with EGFR following their release by MMP-mediated proteolysis (Schenk et al., 2003; Koshikawa et al., 2005), whereas tenascin-C repeats are thought to bind EGFR in the context of the full-length protein (Swindle et al., 2001). Decorin also binds and activates EGFR, although this binding occurs via leucine-rich repeats rather than EGF-like repeats (Iozzo et al., 1999; Santra et al., 2002). The ability of ECM molecules to serve as ligands for growth factor receptors may facilitate a stable signaling environment for the associated cells due to the inability of the ligand to either diffuse or be internalized, thus serving as a long-term pro-migratory and/or pro-proliferative signal (Tran et al., 2004, 2005).

RECEPTORS FOR EXTRACELLULAR MATRIX MOLECULES

In order to establish that ECM molecules themselves directly affect cellular behavior, it was important to identify transmembrane receptors for the specific sequences present on these molecules. As early as 1973, it was observed that, during salivary gland morphogenesis near the sites of glycosaminoglycan deposition, the intracellular microfilaments contracted (Bernfield et al., 1973). These investigators proposed that the ECM could "be involved in regulating microfilament function," suggesting that these molecules can specifically interact with cell surface receptors. It was subsequently shown that various ECM molecules contain specific amino acid motifs that allow them to bind directly to cell surface receptors (Humphries, 1991; Hynes, 1992; Gullberg and Ekblom, 1995). The best characterized motif is the tripeptide RGD, first found in fibronectin (Pierschbacher and Ruoslahti, 1984; Yamada and Kennedy, 1984). Peptides containing this amino acid sequence promote adhesion of cells and inhibit the adhesive properties of fibronectin. This and other amino acid adhesive motifs have been found in laminin, entactin, thrombin, tenascin, fibrinogen, vitronectin, collagens I and VI, bone sialoprotein, and osteopontin (Humphries, 1991).

Integrins, a family of heterodimeric transmembrane proteins composed of α and β subunits, were the first ECM receptors to be identified (Hynes, 1987). At least 18 α and 8 β subunits have been identified so far; they pair with each other in a variety of combinations, giving rise to a large family that recognizes specific sequences on the ECM molecules (Fig. 2.1). Some integrin receptors are very specific, whereas others bind several different epitopes; these may be on the same or different ECM molecules (Fig. 2.1), thus facilitating plasticity and redundancy in specific systems (Hynes, 1992; Desgrosellier and Cheresh, 2010). Although the α and β subunits of integrins are unrelated, there is 30–45% identity within each subunit with the highest divergence in the intracellular domain of the α subunit (Takada et al., 2007). All but one of these subunits (β_4) have large extracellular domains and very small intracellular domains (Wegener and Campbell, 2008). It is important to note that, despite the relatively short length of their cytoplasmic domains, the β subunits remain able to interact with an array of signaling proteins critical in integrin-associated signal transduction (Wegener and Campbell, 2008). The extracellular domain of the α subunits contains four regions that serve as binding sites for divalent cations, which appear to augment ligand binding and increase the strength of the ligand-integrin interactions (Gailit and Ruoslahti, 1988; Pujades et al., 1997; Leitinger et al., 2000).

Although not as extensively studied as the integrins, it has been found that proteoglycans can also serve as receptors for ECM molecules. Members of the syndecan family, CD44, and RHAMM (receptor for hyaluronate mediated motility) are proteoglycan receptors for ECM

FIGURE 2.1

Representative members of the integrin family of ECM receptors and their respective ligands. These heterodimeric receptors are composed of one α and one β subunit, and are capable of binding a variety of ligands, including Ig superfamily cell adhesion molecules, complement factors, and clotting factors in addition to ECM molecules. Cell-cell adhesion is largely mediated through integrin heterodimers containing the β_2 subunits, while cell-matrix adhesion is mediated primarily via integrin heterodimers containing the β_1 and β_3 subunits. In general, the β_1 integrins interact with ligands found in the connective tissue matrix, including laminin, fibronectin, and collagen, whereas the β_3 integrins interact with vascular ligands, including thrombospondin, vitronectin, fibrinogen, and von Willebrand factor.

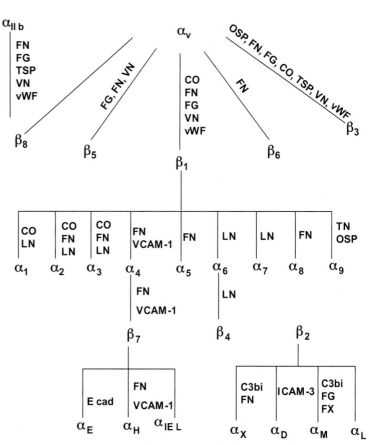

Abbreviations: CO, collagens; C3bi, complement component; FG, fibrinogen; FN, fibronectin; FX, Factor X; ICAM-1, intercellular adhesion molecule-1; ICAM-2, intercellular adhesion molecule-2; ICAM-3, intercellular adhesion molecule-3; LN, laminin; OSP, osteopontin; TN, tenascin; TSP, thrombospondin; VCAM-1, vascular cell adhesion molecule-1; VN, vitronectin; vWF, von Willebrand factor.

molecules (Liu et al., 1998; Slevin et al., 2007; Okina et al., 2009). Syndecans interact with the matrix via chondroitin-, dermatan-, and heparan-sulfate glycosaminoglycans, whose composition varies based upon the specific syndecan family member and the type of tissue in which it is expressed; the differential glycosaminoglycan modifications can alter the binding capacity of particular ligands (Carey, 1997; Granes et al., 1999; Saoncella et al., 1999; Okina et al., 2009). Syndecan proteoglycans also associate with the cytoskeleton, promoting intracellular signaling events and cytoskeletal reorganization through activation of Rho GTPases (Granes et al., 1999; Saoncella et al., 1999). The CD44 receptor also carries chondroitin sulfate and heparan sulfate chains on its extracellular domain (Brown et al., 1991; Ehnis et al., 1996; Tuhkanen et al., 1997), and undergoes tissue-specific splicing and glycosylation to yield multiple isoforms; these may play roles in cell adhesion as well as in ligand binding (Miyake et al., 1990; Peach et al., 1993; Bajorath et al., 1998). CD44 interacts with hyaluronan through an extracellular "link" module; this interaction is thought to switch the link module binding site from a low-affinity conformation to a high-affinity conformation (Banerji et al., 2007). Although hyaluronan is its primary ligand, CD44 interacts with other extracellular matrix molecules, including fibronectin, laminin, collagen IV, and collagen XIV (Jalkanen and Jalkanen, 1992; Ishii et al., 1993, 1994; Ehnis et al., 1996; Mythreye and Blobe, 2009). In contrast to the transmembrane CD44 and syndecan proteoglycans, RHAMM, a cell-associated, non-integral proteoglycan, can also bind extracellular matrix proteins and induce signaling (Hall et al., 1994; Savani et al., 2001). Because RHAMM appears to lack a transmembrane domain, it likely activates intracellular signaling through indirect mechanisms via interactions with transmembrane ECM receptors such as integrins or CD44 (Hamilton et al., 2007; Maxwell et al., 2008).

Cell surface receptors other than integrins or proteoglycans have also been identified as receptors for ECM molecules. A non-integrin splice variant of β-galactosidase, elastin-binding protein (EBP), recognizes the GXXPG sequence of elastin, laminin, fibrillin, and peptides derived from these ECM molecules (Moroy et al., 2009). Together with neuraminidase 1 and cathepsin A, EBP forms a complex known as the elastin/laminin receptor (ELR), which is necessary for elastin deposition (Antonicelli et al., 2009). ELR has been implicated in the signaling downstream of elastin and laminin during mechanotransduction (Spofford and Chilian, 2003). In addition, the neuraminidase subunit desialylates cell-surface growth factor receptors, preventing growth factor-receptor binding and downstream signaling, thereby decreasing cell proliferation (Hinek et al., 2008). Under proteolytic conditions, including wound healing and inflammation, elastin is cleaved to form short peptides. ELR binding to these elastin-derived peptide ligands induces the migration and/or proliferation of keratinocytes, fibroblasts, endothelial cells, and monocytes (Antonicelli et al., 2009). The proliferative and migratory effects of elastin-derived peptides may result from signaling downstream of neuraminidase 1, which can promote ERK1/2 activation (Duca et al., 2007). Another non-integrin receptor, CD36, better known for its function as a scavenger receptor for long chain fatty acids and oxidized LDL, binds thrombospondin, collagen I, and collagen IV (Asch et al., 1993; Febbraio and Silverstein, 2007). CD36-thrombospondin binding activates a variety of signal transduction molecules, ultimately leading to inhibition of angiogenesis via increased endothelial cell apoptosis (Jimenez et al., 2000, 2001; Isenberg et al., 2005; Silverstein and Febbraio, 2007). Furthermore, alternative splice variants of tenascin-C interact with cell-surface annexin II, which may mediate the cellular responses to this particular form (Chung and Erickson, 1994). In addition, ECM molecules have been shown to bind and activate tyrosine kinase receptors, including the EGFR via EGF-like domains (see above) as well as the discoidin domain receptors DDR1 and DDR2. DDR1 and DDR2 function as receptors for various collagens and mediate cell adhesion and signaling events (Vogel et al., 1997; Faraci et al., 2003; Ferri et al., 2004; Leitinger and Hohenester, 2007). The DDR receptors have also been implicated in ECM remodeling, as their overexpression decreases the expression of multiple matrix molecules and their receptors, including collagen, syndecan-1, and integrin α_3, while simultaneously increasing MMP activity (Ferri et al., 2004). Inhibition of signaling by expressing a kinase-dead DDR2 or by treating cells with DDR1/DDR2 soluble extracellular domains resulted in decreased collagen deposition and altered fibrillogenesis, further supporting a role for these receptors in matrix remodeling (Blissett et al., 2009; Flynn et al., 2010).

SIGNAL TRANSDUCTION EVENTS DURING CELL-ECM INTERACTIONS

The interactions between ECM molecules and their receptors as described above can transmit signals directly or indirectly to signaling molecules within the cell, leading to a cascade of events and the coordinated expression of a variety of genes involved in cell adhesion, migration, proliferation, differentiation, and death (Fig. 2.2). There is increasing evidence that cell-ECM interactions, especially through integrins, activate a variety of signaling pathways that can be linked to those specific functions. Some of the signaling events important in these cellular processes are discussed below.

Adhesion and migration

When discussing the importance of cell-matrix adhesions in adhesion and migration, it is important to recall that certain receptors for extracellular matrix molecules, such as the integrins, can participate in both traditional "outside-in" signaling, leading to the activation of intracellular signaling, and "inside-out" signaling, in which intracellular signaling activates the

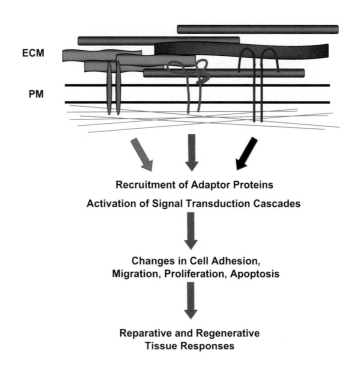

 a/b integrin heterodimer   Non-integrin ECM receptor   Growth factor receptor   Actin

FIGURE 2.2

Schematic diagram of cell-ECM interactions present during the healing and regenerative responses. Such interactions between the ECM receptors and their respective ligands initiate signal transduction cascades culminating in a variety of cellular events important in repair and regeneration, including changes in cellular adhesion and migration and altered rates of proliferation and apoptosis. The presence and/or extent of such changes may influence the balance of repair and regenerative responses to favor one outcome over another; thus, interventions that alter ECM signaling events may shift this balance to favor tissue regeneration and thus decrease scarring.

integrin by increasing its affinity for an ECM molecule. This is further complicated by the fact that integrin activation and ligand binding can, in turn, initiate outside-in signaling. In the following section, the signaling events refer to outside-in signaling unless otherwise specified.

It is now well established that, upon ligand binding, integrins can directly induce biochemical signals inside cells (Takada et al., 2007; Hynes, 2009). The cytoplasmic domain of integrins interacts with the cytoskeleton indirectly through a variety of signaling proteins; ECM signaling through integrins can thus induce changes in cell shape and lead to growth, migration, and/or differentiation (Delon and Brown, 2007; Hynes, 2009). For example, cell migration is promoted when fibronectin binds simultaneously to integrins through its cell-binding domain and to proteoglycan receptors through its heparin-binding domain (Mercurius and Morla, 2001). These receptors interact and colocalize in areas of adhesion where microfilaments associate with the β_1 subunit of the integrin receptor via structural proteins such as talin and α-actinin present in the actin cytoskeleton of the focal adhesions. The cytoplasmic domain of the β_1 subunit also interacts with talin and paxillin, which, in turn, interact with focal adhesion tyrosine kinase (FAK), linking the integrin to this intracellular signaling molecule (Mitra and Schlaepfer, 2006). When the integrin heterodimer interacts with its matrix ligand, FAK becomes autophosphorylated on tyrosine 397, a process that appears to require mechanosensation, as this residue is not phosphorylated when integrins interact with soluble ligand or are clustered via antibodies (Shi and Boettiger, 2003). FAK PY397 then subsequently serves as the binding site for the SH2 domain of the non-receptor tyrosine kinase c-Src, which then phosphorylates FAK at additional sites to enhance FAK activity (Mitra and Schlaepfer, 2006). The FAK/c-Src complex also phosphorylates other components of the focal adhesion plaques, including paxillin, tensin, vinculin, and the protein p130^{cas} (Sefton et al., 1981; Schaller et al., 1999; Volberg et al., 2001; Mitra and Schlaepfer, 2006).

Paxillin has been implicated in the regulation of integrin-mediated signaling events and motility; paxillin-deficient fibroblasts exhibit reduced phosphorylation of signaling molecules downstream of integrin ligation, with a concomitant reduction in cell motility (Lo, 2004). Paxillin phosphorylated via the FAK/c-Src complex interacts with the SH2 domain of the CrkII/DOCK180/ELMO complex; DOCK180, a guanine nucleotide exchange factor (GEF) for Rac1, then activates Rac1 and promotes cell migration (Deakin and Turner, 2008). Phospho-paxillin also appears to activate p190RhoGAP, leading to localized inhibition of RhoA activity (Tsubouchi et al., 2002). The combination of enhanced Rac1 activity and decreased RhoA activity is thought to decrease cell adhesion and promote protrusion formation, thus facilitating cell migration. The specific role of tensin family members in the process of adhesion/de-adhesion during migration is not well understood. Growth factor-induced signaling appears to alter expression of two tensin family members, switching expression from one tensin family member, tensin 3, to another, cten (Katz et al., 2007). Both tensin family members bind to integrin β_1, but, while tensin 3 binds and caps the barbed ends of the actin filament, cten is unable to bind actin (Lo, 2004). As such, switching from tensin 3 to cten promotes actin filament disassembly, which facilitates cell migration (Katz et al., 2007).

Phosphorylation and activation of p130^{cas} promotes its interaction with the adaptor molecules Crk and Nck, which form a scaffold for localized activation of Rac-GTPase and the MAP/JNK kinase pathways, thus facilitating lamellipodia formation and migration (Schlaepfer et al., 1997; Sharma and Mayer, 2008). In addition, it has also been shown that c-Src phosphorylates FAK on tyrosine 925, which serves as a site for binding of Grb2/Sos complex with subsequent activation of Ras and the MAP kinase cascade, which may also be involved in adhesion/de-adhesion and migration (Dedhar, 1999; Ly and Corbett, 2005).

Although FAK and c-Src are best known for their roles in outside-in signaling, as described above, these kinases are also involved in inside-out signaling. FAK promotes integrin activation, cell adhesion to fibronectin, and strengthening of focal adhesions (Michael et al., 2009). These effects appear to require Src binding and/or activity, as a Y397F mutation that prevents FAK autophosphorylation and Src binding at this site also prevents FAK-mediated adhesion (Michael et al., 2009). FAK-induced integrin binding to ECM molecules can then initiate outside-in signaling, leading to more FAK activation, FAK-Src interaction, and downstream signaling that promotes de-adhesion and migration. This suggests a cycle of FAK and Src activity, in which they initially promote de-adhesion and migration, followed by the formation of new adhesions at the leading edge. In support of this FAK/Src cycle of activity, recent data showed the movement of active Src from the focal adhesions to the membrane ruffles at the leading edge during cell migration (Hamadi et al., 2009).

Non-integrin ECM receptors, including proteoglycan receptors, the elastin-laminin receptor, the EGFR, and DDR1, also participate in cell adhesion and migration, although the signaling downstream of receptor-ligand binding is less well known. Syndecans can cooperate with integrin heterodimers to mediate cell adhesion to vitronectin and laminin and induce cell migration (Morgan et al., 2007). Syndecan 4 promotes Rac1 activation in a PKC-dependent manner, and is necessary for Rac1 localization to the leading edge and persistence of directional cell migration (Bass et al., 2007). Syndecan-4-induced PKC, along with integrin-induced signaling, also activates a Rho GAP at the leading edge, inactivating RhoA and further promoting cell migration (Bass et al., 2008). Hyaluronan binding to additional proteoglycan receptors, CD44 and/or RHAMM, promotes endothelial cell adhesion and migration (Savani et al., 2001; Slevin et al., 2007; Gao et al., 2008), while hyaluronan binding to RHAMM induces smooth muscle cell migration in a PI3K- and Rac-dependent manner (Goueffic et al., 2006). RHAMM also participates in fibroblast migration, as shown by defective migration in RHAMM$^{-/-}$ fibroblasts; in these cells, RHAMM is required for surface localization of CD44 and downstream activation of ERK1/2 (Tolg et al., 2006). The elastin-binding protein (EBP) subunit of the elastin-laminin receptor initially promotes cell adhesion by promoting elastin deposition into the extracellular matrix (Duca et al., 2004). However, interaction of the EBP with elastin-derived peptides can promote migration of multiple cell types, including monocytes, keratinocytes, fibroblasts, smooth muscle cells, and endothelial cells (Duca et al., 2004). EBP-elastin peptide binding stimulates cGMP production and activation of protein kinase G (PKG), which appears to be involved in elastin peptide-induced migration of monocytes and macrophages (Kamisato et al., 1997; Uemura and Okamoto, 1997). Activation of EGFR via the EGF-like repeats in tenascin-C decreases fibroblast adhesion to fibronectin, suggesting a role in cell migration (Prieto et al., 1992). Indeed, the EGF-like repeats of thrombospondin promote epithelial cell migration through the activation of EGFR, and may involve the downstream activation of PLCγ (Liu et al., 2009). In contrast, collagen binding to DDR1 appears to inhibit both cell adhesion and migration in kidney epithelial cells; over-expression of DDR1 decreased cell adhesion and migration on collagen, whereas over-expression of a dominant negative version promoted both adhesion and migration (Wang et al., 2006). These effects may be cell type-specific; in fibroblasts, DDR1 appears to promote fibroblast migration by binding to non-muscle myosin IIA heavy chain and promoting myosin filament assembly (Huang et al., 2009).

Proliferation and survival

Extracellular matrix interaction with its receptors can promote cell proliferation and survival, often in conjunction with growth factors or cytokine receptors. Such cooperative effects may occur in a direct manner, as in situations in which the EGF-like repeats of ECM molecules bind and activate growth factor receptors, leading to cell proliferation (Swindle et al., 2001; Tran et al., 2004). However, more is known regarding the importance of indirect cooperative effects, particularly those involved in the anchorage-dependence of cell growth. Anchorage is required for cells to enter S phase; even in the presence of growth factors, cells will not enter the DNA synthesis phase without being anchored to a substrate (Giancotti, 1997; Mainiero et al., 1997; Murgia et al., 1998). In addition, cell detachment from the matrix often promotes apoptosis, a process known as anoikis (Reddig and Juliano, 2005). Thus, adhesion of cells to ECM molecules plays a very important role in regulating cell survival and proliferation. The loss of integrin-mediated adhesion induces the movement of the pro-apoptotic protein Bax from the cytoplasm to the mitochondria, promoting apoptosis (Gilmore et al., 2000). Cell transfection with dominant-negative FAK also promoted Bax translocation and apoptosis, which was, in turn, blocked by overexpression of the p110 subunit of PI3K or Src (Gilmore et al., 2000). These results suggest that, following cell detachment from ECM, the loss of Fak-mediated stimulation of PI3K and Src results in Bax activation and apoptosis, and that Fak activation may promote cell survival by repressing anoikis (Gilmore et al., 2000). Indeed, Fak activation prevents anoikis and promotes survival in fibroblasts and epithelial cells. In fibroblasts, the pro-survival signals downstream of Fak involve p130CAS activation, as dominant negative p130CAS prevents Fak-mediated survival; in epithelial cells, these pro-survival signals involve the activation of Src kinases rather than p130CAS, suggesting that the mechanisms involved in Fak-induced survival are cell-type specific (Zouq et al., 2009). As mentioned above, Rac1 is activated downstream of Fak-induced p130CAS stimulation (Sharma and Mayer, 2008); Rac1 can then promote the activation of the JNK pathway, which also increases cell survival (Almeida et al., 2000). The importance of the Rac/JNK pathway in integrin-mediated proliferation is underscored by studies involving a β_1 integrin cytoplasmic domain mutant, which decreased the activation of the Rac/JNK pathway and also negatively affected fibroblast proliferation and survival; these effects were rescued by the expression of constitutively active

Rac1 (Pozzi et al., 1998). In intestinal epithelial cells, Fak/Src complexes activated by integrin-ECM binding also activate PI3K, leading to the activation of Akt; Akt then alters the ratios of pro- and anti-apoptotic Bcl-2 family members, increasing the levels of the anti-apoptotic Bcl-xl and Mcl-1 and decreasing the levels of pro-apoptotic Bax and Bak, thereby promoting cell survival (Bouchard et al., 2008). Akt can also be activated by another integrin-induced kinase, integrin-linked kinase (ILK), further promoting cell survival (Troussard et al., 2003).

Integrin-ECM binding can promote cell proliferation through multiple signaling pathways, many of which involve the activation of MAP kinase pathways (Guo and Giancotti, 2004). Multiple studies in which integrins are either inhibited or deficient demonstrate that integrin signaling is critical for cell proliferation. For example, studies of mice lacking the $\alpha_1\beta_1$ integrin, which is a primary collagen receptor, showed that the fibroblasts of these mice have reduced proliferation even though they attach normally (Faraldo et al., 2001). In addition, mammary epithelial cells overexpressing a dominant negative β_1 integrin subunit exhibit reduced proliferation (Hirsch et al., 2002). Integrin-ECM interactions can activate Rac1 through the Fak/Src/p130CAS/DOCK180 pathway and thus induce JNK activation, which can stimulate cyclin D expression and cell division (Assoian and Klein, 2008). Src can also activate Rac1 and its downstream signaling through a separate pathway, in which Src-induced PI3K/Akt activates a Rac GEF, Asef-1 (Mainiero et al., 1997; Hirsch et al., 2002; Guo and Giancotti, 2004; Bristow et al., 2009). Other MAP kinases, ERK1/2, can be activated through integrin ligation. Integrin ligation and activation of Src family kinases lead to the recruitment of Shc, an adaptor protein that binds Grb2/Sos and thus activates the Ras/ERK cascade, leading to the phosphorylation of the Elk-1 transcription factor and the expression of early response genes involved in cell cycle progression (Mainiero et al., 1997; Aplin and Juliano, 1999; Roovers et al., 1999; Aplin et al., 2001). Signaling downstream of cell-ECM binding also promotes degradation of cell cycle inhibitors, thus facilitating cell proliferation; indeed, fibronectin-mediated adhesion leads to the degradation of p21 in a Rac1- and Cdc42-dependent manner (Bao et al., 2002).

Integrin-ECM binding also cooperates with growth factor receptor signaling to stimulate cell proliferation (Hynes, 2009). Individual growth factors may require specific integrin-matrix interactions to mediate downstream signaling; for example, bFGF-induced angiogenesis requires $\alpha_v\beta_3$, whereas VEGF-induced angiogenesis requires $\alpha_v\beta_5$ (Hood et al., 2003). Integrins and growth factors can increase the activation of phosphatidylinositol phosphate kinases, thus increasing the levels of phosphatidylinositol bis-phosphate (PIP$_2$). PIP$_2$ then serves as substrate for phospholipase C$_g$ (PLC$_g$), which is activated by growth factors as well as by integrin ligation, ultimately leading to the activation of protein kinase C (PKC) and the promotion of cell proliferation (Cybulsky et al., 1993; Khwaja et al., 1997). Integrin binding to substrate is important for the efficient and prolonged activation of MAPK by growth factors, promoting cyclin D expression and passage through the cell cycle; this may explain, in part, the anchorage dependence of growth factor-mediated proliferation (Miyamoto et al., 1996; Roovers et al., 1999; Assoian and Klein, 2008). Cell adhesion to fibronectin promotes cell proliferation by inducing the autophosphorylation and activation of EGFR (Bill et al., 2004). Inhibition of EGFR blocked some the fibronectin-induced signaling, including the phosphorylation and activation of Shc, ERK2, and Akt, but had no effect on the fibronectin-induced phosphorylation of Fak, Src, or PKC (Bill et al., 2004). By itself, fibronectin could induce Rb phosphorylation, Cdk2 activation, and cyclin D expression, but required EGF to promote cyclin A expression and p27 degradation; as such, signaling induced by both fibronectin and EGF were required to induce cell proliferation (Bill et al., 2004).

There are multiple mechanisms that are involved in cell-matrix adhesion and growth factor receptor signaling, including a direct interaction between integrins and growth factors or growth factor receptors, altered regulation of the integrin or growth factor receptors, and matrix binding to growth factors or growth factor receptors (Desgrosellier and Cheresh, 2010). VEGF appears to bind $\alpha_9\beta_1$ integrin directly, and both $\alpha_9\beta_1$ integrin and VEGFR2 are required for downstream phosphorylation of paxillin and ERK, suggesting that this unique growth factor-integrin interaction may be involved in VEGF-induced proliferation; indeed, inhibition of $\alpha_9\beta_1$ integrin blocked VEGF-induced angiogenesis *in vivo* (Vlahakis et al., 2007). Other growth factors are also able to bind integrins directly; for example, IGF-1 and FGF-2 both interact with $\alpha_v\beta_3$ integrin (Mori et al., 2008; Saegusa et al., 2009). IGF-1 and FGF-2 mutations that prevented binding to $\alpha_v\beta_3$ without impairing their interactions with IGF-1R or FGFR, respectively, abolished their mitogenic and migratory effects (Mori et al., 2008; Saegusa et al., 2009). These results suggest that growth factor-integrin binding may play a critical role in growth factor-mediated cell signaling. Both PDGFRβ and VEGFR2 physically interact with integrin subunits (Borges et al., 2000), and concomitant integrin-mediated cell adhesion further increases both receptor activation and mitogenicity (Schneller et al., 1997; Soldi et al., 1999). Upon cell detachment from the matrix, PDGFR autophosphorylation is decreased and the receptor is internalized and degraded, suggesting a role for integrin-matrix binding in both activation and localization of the receptor (Baron and Schwartz, 2000). In the case of VEGFR2, interaction with integrins participates in VEGF induction of the Ras/ERK pathway in endothelial cells, which is dependent upon both Fak and integrin $\alpha_v\beta_5$ (Hood et al., 2003). EGFR also interacts with integrins following cell adhesion to the matrix in a complex that also contains p130CAS and Src, leading to EGFR phosphorylation (Moro et al., 2002), providing a potential mechanism whereby integrin and EGFR signaling synergize to promote cell proliferation (Bill et al., 2004). In addition, growth factor signaling can activate integrins and thus promote cell-matrix interactions; in the case of $\alpha_v\beta_3$, VEGF promotes $\beta3$ phosphorylation and interaction with VEGFR2 in an Src-dependent manner, increasing phosphorylation of FAK and activation of p38, and culminating in cell adhesion, migration, and proliferation on vitronectin (Masson-Gadais et al., 2003; Mahabeleshwar et al., 2007). $\alpha_v\beta_3$ signaling promotes phosphorylation and activation of VEGFR2 in a reciprocal manner (Mahabeleshwar et al., 2007). Similar results have been shown for EGF family receptors in the activation of integrins in breast carcinoma cells (Adelsman et al., 1999).

Multiple ECM molecules are able to bind to either growth factors or their receptors to regulate their activity (Hynes, 2009). IGF-1 interacts with vitronectin, promoting its signaling through the IGFR (Upton et al., 1999). Both fibronectin and vitronectin bind HGF and induce the formation of integrin-HGF receptor (Met) complexes, promote Met phosphorylation, and increase cell proliferation in an Erk-dependent manner (Rahman et al., 2005). VEGF binding to fibronectin greatly increases its ability to stimulate activation of VEGFR2 and Erk in endothelial cells (Wijelath et al., 2006), and VEGF-B interaction with tenascin-X enhances VEGFR1 activation and endothelial cell proliferation (Ikuta et al., 2001). In addition, VEGF interaction with collagen promotes VEGFR2 interaction with integrin $\beta1$ and increases the duration of VEGFR2 activity (Chen et al., 2010). Several growth factors and cytokines can interact with heparan sulfate proteoglycans, which can either sequester these factors within the matrix, such that they are released upon matrix degradation, or can present them to their receptors. Such proteoglycans have been shown to interact with FGF-2, FGF-10, PDGF, VEGF, and IL-2 through heparan sulfate chains (Whitelock et al., 2008). Binding of FGF family members to heparan sulfate moieties serves to retain FGFs near the source of secretion, protect them from proteolysis, and facilitate FGF binding to FGFR (Moscatelli, 1987; Saksela et al., 1988; Yayon et al., 1991). Interaction with proteoglycans appears to stabilize dimeric and oligomeric forms of FGF and effectively "present" them to the FGFR, promoting receptor activation and cell proliferation (Ornitz et al., 1992; Venkataraman et al., 1996). In addition, heparan sulfate proteoglycans may interact with FGFR directly, physically linking the FGFR to its ligand (Kan et al., 1993). Binding of VEGF to heparan sulfate proteoglycans increases binding to VEGFR, specifically VEGFR2, and promotes cell proliferation (Gitay-Goren et al., 1992; Ono et al., 1999; Whitelock et al., 2008). In addition to growth factor-ECM binding, many growth factor receptors can also interact with ECM molecules, which then promote receptor activation and downstream signaling. For example, laminin and tenascin-C can bind and activate EGFR and associated downstream signaling through their EGF-like domains (Panayotou et al., 1989; Swindle et al., 2001; Iyer et al., 2008). The proteoglycan versican also possesses EGF-like repeats that can bind to EGFR and promote proliferation in fibroblasts (Zhang et al., 1998). Decorin, another matrix proteoglycan, can also bind the EGFR, although the binding occurs through a series of leucine-rich-repeats rather than through EGF-like domains (Santra et al., 2002). In trophoblast cells, decorin binds to VEGFR2 and IGF-1R as well as EGFR, and appears to inhibit migration induced by IGF-1 and proliferation induced by VEGF and EGF (Iacob et al., 2008); decorin may exert these effects through internalization and/or degradation of the receptors, which has been shown for EGFR (Zhu et al., 2005).

Non-integrin ECM receptors have also been implicated in cell proliferation and survival. Endothelial cell proliferation induced by hyaluronan fragments is mediated by RHAMM, and is associated with the phosphorylation of paxillin and ERK (Lokeshwar and Selzer, 2000; Gao et al., 2008). Hyaluronan-induced proliferation in fibroblasts appears to be mediated, at least in part, by CD44, through the downstream activation of Erk and Akt (David-Raoudi et al., 2008). Similarly, binding of low-molecular-weight hyaluronan fragments to CD44 promotes smooth muscle cell proliferation via Erk-mediated increases in cyclin D1 expression (Kothapalli et al., 2008). The binding of elastin-derived peptides to the ELR can promote smooth muscle cell proliferation through the activation of multiple signaling pathways that culminate in the activation of the MAPK cascade and upregulation of cyclins A, E, D1, cdk 2, and cdk4 (Mochizuki et al., 2002). In addition, DDR2, a non-integrin collagen receptor, increases proliferation of fibroblasts and chondrocytes (Labrador et al., 2001; Olaso et al., 2002).

Differentiation

Interaction of cells with ECM molecules, hormones, and growth factors is required to activate genes that are specific for differentiation. In endothelial cells, the interaction of $\alpha_2\beta_1$ integrin with laminin leads to formation of capillary-type structures, whereas the interaction of $\alpha_5\beta_1$ in the same cells with fibronectin results in proliferation (Wary et al., 1998). Similar observations have been made with primary bronchial epithelial cells when they are cultured on collagen matrices (Moghal and Neel, 1998). The formation of endothelial capillary-like tubes also relies upon additional signaling pathways, such as occur upon activation of integrin-linked kinase (ILK); overexpression of this kinase can rescue tube formation in the absence of ECM molecules, while expression of dominant negative ILK prevents tube formation in the presence of ECM and VEGF (Cho et al., 2005; Watanabe et al., 2005). Following the formation of nascent vessels, they must be stabilized by the recruitment and differentiation of pericytes, smooth muscle cells that increase endothelial barrier function and participate in the deposition of a new basement membrane (Hirschi and D'Amore, 1996; Allt and Lawrenson, 2001; Conway et al., 2001; Hellstrom et al., 2001). A conditional β_1 integrin knockout in mural cells decreased pericyte spreading along the endothelium; decreased their expression of smoothelin, a late differentiation marker; and impaired normal pericyte ability to regulate vascular maturation and barrier function (Abraham et al., 2008). As such, β_1 integrin-mediated interactions with ECM appear critical for adhesion and differentiation of these cells.

Other differentiated phenotypes likewise require integrin-mediated signaling events. TGF-β1-mediated myofibroblast differentiation, an event important in both wound healing and liver regeneration, requires adhesion to the EDA domain of fibronectin, as well as the activation of FAK and its associated signaling pathways (Serini et al., 1998; Thannickal et al., 2003; Lygoe et al., 2004; White et al., 2008). Differentiation of keratinocytes is carefully regulated by multiple cell-matrix and cell-cell interactions. Integrin $\alpha_6\beta_4$ interaction with laminin-332, a component of the basement membrane, appears to prevent keratinocyte differentiation, as shown by the enhanced expression of differentiation markers in the epidermis of α_6-deficient

29

Following leukocyte extravasation from the vasculature, the leukocytes are directed to the site of injury by the chemokines, which form relatively stable gradients through interactions with endothelial cell surface proteins and extracellular matrix molecules, thereby promoting directional cell migration through the provisional matrix (Gillitzer and Goebeler, 2001; Patel et al., 2001; Proudfoot et al., 2003). The fibrin-fibronectin meshwork of the provisional matrix serves as substrate for the migration of leukocytes and later keratinocytes during the early stages of healing when inflammation and re-epithelialization are occurring. Leukocyte interactions with ECM molecules via integrin receptors affect many of the functions of these cells, in particular those that lead to cell adhesion and migration or to production of inflammatory mediators. As mentioned above, neutrophils interact with fibronectin and vitronectin, which are present in the provisional matrix. Several types of inflammatory cells interact with fibrinogen, also a major component of the provisional matrix, through integrins $\alpha_M\beta_2$ and $\alpha_X\beta_2$ (Ugarova and Yakubenko, 2001). Integrin $\alpha_M\beta_2$ also binds to urokinase plasminogen activator (uPA), leading to cell adhesion on uPA, migration toward uPA, uPA-mediated plasmin activation, and subsequent degradation of the fibrin clot (Pluskota et al., 2004, 2003). In addition, this integrin also interacts with thrombospondin 4, promoting cell adhesion and migration; signaling induced by this interaction promotes neutrophil secretion of IL-8, activation of the p38 and JNK MAPK pathways, and the respiratory burst downstream of p38 (Pluskota et al., 2005). Monocytes binding to SPARC, which is present in the provisional matrix in small amounts, promotes their production of MMP-1 and MMP-9, which then degrade matrix molecules; this could facilitate both migration through the matrix and matrix turnover (Shankavaram et al., 1997). Another minor component of this matrix is tenascin-C; integrin $\alpha_5\beta_1$-mediated interaction of neutrophils and monocytes to tenascin-C inhibits their migration, and may participate in halting chemotaxis after these cells reach the area where they are needed (Loike et al., 2001).

In addition to the thrombospondin 4-induced IL-8 secretion mentioned above, several other ECM molecules can induce pro-inflammatory cytokine production. In monocytes, fibrin binding to $\alpha_M\beta_2$ promotes IL-1β expression and inhibits the production of its receptor antagonist, IL-1ra (Perez and Roman, 1995). Binding of CD44 to low-molecular-weight hyaluronan stimulates pro-inflammatory cytokine release by tissue macrophages (Hodge-Dufour et al., 1997) and promotes the production of IL-6 and IL-8 by PBMC (Perez and Roman, 1995). Because some inflammatory molecules can be damaging to tissues when produced in excess, the course of inflammation can be affected significantly by the types of ECM encountered by these leukocytes. ECM molecules can also facilitate leukocyte chemotaxis into the inflamed area by binding chemokines, thus creating a stable chemotactic gradient to promote a specific directional migration (Patel et al., 2001; de Paz et al., 2007); mutant chemokines unable to bind glycosaminoglycans were unable to promote chemotaxis *in vivo*, underscoring the importance of ECM binding in leukocyte recruitment (Proudfoot et al., 2003).

Shortly after wounding, activated platelets secrete a variety of growth factors, including EGF and TGFβ, that stimulate the keratinocytes at the wound edge to proliferate and migrate to cover the wounded area, a process known as re-epithelialization (Singer and Clark, 1999). Additional stimulatory factors, including IL-8, FGF, and KGF (keratinocyte growth factor), produced at later time points by neutrophils, macrophages, endothelial cells, and fibroblasts, may maintain the proliferative and pro-migratory signal (Gillitzer and Goebeler, 2001). During the re-epithelialization process, the keratinocytes migrate beneath the provisional extracellular matrix, composed primarily of fibrin and fibronectin, with vitronectin, tenascin, and collagen type III present in lesser amounts (Decline and Rousselle, 2001). Keratinocyte interactions with matrix molecules are mediated by their corresponding integrin receptors and are required for re-epithelialization; re-epithelialization also depends upon the secretion of new ECM proteins (Singer and Clark, 1999). During re-epithelialization, keratinocytes migrate through the provisional matrix and migrate on top of the collagen I and fibronectin in the granulation tissue. Keratinocytes express $\alpha_2\beta_1$, $\alpha_3\beta_1$, $\alpha_5\beta_1$, $\alpha_6\beta_1$, $\alpha_5\beta_4$, and α_v integrin receptors for these ECM molecules, which, in conjunction with various proteases, facilitate their migration to close the wound (Cavani et al., 1993; Juhasz et al., 1993; Gailit et al., 1994; O'Toole, 2001; Li et al., 2004b). The lack of keratinocyte migration on top of the fibrin-based clot may result from their lack of integrin $\alpha_v\beta_3$ expression as well as the ability of fibrin to prevent keratinocyte adhesion to other provisional matrix components, including fibronectin (Kubo et al., 2001). Interestingly, fibrinogen-deficient mice experienced disordered re-epithelialization (Drew et al., 2001). Interaction of keratinocytes with fibrin appears to promote localized plasmin activation and matrix degradation, which is necessary for re-epithelialization (Bugge et al., 1996; Romer et al., 1996; Geer and Andreadis, 2003). As such, the fibrinogen deficiency may prevent plasmin activation and the matrix degradation that is necessary for re-epithelialization. In support of this possibility, mice that are deficient in both fibrinogen and plasmin exhibit normal wound healing (Nikolopoulos et al., 2005; Schneider et al., 2007).

In addition to fibrin/fibrinogen, keratinocytes interact with multiple other matrix components, whose importance in re-epithelialization is demonstrated by studies done in mice lacking these molecules or in human patients with matrix mutations. This keratinocyte migration also requires new laminin deposition, providing a substrate for the migration and proliferation of the keratinocytes that follow (Decline and Rousselle, 2001; Hartwig et al., 2007). Indeed, antibodies against laminin 332 (laminin 5) inhibited keratinocyte migration on fibronectin, collagen I, and collagen IV, and keratinocytes isolated from a human patient with a mutation leading to laminin-332 deficiency also migrated in a disorganized manner (Sullivan et al., 2007). In addition, poorly healing wounds in *db/db* diabetic mice also exhibit decreased expression of laminin 5, which may account for some of the defects seen in diabetic wound healing (Kamei et al., 1998). Cell-ECM interactions are equally important in the closure of other epithelial wounds. Studies examining the sequential deposition of ECM molecules after wounding of retinal pigment epithelial cells showed *de novo* fibronectin deposition 24 hours after wounding, which is followed by deposition of collagen IV and laminin (Hergott et al., 1993; Hoffmann et al., 2005). This sequence of matrix deposition is tightly linked to adhesion and migration of cells to close the wound, and inhibition of integrin-matrix binding using antibodies or cyclic peptides can prevent both cell adhesion and migration, implicating cell-ECM interactions in the observed epithelial closure (Kamei et al., 1998). A similar sequence of events is observed during the repair of airway epithelial cells after mechanical injury (Pilewski et al., 1997; White et al., 1999; Coraux et al., 2008); functional inhibition of fibronectin or various expressed integrins likewise diminished cell migration and healing of this epithelium (Herard et al., 1996; White et al., 1999).

As healing progresses, embryonic-type cellular fibronectin produced by macrophages and fibroblasts in the wound bed contributes to formation of the granulation tissue, a provisional connective tissue containing nascent blood vessels and multiple types of extracellular matrix molecules (Li et al., 2003). This fibronectin serves as substrate for the migration of the keratinocytes (see above), the endothelial cells that form the vasculature of the wound bed, myofibroblasts, and lymphocytes that are chemoattracted to the wound site by a variety of small cytokines (chemokines) secreted by both macrophages and fibroblasts (Greiling and Clark, 1997; Feugate et al., 2002b). These chemokines belong to a large superfamily and have been characterized in humans, other mammals, and avians (Rossi and Zlotnik, 2000; Gillitzer and Goebeler, 2001). Chemokine-mediated chemoattraction of cells involved in granulation tissue formation, in conjunction with the interaction of these cells with ECM via cell surface receptors, results in processes that lead to cell adhesion and migration into the area of the wound to form the granulation tissue (Lukacs and Kunkel, 1998; Martins-Green and Feugate, 1998; Feugate et al., 2002b).

One of the most extensively studied chemokines with functions important in wound healing is IL-8 (Martins-Green and Bissell, 1990; Martins-Green et al., 1992; Martins-Green and

Hanafusa, 1997; Martins-Green and Feugate, 1998; Martins-Green, 2001; Feugate et al., 2002a,b). This has been well illustrated in studies performed using cIL-8/cCAF and chicks as a model system. cIL-8 is stimulated to high levels shortly after wounding in the fibroblasts of the wounded tissue (Martins-Green and Bissell, 1990; Martins-Green et al., 1992), and thrombin, an enzyme involved in coagulation that is activated upon wounding, stimulates these cells to overexpress cIL-8 (Vaingankar and Martins-Green, 1998; Li et al., 2000). This chemokine then chemoattracts monocyte/macrophages and lymphocytes (Martins-Green and Feugate, 1998). We have shown that thrombin can promote further increases in hIL-8 levels by stimulation of hIL-8 expression in THP-1 differentiated macrophages (Zheng and Martins-Green, 2007). Expression of cIL-8 remains elevated during granulation tissue formation due to its secretion by fibroblasts, the endothelial cells of the microvasculature of the wound, and macrophages, as well as from its binding to the interstitial collagens, tenascin, and laminin present in the granulation tissue (Martins-Green and Bissell, 1990; Martins-Green et al., 1992; Martins-Green et al., 1996). Furthermore, both hIL-8 and cIL-8 are angiogenic *in vivo*, and, in the case of cIL-8, the angiogenic portion of the molecule is localized in the C-terminus of the molecule (Martins-Green and Feugate, 1998; Martins-Green and Kelly, 1998). Based on the pattern of expression and functions of IL-8, it appears that this chemokine participates both in inflammation, via chemotaxis for specific leukocytes, and in the formation of the granulation tissue via stimulation of angiogenesis and ECM deposition (Martins-Green and Hanafusa, 1997; Martins-Green, 2001; Feugate et al., 2002b).

Extracellular matrix interactions with endothelial cells are crucial in the cell migration and in the development of blood vessels during granulation tissue formation (Arroyo and Iruela-Arispe, 2010). Human umbilical vein endothelial cells migrate and arrange themselves in tubular structures when cultured for 12 h on a matrix isolated from Engelbreth-Holm-Swarm (EHS) tumors (a basement membrane-like matrix consisting primarily of laminin but also containing collagen IV, proteoglycans, and entactin/nidogen) (Kubota et al., 1988; Grant et al., 1989; Lawley and Kubota, 1989). When these cells are cultured on collagen I, however, tubular structures do not form in this period of time (Kubota et al., 1988); but, if they are grown for a week *inside* collagen gels, giving the endothelial cells time to deposit their own basement membrane, tubes do develop (Montesano et al., 1983; Madri et al., 1988; Bell et al., 2001). The much more rapid tubulogenesis that occurs on EHS suggests that one or more components of the basement membrane plays an important role in the development of the capillary-like structures, a speculation confirmed both in culture and *in vivo* (Sakamoto et al., 1991; Grant et al., 1992). Indeed, preincubation of these endothelial cells with antibodies to laminin, the major component of basement membrane, prevents the formation of tubules *in vitro* (Kubota et al., 1988). Synthetic peptides containing the sequence SIKVAV derived from the A chain of laminin, which interact with integrins $\alpha_6\beta_1$ and $\alpha_3\beta_1$, induce endothelial cell adhesion and elongation and promote angiogenesis (Grant et al., 1992; Freitas et al., 2007). In contrast, peptides containing the sequence YIGSR derived from the laminin B1 chain, which is known to bind the elastin laminin receptor (ELR), promote endothelial tube formation (Grant et al., 1989), although YIGSR peptides block angiogenesis *in vivo* and inhibit endothelial cell migration *in vitro* (Sakamoto et al., 1991; Grant et al., 1992; Dubey et al., 2009). The mechanisms behind the ability of the YIGSR synthetic peptide to yield such different results *in vitro* and *in vivo* may result from competition of this peptide with laminin for ELR binding, as this YIGSR peptide is known to block laminin binding to cells and block migration. If such competition does occur, the binding of the soluble YIGSR peptide to the ELR rather than YIGSR in the normal context of the complete laminin protein may alter downstream signaling events due to changes in the mechanical resistance and ligand presentation afforded by soluble, rather than intact, ligand, as has been suggested for integrin signaling (Stupack and Cheresh, 2002; Desgrosellier and Cheresh, 2010). Regardless of the actual mechanism of action, the fact that soluble receptor-binding regions of ECM molecules may yield results different from those of the intact molecule may be of particular importance during matrix degradation, which releases ECM fragments. For example, matrix-degrading enzymes are activated during angiogenesis to facilitate the migration and invasion of endothelial cells into adjacent tissues and matrix; this matrix degradation may provide angiogenic or anti-angiogenic factors via release from the matrix or by appropriate cleavage of ECM molecules such as laminin (Rundhaug, 2005). *In vivo*, angiogenic sequences or factors could be provided locally, and, when they have served their purpose, inhibition of further action could similarly be initiated by suitable cleavage to generate anti-angiogenic fragments (Arroyo and Iruela-Arispe, 2010). Matrix degradation also participates in angiogenesis by releasing matrix-bound angiogenic factors and by exposing cryptic sites within matrix molecules that promote migration through alterations in integrin binding (Bergers et al., 2000; Xu et al., 2001; Hangai et al., 2002). Therefore, the way matrix molecules are locally cleaved and/or factors are locally released could have important consequences for the formation of the granulation tissue.

Proliferation

Immediately after wounding, the epithelium undergoes changes that lead to wound closure. During this re-epithelialization period, the keratinocytes trailing behind those at the front edge of migration replicate to provide a source of cells to cover the wound. Basement membrane-type ECM still present on the basal surface of these keratinocytes may be important in maintaining this proliferative state. In support of this possibility is the finding that, during normal skin remodeling, fibronectin associated with the basal lamina of epithelia is crucial for maintaining the basal keratinocyte layer in a proliferative state for constant replenishment of the suprabasal layers (Nicholson and Watt, 1991). It has also been shown using a dermal wound model that basement membrane matrices are able to sustain the proliferation of keratinocytes for several days (Dawson et al., 1996). The component of the basement membrane involved in this proliferation may be laminin, as laminin 10/11 can promote keratinocyte proliferation *in vitro* (Pouliot et al., 2002). Specific integrins are critical for keratinocyte proliferation and thus in re-epithelialization. For example, keratinocyte-specific integrin α_9 deficiency resulted in decreased keratinocyte proliferation during wound healing and decreased thickness of the resulting epithelium (Singh et al., 2009). In contrast, specific cell-matrix interactions may prevent excessive proliferation. For example, the keratinocytes of fibrinogen-deficient mice proliferate abnormally during re-epithelialization (Drew et al., 2001). Integrin β_1-deficient keratinocytes exhibit both impaired migration and hyper-proliferation at the wound margin, and re-epithelialized areas frequently detach from the underlying granulation tissue; at least some of these defects may result from defective laminin 332 organization and the prolonged presence of inflammatory cells (Grose et al., 2002).

As re-epithelialization is occurring, the granulation tissue begins to form. This latter tissue is composed of fibroblasts, myofibroblasts, monocytes/macrophages, lymphocytes, endothelial cells of the microvasculature, and ECM molecules, including embryonic fibronectin, hyaluronan, type III collagen, and small amounts of type I collagen (Clark, 1996). These ECM molecules, in conjunction with growth factors released by the platelets and secreted by the cells present in the granulation tissue, provide signals to the cells that lead to their proliferation (Tuan et al., 1996; Hynes, 2009). ECM molecules themselves, including fibronectin and specific fragments of fibronectin, laminin, collagen VI, SPARC/osteonectin, and hyaluronan, have been shown to stimulate fibroblast and endothelial cell proliferation (Bitterman et al., 1983; Panayotou et al., 1989; Atkinson et al., 1996; Grant et al., 1998; Kapila et al., 1998; Ruhl et al., 1999; Sage et al., 2003; David-Raoudi et al., 2008). In the case of laminin, this proliferative activity appears to be mediated by its EGF-like domains (Panayotou et al., 1989), suggesting a potential dependence upon the activation of EGFR (Schenk et al., 2003; Koshikawa et al., 2005). In contrast, ECM molecules and/or peptides derived from their proteolysis can have inhibitory effects on cell proliferation; intact decorin (Sulochana et al., 2005) and SPARC (Funk and Sage, 1991; Chlenski et al., 2005), as well as peptides derived from decorin (Sulochana et al., 2005), SPARC (Sage et al., 2003), collagens XVIII and XV (endostatin)

(O'Reilly et al., 1997; Sasaki et al., 2000), collagen IV (tumstatin) (Hamano et al., 2003), and tenascin-C (Saito et al., 2008) have anti-angiogenic effects due to their inhibition of endothelial cell proliferation. ECM molecules may also cooperate with growth factors in the proliferation of fibroblasts and the development of new blood vessels in the granulation tissue. During this angiogenic process, growth factors such as VEGFs and FGFs associate with ECM molecules and stimulate proliferation of endothelial cells, which then migrate to form the new microvessels (Miao et al., 1996; Ikuta et al., 2000, 2001; Sottile, 2004). Matrix binding can increase growth factor activity and thus promote proliferation. Binding of VEGF to fibronectin or collagen increases VEGFR2 activation, leading to endothelial cell proliferation and angiogenesis (Wijelath et al., 2006; Whitelock et al., 2008; Chen et al., 2010). bFGF interaction with heparan sulfate proteoglycans promotes FGF binding to FGFR, promoting its activation and inducing cell proliferation (Yayon et al., 1991; Ornitz et al., 1992; Venkataraman et al., 1996). Some anti-angiogenic molecules, including thrombospondin and endostatin, may inhibit angiogenesis by competition with these growth factors for ECM binding (Gupta et al., 1999; Reis et al., 2005). Conversely, ECM-growth factor interactions can be inhibitory; for example, VEGF binding of SPARC can inhibit VEGF-induced proliferation (Kupprion et al., 1998). In addition, the proliferation stimulated by growth factors may be dependent upon the presence of specific ECM molecules; for example, TGF-β1 stimulation of fibroblast proliferation is dependent upon fibronectin (Clark et al., 1997).

Differentiation

As healing progresses during the formation of granulation tissue, some of the fibroblasts differentiate into myofibroblasts; they acquire the morphological and biochemical characteristics of smooth muscle cells by expressing α-smooth muscle actin (αSMA) (Desmouliere et al., 2005). Matrix molecules are important in this differentiation process. For example, heparin decreases the proliferation of fibroblasts in culture and induces the expression of α-smooth muscle actin in these cells. *In vivo*, the local application of tumor necrosis factor α leads to the development of granulation tissue, but the presence of cells expressing α-smooth muscle actin was only observed when heparin was also applied (Desmouliere et al., 1992). These results suggest that some of the properties of heparin not related to its anticoagulant effects are important in the induction of α-smooth muscle actin. This function may be related to the ability of heparin and heparin sulfate proteoglycans to bind cytokines and/or growth factors, such as TGFβ, that regulate myofibroblast differentiation (Kim and Mooney, 1998; Kirkland et al., 1998; Menart et al., 2002; Li et al., 2004a). Specific interactions with the extracellular matrix are also important for myofibroblast differentiation; inhibition of the ED-A-containing form of fibronectin or α_v, α_5, or β_1 integrins can block TGF-β1-mediated myofibroblast differentiation (Serini et al., 1998; Lygoe et al., 2004, 2007; White et al., 2008). Hyaluronan participates in the maintenance of differentiated myofibroblasts; inhibition of hyaluronan synthesis decreased expression of α-smooth muscle actin, a marker of myofibroblast differentiation that is critical for cell contraction (Meran et al., 2007; Webber et al., 2009). In addition, cardiac fibroblasts undergo myofibroblast differentiation when plated on collagen VI (Naugle et al., 2005). Interstitial collagens have also been shown to play a role in the acquisition of the myofibroblastic phenotype. When fibroblasts are cultured on relaxed collagen gels or collagen-coated plates, they do not differentiate (Tomasek et al., 1992); however, if they are grown on anchored collagen matrices where the collagen fibers are aligned (much like in the granulation tissue), they show myofibroblast characteristics (Bell et al., 1979; Arora et al., 1999). These observations led to the hypothesis that myofibroblast differentiation is regulated by mechanical tension; more recent studies *in vivo*, during wound healing, and *in vitro* have suggested that this hypothesis is, in fact, correct (Hinz, 2007).

Differentiation of additional cell types, including keratinocytes, endothelial cells, and pericytes, is regulated by cell-matrix interactions. In keratinocytes, certain cell-matrix interactions, including integrin $\alpha_6\beta_4$ binding to laminin 332, prevent terminal differentiation

(Tennenbaum et al., 1996; Rodius et al., 2007). In contrast, binding of the elastin-laminin receptor to elastin-derived peptides induces keratinocyte differentiation (Fujimoto et al., 2000); similarly, hyaluronan fragments appear to induce the differentiation process via CD44 (Bourguignon et al., 2006). Endothelial cells must make specific cell-matrix contacts, including laminin binding via integrin $\alpha_2\beta_1$, to form capillary-like tubes (Wary et al., 1998). *In vivo*, new capillaries must be stabilized by recruiting pericytes whose differentiation also appears to depend upon cell-matrix interactions, as their differentiation and function in vessel maturation was impaired when these cells lacked expression of integrin β_1 (Abraham et al., 2008).

Apoptosis

Many inflammatory cells undergo apoptosis following their activation, and some of these apoptotic events are regulated by ECM molecules. CCL5/RANTES, which both activates T-cells and promotes their apoptosis, must interact with extracellular glycosaminoglycans (GAGs) in order to induce apoptosis; a mutant CCL5 unable to bind GAGs, enzymatic digestion and removal of cell-associated GAGs, and competition with heparin or chondroitin sulfate prevented CCL-5-induced T-cell apoptosis (Murooka et al., 2006). In addition, hylauronan binding to CD44 promotes apoptosis in activated T-cells (Ruffell and Johnson, 2008), and fibronectin may facilitate Fas ligand/Fas-induced apoptosis in T-cells by binding to Fas ligand, which then promotes T-cell activation and apoptosis (Zanin-Zhorov et al., 2003). A specific fibronectin-derived fragment induces caspase activation and apoptosis in monocytes, although the mechanism remains unclear (Natal et al., 2006).

Apoptosis also participates in the wound remodeling phase, as the granulation tissue evolves into scar tissue. As the wound heals, the number of inflammatory cells, fibroblasts, myofibroblasts, endothelial cells, and pericytes decreases dramatically; matrix molecules, especially interstitial collagen, accumulate; and a scar forms (Singer and Clark, 1999). In this remodeling phase of healing, cell death by apoptosis leads to elimination of many cells of various types at once without causing tissue damage. For example, studies using transmission electron microscopy and *in situ* end-labeling of DNA fragments have shown that many myofibroblasts and endothelial cells undergo apoptosis during the remodeling process. In the granulation tissue, the number of cells undergoing apoptosis increases around days 20–25 after injury and this results in a dramatic reduction in cellularity after day 25 (Desmouliere et al., 1995); similar results were noted in cardiac granulation tissue following infarction (Takemura et al., 1998). Moreover, using model systems that mimic regression of granulation tissue, it has been shown that release of mechanical tension triggers apoptosis of human fibroblasts and myofibroblasts (Fluck et al., 1998; Grinnell et al., 1999; Bride et al., 2004). In these models, apoptotic cell death was regulated by interstitial-type collagens in combination with growth factors and mechanical tension and did not require differentiation of the fibroblasts into myofibroblasts, strongly suggesting that contractile collagens determine the susceptibility of fibroblasts of the wound tissue to undergo apoptotic cell death (Fluck et al., 1998; Grinnell et al., 1999). Further studies have also implicated the interactions between thrombospondin-1 and the $\alpha_v\beta_3$ integrin-CD47 complex in the mechanical tension-mediated stimulation of fibroblast apoptosis (Graf et al., 2002). Such apoptosis may be required for resolution of wound healing and the prevention of scarring. Indeed, fibroblast/myofibroblast apoptosis is reduced in keloid and hypertrophic scars, resulting in the excessive matrix accumulation and scarring (van der Veer et al., 2009). In keloid scars, this decreased apoptosis may be due to p53 mutations and/or growth factor receptor overexpression (Ladin et al., 1998; Saed et al., 1998; Messadi et al., 1999; Ishihara et al., 2000; Moulin et al., 2004;); in contrast, it is thought that apoptotic failure in hypertrophic scars results from an overexpression of tissue trans-glutaminase, leading to increased matrix breakdown and decreased collagen contraction (Linge et al., 2005). In addition to cell death by apoptosis, it has also been shown that bronchoalveolar lavage fluid collected during lung remodeling after injury can promote fibroblast cell death by a process that is distinct from that of necrosis or apoptosis (Polunovsky et al., 1993). Although this process of cell death has not been extensively studied, it suggests that there are other processes of programmed cell death that are distinct from apoptosis and occur preferentially in association with wound repair.

CELL-ECM INTERACTIONS DURING REGENERATION

True tissue regeneration following injury rarely occurs in vertebrate species, but it does occur in specific instances, including fetal cutaneous wound healing, liver regeneration, and urodele amphibian limb regeneration. Unlike wound healing in normal adult animals, which is characterized by scarring, fetal cutaneous wounds heal without fibrosis and scar formation, leading to regeneration of the injured area. Similarly, after injury, injured liver very effectively restores both normal function and normal organ size by proliferation and differentiation of pre-existing cell types. The contribution of cell-ECM interactions to regeneration in fetal healing and liver regeneration is discussed below (Fig. 2.3).

Fetal wound healing
ADHESION AND MIGRATION

Scarless fetal wounds have significant differences in cell-ECM interactions in the injured area when compared with scarring adult wounds; these changes occur due to alterations in the composition of the ECM molecules, the rate of their appearance after wounding, and their duration in the wound area. One crucial ECM molecule in fetal wound healing is hyaluronan, which appears to be necessary for the regenerative response; its removal from fetal wounds promotes a healing response more similar to that of adults (Mast et al., 1992), and treatment of normally scarring wounds or wound organ cultures with hyaluronan decreases scarring (Iocono et al., 1998a,b; Hu et al., 2003). Hyaluronan is present at higher levels and for a longer duration in fetal skin wounds compared with adult wounds; the latter may result, in part, from the reduced activity of hyaluronidase in fetal wounds (Krummel et al., 1987; Sawai et al., 1997; West et al., 1997). Fetal fibroblasts also express higher levels of the hyaluronan receptor CD44 (Adolph et al., 1993; Alaish et al., 1994), thus increasing receptor-ligand interactions that promote fibroblast migration (Huang-Lee et al., 1994). Increased fetal hyaluronan may also facilitate fibroblast migration by decreasing or preventing expression of TGF-β1, a factor that inhibits fibroblast migration, increases collagen I deposition, and promotes scar formation (Ignotz and Massague, 1986; Ellis et al., 1992; Hu et al., 2003). In contrast, hyaluronan increases the expression of TGF-β3, a factor highly expressed in fetal skin that promotes

Healing with Scar Formation (Adult Healing)	Healing with Regeneration (Fetal Healing)
↓ Hyaluronic acid, ↑ decorin, presence of ED-A fibronectin	↑ Hyaluronic acid ↓ Decorin
↑ TGF-b1, disorganized collagen deposition	↓ TGF-b1, ↑ collagen organization
↑ Myofibroblast differentiation ↑ contraction	↓ Myofibroblast differentiation ↓ contraction
↑ Scar formation ↓ Regeneration	↓ Scar formation ↑ Regeneration

FIGURE 2.3
A comparison of particular cell-ECM interactions occurring in scar-forming adult healing versus those occurring during regenerative fetal healing. As shown in this diagram, unique subsets of ECM molecules are associated with scarring versus regenerative healing. As such, therapeutic alteration of ECM composition may allow physicians to modulate healing to promote tissue regeneration. Additional therapeutic approaches may be generated upon further investigation into the importance of additional cell-ECM interactions in scarring and regenerative responses.

scarless healing (Chen et al., 2005; David-Raoudi et al., 2008). Another glycosaminoglycan present in large amounts in fetal wounds is chondroitin sulfate, which, like hyaluronan, binds to CD44; as such, chondroitin sulfate may also participate in scarless healing (Whitby and Ferguson, 1991; Coolen et al., 2010).

Tenascin C is expressed at higher levels in fetal skin than in adult skin, and is induced more rapidly and to a greater extent in fetal wounds, thus modulating cell adhesion to fibronectin (Whitby and Ferguson, 1991; Whitby et al., 1991; Coolen et al., 2010). Fibronectin levels also increase more quickly in fetal wounds than in adult wounds (Longaker et al., 1989). This increased expression of tenascin and fibronectin is associated with concomitant increases in the expression of integrins that serve as their receptors. In particular, the $\alpha5$ subunit, $\alpha_v\beta_3$, and $\alpha_v\beta_6$ integrins, which bind fibronectin and/or tenascin, are upregulated in the wounded fetal epithelium (Cass et al., 1998). The combined rapid increases in fibronectin and tenascin, coupled with increased expression of their respective integrin receptors in epithelial cells, are likely important in facilitating cell migration and re-epithelialization in fetal wounds.

In addition, fetal fibroblasts produce more collagen, particularly collagen type III, than adult cells, and the organization of the fibrils in the fetal wound appears normal, while that of the adult wound exhibits an organization indicative of scarring (Hallock et al., 1988; Longaker et al., 1990; Whitby and Ferguson, 1991; Gosiewska et al., 2001; Brink et al., 2009). The changes in the collagen levels and organization in fetal wounds may result from the increased expression in fetal fibroblasts of the collagen receptor DDR1, which is important in collagen expression and organization (Chin et al., 2001). Furthermore, hyaluronan increases collagen synthesis *in vitro*, and may thus contribute to increased collagen deposition in fetal wounds (Mast et al., 1993; David-Raoudi et al., 2008). In spite of the increased collagen production by fetal fibroblasts, the fetal wounds do not exhibit excessive collagen deposition and fibrosis; this may be due to changes in the organization and cross-linking of collagen at the wound site (Lovvorn et al., 1999) or rapid turnover of these ECM components by protease-mediated degradation. For example, levels of uPA and some, though not all, MMPs are increased while the levels of their endogenous inhibitors, PAI-1 and TIMPs, are decreased in fetal wounds, ultimately promoting matrix degradation and turnover (Huang et al., 2002; Peled et al., 2002; Dang et al., 2003; Chen et al., 2007). Hyaluronan fragments can induce MMP-1 and MMP-3 expression and decrease TIMP-1 expression in adult fibroblasts (David-Raoudi et al., 2008), while TGF-$\beta3$ appears to suppress PAI-1 expression in fetal skin (Li et al., 2006). In contrast, the pro-scarring TGF-$\beta1$ decreases MMP-1 levels in fetal wounds, potentially inducing scar formation by preventing matrix turnover (Bullard et al., 1997). Taken together, these data support a role for MMPs, uPA, and plasmin in scarless healing. Not only does the resulting matrix degradation and turnover prevent fibrosis, it also likely facilitates cell migration by reducing matrix density and increases the generation of proteolytic matrix fragments that modulate various stages of wound repair, as mentioned above for laminin and collagen fragments.

PROLIFERATION

As mentioned above, during fetal wound healing, increased levels of hyaluronan are present, and *in vitro* studies indicate that hyaluronan decreases fetal fibroblast proliferation (Mast et al., 1993), although specific hylauronan fragments can induce proliferation of adult fibroblasts (David-Raoudi et al., 2008). In support of a pro-proliferative role of hyaluronan or its fragments, studies comparing fetal wounds with those of newborns and adults showed an increase in fibroblast number in the fetal wounds, and fetal fibroblasts proliferate more rapidly than adult cells (Adzick et al., 1985; Khorramizadeh et al., 1999). Growth factor-induced proliferation and matrix production of fetal fibroblasts may also be altered when compared with that of adult cells. IGF-1, which induces Erk signaling, proliferation, and matrix synthesis in post-natal fibroblasts, induces proliferation to a much lesser extent and fails to induce significant Erk signaling or matrix synthesis in fetal fibroblasts (Rolfe et al., 2007a). While TGF-β1 levels are reduced in fetal wounds when compared with TGF-β3, this factor is, indeed, present, and may function in wound repair (Wilgus, 2007). However, the cellular responses to TGF-β1 differ between fetal and post-natal fibroblasts; TGF-β1 treatment increased PAI-1 to a much greater extent in fetal fibroblasts than in post-natal fibroblasts, increased collagen III synthesis in fetal fibroblasts but not in post-natal fibroblasts, and induced collagen I synthesis in post-natal fibroblasts but not in fetal cells (Rolfe et al., 2007b,c). Furthermore, while TGF-β1 induces proliferation in post-natal fibroblasts, it does not do so in fetal fibroblasts (Moulin et al., 2001; Carre et al., 2010). This may result, at least in part, from the ability of TGF-β1 to induce hyaluronan synthase-2 (HAS-2) expression in fetal, but not post-natal, fibroblasts (Carre et al., 2010); if intact hyaluronan suppresses fetal fibroblast proliferation (Mast et al., 1993), TGF-β1-induced hyaluronan synthesis due to HAS-2 production may prevent the proliferation of these cells.

Another critical event in wound healing is re-epithelialization, which requires both keratinocyte migration and proliferation. Keratinocyte proliferation is decreased in mice lacking CD44 expression in keratinocytes (Kaya et al., 1997), and proliferation is increased in wounds that are treated with modified hyaluronan in a CD44-dependent manner (Kaya et al., 2006), suggesting that interactions between hyaluronan and CD44 may be important for keratinocyte proliferation during healing, and thus for more effective re-epithelialization. This finding may explain, in part, the enhanced rate of healing seen in wounds treated with hyaluronan.

DIFFERENTIATION

Fetal wounds have a decreased number of myofibroblasts, which appear in the wounded site earlier and remain a shorter time than in adult wounds; in fact, one study showed a lack of α-smooth muscle actin-expressing myofibroblasts in the wounds of early-stage fetuses (Estes et al., 1994). This is associated with a general lack of contraction in the fetal wounds themselves (Krummel et al., 1987). Similar results have been observed *in vitro*. Fetal fibroblasts can differentiate into myofibroblasts more rapidly and more transiently in response to TGF-β1, a potent stimulator of myofibroblast differentiation in adult fibroblasts (Rolfe et al., 2007c). In addition, TGF-β1-treated fetal fibroblasts contract less than untreated controls, and do not exhibit increased production of collagen I (Moulin et al., 2001; Rolfe et al., 2007c). Increased levels of hyaluronan present during fetal wound healing may alter the differentiation and/or contractility of myofibroblasts in the wound site; studies *in vitro* have shown that addition of hyaluronan decreases fibroblast contraction of collagen matrices (Huang-Lee et al., 1994). This may be due, in part, to reduced expression of TGF-β1, a major inducer of myofibroblast differentiation and fibrosis. Indeed, incisional adult wounds treated with hyaluronan healed more rapidly with a significant decrease in TGF-β1 levels (Hu et al., 2003). The large amounts of hyaluronan in fetal wounds may thus explain the greatly reduced levels of TGF-β1 in fetal wounds (Nath et al., 1994; Chen et al., 2005). Downregulation of TGF-β1 in adult wounds produces a decrease in scarring similar to that observed with hyaluronan treatment (Choi et al., 1996). Conversely, studies have shown that the addition of TGF-β1 to normally scarless fetal wounds induces a more scarring phenotype, with myofibroblast differentiation, wound contraction, and fibrosis (Lin et al., 1995; Lanning et al., 1999). Thus, hyaluronan-mediated inhibition of TGF-β1 expression may be critical in scarless fetal healing. However, hyaluronan synthesis is necessary for maintenance of the TGF-β1-induced myofibroblastic phenotype in adult cells via regulation of a TGF-β1 autocrine loop, complicating this scenario (Simpson et al., 2009; Webber et al., 2009). Further studies are needed to dissect the interrelationship between hyaluronan and TGF-β1 in myofibroblast differentiation.

The relatively small amount of TGF-β1 present during fetal wound healing may be regulated by inhibitory ECM molecules present in the injured area. One such inhibitor is the proteoglycan fibromodulin, which is capable of binding TGF-β1 and preventing receptor binding, and is expressed to a greater extent in fetal wounds than in adult wounds (Hildebrand et al., 1994; Soo et al., 2000). In addition, adenoviral-mediated overexpression of fibromodulin in adult cutaneous wounds decreased wound levels of TGF-β1 and TGF-β2, increased levels of TGF-β3, and improved wound healing, supporting a role for fibromodulin in scarless healing (Stoff et al., 2007). Another molecule that may alter TGF-β1 activity is decorin, although the function of decorin in modulating TGF-β1 activity is somewhat controversial; some studies indicate that decorin binding decreases TGF-β1 activity (Noble et al., 1992), while others suggest that this interaction either has no effect on TGF-β1 or even actually increases activity (Hausser et al., 1994; Takeuchi et al., 1994). The outcome of decorin-TGF-β1 binding may depend upon the microenvironment, and this has not been extensively studied in fetal wounds. Regardless, decorin levels are decreased in scarless wounds, resulting in decreased decorin-TGF-β1 interactions and altered TGF-β1 activity (Beanes et al., 2001). Decreased activity of this growth factor, combined with low levels of expression in fetal wounds, results in decreased fibrosis, myofibroblast differentiation, and wound contraction, leading to regeneration rather than scarring.

APOPTOSIS

Little is known regarding the apoptotic process in fetal wounds and whether this differs from that of adult wounds. A recent study examined specific indicators of apoptotic induction at very early time points after wounding in both scarless (E15) and scar-forming (E18) fetal mouse wounds (Carter et al., 2009). At these early time points, there was some induction of caspase 7 and PARP cleavage, as well as DNA fragmentation in E15 wounds, with no caspase 7 cleavage, little PARP cleavage, and lower amounts of fragmented DNA in E18 wounds. However, many cells underwent apoptosis in E15 skin as well, so the importance of the increased apoptosis after wounding is unclear; furthermore, the cell types that undergo apoptosis in E15 skin and wounds are not known. Suffice it to say that, as in adult healing, multiple cell types present within the fetal granulation tissue likely disappear via apoptosis. It is also apparent that any myofibroblasts that do differentiate from fetal fibroblasts, either *in vivo* or *in vitro*, disappear rapidly (Estes et al., 1994; Rolfe et al., 2007c), perhaps due to an altered rate of apoptosis in these wounds. If changes in apoptotic efficiency do indeed occur, they may result from the decreased contraction, and thus decreased mechanical tension, in fetal wounds (Krummel et al., 1987; Moulin et al., 2001), as well as altered collagen levels within the collagen matrix (Adzick et al., 1985; Longaker et al., 1990; Lovvorn et al., 1999; Gosiewska et al., 2001). It is also possible that apoptosis is not as critical in the healing of fetal wounds as in adult wounds; leukocyte influx and myofibroblast differentiation appear to be minimal in fetal wounds, and thus may not require large numbers of cells to undergo apoptosis for regeneration to occur (Estes et al., 1994; Harty et al., 2003).

Liver regeneration

ADHESION AND MIGRATION

ECM-cell interactions are also altered during mammalian liver regeneration, leading to changes in adhesion and migration. One major molecule upregulated after liver injury is laminin (Martinez-Hernandez et al., 1991; Kato et al., 1992). Hepatocytes isolated soon after liver injury and plated on laminin attach more efficiently than non-injured hepatocytes, suggesting a concomitant increase in laminin-binding integrins (Carlsson et al., 1981; Kato et al., 1992). Collagen I, III, IV, and V increase in regenerating liver several days after injury. Hepatocytes isolated from this stage of regenerating liver show increased adhesion to collagen, which may indicate increased expression of collagen adhesion receptors (Kato et al., 1992). The increased levels of laminin and collagen IV during regeneration may also promote hepatocyte migration, as both the basal and stimulated migration of hepatocytes is enhanced on laminin and collagen IV relative to other types of ECM (Ma et al., 1999).

Additional ECM molecules upregulated during liver regeneration are fibronectin and collagen I. Together with laminin, fibronectin and collagen I may promote the adhesion and migration of oval cells, liver cells that serve as hepatocyte progenitors. All three matrix molecules are deposited near oval cells in the liver after injury, and deposition precedes oval cell migration, suggesting that these molecules may form a type of "track" upon which the oval cells migrate (van Hul et al., 2009; Zhang et al., 2009). In support of an adhesive and migratory role for fibronectin, the fibronectin-binding molecule CTGF/CCN2 promotes the adhesion and migration of oval cells in an integrin $\alpha_5\beta_1$- and heparan sulfate-dependent manner (Pi et al., 2008). Because integrin $\alpha_5\beta_1$ is a fibronectin receptor, CTGF may promote adhesion and migration via integrin $\alpha_5\beta_1$ through an indirect interaction mediated by fibronectin. In addition, oval cells express CD44 after injury; this finding, coupled with the fact that CTGF-induced adhesion and migration require heparan sulfate, suggests a potential role for hyaluronan in this process (Chiu et al., 2009).

PROLIFERATION

In response to liver injury, hepatocytes and their progenitors proliferate to restore normal liver function and size. After injury, oval cells express CD44, and oval cell proliferation is impaired in CD44-deficient mice, suggesting a role for CD44-hyaluronan interactions in oval cell proliferation (Chiu et al., 2009). *In vitro* studies show that laminin enhances hepatocyte proliferation in general and in response to EGF; thus, the increased laminin present in regenerating tissue may facilitate proliferation (Hirata et al., 1983; Kato et al., 1992). Both the mRNA and the protein levels of plasma fibronectin and its receptor integrin $\alpha_5\beta_1$ increase in regenerating liver following injury (Gluck et al., 1992; Kato et al., 1992; Pujades et al., 1992), which may also increase proliferation. Indeed, intraperitoneal injection of plasma fibronectin further stimulates proliferation in the regenerating liver (Kwon et al., 1990b). Following hepatocyte proliferation after injury, increases in ECM deposition and thus cell-ECM interactions likely inhibit excessive proliferation and also protect the cells from apoptosis. Inhibition of cell-matrix signaling by liver-specific knockout of integrin-linked kinase (ILK) greatly increased hepatocyte proliferation in both injured and non-injured livers, leading to increased liver size but also increased apoptosis, probably through a reduction in survival signals propagated via cell adhesion (Gkretsi et al., 2008; Apte et al., 2009).

The primary growth factor responsible for hepatocyte proliferation is hepatocyte growth factor (HGF); thus, processes that stimulate HGF production and/or release from matrix components will also increase hepatocyte numbers in regenerating liver. Heparan sulfate proteoglycans that are upregulated after injury bind HGF and promote its mitogenic activity (Matsumoto et al., 1993; Kato et al., 1994; Lai et al., 2004). Various proteoglycans are also upregulated after injury, potentially increasing HGF activity in the regenerating liver (Otsu et al., 1992; Gallai et al., 1996). Other ECM molecules are known to bind HGF with low affinity, possibly sequestering HGF in the ECM and preventing its activity (Schuppan et al., 1998). In fact, increased MMP expression or inhibited TIMP expression during regeneration stimulates ECM degradation and hepatocyte proliferation (Mohammed et al., 2005; Hu et al., 2007). This increased proliferation is likely due to the proteolytic processing and release of matrix-bound HGF (Nishio et al., 2003; Mohammed et al., 2005). Increases in MMP production are followed by increased TIMP expression, which may prevent excessive hepatocyte proliferation and/or excessive matrix degradation (Rudolph et al., 1999; Mohammed et al., 2005). HGF, and thus hepatocyte proliferation, can also be activated by plasmin, suggesting a role for plasminogen activators in liver regeneration (Shimizu et al., 2001). After partial hepatectomy in both humans and rats, uPA expression is increased; in the rat, this rapid increase in uPA activity after injury is followed by increases in plasmin activation and fibrinogen cleavage and a rapid loss of fibronectin, laminin, and entactin via proteolysis, although the levels of these latter proteins increase at later stages of healing (Kim et al., 1997; Mangnall et al., 2004). The importance of plasmin activation is underscored by studies in which the livers of uPA and tPA single and double knockout mice or plasminogen knockout mice were injured chemically (Bezerra et al., 1999, 2001). It was found that the plasminogen and uPA single knockouts, as well as the uPA/tPA double knockouts, experienced significant liver regenerative problems accompanied by excessive fibrin and fibronectin, with a lesser effect seen in the tPA knockout. The observed disruption of regeneration may be due to a reduction of hepatocyte proliferation resulting from decreased HGF activity.

DIFFERENTIATION

Myofibroblast differentiation can also occur from the stellate cells of the liver, which can then stimulate excessive ECM deposition, leading to fibrosis and cirrhosis rather than regeneration. Thus, myofibroblast differentiation must be very limited to allow appropriate liver regeneration. Plasma fibronectin levels are increased in the liver regenerating tissue, but are reduced in cirrhotic tissue (Kwon et al., 1990a; Chijiiwa et al., 1994). In addition, myofibroblast differentiation appears to require the ED-A domain of fibronectin (Serini et al., 1998; Kato et al., 2001), which is lacking in plasma fibronectin. These results, when taken together, suggest the possibility that plasma fibronectin may limit myofibroblast differentiation and fibrosis in the liver. This may be particularly important, given the increased quantity and activation of TGF-β1, 2, and 3 in the regenerating liver, which would otherwise promote myofibroblast differentiation and fibrosis (Jakowlew et al., 1991). In contrast, the stellate cell differentiation state may be maintained by the basement membrane, which appears to both maintain the differentiation state of stellate cells and, *in vitro*, promote myofibroblast dedifferentiation back to stellate cells (Friedman et al., 1989; Sohara et al., 2002).

Certain matrix molecules may, themselves, promote myofibroblast differentiation. SPARC, for example, is expressed in fibrotic livers, along with αSMA and collagen I, markers of myofibroblasts, and inhibition of SPARC synthesis via adenoviral-mediated delivery of SPARC antisense RNA reduced liver fibrosis and decreased myofibroblast differentiation (Blazejewski et al., 1997; Camino et al., 2008). Matrix degradation may also prevent liver fibrosis and promote regeneration; PAI-1 is upregulated following pro-fibrotic liver injury, and liver fibrosis is decreased in PAI-1$^{-/-}$ mice (Bergheim et al., 2006). Integrin-mediated signaling through integrin-linked kinase (ILK) appears to be necessary for induction of fibrosis, as adenoviral delivery of shRNA against ILK decreased expression of collagen I, αSMA, TGF-β, and fibronectin, leading to decreased liver fibrosis (Shafiei and Rockey, 2006).

APOPTOSIS

In liver regeneration, prevention of hepatocyte apoptosis is critical for regeneration, while increased apoptotic rates are associated with impaired regeneration. Indeed, extensive cell death following a large liver resection leads to liver failure rather than regeneration (Panis et al., 1997). Liver ischemia-reperfusion injury can also promote apoptosis and liver failure rather than regeneration (Takeda et al., 2002). In the latter case of ischemia-reperfusion injury, prevention of apoptosis can significantly reduce the incidence of liver failure, underscoring the relationship between apoptosis and impaired regeneration or failure (Vilatoba et al., 2005b). The lack of regeneration in such cases is associated with the upregulation of pro-apoptotic gene expression and the downregulation of pro-survival genes (Morita et al., 2002), and may thus be related to the inability of hepatocytes to proliferate under such pro-apoptotic conditions (Iimuro et al., 1998). This hypothesis is supported by studies indicating that apoptosis and liver failure resulting from extensive liver resection or ischemia-reperfusion injury can be largely prevented by treatment conditions that promote cell proliferation (Longo et al., 2005; Vilatoba et al., 2005a). The prevention of apoptosis may thus require ECM molecules that are important in promoting hepatocyte proliferation, including laminin (Hirata et al., 1983; Kato et al., 1992), plasma fibronectin (Kwon et al., 1990b), and HGF-binding proteoglycans (Matsumoto et al., 1993; Kato et al., 1994; Lai et al., 2004). Different MMPs are activated after ischemia-reperfusion injury when compared with forms of injury that regenerate (Cursio et al., 2002), perhaps leading to the degradation of a different profile of ECM proteins; the activation of specific MMPs is thought to promote hepatocyte proliferation by releasing matrix-sequestered HGF (Nishio et al., 2003; Mohammed et al., 2005). The activation of

43

different MMPs and cleavage of different substrates may alter HGF release and subsequent proliferation, leaving these cells more susceptible to apoptosis. This idea is supported by studies in which liver with ischemia-reperfusion injury was treated with an MMP inhibitor, which decreased apoptosis and necrosis in the injured liver (Cursio et al., 2002; Defamie et al., 2008).

Although apoptosis of hepatocytes disrupts the regenerative process, apoptosis of myofibroblastic hepatic stellate cells may be critical in preventing fibrosis and scarring during regeneration (Issa et al., 2001). These myofibroblastic hepatic stellate cells disappear via apoptosis (Saile et al., 1997; Issa et al., 2001) and also potentially by dedifferentiation back to stellate cells (Friedman et al., 1989; Sohara et al., 2002). The apoptosis of these myofibroblastic cells seems to be dependent upon the activation of specific proteases and the subsequent degradation of matrix components. Mice expressing a collagen I gene that is resistant to proteolysis had decreased stellate cell myofibroblast apoptosis and increased fibrosis, and thus impaired regeneration, relative to wild type (Issa et al., 2003). These myofibroblasts also persist in plasminogen-deficient mice and are associated with a general accumulation of non-degraded matrix components (Ng et al., 2001), further supporting a role for matrix degradation in the observed apoptosis. The matrix degradation important in apoptosis also likely involves the activation of MMPs, as inhibition of MMP activity using synthetic inhibitors or TIMP-1 (Murphy et al., 2002; Zhou et al., 2004) prevents apoptosis of myofibroblastic stellate cells *in vitro*, whereas increased MMP-9 activity or inhibition of TIMP-1 promote apoptosis of these cells (Zhou et al., 2004; Roderfeld et al., 2006). In *in vitro* models of cutaneous wound healing, a release of mechanical tension within the collagen matrix (Fluck et al., 1998; Grinnell et al., 1999; Bride et al., 2004) can promote myofibroblast apoptosis. It is possible that a similar release of mechanical tension, perhaps via cleavage of collagen I, is critical for myofibroblast apoptosis in the liver. Proteolysis of ECM components may also contribute to stellate cell apoptosis by abolishing integrin signaling downstream of binding to these components. Experimental disruption of ECM-integrin binding via an RGD-containing peptide (Iwamoto et al., 1999) or various $\alpha_v\beta_3$ antagonists (Zhou et al., 2004) induced stellate cell apoptosis *in vitro*, further supporting a role for integrin-mediated signaling in this apoptotic event.

IMPLICATIONS FOR REGENERATIVE MEDICINE

One primary goal of studies comparing differences in cell-ECM interactions, and thus changes in signaling, that accompany regenerative and non-regenerative healing is to determine which types of interactions promote and which inhibit tissue regeneration (for an example, see Fig. 2.3). After elucidating the functions of particular interactions, it may be possible to increase the regenerative response through (1) the induction of pro-regenerative ECM molecules or signaling events in the wounded area combined with (2) the antagonism of anti-regenerative/scarring interactions or signaling events using specific inhibitors. This discussion of regenerative medicine will focus upon possible strategies to promote regeneration in adult scarring wounds, thus causing adult wounds to more closely resemble fetal scarless wounds. Such an increased regenerative response would be particularly useful in the treatment of wounds that heal abnormally with increased scar formation, such as keloids and hypertrophic scars, ischemic reperfusion injury, and chronic inflammatory responses.

Different types of approaches may be used to increase pro-regenerative ECM levels in the wounded area, including direct application of the molecules themselves, addition of agents that increase their expression, addition of cells producing these types of ECM that have been prepared to minimize immunogenicity, introduction of biomaterials modified to contain adhesive, pro-regenerative regions of these ECM molecules, or wound treatment with inhibitors of their proteolysis. Several different ECM molecules are present at higher levels in fetal wounds than in adult wounds, including hyaluronan, chondroitin sulfate, tenascin, fibronectin, and collagen III (Krummel et al., 1987; Hallock et al., 1988; Longaker et al., 1989; Whitby and Ferguson, 1991; Whitby et al., 1991; Sawai et al., 1997; Coolen et al., 2010), and may play important roles in the regeneration process. Thus, altering the levels of these molecules in a scarring wound may improve regeneration. Some studies have used fetal cells themselves to promote healing and decrease scarring in burn patients; several genes involved in cell-cell and cell-matrix adhesion were upregulated in fetal cells versus adult cells, notably a chondroitin sulfate proteoglycan and CD44, which are thought to be involved in fetal scarless healing (de Buys Roessingh et al., 2006; Ramelet et al., 2009). However, the use of fetal human cells is controversial, and a cell-free system would be less likely to induce an immune response. Preliminary experiments in rat wounds suggest that hyaluronan treatment decreases both the time required for healing and the amount of scar formation (Hu et al., 2003), underscoring the potential for this molecule in therapeutics. It is possible that treatment with tenascin, fibronectin, or collagen III in addition to hyaluronan could yield even more favorable outcomes. Several studies have shown that synthetic, modified versions of hyaluronan or chondroitin sulfate can be further modified by the inclusion of molecules that promote cell adhesion and/or growth factor binding, such as RGD sequences, specific regions of fibronectin or gelatin, heparin, or intact collagen (Serban and Prestwich, 2008), thereby promoting wound healing (Liu et al., 2007). Growth factors that promote tissue repair or regeneration can be added to this "semi-synthetic" biomaterial, where they can interact with heparan or chondroitin sulfate and thus be either effectively presented to their receptors or be released upon degradation of the biomaterial. Alterations in the biomaterial formulation, such as the addition of differing amounts of heparin, can regulate the timing of growth factor release, allowing their release over a relatively long period of time (Cai et al., 2005). As such, this or other biomaterials may be useful for delivery of pro-regenerative ECM molecules and/or growth factors to the injured area, thereby promoting healing and reducing scar formation.

When attempting to promote regeneration, it is also imperative to inhibit events associated with scarring, including excessive ECM deposition, fibrosis, and contraction. During the adult healing process, these scar-associated processes are primarily controlled by the myofibroblast, a differentiated cell type that arises during the adult healing process but that is largely absent throughout fetal wound healing. As such, inhibition of myofibroblast differentiation or function along with the addition of pro-regenerative molecules may facilitate a stronger regenerative response. Inhibition of differentiation could be accomplished by blocking the factors that normally stimulate this process, such as TGF-β1 (Lin et al., 1995; Lanning et al., 1999) and IL-8 (Feugate et al., 2002a), or by preventing fibroblast-ECM interactions that facilitate myofibroblast differentiation, such as ED-A-containing fibronectin (Serini et al., 1998; Kato et al., 2001). Hyaluronan and fibromodulin appear to decrease TGF-β1 levels and activity, respectively; treatment of normally scarring wounds with these matrix components may decrease TGF-β1-mediated scarring (Hildebrand et al., 1994; Soo et al., 2000; Hu et al., 2003). Indeed, treatment of leg ulcers with fetal cells on a collagen scaffold decreased scarring; these fetal cells exhibited increased expression of fibromodulin, which may have interacted with and inhibited TGF-β1 activity, resulting in the decreased scarring (Ramelet et al., 2009). IL-8, on the other hand, is a chemokine that activates G-protein-linked receptors, which are highly amenable to inhibition by small molecules that could be used to reduce the effects of this chemokine on myofibroblast differentiation (Casilli et al., 2005).

In summary, the recent surge in research regarding the ECM molecules themselves and their interactions with particular cells and cell-surface receptors has led to the realization that such interactions are many and complex, and that they are of the utmost importance in determining cell behavior during such events as wound repair and tissue regeneration. As such, the manipulation of specific cell-ECM interactions has the potential to modulate particular aspects of the repair process in order to promote a regenerative response.

References

Abraham, S., Kogata, N., Fassler, R., & Adams, R. H. (2008). Integrin beta1 subunit controls mural cell adhesion, spreading, and blood vessel wall stability. *Circ. Res., 102*, 562–570.

Adams, J. C., & Watt, F. M. (1993). Regulation of development and differentiation by the extracellular matrix. *Development, 117*, 1183–1198.

Adelsman, M. A., McCarthy, J. B., & Shimizu, Y. (1999). Stimulation of beta1-integrin function by epidermal growth factor and heregulin-beta has distinct requirements for erbB2 but a similar dependence on phosphoinositide 3-OH kinase. *Mol. Biol. Cell, 10*, 2861–2878.

Adolph, V. R., Bleacher, J. C., Dillon, P. W., & Krummel, T. M. (1993). Hyaluronate uptake and CD-44 activity in fetal and adult fibroblasts. *J. Surg. Res., 54*, 328–330.

Adzick, N. S., Harrison, M. R., Glick, P. L., Beckstead, J. H., Villa, R. L., Scheuenstuhl, H., et al. (1985). Comparison of fetal, newborn, and adult wound healing by histologic, enzyme-histochemical, and hydroxyproline determinations. *J. Pediatr. Surg., 20*, 315–319.

Akiyama, S. K., Aota, S., & Yamada, K. M. (1995). Function and receptor specificity of a minimal 20 kilodalton cell adhesive fragment of fibronectin. *Cell Adhes. Commun., 3*, 13–25.

Alaish, S. M., Yager, D., Diegelmann, R. F., & Cohen, I. K. (1994). Biology of fetal wound healing: hyaluronate receptor expression in fetal fibroblasts. *J. Pediatr. Surg., 29*, 1040–1043.

Allt, G., & Lawrenson, J. G. (2001). Pericytes: cell biology and pathology. *Cells Tissues Organs, 169*, 1–11.

Almeida, E. A., Ilic, D., Han, Q., Hauck, C. R., Jin, F., Kawakatsu, H., et al. (2000). Matrix survival signaling: from fibronectin via focal adhesion kinase to c-Jun NH(2)-terminal kinase. *J. Cell Biol., 149*, 741–754.

Anderson, S. S., Kim, Y., & Tsilibary, E. C. (1994). Effects of matrix glycation on mesangial cell adhesion, spreading and proliferation. *Kidney Int., 46*, 1359–1367.

Antonicelli, F., Bellon, G., Lorimier, S., & Hornebeck, W. (2009). Role of the elastin receptor complex (S-Gal/Cath-A/Neu-1) in skin repair and regeneration. *Wound Repair Regen., 17*, 631–638.

Aota, S., Nomizu, M., & Yamada, K. M. (1994). The short amino acid sequence Pro-His-Ser-Arg-Asn in human fibronectin enhances cell-adhesive function. *J. Biol. Chem., 269*, 24756–24761.

Aplin, A. E., & Juliano, R. L. (1999). Integrin and cytoskeletal regulation of growth factor signaling to the MAP kinase pathway. *J. Cell Sci., 112*(Pt 5), 695–706.

Aplin, A. E., Stewart, S. A., Assoian, R. K., & Juliano, R. L. (2001). Integrin-mediated adhesion regulates ERK nuclear translocation and phosphorylation of Elk-1. *J. Cell Biol., 153*, 273–282.

Apte, U., Gkretsi, V., Bowen, W. C., Mars, W. M., Luo, J. H., Donthamsetty, S., et al. (2009). Enhanced liver regeneration following changes induced by hepatocyte-specific genetic ablation of integrin-linked kinase. *Hepatology, 50*, 844–851.

Arora, P. D., Narani, N., & McCulloch, C. A. (1999). The compliance of collagen gels regulates transforming growth factor-beta induction of alpha-smooth muscle actin in fibroblasts. *Am. J. Pathol., 154*, 871–882.

Arroyo, A. G., & Iruela-Arispe, M. L. (2010). Extracellular matrix, inflammation, and the angiogenic response. *Cardiovasc. Res, 86*, 226–235.

Asch, A. S., Liu, I., Briccetti, F. M., Barnwell, J. W., Kwakye-Berko, F., Dokun, A., et al. (1993). Analysis of CD36 binding domains: ligand specificity controlled by dephosphorylation of an ectodomain. *Science, 262*, 1436–1440.

Assoian, R. K., & Klein, E. A. (2008). Growth control by intracellular tension and extracellular stiffness. *Trends Cell Biol., 18*, 347–352.

Atkinson, J. C., Ruhl, M., Becker, J., Ackermann, R., & Schuppan, D. (1996). Collagen VI regulates normal and transformed mesenchymal cell proliferation *in vitro*. *Exp. Cell Res., 228*, 283–291.

Avdi, N. J., Nick, J. A., Whitlock, B. B., Billstrom, M. A., Henson, P. M., Johnson, G. L., & Worthen, G. S. (2001). Tumor necrosis factor-alpha activation of the c-Jun N-terminal kinase pathway in human neutrophils. Integrin involvement in a pathway leading from cytoplasmic tyrosine kinases apoptosis. *J. Biol. Chem., 276*, 2189–2199.

Bajorath, J., Greenfield, B., Munro, S. B., Day, A. J., & Aruffo, A. (1998). Identification of CD44 residues important for hyaluronan binding and delineation of the binding site. *J. Biol. Chem., 273*, 338–343.

Banerji, S., Wright, A. J., Noble, M., Mahoney, D. J., Campbell, I. D., Day, A. J., et al. (2007). Structures of the Cd44-hyaluronan complex provide insight into a fundamental carbohydrate-protein interaction. *Nat. Struct. Mol. Biol., 14*, 234–239.

Bao, W., Thullberg, M., Zhang, H., Onischenko, A., & Stromblad, S. (2002). Cell attachment to the extracellular matrix induces proteasomal degradation of p21(CIP1) via Cdc42/Rac1 signaling. *Mol. Cell Biol., 22*, 4587–4597.

Baron, V., & Schwartz, M. (2000). Cell adhesion regulates ubiquitin-mediated degradation of the platelet-derived growth factor receptor beta. *J. Biol. Chem., 275*, 39318–39323.

Bass, M. D., Morgan, M. R., Roach, K. A., Settleman, J., Goryachev, A. B., & Humphries, M. J. (2008). p190RhoGAP is the convergence point of adhesion signals from alpha 5 beta 1 integrin and syndecan-4. *J. Cell Biol., 181,* 1013—1026.

Bass, M. D., Roach, K. A., Morgan, M. R., Mostafavi-Pour, Z., Schoen, T., Muramatsu, T., et al. (2007). Syndecan-4-dependent Rac1 regulation determines directional migration in response to the extracellular matrix. *J. Cell Biol., 177,* 527—538.

Bateman, J. F., Boot-Handford, R. P., & Lamande, S. R. (2009). Genetic diseases of connective tissues: cellular and extracellular effects of ECM mutations. *Nat. Rev. Genet., 10,* 173—183.

Beanes, S. R., Dang, C., Soo, C., Wang, Y., Urata, M., Ting, K., et al. (2001). Down-regulation of decorin, a transforming growth factor-beta modulator, is associated with scarless fetal wound healing. *J. Pediatr. Surg., 36,* 1666—1671.

Bell, E., Ivarsson, B., & Merrill, C. (1979). Production of a tissue-like structure by contraction of collagen lattices by human fibroblasts of different proliferative potential *in vitro. Proc. Natl. Acad. Sci. U.S.A., 76,* 1274—1278.

Bell, S. E., Mavila, A., Salazar, R., Bayless, K. J., Kanagala, S., Maxwell, S. A., et al. (2001). Differential gene expression during capillary morphogenesis in 3D collagen matrices: regulated expression of genes involved in basement membrane matrix assembly, cell cycle progression, cellular differentiation and G-protein signaling. *J. Cell Sci., 114,* 2755—2773.

Bergers, G., Brekken, R., McMahon, G., Vu, T. H., Itoh, T., Tamaki, K., et al. (2000). Matrix metalloproteinase-9 triggers the angiogenic switch during carcinogenesis. *Nat. Cell Biol., 2,* 737—744.

Bergheim, I., Guo, L., Davis, M. A., Duveau, I., & Arteel, G. E. (2006). Critical role of plasminogen activator inhibitor-1 in cholestatic liver injury and fibrosis. *J. Pharmacol. Exp. Ther., 316,* 592—600.

Bernfield, M. R., Banerjee, S. D., & Cohn, R. H. (1972). Dependence of salivary epithelial morphology and branching morphogenesis upon acid mucopolysaccharide-protein (proteoglycan) at the epithelial surface. *J. Cell Biol., 52,* 674—689.

Bernfield, M. R., Cohn, R. H., & Banerjee, S. D. (1973). Glycosaminoglycans and epithelial organ formation. *Amer. Zool., 13,* 1067—1083.

Bezerra, J. A., Bugge, T. H., Melin-Aldana, H., Sabla, G., Kombrinck, K. W., Witte, D. P., et al. (1999). Plasminogen deficiency leads to impaired remodeling after a toxic injury to the liver. *Proc. Natl. Acad. Sci. U.S.A., 96,* 15143—15148.

Bezerra, J. A., Currier, A. R., Melin-Aldana, H., Sabla, G., Bugge, T. H., Kombrinck, K. W., et al. (2001). Plasminogen activators direct reorganization of the liver lobule after acute injury. *Am. J. Pathol., 158,* 921—929.

Bill, H. M., Knudsen, B., Moores, S. L., Muthuswamy, S. K., Rao, V. R., Brugge, J. S., et al. (2004). Epidermal growth factor receptor-dependent regulation of integrin-mediated signaling and cell cycle entry in epithelial cells. *Mol. Cell Biol., 24,* 8586—8599.

Bissell, M. J., Hall, H. G., & Parry, G. (1982). How does the extracellular matrix direct gene expression? *J. Theor. Biol., 99,* 31—68.

Bitterman, P. B., Rennard, S. I., Adelberg, S., & Crystal, R. G. (1983). Role of fibronectin as a growth factor for fibroblasts. *J. Cell Biol., 97,* 1925—1932.

Blazejewski, S., Le Bail, B., Boussarie, L., Blanc, J. F., Malaval, L., Okubo, K., et al. (1997). Osteonectin (SPARC) expression in human liver and in cultured human liver myofibroblasts. *Am. J. Pathol., 151,* 651—657.

Blissett, A. R., Garbellini, D., Calomeni, E. P., Mihai, C., Elton, T. S., & Agarwal, G. (2009). Regulation of collagen fibrillogenesis by cell-surface expression of kinase dead DDR2. *J. Mol. Biol., 385,* 902—911.

Bonewald, L. F. (1999). Regulation and regulatory activities of transforming growth factor beta. *Crit. Rev. Eukaryot. Gene Expr., 9,* 33—44.

Borges, E., Jan, Y., & Ruoslahti, E. (2000). Platelet-derived growth factor receptor beta and vascular endothelial growth factor receptor 2 bind to the beta 3 integrin through its extracellular domain. *J. Biol. Chem., 275,* 39867—39873.

Bouchard, V., Harnois, C., Demers, M. J., Thibodeau, S., Laquerre, V., Gauthier, R., et al. (2008). B1 integrin/Fak/Src signaling in intestinal epithelial crypt cell survival: integration of complex regulatory mechanisms. *Apoptosis, 13,* 531—542.

Boudreau, N., Myers, C., & Bissell, M. J. (1995). From laminin to lamin: regulation of tissue-specific gene expression by the ECM. *Trends Cell Biol., 5,* 1—4.

Bourguignon, L. Y., Ramez, M., Gilad, E., Singleton, P. A., Man, M. Q., Crumrine, D. A., et al. (2006). Hyaluronan-CD44 interaction stimulates keratinocyte differentiation, lamellar body formation/secretion, and permeability barrier homeostasis. *J. Invest. Dermatol., 126,* 1356—1365.

Breitkreutz, D., Mirancea, N., & Nischt, R. (2009). Basement membranes in skin: unique matrix structures with diverse functions? *Histochem. Cell Biol., 132,* 1—10.

Bride, J., Viennet, C., Lucarz-Bietry, A., & Humbert, P. (2004). Indication of fibroblast apoptosis during the maturation of disc-shaped mechanically stressed collagen lattices. *Arch. Dermatol. Res., 295,* 312—317.

47

Brink, H. E., Bernstein, J., & Nicoll, S. B. (2009). Fetal dermal fibroblasts exhibit enhanced growth and collagen production in two- and three-dimensional culture in comparison to adult fibroblasts. *J. Tissue Eng. Regen. Med.*, *3*, 623–633.

Bristow, J. M., Sellers, M. H., Majumdar, D., Anderson, B., Hu, L., & Webb, D. J. (2009). The Rho-family GEF Asef2 activates Rac to modulate adhesion and actin dynamics and thereby regulate cell migration. *J. Cell Sci.*, *122*, 4535–4546.

Brown, T. A., Bouchard, T., St John, T., Wayner, E., & Carter, W. G. (1991). Human keratinocytes express a new CD44 core protein (CD44E) as a heparan-sulfate intrinsic membrane proteoglycan with additional exons. *J. Cell Biol.*, *113*, 207–221.

Bugge, T. H., Kombrinck, K. W., Flick, M. J., Daugherty, C. C., Danton, M. J., & Degen, J. L. (1996). Loss of fibrinogen rescues mice from the pleiotropic effects of plasminogen deficiency. *Cell*, *87*, 709–719.

Bullard, K. M., Cass, D. L., Banda, M. J., & Adzick, N. S. (1997). Transforming growth factor beta-1 decreases interstitial collagenase in healing human fetal skin. *J. Pediatr. Surg.*, *32*, 1023–1027.

Cai, S., Liu, Y., Zheng Shu, X., & Prestwich, G. D. (2005). Injectable glycosaminoglycan hydrogels for controlled release of human basic fibroblast growth factor. *Biomaterials*, *26*, 6054–6067.

Camino, A. M., Atorrasagasti, C., Maccio, D., Prada, F., Salvatierra, E., Rizzo, M., et al. (2008). Adenovirus-mediated inhibition of SPARC attenuates liver fibrosis in rats. *J. Gene Med.*, *10*, 993–1004.

Carey, D. J. (1997). Syndecans: multifunctional cell-surface co-receptors. *Biochem. J.*, *327*(Pt 1), 1–16.

Carlsson, R., Engvall, E., Freeman, A., & Ruoslahti, E. (1981). Laminin and fibronectin in cell adhesion: enhanced adhesion of cells from regenerating liver to laminin. *Proc. Natl. Acad. Sci. U.S.A.*, *78*, 2403–2406.

Carre, A. L., James, A. W., MacLeod, L., Kong, W., Kawai, K., Longaker, M. T., et al. (2010). Interaction of wingless protein (Wnt), transforming growth factor-beta1, and hyaluronan production in fetal and postnatal fibroblasts. *Plast. Reconstr. Surg.*, *125*, 74–88.

Carter, R., Sykes, V., & Lanning, D. (2009). Scarless fetal mouse wound healing may initiate apoptosis through caspase 7 and cleavage of PARP. *J. Surg. Res.*, *156*, 74–79.

Casilli, F., Bianchini, A., Gloaguen, I., Biordi, L., Alesse, E., Festuccia, C., et al. (2005). Inhibition of interleukin-8 (CXCL8/IL-8) responses by repertaxin, a new inhibitor of the chemokine receptors CXCR1 and CXCR2. *Biochem. Pharmacol.*, *69*, 385–394.

Cass, D. L., Bullard, K. M., Sylvester, K. G., Yang, E. Y., Sheppard, D., Herlyn, M., et al. (1998). Epidermal integrin expression is upregulated rapidly in human fetal wound repair. *J. Pediatr. Surg.*, *33*, 312–316.

Cavani, A., Zambruno, G., Marconi, A., Manca, V., Marchetti, M., & Giannetti, A. (1993). Distinctive integrin expression in the newly forming epidermis during wound healing in humans. *J. Invest. Dermatol.*, *101*, 600–604.

Chen, T. T., Luque, A., Lee, S., Anderson, S. M., Segura, T., & Iruela-Arispe, M. L. (2010). Anchorage of VEGF to the extracellular matrix conveys differential signaling responses to endothelial cells. *J. Cell Biol.*, *188*, 595–609.

Chen, W., Fu, X., Ge, S., Sun, T., & Sheng, Z. (2007). Differential expression of matrix metalloproteinases and tissue-derived inhibitors of metalloproteinase in fetal and adult skins. *Int. J. Biochem. Cell Biol.*, *39*, 997–1005.

Chen, W., Fu, X., Ge, S., Sun, T., Zhou, G., Jiang, D., et al. (2005). Ontogeny of expression of transforming growth factor-beta and its receptors and their possible relationship with scarless healing in human fetal skin. *Wound Repair Regen.*, *13*, 68–75.

Chijiiwa, K., Nakano, K., Kameoka, N., Nagai, E., & Tanaka, M. (1994). Proliferating cell nuclear antigen, plasma fibronectin, and liver regeneration rate after seventy percent hepatectomy in normal and cirrhotic rats. *Surgery*, *116*, 544–549.

Chin, G. S., Lee, S., Hsu, M., Liu, W., Kim, W. J., Levinson, H., et al. (2001). Discoidin domain receptors and their ligand, collagen, are temporally regulated in fetal rat fibroblasts *in vitro*. *Plast. Reconstr. Surg.*, *107*, 769–776.

Chiquet-Ehrismann, R., Matsuoka, Y., Hofer, U., Spring, J., Bernasconi, C., & Chiquet, M. (1991). Tenascin variants: differential binding to fibronectin and distinct distribution in cell cultures and tissues. *Cell Regul.*, *2*, 927–938.

Chiu, C. C., Sheu, J. C., Chen, C. H., Lee, C. Z., Chiou, L. L., Chou, S. H., et al. (2009). Global gene expression profiling reveals a key role of CD44 in hepatic oval-cell reaction after 2-AAF/CCl4 injury in rodents. *Histochem. Cell Biol.*, *132*, 479–489.

Chlenski, A., Liu, S., Guerrero, L. J., Yang, Q., Tian, Y., Salwen, H. R., et al. (2005). SPARC expression is associated with impaired tumor growth, inhibited angiogenesis and changes in the extracellular matrix. *Int. J. Cancer*, *118*, 310–316.

Cho, H. J., Youn, S. W., Cheon, S. I., Kim, T. Y., Hur, J., Zhang, S. Y., et al. (2005). Regulation of endothelial cell and endothelial progenitor cell survival and vasculogenesis by integrin-linked kinase. *Arterioscler. Thromb. Vasc. Biol.*, *25*, 1154–1160.

Choi, B. M., Kwak, H. J., Jun, C. D., Park, S. D., Kim, K. Y., Kim, H. R., et al. (1996). Control of scarring in adult wounds using antisense transforming growth factor-beta 1 oligodeoxynucleotides. *Immunol. Cell Biol.*, *74*, 144–150.

Chung, C. Y., & Erickson, H. P. (1994). Cell surface annexin II is a high affinity receptor for the alternatively spliced segment of tenascin-C. *J. Cell Biol., 126,* 539–548.

Clark, R. A., McCoy, G. A., Folkvord, J. M., & McPherson, J. M. (1997). TGF-beta 1 stimulates cultured human fibroblasts to proliferate and produce tissue-like fibroplasia: a fibronectin matrix-dependent event. *J. Cell Physiol., 170,* 69–80.

Clark, R. A. F. (Ed.), (1996). *The Molecular and Cellular Biology of Wound Repair.* New York: Plenum Press.

Conway, E. M., Collen, D., & Carmeliet, P. (2001). Molecular mechanisms of blood vessel growth. *Cardiovasc. Res., 49,* 507–521.

Coolen, N. A., Schouten, K. C., Middelkoop, E., & Ulrich, M. M. (2010). Comparison between human fetal and adult skin. *Arch. Dermatol. Res., 302,* 47–55.

Coraux, C., Roux, J., Jolly, T., & Birembaut, P. (2008). Epithelial cell-extracellular matrix interactions and stem cells in airway epithelial regeneration. *Proc. Am. Thorac. Soc., 5,* 689–694.

Cotman, S. L., Halfter, W., & Cole, G. J. (1999). Identification of extracellular matrix ligands for the heparan sulfate proteoglycan agrin. *Exp. Cell Res., 249,* 54–64.

Cox, E. A., & Huttenlocher, A. (1998). Regulation of integrin-mediated adhesion during cell migration. *Microsc. Res. Tech., 43,* 412–419.

Culav, E. M., Clark, C. H., & Merrilees, M. J. (1999). Connective tissues: matrix composition and its relevance to physical therapy. *Phys. Ther., 79,* 308–319.

Cursio, R., Mari, B., Louis, K., Rostagno, P., Saint-Paul, M. C., Giudicelli, J., et al. (2002). Rat liver injury after normothermic ischemia is prevented by a phosphinic matrix metalloproteinase inhibitor. *Faseb J., 16,* 93–95.

Cybulsky, A. V., Carbonetto, S., Cyr, M. D., McTavish, A. J., & Huang, Q. (1993). Extracellular matrix-stimulated phospholipase activation is mediated by beta 1-integrin. *Am. J. Physiol., 264,* C323–C332.

Dang, C. M., Beanes, S. R., Lee, H., Zhang, X., Soo, C., & Ting, K. (2003). Scarless fetal wounds are associated with an increased matrix metalloproteinase-to-tissue-derived inhibitor of metalloproteinase ratio. *Plast. Reconstr. Surg., 111,* 2273–2285.

Dangerfield, J., Larbi, K. Y., Huang, M. T., Dewar, A., & Nourshargh, S. (2002). PECAM-1 (CD31) homophilic interaction up-regulates alpha6beta1 on transmigrated neutrophils *in vivo* and plays a functional role in the ability of alpha6 integrins to mediate leukocyte migration through the perivascular basement membrane. *J. Exp. Med., 196,* 1201–1211.

David-Raoudi, M., Tranchepain, F., Deschrevel, B., Vincent, J. C., Bogdanowicz, P., Boumediene, K., et al. (2008). Differential effects of hyaluronan and its fragments on fibroblasts: relation to wound healing. *Wound Repair Regen., 16,* 274–287.

Dawson, R. A., Goberdhan, N. J., Freedlander, E., & MacNeil, S. (1996). Influence of extracellular matrix proteins on human keratinocyte attachment, proliferation and transfer to a dermal wound model. *Burns, 22,* 93–100.

de Buys Roessingh, A. S., Hohlfeld, J., Scaletta, C., Hirt-Burri, N., Gerber, S., Hohlfeld, P., et al. (2006). Development, characterization, and use of a fetal skin cell bank for tissue engineering in wound healing. *Cell Transplant., 15,* 823–834.

de Paz, J. L., Moseman, E. A., Noti, C., Polito, L., von Andrian, U. H., & Seeberger, P. H. (2007). Profiling heparin-chemokine interactions using synthetic tools. *ACS Chem. Biol., 2,* 735–744.

Deakin, N. O., & Turner, C. E. (2008). Paxillin comes of age. *J. Cell Sci., 121,* 2435–2444.

Dean, J. W., III, Chandrasekaran, S., & Tanzer, M. L. (1990). A biological role of the carbohydrate moieties of laminin. *J. Biol. Chem., 265,* 12553–12562.

Decline, F., & Rousselle, P. (2001). Keratinocyte migration requires alpha2beta1 integrin-mediated interaction with the laminin 5 gamma2 chain. *J. Cell Sci., 114,* 811–823.

Dedhar, S. (1999). Integrins and signal transduction. *Curr. Opin. Hematol., 6,* 37–43.

Defamie, V., Laurens, M., Patrono, D., Devel, L., Brault, A., Saint-Paul, M. C., et al. (2008). Matrix metalloproteinase inhibition protects rat livers from prolonged cold ischemia-warm reperfusion injury. *Hepatology, 47,* 177–185.

Delon, I., & Brown, N. H. (2007). Integrins and the actin cytoskeleton. *Curr. Opin. Cell Biol., 19,* 43–50.

Desgrosellier, J. S., & Cheresh, D. A. (2010). Integrins in cancer: biological implications and therapeutic opportunities. *Nat. Rev. Cancer, 10,* 9–22.

Desmouliere, A., Chaponnier, C., & Gabbiani, G. (2005). Tissue repair, contraction, and the myofibroblast. *Wound Repair Regen., 13,* 7–12.

Desmouliere, A., Redard, M., Darby, I., & Gabbiani, G. (1995). Apoptosis mediates the decrease in cellularity during the transition between granulation tissue and scar. *Am. J. Pathol., 146,* 56–66.

Desmouliere, A., Rubbia-Brandt, L., Grau, G., & Gabbiani, G. (1992). Heparin induces alpha-smooth muscle actin expression in cultured fibroblasts and in granulation tissue myofibroblasts. *Lab. Invest., 67,* 716–726.

Dhanabal, M., Ramchandran, R., Waterman, M. J., Lu, H., Knebelmann, B., Segal, M., et al. (1999). Endostatin induces endothelial cell apoptosis. *J. Biol. Chem., 274,* 11721–11726.

Dockery, P., Khalid, J., Sarani, S. A., Bulut, H. E., Warren, M. A., Li, T. C., et al. (1998). Changes in basement membrane thickness in the human endometrium during the luteal phase of the menstrual cycle. *Hum. Reprod. Update, 4,* 486–495.

Drew, A. F., Liu, H., Davidson, J. M., Daugherty, C. C., & Degen, J. L. (2001). Wound-healing defects in mice lacking fibrinogen. *Blood, 97,* 3691–3698.

Dubey, P. K., Singodia, D., & Vyas, S. P. (2009). Liposomes modified with YIGSR peptide for tumor targeting. *J. Drug Target, 183,* 73–80.

Duca, L., Blanchevoye, C., Cantarelli, B., Ghoneim, C., Dedieu, S., Delacoux, F., et al. (2007). The elastin receptor complex transduces signals through the catalytic activity of its Neu-1 subunit. *J. Biol. Chem., 282,* 12484–12491.

Duca, L., Floquet, N., Alix, A. J., Haye, B., & Debelle, L. (2004). Elastin as a matrikine. *Crit. Rev. Oncol. Hematol., 49,* 235–244.

Edwards, G., & Streuli, C. (1995). Signalling in extracellular-matrix-mediated control of epithelial cell phenotype. *Biochem. Soc. Trans., 23,* 464–468.

Ehnis, T., Dieterich, W., Bauer, M., Lampe, B., & Schuppan, D. (1996). A chondroitin/dermatan sulfate form of CD44 is a receptor for collagen XIV (undulin). *Exp. Cell Res., 229,* 388–397.

Ekblom, P., Ekblom, M., Fecker, L., Klein, G., Zhang, H. Y., Kadoya, Y., et al. (1994). Role of mesenchymal nidogen for epithelial morphogenesis *in vitro. Development, 120,* 2003–2014.

Ellis, I., Grey, A. M., Schor, A. M., & Schor, S. L. (1992). Antagonistic effects of TGF-beta 1 and MSF on fibroblast migration and hyaluronic acid synthesis. Possible implications for dermal wound healing. *J. Cell Sci., 102*(Pt 3), 447–456.

Estes, J. M., Vande Berg, J. S., Adzick, N. S., MacGillivray, T. E., Desmouliere, A., & Gabbiani, G. (1994). Phenotypic and functional features of myofibroblasts in sheep fetal wounds. *Differentiation, 56,* 173–181.

Faraci, E., Eck, M., Gerstmayer, B., Bosio, A., & Vogel, W. F. (2003). An extracellular matrix-specific microarray allowed the identification of target genes downstream of discoidin domain receptors. *Matrix Biol., 22,* 373–381.

Faraldo, M. M., Deugnier, M. A., Thiery, J. P., & Glukhova, M. A. (2001). Growth defects induced by perturbation of beta1-integrin function in the mammary gland epithelium result from a lack of MAPK activation via the Shc and Akt pathways. *EMBO Rep., 2,* 431–437.

Febbraio, M., & Silverstein, R. L. (2007). CD36: implications in cardiovascular disease. *Int. J. Biochem. Cell Biol., 39,* 2012–2030.

Ferri, N., Carragher, N. O., & Raines, E. W. (2004). Role of discoidin domain receptors 1 and 2 in human smooth muscle cell-mediated collagen remodeling: potential implications in atherosclerosis and lymphangioleiomyomatosis. *Am. J. Pathol., 164,* 1575–1585.

Feugate, J. E., Li, Q., Wong, L., & Martins-Green, M. (2002a). The cxc chemokine cCAF stimulates differentiation of fibroblasts into myofibroblasts and accelerates wound closure. *J. Cell Biol., 156,* 161–172.

Feugate, J. E., Wong, L., Li, Q. J., & Martins-Green, M. (2002b). The CXC chemokine cCAF stimulates precocious deposition of ECM molecules by wound fibroblasts, accelerating development of granulation tissue. *BMC Cell Biol., 3,* 13.

Ffrench-Constant, C., & Hynes, R. O. (1989). Alternative splicing of fibronectin is temporally and spatially regulated in the chicken embryo. *Development, 106,* 375–388.

Flaumenhaft, R., & Rifkin, D. B. (1992). The extracellular regulation of growth factor action. *Mol. Biol. Cell, 3,* 1057–1065.

Fluck, J., Querfeld, C., Cremer, A., Niland, S., Krieg, T., & Sollberg, S. (1998). Normal human primary fibroblasts undergo apoptosis in three-dimensional contractile collagen gels. *J. Invest. Dermatol., 110,* 153–157.

Flynn, L. A., Blissett, A. R., Calomeni, E. P., & Agarwal, G. (2010). Inhibition of collagen fibrillogenesis by cells expressing soluble extracellular domains of DDR1 and DDR2. *J. Mol. Biol., 395,* 533–543.

Francki, A., Motamed, K., McClure, T. D., Kaya, M., Murri, C., Blake, D. J., et al. (2003). SPARC regulates cell cycle progression in mesangial cells via its inhibition of IGF-dependent signaling. *J. Cell Biochem., 88,* 802–811.

Freitas, V. M., Vilas-Boas, V. F., Pimenta, D. C., Loureiro, V., Juliano, M. A., Carvalho, M. R., et al. (2007). SIKVAV, a laminin alpha1-derived peptide, interacts with integrins and increases protease activity of a human salivary gland adenoid cystic carcinoma cell line through the ERK 1/2 signaling pathway. *Am. J. Pathol., 171,* 124–138.

Friedman, S. L., Roll, F. J., Boyles, J., Arenson, D. M., & Bissell, D. M. (1989). Maintenance of differentiated phenotype of cultured rat hepatic lipocytes by basement membrane matrix. *J. Biol. Chem., 264,* 10756–10762.

Fuchs, E., Dowling, J., Segre, J., Lo, S. H., & Yu, Q. C. (1997). Integrators of epidermal growth and differentiation: distinct functions for beta 1 and beta 4 integrins. *Curr. Opin. Genet. Dev., 7,* 672–682.

Fujimoto, N., Tajima, S., & Ishibashi, A. (2000). Elastin peptides induce migration and terminal differentiation of cultured keratinocytes via 67 kDa elastin receptor *in vitro*: 67 kDa elastin receptor is expressed in the keratinocytes eliminating elastic materials in elastosis perforans serpiginosa. *J. Invest. Dermatol., 115,* 633–639.

Fukai, F., Mashimo, M., Akiyama, K., Goto, T., Tanuma, S., & Katayama, T. (1998). Modulation of apoptotic cell death by extracellular matrix proteins and a fibronectin-derived antiadhesive peptide. *Exp. Cell Res., 242*, 92–99.

Funk, S. E., & Sage, E. H. (1991). The Ca2(+)-binding glycoprotein SPARC modulates cell cycle progression in bovine aortic endothelial cells. *Proc. Natl. Acad. Sci. U.S.A., 88*, 2648–2652.

Gailit, J., & Ruoslahti, E. (1988). Regulation of the fibronectin receptor affinity by divalent cations. *J. Biol. Chem., 263*, 12927–12932.

Gailit, J., Welch, M. P., & Clark, R. A. (1994). TGF-beta 1 stimulates expression of keratinocyte integrins during re-epithelialization of cutaneous wounds. *J. Invest. Dermatol., 103*, 221–227.

Gallai, M., Sebestyen, A., Nagy, P., Kovalszky, I., Onody, T., & Thorgeirsson, S. S. (1996). Proteoglycan gene expression in rat liver after partial hepatectomy. *Biochem. Biophys. Res. Commun., 228*, 690–694.

Gao, F., Yang, C. X., Mo, W., Liu, Y. W., & He, Y. Q. (2008). Hyaluronan oligosaccharides are potential stimulators to angiogenesis via RHAMM mediated signal pathway in wound healing. *Clin. Invest. Med., 31*, E106–E116.

Geer, D. J., & Andreadis, S. T. (2003). A novel role of fibrin in epidermal healing: plasminogen-mediated migration and selective detachment of differentiated keratinocytes. *J. Invest. Dermatol., 121*, 1210–1216.

Gharaee-Kermani, M., Hu, B., Phan, S. H., & Gyetko, M. R. (2009). Recent advances in molecular targets and treatment of idiopathic pulmonary fibrosis: focus on TGFbeta signaling and the myofibroblast. *Curr. Med. Chem., 16*, 1400–1417.

Ghert, M. A., Qi, W. N., Erickson, H. P., Block, J. A., & Scully, S. P. (2001). Tenascin-C splice variant adhesive/anti-adhesive effects on chondrosarcoma cell attachment to fibronectin. *Cell Struct. Funct., 26*, 179–187.

Giancotti, F. G. (1997). Integrin signaling: specificity and control of cell survival and cell cycle progression. *Curr. Opin. Cell Biol., 9*, 691–700.

Gillitzer, R., & Goebeler, M. (2001). Chemokines in cutaneous wound healing. *J. Leukoc. Biol., 69*, 513–521.

Gilmore, A. P., Metcalfe, A. D., Romer, L. H., & Streuli, C. H. (2000). Integrin-mediated survival signals regulate the apoptotic function of Bax through its conformation and subcellular localization. *J. Cell Biol., 149*, 431–446.

Gitay-Goren, H., Soker, S., Vlodavsky, I., & Neufeld, G. (1992). The binding of vascular endothelial growth factor to its receptors is dependent on cell surface-associated heparin-like molecules. *J. Biol. Chem., 267*, 6093–6098.

Gkretsi, V., Apte, U., Mars, W. M., Bowen, W. C., Luo, J. H., Yang, Y., et al. (2008). Liver-specific ablation of integrin-linked kinase in mice results in abnormal histology, enhanced cell proliferation, and hepatomegaly. *Hepatology, 48*, 1932–1941.

Gluck, U., Rodriguez Fernandez, J. L., Pankov, R., & Ben-Ze'ev, A. (1992). Regulation of adherens junction protein expression in growth-activated 3T3 cells and in regenerating liver. *Exp. Cell Res., 202*, 477–486.

Gosiewska, A., Yi, C. F., Brown, L. J., Cullen, B., Silcock, D., & Geesin, J. C. (2001). Differential expression and regulation of extracellular matrix-associated genes in fetal and neonatal fibroblasts. *Wound Repair Regen., 9*, 213–222.

Goueffic, Y., Guilluy, C., Guerin, P., Patra, P., Pacaud, P., & Loirand, G. (2006). Hyaluronan induces vascular smooth muscle cell migration through RHAMM-mediated PI3K-dependent Rac activation. *Cardiovasc. Res., 72*, 339–348.

Graf, R., Freyberg, M., Kaiser, D., & Friedl, P. (2002). Mechanosensitive induction of apoptosis in fibroblasts is regulated by thrombospondin-1 and integrin associated protein (CD47). *Apoptosis, 7*, 493–498.

Granes, F., Garcia, R., Casaroli-Marano, R. P., Castel, S., Rocamora, N., Reina, M., et al. (1999). Syndecan-2 induces filopodia by active cdc42Hs. *Exp. Cell Res., 248*, 439–456.

Grant, D. S., Kinsella, J. L., Fridman, R., Auerbach, R., Piasecki, B. A., Yamada, Y., et al. (1992). Interaction of endothelial cells with a laminin A chain peptide (SIKVAV) *in vitro* and induction of angiogenic behavior *in vivo*. *J. Cell Physiol., 153*, 614–625.

Grant, D. S., Tashiro, K., Segui-Real, B., Yamada, Y., Martin, G. R., & Kleinman, H. K. (1989). Two different laminin domains mediate the differentiation of human endothelial cells into capillary-like structures *in vitro*. *Cell, 58*, 933–943.

Grant, M. B., Caballero, S., Bush, D. M., & Spoerri, P. E. (1998). Fibronectin fragments modulate human retinal capillary cell proliferation and migration. *Diabetes, 47*, 1335–1340.

Gratchev, A., Kzhyshkowska, J., Utikal, J., & Goerdt, S. (2005). Interleukin-4 and dexamethasone counterregulate extracellular matrix remodelling and phagocytosis in type-2 macrophages. *Scand. J. Immunol., 61*, 10–17.

Greiling, D., & Clark, R. A. (1997). Fibronectin provides a conduit for fibroblast transmigration from collagenous stroma into fibrin clot provisional matrix. *J. Cell Sci., 110*(Pt 7), 861–870.

Grinnell, F., Zhu, M., Carlson, M. A., & Abrams, J. M. (1999). Release of mechanical tension triggers apoptosis of human fibroblasts in a model of regressing granulation tissue. *Exp. Cell Res., 248*, 608–619.

Grose, R., Hutter, C., Bloch, W., Thorey, I., Watt, F. M., Fassler, R., et al. (2002). A crucial role of beta 1 integrins for keratinocyte migration *in vitro* and during cutaneous wound repair. *Development, 129*, 2303–2315.

Gullberg, D., & Ekblom, P. (1995). Extracellular matrix and its receptors during development. *Int. J. Dev. Biol., 39*, 845–854.

Guo, W., & Giancotti, F. G. (2004). Integrin signalling during tumour progression. *Nat. Rev. Mol. Cell Biol., 5*, 816–826.

Gupta, K., Gupta, P., Wild, R., Ramakrishnan, S., & Hebbel, R. P. (1999). Binding and displacement of vascular endothelial growth factor (VEGF) by thrombospondin: effect on human microvascular endothelial cell proliferation and angiogenesis. *Angiogenesis, 3*, 147–158.

Hall, C. L., Wang, C., Lange, L. A., & Turley, E. A. (1994). Hyaluronan and the hyaluronan receptor RHAMM promote focal adhesion turnover and transient tyrosine kinase activity. *J. Cell Biol., 126*, 575–588.

Hallock, G. G., Rice, D. C., Merkel, J. R., & DiPaolo, B. R. (1988). Analysis of collagen content in the fetal wound. *Ann. Plast. Surg., 21*, 310–315.

Hamadi, A., Deramaudt, T. B., Takeda, K., & Ronde, P. (2009). Src activation and translocation from focal adhesions to membrane ruffles contribute to formation of new adhesion sites. *Cell Mol. Life Sci., 66*, 324–338.

Hamano, Y., Zeisberg, M., Sugimoto, H., Lively, J. C., Maeshima, Y., Yang, C., et al. (2003). Physiological levels of tumstatin, a fragment of collagen IV alpha3 chain, are generated by MMP-9 proteolysis and suppress angiogenesis via alphaV beta3 integrin. *Cancer Cell, 3*, 589–601.

Hamilton, S. R., Fard, S. F., Paiwand, F. F., Tolg, C., Veiseh, M., Wang, C., et al. (2007). The hyaluronan receptors CD44 and Rhamm (CD168) form complexes with ERK1,2 that sustain high basal motility in breast cancer cells. *J. Biol. Chem., 282*, 16667–16680.

Hangai, M., Kitaya, N., Xu, J., Chan, C. K., Kim, J. J., Werb, Z., et al. (2002). Matrix metalloproteinase-9-dependent exposure of a cryptic migratory control site in collagen is required before retinal angiogenesis. *Am. J. Pathol., 161*, 1429–1437.

Hartwig, B., Borm, B., Schneider, H., Arin, M. J., Kirfel, G., & Herzog, V. (2007). Laminin-5-deficient human keratinocytes: defective adhesion results in a saltatory and inefficient mode of migration. *Exp. Cell Res., 313*, 1575–1587.

Harty, M., Neff, A. W., King, M. W., & Mescher, A. L. (2003). Regeneration or scarring: an immunologic perspective. *Dev. Dyn., 226*, 268–279.

Hausser, H., Groning, A., Hasilik, A., Schonherr, E., & Kresse, H. (1994). Selective inactivity of TGF-beta/decorin complexes. *FEBS Lett., 353*, 243–245.

Hay, E. D. (1977). Interaction between the cell surface and extracellular matrix in corneal development. In J. W. Lash & M. M. Burger (Eds.), *Cell and Tissue Interactions* (pp. 115–137). New York: Raven Press.

Hellstrom, M., Gerhardt, H., Kalen, M., Li, X., Eriksson, U., Wolburg, H., et al. (2001). Lack of pericytes leads to endothelial hyperplasia and abnormal vascular morphogenesis. *J. Cell Biol., 153*, 543–553.

Herard, A. L., Pierrot, D., Hinnrasky, J., Kaplan, H., Sheppard, D., Puchelle, E., et al. (1996). Fibronectin and its alpha 5 beta 1-integrin receptor are involved in the wound-repair process of airway epithelium. *Am. J. Physiol., 271*, L726–L733.

Hergott, G. J., Nagai, H., & Kalnins, V. I. (1993). Inhibition of retinal pigment epithelial cell migration and proliferation with monoclonal antibodies against the beta 1 integrin subunit during wound healing in organ culture. *Invest. Ophthalmol. Vis. Sci., 34*, 2761–2768.

Hildebrand, A., Romaris, M., Rasmussen, L. M., Heinegard, D., Twardzik, D. R., Border, W. A., et al. (1994). Interaction of the small interstitial proteoglycans biglycan, decorin and fibromodulin with transforming growth factor beta. *Biochem. J., 302*(Pt 2), 527–534.

Hinek, A., Bodnaruk, T. D., Bunda, S., Wang, Y., & Liu, K. (2008). Neuraminidase-1, a subunit of the cell surface elastin receptor, desialylates and functionally inactivates adjacent receptors interacting with the mitogenic growth factors PDGF-BB and IGF-2. *Am. J. Pathol., 173*, 1042–1056.

Hinz, B. (2007). Formation and function of the myofibroblast during tissue repair. *J. Invest. Dermatol., 127*, 526–537.

Hirata, K., Usui, T., Koshiba, H., Maruyama, Y., Oikawa, I., Freeman, A. E., et al. (1983). Effects of basement membrane matrix on the culture of fetal mouse hepatocytes. *Gann, 74*, 687–692.

Hirsch, E., Barberis, L., Brancaccio, M., Azzolino, O., Xu, D., Kyriakis, J. M., et al. (2002). Defective Rac-mediated proliferation and survival after targeted mutation of the beta1 integrin cytodomain. *J. Cell Biol., 157*, 481–492.

Hirschi, K. K., & D'Amore, P. A. (1996). Pericytes in the microvasculature. *Cardiovasc. Res., 32*, 687–698.

Hodge-Dufour, J., Noble, P. W., Horton, M. R., Bao, C., Wysoka, M., Burdick, M. D., et al. (1997). Induction of IL-12 and chemokines by hyaluronan requires adhesion-dependent priming of resident but not elicited macrophages. *J. Immunol., 159*, 2492–2500.

Hoffman, M., & Monroe, D. M., III. (2001). A cell-based model of hemostasis. *Thromb. Haemost., 85*, 958–965.

Hoffmann, S., He, S., Jin, M., Ehren, M., Wiedemann, P., Ryan, S. J., et al. (2005). A selective cyclic integrin antagonist blocks the integrin receptors alphavbeta3 and alphavbeta5 and inhibits retinal pigment epithelium cell attachment, migration and invasion. *BMC Ophthalmol., 5,* 16.

Hood, J. D., Frausto, R., Kiosses, W. B., Schwartz, M. A., & Cheresh, D. A. (2003). Differential alphav integrin-mediated Ras-ERK signaling during two pathways of angiogenesis. *J. Cell Biol., 162,* 933–943.

Hu, M., Sabelman, E. E., Cao, Y., Chang, J., & Hentz, V. R. (2003). Three-dimensional hyaluronic acid grafts promote healing and reduce scar formation in skin incision wounds. *J. Biomed. Mater. Res. B Appl. Biomater., 67,* 586–592.

Hu, Y. B., Li, D. G., & Lu, H. M. (2007). Modified synthetic siRNA targeting tissue inhibitor of metalloproteinase-2 inhibits hepatic fibrogenesis in rats. *J. Gene Med., 9,* 217–229.

Huang, E. Y., Wu, H., Island, E. R., Chong, S. S., Warburton, D., Anderson, K. D., et al. (2002). Differential expression of urokinase-type plasminogen activator and plasminogen activator inhibitor-1 in early and late gestational mouse skin and skin wounds. *Wound Repair Regen., 10,* 387–396.

Huang, Y., Arora, P., McCulloch, C. A., & Vogel, W. F. (2009). The collagen receptor DDR1 regulates cell spreading and motility by associating with myosin IIA. *J. Cell Sci., 122,* 1637–1646.

Huang-Lee, L. L., Wu, J. H., & Nimni, M. E. (1994). Effects of hyaluronan on collagen fibrillar matrix contraction by fibroblasts. *J. Biomed. Mater. Res., 28,* 123–132.

Humphries, M. J., Mould, A. P., & Yamada, K. M. (1991). Matrix receptors in cell migration. In J. A. McDonald & R. P. Mecham (Eds.), *Receptors for Extracellular Matrix* (pp. 195–253). San Diego: Academic Press.

Hynes, R. O. (1987). Integrins: a family of cell surface receptors. *Cell, 48,* 549–554.

Hynes, R. O. (1992). Integrins: versatility, modulation, and signaling in cell adhesion. *Cell, 69,* 11–25.

Hynes, R. O. (2009). The extracellular matrix: not just pretty fibrils. *Science, 326,* 1216–1219.

Iacob, D., Cai, J., Tsonis, M., Babwah, A., Chakraborty, C., Bhattacharjee, R. N., et al. (2008). Decorin-mediated inhibition of proliferation and migration of the human trophoblast via different tyrosine kinase receptors. *Endocrinology, 149,* 6187–6197.

Ignotz, R. A., & Massague, J. (1986). Transforming growth factor-beta stimulates the expression of fibronectin and collagen and their incorporation into the extracellular matrix. *J. Biol. Chem., 261,* 4337–4345.

Iimuro, Y., Nishiura, T., Hellerbrand, C., Behrns, K. E., Schoonhoven, R., Grisham, J. W., et al. (1998). NFkappaB prevents apoptosis and liver dysfunction during liver regeneration. *J. Clin. Invest., 101,* 802–811.

Ikuta, T., Ariga, H., & Matsumoto, K. (2000). Extracellular matrix tenascin-X in combination with vascular endothelial growth factor B enhances endothelial cell proliferation. *Genes Cells, 5,* 913–927.

Ikuta, T., Ariga, H., & Matsumoto, K. I. (2001). Effect of tenascin-X together with vascular endothelial growth factor A on cell proliferation in cultured embryonic hearts. *Biol. Pharm. Bull., 24,* 1320–1323.

Ingber, D. (1991). Extracellular matrix and cell shape: potential control points for inhibition of angiogenesis. *J. Cell Biochem., 47,* 236–241.

Iocono, J. A., Ehrlich, H. P., Keefer, K. A., & Krummel, T. M. (1998a). Hyaluronan induces scarless repair in mouse limb organ culture. *J. Pediatr. Surg., 33,* 564–567.

Iocono, J. A., Krummel, T. M., Keefer, K. A., Allison, G. M., & Paul, H. (1998b). Repeated additions of hyaluronan alters granulation tissue deposition in sponge implants in mice. *Wound Repair Regen., 6.* 442–428.

Iozzo, R. V., Moscatello, D. K., McQuillan, D. J., & Eichstetter, I. (1999). Decorin is a biological ligand for the epidermal growth factor receptor. *J. Biol. Chem., 274,* 4489–4492.

Isenberg, J. S., Ridnour, L. A., Perruccio, E. M., Espey, M. G., Wink, D. A., & Roberts, D. D. (2005). Thrombospondin-1 inhibits endothelial cell responses to nitric oxide in a cGMP-dependent manner. *Proc. Natl. Acad. Sci. U.S.A., 102,* 13141–13146.

Ishihara, H., Yoshimoto, H., Fujioka, M., Murakami, R., Hirano, A., Fujii, T., et al. (2000). Keloid fibroblasts resist ceramide-induced apoptosis by overexpression of insulin-like growth factor I receptor. *J. Invest. Dermatol., 115,* 1065–1071.

Ishii, S., Ford, R., Thomas, P., Nachman, A., Steele, G., Jr., & Jessup, J. M. (1993). CD44 participates in the adhesion of human colorectal carcinoma cells to laminin and type IV collagen. *Surg. Oncol., 2,* 255–264.

Ishii, S., Steele, G., Jr., Ford, R., Paliotti, G., Thomas, P., Andrews, C., et al. (1994). Normal colonic epithelium adheres to carcinoembryonic antigen and type IV collagen. *Gastroenterology, 106,* 1242–1250.

Issa, R., Williams, E., Trim, N., Kendall, T., Arthur, M. J., Reichen, J., et al. (2001). Apoptosis of hepatic stellate cells: involvement in resolution of biliary fibrosis and regulation by soluble growth factors. *Gut, 48,* 548–557.

Issa, R., Zhou, X., Trim, N., Millward-Sadler, H., Krane, S., Benyon, C., et al. (2003). Mutation in collagen-1 that confers resistance to the action of collagenase results in failure of recovery from CCl4-induced liver fibrosis, persistence of activated hepatic stellate cells, and diminished hepatocyte regeneration. *Faseb J., 17,* 47–49.

Iwamoto, H., Sakai, H., Tada, S., Nakamuta, M., & Nawata, H. (1999). Induction of apoptosis in rat hepatic stellate cells by disruption of integrin-mediated cell adhesion. *J. Lab. Clin. Med., 134,* 83–89.

Iyer, A. K., Tran, K. T., Griffith, L., & Wells, A. (2008). Cell surface restriction of EGFR by a tenascin cytotactin-encoded EGF-like repeat is preferential for motility-related signaling. *J. Cell Physiol., 214*, 504–512.

Jakowlew, S. B., Mead, J. E., Danielpour, D., Wu, J., Roberts, A. B., & Fausto, N. (1991). Transforming growth factor-beta (TGF-beta) isoforms in rat liver regeneration: messenger RNA expression and activation of latent TGF-beta. *Cell Regul., 2*, 535–548.

Jalkanen, S., & Jalkanen, M. (1992). Lymphocyte CD44 binds the COOH-terminal heparin-binding domain of fibronectin. *J. Cell Biol., 116*, 817–825.

Jimenez, B., Volpert, O. V., Crawford, S. E., Febbraio, M., Silverstein, R. L., & Bouck, N. (2000). Signals leading to apoptosis-dependent inhibition of neovascularization by thrombospondin-1. *Nat. Med., 6*, 41–48.

Jimenez, B., Volpert, O. V., Reiher, F., Chang, L., Munoz, A., Karin, M., et al. (2001). c-Jun N-terminal kinase activation is required for the inhibition of neovascularization by thrombospondin-1. *Oncogene, 20*, 3443–3448.

Juhasz, I., Murphy, G. F., Yan, H. C., Herlyn, M., & Albelda, S. M. (1993). Regulation of extracellular matrix proteins and integrin cell substratum adhesion receptors on epithelium during cutaneous human wound healing *in vivo*. *Am. J. Pathol., 143*, 1458–1469.

Juric, V., Chen, C. C., & Lau, L. F. (2009). Fas-mediated apoptosis is regulated by the extracellular matrix protein CCN1 (CYR61) *in vitro* and *in vivo*. *Mol. Cell Biol., 29*, 3266–3279.

Kadl, A., & Leitinger, N. (2005). The role of endothelial cells in the resolution of acute inflammation. *Antioxid. Redox Signal, 7*, 1744–1754.

Kagami, S., Kondo, S., Loster, K., Reutter, W., Urushihara, M., Kitamura, A., et al. (1998). Collagen type I modulates the platelet-derived growth factor (PDGF) regulation of the growth and expression of beta1 integrins by rat mesangial cells. *Biochem. Biophys. Res. Commun., 252*, 728–732.

Kamei, M., Kawasaki, A., & Tano, Y. (1998). Analysis of extracellular matrix synthesis during wound healing of retinal pigment epithelial cells. *Microsc. Res. Tech., 42*, 311–316.

Kamisato, S., Uemura, Y., Takami, N., & Okamoto, K. (1997). Involvement of intracellular cyclic GMP and cyclic GMP-dependent protein kinase in alpha-elastin-induced macrophage chemotaxis. *J. Biochem., 121*, 862–867.

Kan, M., Wang, F., Xu, J., Crabb, J. W., Hou, J., & McKeehan, W. L. (1993). An essential heparin-binding domain in the fibroblast growth factor receptor kinase. *Science, 259*, 1918–1921.

Kapila, Y. L., Lancero, H., & Johnson, P. W. (1998). The response of periodontal ligament cells to fibronectin. *J. Periodontol., 69*, 1008–1019.

Kapila, Y. L., Wang, S., & Johnson, P. W. (1999). Mutations in the heparin binding domain of fibronectin in cooperation with the V region induce decreases in pp125(FAK) levels plus proteoglycan-mediated apoptosis via caspases. *J. Biol. Chem., 274*, 30906–30913.

Kato, R., Kamiya, S., Ueki, M., Yajima, H., Ishii, T., Nakamura, H., et al. (2001). The fibronectin-derived antiadhesive peptides suppress the myofibroblastic conversion of rat hepatic stellate cells. *Exp. Cell Res., 265*, 54–63.

Kato, S., Ishii, T., Hara, H., Sugiura, N., Kimata, K., & Akamatsu, N. (1994). Hepatocyte growth factor immobilized onto culture substrates through heparin and matrigel enhances DNA synthesis in primary rat hepatocytes. *Exp. Cell Res., 211*, 53–58.

Kato, S., Otsu, K., Ohtake, K., Kimura, Y., Yashiro, T., Suzuki, T., et al. (1992). Concurrent changes in sinusoidal expression of laminin and affinity of hepatocytes to laminin during rat liver regeneration. *Exp. Cell Res., 198*, 59–68.

Katz, M., Amit, I., Citri, A., Shay, T., Carvalho, S., Lavi, S., et al. (2007). A reciprocal tensin-3-cten switch mediates EGF-driven mammary cell migration. *Nat. Cell Biol., 9*, 961–969.

Kaya, G., Rodriguez, I., Jorcano, J. L., Vassalli, P., & Stamenkovic, I. (1997). Selective suppression of CD44 in keratinocytes of mice bearing an antisense CD44 transgene driven by a tissue-specific promoter disrupts hyaluronate metabolism in the skin and impairs keratinocyte proliferation. *Genes Dev., 11*, 996–1007.

Kaya, G., Tran, C., Sorg, O., Hotz, R., Grand, D., Carraux, P., et al. (2006). Hyaluronate fragments reverse skin atrophy by a CD44-dependent mechanism. *PLoS Med., 3*, e493.

Kettritz, R., Xu, Y. X., Kerren, T., Quass, P., Klein, J. B., Luft, F. C., et al. (1999). Extracellular matrix regulates apoptosis in human neutrophils. *Kidney Int., 55*, 562–571.

Khorramizadeh, M. R., Tredget, E. E., Telasky, C., Shen, Q., & Ghahary, A. (1999). Aging differentially modulates the expression of collagen and collagenase in dermal fibroblasts. *Mol. Cell Biochem., 194*, 99–108.

Khwaja, A., Rodriguez-Viciana, P., Wennstrom, S., Warne, P. H., & Downward, J. (1997). Matrix adhesion and Ras transformation both activate a phosphoinositide 3-OH kinase and protein kinase B/Akt cellular survival pathway. *Embo J., 16*, 2783–2793.

Kielty, C. M., Wess, T. J., Haston, L., Ashworth, J. L., Sherratt, M. J., & Shuttleworth, C. A. (2002). Fibrillin-rich microfibrils: elastic biopolymers of the extracellular matrix. *J. Muscle Res. Cell Motil., 23*, 581–596.

Kim, B. S., & Mooney, D. J. (1998). Engineering smooth muscle tissue with a predefined structure. *J. Biomed. Mater. Res., 41*, 322–332.

Kim, T. H., Mars, W. M., Stolz, D. B., Petersen, B. E., & Michalopoulos, G. K. (1997). Extracellular matrix remodeling at the early stages of liver regeneration in the rat. *Hepatology, 26*, 896–904.

Kirkland, G., Paizis, K., Wu, L. L., Katerelos, M., & Power, D. A. (1998). Heparin-binding EGF-like growth factor mRNA is upregulated in the peri-infarct region of the remnant kidney model: *in vitro* evidence suggests a regulatory role in myofibroblast transformation. *J. Am. Soc. Nephrol., 9*, 1464–1473.

Koshikawa, N., Minegishi, T., Sharabi, A., Quaranta, V., & Seiki, M. (2005). Membrane-type matrix metal-loproteinase-1 (MT1-MMP) is a processing enzyme for human laminin gamma 2 chain. *J. Biol. Chem., 280*, 88–93.

Kothapalli, D., Flowers, J., Xu, T., Pure, E., & Assoian, R. K. (2008). Differential activation of ERK and Rac mediates the proliferative and anti-proliferative effects of hyaluronan and CD44. *J. Biol. Chem., 283*, 31823–31829.

Krummel, T. M., Nelson, J. M., Diegelmann, R. F., Lindblad, W. J., Salzberg, A. M., et al. (1987). Fetal response to injury in the rabbit. *J. Pediatr. Surg., 22*, 640–644.

Kubo, M., van de Water, L., Plantefaber, L. C., Mosesson, M. W., Simon, M., Tonnesen, M. G., et al. (2001). Fibrinogen and fibrin are anti-adhesive for keratinocytes: a mechanism for fibrin eschar slough during wound repair. *J. Invest. Dermatol., 117*, 1369–1381.

Kubota, Y., Kleinman, H. K., Martin, G. R., & Lawley, T. J. (1988). Role of laminin and basement membrane in the morphological differentiation of human endothelial cells into capillary-like structures. *J. Cell Biol., 107*, 1589–1598.

Kupprion, C., Motamed, K., & Sage, E. H. (1998). SPARC (BM-40, osteonectin) inhibits the mitogenic effect of vascular endothelial growth factor on microvascular endothelial cells. *J. Biol. Chem., 273*, 29635–29640.

Kutz, S. M., Hordines, J., McKeown-Longo, P. J., & Higgins, P. J. (2001). TGF-beta1-induced PAI-1 gene expression requires MEK activity and cell-to-substrate adhesion. *J. Cell Sci., 114*, 3905–3914.

Kwon, A. H., Inada, Y., Uetsuji, S., Yamamura, M., Hioki, K., & Yamamoto, M. (1990a). Response of fibronectin to liver regeneration after hepatectomy. *Hepatology, 11*, 593–598.

Kwon, A. H., Uetsuji, S., Yamamura, M., Hioki, K., & Yamamoto, M. (1990b). Effect of administration of fibronectin or aprotinin on liver regeneration after experimental hepatectomy. *Ann. Surg., 211*, 295–300.

Labrador, J. P., Azcoitia, V., Tuckermann, J., Lin, C., Olaso, E., Manes, S., et al. (2001). The collagen receptor DDR2 regulates proliferation and its elimination leads to dwarfism. *EMBO Rep., 2*, 446–452.

Ladin, D. A., Hou, Z., Patel, D., McPhail, M., Olson, J. C., Saed, G. M., et al. (1998). p53 and apoptosis alterations in keloids and keloid fibroblasts. *Wound Repair Regen., 6*, 28–37.

Lai, J. P., Chien, J. R., Moser, D. R., Staub, J. K., Aderca, I., Montoya, D. P., et al. (2004). hSulf1 Sulfatase promotes apoptosis of hepatocellular cancer cells by decreasing heparin-binding growth factor signaling. *Gastroenterology, 126*, 231–248.

Lamszus, K., Joseph, A., Jin, L., Yao, Y., Chowdhury, S., Fuchs, A., et al. (1996). Scatter factor binds to thrombo-spondin and other extracellular matrix components. *Am. J. Pathol., 149*, 805–819.

Lanning, D. A., Nwomeh, B. C., Montante, S. J., Yager, D. R., Diegelmann, R. F., & Haynes, J. H. (1999). TGF-beta1 alters the healing of cutaneous fetal excisional wounds. *J. Pediatr. Surg., 34*, 695–700.

LaRochelle, W. J., May-Siroff, M., Robbins, K. C., & Aaronson, S. A. (1991). A novel mechanism regulating growth factor association with the cell surface: identification of a PDGF retention domain. *Genes Dev., 5*, 1191–1199.

Laurens, N., Koolwijk, P., & de Maat, M. P. (2006). Fibrin structure and wound healing. *J. Thromb. Haemost., 4*, 932–939.

Laurie, G. W., Horikoshi, S., Killen, P. D., Segui-Real, B., & Yamada, Y. (1989). In situ hybridization reveals temporal and spatial changes in cellular expression of mRNA for a laminin receptor, laminin, and basement membrane (type IV) collagen in the developing kidney. *J. Cell Biol., 109*, 1351–1362.

Lawley, T. J., & Kubota, Y. (1989). Induction of morphologic differentiation of endothelial cells in culture. *J. Invest. Dermatol., 93*, 59S–61S.

Lee, S., Bowrin, K., Hamad, A. R., & Chakravarti, S. (2009). Extracellular matrix lumican deposited on the surface of neutrophils promotes migration by binding to beta2 integrin. *J. Biol. Chem., 284*, 23662–23669.

Leitinger, B., & Hohenester, E. (2007). Mammalian collagen receptors. *Matrix Biol., 26*, 146–155.

Leitinger, B., McDowall, A., Stanley, P., & Hogg, N. (2000). The regulation of integrin function by Ca(2+). *Biochim. Biophys. Acta, 1498*, 91–98.

Li, J., Kleeff, J., Kayed, H., Felix, K., Penzel, R., Buchler, M. W., et al. (2004a). Glypican-1 antisense transfection modulates TGF-beta-dependent signaling in Colo-357 pancreatic cancer cells. *Biochem. Biophys. Res. Commun., 320*, 1148–1155.

Li, J., Zhang, Y. P., & Kirsner, R. S. (2003). Angiogenesis in wound repair: angiogenic growth factors and the extracellular matrix. *Microsc. Res. Tech., 60*, 107–114.

Li, Q. J., Vaingankar, S., Sladek, F. M., & Martins-Green, M. (2000). Novel nuclear target for thrombin: activation of the Elk1 transcription factor leads to chemokine gene expression. *Blood, 96*, 3696–3706.

Li, W., Henry, G., Fan, J., Bandyopadhyay, B., Pang, K., Garner, W., et al. (2004b). Signals that initiate, augment, and provide directionality for human keratinocyte motility. *J. Invest. Dermatol., 123*, 622–633.

Li, W. Y., Huang, E. Y., Dudas, M., Kaartinen, V., Warburton, D., & Tuan, T. L. (2006). Transforming growth factor-beta3 affects plasminogen activator inhibitor-1 expression in fetal mice and modulates fibroblast-mediated collagen gel contraction. *Wound Repair Regen., 14*, 516–525.

Lin, R. Y., Sullivan, K. M., Argenta, P. A., Meuli, M., Lorenz, H. P., & Adzick, N. S. (1995). Exogenous transforming growth factor-beta amplifies its own expression and induces scar formation in a model of human fetal skin repair. *Ann. Surg., 222*, 146–154.

Linge, C., Richardson, J., Vigor, C., Clayton, E., Hardas, B., & Rolfe, K. (2005). Hypertrophic scar cells fail to undergo a form of apoptosis specific to contractile collagen – the role of tissue transglutaminase. *J. Invest. Dermatol., 125*, 72–82.

Liu, A., Garg, P., Yang, S., Gong, P., Pallero, M. A., Annis, D. S., et al. (2009). Epidermal growth factor-like repeats of thrombospondins activate phospholipase Cgamma and increase epithelial cell migration through indirect epidermal growth factor receptor activation. *J. Biol. Chem., 284*, 6389–6402.

Liu, D., Liu, T., Li, R., & Sy, M. S. (1998). Mechanisms regulating the binding activity of CD44 to hyaluronic acid. *Front. Biosci., 3*, d631–d636.

Liu, Y., Cai, S., Shu, X. Z., Shelby, J., & Prestwich, G. D. (2007). Release of basic fibroblast growth factor from a crosslinked glycosaminoglycan hydrogel promotes wound healing. *Wound Repair Regen., 15*, 245–251.

Lo, S. H. (2004). Tensin. *Int. J. Biochem. Cell Biol., 36*, 31–34.

Loike, J. D., Cao, L., Budhu, S., Hoffman, S., & Silverstein, S. C. (2001). Blockade of alpha 5 beta 1 integrins reverses the inhibitory effect of tenascin on chemotaxis of human monocytes and polymorphonuclear leukocytes through three-dimensional gels of extracellular matrix proteins. *J. Immunol., 166*, 7534–7542.

Lokeshwar, V. B., & Selzer, M. G. (2000). Differences in hyaluronic acid-mediated functions and signaling in arterial, microvessel, and vein-derived human endothelial cells. *J. Biol. Chem., 275*, 27641–27649.

Longaker, M. T., Whitby, D. J., Adzick, N. S., Crombleholme, T. M., Langer, J. C., Duncan, B. W., et al. (1990). Studies in fetal wound healing, VI. Second and early third trimester fetal wounds demonstrate rapid collagen deposition without scar formation. *J. Pediatr. Surg., 25*, 63–68, discussion 68–69.

Longaker, M. T., Whitby, D. J., Ferguson, M. W., Harrison, M. R., Crombleholme, T. M., Langer, J. C., et al. (1989). Studies in fetal wound healing: III. Early deposition of fibronectin distinguishes fetal from adult wound healing. *J. Pediatr. Surg., 24*, 799–805.

Longo, C. R., Patel, V. I., Shrikhande, G. V., Scali, S. T., Csizmadia, E., Daniel, S., et al. (2005). A20 protects mice from lethal radical hepatectomy by promoting hepatocyte proliferation via a p21waf1-dependent mechanism. *Hepatology, 42*, 156–164.

Lovvorn, H. N., III, Cheung, D. T., Nimni, M. E., Perelman, N., Estes, J. M., & Adzick, N. S. (1999). Relative distribution and crosslinking of collagen distinguish fetal from adult sheep wound repair. *J. Pediatr. Surg., 34*, 218–223.

Lukacs, N. W., & Kunkel, S. L. (1998). Chemokines and their role in disease. *Int. J. Clin. Lab. Res., 28*, 91–95.

Lundberg, S., Lindholm, J., Lindbom, L., Hellstrom, P. M., & Werr, J. (2006). Integrin alpha2beta1 regulates neutrophil recruitment and inflammatory activity in experimental colitis in mice. *Inflamm. Bowel Dis., 12*, 172–177.

Ly, D. P., & Corbett, S. A. (2005). The integrin alpha5beta1 regulates alphavbeta3-mediated extracellular signal-regulated kinase activation. *J. Surg. Res., 123*, 200–205.

Lygoe, K. A., Norman, J. T., Marshall, J. F., & Lewis, M. P. (2004). AlphaV integrins play an important role in myofibroblast differentiation. *Wound Repair Regen., 12*, 461–470.

Lygoe, K. A., Wall, I., Stephens, P., & Lewis, M. P. (2007). Role of vitronectin and fibronectin receptors in oral mucosal and dermal myofibroblast differentiation. *Biol. Cell, 99*, 601–614.

Ma, T. Y., Kikuchi, M., Sarfeh, I. J., Shimada, H., Hoa, N. T., & Tarnawski, A. S. (1999). Basic fibroblast growth factor stimulates repair of wounded hepatocyte monolayer: modulatory role of protein kinase A and extracellular matrix. *J. Lab. Clin. Med., 134*, 363–371.

Madri, J. A., Pratt, B. M., & Tucker, A. M. (1988). Phenotypic modulation of endothelial cells by transforming growth factor-beta depends upon the composition and organization of the extracellular matrix. *J. Cell Biol., 106*, 1375–1384.

Maeshima, Y., Manfredi, M., Reimer, C., Holthaus, K. A., Hopfer, H., Chandamuri, B. R., et al. (2001). Identification of the anti-angiogenic site within vascular basement membrane-derived tumstatin. *J. Biol. Chem., 276*, 15240–15248.

Mahabeleshwar, G. H., Feng, W., Reddy, K., Plow, E. F., & Byzova, T. V. (2007). Mechanisms of integrin-vascular endothelial growth factor receptor cross-activation in angiogenesis. *Circ. Res., 101*, 570–580.

Mainiero, F., Murgia, C., Wary, K. K., Curatola, A. M., Pepe, A., Blumemberg, M., Westwick, J. K., et al. (1997). The coupling of alpha6beta4 integrin to Ras-MAP kinase pathways mediated by Shc controls keratinocyte proliferation. *Embo J.*, *16*, 2365–2375.

Mangnall, D., Smith, K., Bird, N. C., & Majeed, A. W. (2004). Early increases in plasminogen activator activity following partial hepatectomy in humans. *Comp. Hepatol.*, *3*, 11.

Martinez-Hernandez, A., Delgado, F. M., & Amenta, P. S. (1991). The extracellular matrix in hepatic regeneration. Localization of collagen types I, III, IV, laminin, and fibronectin. *Lab. Invest.*, *64*, 157–166.

Martins-Green, M. (2001). The chicken Chemotactic and Angiogenic Factor (cCAF), a CXC chemokine. *Int. J. Biochem. Cell Biol.*, *33*, 427–432.

Martins-Green, M., Aotaki-Keen, A., Hjelmeland, L. M., & Bissell, M. J. (1992). The 9E3 protein: immunolocalization *in vivo* and evidence for multiple forms in culture. *J. Cell Sci.*, *101*(Pt 3), 701–707.

Martins-Green, M., & Bissell, M. J. (1990). Localization of 9E3/CEF-4 in avian tissues: expression is absent in Rous sarcoma virus-induced tumors but is stimulated by injury. *J. Cell Biol.*, *110*, 581–595.

Martins-Green, M., & Bissell, M. J. (1995). Cell-extracellular matrix interactions in development. *Sems. in Dev. Biol.*, *6*, 149–159.

Martins-Green, M., & Feugate, J. E. (1998). The 9E3/CEF4 gene product is a chemotactic and angiogenic factor that can initiate the wound-healing cascade *in vivo*. *Cytokine*, *10*, 522–535.

Martins-Green, M., & Hanafusa, H. (1997). The 9E3/CEF4 gene and its product the chicken chemotactic and angiogenic factor (cCAF): potential roles in wound healing and tumor development. *Cytokine Growth Factor Rev.*, *8*, 221–232.

Martins-Green, M., & Kelly, T. (1998). The chicken chemotactic and angiogenic factor (9E3 gene product): its angiogenic properties reside in the C-terminus of the molecule. *Cytokine*, *10*, 819–830.

Martins-Green, M., Stoeckle, M., Hampe, A., Wimberly, S., & Hanafusa, H. (1996). The 9E3/CEF4 cytokine: kinetics of secretion, processing by plasmin, and interaction with extracellular matrix. *Cytokine*, *8*, 448–459.

Masson-Gadais, B., Houle, F., Laferriere, J., & Huot, J. (2003). Integrin alphavbeta3, requirement for VEGFR2-mediated activation of SAPK2/p38 and for Hsp90-dependent phosphorylation of focal adhesion kinase in endothelial cells activated by VEGF. *Cell Stress Chaperones*, *8*, 37–52.

Mast, B. A., Diegelmann, R. F., Krummel, T. M., & Cohen, I. K. (1993). Hyaluronic acid modulates proliferation, collagen and protein synthesis of cultured fetal fibroblasts. *Matrix*, *13*, 441–446.

Mast, B. A., Haynes, J. H., Krummel, T. M., Diegelmann, R. F., & Cohen, I. K. (1992). *In vivo* degradation of fetal wound hyaluronic acid results in increased fibroplasia, collagen deposition, and neovascularization. *Plast. Reconstr. Surg.*, *89*, 503–509.

Matsumoto, K., Tajima, H., Okazaki, H., & Nakamura, T. (1993). Heparin as an inducer of hepatocyte growth factor. *J. Biochem (Tokyo)*, *114*, 820–826.

Maxwell, C. A., McCarthy, J., & Turley, E. (2008). Cell-surface and mitotic-spindle RHAMM: moonlighting or dual oncogenic functions? *J. Cell Sci.*, *121*, 925–932.

Menart, V., Fonda, I., Kenig, M., & Porekar, V. G. (2002). Increased *in vitro* cytotoxicity of TNF-alpha analog LK-805 is based on the interaction with cell surface heparan sulfate proteoglycan. *Ann. N.Y. Acad. Sci.*, *973*, 194–206.

Meran, S., Thomas, D., Stephens, P., Martin, J., Bowen, T., Phillips, A., et al. (2007). Involvement of hyaluronan in regulation of fibroblast phenotype. *J. Biol. Chem.*, *282*, 25687–25697.

Mercurius, K. O., & Morla, A. O. (2001). Cell adhesion and signaling on the fibronectin 1st type III repeat; requisite roles for cell surface proteoglycans and integrins. *BMC Cell Biol.*, *2*, 18.

Messadi, D. V., Le, A., Berg, S., Jewett, A., Wen, Z., Kelly, P., et al. (1999). Expression of apoptosis-associated genes by human dermal scar fibroblasts. *Wound Repair Regen.*, *7*, 511–517.

Miao, H. Q., Ishai-Michaeli, R., Atzmon, R., Peretz, T., & Vlodavsky, I. (1996). Sulfate moieties in the subendothelial extracellular matrix are involved in basic fibroblast growth factor sequestration, dimerization, and stimulation of cell proliferation. *J. Biol. Chem.*, *271*, 4879–4886.

Michael, K. E., Dumbauld, D. W., Burns, K. L., Hanks, S. K., & Garcia, A. J. (2009). Focal adhesion kinase modulates cell adhesion strengthening via integrin activation. *Mol. Biol. Cell*, *20*, 2508–2519.

Miralem, T., Steinberg, R., Price, D., & Avraham, H. (2001). VEGF(165) requires extracellular matrix components to induce mitogenic effects and migratory response in breast cancer cells. *Oncogene*, *20*, 5511–5524.

Mitra, S. K., & Schlaepfer, D. D. (2006). Integrin-regulated FAK-Src signaling in normal and cancer cells. *Curr. Opin. Cell Biol.*, *18*, 516–523.

Miyake, K., Underhill, C. B., Lesley, J., & Kincade, P. W. (1990). Hyaluronate can function as a cell adhesion molecule and CD44 participates in hyaluronate recognition. *J. Exp. Med.*, *172*, 69–75.

Miyamoto, S., Teramoto, H., Gutkind, J. S., & Yamada, K. M. (1996). Integrins can collaborate with growth factors for phosphorylation of receptor tyrosine kinases and MAP kinase activation: roles of integrin aggregation and occupancy of receptors. *J. Cell Biol.*, *135*, 1633–1642.

Mochizuki, S., Brassart, B., & Hinek, A. (2002). Signaling pathways transduced through the elastin receptor facilitate proliferation of arterial smooth muscle cells. *J. Biol. Chem., 277,* 44854–44863.

Moghal, N., & Neel, B. G. (1998). Integration of growth factor, extracellular matrix, and retinoid signals during bronchial epithelial cell differentiation. *Mol. Cell Biol., 18,* 6666–6678.

Mohammed, F. F., Pennington, C. J., Kassiri, Z., Rubin, J. S., Soloway, P. D., Ruther, U., et al. (2005). Metalloproteinase inhibitor TIMP-1 affects hepatocyte cell cycle via HGF activation in murine liver regeneration. *Hepatology, 41,* 857–867.

Montesano, R., Orci, L., & Vassalli, P. (1983). *In vitro* rapid organization of endothelial cells into capillary-like networks is promoted by collagen matrices. *J. Cell Biol., 97,* 1648–1652.

Morgan, M. R., Humphries, M. J., & Bass, M. D. (2007). Synergistic control of cell adhesion by integrins and syndecans. *Nat. Rev. Mol. Cell Biol., 8,* 957–969.

Mori, S., Wu, C. Y., Yamaji, S., Saegusa, J., Shi, B., Ma, Z., et al. (2008). Direct binding of integrin alphavbeta3 to FGF1 plays a role in FGF1 signaling. *J. Biol. Chem., 283,* 18066–18075.

Morita, T., Togo, S., Kubota, T., Kamimukai, N., Nishizuka, I., Kobayashi, T., et al. (2002). Mechanism of post-operative liver failure after excessive hepatectomy investigated using a cDNA microarray. *J. Hepatobiliary Pancreat. Surg., 9,* 352–359.

Moro, L., Dolce, L., Cabodi, S., Bergatto, E., Boeri Erba, E., Smeriglio, M., et al. (2002). Integrin-induced epidermal growth factor (EGF) receptor activation requires c-Src and p130Cas and leads to phosphorylation of specific EGF receptor tyrosines. *J. Biol. Chem., 277,* 9405–9414.

Moroy, G., Ostuni, A., Pepe, A., Tamburro, A. M., Alix, A. J., & Hery-Huynh, S. (2009). A proposed interaction mechanism between elastin-derived peptides and the elastin/laminin receptor-binding domain. *Proteins, 76,* 461–476.

Moscatelli, D. (1987). High and low affinity binding sites for basic fibroblast growth factor on cultured cells: absence of a role for low affinity binding in the stimulation of plasminogen activator production by bovine capillary endothelial cells. *J. Cell Physiol., 131,* 123–130.

Mostafavi-Pour, Z., Askari, J. A., Whittard, J. D., & Humphries, M. J. (2001). Identification of a novel heparin-binding site in the alternatively spliced IIICS region of fibronectin: roles of integrins and proteoglycans in cell adhesion to fibronectin splice variants. *Matrix Biol., 20,* 63–73.

Mould, A. P., Askari, J. A., Craig, S. E., Garratt, A. N., Clements, J., & Humphries, M. J. (1994). Integrin alpha 4 beta 1-mediated melanoma cell adhesion and migration on vascular cell adhesion molecule-1 (VCAM-1) and the alternatively spliced IIICS region of fibronectin. *J. Biol. Chem., 269,* 27224–27230.

Moulin, V., Larochelle, S., Langlois, C., Thibault, I., Lopez-Valle, C. A., & Roy, M. (2004). Normal skin wound and hypertrophic scar myofibroblasts have differential responses to apoptotic inductors. *J. Cell Physiol., 198,* 350–358.

Moulin, V., Tam, B. Y., Castilloux, G., Auger, F. A., O'Connor-McCourt, M. D., Philip, A., et al. (2001). Fetal and adult human skin fibroblasts display intrinsic differences in contractile capacity. *J. Cell Physiol., 188,* 211–222.

Murgia, C., Blaikie, P., Kim, N., Dans, M., Petrie, H. T., & Giancotti, F. G. (1998). Cell cycle and adhesion defects in mice carrying a targeted deletion of the integrin beta4 cytoplasmic domain. *Embo J., 17,* 3940–3951.

Murooka, T. T., Wong, M. M., Rahbar, R., Majchrzak-Kita, B., Proudfoot, A. E., & Fish, E. N. (2006). CCL5-CCR5-mediated apoptosis in T cells: requirement for glycosaminoglycan binding and CCL5 aggregation. *J. Biol. Chem., 281,* 25184–25194.

Murphy, F. R., Issa, R., Zhou, X., Ratnarajah, S., Nagase, H., Arthur, M. J., et al. (2002). Inhibition of apoptosis of activated hepatic stellate cells by tissue inhibitor of metalloproteinase-1 is mediated via effects on matrix metalloproteinase inhibition: implications for reversibility of liver fibrosis. *J. Biol. Chem., 277,* 11069–11076.

Mythreye, K., & Blobe, G. C. (2009). Proteoglycan signaling co-receptors: roles in cell adhesion, migration and invasion. *Cell Signal, 21,* 1548–1558.

Natal, C., Oses-Prieto, J. A., Pelacho, B., Iraburu, M. J., & Lopez-Zabalza, M. J. (2006). Regulation of apoptosis by peptides of fibronectin in human monocytes. *Apoptosis, 11,* 209–219.

Nath, R. K., LaRegina, M., Markham, H., Ksander, G. A., & Weeks, P. M. (1994). The expression of transforming growth factor type beta in fetal and adult rabbit skin wounds. *J. Pediatr. Surg., 29,* 416–421.

Nathan, C., & Sporn, M. (1991). Cytokines in context. *J. Cell Biol., 113,* 981–986.

Naugle, J. E., Olson, E. R., Zhang, X., Mase, S. E., Pilati, C. F., et al. (2005). Type VI collagen induces cardiac myofibroblast differentiation: implications for post-infarction remodeling. *Am. J. Physiol. Heart Circ. Physiol, 290,* H323–330.

Ng, V. L., Sabla, G. E., Melin-Aldana, H., Kelley-Loughnane, N., Degen, J. L., & Bezerra, J. A. (2001). Plasminogen deficiency results in poor clearance of non-fibrin matrix and persistent activation of hepatic stellate cells after an acute injury. *J. Hepatol., 35,* 781–789.

Nicholson, L. J., & Watt, F. M. (1991). Decreased expression of fibronectin and the alpha 5 beta 1 integrin during terminal differentiation of human keratinocytes. *J. Cell Sci., 98*(Pt 2), 225–232.

Nikolopoulos, S. N., Blaikie, P., Yoshioka, T., Guo, W., Puri, C., Tacchetti, C., et al. (2005). Targeted deletion of the integrin beta4 signaling domain suppresses laminin-5-dependent nuclear entry of mitogen-activated protein kinases and NF-kappaB, causing defects in epidermal growth and migration. *Mol. Cell Biol.*, 25, 6090–6102.

Nishio, T., Iimuro, Y., Nitta, T., Harada, N., Yoshida, M., Hirose, T., et al. (2003). Increased expression of collagenase in the liver induces hepatocyte proliferation with cytoplasmic accumulation of beta-catenin in the rat. *J. Hepatol.*, 38, 468–475.

Noble, N. A., Harper, J. R., & Border, W. A. (1992). *In vivo* interactions of TGF-beta and extracellular matrix. *Prog. Growth Factor Res.*, 4, 369–382.

Okina, E., Manon-Jensen, T., Whiteford, J. R., & Couchman, J. R. (2009). Syndecan proteoglycan contributions to cytoskeletal organization and contractility. *Scand J. Med. Sci. Sports*, 19, 479–489.

Olaso, E., Labrador, J. P., Wang, L., Ikeda, K., Eng, F. J., Klein, R., et al. (2002). Discoidin domain receptor 2 regulates fibroblast proliferation and migration through the extracellular matrix in association with transcriptional activation of matrix metalloproteinase-2. *J. Biol. Chem.*, 277, 3606–3613.

Ono, K., Hattori, H., Takeshita, S., Kurita, A., & Ishihara, M. (1999). Structural features in heparin that interact with VEGF165 and modulate its biological activity. *Glycobiology*, 9, 705–711.

O'Reilly, M. S., Boehm, T., Shing, Y., Fukai, N., Vasios, G., Lane, W. S., et al. (1997). Endostatin: an endogenous inhibitor of angiogenesis and tumor growth. *Cell*, 88, 277–285.

Ornitz, D. M., Yayon, A., Flanagan, J. G., Svahn, C. M., Levi, E., & Leder, P. (1992). Heparin is required for cell-free binding of basic fibroblast growth factor to a soluble receptor and for mitogenesis in whole cells. *Mol. Cell Biol.*, 12, 240–247.

O'Toole, E. A. (2001). Extracellular matrix and keratinocyte migration. *Clin. Exp. Dermatol.*, 26, 525–530.

Otsu, K., Kato, S., Ohtake, K., & Akamatsu, N. (1992). Alteration of rat liver proteoglycans during regeneration. *Arch. Biochem. Biophys.*, 294, 544–549.

Panayotou, G., End, P., Aumailley, M., Timpl, R., & Engel, J. (1989). Domains of laminin with growth-factor activity. *Cell*, 56, 93–101.

Panis, Y., McMullan, D. M., & Emond, J. C. (1997). Progressive necrosis after hepatectomy and the pathophysiology of liver failure after massive resection. *Surgery*, 121, 142–149.

Passi, A., Sadeghi, P., Kawamura, H., Anand, S., Sato, N., White, L. E., et al. (2004). Hyaluronan suppresses epidermal differentiation in organotypic cultures of rat keratinocytes. *Exp. Cell Res.*, 296, 123–134.

Patel, D. D., Koopmann, W., Imai, T., Whichard, L. P., Yoshie, O., & Krangel, M. S. (2001). Chemokines have diverse abilities to form solid phase gradients. *Clin. Immunol.*, 99, 43–52.

Paulsson, M. (1988). The role of Ca2+ binding in the self-aggregation of laminin-nidogen complexes. *J. Biol. Chem.*, 263, 5425–5430.

Peach, R. J., Hollenbaugh, D., Stamenkovic, I., & Aruffo, A. (1993). Identification of hyaluronic acid binding sites in the extracellular domain of CD44. *J. Cell Biol.*, 122, 257–264.

Peled, Z. M., Phelps, E. D., Updike, D. L., Chang, J., Krummel, T. M., Howard, E. W., et al. (2002). Matrix metalloproteinases and the ontogeny of scarless repair: the other side of the wound healing balance. *Plast. Reconstr. Surg.*, 110, 801–811.

Pérez, R. L., & Roman, J. (1995). Fibrin enhances the expression of IL-1 beta by human peripheral blood mononuclear cells. Implications in pulmonary inflammation. *J. Immunol.*, 154, 1879–1887.

Pi, L., Ding, X., Jorgensen, M., Pan, J. J., Oh, S. H., Pintilie, D., et al. (2008). Connective tissue growth factor with a novel fibronectin binding site promotes cell adhesion and migration during rat oval cell activation. *Hepatology*, 47, 996–1004.

Pierschbacher, M. D., & Ruoslahti, E. (1984). Cell attachment activity of fibronectin can be duplicated by small synthetic fragments of the molecule. *Nature*, 309, 30–33.

Pilewski, J. M., Latoche, J. D., Arcasoy, S. M., & Albelda, S. M. (1997). Expression of integrin cell adhesion receptors during human airway epithelial repair *in vivo*. *Am. J. Physiol.*, 273, L256–L263.

Pluskota, E., Soloviev, D. A., Bdeir, K., Cines, D. B., & Plow, E. F. (2004). Integrin alphaMbeta2 orchestrates and accelerates plasminogen activation and fibrinolysis by neutrophils. *J. Biol. Chem.*, 279, 18063–18072.

Pluskota, E., Soloviev, D. A., & Plow, E. F. (2003). Convergence of the adhesive and fibrinolytic systems: recognition of urokinase by integrin alpha Mbeta 2 as well as by the urokinase receptor regulates cell adhesion and migration. *Blood*, 101, 1582–1590.

Pluskota, E., Stenina, O. I., Krukovets, I., Szpak, D., Topol, E. J., & Plow, E. F. (2005). Mechanism and effect of thrombospondin-4 polymorphisms on neutrophil function. *Blood*, 106, 3970–3978.

Pollock, R. A., & Richardson, W. D. (1992). The alternative-splice isoforms of the PDGF A-chain differ in their ability to associate with the extracellular matrix and to bind heparin *in vitro*. *Growth Factors*, 7, 267–277.

Poltorak, Z., Cohen, T., Sivan, R., Kandelis, Y., Spira, G., Vlodavsky, I., et al. (1997). VEGF145, a secreted vascular endothelial growth factor isoform that binds to extracellular matrix. *J. Biol. Chem.*, 272, 7151–7158.

Polunovsky, V. A., Chen, B., Henke, C., Snover, D., Wendt, C., Ingbar, D. H., et al. (1993). Role of mesenchymal cell death in lung remodeling after injury. *J. Clin. Invest., 92*, 388–397.

Pouliot, N., Saunders, N. A., & Kaur, P. (2002). Laminin 10/11: an alternative adhesive ligand for epidermal keratinocytes with a functional role in promoting proliferation and migration. *Exp. Dermatol., 11*, 387–397.

Pozzi, A., Wary, K. K., Giancotti, F. G., & Gardner, H. A. (1998). Integrin alpha1beta1 mediates a unique collagen-dependent proliferation pathway *in vivo. J. Cell Biol., 142*, 587–594.

Prieto, A. L., Andersson-Fisone, C., & Crossin, K. L. (1992). Characterization of multiple adhesive and counter-adhesive domains in the extracellular matrix protein cytotactin. *J. Cell Biol., 119*, 663–678.

Proudfoot, A. E., Handel, T. M., Johnson, Z., Lau, E. K., LiWang, P., Clark-Lewis, I., et al. (2003). Glycosaminoglycan binding and oligomerization are essential for the *in vivo* activity of certain chemokines. *Proc. Natl. Acad. Sci. U.S.A., 100*, 1885–1890.

Pujades, C., Alon, R., Yauch, R. L., Masumoto, A., Burkly, L. C., Chen, C., et al. (1997). Defining extracellular integrin alpha-chain sites that affect cell adhesion and adhesion strengthening without altering soluble ligand binding. *Mol. Biol. Cell, 8*, 2647–2657.

Pujades, C., Forsberg, E., Enrich, C., & Johansson, S. (1992). Changes in cell surface expression of fibronectin and fibronectin receptor during liver regeneration. *J. Cell Sci., 102*(Pt 4), 815–820.

Rahman, S., Patel, Y., Murray, J., Patel, K. V., Sumathipala, R., Sobel, M., et al. (2005). Novel hepatocyte growth factor (HGF) binding domains on fibronectin and vitronectin coordinate a distinct and amplified Met-integrin induced signalling pathway in endothelial cells. *BMC Cell Biol., 6*, 8.

Ramelet, A. A., Hirt-Burri, N., Raffoul, W., Scaletta, C., Pioletti, D. P., Offord, E., et al. (2009). Chronic wound healing by fetal cell therapy may be explained by differential gene profiling observed in fetal versus old skin cells. *Exp. Gerontol., 44*, 208–218.

Reddig, P. J., & Juliano, R. L. (2005). Clinging to life: cell to matrix adhesion and cell survival. *Cancer Metastasis Rev., 24*, 425–439.

Reis, R. C., Schuppan, D., Barreto, A. C., Bauer, M., Bork, J. P., Hassler, G., et al. (2005). Endostatin competes with bFGF for binding to heparin-like glycosaminoglycans. *Biochem. Biophys. Res. Commun., 333*, 976–983.

Roberts, R., Gallagher, J., Spooncer, E., Allen, T. D., Bloomfield, F., & Dexter, T. M. (1988). Heparan sulphate bound growth factors: a mechanism for stromal cell mediated haemopoiesis. *Nature, 332*, 376–378.

Roderfeld, M., Weiskirchen, R., Wagner, S., Berres, M. L., Henkel, C., Grotzinger, J., et al. (2006). Inhibition of hepatic fibrogenesis by matrix metalloproteinase-9 mutants in mice. *Faseb J., 20*, 444–454.

Rodius, S., Indra, G., Thibault, C., Pfister, V., & Georges-Labouesse, E. (2007). Loss of alpha6 integrins in keratinocytes leads to an increase in TGFbeta and AP1 signaling and in expression of differentiation genes. *J. Cell Physiol., 212*, 439–449.

Rolfe, K. J., Cambrey, A. D., Richardson, J., Irvine, L. M., Grobbelaar, A. O., & Linge, C. (2007a). Dermal fibroblasts derived from fetal and postnatal humans exhibit distinct responses to insulin like growth factors. *BMC Dev. Biol., 7*, 124.

Rolfe, K. J., Irvine, L. M., Grobbelaar, A. O., & Linge, C. (2007b). Differential gene expression in response to transforming growth factor-beta1 by fetal and postnatal dermal fibroblasts. *Wound Repair Regen., 15*, 897–906.

Rolfe, K. J., Richardson, J., Vigor, C., Irvine, L. M., Grobbelaar, A. O., & Linge, C. (2007c). A role for TGF-beta1-induced cellular responses during wound healing of the non-scarring early human fetus? *J. Invest. Dermatol., 127*, 2656–2667.

Romer, J., Bugge, T. H., Pyke, C., Lund, L. R., Flick, M. J., Degen, J. L., et al. (1996). Impaired wound healing in mice with a disrupted plasminogen gene. *Nat. Med., 2*, 287–292.

Roovers, K., Davey, G., Zhu, X., Bottazzi, M. E., & Assoian, R. K. (1999). Alpha5beta1 integrin controls cyclin D1 expression by sustaining mitogen-activated protein kinase activity in growth factor-treated cells. *Mol. Biol. Cell, 10*, 3197–1204.

Rossi, D., & Zlotnik, A. (2000). The biology of chemokines and their receptors. *Annu. Rev. Immunol., 18*, 217–242.

Rudolph, K. L., Trautwein, C., Kubicka, S., Rakemann, T., Bahr, M. J., Sedlaczek, N., et al. (1999). Differential regulation of extracellular matrix synthesis during liver regeneration after partial hepatectomy in rats. *Hepatology, 30*, 1159–1166.

Ruffell, B., & Johnson, P. (2008). Hyaluronan induces cell death in activated T cells through CD44. *J. Immunol., 181*, 7044–7054.

Ruhl, M., Sahin, E., Johannsen, M., Somasundaram, R., Manski, D., Riecken, E. O., et al. (1999). Soluble collagen VI drives serum-starved fibroblasts through S phase and prevents apoptosis via down-regulation of Bax. *J. Biol. Chem., 274*, 34361–34368.

Rundhaug, J. E. (2005). Matrix metalloproteinases and angiogenesis. *J. Cell Mol. Med., 9*, 267–285.

Saed, G. M., Ladin, D., Olson, J., Han, X., Hou, Z., & Fivenson, D. (1998). Analysis of p53 gene mutations in keloids using polymerase chain reaction-based single-strand conformational polymorphism and DNA sequencing. *Arch. Dermatol., 134*, 963−967.

Saegusa, J., Yamaji, S., Ieguchi, K., Wu, C. Y., Lam, K. S., Liu, F. T., et al. (2009). The direct binding of insulin-like growth factor-1 (IGF-1) to integrin alphavbeta3 is involved in IGF-1 signaling. *J. Biol. Chem., 284*, 24106−24114.

Sage, E. H., Reed, M., Funk, S. E., Truong, T., Steadele, M., Puolakkainen, P., et al. (2003). Cleavage of the matricellular protein SPARC by matrix metalloproteinase 3 produces polypeptides that influence angiogenesis. *J. Biol. Chem., 278*, 37849−37857.

Saile, B., Knittel, T., Matthes, N., Schott, P., & Ramadori, G. (1997). CD95/CD95L-mediated apoptosis of the hepatic stellate cell. A mechanism terminating uncontrolled hepatic stellate cell proliferation during hepatic tissue repair. *Am. J. Pathol., 151*, 1265−1272.

Saito, Y., Shiota, Y., Nishisaka, M., Owaki, T., Shimamura, M., & Fukai, F. (2008). Inhibition of angiogenesis by a tenascin-c peptide which is capable of activating beta1-integrins. *Biol. Pharm. Bull., 31*, 1003−1007.

Sakakura, S., Saito, S., & Morikawa, H. (1999). Stimulation of DNA synthesis in trophoblasts and human umbilical vein endothelial cells by hepatocyte growth factor bound to extracellular matrix. *Placenta, 20*, 683−693.

Sakamoto, N., Iwahana, M., Tanaka, N. G., & Osada, Y. (1991). Inhibition of angiogenesis and tumor growth by a synthetic laminin peptide, CDPGYIGSR-NH2. *Cancer Res., 51*, 903−906.

Saksela, O., Moscatelli, D., Sommer, A., & Rifkin, D. B. (1988). Endothelial cell-derived heparan sulfate binds basic fibroblast growth factor and protects it from proteolytic degradation. *J. Cell Biol., 107*, 743−751.

Sanes, J. R., Engvall, E., Butkowski, R., & Hunter, D. D. (1990). Molecular heterogeneity of basal laminae: isoforms of laminin and collagen IV at the neuromuscular junction and elsewhere. *J. Cell Biol., 111*, 1685−1699.

Santra, M., Reed, C. C., & Iozzo, R. V. (2002). Decorin binds to a narrow region of the epidermal growth factor (EGF) receptor, partially overlapping but distinct from the EGF-binding epitope. *J. Biol. Chem., 277*, 35671−35681.

Saoncella, S., Echtermeyer, F., Denhez, F., Nowlen, J. K., Mosher, D. F., Robinson, S. D., et al. (1999). Syndecan-4 signals cooperatively with integrins in a Rho-dependent manner in the assembly of focal adhesions and actin stress fibers. *Proc. Natl. Acad. Sci. U.S.A., 96*, 2805−2810.

Sasaki, T., Larsson, H., Tisi, D., Claesson-Welsh, L., Hohenester, E., & Timpl, R. (2000). Endostatins derived from collagens XV and XVIII differ in structural and binding properties, tissue distribution and anti-angiogenic activity. *J. Mol. Biol., 301*, 1179−1190.

Savani, R. C., Cao, G., Pooler, P. M., Zaman, A., Zhou, Z., & DeLisser, H. M. (2001). Differential involvement of the hyaluronan (HA) receptors CD44 and receptor for HA-mediated motility in endothelial cell function and angiogenesis. *J. Biol. Chem., 276*, 36770−36778.

Sawai, T., Usui, N., Sando, K., Fukui, Y., Kamata, S., Okada, A., et al. (1997). Hyaluronic acid of wound fluid in adult and fetal rabbits. *J. Pediatr. Surg., 32*, 41−43.

Schaller, M. D., Hildebrand, J. D., & Parsons, J. T. (1999). Complex formation with focal adhesion kinase: a mechanism to regulate activity and subcellular localization of Src kinases. *Mol. Biol. Cell, 10*, 3489−3505.

Schenk, S., Hintermann, E., Bilban, M., Koshikawa, N., Hojilla, C., Khokha, R., et al. (2003). Binding to EGF receptor of a laminin-5 EGF-like fragment liberated during MMP-dependent mammary gland involution. *J. Cell Biol., 161*, 197−209.

Schlaepfer, D. D., Broome, M. A., & Hunter, T. (1997). Fibronectin-stimulated signaling from a focal adhesion kinase-c-Src complex: involvement of the Grb2, p130cas, and Nck adaptor proteins. *Mol. Cell Biol., 17*, 1702−1713.

Schneider, H., Muhle, C., & Pacho, F. (2007). Biological function of laminin-5 and pathogenic impact of its deficiency. *Eur. J. Cell Biol., 86*, 701−717.

Schneller, M., Vuori, K., & Ruoslahti, E. (1997). Alphavbeta3 integrin associates with activated insulin and PDGFbeta receptors and potentiates the biological activity of PDGF. *Embo. J., 16*, 5600−5607.

Schonherr, E., & Hausser, H. J. (2000). Extracellular matrix and cytokines: a functional unit. *Dev. Immunol., 7*, 89−101.

Schuppan, D., Schmid, M., Somasundaram, R., Ackermann, R., Ruehl, M., Nakamura, T., et al. (1998). Collagens in the liver extracellular matrix bind hepatocyte growth factor. *Gastroenterology, 114*, 139−152.

Sefton, B. M., Hunter, T., Ball, E. H., & Singer, S. J. (1981). Vinculin: a cytoskeletal target of the transforming protein of Rous sarcoma virus. *Cell, 24*, 165−174.

Serban, M. A., & Prestwich, G. D. (2008). Modular extracellular matrices: solutions for the puzzle. *Methods, 45*, 93−98.

Serini, G., Bochaton-Piallat, M. L., Ropraz, P., Geinoz, A., Borsi, L., Zardi, L., et al. (1998). The fibronectin domain ED-A is crucial for myofibroblastic phenotype induction by transforming growth factor-beta1. *J. Cell Biol., 142*, 873−881.

Shafiei, M. S., & Rockey, D. C. (2006). The role of integrin-linked kinase in liver wound healing. *J. Biol. Chem., 281,* 24863–24872.

Shankavaram, U. T., DeWitt, D. L., Funk, S. E., Sage, E. H., & Wahl, L. M. (1997). Regulation of human monocyte matrix metalloproteinases by SPARC. *J. Cell Physiol., 173,* 327–334.

Sharma, A., & Mayer, B. J. (2008). Phosphorylation of p130Cas initiates Rac activation and membrane ruffling. *BMC Cell Biol., 9,* 50.

Shi, Q., & Boettiger, D. (2003). A novel mode for integrin-mediated signaling: tethering is required for phosphorylation of FAK Y397. *Mol. Biol. Cell, 14,* 4306–4315.

Shimizu, M., Hara, A., Okuno, M., Matsuno, H., Okada, K., Ueshima, S., et al. (2001). Mechanism of retarded liver regeneration in plasminogen activator-deficient mice: impaired activation of hepatocyte growth factor after Fas-mediated massive hepatic apoptosis. *Hepatology, 33,* 569–576.

Silverstein, R. L., & Febbraio, M. (2007). CD36-TSP-HRGP interactions in the regulation of angiogenesis. *Curr. Pharm. Des., 13,* 3559–3567.

Simpson, R. M., Meran, S., Thomas, D., Stephens, P., Bowen, T., Steadman, R., et al. (2009). Age-related changes in pericellular hyaluronan organization leads to impaired dermal fibroblast to myofibroblast differentiation. *Am. J. Pathol., 175,* 1915–1928.

Singer, A. J., & Clark, R. A. (1999). Cutaneous wound healing. *N. Engl. J. Med., 341,* 738–746.

Singh, P., Chen, C., Pal-Ghosh, S., Stepp, M. A., Sheppard, D., & van de Water, L. (2009). Loss of integrin alpha9beta1 results in defects in proliferation, causing poor re-epithelialization during cutaneous wound healing. *J. Invest. Dermatol., 129,* 217–228.

Sixt, M., Hallmann, R., Wendler, O., Scharffetter-Kochanek, K., & Sorokin, L. M. (2001). Cell adhesion and migration properties of beta 2-integrin negative polymorphonuclear granulocytes on defined extracellular matrix molecules. Relevance for leukocyte extravasation. *J. Biol. Chem., 276,* 18878–18887.

Sjaastad, M. D., & Nelson, W. J. (1997). Integrin-mediated calcium signaling and regulation of cell adhesion by intracellular calcium. *Bioessays, 19,* 47–55.

Slevin, M., Krupinski, J., Gaffney, J., Matou, S., West, D., Delisser, H., et al. (2007). Hyaluronan-mediated angiogenesis in vascular disease: uncovering RHAMM and CD44 receptor signaling pathways. *Matrix Biol., 26,* 58–68.

So, C. L., Kaluarachchi, K., Tam, P. P., & Cheah, K. S. (2001). Impact of mutations of cartilage matrix genes on matrix structure, gene activity and chondrogenesis. *Osteoarthritis Cartilage, 9*(Suppl. A), S160–S173.

Sohara, N., Znoyko, I., Levy, M. T., Trojanowska, M., & Reuben, A. (2002). Reversal of activation of human myofibroblast-like cells by culture on a basement membrane-like substrate. *J. Hepatol., 37,* 214–221.

Soldi, R., Mitola, S., Strasly, M., Defilippi, P., Tarone, G., & Bussolino, F. (1999). Role of alphavbeta3 integrin in the activation of vascular endothelial growth factor receptor-2. *Embo J., 18,* 882–892.

Soo, C., Hu, F. Y., Zhang, X., Wang, Y., Beanes, S. R., Lorenz, H. P., et al. (2000). Differential expression of fibromodulin, a transforming growth factor-beta modulator, in fetal skin development and scarless repair. *Am. J. Pathol., 157,* 423–433.

Sottile, J. (2004). Regulation of angiogenesis by extracellular matrix. *Biochim. Biophys. Acta, 1654,* 13–22.

Spofford, C. M., & Chilian, W. M. (2003). Mechanotransduction via the elastin-laminin receptor (ELR) in resistance arteries. *J. Biomech., 36,* 645–652.

Stoff, A., Rivera, A. A., Mathis, J. M., Moore, S. T., Banerjee, N. S., Everts, M., et al. (2007). Effect of adenoviral mediated overexpression of fibromodulin on human dermal fibroblasts and scar formation in full-thickness incisional wounds. *J. Mol. Med., 85,* 481–496.

Streuli, C. H., Schmidhauser, C., Kobrin, M., Bissell, M. J., & Derynck, R. (1993). Extracellular matrix regulates expression of the TGF-beta 1 gene. *J. Cell Biol., 120,* 253–260.

Stupack, D. G., & Cheresh, D. A. (2002). Get a ligand, get a life: integrins, signaling and cell survival. *J. Cell Sci., 115,* 3729–3738.

Stupack, D. G., Puente, X. S., Boutsaboualoy, S., Storgard, C. M., & Cheresh, D. A. (2001). Apoptosis of adherent cells by recruitment of caspase-8 to unligated integrins. *J. Cell Biol., 155,* 459–470.

Sudhakar, A., Sugimoto, H., Yang, C., Lively, J., Zeisberg, M., & Kalluri, R. (2003). Human tumstatin and human endostatin exhibit distinct antiangiogenic activities mediated by alpha v beta 3 and alpha 5 beta 1 integrins. *Proc. Natl. Acad. Sci. U.S.A., 100,* 4766–4771.

Suki, B., Ito, S., Stamenovic, D., Lutchen, K. R., & Ingenito, E. P. (2005). Biomechanics of the lung parenchyma: critical roles of collagen and mechanical forces. *J. Appl. Physiol., 98,* 1892–1899.

Sullivan, S. R., Underwood, R. A., Sigle, R. O., Fukano, Y., Muffley, L. A., Usui, M. L., et al. (2007). Topical application of laminin-332 to diabetic mouse wounds. *J. Dermatol. Sci., 48,* 177–188.

Sulochana, K. N., Fan, H., Jois, S., Subramanian, V., Sun, F., Kini, R. M., et al. (2005). Peptides derived from human decorin leucine-rich repeat 5 inhibit angiogenesis. *J. Biol. Chem., 280,* 27935–27948.

Swindle, C. S., Tran, K. T., Johnson, T. D., Banerjee, P., Mayes, A. M., Griffith, L., et al. (2001). Epidermal growth factor (EGF)-like repeats of human tenascin-C as ligands for EGF receptor. *J. Cell Biol., 154,* 459–468.

Takada, Y., Ye, X., & Simon, S. (2007). The integrins. *Genome Biol., 8,* 215.

Takeda, K., Togo, S., Kunihiro, O., Fujii, Y., Kurosawa, H., Tanaka, K., et al. (2002). Clinicohistological features of liver failure after excessive hepatectomy. *Hepatogastroenterology, 49,* 354–358.

Takemura, G., Ohno, M., Hayakawa, Y., Misao, J., Kanoh, M., Ohno, A., et al. (1998). Role of apoptosis in the disappearance of infiltrated and proliferated interstitial cells after myocardial infarction. *Circ. Res., 82,* 1130–1138.

Takeuchi, Y., Kodama, Y., & Matsumoto, T. (1994). Bone matrix decorin binds transforming growth factor-beta and enhances its bioactivity. *J. Biol. Chem., 269,* 32634–32638.

Tennenbaum, T., Li, L., Belanger, A. J., de Luca, L. M., & Yuspa, S. H. (1996). Selective changes in laminin adhesion and alpha 6 beta 4 integrin regulation are associated with the initial steps in keratinocyte maturation. *Cell Growth Differ., 7,* 615–628.

Thannickal, V. J., Lee, D. Y., White, E. S., Cui, Z., Larios, J. M., Chacon, R., et al. (2003). Myofibroblast differentiation by transforming growth factor-beta1 is dependent on cell adhesion and integrin signaling via focal adhesion kinase. *J. Biol. Chem., 278,* 12384–12389.

Todorovicc, V., Chen, C. C., Hay, N., & Lau, L. F. (2005). The matrix protein CCN1 (CYR61) induces apoptosis in fibroblasts. *J. Cell Biol., 171,* 559–568.

Tolg, C., Hamilton, S. R., Nakrieko, K. A., Kooshesh, F., Walton, P., McCarthy, J. B., et al. (2006). Rhamm−/− fibroblasts are defective in CD44-mediated ERK1,2 motogenic signaling, leading to defective skin wound repair. *J. Cell Biol., 175,* 1017–1028.

Tomasek, J. J., Haaksma, C. J., Eddy, R. J., & Vaughan, M. B. (1992). Fibroblast contraction occurs on release of tension in attached collagen lattices: dependency on an organized actin cytoskeleton and serum. *Anat. Rec., 232,* 359–368.

Torres, V. A., Mielgo, A., Barbero, S., Hsiao, R., Wilkins, J. A., & Stupack, D. G. (2010). Rab5 mediates caspase-8-promoted cell motility and metastasis. *Mol. Biol. Cell, 21,* 369–376.

Tran, K. T., Griffith, L., & Wells, A. (2004). Extracellular matrix signaling through growth factor receptors during wound healing. *Wound Repair Regen., 12,* 262–268.

Tran, K. T., Lamb, P., & Deng, J. S. (2005). Matrikines and matricryptins: implications for cutaneous cancers and skin repair. *J. Dermatol. Sci., 40,* 11–20.

Troussard, A. A., Mawji, N. M., Ong, C., Mui, A., St-Arnaud, R., & Dedhar, S. (2003). Conditional knock-out of integrin-linked kinase demonstrates an essential role in protein kinase B/Akt activation. *J. Biol. Chem., 278,* 22374–22378.

Tsubouchi, A., Sakakura, J., Yagi, R., Mazaki, Y., Schaefer, E., Yano, H., et al. (2002). Localized suppression of RhoA activity by Tyr31/118-phosphorylated paxillin in cell adhesion and migration. *J. Cell Biol., 159,* 673–683.

Tsuda, T., Majumder, K., & Linask, K. K. (1998). Differential expression of flectin in the extracellular matrix and left-right asymmetry in mouse embryonic heart during looping stages. *Dev. Genet., 23,* 203–214.

Tuan, T. L., Song, A., Chang, S., Younai, S., & Nimni, M. E. (1996). *In vitro* fibroplasia: matrix contraction, cell growth, and collagen production of fibroblasts cultured in fibrin gels. *Exp. Cell Res., 223,* 127–134.

Tuhkanen, A. L., Tammi, M., & Tammi, R. (1997). CD44 substituted with heparan sulfate and endo-beta-galactosidase-sensitive oligosaccharides: a major proteoglycan in adult human epidermis. *J. Invest. Dermatol., 109,* 213–218.

Uemura, Y., & Okamoto, K. (1997). Elastin-derived peptide induces monocyte chemotaxis by increasing intracellular cyclic GMP level and activating cyclic GMP dependent protein kinase. *Biochem. Mol. Biol. Int., 41,* 1085–1092.

Ugarova, T. P., & Yakubenko, V. P. (2001). Recognition of fibrinogen by leukocyte integrins. *Ann. N.Y. Acad. Sci., 936,* 368–385.

Upton, Z., Webb, H., Hale, K., Yandell, C. A., McMurtry, J. P., Francis, G. L., et al. (1999). Identification of vitronectin as a novel insulin-like growth factor-II binding protein. *Endocrinology, 140,* 2928–2931.

Vaingankar, S. M., & Martins-Green, M. (1998). Thrombin aivation of the 9E3/CEF4 chemokine involves tyrosine kinases including c-src and the epidermal growth factor receptor. *J. Biol. Chem., 273,* 5226–5234.

van der Veer, W. M., Bloemen, M. C., Ulrich, M. M., Molema, G., van Zuijlen, P. P., Middelkoop, E., et al. (2009). Potential cellular and molecular causes of hypertrophic scar formation. *Burns, 35,* 15–29.

van Hul, N. K., Abarca-Quinones, J., Sempoux, C., Horsmans, Y., & Leclercq, I. A. (2009). Relation between liver progenitor cell expansion and extracellular matrix deposition in a CDE-induced murine model of chronic liver injury. *Hepatology, 49,* 1625–1635.

63

Venkataraman, G., Sasisekharan, V., Herr, A. B., Ornitz, D. M., Waksman, G., Cooney, C. L., et al. (1996). Preferential self-association of basic fibroblast growth factor is stabilized by heparin during receptor dimerization and activation. *Proc. Natl. Acad. Sci. U.S.A., 93,* 845–850.

Vilatoba, M., Eckstein, C., Bilbao, G., Frennete, L., Eckhoff, D. E., & Contreras, J. L. (2005a). 17beta-estradiol differentially activates mitogen-activated protein-kinases and improves survival following reperfusion injury of reduced-size liver in mice. *Transplant Proc., 37,* 399–403.

Vilatoba, M., Eckstein, C., Bilbao, G., Smyth, C. A., Jenkins, S., Thompson, J. A., et al. (2005b). Sodium 4-phenylbutyrate protects against liver ischemia reperfusion injury by inhibition of endoplasmic reticulum-stress mediated apoptosis. *Surgery, 138,* 342–351.

Vlahakis, N. E., Young, B. A., Atakilit, A., Hawkridge, A. E., Issaka, R. B., Boudreau, N., et al. (2007). Integrin alpha9beta1 directly binds to vascular endothelial growth factor (VEGF)-A and contributes to VEGF-A-induced angiogenesis. *J. Biol. Chem., 282,* 15187–15196.

Vlodavsky, I., Miao, H. Q., Medalion, B., Danagher, P., & Ron, D. (1996). Involvement of heparan sulfate and related molecules in sequestration and growth promoting activity of fibroblast growth factor. *Cancer Metastasis Rev., 15,* 177–186.

Vogel, W., Gish, G. D., Alves, F., & Pawson, T. (1997). The discoidin domain receptor tyrosine kinases are activated by collagen. *Mol. Cell, 1,* 13–23.

Volberg, T., Romer, L., Zamir, E., & Geiger, B. (2001). pp60(c-src) and related tyrosine kinases: a role in the assembly and reorganization of matrix adhesions. *J. Cell Sci., 114,* 2279–2289.

Wallner, E. I., Yang, Q., Peterson, D. R., Wada, J., & Kanwar, Y. S. (1998). Relevance of extracellular matrix, its receptors, and cell adhesion molecules in mammalian nephrogenesis. *Am. J. Physiol., 275,* F467–F477.

Wang, C. Z., Su, H. W., Hsu, Y. C., Shen, M. R., & Tang, M. J. (2006). A discoidin domain receptor 1/SHP-2 signaling complex inhibits alpha2beta1-integrin-mediated signal transducers and activators of transcription 1/3 activation and cell migration. *Mol. Biol. Cell, 17,* 2839–2852.

Wary, K. K., Mariotti, A., Zurzolo, C., & Giancotti, F. G. (1998). A requirement for caveolin-1 and associated kinase Fyn in integrin signaling and anchorage-dependent cell growth. *Cell, 94,* 625–634.

Watanabe, M., Fujioka-Kaneko, Y., Kobayashi, H., Kiniwa, M., Kuwano, M., & Basaki, Y. (2005). Involvement of integrin-linked kinase in capillary/tube-like network formation of human vascular endothelial cells. *Biol. Proced. Online, 7,* 41–47.

Webber, J., Meran, S., Steadman, R., & Phillips, A. (2009). Hyaluronan orchestrates transforming growth factor-beta1-dependent maintenance of myofibroblast phenotype. *J. Biol. Chem., 284,* 9083–9092.

Wegener, K. L., & Campbell, I. D. (2008). Transmembrane and cytoplasmic domains in integrin activation and protein-protein interactions (review). *Mol. Membr. Biol., 25,* 376–387.

Werb, Z., & Chin, J. R. (1998). Extracellular matrix remodeling during morphogenesis. *Ann. N.Y. Acad. Sci., 857,* 110–118.

Werr, J., Johansson, J., Eriksson, E. E., Hedqvist, P., Ruoslahti, E., & Lindbom, L. (2000). Integrin alpha(2)beta(1) (VLA-2) is a principal receptor used by neutrophils for locomotion in extravascular tissue. *Blood, 95,* 1804–1809.

Werr, J., Xie, X., Hedqvist, P., Ruoslahti, E., & Lindbom, L. (1998). beta1 integrins are critically involved in neutrophil locomotion in extravascular tissue *in vivo. J. Exp. Med., 187,* 2091–2096.

Wess, T. J., Purslow, P. P., Sherratt, M. J., Ashworth, J., Shuttleworth, C. A., & Kielty, C. M. (1998). Calcium determines the supramolecular organization of fibrillin-rich microfibrils. *J. Cell Biol., 141,* 829–837.

Wessells, N. K., & Cohen, J. H. (1968). Effects of collagenase on developing epithelia *in vitro*: lung, ureteric bud, and pancreas. *Dev. Biol., 18,* 294–309.

West, D. C., Shaw, D. M., Lorenz, P., Adzick, N. S., & Longaker, M. T. (1997). Fibrotic healing of adult and late gestation fetal wounds correlates with increased hyaluronidase activity and removal of hyaluronan. *Int. J. Biochem. Cell Biol., 29,* 201–210.

Whitby, D. J., & Ferguson, M. W. (1991). The extracellular matrix of lip wounds in fetal, neonatal and adult mice. *Development, 112,* 651–668.

Whitby, D. J., Longaker, M. T., Harrison, M. R., Adzick, N. S., & Ferguson, M. W. (1991). Rapid epithelialisation of fetal wounds is associated with the early deposition of tenascin. *J. Cell Sci., 99*(Pt 3), 583–586.

White, E. S., Baralle, F. E., & Muro, A. F. (2008). New insights into form and function of fibronectin splice variants. *J. Pathol., 216,* 1–14.

White, S. R., Dorscheid, D. R., Rabe, K. F., Wojcik, K. R., & Hamann, K. J. (1999). Role of very late adhesion integrins in mediating repair of human airway epithelial cell monolayers after mechanical injury. *Am. J. Respir. Cell Mol. Biol., 20,* 787–796.

Whitelock, J. M., Melrose, J., & Iozzo, R. V. (2008). Diverse cell signaling events modulated by perlecan. *Biochemistry, 47,* 11174–11183.

Wijelath, E. S., Rahman, S., Namekata, M., Murray, J., Nishimura, T., et al. (2006). Heparin-II domain of fibronectin is a vascular endothelial growth factor-binding domain: enhancement of VEGF biological activity by a singular growth factor/matrix protein synergism. *Circ. Res.*, *99*, 853–860.

Wikner, N. E., Elder, J. T., Persichitte, K. A., Mink, P., & Clark, R. A. (1990). Transforming growth factor-beta modulates plasminogen activator activity and plasminogen activator inhibitor type-1 expression in human keratinocytes *in vitro*. *J. Invest. Dermatol.*, *95*, 607–613.

Wilgus, T. A. (2007). Regenerative healing in fetal skin: a review of the literature. *Ostomy. Wound Manage.*, *53*, 16–31, quiz 32–33.

Wondimu, Z., Geberhiwot, T., Ingerpuu, S., Juronen, E., Xie, X., Lindbom, L., et al. (2004). An endothelial laminin isoform, laminin 8 (alpha4beta1gamma1), is secreted by blood neutrophils, promotes neutrophil migration and extravasation, and protects neutrophils from apoptosis. *Blood*, *104*, 1859–1866.

Xu, J., Rodriguez, D., Petitclerc, E., Kim, J. J., Hangai, M., Moon, Y. S., et al. (2001). Proteolytic exposure of a cryptic site within collagen type IV is required for angiogenesis and tumor growth *in vivo*. *J. Cell Biol.*, *154*, 1069–1079.

Xu, T., Bianco, P., Fisher, L. W., Longenecker, G., Smith, E., Goldstein, S., et al. (1998). Targeted disruption of the biglycan gene leads to an osteoporosis-like phenotype in mice. *Nat. Genet.*, *20*, 78–82.

Yamada, K. M., & Kennedy, D. W. (1984). Dualistic nature of adhesive protein function: fibronectin and its biologically active peptide fragments can autoinhibit fibronectin function. *J. Cell Biol.*, *99*, 29–36.

Yayon, A., Klagsbrun, M., Esko, J. D., Leder, P., & Ornitz, D. M. (1991). Cell surface, heparin-like molecules are required for binding of basic fibroblast growth factor to its high affinity receptor. *Cell*, *64*, 841–848.

Young, R. E., Thompson, R. D., Larbi, K. Y., La, M., Roberts, C. E., Shapiro, S. D., et al. (2004). Neutrophil elastase (NE)-deficient mice demonstrate a nonredundant role for NE in neutrophil migration, generation of proinflammatory mediators, and phagocytosis in response to zymosan particles *in vivo*. *J. Immunol.*, *172*, 4493–4502.

Young, R. E., Voisin, M. B., Wang, S., Dangerfield, J., & Nourshargh, S. (2007). Role of neutrophil elastase in LTB4-induced neutrophil transmigration *in vivo* assessed with a specific inhibitor and neutrophil elastase deficient mice. *Br. J. Pharmacol.*, *151*, 628–637.

Zanin-Zhorov, A., Hershkoviz, R., Hecht, I., Cahalon, L., & Lider, O. (2003). Fibronectin-associated Fas ligand rapidly induces opposing and time-dependent effects on the activation and apoptosis of T cells. *J. Immunol.*, *171*, 5882–5889.

Zhang, W., Chen, X. P., Zhang, W. G., Zhang, F., Xiang, S., Dong, H. H., et al. (2009). Hepatic non-parenchymal cells and extracellular matrix participate in oval cell-mediated liver regeneration. *World J. Gastroenterol.*, *15*, 552–560.

Zhang, Y., Cao, L., Yang, B. L., & Yang, B. B. (1998). The G3 domain of versican enhances cell proliferation via epidermial growth factor-like motifs. *J. Biol. Chem.*, *273*, 21342–21351.

Zhao, H., Ross, F. P., & Teitelbaum, S. L. (2005). Unoccupied alpha(v)beta3 integrin regulates osteoclast apoptosis by transmitting a positive death signal. *Mol. Endocrinol.*, *19*, 771–780.

Zhao, Y., Sato, Y., Isaji, T., Fukuda, T., Matsumoto, A., Miyoshi, E., et al. (2008). Branched N-glycans regulate the biological functions of integrins and cadherins. *Febs. J.*, *275*, 1939–1948.

Zheng, L., & Martins-Green, M. (2007). Molecular mechanisms of thrombin-induced interleukin-8 (IL-8/CXCL8) expression in THP-1-derived and primary human macrophages. *J. Leukoc. Biol.*, *82*, 619–629.

Zhou, X., Murphy, F. R., Gehdu, N., Zhang, J., Iredale, J. P., & Benyon, R. C. (2004). Engagement of alphavbeta3 integrin regulates proliferation and apoptosis of hepatic stellate cells. *J. Biol. Chem.*, *279*, 23996–24006.

Zhu, J. X., Goldoni, S., Bix, G., Owens, R. T., McQuillan, D. J., Reed, C. C., et al. (2005). Decorin evokes protracted internalization and degradation of the epidermal growth factor receptor via caveolar endocytosis. *J. Biol. Chem.*, *280*, 32468–32479.

Zhu, Y., McAlinden, A., & Sandell, L. J. (2001). Type IIA procollagen in development of the human intervertebral disc: regulated expression of the NH(2)-propeptide by enzymic processing reveals a unique developmental pathway. *Dev. Dyn.*, *220*, 350–362.

Zouq, N. K., Keeble, J. A., Lindsay, J., Valentijn, A. J., Zhang, L., Mills, D., et al. (2009). FAK engages multiple pathways to maintain survival of fibroblasts and epithelia: differential roles for paxillin and p130Cas. *J. Cell Sci.*, *122*, 357–367.

CHAPTER 3

Mechanisms of Blastema Formation in Regenerating Amphibian Limbs

David L. Stocum, Nandini Rao
Department of Biology and Indiana University Center for Regenerative Biology and
Medicine, Indiana University-Purdue University, Indianapolis, IN, USA

INTRODUCTION

The limbs of larval and adult urodele amphibians are unique among tetrapod vertebrates
in their ability to regenerate from any level of the limb after amputation. Limb regeneration can be divided into two major phases: (1) formation of a blastema that resembles
the early embryonic limb bud and (2) blastema redevelopment, which involves blastema
growth and redifferentiation (Thornton, 1968; Tsonis, 2000; Bryant et al., 2002; Brockes
and Kumar, 2005, 2008; Stocum, 2006; Carlson, 2007 for reviews). Pattern formation, in
which the spatial relationships of the structures to be regenerated are specified, is a process
that spans both phases. Figure 3.1 illustrates these phases of limb regeneration. The ability
to form a blastema after amputation is what distinguishes the limbs of urodeles from
those of anuran amphibians, reptiles, birds, and mammals, and is the primary focus of this
chapter.

Blastema formation is a reverse developmental process realized partly by cell dedifferentiation in tissues local to the amputation plane (Thornton, 1968) and partly by a contribution of muscle stem cells (Morrison et al., 2006). Growth and redifferentiation of the
blastema are similar to embryonic limb bud development, with one major exception:
blastema cell proliferation is dependent on signals supplied by both the apical epidermal
cap (AEC) and the regenerating nerves, whereas the embryonic limb bud relies solely
on signals from the counterpart of the AEC, the apical ectodermal ridge (AER). The
musculoskeletal and dermal tissues of the new limb parts derived from the blastema
redifferentiate in continuity with their parent tissues (Carlson, 1978), and blood vessels
and nerves regenerate by extension from the cut ends of the pre-existing blood vessels and
axons, respectively.

Were we able to understand why some animals such as urodele amphibians are able to form
a regeneration-competent blastema after amputation while others such as adult anurans, birds,
and mammals are not, it might be possible to design chemical approaches to inducing blastema formation in human appendages. At the very least, such knowledge might improve our
ability to deal with non-amputational injuries to musculoskeletal, vascular, and neural tissues.
With this in mind, we review here what is known about blastema formation in the regeneration-competent limbs of urodeles and provide a brief comparison to blastema formation in
the regeneration-deficient anuran, *Xenopus laevis*.

Principles of Regenerative Medicine. DOI: 10.1016/B978-0-12-381422-7.10003-3

FIGURE 3.1
(A) Diagram of phases and stages of regeneration after amputation of a urodele limb. The two black lines indicate the two major phases of regeneration (blastema formation and blastema redevelopment), and the stages of regeneration following amputation (AMP). AB = accumulation blastema; MB = medium bud; LB = late bud; 2FB, 3FB, 4FB = fingerbud stages. The colored lines indicate different subphases of blastema formation and redevelopment.

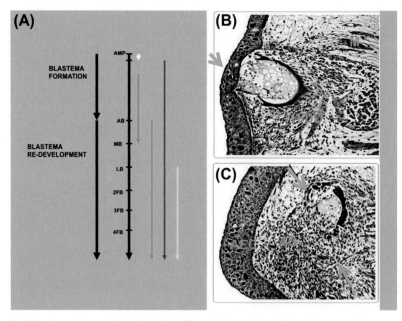

White = hemostasis and re-epithelialization; orange = histolysis and dedifferentiation; green = blastema growth; blue = pattern formation; yellow = redifferentiation. (B) Longitudinal section of regenerating axolotl hindlimb 4 days after amputation through the mid tibia-fibula. Arrow points to the thickening AEC. The cartilage (C), muscle (M), and other tissues are breaking down in a region of histolysis and dedifferentiation (H/DD) under the wound epithelium. 10×, light green and iron hematoxylin stain. (C) Longitudinal section of regenerating axolotl hindlimb 7 days after amputation through the mid tibia-fibula. An accumulation blastema (AB) has formed by the migration of dedifferentiated cells under the AEC. Arrows mark the junction between the accumulation blastema and the still-active region of histolysis and dedifferentiation proximal to it. 10×, light green and iron hematoxylin stain.

68

BLASTEMA FORMATION IN URODELE LIMBS

Blastema formation in regenerating urodele limbs can be subdivided into three overlapping phases: (1) hemostasis and re-epithelialization, (2) histolysis and dedifferentiation, and (3) blastema cell migration and accumulation (Fig. 3.1).

Hemostasis and re-epithelialization

Following limb amputation or after making skin wounds in amphibians, vasoconstriction occurs and a thrombin-catalyzed fibrin clot forms within seconds to protect the wound tissue and provide a temporary matrix from which repair or regeneration is initiated. An epithelium two to three cells thick covers the wound surface within 24 h after amputation, depending on limb size (Thornton, 1968). The basal epidermal cells at the cut edge of the skin migrate as a sheet that is extended by mitosis of cells adjacent to the wound edges (Lash, 1955; Hay and Fischman, 1961; Repesh and Oberpriller, 1978; Mahan and Donaldson, 1986). The fibrin clot contains significant amounts of fibronectin, which the epithelial sheet uses as a substrate for migration (Repesh and Furcht, 1982; Rao et al., 2009). Although structural alterations in the basal epidermal cells are necessary for the migratory movements of wound closure, the migrating cells retain other characteristics of their original state such as intermediate filaments (Repesh and Oberpriller, 1980). Within 2–3 days post-amputation (dpa), the wound epidermis thickens to form the AEC.

The basal cells and gland cells of the wound epidermis/AEC have secretory functions essential for blastema formation, as evidenced by their more extensive endoplasmic reticulum and Golgi network (Singer and Salpeter, 1961). WE3, 4, and 6 are three secretory-related antigens expressed specifically by dermal glands and wound epidermis/AEC (Tassava et al., 1989, 1993;

Castilla and Tassava, 1992; Estrada et al., 1993). Two other antigens, 9G1 (Onda and Tassava, 1991) and NvKII (Ferretti et al., 1991), are also specific to the wound epidermis, but their functions are unknown.

The early wound epidermis has an important function in generating early signals for limb regeneration. Na^+ influx in the amputated newt limb and H^+ efflux in the amputated tail of *Xenopus* tadpoles generate ionic currents across the wound epidermis essential for regeneration. Na^+ influx is via sodium channels (Borgens et al., 1977). H^+ efflux in the amputated tail is driven by a plasma membrane ATPase in the epidermal cells (Adams et al., 2007) and is likely to be important for urodele limb regeneration as well, given that a gene encoding a v-ATPase was the most abundant clone in a suppressive subtraction cDNA library made from 4 dpa regenerating limb tissue in the axolotl (Gorsic et al., 2008). Drug-induced inhibition of either Na^+ or H^+ movements during the first 24 h or so after amputation results in failure of blastema formation (Jenkins et al., 1996; Adams et al., 2007).

Two other early regeneration signals that may be linked to ion flux are nitric oxide (NO) and inositol trisphosphate (IP_3). The enzyme that catalyzes NO synthesis, nitric oxide synthase 1 (NOS1), is strongly upregulated in the wound epidermis of amputated axolotl limbs at 1 dpa (Rao et al., 2009). NO has a wide variety of signaling functions (Lowenstein and Snyder, 1992), is produced by macrophages and neutrophils as a bactericidal agent, and has a role in activating proteases known to be important effectors of histolysis in regenerating limbs. IP_3 and diacylglycerol (DAG) are the products of phosphatidylinositol bisphosphate (PIP_2), which in turn is derived from inositol. IP_3 synthase, a key enzyme for the synthesis of inositol from glucose-6-phosphate, is upregulated during blastema formation in regenerating axolotl limbs (Rao et al., 2009). IP_3 stimulates a rise in cytosolic Ca^{2+} that results in the localization of protein kinase C (PKC) to the plasma membrane, where PKC is activated by DAG and regulates transcription (Lodish et al., 2008). During blastema formation, there is a general downregulation of proteins involved in Ca^{2+} homeostasis, which suggests that IP_3 might signal a rise in cytosolic Ca^{2+} in regenerating limbs to localize PKC to the plasma membrane (Rao et al., 2009). Other studies have shown that IP_3 is generated from PIP_2 within 30 seconds after amputation in newt limbs (Tsonis et al., 1991) and that PKC rises to a peak by the accumulation blastema stage (Oudkhir et al., 1989). Furthermore, beryllium inhibition of IP_3 formation prevents blastema formation (Tsonis et al., 1991). How these early signals are translated into the next phase of blastema formation, histolysis and dedifferentiation, is unknown.

Histolysis and dedifferentiation

Histolysis is the loss of tissue organization resulting from the enzymatic degradation of the extracellular matrix (ECM). Dedifferentiation is the reversal of a given state of differentiation to an earlier state via nuclear reprogramming and loss of specialized structure and function.

All of the tissues subjacent to the wound epidermis undergo intense histolysis (ECM degradation and tissue disorganization) for a distance of 1−2 mm, resulting in the liberation of individual dermal cells, Schwann cells of the peripheral nerves, and skeletal cells from their matrix. Myofibers fragment at their cut ends and break up into mononucleate cells while simultaneously releasing satellite cells (the stem cells that effect muscle regeneration). The liberated cells undergo dedifferentiation to mesenchyme-like cells with large nuclei and sparse cytoplasm that exhibit intense RNA and protein synthesis (Bodemer and Everett, 1959; Bodemer, 1962; Hay and Fischman, 1961; Anton, 1965). Histolysis and dedifferentiation begin within 2−3 days post-amputation in larval urodeles and within 4−5 days in adults, and continue until the medium bud stage of blastema growth (Hay and Fischman, 1961; Thornton, 1968).

MECHANISMS OF HISTOLYSIS

Degradation of tissue ECM is achieved by acid hydrolases and matrix metalloproteinases (MMPs). Acid hydrolases identified in regenerating urodele limbs include cathepsin D, acid phosphatase, β-glucuronidase, carboxyl ester hydrolases, and N-acetyl-glucoaminidase (Schmidt, 1966, 1968 for reviews; Rivera et al., 1981; Ju and Kim, 1998; Park and Kim, 1999). Osteoclasts are abundant in the region of histolysis, where they degrade bone matrix via hydrochloric acid, acid hydrolases, and MMPs. MMPs that are upregulated include MMP-2 and -9 (gelatinases), and MMP-3/10a and b (stromelysins) (Grillo et al., 1968; Dresden and Gross, 1970; Yang and Bryant, 1994; Miyazaki et al., 1996; Ju and Kim, 1998; Park and Kim, 1999; Kato et al., 2003; Vinarsky et al., 2005). The basal layer of the wound epidermis is a source of MMP-3/10a and b in the newt limb, as well as of a novel MMP with low homology to other MMPs (Kato et al., 2003). These MMPs may be responsible for maintaining contact between the wound epidermis and the underlying tissues by preventing reassembly of a basement membrane, and may also diffuse into the underlying tissues. Chondrocytes are the source of MMP-2 and -9 in the newt limb, and these enzymes diffuse outward from the degrading skeletal elements (Kato et al., 2003). The importance of MMPs to histolysis, and the importance of histolysis to the success of regeneration, is underscored by the failure of blastema formation in amputated newt limbs treated with an inhibitor of MMPs (GM6001) (Vinarsky et al., 2005).

Once the accumulation blastema begins to grow, histolysis gradually ceases due to the activity of tissue inhibitors of metalloproteinases (TIMPS) (Stevenson et al., 2006). TIMP1 is upregulated during histolysis, when MMPs are at maximum levels, and exhibits spatial patterns of expression congruent with those of MMPs in the wound epidermis, proximal epidermis, and internal tissues undergoing disorganization.

MECHANISMS OF DEDIFFERENTIATION

Dedifferentiation is a complex process involving changes in transcriptional program to suppress differentiation genes, while activating genes associated with stemness, reduction of cell stress, and remodeling internal structure. Inhibition of the transcriptional shift by actinomycin D does not affect histolysis, but does prevent or retard dedifferentiation, leading to regenerative failure or delay (Carlson, 1969). This suggests that at least part of the proteases involved in histolysis are not regulated at the transcriptional level, but that proteins effecting dedifferentiation are so regulated.

Stemness genes upregulated during blastema formation are *msx1* (Crews et al., 1995; Koshiba et al., 1998; Echeverri and Tanaka, 2005), *nrad* (Shimizu-Nishikawa et al., 2001), *rfrng*, and *notch* (Cadinouche et al., 1999). *Msx1* inhibits myogenesis (Song et al., 1992; Woloshin et al., 1995) and its forced expression in mouse myotubes causes cellularization and reduced expression of muscle regulatory proteins (Odelberg et al., 2000). Inhibition of *msx1* expression in cultured newt myofibers by anti-*msx* morpholinos prevents their cellularization (Kumar et al., 2004). Newt regeneration extract also stimulates mouse myotubes to re-enter the cell cycle, cellularize, and reduce expression of muscle regulatory proteins (McGann et al., 2001). *Nrad* expression is correlated with muscle dedifferentiation (Shimizu-Nishikawa et al., 2001), and *Notch* is a major mediator of stem cell self-renewal (Go et al., 1998; Lundkvist and Lendahl, 2001). Dedifferentiated cells express a more limb bud-like ECM in which the basement membrane is absent, type I collagen synthesis and accumulation are reduced, and fibronectin, tenascin, and hyaluronate accumulate (Toole and Gross, 1971; Gulati et al., 1983; Mescher and Munaim, 1986; Onda et al., 1991; Stocum, 1995 for review).

Nuclear transplantation studies (Burgess, 1967) and transplantation experiments with genetically marked (triploidy, GFP) tissues have shown that blastema cells are not reprogrammed to pluripotency or even multipotency, but are largely constrained to redifferentiate into their parent cell types (Steen, 1968; Kragl et al., 2009). The exception is fibroblasts of the dermis, which after dedifferentiation are able to transdifferentiate at high frequency into cartilage (Steen, 1968; Namenwirth, 1974; Kragl et al., 2009). Regardless of this limited plasticity, it is interesting that three of the six transcription factor genes (*klf4*, *sox2*, *c-myc*) used to reprogram mammalian adult somatic cells to induced pluripotent stem cells (iPSCs) (Takahashi et al., 2007; Yu et al., 2007) are upregulated during blastema formation in regenerating newt limbs, and also during lens regeneration (Maki et al., 2009). The Lin 28 protein, the product of a fourth transcription factor gene used to derive iPSCs (Yu et al., 2007), is upregulated during blastema formation in regenerating axolotl limbs (Rao et al., 2009). Thus, transcription factors that reprogram fibroblasts to iPS cells may also play a role in nuclear reprogramming during limb regeneration, but other factors are clearly in play to ensure that dedifferentiated cells reverse their transcription programs only far enough to maintain a state of "limbness" that can respond to proliferation and patterning signals.

The differential regulation of pathways that protect cells from stress and apoptosis may also play a role in dedifferentiation. Proteomic analysis suggests that reduced metabolic activity, upregulation of pathways that accelerate protein folding or eliminate unfolded proteins (the unfolded protein response, UPR), and differential regulation of apoptotic pathways may largely prevent apoptosis (Rao et al., 2009), which is known to be minimal in regenerating limbs (Mescher et al., 2000; Atkinson et al., 2006). This idea is consistent with other studies on cultured chondrocytes, β cells, and Muller glia cells of the retina showing that cells dedifferentiate as part of a mechanism to combat apoptotic cell stress (see Rao et al., 2009 for discussion).

The details of internal structural remodeling in dedifferentiating cells are poorly understood. Dismantling of phenotypic structure and function is most visible in myofibers, but the molecular details of the process are largely uninvestigated for any limb cell type. Two small molecules, one a trisubstituted purine called myoseverin and the other a disubstituted purine dubbed reversine, have been screened from combinatorial chemical libraries and found to cause cellularization of C2C12 mouse myofibers (Rosania et al., 2000; Chen et al., 2004). Myoseverin disrupts microtubules and upregulates genes for growth factors, immunomodulatory molecules, ECM remodeling proteases, and stress-response genes, consistent with the activation of pathways involved in wound healing and regeneration, but does not activate the whole program of myogenic dedifferentiation (Duckmanton et al., 2005). Reversine treatment of C2C12 myotubes resulted in mononucleate cells that behaved like mesenchymal stem cells (MSCs); i.e. they were able to differentiate *in vitro* into osteoblasts and adipocytes, as well as muscle cells (Anastasia et al., 2006). Myoseverin and reversine will be useful in analyzing the events of structural remodeling, and may have natural counterparts that can be isolated.

The signals that trigger the shift in transcription during dedifferentiation are largely unknown. Degradation of the ECM by proteases would break contacts between ECM molecules and integrin receptors, leading to changes in cell shape and reorganization of the actin cytoskeleton (Juliano and Haskill, 1993). This reorganization might activate the signal transduction pathways that downregulate phenotype-specific transcription programs and upregulate programs characteristic of a less specialized state that allows blastema cell migration and response to proliferation and patterning signals. The molecular characterization of blastema cell surface antigens, transcription factors, and micro-RNAs, and studies of changes in epigenetic marks via chromatin-modifying enzymes, will be crucial for understanding the mechanism of dedifferentiation in regenerating amphibian limbs.

DIFFERENTIAL TISSUE CONTRIBUTIONS TO THE BLASTEMA

Transplantation studies with genetically marked tissues indicate that individual tissues of the limb make differential contributions to the blastema. In the axolotl limb, dermal cells represent 19% and chondrocytes 6% of the cells present at the amputation surface, but contribute 43 and 2% of the blastema cells, respectively (Muneoka et al., 1986). The percentage of blastema cells contributed by periosteum, myofibers and their fibroblasts, and Schwann cells, is not known. Studies on *Pax-7* expression indicate that satellite cells of myofibers make a substantial contribution to the blastema (Morrison et al., 2006). An interesting question is what proportions of the muscle in the regenerated limb parts are derived from dedifferentiated myofibers and satellite cells, and whether these proportions are different for larval and adult urodeles.

CELL CYCLING DURING BLASTEMA FORMATION

Tritiated thymidine (^{3}H-T) labeling studies have shown that cells of amputated urodele limbs initiate cell cycle re-entry coincident with their histolysis and dedifferentiation (Fig. 3.2). The pulse-labeling index reaches 10–30% during the pre-accumulation blastema phase (Mescher and Tassava, 1975; Loyd and Tassava, 1980). However, the mitotic index is very low, between 0.1 and 0.7% (average ~0.4%, or 4/1,000 cells) in both *Ambystoma* larvae (Kelly and Tassava, 1973; Stocum, 1980) and adult newt (Mescher and Tassava, 1975; Mescher, 1976).

Both the labeling and mitotic indices rise as much as 10-fold when the accumulation blastema initiates growth (Fig. 3.2) (Chalkley, 1954; Kelly and Tassava, 1974; Mescher and Tassava, 1975; Loyd and Tassava, 1980; Stocum, 1980). ^{3}H-T pulse labeling studies indicate that the final cycling fraction of blastema cells is between 92 and 96% in larvae and over 90% in adults (Tomlinson et al., 1985; Goldhamer and Tassava, 1987; Tomlinson and Barger, 1987). The mitotic index in the growing blastema is relatively uniform along the proximodistal axis until differentiation sets in, when cells in the proximal region of the blastema withdraw from the cell cycle, creating a distal to proximal gradient of mitosis (Litwiller, 1939; Smith and Crawley, 1977; Stocum 1980).

The low mitotic index prior to establishment of the accumulation blastema suggests that it forms primarily by the accumulation of dedifferentiated cells rather than their mitosis. The cell cycle, measured in regenerating axolotl and adult newt limbs, is 40–53 h in length, with an average of 46 h, and does not vary significantly between larval and adult limbs or between stages of regeneration (McCullough and Tassava, 1976; Maden, 1978; Tassava et al., 1983; Tassava et al., 1987). Mitosis takes up about 1 h of the cycle. The time taken to establish an accumulation blastema, however, is two (small *Ambystoma* larvae) to seven (juvenile axolotl,

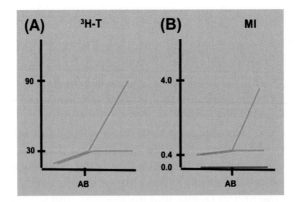

FIGURE 3.2

Diagrammatic representation of changes in ^{3}H-T labeling and mitotic index (MI) during blastema formation and growth, expressed as percentages of total cell number on the ordinate. AB on the abscissa represents the accumulation blastema stage. The growth phase is to the right of the AB. (A) Prior to the accumulation blastema stage, the ^{3}H-T labeling index is the same in control (green line) and in epidermis-free and denervated limbs (both represented by the red line). These indices in deprived limbs fail to rise in concert with the controls during blastema growth and an accumulation blastema does not form. (B) Prior to the accumulation blastema stage, the basal mitotic index of controls (green line) and epidermis-free limbs (red line) are nearly identical, but the MI does not increase with the controls during blastema growth. In contrast, the MI in denervated limbs (blue line) does not achieve the basal level and remains near zero. An accumlation blastema does not form in either denervated or epidermis-free limbs.

adult newt) times the average length of the cell cycle. The fact that cells readily enter the cell cycle during histolysis and dedifferentiation but divide only infrequently suggests that a large proportion of dedifferentiating cells arrest in G_2 (Mescher and Tassava, 1975).

Further indirect evidence for G_2 arrest is the strong upregulation of the ecotropic viral integration factor 5 (EVI5) throughout blastema formation in regenerating axolotl limbs (Rao et al., 2009). EVI5 is a centrosomal protein that accumulates in the nucleus during early G_1 in mammalian cells and prevents them from prematurely entering mitosis by stabilizing Emi1, a protein that inhibits cyclin A degradation by the anaphase-promoting complex/cyclosome (APC/C) (Eldridge et al., 2006). At G_2, Emi1 and EVI5 are phosphorylated by Polo-like kinase 1 (PLK1) and targeted for ubiquitin-driven degradation, allowing the cell to enter mitosis. Thus, high levels of EVI5 during blastema formation may restrain cells from entering mitosis until they are fully dedifferentiated and present in enough numbers to form an accumulation blastema (Rao et al., 2009).

The signals that drive re-entry into the cell cycle have been studied in detail in myofibers of the regenerating newt limb. Cell cycle re-entry in cultured newt and mouse myoblasts and newt myofibers is promoted by a thrombin-activated factor present in the serum of all vertebrates tested thus far, including mammals (Tanaka et al., 1997; Straube and Tanaka, 2006). Mouse myofibers do not respond to this factor. Newt blastema extract promotes DNA synthesis in both newt and mouse myofibers (McGann et al., 2001), suggesting that mouse myofibers lack an essential signal pathway ingredient that is supplied by newt blastema extract, but not by serum. Although the thrombin-activated protein is both necessary and sufficient to stimulate the entry of myonuclei into the cell cycle, it is not sufficient to drive them through mitosis, and myonuclei arrest in G_2. Cell cycle re-entry is independent of myofiber cellularization, since cell cycle-inhibited myofibers implanted into newt limb blastemas break up into mononucleate cells (Velloso et al., 2001). Mitosis, however, does appear to require mononucleate cell status. The mechanism of myofiber fragmentation into single cells is not known, nor is it known whether the thrombin-activated protein is also necessary to drive mononucleate cells such as dedifferentiating chondrocytes and fibroblasts into the cell cycle as well, or whether this is a feature unique to myofibers. Biochemical evidence suggests that the thrombin-activated factor may be a potent growth factor required in very small amounts (Straube and Tanaka, 2006).

WOUND EPIDERMIS, NERVES, AND NON-NEIGHBORING CELL CONTACTS ARE REQUIRED FOR CELL CYCLING

Requirement for wound epidermis and nerves

The wound epidermis of regenerating urodele limbs is invaded by sprouting sensory axons within 2–3 days after amputation, while other sensory axons and motor axons make intimate contact with mesenchyme cells as the blastema forms (Salpeter, 1965; Lentz, 1967). Blastema formation is inhibited when formation of the wound epidermis is prevented by covering the amputation surface with a full-thickness skin flap or inserting the skinned amputated limb tip into the coelom (Goss, 1956; Mescher, 1976), or when the function of the wound epidermis is compromised by UV irradiation of the AEC (Thornton, 1958), or by substituting X-irradiated epidermis for normal epidermis (Lheureux and Carey, 1988). Denervating the limb at the time of amputation also prevents blastema formation (Schotte and Butler, 1944; Singer and Craven, 1948; Powell, 1969). In either case, inhibition of blastema formation is not due to a failure of cells to undergo dedifferentiation, although the number of dedifferentiated cells is fewer in epidermis-free limbs (Singer and Salpeter, 1960). This result is consistent with the role of the wound epidermis in histolysis and with the idea that it is the accumulation of dedifferentiated cells, not mitosis, that is primarily responsible for establishment of the blastema.

Deprivation studies suggest that the wound epidermis and nerves have differential effects on the cell cycle during blastema formation (Fig. 3.2). The ^{3}H-T labeling index is the same as that

73

of controls up to the accumulation blastema stage in both epidermis-free and denervated limbs, suggesting that neither the nerve nor wound epidermis is required for DNA synthesis during this time. Furthermore, the mitotic index in limbs deprived of wound epidermis is the same as controls up to the accumulation blastema stage, indicating that the wound epidermis is also not required for the low basal level of mitosis observed during blastema formation. However, denervated limbs do not achieve the control basal mitotic index and their index remains near zero (Kelly and Tassava, 1973; Tassava et al., 1974; Mescher and Tassava, 1975; Tassava and Mescher, 1976; Maden, 1978; Tassava and McCullough, 1978; Tassava and Garling, 1979). The significant increases in ^{3}H-T labeling and mitotic indices seen after blastema formation in control limbs do not take place in denervated or wound epidermis-free limbs.

Once an accumulation blastema has been established, its differentiation and morphogenesis, but not growth by mitosis, become nerve independent (Schotte and Butler, 1944; Singer and Craven, 1948; Powell, 1969; Maden, 1978; Goldhamer and Tassava, 1987). Blastemas denervated at the medium bud stage form complete but miniature regenerates, illustrating their continued dependence on the nerve for mitosis. Redifferentiation of the early blastema also becomes independent of the AEC, as shown by the ability of epidermis-free medium bud blastemas to form correctly patterned skeletal elements when implanted into dorsal fin tunnels of larval *Abystoma maculatum* (Stocum and Dearlove, 1972). The size of these elements is subnormal, suggesting that blastema cell proliferation has a continuing requirement for the AEC as well. Consistent with this notion, the ^{3}H-thymidine labeling and mitotic indices of epidermis-free newt limb blastemas cultured in the presence of dorsal root ganglia are reduced 3−4-fold (Globus et al., 1980; Smith and Globus 1989). A major difference between the regenerates formed by denervated and epidermis-free blastemas, however, is that the pattern of skeletal elements formed by the latter is more or less distally truncated, depending on the stage at which the epidermis is removed, suggesting that the AEC has a role in proximodistal patterning in addition to proliferation (Stocum and Dearlove, 1972; Stocum, 2006).

Based on these studies, Tassava and Mescher (1975) proposed that the injury of amputation is sufficient to promote entry into the cell cycle and DNA synthesis. The nerve is required for dedifferentiating cells to enter mitosis from G_2, and the wound epidermis is required to maintain post-mitotic cells in the cell cycle and prevent their differentiation. Evidence that the wound epidermis performs this function is that innervated blastemas *in vitro* undergo premature differentiation in the absence of epidermis (Globus et al., 1980). Since hormones, especially insulin, are also critical for regeneration, another model has proposed a tripartite control of proliferation by wound epidermis, nerve, and insulin (Vethamany-Globus et al., 1978).

Molecular factors contributed by wound epidermis and nerves

What are the molecular factors supplied to blastema cells by the wound epidermis and nerves? The wound epidermal factors appear to be members of the fibroblast growth factor (FGF) family. Fgf-1, fgf-2, and fgf-8 are made *in vivo* by the wound epidermis/AEC and fgf-10 by blastema cells (Boilly et al., 1991; Christensen et al., 2001; Han et al., 2001; Giampaoli et al., 2003), and blastema cells express receptors for the FGFs of the AEC (Poulin et al., 1993). Fgf-1 is expressed in the wound epidermis/AEC throughout blastema formation and growth (Giampaoli et al., 2003). Fgf-2 and fgf-8 are expressed at low levels during dedifferentiation, with expression increasing once the accumulation blastema has formed (Christensen et al., 2001; Han et al., 2001; Giampaoli et al., 2003). By contrast, fgf-10 is strongly expressed throughout blastema formation and growth (Christensen et al., 2001). Fgf-1 was shown to elevate the mitotic index of blastema cells cultured in the absence of nerves or AEC (Albert et al., 1987), and fgf-2 to elevate the mitotic index of blastema cells in amputated limbs covered by full-thickness skin (Chew and Cameron, 1983). In other experiments, both fgf-2 and insulin-like growth factor-1 (IGF-I) injected intraperitoneally shortened the time required for formation of the accumulation blastema by amputated limbs (Fahmy and Sicard, 1998).

Many factors that promote blastema cell proliferation *in vitro* have been detected in the nerves of regenerating urodele limbs, including transferrin (Mescher and Kiffmeyer, 1992; Mescher et al., 1997), substance P (Globus and Alles, 1990; Anand et al., 1987), *fgf-2* transcripts (Mullen et al., 1996), and glial growth factor 2 (Ggf-2, Wang et al., 2000). Ggf-2 was reported to rescue regeneration in denervated axolotl limbs when injected intraperitoneally during blastema formation, although the nature of the rescue was not defined (Wang et al., 2000).

Recent experiments, however, suggest that a single protein, the anterior gradient protein (AGP), can substitute for the mitotic function of the nerves in regenerating newt limbs (Kumar et al., 2007). AGP is strongly expressed in the Schwann cells of regenerating newt limbs at 5 and 8 dpa, when initial dedifferentiation is under way (Kumar et al., 2007). Nerve transection at the base of the amputated limb abolishes AGP expression, indicating that it is induced in the Schwann cells by axons. The gene for AGP supports regeneration to digit stages when electroporated into denervated newt limbs at 5 dpa. AGP is a ligand for the blastema cell surface protein Prod1, a member of the Ly6 family of three-finger proteins anchored to the cell surface by a glycosylphosphatidyl inositol (GPI) linkage (da Silva et al., 2002; Brockes and Kumar, 2008). Conditioned medium of Cos7 cells transfected with the AGP gene stimulates BrdU incorporation into cultured blastema cells, and this incorporation is blocked by antibodies to Prod1, suggesting that AGP can act directly on blastema cells through Prod1 to stimulate DNA replication (Kumar et al., 2007).

The function of the wound epidermis may depend on regenerating nerves. The epidermis of a wound made in the skin of an unamputated axolotl limb develops a thickening comparable to the AEC of a regenerating limb, which subsequently regresses. However, if a nerve is deviated into the wound, the thickening is maintained and a blastema-like growth is formed (Endo et al., 2004). This result implies that the initial AEC structure can form independently of the nerve, but that maintenance of AEC structure and function may be nerve-dependent.

Evidence for this possibility comes from two sources. First, AGP expression in the regenerating newt limb shifts from the Schwann sheath to cells of secretory glands subjacent to the wound epidermis by the accumulation blastema stage (Kumar et al., 2007). This shift is nerve dependent, suggesting that the axons reinnervating the wound epidermis induce it to express AGP, which is then supplied to subjacent mesenchymal cells, enabling growth of the blastema. The nerve dependence of mitosis throughout blastema redevelopment implies that this induction is continuous, an idea that might be tested by examining expression patterns of AGP in control and denervated limbs at successively later stages of blastema redevelopment. Second, aneurogenic limbs are AEC-dependent, but nerve-independent for regeneration (Yntema, 1959a,b), and become nerve dependent when reinnervated (Thornton and Thornton, 1970). A similar shift from nerve independence to dependence occurs as nerves invade the differentiating limb bud (Fekete and Brockes, 1987). These shifts again suggest an interaction between nerves and epidermis that renders the limb nerve dependent for regeneration. It would be of interest to investigate the expression of AGP and Prod-1 in regenerating aneurogenic limbs and in reinnervated aneurogenic limbs, as well as regenerating limb buds at various stages of normal development, to help clarify the functional relationship between the AEC and nerves. Another question is whether AGP is able to substitute for the function of the AEC.

Requirement for non-neighboring cell contacts

Amputated urodele limbs will not form a blastema unless cells from non-neighboring positions on the limb circumference interact to sense gaps in structure that need to be filled in by proliferation. This has been shown by experiments in which the normally asymmetrical (anterioposterior, dorsoventral) skin of the newt limb has been made symmetrical by rotating a longitudinal strip of dorsal skin 90° and grafting it around the circumference of the limb, and then amputating through the strip (Lheureux, 1975). The dedifferentiated graft cells all have the same circumferential positional identity and so do not sense any positional gap when they interact, leading to failure of cell cycling and blastema formation. Normal regeneration ensues, however, when short longitudinal skin strips from three or four opposite points of the circumference are rotated and grafted because dedifferentiated cells from these strips have non-neighboring positional identities. Similarly, the cells of blastema-like growths induced by deviating a nerve to limb skin wounds will undergo mitosis only if pieces of skin from opposite circumferential sites cover the wound (Endo et al., 2004). While most investigators consider that the interacting positional identities are those of dermal cells, Campbell and Crews (2008) have proposed that confrontation of epidermal cells from different positional identities is important as well.

Prod-1 has been implicated in recognizing gaps in positional identity between non-neighboring cells (Brockes and Kumar, 2008). Positional identity of blastema cells is associated with a proximodistal gradient of cell adhesivity (Nardi and Stocum, 1983; Crawford and Stocum, 1988; Egar, 1993; Echeverri and Tanaka, 2005; Kragl et al., 2009). Prod-1 is also present in a distal to proximal gradient (da Silva et al., 2002). Antibodies to Prod1, or its removal from the blastema cell surface by phosphatidylinositol-specific phospholipase C (PIPLC), inhibit the recognition of adhesive differentials between distal and proximal blastemas in the Nardi and Stocum (1983) *in vitro* engulfment assay (da Silva et al., 2002). These results suggest that Prod-1 plays a role in recognizing gaps in positional identity that could stimulate mitosis of blastema cells through its ligand, AGP (Brockes and Kumar, 2008).

Blastema cell migration and accumulation

The AEC appears to direct the migration of blastema cells to form the accumulation blastema beneath it. This was shown by experiments in which shifting the position of the AEC laterally caused a corresponding shift in blastema cell accumulation (Thornton, 1960), and transplantation of an additional AEC to the base of the blastema resulted in supernumerary blastema formation (Thornton and Thornton, 1965). Nerve guidance of blastema cells to form eccentric blastemas appeared to be ruled out, since similar experiments on aneurogenic limbs also resulted in eccentric blastema formation (Thornton and Steen, 1962).

The redirected accumulation of blastema cells in these experiments may be due to the migration of the cells on adhesive substrates produced by the eccentric AEC. TGF-β1 is strongly upregulated during blastema formation in amputated axolotl limbs (Hutchison et al., 2007). A target gene of TGF-β1 is fibronectin, a substrate molecule for cell migration that is highly expressed by basal cells of the wound epidermis during blastema formation (Christensen and Tassava, 2000; Rao et al., 2009). Inhibition of TGF-β1 expression by the inhibitor of SMAD phosphorylation, SB-431542, reduces fibronectin expression and results in failure of blastema formation (Hutchison et al., 2007), suggesting that fibronectin provided by the AEC provides directional guidance for blastema cells.

Proximodistal patterning begins during blastema formation

A detailed discussion of pattern formation is beyond the scope of this chapter, but excellent reviews can be found elsewhere (Meinhardt, 1982; Gardiner et al., 1999; Tanaka, 2003; Tamura et al., 2009; Yakushiji et al., 2009). Here we wish to point out just two aspects of regenerate patterning. First, the blastema is a self-organizing system from its inception with regard to proximodistal patterning and morphogenesis (Stocum and Melton, 1977). Second, patterning begins during the phase of blastema formation. Genes specifying the proximodistal axis of the regenerate (and the limb bud), such as *Hoxa-9* and *-13* and *Meis*, are activated even before an accumulation blastema is formed (Gardiner et al., 1999; Mercader et al., 2005). A fascinating problem in limb regeneration is how this self-organization is achieved, particularly how the distal and proximal boundaries of what is to be regenerated are established (Stocum, 2006).

BLASTEMA FORMATION IN *XENOPUS LAEVIS* LIMBS

The amputated limb buds of early anuran tadpoles regenerate prior to their differentiation, but lose the capacity to regenerate at successive proximodistal levels as the limb bud differentiates (Marcucci, 1916; Schotte and Harland, 1943; Dent, 1962). After metamorphosis, most Ranid froglets exhibit zero ability for limb regeneration, whereas some Pipid frogs can regenerate a symmetrical cartilage spike lacking muscle (Stocum, 1995 for review). The events associated with this regenerative deficiency have been best characterized in *Xenopus laevis* (Dent, 1962; Korneluk and Liversage, 1984; Wolfe et al., 2000). The undifferentiated hindlimb buds of *Xenopus* early tadpoles (up to stage 52/53) form a blastema of mesenchymatous cells that regenerates the missing structures in continuity with the structures differentiating proximal to it. The regenerative deficiency of late tadpole and froglet limbs can be traced to impaired histolysis and dedifferentiation, leading to the formation of a poor-quality, non-mesenchymatous fibroblastema.

Formation of the fibroblastema is associated with limited histolysis and dedifferentiation

Following amputation of a froglet limb, the events of hemostasis and re-epithelialization are the same as those in the amputated urodele limb. Histological studies indicate, however, that there is little histolysis and the few cells that are liberated from their ECM appear not to dedifferentiate. Compared to urodeles, the lack of histolysis and dedifferentiation is correlated with an AEC that is thinner (Wolfe et al., 2000; Suzuki et al., 2005, 2006), exhibits increased expression of *inhibitor of differentiation 2* and *3* (*Id2, 3*) genes (Shimizu-Nishikawa et al., 1999), and does not upregulate expression of NOS1 (Rao et al., 2009; Rao et al., in preparation). These observations suggest a lack of production by the wound epidermis of MMPs and/or signals essential for histolysis and dedifferentiation. Stage 52 amputated limbs have been shown to express MMP9 (Carinato et al., 2000), and regenerating nerves of amputated froglet limbs to express neural MMP28 (Werner et al., 2007). We are currently conducting a detailed study in amputated froglet limbs of the types and level of activity of proteases known to be involved in urodele limb histolysis (F. Song et al., in preparation). Metabolism and failure to induce stress response pathways might also play roles in the lack of histolysis. Induction of stress response pathways is requisite for successful regeneration in stage 52 hindlimbs (Pearl et al., 2008), but whether these pathways are induced after amputation of froglet limbs is unknown. Newt blastemas produce large amounts of lactic acid (Schmidt, 1968), which would provide the acidic pH optimum for acid hydrolases. If this acidic environment were absent in amputated *Xenopus* limbs, the activity of such enzymes would be compromised.

In lieu of dedifferentiation, fibroblasts from the periosteum, dermis, and possibly muscle are activated and accumulate between the wound epithelium and cut surface of the bone (Dent, 1962; Korneluk and Liversage, 1984; McLaughlin and Liversage, 1986; Wolfe et al., 2000). These fibroblasts divide to form a "fibroblastema" that goes on to differentiate into the cartilage spike (Fig. 3.3). The spike is symmetrical in the anteroposterior axis due to the failure to activate *sonic hedgehog* (*shh*) expression (Endo et al., 2000; Satoh et al., 2006; Yakushiji et al., 2007). Periosteal fibroblasts also accumulate around the bone shaft proximal to the amputation plane, and differentiate into a cartilage collar that is continuous distally with the spike (Dent, 1962; Wolfe et al., 2000). Histological observations and proteomic data indicate little muscle breakdown (Rao et al., in preparation). Satellite cells are present in the myofibers at the amputation plane but do not become part of the fibroblastema, a situation that can be remedied by transplanting cells that secrete hepatocyte growth factor (HGF) into the blastema (Satoh et al., 2005a). This implies that the factors necessary to attract satellite cells into the

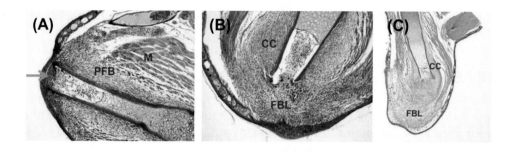

FIGURE 3.3

(A) Longitudinal section of a *Xenopus laevis* froglet limb 5 days after amputation through the mid tarsus of the hindlimb. The arrow points to fibroblasts that have migrated over the cut end of the bones under the wound epithelium. The periosteal fibroblasts (PFB) are proliferating to form a collar around the bones. M = muscle. 10×, hematoxylin and eosin stain. (B) Longitudinal section of a *Xenopus laevis* froglet limb 7 days after amputation through the mid tarsus of the hindlimb. The fibroblasts under the wound epithelium have proliferated to form an accumulation fibroblastema (FBL). The collar of periosteal fibroblasts around the bones is beginning to differentiate into a cartilage collar (CC). M = muscle. 10×, hematoxylin and eosin stain. (C) Longitudinal section of a *Xenopus laevis* froglet limb 12 days after amputation through the mid tarsus of the hindlimb, illustrating the growing fibroblastema (FBL). The base of the fibroblastema is starting to differentiate into the cartilage spike, in continuity with the cartilage collar (CC) differentiating from periosteal fibroblasts. 4×, hematoxylin and eosin stain.

blastema are present in urodeles, although these have not been identified. To further explore the lack of histolysis and dedifferentiation in limbs of *Xenopus* froglets, detailed comparative analyses of acid hydrolases, MMPs, transcription factors, cell surface antigens, and epigenetic factors such as chromatin remodeling enzymes, Polycomb group proteins, and micro-RNAs should be undertaken and compared to similar analyses in axolotl limbs.

Fibroblastema formation in amputated limbs of *Xenopus* froglets appears to have the same epidermal and nerve requirements as blastema formation in urodele limbs. Denervation at the time of amputation (Filoni et al., 1999; Cannata et al., 2001; Suzuki et al., 2005) or prevention of wound epidermis formation (Goss and Holt, 1992) result in failure of fibroblast accumulation, suggesting that the wound epidermis is necessary for fibroblast migration and/or proliferation and that this function of the epidermis requires interaction with nerves. Like the urodele AEC, the wound epidermis of the amputated *Xenopus* froglet limb expresses fgf-8 (Endo et al., 2000; Suzuki et al., 2005). Bone morphogenetic protein (BMP) signaling is crucial for AEC function in amputated stage 52 tadpole limbs (Pearl et al., 2008). BMP is essential for fibroblastema formation in amputated froglet limbs (Beck et al., 2009) and can induce segmentation of the cartilage spike when introduced into the fibroblastema (Satoh et al., 2005b). Studies of the type that have been done on the regenerating urodele limb with regard to cell contribution, DNA labeling, and mitosis during blastema formation have not been done on amputated *Xenopus* limbs. Furthermore, given the importance of Prod-1 and AGP to neural and epidermal function in urodele limb regeneration, the expression pattern of these molecules during fibroblastema formation should be investigated and compared to their patterns during blastema formation in the urodele limb.

Why do *Xenopus* limbs lose their capacity for regeneration as they develop?

While we can correlate the lack of true blastema formation in *Xenopus* with certain deficiencies compared to urodeles, we still do not know the fundamental physiological reasons as to why juvenile and adult urodeles, and early anuran tadpoles, are able to form a regeneration-competent blastema whereas late anuran tadpoles and adults can form only a regeneration-deficient or incompetent blastema. There are several ideas about what underlies these differences between urodeles and anurans.

The first is that the degree of maturity of the immune system determines whether or not a limb can regenerate (Harty et al., 2003; Mescher and Neff, 2005, 2006; Godwin and Brockes, 2006 for reviews). The more developed the immune system, the less capacity for regeneration of complex structures such as limbs. There are two observations that support this idea. First, compared to anurans, urodeles have a less developed immune system that enables them to more easily accept allografts (Cohen, 1971). Second, the immune system of *Xenopus* changes profoundly during development, coincident with loss of limb regenerative capacity. Thus, skin taken from a regeneration-competent early tadpole and cold preserved is rejected when autografted to the donor after metamorphosis (Izutsu and Yoshizato, 1993).

Further support for the idea of an inverse relationship between immune competence and limb regeneration comes from studies of mammalian fetal wounds. Mouse fetal limb buds have some capacity for regeneration (Wanek et al., 1989; Reginelli et al., 1995; Han et al., 2003), and mouse fetal skin regenerates until late in gestation, when it shifts to the adult scarring response to wounding (Martin, 1997; Ferguson and O'Kane, 2004). Skin regeneration in the mouse fetus is correlated with a minimal inflammatory response, reflected in low numbers of platelets and macrophages; a lower ratio of TGF-β1, 2/TGF-β3, and type I/III collagens; lower levels of platelet-derived growth factor (PDGF) and its receptor; and higher levels of hyaluronic acid (HA) and its receptor (Stocum, 2006 for review). Antibodies to TGF-β1, 2 or addition of exogenous TGF-β3 administered early in the course of adult skin repair evoke a more regenerative response (Shah et al., 1995), while hyaluronidase and PDGF administered to fetal skin shift the wound response toward scarring (Haynes et al., 1995; Mast et al., 1995). Skin wounds in antibiotic-maintained PU.1 null mice, which lack macrophages and neutrophils, are repaired by regeneration (Martin et al., 2003).

No studies have investigated the role of changing ratios of growth factors, cytokines, and ECM components in amputated regeneration-competent versus deficient amphibian limbs. For example, do the ratios of TGF-β1 and 2/TGF-β3 and type I/III collagens, and level of PDGF show any correlation with regeneration-competence and deficiency? Likewise, it would be interesting to test whether antibodies to TGF-β3 would retard or inhibit blastema formation in regeneration-competent limbs and whether augmenting TGF-β3 while simultaneously inhibiting TGF-β1, 2 would enhance blastema formation in regeneration-deficient limbs.

A second possibility is that loss of regenerative capacity is not due to the maturity of the immune system but rather to how the changing developmental state of cells alters their response to immune cells. Evidence for this possibility is that the ontogenetic decline in regenerative ability of *Xenopus* limb buds has been shown by transplantation experiments to be the result of intrinsic changes in limb bud cells (Sessions and Bryant, 1988). Furthermore, fetal mouse skin fibroblasts maintain their regenerative response when grafted subcutaneously into adult athymic mice, even though these host mice heal by scarring (Lorenz et al., 1992; Lin et al., 1994) and the skin of early mouse limb buds cultured *in vitro* undergoes the transition from regeneration to scarring in response to wounding in the complete absence of circulating immune cells (Chopra et al., 1997).

The above ideas are based on the assumption that limb regeneration is a reactivation of limb development that is possible over the lifespan of a urodele, but that is progressively suppressed during anuran (and mammalian) development. Thus, if anurans can regenerate their limb buds at early tadpole stages, all the pathways necessary for regeneration must be there, but are inactivated as the limb bud differentiates. This notion assumes, however, that blastema formation in early tadpoles is not simply an extension of normal limb development, but a reverse regenerative process that takes place the same way as it does in the amputated limbs of urodele larvae and adults, something that has not been rigorously proven. It is also possible that urodeles have evolved (or retained) limb regeneration-specific genes not found in other vertebrates that allow their limb cells to undergo dedifferentiation and accumulate as a blastema. If this idea is correct, suites of regeneration-specific genes might have to be engineered into the genome of regeneration-deficient animals to achieve blastema formation.

Clearly, we have a great deal of interesting research ahead in order to understand the secrets of blastema formation and how to apply them to human benefit.

Acknowledgments

Research from this laboratory was supported by the W.M. Keck Foundation and the U.S. Army Research Office (Grant number W911NF07-10176). We thank our colleague Fengyu Song for insightful critiques and advice during the preparation of the manuscript.

References

Adams, D. S., Masi, A., et al. (2007). H+ pump-dependent changes in membrane voltage are an early mechanism necessary and sufficient to induce Xenopus tail regeneration. *Development, 134*(7), 1323–1335.

Albert, P., Boilly, B., et al. (1987). Stimulation in cell culture of mesenchymal cells of newt limb blastemas by EDGF I or II (basic or acidic FGF). *Cell Differ., 21*(1), 63–68.

Anand, P., McGregor, G. P., et al. (1987). Increase of substance P-like immunoreactivity in the peripheral nerve of the axolotl after injury. *Neurosci. Lett., 82*(3), 241–245.

Anastasia, L., Sampaolesi, M., et al. (2006). Reversine-treated fibroblasts acquire myogenic competence in vitro and in regenerating skeletal muscle. *Cell Death Differ., 13*(12), 2042–2051.

Anton, H. J. (1965). The origin of blastema cells and protein synthesis during forelimb regeneration in Triturus. In V. Kiortsis & H. A. L. Trampusch (Eds.), *Regeneration in Animals* (pp. 377–395). Amsterdam: North-Holland Pub. Co.

Atkinson, D. L., Stevenson, T. J., et al. (2006). Cellular electroporation induces dedifferentiation in intact newt limbs. *Dev. Biol., 299*(1), 257–271.

Beck, C. W., Izpisua Belmonte, J. C., et al. (2009). Beyond early development: Xenopus as an emerging model for the study of regenerative mechanisms. *Dev. Dyn., 238*(6), 1226–1248.

Bodemer, C. W. (1962). Distribution of ribonucleic acid in the urodele limb as determined by autoradiographic localization of uridine-H3. *Anat. Rec., 142*, 147–148.

Bodemer, C. W., & Everett, N. B. (1959). Localization of newly synthesized proteins in regenerating newt limbs as determined by radioautographic localization of injected methinine-S35. *Dev. Biol., 1*, 327–342.

Boilly, B., Cavanaugh, K. P., et al. (1991). Acidic fibroblast growth factor is present in regenerating limb blastemas of axolotls and binds specifically to blastema tissues. *Dev. Biol., 145*(2), 302–310.

Borgens, R. B., Vanable, J. W., Jr., et al. (1977). Bioelectricity and regeneration: large currents leave the stumps of regenerating newt limbs. *Proc. Natl. Acad. Sci. U.S.A., 74*(10), 4528–4532.

Brockes, J. P., & Kumar, A. (2005). Appendage regeneration in adult vertebrates and implications for regenerative medicine. *Science, 310*(5756), 1919–1923.

Brockes, J. P., & Kumar, A. (2008). Comparative aspects of animal regeneration. *Annu. Rev. Cell Dev. Biol., 24*, 525–549.

Bryant, S. V., Endo, T., et al. (2002). Vertebrate limb regeneration and the origin of limb stem cells. *Int. J. Dev. Biol., 46*(7), 887–896.

Burgess, A. M. (1967). The developmental potentialities of regeneration blastema cell nuclei as determined by nuclear transplantation. *J. Embryol. Exp. Morphol., 18*(1), 27–41.

Cadinouche, M. Z., Liversage, R. A., et al. (1999). Molecular cloning of the Notophthalmus viridescens radical fringe cDNA and characterization of its expression during forelimb development and adult forelimb regeneration. *Dev. Dyn., 214*(3), 259–268.

Campbell, L. J., & Crews, C. M. (2008). Wound epidermis formation and function in urodele amphibian limb regeneration. *Cell Mol. Life Sci., 65*(1), 73–79.

Cannata, S. M., Bagni, C., et al. (2001). Nerve-independence of limb regeneration in larval Xenopus laevis is correlated to the level of fgf-2 mRNA expression in limb tissues. *Dev. Biol., 231*(2), 436–446.

Carinato, M. E., Walter, B. E., et al. (2000). Xenopus laevis gelatinase B (Xmmp-9): development, regeneration, and wound healing. *Dev. Dyn., 217*(4), 377–387.

Carlson, B. M. (1969). Inhibition of limb regeneration in the axolotl after treatment of the skin with actinomycin D. *Anat. Rec., 163*(3), 389–401.

Carlson, B. M. (1978). Types of morphogenetic phenomena in vertebrate regenerating systems. *Amer. Zool., 18*, 869–882.

Carlson, B. M. (2007). *Principles of Regenerative Biology*. San Diego: Academic Press.

Castilla, M., & Tassava, R. A. (1992). Extraction of the WE3 antigen and comparison of reactivities of mAbs WE3 and WE4 in adult newt regenerate epithelium and body tissues. *Monogr. Dev. Biol., 23*, 116–130.

Chalkley, D. T. (1954). A quantitative histological analysis of forelimb regeneration in Triturus viridescens. *J. Morphol., 94*, 21–70.

Chen, S., Zhang, Q., et al. (2004). Dedifferentiation of lineage-committed cells by a small molecule. *J. Am. Chem. Soc., 126*(2), 410–411.

Chew, K. E., & Cameron, J. A. (1983). Increase in mitotic activity of regenerating axolotl limbs by growth factor-impregnated implants. *J. Exp. Zool., 226*(2), 325–329.

Chopra, V., Blewett, C. V., et al. (1997). Transition from fetal to adult repair occurring in mouse forelimbs maintained in organ culture. *Wound Repair Regen., 5*(1), 47–51.

Christensen, R. N., & Tassava, R. A. (2000). Apical epithelial cap morphology and fibronectin gene expression in regenerating axolotl limbs. *Dev. Dyn., 217*(2), 216–224.

Christensen, R. N., Weinstein, M., et al. (2001). Fibroblast growth factors in regenerating limbs of Ambystoma: cloning and semi-quantitative RT-PCR expression studies. *J. Exp. Zool., 290*, 529–540.

Cohen, N. (1971). Amphibian transplantation reactions: a review. *Amer. Zool., 11*(2), 193–205.

Crawford, K., & Stocum, D. L. (1988). Retinoic acid coordinately proximalizes regenerate pattern and blastema differential affinity in axolotl limbs. *Development, 102*(4), 687–698.

Crews, L., Gates, P. B., et al. (1995). Expression and activity of the newt Msx-1 gene in relation to limb regeneration. *Proc. Biol. Sci., 259*(1355), 161–171.

da Silva, S. M., Gates, P. B., et al. (2002). The newt ortholog of CD59 is implicated in proximodistal identity during amphibian limb regeneration. *Dev. Cell, 3*(4), 547–555.

Dent, J. N. (1962). Limb regeneration in larvae and metamorphosing individuals of the South African clawed toad. *J. Morphol., 110*, 61–77.

Dresden, M. H., & Gross, J. (1970). The collagenolytic enzyme of the regenerating limb of the newt Triturus viridescens. *Dev. Biol., 22*(1), 129–137.

Duckmanton, A., Kumar, A., et al. (2005). A single-cell analysis of myogenic dedifferentiation induced by small molecules. *Chem. Biol., 12*(10), 1117–1126.

Echeverri, K., & Tanaka, E. M. (2005). Proximodistal patterning during limb regeneration. *Dev. Biol., 279*(2), 391–401.

Egar, M. W. (1993). Affinophoresis as a test of axolotl accessory limbs. In J. F. Fallon, P. F. Goetinck, R. O. Kelley & D. L. Stocum (Eds.), *Limb Development and Regeneration, Part B* (pp. 203–211). New York: Wiley-Liss.

Eldridge, A. G., Loktev, A. V., et al. (2006). The evi5 oncogene regulates cyclin accumulation by stabilizing the anaphase-promoting complex inhibitor emi1. *Cell, 124*(2), 367–380.

Endo, T., Bryant, S. V., et al. (2004). A stepwise model system for limb regeneration. *Dev. Biol., 270*(1), 135–145.

Endo, T., Tamura, K., et al. (2000). Analysis of gene expressions during Xenopus forelimb regeneration. *Dev. Biol., 220*(2), 296–306.

Estrada, C. M., Park, C. D., et al. (1993). Monoclonal antibody WE6 identifies an antigen that is up-regulated in the wound epithelium of newts and frogs. In J. F. Fallon, P. F. Goetinck, R. O. Kelley & D. L. Stocum (Eds.), *Limb Development and Regeneration, Part A* (pp. 271–282). New York: Wiley-Liss.

Fahmy, G. H., & Sicard, R. E. (1998). Acceleration of amphibian forelimb regeneration by polypeptide growth factors. *J. Minn. Acad. Sci., 63*, 58–60.

Fekete, D. M., & Brockes, J. P. (1987). A monoclonal antibody detects a difference in the cellular composition of developing and regenerating limbs of newts. *Development, 99*(4), 589–602.

Ferguson, M. W., & O'Kane, S. (2004). Scar-free healing: from embryonic mechanisms to adult therapeutic intervention. *Philos. Trans. R. Soc. Lond. B. Biol. Sci., 359*(1445), 839–850.

Ferretti, P., Brockes, J. P., et al. (1991). A newt type II keratin restricted to normal and regenerating limbs and tails is responsive to retinoic acid. *Development, 111*(2), 497–507.

Filoni, S., Bernardini, S., et al. (1999). Nerve-independence of limb regeneration in larval Xenopus laevis is related to the presence of mitogenic factors in early limb tissues. *J. Exp. Zool., 284*(2), 188–196.

Gardiner, D. M., Carlson, M. R., et al. (1999). Towards a functional analysis of limb regeneration. *Semin. Cell Dev. Biol., 10*(4), 385–393.

Giampaoli, S., Bucci, S., et al. (2003). Expression of FGF2 in the limb blastema of two Salamandridae correlates with their regenerative capability. *Proc. Biol. Sci., 270*(1530), 2197–2205.

Globus, M., & Alles, P. (1990). A search for immunoreactive substance P and other neural peptides in the limb regenerate of the newt Notophthalmus viridescens. *J. Exp. Zool., 254*(2), 165–176.

Globus, M., Vethamany-Globus, S., et al. (1980). Effect of apical epidermal cap on mitotic cycle and cartilage differentiation in regeneration blastemata in the newt, Notophthalmus viridescens. *Dev. Biol., 75*(2), 358–372.

Go, M. J., Eastman, D. S., et al. (1998). Cell proliferation control by Notch signaling in Drosophila development. *Development, 125*(11), 2031–2040.

Godwin, J. W., & Brockes, J. P. (2006). Regeneration, tissue injury and the immune response. *J. Anat., 209*(4), 423–432.

Goldhamer, D. J., & Tassava, R. A. (1987). An analysis of proliferative activity in innervated and denervated forelimb regenerates of the newt, Notophthalmus viridescens. *Development, 100*, 619–628.

Gorsic, M., Majdic, G., et al. (2008). Identification of differentially expressed genes in 4-day axolotl limb blastema by suppression subtractive hybridization. *J. Physiol. Biochem., 64*(1), 37–50.

Goss, R. J. (1956). Regenerative inhibition following limb amputation and immediate insertion into the body cavity. *Anat. Rec., 126*(1), 15–27.

Goss, R. J., & Holt, R. (1992). Epimorphic vs. tissue regeneration in Xenopus forelimbs. *J. Exp. Zool., 261*(4), 451–457.

Grillo, H. C., Lapiere, C. M., et al. (1968). Collagenolytic activity in regenerating forelimbs of the adult newt (Triturus viridescens). *Dev. Biol., 17*(5), 571–583.

Gulati, A. K., Zalewski, A. A., et al. (1983). An immunofluorescent study of the distribution of fibronectin and laminin during limb regeneration in the adult newt. *Dev. Biol., 96*(2), 355–365.

Han, M. J., An, J. Y., et al. (2001). Expression patterns of Fgf-8 during development and limb regeneration of the axolotl. *Dev. Dyn., 220*(1), 40–48.

Han, M., Yang, X., et al. (2003). Digit regeneration is regulated by Msx1 and BMP4 in fetal mice. *Development, 130*(21), 5123–5132.

Harty, M., Neff, A. W., et al. (2003). Regeneration or scarring: an immunologic perspective. *Dev. Dyn., 226*(2), 268–279.

Hay, E. D., & Fischman, D. A. (1961). Origin of the blastema in regenerating limbs of the newt Triturus viridescens. An autoradiographic study using tritiated thymidine to follow cell proliferation and migration. *Dev. Biol., 3*, 26–59.

Haynes, J. H., Mast, B. A., et al. (1995). Exposure to amniotic fluid inhibits closure of open wounds in the fetal rabbit. *Wound Repair Regen., 3*(4), 467–472.

Hutchison, C., Pilote, M., et al. (2007). The axolotl limb: a model for bone development, regeneration and fracture healing. *Bone, 40*(1), 45–56.

Izutsu, Y., & Yoshizato, K. (1993). Metamorphosis-dependent recognition of larval skin as non-self by inbred adult frogs (Xenopus laevis). *J. Exp. Zool., 266*(2), 163–167.

Jenkins, L. S., Duerstock, B. S., et al. (1996). Reduction of the current of injury leaving the amputation inhibits limb regeneration in the red spotted newt. *Dev. Biol., 178*(2), 251–262.

Ju, B. G., & Kim, W. S. (1998). Upregulation of cathepsin D expression in the dedifferentiating salamander limb regenerates and enhancement of its expression by retinoic acid. *Wound Repair Regen., 6*(4), 349–357.

Juliano, R. L., & Haskill, S. (1993). Signal transduction from the extracellular matrix. *J. Cell Biol., 120*(3), 577–585.

Kato, T., Miyazaki, K., et al. (2003). Unique expression patterns of matrix metalloproteinases in regenerating newt limbs. *Dev. Dyn., 226*(2), 366–376.

Kelly, D. J., & Tassava, R. A. (1973). Cell division and ribonucleic acid synthesis during the initiation of limb regeneration in larval axolotls (Ambystoma mexicanum). *J. Exp. Zool., 185*(1), 45–54.

Korneluk, R. G., & Liversage, R. A. (1984). Effects of radius – ulna removal on forelimb regeneration in Xenopus laevis froglets. *J. Embryol. Exp. Morphol., 82*, 9–24.

Koshiba, K., Kuroiwa, A., et al. (1998). Expression of Msx genes in regenerating and developing limbs of axolotl. *J. Exp. Zool., 282*(6), 703–714.

Kragl, M., Knapp, D., et al. (2009). Cells keep a memory of their tissue origin during axolotl limb regeneration. *Nature, 460*(7251), 60–65.

Kumar, A., Godwin, J. W., et al. (2007). Molecular basis for the nerve dependence of limb regeneration in an adult vertebrate. *Science, 318*(5851), 772–777.

Kumar, A., Velloso, C. P., et al. (2004). The regenerative plasticity of isolated urodele myofibers and its dependence on MSX1. *PLoS Biol., 2*(8), E218.

Lash, J. W. (1955). Studies on wound closure in urodeles. *J. Exp. Zool., 128*(1), 13–28.

Lentz, T. L. (1967). Fine structure of nerves in the regenerating limb of the newt Triturus. *Am. J. Anat., 121*(3), 647–669.

Lheureux, E. (1975). Regeneration des membres irradies de *Pleurodeles waltlii* Michah. (Urodele). Influence des qualités et orientations des greffons non irradies. *Wilhelm Roux's Archs Devl. Biol., 176*, 303–327.

Lheureux, E., & Carey, F. (1988). The irradiated epidermis inhibits newt limb regeneration by preventing blastema growth. A histological study. *Biol. Struct. Morphog.*, *1*(2), 49–57.

Lin, R. Y., Sullivan, K. M., et al. (1994). Scarless human fetal skin repair is intrinsic to the fetal fibroblast and occurs in the absence of an inflammatory response. *Wound Repair Regen.*, *2*(4), 297–305.

Litwiller, R. (1939). Mitotic index and size in regenerating amphibian limbs. *J. Exp. Zool.*, *82*, 273–286.

Lodish, H., Berk, A., et al. (2008). *Molecular Cell Biology*. New York: Freeman, W.E. and Co.

Lorenz, H. P., Longaker, M. T., et al. (1992). Scarless wound repair: a human fetal skin model. *Development*, *114*(1), 253–259.

Lowenstein, C. J., & Snyder, S. H. (1992). Nitric oxide, a novel biologic messenger. *Cell*, *70*(5), 705–707.

Loyd, R. M., & Tassava, R. A. (1980). DNA synthesis and mitosis in adult newt limbs following amputation and insertion into the body cavity. *J. Exp. Zool.*, *214*(1), 61–69.

Lundkvist, J., & Lendahl, U. (2001). Notch and the birth of glial cells. *Trends Neurosci.*, *24*(9), 492–494.

Maden, M. (1978). Neurotrophic control of the cell cycle during amphibian limb regeneration. *J. Embryol. Exp. Morphol.*, *48*, 169–175.

Mahan, J. T., & Donaldson, D. J. (1986). Events in the movement of newt epidermal cells across implanted substrates. *J. Exp. Zool.*, *237*(1), 35–44.

Maki, N., Suetsugu-Maki, R., et al. (2009). Expression of stem cell pluripotency factors during regeneration in newts. *Dev. Dyn.*, *238*(6), 1613–1616.

Marcucci, E. (1916). Capacita rigenerativa degliarti nelle larve di Anuri e condizioni che ne determinano la perdita. *Archivio. Zool.*, *9*, 89–117.

Martin, P. (1997). Wound healing — aiming for perfect skin regeneration. *Science*, *276*(5309), 75–81.

Martin, P., d'Souza, D., et al. (2003). Wound healing in the PU.1 null mouse — tissue repair is not dependent on inflammatory cells. *Curr. Biol.*, *13*(13), 1122–1128.

Mast, B. A., Frantz, F. W., et al. (1995). Hyaluronic acid degradation products induce neovascularization and fibroplasia in fetal rabbit wounds. *Wound Repair Regen.*, *3*(1), 66–72.

McCullough, W. D., & Tassava, R. A. (1976). Determination of the blastema cell cycle in regenerating limbs of the larval axolotl, Ambystoma mexicanum. *Ohio J. Sci.*, *76*, 63–65.

McGann, C. J., Odelberg, S. J., et al. (2001). Mammalian myotube dedifferentiation induced by newt regeneration extract. *Proc. Natl. Acad. Sci. U.S.A.*, *98*(24), 13699–13704.

McLaughlin, D. S., & Liversage, R. A. (1986). Effects of denervation and delayed amputation on forelimb regeneration in Xenopus laevis froglets. *Anat. Rec.*, *214*(3), 289–293.

Meinhardt, H. (1982). *Models of Biological Pattern Formation*. London: Academic Press.

Mercader, N., Tanaka, E. M., et al. (2005). Proximodistal identity during vertebrate limb regeneration is regulated by Meis homeodomain proteins. *Development*, *132*(18), 4131–4142.

Mescher, A. L. (1976). Effects on adult newt limb regeneration of partial and complete skin flaps over the amputation surface. *J. Exp. Zool.*, *195*(1), 117–128.

Mescher, A. L., & Neff, A. W. (2006). Limb regeneration in amphibians: immunological considerations. *ScientificWorld Journal*, *6*(Suppl. 1), 1–11.

Mescher, A. L., & Kiffmeyer, W. R. (1992). Axonal release of transferrin in peripheral nerves of axolotls during regeneration. In C. H. Taban & B. Boilly (Eds.), *Keys for Regeneration: Monographs in Developmental Biology, 23* (pp. 100–109). Basel: Karger.

Mescher, A. L., & Munaim, S. I. (1986). Changes in the extracellular matrix and glycosaminoglycan synthesis during the initiation of regeneration in adult newt forelimbs. *Anat. Rec.*, *214*(4), 424–431, 394–395.

Mescher, A. L., & Neff, A. W. (2005). Regenerative capacity and the developing immune system. *Adv. Biochem. Eng. Biotechnol.*, *93*, 39–66.

Mescher, A. L., & Tassava, R. A. (1975). Denervation effects on DNA replication and mitosis during the initiation of limb regeneration in adult newts. *Dev. Biol.*, *44*(1), 187–197.

Mescher, A. L., Connell, E., et al. (1997). Transferrin is necessary and sufficient for the neural effect on growth in amphibian limb regeneration blastemas. *Dev. Growth Differ.*, *39*(6), 677–684.

Mescher, A. L., White, G. W., et al. (2000). Apoptosis in regenerating and denervated, nonregenerating urodele forelimbs. *Wound Repair Regen.*, *8*(2), 110–116.

Miyazaki, K., Uchiyama, K., et al. (1996). Cloning and characterization of cDNAs for matrix metalloproteinases of regenerating newt limbs. *Proc. Natl. Acad. Sci. U.S.A.*, *93*(13), 6819–6824.

Morrison, J. I., Loof, S., et al. (2006). Salamander limb regeneration involves the activation of a multipotent skeletal muscle satellite cell population. *J. Cell Biol.*, *172*(3), 433–440.

Mullen, L. M., Bryant, S. V., et al. (1996). Nerve dependency of regeneration: the role of distal-less and FGF signaling in amphibian limb regeneration. *Development*, *122*(11), 3487–3497.

83

Muneoka, K., Fox, W. F., et al. (1986). Cellular contribution from dermis and cartilage to the regenerating limb blastema in axolotls. *Dev. Biol., 116*(1), 256–260.

Namenwirth, M. (1974). The inheritance of cell differentiation during limb regeneration in the axolotl. *Dev. Biol., 41*(1), 42–56.

Nardi, J. B., & Stocum, D. L. (1983). Surface properties of regenerating limb cells: evidence for gradation along the proximodistal axis. *Differentiation, 25,* 27–31.

Odelberg, S. J., Kollhoff, A., et al. (2000). Dedifferentiation of mammalian myotubes induced by msx1. *Cell, 103*(7), 1099–1109.

Onda, H., & Tassava, R. A. (1991). Expression of the 9G1 antigen in the apical cap of axolotl regenerates requires nerves and mesenchyme. *J. Exp. Zool., 257*(3), 336–349.

Onda, H., Poulin, M. L., et al. (1991). Characterization of a newt tenascin cDNA and localization of tenascin mRNA during newt limb regeneration by in situ hybridization. *Dev. Biol., 148*(1), 219–232.

Oudkhir, M., Martelly, I., et al. (1989). Protein kinase C activity during limb regeneration of amphibians. In V. Kiortsis, S. Koussoulakos & H. Wallace (Eds.), *Recent Trends in Regeneration Research* (pp. 69–79). New York: Plenum Press.

Park, I. S., & Kim, W. S. (1999). Modulation of gelatinase activity correlates with the dedifferentiation profile of regenerating salamander limbs. *Mol. Cells, 9*(2), 119–126.

Pearl, E. J., Barker, D., et al. (2008). Identification of genes associated with regenerative success of Xenopus laevis hindlimbs. *BMC Dev. Biol., 8,* 66.

Poulin, M. L., Patrie, K. M., et al. (1993). Heterogeneity in the expression of fibroblast growth factor receptors during limb regeneration in newts (Notophthalmus viridescens). *Development, 119*(2), 353–361.

Powell, J. A. (1969). Analysis of histogenesis and regenerative ability of denervated forelimb regenerates of Triturus viridescens. *J. Exp. Zool., 170*(2), 125–147.

Rao, N., Jhamb, D., et al. (2009). Proteomic analysis of blastema formation in regenerating axolotl limbs. *BMC Biol., 7,* 83.

Reginelli, A. D., Wang, Y. Q., et al. (1995). Digit tip regeneration correlates with regions of Msx1 (Hox 7) expression in fetal and newborn mice. *Development, 121*(4), 1065–1076.

Repesh, L. A., & Oberpriller, J. C. (1978). Scanning electron microscopy of epidermal cell migration in wound healing during limb regeneration in the adult newt, Notophthalmus viridescens. *Am. J. Anat., 151*(4), 539–555.

Repesh, L. A., & Oberpriller, J. C. (1980). Ultrastructural studies on migrating epidermal cells during the wound healing stage of regeneration in the adult newt, Notophthalmus viridescens. *Am. J. Anat., 159*(2), 187–208.

Repesh, L. A., & Furcht, L. P. (1982). Distribution of fibronectin in regenerating limbs of the adult newt *Notophthalmus viridescens. Differentiation, 22,* 125–131.

Rivera, R., Ortiz, J. R., et al. (1981). N-Acetyl-glucosaminidase activity during limb regeneration in the adult newt. *Differentiation, 19*(1–3), 121–123.

Rosania, G. R., Chang, Y. T., et al. (2000). Myoseverin, a microtubule-binding molecule with novel cellular effects. *Nat. Biotechnol., 18*(3), 304–308.

Salpeter, M. M. (1965). Disposition of nerve fibers in the regenerating limb of the adult newt, Triturus. *J. Morphol., 117*(2), 201–211.

Satoh, A., Endo, T., et al. (2006). Characterization of Xenopus digits and regenerated limbs of the froglet. *Dev. Dyn., 235*(12), 3316–3326.

Satoh, A., Ide, H., et al. (2005a). Muscle formation in regenerating Xenopus froglet limb. *Dev. Dyn., 233*(2), 337–346.

Satoh, A., Suzuki, M., et al. (2005b). Joint development in Xenopus laevis and induction of segmentations in regenerating froglet limb (spike). *Dev. Dyn., 233*(4), 1444–1453.

Schmidt, A. J. (1966). The molecular basis of regeneration: enzymes. In E. R. Kirch, J. P. Marberger, S. R. M. Reynolds & R. J. Winzler (Eds.), *Illinois Monographs in Medical Sciences.* Urbana: University of Illinois Press.

Schmidt, A. J. (1968). *Cellular Biology of Vertebrate Regeneration and Repair.* Chicago: The University of Chicago Press.

Schotte, O. E., & Butler, E. G. (1944). Phases in regeneration of the urodele limb and their dependence upon the nervous system. *J. Exp. Zool., 97*(2), 95–121.

Schotte, O. E., & Harland, M. (1943). Amputation level and regeneration in limbs of late *Rana clamitans* tadpoles. *J. Morphol., 73,* 329–362.

Sessions, S. K., & Bryant, S. V. (1988). Evidence that regenerative ability is an intrinsic property of limb cells in Xenopus. *J. Exp. Zool., 247*(1), 39–44.

Shah, M., Foreman, D. M., et al. (1995). Neutralisation of TGF-beta 1 and TGF-beta 2 or exogenous addition of TGF-beta 3 to cutaneous rat wounds reduces scarring. *J. Cell Sci., 108*(Pt 3), 985–1002.

Shimizu-Nishikawa, K., Tazawa, I., et al. (1999). Expression of helix-loop-helix type negative regulators of differentiation during limb regeneration in urodeles and anurans. *Dev. Growth Differ., 41*(6), 731–743.

Shimizu-Nishikawa, K., Tsuji, S., et al. (2001). Identification and characterization of newt rad (ras associated with diabetes), a gene specifically expressed in regenerating limb muscle. *Dev. Dyn., 220*(1), 74–86.

Singer, M., & Craven, L. (1948). The growth and morphogenesis of the regenerating forelimb of adult Triturus following denervation at various stages of development. *J. Exp. Zool., 108*(2), 279–308.

Singer, M., & Salpeter, M. M. (1961). Regeneration in verebrates: the role of the wound epithelium in vertebrate regeneration. In M. Zarrow (Ed.), *Growth in Living Systems.* New York: Basic Books.

Smith, A. R., & Crawley, A. M. (1977). The pattern of cell division during growth of the blastema of regenerating newt forelimbs. *J. Embryol. Exp. Morphol., 37*(1), 33–48.

Smith, M. J., & Globus, M. (1989). Multiple interactions in juxtaposed monolayers of amphibian neuronal, epidermal, and mesodermal limb blastema cells. In Vitro *Cell Dev. Biol., 25*(9), 849–856.

Song, K., Wang, Y., et al. (1992). Expression of Hox-7.1 in myoblasts inhibits terminal differentiation and induces cell transformation. *Nature, 360*(6403), 477–481.

Steen, T. P. (1968). Stability of chondrocyte differentiation and contribution of muscle to cartilage during limb regeneration in the axolotl (Siredon mexicanum). *J. Exp. Zool., 167*(1), 49–78.

Stevenson, T. J., Vinarsky, V., et al. (2006). Tissue inhibitor of metalloproteinase 1 regulates matrix metalloproteinase activity during newt limb regeneration. *Dev. Dyn., 235*(3), 606–616.

Stocum, D. L. (1980). The relation of mitotic index, cell density, and growth to pattern regulation in regenerating Ambystoma maculatum forelimbs. *J. Exp. Zool., 212*(2), 233–242.

Stocum, D. L. (1995). *Wound Repair, Regeneration and Artificial Tissues.* Austin, TX: RG Landes Co.

Stocum, D. L. (2006). *Regenerative Biology and Medicine.* San Diego: Elsevier Inc.

Stocum, D. L., & Dearlove, G. E. (1972). Epidermal-mesodermal interaction during morphogenesis of the limb regeneration blastema in larval salamanders. *J. Exp. Zool., 181*, 49–61.

Stocum, D. L., & Melton, D. A. (1977). Self-organizational capacity of distally transplanted limb regeneration blastemas in larval salamanders. *J. Exp. Zool., 201*(3), 451–461.

Straube, W. L., & Tanaka, E. M. (2006). Reversibility of the differentiated state: regeneration in amphibians. *Artif. Organs, 30*(10), 743–755.

Suzuki, M., Satoh, A., et al. (2005). Nerve-dependent and -independent events in blastema formation during Xenopus froglet limb regeneration. *Dev. Biol., 286*(1), 361–375.

Suzuki, M., Yakushiji, N., et al. (2006). Limb regeneration in Xenopus laevis froglet. *ScientificWorld Journal, 6* (Suppl. 1), 26–37.

Takahashi, K., Tanabe, K., et al. (2007). Induction of pluripotent stem cells from adult human fibroblasts by defined factors. *Cell, 131*(5), 861–872.

Tamura, K., Ohgo, S., et al. (2009). Limb blastema cell: a stem cell for morphological regeneration. *Dev. Growth Differ., 52*(1), 89–99.

Tanaka, E. M. (2003). Regeneration: if they can do it, why can't we? *Cell, 113*(5), 559–562.

Tanaka, E. M., Gann, A. A., et al. (1997). Newt myotubes reenter the cell cycle by phosphorylation of the retinoblastoma protein. *J. Cell Biol., 136*(1), 155–165.

Tassava, R. A., & Garling, D. J. (1979). Regenerative responses in larval axolotl limbs with skin grafts over the amputation surface. *J. Exp. Zool., 208*(1), 97–110.

Tassava, R. A., & McCullough, W. D. (1978). Neural control of cell cycle events in regenerating salamander limbs. *Amer. Zool., 18*(4), 843–854.

Tassava, R. A., & Mescher, A. L. (1975). The roles of injury, nerves and the wound epithelium during the initiation of amphibian limb regeneration. *Differentiation, 4*, 23–24.

Tassava, R. A., & Mescher, A. L. (1976). Mitotic activity and nucleic acid precursor incorporation in denervated and innervated limb stumps of axolotl larvae. *J. Exp. Zool., 195*(2), 253–262.

Tassava, R. A., Bennett, L. L., et al. (1974). DNA synthesis without mitosis in amputated denervated forelimbs of larval axolotls. *J. Exp. Zool., 190*(1), 111–116.

Tassava, R. A., Castilla, M., et al. (1993). The wound epithelium of regenerating limbs of Pleurodeles waltl and Notophthalmus viridescens: studies with mAbs WE3 and WE4, phalloidin, and DNase 1. *J. Exp. Zool., 267*(2), 180–187.

Tassava, R. A., Goldhamer, D. J., et al. (1987). Cell cycle controls and the role of nerves and the regenerate epithelium in urodele forelimb regeneration: possible modifications of basic concepts. *Biochem. Cell Biol., 65* (8), 739–749.

Tassava, R. A., Tomlinson, B., et al. (1989). Expression of the WE3 antigen in the newt wound epithelium. In V. Kiortsis, S. Koussoulakos & H. Wallace (Eds.), *Recent Trends in Regeneration Research* (pp. 37–49). New York: Plenum.

Tassava, R. A., Treece, D. P., et al. (1983). Effects of partial denervation on the newt blastema cell cycle. *Prog. Clin. Biol. Res., 110*(Pt A), 537–545.

Thornton, C. S. (1958). The inhibition of limb regeneration in urodele larvae by localized irradiation with ultraviolet light. *J. Exp. Zool., 137*(1), 153–179.

Thornton, C. S. (1960). Influence of an eccentric epidermal cap on limb regeneration in Amblystoma larvae. *Dev. Biol., 2,* 551–569.

Thornton, C. S. (1968). Amphibian limb regeneration. In L. Brachet & T. J. King (Eds.), *Advances in Morphogenesis 7* (pp. 205–244). New York: Academic Press.

Thornton, C. S., & Steen, T. P. (1962). Eccentric blastema formation in aneurogenic limbs of Ambystoma larvae following epidermal cap deviation. *Dev. Biol., 5,* 328–343.

Thornton, C. S., & Thornton, M. T. (1965). The regeneration of accessory limb parts following epidermal cap transplantation in urodeles. *Experientia, 21*(3), 146–148.

Thornton, C. S., & Thornton, M. T. (1970). Recuperation of regeneration in denervated limbs of Ambystoma larvae. *J. Exp. Zool., 173*(3), 293–301.

Tomlinson, B. L., & Barger, P. M. (1987). A test of the punctuated-cycling hypothesis in Ambystoma forelimb regenerates: the roles of animal size, limb innervation, and the aneurogenic condition. *Differentiation, 35*(1), 6–15.

Tomlinson, B., Goldhamer, D. J., et al. (1985). Punctuated cell cycling in the regeneration blastema of urodele amphibians: an hypothesis. *Differentiation, 28*(3), 195–199.

Toole, B. P., & Gross, J. (1971). The extracellular matrix of the regenerating newt limb: synthesis and removal of hyaluronate prior to differentiation. *Dev. Biol., 25*(1), 57–77.

Tsonis, P. A. (2000). Regeneration in vertebrates. *Dev. Biol., 221*(2), 273–284.

Tsonis, P. A., English, D., et al. (1991). Increased content of inositol phosphates in amputated limbs of axolotl larvae, and the effect of beryllium. *J. Exp. Zool., 259,* 252–258.

Velloso, C. P., Simon, A., et al. (2001). Mammalian postmitotic nuclei reenter the cell cycle after serum stimulation in newt/mouse hybrid myotubes. *Curr. Biol., 11*(11), 855–858.

Vethamany-Globus, S., Globus, M., et al. (1978). Neural and hormonal stimulation of DNA and protein synthesis in cultured regeneration blastemata in the newt Notophthalmus viridescens. *Dev. Biol., 65*(1), 183–192.

Vinarsky, V., Atkinson, D. L., et al. (2005). Normal newt limb regeneration requires matrix metalloproteinase function. *Dev. Biol., 279*(1), 86–98.

Wanek, N., Muneoka, K., et al. (1989). Evidence for regulation following amputation and tissue grafting in the developing mouse limb. *J. Exp. Zool., 249*(1), 55–61.

Wang, L., Marchionni, M. A., et al. (2000). Cloning and neuronal expression of a type III newt neuregulin and rescue of denervated, nerve-dependent newt limb blastemas by rhGGF2. *J. Neurobiol., 43*(2), 150–158.

Werner, S. R., Mescher, A. L., et al. (2007). Neural MMP-28 expression precedes myelination during development and peripheral nerve repair. *Dev. Dyn., 236*(10), 2852–2864.

Wolfe, A. D., Nye, H. L., et al. (2000). Extent of ossification at the amputation plane is correlated with the decline of blastema formation and regeneration in Xenopus laevis hindlimbs. *Dev. Dyn., 218*(4), 681–697.

Woloshin, P., Song, K., et al. (1995). MSX1 inhibits myoD expression in fibroblast x 10T1/2 cell hybrids. *Cell, 82*(4), 611–620.

Yakushiji, N., Suzuki, M., et al. (2007). Correlation between Shh expression and DNA methylation status of the limb-specific Shh enhancer region during limb regeneration in amphibians. *Dev. Biol., 312*(1), 171–182.

Yakushiji, N., Yokoyama, H., et al. (2009). Repatterning in amphibian limb regeneration: a model for study of genetic and epigenetic control of organ regeneration. *Semin. Cell Dev. Biol., 20*(5), 565–574.

Yang, E. V., & Bryant, S. V. (1994). Developmental regulation of a matrix metalloproteinase during regeneration of axolotl appendages. *Dev. Biol., 166*(2), 696–703.

Yntema, C. L. (1959a). Blastema formation in sparsely innervated and aneurogenic forelimbs of amblystoma larvae. *J. Exp. Zool., 142,* 423–439.

Yntema, C. L. (1959b). Regeneration in sparsely innervated and aneurogenic forelimbs of Amblystoma larvae. *J. Exp. Zool., 140,* 101–123.

Yu, J., Vodyanik, M. A., et al. (2007). Induced pluripotent stem cell lines derived from human somatic cells. *Science, 318*(5858), 1917–1920.

CHAPTER 4

The Molecular Circuitry Underlying Pluripotency in Embryonic Stem Cells and iPS Cells

Harvir Singh, Ali H. Brivanlou
Laboratory of Molecular Embryology, The Rockefeller University, New York, NY, USA

INTRODUCTION

Multiple criteria are currently employed to characterize pluripotent potential including (1) expression of molecular markers and transcription factors known to regulate embryonic stem cell self renewal, (2) absence of molecular and morphological markers defining specific lineages, and (3) the ability to form all three embryonic germ layers including ectoderm, endoderm, and mesoderm upon induction of differentiation *in vitro* and *in vivo*. Upon injection into immunocompromised mice, embryonic stem cells will rapidly form teratomas containing cells from the three germ layers. Ultimately, implantation of ESCs into mouse blastocysts and subsequent contribution of these cells to all tissues of the adult chimeric animal represents one of the most stringent tests of pluripotency.

In this review, we describe the mechanistic details that regulate the maintenance of the pluripotent state at the level of signal transduction and transcription factor control. Particular emphasis is placed on the signaling circuitry regulating human ESC self renewal. Further, we discuss the advent of induced pluripotency, or the reprogramming of somatic cells into embryonic stem cells, and the processes that govern their formation and maintenance.

SIGNALING NETWORKS UNDERLYING PLURIPOTENCY

Initial derivation and maintenance of murine ESCs involved plating cells isolated from the inner cell mass on feeder cells consisting of embryonic fibroblasts and a medium containing serum proteins (Evans et al., 1981; Martin, 1981). The complex mixture of exogenous factors released by fibroblasts into the medium maintains ESCs in their pluripotent state and allows for the undifferentiated self renewal and proliferation of these cells. Upon removal of the feeder cells, or medium conditioned by the feeder cells, ESCs spontaneously differentiate into all three germ layers of the developing organism. A similar protocol allows for the establishment of human ESCs grown on feeder cells or in media conditioned by fibroblasts (Thomson et al., 1998). Despite the complex composition of fibroblast-conditioned medium, which is replete with a variety of unknown factors, several pathways essential for pluripotency have been elucidated. Intriguingly, the signaling molecules that maintain mouse ESCs differ from those necessary for maintenance of human ESCs, indicating a species-specific divergence of signaling circuitry regulating self renewal (Fig. 4.1).

Principles of Regenerative Medicine. DOI: 10.1016/B978-0-12-381422-7.10004-5

87

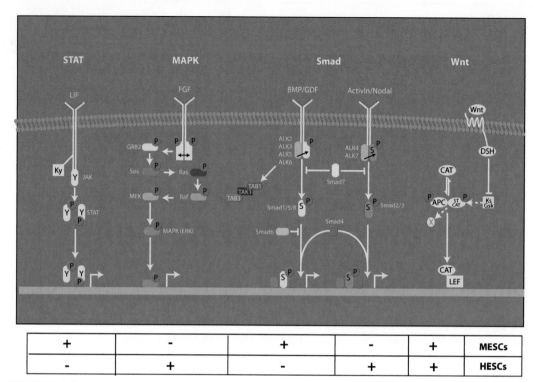

+	-	+	-	+	MESCs
-	+	-	+	+	HESCs

FIGURE 4.1

Signaling circuitry regulating mouse and human embryonic stem cell pluripotency. The Wnt pathway is a highly conserved regulator of pluripotency and is active in both mouse and human ESCs. A species-specific divergence exists for the LIF, BMP, TGF-β, and FGF pathways, respectively. Mouse ESCs require LIF and BMP signals to maintain self-renewal, whereas human ESCs depend on the activity of TGF-β and FGF signals. These pathways ultimately function at multiple levels to maintain the pluripotent state by inhibiting differentiation and feeding into the core transcriptional regulatory circuitry of embryonic stem cells.

LIF and BMP signaling pathways regulate mouse ESC self renewal

Mouse embryonic stem cells require leukemia inhibitory factor (LIF) as well as bone morphogenic proteins (BMP4) to maintain their undifferentiated state (Smith et al., 1988; Ying et al., 2003; Qi et al., 2004). LIF receptor activation leads to receptor dimerization with gp130 subunits and subsequent tyrosine phosphorylation and nuclear localization of the transcriptional activator STAT3 (Heinrich et al., 2003). BMPs are TGFβ superfamily members that bind to Type 1 TGFβ receptors Alk1, Alk2, Alk3, or Alk6. Upon ligand binding, type 1 receptors form heterodimers with type II receptors, which recruit and phosphorylate receptor activated Smads 1, 5, and 8 (R-Smads). Serine/threonine phosphorylation of R-Smads allows association and complex formation with co-Smad 4, which can subsequently enter the nucleus and initiate transcription (Shi and Massagué, 2003; Fig. 4.1).

TGFβ and FGF signaling pathways regulate human ESC self renewal

In stark contrast, human embryonic stem cells require TGFβ/Activin and FGF signaling to self renew and remain undifferentiated (James et al., 2005; Vallier et al., 2005). TGFβ and Activins account for the second branch of the TGFβ superfamily of ligands, and binding to receptors Alk4, Alk5, and Alk7 triggers serine/theonine phosphorylation of the C-terminal region of Smads 2 and 3, which also dimerize with Smad 4 to allow nuclear entry and transcription (Shi and Massagué, 2003). Fibroblast growth factors function through tyrosine receptor dimerization upon ligand binding and subsequent activation of phosphorylation events in the MAP kinase cascade (Chang et al., 2001). Intriguingly, FGF signaling can further phosphorylate both BMP and TGFβ mediated R-Smads at the "linker" domain of the proteins. This phosphorylation has been associated with signal termination as linker phosphorylation allows recognition of Smad proteins by the ubiquitin ligase Smurf1 (Pera et al., 2003; Sapkota et al., 2007).

Polyubiquitination of the Smad proteins by Smurf1 leads to subsequent degradation of the Smad proteins and termination of the signal. Hence, there may exist an intricate balance of antagonistic signaling inputs in the maintenance of human ESC pluripotency. Among their definitive roles in proliferation and survival, FGF signals may additionally act to inhibit differentiation promoting BMP signals in human ESCs by promoting degradation of any active Smad 1/5/8 proteins (Pera et al., 2003). Alternatively, FGF signals may also fine-tune the amount of active TGFβ-mediated Smad 2/3 proteins to produce the proper threshold of activity necessary for maintenance of pluripotency, as an excess of TGFβ/Activin signaling can lead to definitive endoderm formation of ESCs (D'Amour et al., 2005).

Studies demonstrating the necessity of these pathways for the maintenance of self renewal have followed two strategies. First, small molecule inhibition of TGFβ/Activin receptors results in the rapid differentiation of human ESCs even in fibroblast-conditioned medium, illustrating the necessity of TGFβ signals for the maintenance of pluripotency (James et al., 2005). Second, defined medium with select growth factors and cytokines has been developed to substitute fibroblast-conditioned medium, which contains a diverse milieu of undefined components. These studies have revealed that both TGFβ or Activin and FGF-2 at defined concentrations are necessary components for self renewal, and removal of either of these factors results in differentiation of ESCs (Vallier et al., 2005; Ludwig et al., 2006).

Wnt signaling is a conserved regulator of pluripotency across species

Although the aforementioned pathways are mutually exclusive in their ability to maintain mouse or human ESC self renewal respectively, the highly conserved Wnt pathway is necessary for maintenence of pluripotency in both species (Sato et al, 2004; Hao et al., 2006; Ogawa et al., 2006). In the presence of Wnt ligand, a receptor complex forms between receptors Frizzled and LRP5/6. This complex recruits and sequesters Axin and GSK3β, releasing their inhibitory interaction with β-catenin, which is subsequently allowed to accumulate in the nucleus, where it serves as a coactivator for T-Cell-Factor (Tcf) transcription factors to activate Wnt-responsive genes (MacDonald et al., 2009). Functional studies demonstrating the necessity of Wnt signaling have employed small molecule inhibitors of GSK3β, which destabilizes β-catenin. Inhibition of GSK3β results in increased Wnt activity, and cells cultured in the presence of GSK3β inhibitors have increased propensity to maintain their pluripotent state even in differentiation conditions (Sato et al, 2004; Ying et al, 2008). Furthermore, the role of Wnt ligands in supporting stemness has been demonstrated in experiments that show that Wnts secreted by feeder cells or Wnt-conditioned media maintain pluripotency in mouse ESCs (Hao et al., 2006; Ogawa et al., 2006). The necessity of Wnt signals in the maintenance of pluripotency across species highlights its evolutionary significance as a central signaling hub in ESC self-renewal.

SIGNALING PATHWAYS INHIBIT DIFFERENTIATION AND CONVERGE ON CORE TRANSCRIPTIONAL CIRCUITRY TO MAINTAIN PLURIPOTENCY

Conceptually, these signal transduction pathways can promote pluripotency either through direct inhibition of differentiation-promoting genes, direct enhancement of the self renewal transcriptional circuitry, or both. As noted above, LIF and BMP4 are sufficient to maintain pluripotency in mouse ESCs, and autocrine induction of FGF4 expression and subsequent activation of the MAPK cascade propel the cells out of pluripotency and into lineage specification (Kunath et al., 2007). Activation of BMP signaling has been shown to exert a negative effect on the MAPK cascade, thus inhibiting a differentiation-inducing signal (Qi et al., 2004).

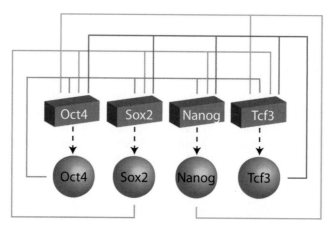

FIGURE 4.2

Core transcriptional circuitry of embryonic stem cells. Four genes, Oct4, Sox2, Nanog, and Tcf3, represent transcription factors crucial for the maintenance of pluripotency. These factors form a self-sustaining autoregulatory loop by binding to each other's promoter regions and activating their transcription.

Furthermore, small molecule inhibition of FGF receptor signaling in the presence of LIF obviates the need for BMP4 or serum (Ying et al., 2008). These results indirectly imply that BMP4 functions to inhibit FGF-induced differentiation in mouse embryonic stem cells. Intriguingly, in human ESCs, which require FGF and TGFβ/Activin signaling for maintenance, BMP4 results in the rapid and efficient differentiation of embryonic stem cells to trophectoderm (Xu et al., 2002). Inhibition of this pathway with the BMP4 antagonist Noggin, in the presence of FGF, preserves the pluripotent state of human embryonic stem cells (Xu et al., 2005). Thus, inhibition of differentiation inducing signals as a mechanism for maintaining pluripotency appears to be a consistent theme across species.

Evidence supporting the latter hypothesis has recently emerged demonstrating a direct interaction of these signaling pathways with the core transcriptional machinery of self renewal, triggering the activation and expression of transcription factors that maintain the pluripotent state including Oct4, Nanog, Sox2, and Tcf3 (Cole et al., 2008; Fig. 4.2). Three of these factors, Oct4, Nanog, and Sox2, coordinately regulate the pluripotency program and are thought to be central to transcriptional regulation of ESC identity because of their essential roles during early development and their ability to maintain the embryonic stem cell state (Nichols et al., 1998; Avilion et al., 2003; Chambers et al., 2003; Mitsui et al., 2003). Disruption of Oct4 in knockout embryos and stem cells results in the inappropriate differentiation of ICM and ES cells to trophectoderm, while Nanog mutants develop into extra-embryonic endoderm (Nichols et al., 1998, Chambers et al., 2003; Mitsui et al., 2003). Sox2 loss-of-function mutants also divert to trophectoderm (Avilion et al., 2003). Intriguingly, the phenotype of mouse ESCs overexpressing Oct4 resembles that of Nanog loss of function, forming embryonic endoderm, whereas cells with Nanog overexpression are highly resistant to differentiation (Niwa et al., 2000; Chambers et al., 2003). Genome-wide analysis has revealed that these three transcription factors form an autoregulatory network by binding to each other's promoter regions and enhancing their own expression (Fig. 4.2; Boyer et al., 2005). Furthermore, these factors regulate the expression of thousands of downstream genes governing aspects of differentiation, cell cycle, and self renewal (Boyer et al., 2005).

Signaling pathways necessary for self renewal have recently been shown to converge upon these transcriptional regulators to induce and maintain their transcription. TGFβ signaling, for example, directly targets and activates transcription of Nanog in human ESCs (Xu et al., 2008). Furthermore, in mouse ESCs, LIF-induced activation of the Jak-Stat3 pathway activates Kruppel transcription factor Klf4 expression, a zinc finger transcription factor that promotes expression of Sox2 and Nanog (Hall et al., 2009; Niwa et al., 2009). Whereas TGFβ and LIF signaling appear to interact with specific components of the core transcriptional circuitry of ESCs, Wnt signaling, remarkably, directly interacts with all of these components. The downstream mediator of Wnt signaling, the TCF transcription factors, binds the promoter regions of Oct4, Nanog, and Sox2, thus activating their expression upon Wnt ligand stimulation (Cole et al., 2008). Interestingly, not only does TCF bind these three pluripotency factors, it also co-occupies promoters across the genome in association with the transcription factors, indicating an intricate role of Wnt signaling integration with the core transcriptional circuitry of pluripotency.

INDUCED PLURIPOTENCY, STOCHASTICITY, AND SIGNALING THRESHOLDS

When Conrad Waddington described his epigenetic landscape for development, scarcely would he have imagined a process in which a complete reversal of fate from differentiated fibroblast to an embryonic state could occur (Waddington, 1957). Yet this complete reversion is exactly what was accomplished in 2006 by Yamanaka and colleagues (Takahashi et al., 2006). By introducing four transcription factors necessary for embryonic stem cell self renewal including Oct4, Sox2, Klf4, and c-Myc into the genome of fibroblasts, some cells underwent complete reprogramming to a state of pluripotency. These induced pluripotent stem cells (iPSCs) possess all the hallmarks of embryonic stem cells in their functional abilities to differentiate into all cell types of an organism (Takahashi et al., 2006, 2007). Importantly, these cells also require the same signaling pathways to maintain their undifferentiated state and respond appropriately to growth factors and cytokines eliciting specific lineages (Vallier et al., 2009).

Surprisingly, Nanog is not one of the primary inducing factors for iPS cells, despite its crucial role in pluripotency. However, as Oct4 and Sox2 form activating autoregulatory loops with each other and Nanog (Fig. 4.2; Boyer et al., 2005), it is conceivable that activation of endogenous Nanog is still necessary for reprogramming to a complete pluripotent state (Hanna et al., 2009). Furthermore, Klf4 and c-Myc can be substituted by Nanog and another transcription factor Lin28, indicating the multiple combinations of transcription factors that exist that can reprogram cells to the same developmental state (Yu et al., 2007).

The process of reprogramming itself is a complicated stochastic process in which epigenetic marks are wiped away and transcriptional circuitry rewired. The process is highly inefficient with an average 0.1–0.2% of cells at most reverting to a pluripotent state. Several small molecules that alter chromatin structure, including DNA methyltransferase inhibitors and histone deacetylase inhibitors, greatly enhance efficiency and can even reduce the number of transcription factors required, highlighting the importance of modulating epigenetic marks in the reprogramming process (Huangfu et al., 2008a,b). Interestingly, enhancement or inhibition of certain signaling pathways can also increase efficiency of reprogramming. As anticipated, increasing Wnt activity via GSK3-β inhibition enhances the reprogramming process (Marson et al., 2008; Silva et al., 2008). Inhibition of TGFβ signaling in mouse fibroblasts can also promote reprogramming by increasing the expression of Nanog and can even replace the transcription factor Sox2 (Ichida et al., 2009). The ability of TGFβ inhibitors to replace reprogramming factors in human fibroblasts has as yet to be demonstrated, although, paradoxically, small molecule inhibition of TGFβ signaling does markedly enhance reprogramming efficiency in human cells (Lin et al., 2009).

Regardless of the cocktail of inhibitors or factors used to reprogram somatic cells to pluripotency, there remain large fluctuations in efficiency and the probability that any given cell will become an iPS cell. Furthermore, as recently observed, not all colonies formed during the reprogramming process are bonafide pluripotent cells (Chan et al., 2009). Rather, some colonies appear to stall in an intermediate state unable to proliferate or to give rise to all cell types of a pluripotent cell. Even among a clonally selected somatic cell population infused with the same copy number of reprogramming factors, heterogeneity abounds, with a minority of cells reprogramming within a few weeks (Hanna et al., 2009). In this study, eventually all cells were able to become pluripotent stem cells over a period of several months; however, the process was highly stochastic and dependent on the rate of cell divisions (Hanna et al., 2009). What causes the aberrant heterogeneity in reprogramming efficiency despite equivalent levels of reprogramming factors?

Part of the answer may lie in the concept of non-genetic heterogeneity and random fluctuations in protein expression levels among a clonal population of cells. Stochastic noise in protein expression or activity, particularly in *in vitro* systems, arises from random fluctuation in the synthesis and breakdown of molecules and is ultimately a representation of thermodynamic principles of chemical reactions (Enver et al., 2009; Huang et al., 2009). These random fluctuations can have rather large functional effects, particularly in lineage specification. For example, embryonic stem cells are known to possess marked heterogeneity in Nanog expression levels (Kalmar et al., 2009). Cells with low Nanog levels might represent a permissive state that allows the initiation of differentiation, whereas cells with high levels might be resistant to the same. Hence, ESCs with high Nanog expression levels may exist in a stable attractor state, whereas those with low levels may define a metastable state and require less energy to proceed towards lineage specification. Indeed, experiments isolating high and low Nanog expressing cells from clonal ESC populations and subsequently exposing them to differentiation conditions reveal that low expressors readily differentiate, whereas high expressors resist lineage commitment (Kalmar et al., 2009). Similarly, in the process of reprogramming somatic cells, heterogeneity in any number of transcription factors, signaling proteins, or epigenetic marks may place a cell in a state either amenable or resistant to manipulation by the reprogramming factors. Dynamic measurements of multiple signaling and biochemical events at the single cell level may elucidate the thresholds at which certain cells reprogram and others fail to do so.

PERSPECTIVES

The molecular basis of pluripotency is a complex coordination of extracellular and environmental factors, and intracellular signal transduction and transcriptional regulation. Over the past few years we have seen significant leaps in our understanding of how signaling cascades converge upon core transcriptional circuitry to coordinate maintenance of pluripotency. Further, understanding of the intricate mechanisms through which signaling pathways create the multitude of tissue lineages remains paramount to understanding basic human development as well as to manipulating and controlling lineage specification for purposes of regenerative medicine. The rapid advent of induced pluripotency via reprogramming of differentiated cells to an embryonic state through cocktails of transcription factors has opened significant doors towards the concept of personalized regenerative medicine. Understanding the fundamental mechanisms of this process may ultimately provide us with unprecedented control to reprogram somatic cells into any desired cell type for the purposes of cell transplantation.

References

Avilion, A. A., Nicolis, S. K., Pevny, L. H., Perez, L., Vivian, N., & Lovell-Badge, R. (2003). Multipotent cell lineages in early mouse development depend on SOX2 function. *Genes Dev.*, *17*(1), 126–140.

Boyer, L. A., Lee, T. I., Cole, M. F., Johnstone, S. E., Levine, S. S., Zucker, J. P., et al. (2005). Core transcriptional regulatory circuitry in human embryonic stem cells. *Cell*, *122*(6), 947–956.

Brivanlou, A. H., & Darnell, J. E., Jr. (2002). Signal transduction and the control of gene expression. *Science*, *295* (5556), 813–818, Review.

Chambers, I., Colby, D., Robertson, M., Nichols, J., Lee, S., Tweedie, S., et al. (2003). Functional expression cloning of Nanog, a pluripotency sustaining factor in embryonic stem cells. *Cell*, *113*(5), 643–655.

Chan, E. M., Ratanasirintrawoot, S., Park, I. H., Manos, P. D., Loh, Y. H., Huo, H., et al. (2009). Live cell imaging distinguishes bona fide human iPS cells from partially reprogrammed cells. *Nat. Biotechnol., 27*(11), 1033–1037.

Chang, L., & Karin, M. (2001). Mammalian MAP kinase signaling cascades. *Nature, 410*(6824), 37–40.

Cole, M. F., Johnstone, S. E., Newman, J. J., Kagey, M. H., & Young, R. A. (2008). Tcf3 is an integral component of the core regulatory circuitry of embryonic stem cells. *Genes Dev., 22*(6), 746–755.

D'Amour, K. A., Agulnick, A. D., Eliazer, S., Kelly, O. G., Kroon, E., & Baetge, E. E. (2005). Efficient differentiation of human embryonic stem cells to definitive endoderm. *Nat. Biotechnol., 23*(12), 1534–1541, Epub 2005 Oct. 28.

Enver, T., Pera, M., Peterson, C., & Andrews, P. W. (2009). Stem cell states, fates, and the rules of attraction. *Cell Stem Cell, 4*(5), 387–397.

Evans, M. J., & Kaufman, M. H. (1981). Establishment in culture of pluripotential cells from mouse embryos. *Nature, 292*, 154–156.

Hall, J., Guo, G., Wray, J., Eyres, I., Nichols, J., Grotewold, L., et al. (2009). Oct4 and LIF/Stat3 additively induce Krüppel factors to sustain embryonic stem cell self-renewal. *Cell Stem Cell, 5*(6), 597–609.

Hanna, J., Saha, K., Pando, B., van Zon, J., Lengner, C. J., Creyghton, M. P., et al. (2009). Direct cell reprogramming is a stochastic process amenable to acceleration. Nature 2009, 462(7273), 595–601.

Hao, J., Li, T. G., Qi, X., Zhao, D. F., & Zhao, G. Q. (2006). WNT/beta-catenin pathway up-regulates Stat3 and converges on LIF to prevent differentiation of mouse embryonic stem cells. *Dev. Biol., 290*(1), 81–91, Epub 2005 Dec. 5.

Heinrich, P. C., Behrmann, I., Haan, S., Hermanns, H. M., Müller-Newen, G., et al. (2003). Principles of interleukin (IL)-6-type cytokine signalling and its regulation. *Biochem. J., 374*(Pt 1), 1–20, Review.

Huang, S. (2009). Non-genetic heterogeneity of cells in development: more than just noise. *Development, 136*(23), 3853–3862.

Huangfu, D., Maehr, R., Guo, W., Eijkelenboom, A., Snitow, M., Chen, A. E., et al. (2008a). Induction of pluripotent stem cells by defined factors is greatly improved by small-molecule compounds. *Nat. Biotechnol., 26*(7), 795–797.

Huangfu, D., Osafune, K., Maehr, R., Guo, W., Eijkelenboom, A., Chen, S., et al. (2008b). Induction of pluripotent stem cells from primary human fibroblasts with only Oct4 and Sox2. *Nat. Biotechnol., 26*(11), 1269–1275.

Ichida, J. K., Blanchard, J., Lam, K., Son, E. Y., Chung, J. E., Egli, D., et al. (2009). A small-molecule inhibitor of tgf-Beta signaling replaces sox2 in reprogramming by inducing nanog. *Cell Stem Cell, 5*(5), 491–503.

James, D., Levine, A. J., Besser, D., & Hemmati-Brivanlou, A. (2005). TGFbeta/activin/nodal signaling is necessary for the maintenance of pluripotency in human embryonic stem cells. *Development, 132*(6), 1273–1282.

Kalmar, T., Lim, C., Hayward, P., Muñoz-Descalzo, S., Nichols, J., Garcia-Ojalvo, J., et al. (2009). Regulated fluctuations in nanog expression mediate cell fate decisions in embryonic stem cells. *PLoS Biol., 7*(7), e1000149.

Kunath, T., Saba-El-Leil, M. K., Almousailleakh, M., Wray, J., Meloche, S., & Smith, A. (2007). FGF stimulation of the Erk1/2 signalling cascade triggers transition of pluripotent embryonic stem cells from self-renewal to lineage commitment. *Development, 134*(16), 2895–2902.

Lin, T., Ambasudhan, R., Yuan, X., Li, W., Hilcove, S., Abujarour, R., et al. (2009). A chemical platform for improved induction of human iPSCs. *Nat. Methods, 6*(11), 805–808, Epub 2009 Oct. 18.

Ludwig, T. E., Levenstein, M. E., Jones, J. M., Berggren, W. T., Mitchen, E. R., Frane, J. L., et al. (2006). Derivation of human embryonic stem cells in defined conditions. *Nat. Biotechnol., 24*(2), 185–187.

MacDonald, B. T., Tamai, K., & He, X. (2009). Wnt/beta-catenin signaling: components, mechanisms, and diseases. *Dev. Cell.*

Maherali, N., & Hochedlinger, K. (2009). Tgfbeta signal inhibition cooperates in the induction of iPSCs and replaces Sox2 and cMyc. *Curr. Biol., 19*(20), 1718–1723.

Marson, A., Foreman, R., Chevalier, B., Bilodeau, S., Kahn, M., Young, R. A., et al. (2008). Wnt signaling promotes reprogramming of somatic cells to pluripotency. *Cell Stem Cell, 3*(2), 132–135.

Martin, G. R. (1981). Isolation of a pluripotent cell line from early mouse embryos cultured in medium conditioned by teratocarcinoma stem cells. *Proc. Natl. Acad. Sci. U.S.A., 78*, 7634–7638.

Mitsui, K., Tokuzawa, Y., Itoh, H., Segawa, K., Murakami, M., Takahashi, K., et al. (2003). The homeoprotein Nanog is required for maintenance of pluripotency in mouse epiblast and ES cells. *Cell, 113*(5), 631–642.

Nichols, J., Zevnik, B., Anastassiadis, K., Niwa, H., Klewe-Nebenius, D., Chambers, I., et al. (2000). Quantitative expression of Oct-3/4 defines differentiation, dedifferentiation or self-renewal of ES cells. *Nat. Genet., 24*(4), 372–376.

Niwa, H., Ogawa, K., Shimosato, D., & Adachi, K. (2009). A parallel circuit of LIF signalling pathways maintains pluripotency of mouse ES cells. *Nature, 460*(7251), 118–122.

Ogawa, K., Nishinakamura, R., Iwamatsu, Y., Shimosato, D., & Niwa, H. (2006). Synergistic action of Wnt and LIF in maintaining pluripotency of mouse ES cells. *Biochem. Biophys. Res. Commun., 343*(1), 159–166, Epub 2006 Mar. 2.

93

Pera, E. M., Ikeda, A., Eivers, E., & de Robertis, E. M. (2003). Integration of IGF, FGF, and anti-BMP signals via Smad1 phosphorylation in neural induction. *Genes Dev., 17*(24), 3023–3308.

Qi, X., Li, T. G., Hao, J., Hu, J., Wang, J., Simmons, H., et al. (2004). BMP4 supports self-renewal of embryonic stem cells by inhibiting mitogen-activated protein kinase pathways. *Proc. Natl. Acad. Sci. U.S.A., 101*(16), 6027–6032.

Rosner, M. H., Vigano, M. A., Ozato, K., Timmons, P. M., Poirier, F., Rigby, P. W., et al. (1990). A POU-domain transcription factor in early stem cells and germ cells of the mammalian embryo. *Nature, 345*(6277), 686–692.

Sapkota, G., Alarcón, C., Spagnoli, F. M., Brivanlou, A. H., & Massagué, J. (2007). Balancing BMP signaling through integrated inputs into the Smad1 linker. *Mol. Cell, 25*(3), 441–454.

Sato, N., Meijer, L., Skaltsounis, L., Greengard, P., & Brivanlou, A. H. (2004). Maintenance of pluripotency in human and mouse embryonic stem cells through activation of Wnt signaling by a pharmacological GSK-3-specific inhibitor. *Nat. Med., 10*(1), 55–63.

Sato, N., Sanjuan, I. M., Heke, M., Uchida, M., Naef, F., & Brivanlou, A. H. (2003). Molecular signature of human embryonic stem cells and its comparison with the mouse. *Dev. Biol., 260*(2), 404–413.

Schöler, H., & Smith, A. (1998). Formation of pluripotent stem cells in the mammalian embryo depends on the POU transcription factor Oct4. *Cell, 95*(3), 379–391.

Shi, Y., & Massagué, J. (2003). Mechanisms of TGF-beta signaling from cell membrane to the nucleus. *Cell, 113*(6), 685–700, Review.

Silva, J., Barrandon, O., Nichols, J., Kawaguchi, J., Theunissen, T. W., & Smith, A. (2008). Promotion of reprogramming to ground state pluripotency by signal inhibition. *PLoS Biol., 6*(10), e253.

Takahashi, K., & Yamanaka, S. (2006). Induction of pluripotent stem cells from mouse embryonic and adult fibroblast cultures by defined factors. *Cell, 126*(4), 663–676.

Takahashi, K., Tanabe, K., Ohnuki, M., Narita, M., Ichisaka, T., Tomoda, K., et al. (2007). Induction of pluripotent stem cells from adult human fibroblasts by defined factors. *Cell, 131*(5), 861–872.

Thomson, J. A., et al. (1998). Embryonic stem cell lines derived from human blastocysts. *Science, 282*, 1145–1147.

Vallier, L., Touboul, T., Brown, S., Cho, C., Bilican, B., Alexander, M., et al. (2009). Signaling pathways controlling pluripotency and early cell fate decisions of human induced pluripotent stem cells. *Stem Cells, 27*(11), 2655–2666.

Waddington, C. (1957). *The Strategy of the Genes*. London: George Allen & Unwin.

Xu, R. H., Chen, X., Li, D. S., Li, R., Addicks, G. C., Glennon, C., et al. (2002). BMP4 initiates human embryonic stem cell differentiation to trophoblast. *Nat. Biotechnol., 20*(12), 1261–1264, Epub 2002 Nov. 11.

Xu, R. H., Peck, R. M., Li, D. S., Feng, X., Ludwig, T., & Thomson, J. A. (2005). Basic FGF and suppression of BMP signaling sustain undifferentiated proliferation of human ES cells. *Nat. Methods, 2*(3), 185–190, Epub 2005 Feb. 17.

Xu, R. H., Sampsell-Barron, T. L., Gu, F., Root, S., Peck, R. M., Pan, G., et al. (2008). NANOG is a direct target of TGFbeta/activin-mediated SMAD signaling in human ESCs. *Cell Stem Cell, 3*(2), 196–206.

Ying, Q. L., Nichols, J., Chambers, I., & Smith, A. (2003). BMP induction of Id proteins suppresses differentiation and sustains embryonic stem cell self-renewal in collaboration with STAT3. *Cell, 115*(3), 281–292.

Ying, Q. L., Wray, J., Nichols, J., Batlle-Morera, L., Doble, B., Woodgett, J., et al. (2008). The ground state of embryonic stem cell self-renewal. *Nature, 453*(7194), 519–523.

Yu, J., Vodyanik, M. A., Smuga-Otto, K., Antosiewicz-Bourget, J., Frane, J. L., Tian, S., et al. (2007). Induced pluripotent stem cell lines derived from human somatic cells. *Science, 318*(5858), 1917–1920.

CHAPTER 5

How Cells Change their Phenotype

Caroline Beth Sangan, David Tosh
Centre for Regenerative Medicine, Department of Biology and Biochemistry,
University of Bath, Claverton Down, Bath, UK

INTRODUCTION

It was long thought that once a cell had acquired a differentiated phenotype it could not be altered, but we now know that this is not the case, and over the past few years a number of well-documented examples have been presented whereby already differentiated cells or tissue-specific stem cells have been shown to alter their phenotype to express functional characteristics of a different tissue. In this chapter, we examine evidence for these examples and comment on the underlying cellular and molecular mechanisms.

DEFINITIONS AND THEORETICAL CONSIDERATIONS

The process of regional specification in embryonic development is now quite well understood. It proceeds hierarchically, starting from the epiblast of the early embryo. Each tissue rudiment is then formed by a sequence of developmental decisions. At each step, a particular combination of transcription factors is activated or repressed in response to an extracellular signal, which may be composed of one or more inducing factors. Different concentrations of the signal or transcription factors will result in the adoption of a different developmental pathway. Hence, each step leads to multiple pathways, a developmental "choice."

We know that it is not necessary to change the activity of hundreds of genes to alter a cell phenotype, because development is controlled by a relatively small number of genes encoding those transcription factors whose activity determines developmental choices between programs of gene expression. These critical genes are sometimes called "master control genes" and the misexpression of these genes is the key to understanding transdifferentiation and metaplasia. At a molecular level, transdifferentiation must arise from the change in expression level of a master gene. "Master" genes therefore determine which part of the body is formed by each region of the embryo. The protein products of the master genes are transcription factors, and their function is to regulate the next level of genes in the hierarchy, which eventually leads to individual tissue types. Overexpressing a "master" gene in another differentiated cell type should therefore be sufficient to induce the cell (or tissue type) type encoded by the "master" gene.

WHY IS IT IMPORTANT TO STUDY TRANSDIFFERENTIATION?

It is important to study the process of transdifferentiation for three reasons. First, for understanding the molecular and cellular basis of embryonic development, as the conversion of one cell type to another generally occurs between cells that arise from neighboring regions of the

Principles of Regenerative Medicine. DOI: 10.1016/B978-0-12-381422-7.10005-7

same germ layer (mesoderm, endoderm, or ectoderm) (Slack, 1986; Tosh and Slack, 2002). Second, transdifferentiation leads to a predisposition towards certain neoplastic transformations, so elucidating the molecular basis of the conversion will also provide information on the processes underlying the development of cancer (Slack, 1986). Third, identification of the key (master) genes responsible for inducing transdifferentiation may be useful in the directed differentiation of stem cells towards therapeutically useful cell types.

In order to demonstrate that transdifferentiation has occurred in a system, Eguchi and Kodama (1993) suggested that two prerequisites be fulfilled. The first involves demonstrating (preferably with molecular evidence) the differentiation state of the two cell types before and after the transdifferentiation event. The second prerequisite involves showing a direct ancestor-descendent relationship between the cells prior to and following transdifferentiation. It is difficult to fulfill these prerequisites under *in vivo* conditions. However, *in vitro* culture systems are more amenable to testing these prerequisites. One of the best-studied *in vitro* models for transdifferentiation is the conversion of pigmented epithelial cells of the retina to lens cells, so-called Wolffian lens regeneration (Eguchi and Kodama, 1993). Developing *in vitro* models for the transdifferentiation of one cell type to another is crucial as it will allow us to define the molecular and cellular mechanisms that distinguish the two cell types involved in the switch.

CONVERSION OF PANCREATIC CELLS TO HEPATOCYTES

The conversion of pancreas to liver is one well-documented example of transdifferentiation. This conversion is not surprising as both organs arise from adjacent regions of the endoderm, and are postulated to arise from bi-potential cells in the foregut endoderm (Deutsch et al., 2001). The appearance of hepatic foci in the pancreas has been naturally observed in the primate, the vervet monkey (Wolfe-Coote et al., 1996), and numerous protocols have been established to induce transdifferentiation in other species, including rats fed a copper-deficient diet (Rao et al., 1986), rats treated with peroxisome proliferators (e.g. ciprofibrate) (Reddy et al., 1984), and transgenic mice overexpressing KGF in pancreatic islets (Krakowski et al., 1999). These *in vivo* models have been extremely valuable in demonstrating the possibility that all three cell types in the pancreas (acinar, endocrine, and ductular) have the potential to transdifferentiate into hepatocytes. Unfortunately, *in vivo* studies are limited in their ability to identify the significant individual changes occurring at the molecular and cellular level.

Consequently, the molecular and cellular basis of transdifferentiation of pancreas to liver has only been investigated in detail recently, via utilization of two *in vitro* models. The first model exploits the pancreatic cell line AR42J. Originally isolated from a carcinoma of an azaserine-treated rat, they are amphicrine cells expressing both exocrine and neuroendocrine properties (e.g. are able to synthesise digestive enzymes and express neurofilaments) (Christophe, 1994). AR42J cells transdifferentiate following treatment with the synthetic glucocorticoid dexamethasone, in a three-step process involving the initial loss of pancreatic markers (e.g. amylase), prior to the gain of fetal liver markers (e.g. alpha-fetoprotein, transferrin) and then finally adult liver markers (e.g. albumin) (Shen et al., 2000). The transdifferentiated hepatocytes function like normal hepatocytes; in particular, they are able to respond to xenobiotics (Tosh et al., 2002; Marek et al., 2003; Lardon et al., 2004). Hepatocyte cell architecture is fundamental to their function; thus, there are also associated morphological changes during transdifferentiation, including changes in cell shape and formation of extensive endoplasmic reticulum and structures resembling bile canaliculi. Lineage experiments based on the expression of green fluorescent protein (GFP) under the exocrine pancreatic elastase promoter were performed in parallel. Some nascent hepatocytes were GFP-positive, indicating that they once had an active elastase promoter, thus validating that the hepatocytes are generated from exocrine cells (Shen et al., 2000).

The second *in vitro* model employed also relies on the addition of dexamethasone to an *ex vivo* culture model for mouse embryonic pancreas. After treatment, liver proteins are expressed; however, it is not clear whether the same cellular and molecular mechanisms are in operation as in the AR42J cells; for example, in a culture model that consists of multiple cell types, the liver-like cells could originate from pancreatic stem cells instead of differentiated cell types.

Due to their close developmental relationship, the pancreas and liver share a similar array of transcription factors. The expression of several liver-enriched transcription factors has been analyzed prior to and following transdifferentiation. Associated with loss of exocrine gene expression and gain of liver gene expression was the induction of transcription factor CCAAT/enhancer binding protein beta (C/EBPβ). Furthermore, transfection of AR42J cells with C/EBPβ is sufficient to transdifferentiate AR42J cells to hepatocytes, while overexpression of liver inhibitory protein (LIP), the dominant negative form of C/EBPβ (which heterodimerizes with full length C/EBPβ) prevents transdifferentiation. *In toto*, C/EBPβ is the key candidate for the "master switch" transcription factor responsible for distinguishing liver and pancreas. *In vivo* data are consistent with this theory as an increase in C/EBPβ is observed in a copper-deficient pancreas; however, it remains to be elucidated whether the increase is due to C/EBPβ's involvement in transdifferentiation or adipogenesis (Tanaka et al., 1997). Interestingly, C/EBP (α and β) are expressed in the early liver rudiment but not in the pancreas (Westmacott et al., 2006), suggesting that C/EBPs may distinguish liver and pancreas during development. A similar upregulation of C/EBPβ along with α-fetoprotein is seen during the dexamethasone-induced transdifferentiation of primary rat pancreatic exocrine cells into hepatocytes (Lardon et al., 2004).

It is also suggested that the transcription factor hepatocyte nuclear factor 4 alpha (HNF4α) may play a role in transdifferentiation, as it is observed translocating into the nuclei during transdifferentiation and previous work indicates HNF4α performs a critical role in regulating liver differentiation, both in development and regeneration (Shen et al., 2003).

TRANSDIFFERENTIATION OF PANCREATIC EXOCRINE TO ENDOCRINE CELLS

The pancreatic acinar cells normally secrete digestive enzymes (e.g. amylase) and are capable of transdifferentiation into endocrine islet insulin-secreting β cells under appropriate conditions. This conversion has been demonstrated by several *in vitro* studies, which involve the culturing of dissociated adult pancreatic acini in the presence of growth factors, such as epidermal growth factor (EGF). Transdifferentiation of exocrine cells to β cells is postulated to operate via EGF signaling, a hypothesis that is corroborated by inhibition of EGF receptor kinase blocking transdifferentiation (Minami et al., 2005) and EGF receptor knockout studies showing impaired β-cell differentiation and islet morphogenesis (Miettinen et al., 2000). Cultures treated with EGF and additional growth factors (e.g. leukemia inhibitory factor (LIF)) exhibit an increase in β-cell mass with the nascent β cells expressing mature phenotypic markers of β cells (for example, GLUT2 and C-peptide 1) and containing insulin-immunoreactive granules. From a functional perspective, the pancreatic β cell is unique in its expression, processing, and secretion of insulin in response to glucose concentrations. Transdifferentiated β cells demonstrate glucose responsiveness, as a four-fold increase in insulin secretion is induced upon glucose stimulation. In addition, when transplanted *in vivo* into alloxan-diabetic mice, these β cells are able to restore normoglycemia, with hyperglycemia recurring upon removal of the graft (Baeyens et al., 2005).

It was confirmed that the β cells originated from exocrine cells and not from a contaminating cell type as, when cultured with nicotinamide, a substance known to prevent acinar exocrine cells from losing their functional characteristics, some transitional cells are identified that were co-positive for amylase and insulin. Similarly analysis by the Cre-loxP-based direct cell lineage tracing system indicates that newly made β cells originate from amylase/elastase-expressing pancreatic acinar cells (Minami et al., 2005).

In order to elucidate the molecular basis of the switch in cell phenotype, Zhou et al. (2008) recently employed an *in vivo* strategy to re-express key developmental transcription factors in adult exocrine cells using adenoviral vectors expressing Pdx1, Ngn3, and MafA. An *in vivo* experiment has the advantage that it allows the induced β cells to reside in their native environment, which may not only enhance survival and maturation but also allow direct comparison with endogenous islet β cells. The transdifferentiated β cells display the appropriate size, shape, and ultrastructure; express functional β-cell genes but not exocrine genes or other endocrine cell-type genes; and can ameliorate hyperglycemia by remodeling local vasculature and secreting insulin. One difference observed is the lack of organization into islet structures, which could ultimately impair function as signaling between β cells is important for enhancing glucose responsiveness. The promotion of exocrine trans-differentiation into endocrine β cells is not reliant on a single factor. The transcription factor Neurogenin 3 is imperative for allowing the genetic switch to an endocrine fate as, in development, it is essential for directing differentiation of pancreatic precursor cells towards the endocrine lineage via regulation of factors further downstream that are required for β-cell differentiation (Gradwohl et al., 2000). Pancreatic duodenal homeobox-1 (Pdx1), which is broadly expressed in all pancreatic cell types during embryonic development, has a different role in the adult pancreas as it is primarily expressed in mature β cells. This is because it has a central role in insulin transcription (binds the A/T-rich elements) leading to activation in conjunction with other transcription factors such as MafA (Ohlsson et al., 1993; Peshavaria et al., 1994). MafA is a β-cell-specific transcription factor again important for insulin activation in mature β cells (Olbrot et al., 2002). In conclusion, the specific combination of these three transcription factors, Ngn3 (Neurog3), Pdx1, and MafA, can reprogram differentiated pancreatic exocrine cells in adult mice into cells that closely resemble β cells (Krakowski et al., 1999).

INDUCED PLURIPOTENT STEM CELLS

The recent discovery of methods for generating induced pluripotent stem cells (iPSCs) has transformed the landscape for stem cell research. iPSCs, closely resembling ESCs, can be created from normal fibroblasts or other cell types by overexpression of specific genes (Takahashi and Yamanaka, 2006; Takahashi et al., 2007; Yu et al., 2007). The holy grail of cell therapy is patient-specific grafts, which would be fully immunocompatible, alleviating the need for post-transplantation immunosuppression. iPSCs can help to achieve this in three ways. First, it is possible to derive iPSCs from individual patients, even those suffering from genetic diseases (Park et al., 2008). Although patient-specific cell culture is currently very expensive, it is possible to envisage considerable technological improvements and cost reductions in the long-term. Second, even if routine patient-specific cell culture is not feasible, the relative ease of making iPSCs suggests that cell banks could be created representing a reasonable match to a large fraction of the population (Daley and Scadden, 2008). Finally, it is possible to make both hepatocytes and hematopoietic stem cells (HSCs) from the same cell line (Kaufman and Thomson, 2002). It has been shown that a graft of HSCs can produce a chimeric bone marrow that is tolerant to subsequent grafts from the same donor. So, it is possible to imagine that an HSC graft could be given to the patient to render them tolerant to the subsequent graft of therapeutically relevant cells (e.g. pancreatic β cells or hepatocytes) made from the same cell line (Kyba and Daley, 2003).

BARRETT'S METAPLASIA

The incidence of esophageal adenocarcinoma (OA) has increased rapidly in the last 30 years and reflects the increasing incidence of Barrett's metaplasia (also referred to as Barrett's esophagus) (Falk, 2002). According to the British Society for Gastroenterology, Barrett's metaplasia is a pathological condition in which the distal esophagus undergoes metaplastic transformation from the normal stratified squamous epithelium (SSQE) to columnar-lined epithelium (CE). Intestinal differentiation is also a feature of Barrett's metaplasia. Although there are four intestinal cell types (enterocytes, goblet cells, enteroendocrine cells, and Paneth cells), Paneth cells are rarely found in histological specimens, prompting the term "incomplete intestinal metaplasia." Barrett's metaplasia generally occurs in the context of chronic gastro-esophageal reflux disease, suggesting a role of reflux components (including bile and acid) (Vaezi and Richter, 1996).

Barrett's metaplasia is the only known precursor for OA and confers an increased risk of 50–100 times that of the normal population. The metaplasia-dysplasia-adenocarcinoma sequence is widely accepted as the pathway for the development of OA (Aldulaimi and Jankowski, 1999; Jankowski et al., 1999). Barrett's metaplasia is the strongest contributory factor, associated with an annual risk of conversion to OA of 0.5–1%. The UK has one of the highest worldwide incidences of Barrett's metaplasia, with an estimated prevalence of 1% and an incidence of esophageal adenocarcinoma two to three times that of Europe or North America (Jankowski and Anderson, 2004; Fitzgerald, 2006).

Current management for Barrett's patients is based on surveillance with the aim of detecting early curable lesions. There are no effective treatments for preventing patients with Barrett's metaplasia from developing adenocarcinoma (Fitzgerald, 2004). The prognosis of esophageal adenocarcinoma is dismal, with a five-year survival rate of less than 10% despite combined treatment with chemotherapy and surgery (Jankowski et al., 2000). Pharmacological treatment is aimed at controlling reflux symptoms, believed to be a contributory factor, but this strategy does not reverse Barrett's metaplasia or eliminate the associated cancer risk (Li et al., 2008). Current UK guidelines are for surveillance endoscopy every two years in those patients where it is considered appropriate. The aim of surveillance is to detect dysplasia but the methods are labor-intensive, costly, and relatively ineffective (Fitzgerald, 2004). A treatment that could eliminate Barrett's metaplasia and its associated cancer risk would have a significant impact on the increasing esophageal cancer figures.

The master switch gene responsible for inducing Barrett's metaplasia is thought to be the *Cdx2* gene, a member of the parahox cluster (Ferrier et al., 2005). Cdx2 is involved in intestinal epithelial differentiation and distinguishes the upper and lower epithelium of the alimentary canal; furthermore, Cdx2 expression has been found to be upregulated in adenocarcinomas of the intestine (de Lott et al., 2005). Experiments have shown that ectopic expression of Cdx2 can induce intestinal metaplasia in the stomach (Silberg et al., 2002). The exact mechanism by which metaplasia is induced in Barrett's is still unclear, and some debate remains as to whether Barrett's may be described as a true transdifferentiation event of the epithelium or simply the metaplasia of esophageal stem cells to intestinal stem cells and subsequent differentiation.

SUMMARY

It is now apparent that transdifferentiation is a biological reality. Whether transdifferentiation really does occur on a day-to-day basis during regeneration or after normal physiological damage has yet to be established. Although some examples of metaplasia and trans-differentiation have been shown to occur *in vivo*, many experiments have been done *in vitro*, and it is not clear whether these changes in phenotype are just tissue-culture phenomena or whether they also occur *in vivo*.

The molecular basis of transdifferentiation is now understood in several cases; for example, the conversion of pancreas to liver and liver to pancreas. These examples generally show a close developmental relationship, perhaps making it easier to determine the genetics of the switch. Understanding the rules for the molecular basis of metaplasia is crucial for rational progress in the area of therapeutic stem-cell transplantation; a technology that is certain to attract considerable attention in the next few years.

References

Aldulaimi, D., & Jankowski, J. (1999). Barrett's esophagus: an overview of the molecular biology. *Dis. Esophagus, 12,* 177–180.

Baeyens, L., de Breuck, S., Lardon, J., Mfopou, J. K., Rooman, I., & Bouwens, L. (2005). *In vitro* generation of insulin-producing beta cells from adult exocrine pancreatic cells. *Diabetologia, 48,* 49–57.

Christophe, J. (1994). Pancreatic tumoral cell line AR42J: an amphicrine model. *Am. J. Physiol., 266,* G963–G971.

Daley, G. Q., & Scadden, D. T. (2008). Prospects for stem cell-based therapy. *Cell, 132,* 544–548.

de Lott, L. B., Morrison, C., Suster, S., Cohn, D. E., & Frankel, W. L. (2005). CDX2 is a useful marker of intestinal-type differentiation: a tissue microarray-based study of 629 tumors from various sites. *Arch. Pathol. Lab. Med., 129,* 1100–1105.

Deutsch, G., Jung, J., Zheng, M., Lóra, J., & Zaret, K. S. (2001). A bipotential precursor population for pancreas and liver within the embryonic endoderm. *Development, 128,* 871–881.

Eguchi, G., & Kodama, R. (1993). Transdifferentiation. *Curr. Opin. Cell Biol., 5,* 1023–1028.

Falk, G. W. (2002). Barrett's esophagus. *Gastroenterology, 122,* 1569–1591.

Ferrier, D. E., Dewar, K., Cook, A., Chang, J. L., Hill-Force, A., & Amemiya, C. (2005). The chordate ParaHox cluster. *Curr. Biol., 15,* R820–R822.

Fitzgerald, R. C. (2004). Review article: Barrett's oesophagus and associated adenocarcinoma – a UK perspective. *Aliment Pharmacol. Ther., 20*(Suppl. 8), 45–49.

Fitzgerald, R. C. (2006). Molecular basis of Barrett's oesophagus and oesophageal adenocarcinoma. *Gut, 55,* 1810–1820.

Gradwohl, G., Dierich, A., LeMeur, M., & Guillemot, F. (2000). Neurogenin3 is required for the development of the four endocrine cell lineages of the pancreas. *Proc. Natl. Acad. Sci. U.S.A., 97,* 1607–1611.

Jankowski, J. A., & Anderson, M. (2004). Review article: management of oesophageal adenocarcinoma – control of acid, bile and inflammation in intervention strategies for Barrett's oesophagus. *Aliment Pharmacol. Ther., 20* (Suppl. 5), 71–80, discussion 95–96.

Jankowski, J. A., et al. (1999). Molecular evolution of the metaplasia-dysplasia-adenocarcinoma sequence in the esophagus. *Am. J. Pathol., 154,* 965–973.

Jankowski, J. A., et al. (2000). Barrett's metaplasia. *Lancet, 356,* 2079–2085.

Kaufman, D. S., & Thomson, J. A. (2002). Human ES cells – haematopoiesis and transplantation strategies. *J. Anat., 200,* 243–248.

Krakowski, M. L., Kritzik, M. R., Jones, E. M., Krahl, T., Lee, J., Arnush, M., et al. (1999). Pancreatic expression of keratinocyte growth factor leads to differentiation of islet hepatocytes and proliferation of duct cells. *Am. J. Pathol., 154,* 683–691.

Kyba, M., & Daley, G. Q. (2003). Hematopoiesis from embryonic stem cells: lessons from and for ontogeny. *Exper. Hematol., 31,* 994–1006.

Lardon, J., de Breuck, S., Rooman, I., van Lommel, L., Kruhøffer, M., Orntoft, T., et al. (2004). Plasticity in the adult rat pancreas: transdifferentiation of exocrine to hepatocyte-like cells in primary culture. *Hepatology, 39,* 1499–1507.

Li, Y. M., et al. (2008). A systematic review and meta-analysis of the treatment for Barrett's esophagus. *Dig. Dis. Sci., 53,* 2837–2846.

Marek, C. J., Cameron, G. A., Elrick, L. J., Hawksworth, G. M., & Wright, M. C. (2003). Generation of hepatocytes expressing functional cytochromes P450 from a pancreatic progenitor cell line *in vitro*. *Biochem. J., 370,* 763–769.

Miettinen, P. J., Huotari, M., Koivisto, T., Ustinov, J., Palgi, J., Rasilainen, S., et al. (2000). Impaired migration and delayed differentiation of pancreatic islet cells in mice lacking EGF-receptors. *Development, 127,* 2617–2627.

Minami, K., Okuno, M., Miyawaki, K., Okumachi, A., Ishizaki, K., Oyama, K., et al. (2005). Lineage tracing and characterization of insulin-secreting cells generated from adult pancreatic acinar cells. *Proc. Natl. Acad. Sci. U.S.A., 102,* 15116–15121.

Ohlsson, H., Karlsson, K., & Edlund, T. (1993). IPF1, a homeodomain-containing transactivator of the insulin gene. *EMBO J., 12,* 4251–4259.

Olbrot, M., Rud, J., Moss, L. G., & Sharma, A. (2002). Identification of beta-cell-specific insulin gene transcription factor RIPE3b1 as mammalian MafA. *Proc. Natl. Acad. Sci. U.S.A., 99,* 6737–6742.

Park, I. H., Arora, N., Huo, H., Maherali, N., Ahfeldt, T., Shimamura, A., et al. (2008). Disease-specific induced pluripotent stem cells. *Cell, 134,* 877–886.

Peshavaria, M., Gamer, L., Henderson, E., Teitelman, G., Wright, C. V., & Stein, R. (1994). XIHbox 8, an endoderm-specific Xenopus homeodomain protein, is closely related to a mammalian insulin gene transcription factor. *Mol. Endocrinol., 8,* 806–816.

Rao, M. S., Subbarao, V., & Reddy, J. K. (1986). Induction of hepatocytes in the pancreas of copper-depleted rats following copper repletion. *Cell Differ., 18,* 109–117.

Reddy, J. K., Rao, M. S., Qureshi, S. A., Reddy, M. K., Scarpelli, D. G., & Lalwani, N. D. (1984). Induction and origin of hepatocytes in rat pancreas. *J. Cell Biol., 98,* 2082–2290.

Shen, C. N., Slack, J. M. W., & Tosh, D. (2000). Molecular basis of transdifferentiation of pancreas to liver. *Nat. Cell Biol., 2,* 879–887.

Shen, C.-N., Horb, M., Slack, J. M. W., & Tosh, D. (2003). Transdifferentiation of pancreas to liver. *Mech. Develop., 120,* 107–116.

Silberg, D. G., Sullivan, J., Kang, E., Swain, G. P., Moffett, J., Sund, N. J., et al. (2002). Cdx2 ectopic expression induces gastric intestinal metaplasia in transgenic mice. *Gastroenterology, 122,* 689–696.

Slack, J. M. W. (1986). Epithelial metaplasia and the second anatomy. *Lancet, 2,* 268–271.

Takahashi, K., & Yamanaka, S. (2006). Induction of pluripotent stem cells from mouse embryonic and adult fibroblast cultures by defined factors. *Cell, 126,* 663–676.

Takahashi, K., Tanabe, K., Ohnuki, M., Narita, M., Ichisaka, T., Tomoda, K., et al. (2007). Induction of pluripotent stem cells from adult human fibroblasts by defined factors. *Cell, 131,* 861–872.

Tanaka, T., Yoshida, N., Kishimoto, T., & Akira, S. (1997). Defective adipocyte differentiation in mice lacking the C/EBPbeta and/or C/EBPdelta gene. *EMBO J., 16,* 7432–7443.

Tosh, D., & Slack, J. M. W. (2002). How cells change their phenotype. *Nat. Rev. Mol. Cell Biol., 3,* 187–194.

Tosh, D., Shen, C. N., & Slack, J. M. W. (2002). Differentiated properties of hepatocytes induced from pancreatic cells. *Hepatology, 36,* 534–543.

Vaezi, M. F., & Richter, J. E. (1996). Role of acid and duodenogastroesophageal reflux in gastro-esophageal reflux disease. *Gastroenterology, 111,* 1192–1199.

Westmacott, A., Burke, Z. D., Oliver, G., Slack, J. M. W., & Tosh, D. (2006). C/EBPα and C/EBPβ are markers of early liver development. *Int. J. Dev. Biol., 50,* 653–657.

Wolfe-Coote, S., Louw, J., Woodroof, C., & Du Toit, D. F. (1996). The non-human primate endocrine pancreas: development, regeneration potential and metaplasia. *Cell Biol. Int., 20,* 95–101.

Yu, J. Y., Vodyanik, M. A., Smuga-Otto, K., Antosiewicz-Bourget, J., Frane, J. L., Tian, S., et al. (2007). Induced pluripotent stem cell lines derived from human somatic cells. *Science, 318,* 1917–1920.

Zhou, Q., Brown, J., Kanarek, A., Rajagopal, J., & Melton, D. A. (2008). *In vivo* reprogramming of adult pancreatic exocrine cells to beta-cells. *Nature, 455,* 627–632.

CHAPTER 6

Scarless Wound Healing

Allison Nauta[*,**], Barrett Larson[*], Michael T. Longaker[*], H. Peter Lorenz[*]

* Hagey Laboratory for Pediatric and Regenerative Medicine, Division of Plastic and Reconstructive Surgery, Department of Surgery, Institute of Stem Cell Biology and Regenerative Medicine, Stanford University School of Medicine, Palo Alto, California, USA; ** Department of Surgery, Georgetown University Hospital, Washington DC, USA

CLINICAL BURDEN

Scarring can affect any tissue or organ in the body, which causes a spectrum of medical problems. For example, a patient undergoing gastrointestinal surgery has bowel scarring, which can cause post-operative bowel obstruction. After traumatic injury or surgery to ligaments and tendons, scarring can cause contracture across joints, which can limit movement and cause functional restriction. Scarring in the nervous system results in loss of function as neuronal connections are destroyed. Scarring in the cornea limits visual acuity. In summary, injury to nearly all tissues results in scarring. The only exceptions in mammals are bone fracture repair and liver repair after partial surgical resection.

Burns and other breaches to skin integrity heal with scarring that can cause functional limitations and restrictions in movement through contractures across joints. Scarring on the face can restrict growth in children and cause ocular and oral dysfunction when around the eyes and mouth, respectively. Approximately 500,000 patients in the USA undergo medical treatment for burn injuries annually, and over one third of patients requiring hospital admission have burns that exceed 10% total body surface area (American Burn Association Burn Incidence Fact Sheet, 2007). Many of these patients are children, a population that is particularly vulnerable to the negative physical and psychological effects of scarring.

Wound healing in healthy adults usually results in a physiologically normal scar, which — though problematic for the reasons discussed above — is preferable to the two extreme outcomes of the repair process: non-healing chronic ulcers and excessive fibroproliferative scarring. Patients with chronic illnesses fail to heal effectively for numerous reasons, including infection, impaired blood flow, severe malnutrition, and inadequate wound care. These patients have become an increasing concern, particularly as the population ages and more healthcare resources are allocated to treat chronic diseases and their associated complications.

The diabetic population is a dramatic example of the chronic wound burden on society. The following statistics, obtained from the CDC's 2008 National Diabetes Fact Sheet, illustrate the magnitude of the burden that diabetic non-healing wounds pose to patients and society: Twenty-three million people in the USA have diabetes, a population that doubled between 1990 and 2005. Diabetes alone is responsible for more than half a million hospital admissions and 28.6 million ambulatory care visits each year. In 2004 alone, 7,100 lower limb non-traumatic amputations were performed in patients with diabetes. Twenty-three percent of all patients with diabetes have foot problems, ranging from numbness to amputations. Twenty-five to fifty percent of all hospital admissions in these patients are for non-healing diabetic

103

Principles of Regenerative Medicine. DOI: 10.1016/B978-0-12-381422-7.10006-9

ulcers, which are the cause of the majority of non-traumatic extremity amputations performed in the USA each year. In 2007, total direct healthcare costs for patients with diabetes were estimated at a staggering $174 billion. In addition, indirect costs, resulting from disability, work loss, and early mortality, totaled $58 billion (National Diabetes Fact Sheet of the National Center for Chronic Disease Prevention and Health Promotion, 2008). These data demonstrate that the diabetic population is rapidly growing, thus requiring greater healthcare resources to manage conditions related to poor wound healing (e.g. Charcot neuroarthropathy, limb ulcerations and infection, and amputations) and the resultant disabilities. Other reasons for chronic non-healing ulcers include peripheral vascular occlusive disease and paraplegia.

On the other extreme, excessive healing is also a burden. Pathological scarring causes hypertrophic scars and keloids. These scarring processes cause functional impairment and symptoms such as burning, itching, and pain. These lesions are difficult to treat medically or with surgery, and no effective uniform treatment exists (Kose and Waseem, 2008).

ADULT SKIN
Anatomy of adult skin

Adult skin is made up of two layers, the epidermis and dermis. The epidermis has five distinct layers, each characterized by the level of keratinocyte maturation. Keratinocytes originate in the basal epidermal layer and migrate to the surface layer over a four-week journey to become soft keratin, which eventually sloughs off.

Epidermal appendages, which are epithelial structures that extend intradermally, are an important source of cells for re-epithelialization. Epidermal appendages include sebaceous glands, sweat glands, apocrine glands, and hair follicles. Appendages can extend deep into the dermis or even through the dermis and into the subcutaneous tissue. The hair follicle is composed of an external outer root sheath attached to the basal lamina that is contiguous with the epidermis. The hair follicle also contains a channel and a hair shaft. Together, the hair follicle and its attached sebaceous gland are called the pilosebaceous unit. The base, or bulb, of the hair follicle contains committed but proliferating progenitor cells and the matrix encasing the dermal papilla, which contains specialized mesenchymal cells. The hair shaft and its channel grow from this region.

Sweat glands — or eccrine glands — produce sweat, which cools the body upon evaporation. The sweat gland contains a coiled intradermal portion that extends into the epidermis by a relatively straight distal duct.

Below the epidermis lie the two layers of the dermis, the more superficial papillary layer and the deeper, more fibrous reticular layer. The papillary dermis is highly vascular, sending capillaries (dermal papillae) superficially into the dermis. The reticular dermis contains densely packed collagen fibers and tends to be less vascular, except where sweat glands and hair follicles run through it. This layer is also rich in elastic fibers and contains some macrophages, fibroblasts, and adipose cells (Cormack, 1997) (Fig. 6.1).

Adult wound healing and scar formation

Adult wound healing is traditionally described as a sequence of temporally overlapping phases: inflammation, proliferation, and remodeling. Disruption of the vascular network within cutaneous wounds results in platelet aggregation and the formation of a fibrin-rich clot, which protects from further extravasation of blood or plasma. Aggregation of platelets initiates the coagulation cascade (Clark, 1996). In addition to providing hemostasis, platelets modulate fibroblast activity through degranulation and secretion of multiple cytokines and growth factors, such as platelet-derived growth factor (PDGF), platelet factor 4 (PF4), and transforming growth factor β1 (TGF-β1). These growth factors and cytokines remain elevated throughout the process of normal wound healing (Moulin et al., 1998; Henry and Garner, 2003).

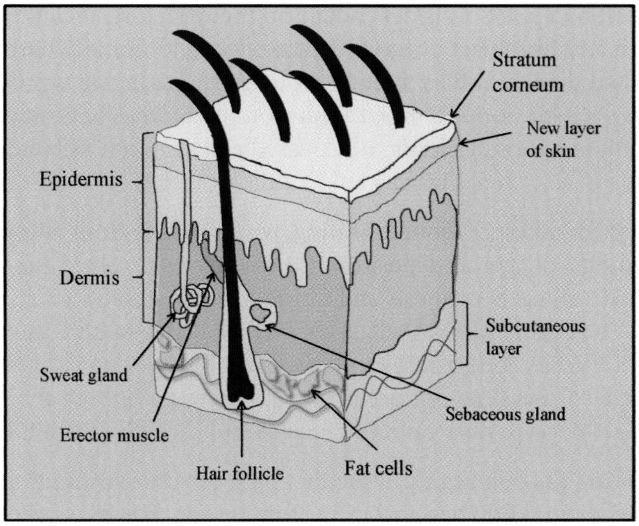

FIGURE 6.1
Adult skin anatomy. Adult skin is dynamic with continuous epidermal cell turnover. Shown are the epidermis, dermis, and subcutaneous layers. Epidermal appendages, such as sweat glands and hair follicles, originate in the epidermis and extend into the dermis and the subcutaneous layer.

Largely under the influence of platelet-derived inflammatory molecules, neutrophils and monocytes initiate their migration to the wound. However, due to the high concentration of neutrophils in circulation, these cells are the first responders to the area of injury and very quickly reach high concentrations, becoming the most dominant influence. Neutrophils primarily produce degradative enzymes and phagocytose foreign and necrotic material, but they also produce vascular endothelial growth factor (VEGF), tumor necrosis factor alpha (TNF-α), interleukin 1 (IL-1), and other growth factors that assist in wound healing. Interestingly, studies show that neutrophil infiltration is not essential to normal healing, demonstrating one of many redundancies in the repair process (Simpson and Ross, 1972).

The level of inflammation depends on the presence or absence of infection. In the presence of infection, neutrophils continue to be active in high concentration, leading to further inflammation and fibrosis (Singer and Clark, 1999). In the absence of infection, neutrophils greatly diminish activity at day 2 or 3, as monocytes increase in number in response to both extravascular and intravascular chemoattractants. Monocytes and macrophages are able to bind to the ECM, which induces phagocytosis and allows for debridement of necrotic cells and fractured structural proteins. During the late inflammatory phase, monocytes transform into tissue macrophages that release cytokines and scavenge dead neutrophils, making macrophages the dominant leukocyte in the wound bed. In contrast to neutrophils, studies on tissue macrophage and monocyte-depleted guinea pigs have demonstrated that macrophages are essential to the normal wound healing process through their stimulation of collagen production, angiogenesis, and re-epithelialization (Leibovich and Ross, 1975). However, similarly to the activity of neutrophils, if macrophages persist, the result is excess scar formation. Under these circumstances, macrophages produce high amounts of cytokines that activate fibroblasts to deposit excessive amounts of collagen (Niessen et al., 1999).

The presence of macrophages in the wound marks the transition between the inflammatory phase and the proliferative phase of wound healing, which begins around day 4 to 5 post-injury in uninfected open wounds. Granulation tissue begins to form and is a loose network of collagen, fibronectin, and hyaluronic acid, embedding a dense population of macrophages, fibroblasts, and neovasculature. During the deposition of granulation tissue, macrophages, fibroblasts, and newly formed blood vessels move into the wound space as a unit (Clark, 1985).

The rate of granulation deposition is dependent on many factors, including the interaction between fibronectin and fibroblast integrin receptors (Xu and Clark, 1996). Fibroblasts in the wound edges and bed deposit collagen and a proteoglycan-rich provisional matrix, a process that is stimulated by TGF-β1 and TGF-β2 in adult wounds. Studies have shown that exogenous administration of these molecules leads to increased collagen and inflammatory cells at the wound site (Roberts et al., 1986; Ogawa et al., 1991).

During the proliferative phase of wound healing, which occurs from approximately day 5 to day 14 post-wounding, collagen is deposited at the wound site. Once a threshold level of collagen is deposited, collagen synthesis and fibroblast accumulation is suppressed by a negative-feedback mechanism (Grinnell, 1994). The balance of collagen synthesis and degradation is controlled by collagenases and tissue inhibitors of metalloproteinases (TIMPs). When this negative feedback does not occur appropriately, pathological scars form with deposition of densely packed, disorganized collagen bundles (Singer and Clark, 1999).

The re-epithelialization process begins in the first 24 hours after wounding, with the goal of creating a protective, natural skin barrier. During this process, basal keratinocytes at the border of the wound — which under normal circumstances are linked together by desmosomes and attached to the ECM — detach from the ECM and migrate laterally to fill the void in the epidermis. Through this process, keratinocytes are exposed to serum for the first time. Keratinocytes are subjected to new and increased levels of inflammatory cytokines and growth factors, which signal their further migration, proliferation, and differentiation (Li et al., 2004).

Neovascularization occurs during the proliferative phase and is influenced by multiple cytokines, as well as circulating endothelial progenitor cells and the ECM (Folkman, 1996). Additionally, the formation of blood vessels is induced by lactic acid, plasminogen activator, collagenases, and low oxygen tension (Singer and Clark, 1999). Apoptotic pathways become active once granulation tissue matures, which stops angiogenesis (Ilan et al., 1998).

The maturation stage of wound healing consists of collagen remodeling, which begins during the second week of healing. At this point, fibroblasts have become myofibroblasts, which are characterized by greater expression of smooth muscle actin. Fibroblasts decrease in number, and the scar tissue becomes less vascular and paler as vessels involute (Montesano and Orci, 1988).

Scar tissue gains tensile strength as collagen cross-links increase during remodeling. However, scar tensile strength will never reach the original strength of unwounded skin. Collagen maturation also involves the replacement of initial, randomly oriented types I and III collagen by predominantly type I collagen, which is organized along the lines of tension. Collagen remodeling is yet another stage during the repair process that can be derailed and cause the creation of a raised and irregular scar (Rahban and Garner, 2003) (Fig. 6.2).

Fibroproliferative scarring

Fibrosis is defined as "the replacement of the normal structural elements of the tissue by distorted, non-functional, and excessive accumulation of scar tissue" (Diegelmann and Evans, 2004). Many medical problems are linked to excessive fibrosis, and a full discussion is outside the scope of this chapter. Keloids and hypertrophic scars are clinical examples of excessive cutaneous fibrosis (Shaffer et al., 2002; Rahban and Garner, 2003).

As previously mentioned, excessive fibroproliferative scarring occurs when the mechanisms of wound healing go into overdrive. Abnormal scar formation is an excess accumulation of an unorganized collagenous extracellular matrix. Although the appearance of scars is often random and unpredictable, there are several factors that influence the severity of scarring. These include not only genetics but also tissue site, sex, race, age, magnitude of injury, and wound contamination. Generally speaking, skin sites with a thicker dermis tend to scar greater

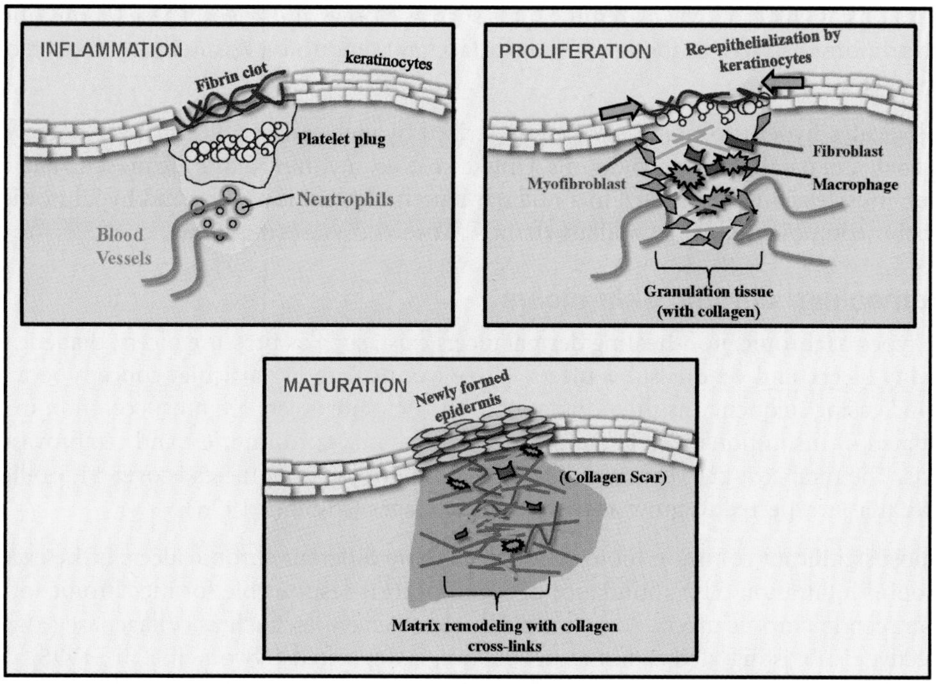

FIGURE 6.2
Adult skin wound healing. Temporally overlapping phases of wound healing. Inflammation: infiltration of neutrophils, followed by monocytes and macrophages. This process is marked by bacterial destruction, phagocytosis, and tissue debridement. Proliferation: coordinated migration of macrophages, fibroblasts, and vascular endothelial cells into the wound bed. Wound contraction and collagen accumulation occurs. Maturation: continued collagen accumulation, cross-linking, and remodeling by cells in the wound bed.

compared to sites with a thinner dermis (all else being equal). Estrogen is believed to promote scarring; as a result, pre-menopausal women often have worse scarring than both post-menopausal women and men. In general, patients with darkly pigmented skin are more prone to thicker scarring, as are young people. Larger, deeper, and more contaminated wounds also tend to produce increased scar formation (Ashcroft et al., 1997a,b,c; Ferguson and O'Kane, 2004).

KELOIDS

Keloids are benign fibrous tumors that develop at sites of skin injury over a period of months to years. The fibrous growth develops a round, smooth surface that extends beyond the area of original injury. These growths can be extremely irritable, though the clinical manifestations can vary from patient to patient. Keloids can be particularly disfiguring because of their nodular appearance, size, and color, which tends to be dark and erythematous (English and Shenefelt, 1999; Niessen et al., 1999; Shaffer et al., 2002; Atiyeh et al., 2005). These lesions can cause pain, burning, and itching and tend not to regress spontaneously. They can continue to slowly grow over many years, with growth correlating to symptoms. The most common areas affected by keloids are upper body sebaceous areas, while the extremities are less commonly involved (Tuan and Nichter, 1998).

Histologically, keloids are characterized by thick, large, closely packed bundles of disorganized collagen. Mucin is deposited focally in the dermis, and hyaluronic acid expression is confined to the thickened, granular/spinous layer of the epidermis (Kose and Waseem, 2008).

HYPERTROPHIC SCARS

The incidence of hypertrophic scars is higher than that of keloids. Hypertrophic scars are often initially erythematous, brownish-red in color, but can become pale with age. Unlike keloids, these lesions often occur over extremity joints such as elbows and knees. These lesions often do not raise more than 4 mm above the skin surface and tend to be less nodular than keloids (Niessen et al., 1999).

Histologically, hypertrophic scars are characterized by collagen bundles that are fine, well organized, and parallel to the epidermis. Unlike keloids, myofibroblasts are present, and alpha smooth muscle actin is expressed in a nodular pattern. Mucin is absent, and hyaluronic acid is a major component of the papillary dermis (Kose and Waseem, 2008).

Underhealing: chronic skin ulcers

Many types of chronic, non-healing dermal ulcers exist, such as pressure ulcers, diabetic lower extremity ulcers, and venous stasis ulcers. These wounds are of particular concern because of their increasing frequency as the population ages. Pressure ulcers are most common in debilitated or institutionalized patients, those with spinal cord injuries, and cerebrovascular infarcts. The total cost per year to care for patients with pressure ulcers is over $1.3 billion, a figure that is expected to grow as the population ages (Allman, 1998).

The most significant common biologic marker for the different chronic ulcers is the excessive neutrophil infiltration. The abundance of neutrophils is responsible for the chronic inflammation seen in chronic ulcers. As neutrophils release enzymes, such as collagenase (MMP 8), connective tissue is digested as fast as new matrix is deposited (Nwomeh et al., 1998, 1999). Neutrophils also release elastase, an enzyme known to destroy the PDGFs and TGF-βs, which are growth factors known to be important for normal wound healing (Yager et al., 1996). The environment of chronic ulcers is also known to contain an abundance of reactive oxygen species that also damage healing tissue (Wenk et al., 2001). Chronic ulcers generally will not heal on their own until the inflammatory response is reduced.

FETAL SKIN
Development of fetal skin

The skin's superficial layer, the epidermis, is derived from surface ectoderm, while the dermis is of mesenchymal origin. The epidermis starts as a single layer of ectodermal cells covering the embryo, which begins to emerge at gestational day 20 in humans (Lane, 1986; Moore and Persaud, 1993). In the second month, a cell division takes place, at which time the periderm (epitrichium) emerges as a thin superficial layer of squamous epithelium overlying the basal germinative layer. Over the next 4 to 8 weeks, the epidermis becomes highly cellular. New cells are produced in the basal germinative layer and are continuously keratinized and shed, which replaces cells of the periderm. These cells are part of the vernix caseosa, a greasy, white film that covers fetal skin. In addition to desquamated cells, the vernix caseosa contains sebum from sebaceous glands (Lane, 1986; Moore and Persaud, 1993). This substance serves as a protective barrier during gestation and facilitates passage through the birth canal at delivery, due to its slippery nature.

Replacement of the periderm continues until the 21st week, at which point the periderm has been replaced by the stratum corneum (Lane, 1986; Moore and Persaud, 1993). Through a series of stages of differentiation, the epidermis stratifies into four layers by the end of the fourth month: the stratus germinativum (derived from the basal layer), the thick spinous layer, the granular layer, and the most superficial stratum corneum. By the time these four layers emerge, interfollicular keratinization has begun, and the epidermis has developed buds that become epidermal appendages. Melanocytes of neural crest origin have invaded the epidermis, synthesizing melanin pigment that can be transmitted to other cells through dendritic processes. By the 21st week, the fetal epidermis has many of the components that will maintain into adulthood (Lane, 1986; Moore and Persaud, 1993). Also, the dermis begins to mature from a thin and cellular to a thick and more fibrous structure. After

24 weeks of gestation and through the neonatal period, the fetal skin matures and thickens to become histologically distinct from its embryonic beginnings (Lane, 1986; Moore and Persaud, 1993).

The fetal scarless repair phenotype

Adult wounds heal with fibrous tissue (scarring), whereas early gestation fetal wounds heal scarlessly. Fetal wounds heal with restoration of normal skin architecture and preservation of tissue strength and function. This observation has been confirmed in multiple animal species, including mice, rats, rabbits, pigs, sheep, and monkeys. The mechanisms responsible for fetal scarless wound healing are intrinsic to fetal tissue and are independent of environmental or systemic factors such as bathing in sterile amniotic fluid, perfusion with fetal serum, or the fetal immune system (Ferguson and Howarth, 1992; Ihara and Motobayashi, 1992; Martin and Lewis, 1992; Longaker et al., 1994). To support this point, studies have shown that human fetal skin transplanted subcutaneously in the dorsolateral flank of athymic mice heals without a scar, further suggesting that the scarless wound phenotype is dependent on characteristics intrinsic to fetal tissue (Lorenz et al., 1992; Adzick and Lorenz, 1994).

The scarless fetal wound repair outcome depends on two factors: the gestational age of the fetus and the size of the wound. Excisional wound healing studies performed on fetal lambs showed that, at a given gestational age, larger wounds healed with an increased incidence of scar formation. Likewise, the frequency of scarring increased with increasing gestational age (Cass et al., 1997). Since the publication of these studies, transitional periods have been found for humans (24 weeks' gestation) (Lorenz et al., 1992), rats (between gestation days 16.5 and 18.5) (Ihara et al., 1990), and mice (Colwell et al., 2006a).

Extensive research has been dedicated to determining what is responsible for the shift to the adult wound healing phenotype. Eventually, instead of depositing bundles of ECM in a normal basket-weave pattern, organisms begin to heal breaches in the skin with collagen scarring composed of large parallel fibers of mainly collagen types I and III.

As fetuses develop and enter into the early period of scar formation, the wound phenotype has been described as a "transition wound." At this point, the repair outcome is tissue that retains the reticular organization of collagen characteristic of normal skin but is devoid of epidermal appendages (Lorenz et al., 1993). The skin does not truly regenerate, but the dermis does not form a scar. This is an intermediate outcome before true scar formation. The transition occurs during the later stages of fetal development.

The fetal ECM was once thought to be inert. However, recent evidence suggests that the ECM is a dynamic structure that plays a pivotal role in cellular signaling and proliferation. The fetal ECM is now known to be a reservoir of growth factors essential to development (Buchanan et al., 2009). The fetal ECM also has a different structural protein composition. For example, the collagen content of the ECM changes as the fetus ages, starting with a relatively high type III to type I collagen ratio and shifting to the adult phenotype in the post-natal period (which tends to have less type III collagen). Another structural difference between fetal and adult ECM is the hyaluronic acid content. Hyaluronic acid, the negatively charged, extremely hydrophilic, non-sulfated glycosaminoglycan of the ECM, has been shown to be in higher concentration in the ECM during rapid growth processes, such as cellular migration and angiogenesis. *In vitro* studies show that hyaluronic acid can cause fibroblasts to increase synthesis of collagen and non-collagen ECM proteins (Mast et al., 1993). During adult repair, hyaluronic acid initially increases dramatically, then decreases from days 5 to 10, after which time the concentration remains at a low level. Interestingly, this hyaluronic acid profile is not the case in the fetal wound ECM, where the hyaluronic acid level remains high. As demonstrated with type III collagen, the ECM hyaluronic acid content decreases from the fetal to the post-natal period (Adzick and Longaker, 1991).

The concentration of other substances such as decorin, fibromodulin, lysyl oxidase, and matrix metalloproteases (MMPs) further set the fetal ECM apart from the adult ECM (Buchanan et al., 2009). These substances are proteoglycan ECM modulators that play a role in the development and maturation of collagen. Lysyl oxidase cross-links collagens, and MMPs degrade collagen. Decorin content and the expression of enzymes such as lysyl oxidase and matrix metalloproteases increase as fetal tissue matures. Fibromodulin modulates collagen fibrillogenesis and has been shown to bind and inactivate the transforming growth factor betas (TGF-βs). The TGF-βs have been implicated in adult wound healing and scar formation. Fibromodulin decreases with gestational age, paralleling the shift from scarless fetal wound healing to scarring adult repair (Soo et al., 2000).

REGENERATIVE HEALING AND SCAR REDUCTION THEORY
Targeting the inflammatory response

Initial research into the mechanisms responsible for scar formation led investigators to focus on the inflammatory phase of wound healing as a target for reducing the incidence and magnitude of scar formation. This choice of direction was based on the observation that regenerative wound healing is replaced by scarring as the immune system in the embryo develops (Martin, 1997). Interestingly, many studies have shown that reduction of inflammation in post-natal skin wounds correlates with reduced scarring (Gawronska-Kozak et al., 2006).

Ashcroft et al. reported one example of reduced inflammation and scarring. Enhanced healing occurred in mice devoid of Smad3, a protein known to transduce TGF-β signals. These mice exhibited more rapid re-epithelialization and decreased inflammation (blunted monocyte activation) (Ashcroft et al., 1999). Martin et al. performed wound healing experiments in PU.1 null mice devoid of functional neutrophils and macrophages. Results showed that these mice healed wounds over a similar time course to their wild-type counterparts but exhibited scar-free healing similar to embryonic wound healing (Martin et al., 2003). These two studies support the contention that the inflammatory response may be deleterious to normal wound repair by contributing to increased fibrosis. Experiments performed on athymic mice (Gawronska-Kozak et al., 2006) and experiments involving antisense downregulation of connexin43, a protein involved in gap junctions and inflammation, support these findings (Qiu et al., 2003; Gawronska-Kozak et al., 2006). Furthermore, other studies have provided evidence that wound inflammatory cells from the circulation produce signals that either directly or indirectly induce collagen deposition and granulation tissue formation, which increase scarring (Martin and Leibovich, 2005).

Although this research points to the inflammatory phase of wound healing as one cause of scar tissue formation, recent studies have provided evidence that the inflammatory phase and scarring might not be as directly linked as previously believed. Cox-2, an enzyme involved in prostaglandin production, is a mediator of inflammation. Two studies show conflicting evidence regarding the effect of Cox-2 inhibition, one study reporting decreased scar formation (Wilgus et al., 2003) and the other claiming no difference in wound healing or scar formation (Blomme et al., 2003). Likewise, a recent study transiently induced neutropenia in mice, which accelerated wound closure but failed to show a difference between collagen content in neutrophil-depleted wounds compared to wild-type controls (Dovi et al., 2003).

In addition to inflammatory cells, other blood-borne cells have been identified as having a role in granulation tissue deposition and scar formation, suggesting that neutrophils and monocytes might not be the only mediators implicated. Fibrocytes are a subpopulation of circulating leukocytes that are thought to be fibroblast-like, expressing both leukocyte markers (e.g. CD34) and ECM proteins (e.g. collagen). These cells increase the intensity of the inflammatory response and, through secretion of cytokines such as PDGF and TGF-β, guide the action of fibroblasts at the wound site. As fibrocytes have increased expression of collagen and decreased expression of CD34 over time, these cells are postulated to mature into fibroblasts at the wound site. Therefore, not only do inflammatory cells influence fibroblasts in a paracrine fashion, they also may differentiate into fibroblasts that are capable of influencing fibrin deposition and collagen scar formation (Quan et al., 2004; Stramer et al., 2007).

Although other possible mediators of scar formation exist, the inflammatory response remains a major target for ongoing research aimed at preventing or reducing the appearance of scar. As Stramer et al. illustrate, many points exist at which interventions could dampen the inflammatory response. The first target could be leukocytes, at any point as they migrate (1) through the vessel wall from the bloodstream, (2) from outside the vessel to the wound, or (3) as they transmit a signal to fibroblasts, inducing the fibrotic response. A second target could be the fibroblasts, and interventions could be designed to block the action of these cells as they respond to leukocyte signaling (Stramer et al., 2007) (Fig. 6.3).

Cytokines and growth factors

TGF-β SUPERFAMILY

By far, most scar-reducing progress has been made in targeting the TGF-β pathways in order to make adult wound healing similar to embryonic healing. The TGF-β superfamily includes TGF-β1, TGF-β2, and TGF-β3, all of which have been shown to influence adult wound healing (Frank et al., 1996). These cytokines are secreted by keratinocytes, fibroblasts, platelets, and macrophages. The TGF-βs influence — through activation and inhibition — the migration of cells such as keratinocytes and fibroblasts to the wound bed. The TGF-β superfamily has also been implicated in matrix remodeling and collagen synthesis (Clark, 1996; Werner and Grose, 2003). TGF-β1 activates myofibroblast differentiation, implicating this pathway in the process of wound contraction and the synthesis of collagen and fibronectin in granulation tissue (Desmouliere et al., 2005).

111

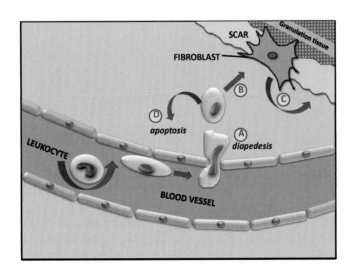

FIGURE 6.3
Inflammatory cell recruitment to the site of tissue damage. Therapeutic intervention aimed at dampening the immune response could target any of the steps along the pathway of inflammatory cell recruitment. (A) Leukocytes in blood vessels adjacent to the site of tissue damage emigrate through the vessel wall by diapedesis and (B) migrate to the site of tissue damage in response to chemotactic signals. Inflammatory cells activate resident fibroblasts and attract other bone marrow-derived cells to the wound, where the repair outcome is (C) scar formation. After acting at the wound site, the activated repair cells either disperse, differentiate, or (D) apoptose, thus ending the repair response.

Investigators have compared the TGF-β isoform profiles of fetal and adult skin, showing that injured fetal epidermis contains a greater amount of TGF-β3, derived from keratinocytes and fibroblasts, and less TGF-β1 and TGF-β2, derived from degranulating platelets, monocytes, and fibroblasts, compared to healing adult skin (Whitby and Ferguson, 1991; Hsu et al., 2001). Since this cytokine profile was discovered, isoforms TGF-β1 and TGF-β2 have generally been thought to be fibrotic, while TGF-β3 is thought to support scarless healing. Discovery of the relative ratios of these isoforms prompted experiments aimed at mimicking the embryonic profile, using antibody neutralization of TGF-β1 and TGF-β2 and treating with exogenous TGF-β3. Shah et al. demonstrated, through a series of experiments on cutaneous rat wounds, that these interventions reduce scar formation (Shah et al., 1995). Although knocking down only TGF-β1 or TGF-β2 had little or no effect on wound healing, subsequent experiments show that antisense RNA knockdown of TGF-β1 reduces scar formation (Choi et al., 1996). Likely, the length of time that TGF-β1 is neutralized over the repair period influences scarring, with longer neutralization needed for greater scar reduction.

CONNECTIVE TISSUE GROWTH FACTOR (CTGF)

CTGF is considered to be profibrotic by a mechanism related to TGF-β. CTGF is a TGF-β target gene that is activated by Smad proteins after TGF-β binds to its receptors. Like TGF-β, CTGF stimulates the deposition of ECM components, including collagen. However, unlike TGF-β, CTGF does not exert any effect on epidermal or inflammatory cells. Thus, CTGF appears to specifically influence ECM deposition at the wound site.

Adult fibroblasts have higher expression of CTGF. Studies show that fetal fibroblasts stimulated by TGF-β show increased expression of CTGF, suggesting scarless fetal repair may be partially a result of lower CTGF expression (Colwell et al., 2006b).

VASCULAR ENDOTHELIAL GROWTH FACTOR (VEGF)

There are four isoforms of VEGF, VEGF A through D. Keratinocytes, fibroblasts, and macrophages produce VEGF, which is thought to be one of the main regulators of angiogenesis and vasculogenesis. VEGF acts through two receptors in endothelial cells, VEGF-R1 and VEGF-R2.

VEGF increases during adult wound healing and has been associated with angiogenesis (Buchanan et al., 2009). However, through studies on fetal rats, Colwell et al. discovered that scarless healing shows an increase in VEGF expression three times higher than what is observed in late-gestation fetal wounds (Colwell et al., 2005). This work suggests that increased VEGF expression is partially responsible for the accelerated wound healing that occurs early in gestation.

FIBROBLAST GROWTH FACTORS (FGFs)

Embryonic wounds contain lower levels of FGFs, growth factors involved in skin morphogenesis (Whitby and Ferguson, 1991). The expression of FGFs, including keratinocyte growth factors 1 and 2, increases as the fetus ages, suggesting that these growth factors are profibrotic (Dang et al., 2003). Many isoforms have been studied, including FGF 5, which doubles in expression at birth, FGF 7, which multiplies more than seven-fold at birth, and FGF 10, which doubles at the transitional period (Buchanan et al., 2009). In general, a downregulation of the FGF isoforms occurs during scarless wound healing, whereas the opposite is true during adult wound healing, suggesting that FGF upregulation is likely partially responsible for scar formation (Buchanan et al., 2009).

PLATELET-DERIVED GROWTH FACTOR (PDGF)

Like FGF, PDGF has been identified as a profibrotic growth factor. Adult wounds contain very high amounts of PDGF, whereas this growth factor is virtually absent in embryonic wounds. One reason may be that platelet degranulation is decreased in embryonic wounds (Whitby and

Ferguson, 1991). Experiments involving the administration of PDGF to fetal wounds show that this growth factor induces scarring through increased inflammation, fibroblast recruitment, and collagen deposition (Haynes et al., 1994).

WNT SIGNALING

Expression of most Wnts increases during skin development and is lost with post-natal development. These glycoproteins are cytokines involved in cell-cell signaling, proliferation, differentiation, and carcinogenesis. With wounding, fetal Wnt expression remains stable at its high basal level, whereas, in adult skin, Wnt signaling increases during repair. These data demonstrate that Wnt is involved in the healing process, but which isoform(s) are specific to scarring remains unknown (Colwell et al., 2006c; Buchanan et al., 2009; Carre et al., 2010).

INTERLEUKINS

The interleukins are a class of cytokines involved in activation of the inflammatory cascade. IL-8 stimulates neovascularization and attracts neutrophils. IL-6 is produced by adult fibroblasts in response to stimulation by PDGF and activates macrophages and stimulates monocyte chemotaxis. With an insult to skin integrity, IL-6 and IL-8 rapidly increase expression (Liechty et al., 2000a, 1998). This elevated expression is maintained over a period of 72 hours during adult repair but is suppressed after 12 hours during scarless fetal repair (Liechty et al., 2000a, 1998). Early fetal fibroblasts express lower levels of both IL-6 and IL-8 than their adult counterparts at baseline and in response to PDGF stimulation. Therefore, these pro-inflammatory cytokines are thought to promote scar formation. Studies show that the administration of IL-6 to fetal wounds induces scarring (Liechty et al., 2000a), which further supports this theory.

IL-10 is thought to be anti-inflammatory based on its antagonism of IL-6 and IL-8. Liechty et al. harvested fetal skin grafts from 15-day gestation IL-10 knockout mice and grafted them to syngeneic adult mice. Incisional wounds on these skin grafts showed scar formation, whereas similar wounds on 15-day gestation wild-type skin grafts on adult wild-type mice healed scarlessly. These results suggest that IL-10 is essential for scarless fetal healing due to its ability to dampen the inflammatory response (Liechty et al., 2000b). In a supporting study, administration of an IL-10 overexpression adenoviral vector reduced inflammation and induced scarless healing in adult mouse wounds (Gordon et al., 2008).

CURRENT THERAPEUTIC INTERVENTIONS

No current commercially available therapy exists that can induce post-natal skin wound regenerative healing. Although many therapeutic interventions are used to reduce scar formation, research has not adequately demonstrated efficacy or safety for many of these treatments secondary to small treatment groups and a lack of well-designed studies. However, the following treatments are used clinically to reduce scarring symptoms and scar formation.

Topical and intralesional corticosteroid injections

Corticosteroids are used commonly to treat symptomatic scars, and triamcinolone is the most common agent used. The mechanism of action is multifactorial. The inflammatory response is globally decreased, which secondarily decreases collagen synthesis and increases collagen degradation. Corticosteroids also inhibit fibroblast proliferation and TGF-β1 and TGF-β2 expression by keratinocytes (Perez et al., 2001; Manuskiatti and Fitzpatrick, 2002; Wu et al., 2006; Stojadinovic et al., 2007). Although 50 to 100% efficacy in symptom improvement has been reported, studies are limited by lack of appropriate controls and poor design (Darzi et al., 1992; Tang, 1992; Manuskiatti and Fitzpatrick, 2002; Reish and Eriksson, 2008).

The use of corticosteroids is limited by reported adverse consequences in 63% of patients. These effects include delayed wound healing, hypopigmentation, dermal atrophy, and scar widening (Maguire, 1965; Manuskiatti and Fitzpatrick, 2002). Based on successful studies combining corticosteroid injections with 5-fluorouracil therapy and laser therapy, polytherapy is the best method to utilize steroids, as lower dosing and fewer adverse effects occur (Manuskiatti and Fitzpatrick, 2002; Alster, 2003; Asilian et al., 2006).

5-Fluorouracil (5-FU)

5-FU has shown the most efficacy in combination with corticosteroids alone or with corticosteroids and laser therapy. 5-FU alone, however, has shown limited efficacy (Fitzpatrick, 1999; Manuskiatti and Fitzpatrick, 2002). The mechanism of action occurs primarily through inhibition of fibroblast proliferation and TGF-β1-induced collagen synthesis (Blumenkranz et al., 1982; Mallick et al., 1985; Wendling et al., 2003). 5-FU may be an efficacious therapy in combination with corticosteroids after all conventional therapies have failed, but this therapy should undergo further controlled studies.

Imiquimod

Imiquimod 5% cream is a topical agent that enhances local production of immune-stimulating cytokines, such as tumor necrosis factor, interleukins, and interferons (Miller et al., 1999). This agent has been used to prevent recurrence of keloids following surgical excision, though clinical trials show mixed results (Cacao et al., 2009). Typically, imiquimod is applied immediately following surgery, followed by daily application for 8 weeks. However, approximately 50% of patients experience hyperpigmentation, and many patients also experience skin irritation at the application site (Berman, 2002; Berman and Kaufman, 2002).

Laser therapy

Pulsed dye laser therapy has been shown to reduce scar erythema, though lack of well-designed controls is a limitation of these studies (Alster, 1994; Alster and Williams, 1995; Reiken et al., 1997; Manuskiatti and Fitzpatrick, 2002). The idea behind targeting fibroproliferative scars with laser treatment comes from the principle that vascularity is partially responsible for the erythematous appearance of scars. Pulsed dye laser therapy produces photothermolysis of the microvasculature, resulting in thrombosis and ischemia; as a result, collagen content decreases (Reiken et al., 1997).

Laser therapy has relatively few adverse effects (hyperpigmentation in 1–24% of patients and transient purpura in some). However, more research to support its efficacy is needed.

Bleomycin

Bleomycin, an antibiotic known to produce antibacterial, antiviral, and antitumor activity, has been demonstrated to improve hypertrophic scars and keloids with intralesional injection (Espana et al., 2001; Saray and Gulec, 2005; Naeini et al., 2006). However, similarly to the therapies discussed above, the studies are limited due to lack of well-designed controls.

Bleomycin is hypothesized to act either through inhibition of lysyl-oxidase or inhibition of TGF-β1, resulting in decreased collagen synthesis (Lee et al., 1991; Hendricks et al., 1993). Adverse effects of this treatment are hyperpigmentation in 75% of patients and dermal atrophy in the skin surrounding the injection site in 10–30% of patients (Bodokh and Brun, 1996).

Silicone gel sheets

Silicone gel sheets are hypothesized to act by hydrating the wound, inhibiting collagen deposition, and downregulating TGF-β2. This therapy has been studied for both treatment and prophylaxis of excessive scarring. Initial studies show conflicting results in terms of efficacy (Quinn, 1987; Ahn et al., 1991; Carney et al., 1994), requiring further study. However, silicone gel sheets will likely continue to be used as a non-invasive treatment with few adverse effects.

Liquid silicone gels are also available, which are applied topically and have no significant adverse effects. Similarly to gel sheets, they lack proven efficacy but likely reduce scar erythema.

Pressure dressings

Despite being in clinical use since the 1970s (Tolhurst, 1977), pressure dressings have not been validated by experimental trials to be efficacious either in prophylaxis or in the treatment of scars (Reish and Eriksson, 2008). These treatments may be efficacious in reducing the appearance of scar if used in polytherapy, but further investigation is warranted. Pressure earrings have been used at sites of earlobe keloid excisions but have not been shown to eliminate recurrence.

Radiation therapy

Radiation therapy is often used as an adjunct to surgical excision in the treatment of keloids and is thought to decrease collagen production by reduction of fibroblast proliferation and neovascular bud formation. Radiation therapy is most effective for recurrent keloids if a single dose is given within 24 h of surgical excision. Radiation treatment decreases recurrence rates after surgical excision from between 45 and 100% to between 16 and 27% (Kovalic and Perez, 1989; Ship et al., 1993; Berman and Bieley, 1996; Ragoowansi et al., 2003).

One limitation of radiation therapy has been in determining a standard dosage, fractionation, time period, and frequency of dosing following surgical procedures. Reish et al. report good results in treatment of recurrent keloids following surgery with 300 to 400 Gy in three to four fractions or 600 Gy in three fractions (Reish and Eriksson, 2008).

Cryotherapy

Cryotherapy has been studied in conjunction with surgical excision to treat keloids and hypertrophic scars. Many of these studies are limited by small sample size and poor controls, but the largest study reported 79.5% response rate with 80% reduction in scar volume (Zouboulis et al., 1993; Har-Shai et al., 2003). Cryotherapy is thought to decrease collagen synthesis and mechanically destroy scar tissue. Side-effects include hypopigmentation and depressed atrophic scar formation (Rusciani et al., 2006). This therapy is an adjunct to surgery, though its long-term efficacy has not been established.

Surgery

Remodeling is a process that can last for one to two years. During this time, scars can lose their dark pigmentation, flatten, and soften, and contractures can lessen. Because scars can often behave in an unpredictable way, surgery is usually reserved until after this period has passed. There are many options for surgical treatment for scarring, including excision with direct closure, local skin flap coverage, or more extensive vascular flap coverage. The aforementioned medical treatments are generally considered prior to, or as an adjunct to, surgical treatment.

FUTURE THERAPEUTIC INTERVENTIONS
TGF-β associated therapies

The first pharmaceutical scar-reducing products are currently being developed. Avotermin (Juvista®) is a recombinant TGF-β3 polypeptide proposed to improve scar appearance with intralesional injection. Phase I and II trials have recently been completed in the UK. According to the company (Renovo), 70% of the wounds treated with avotermin exhibited improvement in scar appearance with statistical significance, as evaluated by both surgeons and laypersons. The drug was additionally found to be safe in the tested population, a group of over 1,500 patients.

Juvidex® is another Renovo product undergoing clinical trials. Juvidex® is a topical formulation of mannose-6-phosphate, an estradiol derivative that inhibits TGF-β1 and TGF-β2 (see www.renovo.com) (Ferguson and O'Kane, 2004). The idea for this formulation stems from research showing that mannose-6-phosphate antagonizes the activation of TGF-β during wound repair, thus decreasing scar (Stevenson et al., 2008).

Targeting gap junctions and connexins

Propagation of cellular signals can occur through many different mechanisms, one of which is the binding of a growth factor or cytokine ligand to a cell surface receptor. Another mechanism is the propagation of a signal from one cell to an adjacent cell through a gap junction. These connections can allow a signal to spread over long distances, as occurs in the heart (Desplantez et al., 2007). The connexin multigene family encodes proteins that aggregate to form inter-cellular channels (Wei et al., 2004). These connections are also important for spreading signals during cutaneous wound healing. Gap junctions are hypothesized to function during wound repair, by transferring injury signals from cell to cell, coordinating the inflammatory response, mediating wound closure, and regulating scar tissue formation in response to injury (Coutinho et al., 2003; Qiu et al., 2003; Zahler et al., 2003; Ehrlich et al., 2006; Gourdie et al., 2006; Mori et al. 2006).

Many connexins are present in the skin, but the most extensively studied connexin is Cx43, which is expressed in both the epidermis and dermis (Qiu et al., 2003). Cx43 has a decreased expression at the wound edge in the first 1 or 2 days post-injury (Goliger and Paul, 1995; Coutinho et al., 2003). During wound repair, increased phosphorylation of Cx43 by protein kinase C occurs at serine368, which may cause decreased gap junctional communication through decrease in unitary channel conductance. This inhibition then initiates the injury-related response by the involved cell (Richards et al., 2004).

By applying Cx43 antisense oligonucleotides to mouse skin wounds, Coutinho et al. were able to demonstrate decreased inflammatory cell infiltration, decreased fibrotic tissue deposition, and accelerated wound healing. These findings were hypothesized to be due to further decreased connexin expression in the epidermis adjacent to the wound (Coutinho et al., 2005). Other studies have shown that transient inhibition of Cx43 decreases scarring after burn injury in wild-type mice and increases re-epithelialization after burn injury in human diabetics (Coutinho et al., 2005; Wang et al., 2007). To further support these data, Cx43 knockouts have accelerated wound closure (Kretz et al., 2004) and decreased collagen type I synthesis in the presence of chemicals that uncouple communication between cells. Interestingly, these treatments did not affect the levels of collagen type I mRNA (Ehrlich et al., 2006). Based on these data, the application of lithium chloride, a substance known to enhance signal propagation through gap junctions, produced the opposite effect: enhancing the deposition of granulation tissue, increasing open wound closure time, and increasing scar (Moyer et al., 2002).

Given the strong correlation between connexin inhibition and improved wound healing, other therapies aimed at blocking signal transduction from cell to cell are currently under investigation. For example, a group at the Medical University of South Carolina synthesized a membrane permanent peptide containing a sequence designed to inhibit interaction of the ZO-1 protein with Cx43. This peptide, now known as ACT1 peptide, decreases the rate of channel organization in gap junctions (Rhett et al., 2008). Through further investigation, researchers have found that this peptide interacts with more than one portion of Cx43, and enhances cutaneous wound healing through decreased inflammation and scarring (Gourdie et al., 2006). The advantage of this novel protein is that Cx43 expression is not altered. Moreover, the expression of other genes is not directly altered, unlike with antisense therapy and gene knockdown modalities.

As with the TGF-β superfamily, several commercial companies are currently attempting to develop connexin-related scar reduction therapies. These therapies include Cx43 antisense-based gene therapy and ACT peptide bioengineering (Rhett et al., 2008), which are in the early stages of testing and will not be available for some time.

Other drugs and biologics

Many possible pathway interventions have been proposed to prevent or reduce scar formation. Some strategies include therapies to increase expression of intrinsic anti-scarring molecules at the wound site. These include fibromodulin, hyaluronic acid, and hepatocyte growth factor (Iocono et al., 1998; Ha et al., 2003; Stoff et al., 2007). Other approaches include treatment with inhibitors of MMP (e.g. GM6001) (Witte et al., 1998), inhibitors of pro-collagen C-proteinase (Fish et al., 2007), and inhibitors of dipeptidyl peptidase IV enzymes (Thielitz et al., 2008), as well as treatment with angiotensin peptides (Rodgers et al., 2003). Adenovirus-p21 overexpression has also been linked to scar reduction (Gu et al., 2005) (Fig. 6.4).

Stem cells

True skin regeneration at sites of injury has not been accomplished by single molecule-specific therapy. Regenerative repair may require cell-based therapy in which multiple cascades of signaling pathways are affected. Stem cell therapy, with the ability to differentiate cells into various cell types, is a promising approach to inducing regenerative repair (Fig. 6.5).

EMBRYONIC STEM CELLS (ES CELLS)

Embryonic stem cells were originally isolated from blastocyst embryos by Thomson et al. in 1998. Embryonic stem cell transplantation into an injured area was hypothesized to regenerate tissue locally by producing differentiated progeny. However, recent evidence presented by Fraidenraich et al. suggests that these cells are more likely "catalysts" that secrete various factors that can then act either locally or systemically (Fraidenraich et al., 2004). The ability of ES cells to regenerate tissue is hypothesized to be due to a necessity *in utero* to correct aberrant development. Because early mistakes have a large effect on development at later stages, it follows that embryonic cells would possess a capacity for regeneration that is more robust than cells found in mature tissue (Heng et al., 2005).

Assuming embryonic stem cells act as a catalyst for regeneration, controversy as to whether transplantation of these cells would be the best way to improve wound healing exists. Chien et al. argue that determination of the cocktail of compounds that ES cells stimulate would be more efficacious (and wrought with less controversy). With that knowledge, recombinant technology could be used to produce these molecules and mimic the regenerative effect of ES cells (Chien et al., 2004). However, many of these molecules are thought to be labile with a short half life *in vivo*. Additionally, the cost of this research would be extraordinary. Therefore, the research focus remains on addressing the obstacles involved in transplanting ES cells.

One obstacle involved in the transplantation of ES cells is the human immune system. In 2008, Wu et al. demonstrated that transplantation of human ESCs to immunocompetent hosts elicits robust humoral and cellular immune responses (Swijnenburg et al., 2008). One way of dealing with the issue of rejection could be transient therapy with immunosuppressive agents with gradual withdrawal (Heng et al., 2005). Other suggestions include encapsulation of ES cells with a biodegradable polymer membrane prior to transplantation (Orive et al., 2003). In both cases, the transplanted cells would ultimately be killed by the host's immune system, but only after regeneration is well under way.

A second obstacle to ES cell transplantation is the potential for teratoma development. Numerous studies have shown that undifferentiated ESCs, when placed in the subcutaneous space of nude mice, form teratomas. In fact, the formation of a teratoma is what defines these cells as pluripotent. In theory, a degree of pre-differentiation prior to transplantation would allow ES cells to better promote tissue regeneration without the risk of teratoma formation. However, the degree of pre-differentiation has not yet been determined and remains an obstacle for both embryonic stem cells and induced pluripotent stem cells (Heng et al., 2005).

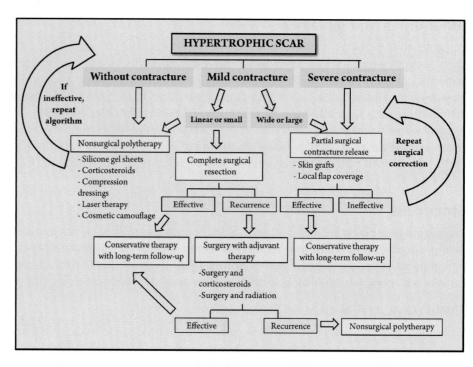

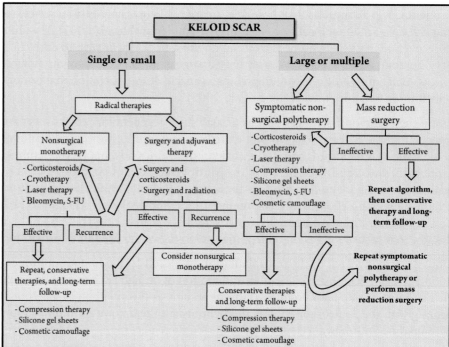

FIGURE 6.4

Scar reduction strategies: algorithms for hypertrophic scars and keloids. (Modified from Ogawa, 2010).

MESENCHYMAL STEM CELLS

Mesenchymal stem cells are non-hematopoeitic bone marrow stromal cells that were initially isolated based on their ability to adhere to plastic culture plates. These cells are unique in that they are capable of differentiating into mesenchymal lineages such as cartilage, fat, muscle, and bone (Chamberlain et al., 2007). MSCs are a heterogeneous group of cells that have had populations isolated not only from the bone marrow but also from adipose tissue and amniotic fluid. Based on their ability to expand *in vivo* and differentiate into multiple tissue types, these cells are thought to be an ideal source of stem cells used for promoting wound

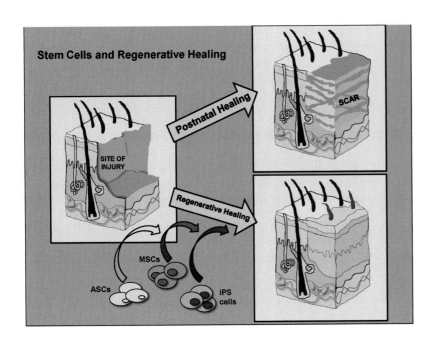

FIGURE 6.5
Stem cells and skin regeneration. The application of stem cells (e.g. mesenchymal stem cells (MSCs), adipose-derived progenitor cells (ASCs), iPS cells) holds great promise as a strategy for inducing regenerative healing in post-natal wounds, which would otherwise heal with scar formation.

119

healing and/or scar-reducing therapies (Zuk et al., 2002; Lee et al., 2004). MSCs could serve as a source of autologous stem cells to be harvested from an adult and transplanted back to the same patient, thereby avoiding rejection and the ethical and moral concerns associated with embryonic stem cell therapies.

Mesenchymal stem cells could affect wound healing and tissue regeneration through many different avenues. These cells are capable of migrating to the site of injury or inflammation, and they may stimulate the proliferation and differentiation of resident progenitor cells, secrete growth factors, participate in remodeling, and modulate the immune and inflammatory responses (Caplan, 2007; Chamberlain et al., 2007; Uccelli et al., 2007).

A wealth of clinical data attests to the safety of bone marrow-derived mesenchymal stem cells, and emerging data support adipose-derived mesenchymal cells as possessing a similar safety profile to bone marrow-derived MSCs (Garcia-Olmo et al., 2005; Fang et al., 2007; Hanson et al., 2010). MSCs could, therefore, be used to affect various pathways involved in wound healing including — but not limited to — inflammation, aging, and cellular senescence. Research using MSCs in wound healing has been encouraging, though limited to mostly small, non-randomized clinical trials (Hanson et al., 2010).

Two examples of human wound healing investigations using MSCs were performed by Falanga et al. and Yoshikawa and colleagues. The first was a small trial using a fibrin glue vehicle in both acute and chronic wounds. Falanga et al. demonstrated that topical application of autologous passage 2 to 10 bone marrow-derived MSCs, combined with fibrin spray, allowed acute surgical wounds and chronic lower extremity ulcers to heal faster. The wound healing speed increased in a manner directly proportional to the number of cells applied (Falanga et al., 2007).

Yoshikawa et al. performed a larger study on patients with various non-healing wounds. This group applied bone marrow-derived MSCs with a dermal replacement to wounds, with or without autologous skin grafts. Results showed accelerated healing in wounds treated with MSCs (Yoshikawa et al., 2008). One limitation of this study is that the cells used were at passage 0, and flow cytometry was not used to characterize the cell types. MSCs are known to represent only 0.001% of nucleated cells in the bone marrow; therefore, the cell population used in these experiments likely contained other cells, such as tissue macrophages, that would also assist in wound healing (Chamberlain et al., 2007).

Further research is needed to characterize MSCs and their niches. As purification techniques improve, the role of MSCs in wound healing will gain clarity. Defining the direct role of MSCs in wound repair, as well as their effects on other cells, will guide their future therapeutic potential.

EPIDERMAL STEM CELLS

As mentioned previously, the epidermis in humans is a dynamic structure undergoing constant renewal. Epidermal turnover is estimated to take place over a 60-day time period in humans, a process that requires a continuous supply of differentiated cells. Epidermal stem cells are thought to have a high capacity for self-renewal, as evidenced by their ability to produce daughter cells that undergo terminal differentiation into keratinocytes (Watt, 1998).

A number of stem cell niches are present in the epidermis. The best-characterized are the interfollicular epidermal stem cells and the hair follicle bulge region, which can resupply each other when damaged. These cells are important sources for re-epithelialization during repair.

Wound closure is not complete until the epidermis is restored. Through clinical observation in burn treatment, scar formation can be reduced when early wound excision and skin grafting is done. This clinical observation suggests that cells intrinsic to the epidermis have regenerative potential. Zhang et al. postulate that epidermal stem cells may be responsible for signals suppressing fibroblast activity after burn injury (Zhang et al., 2009). This hypothesis is supported by previous studies showing that scar tissue contains fewer epidermal stem cells (Zhao et al., 2003).

At this point, the therapeutic potential for epidermal stem cells is largely theoretical, but research will continue to develop at a rapid pace as clinical opportunities remain abundant (e.g. skin grafting for burn victims).

INDUCED PLURIPOTENT STEM CELLS (iPS CELLS)

In 2006, Takahashi and Yamanaka published a landmark paper describing the process of reverting differentiated tissue cells back to a pluripotent state by transduction with specific transcription factors (Oct4, Sox2, Klf4, and c-Myc). Takahashi and Yamanaka were able to reprogram adult murine fibroblasts into ES-like iPS cells, a system that they later used to induce human cells (Takahashi and Yamanaka, 2006; Takahashi et al., 2007). Both murine and human iPS cells resemble and behave like ES cells (Takahashi and Yamanaka, 2006; Maherali et al., 2007; Wernig et al., 2007).

The introduction of iPS cells has been an exciting advance in stem cell technology. The use of iPS cells for tissue regeneration would allow for the use of autologous cells to create patient-specific cell lines for regenerative therapy. iPS technology has the advantage of not being associated with the same ethical or immune rejection concerns as the use of ES cells. However, due to the use of viral vectors (retroviral and lentiviral), the development of iPS cells presents the risk of insertional mutagenesis, leading to uncontrolled genome modification (Pera and Hasegawa, 2008). In response to these concerns, researchers have focused on the development of other reprogramming processes, such as adenoviral, plasmid-based, and recombinant protein-based methods (Okita et al., 2008; Stadtfeld et al., 2008; Zhou et al., 2009). However, all reprogramming factors are known to be oncogenic when overexpressed. Therefore, rigorous investigation into the safety of potential iPS therapies is necessary before their introduction to clinical practice.

PERSPECTIVE

The process of wound repair is highly regulated and complex. Age and systemic influences, such as malnutrition, infection, and chronic disease, may lead to delayed repair, while dysregulation of the mechanisms of wound healing can lead to excessive fibroproliferative scarring. Even when healing proceeds in the usual fashion, the result is the deposition of non-functioning fibrotic tissue in most organs.

Although several decades of research have been dedicated to defining the mechanisms responsible for wound healing, advances have not produced a universally effective or safe method for either preventing or reducing scar formation. Focus on the inflammatory cascade has identified molecules, cytokines, and growth factors that can reduce scarring. Though still in the early stages of discovery, stem cell research offers promising opportunities for improving wound healing and advancing the field of regenerative medicine. Further research in these fields, as well as in the fields of tissue engineering and biomaterials, will provide translational approaches to stem cell research and wound healing.

References

Adzick, N. S., & Longaker, M. T. (1991). Animal models for the study of fetal tissue repair. *J. Surg. Res., 51*, 216–222.

Adzick, N. S., & Lorenz, H. P. (1994). Cells, matrix, growth factors, and the surgeon. The biology of scarless fetal wound repair. *Ann. Surg., 220*, 10–18.

Ahn, S. T., Monafo, W. W., & Mustoe, T. A. (1991). Topical silicone gel for the prevention and treatment of hypertrophic scar. *Arch. Surg., 126*, 499–504.

Allman, R. M. (1998). The impact of pressure ulcers on health care costs and mortality. *Adv. Wound Care, 11*, 2.

Alster, T. (2003). Laser scar revision: comparison study of 585-nm pulsed dye laser with and without intralesional corticosteroids. *Dermatol. Surg., 29*, 25–29.

Alster, T. S. (1994). Improvement of erythematous and hypertrophic scars by the 585-nm flashlamp-pumped pulsed dye laser. *Ann. Plast. Surg., 32*, 186–190.

Alster, T. S., & Williams, C. M. (1995). Treatment of keloid sternotomy scars with 585 nm flashlamp-pumped pulsed-dye laser. *Lancet, 345*, 1198–1200.

American Burn Association. (2007). American Burn Association Burn Incidence Fact Sheet. Chicago. http://www.ameriburn.org/resources_factsheet.php

Ashcroft, G. S., Dodsworth, J., van Boxtel, E., Tarnuzzer, R. W., Horan, M. A., Schultz, G. S., et al. (1997a). Estrogen accelerates cutaneous wound healing associated with an increase in TGF-beta1 levels. *Nat. Med., 3*, 1209–1215.

Ashcroft, G. S., Horan, M. A., & Ferguson, M. W. (1997b). Aging is associated with reduced deposition of specific extracellular matrix components, an upregulation of angiogenesis, and an altered inflammatory response in a murine incisional wound healing model. *J. Invest. Dermatol., 108*, 430–437.

Ashcroft, G. S., Horan, M. A., & Ferguson, M. W. (1997c). The effects of ageing on wound healing: immunolocalisation of growth factors and their receptors in a murine incisional model. *J. Anat., 190*(Pt 3), 351–365.

Ashcroft, G. S., Yang, X., Glick, A. B., Weinstein, M., Letterio, J. L., Mizel, D. E., et al. (1999). Mice lacking Smad3 show accelerated wound healing and an impaired local inflammatory response. *Nat. Cell Biol., 1*, 260–266.

Asilian, A., Darougheh, A., & Shariati, F. (2006). New combination of triamcinolone, 5-fluorouracil, and pulsed-dye laser for treatment of keloid and hypertrophic scars. *Dermatol. Surg., 32*, 907–915.

Atiyeh, B. S., Costagliola, M., & Hayek, S. N. (2005). Keloid or hypertrophic scar: the controversy: review of the literature. *Ann. Plast. Surg., 54*, 676–680.

Berman, B. (2002). Imiquimod: a new immune response modifier for the treatment of external genital warts and other diseases in dermatology. *Int. J. Dermatol., 41*(Suppl. 1), 7–11.

Berman, B., & Bieley, H. C. (1996). Adjunct therapies to surgical management of keloids. *Dermatol. Surg., 22*, 126–130.

Berman, B., & Kaufman, J. (2002). Pilot study of the effect of postoperative imiquimod 5% cream on the recurrence rate of excised keloids. *J. Am. Acad. Dermatol., 47*, S209–S211.

Blomme, E. A., Chinn, K. S., Hardy, M. M., Casler, J. J., Kim, S. H., Opsahl, A. C., et al. (2003). Selective cyclo-oxygenase-2 inhibition does not affect the healing of cutaneous full-thickness incisional wounds in SKH-1 mice. *Br. J. Dermatol., 148*, 211–223.

Blumenkranz, M. S., Ophir, A., Claflin, A. J., & Hajek, A. (1982). Fluorouracil for the treatment of massive peri-retinal proliferation. *Am. J. Ophthalmol., 94*, 458–467.

Bodokh, I., & Brun, P. (1996). Treatment of keloid with intralesional bleomycin. *Ann. Dermatol. Venereol., 123*, 791–794.

Buchanan, E. P., Longaker, M. T., & Lorenz, H. P. (2009). Fetal skin wound healing. *Adv. Clin. Chem., 48*, 137–161.

Cacao, F. M., Tanaka, V., & Messina, M. C. (2009). Failure of imiquimod 5% cream to prevent recurrence of surgically excised trunk keloids. *Dermatol. Surg., 35*, 629–633.

Caplan, A. I. (2007). Adult mesenchymal stem cells for tissue engineering versus regenerative medicine. *J. Cell Physiol., 213*, 341–347.

Carney, S. A., Cason, C. G., Gowar, J. P., Stevenson, J. H., McNee, J., Groves, A. R., et al. (1994). Cica-Care gel sheeting in the management of hypertrophic scarring. *Burns, 20*, 163–167.

Carre, A. L., James, A. W., MacLeod, L., Kong, W., Kawai, K., Longaker, M. T., et al. (2010). Interaction of wingless protein (Wnt), transforming growth factor-beta1, and hyaluronan production in fetal and postnatal fibroblasts. *Plast. Reconstr. Surg., 125*, 74–88.

Cass, D. L., Bullard, K. M., Sylvester, K. G., Yang, E. Y., Longaker, M. T., & Adzick, N. S. (1997). Wound size and gestational age modulate scar formation in fetal wound repair. *J. Pediatr. Surg., 32*, 411–415.

Chamberlain, G., Fox, J., Ashton, B., & Middleton, J. (2007). Concise review: mesenchymal stem cells: their phenotype, differentiation capacity, immunological features, and potential for homing. *Stem Cells, 25*, 2739–2749.

Chien, K. R., Moretti, A., & Laugwitz, K. L. (2004). Development. ES cells to the rescue. *Science, 306*, 239–240.

Choi, B. M., Kwak, H. J., Jun, C. D., Park, S. D., Kim, K. Y., Kim, H. R., et al. (1996). Control of scarring in adult wounds using antisense transforming growth factor-beta 1 oligodeoxynucleotides. *Immunol. Cell Biol., 74*, 144–150.

Clark, R. A. (1985). Cutaneous tissue repair: basic biologic considerations. I. *J. Am. Acad. Dermatol., 13*, 701–725.

Clark, R. A. F. (1996). *The Molecular and Cellular Biology of Wound Repair.* New York: Plenum Press.

Colwell, A. S., Beanes, S. R., Soo, C., Dang, C., Ting, K., Longaker, M. T., et al. (2005). Increased angiogenesis and expression of vascular endothelial growth factor during scarless repair. *Plast. Reconstr. Surg., 115*, 204–212.

Colwell, A. S., Krummel, T. M., Longaker, M. T., & Lorenz, H. P. (2006a). An *in vivo* mouse excisional wound model of scarless healing. *Plast. Reconstr. Surg., 117*, 2292–2296.

Colwell, A. S., Krummel, T. M., Longaker, M. T., & Lorenz, H. P. (2006b). Fetal and adult fibroblasts have similar TGF-beta-mediated, Smad-dependent signaling pathways. *Plast. Reconstr. Surg., 117*, 2277–2283.

Colwell, A. S., Krummel, T. M., Longaker, M. T., & Lorenz, H. P. (2006c). Wnt-4 expression is increased in fibroblasts after TGF-beta1 stimulation and during fetal and postnatal wound repair. *Plast. Reconstr. Surg., 117*, 2297–2301.

Cormack, D. H. (1997). The integumentary system. In R. Anthony (Ed.), *Essential Histology* (pp. 255–268). Philadelphia: Lippincott-Raven Publishers.

Coutinho, P., Qiu, C., Frank, S., Tamber, K., & Becker, D. (2003). Dynamic changes in connexin expression correlate with key events in the wound healing process. *Cell Biol. Int., 27*, 525–541.

Coutinho, P., Qiu, C., Frank, S., Wang, C. M., Brown, T., Green, C. R., et al. (2005). Limiting burn extension by transient inhibition of Connexin43 expression at the site of injury. *Br. J. Plast. Surg., 58*, 658–667.

Dang, C. M., Beanes, S. R., Soo, C., Ting, K., Benhaim, P., Hedrick, M. H., et al. (2003). Decreased expression of fibroblast and keratinocyte growth factor isoforms and receptors during scarless repair. *Plast. Reconstr. Surg., 111*, 1969–1979.

Darzi, M. A., Chowdri, N. A., Kaul, S. K., & Khan, M. (1992). Evaluation of various methods of treating keloids and hypertrophic scars: a 10-year follow-up study. *Br. J. Plast. Surg., 45*, 374–379.

Desmouliere, A., Chaponnier, C., & Gabbiani, G. (2005). Tissue repair, contraction, and the myofibroblast. *Wound Repair Regen., 13*, 7–12.

Desplantez, T., Dupont, E., Severs, N. J., & Weingart, R. (2007). Gap junction channels and cardiac impulse propagation. *J. Membr. Biol., 218*, 13–28.

Diegelmann, R. F., & Evans, M. C. (2004). Wound healing: an overview of acute, fibrotic and delayed healing. *Front. Biosci., 9*, 283–289.

Dovi, J. V., He, L. K., & DiPietro, L. A. (2003). Accelerated wound closure in neutrophil-depleted mice. *J. Leukoc. Biol., 73*, 448–455.

Ehrlich, H. P., Sun, B., Saggers, G. C., & Kromath, F. (2006). Gap junction communications influence upon fibroblast synthesis of Type I collagen and fibronectin. *J. Cell Biochem., 98*, 735–743.

English, R. S., & Shenefelt, P. D. (1999). Keloids and hypertrophic scars. *Dermatol. Surg., 25*, 631–638.

Espana, A., Solano, T., & Quintanilla, E. (2001). Bleomycin in the treatment of keloids and hypertrophic scars by multiple needle punctures. *Dermatol. Surg., 27*, 23–27.

Falanga, V., Iwamoto, S., Chartier, M., Yufit, T., Butmarc, J., Kouttab, N., et al. (2007). Autologous bone marrow-derived cultured mesenchymal stem cells delivered in a fibrin spray accelerate healing in murine and human cutaneous wounds. *Tissue Eng., 13*, 1299–1312.

Fang, B., Song, Y., Liao, L., Zhang, Y., & Zhao, R. C. (2007). Favorable response to human adipose tissue-derived mesenchymal stem cells in steroid-refractory acute graft-versus-host disease. *Transplant. Proc., 39*, 3358–3362.

Ferguson, M. W., & Howarth, G. F. (1992). Marsupial models of scarless fetal wound healing. In N. S. Adzick & M. T. Longaker (Eds.), *Fetal Wound Healing* (pp. 95–124). New York: Elsevier Scientific Press.

Ferguson, M. W., & O'Kane, S. (2004). Scar-free healing: from embryonic mechanisms to adult therapeutic intervention. *Philos. Trans. R. Soc. Lond. B Biol. Sci., 359*, 839–850.

Fish, P. V., Allan, G. A., Bailey, S., Blagg, J., Butt, R., Collis, M. G., et al. (2007). Potent and selective nonpeptidic inhibitors of procollagen C-proteinase. *J. Med. Chem., 50*, 3442–3456.

Fitzpatrick, R. E. (1999). Treatment of inflamed hypertrophic scars using intralesional 5-FU. *Dermatol. Surg., 25*, 224–232.

Folkman, J. (1996). New perspectives in clinical oncology from angiogenesis research. *Eur. J. Cancer., 32A*, 2534–2539.

Fraidenraich, D., Stillwell, E., Romero, E., Wilkes, D., Manova, K., Basson, C. T., et al. (2004). Rescue of cardiac defects in id knockout embryos by injection of embryonic stem cells. *Science, 306*, 247–252.

Frank, S., Madlener, M., & Werner, S. (1996). Transforming growth factors beta1, beta2, and beta3 and their receptors are differentially regulated during normal and impaired wound healing. *J. Biol. Chem., 271*, 10188–10193.

Garcia-Olmo, D., Garcia-Arranz, M., Herreros, D., Pascual, I., Peiro, C., & Rodriguez-Montes, J. A. (2005). A phase I clinical trial of the treatment of Crohn's fistula by adipose mesenchymal stem cell transplantation. *Dis. Colon Rectum., 48*, 1416–1423.

Gawronska-Kozak, B., Bogacki, M., Rim, J. S., Monroe, W. T., & Manuel, J. A. (2006). Scarless skin repair in immunodeficient mice. *Wound Repair Regen., 14*, 265–276.

Goliger, J. A., & Paul, D. L. (1995). Wounding alters epidermal connexin expression and gap junction-mediated intercellular communication. *Mol. Biol. Cell, 6*, 1491–1501.

Gordon, A., Kozin, E. D., Keswani, S. G., Vaikunth, S. S., Katz, A. B., Zoltick, P. W., et al. (2008). Permissive environment in postnatal wounds induced by adenoviral-mediated overexpression of the anti-inflammatory cytokine interleukin-10 prevents scar formation. *Wound Repair Regen., 16*, 70–79.

Gourdie, R. G., Ghatnekar, G. S., O'Quinn, M., Rhett, M. J., Barker, R. J., Zhu, C., et al. (2006). The unstoppable connexin43 carboxyl-terminus: new roles in gap junction organization and wound healing. *Ann. N.Y. Acad. Sci., 1080*, 49–62.

Grinnell, F. (1994). Fibroblasts, myofibroblasts, and wound contraction. *J. Cell Biol., 124*, 401–404.

Gu, D., Atencio, I., Kang, D. W., Looper, L. D., Ahmed, C. M., Levy, A., et al. (2005). Recombinant adenovirus-p21 attenuates proliferative responses associated with excessive scarring. *Wound Repair Regen., 13*, 480–490.

Ha, X., Li, Y., Lao, M., Yuan, B., & Wu, C. T. (2003). Effect of human hepatocyte growth factor on promoting wound healing and preventing scar formation by adenovirus-mediated gene transfer. *Chin. Med. J. (Engl.), 116*, 1029–1033.

Hanson, S. E., Bentz, M. L., & Hematti, P. (2010). Mesenchymal stem cell therapy for nonhealing cutaneous wounds. *Plast. Reconstr. Surg., 125*, 510–516.

Har-Shai, Y., Amar, M., & Sabo, E. (2003). Intralesional cryotherapy for enhancing the involution of hypertrophic scars and keloids. *Plast. Reconstr. Surg., 111*, 1841–1852.

Haynes, J. H., Johnson, D. E., Mast, B. A., Diegelmann, R. F., Salzberg, D. A., Cohen, I. K., et al. (1994). Platelet-derived growth factor induces fetal wound fibrosis. *J. Pediatr. Surg., 29*, 1405–1408.

Hendricks, T., Martens, M. F., Huyben, C. M., & Wobbes, T. (1993). Inhibition of basal and TGF beta-induced fibroblast collagen synthesis by antineoplastic agents. Implications for wound healing. *Br. J. Cancer, 67*, 545–550.

Heng, B. C., Liu, H., & Cao, T. (2005). Transplanted human embryonic stem cells as biological "catalysts" for tissue repair and regeneration. *Med. Hypotheses, 64*, 1085–1088.

Henry, G., & Garner, W. L. (2003). Inflammatory mediators in wound healing. *Surg. Clin. North Am., 83*, 483–507.

Hsu, M., Peled, Z. M., Chin, G. S., Liu, W., & Longaker, M. T. (2001). Ontogeny of expression of transforming growth factor-beta 1 (TGF-beta 1), TGF-beta 3, and TGF-beta receptors I and II in fetal rat fibroblasts and skin. *Plast. Reconstr. Surg., 107*, 1787–1794, discussion 1795–1786.

Ihara, S., & Motobayashi, Y. (1992). Wound closure in foetal rat skin. *Development, 114*, 573–582.

Ihara, S., Motobayashi, Y., Nagao, E., & Kistler, A. (1990). Ontogenetic transition of wound healing pattern in rat skin occurring at the fetal stage. *Development, 110*, 671–680.

Ilan, N., Mahooti, S., & Madri, J. A. (1998). Distinct signal transduction pathways are utilized during the tube formation and survival phases of *in vitro* angiogenesis. *J. Cell Sci., 111*(Pt 24), 3621–3631.

Iocono, J. A., Ehrlich, H. P., Keefer, K. A., & Krummel, T. M. (1998). Hyaluronan induces scarless repair in mouse limb organ culture. *J. Pediatr. Surg., 33*, 564–567.

Kose, O., & Waseem, A. (2008). Keloids and hypertrophic scars: are they two different sides of the same coin? *Dermatol. Surg., 34*, 336–346.

Kovalic, J. J., & Perez, C. A. (1989). Radiation therapy following keloidectomy: a 20-year experience. *Int. J. Radiat. Oncol. Biol Phys., 17*, 77–80.

Kretz, M., Maass, K., & Willecke, K. (2004). Expression and function of connexins in the epidermis, analyzed with transgenic mouse mutants. *Eur. J. Cell Biol., 83*, 647–654.

Lane, A. T. (1986). Human fetal skin development. *Pediatr. Dermatol., 3*, 487–491.

Lee, K. S., Song, J. Y., & Suh, M. H. (1991). Collagen mRNA expression detected by in situ hybridization in keloid tissue. *J. Dermatol. Sci., 2*, 316–323.

Lee, O. K., Kuo, T. K., Chen, W. M., Lee, K. D., Hsieh, S. L., & Chen, T. H. (2004). Isolation of multipotent mesenchymal stem cells from umbilical cord blood. *Blood, 103*, 1669–1675.

Leibovich, S. J., & Ross, R. (1975). The role of the macrophage in wound repair. A study with hydrocortisone and antimacrophage serum. *Am. J. Pathol., 78*, 71–100.

Li, W., Henry, G., Fan, J., Bandyopadhyay, B., Pang, K., Garner, W., et al. (2004). Signals that initiate, augment, and provide directionality for human keratinocyte motility. *J. Invest. Dermatol., 123*, 622–633.

Liechty, K. W., Adzick, N. S., & Crombleholme, T. M. (2000a). Diminished interleukin 6 (IL-6) production during scarless human fetal wound repair. *Cytokine, 12*, 671–676.

Liechty, K. W., Crombleholme, T. M., Cass, D. L., Martin, B., & Adzick, N. S. (1998). Diminished interleukin-8 (IL-8) production in the fetal wound healing response. *J. Surg. Res., 77*, 80–84.

Liechty, K. W., Kim, H. B., Adzick, N. S., & Crombleholme, T. M. (2000b). Fetal wound repair results in scar formation in interleukin-10-deficient mice in a syngeneic murine model of scarless fetal wound repair. *J. Pediatr. Surg., 35*, 866–872, discussion 872–863.

Longaker, M. T., Whitby, D. J., Ferguson, M. W., Lorenz, H. P., Harrison, M. R., & Adzick, N. S. (1994). Adult skin wounds in the fetal environment heal with scar formation. *Ann. Surg., 219*, 65–72.

Lorenz, H. P., Longaker, M. T., Perkocha, L. A., Jennings, R. W., Harrison, M. R., & Adzick, N. S. (1992). Scarless wound repair: a human fetal skin model. *Development, 114*, 253–259.

Lorenz, H. P., Whitby, D. J., Longaker, M. T., & Adzick, N. S. (1993). Fetal wound healing. The ontogeny of scar formation in the non-human primate. *Ann. Surg., 217*, 391–396.

Maguire, H. C., Jr. (1965). Treatment of keloids with triamcinolone acetonide injected intralesionally. *JAMA, 192*, 325–326.

Maherali, N., Sridharan, R., Xie, W., Utikal, J., Eminli, S., Arnold, K., et al. (2007). Directly reprogrammed fibroblasts show global epigenetic remodeling and widespread tissue contribution. *Cell Stem Cell, 1*, 55–70.

Mallick, K. S., Hajek, A. S., & Parrish, R. K., II (1985). Fluorouracil (5-FU) and cytarabine (ara-C) inhibition of corneal epithelial cell and conjunctival fibroblast proliferation. *Arch. Ophthalmol., 103*, 1398–1402.

Manuskiatti, W., & Fitzpatrick, R. E. (2002). Treatment response of keloidal and hypertrophic sternotomy scars: comparison among intralesional corticosteroid, 5-fluorouracil, and 585-nm flashlamp-pumped pulsed-dye laser treatments. *Arch. Dermatol., 138*, 1149–1155.

Martin, P. (1997). Wound healing – aiming for perfect skin regeneration. *Science, 276*, 75–81.

Martin, P., & Leibovich, S. J. (2005). Inflammatory cells during wound repair: the good, the bad and the ugly. *Trends Cell Biol., 15*, 599–607.

Martin, P., & Lewis, J. (1992). Actin cables and epidermal movement in embryonic wound healing. *Nature, 360*, 179–183.

Martin, P., d'Souza, D., Martin, J., Grose, R., Cooper, L., Maki, R., et al. (2003). Wound healing in the PU.1 null mouse – tissue repair is not dependent on inflammatory cells. *Curr. Biol., 13*, 1122–1128.

Mast, B. A., Diegelmann, R. F., Krummel, T. M., & Cohen, I. K. (1993). Hyaluronic acid modulates proliferation, collagen and protein synthesis of cultured fetal fibroblasts. *Matrix, 13*, 441–446.

Miller, R. L., Gerster, J. F., Owens, M. L., Slade, H. B., & Tomai, M. A. (1999). Imiquimod applied topically: a novel immune response modifier and new class of drug. *Int. J. Immunopharmacol., 21*, 1–14.

Montesano, R., & Orci, L. (1988). Transforming growth factor beta stimulates collagen-matrix contraction by fibroblasts: implications for wound healing. *Proc. Natl. Acad. Sci. U.S.A., 85*, 4894–4897.

Moore, K. L., & Persaud, T. V. N. (1993). *The Developing Human: Clinically Oriented Embryology.* Philadelphia: Saunders.

Mori, R., Power, K. T., Wang, C. M., Martin, P., & Becker, D. L. (2006). Acute downregulation of connexin43 at wound sites leads to a reduced inflammatory response, enhanced keratinocyte proliferation and wound fibroblast migration. *J. Cell Sci., 119*, 5193–5203.

Moulin, V., Lawny, F., Barritault, D., & Caruelle, J. P. (1998). Platelet releasate treatment improves skin healing in diabetic rats through endogenous growth factor secretion. *Cell Mol. Biol. (Noisy-le-grand), 44*, 961–971.

Moyer, K. E., Davis, A., Saggers, G. C., Mackay, D. R., & Ehrlich, H. P. (2002). Wound healing: the role of gap junctional communication in rat granulation tissue maturation. *Exp. Mol. Pathol., 72*, 10–16.

Naeini, F. F., Najafian, J., & Ahmadpour, K. (2006). Bleomycin tattooing as a promising therapeutic modality in large keloids and hypertrophic scars. *Dermatol. Surg., 32*, 1023–1029, discussion 1029–1030.

Centers of Disease Control and Prevention. (2008). National Diabetes Fact Sheet of the National Center Chronic Disease Prevention and Health Promotion. Atlanta. http://apps.nccd.cdc.gov/DDTSTRS/FactSheet.aspx

Niessen, F. B., Spauwen, P. H., Schalkwijk, J., & Kon, M. (1999). On the nature of hypertrophic scars and keloids: a review. *Plast. Reconstr. Surg., 104*, 1435–1458.

Nwomeh, B. C., Liang, H. X., Cohen, I. K., & Yager, D. R. (1999). MMP-8 is the predominant collagenase in healing wounds and nonhealing ulcers. *J. Surg. Res., 81*, 189–195.

Nwomeh, B. C., Liang, H. X., Diegelmann, R. F., Cohen, I. K., & Yager, D. R. (1998). Dynamics of the matrix metalloproteinases MMP-1 and MMP-8 in acute open human dermal wounds. *Wound Repair Regen., 6*, 127–134.

Ogawa, R. (2010). The most current algorithms for the treatment and prevention of hypertrophic scars and keloids. *Plast. Reconstr. Surg., 125*, 557–568.

Ogawa, Y., Ksander, G. A., Pratt, B. M., Sawamura, S. J., Ziman, J. M., Gerhardt, C. O., et al. (1991). Differences in the biological activities of transforming growth factor-beta and platelet-derived growth factor *in vivo*. *Growth Factors, 5*, 57–68.

Okita, K., Nakagawa, M., Hyenjong, H., Ichisaka, T., & Yamanaka, S. (2008). Generation of mouse induced pluripotent stem cells without viral vectors. *Science, 322*, 949–953.

Orive, G., Gascon, A. R., Hernandez, R. M., Igartua, M., & Luis Pedraz, J. (2003). Cell microencapsulation technology for biomedical purposes: novel insights and challenges. *Trends Pharmacol. Sci., 24*, 207–210.

Pera, M. F., & Hasegawa, K. (2008). Simpler and safer cell reprogramming. *Nat. Biotechnol., 26*, 59–60.

Perez, P., Page, A., Bravo, A., del Rio, M., Gimenez-Conti, I., Budunova, I., et al. (2001). Altered skin development and impaired proliferative and inflammatory responses in transgenic mice overexpressing the glucocorticoid receptor. *FASEB J., 15*, 2030–2032.

Qiu, C., Coutinho, P., Frank, S., Franke, S., Law, L. Y., Martin, P., et al. (2003). Targeting connexin43 expression accelerates the rate of wound repair. *Curr. Biol., 13*, 1697–1703.

Quan, T. E., Cowper, S., Wu, S. P., Bockenstedt, L. K., & Bucala, R. (2004). Circulating fibrocytes: collagen-secreting cells of the peripheral blood. *Int. J. Biochem. Cell Biol., 36*, 598–606.

Quinn, K. J. (1987). Silicone gel in scar treatment. *Burns Incl. Therm. Inj., 13*(Suppl.), S33–S40.

Ragoowansi, R., Cornes, P. G., Moss, A. L., & Glees, J. P. (2003). Treatment of keloids by surgical excision and immediate postoperative single-fraction radiotherapy. *Plast. Reconstr. Surg., 111*, 1853–1859.

Rahban, S. R., & Garner, W. L. (2003). Fibroproliferative scars. *Clin. Plast. Surg., 30*, 77–89.

Reiken, S. R., Wolfort, S. F., Berthiaume, F., Compton, C., Tompkins, R. G., & Yarmush, M. L. (1997). Control of hypertrophic scar growth using selective photothermolysis. *Lasers Surg. Med., 21*, 7–12.

Reish, R. G., & Eriksson, E. (2008). Scars: a review of emerging and currently available therapies. *Plast. Reconstr. Surg., 122*, 1068–1078.

Rhett, J. M., Ghatnekar, G. S., Palatinus, J. A., O'Quinn, M., Yost, M. J., & Gourdie, R. G. (2008). Novel therapies for scar reduction and regenerative healing of skin wounds. *Trends Biotechnol., 26*, 173–180.

Richards, T. S., Dunn, C. A., Carter, W. G., Usui, M. L., Olerud, J. E., & Lampe, P. D. (2004). Protein kinase C spatially and temporally regulates gap junctional communication during human wound repair via phosphorylation of connexin43 on serine368. *J. Cell Biol., 167*, 555–562.

Roberts, A. B., Sporn, M. B., Assoian, R. K., Smith, J. M., Roche, N. S., Wakefield, L. M., et al. (1986). Transforming growth factor type beta: rapid induction of fibrosis and angiogenesis *in vivo* and stimulation of collagen formation *in vitro*. *Proc. Natl. Acad. Sci. U.S.A., 83*, 4167–4171.

Rodgers, K. E., Roda, N., Felix, J. E., Espinoza, T., Maldonado, S., & diZerega, G. (2003). Histological evaluation of the effects of angiotensin peptides on wound repair in diabetic mice. *Exp. Dermatol., 12*, 784–790.

Rusciani, L., Paradisi, A., Alfano, C., Chiummariello, S., & Rusciani, A. (2006). Cryotherapy in the treatment of keloids. *J. Drugs Dermatol., 5*, 591–595.

Saray, Y., & Gulec, A. T. (2005). Treatment of keloids and hypertrophic scars with dermojet injections of bleomycin: a preliminary study. *Int. J. Dermatol., 44*, 777–784.

Shaffer, J. J., Taylor, S. C., & Cook-Bolden, F. (2002). Keloidal scars: a review with a critical look at therapeutic options. *J. Am. Acad. Dermatol., 46*, S63–S97.

Shah, M., Foreman, D. M., & Ferguson, M. W. (1995). Neutralisation of TGF-beta 1 and TGF-beta 2 or exogenous addition of TGF-beta 3 to cutaneous rat wounds reduces scarring. *J. Cell Sci., 108*(Pt 3), 985–1002.

Ship, A. G., Weiss, P. R., Mincer, F. R., & Wolkstein, W. (1993). Sternal keloids: successful treatment employing surgery and adjunctive radiation. *Ann. Plast. Surg., 31*, 481–487.

Simpson, D. M., & Ross, R. (1972). The neutrophilic leukocyte in wound repair a study with antineutrophil serum. *J. Clin. Invest., 51*, 2009–2023.

Singer, A. J., & Clark, R. A. (1999). Cutaneous wound healing. *N. Engl. J. Med., 341*, 738–746.

Soo, C., Hu, F. Y., Zhang, X., Wang, Y., Beanes, S. R., Lorenz, H. P., et al. (2000). Differential expression of fibromodulin, a transforming growth factor-beta modulator, in fetal skin development and scarless repair. *Am. J. Pathol., 157*, 423–433.

Stadtfeld, M., Nagaya, M., Utikal, J., Weir, G., & Hochedlinger, K. (2008). Induced pluripotent stem cells generated without viral integration. *Science, 322*, 945–949.

Stevenson, S., Nelson, L. D., Sharpe, D. T., & Thornton, M. J. (2008). 17beta-estradiol regulates the secretion of TGF-beta by cultured human dermal fibroblasts. *J. Biomater. Sci. Polym. Ed., 19*, 1097–1109.

Stoff, A., Rivera, A. A., Mathis, J. M., Moore, S. T., Banerjee, N. S., Everts, M., et al. (2007). Effect of adenoviral mediated overexpression of fibromodulin on human dermal fibroblasts and scar formation in full-thickness incisional wounds. *J. Mol. Med., 85*, 481–496.

Stojadinovic, O., Lee, B., Vouthounis, C., Vukelic, S., Pastar, I., Blumenberg, M., et al. (2007). Novel genomic effects of glucocorticoids in epidermal keratinocytes: inhibition of apoptosis, interferon-gamma pathway, and wound healing along with promotion of terminal differentiation. *J. Biol. Chem., 282*, 4021–4034.

Stramer, B. M., Mori, R., & Martin, P. (2007). The inflammation-fibrosis link? A Jekyll and Hyde role for blood cells during wound repair. *J. Invest. Dermatol., 127*, 1009–1017.

Swijnenburg, R. J., Schrepfer, S., Govaert, J. A., Cao, F., Ransohoff, K., Sheikh, A. Y., Haddad, M., et al. (2008). Immunosuppressive therapy mitigates immunological rejection of human embryonic stem cell xenografts. *Proc. Natl. Acad. Sci. U.S.A., 105*, 12991–12996.

Takahashi, K., & Yamanaka, S. (2006). Induction of pluripotent stem cells from mouse embryonic and adult fibroblast cultures by defined factors. *Cell, 126*, 663–676.

Takahashi, K., Tanabe, K., Ohnuki, M., Narita, M., Ichisaka, T., Tomoda, K., et al. (2007). Induction of pluripotent stem cells from adult human fibroblasts by defined factors. *Cell, 131*, 861–872.

Tang, Y. W. (1992). Intra- and postoperative steroid injections for keloids and hypertrophic scars. *Br. J. Plast. Surg., 45*, 371–373.

Thielitz, A., Vetter, R. W., Schultze, B., Wrenger, S., Simeoni, L., Ansorge, S., et al. (2008). Inhibitors of dipeptidyl peptidase IV-like activity mediate antifibrotic effects in normal and keloid-derived skin fibroblasts. *J. Invest. Dermatol., 128*, 855–866.

Thomson, J. A., Itskovitz-Eldor, J., Shapiro, S. S., Waknitz, M. A., Swiergiel, J. J., Marshall, V. S., et al. (1998). Embryonic stem cell lines derived from human blastocysts. *Science, 282*, 1145–1147.

Tolhurst, D. E. (1977). Hypertrophic scarring prevented by pressure: a case report. *Br. J. Plast. Surg., 30*, 218–219.

Tuan, T. L., & Nichter, L. S. (1998). The molecular basis of keloid and hypertrophic scar formation. *Mol. Med. Today, 4*, 19–24.

Uccelli, A., Pistoia, V., & Moretta, L. (2007). Mesenchymal stem cells: a new strategy for immunosuppression? *Trends Immunol., 28*, 219–226.

Wang, C. M., Lincoln, J., Cook, J. E., & Becker, D. L. (2007). Abnormal connexin expression underlies delayed wound healing in diabetic skin. *Diabetes, 56*, 2809–2817.

Watt, F. M. (1998). Epidermal stem cells: markers, patterning and the control of stem cell fate. *Philos. Trans. R. Soc. Lond. B Biol. Sci., 353*, 831–837.

Wei, C. J., Xu, X., & Lo, C. W. (2004). Connexins and cell signaling in development and disease. *Annu. Rev. Cell Dev. Biol., 20*, 811–838.

Wendling, J., Marchand, A., Mauviel, A., & Verrecchia, F. (2003). 5-fluorouracil blocks transforming growth factor-beta-induced alpha 2 type I collagen gene (COL1A2) expression in human fibroblasts via c-Jun NH2-terminal kinase/activator protein-1 activation. *Mol. Pharmacol., 64*, 707–713.

Wenk, J., Foitzik, A., Achterberg, V., Sabiwalsky, A., Dissemond, J., Meewes, C., et al. (2001). Selective pick-up of increased iron by deferoxamine-coupled cellulose abrogates the iron-driven induction of matrix-degrading metalloproteinase 1 and lipid peroxidation in human dermal fibroblasts *in vitro*: a new dressing concept. *J. Invest. Dermatol., 116*, 833–839.

Werner, S., & Grose, R. (2003). Regulation of wound healing by growth factors and cytokines. *Physiol. Rev., 83*, 835–870.

Wernig, M., Meissner, A., Foreman, R., Brambrink, T., Ku, M., Hochedlinger, K., et al. (2007). *In vitro* reprogramming of fibroblasts into a pluripotent ES-cell-like state. *Nature, 448*, 318–324.

Whitby, D. J., & Ferguson, M. W. (1991). Immunohistochemical localization of growth factors in fetal wound healing. *Dev. Biol., 147*, 207–215.

Wilgus, T. A., Vodovotz, Y., Vittadini, E., Clubbs, E. A., & Oberyszyn, T. M. (2003). Reduction of scar formation in full-thickness wounds with topical celecoxib treatment. *Wound Repair Regen., 11*, 25–34.

Witte, M. B., Thornton, F. J., Kiyama, T., Efron, D. T., Schulz, G. S., Moldawer, L. L., & Barbul, A. (1998). Metalloproteinase inhibitors and wound healing: a novel enhancer of wound strength. *Surgery, 124*, 464–470.

Wu, W. S., Wang, F. S., Yang, K. D., Huang, C. C., & Kuo, Y. R. (2006). Dexamethasone induction of keloid regression through effective suppression of VEGF expression and keloid fibroblast proliferation. *J. Invest. Dermatol., 126*, 1264–1271.

Xu, J., & Clark, R. A. (1996). Extracellular matrix alters PDGF regulation of fibroblast integrins. *J. Cell Biol., 132*, 239–249.

Yager, D. R., Zhang, L. Y., Liang, H. X., Diegelmann, R. F., & Cohen, I. K. (1996). Wound fluids from human pressure ulcers contain elevated matrix metalloproteinase levels and activity compared to surgical wound fluids. *J. Invest. Dermatol., 107*, 743–748.

Yoshikawa, T., Mitsuno, H., Nonaka, I., Sen, Y., Kawanishi, K., Inada, Y., et al. (2008). Wound therapy by marrow mesenchymal cell transplantation. *Plast. Reconstr. Surg., 121*, 860–877.

Zahler, S., Hoffmann, A., Gloe, T., & Pohl, U. (2003). Gap-junctional coupling between neutrophils and endothelial cells: a novel modulator of transendothelial migration. *J. Leukoc. Biol., 73*, 118–126.

Zhang, G. Y., Li, X., Chen, X. L., Li, Z. J., Yu, Q., Jiang, L. F., et al. (2009). Contribution of epidermal stem cells to hypertrophic scars pathogenesis. *Med. Hypotheses, 73*, 332–333.

Zhao, Z. L., Fu, X. B., Sun, T. Z., Chen, W., & Sun, X. Q. (2003). Study on the location and the expression characteristics of epidermal stem cells in normal adult skin and scar tissue. *Zhonghua Shao Shang Za Zhi, 19*, 12–14.

Zhou, H., Wu, S., Joo, J. Y., Zhu, S., Han, D. W., Lin, T., et al. (2009). Generation of induced pluripotent stem cells using recombinant proteins. *Cell Stem Cell, 4*, 381–384.

Zouboulis, C. C., Blume, U., Buttner, P., & Orfanos, C. E. (1993). Outcomes of cryosurgery in keloids and hypertrophic scars. A prospective consecutive trial of case series. *Arch. Dermatol., 129*, 1146–1151.

Zuk, P. A., Zhu, M., Ashjian, P., de Ugarte, D. A., Huang, J. I., Mizuno, H., et al. (2002). Human adipose tissue is a source of multipotent stem cells. *Mol. Biol. Cell, 13*, 4279–4295.

CHAPTER 7

Somatic Cloning and Epigenetic Reprogramming in Mammals

Heiner Niemann, Wilfried A. Kues, Andrea Lucas-Hahn, Joseph W. Carnwath
Institute of Farm Animal Genetics, Friedrich-Loeffler-Institut (FLI), Federal Research
Institute for Animal Health, Mariensee, Neustadt, Germany

INTRODUCTION — SHORT HISTORY OF SOMATIC CLONING

More than 50 years ago, Briggs and King (1952) showed that normal hatched tadpoles could be obtained after transplanting the nucleus of a blastula cell into the enucleated egg of the amphibian *Rana pipiens*. However, while cloning with embryonic cells resulted in normal offspring, development became more and more restricted when cells from more differentiated stages of development were employed (Briggs and King, 1952). This led to the hypothesis that the closer the nuclear donor is developmentally to early embryonic stages the more successful nuclear transfer is likely to be. This concept prevailed for many years (Gurdon and Byrne, 2003). Cloning of mammals became possible when equipment became available in the late 1960s and early 1970s that allowed micromanipulation of the small mammalian egg (~100 to 130 μm), which is only one tenth the diameter of an amphibian egg. The first report of cloning an adult mammal was that of Illmensee and Hoppe (1981), who reported the birth of three cloned mice after transfer of nuclei from inner cell mass cells into enucleated zygotes. Unfortunately, these results could not be repeated in other laboratories. Subsequently it was shown that development to blastocysts could only be obtained when the nucleus of a zygote or a two-cell embryo was transferred into an enucleated zygote (McGrath and Solter, 1983) and no development was obtained when donor cell nuclei from later developmental stages were used (McGrath and Solter, 1984). McGrath and Solter (1984) concluded that the cloning of mammals by simple nuclear transfer was biologically impossible, mainly due to the rapid loss of totipotency of the embryonic cells. This conclusion affected research in this field profoundly. The concept that nuclear transfer was only successful when both donor and recipient were at nearly the same developmental stage contrasted with the results of the amphibian experiments, which had demonstrated the use of unfertilized eggs as recipients of somatic donor cell nuclei. However, the contradiction did not withstand the test of time. Willadsen (1986) soon demonstrated the use of blastomeres from cleavage stage mammalian embryos (sheep) for transfer into enucleated oocytes. This formed the basis for successful embryonic cloning in rabbits (Stice and Robl, 1988), mice (Cheong et al., 1993), pigs (Prather et al., 1989), cows (Sims and First, 1994), and monkeys (Meng et al., 1997). Eventually, in 1996, the full potential of somatic cloning in mammals became evident for the first time.

Principles of Regenerative Medicine. DOI: 10.1016/B978-0-12-381422-7.10007-0

Campbell et al. (1996) had success in using cells from an established cell line derived from a day 13 ovine conceptus and maintained *in vitro* for 6–13 passages. These cells had been blocked in a quiescent state by serum starvation prior to fusing them with enucleated sheep oocytes. Transfer of these nuclear transfer-derived embryos to foster mothers resulted in two healthy cloned sheep ("Morag" and "Megan") and formed the basis for the birth of "Dolly," the first mammal cloned from an adult mammary epithelial cell, reported a year later by the same laboratory (Wilmut et al., 1997). "Dolly" launched a worldwide heated ethical debate and sparked a series of science-fiction stories. More than 10 years later, this technology has matured and has become widely accepted as an important tool for research (Wadman, 2007). Initially, scientific progress was slow, but the speed of development has picked up in recent years and the technology is beginning to be used in important agricultural species including cattle, pigs, and horses. At the time of writing, somatic cell nuclear transfer (SCNT) has been successful (i.e. live clones have been obtained) in a total of 16 species, including sheep (Wilmut et al., 1997), cow (Kato et al., 1998), mouse (Wakayama et al., 1998), goat (Baguisi et al., 1999), pig (Polejaeva et al., 2000; Onishi et al., 2000), cat (Shin et al., 2002), rabbit (Chesne et al., 2002), mule (Woods et al., 2003), horse (Galli et al., 2003), rat (Zhou et al., 2003), dog (Lee et al., 2005), ferret (Li et al., 2006), red deer (Berg et al., 2007), buffalo (Shi et al., 2007), gray wolf (Oh et al., 2008), and camel (Wani et al., 2010). The report of a cloned dog (Lee et al., 2005) was questioned in the context of the scandal of South Korean scientist Woo Suk Hwang, whose claims of having derived stem cell lines from human embryos later turned out to be fraudulent. The dog, however, was eventually confirmed as a genuine clone by microsatellite analysis and mitochondrial genotyping (Lee and Park, 2006; Parker et al., 2006).

Worldwide research efforts have been undertaken to unravel the underlying mechanisms for successful somatic nuclear transfer. Initially, one hypothesis for the limited success of SCNT was that clones only arose from a subpopulation of adult stem cells (Hochedlinger and Jaenisch, 2002). However, compelling evidence now shows that differentiated somatic cells can successfully be employed in SCNT. Indeed, the most dramatic epigenetic reprogramming occurs in SCNT when the expression profile of a differentiated cell is abolished and a new embryo-specific expression profile is established that drives embryonic and fetal development (Niemann et al., 2008). This epigenetic reprogramming involves erasure of the gene expression program of the respective donor cell and the re-establishment of the well-orchestrated sequence of expression of the estimated 10,000–12,000 genes that regulate embryonic and fetal development (Kues et al., 2008b). The initial release from somatic cell epigenetic constraints is followed by establishment of post-zygotic expression patterns, X-chromosome inactivation, and adjustment of telomere length (Hochedlinger and Jaenisch, 2003).

Somatic nuclear transfer holds great promise for basic biological research and for various agricultural and biomedical applications. The following is a comprehensive review of the present state of somatic cell nuclear transfer (SCNT)-based cloning, including potential areas of application, with emphasis on the epigenetic reprogramming of the transferred somatic cell nucleus.

TECHNICAL ASPECTS OF SOMATIC NUCLEAR TRANSFER

Common somatic cloning protocols involve the following major technical steps (Figs 7.1, 7.2): (1) collection and enucleation of the recipient oocyte, (2) preparation and subzonal transfer of the donor cell, (3) fusion of the two components, (4) activation of the reconstructed complex, (5) temporary culture of the reconstructed embryo, and (6) transfer to a foster mother or storage in liquid nitrogen.

Collection and enucleation of the recipient oocyte

In many domesticated species, oocytes can be readily obtained from abattoir ovaries. Alternatively, oocytes can be repeatedly collected from live animals by ultrasound-guided

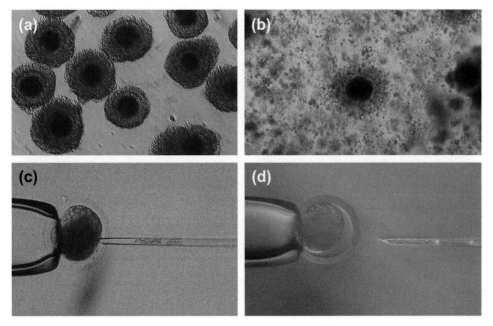

FIGURE 7.1

Sequence of steps in somatic cloning of pigs: *in vitro*-maturation (IVM) and enucleation of porcine oocytes. (a) Porcine cumulus oocyte complexes after isolation from abattoir ovaries. (b) Porcine oocyte after 42 h of IVM; note the expansion of the cumulus cells. (c) Microsurgical removal of the polar body plus adjacent cytoplasm containing the metaphase II chromosomes. (d) Microsurgical enucleation after labeling the DNA with a specific stain; note the fluorescence of the DNA within the cytoplasm indicating the metaphase plate and the polar body located in the enucleation pipette.

aspiration (Oropeza et al., 2007). These immature oocytes are usually at the germinal vesicle (GV) stage and need to be matured *in vitro* but represent a virtually unlimited source of material for cloning experiments. In cattle and pigs, *in vitro* maturation protocols have advanced to the extent that *in vitro*-matured (IVM) oocytes can be used in somatic cloning without major losses in efficiency and are comparable to their *in vivo*-matured counterparts. During the *in vitro* maturation period, the oocytes undergo a complex series of structural and biochemical changes culminating in the metaphase II stage of meiosis, at which point they have acquired the potential to be successfully fertilized and to undergo embryo and fetal development. Compelling evidence indicates that oocytes at the metaphase II stage rather than any other developmental stage are the most appropriate recipients for the production of viable cloned mammalian embryos. These oocytes possess high levels of maturation-promoting

131

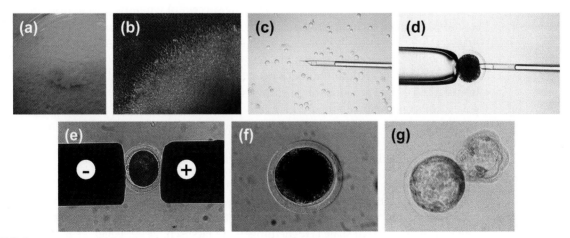

FIGURE 7.2

Sequence of steps in somatic cloning: from donor cell production to cloned blastocysts. (a) Porcine fetus from day 25 after insemination. (b) Outgrowing fibroblasts from minced fetal tissue, cultured as adhesive cells. (c) Isolated fibroblasts ready to be sucked up by the transfer pipette. (d) Transfer of a porcine fetal fibroblast into the perivitelline space of the enucleated recipient oocyte. (e) Fusion of the donor cell with the cytoplast in the electric field; note the great difference in size between donor cell and recipient. (f) Successful fusion of both components within 15 minutes. The donor cell has been completely integrated into the cytoplasm and is not further visible. (g) Cloned porcine blastocyst after 7 days of culture; image taken during the hatching process.

factor (MPF), which is thought to be critical for development of the reconstructed embryo (Miyoshi et al., 2003). Oocytes are enucleated by sucking out their chromosomes with microcapillaries or squeezing out the small portion of oocyte cytoplasm closely apposed to the first polar body, where the metaphase II chromosomes are usually located. The oocyte can be pretreated with a mycotoxin, cytochalasin B, to destabilize its cytoskeleton, but this is washed out immediately after microsurgical removal of the chromosomes. Preliminary evidence suggests that injection of chromatin remodeling factors such as nucleoplasmin or poly-glutamic acid into the oocyte may improve *in vitro* and *in vivo* development of cloned bovine embryos (Betthauser et al., 2006). Significantly higher success rates of bovine cloning were achieved by autologous SCNT, in which a somatic nucleus of the female donor was transferred to its own enucleated oocyte, which had been recovered by ultrasound-guided follicular aspiration (Yang et al., 2006). This higher success rate was explained by reduced epigenetic abnormalities in comparison with allogenic SCNT.

It has also been shown that bovine and murine zygotes can be used as recipient cells for the production of viable cloned offspring (Schurmann et al., 2006; Egli et al., 2007).

Selection, preparation, and subzonal transfer of the donor cell

The entire intact donor cell, i.e. nucleus plus cytoplasm, is isolated from a cell culture dish by trypsin treatment and is inserted under the zona pellucida in intimate contact with the cytoplasmic membrane of the oocyte with the aid of an appropriate micropipette.

A large variety of differentiated somatic cell types, including mammary epithelial cells, cumulus cells, oviductal cells, leucocytes, hepatocytes, granulosa cells, epithelial cells, myocytes, neuronal cells, lymphocytes, immunologically relevant cells, Sertoli cells, germ cells, and, most frequently, fibroblasts, have been successfully used as donors for the production of cloned animals (Brem and Kuhholzer, 2002; Hochedlinger and Jaenisch, 2002; Miyoshi et al., 2003; Eggan et al., 2004; Oback and Wells, 2007). It is unclear which cell type is best for nuclear transfer into oocytes. No differences were found when the efficiency of cloning was compared using various somatic cell types, including those of adult, newborn or fetal, female or male donor cattle (Kato et al., 2000). However, some terminally differentiated, highly specialized cells such as cardiomyocytes cannot be reprogrammed with high enough efficiency with current cloning protocols even when cardiac-specific gene expression was abolished immediately after fusion and activation (Schwarzer et al., 2006). Although initial experience suggested that cloning with adult somatic cells was only successful when cells were from the female reproductive tract, including the mammary epithelium, cumulus, granulosa, or oviductal cells, male mice were eventually cloned from tail-tip cells (Wakayama and Yanagimachi, 1999) and subsequently similar developmental rates were observed for embryos cloned from either male or female nuclei in cattle and mice (Kato et al., 2000; Wakayama and Yanagimachi, 2001). Cells from early passages are most often chosen for somatic cloning, but high rates of development have also been obtained when donor cells from later passages of adult somatic cells were employed (Kubota et al., 2000). Fetal cells, specifically fibroblasts, have frequently been used in somatic cloning experiments with the main agricultural species because they are thought to have less genetic damage and a higher proliferation capacity than adult somatic cells (Kues et al., 2008a).

The successful cloning of mice from terminally differentiated cells such as B and T lymphocytes or neurons demonstrated unequivocally that a fully differentiated nucleus can be returned to a genetically totipotent stage (Hochedlinger and Jaenisch, 2002; Eggan et al., 2004). However, it is still unclear whether the differentiation status of the donor cell is relevant to the success of somatic cloning. Comparative data are available for mice. When testing mouse hematopoietic cells at various stages of differentiation, i.e. hematopoietic stem cells, progenitor cells, and granulocytes, it was reported that cloning efficiency actually increased with differentiation and terminally differentiated post-mitotic granulocytes yielded cloned pups with the greatest efficiency (Sung et al., 2006). However, these results were subsequently challenged and related to specific properties of hematopoietic cells. The endpoint of cloning in mice can be based on the production of ES cells from cloned blastocysts rather than the production of live offspring. Less-differentiated cells were more effective in cloning mice than differentiated cells when measured by ES cell production from cloned blastocysts (Hochedlinger and Jaenisch, 2007). Cloning efficiency, defined as the potential to derive pluripotent ES cells from cloned blastocysts, was <10% with differentiated donor cells, including lymphocytes, natural killer cells, and neurons (Hochedlinger and Jaenisch, 2002; Eggan et al., 2004; Li et al., 2004; Inoue et al., 2005), ~20% with fibroblasts (Wakayama et al., 2005), and reached up to ~50% with embryonic stem cells, embryonic carcinoma cells, or neuronal stem cells (Blelloch et al., 2004, 2006). However, other investigators have not found that undifferentiated cells (neural stem cells) yielded higher rates of cloned mouse production than differentiated donor cells such as Sertoli and cumulus cells (Mizutani et al., 2006). Similarly, bovine muscle cells with different differentiation status, i.e. myotubes, myogenic precursor cells, and muscle fibroblasts, did not give different levels of success in *in vitro* development to blastocysts or in development *in vivo* after transferring the cloned embryos to foster mothers (Green et al., 2007). While viable cloned offspring have been produced from bovine ES-like cells and from porcine fetal and adult somatic stem cells (Saito et al., 2003; Zhu et al., 2004; Hornen et al., 2007), it is not yet clear whether the differentiation status really affects cloning efficiency with somatic donor cells (Oback and Wells, 2007).

In experiments with mice, nuclei from various cancer cells, including leukemia, lymphoma, and breast cancer, could be reprogrammed by nuclear transfer and yielded apparently normal blastocysts. However, ES cells could not be derived from such blastocysts (Hochedlinger et al., 2004). In contrast, ES cells could be derived from blastocysts cloned from melanoma cells, and these ES cells were subsequently able to differentiate into various cell types. Chimeras obtained from these ES cells showed a high incidence of tumor formation (Hochedlinger et al., 2004) suggesting that the tumorigenic potential of the donor cells was not fully erased by the reprogramming process.

Whether donor cells need to be forced into a quiescent state by either serum starvation or treatment with cell cycle inhibitors is still a matter of debate. In most experiments, donor cells are induced to exit the cell cycle by serum starvation, which holds cells at the G0/G1 cell cycle stage (Campbell et al., 1996). Specific cyclin-dependent kinase inhibitors, such as roscovitin, have been reported to increase the efficiency of the cloning process, although final evidence in the form of healthy offspring is lacking (Miyoshi et al., 2003). Nevertheless, unsynchronized somatic donor cells have been successfully used to clone offspring in mice and cattle (Cibelli et al., 1998; Wakayama et al., 1999).

Attempts have been made to improve the success of somatic cloning by altering the epigenetic status of the donor cell by prior treatment with chemicals that affect its DNA methylation and/ or histone acetylation status. Treatment of bovine donor cells with trichostatin A (TSA), a DNA methylation inhibitor, improved *in vitro* development to blastocysts, but *in vivo* development was not tested (Wee et al., 2007). Cloning efficiency was significantly improved in a highly inbred strain of miniature pigs after treating the donor fibroblasts with a specific histone deacetylase inhibitor (Scriptaid). However, this improvement may have been specific for this inbred miniature pig strain, which otherwise could not be successfully cloned (Zhao et al., 2009). It remains to be shown whether treatment of donor cells with either a DNA methylation or histone deacetylase inhibitor ultimately increases yields of normal live cloned offspring. Current evidence indicates that it is primarily the source of the donor cells that seems to have the greatest impact on cloning success. After transferring embryos cloned from six different donor animals, viable calves could be obtained from only two donors (Powell et al., 2004). It is still necessary to develop reliable methods for the selection or modification of donor cells to increase the efficiency of somatic nuclear transfer.

133

Fusion of enucleated oocyte and donor cell

The two components are typically fused by short, high-voltage pulses through the point of contact between the two cells. The fusion pulse opens pores in the membrane and the smaller donor cell is incorporated into the cytoplasm of the recipient oocyte. In mice, instead of electrofusion, a piezoelectric microinjection tool can be used. The donor cell membrane is disrupted and separated from the nucleus by suction into a glass micropipette and the nucleus is then injected into the oocytes' cytoplasm (Wakayama et al., 1998). Once fused, the oocyte containing the donor cell or nucleus is called a reconstructed complex.

Activation of the reconstructed complex

In natural fertilization, the penetrating sperm activates the oocyte by which the egg becomes ready for sustaining embryonic and fetal development. The activation is a complex structural and biochemical process that includes alteration of the pH and the migration of vesicular granulae to the periphery of the oocyte to prevent polyspermy. Activation of nuclear transfer complexes attempts to mimic these events and is achieved either by short electrical pulses or by brief exposure to chemical substances such as ionomycin or dimethylaminopurin (DMAP). Both methods trigger calcium influx and initiation of the cell cycle.

Temporary *in vitro* culture of the reconstructed embryos

Cloned embryos can be cultured *in vitro* to the blastocyst stage (5–7 days) to assess the initial developmental competence prior to transfer into a foster mother. Culture systems for mouse and ruminant embryos are quite advanced and allow the routine production of 30–40% blastocysts from *in vitro* fertilized oocytes isolated from abattoir ovaries (Niemann et al., 2002; Biggers and Summers, 2008). However, cloned embryos appear to require modified culture conditions to obtain maximum development rates *in vitro*. Culture media lacking two of the usual components, EDTA and glutamine, enhanced the *in vitro* development of cloned mouse embryos (Dai et al., 2009). Supplementation of culture media with the antioxidant vitamin E during *in vitro* maturation of oocytes and donor cells enhanced blastocyst formation and decreased DNA damage in cloned bovine embryos (Wongsrikeao et al., 2007). Successful attempts have also been made to increase the developmental potential of cloned embryos by altering their epigenetic status. Supplementation of the culture medium with the deacetylase inhibitor trichostatin (TSA) to increase histone acetylation has met with some success. Supplementation of the culture medium with 50 nM TSA improved *in vitro* and *in vivo* development of cloned porcine embryos while other TSA doses did not improve development (Zhao et al., 2010). The TSA treatment increased acetylation only at specific histone sites in porcine embryos cloned from fibroblasts and the improvement of development was donor cell type-specific, with fibroblast-derived porcine embryos showing better development than embryos cloned from bone marrow cells (Martinez-Diaz et al., 2010). Supplementation of the medium with the histone deacetylase inhibitor valproic acid improved development of cloned miniature-pig embryos and maintained their ability to express the OCT4 gene, which is a critical transcription factor controlling the induction and maintenance of pluripotency (Miyoshi et al., 2010). These results show that exogenous modification of the epigenetic status of cultured cloned embryos may improve the developmental yield for certain types of donor cells.

Transfer to a foster mother or storage in liquid nitrogen ($-196°C$)

Cloned mouse embryos are usually surgically transferred into the oviducts or uterus of pseudopregnant foster mothers. Similarly, in pigs, the activated nuclear transfer complexes are frequently immediately transferred into the oviduct of a recipient animal. Porcine cloning success was significantly improved by a more careful selection of the recipients and by providing a 24 h asynchrony between the pre-ovulatory oviducts of the recipients and the reconstructed embryos (Petersen et al., 2008). Presumably, this gave the embryos additional time to achieve the necessary level of nuclear reprogramming and resulted in pregnancy rates of ~80% and only slightly reduced litter size (Petersen et al., 2008). Bovine embryos at the morula and blastocyst stages can be transferred non-surgically to the uterine horns of synchronized recipients using established embryo transfer procedures or can be frozen with cryopreservation protocols initially developed for *in vitro*-produced embryos. Somatic cells can readily be stored frozen in liquid nitrogen ($-196°C$) and upon thawing can be used as donor cells in SCNT to yield live offspring. Live mice have even been cloned from cells frozen at $-80°C$ or $-20°C$ for a prolonged period of time without cryoprotectant (Li and Mombaerts, 2008; Wakayama et al., 2008). This allows nuclear transfer techniques to be used for "resurrecting" animals and for programs designed to preserve valuable genomic resources.

SUCCESS RATES OF SOMATIC CLONING AND THE QUESTION OF NORMALITY OF CLONED OFFSPRING

The typical success rate (live births) of mammalian somatic nuclear transfer is apparently low: only 1–3% of the transferred embryos. However, for most species there is not much information and what has been published is largely based on early pioneering experiments. For cattle and pigs, larger data sets have been compiled and clearly show that cloning efficiency has steadily improved with greater technical skills and growing knowledge of the underlying mechanisms. Bovine cloning is now achieving a yield of 20–25% live cloned offspring (Panarace et al., 2007; Kues et al., 2008a). Porcine cloning success gives pregnancy rates as high as 80% although average litter size is moderately reduced compared to conventional commercial breeding figures (~6 piglets vs. 9–10 piglets). The yield of live births per embryo transferred is 3–5% (Petersen et al., 2008). Most embryonic losses in cattle occur prior to day 14 of gestation although surviving embryos may show various ultrastructural abnormalities such as secondary lysosomes and vacuoles, fewer mitochondria, polyribosomes, tight junctions, and filamental structures compared with their *in vitro*-produced counterparts (Alexopoulos et al., 2008). When considering these figures, one should be aware that natural embryonic mortality is 35–50% in pigs and cattle.

Pre- and post-natal development can be compromised and a variable proportion of the offspring in cattle, sheep, and mice have shown aberrant developmental patterns and increased perinatal mortality. These abnormalities include a wide range of symptoms, including extended gestation period, oversized offspring, aberrant placenta, cardiovasculatory problems, respiratory defects, immunological deficiencies, problems with tendons, adult obesity, kidney and hepatic malfunctions, deviant behavior, and a higher susceptibility to neonatal diseases. These are summarized as "large offspring syndrome" (LOS) and predominantly have been found in ruminants (sheep, cattle) and mice (Renard et al., 1999; Tamashiro et al., 2000; Ogonuki et al., 2002; Perry and Wakayama, 2002; Rhind et al., 2003; Panarace et al., 2007). The incidence is stochastic and has not been correlated with the aberrant expression of single genes or any specific pathophysiology. A new term, "abnormal offspring syndrome" (AOS), with subclassification according to the outcome of such pregnancies was later proposed to better reflect the broad spectrum of this pathological phenomenon (Farin et al., 2006). The general assumption is that the underlying cause of these pathologies is insufficient or faulty reprogramming of the transferred somatic cell nucleus, which primarily affects the growth and function of the placenta. Although the endometrium has great plasticity, the placental failures in bovine clone pregnancies seem to originate from abnormal embryo/fetal-maternal communication in the peri-implantation period (Bauersachs et al., 2009). Numerous genomic pathways important for metabolism and immune function are significantly altered in bovine clone pregnancies compared to those from commercially bred animals (Mansouri-Attia et al., 2009). Normally, the endometrium can fine-tune its physiological response to the presence of embryos. Bovine and ovine clone pregnancies usually show significantly fewer placentomes in surviving clones and increased caruncle tissue weight, which indicates excessive uterine tissue growth (Lee et al., 2004; Fletcher et al., 2007). Placentome, liver, and kidney overgrowth are predominant features in hydrops syndrome at day 150 of gestation (Lee et al., 2004). Ovine placentomes from clone pregnancies showed more shed trophoblast cells and fetal villous hemorrhaging (Fletcher et al., 2007).

Despite these reports of pathology, a critical survey of the published literature on cloning revealed that most cloned animals are healthy and develop normally (Cibelli et al., 2002). This is consistent with the observation that mammalian development is rather tolerant of minor epigenetic aberrations in the genome. Subtle abnormalities in gene expression do not appear to interfere with the survival of cloned animals (Humpherys et al., 2001). It has become clear that, once cloned offspring have survived the neonatal period and have reached six months of age (cattle, sheep), they are not different from age-matched controls with regard to numerous biochemical blood and urine parameters (Lanza et al., 2001; Chavatte-Palmer et al., 2002), immune status (Lanza et al., 2001), body score (Lanza et al., 2001), somatotrophic axis (Govoni et al., 2002), reproductive parameters (Enright et al., 2002), and yields and composition of milk (Pace et al., 2002). No differences were found in meat and milk composition of bovine clones when compared with age-matched counterparts; all parameters were within the normal range (Kumugai, 2002; Tian et al., 2005; Yang et al., 2007b). Similar results have been reported for cloned pigs (Archer et al., 2003). According to expert committees of the National Academy of Science of the USA, from the Japanese Ministry of Agriculture, Forestry and Fisheries (MAFF), the FDA (2007/2008) and the EFSA (European Food Safety Authority) (2008/2009), there is no scientific basis for questioning the safety of food derived from cloned animals (Rudenko et al., 2007; Fox, 2008). The official documents of the FDA relate to cattle, pigs, and goats; the EFSA document only relates to cattle and pigs. Products from cloned animals and their offspring are expected to enter the market in the next few years, reflecting regional political regulations. Due to the limited experience with somatic cloning, which has only been in general use since 1997, and the relatively long generation intervals in domestic animals, specific effects of cloning on longevity and senescence have not yet been fully assessed. Preliminary data indicate no pathology in second and later generations of cloned mice and cattle (Wakayama et al., 2000; Kubota et al., 2004). Indeed, reiterative cloning of mice for six generations revealed no pathology (Wakayama et al., 2000).

EPIGENETIC REPROGRAMMING
Basic epigenetic mechanisms
DNA METHYLATION

Normal development depends on a precise sequence of changes in the configuration of the chromatin that are primarily related to the acetylation and methylation status of the genomic DNA. These epigenetic modifications control the precise tissue-specific expression of genes. It is estimated that the mammalian genome with its ~25,000 genes contains 30,000–40,000 CpG islands, i.e. areas of between 200 and 2,000 nucleotides where the GC content is >60% and the ratio of CpG dinucleotides is >0.6. These CpG islands are predominantly found in the promoters of housekeeping genes but are also observed in tissue-specific genes (Antequera, 2003). The correct pattern of cytosine methylation in CpG dinucleotides is required for normal mammalian development (Li et al., 1993, Li, 2002). DNA methylation is also thought to play a crucial role in suppressing the activities of parasitic promoters and is thus part of the gene-silencing system in eukaryotic cells (Jones, 1999). Usually, methylation is associated with silencing of a given gene, but an increasing number of genes are found to be activated by methylation, particularly tumor-suppressor genes (Bestor and Tycko, 1996; Jones, 1999, Li, 2002). Epigenetic regulation is critical to achieving the biological complexity of multi-cellular organisms, and the complexity of epigenetic regulation increases with genomic size (Mager and Bartholomei, 2005).

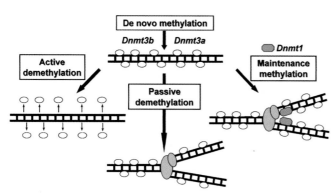

FIGURE 7.3

Methylation and demethylation of DNA (Dnmts). The drawing shows DNA modifications by methylation and the involvement of various DNA-methyltransferases (Dnmts) and their function during methylation, demethylation, and remethylation of a DNA strand.

DNA methylation critically depends on the activity of specific enzymes, the DNA methyltransferases (Dnmts) (Fig. 7.3). DNA-methytransferase1 (Dnmt1) is a maintenance enzyme that is responsible for restoring methylation to hemi-methylated CpG dinucleotides after DNA replication (Bestor, 1992). The oocyte-specific isoform, Dnmt1o, maintains maternal imprints. Dnmt3a and Dnmt3b catalyze *de novo* methylation and are thus critical for establishing DNA methylation during development (Hsieh, 1999; Okano et al., 1999). Dnmt3L co-localizes with Dnmt3a and -b and presumably is involved in establishing specific methylation imprints in the female germline (Bourc'his et al., 2001b). Dnmt activities are linked with histone deacetylases (HDACs), histone methyltransferases (HMTs), and several ATPases and are part of a complex system regulating chromatin structure and thus gene expression (Burgers et al., 2002).

During early mammalian development, reprogramming of the DNA is observed shortly before and shortly after formation of the zygote (Fig. 7.4). Paternal DNA is actively demethylated after fertilization, while the female DNA undergoes passive demethylation in several species, including murine, bovine, porcine, rat, and human zygotes (Mayer et al., 2000; Oswald et al., 2000; Dean et al., 2001; Santos et al., 2002; Beaujean et al., 2004; Xu et al., 2005). Mechanisms of active DNA methylation during pronuclear maturation are highly conserved among mammalian species (Lepikhov et al., 2008). Subsequently, the embryonic DNA is increasingly remethylated at species-specific time points between the two-cell and the blastocyst stages (Fig. 7.4; Dean et al., 2001). These mechanisms ensure that the critical steps of early development, such as timing of first cell division, compaction, blastocyst formation, expansion, and hatching, are regulated by a well-orchestrated succession of gene expression patterns.

137

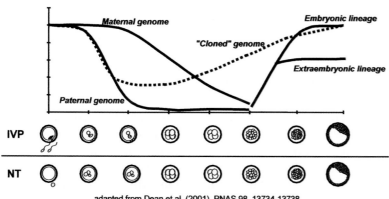

adapted from Dean et al. (2001), PNAS 98, 13734–13738

FIGURE 7.4

Methylation reprogramming of the genome during early bovine development. The paternal genome is rapidly and actively demethylated after fertilization, while the maternal genome becomes passively demethylated over time during cleavage. The embryonic genome is remethylated starting at the morula stage; the two cell lineages of the bovine blastocyst are methylated to different levels. In cloned embryos the methylation pattern may be completely different.

IMPRINTING

Imprinting represents a specific function of DNA methylation. A typical feature of genomic imprinting is that the two alleles of a given gene are expressed differently. Usually one allele, either the maternal or the paternal, is silenced throughout development by covalent addition of methyl groups to cytosine residues in CpG dinucleotides (Constancia et al., 2004). This DNA methylation occurs in imprinting control regions (ICRs) of DNA and is established by the *de novo* methyltransferase Dnmt 3a. A typical feature of imprinted genes is that they are found in clusters and the ICRs exert regional control of gene expression (Reik and Walter, 2001). In the mouse no more than 50, and in humans ~80, imprinted genes have been identified (Dean et al., 2003, Constancia et al., 2004). Imprinting is a genetic mechanism that regulates the demand, provision, and use of resources in mammals, particularly during fetal and neonatal development. Usually genes expressed from the paternally inherited allele increase resource transfer from the mother to the fetus, whereas maternally expressed genes reduce this transfer to secure the mother's well-being (Constancia et al., 2004). Imprints are established during development of germ cells into sperm and eggs. The germ line resets imprints such that mature gametes reflect the sex of a specific germ line due to the sequence of erasure and establishment (Reik and Walter, 2001).

HISTONE MODIFICATIONS

Histones are the main protein component of chromatin and the core histones (H2A, H2B, H3, and H4) form the nucleosome. Covalent post-translational modifications of histones play a crucial role in controlling the capacity of the genome to store, release, and inherit biological information (Fischle et al., 2003). Numerous histone and chromatin-related regulatory options are available, including histone acetylation, phosphorylation and methylation. Binary switches and modification cassettes have been suggested as new concepts to understand the enormous versatility of histone function (Fischle et al., 2003). Specific histone methyltransferases (HMTs) catalyze methylation at specific positions of the nucleosome in mammalian cells. Deacetylation of histones is carried out by isoforms of histone deacetylases (HDACs). Histone acetyltransferases are involved in diverse processes including transcriptional activation, gene silencing, DNA repair, and cell-cycle progression and thus play a critical role in cell growth and development (Carrozza et al., 2003).

Reprogramming can be divided into the pre-zygotic phase, which includes acquisition of genomic imprints and the epigenetic modification of most somatic genes during gametogenesis. X-chromosome inactivation and adjustment of telomere lengths are prominent examples of post-zygotic reprogramming (Hochedlinger and Jaenisch, 2003).

Pre-zygotic reprogramming

IMPRINTED GENE EXPRESSION IN CLONED EMBRYOS AND FETUSES

The majority of imprinted genes are involved in fetal and placental growth and differentiation, which makes them promising candidates for unraveling the developmental aberrations found after somatic nuclear transfer. Disruption of imprinted genes has been observed in cloned mouse embryos (Mann et al., 2003). Knowledge about imprinted genes in bovine development is limited; only one out of eight genes known to be imprinted in mice appeared to be imprinted in bovine blastocysts (Ruddock et al., 2004). The imprinted genes NDN and XIST were found to be aberrantly expressed in cloned bovine embryos compared with their *in vitro*-produced counterparts. This aberrant expression was at least partially associated with histone H4 acetylation at position AcH4K5 (Wee et al., 2006). The normally imprinted H19 gene was expressed bi-allelically in bovine stillborn cloned calves, suggesting that aberrant imprinting is associated with abnormal development (Zhang et al., 2004). In surviving calves, faulty H19 imprinted expression was corrected in the offspring, showing that the program of germ line development was normal (Zhang et al., 2004). Genomic imprinting can be disrupted at the

XIST (X-chromosome inactive specific transcript) locus in cloned fetuses, whereas IGF2 and GTL2 are properly expressed in fetal and placental tissue (Dindot et al., 2004). As in other species, the bovine IGF2 gene is controlled by an extremely complex regulatory mechanism based on multiple promoters, alternative splicing, and genomic imprinting, that can be severely perturbed in cloned fetal, neonatal, and adult tissue (Curchoe et al., 2005). The IGF2 gene is critically involved in fetal and placental development and known to be imprinted in mice (Constancia et al., 2002). A differentially methylated region (DMR) has been discovered in exon 10 of the bovine IGF2 gene and provides a diagnostic tool for in-depth studies of bovine imprinting (Gebert et al., 2006). Using bisulfite sequencing, sex-specific DNA methylation patterns within this DMR in bovine blastocysts produced *in vivo*, by *in vitro* fertilization and culture, by SCNT, and by androgenesis or parthenogenesis were investigated. As expected, in *in vivo* embryos, DNA methylation was removed from this intragenic DMR after fertilization and was partially replaced by the blastocyst stage. DNA methylation was significantly lower in female than in male blastocysts and this sexual dimorphism was maintained in SCNT embryos and can be used as evidence for correct methylation reprogramming (Gebert et al., 2009). Aberrant expression of genes from the insulin-like growth factor (IGF) family was observed in cloned embryos on day 7 and in conceptuses from day 25 (Moore et al., 2007), indicating perturbed imprinting. The SNRPN-imprinted genomic locus was hypomethylated in day 17 cloned fetuses compared to *in vivo*- and *in vitro*-produced controls, indicating faulty reprogramming or maintenance of methylation imprints at this locus (Lucifero et al., 2006). Severe loss of DMR methylation of the SNRPN-imprinted gene was observed in cloned day 17 and day 40 fetuses, and bi-allelic expression was found in all tissues analysed (Suzuki et al., 2009).

Expression of the bovine imprinted genes IGF2, IGF2R, and H19 was aberrant in eight organs of deceased cloned calves. With the exception of IGF2 in muscle, these genes were expressed within the normal range in the tissues of surviving clones (Yang et al., 2005). Thus, the aberrant expression of genes that are normally imprinted may be directly implicated in the higher neonatal mortality in cloned cattle. This assumption is supported by the aberrant expression of other imprinted genes, such as PEG 3, MAOA, XIST, and PEG, in four aborted cloned calves (Liu et al., 2008). Aberrant expression of genes of the IGF family was found in several organs of cloned calves that died shortly after birth when the kidney was most affected (Li et al., 2007). Current data indicate that normal expression of the IGF2 gene and other members of this gene family is critical for normal embryonic and fetal development.

SOMATIC CELL NUCLEAR TRANSFER AND EMBRYONIC GENE EXPRESSION PATTERNS

Somatic cloning typically uses the unfertilized matured oocyte as the recipient cell. Reprogramming must occur within the short interval between the transfer of the donor cell into the oocyte and the initiation of embryonic transcription, the timing of which is species-specific. In the mouse, embryonic transcription begins at the two-cell stage, that of the pig at the four-cell stage, and that of sheep, cattle, and humans at the 8–16-cell stage (Telford et al., 1990; Kues et al., 2008b). Early events of nuclear and nucleolar reprogramming have been studied in bovine SCNT-derived embryos (Oestrup et al., 2009). During the first three hours after SCNT, the chromatin of the transferred nucleus gradually decondensed towards the periphery and the nuclear envelope reformed. Then the somatic cell nucleus gained a pronucleus-like appearance and displayed nucleolar precursor bodies (NPB), suggesting ooplasmic control of development (Oestrup et al., 2009). The effects of somatic cloning on mRNA expression patterns have mostly been analyzed in bovine morula and blastocyst stages and numerous genes related to specific physiological functions have been identified as aberrantly expressed in cloned embryos as compared to their *in vivo*-derived counterparts (see Wrenzycki et al., 2005b). This group includes genes related to stress susceptibility, growth factor signaling, imprinting, trophoblast formation and function, sex chromosome-related mRNA expression, and X-chromosome inactivation (Wrenzycki et al., 2005b). The mRNA expression profile of genes critical for epigenetic reprogramming during early development, including the histone modifiers HDAC2, HAT1, SUV39H1, G9A, and HP1 and the DNA methyltransferases (DNMTs), was significantly altered in cloned bovine blastocysts compared with their *in vivo*-produced counterparts, suggesting widespread epigenetic dysregulation (Nowak-Imialek et al., 2008; Sawai et al., 2010).

Expression of the transcription factor Oct4 within a certain range is crucial for maintaining toti- and pluripotency in early embryos. Oct4 is a transcription factor for a panel of developmentally important genes (Niwa et al., 2000; Pesce and Schöler, 2001). Aberrant spatial expression of Oct4 was found in murine embryos cloned from cumulus cells (Boiani et al., 2002). In a high proportion, up to 40% of cloned mouse embryos, Oct4 regulated genes were found to be aberrantly expressed due to faulty reactivation of Oct4 (Bortvin et al., 2003). These findings indicate that dysregulation of the pluripotent state in embryonic cells can contribute to developmental failure in cloned embryos. Using an Oct4/GFP reporter construct, it was shown that bovine SCNT embryos initiate activation of the Oct4 promoter during the fourth cell cycle. Later in preimplantation development, Oct4 expression differed substantially between individual embryos and was thought to be associated with embryonic developmental potential (Wuensch et al., 2007).

Data from our laboratory have shown that DNMT 1 mRNA expression was significantly increased in cloned bovine embryos compared to *in vivo*-derived controls. Similar observations have been made for DNMT 3a, while DNMT 3b expression did not differ between cloned, *in vitro*-produced and *in vivo*-produced bovine embryos (Wrenzycki and Niemann, 2003). Expression of DNMT 1 and two other chromatin remodeling genes was abnormal in the majority of cloned bovine embryos on day 7 and day 13, suggesting that insufficient nuclear reprogramming caused retarded development. Blastocyst development and DNMT1 expression were to some extent correlated with DNMT1 levels in donor cells, and donor cells in which the DNMT transcription level had been reduced prior to use in SCNT yielded higher rates of development (Giraldo et al., 2008). Mice cloned from cumulus cells show aberrant DNMT 1 localization and expression (Chung et al., 2003).

Using an array assay specific for bovine embryo genomic activation, it was found that endogenous long terminal repeat (LTR) retrotransposons and mitochondrial transcripts were upregulated and transcripts involved in ribosomal protein function were downregulated in cloned bovine embryos at the morula stage. These results demonstrate specific categories of transcripts that are more sensitive to somatic reprogramming and may affect embryo viability more than other gene transcripts (Bui et al., 2009).

These findings suggest perturbation of the normal wave of de- and remethylation in early development, which can be associated with developmental abnormalities in cloned animals. The pattern of aberrations in mRNA expression was extremely variable in embryos derived by *in vitro* production and/or cloning. Embryo production methods thus cause significant up- or downregulation and *de novo* induction or silencing of genes critically involved in embryonic and fetal development (Niemann and Wrenzycki, 2000). Some of the aberrant expression patterns found in cloned blastocysts could be the result of aberrant allocation of cells to the inner cell mass (ICM) and trophectoderm (Koo et al., 2003). But in most cases faulty expression patterns seem to be related to epigenetic errors rather than morphological deviations.

Extended *in vitro* culture of mammalian embryos alone is known to result in aberrations in mRNA expression patterns, affecting imprinted and non-imprinted genes (Young et al., 2001; Wrenzycki et al., 2001a). In the case of cloning, it is difficult to discriminate between the effect of *in vitro* culture and dysregulation due to the cloning process. An analysis using a bovine cDNA microarray with 6,298 unique sequences revealed that the mRNA expression profile of cloned bovine embryos was completely different from that of the donor cells and was surprisingly similar to that of naturally fertilized embryos (Smith et al., 2005), thus confirming previous RT-PCR analyses (Wrenzycki et al., 2001b, 2005a,b). A greater number of genes was differentially expressed in comparisons of artificial insemination (AI) and *in vitro* fertilization (IVF) embryos ($n = 198$) and between nuclear transfer (NT) and IVF embryos ($n = 133$) than between NT and AI embryos ($n = 50$), indicating that cloned embryos had undergone significant nuclear reprogramming at the blastocyst stage (Smith et al., 2005). In this case, it was suggested that aberrations cause effects later in development during organogenesis because small reprogramming errors are magnified downstream in development. Using the bovine genomic Affymetrix microarray, significant differences in the mRNA expression profile were found between bovine embryos cloned from fibroblasts and *in vitro* fertilized and cultured embryos prior to the blastocyst stage. Abnormal OCT4 expression was considered the most critical factor in deteriorated development (Aston et al., 2010). A global gene expression analysis of bovine SCNT-derived blastocysts and cotyledons isolated from cloned pregnancies using the Affymetrix microarray revealed only 28 differentially expressed genes between SCNT and AI-derived blastocysts and 19 differentially expressed cotyledon genes, with none of the differentially expressed genes being common to both groups. Several of the genes were either previously unknown or not well annotated (Aston et al., 2009). Analysis of the mRNA expression profile of day 60 placental tissue revealed several genes that seemed to be associated with embryonic death, including aberrant expression profiles for IGF2, HBA1, HBA2, SPTB, and SPTBN1 in cloned placental material versus conventionally produced tissue (Oishi et al., 2006). Aberrant expression of genes involved in various developmentally important pathways (including NOTCH, hedgehog receptor tyrosine kinase, JAK/STAT, wingless related (WNT), and transforming growth factor-β (TGF-β)) was found in cloned porcine fetuses on day 26, indicating unbalanced regulation of critical pathways with subsequent consequences for embryo survival (Chae et al., 2008).

We have developed the hypothesis that deviations from the normal pattern of mRNA expression that are observed in the early preimplantation embryo persist throughout fetal development up to birth and that the many effects of this period of culture only become manifest later in development (Niemann and Wrenzycki, 2000). Consistent with this hypothesis, genes aberrantly expressed in blastocysts were also aberrantly expressed in the organs of clones that died shortly after birth (Li et al., 2005). This is particularly true for XIST and heat shock protein (HSP) for which aberrant expression patterns had been found in cloned blastocysts (Wrenzycki et al., 2001b, 2002). The recently published comprehensive Affymetrix array analysis of gene expression and transcriptome dynamics of *in vivo* developing bovine embryos serves as a physiological standard for "normal" mRNA expression in preimplantation embryos against which embryos from other production methods and other species can be compared and should thus be useful for improving assisted reproductive technologies, including SCNT cloning (Kues et al., 2008b).

DNA METHYLATION PATTERNS AND HISTONE MODIFICATIONS IN CLONED EMBRYOS AND FETUSES

DNA demethylation is a first step in reprogramming and is essential for Oct4 transcription (Simonsson and Gurdon, 2004). Failure of demethylation is associated with impaired development in cloned mice embryos (Yamazaki et al., 2006). It is critical to assess to what extent the chromatin changes required in the reprogramming of an adult somatic donor nucleus are similar to the changes that take place in gametogenesis and fertilization (Jaenisch and Wilmut, 2001). Indeed, studies in mice suggest that nuclear reprogramming by SCNT utilizes the same chromatin remodeling mechanisms that are active upon fertilization (Chang et al., 2010). Recently, a first attempt was made to describe DNA methylation profiles after SCNT in bovine blastocysts. For the first time, broad demethylation of the genomic DNA in somatic cells upon bovine SCNT was demonstrated (Niemann et al., 2010). A panel of 41 amplicons representing 25 developmentally important genes on 15 chromosomal locations (a total of 1,079 CpG sites) was used to analyze somatic cells from which embryos were cloned and compared this methylation profile with the methylation of SCNT bovine blastocysts, bovine blastocysts produced *in vitro*, and bovine embryos developing *in vivo*. Massive epigenetic reprogramming was demonstrated by reduced levels of methylation in the embryos (Fig. 7.5). Analysis of the 28 most informative amplicons (hotspot loci) revealed subsets of amplicons with methylation patterns that were unique to each class of embryo and may indicate metastable epialleles (Niemann et al., 2010). This subset of amplicons can be used to evaluate blastocyst quality and reprogramming after SCNT.

The abnormalities in cloned fetuses and live offspring cannot simply be due to the source of the donor nuclei. The most likely explanation for the variability is that it reflects the extent of

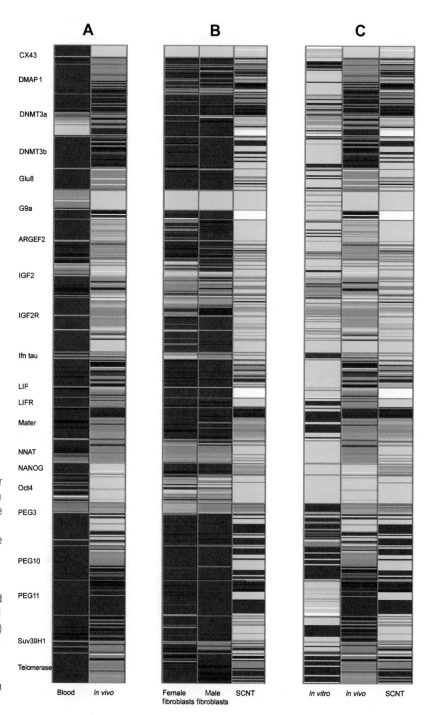

FIGURE 7.5

Differences in methylation for 21 genes that play important roles in early mammalian development (heat map). DNA was derived from peripheral blood mononuclear cells (PBMCs), primary fibroblasts, and bovine embryos that were produced *in vivo* by insemination, *in vitro*, or by somatic cell nuclear transfer (SCNT). Analyzed genes are separated by red lines with each row representing the methylation status of a single CpG. When genes are represented by two amplicons, these are separated by a gray line. Methylation of single CpGs is visualized by a color code ranging from yellow (0% methylation) to green (50% methylation) to blue (100% methylation); white: no CpG information. Differentiated somatic cells are more heavily methylated (blue) while the embryonic samples are less methylated (yellow). Column A shows a comparison of PBMC (blood) and *in vivo* embryos. Column B shows a comparison of SCNT blastocysts with the fibroblasts from which they were produced. Column C shows a comparison of the three types of embryos: *in vitro*, *in vivo*, and SCNT (from Niemann et al., 2010).

failure in genomic reprogramming of the transferred nucleus. Cloned embryos all show aberrant patterns of the global DNA methylation (Kang et al., 2001a,b; Dean et al., 2001). The maintenance of high methylation levels during cleavage is thought to be related to the presence of the somatic form of DNMT, an enzyme brought by the somatic donor cell nucleus into the cloned embryo. This probably interferes with the genome-wide demethylation process that takes place in a normal preimplantation embryo (Reik et al., 2001). Methylation reprogramming is delayed and incomplete in cloned bovine embryos (Bourc'his et al., 2001a). A high degree of variability is observed among individual embryos with regard to methylation levels (Dean et al., 2001). At present it is not fully clear whether the aberrant methylation stems from a defective demethylation of the transferred somatic nucleus or is a consequence of failed nuclear reorganization. Only cloned ovine embryos that show reorganized chromatin appear to survive the early embryonic phase (Beaujean et al., 2004). Attempts to improve the developmental capacity of bovine cloned embryos by either complete or partial erasure of DNA methylation/acetylation of the donor cell by treatment with specific inhibitors prior to use in nuclear transfer have met with limited success (Enright et al., 2003, 2005).

In support of the hypothesis that aberrant mRNA expression patterns persist throughout subsequent development (Niemann and Wrenzycki, 2000), epigenetic analysis revealed that methylation errors produced early in preimplantation development are in fact maintained throughout development and these genome-wide epigenetic aberrations can be identified in cloned bovine fetuses (Cezar et al., 2003). The proportion of methylated cytosine residues is reduced in cloned fetuses compared to *in vivo*-produced controls and survivability of cloned bovine fetuses was found to be closely related to the reduced global DNA methylation status (Cezar et al., 2003). Significant hypermethylation was detected in the liver tissue of cloned bovine fetuses and was found to be correlated with fetal overgrowth (Hiendleder et al., 2004a). These results show that developmental abnormalities can be associated with both hypo- and hypermethylation during fetal bovine development. Significant differences with regard to DNA methylation of the repetitive satellite I sequence were observed between *in vivo*-produced and cloned bovine embryos. The DNA methylation levels of *in vivo*-derived embryos increased from the blastocyst to the elongation stage (day 16 post-insemination) while in cloned conceptuses DNA methylation remained unchanged in the embryonic disc and was significantly reduced in trophectodermal tissue over this time period (Sawai et al., 2010).

Remarkably, the degree of demethylation of repetitive sequences in the donor genome seems to be determined by the recipient ooplasm and not by the donor cell. Ooplasm from different species may have different capacity to demethylate specific genes (Chen et al., 2006). The cytoplasm of the bovine oocyte may be particularly advantageous in this respect. The use of defined sources of highly effective recipient oocytes could render somatic cloning more efficient and could give significant improvements in the cloned phenotype (Hiendleder et al., 2004b).

Post-zygotic reprogramming

X-CHROMOSOME INACTIVATION AFTER SOMATIC CLONING

X-chromosome inactivation is the developmentally regulated process by which one of the two X-chromosomes in female mammals is silenced early in development to provide dosage compensation for X-linked genes. A single X-chromosome is sufficient, as shown in XY males (Lyon, 1961). Although the mechanism of X-chromosome inactivation is not yet fully understood, the paternal X-chromosome is typically inactivated by DNA methylation and remains inactive in placental tissue, while in the embryo proper either the paternal or maternal X-chromosome can be randomly selected on a cell-by-cell basis for inactivation, leading to a mosaic pattern in adult cells (Hajkova and Surani, 2004). Recent findings in the mouse revealed that the paternal imprint in the inner cell mass (ICM), i.e. the pluripotent cells that give rise to the fetus, is erased from the paternal X-chromosome late in preimplantation development followed by random X-inactivation (Mak et al., 2004). The paternal

X-chromosome is partly silent at fertilization and becomes fully inactivated at the two- or four-cell stage (Huynh and Lee, 2003; Okamoto et al., 2004). Female somatic nuclear transfer-derived embryos inherit one active and one inactive X-chromosome from the donor cell. Messenger RNA expression analysis of bovine embryos cloned from adult donor cells at the blastocyst stage revealed a significant upregulation of XIST (X-inactivating specific transcript) compared to *in vitro-* and *in vivo-*derived embryos. Expression of X-chromosome-related genes is delayed in cloned as compared to *in vivo-*derived embryos (Wrenzycki et al., 2002). Premature X-inactivation was observed for the X-chromosome linked inhibitor of apoptosis (XIAP) gene in *in vitro-*produced bovine embryos compared with their *in vivo* counterparts (Knijn et al., 2005). These findings indicate perturbation of X-chromosome inactivation has occurred by the blastocyst stage after somatic cloning or *in vitro* fertilization and culture. In female bovine cloned calves, aberrant expression patterns of X-linked genes and hypo-methylation of XIST in various organs of stillborn calves were observed. Random inactivation of the X-chromosome was found in the placenta of deceased clones but skewed in that of live bovine clones (Xue et al., 2002). This aberrant expression pattern of X-chromosome inactivation initiated in the trophectoderm seems to have resulted from incomplete nuclear reprogramming. Similar findings were obtained in studies of cloned mouse embryos (Eggan et al., 2000).

TELOMERE LENGTH AND SOMATIC CLONING

Telomeres are the natural ends of linear chromosomes and play a crucial role in maintaining the integrity of the entire genome by preventing loss of terminal coding DNA sequences or end-to-end chromosome fusion. Telomeres are composed of repetitive DNA elements and specific DNA proteins, which together form a nucleoprotein complex at the ends of eukaryotic chromosomes (Blackburn, 2001). Although the sequence of these terminal DNA structures varies between organisms, mammalian telomeres are generally composed of a concatamer of short hexamers (5′-TTAGGG-3′). Changes in telomere length are closely related to aging and cancer (de Lange, 2002). As a general rule, some loss of telomeres occurs with each cell division as a result of the incomplete replication of the lagging strand. A specialized reverse transcriptase, the telomerase, is then required to maintain the natural length of telomeric DNA. This ribonucleoprotein enzyme is composed of two essential subunits: the telomerase RNA component (TERC) and the telomerase reverse transcriptase (TERT) component (Nakayama et al., 1998). Telomerase is critically involved in maintaining normal telomere length (Blasco et al., 1999). This enzyme is active in hematopoietic cells, cancer cells, germ cells, and early embryos.

Telomeres of the cloned sheep "Dolly," derived from adult mammary epithelial cells, were found to be shortened when compared to age-matched, naturally bred counterparts and telomere length reduction seemed to be correlated with telomere length of the donor cells (Shiels et al., 1999). Telomeres in sheep clones derived from cultured somatic cells were shortened compared to age-matched controls while offspring derived by sexual reproduction from clones had normal telomere length (Alexander et al., 2007). However, the vast majority of studies reported that telomere length in cloned cattle, pigs, goats and mice, is comparable with age-matched naturally bred controls even when senescent donor cells were used for cloning (see Jiang et al., 2004; Schaetzlein and Betts et al., 2005; Jeon et al., 2005; Rudolph, 2005). Regulation of telomere length is to some extent related to the donor cells employed for cloning. Telomere length in cattle cloned from fibroblasts or muscle cells was similar to that of age-matched controls while clones derived from epithelial cells did not have telomeres restored to normal length (Miyashita et al., 2002). A check point for elongation of telomeres to their species-determined length has been discovered at the morula-to-blastocyst transition in bovine and mouse embryos (Schaetzlein et al., 2004). Telomeres are at the level of the donor cells in cloned morulae (Fig. 7.6), whereas at the blastocyst stage telomeres have been restored to normal length (Fig. 7.7). The telomere elongation process at this particular stage of embryogenesis is telomerase-dependent since it was abrogated in telomerase-deficient mice

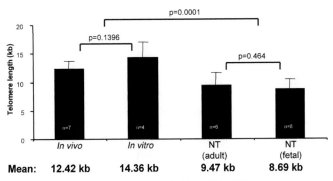

FIGURE 7.6

Telomere length in bovine morulae as determined by qFISH (quantitative fluorescent *in situ* hybridization). Telomeres in morulae produced *in vivo* from superovulated cows or *in vitro* have significantly longer telomeres compared to morulae cloned from either fetal or adult fibroblasts.

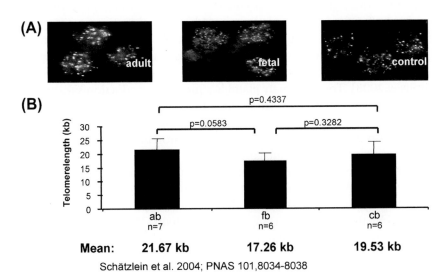

145

FIGURE 7.7

(A) Telomere length in bovine blastocysts as determined by qFISH. (B) The blastocysts cloned from either fetal (fb) or adult (ab) fibroblasts have similar telomere length to the *in vitro*-produced "control" embryos (cb). Telomere length is restored to physiological length at morula/blastocyst transition.

(Schaetzlein et al., 2004). The morula/blastocyst transition is a critical step in preimplantation development leading to first differentiation into two cell lineages: the inner cell mass and the trophoblast, which coincides with dramatic changes in morphology and gene expression (Niemann and Wrenzycki, 2000).

APPLICATION OF SOMATIC NUCLEAR TRANSFER
Reproductive cloning of transgenic animals

SCNT cloning holds great potential in three major areas: reproductive cloning, therapeutic cloning, and basic research (see Table 7.1). SCNT has emerged as a useful methodology for the production of transgenic farm animals and has replaced DNA microinjection of foreign DNA into pronuclei for this purpose. Improved transgenesis is of special relevance to the field of reproductive cloning due to a number of significant advantages over the previously used microinjection technology (Niemann and Kues, 2007). The major advantage is that somatic donor cells can be transfected with various gene constructs and those cells with the most appropriate expression patterns can be selected *in vitro* as donor cells. Even targeted genetic modifications such as a gene knockout by homologous recombination are compatible with

TABLE 7.1 Application Fields for Somatic Cloning

Reproductive cloning	Therapeutic cloning	Basic research
Genetically identical multiplets	Derivation of customized	Toti- and pluripotency
Transgenic animals (transfection,	ES cells	Reprogramming
homologous recombination)	Targeted differentiation	Dedifferentiation
Disease models	Regenerative cells	Redifferentiation
Maintenance of genetic	and tissues (autologous,	Aging
resources	heterologous)	Tumorigenesis
Animal breeding strategies (milk,	Tissue engineering	Epigenetics
meat, etc.)		Telomere biology
		Many other areas

primary cell cultures. The transgenic expression patterns render much more control than was possible with microinjection (Kues and Niemann, 2004).

Pre-eminent areas of application include the production of recombinant pharmaceutically valuable proteins in the mammary glands of transgenic livestock ("pharming") and the generation of transgenic pigs for xenotransplantation research. Gene pharming entails the production of recombinant pharmaceutically active human proteins in transgenic animals. This technology overcomes the limitations of conventional microbial or cell culture-based recombinant-DNA production systems and has advanced to the stage of commercial application (Kind and Schnieke, 2008). The mammary gland is a preferred production site, mainly because of the quantities of protein that can be produced in this organ using mammary gland-specific promoter elements and because GMP (good manufacturing practice) methods have been established for extraction and purification of the resultant proteins from milk. Products derived from the mammary glands of transgenic goats and sheep have progressed through advanced clinical trials and have been approved by regulatory bodies (Kind and Schnieke, 2008). Antithrombin III (ATIII) (ATryn® from GTC-Biotherapeutics, USA) produced in the mammary gland of transgenic goats was approved as a drug by the European Medicines Agency (EMA) in August 2006 and by the FDA in the USA in February 2009. This protein is the first product from a transgenic farm animal to become a registered drug. ATryn® is approved for the treatment of heparin-resistant patients undergoing cardiopulmonary bypass procedures. GTC-Biotherapeutics has also expressed numerous other transgenic proteins in the mammary glands of transgenic goats at concentrations of more than one gram per liter. The enzyme α-glucosidase (Pharming BV) from the milk of transgenic rabbits has orphan drug status and has been successfully used for the treatment of Pompe's disease. Similarly, recombinant C1 inhibitor (Pharming BV) produced in the milk of transgenic rabbits has completed phase III trials and is expected to be approved for use in human medicine in the near future. It is estimated that more than 12 recombinant proteins are currently in different phases of clinical testing (Kind and Schnieke, 2008). The overall global market for recombinant proteins from domestic animals is expected to reach $18.6 billion in 2013.

To close the growing gap between demand and availability of appropriate organs, transplant surgeons are now considering the use of xenografts from domesticated pigs. Overcoming the immunological hurdle for a discordant donor species such as the pig requires the prevention of both hyperacute rejection (HAR) and acute vascular rejection (AVR). The two strategies that have been successfully explored for long-term suppression of the HAR of porcine xenografts are (1) transgenic synthesis of human proteins regulating complement activity (RCAs) in the donor organ and (2) inactivation of the genes producing antigenic structures on the surface of the porcine donor organ. The most important xenotransplantation-relevant antigenic epitope is the α-gal-sugar chain modification of porcine surface proteins, i.e. the α-gal-epitope. Prolonged survival of xenotransplanted porcine organs, where the 1,3-α-galactosyltransferase (α-gal) gene has been knocked out, has been demonstrated. Using α-gal knockout pigs as organ donors and baboons as recipients, six-month survival has been achieved with transplanted hearts and three-month survival has been achieved with kidneys in a few experiments. The current approach to increasing survival time routinely beyond six months is to create donor pigs with multiple transgenes (multi-transgenic pigs) to further suppress the immunological response (see Petersen et al., 2009a). To this end, transgenic pigs expressing either human thrombomodulin (hTM) or human A20 gene (hA20) on top of one or two RCAs have been recently produced to suppress both HAR and the later stage coagulatory disorders observed in experimental porcine-to-primate xenotransplantation (Petersen et al., 2009b; Oropeza et al., 2009). Cloning is the only practical approach to producing multi-transgenic animals for this kind of research as it is the only way to select the genotype precisely. Reproducible survival of porcine xenografts for more than six months in non-human primate recipients is considered to be a necessary precondition to starting clinical trials with human patients (Petersen et al., 2009a).

Typical agricultural applications of transgenesis include improved carcass composition, lactational performance, wool production, enhanced disease resistance, and reduced environmental impact (Niemann and Kues, 2007).

Therapeutic cloning

Therapeutic cloning, whereby patient-specific embryonic stem cells are derived from cloned blastocysts, holds great promise for treatment of many human diseases. Embryonic stem cells have been produced from cloned blastocysts in mice and cattle (Wakayama et al., 2001; Wang et al., 2005), but not yet in humans. The generation of histocompatible tissue by nuclear transplantation has been demonstrated in a bovine model (Lanza et al., 2002). Despite expression of different mitochondrial DNA haptotypes, no rejection responses were observed when cloned renal cells were retransferred to the donor animal (Lanza et al., 2002). Skin grafts between bovine clones with different mitochondrial haplotypes were accepted long-term whereas non-cloned tissues were rejected (Theoret et al., 2006). The feasibility of therapeutic cloning has also been shown in mice, where correction of a genetic defect by cell therapy was demonstrated (Rideout et al., 2002). Mouse ES cells derived from cloned or fertilized blastocysts were similar with regard to their transcriptional profile and differentiation potential and thus have equal value as stem cells (Brambrink et al., 2006). The first preimplantation human embryos were produced from adult fibroblast nuclei; these gave only low blastocyst rates (French et al., 2008). Pre-selection based on the morphology of the first polar body, the perivitelline space, and cytoplasm granula distribution resulted in improved blastocyst yields (Yu et al., 2009). This may be beneficial in the production of human SCNT embryos for therapeutic cloning. The use of animal oocytes (bovine, rabbit) for reprogramming human somatic cells gives the same high level of blastocyst development as human-human SCNT. Nevertheless, the pattern of genomic reprogramming is significantly different between interspecies cloned embryos and intraspecies cloned embryos. Numerous genes were aberrantly expressed in the interspecies cloned embryos (Chung et al., 2009), raising doubts about the wisdom of using animal oocytes to overcome the shortage of human eggs.

Cells cloned from a patient have the advantage that they are accepted by that patient without permanent immune suppression. The production of customized ES cells will be invaluable in human medicine for the treatment of degenerative diseases because no immunosuppressive treatment is required. The concept of "therapeutic cloning" (Fig. 7.8) is fascinating but application in human medicine is still in its infancy. Current knowledge suggests that reprogramming of genes expressed in the inner cell mass, from which ES cells are derived, is rather efficient. Defects in the extraembryonic lineage are a major cause of the low success rate of reproductive cloning, but these would not affect derivation of ES cells (Yang et al., 2007a). However, major practical problems include the limited availability of human oocytes for reprogramming of the donor cells, the low efficiency of somatic nuclear transfer, the difficulty of inserting genetic modifications, the increased risk of oncogenic transformation, and the

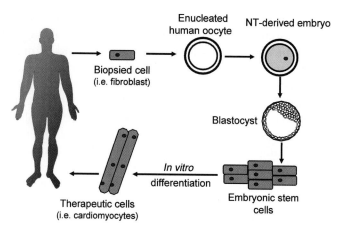

FIGURE 7.8
Principle of therapeutic cloning for the production of autologous cardiomyocytes.

epigenetic instability of embryos and cells derived from somatic cloning (Colman and Kind, 2000; Humpherys et al., 2001). Alternatives to nuclear transfer for reprogramming of somatic cell nuclei for the production of autologous therapeutic cells are being explored (Dennis, 2003). In humans, only preliminary data are available on therapeutic cloning (Cibelli et al., 2001). The papers on human ES cell isolation and cloning (Hwang et al., 2004, 2005) were retracted after discovery of significant fraud (Kennedy, 2006). The long-term goal of therapeutic cloning is to provide data on ES cell growth and differentiation, which may make it possible to stimulate proliferation and differentiation of endogenous stem cells and reparation of sick stocks.

INDUCED PLURIPOTENT STEM CELLS (ιPS)

Recent research has indicated that induced pluripotent stem cells (iPS) may emerge as an alternative for human therapeutic autologous ES cells produced by therapeutic cloning. In a revolutionary experiment, Takahashi and Yamanaka (2006) discovered that the genome of a differentiated somatic cell can be epigenetically reprogrammed to a pluripotent status by the expression of only four transcription factors, resulting in the generation of induced pluripotent stem cells (iPS) that possess pluripotent features equivalent to those of embryonic stem cells. Using viral gene transfer and combinations of Oct4, Sox2, c-myc, Klf4, Nanog, and LIN28, iPS cells have been produced from mice (Okita et al., 2007), humans (Takahashi et al., 2007; Yu et al., 2007), rats (Liao et al. 2009; Li et al., 2009), non-human primates (Liu et al., 2008), and pigs (Esteban et al., 2009; Ezashi et al., 2009; Wu et al., 2009). However, the porcine iPS reported to date have been dependent on the continued expression of the exogenous transcription factors (Esteban et al., 2009; Ezashi et al., 2009; Wu et al., 2009). The underlying mechanisms of the epigenetic reprogramming of somatic cells to iPS cells are not yet well understood, but are probably similar to those required for the epigenetic reprogramming involved in SCNT cloning. On a single cell basis, the overall efficiency of iPS reprogramming is low compared to the reprogramming that occurs in an oocyte, but viral transduction of cultured cells is successful when only one cell in a million is successfully reprogrammed. The use of viral vectors to transduce cells and the use of oncogenes such as c-myc and Klf4 present serious problems for the use of iPSC in regenerative medicine and the production of transgenic animals. Alternative approaches avoid integration of viral sequences in the host genome and reprogramming of somatic cells has been achieved by substituting viral vectors with small molecules (Lin et al., 2009; Li et al., 2009), by using non-integrating adenoviral vectors (Okita et al., 2008; Stadtfeld et al., 2008), and by completely avoiding the use of viruses by delivering the reprogramming factors in the form of proteins (Zhou et al.,

2009). Non-viral gene transfer using transposon technology has also been reported (Yusa et al., 2009). The advantages of using transposons for the generation of iPS cells are enhanced safety, a higher gene integration frequency (similar to the efficiency of viral transduction), and the possibility to remove the transposons from the iPSC genome after the reprogramming process.

CONCLUDING REMARKS

Since the birth of "Dolly," the first cloned mammal, significant progress has been made in increasing the efficiency of somatic cloning. While epigenetic reprogramming is considered to be essential for successful nuclear transfer-based cloning, it is not the only factor affecting cloning efficiency. Additional factors include improved tools for nuclear transfer itself and improvements in reproductive biology and animal husbandry (Hiiragi and Solter, 2005; Petersen et al., 2008). Altogether, it is apparent that there has been a steady increase in the efficiency of somatic mammalian cloning since it was first described in 1997. At the time of writing, cloned animals have been produced in 16 mammalian species. A variety of differentiated somatic cells can be successfully reprogrammed by SCNT, pulling the transferred somatic cell nucleus back from its differentiated status into the totipotent stage of the early embryo. This reprogramming is the most critical factor in the cloning protocol and also in the protocols for producing iPS cells. While the majority of offspring derived from somatic cloning are outwardly normal, cloning is still associated with pathological side-effects summarized as large offspring syndrome, which appear to be the result of incomplete and/or faulty reprogramming. The epigenetic changes essential for successful cloning involve the reversal of differentiation and rebooting the programs found in early preimplantation development that ensure the well-orchestrated gene expression pattern associated with normal embryonic development. DNA methylation and histone modifications seem to be critical for this process. Recent findings have also revealed key roles for small RNAs and proteins with domains that bind methylated DNA and DNA (Law and Jacobsen, 2010). X-chromosome inactivation and telomere length restoration represent additional post-zygotic epigenetic tasks that are important for successful cloning. Identification of the specific factors present in the ooplasm that are necessary for epigenetic reprogramming will give us a better understanding of the underlying mechanisms and will permit improved cloning efficiency. It is now clear that the ectopic expression of four or less transcription factors is sufficient to reprogram differentiated somatic cells into "induced pluripotent stem (iPS) cells." These developments owe their existence to the cloning of "Dolly" and afford a promising route towards autologous therapeutic cells. As a tool in basic research, somatic cloning has opened up the field of regenerative medicine and an expanding universe of epigenetic biology.

Acknowledgments

The authors gratefully acknowledge the valuable support during the course of the experiments on somatic cloning and reprogramming by various members of the Mariensee laboratory, specifically Doris Herrmann, Erika Lemme, Klaus-Gerd Hadeler, Lothar Schindler, Karin Korsawe, Hans-Herrmann Doepke, and Dr. Bjoern Petersen. We thank Susanne Tonks for her expert technical assistance in the production of this manuscript. The financial support of the research on which this review is based through various DFG grants is gratefully acknowledged.

References

Alexander, B., Coppola, G., Perrault, S. D., Peura, T. T., Betts, D. H., & King, W. A. (2007). Telomere length status of somatic cell sheep clones and their offspring. *Mol. Reprod. Dev., 74,* 1525–1537.

Alexopoulos, N. I., Maddox-Hyttel, P., Tveden-Nyborg, P., d'Cruz, N. T., Tecirlioglu, T. R., Cooney, M. A., et al. (2008). Developmental disparity between *in vitro*-produced and somatic cell nuclear transfer bovine days 14 and 21 embryos: Implications for embryonic loss. *Reproduction, 136,* 433–445.

Antequera, F. (2003). Structure, function and evolution of CpG island promoters. *Cell Mol. Life Sci., 60,* 1647–1658.

Archer, G. S., Dindot, S., Friend, T. H., Walker, S., Zaunbrecher, G., Lawhorn, B., et al. (2003). Hierarchical phenotypic and epigenetic variation in cloned swine. *Biol. Reprod., 69,* 430–436.

Aston, K. I., Li, G. P., Hicks, B. A., Sessions, B. R., Davis, A. P., Rickords, L. F., et al. (2010). Abnormal levels of transcript abundance of developmentally important genes in various stages of preimplantation bovine somatic cell nuclear transfer embryos. *Cell Reprogramming, 12*, 23–32.

Aston, K. I., Li, G. P., Hicks, B. A., Sessions, B. R., Davis, A. P., Winger, Q. A., et al. (2009). Global gene expression analysis of bovine somatic cell nuclear transfer blastocysts and cotyledons. *Mol. Reprod. Dev., 76*, 471–482.

Baguisi, A., Behboodi, E., Melican, D. T., Pollock, J. S., Destrempes, M. M., Cammuso, C., et al. (1999). Production of goats by somatic cell nuclear transfer. *Nat. Biotechnol., 17*, 456–461.

Bauersachs, S., Ulbrich, S. E., Zakhartchenko, V., Minten, M., Reichenbach, M., Reichenbach, H. D., et al. (2009). The endometrium responds differently to cloned versus fertilized embryos. *Proc. Natl. Acad. Sci. U.S.A., 106*, 5681–5686.

Beaujean, N., Taylor, J., Gardner, J., Wilmut, I., Meehan, R., & Young, L. (2004). Effect of limited DNA methylation reprogramming in the normal sheep embryo on somatic cell nuclear transfer. *Biol. Reprod., 71*, 185–193.

Berg, D. K., Li, C., Asher, G., Wells, D. N., & Oback, B. (2007). Red deer cloned from antler stem cells and their differentiated progeny. *Biol. Reprod., 77*, 384–394.

Bestor, T. H. (1992). Activation of mammalian DNA methyltransferase by cleavage of a Zn binding regulatory domain. *EMBO J., 11*, 2611–2617.

Bestor, T. H., & Tycko, B. (1996). Creation of genomic methylation patterns. *Nat. Genet., 12*, 363–367.

Betthauser, J. M., Pfister-Genskow, M., Xu, H., Golueke, P. J., Lacson, J. C., Koppang, R. W., et al. (2006). Nucleoplasmin facilitates reprogramming and *in vivo* development of bovine nuclear transfer embryos. *Mol. Reprod. Dev., 73*, 977–986.

Betts, D. H., Perrault, S. D., Petrik, J., Lin, L., Favetta, L. A., Keefer, C. L., et al. (2005). Telomere length analysis in goat clones and their offspring. *Mol. Reprod. Dev., 72*, 461–470.

Biggers, J. D., & Summers, M. C. (2008). Choosing a culture medium: making informed choices. *Fertil. Steril., 90*, 473–483.

Blackburn, E. H. (2001). The telomere and telomerase: nucleic acid-protein complexes acting in a telomere homeostasis system. A review. *Biochemistry (Mosc.), 62*, 1196–1201.

Blasco, M. A., Gasser, S. M., & Lingner, J. (1999). Telomeres and telomerase. *Genes Dev., 13*, 2353–2359.

Blelloch, R. H., Hochedlinger, K., Yamada, Y., Brennan, C., Kim, M., Mintz, B., et al. (2004). Nuclear cloning of embryonal carcinoma cells. *Proc. Natl. Acad. Sci. U.S.A., 101*, 13985–13990.

Blelloch, R., Wang, Z., Meissner, A., Pollard, S., Smith, A., & Jaenisch, R. (2006). Reprogramming efficiency following somatic cell nuclear transfer is influenced by the differentiation and methylation state of the donor nucleus. *Stem Cells, 24*, 2007–2013.

Boiani, M., Eckardt, S., Scholer, H. R., & McLaughlin, K. J. (2002). Oct4 distribution and level in mouse clones: consequences for pluripotency. *Genes Dev., 16*, 1209–1219.

Bortvin, A., Eggan, K., Skaletsky, H., Akutsu, H., Berry, D. L., Yanagimachi, R., et al. (2003). Incomplete reactivation of Oct4-related genes in mouse embryos cloned from somatic nuclei. *Development, 130*, 1673–1680.

Bourc'his, D., Le Bourhis, D., Patin, D., Niveleau, A., Comizzoli, P., Renard, J. P., et al. (2001a). Delayed and incomplete reprogramming of chromosome methylation patterns in bovine cloned embryos. *Curr. Biol., 11*, 1542–1546.

Bourc'his, D., Xu, G. L., Lin, C. S., Bollman, B., & Bestor, T. H. (2001b). Dnmt3L and the establishment of maternal genomic imprints. *Science, 294*, 2536–2539.

Brambrink, T., Hochedlinger, K., Bell, G., & Jaenisch, R. (2006). ES cells derived from cloned and fertilized blastocysts are transcriptionally and functionally indistinguishable. *Proc. Natl. Acad. Sci. U.S.A., 103*, 933–938.

Brem, G., & Kuhholzer, B. (2002). The recent history of somatic cloning in mammals. *Cloning Stem Cells, 4*, 57–63.

Briggs, R., & King, T. J. (1952). Transplantation of living nuclei from blastula cells into enucleated frogs' eggs. *Proc. Natl. Acad. Sci. U.S.A., 38*, 455–463.

Bui, L. C., Evsikov, A. V., Khan, D. R., Archilla, C., Peynot, N., Hénaut, A., et al. (2009). Retrotransposon expression as a defining event of genome reprogramming in fertilized and cloned bovine embryos. *Reprod., 138*, 289–299.

Burgers, W. A., Fuks, F., & Kouzarides, T. (2002). DNA methyltransferases get connected to chromatin. *Trends Genet., 18*, 275–277.

Campbell, K. H., McWhir, J., Ritchie, W. A., & Wilmut, I. (1996). Sheep cloned by nuclear transfer from a cultured cell line. *Nature, 380*, 64–66.

Carrozza, M. J., Utley, R. T., Workman, J. L., & Cote, J. (2003). The diverse functions of histone acetyltransferase complexes. *Trends Genet., 19*, 321–329.

Cezar, G. G., Bartolomei, M. S., Forsberg, E. J., First, N. L., Bishop, M. D., & Eilertsen, K. J. (2003). Genome-wide epigenetic alterations in cloned bovine fetuses. *Biol. Reprod., 68*, 1009–1014.

Chae, J. I., Yu, K., Cho, S. K., Kim, J. H., Koo, D. B., Lee, K. K., et al. (2008). Aberrant expression of developmentally important signaling molecules in cloned porcine extraembryonic tissues. *Proteomics, 8*, 2724–2734.

Chang, C. C., Gao, S., Sung, L. Y., Corry, G. N., Ma, Y., Nagy, Z. P., et al. (2010). Rapid elimination of the histone variant macroH2A from somatic cell heterochromatin after nuclear transfer. *Cell Reprogramming, 12*, 43–53.

Chavatte-Palmer, P., Heyman, Y., Richard, C., Monget, P., LeBourhis, D., Kann, G., et al. (2002). Clinical, hormonal, and hematologic characteristics of bovine calves derived from nuclei from somatic cells. *Biol. Reprod., 66*, 1596–1603.

Chen, T., Zhang, Y. L., Jiang, Y., Liu, J. H., Schatten, H., Chen, D. Y., et al. (2006). Interspecies nuclear transfer reveals that demethylation of specific repetitive sequences is determined by recipient ooplasm but not by donor intrinsic property in cloned embryos. *Mol. Reprod. Dev., 73*, 313–317.

Cheong, H. T., Takahashi, Y., & Kanagawa, H. (1993). Birth of mice after transplantation of early cell-cycle-stage embryonic nuclei into enucleated oocytes. *Biol. Reprod., 48*, 958–963.

Chesne, P., Adenot, P. G., Viglietta, C., Baratte, M., Boulanger, L., & Renard, J. P. (2002). Cloned rabbits produced by nuclear transfer from adult somatic cells. *Nat. Biotechnol., 20*, 366–369.

Chung, Y. G., Ratnam, S., Chaillet, J. R., & Latham, K. E. (2003). Abnormal regulation of DNA methyltransferase expression in cloned mouse embryos. *Biol. Reprod., 69*, 146–153.

Chung, Y., Bishop, C. E., Treff, N. R., Walker, S. J., Sandler, V. M., Becker, S., et al. (2009). Reprogramming of human somatic cells using human and animal oocytes. *Cloning and Stem Cells, 11*, 213–223.

Cibelli, J. B., Campbell, K. H., Seidel, G. E., West, M. D., & Lanza, R. P. (2002). The health profile of cloned animals. *Nat. Biotechnol., 20*, 13–14.

Cibelli, J. B., Kiessling, A. A., Cunniff, K., Richards, C., Lanza, R. P., & West, M. D. (2001). Somatic cell nuclear transfer in humans: pronuclear and early embryonic development. *J. Regener. Med., 2*, 25–31.

Cibelli, J. B., Stice, S. L., Golueke, P. J., Kane, J. J., Jerry, J., Blackwell, C., et al. (1998). Cloned transgenic calves produced from nonquiescent fetal fibroblasts. *Science, 280*, 1256–1258.

Colman, A., & Kind, A. (2000). Therapeutic cloning: concepts and practicalities. *Trends Biotechnol., 18*, 192–196.

Constancia, M., Hemberger, M., Hughes, J., Dean, W., Ferguson-Smith, A., Fundele, R., et al. (2002). Placental-specific IGF-II is a major modulator of placental and fetal growth. *Nature, 417*, 945–948.

Constancia, M., Kelsey, G., & Reik, W. (2004). Resourceful imprinting. *Nature, 432*, 53–57.

Curchoe, C., Zhang, S., Bin, Y., Zhang, X., Yang, L., Feng, D., et al. (2005). Promoter-specific expression of the imprinted IGF2 gene in cattle (Bos taurus). *Biol. Reprod., 73*, 1275–1281.

Dai, X., Hao, J., & Zhou, Q. (2009). A modified culture method significantly improves the development of mouse somatic cell nuclear transfer embryos. *Reproduction, 138*, 301–308.

de Lange, T. (2002). Protection of mammalian telomeres. *Oncogene, 21*, 532–540.

Dean, W., Santos, F., & Reik, W. (2003). Epigenetic reprogramming in early mammalian development and following somatic nuclear transfer. *Semin. Cell Dev. Biol., 14*, 93–100.

Dean, W., Santos, F., Stojkovic, M., Zakhartchenko, V., Walter, J., Wolf, E., et al. (2001). Conservation of methylation reprogramming in mammalian development: aberrant reprogramming in cloned embryos. *Proc. Natl. Acad. Sci. U.S.A., 98*, 13734–13738.

Dennis, C. (2003). Developmental reprogramming: take a cell, any cell.... *Nature, 426*, 490–491.

Dindot, S. V., Farin, P. W., Farin, C. E., Romano, J., Walker, S., Long, C., et al. (2004). Epigenetic and genomic imprinting analysis in nuclear transfer derived Bos gaurus/Bos taurus hybrid fetuses. *Biol. Reprod., 71*, 470–478.

Eggan, K., Akutsu, H., Hochedlinger, K., Rideout, W., III, Yanagimachi, R., & Jaenisch, R. (2000). X-Chromosome inactivation in cloned mouse embryos. *Science, 290*, 1578–1581.

Eggan, K., Baldwin, K., Tackett, M., Osborne, J., Gogos, J., Chess, A., et al. (2004). Mice cloned from olfactory sensory neurons. *Nature, 428*, 44–49.

Egli, D., Rosains, J., Birkhoff, G., & Eggan, K. (2007). Developmental reprogramming after chromosome transfer into mitotic mouse zygotes. *Nature, 447*, 679–685.

Enright, B. P., Kubota, C., Yang, X., & Tian, X. C. (2003). Epigenetic characteristics and development of embryos cloned from donor cells treated by trichostatin A or 5-aza-2′-deoxycytidine. *Biol. Reprod., 69*, 896–901.

Enright, B. P., Sung, L. Y., Chang, C. C., Yang, X., & Tian, X. C. (2005). Methylation and acetylation characteristics of cloned bovine embryos from donor cells treated with 5-aza-2′-deoxycytidine. *Biol. Reprod., 72*, 944–948.

Enright, B. P., Taneja, M., Schreiber, D., Riesen, J., Tian, X. C., Fortune, J. E., et al. (2002). Reproductive characteristics of cloned heifers derived from adult somatic cells. *Biol. Reprod., 66*, 291–296.

Esteban, M., Xu, J., Yang, J., Peng, M., Qin, D., Li, W., et al. (2009). Generation of induced pluripotent stem cell lines from Tibetan miniature pig. *J. Biol. Chem., 284*, 17634–17640.

Ezashi, T., Telugu, B., Alexenko, A., Sachdev, S., Sinha, S., & Roberts, M. (2009). Derivation of induced pluripotent stem cells from pig somatic cells. *Proc. Natl. Acad. Sci. U.S.A., 106*, 10993–10998.

Farin, P., Piedrahita, J. A., & Farin, C. E. (2006). Errors in development of fetuses and placentas from in vitro-produced bovine embryos. *Theriogenology, 65*, 178–191.

151

Fischle, W., Wang, Y., & Allis, C. D. (2003). Binary switches and modification cassettes in histone biology and beyond. *Nature, 425,* 475–479.

Fletcher, C. J., Roberts, C. T., Hartwich, K. M., Walker, S. K., & McMillen, I. C. (2007). Somatic cell nuclear transfer in the sheep induces placental defects that likely precede fetal demise. *Reproduction, 133,* 243–255.

Fox, J. L. (2008). Cloned animals deemed safe to eat, but labelling issues loom. *Nat. Biotech., 26,* 249–250.

French, A. J., Adams, C. A., Anderson, L. S., Kitchen, J. R., Hughes, M. R., & Wood, S. H. (2008). Development of human cloned blastocysts following somatic cell nuclear transfer (SCNT) with adult fibroblasts. *Stem Cells, 26,* 485–493.

Galli, C., Lagutina, I., Crotti, G., Colleoni, S., Turini, P., Ponderato, N., et al. (2003). Pregnancy: a cloned horse born to its dam twin. *Nature, 424,* 635.

Gebert, C., Wrenzycki, C., Herrmann, D., Groger, D., Reinhardt, R., Hajkova, P., et al. (2006). The bovine IGF2 gene is differentially methylated in oocyte and sperm DNA. *Genomics, 88,* 222–229.

Gebert, C., Wrenzycki, C., Herrmann, D., Gröger, D., Thiel, J., Reinhardt, R., et al. (2009). DNA methylation in the *IGF2* intrageneric DMR is re-established in a sex-specific manner in bovine blastocysts after somatic cloning. *Genomics, 94,* 63–69.

Giraldo, A. M., Hylan, D. A., Ballard, C. B., Purpera, M. N., Vaught, T. D., Lynn, J. W., et al. (2008). Effect of epigenetic modifications of donor somatic cells on the subsequent chromatin remodeling of cloned bovine embryos. *Biol. Reprod., 78,* 832–840.

Govoni, K. E., Tian, X. C., Kazmer, G. W., Taneja, M., Enright, B. P., Rivard, A. L., et al. (2002). Age-related changes of the somatotropic axis in cloned Holstein calves. *Biol. Reprod., 66,* 1293–1298.

Green, A. L., Wells, D. N., & Oback, B. (2007). Cattle cloned from increasingly differentiated muscle cells. *Biol. Reprod., 77,* 395–406.

Gurdon, J. B., & Byrne, J. A. (2003). The first half-century of nuclear transplantation. *Proc. Natl. Acad. Sci. U.S.A., 100,* 8048–8052.

Hajkova, P., & Surani, M. A. (2004). Development. Programming the X chromosome. *Science, 303,* 633–634.

Hiendleder, S., Mund, C., Reichenbach, H. D., Wenigerkind, H., Brem, G., Zakhartchenko, V., et al. (2004a). Tissue-specific elevated genomic cytosine methylation levels are associated with an overgrowth phenotype of bovine fetuses derived by *in vitro* techniques. *Biol. Reprod., 71,* 217–223.

Hiendleder, S., Prelle, K., Bruggerhoff, K., Reichenbach, H. D., Wenigerkind, H., Bebbere, D., et al. (2004b). Nuclear-cytoplasmic interactions affect in utero developmental capacity, phenotype, and cellular metabolism of bovine nuclear transfer fetuses. *Biol. Reprod., 70,* 1196–1205.

Hiiragi, T., & Solter, D. (2005). Reprogramming is essential in nuclear transfer. *Mol. Reprod. Dev., 70,* 417–421.

Hochedlinger, K., & Jaenisch, R. (2002). Monoclonal mice generated by nuclear transfer from mature B and T donor cells. *Nature, 415,* 1035–1038.

Hochedlinger, K., & Jaenisch, R. (2003). Nuclear transplantation, embryonic stem cells, and the potential for cell therapy. *N. Engl. J. Med., 349,* 275–286.

Hochedlinger, K., & Jaenisch, R. (2007). On the cloning of animals from terminally differentiated cells. *Nat. Genet., 39,* 136–137.

Hochedlinger, K., Blelloch, R., Brennan, C., Yamada, Y., Kim, M., Chin, L., et al. (2004). Reprogramming of a melanoma genome by nuclear transplantation. *Genes Dev., 18,* 1875–1885.

Hornen, N., Kues, W. A., Carnwath, J. W., Lucas-Hahn, A., Petersen, B., Hassel, P., et al. (2007). Production of viable pigs from fetal somatic stem cells. *Cloning and Stem Cells, 9,* 364–373.

Hsieh, C. L. (1999). *In vivo* activity of murine de novo methyltransferases, Dnmt3a and Dnmt3b. *Mol. Cell Biol., 19,* 8211–8218.

Humpherys, D., Eggan, K., Akutsu, H., Hochedlinger, K., Rideout, W. M., III, Biniszkiewicz, D., et al. (2001). Epigenetic instability in ES cells and cloned mice. *Science, 293,* 95–97.

Huynh, K. D., & Lee, J. T. (2003). Inheritance of a pre-inactivated paternal X chromosome in early mouse embryos. *Nature, 426,* 857–862.

Hwang, W. S., Roh, S. I., Lee, B. C., Kang, S. K., Kwon, D. K., Kim, S., et al. (2005). Patient-specific embryonic stem cells derived from human SCNT blastocysts. *Science, 308,* 1777–1783.

Hwang, W. S., Ryu, Y. J., Park, J. H., Park, E. S., Lee, E. G., Koo, J. M., et al. (2004). Evidence of a pluripotent human embryonic stem cell line derived from a cloned blastocyst. *Science, 303,* 1669–1674.

Illmensee, K., & Hoppe, P. C. (1981). Nuclear transplantation in Mus musculus: developmental potential of nuclei from preimplantation embryos. *Cell, 23,* 9–18.

Inoue, K., Wakao, H., Ogonuki, N., Miki, H., Seino, K.-I., Nambu-Wakao, R., et al. (2005). Generation of cloned mice by direct nuclear transfer from natural killer T cells. *Curr. Biol., 15,* 1114–1118.

Jaenisch, R., & Wilmut, I. (2001). Developmental biology. Don't clone humans! *Science, 291,* 2552.

Jeon, H. Y., Hyun, S. H., Lee, G. S., Kim, H. S., Kim, S., Jeong, Y. W., et al. (2005). The analysis of telomere length and telomerase activity in cloned pigs and cows. *Mol. Reprod. Dev., 71*, 315−320.

Jiang, L., Carter, D. B., Xu, J., Yang, X., Prather, R. S., & Tian, X. C. (2004). Telomere lengths in cloned transgenic pigs. *Biol. Reprod., 70*, 1589−1593.

Jones, P. A. (1999). The DNA methylation paradox. *Trends Genet., 15*, 34−37.

Kang, Y. K., Koo, D. B., Park, J. S., Choi, Y. H., Chung, A. S., Lee, K. K., et al. (2001a). Aberrant methylation of donor genome in cloned bovine embryos. *Nat. Genet., 28*, 173−177.

Kang, Y. K., Koo, D. B., Park, J. S., Choi, Y. H., Kim, H. N., Chang, W. K., et al. (2001b). Typical demethylation events in cloned pig embryos. Clues on species-specific differences in epigenetic reprogramming of a cloned donor genome. *J. Biol. Chem., 276*, 39980−39984.

Kato, Y., Tani, T., & Tsunoda, Y. (2000). Cloning of calves from various somatic cell types of male and female adult, newborn and fetal cows. *J. Reprod. Fertil., 120*, 231−237.

Kato, Y., Tani, T., Sotomaru, Y., Kurokawa, K., Kato, J., Doguchi, H., et al. (1998). Eight calves cloned from somatic cells of a single adult. *Science, 282*, 2095−2098.

Kennedy, D. (2006). Editorial retraction. *Science, 311*, 335.

Kind, A., & Schnieke, A. (2008). Animal pharming, two decades on. *Transgenic Res., 17*, 1025−1033.

Knijn, H. M., Wrenzycki, C., Hendriksen, P. J., Vos, P. L., Zeinstra, E. C., van der Weijden, G. C., et al. (2005). *In vitro* and *in vivo* culture effects on mRNA expression of genes involved in metabolism and apoptosis in bovine embryos. *Reprod. Fertil. Dev., 17*, 775−784.

Koo, D. B., Kang, Y. K., Choi, Y. H., Park, J. S., Kim, H. N., Oh, K. B., et al. (2003). Aberrant allocations of inner cell mass and trophectoderm cells in bovine nuclear transfer blastocysts. *Biol. Reprod., 67*, 487−492.

Kubota, C., Tian, X. C., & Yang, X. (2004). Serial bull cloning by somatic cell nuclear transfer. *Nat. Biotechnol., 22*, 693−694.

Kubota, C., Yamakuchi, H., Todoroki, J., Mizoshita, K., Tabara, N., Barber, M., et al. (2000). Six cloned calves produced from adult fibroblast cells after long-term culture. *Proc. Natl. Acad. Sci. U.S.A., 97*, 990−995.

Kues, W. A., & Niemann, H. (2004). The contribution of farm animals to human health. *Trends Biotechnol., 22*, 286−294.

Kues, W. A., Rath, D., & Niemann, H. (2008a). Reproductive biotechnology in farm animals goes genomics. *CAB Rev., 3*, 1−18.

Kues, W. A., Sudheer, S., Herrmann, D., Carnwath, J. W., Havlicek, S., Besenfelder, U., et al. (2008b). Genome-wide expression profiling reveals distinct clusters of transcriptional regulation during bovine preimplantation development *in vivo. Proc. Natl. Acad. Sci. U.S.A., 105*, 19768−19773.

Kumugai, S. (2002). *Safety of animal foods that utilize cloning technology.* Report to the Japanese Ministry for Agriculture, Forestry and Fishery (MAFF).

Lanza, R. P., Chung, H. Y., Yoo, J. J., Wettstein, P. J., Blackwell, C., Borson, N., et al. (2002). Generation of histocompatible tissues using nuclear transplantation. *Nat. Biotechnol., 20*, 689−696.

Lanza, R. P., Cibelli, J. B., Faber, D., Sweeney, R. W., Henderson, B., Nevala, W., et al. (2001). Cloned cattle can be healthy and normal. *Science, 294*, 1893−1894.

Law, J. A., & Jacobsen, S. E. (2010). Establishing, maintaining and modifying DNA methylation patterns in plants and animals. *Nature Reviews Genetics, 11*, 204−220.

Lee, B. C., Kim, M. K., Jang, G., Oh, H. J., Yuda, F., Kim, H. J., et al. (2005). DNA analysis of a putative dog clone. *Nature, 436*, 641.

Lee, J. B., & Park, C. (2006). Verification that Snuppy is a clone. *Nature, 440*, E2−E3.

Lee, R. S. F., Peterson, A. J., Donnison, M. J., Ravelich, S., Ledgard, A. M., Li, N., et al. (2004). Cloned cattle fetuses with the same nuclear genetics are more variable than contemporary half-siblings resulting from artificial insemination and exhibit fetal and placental growth deregulation even in the first trimester. *Biol. Reprod., 70*, 1−11.

Lepikhov, K., Zakhartchenko, V., Hao, R., Yang, F., Wrenzycki, C., Niemann, H., et al. (2008). Evidence for conserved DNA and histone H3 methylation reprogramming in mouse, bovine and rabbit zygotes. *Epigenetics & Chromatin.* doi:10.1186/1756-8935-1-8.

Li, E. (2002). Chromatin modification and epigenetic reprogramming in mammalian development. *Nature Reviews Genetics, 3*, 662−673.

Li, E., Beard, C., & Jaenisch, R. (1993). Role for DNA methylation in genomic imprinting. *Nature, 366*, 362−365.

Li, J., & Mombaerts, P. (2008). Nuclear transfer-mediated rescue of the nuclear genome of nonviable mouse cells frozen without cryoprotectant. *Biol. Reprod., 79*, 588−593.

Li, J., Ishli, T., Feinstein, P., & Mombaerts, P. (2004). Odorant receptor gene choice is reset by nuclear transfer from mouse olfactory sensory neurons. *Nature, 428*, 393−399.

Li, S., Li, Y., Du, W., Zhang, L., Yu, S., Dai, Y., et al. (2005). Aberrant gene expression in organs of bovine clones that die within two days after birth. *Biol. Reprod., 72*, 258−265.

Li, S., Li, Y., Yu, S., Du, W., Zhang, L., Dai, Y., et al. (2007). Expression of insulin-like growth factors systems in cloned cattle dead within hours after birth. *Mol. Reprod. Dev., 74*, 397–402.

Li, W., Wei, W., Zhu, S., Zhu, J., Shi, Y., Lin, T., et al. (2009). Generation of rat and human induced pluripotent stem cells by combining genetic reprogramming and chemical inhibitors. *Cell Stem Cell, 4*, 16–19.

Li, Z., Sun, X., Chen, J., Liu, X., Wisely, S. M., Zhou, Q., et al. (2006). Cloned ferrets produced by somatic cell nuclear transfer. *Dev. Biol., 293*, 439–448.

Liao, J., Cui, C., Chen, S., Ren, J., Chen, J., Gao, Y., et al. (2009). Generation of induced pluripotent stem cell lines from adult rat cells. *Cell Stem Cell, 4*, 11–15.

Lin, T., Ambasudhan, R., Yuan, X., Li, W., Hilcove, S., Abujarour, R., et al. (2009). A chemical platform for improved induction of human iPSCs. *Nat. Methods, 6*, 805–808.

Liu, H., Zhu, F., Yong, J., Zhang, P., Hou, P., Li, H., et al. (2008). Generation of induced pluripotent stem cells from adult rhesus monkey fibroblasts. *Cell Stem Cell, 4*, 587–590.

Liu, J. H., Yin, S., Xiong, B., Hou, Y., Chen, D. Y., & Sun, Q. Y. (2008). Aberrant DNA methylation imprints in aborted bovine clones. *Mol. Reprod. Dev., 75*, 598–607.

Lucifero, D., Suzuki, J., Bordignon, V., Martel, J., Vigneault, C., Therrien, J., et al. (2006). Bovine *SNRPN* methylation imprint in oocytes and day 17 in vitro-produced and somatic cell nuclear transfer embryos. *Biol. Reprod., 75*, 531–538.

Lyon, M. F. (1961). Gene action in the X-chromosome of the mouse (Mus musculus L.). *Nature, 190*, 372–373.

Mager, J., & Bartholomei, M. S. (2005). Strategies for dissecting epigenetic mechanisms in the mouse. *Nature Genetics, 37*, 1194–1199.

Mak, W., Nesterova, T. B., de Napoles, M., Appanah, R., Yamanaka, S., Otte, A. P., et al. (2004). Reactivation of the paternal X chromosome in early mouse embryos. *Science, 303*, 666–669.

Mann, M. R., Chung, Y. G., Nolen, L. D., Verona, R. I., Latham, K. E., & Bartolomei, M. S. (2003). Disruption of imprinted gene methylation and expression in cloned preimplantation stage mouse embryos. *Biol. Reprod., 69*, 902–914.

Mansouri-Attia, N., Sandra, O., Aubert, J., Degrelle, S., Everts, R. E., Giraud-Delville, C., et al. (2009). Endometrium as an early sensor of *in vitro* embryo manipulation technologies. *Proc. Natl. Acad. Sci. U.S.A., 106*, 5687–5692.

Martinez-Diaz, M. A., Che, L., Albornoz, M., Seneda, M. M., Collis, D., Coutinho, A. R. S., et al. (2010). Pre- and postimplantation development of swine-cloned embryos derived from fibroblasts and bone marrow cells after inhibition of histone deacetylases. *Cell Reprogramming, 12*, 85–94.

Mayer, W., Niveleau, A., Walter, J., Fundele, R., & Haaf, T. (2000). Demethylation of the zygotic paternal genome. *Nature, 403*, 501–502.

McGrath, J., & Solter, D. (1983). Nuclear transplantation in mouse embryos. *J. Exp. Zool., 228*, 355–362.

McGrath, J., & Solter, D. (1984). Inability of mouse blastomere nuclei transferred to enucleated zygotes to support development *in vitro. Science, 226*, 1317–1319.

Meng, L., Ely, J. J., Stouffer, R. L., & Wolf, D. P. (1997). Rhesus monkeys produced by nuclear transfer. *Biol. Reprod., 57*, 454–459.

Miyashita, N., Shiga, K., Yonai, M., Kaneyama, K., Kobayashi, S., Kojima, T., et al. (2002). Remarkable differences in telomere lengths among cloned cattle derived from different cell types. *Biol. Reprod., 66*, 1649–1655.

Miyoshi, K., Mori, H., Mizobe, Y., Akasaka, E., Ozawa, A., Yoshida, M., et al. (2010). Valproic acid enhances *in vitro* development and Oct-3/4 expression of miniature pig somatic cell nuclear transfer embryos. *Cell Reprogramming, 12*, 67–74.

Miyoshi, K., Rzucidlo, S. J., Pratt, S. L., & Stice, S. L. (2003). Improvements in cloning efficiencies may be possible by increasing uniformity in recipient oocytes and donor cells. *Biol. Reprod., 68*, 1079–1086.

Mizutani, E., Ohta, H., Kishigami, S., Thuan, N. V., Hikichi, T., Wakayama, S., et al. (2006). Developmental ability of cloned embryos from neural stem cells. *Reproduction, 132*, 849–857.

Moore, K., Kramer, J. M., Rodriguez-Sallaberry, C. J., Yelich, J. V., & Drost, M. (2007). Insulin-like growth factor (IGF) family genes are aberrantly expressed in bovine conceptuses produced *in vitro* or by nuclear transfer. *Theriogenology, 68*, 717–727.

Nakayama, J., Tahara, H., Tahara, E., Saito, M., Ito, K., Nakamura, H., et al. (1998). Telomerase activation by hTRT in human normal fibroblasts and hepatocellular carcinomas. *Nat. Genet., 18*, 65–68.

Niemann, H., & Kues, W. A. (2007). Transgenic farm animals — an update. *Reprod. Fert. Dev., 19*, 762–770.

Niemann, H., & Wrenzycki, C. (2000). Alterations of expression of developmentally important genes in preimplantation bovine embryos by *in vitro* culture conditions: implications for subsequent development. *Theriogenology, 53*, 21–34.

Niemann, H., Carnwath, J. W., Herrmann, D., Wieczorek, G., Lemme, E., Lucas-Hahn, A., et al. (2010). DNA methylation patterns reflect epigenetic reprogramming in bovine embryos. *Cell Reprogramming, 12*, 33–42.

Niemann, H., Tian, X. C., King, W. A., & Lee, R. S. F. (2008). Epigenetic reprogramming in embryonic and foetal development upon somatic cell nuclear transfer cloning. *Reproduction, 135*, 151–163.

Niemann, H., Wrenzycki, C., Lucas-Hahn, A., Brambrink, T., Kues, W. A., & Carnwath, J. W. (2002). Gene expression patterns in bovine *in vitro*-produced and nuclear transfer-derived embryos and their implications for early development. *Cloning and Stem Cells, 4*, 29–38.

Niwa, H., Miyazaki, J., & Smith, A. G. (2000). Quantitative expression of Oct-3/4 defines differentiation, dedifferentiation or self-renewal of ES cells. *Nat. Genet., 24*, 372–376.

Nowak-Imialek, M., Wrenzycki, C., Herrmann, D., Lucas-Hahn, A., Lagutina, I., Lemme, E., et al. (2008). Messenger RNA expression patterns of histone-associated genes in bovine preimplantation embryos derived from different origins. *Mol. Reprod. Dev., 75*, 731–743.

Oback, B., & Wells, D. N. (2007). Donor cell differentiation, reprogramming, and cloning efficiency: elusive or illusive correlation? *Mol. Reprod. Dev., 74*, 646–654.

Oestrup, O., Petrovica, I., Strejcek, F., Morovic, M., Lucas-Hahn, A., Lemme, E., et al. (2009). Nuclear and nucleolar reprogramming during the first cell cycle in bovine nuclear transfer embryos. *Cloning and Stem Cells, 11*, 367–375.

Ogonuki, N., Inoue, K., Yamamoto, Y., Noguchi, Y., Tanemura, K., Suzuki, O., et al. (2002). Early death of mice cloned from somatic cells. *Nat. Genet., 30*, 253–254.

Oh, H. J., Kim, M. K., Jang, G., Kim, H. J., Hong, S. G., Park, J. E., et al. (2008). Cloning endangered gray wolves (*Canis lupus*) from somatic cells collected postmortem. *Theriogenology, 70*, 638–647.

Oishi, M., Gohma, H., Hashizume, K., Taniguchi, Y., Yasue, H., Takahashi, S., et al. (2006). Early embryonic death — associated changes in genome-wide gene expression profiles in the fetal placenta of the cow carrying somatic nuclear-derived cloned embryo. *Mol. Reprod. Dev., 73*, 404–409.

Okamoto, I., Otte, A. P., Allis, C. D., Reinberg, D., & Heard, E. (2004). Epigenetic dynamics of imprinted X-inactivation during early mouse development. *Science, 303*, 644–649.

Okano, M., Bell, D. W., Haber, D. A., & Li, E. (1999). DNA methyltransferases Dnmt3a and Dnmt3b are essential for de novo methylation and mammalian development. *Cell, 99*, 247–257.

Okita, K., Ichisaka, T., & Yamanaka, S. (2007). Generation of germline-competent induced pluripotent stem cells. *Nature, 448*, 313–317.

Okita, K., Nakagawa, M., Hyenjong, H., Ichisaka, T., & Yamanaka, S. (2008). Generation of mouse induced pluripotent stem cells without viral vectors. *Science, 322*, 949–953.

Onishi, A., Iwamoto, M., Akita, T., Mikawa, S., Takeda, K., Awata, T., et al. (2000). Pig cloning by microinjection of fetal fibroblast nuclei. *Science, 289*, 1188–1190.

Oropeza, A., Hadeler, K. G., & Niemann, H. (2007). Application of ultrasound-guided follicular aspiration (OPU) in prepubertal and adult cattle. *J. Reprod. Dev., 52*, S31–S38.

Oropeza, M., Petersen, B., Carnwath, J. W., Lucas-Hahn, A., Lemme, E., Hassel, P., et al. (2009). Transgenic expression of the human A20 gene in cloned pigs provides protection against apoptotic and inflammatory stimuli. *Xenotransplantation, 16*, 522–534.

Oswald, J., Engemann, S., Lane, N., Mayer, W., Olek, A., Fundele, R., et al. (2000). Active demethylation of the paternal genome in the mouse zygote. *Curr. Biol., 10*, 475–478.

Pace, M. M., Augenstein, M. L., Betthauser, J. M., Childs, L. A., Eilertsen, K. J., Enos, J. M., et al. (2002). Ontogeny of cloned cattle to lactation. *Biol. Reprod., 67*, 334–339.

Panarace, M., Agüero, J. I., Garrote, M., Jauregui, G., Segovia, A., Cané, L., et al. (2007). How healthy are clones and their progeny: 5 years of field experience. *Theriogenology, 67*, 142–151.

Parker, H. G., Kruglyak, L., & Ostrander, E. A. (2006). DNA analysis of a putative dog clone. *Nature, 440*, E1–E2.

Perry, A. C., & Wakayama, T. (2002). Untimely ends and new beginnings in mouse cloning. *Nat. Genet., 30*, 243–244.

Pesce, M., & Schöler, H. R. (2001). Oct-4: gatekeeper in the beginnings of mammalian development. *Stem Cells, 19*, 271–278.

Petersen, B., Carnwath, J. W., & Niemann, H. (2009a). The perspectives for porcine-to-human xenografts. *Comp. Immunol. Microbiol. Infectious Diseases, 32*, 91–105.

Petersen, B., Lucas-Hahn, A., Oropeza, M., Hornen, N., Lemme, E., Hassel, P., et al. (2008). Development and validation of a highly efficient protocol of porcine somatic cloning using preovulatory embryo transfer in peripubertal gilts. *Cloning and Stem Cells, 10*, 355–362.

Petersen, B., Ramackers, W., Tiede, A., Lucas-Hahn, A., Herrmann, D., Barg-Kues, B., et al. (2009b). Pigs transgenic for human thrombomodulin have elevated production of activated protein C. *Xenotransplantation, 16*, 486–495.

Polejaeva, I. A., Chen, S. H., Vaught, T. D., Page, R. L., Mullins, J., Ball, S., et al. (2000). Cloned pigs produced by nuclear transfer from adult somatic cells. *Nature, 407*, 86–90.

Powell, A. M., Talbot, N. C., Wells, K. D., Kerr, D. E., Pursel, V. G., & Wall, R. J. (2004). Cell donor influences success of producing cattle by somatic cell nuclear transfer. *Biol. Reprod., 71,* 210–216.

Prather, R. S., Sims, M. M., & First, N. L. (1989). Nuclear transplantation in early pig embryos. *Biol. Reprod., 41,* 414–418.

Reik, W., & Walter, J. (2001). Genomic imprinting: parental influence on the genome. *Nat. Rev. Genet., 2,* 21–32.

Reik, W., Dean, W., & Walter, J. (2001). Epigenetic reprogramming in mammalian development. *Science, 293,* 1089–1093.

Renard, J. P., Chastant, S., Chesne, P., Richard, C., Marchal, J., Cordonnier, N., et al. (1999). Lymphoid hypoplasia and somatic cloning. *Lancet, 353,* 1489–1491.

Rhind, S. M., King, T. J., Harkness, L. M., Bellamy, C., Wallace, W., DeSousa, P., et al. (2003). Cloned lambs – lessons from pathology. *Nat. Biotechnol., 21,* 744–745.

Rideout, W. M., III, Hochedlinger, K., Kyba, M., Daley, G. Q., & Jaenisch, R. (2002). Correction of a genetic defect by nuclear transplantation and combined cell and gene therapy. *Cell, 109,* 17–27.

Ruddock, N. T., Wilson, K. J., Cooney, M. A., Korfiatis, N. A., Tecirlioglu, R. T., & French, A. J. (2004). Analysis of imprinted messenger RNA expression during bovine preimplantation development. *Biol. Reprod., 70,* 1131–1135.

Rudenko, L., Matheson, J. C., & Sundlof, S. F. (2007). Animal cloning and the FDA – the risk assessment paradigm under public scrutiny. *Nat. Biotech., 25,* 39–43.

Saito, S., Sawai, K., Ugai, H., Moriyasu, S., Minamihashi, A., Yamamoto, Y., et al. (2003). Generation of cloned calves and transgenic chimeric embryos from bovine embryonic stem-like cells. *Biochem. Biophys. Res. Comm., 309,* 104–113.

Santos, F., Hendrich, B., Reik, W., & Dean, W. (2002). Dynamic reprogramming of DNA methylation in the early mouse embryo. *Dev. Biol., 241,* 172–182.

Sawai, K., Takahashi, M., Moriyasu, S., Hirayama, H., Minamihashi, A., Hashizume, T., et al. (2010). Changes in the DNA methylation status of bovine embryos from the blastocyst to elongated stage derived from somatic cell nuclear transfer. *Cell Reprogramming, 12,* 15–22.

Schaetzlein, S., & Rudolph, K. L. (2005). Telomere length regulation during cloning, embryogenesis and ageing. *Reprod. Fertil. Dev., 17,* 85–96.

Schaetzlein, S., Lucas-Hahn, A., Lemme, E., Kues, W. A., Dorsch, M., Manns, M. P., et al. (2004). Telomere length is reset during early mammalian embryogenesis. *Proc. Natl. Acad. Sci. U.S.A., 101,* 8034–8038.

Schurmann, A., Wells, D. N., & Oback, B. (2006). Early zygotes are suitable recipients for bovine somatic nuclear transfer and result in cloned offspring. *Reproduction, 132,* 839–848.

Schwarzer, M., Carnwath, J. W., Lucas-Hahn, A., Lemme, E., Kues, W. A., Wachsmann, B., et al. (2006). Isolation of bovine cardiomyocytes for reprogramming studies based on nuclear transfer. *Cloning and Stem Cells, 8,* 150–158.

Shi, D., Lu, F., Wei, Y., Cui, K., Yang, S., Wei, J., et al. (2007). Buffaloes (*Bubalus bubalis*) cloned by nuclear transfer of somatic cells. *Biol. Reprod., 77,* 285–291.

Shiels, P. G., Kind, A. J., Campbell, K. H., Wilmut, I., Waddington, D., Colman, A., et al. (1999). Analysis of telomere lengths in cloned sheep. *Nature, 399,* 316–317.

Shin, T., Kraemer, D., Pryor, J., Liu, L., Rugila, J., Howe, L., et al. (2002). A cat cloned by nuclear transplantation. *Nature, 415,* 859–860.

Simonsson, S., & Gurdon, J. (2004). DNA demethylation is necessary for the epigenetic reprogramming of somatic cell nuclei. *Nat. Cell Biol., 6,* 984–990.

Sims, M., & First, N. L. (1994). Production of calves by transfer of nuclei from cultured inner cell mass cells. *Proc. Natl. Acad. Sci. U.S.A., 91,* 6143–6147.

Smith, S. L., Everts, R. E., Tian, X. C., Du, F., Sung, L. Y., Rodriguez-Zas, S. L., et al. (2005). Global gene expression profiles reveal significant nuclear reprogramming by the blastocyst stage after cloning. *Proc. Natl. Acad. Sci. U.S.A., 102,* 17582–17587.

Stadtfeld, M., Nagaya, M., Utikal, J., Weir, G., & Hochedlinger, K. (2008). Induced pluripotent stem cells generated without viral integration. *Science, 322,* 945–949.

Stice, S. L., & Robl, J. M. (1988). Nuclear reprogramming in nuclear transplant rabbit embryos. *Biol. Reprod., 39,* 657–664.

Sung, L. Y., Gao, S., Shen, H., Yu, H., Song, Y., Smith, S. L., et al. (2006). Differentiated cells are more efficient than adult stem cells for cloning by somatic cell nuclear transfer. *Nat. Genet., 38,* 1323–1328.

Suzuki, J., Jr., Therrien, J., Filion, F., Lefebvre, R., Goff, A. K., & Smith, L. C. (2009). *In vitro* culture and somatic cell nuclear transfer affect imprinting of SNRPN gene in pre- and post-implantation stages of development in cattle. *BMC Dev. Biol., 9.* doi:10.1186/1471-213X-9-9.

Takahashi, K., & Yamanaka, S. (2006). Induction of pluripotent stem cells from mouse embryonic and adult fibroblast cultures by defined factors. *Cell, 126,* 663–676.

Takahashi, K., Tanabe, K., Ohnuki, M., Narita, M., Ichisaka, T., Tomoda, K., et al. (2007). Induction of pluripotent stem cells from adult human fibroblasts by defined factors. *Cell, 131*, 1—12.

Tamashiro, K. L., Wakayama, T., Blanchard, R. J., Blanchard, D. C., & Yanagimachi, R. (2000). Postnatal growth and behavioral development of mice cloned from adult cumulus cells. *Biol. Reprod., 63*, 328—234.

Telford, N. A., Watson, A. J., & Schultz, G. A. (1990). Transition from maternal to embryonic control in early mammalian development: a comparison of several species. *Mol. Reprod. Dev., 26*, 90—100.

Theoret, C., Dore, M., Mulon, P. Y., Desrochers, A., Viramontes, F., Filion, F., et al. (2006). Short- and longterm skin graft survival in cattle clones with different mitochondrial haplotypes. *Theriogenology, 65*, 1465—1479.

Tian, X. C., Kubota, C., Sakashita, K., Izaike, Y., Okano, R., Tabara, N., et al. (2005). Meat and milk compositions of bovine clones. *Proc. Natl. Acad. Sci. U.S.A., 102*, 6261—6266.

Wadman, M. (2007). Dolly: a decade on. *Nature, 445*, 800—801.

Wakayama, S., Ohta, H., Hikichi, T., Mizutani, E., Iwaki, T., Kanagawa, O., et al. (2008). Production of healthy cloned mice from bodies frozen at −20°C for 16 years. *Proc. Natl. Acad. Sci. U.S.A., 105*, 17318—17322.

Wakayama, S., Ohta, H., Kishigami, S., van Thuan, N., Hikichi, T., Mizutani, E., et al. (2005). Establishment of male and female nuclear transfer embryonic stem cell lines from different mouse strains and tissues. *Biol. Reprod., 72*, 932—936.

Wakayama, T., & Yanagimachi, R. (1999). Cloning of male mice from adult tail-tip cells. *Nat. Genet., 22*, 127—128.

Wakayama, T., & Yanagimachi, R. (2001). Mouse cloning with nucleus donor cells of different age and type. *Mol. Reprod. Dev., 58*, 376—383.

Wakayama, T., Perry, A. C., Zuccotti, M., Johnson, K. R., & Yanagimachi, R. (1998). Full-term development of mice from enucleated oocytes injected with cumulus cell nuclei. *Nature, 394*, 369—374.

Wakayama, T., Rodriguez, I., Perry, A. C., Yanagimachi, R., & Mombaerts, P. (1999). Mice cloned from embryonic stem cells. *Proc. Natl. Acad. Sci. U.S.A., 96*, 14984—14989.

Wakayama, T., Shinkai, Y., Tamashiro, K. L., Niida, H., Blanchard, D. C., Blanchard, R. J., et al. (2000). Cloning of mice to six generations. *Nature, 407*, 318—319.

Wakayama, T., Tabar, V., Rodriguez, I., Perry, A. C., Studer, L., & Mombaerts, P. (2001). Differentiation of embryonic stem cell lines generated from adult somatic cells by nuclear transfer. *Science, 292*, 740—743.

Wang, L., Duan, E., Sung, L. Y., Jeong, B. S., Yang, X., & Tian, X. C. (2005). Generation and characterization of pluripotent cells from cloned bovine embryos. *Biol. Reprod., 73*, 149—155.

Wani, N. A., Wernery, U., Hassan, F. A. H., Wernery, R., & Skidmore, J. A. (2010). Production of the first cloned camel by somatic cell nuclear transfer. *Biol. Reprod., 82*, 373—379.

Wee, G., Koo, D. B., Song, B. S., Kim, J. S., Kang, M. J., Moon, S. J., et al. (2006). Inheritable histone H4 acetylation of somatic chromatins in cloned embryos. *J. Biol. Chem., 281*, 6048—6057.

Wee, G., Shim, J. J., Koo, D. B., Chae, J. I., Lee, K. K., & Han, Y. M. (2007). Epigenetic alteration of the donor cells does not recapitulate the reprogramming of DNA methylation in cloned embryos. *Reproduction, 134*, 781—787.

Willadsen, S. M. (1986). Nuclear transplantation in sheep embryos. *Nature, 320*, 63—65.

Wilmut, I., Schnieke, A. E., McWhir, J., Kind, A. J., & Campbell, K. H. (1997). Viable offspring derived from fetal and adult mammalian cells. *Nature, 385*, 810—813.

Wongsrikeao, P., Nagai, T., Agung, B., Taniguchi, M., Kunishi, M., Suto, S., et al. (2007). Improvement of transgenic cloning efficiencies by culturing recipient oocytes and donor cells with antioxidant vitamins in cattle. *Mol. Reprod. Dev., 74*, 694—702.

Woods, G. L., White, K. L., Vanderwall, D. K., Li, G. P., Aston, K. I., Bunch, T. D., et al. (2003). A mule cloned from fetal cells by nuclear transfer. *Science, 301*, 1063.

Wrenzycki, C., & Niemann, H. (2003). Epigenetic reprogramming in early embryonic development: effects of in-vitro production and somatic nuclear transfer. *Reprod. Biomed. Online, 7*, 649—656.

Wrenzycki, C., Herrmann, D., Keskintepe, L., Martins, A., Jr., Sirisathien, S., Brackett, B., et al. (2001a). Effects of culture system and protein supplementation on mRNA expression in pre-implantation bovine embryos. *Hum. Reprod., 16*, 893—901.

Wrenzycki, C., Herrmann, D., Lucas-Hahn, A., Gebert, C., Korsawe, K., Lemme, E., et al. (2005a). Epigenetic reprogramming throughout preimplantation development and consequences for assisted reproductive technologies. *Birth Defects Res. C Embryo Today, 75*, 1—9.

Wrenzycki, C., Herrmann, D., Lucas-Hahn, A., Korsawe, K., Lemme, E., & Niemann, H. (2005b). Messenger RNA expression patterns in bovine embryos derived from *in vitro* procedures and their implications for development. *Reprod. Fertil. Dev., 17*, 23—35.

Wrenzycki, C., Lucas-Hahn, A., Herrmann, D., Lemme, E., Korsawe, K., & Niemann, H. (2002). *In vitro* production and nuclear transfer affect dosage compensation of the X-linked gene transcripts G6PD, PGK, and Xist in preimplantation bovine embryos. *Biol. Reprod., 66*, 127—134.

Wrenzycki, C., Wells, D., Herrmann, D., Miller, A., Oliver, J., Tervit, R., et al. (2001b). Nuclear transfer protocol affects messenger RNA expression patterns in cloned bovine blastocysts. *Biol. Reprod., 65*, 309—317.

157

Wu, Z., Chen, J., Ren, J., Bao, L., Liao, J., Cui, C., et al. (2009). Generation of pig-induced pluripotent stem cells with a drug-inducible system. *J. Mol. Cell Biol., 1,* 46–54.

Wuensch, A., Habermann, F. A., Kurosaka, S., Klose, R., Zakhartchenko, V., Reichenbach, H. D., et al. (2007). Quantitative monitoring of pluripotency gene activation after somatic cloning in cattle. *Biol. Reprod., 76,* 983–991.

Xu, Y., Zhang, J. J., Grifo, J. A., & Krey, L. C. (2005). DNA methylation patterns in human tripronucleate zygotes. *Mol. Hum. Reprod., 11,* 167–171.

Xue, F., Tian, X. C., Du, F., Kubota, C., Taneja, M., Dinnyes, A., et al. (2002). Aberrant patterns of X chromosome inactivation in bovine clones. *Nat. Genet., 31,* 216–220.

Yamazaki, Y., Fujita, T. C., Low, E. W., Alarcón, V. B., Yanagimachi, R., & Marikawa, Y. (2006). Gradual DNA demethylation of the *Oct4* promoter in cloned mouse embryos. *Mol. Reprod. Dev., 73,* 180–188.

Yang, L., Chavatte-Palmer, P., Kubota, C., O'Neill, M., Hoagland, T., Renard, J. P., et al. (2005). Expression of imprinted genes is aberrant in deceased newborn cloned calves and relatively normal in surviving adult clones. *Mol. Reprod. Dev., 71,* 431–438.

Yang, X., Li, H., Ma, Q., Yan, J., Zhao, J., Li, H., et al. (2006). Improved efficiency of bovine cloning by autologous somatic cell nuclear transfer. *Reprod., 132,* 733–739.

Yang, X., Smith, S. L., Tian, X. C., Lewin, H. A., Renard, J. P., & Wakayama, T. (2007a). Nuclear reprogramming of cloned embryos and its implications for therapeutic cloning. *Nat. Genet., 39,* 295–302.

Yang, X., Tian, X. C., Kubota, C., Page, R., Xu, J., Cibelli, J., et al. (2007b). Risk assessment of meat and milk from cloned animals. *Nat. Biotech., 25,* 77–83.

Young, L. E., Fernandes, K., McEvoy, T. G., Butterwith, S. C., Gutierrez, C. G., Carolan, C., et al. (2001). Epigenetic change in IGF2R is associated with fetal overgrowth after sheep embryo culture. *Nat. Genet., 27,* 153–154.

Yu, J., Vodyanik, M., Smuga-Otto, K., Antosiewicz-Bourget, J., Frane, J., Tian, S., et al. (2007). Induced pluripotent stem cell lines derived from human somatic cells. *Science, 318,* 1917–1920.

Yu, Y., Mai, Q., Chen, X., Wang, L., Gao, L., Zhou, C., et al. (2009). Assessment of the developmental competence of human somatic cell nuclear transfer embryos by oocyte morphology classification. *Hum. Reprod., 24,* 649–657.

Yusa, K., Rad, R., Takeda, J., & Bradley, A. (2009). Generation of transgene-free induced pluripotent mouse stem cells by the piggyBac transposon. *Nat. Methods, 6,* 363–369.

Zhang, S., Kubota, C., Yang, L., Zhang, Y., Page, R., O'Neill, M., et al. (2004). Genomic imprinting of H19 in naturally reproduced and cloned cattle. *Biol. Reprod., 71,* 1540–1544.

Zhao, J., Hao, Y., Ross, J. W., Spate, L. D., Walters, E. M., Samuel, M. S., et al. (2010). Histone deacetylase inhibitors improve *in vitro* and *in vivo* developmental competence of somatic cell nuclear transfer porcine embryos. *Cell Reprogramming, 12,* 75–83.

Zhao, J., Ross, J. W., Hao, Y., Spate, L. D., Walters, E. M., Samuel, M. S., et al. (2009). Significant improvement in cloning efficiency of an inbred miniature pig by histone deacetylase inhibitor treatment after somatic cell nuclear transfer. *Biol. Reprod., 81,* 525–530.

Zhou, H., Wu, S., Joo, J., Zhu, S., Han, D., Lin, T., Trauger, S., et al. (2009). Generation of induced pluripotent stem cells using recombinant proteins. *Cell Stem Cell, 4,* 1–4.

Zhou, Q., Renard, J.-P., Friec, G., Brochard, V., Beaujean, N., Cherifi, Y., et al. (2003). Generation of fertile cloned rats using controlled timing of oocyte activation. *Science, 302,* 1179.

Zhu, H., Craig, J. A., Dyce, P. W., Sunnen, N., & Li, J. (2004). Embryos derived from porcine skin-derived stem cells exhibit enhanced preimplantation development. *Biol. Reprod., 71,* 1890–1897.

CHAPTER 8

Engineered Proteins for Controlling Gene Expression

Charles A. Gersbach
Department of Biomedical Engineering, Duke University, Hudson Hall, Durham, NC, USA

INTRODUCTION

Regenerative medicine is focused on biologic approaches to repairing, restoring, or replacing damaged or diseased tissues (Atala, 2009). Typically, this involves using cells for engineering a living tissue substitute or implantation into the target tissue in the patient. Ideally, these cells are isolated from the patient in order to minimize immune responses or the possibility of disease transmission. However, it is often not possible to harvest the necessary cell type directly from the patient. For example, cardiomyocytes, osteoblasts, β cells, and dopaminergic neurons are the necessary cell types for treating damaged heart tissue, bone defects, diabetes, and Parkinson's disease, respectively. Because these cells are not readily accessible from patients with these complications or diseases, researchers have explored the possibility of directing the differentiation of a more readily available cell source into the cell type of interest. These cell sources could include adult stem cells or progenitor cells, such as bone marrow-derived mesenchymal stem cells, blood-derived hematopoietic stem cells, muscle-derived stem cells, or adipose-derived stem cells. Alternatively, lineage-committed adult cell types, such as skin-derived fibroblasts or myoblasts from skeletal muscle, can be reprogrammed, or trans-differentiated, into a new cell type for regenerative medicine (Gurdon and Melton, 2008; Muller et al., 2009).

Approaches for directing cells into specific lineages or converting from one lineage into another are diverse. Most frequently, cells are treated with soluble factors that activate cellular signaling pathways that lead to cellular differentiation. These soluble factors could include small molecule drugs, growth factors, cytokines, or other engineered proteins including peptides or antibodies (Lutolf and Hubell, 2005; Phelps and Garcia, 2009). Alternatively, these same signaling pathways may be activated through material properties of the substrate on which the cells are cultured or implanted. These properties include surface chemistry, conformation and density of adsorbed proteins, stiffness, and micro- or nano-architecture (Rehfeldt et al., 2007; Lutolf et al., 2009). Finally, the signaling pathways may also be stimulated by physical stress (Setton and Chen, 2006; Chiu et al., 2009; Davies, 2009), such as shear flow, or electrical stimuli (Aaron et al., 2004; Gordon, 2007). All of these methods of directing cell differentiation are based on mimicking the natural stimuli that cells encounter during normal developmental processes and organogenesis. Importantly, the signaling pathways that are activated by these stimuli ultimately converge in the nucleus, where changes in cellular gene expression lead to long-term effects on cell fate and lineage commitment in response to activation and repression of specific gene networks.

Principles of Regenerative Medicine. DOI: 10.1016/B978-0-12-381422-7.10008-2

159

Over the last 20 years, the molecular mechanisms of these signaling pathways and the critical regulatory components of the lineage-specific gene networks have been elucidated. Consequently, a new area of research has emerged focusing on directly coordinating these networks with the molecular machinery that normally performs this function in cells — transcription factors — in contrast to the indirect extracellular stimuli described above. The rationale for this work is that by directly controlling gene networks at the level of transcription it may be possible to achieve enhanced levels of specificity, potency, and control of cell differentiation. This chapter will describe the various efforts in this area, including the use of natural transcriptional regulators, enhancement of these regulators through molecular engineering, and the engineering of entirely synthetic transcription factors for targeted gene regulation.

GENETIC REPROGRAMMING AND THE REGULATION OF GENE NETWORKS

The control of cell differentiation, tissue development and repair, and organism function largely occurs through regulation of gene expression. Each of the somatic cells of the human body contains identical genomes of the same set of >20,000 genes, as well as other non-coding regulatory elements (Lander et al., 2001; Venter et al., 2001). Each of >200 distinct cell types in the human body is defined by how this large set of genes is differentially regulated. Many genes are common to fundamental cellular processes and therefore are shared among many, if not all, cell types. However, certain sets of genes, or gene networks, are specific to a particular cell type. For example, there are specific gene networks that correspond to cells that make muscle, bone, or blood vessels. Similarly, there are specific gene networks that regulate stem cell pluripotency. These gene networks are primarily activated and repressed by cellular proteins called transcription factors that bind to DNA nearby the genes that comprise these networks. Although many transcription factors may belong to a gene network, there is often a "master regulatory factor" or combination of factors that is capable of activating the complete gene network under specific conditions (Fig. 8.1). This master factor can regulate many different types of genes in this network, including target genes that encode the proteins responsible for making a particular tissue, or secondary transcription factors that also regulate these target genes but are considered to act downstream of the master factor in the pathway. Importantly, these gene networks typically are not organized in a linear manner and there are numerous

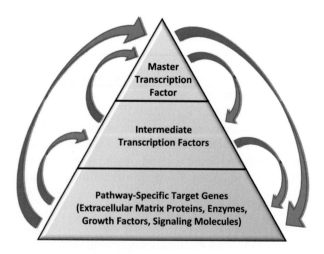

FIGURE 8.1
Lineage-specific gene networks can be conceptualized as a pyramid, with the master regulatory factor for that gene network at the top. This master factor regulates the expression of numerous target genes, including genes for intermediate transcription factors and genes necessary for cell differentiation and tissue formation. Numerous examples of redundancy, positive and negative feedback, and feedforward loops between these classes of genes are the basis for complex and non-linear network behaviors.

mechanisms for positive and negative feedback and feedforward signaling between the master transcription factor, the secondary transcription factors, and the terminal target genes (Fig. 8.1). Consequently, it is often controversial or difficult to experimentally determine which transcription factor, if any, in the network is the "master factor."

The identification of several potential master transcription factors has led to the development of genetic reprogramming as a means for controlling cell behavior and lineage commitment (Pomerantz and Blau, 2004; Gurdon and Melton, 2008; Muller et al., 2009). Genetic reprogramming is based on the hypothesis that any gene network can be activated in any cell type by the corresponding master transcription factor. For example, a skin fibroblast could be reprogrammed into a skeletal myoblast, osteoblast, cardiomyocyte, or neuron by activation of the appropriate gene networks. The concept of genetic reprogramming has been validated experimentally by genetically engineering cells to overexpress master transcription factors. A list of putative master transcription factors, their corresponding cell lineage, their potential applications in regenerative medicine, and representative publications demonstrating this approach is presented in Table 8.1.

The principle of genetic reprogramming was first demonstrated experimentally through the success of somatic cell nuclear transfer (SCNT). In SCNT, the nucleus is removed from an oocyte and replaced with the nucleus from a differentiated cell. Under appropriate conditions, some of the cells treated in this manner are capable of undergoing full organismal development. The SCNT technology was originally demonstrated in frogs (Briggs and King, 1952) but was later extended to mammals, including the widely publicized cloning of Dolly the sheep (Wilmut et al., 1997; Kato et al., 1998; Wakayama et al., 1998; Baguisi et al., 1999; Byrne et al., 2007). This work showed that the enucleated oocyte contains all of the necessary factors, in the form of cytoplasmic proteins and mRNA molecules, to activate the gene networks necessary for pluripotency. Presumably, some of these unknown molecules are transcription factors that

TABLE 8.1 **Master Regulatory Transcription Factors and Corresponding Cell Types and Therapeutic Applications**

Transcription factor	Cell/tissue type	Therapeutic applications	Representative publications
Oct4, Sox2, Klf4, Nanog	Pluripotent stem cells	Regenerative medicine, drug discovery	Takahashi et al. (2006); Wernig et al. (2007); Okita et al. (2007); Maherali et al. (2007); Park et al. (2008)
MyoD	Myoblast	Muscle regeneration, muscular dystrophy	Weintraub et al. (1989); Murry et al. (1996); Goudenege et al. (2009)
Runx2	Osteoblast	Bone regeneration, osteoporosis, osteogenesis imperfecta	Ducy et al. (1997); Byers et al. (2002); Gersbach et al. (2004b); Zheng et al. (2004); Zhao et al. (2007)
Hif1α	Angiogenesis	Wound healing	Vincent et al. (2000); Pajusola et al. (2005); Rajagopalan et al. (2007); Botusan et al. (2008); Jiang et al. (2008); Kajiwara et al. (2009); Huang et al. (2009)
Gata4, Tbx5, Nkx2-5	Cardiomyocytes/ endothelium	Repairing myocardium and vasculature	Bian et al. (2007); Yamada et al. (2007); David et al. (2009); Takeuchi and Bruneau (2009); Ferdous et al. (2009)
Pdx1, Ngn3	β-Cells	Diabetes	Koya et al. (2008); Yechoor et al. (2009)
Pitx3, Nurr1	Dopaminergic neurons	Parkinson's disease	Kim et al. (2003); Kim et al. (2006); Andersson et al. (2007); Li et al. (2007); Chung et al. (2005); Yang et al. (2008)
Ascl1	Oligodendrocytes	Multiple sclerosis, epilepsy	Jessberger et al. (2008)
Sox9	Chondrocyte	Cartilage regeneration, arthritis	Paul et al. (2003)
p53	DNA repair	Cancer	Clayman et al. (1995); Peng (2005); Ventura et al. (2007); Martins et al. (2006)

travel into the nucleus of the differentiated cell type and reprogram the gene expression profile. These experiments provided the first evidence that cell differentiation is not a unidirectional path and motivated the search for a minimal set of factors that are necessary for genetic reprogramming.

Yamanaka and colleagues completed the most monumental advance to date in the search for these reprogramming factors through their discovery of induced pluripotent stem cells (iPSCs) (Jaenisch and Young, 2008; Yamanaka, 2009). They began with 24 candidate transcription factors with known roles in regulating the gene network associated with stem cell pluripotency (Takahashi and Yamanaka, 2006). By testing various combinations of these factors for the ability to regulate genes associated with stem cell pluripotency, they identified a specific set of four factors that could induce pluripotency in mouse fibroblasts (Fig. 8.2). Subsequently, there have been numerous studies dedicated to advancing this technology, including the identification and characterization of alternative sets of transcription factors and substitutes for transcription factors that are capable of generating iPSCs (Jaenisch and Young, 2008; Yamanaka, 2009). It is now generally accepted that adult cell types can be used to generate any cell type in the human body by reverting into a pluripotent state through genetic reprogramming and subsequently differentiating into an alternative lineage of interest. The iPSC technology is covered in detail elsewhere in this book, and is presented briefly here to highlight arguably the most significant example of manipulating gene expression for regenerative medicine in contemporary research.

The successful reversion of a differentiated cell into a pluripotent cell capable of generating a whole organism through SCNT also suggested that it should be possible to convert differentiated cells between lineages by activating and suppressing the appropriate gene networks. This concept of direct reprogramming was originally demonstrated following the discovery of MyoD, the master transcriptional regulator of the skeletal muscle gene network (Berkes and Tapscott, 2005). When MyoD was overexpressed in differentiated fibroblasts, muscle-specific

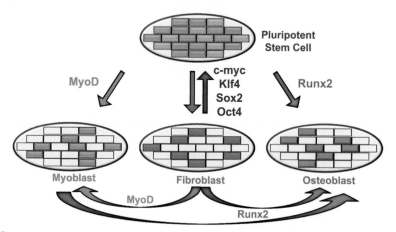

FIGURE 8.2

The concept of genetic reprogramming is founded on the fact that all of the somatic cells that make up various tissues contain the same set of genes. Different cell types form specific tissues based on how this set of genes is differentially regulated by transcription factors that activate gene networks. For example, MyoD and Runx2 are transcription factors that coordinate the gene networks corresponding to cell differentiation into skeletal myoblasts and osteoblasts, respectively, during the natural course of organism development. Genetic reprogramming occurs when gene networks within a cell are repressed or activated in order to convert one cell type into another. For example, fibroblasts have been reprogrammed into skeletal myoblasts by the overexpression of MyoD (Davis et al., 1987; Weintraub et al., 1989; Choi et al., 1990) or into pluripotent stem cells by the combined overexpression of Oct4, Sox2, Klf4, and c-myc (Takahashi and Yamanaka, 2006; Okita et al., 2007; Wernig et al., 2007). Alternatively, skeletal myoblasts and fibroblasts have been reprogrammed into osteoblasts by overexpression of Runx2 (Ducy et al., 1997; Byers et al., 2002; Gersbach et al., 2004b).

gene expression was induced and these cells converted into myoblasts capable of fusing into multinucleated myotubes (Davis et al., 1987; Weintraub et al., 1989; Choi et al., 1990; Fig. 8.2). This represents one of the earliest examples of induced transdifferentiation via genetic reprogramming by a defined master regulator of gene expression. Building on this work, cells have been genetically engineered with MyoD to simulate myoblast differentiation for several applications relevant to regenerative medicine, including cell-based treatments for muscular dystrophy and myocardial infarction (Murry et al., 1996; Chaouch et al., 2009; Goudenege et al., 2009).

Another successful example of direct genetic reprogramming is the stimulation of trans-differentiation into an osteoblastic phenotype by the osteoblast-specific transcription factor Runx2. Runx2 regulates the gene network responsible for bone formation, and knockout of Runx2 alleles in mice leads to the complete absence of mineralized tissue formation (Ducy et al., 1997; Komori et al., 1997). Forced expression of Runx2 leads to genetic reprogramming of several cell types into an osteoblastic lineage, including multipotent progenitor cells (Ducy et al., 1997; Byers et al., 2002; Yang et al., 2003; Byers and Garcia, 2004), skeletal myoblasts (Gersbach et al., 2004a,b, 2006), and fibroblasts (Ducy et al., 1997; Byers et al., 2002; Phillips et al., 2006a,b, 2007a). These successes have led to the application of genetic engineering with Runx2 to generate mineralized tissues *in vitro* and repair bone defects *in vivo* (Yang et al., 2003; Byers et al., 2004, 2006; Zheng et al., 2004; Zhao et al., 2005, 2007; Gersbach et al., 2006, 2007; Phillips et al., 2006b, 2007a, 2008; Itaka et al., 2007; Bhat et al., 2008; Zhang et al., 2010).

Many other transcription factors have also been identified as regulators of gene networks associated with cell types central to the goals of regenerative medicine (Table 8.1). For example, master transcription factors have been used to induce cell differentiation into β-cells (Koya et al., 2008; Zhou et al., 2008), cardiomyocytes (Bian et al., 2007; Yamada et al., 2007; David et al., 2009; Takeuchi and Bruneau, 2009), endothelial cells (Ferdous et al., 2009), neurons (Kim et al., 2003, 2006; Chung et al., 2005; Andersson et al., 2007; Li et al., 2007; Yang et al., 2008; Flames and Hobert, 2009; Vierbuchen et al., 2010), oligodendrocytes (Jessberger et al., 2008), and chondrocytes (Paul et al., 2003), as well as the formation of new blood vessels (Vincent et al., 2000; Trentin et al., 2006; Rajagopalan et al., 2007; Rey et al., 2009; Sarkar et al., 2009) and tumor suppression (Clayman et al., 1995; Peng, 2005; Martins et al., 2006; Ventura et al., 2007). The identification of numerous master transcription factors that regulate gene networks corresponding to a wide variety of cell types suggests that genetic reprogramming is a promising strategy for directing cell differentiation for applications in regenerative medicine. Furthermore, reprogramming with these factors represents an interesting approach to understanding cell differentiation and lineage commitment, including the identification of drug targets critical to these processes.

MOLECULAR ENGINEERING OF NATURAL TRANSCRIPTION FACTORS

As described above, there are many examples of successful direct genetic reprogramming with the natural transcription factor that corresponds to a specific cell lineage. However, for many applications, the natural ability of the transcription factor to activate a gene network is insufficient to produce the desired effect. These applications require increased potency or control in regards to transactivation activity of the particular factor. To address this need, many transcription factors have been engineered into forms with enhanced or controllable activity (Table 8.2). This can be achieved by mutating critical residues of the protein, altering or removing destabilizing regions of the protein, or fusing the factor to another protein domain that enhances transcriptional activation. This approach can be best exemplified by the molecular engineering of the angiogenic transcription factor HIF-1α.

TABLE 8.2 Representative Engineered Modifications of Natural Transcription Factors

Transcription factor	Engineered modification	Publications
Runx2	Point mutations that prevent inhibitory serine phosphorylation	Phillips et al. (2006a)
MyoD	Fusion to inducible steroid receptor domain	Hollenberg et al. (1993)
Hif-1α	Fusion to VP16 constitutive transactivation domain, point mutations to prevent ubiquitination, truncation to remove destabilizing domain	Vincent et al. (2000); Kelly et al. (2003); Pajusola et al. (2005)
Oct4	Point mutations that mimic serine phosphorylation	Saxe et al. (2009)

Early studies dedicated to elucidating the mechanism by which erythropoietin is induced in response to hypoxia led to the discovery of HIF-1 as the primary transcriptional mediator of cellular oxygen sensing (Semenza and Wang, 1992). It is now clear that HIF-1 regulates the expression of hundreds of genes in response to hypoxia, many of which are also transcription factors (Manalo et al., 2005). This supports the role of HIF-1 as a master orchestrator of the complex network of spatial and temporal signals that lead to new blood vessel formation (Hirota and Semenza, 2006). HIF-1 is a heterodimeric transcription factor composed of two basic helix-loop-helix (bHLH) proteins, HIF-1α and HIF-1β (Wang et al., 1995). HIF-1β is constitutively expressed in an active form in the nucleus of oxygen-sensing cells. In contrast, HIF-1α is highly inducible by hypoxia, primarily through post-translational regulation (Jiang et al., 1996). Under normoxic conditions, HIF-1α is rapidly degraded through hydroxylation of proline residues in the N- and C-terminal oxygen-dependent degradation domains (NODDD and CODDD) (Fig. 8.3) (Salceda and Caro, 1997; Huang et al., 1998; Maxwell et al., 1999; Schofield and Ratcliffe, 2004). This post-translational modification is mediated by a family of three HIF-1α prolyl hydroxylases that use oxygen as a substrate such that enzymatic activity is tightly regulated by oxygen concentration (Epstein et al., 2001; McNeill et al., 2002; Baek et al., 2005). Hydroxylated proline residues are recognized by the von Hippel Lindau tumor suppressor (VHL), which targets HIF-1α for proteosomal proteolysis via ubiquitin ligation (Maxwell et al., 1999). In the hypoxic environment, the HIF-1α prolyl hydroxylases are inactive,

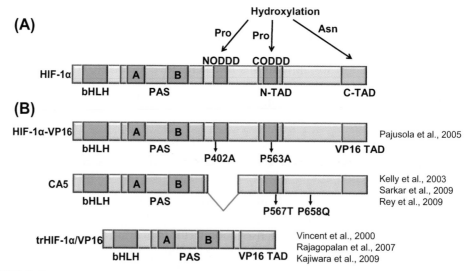

FIGURE 8.3

Structure of (A) HIF-1α and (B) variants of HIF-1α used in clinical and preclinical studies. Core elements of the HIF-1α protein include the basic helix-loop-helix DNA-binding domain and PAS domain. The natural protein contains N- and C-terminal oxygen-dependent degradation domains (ODDDs) and transcriptional activation domains (TADs). These domains may be substituted with a constitutively active VP16 TAD. Destabilizing proline and asparagine residues may also be mutated to enhance protein stability.

leading to dehydroxylation of proline residues and increased levels of a stabilized HIF-1α protein. HIF-1α also contains two distinct transcriptional activation domains (TADs) (Fig. 8.3). Asparaginyl hydroxylation within the C-terminal activation domain blocks interaction with the HIF-1 co-activator p300 (Arany et al., 1996; Lando et al., 2002; Schofield and Ratcliffe, 2004). In hypoxic environments, the stabilized HIF-1 heterodimer regulates the expression of a variety of angiogenic growth factors, including VEGF, PDGF, PLGF, angiopoietin 1, and angiopoietin 2, as well as their receptors (Hirota and Semenza, 2006). Other genes regulated by HIF-1 included factors involved in matrix metabolism, including MMPs, plasminogen activator receptors and inhibitors, and procollagen prolyl hydroxylase (Hirota and Semenza, 2006). Global gene analysis of arterial endothelial cells suggests that nearly 2% of all human genes may be directly or indirectly regulated by HIF-1 (Manalo et al., 2005). The number and variety of HIF-1 target genes that are critical to angiogenesis has stimulated the investigation of HIF-1 as a provascular therapeutic.

The efficacy of gene therapies with HIF-1α to stimulate angiogenesis has been demonstrated in numerous preclinical studies and one clinical study. All of these studies have used engineered forms of HIF-1α in which the coding sequence was modified to stabilize the protein and prevent degradation. In some cases, the transactivation domain from the herpes simplex virus VP16 was added to the protein to create a constitutively active HIF-1α. An early study demonstrated that delivery of a plasmid encoding a truncated HIF-1α fused to the VP16 domain (trHIF-1α/VP16; Fig. 8.3) to the ischemic hind limbs of rabbits led to significant improvements in calf blood pressure ratio, angiographic score, resting and maximal regional blood flow, and capillary density (Vincent et al., 2000). This study was the basis for a subsequent phase I dose-identification clinical trial in no-option patients with critical limb ischemia. This trial showed that adenoviral delivery of HIF-1α or trHIF-1α/VP16 was well tolerated and provided encouraging evidence of efficacy (Rajagopalan et al., 2007). A phase II trial is under way. This form of HIF-1α has also been shown to reduce infarct size and enhance neovascularization following plasmid DNA delivery to an acute myocardial infarction (Shyu et al., 2002). Alternatively, a truncated form of HIF-1α with mutations to destabilizing proline residues (CA5; Fig. 8.3) has been used by adenoviral delivery to induce angiogenesis in non-ischemic tissues (Kelly et al., 2003), improve perfusion and arterial remodeling in an endovascular model of limb ischemia (Patel et al., 2005), and treat critical limb ischemia in mice (Rey et al., 2009; Sarkar et al., 2009). A variety of additional preclinical results that support the gene delivery of various forms of HIF-1α in multiple small animal models of ischemia have been published (Jiang et al., 2008; Tal et al., 2008; Huang et al., 2009; Kajiwara et al., 2009). Notably, several of these studies demonstrate enhanced efficacy of HIF-1α relative to VEGF treatment (Pajusola et al., 2005; Trentin et al., 2006). Collectively, this work has validated the rationale of HIF-1α-based angiogenic gene therapy and shown the utility of modifying the natural HIF-1α protein sequence to enhance activity.

Although molecular engineering of HIF-1α has been examined the most extensively, there are a variety of other examples of transcription factor engineering to enhance or control protein activity. Fusion proteins of the myogenic factor MyoD and hormone-binding domains of steroid receptors have been created in order to control the myogenic activity of MyoD with hormone treatment (Hollenberg et al., 1993). Point mutations to the osteogenic transcription factor Runx2 have been identified that mimic post-translational modifications responsible for regulating Runx2 activity (Phillips et al., 2006a). Overexpression of the Runx2 variant containing these mutations in dermal fibroblasts led to enhanced osteoblastic gene expression and mineralized tissue formation relative to the wild-type sequence as a result of bypassing these regulatory mechanisms. Similar mutations have been identified that modulate the activity of the Oct4 transcription factor which regulates the gene network responsible for stem cell pluripotency (Saxe et al., 2009).

Collectively, these varied examples of enhancing the properties of transcription factors represent a general approach to refining the potency and control of reprogramming gene networks. Given the recent advances in genetic reprogramming to create new cell sources for regenerative medicine, it is likely that these approaches will be highlighted and expanded in the near future.

SYNTHETIC TRANSCRIPTION FACTORS FOR TARGETED GENE REGULATION

The effectiveness of directly coordinating changes in gene expression as a means to control cell behavior for regenerative medicine and gene therapy has led to an interest in engineering artificial regulators of specific target genes. For many applications, it may be desirable to control the expression of a specific gene, rather than entire gene networks. For this purpose, scientists have used naturally occurring DNA-binding molecules, including Cys2-His2 zinc finger proteins, as a guide for engineering artificial transcription factors. The Cys2-His2 zinc finger domain is the most common DNA-binding motif in the human proteome (Lander et al., 2001; Venter et al., 2001) and consists of a ββα configuration, where the α-helix projects into the major groove of DNA (Pavletich and Pabo, 1991) (Fig. 8.4A). Structural analysis demonstrates that although a single zinc finger contains approximately 30 amino acids, the domain typically functions by binding three consecutive base pairs of DNA via interactions of a single amino acid side chain per base pair (Elrod-Erickson et al., 1996; Pavletich and Pabo, 1991). The specificity of particular zinc fingers for the 64 different nucleotide triplets has been examined extensively through site-directed mutagenesis and rational design (Desjarlais and Berg, 1992; Nardelli et al., 1992) or the selection of large combinatorial libraries by phage display (Choo and Klug, 1994; Jamieson et al., 1994; Rebar and Pabo, 1994; Wu et al., 1995; Greisman and Pabo, 1997). As a result of this work, synthetic zinc finger domains have been isolated that bind to almost all of the possible nucleotide triplets (Segal et al., 1999; Dreier et al., 2001, 2005). Significantly, the modular structure of zinc finger motifs permits the conjunction of several domains in series, allowing for the recognition and targeting of extended sequences in multiples of three nucleotides (Beerli and Barbas, 2002; Segal et al., 2003). As a result, zinc finger protein can be designed to bind with high affinity and specificity to any target site in a cellular genome. These DNA-binding domains can then be combined with effector domains to create functional molecules that act at targeted genomic locations (Fig. 8.4B). Established effector domains include activating (Seipel et al., 1992), repressing (Hanna-Rose and Hansen, 1996), and inducible (Beerli et al., 2000) motifs for regulating

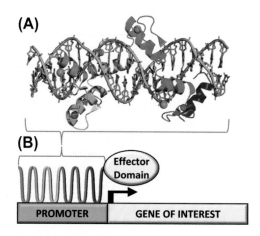

FIGURE 8.4

Engineered zinc finger proteins for targeted gene regulation. (A) Structure of the engineered six-finger zinc finger protein Aart (Segal et al., 2006). Each finger is represented with a different color. (B) Individual zinc finger domains can be linked together to recognize target sequences in the genome with high specificity and affinity. When fused to effector domains, such as transcriptional activators or repressors, these proteins become functional artificial transcription factors.

transcription, nucleases for gene modification (Kim and Chandrasegaran, 1994; Porteus and Baltimore, 2003; Urnov et al., 2005), methylases for gene silencing (Snowden et al., 2002; Nomura and Barbas, 2007), integrases to direct chromosomal integration of viral DNA (Tan et al., 2004, 2006), and recombinases for rearranging gene sequences (Gordley et al., 2007, 2009). The control of DNA-binding specificity and the range of effector domain functionalities have created considerable enthusiasm for engineered zinc finger proteins as tools for the study and treatment of a vast range of pathologies.

Artificial transcription factors based on zinc finger proteins have been engineered to regulate a variety of genes relevant to regenerative medicine (Blancafort and Beltran, 2008). The most notable example to date is a zinc finger transcription factor designed to regulate the gene for vascular endothelial growth factor (VEGF). VEGF is known to stimulate the formation of new blood vessels necessary for wound healing and the repair of injured cardiovascular tissues. Consequently, VEGF has been pursued as a candidate for proangiogenic therapies. However, results to date have shown that direct delivery of a single VEGF isoform results in the formation of new blood vessels that are leaky, poorly interconnected, and generally have a structure that does not mirror normal vasculature (Ehrbar et al., 2004; Phelps and Garcia, 2009). Studies have shown that the presence of multiple VEGF isoforms in the correct ratio is a critical factor in proper blood vessel formation (Whitlock et al., 2004; Amano et al., 2005). Therefore, Rebar and colleagues designed an artificial zinc finger transcription factor that regulates the endogenous VEGF promoter and gene sequence (Liu et al., 2001; Rebar et al., 2002). By inducing expression from the endogenous VEGF gene, all of the natural mechanisms of VEGF regulation, including mRNA splicing and isoform generation, were retained. This transcription factor was shown to have an enhanced capacity for angiogenesis and wound healing relative to the most common VEGF isoform (Rebar et al., 2002). The therapeutic efficacy of artificial transcription factors regulating the VEGF gene has also been validated in models of hind limb ischemia and diabetic neuropathy (Dai et al., 2004; Price et al., 2006; Yu et al., 2006). As a result of these successes, this artificial transcription factor has moved into clinical trials for a variety of indications (Rebar, 2004; Klug, 2005).

Artificial zinc finger transcription factors have also been engineered to target a variety of other genes relevant to regenerative medicine (Table 8.3). For example, upregulation of the utrophin gene can be used as a substitute for dystrophin expression, which is lost in Duchenne muscular dystrophy as a result of mutation to the dystrophin gene. Therefore, artificial zinc finger transcription factors have been designed to regulate the utrophin promoter and induce utrophin expression in target cells (Corbi et al., 2000; Onori et al.,

TABLE 8.3 Selected Artificial Zinc Finger Transcription Factors Relevant to Regenerative Medicine

Target gene	Application	Publications
VEGF	Tissue ischemia, cardiovascular disease, diabetic neuropathy	Liu et al. (2001); Rebar et al. (2002); Dai et al. (2004); Yu et al. (2006); Price et al. (2006)
Utrophin	Duchenne muscular dystrophy	Corbi et al. (2000); Onori et al. (2007); Lu et al. (2008); Desantis et al. (2009); di Certo et al. (2010)
γ-Globin	Sickle cell disease	Blau et al. (2005); Graslund et al. (2005); Wilber et al. (2010)
Erythropoietin	Red blood cell production	Zhang et al. (2000)
Oct4	Embryonic stem cell differentiation	Bartsevich et al. (2003)
HIV	Repressing viral replication	Reynolds et al. (2003); Segal et al. (2004); Eberhardy et al. (2006)
Mediators of drug resistance	Sensitizing tumor cells to chemotherapy	Blancafort et al. (2005)
Maspin	Suppressing tumor growth	Beltran et al. (2007); Beltran et al. (2008)

2007; Lu et al., 2008; Desantis et al., 2009). These transcription factors can alleviate disease symptoms in animal models of Duchenne muscular dystrophy (Mattei et al., 2007; Lu et al., 2008; di Certo et al., 2010). Similarly, artificial zinc finger transcription factors have been generated to activate the γ-globin gene as a functional substitute for β-globin, which is lost in sickle-cell disease (Blau et al., 2005; Graslund et al., 2005; Tschulena et al., 2009; Wilber et al., 2010). Artificial zinc finger transcription factors have also been engineered to repress replication of the HIV genome (Reynolds et al., 2003; Segal et al., 2004; Eberhardy et al., 2006) and regulate oncogenes (Beerli et al., 1998, 2000; Blancafort et al., 2005; Lund et al., 2005), tumor suppressors (Beltran et al., 2007, 2008), molecules involved in cell-cell adhesion (Blancafort et al., 2003; Magnenat et al., 2004), and regulators of adipogenesis (Ren et al., 2002), erythropoiesis (Zhang et al., 2000), and stem cell pluripotency (Bartsevich et al., 2003).

The diversity of genes that have been targeted for activation or repression by engineered zinc finger transcription factors is convincing evidence of the robustness of this approach. Given that any gene in the human genome can be regulated by these factors, including silenced genes (Beltran et al., 2008), there are a great variety of means by which this approach might be used for regenerative medicine. For example, artificial transcription factors could be designed to target genes related to directing cell differentiation into specific lineages or used to regulate therapeutic molecules, such as growth factors and cytokines. Therefore, this strategy of protein engineering for targeted gene regulation is a powerful approach to repairing damaged tissues.

DELIVERY AND REGULATION

A variety of methods are available for delivering transcriptional regulators to cells and controlling their activity inside the cell. Typically the gene sequences for the transcription factors are delivered and the transcription factors are expressed inside the cell. Consequently, all of the benefits and limitations of various gene delivery vehicles, including plasmid DNA and viral vectors, are applicable to these approaches (Gersbach et al., 2007; Phillips et al., 2007b). Additionally, the transcription factors have been expressed and purified from bacteria as fusions to cell-penetrating peptides (Joliot and Prochiantz, 2004; Gump and Dowdy, 2007). These purified proteins can then be added directly to cell culture, cross the cell membrane, and enter the nucleus to coordinate changes in gene expression. This approach has been validated both *in vitro* and *in vivo* for a variety of applications relevant to regenerative medicine, including stimulating angiogenesis (Tachikawa et al., 2004; Yun et al., 2008), promoting β-cell regeneration (Koya et al., 2008), and generating iPSCs (Kim et al., 2009; Zhou et al., 2009).

Strategies for controlling transcription factor activity inside the cell are critical for ensuring safety and efficacy of tissue regeneration. Expression of the transcription factors can be controlled by regulating the gene with an inducible promoter (Kelm et al., 2004; Weber and Fussenegger, 2004). These systems permit the control of transgene expression through the administration of antibiotics, hormone analogues, quorum-sensing messengers, or secondary metabolites to genetically engineered cells *in vitro* or *in vivo*. These systems have been used in a variety of contexts for regulating cell differentiation and tissue regeneration (Gersbach et al., 2006, 2007). Alternatively, transcription factors can be regulated at the protein level by linking them to steroid receptors as a fusion protein. In these systems, the steroid receptors undergo conformation changes, such as dimerization, upon the addition of a drug that leads to reconstitution of protein activity. This approach has been used to regulate the activity of natural transcription factors (Hollenberg et al., 1993) and engineered zinc finger transcription factors (Beerli et al., 2000; Pollock et al., 2002; Magnenat et al., 2008). The multiple levels of regulation afforded by these genetically engineered systems allow for finely tuned control of gene expression in a variety of contexts.

CONCLUSION

Recent progress in cell and molecular biology has clearly demonstrated the role of gene expression in determining disease states and tissue regeneration. In parallel, advances in protein and genetic engineering have provided scientists with the methods necessary for directly reprogramming or modulating the gene expression networks that are central to these processes. Consequently, natural and engineered proteins that regulate gene expression are serving a central role in many regenerative medicine strategies. Highlighted by the discovery of iPSCs, genetic reprogramming with master regulatory factors is a general approach that can be used to coordinate a variety of gene networks critical to cell differentiation. Alternatively, specific genes can be individually regulated by artificial transcription factors, such as engineered zinc finger proteins. Collectively, these methods for gene regulation constitute a unique and powerful set of tools to address the persistent challenges of controlling cell behavior for regenerative medicine.

References

Aaron, R. K., Boyan, B. D., Ciombor, D. M., Schwartz, Z., & Simon, B. J. (2004). Stimulation of growth factor synthesis by electric and electromagnetic fields. *Clin. Orthop. Relat. Res.*, 30—37.

Amano, H., Hackett, N. R., Kaner, R. J., Whitlock, P., Rosengart, T. K., & Crystal, R. G. (2005). Alteration of splicing signals in a genomic/cDNA hybrid VEGF gene to modify the ratio of expressed VEGF isoforms enhances safety of angiogenic gene therapy. *Mol. Ther.*, *12*, 716—724.

Andersson, E. K., Irvin, D. K., Ahlsio, J., & Parmar, M. (2007). Ngn2 and Nurr1 act in synergy to induce midbrain dopaminergic neurons from expanded neural stem and progenitor cells. *Exp. Cell Res.*, *313*, 1172—1180.

Arany, Z., Huang, L. E., Eckner, R., Bhattacharya, S., Jiang, C., Goldberg, M. A., et al. (1996). An essential role for p300/CBP in the cellular response to hypoxia. *Proc. Natl. Acad. Sci. U.S.A.*, *93*, 12969—12973.

Atala, A. (2009). Engineering organs. *Curr. Opin. Biotechnol.*, *20*, 575—592.

Baek, J. H., Mahon, P. C., Oh, J., Kelly, B., Krishnamachary, B., Pearson, M., et al. (2005). OS-9 interacts with hypoxia-inducible factor 1alpha and prolyl hydroxylases to promote oxygen-dependent degradation of HIF-1alpha. *Mol. Cell*, *17*, 503—512.

Baguisi, A., Behboodi, E., Melican, D. T., Pollock, J. S., Destrempes, M. M., Cammuso, C., et al. (1999). Production of goats by somatic cell nuclear transfer. *Nat. Biotechnol.*, *17*, 456—461.

Bartsevich, V. V., Miller, J. C., Case, C. C., & Pabo, C. O. (2003). Engineered zinc finger proteins for controlling stem cell fate. *Stem Cells*, *21*, 632—637.

Beerli, R. R., & Barbas, C. F., III (2002). Engineering polydactyl zinc-finger transcription factors. *Nat. Biotechnol.*, *20*, 135—141.

Beerli, R. R., Dreier, B., & Barbas, C. F., III. (2000). Positive and negative regulation of endogenous genes by designed transcription factors. *Proc. Natl. Acad. Sci. U.S.A.*, *97*, 1495—1500.

Beerli, R. R., Schopfer, U., Dreier, B., & Barbas, C. F., III. (2000). Chemically regulated zinc finger transcription factors. *J. Biol. Chem.*, *275*, 32617—32627.

Beerli, R. R., Segal, D. J., Dreier, B., & Barbas, C. F., III. (1998). Toward controlling gene expression at will: specific regulation of the erbB-2/HER-2 promoter by using polydactyl zinc finger proteins constructed from modular building blocks. *Proc. Natl. Acad. Sci. U.S.A.*, *95*, 14628—14633.

Beltran, A., Parikh, S., Liu, Y., Cuevas, B. D., Johnson, G. L., Futscher, B. W., et al. (2007). Re-activation of a dormant tumor suppressor gene maspin by designed transcription factors. *Oncogene*, *26*, 2791—2798.

Beltran, A. S., Sun, X., Lizardi, P. M., & Blancafort, P. (2008). Reprogramming epigenetic silencing: artificial transcription factors synergize with chromatin remodeling drugs to reactivate the tumor suppressor mammary serine protease inhibitor. *Mol. Cancer Ther.*, *7*, 1080—1090.

Berkes, C. A., & Tapscott, S. J. (2005). MyoD and the transcriptional control of myogenesis. *Semin. Cell Dev. Biol.*, *16*, 585—595.

Bhat, B. M., Robinson, J. A., Coleburn, V. E., Zhao, W., & Kharode, Y. (2008). Evidence of in vivo osteoinduction in adult rat bone by adeno-Runx2 intra-femoral delivery. *J. Cell Biochem.*, *103*, 1912—1924.

Bian, J., Popovic, Z. B., Benejam, C., Kiedrowski, M., Rodriguez, L. L., & Penn, M. S. (2007). Effect of cell-based intercellular delivery of transcription factor GATA4 on ischemic cardiomyopathy. *Circ. Res.*, *100*, 1626—1633.

Blancafort, P., & Beltran, A. S. (2008). Rational design, selection and specificity of artificial transcription factors (ATFs): the influence of chromatin in target gene regulation. *Comb. Chem. High Throughput Screen*, *11*, 146—158.

Blancafort, P., Chen, E. I., Gonzalez, B., Bergquist, S., Zijlstra, A., Guthy, D., et al. (2005). Genetic reprogramming of tumor cells by zinc finger transcription factors. *Proc. Natl. Acad. Sci. U.S.A.*, *102*, 11716—11721.

Blancafort, P., Magnenat, L., & Barbas, C. F., III. (2003). Scanning the human genome with combinatorial transcription factor libraries. *Nat. Biotechnol., 21,* 269–274.

Blau, C. A., Barbas, C. F., III., Bomhoff, A. L., Neades, R., Yan, J., Navas, P. A., et al. (2005). γ-Globin gene expression in chemical inducer of dimerization (CID)-dependent multipotential cells established from human β-globin locus yeast artificial chromosome (β-YAC) transgenic mice. *J. Biol. Chem., 280,* 36642–36647.

Botusan, I. R., Sunkari, V. G., Savu, O., Catrina, A. I., Grünler, J., Lindberg, S., et al. (2008). Stabilization of HIF-1alpha is critical to improve wound healing in diabetic mice. *Proc. Natl. Acad. Sci. U.S.A., 105*(49), 19426–19431.

Briggs, R., & King, T. J. (1952). Transplantation of living nuclei from blastula cells into enucleated frogs' eggs. *Proc. Natl. Acad. Sci. U.S.A., 38,* 455–463.

Byers, B. A., & Garcia, A. J. (2004). Exogenous Runx2 expression enhances in vitro osteoblastic differentiation and mineralization in primary bone marrow stromal cells. *Tissue Eng., 10,* 1623–1632.

Byers, B. A., Guldberg, R. E., & Garcia, A. J. (2004). Synergy between genetic and tissue engineering: Runx2 overexpression and in vitro construct development enhance in vivo mineralization. *Tissue Eng., 10,* 1757–1766.

Byers, B. A., Guldberg, R. E., Hutmacher, D. W., & Garcia, A. J. (2006). Effects of Runx2 genetic engineering and in vitro maturation of tissue-engineered constructs on the repair of critical size bone defects. *J. Biomed. Mater. Res. A, 76,* 646–655.

Byers, B. A., Pavlath, G. K., Murphy, T. J., Karsenty, G., & Garcia, A. J. (2002). Cell-type-dependent up-regulation of in vitro mineralization after overexpression of the osteoblast-specific transcription factor Runx2/Cbfal. *J. Bone Miner. Res., 17,* 1931–1944.

Byrne, J. A., Pedersen, D. A., Clepper, L. L., Nelson, M., Sanger, W. G., Gokhale, S., et al. (2007). Producing primate embryonic stem cells by somatic cell nuclear transfer. *Nature, 450,* 497–502.

Chaouch, S., Mouly, V., Goyenvalle, A., Vulin, A., Mamchaoui, K., Negroni, E., et al. (2009). Immortalized skin fibroblasts expressing conditional MyoD as a renewable and reliable source of converted human muscle cells to assess therapeutic strategies for muscular dystrophies: validation of an exon-skipping approach to restore dystrophin in duchenne muscular dystrophy cells. *Hum. Gene Ther., 20,* 784–790.

Chiu, J. J., Usami, S., & Chien, S. (2009). Vascular endothelial responses to altered shear stress: pathologic implications for atherosclerosis. *Ann. Med., 41,* 19–28.

Choi, J., Costa, M. L., Mermelstein, C. S., Chagas, C., Holtzer, S., & Holtzer, H. (1990). MyoD converts primary dermal fibroblasts, chondroblasts, smooth muscle, and retinal pigmented epithelial cells into striated mononucleated myoblasts and multinucleated myotubes. *Proc. Natl. Acad. Sci. U.S.A., 87,* 7988–7992.

Choo, Y., & Klug, A. (1994). Toward a code for the interactions of zinc fingers with DNA: selection of randomized fingers displayed on phage. *Proc. Natl. Acad. Sci. U.S.A., 91,* 11163–11167.

Chung, S., Hedlund, E., Hwang, M., Kim, D. W., Shin, B. S., Hwang, D. Y., et al. (2005). The homeodomain transcription factor Pitx3 facilitates differentiation of mouse embryonic stem cells into AHD2-expressing dopaminergic neurons. *Mol. Cell Neurosci., 28,* 241–252.

Clayman, G. L., el-Naggar, A. K., Roth, J. A., Zhang, W. W., Goepfert, H., Taylor, D. L., et al. (1995). In vivo molecular therapy with p53 adenovirus for microscopic residual head and neck squamous carcinoma. *Cancer Res., 55,* 1–6.

Corbi, N., Libri, V., Fanciulli, M., Tinsley, J. M., Davies, K. E., & Passananti, C. (2000). The artificial zinc finger coding gene "Jazz" binds the utrophin promoter and activates transcription. *Gene Ther., 7,* 1076–1083.

Dai, Q., Huang, J., Klitzman, B., Dong, C., Goldschmidt-Clermont, P. J., March, K. L., et al. (2004). Engineered zinc finger-activating vascular endothelial growth factor transcription factor plasmid DNA induces therapeutic angiogenesis in rabbits with hindlimb ischemia. *Circulation, 110,* 2467–2475.

David, R., Stieber, J., Fischer, E., Brunner, S., Brenner, C., Pfeiler, S., et al. (2009). Forward programming of pluripotent stem cells towards distinct cardiovascular cell types. *Cardiovasc. Res., 84,* 263–272.

Davies, P. F. (2009). Hemodynamic shear stress and the endothelium in cardiovascular pathophysiology. *Nat. Clin. Pract. Cardiovasc. Med., 6,* 16–26.

Davis, R. L., Weintraub, H., & Lassar, A. B. (1987). Expression of a single transfected cDNA converts fibroblasts to myoblasts. *Cell, 51,* 987–1000.

Desantis, A., Onori, A., di Certo, M. G., Mattei, E., Fanciulli, M., Passananti, C., et al. (2009). Novel activation domain derived from Che-1 cofactor coupled with the artificial protein Jazz drives utrophin upregulation. *Neuromuscul. Disord., 19,* 158–162.

Desjarlais, J. R., & Berg, J. M. (1992). Toward rules relating zinc finger protein sequences and DNA binding site preferences. *Proc. Natl. Acad. Sci. U.S.A., 89,* 7345–7349.

di Certo, M.G., Corbi, N., Strimpakos, G., Onori, A., Luvisetto, S., Severini, C., et al. (2010). The artificial gene Jazz, a transcriptional regulator of utrophin, corrects the dystrophic pathology in mdx mice. *Hum. Mol. Genet., 19,* 752–760.

Dreier, B., Beerli, R. R., Segal, D. J., Flippin, J. D., & Barbas, C. F., III. (2001). Development of zinc finger domains for recognition of the 5′-ANN-3′ family of DNA sequences and their use in the construction of artificial transcription factors. *J. Biol. Chem., 276*, 29466–29478.

Dreier, B., Fuller, R. P., Segal, D. J., Lund, C. V., Blancafort, P., Huber, A., et al. (2005). Development of zinc finger domains for recognition of the 5′-CNN-3′ family DNA sequences and their use in the construction of artificial transcription factors. *J. Biol. Chem., 280*, 35588–35597.

Ducy, P., Zhang, R., Geoffroy, V., Ridall, A. L., & Karsenty, G. (1997). Osf2/Cbfa1: a transcriptional activator of osteoblast differentiation. *Cell, 89*, 747–754.

Eberhardy, S. R., Goncalves, J., Coelho, S., Segal, D. J., Berkhout, B., & Barbas, C. F., III. (2006). Inhibition of human immunodeficiency virus type 1 replication with artificial transcription factors targeting the highly conserved primer-binding site. *J. Virol., 80*, 2873–2883.

Ehrbar, M., Djonov, V. G., Schnell, C., Tschanz, S. A., Martiny-Baron, G., Schenk, U., et al. (2004). Cell-demanded liberation of VEGF121 from fibrin implants induces local and controlled blood vessel growth. *Circ. Res., 94*, 1124–1132.

Elrod-Erickson, M., Rould, M. A., Nekludova, L., & Pabo, C. O. (1996). Zif268 protein-DNA complex refined at 1.6 A: a model system for understanding zinc finger-DNA interactions. *Structure, 4*, 1171–1180.

Epstein, A. C., Gleadle, J. M., McNeill, L. A., Hewitson, K. S., O'Rourke, J., Mole, D. R., et al. (2001). C. elegans EGL-9 and mammalian homologs define a family of dioxygenases that regulate HIF by prolyl hydroxylation. *Cell, 107*, 43–54.

Ferdous, A., Caprioli, A., Iacovino, M., Martin, C. M., Morris, J., Richardson, J. A., et al. (2009). Nkx2-5 trans-activates the Ets-related protein 71 gene and specifies an endothelial/endocardial fate in the developing embryo. *Proc. Natl. Acad. Sci. U.S.A., 106*, 814–819.

Flames, N., & Hobert, O. (2009). Gene regulatory logic of dopamine neuron differentiation. *Nature, 458*, 885–889.

Gersbach, C. A., Byers, B. A., Pavlath, G. K., & Garcia, A. J. (2004b). Runx2/Cbfa1 stimulates trans-differentiation of primary skeletal myoblasts into a mineralizing osteoblastic phenotype. *Exp. Cell Res., 300*, 406–417.

Gersbach, C. A., Byers, B. A., Pavlath, G. K., Guldberg, R. E., & Garcia, A. J. (2004a). Runx2/Cbfa1-genetically engineered skeletal myoblasts mineralize collagen scaffolds in vitro. *Biotechnol. Bioeng., 88*, 369–378.

Gersbach, C. A., Guldberg, R. E., & Garcia, A. J. (2007). In vitro and in vivo osteoblastic differentiation of BMP-2- and Runx2-engineered skeletal myoblasts. *J. Cell Biochem., 100*, 1324–1336.

Gersbach, C. A., Le Doux, J. M., Guldberg, R. E., & Garcia, A. J. (2006). Inducible regulation of Runx2-stimulated osteogenesis. *Gene Ther., 13*, 873–882.

Gersbach, C. A., Phillips, J. E., & Garcia, A. J. (2007). Genetic engineering for skeletal regenerative medicine. *Annu. Rev. Biomed. Eng., 9*, 87–119.

Gordley, R. M., Gersbach, C. A., & Barbas, C. F., III. (2009). Synthesis of programmable integrases. *Proc. Natl. Acad. Sci. U.S.A., 106*, 5053–5058.

Gordley, R. M., Smith, J. D., Graslund, T., & Barbas, C. F., III. (2007). Evolution of programmable zinc finger-recombinases with activity in human cells. *J. Mol. Biol., 367*, 802–813.

Gordon, G. A. (2007). Designed electromagnetic pulsed therapy: clinical applications. *J. Cell Physiol., 212*, 579–582.

Goudenege, S., Pisani, D. F., Wdziekonski, B., di Santo, J. P., Bagnis, C., Dani, C., et al. (2009). Enhancement of myogenic and muscle repair capacities of human adipose-derived stem cells with forced expression of MyoD. *Mol. Ther., 17*, 1064–1072.

Graslund, T., Li, X., Magnenat, L., Popkov, M., & Barbas, C. F., III. (2005). Exploring strategies for the design of artificial transcription factors: targeting sites proximal to known regulatory regions for the induction of gamma-globin expression and the treatment of sickle cell disease. *J. Biol. Chem., 280*, 3707–3714.

Greisman, H. A., & Pabo, C. O. (1997). A general strategy for selecting high-affinity zinc finger proteins for diverse DNA target sites. *Science, 275*, 657–661.

Gump, J. M., & Dowdy, S. F. (2007). TAT transduction: the molecular mechanism and therapeutic prospects. *Trends Mol. Med., 13*, 443–448.

Gurdon, J. B., & Melton, D. A. (2008). Nuclear reprogramming in cells. *Science, 322*, 1811–1815.

Hanna-Rose, W., & Hansen, U. (1996). Active repression mechanisms of eukaryotic transcription repressors. *Trends Genet., 12*, 229–234.

Hirota, K., & Semenza, G. L. (2006). Regulation of angiogenesis by hypoxia-inducible factor 1. *Crit. Rev. Oncol. Hematol., 59*, 15–26.

Hollenberg, S. M., Cheng, P. F., & Weintraub, H. (1993). Use of a conditional MyoD transcription factor in studies of MyoD trans-activation and muscle determination. *Proc. Natl. Acad. Sci. U.S.A., 90*, 8028–8032.

Huang, L. E., Gu, J., Schau, M., & Bunn, H. F. (1998). Regulation of hypoxia-inducible factor 1alpha is mediated by an O2-dependent degradation domain via the ubiquitin-proteasome pathway. *Proc. Natl. Acad. Sci. U.S.A., 95*, 7987–7992.

171

Huang, M., Chen, Z., Hu, S., Jia, F., Li, Z., Hoyt, G., et al. (2009). Novel minicircle vector for gene therapy in murine myocardial infarction. *Circulation, 120*, S230−S237.

Itaka, K., Ohba, S., Miyata, K., Kawaguchi, H., Nakamura, K., Takato, T., et al. (2007). Bone regeneration by regulated in vivo gene transfer using biocompatible polyplex nanomicelles. *Mol. Ther., 15*, 1655−1662.

Jaenisch, R., & Young, R. (2008). Stem cells, the molecular circuitry of pluripotency and nuclear reprogramming. *Cell, 132*, 567−582.

Jamieson, A. C., Kim, S. H., & Wells, J. A. (1994). In vitro selection of zinc fingers with altered DNA-binding specificity. *Biochemistry, 33*, 5689−5695.

Jessberger, S., Toni, N., Clemenson, G. D., Jr., Ray, J., & Gage, F. H. (2008). Directed differentiation of hippocampal stem/progenitor cells in the adult brain. *Nat. Neurosci., 11*, 888−893.

Jiang, B. H., Semenza, G. L., Bauer, C., & Marti, H. H. (1996). Hypoxia-inducible factor 1 levels vary exponentially over a physiologically relevant range of O2 tension. *Am. J. Physiol., 271*, C1172−C1180.

Jiang, M., Wang, B., Wang, C., He, B., Fan, H., Guo, T. B., et al. (2008). Angiogenesis by transplantation of HIF-1 alpha modified EPCs into ischemic limbs. *J. Cell Biochem., 103*, 321−334.

Joliot, A., & Prochiantz, A. (2004). Transduction peptides: from technology to physiology. *Nat. Cell Biol., 6*, 189−196.

Kajiwara, H., Luo, Z., Belanger, A. J., Urabe, A., Vincent, K. A., Akita, G. Y., et al. (2009). A hypoxic inducible factor-1 alpha hybrid enhances collateral development and reduces vascular leakage in diabetic rats. *J. Gene Med., 11*, 390−400.

Kato, Y., Tani, T., Sotomaru, Y., Kurokawa, K., Kato, J., Doguchi, H., et al. (1998). Eight calves cloned from somatic cells of a single adult. *Science, 282*, 2095−2098.

Kelly, B. D., Hackett, S. F., Hirota, K., Oshima, Y., Cai, Z., Berg-Dixon, S., et al. (2003). Cell type-specific regulation of angiogenic growth factor gene expression and induction of angiogenesis in nonischemic tissue by a constitutively active form of hypoxia-inducible factor 1. *Circ. Res., 93*, 1074−1081.

Kelm, J. M., Kramer, B. P., Gonzalez-Nicolini, V., Ley, B., & Fussenegger, M. (2004). Synergies of microtissue design, viral transduction and adjustable transgene expression for regenerative medicine. *Biotechnol. Appl. Biochem., 39*, 3−16.

Kim, D., Kim, C. H., Moon, J. I., Chung, Y. G., Chang, M. Y., Han, B. S., et al. (2009). Generation of human induced pluripotent stem cells by direct delivery of reprogramming proteins. *Cell Stem Cell, 4*, 472−476.

Kim, D. W., Chung, S., Hwang, M., Ferree, A., Tsai, H. C., Park, J. J., et al. (2006). Stromal cell-derived inducing activity, Nurr1, and signaling molecules synergistically induce dopaminergic neurons from mouse embryonic stem cells. *Stem Cells, 24*, 557−567.

Kim, J. Y., Koh, H. C., Lee, J. Y., Chang, M. Y., Kim, Y. C., Chung, H. Y., et al. (2003). Dopaminergic neuronal differentiation from rat embryonic neural precursors by Nurr1 overexpression. *J. Neurochem., 85*, 1443−1454.

Kim, Y. G., & Chandrasegaran, S. (1994). Chimeric restriction endonuclease. *Proc. Natl. Acad. Sci. U.S.A., 91*, 883−887.

Klug, A. (2005). Towards therapeutic applications of engineered zinc finger proteins. *FEBS Lett., 579*, 892−894.

Komori, T., Yagi, H., Nomura, S., Yamaguchi, A., Sasaki, K., Deguchi, K., et al. (1997). Targeted disruption of Cbfa1 results in a complete lack of bone formation owing to maturational arrest of osteoblasts. *Cell, 89*, 755−764.

Koya, V., Lu, S., Sun, Y. P., Purich, D. L., Atkinson, M. A., Li, S. W., et al. (2008). Reversal of streptozotocin-induced diabetes in mice by cellular transduction with recombinant pancreatic transcription factor pancreatic duodenal homeobox-1: a novel protein transduction domain-based therapy. *Diabetes, 57*, 757−769.

Lander, E. S., Linton, L. M., Birren, B., Nusbaum, C., Zody, M. C., Baldwin, J., et al. (2001). Initial sequencing and analysis of the human genome. *Nature, 409*, 860−921.

Lando, D., Peet, D. J., Whelan, D. A., Gorman, J. J., & Whitelaw, M. L. (2002). Asparagine hydroxylation of the HIF transactivation domain a hypoxic switch. *Science, 295*, 858−861.

Li, Q. J., Tang, Y. M., Liu, J., Zhou, D. Y., Li, X. P., Xiao, S. H., et al. (2007). Treatment of Parkinson disease with C17.2 neural stem cells overexpressing NURR1 with a recombinant republic-deficit adenovirus containing the NURR1 gene. *Synapse, 61*, 971−977.

Liu, P. Q., Rebar, E. J., Zhang, L., Liu, Q., Jamieson, A. C., Liang, Y., et al. (2001). Regulation of an endogenous locus using a panel of designed zinc finger proteins targeted to accessible chromatin regions. Activation of vascular endothelial growth factor A. *J. Biol. Chem., 276*, 11323−11334.

Lu, Y., Tian, C., Danialou, G., Gilbert, R., Petrof, B. J., Karpati, G., et al. (2008). Targeting artificial transcription factors to the utrophin A promoter: effects on dystrophic pathology and muscle function. *J. Biol. Chem., 283*, 34720−34727.

Lund, C. V., Popkov, M., Magnenat, L., & Barbas, C. F., III. (2005). Zinc finger transcription factors designed for bispecific coregulation of ErbB2 and ErbB3 receptors: insights into ErbB receptor biology. *Mol. Cell Biol., 25*, 9082−9091.

Lutolf, M. P., & Hubbell, J. A. (2005). Synthetic biomaterials as instructive extracellular microenvironments for morphogenesis in tissue engineering. *Nat. Biotechnol., 23*, 47−55.

Lutolf, M. P., Gilbert, P. M., & Blau, H. M. (2009). Designing materials to direct stem-cell fate. *Nature, 462*, 433–441.

Magnenat, L., Blancafort, P., & Barbas, C. F., III. (2004). In vivo selection of combinatorial libraries and designed affinity maturation of polydactyl zinc finger transcription factors for ICAM-1 provides new insights into gene regulation. *J. Mol. Biol., 341*, 635–649.

Magnenat, L., Schwimmer, L. J., & Barbas, C. F., III. (2008). Drug-inducible and simultaneous regulation of endogenous genes by single-chain nuclear receptor-based zinc-finger transcription factor gene switches. *Gene Ther., 15*, 1223–1232.

Maherali, N., Sridharan, R., Xie, W., Utikal, J., Eminli, S., Arnold, K., et al. (2007). Directly reprogrammed fibroblasts show global epigenetic remodeling and widespread tissue contribution. *Cell Stem Cell, 1*(1), 55–70.

Manalo, D. J., Rowan, A., Lavoie, T., Natarajan, L., Kelly, B. D., Ye, S. Q., et al. (2005). Transcriptional regulation of vascular endothelial cell responses to hypoxia by HIF-1. *Blood, 105*, 659–669.

Martins, C. P., Brown-Swigart, L., & Evan, G. I. (2006). Modeling the therapeutic efficacy of p53 restoration in tumors. *Cell, 127*, 1323–1334.

Mattei, E., Corbi, N., di Certo, M. G., Strimpakos, G., Severini, C., Onori, A., et al. (2007). Utrophin up-regulation by an artificial transcription factor in transgenic mice. *PLoS One, 2*, e774.

Maxwell, P. H., Wiesener, M. S., Chang, G. W., Clifford, S. C., Vaux, E. C., Cockman, M. E., et al. (1999). The tumour suppressor protein VHL targets hypoxia-inducible factors for oxygen-dependent proteolysis. *Nature, 399*, 271–275.

McNeill, L. A., Hewitson, K. S., Gleadle, J. M., Horsfall, L. E., Oldham, N. J., Maxwell, P. H., et al. (2002). The use of dioxygen by HIF prolyl hydroxylase (PHD1). *Bioorg. Med. Chem. Lett., 12*, 1547–1550.

Muller, L. U., Daley, G. Q., & Williams, D. A. (2009). Upping the ante: recent advances in direct reprogramming. *Mol. Ther., 17*, 947–953.

Murry, C. E., Kay, M. A., Bartosek, T., Hauschka, S. D., & Schwartz, S. M. (1996). Muscle differentiation during repair of myocardial necrosis in rats via gene transfer with MyoD. *J. Clin. Invest., 98*, 2209–2217.

Nardelli, J., Gibson, T., & Charnay, P. (1992). Zinc finger-DNA recognition: analysis of base specificity by site-directed mutagenesis. *Nucleic Acids Res., 20*, 4137–4144.

Nomura, W., & Barbas, C. F., III. (2007). In vivo site-specific DNA methylation with a designed sequence-enabled DNA methylase. *J. Am. Chem. Soc., 129*, 8676–8677.

Okita, K., Ichisaka, T., & Yamanaka, S. (2007). Generation of germline-competent induced pluripotent stem cells. *Nature, 448*, 313–317.

Onori, A., Desantis, A., Buontempo, S., di Certo, M. G., Fanciulli, M., Salvatori, L., et al. (2007). The artificial 4-zinc-finger protein Bagly binds human utrophin promoter A at the endogenous chromosomal site and activates transcription. *Biochem. Cell Biol., 85*, 358–365.

Pajusola, K., Kunnapuu, J., Vuorikoski, S., Soronen, J., Andre, H., Pereira, T., et al. (2005). Stabilized HIF-1alpha is superior to VEGF for angiogenesis in skeletal muscle via adeno-associated virus gene transfer. *FASEB J., 19*, 1365–1367.

Park, I. H., Arora, N., Huo, H., Maherali, N., Ahfeldt, T., Shimamura, A., et al. (2008). Disease-specific induced pluripotent stem cells. *Cell, 134*(5), 877–886.

Patel, T. H., Kimura, H., Weiss, C. R., Semenza, G. L., & Hofmann, L. V. (2005). Constitutively active HIF-1alpha improves perfusion and arterial remodeling in an endovascular model of limb ischemia. *Cardiovasc. Res., 68*, 144–154.

Paul, R., Haydon, R. C., Cheng, H., Ishikawa, A., Nenadovich, N., Jiang, W., et al. (2003). Potential use of Sox9 gene therapy for intervertebral degenerative disc disease. *Spine (Phila. Pa 1976), 28*, 755–763.

Pavletich, N. P., & Pabo, C. O. (1991). Zinc finger-DNA recognition: crystal structure of a Zif268-DNA complex at 2.1 A. *Science, 252*, 809–817.

Peng, Z. (2005). Current status of gendicine in China: recombinant human Ad-p53 agent for treatment of cancers. *Hum. Gene Ther., 16*, 1016–1027.

Phelps, E. A., & Garcia, A. J. (2009). Update on therapeutic vascularization strategies. *Regen. Med., 4*, 65–80.

Phillips, J. E., Burns, K. L., Le Doux, J. M., Guldberg, R. E., & Garcia, A. J. (2008). Engineering graded tissue interfaces. *Proc. Natl. Acad. Sci. U.S.A., 105*, 12170–12175.

Phillips, J. E., Gersbach, C. A., & Garcia, A. J. (2007b). Virus-based gene therapy strategies for bone regeneration. *Biomaterials, 28*, 211–229.

Phillips, J. E., Gersbach, C. A., Wojtowicz, A. M., & Garcia, A. J. (2006a). Glucocorticoid-induced osteogenesis is negatively regulated by Runx2/Cbfa1 serine phosphorylation. *J. Cell Sci., 119*, 581–591.

Phillips, J. E., Guldberg, R. E., & Garcia, A. J. (2007a). Dermal fibroblasts genetically modified to express Runx2/Cbfa1 as a mineralizing cell source for bone tissue engineering. *Tissue Eng., 13*, 2029–2040.

Phillips, J. E., Hutmacher, D. W., Guldberg, R. E., & Garcia, A. J. (2006b). Mineralization capacity of Runx2/Cbfa1-genetically engineered fibroblasts is scaffold dependent. *Biomaterials, 27*, 5535–5545.

Pollock, R., Giel, M., Linher, K., & Clackson, T. (2002). Regulation of endogenous gene expression with a small-molecule dimerizer. *Nat. Biotechnol., 20*, 729–733.

Pomerantz, J., & Blau, H. M. (2004). Nuclear reprogramming: a key to stem cell function in regenerative medicine. *Nat. Cell Biol., 6*, 810–816.

Porteus, M. H., & Baltimore, D. (2003). Chimeric nucleases stimulate gene targeting in human cells. *Science, 300*, 763.

Price, S. A., Dent, C., Duran-Jimenez, B., Liang, Y., Zhang, L., Rebar, E. J., et al. (2006). Gene transfer of an engineered transcription factor promoting expression of VEGF-A protects against experimental diabetic neuropathy. *Diabetes, 55*, 1847–1854.

Rajagopalan, S., Olin, J., Deitcher, S., Pieczek, A., Laird, J., Grossman, P. M., et al. (2007). Use of a constitutively active hypoxia-inducible factor-1alpha transgene as a therapeutic strategy in no-option critical limb ischemia patients: phase I dose-escalation experience. *Circulation, 115*, 1234–1243.

Rebar, E. J. (2004). Development of pro-angiogenic engineered transcription factors for the treatment of cardiovascular disease. *Expert Opin. Investig. Drugs, 13*, 829–839.

Rebar, E. J., & Pabo, C. O. (1994). Zinc finger phage: affinity selection of fingers with new DNA-binding specificities. *Science, 263*, 671–673.

Rebar, E. J., Huang, Y., Hickey, R., Nath, A. K., Meoli, D., Nath, S., et al. (2002). Induction of angiogenesis in a mouse model using engineered transcription factors. *Nat. Med., 8*, 1427–1432.

Rehfeldt, F., Engler, A. J., Eckhardt, A., Ahmed, F., & Discher, D. E. (2007). Cell responses to the mechanochemical microenvironment – implications for regenerative medicine and drug delivery. *Adv. Drug Deliv. Rev., 59*, 1329–1339.

Ren, D., Collingwood, T. N., Rebar, E. J., Wolffe, A. P., & Camp, H. S. (2002). PPARgamma knockdown by engineered transcription factors: exogenous PPARgamma2 but not PPARgamma1 reactivates adipogenesis. *Genes Dev., 16*, 27–32.

Rey, S., Lee, K., Wang, C. J., Gupta, K., Chen, S., McMillan, A., et al. (2009). Synergistic effect of HIF-1alpha gene therapy and HIF-1-activated bone marrow-derived angiogenic cells in a mouse model of limb ischemia. *Proc. Natl. Acad. Sci. U.S.A., 106*, 20399–20404.

Reynolds, L., Ullman, C., Moore, M., Isalan, M., West, M. J., Clapham, P., et al. (2003). Repression of the HIV-1 5′ LTR promoter and inhibition of HIV-1 replication by using engineered zinc-finger transcription factors. *Proc. Natl. Acad. Sci. U.S.A., 100*, 1615–1620.

Salceda, S., & Caro, J. (1997). Hypoxia-inducible factor 1alpha (HIF-1alpha) protein is rapidly degraded by the ubiquitin-proteasome system under normoxic conditions. Its stabilization by hypoxia depends on redox-induced changes. *J. Biol. Chem., 272*, 22642–22647.

Sarkar, K., Fox-Talbot, K., Steenbergen, C., Bosch-Marce, M., & Semenza, G. L. (2009). Adenoviral transfer of HIF-1alpha enhances vascular responses to critical limb ischemia in diabetic mice. *Proc. Natl. Acad. Sci. U.S.A., 106*, 18769–18774.

Saxe, J. P., Tomilin, A., Scholer, H. R., Plath, K., & Huang, J. (2009). Post-translational regulation of Oct4 transcriptional activity. *PLoS One, 4*, e4467.

Schofield, C. J., & Ratcliffe, P. J. (2004). Oxygen sensing by HIF hydroxylases. *Nat. Rev. Mol. Cell Biol., 5*, 343–354.

Segal, D. J., Beerli, R. R., Blancafort, P., Dreier, B., Effertz, K., Huber, A., et al. (2003). Evaluation of a modular strategy for the construction of novel polydactyl zinc finger DNA-binding proteins. *Biochemistry, 42*, 2137–2148.

Segal, D. J., Crotty, J. W., Bhakta, M. S., Barbas, C. F., III., & Horton, N. C. (2006). Structure of Aart, a designed six-finger zinc finger peptide, bound to DNA. *J. Mol. Biol., 363*, 405–421.

Segal, D. J., Dreier, B., Beerli, R. R., & Barbas, C. F., III. (1999). Toward controlling gene expression at will: selection and design of zinc finger domains recognizing each of the 5′–GNN-3′ DNA target sequences. *Proc. Natl. Acad. Sci. U.S.A., 96*, 2758–2763.

Segal, D. J., Goncalves, J., Eberhardy, S., Swan, C. H., Torbett, B. E., Li, X., et al. (2004). Attenuation of HIV-1 replication in primary human cells with a designed zinc finger transcription factor. *J. Biol Chem., 279*, 14509–14519.

Seipel, K., Georgiev, O., & Schaffner, W. (1992). Different activation domains stimulate transcription from remote ("enhancer") and proximal ("promoter") positions. *EMBO J., 11*, 4961–4968.

Semenza, G. L., & Wang, G. L. (1992). A nuclear factor induced by hypoxia via de novo protein synthesis binds to the human erythropoietin gene enhancer at a site required for transcriptional activation. *Mol. Cell Biol., 12*, 5447–5454.

Setton, L. A., & Chen, J. (2006). Mechanobiology of the intervertebral disc and relevance to disc degeneration. *J. Bone Joint Surg. Am., 88*(Suppl. 2), 52–57.

Shyu, K. G., Wang, M. T., Wang, B. W., Chang, C. C., Leu, J. G., Kuan, P., et al. (2002). Intramyocardial injection of naked DNA encoding HIF-1alpha/VP16 hybrid to enhance angiogenesis in an acute myocardial infarction model in the rat. *Cardiovasc. Res., 54*, 576–583.

174

Snowden, A. W., Gregory, P. D., Case, C. C., & Pabo, C. O. (2002). Gene-specific targeting of H3K9 methylation is sufficient for initiating repression in vivo. *Curr. Biol., 12*, 2159–2166.

Tachikawa, K., Schroder, O., Frey, G., Briggs, S. P., & Sera, T. (2004). Regulation of the endogenous VEGF-A gene by exogenous designed regulatory proteins. *Proc. Natl. Acad. Sci. U.S.A., 101*, 15225–15230.

Takahashi, K., & Yamanaka, S. (2006). Induction of pluripotent stem cells from mouse embryonic and adult fibroblast cultures by defined factors. *Cell, 126*, 663–676.

Takeuchi, J. K., & Bruneau, B. G. (2009). Directed transdifferentiation of mouse mesoderm to heart tissue by defined factors. *Nature, 459*, 708–711.

Tal, R., Shaish, A., Rofe, K., Feige, E., Varda-Bloom, N., Afek, A., et al. (2008). Endothelial-targeted gene transfer of hypoxia-inducible factor-1alpha augments ischemic neovascularization following systemic administration. *Mol. Ther., 16*, 1927–1936.

Tan, W., Dong, Z., Wilkinson, T. A., Barbas, C. F., III., & Chow, S. A. (2006). Human immunodeficiency virus type 1 incorporated with fusion proteins consisting of integrase and the designed polydactyl zinc finger protein E2C can bias integration of viral DNA into a predetermined chromosomal region in human cells. *J. Virol., 80*, 1939–1948.

Tan, W., Zhu, K., Segal, D. J., Barbas, C. F., III., & Chow, S. A. (2004). Fusion proteins consisting of human immunodeficiency virus type 1 integrase and the designed polydactyl zinc finger protein E2C direct integration of viral DNA into specific sites. *J. Virol., 78*, 1301–1313.

Trentin, D., Hall, H., Wechsler, S., & Hubbell, J. A. (2006). Peptide-matrix-mediated gene transfer of an oxygen-insensitive hypoxia-inducible factor-1alpha variant for local induction of angiogenesis. *Proc. Natl. Acad. Sci. U.S.A., 103*, 2506–2511.

Tschulena, U., Peterson, K. R., Gonzalez, B., Fedosyuk, H., & Barbas, C. F., III. (2009). Positive selection of DNA-protein interactions in mammalian cells through phenotypic coupling with retrovirus production. *Nat. Struct. Mol. Biol., 16*(11), 1195–1199.

Urnov, F. D., Miller, J. C., Lee, Y. L., Beausejour, C. M., Rock, J. M., Augustus, S., et al. (2005). Highly efficient endogenous human gene correction using designed zinc-finger nucleases. *Nature, 435*, 646–651.

Venter, J. C., Adams, M. D., Myers, E. W., Li, P. W., Mural, R. J., Sutton, G. G., et al. (2001). The sequence of the human genome. *Science, 291*, 1304–1351.

Ventura, A., Kirsch, D. G., McLaughlin, M. E., Tuveson, D. A., Grimm, J., Lintault, L., et al. (2007). Restoration of p53 function leads to tumour regression in vivo. *Nature, 445*, 661–665.

Vierbuchen, T., Ostermeier, A., Pang, Z. P., Kokubu, Y., Sudhof, T. C., & Wernig, M. (2010). Direct conversion of fibroblasts to functional neurons by defined factors. *Nature, 463*, 1035–1041.

Vincent, K. A., Shyu, K. G., Luo, Y., Magner, M., Tio, R. A., Jiang, C., et al. (2000). Angiogenesis is induced in a rabbit model of hindlimb ischemia by naked DNA encoding an HIF-1alpha/VP16 hybrid transcription factor. *Circulation, 102*, 2255–2261.

Wakayama, T., Perry, A. C., Zuccotti, M., Johnson, K. R., & Yanagimachi, R. (1998). Full-term development of mice from enucleated oocytes injected with cumulus cell nuclei. *Nature, 394*, 369–374.

Wang, G. L., Jiang, B. H., Rue, E. A., & Semenza, G. L. (1995). Hypoxia-inducible factor 1 is a basic-helix-loop-helix-PAS heterodimer regulated by cellular O2 tension. *Proc. Natl. Acad. Sci. U.S.A., 92*, 5510–5514.

Weber, W., & Fussenegger, M. (2004). Approaches for trigger-inducible viral transgene regulation in gene-based tissue engineering. *Curr. Opin. Biotechnol., 15*, 383–391.

Weintraub, H., Tapscott, S. J., Davis, R. L., Thayer, M. J., Adam, M. A., Lassar, A. B., et al. (1989). Activation of muscle-specific genes in pigment, nerve, fat, liver, and fibroblast cell lines by forced expression of MyoD. *Proc. Natl. Acad. Sci. U.S.A., 86*, 5434–5438.

Wernig, M., Meissner, A., Foreman, R., Brambrink, T., Ku, M., Hochedlinger, K., et al. (2007). In vitro reprogramming of fibroblasts into a pluripotent ES-cell-like state. *Nature, 448*, 318–324.

Whitlock, P. R., Hackett, N. R., Leopold, P. L., Rosengart, T. K., & Crystal, R. G. (2004). Adenovirus-mediated transfer of a minigene expressing multiple isoforms of VEGF is more effective at inducing angiogenesis than comparable vectors expressing individual VEGF cDNAs. *Mol. Ther., 9*, 67–75.

Wilber, A., Tschulena, U., Hargrove, P. W., Kim, Y. S., Persons, D. A., Barbas, C. F., III., et al. (2010). A zinc-finger transcriptional activator designed to interact with the gamma-globin gene promoters enhances fetal hemoglobin production in primary human adult erythroblasts. *Blood, 115*, 3033–3041.

Wilmut, I., Schnieke, A. E., McWhir, J., Kind, A. J., & Campbell, K. H. (1997). Viable offspring derived from fetal and adult mammalian cells. *Nature, 385*, 810–813.

Wu, H., Yang, W. P., & Barbas, C. F., III. (1995). Building zinc fingers by selection: toward a therapeutic application. *Proc. Natl. Acad. Sci. U.S.A., 92*, 344–348.

Yamada, Y., Sakurada, K., Takeda, Y., Gojo, S., & Umezawa, A. (2007). Single-cell-derived mesenchymal stem cells overexpressing Csx/Nkx2.5 and GATA4 undergo the stochastic cardiomyogenic fate and behave like transient amplifying cells. *Exp. Cell Res., 313*, 698–706.

Yamanaka, S. (2009). A fresh look at iPS cells. *Cell, 137*, 13–17.

Yang, D., Peng, C., Li, X., Fan, X., Li, L., Ming, M., et al. (2008). Pitx3-transfected astrocytes secrete brain-derived neurotrophic factor and glial cell line-derived neurotrophic factor and protect dopamine neurons in mesencephalon cultures. *J. Neurosci. Res., 86*, 3393–3400.

Yang, S., Wei, D., Wang, D., Phimphilai, M., Krebsbach, P. H., & Franceschi, R. T. (2003). In vitro and in vivo synergistic interactions between the Runx2/Cbfa1 transcription factor and bone morphogenetic protein-2 in stimulating osteoblast differentiation. *J. Bone Miner. Res., 18*, 705–715.

Yechoor, V., Liu, V., Espiritu, C., Paul, A., Oka, K., Kojima, H., et al. (2009). Neurogenin3 is sufficient for trans-determination of hepatic progenitor cells into neo-islets in vivo but not transdifferentiation of hepatocytes. *Dev. Cell, 16*(3), 358–373.

Yu, J., Lei, L., Liang, Y., Hinh, L., Hickey, R. P., Huang, Y., et al. (2006). An engineered VEGF-activating zinc finger protein transcription factor improves blood flow and limb salvage in advanced-age mice. *FASEB J., 20*, 479–481.

Yun, C. O., Shin, H. C., Kim, T. D., Yoon, W. H., Kang, Y. A., Kwon, H. S., et al. (2008). Transduction of artificial transcriptional regulatory proteins into human cells. *Nucleic Acids Res., 36*, e103.

Zhang, L., Spratt, S. K., Liu, Q., Johnstone, B., Qi, H., Raschke, E. E., et al. (2000). Synthetic zinc finger transcription factor action at an endogenous chromosomal site. Activation of the human erythropoietin gene. *J. Biol. Chem., 275*, 33850–33860.

Zhang, Y., Deng, X., Scheller, E. L., Kwon, T. G., Lahann, J., Franceschi, R. T., et al. (2010). The effects of Runx2 immobilization on poly (epsilon-caprolactone) on osteoblast differentiation of bone marrow stromal cells in vitro. *Biomaterials, 31*(12), 3231–3236.

Zhao, Z., Wang, Z., Ge, C., Krebsbach, P., & Franceschi, R. T. (2007). Healing cranial defects with AdRunx2-transduced marrow stromal cells. *J. Dent. Res., 86*, 1207–1211.

Zhao, Z., Zhao, M., Xiao, G., & Franceschi, R. T. (2005). Gene transfer of the Runx2 transcription factor enhances osteogenic activity of bone marrow stromal cells in vitro and in vivo. *Mol. Ther., 12*, 247–253.

Zheng, H., Guo, Z., Ma, Q., Jia, H., & Dang, G. (2004). Cbfa1/osf2 transduced bone marrow stromal cells facilitate bone formation in vitro and in vivo. *Calcif. Tissue Int., 74*, 194–203.

Zhou, H., Wu, S., Joo, J. Y., Zhu, S., Han, D. W., Lin, T., et al. (2009). Generation of induced pluripotent stem cells using recombinant proteins. *Cell Stem Cell, 4*, 381–384.

Zhou, Q., Brown, J., Kanarek, A., Rajagopal, J., & Melton, D. A. (2008). In vivo reprogramming of adult pancreatic exocrine cells to beta-cells. *Nature, 455*, 627–632.

PART 2

Cells and Tissue Development

Genetic Approaches in Human Embryonic Stem Cells and their Derivatives: Prospects for Regenerative Medicine

Junfeng Ji, Bonan Zhong, Mickie Bhatia
Stem Cell and Cancer Research Institute, Michael G. DeGroote School of Medicine; and Department of Biochemistry and Biomedical Studies, McMaster University, Hamilton, Ontario, Canada

INTRODUCTION

Human embryonic stem cells (hESCs) were first derived from the inner cell mass of blastocyst-stage embryos in 1998 (Thomson et al., 1998). Isolation of hESCs opened up exciting new opportunities to study human development that is inaccessible *in vivo* and develop cell replacement approaches to the treatment of a broad range of diseases based on two unique properties: (1) *self-renewal capacity*: hESCs are able to proliferate for extended periods of time while maintaining their undifferentiated state and normal karyotypes in the proper culture conditions *in vitro* and (2) *broad developmental potential*: hESCs are pluripotent cells that can give rise to cell types representing ectodermal, mesodermal, and endodermal germ layers as assessed by *in vitro* formation of embryonic bodies (EBs) and *in vivo* teratoma assay (Itskovitz-Eldor et al., 2000; Schuldiner et al., 2000; Dvash et al., 2004).

Despite the promising prospect of hESCs as an invaluable system to model human development *in vitro* and as an unlimited source of cells for transplantation for a broad spectrum of human disease, the emerging hESCs field is still in its infancy and fundamental questions regarding the biology of hESCs remain to be addressed. Optimization of culture conditions to maintain hESCs in the undifferentiated state for a prolonged time *in vitro* is the first crucial step prior to any means of exploring the therapeutic potential of hESCs, the success of which requires a thorough understanding of molecular pathways regulating the self-renewal, pluripotency, apoptosis, and differentiation of hESCs. Moreover, only upon elucidation of cellular and molecular events dictating lineage specification and commitment of hESCs that faithfully recapitulate early human development will it be feasible to develop protocols to efficiently differentiate hESCs into diverse cell lineages potentially used for transplantation in the clinic. Genetic approaches to manipulating mouse embryonic stem cells (mESCs) in studies during the past 20 years have provided invaluable insights into the understanding of molecular signals governing pluripotency and specification of mESCs

Principles of Regenerative Medicine. DOI: 10.1016/B978-0-12-381422-7.10009-4

(Boiani and Scholer, 2005). To date, there is mounting evidence demonstrating that genetic manipulations such as homologous recombination, RNA interference (RNAi), over-expression of genes by transient transfection, and stable viral infection are applicable to hESCs and their derivatives, which will allow us to investigate the genetic programming regulating pluripotency maintenance versus differentiation of hESCs into diverse lineages (Gropp et al., 2003; Zwaka and Thomson, 2003; Menendez et al., 2004; Zaehres et al., 2005). In this chapter, we will review current protocols to maintain hESCs, genetic approaches to modifying undifferentiated hESCs, differentiation of hESCs into multiple lineages and transplantation of their derivatives, and genetic manipulation of hESC-derived progenies, and discuss the potential applications of genetic modifications of hESCs and their derivatives in the context of regenerative medicine.

MAINTAINING UNDIFFERENTIATED hESCs

hESCs were originally established and maintained by co-culture with mouse embryonic fibro-blast (MEF) feeder layer (Thomson et al., 1998). In an attempt to free hESCs from animal feeder layer, researchers have successfully used human feeder cells to derive and grow hESCs (Richards et al., 2002). Xu and colleagues went one step further to show that hESCs can be maintained in feeder-free condition where hESCs are cultured on Matrigel, laminin, or fibronectin in media conditioned by MEFs (Xu et al., 2001). However, culturing hESCs on either feeder cells or in conditioned media from supportive feeder cells adds additional difficulties to the maintenance and propagation of hESCs, because preparing feeder layer or feeder layer-conditioned media is time consuming in that feeder cells such as MEFs undergo senescence after approximately five passages and different batches vary significantly in their ability to support hESC growth. Moreover, the presence of xenogeneic components derived from MEFs or their conditioned media in hESC culture harbors a potential risk for transmission of animal pathogens into humans if cells derived in such conditions are used for cell replacement therapies in the clinic.

Recently, four groups have made significant progress in eliminating animal product from hESC culture (Amit et al., 2004; Wang et al., 2005a; Xu et al., 2005a,b). Amit et al. reported a feeder layer-free system where hESCs were cultured on fibronectin-coated plate in media supple-mented with 15% serum replacement (SR), a combination of growth factors including basic fibroblast growth factor (bFGF), leukemia inhibitory factor (LIF), and transforming growth factor beta 1 (TGF-β1) (Amit et al., 2004). Xu and colleagues have successfully sustained undifferentiated proliferation of hESCs on Matrigel in unconditioned media supplemented with 20% SR plus a high dose of bFGF (40 ng/ml) and bone morphogenetic protein (BMP) antagonist noggin (Xu et al., 2005b). Similarly, Wang et al. have been able to maintain hESCs by culturing them on Matrigel in media supplemented with 20% SR and a high dose of bFGF (36 ng/ml) alone (Wang et al., 2005a). Finally, Xu et al. demonstrated that Matrigel and SR supplemented with bFGF alone or in combination with other factors such as stem cell factor (SCF) or fetal liver tyrosine kinase 3 ligand (Flt3L) were able to maintain the growth of hESCs. Although all the above groups used SR and/or Matrigel to substitute for MEFs or their conditioned media to support hESCs, both SR and Matrigel are undefined and still contain animal-derived product. Subsequent to the reports, two groups have further demonstrated the successful derivation and growth of hESCs in defined culture conditions that solely consist of human materials (Lu et al., 2006; Ludwig et al., 2006). Ludwig and colleagues reported the generation of two new hESC lines in TeSR1 media that are composed of DMEM/F12 base supplemented with human serum albumin, vitamins, antioxidants, trace minerals, specific lipids, and growth factors of human origin including bFGF, LiCl, gamma-aminobutyric acid (GABA), pipecolic acid, and TGF-β (Ludwig et al., 2006). Derivation of hESC lines in TeSR1 also requires a combination of collagen, fibronectin, laminin, and vibronectin as supporting matrices, along with pH (7.2), osmolarity (350 nanoosmoles), and gas atmosphere (10% CO_2/5% O_2). Lu et al. developed a less complex hESC cocktail (hESCO) containing bFGF, Wnt3a, a proliferation-inducing ligand (April), B-cell-activating factor belonging to TNF

(BAFF), albumin, cholesterol, insulin, and transferin to support the self-renewal of hESCs (Lu et al., 2006). However, both of the two studies used incompletely defined albumin derived from human sources in their culture conditions, which may introduce human pathogens into the hESC culture to comprise their potential application in the clinic. In addition, one new hESC line derived in TeSR1 media, although originally normal, developed genetic abnormality as previously observed (Draper et al., 2004) after a relatively long-term culture *in vitro* (Ludwig et al., 2006).

Therefore, other than the requirement to eliminate feeder cells, animal product, and undefined components from hESC culture, an optimal culture condition for the growth of hESC must be able to prevent spontaneous differentiation and maintain genomic stability in the long-term culture. Maintained in the existing conditions, hESC culture consists of morphologically heterogeneous populations of cells in which a subset of fibroblast-like cells that are spontaneously differentiated from hESCs usually surrounds colonies. Although hESC-derived fibroblast-like cells have been used as a feeder layer to support the growth of hESCs (Yoo et al., 2005), the cellular and molecular identity and heterogeneity of hESC-derived fibroblasts related to the proliferation propensity and developmental potential between individual colonies within hESC culture remain to be determined. Furthermore, during long-term hESC culture in suboptimal conditions, hESCs have been shown to progressively adapt to the culture and select for clones with alterations in survival and proliferation capacity (Enver et al., 2005). Maitra et al. reported that eight of nine late-passage hESC lines acquired genetic and epigenetic abnormalities implicated in human cancer development (Maitra et al., 2005). In an attempt to develop measures to ensure the genetic normality of hESCs, a recent study has established differential expression of CD30, a member of the tumor necrosis factor receptor superfamily, in transformed versus normal hESC lines, implying that CD30 may serve as a biomarker for transformed hESCs (Herszfeld et al., 2006). However, examination of CD30 expression must be extended to a larger array of normal hESC lines and their variants with subtle genetic alterations. Determining the cellular and molecular bases of heterogeneity and transformation due to spontaneous differentiation and adaptation is important for devising improved culture conditions that minimize the selective advantage of variant cells and therefore help to maintain genetically normal cells suitable for therapeutic applications. Molecular dissection of signals dictating pluripotency and specification of hESCs by means of genetic manipulation will facilitate the optimization of culture conditions to maintain and specify hESCs.

GENETIC APPROACHES TO MANIPULATING hESCs
Gene regulation
KNOCK-IN/KNOCKOUT

Traditionally, knock-in/knockout technologies based on homologous recombination are the most widely used methods to study gene function in most mammals. Homologous recombination in hESCs is important for modifying specific hESC-derived tissues for therapeutic applications in transplantation medicine. *In vitro* studies of hESCs involved in understanding the pathogenesis of gene disorder diseases such as Wiskott-Aldrich syndrome or cancer also need the loss-and-gain methods. Although homologous recombination was efficient in generating mESC-derived mutant and knockout mice (Joyner, 2000), it is difficult to apply it to hESCs. First, compared to their murine counterparts, hESCs cannot be cloned efficiently from single cells, making it difficult to screen for rare recombination events. Second, since hESCs (14 μm) are larger than mESCs (8 μm), the transfection strategies between humans and mESCs are different. Based on an electroporation method, the first homologous recombination in hESCs succeeded in generating the hypoxanthine phosphoribosyltranferase-1 (HPRT-1) knockout mutant and the *oct-4* knock-in mutant (Zwaka and Thomson, 2003). The transfection rate was $5.6 \times 10^{2^5}$ and the frequency of homologous recombination itself in hESCs was comparable to that in mESCs (2–40% and 2.7–86%, respectively) (Mountford et al., 1994).

KNOCKDOWN

In 1998, the same year that hESCs were derived, RNAi was discovered in *Caenorhabditis elegans* and has since been intensively investigated (Fire et al., 1998). The first application of RNAi in hESC was achieved in hESCs six years later; *oct-4*, the important gene keeping hESCs in an undifferentiated state, was efficiently knocked down (Hay et al., 2004; Matin et al., 2004; Zaehres et al., 2005). RNAi is a mechanism of post-transcription silencing that degrades mRNA transcripts through homologous short RNA species in two steps: (1) double-stranded RNAs (dsRNAs) larger than 30 bp are recognized by the highly conserved RNAse III nuclease, named Dicer, and cleaved into <21–24 nucleotides of small interfering RNAs (siRNAs) and (2) siRNAs are recruited into "RNA-induced silencing complex" (RISC), which is a multi-protein complex (with endogenous RNase activity) that induces endonucleolytic cleavage of the target mRNA (recognized by hybridization with the RISC-bound siRNA antisense strand). While a lower degree of sequence complementary to the target mRNA only leads the RISC to interfere with the translational machinery, leaving mRNA intact. Previous studies have found that, in mammalian cells, dsRNAs larger than 30 bp (usually ranging from 500 to 1,000 bp) can trigger an interferon response by activating the dsRNA-dependent kinase (PKR), resulting in a non-specific global inhibition of protein translation and mRNA/rRNA hydrolysis (Kumar and Carmichael, 1998). Synthetic siRNA or short hairpin RNA (shRNA) can be exogenously delivered into cells to induce RNAi of target genes specifically without the activation of interferon response, which made RNAi applicable to the manipulation of genes in the hESCs study (Amarzguioui et al., 2005).

The screening of RNAi libraries is very useful in identifying novel gene functions, especially in the study of hESC differentiation. Libraries of synthetic shRNAs or siRNAs against a specified gene have been reported (Berns et al., 2004). One or multiple siRNAs could be delivered into the target cells with various transfection methods to increase the chance of successful repression. Because of the transient transfection, this method does not offer long-term stability. However, it also reduces the chances that potential inhibition of unknown genes may occur in the long-term assay. In order to achieve long-term therapeutic aims, more stable knockdown is required; for instance, expression of miRNAs, siRNAs, or shRNA *in vivo* may rely on the chromosomal integration of viral vectors containing homologous and complementary DNA sequence under the control of RNApol II or RNApol III promoters (Denti et al., 2004; Stegmeier et al., 2005). Among all these promoters, H1 and U6 promoters were mostly widely used to drive shRNAs/siRNAs (Tiscornia et al., 2003; Kaeser et al., 2004; Schomber et al., 2004; Zaehres et al., 2005), while vectors containing U6 promoter gave a higher frequency of interferon response induction than comparable H1 promoter-containing vectors. To avoid interferon induction by U6 promoter-driven vectors, Pebernard and colleagues recommended preserving the wild-type sequence around the transcription start site, in particular a C/G sequence at positions $-1/+1$ (Pebernard and Iggo, 2004). In addition, the promoters could be constructed under various chemical-regulated transcription systems to achieve inducible expression, which offers the option of inhibiting gene expression at certain steps during hESC differentiation (Wiznerowicz and Trono, 2003; Gupta et al., 2004; Higuchi et al., 2004; Tiscornia et al., 2004; Szulc et al., 2006).

Because the sequence of shRNA plays a critical role in the efficiency of gene knockdown, several factors need to be considered during the designing of shRNA: the length of sequences should be within 19–23 bp; the first 75–100 nucleotides (possible protein binding site) of target mRNA should be avoided; G/C component should be within 30–50%; there should be low internal stability at 5′ antisense sequence; there should be high internal stability at 5′ sense sequence; there should be absence of internal repeats or palindromes; 3′ end of sense and antisense sequences should have two "U"; G/C, A, U, and A are preferred at the 1st, 3rd, 10th, and 19th nucleotide positions respectively in the sense sequence; and the appearance of G/C and G at the 19th and 13th positions should be avoided (Bantounas et al., 2004; Gilmore et al., 2004; Pebernard and Iggo, 2004). The most popularly used sequence, CAAGAGA, was designed as the nucleotides loop to link the sense and antisense sequence (Brummelkamp et al., 2002; Anderson et al., 2003; Kunath et al., 2003). Alternatively, because the nucleotide size of shRNA or siRNA is usually very small and hard to be examined during siRNA vector cloning based on enzyme-digestion, using a restriction enzyme-recognizing sequence as the nucleotide loop will be more convenient to screen the positive shRNA-inserted clones.

There are several advantages of RNAi over "antisense oligonucleotides" and "knockout" strategies. To silence the same gene, siRNA strategy was much more efficient than antisense oligonucleotides, as well as having higher stability and less toxic side-effects (Miyagishi et al., 2003). Whereas the "knockout" strategy is time- and cost-consuming, RNAi can achieve a gene knockdown in hESCs within several months. By lowering the expression level of one gene instead of completely eliciting it, RNAi allows a molecular "turning dial." However, knock-down based on RNAi cannot simply replace traditional knockout techniques, but works as a relatively complementary tool. Recently, Persengiev and colleagues detected changes of 1,000 genes during introduction of siRNA against a non-existing gene (Persengiev et al., 2004). This indicates that it will be necessary to pay more attention to the issue of off-target effects versus target-specific effects in future studies.

Transfection

CHEMICAL TRANSFECTION

Synthetic chemicals such as cationic lipids have been extensively used for the delivery of DNA into hESCs. These chemical approaches are based on the neutralization of cationic lipids to negatively charged DNA followed by the formation of DNA/lipid complexes, which possess an excess of positive charges. These complexes bind to the negatively charged membranes of hESCs and are subsequently taken by the cells through endocytosis. Although various optimizations have been compared by combining different chemicals, such as ExGen500 (Fermentas) (Eiges et al., 2001; Matin et al., 2004), calcium phosphate (Darr et al., 2006), FuGENE (Boehringer Mannheim) (Liu et al., 2004), LipofectAMINE Plus (Life Technologies) (Vallier et al., 2004), or Lipofectamine 2000 (Invitrogen) (Hay et al., 2004), with different media, concentrations of DNAs, and cells, the transfection efficiency was still not promising. The inefficient performance of chemical methods may be due to cell cycle phase, the degradation of DNA caused by phagocytosis, or other unidentified factors.

PHYSICAL TRANSFECTION

Oligo delivery through electroporation is based on transient permeabilization of cell membrane via reversible formation of pores. Electrophoretic and electro-osmotic forces drive DNA through the destabilized cell membrane. Pre-stimulation of target cells by cytokines has varying transfection results from different groups (Wu et al., 2001; Weissinger et al., 2003). This may be due to the different use of plasmids and various electroporation conditions. Lots of evidence indicates that electroporation is an efficient gene delivery method in mESCs, hematopoietic stem cells (HSCs), and hESCs (Kunath et al., 2003; Oliveira and Goodell, 2003; Fathi et al., 2006), and CD34+ HSCs were relatively tolerant to electric forces and exhibited a higher cell survival rate after transfection compared to other primary cells. The death of the electroporated cells is caused by colloidal-osmotic swelling of cells as well as the uptake of exogenous DNA, which triggers apoptosis. The optimized protocol showed obvious greater post-electroporation viability when the hESCs were electroporated in clumps and plated out at high densities in isotonic, protein-rich medium instead of phosphate-buffered saline (PBS) (Zwaka and Thomson, 2003).

The more recent emergence of "nucleofection" has yielded acceptable cell survival rates (>70%), and 66% of the surviving cells showed transgene expression 24 h after nucleofection (Siemen et al., 2005; Levetzow et al., 2006). As the oilgo is delivered into the nucleus, the transfection rate is comparable to those of retroviral systems. Thus, this method is promising for wider application in the near future. Some other methods such as molecular vibration-mediated transfection and microinjection had high gene transfer rates (up to 100%); these one-step efficient procedures have attracted more attention in stem cell research (Capecchi, 1980; Wakayama et al., 2001; Song et al., 2004).

Overall, physical methods of transfection are more efficient methods for plasmid DNA delivery, are free from biocontamination, and raise fewer concerns about immune reaction. These physical transfection methods have low cost, ease of handling, and is highly reproducible, but most importantly it is biosafe. However, transient transgene expression in hESC colonies is difficult to retain for longer than five passages (Vallier et al., 2004). To achieve long-term transgene expression, especially in the fast-replicating cells, viral vector delivery may be needed.

Viral transduction

RETROVIRAL VECTOR

In the past two decades, retroviral vectors have been used for stable gene transfer into mammalian cells (Cone and Mulligan, 1984). The first vectors studied in a clinical trial (adenosine deaminase deficiency) were also retroviral vectors (Anderson, 1990). In 2000, the first successful treatment of a genetic disease relied on retroviral vectors, demonstrating the concept of gene therapy (Cavazzana-Calvo et al., 2000). The most popularly used retroviral vectors were those derived from the Moloney murine leukemia virus, which was also widely reported in the transduction of HSCs for gene therapy. Relative simplicity of their genomes, ease and safety of use, and the ability of integrating into the cell genome resulting in long-term transgene expression render them ideal vectors for genetic alteration. Stem cells in general, especially HSCs, constitute the best targets for retroviral vector-mediated gene transfer. Transgenes could be expressed long-term *in vivo* and may give rise to a large progeny of gene-modified mature cells during the continuous amplification process.

Retroviral vectors are derived from retroviruses. This family consists of seven genera: alpha-retrovirus, betaretrovirus, gammaretrovirus, deltaretrovirus, epsilonretrovirus, lentivirus, and spumavirus. The first five genera were previously classified as oncoretrovirus. Strictly speaking, vectors based on lentivirus or spumavirus are also retroviral vectors. However, the name retroviral vector is often used to refer to vectors based on murine leukemia virus or other oncoretrovirus. All retroviruses share some common features: lipid-enveloped particles containing two identical copies of liner single-stranded RNA; dependence on a specific cell membrane receptor for viral entry; and the RNA is reverse transcribed and integrates randomly into the target cell genome upon infection. All retroviral vectors contain long terminal repeats at the 5′ and 3′ ends (5′LTR and 3′LTR), a packaging signal located 3′ of the 5′LTR(ψ), and the three groups of structural genes, *gag*, *pol*, and *env*, coding for the capsid proteins, reverse transcriptase and integrase, and envelop proteins, respectively. For the production of retroviral vectors, the complete coding region of the *pol* and *env* genes and the majority coding region of the *gag* are removed, leaving a backbone of the 5′ and 3′ LTRs, part of the *gag* coding region, and the packaging signal (ψ). The transgene is constructed between the LTRs, and the resulting RNA transcript can be packaged into a virus with co-transfection of other separate packaging vectors (coding gag/pol, env proteins) within a cell.

Some features of retrovirus have been problematic in retroviral vector design. First, cells not expressing the appropriate receptor are resistant to certain retroviruses, which limits the application of retroviral vectors for host transduction. To obtain a broad host range, retroviral vectors have been pseudotyped with amphoteric envelope, gibbon ape leukemia virus (GALV) envelope (transduction in hESC-derived CD45^{neg}PFV hemogenic precursors), or vesicular stomatitis virus glycoprotein (VSV-G), by which retroviruses were able to be transduced into even non-mammalian cells derived from fish, *Xenopus*, mosquito, and

Lepidoptera (Burns et al., 1993; Menendez et al., 2004). The VSV-G envelope is also useful to stabilize retroviruses during viral particle concentration by ultracentrifugation. However, the expression of the VSV-G is toxic to cells, resulting in only transient production of vectors in the producer cell line. Therefore, the conditional expression system of VSV-G in the retroviral vector has been developed (Yang et al., 1995). Second, the nuclear membrane is a physical barrier for most retroviruses to migrate their transcribed dsDNA into the cell nucleus. Therefore, targets of most retroviral vectors, such as those based on murine leukemia virus, are limited to actively dividing cells (Miller et al., 1990). To disrupt the nuclear membrane, addition of a variety of stimulatory cytokines to introduce cycling in the HSC population is usually applied before retrovirus infection. Third, retroviral regulatory elements are repressed in ESCs and HSCs, and this makes long-term expression mediated by integrated retroviral vector difficult to achieve. Short-term silencing of recombinant genes is due to the binding of *trans*-acting transcriptional repressor on a specific region within the promoter of retroviral vector (Gautsch, 1980). Modification of the sequences in LTR to decrease the affinity of negative regulators has been applied to solve this problem (Laker et al., 1998). By engineering the regulatory regions, generation of novel retroviral vectors was reported, for example Friend mink cell focus-forming virus/murine ES cell virus hybrid vectors (FMEV), and higher expression levels of transgene than conventional retroviral vectors were observed in HSCs (Baum et al., 1995). In contrast, long-term silencing of the target gene is often observed in retroviral vectors based on murine stem cell virus. Because of the high *cis*-acting methylation activity of ES cells, effective DNA methylation leads to the silencing of integrated retroviral vectors, though this was not detected within differentiated cells showing low methylation activity. Alteration of the *cis* elements in LTR could decrease the DNA methylation and increase transgene expression in embryonic carcinoma cells (Challita et al., 1995). From the cells perspective, disruption of the methyltransferase gene *Dmnt1* to alter the endogenous level of DNA methylation in target ESCs may lead to another potential solution. As a result of the multiple defects of retroviral vectors, lentivirus-based vectors are more attractive in the genetic research of hESCs.

LENTIVIRAL VECTOR

Lentivirus is one genus of retrovirus and includes the human immunodeficiency virus (HIV) type 1. Principally, lentiviral vectors are derived from lentiviruses in a similar way to retroviral vectors. Some features of lentiviruses make lentiviral vectors better alternatives for gene regulation within the hESCs. Because their pre-integration complex can get through the intact membrane of the nucleus within the target cell, lentiviruses can infect both dividing and non-dividing cells or terminally differentiated cells such as macrophages, retinal photoreceptors, and liver cells (Naldini et al., 1996). Lentiviral vectors are also promising gene transfer vehicles for HSCs, which reside almost exclusively in the G_0/G_1 phase of the cell cycle (Cheshier et al., 1999). The only cells lentiviruses cannot gain access to are quiescent cells in the G_0 state, which block the reverse transcription step (Amado and Chen, 1999). Lentiviruses can stably change the gene expression within hESCs for up to six months and are more resistant to transcriptional silencing (Pfeifer et al., 2002). High expression level of enhanced green fluorescent protein (eGFP) was achieved both in undifferentiated hESCs and their derivatives (Gropp et al., 2003). Overexpression of different genes, for instance *oct-4*, *nanog*, and *eGFP*, has been reported under the control of various promoters, such as human cytomegalovirus (CMV) immediate early region enhancer-promoter, the composite CAG promoter (consisting of the CMV immediate early enhancer and the chicken β-actin promoter), human phosphoglycerate kinase 1(PGK) promoter, human elongation factor 1α (EF1α) promoter, and ubiquitin (Ub) promoter (Ramezani et al., 2000; Salmon et al., 2000; Luther-Wyrsch et al., 2001; Gropp et al., 2003; Ma et al., 2003). Among these promoters, the CMV promoter does not perform well in HSCs (Boshart et al., 1985). Moreover, it is often subject to extinction of expression and silencing *in vivo* (Kay et al., 1992). In comparison, EF1α promoter was the most popularly used and showed consistently better performance.

Single transgene expression can shorten the length of lentiviral vector, leading to relatively higher transduction efficiency of the recombinant lentivirus in the hESCs. However, screening of the positively transduced cells from the polyclonal population cannot be achieved unless the overexpressed gene encodes a fluorescent or membrane protein, or an antibiotics cassette. Instead, to express two recombinant genes and for one of them to work as an integration reporter, internal ribosome entry sites (IRES) and double-promoters have been extensively studied in lentiviral vector design. IRES are sequences that can recruit ribosomes and allow cap-independent translation, which can link two coding sequences in one bicistronic vector and allow the translation of both proteins in hESCs. The expression level of target gene by bicistronic vectors could be higher than that by single gene vectors; however, the percentage of positively transduced cells was relatively lower (Ben-Dor et al., 2006). Besides, the expression of downstream gene to IRES may inconsistently depend on the sequence of its upstream gene in an unpredictable manner (Yu et al., 2003). In comparison, lentiviral vectors containing double-promoters allow expression of reporter gene and target gene independently as well as the permission of transgene expression under tissue-specific promoter.

Gene regulation based on the bacterial tetracycline repressor/operator (tetR/tetO) system has been applied to lentiviral vector design. To make the expression of a transgene inducible, the tetO cassette is inserted upstream of the transgene promoter and the tetR cassette can be transcribed either by the same gene expression vector or by a separate vector within the same hESC, binding to the tetO and inhibiting gene expression. Conditional gene expression can be achieved when tetracycline or doxycycline is added to the cells, releasing the tetR binding and turning on the promoter (Szulc et al., 2006).

Accompanied with various benefits using lentiviral vectors in hESCs, the obvious concern was due to biosafety issues. The lentiviral vectors based on HIV could self-replicate and could be produced during manufacture of the vectors in the packaging cells by a process of recombination. Also, a self-replicating infectious vector may transform hESC into a cancer stem cell by chromosome integration and activation of a neighboring proto-oncogene. Therefore, a number of modifications and changes were made over time, leading to the safe production of high-titer lentiviral vector preparations. In addition to the structural *gag*, *pol*, and *env* genes common to all retroviruses, more complex lentiviruses contain two regulatory genes, *tat* and *rev*, crucial for viral replication, and four accessory genes, *vif*, *vpr*, *vpu*, and *nef*, which are not critical for viral growth *in vitro* but are essential for *in vivo* replication and pathogenesis. The Tat protein regulates the promoter activity of the 5 BMPs' LTR and is necessary for the transcription from the 5′ LTR. The Rev protein regulates gene expression at post-transcription level. It promotes the transport of unspliced and singly spliced viral transcripts into cytoplasma, allowing the production of the late viral proteins. The Tat and Rev are necessary for efficient *gag* and *pol* expression and new viral particle production. Understanding the functions of these genes leads to a 10-year path of lentiviral vector design.

The first generation of HIV-derived vectors was produced transiently by transfection of plasmids coding for the packaging functions and the transgene plasmid into a suitable cell line mostly derived from 293 cells (Naldini et al., 1996). The ψ sequences and the *env* gene were removed from the HIV genome, the 5′ LTR was replaced by heterologous promoter, and the 3′ LTR was replaced by a polyadenylation signal. The envelope was replaced by another virus, and was most often VSV-G (Burns et al., 1993).

In the second generation, to attenuate the virulence of the virus, all four accessory genes were removed and the HIV-derived packaging component was reduced to the *gag*, *pol*, *tat*, and *rev* genes of HIV-1 in the second version of the system (Zufferey et al., 1997). However, viruses can still be produced *in vitro*.

In the third generation, constitutively active promoter sequences replaced part of the U3 region in the 5′ LTR in the transgene vector. The activity of the 5′ LTR during vector production became independent of the *tat* gene, which could be completely removed from the packaging construct. The *rev* gene, necessary for the *gag/pol* expression, was separately cloned into another plasmid to minimize the likelihood of recombination. In addition, a 299 bp deletion in the 3' LTR blocked the function of enhancer and promoter, resulting in the self-inactivation (SIN) of the provirus in the infected cells and minimizing the risk of insertional oncogenesis. Therefore, an internal promoter is needed for SIN vectors to drive transgene expression, allowing the use of tissue-specific or inducible promoters. The resulting gene delivery system, which conserves only three genes (*rev, gag, pol*) of HIV-1 and relies on four separate transcriptional units for the production of transducing particles, offers significant advantages for its predicted biosafety.

Other modifications of lentiviral vectors were performed to satisfy different expression requirements. To enhance the susceptibility to infection, the central polypurine tract (cPPT) is often included in the transgene vectors. Insertion of the woodchuck hepatitis virus post-transcriptional regulatory element (WPRE) was previously found to enhance transgene expression (Zufferey et al., 1999). However, inclusion of WPRE from certain lentiviral vectors showed lower transgene expression in human HSCs KG1a cell line (Ramezani et al., 2000). Besides stable gene expression, mutation of integrase protein itself and the integrase recognition sequences (*att*) in the lentiviral LTR could disable the integration of lentiviral vector and permitted transient gene expression (Nightingale et al., 2006). To lower the possibility of integration by LTR during lentiviral vector construction, *E. coli* Stbl3 and *E. coli* Stbl2 strains (Invitrogen) instead of DH5 α were developed, and optimization of culturing temperature under 30°C instead of 37°C reduced the possibility of LTR recombination.

ADENOVIRAL VECTORS AND ADENO-ASSOCIATED VIRAL VECTORS

Adenoviruses are a group of non-pathogenic viruses that contain a linear double-stranded DNA genome without envelope. They have been developed as gene delivery vehicles due to the ability to infect non-dividing cells. Adenoviral vectors do not integrate into the genome of host cells providing a transient expression of the transgene. Adenoviruses are capable of transducing cells *in vivo* taking up to 30 kb exogenous DNA, and adenovirus-associated viruses can express 4.8 kb transgene (Tatsis and Ertl, 2004; Volpers and Kochanek, 2004). Co-infection with helper viruses such as herpes simplex virus is required for adeno-associated viral vectors, which still need to be optimized to achieve productive infection. Adenovirus-derived vectors have been successfully used in mESC studies (Mitani et al., 1995; Kawabata et al., 2005), and their applications as homologous recombination and gene transfer vehicles in the hESCs and/or their differentiating progenies are under investigation (Ohbayashi et al., 2005; Stone et al., 2005).

DIFFERENTIATION OF hESCs INTO TISSUE-SPECIFIC LINEAGES AND TRANSPLANTATION OF hESC-DERIVED CELLS

To date, a large number of methods and protocols to drive the differentiation of hESCs into a broad spectrum of tissue-specific lineages *in vitro* representing three germ layers have been documented. However, hESC-based regenerative medicine largely relies on the generation of transplantable progenies from hESCs that will function *in vivo*. Therefore, in addition to identifying tissue-specific lineages derived from hESCs by morphological and phenotypic criteria and *in vitro* functional assays, hESC-derived progenies have to be functionally evaluated *in vivo* by transplantation into appropriate animal models. In this chapter, we review the approaches to generating diverse cell lineages from hESCs that have been functionally assessed *in vivo* by transplantation assays.

Mesodermal derivatives and their transplantation

Mesodermal derivatives, including hematopoietic, vascular, and cardiac differentiation from hESCs, have been well characterized in great detail. Derivation of hematopoietic cells from hESCs is not only important for studying hematopoietic development in humans but is also opening exciting opportunities to create an alternative cell source in addition to cord blood and bone marrow for transplantation in the clinic. Different methods have been used to induce hematopoietic differentiation from hESCs *in vitro*. The first report on derivation of hematopoietic cells from hESCs employed co-culture of hESCs with murine bone marrow cell line S17 or the yolk sac endothelial cell line C166 (Kaufman et al., 2001). An improvement in the production of CD34+ hematopoietic progenitor cells was then achieved by co-culturing hESCs with OP9 stromal cells, a bone marrow stromal cell line created from mice deficient in macrophage colony stimulating factor (M-CSF) (Vodyanik et al., 2005). Nevertheless, hematopoietic differentiation by the co-culture system is inefficient and hematopoietic cells derived from the system lack the expression of pan-leukocyte marker CD45. Our group has recently demonstrated that a combination of hematopoietic cytokines and BMP-4 efficiently augments hematopoietic differentiation from hEBs (Chadwick et al., 2003; Cerdan et al., 2004), and identified a rare subpopulation of cells lacking CD45 but expressing PECAM-1, Flk-1, and VE-Cadherin (termed CD45[neg]PFV precursors) that are exclusively responsible for hematopoietic cell fate (Wang et al., 2004). The function of hematopoietic cells derived by either stromal co-culture or EB formation system has been evaluated *in vivo* by xeno-transplantation repopulation assays that have been instrumental in measuring human somatic HSCs (Dick et al., 1997). However, generation of *in vivo* repopulating hematopoietic cells from hESCs has been proven to be difficult. Our laboratory has recently demonstrated that CD45+ cells isolated from EBs cannot be successfully intravenously transplanted into immunocompromised mice due to the rapid aggregation upon exposure to mouse serum, and that the levels of reconstitution were still very low despite direct intra-femoral injection of hESC-derived hematopoietic cells to bypass the circulation and allow mice to survive (Wang et al., 2005b). Moreover, CD45[neg]PFV precursors or their derived hematopoietic cells were unable to engraft even after transplantation into the liver of newborn immunocom-promised mice (unpublished data), an assay more amenable to readout repopulating hematopoietic cells (Yoder et al., 1997). In addition to our studies, sorted CD34+lineage[−] cells or unsorted cells from hESCs differentiated on S17 stromal cells have recently been shown to engraft, but at a very low level, after transplantation into fetal sheep or adult non-obese severe combined immunodeficient NOD/SCID mice, respectively (Narayan et al., 2006; Tian et al., 2006). Taken together, these studies suggest that full understanding of molecular and cellular events dictating hematopoiesis from hESCs is required to improve means of generating HSCs with potent repopulating ability from hESCs.

Initiation of vascular development has been shown to be closely associated with the emergence of hematopoiesis, and a common precursor termed "hemangioblast" with both vascular and hematopoietic potential has been identified during hematopoietic differentiation of mESCs and in the primitive streak of the mouse embryo (Choi et al., 1998; Huber et al., 2004). In humans, our laboratory has recently identified a subpopulation of primitive endothelium-like cells termed CD45[neg]PFV precursors with hemangioblast properties during EB differentiation of hESCs in the presence of exogenous hematopoietic cytokines and BMP-4 (Wang et al., 2004). Cells expressing PECAM1/CD31, a marker associated with cells capable of early hematopoietic potential in the human embryo (Oberlin et al., 2002), first emerged at day 3 and significantly increased at day 7 through day 10 of EB development. An isolated subpopulation of CD45[neg]PFV precursors contained single cells with both hematopoietic and endothelial capacity. After 7 days in culture condition conducive to endothelial maturation, the cells not only strongly expressed CD31, VE-cadherin, and mature endothelium markers vWF and eNOS, but also possessed low-density lipoprotein (LDL) uptake capacity (Wang et al., 2004). However, the *in vivo* function of hESC-derived endothelial cells from our system has not been assessed. Levenberg et al. reported the first study to characterize differentiation of hESCs into endothelial cells during spontaneous EB differentiation without adding any exogenous growth factors by functionally evaluating hESC-derived endothelial cells both *in vitro* and *in vivo* (Levenberg et al., 2002). Although the efficiency of endothelial differentiation is relatively low in the spontaneous system as opposed to our system, their differentiation kinetics are similar in that the expression of CD31, VE-cadherin, and CD34 appeared at days 3—5 and reached a maximum of about 2% at days 13—15 during EB differentiation. CD31+ cells isolated from day 13 EBs displayed endothelium characteristics by expressing endothelium-specific markers VE-cadherin and vWF, taking up acetylated LDL (ac-LDL) and forming tube-like structures (Levenberg et al., 2002). Furthermore, hESC-derived CD31+ cells were able to form functional blood-carrying microvessels after transplantation into SCID mice (Levenberg et al., 2002). A recent study from the same group has further shown that hESC-derived endothelial cells are able to vascularize skeletal muscle tissue construct using a three-dimensional multiculture system *in vitro* (Levenberg et al., 2005). More significantly, pre-endothelialization of the construct, by promoting implant vascularization, can improve blood perfusion to the implant and implant survival *in vivo* (Levenberg et al., 2005). In summary, these studies demonstrate that endothelial differentiation of hESCs likely recapitulates vasculogenesis during human development and hESC-derived endothelial cells are able to vascularize tissue construct *in vitro* and implant *in vivo*. However, it remains to further determine potential therapeutic implications of embryonic endothelial cells generated from hESCs for treatment of vascular disease and repair of ischemic tissues.

Methods from different laboratories to induce cardiac differentiation from hESCs have also been demonstrated (Kehat et al., 2001; Xu et al., 2002; Mummery et al., 2003). During spontaneous EB differentiation of hESCs, 8% of EBs contained contracting cardiomyocytes that displayed structural, phenotypic, and functional properties of early-state cardiomyocytes (Kehat et al., 2001). Treatment of cells with 5-aza-2′-deoxycytidine increased cardiomyocyte differentiation in a time-dependent and concentration-dependent manner and Percoll density centrifugation could achieve a population containing 70% cardiomyocytes (Xu et al., 2002). In addition to spontaneous differentiation, co-culture of hESCs with visceral endoderm-like cell line, END-2, has also been shown to induce cardiac differentiation of hESCs (Mummery et al., 2003). The induction events for cardiac development in the hESCs remain to be further defined in detail as cardiomyocytes are generated in serum-containing conditions in most studies. Recently, hESC-derived cardiomyocytes have been functionally tested in a swine model of complete atrioventricular block as a "biologic pacemaker" for the treatment of bradycardia; the transplanted cells survived, integrated, and successfully paced the ventricle with complete heart block (Kehat et al., 2004). However, long-term pacemaking function of grafted hESC-derived cardiomyocytes was not evaluated in the study, which also raises the concern that transplanted cells could serve as a nidus for arrhythmia.

Ectodermal derivatives and their transplantation

Most studies on derivation of ectodermal lineages from hESCs have focused on neuroectoderm and neural cells, aiming to create an unlimited source of neural cells for transplantation therapies. Differentiation of hESCs into neural lineages has been induced using different methods (Carpenter et al., 2001; Reubinoff et al., 2001; Zhang et al., 2001). hESC-derived neural progenitors that could differentiate into three neural lineages — mature neurons, astrocytes, and oligodendrocytes *in vitro* — have been transplanted into neonatal mouse brain, where they are incorporated into host brain parenchyma, migrated along established brain migratory tracks, and differentiated into progeny of three neural lineages *in vivo* (Reubinoff et al., 2001; Zhang et al., 2001). Furthermore, enriched population of neural progenitors from hESCs that were grafted into the striatum of Parkinsonian rats induced partial behavioral recovery (Ben-Hur et al., 2004). The functional improvement is likely due to release of neurotropic factors from the graft to promote survival of impaired endogenous dopamine neurons as hESC-derived neural progenitors could not acquire dopaminergic fate in the host tissue. Despite recent availability of protocols to generate specific dopaminergic neurons from hESCs (Park et al., 2004; Perrier et al., 2004; Schulz et al., 2004; Zeng et al., 2004),

189

only one of the studies has examined the *in vivo* functions of hESC-derived dopamine neurons after transplantation into the striatum of 6-hydroxydopamine-treated rats and the significance of the study is unclear because only a few dopaminergic neurons survived 5 weeks after transplantation and no functional improvement has been demonstrated (Zeng et al., 2004). Future studies are required to determine the appropriate cell type for transplantation therapies by functionally evaluating hESC-derived dopamine neurons in comparison to neural progenitors in animal models of Parkinson's disease. In addition to dopamine neurons, other specific neuronal subtypes, such as motoneurons, which have also been recently generated from hESCs (Li et al., 2005), have to be functionally assessed in animal models of spinal cord injuries and motoneuronal degeneration.

Endodermal derivatives and their transplantation

In contrast to mesodermal and ectodermal differentiation of hESCs, specification of hESCs into endodermal lineages, specifically insulin-producing cells, is less studied. Although differentiation of hESCs into insulin-producing cells has been demonstrated by either spontaneous system, exposure to inducing factors, or overexpression of Pdx1 or Foxa2 (important transcription factors involved in pancreatic development (Assady et al., 2001; Segev et al., 2004; Brolen et al., 2005; Lavon et al., 2006)), the frequency of these cells generated in the current differentiation conditions is too low to allow detailed characterization and functional analysis.

GENETIC MODIFICATIONS OF hESC-DERIVED PROGENIES

Successful derivation of diverse tissue-specific lineages from hESCs sets the stage to genetically manipulate hESC-derived progenies. However, in sharp contrast to the broad applications of genetic modifications to undifferentiated hESCs, very few studies have investigated genetic manipulations of specific lineages derived from hESCs, possibly due to the difficulties in prospectively isolating a low frequency of lineage-specific progenies from the bulk population to allow detailed studies. To date, hESC-derived hematopoietic cells are the only cell type to which retrovirus-based gene transfer has been successfully applied (Menendez et al., 2004). Our laboratory has recently characterized and optimized a GALV-pseudotyped retroviral gene transfer strategy to stably transduce the hematopoietic progenitor cells derived from CD45^{neg}PFV hemogenic precursors that were prospectively isolated from hEBs (Menendez et al., 2004). We achieved >25% transduction efficiency using GALV-pseudotyped retrovirus into CD45^{neg}PFV precursors-derived hematopoietic cells and a proportion of transduced cells co expressed CD34 and were able to give rise to a hematopoietic colony-forming unit (Menendez et al., 2004). These studies are expected to provide a method to examine the functional effects of ectopic expression of candidate genes that may regulate primitive human hematopoietic development. Using the GALV-pseudotyped retroviral gene delivery method, we have very recently evaluated the role of HoxB4 overexpression in CD45^{neg}PFV precursors derived from hESCs (Wang et al., 2005b). In contrast to the generation of repopulating hematopoietic cells from mESCs by overexpressing HoxB4 in mESC-derived hematopoietic progenitors, ectopic expression of HoxB4 in hESC-derived hematopoietic cells does not confer engraftment potential (Kyba et al., 2002; Wang et al., 2005c). Overexpression and knockdown of genes associated with lineage development in hESC-derived progenies is critical to further understand lineage specification and commitment from hESCs.

POTENTIAL APPLICATIONS OF GENETICALLY MANIPULATED hESCs AND THEIR DERIVATIVES
Augmenting differentiation of hESCs into specific lineages

Once formed as EBs in serum-containing medium, hESCs will spontaneously differentiate into diverse lineages representing three germ layers, but at very low levels. Although many studies have demonstrated that adding growth factors or morphogens related to lineage development into the medium can significantly increase the differentiation of hESCs into specific lineages, the frequencies of lineage-specific cells are, in general, still low (Chadwick et al., 2003). In the setting of hematopoietic differentiation, our group has observed that 10–20% of EBs at days 10–13 still contained Oct-4-positive cells (unpublished observation), suggesting that the differentiation processes of cells within the EBs are not synchronized and some cells are reluctant to respond to differentiation clues in the culture. A very recent genetic mapping study has suggested that pluripotency-associated transcription factors Oct-4, Nanog, and Sox2 repress a set of developmental regulators of lineage specification to maintain the pluripotent status of hESCs (Lee et al., 2006). Therefore, RNAi-based genetic knockdown of Oct-4, Nanog, or Sox2 is expected to release the repression of differentiation and thereby facilitate the generation of tissue-specific progenies from hESCs with the induction of proper growth factors along the pathways of lineage development. Indeed, Oct-4 knockdown in hESCs has been shown to induce endoderm differentiation (Hay et al., 2004). On the other hand, enforced expression of lineage-specific genes in undifferentiated hESCs will likely promote the differentiation of hESCs into specific lineages. In the context of hematopoietic differentiation, overexpression of HoxB4, a transcription factor involved in hematopoietic development and self-renewal of HSCs, in undifferentiated hESCs by lipofection promotes a 6–20-fold increase in the frequency of hematopoietic cells derived from hESCs (Bowles et al., 2006). In line with the augmenting effect of constitutive expression of HoxB4 on the hematopoietic differentiation of hESCs, our group has observed that the mRNA expression profile of HoxB4 during EB differentiation is temporally correlated with hematopoietic development from hESCs (unpublished observation). A very recent study has evaluated the effect of transfection-based overexpression of Foxa2 and Pdx1, transcription factors involved in different phases of early endoderm and pancreatic development, on the differentiation of hESCs into pancreatic cells (Lavon et al., 2006). In contrast to the insignificant effect of overexpression of Foxa2 on the differentiation of hESCs into endoderm lineage, constitutive expression of Pdx1 promoted the differentiation of hESCs toward insulin cells, as shown by induced expression of most transcription factors involved in pancreatic development (Lavon et al., 2006). However, expression of insulin gene was not induced by enforced Pdx1 expression, suggesting that differentiation signals that can further drive the specification into insulin cells is still missing in spite of constitutive expression of Pdx1. Future studies are required to investigate introduction of inducible gene expression system into hESCs, which will allow us to study the role of lineage-specific genes in lineage development from hESCs at specific stages of hESC differentiation.

Lineage tracking and purification

In order to better understand temporal differentiation and spatial organization of specific lineages from hESCs, it is important to trace lineage specification and commitment within heterogeneous populations of cells during EB differentiation. Introduction of reporter/selection genes under the control of lineage-specific promoters will allow us to monitor the differentiation of hESCs toward specific lineages. Furthermore, it offers us the feasibility to select and purify specific lineages and eliminate undesirable cells from the bulk population based on reporter gene expression, which is critical for the potential use of these hESC-derived lineages in cell-based therapies, since any potential contamination by undifferentiated hESCs will likely result in the development of teratomas. Eiges et al. and Gerrard et al. introduced eGFP reporter gene under the control of ESC-enriched gene murine Rex1 or Oct-4 promoter into hESCs to select the undifferentiated hESCs from their spontaneously differentiated derivatives in the culture (Eiges et al., 2001; Gerrard et al., 2005). Lavon et al. have very recently traced the differentiation of hESCs into pancreatic cells by generating and differentiating hESC lines carrying eGFP reporter gene under the control of insulin promoter or Pdx1 promoter (Lavon et al., 2006). These studies paved the way for future endeavors to examine the molecular and cellular mechanisms governing lineage specification, which in turn will provide insight into better generation of lineage-specific cells from hESCs.

Modifying the immunogenicity of hESCs and their derivatives

hESC-derived tissue-specific progenies represent a promising source for the potential transplantation of therapies to a broad spectrum of diseases in the clinic. However, immune response launched by the host immune system to the graft may comprise the therapeutic potential of derivatives from hESCs. Although we and others have demonstrated that hESCs and their derivatives after a short period of differentiation *in vitro* express low levels of major histocompatibility complex (MHC) class I and are less susceptible to immune rejection than adult cells (Li et al., 2004; Drukker et al., 2006), it remains unclear whether hESC-derived cells differentiated to a fully functional adult phenotype after successful engraftment will still possess immuno-privileged properties to permanently evade immune rejection. To overcome potential immune rejection, a few approaches have been proposed, which include somatic cell nuclear transfer to create hESC lines with identical MHC to that of host tissue, collection of hESC banks representing the broadest diversity of MHC polymorphisms, and induction of a state of immune tolerance to an hESC line using tolerogenic HSCs derived from it. Though promising, the feasibility of these strategies remains to be validated. Alternatively, strategies to genetically modify the immunogenicity of hESCs and their derivatives by targeting genes that encode and control the cell surface expression of MHC classes I and II molecules provide another theoretical means to circumvent the immune barrier. The deletion of both classes of MHC molecule has been achieved in mESCs by disruption of the genes critical for the correct assembly and membrane expression of MHC classes I and II (Zijlstra et al., 1990; Grusby et al., 1991). Although grafts deficient in the expression of either MHC class I or II target molecules do not completely avoid rejection by immunologically intact allogeneic hosts, MHC class I-deficient grafts are rejected more slowly than grafts from normal mice. Genetic modifications of similar target genes for MHC class I expression in hESCs and their derivatives remain to be fully explored in future studies, given the applicability of multiple genetic tools to manipulate hESCs and their progenies.

CONCLUSION

Derivation of hESCs opens up a new era for human development biology and regenerative medicine. The almost one decade of research to date has made considerable progress in defining culture conditions to grow hESCs and developing protocols to differentiate hESCs into tissue-specific lineages. However, a formulated culture condition completely devoid of animal component and uncharacterized serum elements to maintain hESCs remains to be further optimized. Moreover, efficient generation of specialized derivatives from hESCs that are able to function *in vivo* after transplantation into animal models has not been achieved so far. Realization of hESCs as a model system to study human development and unlimited source for regenerative medicine relies on the dissection of molecular and cellular mechanisms dictating the pluripotency, self-renewal, and lineage specification of hESCs. Genetic manipulations of hESCs and their derivatives are anticipated to provide invaluable insight into the understanding of fundamental biology of hESCs, which in turn will be instrumental in the optimization of protocols to either maintain hESCs or specify hESCs into functional tissue-specific lineages with potential use in the clinic.

Acknowledgments

We thank Dr. Marc Bosse in the Bhatia laboratory for his critical comments and insights during the preparation of this review.

References

Amado, R. G., & Chen, I. S. (1999). Lentiviral vectors — the promise of gene therapy within reach? *Science, 285,* 674—676.

Amarzguioui, M., Rossi, J. J., & Kim, D. (2005). Approaches for chemically synthesized siRNA and vector-mediated RNAi. *FEBS Lett., 579,* 5974—5981.

Amit, M., Shariki, C., Margulets, V., & Itskovitz-Eldor, J. (2004). Feeder layer- and serum-free culture of human embryonic stem cells. *Biol. Reprod., 70,* 837—845.

Anderson, J., Banerjea, A., Planelles, V., & Akkina, R. (2003). Potent suppression of HIV type 1 infection by a short hairpin anti-CXCR4 siRNA. *AIDS Res. Hum. Retrov., 19,* 699—706.

Anderson, W. F. (1990). The beginning. *Hum. Gene. Ther., 1,* 371—372.

Assady, S., Maor, G., Amit, M., Itskovitz-Eldor, J., Skorecki, K. L., & Tzukerman, M. (2001). Insulin production by human embryonic stem cells. *Diabetes, 50,* 1691—1697.

Bantounas, I., Phylactou, L. A., & Uney, J. B. (2004). RNA interference and the use of small interfering RNA to study gene function in mammalian systems. *J. Mol. Endocrinol., 33,* 545—557.

Baum, C., Hegewisch-Becker, S., Eckert, H. G., Stocking, C., & Ostertag, W. (1995). Novel retroviral vectors for efficient expression of the multidrug resistance (mdr-1) gene in early hematopoietic cells. *J. Virol., 69,* 7541—7547.

Ben-Dor, I., Itsykson, P., Goldenberg, D., Galun, E., & Reubinoff, B. E. (2006). Lentiviral vectors harboring a dual-gene system allow high and homogeneous transgene expression in selected polyclonal human embryonic stem cells. *Mol. Ther., 14,* 255—267.

Ben-Hur, T., Idelson, M., Khaner, H., Pera, M., Reinhartz, E., Itzik, A., et al. (2004). Transplantation of human embryonic stem cell-derived neural progenitors improves behavioral deficit in Parkinsonian rats. *Stem Cells, 22,* 1246—1255.

Berns, K., Hijmans, E. M., Mullenders, J., Brummelkamp, T. R., Velds, A., Heimerikx, M., et al. (2004). A large-scale RNAi screen in human cells identifies new components of the p53 pathway. *Nature, 428,* 431—437.

Boiani, M., & Scholer, H. R. (2005). Regulatory networks in embryo-derived pluripotent stem cells. *Nat. Rev. Mol. Cell. Biol., 6,* 872—884.

Boshart, M., Weber, F., Jahn, G., Dorsch-Hasler, K., Fleckenstein, B., & Schaffner, W. (1985). A very strong enhancer is located upstream of an immediate early gene of human cytomegalovirus. *Cell, 41,* 521—530.

Bowles, K. M., Vallier, L., Smith, J. R., Alexander, M. R., & Pedersen, R. A. (2006). HOXB4 overexpression promotes hematopoietic development by human embryonic stem cells. *Stem Cells, 24,* 1359—1369.

Brolen, G. K., Heins, N., Edsbagge, J., & Semb, H. (2005). Signals from the embryonic mouse pancreas induce differentiation of human embryonic stem cells into insulin-producing beta-cell-like cells. *Diabetes, 54,* 2867—2874.

Brummelkamp, T. R., Bernards, R., & Agami, R. (2002). A system for stable expression of short interfering RNAs in mammalian cells. *Science, 296,* 550—553.

Burns, J. C., Friedmann, T., Driever, W., Burrascano, M., & Yee, J. K. (1993). Vesicular stomatitis virus G glycoprotein pseudotyped retroviral vectors: concentration to very high titer and efficient gene transfer into mammalian and nonmammalian cells. *Proc. Natl. Acad. Sci. U.S.A., 90,* 8033—8037.

Capecchi, M. R. (1980). High efficiency transformation by direct microinjection of DNA into cultured mammalian cells. *Cell, 22,* 479—488.

Carpenter, M. K., Inokuma, M. S., Denham, J., Mujtaba, T., Chiu, C. P., & Rao, M. S. (2001). Enrichment of neurons and neural precursors from human embryonic stem cells. *Exp. Neurol., 172,* 383—397.

Cavazzana-Calvo, M., Hacein-Bey, S., de Saint Basile, G., Gross, F., Yvon, E., Nusbaum, P., et al. (2000). Gene therapy of human severe combined immunodeficiency (SCID)-X1 disease. *Science, 288,* 669—672.

Cerdan, C., Rouleau, A., & Bhatia, M. (2004). VEGF-A165 augments erythropoietic development from human embryonic stem cells. *Blood, 103,* 2504—2512.

Chadwick, K., Wang, L., Li, L., Menendez, P., Murdoch, B., Rouleau, A., et al. (2003). Cytokines and BMP-4 promote hematopoietic differentiation of human embryonic stem cells. *Blood, 102,* 906—915.

Challita, P. M., Skelton, D., el-Khoueiry, A., Yu, X. J., Weinberg, K., & Kohn, D. B. (1995). Multiple modifications in *cis* elements of the long terminal repeat of retroviral vectors lead to increased expression and decreased DNA methylation in embryonic carcinoma cells. *J. Virol., 69,* 748—755.

Cheshier, S. H., Morrison, S. J., Liao, X., & Weissman, I. L. (1999). *In vivo* proliferation and cell cycle kinetics of long-term self-renewing hematopoietic stem cells. *Proc. Natl. Acad. Sci. U.S.A., 96,* 3120—3125.

Choi, K., Kennedy, M., Kazarov, A., Papadimitriou, J. C., & Keller, G. (1998). A common precursor for hemato-poietic and endothelial cells. *Development, 125,* 725—732.

Cone, R. D., & Mulligan, R. C. (1984). High-efficiency gene transfer into mammalian cells: generation of helper-free recombinant retrovirus with broad mammalian host range. *Proc. Natl. Acad. Sci. U.S.A., 81,* 6349—6353.

Darr, H., Mayshar, Y., & Benvenisty, N. (2006). Overexpression of NANOG in human ES cells enables feeder-free growth while inducing primitive ectoderm features. *Development, 133,* 1193–1201.

Denti, M. A., Rosa, A., Sthandier, O., de Angelis, F. G., & Bozzoni, I. (2004). A new vector, based on the PolII promoter of the U1 snRNA gene, for the expression of siRNAs in mammalian cells. *Mol. Ther., 10,* 191–199.

Dick, J. E., Bhatia, M., Gan, O., Kapp, U., & Wang, J. C. (1997). Assay of human stem cells by repopulation of NOD/SCID mice. *Stem Cells, 15*(Suppl. 1), 199–203.

Draper, J. S., Smith, K., Gokhale, P., Moore, H. D., Maltby, E., Johnson, J., et al. (2004). Recurrent gain of chromosomes 17q and 12 in cultured human embryonic stem cells. *Nat. Biotechnol., 22,* 53–54.

Drukker, M., Katchman, H., Katz, G., Even-Tov Friedman, S., Shezen, E., Hornstein, E., et al. (2006). Human embryonic stem cells and their differentiated derivatives are less susceptible to immune rejection than adult cells. *Stem Cells, 24,* 221–229.

Dvash, T., Mayshar, Y., Darr, H., McElhaney, M., Barker, D., Yanuka, O., et al. (2004). Temporal gene expression during differentiation of human embryonic stem cells and embryoid bodies. *Hum. Reprod., 19,* 2875–2883.

Eiges, R., Schuldiner, M., Drukker, M., Yanuka, O., Itskovitz-Eldor, J., et al. (2001). Establishment of human embryonic stem cell-transfected clones carrying a marker for undifferentiated cells. *Curr. Biol., 11,* 514–518.

Enver, T., Soneji, S., Joshi, C., Brown, J., Iborra, F., Orntoft, T., et al. (2005). Cellular differentiation hierarchies in normal and culture-adapted human embryonic stem cells. *Hum. Mol. Genet., 14,* 3129–3140.

Fathi, F., Tiraihi, T., Mowla, S. J., & Movahedin, M. (2006). Transfection of CCE mouse embryonic stem cells with EGFP and BDNF genes by the electroporation method. *Rejuv. Res., 9,* 26–30.

Fire, A., Xu, S., Montgomery, M. K., Kostas, S. A., Driver, S. E., & Mello, C. C. (1998). Potent and specific genetic interference by double-stranded RNA in Caenorhabditis elegans. *Nature, 391,* 806–811.

Gautsch, J. W. (1980). Embryonal carcinoma stem cells lack a function required for virus replication. *Nature, 285,* 110–112.

Gerrard, L., Zhao, D., Clark, A. J., & Cui, W. (2005). Stably transfected human embryonic stem cell clones express OCT4-specific green fluorescent protein and maintain self-renewal and pluripotency. *Stem Cells, 23,* 124–133.

Gilmore, I. R., Fox, S. P., Hollins, A. J., Sohail, M., & Akhtar, S. (2004). The design and exogenous delivery of siRNA for post-transcriptional gene silencing. *J. Drug Target., 12,* 315–340.

Gropp, M., Itsykson, P., Singer, O., Ben-Hur, T., Reinhartz, E., Galun, E., et al. (2003). Stable genetic modification of human embryonic stem cells by lentiviral vectors. *Mol. Ther., 7,* 281–287.

Grusby, M. J., Johnson, R. S., Papaioannou, V. E., & Glimcher, L. H. (1991). Depletion of CD41 T cells in major histocompatibility complex class II-deficient mice. *Science, 253,* 1417–1420.

Gupta, S., Schoer, R. A., Egan, J. E., Hannon, G. J., & Mittal, V. (2004). Inducible, reversible, and stable RNA interference in mammalian cells. *Proc. Natl. Acad. Sci. U.S.A., 101,* 1927–1932.

Hay, D. C., Sutherland, L., Clark, J., & Burdon, T. (2004). Oct-4 knockdown induces similar patterns of endoderm and trophoblast differentiation markers in human and mouse embryonic stem cells. *Stem Cells, 22,* 225–235.

Herszfeld, D., Wolvetang, E., Langton-Bunker, E., Chung, T. L., Filipczyk, A. A., Houssami, S., et al. (2006). CD30 is a survival factor and a biomarker for transformed human pluripotent stem cells. *Nat. Biotechnol., 24,* 351–357.

Higuchi, M., Tsutsumi, R., Higashi, H., & Hatakeyama, M. (2004). Conditional gene silencing utilizing the lac repressor reveals a role of SHP-2 in cagA-positive Helicobacter pylori pathogenicity. *Cancer Sci., 95,* 442–447.

Huber, T. L., Kouskoff, V., Fehling, H. J., Palis, J., & Keller, G. (2004). Haemangioblast commitment is initiated in the primitive streak of the mouse embryo. *Nature, 432,* 625–630.

Itskovitz-Eldor, J., Schuldiner, M., Karsenti, D., Eden, A., Yanuka, O., Amit, M., et al. (2000). Differentiation of human embryonic stem cells into embryoid bodies comprising the three embryonic germ layers. *Mol. Med., 6,* 88–95.

Joyner, A. L. (2000). *Gene Targeting: A Practical Approach.* Oxford: Oxford University Press.

Kaeser, M. D., Pebernard, S., & Iggo, R. D. (2004). Regulation of p53 stability and function in HCT116 colon cancer cells. *J. Biol. Chem., 279,* 7598–7605.

Kaufman, D. S., Hanson, E. T., Lewis, R. L., Auerbach, R., & Thomson, J. A. (2001). Hematopoietic colony-forming cells derived from human embryonic stem cells. *Proc. Natl. Acad. Sci. U.S.A., 98,* 10716–10721.

Kawabata, K., Sakurai, F., Yamaguchi, T., Hayakawa, T., Mizuguchi, H., Mitani, K., et al. (2005). Efficient gene transfer into mouse embryonic stem cells with adenovirus vectors. *Mol. Ther., 12,* 547–554.

Kay, M. A., Baley, P., Rothenberg, S., Leland, F., Fleming, L., Ponder, K. P., et al. (1992). Expression of human alpha 1-antitrypsin in dogs after autologous transplantation of retroviral transduced hepatocytes. *Proc. Natl. Acad. Sci. U.S.A., 89,* 89–93.

Kehat, I., Kenyagin-Karsenti, D., Snir, M., Segev, H., Amit, M., Gepstein, A., et al. (2001). Human embryonic stem cells can differentiate into myocytes with structural and functional properties of cardiomyocytes. *J. Clin. Invest., 108,* 407–414.

Kehat, I., Khimovich, L., Caspi, O., Gepstein, A., Shofti, R., Arbel, G., et al. (2004). Electromechanical integration of cardiomyocytes derived from human embryonic stem cells. *Nat. Biotechnol., 22*, 1282–1289.

Kumar, M., & Carmichael, G. G. (1998). Antisense RNA: function and fate of duplex RNA in cells of higher eukaryotes. *Microbiol. Mol. Biol. Rev., 62*, 1415–1434.

Kunath, T., Gish, G., Lickert, H., Jones, N., Pawson, T., & Rossant, J. (2003). Transgenic RNA interference in ES cell-derived embryos recapitulates a genetic null phenotype. *Nat. Biotechnol., 21*, 559–561.

Kyba, M., Perlingeiro, R. C., & Daley, G. Q. (2002). HoxB4 confers definitive lymphoid-myeloid engraftment potential on embryonic stem cell and yolk sac hematopoietic progenitors. *Cell, 109*, 29–37.

Laker, C., Meyer, J., Schopen, A., Friel, J., Heberlein, C., Ostertag, W., et al. (1998). Host *cis*-mediated extinction of a retrovirus permissive for expression in embryonal stem cells during differentiation. *J. Virol., 72*, 339–348.

Lavon, N., Yanuka, O., & Benvenisty, N. (2006). The effect of over expression of Pdx1 and Foxa2 on the differentiation of human embryonic stem cells into pancreatic cells. *Stem Cells, 24*, 1923–1930.

Lee, T. I., Jenner, R. G., Boyer, L. A., Guenther, M. G., Levine, S. S., Kumar, R. M., et al. (2006). Control of developmental regulators by Polycomb in human embryonic stem cells. *Cell, 125*, 301–313.

Levenberg, S., Golub, J. S., Amit, M., Itskovitz-Eldor, J., & Langer, R. (2002). Endothelial cells derived from human embryonic stem cells. *Proc. Natl. Acad. Sci. U.S.A., 99*, 4391–4396.

Levenberg, S., Rouwkema, J., Macdonald, M., Garfein, E. S., Kohane, D. S., Darland, D. C., et al. (2005). Engineering vascularized skeletal muscle tissue. *Nat. Biotechnol., 23*, 879–884.

Levetzow, G. V., Spanholtz, J., Beckmann, J., Fischer, J., Kogler, G., Wernet, P., et al. (2006). Nucleofection, an efficient nonviral method to transfer genes into human hematopoietic stem and progenitor cells. *Stem Cells Dev., 15*, 278–285.

Li, L., Baroja, M. L., Majumdar, A., Chadwick, K., Rouleau, A., Gallacher, L., et al. (2004). Human embryonic stem cells possess immune-privileged properties. *Stem Cells, 22*, 448–456.

Li, X. J., Du, Z. W., Zarnowska, E. D., Pankratz, M., Hansen, L. O., Pearce, R. A., et al. (2005). Specification of motoneurons from human embryonic stem cells. *Nat. Biotechnol., 23*, 215–221.

Liu, Y. P., Dovzhenko, O. V., Garthwaite, M. A., Dambaeva, S. V., Durning, M., Pollastrini, L. M., et al. (2004). Maintenance of pluripotency in human embryonic stem cells stably over-expressing enhanced green fluorescent protein. *Stem Cells Dev., 13*, 636–645.

Lu, J., Hou, R., Booth, C. J., Yang, S. H., & Snyder, M. (2006). Defined culture conditions of human embryonic stem cells. *Proc. Natl. Acad. Sci. U.S.A., 103*, 5688–5693.

Ludwig, T. E., Levenstein, M. E., Jones, J. M., Berggren, W. T., Mitchen, E. R., Frane, J. L., et al. (2006). Derivation of human embryonic stem cells in defined conditions. *Nat. Biotechnol., 24*, 185–187.

Luther-Wyrsch, A., Costello, E., Thali, M., Buetti, E., Nissen, C., Surbek, D., et al. (2001). Stable transduction with lentiviral vectors and amplification of immature hematopoietic progenitors from cord blood of preterm human fetuses. *Hum. Gene. Ther., 12*, 377–389.

Ma, Y., Ramezani, A., Lewis, R., Hawley, R. G., & Thomson, J. A. (2003). High-level sustained transgene expression in human embryonic stem cells using lentiviral vectors. *Stem Cells, 21*, 111–117.

Maitra, A., Arking, D. E., Shivapurkar, N., Ikeda, M., Stastny, V., Kassauei, K., et al. (2005). Genomic alterations in cultured human embryonic stem cells. *Nat. Genet., 37*, 1099–1103.

Matin, M. M., Walsh, J. R., Gokhale, P. J., Draper, J. S., Bahrami, A. R., Morton, I., et al. (2004). Specific knockdown of Oct4 and beta2-microglobulin expression by RNA interference in human embryonic stem cells and embryonic carcinoma cells. *Stem Cells, 22*, 659–668.

Menendez, P., Wang, L., Chadwick, K., Li, L., & Bhatia, M. (2004). Retroviral transduction of hematopoietic cells differentiated from human embryonic stem cell-derived CD45(neg)PFV hemogenic precursors. *Mol. Ther., 10*, 1109–1120.

Miller, D. G., Adam, M. A., & Miller, A. D. (1990). Gene transfer by retrovirus vectors occurs only in cells that are actively replicating at the time of infection. *Mol. Cell. Biol., 10*, 4239–4242.

Mitani, K., Wakamiya, M., Hasty, P., Graham, F. L., Bradley, A., & Caskey, C. T. (1995). Gene targeting in mouse embryonic stem cells with an adenoviral vector. *Somat. Cell Mol. Genet., 21*, 221–231.

Miyagishi, M., Hayashi, M., & Taira, K. (2003). Comparison of the suppressive effects of antisense oligonucleotides and siRNAs directed against the same targets in mammalian cells. *Antisense Nucleic Acid Drug Dev., 13*, 1–7.

Mountford, P., Zevnik, B., Duwel, A., Nichols, J., Li, M., Dani, C., et al. (1994). Dicistronic targeting constructs: reporters and modifiers of mammalian gene expression. *Proc. Natl. Acad. Sci. U.S.A., 91*, 4303–4307.

Mummery, C., Ward-van Oostwaard, D., Doevendans, P., Spijker, R., van den Brink, S., Hassink, R., et al. (2003). Differentiation of human embryonic stem cells to cardiomyocytes: role of coculture with visceral endoderm-like cells. *Circulation, 107*, 2733–2740.

Naldini, L., Blomer, U., Gallay, P., Ory, D., Mulligan, R., Gage, F. H., et al. (1996). *In vivo* gene delivery and stable transduction of nondividing cells by a lentiviral vector. *Science, 272*, 263–267.

Narayan, A. D., Chase, J. L., Lewis, R. L., Tian, X., Kaufman, D. S., Thomson, J. A., et al. (2006). Human embryonic stem cell-derived hematopoietic cells are capable of engrafting primary as well as secondary fetal sheep recipients. *Blood, 107,* 2180—2183.

Nightingale, S. J., Hollis, R. P., Pepper, K. A., Petersen, D., Yu, X. J., Yang, C., et al. (2006). Transient gene expression by nonintegrating lentiviral vectors. *Mol. Ther., 13,* 1121—1132.

Oberlin, E., Tavian, M., Blazsek, I., & Peault, B. (2002). Blood-forming potential of vascular endothelium in the human embryo. *Development, 129,* 4147—4157.

Ohbayashi, F., Balamotis, M. A., Kishimoto, A., Aizawa, E., Diaz, A., Hasty, P., et al. (2005). Correction of chromosomal mutation and random integration in embryonic stem cells with helper-dependent adenoviral vectors. *Proc. Natl. Acad. Sci. U.S.A., 102,* 13628—13633.

Oliveira, D. M., & Goodell, M. A. (2003). Transient RNA interference in hematopoietic progenitors with functional consequences. *Genesis, 36,* 203—208.

Park, S., Lee, K. S., Lee, Y. J., Shin, H. A., Cho, H. Y., Wang, K. C., et al. (2004). Generation of dopaminergic neurons *in vitro* from human embryonic stem cells treated with neurotrophic factors. *Neurosci. Lett., 359,* 99—103.

Pebernard, S., & Iggo, R. D. (2004). Determinants of interferon-stimulated gene induction by RNAi vectors. *Differentiation, 72,* 103—111.

Perrier, A. L., Tabar, V., Barberi, T., Rubio, M. E., Bruses, J., Topf, N., et al. (2004). Derivation of midbrain dopamine neurons from human embryonic stem cells. *Proc. Natl. Acad. Sci. U.S.A., 101,* 12543—12548.

Persengiev, S. P., Zhu, X., & Green, M. R. (2004). Nonspecific, concentration-dependent stimulation and repression of mammalian gene expression by small interfering RNAs (siRNAs). *RNA, 10,* 12—18.

Pfeifer, A., Ikawa, M., Dayn, Y., & Verma, I. M. (2002). Transgenesis by lentiviral vectors: lack of gene silencing in mammalian embryonic stem cells and preimplantation embryos. *Proc. Natl. Acad. Sci. U.S.A., 99,* 2140—2145.

Ramezani, A., Hawley, T. S., & Hawley, R. G. (2000). Lentiviral vectors for enhanced gene expression in human hematopoietic cells. *Mol. Ther., 2,* 458—469.

Reubinoff, B. E., Itsykson, P., Turetsky, T., Pera, M. F., Reinhartz, E., Itzik, A., et al. (2001). Neural progenitors from human embryonic stem cells. *Nat. Biotechnol., 19,* 1134—1140.

Richards, M., Fong, C. Y., Chan, W. K., Wong, P. C., & Bongso, A. (2002). Human feeders support prolonged undifferentiated growth of human inner cell masses and embryonic stem cells. *Nat. Biotechnol., 20,* 933—936.

Salmon, P., Kindler, V., Ducrey, O., Chapuis, B., Zubler, R. H., & Trono, D. (2000). High-level transgene expression in human hematopoietic progenitors and differentiated blood lineages after transduction with improved lentiviral vectors. *Blood, 96,* 3392—3398.

Schomber, T., Kalberer, C. P., Wodnar-Filipowicz, A., & Skoda, R. C. (2004). Gene silencing by lentivirus-mediated delivery of siRNA in human CD341 cells. *Blood, 103,* 4511—4513.

Schuldiner, M., Yanuka, O., Itskovitz-Eldor, J., Melton, D. A., & Benvenisty, N. (2000). Effects of eight growth factors on the differentiation of cells derived from human embryonic stem cells. *Proc. Natl. Acad. Sci. U.S.A., 97,* 11307—11312.

Schulz, T. C., Noggle, S. A., Palmarini, G. M., Weiler, D. A., Lyons, I. G., Pensa, K. A., et al. (2004). Differentiation of human embryonic stem cells to dopaminergic neurons in serum-free suspension culture. *Stem Cells, 22,* 1218—1238.

Segev, H., Fishman, B., Ziskind, A., Shulman, M., & Itskovitz-Eldor, J. (2004). Differentiation of human embryonic stem cells into insulin-producing clusters. *Stem Cells, 22,* 265—274.

Siemen, H., Nix, M., Endl, E., Koch, P., Itskovitz-Eldor, J., & Brustle, O. (2005). Nucleofection of human embryonic stem cells. *Stem Cells Dev., 14,* 378—383.

Song, L., Chau, L., Sakamoto, Y., Nakashima, J., Koide, M., & Tuan, R. S. (2004). Electric field-induced molecular vibration for noninvasive, high-efficiency DNA transfection. *Mol. Ther., 9,* 607—616.

Stegmeier, F., Hu, G., Rickles, R. J., Hannon, G. J., & Elledge, S. J. (2005). A lentiviral microRNA-based system for single-copy polymerase II-regulated RNA interference in mammalian cells. *Proc. Natl. Acad. Sci. U.S.A., 102,* 13212—13217.

Stone, D., Ni, S., Li, Z. Y., Gaggar, A., DiPaolo, N., Feng, Q., et al. (2005). Development and assessment of human adenovirus type 11 as a gene transfer vector. *J. Virol., 79,* 5090—5104.

Szulc, J., Wiznerowicz, M., Sauvain, M. O., Trono, D., & Aebischer, P. (2006). A versatile tool for conditional gene expression and knockdown. *Nat. Meth., 3*(2), 109—116.

Tatsis, N., & Ertl, H. C. (2004). Adenoviruses as vaccine vectors. *Mol. Ther., 10,* 616—629.

Thomson, J. A., Itskovitz-Eldor, J., Shapiro, S. S., Waknitz, M. A., Swiergiel, J. J., Marshall, V. S., et al. (1998). Embryonic stem cell lines derived from human blastocysts. *Science, 282,* 1145—1147.

Tian, X., Woll, P. S., Morris, J. K., Linehan, J. L., & Kaufman, D. S. (2006). Hematopoietic engraftment of human embryonic stem cell-derived cells is regulated by recipient innate immunity. *Stem Cells, 24,* 1370—1380.

Tiscornia, G., Singer, O., Ikawa, M., & Verma, I. M. (2003). A general method for gene knockdown in mice by using lentiviral vectors expressing small interfering RNA. *Proc. Natl. Acad. Sci. U.S.A., 100*, 1844−1848.

Tiscornia, G., Tergaonkar, V., Galimi, F., & Verma, I. M. (2004). CRE recombinase-inducible RNA interference mediated by lentiviral vectors. *Proc. Natl. Acad. Sci. U.S.A., 101*, 7347−7351.

Vallier, L., Rugg-Gunn, P. J., Bouhon, I. A., Andersson, F. K., Sadler, A. J., & Pedersen, R. A. (2004). Enhancing and diminishing gene function in human embryonic stem cells. *Stem Cells, 22*, 2−11.

Vodyanik, M. A., Bork, J. A., Thomson, J. A., & Slukvin, I. I. (2005). Human embryonic stem cell-derived CD34+ cells: efficient production in the coculture with OP9 stromal cells and analysis of lymphohematopoietic potential. *Blood, 105*, 617−626.

Volpers, C., & Kochanek, S. (2004). Adenoviral vectors for gene transfer and therapy. *J. Gene. Med., 6*(Suppl. 1), S164−S171.

Wakayama, T., Tabar, V., Rodriguez, I., Perry, A. C., Studer, L., & Mombaerts, P. (2001). Differentiation of embryonic stem cell lines generated from adult somatic cells by nuclear transfer. *Science, 292*, 740−743.

Wang, L., Li, L., Menendez, P., Cerdan, C., & Bhatia, M. (2005a). Human embryonic stem cells maintained in the absence of mouse embryonic fibroblasts or conditioned media are capable of hematopoietic development. *Blood, 105*, 4598−4603.

Wang, L., Li, L., Shojaei, F., Levac, K., Cerdan, C., Menendez, P., et al. (2004). Endothelial and hematopoietic cell fate of human embryonic stem cells originates from primitive endothelium with hemangioblastic properties. *Immunity, 21*, 31−41.

Wang, L., Menendez, P., Shojaei, F., Li, L., Mazurier, F., Dick, J. E., et al. (2005b). Generation of hematopoietic repopulating cells from human embryonic stem cells independent of ectopic HOXB4 expression. *J. Exp. Med., 201*, 1603−1614.

Wang, Y., Yates, F., Naveiras, O., Ernst, P., & Daley, G. Q. (2005c). Embryonic stem cell-derived hematopoietic stem cells. *Proc. Natl. Acad. Sci. U.S.A., 102*, 19081−19086.

Weissinger, F., Reimer, P., Waessa, T., Buchhofer, S., Schertlin, T., Kunzmann, V., et al. (2003). Gene transfer in purified human hematopoietic peripheral-blood stem cells by means of electroporation without prestimulation. *J. Lab. Clin. Med., 141*, 138−149.

Wiznerowicz, M., & Trono, D. (2003). Conditional suppression of cellular genes: lentivirus vector-mediated drug-inducible RNA interference. *J. Virol., 77*, 8957−8961.

Wu, M. H., Liebowitz, D. N., Smith, S. L., Williams, S. F., & Dolan, M. E. (2001). Efficient expression of foreign genes in human CD34(+) hematopoietic precursor cells using electroporation. *Gene. Ther., 8*, 384−390.

Xu, C., Inokuma, M. S., Denham, J., Golds, K., Kundu, P., Gold, J. D., et al. (2001). Feeder-free growth of undifferentiated human embryonic stem cells. *Nat. Biotechnol., 19*, 971−974.

Xu, C., Police, S., Rao, N., & Carpenter, M. K. (2002). Characterization and enrichment of cardiomyocytes derived from human embryonic stem cells. *Circ. Res., 91*, 501−508.

Xu, C., Rosler, E., Jiang, J., Lebkowski, J. S., Gold, J. D., O'Sullivan, C., et al. (2005a). Basic fibroblast growth factor supports undifferentiated human embryonic stem cell growth without conditioned medium. *Stem Cells, 23*, 315−323.

Xu, R. H., Peck, R. M., Li, D. S., Feng, X., Ludwig, T., & Thomson, J. A. (2005b). Basic FGF and suppression of BMP signaling sustain undifferentiated proliferation of human ES cells. *Nat. Meth., 2*, 185−190.

Yang, Y., Vanin, E. F., Whitt, M. A., Fornerod, M., Zwart, R., Schneiderman, R. D., et al. (1995). Inducible, high-level production of infectious murine leukemia retroviral vector particles pseudotyped with vesicular stomatitis virus G envelope protein. *Hum. Gene. Ther., 6*, 1203−1213.

Yoder, M. C., Hiatt, K., & Mukherjee, P. (1997). *In vivo* repopulating hematopoietic stem cells are present in the murine yolk sac at day 9.0 postcoitus. *Proc. Natl. Acad. Sci. U.S.A., 94*, 6776−6780.

Yoo, S. J., Yoon, B. S., Kim, J. M., Song, J. M., Roh, S., You, S., et al. (2005). Efficient culture system for human embryonic stem cells using autologous human embryonic stem cell-derived feeder cells. *Exp. Mol. Med., 37*, 399−407.

Yu, X., Zhan, X., d'Costa, J., Tanavde, V. M., Ye, Z., Peng, T., et al. (2003). Lentiviral vectors with two independent internal promoters transfer high-level expression of multiple transgenes to human hematopoietic stem-progenitor cells. *Mol. Ther., 7*, 827−838.

Zaehres, H., Lensch, M. W., Daheron, L., Stewart, S. A., Itskovitz-Eldor, J., & Daley, G. Q. (2005). High-efficiency RNA interference in human embryonic stem cells. *Stem Cells, 23*, 299−305.

Zeng, X., Cai, J., Chen, J., Luo, Y., You, Z. B., Fotter, E., et al. (2004). Dopaminergic differentiation of human embryonic stem cells. *Stem Cells, 22*, 925−940.

Zhang, S. C., Wernig, M., Duncan, I. D., Brustle, O., & Thomson, J. A. (2001). *In vitro* differentiation of transplantable neural precursors from human embryonic stem cells. *Nat. Biotechnol., 19*, 1129−1133.

Zijlstra, M., Bix, M., Simister, N. E., Loring, J. M., Raulet, D. H., & Jaenisch, R. (1990). Beta 2-microglobulin deficient mice lack CD4-81 cytolytic T cells. *Nature, 344*, 742−746.

Zufferey, R., Donello, J. E., Trono, D., & Hope, T. J. (1999). Woodchuck hepatitis virus posttranscriptional regulatory element enhances expression of transgenes delivered by retroviral vectors. *J. Virol., 73*, 2886–2892.

Zufferey, R., Nagy, D., Mandel, R. J., Naldini, L., & Trono, D. (1997). Multiply attenuated lentiviral vector achieves efficient gene delivery *in vivo. Nat. Biotechnol., 15*, 871–875.

Zwaka, T. P., & Thomson, J. A. (2003). Homologous recombination in human embryonic stem cells. *Nat. Biotechnol., 21*, 319–321.

CHAPTER 10

Embryonic Stem Cells: Derivation and Properties

Junying Yu[*], **James A. Thomson**[**,***,†,‡]

[*] Cellular Dynamics International, Inc., Science Drive, Madison, WI, USA
[**] National Primate Research Center, University of Wisconsin Graduate School, Madison, WI, USA
[***] WiCell Research Institute, Madison, WI, USA
[†] Department of Anatomy, University of Wisconsin Medical School, Madison, WI, USA
[‡] Genome Center of Wisconsin, University of Wisconsin-Madison, Madison, WI, USA

INTRODUCTION

Embryonic stem (ES) cells are derived from early embryos, and are capable of indefinite self-renewal *in vitro* while maintaining the potential to develop into all cell types of the body — they are pluripotent. With these remarkable features, ES cells hold great promise in both regenerative medicine and basic biological research. In this chapter, we will discuss how embryonic stem cells are derived and what is known about the mechanisms that allow these cells to maintain their pluripotency while proliferating *in vitro*.

DERIVATION OF EMBRYONIC STEM CELLS
Embryonic carcinoma cells

Teratocarcinoma is a form of malignant germ cell tumor that occurs in both animals and humans. These tumors comprise an undifferentiated embryonal carcinoma (EC) component and differentiated derivatives that can include all three germ layers. Although teratocarcinomas had been known as medical curiosities for centuries (Wheeler, 1983), it was the discovery that male mice of strain 129 had a high incidence of testicular teratocarcinomas (Stevens and Little, 1954) that made these tumors more routinely amenable to experimental analysis. Because their growth is sustained by a persistent EC cell component, teratocarcinomas can be serially transplanted between mice. In 1964, Kleinsmith and Pierce demonstrated that a single EC cell was capable of both self-renewal and multilineage differentiation, and this formal demonstration of a pluripotent stem cell provided the intellectual framework for both mouse and human ES cells.

The first mouse EC cell lines were established in the early 1970s (Kahan and Ephrussi, 1970; Evans, 1972). EC cells exhibit similar antigen and protein expression to the cells present in the inner cell mass (ICM) (Klavins et al., 1971; Comoglio et al., 1975; Gachelin et al., 1977; Solter and Knowles, 1978; Calarco and Banka, 1979; Howe et al., 1980; Henderson et al., 2002), and this led to the notion that EC cells are the counterpart of pluripotent cells present in the ICM (Martin, 1980; Rossant and Papaioannou, 1984). When injected into mouse blastocysts, some

EC cell lines are able to contribute to various somatic cell types (Brinster, 1974; Mintz and Illmensee, 1975; Papaioannou et al., 1975; Illmensee and Mintz, 1976), but most EC cell lines have limited developmental potential and contribute poorly to chimeric mice, probably reflecting genetic changes acquired during teratocarcinoma formation (Atkin et al., 1974; McBurney, 1976; Bronson et al., 1980; Zeuthen et al., 1980). Mutations that confer growth advantages to EC cells are likely to accumulate during tumorigenesis, and EC cells in chimeras can result in tumor formation (Papaioannou et al., 1978). As a result, there are limitations in the application of EC cells to both regenerative medicine and research in basic developmental biology.

Following fertilization, as the one-cell embryo migrates down the oviduct, it undergoes a series of cleavage divisions resulting in a morula. During blastocyst formation, the outer cell layer of the morula delaminates from the rest of the embryo to form the trophectoderm. The ICM of the blastocyst gives rise to all the fetal tissues (ectoderm, mesoderm, and endoderm) and some extraembryonic tissues, and the trophectoderm gives rise to the trophoblast. Although the early ICM can contribute to the trophoblast, the late ICM does not (Winkel and Pedersen, 1988), suggesting there is some restriction in developmental potential at this stage. In normal embryos, the pluripotent cells of the embryo have a transient existence, as these cells quickly give rise to other non-pluripotent cells through the normal developmental program. Thus, the pluripotent cells of the intact embryo really function *in vivo* as precursor cells and not as stem cells. However, if early mouse embryos are transferred to extrauterine sites, such as the kidney or testis capsules of adult mice, they can develop into teratocarcinomas that include pluripotent stem (EC) cells (Solter et al., 1970; Stevens, 1970). These ectopic transplantation experiments result in teratocarcinomas at high frequencies, even in strains that do not spontaneously have elevated incidence of germ cell tumors, suggesting that this process is not the result of rare neoplastic transformation events. These key transplantation experiments led to the search for culture conditions that would allow the *in vitro* derivation of pluripotent stem cells directly from the embryo, without the intermediate need to form teratocarcinomas *in vivo*.

Derivation of embryonic stem cells

In 1981, pluripotent embryonic stem (ES) cell lines were derived directly from the ICM of mouse blastocysts using culture conditions previously developed for mouse EC cells (Evans and Kaufman, 1981; Martin, 1981). ES cell cultures derived from a single cell could differentiate into a wide variety of cell types, or could form teratocarcinomas when injected into mice (Martin, 1981). Unlike EC cells, however, these karyotypically normal cells contributed at a high frequency to a variety of tissues in chimeras, including germ cells, and thus provided a practical way to introduce modifications to the mouse germ line (Bradley et al., 1984).

The efficiency in mouse ES cell derivation is influenced by genetic background. For example, ES cells can be easily derived from the inbred 129/ter-Sv strain, but less efficiently from C57BL/6 and other mouse strains (Ledermann and Burki, 1991; Kitani et al., 1996), and these strain differences somewhat correspond with the propensity of mice of different strains to develop teratocarcinomas. These observations suggested that genetic and/or epigenetic components play an important role in the derivation of mouse ES cells. On the other hand, the efficiency of teratocarcinoma formation induced through extrauterine mouse embryo transplantations appears to be somewhat less strain-dependent (Damjanov et al., 1983). This indicates that the difference in the efficiency of ES cell derivation from different mouse strains might be due to suboptimal culture conditions. Indeed, mouse ES cells can be derived from some non-permissive strains using modified protocols; e.g. dual inhibition of differentiation-inducing signaling from mitogen-activated protein kinase and glycogen synthase kinase-3 (GSK3) enabled the efficient derivation of germ line-competent ES cells from non-obese diabetic mice (McWhir et al., 1996; Brook and Gardner, 1997; Nichols et al., 2009).

ES cell lines are generally derived from the culture of the ICM, but this does not mean that ES cells are the *in vitro* equivalent to ICM cells, or even that ICM cells are the immediate precursor to ES cells. It is possible that, during culture, ICM cells give rise to other cells that serve as the immediate precursors. Some experiments suggest that ES cells more closely resemble cells from the primitive ectoderm, the cell layer derived from the ICM after delamination of the primitive endoderm. Isolated primitive ectoderm from the mouse gives rise to ES cell lines at a high frequency and allows the isolation of ES cell lines from mouse strains that had previously been refractory to ES cell isolation (Brook and Gardner, 1997). Indeed, single primitive ectoderm cells can give rise to ES cell lines at a reasonable frequency, something not possible with early ICM cells (Brook and Gardner, 1997). Although these experiments do suggest that ES cells are more closely related to primitive ectoderm than to ICM, they do not reveal whether ES cells more closely resemble primitive ectoderm or another cell type (for example, very early germ cells) derived from it *in vitro* (Zwaka and Thomson, 2005). As no pluripotent cell in the intact embryo undergoes long-term self-renewal, ES cells are in some ways tissue culture artifacts. It is surprising that even more than 20 years after their derivation, the origin of these cells is not completely understood. Given the dramatic improvement in molecular techniques since the initial derivation in the 1980s, there is considerable value in reexamining the origin of ES cells to better understand the control of their proliferative pluripotent state (Zwaka and Thomson, 2005).

In addition to derivation from the ICM and isolated primitive ectoderm, mouse ES cells have also been derived from morula-stage embryos and even from individual blastomeres (Eistetter, 1989; Delhaise et al., 1996; Tesar, 2005; Chung et al., 2006). Again, although the ES cell lines were derived from morula, there may well be a progression of intermediate states during the derivation process. The frequencies of success were lower when starting with morula or blastomeres, but these results do suggest that it might be possible to derive human ES cells without the destruction of an embryo. Such cell lines could prove useful to the child resulting from the transfer of a biopsied embryo, as they would be genetically matched to the child.

Derivation of human embryonic stem cells

In 1978 the first baby was born from an embryo fertilized *in vitro* (Steptoe and Edwards, 1978) and, without this event, the derivation of human ES cells would not have been possible. Although there were attempts to derive human ES cells as early as the 1980s, species-specific differences and suboptimal human embryo culture media delayed their successful isolation until 1998 (Thomson et al., 1998). For example, the culture of isolated ICMs from human blastocysts was reported (Bongso et al., 1994), but stable undifferentiated cell lines were not produced in medium supplemented with leukemia inhibitory factor (LIF) in the presence of feeder layers, conditions that allow the isolation of mouse ES cells. In the mid-1990s, ES cell lines were derived from two non-human primates: the rhesus monkey and the common marmoset (Thomson et al., 1995, 1996). Experience with these ES cell lines and concomitant improvements in culture conditions for human IVF embryos (Gardner et al., 1998) resulted in the successful derivation of human ES cell lines (Thomson et al., 1998). These human ES cells had normal karyotypes and, even after prolonged undifferentiated proliferation, maintained the developmental potential to contribute to advanced derivatives of all three germ layers.

To date, more than 120 human ES cell lines have been established worldwide (Stojkovic et al., 2004b). Although most were derived from isolated ICMs, some were derived from morulae or later blastocyst stage embryos (Stojkovic et al., 2004a; Strelchenko et al., 2004). It is not yet known whether ES cells derived from these different developmental stages have any consistent differences or whether they are developmentally equivalent. Human ES cell lines have also been derived from embryos carrying various disease-associated genetic changes, which provide new *in vitro* models of disease (Verlinsky et al., 2005).

CULTURE OF EMBRYONIC STEM CELLS
Culture of mouse embryonic stem cells

Mitotically inactivated feeder layers were first used to support difficult-to-culture epithelial cells (Puck et al., 1956), and were later successfully adapted for the culture of mouse EC cells (Martin and Evans, 1975; Martin et al., 1977) and mouse ES cells (Evans and Kaufman, 1981; Martin, 1981). Medium that is "conditioned" by co-culture with fibroblasts sustains EC cells (Smith and Hooper, 1983). Fractionation of conditioned medium led to the identification of a cytokine, leukemia inhibitory factor (LIF), that sustains ES cells (Smith et al., 1988; Williams et al., 1988). LIF and its related cytokines act via the gp130 receptor (Yoshida et al., 1994). Binding of LIF induces dimerization of LIF/gp130 receptors, which in turn activates the latent transcription factor STAT3 (Lutticken et al., 1994; Wegenka et al., 1994) and ERK mitogen-activated protein kinase (MAPK) cascade (Takahashi-Tezuka et al., 1998). STAT3 activation is sufficient for LIF-mediated self-renewal of mouse ES cells in the presence of serum (Matsuda et al., 1999). In contrast, suppression of the ERK pathway promotes ES cell proliferation (Burdon et al., 1999). In serum-free medium, LIF alone is insufficient to prevent mouse ES cell differentiation but, in combination with BMP (bone morphogenetic protein, a member of the TGFβ superfamily), mouse ES cells are sustained (Ying et al., 2003a). BMPs induce expression of Id (inhibitor of differentiation) proteins and inhibit the ERK and p38 MAPK pathways, thus attenuating the pro-differentiation activation of ERK MAPK pathway by LIF. These earlier works suggest the dependence on the extrinsic stimuli for the self-renewal of mouse ES cells, which was brought into question by recent studies. Inhibition of the ERK cascade (e.g. SU5402 and PD184352 or PD0325901) and GSK3 (CHIR99021) was sufficient to support the derivation, proliferation, and pluripotency of mouse ES cells; i.e. mouse ES cells do not rely on the extrinsic signals for self-renewal (Ying et al., 2008). Indeed, such conditions not only enabled the efficient derivation of ES cells from previously non-permissive mouse strains (Nichols et al., 2009), but also from refractory species (Buehr et al., 2008; Li et al., 2008).

Culture of human embryonic stem cells

Mitotically inactivated fibroblast feeder layers and serum-containing medium were used in the initial derivation of human ES cells, essentially the same conditions used for the derivation of mouse ES cells prior to the identification of LIF (Thomson et al., 1998; Reubinoff et al., 2000). However, it now appears largely to be a lucky coincidence that fibroblast feeder layers support both mouse and human ES cells, as the specific factors identified to date that sustain mouse ES cells do not support human ES cells. LIF and its related cytokines fail to support human or non-human primate ES cells in serum-containing media that supports mouse ES cells (Thomson et al., 1998; Daheron et al., 2004; Humphrey et al., 2004; Sumi et al., 2004), and BMPs, when added to human ES cells, cause rapid differentiation in conditions that would otherwise support their self-renewal (Xu et al., 2002; Pera et al., 2004). Indeed, the LIF/STAT3 pathway has yet to be shown to have any relevance to the self-renewal of human ES cells (Thomson et al., 1998; Daheron et al., 2004; Humphrey et al., 2004).

In contrast to mouse ES cells, FGF signaling appears to be of central importance in the self-renewal of human ES cells. Basic FGF (bFGF or FGF2) allows the clonal growth of human ES cells on fibroblasts in the presence of a commercially available serum replacement (Amit et al., 2000; Xu et al., 2001). At higher concentrations, bFGF allows feeder-independent growth of human ES cells cultured in the same serum replacement (Wang et al., 2005; Xu et al., 2005a,b). The mechanism through which these high concentrations of bFGF exert their functions is incompletely known, although one of the effects is suppression of BMP signaling (Xu et al., 2005b). Serum and the serum replacement currently used have significant BMP-like activity, which is sufficient to induce differentiation of human ES cells, and conditioning this medium on fibroblasts reduces this activity (Xu et al., 2005b). At moderate concentrations of bFGF (40 ng/ml), the addition of noggin or other inhibitors of BMP signaling significantly decreases background differentiation of human ES cells. At higher concentrations (100 ng/ml), bFGF itself suppresses BMP signaling in human ES cells to levels comparable to those observed in fibroblast-conditioned medium, and the addition of noggin is no longer needed for feeder-independent growth (Xu et al., 2005b). As more defined culture conditions are developed for human ES cells that lack serum products containing BMP activity, it is not yet clear how important the suppression of the BMP pathway will be, unless there is significant production of BMPs by the ES cells themselves. Also, the effects of BMP signaling could change depending on context. Even in mouse ES cells, BMPs are inducers of differentiation unless they are presented in combination with LIF, and it is entirely possible that, in a different signaling context, the effects of BMPs on human ES cells could change.

Suppression of BMP activity by itself is insufficient to maintain human ES cells (Xu et al., 2005b); thus, bFGF must be serving other signaling functions. Human ES cells themselves produce FGFs, and, in high-density cultures either on fibroblasts or in fibroblast-conditioned medium, it is not necessary to add FGFs. However, chemical inhibitors of FGF receptor-mediated phosphorylation cause differentiation of human ES cells under these standard culture conditions (Dvorak et al., 2005). The required downstream events are not yet well worked out, but some evidence implicates activation of the ERK pathway (Kang et al., 2005).

Although FGF signaling appears to have a central role in the self-renewal of human ES cells, other pathways have also been implicated. When combined with low to moderate levels of FGFs, TGFβ/Activin/Nodal signaling has a positive effect on the undifferentiated proliferation of human ES cells (Amit et al., 2004; Beattie et al., 2005; James et al., 2005; Vallier et al., 2005), and inhibition of this pathway leads to differentiation (James et al., 2005; Vallier et al., 2005). However, one of the effects of inhibiting the TGFβ/Activin/Nodal pathway is a stimulation of the BMP pathway (James et al., 2005), which in itself would be sufficient to induce differentiation. Thus, it is not yet clear whether TGFβ/Activin/Nodal signaling has a role in human ES cell self-renewal independent of its effects on BMP signaling. Further studies directly inhibiting the BMP pathway in the context of inhibition or stimulation of the TGFβ/Activin/Nodal are needed to resolve this issue.

The molecular components of the Wnt pathway are well represented in human ES cells (Sperger et al., 2003). In short-term cultures, activation of Wnt signaling by a pharmacological GSK-3-specfic inhibitor (BIO) has been reported to have a positive effect on human ES cell self-renewal (Sato et al., 2004), but, in a different study, inhibition of Wnt signaling or stimulation of Wnt signaling by the addition of recombinant Wnt proteins showed no effect on the maintenance of human ES cells (Dravid et al., 2005). It is possible that the positive observed effect of BIO on human ES cells is mediated through other pathways (James et al., 2005).

For human ES cells to be used in a clinical setting, it would be useful for these cells to be derived and maintained in conditions that are free of animal products. For example, human ES cells derived with mouse embryonic fibroblasts were shown to be contaminated with immunogenic non-human sialic acid, which would cause an immune reaction if the cells were used in human patients (Martin et al., 2005). Towards this goal, protein matrices including laminin and fibronectin, and different types of human feeder cells, were developed to sustain human ES cells (Xu et al., 2001; Amit et al., 2003; Richards et al., 2003). New human ES cell lines have been derived in the absence of feeder cells, but in the presence of a mouse-derived matrix and a bovine-derived serum replacement product (Klimanskaya et al., 2005). Existing human ES cell lines have been grown in defined serum-free medium that included sphingosine-1-phosphate (S1P) and PDGF (Pebay et al., 2005), but this medium does not eliminate the need for feeder layers. Existing human ES cell lines have also been adapted to feeder-free conditions in which none of the protein components are animal-derived, but it is not yet known whether these specific conditions will allow derivation of new lines (Li et al., 2005). Recent improvements in human ES cell culture have enabled the commercial development of completely defined, feeder-free culture conditions such as mTeSR1 and STEMPRO® hESC SFM

(Ludwig et al., 2006; Wang et al., 2007). Such conditions allow the derivation of new cell lines that will be more directly applicable to therapeutic purposes.

During extended culture, genetic changes can accumulate in human ES cells (Draper et al., 2004; Maitra et al., 2005). The status of imprinted genes appears to be relatively stable in human ES cells, but can also change (Rugg-Gunn et al., 2005). Such genetic and epigenetic alterations present a challenge that must be appropriately managed if human ES cells are to be used in cell replacement therapy. The rates at which these changes accumulate in culture likely depend on the culture system used and the particular selective pressures applied. For example, in all current culture conditions, the cloning efficiency of human ES cells is poor: typically 1% or less (Amit et al., 2000). If cells are dispersed into a suspension of single cells, there is a tremendous selective pressure for cells that clone at a higher efficiency, and indeed such an increase in cloning efficiency is observed in karyotypically abnormal cells (Enver et al., 2005). Enzymatic methods of passaging ES cells can allow long-term passage without karyotypic changes if the clump size is carefully controlled (Amit et al., 2000), but, if such methods are used to disperse cells to single cell suspensions or small clumps, karyotypic changes are more frequent (Cowan et al., 2004). This is a likely explanation for why mechanical splitting of individual colonies allows such long-term karyotypic stability (Buzzard et al., 2004). Understanding the rates at which genetic changes occur and the selective pressures that allow them to overgrow a culture in different culture conditions will be critical to the large-scale expansion and clinical use of human ES cells. For example, ROCK inhibitors could significantly improve the survival of dissociated human ES cells (Watanabe et al., 2007). Inclusion of these small molecules could potentially minimize the selection pressure and facilitate the development of large-scale human ES cell culture.

204

DEVELOPMENTAL POTENTIAL OF EMBRYONIC STEM CELLS
Differentiation of embryonic stem cells

Since ES cells have the ability to differentiate into clinically relevant cell types such as dopamine neurons, cardiomyocytes, and β cells, there is tremendous interest in using these cells both in basic biological research and in transplantation medicine. Both uses demand a great deal of control over lineage allocation and expansion. There are several experimental approaches to demonstrating the developmental potential of embryonic stem cells and to directing their differentiation to specific lineages. These approaches range in complexity and experimental control from allowing the ES cells to respond to normal developmental cues in a chimera within an intact embryo, to the addition of defined growth factors to a monolayer culture.

Mouse ES cells reintroduced into blastocysts participate in normal embryogenesis, even after prolonged culture and extensive manipulation *in vitro*. In such chimeras, the progeny of ES cells contribute to both somatic tissues and germ cells (Bradley et al., 1984). When ES cells are introduced into tetraploid blastocysts, mice entirely derived from ES cells can be produced, as the tetraploid component is outcompeted in the ICM-derived somatic tissues (Nagy et al., 1993; Ueda et al., 1995). Although mice entirely derived from ES cells can be generated, signals from the ICM of the blastocyst are likely necessary for mouse ES cells to contribute to offspring, as fetal development has not been reported when the ICM is completely replaced with ES cells.

ES cells injected into syngeneic or immunocompromised adult mice form teratomas that contain differentiated derivatives of all three germ layers (ectoderm, mesoderm, and endoderm) (Martin, 1981). This property is similar to both early embryos and EC cells, and is an approach now routinely used to demonstrate the pluripotency of human ES cells (Thomson et al., 1998). Very complex structures resembling neural tube, gut, teeth, and hair form in these teratomas in a very consistent temporal pattern, and these teratomas do offer an experimental model to study the development of these structures in human material, but the environment of differentiation is complex and difficult to manipulate.

Aggregates of EC cells or ES cells cultured in conditions that prevent their attachment form cystic "embryoid bodies" (Martin and Evans, 1975; Martin et al., 1977) that recapitulate some of the events of early development. Differentiated derivatives of all three germ layers form in these structures, and for ES cells the temporal events occurring mimic *in vivo* embryogenesis. The formation of embryoid bodies has been used, for example, to produce neural cells (Bain et al., 1995; Zhang et al., 2001), cardiomyocyte (Klug et al., 1996; He et al., 2003), hematopoietic precursors (Keller et al., 1993; Chadwick et al., 2003), β-like cells (Assady et al., 2001; Lumelsky et al., 2001), hepatocytes (Hamazaki et al., 2001; Rambhatla et al., 2003), and germ cells (Hubner et al., 2003; Toyooka et al., 2003; Geijsen et al., 2004). The formation of a three-dimensional structure in EBs is useful to promote certain developmental events, but the complicated cell-cell interaction makes it difficult to elucidate the essential signaling pathways involved.

A somewhat more controlled method to differentiate ES cells is to co-culture them with differentiated cells that induce their differentiation to specific lineages. For example, MS5, S2, and PA6 stromal cells have been used to derive dopamine neurons from human ES cells (Perrier et al., 2004; Zeng et al., 2004); bone marrow stromal cell lines S17 and OP9 support efficient hematopoietic differentiation (Kaufman et al., 2001; Vodyanik et al., 2005). The inducing activity provided by such stromal cells, while efficient in directing ES cell differentiation, contains many unknown factors, and such activity can change both between and within cell lines as a function of culture conditions.

An even more controlled method is differentiation in monolayers on defined matrices in the presence of specific growth factors. Both mouse and human ES cells differentiate into neuroectodermal precursors in monolayer culture (Ying et al., 2003b; Gerrard et al., 2005), and human ES cells can be efficiently induced to differentiate into trophoblasts with addition of BMPs (Xu et al., 2002). This method eliminates many unknown factors provided by either EBs or stromal cells, thus allowing precise analysis of specific factors on the differentiation of ES cells into lineages of choice. With improved understanding of regulatory events governing germ layer and cell lineage specifications, more cell types will likely be derived from ES cells in increasingly defined conditions.

Molecular control of pluripotency

We remain remarkably ignorant about why one cell is pluripotent and another is not, although some of the key players important to maintaining this remarkable state have been identified. Oct4, a member of the POU family of transcription factors, is essential for both the derivation and maintenance of ES cells (Pesce et al., 1998). The expression of Oct4 in the mouse is restricted to early embryos and germ cells (Scholer et al., 1989; Okamoto et al., 1990), and homozygous deletion of this gene causes a failure in the formation of the ICM (Nichols et al., 1998). For mouse ES cells to remain undifferentiated, the expression of Oct4 must be maintained within a critical range. Overexpression of this protein causes differentiation into endoderm and mesoderm, while decreased expression leads to differentiation into trophoblast (Niwa et al., 2000). The expression of Oct4 is also a hallmark of human ES cells (Hansis et al., 2000), and its downregulation also leads to differentiation and expression of trophoblast markers (Matin et al., 2004).

Another transcription factor important for the pluripotency of ES cells is Nanog (Chambers et al., 2003; Mitsui et al., 2003). Similar to Oct4, the expression of Nanog decreases rapidly as ES cells differentiate. However, unlike Oct4, overexpression of this protein in mouse ES cells allows their self-renewal to be independent of LIF/STAT3, though Nanog appears not to be a direct downstream target of the LIF/STAT3 pathway (Chambers et al., 2003). Moreover, increased Nanog expression stimulates the activation of pluripotent genes from the somatic genome in cell-cell fusion models (Silva et al., 2006). In human ES cells, the expression of NANOG was directly activated by the TGFβ/activin-mediated SMAD signaling (Xu et al.,

2008), and its overexpression enabled feeder-free growth (Darr et al., 2006). In both mouse and human ES cells, reduced expression of Nanog causes differentiation into extraembryonic lineages (Chambers et al., 2003; Mitsui et al., 2003; Hyslop et al., 2005). Interestingly, although they are prone to differentiating, mouse ES cells can self-renew indefinitely and contribute to multilineages in chimaeras in the absence of Nanog (Chambers et al., 2007). The function of Nanog in ES cells, thus, is more likely involved in the stabilization of the pluripotent state, while dispensable for its establishment.

The expression of genes enriched in ES cells has been extensively studied by several groups (see, e.g., Rao and Stice, 2004, and references therein), and includes, for example, transcription factors Sox2 and *foxd3*, RNA-binding protein Esg-1 (Dppa5), and *de novo* DNA methyl-transferase 3b. Deletion of some of them in mice does demonstrate a critical function in early development (Table 10.1). ES cells also express high levels of genes involved in protein synthesis and mRNA processing (Richards et al., 2004), and non-coding RNAs unique to ES cells (Suh et al., 2004). A surprisingly high percentage of genes enriched in ES cells have unknown functions (Tanaka et al., 2002; Robson, 2004, and references therein).

206

TABLE 10.1

Genes	Protein Features and Functions	References
Sox2	HMG-box transcription factor; interacts with Oct4 to regulate transcription; Sox2-/- mouse embryos died shortly after implantation with loss of epiblast at ~ E6.0	Avilion et al., 2003
FOXD3	Forkhead family transcription factor; FoxD3-/- mouse embryos died shortly after implantation with loss of epiblast (~E6.5); no FoxD3-/- ES cells can be established	Hanna et al., 2002
Rex-1(Zfp-42)	Zinc-finger transcription factor; direct target of Oct4; Rex-1-/- EC cells failed to differentiate into primitive and visceral endoderm	Rosfjord and Rizzino, 1994; Thompson and Gudas, 2002
Gbx2(Stra7)	Homeobox-containing transcription factor; Gbx-/- embryos displayed defects in neural crest cell patterning and pharyngeal arch artery	Byrd and Meyers, 2005
Sall1	Potent zinc-finger transcription repressor; heterozygous mutations in humans cause Townes-Brocks syndrome; Sall1-/- mice died perinatally	Kohlhase et al., 1998; Nishinakamura et al., 2001; Kiefer et al., 2002
Sall2	Homolog of Sall1; Sall-/- mice showed no phenotype	Sato et al., 2003
Hoxa11	Transcription factor; Hoxa11-/- mice showed defects in male and female fertility	Hsieh-Li et al., 1995
UTF1	Transcriptional coactivator; stimulate ES cell proliferation	Nishimoto et al., 2005
TERT	Reverse transcriptase (catalytic component of telomerase)	Liu et al., 2000
TERF1	Telomere repeat-binding factor 1; TERF1-/- mouse embryos died at E5-6 with severe growth defect in ICM	Karlseder et al., 2003
TERF2	Telomere repeat-binding factor 2	Sakaguchi et al., 1998
DNMT3b	*De novo* DNA methyltransferase; required for methylation of centrimeric minor satellite repeats; DNMT3ß-/- embryos died before birth	Okano et al., 1999
DNMT3a	*De novo* DNA methyltransferase; DNMT3a-/- mice died at the age of 4 weeks	Okano et al., 1999
Dppa2	Putative DNA binding motif SAP	Bortvin et al., 2003
Dppa3 (PGC7, Stella)	Putative DNA binding motif SAP	Saitou et al., 2002; Sato et al., 2002; Bortvin et al., 2003; Bowles et al., 2003

Continued

TABLE 10.1 continued

Genes	Protein Features and Functions	References
Dppa4 (FLJ10713)	Putative DNA binding motif SAP	Bortvin et al., 2003; Sperger et al., 2003
Dppa5 (Ph34, Esg-1)	Similar to KH RNA-binding motif	Astigiano et al., 1991; Tanaka et al., 2002
ECAT11(FLJ10884)	Conserved transposase 22 domain	Sperger et al., 2003

A recent genome-wide location analysis of human ES cells showed that Oct4 and Nanog, along with Sox2, co-occupy the promoters of a high number of genes, many of which are transcription factors such as Oct4, Nanog, and Sox2 (Boyer et al., 2005). These three proteins, in addition to regulating their own transcription as previously shown (Catena et al., 2004; Kuroda et al., 2005; Okumura-Nakanishi et al., 2005; Rodda et al., 2005), could also activate or repress the expression of many other genes. These genome-wide approaches hold great promise in elucidating the networks that control the pluripotent state.

CONCLUSION

Progress in developmental biology has been dramatic over the last few decades, and one of the legacies of the derivation of human ES cells is that they provide a compelling link between that progress and the understanding and treatment of human disease. The derivation of mouse ES cells in 1981 and subsequent development of homologous recombination revolutionized mammalian developmental biology, as it allowed the very specific modification of the mouse genome to test gene function. Yet, although the use of mouse ES cells as an *in vitro* model of differentiation was established soon after their initial derivation, it was only after the derivation of human ES cells in 1998, and their potential use in transplantation medicine was immediately appreciated, that there was an explosion of interest in the *in vitro*, lineage-specific differentiation of ES cells. Significant progress has been made in lineage-specific differentiation of human ES cells, and progress in this area is accelerating as new groups are now rapidly entering this field. An understanding of the basic mechanism controlling germ layer and lineage specification is rapidly unfolding through the interplay of knockout mice, *in vitro* differentiation of ES cells, and conserved mechanisms identified in other model organisms.

The basic biology of pluripotency is another area of research that the isolation of human ES cells rekindled. Even though significant differences exist between mouse and human ES cells, they share many key genes involved in pluripotency, such as Oct4 and Nanog. Global gene expression analysis of mouse and human ES cells has revealed the existence of many novel genes unique to ES cells, but the challenge remains in identifying functions of those genes and coming to understand how the proliferative, pluripotent state is established and maintained. Indeed, although certain genes have been identified that are required to maintain the pluripotent state, it remains a central problem in biology to understand why one cell can form anything in the body and another cannot. Such a basic understanding has implications for regenerative medicine that go far beyond the use of ES cells in transplantation, and may lead to methods of causing tissues to regenerate that fail to do so naturally. The derivation of ES cell-like induced pluripotent stem cells from differentiated somatic cells with a small set of transgenes is a first groundbreaking step in this direction (Yu et al., 2007; Takahashi et al., 2006, 2007).

References

Amit, M., Carpenter, M. K., Inokuma, M. S., Chiu, C. P., Harris, C. P., Waknitz, M. A., et al. (2000). Clonally derived human embryonic stem cell lines maintain pluripotency and proliferative potential for prolonged periods of culture. *Dev. Biol., 227,* 271–278.

Amit, M., Margulets, V., Segev, H., Shariki, K., Laevsky, I., Coleman, R., et al. (2003). Human feeder layers for human embryonic stem cells. *Biol. Reprod., 68,* 2150–2156.

Amit, M., Shariki, C., Margulets, V., & Itskovitz-Eldor, J. (2004). Feeder layer- and serum-free culture of human embryonic stem cells. *Biol. Reprod., 70*, 837–845.

Assady, S., Maor, G., Amit, M., Itskovitz-Eldor, J., Skorecki, K. L., & Tzukerman, M. (2001). Insulin production by human embryonic stem cells. *Diabetes, 50*, 1691–1697.

Astigiano, S., Barkai, U., Abarzua, P., Tan, S. C., Harper, M. I., & Sherman, M. I. (1991). Changes in gene expression following exposure of nulli-SCCl murine embryonal carcinoma cells to inducers of differentiation: characterization of a down-regulated mRNA. *Differentiation, 46*, 61–67.

Atkin, N. B., Baker, M. C., Robinson, R., & Gaze, S. E. (1974). Chromosome studies on 14 near-diploid carcinomas of the ovary. *Eur. J. Cancer, 10*, 144–146.

Bain, G., Kitchens, D., Yao, M., Huettner, J. E., & Gottlieb, D. I. (1995). Embryonic stem cells express neuronal properties in vitro. *Dev. Biol., 168*, 342–357.

Beattie, G. M., Lopez, A. D., Bucay, N., Hinton, A., Firpo, M. T., King, C. C., et al. (2005). Activin A maintains pluripotency of human embryonic stem cells in the absence of feeder layers. *Stem Cells, 23*, 489–495.

Bongso, A., Fong, C. Y., Ng, S. C., & Ratnam, S. (1994). Isolation and culture of inner cell mass cells from human blastocysts. *Hum. Reprod., 9*, 2110–2117.

Bowles, J., Teasdale, R. P., James, K., & Koopman, P. (2003). Dppa3 is a marker of pluripotency and has a human homologue that is expressed in germ cell tumours. *Cytogenet. Genome Res., 101*, 261–265.

Boyer, L. A., Lee, T. I., Cole, M. F., Johnstone, S. E., Levine, S. S., Zucker, J. P., et al. (2005). Core transcriptional regulatory circuitry in human embryonic stem cells. *Cell, 122*, 947–956.

Bradley, A., Evans, M., Kaufman, M. H., & Robertson, E. (1984). Formation of germ-line chimaeras from embryo-derived teratocarcinoma cell lines. *Nature, 309*, 255–256.

Brinster, R. L. (1974). The effect of cells transferred into the mouse blastocyst on subsequent development. *J. Exp. Med., 140*, 1049–1056.

Bronson, D. L., Andrews, P. W., Solter, D., Cervenka, J., Lange, P. H., & Fraley, E. E. (1980). Cell line derived from a metastasis of a human testicular germ cell tumor. *Cancer Res., 40*, 2500–2506.

Brook, F. A., & Gardner, R. L. (1997). The origin and efficient derivation of embryonic stem cells in the mouse. *Proc. Natl. Acad. Sci. U.S.A., 94*, 5709–5712.

Buehr, M., Meek, S., Blair, K., Yang, J., Ure, J., Silva, J., et al. (2008). Capture of authentic embryonic stem cells from rat blastocysts. *Cell, 135*, 1287–1298.

Buehr, M., Nichols, J., Stenhouse, F., Mountford, P., Greenhalgh, C. J., Kantachuvesiri, S., et al. (2003). Rapid loss of Oct-4 and pluripotency in cultured rodent blastocysts and derivative cell lines. *Biol. Reprod., 68*, 222–229.

Burdon, T., Stracey, C., Chambers, I., Nichols, J., & Smith, A. (1999). Suppression of SHP-2 and ERK signalling promotes self-renewal of mouse embryonic stem cells. *Dev. Biol., 210*, 30–43.

Buzzard, J. J., Gough, N. M., Crook, J. M., & Colman, A. (2004). Karyotype of human ES cells during extended culture. *Nat. Biotechnol., 22*, 381–382, author reply 382.

Byrd, N. A., & Meyers, E. N. (2005). Loss of Gbx2 results in neural crest cell patterning and pharyngeal arch artery defects in the mouse embryo. *Dev. Biol., 284*, 233–245.

Calarco, P. G., & Banka, C. L. (1979). Cell surface antigens of preimplantation mouse embryos detected by an antiserum to an embryonal carcinoma cell line. *Biol. Reprod., 20*, 699–704.

Catena, R., Tiveron, C., Ronchi, A., Porta, S., Ferri, A., Tatangelo, L., et al. (2004). Conserved POU binding DNA sites in the Sox2 upstream enhancer regulate gene expression in embryonic and neural stem cells. *J. Biol. Chem., 279*, 41846–41857.

Chadwick, K., Wang, L., Li, L., Menendez, P., Murdoch, B., Rouleau, A., et al. (2003). Cytokines and BMP-4 promote hematopoietic differentiation of human embryonic stem cells. *Blood, 102*, 906–915.

Chambers, I., Colby, D., Robertson, M., Nichols, J., Lee, S., Tweedie, S., et al. (2003). Functional expression cloning of Nanog, a pluripotency sustaining factor in embryonic stem cells. *Cell, 113*, 643–655.

Chambers, I., Silva, J., Colby, D., Nichols, J., Nijmeijer, B., Robertson, M., et al. (2007). Nanog safeguards pluripotency and mediates germline development. *Nature, 450*, 1230–1234.

Chung, Y., Klimanskaya, I., Becker, S., Marh, J., Lu, S. J., & Johnson, J. (2006). Embryonic and extraembryonic stem cell lines derived from single mouse blastomeres. *Nature* JN 12, *439*(7073), 216–219.

Chung, Y., Klimanskaya, I., Becker, S., Marh, J., Lu, S. J., Johnson, J., et al. (2006). Embryonic and extraembryonic stem cell lines derived from single mouse blastomeres. *Nature, 439*, 216–219.

Comoglio, P. M., Bertini, M., & Forni, G. (1975). Evidence for a membrane carrier molecule common to embryonal and tumour-specific antigenic determinants expressed by a mouse transplantable tumour. *Immunology, 29*, 353–364.

Cowan, C. A., Klimanskaya, I., McMahon, J., Atienza, J., Witmyer, J., Zucker, J. P., et al. (2004). Derivation of embryonic stem-cell lines from human blastocysts. *N. Engl. J. Med., 350*, 1353–1356.

Daheron, L., Opitz, S. L., Zaehres, H., Lensch, W. M., Andrews, P. W., Itskovitz-Eldor, J., & Daley, G. Q. (2004). LIF/STAT3 signaling fails to maintain self-renewal of human embryonic stem cells. *Stem Cells, 22*, 770–778.

Damjanov, I., Bagasra, O., & Solter, D. (1983). Genetic and epigenetic factors regulate the evolving malignancy of embryo-derived teratomas. *Cold Spring Harbor Conference, Cell Proliferation, 10*, 501–517.

Darr, H., Mayshar, Y., & Benvenisty, N. (2006). Overexpression of NANOG in human ES cells enables feeder-free growth while inducing primitive ectoderm features. *Development, 133*, 1193–1201.

Delhaise, F., Bralion, V., Schuurbiers, N., & Dessy, F. (1996). Establishment of an embryonic stem cell line from 8-cell stage mouse embryos. *Eur. J. Morphol., 34*, 237–243.

Draper, J. S., Smith, K., Gokhale, P., Moore, H. D., Maltby, E., Johnson, J., et al. (2004). Recurrent gain of chromosomes 17q and 12 in cultured human embryonic stem cells. *Nat. Biotechnol., 22*, 53–54.

Dravid, G., Ye, Z., Hammond, H., Chen, G., Pyle, A., Donovan, P., et al. (2005). Defining the role of Wnt/β-catenin signaling in the survival, proliferation and self-renewal of human embryonic stem cells. *Stem Cells, 23*, 1489–1501.

Dvorak, P., Dvorakova, D., Koskova, S., Vodinska, M., Najvirtova, M., Krekac, D., et al. (2005). Expression and potential role of fibroblast growth factor 2 and its receptors in human embryonic stem cells. *Stem Cells, 23*, 1200–1211.

Eistetter, H. (1989). Pluripotent embryonal stem cell lines can be established from disaggregated mouse morulae. *Dev. Growth Differ., 31*, 275–282.

Enver, T., Soneji, S., Joshi, C., Brown, J., Iborra, F., Orntoft, T., et al. (2005). Cellular differentiation hierarchies in normal and culture adapted human embryonic stem cells. *Hum. Mol. Genet., 14*(21), 3129–3140.

Evans, M. J. (1972). The isolation and properties of a clonal tissue culture strain of pluripotent mouse teratoma cells. *J. Embryol. Exp. Morphol., 28*, 163–176.

Evans, M. J., & Kaufman, M. H. (1981). Establishment in culture of pluripotential cells from mouse embryos. *Nature, 292*, 154–156.

Gachelin, G., Kemler, R., Kelly, F., & Jacob, F. (1977). PCC4, a new cell surface antigen common to multipotential embryonal carcinoma cells, spermatozoa, and mouse early embryos. *Dev. Biol., 57*, 199–209.

Gardner, D. K., Vella, P., Lane, M., Wagley, L., Schlenker, T., & Schoolcraft, W. B. (1998). Culture and transfer of human blastocysts increases implantation rates and reduces the need for multiple embryo transfers. *Fertil. Steril., 69*, 84–88.

Geijsen, N., Horoschak, M., Kim, K., Gribnau, J., Eggan, K., & Daley, G. Q. (2004). Derivation of embryonic germ cells and male gametes from embryonic stem cells. *Nature, 427*, 148–154.

Gerrard, L., Rodgers, L., & Cui, W. (2005). Differentiation of human embryonic stem cells to neural lineages in adherent culture by blocking BMP signaling. *Stem Cells, 23*(9), 1234–1241.

Hamazaki, T., Iiboshi, Y., Oka, M., Papst, P. J., Meacham, A. M., Zon, L. I., et al. (2001). Hepatic maturation in differentiating embryonic stem cells in vitro. *FEBS Lett., 497*, 15–19.

Hanna, L. A., Foreman, R. K., Tarasenko, I. A., Kessler, D. S., & Labosky, P. A. (2002). Requirement for Foxd3 in maintaining pluripotent cells of the early mouse embryo. *Genes Dev., 16*, 2650–2661.

Hansis, C., Grifo, J. A., & Krey, L. C. (2000). Oct-4 expression in inner cell mass and trophectoderm of human blastocysts. *Mol. Hum. Reprod., 6*, 999–1004.

He, J. Q., Ma, Y., Lee, Y., Thomson, J. A., & Kamp, T. J. (2003). Human embryonic stem cells develop into multiple types of cardiac myocytes: action potential characterization. *Circ. Res., 93*, 32–39.

Henderson, J. K., Draper, J. S., Baillie, H. S., Fishel, S., Thomson, J. A., Moore, H., et al. (2002). Preimplantation human embryos and embryonic stem cells show comparable expression of stage-specific embryonic antigens. *Stem Cells, 20*, 329–337.

Howe, C. C., Gmur, R., & Solter, D. (1980). Cytoplasmic and nuclear protein synthesis during in vitro differentiation of murine ICM and embryonal carcinoma cells. *Dev. Biol., 74*, 351–363.

Hsieh-Li, H. M., Witte, D. P., Weinstein, M., Branford, W., Li, H., Small, K., et al. (1995). Hoxa 11 structure, extensive antisense transcription, and function in male and female fertility. *Development, 121*, 1373–1385.

Hubner, K., Fuhrmann, G., Christenson, L. K., Kehler, J., Reinbold, R., de la Fuente, R., et al. (2003). Derivation of oocytes from mouse embryonic stem cells. *Science, 300*, 1251–1256.

Humphrey, R. K., Beattie, G. M., Lopez, A. D., Bucay, N., King, C. C., Firpo, M. T., et al. (2004). Maintenance of pluripotency in human embryonic stem cells is STAT3 independent. *Stem Cells, 22*, 522–530.

Hyslop, L., Stojkovic, M., Armstrong, L., Walter, T., Stojkovic, P., Przyborski, S., et al. (2005). Downregulation of NANOG induces differentiation of human embryonic stem cells to extraembryonic lineages. *Stem Cells, 23*, 1035–1043.

Illmensee, K., & Mintz, B. (1976). Totipotency and normal differentiation of single teratocarcinoma cells cloned by injection into blastocysts. *Proc. Natl. Acad. Sci. U.S.A., 73*, 549–553.

James, D., Levine, A. J., Besser, D., & Hemmati-Brivanlou, A. (2005). TGFbeta/activin/nodal signaling is necessary for the maintenance of pluripotency in human embryonic stem cells. *Development, 132*, 1273–1282.

Kahan, B. W., & Ephrussi, B. (1970). Developmental potentialities of clonal in vitro cultures of mouse testicular teratoma. *J. Natl. Cancer Inst., 44*, 1015–1036.

Kang, H. B., Kim, J. S., Kwon, H. J., Nam, K. H., Youn, H. S., Sok, D. E., et al. (2005). Basic fibroblast growth factor activates ERK and induces c-fos in human embryonic stem cell line MizhES1. *Stem Cells Dev., 14*, 395–401.

Karlseder, J., Kachatrian, L., Takai, H., Mercer, K., Hingorani, S., Jacks, T., et al. (2003). Targeted deletion reveals an essential function for the telomere length regulator Trf1. *Mol. Cell Biol., 23*, 6533–6541.

Kaufman, D. S., Hanson, E. T., Lewis, R. L., Auerbach, R., & Thomson, J. A. (2001). Hematopoietic colony-forming cells derived from human embryonic stem cells. *Proc. Natl. Acad. Sci. U.S.A., 98*, 10716–10721.

Keller, G., Kennedy, M., Papayannopoulou, T., & Wiles, M. V. (1993). Hematopoietic commitment during embryonic stem cell differentiation in culture. *Mol. Cell Biol., 13*, 473–486.

Kiefer, S. M., McDill, B. W., Yang, J., & Rauchman, M. (2002). Murine Sall1 represses transcription by recruiting a histone deacetylase complex. *J. Biol. Chem., 277*, 14869–14876.

Kitani, H., Takagi, N., Atsumi, T., Kawakura, K., Imamura, K., Goto, S., et al. (1996). Isolation of a germline-transmissible embryonic stem (ES) cell line from C3H/He mice. *Zoolog. Sci., 13*, 865–871.

Klavins, J. V., Mesa-Tejada, R., & Weiss, M. (1971). Human carcinoma antigens cross reacting with anti-embryonic antibodies. *Nat. New Biol., 234*, 153–154.

Kleinsmith, L. J., & Pierce, G. B., Jr. (1964). Multipotentiality of Single Embryonal Carcinoma Cells. *Cancer Res., 24*, 1544–1551.

Klimanskaya, I., Chung, Y., Meisner, L., Johnson, J., West, M. D., & Lanza, R. (2005). Human embryonic stem cells derived without feeder cells. *Lancet, 365*, 1636–1641.

Klug, M. G., Soonpaa, M. H., Koh, G. Y., & Field, L. J. (1996). Genetically selected cardiomyocytes from differentiating embronic stem cells form stable intracardiac grafts. *J. Clin. Invest., 98*, 216–224.

Kohlhase, J., Wischermann, A., Reichenbach, H., Froster, U., & Engel, W. (1998). Mutations in the SALL1 putative transcription factor gene cause Townes-Brocks syndrome. *Nat. Genet., 18*, 81–83.

Kuroda, T., Tada, M., Kubota, H., Kimura, H., Hatano, S. Y., Suemori, H., et al. (2005). Octamer and Sox elements are required for transcriptional cis regulation of Nanog gene expression. *Mol. Cell Biol., 25*, 2475–2485.

Ledermann, B., & Burki, K. (1991). Establishment of a germ-line competent C57BL/6 embryonic stem cell line. *Exp. Cell Res., 197*, 254–258.

Li, P., Tong, C., Mehrian-Shai, R., Jia, L., Wu, N., Yan, Y., et al. (2008). Germline competent embryonic stem cells derived from rat blastocysts. *Cell, 135*, 1299–1310.

Li, Y., Powell, S., Brunette, E., Lebkowski, J., & Mandalam, R. (2005). Expansion of human embryonic stem cells in defined serum-free medium devoid of animal-derived products. *Biotechnol. Bioeng., 91*, 688–698.

Liu, Y., Snow, B. E., Hande, M. P., Yeung, D., Erdmann, N. J., Wakeham, A., et al. (2000). The telomerase reverse transcriptase is limiting and necessary for telomerase function in vivo. *Curr. Biol., 10*, 1459–1462.

Ludwig, T. E., Levenstein, M. E., Jones, J. M., Berggren, W. T., Mitchen, E. R., Frane, J. L., et al. (2006). Derivation of human embryonic stem cells in defined conditions. *Nat. Biotechnol., 24*, 185–187.

Lumelsky, N., Blondel, O., Laeng, P., Velasco, I., Ravin, R., & McKay, R. (2001). Differentiation of embryonic stem cells to insulin-secreting structures similar to pancreatic islets. *Science, 292*, 1389–1394.

Lutticken, C., Wegenka, U. M., Yuan, J., Buschmann, J., Schindler, C., Ziemiecki, A., et al. (1994). Association of transcription factor APRF and protein kinase Jak1 with the interleukin-6 signal transducer gp130. *Science, 263*, 89–92.

Maitra, A., Arking, D. E., Shivapurkar, N., Ikeda, M., Stastny, V., Kassauei, K., et al. (2005). Genomic alterations in cultured human embryonic stem cells. *Nat. Genet., 37*(10), 1099–1103.

Martin, G. R. (1980). Teratocarcinomas and mammalian embryogenesis. *Science, 209*, 768–776.

Martin, G. R. (1981). Isolation of a pluripotent cell line from early mouse embryos cultured in medium conditioned by teratocarcinoma stem cells. *Proc. Natl. Acad. Sci. U.S.A., 78*, 7634–7638.

Martin, G. R., & Evans, M. J. (1975). Differentiation of clonal lines of teratocarcinoma cells: formation of embryoid bodies in vitro. *Proc. Natl. Acad. Sci. U.S.A., 72*, 1441–1445.

Martin, G. R., Wiley, L. M., & Damjanov, I. (1977). The development of cystic embryoid bodies in vitro from clonal teratocarcinoma stem cells. *Dev. Biol., 61*, 230–244.

Martin, M. J., Muotri, A., Gage, F., & Varki, A. (2005). Human embryonic stem cells express an immunogenic nonhuman sialic acid. *Nat. Med., 11*, 228–232.

Matin, M. M., Walsh, J. R., Gokhale, P. J., Draper, J. S., Bahrami, A. R., Morton, I., et al. (2004). Specific knockdown of Oct4 and beta2-microglobulin expression by RNA interference in human embryonic stem cells and embryonic carcinoma cells. *Stem Cells, 22*, 659–668.

Matsuda, T., Nakamura, T., Nakao, K., Arai, T., Katsuki, M., Heike, T., et al. (1999). STAT3 activation is sufficient to maintain an undifferentiated state of mouse embryonic stem cells. *Embo J., 18*, 4261–4429.

McBurney, M. W. (1976). Clonal lines of teratocarcinoma cells in vitro: differentiation and cytogenetic characteristics. *J. Cell Physiol.*, *89*, 441−455.

McWhir, J., Schnieke, A. E., Ansell, R., Wallace, H., Colman, A., Scott, A. R., et al. (1996). Selective ablation of differentiated cells permits isolation of embryonic stem cell lines from murine embryos with a non-permissive genetic background. *Nat. Genet.*, *14*, 223−226.

Mintz, B., & Illmensee, K. (1975). Normal genetically mosaic mice produced from malignant teratocarcinoma cells. *Proc. Natl. Acad. Sci. U.S.A.*, *72*, 3585−3589.

Mitsui, K., Tokuzawa, Y., Itoh, H., Segawa, K., Murakami, M., Takahashi, K., et al. (2003). The homeoprotein Nanog is required for maintenance of pluripotency in mouse epiblast and ES cells. *Cell*, *113*, 631−642.

Nagy, A., Rossant, J., Nagy, R., Abramow-Newerly, W., & Roder, J. C. (1993). Derivation of completely cell culture-derived mice from early-passage embryonic stem cells. *Proc. Natl. Acad. Sci. U.S.A.*, *90*, 8424−8428.

Nichols, J., Zevnik, B., Anastassiadis, K., Niwa, H., Klewe-Nebenius, D., Chambers, I., et al. (1998). Formation of pluripotent stem cells in the mammalian embryo depends on the POU transcription factor Oct4. *Cell*, *95*, 379−391.

Nichols, J., Jones, K., Phillips, J. M., Newland, S. A., Roode, M., Mansfield, W., et al. (2009). Validated germline-competent embryonic stem cell lines from nonobese diabetic mice. *Nat. Med.*, *15*, 814−818.

Nishimoto, M., Miyagi, S., Yamagishi, T., Sakaguchi, T., Niwa, H., Muramatsu, M., et al. (2005). Oct-3/4 maintains the proliferative embryonic stem cell state via specific binding to a variant octamer sequence in the regulatory region of the UTF1 locus. *Mol. Cell Biol.*, *25*, 5084−5094.

Nishinakamura, R., Matsumoto, Y., Nakao, K., Nakamura, K., Sato, A., Copeland, N. G., et al. (2001). Murine homolog of SALL1 is essential for ureteric bud invasion in kidney development. *Development*, *128*, 3105−3115.

Niwa, H., Miyazaki, J., & Smith, A. G. (2000). Quantitative expression of Oct-3/4 defines differentiation, dedifferentiation or self-renewal of ES cells. *Nat. Genet.*, *24*, 372−376.

Okamoto, K., Okazawa, H., Okuda, A., Sakai, M., Muramatsu, M., & Hamada, H. (1990). A novel octamer binding transcription factor is differentially expressed in mouse embryonic cells. *Cell*, *60*, 461−472.

Okano, M., Bell, D. W., Haber, D. A., & Li, E. (1999). DNA methyltransferases Dnmt3a and Dnmt3b are essential for de novo methylation and mammalian development. *Cell*, *99*, 247−257.

Okumura-Nakanishi, S., Saito, M., Niwa, H., & Ishikawa, F. (2005). Oct-3/4 and Sox2 regulate Oct-3/4 gene in embryonic stem cells. *J. Biol. Chem.*, *280*, 5307−5317.

Papaioannou, V. E., Gardner, R. L., McBurney, M. W., Babinet, C., & Evans, M. J. (1978). Participation of cultured teratocarcinoma cells in mouse embryogenesis. *J. Embryol. Exp. Morphol.*, *44*, 93−104.

Papaioannou, V. E., McBurney, M. W., Gardner, R. L., & Evans, M. J. (1975). Fate of teratocarcinoma cells injected into early mouse embryos. *Nature*, *258*, 70−73.

Pebay, A., Wong, R. C., Pitson, S. M., Wolvetang, E., Peh, G. S., Filipczyk, A., et al. (2005). Essential roles of sphingosine-1-phosphate and platelet-derived growth factor in the maintenance of human embryonic stem cells. *Stem Cells*, *23*(10), 1541−1548.

Pera, M. F., Andrade, J., Houssami, S., Reubinoff, B., Trounson, A., Stanley, E. G., et al. (2004). Regulation of human embryonic stem cell differentiation by BMP-2 and its antagonist noggin. *J. Cell Sci.*, *117*, 1269−1280.

Perrier, A. L., Tabar, V., Barberi, T., Rubio, M. E., Bruses, J., Topf, N., et al. (2004). Derivation of midbrain dopamine neurons from human embryonic stem cells. *Proc. Natl. Acad. Sci. U.S.A.*, *101*, 12543−12548.

Pesce, M., Gross, M. K., & Scholer, H. R. (1998). In line with our ancestors: Oct-4 and the mammalian germ. *Bioessays*, *20*, 722−732.

Puck, T. T., Marcus, P. I., & Cieciura, S. J. (1956). Clonal growth of mammalian cells in vitro; growth characteristics of colonies from single HeLa cells with and without a feeder layer. *J. Exp. Med.*, *103*, 273−283.

Rambhatla, L., Chiu, C. P., Kundu, P., Peng, Y., & Carpenter, M. K. (2003). Generation of hepatocyte-like cells from human embryonic stem cells. *Cell Transplant.*, *12*, 1−11.

Rao, R. R., & Stice, S. L. (2004). Gene expression profiling of embryonic stem cells leads to greater understanding of pluripotency and early developmental events. *Biol. Reprod.*, *71*, 1772−1778.

Reubinoff, B. E., Pera, M. F., Fong, C. Y., Trounson, A., & Bongso, A. (2000). Embryonic stem cell lines from human blastocysts: somatic differentiation in vitro. *Nat. Biotechnol.*, *18*, 399−404.

Richards, M., Tan, S., Fong, C. Y., Biswas, A., Chan, W. K., & Bongso, A. (2003). Comparative evaluation of various human feeders for prolonged undifferentiated growth of human embryonic stem cells. *Stem Cells*, *21*, 546−556.

Richards, M., Tan, S. P., Tan, J. H., Chan, W. K., & Bongso, A. (2004). The transcriptome profile of human embryonic stem cells as defined by SAGE. *Stem Cells*, *22*, 51−64.

Robson, P. (2004). The maturing of the human embryonic stem cell transcriptome profile. *Trends Biotechnol.*, *22*, 609−612.

Rodda, D. J., Chew, J. L., Lim, L. H., Loh, Y. H., Wang, B., Ng, H. H., et al. (2005). Transcriptional regulation of nanog by OCT4 and SOX2. *J. Biol. Chem.*, *280*, 24731−24737.

211

Rosfjord, E., & Rizzino, A. (1994). The octamer motif present in the Rex-1 promoter binds Oct-1 and Oct-3 expressed by EC cells and ES cells. *Biochem. Biophys. Res. Commun., 203*, 1795–1802.

Rossant, J., & Papaioannou, V. E. (1984). The relationship between embryonic, embryonal carcinoma and embryo-derived stem cells. *Cell Differ., 15*, 155–161.

Rugg-Gunn, P. J., Ferguson-Smith, A. C., & Pedersen, R. A. (2005). Epigenetic status of human embryonic stem cells. *Nat. Genet., 37*, 585–587.

Saitou, M., Barton, S. C., & Surani, M. A. (2002). A molecular programme for the specification of germ cell fate in mice. *Nature, 418*, 293–300.

Sakaguchi, A. Y., Padalecki, S. S., Mattern, V., Rodriguez, A., Leach, R. J., McGill, J. R., et al. (1998). Chromosomal sublocalization of the transcribed human telomere repeat binding factor 2 gene and comparative mapping in the mouse. *Somat. Cell. Mol. Genet., 24*, 157–163.

Sato, A., Matsumoto, Y., Koide, U., Kataoka, Y., Yoshida, N., Yokota, T., et al. (2003). Zinc finger protein sall2 is not essential for embryonic and kidney development. *Mol. Cell Biol., 23*, 62–69.

Sato, N., Meijer, L., Skaltsounis, L., Greengard, P., & Brivanlou, A. H. (2004). Maintenance of pluripotency in human and mouse embryonic stem cells through activation of Wnt signaling by a pharmacological GSK-3-specific inhibitor. *Nat. Med., 10*, 55–63.

Scholer, H. R., Balling, R., Hatzopoulos, A. K., Suzuki, N., & Gruss, P. (1989). Octamer binding proteins confer transcriptional activity in early mouse embryogenesis. *Embo J., 8*, 2551–2557.

Silva, J., Chambers, I., Pollard, S., & Smith, A. (2006). Nanog promotes transfer of pluripotency after cell fusion. *Nature, 441*, 997–1001.

Smith, A. G., Heath, J. K., Donaldson, D. D., Wong, G. G., Moreau, J., Stahl, M., et al. (1988). Inhibition of pluripotential embryonic stem cell differentiation by purified polypeptides. *Nature, 336*, 688–690.

Smith, T. A., & Hooper, M. L. (1983). Medium conditioned by feeder cells inhibits the differentiation of embryonal carcinoma cultures. *Exp. Cell Res., 145*, 458–462.

Solter, D., & Knowles, B. B. (1978). Monoclonal antibody defining a stage-specific mouse embryonic antigen (SSEA-1). *Proc. Natl. Acad. Sci. U.S.A., 75*, 5565–5569.

Solter, D., Skreb, N., & Damjanov, I. (1970). Extrauterine growth of mouse egg-cylinders results in malignant teratoma. *Nature, 227*, 503–504.

Sperger, J. M., Chen, X., Draper, J. S., Antosiewicz, J. E., Chon, C. H., Jones, S. B., et al. (2003). Gene expression patterns in human embryonic stem cells and human pluripotent germ cell tumors. *Proc. Natl. Acad. Sci. U.S.A., 100*, 13350–13355.

Steptoe, P. C., & Edwards, R. G. (1978). Birth after the reimplantation of a human embryo. *Lancet, 2*, 366.

Stevens, L. C. (1970). The development of transplantable teratocarcinomas from intratesticular grafts of pre- and postimplantation mouse embryos. *Dev. Biol., 21*, 364–382.

Stevens, L. C., & Little, C. C. (1954). Spontaneous testicular teratomas in an inbred strain of mice. *Proc. Natl. Acad. Sci. U.S.A., 40*, 1080–1087.

Stojkovic, M., Lako, M., Stojkovic, P., Stewart, R., Przyborski, S., Armstrong, L., et al. (2004a). Derivation of human embryonic stem cells from day-8 blastocysts recovered after three-step in vitro culture. *Stem Cells, 22*, 790–797.

Stojkovic, M., Lako, M., Strachan, T., & Murdoch, A. (2004b). Derivation, growth and applications of human embryonic stem cells. *Reproduction, 128*, 259–267.

Strelchenko, N., Verlinsky, O., Kukharenko, V., & Verlinsky, Y. (2004). Morula-derived human embryonic stem cells. *Reprod. Biomed. Online, 9*, 623–629.

Suh, M. R., Lee, Y., Kim, J. Y., Kim, S. K., Moon, S. H., Lee, J. Y., et al. (2004). Human embryonic stem cells express a unique set of microRNAs. *Dev. Biol., 270*, 488–498.

Sumi, T., Fujimoto, Y., Nakatsuji, N., & Suemori, H. (2004). STAT3 is dispensable for maintenance of self-renewal in nonhuman primate embryonic stem cells. *Stem Cells, 22*, 861–872.

Takahashi-Tezuka, M., Yoshida, Y., Fukada, T., Ohtani, T., Yamanaka, Y., Nishida, K., et al. (1998). Gab1 acts as an adapter molecule linking the cytokine receptor gp130 to ERK mitogen-activated protein kinase. *Mol. Cell Biol., 18*, 4109–4117.

Takahashi, K., & Yamanaka, S. (2006). Induction of pluripotent stem cells from mouse embryonic and adult fibroblast cultures by defined factors. *Cell, 126*, 663–676.

Takahashi, K., Tanabe, K., Ohnuki, M., Narita, M., Ichisaka, T., Tomoda, K., et al. (2007). Induction of pluripotent stem cells from adult human fibroblasts by defined factors. *Cell, 131*, 861–872.

Tanaka, T. S., Kunath, T., Kimber, W. L., Jaradat, S. A., Stagg, C. A., Usuda, M., et al. (2002). Gene expression profiling of embryo-derived stem cells reveals candidate genes associated with pluripotency and lineage specificity. *Genome Res., 12*, 1921–1928.

Tesar, P. J. (2005). Derivation of germ-line-competent embryonic stem cell lines from preblastocyst mouse embryos. *Proc. Natl. Acad. Sci. U.S.A., 102*, 8239–8244.

Thomson, J. A., Itskovitz-Eldor, J., Shapiro, S. S., Waknitz, M. A., Swiergiel, J. J., Marshall, V. S., et al. (1998). Embryonic stem cell lines derived from human blastocysts. *Science, 282*, 1145–1147.

Thomson, J. A., Kalishman, J., Golos, T. G., Durning, M., Harris, C. P., Becker, R. A., et al. (1995). Isolation of a primate embryonic stem cell line. *Proc. Natl. Acad. Sci. U.S.A., 92*, 7844–7848.

Thomson, J. A., Kalishman, J., Golos, T. G., Durning, M., Harris, C. P., & Hearn, J. P. (1996). Pluripotent cell lines derived from common marmoset (Callithrix jacchus) blastocysts. *Biol. Reprod., 55*, 254–259.

Thompson, J. R., & Gudas, L. J. (2002). Retinoic acid induces parietal endoderm but not primitive endoderm and visceral endoderm differentiation in F9 teratocarcinoma stem cells with a targeted deletion of the Rex-1 (Zfp-42) gene. *Mol Cell Endocrinol, 195*, 119–133.

Toyooka, Y., Tsunekawa, N., Akasu, R., & Noce, T. (2003). Embryonic stem cells can form germ cells in vitro. *Proc. Natl. Acad. Sci. U.S.A., 100*, 11457–11462.

Ueda, O., Jishage, K., Kamada, N., Uchida, S., & Suzuki, H. (1995). Production of mice entirely derived from embryonic stem (ES) cell with many passages by coculture of ES cells with cytochalasin B induced tetraploid embryos. *Exp. Anim., 44*, 205–210.

Vallier, L., Alexander, M., & Pedersen, R. A. (2005). Activin/Nodal and FGF pathways cooperate to maintain pluripotency of human embryonic stem cells. *J. Cell Sci., 118*, 4495–4509.

Verlinsky, Y., Strelchenko, N., Kukharenko, V., Rechitsky, S., Verlinsky, O., Galat, V., et al. (2005). Human embryonic stem cell lines with genetic disorders. *Reprod. Biomed. Online, 10*, 105–110.

Vodyanik, M. A., Bork, J. A., Thomson, J. A., & Slukvin, I. I. (2005). Human embryonic stem cell-derived CD34+ cells: efficient production in the coculture with OP9 stromal cells and analysis of lymphohematopoietic potential. *Blood, 105*, 617–626.

Wang, L., Li, L., Menendez, P., Cerdan, C., & Bhatia, M. (2005). Human embryonic stem cells maintained in the absence of mouse embryonic fibroblasts or conditioned media are capable of hematopoietic development. *Blood, 105*, 4598–4603.

Wang, L., Schulz, T. C., Sherrer, E. S., Dauphin, D. S., Shin, S., Nelson, A. M., et al. (2007). Self-renewal of human embryonic stem cells requires insulin-like growth factor-1 receptor and ERBB2 receptor signaling. *Blood, 110*, 4111–4119.

Watanabe, K., Ueno, M., Kamiya, D., Nishiyama, A., Matsumura, M., Wataya, T., et al. (2007). A ROCK inhibitor permits survival of dissociated human embryonic stem cells. *Nat. Biotechnol., 25*, 681–686.

Wegenka, U. M., Lutticken, C., Buschmann, J., Yuan, J., Lottspeich, F., Muller-Esterl, W., et al. (1994). The interleukin-6-activated acute-phase response factor is antigenically and functionally related to members of the signal transducer and activator of transcription (STAT) family. *Mol. Cell Biol., 14*, 3186–3196.

Wheeler, J. E. (1983). *The Human Teratomas: Experimental and Clinical Biology.* Clifton, NJ: Humana Press. pp. 1–22.

Williams, R. L., Hilton, D. J., Pease, S., Willson, T. A., Stewart, C. L., Gearing, D. P., et al. (1988). Myeloid leukaemia inhibitory factor maintains the developmental potential of embryonic stem cells. *Nature, 336*, 684–687.

Winkel, G. K., & Pedersen, R. A. (1988). Fate of the inner cell mass in mouse embryos as studied by microinjection of lineage tracers. *Dev. Biol., 127*, 143–156.

Xu, C., Inokuma, M. S., Denham, J., Golds, K., Kundu, P., Gold, J. D., et al. (2001). Feeder-free growth of undifferentiated human embryonic stem cells. *Nat. Biotechnol., 19*, 971–974.

Xu, C., Rosler, E., Jiang, J., Lebkowski, J. S., Gold, J. D., O'Sullivan, C., et al. (2005a). Basic fibroblast growth factor supports undifferentiated human embryonic stem cell growth without conditioned medium. *Stem Cells, 23*, 315–323.

Xu, R. H., Chen, X., Li, D. S., Li, R., Addicks, G. C., Glennon, C., et al. (2002). BMP4 initiates human embryonic stem cell differentiation to trophoblast. *Nat. Biotechnol., 20*, 1261–1264.

Xu, R. H., Peck, R. M., Li, D. S., Feng, X., Ludwig, T., & Thomson, J. A. (2005b). Basic FGF and suppression of BMP signaling sustain undifferentiated proliferation of human ES cells. *Nat. Methods, 2*, 185–190.

Xu, R. H., Sampsell-Barron, T. L., Gu, F., Root, S., Peck, R. M., Pan, G., et al. (2008). NANOG is a direct target of TGFbeta/activin-mediated SMAD signaling in human ESCs. *Cell Stem Cell, 3*, 196–206.

Ying, Q. L., Nichols, J., Chambers, I., & Smith, A. (2003a). BMP induction of Id proteins suppresses differentiation and sustains embryonic stem cell self-renewal in collaboration with STAT3. *Cell, 115*, 281–292.

Ying, Q. L., Stavridis, M., Griffiths, D., Li, M., & Smith, A. (2003b). Conversion of embryonic stem cells into neuroectodermal precursors in adherent monoculture. *Nat. Biotechnol., 21*, 183–186.

Ying, Q. L., Wray, J., Nichols, J., Batlle-Morera, L., Doble, B., Woodgett, J., et al. (2008). The ground state of embryonic stem cell self-renewal. *Nature, 453*, 519–523.

Yoshida, K., Chambers, I., Nichols, J., Smith, A., Saito, M., Yasukawa, K., et al. (1994). Maintenance of the pluripotential phenotype of embryonic stem cells through direct activation of gp130 signalling pathways. *Mech. Dev., 45*, 163–171.

Yu, J., Vodyanik, M. A., Smuga-Otto, K., Antosiewicz-Bourget, J., Frane, J. L., Tian, S., et al. (2007). Induced pluripotent stem cell lines derived from human somatic cells. *Science, 318,* 1917–1920.

Zeng, X., Cai, J., Chen, J., Luo, Y., You, Z. B., Fotter, E., et al. (2004). Dopaminergic differentiation of human embryonic stem cells. *Stem Cells, 22,* 925–940.

Zeuthen, J., Norgaard, J. O., Avner, P., Fellous, M., Wartiovaara, J., Vaheri, A., et al. (1980). Characterization of a human ovarian teratocarcinoma-derived cell line. *Int. J. Cancer, 25,* 19–32.

Zhang, S. C., Wernig, M., Duncan, I. D., Brustle, O., & Thomson, J. A. (2001). In vitro differentiation of transplantable neural precursors from human embryonic stem cells. *Nat. Biotechnol., 19,* 1129–1133.

Zwaka, T. P., & Thomson, J. A. (2005). A germ cell origin of embryonic stem cells? *Development, 132,* 227–233.

CHAPTER 11

Alternative Sources of Human Embryonic Stem Cells

Svetlana Gavrilov[*], Virginia E. Papaioannou[*], Donald W. Landry[**]
* Department of Genetics and Development, College of Physicians and Surgeons of Columbia University, New York, NY, USA
** Department of Medicine, College of Physicians and Surgeons of Columbia University, New York, NY, USA

INTRODUCTION

Human embryonic stem (ES) cells are conventionally derived from viable preimplantation embryos produced by *in vitro* fertilization (IVF) (Thomson et al., 1998). The derivation of human ES cells is considered ethically controversial due to the typical destruction of an embryo during this process (Landry and Zucker, 2004; Green, 2007; McLaren, 2007; Gavrilov et al., 2009a). A human embryo constitutes an object of moral concern (Guenin, 2004) due to its identity as a human being at the embryonic stage of development. In biological terms, a human embryo has a distinct, unique, and unambiguous status due to this identity. However, the political and moral status of human embryos is in a state of flux. While there is universal opposition to reproductive cloning of humans by any method, there is diversity in public views toward the use of human embryos for derivation of human ES cells and, subsequently, potential therapies derived from them (Einsiedel et al., 2009; Peddie et al., 2009). Ethical and cultural imperatives to respect human dignity from the moment of fertilization conflict with a utilitarian desire to relieve human suffering even at the expense of embryonic human life. These conflicting perspectives have fueled an intense debate and have influenced legislative regulation of stem cell research in the USA and internationally (Landry and Zucker, 2004; Green, 2007; McLaren, 2007; Gavrilov et al., 2009a; ISSCR, 2010; NIH, 2010). At the time of writing, US stem cell research policy is regulated on the federal level by the Dickey amendment and President Obama's executive order 13505, and additionally by individual state laws (see Box 11.1) (NIH, 2010). The use of federal funding for derivation of new human ES cells that would entail the destruction of human embryos is forbidden. Also, in many European countries (Austria, Germany, Ireland, Italy, Lithuania, Norway, Poland, and Slovakia), the derivation of human ES cells from surplus embryos is not allowed (ISSCR, 2010). As stem cell biology is at the research forefront, legislative acts change rapidly. (For up-to-date legislative regulation of human ES cell research refer to links provided in Box 11.2 (ISSCR, 2010; NIH, 2010).)

Another consideration is the constant demand for deriving new human ES lines for both basic and clinical applications due to the loss of genetic and epigenetic stability arising during human ES cell culture and manipulation (Cowan et al., 2004; Maitra et al., 2005; Allegrucci and Young, 2007; Rugg-Gunn et al., 2007). Many of the currently available human ES cell lines

215

> ## BOX 11.1 BRIEF OVERVIEW OF CURRENT US FEDERAL STEM CELL POLICY
>
> At the time of writing, US policy on stem cell research is shaped by the following legislative act and executive order:
>
> - The **"Dickey amendment,"** a rider issued in **1996** that framed all subsequent political discussions regarding hESC research. The amendment stated that no federal funding may be employed for (1) the creation of a human embryo or embryos for research purposes or (2) research in which a human embryo or embryos are destroyed, discarded, or knowingly subjected to risk of injury or death (beyond that permitted for fetuses *in utero* under the Public Health Service Act).
> - **Executive Order (EO) 13505,** which removed barriers to responsible scientific research involving human stem cells. This EO was issued by **President Obama** on **March 9, 2009** and it states that the Secretary of Health and Human Services, through the director of NIH, may support and conduct responsible, scientifically worthy human stem cell research, including human stem cell research, to the extent permitted by law. In addition, this EO revoked two items issued by President George W. Bush: (1) a presidential statement that permitted work only on human ES cell lines generated prior to August 9, 2001 and (2) EO 13435 that favored all research on stem cells without harming a human embryo.

> ## BOX 11.2 USEFUL LINKS AND RESOURCES FOR UP-TO-DATE INFORMATION ON CURRENT LEGISLATION IN THE USA AND INTERNATIONALLY
>
> **National Institutes of Health (NIH) Stem Cell Information webpage**: contains relevant information on current US stem cell policy, NIH Stem Cell Registry with a list of eligible lines for NIH funding. The page also contains public comments on draft NIH human stem cell guidelines that supplement EO 13505. **http://stemcells.nih.gov/index.asp**
>
> **International Society for Stem Cell Research webpage**: contains comprehensive information on international legislation on human embryonic stem cell research; periodically updated. **http://www.isscr.org/**

were exposed to animal material during derivation or culture (Gavrilov et al., 2009a; Skottman et al., 2006). It is currently acceptable to expose human ES cell lines to products of human origin, but it remains the ultimate goal to pursue human ES cell derivation under stringent xeno-free conditions for eventual clinical use (Gavrilov et al., 2009a; Skottman et al., 2006).

The debate on embryo-destructive derivation of ES cells often focuses on the moral sensibilities of investigators and their desires for research unfettered by ethical considerations. However, the goal of human ES cell research is to find therapies that would ease human suffering from debilitating illness or injury (Klimanskaya et al., 2008; Gavrilov et al., 2009a; Leeb et al., 2009). In the latter context, the sensibilities of many millions of the populace — the intended beneficiaries of this work — should be instructive. As a result, a variety of different derivation strategies have been proposed (see Fig. 11.1) to avoid the use of an embryo as a source of human stem cells (detailed information can be found in appropriate chapters of this book or elsewhere) (Green, 2007; Gavrilov et al., 2009a). In this chapter we will discuss two alternative approaches to yielding genetically unmodified human ES cells that do not interfere with the developmental potential of human embryos: single blastomere biopsy and organismically dead embryos (Fig. 11.1) (Gavrilov et al., 2009a).

SINGLE BLASTOMERE BIOPSY

Single blastomere biopsy (SBB) for the purpose of deriving ES cells was developed by Lanza and colleagues (Chung et al., 2006, 2008; Klimanskaya et al., 2006, 2007). Human

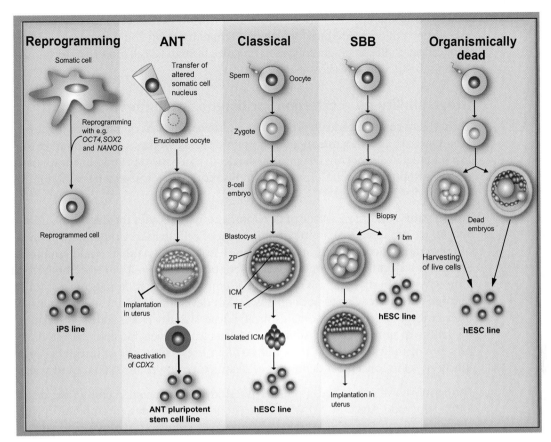

FIGURE 11.1

Classical and alternative strategies for the generation of human stem cells by reprogramming with exogenous genes (iPS), transfer of a genetically altered somatic cell nucleus into an oocyte (ANT), the classical derivation of hESCs from blastocyst culture, derivation of hESCs from a biopsied single blastomere (SBB), and derivation from organismically dead embryos. bm = blastomere; ICM = inner cell mass; iPS = induced pluripotent stem cells; TE = trophectoderm; ZP = zona pelucida *(reproduced with permission from Gavrilov et al., 1999a).*

217

ES cells are created from a single blastomere that is removed from the embryo (Klimanskaya et al., 2006, 2007) utilizing a technique that was originally developed for preimplantation genetic diagnosis (PGD) (Staessen et al., 2004; Verlinsky et al., 2004; Ogilvie et al., 2005; Gavrilov et al., 2009a). This procedure bypasses the ethical issue of embryo destruction, as biopsied embryos continue developing and reach the blastocyst stage and beyond, as demonstrated by more than a decade of experience with PGD (Verlinsky et al., 2004; Gavrilov et al., 2009a). SBB of both murine and human eight-cell stage embryos has been used successfully as a source of material to derive ES cell lines (see Fig. 11.1) (Chung et al., 2006, 2008; Klimanskaya et al., 2006, 2007; Gavrilov et al., 2009a). The risk associated with embryo biopsy (American Society for Reproductive Medicine, 2007) is accepted by patients as part of the PGD procedure, but it would be considered unjustified in a research setting in the absence of a clinical indication (Gavrilov et al., 2009a). In addition, US regulations forbid research on an embryo that imposes greater than minimal risk, unless the research is for the direct benefit of the fetus (Box 11.1) (Department of Health and Human Services, 2010). To date, none of the human ES cell lines derived by SBB have been approved for NIH funding (NIH, 2010).

ORGANISMICALLY DEAD EMBRYOS

Our group proposed the derivation of human ES cells from irreversibly arrested, non-viable human embryos that have died, despite best efforts, during the course of IVF for reproductive purposes (Gavrilov et al., 2009a). This proposal to harvest live cells from dead embryos is analogous to the harvesting of essential organs from deceased donors. We suggested that the established ethical guidelines for essential organ donation could be employed for the clinical application of this paradigm for generating new human ES cell lines (Landry and Zucker, 2004; Landry et al., 2006; Gavrilov et al., 2009a,b).

Irreversibility as a criterion for diagnosing embryonic death

The modern concept of death is based on an irreversible loss of integrated organismic function (Landry et al., 2006; Egonsson, 2009). Brain death is used as a reliable marker for irreversible loss of integrated function. Diagnosing the death of a patient prior to the death of that patient's tissues is important for the appropriate application of medical resources and for the possibility of organ donation.

To apply this concept to a stage of development that precedes the development of the nervous system, we proposed that an irreversible arrest of cell division would mark an irreversible loss of integrated function. Thus, it was necessary to find criteria that would establish irreversible cessation of normal embryonic development before every cell of the embryo has died. Through retrospective analysis of early-stage embryos that had been generated for reproductive purpose but rejected due to poor quality and/or developmental arrest, we showed that many of these embryos were, in fact, organismically dead (Landry et al., 2006). Our data showed that the failure of normal cell division for 48 hours was irreversible and, despite the possible presence of individual living cells, indicated an irreversible loss of integrated organismic function – the conceptual definition of death (Gavrilov et al., 2009a; Landry et al., 2006).

Furthermore, we conducted a prospective study to characterize embryonic death (Green, 2007; Hipp and Atala, 2008) where the progression of arrested embryos, including abnormal blastocysts, was examined in extended culture (Gavrilov et al., 2009b). Our data demonstrated that developmental arrest observed in some human embryos by embryonic day 6 (ED6) following IVF cannot be reversed by extended culture in conditions suitable for preimplantation embryos, as we saw no morphological changes indicative of developmental progression in the majority of embryos and observed no unequivocal instances of further cell divisions (Gavrilov et al., 2009b). Moreover, these observations are in line with standard IVF practice, which dictates that such embryos should not be transferred or cryopreserved because they are known not to produce live offspring (Cummins et al., 1986; Puissant et al., 1987; Bolton et al., 1989; Erenus et al., 1991; Staessen et al., 1992; Steer et al., 1992; Giorgetti et al., 1995; Ziebe et al., 1997; Gavrilov et al., 2009b). In an attempt to correlate morphology with cell number, we categorized the embryos at ED6 on the basis of gross morphology (Fig. 11.2) (Gavrilov et al., 2009b). We showed that morphological categorization was of limited value in predicting cell number. Nevertheless, the higher cell number associated with cavitation might predict greater potential for success of human ES cell derivation (Gavrilov et al., 2009b). In addition, we determined the proportion of living and non-living cells in non-viable ED6 human embryos (Fig. 11.2) and showed that the majority of irreversibly arrested embryos contain a high proportion of vital cells regardless of the stage of arrest, indicating that harvesting cells and deriving hESC from such non-viable embryos should be feasible (Gavrilov et al., 2009b).

Human ES cell lines derived from irreversibly arrested, non-viable embryos

In fact, the proof of principle for this alternative method has been obtained as, to date, 14 human ES cell lines have been successfully derived from non-viable embryos that were irreversibly arrested by our criteria (Table 11.1) (Zhang et al., 2006; Lerou et al., 2008; Gavrilov et al., submitted). The first cell line (hES-NCL9) was derived by Stojkovic and colleagues from 132 arrested embryos (Zhang et al., 2006). Subsequently, Daley and colleagues derived 11 lines from 413 poor-quality embryos rejected for clinical use (Lerou et al., 2008). Additionally, our

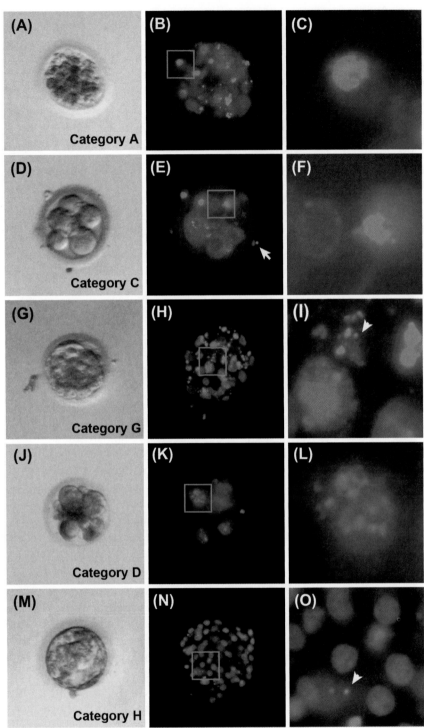

FIGURE 11.2
Morphology and differential propidium iodide/Hoechst fluorescent nuclear staining of non-viable embryos at ED6.
Brightfield images (A, D, G, J, M) with corresponding fluorescence images (B, E, H, K, N), and enlarged details (C, F, I, L, O) as indicated by the green squares. (A—C) Category A embryo showing degeneration at ED6. All nuclei, including nuclear fragments, are pink, indicating that there are no living cells in the embryo. Detail shows pink nucleus from a dead cell. (D—F) Category C embryo with living and dead cells indicated by the blue and pink nuclei, respectively. Detail shows nuclei from one living and one dead cell. Arrow in E indicates a sperm nucleus outside the ZP. (G—I) Category G embryo with living and dead cells as well as fragmented nuclei. Detail shows intact and fragmented nuclei. (J—L) Category D embryo with all live cells. Detail shows blue fragmented nucleus. (M—O) Category H embryo with many living and a few dead cells. Arrowheads in I and O indicate nuclear fragments *(reproduced with permission from Gavrilov et al., 1999b)*.

TABLE 11.1 List of hESC Lines Derived from Non-viable Organismically Dead Embryos

Cell line name	Type of embryo	Karyotype	Stem cell markers	EB assay	Teratoma	Eligible for NIH funding?	Reference
hES-NCL9	Day 6–7 late arrested embryo (16–24 cells)	46 XX	yes	yes	yes	ND	Zhang et al., 2006
CHB-1	Day 3 PQE	46 XY	yes	NR	yes	yes	Lerou et al., 2008
CHB-2	Day 5 PQE	46 XX	yes	NR	yes	yes	Lerou et al., 2008
CHB-3	Day 5 PQE	46 XX	yes	NR	yes	yes	Lerou et al., 2008
CHB-4	Day 5 PQE	46 XY	yes	NR	yes	yes	Lerou et al., 2008
CHB-5	Day 5 PQE	46 XX	yes	NR	yes	yes	Lerou et al., 2008
CHB-6	Day 5 PQE	46 XX	yes	NR	yes	yes	Lerou et al., 2008
CHB-8	Day 5 PQE	46 XX	yes	NR	yes	yes	Lerou et al., 2008
CHB-9	Day 5 PQE	46 XY	yes	NR	yes	yes	Lerou et al., 2008
CHB-10	Day 5 PQE	46 XY	yes	NR	yes	yes	Lerou et al., 2008
CHB-11	Day 5 PQE	46 XX	yes	NR	yes	yes	Lerou et al., 2008
CHB-12	Day 5 PQE	46 XX	yes	NR	yes	yes	Lerou et al., 2008
CU1	Day 6 arrested poor blastocyst	46 XX	yes	yes	ND	ND	Gavrilov et al., submitted
CU2	Day 6 arrested early blastocyst	46 XX*	yes	yes	ND	ND	Gavrilov et al., submitted

ND = not determined; NR = not reported; PQE = poor quality embryo
*Putative normal karyotype – possible low level of mosaicism

group has derived two human ES lines: CU1 and CU2 from 159 ED6 irreversibly arrested, non-viable human embryos (Gavrilov et al., submitted). Although many arrested embryos might be expected to be aneuploid (Hardy et al., 1989; Magli et al., 2000; Sandalinas et al., 2001; Findikli et al., 2004; Munne et al., 2007), all 14 hESC lines were karyotypically normal and, additionally, pluripotency and differentiation potential were demonstrated *in vitro* and/or *in vivo* (Zhang et al., 2006; Lerou et al., 2008; Gavrilov et al., submitted).

Morphological criteria for predicting the capacity of irreversibly arrested, non-viable human embryos to give rise to a human ES cell line

In order to define morphological criteria that could be used to predict the capacity of discarded, irreversibly arrested, non-viable embryos to give rise to a human ES cell line, we carried out a retrospective analysis of the morphological progression from ED5 to ED6 in 2,480 embryos that were rejected for clinical use (Gavrilov et al., submitted). Embryos were given a morphological category, commonly used for clinical grading as per standard IVF practice (e.g. single-celled embryo, multicell, morula, blastocyst, etc.). If an embryo had reached the blastocyst stage (i.e. showing advanced cavitation), it was given an overall grade of good, fair, or poor and, additionally, scored for inner cell mass and trophectoderm quality. Our analysis showed that non-viable embryos defined as poor do not improve with extended *in vitro* culture and yet retain the capacity to yield human ES cell lines despite arrested development (Gavrilov et al., submitted). We have postulated that, if derivation efforts are targeted on this subgroup, derivation success rate could be increased and production of new hESC lines brought closer to clinical application (Gavrilov et al., submitted).

CONCLUSION

Derivation of human ES cells from organismically dead embryos is a unique approach because it defines a common ground in the human ES debate. Harvesting live cells from dead human embryos has the likelihood of being accepted by the staunchest opponents of embryo-destructive ES derivation. The ES cells generated by this approach appear suitable for clinical research. Thus far, 11 human ES lines derived by Daley and colleagues have been included in the NIH stem cell registry and are available for research with NIH funding (NIH, 2010). Human ES lines generated from organismically dead embryos are of equal quality when compared with lines derived by the classical, ICM-derivation approach, but further characterization of these lines is needed (Gavrilov et al., 2009a).

During routine IVF procedures large proportions of embryos fail to develop properly (Alikani et al., 2000; Magli et al., 2001; Munne et al., 2007) and are discarded as being unsuitable for clinical use (Gavrilov et al., 2009a,b). Despite the low efficiency of isolation of human ES cells from organismically dead embryos, large-scale derivation is not limited since in the USA alone nearly half a million such embryos are generated yearly as a by-product of assisted reproductive technologies (Gavrilov et al., 2009a,b). The prospect for thousands of human ES cell lines generated by this method and deposited into stem cell banks renders clinical applications based on HLA (human leukocyte antigen) matching feasible.

References

American Society for Reproductive Medicine. (2007). Preimplantation genetic testing: a Practice Committee opinion. *Fertil. Steril.,* 88, 1497–1504.

Alikani, M., Calderon, G., Tomkin, G., Garrisi, J., Kokot, M., & Cohen, J. (2000). Cleavage anomalies in early human embryos and survival after prolonged culture in-vitro. *Hum. Reprod.,* 15, 2634–22643.

Allegrucci, C., & Young, L. E. (2007). Differences between human embryonic stem cell lines. *Hum. Reprod. Update,* 13, 103–120.

Bolton, V. N., Hawes, S. M., Taylor, C. T., & Parsons, J. H. (1989). Development of spare human preimplantation embryos in vitro: an analysis of the correlations among gross morphology, cleavage rates, and development to the blastocyst. *J. In Vitro Fert. Embryo. Transf.,* 6, 30–35.

Chung, Y., Klimanskaya, I., Becker, S., Li, T., Maserati, M., Lu, S. J., et al. (2008). Human embryonic stem cell lines generated without embryo destruction. *Cell Stem Cell,* 2, 113–117.

Chung, Y., Klimanskaya, I., Becker, S., Marh, J., Lu, S. J., Johnson, J., et al. (2006). Embryonic and extraembryonic stem cell lines derived from single mouse blastomeres. *Nature,* 439, 216–219.

Cowan, C. A., Klimanskaya, I., McMahon, J., Atienza, J., Witmyer, J., Zucker, J. P., et al. (2004). Derivation of embryonic stem-cell lines from human blastocysts. *N. Engl. J. Med.,* 350, 1353–1356.

Cummins, J. M., Breen, T. M., Harrison, K. L., Shaw, J. M., Wilson, L. M., & Hennessey, J. F. (1986). A formula for scoring human embryo growth rates in in vitro fertilization: its value in predicting pregnancy and in comparison with visual estimates of embryo quality. *J. In Vitro Fert. Embryo Transf.,* 3, 284–295.

Department of Health And Human Services. (2010). §46.204. *Research involving pregnant women or fetuses, Vol. 46.*

Egonsson, D. (2009). Death and irreversibility. *Rev. Neurosci.,* 20, 275–281.

Einsiedel, E., Premji, S., Geransar, R., Orton, N. C., Thavaratnam, T., & Bennett, L. K. (2009). Diversity in public views toward stem cell sources and policies. *Stem Cell Rev.,* 5, 102–107.

Erenus, M., Zouves, C., Rajamahendran, P., Leung, S., Fluker, M., & Gomel, V. (1991). The effect of embryo quality on subsequent pregnancy rates after in vitro fertilization. *Fertil. Steril.,* 56, 707–710.

Findikli, N., Kahraman, S., Kumtepe, Y., Donmez, E., Benkhalifa, M., Biricik, A., et al. (2004). Assessment of DNA fragmentation and aneuploidy on poor quality human embryos. *Reprod. Biomed. Online,* 8, 196–206.

Gavrilov, S., Marolt, D., Douglas, N. C., Prosser, R. W., Khalid, I., Sauer, M. V., et al. Derivation of two new human embryonic stem cell (hESC) lines from irreversibly-arrested, non-viable human embryos. Submitted.

Gavrilov, S., Papaioannou, V. E., & Landry, D. W. (2009a). Alternative strategies for the derivation of human embryonic stem cell lines and the role of dead embryos. *Curr. Stem Cell Res. Ther.,* 4, 81–86.

Gavrilov, S., Prosser, R. W., Khalid, I., MacDonald, J., Sauer, M. V., Landry, D. W., et al. (2009b). Non-viable human embryos as a source of viable cells for embryonic stem cell derivation. *Reprod. Biomed. Online,* 18, 301–308.

Giorgetti, C., Terriou, P., Auquier, P., Hans, E., Spach, J. L., Salzmann, J., et al. (1995). Embryo score to predict implantation after in-vitro fertilization: based on 957 single embryo transfers. *Hum. Reprod.,* 10, 2427–2431.

Green, R. M. (2007). Can we develop ethically universal embryonic stem-cell lines? *Nat. Rev. Genet.,* 8, 480–485.

Guenin, L. M. (2004). The morality of unenabled embryo use — arguments that work and arguments that don't. *Mayo Clin. Proc.,* 79, 801–808.

Hardy, K., Handyside, A. H., & Winston, R. M. (1989). The human blastocyst: cell number, death and allocation during late preimplantation development in vitro. *Development,* 107, 597–604.

Hipp, J., & Atala, A. (2008). Sources of stem cells for regenerative medicine. *Stem Cell Rev.,* 4, 3–11.

ISSCR (2010). Vol. 2010. http://www.isscr.org.

Klimanskaya, I., Chung, Y., Becker, S., Lu, S. J., & Lanza, R. (2006). Human embryonic stem cell lines derived from single blastomeres. *Nature, 444*, 481–485.

Klimanskaya, I., Chung, Y., Becker, S., Lu, S. J., & Lanza, R. (2007). Derivation of human embryonic stem cells from single blastomeres. *Nat. Protoc., 2*, 1963–1972.

Klimanskaya, I., Rosenthal, N., & Lanza, R. (2008). Derive and conquer: sourcing and differentiating stem cells for therapeutic applications. *Nat. Rev. Drug Discov., 7*, 131–142.

Landry, D. W., & Zucker, H. A. (2004). Embryonic death and the creation of human embryonic stem cells. *J. Clin. Invest., 114*, 1184–1186.

Landry, D. W., Zucker, H. A., Sauer, M. V., Reznik, M., & Wiebe, L. (2006). Hypocellularity and absence of compaction as criteria for embryonic death. *Regen. Med., 1*, 367–371.

Leeb, C., Jurga, M., McGuckin, C., Moriggl, R., & Kenner, L. (2009). Promising new sources for pluripotent stem cells. *Stem Cell Rev, 6*(1), 15–26.

Lerou, P. H., Yabuuchi, A., Huo, H., Takeuchi, A., Shea, J., Cimini, T., et al. (2008). Human embryonic stem cell derivation from poor-quality embryos. *Nat. Biotechnol, 26*(2), 212–214.

Magli, M. C., Gianaroli, L., & Ferraretti, A. P. (2001). Chromosomal abnormalities in embryos. *Mol. Cell Endocrinol., 183*(Suppl. 1), S29–S34.

Magli, M. C., Jones, G. M., Gras, L., Gianaroli, L., Korman, I., & Trounson, A. O. (2000). Chromosome mosaicism in day 3 aneuploid embryos that develop to morphologically normal blastocysts in vitro. *Hum. Reprod., 15*, 1781–1786.

Maitra, A., Arking, D. E., Shivapurkar, N., Ikeda, M., Stastny, V., Kassauei, K., et al. (2005). Genomic alterations in cultured human embryonic stem cells. *Nat. Genet., 37*, 1099–1103.

McLaren, A. (2007). A scientist's view of the ethics of human embryonic stem cell research. *Cell Stem Cell, 1*, 23–26.

Munne, S., Chen, S., Colls, P., Garrisi, J., Zheng, X., Cekleniak, N., et al. (2007). Maternal age, morphology, development and chromosome abnormalities in over 6000 cleavage-stage embryos. *Reprod. Biomed. Online, 14*, 628–634.

NIH. (2010). *Stem Cell Information, Vol. 2010.* http://stemcells.nih.gov/index.asp.

Ogilvie, C. M., Braude, P. R., & Scriven, P. N. (2005). Preimplantation genetic diagnosis – an overview. *J. Histochem. Cytochem., 53*, 255–260.

Peddie, V. L., Porter, M., Counsell, C., Caie, L., Pearson, D., & Bhattacharya, S. (2009). "Not taken in by media hype": how potential donors, recipients and members of the general public perceive stem cell research. *Hum. Reprod., 24*, 1106–1113.

Puissant, F., van Rysselberge, M., Barlow, P., Deweze, J., & Leroy, F. (1987). Embryo scoring as a prognostic tool in IVF treatment. *Hum. Reprod., 2*, 705–708.

Rugg-Gunn, P. J., Ferguson-Smith, A. C., & Pedersen, R. A. (2007). Status of genomic imprinting in human embryonic stem cells as revealed by a large cohort of independently derived and maintained lines. *Hum. Mol. Genet., 16 Spec. No. 2*, R243–R251.

Sandalinas, M., Sadowy, S., Alikani, M., Calderon, G., Cohen, J., & Munne, S. (2001). Developmental ability of chromosomally abnormal human embryos to develop to the blastocyst stage. *Hum. Reprod., 16*, 1954–1958.

Skottman, H., Dilber, M. S., & Hovatta, O. (2006). The derivation of clinical-grade human embryonic stem cell lines. *FEBS Lett., 580*, 2875–2878.

Staessen, C., Camus, M., Bollen, N., Devroey, P., & van Steirteghem, A. C. (1992). The relationship between embryo quality and the occurrence of multiple pregnancies. *Fertil. Steril., 57*, 626–630.

Staessen, C., Platteau, P., van Assche, E., Michiels, A., Tournaye, H., Camus, M., et al. (2004). Comparison of blastocyst transfer with or without preimplantation genetic diagnosis for aneuploidy screening in couples with advanced maternal age: a prospective randomized controlled trial. *Hum. Reprod., 19*, 2849–2858.

Steer, C. V., Mills, C. L., Tan, S. L., Campbell, S., & Edwards, R. G. (1992). The cumulative embryo score: a predictive embryo scoring technique to select the optimal number of embryos to transfer in an in-vitro fertilization and embryo transfer programme. *Hum. Reprod., 7*, 117–119.

Thomson, J. A., Itskovitz-Eldor, J., Shapiro, S. S., Waknitz, M. A., Swiergiel, J. J., Marshall, V. S., et al. (1998). Embryonic stem cell lines derived from human blastocysts. *Science, 282*, 1145–1147.

Verlinsky, Y., Cohen, J., Munne, S., Gianaroli, L., Simpson, J. L., Ferraretti, A. P., et al. (2004). Over a decade of experience with preimplantation genetic diagnosis: a multicenter report. *Fertil. Steril., 82*, 292–294.

Zhang, X., Stojkovic, P., Przyborski, S., Cooke, M., Armstrong, L., Lako, M., et al. (2006). Derivation of human embryonic stem cells from developing and arrested embryos. *Stem Cells, 24*, 2669–2676.

Ziebe, S., Petersen, K., Lindenberg, S., Andersen, A. G., Gabrielsen, A., & Andersen, A. N. (1997). Embryo morphology or cleavage stage: how to select the best embryos for transfer after in-vitro fertilization. *Hum. Reprod., 12*, 1545–1549.

CHAPTER 12

Stem Cells from Amniotic Fluid

Mara Cananzi[*,**], **Anthony Atala**[***], **Paolo de Coppi**[*,**,***]
* Surgery Unit, UCL Institute of Child Health and Great Ormond Street Hospital, London, UK
** Department of Paediatrics, University of Padua, Padua, Italy
*** Wake Forest Institute for Regenerative Medicine, Winston Salem, NC, USA

INTRODUCTION

In this chapter, we provide an overview of the potential advantages and disadvantages of different stem and progenitor cell populations identified to date in amniotic fluid, along with their properties and potential clinical applications.

In the last ten years, placenta, fetal membranes (i.e. amnion and chorion), and amniotic fluid have been extensively investigated as a potential non-controversial source of stem cells. They are usually discarded after delivery and are accessible during pregnancy through amniocentesis and chorionic villus sampling (Marcus and Woodbury, 2008). Several populations of cells with multilineage differentiation potential and immunomodulatory properties have been isolated from the human placenta and fetal membranes; they have been classified by an international workshop (Parolini et al., 2007) as human amniotic epithelial cells (hAECs) (Tamagawa et al., 2004; Miki et al., 2005; Miki and Strom, 2006; Kim et al., 2007a; Marcus et al., 2008), human amniotic mesenchymal stromal cells (hAMSCs) (Alviano et al., 2007; Soncini et al., 2007), human chorionic mesenchymal stromal cells (hCMSCs) (Igura et al., 2004; In 't Anker et al., 2004), and human chorionic trophoblastic cells (hCTCs). In the amniotic fluid (AF), two main populations of stem cells have been isolated so far: (1) amniotic fluid mesenchymal stem cells (AFMSCs) and (2) amniotic fluid stem (AFS) cells. Although only recently described, these cells may, given the easier accessibility of the AF in comparison to other extra-embryonic tissues, hold much promise in regenerative medicine.

AMNIOTIC FLUID: FUNCTION, ORIGIN, AND COMPOSITION

The AF is the clear, watery liquid that surrounds the growing fetus within the amniotic cavity. It allows the fetus to freely grow and move inside the uterus, protects it from outside injuries by cushioning sudden blows or movements by maintaining consistent pressure and temperature, and acts as a vehicle for the exchange of body chemicals with the mother (Riboldi and Simon, 2009; Underwood et al., 2005).

In humans, the AF starts to appear at the beginning of the second week of gestation as a small film of liquid between the cells of the epiblast. Between days 8 and 10 after fertilization, this fluid gradually expands and separates the epiblast (i.e. the future embryo) from the amnioblasts (i.e. the future amnion), thus forming the amniotic cavity (Miki and Strom, 2006). Thereafter, it progressively increases in volume, completely surrounding the embryo after the fourth week of pregnancy. Over the course of gestation, AF volume markedly changes from

223

Principles of Regenerative Medicine. DOI: 10.1016/B978-0-12-381422-7.10012-4

20 ml in the seventh week to 600 ml in the 25th week, 1,000 ml in the 34th week, and 800 ml at birth. During the first half of gestation, the AF results from active sodium and chloride transport across the amniotic membrane and the non-keratinized fetal skin, with concomitant passive movement of water (Brace and Resnik, 1999). In the second half of gestation, the AF is constituted by fetal urine, gastrointestinal excretions, respiratory secretions, and substances exchanged through the sac membranes (Mescher et al., 1975; Lotgering and Wallenburg, 1986; Muller et al., 1994; Fauza, 2004).

The AF is primarily composed of water and electrolytes (98–99%) but also contains chemical substances (e.g. glucose, lipids, proteins, hormones, and enzymes), suspended materials (e.g. vernix caseosa, lanugo hair, and meconium), and cells. AF cells derive both from extra-embryonic structures (i.e. placenta and fetal membranes) and from embryonic and fetal tissues (Thakar et al., 1982; Gosden, 1983). Although AF cells are known to express markers of all three germ layers (Cremer et al., 1981), their exact origin still represents a matter of discussion; the consensus is that they mainly consist of cells shed in the amniotic cavity from the developing skin, respiratory apparatus, and urinary and gastrointestinal tracts (Milunsky, 1979; von Koskull et al., 1984; Fauza, 2004). AF cells display a broad range of morphologies and behaviors varying with gestational age and fetal development (Hoehn and Salk, 1982). In normal conditions, the number of AF cells increases with advancing gestation; if a fetal disease is present, AF cell counts can be either dramatically reduced (e.g. intrauterine death, urogenital atresia) or abnormally elevated (e.g. anencephaly, spina bifida, exomphalos) (Nelson, 1973; Gosden and Brock, 1978). Based on their morphological and growth characteristics, viable adherent cells from the AF are classified into three main groups: epithelioid (33.7%), amniotic fluid (60.8%), and fibroblastic type (5.5%) (Hoehn et al., 1975). In the event of fetal abnormalities, other types of cells can be found in the AF, e.g. neural cells in the presence of neural tube defects and peritoneal cells in the case of abdominal wall malformations (Gosden et al., 1978; Aula et al., 1980; von Koskull et al., 1981).

The majority of cells present in the AF are terminally differentiated and have limited proliferative capabilities (Gosden et al., 1978; Siegel et al., 2007). In the 1990s, however, two groups demonstrated the presence in the AF of small subsets of cells harboring a proliferation and differentiation potential. First, Torricelli reported the presence of hematopoietic progenitors in the AF collected before the 12th week of gestation (Torricelli et al., 1993). Then Streubel was able to differentiate AF cells into myocytes, thus suggesting the presence in the AF of non-hematopoietic precursors (Streubel et al., 1996). These results initiated a new interest in the AF as an alternative source of cells for therapeutic applications.

AMNIOTIC FLUID MESENCHYMAL STEM CELLS

Mesenchymal stem cells (MSCs) represent a population of multipotent stem cells able to differentiate towards mesoderm-derived lineages (i.e. adipogenic, chondrogenic, myogenic, and osteogenic) (Pittenger et al., 1999). Initially identified in adult bone marrow, where they represent 0.001–0.01% of total nucleated cells (Owen and Friedenstein, 1988), MSCs have since been isolated from several adult (e.g. adipose tissue, skeletal muscle, liver, brain), fetal (i.e. bone marrow, liver, blood), and extra-embryonic tissues (i.e. placenta, amnion) (Porada et al., 2006).

The presence of a subpopulation of AF cells with mesenchymal features, able to proliferate *in vitro* more rapidly than comparable fetal and adult cells, was described for the first time in 2001 (Kaviani et al., 2001). In 2003, In 't Anker demonstrated that the AF can be an abundant source of fetal cells that exhibit a phenotype and a multilineage differentiation potential similar to that of bone marrow-derived MSCs; these cells were named AF mesenchymal stem cells (AFMSCs) (In 't Anker et al., 2003). Soon after this paper, other groups independently confirmed similar results.

Isolation and culture

AFMSCs can be easily obtained: in humans, from small volumes (2−5 ml) of second and third trimester AF (Tsai et al., 2004; You et al., 2009), where their percentage is estimated to be 0.9−1.5% of the total AF cells (Roubelakis et al., 2007), and in rodents, from the AF collected during the second or third week of pregnancy (de Coppi et al., 2007a; Nadri and Soleimani, 2008). Various protocols have been proposed for their isolation; all are based on the expansion of unselected populations of AF cells in serum-rich conditions without feeder layers, allowing cell selection by culture conditions. The success rate of the isolation of AFMSCs is reported by different authors to be 100% (Tsai et al., 2004; Nadri and Soleimani, 2008). AFMSCs grow in basic medium containing fetal bovine serum (20%) and fibroblast growth factor (5 ng/ml). Importantly, it has been very recently shown that human AFMSCs can be also cultured in the absence of animal serum without losing their properties (Kunisaki et al., 2007); this finding is a fundamental prerequisite for the beginning of clinical trials in humans.

Characterization

The fetal versus maternal origin of AFMSCs has been investigated by different authors. Molecular HLA typing and amplification of the SRY gene in AF samples collected from male fetuses (In 't Anker et al., 2003; Roubelakis et al., 2007) demonstrated the exclusive fetal derivation of these cells. However, whether AFMSCs originate from the fetus or from the fetal portion of extra-embryonic tissues is still a matter of debate (Kunisaki et al., 2007).

AFMSCs display a uniform spindle-shaped fibroblast-like morphology similar to that of other MSC populations and expand rapidly in culture (Tsai et al., 2007). Human cells derived from a single 2 ml AF sample can increase up to 180×10^6 cells within four weeks (three passages) and, as demonstrated by growth kinetics assays, possess a greater proliferative potential (average doubling time 25−38 hours) in comparison to that of bone marrow-derived MSCs (average doubling time 30−90 hours) (In 't Anker et al., 2003; Roubelakis et al., 2007; Nadri and Soleimani, 2008; Sessarego et al., 2008). Moreover, AFMSCs' clonogenic potential has been proved to exceed that of MSCs isolated from bone marrow (86 ± 4.3 vs. 70 ± 5.1 colonies) (Nadri and Soleimani, 2008). Despite their high proliferation rate, AFMSCs retain a normal karyotype and do not display tumorigenic potential even after extensive expansion in culture (Roubelakis et al., 2007; Sessarego et al., 2008).

Analysis of AFMSC transcriptome demonstrated that: (1) AFMSCs' gene expression profile, as well as that of other MSC populations, remains stable between passages in culture, enduring cryopreservation and thawing well; (2) AFMSCs share with MSCs derived from other sources a core set of genes involved in extracellular matrix remodeling, cytoskeletal organization, chemokine regulation, plasmin activation, TGF-β and Wnt signaling pathways; (3) in comparison to other MSCs, AFMSCs show a unique gene expression signature that consists of the upregulation of genes involved in signal transduction pathways (e.g. HHAT, F2R, F2RL) and in uterine maturation and contraction (e.g. OXTR, PLA2G10), thus suggesting a role of AFMSCs in modulating the interactions between the fetus and the uterus during pregnancy (Tsai et al., 2007).

The cell-surface antigenic profile of human AFMSCs has been determined through flow cytometry by different investigators (Table 12.1). Cultured human AFMSCs are positive for mesenchymal markers (i.e. CD90, CD73, CD105, CD166), for several adhesion molecules (i.e. CD29, CD44, CD49e, CD54), and for antigens belonging to the major histocompatibility complex I (MHC-I). They are negative for hematopoietic and endothelial markers (e.g. CD45, CD34, CD14, CD133, CD31).

AFMSCs exhibit a broad differentiation potential towards mesenchymal lineages. Under specific *in vitro* inducing conditions, they are able to differentiate towards the adipogenic, osteogenic, and chondrogenic lineage (In 't Anker et al., 2003; Tsai et al., 2007; Nadri and Soleimani, 2008).

225

TABLE 12.1 Immunophenotype of Culture-expanded Second and Third Trimester Human AFMSC: Results by Different Groups

Markers	Antigen	CD no.	You et al., 2009	Roubelakis et al., 2007	Tsai et al., 2004	In 't Anker et al., 2003
Mesenchymal	SH2, SH3, SH4	CD73	+	+	+	+
	Thy1	CD90	+	+	+	+
	Endoglin	CD105	+	+	+	+
	SB10/ALCAM	CD166	nt	+	nt	+
Endothelial and hematopoietic	LCA	CD14	nt	-	nt	-
	gp105-120	CD34	nt	-	-	-
	LPS-R	CD45	-	-	-	-
	Prominin-1	CD133	nt	-	nt	nt
Integrins	β1-integrin	CD29	+	+	+	nt
	β3-integrin	CD61	-	nt	nt	nt
	α4-integrin	CD49d	nt	-	nt	-
	α5-integrin	CD49e	nt	+	nt	+
	LFA-1	CD11a	nt	+	nt	-
Selectins	E-Selectin	CD62E	nt	+	nt	-
	P-selectin	CD62P	nt	+	nt	-
Ig-superfamily	PECAM-1	CD31	-	+	-	-
	ICAM-1	CD54	nt	+	nt	+
	ICAM-3	CD50	nt	-	nt	
	VCAM-1	CD106	nt	+	nt	-
	HCAM-1	CD44	nt	+	+	+
MHC	I (HLA-ABC)	none	nt	+	+	+
	II (HLA-DR,DP,DQ)	none	nt	nt	-	-

nt = not tested.

Despite not being pluripotent, AFMSCs can be efficiently reprogrammed into pluripotent stem cells (iPS) via retroviral transduction of defined transcription factors (Oct4, Sox2, Klf-4, c-Myc). Strikingly, AFMSC reprogramming capacity is significantly higher (100-fold) and much quicker (6 days vs. 16–30 days) in comparison to that of somatic cells such as skin fibroblasts. As iPS derived from adult cells, AF-derived iPS generate embryoid bodies (EBs) and differentiate towards all three germ layers *in vitro*, and *in vivo* form teratomas when injected into SCID mice (Li et al., 2009).

Preclinical studies

After AFMSC identification, various studies investigated their therapeutic potential in different experimental settings. Different groups demonstrated that AFMSCs are able not only to express cardiac and endothelial specific markers under specific culture conditions, but also to integrate into normal and ischemic cardiac tissue, where they differentiate into cardiomyocytes and endothelial cells (Zhao et al., 2005; Iop et al., 2008; Yeh et al., 2010; Zhang et al., 2010). In a rat model of bladder cryo-injury, AFMSCs show the ability to differentiate into smooth muscle and to prevent the compensatory hypertrophy of surviving smooth muscle cells (de Coppi et al., 2007a).

AFMSCs can be a suitable cell source for tissue engineering of congenital malformations. In an ovine model of diaphragmatic hernia, repair of the muscle deficit using grafts engineered with autologous mesenchymal amniocytes leads to better structural and functional results in comparison to equivalent fetal myoblast-based and acellular implants (Fuchs et al., 2004; Kunisaki et al., 2006a). Engineered cartilaginous grafts have been derived from AFMSCs grown on biodegradable meshes in serum-free chondrogenic conditions for at least 12 weeks; these grafts have been successfully used to repair tracheal defects in foetal lambs when implanted *in utero* (Kunisaki et al., 2006b). The surgical implantation of AFMSCs seeded on nanofibrous scaffolds and predifferentiated *in vitro* towards the osteogenic lineage into a leporine model of sternal defect leads to a complete bone repair in 2 months' time (Steigman et al., 2009).

Intriguingly, recent studies suggest that AFMSCs can harbor trophic and protective effects in the central and peripheral nervous systems. Pan showed that AFMSCs facilitate peripheral nerve regeneration after injury and hypothesized that this can be determined by cell secretion of neurotrophic factors (Pan et al., 2006, 2007; Chen et al., 2009). After transplantation into the striatum, AFMSCs are capable of surviving and integrating in the rat adult brain and migrating towards areas of ischemic damage (Cipriani et al., 2007). Moreover, the intra-ventricular administration of AFMSCs in mice with focal cerebral ischemia-reperfusion injuries significantly reverses neurological deficits in the treated animals (Rehni et al., 2007).

Remarkably, it has also been observed that AFMSCs present *in vitro* an immunosuppressive effect similar to that of bone marrow-derived MSCs (Uccelli et al., 2007). Following stimulation of peripheral blood mononuclear cells with anti-CD3, anti-CD28, or phytohemagglutinin, irradiated AFMSCs determine a significant inhibition of T-cell proliferation with dose-dependent kinetics (Sessarego et al., 2008).

AMNIOTIC FLUID STEM CELLS

The first evidence that the AF could contain pluripotent stem cells was provided in 2003 when Prusa described the presence of a distinct subpopulation of proliferating AF cells (0.1–0.5% of the cells present in the AF) expressing the pluripotency marker Oct4 at both transcriptional and proteic levels (Prusa et al., 2003). Oct4 (i.e. octamer binding transcription factor 4) is a nuclear transcription factor that plays a critical role in maintaining ES cell differentiation potential and capacity of self-renewal (Schöler et al., 1989; Nichols et al., 1998; Niwa et al., 2000). Other than by ES cells, Oct4 is specifically expressed by germ cells, where its inactivation results in apoptosis, and by embryonal carcinoma cells and tumors of germ cell origin, where it acts as an oncogenic fate determinant (Donovan, 2001; Pesce and Schöler, 2001; Gidekel et al., 2003; Looijenga et al., 2003). While its role in stem cells of fetal origin has not been completely addressed, it has been recently demonstrated that Oct4 is neither expressed nor required by somatic stem cells or progenitors (Berg and Goodell, 2007; Lengner et al., 2007; Liedtke et al., 2007).

After Prusa, different groups confirmed the expression of Oct4 and of its transcriptional targets (e.g. Rex-1) in the AF (Bossolasco et al., 2006; Stefanidis et al., 2007). Remarkably, Karlmark transfected human AF cells with the green fluorescent protein gene under either the Oct4 or the Rex-1 promoter and established that some AF cells were able to activate these promoters (Karlmark et al., 2005). Several authors subsequently reported the possibility of harvesting AF cells displaying features of pluripotent stem cells (Kim et al., 2007b; Tsai et al., 2006). Thereafter, the presence of a cell population able to generate clonal cell lines capable of differentiating into lineages representative of all three embryonic germ layers was definitively demonstrated (de Coppi et al., 2007b). These cells, named AF stem (AFS) cells, are characterized by the expression of the surface antigen c-kit (CD117), which is the type III tyrosine kinase receptor of the stem cell factor (Zsebo et al., 1990).

Isolation and culture

The proportion of c-kit+ cells in the amniotic fluid varies over the course of gestation, roughly describing a Gaussian curve; they appear at very early time points in gestation (i.e. at 7 weeks of amenorrhea in humans and at E9.5 in mice) and present a peak at midgestation equal to 90×10^4 cells/fetus at 20 weeks of pregnancy in humans and to 10,000 cells/fetus at E12.5 in mice (Ditadi et al., 2009). Human AFS cells can be derived either from small volumes (5 ml) of second trimester AF (14–22 weeks of gestation) or from confluent back-up amniocentesis cultures. Murine AFS cells are obtainable from the AF collected during the second week of gestation (E11.5–14.5) (de Coppi et al., 2007b; Kim et al., 2007b; Siegel et al., 2009b; Tsai et al., 2006). AFS cell isolation is based on a two-step protocol consisting of the prior immunological selection of c-kit positive cells from the AF (approximately 1% of total AF cells) and of the subsequent expansion of these cells in culture (de Coppi et al., 2007b; Kolambar et al., 2007; Perin et al., 2007; Chen et al., 2009; Siegel et al. 2009b; Valli et al., 2009). Isolated AFS cells can be expanded in feeder layer-free, serum-rich conditions without evidence of spontaneous differentiation *in vitro*. Cells are cultured in basic medium containing 15% of fetal bovine serum and Chang supplement (de Coppi et al., 2007b; Valli et al., 2009).

Characterization

Karyotype analysis of human AFS cells deriving from pregnancies in which the fetus was male revealed the fetal origin of these cells (de Coppi et al., 2007b).

AFS cells proliferate well during *ex vivo* expansion. When cultivated, they display a spectrum of morphologies ranging from a fibroblast-like to an oval-round shape (Fig. 12.1a). As demonstrated by different authors, AFS cells possess a great clonogenic potential (de Coppi et al., 2007b; Tsai et al., 2006). Clonal AFS cell lines expand rapidly in culture (doubling time = 36 h) and, more interestingly, maintain a constant telomere length (20 kbp) between early and late passages (Fig. 12.1b). Almost all clonal AFS cell lines express markers of a pluripotent undifferentiated state: Oct4 and NANOG (Tsai et al., 2006; Chambers et al., 2007; de Coppi et al., 2007b; Chen et al., 2009; Valli et al., 2009). However, they have been proved not to form tumors when injected in severe combined immunodeficient (SCID) mice (de Coppi et al., 2007b).

The cell-surface antigenic profile of AFS cells has been determined through flow cytometry by different investigators (Table 12.2). Cultured human AFS cells are positive for ES cell (e.g. SSEA-4) and mesenchymal markers (e.g. CD73, CD90, CD105), for several adhesion molecules (e.g. CD29, CD44), and for antigens belonging to the MHC-I. They are negative for hematopoietic and endothelial markers (e.g. CD14, CD34, CD45, CD133, CD31) and for antigens belonging to the major histocompatibility complex II (MHC-II).

As stability of cell lines is a fundamental prerequisite for basic and translational research, AFS cell capacity of maintaining their baseline characteristics over passages has been evaluated based on multiple parameters. Despite their high proliferation rate, AFS cells and derived clonal lines show a homogeneous, diploid DNA content without evidence of chromosomal rearrangement even after expansion to 250 population doublings (de Coppi et al., 2007b; Chen et al., 2009) (Fig. 12.1C). Moreover, AFS cells maintain constant morphology, doubling time, apoptosis rate, cell cycle distribution, and marker expression (e.g. Oct4, CD117, CD29, CD44) up to 25 passages (Chen et al., 2009; Valli et al., 2009). During *in vitro* expansion,

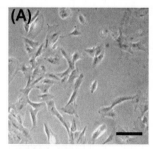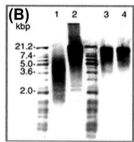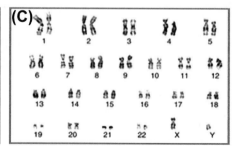

FIGURE 12.1

(A) Human AFS cells mainly display a spindle-shaped morphology during *in vitro* cultivation under feeder layer-free, serum-rich conditions. (B—C) Clonal human AFS cell lines retain long telomeres and a normal karyotype after more than 250 cell divisions. (B) Conserved telomere length of AFS cells between early passage (20 population doublings, lane 3) and late passage (250 population doublings, lane 4). Short length (lane 1) and high length (lane 2) telomere standards provided in the assay kit. (C) Giemsa band karyogram showing chromosomes of late passage (250 population doublings) cells. (Adapted from de Coppi et al. (2007b).

TABLE 12.2 Surface Markers Expressed by Human c-kit+ AF Stem Cells: Results by Different Groups

Markers	Antigen	CD no.	Ditadi et al., 2009	De Coppi et al., 2007b	Kim et al., 2007	Tsai et al., 2006
ES cells	SSEA-3	none	nt	-	+	nt
	SSEA-4	none	nt	+	+	nt
	Tra-1-60	none	nt	-	+	nt
	Tra-1-81	none	nt	-	nt	nt
Mesenchymal	SH2, SH3, SH4	CD73	nt	+	nt	+
	Thy1	CD90	+	+	nt	+
	Endoglin	CD105	nt	+	nt	+
Endothelial and hematopoieic	LCA	CD14	nt	nt	nt	-
	gp105-120	CD34	-	-	nt	-
	LPS-R	CD45	+	-	nt	nt
	Prominin-1	CD133	-	-	nt	nt
Integrins	β1-integrin	CD29	nt	+	nt	+
Ig-superfamily	PECAM-1	CD31	nt	nt	+	nt
	ICAM-1	CD54	nt	nt	+	nt
	VCAM-1	CD106	nt	nt	+	nt
	HCAM-1	CD44	+	+	+	+
MHC	I (HLA-ABC)	None	+	+	+	+
	II (HLA-DR,DP,DQ)	none	-	-	-	-

nt = not tested.

however, cell volume tends to increase and significant fluctuations of proteins involved in different networks (i.e. signaling, antioxidant, proteasomal, cytoskeleton, connective tissue, and chaperone proteins) can be observed using a gel-based proteomic approach (Chen et al., 2009); the significance of these modifications warrants further investigations but needs to be taken into consideration when interpreting experiments run over several passages and comparing results from different groups.

AFS cells and, more importantly, derived clonal cell lines are able to differentiate towards tissues representative of all three embryonic germ layers, both spontaneously, when cultured in suspension to form EBs, and when grown in specific differentiation conditions.

EBs consist of three-dimensional aggregates of ES cells, which recapitulate the first steps of early mammalian embryogenesis (Itskovitz-Eldor et al., 2000; Koike et al., 2007; Ungrin et al., 2008). As ES cells, when cultured in suspension and without anti-differentiation factors, AFS cells harbor the potential to form EBs with high efficiency: the incidence of EB formation (i.e. percentage of number of EB recovered from 15 hanging drops) is estimated to be around 28% for AFS cell lines and around 67% for AFS cell clonal lines. Similarly to ES cells, EB generation by AFS cells is regulated by the mTor (i.e. mammalian target of rapamycin) pathway and is accompanied by a decrease of Oct4 and Nodal expression and by an induction of endodermal (GATA4), mesodermal (Brachyury, HBE1), and ectodermal (Nestin, Pax6) markers (Siegel et al., 2009a; Valli et al., 2009).

In specific mesenchymal differentiation conditions, AFS cells express molecular markers of adipose, bone, muscle, and endothelial differentiated cells (e.g. LPL, desmin, osteocalcin, and V-CAM1). In the adipogenic, chondrogenic, and osteogenic medium, AFS cells respectively develop intracellular lipid droplets, secrete glycosaminoglycans, and produce mineralized calcium (Kim et al., 2007b; Tsai et al., 2006). In conditions inducing cell differentiation towards the hepatic lineage, AFS cells express hepatocyte-specific transcripts (e.g. albumin, alpha-fetoprotein, multidrug resistance membrane transporter 1) and acquire the liver-specific function of urea secretion (Fig. 12.2A) (de Coppi et al., 2007b). In neuronal conditions, AFS cells are capable of entering the neuroectodermal lineage. After induction, they express

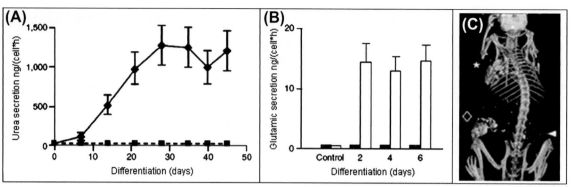

FIGURE 12.2

AFS cells differentiation into lineages representative of the three embryonic germ layers. (A) Hepatogenic differentiation: urea secretion by human AFS cells before (rectangles) and after (diamonds) hepatogenic *in vitro* differentiation. (B) Neurogenic differentiation: secretion of neurotransmitter glutamic acid in response to potassium ions. (C) Osteogenic differentiation: mouse micro CT scan 18 weeks after implantation of printed constructs of engineered bone from human AFS cells; arrow head: region of implantation of control scaffold without AFS cells; rhombus: scaffolds seeded with AFS cells. Adapted from de Coppi et al. (2007b).

neuronal markers (e.g. GIRK potassium channels), exhibit barium-sensitive potassium current, and release glutamate after stimulation (Fig. 12.2b). Ongoing studies are investigating AFS cell capacity to yield mature, functional neurons (Santos et al., 2008; Toselli et al., 2008; Donaldson et al., 2009).

AFS cells can be easily manipulated *in vitro*. They can be transduced with viral vectors more efficiently than adult MSCs, and, after infection, maintain their antigenic profile and the ability to differentiate into different lineages (Grisafi et al., 2008). AFS cells labeled with super-paramagnetic micrometer-sized iron oxide particles (MPIOs) retain their potency and can be non-invasively tracked by MRI for at least four weeks after injection *in vivo* (Delo et al., 2008).

Preclinical studies

Despite the very recent identification of AFS cells, several reports have investigated their potential applications in different settings.

BONE

Critical-sized segmental bone defects are one of the most challenging problems faced by orthopedics surgeons. Autologous and heterologous bone grafting are limited respectively by the small amount of tissue available for transplantation and by high refracture rates (Salgado et al., 2006; Beardi et al., 2008; Muscolo et al., 2009). Tissue engineering strategies that combine biodegradable scaffolds with stem cells capable of osteogenesis have been indicated as promising alternatives to bone grafting (Bianco and Robey, 2001); however, bone regeneration through cell-based therapies has been limited so far by the insufficient availability of osteogenic cells (Peister et al., 2009).

The potential of AFS cells to synthesize mineralized extracellular matrix within porous scaffolds has been investigated by different groups. After exposure to osteogenic conditions in static two-dimensional cultures, AFS cells differentiate into functional osteoblasts (i.e. activate the expression of osteogenic genes such as Runx2, Osx, Bsp, Opn, and Ocn, and produce alkaline phosphatase) and form dense layers of mineralized matrix (de Coppi et al., 2007b; Peister et al., 2009; Sun, 2010). As demonstrated by clonogenic mineralization assays, 85% of AFS cells versus 50% of MSCs are capable of forming osteogenic colonies (Sakaguchi et al., 2004; Morito et al., 2008; Peister et al., 2009). When seeded into three-dimensional biodegradable scaffolds and stimulated by osteogenic supplements (i.e. rhBMP-7 or dexamethasone), AFS cells remain highly viable up to several months in culture and produce extensive mineralization throughout the entire volume of the scaffold (de Coppi et al., 2007b; Peister et al., 2009; Sun 2010). *In vivo*, when subcutaneously injected into nude rodents, predifferentiated AFS cell-scaffold constructs are able to generate ectopic bone structures in four weeks' time (de Coppi et al., 2007b; Peister et al., 2009; Sun 2010) (Fig. 12.2C). AFS cells embedded in scaffolds, however, are not able to mineralize *in vivo* at ectopic sites unless previously predifferentiated *in vitro* (Peister et al., 2009). These studies demonstrate the potential of AFS cells to produce three-dimensional mineralized bioengineered constructs and suggest that AFS cells may be an effective cell source for functional repair of large bone defects. Further studies are needed to explore AFS cell osteogenic potential when injected into sites of bone injury.

CARTILAGE

Enhancing the regeneration potential of hyaline cartilage is one of the most significant challenges for treating damaged cartilage (Deans and Elisseeff, 2009; Koelling and Miosge, 2009).

The capacity of AFS cells to differentiate into functional chondrocytes has been tested *in vitro*. Human AFS cells treated with TGF-β1 have been proven to produce significant amounts of cartilaginous matrix (i.e. sulfated glycosaminoglycans and type II collagen) both in pellet and alginate hydrogel cultures (Kolambar et al., 2007).

SKELETAL MUSCLE

Stem cell therapy is an attractive method to treat muscular degenerative diseases because only a small number of cells, together with a stimulatory signal for expansion, are required to obtain a therapeutic effect (Price et al., 2007). The identification of a stem cell population providing efficient muscle regeneration is critical for the progression of cell therapy for muscle diseases (Farini et al., 2009).

AFS cell capacity of differentiating into the myogenic lineage has recently started to be explored. Under the influence of specific induction media containing 5-Aza-2′-deoxycytidine, AFS cells are able to express myogenic-associated markers such as Mrf4, Myo-D, and desmin both at a molecular and proteic level (de Coppi et al., 2007b; Gekas et al., 2010). However, when transplanted undifferentiated into damaged skeletal muscles of SCID mice, despite displaying a good tissue engraftment AFS cells did not differentiate towards the myogenic lineage (Gekas et al., 2010). Further studies are needed to confirm the results of this single report.

HEART

Cardiovascular diseases are the first cause of mortality in developed countries despite advances in pharmacological, interventional, and surgical therapies (Walther et al., 2009). Cell transplantation is an attractive strategy to replace endogenous cardiomyocytes lost by myocardial infarction. Fetal and neonatal cardiomyocites are the ideal cells for cardiac regeneration as they have been shown to integrate structurally and functionally into the myocardium after transplantation (Yao et al., 2003; Ott et al., 2008). However, their application is limited by the ethical restrictions involved in the use of fetal and neonatal cardiac tissues (Dai and Kloner, 2007).

Chiavegato et al. investigated human AFS cell plasticity towards the cardiac lineage. Undifferentiated AFS cells express cardiac transcription factors at a molecular level (i.e. Nkx2.5 and GATA-4 mRNA) but do not produce any myocardial differentiation marker. Under *in vitro* cardiovascular inducing conditions (i.e. co-culture with neonatal rat cardiomyocytes), AFS cells express differentiated cardiomyocyte markers such as cTnI, indicating that an *in vitro* cardiomyogenic-like medium can lead to a spontaneous differentiation of AFS cells into cardiomyocyte-like cells. *In vivo*, when xenotransplanted in the hearts of immunodeficient rats 20 minutes after creating a myocardial infarction, AFS cell differentiation capabilities were impaired by cell immune rejection (Chiavegato et al., 2007). More recently, we have proved that we could activate the myocardial gene program in GFP-positive rat AFS (GFP-rAFS cells) by co-culture with rCMs. The differentiation attained via a paracrine/contact action was confirmed using immunofluorescence, RT-PCR, and single-cell electrophysiological tests. Moreover, despite only a small number of Endorem-labeled GFP-rAFS, cells acquired an endothelial or smooth muscle phenotype and to a lesser extent CMs in an allogeneic acute myocardial infarction (AMI) context, and there was still improvement of ejection fraction as measured by magnetic resonance imaging (MRI) three weeks after injection (Bollini et al., submitted). This could be partially due to a paracrine action perhaps mediated by the secretion of thymosin β4 (Bollini et al., submitted).

HEMATOPOIETIC SYSTEM

Hematopoietic stem cells (HSCs) lie at the top of hematopoietic ontogeny and, if engrafted in the right niche, can theoretically reconstitute the organism's entire blood supply. Thus, the generation of autologous HSCs from pluripotent, patient-specific stem cells offers real promise for cell-therapy of both genetic and malignant blood disorders (Kim and Daley, 2009).

The hematopoietic potential of c-kit+ hematopoietic lineage negative cells present in the amniotic fluid (AFKL cells) has been recently explored (Ditadi et al., 2009). *In vitro*, human and murine AFKL cells exhibit strong multilineage hematopoietic potential. Cultured in semisolid medium, these cells are able to generate erythroid, myeloid, and lymphoid colonies. Moreover, murine cells exhibit the same clonogenic potential (0.03%) as hematopoietic progenitors present in the liver at the same stage of development. *In vivo*, mouse AFKL cells (i.e. 2×10^4 cells intravenously injected) are able to generate all three hematopoietic lineages after primary and secondary transplantation into immunocompromised hosts (i.e. sublethally irradiated Rag$^{-/-}$ mice), demonstrating their ability to self-renew. These results clearly show that c-kit+ cells present in the amniotic fluid have true hematopoietic potential both *in vitro* and *in vivo*.

KIDNEY

The incidence and prevalence of end stage renal disease (ESRD) continues to increase worldwide. Although renal transplantation represents a good treatment option, the shortage of compatible organs remains a critical issue for patients affected by ESRD. Therefore, the possibility of developing stem cell-based therapies for both glomerular and tubular repair has received intensive investigation in recent years (Bussolati et al., 2009). Different stem cell types have shown some potential in the generation of functional nephrons (Gupta et al., 2002; Bussolati et al., 2005; Kramer et al., 2006; Bruce et al., 2007; Morigi et al., 2008, 2010; Bruno et al., 2009) but the most appropriate cell type for transplantation is still to be established (Murray et al., 2007).

The potential of AFS cells in contributing to kidney development has been recently explored. Using a mesenchymal/epithelial differentiation protocol previously applied to demonstrate the renal differentiation potential of kidney stem cells (Bussolati et al., 2005), Siegel demonstrated that AFS cells and clonal-derived cell lines can differentiate towards the renal lineage; AFS cells sequentially grown in a mesenchymal differentiation medium containing EGF and PDGF-BB, and in an epithelial differentiation medium containing HGF and FGF4, reduce the expression of pluripotency markers (i.e. Oct4 and c-Kit) and switch on the expression of epithelial (i.e. CD51, ZO-1) and podocyte markers (i.e. CD2AP, NPHS2) (Siegel et al., 2009). AFS cells have also been shown to contribute to the development of primordial kidney structures during *in vitro* organogenesis; undifferentiated human AFS cells injected into a mouse embryonic kidney cultured *ex vivo* are able to integrate in the renal tissue, participate in all steps of nephrogenesis, and express molecular markers of early kidney differentiation such as ZO-1, claudin, and GDNF (Perin et al., 2007; Giuliani et al., 2008). Finally, very recent *in vivo* experiments show that AFS cells directly injected into damaged kidneys are able to survive, integrate into tubular structures, express mature kidney markers, and restore renal

232

function (Perin, 2010). These studies demonstrate the nephrogenic potential of AFS cells and warrant further investigation of their potential use for cell-based kidney therapies.

LUNG

Chronic lung diseases are common and debilitating; medical therapies have restricted efficacy and lung transplantation is often the only effective treatment (Loebinger, 2008). The use of stem cells for lung repair and regeneration after injury holds promise as a potential therapeutic approach for many lung diseases; however, current studies are still in their infancy (Weiss, 2008).

AFS cell ability to integrate into the lung and to differentiate into pulmonary lineages has been elegantly investigated in different experimental models of lung damage and development. *In vitro*, human AFS cells injected into mouse embryonic lung explants engraft into the epithelium and into the mesenchyme and express the early pulmonary differentiation marker TFF1 (Carraro et al., 2008). *In vivo*, in the absence of lung damage, systemically administered AFS cells show the capacity to home to the lung but not to differentiate into specialized cells; while, in the presence of lung injury, AFS cells not only exhibit a strong tissue engraftment but also express specific alveolar and bronchiolar epithelial markers (e.g. TFF1, SPC, CC10). Remarkably, cell fusion fenomena were elegantly excluded and long-term experiments confirmed the absence of tumor formation in the treated animals up to 7 months after AFS cell injection (Carraro et al., 2008).

INTESTINE

To date, very few studies have considered the employment of stem cells in gastroenterological diseases. Although still at initial stages and associated with numerous problems, ever-increasing experimental evidence supports the intriguing hypothesis that stem cells may be possible candidates to treat and/or prevent intestinal diseases (Khalil et al., 2007; Srivastava et al., 2007; Hotta et al., 2009; Panés and Salas, 2009).

In a study evaluating AFS cell transplantation into healthy newborn rats, Ghionzoli demonstrated that, after intraperitoneal injection, AFS cells (1) diffuse systemically within a few hours from their administration in 90% of the animals, (2) engraft in several organs of the abdominal and thoracic compartment and (3) localize preferentially in the intestine colonizing the gut in 60% of the animals (Ghionzoli et al., 2009). Preliminary *in vivo* experiments investigating the role of AFS cells in a neonatal rat model of necrotizing enterocolitis show that intraperitoneal-injected AFS cells are able not only to integrate into all gut layers but also to reduce bowel damage, improve rat clinical status, and lengthen animal survival (Zani et al., 2009).

CONCLUSIONS

Many stem cell populations (e.g. embryonic, adult, and fetal stem cells) as well as methods for generating pluripotent cells (e.g. nuclear reprogramming) have been described to date. All of them carry specific advantages and disadvantages and, at present, it has yet to be established which type of stem cell represents the best candidate for cell therapy. However, although it is likely that one cell type may be better than another, depending on the clinical scenario, the recent discovery of easily accessible cells of fetal derivation, not burdened by ethical concerns, in the AF has the potential to open new horizons in regenerative medicine. Amniocentesis, in fact, is routinely performed for the antenatal diagnosis of genetic diseases and its safety has been established by several studies documenting an extremely low overall fetal loss rate (0.06–0.83%) related to this procedure (Caughey et al., 2006; Eddleman et al., 2006). Moreover, stem cells can be obtained from AF samples without interfering with diagnostic procedures.

Two stem cell populations have been isolated from the AF so far (i.e. AFMSCs and AFS cells) and both can be used as primary (not transformed or immortalized) cells without further technical manipulations. AFMSCs exhibit typical MSC characteristics: fibroblastic-like morphology, clonogenic capacity, multilineage differentiation potential, immunosuppressive properties, and expression of a mesenchymal gene expression profile and of a mesenchymal set of surface antigens. However, ahead of other MSC sources, AFMSCs are easier to isolate and show better proliferation capacities. The harvest of bone marrow remains, in fact, a highly invasive and painful procedure, and the number, the proliferation, and the differentiation potential of these cells decline with increasing age (D'Ippolito et al., 1999; Kern et al., 2006). Similarly, UCB-derived MSCs exist at a low percentage and expand slowly in culture (Bieback et al., 2004).

AFS cells, on the other hand, represent a novel class of pluripotent stem cells with intermediate characteristics between ES cells and AS cells (Siegel et al., 2007; Bajada et al., 2008). They express both embryonic and mesenchymal stem cell markers, are able to differentiate into lineages representative of all embryonic germ layers, and do not form tumors after implantation *in vivo*. However, AFS cells have only recently identified and many questions need to be answered concerning their origin, epigenetic state, immunological reactivity, and regeneration and differentiation potential *in vivo*. AFS cells, in fact, may not differentiate as promptly as ES cells and their lack of tumorigenesis can be argued against their pluripotency.

Although further studies are needed to better understand their biologic properties and to define their therapeutic potential, stem cells present in the AF appear to be promising candidates for cell therapy and tissue engineering. In particular, they represent an attractive source for the treatment of perinatal disorders such as congenital malformations (e.g. congenital diaphragmatic hernia) and acquired neonatal diseases requiring tissue repair/ regeneration (e.g. necrotizing enterocolitis). In a future clinical scenario, AF cells collected during a routinely performed amniocentesis could be banked and, in case of need, subsequently expanded in culture or engineered in acellular grafts (Kunisaki et al., 2007; Siegel et al., 2007). In this way, affected children could benefit from having autologous expanded/engineered cells ready for implantation either before birth or in the neonatal period.

References

Alviano, F., Fossati, V., Marchionni, C., et al. (2007). Term amniotic membrane is a high throughput source for multipotent mesenchymal stem cells with the ability to differentiate into endothelial cells *in vitro*. *BMC Developmental Biology, 7*, 11.

Andrade, C. F., Wong, A. P., Waddell, T. K., et al. (2007). Cell-based tissue engineering for lung regeneration. *American Journal of Physiology. Lung Cellular and Molecular Physiology, 292*(2), L510−L518.

Aula, P., von Koskull, H., Teramo, K., et al. (1980). Glial origin of rapidly adhering amniotic fluid cells. *British Medical Journal, 281*(6253), 1456−1457.

Bajada, S., Mazakova, I., Richardson, J. B., et al. (2008). Updates on stem cells and their applications in regenerative medicine. *Journal of Tissue Engineering and Regenerative Medicine, 2*(4), 169−183.

Beardi, J., Hessmann, M., Hansen, M., et al. (2008). Operative treatment of tibial shaft fractures: a comparison of different methods of primary stabilisation. *Archives of Orthopaedic and Trauma Surgery, 128*(7), 709−715.

Berg, J. S., & Goodell, M. A. (2007). An argument against a role for Oct4 in somatic stem cells. *Cell Stem Cell, 1*(4), 359−360.

Bianco, P., & Robey, P. G. (2001). Stem cells in tissue engineering. *Nature, 414*(6859), 118−121.

Bieback, K., Kern, S., Klüter, H., et al. (2004). Critical parameters for the isolation of mesenchymal stem cells from umbilical cord blood. *Stem Cells, 22*(4), 625−634.

Bollini, S., Pozzobon, M., Nobles, M., et al. Amniotic fluid stem cells can acquire a cardiomyogenic phenotype *in vitro* but their survival *in vivo* is limited when used in an allogeneic setting. (Submitted.)

Bossolasco, P., Montemurro, T., Cova, L., et al. (2006). Molecular and phenotypic characterization of human AF cells and their differentiation potential. *Cell Research, 16*(4), 329−336.

Brace, R. A., & Resnik, R. (1999). Dynamics and disorders of amniotic fluid. In R. K. Creasy & R. Renik (Eds.), *Maternal Fetal Medicine* (pp. 623−643). Philadelphia: Saunders.

Bruce, S. J., Rea, R. W., Steptoe, A. L., et al. (2007). *In vitro* differentiation of murine embryonic stem cells toward a renal lineage. *Differentiation, 75*(5), 337−349.

Bruno, S., Bussolati, B., Grange, C., et al. (2009). Isolation and characterization of resident mesenchymal stem cells in human glomeruli. *Stem Cells and Development, 18*(6), 867–880.

Bussolati, B., Bruno, S., Grange, C., et al. (2005). Isolation of renal progenitor cells from adult human kidney. *American Journal of Pathology, 166*(2), 545–555.

Bussolati, B., Hauser, P. V., Carvalhosa, R., et al. (2009). Contribution of stem cells to kidney repair. *Current Stem Cell Research & Therapy, 4*(1), 2–8.

Cananzi, M., Atala, A., & de Coppi, P. (2009). Stem cells derived from amniotic fluid: new potentials in regenerative medicine. *Reproductive Biomedicine Online, 18*(Suppl. 1), 17–27.

Carraro, G., Perin, L., Sedrakyan, S., et al. (2008). Human amniotic fluid stem cells can integrate and differentiate into epithelial lung lineages. *Stem Cells.* (Epub ahead of print).

Chambers, I., Silva, J., Colby, D., et al. (2007). Nanog safeguards pluripotency and mediates germline development. *Nature, 450*(7173), 1230–1234.

Chen, W. Q., Siegel, N., Li, L., et al. (2009). Variations of protein levels in human amniotic fluid stem cells CD117/2 over passages 5–25. *Journal of Proteome Research, 8*(11), 5285–5295.

Chiavegato, A., Bollini, S., Pozzobon, M., et al. (2007). Human AF-derived stem cells are rejected after transplantation in the myocardium of normal, ischemic, immuno-suppressed or immuno-deficient rat. *Journal of Molecular and Cellular Cardiology, 42*(4), 746–759.

Cipriani, S., Bonini, D., Marchina, E., et al. (2007). Mesenchymal cells from human AF survive and migrate after transplantation into adult rat brain. *Cell Biology International, 31*(8), 845–850.

Caughey, A. B., Hopkins, L. M., & Norton, M. E. (2006). Chorionic villus sampling compared with amniocentesis and the difference in the rate of pregnancy loss. *Obstetrics and Gynecology, 108*, 612–616.

Cheng, F. C., Tai, M. H., Sheu, M. L., et al. (2009). Enhancement of regeneration with glia cell line-derived neurotrophic factor-transduced human amniotic fluid mesenchymal stem cells after sciatic nerve crush injury. *Journal of Neurosurgery.* (Epub ahead of print).

Cremer, M., Treiss, I., Cremer, T., et al. (1981). Characterization of cells of AFs by immunological identification of intermediate-sized filaments: presence of cells of different tissue origin. *Human Genetics, 59*(4), 373–379.

d'Ippolito, G., Schiller, P. C., Ricordi, C., et al. (1999). Age-related osteogenic potential of mesenchymal stromal stem cells from human vertebral bone marrow. *Journal of Bone and Mineral Research, 14*, 1115–1122.

Dai, W., & Kloner, R. A. (2007). Myocardial regeneration by human amniotic fluid stem cells: challenges to be overcome. *Journal of Molecular Cellular Cardiology, 42*(4), 730–732.

Deans, T. L., & Elisseeff, J. H. (2009). Stem cells in musculoskeletal engineered tissue. *Current Opinion in Biotechnology, 20*(5), 537–544.

de Coppi, P., Callegari, A., Chiavegato, A., et al. (2007a). AF and bone marrow derived mesenchymal stem cells can be converted to smooth muscle cells in the cryo-injured rat bladder and prevent compensatory hypertrophy of surviving smooth muscle cells. *Journal of Urology, 177*(1), 369–376.

de Coppi, P., Bartsch, G., Jr., Siddiqui, M. M., et al. (2007b). Isolation of amniotic stem cell lines with potential for therapy. *Nature Biotechnology, 25*(1), 100–106.

Delo, D. M., Olson, J., Baptista, P. M., et al. (2008). Non-invasive longitudinal tracking of human amniotic fluid stem cells in the mouse heart. *Stem Cells and Development, 17*(6), 1185–1194.

Ditadi, A., de Coppi, P., Picone, O., et al. (2008). Human and murine amniotic fluidic-Kit+Lin-cells display hematopoietic activity. *Blood, 113*(17), 3953–3960.

Donaldson, A. E., Cai, J., Yang, M., et al. (2009). Human amniotic fluid stem cells do not differentiate into dopamine neurons *in vitro* or after transplantation *in vivo*. *Stem Cells and Development, 18*(7), 1003–1012.

Donovan, P. J. (2001). High Oct-ane fuel powers the stem cell. *Nature Genetics, 29*(3), 246–247.

Eddleman, K. A., Malone, F. D., Sullivan, L., et al. (2006). Pregnancy loss rates after midtrimester amniocentesis. *Obstetrics and Gynecology, 108*(5), 1067–1072.

Farini, A., Razini, P., Erratico, S., et al. (2009). Cell based therapy for Duchenne muscular dystrophy. *Journal of Cell Physiology, 221*(3), 526–534.

Fauza, D. (2004). AF and placental stem cells. *Best Practice and Research Clinical Obstetrics and Gynaecology, 18*(6), 877–891.

Fuchs, J. R., Kaviani, A., Oh, J. T., et al. (2004). Diaphragmatic reconstruction with autologous tendon engineered from mesenchymal amniocytes. *Journal of Pediatric Surgery, 39*(6), 834–838.

Gang, E. J., Bosnakovski, D., Figueiredo, C. A., et al. (2007). SSEA–4 identifies mesenchymal stem cells from bone marrow. *Blood, 109*(4), 1743–1751.

Gao, J., Coggeshall, R. E., Chung, J. M., et al. (2007). Functional motoneurons develop from human neural stem cell transplants in adult rats. *Neuroreport, 18*(6), 565–569.

Gekas, J., Walther, G., Skuk, D., et al. (2010). *In vitro* and *in vivo* study of human amniotic fluid-derived stem cell differentiation into myogenic lineage. *Clinical and Experimental Medicine, 10*(1), 1–6.

Ghionzoli, M., Cananzi, M., Zani, A., et al. (2009). Amniotic fluid stem cell migration after intraperitoneal injection in pup rats: implication for therapy. *Pediatric Surgical International, 26*(1), 79–84.

Gidekel, S., Pizov, G., Bergman, Y., et al. (2003). Oct-3/4 is a dose-dependent oncogenic fate determinant. *Cancer Cell, 4*(5), 361–370.

Giuliani, S., Perin, L., Sedrakyan, S., et al. (2008). Ex vivo whole embryonic kidney culture: a novel method for research in development, regeneration and transplantation. *Journal of Urology, 179*(1), 365–370.

Gosden, C. M. (1983). AF cell types and culture. *British Medical Bulletin, 39*(4), 348–354.

Gosden, C., & Brock, D. J. (1978). Combined use of alphafetoprotein and amniotic fluid cell morphology in early prenatal diagnosis of fetal abnormalities. *Journal of Medical Genetics, 15*(4), 262–270.

Grisafi, D., Piccoli, M., Pozzobon, M., et al. (2008). High transduction efficiency of human amniotic fluid stem cells mediated by adenovirus vectors. *Stem Cells and Development, 17*(5), 953–962.

Gupta, S., Verfaillie, C., Chmielewski, D., et al. (2006). Isolation and characterization of kidney-derived stem cells. *Journal of the American Society of Nephrology, 7*(11), 3028–3040.

Gupta, S., Verfaillie, C., Chmielewski, D., et al. (2002). A role for extrarenal cells in the regeneration following acute renal failure. *Kidney International, 62*(4), 1285–1290.

Hoehn, H., Bryant, E. M., Karp, L. E., et al. (1975). Cultivated cells from diagnostic amniocentesis in second trimester pregnancies. II. Cytogenetic parameters as functions of clonal type and preparative technique. *Clinical Genetics, 7*(1), 29–36.

Hoehn, H., & Salk, D. (1982). Morphological and biochemical heterogeneity of amniotic fluid cells in culture. *Methods in Cell Biology, 26*, 11–34.

Hotta, R., Natarajan, D., & Thapar, N. (2009). Potential of cell therapy to treat pediatric motility disorders. *Seminars in Pediatric Surgery, 18*(4), 263–273.

Igura, K., Zhang, X., Takahashi, K., et al. (2004). Isolation and characterization of mesenchymal progenitor cells from chorionic villi of human placenta. *Cytotherapy, 6*(6), 543–553.

In 't Anker, P. S., Scherjon, S. A., Kleijburg-van der Keur, C., et al. (2003). AF as a novel source of mesenchymal stem cells for therapeutic transplantation. *Blood, 102*(4), 1548–1549.

In 't Anker, P. S., Scherjon, S. A., Kleijburg-van der Keur, C., et al. (2004). Isolation of mesenchymal stem cells of fetal or maternal origin from human placenta. *Stem Cells, 22*(7), 1338–1345.

Iop, L., Chiavegato, A., Callegari, A., et al. (2008). Different cardiovascular potential of adult- and fetal-type mesenchymal stem cells in a rat model of heart cryoinjury. *Cell Transplantation, 17*(6), 679–694.

Itskovitz-Eldor, J., Schuldiner, M., Karsenti, D., et al. (2000). Differentiation of human embryonic stem cells into embryoid bodies compromising the three embryonic germ layers. *Molecular Medicine, 6*(2), 88–95.

Karlmark, K. R., Freilinger, A., Marton, E., et al. (2005). Activation of ectopic Oct4 and Rex-1 promoters in human AF cells. *International Journal of Molecular Medicine, 16*(6), 987–992.

Kaviani, A., Perry, T. E., Dzakovic, A., et al. (2001). The AF as a source of cells for fetal tissue engineering. *Journal of Pediatric Surgery, 36*(11), 1662–1665.

Kern, S., Eichler, H., Stoeve, J., et al. (2006). Comparative analysis of mesenchymal stem cells from bone marrow, umbilical cord blood, or adipose tissue. *Stem Cells, 24*(5), 1294–1301.

Khalil, P. N., Weiler, V., Nelson, P. J., et al. (2007). Nonmyeloablative stem cell therapy enhances microcirculation and tissue regeneration in murine inflammatory bowel disease. *Gastroenterology, 132*(3), 944–954.

Kim, C. F., Jackson, E. L., Woolfenden, A. E., et al. (2005). Identification of bronchioalveolar stem cells in normal lung and lung cancer. *Cell, 121*, 823–835.

Kim, J., Kang, H. M., Kim, H., et al. (2007a). Ex vivo characteristics of human amniotic membrane-derived stem cells. *Cloning Stem Cells, 9*(4), 581–594.

Kim, J., Lee, Y., Kim, H., et al. (2007b). Human AF-derived stem cells have characteristics of multipotent stem cells. *Cell Proliferation, 40*(1), 75–90.

Kim, P. G., & Daley, G. Q. (2009). Application of induced pluripotent stem cells to hematologic disease. *Cytotherapy, 11*(8), 980–989.

Koelling, S., & Miosge, N. (2009). Stem cell therapy for cartilage regeneration in osteoarthritis. *Expert Opinion on Biological Therapy, 9*(11), 1399–1405.

Koike, M., Sakaki, S., Amano, Y., et al. (2007). Characterization of embryoid bodies of mouse embryonic stem cells formed under various culture conditions and estimation of differentiation status of such bodies. *Journal of Bioscience and Bioengineering, 104*(4), 294–299.

Kolambkar, Y. M., Peister, A., Soker, S., et al. (2007). Chondrogenic differentiation of AF-derived stem cells. *Journal of Molecular Histology, 38*(5), 405–413.

Kramer, J., Steinhoff, J., Klinger, M., et al. (2006). Cells differentiated from mouse embryonic stem cells via embryoid bodies express renal marker molecules. *Differentiation, 74*(2–3), 91–104.

236

Kunisaki, S. M., Fuchs, J. R., Kaviani, A., et al. (2006a). Diaphragmatic repair through fetal tissue engineering: a comparison between mesenchymal amniocyte- and myoblast-based constructs. *Journal of Pediatric Surgery, 41* (1), 34–39.

Kunisaki, S. M., Freedman, D. A., & Fauza, D. O. (2006b). Fetal tracheal reconstruction with cartilaginous grafts engineered from mesenchymal amniocytes. *Journal of Pediatric Surgery, 41*(4), 675–682.

Kunisaki, S. M., Armant, M., Kao, G. S., et al. (2007). Tissue engineering from human mesenchymal amniocytes: a prelude to clinical trials. *Journal of Pediatric Surgery, 42*(6), 974–979.

Lengner, C. J., Camargo, F. D., Hochedlinger, K., et al. (2007). Oct4 expression is not required for mouse somatic stem cell self-renewal. *Cell Stem Cell, 1*(4), 403–415.

Li, C., Zhou, J., Shi, G., et al. (2009). Pluripotency can be rapidly and efficiently induced in human amniotic fluid-derived cells. *Human Molecular Genetics, 18*(22), 4340–4349.

Liedtke, S., Enczmann, J., Waclawczyk, S., et al. (2007). Oct4 and its pseudogenes confuse stem cell research. *Cell Stem Cell, 1*(4), 364–366.

Looijenga, L. H., Stoop, H., de Leeuw, H. P., et al. (2003). POU5F1 (OCT3/4) identifies cells with pluripotent potential in human germ cell tumors. *Cancer Research, 63*(9), 2244–2250.

Lotgering, F. K., & Wallenburg, H. C. (1986). Mechanisms of production and clearance of AF. *Seminars in Perinatology, 10*(2), 94–102.

Marcus, A. J., & Woodbury, D. (2008). Fetal stem cells from extra-embryonic tissues: do not discard. *Journal of Cellular and Molecular Medicine, 12*(3), 730–742.

Marcus, A. J., Coyne, T. M., Rauch, J., et al. (2008). Isolation, characterization, and differentiation of stem cells derived from the rat amniotic membrane. *Differentiation, 76*(2), 130–144.

Mescher, E. J., Platzker, A. C., Ballard, P. L., et al. (1975). Ontogeny of tracheal fluid, pulmonary surfactant, and plasma corticoids in the fetal lamb. *Journal of Applied Physiology, 39*(6), 1017–1021.

Miki, T., Lehmann, T., Cai, H., et al. (2005). Stem cell characteristics of amniotic epithelial cells. *Stem Cells, 23*(10), 1549–1559.

Miki, T., & Strom, S. C. (2006). Amnion-derived pluripotent/multipotent stem cells. *Stem Cell Reviews, 2*(2), 133–142.

Milunsky, A. (1979). *Genetic Disorder of the Fetus.* New York: Plenum Press. pp. 75–84.

Morigi, M., Introna, M., Imberti, B., et al. (2008). Human bone marrow mesenchymal stem cells accelerate recovery of acute renal injury and prolong survival in mice. *Stem Cells, 26*(8), 2075–2082.

Morigi, M., Rota, C., Montemurro, T., et al. (2010). Life-sparing effect of human cord-blood mesenchymal stem cells in experimental acute kidney injury. *Stem Cells.* (Epub ahead of print).

Morito, T., Muneta, T., Hara, K., et al. (2008). Synovial fluid-derived mesenchymal stem cells increase after intra-articular ligament injury in humans. *Rheumatology, 47*(8), 1137–1143.

Muller, F., Dommergues, M., Ville, Y., et al. (1994). Amniotic fluid digestive enzymes: diagnostic value in fetal gastrointestinal obstructions. *Prenatal Diagnosis, 14*(10), 973–979.

Murray, P., Camussi, G., Davies, J. A., et al. (2007). The KIDSTEM European Research Training Network: developing a stem cell based therapy to replace nephrons lost through reflux nephropathy. *Organogenesis, 3*(1), 2–5.

Muscolo, D. L., Ayerza, M. A., Farfalli, G., et al. (2009). Proximal tibia osteoarticular allografts in tumor limb salvage surgery. *Clinical Orthopaedics and Related Research.* (Epub ahead of print).

Nadri, S., & Soleimani, M. (2008). Comparative analysis of mesenchymal stromal cells from murine bone marrow and amniotic fluid. *Cytotherapy, 9*(8), 729–737.

Nelson, M. M. (1973). Amniotic fluid cell culture and chromosome studies. In A. E. H. Emery (Ed.), *Antenatal Diagnosis of Genetic Disease* (pp. 69–81). Churchill Livingstone Edinburgh.

Nichols, J., Zevnik, B., Anastassiadis, K., et al. (1998). Formation of pluripotent stem cells in the mammalian embryo depends on the POU transcription factor Oct4. *Cell, 95*(3), 379–391.

Niwa, H., Miyazaki, J., & Smith, A. G. (2000). Quantitative expression of Oct-3/4 defines differentiation, dedifferentiation or self-renewal of ES cells. *Nature Genetics, 24*(4), 372–376.

Ott, H. C., Matthiesen, T. S., Goh, S. K., et al. (2008). Perfusion-decellularized matrix: using nature's platform to engineer a bioartificial heart. *Nature Medicine, 14*(2), 213–221.

Owen, M., & Friedenstein, A. J. (1988). Stromal stem cells: marrow-derived osteogenic precursors. *Ciba Foundation Symposium, 136,* 42–60.

Pan, H. C., Yang, D. Y., Chiu, Y. T., et al. (2006). Enhanced regeneration in injured sciatic nerve by human amniotic mesenchymal stem cell. *Journal of Clinical Neuroscience, 13*(5), 570–575.

Pan, H. C., Cheng, F. C., Chen, C. J., et al. (2007). Post-injury regeneration in rat sciatic nerve facilitated by neurotrophic factors secreted by AF mesenchymal stem cells. *Journal of Clinical Neuroscience, 14*(11), 1089–1098.

Panés, J., & Salas, A. (2009). Mechanisms underlying the beneficial effects of stem cell therapies for inflammatory bowel diseases. *Gut, 58*(7), 898–900.

Parolini, O., Alviano, F., Bagnara, G. P., et al. (2007). Concise review: isolation and characterization of cells from human term placenta: outcome of the first international Workshop on Placenta Derived Stem Cells. *Stem Cells, 26*(2), 300–311.

Peister, A., Deutsch, E. R., Kolambkar, Y., et al. (2009). Amniotic fluid stem cells produce robust mineral deposits on biodegradable scaffolds. *Tissue Engineering Part A, 15*(10), 3129–3123.

Perin, L., Giuliani, S., Jin, D., et al. (2007). Renal differentiation of AF stem cells. *Cell Proliferation, 40*(6), 936–948.

Perin, L., Giuliani, S., Sedrakyan, S., et al. (2008). Stem cell and regenerative science applications in the development of bioengineering of renal tissue. *Pediatric Research, 63*(5), 467–471.

Pesce, M., & Schöler, H. R. (2001). Oct4: gatekeeper in the beginnings of mammalian development. *Stem Cells, 19* (4), 271–278.

Pittenger, M. F., Mackay, A. M., Beck, S. C., et al. (1999). Multilineage potential of adult human mesenchymal stem cells. *Science, 284*(5411), 143–147.

Porada, C. D., Zanjani, E. D., & Almeida-Porad, G. (2006). Adult mesenchymal stem cells: a pluripotent population with multiple applications. *Current Stem Cell Research Therapy, 1*(3), 365–369.

Price, F. D., Kuroda, K., & Rudnicki, M. A. (2007). Stem cell based therapies to treat muscular dystrophy. *Biochimica and Biophysica Acta, 1772*(2), 272–283.

Prusa, A. R., Marton, E., Rosner, M., et al. (2003). Oct4-expressing cells in human AF: a new source for stem cell research? *Human Reproduction, 18*(7), 1489–1493.

Rehni, A. K., Singh, N., Jaggi, A. S., et al. (2007). AF derived stem cells ameliorate focal cerebral ischaemia-reperfusion injury induced behavioural deficits in mice. *Behavioural Brain Research, 183*(1), 95–100.

Riboldi, M., & Simon, C. (2009). Extraembryonic tissues as a source of stem cells. *Gynecological Endocrinology, 25*(6), 351–355.

Roubelakis, M. G., Pappa, K. I., Bitsika, V., et al. (2007). Molecular and proteomic characterization of human mesenchymal stem cells derived from amniotic fluid: comparison to bone marrow mesenchymal stem cells. *Stem Cells and Development, 16*(6), 931–952.

Sakaguchi, Y., Sekiya, I., Yagishita, K., et al. (2004). Suspended cells from trabecular bone by collagenase digestion become virtually identical to mesenchymal stem cells obtained from marrow aspirates. *Blood, 104*(9), 2728–2735.

Salgado, A. J., Oliveira, J. T., Pedro, A. J., et al. (2006). Adult stem cells in bone and cartilage tissue engineering. *Current Stem Cell Research & Therapy, 1*(3), 345–364.

Santos, C. C., Furth, M. E., Snyder, E. Y., et al. (2008). Response to Do amniotic fluid-derived stem cells differentiate into neurons *in vitro*? *Nature Biotechnology, 26*(3), 270–271.

Schöler, H. R., Balling, R., Hatzopoulos, A. K., et al. (1989). Octamer binding proteins confer transcriptional activity in early mouse embryogenesis. *The EMBO Journal, 8*(9), 2551–2557.

Seaberg, R. M., Smukler, S. R., Kieffer, T. J., et al. (2004). Clonal identification of multipotent precursors from adult mouse pancreas that generate neural and pancreatic lineages. *Nature Biotechnology, 22*, 1115–1124.

Sessarego, N., Parodi, A., Podestà, M., et al. (2008). Multipotent mesenchymal stromal cells from amniotic fluid: solid perspectives for clinical application. *Haematologica, 93*(3), 339–346.

Siegel, N., Rosner, M., Hanneder, M., et al. (2007). Stem cells in amniotic fluid as new tools to study human genetic diseases. *Stem Cell Reviews, 3*(4), 256–264.

Siegel, N., Valli, A., Fuchs, C., et al. (2009a). Expression of mTOR pathway proteins in human amniotic fluid stem cells. *International Journal of Molecular Medicine, 23*(6), 779–784.

Siegel, N., Valli, A., Fuchs, C., et al. (2009b). Induction of mesenchymal/epithelial marker expression in human amniotic fluid stem cells. *Reproductive Biomedicine Online, 19*(6), 838–846.

Soncini, M., Vertua, E., Gibelli, L., et al. (2007). Isolation and characterization of mesenchymal cells from human fetal membranes. *Journal of Tissue Engineering and Regenerative Medicine, 1*(4), 296–305.

Stefanidis, K., Loutradis, D., Koumbi, L., et al. (2007). Deleted in Azoospermia-Like (DAZL) gene-expressing cells in human AF: a new source for germ cells research? *Fertility and Sterility.* (Epub ahead of print).

Steigman, S. A., Ahmed, A., Shanti, R. M., et al. (2009). Sternal repair with bone grafts engineered from amniotic mesenchymal stem cells. *Journal of Pediatric Surgery, 44*(6), 1120–1126.

Streubel, B., Martucci-Ivessa, G., Fleck, T., et al. (1996). *In vitro* transformation of amniotic cells to muscle cells — background and outlook. *Wiener medizinische Wochenschrift, 146*(9–10), 216–217.

Srivastava, A. S., Feng, Z., Mishra, R., et al. (2007). Embryonic stem cells ameliorate piroxicam-induced colitis in IL10-/- KO mice. *Biochemical Biophysical Research Communication, 361*(4), 953–959.

Tamagawa, T., Ishiwata, I., & Saito, S. (2004). Establishment and characterization of a pluripotent stem cell line derived from human amniotic membranes and initiation of germ layers *in vitro*. *Human Cell, 17*(3), 125–130.

Thakar, N., Priest, R. E., & Priest, J. H. (1982). Estrogen production by cultured amniotic fluid cells. *Clinical Research, 30*, 888A.

Torricelli, F., Brizzi, L., Bernabei, P. A., et al. (1993). Identification of hematopoietic progenitor cells in human amniotic fluid before the 12th week of gestation. *Italian Journal of Anatomy and Embryology, 98*(2), 119—126.

Toselli, M., Cerbai, E., Rossi, F., et al. (2008). Do amniotic fluid-derived stem cells differentiate into neurons *in vitro*? *Nature Biotechnology, 26*(3), 269—270.

Tsai, M. S., Lee, J. L., Chang, Y. J., et al. (2004). Isolation of human multipotent mesenchymal stem cells from second-trimester AF using a novel two-stage culture protocol. *Human Reproduction, 19*(6), 1450—1456.

Tsai, M. S., Hwang, S. M., Tsai, Y. L., et al. (2006). Clonal AF-derived stem cells express characteristics of both mesenchymal and neural stem cells. *Biology of Reproduction, 74*(3), 545—551.

Tsai, M. S., Hwang, S. M., Chen, K. D., et al. (2007). Functional network analysis of the transcriptomes of mesenchymal stem cells derived from amniotic fluid, amniotic membrane, cord blood, and bone marrow. *Stem Cells, 25*(10), 2511—2523.

Uccelli, A., Pistoia, V., & Moretta, L. (2007). Mesenchymal stem cells: a new strategy for immunosuppression? *Trends in Immunology, 28*, 219—226.

Underwood, M. A., Gilbert, W. M., & Sherman, M. P. (2005). AF: not just fetal urine anymore. *Journal of Perinatology, 25*(5), 341—348.

Ungrin, M. D., Joshi, C., Nica, A., et al. (2008). Reproducible, ultra high-throughput formation of multicellular organization from single cell suspension-derived human embryonic stem cell aggregates. *PLoS One, 3*(2), e1565.

Valli, A., Rosner, M., Fuchs, C., et al. (2009). Embryoid body formation of human amniotic fluid stem cells depends on mTOR. *Oncogene, 29*(7), 966—977.

von Koskull, H., Virtanen, I., Lehto, V. P., et al. (1981). Glial and neuronal cells in amniotic fluid of anencephalic pregnancies. *Prenatal Diagnosis, 1*(4), 259—267.

von Koskull, H., Aula, P., Trejdosiewicz, L. K., et al. (1984). Identification of cells from fetal bladder epithelium in human AF. *Human Genetics, 65*(3), 262—267.

Walther, G., Gekas, J., & Bertrand, O. F. (2009). Amniotic stem cells for cellular cardiomyoplasty: promises and premises. *Catheterization and Cardiovascular Interventions, 73*(7), 917—924.

Weiss, D. J. (2008). Stem cells and cell therapies for cystic fibrosis and other lung diseases. *Pulmonary Pharmacology and Therapeutics, 21*(4), 588—594.

Wolbank, S., Peterbauer, A., Fahrner, M., et al. Dose-dependent immunomodulatory effect of human stem cells from amniotic membrane: a comparison with human mesenchymal stem cells from adipose tissue. *Tissue Engineering, 13*(6), 1173—1183

Yao, M., Dieterle, T., Hale, S. L., et al. (2003). Long-term outcome of fetal cell transplantation on postinfarction ventricular remodeling and function. *Journal of Molecular and Cellular Cardiology, 35*(6), 661—670.

Yeh, Y. C., Wei, H. J., Lee, W. Y., et al. (2010). Cellular cardiomyoplasty with human amniotic fluid stem cells: *in vitro* and *in vivo* studies. *Tissue Engineering Part A.* (Epub ahead of print).

You, Q., Tong, X., Guan, Y., et al. (2009). The biological characteristics of human third trimester amniotic fluid stem cells. *Journal of International Medical Research, 37*(1), 105—112.

Zani, A., Cananzi, M., Eaton, S., et al. (2009). Stem cells as a potential treatment of necrotizing enterocolitis. *Journal of Pediatric Surgery, 44*(3), 659—660.

Zhang, P., Baxter, J., Vinod, K., et al. (2010). Endothelial differentiation of amniotic fluid-derived stem cells: synergism of biochemical and shear force stimuli. *Stem Cells Development, 18*(9), 1299—1308.

Zhao, P., Ise, H., Hongo, M., et al. (2005). Human amniotic mesenchymal cells have some characteristics of cardiomyocytes. *Transplantation, 79*(5), 528—535.

Zsebo, K. M., Williams, D. A., Geissler, E. N., et al. (1990). Stem cell factor is encoded at the Sl locus of the mouse and is the ligand for the c-kit tyrosine kinase receptor. *Cell, 63*(1), 213—224.

239

CHAPTER 13

Induced Pluripotent Stem Cells

Keisuke Okita[*], **Shinya Yamanaka**[*,**,***,†]
*Center for iPS Cell Research and Application (CiRA), Institute for Integrated Cell-Material Sciences, Kyoto University, Kyoto, Japan
**Department of Stem Cell Biology, Institute for Frontier Medical Sciences, Kyoto University, Kyoto, Japan
***Yamanaka iPS Cell Special Project, Japan Science and Technology Agency, Kawaguchi, Japan
†Gladstone Institute of Cardiovascular Disease, San Francisco, CA, USA

INTRODUCTION

Reprogramming of somatic cells has been extensively investigated. Successful studies yielded the generation of cloned animals from frog (Gurdon, 1962) and sheep (Wilmut et al., 1997) somatic cells. The somatic cells were fused with enucleated oocyte in those studies, which indicated the existence of a reprogramming factor in the oocyte. ES cells, which are derived from early embryonic tissue, have similar activity, and they can reprogram somatic cells by cell fusion (Tada et al., 2001). Reprogramming with defined factors based on those results was reported in 2006 (Takahashi and Yamanaka, 2006). Takahashi et al. introduced four defined transcription factors (Oct3/4, Sox2, Klf4, and c-Myc), which are expressed abundantly in ES cells into mouse fibroblasts, and obtained pluripotent stem cells. These artificial cells were termed induced pluripotent stem (iPS) cells. The iPS cells have similar morphology, proliferation, and gene expression profile to those of ES cells. The mouse iPS cells can be transferred into early embryos, where they contribute to tissue development, make adult chimeric mice, and are transmitted through the germ line to the next generation (Maherali et al., 2007; Okita et al., 2007; Wernig et al., 2007). Establishment of iPS cells from human somatic cells was reported in 2007 (Takahashi et al., 2007; Yu et al., 2007b), and since then iPS cells have been generated in several animals, including rats (Liao et al., 2009; Li et al., 2009b), monkeys (Liu et al., 2008), pigs (Esteban et al., 2009a), and dogs (Shimada et al., 2010). The *in vitro* reprogrammed cells have been attracting a lot of attention because they could supply patient-specific pluripotent stem cells for use in many fields, such as the study of disease pathogenesis, drug discovery, toxicology, and even cell transplantation therapy in the future. This chapter will summarize the recent research and future problems associated with iPS cells.

GENERATION OF iPS CELLS
Reprogramming factors

iPS cells are established by the forced expression of several transgenes. The classic mixture is Oct3/4, Sox2, Klf4, and c-Myc (Takahashi and Yamanaka, 2006). This mixture can reprogram mouse, human, rat, monkey, and dog somatic cells. All of these factors have transcriptional activity, and Oct3/4, Sox2, and Klf4 regulate many ES-cell-specific genes in combination (Jiang

Principles of Regenerative Medicine. DOI: 10.1016/B978-0-12-381422-7.10013-6

et al., 2008). These factors also regulate their own expression. There are families of genes for Oct3/4, Sox2, and Klf4, and some of them can induce iPS cells. For example, Sox2 can be replaced with Sox1, Sox3, Sox7, Sox15, Sox17, or Sox18, and Klf4 with Klf2 (Nakagawa et al., 2008). Comparison of the target genes among reprogramming factors and the family genes might be useful to understand the molecular mechanisms underlying iPS cell formation. Other combinations, such as Oct3/4, Sox2, Nanog, and Lin28, have been reported for the generation of human iPS cells (Yu et al., 2007b). Nanog is one of the most important transcription factors for stabilization of pluripotent state in mouse ES cells (Chambers et al., 2003; Mitsui et al., 2003). It also makes a transcriptional circuit with Oct3/4, Sox2, and Klf4 (Jiang et al., 2008). Oct3/4, Sox2, and Nanog bind and upregulate ES-cell-specific genes such as STAT3 and ZIC3 with RNA polymerase II (Boyer et al., 2005; Lee et al., 2006). On the other hand, they also localize to developmental regulator genes, such as PAX6 and ATBF1, with SUZ12, where they work as suppressors. The forced expression of some core components of ES cells would induce ES-cell-like transcription networks in somatic cells and change their state. c-Myc is associated with many aspects of reprogramming, but its precise function is unclear. The process of iPS induction is thought to have some stochastic events dependent on cell proliferation, such as passive DNA demethylation. The expression of c-Myc blocks cell senescence, accelerates proliferation of fibroblasts, and leads to enhancement of iPS induction. c-Myc binds to more than 4,000 sites of the genome (Li et al., 2003); therefore, it could loosen tightly packed chromosomes in somatic cells and increase the accessibility of other transcription factors to the genome during iPS induction. Overexpression of c-Myc itself also shifts the gene expression profile of mouse embryonic fibroblasts (MEFs) towards pluripotent cells (Sridharan et al., 2009). LIN28 is an RNA-binding protein and negatively regulates Let7 microRNA (miRNA) families (Heo et al., 2009; Viswanathan et al., 2009). Let-7 promotes differentiation of breast cancer cells and inhibits their proliferation (Yu et al., 2007a). Therefore, LIN28 seems to indirectly enhance reprogramming efficiency through Let7 families. A combination of extra factors used in the induction can improve the reprogramming efficiency and quality. The addition of transcription factors, such as ESRRB26 (Feng et al., 2009), UTF127 (Zhao et al., 2008), and SALL428 (Tsubooka et al., 2009), increased the efficiency. All of these factors are expressed in ES cells, and are involved in the formation of an ES-like transcriptional network. Han et al. reported that Tbx3 significantly improves the quality and the germ line competency of mouse iPS cells (Han et al., 2010). Some variations of inducing factors have been reported for iPS generation. The factor(s) in the reprogramming cocktail can be reduced if the somatic cells have sufficient endogenous expression of either of the reprogramming factor(s). For example, neural precursor cells express endogenous SOX2, KLF4, and c-MYC, and they only need OCT3/4 transgenes for iPS cell induction (Kim et al., 2009c).

The acceleration of cell proliferation and the inhibition of senescence by the suppression of the p53 and p21 pathways can also dramatically increase the efficiency (Hong et al., 2009; Kawamura et al., 2009; Li et al., 2009a; Marion et al., 2009; Utikal et al., 2009). An increase in the number of cells under induction results in high iPS colony formation because the reprogramming process includes stochastic events. Hanna et al. showed that the suppression of p53 increases reprogramming efficiency predominantly through acceleration of cell division (Hanna et al., 2009). On the other hand, the addition of Nanog to the reprogramming factor upregulated the net reprogramming efficiency, in a cell-division-independent manner. However, the suppression of the p53 and p21 pathways increases the genomic instability of iPS cells. Therefore, the permanent suppression of the pathway should be avoided because it would lower the quality of iPS cells. The transient suppression of inhibitors or siRNAs could be useful for the enhancement of reprogramming.

iPS cell induction takes at least one week in the mouse and two weeks in humans. On the other hand, reprogramming by fusion of ES cells occurs very rapidly. The activation of endogenous Oct3/4 promoter of somatic cell nuclei is observed within 2 days (Tada et al., 2001). Although transgene expression in iPS cells requires a few days after vector transduction, the reprogramming of iPS cells seems to take much more time than that of cell fusion. ES cells must have other factor(s) that facilitate the reprogramming. Reprogramming events occur naturally *in vivo* during early developmental stages. The fertilized eggs erase almost all epigenetic status except imprinting before blastocyst formation, and they rebuild it as differentiation proceeds. The eggs also have high reprogramming activity, since they can produce a cloned animal after enucleation and fusion with somatic cells (Wakayama et al., 1998). Although the mechanism remains elusive, cloning might provide helpful hints to improve the generation of iPS cells. However, cloned mice have some abnormalities, such as a large placenta and a tendency to gain excess weight. There may be some limitation in the artificial reprogramming that must be considered.

Transduction methods

iPS cells were originally established by the delivery of transgenes by MMLV (Moloney murine leukemia virus)-based retroviral vectors (Morita et al., 2000). A retrovirus can robustly infect mouse fibroblasts and introduce its RNA genome into the host genome by reverse transcriptase. Therefore, the iPS cells integrate numerous transgenes, which thereby enable constant transgene expression. The inactivation of the retroviral promoter by DNA methylation is observed in ES cells as well as in iPS cells (Okita et al., 2007). Therefore, the expression of retroviral transgenes is gradually suppressed during the reprogramming process, and the silencing is complete when the cells become iPS cells. This automatic silencing mechanism is thought to provide effective reprogramming in somatic cells. However, the exogenous sequences remain in the genome of iPS cells and the alteration of genomic organization could induce some abnormalities. In particular, c-Myc, one of the reprogramming factors, is a proto-oncogene, and its reactivation could give rise to transgene-derived tumor formation (Okita et al., 2007). There have been improvements in the transduction method for making safe iPS cells. Elimination of the c-Myc transgene for iPS cell induction is one important approach. Human and mouse iPS cells can be established from fibroblasts with only Oct3/4, Sox2, and Klf4, although the efficiency is significantly lower (Nakagawa et al., 2008). Mouse iPS cells without c-Myc do not show enhanced tumor formation during the observation period (6 months) in comparison to control mice. Another approach is to reduce the number of integration sites by attaching the reprogramming factors with IRES or 2A self-cleavage peptide and putting them into a single vector. This reprogramming cassette was used with a lentivirus system containing a loxP sequence and produced iPS cells with only single insertions (Sommer et al., 2009a,b; Soldner et al., 2009). The expression of Cre recombinase successfully cuts out the cassette, although a truncated LTR remains in the iPS genome. The elimination of transgenes from the genome avoids the leaky expression of reprogramming factors, and improves the gene expression profile and the differentiation potential of iPS cells (Sommer et al., 2009a). A transposon system has also been used for iPS induction (Kaji et al., 2009; Woltjen et al., 2009). A plasmid-based transposon vector with a reprogramming cassette can integrate into host genome with transposase. The re-expression of the transposase after establishment of iPS cells recognizes the terminal repeat of integrated transposon vector, and excises it from genome. The excision of the transposon does not leave a footprint in most cases, so it maintains the original endogenous sequences. Non-integration methods were also reported with viral vectors (adenovirus (Stadtfeld et al., 2008a; Zhou and Freed, 2009) and sendaivirus (Fusaki et al., 2009)), DNA vectors (plasmid (Okita et al., 2008), episomal plasmid (Yu et al., 2009), and minicircle vector (Jia et al., 2010)), and direct protein delivery (Zhou et al., 2009; Kim et al., 2009a). Although the induction efficiency of iPS cells with these methods is still low, they could become future standard methods.

CULTURE CONDITIONS AND CELL SIGNALING

Culture conditions and cell signaling have a great influence on iPS generation. iPS cells are cultured in medium optimized for ES cells. Leukemia inhibitory factor and basic fibroblast growth factor are important for mouse and human ES cell maintenance, respectively. However, the roles of these cytokines in the induction process are still unclear. Wnt signaling supports self-renewal of ES cells (Ying et al., 2008). The Wnt3a signal is mediated by glycogen synthase kinase (GSK) 3-β. Without the Wnt signal, GSK3-β inactivates target genes, such as β-catenin and c-Myc, by phosphorylation and proteasome-mediated degradation. Hence, the inhibition of GSK3-β with a chemical drug, such as CHIR99021, results in activation of Wnt signaling. Addition of Wnt3a (Marson et al., 2008a) or CHIR99021 (Li et al., 2009a) enhances the reprogramming efficiency. Kenpaullone is an inhibitor whose targets are GSK3-β as well as CDKs, and can replace Klf4 in reprogramming induction from MEF with Oct3/4, Sox2, and c-Myc (Lyssiotis et al., 2009). Although more specific GSK3-β inhibitors, such as CHIR99021, or CDK inhibitor, purvalanol A, were unable to generate mouse iPS cells with the same combination of transcription factors and Kenpaullone itself did not increase endogenous Klf4 expression, the function of Kenpaullone is still elusive. Importantly, Li et al. found that the combination of CHIR99021 and Parnate, an inhibitor of lysine-specific demethylase 1, can generate iPS cells from human primary keratinocytes with only Oct3/4 and Klf4 (Li et al., 2009c). The addition of vitamin C enhances iPS cell generation from both mouse and human somatic cells (Esteban et al., 2009b). Vitamin C works at least in part by alleviating cell senescence.

O_2 tension is also an important factor for stem cell maintenance and differentiation. For instance, low O_2 tension promotes the survival of neural crest cells and hematopoietic stem cells, and prevents differentiation of human ES cells. Yoshida et al. found up to four-fold enhancement of the reprogramming efficiency when the iPS induction was performed in hypoxic conditions (5% O_2), in both mouse and human fibroblasts (Yoshida et al., 2009).

Cell source

iPS cells were first established from primary mouse fibroblast culture. Their origin was thought to be some tissue stem cells included in the culture since the efficiency of iPS cell induction was very low (less than 0.1%). Aoi et al. showed that mouse iPS cells can be established from mouse hepatocytes and stomach epithelial cells and linage tracing experiments showed that most hepatocyte-derived iPS cells were from albumin-positive cells (Aoi et al., 2008). Mouse iPS cells were also established from pancreatic islet β cells (Stadtfeld et al., 2008b). Therefore, the origin of iPS cells is not only tissue stem cells but also differentiated somatic cells. Human iPS cells have been established from various tissues, including fibroblasts (adult and embryo) (Takahashi et al., 2007; Yu et al., 2007b), adult keratinocyte (Aasen et al., 2008), adipose tissue (Sun et al., 2009), peripheral blood (Loh et al., 2009), cord blood (Giorgetti et al., 2009), amniotic fluid-derived cells (Ye et al., 2009), and neural precursor cells (Kim et al., 2009c). Hence, all somatic cells are thought to have the ability to yield iPS cells, although they show differential efficiency. However, it is unclear whether iPS cells from different cell sources have the same potential. Mouse iPS cells derived by the current reprogramming method from different tissues apparently have divergent characteristics. Miura et al. compared neural differentiation potential and safety of mouse iPS cells derived from MEF, tail-tip fibroblasts (TTFs), and hepatocytes (Miura et al., 2009). Most iPS clones form a neural sphere under *in vitro* directed differentiation conditions. The neural sphere contains neural precursor cells that can produce three neuronal cell types, neuron, astrocyte, and oligodendrocyte. The neurosphere from ES cells could contribute the neural tissues when transplanted into the mouse brain. However, the neurospheres prepared from TTF-derived iPS cells tended to form teratomas after transplantation into mouse brain. Teratoma formation has been reported in the transplantation of neurospheres formed from ES cells containing undifferentiated cells the remained after the differentiation process (Dihne et al., 2006). The population of undifferentiated cells is rare in neurospheres from MEF-derived iPS cells and ES cells, but is obvious in those from TTF-derived iPS cells (up to 20%) (Miura et al., 2009). The study revealed that the existence of undifferentiated cells varies depending on the cell source. Accessibility to a cell source is another important point in the selection of tissues, especially for induction of human iPS cells. Human iPS cells can be established from neural precursor cells with only OCT3/4 transgenes (Kim et al., 2009c); however, constant acquisition of the neural tissue is difficult.

MOLECULAR MECHANISMS IN iPS CELL INDUCTION

Epigenetics

The generation of iPS cells includes epigenetic alterations. DNA methylation status and histone modifications of promoter regions including Nanog, Oct3/4, Sox2, and Fbxo15 achieve an ES-like state after reprogramming. The addition of a histone deacetylase (HDAC) inhibitor, valproic acid (VPA), improves the reprogramming efficiency in both mouse (Huangfu et al., 2008a) and human (Huangfu et al., 2008b) fibroblasts. Other HDAC inhibitors, such as suberoylanilide hydroxamic acid and trichostatin A, also work in mouse fibroblasts (Huangfu et al., 2008a). Inhibitors of DNA methyltransferase, such as 5′-azacytidine and RG108 (Shi et al., 2008a), and BIX-01294 (Shi et al., 2008b) for G9a histone methyltransferase increased reprogramming efficiency. These results supported the hypothesis that the process of iPS generation involves epigenetic changes. Some of the inhibitors could abolish the use of one or two reprogramming factor(s). For example, VPA treatment of human fibroblasts enables reprogramming with only two factors, Oct4 and Sox2, and eliminates the oncogenic c-Myc or Klf4 (Huangfu et al., 2008b). However, it is doubtful whether these drugs fill in the exact function of reprogramming genes; rather, they seemed to enhance the induction efficiency that allows the reduction of reprogramming factor(s).

iPS induction requires the establishment of an ES-like transcription factor circuit in somatic cells. In fact, iPS cells have the same expression profile as ES cells; however, they have differences in epigenetic modifications, especially in genes not involved in pluripotency. Cell differentiation is a process of limitation of the differentiation potential by epigenetic modification. Each type of somatic cell has its specific epigenetics by which cells are able to stabilize their state. The forced expression of reprogramming factors can affect several downstream genes in somatic cells and alter their epigenetic modifications. However, it is difficult to think that the factors control all genes throughout the genome. In fact, genome-wide analysis showed similar DNA methylation patterns of iPS and ES cells, but they also detected differentially methylated regions between iPS and ES cells (Ball et al., 2009; Deng et al., 2009; Doi et al., 2009). The uncontrolled genes would keep their epigenetic profiles even in iPS cells. This could influence the differentiation potential of iPS cells. For example, the methylation status of the enhancer binding site in the astrocyte gene, GFAP, controls the differentiation fate of neuronal precursor by changing the binding activity for an enhancer, STAT3 (Takizawa et al., 2001). Without such methylation they instead tended to become astrocytes, whereas in the presence of methylation they tended to demonstrate neuronal differentiation.

MicroRNAs

miRNAs are small single-stranded RNAs (around 22 nt) that directly interact with target mRNAs through complementary base-pairing and inhibit the expression of the target genes. miRNAs also work at the transcriptional level. miRNAs are generated as long RNA sequences and are digested to the short mature form by Dicer. miRNAs are involved in many features of cell properties, such as proliferation, apoptosis, and differentiation, by fine-tuning gene expression. ES cells have the characteristic expression of miRNAs, and iPS cells also showed a similar expression profile. Over 70% of mRNAs in mouse ES cells are the miR-290 cluster, which contributes to the ES cell-specific rapid cell cycle progression (Marson et al., 2008b; Wang et al., 2008). The cluster includes miR-291-3p, miR-292-3p, miR-293, miR-294, and miR-295. miR-291-3p, miR-294, or miR-295 increases the reprogramming efficiency from MEF with Oct4, Sox2, and Klf4 (Judson et al., 2009). They appear to be downstream targets of c-Myc, because the miRNAs did not enhance reprogramming efficiency in the presence of c-Myc transgene, and c-Myc binds the promoter region of the cluster. The three miRNAs share a conserved seed sequence, which mainly specifies target genes, suggesting they work through common targets. LIN28 is a negative regulator of Let7 miRNA families. Lin28 induced uridylation of immature let7 RNA by a non-canonical poly (A) polymerase, TUTase4, and it leads degradation of the RNA (Heo et al., 2008, 2009). Lin28 gradually decreases during ES cell differentiation, and mature let7 family miRNAs accumulate with inverse correlation. The addition of Lin28 enhances the reprogramming efficiency from both human and mouse fibroblasts (Liao et al., 2008; Hanna et al., 2009). A detailed analysis performed by Hanna et al. showed that Lin28 accelerates the efficiency in a cell cycle-dependent manner. This is consistent with the concept that the targets of mature let7 include oncogenic genes, such as K-Ras and c-Myc (Kim et al., 2009b; Oh et al., 2010). Lin28 facilitates the expression of Oct4 at the post-transcriptional level by direct binding to its mRNA (Qiu et al., 2010).

RECAPITULATION OF DISEASE ONTOLOGY AND DRUG SCREENING

Patient-derived iPS cells are useful in understanding disease ontology. The iPS cells have the same genomic information as the patient. Many iPS cells have been established from somatic cells obtained from patients with adenosine deaminase deficiency-related severe combined immunodeficiency, Duchenne and Becker muscular dystrophy (Park et al., 2008), and amyotrophic lateral sclerosis (Dimos et al., 2008). Ebert et al. established iPS cells from skin fibroblast of a spinal muscular atrophy (SMA) patient (Ebert et al., 2009). SMA is an auto-somal recessive genetic disorder that is characterized by degeneration of motor neurons following progressive muscular atrophy. The most common cause of SMA is a mutation of the survival motor neuron 1 (SMN1) gene, and it significantly reduces the level of protein expression. The motor neurons generated from the patient's iPS cells can recapitulate the disease ontology as they show reduced SMN expression in comparison to those derived from the child's unaffected mother (Ebert et al., 2009). Treatment of VPA or tobramycin increases SMN expression. Importantly, the same treatment also worked in the motor neurons prepared from the patient's iPS cells. The results indicate that iPS cells could provide a useful screening system for identification of a specific drug from thousands of candidate compounds. The patient-derived iPS cells will also be used to find developing drugs that would be harmful to the human body. Some compounds work on target tissue, but have severe side-effects. Long QT syndrome is an inborn heart defect that shows characteristic prolongation of the QT interval on electrocardiogram, increases the risk of irregular heartbeat, and threatens life. It occurs only after drug administration in some individuals. Cardiomyocytes established from the patients of long QT syndrome via iPS cells could therefore be used to identify any possible toxic side effect of candidate compounds before starting clinical trials.

Most diseases do not have a simple cause; they are the total sum of genetic/epigenetic issues, environment, aging, etc., in a complicated relationship between several cell types in the body. It is therefore necessary to establish a way to recapitulate late-onset disease and environmental effects *in vitro* or in an animal model.

iPS CELL BANKING

It will require time to establish a clinical grade of useful cells from a patient's own somatic cells. The applicability and safety of cell type must be assessed. The clinical applications of iPS cells must also be considered from an economic point of view. Complete tailor-made iPS cell therapy would cost too much to apply to a large number of people. Therefore, a banking system should be established for iPS cells. iPS cells having various HLA haplotypes should be collected to avoid immune rejection. Experience with organ transplantation has revealed that the HLA class I molecules, HLA-A and HLA-B, and the class II molecule, HLA-DR, are the most important HLA molecules to match. Therefore, the HLA matching of these loci reduces the incidence of acute rejection and improves transplant survival. Estimations of stem cell bank size have been calculated in Japanese (Nakajima et al., 2006) and UK (Taylor et al., 2005) populations. The random establishment of 170 lines of iPS cells would provide donor lines for 80% of patients with a single mismatch at one of three HLA loci (HLA-A, -B, and -DR) or better match among the Japanese. A comparable bank of 150 lines could provide an acceptable or better match for 84.9% of the UK population. Importantly, a bank size of only 50 lines could provide a three-locus match in 90.7% of the Japanese population, if iPS cells are established from HLA homozygous cells (Nakatsuji et al., 2008). Screening an HLA-type database of 24,000 individuals would be required to identify at least one homozygote for each of 50 different HLA haplotypes. This could be possible if the iPS banks cooperate with other banks, in the same manner as do cord blood banks and bone marrow banks.

SAFETY CONCERNS FOR MEDICAL APPLICATION

The safety issue is a most important point for clinical application of iPS cells. Each culture of iPS cells would have different properties for differentiation and safety. Human iPS cells can be generated from several cell types with different combinations of reprogramming factors by various transduction methods, as described above. Still, no one knows the best way to obtain fully reprogrammed safe iPS cells. Chimeric mice assay has revealed that genomic integration of c-Myc transgene is associated with a high risk of tumor formation and should be avoided (Okita et al., 2007). The integration of Oct3/4, Sox2, and Klf4 seems to have no/little effect on tumorigenesis (Nakagawa et al., 2008). However, the overexpression of Oct3/4 (Hochedlinger et al., 2005) and Klf4 (Rowland and Peeper, 2006) causes tumor formation, and various human tumors express OCT3/4, SOX2, and KLF4. Furthermore, the retroviral insertion to the genome itself may disturb endogenous gene structure and increase tumor risks (Hacein-Bey-Abina et al., 2003). However, there are from one to 40 genomic integration sites of retro- and lentivirus in iPS cells and a PCR-based analysis detects all the integration sites (Aoi et al., 2008; Varas et al., 2008). Therefore, it is possible to estimate the risk beforehand. Non-integration methods have been established, but they have low induction efficiency of iPS cells, which suggests they yield reprogrammed iPS cells of lower quality than integration methods. This might be improved by using better combinations of reprogramming factors and choosing a better cell source. The retroviral induction method might be selected after a careful risk assessment if it induces better reprogramming than other transient or non-integration methods.

Residual undifferentiated cells are a common problem when using stem cells for cell transplantation therapy. As described above, most mouse iPS cells can differentiate into neurospheres; however, a small portion of cells remains in an undifferentiated state in the sphere, thereby giving rise to tumor formation when transplanted (Miura et al., 2009). An effective protocol to eliminate undifferentiated cells should be established, such as improvement of the differentiation protocol and sorting by flow cytometry.

MEDICAL APPLICATION

Mouse iPS cells have been applied for treatment of a humanized sickle cell anemia mouse model (Hanna et al., 2007). Homozygous mice for mutant human β-globin genes show characteristic symptoms including severe anemia due to erythrocyte sickling, splenic infarcts, urine concentration defects, and poor health. The iPS cells established from the mouse have the same genomic mutation. The mutation was corrected by homologous recombination with a non-mutated construct. The rescued iPS cells were differentiated into hematopoietic progenitors and transplanted into a "patient" mouse. The study provided proof of principle for application of iPS cells in combination with gene repair for cell therapy. Efficient gene correction methods have been established in human pluripotent stem cells. Suzuki et al. showed the homologous recombination in human ES cells with helper-dependent adenoviral vectors (Suzuki et al., 2008). The homologous recombination was also performed using zinc finger nuclease-mediated genome editing in both human ES and iPS cells (Hockemeyer et al.,

2009). Human disease-corrected iPS cells have been established from Fanconi anemia patients by lentiviral delivery of a normal gene (Raya et al., 2009). Duchenne muscular dystrophy is caused by defect of the Dystrophine gene, which has an extremely large size of 2.4 Mbp. iPS cells from a DMD patient were transferred and corrected with the human-artificial-chromosome-encoding Dystrophine gene (Kazuki et al., 2009). These techniques could supply patient-specific but gene-corrected iPS cells.

DIRECT FATE SWITCH

The establishment of iPS cells introduced a new paradigm: that forced expression of master genes can alter the cell state. This contributes to the study of direct reprogramming from one somatic cell type into another cell type without mediation of stem cells. Zhou et al. screened more than 1,100 transcription factors and chose nine candidates for β-cell induction (Zhou et al., 2008). They injected a combination of these genes by adenovirus vectors into the pancreata of mice. Insulin-positive cells developed in one month with the mixture of Ngn3, Pdx1, and Mafa. The cells were derived from pancreatic exocrine cells and closely resembled normal β-cells. These cells could ameliorate hyperglycemia by remodeling local vasculature and secreting insulin in mice rendered diabetic by streptozotocin injection. Another example is the conversion of mouse fibroblasts into neurons by induction of three transcription factors, Ascl1, Brn2, and Myt1l (Vierbuchen et al., 2010). The induced neuronal (iN) cells expressed several neuronal markers, generated action potentials, and formed functional synapses. The iN cells are useful for neurological disease modeling and regenerative medicine. Although further study is required, these approaches could therefore supply an alternative method to make specific differentiated cells from a patient's somatic cells or iPS cells.

CONCLUSION

iPS cells have tremendous potential to supply patient-specific pluripotent stem cells for use in the study of disease pathogenesis, drug discovery, toxicology, and cell transplantation therapy. Several lines of evidence support the finding that iPS cells are very similar, but not identical, to ES cells. However, there is insufficient data to definitively determine whether or not this difference is critical. Mouse and rat iPS cells can contribute to chimeric animals after injection into blastocysts. Direct and detailed comparison between iPS cells and ES cells is required. The establishment of iPS cells would also apply not only to the medical field but also to the elucidation of the control mechanisms of stem cells and the development of efficient differentiation protocols. Studies of disease pathogenesis and drug discovery have already been launched, and the results could provide relief to countless people throughout the world. The application of iPS cells to human disease will take time. In addition, both the research and medical application of human iPS cells will also be subject to a wide range of laws and research ethical policies.

References

Aasen, T., Raya, A., Barrero, M. J., Garreta, E., Consiglio, A., Gonzalez, F., et al. (2008). Efficient and rapid generation of induced pluripotent stem cells from human keratinocytes. *Nat. Biotechnol., 26,* 1276–1284.

Aoi, T., Yae, K., Nakagawa, M., Ichisaka, T., Okita, K., Takahashi, K., et al. (2008). Generation of pluripotent stem cells from adult mouse liver and stomach cells. *Science, 321,* 699–702.

Ball, M. P., Li, J. B., Gao, Y., Lee, J. H., LeProust, E. M., Park, I. H., et al. (2009). Targeted and genome-scale strategies reveal gene-body methylation signatures in human cells. *Nat. Biotechnol., 27,* 361–368.

Boyer, L. A., Lee, T. I., Cole, M. F., Johnstone, S. E., Levine, S. S., Zucker, J. P., et al. (2005). Core transcriptional regulatory circuitry in human embryonic stem cells. *Cell, 122,* 947–956.

Chambers, I., Colby, D., Robertson, M., Nichols, J., Lee, S., Tweedie, S., et al. (2003). Functional expression cloning of nanog, a pluripotency sustaining factor in embryonic stem cells. *Cell, 113,* 643–655.

Deng, J., Shoemaker, R., Xie, B., Gore, A., LeProust, E. M., Antosiewicz-Bourget, J., et al. (2009). Targeted bisulfite sequencing reveals changes in DNA methylation associated with nuclear reprogramming. *Nat. Biotechnol.* (Advanced online publication).

Dihne, M., Bernreuther, C., Hagel, C., Wesche, K. O., & Schachner, M. (2006). Embryonic stem cell-derived neuronally committed precursor cells with reduced teratoma formation after transplantation into the lesioned adult mouse brain. *Stem Cells, 24*, 1458–1466.

Dimos, J. T., Rodolfa, K. T., Niakan, K. K., Weisenthal, L. M., Mitsumoto, H., Chung, W., et al. (2008). Induced pluripotent stem cells generated from patients with ALS can be differentiated into motor neurons. *Science, 321*, 1218–1221.

Doi, A., Park, I. H., Wen, B., Murakami, P., Aryee, M. J., Irizarry, R., et al. (2009). Differential methylation of tissue- and cancer-specific CpG island shores distinguishes human induced pluripotent stem cells, embryonic stem cells and fibroblasts. *Nat. Genet., 41*, 1350–1353.

Ebert, A. D., Yu, J., Rose, F. F., Jr., Mattis, V. B., Lorson, C. L., Thomson, J. A., et al. (2009). Induced pluripotent stem cells from a spinal muscular atrophy patient. *Nature, 457*, 277–280.

Esteban, M. A., Wang, T., Qin, B., Yang, J., Qin, D., Cai, J., et al. (2009a). Vitamin C enhances the generation of mouse and human induced pluripotent stem cells. *Cell Stem Cell.*

Esteban, M. A., Xu, J., Yang, J., Peng, M., Qin, D., Li, W., et al. (2009b). Generation of induced pluripotent stem cell lines from Tibetan miniature pig. *J. Biol. Chem., 284*, 17634–17640.

Feng, B., Jiang, J., Kraus, P., Ng, J. H., Heng, J. C., Chan, Y. S., et al. (2009). Reprogramming of fibroblasts into induced pluripotent stem cells with orphan nuclear receptor Esrrb. *Nat. Cell Biol., 11*, 197–203.

Fusaki, N., Ban, H., Nishiyama, A., Saeki, K., & Hasegawa, M. (2009). Efficient induction of transgene-free human pluripotent stem cells using a vector based on Sendai virus, an RNA virus that does not integrate into the host genome. *Proc. Jpn. Acad., Ser. B, 85*, 15.

Giorgetti, A., Montserrat, N., Aasen, T., Gonzalez, F., Rodriguez-Piza, I., Vassena, R., et al. (2009). Generation of induced pluripotent stem cells from human cord blood using OCT4 and SOX2. *Cell Stem Cell, 5*, 353–357.

Gurdon, J. B. (1962). The developmental capacity of nuclei taken from intestinal epithelium cells of feeding tadpoles. *J. Embryol. Exp. Morphol., 10*, 622–640.

Hacein-Bey-Abina, S., von Kalle, C., Schmidt, M., McCormack, M. P., Wulffraat, N., Leboulch, P., et al. (2003). LMO2-associated clonal T cell proliferation in two patients after gene therapy for SCID-X1. *Science, 302*, 415–419.

Han, J., Yuan, P., Yang, H., Zhang, J., Soh, B.S., Li, P., et al. (2010) Tbx3 improves the germ-line competency of induced pluripotent stem cells. *Nature, 463*, 1096–1100.

Hanna, J., Saha, K., Pando, B., van Zon, J., Lengner, C. J., Creyghton, M. P., et al. (2009). Direct cell reprogramming is a stochastic process amenable to acceleration. *Nature, 462*, 595–601.

Hanna, J., Wernig, M., Markoulaki, S., Sun, C. W., Meissner, A., Cassady, J. P., et al. (2007). Treatment of sickle cell anemia mouse model with iPS cells generated from autologous skin. *Science, 318*, 1920–1923.

Heo, I., Joo, C., Cho, J., Ha, M., Han, J., & Kim, V. N. (2008). Lin28 mediates the terminal uridylation of let-7 precursor MicroRNA. *Mol. Cell, 32*, 276–284.

Heo, I., Joo, C., Kim, Y. K., Ha, M., Yoon, M. J., Cho, J., et al. (2009). TUT4 in concert with Lin28 suppresses microRNA biogenesis through pre-microRNA uridylation. *Cell, 138*, 696–708.

Hochedlinger, K., Yamada, Y., Beard, C., & Jaenisch, R. (2005). Ectopic expression of oct-4 blocks progenitor-cell differentiation and causes dysplasia in epithelial tissues. *Cell, 121*, 465–477.

Hockemeyer, D., Soldner, F., Beard, C., Gao, Q., Mitalipova, M., DeKelver, R. C., et al. (2009). Efficient targeting of expressed and silent genes in human ESCs and iPSCs using zinc-finger nucleases. *Nat. Biotechnol., 27*, 851–857.

Hong, H., Takahashi, K., Ichisaka, T., Aoi, T., Kanagawa, O., Nakagawa, M., et al. (2009). Suppression of induced pluripotent stem cell generation by the p53-p21 pathway. *Nature, 460*, 1132–1135.

Huangfu, D., Maehr, R., Guo, W., Eijkelenboom, A., Snitow, M., Chen, A. E., et al. (2008a). Induction of pluripotent stem cells by defined factors is greatly improved by small-molecule compounds. *Nat. Biotechnol., 26*, 795–797.

Huangfu, D., Osafune, K., Maehr, R., Guo, W., Eijkelenboom, A., Chen, S., et al. (2008b). Induction of pluripotent stem cells from primary human fibroblasts with only Oct4 and Sox2. *Nat. Biotechnol., 26*, 1269–1275.

Jia, F., Wilson, K. D., Sun, N., Gupta, D. M., Huang, M., Li, Z., Panetta, N. J., et al. (2010). A nonviral minicircle vector for deriving human iPS cells. *Nat. Methods, 7*, 197–199.

Jiang, J., Chan, Y. S., Loh, Y. H., Cai, J., Tong, G. Q., Lim, C. A., et al. (2008). A core Klf circuitry regulates self-renewal of embryonic stem cells. *Nat. Cell Biol., 10*, 353–360.

Judson, R. L., Babiarz, J. E., Venere, M., & Blelloch, R. (2009). Embryonic stem cell-specific microRNAs promote induced pluripotency. *Nat. Biotechnol.* (Advanced online publication).

Kaji, K., Norrby, K., Paca, A., Mileikovsky, M., Mohseni, P., & Woltjen, K. (2009). Virus-free induction of pluripotency and subsequent excision of reprogramming factors. *Nature, 458*, 771–775.

Kawamura, T., Suzuki, J., Wang, Y. V., Menendez, S., Morera, L. B., Raya, A., et al. (2009). Linking the p53 tumour suppressor pathway to somatic cell reprogramming. *Nature, 460*, 1140–1144.

Kazuki, Y., Hiratsuka, M., Takiguchi, M., Osaki, M., Kajitani, N., Hoshiya, H., et al. (2009). Complete genetic correction of iPS cells from Duchenne muscular dystrophy. *Mol Ther, 18*, 386–393.

Kim, D. H., Kim, C. H., Moon, J. I., Chung, Y. G., Chang, M. Y., Han, B. S., et al. (2009a). Generation of human induced pluripotent stem cells by direct delivery of reprogramming proteins. *Cell Stem Cell* (Advanced online publication).

Kim, H. H., Kuwano, Y., Srikantan, S., Lee, E. K., Martindale, J. L., & Gorospe, M. (2009b). HuR recruits let-7/RISC to repress c-Myc expression. *Genes Dev., 23,* 1743–1748.

Kim, J. B., Sebastiano, V., Wu, G., Arauzo-Bravo, M. J., Sasse, P., Gentile, L., et al. (2009c). Oct4-induced pluripotency in adult neural stem cells. *Cell, 136,* 411–419.

Lee, T. I., Jenner, R. G., Boyer, L. A., Guenther, M. G., Levine, S. S., Kumar, R. M., et al. (2006). Control of developmental regulators by Polycomb in human embryonic stem cells. *Cell, 125,* 301–313.

Li, H., Collado, M., Villasante, A., Strati, K., Ortega, S., Canamero, M., et al. (2009a). The Ink4/Arf locus is a barrier for iPS cell reprogramming. *Nature, 460,* 1136–1139.

Li, W., Wei, W., Zhu, S., Zhu, J., Shi, Y., Lin, T., et al. (2009b). Generation of rat and human induced pluripotent stem cells by combining genetic reprogramming and chemical inhibitors. *Cell Stem Cell, 4,* 16–19.

Li, W., Zhou, H., Abujarour, R., Zhu, S., Joo, J. Y., Lin, T., et al. (2009c). Generation of human induced pluripotent stem cells in the absence of exogenous Sox2. *Stem Cells, 27,* 2992–3000.

Li, Z., van Calcar, S., Qu, C., Cavenee, W. K., Zhang, M. Q., & Ren, B. (2003). A global transcriptional regulatory role for c-Myc in Burkitt's lymphoma cells. *Proc. Natl. Acad. Sci. U.S.A., 100,* 8164–8169.

Liao, J., Cui, C., Chen, S., Ren, J., Chen, J., Gao, Y., et al. (2009). Generation of induced pluripotent stem cell lines from adult rat cells. *Cell Stem Cell, 4,* 11–15.

Liao, J., Wu, Z., Wang, Y., Cheng, L., Cui, C., Gao, Y., et al. (2008). Enhanced efficiency of generating induced pluripotent stem (iPS) cells from human somatic cells by a combination of six transcription factors. *Cell Res., 18,* 600–603.

Liu, H., Zhu, F., Yong, J., Zhang, P., Hou, P., Li, H., et al. (2008). Generation of induced pluripotent stem cells from adult rhesus monkey fibroblasts. *Cell Stem Cell, 3,* 587–590.

Loh, Y. H., Agarwal, S., Park, I. H., Urbach, A., Huo, H., Heffner, G. C., et al. (2009). Generation of induced pluripotent stem cells from human blood. *Blood* (Advanced online publication).

Lyssiotis, C. A., Foreman, R. K., Staerk, J., Garcia, M., Mathur, D., Markoulaki, S., et al. (2009). Reprogramming of murine fibroblasts to induced pluripotent stem cells with chemical complementation of Klf4. *Proc. Natl. Acad. Sci. U.S.A., 106,* 8912–8917.

Maherali, N., Sridharan, R., Xie, W., Utikal, J., Eminli, S., Arnold, K., et al. (2007). Directly reprogrammed fibroblasts show global epigenetic remodelling and widespread tissue contribution. *Cell Stem Cell, 1,* 55–70.

Marion, R. M., Strati, K., Li, H., Murga, M., Blanco, R., Ortega, S., et al. (2009). A p53-mediated DNA damage response limits reprogramming to ensure iPS cell genomic integrity. *Nature, 460,* 1149–1153.

Marson, A., Foreman, R., Chevalier, B., Bilodeau, S., Kahn, M., Young, R. A., et al. (2008a). Wnt signaling promotes reprogramming of somatic cells to pluripotency. *Cell Stem Cell, 3,* 132–135.

Marson, A., Levine, S. S., Cole, M. F., Frampton, G. M., Brambrink, T., Johnstone, S., et al. (2008b). Connecting microRNA genes to the core transcriptional regulatory circuitry of embryonic stem cells. *Cell, 134,* 521–533.

Mitsui, K., Tokuzawa, Y., Itoh, H., Segawa, K., Murakami, M., Takahashi, K., et al. (2003). The homeoprotein Nanog is required for maintenance of pluripotency in mouse epiblast and ES cells. *Cell, 113,* 631–642.

Miura, K., Okada, Y., Aoi, T., Okada, A., Takahashi, K., Okita, K., et al. (2009). Variation in the safety of induced pluripotent stem cell lines. *Nat. Biotechnol, 27,* 743–745.

Morita, S., Kojima, T., & Kitamura, T. (2000). Plat-E: an efficient and stable system for transient packaging of retroviruses. *Gene Ther., 7,* 1063–1066.

Nakagawa, M., Koyanagi, M., Tanabe, K., Takahashi, K., Ichisaka, T., Aoi, T., et al. (2008). Generation of induced pluripotent stem cells without Myc from mouse and human fibroblasts. *Nat. Biotechnol., 26,* 101–106.

Nakajima, F., Tokunaga, K., & Nakatsuji, N. (2006). HLA matching estimations in a hypothetical bank of human embryonic stem cell lines in the Japanese population for use in cell transplantation therapy. *Stem Cells, 25,* 983–985.

Nakatsuji, N., Nakajima, F., & Tokunaga, K. (2008). HLA-haplotype banking and iPS cells. *Nat. Biotechnol., 26,* 739–740.

Oh, J. S., Kim, J. J., Byun, J. Y., & Kim, I. A. (2010). Lin28-let7 modulates radiosensitivity of human cancer cells with activation of K-Ras. *Int. J. Radiat. Oncol. Biol. Phys., 76,* 5–8.

Okita, K., Ichisaka, T., & Yamanaka, S. (2007). Generation of germ-line competent induced pluripotent stem cells. *Nature, 448,* 313–317.

Okita, K., Nakagawa, M., Hyenjong, H., Ichisaka, T., & Yamanaka, S. (2008). Generation of mouse induced pluripotent stem cells without viral vectors. *Science, 322,* 949–953.

Park, I. H., Arora, N., Huo, H., Maherali, N., Ahfeldt, T., Shimamura, A., et al. (2008). Disease-specific induced pluripotent stem cells. *Cell, 134,* 877–886.

Qiu, C., Ma, Y., Wang, J., Peng, S., & Huang, Y. (2010). Lin28-mediated post-transcriptional regulation of Oct4 expression in human embryonic stem cells. *Nucleic Acids Res.*, 38, 1240–1248.

Raya, A., Rodriguez-Piza, I., Guenechea, G., Vassena, R., Navarro, S., Barrero, M. J., et al. (2009). Disease-corrected haematopoietic progenitors from Fanconi anaemia induced pluripotent stem cells. *Nature*, 460, 53–59.

Rowland, B. D., & Peeper, D. S. (2006). KLF4, p21 and context-dependent opposing forces in cancer. *Nat. Rev. Cancer*, 6, 11–23.

Shi, Y., Desponts, C., Do, J. T., Hahm, H. S., Scholer, H. R., & Ding, S. (2008a). Induction of pluripotent stem cells from mouse embryonic fibroblasts by Oct4 and Klf4 with small-molecule compounds. *Cell Stem Cell*, 3, 568–574.

Shi, Y., Do, J. T., Desponts, C., Hahm, H. S., Scholer, H. R., & Ding, S. (2008b). A combined chemical and genetic approach for the generation of induced pluripotent stem cells. *Cell Stem Cell*, 2, 525–528.

Shimada, H., Nakada, A., Hashimoto, Y., Shigeno, K., Shionoya, Y., & Nakamura, T. (2010). Generation of canine induced pluripotent stem cells by retroviral transduction and chemical inhibitors. *Mol. Reprod. Dev.*, 77, 2.

Soldner, F., Hockemeyer, D., Beard, C., Gao, Q., Bell, G. W., Cook, E. G., et al. (2009). Parkinson's disease patient-derived induced pluripotent stem cells free of viral reprogramming factors. *Cell*, 136, 964–977.

Sommer, C. A., Gianotti Sommer, A., Longmire, T. A., Christodoulou, C., Thomas, D. D., Gostissa, M., et al. (2009a). Excision of reprogramming transgenes improves the differentiation potential of iPS cells generated with a single excisable vector. *Stem Cells*, 28, 64–74.

Sommer, C. A., Stadtfeld, M., Murphy, G. J., Hochedlinger, K., Kotton, D. N., & Mostoslavsky, G. (2009b). Induced pluripotent stem cell generation using a single lentiviral stem cell cassette. *Stem Cells*, 27, 543–549.

Sridharan, R., Tchieu, J., Mason, M. J., Yachechko, R., Kuoy, E., Horvath, S., et al. (2009). Role of the murine reprogramming factors in the induction of pluripotency. *Cell*, 136, 364–377.

Stadtfeld, M., Brennand, K., & Hochedlinger, K. (2008a). Reprogramming of pancreatic beta cells into induced pluripotent stem cells. *Curr. Biol.*, 18, 890–894.

Stadtfeld, M., Nagaya, M., Utikal, J., Weir, G., & Hochedlinger, K. (2008b). Induced pluripotent stem cells generated without viral integration. *Science*, 322, 945–949.

Sun, N., Panetta, N. J., Gupta, D. M., Wilson, K. D., Lee, A., Jia, F., et al. (2009). Feeder-free derivation of induced pluripotent stem cells from adult human adipose stem cells. *Proc. Natl. Acad. Sci. U.S.A.*, 106, 15720–15725.

Suzuki, K., Mitsui, K., Aizawa, E., Hasegawa, K., Kawase, E., Yamagishi, T., et al. (2008). Highly efficient transient gene expression and gene targeting in primate embryonic stem cells with helper-dependent adenoviral vectors. *Proc. Natl. Acad. Sci. U.S.A.*, 105, 13781–13786.

Tada, M., Takahama, Y., Abe, K., Nakatsuji, N., & Tada, T. (2001). Nuclear reprogramming of somatic cells by in vitro hybridization with ES cells. *Curr. Biol.*, 11, 1553–1558.

Takahashi, K., & Yamanaka, S. (2006). Induction of pluripotent stem cells from mouse embryonic and adult fibroblast cultures by defined factors. *Cell*, 126, 663–676.

Takahashi, K., Tanabe, K., Ohnuki, M., Narita, M., Ichisaka, T., Tomoda, K., et al. (2007). Induction of pluripotent stem cells from adult human fibroblasts by defined factors. *Cell*, 131, 861–872.

Takizawa, T., Nakashima, K., Namihira, M., Ochiai, W., Uemura, A., Yanagisawa, M., et al. (2001). DNA methylation is a critical cell-intrinsic determinant of astrocyte differentiation in the fetal brain. *Dev. Cell*, 1, 749–758.

Taylor, C. J., Bolton, E. M., Pocock, S., Sharples, L. D., Pedersen, R. A., & Bradley, J. A. (2005). Banking on human embryonic stem cells: estimating the number of donor cell lines needed for HLA matching. *Lancet*, 366, 2019–2025.

Tsubooka, N., Ichisaka, T., Okita, K., Takahashi, K., Nakagawa, M., & Yamanaka, S. (2009). Roles of Sall4 in the generation of pluripotent stem cells from blastocysts and fibroblasts. *Genes Cells*, 14, 683–694.

Utikal, J., Polo, J. M., Stadtfeld, M., Maherali, N., Kulalert, W., Walsh, R. M., et al. (2009). Immortalization eliminates a roadblock during cellular reprogramming into iPS cells. *Nature*, 460, 1145–1148.

Varas, F., Stadtfeld, M., de Andres-Aguayo, L., Maherali, N., di Tullio, A., Pantano, L., et al. (2008). Fibroblast derived induced pluripotent stem cells show no common retroviral vector insertions. *Stem Cells*, 27, 300–306.

Vierbuchen, T., Ostermeier, A., Pang, Z. P., Kokubu, Y., Sudhof, T. C., & Wernig, M. (2010). Direct conversion of fibroblasts to functional neurons by defined factors. *Nature*, 463, 1035–1045.

Viswanathan, S. R., Powers, J. T., Einhorn, W., Hoshida, Y., Ng, T. L., Toffanin, S., et al. (2009). Lin28 promotes transformation and is associated with advanced human malignancies. *Nat. Genet.*, 41, 843–848.

Wakayama, T., Perry, A. C., Zuccotti, M., Johnson, K. R., & Yanagimachi, R. (1998). Full-term development of mice from enucleated oocytes injected with cumulus cell nuclei. *Nature*, 394, 369–374.

Wang, Y., Baskerville, S., Shenoy, A., Babiarz, J. E., Baehner, L., & Blelloch, R. (2008). Embryonic stem cell-specific microRNAs regulate the G1-S transition and promote rapid proliferation. *Nat. Genet.*, 40, 1478–1483.

Wernig, M., Meissner, A., Foreman, R., Brambrink, T., Ku, M., Hochedlinger, K., et al. (2007). In vitro reprogramming of fibroblasts into a pluripotent ES cell-like state. *Nature*, 448, 318–324.

Wilmut, I., Schnieke, A. E., McWhir, J., Kind, A. J., & Campbell, K. H. (1997). Viable offspring derived from fetal and adult mammalian cells. *Nature, 385*, 810–813.

Woltjen, K., Michael, I. P., Mohseni, P., Desai, R., Mileikovsky, M., Hamalainen, R., et al. (2009). piggyBac transposition reprograms fibroblasts to induced pluripotent stem cells. *Nature, 458*, 766–770.

Ye, L., Chang, J. C., Lin, C., Sun, X., Yu, J., & Kan, Y. W. (2009). Induced pluripotent stem cells offer new approach to therapy in thalassemia and sickle cell anemia and option in prenatal diagnosis in genetic diseases. *Proc. Natl. Acad. Sci. U.S.A., 106*, 9826–9830.

Ying, Q. L., Wray, J., Nichols, J., Batlle-Morera, L., Doble, B., Woodgett, J., et al. (2008). The ground state of embryonic stem cell self-renewal. *Nature, 453*, 519–523.

Yoshida, Y., Takahashi, K., Okita, K., Ichisaka, T., & Yamanaka, S. (2009). Hypoxia enhances the generation of induced pluripotent stem cells. *Cell Stem Cell, 5*, 237–241.

Yu, F., Yao, H., Zhu, P., Zhang, X., Pan, Q., Gong, C., et al. (2007a). let-7 regulates self renewal and tumorigenicity of breast cancer cells. *Cell, 131*, 1109–1123.

Yu, J., Hu, K., Smuga-Otto, K., Tian, S., Stewart, R., Slukvin, I. I., et al. (2009). Human induced pluripotent stem cells free of vector and transgene sequences. *Science, 324*, 797–801.

Yu, J., Vodyanik, M. A., Smuga-Otto, K., Antosiewicz-Bourget, J., Frane, J. L., Tian, S., et al. (2007b). Induced pluripotent stem cell lines derived from human somatic cells. *Science, 318*, 1917–1920.

Zhao, Y., Yin, X., Qin, H., Zhu, F., Liu, H., Yang, W., et al. (2008). Two supporting factors greatly improve the efficiency of human iPSC generation. *Cell Stem Cell, 3*, 475–479.

Zhou, H., Wu, S., Joo, J. Y., Zhu, S., Han, D. W., Lin, T., et al. (2009). Generation of induced pluripotent stem cells using recombinant proteins. *Cell Stem Cell.* In press, doi:10.1016/j.stem.2009.04.005.

Zhou, Q., Brown, J., Kanarek, A., Rajagopal, J., & Melton, D. A. (2008). In vivo reprogramming of adult pancreatic exocrine cells to beta-cells. *Nature, 455*, 627–632.

Zhou, W., & Freed, C. R. (2009). Adenoviral gene delivery can reprogram human fibroblasts to induced pluripotent stem cells. *Stem Cells, 27*, 2667–2674.

CHAPTER 14

MSCs in Regenerative Medicine

Arnold I. Caplan
Professor of Biology, Director, Skeletal Research Center, Case Western Reserve
University, Euclid Avenue, Cleveland, OH, USA

INTRODUCTION AND HISTORY

The body continuously changes, and this is controlled primarily by genetic factors. The same genomic program that brings the fertilized egg through a series of multiplication and differentiation changes to bring about the birth of a complete, multi-tissued organism also controls the continuous changes through neonatal, juvenile, and teen stages, and all of adulthood. Ten-30-, 50-, 70- and 90-year-olds exemplify this continuous genetic and distinctive process of change. Importantly, the later stage of this process, referred to as aging, is not a disease state, but rather part of this genomically controlled continuum. The central feature of change is that progenitor cells divide and their progeny differentiate in a sequence of site-specific changes to both expand the dimensions of tissues and *replace* cells that naturally expire. Every cell in the body has a lifespan measured in minutes, weeks, or, in some rare cases, years. With only a few exceptions, the end-stage differentiated cells die within a fixed timeframe. The progenitor cells for that expired cell must replace the expired cells; the rate of replacement controls whether the tissue will increase in size, be maintained, or experience atrophy as seen in old age.

Within this view, I proposed the scheme pictured in Figure 14.1 recognizing that bone marrow contained osteochondral progenitors, which had been used for decades to repair bones (Chutro, 1918; Caplan, 1988; Connolly et al., 1991). Figure 14.1 was established in schematic form to mimic what was known about the lineages of hematopoiesis (Orkin and Zon, 2008), except that in 1988 we knew the most about the lineage progression on the left of Figure 14.1 (Bruder and Caplan, 1990) and nothing about the lineages on the right. Additionally, Professor Maureen Owen, in reviewing the prior experiments from Dr. A. Friedenstein's lab in Moscow and her own findings, also proposed the existence of osteochondral progenitors in a logic mirroring hematopoiesis (Owen and Friedenstein, 1988). It should be emphasized that the dogma of that era (the 1980s and early 1990s) was that there was only one progenitor or stem cell, the hematopoietic stem cell (HSC), in adult organisms. The full scope of Figure 14.1 was clearly not envisioned in 1988 and the term mesenchymal stem cell was considered by some to be provocative at best and by others to be outlandish. Because the focus of Figure 14.1 was on the differentiation lineages, all of the research of that era focused on tissue engineering strategies to rebuild damaged tissues, with most experiments/transplantations formulated in the orthopedic sector with bone (Bruder et al., 1994; Jaiswal et al., 1997), cartilage (Wakitani et al., 1994; Yoo et al., 1998), muscle (Wakitani et al., 1995; Saito et al., 1996), and tendon (Young et al., 1998) being prominent.

Because the MSCs were isolated from marrow, it was hypothesized that these cells were the progenitors for the bone marrow stroma (the highly differentiated tissue that supports all of

Principles of Regenerative Medicine. DOI: 10.1016/B978-0-12-381422-7.10014-8

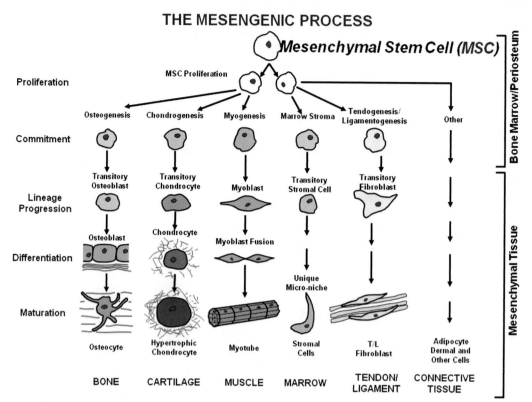

FIGURE 14.1

The mesengenic process. This figure was generated in the late 1980s and proposed that an MSC existed in bone marrow and that its progeny could be induced to enter one of several mesenchymal lineage pathways (Owen and Friedenstein, 1988; Bruder and Caplan, 1990; Bruder et al., 1994; Wakitani et al., 1994; Wakitani et al., 1995; Saito et al., 1996; Jaiswal et al., 1997; Yoo et al., 1998; Young et al., 1998; Reese et al., 1999). The left side of the diagram was already experimentally established (Bruder and Caplan, 1990) while very little was known about the lineages on the right. The lineage format was constructed from what was known about the hematopoietic lineage pathway (Orkin and Zon, 2008).

hematopoiesis) (Reese et al., 1999). Although the majority of preclinical models using MSCs were in the orthopaedic sector (bone, cartilage, tendon) in the early 1990s, the first clinical trial using hMSCs conducted by my colleagues in hematology-oncology was to supplement bone marrow transplantations with culture-expanded hMSCs for cancer patients and eventually for patients with gene defects in mesenchymal-linked processes or tissues (Lazarus et al., 1995; Koc et al., 1999; Maitra et al., 2004a). The logic of that era was that the MSCs, like the HSCs, would "home" back to the marrow when infused into the bloodstream; once home, the MSCs would rebuild or enhance the marrow scaffold to accelerate the engraftment of the HSCs and stimulate the recovery of the blood cell-forming capacity of the marrow by directly fabricating microenvironments for each of the distinct lineage pathways for hematopoiesis. Indeed, the early clinical evidence showed that the added MSCs improved both the kinetics and outcomes of hematopoietic recovery in bone marrow-transplanted cancer patients (Koc et al., 2000).

This involvement of the hematologists-oncologists led to the realization that hMSCs were immuno-modulators and, thus, allogeneic MSCs were not interrogated by the host's immune system (Le Blanc et al., 2003; Maitra et al., 2004b; Sundin et al., 2007). This allows allogeneic hMSCs to be infused into the bloodstream as a delivery modality with the cells engrafting in tissue areas of inflammation or vascular damage. As discussed below, this also allows human MSCs to be used in rodent models of MS (Bonfield et al., 2010a,b), inflammatory bowel disease (Ko et al., 2010), graft versus host disease (GvHD) (Le Blanc et al., 2008), etc., without rapid and intense immuno-rejection.

NEW INSIGHT

When Steven Haynesworth and I first started working with hMSCs, we developed monoclonal antibodies to cell-surface antigens on these cells called SH2, SH3, and SH4 (Haynesworth et al., 1992). Subsequently, we showed that SH2 was a unique antigen on endoglin (also known as CD105) (Barry et al., 1999) and SH3=SH4=CD73 (Barry et al., 2001). The initial screens for these monoclonal antibodies were to their binding to hMSCs in culture and then to frozen sections of marrow plugs. It was not until years later that we realized that SH2-positive cells co-localized with the external aspects of small blood vessels in those marrow sections. Through the detailed research from B. Peault's (Crisan et al., 2008) and I. Bianco's (Sacchetti et al., 2007) laboratories and others (Hirschi and D'Amore, 1996), we now understand that MSCs are identical to perivascular cells, referred to here as pericytes. Indeed, I have hypothesized (Caplan, 2008) that *all* MSCs are pericytes (the reverse is not correct in that some pericytes are not MSCs). In this context, over ten years ago we published a paper studying human skin showing SH2-positive cells sitting on blood vessels (Fleming et al., 1998). For reasons still not clear, only a small proportion of such cells stained positive in sections of skin, with the most numerous positive images in specimens from young donors.

The current literature documents that MSCs can be isolated from marrow (Brighton et al., 1992), fat (Krampera et al., 2007; Bieback, 2009), muscle (Lee et al., 2000), skin (Toma et al., 2001), periosteum (Nakahara et al., 1991), tendon (Salingcarnboriboon et al., 2003), neural tissues (Covas et al., 2008), etc. These tissues all have blood vessels in common and such vessels have perivascular cells on the ablumenal surface. The positioning of such pericytes is controlled by a number of factors, but the prominent component is the PDGF receptor (Gerhardt and Semb, 2008). Thus, PDGF not only acts as a powerful chemo-attractant and mitogen for mesenchymal cells, but its receptors function to stabilize the interactions between perivascular cells and vascular endothelial cells. I would suggest that, in tissue domains of inflammation or blood vessel damage, the pericytes are liberated from their ablumenal locations and apparently function as activated MSCs.

Based on the clinically relevant effects of activated MSCs, we have suggested that the therapeutic capabilities of MSCs (i.e. liberated pericytes) reflect their biologic functionality at sites of tissue injury or inflammation. These therapeutic activities involve the modulation of the local immune cells to inhibit their surveillance of the damaged tissue, thus inhibiting the initiation of autoimmune activities (Da Silva Meirelles et al., 2008). Furthermore, the activated MSCs produce trophic effects by secreting a spectrum of bioactive molecules that inhibit apoptosis (especially in areas of ischemia), prevent scar formation, and stimulate angiogenesis by secreting VEGF to attract endothelial cells to form new blood vessels; by forming perivascular contacts stabilizing such newly formed vessels; and, last, by secreting mitogens that directly affect tissue intrinsic progenitors (Caplan and Dennis, 2006) (see Fig. 14.2). These effects have resulted in clinically relevant therapies, as discussed below and outlined in Table 14.1.

Given the proposed pericyte:MSC relationship, it is now easy to understand why large numbers of MSCs can be isolated from adipose or muscle tissue. Indeed, although marrow MSCs are the most widely studied, the data from fat and muscle appear to be complementary (Bosch et al., 2000; Garcia-Olmo et al., 2005). The fact that fat contains a 300- to 500-fold greater number of MSCs per milliliter of material accounts for the assertion that adipose-derived MSCs can be obtained in sufficient numbers without *in vitro* expansion (Gimble et al., 2007). Simply, liposuctioned fat is dispersed in a homogeneous slurry and incubated with digestive enzymes such as collagenase to free the pericytes from their capillary endothelial cell attachments, and the liberated cells are centrifuged following saline washes (Bieback et al., 2008). The adipocytes float and the pelleted cells, called the stromal vascular fraction (SVF), contain mostly MSCs with some contaminating endothelial cells and monocytes (Iwashima et al., 2009). For autologous cell-based MSC therapy, the SVF has been used directly with clear

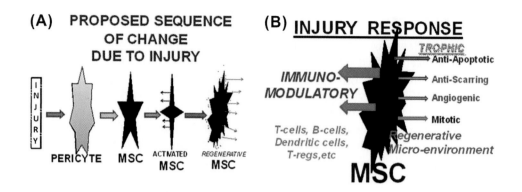

FIGURE 14.2

MSCs are immunomodulatory and trophic. (A) The sequential activation as a response to injury affects pericytes and liberates them from functional contact with blood vessels to become functional MSCs. These MSCs become immuno-activated and secrete factors that organize a regenerative microenvironment. (B) The bioactive molecules secreted by activated MSCs are immunomodulatory and affect a variety of immuno-related cells (Hirschi and D'Amore, 1996; Le Blanc et al., 2003; Maitra et al., 2004b; Caplan, 2008). Other secreted molecules establish a regenerative microenvironment by establishing a powerful trophic field (Caplan and Dennis, 2006; Da Silva Meirelles et al., 2008).

TABLE 14.1 Therapeutic Effects of MSCs

Tissue	Injury/Disease	Reference
Bone marrow	GvHD	Frassoni et al., 2002; Le Blanc et al., 2008
Brain	Stroke/MS/ALS	Mahmoud et al., 2003; Estes et al., 2006; Riordan et al., 2009;
Spinal cord	Cut/Contusion	Chopp and Li, 2002; Neuhuber et al., 2005
Bowel	Inflammatory bowel disease	Garcia-Olmo et al., 2005; Ko et al., 2010
Tendon	Tendonitis	Black et al., 2008, 2007
Muscle	Defect/MD	Bosch et al., 2000; Lee et al., 2000
Pancreas	Diabetes	Farge et al., 2008
Lung	Asthma/CF	Solchaga et al., 2005; Bonfield et al., 2010ab
Heart	AMI, chronic	Toma et al., 2002; Pittenger and Martin, 2004

AMI = acute myocardial infraction; ALS = amyotrophic lateral sclerosis; CF = cystic fibrosis; GvHD = graft versus host disease; MD = Muscular dystrophy; MS = multiple sclerosis.

therapeutic effects (Fauza, 2004; Riordan et al., 2009). Likewise, the SVF represents a relatively pure starting suspension for the culture attachment of MSCs and their further expansion.

Again, for emphasis, relatively large numbers of MSCs can be obtained from muscle (Lee et al., 2000), placenta (Fauza, 2004), and other highly vascularized tissues (de Bari et al., 2001). From unpublished studies in my lab (Sung, Lennon, and Caplan), I would suggest that MSCs are absent in cord blood, but plentiful in the perivascular fraction of the cord. I would further assert that, if MSCs have been identified in cord blood, they have been found as an artifact of collection and are dislodged from their natural habitat (by the needle insertion) during the cord blood retrieval.

ALL MSCs ARE NOT CREATED EQUAL

Since I have asserted that *all* MSCs are pericytes, it also follows that MSCs from different tissue sources must be different since they reside in different tissue microenvironments. We have published various assays to test for the capacity of MSCs (primarily from marrow) to differentiate into bone (Jaiswal et al., 1997), cartilage (Yoo et al., 1998), muscle (Wakitani et al., 1995), and marrow stroma (Majumdar et al., 1998). In many cases, these assays have been optimized for marrow MSCs (Solchaga et al., 2005), but not for MSCs from other tissues. For example, dexamethasone (dex) stimulates human MSCs to differentiate into osteoblasts but induces mouse MSCs to become adipocytes (Meirelles and Nardi, 2003). For cartilage induction, human MSCs from marrow require TGF-β in defined medium (Yoo et al., 1998). Fat-derived human MSCs do not respond to TGB-β alone, but require the addition of BMP-6 with TGF-β for optimal chondrogenic differentiation (Estes et al., 2006). Currently, the relationship between the capacity of MSCs to be induced into specific and multiple phenotypic pathways and their intrinsic secretory profile is unknown. It is clear that MSCs from a variety of tissues are multipotent, but their capacity to respond to certain signaling molecules and their ability to differentiate are apparently independent of their source or their therapeutic or secretory potential. MSCs are presumably conditioned by the tissue microenvironment from which they were purified and, thus, all MSCs are not equivalent.

CLINICALLY RELEVANT THERAPIES USING MSCs

Often the FDA requires animal trials as preclinical proof-of-concept for the use of a drug or device. It must be stated that there are no animal models that exactly match the human disease condition. Indeed, in most cases of animal models of cell-based therapies, autologous animal cells have been preferentially used (Wakitani et al., 1994; Young et al., 1998; Lee et al., 2000; Salingcarnboriboon et al., 2003). The reason for including this section on preclinical studies here is to set forth the unique example that *human* (h) MSCs are being used in animal models to cure their disease states (see, for example, Bonfield et al., 2010a,b and Bai et al., 2009). This, in itself, is a quite remarkable property of hMSCs in that the animals' immune systems do not overtly reject the human cells while the (usually, second passage) hMSCs orchestrate the cures.

Multiple sclerosis (MS)

EXPERIMENTAL AUTOIMMUNE ENCEPHALITIS (EAE) MODEL

A mouse model for MS involves exposing immuno-competent animals to molecules from myelin in the presence of adjuvant that causes the animals to mount an immune reaction to these molecules, which in turn causes a series of demyelinating events mimicking human MS. Two models have been used by us: severe, monophasic MS and recurring-remitting MS. In both cases, hMSCs cure the animal following a single tail-vein injection of 10^6 hMSCs (Bai et al., 2009). We have separately shown that molecules secreted by hMSCs preferentially cause central nervous system neural stem cells to differentiate into oligodendrocytes, the cells that wrap nerve axons with myelin (Bai et al., 2007). The end result is a cured animal that exhibits no signs of MS.

Asthma

OVALBUMIN MODEL

Mice were injected with the chicken protein ovalbumin in adjuvant to immune-sensitize the animal. After 2 weeks, ovalbumin was atomized daily or every other day into the lungs of these sensitized mice, causing an intense inflammatory response. At 3 or 7 days or 11 weeks following the initial exposure of the lungs to ovalbumin followed by daily or every-other-day exposure to ovalbumin, 5–10 mice were tail-vein injected with 10^6 hMSCs (Solchaga et al., 2005; Bonfield et al., 2010a,b). The ovalbumin treatment was continued for 4 days or 1 week following the injection of hMSCs with animal sacrifice at 1, 4, 8 or 12 weeks after starting the initial ovalbumin exposure to the lungs. The hMSCs were compared to saline-injected animals. In the saline-injected ovalbumin-atomized mice, the lung tissue was hugely inflamed and, at the longer time points, scar tissue was apparent. In the hMSC-injected mice, the lungs looked normal with the re-establishment of intact endothelial lining and the absence of inflammation (Solchaga et al., 2005; Bonfield et al., 2010a,b). Experiments are now under way to determine whether the scar tissue is replaced by normal tissue following multiple hMSC exposures (T. Bonfield, A. Caplan, personal communications). Again, human MSCs facilitate the curing of this asthma-like condition.

Inflammatory bowel disease

Mice can be exposed to DDS, which causes severe inflammatory bowel disease that is lethal. When hMSCs are targeted to the sites of inflammation using a proprietary cell-docking system, 80–90% of the animals are cured (Ko et al., 2010). No other single treatment can accomplish this cure. Whether the enhanced targeting is required for human subjects is not known, although Phase I and II clinical trial data appear to indicate that substantial improvement in Crohn's patients can be experienced by infusion of hMSCs alone, with rather high doses required (Osiris Therapeutics, Inc., www.osiristx.com).

Stroke and acute myocardial infarct (AMI)

By occluding a major artery leading to the brain (Li et al., 2002) or nurturing the heart (Penn and Khalil, 2008), mouse models of stroke or AMI have been created. From 2 to 7 days after creating these models of severe ischemia, hMSCs are injected via the tail vein (Li et al., 2002; Penn and Khalil, 2008). The animals are monitored for 5–10 weeks and those that receive hMSC therapy return to full and normal function. The therapeutic effects result from the trophic activity of those hMSCs that home to sites of vascular injury and inflammation.

Diabetes

In both mice and rats, the immune destruction of pancreatic islet cells can be initiated by exposing the animal to streptozotocin. By monitoring systemic secretion of insulin, a continuous, autoimmune-mediated elimination of islets and, thus, the diminution of insulin secretion can be observed. By using a single intravenous injection of human MSCs, the decline of insulin secretion can be halted and the eventual regeneration of islet cells can be observed (Farge et al., 2008).

CLINICAL TRIALS

Worldwide, over 85 separate clinical trials are being conducted using MSCs for various clinical maladies with most being conducted outside the USA. Such clinical trials listings do not include infusions of MSCs occurring in clinics and hospitals operating within the medical tourism industry. Indeed, the use of MSCs, both autologous (SVF and culture-expanded MSCs) and allogeneic MSCs from marrow and other sources, form an expanding sector of the medical tourism markets in the Caribbean, Central and South America, and Asia. Most, if not all, of these MSC uses are in open-label, non-randomized, non-placebo-controlled studies. Although one can find fault with such studies, they are providing patients with access to potential therapeutics long before MSCs will be officially approved for use by the regulatory agencies. Since many drugs are used "off-label" by medical practitioners, it is tempting to be enthusiastic about such treatments, especially using autologous SVF. In this regard, somewhere between 3,000 and 30,000 individuals have been infused with various preparations of MSCs without any reported adverse reactions. Certainly, if one million patients are infused with MSCs, we should expect some adverse events. In this regard, whether in sanctioned clinical trials or in other situations, the medical community must quickly and completely be informed about any and all adverse events so we can understand the problems and limitations of these new therapies.

THE NEW MSCs

Over 20 years ago, when we developed the isolation and culturing technology and assays for hMSCs (Caplan and Haynesworth, 1993; Caplan, 1994), we envisioned using these cells in tissue-engineering applications in orthopedics. We started Osiris Therapeutics, Inc. as a "BioOrthopaedic" company, which interestingly followed our academic lead by furthering our initial investigator-initiated clinical trials in the augmentation of bone marrow transplantations for cancer patients (Lazarus et al., 1995; Koc et al., 1999, 2000; Le Blanc et al., 2003; Maitra et al., 2004a,b;). Within the first 10 years of Osiris, it became clear that the MSCs had a unique immuno-modulatory capacity (Le Blanc et al., 2003; Maitra et al., 2004b). This led to the use of MSCs in GvHD and the new management of Osiris brought us into the current wide range of use of MSCs as briefly outlined above. For the record, it must be clearly stated that we set up the correct animal models for the wrong reasons. For example, the AMI model was established because it was hypothesized that MSCs would differentiate into cardiac myocytes and, thus, fix the damaged heart (Toma et al., 2002). Early reports documented such differentiation by showing the labeled MSCs docked in ischemic heart tissue and the label was seen in/on cells with markers for cardiac myocytes (Pittenger and Martin, 2004). This was most probably due to the very low frequency of fusion between the labeled MSCs and resident cardiac myocytes. We now more clearly understand that the MSCs have a complex, multi-component trophic effect that establishes a protective and regenerative microenvironment that allows the heart to minimize the damage due to blood vessel blockage, as shown schematically in Figure 14.2.

The "new hypothesis" regarding MSCs is that they are site-regulated pumps for bioactive molecules that are both immuno-modulatory and trophic. As outlined in recent review articles (Da Silva Meirelles et al., 2008, 2009; Caplan, 2009), tens to hundreds of molecules are provided and modulated at each site of injury or inflammation. Such modulation is probably site-specific and also controlled by the genetic mark-up of both the host and donor. The corollary is that it may be that patients exhibiting autoimmune diseases may have defects in their MSCs with regard to specific site-regulated responses. Such defects may not be manifested by the absences of key components, but rather may be a consequence of low levels of key components. Such low levels may not be detrimental until a specific age or stress condition is experienced. Likewise, the reported decreases in MSCs with age may be controlled by the age-related decrease in tissue vascular density that is seen with age or disease state, such as diabetes. Given the relationship between pericytes and MSCs (Hirschi and D'Amore, 1996; Sacchetti et al., 2007; Caplan, 2008; Crisan et al., 2008;), the central issue is to determine the cause of vascular density decreases: is this controlled by the MSC/pericytes, the endothelial cells, or the tissues hosting the vasculature?

The "new" MSC, the pericyte, opens a new window onto a better understanding of both a variety of diseases and an understanding of the development and maintenance capacities of a variety of tissues. The new treatment protocols using both culture-expanded and freshly isolated uncultured MSCs (i.e. SVF) open the door for new therapies and new medical horizons. The "old" MSCs can still be used for a variety of orthopedic indications, but they will now be used with new insights and new hypotheses. The second decade of this century will usher in the "new medicine" and new treatment protocols using MSCs. The next technical hurdle will be to learn how to more efficiently orchestrate the docking or targeting (Dennis et al., 2004; Sackstein et al., 2008; Ko et al., 2010) of systemically introduced MSCs to effect efficient therapeutic outcomes.

References

Bai, L., Caplan, A. I., Lennon, D. L., & Miller, R. H. (2007). Human mesenchymal stem cells signals regulate neural stem cell fate. *Neurochem. Res.*, *32*, 353–362.

Bai, L., Lennon, D. P., Eaton, V., Maier, K., Caplan, A. I., Miller, S. D., et al. (2009). Human bone marrow-derived mesenchymal stem cells induce Th2-polarized immune response and promote endogenous repair in animal models of multiple sclerosis. *Glia*, *57*, 1192–1203.

Barry, F., Boynton, R., Murphy, M., & Zaia, J. (2001). The SH-3 and SH-4 antibodies recognize distinct epitopes on CD73 from human mesenchymal stem cells. *Biochem. Biophys. Res. Commun.*, *28*, 519–524.

Barry, F. P., Boynton, R. E., Haynesworth, S., Murphy, J. M., & Zaia, J. (1999). The monoclonal antibody SH-2, raised against human mesenchymal stem cells, recognizes an epitope of endoglin (CD105). *Biochem. Biophys. Res. Commun.*, *265*, 134–139.

Bieback, K. (2009). Fatty tissue: not all bad? Optimally cultured adipose tissue-derived stromal cells improve experimentally-induced ischemia. *Stem Cells Dev., 18,* 531–532.

Bieback, K., Schallmoser, K., Klutera, H., & Strunk, D. (2008). Clinical protocols for the isolation and expansion of mesenchymal stromal cells. *Transfus. Med. Hemother., 35,* 286–294.

Black, L. L., Gaynor, J., Adams, C., Dhupa, S., Sams, A. E., Taylor, R., et al. (2008). Effect of intra-articular injection of autologous adipose-derived mesenchymal stem and regenerative cells on clinical signs of chronic osteoarthritis of the elbow joint in dogs. *Vet. Ther., 9,* 192–299.

Black, L. L., Gaynor, J., Gahring, D., Adams, C., Aron, D., Harman, S., et al. (2007). Effect of adipose-derived mesenchymal stem and regenerative cells on lameness in dogs with chronic osteoarthritis of the coxofemoral joints: a randomized, double-blinded, multicenter, controlled trial. *Vet. Ther., 8,* 272–284.

Bonfield, T. L., Koloze, M., Lennon, D., Zuchowski, B., Yang, S. E., & Caplan, A. I. (2010a). Acute asthma: an in vivo model for human mesenchymal stem cell efficacy. *J. Immunol. Methods.* (Submitted).

Bonfield, T. L., Koloze, M., Lennon, D., Zuchowski, B., Yang, S. E., & Caplan, A. I. (2010b). Human mesenchymal stem cells suppress chronic airway inflammation in the murine ovalbumin asthma model. *Am. J. Physiology Lung.* (In revision).

Bosch, P., Musgrave, D. S., Lee, J. Y., Cummins, J., Shuler, F., Ghivizzani, S. C., et al. (2000). Osteoprogenitor cells within skeletal muscle. *J. Orthop. Res., 18,* 933–944.

Brighton, C. T., Lorich, D. G., Kupcha, R., et al. (1992). The pericyte as a possible osteoblast progenitor-cell. *Clin. Orthop. Relat. Res., 276,* 287–299.

Bruder, S. P., & Caplan, A. I. (1990). Osteogenic cell lineage analysis is facilitated by organ culture of embryonic chick periosteum. *Dev. Biol., 141,* 319–329.

Bruder, S. P., Fink, D. J., & Caplan, A. I. (1994). Mesenchymal stem cells in bone development, bone repair, and skeletal regeneration. *J. Cell. Biochem., 56,* 283–294.

Caplan, A. I. (1994). The mesengenic process. *Clin. Plast. Surg., 21,* 429–435.

Caplan, A. I. (2008). All MSCs are pericytes? *Cell Stem Cell, 3,* 229–230.

Caplan, A. I. (2009). Why are MSCs therapeutic? New data: new insight. *J. Pathol., 217,* 318–324.

Caplan, A. I., & Dennis, J. E. (2006). Mesenchymal stem cells as trophic mediators. *J. Cell. Biochem., 98,* 1076–1084.

Caplan, A. I., & Haynesworth, S. E. (1993). Method for enhancing the implantation and differentiation of marrow-derived mesenchymal cells. *Patent No., 5,* 197–985.

Caplan, A. I. (1988). Biomaterials and bone repair. *Biomat, 87,* 15–24.

Chopp, M., & Li, Y. (2002). Treatment of neural injury with marrow stromal cells. *Lancet Neurol., 1,* 92–100.

Chutro, P. (1918). Greffeosseuse osseuse du tibia. *Bulletins et Mémoires de la Société de Chirurgie de Paris, 44,* 570.

Connolly, J. F., Guse, R., Tiedeman, J., & Dehne, R. (1991). Autologous marrow injection as a substitute for operative grafting of tibial nonunions. *Clin. Orth. Rel. Res., 266,* 259–269.

Covas, D. T., Panepucci, R. A., Fontes, A. M., et al. (2008). Multipotent mesenchymal stromal cells obtained from diverse human tissues share functional properties and gene-expression profile with CD146(+) perivascular cells and fibroblasts. *Exp. Hematol., 36,* 642–654.

Crisan, M., Yap, S., Casteilla, L., Chen, C., Corselli, M., Park, T. S., et al. (2008). A perivascular origin for mesenchymal stem cells in multiple human organs. *Cell Stem Cell, 3,* 301–313.

da Silva Meirelles, L., Fontes, A. M., Covas, D. T., & Caplan, A. I. (2009). Mechanisms involved in the therapeutic properties of mesenchymal stem cells. *Cytokine Growth Factor Rev., 20,* 419–427.

da Silva Meirelles, L., Caplan, A. I., & Nardi, N. B. (2008). In search of the in vivo identity of mesenchymal stem cells. *Stem Cells, 26,* 2287–2299.

de Bari, C., Dell'Accio, F., Tylzanowski, P., et al. (2001). Multipotent mesenchymal stem cells from adult human synovial membrane. *Arthritis Rheum., 44,* 1928–1942.

Dennis, J. E., Cohen, N., Goldberg, V. M., & Caplan, A. I. (2004). Targeted delivery of progenitor cells for cartilage repair. *J. Orthop. Res., 22,* 735–741.

Estes, B. T., Wu, A. W., & Guilak, F. (2006). Potent induction of chondrocytic differentiation of human adipose-derived adult stem cells by bone morphogenetic protein 6. *Arthritis Rheum., 54,* 1222–1232.

Farge, D., Vija, L., Gautier, J. F., Vexiau, P., Dumitrache, C., Verrecchia, F., et al. (2008). Mesenchymal stem cells — stem cell therapy perspectives for type 1 diabetes. *Proc. Rom. Acad., 2,* 59–70.

Fauza, D. (2004). Amniotic fluid and placental stem cells. *Best Pract. Res. Cl. Ob., 18,* 877–891.

Fleming, J. E., Jr., Cassiede, P., Baber, M., Haynesworth, S. E., & Caplan, A. I. (1998). A monoclonal antibody against adult marrow-derived mesenchymal cells recognizes developing vasculature in embryonic human skin. *Dev. Dyn., 212,* 119–132.

Frassoni, F., Labopin, M., Gluckman, E., Rocha, V., Bruno, B., Lazarus, H. M., et al. (2002). Expanded MSCs coinfused with HLA identical hemopoietic stem cell transplants, reduce acute and chronic graft versus host disease: a matched pair analysis. *Bone Marrow Transplant., 29,* S2–S75.

Garcia-Olmo, D., Garcia-Arranz, M., Herreros, D., Pascual, I., Peiro, C., & Rodriguez-Montes, J. A. (2005). A phase I clinical trial of the treatment of Crohn's fistula by adipose mesenchymal stem cell transplantation. *Dis. Colon Rectum.*, *48*, 1416–1423.

Gerhardt, H., & Semb, H. (2008). Pericytes: gatekeepers in tumour cell metastasis? *J. Mol. Med.*, *86*, 135–144.

Gimble, J. M., Katz, A. J., & Bunnell, B. A. (2007). Adipose-derived stem cells for regenerative medicine. *Circ. Res.*, *100*, 1249–1260.

Haynesworth, S. E., Baber, M. A., & Caplan, A. I. (1992). Cell surface antigens on human marrow-derived mesenchymal cells are detected by monoclonal antibodies. *Bone*, *13*, 69–80.

Hirschi, K., & D'Amore, P. A. (1996). Pericytes in the microvasculature. *Cardiovasc. Res.*, *32*, 687–698.

Iwashima, S., Ozaki, T., Maruyama, S., Saka, Y., Kobori, M., Omae, K., et al. (2009). Novel culture system of mesenchymal stromal cells from human subcutaneous adipose tissue. *Stem Cells Dev.*, *18*, 533–544.

Jaiswal, N., Haynesworth, S. E., Caplan, A. I., & Bruder, S. P. (1997). Osteogenic differentiation of purified, culture-expanded human mesenchymal stem cells in vitro. *J. Cell. Biochem.*, *64*, 295–312.

Ko, K., Kim, B. G., Awadallah, A., Mikulan, J., Lin, P., Letterio, J. J., et al. (2010). Targeting improves MSC treatment of inflammatory bowel disease. *Mol. Ther.* (Submitted).

Koc, O. N., Gerson, S. L., Cooper, B. W., Dyhouse, S. M., Haynesworth, S. E., Caplan, A. I., et al. (2000). Rapid hematopoietic recovery after co-infusion of autologous blood stem cells and culture expanded marrow mesenchymal stem cells in advanced breast cancer patients receiving high dose chemotherapy. *J. Clin. Oncology*, *18*, 307–316.

Koc, O. N., Peters, C., Aubourg, P., Raghavan, S., Dyhouse, S., DeGasperi, R., et al. (1999). Bone marrow derived mesenchymal stem cells remain host-derived despite successful hematopoietic engraftment after allogeneic transplantation in patients with lysosomal and peroxisomal storage diseases. *Exp. Hematol.*, *27*, 1675–1681.

Krampera, M., Marconi, S., Pasini, A., et al. (2007). Induction of neural-like differentiation in human mesenchymal stem cells derived from bone marrow, fat, spleen and thymus. *Bone*, *40*, 382–390.

Lazarus, H. M., Haynesworth, S. E., Gerson, S. L., Rosenthal, N., & Caplan, A. I. (1995). Ex-vivo expansion and subsequent infusion of human bone marrow-derived stromal progenitor cells (mesenchymal progenitor cells) [MPCs]: implications for therapeutic use. *Bone Marrow Transpl.*, *16*, 557–564.

le Blanc, K., Frassoni, F., Ball, L., Locatelli, F., Roelofs, H., Lewis, I., et al. (2008). Mesenchymal stem cells for treatment of steroid-resistant, severe, acute graft-versus-host disease: a phase II study. *Lancet*, *371*, 1579–1586.

le Blanc, K., Tammit, L. L., Sundberg, B., Haynesworth, S. E., & Ringden, O. (2003). Mesenchymal stem cells inhibit and stimulate mixed lymphocyte cultures and mitogenic responses independently of the major histocompatibility complex. *Scand. J. Immunol.*, *57*, 11–20.

Lee, J. Y., Qu-Petersen, Z., Cao, B., et al. (2000). Clonal isolation of muscle-derived cells capable of enhancing muscle regeneration and bone healing. *J. Cell. Biol.*, *150*, 1085–1100.

Li, Y., Chen, J., Chen, X. G., Wang, L., Guatam, S. C., Xu, Y. X., et al. (2002). Human marrow stromal cell therapy for stroke in rat: neurotrophins and functional recovery. *Neurology*, *59*, 514.

Mahmoud, A., Lu, D., Lu, M., & Chopp, M. (2003). Treatment of traumatic brain injury in adult rats with intra-venous administration of human bone marrow stromal cells. *Neurosurgery*, *53*, 697–703.

Maitra, B., Szekely, E., Gjini, K., Laughlin, M. J., Dennis, J., Haynesworth, S. E., et al. (2004a). Human mesenchymal stem cells support unrelated donor hematopoietic stem cells and suppress T cell activation. *Bone Marrow Transpl.*, *33*, 597–604.

Maitra, B., Szekely, E., Gjini, K., Laughlin, M. J., Dennis, J., Haynesworth, S. E., et al. (2004b). Human mesenchymal stem cells support unrelated donor hematopoietic stem cells and suppress T-cell activation. *Bone Marrow Transplant.*, *33*, 597–604.

Majumdar, M. K., Thiede, M. A., Mosca, J. D., Moorman, M., & Gerson, S. L. (1998). Phenotypic and functional comparison of cultures of marrow-derived mesenchymal stem cells (MSCs) and stromal cells. *J. Cell. Physiol.*, *176*, 57–66.

Meirelles, L. S., & Nardi, N. B. (2003). Murine marrow-derived mesenchymal stem cell: isolation, in vitro expansion, and characterization. *Br. J. Haematol.*, *123*, 702–711.

Nakahara, H., Goldberg, V. M., & Caplan, A. I. (1991). Culture-expanded human periosteal-derived cells exhibit osteochondral potential in vivo. *J. Ortho. Res.*, *9*, 465–476.

Neuhuber, B., Himes, B. T., Shumsky, J. S., Gallo, G., & Fischer, I. (2005). Axon growth and recovery of function supported by human bone marrow stromal cells in the injured spinal cord exhibit donor variations. *Brain Res.*, *1035*, 73–75.

Orkin, S. H., & Zon, L. I. (2008). Hematopoiesis: an evolving paradigm for stem cell biology. *Cell*, *132*(4), 631–644.

Owen, M., & Friedenstein, A. J. (1988). Stromal stem cells: marrow-derived osteogenic precursors. *Ciba Found. Symp.*, *136*, 42–60.

Penn, M. S., & Khalil, M. K. (2008). Exploitation of stem cell homing for gene delivery. *Expert Opin. Biol. Ther.*, *8*, 17—23.

Pittenger, M. F., & Martin, B. J. (2004). Mesenchymal stem cells and their potential as cardiac therapeutics. *Circ. Res.*, *95*, 9—20.

Reese, J. S., Koc, O. N., & Gerson, S. L. (1999). Human mesenchymal stem cells provide stromal support for efficient CD34+ transduction. *J. Hematoher. Stem Cell Res.*, *8*, 515—523.

Riordan, N. H., Ichim, T. E., Min, W. P., Wang, H., Solano, F., Lara, F., et al. (2009). Non-expanded adipose stromal vascular fraction cell therapy for multiple sclerosis. *J. Transl. Med.*, *7*, 29.

Sacchetti, B., Funari, A., Michienzi, S., di Cesare, S., Piersanti, S., Saggio, I., et al. (2007). Self-renewing osteopro-genitors in bone marrow sinusoids can organize a hematopoietic microenvironment. *Cell*, *131*, 324.

Sackstein, R., Merzaban, J. S., Cain, D. W., Dagia, N. M., Spencer, J. A., Lin, C. P., et al. (2008). Ex vivo glycan engineering of CD44 programs human multipotent mesenchymal stromal cell trafficking to bone. *Nat. Med.*, *14*, 181—187.

Saito, T., Dennis, J. E., Lennon, D. P., Young, R. G., & Caplan, A. I. (1996). Myogenic expression of mesenchymal stem cells within myotubes of MDX mice in vitro and in vivo. *Tissue Eng.*, *1*, 327—344.

Salingcarnboriboon, R., Yoshitake, H., Tsuji, K., et al. (2003). Establishment of tendon-derived cell lines exhibiting pluripotent mesenchymal stem cell-like property. *Exp. Cell. Res.*, *287*, 289—300.

Solchaga, L. A., Penick, K., Porter, J. D., Goldberg, V. M., Caplan, A. I., & Welter, J. F. (2005). FGF-2 enhances the mitotic and chondrogenic potentials of human adult bone marrow-derived mesenchymal stem cells. *J. Cell. Physiol.*, *203*, 398—409.

Sundin, M., Ringden, O., Sundberg, B., Nava, S., Gotherstrom, C., & LeBlanc, K. (2007). No alloantibodies against mesenchymal stromal cells, but presence of anti-fetal calf serum antibodies, after transplantation in allogeneic hematopoietic stem cell recipients. *Haematologica*, *92*, 1208—1215.

Toma, C., Pittenger, M. F., Cahil, K. S., Byrne, B. J., & Kessler, P. D. (2002). Human mesenchymal stem cells differentiate to a cardiomyocyte phenotype in the adult murine heart. *Circulation*, *105*, 93—98.

Toma, J. G., Akhavan, M., Fernandes, K. J., et al. (2001). Isolation of multipotent adult stem cells from the dermis of mammalian skin. *Nat. Cell Biol.*, *3*, 778—784.

Wakitani, S., Goto, T., Pineda, S. J., Young, R. G., Mansour, J. M., Goldberg, V. M., et al. (1994). Mesenchymal cell-based repair of large full-thickness defects of articular cartilage and underlying bone. *J. Bone Joint Surg.*, *76*, 579—592.

Wakitani, S., Saito, T., & Caplan, A. I. (1995). Myogenic cells derived from rat bone marrow mesenchymal stem cells exposed to 5-azacytidine. *Muscle Nerve*, *18*, 1417—1426.

Yoo, J. U., Barthel, T. S., Nishimura, K., Solchaga, L. A., Caplan, A. I., Goldberg, V. M., et al. (1998). The chondrogenic potential of human bone-marrow-derived mesenchymal progenitor cells. *J. Bone and Joint Surg.*, *80*, 1745—1757.

Young, R. G., Butler, D. L., Weber, W., Gordon, S. L., Fink, D. J., & Caplan, A. I. (1998). The use of mesenchymal stem cells in achilles tendon repair. *J. Orthop. Res.*, *16*, 406—413.

Multipotent Adult Progenitor Cells

Philip Roelandt, Catherine M. Verfaillie
Interdepartmental Stem Cell Institute Leuven, Catholic University Leuven, Onderwijs & Navorsing, Herestraat Leuven, Belgium

PLURIPOTENT STEM CELLS: EMBRYONIC STEM CELLS

Embryonic stem cells (ESCs) are pluripotent stem cells as they can be propagated indefinitely, and differentiate into cells of all three germ layers (endoderm, mesoderm, and ectoderm), shown by teratoma and embryoid body (EB) formation. Following blastocyst injection, mouse ESCs contribute to all somatic and germ-line lineages. ESCs are derived from the inner cell mass (ICM) of the blastocyst and are true pluripotent stem cells. Mouse ESCs express the cell surface antigen SSEA1 and human ESC SSEA4, and both are characterized by the expression of a number of relative ESC-specific genes, including the transcription factors (TFs) *Oct4* (Schöler et al., 1989), *Rex1* (Ben-Shushan et al., 1998), *Nanog* (Chambers et al., 2003; Mitsui et al., 2003), and *Sox2* (Avilion et al., 2003). *Oct4* is expressed in the pre-gastrulation embryo, primordial germ cells, the ICM, and germ cells (Schöler et al., 1989; Rosner et al., 1990). While normal expression levels of *Oct4* maintain mouse ESC self-renewal, a decrease in expression to <50% leads to spontaneous trophectoderm differentiation, and an increase to levels >200% to primitive endoderm differentiation (Niwa et al., 2000). *Oct4* promotes self-renewal by promoting transcription of genes such as *Oct4* (Boyer et al., 2005) and *Sox2* (Catena et al., 2004), and repressing genes such as *Hand1* and *Cdx2* that promote trophectoderm differentiation (Niwa et al., 2000). In the past years more and more is known about what regulates the expression of *Oct4*. Initial studies have shown that *Sall4* (Zhang et al., 2006), *Epas1* (*Hif-2α*) (Covello et al., 2006), *SF1*, and *RAR* (Botquin et al., 1998) activate the *Oct4* promoter, while *Tcf3* suppresses *Oct4* transcription (Cole et al., 2008). More recently, DNA methylation of regulatory enhancers by *Dnmt3a* and *Dnmt3b* were found to drive the *Oct4* expression (Li et al., 2007). A number of orphan receptors were identified that can either be suppressive (for example, *GCNF* and *COUP-TFII*) or stimulatory (*Nr5a2*) (Kellner and Kikyo, 2010). The homeoprotein *Nanog* is found to be equally essential for early mouse development and ESC propagation. *Nanog* prevents ICM cells from differentiating into extra-embryonic endoderm by inhibiting genes such as *Gata4* and *Gata6* that promote primitive endoderm differentiation. Older studies suggested that *Nanog*$^{-/-}$ mice do not develop an epiblast, and *Nanog*$^{-/-}$ ESCs differentiate into mesoderm and endoderm (Chambers et al., 2003; Mitsui et al., 2003). More recently it has been demonstrated that *Nanog*$^{-/-}$ cells are blocked in a transitional pre-pluripotent stage and eventually will develop into trophoblast rather than mesendoderm or will undergo apoptosis (Silva et al., 2009). Forced expression of *Nanog* in ESC results in LIF-independent proliferation, demonstrating its important role in maintaining ESC pluripotency (Chambers et al., 2003; Mitsui et al., 2003).

263

Principles of Regenerative Medicine. DOI: 10.1016/B978-0-12-381422-7.10015-X

Intricate TF binding networks involving *Oct4*, *Sox2*, and *Nanog* are crucial for global transcriptional activation and repression in ESCs. Using ChiP on ChiP assays, unique and overlapping promoter binding sites have been identified for *Oct4*, *Sox2*, and *Nanog*, which serve as positive or negative regulators of transcription (Boyer et al., 2005). These interactions are controlled by feed-forward loops, where initial regulators control other regulators with the option of converging and controlling downstream target genes. Others have used proteomics to identify *Nanog* partners (Wang et al., 2006). This technique identified *Nanog*-bound genes such as *Oct4*, as well as other TFs including *Sall1* and *Sall4*.

POST-NATAL TISSUE-SPECIFIC STEM CELLS: ARE SOME MORE THAN MULTIPOTENT?

During gastrulation, the pluripotent cells in the ICM become restricted first to a specific germ layer and then to a specific tissue. The latter persist throughout adult life, and are termed multipotent stem cells.

Already, since the late 1990s, studies have suggested that classical adult stem cells, thought to be multipotent, may actually be more pluripotent, as adult stem cells from a given tissue were reported to be able to become, under some circumstances, a cell of an unexpected tissue. Reports describing stem cell plasticity initially caused great excitement, as they challenged the concept that adult stem cells function solely to maintain the tissue of origin, suggesting that they might therefore provide a source of easily accessible cells not marred by ethical considerations and they could be used to treat a number of degenerative and genetic diseases. For instance, hematopoietic stem cells (HSCs) have been reported to differentiate into a variety of cell types of endoderm (lung epithelium, intestinal epithelium, kidney epithelium, endocrine pancreas, liver, bile ducts) (Petersen et al., 1999; Lagasse et al., 2000; Theise et al., 2000; Krause et al., 2001; Wagers et al., 2002; Alvarez-Dolado et al., 2003; Ianus et al., 2003; Kale et al., 2003; Vassilopoulos et al., 2003; Wang et al., 2003), ectoderm (epidermis and neural cells) (Brazelton et al., 2000; Mezey et al., 2000; Krause et al., 2001; Priller et al., 2001; Wagers et al., 2002; Alvarez-Dolado et al., 2003; Weimann et al., 2003a,b), as well as into mesoderm derivatives other than blood cells (skeletal and cardiac muscle, endothelium) (Ferrari et al., 1998; Gussoni et al., 1999; Orlic et al., 2001a,b; Jackson et al., 2001; Grant et al., 2002; LaBarge and Blau, 2002; Camargo et al., 2003; Corbel et al., 2003; Balsam et al., 2004; Murry et al., 2004; Kajstura et al., 2005).

However, after the initial series of optimistic reports, a number of reports appeared that challenge the initial observation, or provide alternative explanations to the claim of greater potency of adult stem cells. For instance, there is evidence that stem cells, such as HSCs, may not only reside in the bone marrow (BM) but can also be present in other tissues (Jackson et al., 1999; Kawada and Ogawa, 2001; McKinney-Freeman et al., 2002).

A second explanation for the perceived plasticity of chiefly hematopoietic cells is fusion between the hematopoietic cells and certain host cells *in vivo*, a phenomenon known from hybridoma cell production and also shown to occur *in vitro* between hematopoietic cells or neurospheres and ESCs (Terada et al., 2002; Ying et al., 2002). Indeed, a number of studies described fusion between cells of hematopoietic origin and hepatocytes, cardiomyocytes, skeletal muscle cells, and Purkinje cells in the brain (Wagers et al., 2002; Alvarez-Dolado et al., 2003; Weimann et al., 2003a; Balsam et al., 2004; Doyonnas et al., 2004). In many instances, the nucleus of the donor cell becomes partially reprogrammed with suppression of the hematopoietic program and activation of genes from which the donor cell fused (Wang et al., 2003; Weimann et al., 2003b; Cossu, 2004). Others have presented relatively convincing evidence that not all apparent plasticity is due to cell fusion, including differentiation of hematopoietic cells to lung epithelial cells (Harris et al., 2004), and neuronal lineage cells into endothelial cells (Wurmser et al., 2004). However, the efficiency with which one stem cell appears to acquire the phenotype of a tissue cell different from the tissue of origin, whether via fusion or direct, is limited, and it remains to be determined whether this would have clinical relevance.

The two remaining possible explanations for the apparent ability of some adult stem cells to generate cells of a tissue lineage different from the tissue of origin are that stem cells with more pluripotent characteristics persist into adulthood, or that adult stem cells can be reprogrammed via a process of dedifferentiation and redifferentiation to another lineage, or via a process of transdifferentiation.

CAN PLURIPOTENCY BE ACQUIRED?

In 2007, Takahashi et al. demonstrated that mouse adult fibroblasts can be reprogrammed towards cells with all ESC characteristics, so-called induced pluripotent stem cells (iPSCs), by the introduction of four transcription factors known to be expressed in ESCs (*Oct4*, *Sox2*, *Klf4*, and *c-Myc*), and selecting for cells that start to express endogenous *Nanog* or *Oct4* (Takahashi et al., 2007). Transfection with *Oct4*, *Sox2*, and *Klf4* drives somatic cells to a *Nanog*⁻ pre-pluripotent stage and acquisition of *Nanog* is mandatory to gain full reprogramming to pluripotent cells (Silva et al., 2009). This provides proof of the principle that adult somatic cells can be reprogrammed.

Since the initial description, many groups have created iPSCs from cell types and species other than mouse fibroblasts (Kim et al., 2009; Utikal et al., 2009; Yan et al., 2009). Moreover, a number of similar protocols have been generated by replacing one or more of the initial transcription factors with less or other transcription factors (*Nanog* and *Lin28* (Yu et al., 2007)), nuclear orphan receptors (*Esrrb* (Feng et al., 2009), *Nr5a2* (Heng et al., 2010)), or small molecules (Shi et al., 2008).

In 1993, spermatogonial stem cells were isolated for the first time from mouse testis, representing only 0.03% of all germ cells (Tegelenbosch and de Rooij, 1993). These spermatogonial stem cells can be transformed into ES-like cells easily *in vitro* by growing them on feeder layers or by the addition of LIF to the culture (de Rooij and Mizrak, 2008). The four transcription genes used for reprogramming are already present at low levels in spermatogonial stem cells (Kanatsu-Shinohara et al., 2008), but not *Nanog* (Kanatsu-Shinohara et al., 2004), representing the pre-pluripotent status mentioned above. Upon transition to ES-like cells, *Nanog* and *Sox2* are highly upregulated while typical spermatogonial genes are downregulated (Seandel et al., 2007).

Since 2001, a number of papers have reported that cells with greater potency can be isolated from other tissues than testis. These include the isolation of SKPs (skin-derived progenitors) (Toma et al., 2001), PMPs (pancreas-derived multipotent precursors) (Seaberg et al., 2004), and hFLMPCs (human fetal liver multipotent progenitor cells) (Dan et al., 2006) that can differentiate into cells of two germ layers. We isolated stem cells with increased pluripotency from the BM of mouse and rat (Reyes and Verfaillie, 2001; Jiang et al., 2002b; Zeng et al., 2006), termed multipotent adult progenitor cells (MAPCs). Since the initial description of MAPCs, a number of other cell populations isolated by culture of BM, umbilical cord blood, placental tissue, and amniotic fluid have been described that have the ability to differentiate into cells of the three germ layers. They have been named marrow-isolated adult multilineage inducible cells (MIAMI cells) (d'Ippolito et al., 2004), human bone marrow-derived stem cells (hBMSCs) (Yoon et al., 2005), unrestricted somatic stem cells (USSCs) (Kogler et al., 2004), fetal stem cells from somatic tissue (FSSCs) (Kues et al., 2005), very small embryonic-like cells (VSELs) (Kucia et al., 2005, 2006b), pre-mesenchymal stem cells (pre-MSCs) (Anjos-Afonso and Bonnet, 2007), multipotent adult stem cells (MASCs) (Beltrami et al., 2007), and amniotic fluid stem cells (AFSs) (de Coppi et al., 2007). Although the phenotype differs somewhat between these different cell populations, they have in common that they can be expanded extensively *in vitro*; that some of them reportedly express stem-cell specific genes such as Oct4;

and that they can differentiate *in vitro* to cells with at least some features of mesoderm, endoderm, and ectoderm. However, not all studies show this at the single-cell level, and the proof of differentiation differs between publications. Moreover, few if any of the studies have shown that the more potent cells can also regenerate a tissue *in vivo*.

ISOLATION OF RODENT MAPCs

In 2001 and 2002, we described the isolation of MAPCs from BM of human, mouse, and rat. Rodent MAPCs can be expanded *in vitro* without obvious senescence, and can at the single-cell level give rise to cells of mesoderm, endoderm, and ectoderm *in vitro*. We also demonstrated that a Rosa26 mouse-derived MAPC line contributed to many somatic tissues of the mouse when injected into the blastocyst (Jiang et al., 2002a).

Since the initial description of MAPC isolation, we have made changes to the culture method (Subramanian et al., 2010). MAPC isolation is now performed under hypoxic conditions: bone marrow cells are plated at relatively high density on fibronectin-coated plates in 5% O_2 and 6% CO_2. After approximately one month, cells are passed through a Myltenii column to remove CD45+ cells and Ter119+ cells, and cells subcloned at 5 cells/well. Clones are identified based on morphology and *Oct4* mRNA levels (q-RT-PCR), and expanded (Subramanian et al., 2010). This has led to the isolation of MAPCs that have significantly higher levels of Oct4. In addition, 90% of MAPCs thus isolated and maintained express Oct4 protein in the nucleus. The phenotype of mouse MAPC is B220, CD3, CD15, CD31, CD34, CD44, CD45, CD105, Thy1.1, Sca-1, E-cadherin, MHC classes I and II negative, epithelial cell adhesion molecule (EpCAM) low, and c-Kit, VLA-6, and CD9 positive. For rat MAPC, the phenotype is CD44, CD45, MHC classes I and II negative, but CD31 positive. To generate single cell-derived populations of MAPC, we subclone established MAPC lines at 0.8 cells/well. Such subcloning is not usually possible at the initial subcloning step, but has 30% efficiency when cells initially subcloned at 5 cells/well are subsequently subcloned at 0.8 cells/well.

Transcriptome analysis demonstrates that rodent MAPCs differ significantly from MSCs, but also differ significantly from ESCs (Ulloa-Montoya et al., 2007). Rodent MAPCs express a number of genes identified to be ESC-specific (ES cell-associated transcripts or ECATs) (Mitsui et al., 2003), including *Oct4*, *Rex1*, and eight other genes, but they do not express Nanog and Sox2, as well as eight other ECATs. Of note, rMAPCs also express gene characteristics for primitive endoderm, such as *Sox7*, *Sox17*, *Gata4*, *Gata6*, *Foxa2*, and *Hnf1β* (Ulloa-Montoya et al., 2007).

ISOLATION OF HUMAN MAPCs

Like rodent MAPCs, human MAPCs are isolated from BM and, like rodent MAPCs, they can be expanded extensively, although they do undergo eventual senescence. The cell surface is CD31, CD34, CD36, CD44, CD45, HLA class I, HLA-DR, c-Kit, Tie, VE-cadherin, VCAM, and ICAM-1 negative. Human MAPCs express very low levels of β2-microglobulin, AC133, Flk1, and Flt1, and high levels of CD13 and CD49b. Like rodent MAPCs, transcriptome studies have shown that human MAPCs differ from human MSCs and human ESCs; unlike rodent MAPCs, however, human MAPCs do not express significant levels of *Oct3a* (Reyes et al., 2002).

Differentiation ability of MAPCs *in vitro*

Rodent and human MAPCs differentiate to mesenchymal-type cells such as smooth muscle cells, osteoblasts, chondroblasts, and adipocytes (Reyes and Verfaillie, 2001; Zeng et al., 2006; Carmeliet et al., 2001; Ross et al., 2006), as well as to endothelial cells *in vitro* and *in vivo* (Reyes and Verfaillie, 2001; Jiang et al., 2002a; Reyes et al., 2002; Luttun et al., 2006; Zeng et al., 2006; Aranguren et al., 2008).

Since the initial description of differentiation of MAPCs to hepatocyte-like cells (Jiang et al., 2002a; Schwartz et al., 2002), we have developed a differentiation protocol that induces a more robust acquisition of phenotypic and functional characteristics of hepatocytes from rodent MAPCs (Roelandt et al., 2010a,b). However, the differentiation ability of human MAPCs using this new protocol is not enhanced compared with what we described in 2003. These culture conditions consist of initial induction of endoderm using Wnt3 and Activin-A; induction of hepatic endoderm using sequentially the mesodermal-derived cytokines BMP4 and FGF2 followed by FGF1, FGF4, and FGF8; and finally hepatocyte growth factor (HGF) and follistatin. This yields a mixed population of cells wherein a fraction expresses mature liver markers and has several functional characteristics of hepatocytes including albumin secretion, conversion of ammonia to urea, glycogen storage, bilirubin conjugation, and inducible cytochrome P450 activity. With minor adjustments, the protocol can also be applied to induce differentiation of mouse and human ESCs towards functional hepatocyte-like cells.

Engraftment of MAPCs *in vivo*

When mouse MAPCs were grafted intravenously, we found hematopoietic reconstitution (Jiang et al., 2002a). This study was further elaborated on in 2007, when two independent lines of MAPCs were grafted in sublethally irradiated NOD-SCID mice also treated with anti-NK antibodies: Tolar et al. demonstrated that engraftment of MAPCs that are MHC class-I-negative is inhibited by natural killer (NK) activity (Tolar et al., 2006). We found multilineage hematopoietic reconstitution in 75% of animals, without evidence of fusion in the hematopoietic progeny. MAPC-derived KLS cells from primary recipients could rescue secondary C57/Bl6 mice from lethal irradiation and establish long-term hematopoiesis. MAPC-derived progeny cells that are CD45-negative can be found in multiple organs, although differentiation in a tissue-specific manner was not seen. In 2008, we demonstrated that both human and murine MAPCs improve both blood flow and function of ischemic limb in mice (Aranguren et al., 2008), via chiefly trophic effects, although some direct contribution to endothelial cells and skeletal muscle was observed. Likewise, when injected in the heart following left anterior descendant artery occlusion, we and others have shown that murine and swine MAPCs improve cardiac function in comparison with other cell populations, such as MEFs (Zeng et al., 2006; Pelacho et al., 2007a,b,c; Tolar et al., 2007), and this again via trophic effect on cardiac cell survival and function, as well as angiogenesis.

Like MSCs, murine, rat, and human MAPCs have extensive immunomodulatory functions and can decrease T-cell-mediated immune reactions. In some of the studies this was found following systemic injection, whereas in other studies this was only observed by local injection (Ting et al., 2008; Highfill et al., 2009; Kovacsovics-Bankowski et al., 2009; Luyckx et al., 2010).

Contribution of rodent MAPCs to chimeras

Although the initial MAPC line described in *Nature* (Jiang et al. 2002a) contributed to chimeras, subsequent lines do not contribute in a significant manner. As the cells isolated under the new culture conditions have a primitive endoderm phenotype, for example Xen-P cells (Debeb et al., 2009), and the latter have been shown to contribute to the visceral endoderm (Galat et al., 2009), we are currently evaluating the contribution of MAPCs to the yolk sac.

The mechanisms underlying greater potency of MAPCs and similar adult stem cells with greater potency

One question that has not been answered is whether the cell populations described above (SKPs, PMPs, hFLMPCs, MAPCs, MIAMI cells, hBMSCs, USSCs, FSSCs, AFS, MASCs, VSELs, and pre-MSCs) exist *in vivo* or are created in culture as the result of dedifferentiation. From all the cells described, SKPs have recently been isolated directly from skin without the intervening culture step. Toma et al. showed that SKPs can also be derived freshly, without the preceding culture, from fetal mice as well as from adult mice, where they appear to reside in a niche in the hair papillae and whisker follicles (Fernandes et al., 2004). Anjos-Afonso and Bonnet found the SSEA1$^+$ pre-MSCs that express high levels of *Oct4* and can be expanded under MAPC conditions to generate cells capable of differentiating to the mesodermal, endodermal, and ectodermal lineage, and can contribute to hematopoiesis when grafted *in vivo*, can be isolated from mesenchymal cultures at passage 1 (Anjos-Afonso and Bonnet, 2007). In contrast to MAPCs, the cells isolated by Anjos-Afonso also expressed *Nanog* and *Sox2*. In addition, Kucia et al. demonstrated that a homogeneous population of rare Sca-1+ Lin-, CD45-cells can be selected directly from the BM of mice and humans (Kucia et al., 2006a). These VSELs express — like the cells identified by Anjos-Afonso and Bonnet (2007) and like ESCs — SSEA1, *Oct4*, *Nanog*, and *Rex1*. The latter two studies suggest that rare cells exist in murine and human marrow with phenotypic features of MAPCs, MIAMI cells, hBMSCs, USSCs, AFS, or FSSCs. Whether the differentiation ability ascribed to MAPCs and similar cells (Jiang et al., 2002a; d'Ippolito et al., 2004; Kogler et al., 2004; Kues et al., 2005; Yoon et al., 2005; Anjos-Afonso and Bonnet, 2007) is already present in the primary selected, uncultured BM cells isolated by Anjos-Afonso and Bonnet (2007) and Kucia et al. (2006b), hence representing cells with greater potency persisting *in vivo* into post-natal life, or whether the differentiation ability is acquired once cells are culture-expanded *in vitro*, therefore representing dedifferentiation of a rare *Oct4*$^+$ cell, is not known.

The question as to whether MAPCs, and similar cells, exist as such is not only of academic importance; the answer may have profound biological implications as well as potential clinical applications. *In vitro*-generated cells have tremendous potential clinical usefulness, as long as the cells can be generated in an efficient and reliable manner. If MAPCs exist as such *in vivo*, it may one day be possible to manipulate their function *in vivo*, without the need for *in vitro* manipulation. Hence, future studies should aim to determine whether MAPCs and similar cells exist *in vivo*, and if so what the optimal method of isolation and *in vitro* expansion is; and whether they could be mobilized and/or activated *in vivo*. If the answer is "no," then it will be of the utmost importance to determine which cell population in a given tissue generates cells with greater potency *in vitro*, and develop strategies to select the precursor and induce with great efficiency the phenotype *in vitro*.

Acknowledgments

We acknowledge the support of the FWO (Odysseus fund) and the KUL COE funding. PR has a doctoral grant from IWT.

References

Alvarez-Dolado, M., Pardal, R., Garcia-Verdugo, J. M., Fike, J. R., Lee, H. O., Pfeffer, K., et al. (2003). Fusion of bone-marrow-derived cells with Purkinje neurons, cardiomyocytes and hepatocytes. *Nature, 425*, 968–973.

Anjos-Afonso, F., & Bonnet, D. (2007). Non-hematopoietic/endothelial SSEA-1+ cells defines the most primitive progenitors in the adult murine bone marrow mesenchymal compartment. *Blood, 109*, 1298–1306.

Aranguren, X. L., McCue, J. D., Hendrickx, B., Zhu, X.-H., Du, F., Chen, E., et al. (2008). Multipotent adult progenitor cells sustain function of ischemic limbs in mice. *J. Clin. Invest., 118*, 505–514.

Avilion, A. A., Nicolis, S. K., Pevny, L. H., Perez, L., Vivian, N., & Lovell-Badge, R. (2003). Multipotent cell lineages in early mouse development depend on SOX2 function. *Genes Dev., 17*, 126–140.

Balsam, L. B., Wagers, A. J., Christensen, J. L., Kofidis, T., Weissman, I. L., & Robbins, R. C. (2004). Haematopoietic stem cells adopt mature haematopoietic fates in ischaemic myocardium. *Nature, 428*, 668–673.

Beltrami, A. P., Cesselli, D., Bergamin, N., Marcon, P., Rigo, S., Puppato, E., et al. (2007). Multipotent cells can be generated in vitro from several adult human organs (heart, liver, and bone marrow). *Blood, 110*, 3438–3446.

Ben-Shushan, E., Thompson, J. R., Gudas, L. J., & Bergman, Y. (1998). Rex-1, a gene encoding a transcription factor expressed in the early embryo, is regulated via Oct-3/4 and Oct-6 binding to an octamer site and a novel protein, Rox-1, binding to an adjacent site. *Mol. Cell Biol., 18*, 1866–1878.

Botquin, V., Hess, H., Fuhrmann, G., Anastassiadis, C., Gross, M. K., Vriend, G., et al. (1998). New POU dimer configuration mediates antagonistic control of an osteopontin preimplantation enhancer by Oct-4 and Sox-2. *Genes Dev., 12*, 2073–2090.

Boyer, L. A., Lee, T. I., Cole, M. F., Johnstone, S. E., Levine, S. S., Zucker, J. P., et al. (2005). Core transcriptional regulatory circuitry in human embryonic stem cells. *Cell, 122*, 947–956.

Brazelton, T. R., Rossi, F. M. V., Keshet, G. I., & Blau, H. M. (2000). From marrow to brain: expression of neuronal phenotypes in adult mice. *Science, 290*, 1775–1779.

Camargo, F. D., Green, R., Capetanaki, Y., Jackson, K. A., & Goodell, M. A. (2003). Single hematopoietic stem cells generate skeletal muscle through myeloid intermediates. *Nat. Med., 9*, 1520–1527.

Carmeliet, P., Moons, L., Luttun, A., Vincenti, V., Compernolle, V., de Mol, M., et al. (2001). Synergism between vascular endothelial growth factor and placental growth factor contributes to angiogenesis and plasma extravasation in pathological conditions. *Nat. Med., 7*, 575–583.

Catena, R., Tiveron, C., Ronchi, A., Porta, S., Ferri, A., Tatangelo, L., et al. (2004). Conserved POU binding DNA sites in the Sox2 upstream enhancer regulate gene expression in embryonic and neural stem cells. *J. Biol. Chem., 279*, 41846–41857.

Chambers, I., Colby, D., Robertson, M., Nichols, J., Lee, S., Tweedie, S., et al. (2003). Functional expression cloning of Nanog, a pluripotency sustaining factor in embryonic stem cells. *Cell, 113*, 643–655.

Cole, M. F., Johnstone, S. E., Newman, J. J., Kagey, M. H., & Young, R. A. (2008). Tcf3 is an integral component of the core regulatory circuitry of embryonic stem cells. *Genes Dev., 22*, 746–755.

Corbel, S. Y., Lee, A., Yi, L., Duenas, J., Brazelton, T. R., Blau, H. M., et al. (2003). Contribution of hematopoietic stem cells to skeletal muscle. *Nat. Med., 9*, 1528–1532.

Cossu, G. (2004). Fusion of bone marrow-derived stem cells with striated muscle may not be sufficient to activate muscle genes. *J. Clin. Invest., 114*, 1540–1543.

Covello, K. L., Kehler, J., Yu, H., Gordan, J. D., Arsham, A. M., Hu, C.-J., et al. (2006). HIF-2alpha regulates Oct-4: effects of hypoxia on stem cell function, embryonic development, and tumor growth. *Genes Dev., 20*, 557–570.

Dan, Y. Y., Riehle, K. J., Lazaro, C., Teoh, N., Haque, J., Campbell, J. S., et al. (2006). Isolation of multipotent progenitor cells from human fetal liver capable of differentiating into liver and mesenchymal lineages. *Proc. Natl. Acad. Sci. U.S.A., 103*, 9912–9917.

de Coppi, P., Bartsch, G., Siddiqui, M. M., Xu, T., Santos, C. C., Perin, L., et al. (2007). Isolation of amniotic stem cell lines with potential for therapy. *Nat. Biotech., 25*, 100–106.

de Rooij, D. G., & Mizrak, S. C. (2008). Deriving multipotent stem cells from mouse spermatogonial stem cells: a new tool for developmental and clinical research. *Development, 135*, 2207–2213.

Debeb, B. G., Galat, V., Epple-Farmer, J., Iannaccone, S., Woodward, W. A., Bader, M., et al. (2009). Isolation of Oct4-expressing extraembryonic endoderm precursor cell lines. *PLoS ONE, 4*, e7216.

d'Ippolito, G., Diabira, S., Howard, G. A., Menei, P., Roos, B. A., & Schiller, P. C. (2004). Marrow-isolated adult multilineage inducible (MIAMI) cells, a unique population of postnatal young and old human cells with extensive expansion and differentiation potential. *J. Cell Sci., 117*, 2971–2981.

Doyonnas, R., LaBarge, M. A., Sacco, A., Charlton, C., & Blau, H. M. (2004). Hematopoietic contribution to skeletal muscle regeneration by myelomonocytic precursors. *Proc. Natl. Acad. Sci. U.S.A., 101*, 13507–13512.

Feng, B., Jiang, J., Kraus, P., Ng, J. H., Heng, J. C., Chan, Y. S., et al. (2009). Reprogramming of fibroblasts into induced pluripotent stem cells with orphan nuclear receptor Esrrb. *Nat. Cell Biol., 11*, 197–203.

Fernandes, K. J. L., McKenzie, I. A., Mill, P., Smith, K. M., Akhavan, M., Barnabe-Heider, F., et al. (2004). A dermal niche for multipotent adult skin-derived precursor cells. *Nat. Cell Biol., 6*, 1082–1093.

Ferrari, G., Cusella-de Angelis, G., Coletta, M., Paolucci, E., Stornaiuolo, A., Cossu, G., et al. (1998). Muscle regeneration by bone marrow-derived myogenic progenitors. *Science, 279*, 1528–1530.

Galat, V., Binas, B., Iannaccone, S., Postovit, L.-M., Debeb, B., & Iannaccone, P. (2009). Developmental potential of rat extraembryonic stem cells. *Stem Cells Dev., 18*, 1309–1318.

Grant, M. B., May, W. S., Caballero, S., Brown, G. A. J., Guthrie, S. M., Mames, R. N., et al. (2002). Adult hematopoietic stem cells provide functional hemangioblast activity during retinal neovascularization. *Nat. Med., 8*, 607–612.

Gussoni, E., Soneoka, Y., Strickland, C. D., Buzney, E. A., Khan, M. K., Flint, A. F., et al. (1999). Dystrophin expression in the mdx mouse restored by stem cell transplantation. *Nature, 401*, 390–394.

Harris, R. G., Herzog, E. L., Bruscia, E. M., Grove, J. E., van Arnam, J. S., & Krause, D. S. (2004). Lack of a fusion requirement for development of bone marrow-derived epithelia. *Science, 305*, 90–93.

Heng, J. C., Feng, B., Han, J., Jiang, J., Kraus, P., Ng, J. H., et al. (2010). The nuclear receptor Nr5a2 can replace Oct4 in the reprogramming of murine somatic cells to pluripotent cells. *Cell Stem Cell, 6*, 167–174.

Highfill, S. L., Kelly, R. M., O'Shaughnessy, M. J., Zhou, Q., Xia, L., Panoskaltsis-Mortari, A., et al. (2009). Multipotent adult progenitor cells can suppress graft-versus-host disease via prostaglandin E2 synthesis and only if localized to sites of allopriming. *Blood, 114*, 693–701.

Ianus, A., Holz, G. G., Theise, N. D., & Hussain, M. A. (2003). In vivo derivation of glucose-competent pancreatic endocrine cells from bone marrow without evidence of cell fusion. *J. Clin. Invest., 111*, 843–850.

Jackson, K. A., Majka, S. M., Wang, H., Pocius, J., Hartley, C. J., Majesky, M. W., et al. (2001). Regeneration of ischemic cardiac muscle and vascular endothelium by adult stem cells. *J. Clin. Invest., 107*, 1395–1402.

269

Jackson, K. A., Mi, T., & Goodell, M. A. (1999). Hematopoietic potential of stem cells isolated from murine skeletal muscle. *Proc. Natl. Acad. Sci. U.S.A.*, *96*, 14482−14486.

Jiang, Y., Jahagirdar, B. N., Reinhardt, R. L., Schwartz, R. E., Keene, C. D., Ortiz-Gonzalez, X. R., et al. (2002a). Pluripotency of mesenchymal stem cells derived from adult marrow. *Nature*, *418*, 41−49.

Jiang, Y., Vaessen, B., Lenvik, T., Blackstad, M., Reyes, M., & Verfaillie, C. M. (2002b). Multipotent progenitor cells can be isolated from postnatal murine bone marrow, muscle, and brain. *Exp. Hematol.*, *30*, 896−904.

Kajstura, J., Rota, M., Whang, B., Cascapera, S., Hosoda, T., Bearzi, C., et al. (2005). Bone marrow cells differentiate in cardiac cell lineages after infarction independently of cell fusion. *Circ. Res.*, *96*, 127−137.

Kale, S., Karihaloo, A., Clark, P. R., Kashgarian, M., Krause, D. S., & Cantley, L. G. (2003). Bone marrow stem cells contribute to repair of the ischemically injured renal tubule. *J. Clin. Invest.*, *112*, 42−49.

Kanatsu-Shinohara, M., Inoue, K., Lee, J., Yoshimoto, M., Ogonuki, N., Miki, H., et al. (2004). Generation of pluripotent stem cells from neonatal mouse testis. *Cell*, *119*, 1001−1012.

Kanatsu-Shinohara, M., Lee, J., Inoue, K., Ogonuki, N., Miki, H., Toyokuni, S., et al. (2008). Pluripotency of a single spermatogonial stem cell in mice. *Biol. Reprod.*, *78*, 681−687.

Kawada, H., & Ogawa, M. (2001). Bone marrow origin of hematopoietic progenitors and stem cells in murine muscle. *Blood*, *98*, 2008−2013.

Kellner, S., & Kikyo, N. (2010). Transcriptional regulation of the Oct4 gene, a master gene for pluripotency. *Histol. Histopathol.*, *25*, 405−412.

Kim, J. B., Zaehres, H., Arauzo-Bravo, M. J., & Schöler, H. R. (2009). Generation of induced pluripotent stem cells from neural stem cells. *Nat. Protoc.*, *4*, 1464−1470.

Kogler, G., Sensken, S., Airey, J. A., Trapp, T., Muschen, M., Feldhahn, N., et al. (2004). A new human somatic stem cell from placental cord blood with intrinsic pluripotent differentiation potential. *J. Exp. Med.*, *200*, 123−135.

Kovacsovics-Bankowski, M., Streeter, P. R., Mauch, K. A., Frey, M. R., Raber, A., van't Hof, W., Deans, R., et al. (2009). Clinical scale expanded adult pluripotent stem cells prevent graft-versus-host disease. *Cell Immunol.*, *255*, 55−60.

Krause, D. S., Theise, N. D., Collector, M. I., Henegariu, O., Hwang, S., Gardner, R., et al. (2001). Multi-organ, multi-lineage engraftment by a single bone marrow-derived stem cell. *Cell*, *105*, 369−377.

Kucia, M., Halasa, M., Wysoczynski, M., Baskiewicz-Masiuk, M., Moldenhawer, S., Zuba-Surma, E., et al. (2006a). Morphological and molecular characterization of novel population of CXCR4+ SSEA-4+ Oct-4+ very small embryonic-like cells purified from human cord blood − preliminary report. *Leukemia*, *21*, 297−303.

Kucia, M., Ratajczak, J., & Ratajczak, M. Z. (2005). Bone marrow as a source of circulating CXCR4+ tissue-committed stem cells. *Biol. Cell*, *97*, 133−146.

Kucia, M., Reca, R., Campbell, F. R., Zuba-Surma, E., Majka, M., Ratajczak, J., et al. (2006b). A population of very small embryonic-like (VSEL) CXCR4+ SSEA-1+ Oct-4+ stem cells identified in adult bone marrow. *Leukemia*, *20*, 857−869.

Kues, W. A., Carnwath, J. W., & Niemann, H. (2005). From fibroblasts and stem cells: implications for cell therapies and somatic cloning. *Reprod. Fertil. Dev.*, *17*, 125−134.

LaBarge, M. A., & Blau, H. M. (2002). Biological progression from adult bone marrow to mononucleate muscle stem cell to multinucleate muscle fiber in response to injury. *Cell*, *111*, 589−601.

Lagasse, E., Connors, H., Al-Dhalimy, M., Reitsma, M., Dohse, M., Osborne, L., et al. (2000). Purified hematopoietic stem cells can differentiate into hepatocytes in vivo. *Nat. Med.*, *6*, 1229−1234.

Li, J. Y., Pu, M. T., Hirasawa, R., Li, B. Z., Huang, Y. N., Zeng, R., et al. (2007). Synergistic function of DNA methyltransferases Dnmt3a and Dnmt3b in the methylation of Oct4 and Nanog. *Mol. Cell Biol.*, *27*, 8748−8759.

Luttun, A., Ross, J. J., Verfaillie, C., Aranguren, X. L., & Prosper, F. (2006). Differentiation of multipotent adult progenitor cells into functional endothelial and smooth muscle cells. *Curr. Protoc. Immunol.*, *75*, 22F.9.1−22F.9.31.

Luyckx, A., de Somer, L., Rutgeerts, O., Waer, M., Verfaillie, C. M., van Gool, S., et al. (2010). Mouse MAPC-mediated immunomodulation: cell-line dependent variation. *Exp. Hematol.*, *38*, 1−2.

McKinney-Freeman, S. L., Jackson, K. A., Camargo, F. D., Ferrari, G., Mavilio, F., & Goodell, M. A. (2002). Muscle-derived hematopoietic stem cells are hematopoietic in origin. *Proc. Natl. Acad. Sci. U.S.A.*, *99*, 1341−1346.

Mezey, E., Chandross, K. J., Harta, G., Maki, R. A., & McKercher, S. R. (2000). Turning blood into brain: cells bearing neuronal antigens generated in vivo from bone marrow. *Science*, *290*, 1779−1782.

Mitsui, K., Tokuzawa, Y., Itoh, H., Segawa, K., Murakami, M., Takahashi, K., et al. (2003). The homeoprotein Nanog is required for maintenance of pluripotency in mouse epiblast and ES cells. *Cell*, *113*, 631−642.

Murry, C. E., Soonpaa, M. H., Reinecke, H., Nakajima, H., Nakajima, H. O., Rubart, M., et al. (2004). Haemato-poietic stem cells do not transdifferentiate into cardiac myocytes in myocardial infarcts. *Nature*, *428*, 664−668.

Niwa, H., Miyazaki, J., & Smith, A. G. (2000). Quantitative expression of Oct-3/4 defines differentiation, dedif-ferentiation or self-renewal of ES cells. *Nat. Genet.*, *24*, 372−376.

Orlic, D., Kajstura, J., Chimenti, S., Jakoniuk, I., Anderson, S. M., Li, B., et al. (2001a). Bone marrow cells regenerate infarcted myocardium. *Nature, 410,* 701−705.

Orlic, D., Kajstura, J., Chimenti, S., Limana, F., Jakoniuk, I., Quaini, F., et al. (2001b). Mobilized bone marrow cells repair the infarcted heart, improving function and survival. *Proc. Natl. Acad. Sci. U.S.A., 98,* 10344−10349.

Pelacho, B., Aranguren, X. L., Mazo, M., Abizanda, G., Gavira, J. J., Clavel, C., et al. (2007a). Plasticity and cardiovascular applications of multipotent adult progenitor cells. *Nat. Clin. Pract. Cardiovasc. Med., 4*(Suppl. 1), S15−S20.

Pelacho, B., Luttun, A., Aranguren, X. L., Verfaillie, C. M., & Prosper, F. (2007b). Therapeutic potential of adult progenitor cells in cardiovascular disease. *Expert Opin. Biol. Ther., 7,* 1153−1165.

Pelacho, B., Nakamura, Y., Zhang, J., Ross, J., Heremans, Y., Nelson-Holte, M., et al. (2007c). Multipotent adult progenitor cell transplantation increases vascularity and improves left ventricular function after myocardial infarction. *J. Tissue Eng. Regen. Med., 1,* 51−59.

Petersen, B. E., Bowen, W. C., Patrene, K. D., Mars, W. M., Sullivan, A. K., Murase, N., et al. (1999). Bone marrow as a potential source of hepatic oval cells. *Science, 284,* 1168−1170.

Priller, J., Persons, D. A., Klett, F. F., Kempermann, G., Kreutzberg, G. W., & Dirnagl, U. (2001). Neogenesis of cerebellar Purkinje neurons from gene-marked bone marrow cells in vivo. *J. Cell Biol., 155,* 733−738.

Reyes, M., & Verfaillie, C. M. (2001). Characterization of multipotent adult progenitor cells, a subpopulation of mesenchymal stem cells. *Ann. N.Y. Acad. Sci., 938,* 231−235.

Reyes, M., Dudek, A., Jahagirdar, B., Koodie, L., Marker, P. H., & Verfaillie, C. M. (2002). Origin of endothelial progenitors in human postnatal bone marrow. *J. Clin. Invest., 109,* 337−346.

Roelandt, P., Pauwelyn, K. A., Sancho-Bru, P., Subramanian, K., Bose, B., Ordovas, L., et al. (2010a). Human embryonic and rat adult stem cells with primitive endoderm-like phenotype can be fated to definitive endoderm, and finally functional hepatocyte-like cells. *Plos ONE, 5,* e12101.

Roelandt, P., Sancho-Bru, P., Pauwelyn, K., & Verfaillie, C. (2010b). Differentiation of rat multipotent adult progenitor cells to functional hepatocyte-like cells by mimicking embryonic liver development. *Nat. Protoc., 5,* 1324−1336.

Rosner, M. H., Vigano, M. A., Ozato, K., Timmons, P. M., Poirier, F., Rigby, P. W., et al. (1990). A POU-domain transcription factor in early stem cells and germ cells of the mammalian embryo. *Nature, 345,* 686−692.

Ross, J. J., Hong, Z., Willenbring, B., Zeng, L., Isenberg, B., Lee, E. H., et al. (2006). Cytokine-induced differentiation of multipotent adult progenitor cells into functional smooth muscle cells. *J. Clin. Invest., 116,* 3139−3149.

Schöler, H. R., Balling, R., Hatzopoulos, A. K., Suzuki, N., & Gruss, P. (1989). Octamer binding proteins confer transcriptional activity in early mouse embryogenesis. *EMBO J., 8,* 2551−2557.

Schwartz, R. E., Reyes, M., Koodie, L., Jiang, Y., Blackstad, M., Lund, T., et al. (2002). Multipotent adult progenitor cells from bone marrow differentiate into functional hepatocyte-like cells. *J. Clin. Invest., 109,* 1291−1302.

Seaberg, R. M., Smukler, S. R., Kieffer, T. J., Enikolopov, G., Asghar, Z., Wheeler, M. B., et al. (2004). Clonal identification of multipotent precursors from adult mouse pancreas that generate neural and pancreatic lineages. *Nat. Biotechnol., 22,* 1115−1124.

Seandel, M., James, D., Shmelkov, S. V., Falciatori, I., Kim, J., Chavala, S., et al. (2007). Generation of functional multipotent adult stem cells from GPR125+ germline progenitors. *Nature, 449,* 346−350.

Shi, Y., Desponts, C., Do, J. T., Hahm, H. S., Schöler, H. R., & Ding, S. (2008). Induction of pluripotent stem cells from mouse embryonic fibroblasts by Oct4 and Klf4 with small-molecule compounds. *Cell Stem Cell, 3,* 568−574.

Silva, J., Nichols, J., Theunissen, T. W., Guo, G., van Oosten, A. L., Barrandon, O., et al. (2009). Nanog is the gateway to the pluripotent ground state. *Cell, 138,* 722−737.

Subramanian, K., Geraerts, M., Pauwelyn, K. A., Park, Y., Owens, D. J., Muijtjens, M., et al. (2010). Isolation procedure and characterization of multipotent adult progenitor cells (MAPC) from rat bone marrow. In S. Ding (Ed.), *Cellular Programming and Reprogramming* (pp. 55−78). New York: Humana Press.

Takahashi, K., Tanabe, K., Ohnuki, M., Narita, M., Ichisaka, T., Tomoda, K., et al. (2007). Induction of pluripotent stem cells from adult human fibroblasts by defined factors. *Cell, 131,* 861−872.

Tegelenbosch, R. A., & de Rooij, D. G. (1993). A quantitative study of spermatogonial multiplication and stem cell renewal in the C3H/101 F1 hybrid mouse. *Mutat. Res., 290,* 193−200.

Terada, N., Hamazaki, T., Oka, M., Hoki, M., Mastalerz, D. M., Nakano, Y., et al. (2002). Bone marrow cells adopt the phenotype of other cells by spontaneous cell fusion. *Nature, 416,* 542−545.

Theise, N. D., Badve, S., Saxena, R., Henegariu, O., Sell, S., Crawford, J. M., et al. (2000). Derivation of hepatocytes from bone marrow cells in mice after radiation-induced myeloablation. *Hepatology, 31,* 235−240.

Ting, A. E., Mays, R. W., Frey, M. R., Hof, W. V., Medicetty, S., & Deans, R. (2008). Therapeutic pathways of adult stem cell repair. *Crit. Rev. Oncol. Hematol., 65,* 81−93.

Tolar, J., O'Shaughnessy, M. J., Panoskaltsis-Mortari, A., McElmurry, R. T., Bell, S., Riddle, M., et al. (2006). Host factors that impact the biodistribution and persistence of multipotent adult progenitor cells. *Blood, 107,* 4182−4188.

Tolar, J., Wang, X., Braunlin, E., McElmurry, R. T., Nakamura, Y., Bell, S., et al. (2007). The host immune response is essential for the beneficial effect of adult stem cells after myocardial ischemia. *Exp. Hematol., 35*, 682–690.

Toma, J. G., Akhavan, M., Fernandes, K. J., Barnabe-Heider, F., Sadikot, A., Kaplan, D. R., et al. (2001). Isolation of multipotent adult stem cells from the dermis of mammalian skin. *Nat. Cell Biol., 3*, 778–784.

Ulloa-Montoya, F., Kidder, B., Pauwelyn, K., Chase, L., Luttun, A., Crabbe, A., et al. (2007). Comparative transcriptome analysis of embryonic and adult stem cells with extended and limited differentiation capacity. *Genome Biol., 8*, R163.

Utikal, J., Maherali, N., Kulalert, W., & Hochedlinger, K. (2009). Sox2 is dispensable for the reprogramming of melanocytes and melanoma cells into induced pluripotent stem cells. *J. Cell Sci., 122*, 3502–3510.

Vassilopoulos, G., Wang, P.-R., & Russell, D. W. (2003). Transplanted bone marrow regenerates liver by cell fusion. *Nature, 422*, 901–904.

Wagers, A. J., Sherwood, R. I., Christensen, J. L., & Weissman, I. L. (2002). Little evidence for developmental plasticity of adult hematopoietic stem cells. *Science, 297*, 2256–2259.

Wang, J., Rao, S., Chu, J., Shen, X., Levasseur, D. N., Theunissen, T. W., et al. (2006). A protein interaction network for pluripotency of embryonic stem cells. *Nature, 444*, 364–368.

Wang, X., Willenbring, H., Akkari, Y., Torimaru, Y., Foster, M., Al-Dhalimy, M., et al. (2003). Cell fusion is the principal source of bone-marrow-derived hepatocytes. *Nature, 422*, 897–901.

Weimann, J. M., Charlton, C. A., Brazelton, T. R., Hackman, R. C., & Blau, H. M. (2003a). Contribution of transplanted bone marrow cells to Purkinje neurons in human adult brains. *Proc. Natl. Acad. Sci. U.S.A., 100*, 2088–2093.

Weimann, J. M., Johansson, C. B., Trejo, A., & Blau, H. M. (2003b). Stable reprogrammed heterokaryons form spontaneously in Purkinje neurons after bone marrow transplant. *Nat. Cell Biol., 5*, 959–966.

Wurmser, A. E., Nakashima, K., Summers, R. G., Toni, N., d'Amour, K. A., Lie, D. C., et al. (2004). Cell fusion-independent differentiation of neural stem cells to the endothelial lineage. *Nature, 430*, 350–356.

Yan, X., Qin, H., Qu, C., Tuan, R. S., Shi, S., & Huang, G. T. (2009). iPS cells reprogrammed from mesenchymal-like stem/progenitor cells of dental tissue origin. *Stem Cells Dev, 19*, 469–480.

Ying, Q.-L., Nichols, J., Evans, E. P., & Smith, A. G. (2002). Changing potency by spontaneous fusion. *Nature, 426*, 545–548.

Yoon, Y. S., Wecker, A., Heyd, L., Park, J. S., Tkebuchava, T., Kusano, K., et al. (2005). Clonally expanded novel multipotent stem cells from human bone marrow regenerate myocardium after myocardial infarction. *J. Clin. Invest., 115*, 326–338.

Yu, J., Vodyanik, M. A., Smuga-Otto, K., Antosiewicz-Bourget, J., Frane, J. L., Tian, S., et al. (2007). Induced pluripotent stem cell lines derived from human somatic cells. *Science, 318*, 1917–1920.

Zeng, L., Rahrmann, E., Hu, Q., Lund, T., Sandquist, L., Felten, M., et al. (2006). Multipotent adult progenitor cells from swine bone marrow. *Stem Cells, 24*, 2355–2366.

Zhang, J., Tam, W. L., Tong, G. Q., Wu, Q., Chan, H. Y., Soh, B. S., et al. (2006). Sall4 modulates embryonic stem cell pluripotency and early embryonic development by the transcriptional regulation of Pou5f1. *Nat. Cell Biol., 8*, 1114–1123.

CHAPTER 16

Hematopoietic Stem Cell Properties, Markers, and Therapeutics

Kuanyin K. Lin, Grant A. Challen, Margaret A. Goodell
Center for Cell and Gene Therapy, Stem Cell and Regenerative Medicine Center, Department of Pathology and Immunology, Baylor College of Medicine, Houston, TX, USA

INTRODUCTION

Hematopoietic stem cells (HSCs), which primarily reside in bone marrow, maintain blood formation and replenish themselves throughout the adult's lifespan. The activity of bone marrow HSCs was discovered half a century ago when Ford et al. identified a robust contribution of donor bone marrow cells in lethally irradiated recipient mice (Ford et al., 1956). After three decades of work, the contribution of donor hematopoietic cells in recipients had been demonstrated to originate from a few "clones," suggesting the existence of HSCs (Becker et al., 1963; Lemischka et al., 1986). However, the isolation of HSCs was not achieved until 1988, when Weissman and his colleagues enriched HSCs from the murine bone marrow using a fluorescent-activated cell sorter (Spangrude et al., 1988). Since these seminal studies, researchers have been able to demonstrate that HSCs possess stem cell properties including the ability to give rise to daughter HSCs (self-renewal) as well as to repopulate all of the hematopoietic lineages (differentiation, Fig. 16.1).

DEVELOPMENTAL ORIGIN OF HEMATOPOIESIS

Developmental origins of emerging hematopoietic cells have been characterized in animal models such as mice (*Mus musculus*) and zebrafish (*Danio rerio*). During mouse embryonic development, two waves of hematopoiesis occur, primitive hematopoiesis and definitive hematopoiesis, which respectively give rise to embryonic and adult hematopoietic cells. Primitive hematopoiesis begins at day 7 of gestation in the mouse yolk sac and generates embryonic primitive erythroblasts (EryPs) (reviewed in Lensch and Daley, 2004). However, the hematopoietic precursors from yolk sac are not able to reconstitute lethally irradiated adult recipients (Medvinsky et al., 1993; Muller et al., 1994), which is the gold standard for demonstrating functional hematopoietic stem cell activity. The second wave of hematopoiesis, definitive or adult hematopoiesis, arises around day 10 of gestation in the aorta-gonad-mesonephros (AGM) region (Muller et al., 1994; Medvinsky and Dzierzak, 1996). The definitive embryonic HSCs are able to self-renew and give rise to mature hematopoietic lineages in adults. These cells seed the bone marrow (BM), where HSCs contribute to blood formation throughout the lifespan of the adult (Lensch and Daley, 2004).

Principles of Regenerative Medicine. DOI: 10.1016/B978-0-12-381422-7.10016-1

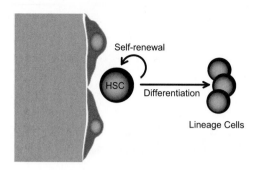

FIGURE 16.1
Self-renewal and differentiation of HSCs. The HSC-to-niche interaction influences the two definitive properties of HSCs. HSCs can expand to create more HSCs (self-renew) and they can regenerate the hematopoietic system (differentiate).

Zebrafish propagate quickly and are readily genetically manipulated, both of which facts make zebrafish a good model organism for genetic screening. In addition, the transparent larvae of zebrafish make the developing blood cells easily visible under the microscope, and allow the identification of molecules essential for early hematopoiesis. Zebrafish embryos with different defective development phenotypes in hematopoiesis can be categorized based on the blood morphologies, and mapped for genetic mutations causing phenotypes (Driever et al., 1996; Haffter et al., 1996; Weinstein et al., 1996). In zebrafish (*Danio rerio*), there are equivalent sites where primitive and definitive hematopoiesis occur. The primitive (embryo) hematopoiesis is found in the caudal/posterior lateral plate mesoderm (LPM) (Herbomel et al., 1999), later forming intermediate cell mass (ICM), which is equivalent to the mammalian yolk sac (Carradice and Lieschke, 2008). The second definitive wave of hematopoiesis occurs later at the ventral wall of the dorsal-aorta region, the equivalent of the mammalian AGM, and gives rise to multilineage blood cells (Carradice and Lieschke, 2008). In general, because it is more feasible to undertake large-scale screening in zebrafish, it has been an invaluable model organism to discover novel genes in hematopoietic development.

Several studies have identified molecules required for primitive and definitive hematopoiesis in mice, bringing insight to HSC ontogeny and shedding light on mechanisms that regulate HSC self-renewal and differentiation (reviewed in Lensch and Daley, 2004; Medvinsky and Dzierzak, 1998). Deficiency in some of these genes is found to cause embryonic anemia due to inefficient hematopoiesis, indicating that they are essential for HSC formation. For example, mutants of Runx1, a member of the runt transcription factor family, exhibit normal primitive hematopoiesis in the yolk sac but lack of hematopoietic clusters in the intra-aorta region at E10.5 of the AGM, and are embryonic lethal at E12.5 with anemic fetal liver, a temporary site of primitive hematopoiesis between E11 and E14 (Okuda et al., 1996; North et al., 1999). The evidence indicates that Runx1 is indispensable for definitive hematopoiesis but not primitive hematopoiesis. Flk-1 (vascular endothelial growth factor receptor-2) null mice are also embryo lethal (at E8.5–E9.5) with defects in forming blood clusters and in developing vascular network in the yolk sac region (Shalaby et al., 1995; Sakurai et al., 2005). Likewise, Scl/Tal1 null mice are found to be embryonic lethal (at E9.5–E11.5), and lack yolk sac vitelline vessels and primitive hematopoiesis (Robb et al., 1995; Shivdasani et al., 1995). Scl/Tal1 null cells also fail to contribute to definitive hematopoiesis of both the AGM and fetal liver in chimeric mice (Porcher et al., 1996; Robb et al., 1996), suggesting critical roles of Scl/Tal1 in both primitive and definitive hematopoiesis.

FUNCTIONAL CHARACTERISTICS OF HSCs

Unlike fetal HSCs, adult HSCs are relatively dormant under homeostasis, but can extensively proliferate when they encounter regenerative stresses. Adult HSCs comprise only ~0.02% of whole bone marrow cells but possess abilities to self-renew and differentiate (hematopoiesis) to replenish the whole hematopoietic system. As few as a single HSC is sufficient to establish long-term multilineage engraftment (Osawa et al., 1996; Camargo et al., 2005), which would only be possible through a self-renewal process. Self-renewal, a signature process of all stem cells, is the process by which one stem cell is able to give rise to at least one daughter stem cell via cell division. While it is unclear how the cell fate determination of HSCs is facilitated during cell division, self-renewal is defined for the daughter HSCs to inherit the ability to regenerate another HSC once divided, to repopulate multiple lineages during hematopoiesis, and in most cases to regain their dormant cell cycle status.

During adult hematopoiesis, BM-HSCs generate both lymphoid and myeloid cells (Fig. 16.2). Lymphoid cells are comprised of primarily T-cells, B cells, and natural killer (NK) cells. Myeloid cells include granulocytes, macrophages, megakaryocytes, and erythrocytes. Hematopoiesis is a gradual differentiation process that involves multiple decision points beginning with HSCs and ending with terminally differentiated lineages (Fig. 16.2). This concept of stepwise hematopoiesis has led to the identification of several differentiation intermediates. From this concept, Morrison and Weissman described two populations within bone marrow that possess transient engraftment ability when transplanted into lethally irradiated mice. These two populations are considered short-term HSCs (ST-HSCs, Mac-1^{lo}CD4^{-}) and multipotent progenitor (MPP, Mac-1^{lo}CD4^{lo}), which are distinguished from long-term repopulating HSCs (Morrison and Weissman, 1994; Morrison et al., 1997). Weissman and his colleagues also first identified common lymphoid progenitors (CLPs) within bone marrow that specifically give rise to lymphoid lineages (T-cells, B cells, and NK cells) (Kondo et al., 1997), and common myeloid progenitors (CMPs), which give rise to granulocytes/macrophages and megakaryocytes/erythrocytes colonies in methylcellulose cell culture and in lethally irradiated recipients (Akashi et al., 2000). More recently, Adolfsson et al. have revised the role of MPPs as lymphoid-primed multipotent progenitors (LMPPs). They discovered that MPPs (now LMPPs) preferentially differentiate into lymphoid lineages, but retain some myeloid development capacity (dashed line in Fig. 16.2), restricted to granulocytes and macrophages (Adolfsson et al., 2005). The CMP is the major generator of all myeloid cells, including megakaryocytes/erythrocytes lineages. In summary, these studies have suggested a hematopoietic hierarchy in which long-term HSCs give rise to ST-HSCs that differentiate into CMPs and CLPs. The CMPs and CLPs then generate the myeloid and lymphoid lineages (Fig. 16.2). Although the differentiation pathway in humans is not as well established as in rodent models, transplantation of HSCs has been utilized for decades to treat patients with

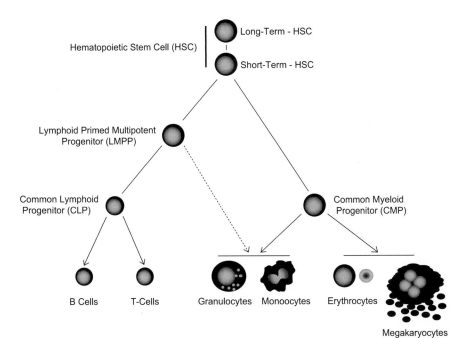

FIGURE 16.2

HSC differentiation. HSCs regenerate the hematopoietic system, which is comprised of a myeloid and lymphoid branch, ultimately creating all the cells that comprise the blood. The multipotent progenitor (MPP), previously thought of as a bi-potential progenitor, is now identified as a lymphoid-primed progenitor (LMPP).

Hematopoietic Stem Cell (HSC)
Long-Term - HSC
Short-Term - HSC
Lymphoid Primed Multipotent Progenitor (LMPP)
Common Lymphoid Progenitor (CLP)
Common Myeloid Progenitor (CMP)
B Cells
T-Cells
Granulocytes
Monoocytes
Erythrocytes
Megakaryocytes

hematopoietic diseases, demonstrating the repopulation ability of human HSCs to reconstitute the entire hematopoietic system.

HSC NICHE

In adults, HSCs reside in the bone marrow cavity, closely associated with surrounding stromal cells. There is mounting evidence suggesting that the most primitive HSCs localize to the interior surface of bone (periosteum/endosteum border) on the basis of colony-forming assays (Lord and Hendry, 1972) and Brd-U label retention (Zhang et al., 2003), bringing them within close contact with osteoblasts. Murine osteoblasts have long been thought to provide essential cues for HSCs, as they (or their transformed counterparts) express various cytokines known to influence hematopoiesis, including but not limited to G/M/GM-CSFs, IL-1, IL-6, SDF-1, and VEGF (reviewed in Taichman, 2005). In addition to expressing HSC-modulating cytokines, genetic evidence from mice has demonstrated that expanding the number of osteoblasts within the bone marrow increases the relative percentage of HSCs (Calvi et al., 2003; Zhang et al., 2003), and genetically ablating osteoblasts results in the failure of bone marrow hematopoiesis (Visnjic et al., 2004), suggesting that osteoblasts provide a direct physical niche for HSCs that maintains their self-renewing capacity via various cell surface molecules (Fig. 16.3). Researchers have also provided evidence of a second HSC niche provided by sinusoidal endothelial cells resident in the bone marrow (Kiel et al., 2005).

Molecules including N-cadherin, Notch-1, Tie2, and CXCR-4 have all been implicated in the HSC-to-niche interface. N-cadherin, a Ca^{2+} dependent homophilic adhesion molecule expressed on osteoblasts, exhibits an asymmetrical localization on HSCs, as determined by fluorescence microscopy (Zhang et al., 2003), and is found expressed on a fraction (10%) of hematopoietic progenitors that include the HSCs, thus suggesting that it may be involved in self-renewal or niche retention. However, it remains to be seen whether or not N-cadherin is required for HSC-to-niche interaction, by examining whether there is a functional difference between N-cadherin$^+$ and N-cadherin$^-$ HSC *in vivo*.

Several studies have demonstrated that Notch-1 is expressed on HSCs and its activation by incubating with Jagged-1-expressing cells or constitutive Notch-1 signaling (Varnum-Finney et al., 2000) results in *in vitro* expansion of self-renewing HSCs with normal homeostasis while transplanting into lethally irradiated mice. In contrast to the evidence discovered from *in vitro* Notch1 stimulation, targeted disruption of Jagged1 and Notch-1 in mice does not result in reconstitution or self-renewal defects *in vivo* (Mancini et al., 2005). Therefore, Notch-1

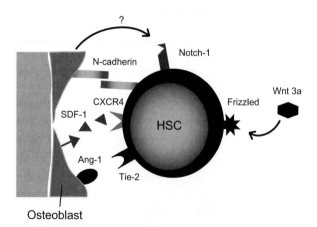

FIGURE 16.3
Hypothetical HSC-to-niche interactions. Through several cell surface molecules (markers) and cytokines, the HSC is thought to directly or indrectly interact with osteoblast cells. These are a few of the molecules that provide instructions for self-renewal and differentiation to the HSC.

stimulation may regulate HSC cell fate decision toward self-renewal during *in vitro* cell culture, although it is not essential for HSC homeostasis *in vivo*. Hence, stimulation of the signaling pathway is of interest in HSC expansion.

A second such receptor-ligand interaction is that of Tie2, a tyrosine kinase receptor that is expressed on HSCs, and its ligand angiopoietin-1 (Ang-1), expressed by osteoblasts. These two molecules were demonstrated to be important retention factors for HSCs (Arai et al., 2004). Incubation of HSCs in Ang-1 or overexpression of Tie-2 in transduced bone marrow cells resulted in expansion of the quiescent portion of the HSC compartment. A loss of self-renewal or rapid differentiation continues to be a hurdle for *in vitro* expansion of HSCs. Exploiting Tie2 signaling with soluble Ang-1 may be a promising method for expanding long-term, quiescent HSC cultures.

One such receptor that has been employed in the clinic is the chemokine receptor, CXCR-4, expressed on both human (Viardot et al., 1998) and mouse (Wright et al., 2002) HSCs. In addition, SDF-1, the ligand for CXCR-4, is a well-established HSC homing factor expressed by osteoblasts and bone marrow fibroblasts (Ponomaryov et al., 2000). Antibodies against SDF-1 or CXCR-4 block human HSC engraftment in the non-obese severe combined immunodeficient (NOD/SCID) mouse model, and SDF-1 enhances transwell migration of human HSCs (Peled et al., 1999). Therefore, the CXCR-4/SDF-1 axis is thought to provide a BM homing and retention mechanism for HSCs *in vivo*. In a secondary transplantation assay, treatment of stem cell factor (SCF) and IL-6 enhanced HSC engraftment correlating with elevated CXCR4 expression and increased *in vitro* migration activity to SDF-1 (Peled et al., 1999). However, they did not distinguish between a rescued migratory defect and enhanced self-renewal as the cause for increased engraftment *in vivo*. Therefore, it remains less clear whether CXCR-4 plays a role in HSC self-renewal and *in vitro* expansion of HSCs.

In summary, over the past decade it has become increasingly clear that components such as N-cadherin, Tie2/Ang-1, Notch-1, and CXCR-4/SDF-1 may be required in order to establish an *ex vivo* niche for HSC expansion and self-renewal; unfortunately, research has yet to elucidate the appropriate cocktail of soluble factors, cytokines, and/or niche support cells needed to stimulate faithful and prolonged *in vitro* HSC self-renewal and expansion.

EXPERIMENTAL MODELS TO CHARACTERIZE HEMATOPOIETIC STEM CELLS IN VERTEBRATES

Combining animal models with gene modifications aids the understanding of molecular mechanisms underlying behaviors of hematopoietic stem cells *in vivo*. In the murine model, bone marrow transplantation is a widely used approach to test the self-renewal capacity of adult HSCs. In a bone marrow transplantation, self-renewing HSCs are given the most stringent stress in which they have to regenerate another set of self-renewing HSCs as well as give rise to blood progeny that re-establishes homeostasis of the entire hematopoietic system. In such assay, the test cells (donors) are intravenously injected into lethally irradiated mice (recipients) and subjected to the analysis of engraftment for short-term (4−6 weeks post-transplantation) and long-term (8−24 weeks post-transplantation) engraftment. To analyze the donor-derived progeny, it is vital to "mark" the donor cells in the recipient mice. Retroviral integration was an earlier approach utilized to demonstrate clonal expansion of HSCs *in vivo* in a quantitative manner (Lemischka et al., 1986). In recent decades, congenic mice that bear equivalent genetic background but with different allelic variants of CD45, a tyrosine phosphatase commonly expressed in blood, have been mostly used (Spangrude et al., 1988). In this CD45-congenic-mice system, a population of donor cells that bears the wild-type CD45 allele, CD45.2, is transplanted into lethally irradiated recipient mice expressing the other CD45 allele, named CD45.1 (Fig. 16.4). The contribution of donor stem cells can then be measured by quantifying the percentage of CD45.1 cells in the hematopoietic organs of the hosts (recipient mice) with flow cytometry. An additional advantage of this

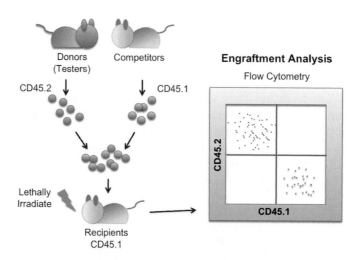

FIGURE 16.4

Competitive transplantation analysis using CD45.1/CD45.2 congenic mice. Bone marrow transplantation using congenic mouse strains is common. Pairs of CD45.2 and CD45.1 mice can be utilized in one competitive transplantation assay. For instance, tester bone marrows from the donor mice that bear CD45.2 are transplanted into lethally irradiated recipient mice that express CD45.1 along with competitor bone marrow cells that express the same CD45 allele as recipient mice; that is, CD45.1 here. Engraftment of donor-derived cells in the hematopoietic organs of recipients can then be distinguished by monoclonal antibodies specific to each allele and analyzed with flow cytometry.

system is to allow repurification of HSCs from the recipients for further analysis such as serial transplantation assay that analyzes the stem cell activity of regenerated daughter HSCs. In general, bone marrow transplantation and serial bone marrow transplantation are the gold-standard assays to test the self-renewal capacity of HSCs *in vivo* by investigating the regeneration of hematopoietic compartments in the recipients, including HSCs, progenitors, lymphoid cells (B and T-cells), and myeloid cells (Gr-1 and Mac-1 cells).

In addition to mouse models, the zebrafish is an established model organism for studying hematopoietic stem cells in vertebrates, especially on large scales. More recently, the kidney marrow transplantation model has allowed the study of the role of genes in adult (definitive) hematopoiesis (Traver et al., 2003). Moreover, FACS sorting, which is routinely used to purify HSCs in mammals, has also been attempted in zebrafish (Traver et al., 2003). Although purification of hematopoietic stem cells in zebrafish lacks appropriate surface markers, the power of being able to combine imaging tools with easily manipulated genetic approaches makes the zebrafish a unique tool to screen for genes essential for hematopoiesis.

PURIFICATION MARKERS FOR HSCs

One of the major technological breakthroughs that have greatly facilitated the field of HSC research was the coupling of fluorescently conjugated monoclonal antibodies with multi-parameter flow cytometry. Improvement in these technologies has yielded a vast amount of information about the phenotype of mouse HSCs, particularly in terms of their cell-surface marker profile. The identification and purification of LT-HSCs relies on combinations of cell surface markers, with the presence or absence of specific antigens allowing discrimination of these cells from other bone marrow cell types including the immediately downstream ST-HSCs and progenitor cell compartments.

In humans, HSCs are defined on the basis of positive expression of CD34 and negative expression of CD38 ($CD34^+$ and $CD38^-$) (Bhatia et al., 1998). In mice, almost all HSC purification strategies rely on the basis of lack of expression of markers of mature blood cells ($Lineage^-$) and positive selection for the canonical HSC markers Sca-1 and c-Kit such that the combination of $c\text{-}Kit^+$ $Sca\text{-}1^+$ $Lineage^-$ is termed "KSL." But the KSL fraction of bone marrow is still very heterogeneous, comprising all the various hematopoietic progenitor cell populations in addition to the LT-HSCs. Further enrichment of the LT-HSCs within the KSL fraction can be achieved by positive selection for CD201/EPCR (Balazs, 2006) and exclusion of cells expressing CD34 (Osawa et al., 1996), CD48 (Kiel, 2005), CD49b (Benveniste, 2010) and Flk2 (Christensen and Weissman, 2001). Thus far, the only marker that seems to subfractionate LT-HSCs into functional categories is CD150 (Kent et al., 2009; Challen et al., 2010), with CD150$^+$ LT-HSCs showing greater propensity for myeloid production and CD150$^-$ LT-HSCs being biased towards lymphoid cell generation. In addition to cell surface markers, HSCs can also be purified according to their characteristic staining patterns with vital dyes such as Hoechst 33342 and Rhodamine 123 such that the HSCs reside in a distinct population termed the "side population" or "SP" (Goodell et al., 1996). However, the identification of more markers for HSC purification has produced a double-edged sword, simultaneously presenting researchers with more options for HSC purification but raising the question of which markers are the best or most appropriate for a particular experiment. Confounding the issue is the fact that several recent studies have raised the possibility that the hematopoietic system is not maintained solely by a uniform population of LT-HSCs, but rather distinct pools of HSC subtypes that contribute to various aspects of hematopoietic cell generation (Muller-Sieburg and Sieburg, 2006; Dykstra et al., 2007). This view has been further supported by studies utilizing single-cell HSC transplantation based on combinations of markers to identify lineage-biased HSCs with different propensities for self-renewal, proliferation, and myeloid versus lymphoid differentiation (Kent et al., 2009; Challen et al., 2010; Morita et al., 2010).

NATURALLY OCCURRING STIMULI FOR HSCs

Stresses that lead to fluctuation in hematopoietic homeostasis may stimulate HSCs to proliferate in order to replenish hematopoietic lineages. Collective evidence of naturally occurring stresses such as infection and aging have provided invaluable clinical implications. HSCs are found to respond to infection stress through the influence of inflammation cytokines (Essers et al., 2009; Baldridge et al., 2010). Genetic evidence suggests that regulators in interferon pathways are essential for HSC self-renewal, as lacking receptors to inflammation cytokines such as interferons impacts the self-renewal ability in HSCs (Feng et al., 2008; Essers et al., 2009; Sato et al., 2009; Baldridge et al., 2010).

Another naturally occurring stimuli for HSCs is aging. The aging stresses to HSCs are thought to be mild and constant, and the effect of aging on HSCs is not solely cell-autonomous. Environmental effects of aging were found to play a profound role in stem cell activity of HSCs, such that conditioning young HSCs with an old environment drastically decreases the engraftment ability of young HSCs, while conditioning old HSCs with a young environment improves the engraftment ability of old HSCs (Mayack et al., 2010). When aged, HSCs present different immunophenotypes such that there are more side population-low cells (Chambers et al., 2007). With current purification markers, aged HSCs are found to be different from young HSCs as aged HSCs are more myeloid-biased (Sudo et al., 2000; Rossi et al., 2005), and less capable of regenerating blood progeny (Sudo et al., 2000) and of engrafting upon transplantation (Morrison et al., 1996; Chambers et al., 2007). Moreover, gene expression pattern changes in aged HSCs suggest epigenetic mechanisms that at least partly explain the functional changes (Rossi et al., 2005; Chambers et al., 2007).

DEVELOPMENTAL PATHWAYS AND SIGNALING PATHWAYS UNDERLIE HSC SELF-RENEWAL

Pathways that trigger cell fate decisions during early development of vertebrates and invertebrates have been demonstrated to be involved in dictating cell fate decisions of HSCs during cell division. Overexpression of constitutively activated Notch1 in HSCs immortalizes HSCs in long-term *in vitro* culture. A single Notch1-transduced HSC clone is able to undergo multilineage repopulation *in vivo* (Varnum-Finney et al., 2000). The Wnt signaling pathway has also been implicated in HSC self-renewal and recently extensively studied. In particular, roles of canonical Wnt pathways in HSC self-renewal have had somewhat controversial results. Extrinsic stimulation of Wnt3a or overexpression of beta-catenin in combination with the presence of Bcl2 lead to HSCs with a high repopulating activity after a long period of *in vitro* cell culture while ectopic expression of Axin, an inhibitor of the Wnt pathway, decreased HSC proliferation *in vitro* and reconstitution function *in vivo* (Reya et al., 2003). Lack of a canonical Wnt ligand, Wnt3a, was found to lead to impaired HSC self-renewal in fetal liver, a developing hematopoietic organ (Luis et al., 2009), while endosteal (osteoblast)-specific overexpression of a pan-canonical Wnt inhibitor, DKK1, led to impaired HSC self-renewal in the context of transplantation and a proliferating phenotype in HSCs (Fleming et al., 2008). The evidence above indicates an essential role of canonical Wnt signaling in promoting HSC self-renewal. On the other hand, contrary evidence had shown otherwise, such that the canonical Wnt signaling pathway works against HSC self-renewal. It has been shown that culturing HSCs with canonical ligand, Wnt3a, promotes cell proliferation and decreases repopulation ability (Nemeth et al., 2007). In addition, constitutive activation of the central regulator of canonical Wnt pathway, beta-catenin, leads to impaired hemaopoietic stem cell function including self-renewal and myeloid differentiation, indicating canonical Wnt pathway may negatively impact HSC self-renewal. These contradictory findings have brought several suggestions that the impact of Wnt pathway on HSCs is dependent on the context of the niche environment and may be dose-dependent. Clearly, this important pathway merits further study in HSCs.

Stress such as inflammation that triggers JAK/STAT pathways has also been implicated in the initiation of HSC regeneration. Among the cytokines that activate JAK/STAT, interferons have been shown to activate HSC proliferation (Essers et al., 2009). Regulators in the JAK/STAT pathways such as interferon regulator factor 2 (Irf2) (Sato et al., 2009) and p47 GTPase (Lrg47/Irgm1) (Feng et al., 2008) were also shown to be important to maintain HSC quiescence and to preserve the stem cell activity. The evidence indicates that several molecular mechanisms exist to maintain balance between HSC proliferation and quiescence when HSCs encounter proliferation stresses.

One other signaling pathway recently found to be essential to promote HSC self-renewal is the AKT/FOXO pathway. The AKT/FOXO pathway regulates various aspects of cell function including cell proliferation, apoptosis, and anti-oxidant stress. Players in this pathway such as PTEN (phosphatase and tensin homolog), FOXO1 (Forkhead O1), FOXO3 (Forkhead O3), and FOXO4 (Forkhead O4), have been found to positively impact HSC self-renewal. In HSCs, loss of PTEN, a negative regulator of the PI3K/AKT pathway that inhibits the phosphorylation of phosphoinositols-3, 4-P_2 and subsequently prevents the activation of AKT (Manning and Cantley, 2007), results in a myeloproliferation phenotype and defective HSC self-renewal. Consistent with what was found in the PTEN$^{-/-}$ HSCs, studies that characterized HSCs lacking the downstream effectors, the FOXO (Forkhead O) family members, have shown impaired self-renewal coincident with properties such as overproliferation and increased reactive oxygen species (ROS) (Miyamoto et al., 2007, 2008; Tothova et al., 2007). Taken together, these studies suggest that molecules that negatively regulate the AKT/FOXO pathway are involved in promoting HSC quiescence when they encounter stress, and thereby preserve their self-renewal capacity.

Collectively, it may have seemed that molecules in a pathway in favor of proliferation facilitate HSC differentiation, while molecules in favor of cell quiescence maintain HSC self-renewal.

IN VITRO DIFFERENTIATION OF HEMATOPOIETIC LINEAGES

In vitro cell culture of blood precursors has been established to quantify the differentiation ability of hematopoietic precursors. The *in vitro* culture conditions were established with the goal of mimicking the *in vivo* growth stimuli and maturation signals. A functional assay of hematopoietic precursors, the colony-forming unit in cell culture (CFU-C), was first established in the 1980s and measured the generation of myeloid and erythroid cells (Metcalf, 1989). *In vitro* differentiation of lymphocytes was later found to require microenvironments that are provided by co-cultured cells. *In vitro* co-culture of stromal cell line with B-cell precursors (Cumano et al., 1990), and fetal thymic organ culture (FTOC) to generate thymocytes (Robinson and Owen, 1978), have been aimed at recapitulating the *in vivo* environment and identifying regulators of lymphocyte differentiation. Moreover, *in vitro* assays such as cobblestone area forming cells (CAFC) and long-term culture initiating cells (LT-CIC) were developed to detect earlier precursors, including the HSCs. A more comprehensive outline of *in vitro* HSC differentiation assays has been described by Ramos et al. (2003). In addition, differentiation of murine and human hematopoietic progenitor cells *in vitro* has been utilized to modulate immune responses. The best example is terminally differentiated dendritic cells, one of the professional antigen presenting cells in the immune system. Antigen (Ag)-pulsed dendritic cells (DCs) have been utilized to modulate Ag-specific immune response. In clinical trials of immunotherapy, these *in vitro*-differentiated DCs are able to stimulate tumor antigen-specific immune responses and to induce tolerance in autoimmune diseases (Figdor et al., 2004).

IN VITRO EXPANSION OF SELF-RENEWING HSCs

Expansion of HSCs *in vitro* has been the most difficult challenge for decades due to a decline in repopulation capacity of HSCs in long-term *ex vivo* culture. There are currently two main strategies for *ex vivo* HSC expansion: HSC-stromal cell co-cultivation and HSC suspension culture. BM stromal cells support HSC maintenance, measured by repopulating ability, in the absence of additional growth factors (Fraser et al., 1990, 1992). BM stromal cells are thought to mimic the microenvironment of the HSC niche. To identify the molecules in these stromal cells that retain HSC function, genome-wide studies of cloned stromal cells have been reported (Moore, 2004), providing an emerging picture that underlies the HSC-to-niche interaction, which we will discuss below. For suspension culture, growth factor cocktails have been utilized in an attempt to expand human and murine HSCs (Sauvageau et al., 2004), albeit with low recovery rate of repopulating HSCs. The differentiation of HSCs during *in vitro* culture has drawn into question whether or not HSC self-renewal occurs during cell expansion. To address the question, Glimm and Eaves labeled HSCs with a fluorescent membrane-specific dye, carboxyfluorescein diacetate succinimidyl (CFSE), to track the proliferation history. By transplanting cells that had divided from an *in vitro* cell suspension culture, they discovered that human HSCs were still able to give rise to multilineage repopulation after a low number of cell divisions (Glimm and Eaves, 1999). Additionally, Nakauchi and his colleagues have cultured highly purified murine single-cell HSCs to track cell division as well as cell fate. They have been able to generate limited self-renewing HSCs under the influence of various combinations of cytokines (Ema et al., 2000). However, after more than two cell divisions *in vitro*, the HSCs greatly lost their repopulating activity.

CONCLUSIONS

The hematopoietic stem cell field continues to be at the forefront of regenerative medicine, with therapeutic potential to cure a wide range of diseases. Developmental origin, self-renewal, differentiation, molecular signature, and therapeutic potential of HSCs are currently being established. The next horizon for the HSC field is to create an *ex vivo* niche. Identifying the essential support cells and their secreted cytokines required for HSC self-renewal and expansion would open the gateway for unfettered genetic modification of HSCs to aid the continued efforts in gene therapy. Equally important, HSC expansion could drastically reduce the numbers of HSCs needed to provide adequate reconstitution in human BM transplant therapy. Therefore, further characterization of components that contribute to the regulation of HSC self-renewal is needed in order to coordinate HSC expansion.

References

Adolfsson, J., Mansson, R., Buza-Vidas, N., Hultquist, A., Liuba, K., Jensen, C. T., et al. (2005). Identification of Flt3+ lympho-myeloid stem cells lacking erythro-megakaryocytic potential a revised road map for adult blood lineage commitment. *Cell, 121,* 295–306.

Akashi, K., Traver, D., Miyamoto, T., & Weissman, I. L. (2000). A clonogenic common myeloid progenitor that gives rise to all myeloid lineages. *Nature, 404,* 193–197.

Arai, F., Hirao, A., Ohmura, M., Sato, H., Matsuoka, S., Takubo, K., et al. (2004). Tie2/angiopoietin-1 signaling regulates hematopoietic stem cell quiescence in the bone marrow niche. *Cell, 118,* 149–161.

Balazs, A. B., Fabian, A. J., Esmon, C. T., & Mulligan, R. C. (2006). Endothelial protein C receptor (CD201) explicitly identifies hematopoietic stem cells in murine bone marrow. *Blood, 107,* 2317–2321.

Baldridge, M. T., King, K. Y., Boles, N. C., Weksberg, D. C., & Goodell, M. A. (2010). Quiescent haematopoietic stem cells are activated by IFN-gamma in response to chronic infection. *Nature, 465,* 793–797.

Becker, A. J., Mc, C. E., & Till, J. E. (1963). Cytological demonstration of the clonal nature of spleen colonies derived from transplanted mouse marrow cells. *Nature, 197,* 452–454.

Benveniste, P., Frelin, C., Janmohamed, S., Barbara, M., Herrington, R., Hyam, D., et al. (2010). Intermediate-term hematopoietic stem cells with extended but time-limited reconstitution potential. *Cell stem cell, 6,* 48–58.

Bhatia, M., Bonnet, D., Murdoch, B., Gan, O. I., & Dick, J. E. (1998). A newly discovered class of human hematopoietic cells with SCID-repopulating activity. *Nat. Med., 4,* 1038–1045.

Calvi, L. M., Adams, G. B., Weibrecht, K. W., Weber, J. M., Olson, D. P., Knight, M. C., et al. (2003). Osteoblastic cells regulate the haematopoietic stem cell niche. *Nature, 425,* 841–846.

Camargo, F. D., Chambers, S. M., Drew, E., McNagny, K. M., & Goodell, M. A. (2005). Hematopoietic stem cells do not engraft with absolute efficiencies. *Blood.*

Carradice, D., & Lieschke, G. J. (2008). Zebrafish in hematology: sushi or science? *Blood, 111,* 3331–3342.

Challen, G. A., Boles, N. C., Chambers, S. M., & Goodell, M. A. (2010). Distinct hematopoietic stem cell subtypes are differentially regulated by TGF-beta1. *Cell stem cell, 6,* 265–278.

Chambers, S. M., Shaw, C. A., Gatza, C., Fisk, C. J., Donehower, L. A., & Goodell, M. A. (2007). Aging hematopoietic stem cells decline in function and exhibit epigenetic dysregulation. *PLoS Biology, 5,* e201.

Christensen, J. L., & Weissman, I. L. (2001). Flk-2 is a marker in hematopoietic stem cell differentiation: a simple method to isolate long-term stem cells. *Proc. Natl. Acad. Sci. U.S.A., 98,* 14541–14546.

Cumano, A., Dorshkind, K., Gillis, S., & Paige, C. J. (1990). The influence of S17 stromal cells and interleukin 7 on B cell development. *Eur. J. Immunol., 20,* 2183–2189.

Driever, W., Solnica-Krezel, L., Schier, A. F., Neuhauss, S. C., Malicki, J., Stemple, D. L., et al. (1996). A genetic screen for mutations affecting embryogenesis in zebrafish. *Development, 123,* 37–46.

Dykstra, B., Kent, D., Bowie, M., McCaffrey, L., Hamilton, M., Lyons, K., et al. (2007). Long-term propagation of distinct hematopoietic differentiation programs in vivo. *Cell stem cell, 1,* 218–229.

Ema, H., Takano, H., Sudo, K., & Nakauchi, H. (2000). In vitro self-renewal division of hematopoietic stem cells. *J. Exp. Med., 192,* 1281–1288.

Essers, M. A., Offner, S., Blanco-Bose, W. E., Waibler, Z., Kalinke, U., Duchosal, M. A., et al. (2009). IFNalpha activates dormant haematopoietic stem cells in vivo. *Nature, 458,* 904–908.

Feng, C. G., Weksberg, D. C., Taylor, G. A., Sher, A., & Goodell, M. A. (2008). The p47 GTPase Lrg-47 (Irgm1) links host defense and hematopoietic stem cell proliferation. *Cell Stem Cell, 2,* 83–89.

Figdor, C. G., de Vries, I. J., Lesterhuis, W. J., & Melief, C. J. (2004). Dendritic cell immunotherapy: mapping the way. *Nat. Med., 10,* 475–480.

Fleming, H. E., Janzen, V., Lo Celso, C., Guo, J., Leahy, K. M., Kronenberg, H. M., et al. (2008). Wnt signaling in the niche enforces hematopoietic stem cell quiescence and is necessary to preserve self-renewal in vivo. *Cell Stem Cell, 2,* 274–283.

Ford, C. E., Hamerton, J. L., Barnes, D. W., & Loutit, J. F. (1956). Cytological identification of radiation-chimaeras. *Nature, 177,* 452–454.

Fraser, C. C., Eaves, C. J., Szilvassy, S. J., & Humphries, R. K. (1990). Expansion in vitro of retrovirally marked totipotent hematopoietic stem cells. *Blood, 76,* 1071–1076.

Fraser, C. C., Szilvassy, S. J., Eaves, C. J., & Humphries, R. K. (1992). Proliferation of totipotent hematopoietic stem cells in vitro with retention of long-term competitive in vivo reconstituting ability. *Proc. Natl. Acad. Sci. U.S.A., 89,* 1968–1972.

Glimm, H., & Eaves, C. J. (1999). Direct evidence for multiple self-renewal divisions of human in vivo repopulating hematopoietic cells in short-term culture. *Blood, 94,* 2161–2168.

Goodell, M. A., Brose, K., Paradis, G., Conner, A. S., & Mulligan, R. C. (1996). Isolation and functional properties of murine hematopoietic stem cells that are replicating in vivo. *J. Exp. Med., 183,* 1797–1806.

Haffter, P., Granato, M., Brand, M., Mullins, M. C., Hammerschmidt, M., Kane, D. A., et al. (1996). The identification of genes with unique and essential functions in the development of the zebrafish, *Danio rerio*. *Development, 123*, 1–36.

Herbomel, P., Thisse, B., & Thisse, C. (1999). Ontogeny and behaviour of early macrophages in the zebrafish embryo. *Development, 126*, 3735–3745.

Kent, D. G., Copley, M. R., Benz, C., Wohrer, S., Dykstra, B. J., Ma, E., et al. (2009). Prospective isolation and molecular characterization of hematopoietic stem cells with durable self-renewal potential. *Blood, 113*, 6342–6350.

Kiel, M. J., Yilmaz, O. H., Iwashita, T., Yilmaz, O. H., Terhorst, C., & Morrison, S. J. (2005). SLAM family receptors distinguish hematopoietic stem and progenitor cells and reveal endothelial niches for stem cells. *Cell, 121*, 1109–1121.

Kondo, M., Weissman, I. L., & Akashi, K. (1997). Identification of clonogenic common lymphoid progenitors in mouse bone marrow. *Cell, 91*, 661–672.

Lemischka, I. R., Raulet, D. H., & Mulligan, R. C. (1986). Developmental potential and dynamic behavior of hematopoietic stem cells. *Cell, 45*, 917–927.

Lensch, M. W., & Daley, G. Q. (2004). Origins of mammalian hematopoiesis: in vivo paradigms and in vitro models. *Curr. Top. Dev. Biol., 60*, 127–196.

Lord, B. I., & Hendry, J. H. (1972). The distribution of haemopoietic colony-forming units in the mouse femur, and its modification by x rays. *Br. J. Radiol., 45*, 110–115.

Luis, T. C., Weerkamp, F., Naber, B. A., Baert, M. R., de Haas, E. F., Nikolic, T., et al. (2009). Wnt3a deficiency irreversibly impairs hematopoietic stem cell self-renewal and leads to defects in progenitor cell differentiation. *Blood, 113*, 546–554.

Mancini, S. J., Mantei, N., Dumortier, A., Suter, U., Macdonald, H. R., & Radtke, F. (2005). Jagged1-dependent Notch signaling is dispensable for hematopoietic stem cell self-renewal and differentiation. *Blood, 105*, 2340–2342.

Manning, B. D., & Cantley, L. C. (2007). AKT/PKB signaling: navigating downstream. *Cell, 129*, 1261–1274.

Mayack, S.R., Shadrach, J.L., Kim, F.S. and Wagers, A.J. (2010). Systemic signals regulate ageing and rejuvenation of blood stem cell niches. *Nature 463*, 495-500

Medvinsky, A., & Dzierzak, E. (1996). Definitive hematopoiesis is autonomously initiated by the AGM region. *Cell, 86*, 897–906.

Medvinsky, A. L., & Dzierzak, E. A. (1998). Development of the definitive hematopoietic hierarchy in the mouse. *Dev. Comp. Immunol., 22*, 289–301.

Medvinsky, A. L., Samoylina, N. L., Muller, A. M., & Dzierzak, E. A. (1993). An early pre-liver intraembryonic source of CFU-S in the developing mouse. *Nature, 364*, 64–67.

Metcalf, D. (1989). The molecular control of cell division, differentiation commitment and maturation in haemopoietic cells. *Nature, 339*, 27–30.

Miyamoto, K., Araki, K. Y., Naka, K., Arai, F., Takubo, K., Yamazaki, S., et al. (2007). Foxo3a is essential for maintenance of the hematopoietic stem cell pool. *Cell Stem Cell, 1*, 101–112.

Miyamoto, K., Miyamoto, T., Kato, R., Yoshimura, A., Motoyama, N., & Suda, T. (2008). FoxO3a regulates hematopoietic homeostasis through a negative feedback pathway in conditions of stress or aging. *Blood, 112*, 4485–4493.

Moore, K. A. (2004). Recent advances in defining the hematopoietic stem cell niche. *Curr. Opin. Hematol., 11*, 107–111.

Morita, Y., Ema, H., & Nakauchi, H. (2010). Heterogeneity and hierarchy within the most primitive hematopoietic stem cell compartment. *J. Exp. Med., 207*, 1173–1182.

Morrison, S. J., Wandycz, A. M., Akashi, K., Globerson, A., & Weissman, I. L. (1996). The aging of hematopoietic stem cells. *Nature Medicine, 2*, 1011–1016.

Morrison, S. J., Wandycz, A. M., Hemmati, H. D., Wright, D. E., & Weissman, I. L. (1997). Identification of a lineage of multipotent hematopoietic progenitors. *Development, 124*, 1929–1939.

Morrison, S. J., & Weissman, I. L. (1994). The long-term repopulating subset of hematopoietic stem cells is deterministic and isolatable by phenotype. *Immunity, 1*, 661–673.

Muller, A. M., Medvinsky, A., Strouboulis, J., Grosveld, F., & Dzierzak, E. (1994). Development of hematopoietic stem cell activity in the mouse embryo. *Immunity, 1*, 291–301.

Muller-Sieburg, C. E., & Sieburg, H. B. (2006). Clonal diversity of the stem cell compartment. *Curr. Opin. Hematol., 13*, 243–248.

Nemeth, M. J., Topol, L., Anderson, S. M., Yang, Y., & Bodine, D. M. (2007). Wnt5a inhibits canonical Wnt signaling in hematopoietic stem cells and enhances repopulation. *Proc. Natl. Acad. Sci. U.S.A., 104*, 15436–15441.

North, T., Gu, T. L., Stacy, T., Wang, Q., Howard, L., Binder, M., et al. (1999). Cbfa2 is required for the formation of intra-aortic hematopoietic clusters. *Development, 126*, 2563–2575.

Okuda, T., van Deursen, J., Hiebert, S. W., Grosveld, G., & Downing, J. R. (1996). AML1, the target of multiple chromosomal translocations in human leukemia, is essential for normal fetal liver hematopoiesis. *Cell, 84*, 321–330.

Osawa, M., Hanada, K., Hamada, H., & Nakauchi, H. (1996). Long-term lymphohematopoietic reconstitution by a single CD34-low/negative hematopoietic stem cell. *Science, 273*, 242–245.

Peled, A., Petit, I., Kollet, O., Magid, M., Ponomaryov, T., Byk, T., et al. (1999). Dependence of human stem cell engraftment and repopulation of NOD/SCID mice on CXCR4. *Science, 283*, 845–848.

Ponomaryov, T., Peled, A., Petit, I., Taichman, R. S., Habler, L., Sandbank, J., et al. (2000). Induction of the chemokine stromal-derived factor-1 following DNA damage improves human stem cell function. *J. Clin. Invest., 106*, 1331–1339.

Porcher, C., Swat, W., Rockwell, K., Fujiwara, Y., Alt, F. W., & Orkin, S. H. (1996). The T cell leukemia oncoprotein SCL/tal-1 is essential for development of all hematopoietic lineages. *Cell, 86*, 47–57.

Ramos, C. A., Venezia, T. A., Camargo, F. A., & Goodell, M. A. (2003). Techniques for the study of adult stem cells: be fruitful and multiply. *Biotechniques, 34*, 572–591.

Reya, T., Duncan, A. W., Ailles, L., Domen, J., Scherer, D. C., Willert, K., et al. (2003). A role for Wnt signalling in self-renewal of haematopoietic stem cells. *Nature, 423*, 409–414.

Robb, L., Elwood, N. J., Elefanty, A. G., Kontgen, F., Li, R., Barnett, L. D., et al. (1996). The scl gene product is required for the generation of all hematopoietic lineages in the adult mouse. *Embo J., 15*, 4123–4129.

Robb, L., Lyons, I., Li, R., Hartley, L., Kontgen, F., Harvey, R. P., et al. (1995). Absence of yolk sac hematopoiesis from mice with a targeted disruption of the scl gene. *Proc. Natl. Acad. Sci. U.S.A., 92*, 7075–7079.

Robinson, J. H., & Owen, J. J. (1978). Transplantation tolerance induced in foetal mouse thymus in vitro. *Nature, 271*, 758–760.

Rossi, D. J., Bryder, D., Zahn, J. M., Ahlenius, H., Sonu, R., Wagers, A. J., et al. (2005). Cell intrinsic alterations underlie hematopoietic stem cell aging. *Proc. Natl. Acad. Sci. U.S.A., 102*, 9194–9199.

Sakurai, Y., Ohgimoto, K., Kataoka, Y., Yoshida, N., & Shibuya, M. (2005). Essential role of Flk-1 (VEGF receptor 2) tyrosine residue 1173 in vasculogenesis in mice. *Proc. Natl. Acad. Sci. U.S.A., 102*, 1076–1081.

Sato, T., Onai, N., Yoshihara, H., Arai, F., Suda, T., & Ohteki, T. (2009). Interferon regulatory factor-2 protects quiescent hematopoietic stem cells from type I interferon-dependent exhaustion. *Nat. Med., 15*, 696–700.

Sauvageau, G., Iscove, N. N., & Humphries, R. K. (2004). In vitro and in vivo expansion of hematopoietic stem cells. *Oncogene, 23*, 7223–7232.

Shalaby, F., Rossant, J., Yamaguchi, T. P., Gertsenstein, M., Wu, X. F., Breitman, M. L., et al. (1995). Failure of blood-island formation and vasculogenesis in Flk-1-deficient mice. *Nature, 376*, 62–66.

Shivdasani, R. A., Mayer, E. L., & Orkin, S. H. (1995). Absence of blood formation in mice lacking the T-cell leukaemia oncoprotein tal-1/SCL. *Nature, 373*, 432–434.

Spangrude, G. J., Heimfeld, S., & Weissman, I. L. (1988). Purification and characterization of mouse hematopoietic stem cells. *Science (New York, NY), 241*, 58–62.

Sudo, K., Ema, H., Morita, Y., & Nakauchi, H. (2000). Age-associated characteristics of murine hematopoietic stem cells. *J. Exp. Med., 192*, 1273–1280.

Taichman, R. S. (2005). Blood and bone: two tissues whose fates are intertwined to create the hematopoietic stem-cell niche. *Blood, 105*, 2631–2639.

Tothova, Z., Kollipara, R., Huntly, B. J., Lee, B. H., Castrillon, D. H., Cullen, D. E., et al. (2007). FoxOs are critical mediators of hematopoietic stem cell resistance to physiologic oxidative stress. *Cell, 128*, 325–339.

Traver, D., Paw, B. H., Poss, K. D., Penberthy, W. T., Lin, S., & Zon, L. I. (2003). Transplantation and in vivo imaging of multilineage engraftment in zebrafish bloodless mutants. *Nat. Immunol., 4*, 1238–1246.

Varnum-Finney, B., Xu, L., Brashem-Stein, C., Nourigat, C., Flowers, D., Bakkour, S., et al. (2000). Pluripotent, cytokine-dependent, hematopoietic stem cells are immortalized by constitutive Notch1 signaling. *Nat. Med., 6*, 1278–1281.

Viardot, A., Kronenwett, R., Deichmann, M., & Haas, R. (1998). The human immunodeficiency virus (HIV)-type 1 coreceptor CXCR-4 (fusin) is preferentially expressed on the more immature CD34+ hematopoietic stem cells. *Ann. Hematol., 77*, 193–197.

Visnjic, D., Kalajzic, Z., Rowe, D. W., Katavic, V., Lorenzo, J., & Aguila, H. L. (2004). Hematopoiesis is severely altered in mice with an induced osteoblast deficiency. *Blood, 103*, 3258–3264.

Weinstein, B. M., Schier, A. F., Abdelilah, S., Malicki, J., Solnica-Krezel, L., Stemple, D. L., et al. (1996). Hematopoietic mutations in the zebrafish. *Development, 123*, 303–309.

Wright, D. E., Bowman, E. P., Wagers, A. J., Butcher, E. C., & Weissman, I. L. (2002). Hematopoietic stem cells are uniquely selective in their migratory response to chemokines. *J. Exp. Med., 195*, 1145–1154.

Zhang, J., Niu, C., Ye, L., Huang, H., He, X., Tong, W. G., et al. (2003). Identification of the haematopoietic stem cell niche and control of the niche size. *Nature, 425*, 836–841.

CHAPTER 17

Mesenchymal Stem Cells

Zulma Gazit[*,**], **Gadi Pelled**[*,**], **Dima Sheyn**[*], **Nadav Kimelman**[*], **Dan Gazit**[*,**]
* Skeletal Biotechnology Laboratory, Hebrew University—Hadassah Faculty of Dental Medicine, Jerusalem, Israel;
** Department of Surgery and Cedars-Sinai Regenerative Medicine Institute (CS-RMI), Cedars-Sinai Medical Center, Los Angeles, CA, USA

THE DEFINITION OF MSCs

BM was the first tissue described as a source of plastic-adherent, fibroblast-like cells that develops colony-forming unit fibroblastic (CFU-F) when seeded in tissue culture plates (Friedenstein et al., 1982, 1987). These cells, originally designated stromal cells, elicited much attention, and the main goal of thousands of studies conducted using these cells was to find an ultimate pure cell population that could be further utilized for regenerative purposes. In these studies, cells were isolated using several methods that will be discussed later in this chapter and were given names such as MSCs (mesenchymal stem cells), mesenchymal progenitors, stromal stem cells, among others. Lately, a committee of the International Society for Cytotherapy suggested the name "multipotent mesenchymal stromal cells" (Dominici et al., 2006). However, most scientists have been referring to them simply as "MSCs."

The precise definition of these cells remains a matter of debate. Nevertheless, to date MSCs are widely defined as a plastic-adherent cell population that can be directed to differentiate *in vitro* into cells of osteogenic, chondrogenic, adipogenic, myogenic, and other lineages (Pittenger et al., 1999; Javazon et al., 2004; Prockop, 2009). As part of their stem cell nature, MSCs proliferate and give rise to daughter cells that have the same pattern of gene expression and phenotype and, therefore, maintain the "stemness" of the original cells. Self-renewal and differentiation potential are two criteria that define MSCs as real stem cells; however, these characteristics have only been proved after *in vitro* manipulation, in bulk and at single-cell level, and there is no clear description of the characteristics displayed by unmanipulated MSCs *in vivo* (Javazon et al., 2004; Yoshimura et al., 2006; Lee et al., 2010a).

In contrast to other stem cells such as hematopoietic stem cells (HSCs), which are identified by the expression of the CD34 surface marker, MSCs lack a unique marker. The CD105 surface antigen (endoglin) has been recently used to isolate hMSCs (human mesenchymal stem cells) from BM and such an approach enabled the characterization of freshly isolated hMSCs before culture. A distinct expression of certain surface antigens such as CD45 and CD31 was demonstrated in freshly isolated hMSCs and the expression of these molecules was lower in culture-expanded hMSCs (Aslan et al., 2006b). These data suggest, again, the alterations that hMSCs may undergo during culture (Boquest et al., 2005).

In several studies, cultured MSCs have been characterized either by using cell surface antigens or by examining the cells' differentiation potential. Lately, the Mesenchymal and Tissue Stem Cell Committee of the International Society for Cellular Therapy proposed minimal criteria to define human MSCs: (1) MSCs must be plastic-adherent when maintained in standard culture conditions and form CFU-Fs, (2) MSCs must express CD105, CD73, and CD90, and lack

285

expression of CD45, CD34, CD14 or CD11b, CD79alpha or CD19, and HLA-DR surface molecules, and (3) MSCs must differentiate to osteoblasts, adipocytes, and chondroblasts *in vitro* (Dominici et al., 2006)

THE STEM CELL NATURE OF MSCs

Stem cells are defined by their ability to self-renew and by their potential to undergo differentiation into functional cells under the right conditions. As detailed below, MSCs exhibit the potential to differentiate into the osteogenic, chondrogenic, adipogenic, tenogenic, myogenic, or stromal lineages.

The ongoing public discussion regarding whether MSCs are strictly stem cells requires a revision of the definition of stem cells, as MSCs apply to a wide cluster of non-hematopoietic stem-like cells isolated from mesenchymal tissues such as bone marrow, adipose, amniotic fluid, and blood vessels. The central question would be whether they might be differentiated into cells of other than a mesenchymal nature. Researchers have reported that MSCs from bone marrow and other tissues can be differentiated into epithelial, endothelial, and neural cells (Spees et al., 2003; Greco and Rameshwar, 2007; Yue et al., 2008). As stated above, there is a consensus on specific MSC markers, but a unique marker of "stemness" and multipotentiality has not yet been defined, since culture-expanded MSCs may lose some of these markers and acquire others, which are non-specific, but cells retain their multipotentiality (Jones and McGonagle, 2008). The molecular signature and *in vivo* distribution status of MSCs remain unknown and, as such, subject to investigation, even though *ex vivo*-expanded MSCs have been widely used in numerous studies (Prockop, 2007; Kubo et al., 2009; Pricola et al., 2009).

In local models, direct injection of hMSCs into the brain tissue of rats resulted in the cells' long-term engraftment and subsequent migration along pathways similar to those used by neural stem cells (Azizi et al., 1998). The results of these studies demonstrate the multilineage differentiation potential of BM-derived adult MSCs and aid in defining them as suitable candidates for the regeneration of several mesenchymal tissues.

WHICH TISSUES CONTAIN MSCs?

The embryonic origin of MSCs is still unclear; however, some findings indicate a possible origin of MSCs in a supporting layer of the dorsal aorta in the aorto-gonadal-mesonephric region (Cortes et al., 1999; Marshall et al., 1999). Consistent with these findings, MSC-like cells were found circulating within early human blood (Campagnoli et al., 2001). In adults, MSCs appear to be "resident" stem cells in many tissues, and they function in the normal turnover of these tissues. When tissue repair is required, these cells can be stimulated to proliferate and differentiate.

The most studied MSCs compose the stroma-supportive system of BM along with endothelial cells and adipocytes (Bianco et al., 2001). MSC population was also found in the BM of the craniofacial complex (Steinhardt et al., 2008). However, many studies have demonstrated the presence of MSCs or MSC-like cells within other tissues such as adipose tissue (ASCs) (Zuk et al., 2001, 2002), dermal tissue (Bartsch et al., 2005), intervertebral disc (Risbud et al., 2007b), amniotic fluid (de Coppi et al., 2007), various dental tissues (Huang et al., 2009), human placenta (Parolini et al., 2008), cord blood (Hutson et al., 2005), and peripheral blood, although the latter finding is still controversial (Fernandez et al., 1997; Conrad et al., 2002).

ASCs are quite similar to BM-derived MSCs morphologically and immunophenotypically; however, ASCs form more CFU-Fs when plated in culture (Kern et al., 2006). Adipose tissue is an attractive source of MSCs for regenerative medical purposes: it is relatively easy to obtain, can be collected with the use of local anesthesia, and is associated with minimal discomfort and risks (Mizuno and Hyakusoku, 2003).

MSC ISOLATION TECHNIQUES

Application of MSCs requires their isolation and directing the differentiation of these cells into the appropriate lineage. Since the 1980s (Friedenstein et al., 1982, 1987), a density gradient has been used to separate mononuclear cells (MNCs) and red blood cells in the BM. The MNCs are then collected and seeded in medium containing 10% fetal bovine serum (FBS) at a density of $10-15 \times 10^5$ cells/cm^2 growth area (Pittenger et al., 1999). Adherent spindle-shaped cells appeared within 48 h after the initial seeding, and the estimated percentage of MNCs ranges from 0.001 to 0.01%.

Adipose tissue-derived stem cells (ASCs) can be isolated also from adipose tissue after enzymatic treatment with collagenase. Then, a stromal vascular fraction (SVF) is obtained that parallels the MNC fraction in BM. This fraction is collected while the adipocytes-containing fraction is removed during the first steps of centrifugation due to its high content of fatty acids. Plastic-adherent cells within the SVF were shown to have a high potential for *in vitro* expansion and for differentiation into several mesenchymal lineages (Zuk et al., 2001, 2002; Katz et al., 2005).

The major disadvantages of these methods are the presence of adherent cells of hematopoietic origin within the cultures during the first days and the need for *in vitro* culturing and expansion. The solution to these downsides will include isolation of cells based on intrinsic properties of MSCs avoiding culturing and the generation of immortalized cell lines.

Immunoisolation is a method to isolate non-cultured MSCs based on cell surface markers. Several studies employed the positive selection technique by immunoisolating MSCs with antibodies directed against the endoglin (CD105) (Aslan et al., 2006b), Stro-1 (Gronthos and Simmons, 1995; Gronthos et al., 2003), CD146 (Sorrentino et al., 2008), and MSC markers. Furthermore, immunodepletion is a "negative selection" approach, in which the MSC population is enriched by washing out the cells labeled with antibodies, mostly directed against hematopoietic markers (Phinney, 2008). Recently, more specific and pure populations were isolated utilizing a combination of immunoisolation and immunodepletion based on different surface markers (Kastrinaki et al., 2008).

Roda and collegues recently developed another technique for non-cultured MSC isolation that does not rely on surface marker, but on biophysical properties that cells acquire when in suspension under fluidic conditions (Roda et al., 2009b); this was further described in a detailed protocol (Roda et al., 2009a).

IMMUNOMODULATORY EFFECTS OF MSCs

Several studies have shown that MSCs escape immune recognition and inhibit immune responses (Noel et al., 2007). The modulation of the immune system was detected in both BM-MSCs and ASCs (Niemeyer et al., 2007). This property of MSCs facilitates clinical use of MSCs in an allogeneic maner in diverse regenerative medicine approaches, for example liver transplantation (Popp et al., 2009).

A variety of suggested mechanisms explicate how MSCs prevent allogeneic rejection among different species (Ren et al., 2009), such as weak immunogenicity, interference in the maturation and function of dendritic cells (DCs), abolishment of T-cell proliferation, or interaction with natural killer (NK) cells in cell-to-cell contact or through the release of soluble secreted factors. Although there have been discrepancies, probably due to the different implemented experimental systems, the majority of the reports have indicated no or low expression of MHC class II proteins (Majumdar et al., 2003; Gotherstrom et al., 2004). Beyth et al. (2005) have provided evidence for the interference in the maturation of DCs: it was demonstrated that, although hMSCs are able to promote antigen-induced activation of purified T-cells, an addition of antigen-presenting cells (APCs) — monocytes or DCs — to cultures inhibited, in a contact-dependent manner, the T-cell responses. This inhibition could be partially over-ridden by the addition of factors that promote APC maturation. These data have been supported by findings of co-culture experiments, in which Zhang et al. (2004) showed that both MSCs and their supernatants interfered with the endocytosis of DCs and decreased their capacity to secrete (interleukin) IL-12 and activate alloreactive T-cells. Similar conclusions have been reported by Aggarwal and Pittenger (2005), who in co-cultures of hMSCs and DCs demonstrated decreased tumor necrosis factor secretion in mature type I DCs and increased secretion of IL-10.

Numerous groups support the direct interaction of MSCs and T-cells, either by cell contact or by the release of soluble factors. Rasmusson et al. made the distinction between T-cell stimulation in culture by mitogen and alloantigens. They found that MSCs increased the levels of IL-2 and the IL-2-soluble receptor, as well as that of IL-10 in MLCs. None of these factors are constitutively secreted by MSCs (Beyth et al., 2005; Rasmusson et al., 2005). When peripheral blood lymphocytes were stimulated with phytohemagglutinin (PHA), decreases in levels of IL-2 and the IL-2 soluble receptor were observed, whereas IL-10 levels were not affected. Moreover, the addition of a prostaglandin inhibitor, indomethacin, restored the inhibition induced by MSCs in PHA cultures, but did not influence MLCs (Rasmusson et al., 2005). Di Nicola et al. identified TGFβ1 and HGF as mediators of MSC effects on T-lymphocyte-suppressed proliferation by using neutralizing monoclonal antibodies. They demonstrated that cellular stimuli were effective as well as non-specific mitogens, and that T-cell inhibition is conducted by soluble factors, as shown by transwell experiments, in which cell-to-cell contact between MSCs and effector cells was avoided (di Nicola et al., 2002).

Alternatively, Sotiropoulou et al. found that MSCs alter the phenotype of NK cells and suppress proliferation and cytokine secretion. Some of these effects were mediated by soluble factors including TGFβ1 and PGE-2 (Sotiropoulou et al., 2006). Others reported no involvement in T-cell inhibition by MSCs (Djouad et al., 2003). The upregulation of PGE-2 in co-cultures has been observed as well, although the role of PGE-2 in the downregulation of MLCs diverged from that mentioned above (Tse et al., 2003; Rasmusson et al., 2005).

Overall, the way by which MSC avoid detection by the immune system is not thoroughly elucidated yet; still, novel mechanisms might be revealed as additional soluble factors and cells are actively under research.

SKELETAL TISSUE REGENERATION BY MSCs
Bone

Bone fractures and small defects usually regenerate and heal without the need for surgical intervention. Yet, in certain conditions, tissue loss is too extensive and complete spontaneous healing cannot be achieved. This is the case for non-union fractures and other critical-size defects that might occur in long bones, the spinal column, or the craniofacial complex. In addition, certain procedures, such as spine fusion, require neo-formation of bone in sites where osteogenesis does not physiologically occur.

Numerous studies have attempted to demonstrate the feasibility of MSC-mediated bone regeneration. In general, MSCs can either be systemically administered using intravenous (iv) injection or directly implanted in the bone defect site. The systemic approach assumes that MSCs have the capability of migrating across the endothelium and homing to injured tissues in a manner similar to the migration of leukocytes to sites of inflammation. This phenomenon has been shown in different experimental models including injuries to heart, brain, liver, and lungs (Chamberlain et al., 2007). Several studies have also shown that MSCs home to sites of bone fractures (Devine et al., 2002; Shirley et al., 2005; Kumar and Ponnazhagan, 2007; Kitaori et al., 2009; Kumar et al., 2010) or to bones with impaired development, as in several patients of osteogenesis imperfecta treated with allogeneic MSCs (Horwitz et al., 2002). Yet, although the systemic approach is attractive for clinical use, it is still unknown what percentage of the injected cells will eventually engraft at the injured tissue. It has been shown that, shortly after iv injection, the MSCs are entrapped in the lungs and are probably released to the circulation a few days later (Kumar et al., 2010). Thus, the direct implantation approach aims at concentrating a high number of MSCs at the site of the injury without the risk of cell migration to other sites in the body.

Undifferentiated MSCs tend to form a non-specific connective tissue even in bone defects (Moutsatsos et al., 2001; Turgeman et al., 2001); therefore, it is essential to either induce osteogenic differentiation of the cells *in vitro* prior to implantation, or to seed them onto an osteoinductive and osteoconductive scaffold, which is usually composed of hydroxyapatite and β-tricalcium phosphate. The potential of MSC-loaded osteoinductive scaffolds to repair segmental defects in long bones has been shown in a number of animal models (Bruder et al., 1998a,b). Using a similar approach, spine fusion was achieved in large animals including rabbits, sheep, and rhesus monkeys (Kon et al., 2000; Arinzeh et al., 2003; Cinotti et al., 2004). Following this solid experimental proof of principle, Quarto et al. attempted to use this tissue-engineering method in the treatment of three human patients who suffered a bone loss of 4—7 cm in long bones (Quarto et al., 2001). A good integration of the implants was evident 2 months post-surgery. The patients recovered function in 6—7 months after surgery (one half to one third of the time needed for recovery using "conventional" bone grafts) and no special problems were recorded over a six-year follow-up period (Mastrogiacomo et al., 2005). Since then, several reports have described the use of this approach for bone regeneration in human patients in different sites including the jaws (Shayesteh et al., 2008), spine (Gan et al., 2008), and femoral head (Kawate et al., 2006).

The downside of using hydroxyapatite scaffolds is their slow resorption rate *in vivo*. In fact, a large portion of these scaffolds does not resorb even after a few years (Mastrogiacomo et al., 2005), thus preventing complete bone regeneration. An alternative approach could be the combination of MSCs and an osteogenic factor such as bone morphogenetic protein (BMP)-2. BMP-2 can be incorporated into a scaffold during its preparation and then combined with MSCs (Kim et al., 2007; Na et al., 2007). In this manner, BMP-2 is slowly released from the scaffold upon implantation, and its release is in correlation with the degradation rate of the scaffold itself. It is assumed that BMP-2 induces the osteogenic differentiation of the implanted MSCs and resident MSCs at the site of implantation. The shortcoming of this strategy lies in the short half-life of BMP (Johnson et al., 2009a). Thus, the effect of BMP-2 in this system could be limited.

A different approach, which combines MSCs and a continuous secretion of an osteogenic protein at the fractures site, is known as MSC-based gene therapy. This method requires the genetic modification of MSCs to overexpress a transgene encoding for an osteogenic gene. BMP-2 has been widely used for this purpose (Gazit et al., 1999; Moutsatsos et al., 2001; Turgeman et al., 2001), and also other members of the BMP family, such as BMP-4, BMP-6, and BMP-9 (Chen et al., 2002; Dumont et al., 2002; Gysin et al., 2002; Peng et al., 2002; Wright et al., 2002; Aslan et al., 2006a; Sheyn et al., 2008). There are several advantages to this approach of tissue regeneration. First, the implanted MSCs secrete physiological quantities of the osteogenic factor, over a period of time (Moutsatsos et al., 2001; Aslan et al., 2006a). Second, MSCs tend to migrate to the fracture edges and induce an organized pattern of fracture repair, when compared to BMP-2 treatment, which induces the formation of scattered foci of ossification instead (Gazit et al., 1999). Third, due to a continuous secretion of the osteogenic factor, an autocrine-paracrine effect is exerted inducing the osteogenic differentiation of the implanted MSCs and resident stem cells in the surrounding tissue (Moutsatsos et al., 2001). It is important to note that, when new bone generated by BMP-2-engineered MSCs was analyzed for its chemical, structural, and nanobiomechanical properties, it showed remarkably similar values to its natural counterpart (Pelled et al., 2007; Tai et al., 2008).

Cartilage

Regeneration of damaged cartilage presents a great challenge for orthopedic medicine, because articular cartilage has very limited capacity for effective repair. Adult MSCs have the potential to proliferate and differentiate into chondrocytes; they can therefore be considered ideal candidates for cartilage tissue repair. Several attempts have been made to implant cells in cartilage defects. The first attempt was to culture autologous chondrocytes and implant them in a cartilage defect in patients younger than 50 years of age who were believed to have healthy chondrocytes (Brittberg et al., 1994). It appeared, however, that chondrocytes could only achieve limited success in regenerating cartilage defects (Liu et al., 2002). It was also shown that chondrocytes loaded onto a polymeric carrier underwent apoptosis, which limited their therapeutic potential (Gille et al., 2002). These results prompted research into autologous pluripotent cells with chondrocyte-differentiating capacities (Caplan et al., 1997). Evidence that MSCs can produce cartilage regeneration has been controversial. Findings of some studies indicate that MSCs fail to produce full regeneration over long time periods (Tatebe et al., 2005). MSCs have also been found to have limited success in forming long-lasting cartilage tissue (Wakitani et al., 2002; Wakitani and Yamamoto, 2002). Other studies, in which sheep, pig, and rabbit models were used, have demonstrated the feasibility of using biodegradable scaffolds seeded with MSCs for articular cartilage repair (Im et al., 2001; Li et al., 2009b).

Genetically modified MSCs have also been used in an attempt at cartilage formation; however, only a few genes have been shown to induce chondrogenic differentiation in these cells. Kawamura et al. (2005) and Palmer et al. (2005) have shown that, when infected with adeno-TGFβ but not with adeno-IGF-1, MSCs differentiated into chondrocytes *in vitro*. A combination of IGF-1 and TGFβ or BMP-2 gene delivery to MSCs led to enhanced chondrogenesis *in vitro*, however, with the expression of collagen X, a marker of hypertrophic cartilage (Steinert et al., 2009). Successful induction of MSC chondrogenic differentiation *in vivo* was achieved using the overexpression of Brachyury transcription factor (Hoffmann et al., 2002). Brachyury-expressing MSCs secreted collagen II, but not collagen X, *in vitro* and *in vivo*. Moreover, the implantation of these cells in ectopic sites *in vivo* has led to the formation of a chondrogenic tissue composed of proliferative chondrocytes. Interestingly, the engineered chondrogenic tissue generated *in vivo* was resistant to the destructive effect of rheumatoid arthritis synovial fibroblasts (Dinser et al., 2009).

Tendon

Although of low occurrence, tendon and ligament lesions (especially rotator cuff, Achilles' tendon, and patellar tendon defects) are among the most common soft-tissue injuries (Juncosa-Melvin et al., 2006). Repairing these defects is not a simple task, and indeed the available surgical treatments are not satisfactory (Wang et al., 2005). The *in vitro* differentiation of MSCs into tendon or ligament cells has only been shown in a few studies either by application of exogenous forces on the scaffold on which the cells are grown (Altman et al., 2002) or by the use of a specific scaffold made of hyaluronic acid, which induces ligament differentiation of hMSCs (Cristino et al., 2005). There is no evidence that MSCs that have differentiated *in vitro* into tendon or ligament cells can indeed repair those tissues *in vivo*.

One possible treatment for *in vivo* tendon repair involves the implantation of non-differentiated MSCs that have been seeded onto various biodegradable scaffolds. So far there have been contradictory reports in the literature. It has been shown that the implantation of autologous MSCs in rabbit Achilles' tendon defects improves the physical properties of the damaged tendon when compared with tendons treated only with hydrogel, scaffold, or sutures (Juncosa-Melvin et al., 2006), yet this effect could be detected only for a few weeks post-surgery (Chong et al., 2007). Dressler et al. have also observed that MSCs obtained from older animals are able to induce tendon repair in young ones (Dressler et al., 2005). A recent publication found no added value for the implantation of MSCs in a rat rotator cuff tear (Gulotta et al.,

2009). One adverse effect discovered in some of these studies was the formation of ectopic bone within tendons implanted with MSCs (Harris et al., 2004). Awad et al. (1999) have also posited that there is no morphometric difference between tendons implanted with MSCs and ones implanted with collagen gel.

MSCs genetically modified to overexpress the Smad8 and BMP-2 cDNAs were shown to differentiate to tenocyte-like cells *in vitro* and *in vivo*. In addition, when implanted into a 3 mm defect in a rat's Achilles' tendon defect, complete regeneration was achieved, as demonstrated by double-quantum filtered magnetic resonance (MR) and histology (Hoffmann et al., 2006). So far this has been the only report of genetically engineered MSCs used for tendon or ligament regeneration.

Intervertebral disc

Regeneration of an intervertebral disc (IVD) poses great challenges for stem cell therapy due to the hostile environment in which implanted cells must survive. The IVD is avascular and hypoxic; in the rabbit IVD, the nearest blood vessel can be 5—8 mm away from cells at the disc center (Gan et al., 2003). The disc cells (mainly nucleus pulposus (NP) cells) use anaerobic metabolism to generate energy (Gan et al., 2003; Roughley, 2004; Risbud et al. 2007a). As a result, lactic acid (the main product of glycolysis) can accumulate, resulting in a low pH environment (Roughley, 2004).

Studies have outlined two strategies for stem cell-mediated IVD regeneration. The first, which is indicated for early disc degeneration, is to regenerate only the NP. This could be achieved by direct injection of MSCs, similarly to what is done in discography procedures in the clinic today. Several works have shown that MSCs differentiate to NP-like cells when co-cultured with NP cells (Wei et al., 2009) injected into a disc organ culture (le Maitre et al., 2009) or cultured in specific scaffolds (Richardson et al., 2008). It has been shown that low pH levels that exist in degenerated discs might have a significant effect on MSC proliferation and differentiation (Wuertz et al., 2009). Nevertheless, studies in rodents, canines, and rabbits showed that MSCs could survive in "nucleotomized" discs for several weeks (up to 48), enhance extracellular matrix production, and increase disc height (Crevensten et al., 2004; Bertram et al., 2005; Sakai et al., 2005, 2006; Hiyama et al., 2008; Yang et al., 2009). A comprehensive biomechanical comparison between native and engineered tissues should be performed to evaluate the ability of this approach to generate functional NP tissue. In addition, care should be taken when choosing the right needle for stem cell injection, since its diameter might have an effect on the damage caused to the disc (Elliott et al., 2008). Moreover, the injected cells might leak out via the entry site of the needle, as shown in a pig model (Omlor et al., 2009).

A second strategy for IVD regeneration relates to the complete regeneration of the IVD, which could be relevant for a late-stage disease. This is a more challenging tissue-engineering goal, which will require the combination of designated scaffolds and inducing factors that will regenerate both the NP and the annulus fibrosus (Nesti et al., 2008).

Interestingly, there is evidence showing the presence of resident MSCs in the degenerated IVD (Risbud et al., 2007a), which might imply that a potential therapy could include the activation of these cells in order to regenerate the disc.

NON-SKELETAL TISSUE REGENERATION BY MSCs

During the mid-1990s, the first two reports demonstrating non-skeletal differentiation potential of MSCs were presented (Okuyama et al., 1995; Wakitani et al., 1995). These reports were validated a few years later (Liechty et al., 2000; Fukuda, 2001, 2002). Since then, MSCs have been used as regenerators of heart, skeletal muscle, nerve, liver, kidney, lungs, and pancreas (Burt et al., 2002; Lardon et al., 2002; Bonafe et al., 2003; Dabeva and Shafritz, 2003;

Abedin et al., 2004; Kim et al., 2004; Jain et al., 2005; Sonoyama et al., 2005; Goncalves et al., 2006; Lee et al., 2009; Zubko and Frishman, 2009; Reinders et al., 2010).

The use of pluripotent stem cells to regenerate damaged heart tissue is being advocated as the new treatment for heart failure secondary to heart disease or severe myocardial infarction. Promising results at the research stage have now led to the challenge of applying stem cell technology in the clinical setting (Fukuda, 2003a,b; Itescu et al., 2003; Orlic, 2003; Amado et al., 2005; Bayes-Genis et al., 2005; Fazel et al., 2005; Fukuda, 2005; Jain et al., 2005; Siepe et al., 2005; Smits et al., 2005; Wojakowski and Tendera, 2005; Yamaguchi et al., 2005; Yoon et al., 2005; Minguell and Erices, 2006).

Cardiomyocytes generated from MSCs were able to stay differentiated after being transplanted into the adult murine heart (Makino et al., 1999; Toma et al., 2002). Transplantation of MSCs improved cardiac function in animal models (Mangi et al., 2003), possibly through induction of myogenesis and angiogenesis and inhibition of myocardial fibrosis by the cells' ability to supply angiogenic, anti-apoptotic, and mitogenic factors (Nagaya et al., 2005; Pons et al., 2009; Zisa et al., 2009).

In order to achieve clinical application, studies were conducted in pre-clinical large animal models with encouraging results (Potapova et al., 2008; Qi et al., 2009; Wolf et al., 2009). Methods for introducing the cells to the tissue (Martens et al., 2009) and monitoring their homing, survival, and post-implantation effect were developed (Rosen et al., 2007; Chacko et al., 2009; Qi et al., 2009). New sources from which stem cells can be isolated for cardiac applications were identified (Imanishi et al., 2009; Madonna et al., 2009; Okura et al., 2009) and, in recent research, the effect of immunosuppressive drugs that will probably be used in implanted patients was tested on MSC activity (Song et al., 2010). Finally, clinical studies have been undertaken and report excellent results: while the use/application of mesenchymal stem cells is now studied in small scale, mainly for safety and feasibility (Joggerst and Hatzopoulos, 2009; Trivedi et al., 2010), the use of bone marrow mononuclear cells has been extensively studied and showed a significant reduction in subsequent cardiovascular events (Joggerst and Hatzopoulos, 2009; Krause et al., 2009).

MSCs have been shown to promote neuron survival and limit the severity of neurological impairment in animal models of traumatic brain injury (Lu et al., 2001; Mahmood et al., 2003) and stroke (Chen et al., 2001; Zhao et al., 2002). Direct implantation of MSCs, either native or genetically engineered, into the spinal column has also been shown to promote functional recovery following spinal cord injury (Chopp et al., 2000; Akiyama et al., 2002; Hofstetter et al., 2002; Gu et al., 2009; Sasaki et al., 2009). Many pre-clinical animal studies have been conducted recently, with promising results that show MSC migration, differentiation, and regeneration effects in the brain (Dharmasaroja, 2009; Cova et al., 2010). MSCs were able to reduce neuropathic pain (Siniscalco et al., 2010), confer neuroprotection in a rat model for glaucoma (Johnson et al., 2009b), induce neuronal regeneration after neonatal ischemic brain injury (Pimentel-Coelho et al., 2009; van Velthoven et al., 2009; Lee et al., 2010b), and treat depression (Tfilin et al., 2009), Parkinson's disease (Chao et al., 2009; Glavaski-Joksimovic et al., 2009), and even epilepsy (Li et al., 2009a). The neuroprotective effects of MSCs are thought to result in part from their ability to replace diseased or damaged neurons via cellular differentiation (Black and Woodbury, 2001; Crigler et al., 2006) as well as by induction of neurogenesis, angiogenesis, synapse formation, activation of endogenous restorative processes (Dharmasaroja, 2009; Wang et al., 2009a; Wilkins et al., 2009; Kim et al., 2010), and modulation of inflammatory response (Walker et al., 2009). Still, few clinical studies have been performed that have demonstrated the safety and regenerative effect of MSCs (Bang et al., 2005; Lee et al., 2008).

As the prevalence of diabetes increases, new treatment avenues are being sought and MSCs have been identified as prime candidates. Scientists have been able to obtain islet-like functional cells through differentiation of both animal and human MSCs from BM by modifying the cell culture environment (Chen et al., 2004; Choi et al., 2005; Moriscot et al., 2005; Chandra et al., 2009; Chang et al., 2009; Xu et al., 2009b). Lately, MSCs were also identified in pancreatic tissue cultures (Sordi et al., 2010); in fact, a combined implantation of MSCs with islet graft improved the graft's function significantly (Figliuzzi et al., 2009; Solari et al., 2009). Moreover, MSC systemic and local administration was found to be effective in different reports: reversion of hyperglycemia, reduction of albuminuria, and regeneration of beta-pancreatic islets on mouse and rat models of diabetes (Dong et al., 2008; Ezquer et al., 2008; Lin et al., 2009; Zhang et al., 2009); amelioration of diabetic nephropathy in rats and mice (Ezquer et al., 2008; Zhou et al., 2009); and even improvement of diabetic poly-neuropathy in rats (Shibata et al., 2008). Large animal models such as dogs and pigs confirm the feasibility of MSC-based therapy for diabetes (Chang et al., 2008; Zhu et al., 2009). The therapeutic effect of MSCs in diabetes models may be due to their immunomodulatory capacity (Ding et al., 2009; Fiorina et al., 2009; Madec et al., 2009; Volarevic et al., 2009) and paracrine activity (Xu et al., 2009a), apart from their direct cell differentiation.

Schwartz and associates reported for the first time that multipotent adult progenitor cells (MAPCs) could differentiate into functional hepatocyte-like cells (Schwartz et al., 2002). Since then, many studies have demonstrated hepatic differentiation of MSCs (Wang et al., 2004; Kang et al., 2005; Lange et al., 2005; Saulnier et al., 2009) and their applications in pre-clinical models (Sato et al., 2005; Oyagi et al., 2006; Yu et al., 2007; Wang et al., 2009b; Pulavendran et al., 2010) and even in a phase-1 clinical trial (Mohamadnejad et al., 2007).

In summary, even for non-skeletal tissue injuries and diseases, MSCs are seen as serious candidates that indeed possess the potential for future treatment of choice.

CONCLUSIONS

293

MSCs constitute a unique population of adult stem cells that hold great promise for various tissue-engineering applications. These cells can readily be isolated from various sites in the human body, especially from BM and adipose tissues. Established protocols exist for the induction of specific differentiation patterns of MSCs into different committed cells, most notably into osteoblasts, chondrocytes, and adipocytes. So far it has been demonstrated that the use of genetically modified MSCs, overexpression of various therapeutic transgenes, is a powerful tool in the induction of differentiation and in the promotion of tissue regener-ation *in vivo*. Novel technologies, which utilize electroporation-based systems, allow for the safe and efficient gene delivery into MSCs and bypass the need for using non-safe viral vectors. It has been shown that the ultrastructural, chemical, and nanobiomechanical properties of engineered bone derived from MSCs were similar to those of native origin. The conventional method of MSC isolation using plastic adherence has been shown to be costly and might reduce the stemness of the cells. Therefore, an attractive alternative has been developed that includes the immediate use of immuno-isolated, non-cultured MSCs for *in vivo* implantation. Future challenges require the identification of an optimal scaffold for MSC implantation *in vivo* and, finally, the development of a preservation method for future reuse of autologous cells. Non-invasive imaging will continue to play an important role in analyzing the power of MSCs to regenerate tissues in various defect models. Overcoming these hurdles will no doubt make MSCs the optimal tool for biological tissue replacement within this century.

Acknowledgments

Funding derived from NIH (R01DE019902, RO3AR057143, R01AR056694, R43AR057587-01), CIRM (RT1-01027), and Israel Science Foundation (ISF) grants. We thank Olga Mizrahi, Amir Lavi, Shimon Benjamin, and Ilan Kallai, graduate students, for their bibliographic assistance and enthusiastic help.

References

Abedin, M., Tintut, Y., & Demer, L. L. (2004). Mesenchymal stem cells and the artery wall. *Circ. Res., 95*(7), 671–676.

Aggarwal, S., & Pittenger, M. F. (2005). Human mesenchymal stem cells modulate allogeneic immune cell responses. *Blood, 105*(4), 1815–1822.

Akiyama, Y., Radtke, C., & Kocsis, J. D. (2002). Remyelination of the rat spinal cord by transplantation of identified bone marrow stromal cells. *J. Neurosci., 22*(15), 6623–6630.

Altman, G. H., Horan, R. L., Martin, I., Farhadi, J., Stark, P. R., Volloch, V., et al. (2002). Cell differentiation by mechanical stress. *Faseb J., 16*(2), 270–272.

Amado, L. C., Saliaris, A. P., Schuleri, K. H., St John, M., Xie, J. S., Cattaneo, S., et al. (2005). Cardiac repair with intramyocardial injection of allogeneic mesenchymal stem cells after myocardial infarction. *Proc. Natl. Acad. Sci. U.S.A., 102*(32), 11474–11479.

Arinzeh, T. L., Peter, S. J., Archambault, M. P., van den Bos, C., Gordon, S., Kraus, K., et al. (2003). Allogeneic mesenchymal stem cells regenerate bone in a critical-sized canine segmental defect. *J. Bone Joint Surg. Am., 85-A* (10), 1927–1935.

Aslan, H., Zilberman, Y., Arbeli, V., Sheyn, D., Matan, Y., Liebergall, M., et al. (2006a). Nucleofection-based ex vivo nonviral gene delivery to human stem cells as a platform for tissue regeneration. *Tissue Eng., 12*(4), 877–889.

Aslan, H., Zilberman, Y., Kandel, L., Liebergall, M., Oskouian, R. J., Gazit, D., et al. (2006b). Osteogenic differentiation of noncultured immunoisolated bone marrow-derived CD105+ cells. *Stem Cells, 24*(7), 1728–1737.

Awad, H. A., Butler, D. L., Boivin, G. P., Smith, F. N., Malaviya, P., Huibregtse, B., et al. (1999). Autologous mesenchymal stem cell-mediated repair of tendon. *Tissue Eng., 5*(3), 267–277.

Azizi, S. A., Stokes, D., Augelli, B. J., DiGirolamo, C., & Prockop, D. J. (1998). Engraftment and migration of human bone marrow stromal cells implanted in the brains of albino rats – similarities to astrocyte grafts. *Proc. Natl. Acad. Sci. U.S.A., 95*(7), 3908–3913.

Bang, O. Y., Lee, J. S., Lee, P. H., & Lee, G. (2005). Autologous mesenchymal stem cell transplantation in stroke patients. *Ann. Neurol., 57*(6), 874–882.

Bartsch, G., Yoo, J. J., de Coppi, P., Siddiqui, M. M., Schuch, G., Pohl, H. G., et al. (2005). Propagation, expansion, and multilineage differentiation of human somatic stem cells from dermal progenitors. *Stem Cells Dev., 14*(3), 337–348.

Bayes-Genis, A., Roura, S., Soler-Botija, C., Farre, J., Hove-Madsen, L., Llach, A., et al. (2005). Identification of cardiomyogenic lineage markers in untreated human bone marrow-derived mesenchymal stem cells. *Transplant Proc., 37*(9), 4077–4079.

Bertram, H., Kroeber, M., Wang, H., Unglaub, F., Guehring, T., Carstens, C., et al. (2005). Matrix-assisted cell transfer for intervertebral disc cell therapy. *Biochem. Biophys. Res. Commun., 331*(4), 1185–1192.

Beyth, S., Borovsky, Z., Mevorach, D., Liebergall, M., Gazit, Z., Aslan, H., et al. (2005). Human mesenchymal stem cells alter antigen-presenting cell maturation and induce T-cell unresponsiveness. *Blood, 105*(5), 2214–2219.

Bianco, P., Riminucci, M., Gronthos, S., & Robey, P. G. (2001). Bone marrow stromal stem cells: nature, biology, and potential applications. *Stem Cells, 19*(3), 180–192.

Black, I. B., & Woodbury, D. (2001). Adult rat and human bone marrow stromal stem cells differentiate into neurons. *Blood Cells Mol. Dis., 27*(3), 632–636.

Bonafe, F., Muscari, C., Guarnieri, C., & Caldarera, C. M. (2003). Regeneration of infarcted cardiac tissue: the route of stem cells. *Ital. Heart J. Suppl., 4*(4), 299–305.

Boquest, A. C., Shahdadfar, A., Fronsdal, K., Sigurjonsson, O., Tunheim, S. H., Collas, P., et al. (2005). Isolation and transcription profiling of purified uncultured human stromal stem cells: alteration of gene expression after in vitro cell culture. *Mol. Biol. Cell, 16*(3), 1131–1141.

Brittberg, M., Lindahl, A., Nilsson, A., Ohlsson, C., Isaksson, O., & Peterson, L. (1994). Treatment of deep cartilage defects in the knee with autologous chondrocyte transplantation. *N. Engl. J. Med., 331*(14), 889–895.

Bruder, S. P., Kraus, K. H., Goldberg, V. M., & Kadiyala, S. (1998a). The effect of implants loaded with autologous mesenchymal stem cells on the healing of canine segmental bone defects. *J. Bone Joint Surg. Am., 80*(7), 985–996.

Bruder, S. P., Kurth, A. A., Shea, M., Hayes, W. C., Jaiswal, N., & Kadiyala, S. (1998b). Bone regeneration by implantation of purified, culture-expanded human mesenchymal stem cells. *J. Orthop. Res., 16*(2), 155–162.

Burt, R. K., Oyama, Y., Traynor, A., & Kenyon, N. S. (2002). Hematopoietic stem cell therapy for type 1 diabetes: induction of tolerance and islet cell neogenesis. *Autoimmun. Rev., 1*(3), 133–138.

Campagnoli, C., Roberts, I. A., Kumar, S., Bennett, P. R., Bellantuono, I., & Fisk, N. M. (2001). Identification of mesenchymal stem/progenitor cells in human first-trimester fetal blood, liver, and bone marrow. *Blood, 98*(8), 2396–2402.

Caplan, A. I., Elyaderani, M., Mochizuki, Y., Wakitani, S., & Goldberg, V. M. (1997). Principles of cartilage repair and regeneration. *Clin. Orthop. Relat. Res., 342*, 254–269.

Chacko, S. M., Khan, M., Kuppusamy, M. L., Pandian, R. P., Varadharaj, S., Selvendiran, K., et al. (2009). Myocardial oxygenation and functional recovery in infarct rat hearts transplanted with mesenchymal stem cells. *Am. J. Physiol. Heart Circ. Physiol., 296*(5), H1263−H1273.

Chamberlain, G., Fox, J., Ashton, B., & Middleton, J. (2007). Concise review: mesenchymal stem cells: their phenotype, differentiation capacity, immunological features, and potential for homing. *Stem Cells, 25*(11), 2739−2749.

Chandra, V., Swetha, G., Phadnis, S., Nair, P. D., & Bhonde, R. R. (2009). Generation of pancreatic hormone-expressing islet-like cell aggregates from murine adipose tissue-derived stem cells. *Stem Cells, 27*(8), 1941−1953.

Chang, C., Niu, D., Zhou, H., Zhang, Y., Li, F., & Gong, F. (2008). Mesenchymal stroma cells improve hyperglycemia and insulin deficiency in the diabetic porcine pancreatic microenvironment. *Cytotherapy, 10*(8), 796−805.

Chang, C., Wang, X., Niu, D., Zhang, Z., Zhao, H., & Gong, F. (2009). Mesenchymal stem cells adopt beta-cell fate upon diabetic pancreatic microenvironment. *Pancreas, 38*(3), 275−281.

Chao, Y. X., He, B. P., & Tay, S. S. (2009). Mesenchymal stem cell transplantation attenuates blood brain barrier damage and neuroinflammation and protects dopaminergic neurons against MPTP toxicity in the substantia nigra in a model of Parkinson's disease. *J. Neuroimmunol., 216*(1−2), 39−50.

Chen, J., Li, Y., Wang, L., Lu, M., Zhang, X., & Chopp, M. (2001). Therapeutic benefit of intracerebral transplantation of bone marrow stromal cells after cerebral ischemia in rats. *J. Neurol. Sci., 189*(1−2), 49−57.

Chen, L. B., Jiang, X. B., & Yang, L. (2004). Differentiation of rat marrow mesenchymal stem cells into pancreatic islet beta-cells. *World J. Gastroenterol., 10*(20), 3016−3020.

Chen, Y., Cheung, K. M., Kung, H. F., Leong, J. C., Lu, W. W., & Luk, K. D. (2002). In vivo new bone formation by direct transfer of adenoviral-mediated bone morphogenetic protein-4 gene. *Biochem. Biophys. Res. Commun., 298* (1), 121−127.

Choi, K. S., Shin, J. S., Lee, J. J., Kim, Y. S., Kim, S. B., & Kim, C. W. (2005). In vitro trans-differentiation of rat mesenchymal cells into insulin-producing cells by rat pancreatic extract. *Biochem. Biophys. Res. Commun., 330* (4), 1299−1305.

Chong, A. K., Ang, A. D., Goh, J. C., Hui, J. H., Lim, A. Y., Lee, E. H., et al. (2007). Bone marrow-derived mesenchymal stem cells influence early tendon-healing in a rabbit achilles tendon model. *J. Bone Joint Surg. Am., 89*(1), 74−81.

Chopp, M., Zhang, X. H., Li, Y., Wang, L., Chen, J., Lu, D., et al. (2000). Spinal cord injury in rat: treatment with bone marrow stromal cell transplantation. *Neuroreport, 11*(13), 3001−3005.

Cinotti, G., Patti, A. M., Vulcano, A., Della Rocca, C., Polveroni, G., Giannicola, G., et al. (2004). Experimental posterolateral spinal fusion with porous ceramics and mesenchymal stem cells. *J. Bone Joint Surg. Br., 86*(1), 135−142.

Conrad, C., Gottgens, B., Kinston, S., Ellwart, J., & Huss, R. (2002). GATA transcription in a small rhodamine 123 (low)CD34(+) subpopulation of a peripheral blood-derived CD34(−)CD105(+) mesenchymal cell line. *Exp. Hematol., 30*(8), 887−895.

Cortes, F., Deschaseaux, F., Uchida, N., Labastie, M. C., Friera, A. M., He, D., et al. (1999). HCA, an immuno-globulin-like adhesion molecule present on the earliest human hematopoietic precursor cells, is also expressed by stromal cells in blood-forming tissues. *Blood, 93*(3), 826−837.

Cova, L., Armentero, M. T., Zennaro, E., Calzarossa, C., Bossolasco, P., Busca, G., et al. (2010). Multiple neurogenic and neurorescue effects of human mesenchymal stem cell after transplantation in an experimental model of Parkinson's disease. *Brain Res., 1311*, 12−27.

Crevensten, G., Walsh, A. J., Ananthakrishnan, D., Page, P., Wahba, G. M., Lotz, J. C., et al. (2004). Intervertebral disc cell therapy for regeneration: mesenchymal stem cell implantation in rat intervertebral discs. *Ann. Biomed. Eng., 32*(3), 430−434.

Crigler, L., Robey, R. C., Asawachaicharn, A., Gaupp, D., & Phinney, D. G. (2006). Human mesenchymal stem cell subpopulations express a variety of neuro-regulatory molecules and promote neuronal cell survival and neuritogenesis. *Exp. Neurol., 198*(1), 54−64.

Cristino, S., Grassi, F., Toneguzzi, S., Piacentini, A., Grigolo, B., Santi, S., et al. (2005). Analysis of mesenchymal stem cells grown on a three-dimensional HYAFF 11-based prototype ligament scaffold. *J. Biomed. Mater. Res. A, 73*(3), 275−283.

Dabeva, M. D., & Shafritz, D. A. (2003). Hepatic stem cells and liver repopulation. *Semin Liver Dis., 23*(4), 349−362.

de Coppi, P., Bartsch, G., Jr., Siddiqui, M. M., Xu, T., Santos, C. C., Perin, L., et al. (2007). Isolation of amniotic stem cell lines with potential for therapy. *Nat. Biotechnol., 25*(1), 100−106.

Devine, M. J., Mierisch, C. M., Jang, E., Anderson, P. C., & Balian, G. (2002). Transplanted bone marrow cells localize to fracture callus in a mouse model. *J. Orth. Res., 20*(6), 1232−1239.

Dharmasaroja, P. (2009). Bone marrow-derived mesenchymal stem cells for the treatment of ischemic stroke. *J. Clin. Neurosci., 16*(1), 12−20.

295

di Nicola, M., Carlo-Stella, C., Magni, M., Milanesi, M., Longoni, P. D., Matteucci, P., et al. (2002). Human bone marrow stromal cells suppress T-lymphocyte proliferation induced by cellular or nonspecific mitogenic stimuli. *Blood, 99*(10), 3838–3843.

Ding, Y., Xu, D., Feng, G., Bushell, A., Muschel, R. J., & Wood, K. J. (2009). Mesenchymal stem cells prevent the rejection of fully allogenic islet grafts by the immunosuppressive activity of matrix metalloproteinase-2 and -9. *Diabetes, 58*(8), 1797–1806.

Dinser, R., Pelled, G., Muller-Ladner, U., Gazit, D., & Neumann, E. (2009). Expression of Brachyury in mesenchymal progenitor cells leads to cartilage-like tissue that is resistant to the destructive effect of rheumatoid arthritis synovial fibroblasts. *J. Tissue Eng. Regen. Med., 3*(2), 124–128.

Djouad, F., Plence, P., Bony, C., Tropel, P., Apparailly, F., Sany, J., et al. (2003). Immunosuppressive effect of mesenchymal stem cells favors tumor growth in allogeneic animals. *Blood, 102*(10), 3837–3844.

Dominici, M., le Blanc, K., Mueller, I., Slaper-Cortenbach, I., Marini, F., Krause, D., et al. (2006). Minimal criteria for defining multipotent mesenchymal stromal cells. The International Society for Cellular Therapy position statement. *Cytotherapy, 8*(4), 315–317.

Dong, Q. Y., Chen, L., Gao, G. Q., Wang, L., Song, J., Chen, B., et al. (2008). Allogeneic diabetic mesenchymal stem cells transplantation in streptozotocin-induced diabetic rat. *Clin. Invest. Med., 31*(6), E328–E337.

Dressler, M. R., Butler, D. L., & Boivin, G. P. (2005). Effects of age on the repair ability of mesenchymal stem cells in rabbit tendon. *J. Orthop. Res., 23*(2), 287–293.

Dumont, R. J., Dayoub, H., Li, J. Z., Dumont, A. S., Kallmes, D. F., Hankins, G. R., et al. (2002). Ex vivo bone morphogenetic protein-9 gene therapy using human mesenchymal stem cells induces spinal fusion in rodents. *Neurosurgery, 51*(5), 1239–1244, discussion 1244–1235.

Elliott, D. M., Yerramalli, C. S., Beckstein, J. C., Boxberger, J. I., Johannessen, W., & Vresilovic, E. J. (2008). The effect of relative needle diameter in puncture and sham injection animal models of degeneration. *Spine (Phila. Pa. 1976), 33*(6), 588–596.

Ezquer, F. E., Ezquer, M. E., Parrau, D. B., Carpio, D., Yanez, A. J., & Conget, P. A. (2008). Systemic administration of multipotent mesenchymal stromal cells reverts hyperglycemia and prevents nephropathy in type 1 diabetic mice. *Biol. Blood Marrow Transplant., 14*(6), 631–640.

Fazel, S., Chen, L., Weisel, R. D., Angoulvant, D., Seneviratne, C., Fazel, A., et al. (2005). Cell transplantation preserves cardiac function after infarction by infarct stabilization: augmentation by stem cell factor. *J. Thorac. Cardiovasc. Surg., 130*(5), 1310.

Fernandez, M., Simon, V., Herrera, G., Cao, C., del Favero, H., & Minguell, J. J. (1997). Detection of stromal cells in peripheral blood progenitor cell collections from breast cancer patients. *Bone Marrow Transplant., 20*(4), 265–271.

Figliuzzi, M., Cornolti, R., Perico, N., Rota, C., Morigi, M., Remuzzi, G., et al. (2009). Bone marrow-derived mesenchymal stem cells improve islet graft function in diabetic rats. *Transplant. Proc., 41*(5), 1797–1800.

Fiorina, P., Jurewicz, M., Augello, A., Vergani, A., Dada, S., la Rosa, S., et al. (2009). Immunomodulatory function of bone marrow-derived mesenchymal stem cells in experimental autoimmune type 1 diabetes. *J. Immunol., 183*(2), 993–1004.

Friedenstein, A. J., Chailakhyan, R. K., & Gerasimov, U. V. (1987). Bone marrow osteogenic stem cells: in vitro cultivation and transplantation in diffusion chambers. *Cell Tissue Kinet., 20*(3), 263–272.

Friedenstein, A. J., Latzinik, N. W., Grosheva, A. G., & Gorskaya, U. F. (1982). Marrow microenvironment transfer by heterotopic transplantation of freshly isolated and cultured cells in porous sponges. *Exp. Hematol., 10*(2), 217–227.

Fukuda, K. (2001). Development of regenerative cardiomyocytes from mesenchymal stem cells for cardiovascular tissue engineering. *Artif. Organs, 25*(3), 187–193.

Fukuda, K. (2002). Molecular characterization of regenerated cardiomyocytes derived from adult mesenchymal stem cells. *Congenit. Anom. (Kyoto), 42*(1), 1–9.

Fukuda, K. (2003a). Application of mesenchymal stem cells for the regeneration of cardiomyocyte and its use for cell transplantation therapy. *Hum. Cell, 16*(3), 83–94.

Fukuda, K. (2003b). Use of adult marrow mesenchymal stem cells for regeneration of cardiomyocytes. *Bone Marrow Transplant., 32*(Suppl. 1), S25–S27.

Fukuda, K. (2005). Progress in myocardial regeneration and cell transplantation. *Circ. J., 69*(12), 1431–1446.

Gan, J. C., Ducheyne, P., Vresilovic, E. J., Swaim, W., & Shapiro, I. M. (2003). Intervertebral disc tissue engineering I: characterization of the nucleus pulposus. *Clin. Orthop. Relat. Res., 411*, 305–314.

Gan, Y., Dai, K., Zhang, P., Tang, T., Zhu, Z., & Lu, J. (2008). The clinical use of enriched bone marrow stem cells combined with porous beta-tricalcium phosphate in posterior spinal fusion. *Biomaterials, 29*(29), 3973–3982.

Gazit, D., Turgeman, G., Kelley, P., Wang, E., Jalenak, M., Zilberman, Y., et al. (1999). Engineered pluripotent mesenchymal cells integrate and differentiate in regenerating bone: a novel cell-mediated gene therapy. *J. Gene Med., 1*(2), 121–133.

Gille, J., Ehlers, E. M., Okroi, M., Russlies, M., & Behrens, P. (2002). Apoptotic chondrocyte death in cell-matrix biocomposites used in autologous chondrocyte transplantation. *Ann. Anat., 184*(4), 325–332.

Glavaski-Joksimovic, A., Virag, T., Chang, Q. A., West, N. C., Mangatu, T. A., McGrogan, M. P., et al. (2009). Reversal of dopaminergic degeneration in a parkinsonian rat following micrografting of human bone marrow-derived neural progenitors. *Cell Transplant., 18*(7), 801–814.

Goncalves, M. A., de Vries, A. A., Holkers, M., van de Watering, M. J., van der Velde, I., van Nierop, G. P., et al. (2006). Human mesenchymal stem cells ectopically expressing full-length dystrophin can complement Duchenne muscular dystrophy myotubes by cell fusion. *Hum. Mol. Genet., 15*(2), 213–221.

Gotherstrom, C., Ringden, O., Tammik, C., Zetterberg, E., Westgren, M., & le Blanc, K. (2004). Immunologic properties of human fetal mesenchymal stem cells. *Am. J. Obstet. Gynecol., 190*(1), 239–245.

Greco, S. J., & Rameshwar, P. (2007). Enhancing effect of IL-1alpha on neurogenesis from adult human mesenchymal stem cells: implication for inflammatory mediators in regenerative medicine. *J. Immunol., 179*(5), 3342–3350.

Gronthos, S., & Simmons, P. J. (1995). The growth factor requirements of STRO-1-positive human bone marrow stromal precursors under serum-deprived conditions in vitro. *Blood, 85*(4), 929–940.

Gronthos, S., Zannettino, A. C., Hay, S. J., Shi, S., Graves, S. E., Kortesidis, A., et al. (2003). Molecular and cellular characterisation of highly purified stromal stem cells derived from human bone marrow. *J. Cell Sci., 116*(Pt 9), 1827–1835.

Gu, W., Zhang, F., Xue, Q., Ma, Z., Lu, P., & Yu, B. (2009). Transplantation of bone marrow mesenchymal stem cells reduces lesion volume and induces axonal regrowth of injured spinal cord. *Neuropathology, 30*(3), 205–217.

Gulotta, L. V., Kovacevic, D., Ehteshami, J. R., Dagher, E., Packer, J. D., & Rodeo, S. A. (2009). Application of bone marrow-derived mesenchymal stem cells in a rotator cuff repair model. *Am. J. Sports Med., 37*(11), 2126–2133.

Gysin, R., Wergedal, J. E., Sheng, M. H., Kasukawa, Y., Miyakoshi, N., Chen, S. T., et al. (2002). Ex vivo gene therapy with stromal cells transduced with a retroviral vector containing the BMP4 gene completely heals critical size calvarial defect in rats. *Gene Ther., 9*(15), 991–999.

Harris, M. T., Butler, D. L., Boivin, G. P., Florer, J. B., Schantz, E. J., & Wenstrup, R. J. (2004). Mesenchymal stem cells used for rabbit tendon repair can form ectopic bone and express alkaline phosphatase activity in constructs. *J. Orthop. Res., 22*(5), 998–1003.

Hiyama, A., Mochida, J., Iwashina, T., Omi, H., Watanabe, T., Serigano, K., et al. (2008). Transplantation of mesenchymal stem cells in a canine disc degeneration model. *J. Orthop. Res., 26*(5), 589–600.

Hoffmann, A., Czichos, S., Kaps, C., Bachner, D., Mayer, H., Kurkalli, B. G., et al. (2002). The T-box transcription factor Brachyury mediates cartilage development in mesenchymal stem cell line C3H10T1/2. *J. Cell Sci., 115* (Pt 4), 769–781.

Hoffmann, A., Pelled, G., Turgeman, G., Eberle, P., Zilberman, Y., Shinar, H., et al. (2006). Neotendon formation induced by manipulation of the Smad8 signalling pathway in mesenchymal stem cells. *J. Clin. Invest., 116*(4), 940–952.

Hofstetter, C. P., Schwarz, E. J., Hess, D., Widenfalk, J., el Manira, A., Prockop, D. J., et al. (2002). Marrow stromal cells form guiding strands in the injured spinal cord and promote recovery. *Proc. Natl. Acad. Sci. U.S.A., 99*(4), 2199–2204.

Horwitz, E. M., Gordon, P. L., Koo, W. K. K., Marx, J. C., Neel, M. D., McNall, R. Y., et al. (2002). Isolated allogeneic bone marrow-derived mesenchymal cells engraft and stimulate growth in children with osteogenesis imperfecta: implications for cell therapy of bone. *Proc. Natl. Acad. Sci. U.S.A., 99*(13), 8932–8937.

Huang, G. T., Gronthos, S., & Shi, S. (2009). Mesenchymal stem cells derived from dental tissues vs. those from other sources: their biology and role in regenerative medicine. *J. Dent. Res., 88*(9), 792–806.

Hutson, E. L., Boyer, S., & Genever, P. G. (2005). Rapid isolation, expansion, and differentiation of osteoprogenitors from full-term umbilical cord blood. *Tissue Eng., 11*(9–10), 1407–1420.

Im, G. I., Kim, D. Y., Shin, J. H., Hyun, C. W., & Cho, W. H. (2001). Repair of cartilage defect in the rabbit with cultured mesenchymal stem cells from bone marrow. *J. Bone Joint Surg. Br., 83*(2), 289–294.

Imanishi, Y., Miyagawa, S., Kitagawa-Sakakida, S., Taketani, S., Sekiya, N., & Sawa, Y. (2009). Impact of synovial membrane-derived stem cell transplantation in a rat model of myocardial infarction. *J. Artif. Organs, 12*(3), 187–193.

Itescu, S., Schuster, M. D., & Kocher, A. A. (2003). New directions in strategies using cell therapy for heart disease. *J. Mol. Med., 81*(5), 288–296.

Jain, M., Pfister, O., Hajjar, R. J., & Liao, R. (2005). Mesenchymal stem cells in the infarcted heart. *Coron. Artery Dis., 16*(2), 93–97.

Javazon, E. H., Beggs, K. J., & Flake, A. W. (2004). Mesenchymal stem cells: paradoxes of passaging. *Exp. Hematol., 32* (5), 414–425.

Joggerst, S. J., & Hatzopoulos, A. K. (2009). Stem cell therapy for cardiac repair: benefits and barriers. *Expert Rev. Mol. Med., 11*, e20.

Johnson, M. R., Lee, H. J., Bellamkonda, R. V., & Guldberg, R. E. (2009a). Sustained release of BMP-2 in a lipid-based microtube vehicle. *Acta Biomater., 5*(1), 23–28.

Johnson, T. V., Bull, N. D., Hunt, D. P., Marina, N., Tomarev, S. I., & Martin, K. R. (2009b). Local mesenchymal stem cell transplantation confers neuroprotection in experimental glaucoma. *Invest. Ophthalmol. Vis. Sci, 51*(4), 2051–2059.

Jones, E., & McGonagle, D. (2008). Human bone marrow mesenchymal stem cells in vivo. *Rheumatology (Oxford), 47*(2), 126–131.

Juncosa-Melvin, N., Boivin, G. P., Galloway, M. T., Gooch, C., West, J. R., & Butler, D. L. (2006). Effects of cell-to-collagen ratio in stem cell-seeded constructs for Achilles tendon repair. *Tissue Eng., 12*(4), 681–689.

Kang, X. Q., Zang, W. J., Song, T. S., Xu, X. L., Yu, X. J., Li, D. L., et al. (2005). Rat bone marrow mesenchymal stem cells differentiate into hepatocytes in vitro. *World J. Gastroenterol., 11*(22), 3479–3484.

Kastrinaki, M. C., Andreakou, I., Charbord, P., & Papadaki, H. A. (2008). Isolation of human bone marrow mesenchymal stem cells using different membrane markers: comparison of colony/cloning efficiency, differentiation potential, and molecular profile. *Tissue Eng. Part C Methods, 14*(4), 333–339.

Katz, A. J., Tholpady, A., Tholpady, S. S., Shang, H., & Ogle, R. C. (2005). Cell surface and transcriptional characterization of human adipose-derived adherent stromal (hADAS) cells. *Stem Cells, 23*(3), 412–423.

Kawamura, K., Chu, C. R., Sobajima, S., Robbins, P. D., Fu, F. H., Izzo, N. J., et al. (2005). Adenoviral-mediated transfer of TGF-beta1 but not IGF-1 induces chondrogenic differentiation of human mesenchymal stem cells in pellet cultures. *Exp. Hematol., 33*(8), 865–872.

Kawate, K., Yajima, H., Ohgushi, H., Kotobuki, N., Sugimoto, K., Ohmura, T., et al. (2006). Tissue-engineered approach for the treatment of steroid-induced osteonecrosis of the femoral head: transplantation of autologous mesenchymal stem cells cultured with beta-tricalcium phosphate ceramics and free vascularized fibula. *Artif. Organs, 30*(12), 960–962.

Kern, S., Eichler, H., Stoeve, J., Kluter, H., & Bieback, K. (2006). Comparative analysis of mesenchymal stem cells from bone marrow, umbilical cord blood, or adipose tissue. *Stem Cells, 24*(5), 1294–1301.

Kim, H. J., Lee, J. H., & Kim, S. H. (2010). Therapeutic effects of human mesenchymal stem cells on traumatic brain injury in rats: secretion of neurotrophic factors and inhibition of apoptosis. *J. Neurotrauma, 27*(1), 131–138.

Kim, J., Kim, I. S., Cho, T. H., Lee, K. B., Hwang, S. J., Tae, G., et al. (2007). Bone regeneration using hyaluronic acid-based hydrogel with bone morphogenic protein-2 and human mesenchymal stem cells. *Biomaterials, 28*(10), 1830–1837.

Kim, S. Y., Lee, S. H., Kim, B. M., Kim, E. H., Min, B. H., Bendayan, M., et al. (2004). Activation of nestin-positive duct stem (NPDS) cells in pancreas upon neogenic motivation and possible cytodifferentiation into insulin-secreting cells from NPDS cells. *Dev. Dyn., 230*(1), 1–11.

Kitaori, T., Ito, H., Schwarz, E. A., Tsutsumi, R., Yoshitomi, H., Oishi, S., et al. (2009). Stromal cell-derived factor 1/CXCR4 signaling is critical for the recruitment of mesenchymal stem cells to the fracture site during skeletal repair in a mouse model. *Arthritis and Rheumatism, 60*(3), 813–823.

Kon, E., Muraglia, A., Corsi, A., Bianco, P., Marcacci, M., Martin, I., et al. (2000). Autologous bone marrow stromal cells loaded onto porous hydroxyapatite ceramic accelerate bone repair in critical-size defects of sheep long bones. *J. Biomed. Mater. Res., 49*(3), 328–337.

Krause, K., Jaquet, K., Schneider, C., Haupt, S., Lioznov, M. V., Otte, K. M., et al. (2009). Percutaneous intramyocardial stem cell injection in patients with acute myocardial infarction: first-in-man study. *Heart, 95*(14), 1145–1152.

Kubo, H., Shimizu, M., Taya, Y., Kawamoto, T., Michida, M., Kaneko, E., et al. (2009). Identification of mesenchymal stem cell (MSC)-transcription factors by microarray and knockdown analyses, and signature molecule-marked MSC in bone marrow by immunohistochemistry. *Genes Cells, 14*(3), 407–424.

Kumar, S., & Ponnazhagan, S. (2007). Bone homing of mesenchymal stem cells by ectopic alpha 4 integrin expression. *Faseb J., 21*(14), 3917–3927.

Kumar, S., Wan, C., Ramaswamy, G., Clemens, T., & Ponnazhagan, S. (2010). Mesenchymal stem cells expressing osteogenic and angiogenic factors synergistically enhance bone formation in a mouse model of segmental bone defect. *Mol. Ther.* In press.

Lange, C., Bassler, P., Lioznov, M. V., Bruns, H., Kluth, D., Zander, A. R., et al. (2005). Liver-specific gene expression in mesenchymal stem cells is induced by liver cells. *World J. Gastroenterol., 11*(29), 4497–4504.

Lardon, J., Rooman, I., & Bouwens, L. (2002). Nestin expression in pancreatic stellate cells and angiogenic endothelial cells. *Histochem. Cell. Biol., 117*(6), 535–540.

Le Maitre, C. L., Baird, P., Freemont, A. J., & Hoyland, J. A. (2009). An in vitro study investigating the survival and phenotype of mesenchymal stem cells following injection into nucleus pulposus tissue. *Arthritis Res. Ther., 11*(1), R20.

Lee, C. C., Christensen, J. E., Yoder, M. C., & Tarantal, A. F. (2010a). Clonal analysis and hierarchy of human bone marrow mesenchymal stem and progenitor cells. *Exp. Hematol., 38*(1), 46–54.

Lee, J. A., Kim, B. I., Jo, C. H., Choi, C. W., Kim, E. K., Kim, H. S., et al. (2010b). Mesenchymal stem-cell transplantation for hypoxic-ischemic brain injury in neonatal rat model. *Pediatr. Res., 67*(1), 42–46.

Lee, J. W., Gupta, N., Serikov, V., & Matthay, M. A. (2009). Potential application of mesenchymal stem cells in acute lung injury. *Expert Opin. Biol. Ther., 9*(10), 1259–1270.

Lee, P. H., Kim, J. W., Bang, O. Y., Ahn, Y. H., Joo, I. S., & Huh, K. (2008). Autologous mesenchymal stem cell therapy delays the progression of neurological deficits in patients with multiple system atrophy. *Clin. Pharmacol. Ther., 83*(5), 723–730.

Li, T., Ren, G., Kaplan, D. L., & Boison, D. (2009a). Human mesenchymal stem cell grafts engineered to release adenosine reduce chronic seizures in a mouse model of CA3-selective epileptogenesis. *Epilepsy Res., 84*(2–3), 238–241.

Li, W. J., Chiang, H., Kuo, T. F., Lee, H. S., Jiang, C. C., & Tuan, R. S. (2009b). Evaluation of articular cartilage repair using biodegradable nanofibrous scaffolds in a swine model: a pilot study. *J. Tissue Eng. Regen. Med., 3*(1), 1–10.

Liechty, K. W., MacKenzie, T. C., Shaaban, A. F., Radu, A., Moseley, A. M., Deans, R., et al. (2000). Human mesenchymal stem cells engraft and demonstrate site-specific differentiation after in utero transplantation in sheep. *Nat. Med., 6*(11), 1282–1286.

Lin, P., Chen, L., Yang, N., Sun, Y., & Xu, Y. X. (2009). Evaluation of stem cell differentiation in diabetic rats transplanted with bone marrow mesenchymal stem cells. *Transplant Proc., 41*(5), 1891–1893.

Liu, Y., Chen, F., Liu, W., Cui, L., Shang, Q., Xia, W., et al. (2002). Repairing large porcine full-thickness defects of articular cartilage using autologous chondrocyte-engineered cartilage. *Tissue Eng., 8*(4), 709–721.

Lu, D., Mahmood, A., Wang, L., Li, Y., Lu, M., & Chopp, M. (2001). Adult bone marrow stromal cells administered intravenously to rats after traumatic brain injury migrate into brain and improve neurological outcome. *Neuroreport, 12*(3), 559–563.

Madec, A. M., Mallone, R., Afonso, G., Abou Mrad, E., Mesnier, Eljaafari, A., & Thivolet, C. (2009). Mesenchymal stem cells protect NOD mice from diabetes by inducing regulatory T cells. *Diabetologia, 52*(7), 1391–1399.

Madonna, R., Geng, Y. J., & de Caterina, R. (2009). Adipose tissue-derived stem cells: characterization and potential for cardiovascular repair. *Arterioscler. Thromb. Vasc. Biol., 29*(11), 1723–1729.

Mahmood, A., Lu, D., Lu, M., & Chopp, M. (2003). Treatment of traumatic brain injury in adult rats with intravenous administration of human bone marrow stromal cells. *Neurosurgery, 53*(3), 697–702, discussion 702–703.

Majumdar, M. K., Keane-Moore, M., Buyaner, D., Hardy, W. B., Moorman, M. A., McIntosh, K. R., et al. (2003). Characterization and functionality of cell surface molecules on human mesenchymal stem cells. *J. Biomed. Sci., 10*(2), 228–241.

Makino, S., Fukuda, K., Miyoshi, S., Konishi, F., Kodama, H., Pan, J., et al. (1999). Cardiomyocytes can be generated from marrow stromal cells in vitro. *J. Clin. Invest., 103*(5), 697–705.

Mangi, A. A., Noiseux, N., Kong, D., He, H., Rezvani, M., Ingwall, J. S., et al. (2003). Mesenchymal stem cells modified with Akt prevent remodeling and restore performance of infarcted hearts. *Nat. Med., 9*(9), 1195–1201.

Marshall, C. J., Moore, R. L., Thorogood, P., Brickell, P. M., Kinnon, C., & Thrasher, A. J. (1999). Detailed characterization of the human aorta-gonad-mesonephros region reveals morphological polarity resembling a hematopoietic stromal layer. *Dev. Dyn., 215*(2), 139–147.

Martens, T. P., Godier, A. F., Parks, J. J., Wan, L. Q., Koeckert, M. S., Eng, G. M., et al. (2009). Percutaneous cell delivery into the heart using hydrogels polymerizing in situ. *Cell Transplant., 18*(3), 297–304.

Mastrogiacomo, M., Muraglia, A., Komlev, V., Peyrin, F., Rustichelli, F., Crovace, A., et al. (2005). Tissue engineering of bone: search for a better scaffold. *Orthod. Craniofac. Res., 8*(4), 277–284.

Minguell, J. J., & Erices, A. (2006). Mesenchymal stem cells and the treatment of cardiac disease. *Exp. Biol. Med. (Maywood), 231*(1), 39–49.

Mizuno, H., & Hyakusoku, H. (2003). Mesengenic potential and future clinical perspective of human processed lipoaspirate cells. *J. Nippon Med. Sch., 70*(4), 300–306.

Mohamadnejad, M., Alimoghaddam, K., Mohyeddin-Bonab, M., Bagheri, M., Bashtar, M., Ghanaati, H., et al. (2007). Phase 1 trial of autologous bone marrow mesenchymal stem cell transplantation in patients with decompensated liver cirrhosis. *Arch. Iran. Med., 10*(4), 459–466.

Moriscot, C., de Fraipont, F., Richard, M. J., Marchand, M., Savatier, P., Bosco, D., et al. (2005). Human bone marrow mesenchymal stem cells can express insulin and key transcription factors of the endocrine pancreas developmental pathway upon genetic and/or microenvironmental manipulation in vitro. *Stem Cells, 23*(4), 594–603.

Moutsatsos, I. K., Turgeman, G., Zhou, S., Kurkalli, B. G., Pelled, G., Tzur, L., et al. (2001). Exogenously regulated stem cell-mediated gene therapy for bone regeneration. *Mol. Ther., 3*(4), 449–461.

Na, K., Kim, S. W., Sun, B. K., Woo, D. G., Yang, H. N., Chung, H. M., et al. (2007). Osteogenic differentiation of rabbit mesenchymal stem cells in thermo-reversible hydrogel constructs containing hydroxyapatite and bone morphogenic protein-2 (BMP-2). *Biomaterials, 28*(16), 2631–2637.

Nagaya, N., Kangawa, K., Itoh, T., Iwase, T., Murakami, S., Miyahara, Y., et al. (2005). Transplantation of mesenchymal stem cells improves cardiac function in a rat model of dilated cardiomyopathy. *Circulation, 112*(8), 1128–1135.

Nesti, L. J., Li, W. J., Shanti, R. M., Jiang, Y. J., Jackson, W., Freedman, B. A., et al. (2008). Intervertebral disc tissue engineering using a novel hyaluronic acid-nanofibrous scaffold (HANFS) amalgam. *Tissue Eng. Part A, 14*(9), 1527–1537.

Niemeyer, P., Kornacker, M., Mehlhorn, A., Seckinger, A., Vohrer, J., Schmal, H., et al. (2007). Comparison of immunological properties of bone marrow stromal cells and adipose tissue-derived stem cells before and after osteogenic differentiation in vitro. *Tissue Eng., 13*(1), 111–121.

Noel, D., Djouad, F., Bouffi, C., Mrugala, D., & Jorgensen, C. (2007). Multipotent mesenchymal stromal cells and immune tolerance. *Leuk. Lymphoma, 48*(7), 1283–1289.

Okura, H., Matsuyama, A., Lee, C. M., Saga, A., Kakuta-Yamamoto, A., Nagao, A., et al. (2009). Cardiomyoblast-like cells differentiated from human adipose tissue-derived mesenchymal stem cells improve left ventricular dysfunction and survival in a rat myocardial infarction model. *Tissue Eng. Part C Methods, 16*(3), 417–425.

Okuyama, R., Yanai, N., & Obinata, M. (1995). Differentiation capacity toward mesenchymal cell lineages of bone marrow stromal cells established from temperature-sensitive SV40 T-antigen gene transgenic mouse. *Exp. Cell Res., 218*(2), 424–429.

Omlor, G. W., Bertram, H., Kleinschmidt, K., Fischer, J., Brohm, K., Guehring, T., et al. (2009). Methods to monitor distribution and metabolic activity of mesenchymal stem cells following in vivo injection into nucleotomized porcine intervertebral discs. *Eur. Spine J., 19*(4), 601–612.

Orlic, D. (2003). Adult bone marrow stem cells regenerate myocardium in ischemic heart disease. *Ann. N.Y. Acad. Sci., 996*, 152–157.

Oyagi, S., Hirose, M., Kojima, M., Okuyama, M., Kawase, M., Nakamura, T., et al. (2006). Therapeutic effect of transplanting HGF-treated bone marrow mesenchymal cells into CCl4-injured rats. *J. Hepatol., 44*(4), 742–748.

Palmer, G. D., Steinert, A., Pascher, A., Gouze, E., Gouze, J. N., Betz, O., et al. (2005). Gene-induced chondrogenesis of primary mesenchymal stem cells in vitro. *Mol. Ther., 12*(2), 219–228.

Parolini, O., Alviano, F., Bagnara, G. P., et al. (2008). Concise review: isolation and characterization of cells from human term placenta: outcome of the first international Workshop on Placenta Derived Stem Cells. *Stem Cells, 26*(2), 300–311.

Pelled, G., Tai, K., Sheyn, D., Zilberman, Y., Kumbar, S., Nair, L. S., et al. (2007). Structural and nanoindentation studies of stem cell-based tissue-engineered bone. *J. Biomech., 40*(2), 399–411.

Peng, H., Wright, V., Usas, A., Gearhart, B., Shen, H. C., Cummins, J., et al. (2002). Synergistic enhancement of bone formation and healing by stem cell-expressed VEGF and bone morphogenetic protein-4. *J. Clin Invest., 110*(6), 751–759.

Phinney, D. G. (2008). Isolation of mesenchymal stem cells from murine bone marrow by immunodepletion. *Methods Mol. Biol., 449*, 171–186.

Pimentel-Coelho, P. M., & Mendez-Otero, R. (2009). Cell therapy for neonatal hypoxic-ischemic encephalopathy. *Stem Cells Dev., 19*(3), 299–310.

Pittenger, M. F., Mackay, A. M., Beck, S. C., Jaiswal, R. K., Douglas, R., Mosca, J. D., et al. (1999). Multilineage potential of adult human mesenchymal stem cells. *Science, 284*(5411), 143–147.

Pons, J., Huang, Y., Takagawa, J., Arakawa-Hoyt, J., Ye, J., Grossman, W., et al. (2009). Combining angiogenic gene and stem cell therapies for myocardial infarction. *J. Gene Med., 11*(9), 743–753.

Popp, F. C., Renner, P., Eggenhofer, E., Slowik, P., Geissler, E. K., Piso, P., et al. (2009). Mesenchymal stem cells as immunomodulators after liver transplantation. *Liver Transpl., 15*(10), 1192–1198.

Potapova, I. A., Doronin, S. V., Kelly, D. J., Rosen, A. B., Schuldt, A. J., Lu, Z., et al. (2008). Enhanced recovery of mechanical function in the canine heart by seeding an extracellular matrix patch with mesenchymal stem cells committed to a cardiac lineage. *Am. J. Physiol. Heart Circ. Physiol., 295*(6), H2257–H2263.

Pricola, K. L., Kuhn, N. Z., Haleem-Smith, H., Song, Y., & Tuan, R. S. (2009). Interleukin-6 maintains bone marrow-derived mesenchymal stem cell stemness by an ERK1/2-dependent mechanism. *J. Cell. Biochem., 108*(3), 577–588.

Prockop, D. J. (2007). "Stemness" does not explain the repair of many tissues by mesenchymal stem/multipotent stromal cells (MSCs). *Clin. Pharmacol. Ther., 82*(3), 241–243.

Prockop, D. J. (2009). Repair of tissues by adult stem/progenitor cells (MSCs), controversies, myths, and changing paradigms. *Mol. Ther., 17*(6), 939–946.

Pulavendran, S., Vignesh, J., & Rose, C. (2010). Differential anti-inflammatory and anti-fibrotic activity of transplanted mesenchymal vs. hematopoietic stem cells in carbon tetrachloride-induced liver injury in mice. *Int Immunopharmacol., 10*(4), 513–519.

Qi, C. M., Ma, G. S., Liu, N. F., Shen, C. X., Chen, Z., Liu, X. J., et al. (2009). Identification and differentiation of magnetically labeled mesenchymal stem cells in vivo in swines with myocardial infarction. *Int. J. Cardiol., 131*(3), 417–419.

Quarto, R., Mastrogiacomo, M., Cancedda, R., Kutepov, S. M., Mukhachev, V., Lavroukov, A., et al. (2001). Repair of large bone defects with the use of autologous bone marrow stromal cells. *N. Engl. J. Med., 344*(5), 385−386.

Rasmusson, I., Ringden, O., Sundberg, B., & le Blanc, K. (2005). Mesenchymal stem cells inhibit lymphocyte proliferation by mitogens and alloantigens by different mechanisms. *Exp. Cell Res., 305*(1), 33−41.

Reinders, M. E., Fibbe, W. E., & Rabelink, T. J. (2010). Multipotent mesenchymal stromal cell therapy in renal disease and kidney transplantation. *Nephrol. Dial. Transplant., 25*(1), 17−24.

Ren, G., Su, J., Zhang, L., Zhao, X., Ling, W., L'Huillie, A., et al. (2009). Species variation in the mechanisms of mesenchymal stem cell-mediated immunosuppression. *Stem Cells, 27*(8), 1954−1962.

Richardson, S. M., Hughes, N., Hunt, J. A., Freemont, A. J., & Hoyland, J. A. (2008). Human mesenchymal stem cell differentiation to NP-like cells in chitosan-glycerophosphate hydrogels. *Biomaterials, 29*(1), 85−93.

Risbud, M. V., Guttapalli, A., Tsai, T. T., Lee, J. Y., Danielson, K. G., Vaccaro, A. R., et al. (2007a). Evidence for skeletal progenitor cells in the degenerate human intervertebral disc. *Spine (Phila. Pa. 1976), 32*(23), 2537−2544.

Risbud, M. V., Guttapalli, A., Tsai, T. T., Lee, J. Y., Danielson, K. G., Vaccaro, A. R., et al. (2007b). Evidence for skeletal progenitor cells in the degenerate human intervertebral disc. *Spine, 32*(23), 2537−2544.

Risbud, M.V., Schipani, E. & Shapiro, I.M. Hypoxic regulation of nucleus pulposus cell survival. From Niche to Notch. *Am. J. Pathol., 176*(4), 1577−1578.

Roda, B., Lanzoni, G., Alviano, F., Zattoni, A., Costa, R., di Carlo, A., et al. (2009a). A novel stem cell tag-less sorting method. *Stem Cell Rev., 5*(4), 420−427.

Roda, B., Reschiglian, P., Zattoni, A., Alviano, F., Lanzoni, G., Costa, R., et al. (2009b). A tag-less method of sorting stem cells from clinical specimens and separating mesenchymal from epithelial progenitor cells. *Cytometry B Clin. Cytom., 76B*(4), 285−290.

Rosen, A. B., Kelly, D. J., Schuldt, A. J., Lu, J., Potapova, I. A., Doronin, S. V., et al. (2007). Finding fluorescent needles in the cardiac haystack: tracking human mesenchymal stem cells labeled with quantum dots for quantitative in vivo three-dimensional fluorescence analysis. *Stem Cells, 25*(8), 2128−2138.

Roughley, P. J. (2004). Biology of intervertebral disc aging and degeneration: involvement of the extracellular matrix. *Spine (Phila. Pa. 1976), 29*(23), 2691−2699.

Sakai, D., Mochida, J., Iwashina, T., Hiyama, A., Omi, H., Imai, M., et al. (2006). Regenerative effects of transplanting mesenchymal stem cells embedded in atelocollagen to the degenerated intervertebral disc. *Biomaterials, 27*(3), 335−345.

Sakai, D., Mochida, J., Iwashina, T., Watanabe, T., Nakai, T., Ando, K., et al. (2005). Differentiation of mesenchymal stem cells transplanted to a rabbit degenerative disc model: potential and limitations for stem cell therapy in disc regeneration. *Spine (Phila. Pa. 1976), 30*(21), 2379−2387.

Sasaki, M., Radtke, C., Tan, A. M., Zhao, P., Hamada, H., Houkin, K., et al. (2009). BDNF-hypersecreting human mesenchymal stem cells promote functional recovery, axonal sprouting, and protection of corticospinal neurons after spinal cord injury. *J. Neurosci., 29*(47), 14932−14941.

Sato, Y., Araki, H., Kato, J., Nakamura, K., Kawano, Y., Kobune, M., et al. (2005). Human mesenchymal stem cells xenografted directly to rat liver are differentiated into human hepatocytes without fusion. *Blood, 106*(2), 756−763.

Saulnier, N., Lattanzi, W., Puglisi, M. A., Pani, G., Barba, M., Piscaglia, A. C., et al. (2009). Mesenchymal stromal cells multipotency and plasticity: induction toward the hepatic lineage. *Eur. Rev. Med. Pharmacol. Sci., 13*(Suppl. 1), 71−78.

Schwartz, R. E., Reyes, M., Koodie, L., Jiang, Y., Blackstad, M., Lund, T., et al. (2002). Multipotent adult progenitor cells from bone marrow differentiate into functional hepatocyte-like cells. *J. Clin. Invest., 109*(10), 1291−1302.

Shayesteh, Y. S., Khojasteh, A., Soleimani, M., Alikhasi, M., Khoshzaban, A., et al. (2008). Sinus augmentation using human mesenchymal stem cells loaded into a beta-tricalcium phosphate/hydroxyapatite scaffold. *Oral Surg. Oral Med. Oral Pathol. Oral Radiol. Endod., 106*(2), 203−209.

Sheyn, D., Pelled, G., Zilberman, Y., Talasazan, F., Frank, J. M., Gazit, D., et al. (2008). Nonvirally engineered porcine adipose tissue-derived stem cells: use in posterior spinal fusion. *Stem Cells, 26*(4), 1056−1064.

Shibata, T., Naruse, K., Kamiya, H., Kozakae, M., Kondo, M., Yasuda, Y., et al. (2008). Transplantation of bone marrow-derived mesenchymal stem cells improves diabetic polyneuropathy in rats. *Diabetes, 57*(11), 3099−3107.

Shirley, D., Marsh, D., Jordan, G., McQuaid, S., & Li, G. (2005). Systemic recruitment of osteoblastic cells in fracture healing. *J. Orth. Res., 23*(5), 1013−1021.

Siepe, M., Heilmann, C., von Samson, P., Menasche, P., & Beyersdorf, F. (2005). Stem cell research and cell transplantation for myocardial regeneration. *Eur. J. Cardiothorac. Surg., 28*(2), 318−324.

Siniscalco, D., Giordano, C., Galderisi, U., Luongo, L., Alessio, N., di Bernardo, G., et al. (2010). Intra-brain microinjection of human mesenchymal stem cells decreases allodynia in neuropathic mice. *Cell. Mol. Life Sci., 67*(4), 655−669.

Smits, A. M., van Vliet, P., Hassink, R. J., Goumans, M. J., & Doevendans, P. A. (2005). The role of stem cells in cardiac regeneration. *J. Cell. Mol. Med.*, *9*(1), 25–36.

Solari, M. G., Srinivasan, S., Boumaza, I., Unadkat, J., Harb, G., Garcia-Ocana, A., et al. (2009). Marginal mass islet transplantation with autologous mesenchymal stem cells promotes long-term islet allograft survival and sustained normoglycemia. *J. Autoimmun.*, *32*(2), 116–124.

Song, H., Cha, M. J., Song, B. W., Kim, I. K., Chang, W., Lim, S., et al. (2010). Reactive oxygen species inhibit adhesion of mesenchymal stem cells implanted into ischemic myocardium via interference of focal adhesion complex. *Stem Cells.*, *28*(3), 555–563.

Sonoyama, W., Coppe, C., Gronthos, S., & Shi, S. (2005). Skeletal stem cells in regenerative medicine. *Curr. Top Dev. Biol.*, *67*, 305–323.

Sordi, V., Melzi, R., Mercalli, A., Formicola, R., Doglioni, C., Tiboni, F., et al. (2010). Mesenchymal cells appearing in pancreatic tissue culture are bone marrow-derived stem cells with the capacity to improve transplanted islet function. *Stem Cells*, *28*(1), 140–151.

Sorrentino, A., Ferracin, M., Castelli, G., Biffoni, M., Tomaselli, G., Baiocchi, M., et al. (2008). Isolation and characterization of CD146+ multipotent mesenchymal stromal cells. *Exp. Hematol.*, *36*(8), 1035–1046.

Sotiropoulou, P. A., Perez, S. A., Gritzapis, A. D., Baxevanis, C. N., & Papamichail, M. (2006). Interactions between human mesenchymal stem cells and natural killer cells. *Stem Cells*, *24*(1), 74–85.

Spees, J. L., Olson, S. D., Ylostalo, J., Lynch, P. J., Smith, J., Perry, A., et al. (2003). Differentiation, cell fusion, and nuclear fusion during ex vivo repair of epithelium by human adult stem cells from bone marrow stroma. *Proc. Natl. Acad. Sci. U.S.A.*, *100*(5), 2397–2402.

Steinert, A. F., Palmer, G. D., Pilapil, C., Noth, U., Evans, C. H., & Ghivizzani, S. C. (2009). Enhanced in vitro chondrogenesis of primary mesenchymal stem cells by combined gene transfer. *Tissue Eng. Part A*, *15*(5), 1127–1139.

Steinhardt, Y., Aslan, H., Regev, E., Zilberman, Y., Kallai, I., Gazit, D., et al. (2008). Maxillofacial-derived stem cells regenerate critical mandibular bone defect. *Tissue Eng. Part A*, *14*(11), 1763–1773.

Tai, K., Pelled, G., Sheyn, D., Bershteyn, A., Han, L., Kallai, I., et al. (2008). Nanobiomechanics of repair bone regenerated by genetically modified mesenchymal stem cells. *Tissue Eng. Part A*, *14*(10), 1709–1720.

Tatebe, M., Nakamura, R., Kagami, H., Okada, K., & Ueda, M. (2005). Differentiation of transplanted mesenchymal stem cells in a large osteochondral defect in rabbit. *Cytotherapy*, *7*(6), 520–530.

Tfilin, M., Sudai, E., Merenlender, A., Gispan, I., Yadid, G., & Turgeman, G. (2009). Mesenchymal stem cells increase hippocampal neurogenesis and counteract depressive-like behavior. *Mol. Psychiatry*. Epub ahead of print.

Toma, C., Pittenger, M. F., Cahill, K. S., Byrne, B. J., & Kessler, P. D. (2002). Human mesenchymal stem cells differentiate to a cardiomyocyte phenotype in the adult murine heart. *Circulation*, *105*(1), 93–98.

Trivedi, P. S., Tray, N. J., Nguyen, T. D., Nigam, N., & Gallicano, G. I. (2010). Mesenchymal stem cell therapy for treatment of cardiovascular disease: helping people sooner or later. *Stem Cells Dev.*, *19*(7), 1109–1120.

Tse, W. T., Pendleton, J. D., Beyer, W. M., Egalka, M. C., & Guinan, E. C. (2003). Suppression of allogeneic T-cell proliferation by human marrow stromal cells: implications in transplantation. *Transplantation*, *75*(3), 389–397.

Turgeman, G., Pittman, D. D., Muller, R., Kurkalli, B. G., Zhou, S., Pelled, G., et al. (2001). Engineered human mesenchymal stem cells: a novel platform for skeletal cell mediated gene therapy. *J. Gene Med.*, *3*(3), 240–251.

van Velthoven, C. T., Kavelaars, A., van Bel, F., & Heijnen, C. J. (2009). Mesenchymal stem cell treatment after neonatal hypoxic-ischemic brain injury improves behavioral outcome and induces neuronal and oligodendrocyte regeneration. *Brain Behav. Immun.*, *24*(3), 387–389.

Volarevic, V., al-Qahtani, A., Arsenijevic, N., Pajovic, S., & Lukic, M. L. (2009). Interleukin-1 receptor antagonist (IL-1Ra) and IL-1Ra producing mesenchymal stem cells as modulators of diabetogenesis. *Autoimmunity*, *43*(4), 255–263.

Wakitani, S., Imoto, K., Yamamoto, T., Saito, M., Murata, N., & Yoneda, M. (2002). Human autologous culture expanded bone marrow mesenchymal cell transplantation for repair of cartilage defects in osteoarthritic knees. *Osteoarthritis Cartilage*, *10*(3), 199–206.

Wakitani, S., Saito, T., & Caplan, A. I. (1995). Myogenic cells derived from rat bone marrow mesenchymal stem cells exposed to 5-azacytidine. *Muscle Nerve*, *18*(12), 1417–1426.

Wakitani, S., & Yamamoto, T. (2002). Response of the donor and recipient cells in mesenchymal cell transplantation to cartilage defect. *Microsc. Res. Tech.*, *58*(1), 14–18.

Walker, P. A., Harting, M. T., Jimenez, F., Shah, S. K., Pati, S., Dash, P. K., et al. (2009). Direct intrathecal implantation of mesenchymal stromal cells leads to enhanced neuroprotection via an NFkappaB mediated increase in Interleukin 6 (IL-6) production. *Stem Cells Dev.*, *19*(6), 867–876.

Wang, J., Ding, F., Gu, Y., Liu, J., & Gu, X. (2009a). Bone marrow mesenchymal stem cells promote cell proliferation and neurotrophic function of Schwann cells in vitro and in vivo. *Brain Res.*, *1262*, 7–15.

Wang, P. P., Wang, J. H., Yan, Z. P., Hu, M. Y., Lau, G. K., Fan, S. T., et al. (2004). Expression of hepatocyte-like phenotypes in bone marrow stromal cells after HGF induction. *Biochem. Biophys. Res. Commun., 320*(3), 712–716.

Wang, Q. W., Chen, Z. L., & Piao, Y. J. (2005). Mesenchymal stem cells differentiate into tenocytes by bone morphogenetic protein (BMP) 12 gene transfer. *J. Biosci. Bioeng., 100*(4), 418–422.

Wang, Y., Zhang, A., Ye, Z., Xie, H., & Zheng, S. (2009b). Bone marrow-derived mesenchymal stem cells inhibit acute rejection of rat liver allografts in association with regulatory T-cell expansion. *Transplant Proc., 41*(10), 4352–4356.

Wei, A., Chung, S. A., Tao, H., Brisby, H., Lin, Z., Shen, B., et al. (2009). Differentiation of rodent bone marrow mesenchymal stem cells into intervertebral disc-like cells following coculture with rat disc tissue. *Tissue Eng. Part A, 15*(9), 2581–2595.

Wilkins, A., Kemp, K., Ginty, M., Hares, K., Mallam, E., & Scolding, N. (2009). Human bone marrow-derived mesenchymal stem cells secrete brain-derived neurotrophic factor which promotes neuronal survival in vitro. *Stem Cell Res.* Epub ahead of print.

Wojakowski, W., & Tendera, M. (2005). Mobilization of bone marrow-derived progenitor cells in acute coronary syndromes. *Folia Histochem. Cytobiol., 43*(4), 229–232.

Wolf, D., Reinhard, A., Seckinger, A., Gross, L., Katus, H. A., & Hansen, A. (2009). Regenerative capacity of intravenous autologous, allogeneic and human mesenchymal stem cells in the infarcted pig myocardium — complicated by myocardial tumor formation. *Scand. Cardiovasc. J., 43*(1), 39–45.

Wright, V., Peng, H., Usas, A., Young, B., Gearhart, B., Cummins, J., et al. (2002). BMP4-expressing muscle-derived stem cells differentiate into osteogenic lineage and improve bone healing in immunocompetent mice. *Mol. Ther., 6*(2), 169–178.

Wuertz, K., Godburn, K., & Iatridis, J. C. (2009). MSC response to pH levels found in degenerating intervertebral discs. *Biochem. Biophys. Res. Commun., 379*(4), 824–829.

Xu, R. X., Chen, X., Chen, J. H., Han, Y., & Han, B. M. (2009a). Mesenchymal stem cells promote cardiomyocyte hypertrophy in vitro through hypoxia-induced paracrine mechanisms. *Clin. Exp. Pharmacol. Physiol., 36*(2), 176–180.

Xu, Y. X., Chen, L., Hou, W. K., Lin, P., Sun, L., Sun, Y., et al. (2009b). Mesenchymal stem cells treated with rat pancreatic extract secrete cytokines that improve the glycometabolism of diabetic rats. *Transplant Proc., 41*(5), 1878–1884.

Yamaguchi, Y., Kubo, T., Murakami, T., Takahashi, M., Hakamata, Y., Kobayashi, E., et al. (2005). Bone marrow cells differentiate into wound myofibroblasts and accelerate the healing of wounds with exposed bones when combined with an occlusive dressing. *Br. J. Dermatol., 152*(4), 616–622.

Yang, F., Leung, V. Y., Luk, K. D., Chan, D., & Cheung, K. M. (2009). Mesenchymal stem cells arrest intervertebral disc degeneration through chondrocytic differentiation and stimulation of endogenous cells. *Mol. Ther., 17*(11), 1959–1966.

Yoon, Y. S., Lee, N., & Scadova, H. (2005). Myocardial regeneration with bone-marrow-derived stem cells. *Biol. Cell, 97*(4), 253–263.

Yoshimura, K., Shigeura, T., Matsumoto, D., Sato, T., Takaki, Y., Aiba-Kojima, E., et al. (2006). Characterization of freshly isolated and cultured cells derived from the fatty and fluid portions of liposuction aspirates. *J. Cell Physiol., 208*(1), 64–76.

Yu, Y., Yao, A. H., Chen, N., Pu, L. Y., Fan, Y., Lv, L., et al. (2007). Mesenchymal stem cells over-expressing hepatocyte growth factor improve small-for-size liver grafts regeneration. *Mol. Ther., 15*(7), 1582–1589.

Yue, W. M., Liu, W., Bi, Y. W., He, X. P., Sun, W. Y., Pang, X. Y., et al. (2008). Mesenchymal stem cells differentiate into an endothelial phenotype, reduce neointimal formation, and enhance endothelial function in a rat vein grafting model. *Stem Cells Dev., 17*(4), 785–793.

Zhang, W., Ge, W., Li, C., You, S., Liao, L., Han, Q., et al. (2004). Effects of mesenchymal stem cells on differentiation, maturation, and function of human monocyte-derived dendritic cells. *Stem Cells Dev., 13*(3), 263–271.

Zhang, Y. H., Wang, H. F., Liu, W., Wei, B., Bing, L. J., & Gao, Y. M. (2009). Insulin-producing cells derived from rat bone marrow and their autologous transplantation in the duodenal wall for treating diabetes. *Anat. Rec. (Hoboken), 292*(5), 728–735.

Zhao, L. R., Duan, W. M., Reyes, M., Keene, C. D., Verfaillie, C. M., & Low, W. C. (2002). Human bone marrow stem cells exhibit neural phenotypes and ameliorate neurological deficits after grafting into the ischemic brain of rats. *Exp. Neurol., 174*(1), 11–20.

Zhou, H., Tian, H. M., Long, Y., Zhang, X. X., Zhong, L., Deng, L., et al. (2009). Mesenchymal stem cells transplantation mildly ameliorates experimental diabetic nephropathy in rats. *Chin. Med. J. (Engl.), 122*(21), 2573–2579.

Zhu, S., Lu, Y., Zhu, J., Xu, J., Huang, H., Zhu, M., et al. (2009). Effects of intrahepatic bone-derived mesenchymal stem cells autotransplantation on the diabetic beagle dogs. *J. Surg. Res.* Epub ahead of print.

303

Zisa, D., Shabbir, A., Suzuki, G., & Lee, T. (2009). Vascular endothelial growth factor (VEGF) as a key therapeutic trophic factor in bone marrow mesenchymal stem cell-mediated cardiac repair. *Biochem. Biophys. Res. Commun.*, *390*(3), 834–838.

Zubko, R., & Frishman, W. (2009). Stem cell therapy for the kidney? *Am. J. Ther.*, *16*(3), 247–256.

Zuk, P. A., Zhu, M., Ashjian, P., de Ugarte, D. A., Huang, J. I., Mizuno, H., et al. (2002). Human adipose tissue is a source of multipotent stem cells. *Mol. Biol. Cell*, *13*(12), 4279–4295.

Zuk, P. A., Zhu, M., Mizuno, H., Huang, J., Futrell, J. W., Katz, A. J., et al. (2001). Multilineage cells from human adipose tissue: implications for cell-based therapies. *Tissue Eng.*, *7*(2), 211–228.

Cell Therapy of Liver Disease: From Hepatocytes to Stem Cells

Stephen C. Strom[*], **Ewa C.S. Ellis**[**]
* Department of Pathology, University of Pittsburgh, PA, USA
** Department of Clinical Science, Intervention and Technology, Division of Transplantation, Liver Cell Lab., Karolinska Institute, Stockholm, Sweden

INTRODUCTION

The concept of regenerative medicine implies that the clinician works with the innate healing and regenerative process of the body to effect an improvement in a patient's health. Perhaps more than with any other organ, the liver offers the greatest opportunity for regenerative medicine. This is because, unlike most other tissues, the liver has the capacity to regenerate following massive chemical or physical insult and tissue loss (Michalopoulos and deFrances, 1997). Our very existence may well rely on the ability to regenerate liver mass. The liver is an incredibly complex organ that performs quite diverse biological functions, from glycogen storage and catabolism to maintaining blood sugar levels, to the production and secretion of critical plasma proteins including albumin, clotting factors, and protease inhibitors. In addition, the liver is the major site in the body for the metabolism and excretion of hormones, metabolic waste products such as ammonia as well as exogenous compounds such as toxins, drugs, and a variety of other compounds to which we are exposed through diet and environment. These processes are so critical to survival that the loss of any of these functions has serious and often lethal consequences for the individual.

Until recently, the only option for treating chronic liver disease or metabolic defects in liver function has been whole organ transplantation. Recently, hepatocyte transplantation has been performed. Although still an experimental therapy, there are some potential advantages for a cell therapy approach to treat liver disease. Some of the advantages of and problems with the current treatments for liver disease are listed in Table 18.1.

Despite the unquestioned success of this technique, orthotopic liver transplantation (OLT) requires major surgery and has a significantly long recovery period. The financial costs associated with OLT and lifelong immunosuppression are considerable. There is a high incidence of complications from the surgical procedure and the concomitant immunosuppression that is required following the organ transplant. Complications can range from simple infections to renal failure, hyperlipidemia, and an increased incidence of skin and other types of cancers following long-term immunosuppression. As with all other organs, the number of liver donors does not nearly equal the number of patients on the waiting list. Patients may wait two or more years for a liver transplant, and there is a death rate of greater than 10% per year of patients on the waiting list. Timing is critical for whole organ transplant. An ABO-compatible liver donor

305

Principles of Regenerative Medicine. DOI: 10.1016/B978-0-12-381422-7.10018-5

TABLE 18.1 Current Treatments for Liver Disease

Orthotopic liver transplantation

Major and expensive surgery
Extensive recovery period
High incidence of complications
Expensive maintenance therapy
Shortage of donor organs
Timing is critical

Hepatocyte transplantation

Less invasive and less costly procedure
Complications fewer and less severe
Timing of procedure is easier
Alternative cell sources
Patient retains native liver
Graft loss is not necessarily lethal
Option remains for whole organ transplant

must be available when a patient requires the transplant. Some of the limitations associated with whole organ transplants are addressed with hepatocyte transplants (Table 18.1). Hepatocyte transplants do not require major surgical procedures as they are performed by infusion of cells into the blood supply to an organ such as the liver or spleen. Thus, hepatocyte transplants are less invasive and less costly procedures. Because major surgery is not required there are fewer complications associated with the procedure.

Since cell infusions are minor procedures, there is essentially no recovery period needed. If patients were healthy prior to the procedures, for example a stable metabolic disease patient, they would likely feel no adverse effects from the procedure other than from the placement of a catheter. Hepatocytes can be banked and cryopreserved, so, theoretically, cells could be available at any time for a patient transplant. The timing of a hepatocyte transplant depends on the status of the patient rather than on the availability of a suitable organ. Currently, the source of hepatocytes for hepatocyte transplants is mainly discarded organs not suitable for whole organ transplant (Nakazawa et al., 2002). Currently, there are not enough hepatocytes to transplant all recipients who would likely benefit from the procedure. However, some inventive new ideas have been proposed, such as the use of segment IV, which can be made available from a split-liver procedure (Mitry et al., 2004) to make more hepatocytes available for transplants. Alternative sources of hepatocytes could also be available in the future. Although many options are discussed, the most prominent sources are xenotransplants from pigs or other species, immortalized hepatocytes, and most recently stem cell-derived hepatocytes (Strom and Fisher, 2003). Future developments in these areas may make the number of cells available for hepatocyte transplants virtually unlimited.

A significant benefit of hepatocyte transplantation is that the patient retains their native liver. In cases of cell transplants for metabolic disease, the patient's native liver still performs all of the liver functions with the exception of the function that initiates the disease. Patients with ornithine transcarbamylase deficiency (OTC) have mutation in an enzyme involved in the urea cycle that prevents the metabolism and elimination of ammonia. Although the native liver is not proficient in ammonia metabolism, it is still capable of performing other liver functions including the secretion of clotting factors, albumin, drug metabolism, and all other metabolic and synthetic processes. A cell transplant need only support the ammonia metabolism for the patient, and will not be required to provide complete liver support. Because all liver functions are not dependent on donor cells, loss of the cell graft or failure of the cells to function properly will not necessarily be life threatening, especially for a stable metabolic disease patient. Finally, a whole organ transplant always remains as an option for the cell transplant patient. Even if the cell transplant fails to function or is rejected, to do nothing as part of the cell transplant procedure would likely interfere with a subsequent whole organ transplant. Fisher et al. (1998) reported that prior hepatocyte transplantation did not sensitize the cell transplant recipient to either the donor cells or to an eventual liver graft. Thus, despite sometimes transplanting hepatocytes directly into an immunological response organ, the spleen, no immunological reactions are initiated that are deleterious to the cell transplant or an eventual whole organ transplant.

There are potential disadvantages of hepatocyte transplants as well. First, there are no reports of long-term complete corrections of metabolic liver disease in patients following cell transplantation alone. Because it is a new field, much additional experimentation will be required to determine the full efficacy of cell therapy of liver disease and the length of time for which the cell graft will function. Also, like whole organ transplants, it is believed that cell transplant recipients will require the administration of immunosuppressive drugs. It is likely that lower doses of the drugs will be needed to prevent rejection of cell transplants than are required for whole organ transplants. Because of this, fewer and less severe side-effects from immunosuppressive drugs would be expected, but definitive studies are lacking.

BACKGROUND STUDIES
Choice of sites for hepatocyte transplantation

Hepatocyte transplants have been conducted for over 20 years. A number of good reviews are available for details of the experiments and the original references, which may be omitted in this review (Strom et al., 1999, 2006; Malhi and Gupta, 2001; Ohashi et al., 2001; Fox, 2002; Fox and Roy-Chowdhury, 2004). The large numbers of preclinical studies conducted on hepatocyte transplants firmly establish that the transplants are safe and effective. The most common sites for the transplantation of hepatocytes are the spleen and the liver; however, transplants to the peritoneal cavity, stomach, or omentum have been reported. Long-term survival of the cells is readily measured following transplants into the spleen or liver. The majority of cells transplanted into the peritoneal cavity intraperitoneal (IP) are rapidly lost. Following IP transplants, only those cells that nidate near blood vessels and can attract sufficient nutrition survive long-term. Despite the ease of the procedure, IP transplants of hepatocytes have only limited efficacy. Transplants of hepatocytes to the spleen or the liver have been shown to function for the lifetime of the recipient (Mito et al., 1979; Gupta et al., 1991; Ponder et al., 1991; Holzman et al., 1993). Studies by Mito and co-workers clearly show long-term survival of hepatocytes and that over time the spleen of an animal can be "hepatized" to where 80% of the mass of the organ can be replaced with hepatocytes (Mito et al., 1978, 1979; Kusano et al., 1981, 1992; Kusano and Mito, 1982).

The concept of establishing ectopic liver function in the spleen is similar in theory to the bioartificial liver (BAL). In the BAL, the hepatocytes are seeded into and maintained in some form of an extracorporal device. The patient's blood or plasma is pumped to the device, where it interacts with the hepatocytes across membrane barriers and is then returned to the patient by a second series of pumps. There are reports that BAL can provide short-term synthetic and metabolic support (Gerlach et al., 2003; Demetriou et al., 2004). The ease of transplant of hepatocytes and the abundance of the patient's own natural basement membrane components coupled with the naturally high blood flow make the spleen a useful site for the establishment of short- or long-term ectopic liver function. It is likely that hepatocyte transplants will be easier, cheaper, and more efficient, and will provide the same, or better, level of support as extracorporal devices.

For transplants into the liver, the preferred route for administration of cells is via the portal vein. Cells are infused into the blood supply that feeds the liver and the hepatocytes are distributed to the different lobes in proportion to the blood flow they receive from the portal vein. Portal vein injections are difficult in small animals, so an alternative method is used in these studies. Hepatocytes are injected directly into the splenic pulp. The proportion of the cells that remains in the spleen is determined by the extent to which the outflow through splenic veins is impeded. In the studies of Mito et al. (1979), where the spleen was "hepatized", the authors briefly occlude the splenic outflow, which helps retain the cells in the spleen. Alternatively, when the spleen is used as a method to affect a portal vein injection, the splenic veins are left open. It was reported that up to 52% of the cells injected into the spleen traverse to the liver via the splenic and portal veins within a few minutes (Gupta et al., 1991; Ponder et al., 1991).

INTEGRATION OF HEPATOCYTES FOLLOWING TRANSPLANTATION

Integration of hepatocytes into recipient liver is a complex process that requires the interaction of donor and native hepatocytes to form an integrated tissue. The process may be considered in four steps (Table 18.2) (Gupta et al., 1995, 1999b, 2000; Koenig et al., 2005). Although they are presented as separate, there is considerable overlap of the steps in both time and space. Some of the most spectacular photographs of the entire process are provided by Koenig et al. (2005). Following infusion into the portal vein, hepatocytes must traverse the endothelium to escape the vascular system. Although the liver has fenestrated endothelium, under normal conditions the pores, which are in the range of 150 nm, are far too small to provide a simple transit of parenchymal hepatocytes, which range in size from 20 to 50 μm. Infusions of hepatocytes quickly fill the portal veins and embolize secondary and tertiary portal radicals (Gupta et al., 1999a). Portal pressures increase as flow is restricted by hepatocyte plugs in the portal veins. Venograms that were normal prior to cell transplantation become markedly attenuated and show greater filling of vessels proximal to the portal vein, including the mesenteric and splenic vein. If the number of hepatocytes transplanted is in the range of 5% of the total number of hepatocytes in the native liver, the portal hypertension is transient and resolves within minutes to hours.

A proportion of transplanted cells begins to fill sinusoidal spaces and the space of Disse as the endothelium in the region of the transplanted cells begins to degenerate. It is likely that both physical and humoral (growth factors, cytokines) factors are involved in this process. Microscopic analysis of tissue sections reveals that endothelium is breached in many places and donor hepatocytes leave the portal veins in regions where endothelium is incomplete and broken. Reports suggest that most of the hepatocytes that eventually integrate into the recipient liver will have traversed the endothelial barrier by 24 h post-transplant. Cells that remain in the portal vessels are eventually removed by macrophages between 16 and 24 h post-transplant. Other reports suggest that cells may continue to integrate into parenchyma for 2–3 days following transplantation (Shani-Peretz et al., 2005). Transient hypoxia in the region of the occluded vessels leads to changes in the endothelium as well as both recipient and donor hepatocytes. Endothelium and donor and native hepatocytes all express vascular endothelial growth factor (VEGF) in the areas of hepatocyte integration (Gupta et al., 1999b; Shani-Peretz et al., 2005), a factor known to be induced by hypoxia. It is interesting that VEGF was previously known as vascular permeability factor (VPF). Expression and secretion of VEGF/VPF, a potent angiogenesis factor, is thought to contribute to the reformation of new sinusoids and restoration of the endothelial barrier following cell transplantation.

TABLE 18.2 Integration of Donor Hepatocytes into Native Liver following Transplantation

Filling vascular spaces with donor cells
Disruption of the sinusoidal endothelium
Donor cell integration in host parenchyma
Remodeling of liver via modulation of extracellular matrix

Passage through the endothelial barrier allows donor hepatocytes to become integrated into recipient parenchyma. Full integration of donor hepatocytes and restoration of full hepatic function is difficult to ascertain. However, careful studies of the expression of antigens and activities localized to specific membrane fractions clearly demonstrate that donor hepatocytes fully integrate into the hepatic plate of native liver, and for hybrid structures between native and donor cells, within 3–5 days following transplantation. The antibody to CD26 recognizes the dipeptidylpeptidase IV (DPPIV) antigen, which is localized to the basolateral membrane of hepatocytes. Antibodies to connexin 32 can be used to visualize gap junctions between adjacent hepatocytes. Likewise, canicular ATPase activity can be used to identify bile cannicular regions between adjacent hepatocytes. The proper localization of these different antigens and activities requires that the hepatocyte be fully integrated into the hepatic plate and polarized. By 3–7 days post-transplant, hybrid structures could be visualized in recipient liver containing both donor (DPPIV) hepatocytes and recipient ATPase activity (Gupta et al., 1995) or donor DPPIV co-localized with connexin 32 (Koenig et al., 2005). Both studies clearly demonstrate proper integration of donor hepatocytes as well as the re-establishment of intracellular communication (connexin 32) between donor and recipient hepatocytes. Hybrid structures between donor and recipient hepatocytes were shown to be functional by the transport and excretion of a fluorescent conjugated bile acid (Gupta et al., 1995). Hepatic transport of indocyanine and sulfobromothalein into the bile following hepatocyte transplantation was also reported by Hamaguchi et al. (1994). Hepatocyte transplants were conducted on Eizai-hyperbilirubinemic rats. These animals have a defect in multidrug resistance protein2 (MRP2), which prevents the normal transport of bile acid conjugates and their excretion into bile. This is a relevant animal model of metabolic disease as the condition is similar to Dubin-Johnson syndrome in humans. The correction of this transport defect by hepatocyte transplantation is definitive proof of the complete functional integration of donor hepatocytes into recipient liver.

As part of the integration process, there is significant remodeling of the hepatic parenchyma. Koenig et al. (2005) have reported the activation and release of matrix metaloprotease-2 (MMP-2) in the immediate area of donor cells. It is not clear whether the proteases are produced by the donor or recipients cells or even which cell type is the source of the protease, but the degradation of extracellular matrix components helps to create space for the donor cells. Expression of MMP-2 was detected in and surrounding foci of proliferating donor hepatocytes two months following cell transplantation. Increased production and release of MMP-2 were also observed at the growth edge of nodules of fetal rat hepatocytes proliferating in adult liver following transplantation (Oertel et al., 2006). While all of the components of the process are not completely understood, it is clear that hepatocytes can be transplanted into the vascular supply of the liver, breach the endothelial barrier, remodel and integrate into hepatic parenchyma, and establish communication with adjacent cells and the biliary tree all within 3–5 days in a process of remodeling that completely retains normal host hepatic architecture.

CLINICAL HEPATOCYTE TRANSPLANTATION

Hepatocyte transplantation has been employed in clinics in three types of procedures (Table 18.3). Cell transplants have been used to provide short-term liver support to patients who were dying of their disease before a suitable organ could be found. As these patients are already listed for a whole organ transplant, the hepatocyte infusion used is sometimes referred to as a "bridge" to transplant. A second use for hepatocyte transplants grew out of the attempts to

TABLE 18.3 Opportunities for Hepatocyte Transplantation

"Bridge" for patients to whole organ transplantation
Cell support for acute liver failure
"Cell therapy" for metabolic disease

bridge people to OLT. It was discovered that some of the patients receiving hepatocyte transplants recovered completely following the hepatocyte transplants and no longer required whole organ transplant. The third general use for hepatocyte transplants is for the correction of metabolic liver disease. Each technique will be discussed separately.

HEPATOCYTE BRIDGE

With the bridge technique, hepatocytes are provided to a patient in acute liver failure or experiencing acute decompensation following chronic liver disease. The majority of these patients are already listed for OLT, and they are in danger of dying before a suitable organ can be found. Hepatocyte transplants have been conducted on these patients in an effort to keep them alive long enough to receive OLT. The primary goal of the bridge transplant is not to prevent whole organ transplant, but rather to support and sustain the patient until an organ becomes available. Preclinical studies with several different models of acute or chronic liver failure have demonstrated that hepatocyte transplantation can support liver function and improve survival (Sutherland et al., 1977; Sommer et al., 1979; Makowka et al., 1981; Demetriou et al., 1988; Mito et al., 1993; Takeshita et al., 1993; Arkadopoulos et al., 1998b; Kobayashi et al., 2000; Ahmad et al., 2002; Aoki et al., 2005). The results with human hepatocyte transplantation in the clinics also show an increase in the survival of patients following hepatocyte transplantation. There are now several reports and review articles that provide details of the patients and the transplant procedures (Habibullah et al., 1994; Strom et al., 1997a,b, 1999, 2006; Bilir et al., 2000; Ohashi et al., 2001; Soriano, 2002; Fox and Roy-Chowdhury, 2004; Fisher and Strom, 2006). The results indicate that there is a 65% survival rate for patients receiving hepatocyte transplants. Although randomized control studies could not be conducted, the preliminary results with approximately 25 patients indicate a survival advantage to those patients receiving cell transplants. In addition to increase survival, there are consistent reports that clinical parameters such as ammonia levels, intracranial pressures, and cerebral blood flow are improved following hepatocyte transplantation (Strom et al., 1997a, b, 1999; Soriano et al., 1998; Bilir et al., 2000; Fisher, 2004; Fisher and Strom, 2006). These results indicate that desperately ill patients who receive hepatocyte transplants are more likely to survive long enough to receive OLT than the non-transplant controls.

Most of the patients who would be candidates for the hepatocyte bridge technique suffer from chronic liver disease and have advanced cirrhosis. Because of the cirrhotic changes in the liver and the accompanying portal hypertension, hepatocytes were not transplanted into the liver (portal vein) in most of the clinical studies. Preclinical studies were conducted where cirrhosis was induced in rats by the administration of phenobarbital and carbon tetrachloride (Gupta et al., 1993). When hepatocytes were subsequently transplanted into animals with increased portal pressures and cirrhosis, there was significantly greater intrapulmonary translocation of donor cells, presumably because of portosystemic shunting. These results suggest that serious complications could arise if portal infusion of hepatocytes were conducted on cirrhotic patients with portal hypertension. Indeed, shunting of transplanted hepatocytes to pulmonary vascular beds has been reported in one clinical study (Bilir et al., 2000). To avoid this possible complication, Fisher et al. recommend that hepatocytes be transplanted into the spleen in cirrhotic patients via the splenic artery (Strom et al., 1997b; Fisher and Strom, 2006). Despite the obvious success of the splenic artery route for hepatocyte transplantation, a recent report suggests that transplantation of hepatocytes by direct splenic puncture results in superior engraftment and fewer serious complications, although long-term engraftment was not studied (Nagata et al., 2003b). Although the method for splenic delivery of cells may not be settled, it is clear that, in cases where physical and/or anatomic abnormalities are present in the native liver, the preferred route for hepatocyte transplantation is to an ectopic site, the spleen.

The promising results reported to date suggest that hepatocyte transplantation is beneficial to patients suffering from severe hepatic insufficiency while awaiting OLT. A logical extension of these results might be for the use of hepatocyte transplants earlier in the process. Rather than wait until the patient is near death and with no immediate prospect for a whole organ transplant, a more pre-emptive approach might be warranted. Hepatocyte transplants could be performed when patients awaiting OLT become unstable. This would presumably stabilize the patient and avoid or at least delay more serious complications of liver failure. Early intervention might avoid more costly hospitalization and other treatments.

HEPATOCYTE TRANSPLANTATION IN ACUTE LIVER FAILURE

As described above, hepatocyte transplants have been used as a bridge to OLT. Most of the patients who have been referred for bridge transplants suffered from chronic liver disease and had cirrhotic changes in liver architecture. There is a subgroup of patients referred for OLT who experience acute liver failure. In these patients there is massive loss of hepatocytes over a short period of time, leading to hepatic insufficiency. Except for the dramatic loss of hepatocytes, there is no long-standing pathological change in liver architecture. Since the liver has the capacity for robust regeneration following loss of liver mass (Michalopoulos and deFrances, 1997), there is considerable interest in trying to correct acute liver failure with hepatocyte transplantation. The hypothesis is similar to the bridge technique, where hepatocyte transplantation is used to provide support at a time of critical and otherwise lethal liver failure. The expectation is that, if the patient survives the acute loss of tissue mass, their native liver will regenerate. If the native liver regenerates, there will no longer be a need for OLT. An exogenous source of hepatocytes by transplantation would provide support of liver function to prevent lethal hepatic failure. Both donor and native hepatocytes would be expected to participate in the regeneration response. Once the native liver has been fully restored, there might not be a need for donor-derived hepatocytes. If the chimeric liver generated following the transplant is composed predominantly of native hepatocytes, the patient could be safely removed from immunosuppressive therapy. In this manner, the patient receives what amounts to a temporary liver cell transplant. If cell therapy is sufficient, the patient will be spared whole organ transplantation and lifelong immunosuppression. Several preclinical studies support the hypothesis that hepatocyte transplantation can provide sufficient liver function to maintain an animal experiencing acute liver failure. Studies have shown that hepatocyte transplants dramatically improve survival of animals with acute liver failure induced by D-galactosamine (Sutherland et al., 1977; Sommer et al., 1979; Makowka et al., 1981; Baumgartner et al., 1983), 90% hepatectomy (Cuervas-Mons et al., 1984; Demetriou et al., 1988; Mito et al., 1993; Kobayashi et al., 2000), or ischemic liver injury (Takeshita et al., 1993; Arkadopoulos et al., 1998a).

There are now reports of reversal of acute liver failure in four patients following hepatocyte transplantation (Fisher et al., 2000; Soriano, 2002; Fisher and Strom, 2006; Ott et al., 2006; Strom et al., 2006). The causes of acute liver failure ranged from hepatitis B-induced liver failure to acetaminophen intoxication, to liver toxicity following eating poisonous mushrooms to liver failure of unknown etiology in a pediatric patient. In each case patients presented with classic symptoms of acute liver failure, and most were immediately listed for OLT. The number of cells transplanted varied between different procedures but ranged from approximately 1 to 5 billion total viable cells. In all cases cells were transplanted into the portal vein to get a direct transplant into the liver. In general, patients were given fresh frozen plasma prior to placement of the catheter to prevent bleeding. The results presented by Fisher et al. (2000) are typical of the response to hepatocyte transplantation. There is usually a rapid fall in ammonia levels following the transplant. Circulating levels of clotting factors stabilize following the transplant and then slowly increase over the next 2 weeks. Fisher et al. report that Factor VII levels were 1% of normal prior to transplant and increased to 25% by 7 days and 64% of normal by week 2 post-cell transplant. The recovery of the clotting factors is usually rapid enough that, following the cell transplant, no additional fresh frozen plasma is required.

Patients are generally discharged within 2–4 weeks and are judged to experience a complete recovery. The cell transplant recipients ranged in age from three to 64 years in age, indicating that even older patients have sufficient regenerative capacity to be supported by hepatocyte transplantation.

As is observed with donor tissue allografts, hepatocyte allografts produce and secrete human leukocyte antigen-I (sHLA-I) immediately upon implantation. If there is a mismatch between the donor and recipient, the donor-specific sHLA-I can be detected in the circulation and quantified by enzyme-linked immunosorbent assay (ELISA). Donor-specific HLA class I alleles can be identified and quantified by polymerase chain reaction (PCR) analysis of tissue samples taken at biopsy. When it is determined that the preponderance of cells in the patient's liver are native, the patient can slowly be removed from immunosuppressive therapy as was described by Fisher et al. (2000). In the cases described to date, the patients recovered completely from liver failure following hepatocyte transplantation without serious adverse consequences and without whole organ transplant and lifelong immunosuppression. Although the numbers of patients are small, the treatment of acute liver failure by hepatocyte transplant has some significant advantages that make further investigation of this novel therapy appropriate (Table 18.3).

HEPATOCYTE TRANSPLANTATION FOR METABOLIC LIVER DISEASE

A common indication for whole organ transplantation in pediatric patients is metabolic liver disease. In these cases, there is usually a genetic defect in an enzyme or protein that is produced in the liver that inactivates a critical liver function. Although all other liver functions are generally normal, the liver is removed and replaced with a liver that can perform the missing function. Because there is usually only one genetic defect associated with each metabolic liver disease, a gene therapy approach to correct the defect would seem appropriate. Unfortunately, gene therapy has met with considerable problems that have prevented successful use of this experimental technique. Hepatocyte transplantation has been used in attempts to correct the metabolic defects associated with several types of metabolic liver disease (Table 18.4).

In an approach similar to gene therapy, with hepatocyte transplants one tries to seed the patient's liver with cells that are proficient in the enzyme or function missing in the native liver. The goal is to repopulate the liver of the transplant recipient with sufficient numbers of hepatocytes to provide the missing liver function by donor cells.

Large numbers of hepatocytes cannot be infused into the portal system because of the problems with embolism of the portal veins and portal hypertension. Generally, we infuse

TABLE 18.4 Clinical Transplants for Metabolic Liver Disease

Familial hypercholesterolemia
Crigler-Najjar
Ornithine transcarbamylase deficiency
Arginosuccinate lyase deficiency
Citrullinemia
Factor VII deficiency
Glycogen storage disease, Type 1a and 1b
Infantile Refsum disease
Progressive familial intrahepatic cholestasis
Alpha-1 antitrypsin deficiency
Carbamoylphosphate synthase deficiency
Phenylketonuria

approximately 2×10^8 cells/kg body weight of the recipient (Fox et al., 1998; Horslen et al., 2003). Infusions of these cell numbers have not resulted in any long-term complications. There is always a transient increase in portal pressures that resolves within hours (Strom et al., 1997a; Fox et al., 1998; Bohnen et al., 2000; Soriano, 2002; Horslen et al., 2003; Sokal et al., 2003; Horslen and Fox, 2004). While quite experimental, this number was arrived at by an extrapolation from preclinical studies with non-human primates. Grossman et al. (1992) reported that the infusion of between 1 and 2×10^8 cells/kg into baboons who had previously received a left or right lobectomy was accomplished without serious complications and with only transient increases in portal pressures. Because only a few percent of liver mass can be transplanted at any one time, single hepatocyte transplants cannot be expected to replace a large percentage of liver with donor cells. For this reason, the metabolic diseases that are candidates for cell transplants are those in which the restoration of 10% or less of total liver function or activity is likely to correct the disease. The liver has highly redundant functions. Thus, it is recognized that 10% of a normal amount of gene product or enzyme activity would likely correct the symptoms of most metabolic liver diseases. Exceptions exist, such as hypercholesterolemia, where more than 50% replacement of liver with donor cells would likely be needed to correct circulating low-density lipoprotein levels. However, for most metabolic liver diseases and all of those listed in Table 18.4, it is believed that the replacement of the liver with 10% donor hepatocytes would either be completely corrective or at least ameliorate most of the symptoms of the disease.

In general, hepatocyte transplants work best when the donor cells have a selective growth advantage. There are a number of animal models of liver disease where the native hepatocytes show an increased death rate as compared to normal liver (Sandgren et al., 1991; Rhim et al., 1994; Overturf et al., 1996; de Vree et al., 2000). In these situations, when cells without the defect are transplanted into the diseased liver, the donor cells have a strong and selective growth advantage as compared to the native hepatocytes. Over time, the liver may become nearly completely replaced with donor cells. In certain human diseases, there might be sufficient selective pressure to strongly favor the replacement of large parts of the liver with donor cells. Such diseases include tyrosinemia Type 1, Wilson's disease (Irani et al., 2001), progressive familial intrahepatic cholestasis (PFIC) (de Vree et al., 2000), and alpha-1 anti-trypsin deficiency (A1AT) (Rudnick and Perlmutter, 2005). In these diseases, integration of only a small proportion of liver mass by hepatocyte transplantation would likely be necessary because the donor cells would be expected to continue to proliferate in the host liver, and over time replace the diseased cells. Although there are clear examples of this in studies of transplants of laboratory animals, there are no studies with human patients showing comparable results.

Most metabolic diseases such as Crigler-Najjar (CN), OTC deficiency, and all of those diseases listed in Table 18.4 would not be expected to show such selective growth pressure for donor cells. For diseases such as these, multiple transplants over time will be required to populate the liver with 10% donor cells (Rozga et al., 1995).

A large number of studies with different animal models have shown the efficacy of hepatocyte transplantation to correct metabolic liver disease (reviewed in Malhi and Gupta, 2001 and Strom et al., 2006). Metabolic defects in bilirubin metabolism (Matas et al., 1976; Groth et al., 1977; Vroemen et al., 1986; Demetriou et al., 1988; Moscioni et al., 1989; Holzman et al., 1993; Hamaguchi et al., 1994), albumin secretion (Mito et al., 1979; Kusano and Mito, 1982; Demetriou et al., 1993; Rozga et al., 1995; Moscioni et al., 1996; Oren et al., 1999), ascorbic acid production (Onodera et al., 1995; Nakazawa et al., 1996), tyrosinemia Type 1 (Overturf et al., 1996), copper excretion (Yoshida et al., 1996; Irani et al., 2001; Allen et al., 2004), PFIC (de Vree et al., 2000), as well as other defects in biliary transport similar to Dubin-Johnson syndrome in humans (Hamaguchi et al., 1994) have been shown to be amenable to correction by hepatocyte transplantation. These encouraging results suggested that similar defects in human patients could be corrected by hepatocyte transplantation. The diseases listed in Table 18.4 have been the focus of human trials of hepatocyte transplants.

Hepatocyte transplants were previously shown to result in a rapid correction of ammonia levels (Strom et al., 1997b, 1999; Bilir et al., 2000; Soriano, 2002). For this reason, urea cycle defects that result in life-threatening hyperammonemia were the first metabolic disease target for hepatocyte transplants (Strom et al., 1997b; Bohnen et al., 2000). In the initial study, 1 billion viable cells were transplanted into the portal vein of a five-year-old recipient. Portal pressures increased from 11 cm of water prior to cell transplant to 19 cm immediately following the cell infusion, but recovered rapidly. The patient's ammonia levels normalized without medical intervention within 48 h of cell infusion and his glutamine levels returned to normal. Although OTC activity was undetectable prior to cell transplant, measurable OTC activity was detected in a biopsy performed at 28 days. In these studies, 10% of the cells were labeled with indium[111] prior to infusion into the patient to monitor distribution of the cells. Quantitative analysis of the scientigraphic images showed an average distribution ratio of liver:spleen of 9.5:1. Measurements made prior to cell infusion indicated that free indium was released from hepatocytes at a rate of 10% per hour, and free indium is rapidly cleared from circulation by reticuloendothelial systems such as the spleen. Thus, most of the tracer in the spleen following cell infusion was thought to be free indium, not hepatocytes. Pulmonary radiotracer uptake was consistent with background counts, indicating the absence of porto-systemic shunting despite the modest increase in portal pressures observed at the time of transplant. This first transplant for metabolic liver disease indicated that hepatocyte transplantation into the portal vein could be conducted safely in patients with no significant liver pathology with only a moderate and reversible increase in portal pressures. From the rapid normalization of ammonia levels following hepatocyte transplant, it was concluded that cell transplantation can partially correct the hyperammonemia associated with the disease. Subsequent studies have verified that partial corrections of ammonia levels are possible by cell transplants alone (Horslen et al., 2003; Dhawan et al., 2004; Stephenne et al., 2005; Meyburg et al., 2009). While complete corrections of OTC deficiency have not been accomplished, these studies indicate that cell transplants provide much-needed metabolic control of ammonia levels. Even in the absence of complete correction, liver cell transplantation should be considered as a bridge to whole organ transplantation for OTC patients to prevent the neurological problems associated with uncontrolled hyperammonemia (Bohnen et al., 2000; Stephenne et al., 2005; Meyburg et al., 2009).

A number of groups have attempted to correct CN syndrome, Type 1 with hepatocyte transplants. The first case was in many ways typical of the results obtained by other groups and will be discussed in greater detail (Fox et al., 1998). This disease is caused by a defect in the enzyme that is responsible for the conjugation and eventual excretion of bilirubin. The absence of the enzyme results in severe hyperbilirubinemia, which can lead to central nervous system (CNS) toxicity including kernicterus. Following the transplantation of approximately 7.5 billion cells into the liver of a 10-year-old female, there was a slow and continuous decrease in circulating bilirubin levels over the first 30–40 days, and bilirubin conjugates were readily detected in the bile. Overall, there was approximately a 60–65% decrease in bilirubin levels as compared to pretransplant levels. Because the bilirubin conjugates could only be produced by the donor cells, their detection in the bile demonstrated the robust biochemical function of the transplanted cells and established that donor hepatocytes integrated into the hepatic parenchyma and quickly established connections with the recipient's biliary tree.

Several important findings were gained from this transplant. First, large numbers of hepatocytes could be safely transplanted into the portal vein without complication. Although the total numbers of hepatocytes in liver are difficult to assess, a transplant of 7.5 billion cells represents an estimated 3.5–7.5% of the liver mass, which was transplanted without complication over approximately a 15 h period. Second, the apparent engraftment and function of hepatocytes in the clinical trials seems to exceed that found in previous animal studies. The transplantation of 3.5–7.5% of liver mass resulted in the restoration of approximately 5% of a normal amount of bilirubin conjugation capacity in the liver. Third, a long-term correction in bilirubin levels was observed. This patient was followed for more than 1.5 years. Fourth, single transplants of hepatocytes are effective in creating partial corrections of the disease, but, given the limitation of transplanting 2×10^8 cells/kg body weight, one cannot transplant sufficient numbers of hepatocytes to achieve a complete correction of metabolic liver disease with one transplant. It is estimated that complete corrections would require 2–4 transplants if each were as successful and efficient as the first. Finally, this was the first unequivocal demonstration of the long-term success of hepatocyte transplantation. Although patients were bridged to transplant and clinical parameters such as ammonia levels rapidly changed following transplantation, many of the previous patients underwent subsequent OLT the long-term metabolic function of the transplanted cells was difficult to assess.

These studies firmly established that hepatocyte transplants were an effective means of correcting metabolic liver disease. The results of hepatocyte transplants of other patients with CN largely confirm those seen with the first patient (Dhawan et al., 2004; Ambrosino et al., 2005).

Muraca et al. (2002) and Lee et al. (2007) reported partial correction of glycogen storage disease, Type 1 following hepatocyte transplantation. Improvement was documented by the patient's ability to maintain blood glucose between meals as well as sustained and higher glucose levels with meals. Sokal et al. (2003) employed hepatocyte transplants to achieve a partial correction of infantile Refsum disease an autosomal recessive inborn error in peroxisome metabolism of very long chain fatty acid metabolism, bile acid, and pipecolic acid. The authors reported improvement in fatty acids metabolism and a reduction in circulating pipecolic acid and bile salt levels. An overall improvement in the health of the patient was evidenced by the report of significant increase in muscle strength and weight gain. Dhawan et al. (2004) reported that hepatocyte transplantation partially corrected a severe deficiency in the production and secretion of coagulation Factor VII. Following cell transplant, exogenous Factor VII requirement was reduced to 20% of that needed prior to the cell transplant. Recently, Stephenne et al. (2006) reported the complete correction of a 3.5-year-old female patient with neonatal onset arginosuccinate lyase (ASL) deficiency. Like OTC deficiency, ASL patients are at risk of brain damage from hyperammonemia. The patient received three sequential hepatocyte transplants over a 5-month period. Both freshly isolated and previously cryopreserved hepatocytes were used. At 1 year post-transplant the patient displayed 3% of normal ASL activity in hepatic biopsy samples. Engraftment of donor cells could be demonstrated by fluorescence *in situ* hybridization for Y chromosome. These results confirm that hepatocyte transplantation can achieve sustained engraftment of donor cells and sustained metabolic and clinical control.

HEPATOCYTE TRANSPLANTATION NOVEL USES, CHALLENGES, AND FUTURE DIRECTIONS
Hepatocyte transplants for non-organ transplant candidates

Most of the patients who have received a hepatocyte transplant were already listed for a whole organ transplant. The need for liver support is not limited to this group. There are large numbers of patients for whom OLT is not an option. Patients in this group could include alcoholic cirrhotic patients who have not met the required abstinence period, patients with acute liver failure resulting from suicide attempts, and cancer patients. Early case reports suggested that hepatocyte transplants into the spleen could be useful to restore liver function to end-stage cirrhotic patients (Strom et al., 1999). Although both of the patients in the reported study eventually died of concomitant renal failure that was left untreated, the patients were sufficiently improved following the cell transplants that they were able to be discharged from the hospital. Fox and co-workers created an animal model to study the efficacy of hepatocyte transplants to support liver function in cirrhosis in a more controlled setting. Their studies clearly demonstrated that hepatocyte transplants significantly improve liver function and survival of rats experiencing chronic liver failure following repeated injections of carbon tetrachloride (Ahmad et al., 2002; Cai et al., 2002; Nagata et al., 2003a). With millions of patients currently infected with hepatitis viruses, there is clearly a need for additional means to support liver function in these patients. Not withstanding the difficulties of such clinical studies in cirrhotic patients, cell transplantation should be thoroughly evaluated as a possible support therapy.

In addition to cirrhotic patients who may not be candidates for OLT, there are metabolic liver diseases such as phenylketonuria (PKU) that are not currently referred for OLT. Although some still believe that diseases such as PKU can be adequately controlled by diet, there is evidence of continued and progressive mental deterioration in most patients treated with diet alone. It is likely that cell therapy with hepatocytes would improve control of phenylalanine levels in these patients. Severely affected PKU patients and those not controlled well by diet alone should be given serious consideration for inclusion in hepatocyte transplant protocols, as it seems that the benefits would likely outweigh the risks for these individuals (Harding, 2008).

An important factor preventing the use of hepatocyte transplants in additional medical centers is the limited availability of hepatocytes. The normal source of cells for hepatocyte transplants is livers with greater than 50% steatosis, vascular plaques, or other factors that render the tissue unsuitable for whole organ transplantation (Strom et al., 1997a,b; Fox et al., 1998; Bilir et al., 2000; Fisher et al., 2000; Muraca et al., 2002; Nakazawa et al., 2002; Soriano, 2002; Horslen et al., 2003; Mitry et al., 2003; Strom and Fisher, 2003; Ott et al., 2004; Strom et al., 2006). Better utilization of existing liver tissue could increase the numbers of hepatocytes available immediately. In the USA there are no regulations requiring that donor organs be allocated to transplantation research centers for hepatocyte isolation, and relatively few organs go to centers where hepatocyte transplant is a possibility. Most of the organs not used for whole organ transplant are provided to commercial firms where hepatocytes are isolated for resale or for in-house metabolism and toxicology studies. While most uses of donor liver tissue have merit, simple allocation procedures could be instituted to route the organs to transplant centers for initial review and selection of the most suitable cases for cell isolation. Split-liver procedures have made it possible to use caudate lobe and segment IV for hepatocyte isolation. Depending on the surgical procedure, these portions of liver tissue may remain untransplanted and have been shown to be useful for hepatocyte isolation (Mitry et al., 2004). Although currently quite hypothetical, most livers that are currently transplanted could be split. A portion such as the left lateral segment could be made available for cell isolation while the remaining liver tissue is utilized as a tissue graft. Because hepatocyte transplantation is not currently the standard of care, such proposals are not presently feasible. However, if the efficacy of hepatocyte transplants was firmly established, the risk and the extra time needed for the split procedure would be outweighed by the benefit of the cell transplants. Cell transplants rather than OLT could free up the organs that are now used for acute liver failure and metabolic disease patients.

Methods to improve engraftment and repopulation

Hepatocyte transplants will not be able to progress past the small numbers of patients currently being transplanted until sufficient numbers of hepatocytes become available or engraftment and repopulation are significantly improved (Strom and Fisher, 2003). It has become evident that pretreatment of the native liver of the transplant recipient to induce regeneration and proliferation of donor hepatocytes may be needed prior to hepatocyte transplantation. Most of the pretreatment conditioning regimens used in studies with experimental animals are too hazardous to be applied in a clinical setting. The two most common approaches that can be applied in the clinic that have been suggested are portal embolization and hepatic irradiation (Guha et al. 2001a,b; Dagher et al. 2006; Weber et al. 2009). The theories and literature on these techniques are widely available and will not be presented here. There is another technique that is commonly used in experimental animals that has not been given serious consideration as a pretreatment to hepatocyte transplantation. Partial hepatic resection, more commonly called partial hepatectomy (PH), has been considered too risky for clinical application. Due to improved techniques and instruments and increased activities at experienced centers, liver resection and living donor liver transplantation are now common procedures. With today's advanced surgical techniques and in the hands of surgeons with considerable experience with reduced grafts, split livers, and partial liver resection, this surgical procedure would likely be as safe as portal embolization and hepatic irradiation and should also be considered as a possible pretreatment for hepatocyte transplantation. A partial resection to induce liver regeneration would be a much simpler and safer surgical procedure than those routinely performed in the clinic today. Partial liver resections are most commonly performed to remove malignancies or during living donor transplant procedures. When performing a surgical procedure for tumor removal, the amount of tissue and the location of the procedure are dictated by the location of the tumor(s). Likewise, in the case of living donor liver graft removal, the surgeons need to consider preservation of vessels as well as minimize ischemia injury to the graft during surgery.

Removal of liver tissue to induce liver regeneration will be much safer as the only concern for the surgeon will be safety of the patient. Since the resected liver tissue will not be used as a tissue graft, the surgeon will not need to be concerned about the vessels in the resected tissue. The amount and exact location of the tissue to be removed can be chosen by the surgeon and the procedure performed in the safest, fastest, and easiest way. Although there are risks associated with both the surgery and anesthesia, these risks would likely be much lower than for living donor liver transplantation. As an example, in a recent report of 100 donor resections from one center, no life-threatening complications occurred (Fernandes et al., 2010). Although for a different reason, hepatectomy prior to hepatocyte transplantation has already been done in a series of patients transplanted for familial hypercholesterolemia in 1992–1994. These patients underwent left lateral hepatectomy to harvest tissue for hepatocyte isolation and subsequent retroviral transduction of the LDL receptor (Grossman et al., 1995). Transplantation of transduced autologous hepatocytes was performed on day 3 post-operation. The surgical safety of this procedure was thoroughly studied and reported without any major complications (Raper et al., 1996). A point that remains unanswered is to what extent hepatectomy will generate a sufficient signal to improve the engraftment or proliferation of transplanted hepatocytes. The timing of the transplantation after hepatectomy might also be important. Efimova et al. measured serum growth factors in healthy individuals after living related liver donation and showed that HGF increased 12-fold at 2 h post-operation and thereafter stabilized at a three-fold higher level compared to pre-operation for an additional 5 days (Efimova et al., 2005). Other growth factors such as VEGF and EGF did not change significantly and TGF-alpha was not detected at all. These data suggest that partial hepatic resection could provide a significant stimulus to donor hepatocytes, and that the effect would last at least 5 days. Taken together, these data suggest that partial hepatic resection could be a safe and effective pretreatment to hepatocyte transplantation.

In addition, the tissue removed as a pretreatment could be used for cell isolation. One could also, at least in theory, envision use of tissue removed from a patient with a metabolic disease for cell isolation and subsequent domino transplantation to a patient with a different metabolic disease.

Stem cells and alternative cell sources for liver therapy

In addition to attempts to improve engraftment and repopulation of the liver, alternative cell sources for hepatocytes have been proposed. Xenotransplants (Nagata et al., 2003a), immortalized human hepatocytes (Kobayashi et al., 2000; Cai et al., 2002; Wege et al.,

2003a,b), stem cell-derived hepatocytes (Avital et al., 2002; Miki et al., 2002, 2005; Davila et al., 2004; Ruhnke et al., 2005), and fetal hepatocytes have been proposed as alternative sources of cells for clinical transplants. To date, no alternative cell source has been found that meets all of the requirements for safety and efficacy. There is currently great interest in stem cell-derived hepatocytes and the possibility that they might become a future source of cells for clinical transplantation.

Proponents for the use of stem cells suggest that, because of their wide availability and small size, stem cells could be a feasible and efficient alternative to hepatocytes for cellular therapy (Dolle et al., 2010). While it is true that stem cells, once generated, should be available in sufficient numbers for transplant protocols, it is not clear that the small size of the stem cells would actually favor engraftment and repopulation of target organs. It is pleasing to speculate that the relatively low levels of engraftment of mature hepatocytes following transplant is due to their large diameter, and that the 50—90% of transplanted hepatocytes do not engraft because they temporarily obstruct and get trapped in portal veins or hepatic sinusoids. However, obstruction of the portal vessels and hepatic sinusoids and transient increases in portal pressure may actually be a necessary step in engraftment into the liver parenchyma. Smaller, stem cell-derived hepatocyte-like cells may actually engraft less effectively than mature hepatocytes. When examined in transplant models, smaller hepatocytes or hepatocyte-like cells were usually less effective than larger, or more mature, hepatocytes (Overturf et al., 1999; Weglarz, et al. 2000; Strom et al., 2006; Utoh et al., 2008). Sharma et al. directly examined and compared the engraftment and proliferation of mouse ES-derived hepatocyte-like cells and mature hepatocytes in the $FAH^{-/-}$ mouse (Sharma et al., 2008). The $FAH^{-/-}$ is a robust model of metabolic liver disease where transplanted donor ($FAH^{+/+}$) hepatocytes are under strong positive growth selection leading to rapid and effective repopulation of the diseased mouse liver. Mouse ES-derived cells with hepatic features were found to engraft less efficiently when compared to mature hepatocytes and showed very limited capacity for repopulation and tissue formation. Of the cell types most often suggested to become a source of cells for clinical transplants, embryonic stem cell (ES) and induced pluripotent stem cell (iPS) may hold the greatest promise for future therapy. However, to move this potential therapy to the clinics, two significant roadblocks would have to be overcome: efficient and effective hepatic differentiation of the stem cells and removal of the tumorigenic potential of the transplanted cells. To date, neither condition has been met. No published protocols are efficient or effective enough to produce large numbers of mature human hepatocytes that could be immediately used for transplants. The problem of tumor formation from cells in the population that did not undergo hepatic differentiation and the possibility that differentiated hepatocyte-like cells could regress to undifferentiated stem cells following transplantation will have to be overcome before either ES or iPS cells could be considered for clinical protocols. Also, as stated above, ES-derived hepatocyte-like cells showed limited capacity for liver tissue formation following transplantation when compared to mature hepatocytes (Sharma et al., 2008); thus, much more basic research will be required before ES or iPS-derived hepatocytes are ready for the clinics. Liver stem cells are covered in Chapter 20 and are not discussed here.

Cell types that are currently in clinical practice and could potentially be available for cellular therapy in the near future are those from bone marrow and mesenchymal stromal cells (MSCs). Following the initial publication of Petersen et al., there was great excitement at the possibility that bone marrow cells might serve as a source of hepatocytes for the correction of liver disease (Petersen et al., 1999). Subsequent detailed work has suggested that cell fusion is the principal source of bone marrow-derived hepatocytes observed in the experimental model (Wang et al., 2003). Although there is still some controversy over this issue, the bulk of the more recent data suggests that bone marrow is not the source of the progenitor cells in the liver (Menthena et al., 2004; Thorgeirsson and Grisham, 2006) and there is little evidence for the conversion of hematopoietic cells to hepatocytes, *in vivo*, in experiments with animals (Cantz et al., 2004; Yamaguchi et al., 2006) or in a clinical setting (Fogt et al., 2002). When the presence of X and Y chromosomes was analyzed in liver biopsies taken from sex-mismatched liver transplant recipients (eight female to male and five male to female) the recipient-specific sex chromosome pattern was only detected in the inflammatory cells, and not hepatocytes (Fogt et al., 2002). This study had a transplant-to-biopsy interval of 4.5 years (range 1.2–12 years), and the authors concluded that recipient engraftment of stem hematopoietic cells is an infrequent feature in long-term grafts. Thus, bone marrow cells may not be a relevant source of hepatocytes for the treatment of liver disease.

MSCs isolated from a variety of tissues including cord blood, skin, and human liver have been proposed as a source of hepatocytes for transplantation. There are now numerous reports of mesenchymal cells adopting hepatic features when cultures are placed under specific conditions or upon transplantation, *in vivo*. Several groups have now reported the expression of hepatic genes and proteins normally expressed in the liver such as albumin, α-1 antitrypsin, α-fetoprotein, fibrinogen, glycogen, and even some more mature hepatic markers such as drug metabolizing genes including CYP3A4 (Duret et al., 2007; Najimi et al., 2007; Lysy et al., 2007; Campard et al., 2008). In all cases, the levels of expression of hepatic genes and their functions, when measured, were quite low when compared to authentic human hepatocytes. The utility of MSCs as a source of hepatocytes may depend on their ability to differentiate to cells with a mature hepatic phenotype upon transplantation. When examined carefully, following transplantation into the mouse liver, human cord blood mononuclear cells gave rise to small clusters of hepatocyte-like cells that expressed human albumin and Hep Par, a marker protein found in hepatocytes; however, the cells also expressed mouse cytokeratin 18, suggesting that the clusters of hepatocyte-like cells were the result of cell fusion with endogenous mouse hepatocytes (Sharma et al., 2005). At the present time, there is no convincing evidence that MSCs can differentiate to cells with a broad range of mature hepatic functions.

Although there is little substantial evidence that bone marrow or MSCs from different sources form mature hepatocytes, *in vitro* or *in vivo*, there is growing evidence that these cells may improve liver function when they are infused into patients with cirrhosis. A remarkable paper by Sakaida et al. reported that the intravenous transplantation of bone marrow cells reduced liver fibrosis in a model of cirrhosis induced by treating mice with CCL_4 (Sakaida et al., 2004), which eventually led to the proposal to use autologous bone marrow therapy for liver cirrhosis (Sakaida et al., 2005). Several groups soon began phase I safety and feasibility studies with bone marrow cell infusions in cirrhotic patients. Most of the studies have been uncontrolled investigations of the infusion of bone marrow-derived mononuclear cells (Terai et al., 2006; Lyra et al., 2007) or CD34-selected cells (Gaia et al., 2006; Gordon et al., 2006; Yannaki et al., 2006; Mohamadnejad et al., 2007a; Pai et al., 2008). In most of the studies G-CSF was used to mobilize $CD34^+$ cells. Cells were delivered through a peripheral vein or were infused directly through a hepatic artery. The most common findings were a slight improvement in liver function as measured by a small decrease in bilirubin levels which was usually accompanied by a small increase in serum albumin levels. Improvements in the Child-Pugh and/or the MELD scores were also frequently reported (Gaia et al., 2006; Yannaki et al., 2006; Mohamadnejad et al., 2007a; Pai et al., 2008). In one study, bone marrow-derived MSCs rather than $CD34^+$ or unfractionated bone marrow mononuclear cells were infused via a peripheral vein with similar results (Mohamadnejad et al., 2007b). It is important to note that the authors observed an increase in mortality and other complications if the MSCs were infused into the hepatic artery of patients with decompensated cirrhosis rather than via a peripheral vein.

In the only controlled trial reported, a minimum of 1×10^8 mononuclear cells from bone marrow aspirates (without G-CSF pretreatment) were infused through the hepatic artery of patients with cirrhosis (Lyra et al., 2010). Fifteen patients were randomized to each arm of the study. The results indicated that the Child-Pugh score improved in the cell therapy group relative to the controls. The MELD score remained stable in patients receiving the cell therapy, while in the control group the MELD score increased. Serum bilirubin levels were also

319

improved in the treated group. The improvements noted in the different endpoints were only significant for 90 days. The clinical findings suggest that slight improvements in hepatic function as measured by bilirubin and albumin levels, NELD, or Child-Pugh scores are obtained following the transplantation of bone marrow mononuclear cells or partially puri-fied CD34$^+$ cells. It is comforting that similar findings were obtained with different protocols and by different groups. What is not clear from these initial studies is the most useful cell type to transplant and the best route of delivery. Perhaps sustained improvements in liver function could be obtained if these parameters were optimized.

CONCLUSION

With the exception of bone marrow or bone marrow-derived stem cells, other stem cell sources have not been employed for the treatment of liver disease. Much more research will have to be conducted before cell types such as ES or iPS can be approved for clinical trials. At the current time, adult stem cells do not seem to show sufficient engraftment, proliferation, and differ-entiation to hepatocytes to warrant clinical trials with any of the cell types. For now, authentic hepatocytes remain the preferred cell type for the treatment of liver disease. Future work with hepatocyte-based therapy will need to focus on the improvement of engraftment and/or proliferation of donor cells post-transplant. Even 2—4-fold increases in liver repopulation by hepatocytes over the levels obtained with current transplant procedures could lead to substantial improvement in the clinical outcome of patients with liver-based metabolic diseases. It is likely that the incorporation of preconditioning regimens with hepatic resection, ischemia/reperfusion injury, and/or radiation-induced blockage of the growth of native liver will provide the selective growth advantage to the donor cells required to attain the levels of liver repopulation required to normalize the alterations observed in metabolic disease patients. Since these types of studies are currently being planned at medical centers around the world, the efficacy of these modified protocols should soon be apparent.

Hepatocyte transplantation studies conducted in animal models of liver failure and liver-based metabolic disease have proven safe and effective means to provide short- or long-term synthetic and metabolic support of liver function. For certain organ transplant candidates such as those with metabolic liver disease, cell transplantation alone could provide relief of the clinical symptoms. Cell transplant studies in patients with acute or chronic liver failure or genetic defects in liver function clearly demonstrate the efficacy of hepatocyte transplantation to treat liver disease. In virtually all cases, a clinical improvement in the condition of the patient could be documented. No serious complications of hepatocyte transplant have been reported. Although all of the initial reports concerning hepatocyte transplants are encouraging, it must be realized that there are still no reports of long-term and complete corrections of any metabolic disease in patients. The recent report of a complete correction of a patient with a urea cycle defect is most encouraging; however, the length of time that human hepatocytes will function following transplantation has not been determined. Studies in animal models of liver disease have documented that donor hepatocytes transplanted into the spleen or the liver function for the lifetime of the recipient and participate in normal regenerative events. Although it is likely that human hepatocyte transplantation will result in lifelong and normal function of donor cells, this needs to be clearly demonstrated in a clinical study.

Future work will have to be conducted to establish optimal transplant and immunosuppres-sion protocols to minimize complications and maximize engraftment and function. A major problem for clinical hepatocyte transplant is the inability to track donor cells following transplantation. Except for the short-term tracking of hepatocytes pre-labeled with radioactive substances such as indium[111] (Bohnen et al., 2000), and following differences between donor and recipient-secreted HLA (Fisher et al., 2000), there are no reports of quantitative and facile methods to detect donor cells. Relatively non-invasive methods will be needed to optimize transplant and immunosuppressive protocols as well as for day-to-day monitoring of the cell graft. None of the problems cited here seem insurmountable. There are now reports of successful hepatocyte transplants from laboratories in many different countries. The cooperative spirit that has developed between the investigators at the different transplant centers should benefit the research field and especially the future recipients of hepatocyte transplants.

References

Ahmad, T. A., Eguchi, S., Yanaga, K., Miyamoto, S., Kamohara, Y., Fujioka, H., et al. (2002). Role of intrasplenic hepatocyte transplantation in improving survival and liver regeneration after hepatic resection in cirrhotic rats. *Cell Transplant.*, *11*, 399–402.

Allen, K. J., Cheah, D. M., Wright, P. F., Gazeas, S., Pettigrew-Buck, N. E., Deal, Y. H., et al. (2004). Liver cell transplantation leads to repopulation and functional correction in a mouse model of Wilson's disease. *J. Gastroenterol. Hepatol.*, *19*, 1283–1290.

Ambrosino, G., Varotto, S., Strom, S. C., Guariso, G., Franchin, E., Miotto, D., et al. (2005). Isolated hepatocyte transplantation for Crigler-Najjar syndrome type 1. *Cell Transplant.*, *14*, 151–157.

Aoki, T., Jin, Z., Nishino, N., Kato, H., Shimizu, Y., Niiya, T., et al. (2005). Intrasplenic transplantation of encapsulated hepatocytes decreases mortality and improves liver functions in fulminant hepatic failure from 90% partial hepatectomy in rats. *Transplantation*, *79*, 783–790.

Arkadopoulos, N., Chen, S. C., Khalili, T. M., Detry, O., Hewitt, W. R., Lilja, H., et al. (1998a). Transplantation of hepatocytes for prevention of intracranial hypertension in pigs with ischemic liver failure. *Cell Transplant.*, *7*, 357–363.

Arkadopoulos, N., Lilja, H., Suh, K. S., Demetriou, A. A., & Rozga, J. (1998b). Intrasplenic transplantation of allogeneic hepatocytes prolongs survival in anhepatic rats. *Hepatology*, *28*, 1365–1370.

Avital, I., Feraresso, C., Aoki, T., Hui, T., Rozga, J., Demetriou, A., et al. (2002). Bone marrow-derived liver stem cell and mature hepatocyte engraftment in livers undergoing rejection. *Surgery*, *132*, 384–390.

Baumgartner, D., LaPlante-O'Neill, P. M., Sutherland, D. E., & Najarian, J. S. (1983). Effects of intrasplenic injection of hepatocytes, hepatocyte fragments and hepatocyte culture supernatants on D-galactosamine-induced liver failure in rats. *Eur. Surg. Res.*, *15*, 129–135.

Bilir, B. M., Guinette, D., Karrer, F., Kumpe, D. A., Krysl, J., Stephens, J., et al. (2000). Hepatocyte transplantation in acute liver failure. *Liver Transplant.*, *6*, 32–40.

Bohnen, N. I., Charron, M., Reyes, J., Rubinstein, W., Strom, S. C., Swanson, D., et al. (2000). Use of indium-111-labeled hepatocytes to determine the biodistribution of transplanted hepatocytes through portal vein infusion. *Clin. Nucl. Med.*, *25*, 447–450.

Cai, J., Ito, M., Nagata, H., Westerman, K. A., Lafleur, D., Chowdhury, J. R., et al. (2002). Treatment of liver failure in rats with end-stage cirrhosis by transplantation of immortalized hepatocytes. *Hepatology*, *36*, 386–394.

Campard, D., Lysy, P. A., Najimi, M., & Sokal, E. M. (2008). Native umbilical cord matrix stem cells express hepatic markers and differentiate into hepatocyte-like cells. *Gastroenterology*, *134*, 833–848.

Cantz, T., Sharma, A. D., Jochheim-Richter, A., Arseniev, L., Klein, C., Manns, M. P., et al. (2004). Reevaluation of bone marrow-derived cells as a source for hepatocyte regeneration. *Cell Transplant.*, *13*, 659–666.

Cuervas-Mons, V., Cienfuegos, J. A., Maganto, P., Golitsin, A., Eroles, G., Castillo-Olivares, J., et al. (1984). Time-related efficacy of liver cell isografts in fulminant hepatic failure. *Transplantation*, *38*, 23–25.

Dagher, I., Boudechiche, L., Branger, J., Coulomb-Lhermine, A., Parouchev, A., Sentilhes, L., et al. (2006). Efficient hepatocyte engraftment in a nonhuman primate model after partial portal vein embolization. *Transplantation*, *82*, 1067–1073.

Davila, J. C., Cezar, G. G., Thiede, M., Strom, S., Miki, T., & Trosko, J. (2004). Use and application of stem cells in toxicology. *Toxicol. Sci.*, *79*, 214–223.

de Vree, J. M., Ottenhoff, R., Bosma, P. J., Smith, A. J., Aten, J., & Oude Elferink, R. P. (2000). Correction of liver disease by hepatocyte transplantation in a mouse model of progressive familial intrahepatic cholestasis. *Gastroenterology*, *119*, 1720–1730.

Demetriou, A. A., Brown, R. S., Jr., Busuttil, R. W., Fair, J., McGuire, B. M., Rosenthal, P., Am Esch, J. S., II, et al. (2004). Prospective, randomized, multicenter, controlled trial of a bioartificial liver in treating acute liver failure. *Ann. Surg.*, *239*, 660–667, discussion 667–670.

Demetriou, A. A., Holzman, M., Moscioni, A. D., & Rozga, J. (1993). Hepatic cell transplantation. *Adv. Vet. Sci. Comp. Med.*, *37*, 313–332.

Demetriou, A. A., Reisner, A., Sanchez, J., Levenson, S. M., Moscioni, A. D., & Chowdhury, J. R. (1988). Transplantation of microcarrier-attached hepatocytes into 90% partially hepatectomized rats. *Hepatology*, *8*, 1006–1009.

Dhawan, A., Mitry, R. R., Hughes, R. D., Lehec, S., Terry, C., Bansal, S., et al. (2004). Hepatocyte transplantation for inherited factor VII deficiency. *Transplantation, 78,* 1812–1814.

Dolle, L., Best, J., Mei, J., Al Battah, F., Reynaert, H., van Grunsven, L. A., et al. (2010). The quest for liver progenitor cells: a practical point of view. *J. Hepatol., 52,* 117–129.

Duret, C., Gerbal-Chaloin, S., Ramos, J., Fabre, J. M., Jacquet, E., Navarro, F., et al. (2007). Isolation, characterization, and differentiation to hepatocyte-like cells of nonparenchymal epithelial cells from adult human liver. *Stem Cells, 25,* 1779–1790.

Efimova, E. A., Glanemann, M., Nussler, A. K., Schumacher, G., Settmacher, U., Jonas, S., et al. (2005). Changes in serum levels of growth factors in healthy individuals after living related liver donation. *Transplant. Proc., 37,* 1074–1075.

Fernandes, R., Pacheco-Moreira, L. F., Enne, M., Steinbruck, K., Alves, J. A., Filho, G. D., et al. (2010). Surgical complications in 100 donor hepatectomies for living donor liver transplantation in a single Brazilian center. *Transplant. Proc., 42,* 421–423.

Fisher, R. A., & Strom, S. C. (2006). Human hepatocyte transplantation: worldwide results. *Transplantation, 82,* 441–449.

Fisher, R. A. (2004). Adult human hepatocyte transplantation. *7th International Congress of Cell Transplantation Society, Boston,* 304.

Fisher, R. A., Bu, D., Thompson, M., Tisnado, J., Prasad, U., Sterling, R., et al. (2000). Defining hepatocellular chimerism in a liver failure patient bridged with hepatocyte infusion. *Transplantation, 69,* 303–307.

Fisher, R. A., Kimball, P. M., Thompson, M., Saggi, B., Wolfe, L., & Posner, M. (1998). An immunologic study of liver failure in patients bridged with hepatocyte infusions. *Transplantation, 65,* S45.

Fogt, F., Beyser, K. H., Poremba, C., Zimmerman, R. L., Khettry, U., & Ruschoff, J. (2002). Recipient-derived hepatocytes in liver transplants: a rare event in sex-mismatched transplants. *Hepatology, 36,* 173–176.

Fox, I. J. (2002). Transplantation into and inside the liver. *Hepatology, 36,* 249–251.

Fox, I. J., & Roy-Chowdhury, J. (2004). Hepatocyte transplantation. *J. Hepatol., 40,* 878–886.

Fox, I. J., Chowdhury, J. R., Kaufman, S. S., Goertzen, T. C., Chowdhury, N. R., Warkentin, P. I., et al. (1998). Treatment of the Crigler-Najjar syndrome type I with hepatocyte transplantation. *N. Engl. J. Med., 338,* 1422–1426.

Gaia, S., Smedile, A., Omede, P., Olivero, A., Sanavio, F., Balzola, F., et al. (2006). Feasibility and safety of G-CSF administration to induce bone marrow-derived cells mobilization in patients with end stage liver disease. *J. Hepatol., 45,* 13–19.

Gerlach, J. C., Mutig, K., Sauer, I. M., Schrade, P., Efimova, E., Mieder, T., et al. (2003). Use of primary human liver cells originating from discarded grafts in a bioreactor for liver support therapy and the prospects of culturing adult liver stem cells in bioreactors: a morphologic study. *Transplantation, 76,* 781–786.

Gordon, M. Y., Levicar, N., Pai, M., Bachellier, P., Dimarakis, I., Al-Allaf, F., et al. (2006). Characterization and clinical application of human CD34+ stem/progenitor cell populations mobilized into the blood by granulocyte colony-stimulating factor. *Stem Cells, 24,* 1822–1830.

Grossman, M., Rader, D. J., Muller, D. W., Kolansky, D. M., Kozarsky, K., Clark, B. J., III, et al. (1995). A pilot study of ex vivo gene therapy for homozygous familial hypercholesterolaemia. *Nat. Med., 1,* 1148–1154.

Grossman, M., Raper, S. E., & Wilson, J. M. (1992). Transplantation of genetically modified autologous hepatocytes into nonhuman primates: feasibility and short-term toxicity. *Hum. Gene. Ther., 3,* 501–510.

Groth, C. G., Arborgh, B., Bjorken, C., Sundberg, B., & Lundgren, G. (1977). Correction of hyperbilirubinemia in the glucuronyltransferase-deficient rat by intraportal hepatocyte transplantation. *Transplant. Proc., 9,* 313–316.

Guha, C., Deb, N. J., Sappal, B. S., Ghosh, S. S., Roy-Chowdhury, N., & Roy-Chowdhury, J. (2001b). Amplification of engrafted hepatocytes by preparative manipulation of the host liver. *Artif. Organs, 25,* 522–528.

Guha, C., Parashar, B., Deb, N. J., Sharma, A., Gorla, G. R., Alfieri, A., et al. (2001a). Liver irradiation: a potential preparative regimen for hepatocyte transplantation. *Int. J. Radiat. Oncol. Biol. Phys., 49,* 451–457.

Gupta, S., Aragona, E., Vemuru, R. P., Bhargava, K. K., Burk, R. D., & Chowdhury, J. R. (1991). Permanent engraftment and function of hepatocytes delivered to the liver: implications for gene therapy and liver repopulation. *Hepatology, 14,* 144–149.

Gupta, S., Rajvanshi, P., & Lee, C. D. (1995). Integration of transplanted hepatocytes into host liver plates demonstrated with dipeptidyl peptidase IV-deficient rats. *Proc. Natl. Acad. Sci. U.S.A., 92,* 5860–5864.

Gupta, S., Rajvanshi, P., Aragona, E., Lee, C. D., Yerneni, P. R., & Burk, R. D. (1999a). Transplanted hepatocytes proliferate differently after CCl_4 treatment and hepatocyte growth factor infusion. *Am. J. Physiol., 276,* G629–G638.

Gupta, S., Rajvanshi, P., Malhi, H., Slehria, S., Sokhi, R. P., Vasa, S. R., et al. (2000). Cell transplantation causes loss of gap junctions and activates GGT expression permanently in host liver. *Am. J. Physiol. Gastrointest. Liver Physiol., 279,* G815–G826.

Gupta, S., Rajvanshi, P., Sokhi, R., Slehria, S., Yam, A., Kerr, A., et al. (1999b). Entry and integration of transplanted hepatocytes in rat liver plates occur by disruption of hepatic sinusoidal endothelium. *Hepatology, 29*, 509–519.

Gupta, S., Yerneni, P. R., Vemuru, R. P., Lee, C. D., Yellin, E. L., & Bhargava, K. K. (1993). Studies on the safety of intrasplenic hepatocyte transplantation: relevance to ex vivo gene therapy and liver repopulation in acute hepatic failure. *Hum. Gene. Ther., 4*, 249–257.

Habibullah, C. M., Syed, I. H., Qamar, A., & Taher-Uz, Z. (1994). Human fetal hepatocyte transplantation in patients with fulminant hepatic failure. *Transplantation, 58*, 951–952.

Hamaguchi, H., Yamaguchi, Y., Goto, M., Misumi, M., Hisama, N., Miyanari, N., et al. (1994). Hepatic biliary transport after hepatocyte transplantation in Eizai hyperbilirubinemic rats. *Hepatology, 20*, 220–224.

Harding, C. (2008). Progress towards cell-directed therapy for Phenylketonuria. *Clin. Genet., 74*, 97–104.

Holzman, M. D., Rozga, J., Neuzil, D. F., Griffin, D., Moscioni, A. D., & Demetriou, A. A. (1993). Selective intra-portal hepatocyte transplantation in analbuminemic and Gunn rats. *Transplantation, 55*, 1213–1219.

Horslen, S. P., & Fox, I. J. (2004). Hepatocyte transplantation. *Transplantation, 77*, 1481–1486.

Horslen, S. P., McCowan, T. C., Goertzen, T. C., Warkentin, P. I., Cai, H. B., Strom, S. C., et al. (2003). Isolated hepatocyte transplantation in an infant with a severe urea cycle disorder. *Pediatrics, 111*, 1262–1267.

Irani, A. N., Malhi, H., Slehria, S., Gorla, G. R., Volenberg, I., Schilsky, M. L., et al. (2001). Correction of liver disease following transplantation of normal rat hepatocytes into Long-Evans Cinnamon rats modeling Wilson's disease. *Mol. Ther., 3*, 302–309.

Kobayashi, N., Fujiwara, T., Westerman, K. A., Inoue, Y., Sakaguchi, M., Noguchi, H., et al. (2000). Prevention of acute liver failure in rats with reversibly immortalized human hepatocytes. *Science, 287*, 1258–1262.

Koenig, S., Stoesser, C., Krause, P., Becker, H., & Markus, P. M. (2005). Liver repopulation after hepatocellular transplantation: integration and interaction of transplanted hepatocytes in the host. *Cell Transplant., 14*, 31–40.

Kusano, M., & Mito, M. (1982). Observations on the fine structure of long-survived isolated hepatocytes inoculated into rat spleen. *Gastroenterology, 82*, 616–628.

Kusano, M., Ebata, H., Onishi, T., Saito, T., & Mito, M. (1981). Transplantation of cryopreserved isolated hepatocytes into the rat spleen. *Transplant. Proc., 13*, 848–854.

Kusano, M., Sawa, M., Jiang, B., Kino, S., Itoh, K., Sakata, H., et al. (1992). Proliferation and differentiation of fetal liver cells transplanted into rat spleen. *Transplant. Proc., 24*, 2960–2961.

Lee, K. W., Lee, J. H., Shin, S. W., Kim, S. J., Joh, J. W., Lee, D. H., et al. (2007). Hepatocyte transplantation for glycogen storage disease type Ib. *Cell Transplant., 16*, 629–637.

Lyra, A. C., Soares, M. B., da Silva, L. F., Braga, E. L., Oliveira, S. A., Fortes, M. F., et al. (2010). Infusion of autologous bone marrow mononuclear cells through hepatic artery results in a short-term improvement of liver function in patients with chronic liver disease: a pilot randomized controlled study. *Eur. J. Gastroenterol. Hepatol., 22*, 33–42.

Lyra, A. C., Soares, M. B., da Silva, L. F., Fortes, M. F., Silva, A. G., Mota, A. C., et al. (2007). Feasibility and safety of autologous bone marrow mononuclear cell transplantation in patients with advanced chronic liver disease. *World J. Gastroenterol., 13*, 1067–1073.

Lysy, P. A., Smets, F., Sibille, C., Najimi, M., & Sokal, E. M. (2007). Human skin fibroblasts: from mesodermal to hepatocyte-like differentiation. *Hepatology, 46*, 1574–1585.

Makowka, L., Rotstein, L. E., Falk, R. E., Falk, J. A., Zuk, R., Langer, B., et al. (1981). Studies into the mechanism of reversal of experimental acute hepatic failure by hepatocyte transplantation. 1. *Can. J. Surg., 24*, 39–44.

Malhi, H., & Gupta, S. (2001). Hepatocyte transplantation: new horizons and challenges. *J. Hepatobiliary Pancreat. Surg., 8*, 40–50.

Matas, A. J., Sutherland, D. E., Steffes, M. W., Mauer, S. M., Sowe, A., Simmons, R. L., et al. (1976). Hepatocellular transplantation for metabolic deficiencies: decrease of plasms bilirubin in Gunn rats. *Science, 192*, 892–894.

Menthena, A., Deb, N., Oertel, M., Grozdanov, P. N., Sandhu, J., Shah, S., et al. (2004). Bone marrow progenitors are not the source of expanding oval cells in injured liver. *Stem Cells, 22*, 1049–1061.

Meyburg, J., Das, A. M., Hoerster, F., Lindner, M., Kriegbaum, H., Engelmann, G., et al. (2009). One liver for four children: first clinical series of liver cell transplantation for severe neonatal urea cycle defects. *Transplantation, 87*, 636–641.

Michalopoulos, G. K., & deFrances, M. C. (1997). Liver regeneration. *Science, 276*, 60–66.

Miki, T., Cai, H., Lehmann, T., & Strom, S. (2002). Production of hepatocytes from human amniotic stem cells. *Hepatology, 36*, 171A.

Miki, T., Lehmann, T., Cai, H., Stolz, D. B., & Strom, S. C. (2005). Stem cell characteristics of amniotic epithelial cells. *Stem Cells, 23*, 1549–1559.

Mito, M., Ebata, H., Kusano, M., Onishi, T., Saito, T., & Sakamoto, S. (1979). Morphology and function of isolated hepatocytes transplanted into rat spleen. *Transplantation, 28*, 499–505.

Mito, M., Kusano, M., & Sawa, M. (1993). Hepatocyte transplantation for hepatic failure. *Transplant. Rev., 7*, 35.

323

Mito, M., Kusano, M., Onishi, T., Saito, T., & Ebata, H. (1978). Hepatocellular transplantation — morphological study on hepatocytes transplanted into rat spleen. *Gastroenterol. Jpn, 13*, 480–490.

Mitry, R. R., Dhawan, A., Hughes, R. D., Bansal, S., Lehec, S., Terry, C., et al. (2004). One liver, three recipients: segment IV from split-liver procedures as a source of hepatocytes for cell transplantation. *Transplantation, 77*, 1614–1616.

Mitry, R. R., Hughes, R. D., Aw, M. M., Terry, C., Mieli-Vergani, G., Girlanda, R., et al. (2003). Human hepatocyte isolation and relationship of cell viability to early graft function. *Cell Transplant., 12*, 69–74.

Mohamadnejad, M., Namiri, M., Bagheri, M., Hashemi, S. M., Ghanaati, H., Zare Mehrjardi, N., et al. (2007a). Phase 1 human trial of autologous bone marrow-hematopoietic stem cell transplantation in patients with decompensated cirrhosis. *World J. Gastroenterol., 13*, 3359–3363.

Mohamadnejad, M., Alimoghaddam, K., Mohyeddin-Bonab, M., Bagheri, M., Bashtar, M., Ghanaati, H., et al. (2007b). Phase 1 trial of autologous bone marrow mesenchymal stem cell transplantation in patients with decompensated liver cirrhosis. *Arch. Iran. Med., 10*, 459–466.

Moscioni, A. D., Roy-Chowdhury, J., Barbour, R., Brown, L. L., Roy-Chowdhury, N., Competiello, L. S., et al. (1989). Human liver cell transplantation. Prolonged function in athymic-Gunn and athymic-analbuminemic hybrid rats. *Gastroenterology, 96*, 1546–1551.

Moscioni, A. D., Rozga, J., Chen, S., Naim, A., Scott, H. S., & Demetriou, A. A. (1996). Long-term correction of albumin levels in the Nagase analbuminemic rat: repopulation of the liver by transplanted normal hepatocytes under a regeneration response. *Cell Transplant., 5*, 499–503.

Muraca, M., Gerunda, G., Neri, D., Vilei, M. T., Granato, A., Feltracco, P., et al. (2002). Hepatocyte transplantation as a treatment for glycogen storage disease type 1a. *Lancet, 359*, 317–318.

Nagata, H., Ito, M., Cai, J., Edge, A. S., Platt, J. L., & Fox, I. J. (2003a). Treatment of cirrhosis and liver failure in rats by hepatocyte xenotransplantation. *Gastroenterology, 124*, 422–431.

Nagata, H., Ito, M., Shirota, C., Edge, A., McCowan, T. C., & Fox, I. J. (2003b). Route of hepatocyte delivery affects hepatocyte engraftment in the spleen. *Transplantation, 76*, 732–734.

Najimi, M., Khuu, D. N., Lysy, P. A., Jazouli, N., Abarca, J., Sempoux, C., et al. (2007). Adult-derived human liver mesenchymal-like cells as a potential progenitor reservoir of hepatocytes? *Cell Transplant., 16*, 717–728.

Nakazawa, F., Cai, H., Miki, T., Dorko, K., Abdelmeguid, A., Walldorf, J., et al. (2002). Human hepatocyte isolation from cadaver donor liver. In *Proceedings of Falk Symposium, Hepatocyte Transplantation, Vol 126* (pp. 147–158). Lancaster: Kluwer Academic Publishers.

Nakazawa, F., Onodera, K., Kato, K., Sawa, M., Kino, Y., Imai, M., et al. (1996). Multilocational hepatocyte transplantation for treatment of congenital ascorbic acid deficiency rats. *Cell Transplant., 5*, S23–S25.

Oertel, M., Menthena, A., Dabeva, M. D., & Shafritz, D. A. (2006). Cell competition leads to a high level of normal liver reconstitution by transplanted fetal liver stem/progenitor cells. *Gastroenterology, 130*, 507–520, quiz 590.

Ohashi, K., Park, F., & Kay, M. A. (2001). Hepatocyte transplantation: clinical and experimental application. *J. Mol. Med., 79*, 617–630.

Onodera, K., Kasai, S., Kato, K., Nakazawa, F., & Mito, M. (1995). Long-term effect of intrasplenic hepatocyte transplantation in congenitally ascorbic acid biosynthetic enzyme-deficient rats. *Cell Transplant., 4*(Suppl. 1), S41–S43.

Oren, R., Dabeva, M. D., Petkov, P. M., Hurston, E., Laconi, E., & Shafritz, D. A. (1999). Restoration of serum albumin levels in nagase analbuminemic rats by hepatocyte transplantation. *Hepatology, 29*, 75–81.

Ott, M. C., Barthold, M., Alexandrova, K., Griesel, C., Shchneider, A., Attaran, M., et al. (2004). Isolation of human hepatocytes from donor organs under cgmp conditions and clinical application in patients with liver disease. *7th International Congress of Cell Transplantation Society, Boston, 142*.

Ott, M., Schneider, A., Attaran, M., & Manns, M. P. (2006). Transplantation of hepatocytes in liver failure. *Dtsch. Med. Wochenschr., 131*, 888–891.

Overturf, K., Al-Dhalimy, M., Finegold, M., & Grompe, M. (1999). The repopulation potential of hepatocyte populations differing in size and prior mitotic expansion. *Am. J. Pathol., 155*, 2135–2143.

Overturf, K., Al-Dhalimy, M., Tanguay, R., Brantly, M., Ou, C. N., Finegold, M., et al. (1996). Hepatocytes corrected by gene therapy are selected *in vivo* in a murine model of hereditary tyrosinaemia type I. *Nat. Genet., 12*, 266–273.

Pai, M., Zacharoulis, D., Milicevic, M. N., Helmy, S., Jiao, L. R., Levicar, N., et al. (2008). Autologous infusion of expanded mobilized adult bone marrow-derived CD34+ cells into patients with alcoholic liver cirrhosis. *Am. J. Gastroenterol., 103*, 1952–1958.

Petersen, B. E., Bowen, W. C., Patrene, K. D., Mars, W. M., Sullivan, A. K., Murase, N., et al. (1999). Bone marrow as a potential source of hepatic oval cells. *Science, 284*, 1168–1170.

Ponder, K. P., Gupta, S., Leland, F., Darlington, G., Finegold, M., DeMayo, J., et al. (1991). Mouse hepatocytes migrate to liver parenchyma and function indefinitely after intrasplenic transplantation. *Proc. Natl. Acad. Sci. U.S.A., 88*, 1217–1221.

Raper, S. E., Grossman, M., Rader, D. J., Thoene, J. G., Clark, B. J., III, Kolansky, D. M., et al. (1996). Safety and feasibility of liver-directed ex vivo gene therapy for homozygous familial hypercholesterolemia. *Ann. Surg., 223*, 116–126.

Rhim, J. A., Sandgren, E. P., Degen, J. L., Palmiter, R. D., & Brinster, R. L. (1994). Replacement of diseased mouse liver by hepatic cell transplantation. *Science, 263*, 1149–1152.

Rozga, J., Holzman, M., Moscioni, A. D., Fujioka, H., Morsiani, E., & Demetriou, A. A. (1995). Repeated intraportal hepatocyte transplantation in analbuminemic rats. *Cell Transplant., 4*, 237–243.

Rudnick, D. A., & Perlmutter, D. H. (2005). Alpha-1-antitrypsin deficiency: a new paradigm for hepatocellular carcinoma in genetic liver disease. *Hepatology, 42*, 514–521.

Ruhnke, M., Nussler, A. K., Ungefroren, H., Hengstler, J. G., Kremer, B., Hoeckh, W., et al. (2005). Human monocyte-derived neohepatocytes: a promising alternative to primary human hepatocytes for autologous cell therapy. *Transplantation, 79*, 1097–1103.

Sakaida, I., Terai, S., & Okita, K. (2005). Use of bone marrow cells for the development of cellular therapy in liver diseases. *Hepatol. Res., 31*, 195–196.

Sakaida, I., Terai, S., Yamamoto, N., Aoyama, K., Ishikawa, T., Nishina, H., et al. (2004). Transplantation of bone marrow cells reduces CCl4-induced liver fibrosis in mice. *Hepatology, 40*, 1304–1311.

Sandgren, E. P., Palmiter, R. D., Heckel, J. L., Daugherty, C. C., Brinster, R. L., & Degen, J. L. (1991). Complete hepatic regeneration after somatic deletion of an albumin-plasminogen activator transgene. *Cell, 66*, 245–256.

Shani-Peretz, H., Tsiperson, V., Shoshani, G., Veitzman, E., Neufeld, G., & Baruch, Y. (2005). HVEGF165 increases survival of transplanted hepatocytes within portal radicals: suggested mechanism for early cell engraftment. *Cell Transplant., 14*, 49–57.

Sharma, A. D., Cantz, T., Richter, R., Eckert, K., Henschler, R., Wilkens, L., et al. (2005). Human cord blood stem cells generate human cytokeratin 18-negative hepatocyte-like cells in injured mouse liver. *Am. J. Pathol., 167*, 555–564.

Sharma, A. D., Cantz, T., Vogel, A., Schambach, A., Haridass, D., Iken, M., et al. (2008). Murine embryonic stem cells-derived hepatocyte progenitors cells engraft in recipient livers with limited capacity of liver tissue formation. *Cell Transplant., 17*(3), 313–323.

Sokal, E. M., Smets, F., Bourgois, A., van Maldergem, L., Buts, J. P., Reding, R., et al. (2003). Hepatocyte transplantation in a 4-year-old girl with peroxisomal biogenesis disease: technique, safety, and metabolic follow-up. *Transplantation, 76*, 735–738.

Sommer, B. G., Sutherland, D. E., Matas, A. J., Simmons, R. L., & Najarian, J. S. (1979). Hepatocellular transplantation for treatment of D-galactosamine-induced acute liver failure in rats. *Transplant. Proc., 11*, 578–584.

Soriano, H. E. (2002). Liver cell transplantation: human applications in adults and children. In *Proceedings of Falk Symposium, Hepatocyte Transplantation, Vol 126* (pp. 99–115). Lancaster: Kluwer Academic Publishers.

Soriano, H. E., Kang, D. C., Finegold, M. J., Hicks, M. J., Wang, N. D., Harrison, W., et al. (1998). Lack of C/EBP alpha gene expression results in increased DNA synthesis and an increased frequency of immortalization of freshly isolated mice [correction of rat] hepatocytes. *Hepatology, 27*, 392–401.

Stephenne, X., Najimi, M., Sibille, C., Nassogne, M. C., Smets, F., & Sokal, E. M. (2006). Sustained engraftment and tissue enzyme activity after liver cell transplantation for argininosuccinate lyase deficiency. *Gastroenterology, 130*, 1317–1323.

Stephenne, X., Najimi, M., Smets, F., Reding, R., de Ville de Goyet, J., & Sokal, E. M. (2005). Cryopreserved liver cell transplantation controls ornithine transcarbamylase deficient patient while awaiting liver transplantation. *Am. J. Transplant., 5*, 2058–2061.

Strom, S., & Fisher, R. (2003). Hepatocyte transplantation: new possibilities for therapy. *Gastroenterology, 124*, 568–571.

Strom, S. C., Chowdhury, J. R., & Fox, I. J. (1999). Hepatocyte transplantation for the treatment of human disease. *Semin. Liver Dis., 19*, 39–48.

Strom, S. C., Fisher, R. A., Rubinstein, W. S., Barranger, J. A., Towbin, R. B., Charron, M., et al. (1997a). Transplantation of human hepatocytes. *Transplant. Proc., 29*, 2103–2106.

Strom, S. C., Fisher, R. A., Thompson, M. T., Sanyal, A. J., Cole, P. E., Ham, J. M., et al. (1997b). Hepatocyte transplantation as a bridge to orthotopic liver transplantation in terminal liver failure. *Transplantation, 63*, 559–569.

Strom, S., Bruzzone, P., Cai, H., Ellis, E., Lehmann, T., Mitamura, K., et al. (2006). Hepatocyte transplantation: clinical experience and potential for future use. *Cell Transplant., 15*, S105–S110.

Sutherland, D. E., Numata, M., Matas, A. J., Simmons, R. L., & Najarian, J. S. (1977). Hepatocellular transplantation in acute liver failure. *Surgery, 82*, 124–132.

Takeshita, K., Ishibashi, H., Suzuki, M., & Kodama, M. (1993). Hepatocellular transplantation for metabolic support in experimental acute ischemic liver failure in rats. *Cell Transplant., 2*, 319–324.

325

Terai, S., Ishikawa, T., Omori, K., Aoyama, K., Marumoto, Y., Urata, Y., et al. (2006). Improved liver function in patients with liver cirrhosis after autologous bone marrow cell infusion therapy. *Stem Cells, 24*, 2292–2298.

Thorgeirsson, S. S., & Grisham, J. W. (2006). Hematopoietic cells as hepatocyte stem cells: a critical review of the evidence. *Hepatology, 43*, 2–8.

Utoh, R., Tateno, C., Yamasaki, C., Hiraga, N., Kataoka, M., Shimada, T., et al. (2008). Susceptibility of chimeric mice with livers repopulated by serially subcultured human hepatocytes to hepatitis B virus. *Hepatology, 47*, 435–446.

Vroemen, J. P., Buurman, W. A., Heirwegh, K. P., van der Linden, C. J., & Kootstra, G. (1986). Hepatocyte transplantation for enzyme deficiency disease in congenic rats. *Transplantation, 42*, 130–135.

Wang, X., Willenbring, H., Akkari, Y., Torimaru, Y., Foster, M., Al-Dhalimy, M., et al. (2003). Cell fusion is the principal source of bone-marrow-derived hepatocytes. *Nature, 422*, 897–901.

Weber, A., Groyer-Picard, M. T., & Dagher, I. (2009). Hepatocyte transplantation techniques: large animal models. *Methods Mol. Biol., 481*, 83–96.

Wege, H., Chui, M. S., Le, H. T., Strom, S., & Zern, M. A. (2003a). *In vitro* expansion of human hepatocytes is restricted by telomere-dependent replicative aging. *Cell Transplant., 12*, 897–906.

Wege, H., Le, H. T., Chui, M. S., Liu, L., Wu, J., Giri, R., et al. (2003b). Telomerase reconstitution immortalizes human fetal hepatocytes without disrupting their differentiation potential. *Gastroenterology, 124*, 432–444.

Weglarz, T. C., Degen, J. L., & Sandgren, E. P. (2000). Hepatocyte transplantation into diseased mouse liver. Kinetics of parenchymal repopulation and identification of the proliferative capacity of tetraploid and octaploid hepatocytes. *Am. J. Pathol., 157*, 1963–1974.

Yamaguchi, K., Itoh, K., Masuda, T., Umemura, A., Baum, C., Itoh, Y., et al. (2006). In vivo selection of transduced hematopoietic stem cells and little evidence of their conversion into hepatocytes in vivo. *J. Hepatol.*

Yannaki, E., Anagnostopoulos, A., Kapetanos, D., Xagorari, A., Iordanidis, F., Batsis, I., et al. (2006). Lasting amelioration in the clinical course of decompensated alcoholic cirrhosis with boost infusions of mobilized peripheral blood stem cells. *Exp. Hematol., 34*, 1583–1587.

Yoshida, Y., Tokusashi, Y., Lee, G. H., & Ogawa, K. (1996). Intrahepatic transplantation of normal hepatocytes prevents Wilson's disease in Long-Evans cinnamon rats. *Gastroenterology, 111*, 1654–1660.

Cardiac Stem Cells: Biology and Therapeutic Applications

Sarah Selem, Konstantinos E. Hatzistergos, Joshua M. Hare
University of Miami, Miller School of Medicine, Interdisciplinary Stem Cell Institute, Miami, FL, USA

The heart, previously considered a prototypic terminally differentiated organ, is now known to contain reservoirs or compartments of precursor cells. Both post-natal persistence of precursor cells that govern heart formation and adult stem cell niches are described, although the former mechanism appears to be of limited significance. The exact nature and role of adult cell niches continues to be debated, as does the degree to which cardiomyocytes turn over in the adult. There continues to be growing support for a new paradigm of cardiac biology in which myocyte homeostasis occurs throughout life. In this paradigm, diseases of the heart can now be viewed as disruptions in the balance between physiologic myocyte loss and replacement. Moreover, a new therapeutic possibility has emerged whereby cardiac regenerative mechanisms may be deployed in the form of cell therapeutics that either replace lost cells primarily or promote endogenous cellular repair. In this chapter, we review the biology of cardiac stem cells and the growing body of knowledge regarding cell-based therapeutics.

327

MAMMALIAN CARDIOGENESIS: EVIDENCE FOR PROGRESSIVE LINEAGE RESTRICTION

Mammalian cardiogenesis involves proper synchrony of function among a diverse population of cells that comprise the heart: atrial/ventricular cardiomyocytes, smooth muscle cells, endothelial cells, epicardial cells, conduction system cells, valvular components, and connective tissue (Garry and Olson, 2006; Lam et al., 2009). The differentiation of these cell lineages is dependent on spatially and temporally controlled developmental steps (Domian et al., 2009; Yi et al., 2010). There is increasing evidence that the heart, like blood, may develop from a single progenitor cell by progressive lineage restriction (Wu et al., 2006; Yang et al., 2008). Three cardiac anlagen — the cardiogenic mesoderm, cardiac neural crest, and proepicardial organ (Yi et al., 2010) — are the principal sources for the cardiac precursors that give rise to progenitor cells. Essentially, these are responsible for the development of the various cardiac structures (Lam et al., 2009). The role of cardiac neural crest precursors is ultimately to give rise to the vascular smooth muscle of the aortic arch, ductus arteriosus, and great vessel, as well as to contribute to the cardiac autonomic nervous system. The proepicardial cells give rise to the smooth muscle cells of the coronary vessels and other epicardial cells (Domian et al., 2009; Yi et al., 2010). Finally, cardiogenic mesodermal precursor(s) are responsible for the formation of the cardiomyocytes and are most likely to be the source of reservoirs of precursor cells post-natally.

Principles of Regenerative Medicine. DOI: 10.1016/B978-0-12-381422-7.10019-7

Cardiogenesis begins during gastrulation when precursor cells derived from the mesoderm are induced by adjacent tissue to migrate away from the primitive streak to the lateral-cranial parts of the embryo and form the "cardiogenic regions," including the cardiogenic mesoderm (Srivastava and Olson, 2000). During this migration, the precursor cells from the cardiogenic mesoderm downregulate the T-box transcription factor Brachyury (Bry), which is the earliest mesodermal marker, and upregulate the mesoderm posterior 1 (Mesp1) gene (Fig. 19.1). All cardiac precursors express Mesp1, but the expression of this gene is downregulated upon formation of the cardiac crescent. At this stage in development, progenitors irreversibly commit to the cardiac lineage, expressing LIM-homeodomain Islet1 (Isl1) transcription factor, NK2 transcription related, locus 5 (Nkx2-5), and VEGF receptor-2 fetal liver kinase 1(Flk-1) (Fig. 19.1) (Lam et al., 2009).

Recent studies have demonstrated that the cardiogenic mesoderm maintains two multipotent progenitor cell populations, known as the first heart field (FHF) and second heart field (SHF), which give rise to various cardiac structures. The FHF originates in the anterior splanchnic mesoderm and forms the cardiac crescent, from which the cells migrate medially to form the linear heart tube. Ultimately, the FHF contributes to left ventricle and atria formation (Kelly et al., 2001; Cai et al., 2003; bu-Issa et al., 2004). The SHF originates in the pharyngeal mesoderm medial to the cardiac crescent and lies anterior and dorsal to the linear heart tube. Ultimately, the SHF contributes to right ventricle, outflow tract, and atrial tissue formation, accounting for two thirds of the cells in the heart (Buckingham et al., 2005). While not all genetic markers unique to the FHF and SHF have been identified, some transcription factors and signaling molecules are known to characterize each population. Both are marked by Nkx2-5, but the FHF progenitors are distinguished by T-box transcription factor Tbx5 and bHLH transcription factor Hand1, whereas the SHF progenitors are distinguished by Hand2, Isl1, and Fgf10 (Kelly et al., 2001; Cai et al., 2003). Although a retrospective clonal analysis in the mouse embryo suggested the FHF and SHF progenitors originate from a common precursor (Meilhac et al., 2004), further investigations must be pursued in order to determine whether the FHF and SHF progenitors stem from a single precursor or, perhaps, a subset of precursors (Yi et al., 2010).

Recently, multipotent Isl1[+] progenitors have been isolated and expanded from embryonic and post-natal hearts, and have been shown to differentiate into cardiomyocytes, endothelial cells, and smooth muscle cells, among others (Moretti et al., 2006; Domian et al., 2009). However, it

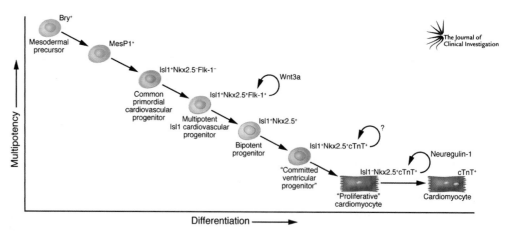

FIGURE 19.1

Proposed pathway for the differentiation of cardiomyocytes along the Isl1 lineage pathway. Differentiation proceeds from a mesodermal precursor cell to a common primordial cardiovascular progenitor that gives off a multipotent Isl1[+] cardiovascular progenitor. Further differentiation proceeds via a committed ventricular progenitor to a terminally differentiated cardiomyocyte. cTnT, cardiac troponin T; MesP1, mesoderm posterior 1. Reproduced from Yi et al., 2010.

is still uncertain whether Isl1 expression is restricted to the SHF or whether it is transiently expressed in FHF progenitors. In argument for the former, it has been demonstrated that homozygous Isl1 mutant mice are growth retarded and die at about E10.5–11. Histological analysis revealed a failure in these hearts to properly undergo looping or even form the right ventricle and outflow tract, such that the Isl1-null hearts are uni-ventricular (Cai et al., 2003). A Cre-loxP strategy of lineage tracing complemented this evidence, revealing that Isl1 was expressed in cells of the outflow tract, right ventricle, as well as part of the atria and the inner curvature of the left ventricle (Buckingham et al., 2005). Some experimental evidence, however, favors the latter notion, that Isl1 is transiently expressed in FHF progenitors. Another lineage-tracing experiment used an Isl1-Cre knock-in mouse line demonstrating that most of the cells in the left ventricle express β-galactosidase (β-gal) (Srivastava and Olson, 2000; Park et al., 2006), a protein that is also found in the outflow tract and right ventricle, both of which originate from the SHF, implying that Isl1 may be expressed by the FHF (Kelly et al., 2001; Brade et al., 2007). Other studies revealed that Is1 protein, unlike Isl1 mRNA, is expressed throughout the anterior intra-embryonic coelemic walls and proximal head mesenchyme (Prall et al., 2007). These regions originate from both FHF and SHF progenitors. Moreover, the cardiac crescent in *Xenopus*, which is derived from the FHF in amphibians as well as mammals, was revealed to coexpress Isl1 and Nkx2-5 (Brade et al., 2007). If Isl1 is expressed by the FHF, it would have major implications for treating ischemic cardiomyopathies.

While it is not clear whether the FHF and SHF progenitors have originated by lineage restriction, Moretti and colleagues have characterized a hierarchy of multipotent Isl1$^+$ cardiovascular progenitors (MICPs) that give rise to endothelial cells, smooth muscle cells, conduction system cells, and ventricular myocytes by way of lineage restriction (Moretti et al., 2006).

As mentioned above, the FHF and SHF are governed by both shared and distinct genetic programs. Both populations are regulated by zinc finger-containing transcription factors GATA4, 5, and 6 (Charron and Nemer, 1999), as well as Nkx2-5. However, studies with mice have revealed that GATA, Nkx2-5, Isl1, and Forkhead box H1 (Foxh1), also expressed in the SHF (von Both et al., 2004), activate expression of myocyte enhancer factor 2C (Mef2c) in the SHF by regulating two enhancer regions (Dodou et al., 2004). Moreover, Mef2c directly activates expression of the SET-domain protein Smyd1, which regulates Hand2 expression during development of the SHF (Phan et al., 2005). The Forkhead family of factors, Foxa2, Foxc1, and Foxc2, have also been revealed to reinforce the Isl1-GATA-Mef2c pathway by binding and activating a Tbx1 enhancer (Maeda et al., 2006), which then activates Fgf8 (Hu et al., 2004). As a result, Fgf8 loses its function in the SHF, causing downregulation of Isl1 in the pharyngeal mesoderm and outflow tract upon their formation (Park et al., 2006). Overall, NKx2-5 and GATA transcription factors mark the FHF progenitor population and Isl1, Foxh1, and Mef2c mark the specialization of the SHF progenitors. However, the transcriptional control that comes with the lineage segregation is still uncertain.

Prior to isolating Isl1$^+$ progenitors, other cardiac progenitors had been isolated and partially characterized, including cells expressing the receptor for stem cell factor (c-kit) (Beltrami et al., 2003), cells expressing Sca-1 but not expressing c-kit (Oh et al., 2003), and cells expressing transport protein Abcg2 (called "side population" or SP cells) (Martin et al., 2004). Importantly, these precursor populations can be identified in the adult myocardium.

C-kit$^+$ progenitor cells

Beltrami and colleagues isolated c-kit$^+$ progenitor cells (stem cell factor (SCF), ligand) from adult rat myocardium (Beltrami et al., 2003) (Fig. 19.2). They primarily reside throughout the ventricular and atrial myocardium, and have a higher density in the ventricular apex. These progenitors do not express transcription factors or membrane and cytoplasmic proteins that may characterize them as bone marrow, neural, skeletal, skeletal muscle, or cardiac cells; that is, they are lineage negative (lin$^-$). C-kit$^+$/lin$^-$ cells proved to be self-renewing, clonogenic,

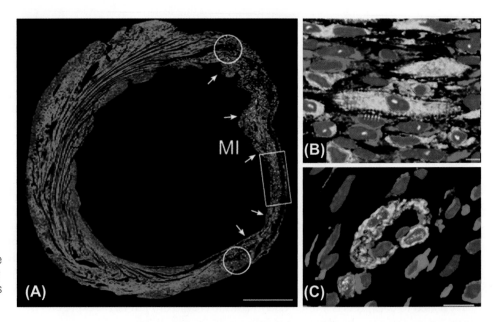

FIGURE 19.2
Bone marrow-derived Lin(−)/C-kit(+) stem cells in an infarcted mouse heart (A) differentiate into new cardiac myocytes (B, yellow and/or white dots) and coronary vessels (C, yellow). Implantation of green fluorescence protein (GFP)-tagged stem cells (yellow color) of male origin (Y chromosome, white dots in B) in female animals is the most widely used strategy to accurately determine the fate of the exogenous cells *in vivo*.

and multipotent, giving rise to cardiomyocytes, endothelial cells, and smooth muscle cells (Anversa et al., 2006). It may, however, be difficult to characterize these cells with regard to transcriptional expression as they are heterogeneous in nature, as indicated by the findings that around 10% of the population express various early myocardial lineage transcription factors (Beltrami et al., 2003). C-kit$^+$ cells have been expanded from rodent, canine, porcine, and human hearts and transplanted into the infarcted ventricle, resulting in the multilineage differentiation of cells that replaced necrotic tissue with functional myocardium (Beltrami et al., 2003; Bearzi et al., 2007; Hatzistergos et al., 2010). These cells have also been injected after ischemia reperfusion injury and been shown to limit infarct size and reduce ventricular remodeling, promoting cardiac functioning. While their role in cardiogenesis remains to be fully investigated, cardiac c-kit cells are described in prenatal hearts (Tallini et al., 2009). Further research is required to fully understand how c-kit cells are related to embryonic cardiac precursors and other progenitors, including their precise status within the hierarchy of cardiac progenitors.

Sca-1$^+$ progenitor cells

Oh and colleagues isolated Sca-1$^+$ progenitor cells from the adult mouse heart (Oh et al., 2003). These are approximately 100- to 700-fold more frequent than c-kit$^+$ cells (Beltrami et al., 2003). These cells were shown to express cardiac marker Nkx2-5 in response to DNA demethylation with 5'-azacytidine; after four weeks under this treatment, the cells differentiated into cardiomyocytes. However, Cre recombinase techniques were used to reveal that the seemingly differentiated cells were due to fusion in approximately 50% of cases (Oh et al., 2003). Interestingly, Matsuura's group reported isolation of Sca-1$^+$ cells from adult murine hearts, of which 1% differentiated into beating cardiomyocytes (Matsuura et al., 2004).

In vivo studies revealed that these cells engrafted in the infracted myocardium of rat heart 2 weeks post-injection. Overall, the self-renewal, clonogenic, multipotent properties of Sca-1$^+$ cells remain obscure. As with c-kit cells, the developmental origin of these progenitors and their role in cardiogenesis require further exploration. In terms of translation to humans, challenges exist as sca-1 or a homologue are notably absent in humans.

Side population cells

Side population cells are a subset of Sca-1$^+$ cells isolated by Garry and colleagues. The cells are characterized by their exclusion of dyes such as Hoechst33342 and Rhodamine 123, and expression of ATP-binding cassette transporter, Abcg2. Cells positive for MDR1-Abcg2 in the bone marrow give rise to the myeloid, lymphoid, and erythroid cell lineages (Bunting, 2002). Those from the skeletal muscle have been found to regenerate muscle fibers. Moreover, cardiac side population cells were shown to differentiate into cardiomyocytes, endothelial cells, or smooth muscle cells in infarcted rat heart (4.4%, 6.7%, and 29% of total CSP-derived cells) (Matsuura et al., 2004). This heterogeneous population of cells, however, is one of the least characterized as their self-renewal, clonogenic, and multipotent properties have not been established. Moreover, their effects *in vivo*, in terms of migration, proliferation, engraftment, and subsequent cardiac function, have not been determined.

CELL-BASED THERAPEUTICS FOR HEART DISEASE

The most important unmet need in cardiovascular medicine is that of a regenerative therapy. Although the heart has regenerative capacity, it is limited, and ischemic and other types of cardiac injury leave permanent injury and impairment to heart function which in turn produces major burdens of morbidity, mortality, and healthcare costs. As such, there is major impetus to translate the new knowledge of cardiac stem cell biology to the therapeutic arena (Fig. 19.3). There have been attempts at cell-based therapy using both cardiopoietic cells — either cardiac stem cells or ESC/iPS strategies — and cells derived from other body sites, most notably bone marrow (Abdel-Latif et al., 2007).

Although mortality rates from ischemic heart disease are falling, paradoxically the incidence of heart failure (HF) is a growing cause of morbidity and mortality worldwide because of the improved short-term survival from myocardial infarction (MI) (Mosterd and Hoes, 2007). In the USA, it is estimated that the lifetime risk of developing HF is one in five (Mosterd and Hoes, 2007). Despite the significant advances in the management of HF during the last decade, one-year mortality rates remain approximately 20%, with even higher rates among patients hospitalized for HF (Mosterd and Hoes, 2007). Coronary heart disease is the predominant cause of HF in developed countries (Mosterd and Hoes, 2007). The pathophysiologic underpinning of this phenomenon is ventricular remodeling, which ensues following MI, and regenerative therapies therefore are targeted against preventing or reversing the remodeling process.

To date, the majority of trials have employed bone marrow-derived strategies to treat patients following acute myocardial infarction, with the goal of preventing remodeling largely through

331

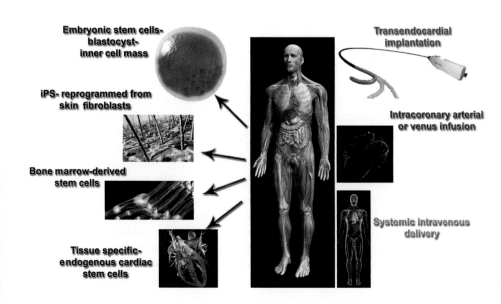

FIGURE 19.3

Stem cell-based therapeutic strategies for cardiac regeneration. The sources for stem cells with cardiac reparative capacities are numerous, including embryonic, bone marrow-derived, and cardiac-specific stem cells. After their expansion into therapeutic quantities, they can be transplanted into the patient using direct transendocardial implantation, intracoronary infusions, or subcutaneous intravenous delivery.

ameliorating infarct size. More recently, strategies have begun to be tested for the reversal of remodeling in patients with heart failure and LV dysfunction.

The basic premise underlying early attempts to repair cardiac injury is that of myocyte terminal differentiation. Early and ongoing attempts, therefore, are designed around cell replacement with cells capable of differentiation. An early attempt at cell-based therapeutic myocardial replacement was reported in 1993, when Koh and co-workers used murine cardiomyocyte-like tumor cells to create intracardiac grafts (Koh et al., 1993). By genetically manipulating murine hearts to overexpress oncogenes such as the simian virus 40 large T antigen (T-Ag), the investigators created a tumorigenic cardiomyocyte cell line with the capacity for proliferation both *in vitro* and *in vivo*. Implantation of these cells into healthy mouse hearts was accompanied by long-term engraftment in ~50% of the recipients without affecting their heart function, introducing the notion of using cells to replace injured or lost cardiomyocytes. This experiment opened a new field of investigation that has led to the exploration of multiple types and sources of cells as potential cardiac therapeutic agents. In 1998, Anversa and co-workers (Anversa and Kajstura, 1998) reported evidence of mitosis in adult cardiomyocytes, suggesting cellular renewal throughout adult life. This discovery instigated the notion that myocyte renewal may not be attributed only to cardiomyocyte proliferation *per se*, but also to homing and differentiation of endogenous stem cells. To address this idea, Orlic and co-workers (Orlic et al., 2001) demonstrated that transplantation of bone marrow (BM)-derived lineage-negative [Lin(−)/C-kit(+)] stem cells into the infarcted mouse heart caused differentiation of the bone marrow cells into cardiac myocytes and vessels and substantial recovery of cardiac function (Fig. 19.2).

Cardiac engraftment of cells from distant sites was demonstrated in 2002 by Quaini et al. (Quaini et al., 2002) in studies of sex-mismatched heart transplants. In these studies, male cardiomyocytes were demonstrated in female hearts transplanted into male patients. More importantly, the investigators also detected a subpopulation of cardiac precursor cells of donor origin, suggesting for the first time that the heart could contain its own cardiac stem cell population. These observations opened up the field of adult stem cell-based therapy and led to a plethora of studies utilizing multiple sources of cells to stimulate post-injury cardiac repair. Based on the Orlic observations, bone marrow was preferentially used as a cell source, prompting a decade-long quest to establish the clinical value and mechanism of action of bone marrow-derived cell-based therapy for heart disease. This quest, however, is not without controversy, and other experimental observations have challenged the hypothesis of cell transdifferentiation as a dominant mechanism of action for cell-based cardiac repair (Guan and Hasenfuss, 2007).

MECHANISMS OF ACTION

Currently there are several contemplated mechanisms of actions underlying successful cell-based therapeutics, each with varying degrees of experimental support (Fig. 19.4). These include differentiation, paracrine signaling, fusion, and cell autonomous niche reconstitution (Mazhari and Hare, 2007; Hatzistergos et al., 2010). *Ex vivo* culture of adult and embryonic stem cells under specific conditions such as the hanging drop technique, stimulation by biochemical compounds, or co-culture with cardiac myocytes have demonstrated the capacity of stem cells to differentiate into beating cardiomyocytes and vascular lineages (Christoforou and Gearhart, 2007). However, many experimental studies suggest that this mechanism is unlikely to account solely for the cardiac repair observed in response to cell-based therapy, since the therapeutic outcomes appear to be in excess of documented levels of cell engraftment and differentiation (Segers and Lee, 2008). Whether the currently employed techniques (i.e. labeling of stem cells with reporter genes, magnetic particles, etc.) are sensitive enough to accurately trace the fate of the implanted cells throughout time is as yet uncertain. Nonetheless, the majority of these studies agree that the exogenously administered stem cells positively

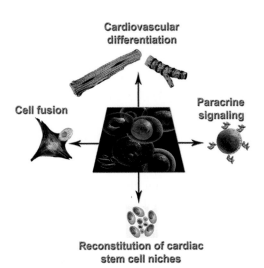

FIGURE 19.4
Mechanisms of cardiac repair in cellular cardiomyoplasty. The transplanted grafts have the capacity for trilineage differentiation into cardiac myocytes, endothelial cells, and vascular smooth muscle cells. Fusion with adjoining host cells and paracrine signaling are also critical and stimulate mechanisms for survival and proliferation of the host cells, as well as the mobilization of endogenous stem cells. An intriguing novel hypothesis is that of stem cell niche reconstitution following mesenchymal stem cell transplantation.

regulate the host's cardiac milieu; by fusing with native cells (a very low frequency event) or secreting several cytokines and growth factors, transplanted stem cells can promote angiogenesis and cell survival (Mangi et al., 2003; Nygren et al., 2004; Mirotsou et al., 2007). This concept is further advanced by the discovery of endogenous stem cells found in cardiac niches. The presence of endogenous stem cells suggests a broader cell autonomous mechanism of action for successful cell-based therapy (Mazhari and Hare, 2007; Nakanishi et al., 2008; Hatzistergos et al., 2010).

CLINICAL TRIALS

The field of stem cell research has gained enormous attention during the last decade, and experimental work has employed both embryonic and adult stem cells to treat heart disease. While a consensus has emerged that there is some functional merit to cell-based therapies for heart disease, underlying mechanisms of action remain controversial. In addition to this central issue, other key issues that need to be settled include the role of host factors in cell functionality (Kissel et al., 2007) and the exciting possibility of using allogeneic grafts because of the unique immunoprivileged properties of some cell types. Despite much controversy, substantial work in the clinical arena with trials of growing size and sophistication have produced a major database of safety and efficacy data that is paving the way forward for future clinical development. Below we review the developments for each cell-based stategy.

Cardiopoietic stem cells

EMBRYONIC STEM CELLS

Murine embryonic stem cells (ESCs) were first identified in 1981 by Evans and Kaufman (1981) and Martin (1981). These cells arise from the inner cell mass of late mice blastocysts and, because of the capacity to differentiate into cell types of all three germ layers including cardiomyocytes, represent a prototypic pluripotent cell. From a practical standpoint, ESCs, because they are pluripotent, have a high probability of causing teratogenicity. However, several groups have employed selective predifferentiation strategies to enhance cardiopoiesis and reduce the risk of teratoma formation (Behfar et al., 2007; Caspi et al., 2007; Christoforou et al., 2008). Animal studies suggest that these ESC-derived committed cells have the capacity to improve myocardial function and structure after MI through generation of new cardio-myocytes in the infarcted area (Christoforou et al., 2010). Recently, embryonic cell-derived endothelial cells were described that also improve myocardial contractility in mice following MI through stimulation of angiogenesis. As with cardiogenic cells, teratomas did not form after administration of these cells (Li et al., 2007).

Another strategy to obtain pluripotent cells involves adult cell genetic reprogramming, so-called induced pluripotent stem cells (iPS). In two pioneering studies conducted by Yu et al. (2007) and Takahashi et al. (2007), adult human skin fibroblasts were reprogrammed into pluripotent embryonic-like stem cells (iPS) by transfection with stem cell-related genes such as Lin28, c-myc, oct 4, sox 2, klf-4, and Nanog. Importantly, since viral transfection techniques, especially when combined with the induction of the oncogene c-myc into the host genome, are accompanied with a high risk for tumor development, virus-free iPS can be generated by using triplets of the above genes with or without using this specific oncogene (Okita et al., 2008). This approach has stimulated enormous enthusiasm given the potential to develop pluripotent cells without using human embryos, offering substantial availability of the cells. Not the least of the advantages of this approach is the prospect of developing host-tailored stem cells that could escape immune rejection, or the development of pluripotent cell lines from hosts with genetic diseases providing an optimal *in vitro* experimental system. To date, EPS or iPS cells have not entered the clinic, although a trial for patients with spinal cord injury is reported to be initiated shortly (Cyranoski, 2008).

ADULT STEM CELLS

The field of cardiac stem cell therapy has been substantially advanced through the discovery that adult stem cells have the capacity to (trans)-differentiate into lineages other than the tissue of origin. This stem cell plasticity allows the use of stem cells isolated from a variety of easily accessible sources such as bone marrow (BM), peripheral blood, fat, umbilical cord, or even testis to be used for cell-based repair of damaged organs. To date, human trials have shown BM-derived mononuclear cells (BMMNCs) and mesenchymal stem cells (MSCs) as the most promising candidates for treating heart disease (Burt et al., 2008), while tissue-specific cardiac stem cells (CSCs) are currently entering trials and potentially offer great promise (http://clinicaltrials.gov/ct2/show/NCT00474461).

BM STEM CELLS

Because of its various well-defined stem cell compartments and its ease of access, whole BM and BM-derived mononuclear stem cells (BMMNCs) are to date the most widely studied type of cell for cellular cardiomyoplasty. Using different cell-surface markers, BMMNCs can be fractionated to hematopoietic (HSCs) or non-hematopoietic stem cells. The latter includes a number of distinct subtypes named as side population (SPs) (Jackson et al., 2001), endothelial progenitor cells (EPCs) (Asahara et al., 1997), MSCs (Zimmet and Hare, 2005), multipotent adult progenitor cells (MAPCs) (Jiang et al., 2002), multilineage inducible (MIAMI) cells (D'Ippolito et al., 2006) cells and very small embryonic like (VSEL), stem cells (Kucia et al., 2006).

BMMNCs

Numerous experimental and clinical studies have tested BMMNCs for a range of therapeutic strategies involving their transplantation and/or their mobilization to sites of cardiac injury. The totality of evidence of trials of BM cells and derivatives supports both the safety and provisional efficacy of this approach. Three meta-analyses evaluating data from approximately 18 trials and close to 1,000 patients (Abdel-Latif et al., 2007; Burt et al., 2008; Martin-Rendon et al., 2008) conclude that BM cell-based therapies contribute to modest improvements in cardiac function by reducing infarct size, preserving LV dimensions, and increasing ejection fraction by 2–3% within 6 months after transplantation. Long-term follow-up data derived from the BOOST and TOPCARE-AMI studies have documented that the therapeutic results are sustained up to 5 years post-transplantation (Dimmeler et al., 2008). One of the most exciting observations derives from the DSMB data of the REPAIR-AMI study (Schachinger et al., 2006). In this study, 204 patients with acute myocardial infarction underwent successful reperfusion of the culprit coronary vessel(s), and 3–7 days later were randomized to receive intracoronary infusion of autologous BMMNCs or placebo. By four months, patients who had received the cells showed a significantly improved LVEF compared to the placebo, with the ones having larger infarcts (baseline LVEF < 48.9%) being more responsive to the therapy than the others. Importantly, 1-year follow-up data from REPAIR-AMI demonstrate improved event-free survival (death, recurrence of MI, revascularization, or rehospitalization for heart failure) of the BMMNC-treated patients compared to the placebo (Fig. 19.5). In addition to the post-MI setting, several trials have employed BMMNCs for patients with established LV dysfunction and/or heart failure due to either ischemic or non-ischemic causes; data are sufficiently promising to warrant further study. Based upon the totality of evidence, BMMNCs are poised to enter into pivotal clinical trials (Abdel-Latif et al., 2007; Martin-Rendon et al., 2008). In addition, ongoing trials are being conducted to refine the key issues of dose, timing, and host-specific impairments in autologous cells (Schachinger et al., 2006).

ENDOTHELIAL PROGENITOR CELLS (EPCs)

This subset of hematopoietic stem cells can be isolated from BMMNCs as well as peripheral blood mononuclear cells, based on the expression of HSC surface markers such as CD34,

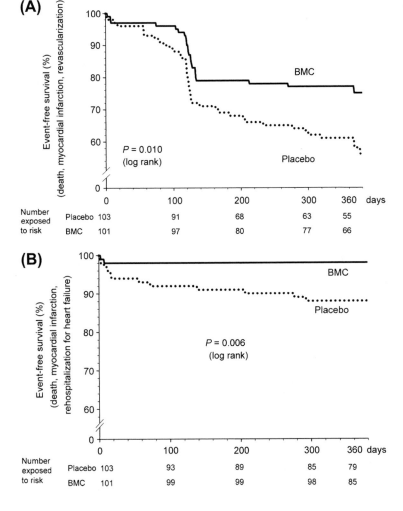

FIGURE 19.5

Kaplan Meier event-free survival analysis of the REPAIR-AMI trial illustrates that patients who had received BMMNCs had significantly reduced frequencies of (A) death, myocardial infarction, or revascularization (combined end points) and (B) death, myocardial infarction, or rehospitalization (combined end points), compared to the patients treated with placebo. Reproduced from Assmus et al., 2010.

CD133, and the vascular endothelial growth factor receptor 2 (VEGF-R2 or KDR). The main mechanism of action of these cells is the formation of new vessels in the infarcted myocardium; however, little evidence exists for their *in vivo* transdifferentiation into new cardiac myocytes (Segers and Lee, 2008). In rats with AMI, intravenously injected EPC stimulated development of collateral vessels from pre-existing vessels as well as *de novo* capillary formation (Kocher et al., 2001). This was associated with decreased apoptosis of myocytes in the borderline zone; reduced fibrosis and scar formation, resulting in prevention of LV remodeling; and improvement in myocardial function (Kocher et al., 2001). It was also reported that infusion of EPC in the infarct-related arteries improves vasomotor function, an effect that could contribute to improved myocardial function (Erbs et al., 2007). In patients with old MI and chronic coronary total occlusion, intracoronary infusion of EPC after recanalization of the occluded artery improved myocardial perfusion, reduced infarct size, and ameliorated myocardial function (Erbs et al., 2005).

However, a clinical trial that was comparing the effects of G-CSF and PBMCs as an alternative approach to recruit EPCs at the sites of myocardial infarction was terminated prematurely due to the potential adverse reaction of increased restenosis (Kang et al., 2004). Lately, approaches that involve EPC therapies combined with gene therapies or even the genetic manipulation of EPCs before transplantation have emerged in an attempt to minimize side-effects and improve outcome (Roncalli et al., 2008).

MESENCHYMAL STEM CELLS (MSCs)

The MSC, an adult stem cell with self-replication and differentiation capacity, represents a promising adult stem cell for regenerative medicine. MSCs can be isolated from a variety of tissues such as adipose, umbilical cord/umbilical cord blood, and BM, although whether they all share common cardiopoietic and immunomodulatory properties is still not clear. MSCs lack hematopoietic lineage markers such as CD14, CD34, and CD45 and express specific stromal cell-surface markers such as Stro1, CD105, CD90, and CD71. They adhere to plastic surfaces and grow as cell monolayers without losing their stem cell phenotype. Furthermore, they have reduced expression levels of MHC class-I molecule and lack MHC class-II (although interferon-γ will induce MHC class-II) and co-stimulatory molecules CD80 (B7-1), CD86 (B7-2), and CD40. MSCs are therefore the prototypic immunoprivileged cell-based therapy and have been tested in phase I double-blind randomized clinical trials as an allogeneic graft (Hare et al., 2009).

The mechanism of action of MSCs as a cardiac regenerative agent appears to be multi-factorial. While definite evidence of their *in vivo* transdifferentiation into cardiovascular elements is reported, the degree to which they differentiate does not explain in full their substantial cardiac reparative properties (Quevedo et al., 2009; Schuleri et al., 2009; Hatzistergos et al., 2010). Indeed, MSCs appear to have additional powerful effects mediated by secreted factors and cytokines that evoke the therapeutic response (Mirotsou et al., 2007). However, recent data document that MSCs facilitate cardiac regeneration through mechanisms that involve both differentiation and paracrine stimulation of innate repair pathways (Mazhari and Hare, 2007; Hatzistergos et al., 2010). Particularly, MSCs seem to have a unique capacity to gain control over the endogenous c-kit+ cardiac precursors cell content and establish the necessary cardiopoietic cues that instruct the latter to massively regenerate a myocardial scar (Hatzistergos et al., 2010).

In various animal models of AMI, intramyocardial injection of MSCs prevents ventricular remodeling by reducing scar formation, leading to a net improvement in myocardial function (Fig. 19.6) (Quevedo et al., 2009; Schuleri et al., 2009). However, it should also be mentioned that some studies in mice with AMI failed to show a sustained benefit of MSC therapy despite an early benefit (Dai et al., 2005; Meyer et al., 2006). In patients with AMI, intracoronary MSC infusion improved myocardial perfusion and function and reduced LV end-systolic and end-diastolic volumes (Chen et al., 2004). In another small study, combined intracoronary

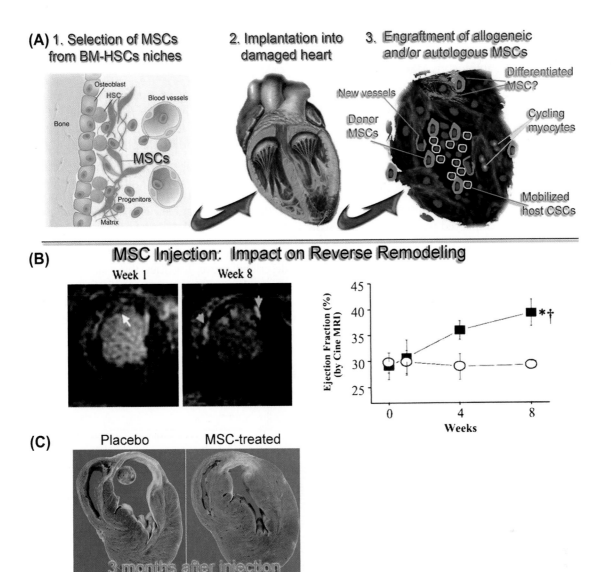

(A) 1. Selection of MSCs from BM-HSCs niches

2. Implantation into damaged heart

3. Engraftment of allogeneic and/or autologous MSCs

(B) MSC Injection: Impact on Reverse Remodeling

(C) Placebo — MSC-treated

FIGURE 19.6
(A) MSCs can provide a safe allogeneic source for cell-based therapies. Their mechanisms of action are believed to be multifaceted since activation of innate (stimulation of angiogenesis, cardiomyocyte proliferation, CSC mobilization) and exogenous (differentiation onto cardiovascular lineages) repair pathways have been reported. (B) Prevention of remodeling in the porcine heart. Cardiac MRI and MDCT document the development of a sub-endocardial rim following MSC transplantation in the damaged zones of infarcted hearts. The newly formed tissue rendered an ~50% decrease in infarct size and restoration of the heart function. (C) Reverse remodeling induced by MSC injection in the porcine heart.

administration of MSC and EPC improved perfusion and contractility of the infarcted area (Katritsis et al., 2005). Even though early uncontrolled animal studies suggested that intra-coronary injection of MSCs could induce coronary artery occlusion and MI (Vulliet et al., 2004), this was not observed in humans (Chen et al., 2004). In a rat model of dilated cardiomyopathy, intramyocardial injection of MSC exerted antifibrotic effects and improved myocardial perfusion and function (Nagaya et al., 2005). A number of strategies have been developed in order to enhance the efficacy of MSCs. Studies in animals showed that *ex vivo* modification of MSCs resulting in overexpression of anti-apoptotic genes augments their regenerative potential (Mangi et al., 2003). MSC treatment appears to be safe (Amado et al., 2005, 2006; Lim et al., 2006; Hu et al., 2007). Importantly, allogeneic MSCs are not rejected (Amado et al., 2005), suggesting that MSCs may obviate the need to harvest bone marrow from patients (Amado et al., 2005). Furthermore, EPCs and BMMNCs from patients with established CHD or cardiovascular risk factors show impaired proliferating and migratory capacity

(Vasa et al., 2001; Heeschen et al., 2004). Therefore, treatment of patients with CHD with MSCs obtained from healthy patients may be equally safe and more advantageous than using autologous MSCs.

The most definitive clinical study of allogeneic MSCs was a 53-patient double-blind placebo-controlled trial of MSCs administered intravenously within 10 days after acute MI (Hare et al., 2009). This phase I study demonstrates acute and long-term safety of the approach and provides provocative data supporting the conduct of additional phase II studies (Hare et al., 2009). While this study was primarily designed to test safety, a phase I study, four domains of pre-specified safety monitoring supported an improved outcome in the cell-treated patients. These included a reduction in malignant ventricular arrhythmias (Fig. 19.7), improved pulmonary function, an improved EF in the subset of patients with anterior MI, and finally an improved patient well-being score at 6 months.

The total database of studies of MSC therapy for acute MI now includes three proofs of concept and the phase I clinical trial described above. Together, these studies illustrate that MSCs can be successfully used either as autologous or allogeneic grafts for treating heart disease (Chen et al., 2004, 2006; Katritsis et al., 2005). While safety has served as the primary end-point thus far, an overall provisional efficacy of MSCs on reversing heart disease has also

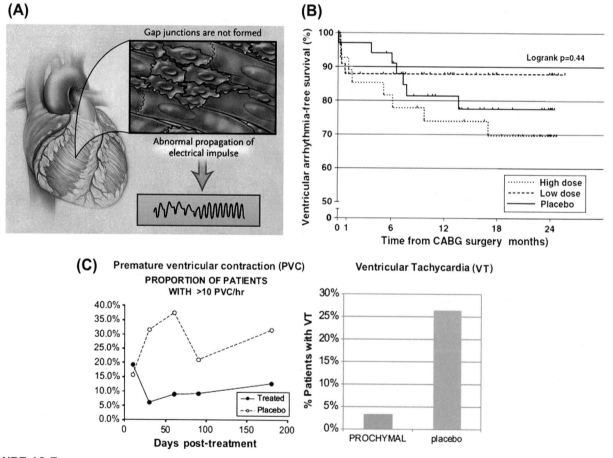

FIGURE 19.7

(A) Experimental studies have shown that skeletal fibroblasts engraft and survive in the damaged myocardium. However, these contractile cells do not express gap junctional proteins such as Connexin-43 and fail to couple with the host myocytes. As a result, the transplanted cells cannot propagate the conduction signals, giving rise to an arrhythmogenic substrate. (B) In the MAGIC trial, patients with ischemic cardiomyopathy who received skeletal myoblasts were more prone to developing arrhythmias compared to the placebo-treated group. CABG = coronary artery bypass grafting. (C) MSCs, in contrast, have an anti-arrhythmic effect.

been strongly suggested. All four studies report significant improvements in cardiac function accompanied by a substantial reduction in scar size, whereas, in addition, 42% of the patients that were treated with an allogeneic "off-the-shelf" MSC preparation showed improvement in their overall condition compared to 11% of patients who had received the placebo and also improved. As a result, MSCs are now in phase II trials for acute MI and there are a number of phase I/II clinical trials under way for ischemic cardiomyopathy, supported by robust preclinical data (Fig. 19.6C, clinicaltrials.gov).

MYOBLASTS

Skeletal myoblasts were the first contractile cell type transplanted in the infarcted heart with the goal of restoring cardiac function (Yoon et al., 1995). These cells can be purified from the skeletal muscle of the patient and, after expansion into therapeutic quantities, can be transplanted into the myocardium. Transplantation of these cells to the heart is accomplished either surgically or by catheter delivery system (Heldman and Hare, 2007). There have been two large well-conducted phase I/II clinical studies (Menasche, 2008). The MAGIC trial revealed that there was a dose-dependent attenuation in LV remodeling that, however, was not accompanied by functional improvements in cardiac function (Fig. 19.7). In addition, there are ongoing concerns that skeletal myoblast can precipitate arrhythmias (Menasche et al., 2008).

In experimental models of AMI, intramyocardial injection of myoblasts preserves myocardial function and abrogates the remodeling process (Ghostine et al., 2002). Myoblasts can differentiate into slow-twitch myotubes in the infarcted area, which can contribute to myocardial systole (Ghostine et al., 2002). It has been suggested that skeletal muscle-derived stem cells have greater potential for myocyte regeneration than myoblasts and can additionally stimulate innate angiogenesis. This cell population appears to be more effective in improving myocardial perfusion and contractility and attenuating remodeling in animal models of MI (Oshima et al., 2005). The failure of myoblasts to improve cardiac function in humans has been attributed to their inability to differentiate into cardiac myocytes and the *in situ* development of dysfunctional electrical coupling with resident cardiomyocytes (Fig. 19.7) (Reinecke et al., 2002). Recent studies are focused towards the identification and characterization of a more cardiogenic skeletal muscle-derived cell population that may improve cardiac repair (Okada et al., 2008).

CARDIAC STEM CELLS

CSCs are tissue-specific stem cells that reside within the heart itself (Fig. 19.8). Cardiac progenitors were first reported in 2002, when Hierlihy et al. (2002) detected a robust side population of cells (SP cells) in the post-natal murine heart that expressed the ATP-binding cassette transporter Abcg2 and extruded Hoechst dye. These cells represented ∼1% of total cardiac cells and differentiated into cardiac myocytes *in vitro*. Following this observation, in 2003 two different groups, Beltrami et al. (2003) and Oh et al. (Oh et al., 2003; Matsuura et al., 2004), reported the isolation and characterization of two novel cardiac stem cells (CSCs) from the murine heart. These resident stem cells have been reported to correspond to from 0.01 to 2% (or ∼1 CSC for every 13,000 cardiomyocytes) of the total cell population of the human heart and are mostly recognized according to the expression of three cell-surface markers: C-kit (the receptor for stem cell factor (SCF)), MDR-1 (multidrug resistance protein-1), and/or Sca-1 (stem cell antigen-1). CSCs are self-renewing, clonogenic, multipotent, and are able to differentiate both *in vitro* and *in vivo* into myogenic, endothelial, and vascular smooth muscle lineages. Two different methods for the isolation of human CSCs have been reported, but whether the purified cells share the same properties is yet unknown. The first method involves the homogenization of relatively large amounts of cardiac tissue (∼30−60 mg for successful isolation) and subsequent antibody-based selection of CSCs (Bearzi et al., 2007). It is apparent that the applicability of this method is limited only in patients that undergo major cardiac interventions such as CABG, LVAD placement, or heart transplantation. The second method

339

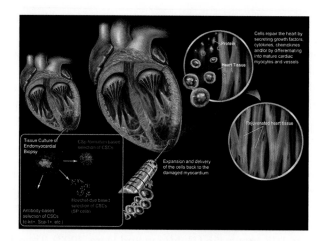

FIGURE 19.8

Cardiac stem cells represent a heterogeneous population of myogenic and vasculogenic progenitor cells. They can be purified from the heart tissue based on the expression of surface molecules (such as c-kit, sca-1, abcg2, and MDR1), their ability to extrude Hoechst dye, or based on a novel identified property to develop cardiospheres. The exact differences and cardiopoietic potentials between these different CSC populations are not fully understood. Modified from http://www.mirm.pitt.edu/news/article.asp?qEmpID=110

has been adapted from the field of neurological sciences and involves the culture of a single biopsy, from which cardiac stem cells are selected with antibodies as a subpopulation of the outgrowing cells. Alternatively, the CSCs can be selected without the use of antibodies, based on their property of forming cardiospheres (Messina et al., 2004). The discovery of cardiac stem cells represents a major biological discovery furthering understanding of cardiac pathophysiology and facilitating cardiac cell-based therapeutics. Interestingly, in the post-natal senescent human heart, CSCs are found to reside in structures with the properties of stem cell niches. These niches are structurally and functionally similar to stem cell niches found in highly regenerating tissues such as bone marrow, gut, and hair follicles. In cardiac niches, CSCs are regulated by the surrounding cellular and non-cellular constituents so as to maintain homeostasis of both the myocardium and the niche population throughout their lifespan. CSCs undergo either symmetric (one CSC gives rise to two CSCs) or asymmetric (one CSC gives rise to one CSC and one committed cell (i.e cardiomyocyte precursor)) division.

Following myocardial damage, CSC niches are also damaged and replaced by scarred tissue, thereby restricting the capacity of the heart to heal itself (Mazhari and Hare, 2007). Another important concept that arises from the description of CSC niches is that of host-related dysfunction of CSCs and/or niches due to comorbid diseases or aging (Anversa et al., 2006).

C-kit$^+$ CSCs

C-kit$^+$ CSCs represent a highly promising candidate for cardiac-specific stem cell lineages. This cell type is extensively described in multiple species ranging from rodents to large animals to humans. Endogenous cardiac repair mechanisms involve the mobilization of c-kit$^+$ CSCs to the areas of cardiac injury soon after infarction (Fransioli et al., 2008). In addition, animal studies document how implantation of cardiac stem cells in rodent myocardium can reduce infarct size and improve cardiac function through their extensive differentiation into new cardiac muscle and vasculature (Beltrami et al., 2003). Based on these findings, c-kit$^+$ CSCs are the first cardiac-specific stem cell population to be approved for human testing in a phase I clinical trial (www.clinicaltrials.gov, NCT00474461).

Other CSCs

Sca-1$^+$ CSCs are an alternative adult CSC. Evaluation of the corresponding human cell is limited by the absence of the Sca-1 antigen in humans (Holmes and Stanford, 2007). The

Islet-1$^+$ CSCs have been isolated only from embryonic and very young murine cardiac tissues that do not exceed 8 days of age (Cai et al., 2003; Barile et al., 2007), indicating that they possibly represent cell remnants from embryonic development. Isl-1 cardioblasts have not yet been isolated from humans, but have been developed in the lab from bioengineered ESCs and iPS human stem cell lines (Bu et al., 2009; Moretti et al., 2010). Therefore, our knowledge of their reparative capacities following cardiac injury is very limited, and more studies are needed in order to assess whether their use can comprise a good therapeutic strategy. Abcg2$^+$ side population cells are a well-characterized cardiac precursor cell population in rodent hearts (Hierlihy et al., 2002; Pfister et al., 2008). However, although immunohistological studies have documented the existence of MDR1$^+$ and Abcg2$^+$ cells in the post-natal human heart (Quaini et al., 2002; Meissner et al., 2006), the isolation and expansion of these cells into therapeutic quantities is yet to be reported.

METHODS FOR EXPANSION OF CARDIAC STEM CELLS

Two general mechanisms have been employed to isolate and expand cardiac stem cells. These are antigen panning techniques to identify cells such as c-kit, sca-1, or abcg-2 or direct cell amplification. In the latter case, cells can be readily amplified from cardiac explants. These have been termed cardiospheres (CSs) or cardiac explants-derived precursor cells (CEDPC).

Cardiosphere-forming cells

There have been several attempts to culture cells from the adult heart. Messina and colleagues reported on "cardiospheres," structures akin to "neurospheres." CS are self-aggregating structures arising from cultured cardiac cells, and represent a heterogeneous population, possessing cardiopoietic properties *in vitro* and *in vivo* (Smith et al., 2007; Takehara et al., 2008). CSs can be derived from human biopsies and are reported to contain c-kit$^+$, Sca-1$^+$, and Flk1$^+$ cells. When co-cultured with neonatal rat cardiomyocytes, they transdifferentiate into cardiomyocytes, demonstrating calcium transients synchronous among the myocytes as well as spontaneous action potentials. When injected into infarcted rat hearts, ventricular function improved. In a recent study by Johnston et al., CSCs were injected into a porcine model of MI and were shown to abbreviate but not reverse progressive cardiac remodeling (Johnston et al., 2009).

Cardiospheres represent a potential therapeutic opportunity because of their ability to expand potential cardiac precursor cells from smaller amounts of myocardial tissue, such as a cardiac biopsy. Cardiospheres are incompletely characterized, and whether they offer superior regenerative capacities to BM-derived stem cells will require formal testing. Whether the cells comprising CSs offer any significant advantage is unclear and other methods for CEDPC isolation have been reported. CS-derived cells (CDCs) have entered clinical trials (Johnston et al., 2009).

CONCLUSIONS

The past decade has witnessed the rapid development of mechanistic and clinical trial support for the notion of a new paradigm in treatment for heart disease based upon cellular therapeutics. Several key insights have emerged supporting this paradigm. They include: (1) the discovery that the heart has capacity for self-renewal and harbors reservoirs of precursor cells that can be tapped or manipulated for therapeutic benefit; (2) remote cell sources, notably bone marrow, also contain cellular constituents with profound therapeutic potential; (3) cell-based therapies have a remarkable safety profile and can be delivered by diverse methodologies that range from intravenous administration in acute MI to directed catheter-based injection systems in chronic heart failure. There is an emerging database of clinical trials and fundamental scientific enquiry that provides a foundation for this strategy and holds promise for a treatment strategy aimed at a key pathophysiologic target in heart disease, that of ventricular remodeling.

References

Abdel-Latif, A., Bolli, R., Tleyjeh, I. M., et al. (2007). Adult bone marrow-derived cells for cardiac repair: a systematic review and meta-analysis. *Arch. Intern. Med., 167*, 989–997.

Amado, L. C., Saliaris, A. P., Schuleri, K. H., et al. (2005). Cardiac repair with intramyocardial injection of allogeneic mesenchymal stem cells after myocardial infarction. *Proc. Natl. Acad. Sci. U.S.A., 102*, 11474–11479.

Amado, L. C., Schuleri, K. H., Saliaris, A. P., et al. (2006). Multimodality noninvasive imaging demonstrates in vivo cardiac regeneration after mesenchymal stem cell therapy. *J. Am. Coll. Cardiol., 48*, 2116–2124.

Anversa, P., & Kajstura, J. (1998). Ventricular myocytes are not terminally differentiated in the adult mammalian heart. *Circ. Res., 83*, 1–14.

Anversa, P., Kajstura, J., Leri, A., et al. (2006). Life and death of cardiac stem cells: a paradigm shift in cardiac biology. *Circulation, 113*, 1451–1463.

Asahara, T., Murohara, T., Sullivan, A., et al. (1997). Isolation of putative progenitor endothelial cells for angiogenesis. *Science, 275*, 964–967.

Assmus, B., Rolf, A., Erbs, S., Elsässer, A., Haberbosch, W., Hambrecht, R., et al. (2010). Clinical outcome 2 years after intracoronary administration of bone marrow-derived progenitor cells in acute myocardial infarction. *Circ. Heart Fail., 3*, 89–96.

Barile, L., Messina, E., Giacomello, A., et al. (2007). Endogenous cardiac stem cells. *Prog. Cardiovasc. Dis., 50*, 31–48.

Bearzi, C., Rota, M., Hosoda, T., et al. (2007). Human cardiac stem cells. *Proc. Natl. Acad. Sci. U.S.A., 104*, 14068–14073.

Behfar, A., Perez-Terzic, C., Faustino, R. S., et al. (2007). Cardiopoietic programming of embryonic stem cells for tumor-free heart repair. *J. Exp. Med., 204*, 405–420.

Beltrami, A. P., Barlucchi, L., Torella, D., et al. (2003). Adult cardiac stem cells are multipotent and support myocardial regeneration. *Cell, 114*, 763–776.

Brade, T., Gessert, S., Kuhl, M., et al. (2007). The amphibian second heart field: Xenopus islet-1 is required for cardiovascular development. *Dev. Biol., 311*, 297–310.

Buckingham, M., Meilhac, S., & Zaffran, S. (2005). Building the mammalian heart from two sources of myocardial cells. *Nat. Rev. Genet., 6*, 826–835.

Bu, L., Jiang, X., Martin-Puig, S., Caron, L., Zhu, S., Shao, Y., et al. (2009). Human ISL1 heart progenitors generate diverse multipotent cardiovascular cell lineages. *Nature, 460*(7251), 113–117.

bu-Issa, R., Waldo, K., & Kirby, M. L. (2004). Heart fields: one, two or more? *Dev. Biol., 272*, 281–285.

Bunting, K. D. (2002). ABC transporters as phenotypic markers and functional regulators of stem cells. *Stem Cells, 20*, 11–20.

Burt, R. K., Loh, Y., Pearce, W., et al. (2008). Clinical applications of blood-derived and marrow-derived stem cells for nonmalignant diseases. *JAMA, 299*, 925–936.

Cai, C. L., Liang, X., Shi, Y., et al. (2003). Isl1 identifies a cardiac progenitor population that proliferates prior to differentiation and contributes a majority of cells to the heart. *Dev. Cell, 5*, 877–889.

Caspi, O., Huber, I., Kehat, I., et al. (2007). Transplantation of human embryonic stem cell-derived cardiomyocytes improves myocardial performance in infarcted rat hearts. *J. Am. Coll. Cardiol., 50*, 1884–1893.

Charron, F., & Nemer, M. (1999). GATA transcription factors and cardiac development. *Semin. Cell Dev. Biol., 10*, 85–91.

Chen, S., Liu, Z., Tian, N., et al. (2006). Intracoronary transplantation of autologous bone marrow mesenchymal stem cells for ischemic cardiomyopathy due to isolated chronic occluded left anterior descending artery. *J. Invasive Cardiol., 18*, 552–556.

Chen, S. L., Fang, W. W., Ye, F., et al. (2004). Effect on left ventricular function of intracoronary transplantation of autologous bone marrow mesenchymal stem cell in patients with acute myocardial infarction. *Am. J. Cardiol., 94*, 92–95.

Christoforou, N., & Gearhart, J. D. (2007). Stem cells and their potential in cell-based cardiac therapies. *Prog. Cardiovasc. Dis., 49*, 396–413.

Christoforou, N., Miller, R. A., Hill, C. M., et al. (2008). Mouse ES cell-derived cardiac precursor cells are multipotent and facilitate identification of novel cardiac genes. *J. Clin. Invest, 120*, 20–28.

Christoforou, N., Oskouei, B. N., Esteso, P., Hill, C. M., Zimmet, J. M., Bian, W., et al. (2010). Implantation of mouse embryonic stem cell-derived cardiac progenitor cells preserves function of infarcted murine hearts. *PLoS One, 5*, e11536.

Cyranoski, D. (2008). Stem cells: 5 things to know before jumping on the iPS bandwagon. *Nature, 452,* 406—408.

Dai, W., Hale, S.L., Martin, B.J., et al. (2005). Allogeneic mesenchymal stem cell transplantation in postinfarcted rat myocardium: short- and long-term effects. *Circulation, 112,* 214—223.

Dimmeler, S., Burchfield, J., & Zeiher, A. M. (2008). Cell-based therapy of myocardial infarction. *Arterioscler. Thromb. Vasc. Biol., 28,* 208—216.

D'Ippolito, G., Howard, G. A., Roos, B. A., et al. (2006). Isolation and characterization of marrow-isolated adult multilineage inducible (MIAMI) cells. *Exp. Hematol., 34,* 1608—1610.

Dodou, E., Verzi, M. P., Anderson, J. P., et al. (2004). Mef2c is a direct transcriptional target of ISL1 and GATA factors in the anterior heart field during mouse embryonic development. *Development, 131,* 3931—3942.

Domian, I. J., Chiravuri, M., van der Meer, P., et al. (2009). Generation of functional ventricular heart muscle from mouse ventricular progenitor cells. *Science, 326,* 426—429.

Erbs, S., Linke, A., Adams, V., et al. (2005). Transplantation of blood-derived progenitor cells after recanalization of chronic coronary artery occlusion: first randomized and placebo-controlled study. *Circ. Res., 97,* 756—762.

Erbs, S., Linke, A., Schachinger, V., et al. (2007). Restoration of microvascular function in the infarct-related artery by intracoronary transplantation of bone marrow progenitor cells in patients with acute myocardial infarction: the Doppler Substudy of the Reinfusion of Enriched Progenitor Cells and Infarct Remodeling in Acute Myocardial Infarction (REPAIR-AMI) trial. *Circulation, 116,* 366—374.

Evans, M. J., & Kaufman, M. H. (1981). Establishment in culture of pluripotential cells from mouse embryos. *Nature, 292,* 154—156.

Fransioli, J., Bailey, B., Gude, N. A., et al. (2008). Evolution of the c-kit-positive cell response to pathological challenge in the myocardium. *Stem Cells, 26,* 1315—1324.

Garry, D. J., & Olson, E. N. (2006). A common progenitor at the heart of development. *Cell, 127,* 1101—1104.

Ghostine, S., Carrion, C., Souza, L. C., et al. (2002). Long-term efficacy of myoblast transplantation on regional structure and function after myocardial infarction. *Circulation, 106,* I131—I136.

Guan, K., & Hasenfuss, G. (2007). Do stem cells in the heart truly differentiate into cardiomyocytes? *J. Mol. Cell Cardiol., 43,* 377—387.

Hare, J. M., Traverse, J. H., Henry, T. D., et al. (2009). A randomized, double-blind, placebo-controlled, dose-escalation study of intravenous adult human mesenchymal stem cells (prochymal) after acute myocardial infarction. *J. Am. Coll. Cardiol., 54,* 2277—2286.

Hatzistergos, K. E., Quevedo, H., Oskouei, B. N., Hu, Q., Feigenbaum, G. S., Margitich, I. S., et al. (2010). Bone marrow mesenchymal stem cells stimulate cardiac stem cell proliferation and differentiation. *Circ. Res., 107*(7), 913—922.

Heeschen, C., Lehmann, R., Honold, J., et al. (2004). Profoundly reduced neovascularization capacity of bone marrow mononuclear cells derived from patients with chronic ischemic heart disease. *Circulation, 109,* 1615—1622.

Heldman, A. W., & Hare, J. M. (2007). Cell therapy for myocardial infarction: special delivery. *J. Mol. Cell Cardiol, 44,* 473—476.

Hierlihy, A. M., Seale, P., Lobe, C. G., et al. (2002). The post-natal heart contains a myocardial stem cell population. *FEBS Lett., 530,* 239—243.

Holmes, C., & Stanford, W. L. (2007). Concise review: stem cell antigen-1: expression, function, and enigma. *Stem Cells, 25,* 1339—1347.

Hu, T., Yamagishi, H., Maeda, J., et al. (2004). Tbx1 regulates fibroblast growth factors in the anterior heart field through a reinforcing autoregulatory loop involving forkhead transcription factors. *Development, 131,* 5491—5502.

Hu, X., Wang, J., Chen, J., et al. (2007). Optimal temporal delivery of bone marrow mesenchymal stem cells in rats with myocardial infarction. *Eur. J. Cardiothorac. Surg., 31,* 438—443.

Jackson, K. A., Majka, S. M., Wang, H., et al. (2001). Regeneration of ischemic cardiac muscle and vascular endothelium by adult stem cells. *J. Clin. Invest., 107,* 1395—1402.

Jiang, Y., Jahagirdar, B. N., Reinhardt, R. L., et al. (2002). Pluripotency of mesenchymal stem cells derived from adult marrow. *Nature, 418,* 41—49.

Johnston, P. V., Sasano, T., Mills, K., et al. (2009). Engraftment, differentiation, and functional benefits of autologous cardiosphere-derived cells in porcine ischemic cardiomyopathy. *Circulation, 120,* 1075—1083.

Kang, H. J., Kim, H. S., Zhang, S. Y., et al. (2004). Effects of intracoronary infusion of peripheral blood stem-cells mobilised with granulocyte-colony stimulating factor on left ventricular systolic function and restenosis after coronary stenting in myocardial infarction: the MAGIC cell randomised clinical trial. *Lancet, 363,* 751—756.

Katritsis, D. G., Sotiropoulou, P. A., Karvouni, E., et al. (2005). Transcoronary transplantation of autologous mesenchymal stem cells and endothelial progenitors into infarcted human myocardium. *Catheter Cardiovasc. Interv., 65,* 321—329.

Kelly, R. G., Brown, N. A., & Buckingham, M. E. (2001). The arterial pole of the mouse heart forms from Fgf10-expressing cells in pharyngeal mesoderm. *Dev. Cell, 1,* 435–440.

Kissel, C. K., Lehmann, R., Assmus, B., et al. (2007). Selective functional exhaustion of hematopoietic progenitor cells in the bone marrow of patients with postinfarction heart failure. *J. Am. Coll. Cardiol., 49,* 2341–2349.

Kocher, A. A., Schuster, M. D., Szabolcs, M. J., et al. (2001). Neovascularization of ischemic myocardium by human bone-marrow-derived angioblasts prevents cardiomyocyte apoptosis, reduces remodeling and improves cardiac function. *Nat. Med., 7,* 430–436.

Koh, G. Y., Soonpaa, M. H., Klug, M. G., et al. (1993). Long-term survival of AT-1 cardiomyocyte grafts in syngeneic myocardium. *Am. J. Physiol., 264,* H1727–H1733.

Kucia, M., Reca, R., Campbell, F. R., et al. (2006). A population of very small embryonic-like (VSEL) CXCR4(+) SSEA-1(+)Oct-4+ stem cells identified in adult bone marrow. *Leukemia, 20,* 857–869.

Lam, J. T., Moretti, A., & Laugwitz, K. L. (2009). Multipotent progenitor cells in regenerative cardiovascular medicine. *Pediatr. Cardiol., 30,* 690–698.

Li, Z., Wu, J. C., Sheikh, A. Y., et al. (2007). Differentiation, survival, and function of embryonic stem cell derived endothelial cells for ischemic heart disease. *Circulation, 116,* I46–I54.

Lim, S. Y., Kim, Y. S., Ahn, Y., et al. (2006). The effects of mesenchymal stem cells transduced with Akt in a porcine myocardial infarction model. *Cardiovasc. Res., 70,* 530–542.

Maeda, J., Yamagishi, H., McAnally, J., et al. (2006). Tbx1 is regulated by forkhead proteins in the secondary heart field. *Dev. Dyn., 235,* 701–710.

Mangi, A. A., Noiseux, N., Kong, D., et al. (2003). Mesenchymal stem cells modified with Akt prevent remodeling and restore performance of infarcted hearts. *Nat. Med., 9,* 1195–1201.

Martin, C. M., Meeson, A. P., Robertson, S. M., et al. (2004). Persistent expression of the ATP-binding cassette transporter, Abcg2, identifies cardiac SP cells in the developing and adult heart. *Dev. Biol., 265,* 262–275.

Martin, G. R. (1981). Isolation of a pluripotent cell line from early mouse embryos cultured in medium conditioned by teratocarcinoma stem cells. *Proc. Natl. Acad. Sci. U.S.A., 78,* 7634–7638.

Martin-Rendon, E., Brunskill, S. J., Hyde, C. J., et al. (2008). Autologous bone marrow stem cells to treat acute myocardial infarction: a systematic review. *Eur. Heart J., 15,* 1807–1818.

Matsuura, K., Nagai, T., Nishigaki, N., et al. (2004). Adult cardiac Sca-1-positive cells differentiate into beating cardiomyocytes. *J. Biol. Chem., 279,* 11384–11391.

Mazhari, R., & Hare, J. M. (2007). Mechanisms of action of mesenchymal stem cells in cardiac repair: potential influences on the cardiac stem cell niche. *Nat. Clin. Pract. Cardiovasc. Med., 4*(Suppl. 1), S21–S26.

Meilhac, S. M., Esner, M., Kelly, R. G., et al. (2004). The clonal origin of myocardial cells in different regions of the embryonic mouse heart. *Dev. Cell, 6,* 685–698.

Meissner, K., Heydrich, B., Jedlitschky, G., et al. (2006). The ATP-binding cassette transporter ABCG2 (BCRP), a marker for side population stem cells, is expressed in human heart. *J. Histochem. Cytochem., 54,* 215–221.

Menasche, P., Alfieri, O., Janssens, S., et al. (2008). The Myoblast Autologous Grafting in Ischemic Cardiomyopathy (MAGIC) trial: first randomized placebo-controlled study of myoblast transplantation. *Circulation, 117,* 1189–1200.

Menasche, P. (2008). Skeletal myoblasts for cardiac repair: Act II? *J. Am. Coll. Cardiol., 52,* 1881–1883.

Messina, E., De, A. L., Frati, G., et al. (2004). Isolation and expansion of adult cardiac stem cells from human and murine heart. *Circ. Res., 95,* 911–921.

Meyer, G. P., Wollert, K. C., Lotz, J., et al. (2006). Intracoronary bone marrow cell transfer after myocardial infarction: eighteen months' follow-up data from the randomized, controlled BOOST (BOne marrOw transfer to enhance ST-elevation infarct regeneration) trial. *Circulation, 113,* 1287–1294.

Mirotsou, M., Zhang, Z., Deb, A., et al. (2007). Secreted frizzled related protein 2 (Sfrp2) is the key Akt-mesenchymal stem cell-released paracrine factor mediating myocardial survival and repair. *Proc. Natl. Acad. Sci. U.S.A., 104,* 1643–1648.

Moretti, A., Bellin, M., Jung, C. B., Thies, T. M., Takashima, Y., Bernshausen, A., et al. (2010). Mouse and human induced pluripotent stem cells as a source for multipotent Isl1+ cardiovascular progenitors. *FASEB J., 24*(3), 700–711.

Moretti, A., Caron, L., Nakano, A., et al. (2006). Multipotent embryonic isl1+ progenitor cells lead to cardiac, smooth muscle, and endothelial cell diversification. *Cell, 127,* 1151–1165.

Mosterd, A., & Hoes, A. W. (2007). Clinical epidemiology of heart failure. *Heart, 93,* 1137–1146.

Nagaya, N., Kangawa, K., Itoh, T., et al. (2005). Transplantation of mesenchymal stem cells improves cardiac function in a rat model of dilated cardiomyopathy. *Circulation, 112,* 1128–1135.

Nakanishi, C., Yamagishi, M., Yamahara, K., et al. (2008). Activation of cardiac progenitor cells through paracrine effects of mesenchymal stem cells. *Biochem. Biophys. Res. Commun., 374,* 11–16.

Nygren, J. M., Jovinge, S., Breitbach, M., et al. (2004). Bone marrow-derived hematopoietic cells generate cardiomyocytes at a low frequency through cell fusion, but not transdifferentiation. *Nat. Med., 10*, 494–501.

Oh, H., Bradfute, S. B., Gallardo, T. D., et al. (2003). Cardiac progenitor cells from adult myocardium: homing, differentiation, and fusion after infarction. *Proc. Natl. Acad. Sci. U.S.A., 100*, 12313–12318.

Okada, M., Payne, T. R., Zheng, B., et al. (2008). Myogenic endothelial cells purified from human skeletal muscle improve cardiac function after transplantation into infarcted myocardium. *J. Am. Coll. Cardiol., 52*, 1869–1880.

Okita, K., Nakagawa, M., Hyenjong, H., et al. (2008). Generation of mouse induced pluripotent stem cells without viral vectors. *Science, 322*, 949–953.

Orlic, D., Kajstura, J., Chimenti, S., et al. (2001). Bone marrow cells regenerate infarcted myocardium. *Nature, 410*, 701–705.

Oshima, H., Payne, T. R., Urish, K. L., et al. (2005). Differential myocardial infarct repair with muscle stem cells compared to myoblasts. *Mol. Ther., 12*, 1130–1141.

Park, E. J., Ogden, L. A., Talbot, A., et al. (2006). Required, tissue-specific roles for Fgf8 in outflow tract formation and remodeling. *Development, 133*, 2419–2433.

Pfister, O., Oikonomopoulos, A., Sereti, K. I., et al. (2008). Role of the ATP-binding cassette transporter Abcg2 in the phenotype and function of cardiac side population cells. *Circ. Res., 103*, 825–835.

Phan, D., Rasmussen, T. L., Nakagawa, O., et al. (2005). BOP, a regulator of right ventricular heart development, is a direct transcriptional target of MEF2C in the developing heart. *Development, 132*, 2669–2678.

Prall, O. W., Menon, M. K., Solloway, M. J., et al. (2007). An Nkx2-5/Bmp2/Smad1 negative feedback loop controls heart progenitor specification and proliferation. *Cell, 128*, 947–959.

Quaini, F., Urbanek, K., Beltrami, A. P., et al. (2002). Chimerism of the transplanted heart. *N. Engl. J. Med., 346*, 5–15.

Quevedo, H.C., Hatzistergos, K.E., Oskouei, B.N., et al. (2009). Allogeneic mesenchymal stem cells restore cardiac function in chronic ischemic cardiomyopathy via trilineage differentiating capacity. *Proc. Natl. Acad. Sci. U.S.A., 106*, 14022–14027.

Reinecke, H., Poppa, V., & Murry, C. E. (2002). Skeletal muscle stem cells do not transdifferentiate into cardiomyocytes after cardiac grafting. *J. Mol. Cell Cardiol., 34*, 241–249.

Roncalli, J., Tongers, J., Renault, M. A., et al. (2008). Biological approaches to ischemic tissue repair: gene- and cell-based strategies. *Expert Rev. Cardiovasc. Ther., 6*, 653–668.

Schachinger, V., Erbs, S., Elsasser, A., et al. (2006). Intracoronary bone marrow-derived progenitor cells in acute myocardial infarction. *N. Engl. J. Med., 355*, 1210–1221.

Schuleri, K. H., Feigenbaum, G. S., Centola, M., et al. (2009). Autologous mesenchymal stem cells produce reverse remodelling in chronic ischaemic cardiomyopathy. *Eur. Heart J., 30*, 2722–2732.

Segers, V. F., & Lee, R. T. (2008). Stem-cell therapy for cardiac disease. *Nature, 451*, 937–942.

Smith, R. R., Barile, L., Cho, H. C., et al. (2007). Regenerative potential of cardiosphere-derived cells expanded from percutaneous endomyocardial biopsy specimens. *Circulation, 115*, 896–908.

Srivastava, D., & Olson, E. N. (2000). A genetic blueprint for cardiac development. *Nature, 407*, 221–226.

Takahashi, K., Tanabe, K., Ohnuki, M., et al. (2007). Induction of pluripotent stem cells from adult human fibroblasts by defined factors. *Cell, 131*, 861–872.

Takehara, N., Tsutsumi, Y., Tateishi, K., et al. (2008). Controlled delivery of basic fibroblast growth factor promotes human cardiosphere-derived cell engraftment to enhance cardiac repair for chronic myocardial infarction. *J. Am. Coll. Cardiol., 52*, 1858–1865.

Tallini, Y. N., Greene, K. S., Craven, M., et al. (2009). C-kit expression identifies cardiovascular precursors in the neonatal heart. *Proc. Natl. Acad. Sci. U.S.A., 106*, 1808–1813.

Vasa, M., Fichtlscherer, S., Aicher, A., et al. (2001). Number and migratory activity of circulating endothelial progenitor cells inversely correlate with risk factors for coronary artery disease. *Circ. Res., 89*, E1–E7.

von Both, I., Silvestri, C., Erdemir, T., et al. (2004). Foxh1 is essential for development of the anterior heart field. *Dev. Cell, 7*, 331–345.

Vulliet, P. R., Greeley, M., Halloran, S. M., et al. (2004). Intra-coronary arterial injection of mesenchymal stromal cells and microinfarction in dogs. *Lancet, 363*, 783–784.

Wu, S. M., Fujiwara, Y., Cibulsky, S. M., et al. (2006). Developmental origin of a bipotential myocardial and smooth muscle cell precursor in the mammalian heart. *Cell, 127*, 1137–1150.

Yang, L., Soonpaa, M. H., Adler, E. D., et al. (2008). Human cardiovascular progenitor cells develop from a KDR+ embryonic-stem-cell-derived population. *Nature, 453*, 524–528.

Yi, B. A., Wernet, O., & Chien, K. R. (2010). Pregenerative medicine: developmental paradigms in the biology of cardiovascular regeneration. *J. Clin. Invest., 120*, 20–28.

Yoon, P. D., Kao, R. L., & Magovern, G. J. (1995). Myocardial regeneration. Transplanting satellite cells into damaged myocardium. *Tex. Heart Inst. J., 22*, 119–125.

Yu, J., Vodyanik, M. A., Smuga-Otto, K., et al. (2007). Induced pluripotent stem cell lines derived from human somatic cells. *Science, 318*, 1917–1920.

Zimmet, J. M., & Hare, J. M. (2005). Emerging role for bone marrow derived mesenchymal stem cells in myocardial regenerative therapy. *Basic Res. Cardiol., 100*, 471–481.

CHAPTER 20

Skeletal Muscle Stem Cells

Benjamin D. Cosgrove, Helen M. Blau
Baxter Laboratory for Stem Cell Biology, Stanford University School of Medicine,
Campus Drive, Center for Clinical Sciences Research, Stanford, CA, USA

INTRODUCTION

Skeletal muscle is composed of bundles of muscle fibers (myofibers) that contract to generate force and movement following excitatory signals. Myofibers are terminally differentiated post-mitotic syncytial cells that contain hundreds to thousands of peripherally localized nuclei and contractile myofibrils within a large shared cytoplasm (Fig. 20.1A). Post-natal skeletal muscle growth, maintenance, and regeneration are dependent on a population of tissue-specific muscle stem cells (MuSCs) present in skeletal muscle. Satellite cells, mononuclear cells that reside in between the myofiber plasma membrane and basal lamina surrounding each myofiber (Mauro, 1961), are the muscle stem cells that most prominently contribute to physiological skeletal muscle regeneration and have been most extensively molecularly and functionally characterized (Fig. 20.1B). Other cell sources with putative muscle stem cell properties include muscle interstitial cells, muscle-derived stem cells, bone marrow-derived hematopoietic progenitors, blood vessel-associated mesoangioblasts and pericytes, and mesenchymal stem cells (Tedesco et al., 2010). Although these non-satellite cell types may not contribute substantially to physiological muscle regeneration, some appear to have advantageous characteristics over satellite cells for regenerative muscle cell therapies.

Normal adult skeletal muscle tissue is relatively stable in homeostatic conditions, with infrequent turnover of muscle cells (only ~1−2% of myonuclei replaced weekly) compared to other regenerative tissues such as the blood and skin (Decary et al., 1996; Schmalbruch and Lewis, 2000). Healthy adult skeletal muscle has a remarkable ability to regenerate in response to injury (Carlson, 1973). In the initial phase of skeletal muscle regeneration, muscle damage and disruption of myofiber integrity lead to an inflammatory response and an infiltration of inflammatory cells (Grounds, 1991). Subsequently, satellite cells are rapidly activated and then proliferate to produce committed muscle progenitor cells (myoblasts) without depleting MuSCs in the satellite cell pool. Myogenic progenitor cells then expand and subsequently fuse with existing myofibers or each other to form *de novo* myofibers, with characteristic centrally located nuclei. By increasing myofiber content, the cells that fuse into myofibers replenish skeletal muscle mass and contractile function.

Inherited muscular dystrophies and non-inherited muscle wasting diseases are characterized by extensive muscle degeneration, and thus are candidates for cell-based regenerative therapies. Heritable muscular dystrophies lead to progressive, and often fatal, muscle wasting. The most common form of muscular dystrophy is Duchenne muscular dystrophy (DMD), a severe X-linked recessive disorder that affects 1 in 3,500 males (Emery, 2002). DMD is caused by

Principles of Regenerative Medicine. DOI: 10.1016/B978-0-12-381422-7.10020-3

mutations in the dystrophin gene, which encodes a protein component of the membrane-spanning dystroglycan complex linking the myofiber cytoskeleton to extracellular matrix proteins of the basal lamina. Dystrophin mutation results in a defective structural complex that, in response to normal shear forces of muscle contraction, leads to failure of myofiber membrane integrity, loss of myofiber function, and widespread muscle degeneration and necrosis. Clinical symptoms of DMD in early childhood include muscle weakness and limited mobility. The regenerative potential of skeletal muscle is progressively depleted, in part due to the reduced proliferative capacity of myogenic progenitor cells caused by the excessive demands of constant degeneration (Webster and Blau, 1990; Blau et al., 1993; Heslop et al., 2000). By young adulthood, symptoms include severe muscle deterioration, increased connective and fibrotic tissue deposition, loss of mobility, and paralysis, ultimately resulting in death due to cardiorespiratory failure.

Treatments for muscular dystrophies and muscle wasting conditions, such as aging-associated sarcopenia and pathology-associated cachexia, must provide therapy to defective post-mitotic myofibers, which are either isolated to specific muscle groups or are present throughout the body depending on the disease context (Jones et al., 2009; Saini et al., 2009). Although numerous pharmacologic and genetic therapies have been developed to support muscle regeneration for these diseases, they have met with little clinical success, especially for the most devastating degenerative conditions (Bogdanovich et al., 2004; Muir and Chamberlain, 2009). Due to these disappointments, increased focus has been placed on the development of muscle stem cell-based interventions for efficacious long-term regenerative therapies. The prospects for clinical muscle stem cell therapies are supported by recent advances demonstrating that MuSCs with remarkable regenerative potential can be prospectively isolated and transplanted in pre-clinical animal models (Tedesco et al., 2010). In addition, there has been substantial progress in elucidating the molecular signals and myogenic gene regulatory mechanisms that orchestrate the regenerative contributions of MuSCs (Cosgrove et al., 2009). In this chapter, we describe the identification of satellite cells as an endogenous source of muscle stem cells that contribute to muscle regeneration. Advances in the regulation of satellite cell activation and self-renewal by biochemical and biophysical components of their niche microenvironment have been substantial. We evaluate the potential of satellite cells and other cell sources with myogenic regenerative properties for cell-based therapy for human muscle diseases. We also describe how technological advances offer promise for advancements in stimulating endogenous stem cells in muscle wasting disorders or for cell therapy of heritable muscular dystrophies.

IDENTIFICATION OF SATELLITE CELLS AS SKELETAL MUSCLE STEM CELLS
Requirements of muscle stem cells

The requirements of a tissue-specific stem cell are (1) to proliferate in a self-renewing manner that yields at least one daughter cell retaining its stem cell identity and (2) to generate progeny capable of differentiating into all of the specialized cells in a given tissue (Fig. 20.2). For some tissues, this requirement implies that a stem cell gives rise to a set of transient progenitors that can produce the complete diversity of terminally differentiated cells that make up the tissue. For example, a single hematopoietic stem cell (HSC) must be capable of replacing the entire cellular blood compartment of the body including all of its manifold lymphoid, myeloid, and erythroid progeny. In contrast, MuSCs have a simpler task, as skeletal muscle tissue is defined by a single terminally differentiated cell type, the multinucleated myofiber. Consequently, a *bona fide* MuSC must be a self-renewing unipotent cell capable of generating progeny that fuse with existing myofibers or generate new myofibers. The requirement of MuSCs to self-renew implies that MuSC division gives rise to specialized progenitors in a manner that does not deplete the stem cell pool. This property distinguishes MuSCs from their myoblast progeny, which are unable to undergo extensive division *in vivo* without further myogenic differentiation causing their depletion. The long-term self-renewal characteristics of MuSCs confer the ability to contribute to myogenesis throughout the lifespan of an individual. Moreover, these fundamental characteristics of MuSCs fulfill essential requirements of an ongoing, continuous therapeutic cell source for long-term skeletal muscle regenerative medicine applications. In this section, we summarize the evidence that satellite cells are an intrinsic skeletal muscle stem cell population, describe their molecular characteristics, and highlight how they can be prospectively isolated and transplanted.

Identification of muscle satellite cells

Almost 50 years ago, Mauro predicted that there were muscle stem cells, which he defined as satellite cells, based on their anatomical location in electron microscopy studies of the peripheral region of frog tibialis anticus muscle (Mauro, 1961). Mauro described satellite cells as mononucleated cells with a high nucleus-to-cytoplasm ratio residing in an anatomical compartment between the myofiber basal lamina and its plasma membrane (Mauro, 1961) (Fig. 20.1A). Prior to this discovery, the cell source for mammalian muscle regeneration was actively questioned. Competing hypotheses proposed that a mononuclear cell reservoir existed within mammalian muscle or that nuclei from damaged myofibers became recellularized as mononuclear cells and then fused to form *de novo* myofibers, as occurs in newts and axolotls. Shortly after Mauro's landmark findings, satellite cells were identified by electron microscopy in other skeletal muscles of the frog and rat. Studies using radiolabeled thymidine incorporation to assess satellite cell proliferation demonstrated that only a small fraction of satellite cells actively proliferate under normal conditions, but, following whole muscle transplantation, a large fraction of satellite cells and myofiber nuclei incorporate thymidine (Mauro, 1979; Bischoff, 1986). Thus, satellite cells, like many tissue-resident stem cells, are not consistently proliferating. When individual myofibers are physically dissociated from muscle and placed in culture, satellite cells are activated and migrate off the myofiber to give rise to proliferating myoblasts (progenitors), which eventually form colonies that differentiate and fuse to form multinucleated syncytial cells (myotubes) resembling myofibers (Bischoff, 1974). These initial observations suggested that satellite cells might serve as a resident cell source for mammalian skeletal muscle regeneration after injury. However, proof of the stem cell function of satellite cells was not obtained until four decades from the time of their discovery by Mauro (Collins et al., 2005; Montarras et al., 2005; Cerletti et al., 2008; Sacco et al., 2008).

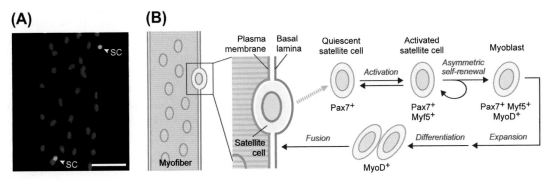

FIGURE 20.1

Satellite cells, a resident muscle stem cell population, activate and self-renew to contribute to muscle regeneration. (A) Confocal microscopy projection image of an isolated mouse myofiber, demonstrating Pax7 transcription factor expression (green) in satellite cells (SC), the myofiber basal lamina (red), and myonuclei (blue). Scale bar, 50 μm. Image courtesy of Rose Tran. (B) Myofibers are terminally differentiated, post-mitotic multinucleated cells that provide the contractile function of skeletal muscle. Satellite cells are mononucleated cells that reside in an anatomical compartment between the myofiber plasma membrane and basal lamina, which serves as a stem cell niche microenvironment. Satellite cells are largely quiescent during homeostasis, but are activated following muscle injury or degeneration. Activated satellite cells undergo proliferative self-renewal to produce differentiated myoblasts (myogenic progenitor cells) while retaining a pool of satellite cells with muscle stem cell properties. Myoblasts further differentiate and fuse into myofibers to repair injured tissue and replenish contractile function.

Molecular and functional characteristics of satellite cells

Since the discovery of satellite cells, our understanding of the molecular and functional characteristics of these cells and their role in muscle regeneration has vastly advanced (Fig. 20.1B). Although definitively identified by their anatomical location, satellite cells are molecularly characterized by (1) the expression of the paired-box transcription factor Pax7 and (2) the absence of transcription factors that govern further myogenic specialization, such as MyoD and myogenin (Cornelison and Wold, 1997; Seale et al., 2000; Olguin and Olwin, 2004; Oustanina et al., 2004; Zammit et al., 2004; Kuang et al., 2006; Relaix et al., 2006; Lepper et al., 2009). In skeletal muscles, Pax7 is essential for satellite cell viability and self-renewal in young but not adult mice (Oustanina et al., 2004; Relaix et al., 2006; Lepper et al., 2009). In the diaphragm muscle, the paired box transcription factor Pax3 is constitutively expressed in satellite cells but is only transiently expressed in hindlimb muscles, and its role as an essential factor in satellite cell maintenance is not clear (Montarras et al., 2005; Boutet et al., 2007; Lepper et al., 2009). In the heart, no definitive satellite cell population has been identified to date. We will therefore restrict our discussion to the satellite cells of skeletal muscles.

In normal adult skeletal muscle, satellite cells are largely mitotically and metabolically quiescent. Following injury or other proliferation signals, satellite cells become "activated" and have strongly upregulated expression of the myogenic regulatory factor Myf5. The myoblast progeny of activated satellite cells, specialized myogenic progenitors, maintain expression of Pax7 and Myf5 and upregulate the expression of MyoD, which is necessary for their differentiation into fusion-competent myocytes (Sabourin et al., 1999; Cornelison et al., 2000). Fusion of these progenitors into existing or *de novo* myofibers is associated with a downregulation of Pax7 and Myf5 expression and the onset of myogenin and myosin heavy chain (MHC) expression (Venuti et al., 1995; Knapp et al., 2006). Notably, the pattern of expression of these satellite cell and myogenic differentiation genes differs among skeletal muscle tissues (e.g. head, diaphragm, and limb) and even between satellite cells within the same myofiber, suggesting that satellite cells and their progeny have heterogeneous gene expression and stem cell phenotypes (Kuang et al., 2008).

Numerous cell surface markers have been identified that aid in prospectively isolating satellite cells by fluorescence-activated cell sorting (FACS). Surface markers expressed on some fractions of mouse satellite cells include α7-integrin (Sherwood et al., 2004), β1-integrin (Kuang et al., 2007; Cerletti et al., 2008), CD34 (Beauchamp et al., 2000), CXCR4 (Cerletti et al., 2008), syndecan-3/4 (Cornelison et al., 2001), M-cadherin (Irintchev et al., 1994; Beauchamp et al., 2000), neural cell adhesion molecule (NCAM) (Irintchev et al., 1994; Bosnakovski et al., 2008; Capkovic et al., 2008), c-met (Cornelison and Wold, 1997), ABGC2 (Tanaka et al., 2009), and the unknown antigen for the SM/C-2.6 monoclonal antibody (Fukada et al., 2004). Other mouse satellite cell prospective isolation strategies use transgenic fluorescent protein reporters under the control of promoter elements of satellite cell-associated transcription factors such as *Pax3-GFP* (Montarras et al., 2005), *Pax7-ZsGreen* (Bosnakovski et al., 2008), and *Myf5-Cre/ROSA26-YFP* (Kuang et al., 2007). Since none of these markers definitively and exclusively mark all mouse satellite cells, prospective isolation strategies routinely utilize a combination of markers (including the absence of other lineage markers such as CD45 and Sca1) to obtain a highly enriched cell population that likely represents only a subpopulation of satellite cells (Cerletti et al., 2008; Sacco et al., 2008). The identification of human satellite cells is an ongoing area of investigation, as the cell surface markers found on putative human satellite cells do not completely correspond to those on well-defined mouse satellite cells. For example, unlike mouse satellite cells, human satellite cells are not CD34$^+$ (Peault et al., 2007). It has been reported that CD56$^+$ (NCAM$^+$) CD34$^-$ cells prospectively isolated from enzymatically digested human muscle tissue represent a population of highly myogenic and non-adipogenic cells, which might comprise a subset of human satellite cells (Pisani et al., 2010).

Satellite cells are muscle stem cells

The definitive demonstration that satellite cells are muscle stem cells entailed a functional assay, first established by Partridge, Buckingham, and colleagues, that revealed their remarkable potential to regenerate damaged mouse muscle tissues following transplantation (Collins et al., 2005; Montarras et al., 2005). The initial proof derived from studies of genetically labeled single myofibers that were isolated together with their resident satellite cells and then transplanted into damaged hosts (Collins et al., 2005). Subsequently, genetically labeled satellite cells isolated from FACS sorting of enzymatically digested muscle tissue or from mechanical trituration of isolated myofibers were transplanted by intramuscular injection into recipient muscle of regeneration-deficient mice (Montarras et al., 2005; Cerletti et al., 2008; Sacco et al., 2008). In such assays, donor cells are typically isolated from transgenic mice that constitutively and ubiquitously express reporter genes such as green fluorescent protein (GFP) or β-galactosidase to allow for assessment of contributions of transplanted cells to the satellite cell and myofiber populations in the recipient mice. Recipient mice are generally injured acutely either with a chemical toxin such as notexin or direct tissue injury (e.g. freeze-probe application). Alternatively, constitutive degeneration due to a heritable muscle degeneration phenotype is studied to assess the ability of the stem cells to meet a regenerative demand. A commonly used model of Duchenne muscular dystrophy is the *mdx* mouse, which, like DMD patients, bears a mutation in the dystrophin gene (Bulfield et al., 1984), although for unknown reasons its dystrophic phenotype is very mild, by contrast with patients who die within the third decade of life. Recipient muscles are often irradiated pre-transplantation to limit the contribution and competition by endogenous cell populations. Through these approaches, multiple investigators have demonstrated that transplanted satellite cells robustly contribute to the regeneration of recipient myofibers and, remarkably, are found occupying myofiber membrane-defined satellite cell niche compartments, suggesting that they can home to the niche and replenish lost satellite cells long-term (Sherwood et al., 2004; Collins et al., 2005; Montarras et al., 2005; Kuang et al., 2007; Cerletti et al., 2008; Sacco et al., 2008).

A definitive demonstration that a cell is a muscle stem cell derives from definitions of classically studied stem cells in *Drosophila* and mammalian blood (hematopoietic stem cells). Accordingly, a *single* muscle stem cell (MuSC) must be able to give rise to progeny that can differentiate into both mature myofibers and additional stem cells, in a process known as self-renewal. The aforementioned studies confirmed that satellite cells are potent contributors to muscle regeneration but did not provide definitive evidence of the single cell criterion required of a true MuSC, as they involved the transplantation of hundreds to thousands of cells and it was therefore not possible to ascribe the findings to a single transplanted cell. Recently, Blau and colleagues clearly demonstrated that this criterion is satisfied by MuSCs by injecting single FACS-isolated CD34[+] α7-integrin[+] cells from a double-transgenic *GFP/luciferase* mouse into an irradiated recipient mouse (Sacco et al., 2008). These laborious studies involved transplanting single cells into the irradiated limbs of hundreds of mice and would have been exceedingly difficult without rapid, quantitative assessment of MuSC contribution to muscle tissues over time provided by bioluminescence imaging (as described later in this chapter). These studies showed that progeny from singly transplanted cells both contributed to myofibers and generated multiple mononuclear Pax7[+] cells in canonical satellite cell positions, demonstrating that a single prospectively isolated satellite cell is capable of both self-renewal and complete myogenic differentiation.

Non-invasive bioluminescence imaging (BLI) not only enabled the demonstration that a single satellite cell fulfilled the stringent criteria of a muscle stem cell, but should find useful application in studies of numerous transplantable stem cell types and evaluation of their regenerative potential. As one demonstration of the utility of this BLI technology in stem cell research, the single cell transplant study described above allowed the identification of 4% of the total 144 mice injected with single muscle stem cells that exhibited detectable engraftment, which would

FIGURE 20.2

Non-invasive bioluminescence imaging (BLI) reveals the dynamic and quantitative contributions of transplanted luciferase-expressing cell populations to muscle regeneration. (A) BLI image of prospectively isolated muscle stem cells (MuSCs) transplanted by intra-muscular injection into the hindlimb muscle of a recipient NOD/SCID mouse, at four weeks post-transplant. Image courtesy of Penney Gilbert. (B) Scheme of a theoretical comparison of the engraftment and proliferation of transplanted muscle stem cells (MuSCs), myoblasts, and a tumorigenic cell line (at ~100 luciferase-expressing cells per type) as assessed by BLI post-transplantation. BLI quantitatively correlates with luciferase-expressing cell number and allows for a dynamic assessment of transplanted cell behavior without requiring sacrifice, as is necessary for traditional histological analyses. Transplanted MuSCs engraft (above the BLI signal level corresponding to detection threshold) and produce progeny that contribute to muscle regeneration. The number of luciferase-expressing progeny reaches a plateau as they cease proliferating and differentiate into mature myofibers. In contrast, transplanted tumorigenic cells would exponentially proliferate without reaching a homeostatic plateau. Transplanted myoblasts have very poor viability and do not robustly engraft or expand post-transplantation. BLI will be instrumental to quantitative, dynamic comparisons of MuSCs isolated by different sorting criteria, maintained in different culture conditions, and transplanted into different injury models.

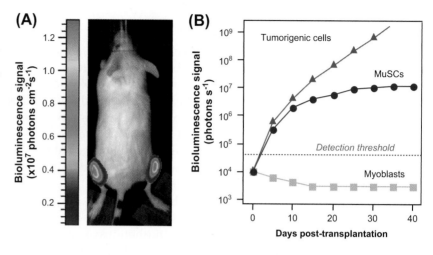

have been a heroic task using classical histological methods. BLI entails transplanting luciferase-expressing satellite cells and allows the dynamic monitoring of the cells as they contribute to tissue regeneration, providing a quantitative time-course (Fig. 20.2). BLI is quantitative as the signal obtained correlates with transplanted cell number. BLI also allows a temporal assessment of stem cell behavior as the same mice can be imaged repeatedly over time, whereas classical histological analyses require sacrificing the mice. BLI revealed that the progeny of transplanted satellite cells divide exponentially and then cease to divide reaching a plateau upon achieving homeostasis, when no more stem cells are required (Sacco et al., 2008). By contrast, 90% of progenitor cells without self-renewal capacity (myoblasts) die upon injection and no increase in numbers is observed over time, explaining in part why clinical trials with these cells have been disappointing (Karpati, 1990; Gussoni et al., 1997; Peault et al., 2007). Furthermore, it is possible to assess whether stem cells are transformed, as the exponential proliferation of tumorigenic cells does not cease and a plateau is never achieved (Contag and Ross, 2002). This projected outcome for a comparision between muscle stem cells, myogenic progenitors, and tumorigenic cells is depicted in the scheme in Figure 20.2B. For studies of muscle regeneration, BLI will be instrumental in rigorous quantitative and temporal comparisons of MuSCs isolated by different sorting criteria, maintained in different culture environments, and transplanted in different mouse models of acute muscle injury, and in different mouse models of chronic and acute muscle injuries typical of human muscle diseases. Given its quantitative, dynamic, and high-throughput monitoring capabilities, BLI should prove useful in future muscle stem cell assays and preclinical regenerative medicine evaluations in general.

REGULATION OF SATELLITE CELL BEHAVIORS BY NICHE COMPONENTS

Stem cell niches

Tissue-specific stem cells commonly reside in instructive microenvironments, or "niches." First proposed by Schofield in 1978, stem cell niches have been extensively characterized for *Drosophila melanogaster* and *Caenorhabditis elegans* germ cells (Fuller and Spradling, 2007; Kimble and Crittenden, 2007) and characterized to some extent in some mammalian tissues, including the hematopoietic system, epidermis, intestinal epithelium, brain, testis, and skeletal muscle (Fuchs et al., 2004; Jones and Wagers, 2008). These complex microenvironments consist of specialized combinations of cellular (stem and supporting cells), biochemical (soluble and extracellular matrix factors), and biophysical components. Niches are thought to dynamically regulate stem cell behavior to ensure and stabilize a quiescent, slowly dividing stem cell population during adult tissue homeostasis. In response to tissue injury, niche disruption signals the activation and self-renewal of stem cells leading to their extensive proliferation and release of their progeny from the niche microenvironment.

Molecular components of the satellite cell niche

The satellite cell niche is a membrane-enclosed compartment between the myofiber plasma membrane and the basal lamina that surrounds the myofiber (Mauro, 1961; Kuang et al., 2008). In this niche, satellite cells are subjected to an asymmetric arrangement of certain niche components, with myofiber signals on their apical surface and basal lamina signals on their basal surface. The characterization of the biochemical and biophysical nature of the satellite cell niche remains incomplete. However, it is assumed that a "bipolar" distribution of niche signals, as in the case of other tissue-specific stem cell niches, may be instrumental for maintaining stem cell polarity and asymmetric self-renewal (Fuchs et al., 2004; Kuang et al., 2008).

Some of the molecular and biophysical components of the satellite cell niche in mice have been identified, although their roles in governing satellite cell quiescence, activation, migration, and self-renewal have not yet been fully elucidated. The myofiber basal lamina contains numerous extracellular matrix components, including type IV collagen, laminin, fibronectin, entactin, and other proteoglycans and glycoproteins (Sanes, 2003; Cosgrove et al., 2009). Satellite cells express α7/β1-integrin heterodimers on their basal surfaces to adhere to laminin and other basal lamina components (Burkin and Kaufman, 1999; Sanes, 2003). Growth factors, including basic fibroblast growth factor (bFGF), hepatocyte growth factor (HGF), epidermal growth factor (EGF), insulin-like growth factor-1 (IGF-1), and Wnt glycoproteins (DiMario et al., 1989; Tatsumi et al., 1998; Machida and Booth, 2004; Golding et al., 2007; Brack et al., 2008; le Grand et al., 2009) are secreted from systemic and local (satellite cell and myofiber) sources and stimulate satellite cell survival, activation, and proliferation. These growth factors can bind to proteoglycans in the myofiber basal lamina on the satellite cell surface itself, or both, to provide a niche repository of signaling molecules (Olwin and Rapraeger, 1992; Tatsumi et al., 1998; Cornelison et al., 2001; Jenniskens et al., 2006; Langsdorf et al., 2007). Myofiber secretion of SDF-1 binds the receptor CXCR4 on satellite cells and activates a migratory response (Ratajczak et al., 2003; Sherwood et al., 2004). M-cadherin presented on the myofiber plasma membrane mediates myofiber-satellite cell adhesion and has been hypothesized to facilitate fusion of myogenic progenitors into myofibers (Irintchev et al., 1994). Moreover, satellite cells express factors, such as ligands for the Notch receptor family, that influence their own behaviors, including self-renewal, through autocrine and juxtacrine signaling (Conboy and Rando, 2002; Kuang et al., 2007). Factors arising from cells outside the myofiber-defined niche can also influence satellite cells; these include the TGF-β family member myostatin (McCroskery et al., 2003) and Wnt3a (Brack et al., 2007), which are both produced by distant cells, arrive via the circulation, and diffuse through the basal lamina to promote satellite cell and myogenic progenitor differentiation. Circulating signaling factors can gain access to satellite cells, due to the close proximity of their niches to both muscle microvasculature (Christov et al., 2007) and neuromuscular junctions (Kelly, 1978). Further elucidation of how satellite cell behaviors are regulated in normal muscle homeostasis and regeneration and dysregulated in degenerative muscle conditions will be instrumental for the development of methods for the maintenance and expansion of MuSCs in culture and therapies directed towards endogenous MuSCs or using transplanted MuSCs.

Biophysical properties of the satellite cell niche

Biophysical cues in stem cell niches include extracellular matrix mechanical (e.g. elasticity) and topographical properties (Lutolf et al., 2009a; Reilly and Engler, 2010). Cells can respond to extracellular biophysical cues through a variety of mechanotransduction mechanisms (Lopez et al., 2008; Geiger et al., 2009). Biophysical cues provided by the extracellular microenvironment could affect the behaviors of stem cells, including satellite cells, within their niches (Lopez et al., 2008; Guilak et al., 2009). The most well-studied biophysical property of tissue microenvironments is the elastic stiffness, which describes the deformability of a tissue under stretching and is ~12 kPa in healthy skeletal muscle (Collinsworth et al., 2002; Engler et al., 2004). The elastic stiffness is increased in aged and diseased skeletal muscle tissue, likely due to the increased deposition of extracellular matrix (fibrosis) in muscle tissues in these conditions (Stedman et al., 1991; Engler et al., 2004; Rosant et al., 2007; Gao et al., 2008). This increase in elastic modulus is exemplified by the elevated rigidity (>18 kPa) observed in the *mdx* mouse model of Duchenne muscular dystrophy after four months of age (Stedman et al., 1991; Engler et al., 2004). Recent findings demonstrate that even subtle changes in extracellular stiffness can potently affect the ability of myoblasts and mesenchymal stem cells to undergo myogenic differentiation in cell culture (Engler et al., 2004, 2006; Boonen et al., 2009). These findings suggest that alterations in the elasticity or rigidity of skeletal muscle tissue may have profound effects on the contributions of satellite cells and myogenic progenitors to muscle regeneration.

Satellite cell self-renewal mechanisms

The mechanisms by which specific niche components actively regulate satellite cell quiescence, activation, and self-renewal have begun to be elucidated. Stem cell self-renewal can be maintained by two proliferation mechanisms: (1) asymmetric self-renewal in which a stem cell divides into one differentiated cell and one stem cell and (2) symmetric self-renewal in which a stem cell divides into two equivalent stem cells (Fig. 20.3). Asymmetric self-renewal is thought to be sufficient to meet the demands of normal cell turnover under homeostatic conditions, but symmetric self-renewal is presumably necessary to yield an expansion of the stem cell pool under demands resulting from injury or disease (Morrison and Kimble, 2006). Many of the insights into murine satellite cell self-renewal mechanisms have been elucidated using lineage tracing in transgenic *Myf5-Cre/ROSA26-YFP* mice (Kuang et al., 2007; le Grand et al., 2009). In these mice, "quiescent" satellite cells are YFP⁻ due to the lack of Myf5 expression and "activated" satellite cells are YFP⁺ following the onset of Myf5 expression. Using transgenic lineage tracing of satellite cells on isolated intact myofibers, Rudnicki and colleagues have reported that Notch signaling, stimulated by Delta ligands on adjacent myofibers, may drive the asymmetric division of "quiescent" Pax7⁺ Myf5⁻ satellite cells into one Pax7⁺ Myf5⁻ satellite cell and one Pax7⁺ Myf5⁺ satellite cell (Kuang et al., 2007). These investigators further reported that non-canonical Wnt7a signaling regulates symmetric division leading one satellite cell giving rise to two Pax7⁺ Myf5⁻ satellite cells (le Grand et al., 2009). An additional potential mechanism of satellite cell self-renewal is reversion from a committed myogenic progenitor back into a quiescent satellite cell. Zammit and colleagues have observed that a rare subset of myofiber-derived Pax7⁺ MyoD⁺ myogenic progenitors do not further differentiate into myogenin⁺ cells but instead lose MyoD expression and revert to

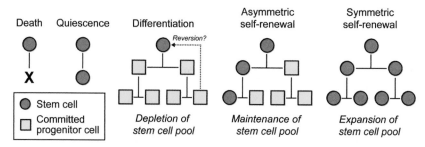

FIGURE 20.3

Fates of tissue-specific stem cells. The stem cell pool is depleted through death and differentiation, and is maintained through quiescence. Asymmetric stem cell self-renewal, in which a stem cell divides into one differentiated cell and one stem cell, meets the demands of normal cell turnover under homeostatic conditions. Symmetric stem cell self-renewal, in which a stem cell divides into two equivalent stem cells, yields an expansion of the stem cell pool. An additional potential mechanism of stem cell self-renewal is the reversion of a differentiated cell back into a stem cell.

a quiescent Pax7$^+$ MyoD$^-$ satellite cell (Zammit et al., 2004), similarly to reports of reversion from progenitor to stem cell status in *Drosophila* testis (Yamashita et al., 2005). This reversion may be mediated by the upregulation of Sprouty1, a negative regulator of growth factor-activated receptor tyrosine kinase signaling, whose expression is associated with the re-establishment of satellite cell quiescence following muscle regeneration (Shea et al., 2010). These reports suggest that the maintenance and expansion of the satellite cell pool are governed by self-renewal mechanisms that, although incompletely understood, are likely regulated by the active presentation of specific niche-associated extrinsic signals in a dose-dependent, temporally and spatially controlled manner (Brack et al., 2008). Further studies that replicate or interfere with these niche signals should facilitate *ex vivo* maintenance and expansion, transplantation, and endogenous activation of MuSCs for regenerative medicine approaches.

AGING OF SKELETAL MUSCLE STEM CELLS AND THEIR NICHE

In the course of aging in mice and humans, skeletal muscle function and regenerative potential both diminish and are susceptible to muscle wasting disease (e.g. sarcopenia) increases (Grounds, 1998; Gopinath and Rando, 2008). Impeded regeneration in aged muscle appears to be partially due to declining satellite cell functionality. Since number and density of satellite cells do not conclusively change with aging in mice and humans, intrinsic and extrinsic factors that regulate satellite cells have been examined for aging-related changes that might explain the observed regenerative decline (Brack and Rando, 2007). Satellite cells in aged muscle tissue exhibit a defective response to activating and proliferative signals (Lipton and Schultz, 1979; Conboy et al., 2003). To examine whether the defective behavior of aged satellite cells is cell-intrinsic or cell-extrinsic, researchers have utilized (1) surgical joining of young and aged mice to establish heterochronic parabiotic pairings with a common circulatory system and (2) transplantation of prospectively isolated satellite cells to young and to aged mice. Exposure of aged satellite cells to the young circulatory environment through parabiosis effectively, but not fully, rejuvenated their ability to contribute to muscle regeneration (Conboy et al., 2005). Correspondingly, satellite cells isolated from young and aged mice show a similar ability to engraft when transplanted into young *mdx* mouse recipients (Collins et al., 2007). Conversely, muscle regeneration in young mice exposed to an aged circulation (Brack et al., 2007) and transplantation of young satellite cells into aged *mdx* recipient mice (Boldrin et al., 2009) are defective compared to young parabiotic and recipient controls, respectively. These remarkable findings indicate that regeneration defects in aged muscle are not wholly due to unalterable intrinsic changes in satellite cells, but instead are significantly dependent on extrinsic factors influencing satellite cell function.

The above studies of young and old satellite cells have sparked a closer focus on the changes in local and systemic extrinsic satellite cell regulatory factors due to aging. The basal lamina of aged myofibers contains increased extracellular matrix (ECM) deposition with an altered protein composition (Snow, 1977; Scime et al., 2010). These changes result in a more elastically rigid basal lamina (Rosant et al., 2007; Gao et al., 2008) and are thought to diminish the growth factor-binding capacity of the basal lamina, possibly influencing the composition of sequestered growth factors present in the satellite cell niche (Alexakis et al., 2007). Moreover, regenerative defects in aged mice have been attributed to diminished activation of the Notch pathway in satellite cells due to decreased expression of the Notch ligand Delta-1 on adjacent myofiber membranes (Conboy et al., 2003). Furthermore, aged mice have increased circulating concentrations of Wnt3a and TGF-β1. Wnt3a and TGF-β1 impede with effective muscle regeneration through antagonism of Notch-stimulated satellite cell proliferation and induction of premature and aberrant differentiation of myogenic progenitors (Brack et al., 2007; Carlson et al., 2008, 2009). These studies highlight the role of Notch activation for adequate MuSC function and the changes that accompany aging that limit its activation. Although these findings do not exclude satellite cell-intrinsic changes as causes for the diminished regenerative potential of aged muscle, they clearly show that regenerative defects in aging can be significantly attributed to changes in extrinsic influences. Indeed, interventions targeted to the Notch, Wnt3a, and TGF-β1 pathways have already demonstrated effective improvements in muscle regeneration in aged mice. These results strongly suggest that targeting these and other extrinsic regulatory factors could prove beneficial for treating aging-associated muscle degenerative diseases.

CHALLENGES IN THE USE OF SATELLITE CELLS IN REGENERATIVE MEDICINE

Although murine satellite cells satisfy the definition of muscle stem cells and can be prospectively isolated with reasonably high purity, there remain significant unmet challenges for the use of human satellite cells in transplant-based clinical regenerative medicine applications. A first challenge is that markers for the prospective isolation of human satellite cells are poorly correlated with those on mouse satellite cells and have not yet been refined to isolate a highly enriched human myogenic cell population. Further evaluation of human satellite cell surface markers and characterization of MuSC phenotypes is necessary for use of prospective satellite cell isolation in clinical approaches. A second challenge is that satellite cell and myoblast transplantations require intramuscular injections to deliver cells efficiently to recipient muscle tissue. To date, these cell types are unable to access muscle tissue when delivered systemically through the vasculature (Dellavalle et al., 2007; Price et al., 2007). This limitation makes the use of satellite cell transplantations in conditions that require regenerative therapy to all muscles or muscles that are difficult to access, such as the diaphragm, impractical.

A third challenge to preclinical and clinical studies is that murine and human satellite cells are both rare (~5% of all muscle nuclei) and currently cannot be maintained for more than a few days or expanded in culture, therefore limiting the numbers of cells available for transplantation. To overcome these cell number limitations, researchers and clinicians have long proposed using human myoblasts as a source of cells for muscle therapeutic applications. Cultured myoblasts are a population of rapidly proliferating cells generated from enzymatically digested muscle tissue or isolated satellite cells (Rando and Blau, 1994). Given that they can be maintained and expanded in culture and retain myogenic differentiation and fusion potential, myoblasts provide an attractive source for cell-based therapies of degenerative human muscle diseases. In clinical trials in the 1980s, transplanted human myoblasts were shown to fuse with endogenous muscle fibers and led to the production of functional dystrophin in the myofibers of DMD patients (Peault et al., 2007; Gussoni et al., 1997; Karpati, 1990; Cossu and Sampaolesi, 2004). However, the myoblasts did not significantly migrate beyond the site of injection and ultimately failed to provide long-term therapeutic benefit. These clinical findings have been recently corroborated by comparisons of satellite cell and myoblast transplantations into damaged mouse muscles, which show that maintenance of satellite cells for more than a few days in standard culture platforms results in conversion to a myoblast phenotype with a dramatically diminished self-renewal potential (Montarras et al., 2005; Sacco et al., 2008). Clearly, improvements in culture technologies are necessary to allow for the maintenance and, ideally, expansion of satellite cells without loss of their stem cell phenotype. This is especially necessary for autologous satellite cells bearing genetic defects, such as those present in DMD patients, that require *ex vivo* gene therapy before transplantation (Blau and Springer, 1995).

IMPROVED CULTURE TECHNOLOGIES FOR SATELLITE CELL STUDIES

To date, biomaterials approaches have almost exclusively been implemented for the *in vitro* engineering of differentiated, functional muscle tissue rather than to support MuSC self-renewal and expansion (Levenberg et al., 2005; Eberli et al., 2009). Given the sensitivity of satellite cells to both biochemical and biophysical cues, effective *in vitro* maintenance and expansion of satellite cells may require engineered culture platforms that not only contain essential niche proteins (Kuang et al., 2008) but also accurately replicate the biophysical properties such as elastic stiffness of the satellite cell niche in healthy skeletal muscle (Collinsworth et al., 2002; Engler et al., 2004). Advances in biomaterials technologies facilitate the precise control of both biochemical and biophysical cues in artificial microenvironments that could potentially replicate these essential components of the physiological satellite cell niche (Cosgrove et al., 2009; Lutolf et al., 2009a). Poly(ethylene glycol) (PEG)-based hydrogels can be engineered with both specified tethered ligand (adhesion or growth factor) presentation and tunable elastic stiffness, and thus allow independent control of specific biochemical and biophysical components of *in vitro* stem cell niche mimics (Lutolf and Hubbell, 2005; Lutolf et al., 2009a). PEG hydrogels are resistant to non-specific protein adsorption, limiting cell adhesion in the absence of covalently tethered protein ligands. This property allows for the engineering of PEG substrates containing specific adhesion ligands or growth factors presented in a "tethered," mode which is more akin to their physiological context than is the standard culture "soluble" presentation mode (Cosgrove et al., 2009). Lutolf, Blau, and colleagues have recently demonstrated that a PEG hydrogel platform containing arrayed "microwells" supports the identification of factors governing clonal HSC expansion (Lutolf et al., 2009b). In theory, such a platform could be employed in high-throughput screens of a library of niche ligands to identify novel therapeutic factors. Similar approaches to studying healthy and diseased satellite cells in a high-throughput, molecularly specified manner *in vitro* should prove fruitful for the maintenance and expansion of satellite cells as well as for the identification of novel treatments aimed at modulating satellite cells within their endogenous niche.

ALTERNATIVE MUSCLE STEM CELL SOURCES

A number of other putative muscle stem cell sources have been characterized that could potentially overcome some of the limitations in using satellite cells in clinical regenerative medicine applications. Non-satellite cells with putative muscle stem cell and myogenic progenitor properties include muscle interstitial cells (Asakura et al., 2002; Tamaki et al., 2002; Mitchell et al., 2010), muscle-derived stem cells (Peng and Huard, 2004; Peault et al., 2007), mesenchymal stem cells (Dezawa et al., 2005), bone marrow-derived hematopoietic progenitors (Ferrari et al., 1998; LaBarge and Blau, 2002; Camargo et al., 2003; Corbel et al., 2003;), blood vessel-associated mesoangioblasts (Sampaolesi et al., 2003) and pericytes (Dellavalle et al., 2007), PW1[+] muscle interstitial cells, and myogenic cells derived from induced pluripotent (iPS) cells (Mizuno et al., 2010). In this section, the functional characteristics and methods used to isolate and transplant these non-satellite cells are discussed. Readers are directed to an excellent review (Tedesco et al., 2010) for further details and additional putative myogenic cell sources not discussed here due to space limitations. Although many of these cell sources have promising characteristics for cell therapy applications, their roles in normal development and muscle regeneration and their anatomical locations have not been fully documented, nor have they been shown to satisfy the rigorous requirements that define muscle stem cells. Rigorous evaluation of these alternative cell populations through single-cell transplantation studies is necessary to establish whether they are truly muscle stem cells.

Non-satellite muscle cells

Muscle side-population (SP) cells have been found in the interstitial space of skeletal muscle (Tamaki et al., 2002) that can engraft in the satellite cell niche following intravenous tail-vein injection (Asakura et al., 2002), suggesting they could comprise an endogenous precursor to satellite cells. Muscle side-population cells are isolated based on their exclusion of Hoechst 33342 dye by elevated activity of the drug efflux pump ABCG2, which is upregulated in multiple stem cell populations (Asakura et al., 2002; Muskiewicz et al., 2005). Their very low engraftment efficiency and inability to be expanded in culture confounds their utility in cell therapy applications. An additional, but not necessarily distinct, population of muscle-resident cells that retains myogenic potential is PW1$^+$ Pax7$^-$ interstitial cells (PICs). Sassoon and colleagues have proposed that PICs may serve as physiological precursors to satellite cells (Mitchell et al., 2010).

Non-muscle cells

Adult bone marrow contains progenitors capable of contributing to myogenic differentiation following transplantation (Ferrari et al., 1998; LaBarge and Blau, 2002). Given the very low efficiency of engraftment following whole bone marrow transplantation, rare subsets of bone marrow cells have been evaluated for myogenic potential. These experiments have shown that CD45 expression marks populations of hematopoietic stem cells that retain the myogenic potential of bone marrow (Camargo et al., 2003; Corbel et al., 2003). Though HSCs can be delivered via arterial circulation, the very limited contribution efficiency of transplanted HSCs to muscle regeneration raises questions regarding their utility as a muscle cell therapy source.

Mesoangioblasts are blood vessel-associated multipotent mesodermal progenitors that can be isolated from fetal muscle biopsy tissue fragments containing small blood vessels (Sampaolesi et al., 2003). Mesoangioblasts can be expanded in culture and effectively contribute to muscle regeneration throughout the body, without tumor formation, following intra-arterial delivery due to their ability to extravasate from the vasculature. The reliance on fetal tissue for the isolation of mesoangioblasts can be overcome through the use of an alternative but related population of blood vessel-associated cells called pericytes that are present in adult humans and have myogenic potential (Dellavalle et al., 2007). The advantageous characteristics of mesoangioblasts have motivated their extensive preclinical evaluation, including in the golden retriever muscular dystrophy model, which closely mimics human DMD (Sampaolesi et al., 2006). Promising outcomes in these preclinical models have made mesoangioblasts the leading candidate for cell therapy for muscular dystrophy patients even though they have not been demonstrated to be a *bona fide* muscle stem cell population nor shown to be a significant contributor to physiological muscle regeneration.

CONCLUSIONS

Muscle satellite cells provide essential contributions to physiological post-natal skeletal muscle regeneration. Although satellite cells are muscle stem cells and their study offers critical insights into the process of muscle regeneration, their use in cell transplantations is currently restricted by their inability to effectively contribute to myogenesis following systemic delivery and current limits on their maintenance and expansion in culture. Recent technological advances, such as bioengineered cell culture platforms, offer great promise to overcome these limitations by identifying novel therapeutic factors for stimulating the regenerative contributions of endogenous satellite cells and improving satellite cell transplantation in preclinical models. In addition, novel non-invasive imaging modalities such as BLI enable quantitative assessment of regeneration over time. Additionally, advances in the isolation of human satellite cells with muscle stem cell properties and the validation of alternative sources of human muscle stem cells amenable to systemic delivery may lead to the clinical realization of much-needed cell therapies for prevalent muscular dystrophies and muscle wasting conditions.

Acknowledgments

We thank Penney Gilbert, Alessandra Sacco, and Mara Damian for helpful discussions. B.D.C. is financially supported by NIH postdoctoral training grant 5R25CA118681. H.M.B. is financially supported by NIH grants HL096113, AG009521, AG020961, U01 HL100397, and RAR059365Z; JDRF grant 34-2008-623; MDA grant 4320; LLS grant TR6025-09; CIRM grants RT1-01001 and RB1-02192; and the Baxter Foundation.

References

Alexakis, C., Partridge, T., & Bou-Gharios, G. (2007). Implication of the satellite cell in dystrophic muscle fibrosis: a self-perpetuating mechanism of collagen overproduction. *Am. J. Physiol. Cell Physiol., 293*, C661–C669.

Asakura, A., Seale, P., Girgis-Gabardo, A., & Rudnicki, M. A. (2002). Myogenic specification of side population cells in skeletal muscle. *J. Cell Biol., 159*, 123–134.

Beauchamp, J. R., Heslop, L., Yu, D. S., Tajbakhsh, S., Kelly, R. G., Wernig, A., et al. (2000). Expression of CD34 and Myf5 defines the majority of quiescent adult skeletal muscle satellite cells. *J. Cell Biol., 151*, 1221–1234.

Bischoff, R. (1974). Enzymatic liberation of myogenic cells from adult rat muscle. *Anat. Rec., 180*, 645–661.

Bischoff, R. (1986). Proliferation of muscle satellite cells on intact myofibers in culture. *Dev. Biol., 115*, 129–139.

Blau, H. M., & Springer, M. L. (1995). Muscle-mediated gene therapy. *N. Engl. J. Med., 333*, 1554–1556.

Blau, H. M., Webster, C., & Pavlath, G. K. (1983). Defective myoblasts identified in Duchenne muscular dystrophy. *Proc. Natl. Acad. Sci. U.S.A., 80*, 4856–4860.

Bogdanovich, S., Perkins, K. J., Krag, T. O., & Khurana, T. S. (2004). Therapeutics for Duchenne muscular dystrophy: current approaches and future directions. *J. Mol. Med., 82*, 102–115.

Boldrin, L., Zammit, P. S., Muntoni, F., & Morgan, J. E. (2009). Mature adult dystrophic mouse muscle environment does not impede efficient engrafted satellite cell regeneration and self-renewal. *Stem Cells, 27*, 2478–2487.

Boonen, K. J., Rosaria-Chak, K. Y., Baaijens, F. P., van der Schaft, D. W., & Post, M. J. (2009). Essential environmental cues from the satellite cell niche: optimizing proliferation and differentiation. *Am. J. Physiol. Cell Physiol., 296*, C1338–C1345.

Bosnakovski, D., Xu, Z., Li, W., Thet, S., Cleaver, O., Perlingeiro, R. C., et al. (2008). Prospective isolation of skeletal muscle stem cells with a Pax7 reporter. *Stem Cells, 26*, 3194–3204.

Boutet, S. C., Disatnik, M. H., Chan, L. S., Iori, K., & Rando, T. A. (2007). Regulation of Pax3 by proteasomal degradation of monoubiquitinated protein in skeletal muscle progenitors. *Cell, 130*, 349–362.

Brack, A. S., & Rando, T. A. (2007). Intrinsic changes and extrinsic influences of myogenic stem cell function during aging. *Stem Cell Rev., 3*, 226–237.

Brack, A. S., Conboy, I. M., Conboy, M. J., Shen, J., & Rando, T. A. (2008). A temporal switch from notch to Wnt signaling in muscle stem cells is necessary for normal adult myogenesis. *Cell Stem Cell, 2*, 50–59.

Brack, A. S., Conboy, M. J., Roy, S., Lee, M., Kuo, C. J., Keller, C., et al. (2007). Increased Wnt signaling during aging alters muscle stem cell fate and increases fibrosis. *Science, 317*, 807–810.

Bulfield, G., Siller, W. G., Wight, P. A., & Moore, K. J. (1984). X chromosome-linked muscular dystrophy (mdx) in the mouse. *Proc. Natl. Acad. Sci. U.S.A., 81*, 1189–1192.

Burkin, D. J., & Kaufman, S. J. (1999). The alpha7beta1 integrin in muscle development and disease. *Cell Tissue Res., 296*, 183–190.

Camargo, F. D., Green, R., Capetanaki, Y., Jackson, K. A., & Goodell, M. A. (2003). Single hematopoietic stem cells generate skeletal muscle through myeloid intermediates. *Nat. Med., 9*, 1520–1527.

Capkovic, K. L., Stevenson, S., Johnson, M. C., Thelen, J. J., & Cornelison, D. D. (2008). Neural cell adhesion molecule (NCAM) marks adult myogenic cells committed to differentiation. *Exp. Cell Res., 314*, 1553–1565.

Carlson, B. M. (1973). The regeneration of skeletal muscle. A review. *Am. J. Anat., 137*, 119–149.

Carlson, M. E., Conboy, M. J., Hsu, M., Barchas, L., Jeong, J., Agrawal, A., et al. (2009). Relative roles of TGF-beta1 and Wnt in the systemic regulation and aging of satellite cell responses. *Aging Cell, 8*, 676–689.

Carlson, M. E., Hsu, M., & Conboy, I. M. (2008). Imbalance between pSmad3 and Notch induces CDK inhibitors in old muscle stem cells. *Nature, 454*, 528–532.

Cerletti, M., Jurga, S., Witczak, C. A., Hirshman, M. F., Shadrach, J. L., Goodyear, L. J., et al. (2008). Highly efficient, functional engraftment of skeletal muscle stem cells in dystrophic muscles. *Cell, 134*, 37–47.

Christov, C., Chretien, F., Abou-Khalil, R., Bassez, G., Vallet, G., Authier, F. J., et al. (2007). Muscle satellite cells and endothelial cells: close neighbors and privileged partners. *Mol. Biol. Cell, 18*, 1397–1409.

Collins, C. A., Olsen, I., Zammit, P. S., Heslop, L., Petrie, A., Partridge, T. A., et al. (2005). Stem cell function, self-renewal, and behavioral heterogeneity of cells from the adult muscle satellite cell niche. *Cell, 122*, 289–301.

Collins, C. A., Zammit, P. S., Ruiz, A. P., Morgan, J. E., & Partridge, T. A. (2007). A population of myogenic stem cells that survives skeletal muscle aging. *Stem Cells, 25*, 885–894.

Collinsworth, A. M., Zhang, S., Kraus, W. E., & Truskey, G. A. (2002). Apparent elastic modulus and hysteresis of skeletal muscle cells throughout differentiation. *Am. J. Physiol. Cell Physiol., 283*, C1219–C1227.

Conboy, I. M., & Rando, T. A. (2002). The regulation of Notch signaling controls satellite cell activation and cell fate determination in postnatal myogenesis. *Dev. Cell, 3*, 397–409.

Conboy, I. M., Conboy, M. J., Smythe, G. M., & Rando, T. A. (2003). Notch-mediated restoration of regenerative potential to aged muscle. *Science, 302*, 1575–1577.

359

Conboy, I. M., Conboy, M. J., Wagers, A. J., Girma, E. R., Weissman, I. L., & Rando, T. A. (2005). Rejuvenation of aged progenitor cells by exposure to a young systemic environment. *Nature, 433*, 760–764.

Contag, C. H., & Ross, B. D. (2002). It's not just about anatomy: in vivo bioluminescence imaging as an eyepiece into biology. *J. Magn. Reson. Imaging, 16*, 378–387.

Corbel, S. Y., Lee, A., Yi, L., Duenas, J., Brazelton, T. R., Blau, H. M., et al. (2003). Contribution of hematopoietic stem cells to skeletal muscle. *Nat. Med., 9*, 1528–1532.

Cornelison, D. D., & Wold, B. J. (1997). Single-cell analysis of regulatory gene expression in quiescent and activated mouse skeletal muscle satellite cells. *Dev. Biol., 191*, 270–283.

Cornelison, D. D., Filla, M. S., Stanley, H. M., Rapraeger, A. C., & Olwin, B. B. (2001). Syndecan-3 and syndecan-4 specifically mark skeletal muscle satellite cells and are implicated in satellite cell maintenance and muscle regeneration. *Dev. Biol., 239*, 79–94.

Cornelison, D. D., Olwin, B. B., Rudnicki, M. A., & Wold, B. J. (2000). MyoD(−/−) satellite cells in single-fiber culture are differentiation defective and MRF4 deficient. *Dev. Biol., 224*, 122–137.

Cosgrove, B. D., Sacco, A., Gilbert, P. M., & Blau, H. M. (2009). A home away from home: challenges and opportunities in engineering in vitro muscle satellite cell niches. *Differentiation, 78*, 185–194.

Cossu, G., & Sampaolesi, M. (2004). New therapies for muscular dystrophy: cautious optimism. *Trends Mol. Med., 10*, 516–520.

Decary, S., Mouly, V., & Butler-Browne, G. S. (1996). Telomere length as a tool to monitor satellite cell amplification for cell-mediated gene therapy. *Hum. Gene Ther., 7*, 1347–1350.

Dellavalle, A., Sampaolesi, M., Tonlorenzi, R., Tagliafico, E., Sacchetti, B., Perani, L., et al. (2007). Pericytes of human skeletal muscle are myogenic precursors distinct from satellite cells. *Nat. Cell Biol., 9*, 255–267.

Dezawa, M., Ishikawa, H., Itokazu, Y., Yoshihara, T., Hoshino, M., Takeda, S., et al. (2005). Bone marrow stromal cells generate muscle cells and repair muscle degeneration. *Science, 309*, 314–317.

DiMario, J., Buffinger, N., Yamada, S., & Strohman, R. C. (1989). Fibroblast growth factor in the extracellular matrix of dystrophic (mdx) mouse muscle. *Science, 244*, 688–690.

Eberli, D., Soker, S., Atala, A., & Yoo, J. J. (2009). Optimization of human skeletal muscle precursor cell culture and myofiber formation in vitro. *Methods, 47*, 98–103.

Emery, A. E. (2002). The muscular dystrophies. *Lancet, 359*, 687–695.

Engler, A. J., Griffin, M. A., Sen, S., Bonnemann, C. G., Sweeney, H. L., & Discher, D. E. (2004). Myotubes differentiate optimally on substrates with tissue-like stiffness: pathological implications for soft or stiff microenvironments. *J. Cell Biol., 166*, 877–887.

Engler, A. J., Sen, S., Sweeney, H. L., & Discher, D. E. (2006). Matrix elasticity directs stem cell lineage specification. *Cell, 126*, 677–689.

Ferrari, G., Cusella-de Angelis, G., Coletta, M., Paolucci, E., Stornaiuolo, A., Cossu, G., et al. (1998). Muscle regeneration by bone marrow-derived myogenic progenitors. *Science, 279*, 1528–1530.

Fuchs, E., Tumbar, T., & Guasch, G. (2004). Socializing with the neighbors: stem cells and their niche. *Cell, 116*, 769–778.

Fukada, S., Higuchi, S., Segawa, M., Koda, K., Yamamoto, Y., Tsujikawa, K., et al. (2004). Purification and cell-surface marker characterization of quiescent satellite cells from murine skeletal muscle by a novel monoclonal antibody. *Exp. Cell Res., 296*, 245–255.

Fuller, M. T., & Spradling, A. C. (2007). Male and female Drosophila germline stem cells: two versions of immortality. *Science, 316*, 402–404.

Gao, Y., Kostrominova, T. Y., Faulkner, J. A., & Wineman, A. S. (2008). Age-related changes in the mechanical properties of the epimysium in skeletal muscles of rats. *J. Biomech., 41*, 465–469.

Geiger, B., Spatz, J. P., & Bershadsky, A. D. (2009). Environmental sensing through focal adhesions. *Nat. Rev. Mol. Cell Biol., 10*, 21–33.

Golding, J. P., Calderbank, E., Partridge, T. A., & Beauchamp, J. R. (2007). Skeletal muscle stem cells express anti-apoptotic ErbB receptors during activation from quiescence. *Exp. Cell Res., 313*, 341–356.

Gopinath, S. D., & Rando, T. A. (2008). Stem cell review series: aging of the skeletal muscle stem cell niche. *Aging Cell, 7*, 590–598.

Grounds, M. D. (1991). Towards understanding skeletal muscle regeneration. *Pathol. Res. Pract., 187*, 1–22.

Grounds, M. D. (1998). Age-associated changes in the response of skeletal muscle cells to exercise and regeneration. *Ann. N.Y. Acad. Sci., 854*, 78–91.

Guilak, F., Cohen, D. M., Estes, B. T., Gimble, J. M., Liedtke, W., & Chen, C. S. (2009). Control of stem cell fate by physical interactions with the extracellular matrix. *Cell Stem Cell, 5*, 17–26.

Gussoni, E., Blau, H. M., & Kunkel, L. M. (1997). The fate of individual myoblasts after transplantation into muscles of DMD patients. *Nat. Med., 3*, 970–977.

Heslop, L., Morgan, J. E., & Partridge, T. A. (2000). Evidence for a myogenic stem cell that is exhausted in dystrophic muscle. *J. Cell Sci.*, *113*(Pt 12), 2299–2308.

Irintchev, A., Zeschnigk, M., Starzinski-Powitz, A., & Wernig, A. (1994). Expression pattern of M-cadherin in normal, denervated, and regenerating mouse muscles. *Dev. Dyn.*, *199*, 326–337.

Jenniskens, G. J., Veerkamp, J. H., & van Kuppevelt, T. H. (2006). Heparan sulfates in skeletal muscle development and physiology. *J. Cell Physiol.*, *206*, 283–294.

Jones, D. L., & Wagers, A. J. (2008). No place like home: anatomy and function of the stem cell niche. *Nat. Rev. Mol. Cell Biol.*, *9*, 11–21.

Jones, T. E., Stephenson, K. W., King, J. G., Knight, K. R., Marshall, T. L., & Scott, W. B. (2009). Sarcopenia — mechanisms and treatments. *J. Geriatr. Phys. Ther.*, *32*, 39–45.

Karpati, G. (1990). The principles and practice of myoblast transfer. *Adv. Exp. Med. Biol.*, *280*, 69–74.

Kelly, A. M. (1978). Perisynaptic satellite cells in the developing and mature rat soleus muscle. *Anat. Rec.*, *190*, 891–903.

Kimble, J., & Crittenden, S. L. (2007). Controls of germline stem cells, entry into meiosis, and the sperm/oocyte decision in Caenorhabditis elegans. *Annu. Rev. Cell Dev. Biol.*, *23*, 405–433.

Knapp, J. R., Davie, J. K., Myer, A., Meadows, E., Olson, E. N., & Klein, W. H. (2006). Loss of myogenin in postnatal life leads to normal skeletal muscle but reduced body size. *Development*, *133*, 601–610.

Kuang, S., Charge, S. B., Seale, P., Huh, M., & Rudnicki, M. A. (2006). Distinct roles for Pax7 and Pax3 in adult regenerative myogenesis. *J. Cell Biol.*, *172*, 103–113.

Kuang, S., Gillespie, M. A., & Rudnicki, M. A. (2008). Niche regulation of muscle satellite cell self-renewal and differentiation. *Cell Stem Cell*, *2*, 22–31.

Kuang, S., Kuroda, K., le Grand, F., & Rudnicki, M. A. (2007). Asymmetric self-renewal and commitment of satellite stem cells in muscle. *Cell*, *129*, 999–1010.

LaBarge, M. A., & Blau, H. M. (2002). Biological progression from adult bone marrow to mononucleate muscle stem cell to multinucleate muscle fiber in response to injury. *Cell*, *111*, 589–601.

Langsdorf, A., Do, A. T., Kusche-Gullberg, M., Emerson, C. P., Jr., & Ai, X. (2007). Sulfs are regulators of growth factor signaling for satellite cell differentiation and muscle regeneration. *Dev. Biol.*, *311*, 464–477.

le Grand, F., Jones, A. E., Seale, V., Scime, A., & Rudnicki, M. A. (2009). Wnt7a activates the planar cell polarity pathway to drive the symmetric expansion of satellite stem cells. *Cell Stem Cell*, *4*, 535–547.

Lepper, C., Conway, S. J., & Fan, C. M. (2009). Adult satellite cells and embryonic muscle progenitors have distinct genetic requirements. *Nature*, *460*, 627–631.

Levenberg, S., Rouwkema, J., Macdonald, M., Garfein, E. S., Kohane, D. S., Darland, D. C., et al. (2005). Engineering vascularized skeletal muscle tissue. *Nat. Biotechnol.*, *23*, 879–884.

Lipton, B. H., & Schultz, E. (1979). Developmental fate of skeletal muscle satellite cells. *Science*, *205*, 1292–1294.

Lopez, J. I., Mouw, J. K., & Weaver, V. M. (2008). Biomechanical regulation of cell orientation and fate. *Oncogene*, *27*, 6981–6993.

Lutolf, M. P., & Hubbell, J. A. (2005). Synthetic biomaterials as instructive extracellular microenvironments for morphogenesis in tissue engineering. *Nat. Biotechnol.*, *23*, 47–55.

Lutolf, M. P., Gilbert, P. M., & Blau, H. M. (2009a). Designing materials to direct stem-cell fate. *Nature*, *462*, 433–441.

Lutolf, M. P., Doyonnas, R., Havenstrite, K., Koleckar, K., & Blau, H. M. (2009b). Perturbation of single hematopoietic stem cell fates in artificial niches. *Integ. Biol.*, *1*, 59–69.

Machida, S., & Booth, F. W. (2004). Insulin-like growth factor 1 and muscle growth: implication for satellite cell proliferation. *Proc. Nutr. Soc.*, *63*, 337–340.

Mauro, A. (1961). Satellite cell of skeletal muscle fibers. *J. Biophys. Biochem. Cytol.*, *9*, 493–495.

Mauro, A. (1979). *Muscle Regeneration*. New York: Raven Press.

McCroskery, S., Thomas, M., Maxwell, L., Sharma, M., & Kambadur, R. (2003). Myostatin negatively regulates satellite cell activation and self-renewal. *J. Cell Biol.*, *162*, 1135–1147.

Mitchell, K. J., Pannerec, A., Cadot, B., Parlakian, A., Besson, V., Gomes, E. R., et al. (2010). Identification and characterization of a non-satellite cell muscle resident progenitor during postnatal development. *Nat. Cell Biol.*, *12*, 257–266.

Mizuno, Y., Chang, H., Umeda, K., Niwa, A., Iwasa, T., Awaya, T., et al. (2010). Generation of skeletal muscle stem/ progenitor cells from murine induced pluripotent stem cells. *Faseb J.*, *24*, 2245–2253.

Montarras, D., Morgan, J., Collins, C., Relaix, F., Zaffran, S., Cumano, A., Partridge, T., & Buckingham, M. (2005). Direct isolation of satellite cells for skeletal muscle regeneration. *Science*, *309*, 2064–2067.

Morrison, S. J., & Kimble, J. (2006). Asymmetric and symmetric stem-cell divisions in development and cancer. *Nature*, *441*, 1068–1074.

Muir, L. A., & Chamberlain, J. S. (2009). Emerging strategies for cell and gene therapy of the muscular dystrophies. *Expert Rev. Mol. Med., 11*, e18.

Muskiewicz, K. R., Frank, N. Y., Flint, A. F., & Gussoni, E. (2005). Myogenic potential of muscle side and main population cells after intravenous injection into sub-lethally irradiated mdx mice. *J. Histochem. Cytochem., 53*, 861–873.

Olguin, H. C., & Olwin, B. B. (2004). Pax-7 up-regulation inhibits myogenesis and cell cycle progression in satellite cells: a potential mechanism for self-renewal. *Dev. Biol., 275*, 375–388.

Olwin, B. B., & Rapraeger, A. (1992). Repression of myogenic differentiation by aFGF, bFGF, and K-FGF is dependent on cellular heparan sulfate. *J. Cell Biol., 118*, 631–639.

Oustanina, S., Hause, G., & Braun, T. (2004). Pax7 directs postnatal renewal and propagation of myogenic satellite cells but not their specification. *Embo J., 23*, 3430–3439.

Peault, B., Rudnicki, M., Torrente, Y., Cossu, G., Tremblay, J. P., Partridge, T., et al. (2007). Stem and progenitor cells in skeletal muscle development, maintenance, and therapy. *Mol. Ther., 15*, 867–877.

Peng, H., & Huard, J. (2004). Muscle-derived stem cells for musculoskeletal tissue regeneration and repair. *Transpl. Immunol., 12*, 311–319.

Pisani, D. F., Dechesne, C. A., Sacconi, S., Delplace, S., Belmonte, N., Cochet, O., et al. (2010). Isolation of a highly myogenic CD34-negative subset of human skeletal muscle cells free of adipogenic potential. *Stem Cells, 28*, 753–764.

Price, F. D., Kuroda, K., & Rudnicki, M. A. (2007). Stem cell based therapies to treat muscular dystrophy. *Biochim. Biophys. Acta, 1772*, 272–283.

Rando, T. A., & Blau, H. M. (1994). Primary mouse myoblast purification, characterization, and transplantation for cell-mediated gene therapy. *J. Cell Biol., 125*, 1275–1287.

Ratajczak, M. Z., Majka, M., Kucia, M., Drukala, J., Pietrzkowski, Z., Peiper, S., et al. (2003). Expression of functional CXCR4 by muscle satellite cells and secretion of SDF-1 by muscle-derived fibroblasts is associated with the presence of both muscle progenitors in bone marrow and hematopoietic stem/progenitor cells in muscles. *Stem Cells, 21*, 363–371.

Reilly, G. C., & Engler, A. J. (2010). Intrinsic extracellular matrix properties regulate stem cell differentiation. *J. Biomech., 43*, 55–62.

Relaix, F., Montarras, D., Zaffran, S., Gayraud-Morel, B., Rocancourt, D., Tajbakhsh, S., et al. (2006). Pax3 and Pax7 have distinct and overlapping functions in adult muscle progenitor cells. *J. Cell Biol., 172*, 91–102.

Rosant, C., Nagel, M. D., & Perot, C. (2007). Aging affects passive stiffness and spindle function of the rat soleus muscle. *Exp. Gerontol., 42*, 301–308.

Sabourin, L. A., Girgis-Gabardo, A., Seale, P., Asakura, A., & Rudnicki, M. A. (1999). Reduced differentiation potential of primary MyoD−/− myogenic cells derived from adult skeletal muscle. *J. Cell Biol., 144*, 631–643.

Sacco, A., Doyonnas, R., Kraft, P., Vitorovic, S., & Blau, H. M. (2008). Self-renewal and expansion of single transplanted muscle stem cells. *Nature, 456*, 502–506.

Saini, A., Faulkner, S., Al-Shanti, N., & Stewart, C. (2009). Powerful signals for weak muscles. *Ageing Res. Rev., 8*, 251–267.

Sampaolesi, M., Blot, S., d'Antona, G., Granger, N., Tonlorenzi, R., Innocenzi, A., et al. (2006). Mesoangioblast stem cells ameliorate muscle function in dystrophic dogs. *Nature, 444*, 574–579.

Sampaolesi, M., Torrente, Y., Innocenzi, A., Tonlorenzi, R., d'Antona, G., Pellegrino, M. A., et al. (2003). Cell therapy of alpha-sarcoglycan null dystrophic mice through intra-arterial delivery of mesoangioblasts. *Science, 301*, 487–492.

Sanes, J. R. (2003). The basement membrane/basal lamina of skeletal muscle. *J. Biol. Chem., 278*, 12601–12604.

Schmalbruch, H., & Lewis, D. M. (2000). Dynamics of nuclei of muscle fibers and connective tissue cells in normal and denervated rat muscles. *Muscle Nerve, 23*, 617–626.

Schofield, R. (1978). The relationship between the spleen colony-forming cell and the haemopoietic stem cell. *Blood Cells, 4*, 7–25.

Scime, A., Desrosiers, J., Trensz, F., Palidwor, G. A., Caron, A. Z., Andrade-Navarro, M. A., et al. (2010). Transcriptional profiling of skeletal muscle reveals factors that are necessary to maintain satellite cell integrity during ageing. *Mech. Ageing Dev., 131*, 9–20.

Seale, P., Sabourin, L. A., Girgis-Gabardo, A., Mansouri, A., Gruss, P., & Rudnicki, M. A. (2000). Pax7 is required for the specification of myogenic satellite cells. *Cell, 102*, 777–786.

Shea, K. L., Xiang, W., LaPorta, V. S., Licht, J. D., Keller, C., Basson, M. A., et al. (2010). Sprouty1 regulates reversible quiescence of a self-renewing adult muscle stem cell pool during regeneration. *Cell Stem Cell, 6*, 117–129.

Sherwood, R. I., Christensen, J. L., Conboy, I. M., Conboy, M. J., Rando, T. A., Weissman, I. L., et al. (2004). Isolation of adult mouse myogenic progenitors: functional heterogeneity of cells within and engrafting skeletal muscle. *Cell, 119*, 543–554.

Snow, M. H. (1977). The effects of aging on satellite cells in skeletal muscles of mice and rats. *Cell Tissue Res., 185,* 399–408.

Stedman, H. H., Sweeney, H. L., Shrager, J. B., Maguire, H. C., Panettieri, R. A., Petrof, B., et al. (1991). The mdx mouse diaphragm reproduces the degenerative changes of Duchenne muscular dystrophy. *Nature, 352,* 536–539.

Tamaki, T., Akatsuka, A., Ando, K., Nakamura, Y., Matsuzawa, H., Hotta, T., et al. (2002). Identification of myogenic-endothelial progenitor cells in the interstitial spaces of skeletal muscle. *J. Cell Biol., 157,* 571–577.

Tanaka, K. K., Hall, J. K., Troy, A. A., Cornelison, D. D., Majka, S. M., & Olwin, B. B. (2009). Syndecan-4-expressing muscle progenitor cells in the SP engraft as satellite cells during muscle regeneration. *Cell Stem Cell, 4,* 217–225.

Tatsumi, R., Anderson, J. E., Nevoret, C. J., Halevy, O., & Allen, R. E. (1998). HGF/SF is present in normal adult skeletal muscle and is capable of activating satellite cells. *Dev. Biol., 194,* 114–128.

Tedesco, F. S., Dellavalle, A., Diaz-Manera, J., Messina, G., & Cossu, G. (2010). Repairing skeletal muscle: regenerative potential of skeletal muscle stem cells. *J. Clin. Invest., 120,* 11–19.

Venuti, J. M., Morris, J. H., Vivian, J. L., Olson, E. N., & Klein, W. H. (1995). Myogenin is required for late but not early aspects of myogenesis during mouse development. *J. Cell Biol., 128,* 563–576.

Webster, C., & Blau, H. M. (1990). Accelerated age-related decline in replicative life-span of Duchenne muscular dystrophy myoblasts: implications for cell and gene therapy. *Somat. Cell Mol. Genet., 16,* 557–565.

Yamashita, Y. M., Fuller, M. T., & Jones, D. L. (2005). Signaling in stem cell niches: lessons from the Drosophila germline. *J. Cell Sci., 118,* 665–672.

Zammit, P. S., Golding, J. P., Nagata, Y., Hudon, V., Partridge, T. A., & Beauchamp, J. R. (2004). Muscle satellite cells adopt divergent fates: a mechanism for self-renewal? *J. Cell Biol., 166,* 347–357.

CHAPTER 21

Stem Cells Derived from Fat

Adam J. Katz[*], **Alexander F. Mericli**[**]
* Department of Plastic Surgery, Department of Biomedical Engineering, Laboratory of Applied Developmental Plasticity, University of Virginia Health System, Virginia, USA
** Department of Plastic Surgery, University of Virginia Health System, Virginia, USA

INTRODUCTION

Regenerative medicine combines the disciplines of medicine, engineering, and biology, and promises a paradigm shift in the treatment of many diseases. The keystone of regenerative medicine and tissue engineering is the stem cell. Use of embryonic stem cells remains controversial; therefore, research efforts involving isolation of multipotent cells from adult tissue is playing an increasingly important role. Adult stem cells have been identified in several different types of tissue, including bone marrow, blood, nervous tissue, skeletal muscle, gut, and adipose. The literature is now replete with evidence that adipose tissue (i.e. fat tissue) contains a readily available, abundant, and expendable source of adult stem cells that can be directed towards several different lineages (Halvorsen et al., 2001; Zuk et al., 2001, 2002; Erickson et al., 2002; Mizuno et al., 2002; Safford et al., 2002; Rehman et al., 2004; Planat-Bénard et al., 2004a; Banas et al., 2007; Lee et al., 2008; Neupane et al., 2008). Approximately 400,000 liposuction surgeries are performed in the USA each year, routinely yielding volumes of up to 3 L of valuable tissue (Katz et al., 1999). One gram of adipose tissue yields approximately 5,000 stem cells, which is 500 times greater than the number of stem cells isolated from 1 g of bone marrow (Fraser et al., 2006). As such, adipose tissue is a promising, readily available, and rich source of adult stem cells.

The process of isolating ASCs from adipose tissue is derived from the initial description by Rodbell in the 1960s (Rodbell, 1966). The tissue is first minced, then washed and dissociated by collagenase. The resulting slurry is then centrifuged, thereby yielding the pelleted stromal vascular fraction (SVF), which can be further washed and/or filtered. The ASC population is then further selected and/or enriched based on cell adherence to tissue culture plastic (Fig. 21.1).

All fat is not the same, and therefore the anatomic site from which fat is harvested, as well as donor characteristics such as age, sex, and medical co-morbidities, may be of particular importance when considering ASC yield, plasticity, and potency. There are two main types of fat within the human body — brown adipose tissue (BAT) and white adipose tissue (WAT). BAT is mainly found surrounding the viscera in infants; in adults it has been identified in the neck, supraclavicular, paraaortic, paravertebral, and suprarenal areas (Nedergaard et al., 2007). BAT serves primarily to generate heat via a specific uncoupling mechanism identified in the cells' mitochondrial electron transport chain. WAT is found subcutaneously, is highly vascularized, and contains the ASC population; its primary function is to serve as an energy reserve but it is

Principles of Regenerative Medicine. DOI: 10.1016/B978-0-12-381422-7.10021-5

366

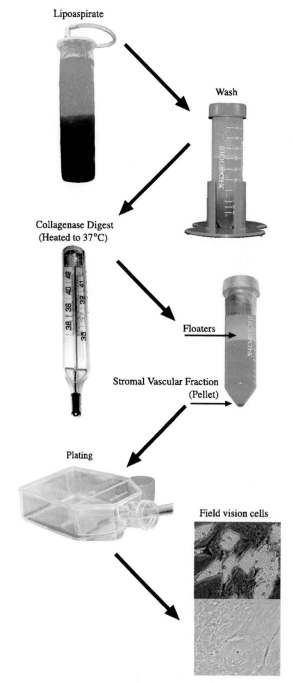

FIGURE 21.1
Processing of lipoaspirate and isolation of adipose derived stem cells. *(From Gimble et al., 2007, with permission.)*

now recognized as a dynamic and highly bioactive tissue with important roles in systemic inflammation and other processes (Trayhurn and Beattie, 2001; Tang et al., 2008). Major depots of subcutaneous WAT in humans include abdominal, hip, thigh, and gluteal locations. Studies indicate that the density of ASCs varies depending on the anatomic location, age, sex and co-morbidities of the donor (Prunet-Marcassus et al., 2006; DiMuzio and Tulenko, 2007).

The definitive cell surface identity of ASCs is problematic, as surface protein expression has been shown to change depending on passage and cell density, among other variables.

TABLE 21.1 Abridged List of ASC Surface Proteins

CD Antigen	Protein Name	References
CD9	Tetraspan protein	Kim et al., 2007b
CD11b	Integrin α_M	Gronthos et al., 2001
CD13	Aminopeptidase N	Mitchell et al., 2006; Yañez et al., 2006
CD29	Integrins $\beta1$	Gronthos et al., 2001
CD44	H-CAM	Boquest et al., 2005; Mitchell et al., 2006
CD49a-e	Integrin α_{1-5}	Boquest et al., 2005
CD55	DAF	Yañez et al., 2006
CD61	gpIIIa	Boquest et al., 2005
CD62e	E-Selectin	Boquest et al., 2005
CD62l	L-Selectin	Boquest et al., 2005
CD62p	P-Selectin	Boquest et al., 2005
CD63	LAMP-3	Boquest et al., 2005
CD73	Ecto-5'-nucleotidase	Boquest et al., 2005; Mitchell et al., 2006
CD90	Thy-1	Boquest et al, 2005
CD105	Endoglin	Mitchell et al., 2006
CD106	VCAM-1	Yañez et al., 2006
CD144	VE-Cadherin	Mitchell et al., 2006
CD166	ALCAM	Mitchell et al., 2006

However, after two or more passages, ASCs express relatively characteristic and reproducible surface and adhesion molecules, cytoskeletal elements, and surface enzymes. For example, ASCs consistently express the tetraspan protein (CD9), several integrins (CD29, CD49a-e), and CD-62, -105, -106, and -166, among others (Table 21.1). Overall, the ASC immunophenotype largely resembles that reported for other adult mesenchymal stem cells — such as bone marrow-derived and skeletal muscle stem cells — with only a few differences (Gronthos et al., 2001, 2003; Zuk et al., 2001, 2002; Katz et al., 2005). Ultimately, cell populations are best characterized by biological activity and potency; still, it would be useful to consolidate and continuously update the various ASC-associated surface protein markers that are reported in the literature. To this end, our team is in the process of compiling an online, searchable database denoting all reported ASC surface protein data and related information. The database will contain links to the original article that published the surface protein data and will be easily accessible through the Department of Plastic Surgery website at the University of Virginia. An abridged version of the ASC surface immunophenotype database is found in Table 21.1.

ASCs have been shown to demonstrate phenotypes consistent with several different mesodermally, endodermally, and ectodermally derived mature cell types both *in vitro* and *in vivo* (Table 21.2). While these developmental findings are remarkable and continue to evolve, there remain many unanswered questions and perhaps some appropriate residual skepticism related to the "applied" *in vivo* developmental plasticity and functional integration of adult stem cells. Part of this relates to limitations of terminology, standardization, and existing technologies, and also to the human need to characterize the elusive nature of dynamic living systems with static labels and/or pathways. For example, establishing a scientific consensus as to what "minimal threshold" definitively describes a particular phenotype/cell lineage is elusive to say the least, and the "minimal essential threshold" continues to evolve as the field matures and realizes that there are few, if any, genes/proteins that are truly "lineage specific." While these issues are of great interest and importance for basic scientists and developmental biologists, the ultimate benchmark of these cells for many in the field remains the reproducible and

TABLE 21.2 Summary of Differentiation Capability of ASCs and Lineage-specific Induction Factors

Cell Lineage	Inductive Factors	Phenotype	References
Adipocyte	Dexamethasone, isobutyl methylxanthine, indomethacin, insulin, thiazolidinedione	Oil Red-O Staining	Halvorsen et al., 2001; Zuk et al., 2001; Neupane et al., 2008
Cardiomyocyte	Transferrin, IL-3,IL-6,VEGF, culture on laminin, TGF-β	Cardiac myosin heavy chain, troponin I, α-sarcomeric actin, spontaneous contraction	Rangappa et al., 2003; Planat-Bénard et al., 2004a; Miyahara et al., 2006; Song et al., 2007
Chrondrocyte	Ascorbic acid,BMP-6, dexamethasone, insulin, hydrostatic pressure, TGF-β	Toluidine Blue,collagen II and X production	Zuk et al., 2001; Erickson et al., 2002
Endothelial	Proprietary medium containing multiple growth factors (EGM-2; Cambrex)	CD31 and von Willeband factor expression, tube formation in matrigel, incorporation into microvasculature	Miranville et al., 2004; Rehman et al., 2004; Planat-Benard et al., 2004b
Hepatocyte	HGF, FGF-1, FGF-4	Urea synthesis, maintain glycogen stores, hepatocyte mRNA markers	Banas et al., 2007
Myocyte	Dexamethasone, horse serum	Multi-nucleation, skeletal muscle myosin heavy-chain II, MyoD1 expression	Zuk et al., 2001; Mizuno et al., 2002
Neuronal-like	Butylated hydroxyanisole, valproic acid, insulin	Nestin, NeuN, intermediate filament, MAP2, b-III tubulin, glutamate receptor subunits NR1 and NR2 expression, electrophysiologic properties	Safford et al., 2002; Safford et al., 2004
Osteoblast	Ascorbic acid, BMP-2, dexamethasone, 1,25 dihydroxy vitamin D3	Alizarin red, von Kassa stain, collagen I, alkaline phosphatase, osteopontin, osteonectin, osteocalcin	Halvorsen et al., 2001; Zuk et al., 2001; Hicok et al., 2004; Neupane et al., 2008;
Pancreatic	Nictinamide, activin A, exendin-4, HGF, pentagastrin, cytosolic extract from regenerating pancrease	Insulin secretion, glucogon, somatostatin	Lee et al., 2008
Smooth muscle	Angiotensin II, shingosylphosphorylcholine, TGF-β	Calponin, caldesmon, myosin heavy chain expression; contractile behavior	Harris et al., 2009

For a more thorough description of the demonstrated *in vitro* and *in vivo* differentiation capabilities of ASCs, please see the following articles listed in this chapter's references: Parker and Katz, 2006; Gimble et al., 2007; Tholpady et al., 2009; Bailey et al., 2010.
Modified and updated from Bailey et al., 2010, with permission.

predictable therapeutic safety and efficacy in the clinical setting, regardless of the mechanism of action.

Clearly, there is still a great deal of work that needs to be done to further confirm, expand, and understand the essence of adult stem cell plasticity. At present, there is limited evidence for definitive *in vivo* differentiation, integration, and *de novo* tissue (re)generation by MSCs of various origins. Initially, this "building block" theory was proposed as a primary mechanism of action for stem cells, wherein undifferentiated cells would migrate towards or be delivered to a diseased or injured tissue and repopulate the tissue/organ through proliferation and differentiation (Kim et al., 2009a). However, this thinking is now being challenged by a growing body of data and experience that suggest that the survival of engrafted cells is too low to justify (explain) therapeutic benefit in most cases (Uemura et al., 2006). Furthermore, acute functional improvement within days makes it difficult to fully explain this mechanism of tissue repair and regeneration (Wang et al., 2006; Crisostomo et al., 2007).

Several groups have now demonstrated that much of the functional improvement and attenuation of injury afforded by stem cells can be repeated by treatment with conditioned medium alone (Patel et al., 2007). In fact, recent studies have revealed that ASCs exert their role in cardiac repair not only through putative differentiation but also (and quite possibly predominantly) through paracrine effects via secretion of a variety of cytokines and chemokines, such as VEGF, TGF-β, and HGF (Rehman et al., 2004; Nakagami et al., 2005; Moon et al., 2006). Additionally, ASCs secrete anti-inflammatory factors (such as IL-10, IL-8, MCP-1, and others) and possess immunosuppressive capabilities, having been shown to inhibit macrophage function and suppress T-helper cell activation (Puissant et al., 2005; Gonzalez-Rey et al., 2010). Thus, it can be deduced that ASCs may exert their beneficial *in vivo* effects via complex paracrine and anti-inflammatory mechanisms in addition to, or instead of, a building-block function.

As we continue to learn more about the function and therapeutic mechanisms of adipose-derived stem cells, regenerative medicine moves closer to the bedside. One way of thinking about the application of ASCs to treat human pathology can best be described as an "adipose therapeutic spectrum" (Fig. 21.2). The spectrum encapsulates current studies and the future application of ASCs to treat a broad range of human diseases. Therapeutic strategies that are further to the right on the spectrum reflect increasingly complex scientific, manufacturing, and/or regulatory hurdles. The remainder of this chapter will explore the therapeutic application of ASCs within the context of the adipose therapeutic spectrum and discuss our current understanding of adipose-derived stem cells as they apply to regenerative medicine.

SYSTEMIC-PARENTERAL APPLICATION OF ASCs

The systemic administration of ASCs as cell suspensions has great translational appeal as it provides an easy, minimally invasive treatment modality. However, it is reliant on the specific and efficient homing of cells to the site(s) of interest/injury. When injected intravenously, ASCs engraft in several different organs, including the liver, lungs, heart, and brain. It is unclear, however, whether this phenomenon is due to an innate stem-cell homing mechanism, the result of the increased perfusion that occurs after damage to an organ, or merely due to the cells being predominantly removed on first pass through the "filter organs." Chemokine (chemotactic cytokine) receptors and ligands are essential components involved in the trafficking, adhesion, and migration of leukocytes into sites of injury/inflammation, and it has recently been shown that ASCs express some of these same signaling molecules (Table 21.1) (Boquest

369

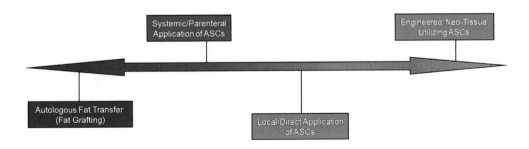

FIGURE 21.2
Adipose therapeutic spectrum. As the spectrum progresses from left to right, therapeutic strategies increase in scientific, manufacturing, and regulatory complexity.

et al., 2005; Amos et al., 2008; Bailey et al., 2010). Therefore, parenterally injected ASCs may arrive at sites of injury similar to the transendothelial migration and diapodesis mechanism of leukocytes. Several groups have been studying the expression of chemokines in mesenchymal stem cells, including ASCs, although results have been variable (Sordi et al., 2005; Honczarenko et al., 2006). These apparently conflicting data may simply reflect phenotypic and biological differences (e.g. cell activation) in response to a variety of culture conditions (e.g. media supplements, cell density). The fact that they express a variety of chemokine receptors may also suggest that they have the potential to home to multiple different tissues. Challenges for the future relate to directing cell homing to achieve a higher efficiency of cell delivery at a site of need, thereby increasing chances for therapeutic efficacy while minimizing cell dose and the chances for co-morbidities. Various strategic approaches exist for such objectives, including the pre-sorting of specific cell subpopulations according to their ability to express receptors such as P-selectin and/or to bind a target substrate.

From a clinical perspective, the majority of work regarding the parenteral delivery of ASCs has been in reference to cardiovascular disease, with the hope of mitigating or repairing ischemic myocardium and/or damaged vessels. Several groups have demonstrated the ability of ASCs to differentiate towards the cardiomyocyte lineage *in vitro*, including the adoption of cardiomyocyte morphology, spontaneous contraction, functional responsiveness to adrenergic and cholinergic agonists, and positive immunostaining for myosin heavy chain, alpha-actinin, and troponin-I (Rangappa et al., 2003; Planat-Bénard et al., 2004a; Parker and Katz, 2006). Valina and colleagues (2007) have shown that the trans-catheter intracoronary administration of ASCs into acutely infarcted myocardium allows ASCs to differentiate into endothelial and vascular smooth muscle cells, resulting in an improvement of left ventricular function, wall thickness, myocardial remodeling, and perfusion in a porcine model. These results were found to be similar in efficacy to those achieved by intracoronary administration of bone-marrow-derived stem cells. Kim et al. (2009b) have demonstrated that intravenously administered ASCs home to the site of radiofrequency catheter ablated canine atrium, engraft in myocardium, and subsequently adopt a cardiomyocyte phenotype. Similarly, with respect to peripheral vascular disease, after ligation of the femoral artery in mice, iv injection of ASCs has also been associated with salvaged perfusion to the lower extremity (Fig. 21.3) (Cao et al., 2005). Several mechanisms have been proposed to explain the recovery of ischemic tissues after ASC delivery, such as secretion of angiogenic and vasculogenic growth factors like VEGF and hepatocyte growth factor (HGF), and/or the differentiation of ASCs into cardiomyocytes (Rangappa et al., 2003; Planat-Bénard et al., 2004a; Miyahara et al., 2006), smooth muscle cells, and/or endothelial cells (Miranville et al., 2004; Planat-Bénard et al., 2004b; Harris et al., 2009).

ASCs have also been delivered intravenously in animal models of several different experimentally created neurologic diseases, such as embolic stroke, hemorrhagic stroke, autoimmune encephalitis, olfactory dysfunction, and spinal cord injury (Kang et al., 2003, 2006; Kim et al., 2007a, 2009c; Constantin et al., 2009). Similarly, systemic ASC therapy has been shown to favorably impact a number of other diverse preclinical models of human disease, including Duchenne muscular dystrophy, urinary stress incontinence, and hepatic injury (Liu et al., 2007; Lin et al., 2010; Kim et al., 2003).

At present, at least two human clinical studies (initiated in 2007) are active that involve the systemic delivery of ASCs for the treatment of myocardial ischemia: APOLLO and PRECISE (Randomized Clinical Trial of Adipose-Derived Stem Cells in the Treatment of Pts With ST-Elevation Myocardial Infarction, 2007; Randomized Clinical Trial of Adipose-Derived Stem Cells in Treatment of Non Revascularizable Ischemic Myocardium, 2007). The APOLLO (AdiPOse-derived stem ceLLs in the treatment of patients with st-elevation myOcardial infarction), and the PRECISE (adiPose-deRived stEm and regenerative Cells In the treatment of patients with non-revaScularizable ischEmic myocardium) studies are both prospective, double-blind, randomized, placebo-controlled phase I trials, currently in the recruiting phase.

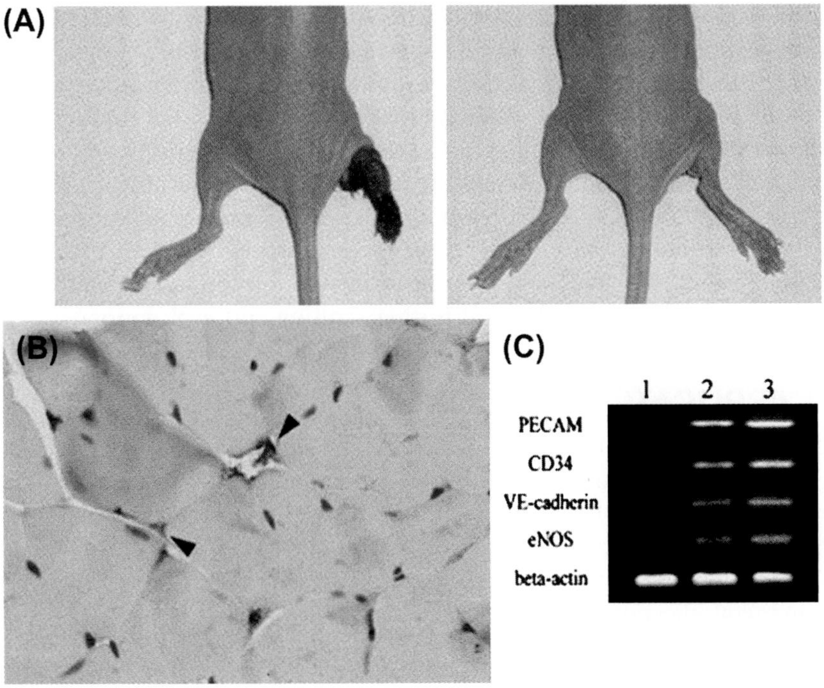

FIGURE 21.3
Culture-expanded ASCs differentiated into endothelial-like cells after intravascular injection following ligation of the femoral artery (mouse ischemic hindlimb model). (A) Representative photographs of control medium (left) and ASC-treated (right) ischemic hindlimbs. (B) Immunohistochemistry staining of human endothelial cells in ischemic hindlimb muscle 14 days after ischemia. Representative photomicrographs of immunohistochemistry using antibody specifically against human CD34 isoform (cluster differentiation protein normally found on endothelial cells). (C) Results of RT-PCR specific for human PECAM, CD34, VE-cadherin, and eNOS. Lane 1 shows non-treated mice; lane 2 shows ASC-treated mice; and lane 3 shows HUVECs as positive control. Reproduced from Cao et al., 2005, with permission.

For the APOLLO trial, inclusion criteria are: clinical symptoms consistent with acute myocardial infarction for a minimum of 2 and a maximum of 12 h from onset to percutaneous coronary intervention (PCI), and unresponsive to nitroglycerin; a successful revascularization of the culprit lesion in the major epicardial vessel; an area of hypokinesia or akinesia corresponding to the culprit lesion, as determined by left ventriculography at the time of primary PCI; a mild to moderate left ventricular dysfunction, reflected by a left ventricular ejection fraction in a range between 30 and 50% by left ventriculography at the time of successful revascularization; and the ability to undergo liposuction. For the PRECISE trial, inclusion criteria are: coronary artery disease not amenable to any type of revascularization (percutaneous or surgical) in the target area; hemodynamic stability; and the ability to undergo liposuction. Of note, both of these trials involve the "point-of-care" isolation and delivery of "fresh" SVF cells from adipose tissue − as opposed to cells further enriched by plating and adherence to tissue culture plastic. This approach may be motivated as much by regulatory considerations as by scientific precedence, as the vast majority of preclinical literature involves the use of culture-expanded cell populations. Recently, however, several groups have reported on the study/use of fresh SVF cells, potentially signaling greater interest in this therapeutic strategy (Boquest et al., 2005; Yoshimura et al., 2006). Compared to culture-expanded cells, fresh SVF isolates are significantly more heterogeneous in nature, consisting of up to 50% leukocytes along with stromal, stem, and endothelial cells and related lineages (Sengenès et al., 2005; Yoshimura et al., 2006).

Safety concerns pertaining to intravascular administration of stem cells do exist. When injecting cells into the circulation, the most worrisome adverse event is the inadvertent creation of emboli. Furlani and colleagues (2009) recently studied the safety of systemic administration of human mesenchymal stem cells in mice. After histopathologic analysis, the authors concluded that up to 40% of animals died post-injection from pulmonary embolism. Interestingly, however, other authors have recently found that these stem cell emboli are not necessarily harmful, but instead may actually contribute to and underlie the cell's therapeutic benefit − particularly in regard to the treatment of cardiac disease. More specifically, Lee et al. (2009a) found that MSCs injected intravenously quickly entrap in the pulmonary vasculature and thereafter increase translation of the anti-inflammatory protein, TSG-6. Of note, TSG-6 alone (i.e. *without* cell therapy) was shown to decrease the inflammatory response, reduce myocardial infarct size, and improve cardiac function similar to that observed with cell therapy. Therefore, MSCs may entrap in pulmonary tissue as microemboli and produce anti-inflammatory cytokines, which then are carried through the pulmonary vasculature and into the heart, inducing a paracrine effect, repairing and restoring cardiac function. This would help to explain findings of functional organ improvement in the face of minimal cell engraftment in the target organ/tissue. At the very least, these findings reveal a completely new perspective on how regenerative cell therapies may produce therapeutic benefit, provide insights into potential safety and dosing concerns, and argue for alternative mechanisms that warrant further scrutiny and exploration.

LOCAL/DIRECT APPLICATION OF ASCs

The direct, local delivery of ASCs to a site of intended tissue regeneration/repair has been demonstrated successfully for a wide variety of tissue types and pathology. Direct application potentially minimizes the importance of a chemotactic homing mechanism, as the cells are applied directly to the site of tissue injury. Furthermore, this method circumvents the potential issue of pulmonary embolism and unintended hepatic or splenic engraftment that has been demonstrated after parenteral injection.

Several examples exist in the literature involving local delivery of cell to myocardium. Katz et al. (in press) demonstrated the engraftment of culture-expanded human ASCs delivered by direct injection into infarcted myocardium using an immunocompromised murine model of acute reperfused myocardial infarction. Engrafted ASCs exhibited trends toward enhanced cardiac structure and function compared with historical controls. In another study, mice were subjected to direct myocardial injection of ASCs after a 30-minute occlusion of the left anterior descending artery (Cai et al., 2008). At 1 month, they found significant improvement in stroke volume, cardiac output, and ejection fraction in the experimental group. In a study by Miyahara and colleagues (Miyahara et al., 2006), a patch-like sheet of monolayer-cultured ASCs and ECM was overlaid onto scarred myocardium 4 weeks after coronary ligation in rats. The engrafted sheet gradually grew to form a thick stratum that included newly formed vessels, undifferentiated cells, and several cardiomyocytes, resulting in increased left ventricular maximum diastolic relaxation rate and decreased left ventricular end-diastolic pressure.

With respect to peripheral vascular pathology, many groups have demonstrated revascularization after direct intramuscular injection of ASCs in the mouse ischemic hindlimb model. In this model, the mouse femoral artery is ligated, ultimately producing ischemic necrosis in the control animals/limbs. After direct intramuscular injection of ASCs into the affected limb, however, perfusion is restored through a combination of angiogenesis and *de novo* vasculogenesis (Planat-Bénard et al., 2004b). Likely via a related mechanism of action, Lu et al. (2008) demonstrated that, when ASCs are injected into the pedicle of random-pattern skin flaps in mice, there is greater flap viability and increased capillary density compared to controls. Several groups have also demonstrated favorable results using local delivery of ASCs for the treatment of neurological injury and disease, including experimentally created Huntington's disease, paraplegia, traumatic brain injury, and peripheral nerve regeneration (Okonkow et al., 2005; Lee et al., 2009b; Ryu et al., 2009; Santiago et al., 2009).

Bone, cartilage, tendon, and ligamentous tissue are well suited for tissue-engineering approaches via direct, localized application of ASCs. The *in vitro* chondrogenic potential of ASCs has been evaluated by seeding into agarose, alginate, or gelatin three-dimensions scaffolds, or by exposure to TGFβ, dexamethasone and ascorbate supplemented media (Lin et al., 2005). Successful differentiation is marked by expression of typical gene and chondrogenic extracellular matrix proteins such as sox-9, collagen type II, and chondroitin sulfate, among others. *In vivo*, ASCs have been demonstrated to be able to heal cartilaginous defects. Dragoo et al. (2007) showed formation of new hyaline cartilage after ASCs, precultured in FGF2 and TGFβ and inserted into a fibrin scaffold, were placed in a rabbit chondral defect.

Osteogenic induction of ASCs is well documented in the literature and *in vivo* applications often involve seeding into scaffolds with or without the prior osteo-induction of the cells *in vitro* (Table 21.2) (Tapp et al., 2009). In an early study using hydroxyapatite/tricalcium phosphate scaffolds, osteoid formation was present in 80% of SCID mice subcutaneous implants loaded with ASCs, but absent in cell-free implants (Hicok et al., 2004). Femoral bone defects were healed by ASCs genetically modified to overexpress BMP-2 and loaded onto a collagen-ceramic scaffold (Peterson et al., 2005; Dudas et al., 2006; Yoon et al., 2007). Early work also shows great promise in the repair of critical-sized calvarial defects using ASCs (Dudas et al., 2006; Yoon et al., 2007).

Repair of intervertebral disks and tendons using stem cells is of great interest given the inability of these tissues to regenerate and heal effectively. ASCs were recently found to be successful in regenerating experimentally damaged canine intervertebral disks, with treated disks demonstrating greater type II collagen and cell density (Ganey et al., 2009). Although there are no published studies to date using ASCs for tendon repair, several studies have shown that bone marrow-derived MSCs *are* able to enhance tendon healing (Awad et al., 1999; Juncosa-Melvin et al., 2005, 2006). Given the extensive documented similarities between ASCs and bone marrow MSCs, it is not unreasonable to postulate that ASCs may be able to improve *in vivo* tendon healing through similar mechanisms.

Still other work suggests that the direct application of ASCs to soft tissue wounds may accelerate healing through a paracrine or other mechanism. Nambu et al. (2009) demonstrated that, when ASCs combined with a collagen/silicon scaffold are applied to a wound in a healing-impaired diabetic mouse, there is greater granulation tissue, capillary density, and epithelialization compared to controls. In a similar study, Amos et al. (2009) showed that ASCs applied to wounds in diabetic mice result in a faster rate of healing when the ASCs are delivered as multicellular aggregates compared to when the stem cells are delivered as a cell suspension. Furthermore, ASCs formulated as three-dimensional cellular aggregates produce significantly more extracellular matrix proteins and secrete higher levels of bioactive factors compared to monolayer culture (Amos et al., 2009). These results emphasize the importance of cell culture, formulation, and delivery mechanism in ASC biology and therapeutic effect.

A clinical application receiving great interest and effort at present involves the localized delivery of ASCs to reconstruct and/or augment soft tissue. Large soft tissue defects are a common problem after burns, trauma, and oncologic resection, such as mastectomy. In addition, many patients seek to smooth cutaneous wrinkles and augment naturally occurring adipose tissue reserves via aesthetic surgery. Unfortunately, adipose tissue (mature adipocytes in particular) is relatively fragile and prone to ischemia, making autologous fat transfer more art than science. In order to develop more reproducible, scientifically based methods for soft tissue reconstruction, several laboratories have investigated the possibility of creating tissue-engineered cell-seeded scaffolds for the generation of *de novo* adipose tissue. Initial successes have been described using these cells to seed artificial scaffolds subsequently implanted subcutaneously in mice and rats (Patrick et al., 1999; von Heimburg et al., 2001). However, reconstruction of large volume defects remains a challenge with this approach, as the constructs rely on diffusion for cell/tissue survival rather than on a pre-existing microvascular network, or one that forms *in vivo* within the requisite timeframe.

In a related approach to soft tissue reconstruction/augmentation that has now reached the clinic, ASCs are combined with intact adipose tissue fragments in a technique termed "cell-assisted lipotransfer" (Yoshimura et al., 2008). In this series of 40 patients, adipose tissue was first harvested by suction lipectomy, and the SVF was isolated and concentrated from a *portion* of the harvested tissue sample. The SVF cells were then combined with the remaining "intact" adipose tissue portion to generate stem cell-enriched adipose tissue grafts. This cell-enriched autologous fat graft was then injected into the breast. At two months post-operative, the resultant breast mound volume was increased 100−200% without any signs of resorption (Fig. 21.4). The authors conclude that this increased volume is due to the stem cell population within the SVF differentiating towards adipose tissue phenotypes and/or enhancing revascularization and "take" of the intact tissue pieces in the grafts. Patients in the study were overwhelmingly pleased given the more natural appearance and consistency of the tissue compared to conventionally augmented breasts (i.e. silicone or saline implants). The long-term safety and efficacy of this approach remains to be seen, especially as compared to traditional fat grafting techniques; however, emerging preclinical and clinical data currently suggest a therapeutic advantage. Ideally, future randomized controlled clinical trials will compare the efficacy of these two approaches to soft tissue augmentation.

Finally, a few groups have utilized direct application of ASCs in the clinical setting for other therapeutic objectives. Yamamoto and colleagues (in press) injected autologous ASCs in a peri-urethral distribution in two patients who were afflicted by stress urinary incontinence, recalcitrant to current treatments. This group reported decreased urine leakage volume, decreased frequency and amount of incontinence, and increased maximum urethral closing pressure. The authors theorize that the mechanism of action was multi-factorial, including a bulking effect, differentiation of the ASCs into contractile cells, and improved vascularity to the sphincter secondary to ASC-related angiogenesis. Garcia-Olmo and colleagues (2009)

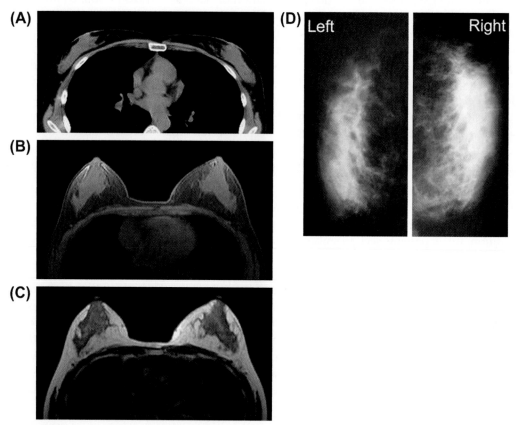

FIGURE 21.4
Radiologic views showing the chest of a patient who underwent cell-assisted lipotransfer for cosmetic breast augmentation. (A) A preoperative computed tomography (CT) image in the horizontal plane at the level of the nipples. (B and C) Horizontal images by magnetic resonance imaging (MRI) 12 months after surgery: (B) T1-image; (C) T2-image. The adipose tissue is augmented around and under the mammary glands. A small cyst (<10 mm) appears in the fatty layer under the right mammary gland. (D) Mammograms at 12 months show no abnormal calcifications. Reproduced from Yoshimura et al., 2008, with permission.

reported their phase II clinical study results evaluating the use of ASCs in healing Crohn's fistulae. Patients with Crohn's fistulae were randomized into two groups: one group to receive intralesional fibrin glue and the other to receive intralesional fibrin glue plus 20 million ASCs. In the ASC group, the authors found a statistically significant higher rate of healing and better "physical component" quality of life score. The authors theorize that the ASCs help to heal Crohn's fistulae by reducing inflammation and inducing immunosuppressive activity. To further support this theory, the authors note that peri-anal sepsis was observed more frequently in the fibrin-glue-only group, potentially because of the anti-inflammatory effect of the ASCs.

ENGINEERED NEO-TISSUE

In the long-term, ASCs may prove to be a valuable cell source for fabricating a variety of *ex vivo* engineered tissue constructs. To date, investigators have successfully regenerated bone, bladder, trachea, and skin, among other types of tissue, both *in vitro* and in human subjects, using ASCs as the cellular foundation. Jack et al. (2009) created a three-dimensional synthetic bladder mold and then seeded it with smooth muscle-like cells derived from human ASCs. These neo-bladders were then implanted into rats. At 12 weeks, the neo-bladders exhibited compliance and capacity similar to *de novo* bladder as well as an increased number of viable smooth muscle cells compared to controls. ASCs applied to a collagen scaffold and implanted in a rat tracheal defect demonstrated accelerated epithelialization and improved angiogenesis compared to controls, and provided an adequate airway (Suzuki et al., 2008). The ability to tissue engineer skin would be of enormous benefit to patients suffering from thermal injury. In one recent study, ASCs were successfully used in place of dermal fibroblasts for reconstruction of the neo-skin's dermal layer. Additionally, the authors concluded that using ASCs for the creation of neo-skin resulted in a superior neo-organ because only the samples containing ASCs were found to possess a histologically distinct hypodermis (Trottier et al., 2008).

375

Reconstruction of large bony defects is complicated and often associated with significant donor site morbidity. As such, there is great interest in using ASCs to improve management of these defects. In Finland, a 65-year-old male underwent maxillary reconstruction using ASCs to generate a hemi-maxilla (Mesimäki et al., 2009). This patient had undergone a hemi-maxillectomy 28 months previously due to a large recurrent keratocyst. This reconstruction took three separate surgeries, beginning with fat and autologous serum harvesting. Over a 14-day period, ASCs were expanded *in vitro* using the patient's autologous serum and DMEM. A three-dimensional scaffold was assembled using β-TCP (tri-calcium phosphate) granules incubated in rhBMP-2-containing media. The rhBMP-2-containing media was discarded and the ASCs were mixed with the rhBMP-2-conditioned β-TCP for 48 h. It is important to note that neither the ASCs nor the patient were ever directly exposed to rhBMP-2. In the second operation, a preformed titanium cage filled with the ASC/β-TCP solution was implanted within the patient's rectus abdominis muscle. The implanted construct was followed radiologically and clinically over a period of eight months while ossification and vascularization occurred. In the third and final operation, the rectus abdominis containing the neo-maxilla was elevated and inset as a free flap in order to reconstruct the patient's hemimaxilla. Four months after flap inset, osseo-integrated dental implants were placed in the neo-maxilla and 1 year after reconstruction there were no signs of bone resorption or neo-maxilla breakdown (Fig. 21.5).

Other potential — albeit completely experimental — methods of producing engineered neo-tissue constructs using ASCs include techniques involving organ printing and/or explanted microcirculatory beds (EMBs) (Mironov et al., 2009). Tissue/organ printing may enable large-scale industrial robotic fabrication of living human organ constructs with a "built-in" perfusable intraorgan branched vascular tree (Mironov et al., 2009). These vascularized organ-constructs could then be perfused with a solution containing predifferentiated ASCs, which would subsequently engraft and further differentiate, creating a functioning,

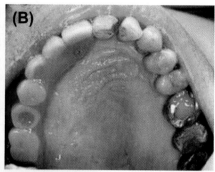

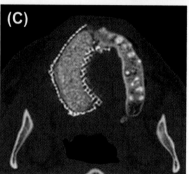

FIGURE 21.5

Images depicting the use of ASCs for the reconstruction of a hemimaxillectomy defect. (A) Pre-reconstruction 28 months after removal of keratocyst by hemimaxillectomy. (B) Final result 1 year after bone and soft tissue reconstruction, with temporary dental implant rehabilitation. (C) Axial CT scan demonstrates the shape and normal bone density of the neo-maxilla. Reproduced from Mesimäki et al., 2009, with permission.

physiologic neo-organ. Similarly, Chang et al. (2009) demonstrated that mouse EMBs (adipofasciocutaneous-free flaps) could be perfused *ex vivo* with a solution containing human ASCs and that a certain percentage of the ASCs in the perfusate could engraft and survive in the EMBs. However, the ability to regenerate a functional, physiologic, vascularized, three-dimensional organ remains an elusive goal for tissue engineers.

ASCs AND CARCINOGENESIS

The remarkable pluripotency of adult stem cells has raised concern regarding the cells' possible contribution to neoplastic disease. The theoretical possibility exists that adult stem cells administered to a patient with the intent to treat disease or restore form could in fact inadvertently introduce or encourage malignancy. Non-adipose mesenchymal stem cells have been demonstrated in some studies to contribute to increased tumor cell viability and improved tumor blood supply; however, other studies have shown either no effect on tumor growth or a *prohibitive* effect on tumor cell proliferation (Zhu et al., 2006; Zhang and Zhang, 2009). Similar conflicting evidence exists regarding the study of ASCs. Sun et al. (2009) did not appreciate any trophic effect of ASCs on tumor growth or cell proliferation, whereas Muehlberg and colleagues (2009) found that co-culture of ASCs with a breast cancer cell line resulted in an increased production of stromal cell-derived factor 1, which acts in a paracrine fashion on the cancer cells to enhance their motility, invasion, and metastasis. Furthermore, ASCs were shown to home to the tumor site when injected locally or intravenously and to incorporate into the tumor's blood vessels as differentiated endothelial cells (Muehlberg et al., 2009).

To date, there is a large and growing number of humans that have received MSC therapies (both bone marrow- and adipose-derived cells; both allogeneic and autologous cells) for a variety of clinical indications, by a variety of delivery routes. As of this writing, we are not aware of any reported oncologic-related adverse events associated with these studies/therapies. Still, there is no doubt that the (putative) oncologic risks associated with regenerative cell therapies deserve focused and continued thorough evaluation. And the mantra of "first do no harm" must guide a clinician's activities, no matter how noble the intent. On the other hand, worthy efforts to advance and translate the field of regenerative medicine will inevitably be associated with risks inherent to the new and unknown — even in light of thorough and meticulous science, and the extensive education and comprehensive and ethical consent of courageous, sometimes desperate patients.

CLINICAL TRIALS

A search for "adipose-derived stem cells" on clinicaltrials.gov results in 17 currently active or recently concluded clinical trials involving adipose-derived cells. Studies listed include using ASCs to treat lipodystrophy, diabetes mellitus type I and II, fecal incontinence, Crohn's and non-Crohn's fistulae, ST elevation myocardial infarction, myocardial ischemia, breast deformity after lumpectomy, treatment of depressed scars, and hepatic regeneration, to mention a few. Because of the regulations instituted by the United States Food and Drug Administration, nearly all trials are located in countries other than the USA.

CONCLUSION

Human adipose tissue is an abundant and accessible source of adult stem cells. ASCs have potential applications in the treatment of acute and chronic musculoskeletal disorders, cardiovascular disease, bony and soft tissue reconstruction, cosmetic surgery, central nervous system injury, and other conditions. However, if ASCs can be used to routinely treat patients in the USA, they will need to be scrutinized by the Food and Drug Administration for both safety and efficacy. The manufacture of human ASCs must demonstrate *in vitro* quality assurances that maximize the safety of the finished product. Similarly to any new biologic, rigorous *in vivo* animal trials will need to be instituted in order to determine safety and efficacy before human clinical trials can be initiated. And many challenges exist related to the scaled production of clinical-grade cell therapies, not to mention packaging, inventory, and shipping considerations.

ASCs are a multipotent group of cells with demonstrated successful application across a therapeutic spectrum, involving both bench and bedside research. Research has demonstrated the successful ability of these cells to adopt phenotypes consistent with several different mature cell types; studies have shown they possess extensive paracrine signaling capacity with the ability to impact and alter inflammatory, immunomodulatory, and angiogenic responses, to name a few; furthermore, they can be therapeutically delivered through several different approaches depending on the pathology and the intended target organ. ASCs will likely become a cellular keystone in the field of regenerative medicine and improve our ability to treat countless diseases and disorders.

References

Amos, P. J., Bailey, A. M., Shang, H., Katz, A. J., Lawrence, M. B., & Peirce, S. M. (2008). Functional binding of human adipose-derived stromal cells: effects of extraction method and hypoxia pretreatment. *Ann. Plast. Surg.*, *60*(4), 437—444.

Amos, P. J., Kapur, S. K., Shang, H., Stapor, P. C., Bekiranov, S., Rodeheaver, G. T., et al. (2009). Human adipose stromal cells (ASCs) accelerate diabetic wound healing: impact of cell formulation and delivery. *Tissue Eng. Part A.* In press.

Awad, H. A., Butler, D. L., Boivin, G. P., Smith, F. N., Malaviya, P., Huibregtse, B., et al. (1999). Autologous mesenchymal stem cell-mediated repair of tendon. *Tissue Eng.*, *5*(3), 267—277.

Bailey, A. M., Kapur, S., & Katz, A. J. (2010). Characterization of adipose-derived stem cells: an update. *Curr. Stem Cell Res. Ther.*, *5*, 1—9.

Banas, A., Teratani, T., Yamamoto, Y., Tokuhara, M., Takeshita, F., Quinn, G., et al. (2007). Adipose tissue-derived mesenchymal stem cells as a source of human hepatocytes. *Hepatology*, *46*, 219—228.

Boquest, A. C., Shahdadfar, A., Frønsdal, K., Sigurjonsson, O., Tunheim, S. H., Collas, P., et al. (2005). Isolation and transcription profiling of purified uncultured human stromal stem cells: alteration of gene expression after in vitro cell culture. *Mol. Biol. Cell*, *16*(3), 1131—1141.

Cai, L., Johnstone, B. H., Cook, T. G., Tan, J., Fishbein, M. C., Chen, P. S., et al. (2008). IFATS series: human adipose tissue-derived stem cells induce angiogenesis and nerve sprouting following myocardial infarction, in conjunction with potent preservation of cardiac function. *Stem Cells, 27*, 230–237.

Cao, Y., Sun, Z., Liao, L., Meng, Y., Han, Q., & Zhao, R. C. (2005). Human adipose tissue-derived stem cells differentiate into endothelial cells in vitro and improve postnatal neovascularization in vivo. *Biochem. Biophys. Res. Commun., 332*(2), 370–379.

Chang, E. I., Bonillas, R. G., El-ftesi, S., Chang, E. I., Ceradini, D. J., Vial, I. N., et al. (2009). Tissue engineering using autologous microcirculatory beds as vascularized bioscaffolds. *FASEB J., 23*(3), 906–901.

Constantin, G., Marconi, S., Rossi, B., Angiari, S., Calderan, L., Anghileri, E., et al. (2009). Adipose-derived mesenchymal stem cells ameliorate chronic experimental autoimmune encephalomyelitis. *Stem Cells, 27*(10), 2624–2635.

Crisostomo, P. R., Markel, T. A., Wang, M., & Meldrum, D. R. (2007). In the adult mesenchymal stem cell population, source gender is a biologically relevant aspect of protective power. *Surgery, 142*(2), 215–221.

DiMuzio, P., & Tulenko, T. (2007). Tissue engineering applications to vascular bypass graft development: the use of adipose-derived stem cells. *J. Vasc. Surg., 45*(Suppl. A), A99–A103.

Dragoo, J. L., Carlson, G., McCormick, F., Khan-Farooqi, H., Zhu, M., Zuk, P. A., et al. (2007). Healing full-thickness cartilage defects using adipose-derived stem cells. *Tissue Eng., 13*, 1615–1621.

Dudas, J. R., Marra, K. G., Cooper, G. M., Penascino, V. M., Mooney, M. P., Jiang, S., et al. (2006). The osteogenic potential of adipose-derived stem cells for the repair of rabbit calvarial defects. *Ann. Plast. Surg., 56*(5), 543–548.

Erickson, G. R., Gimble, J. M., Franklin, D. M., Rice, H. E., Awad, H., & Guilak, F. (2002). Chondrogenic potential of adipose tissue-derived stromal cells *in vitro* and *in vivo*. *Biochem. Biophys. Res. Commun., 290*, 763–769.

Fraser, J. K., Wulur, I., Alfonso, Z., & Hedrick, M. H. (2006). Fat tissue: an underappreciated source of stem cells for biotechnology. *Trends Biotechnol., 24*, 150–154.

Furlani, D., Ugurlucan, M., Ong, L., Bieback, K., Pittermann, E., Westien, I., et al. (2009). Is the intravascular administration of mesenchymal stem cells safe? Mesenchymal stem cells and intravital microscopy. *Microvasc. Res., 77*(3), 370–376.

Ganey, T., Hutton, W. C., Moseley, T., Hedrick, M., & Meisel, H. J. (2009). Intervertebral disc repair using adipose tissue-derived stem and regenerative cells: experiments in a canine model. *Spine, 34*(21), 2297–2304.

Garcia-Olmo, D., Herreros, D., Pascual, I., Pascual, J. A., Del-Valle, E., Zorrilla, J., et al. (2009). Expanded adipose-derived stem cells for the treatment of complex perianal fistula: a phase II clinical trial. *Dis. Colon Rectum, 52*(1), 79–86.

Gimble, J. M., Katz, A. J., & Bunnell, B. A. (2007). Adipose-derived stem cells for regenerative medicine. *Circ. Res., 100*(9), 1249–1260.

Gonzalez-Rey, E., Gonzalez, M. A., Varela, N., O'Valle, F., Hernandez-Cortes, P., Rico, L., et al. (2010). Human adipose-derived mesenchymal stem cells reduce inflammatory and T cell responses and induce regulatory T cells in vitro in rheumatoid arthritis. *Ann. Rheum. Dis., 69*(1), 241–248.

Gronthos, S., Franklin, D. M., Leddy, H. A., Robey, P. G., Storms, R. W., & Gimble, J. M. (2001). Surface protein characterization of human adipose tissue-derived stromal cells. *J. Cell Physiol., 189*, 54–63.

Gronthos, S., Zannettino, A. C., Hay, S. J., Shi, S., Graves, S. E., Kortesidis, A., et al. (2003). Molecular and cellular characterisation of highly purified stromal stem cells derived from human bone marrow. *Cell Sci., 116*(Pt 9), 1827–1835.

Halvorsen, Y. D., Franklin, D., Bond, A. L., Hitt, D. C., Auchter, C., Boskey, A. L., et al. (2001). Extracellular matrix mineralization and osteoblast gene expression by human adipose tissue-derived stromal cells. *Tissue Eng., 7*(6), 729–741.

Harris, L. J., Abdollahi, H., Zhang, P., McIlhenny, S., Tulenko, T. N., & Dimuzio, P. J. (2009). Differentiation of adult stem cells into smooth muscle for vascular tissue engineering. *J. Surg. Res.* In press.

Hicok, K. C., du Laney, T. V., Zhou, Y. S., Halvorsen, Y. D., Hitt, D. C., Cooper, L. F., et al. (2004). Human adipose-derived adult stem cells produce osteoid in vivo. *Tissue Eng., 10*(3–4), 371–380.

Honczarenko, M., Le, Y., Swierkowski, M., et al. (2006). Human bone marrow stromal cells express a distinct set of biologically functional chemokine receptors. *Stem Cells, 24*, 1030–1041.

Jack, G. S., Zhang, R., Lee, M., Xu, Y., Wu, B. M., & Rodríguez, L. V. (2009). Urinary bladder smooth muscle engineered from adipose stem cells and a three dimensional synthetic composite. *Biomaterials, 30*(19), 3259–3270.

Juncosa-Melvin, N., Boivin, G. P., Galloway, M. T., Gooch, C., West, J. R., Sklenka, A. M., et al. (2005). Effects of cell-to-collagen ratio in mesenchymal stem cell-seeded implants on tendon repair biomechanics and histology. *Tissue Eng., 11*(3–4), 448–457.

Juncosa-Melvin, N., Boivin, G. P., Gooch, C., Galloway, M. T., West, J. R., Dunn, M. G., et al. (2006). The effect of autologous mesenchymal stem cells on the biomechanics and histology of gel-collagen sponge constructs used for rabbit patellar tendon repair. *Tissue Eng., 12*(2), 369–379.

378

Kang, S. K., Lee, D. H., Bae, Y. C., Kim, H. K., Baik, S. Y., & Jung, J. S. (2003). Improvement of neurological deficits by intracerebral transplantation of human adipose tissue-derived stromal cells after cerebral ischemia in rats. *Exp. Neurol., 183*(2), 355–366.

Kang, S. K., Shin, M. J., Jung, J. S., Kim, Y. G., & Kim, C. H. (2006). Autologous adipose tissue-derived stromal cells for treatment of spinal cord injury. *Stem Cells Dev., 15*(4), 583–594.

Katz, A. J., Llull, R., Hedrick, M. H., & Futrell, J. W. (1999). Emerging approaches to the tissue engineering of fat. *Clin. Plast. Surg., 26*(4), 587–603.

Katz, A. J., Tholpady, A., Tholpady, S. S., Shang, H., & Ogle, R. C. (2005). Cell surface and transcriptional characterization of human adipose-derived adherent stromal (hADAS) cells. *Stem Cells, 23*(3), 412–423.

Katz, A.J., Zang, Z., Shang, H., Chamberlain, A.T., Berr, S.S., Roy, R.J., et al. Serial MRI assessment of human adipose-derived stem cells (HASCS) in a murine model of reperfused myocardial infarction. *Adipocytes*. In press.

Kim, D. H., Je, C. M., Sin, J. Y., & Jung, J. S. (2003). Effect of partial hepatectomy on in vivo engraftment after intravenous administration of human adipose tissue stromal cells in mouse. *Microsurgery, 23*(5), 424–431.

Kim, J. M., Lee, S. T., Chu, K., Jung, K. H., Song, E. C., Kim, S. J., et al. (2007a). Systemic transplantation of human adipose stem cells attenuated cerebral inflammation and degeneration in a hemorrhagic stroke model. *Brain Res., 1183*, 43–50.

Kim, Y. J., Yu, J. M., Joo, H. J., Kim, H. K., Cho, H. H., Bae, Y. C., et al. (2007b). Role of CD9 in proliferation and proangiogenic action of human adipose-derived mesenchymal stem cells. *Pflugers Arch., 455*(2), 283–296.

Kim, W. S., Park, B. S., & Sung, J. H. (2009a). The wound-healing and antioxidant effects of adipose-derived stem cells. *Expert Opin. Biol. Ther., 9*(7), 879–887.

Kim, U., Shin, D. G., Park, J. S., Kim, Y. J., Park, S. I., Moon, Y. M., et al. (2009b). Homing of adipose-derived stem cells to radiofrequency catheter ablated canine atrium and differentiation into cardiomyocyte-like cells. *Int. J. Cardiol., 14*, 19–38.

Kim, Y. M., Choi, Y. S., Choi, J. W., Park, Y. H., Koo, B. S., Roh, H. J., & Rha, K. S. (2009c). Effects of systemic transplantation of adipose tissue-derived stem cells on olfactory epithelium regeneration. *Laryngoscope, 119*(5), 993–999.

Lee, J., Han, D. J., & Kim, S. C. (2008). *In vitro* differentiation of human adipose tissue-derived stem cells into cells with pancreatic phenotype by regenerating pancreas extract. *Biochem. Biophys. Res. Commun., 375*, 547–551.

Lee, R. H., Pulin, A. A., Seo, M. J., Kota, D. J., Ylostalo, J., Larson, B. L., et al. (2009a). Intravenous hMSCs improve myocardial infarction in mice because cells embolized in lung are activated to secrete the anti-inflammatory protein TSG-6. *Cell Stem Cell, 5*(1), 54–63.

Lee, S. T., Chu, K., Jung, K. H., Im, W. S., Park, J. E., Lim, H. C., et al. (2009b). Slowed progression in models of Huntington disease by adipose stem cell transplantation. *Ann. Neurol., 66*(5), 671–681.

Lin, G., Wang, G., Banie, L., Ning, H., Shindel, A. W., Fandel, T. M., et al. (2010). Treatment of stress urinary incontinence with adipose tissue-derived stem cells. *Cytotherapy, 12*(1), 88–95.

Lin, Y., Luo, E., Chen, X., Liu, L., Qiao, J., Yan, Z., et al. (2005). Molecular and cellular characterization during chondrogenic differentiation of adipose tissue-derived stromal cells in vitro and cartilage formation in vivo. *J. Cell Mol. Med., 9*, 929–939.

Liu, Y., Yan, X., Sun, Z., Chen, B., Han, Q., Li, J., et al. (2007). Flk-1+ adipose-derived mesenchymal stem cells differentiate into skeletal muscle satellite cells and ameliorate muscular dystrophy in mdx mice. *Stem Cells Dev., 16*(5), 695–706.

Lu, F., Mizuno, H., Uysal, C. A., Cai, X., Ogawa, R., & Hyakusoku, H. (2008). Improved viability of random pattern skin flaps through the use of adipose-derived stem cells. *Plast. Reconstr. Surg., 121*(1), 50–58.

Mesimäki, K., Lindroos, B., Törnwall, J., Mauno, J., Lindqvist, C., Kontio, R., et al. (2009). Novel maxillary reconstruction with ectopic bone formation by GMP adipose stem cells. *Int. J. Oral Maxillofac. Surg., 38*(3), 201–209.

Miranville, A., Heeschen, C., Sengenès, C., Curat, C. A., Busse, R., & Bouloumié, A. (2004). Improvement of postnatal neovascularization by human adipose tissue-derived stem cells. *Circulation, 110*(3), 349–355.

Mironov, V., Visconti, R. P., Kasyanov, V., Forgacs, G., Drake, C. J., & Markwald, R. R. (2009). Organ printing: tissue spheroids as building blocks. *Biomaterials, 30*(12), 2164–2174.

Mitchell, J. B., McIntosh, K., Zvonic, S., Garrett, S., Floyd, Z. E., Kloster, A., et al. (2006). Immunophenotype of human adipose-derived cells: temporal changes in stromal-associated and stem cell-associated markers. *Stem Cells, 24*(2), 376–385.

Miyahara, Y., Nagaya, N., Kataoka, M., Yanagawa, B., Tanaka, K., Hao, H., et al. (2006). Monolayered mesenchymal stem cells repair scarred myocardium after myocardial infarction. *Nat. Med., 12*(4), 459–465.

Mizuno, H., Zuk, P. A., Zhu, M., Lorenz, H. P., Benhaim, P., & Hedrick, M. H. (2002). Myogenic differentiation by human processed lipoaspirate cells. *Plast. Reconstr. Surg., 109*, 199–209.

Moon, M. H., Kim, S. Y., Kim, Y. J., Kim, S. J., Lee, J. B., Bae, Y. C., et al. (2006). Human adipose tissue-derived mesenchymal stem cells improve postnatal neovascularization in a mouse model of hindlimb ischemia. *Cell Physiol. Biochem., 17*, 279–290.

Muehlberg, F. L., Song, Y. H., Krohn, A., Pinilla, S. P., Droll, L. H., Leng, X., et al. (2009). Tissue-resident stem cells promote breast cancer growth and metastasis. *Carcinogenesis, 30*(4), 589–597.

Nakagami, H., Maeda, K., Morishita, R., Iguchi, S., Nishikawa, T., Takami, Y., et al. (2005). Novel autologous cell therapy in ischemic limb disease through growth factor secretion by cultured adipose tissue-derived stromal cells. *Arterioscler. Thromb. Vasc. Biol., 25*, 2542–2547.

Nambu, M., Kishimoto, S., Nakamura, S., Mizuno, H., Yanagibayashi, S., Yamamoto, N., et al. (2009). Accelerated wound healing in healing-impaired db/db mice by autologous adipose tissue-derived stromal cells combined with atelocollagen matrix. *Ann. Plast. Surg., 62*(3), 317–321.

Nedergaard, J., Bengtsson, T., & Cannon, B. (2007). Unexpected evidence for active brown adipose tissue in adult humans. *Am. J. Physiol. Endocrinol. Metab., 293*, E444–E452.

Neupane, M., Chang, C. C., Kiupel, M., & Yuzbasiyan-Gurkan, V. (2008). Isolation and characterization of canine adipose-derived mesenchymal stem cells. *Tissue Eng. Part A, 14*, 1007–1015.

Okonkow, D. O., Katz, A. J., Lee, K. S., & Jane, J. A. (2005). Repopulation of injured brain with human adipo-derived stem cells after traumatic brain injury. *J. Am. Coll. Surg., 201*(3), S47.

Parker, A.M., & Katz, A.J., (2006). Adipose-derived stem cells for the regeneration of damaged tissues. *Expert Opin. Biol. Ther., 6*(6), 567–578.

Patel, K. M., Crisostomo, P., Lahm, T., Markel, T., Herring, C., Wang, M., et al. (2007). Mesenchymal stem cells attenuate hypoxic pulmonary vasoconstriction by a paracrine mechanism. *J. Surg. Res., 143*(2), 281–285.

Patrick, C. W., Jr., Chauvin, P. B., Hobley, J., & Reece, G. P. (1999). Preadipocyte seeded PLGA scaffolds for adipose tissue engineering. *Tissue Eng., 5*(2), 139–151.

Peterson, B., Zhang, J., Iglesias, R., Kabo, M., Hedrick, M., Benhaim, P., et al. (2005). Healing of critically sized femoral defects, using genetically modified mesenchymal stem cells from human adipose tissue. *Tissue Eng., 11*(1–2), 120–129.

Planat-Bénard, V., Menard, C., André, M., Puceat, M., Perez, A., Garcia-Verdugo, J. M., et al. (2004a). Spontaneous cardiomyocyte differentiation from adipose tissue stroma cells. *Circ. Res., 94*(2), 223–229.

Planat-Bénard, V., Silvestre, J. S., Cousin, B., André, M., Nibbelink, M., Tamarat, R., et al. (2004b). Plasticity of human adipose lineage cells toward endothelial cells: physiological and therapeutic perspectives. *Circulation, 109*(5), 656–663.

Prunet-Marcassus, B., Cousin, B., Caton, D., André, M., Pénicaud, L., & Casteilla, L. (2006). From heterogeneity to plasticity in adipose tissues: site-specific differences. *Exp. Cell Res., 312*(6), 727–736.

Puissant, B., Barreau, C., Bourin, P., Clavel, C., Corre, J., Bousquet, C., et al. (2005). Immunomodulatory effect of human adipose tissue-derived adult stem cells: comparison with bone marrow mesenchymal stem cells. *Br. J. Haematol., 129*(1), 118–129.

Randomized Clinical Trial of Adipose-Derived Stem Cells in the Treatment of Pts with ST-Elevation Myocardial Infarction (2007). Available at: http://clinicaltrials.gov/ct2/show/NCT00442806. Accessed February 25, 2010.

Randomized Clinical Trial of Adipose-Derived Stem Cells in Treatment of Non Revascularizable Ischemic Myocardium (2007). Available at: http://clinicaltrials.gov/ct2/show/NCT00426868. Accessed February 25, 2010.

Rangappa, S., Fen, C., Lee, E.H., Bongso, A., & and Sim, E.K. (2003). Transformation of adult mesenchymal stem cells isolated from the fatty tissue into cardiomyocytes. *Ann. Thorac. Surg., 75*(3), 775–779.

Rehman, J., Traktuev, D., Li, J., Merfeld-Clauss, S., Temm-Grove, C. J., Bovenkerk, J. E., et al. (2004). Secretion of angiogenic and antiapoptotic factors by human adipose stromal cells. *Circulation, 109*(10), 1292–1298.

Rodbell, M. (1966). Metabolism of isolated fat cells. II. The similar effects of phospholipase c (clostridium perfringens alpha toxin) and of insulin on glucose and amino acid metabolism. *J. Biol. Chem., 241*, 130–139.

Ryu, H. H., Lim, J. H., Byeon, Y. E., Park, J. R., Seo, M. S., Lee, Y. W., et al. (2009). Functional recovery and neural differentiation after transplantation of allogenic adipose-derived stem cells in a canine model of acute spinal cord injury. *J. Vet. Sci., 10*(4), 273–284.

Safford, K. M., Hicok, K. C., Safford, S. D., et al. (2002). Neurogenic differentiation of murine and human adipose-derived stromal cells. *Biochem. Biophys. Res. Commun., 294*, 371–379.

Safford, K. M., Safford, S. D., Gimble, J. M., Shetty, A. K., & Rice, H. E. (2004). Characterization of neuronal/glial differentiation of murine adipose-derived adult stromal cells. *Exp. Neurol., 187*, 319–328.

Santiago, L. Y., Clavijo-Alvarez, J., Brayfield, C., Rubin, J. P., & Marra, K. G. (2009). Delivery of adipose-derived precursor cells for peripheral nerve repair. *Cell Transplant., 18*(2), 145–158.

Sengenès, C., Lolmède, K., Zakaroff-Girard, A., Busse, R., & Bouloumié, A. (2005). Preadipocytes in the human subcutaneous adipose tissue display distinct features from the adult mesenchymal and hematopoietic stem cells. *J. Cell Physiol., 205*(1), 114–122.

Song, Y. H., Gehmert, S., Sadat, S., et al. (2007). VEGF is critical for spontaneous differentiation of stem cells into cardiomyocytes. *Biochem. Biophys. Res. Commun., 354*, 999–1003.

Sordi, V., Malosio, M. L., Marchesi, F., et al. (2005). Bone marrow mesenchymal stem cells express a restricted set of functionally active chemokine receptors capable of promoting migration to pancreatic islets. *Blood, 106,* 419–427.

Sun, B., Roh, K. H., Park, J. R., Lee, S. R., Park, S. B., Jung, J. W., et al. (2009). Therapeutic potential of mesenchymal stromal cells in a mouse breast cancer metastasis model. *Cytotherapy, 11*(3), 289–298.

Suzuki, T., Kobayashi, K., Tada, Y., Suzuki, Y., Wada, I., Nakamura, T., et al. (2008). Regeneration of the trachea using a bioengineered scaffold with adipose-derived stem cells. *Ann. Otol. Rhinol. Laryngol., 117*(6), 453–463.

Tang, W., Zeve, D., Suh, J. M., Bosnakovski, D., Kyba, M., Hammer, R. E., et al. (2008). White fat progenitor cells reside in the adipose vasculature. *Science, 322,* 583–586.

Tapp, H., Hanley, E. N., Jr., Patt, J. C., & Gruber, H. E. (2009). Adipose-derived stem cells: characterization and current application in orthopaedic tissue repair. *Exp. Biol. Med. (Maywood), 234*(1), 1–9.

Tholpady, S. S., Ogle, R. C., & Katz, A. J. (2009). Adipose stem cells and solid organ transplantation. *Curr. Opin. Organ Transplant., 14*(1), 51–55.

Trayhurn, P., & Beattie, J. H. (2001). Physiological role of adipose tissue: white adipose tissue as an endocrine and secretory organ. *Proc. Nutr. Soc., 60,* 329–339.

Trottier, V., Marceau-Fortier, G., Germain, L., Vincent, C., & Fradette, J. (2008). IFATS collection: using human adipose-derived stem/stromal cells for the production of new skin substitutes. *Stem Cells, 26*(10), 2713–2723.

Uemura, R., Xu, M., Ahmad, N., & Ashraf, M. (2006). Bone marrow stem cells prevent left ventricular remodeling of ischemic heart through paracrine signaling. *Circ. Res., 98*(11), 1414–1421.

Valina, C., Pinkernell, K., Song, Y. H., Bai, X., Sadat, S., Campeau, R. J., et al. (2007). Intracoronary administration of autologous adipose tissue-derived stem cells improves left ventricular function, perfusion, and remodelling after acute myocardial infarction. *Eur. Heart J., 28*(21), 2667–2677.

von Heimburg, D., Zachariah, S., Low, A., & Pallua, N. (2001). Influence of different biodegradable carriers on the in vivo behavior of human adipose precursor cells. *Plast. Reconstr. Surg., 108*(2), 411–420.

Wang, M., Tsai, B. M., Crisostomo, P. R., & Meldrum, D. R. (2006). Pretreatment with adult progenitor cells improves recovery and decreases native myocardial proinflammatory signaling after ischemia. *Shock, 25*(5), 454–459.

Yamamoto, T., Gotoh, M., Hattori, R., Toriyama, K., Kamei, Y., Iwaguro, H., et al. Periurethral injection of autologous adipose-derived stem cells for the treatment of stress urinary incontinence in patients undergoing radical prostatectomy: report of two initial cases. *Int. J. Urol.* In press.

Yañez, R., Lamana, M. L., García-Castro, J., Colmenero, I., Ramírez, M., & Bueren, J. A. (2006). Adipose tissue-derived mesenchymal stem cells have in vivo immunosuppressive properties applicable for the control of the graft-versus-host disease. *Stem Cells, 24*(11), 2582–2591.

Yoon, E., Dhar, S., Chun, D. E., Gharibjanian, N. A., & Evans, G. R. (2007). In vivo osteogenic potential of human adipose-derived stem cells/poly lactide-co-glycolic acid constructs for bone regeneration in a rat critical-sized calvarial defect model. *Tissue Eng., 13*(3), 619–627.

Yoshimura, K., Sato, K., Aoi, N., Kurita, M., Hirohi, T., & Harii, K. (2008). Cell-assisted lipotransfer for cosmetic breast augmentation: supportive use of adipose-derived stem/stromal cells. *Aesthetic Plast. Surg., 32*(1), 48–55.

Yoshimura, K., Shigeura, T., Matsumoto, D., Sato, T., Takaki, Y., Aiba-Kojima, E., et al. (2006). Characterization of freshly isolated and cultured cells derived from the fatty and fluid portions of liposuction aspirates. *J. Cell Physiol., 208*(1), 64–76.

Zhang, H. M., & Zhang, L. S. (2009). Influence of human bone marrow mesenchymal stem cells on proliferation of chronic myeloid leukemia cells. *Ai Zheng., 28*(1), 29–32.

Zhu, W., Xu, W., Jiang, R., Qian, H., Chen, M., Hu, J., et al. (2006). Mesenchymal stem cells derived from bone marrow favor tumor cell growth in vivo. *Exp. Mol. Pathol., 80*(3), 267–274.

Zuk, P. A., Zhu, M., Ashjian, P., de Ugarte, D. A., Huang, J. I., Mizuno, H., et al. (2002). Human adipose tissue is a source of multipotent stem cells. *Mol. Biol. Cell, 13*(12), 4279–4295.

Zuk, P. A., Zhu, M., Mizuno, H., Huang, J., Futrell, J. W., Katz, A. J., et al. (2001). Multilineage cells from human adipose tissue: implications for cell-based therapies. *Tissue Eng., 7*(2), 211–228.

CHAPTER 22

Peripheral Blood Stem Cells

Zhan Wang, Gunter Schuch, J. Koudy Williams, Shay Soker
Institute for Regenerative Medicine, Wake Forest University School of Medicine, Medical Center Blvd, Winston-Salem, NC, USA

INTRODUCTION

Adult stem and progenitor cells have been isolated from a wide variety of sources. Lineage-committed progenitors are found in a number of tissues including skin, fat, muscle, heart, brain, liver, pancreas, bladder, and more. Adults have another source of stem and progenitor cells that are not restricted to a specific tissue. These "universal" stem and progenitor cells are found circulating in peripheral blood, allowing them to reach and integrate into all tissues.

The bone marrow is, most likely, the source of peripheral blood stem and progenitor cells. Hemangioblasts are the embryonic precursors of hematopoietic stem cells (HSCs), giving rise to committed hematopoietic progenitors such as lymphoids, thymocytes, myeloids, granulocites-monocytes, megakaryocytes-erythrocytes, and mast cells. These progenitor cells complete their differentiation in the bone marrow, peripheral blood, and thymus and in the target tissues. Extensive research in hematology/oncology has resulted in the identification of a wide variety of cell-surface markers that allow the characterization and isolation of HSCs at different stages of their differentiation. Initially, adult bone marrow mesenchymal cells (MCSs) were isolated, expanded *in vitro*, and examined for their multi lineage differentiation potentials. These early studies were followed by extensive research on bone marrow-derived multipotent adult progenitor cells (MAPCs). This special cell population can proliferate long-term without senescence and can differentiate to multiple lineages *in vitro* and contribute to the regeneration of several tissues *in vivo* (Verfaillie, 2005). Like HSCs, MSCs may leave the bone marrow environment and be found in peripheral blood. Identification and isolation of MSCs is based on differential expression of cell-surface markers that distinguish them from circulating HSCs. Endothelial progenitor cells (EPCs) are probably derived from the same hemangioblast precursors of HSCs but they take a separate path of differentiation in the bone marrow. The identification of circulating EPCs suggested that the process of vasculogenesis, previously believed to be restricted to the embryonic stages, continues into adulthood. The circulating EPCs have specific cell-surface markers that are not found on mature ECs and that are lost when the EPCs differentiate to ECs.

This chapter will briefly review the types and source of stem cells in peripheral blood, their specific cell-surface markers, and factors that change their abundance in peripheral blood. We will focus on the isolation and *in vitro* expansion of peripheral blood-derived MCSs and EPCs and describe their therapeutic applications for regenerative medicine. We will further describe the role of peripheral blood-derived stem cells in normal and pathological processes. Although much information was gathered in the past on the identification of different populations of peripheral blood stem cells, their clinical potential for therapy is just now being explored.

Principles of Regenerative Medicine. DOI: 10.1016/B978-0-12-381422-7.10022-7

Peripheral blood is readily obtainable and is a viable source of cells for regenerative medicine, and therefore deserves special attention.

TYPES AND SOURCE OF STEM CELLS IN THE PERIPHERAL BLOOD

It is well documented that the bone marrow is the source of cells in peripheral blood. Hematopoietic stem cells (HSCs) are characteristically quiescent, multipotent cells with the capacity for both self-renewal and differentiation. After development in the fetus, HSCs reside in adult bone marrow and serve to replenish lymphoid, megakaryocytic, erythroid, and myeloid hematopoietic lineages throughout adulthood. Observations that systemically administrated mesenchymal stem cells (MSCs) could home back to the bone marrow suggested that MSCs may also reside in bone marrow. Results of our recent studies indicate that mature cells, such as endothelial cells (ECs), may enter the circulation (Beaudry et al., 2005). A different study tested the fate of muscle progenitor cells introduced into the circulation of lethally irradiated recipient mice together with distinguishable bone marrow cells. All recipients showed high-level engraftment of muscle-derived cells representing all major adult blood lineages (Goodell et al., 2001). Collectively, these results indicate that there is a constant exchange of cells from the bone marrow to peripheral blood. On the other hand, bone marrow transplantation studies have indicated that this process may be reversed and cells from peripheral blood may repopulate the bone marrow.

Mobilization of bone marrow cells

Stem cell numbers in peripheral blood are very low compared to those in the bone marrow. Although stem cells can be collected by apharesis, this requires the processing of large volumes of blood. Amplification of peripheral blood stem cells can facilitate collection and allows for rescuing autologous stem cells from the bone marrow. Mobilization of HSCs from bone marrow into peripheral blood can be achieved by hematopoietic growth factors. Recombinant human granulocyte (G)- and granulocyte-macrophage (GM) colony-stimulating factor (CSF) have been used as stimulators of hematopoiesis. Results of studies indicate higher numbers of circulating progenitor cells in patients receiving G-CSF or GM-CSF (Baumann et al., 1993; Gianni et al., 1990; Kawano et al., 1993). In fact, transplantation of G-CSF-mobilized stem cells harvested from peripheral blood is replacing bone marrow biopsy, the method of choice for collection of stem cells for autologous bone marrow transplantation. However, it is important to find better mobilizing techniques to provide more efficient harvesting and faster hematopoietic recovery. Recently, elegant studies were designed to prove the role of angiogenic factors in EPC mobilization. Rafii and colleagues reported that mobilization of HSCs and EPCs from bone marrow is mediated through activation of metalloproteinases and adhesion molecules (Eriksson and Alitalo, 2002; Hattori et al., 2002; Heissig et al., 2002; Rafii and Lyden, 2003). In the bone marrow, vascular endothelial growth factor (VEGF) and placental growth factor (PlGF) induce MMP-9 expression. Activation of MMP-9 results in the release of stem cell-active soluble kit ligand, which mobilizes quiescent HSCs and EPCs to the vascular zone, where they are released to the circulation. The results of these studies indicate that co-mobilization of EPCs and HSCs contributes to the revascularization processes.

ENDOTHELIAL PROGENITOR CELLS

Initial evidence that EPCs can be detected in peripheral blood came from research conducted mainly by the groups of Isner and Asahara in Boston and Rafii in New York (Asahara et al., 1997; Rafii et al., 1995b). They showed that cells with EC characteristics can be isolated from peripheral blood and expanded *in vitro*. They and others have shown that the numbers of EPCs in peripheral blood are significantly increased as a result of acute vascular injuries, angiogenic stimuli, estrogen, and nitric oxide synthase, but reduced by certain chronic disease states (e.g. coronary artery disease) (Gill et al., 2001). Circulating EPCs originate primarily from the bone marrow and can be identified by differential expression of hematopoietic and endothelial cell markers. This is important because HSCs and EPCs probably share a common precursor, the hemangioblasts (Hirschi and Goodell, 2001). Hemangioblasts reside mainly in the bone marrow and differentiate into HSCs and angioblasts. This process occurs mainly during early embryogenesis but was shown to exist in adults (Gill et al., 2001; Hattori et al., 2001, 2002). Angioblasts will give rise to EPCs that upon stimulation with angiogenic factors such as VEGF and PlGF are mobilized from bone marrow to peripheral blood (Gill et al., 2001; Heissig et al., 2002; Rafii et al., 2002a; Shintani et al., 2001). Once in peripheral blood, EPCs can be recruited to sites of active neovascularization, as seen in wounds, diabetic retinopathy, and tumors (review in Rafii and Lyden, 2003). The role of EPCs in physiological and pathological neovascularization and their therapeutic applications are described below.

Identification and isolation of EPCs

Marrow and peripheral blood cells expressing CD34 can give rise to EPCs (Asahara et al., 1997; Shi et al., 1998; Bhattacharya et al., 2000; Peichev et al., 2000; Dimmeler et al., 2001). Although CD34 is commonly used to isolate EPCs, CD34 expression is also shared by HSCs and MSCs and cannot be used to distinguish between these populations. Likewise, VEGF receptor 2 (human KDR and mouse Flk-1), which is used to identify EPCs, is expressed also on HSCs (Asahara et al., 1997; Isner and Asahara, 1999). In humans, CD133 (AC133) is used to distinguish EPCs from mature ECs, since CD133 is not expressed by mature ECs (Peichev et al., 2000; Rafii et al., 2002b). CD133 is a stem cell marker with as yet unrecognized functions (Rafii, 2000). Additionally, Hebbel and colleagues have used P1H12 antibodies that recognize CD146 (MUC18) on circulating ECs in peripheral blood but not on monocytes, granulocytes, platelets, megacaryocytes, or T or B lymphocytes (Solovey et al., 1997, 2001; Sodian et al., 2000). Other markers common to progenitor and mature ECs are the cell surface receptors KDR and Tie2 (Rafii and Lyden, 2003; Asahara and Kawamoto, 2004; Ishikawa and Asahara, 2004). Purified populations of CD133$^+$/KDR$^+$ EPCs proliferate *in vitro* in an anchorage-independent manner and can be induced to differentiate into mature adherent ECs (Rafii and Lyden, 2003). It is thought that CD133$^+$/KDR$^+$ EPCs are a population of immature ECs that are mobilized from the bone marrow to participate in neovascularization. As myelomonocytic cells have lost surface expression of CD133, this marker also provides an effective means to distinguish true EPCs from cells of myelomonocytic origin. Yet, recent studies showed that cells expressing CD14, considered as a typical monocytic lineage marker, can give rise to ECs (Kim et al., 2005; Romagnani et al., 2005). Collectively, these studies suggest that identification of circulating EPCs may be achieved using different markers that may define subpopulations of EPCs based on their differentiation stage and origin (Table 22.1).

385

TABLE 22.1 Cell-surface Makers Expressed on Progenitor and Mature Endothelial Cells

	EPC/ECFC	Vessel wall-derived CEC
Proliferative capacity:	High	Limited
Proposed source and mobilization	Bone marrow release and proliferation to CSF, VEGF, and other stimuli	EC damage, VEGF decrease, and apoptosis
Marker		
VEGFR-2 (KDR)	+	+
CD34	+	+
CD31 (PECAM)	+/−	+
CD133 (AC133)	+	−
CD146 (P1H12, MUC18)	−	+

The number of EPCs in bone marrow is very low ($<$10 per 10×10^5 mononuclear cells) and the reported numbers vary a great deal based on which identifying markers are used among the different studies. For practical applications, EPC fraction may be enriched using cell-surface markers such as CD34, CD133, and KDR (Asahara et al., 1997; Shi et al., 1998; Ishikawa and Asahara, 2004). Stromal cell-derived factor-1 stimulates the mobilization of EPCs via an enhancement of protein kinase B (Akt) and endothelial nitric oxide synthase activity (Hiasa et al., 2004). Interestingly, VEGF has been found to promote endothelial cells to express SDF-1 (also known as CXCL12) and CXCR4 (the SDF-1 receptor) (Kryczek et al., 2005); on the other hand, SDF-1 has the potency to induce the expression of VEGF (Salcedo et al., 1999). Thus, in the bone marrow vascular niche, the expression of VEGF and endothelial's SDF-1 possibly modulate the microenviroment through proteolytic enzymes (MMP2, MMP9) and controlling EPC mobilization. In addition to the above factors, recent data indicate that erythropoietin (Epo) (Aicher et al., 2003), platelet-derived growth factor (PDGF) (Amado et al., 2005), and nitric oxide (Aicher et al., 2003) stimulate EPC mobilization as well.

One functional assay capitalizes on *in vitro* growth kinetics to discriminate bone marrow-derived EPCs and circulating ECs from vessel wall-derived mature ECs (Rafii and Lyden, 2003). In this assay, the isolated cells are incubated with VEGF, basic fibroblast growth factor (bFGF), insulin-like growth factor (IGF), and fibronectin or collagen. EC colonies that appear early are derived from the recipient vessel wall circulating ECs, whereas late-outgrowth cells or colonies originate mainly from bone marrow-derived EPCs. Lin et al. (Bader et al., 2000) and Gulati et al. (2003) have reported that early EPCs (7-day cultures) are derived from CD14$^+$ cells (monocytes), while late outgrowing EPCs (4−6 weeks of culture) are derived from CD14$^-$ cells that bear endothelial markers (VE-cardherin$^+$Flk-1$^+$vWF$^+$CD36$^+$CD146$^+$). In addition, Hur et al. (2004) cultured total mononuclear cells obtained from human peripheral blood and also reported two EPC types. Some cells organized into clusters as early as 3 days after plating, and these were called early EPCs. These cells had limited proliferative capacity and disappeared after 4 weeks of culture. In contrast, another group of cells appeared later, at 2 to 3 weeks after plating. These cells demonstrated endothelial cell-like cobblestone morphology and were called late EPCs. Late EPCs proliferated robustly, and were positive for VE-cadherin, Flt-1, and KDR, but negative for CD45. The late EPCs produced nitric oxide and formed capillary tubes. Yoder's group (Case et al., 2007) reported late outgrowth of cells from umbilical cord or circulating blood mononuclear cells that they termed endothelial colony-forming cells (ECFCs). These cells appeared at 14−21 days after plating and formed adherent colonies with cobblestone morphology. ECFCs expressed the cell-surface antigens CD31, CD105, CD144, CD146, vWF, and KDR and took up aceylated low-density lipoprotein (AcLDL). ECFCs did not express hematopoietic or monocyte/macrophage cell-surface antigens such as CD14, CD45, or CD115. Whether isolated from cord or adult peripheral blood, ECFCs display clonal proliferative potential and relatively high levels of telomerase (Ingram et al., 2005). Taken together, the recent consensus is that late-outgrowth endothelial colonies may be considered as angioblast-like EPCs.

The cell-surface antigen CD31 (also known as PECAM-1) has been used as a marker for endothelial cells for many years (Muller et al., 1989; Albelda et al., 1990; Newman et al., 1990). Recently, it was suggested that CD31 could also be used as a marker for EPCs (Melero-Martin et al., 2007, 2008). However, since it has been shown that CD31$^+$ are fully differentiated ECs, it could be argued that CD31 may not be a suitable marker for identification of EPCs, since it may not distinguish between progenitor and fully differentiated endothelial cells. However, in these studies, CD31-selected EPCs had a typical EC/EPC phenotype, proliferated robustly and integrated into nascent vasculature. The close phenotypic similarity between mature and progenitor ECs was highlighted by Ingram et al. (2005). These researchers used single cell colony forming assays to demonstrate that populations of both human umbilical vein endothelial cells (HUVECs) and human aortic endothelial cells (HAECs), which were thought to be fully differentiated endothelial cells, contain a subpopulation of EPCs with different clonogenic and proliferative potential. This work suggested that HUVECs and HAECs are not homogeneous populations and appear to contain some fully differentiated ECs that are unable to proliferate as well as other cell fractions that demonstrate more "stemness," as shown by their robust ability to proliferate and form colonies. It is not yet clear why HUVEC and HAEC populations demonstrate heterogeneity in *ex vivo* culture conditions, but these findings may suggest that culturing and sorting for EPCs may help to standardize experimental data obtained from different research groups.

Isolation of EPCs is conventionally based on FACS or immuno-magnetic techniques, which are complex, time-consuming procedures and need experienced operators. New devices could be engineered to simplify operation and shorten the time for obtaining EPCs. Miniaturization of EPC isolation procedures by using similar strategies, adopted from microfluidic lab-on-a-chip devices, may significantly facilitate EPC isolation (Li et al., 2005; Yager et al., 2006). Plouffe et al. (2009) developed a microfluidics chamber, coated with antibodies against CD34, VEGFR-2, CD31, CD146, or CD45, to capture EPCs. When cells were flowing through the "EPC chip," it demonstrated specific affinity to EPCs. Although this prototype device can only be coated with one antibody on one single chamber, it is potentially one alternative option for easy and rapid separation of EPCs. Tillman et al. (2009) developed an extracorporeal cellular affinity (ECA) column that can recover CD133-expressing progenitor cells with high efficiency. In a sheep model, 1.8 L of blood were passed through a Sepharose-based column with affinity for CD133, and unbound cells and plasma were returned to the animal. The results show that this process has a minimal effect on the hematologic and physiologic parameters of the animal and EPC recovery was over 600-fold more efficient than conventional density centrifugation from a peripheral blood specimen. This technology may facilitate the generation of large numbers of progenitor-derived cells for clinical therapies and reduce the time required to attain clinically relevant cell numbers while minimizing loss of other important cell types to the donor.

387

In vitro expansion of EPCs

The relative paucity of EPCs in the circulation likely contributes to the difficulty associated with EPC culture (Ingram et al., 2005). Density-gradient separation procedures, designed for separating mononuclear cells (MNCs) from blood, may also contribute to this problem. Thus, successful culture of EPCs usually requires a relatively large amount (50−100 ml) of peripheral or cord blood and, still, very few CFUs or ECFCs can be obtained (see Table 1 in Case et al., 2007). The mononuclear fraction is placed in fibronectin-coated plates containing endothelial basal medium that contains angiogenic growth factors such as VEGF and bFGF. Other growth factors such as EGF and IGF contribute to cell growth but not differentiation. In one of their earlier studies, Asahara et al. showed that VEGF and not bFGF is important for EPC differentiation (Asahara et al., 1999b) and bFGF may be used by the differentiated ECs for subsequent proliferation. Inclusion of angiogenic factors in the media helps to prevent "contamination" by other cell types including lymphocytes, macrophages, and dendritic cells. VEGF appears to inhibit dendritic cell maturation from CD34[+] MNC fraction (Gabrilovich et al., 1996, 1998, 1999). Within 7−10 days of culture in fibronectin or collagen-coated dishes, colonies with spindle-shaped cells appear in the dish. These are "slow growing" cells defined as "late outgrowth" EPCs. They differ from the mature circulating ECs that readily proliferate *in vitro* (Gill et al., 2001). The whole blood culture method may be used to increase the likelihood of obtaining EPC cultures. This approach does not require MNC separation procedures, and this may minimize cell damage caused by density-gradient separation. Recently, Reinisch et al. reported that whole blood culture yielded nearly eight-fold more ECFC colonies compared with density gradient separation (Reinisch et al., 2009). The whole blood culture approach may simplify EPC culture procedures and increase the possibility that EPCs will be used for clinical diagnostic and therapeutic purposes.

Commercially available media kits were introduced to facilitate EPC growth. These kits provide reagents and media for culturing EPCs from peripheral blood. However, these culture kits may not be optimal for EPC expansion and differentiation. Hur et al. (2004) reported that cells isolated with such kits displayed limited proliferative capacity and did not integrate into nascent vasculature (Sharpe et al., 2006). Further studies revealed that monocytes and T-cells can contribute to CFU-ECs (Case et al., 2007; Rohde et al., 2007) and therefore it is reasonable to argue that some of these kits are not suitable for EPC culture and assays. Instead, ECFCs (Case et al., 2007) should be cultured in EGM-2 culture medium, which was developed from MCDB131 medium in the 1980s (Knedler and Ham, 1987; Ehringer et al., 2000). ECFCs demonstrate late colony forming, robust proliferation, endothelial morphology, expression of EPC/EC-related markers, and the ability to integrate into nascent vasculature. Regardless of the EPC source, the cells assume a typical flat EC morphology after 2−3 weeks and present mature EC markers such as CD31, VE-cadherin, and CD146 (P1H12). They metabolize acetylated low-density lipoprotein (acLDL), bind Ulex Europaeus agglutinin 1(UEA-1), and produce nitric oxide (NO), consistent with EC properties. Proper characterization of EPC-derived ECs requires the analysis of a combination of cell surface markers.

The role of EPCs in physiological and pathological neovascularization

Blood vessels form by two processes: (1) angiogenesis — the sprouting of capillaries from pre-existing blood vessels; and (2) vasculogenesis — the *in situ* assembly of capillaries from undifferentiated ECs. Vasculogenesis takes place mostly during the early stages of embryo-genesis (Folkman and d'Amore, 1996; Yancopoulos et al., 1998). Vascular channels in the yolk sac originate from the mesoderm by differentiation of angioblasts, which subsequently generate primitive blood vessels (Breier et al., 1997). The early findings that EPCs can partic-ipate in angiogenic processes indicate that post-natal neovascularization does not rely only on sprouting from pre-existing blood vessels (angiogenesis), but may be assisted by EPCs via post-natal vasculogenesis (Asahara et al., 1999a,b; Takahashi et al., 1999; Young et al., 1999).

Bone marrow-derived EPCs contribute to adult tissue neovascularization in several models including wound healing and cornea and tumor angiogenesis (Asahara et al., 1999a; Rafii et al., 2002c). Bone marrow-derived EPCs could be detected in normal organs including spleen, lung, liver, intestine, skin, hindlimb muscle, ovary, and uterus, indicating their participation in main-tenance of physiological neovascularization (Asahara et al., 1999a). Hormonally induced ovulation cycles were also associated with localization of bone marrow-derived EPCs in the corpus lutea and in the uterus endometrium and stroma. These findings indicate that EPCs contribute to physiological neovascularization associated with post-natal regenerative processes.

A recent study examined the presence of endothelial, smooth muscle, and Schwann cell chimerism in patients with sex-mismatched (female-to-male) heart transplants (Minami et al., 2005). The Y chromosome was used to determine chimerism. Biopsy specimens taken at increasing times after heart transplantation showed that endothelial cells had the highest degree of chimerism (24.3%), Schwann cells showed the next highest chimerism (11.2%), and vascular smooth muscle cells the lowest (3.4%). Results of this study indicate that circulating progenitor cells are capable of repopulating most major cell types in the heart, but they do so with varying frequency. The signals for endothelial progenitor recruitment occur early and could relate to the injury during the surgery.

Cumulative evidences show that bone marrow cells may improve ischemic myocardium function by paracrine stimulation of angiogenesis rather than differentiation into contractile cardiomyocytes (Korf-Klingebiel et al., 2008). In addition to directly and indirectly contrib-uting to neovascularization, EPCs probably can also provide paracrine survival signals to cardiomyocytes (Narmoneva et al., 2004; Rubart and Field, 2006; Young et al., 2007). Clinical studies also suggested that cell-based therapy with EPCs can improve myocardial function (Abdel-Latif et al., 2007).

In parallel, EPCs were found incorporated into the vasculature of pathological lesions such as atherosclerotic plaques, tumors, the retina, and ischemic brain tissue. Vascular smooth muscle cell (SMC) proliferation results in neointimal hyperplasia and the development of restenosis. Bone marrow-derived SMCs can integrate into the hyperplastic neointima and atherosclerotic plaques (Luttun et al., 2002; Sata et al., 2002). Evidence for the contribution of bone marrow-derived MSCs to human atherosclerotic plaques originated from a study showing donor-derived neointimal cells within the plaques (Caplice et al., 2003). Also, decreased EPCs in the circulation have been correlated with a higher risk of cardiovascular complications (Hill et al., 2003). It was hypothesized that lower levels of peripheral blood-EPCs were associated with an impaired capacity to repair the damaged vessels, but the pathophysiological role of bone marrow-derived EPCs remains unclear. Recruitment of peripheral blood-EPCs to damaged or diseased tissues is dependent on the underlying pathology and is probably due to the release of specific growth factors and chemokines by these tissues (Hillebrands et al., 2001, 2002). Abnormal retinal neovascularization contributes to the pathogenesis of proliferative retinopathy in diabetes and age-related prematurity and macular degeneration. Bone marrow-derived hemangioblasts were shown to contribute to retinal neovascularization in models of proliferative retinopathy (Grant et al., 2002; Otani et al., 2002). This study documented the incorporation of EPCs into mature endothelium of the retinal blood vessels. Cerebral infarction is associated with neovascularization of the ischemic zone and new vessel growth. Bone marrow transplantation studies showed that EPCs could be detected in the neovessels at the repair sites after 3 days (Hess et al., 2002; Zhang et al., 2002).

Compelling evidence for the role of EPCs in tumor vascularization come from a study by Lyden and colleagues using an angiogenesis-defective mouse model. Mice lacking both alleles of Id1 ($id1^{-/-}$) and Id3 ($id3^{-/-}$) died by embryonic day 13.5 and exhibited massive vascular malformation (Lyden et al., 1999). The $Id3^{-/-}/id1^{+/-}$ mice survived but could not support the growth of several tumor types due to insufficient tumor vascularization. However, transplantation of $id3^{-/-}/id1^{+/-}$ mutant mice with bone marrow from wild-type mice gave rise to tumors that were indistinguishable from tumors grown on wild-type mice (Lyden et al., 2001). Furthermore, 90% of the tumor vessels contained bone marrow-derived ECs, indicating the contribution of EPCs to tumor neovascularization. VEGF treatment failed to elevate the number of EPCs in $id3^{-/-}/id1^{+/-}$ mutant mice but not in $id3^{-/-}/id1^{+/-}$ transplanted with wild-type bone marrow. Further evidence is provided by a model in which transplantation of human bone marrow-derived MAPCs into tumor xenograft-bearing mice resulted in the incorporation of human cells as 40% of the tumor vessel endothelium, indicating the importance of circulating endothelial cells (CECs) for tumor neovascularization (Reyes et al., 2002). Different tumors secrete different types and concentrations of angiogenic factors that may have a different capability to induce mobilization of EPCs. Although a formal correlation between tumor type/stage/size and number of EPCs has not been established in human cancer, some tumor types may be more dependent than others on CECs as a source of endothelium (Rafii et al., 2002b).

Taken together, the results of these studies indicate that EPCs' contribution to neovascularization is not restricted to normal healing processes and they contribute significantly to several pathological processes.

MESENCHYMAL STEM CELLS (MSCs)

Mesenchymal stem cells (MSCs) are multipotent cells that can differentiate into various mesenchymal lineages including bone, cartilage, fat, and muscle. MSCs were initially found in adult bone marrow (Friedenstein et al., 1987; Caplan, 1991), and were first identified as osteogenic progenitors capable of forming bone-like structures *in vitro* (Friedenstein, 1976; Owen, 1988). These early studies suggested that bone marrow MSCs are also fat cell progenitors (Caplan, 1994). Further studies report that MSCs may be found in every mesenchymal tissue that has regeneration capacity. In addition to bone marrow, MSCs were isolated from muscle, fat, skin, cartilage, bone, and blood vessels (Peng and Huard, 2003; Bartsch et al., 2005). MSCs have some of the basic properties of stem cells including self-renewal, multi lineage differentiation capacity, clonality, and the ability to regenerate tissues *in vivo* (Roufosse et al., 2004; Verfaillie, 2002a,b). In addition, Verfaillie and colleagues have shown that adult bone marrow MSCs proliferate for many passages without senescence. They analyzed telomere length in these cells and showed that it was longer than in neutrophils and lymphocytes and was not different between young or old donors (Reyes and Verfaillie, 2001). Their results indicated that bone marrow MSCs have high telomerase activity *in vivo* and came from a population of quiescent cells.

Identification and isolation

Because of the multiple sources and methods of isolation of MSCs, their identifying markers vary between studies. Some of the "classical" markers of bone marrow-derived MSCs include CD34, CD44, CD45, c-Kit, Sca-1 (murine), CD133 (human), and CD105 (Thy-1), and higher concentrations of CD13 and stage-specific antigen I (SSEA-1) (Jiang et al., 2002). As stated above, MSCs were isolated from multiple sources but only a few studies have analyzed their presence in peripheral blood. Systemic infusion of MSCs showed that they may be engrafted in various mesenchymal tissues. These results suggest that MSCs may be present in peripheral blood. In fact, MSCs were isolated from peripheral blood of cancer patients given G-CSF and GM-CSF. The cells were grown *in vitro* and had a fibroblast-like phenotype (Fernandez et al., 1997). The cells were negative for hematopoietic markers and CD34, but expressed CD105, SH3, I-CAM, and V-CAM. MSCs were also isolated from normal human peripheral blood without "mobilization" (Zvaifler et al., 2000). The cells were isolated by gradient centrifugation and plated in growth media. After 2 weeks, adherent fibroblast-like cells appear in the culture. These cells were positive for CD105, Stro-1, vimentin, and BMP receptors, but were negative for CD34. Taken together, these results indicate that a small population of MSCs exists in peripheral blood. These cells are difficult to isolate, but may be identified by their morphology and the expression of a subset of MSC markers.

In vitro expansion

Peripheral blood-derived MSCs are obtained through density centrifugation using Histopaque™ or Ficoll™. There are several factors that are important for successful maintenance of MSCs including cell density, pH of the medium, source of sera, and the type of culture dishes. Human MSCs require densities of 1,500–3,000 cells/cm^2 in order to prevent spontaneous differentiation at higher cell densities. The basal media may be DMEM or α-MEM with 10% fetal serum (Kuznetsov et al., 2001). Collectively, the methods used for MSC expansion *in vitro* do not differ from those used to expand bone marrow-derived MSCs. Reviews from Verfaillie and Caplan describe these methods in detail (Caplan and Bruder, 2001; Verfaillie et al., 2003). Following expansion, MSCs can be differentiated *in vitro* into the mesenchymal lineages and tested *in vivo*. Interestingly, marrow-derived MSCs were induced to differentiate into cells with functional properties of endothelial cells (Reyes et al., 2002), hepatocytes (Schwartz et al., 2002), and neuroectodermal cells (Jiang et al., 2003). *In vitro*-differentiated cells may be used for future therapeutic applications. However, we need to define the appropriate phenotype and functional properties of the differentiated cells before they can be used clinically.

THERAPEUTIC APPLICATIONS OF PERIPHERAL BLOOD STEM CELLS

The physiological role of MSCs in tissue regeneration prompted researchers to evaluate their use in therapeutic applications. The ethical discussions regarding embryonic stem cells underscore the need to explore the clinical applications of adult stem cells, including MSCs. MSCs were first tested in several animal models and have recently been used in clinical studies.

Although the results of the animal experiments are promising, the mechanisms behind the regenerative potential of peripheral blood MSCs are not fully understood. The therapeutic applications can be divided into three groups: (1) tissue engineering, (2) cell delivery applications, and (3) MSCs as a vehicle for gene therapy (review in Rafii and Lyden, 2003). The main advantage of MSCs for clinical use is their presence in peripheral blood. However, as discussed above, further work is needed to evaluate their culture and expansion properties.

EPCs

In many cases, organ and tissue regeneration require re-establishment of the vascular network. There are two possible sources of endothelialization: (1) mature endothelial cells that migrate from pre-existing vessels (Hanahan and Folkman, 1996) and (2) circulating EPCs from peripheral blood (Shi et al., 1998; Peichev et al., 2000; Rafii, 2000). Cultured EPCs offer a robust cell source for tissue engineering and cell delivery applications. EPCs can be obtained from the same patient to avoid immune rejection. Although EPCs were shown to contribute to tissue revascularization, their function in a clinical setting has not been established. The use of EPCs for tissue engineering requires *ex vivo* expansion that is not optimal for clinical use because of animal products and inadequate tissue culture environment.

Two concerns must be addressed before EPCs can be used for therapeutic applications. First, there is the possibility that, when allogeneic cells are implanted, a graft-versus-host immune reaction, caused by a residual T-cell fraction derived from CD34$^+$ cells, could occur. Studies showed that bone marrow-derived CD34$^+$ cells can be differentiated into T-cells both *in vitro* and *in vivo* (DiGiusto et al., 1994). Currently, CD34 is one of the most widely used surface makers used for EPC isolation. It is important to use CD34 in combination with other surface markers to identify the subsets of EPCs that will be suitable for clinical use and reduce the risk of subsequent T-cell differentiation. Second, in order to develop cells for clinical use, it is necessary to remove all animal substances, such as serum, from the culture environment, because they might be pathogenic or immunogenic. For clinical cell therapy applications, EPC culture conditions must be free of animal products to meet GMP standards. Efforts were made to develop clinical-grade human embryonic stem cell lines and culture conditions (review in Unger et al., 2008). Reinisch et al. recently published results of clinically approved culture conditions to expand peripheral blood EPCs (Reinisch et al., 2009). By using pooled human platelet lysate (pHPL) to replace FBS in the culture environment, they reported successful recovery of EPCs (ECFCs) from peripheral blood. The ECFCs were characterized by robust proliferative potential (more than 30 population doublings), normal karyotype, and vascular network-forming ability.

TISSUE ENGINEERING

Vascular diseases are the leading causes of morbidity and mortality in the USA each year (Ross, 1993). Over 500,000 coronary bypass grafts and 50,000 peripheral bypass grafts are performed annually in the USA (www.americanheart.org) (British Cardiac Society, 1991). However, up to 30% of the patients who require arterial bypass surgery lack suitable or sufficient amounts of suitable autologous conduits such as small-caliber arteries or saphenous veins (Edwards et al., 1966; Motwani and Topol, 1998; Pomposelli et al., 1998). Synthetic grafts, such as poly-tetrafluoroethylene (PTFE) or Dacron (polyethylene terephthalate fiber), have been used successfully to bypass large-caliber, high-flow blood vessels. However, these grafts invariably fail when used to bypass small-caliber, low-flow blood vessels due to increased thrombogenicity and accelerated intimal thickening leading to early graft stenosis and occlusion (Stephen et al., 1977; O'Donnell et al., 1984; Sayers et al., 1998; Ao et al., 2000). It has been shown that a confluent endothelial cell (EC) monolayer on small-caliber prosthetic grafts may provide immediate protection from thrombus formation following implantation (Furchgott and Zawadzki, 1980; Cybulsky and Gimbrone, 1991; Seifalian et al., 2002). However, the use of allogeneic endothelial cells is limited by rejection, whereas the use of autologous human endothelial cells for the construction of vascular grafts has not been widely explored. The idea of using EPCs to seed the lumen of engineered blood vessels came from the observations that MSCs contributed to the lining of vascular grafts *in vivo* (Shi et al., 1998; Bhattacharya et al., 2000). We have shown that EPCs might be an ideal source of autologous ECs for seeding small-diameter grafts, eliminating the need to remove native vessel from which to culture ECs. By seeding EPC-derived ECs onto a scaffold, a non-thrombogenic barrier between blood and vessel wall is created, thereby promoting patency *in vivo*. EPC-seeded collagen matrices derived from decellularized porcine arteries were used for carotid artery reconstruction in sheep (Kaushal et al., 2001). These bioengineered arteries remained patent for more than 4 months, whereas control grafts without autologous ECs occluded within 15 days. Thus, functional vessels can be engineered using decellularized arteries and EPCs. Moreover, we have shown that these bioengineered blood vessels, after a brief period of healing *in vivo*, develop a fully cellularized wall of three distinct layers analogous to normal adventitia, media, and intima. Although these are exciting results, bioengineered grafts will need to be constructed in a mechanically relevant environment.

In vitro engineering of blood vessels should mimic the flow conditions that exist *in vivo* in order to enhance tissue formation. Neram et al. have shown that local blood flow properties induce changes in endothelial cell morphology and orientation (Nerem et al., 1981; Nerem, 1984). Further studies showed that the levels and duration of shear stress induced different changes in EC morphology, proliferation, and differentiation (Sprague et al., 1987; Levesque et al., 1990). EPCs cultured under biologic-like shear conditions expressed higher levels of VE-cadherin than those cultured under static conditions (Yamamoto et al., 2003).

A recent study assessed the use of EPCs for bioengineered heart valves. Two endothelial cell types, valve-derived mature ECs and EPCs from peripheral blood, were used (Dvorin et al., 2003). The study showed that both sources of endothelial cells, when seeded on PGA/P4HB scaffolds, proliferate in response to VEGF. The EPCs could be induced to transdifferentiate to a mesenchymal phenotype on PGA/P4HB in response to TGF-β1. These results indicate that EPCs can respond to soluble signals that induce events that occur during valvulogenesis (Fig. 22.1).

One problem with all of these studies is that heterogeneous cell populations are being expanded for seeding onto vascular scaffolds. As mentioned previously, one solution is to isolate MSCs and to differentiate them to EPCs. Another general problem with these bioengineered vascular grafts is immediate availability. For instance, when an emergency bypass needs to be performed, growth of an artificial vessel and preparation for implantation would

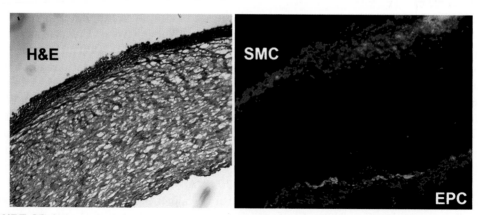

FIGURE 22.1
Acellular porcine arterial segment (stained with hematoxyline and eosin (H&E)) and seeded with peripheral blood-derived SMC (dyed red with PKH26) and EPC (dyed green with PKH27).

take too much time if autologous cells are to be implemented. Alternatively, these bioengineered grafts could be seeded with stem cells that were differentiated into ECs.

TISSUE REGENERATION

Several studies have suggested that EPCs participate in the vascular healing process, in part by recruitment of EPCs to the regenerated site (Asahara et al., 1997; Takahashi et al., 1999). Genetically labeled EPCs were detected in ischemic limbs of mice and shown to accelerate the revascularization process. Administration of cytokines such as G-CSF and GM-SCF appear to enhance mobilization of EPCs and revascularization. In humans, EPCs contributed to wound healing of patients implanted with a left ventricular-assisted device (Rafii et al., 1995a). The EPCs adhered to the device and formed a non-thrombogenic surface. These studies suggested that EPCs may be recruited to assist endothelialization and served as the basis for preclinical and clinical studies as described below.

Melero-Martin et al. (2008) reported that, when cord blood EPCs and MSCs were injected subcutaneously into nude mice, they formed vascular structures that were stable for 4 weeks. The vascular network had endothelialized lumen and pericytes expressing smooth muscle actin. Au et al. (2008) engineered blood vessels from HUVECs and human MSCs that remained stable and functional for more than 130 days *in vivo*. These findings could guide future practices in tissue engineering and regenerative medicine, allowing physicians to form stable and long-lasting vasculature for engineered tissue.

Given the morbidity associated with limb ischemia, EPCs may be used for vascular therapy as an alternative to bypass approaches. In preclinical studies, introduction of bone marrow-derived EPCs significantly improved collateral vessel formation and minimized limb ischemia (Asahara et al., 1999b; Takahashi et al., 1999; Kalka et al., 2000b). In patients suffering from peripheral arterial disease, injection of autologous whole bone marrow mononuclear cells into ischemic gastronemius muscle resulted in restoration of limb function (Tateishi-Yuyama et al., 2002). The improvement in muscle perfusion suggested that it was due to the presence of EPCs in the cell preparation. However, it remains to be determined whether the improvement was due in part to the introduction of myelomonocytic cells.

Bone marrow-derived MSCs were recently shown to contribute to myocardial regeneration and revascularization. In nude rats that underwent myocardial infarction, cytokine-mobilized EPCs homed to the infarcted tissue and contributed to neoangiogenesis (Orlic et al., 2001a). In similar studies, bone marrow-derived MSCs were injected into the infarcted border and were shown to differentiate into myocardial cells and ECs (Jackson et al., 2001; Kocher et al., 2001; Orlic et al., 2001b). In most studies, direct introduction of these cells into an active angiogenic site, such as infarcted or ischemic myocardium, was essential for successful incorporation of the cells and improvement of cardiac function.

Acute myocardial infarction or chronic ischemic heart disease result in the loss of cardiomyocytes and vasculature. Several animal studies have shown that introduction of autologous bone marrow MSCs contributes to neoangiogenesis in the ischemic myocardium (Rafii and Lyden, 2003). In patients, whole autologous bone marrow mononuclear cells were delivered into the coronary arteries feeding the infarcted and ischemic tissue (Rafii and Lyden, 2003). In all of these studies, there was improved cardiac perfusion and left ventricular function, suggesting that delivery of autologous progenitor cells is feasible and safe, and may have a short-term therapeutic benefit. However, follow-up studies in animals and humans detected only a few bone marrow-derived cells in the regenerated vascular network, suggesting that only a small portion of the cells may contribute to revascularization.

Despite the excitement generated by these initial observational clinical trials, it remains to be determined in double-blind placebo-controlled randomized clinical trials whether this cellular therapy approach will result in any long-standing cardiac benefits. Importantly, it remains unclear whether any long-term toxicity exists with this therapy. Such toxicity may result if myeloid cells are incorporated into regenerating myocardium and generate non-cardiac or fibrotic tissues. Therefore, progenitor cells that have been predifferentiated into EPCs should be used with caution and long-term monitoring.

Mesenchymal stem cells

In the case of MSCs, the lineage-committed cells can generate a variety of specialized mesenchymal tissues including bone, cartilage, muscle, marrow stroma, tendon, ligament, fat, and a variety of other connective tissues (Caplan, 1994). As such, MSCs may have a dramatic impact on the overall health status of individuals by controlling the body's capacity to naturally remodel, repair, and upon demand rejuvenate various tissues. In human clinical research, initial efforts are focused on applications of MSC-based tissue repair using cell delivery approaches. The most logical application of MSCs is to regenerate non-union bone defects. A number of studies showed that MSCs from animals and humans, delivered in a porous, calcium phosphate vehicle, were able to regenerate bone tissue (Bruder et al., 1994, 1998; Jaiswal et al., 2000). Additionally, these cells may be beneficial for cartilage research. The cartilage is a tissue that cannot repair itself in adults. MSCs have been applied in hyaluronan scaffolds for cartilage tissue repair with good results and are now in clinical trials (Solchaga et al., 1999, 2000). Bone marrow-derived MSCs have also been used for muscle repair and fuse with the host myotubes and form functional muscle fibers (Shake et al., 2002; Toma et al., 2002). Systemic delivery of bone marrow-derived MSCs showed that they can home back to the bone marrow. This observation prompted clinical studies to use MSCs to restore the bone marrow in patients undergoing radiation and chemotherapy-mediated myeloablation (Koc et al., 2000; Lazarus et al., 1995).

Although a subset of MSCs was reported to differentiate into cardiomyocytes under specific conditions *in vitro*, it is still controversial whether MSCs can differentiate into cardiomyocytes (Caplan and Dennis, 2006; Wollert and Drexler, 2005). *In vivo* MSCs differentiating into cardiomyocytes have also been observed, but at an extremely low rate (Amado et al., 2005; Miyahara et al., 2006). On the other hand, like EPCs, MSCs can provide paracrine factors to support injured myocardium. This could be the major mechanism for the beneficial effects of these cells (Gnecchi et al., 2005; Caplan and Dennis, 2006). So far, most clinical studies using bone marrow MNCs showed no or small (but possibly clinically important) improvement in cardiac function (Abdel-Latif et al., 2007) and the functional improvement was considered related to paracrine instead of mesenchymal cell differentiation.

The use of peripheral blood stem cells for gene therapy

Gene and cell therapy have been proposed for regenerative medicine and tested in a number of clinical trials. Genetically modified MSCs offer a unique approach as cells with growth potential may represent a useful tool for tissue engineering and cell therapy. A detailed knowledge of vector delivery systems is critical for practical applications. One of the most popular vectors used for gene delivery to progenitor cells is replication-deficient adenovirus (Ad). Ad vectors offer two important advantages that make them ideal for gene therapy. First, they can efficiently infect non-dividing cells, which is important for MSCs that live primarily in the G0/G1 phase of the cell cycle (Hawley, 2001; Alessandri et al., 2004). Second, the Ad vector can offer transient expression of the recombinant gene for a time period of approximately 3 weeks (Iwaguro et al., 2002). However, Ad vectors have been shown to elicit an unwanted inflammatory response. Genetically modified stem cells have been explored in a number of studies to regenerate bone and cartilage or for neovascularization (Grande et al., 2003; Kondoh et al., 2004; Shen et al., 2004). The most common genes used in these studies are growth factors such as VEGF. VEGF, as mentioned above, is a potent angiogenic factor that supports the differentiation of MSCs along endothelial lineages. In order to enhance vascularization of engineered muscle tissue, we have transfected primary cultures of rat myoblasts with a plasmid encoding VEGF and green fluorescence protein (GFP). Cells expressing GFP were selected by a fluorescent-activated cells sorter (FACS) and injected, mixed with gelatin, into the subcutaneous space of immune-deficient mice (de Coppi et al., 2005). Tissue volumes of VEGF-transfected cells increased during 21 days and tripled their size. In contrast, the volume of tissues containing cells that were transfected with control plasmid gradually decreased and the tissues were minimally visible after 21 days. Immunohistochemical analysis of VEGF-expressing tissue with anti-von Willebrand factor revealed typical muscle formation and a developed vascular network. VEGF gene transfer to stem cells has been used by *in situ* neovascularization and angiogenesis in order to salvage ischemic limbs (Iwaguro et al., 2002; Kalka et al., 2000a,c). Other studies looked at the combinations of growth factors to mimic the environment of vascular development. Both bFGF and angiopoietin-1 have been transfected with VEGF into progenitor cells to induce the development of mature blood vessels including the medial and outer adventitial layers (Kondoh et al., 2004). This approach also succeeded in reducing the VEGF-mediated permeability and fluid leakage of the new vessels. The future of stem cell-mediated gene therapy is dependent on the resolution of some key questions. The efficiency of gene transfer needs to be close to 100% to ensure that unmodified cells do not interfere with the regenerative process. The most feasible stem cell source needs to be used for successful clinical applications. Finally, the mode of cell delivery, systemic or local injection, needs to be decided. Regardless of the solution to each of these questions, stem cell-based therapies will benefit enormously from gene modification.

CONCLUSIONS AND FUTURE DIRECTIONS

The bone marrow is probably the source of peripheral blood stem and progenitor cells. Hemangioblasts are the embryonic precursors of hematopoietic stem cells (HSCs), giving rise to committed hematopoietic progenitors. The bone marrow is also a source of other progenitor and stem cells, the mesenchymal stem cells (MSCs), which can be expanded *in vitro* and have multi lineage differentiation potentials. Numerous studies, described here, have shown that there is a constant exchange of cells from the bone marrow to peripheral blood. On the other hand, bone marrow transplantation studies have indicated that this process may be reversed and cells from peripheral blood may repopulate the bone marrow. Future success in applying adult peripheral blood-derived stem cells for clinical applications will depend on development of strategies to mobilize, isolate, expand, differentiate, and deliver these cells. For example, EPCs may be isolated from peripheral blood and used for therapeutic angiogenesis directly or after a period of *ex vivo* expansion. Understanding the signals involved in the recruitment of these cells to the regenerating tissues will play a crucial role in optimizing this technology for clinical use. The studies summarized here provide support for the presence of stem cells in peripheral blood and mechanisms by which they can be mobilized from bone marrow in order to increase their numbers in blood. Although various attempts have been made to use peripheral blood-derived stem cells in humans, and some encouraging results have been obtained, standard clinical use of these techniques must await further validation and long-term toxicity evaluations.

References

Abdel-Latif, A., Bolli, R., et al. (2007). Adult bone marrow-derived cells for cardiac repair: a systematic review and meta-analysis. *Arch. Intern. Med., 167*, 989—997.

Aicher, A., Heeschen, C., et al. (2003). Essential role of endothelial nitric oxide synthase for mobilization of stem and progenitor cells. *Nat. Med., 9*, 1370—1376.

Albelda, S. M., Oliver, P. D., et al. (1990). EndoCAM: a novel endothelial cell-cell adhesion molecule. *J. Cell Biol., 110*, 1227—1237.

Alessandri, G., Emanueli, C., et al. (2004). Genetically engineered stem cell therapy for tissue regeneration. *Ann. N.Y. Acad. Sci., 1015*, 271—284.

Amado, L. C., Saliaris, A. P., et al. (2005). Cardiac repair with intramyocardial injection of allogeneic mesenchymal stem cells after myocardial infarction. *Proc. Natl. Acad. Sci. U.S.A., 102*, 11474—11479.

Ao, P. Y., Hawthorne, W. J., et al. (2000). Development of intimal hyperplasia in six different vascular prostheses. *Eur. J. Vasc. Endovasc. Surg., 20,* 241–249.

Asahara, T., & Kawamoto, A. (2004). Endothelial progenitor cells for postnatal vasculogenesis. *Am. J. Physiol. Cell Physiol., 287,* C572–C579.

Asahara, T., Masuda, H., et al. (1999a). Bone marrow origin of endothelial progenitor cells responsible for postnatal vasculogenesis in physiological and pathological neovascularization. *Circ. Res., 85,* 221–228.

Asahara, T., Murohara, T., et al. (1997). Isolation of putative progenitor endothelial cells for angiogenesis. *Science, 275,* 964–967.

Asahara, T., Takahashi, T., et al. (1999b). VEGF contributes to postnatal neovascularization by mobilizing bone marrow-derived endothelial progenitor cells. *Embo J., 18,* 3964–3972.

Au, P., Tam, J., et al. (2008). Bone marrow-derived mesenchymal stem cells facilitate engineering of long-lasting functional vasculature. *Blood, 111,* 4551–4558.

Bader, A., Steinhoff, G., et al. (2000). Engineering of human vascular aortic tissue based on a xenogeneic starter matrix. *Transplantation, 70,* 7–14.

Bartsch, G., Yoo, J. J., et al. (2005). Propagation, expansion, and multilineage differentiation of human somatic stem cells from dermal progenitors. *Stem Cells Dev., 14,* 337–348.

Baumann, I., Testa, N. G., et al. (1993). Haemopoietic cells mobilised into the circulation by lenograstim as alternative to bone marrow for allogeneic transplants. *Lancet, 341,* 369.

Beaudry, P., Force, J., et al. (2005). Differential effects of vascular endothelial growth factor receptor-2 inhibitor ZD6474 on circulating endothelial progenitors and mature circulating endothelial cells: implications for use as a surrogate marker of antiangiogenic activity. *Clin. Cancer Res., 11,* 3514–3522.

Bhattacharya, V., McSweeney, P. A., et al. (2000). Enhanced endothelialization and microvessel formation in polyester grafts seeded with CD34(+) bone marrow cells. *Blood, 95,* 581–585.

Breier, G., Damert, A., et al. (1997). Angiogenesis in embryos and ischemic diseases. *Thromb. Haemost., 78,* 678–683.

British Cardiac Society. (1991). Report of a working party of the British Cardiac Society: coronary angioplasty in the United Kingdom. *Br. Heart J., 66,* 325.

Bruder, S. P., Fink, D. J., et al. (1994). Mesenchymal stem cells in bone development, bone repair, and skeletal regeneration therapy. *J. Cell Biochem., 56,* 283–294.

Bruder, S. P., Kurth, A. A., et al. (1998). Bone regeneration by implantation of purified, culture-expanded human mesenchymal stem cells. *J. Orthop. Res., 16,* 155–162.

Caplan, A. I. (1991). Mesenchymal stem cells. *J. Orthop. Res., 9,* 641–650.

Caplan, A. I. (1994). The mesengenic process. *Clin. Plast. Surg., 21,* 429–435.

Caplan, A. I., & Bruder, S. P. (2001). Mesenchymal stem cells: building blocks for molecular medicine in the 21st century. *Trends Mol. Med., 7,* 259–264.

Caplan, A. I., & Dennis, J. E. (2006). Mesenchymal stem cells as trophic mediators. *J. Cell Biochem., 98,* 1076–1084.

Caplice, N. M., Bunch, T. J., et al. (2003). Smooth muscle cells in human coronary atherosclerosis can originate from cells administered at marrow transplantation. *Proc. Natl. Acad. Sci. U.S.A., 100,* 4754–4759.

Case, J., Mead, L. E., et al. (2007). Human CD34+AC133+VEGFR-2+ cells are not endothelial progenitor cells but distinct, primitive hematopoietic progenitors. *Exp. Hematol., 35,* 1109–1118.

Cybulsky, M. I., & Gimbrone, M. A., Jr. (1991). Endothelial expression of a mononuclear leukocyte adhesion molecule during atherogenesis. *Science, 251,* 788–791.

de Coppi, P., Delo, D., et al. (2005). Angiogenic gene-modified muscle cells for enhancement of tissue formation. *Tissue Eng., 11,* 1034–1044.

DiGiusto, D., Chen, S., et al. (1994). Human fetal bone marrow early progenitors for T, B, and myeloid cells are found exclusively in the population expressing high levels of CD34. *Blood, 84,* 421–432.

Dimmeler, S., Aicher, A., et al. (2001). HMG-CoA reductase inhibitors (statins) increase endothelial progenitor cells via the PI 3-kinase/Akt pathway. *J. Clin. Invest., 108,* 391–397.

Dvorin, E. L., Wylie-Sears, J., et al. (2003). Quantitative evaluation of endothelial progenitors and cardiac valve endothelial cells: proliferation and differentiation on poly-glycolic acid/poly-4-hydroxybutyrate scaffold in response to vascular endothelial growth factor and transforming growth factor beta1. *Tissue Eng., 9,* 487–493.

Edwards, W. S., Holdefer, W. F., et al. (1966). The importance of proper caliber of lumen in femoral-popliteal artery reconstruction. *Surg. Gynecol. Obstet., 122,* 37–40.

Ehringer, W. D., Wang, O. L., et al. (2000). Bradykinin and alpha-thrombin increase human umbilical vein endothelial macromolecular permeability by different mechanisms. *Inflammation, 24,* 175–193.

Eriksson, U., & Alitalo, K. (2002). VEGF receptor 1 stimulates stem-cell recruitment and new hope for angiogenesis therapies. *Nat. Med., 8,* 775–777.

Fernandez, M., Simon, V., et al. (1997). Detection of stromal cells in peripheral blood progenitor cell collections from breast cancer patients. *Bone Marrow Transplant., 20*, 265–271.

Folkman, J., & d'Amore, P. A. (1996). Blood vessel formation: what is its molecular basis? *Cell, 87*, 1153–1155.

Friedenstein, A. J. (1976). Precursor cells of mechanocytes. *Int. Rev. Cytol., 47*, 327–359.

Friedenstein, A. J., Chailakhyan, R. K., et al. (1987). Bone marrow osteogenic stem cells: in vitro cultivation and transplantation in diffusion chambers. *Cell Tissue Kinet., 20*, 263–272.

Furchgott, R. F., & Zawadzki, J. V. (1980). The obligatory role of endothelial cells in the relaxation of arterial smooth muscle by acetylcholine. *Nature, 288*, 373–376.

Gabrilovich, D., Ishida, T., et al. (1998). Vascular endothelial growth factor inhibits the development of dendritic cells and dramatically affects the differentiation of multiple hematopoietic lineages in vivo. *Blood, 92*, 4150–4166.

Gabrilovich, D. I., Chen, H. L., et al. (1996). Production of vascular endothelial growth factor by human tumors inhibits the functional maturation of dendritic cells. *Nat. Med., 2*, 1096–1103.

Gabrilovich, D. I., Ishida, T., et al. (1999). Antibodies to vascular endothelial growth factor enhance the efficacy of cancer immunotherapy by improving endogenous dendritic cell function. *Clin. Cancer Res., 5*, 2963–2970.

Gianni, A. M., Tarella, C., et al. (1990). Durable and complete hematopoietic reconstitution after autografting of rhGM-CSF exposed peripheral blood progenitor cells. *Bone Marrow Transplant., 6*, 143–145.

Gill, M., Dias, S., et al. (2001). Vascular trauma induces rapid but transient mobilization of VEGFR2(+)AC133(+) endothelial precursor cells. *Circ. Res., 88*, 167–174.

Gnecchi, M., He, H., et al. (2005). Paracrine action accounts for marked protection of ischemic heart by Akt-modified mesenchymal stem cells. *Nat. Med., 11*, 367–368.

Goodell, M. A., Jackson, K. A., et al. (2001). Stem cell plasticity in muscle and bone marrow. *Ann. N.Y. Acad. Sci., 938*, 208–218, discussion 218–220.

Grande, D. A., Mason, J., et al. (2003). Stem cells as platforms for delivery of genes to enhance cartilage repair. *J. Bone Joint Surg. Am., 85-A*(Suppl. 2), 111–116.

Grant, M. B., May, W. S., et al. (2002). Adult hematopoietic stem cells provide functional hemangioblast activity during retinal neovascularization. *Nat. Med., 8*, 607–612.

Gulati, R., Jevremovic, D., et al. (2003). Diverse origin and function of cells with endothelial phenotype obtained from adult human blood. *Circ. Res., 93*, 1023–1025.

Hanahan, D., & Folkman, J. (1996). Patterns and emerging mechanisms of the angiogenic switch during tumorigenesis. *Cell, 86*, 353–364.

Hattori, K., Dias, S., et al. (2001). Vascular endothelial growth factor and angiopoietin-1 stimulate postnatal hematopoiesis by recruitment of vasculogenic and hematopoietic stem cells. *J. Exp. Med., 193*, 1005–1014.

Hattori, K., Heissig, B., et al. (2002). Placental growth factor reconstitutes hematopoiesis by recruiting VEGFR1(+) stem cells from bone-marrow microenvironment. *Nat. Med., 8*, 841–849.

Hawley, R. G. (2001). Progress toward vector design for hematopoietic stem cell gene therapy. *Curr. Gene Ther., 1*, 1–17.

Heissig, B., Hattori, K., et al. (2002). Recruitment of stem and progenitor cells from the bone marrow niche requires MMP-9 mediated release of kit-ligand. *Cell, 109*, 625–637.

Hess, D. C., Hill, W. D., et al. (2002). Blood into brain after stroke. *Trends Mol. Med., 8*, 452–453.

Hiasa, K., Ishibashi, M., et al. (2004). Gene transfer of stromal cell-derived factor-1alpha enhances ischemic vasculogenesis and angiogenesis via vascular endothelial growth factor/endothelial nitric oxide synthase-related pathway: next-generation chemokine therapy for therapeutic neovascularization. *Circulation, 109*, 2454–2461.

Hill, J. M., Zalos, G., et al. (2003). Circulating endothelial progenitor cells, vascular function, and cardiovascular risk. *N. Engl. J. Med., 348*, 593–600.

Hillebrands, J. L., Klatter, F. A., et al. (2001). Origin of neointimal endothelium and alpha-actin-positive smooth muscle cells in transplant arteriosclerosis. *J. Clin. Invest., 107*, 1411–1422.

Hillebrands, J. L., Klatter, F. A., et al. (2002). Bone marrow does not contribute substantially to endothelial-cell replacement in transplant arteriosclerosis. *Nat. Med., 8*, 194–195.

Hirschi, K., & Goodell, M. (2001). Common origins of blood and blood vessels in adults? *Differentiation, 68*, 186–192.

Hur, J., Yoon, C. H., et al. (2004). Characterization of two types of endothelial progenitor cells and their different contributions to neovasculogenesis. *Arterioscler. Thromb. Vasc. Biol., 24*, 288–293.

Ingram, D. A., Mead, L. E., et al. (2005). Vessel wall-derived endothelial cells rapidly proliferate because they contain a complete hierarchy of endothelial progenitor cells. *Blood, 105*, 2783–2786.

Ishikawa, M., & Asahara, T. (2004). Endothelial progenitor cell culture for vascular regeneration. *Stem Cells Dev., 13*, 344–349.

Isner, J. M., & Asahara, T. (1999). Angiogenesis and vasculogenesis as therapeutic strategies for postnatal neovascularization. *J. Clin. Invest., 103*, 1231–1236.

Iwaguro, H., Yamaguchi, J., et al. (2002). Endothelial progenitor cell vascular endothelial growth factor gene transfer for vascular regeneration. *Circulation, 105*, 732–738.

Jackson, K. A., Majka, S. M., et al. (2001). Regeneration of ischemic cardiac muscle and vascular endothelium by adult stem cells. *J. Clin. Invest., 107*, 1395–1402.

Jaiswal, R. K., Jaiswal, N., et al. (2000). Adult human mesenchymal stem cell differentiation to the osteogenic or adipogenic lineage is regulated by mitogen-activated protein kinase. *J. Biol. Chem., 275*, 9645–9652.

Jiang, Y., Henderson, D., et al. (2003). Neuroectodermal differentiation from mouse multipotent adult progenitor cells. *Proc. Natl. Acad. Sci. U.S.A., 100*(Suppl. 1), 11854–11860.

Jiang, Y., Jahagirdar, B. N., et al. (2002). Pluripotency of mesenchymal stem cells derived from adult marrow. *Nature, 418*, 41–49.

Kalka, C., Masuda, H., et al. (2000a). Vascular endothelial growth factor(165) gene transfer augments circulating endothelial progenitor cells in human subjects. *Circ. Res., 86*, 1198–1202.

Kalka, C., Masuda, H., et al. (2000b). Transplantation of ex vivo expanded endothelial progenitor cells for therapeutic neovascularization. *Proc. Natl. Acad. Sci. U.S.A., 97*, 3422–3427.

Kalka, C., Tehrani, H., et al. (2000c). VEGF gene transfer mobilizes endothelial progenitor cells in patients with inoperable coronary disease. *Ann. Thorac. Surg., 70*, 829–834.

Kaushal, S., Amiel, G. E., et al. (2001). Functional small-diameter neovessels created using endothelial progenitor cells expanded ex vivo. *Nat. Med., 7*, 1035–1040.

Kawano, Y., Takaue, Y., et al. (1993). Effects of progenitor cell dose and preleukapheresis use of human recombinant granulocyte colony-stimulating factor on the recovery of hematopoiesis after blood stem cell autografting in children. *Exp. Hematol., 21*, 103–108.

Kim, S. Y., Park, S. Y., et al. (2005). Differentiation of endothelial cells from human umbilical cord blood AC133−CD14+ cells. *Ann. Hematol., 84*, 417–422.

Knedler, A., & Ham, R. G. (1987). Optimized medium for clonal growth of human microvascular endothelial cells with minimal serum. *In Vitro Cell Dev. Biol., 23*, 481–491.

Koc, O. N., Gerson, S. L., et al. (2000). Rapid hematopoietic recovery after coinfusion of autologous-blood stem cells and culture-expanded marrow mesenchymal stem cells in advanced breast cancer patients receiving high-dose chemotherapy. *J. Clin. Oncol., 18*, 307–316.

Kocher, A. A., Schuster, M. D., et al. (2001). Neovascularization of ischemic myocardium by human bone-marrow-derived angioblasts prevents cardiomyocyte apoptosis, reduces remodeling and improves cardiac function. *Nat. Med., 7*, 430–436.

Kondoh, K., Koyama, H., et al. (2004). Conduction performance of collateral vessels induced by vascular endothelial growth factor or basic fibroblast growth factor. *Cardiovasc. Res., 61*, 132–142.

Korf-Klingebiel, M., Kempf, T., et al. (2008). Bone marrow cells are a rich source of growth factors and cytokines: implications for cell therapy trials after myocardial infarction. *Eur. Heart J., 29*, 2851–2858.

Kryczek, I., Lange, A., et al. (2005). CXCL12 and vascular endothelial growth factor synergistically induce neoangiogenesis in human ovarian cancers. *Cancer Res., 65*, 465–472.

Kuznetsov, S. A., Mankani, M. H., et al. (2001). Circulating skeletal stem cells. *J. Cell Biol., 153*, 1133–1140.

Lazarus, H. M., Haynesworth, S. E., et al. (1995). Ex vivo expansion and subsequent infusion of human bone marrow-derived stromal progenitor cells (mesenchymal progenitor cells): implications for therapeutic use. *Bone Marrow Transplant., 16*, 557–564.

Levesque, M. J., Nerem, R. M., et al. (1990). Vascular endothelial cell proliferation in culture and the influence of flow. *Biomaterials, 11*, 702–707.

Li, X., Tjwa, M., et al. (2005). Revascularization of ischemic tissues by PDGF-CC via effects on endothelial cells and their progenitors. *J. Clin. Invest., 115*, 118–127.

Luttun, A., Tjwa, M., et al. (2002). Revascularization of ischemic tissues by PlGF treatment, and inhibition of tumor angiogenesis, arthritis and atherosclerosis by anti-Flt1. *Nat. Med., 8*, 831–840.

Lyden, D., Hattori, K., et al. (2001). Impaired recruitment of bone-marrow-derived endothelial and hematopoietic precursor cells blocks tumor angiogenesis and growth. *Nat. Med., 7*, 1194–1201.

Lyden, D., Young, A. Z., et al. (1999). Id1 and Id3 are required for neurogenesis, angiogenesis and vascularization of tumour xenografts. *Nature, 401*, 670–677.

Melero-Martin, J. M., de Obaldia, M. E., et al. (2008). Engineering robust and functional vascular networks in vivo with human adult and cord blood-derived progenitor cells. *Circ. Res., 103*, 194–202.

Melero-Martin, J. M., Khan, Z. A., et al. (2007). In vivo vasculogenic potential of human blood-derived endothelial progenitor cells. *Blood, 109*, 4761–4768.

Minami, E., Laflamme, M. A., et al. (2005). Extracardiac progenitor cells repopulate most major cell types in the transplanted human heart. *Circulation, 112*, 2951–2958.

Miyahara, Y., Nagaya, N., et al. (2006). Monolayered mesenchymal stem cells repair scarred myocardium after myocardial infarction. *Nat. Med., 12*, 459–465.

Motwani, J. G., & Topol, E. J. (1998). Aortocoronary saphenous vein graft disease: pathogenesis, predisposition, and prevention. *Circulation, 97*, 916–931.

Muller, W. A., Ratti, C. M., et al. (1989). A human endothelial cell-restricted, externally disposed plasmalemmal protein enriched in intercellular junctions. *J. Exp. Med., 170*, 399–414.

Narmoneva, D. A., Vukmirovic, R., et al. (2004). Endothelial cells promote cardiac myocyte survival and spatial reorganization: implications for cardiac regeneration. *Circulation, 110*, 962–968.

Nerem, R. M. (1984). Atherogenesis: hemodynamics, vascular geometry, and the endothelium. *Biorheology, 21*, 565–569.

Nerem, R. M., Levesque, M. J., et al. (1981). Vascular endothelial morphology as an indicator of the pattern of blood flow. *J. Biomech. Eng., 103*, 172–176.

Newman, P. J., Berndt, M. C., et al. (1990). PECAM-1 (CD31) cloning and relation to adhesion molecules of the immunoglobulin gene superfamily. *Science, 247*, 1219–1222.

O'Donnell, T. F., Jr., Mackey, W., et al. (1984). Correlation of operative findings with angiographic and noninvasive hemodynamic factors associated with failure of polytetrafluoroethylene grafts. *J. Vasc. Surg., 1*, 136–148.

Orlic, D., Kajstura, J., et al. (2001a). Transplanted adult bone marrow cells repair myocardial infarcts in mice. *Ann. N.Y. Acad. Sci., 938*, 221–229, discussion 229–230.

Orlic, D., Kajstura, J., et al. (2001b). Mobilized bone marrow cells repair the infarcted heart, improving function and survival. *Proc. Natl. Acad. Sci. U.S.A., 98*, 10344–10349.

Otani, A., Kinder, K., et al. (2002). Bone marrow-derived stem cells target retinal astrocytes and can promote or inhibit retinal angiogenesis. *Nat. Med., 8*, 1004–1010.

Owen, M. (1988). Marrow stromal stem cells. *J. Cell Sci. Suppl., 10*, 63–76.

Peichev, M., Naiyer, A. J., et al. (2000). Expression of VEGFR-2 and AC133 by circulating human CD34(+) cells identifies a population of functional endothelial precursors. *Blood, 95*, 952–958.

Peng, H., & Huard, J. (2003). Stem cells in the treatment of muscle and connective tissue diseases. *Curr. Opin. Pharmacol., 3*, 329–333.

Plouffe, B. D., Kniazeva, T., et al. (2009). Development of microfluidics as endothelial progenitor cell capture technology for cardiovascular tissue engineering and diagnostic medicine. *FASEB J., 23*, 3309–3314.

Pomposelli, F. B., Jr., Arora, S., et al. (1998). Lower extremity arterial reconstruction in the very elderly: successful outcome preserves not only the limb but also residential status and ambulatory function. *J. Vasc. Surg., 28*, 215–225.

Rafii, S. (2000). Circulating endothelial precursors: mystery, reality, and promise. *J. Clin. Invest., 105*, 17–19.

Rafii, S., Heissig, B., et al. (2002a). Efficient mobilization and recruitment of marrow-derived endothelial and hematopoietic stem cells by adenoviral vectors expressing angiogenic factors. *Gene Ther., 9*, 631–641.

Rafii, S., & Lyden, D. (2003). Therapeutic stem and progenitor cell transplantation for organ vascularization and regeneration. *Nat. Med., 9*, 702–712.

Rafii, S., Lyden, D., et al. (2002b). Vascular and haematopoietic stem cells: novel targets for anti-angiogenesis therapy? *Nat. Rev. Cancer, 2*, 826–835.

Rafii, S., Meeus, S., et al. (2002c). Contribution of marrow-derived progenitors to vascular and cardiac regeneration. *Semin. Cell Dev. Biol., 13*, 61–67.

Rafii, S., Oz, M. C., et al. (1995a). Characterization of hematopoietic cells arising on the textured surface of left ventricular assist devices. *Ann. Thorac. Surg., 60*, 1627–1632.

Rafii, S., Shapiro, F., et al. (1995b). Human bone marrow microvascular endothelial cells support long-term proliferation and differentiation of myeloid and megakaryocytic progenitors. *Blood, 86*, 3353–3363.

Reinisch, A., Hofmann, N. A., et al. (2009). Humanized large-scale expanded endothelial colony-forming cells function in vitro and in vivo. *Blood, 113*, 6716–6725.

Reyes, M., Dudek, A., et al. (2002). Origin of endothelial progenitors in human postnatal bone marrow. *J. Clin. Invest., 109*, 337–346.

Reyes, M., & Verfaillie, C. M. (2001). Characterization of multipotent adult progenitor cells, a subpopulation of mesenchymal stem cells. *Ann. N.Y. Acad. Sci., 938*, 231–233, discussion 233–235.

Rohde, E., Bartmann, C., et al. (2007). Immune cells mimic the morphology of endothelial progenitor colonies in vitro. *Stem Cells, 25*, 1746–1752.

Romagnani, P., Annunziato, F., et al. (2005). CD14+CD34 low cells with stem cell phenotypic and functional features are the major source of circulating endothelial progenitors. *Circ. Res., 97*, 314–322.

399

Ross, R. (1993). The pathogenesis of atherosclerosis: a perspective for the 1990s. *Nature, 362,* 801–809.

Roufosse, C. A., Direkze, N. C., et al. (2004). Circulating mesenchymal stem cells. *Int. J. Biochem. Cell Biol., 36,* 585–597.

Rubart, M., & Field, L. J. (2006). Cardiac regeneration: repopulating the heart. *Annu. Rev. Physiol., 68,* 29–49.

Salcedo, R., Wasserman, K., et al. (1999). Vascular endothelial growth factor and basic fibroblast growth factor induce expression of CXCR4 on human endothelial cells: in vivo neovascularization induced by stromal-derived factor-1alpha. *Am. J. Pathol., 154,* 1125–1135.

Sata, M., Saiura, A., et al. (2002). Hematopoietic stem cells differentiate into vascular cells that participate in the pathogenesis of atherosclerosis. *Nat. Med., 8,* 403–409.

Sayers, R. D., Raptis, S., et al. (1998). Long-term results of femorotibial bypass with vein or polytetrafluoroethylene. *Br. J. Surg., 85,* 934–938.

Schwartz, R. E., Reyes, M., et al. (2002). Multipotent adult progenitor cells from bone marrow differentiate into functional hepatocyte-like cells. *J. Clin. Invest., 109,* 1291–1302.

Seifalian, A. M., Tiwari, A., et al. (2002). Improving the clinical patency of prosthetic vascular and coronary bypass grafts: the role of seeding and tissue engineering. *Artif. Organs, 26,* 307–320.

Shake, J. G., Gruber, P. J., et al. (2002). Mesenchymal stem cell implantation in a swine myocardial infarct model: engraftment and functional effects. *Ann. Thorac. Surg., 73,* 1919–1925, discussion 1926.

Sharpe, E. E., III, Teleron, A. A., et al. (2006). The origin and in vivo significance of murine and human culture-expanded endothelial progenitor cells. *Am. J. Pathol., 168,* 1710–1721.

Shen, H. C., Peng, H., et al. (2004). Ex vivo gene therapy-induced endochondral bone formation: comparison of muscle-derived stem cells and different subpopulations of primary muscle-derived cells. *Bone, 34,* 982–992.

Shi, Q., Rafii, S., et al. (1998). Evidence for circulating bone marrow-derived endothelial cells. *Blood, 92,* 362–367.

Shintani, S., Murohara, T., et al. (2001). Mobilization of endothelial progenitor cells in patients with acute myocardial infarction. *Circulation, 103,* 2776–2779.

Sodian, R., Hoerstrup, S. P., et al. (2000). Early in vivo experience with tissue-engineered trileaflet heart valves. *Circulation, 102,* III22–III29.

Solchaga, L. A., Dennis, J. E., et al. (1999). Hyaluronic acid-based polymers as cell carriers for tissue-engineered repair of bone and cartilage. *J. Orthop. Res., 17,* 205–213.

Solchaga, L. A., Yoo, J. U., et al. (2000). Hyaluronan-based polymers in the treatment of osteochondral defects. *J. Orthop. Res., 18,* 773–780.

Solovey, A., Lin, Y., et al. (1997). Circulating activated endothelial cells in sickle cell anemia. *N. Engl. J. Med., 337,* 1584–1590.

Solovey, A. N., Gui, L., et al. (2001). Identification and functional assessment of endothelial P1H12. *J. Lab. Clin. Med., 138,* 322–331.

Sprague, E. A., Steinbach, B. L., et al. (1987). Influence of a laminar steady-state fluid-imposed wall shear stress on the binding, internalization, and degradation of low-density lipoproteins by cultured arterial endothelium. *Circulation, 76,* 648–656.

Stephen, M., Loewenthal, J., et al. (1977). Autogenous veins and velour dacron in femoropopliteal arterial bypass. *Surgery, 81,* 314–318.

Takahashi, T., Kalka, C., et al. (1999). Ischemia- and cytokine-induced mobilization of bone marrow-derived endothelial progenitor cells for neovascularization. *Nat. Med., 5,* 434–438.

Tateishi-Yuyama, E., Matsubara, H., et al. (2002). Therapeutic angiogenesis for patients with limb ischaemia by autologous transplantation of bone-marrow cells: a pilot study and a randomised controlled trial. *Lancet, 360,* 427–435.

Tillman, B. W., Yazdani, S. K., et al. (2009). Efficient recovery of endothelial progenitors for clinical translation. *Tissue Eng., 15,* 213–221.

Toma, C., Pittenger, M. F., et al. (2002). Human mesenchymal stem cells differentiate to a cardiomyocyte phenotype in the adult murine heart. *Circulation, 105,* 93–98.

Unger, C., Skottman, H., et al. (2008). Good manufacturing practice and clinical-grade human embryonic stem cell lines. *Hum. Mol. Genet., 17,* R48–R53.

Verfaillie, C. M. (2002a). Adult stem cells: assessing the case for pluripotency. *Trends Cell Biol., 12,* 502–508.

Verfaillie, C. M. (2002b). Hematopoietic stem cells for transplantation. *Nat. Immunol., 3,* 314–317.

Verfaillie, C. M. (2005). Multipotent adult progenitor cells: an update. *Novartis Found. Symp., 265,* 55–61.

Verfaillie, C. M., Schwartz, R., et al. (2003). Unexpected potential of adult stem cells. *Ann. N.Y. Acad. Sci., 996,* 231–234.

Wollert, K. C., & Drexler, H. (2005). Clinical applications of stem cells for the heart. *Circ. Res., 96,* 151–163.

Yager, P., Edwards, T., et al. (2006). Microfluidic diagnostic technologies for global public health. *Nature, 442,* 412–418.

Yamamoto, K., Takahashi, T., et al. (2003). Proliferation, differentiation, and tube formation by endothelial progenitor cells in response to shear stress. *J. Appl. Physiol., 95,* 2081–2088.

Yancopoulos, G. D., Klagsbrun, M., et al. (1998). Vasculogenesis, angiogenesis, and growth factors: ephrins enter the fray at the border. *Cell, 93,* 661–664.

Young, M. R., Kolesiak, K., et al. (1999). Chemoattraction of femoral CD34+ progenitor cells by tumor-derived vascular endothelial cell growth factor. *Clin. Exp. Metastasis, 17,* 881–888.

Young, P. P., Vaughan, D. E., et al. (2007). Biologic properties of endothelial progenitor cells and their potential for cell therapy. *Prog. Cardiovasc. Dis., 49,* 421–429.

Zhang, Z. G., Zhang, L., et al. (2002). Bone marrow-derived endothelial progenitor cells participate in cerebral neovascularization after focal cerebral ischemia in the adult mouse. *Circ. Res., 90,* 284–288.

Zvaifler, N. J., Marinova-Mutafchieva, L., et al. (2000). Mesenchymal precursor cells in the blood of normal individuals. *Arthritis Res., 2,* 477–488.

CHAPTER 23

Islet Cell Therapy and Pancreatic Stem Cells

Juan Domínguez-Bendala, Antonello Pileggi, Camillo Ricordi
Diabetes Research Institute, Cell Transplant Center, University of Miami, Miami, FL, USA

INTRODUCTION

Replacement of insulin-producing cell function represents an appealing approach for the treatment of diabetes mellitus, a condition characterized by loss of β-cell mass and/or function (Ricordi, 2003; Ricordi and Strom, 2004) consequent to autoimmunity (Type 1 diabetes mellitus (T1DM)), metabolic disorders (i.e. cystic fibrosis, hemochromatosis, and liver cirrhosis), surgery (i.e. iatrogenic diabetes following pancreatectomy for relapsing, chronic pancreatitis), or β-cell dysfunction secondary to insulin resistance and hyperinsulinism (Type 2 diabetes mellitus (T2DM)) (Pileggi et al., 2004). Exogenous insulin injections have represented a life-saving treatment in T1DM, changing the natural history of diabetes, and remarkable progress has been achieved in recent years in the management of glycemic control by combining diet, exercise, and improved exogenous insulin treatment options. However, this approach requires continuous adjustments in insulin administrations with significant challenges in attaining tight glycemic control in the absence of severe hypoglycemic episodes. Tight metabolic control with avoidance of wide glycemic excursions is necessary to decrease the risk of development and/or progression of the chronic complications that can negatively impact the quality of life and life expectancy of patients with diabetes. Hundreds of thousands of endocrine cell clusters, from <50 to >500 microns in diameter (islets of Langerhans), are scattered into the pancreatic tissue, representing approximately 1−2% of the entire organ. The islets are "micro-organs" with a unique cytoarchitecture, composed of heterogeneous cell subsets specializing in the production and secretion of endocrine hormones (α-cells for glucagon; β-cells for insulin; δ-cells for somatostatin; PP-cells for pancreatic polypeptide) that are essential for the regulation of glucose homeostasis in the blood (Bosco et al., 2010; Brissova et al., 2005; Cabrera et al., 2006). Complex interactions between the cell subsets composing the islets, their innervation, and the rich vascular bed result in "real-time" secretion of endocrine hormones that maintain glycemic values within physiologic ranges. Better understanding of pancreatic islet cell ontogeny, biology, and physiology will be of assistance in developing efficient protocols for cellular therapies for the restoration of metabolic control in patients with diabetes.

Considerable progress has been achieved in the last three decades in the field of β-cell replacement therapy, either by transplantation of the pancreas as a vascularized organ, or by infusion of islet cell products. The encouraging results of recent clinical trials support the value of this approach, which has been shown to improve both quality of life and metabolic control in patients with T1DM following intrahepatic islet transplantation (Ryan et al., 2001, 2002, 2005a; Froud et al., 2005; Poggioli et al., 2006; Pileggi et al., 2006a). Current challenges to the widespread application of β-cell replacement therapies include the shortage of transplantable

403

Principles of Regenerative Medicine. DOI: 10.1016/B978-0-12-381422-7.10023-9

tissue and the need for more effective and safer immune interventions that favor long-term graft function (Ricordi et al., 2005; Pileggi et al., 2006a,b). Ultimately, successful strategies for immunoisolation (Halle' et al., 2009), tissue engineering with targeted immunomodulation (Bocca et al., 2009), or the development of safe protocols for the induction of immune tolerance to avoid the need for life-long immunosuppression of the recipients will be necessary for the widespread applicability of islet cell therapy (Pileggi et al., 2006b; Mineo et al., 2009). In fact, the current requirements for life-long immunosuppression of the recipients severely limit the current indications for islet transplantation to the most severe cases of T1DM or in patients already undergoing organ transplantation and therefore already undergoing immunosuppressive treatment (Mineo et al., 2009). When islet transplantation is possible without chronic recipient immunosuppression, current sources of donor pancreata will clearly be insufficient to meet the demand. This is why it is so critical to continue to work towards the identification of alternative sources of insulin-producing cells. Encouraging data are emerging in the field of islet cell neogenesis and stem cell research (Ricordi and Edlund, 2008) that justify a cautious optimism for the years to come.

This chapter will review the current status, challenges, and perspectives in clinical islet transplantation for treatment of diabetes and the progress of selected areas of stem cell and β-cell regeneration.

BENEFITS OF β-CELL REPLACEMENT THERAPY

Transplantation of β-cells is currently performed as vascularized pancreas or isolated islet cell grafts. Both procedures can result in improved glycemic control in patients with diabetes (Pileggi et al., 2006a). Transplantation of pancreatic islets offers substantial advantage over whole pancreas transplantation because of the lower risks for procedure-related complications and the possibility of preconditioning the graft *in vitro* prior to implantation (Mineo et al., 2009).

Islets are isolated from the donor pancreas by a mechanically enhanced, enzymatic digestion process that allows for the physical dissociation of pancreatic tissue into small fragments and liberation of the endocrine cell clusters with preserved integrity (Ricordi et al., 1988). The dissociation phase is followed by purification on density gradients that enriches for fractions with higher endocrine cell clusters ($\sim$2% of the whole pancreatic tissue) while minimizing contamination with non-endocrine tissue (Alejandro et al., 1990; Ichii et al., 2005b). After isolation and culture, islets are ready to be transplanted and fractions with different degrees of purity are prepared into infusion bags or shipped to distant transplant centers (Ichii et al., 2007).

Islet transplantation is performed using minimally invasive interventional radiology techniques consisting of percutaneous, trans-hepatic cannulation of the portal vein and infusion of the islets by gravity (Alejandro and Mintz, 1988; Baidal et al., 2003; Froud et al., 2004; Pileggi et al., 2004). After infusion, the islet cell clusters remain trapped at the pre-sinusoidal level of the portal tree. The purification procedure substantially reduces the volume of tissue to be infused, therefore minimizing the risk of portal thrombosis and portal hypertension consequent to the intrahepatic embolization of the islet grafts (Froud et al., 2004, 2005), which was observed in a few cases when unpurified islet preparations or inadequate islet infusion techniques were used.

Islet transplantation is indicated in patients who have lost insulin-producing cell function. Recent clinical trials have shown the importance of intensive insulin treatment to obtain tight glycemic control, and its ability to prevent or delay the dreadful complications of unstable glycemic control, including neuropathy, vasculopathy, and nephropathy (DCCTRG, 1993). Unfortunately, intensive insulin treatment cannot maintain glycemic values within normal ranges throughout the day and is associated with an increased risk of severe hypoglycemia, at times fatal. Restoration of islet β-cell function is a highly desirable goal for patients with T1DM as it can provide a more physiological glycemic metabolic control than exogenous insulin.

Transplantation of autologous islets (autotransplantation) is generally performed to prevent iatrogenic diabetes in patients undergoing pancreatectomy due to severe pain for chronic, relapsing pancreatitis, for traumatic injury, or for non-enucleable benign neoplasm of the pancreas (Robertson et al., 2001; Oberholzer et al., 2003; Garraway et al., 2009;). The islets are isolated from the recipient's pancreas after total pancreatectomy and then transplanted into his/her own liver.

Transplantation of allogeneic islets (obtained from the pancreas of deceased multi-organ donors) is generally performed in patients with T1DM for whom loss of pancreatic β-cells in the pancreatic islets is due to an autoimmune process (Ricordi, 2003; Pileggi et al., 2004). The transplant is indicated in non-uremic, C-peptide-negative patients with unstable diabetes complicated by severe hypoglycemia and performed as solitary islet transplantation (islet transplantation alone (ITA)) and in patients with end-stage renal disease (ESRD) receiving a kidney graft before (islet after kidney (IAK)) or at the time of (simultaneous islet-kidney (SIK)) islet transportation (Ricordi, 2003; Ricordi and Strom, 2004; Shapiro et al., 2000). Allogeneic islet transplantation has been also performed in patients with diabetes associated with metabolic diseases (i.e. cystic fibrosis, hemochromatosis, and liver cirrhosis) and surgical removal of the pancreas (for trauma or benign abdominal diseases) in combination with liver, lung, or clustered abdominal organs (Tzakis et al., 1990, 1996; Brunicardi et al., 1995; Ricordi et al., 1997; Tschopp et al., 1997; Angelico et al., 1999; Jindal et al., 2010).

After transplantation of pancreatic islets, dramatic improvement of metabolic control is generally observed with reduction of mean amplitude of glycemic excursions and of insulin requirements, normalization of glycated hemoglobin, and absence of severe hypoglycemia (Alejandro et al., 1997; Ryan et al., 2002, 2004, 2005a,b; Geiger et al., 2003; Froud et al., 2005). Insulin independence is achieved when a sufficient islet mass is implanted, a goal generally obtained using islets obtained from one or more donor pancreata (Shapiro et al., 2000; Markmann et al., 2003; Pileggi et al., 2004; Froud et al., 2005; Hering et al., 2005).

Long-term graft function has been reported after islet autotransplantation (Robertson et al., 2001) and allogeneic islet transplantation (Carroll et al., 1995; Alejandro et al., 1997; Pileggi et al., 2004; Froud et al., 2005; Ryan et al., 2005a; Berney et al., 2009; Vantyghem et al., 2009), with improved metabolic control and absence of severe hypoglycemia even in patients under exogenous insulin treatment (Mineo et al., 2009). Recent clinical trials of allogeneic islet transplantation have shown that insulin independence can be obtained in approximately 80% of patients at 1 year, but progressive graft dysfunction has been observed over time, despite sustained C-peptide production and good metabolic control with reintroduction of exogenous insulin (Froud et al., 2005; Ryan et al., 2005a; Pileggi et al., 2006a; Alejandro et al., 2008; Mineo et al., 2009). Sustained insulin independence in 50−60% of the patients has been reported using conventional immunosuppressive regimens (Vantyghem et al., 2009) and using novel protocols based on lymphodepleting agents in combination with maintenance immunosuppressive protocols minimizing β-cell toxicity (Bellin et al., 2008; Faradji et al., 2008; Froud et al., 2008a). These encouraging results indicate that ITA can attain long-term insulin independence rates that are comparable to those of pancreas transplant alone (56% at 5 years) (Gruessner et al., 2004) and justify the need for reassessment of islet transplantation's place as a clinical option for β-cell replacement.

The benefits of replacing β-cell function by islet transplantation include a dramatic improvement of the quality of life associated with the enhanced glycemic control and reduced fear of severe hypoglycemia (Johnson et al., 2004; Barshes et al., 2005; Poggioli et al., 2006; Cure et al., 2008b; Tharavanij et al., 2008). The positive effects of islet transplantation are maintained even in patients experiencing partial graft dysfunction and requiring reintroduction of exogenous insulin, until measurable C-peptide persists (Alejandro et al., 1997; Pileggi et al., 2004). Following islet transplantation, partial restoration of physiological β-cell response to secreta-gogues is observed with improved first-phase insulin secretion upon intravenous stimulation and increased overall C-peptide levels following oral challenge (Rickels et al., 2005a, 2007). In addition, glucagon response to hypoglycemia is partially restored and paralleled by recovery of sympatho-adrenal responses and by re-establishment of a normal glucagon response to hyperglycemia with reduction of hepatic glucose output (Paty et al., 2002; Rickels et al., 2005b). Collectively, these effects may contribute to the improvement of metabolic control and regaining of hypoglycemia awareness in islet transplant recipients, a phenomenon that persists for some time even after complete loss of graft function (Leitao et al., 2008).

The impact of the restoration of β-cell function by islet transplantation on diabetes complication is currently under evaluation, and some of the encouraging preliminary reports are based on non-randomized studies in relatively small series of cases. Benefits of islet transplantation include improved survival and function of renal allografts (Fiorina et al., 2003b, 2005b), improvement of vasculopathy (Fiorina et al., 2003a), better cardiovascular function (Fiorina et al., 2005a,b) in IAK recipients, and stabilization of diabetic retinopathy and neuropathy in recipients of ITA (Lee et al., 2005; Venturini et al., 2006; del Carro et al., 2007; Thompson et al., 2008).

The transplantation procedure has been associated with a relatively low incidence of side-effects to date (Goss et al., 2003; Markmann et al., 2003; Owen et al., 2003; Frank et al., 2004; Froud et al., 2004, 2005; Hafiz et al., 2005; Ryan et al., 2005a; Venturini et al., 2005). Expected untoward complications of the immunosuppressive drugs utilized to prevent graft rejection have been described in recent clinical trials (Hirshberg et al., 2003; Cure et al., 2004, 2006, 2008a; Frank et al., 2004; Andres et al., 2005; Froud et al., 2005; Hafiz et al., 2005; Molinari et al., 2005; Ryan et al., 2005a; Senior et al., 2005; Ponte et al., 2007a; Alejandro et al., 2008; Mineo et al., 2009); these are similar to those observed for other organs and tissues.

CURRENT LIMITATIONS OF β-CELL REPLACEMENT THERAPIES

Hurdles to the widespread application of β-cell replacement therapy based on the transplantation of allogeneic islets include the relatively high numbers of islets required to achieve insulin independence, due to the shortage of deceased donor organs available for transplantation (Pileggi et al., 2006b; Mineo et al., 2009). Several factors negatively impact the yield and quality of islets obtained from a donor pancreas (Pileggi et al., 2006b, 2009). While improved donor management after brain death and refined organ procurement (Lee et al., 2004; Ponte et al., 2007b) and preservation (Kuroda et al., 1988; Matsumoto et al., 1996; Fraker et al., 2002) techniques have allowed for better results in recent years, expansion of the donor pool to marginal donors (Ricordi et al., 2003; Tsujimura et al., 2004a,b) and donation after cardiac arrest (Goto et al., 2005; Matsumoto and Tanaka, 2005) appear promising avenues to increase organ utilization and obtain adequate (both qualitatively and quantitatively) islet cells from a single donor pancreas for transplantation. Unfortunately, a large number of organs suitable for transplantation are underutilized (Krieger et al., 2003), indicating the need for improved management of potential pancreas donors and organ recovery policies to increase organ availability. An appealing alternative to overcome donor organ shortage is the use of living donor organs (namely, distal pancreatectomy) as a source of islets (Matsumoto et al., 2005), although for a large-scale application of this approach a thorough evaluation of risks/benefits for both donors and recipients should be undertaken to avoid onset of insulin-requiring diabetes in the pancreas segment donor later in life (Robertson, 2004) and prevent loss of transplanted islets in the recipients due to the lack of safe and non-diabetogenic immunosuppressive/tolerogenic protocols at the present time.

Steady improvements in islet cell processing, purification, and culture have been implemented in recent years (Ricordi et al., 1988; Alejandro et al., 1990; Lakey et al., 1999; Ichii et al., 2005b, 2006; Barbaro et al., 2007; Ponte et al., 2007b; Gangemi et al., 2008) that have allowed for the recovery of better islet yields from a single donor pancreas, thereby maximizing organ utilization for islet transplantation. Additionally, active research is ongoing toward the definition of sensitive predictive tests of islet potency (Street et al., 2004; Ichii et al., 2005a; Fraker et al., 2006) that could discriminate preparations yielding adequate islets for transplant from those that would not, as they could contribute to improve islet transplantation outcomes.

Islet transplantation is regulated by the Food and Drug Administration under the category Investigational New Drug (IND) (Wonnacott, 2005). Implementation of current good manufacture practice (cGMP), availability of specific infrastructures, and of dedicated personnel are required to yield high-quality standards and consistency in islet cell quality for transplantation (Weber, 2004). These requirements impose a remarkable economic burden on clinical islet transplantation programs (Markmann et al., 2003; Guignard et al., 2004). The creation of "regional" human islet cell processing facilities that can provide cGMP quality islet cell products for research and clinical transplant applications may represent a viable option to improve the consistency and quality of the final islet cell products, while containing the costs (Oberholzer et al., 2000; Goss et al., 2002, 2003, 2004; Lee et al., 2004; Kempf et al., 2005; Ichii et al., 2007).

The relatively high islet numbers required for successful post-transplant outcome also depend on the quality of the islet cell product infused into the recipients and the impaired engraftment of a relatively large proportion of islets due to the generation of inflammation in the liver microenvironment (Pileggi et al., 2001, 2006b). Inflammation and hypoxia (due to lack of vascularization in the early period of post-implantation) could contribute to functional impairment and/or islet cell death early after islet transplantation. Engraftment of a suboptimal islet mass may also result in graft dysfunction due to metabolic exhaustion (Froud et al., 2005) that could be further worsened by the relatively hyperglycemic liver environment and toxic levels of immunosuppressive drugs in the hepatic vascular district (Desai et al., 2003; Shapiro et al., 2005; Pileggi et al., 2006a).

Implementation of therapeutic interventions aimed at preserving islet cell functional potency (i.e. incretin mimetics) has proven effective in improving and prolonging islet graft function in recent clinical trials. In particular, off-label use of Exenatide in islet transplant recipients has been associated with the achievement of insulin independence with a reduced number of islets (Ghofaili et al., 2007; Gangemi et al., 2008), recovery/improvement of graft function after the onset of dysfunction (Froud et al., 2008b), improved longevity of insulin independence after delayed supplemental islet infusion (Faradji et al., 2008), as well as enhanced insulin secretion during metabolic tests (Ghofaili et al., 2007; Faradji et al., 2008, 2009; Gangemi et al., 2008). These observations suggest the potential window of opportunity in treatments can be beneficial and lead to a more reproducible success of single-donor islet transplantation in future clinical trials.

Other important hurdles of β-cell replacement therapies include the need for life-long immunosuppression to prevent rejection and autoimmunity recurrence in the case of T1DM (Vendrame et al., 2010). The negative effects of calcineurin inhibitors (CNI) and mTor-inhibitors on renal function have been recognized (Rangan, 2006; Williams and Haragsim, 2006). The potential negative impact of these drugs on the progression of diabetic nephropathy in non-uremic subjects needs to be fully appreciated, even though, in the context of islet transplantation, decline of renal function is questionable (Maffi et al., 2007; Senior et al., 2007), particularly based on the observation of stable renal function and lack of worsening diabetic nephropathy at long-term follow-up (Fung et al., 2007; Thompson et al., 2008; Warnock et al., 2008; Cure et al., 2008b; Leitão et al., 2009b) thanks to strict selection of islet transplant candidates without previous renal dysfunction (i.e. micro-albuminuria and low estimated glomerular filtration rates) and timely implementation of nephroprotective and anti-hypertensive therapies (angiotensin-converting enzyme inhibitors and/or angiotensin-receptor blockers) (Leitão et al., 2009b). Immunosuppressive protocols void of nephrotoxicity are highly desirable, and indeed ongoing clinical trials are showing good results in patients undergoing conversion of either CNI or mTOR-inhibitors to mychophenolic acid (MPA) therapy with maintenance of good renal and islet function (Cure et al., 2008b).

Furthermore, direct β-cell toxicity and functional impairment consequent to exposure to CNI have been widely recognized (Ricordi et al., 1991; Fernandez et al., 1999), with increasing experimental evidence supporting the anti-proliferative effects of mTOR-inhibitors and CNI that may result in impaired islet engraftment (i.e. altered neovascularization and tissue repair/remodeling) and/or reduced β-cell self-renewal (Nir et al., 2007; Zahr et al., 2007). Additionally, increased lipid levels are commonly associated with immunosuppression (mainly mTOR-inhibitors) and may result in β-cell lipotoxicity contributing to loss of functional β-cell mass over time (Hafiz et al., 2005; Leitão et al., 2009a).

Maintenance of adequate immunosuppression in islet transplant recipients can prevent the development of alloantibodies and neutralize their potentially negative impact on graft survival, even in the presence of a low degree of panel reactive alloantibodies (PRA) pre-transplant (Cardani et al., 2007). Nevertheless, development of donor- and non-donor-specific alloantibodies may occur after drug dose reduction (i.e. for medical reasons), while it invariably develops when immunosuppression is withdrawn (Cardani et al., 2007). The significance of this phenomenon of allosensitization and its potential impact on long-term islet graft function or subsequent allografts has not yet been established, even though there is a concern for potentially limiting future therapeutic options (i.e. subsequent islet, pancreas, and/or renal transplantation for ESRD) (Cardani et al., 2007). Possible strategies to overcome this issue may consist of selecting subjects with slow progression of diabetic nephropathy who are unlikely to develop ESRD; attempting more stringent HLA matching may contribute to reducing the risk of allosensitization in islet transplant recipients (Leitão et al., 2009b), as well as maximizing the success of single-donor islet transplantation. It is conceivable that development of immunosuppression-weaning protocols may be of assistance in reducing the risk of sensitization.

The steady improvement in islet cell processing will be of assistance in optimizing both yields and quality of islets from donor pancreata, thereby contributing to increasing the number of transplants in the years to come. It is conceivable that this approach will not suffice in overcoming the increasing demand of islet grafts due to the disproportionate pancreas donor-to-recipient ratio: there will be a large number of patients with insulin-requiring diabetes who would benefit from restoration of β-cell function and not sufficient human pancreata for processing. Alternative sources of insulin-producing β-cells (from either allogeneic or xenogeneic donors) or induction of self-regeneration of the patient's own β-cells (in combination with adequate immunomodulation to prevent recurrence of autoimmunity) (Ogawa et al., 2004) may help in achieving the desired metabolic control in the near future (Ricordi et al., 2005).

For β-cell replacement therapies to become the treatment of choice for patients with diabetes, successful restoration of metabolic function should be achieved long-term. For this reason, implementing a sequential, integrated approach that combines strategies aiming at improving β-cell mass together with those focusing on the modulation of the immune response (i.e. preventing rejection and recurrence of autoimmunity) could represent an essential element towards definition of successful therapeutic strategies (Ricordi and Strom, 2004; Ricordi et al., 2005; Vendrame et al., 2010). Promising data on the induction of donor-specific unresponsiveness and tolerance to transplanted tissues in experimental models justify optimism for the near future, and may allow the achievement of long-term function of transplanted insulin-producing cells in the absence of rejection and recurrence of autoimmunity without the need for chronic immunosuppression in the clinical setting (Inverardi and Ricordi, 2001; Inverardi et al., 2004; Ricordi and Strom, 2004; Mineo et al., 2009).

ALTERNATIVE SOURCES OF INSULIN-PRODUCING CELLS: STEM CELLS AND β-CELL REGENERATION

Stem cells could be defined as undifferentiated cells that have the ability to proliferate while retaining the potential to fully mature into other cell types. The extent to which stem cells can be induced to proliferate or differentiate depends on their origin and stage of development. Arguably, the most powerful stem cells available are embryonic stem (ES) cells. These cells, which are obtained from the inner cell mass (ICM) of the blastocyst, can be maintained indefinitely in an undifferentiated, proliferative stage *in vitro* (Thomson et al., 1998). When transplanted into immunodeficient animals or otherwise induced to spontaneously differentiate, they can give rise to cells of all three embryonal layers (endoderm, ectoderm, and mesoderm). The prospects of turning human ES (huES) cells into islet cell types are therefore substantiated, but not exempt from safety and ethical concerns.

Stem cells of fetal origin may still retain some degree of multilineage differentiation, as well as the potential to proliferate *in vitro*. Despite their embryonic origin, these cells should not be confused with the blastocyst-derived ES cells. In fact, in many respects, these cells are more akin to adult cell types than to ES cells. This, together with the controversy surrounding their procurement, makes them unlikely candidates to become an alternative source of islets in the near future.

Expansion of fully differentiated, adult β-cells has been reported *in vitro*. However, the induction of β-cell proliferation has been generally associated with loss of mature cell phenotype and of functional competence that is only partially recovered after redifferentiation. Many adult tissues also have stem cells involved in their physiologic maintenance and repair mechanisms. Whether the adult pancreas contains endocrine stem cells or not is still the subject of heated debate. In general, tissue-specific stem cells are elusive and difficult to culture *in vitro*, and their differentiation potential is much more restricted than that of ES cells. One possible exception to this rule are the bone marrow (BM)-derived multipotent adult progenitor cells (MAPCs) described by Verfaillie and colleagues (Jiang et al., 2002). These cells have been shown to proliferate extensively *in vitro* and are able to give rise to the three embryonal layers when injected into mouse blastyocysts. However, their routine isolation and culture are still far from mainstream, as they have proven more challenging than working with ES cells.

An additional potential approach to obtain insulin-producing cells is transdifferentiation of adult cells (i.e. hepatocytes) under selected conditions both *in vivo* and *ex vivo*.

PANCREATIC DEVELOPMENT

Research conducted over the last decade has outlined a basic "roadmap" of the major molecular events that shape islet development (Fig. 23.1) (Edlund, 2002). Critical developmental milestones are: (1) generation of endoderm/gut endothelium; (2) pancreatic differentiation; (3) endocrine specification; and (4) β-cell differentiation. Transition between each of these stages of development appears to be catalyzed by a surprisingly small number of transcription factors, which are highly conserved between mouse and human.

Generation of endoderm/gut endothelium

ES cells are an artificially frozen snapshot of the inner cell mass (ICM) cells found at the blastocyst stage (embryonic day (e) 3.5). Expression of genes such as *telomerase*, *Oct3/4*, and *Nanog* make these cells immortal and pluripotent under defined conditions *in vitro* (Thomson et al., 1998; Thomson and Odorico, 2000; Boyer et al., 2005). Subsequent differentiation will be marked by the permanent downregulation of these genes. Visceral endoderm and epiblast, respectively, constitute the outer and inner layers of the ICM immediately before gastrulation. The visceral endoderm will become part of the yolk sac,

409

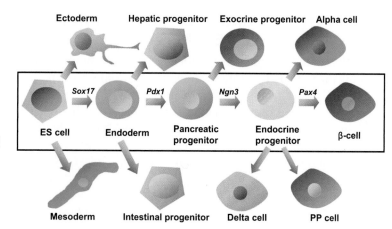

FIGURE 23.1
Schematic representation of the differentiation pathway from ES cells to β-cells. Genes whose expression is necessary for the transition between each step are indicated in italics.

without contribution to the embryo proper. In contrast, the definitive endoderm is formed during gastrulation when epiblast cells leave the ICM through the primitive streak. There is an intermediate stage in definitive endoderm formation, called mesendoderm. Although visceral and definitive endoderm are similar, mesendoderm-specific genes such as *Gsc* and *Bry* do not appear during visceral endoderm differentiation (Kispert and Herrmann, 1994; Tam et al., 2003; Kubo et al., 2004; Yasunaga et al., 2005), and therefore can be used to identify true definitive endoderm (de Santa Barbara et al., 2003). The anterior part of the definitive endoderm will evolve into the foregut, from which pancreas, liver, and lungs will eventually bud out. The posterior definitive endoderm, on the other hand, becomes the midgut and hindgut, which will differentiate into large and small intestine. Graded *Nodal* signaling is responsible for the initial patterning of the primitive gut endothelium. Many genes have been associated with the formation of true endoderm, including *Foxa2*, *Mixl1*, *GATA-4*, and several members of the *Sox* family, chiefly *Sox17* (Hudson et al., 1997; Kanai-Azuma et al., 2002; Yasunaga et al., 2005).

Pancreatic differentiation

There is a cross-communication between the gut endoderm and the surrounding mesoderm, mediated by *Shh* signaling. *Shh* is highly expressed throughout the gut endothelium, but is downregulated in a *Pdx1*-positive region that will later become the pancreas at e8. Both *Shh* repression and *Pdx1* activation are defining events of pancreatic specification. *Pdx1* knockouts are born without pancreas (Jonsson et al., 1994). Chemical inhibition of *Shh* enhances pancreatic differentiation at the expense of intestine (Hebrok et al., 1998, 2000; Kim and Melton, 1998; Ramalho-Santos et al., 2000). Conversely, ectopic expression of *Shh* under the control of the *Pdx1* promoter induces intestinal fates at the expense of the pancreas (Apelqvist et al., 1997).

Endocrine specification

Endocrine differentiation occurs through a lateral inhibition process, mediated by *Notch* signaling. Cells in which the *Notch* receptor is activated by the ligands *delta* or *serrate* express high levels of *HES-1*, which in turn represses the pro-endocrine gene *Ngn3*. Lower levels of *Notch* signaling may randomly occur in individual cells, where *HES-1* expression will not be upregulated. In the absence of its repressor, *Ngn3* will be expressed robustly, and the cell will adopt a pro-endocrine fate (Apelqvist et al., 1999; Gradwohl et al., 2000; Jensen et al., 2000). The differentiation into each of the five endocrine cell types within the islet (α-, β-, δ-, PP-, and ε-cells) is preferentially observed at specific time points during embryogenesis, suggesting that *Ngn3*-positive cells adapt their responses to an evolving milieu of signals (Domínguez-Bendala et al., 2005).

β-Cell differentiation

Little is known about the extracellular signals that drive β-cell specification from *Ngn3*-positive progenitors. Animals lacking *Nkx6.1* (Sander et al., 2000) and *Nkx2.2* (Sussel et al., 1998) have defects in β-cell formation. However, several observations point to *Pax4* as the main hallmark of β-cell differentiation: (1) the knockout of this gene results in the total absence of β-cells (Sosa-Pineda et al., 1997), but not α-cells; (2) its expression peaks between e13.5 and e15.5, which coincides with the period of maximal differentiation of β-cell precursors (Sosa-Pineda, 2004); and (3) shortly after endocrine specification, *Ngn3* co-localizes with *Pax4*, which suggests that the latter may be one of the targets of the former (Wang et al., 2004). Recent evidence indicates that *Pax4* and *Arx* are mutually repressed, and that the balance between the two determines α- (*Arx*) or β-cell (*Pax4*) specification from *Ngn3+* progenitors (Collombat et al., 2003, 2005).

ISLET NEOGENESIS FROM ES CELLS

Ideally, the "education" of human ES (huES) cells along the β-cell lineage would require the exact recapitulation of the differentiation steps described earlier. If we could identify the instructive extracellular signals that naturally drive this process, such signals could then be added in the proper sequence to the culture medium in the hope that the cells would respond accordingly (Fig. 23.1). However, our understanding of the fine regulation of extracellular signaling is still somewhat limited at the present time. In fact, the combined action of signals such as *FGF*, *Nodal*, *Hedgehog*, *Notch*, *BMP/TGF-β*, and *Wnt* is responsible for the patterning and development of most organs. During development, cells respond differentially to environmental cues, depending on their exact location, their interaction with other developing tissues, and time. Fine gradients of *Nodal* (for endoderm/gut endothelium specification), *FGF*, and *Shh* (for pancreatic differentiation), as well as random cell-to-cell interactions in the *Notch* pathway, are examples of the complex differentiation mechanisms that we are just starting to unravel.

Given these limitations, it is not surprising that most attempts at generating β-cells from ES cells have been unsuccessful so far. The observation by Assady and colleagues that spontaneous *in vitro* differentiation of huES cells resulted in the scattered appearance of insulin-producing cells (Assady et al., 2001) merely confirmed the well-known fact that these cells have unlimited differentiation potential. Protocols for the efficient differentiation of β-cells were still necessary, and several groups were set up to develop them. Lumelsky and colleagues, for instance, described a five-step method to generate islet-like cells from murine ES cells (Lumelsky et al., 2001) based on the derivation of cells positive for the intermediate filament protein *Nestin*, a known marker of neuroectodermal and mesodermal fates. Further analyses on these cells demonstrated that they were not true pancreatic endocrine cells, but rather neuroectodermal derivatives that absorbed insulin from the culture medium (Hansson et al., 2004). Further refinements on this method have led to somewhat improved results, although the amount of insulin expressed by these cells is still quite reduced compared to that of mature β-cells (Fujikawa et al., 2005; Baharvand et al., 2006). Using a genetic engineering approach, Soria and colleagues (2000) generated murine ES cell lines where a selectable marker (neomycin, which confers resistance to the drug G418) was placed under the control of the insulin promoter. Thus, when allowed to spontaneously differentiate, G418 selection yielded insulin-producing clones. Although elegant, this method requires the introduction of foreign genes. Also, it must be taken into account that insulin expression is not a very stringent criterion for the selection of β-cells, as many other tissues do express it. Indeed, the same authors later confirmed the ectodermal identity of some of the selected clones (Roche et al., 2005).

The most exciting developments in the field of ES cell differentiation have been the result of a seemingly less ambitious approach. Instead of attempting the direct differentiation of ES cells into insulin-producing cells, several groups have focused on the key first step of the process, namely endoderm specification. The difficulty of this enterprise is highlighted by the fact that standard culture conditions strongly favor ectoderm and mesoderm over endoderm specification (hence the proliferation of ectoderm-based differentiation protocols). Also, early attempts to generate endoderm could not direct ES cells specifically towards definitive, rather than visceral, endoderm. Kubo and colleagues were the first to report the generation of definitive endoderm from murine ES cells, albeit at a low frequency (Kubo et al., 2004). Far more striking results were successively described by Tada and colleagues (2005), also in mouse ES cells, and d'Amour and collaborators in huES cells (d'Amour et al., 2005). The latter was based on the addition of high concentrations of *Activin A*, a *TGF-β*-related agonist of *Nodal*, in low-serum conditions.

Since endoderm specification had been widely regarded as the main obstacle towards pancreatic differentiation, an accelerated progress of these lines of research was to be expected. The first solid report on β-cell differentiation from human ES cells was published shortly after the seminal paper on definitive endoderm specification, by the same group (d'Amour et al., 2006). The basis of this protocol was the sequential addition of factors known to be involved in the progression of pancreatic development. The resulting insulin-producing cells were similar to true β-cells, with the significant exception that their glucose-regulated insulin secretion mechanisms appeared to be severely impaired both *in vitro* and *in vivo*. Also, the efficiency of the method was very low, with only 5–7% of the total population expressing insulin at detectable levels. Subsequent modifications of the approach involved the transplantation of ES cell-derived pancreatic progenitors under the kidney capsule of recipient mice, with the idea that a natural microenvironment might be more permissive for the functional maturation of these cells. Such a strategy turned out to be successful, as evidenced by the observation that mice treated this way became resistant to streptozotocin-induced diabetes (human β-cells, unlike their murine counterparts, are naturally resistant to this chemical). However, a caveat of this experimental procedure was that up to 20% of the transplanted mice developed teratogenic lesions due to the carryover of cells that were not fully differentiated (Kroon et al., 2008). It is anticipated that progressive advances in the efficiency and safety of this protocol may lead to clinical trials in the not too distant future.

A more recent development is now widely recognized as a potential game-changer in the field of regenerative medicine. We refer to the development of nuclear reprogramming techniques for the generation of induced pluripotent stem (iPS) cells. Two back-to-back reports in 2007 (Takahashi et al., 2007; Yu et al., 2007) described a simple method to dedifferentiate fully mature somatic cells into iPS cells, which were virtually indistinguishable from ES cells. The method entailed the simultaneous transduction of four key master regulators of pluripotency by means of retroviral vectors. Successfully reprogrammed cells could be selected from the rest of the population based on growth and morphological criteria. Subsequent refinements in the technique have resulted in a reduction in the number of reprogramming factors, to the point that now Oct3/4 is considered the only indispensable one (Eminli et al., 2008). Transducible proteins (Kim et al., 2009; Zhou, et al., 2009), episomal constructs (Yu et al., 2009), and even small molecules with reprogramming properties can now be used in lieu of viral vectors (Shi et al., 2008), thus effectively improving the clinical prospects of these cells. The advent of this technology has opened the door to personalized medicine strategies in which a skin biopsy of the patient would theoretically suffice to obtain a genetically matched line of embryonic-like cells with the potential to become β-cells (Park et al., 2008; Tateishi et al., 2008; Maehr et al., 2009). The implications of such possibilities are discussed at the end of this chapter.

ISLET NEOGENESIS FROM ADULT STEM CELLS

The ability of adult pancreatic islet cells to retain regenerative potential during adulthood has been recognized. Several experimental models such as partial pancreatectomy (Bonner-Weir et al., 1983; Lee et al., 1989b; Peshavaria et al., 2006), cellophane wrapping of the pancreas (Rafaeloff et al., 1992; Rosenberg et al., 1983, 1989), duct ligation (Rafaeloff et al., 1992; Wang et al., 1995; Peters et al., 2005; Bonner-Weir et al., 2008; Xu et al., 2008), or treatment with the β-cell toxin streptozotocin (Guz et al., 2001), as well as physiological conditions such as pregnancy (Brelje and Sorenson, 1991; Nielsen et al., 1992; Parsons et al., 1992; Brelje et al., 1993; Sorenson and Brelje, 1997; Bernard-Kargar and Ktorza, 2001; Amaral et al., 2004; Johansson et al., 2006; Butler et al., 2007; Zahr et al., 2007) and perhaps long-standing T1DM (Meier et al., 2005), confirm that insulin-producing cells can regenerate in adult life. However, the quest for endocrine pancreatic stem cells has been an elusive one (Fig. 23.2). Numerous observations suggest that these cells may reside in the ductal epithelium. Aside from countless microphotographic snapshots showing insulin-positive cells that appear to sprout from the pancreatic ducts, cultured ductal cells respond to various stimuli *in vitro* by expressing several β-cell markers and even secreting low levels of insulin (Bonner-Weir et al., 1993, 2000, 2004; Bonner-Weir and Weir, 2005). Other groups have identified Nestin-positive cells within the pancreas with a remarkable ability to expand, although their ability to emulate β-cells upon differentiation was less impressive (Peters et al., 2005; Zulewski et al., 2001). More recently, Gershengorn and colleagues demonstrated that adult islets can undergo a process similar to a reversible epithelial-to-mesenchymal transition *in vitro* (Gershengorn et al., 2004). Upon "dedifferentiation," these cells could be expanded by a factor of 10^{12}, which is well within the realm of clinical applicability. However, when "redifferentiated," these putative β-cells expressed a mere 0.02% of the amount of insulin found in mature islets. A variation on this protocol resulted in enhanced insulin production (Ouziel-Yahalom et al., 2006), but the ability of these cells to proliferate was much more modest. Finally, it has been proposed that acinar tissue may also contain endocrine stem cells (Hao et al., 2006). In this case, trans-differentiation was almost negligible, and no effort was made at characterizing either the neogenic insulin-positive cells or their putative progenitors.

In short, thus far nobody has been able to present conclusive evidence that adult stem cells can generate genuine β-cells *in vitro*. Most of these cellular by-products are, at best, oddities that coexpress markers found in many diverse cell types. Concerns that these cells may just be culture artifacts are justified, and were further fueled when Dor and collaborators (2004)

413

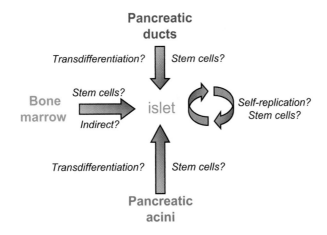

FIGURE 23.2

Islet regeneration during adulthood may occur through several mechanisms. Yet-unidentified endocrine stem cells within the islet may be responsible for β-cell turnover, although data obtained in a mouse model suggest that islet self-maintenance is preferentially due to replication of existing β-cells. Other investigators hypothesize that islets can be regenerated from ductal or acinar tissue, although it is not clear yet whether this phenomenon would be mediated by putative stem cells or by transdifferentiation. Finally, the BM has also been proposed as a reservoir of β-cell progenitors. Recent evidence, however, suggests that the regenerative capacity of migrating BM cells might rather be due to their *in situ* differentiation into supporting endothelial cell types.

suggested that adult β-cells regenerate by replication rather than differentiation. Lineage-tracing experiments conducted in rodents convincingly demonstrated that the regeneration and normal turnover of islets occurs preferentially by division of existing β-cells. This report did not rule out the possibility that stem cells may still exist in the pancreas, but their importance was suddenly − and dramatically − reduced.

Although the burden of proof is now on those who defend the existence and biological significance of pancreatic stem cells, the jury is still out. For instance, it has been shown that one particular model of adult pancreatic regeneration, namely partial duct ligation, is able to induce the appearance of Ngn3+ cells with the ability to replenish β-cell mass (Xu et al., 2008). Another consideration is that human β-cells replicate at a much lower rate than their mouse counterparts. From this perspective, it is conceivable that the adult human pancreas may have evolved different mechanisms for normal β-cell turnover and damage-induced regeneration. The impossibility of conducting lineage-tracing experiments in humans will keep this controversy alive for years to come.

TRANSDIFFERENTIATION

Adult, differentiated cells from specific tissues can turn into completely different cell types in certain conditions. This phenomenon has been termed "transdifferentiation." We will examine here two cell substrates (namely, bone marrow and liver) that have shown some promise at transdifferentiating into pancreatic cell types.

Stem cells derived from the bone marrow (BM) have been associated with numerous examples of tissue repair and regeneration *in vivo*. It has been documented that transplanted BM cells can migrate from their niche to various tissues in response to injury (Ferrari et al., 1998; Lagasse et al., 2000; Wu et al., 2007). In some cases, this migration was accompanied by a significant regeneration of the damaged tissue, which led to the hypothesis that some cell types within the BM may have either ES cell-like properties or the ability to transdifferentiate. However, as was confirmed in a model of liver disease (Grompe, 2003, 2005; Willenbring et al., 2004), the regenerative effect can also be due to the fusion of the BM cells with cells of the target tissue. An early study showed that up to 3% of islet β-cells were of donor origin 1 month after BM cell transplantation, without evidence of cell fusion (Ianus et al., 2003). The conclusions of this report, however, were subsequently contested by Lechner and colleagues, (Lechner and Habener, 2003; Lechner et al., 2004), who could not find any significant contribution of the BM to islets either in healthy mice or in models of pancreatic injury. Furthermore, Kang and collaborators reported that, while BM cell transplantation was enough to prevent diabetes onset in NOD mice, there was little or no involvement of the BM cells in islet cell regeneration once the disease was overt (Kang et al., 2005). Still yet another report presented evidence that donor BM cells *do* promote islet regeneration in a mouse model of diabetes (Mathews et al., 2004). Interestingly, the authors of this study did not find any evidence of transdifferentiation of BM cells into β-cells: the beneficial effect was rather due to the recruitment of donor-derived endothelial cells to the injured islets, where they induced their regeneration.

There is also a wealth of observations indicating that liver and pancreas are especially susceptible to interconversion. Many invertebrates have a single organ that comprises both hepatic and pancreatic functions, which suggests that the separation of these two organs is a relatively late evolutionary event. Indeed, both originate from common endodermal progenitors in the early foregut of vertebrate embryos (Zaret, 1998; Jung et al., 1999; Deutsch et al., 2001; Lemaigre and Zaret, 2004; Tremblay and Zaret, 2005). In general, hepatocytes and β-cells share not only many developmental features but also similar molecular machinery for glucose sensing and secretion (Kim and Ahn, 2004; Nordlie et al., 1999). Many studies confirm that interconversion of liver and pancreas occurs under a variety of experimental conditions, as well as in certain diseases (Rao et al., 1986, 1988; Lee et al., 1989a; Wolf et al., 1990; Rao and Reddy, 1995; Shen et al., 2000).

Based on the above evidence, Ferber and colleagues (2000) set out to demonstrate that ectopic expression of *Pdx1* in liver cells could induce transdifferentiation into pancreatic cell types. Using an adenoviral vector, a *Pdx1* cassette was delivered to the livers of recipient mice, where normally silent, β-cell-specific genes were activated. Subsequent experiments indicated that endogenous Pdx1 expression persisted beyond the time the adenoviral cassette was expected to be active. This was strongly suggestive that the exogenous gene had primed endogenous genetic networks that made the transdifferentiation process permanent (Ber et al., 2003).

Even more conclusive results were later reported by Slack and collaborators, who showed that large portions of the liver could be completely transdifferentiated into pancreas in transgenic frogs where a *Pdx1-VP16* fusion cassette was expressed under the control of the liver-specific promoter *TTR* (Horb et al., 2003; Li et al., 2005). The rationale for the use of *VP16*, a potent transcriptional transactivator from the herpes simplex virus (Sadowski et al., 1988; Triezenberg et al., 1988), is that non-pancreatic cells may lack the appropriate molecular partners for *Pdx1* to exert its biological function. Indeed, no transdifferentiation was observed when *Pdx1*, without *VP16*, was used. This observation suggests that *Pdx1* is necessary but not sufficient to promote true pancreatic differentiation from the liver.

Other noteworthy reprogramming efforts include the recent discovery that a triad of factors, namely Pdx1, Ngn3, and MafA, can transdifferentiate adult pancreatic exocrine tissue along the β-cell lineage *in vivo*. In this report, Zhou et al. (2008) injected the pancreatic parenchyma of recipient mice with a combination of adenoviruses harboring the above transcription factors under the control of constitutive promoters. New insulin-producing cells could be detected within the exocrine tissue within 3 days of the injection, and their number kept expanding for up to 3 months. When diabetic mice received the treatment, they responded with a significant and persistent reduction in blood sugar levels, although full glycemic control was never attained. This may be due to the fact that new β-cells failed to aggregate in clusters, remaining instead dispersed throughout the acinar parenchyma (Konstantinova et al., 2007).

WHAT THE FUTURE MAY HOLD

Steady progress in the field of β-cell replacement has made islet cell transplantation a therapeutic reality for patients with the most severe forms of diabetes. The benefits of this approach both in terms of metabolic control and quality of life after islet transplantation support the advantage of restoring β-cell function, compared to exogenous insulin treatment. The pace of stem cell research over the last decade has also been significant. Diseases thus far considered incurable now seem within the reach of our ever increasing therapeutic arsenal. Stem cells, be they of embryonic or adult origin, may provide in the future an unlimited supply of insulin-producing cells for treatment of diabetes. It is important, however, not to lose perspective on the many challenges ahead. First, no protocol for the *efficient* derivation of fully competent β-cells from stem cells has been described yet. In our opinion, the most promising approaches are based on the generation of true endoderm from ES or iPS cells, but this would be just the first of several steps. Terminal differentiation of β-cells may require further advances in our ability to mimic their unique biological niche, which is known to be highly oxygenated through extensive vascularization. Another important consideration is safety. While islet transplantation is generally considered a safe procedure, ES cell-based approaches may require additional precautions to prevent the formation of tumors by carryover of undifferentiated cells. The same considerations may apply to protocols aiming at *in vivo* β-cell regeneration in the native pancreas, since stimulation of β-cell proliferation may be associated with increased risk of hyperplasia or neoplastic transformation (e.g. nesidioblastosis or insulinomas).

Solving the problem of supply is just one component of the puzzle. At this point there is a widespread consensus that a future cure is likely to combine a stem cell component to replenish the β-cell mass and targeted interventions to induce immune tolerance to the transplanted cells. Regardless of the source of these cells, a major decision to be made is whether the new β-cells should be immunologically matched (autologous; obtained from the same individual) or mismatched (allogeneic; obtained from related or unrelated donors) to the patient. Unlike for T2DM, where the use of autologous cells may be suited to the restoration of a functional β-cell mass, the answer is not obvious in the case of T1DM, where destruction of β-cells is mediated by strong autoimmune responses that are known to remain active when new β-cells are transplanted. That was the case when twin-to-twin pancreatic transplantation resulted in a fast re-enactment of T1DM shortly after the initial reversal (Sutherland et al., 1984, 1989; Sibley et al., 1985; Santamaria et al., 1992). This is the reason why in the current clinical setting of islet and pancreas transplantation some degree of histocompatibility mismatching is not considered a disadvantage. Despite genetic disparity between donor and recipient, recurrence of autoimmunity after allogeneic islet or pancreas transplantation has been described (Sibley and Sutherland, 1987; Bosi et al., 2001; WHITG, 2006; Huurman et al., 2008; Monti et al., 2008; Vendrame et al., 2010), and the magnitude of this phenomenon remains to be fully determined (~5%?) (Vendrame et al., 2010).

It could be argued that the choice of autologous cells could have the advantage of taking allorejection out of the picture, leaving autoimmunity as the only remaining challenge to be addressed. More targeted interventions to restore self-tolerance could therefore lead to a successful replacement strategy. On the other hand, stopping the onset of the disease in its tracks could be sufficient to "cure" the disease if β-cells could regenerate from native precursors or if a residual β-cell mass could effectively replicate. The question of whether autologous is better than allogeneic in the context of future stem cell therapies for T1DM needs to be carefully considered. However, it is possible that novel technologies of immunoisolation, such as those utilizing new generation nano-scale conformal coating techniques, could be equally effective in allowing for successful transplantation of either autologous or allogeneic islets. This would render the entire debate a moot point, limiting the decision-making process to which source of insulin-producing cells will allow the treatment of the highest number of subjects in the safest and most cost-effective way.

CONCLUDING REMARKS

Stem cell research bears an invaluable potential for treatment of diabetes and many other conditions. The enormous potential impact of stem cell-derived therapies in future medical practices warrants a renewed investment of resources in this field of investigation in the context of academic institutions, under strict ethical and regulatory oversight. In this context, support of stem cell research by government agencies would allow for a faster, regulated, and safer advancement of this field. Notwithstanding the challenges, it appears that the prospect of defeating insulin-requiring diabetes is now within reach, and that successful therapeutic strategies can be developed as a result of a multidisciplinary, integrated approach.

Acknowledgments

This work was supported by the agencies listed below. National Institutes of Health: National Center for Research Resources − Islet Cell Resources (5U42RR016603-08S1; M01RR16587); National Institute of Diabetes and Digestive and Kidney Diseases (1DP2DK083096-01; 5R01DK059993-04; 1R21DK076098-01; 1U01 DK70460-02; 5R01DK25802-24; 5R01DK56953-05; 5R01DK55347; 5R01DK056953; R01DK025802; 1R21HD060195-01; 1R43DK083832); National Institute of Biomedical Imaging and Bioengineering (1R01 EB008009-02); Juvenile Diabetes Research Foundation International (4-2000-946; 4-2000-947; 4-2004-361; 4-2008-811; 17-2010-5); the Helmsley Charitable Trust; American Diabetes Association; State of Florida; a contract for support of this research, sponsored by Congressman Bill Young and funded by a special congressional out of the Navy Bureau of Medicine and Surgery, is currently managed by the Naval Health Research Center, San Diego, CA; Converge Biotech, Inc.; Biorep Technologies, Inc.; and the continuous support of the Diabetes Research Institute Foundation (www.diabetesresearch.org).

References

Alejandro, R., Barton, F. B., Hering, B. J., & Wease, S. (2008). 2008 Update from the Collaborative Islet Transplant Registry. *Transplantation, 86,* 1783–1788.

Alejandro, R., Lehmann, R., Ricordi, C., Kenyon, N. S., Angelico, M. C., Burke, G., et al. (1997). Long-term function (6 years) of islet allografts in type 1 diabetes. *Diabetes, 46,* 1983–1989.

Alejandro, R., & Mintz, D. H. (1988). Experimental and clinical methods of islet transplantation. In R. van Schilfgaarde, & M. A. Hardy (Eds.), *Transplantation of the Endocrine Pancreas in Diabetes Mellitus* (pp. 217–223). Amsterdam: Elsevier Science.

Alejandro, R., Strasser, S., Zucker, P. F., & Mintz, D. H. (1990). Isolation of pancreatic islets from dogs. Semi-automated purification on albumin gradients. *Transplantation, 50,* 207–210.

Amaral, M. E., Cunha, D. A., Anhe, G. F., Ueno, M., Carneiro, E. M., Velloso, L. A., et al. (2004). Participation of prolactin receptors and phosphatidylinositol 3-kinase and MAP kinase pathways in the increase in pancreatic islet mass and sensitivity to glucose during pregnancy. *J. Endocrinol., 183,* 469–476.

Andres, A., Toso, C., Morel, P., Demuylder-Mischler, S., Bosco, D., Baertschiger, R., et al. (2005). Impairment of renal function after islet transplant alone or islet-after-kidney transplantation using a sirolimus/tacrolimus-based immunosuppressive regimen. *Transpl. Int., 18,* 1226–1230.

Angelico, M. C., Alejandro, R., Nery, J., Webb, M., Bottino, R., Kong, S. S., et al. (1999). Transplantation of islets of Langerhans in patients with insulin-requiring diabetes mellitus undergoing orthotopic liver transplantation — the Miami experience. *J. Mol. Med., 77,* 144–147.

Apelqvist, A., Ahlgren, U., & Edlund, H. (1997). Sonic hedgehog directs specialised mesoderm differentiation in the intestine and pancreas. *Curr. Biol., 7,* 801–804.

Apelqvist, A., Li, H., Sommer, L., Beatus, P., Anderson, D. J., Honjo, T., et al. (1999). Notch signalling controls pancreatic cell differentiation. *Nature, 400,* 877–881.

Assady, S., Maor, G., Amit, M., Itskovitz-Eldor, J., Skorecki, K. L., & Tzukerman, M. (2001). Insulin production by human embryonic stem cells. *Diabetes, 50,* 1691–1697.

Baharvand, H., Jafary, H., Massumi, M., & Ashtiani, S. K. (2006). Generation of insulin-secreting cells from human embryonic stem cells. *Dev. Growth Differ., 48,* 323–332.

Baidal, D. A., Froud, T., Ferreira, J. V., Khan, A., Alejandro, R., & Ricordi, C. (2003). The bag method for islet cell infusion. *Cell Transplant., 12,* 809–813.

Barbaro, B., Salehi, P., Wang, Y., Qi, M., Gangemi, A., Kuechle, J., et al. (2007). Improved human pancreatic islet purification with the refined UIC-UB density gradient. *Transplantation, 84,* 1200–1203.

Barshes, N. R., Vanatta, J. M., Mote, A., Lee, T. C., Schock, A. P., Balkrishnan, R., et al. (2005). Health-related quality of life after pancreatic islet transplantation: a longitudinal study. *Transplantation, 79,* 1727–1730.

Bellin, M. D., Kandaswamy, R., Parkey, J., Zhang, H. J., Liu, B., Ihm, S. H., et al. (2008). Prolonged insulin independence after islet allotransplants in recipients with type 1 diabetes. *Am. J. Transplant., 8,* 2463–2470.

Ber, I., Shternhall, K., Perl, S., Ohanuna, Z., Goldberg, I., Barshack, I., et al. (2003). Functional, persistent, and extended liver to pancreas transdifferentiation. *J. Biol. Chem., 278,* 31950–31957.

Bernard-Kargar, C., & Ktorza, A. (2001). Endocrine pancreas plasticity under physiological and pathological conditions. *Diabetes, 50*(Suppl. 1), S30–S35.

Berney, T., Ferrari-Lacraz, S., Buhler, L., Oberholzer, J., Marangon, N., Philippe, J., et al. (2009). Long-term insulin-independence after allogeneic islet transplantation for type 1 diabetes: over the 10-year mark. *Am. J. Transplant., 9,* 419–423.

Bocca, N., Buchwald, P., Molano, R. D., Marzorati, S., Stabler, C., Pileggi, A., et al. (2009). Biohybrid devices and local immunomodulation as opposed to systemic immunosuppression. In J. P. Hallé, P. de Vos, & L. Rosenberg (Eds.), *The Bioartificial Pancreas and Other Biohybrid Therapies* (pp. 309–328). Kerala, India: Research Signpost.

Bonner-Weir, S., Baxter, L. A., Schuppin, G. T., & Smith, F. E. (1993). A second pathway for regeneration of adult exocrine and endocrine pancreas. A possible recapitulation of embryonic development. *Diabetes, 42,* 1715–1720.

Bonner-Weir, S., Inada, A., Yatoh, S., Li, W. C., Aye, T., Toschi, E., et al. (2008). Transdifferentiation of pancreatic ductal cells to endocrine beta-cells. *Biochem. Soc. Trans., 36,* 353–356.

Bonner-Weir, S., Taneja, M., Weir, G. C., Tatarkiewicz, K., Song, K. H., Sharma, A., et al. (2000). In vitro cultivation of human islets from expanded ductal tissue. *Proc. Natl. Acad. Sci. U.S.A., 97,* 7999–8004.

Bonner-Weir, S., Toschi, E., Inada, A., Reitz, P., Fonseca, S. Y., Aye, T., et al. (2004). The pancreatic ductal epithelium serves as a potential pool of progenitor cells. *Pediatr. Diabetes, 5*(Suppl. 2), 16–22.

Bonner-Weir, S., Trent, D. F., & Weir, G. C. (1983). Partial pancreatectomy in the rat and subsequent defect in glucose-induced insulin release. *J. Clin. Invest., 71,* 1544–1553.

Bonner-Weir, S., & Weir, G. C. (2005). New sources of pancreatic beta-cells. *Nature Biotech., 23,* 857–861.

417

Bosco, D., Armanet, M., Morel, P., Niclauss, N., Sgroi, A., Muller, Y. D., et al. (2010). Unique arrangement of alpha- and beta-cells in human islets of Langerhans. *Diabetes, 59*, 1202–1210.

Bosi, E., Braghi, S., Maffi, P., Scirpoli, M., Bertuzzi, F., Pozza, G., et al. (2001). Autoantibody response to islet transplantation in type 1 diabetes. *Diabetes, 50*, 2464–2471.

Boyer, L. A., Lee, T. I., Cole, M. F., Johnstone, S. E., Levine, S. S., Zucker, J. P., et al. (2005). Core transcriptional regulatory circuitry in human embryonic stem cells. *Cell, 122*, 947–956.

Brelje, T. C., Scharp, D. W., Lacy, P. E., Ogren, L., Talamantes, F., Robertson, M., et al. (1993). Effect of homologous placental lactogens, prolactins, and growth hormones on islet B-cell division and insulin secretion in rat, mouse, and human islets: implication for placental lactogen regulation of islet function during pregnancy. *Endocrinology, 132*, 879–887.

Brelje, T. C., & Sorenson, R. L. (1991). Role of prolactin versus growth hormone on islet B-cell proliferation in vitro: implications for pregnancy. *Endocrinology, 128*, 45–57.

Brissova, M., Fowler, M. J., Nicholson, W. E., Chu, A., Hirshberg, B., Harlan, D. M., et al. (2005). Assessment of human pancreatic islet architecture and composition by laser scanning confocal microscopy. *J. Histochem. Cytochem., 53*, 1087–1097.

Brunicardi, F. C., Atiya, A., Stock, P., Kenmochi, T., Une, S., Benhamou, P. Y., et al. (1995). Clinical islet transplantation experience of the University of California Islet Transplant Consortium. *Surgery, 118*, 967–971, discussion 967–971.

Butler, P. C., Meier, J. J., Butler, A. E., & Bhushan, A. (2007). The replication of beta cells in normal physiology, in disease and for therapy. *Nat. Clin. Pract. Endocrinol. Metab., 3*, 758–768.

Cabrera, O., Berman, D. M., Kenyon, N. S., Ricordi, C., Berggren, P. O., & Caicedo, A. (2006). The unique cytoarchitecture of human pancreatic islets has implications for islet cell function. *Proc. Natl. Acad. Sci. U.S.A., 103*, 2334–2339.

Cardani, R., Pileggi, A., Ricordi, C., Gomez, C., Baidal, D. A., Ponte, G. G., et al. (2007). Allosensitization of islet allograft recipients. *Transplantation, 84*, 1413–1427.

Carroll, P. B., Rilo, H. L., Alejandro, R., Zeng, Y., Khan, R., Fontes, P., et al. (1995). Long-term (>3-year) insulin independence in a patient with pancreatic islet cell transplantation following upper abdominal exenteration and liver replacement for fibrolamellar hepatocellular carcinoma. *Transplantation, 59*, 875–879.

Collombat, P., Hecksher-Sorensen, J., Broccoli, V., Krull, J., Ponte, I., Mundiger, T., et al. (2005). The simultaneous loss of Arx and Pax4 genes promotes a somatostatin-producing cell fate specification at the expense of the alpha- and beta-cell lineages in the mouse endocrine pancreas. *Development, 132*, 2969–2980.

Collombat, P., Mansouri, A., Hecksher-Sorensen, J., Serup, P., Krull, J., Gradwohl, G., et al. (2003). Opposing actions of Arx and Pax4 in endocrine pancreas development. *Genes Dev., 17*, 2591–2603.

Cure, P., Froud, T., Leitao, C. B., Pileggi, A., Tharavanij, T., Bernetti, K., et al. (2008a). Late Epstein Barr virus reactivation in islet after kidney transplantation. *Transplantation, 86*, 1324–1325.

Cure, P., Pileggi, A., Faradji, R. N., Baidal, D. A., Froud, T., Selvaggi, G., et al. (2006). Cytomegalovirus infection in a recipient of solitary allogeneic islets. *Am. J. Transplant., 6*, 1089–1090.

Cure, P., Pileggi, A., Froud, T., Messinger, S., Faradji, R. N., Baidal, D. A., et al. (2008b). Improved metabolic control and quality of life in seven patients with type 1 diabetes following islet after kidney transplantation. *Transplantation, 85*, 801–812.

Cure, P., Pileggi, A., Froud, T., Norris, P. M., Baidal, D. A., Cornejo, A., et al. (2004). Alterations of the female reproductive system in recipients of islet grafts. *Transplantation, 78*, 1576–1581.

d'Amour, K. A., Agulnick, A. D., Eliazer, S., Kelly, O. G., Kroon, E., & Baetge, E. E. (2005). Efficient differentiation of human embryonic stem cells to definitive endoderm. *Nat. Biotechnol., 23*, 1534–1541.

d'Amour, K. A., Bang, A. G., Eliazer, S., Kelly, O. G., Agulnick, A. D., Smart, N. G., et al. (2006). Production of pancreatic hormone — expressing endocrine cells from human embryonic stem cells. *Nat. Biotechnol, 24*, 1392–1401.

DCCTRG. (1993). The effect of intensive treatment of diabetes on the development and progression of long-term complications in insulin-dependent diabetes mellitus. The Diabetes Control and Complications Trial Research Group. *N. Engl. J. Med., 329*, 977–986.

de Santa Barbara, P., van den Brink, G. R., & Roberts, D. J. (2003). Development and differentiation of the intestinal epithelium. *Cell Mol. Life Sci., 60*, 1322–1332.

del Carro, U., Fiorina, P., Amadio, S., de Toni Franceschini, L., Petrelli, A., Menini, S., et al. (2007). Evaluation of polyneuropathy markers in type 1 diabetic kidney transplant patients and effects of islet transplantation: neurophysiological and skin biopsy longitudinal analysis. *Diabetes Care, 30*, 3063–3069.

Desai, N. M., Goss, J. A., Deng, S., Wolf, B. A., Markmann, E., Palanjian, M., et al. (2003). Elevated portal vein drug levels of sirolimus and tacrolimus in islet transplant recipients: local immunosuppression or islet toxicity? *Transplantation, 76*, 1623–1625.

Deutsch, G., Jung, J., Zheng, M., Lora, J., & Zaret, K. S. (2001). A bipotential precursor population for pancreas and liver within the embryonic endoderm. *Development, 128*, 871–881.

Domínguez-Bendala, J., Klein, D., Ribeiro, M., Ricordi, C., Inverardi, L., Pastori, R., et al. (2005). TAT-mediated neurogenin 3 protein transduction stimulates pancreatic endocrine differentiation in vitro. *Diabetes, 54*, 720–726.

Dor, Y., Brown, J., Martinez, O. I., & Melton, D. A. (2004). Adult pancreatic beta-cells are formed by self-duplication rather than stem-cell differentiation. *Nature, 429*, 41–46.

Edlund, H. (2002). Pancreatic organogenesis – developmental mechanisms and implications for therapy. *Nat. Rev. Genet., 3*, 524–532.

Eminli, S., Utikal, J., Arnold, K., Jaenisch, R., & Hochedlinger, K. (2008). Reprogramming of neural progenitor cells into induced pluripotent stem cells in the absence of exogenous Sox2 expression. *Stem Cells, 26*, 2467–2474.

Faradji, R. N., Froud, T., Messinger, S., Monroy, K., Pileggi, A., Mineo, D., et al. (2009). Long-term metabolic and hormonal effects of exenatide on islet transplant recipients with allograft dysfunction. *Cell Transplant., 18*, 1247–1259.

Faradji, R. N., Tharavanij, T., Messinger, S., Froud, T., Pileggi, A., Monroy, K., et al. (2008). Long-term insulin independence and improvement in insulin secretion after supplemental islet infusion under exenatide and etanercept. *Transplantation, 86*, 1658–1665.

Ferber, S., Halkin, A., Cohen, H., Ber, I., Einav, Y., Goldberg, I., et al. (2000). Pancreatic and duodenal homeobox gene 1 induces expression of insulin genes in liver and ameliorates streptozotocin-induced hyperglycemia. *Nat. Med., 6*, 568–572.

Fernandez, L. A., Lehmann, R., Luzi, L., Battezzati, A., Angelico, M. C., Ricordi, C., et al. (1999). The effects of maintenance doses of FK506 versus cyclosporin A on glucose and lipid metabolism after orthotopic liver transplantation. *Transplantation, 68*, 1532–1541.

Ferrari, G., Cusella-de Angelis, G., Coletta, M., Paolucci, E., Stornaiuolo, A., Cossu, G., et al. (1998). Muscle regeneration by bone marrow-derived myogenic progenitors. *Science, 279*, 1528–1530.

Fiorina, P., Folli, F., Maffi, P., Placidi, C., Venturini, M., Finzi, G., et al. (2003a). Islet transplantation improves vascular diabetic complications in patients with diabetes who underwent kidney transplantation: a comparison between kidney-pancreas and kidney-alone transplantation. *Transplantation, 75*, 1296–1301.

Fiorina, P., Folli, F., Zerbini, G., Maffi, P., Gremizzi, C., di Carlo, V., et al. (2003b). Islet transplantation is associated with improvement of renal function among uremic patients with type I diabetes mellitus and kidney transplants. *J. Am. Soc. Nephrol., 14*, 2150–2158.

Fiorina, P., Gremizzi, C., Maffi, P., Caldara, R., Tavano, D., Monti, L., et al. (2005a). Islet transplantation is associated with an improvement of cardiovascular function in type 1 diabetic kidney transplant patients. *Diabetes Care, 28*, 1358–1365.

Fiorina, P., Venturini, M., Folli, F., Losio, C., Maffi, P., Placidi, C., et al. (2005b). Natural history of kidney graft survival, hypertrophy, and vascular function in end-stage renal disease type 1 diabetic kidney-transplanted patients: beneficial impact of pancreas and successful islet cotransplantation. *Diabetes Care, 28*, 1303–1310.

Fraker, C., Timmins, M. R., Guarino, R. D., Haaland, P. D., Ichii, H., Molano, D., et al. (2006). The use of the BD oxygen biosensor system to assess isolated human islets of Langerhans: oxygen consumption as a potential measure of islet potency. *Cell Transplant., 15*, 745–758.

Fraker, C. A., Alejandro, R., & Ricordi, C. (2002). Use of oxygenated perfluorocarbon toward making every pancreas count. *Transplantation, 74*, 1811–1812.

Frank, A., Deng, S., Huang, X., Velidedeoglu, E., Bae, Y. S., Liu, C., et al. (2004). Transplantation for type I diabetes: comparison of vascularized whole-organ pancreas with isolated pancreatic islets. *Ann. Surg., 240*, 631–643.

Froud, T., Baidal, D. A., Faradji, R., Cure, P., Mineo, D., Selvaggi, G., et al. (2008a). Islet transplantation with alemtuzumab induction and calcineurin-free maintenance immunosuppression results in improved short- and long-term outcomes. *Transplantation, 86*, 1695–1701.

Froud, T., Faradji, R., Pileggi, A., Messinger, S., Baidal, D., Ponte, G., et al. (2008b). The use of exenatide in islet transplant recipients with chronic allograft dysfunction: safety, efficacy, and metabolic effects. *Transplantation, 86*, 36–45.

Froud, T., Ricordi, C., Baidal, D. A., Hafiz, M. M., Ponte, G., Cure, P., et al. (2005). Islet transplantation in type 1 diabetes mellitus using cultured islets and steroid-free immunosuppression: Miami experience. *Am. J. Transplant., 5*, 2037–2046.

Froud, T., Yrizarry, J. M., Alejandro, R., & Ricordi, C. (2004). Use of D-STAT to prevent bleeding following percutaneous transhepatic intraportal islet transplantation. *Cell Transplant., 13*, 55–59.

Fujikawa, T., Oh, S. H., Pi, L., Hatch, H. M., Shupe, T., & Petersen, B. E. (2005). Teratoma formation leads to failure of treatment for type I diabetes using embryonic stem cell-derived insulin-producing cells. *Am. J. Pathol., 166*, 1781–1791.

Fung, M. A., Warnock, G. L., Ao, Z., Keown, P., Meloche, M., Shapiro, R. J., et al. (2007). The effect of medical therapy and islet cell transplantation on diabetic nephropathy: an interim report. *Transplantation, 84*, 17—22.

Gangemi, A., Salehi, P., Hatipoglu, B., Martellotto, J., Barbaro, B., Kuechle, J. B., et al. (2008). Islet transplantation for brittle type 1 diabetes: the UIC protocol. *Am. J. Transplant., 8*, 1250—1261.

Garraway, N. R., Dean, S., Buczkowski, A., Brown, D. R., Scudamore, C. H., Meloche, M., et al. (2009). Islet autotransplantation after distal pancreatectomy for pancreatic trauma. *J. Trauma, 67*, E187—189.

Geiger, M., Ferreira, J., Baidal, D., Froud, T., Hafiz, M., Rothenberg, L., et al. (2003). Use of continuous glucose monitoring system in the evaluation of metabolic control in patients with type I diabetes mellitus after islet cell transplant. *Diabetes, 52*. A94—A94.

Gershengorn, M. C., Hardikar, A. A., Wei, C., Geras-Raaka, E., Marcus-Samuels, B., & Raaka, B. M. (2004). Epithelial-to-mesenchymal transition generates proliferative human islet precursor cells. *Science, 306*, 2261—2264.

Ghofaili, K., Fung, M., Ao, Z., Meloche, M., Shapiro, R., Warnock, G., et al. (2007). Effect of exenatide on beta cell function after islet transplantation in type 1 diabetes. *Transplantation, 83*, 24—28.

Goss, J. A., Goodpastor, S. E., Brunicardi, F. C., Barth, M. H., Soltes, G. D., Garber, A. J., et al. (2004). Development of a human pancreatic islet-transplant program through a collaborative relationship with a remote islet-isolation center. *Transplantation, 77*, 462—466.

Goss, J. A., Schock, A. P., Brunicardi, F. C., Goodpastor, S. E., Garber, A. J., Soltes, G., et al. (2002). Achievement of insulin independence in three consecutive type-1 diabetic patients via pancreatic islet transplantation using islets isolated at a remote islet isolation center. *Transplantation, 74*, 1761—1766.

Goss, J. A., Soltes, G., Goodpastor, S. E., Barth, M., Lam, R., Brunicardi, F. C., et al. (2003). Pancreatic islet transplantation: the radiographic approach. *Transplantation, 76*, 199—203.

Goto, T., Tanioka, Y., Sakai, T., Matsumoto, I., Kakinoki, K., Tanaka, T., et al. (2005). Successful islet transplantation from a single pancreas harvested from a young, low-BMI, non-heart-beating cadaver. *Transplant. Proc., 37*, 3430—3432.

Gradwohl, G., Dierich, A., LeMeur, M., & Guillemot, F. (2000). Neurogenin3 is required for the development of the four endocrine cell lineages of the pancreas. *Proc. Natl. Acad. Sci. U.S.A., 97*, 1607—1611.

Grompe, M. (2003). The role of bone marrow stem cells in liver regeneration. *Semin. Liver Dis., 23*, 363—372.

Grompe, M. (2005). Bone marrow-derived hepatocytes. *Novartis Found. Symp., 265*, 20—27, discussion 28—34, 92—97.

Gruessner, R. W., Sutherland, D. E., & Gruessner, A. C. (2004). Mortality assessment for pancreas transplants. *Am. J. Transplant., 4*, 2018—2026.

Guignard, A. P., Oberholzer, J., Benhamou, P. Y., Touzet, S., Bucher, P., Penfornis, A., et al. (2004). Cost analysis of human islet transplantation for the treatment of type 1 diabetes in the Swiss-French Consortium GRAGIL. *Diabetes Care, 27*, 895—900.

Guz, Y., Nasir, I., & Teitelman, G. (2001). Regeneration of pancreatic beta cells from intra-islet precursor cells in an experimental model of diabetes. *Endocrinology, 142*, 4956—4968.

Hafiz, M. M., Faradji, R. N., Froud, T., Pileggi, A., Baidal, D. A., Cure, P., et al. (2005). Immunosuppression and procedure-related complications in 26 patients with type 1 diabetes mellitus receiving allogeneic islet cell transplantation. *Transplantation, 80*, 1718—1728.

Halle', J.P., de Vos, & Rosenberg, L. (2009). *The Bioartificial Pancreas and other Biohybrid therapies.* Kerala; India: Transworld Research Network.

Hansson, M., Tonning, A., Frandsen, U., Petri, A., Rajagopal, J., Englund, M. C., et al. (2004). Artifactual insulin release from differentiated embryonic stem cells. *Diabetes, 53*, 2603—2609.

Hao, E., Tyrberg, B., Itkin-Ansari, P., Lakey, J. R., Geron, I., Monosov, E. Z., et al. (2006). Beta-cell differentiation from nonendocrine epithelial cells of the adult human pancreas. *Nat. Med., 12*, 310—316.

Hebrok, M., Kim, S. K., & Melton, D. A. (1998). Notochord repression of endodermal Sonic hedgehog permits pancreas development. *Genes Dev., 12*, 1705—1713.

Hebrok, M., Kim, S. K., St Jacques, B., McMahon, A. P., & Melton, D. A. (2000). Regulation of pancreas development by hedgehog signaling. *Development, 127*, 4905—4913.

Hering, B. J., Kandaswamy, R., Ansite, J. D., Eckman, P. M., Nakano, M., Sawada, T., et al. (2005). Single-donor, marginal-dose islet transplantation in patients with type 1 diabetes. *Jama, 293*, 830—835.

Hirshberg, B., Rother, K. I., Digon, B. J., III, Lee, J., Gaglia, J. L., Hines, K., et al. (2003). Benefits and risks of solitary islet transplantation for type 1 diabetes using steroid-sparing immunosuppression: the National Institutes of Health experience. *Diabetes Care, 26*, 3288—3295.

Horb, M. E., Shen, C. N., Tosh, D., & Slack, J. M. (2003). Experimental conversion of liver to pancreas. *Curr. Biol., 13*, 105—115.

Hudson, C., Clements, D., Friday, R. V., Stott, D., & Woodland, H. R. (1997). Xsox17alpha and -beta mediate endoderm formation in Xenopus. *Cell, 91*, 397—405.

Huurman, V. A., Hilbrands, R., Pinkse, G. G., Gillard, P., Duinkerken, G., van de Linde, P., et al. (2008). Cellular islet autoimmunity associates with clinical outcome of islet cell transplantation. *PLoS ONE, 3,* e2435.

Ianus, A., Holz, G. G., Theise, N. D., & Hussain, M. A. (2003). In vivo derivation of glucose-competent pancreatic endocrine cells from bone marrow without evidence of cell fusion. *J. Clin. Invest., 111,* 843–850.

Ichii, H., Inverardi, L., Pileggi, A., Molano, R. D., Cabrera, O., Caicedo, A., et al. (2005a). A novel method for the assessment of cellular composition and beta-cell viability in human islet preparations. *Am. J. Transplant., 5,* 1635–1645.

Ichii, H., Pileggi, A., Molano, R. D., Baidal, D. A., Khan, A., Kuroda, Y., et al. (2005b). Rescue purification maximizes the use of human islet preparations for transplantation. *Am. J. Transplant., 5,* 21–30.

Ichii, H., Sakuma, Y., Pileggi, A., Fraker, C., Alvarez, A., Montelongo, J., et al. (2007). Shipment of human islets for transplantation. *Am. J. Transplant., 7,* 1010–1020.

Ichii, H., Wang, X., Messinger, S., Alvarez, A., Fraker, C., Khan, A., et al. (2006). Improved human islet isolation using nicotinamide. *Am. J. Transplant., 6,* 2060–2068.

Inverardi, L., Linetsky, E., Pileggi, A., Molano, R. D., Serafini, A., Paganelli, G., et al. (2004). Targeted bone marrow radioablation with 153Samarium-lexidronam promotes allogeneic hematopoietic chimerism and donor-specific immunologic hyporesponsiveness. *Transplantation, 77,* 647–655.

Inverardi, L., & Ricordi, C. (2001). Tolerance and pancreatic islet transplantation. *Philos. Trans. R. Soc. Lond. B. Biol. Sci., 356,* 759–765.

Jensen, J., Pedersen, E. E., Galante, P., Hald, J., Heller, R. S., Ishibashi, M., et al. (2000). Control of endodermal endocrine development by Hes-1. *Nat. Genet., 24,* 36–44.

Jiang, Y., Jahagirdar, B. N., Reinhardt, R. L., Schwartz, R. E., Keene, C. D., Ortiz-Gonzalez, X. R., et al. (2002). Pluripotency of mesenchymal stem cells derived from adult marrow. *Nature, 418,* 41–49.

Jindal, R. M., Ricordi, C., & Shrives, C. D. (2010). Autologous islet transportation for severe trauma. *N. Eng. J. Med., 362,* 1550.

Johansson, M., Mattsson, G., Andersson, A., Jansson, L., & Carlsson, P. O. (2006). Islet endothelial cells and pancreatic β-cell proliferation: studies in vitro and during pregnancy in adult rats. *Endocrinology, 147*(5), 2315–2324.

Johnson, J. A., Kotovych, M., Ryan, E. A., & Shapiro, A. M. (2004). Reduced fear of hypoglycemia in successful islet transplantation. *Diabetes Care, 27,* 624–625.

Jonsson, J., Carlsson, L., Edlund, T., & Edlund, H. (1994). Insulin-promoter-factor 1 is required for pancreas development in mice. *Nature, 371,* 606–609.

Jung, J., Zheng, M., Goldfarb, M., & Zaret, K. S. (1999). Initiation of mammalian liver development from endoderm by fibroblast growth factors. *Science, 284,* 1998–2003.

Kanai-Azuma, M., Kanai, Y., Gad, J. M., Tajima, Y., Taya, C., Kurohmaru, M., et al. (2002). Depletion of definitive gut endoderm in Sox17-null mutant mice. *Development, 129,* 2367–2379.

Kang, E. M., Zickler, P. P., Burns, S., Langemeijer, S. M., Brenner, S., Phang, O. A., et al. (2005). Hematopoietic stem cell transplantation prevents diabetes in NOD mice but does not contribute to significant islet cell regeneration once disease is established. *Exp. Hematol., 33,* 699–705.

Kempf, M. C., Andres, A., Morel, P., Benhamou, P. Y., Bayle, F., Kessler, L., et al. (2005). Logistics and transplant coordination activity in the GRAGIL Swiss-French multicenter network of islet transplantation. *Transplantation, 79,* 1200–1205.

Kim, D., Kim, C. H., Moon, J. I., Chung, Y. G., Chang, M. Y., Han, B. S., et al. (2009). Generation of human induced pluripotent stem cells by direct delivery of reprogramming proteins. *Cell Stem Cell, 4,* 472–476.

Kim, H. I., & Ahn, Y. H. (2004). Role of peroxisome proliferator-activated receptor-gamma in the glucose-sensing apparatus of liver and beta-cells. *Diabetes, 53*(Suppl. 1), S60–S65.

Kim, S. K., & Melton, D. A. (1998). Pancreas development is promoted by cyclopamine, a hedgehog signaling inhibitor. *Proc. Natl. Acad. Sci. U.S.A., 95,* 13036–13041.

Kispert, A., & Herrmann, B. G. (1994). Immunohistochemical analysis of the Brachyury protein in wild-type and mutant mouse embryos. *Dev. Biol., 161,* 179–193.

Konstantinova, I., Nikolova, G., Ohara-Imaizumi, M., Meda, P., Kucera, T., Zarbalis, K., et al. (2007). EphA-Ephrin-A-mediated beta cell communication regulates insulin secretion from pancreatic islets. *Cell, 129,* 359–370.

Krieger, N. R., Odorico, J. S., Heisey, D. M., d'Alessandro, A. M., Knechtle, S. J., Pirsch, J. D., et al. (2003). Underutilization of pancreas donors. *Transplantation, 75,* 1271–1276.

Kroon, E., Martinson, L. A., Kadoya, K., Bang, A. G., Kelly, O. G., Eliazer, S., et al. (2008). Pancreatic endoderm derived from human embryonic stem cells generates glucose-responsive insulin-secreting cells in vivo. *Nat. Biotechnol., 26,* 443–452.

Kubo, A., Shinozaki, K., Shannon, J. M., Kouskoff, V., Kennedy, M., Woo, S., et al. (2004). Development of definitive endoderm from embryonic stem cells in culture. *Development, 131,* 1651–1662.

421

Kuroda, Y., Kawamura, T., Suzuki, Y., Fujiwara, H., Yamamoto, K., & Saitoh, Y. (1988). A new, simple method for cold storage of the pancreas using perfluorochemical. *Transplantation, 46,* 457–460.

Lagasse, E., Connors, H., Al-Dhalimy, M., Reitsma, M., Dohse, M., Osborne, L., et al. (2000). Purified hematopoietic stem cells can differentiate into hepatocytes in vivo. *Nat. Med., 6,* 1229–1234.

Lakey, J. R., Warnock, G. L., Shapiro, A. M., Korbutt, G. S., Ao, Z., Kneteman, N. M., et al. (1999). Intraductal collagenase delivery into the human pancreas using syringe loading or controlled perfusion. *Cell Transplant., 8,* 285–292.

Lechner, A., & Habener, J. F. (2003). Bone marrow stem cells find a path to the pancreas. *Nat. Biotechnol., 21,* 755–756.

Lechner, A., Yang, Y. G., Blacken, R. A., Wang, L., Nolan, A. L., & Habener, J. F. (2004). No evidence for significant transdifferentiation of bone marrow into pancreatic beta-cells in vivo. *Diabetes, 53,* 616–623.

Lee, B. C., Hendricks, J. D., & Bailey, G. S. (1989a). Metaplastic pancreatic cells in liver tumors induced by diethylnitrosamine. *Exp. Mol. Pathol., 50,* 104–113.

Lee, H. C., Bonner-Weir, S., Weir, G. C., & Leahy, J. L. (1989b). Compensatory adaption to partial pancreatectomy in the rat. *Endocrinology, 124,* 1571–1575.

Lee, T. C., Barshes, N. R., Brunicardi, F. C., Alejandro, R., Ricordi, C., Nguyen, L., et al. (2004). Procurement of the human pancreas for pancreatic islet transplantation. *Transplantation, 78,* 481–483.

Lee, T. C., Barshes, N. R., O'Mahony, C. A., Nguyen, L., Brunicardi, F. C., Ricordi, C., et al. (2005). The effect of pancreatic islet transplantation on progression of diabetic retinopathy and neuropathy. *Transplantation Proceedings, 37,* 2263–2265.

Leitão, C. B., Bernetti, K., Tharavanij, T., Cure, P., Lauriola, V., Berggren, P. O., et al. (2009a). Lipotoxicity and decreased islet graft survival. *Diabetes Care, 33*(3), 658–660.

Leitão, C. B., Cure, P., Messinger, S., Pileggi, A., Lenz, O., Froud, T., et al. (2009b). Stable renal function after islet transplantation: importance of patient selection and aggressive clinical management. *Transplantation, 87,* 681–688.

Leitão, C. B., Tharavanij, T., Cure, P., Pileggi, A., Baidal, D. A., Ricordi, C., et al. (2008). Restoration of hypoglycemia awareness after islet transplantation. *Diabetes Care, 31,* 2113–2115.

Lemaigre, F., & Zaret, K. S. (2004). Liver development update: new embryo models, cell lineage control, and morphogenesis. *Curr. Opin. Genet. Dev., 14,* 582–590.

Li, W. C., Horb, M. E., Tosh, D., & Slack, J. M. (2005). In vitro transdifferentiation of hepatoma cells into functional pancreatic cells. *Mech. Dev., 122,* 835–847.

Lumelsky, N., Blondel, O., Laeng, P., Velasco, I., Ravin, R., & McKay, R. (2001). Differentiation of embryonic stem cells to insulin-secreting structures similar to pancreatic islets. *Science, 292,* 1389–1394.

Maehr, R., Chen, S., Snitow, M., Ludwig, T., Yagasaki, L., Goland, R., et al. (2009). Generation of pluripotent stem cells from patients with type 1 diabetes. *Proc. Natl. Acad. Sci. U.S.A., 106,* 15768–15773.

Maffi, P., Bertuzzi, F., de Taddeo, F., Magistretti, P., Nano, R., Fiorina, P., et al. (2007). Kidney function after islet transplant alone in type 1 diabetes: impact of immunosuppressive therapy on progression of diabetic nephropathy. *Diabetes Care, 30,* 1150–1155.

Markmann, J. F., Deng, S., Huang, X., Desai, N. M., Velidedeoglu, E. H., Lui, C., et al. (2003). Insulin independence following isolated islet transplantation and single islet infusions. *Ann. Surg., 237,* 741–749, discussion 749–750.

Mathews, V., Hanson, P. T., Ford, E., Fujita, J., Polonsky, K. S., & Graubert, T. A. (2004). Recruitment of bone marrow-derived endothelial cells to sites of pancreatic beta-cell injury. *Diabetes, 53,* 91–98.

Matsumoto, S., Kuroda, Y., Fujita, H., Tanioka, Y., Sakai, T., Hamano, M., et al. (1996). Extending the margin of safety of preservation period for resuscitation of ischemically damaged pancreas during preservation using the two-layer (University of Wisconsin solution/perfluorochemical) method at 20 degrees C with thromboxane A2 synthesis inhibitor OKY046. *Transplantation, 62,* 879–883.

Matsumoto, S., Okitsu, T., Iwanaga, Y., Noguchi, H., Nagata, H., Yonekawa, Y., et al. (2005). Insulin independence after living-donor distal pancreatectomy and islet allotransplantation. *Lancet, 365,* 1642–1644.

Matsumoto, S., & Tanaka, K. (2005). Pancreatic islet cell transplantation using non-heart-beating donors (NHBDs). *J. Hepatobiliary Pancreat. Surg., 12,* 227–230.

Meier, J. J., Bhushan, A., Butler, A. E., Rizza, R. A., & Butler, P. C. (2005). Sustained beta cell apoptosis in patients with long-standing type 1 diabetes: indirect evidence for islet regeneration? *Diabetologia, 48,* 2221–2228.

Mineo, D., Pileggi, A., Alejandro, R., & Ricordi, C. (2009). Point: steady progress and current challenges in clinical islet transplantation. *Diabetes Care, 32,* 1563–1569.

Molinari, M., Al-Saif, F., Ryan, E. A., Lakey, J. R. T., Senior, P. A., Paty, B. W., et al. (2005). Sirolimus-induced ulceration of the small bowel in islet transplant recipients: report of two cases. *Am. J. Transplant., 5,* 2799–2804.

Monti, P., Scirpoli, M., Maffi, P., Ghidoli, N., de Taddeo, F., Bertuzzi, F., et al. (2008). Islet transplantation in patients with autoimmune diabetes induces homeostatic cytokines that expand autoreactive memory T cells. *J. Clin. Invest.*, *118*, 1806–1814.

Nielsen, J. H., Moldrup, A., Billestrup, N., Petersen, E. D., Allevato, G., & Stahl, M. (1992). The role of growth hormone and prolactin in beta cell growth and regeneration. *Adv. Exp. Med. Biol.*, *321*, 9–17, discussion 19–20.

Nir, T., Melton, D. A., & Dor, Y. (2007). Recovery from diabetes in mice by beta cell regeneration. *J. Clin. Invest.*, *117*, 2553–2561.

Nordlie, R. C., Foster, J. D., & Lange, A. J. (1999). Regulation of glucose production by the liver. *Annu. Rev. Nutr.*, *19*, 379–406.

Oberholzer, J., Mathe, Z., Bucher, P., Triponez, F., Bosco, D., Fournier, B., et al. (2003). Islet autotransplantation after left pancreatectomy for non-enucleable insulinoma. *Am. J. Transplant.*, *3*, 1302–1307.

Oberholzer, J., Triponez, F., Mage, R., Andereggen, E., Buhler, L., Cretin, N., et al. (2000). Human islet transplantation: lessons from 13 autologous and 13 allogeneic transplantations. *Transplantation*, *69*, 1115–1123.

Ogawa, N., List, J. F., Habener, J. F., & Maki, T. (2004). Cure of overt diabetes in NOD mice by transient treatment with anti-lymphocyte serum and exendin-4. *Diabetes*, *53*, 1700–1705.

Ouziel-Yahalom, L., Zalzman, M., Anker-Kitai, L., Knoller, S., Bar, Y., Glandt, M., et al. (2006). Expansion and redifferentiation of adult human pancreatic islet cells. *Biochem. Biophys. Res. Commun.*, *341*, 291–298.

Owen, R. J., Ryan, E. A., O'Kelly, K., Lakey, J. R., McCarthy, M. C., Paty, B. W., et al. (2003). Percutaneous trans-hepatic pancreatic islet cell transplantation in type 1 diabetes mellitus: radiologic aspects. *Radiology*, *229*, 165–170.

Park, I. H., Arora, N., Huo, H., Maherali, N., Ahfeldt, T., Shimamura, A., et al. (2008). Disease-specific induced pluripotent stem cells. *Cell*, *134*, 877–886.

Parsons, J. A., Brelje, T. C., & Sorenson, R. L. (1992). Adaptation of islets of Langerhans to pregnancy: increased islet cell proliferation and insulin secretion correlates with the onset of placental lactogen secretion. *Endocrinology*, *130*, 1459–1466.

Paty, B. W., Ryan, E. A., Shapiro, A. M., Lakey, J. R., & Robertson, R. P. (2002). Intrahepatic islet transplantation in type 1 diabetic patients does not restore hypoglycemic hormonal counterregulation or symptom recognition after insulin independence. *Diabetes*, *51*, 3428–3434.

Peshavaria, M., Larmie, B. L., Lausier, J., Satish, B., Habibovic, A., Roskens, V., et al. (2006). Regulation of pancreatic beta-cell regeneration in the normoglycemic 60% partial-pancreatectomy mouse. *Diabetes*, *55*, 3289–3298.

Peters, K., Panienka, R., Li, J., Kloppel, G., & Wang, R. (2005). Expression of stem cell markers and transcription factors during the remodeling of the rat pancreas after duct ligation. *Virchows Arch.*, *446*, 56–63.

Pileggi, A., Alejandro, R., & Ricordi, C. (2006a). Islet transplantation: steady progress and current challenges. *Curr. Opin. Organ Transplant.*, *11*, 7–13.

Pileggi, A., Cobianchi, L., Inverardi, L., & Ricordi, C. (2006b). Overcoming the challenges now limiting islet transplantation: a sequential, integrated approach. *Ann. N.Y. Acad. Sci.*, *1079*, 383–398.

Pileggi, A., Ribeiro, M. M., Hogan, A. R., Molano, R. D., Cobianchi, L., Ichii, H., et al. (2009). Impact of pancreatic cold preservation on rat islet recovery and function. *Transplantation*, *87*, 1442–1450.

Pileggi, A., Ricordi, C., Alessiani, M., & Inverardi, L. (2001). Factors influencing Islet of Langerhans graft function and monitoring. *Clin. Chim. Acta*, *310*, 3–16.

Pileggi, A., Ricordi, C., Kenyon, N. S., Froud, T., Baidal, D. A., Kahn, A., et al. (2004). Twenty years of clinical islet transplantation at the Diabetes Research Institute – University of Miami. *Clinical Transplants*, 177–204.

Poggioli, R., Faradji, R. N., Ponte, G., Betancourt, A., Messinger, S., Baidal, D. A., et al. (2006). Quality of life after islet transplantation. *Am. J. Transplant.*, *6*, 371–378.

Ponte, G. M., Baidal, D. A., Romanelli, P., Faradji, R. N., Poggioli, R., Cure, P., et al. (2007a). Resolution of severe atopic dermatitis after tacrolimus withdrawal. *Cell Transplant.*, *16*, 23–30.

Ponte, G. M., Pileggi, A., Messinger, S., Alejandro, A., Ichii, H., Baidal, D. A., et al. (2007b). Toward maximizing the success rates of human islet isolation: influence of donor and isolation factors. *Cell Transplant*, *16*, 595–607.

Rafaeloff, R., Rosenberg, L., & Vinik, A. I. (1992). Expression of growth factors in a pancreatic islet regeneration model. *Adv. Exp. Med. Biol.*, *321*, 133–140, discussion 141.

Ramalho-Santos, M., Melton, D. A., & McMahon, A. P. (2000). Hedgehog signals regulate multiple aspects of gastrointestinal development. *Development*, *127*, 2763–2772.

Rangan, G. K. (2006). Sirolimus-associated proteinuria and renal dysfunction. *Drug Saf.*, *29*, 1153–1161.

Rao, M. S., Dwivedi, R. S., Subbarao, V., Usman, M. I., Scarpelli, D. G., Nemali, M. R., et al. (1988). Almost total conversion of pancreas to liver in the adult rat: a reliable model to study transdifferentiation. *Biochem. Biophys. Res. Commun.*, *156*, 131–136.

Rao, M. S., & Reddy, J. K. (1995). Hepatic transdifferentiation in the pancreas. *Semin. Cell Biol.*, *6*, 151–156.

Rao, M. S., Subbarao, V., & Reddy, J. K. (1986). Induction of hepatocytes in the pancreas of copper-depleted rats following copper repletion. *Cell Differ., 18*, 109–117.

Rickels, M. R., Naji, A., & Teff, K. L. (2007). Acute insulin responses to glucose and arginine as predictors of beta-cell secretory capacity in human islet transplantation. *Transplantation, 84*, 1357–1360.

Rickels, M. R., Schutta, M. H., Markmann, J. F., Barker, C. F., Naji, A., & Teff, K. L. (2005a). β-Cell function following human islet transplantation for type 1 diabetes. *Diabetes, 54*, 100–106.

Rickels, M. R., Schutta, M. H., Mueller, R., Markmann, J. F., Barker, C. F., Naji, A., et al. (2005b). Islet cell hormonal responses to hypoglycemia after human islet transplantation for type 1 diabetes. *Diabetes, 54*, 3205–3211.

Ricordi, C. (2003). Islet transplantation: a brave new world. *Diabetes, 52*, 1595–1603.

Ricordi, C., Alejandro, R., Angelico, M. C., Fernandez, L. A., Nery, J., Webb, M., et al. (1997). Human islet allografts in patients with type 2 diabetes undergoing liver transplantation. *Transplantation, 63*, 473–475.

Ricordi, C., & Edlund, H. (2008). Toward a renewable source of pancreatic beta-cells. *Nat. Biotechnol., 26*, 397–398.

Ricordi, C., Fraker, C., Szust, J., Al-Abdullah, I., Poggioli, R., Kirlew, T., et al. (2003). Improved human islet isolation outcome from marginal donors following addition of oxygenated perfluorocarbon to the cold-storage solution. *Transplantation, 75*, 1524–1527.

Ricordi, C., Inverardi, L., Kenyon, N. S., Goss, J., Bertuzzi, F., & Alejandro, R. (2005). Requirements for success in clinical islet transplantation. *Transplantation, 79*, 1298–1300.

Ricordi, C., Lacy, P. E., Finke, E. H., Olack, B. J., & Scharp, D. W. (1988). Automated method for isolation of human pancreatic islets. *Diabetes, 37*, 413–420.

Ricordi, C., & Strom, T. B. (2004). Clinical islet transplantation: advances and immunological challenges. *Nat. Rev. Immunol., 4*, 259–268.

Ricordi, C., Zeng, Y. J., Alejandro, R., Tzakis, A., Venkataramanan, R., Fung, J., et al. (1991). In vivo effect of FK506 on human pancreatic islets. *Transplantation, 52*, 519–522.

Robertson, R. P. (2004). Consequences on beta-cell function and reserve after long-term pancreas transplantation. *Diabetes, 53*, 633–644.

Robertson, R. P., Lanz, K. J., Sutherland, D. E., & Kendall, D. M. (2001). Prevention of diabetes for up to 13 years by autoislet transplantation after pancreatectomy for chronic pancreatitis. *Diabetes, 50*, 47–50.

Roche, E., Sepulcre, P., Reig, J. A., Santana, A., & Soria, B. (2005). Ectodermal commitment of insulin-producing cells derived from mouse embryonic stem cells. *Faseb J., 19*, 1341–1343.

Rosenberg, L., Brown, R. A., & Duguid, W. P. (1983). A new approach to the induction of duct epithelial hyperplasia and nesidioblastosis by cellophane wrapping of the hamster pancreas. *J. Surg. Res., 35*, 63–72.

Rosenberg, L., Duguid, W. P., & Vinik, A. I. (1989). The effect of cellophane wrapping of the pancreas in the Syrian golden hamster: autoradiographic observations. *Pancreas, 4*, 31–37.

Ryan, E. A., Lakey, J. R., Paty, B. W., Imes, S., Korbutt, G. S., Kneteman, N. M., et al. (2002). Successful islet transplantation: continued insulin reserve provides long-term glycemic control. *Diabetes, 51*, 2148–2157.

Ryan, E. A., Lakey, J. R., Rajotte, R. V., Korbutt, G. S., Kin, T., Imes, S., et al. (2001). Clinical outcomes and insulin secretion after islet transplantation with the Edmonton protocol. *Diabetes, 50*, 710–719.

Ryan, E. A., Paty, B. W., Senior, P. A., Bigam, D., Alfadhli, E., Kneteman, N. M., et al. (2005a). Five-year follow-up after clinical islet transplantation. *Diabetes, 54*, 2060–2069.

Ryan, E. A., Paty, B. W., Senior, P. A., Lakey, J. R., Bigam, D., & Shapiro, A. M. (2005b). β-Score: an assessment of β-cell function after islet transplantation. *Diabetes Care, 28*, 343–347.

Ryan, E. A., Shandro, T., Green, K., Paty, B. W., Senior, P. A., Bigam, D., et al. (2004). Assessment of the severity of hypoglycemia and glycemic lability in type 1 diabetic subjects undergoing islet transplantation. *Diabetes, 53*, 955–962.

Sadowski, I., Ma, J., Triezenberg, S., & Ptashne, M. (1988). GAL4-VP16 is an unusually potent transcriptional activator. *Nature, 335*, 563–564.

Sander, M., Sussel, L., Conners, J., Scheel, D., Kalamaras, J., Dela Cruz, F., et al. (2000). Homeobox gene Nkx6.1 lies downstream of Nkx2.2 in the major pathway of beta-cell formation in the pancreas. *Development, 127*, 5533–5540.

Santamaria, P., Nakhleh, R. E., Sutherland, D. E., & Barbosa, J. J. (1992). Characterization of T lymphocytes infiltrating human pancreas allograft affected by isletitis and recurrent diabetes. *Diabetes, 41*, 53–61.

Senior, P. A., Paty, B. W., Cockfield, S. M., Ryan, E. A., & Shapiro, A. M. (2005). Proteinuria developing after clinical islet transplantation resolves with sirolimus withdrawal and increased tacrolimus dosing. *Am. J. Transplant., 5*, 2318–2323.

Senior, P. A., Zeman, M., Paty, B. W., Ryan, E. A., & Shapiro, A. M. (2007). Changes in renal function after clinical islet transplantation: four-year observational study. *Am. J. Transplant., 7*, 91–98.

Shapiro, A. M., Gallant, H. L., Hao, E. G., Lakey, J. R., McCready, T., Rajotte, R. V., et al. (2005). The portal immunosuppressive storm: relevance to islet transplantation? *Ther. Drug Monit., 27*, 35–37.

Shapiro, A. M., Lakey, J. R., Ryan, E. A., Korbutt, G. S., Toth, E., Warnock, G. L., et al. (2000). Islet transplantation in seven patients with type 1 diabetes mellitus using a glucocorticoid-free immunosuppressive regimen. *N. Engl. J. Med., 343*, 230–238.

Shen, C. N., Slack, J. M., & Tosh, D. (2000). Molecular basis of transdifferentiation of pancreas to liver. *Nat. Cell Biol., 2*, 879–887.

Shi, Y., Desponts, C., Do, J. T., Hahm, H. S., Scholer, H. R., & Ding, S. (2008). Induction of pluripotent stem cells from mouse embryonic fibroblasts by Oct4 and Klf4 with small-molecule compounds. *Cell Stem Cell, 3*, 568–574.

Sibley, R. K., & Sutherland, D. E. (1987). Pancreas transplantation. An immunohistologic and histopathologic examination of 100 grafts. *Am. J. Pathol., 128*, 151–170.

Sibley, R. K., Sutherland, D. E., Goetz, F., & Michael, A. F. (1985). Recurrent diabetes mellitus in the pancreas iso- and allograft. A light and electron microscopic and immunohistochemical analysis of four cases. *Lab. Invest., 53*, 132–144.

Sorenson, R. L., & Brelje, T. C. (1997). Adaptation of islets of Langerhans to pregnancy: beta-cell growth, enhanced insulin secretion and the role of lactogenic hormones. *Horm. Metab. Res., 29*, 301–307.

Soria, B., Roche, E., Berna, G., Leon-Quinto, T., Reig, J. A., & Martin, F. (2000). Insulin-secreting cells derived from embryonic stem cells normalize glycemia in streptozotocin-induced diabetic mice. *Diabetes, 49*, 157–162.

Sosa-Pineda, B. (2004). The gene Pax4 is an essential regulator of pancreatic beta-cell development. *Mol. Cells, 18*, 289–294.

Sosa-Pineda, B., Chowdhury, K., Torres, M., Oliver, G., & Gruss, P. (1997). The Pax4 gene is essential for differentiation of insulin-producing beta cells in the mammalian pancreas. *Nature, 386*, 399–402.

Street, C. N., Lakey, J. R., Shapiro, A. M., Imes, S., Rajotte, R. V., Ryan, E. A., et al. (2004). Islet graft assessment in the Edmonton Protocol: implications for predicting long-term clinical outcome. *Diabetes, 53*, 3107–3114.

Sussel, L., Kalamaras, J., Hartigan-O'Connor, D. J., Meneses, J. J., Pedersen, R. A., Rubenstein, J. L., et al. (1998). Mice lacking the homeodomain transcription factor Nkx2.2 have diabetes due to arrested differentiation of pancreatic beta cells. *Development, 125*, 2213–2221.

Sutherland, D. E., Goetz, F. C., & Sibley, R. K. (1989). Recurrence of disease in pancreas transplants. *Diabetes, 38* (Suppl. 1), 85–87.

Sutherland, D. E., Sibley, R., Xu, X. Z., Michael, A., Srikanta, A. M., Taub, F., et al. (1984). Twin-to-twin pancreas transplantation: reversal and reenactment of the pathogenesis of type I diabetes. *Trans. Assoc. Am. Physicians, 97*, 80–87.

Tada, S., Era, T., Furusawa, C., Sakurai, H., Nishikawa, S., Kinoshita, M., et al. (2005). Characterization of mesendoderm: a diverging point of the definitive endoderm and mesoderm in embryonic stem cell differentiation culture. *Development, 132*, 4363–4374.

Takahashi, K., Tanabe, K., Ohnuki, M., Narita, M., Ichisaka, T., Tomoda, K., et al. (2007). Induction of pluripotent stem cells from adult human fibroblasts by defined factors. *Cell, 131*, 861–872.

Tam, P. P., Kanai-Azuma, M., & Kanai, Y. (2003). Early endoderm development in vertebrates: lineage differentiation and morphogenetic function. *Curr. Opin. Genet. Dev., 13*, 393–400.

Tateishi, K., He, J., Taranova, O., Liang, G., d'Alessio, A. C., & Zhang, Y. (2008). Generation of insulin-secreting islet-like clusters from human skin fibroblasts. *J. Biol. Chem, 283*(46), 31601–31607.

Tharavanij, T., Betancourt, A., Messinger, S., Cure, P., Leitao, C. B., Baidal, D. A., et al. (2008). Improved long-term health-related quality of life after islet transplantation. *Transplantation, 86*, 1161–1167.

Thompson, D. M., Begg, I. S., Harris, C., Ao, Z., Fung, M. A., Meloche, R. M., et al. (2008). Reduced progression of diabetic retinopathy after islet cell transplantation compared with intensive medical therapy. *Transplantation, 85*, 1400–1405.

Thomson, J. A., Itskovitz-Eldor, J., Shapiro, S. S., Waknitz, M. A., Swiergiel, J. J., Marshall, V. S., et al. (1998). Embryonic stem cell lines derived from human blastocysts. *Science, 282*, 1145–1147.

Thomson, J. A., & Odorico, J. S. (2000). Human embryonic stem cell and embryonic germ cell lines. *Trends Biotechnol., 18*, 53–57.

Tremblay, K. D., & Zaret, K. S. (2005). Distinct populations of endoderm cells converge to generate the embryonic liver bud and ventral foregut tissues. *Dev. Biol., 280*, 87–99.

Triezenberg, S. J., Kingsbury, R. C., & McKnight, S. L. (1988). Functional dissection of VP16, the trans-activator of herpes simplex virus immediate early gene expression. *Genes Dev., 2*, 718–729.

Tschopp, J. M., Brutsche, M. H., Frey, J. G., Spiliopoulos, A., Nicod, L., Rochat, T., et al. (1997). End-stage cystic fibrosis: improved diabetes control 2 years after successful isolated pancreatic cell and double-lung transplantation. *Chest, 112*, 1685–1687.

Tsujimura, T., Kuroda, Y., Avila, J. G., Kin, T., Oberholzer, J., Shapiro, A. M., et al. (2004a). Influence of pancreas preservation on human islet isolation outcomes: impact of the two-layer method. *Transplantation, 78*, 96–100.

Tsujimura, T., Kuroda, Y., Churchill, T. A., Avila, J. G., Kin, T., Shapiro, A. M., et al. (2004b). Short-term storage of the ischemically damaged human pancreas by the two-layer method prior to islet isolation. *Cell Transplant.*, *13*, 67–73.

Tzakis, A., Webb, M., Nery, J., Rogers, A., Koutouby, R., Ruiz, P., et al. (1996). Experience with intestinal transplantation at the University of Miami. *Transplant Proc.*, *28*, 2748–2749.

Tzakis, A. G., Ricordi, C., Alejandro, R., Zeng, Y., Fung, J. J., Todo, S., et al. (1990). Pancreatic islet transplantation after upper abdominal exenteration and liver replacement. *Lancet*, *336*, 402–405.

Vantyghem, M. C., Kerr-Conte, J., Arnalsteen, L., Sergent, G., Defrance, F., Gmyr, V., et al. (2009). Primary graft function, metabolic control, and graft survival after islet transplantation. *Diabetes Care*, *32*, 1473–1478.

Vendrame, F., Pileggi, A., Laughlin, E., Allende, G., Martin-Pagola, A., Molano, R. D., et al. (2010). Recurrence of type 1 diabetes after simultaneous pancreas-kidney transplantation, despite immunosuppression, associated with autoantibodies and pathogenic autoreactive CD4 T-cells. *Diabetes*, *59*.

Venturini, M., Angeli, E., Maffi, P., Fiorina, P., Bertuzzi, F., Salvioni, M., et al. (2005). Technique, complications, and therapeutic efficacy of percutaneous transplantation of human pancreatic islet cells in type 1 diabetes: the role of US. *Radiology*, *234*, 617–624.

Venturini, M., Fiorina, P., Maffi, P., Losio, C., Vergani, A., Secchi, A., et al. (2006). Early increase of retinal arterial and venous blood flow velocities at color Doppler imaging in brittle type 1 diabetes after islet transplant alone. *Transplantation*, *81*, 1274–1277.

Wang, J., Elghazi, L., Parker, S. E., Kizilocak, H., Asano, M., Sussel, L., et al. (2004). The concerted activities of Pax4 and Nkx2.2 are essential to initiate pancreatic beta-cell differentiation. *Dev. Biol.*, *266*, 178–189.

Wang, R. N., Kloppel, G., & Bouwens, L. (1995). Duct- to islet-cell differentiation and islet growth in the pancreas of duct-ligated adult rats. *Diabetologia*, *38*, 1405–1411.

Warnock, G. L., Thompson, D. M., Meloche, R. M., Shapiro, R. J., Ao, Z., Keown, P., et al. (2008). A multi-year analysis of islet transplantation compared with intensive medical therapy on progression of complications in type 1 diabetes. *Transplantation*, *86*, 1762–1766.

Weber, D. J. (2004). FDA regulation of allogeneic islets as a biological product. *Cell Biochem. Biophys.*, *40*, 19–22.

WHITG. (2006). Autoimmunity after islet-cell allotransplantation (Worcester Human Islet Transplantation Group). *N. Engl. J. Med.*, *355*, 1397–1399.

Willenbring, H., Bailey, A. S., Foster, M., Akkari, Y., Dorrell, C., Olson, S., et al. (2004). Myelomonocytic cells are sufficient for therapeutic cell fusion in liver. *Nat. Med.*, *10*, 744–748.

Williams, D., & Haragsim, L. (2006). Calcineurin nephrotoxicity. *Adv. Chron. Kidney Disease*, *13*, 47–55.

Wolf, H. K., Burchette, J. L., Jr., Garcia, J. A., & Michalopoulos, G. (1990). Exocrine pancreatic tissue in human liver: a metaplastic process? *Am. J. Surg. Pathol.*, *14*, 590–595.

Wonnacott, K. (2005). Update on regulatory issues in pancreatic islet transplantation. *Am. J. Ther.*, *12*, 600–604.

Wu, Y., Wang, J., Scott, P. G., & Tredget, E. E. (2007). Bone marrow-derived stem cells in wound healing: a review. *Wound Repair Regen.*, *15*(Suppl. 1), S18–S26.

Xu, X., d'Hoker, J., Stange, G., Bonne, S., de Leu, N., Xiao, X., et al. (2008). Beta cells can be generated from endogenous progenitors in injured adult mouse pancreas. *Cell*, *132*, 197–207.

Yasunaga, M., Tada, S., Torikai-Nishikawa, S., Nakano, Y., Okada, M., Jakt, L. M., et al. (2005). Induction and monitoring of definitive and visceral endoderm differentiation of mouse ES cells. *Nat. Biotechnol.*, *23*, 1542–1550.

Yu, J., Hu, K., Smuga-Otto, K., Tian, S., Stewart, R., Slukvin, I. I., et al. (2009). Human induced pluripotent stem cells free of vector and transgene sequences. *Science*, *324*(5928), 797–801.

Yu, J., Vodyanik, M. A., Smuga-Otto, K., Antosiewicz-Bourget, J., Frane, J. L., Tian, S., et al. (2007). Induced pluripotent stem cell lines derived from human somatic cells. *Science*, *318*, 1917–1920.

Zahr, E., Molano, R. D., Pileggi, A., Ichii, H., Jose, S. S., Bocca, N., et al. (2007). Rapamycin impairs in vivo proliferation of islet beta-cells. *Transplantation*, *84*, 1576–1583.

Zaret, K. (1998). Early liver differentiation: genetic potentiation and multilevel growth control. *Curr. Opin. Genet. Dev.*, *8*, 526–531.

Zhou, H., Wu, S., Young Joo, J., Zhu, S., Wook Han, D., Lin, T., et al. (2009). Generation of induced pluripotent stem cells using recombinant proteins. *Cell Stem Cell*. doi:10.1016/j.stem.2009.05.005.

Zhou, Q., Brown, J., Kanarek, A., Rajagopal, J., & Melton, D. A. (2008). In vivo reprogramming of adult pancreatic exocrine cells to beta-cells. *Nature*, *455*, 627–632.

Zulewski, H., Abraham, E. J., Gerlach, M. J., Daniel, P. B., Moritz, W., Muller, B., et al. (2001). Multipotential nestin-positive stem cells isolated from adult pancreatic islets differentiate ex vivo into pancreatic endocrine, exocrine, and hepatic phenotypes. *Diabetes*, *50*, 521–533.

426

CHAPTER 24

Regenerative Medicine for Diseases of the Retina

Deepak A. Lamba[*], **Thomas A. Reh**[†]

[*] Department of Ophthalmology, University of Washington, Seattle, WA, USA
[†] Department of Biological Structure, University of Washington, Seattle, WA, USA

INTRODUCTION

The vertebrate retina is subject to a variety of degenerative conditions. Glaucoma, diabetes, macular degeneration, and retinitis pigmentosa are among the more common conditions that lead to loss of one or more of the retinal cell types and frequently result in partial or complete blindness. While some of these disorders have treatments that can slow the progression of visual loss, in most cases there will be an untreatable visual impairment. The development of effective cell therapies is thus a goal of many individuals working in the visual sciences, and there have been steady advances using a variety of approaches towards this end. Some of these approaches have relied on the interesting fact that, in many non-mammalian vertebrates, the retina can spontaneously repair itself to a truly remarkable degree. In the early days of regeneration research, investigators used the eye as one of the key model systems to study the phenomenon of regeneration. In this chapter, we will review (1) the basic developmental biology of the eye, describing the relationships between retinal stem cells and progenitors during development, (2) the sources of retinal stem cells and progenitors in mature animals that mediate retinal regeneration, and (3) the derivation of retinal progenitor cells from embryonic stem cells (ESCs) and induced pluripotent stem cells (iPSCs). This review is not meant to be exhaustive, but we hope it will illustrate the main currents of research in this field.

EYE FIELD TRANSCRIPTION FACTORS: SPECIFICATION OF THE EYE

The vertebrate retina arises from the ventral diencephalon of the neural tube (Fig. 24.1). Paired evaginations, known as the optic vesicles, emerge from the anterior region of the neural plate. The optic vesicle cells express a unique complement of transcription factors, termed eye field transcription factors (EFTFs), that set them apart from the surrounding regions of the neural plate (see below). The optic vesicle cells undergo extensive proliferation over the next phases of retinal development, and will ultimately generate all the various cell types of the neural retina, as well as several non-retinal ocular structures, such as the ciliary epithelium, the pigmented epithelium, and the iris.

The presumptive eye-forming region of the embryo, or eye field, was first defined by the transplantation experiments of Hans Spemann as the anterior region of the neural plate (Fig. 24.2). More recently it has been possible to identify the same region by monitoring expression of a group of transcription factors called EFTFs. The EFTFs expressed early in eye field specification include Rx, Pax6, Six3, Lhx2, and Optx2 (Six6) (Fig. 24.2C). The eye field forms late in gastrulation, at the anterior end of the neural plate in the diencephalic region of

Principles of Regenerative Medicine. DOI: 10.1016/B978-0-12-381422-7.10024-0

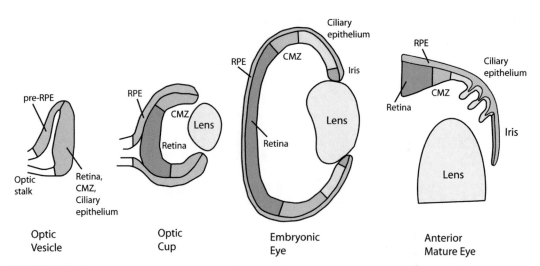

Optic   Optic   Embryonic   Anterior
Vesicle   Cup   Eye   Mature Eye

FIGURE 24.1

The development of the various parts of the eye that are derived from the neural tube. In the first stage of eye development, the *optic vesicle* appears as an evagination from the diencephalons of the neural tube. Even at this early stage of development the vesicle is patterned into a presumptive RPE domain (gray) and a presumptive neural retina domain (red). The vesicle then becomes the *optic cup* as it folds in on itself, and the lens pinches off from the overlying ectoderm. At the optic cup stage the first neurons emerge in the central retinal domain (blue), but not in the more peripheral regions. In the next stage, the *embryonic eye* begins to show distinct gene expression patterns in the more anterior (peripheral) regions, which will become the epithelial part of the ciliary body (ciliary epithelium—yellow) and the iris (green). Note that the iris and ciliary epithelium each have two layers: a pigmented and a non-pigmented layer. The pigmented layer of each region is continuous with the RPE, whereas the non-pigmented epithelial layer of both the iris and the ciliary epithelium is continuous with the retina. The ciliary marginal zone (CMZ), which contains the persistent progenitors/retinal stem cells in non-mammalian vertebrates, arises at the junction between the ciliary epithelium and the retina (red) and may be similar to the very early optic vesicle cells. A small part of the anterior eye is shown at the right of the diagram to show the eventual relationships among the various domains in the *mature eye*.

the forebrain. The eye field initially extends across the midline, as a single domain. This single field is subsequently split into two lateral domains, due to the repression of EFTFs by sonic hedgehog (Shh), which is released from the prechordal mesoderm at the midline (Li et al., 1997). The EFTFs are essential for eye development; mutations in each of these genes are associated with either anophthalmia (no eye) or microphthalmia (small eyes) (Hill et al., 1991; Mathers et al., 1997; Porter et al., 1997; Carl et al., 2002; Zuber et al., 2003).

Prior to the development of the eye field, the anterior end of the nervous system becomes distinct from the posterior, and Otx2 transcription factor (a member of the orthodenticle family) is critical in the control of this distinction (Simeone et al., 1993). Otx2 is down-regulated in the eye field when a related transcription factor, Rx, is expressed (Andreazzoli et al., 1999). Otx2 expression persists in the periphery of the eye field and becomes restricted to the pigment epithelium, though it is re-expressed in the photoreceptors and bipolar cells of the neural retina (Bovolenta et al., 1997). Otx2$^{-/-}$ mutants do not form any structures anterior to rhombomere 3 (Matsuo et al., 1995), including the eye fields.

Among the first EFTFs to be expressed in the eye field is Rx/Rax, a paired-like homeobox transcription factor. Rx expression begins in areas that will give rise to the ventral forebrain and optic vesicles. Once the optic vesicles form, Rx expression is restricted to the ventral diencephalon and the optic vesicles and is eventually restricted to the developing retina (Furukawa et al., 1997). Homozygous null mutations of the Rx gene in mouse result in anophthalmia, with no eye development after the optic vesicle stage (Mathers et al., 1997). The Rx-deficient mice also lack expression of other EFTFs such as Pax6 and Six3, indicating that Rx may have a role in inducing these genes (Zhang et al., 2000). A similar anophthalmia phenotype was observed in loss of function experiments in *Xenopus* embryos using morpholino

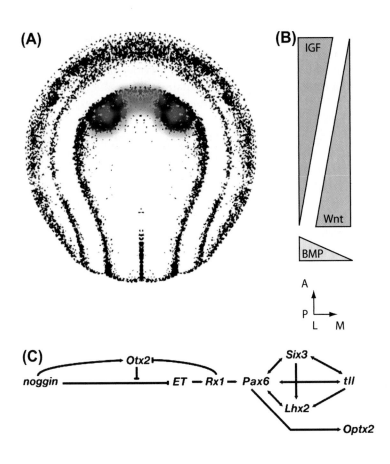

(A)

(B)

(C)

FIGURE 24.2
Formation of the eye field. (A) Drawing of the presumptive eye field as deduced from Hans Spemann's transplantation experiments. The eye field starts as a single region (marked in purple) at the anterior end of neural ectoderm and then splits into two by the midline Shh signaling. (B) The key signaling pathways involved in formation of the eye field include BMP, Wnt, and IGF, such that this region forms in the area of low Wnt, low BMP, and high IGF levels. (C) This signaling results in an induction of a group of transcription factors called the eye field transcription factors and includes Rx, Pax6, Six3, Lhx2, and Six6 (Optx2). Reproduced from Zuber at al., 2003.

oligonucleotides against the *Xenopus* homolog to Rx (Andreazzoli et al., 2003). A mutation in the Rx gene in humans has been identified in a patient suffering from anophthalmia and sclerocornea (Voronina et al., 2004). Overexpression of Rx in *Xenopus* embryos results in hyperproliferation of the neural retina and retinal pigment epithelium as well as formation of ectopic retinal tissue (Mathers et al., 1997). Similar results were obtained in mis-expression studies carried out in zebrafish (Chuang and Raymond, 2001).

The most well-studied eye field transcription factor is Pax6. It has been proposed that Pax6 is the master regulatory gene in eye development. It belongs to the family of paired box homeodomain genes and has been highly conserved across species. Pax6 is expressed in the anterior neural plate at the end of gastrulation and continues to be expressed in the optic vesicles and their derivatives. Its expression persists throughout optic development and ultimately into adult animals in ganglion, horizontal, and amacrine cells of the neural retina (Grindley et al., 1995; de Melo et al., 2003). Mutations in the Pax6 gene result in a variety of phenotypes, depending on the gene dosage. Homozygous mutations that cause a loss of all Pax6 expression result in anophthalmia in mice and rats (Matsuo et al., 1993; Grindley et al., 1995). However, Pax6 mutant mouse embryos have normal Rx expression, suggesting that Pax6 is downstream of Rx (Zhang et al., 2000). Mis-expression studies with Pax6 have been carried out in *Drosophila* (Halder et al., 1995) and *Xenopus* (Chow et al., 1999) and in both species this induces ectopic eye tissue. Overexpression of Pax6 in *Xenopus* results in multiple ectopic eyes all along the dorsal CNS along with ectopic expression of other EFTFs including Rx in these areas. This suggests that Pax6 also has a role in the induction of Rx. The ectopic eyes display a similar morphology to the normal eye, having both a neural retina and a lens. Thus, loss of function studies, as well as mis-expression studies, lend support to the idea that Pax6 is a critical regulatory gene during eye development (see Zuber and Harris, 2006 for a review).

In addition to Rx and Pax6, there are several other members of the EFTFs. Lhx2 is an eye field transcription factor belonging to the family of Lim-homeodomain genes. It is expressed in the optic vesicles just before completion of gastrulation (Xu et al., 1993; Porter et al., 1997). Lhx2 null mutants fail to form eyes (Porter et al., 1997). Developmentally, eye formation gets stalled at the optic vesicle stage and the optic cup and lens do not form. Analysis of Pax6 expression in these mice shows a normal pattern of Pax6 in the optic vesicle, and so Lhx2 may lie downstream of Pax6. Overexpression of Lhx2 in *Xenopus* embryo results both in large eyes and ectopic retinal tissue (Zuber et al., 2003). Six3 belongs to the Six-Homeodomain family of genes. Six3 appears in the region of the presumptive eye field around the same time as Pax6 (Oliver et al., 1995; Bovolenta et al., 1998; Loosli et al., 1998). Six3 inactivation in medaka fish has been shown to result in anophthalmia and forebrain agenesis (Carl et al., 2002). Overexpression of Six3 in medaka fish results in multiple eye-like structures that express other EFTFs (Loosli et al., 1999) while in zebrafish it results in enlargement of the optic stalk (Kobayashi et al., 1998). Optx2 (Six6, Six9) also belongs to the Six-Homeodomain family of genes. It is expressed from the optic vesicle stage of eye development onwards (Toy et al., 1998; Jean et al., 1999; Lopez-Rios et al., 1999; Toy and Sundin, 1999). Overexpression of Optx2 in *Xenopus* embryos results in a large expansion of the retinal domain as well as hyperproliferation of cultured retinal progenitors (Zuber et al., 1999; Bernier et al., 2000).

The signaling factors that pattern the expression of the EFTFs have been the subject of a great deal of investigation in recent years (Esteve and Bovolenta, 2006). It has been appreciated for several years that the IGF (insulin-like growth factor) pathway is important in the development of anterior structures in the embryo, including the eyes. In *Xenopus*, overactivation of the IGF pathway early in embryogenesis causes an increase in the development of anterior fates; ectopic eyes were particularly striking (Pera et al., 2001; Richard-Parpaillon et al., 2002; Eivers et al., 2004), and several EFTFs are upregulated. These experimental studies are consistent with the observed greater levels of expression of IGF and receptors in the anterior neural plate (Richard-Parpaillon et al., 2002). While these early studies demonstrated IGF could promote all anterior fates, more recent studies have shown that differences in downstream components of the IGF pathway may lead to the distinction between the eye field and the adjacent telencephalon (Wu et al., 2006). Activation of the IGF receptor results in at least two types of second messenger signaling within the cell, the MAPK pathway and the PI3K/Akt pathway. Differential activation of these two second messengers may underlie the distinction between the eye field and the telencephalon. A recently described potentiator of the Akt pathway, Kermit2, specifically promotes EFTFs when overexpressed, and has little effect on telencephalon markers. In addition, loss of Kermit2 function leads to an inhibition in eye formation, which can be rescued by expression of a constitutively active PI3 kinase. By contrast, loss of Kermit2 does not affect the activation of the MAPK pathway by IGF, or the ability of IGF to promote other anterior fates, like the telencephalon.

The mechanism by which activation of the Akt pathway by IGF leads to expression of EFTFs involves its interactions with another key developmental signaling pathway, the Wnt pathway. Richard-Parpaillon et al. (2002) found that the overexpression of IGF1 in *Xenopus* ectodermal explants inhibited canonical Wnt signaling, as assayed by beta-catenin and Wnt reporter constructs. Moreover, IGF-1 overexpression rescued eye formation in embryos "posteriorized" by Wnt8 overexpression. Several Wnt ligands, frizzled receptors, and downstream signaling components of this pathway are expressed in or around the domain of EFTF expression. Wnt1, Wnt8, Wnt11, and Wnt4 are expressed posterior to the eye field, while the Wnt receptor, Frizzled-5, is expressed by the cells of the eye field. Experimental manipulations in Wnt signaling in a variety of species indicate that antagonism of canonical Wnt signaling is required for eye field formation. Wnt/beta-catenin signaling activates genes expressed in the posterior diencephalon and suppresses the EFTFs (Cavodeassi et al., 2005), while activation of the noncanonical Wnt pathway (Cavodeassi et al., 2005; Maurus et al., 2005), likely through Wnt4 or Wnt11 and Frizzled-5, promotes EFTF expression.

The above studies support a role for Wnt signaling in creating the boundary between the eye field and the posterior diencephalon; canonical Wnt signaling is necessary for the posterior diencephalon, while the inhibition of this pathway promotes EFTF expression (Fig. 24.2B). As described above, canonical Wnt signaling is inhibited by both IGF and non-canonical Wnt signaling. Consistent with this model, the Wnt inhibitor, Sfrp1, is also expressed in the eye field. Morpholino knockdown of this gene in fish also leads to a reduction in the expression domain of the EFTFs (Esteve et al., 2004), adding further evidence for this model. Although most of the studies analyzing the factors that regulate eye field expression have been carried out in frogs or fish, recent studies have provided evidence that similar mechanisms are at work in mammals as well. Fz5 is expressed the embryonic eye field in mice (Liu et al., 2008) and conditional deletion of this gene leads to a smaller eye size at E10.5, consistent with a role in early eye field patterning, though there are many other defects in retinal development as well.

There are still many questions concerning this emerging model for patterning the eye field. There appears to be a critical role for Ephrin signaling in this process, and interactions between FGF signaling and EphrinB1 are important for coordinating key morphogenetic movements that underlie the formation of the eye fields (Moore et al., 2004). Increasing EphrinB1 causes more cells to move into the eye field and a resulting expansion of the eyes, while blocking this pathway leads to reduction in eye size. Surprisingly, EphrinB1 signaling also feeds into the Wnt pathway (Lee et al., 2006); overexpression of the Wnt pathway component, disheveled, can rescue the phenotype caused by the loss of EphrinB1, and this rescue effect is dependent on other components of the planar cell polarity (PCP) pathway – Egl-10 and pleckstrin. Whether the activation of the PCP pathway through EphrinB1 also causes EFTF expression through the inhibition of the canonical Wnt pathway, and how this relates to the changes in migration of these cells, are questions that will require further investigation. A recent, more puzzling, study shows that EFTF expression is regulated by purine-mediated signaling (Masse et al., 2007). Overexpression of an enzyme (E-NTPDase2) that converts ATP to ADP causes ectopic eyes in *Xenopus* embryos very similar to those observed with IGF overexpression. This effect is mediated at least partly by the purinergic receptor, P2Y1, since knockdown (or overexpression) of P2Y1 along with E-NTPDase2 caused a synergistic loss (or increase) of Pax6 and Rx1. How purinergic signaling interacts with the other patterning systems is currently not clear, but should provide a fascinating new direction for future investigations.

Retinal progenitors: from optic cup to retina

The next phase of retinal development involves a massive proliferation of a group of cells that occupy the structure known as the optic cup. The optic cup forms from an involution of the optic vesicle, and as noted above mutations in the EFTFs prevent the eye from progressing to this stage, or much beyond it. The cells of the optic cup resemble neural progenitors from other regions of the CNS. They have a simple bipolar morphology, span the width of the neuro-epithelium, undergo mitosis at the scleral (ventricular) surface, and progress through stages of interkinetic nuclear migration during S-phase (Fig. 24.3). These cells were once thought to be homogeneous, though more recently they have been shown to have distinct patterns of gene expression. Ever since the first birthdating studies of Sidman (1961), it has been consistently found that the different types of retinal neurons are generated in a sequence, with ganglion cells, cone photoreceptors, amacrine cells, and horizontal cells generated during early stages of development, and most rod photoreceptors, bipolar cells, and Müller glia generated in the latter half of the period of retinogenesis (Fig. 24.3). Clonal analysis of the progeny of these cells shows that they can give rise to all the different types of retinal neurons, and that the clones have mixed neuronal and glial lineages (Holt et al., 1988; Turner et al., 1990 for a review; Cepko et al., 1996). Clonal analysis of retinal progenitor cells has also demonstrated a wide variety of clone sizes (Fekete et al., 1994), with some clones containing thousands of progeny.

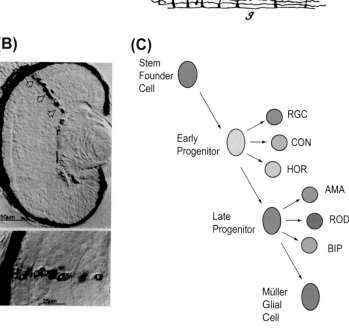

432

FIGURE 24.3

Optic cup to neural retina. (A)The central regions of the optic cup contain a mix of mitotically active progenitor cells (d,e) at various stages of the cell cycle and differentiating ganglion cells (a,b,c) and cones (f). (B) Labeling the progeny of the mitotic cells at early stages of retinal development gives clones that contain all the cell types of the differentiated retina (g = ganglion cell; a = amacrine cell; m = Müller glia; b = bipolar; p = photoreceptor (rod or cone)). (C) Proposed model of retinal development. Birthdating studies in a variety of vertebrates indicate that there is a sequence to the production of the different retinal cell types, and that the progenitors have qualitatively different properties when isolated from early stages of development vs. late stages of development. Müller glia are likely to be the last cell type generated in most species.

The mechanisms that direct progenitor cells to different fates have been the subject of much investigation in the retina. While a thorough discussion of these findings is beyond the scope of this chapter, there have been two basic hypotheses proposed. First, it has been proposed that retinal progenitor cells undergo a progressive change during development that constrains them to a smaller range of fates (Reh and Cagan, 1994; Brzezinski and Reh, 2010). This implies that there is some sort of "molecular clock" keeping track of the developmental stage. The conserved birth order of the different classes of neurons can then be explained in that those cells that become post-mitotic at a particular stage of development are constrained to a specific cell fate. An alternative model is that a changing environment directs the cells to progressively later fates, but the progenitor cells themselves remain competent to generate all retinal cell types throughout the period of retinogenesis (James et al., 2003). There is experimental support for both models, and both the environment and intrinsic state of the cell are likely to be important factors in determining its ultimate fate (Reh and Cagan, 1994; Livesey and Cepko, 2001 for a review).

Are all retinal progenitor cells the same? Are there differences between the retinal progenitors and a more primitive retinal stem cell or founder cell? Several lines of evidence suggest that progenitors are not all identical. Proneural bHLH gene expression profiles appear to differ among the progenitors. For example, the bHLH transcription factor Ascl1 (also known as Mash1 or Cash1) is expressed in only a subset of retinal progenitors (Jasoni et al., 1994; Jasoni and Reh, 1996). Other proneural genes, e.g. neurogenin2, are expressed in subsets of progenitors as well (Marquardt et al., 2001; Nelson and Reh, 2008). Another transcription factor, Foxn4, is also expressed in only a subset of retinal progenitors; this gene is thought to specifically bias progenitors to generate either amacrine or horizontal cells (Li et al., 2004). Retinal progenitors can also be distinguished by their response to growth factors and intracellular signaling. Progenitor cells isolated from late stages of embryonic development or from neonatal retina are induced to differentiate by treatments that raise cAMP; progenitors isolated from early stages of embryogenesis have the opposite response: their proliferation is stimulated by increasing intracellular cAMP (Taylor and Reh, 1990). Progenitors isolated from the early embryonic retina are stimulated to proliferate by FGF, but are only minimally responsive to EGF or TGFα (Anchan et al., 1991; Lillien and Cepko, 1992; Anchan and Reh, 1995). At later embryonic stages, and in the post-natal retina, the progenitors acquire a robust response to EGF. The evidence that all retinal progenitors are not identical is somewhat at variance with the fact that lineage studies have not demonstrated distinct subpopulations of progenitors that generate specific cell classes. One possibility is that the different types of progenitor cells can interconvert among themselves. More generally in the developing CNS, it is thought that neural stem cells can convert from being FGF-responsive to being EGF-responsive (Ciccolini and Svendsen, 1998).

In summary, the multipotent progenitors make up the majority of mitotically active cells in the embryonic retina. At early stages of retinal development, these cells are competent to generate the entire complement of retinal neurons and glia; however, at later stages of development their progeny become restricted to rod photoreceptors, bipolar cells, and Müller glia. Although these cells are typically referred to as multipotent progenitors, those isolated from the early stages of retinogenesis could also be considered retinal stem cells because (1) they generate all retinal cell types; (2) they can generate very large clones; and (3) many of their divisions are symmetric. On the other hand, in mammals, the retinal stem/progenitor cells that are present in the developing retina do not continue to generate new retinal neurons throughout life, and so in this sense they do not appear to be neural stem cells in the retina of mammals like those in the hippocampus or subventricular zone. Nevertheless, the adult retina of some vertebrates continues to add new neurons and glia at the peripheral margin, and thus true retinal stem cells exist in these species. Presumably these cells were derived from a population of similar cells in the developing retina, but at this point there is no definitive way to distinguish the stem cells from the progenitors during retinogenesis.

Retinal stem cells and persistent progenitors in adult vertebrates: the ciliary marginal zone (CMZ)

The development of the amphibian, fish, or avian retina is not complete after the embryonic or neonatal period. In these animals, the retina continues to add new neurons into adulthood. This process is most apparent in teleost fish, which show a dramatic growth of the eye during their lifetime, of up to 100-fold. New retinal neurons are generated from a zone of cells at the peripheral margin of the retina, where it joins with the ciliary epithelium. These cells form a ring around the ciliary margin of the retina called the ciliary marginal zone (or CMZ) (Hollyfield, 1968). The CMZ cells of non-mammalian vertebrates resemble the early progenitor cells of the eye, and possibly even the "founder" cells of the optic vesicle. In fact, most of the retina of the mature frog (Reh and Constantine-Paton, 1983) and fish is generated by the CMZ cells. Wetts and Fraser (1988) carried out lineage tracing studies of these cells, similar to those done in embryos. They found that these cells can give rise to clones that contain all types of retinal neurons, like those of the embryonic retina. Therefore, it is likely that the CMZ contains a population of true retinal stem cells.

Recent molecular analysis of this region in frogs and chicks has shown that CMZ cells express most, if not all, of the EFTFs (Perron et al., 1998; Fischer and Reh, 2000). The CMZ cells also express bHLH transcription factors, such as Ngn2 and Ascl1 (Perron et al., 1998), and at least some of the CMZ cells respond to the same mitogenic growth factors as the embryonic progenitors (Mack and Fernald, 1993; Fischer and Reh, 2000; Moshiri et al., 2005). The CMZ is highly productive in fish and some amphibians, but in birds it is greatly reduced, and is absent in mammals (Fig. 24.4). Although the CMZ is robust in fish and amphibians, it is not known what percentage of the cells in this zone represents true retinal stem cells and what proportion of them are progenitors. In birds, most of the retina is generated during embryonic development and only a small number of retinal neurons are generated by the CMZ (Prada et al., 1991). It is not known whether this zone persists throughout the lifetime of the bird, but new retinal neurons are generated at the peripheral edge of the retina in chickens up to 1 month of

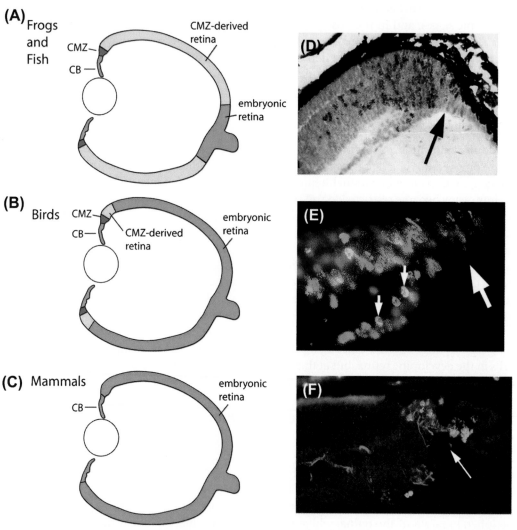

FIGURE 24.4

The CMZ of non-mammalian vertebrates contains retinal stem cells. (A—C) Diagrammatic representations of the regions of the retina generated in embryonic development (blue) or by the CMZ (yellow). In frogs and fish, most of the retina of the adult animal is generated by the CMZ, whereas in chicks only a small region is generated by the CMZ and in mammals this zone is absent. (D) The frog CMZ cells labeled with H3-thymidine. Arrow points to the anterior-most point of labeling, where the CMZ joins with the ciliary epithelium (from Reh and Constantine-Paton, 1983). (E) Chick CMZ labeled with BrdU and Islet1, to show new neurogenesis (small arrows: double labeled ganglion cells), as well as the point where the CMZ joins with the ciliary epithelium (large arrow) (from Fischer and Reh, 2000). (F) Nestin-BrdU double labeling shows a CMZ-like zone (arrow) in a mouse that is haplo-insufficient for the patched Shh receptor (from Moshiri and Reh, 2004).

age (Fischer and Reh, 2000), and in the quail eye for up to a year (Kubota et al., 2002). In addition to their potential to generate new retinal neurons, the chicken CMZ cells express many of the EFTFs, including Pax6 and Chx10 (Fischer and Reh, 2000).

The CMZ is greatly reduced or absent in the mammalian eye. Several groups have analyzed various species for evidence of ongoing proliferation in the margin of the retina, near its junction with the ciliary epithelium, but no mitotic cells are present in normal mice, rats, or macaques (Ahmad et al., 2000; Kubota et al., 2002; Moshiri and Reh, 2004). However, there is evidence that this zone may be repressed in the mammalian retina. Moshiri and Reh (2004) analyzed mice with a single functional allele of the Patched (Ptch) gene, a negative regulator of Shh signaling. They found a small number of proliferating cells at the retinal margin of these mice, into adulthood (Fig. 24.4F). Moreover, when the Ptch +/− mice were bred onto a background with photoreceptor degeneration, the proliferation in this zone was increased. This is reminiscent of the response to retinal damage observed in the CMZ cells of lower vertebrates, and suggests that the CMZ-like zone in Ptch +/− mice has much in common with the CMZ of frogs and fish. In addition, recent studies have found that proliferation can be stimulated after the progenitor cells have normally withdrawn from the cell cycle in the neonatal mammalian retina by the injection of specific growth factors (Close et al., 2005, 2006; Zhao et al., 2005), suggesting that the proliferation at the retinal margin may be suppressed in the mammalian retina by factors in their microenvironment.

TRANSDIFFERENTIATION AND RETINAL REGENERATION
The pigmented epithelium

One of the most striking examples of regeneration in vertebrates is the regeneration of the newt eye. These animals are capable of remarkable regeneration of a variety of tissues, and the eye is no exception. The neural retina can be completely removed in these animals, and within five weeks it is restored and the animal can respond to a visual stimulus. Regeneration of the retina in newts, and many other amphibians, occurs through a highly stereotypic process: shortly after the retina is removed, the adjacent pigmented epithelial tissue re-enters the cell cycle (Stone, 1950; Reyer, 1983; Stroeva and Mitashov, 1983). The proliferating pigmented epithelial cells lose their pigmentation and begin to express markers of retinal progenitors; this process was one of the first examples of transdifferentiation (Okada, 1980). The dedifferentiated pigment epithelial cells go on to generate new retinal neurons in a manner that resembles normal retinal histogenesis (Reh et al., 1987; Sakaguchi et al., 2003). Over a period of just a few weeks, the developmental process is recapitulated and the new retinal ganglion cells regrow connections with the brain.

Most of the details of the cellular transformations that occur during retinal regeneration in amphibians have been well established for many years; however, only recently have there been studies into the molecular mechanisms underlying this process. The use of molecular markers has established that the dedifferentiating pigment cells progress through a stage in which they resemble retinal progenitors (Reh, 1987; Sakami et al., 2005); however, it is possible that these cells go through a stage where they resemble stem or "founder" cells, because the retinal pigment epithelium (RPE) cells can regenerate the entire retina in some species, up to four complete times (Stone, 1957; Stone and Steinitz, 1957). A similar process of dedifferentiation of the pigmented epithelial cells occurs in embryonic chicks and mammals, and this also leads to retinal regeneration. However, the ability of RPE cells to transdifferentiate into retinal stem or progenitor cells is present only in the early stages of eye development (Park and Hollenberg, 1993; Pittack et al., 1997; Zhao et al., 2002). A key stimulus for retinal regeneration from the RPE in both amphibians and chick embryos is FGF. When added to cultures of pigment cells or *in vivo*, this factor stimulates the RPE cells to adopt a retinal progenitor identity (Park and Hollenberg, 1989, 1991; Pittack et al., 1991; Sakaguchi et al., 1997), and new laminated retina is generated. Recent evidence also indicates that Shh also plays a critical role in the process of RPE transdifferentiation (Spence et al., 2004).

The ciliary epithelium

The ciliary body has also been proposed to harbor retinal stem cells or progenitors. This region of the eye is made up of derivatives of both the neural tube (the pigmented and non-pigmented ciliary epithelia; Fig. 24.1) and the neural crest. The non-pigmented ciliary epithelium is the anterior extension of the neural retina, while the pigmented ciliary epithelium is the anterior-most extension of the pigmented epithelium. It is likely that both regions have potential to generate neurons, at least in some animals. A few years ago, we found that intraocular injections of growth factors (insulin, FGF2, and EGF) stimulated the proliferation and ultimate neuronal differentiation of cells within the ciliary epithelium (Fischer and Reh, 2003). Like the CMZ, these cells also express the EFTFs, Chx10, and Pax6. The neurons that develop in this region following growth factor treatments resemble amacrine cells, ganglion cells, and Müller glia, but they do not express markers of bipolar cells or photoreceptors.

The mammalian ciliary epithelium might also have some ability to generate neurons. Fischer and Reh (2001) described cells with neuronal and proliferation markers in the non-pigmented ciliary epithelium of the mature Macaque eye (Fischer et al., 2001). In addition, cells within the iris, the most anterior derivative of the primitive ocular neuroepithelium, are able to express photoreceptor genes when transfected with Crx (Haruta et al., 2001). Several groups have found that cells expressing neuronal markers can be generated from dissociated cell cultures of either the pigmented or non-pigmented cells of the ciliary epithelium (Ahmad et al., 2000; Tropepe et al., 2000; Das et al., 2005; Inoue et al., 2006). Tropepe et al. (2000) found that a small subpopulation of the pigmented cells form neurospheres and can be passaged to form new spheres. As a result of these characteristics, the cells have been termed retinal stem cells (Ahmad et al., 2000; Tropepe et al., 2000). Human eyes also contain these cells (Coles et al., 2004) and they can be grown *in vitro* for extended periods of time, and expanded and transplanted. However, although it has been claimed that these sphere-forming cells represent mammalian retinal stem cells, this is controversial. It has become apparent that the potential to form spheres is not a very reliable assay for stem cells in general. Moreover, a recent report by Dyer and colleagues demonstrated that the cells in the spheres do not lose their ciliary epithelial identity and only very low levels of neuronal genes are expressed (Cicero et al., 2009). Thus, they conclude that ciliary epithelial cells can be grown as spheres, but these are not retinal stem cells or even retinal progenitors. Therefore, at this time, it is not known whether these cells will be useful for reconstructing functional retinal circuits, and more work needs to be done to assess the potential of these cells. The relationship between the sphere-forming pigmented cells and the true retinal stem cells present in the CMZ of fish and frogs is also not clear, since the latter are not thought to be pigmented.

Intrinsic stem cells, rod precursors, and Müller glia

Müller glia are the primary glial cells intrinsic to the retina and the only glia generated by the multipotent retinal progenitors. Teleost fish also have a remarkable ability to regenerate after damage, not from the pigmented epithelium, but rather from the Müller glia (Raymond and Hitchcock, 1997). After experimental damage, the Müller glia in the fish retina undergo a robust proliferative response, upregulate genes normally present in retinal progenitors, and have the capacity to regenerate any of the different types of neurons (Fausett and Goldman, 2006; Bernardos et al., 2007; Fimbel et al., 2007). Some of the critical molecules necessary for regeneration in the fish retina have been recently defined. The proneural bHLH protein Achaete-scute homolog 1a (Ascl1a) is upregulated in the Müller glia soon after damage, and, if this gene is knocked down, the process of retinal regeneration fails (Fausett et al., 2008). Other neural progenitor genes, including those encoding Olig2, Notch1, and Pax6, are also upregulated in the Müller glia after damage, suggesting that Müller glia acquire the retinal progenitor phenotype in response to damage (Raymond et al., 2006; Yurco and Cameron, 2007; Thummel et al., 2008; Qin et al., 2009). Signaling molecules and growth factors have also been shown to be important in retinal regeneration in fish; midkine-a and -b and ciliary neurotrophic factor (CNTF) are important for fish retinal regeneration (Calinescu et al., 2009; Kassen et al., 2009). To date, at least four genes have been shown to be necessary for regeneration in the fish retina — encoding Ascl1a, PCNA, hspd1, and mps1.

The Müller glia of post-hatch chicks also respond to neurotoxic damage in the retina by re-entering the mitotic cell cycle (Fischer and Reh, 2001, 2002). In undamaged retina, the Müller glia do not proliferate; however, after retinal damage, Müller glial proliferation is extensive. Some of the proliferating Müller cells express progenitor genes; they upregulate the homeobox domain protein Chx10 and the paired box protein Pax6 and they express Cash1/Ascl1a, FoxN4, Notch1, Dll1, and Hes5 (Fischer and Reh, 2001; Hayes et al., 2007). These results suggest that the Müller glia can enter a progenitor state in chicks similar to that which occurs in the fish. However, although the majority of Müller glia re-enter the cell cycle after NMDA damage, only a subset express these markers of progenitors, and an even smaller number of these progenitor-like cells differentiate into cells that express neuronal markers. In the weeks following the NMDA treatment, the progeny of the Müller glia express markers of amacrine cells (calretinin, and the RNA-binding proteins HuC/D), bipolar cells (Islet1), and rare ganglion cells (Brn3; neurofilament). Most of the progeny of the Müller glia remain as Müller glia or persist as progenitor-like cells, and continue to express progenitor markers such as Chx10 and Pax6, but do not progress to neuronal differentiation.

The regenerative response of the post-hatch chick retina can be contrasted with that of the fish. Although the initial stages in the process appear quite similar, and retinal damage in both species leads to Müller cell proliferation and expression of neural progenitor genes, there are key differences. First, in fish, Müller glial-derived progenitors divide multiple times after damage, while, in the chick, the Müller glia appear to divide only once. Second, in the chick, only a small percentage of the progeny of the Müller glia differentiate into neurons; in fish, most seem to do so. The regenerative response of mammalian Müller glia to retinal injury is even more limited than that of the bird. Mammalian Müller glia become reactive and hypertrophic, but few re-enter the mitotic cycle after retinal damage (Dyer and Cepko, 2000). Nevertheless, several studies have found that a small number of Müller glia will re-enter the mitotic cycle after specific experimental paradigms (Ooto et al., 2004; Wan et al., 2008). In addition, this process can be stimulated by treating the retinas with mitogens after the damage (Close et al., 2006; Osakada et al., 2007; Karl et al., 2008; Takeda et al., 2008). As noted above, many of the proliferating Müller glia in fish and chicks express markers of progenitor cells after retinal damage, and the ability of Müller glia to regenerate new retinal neurons correlates with their adoption of a retinal progenitor pattern of gene expression. To determine whether aspects of the progenitor developmental program are reinitiated in mammalian Müller cells after damage, Karl et al. (2008) analyzed a panel of genes by RT-PCR in mouse retinas that had received NMDA-induced retinal damage and growth factor treatment. They found that Pax6 is upregulated in the Müller glia after NMDA damage, as are components of the Notch pathway, Dll1 and Notch1 (Das et al., 2006; Karl et al., 2008). Other groups have also shown upregulation of additional progenitor genes in Müller glia after other types of retinal injury (Osakada et al., 2007; Wan et al., 2008). Even in human Müller glia, it appears that at least some progenitor genes are expressed when the cells are grown in dissociated cell culture (Bull et al., 2008).

Taken together, these studies indicate that mammalian Müller glia can upregulate progenitor genes following retinal damage. However, the evidence is not as clear that these "dedifferentiated" Müller glia can regenerate new neurons in the mammalian retina. The first study to report evidence for neuronal regeneration in the mammalian retinas was that of Ooto et al. (2004). They reported that at least some of the BrdU+ Müller glia present at 2–3 days after

437

retinal injury in adult rats went on to express markers of bipolar cells and photoreceptors 2–3 weeks later; interestingly, they did not report that Müller glia could give rise to amacrine cells. Karl et al. (2008) found that the combination of NMDA and mitogen treatments promoted Müller glial proliferation and subsequent differentiation of some of their progeny into BrdU+ cells that express markers of amacrine cells, including Calretinin, NeuN, Pax6, Prox1, and GAD67-GFP (Fig. 24.5). These studies were carried out with GAD67-GFP-expressing mice, and so it was possible to verify that the BrdU+ cells had both laminar position and morphology consistent with an amacrine cell identity. The presence of amacrine cell regeneration in the NMDA-treated mouse retinas is consistent with the fact that amacrine cells and ganglion cells are the cells destroyed by this treatment, suggesting that there is some specificity to this process. In addition, one of the transcription factors upregulated in Müller glia after retinal injury in mice and post-hatch chicks is FoxN4, a transcription factor necessary for the production of amacrine cells during retinal histogenesis (Hayes et al., 2007; Karl et al., 2008). Other groups

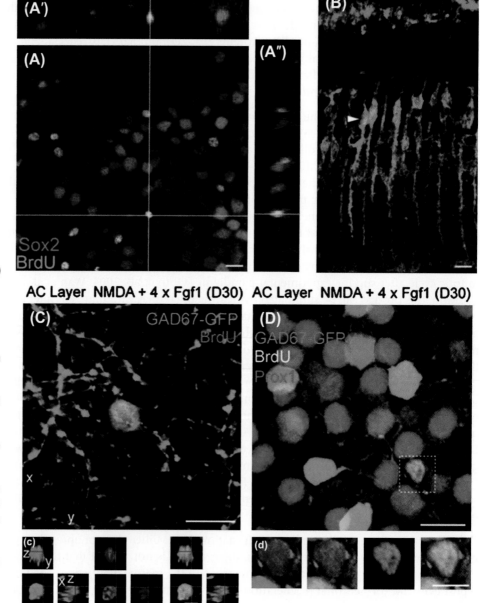

FIGURE 24.5

Regeneration in mammalian retina from the resident Müller glia. (A–A″) Confocal image of retinal flatmount showing BrdU+ (green) and Sox2+ (red) double-labeled cells at depths corresponding to Müller glia as early as 4 h after induction of proliferation following NMDA damage and EGF stimulation. (B) Confocal image of a retinal section showing Sox2 expression (red, arrow) in the GFAP-expressing Müller glia. (C–D) GAD67-GFP mice express GFP in immature neurons as well as a subset of mature retinal ganglion cells (RGCs), horizontal cells, and all GABAergic amacrine cells in adult mouse retina. NMDA followed by 4 FGF1+insulin injections of GAD67-GFP mouse retinas leads to BrdU+ cells in the amacrine cell layer (C and D). (D) At D30 some BrdU+ GAD67-GFP+ cells are also labeled for Prox1, marker of amacrine and horizontal cells in the retina. (c–d) represent confocal x–z and y–z confocal views showing co-localization of the markers and GFP. Reproduced from Karl et al., 2008.

have used additional types of damage to study the potential for retinal regeneration from Müller glia in mice and rats, e.g. Wnt3a (Osakada et al., 2007), MNU damage (Wan et al., 2007, 2008), and alpha-amino adipic acid (Takeda et al., 2008). All these treatments stimulate Müller glial proliferation, and some of the BrdU+ cells in each study were co-labeled with markers for different types of retinal neurons, and particularly rod photoreceptors. Although there are many differences in the specifics of each study, taken together they suggest that a subset of the Müller glia in the rodent retina can regenerate at least small numbers of amacrine cells and photoreceptors.

RETINAL NEURONS FROM EMBRYONIC STEM CELLS AND INDUCED PLURIPOTENT STEM CELLS

As discussed in the previous section, the retinas of non-mammalian vertebrates have a variety of different strategies for repair, and these typically involve dedifferentiation of existing retinal progenitor cells; however, this potential is greatly reduced in mammalian retinas. A number of investigators have therefore attempted to transplant fetal retinal progenitor cells in various models of retinal degeneration, including a current Phase 1 clinical trial, with some success (Lund et al., 2003; Radtke et al., 2008). However, it is difficult to imagine that primary human fetal retinal progenitors will ever be readily accessible, as fetal tissue has been limiting in other cell-based strategies elsewhere in the nervous system. Thus, several investigators are developing methods to direct human embryonic stem cells and more recently human-induced pluripotent stem cells to a retinal progenitor and a photoreceptor cell identity. In this section we will review the progress made to date. Embryonic stem cells, derived from the inner cell mass of the blastocyst, can self-renew indefinitely under appropriate culture conditions (Thomson et al., 1998) and their ability to differentiate into most, if not all, cells in the body makes them an attractive alternative to endogenous retinal progenitor cells for tissue engineering. More recently, researchers have been able to successfully reprogram fully differentiated adult cells back to a truly pluripotent state. These cells, termed induced pluripotent stem cells, could help eliminate some of the ethical concerns associated with the use of human ES cells.

Several groups developed protocols for inducing mouse embryonic stem cells into a retinal differentiation pathway (Zhao et al., 2002; Hirano et al., 2003; Meyer et al., 2004; Tabata et al., 2004; Ikeda et al., 2005; Sugie et al., 2005; Aoki et al., 2006). Some of the early work on neural induction involved use of retinoic acid (Bain et al., 1995, 1996; Fraichard et al., 1995) or basic fibroblast growth factor (FGF-2) and a combination of insulin, transferrin, selenium, and fibronectin (ITSFn) (Okabe et al., 1996; Lee et al., 2000). This approach has yielded a high proportion of neuroepithelial cells, which can then be induced to differentiate into neurons and glia. Although these early attempts did not demonstrate definitive retinal markers in the ES cultures, placing the neurally induced ES cells into a neurogenic micro-environment, either *in vitro* (e.g. dissociated newborn rat retinal cells (Zhao et al., 2002) or embryonic chicken retinal tissue (Sugie et al., 2005)) or *in vivo* (e.g. in a degenerating retina (Meyer et al., 2004)) provided some stimulus for further retinal differentiation. Upon co-culture with neurogenic retinal tissue, some cells even expressed markers of photoreceptor precursors such as Crx and Nrl, but rarely differentiated photoreceptor markers such as rhodopsin and interphotoreceptor retinoid-binding protein (IRBP). When neurally induced ES cells were transplanted to a degenerating retina, Meyer et al. (2004) found that the cells penetrated the retina and acquired neuronal morphology; however, the cells did not express definitive photoreceptor markers, though they did seem to promote better survival of the remaining host photoreceptors (Meyer et al., 2004).

Other approaches have also shown some promise in directed differentiation of mouse ES cells to a retinal identity. Culturing ESC-derived embryoid bodies in combination with the stromal cell line PA6 produces retina and pigmented epithelial cells (Kawasaki et al., 2000; Hirano et al., 2003; Yoshizaki et al., 2004; Aoki et al., 2006). Driving retinal differentiation by

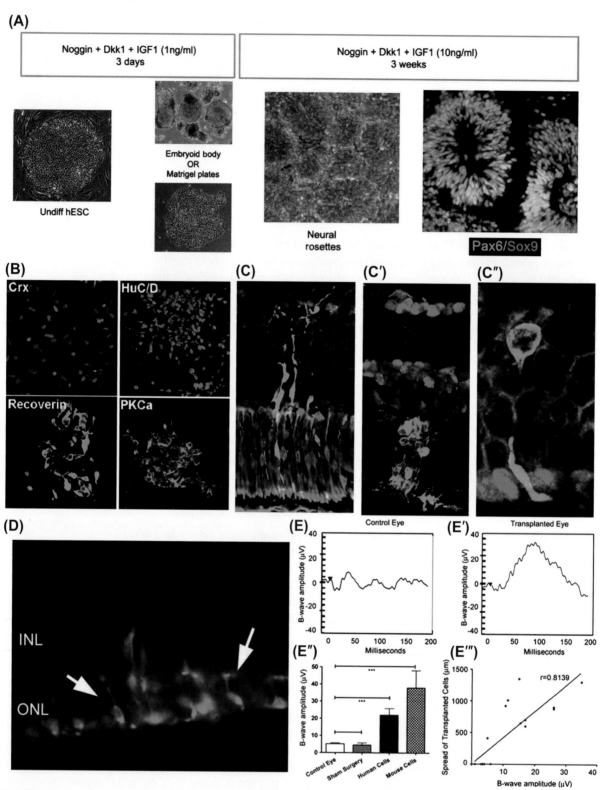

440

FIGURE 24.6

Embryonic stem cells for retinal repair. (A) Schematic of the steps involved in retinal induction of undifferentiated human ES and iPS cells. Lamba and Reh published a two-step protocol with or without embryoid body formation using a combination of BMP inhibitor noggin, Wnt inhibitor dkk1, and IGF-1. This protocol results in the formation of neural rosettes by 2—3 weeks and these rosettes stain for retinal progenitor markers Pax6 and Sox9 (from Lamba et al., 2010). (B) Upon further differentiation the cells express markers of differentiated retinal neurons such as Crx and recoverin (photoreceptors), Hu C/D (ganglion and amacrine cells), and PKC-a (bipolar cells). (C—C″) These cells (infected with a GFP-expressing lentivirus) when transplanted in normal mouse eyes either at birth (C) or in adults (C′—C″) migrate and integrate into the appropriate retinal layers and show the typical photoreceptor morphology (C″). (D—E) When these human ESC-derived retinal cells were transplanted into blind Crx−/− mice, a model of LCA, the cells again migrated into the host retina (D) and had the ability to restore light-response as measured by ERG (E—E″). The ERG b-wave restoration was found to be dependent on integration efficiency (E‴). Reproduced from Lamba et al., 2009.

overexpressing eye field transcription factors such as Rx (Tabata et al., 2004) is also effective. However, studies that have used specific combinations of signaling factors have so far had the most success. Masayo Takahashi's group (Ikeda et al., 2005) used a combination of three factors, lefty-A, Dkk1, and Activin A, to direct mouse ES cells to a retinal fate. Lefty-A acts as a neural inducer by inhibiting BMP signaling (Meno et al., 1997); Dkk1 induces anterior neural fates by inhibiting Wnt signaling (del Barco Barrantes et al., 2003); Activin A directs optic vesicle cells to a pigmented epithelial fate, and in later stages of retinal development promotes differentiation of progenitors into photoreceptors (Davis et al., 2000; Fuhrmann et al., 2000). This protocol resulted in almost 30% of all cells expressing Pax6 and Rx, two key EFTFs (see above). Upon co-culture of these cells with re-aggregated adult retinal neurons, a large proportion of these cells expressed rhodopsin and recoverin, markers of photoreceptors. Transplantation of these cells onto retinal explants resulted in their integration into host retina *in vitro*.

The combination of key developmental signaling factors also proved critical for the design of a protocol for the efficient generation of retinal cells from human ES cells. Lamba et al. (2006) reported that human ES cells can be efficiently directed to a retinal cell fate using a combination of a BMP inhibitor (Noggin), a Wnt inhibitor (Dkk1), and insulin-like growth factor (IGF-1) (Lamba et al., 2006). This protocol resulted in highly efficient directed differentiation of the ES cells to the retinal progenitor fate (Fig. 24.6). The cells differentiated under this protocol expressed all the key eye field transcription factors, and early markers of photoreceptor differentiation such as Crx, Nrl, and recoverin. Photoreceptor differentiation was found to be enhanced when the human ESC-derived retinal cells were co-cultured with explants of adult retinas from wild-type and degenerated retinas from wild-type and retinal degeneration mice, suggesting that micro-environmental cues for full photoreceptor differentiation may be missing from the ES cells alone. A similar protocol was described by Osakada et al., using alternate antagonists of Wnt and BMP/nodal pathways (Osakada et al., 2008). Taken together, these reports indicate that eye field induction in human and mouse ES cells is likely quite similar to the normal developmental process, in which Wnt antagonists and BMP antagonists combine to produce anterior neural tissue (see above). Although many of the different markers of the eye field are also expressed in other anterior neural tissue, Lamba et al. (2006) found that markers of cerebral cortex, such as Emx, were not induced by this protocol. More importantly, photoreceptors are only generated by retinal progenitors, and both the Lamba and Osakada protocols reported the early stages of photoreceptor differentiation in the cultures. More recently it has been shown that combinations of small molecules that modulate these same pathways can also be substituted in the retinal differentiation protocols (Osakada et al., 2009). A very different protocol has been developed by Meyer et al. (2009); their protocol relies more on the identification and selection of the retinal progenitors from other neural and non-neural cells in the differentiating ES cultures; however, by this method, the authors were able to show that the differentiation of human ES cells to retina follows the same stages *in vitro* as are known to occur *in vivo* (Meyer et al., 2009).

As mentioned above, human iPS cells could be an alternative to ES cells. These cells circumvent some of the ethical issues of ES cells, in that destruction of embryos is not necessary for their derivation. Another advantage is that iPS cells can be generated from adult human fibroblasts; this allows investigators to generate pluripotent cells from patients suffering from specific inherited disorders for generation of disease models and/or patient-specific cell replacement therapy. Several groups have recently demonstrated that iPS cell lines can be directed to retinal cell fates using the protocols that were developed for ES cells (Hirami et al., 2009; Meyer et al., 2009; Lamba et al., 2010). The cells develop many characteristics of immature photoreceptors *in vitro*, and, following transplantation to normal mouse retinas, they can incorporate into the appropriate layer (Lamba et al., 2010). Although the efficiency of retinal differentiation may vary with specific cell lines and the methods used to generate pluripotency, in some cases these cells have proven to be equal or even better than many ES lines in the retinal differentiation protocols.

One of the key assays used to assess the effectiveness of directed differentiation is the ability of the differentiated cells to integrate appropriately following transplantation. For cell-based therapies, this is essentially a proof of principle. Banin et al. were the first to show that ESC-derived retinal cells can survive following transplantation to the eye (Banin et al., 2006); however, the protocol used for retinal differentiation in this study was not very efficient, and there was little evidence for definitive retinal cell differentiation markers in the transplanted cells. The first demonstration that human ESC-derived photoreceptors can functionally restore a light response to congenitally blind mice was that of Lamba et al. (2009). ES cells directed to the retinal fate were labeled with a GFP-expressing virus and then transplanted to the retinas of either normal mice or mice with a specific genetic retinal defect (i.e. Crx−/−) (Lamba et al., 2009). In the Crx-deficient mice, a defect also present in humans in one form of Leber's Congenital Amaurosis (LCA), the photoreceptors fail to express the genes necessary for photo-transduction and eventually degenerate. When the human ESC-derived retinal cells were transplanted to the subretinal space of these animals, the cells had the ability to integrate in the degenerated environment, and even restore some light response to the mice (Fig. 24.6D−E). These data suggest that, in principle, human ES cells might be useful for photoreceptor replacement therapies.

If cell replacement is ever to become a viable approach to restoration of vision in patients with retinal degenerations, there are a number of critical issues that must be addressed. Foremost is the issue of safety, and the possibility of teratoma formation is a concern in ocular use of ES cells as it is in other tissues. In addition, several studies have suggested that the cells that integrate best into host retina after transplantation are the recently "born" photoreceptors; i.e. those rods and cones that have been generated from the retinal progenitor within a few days of harvest for transplantation (MacLaren et al., 2006). One approach to addressing both of these considerations would be to purify recently generated photoreceptors from the hES cells after a directed differentiation protocol. Recently, Lamba et al. (2010) have reported that hES cells, directed to retinal fates, can be infected with lentivirus that contains a photoreceptor-specific promoter driving GFP to identify the photoreceptors in the live cultures. These cells can then be purified to over 90% using fluorescent activated-cell sorting (FACS). Subsequent trans-plantation of the FACS-purified photoreceptors shows that they can integrate into mouse retina, and, since the photoreceptors are a post-mitotic population, the risk of teratoma is almost completely eliminated.

CONCLUSION

Retinal diseases that cause blindness through the loss of one or more retinal neuron type are becoming increasingly common in the population. The ability to treat blindness by cell replacement therapy would therefore be a useful addition to the research efforts in the prevention and treatment of blindness. The retina has been a classic model for regeneration studies, particularly in lower vertebrates, and focused efforts to uncover similar mechanisms in mammals are meeting with some success. While an effective cell-based therapy is still many years away, there are several promising approaches: (1) stimulation of endogenous repair; (2) harvest, *in vitro* expansion, and transplantation of adult stem cells from the eye; and (3) direction of human ES and iPS cells to a retinal identity for transplantation.

Towards these goals, it is clear that a number of key pieces of biology need to be better understood. For example, at the present time we cannot distinguish between stem cells and progenitors at any stage of development or in adult animals. While there are some assays that claim to distinguish between these two potential types of cells in other regions of the nervous system, these typically rely on the fact that stem cells are multipotent and self-renewing. In the retina, lineage analysis during development has shown that the majority of dividing cells are multipotent and can make a variety of different clone sizes, some reaching many thousands of cells. Moreover, both symmetric (e.g. two dividing cells) and asymmetric (dividing cell and neuron) divisions are equally common. Thus, at present *in vitro* assays that show multi-potentiality or self-renewal based on subcloning procedures cannot discriminate among the different types of dividing cells in the retina, as they have been used to do in other areas of the CNS.

We also do not understand the process of dedifferentiation or cellular plasticity. RPE can dedifferentiate into stem or progenitor cells that generate an entire new retina in some species. In mammals, pigmented cells can lose their pigmentation *in vitro* and go on to express a variety of proteins normally present only in the neural retina; however, the cells do not appear to recapitulate the entire program of regeneration, and most of the cells generated in these cultures do not resemble neurons morphologically or functionally. Are there intrinsic limitations to the potential of these cells to generate true functional neurons in mammals, or are necessary factors in the local microenvironment not present in the damaged mammalian retina?

After either neurotoxic or surgical damage, both fish and birds can generate new neurons from intrinsic sources. In fish, the intrinsic retinal source of regeneration may include rod progenitors, an intrinsic stem cell and/or Müller glia. In the case of the chick retina, the Müller glia appear to be the only source of the new neurons. However, there is a large difference in the regenerative responses between these two animals. In the fish, the regeneration is nearly perfect, whereas in the bird most of the proliferating glia do not go on to make new neurons, but rather remain in an undifferentiated state. What is the block to efficient regeneration in the bird? Further studies of the factors that regulate Müller glial proliferation after damage in both chicks and mammals may lead to clues for stimulating the process of regeneration. Moreover, gene expression profiles between Müller glia and retinal progenitors may also lead to a better understanding of the process of dedifferentiation in Müller glia that precedes neuronal regeneration.

Lastly, the future of retinal repair may well require the transplantation of retinal cells that have been generated from progenitors or stem cells *in vitro*. While the work on human embryonic stem cells is proceeding at a rapid pace, there are still some fundamental questions that will need to be resolved before a cell-based transplantation therapy can become a reality. The transplantation studies using fetal cells that have been carried out over the past two decades suggest that survival and integration of the transplanted cells may be two key barriers to functional restoration of degenerated retina. Moreover, the *in vitro* expansion and appropriate cell type differentiation of retinal progenitors, derived either from embryonic stem cells or an adult retinal cell, will require a better understanding of the factors that normally control retinal cell fate during development. We have come a long way in our understanding of the phenomenon of retinal regeneration, far enough to appreciate the enormity of the task ahead for the translation of this knowledge to clinical practice.

References

Ahmad, I., Tang, L., & Pham, H. (2000). Identification of neural progenitors in the adult mammalian eye. *Biochem. Biophys. Res. Commun., 270*, 517–521.

Anchan, R. M., & Reh, T. A. (1995). Transforming growth factor-beta-3 is mitogenic for rat retinal progenitor cells in vitro. *J. Neurobiol., 28*, 133–145.

Anchan, R. M., Reh, T. A., Angello, J., Balliet, A., & Walker, M. (1991). EGF and TGF-alpha stimulate retinal neuroepithelial cell proliferation in vitro. *Neuron, 6*, 923–936.

Andreazzoli, M., Gestri, G., Angeloni, D., Menna, E., & Barsacchi, G. (1999). Role of Xrx1 in Xenopus eye and anterior brain development. *Development, 126*, 2451–2460.

Andreazzoli, M., Gestri, G., Cremisi, F., Casarosa, S., Dawid, I. B., & Barsacchi, G. (2003). Xrx1 controls proliferation and neurogenesis in Xenopus anterior neural plate. *Development, 130*, 5143–5154.

Aoki, H., Hara, A., Nakagawa, S., Motohashi, T., Hirano, M., Takahashi, Y., et al. (2006). Embryonic stem cells that differentiate into RPE cell precursors in vitro develop into RPE cell monolayers in vivo. *Exp. Eye Res., 82*, 265–274.

Bain, G., Kitchens, D., Yao, M., Huettner, J. E., & Gottlieb, D. I. (1995). Embryonic stem cells express neuronal properties in vitro. *Dev. Biol.*, *168*, 342–357.

Bain, G., Ray, W. J., Yao, M., & Gottlieb, D. I. (1996). Retinoic acid promotes neural and represses mesodermal gene expression in mouse embryonic stem cells in culture. *Biochem. Biophys. Res. Commun.*, *223*, 691–694.

Banin, E., Obolensky, A., Idelson, M., Hemo, I., Reinhardtz, E., Pikarsky, E., et al. (2006). Retinal incorporation and differentiation of neural precursors derived from human embryonic stem cells. *Stem Cells*, *24*, 246–257.

Bernardos, R. L., Barthel, L. K., Meyers, J. R., & Raymond, P. A. (2007). Late-stage neuronal progenitors in the retina are radial Muller glia that function as retinal stem cells. *J. Neurosci.*, *27*, 7028–7040.

Bernier, G., Panitz, F., Zhou, X., Hollemann, T., Gruss, P., & Pieler, T. (2000). Expanded retina territory by midbrain transformation upon overexpression of Six6 (Optx2) in Xenopus embryos. *Mech. Dev.*, *93*, 59–69.

Bovolenta, P., Mallamaci, A., Briata, P., Corte, G., & Boncinelli, E. (1997). Implication of OTX2 in pigment epithelium determination and neural retina differentiation. *J. Neurosci.*, *17*, 4243–4252.

Bovolenta, P., Mallamaci, A., Puelles, L., & Boncinelli, E. (1998). Expression pattern of cSix3, a member of the Six/sine oculis family of transcription factors. *Mech. Dev.*, *70*, 201–203.

Bull, N. D., Limb, G. A., & Martin, K. R. (2008). Human Muller stem cell (MIO-M1) transplantation in a rat model of glaucoma: survival, differentiation, and integration. *Invest. Ophthalmol. Vis. Sci.*, *49*, 3449–3456.

Calinescu, A. A., Vihtelic, T. S., Hyde, D. R., & Hitchcock, P. F. (2009). Cellular expression of midkine-a and midkine-b during retinal development and photoreceptor regeneration in zebrafish. *J. Comp. Neurol.*, *514*, 1–10.

Carl, M., Loosli, F., & Wittbrodt, J. (2002). Six3 inactivation reveals its essential role for the formation and patterning of the vertebrate eye. *Development*, *129*, 4057–4063.

Cavodeassi, F., Carreira-Barbosa, F., Young, R. M., Concha, M. L., Allende, M. L., Houart, C., et al. (2005). Early stages of zebrafish eye formation require the coordinated activity of Wnt11, Fz5, and the Wnt/beta-catenin pathway. *Neuron*, *47*, 43–56.

Cepko, C. L., Austin, C. P., Yang, X., Alexiades, M., & Ezzeddine, D. (1996). Cell fate determination in the vertebrate retina. *Proc. Natl. Acad. Sci. U.S.A.*, *93*, 589–595.

Chow, R. L., Altmann, C. R., Lang, R. A., & Hemmati-Brivanlou, A. (1999). Pax6 induces ectopic eyes in a vertebrate. *Development*, *126*, 4213–4222.

Chuang, J. C., & Raymond, P. A. (2001). Zebrafish genes rx1 and rx2 help define the region of forebrain that gives rise to retina. *Dev. Biol.*, *231*, 13–30.

Ciccolini, F., & Svendsen, C. N. (1998). Fibroblast growth factor 2 (FGF-2) promotes acquisition of epidermal growth factor (EGF) responsiveness in mouse striatal precursor cells: identification of neural precursors responding to both EGF and FGF-2. *J. Neurosci.*, *18*, 7869–7880.

Cicero, S. A., Johnson, D., Reyntjens, S., Frase, S., Connell, S., Chow, L. M., et al. (2009). Cells previously identified as retinal stem cells are pigmented ciliary epithelial cells. *Proc. Natl. Acad. Sci. U.S.A.*, *106*, 6685–6690.

Close, J. L., Gumuscu, B., & Reh, T. A. (2005). Retinal neurons regulate proliferation of postnatal progenitors and Muller glia in the rat retina via TGF beta signaling. *Development*, *132*, 3015–3026.

Close, J. L., Liu, J., Gumuscu, B., & Reh, T. A. (2006). Epidermal growth factor receptor expression regulates proliferation in the postnatal rat retina. *Glia*, *54*, 94–104.

Coles, B. L., Angenieux, B., Inoue, T., del Rio-Tsonis, K., Spence, J. R., McInnes, R. R., et al. (2004). Facile isolation and the characterization of human retinal stem cells. *Proc. Natl. Acad. Sci. U.S.A.*, *101*, 15772–15777.

Das, A. V., James, J., Rahnenfuhrer, J., Thoreson, W. B., Bhattacharya, S., Zhao, X., et al. (2005). Retinal properties and potential of the adult mammalian ciliary epithelium stem cells. *Vision Res.*, *45*, 1653–1666.

Das, A. V., Mallya, K. B., Zhao, X., Ahmad, F., Bhattacharya, S., Thoreson, W. B., et al. (2006). Neural stem cell properties of Muller glia in the mammalian retina: regulation by Notch and Wnt signaling. *Dev. Biol.*, *299*, 283–302.

Davis, A. A., Matzuk, M. M., & Reh, T. A. (2000). Activin A promotes progenitor differentiation into photoreceptors in rodent retina. *Mol. Cell Neurosci.*, *15*, 11–21.

de Melo, J., Qiu, X., Du, G., Cristante, L., & Eisenstat, D. D. (2003). Dlx1, Dlx2, Pax6, Brn3b, and Chx10 homeobox gene expression defines the retinal ganglion and inner nuclear layers of the developing and adult mouse retina. *J. Comp. Neurol.*, *461*, 187–204.

del Barco Barrantes, I., Davidson, G., Grone, H. J., Westphal, H., & Niehrs, C. (2003). Dkk1 and noggin cooperate in mammalian head induction. *Genes Dev.*, *17*, 2239–2244.

Dyer, M. A., & Cepko, C. L. (2000). Control of Muller glial cell proliferation and activation following retinal injury. *Nat. Neurosci.*, *3*, 873–880.

Eivers, E., McCarthy, K., Glynn, C., Nolan, C. M., & Byrnes, L. (2004). Insulin-like growth factor (IGF) signalling is required for early dorso-anterior development of the zebrafish embryo. *Int. J. Dev. Biol.*, *48*, 1131–1140.

Esteve, P., & Bovolenta, P. (2006). Secreted inducers in vertebrate eye development: more functions for old morphogens. *Curr. Opin. Neurobiol.*, *16*, 13–19.

Esteve, P., Lopez-Rios, J., & Bovolenta, P. (2004). SFRP1 is required for the proper establishment of the eye field in the medaka fish. *Mech. Dev., 121*, 687–701.

Fausett, B. V., & Goldman, D. (2006). A role for alpha1 tubulin-expressing Muller glia in regeneration of the injured zebrafish retina. *J. Neurosci., 26*, 6303–6313.

Fausett, B. V., Gumerson, J. D., & Goldman, D. (2008). The proneural basic helix-loop-helix gene ascl1a is required for retina regeneration. *J. Neurosci., 28*, 1109–1117.

Fekete, D. M., Perez-Miguelsanz, J., Ryder, E. F., & Cepko, C. L. (1994). Clonal analysis in the chicken retina reveals tangential dispersion of clonally related cells. *Dev. Biol., 166*, 666–682.

Fimbel, S. M., Montgomery, J. E., Burket, C. T., & Hyde, D. R. (2007). Regeneration of inner retinal neurons after intravitreal injection of ouabain in zebrafish. *J. Neurosci., 27*, 1712–1724.

Fischer, A. J., Hendrickson, A., & Reh, T. A. (2001). Immunocytochemical characterization of cysts in the peripheral retina and pars plana of the adult primate. *Invest. Ophthalmol. Vis. Sci., 42*, 3256–3263.

Fischer, A. J., & Reh, T. A. (2000). Identification of a proliferating marginal zone of retinal progenitors in postnatal chickens. *Dev. Biol., 220*, 197–210.

Fischer, A. J., & Reh, T. A. (2001). Muller glia are a potential source of neural regeneration in the postnatal chicken retina. *Nat. Neurosci., 4*, 247–252.

Fischer, A. J., & Reh, T. A. (2002). Exogenous growth factors stimulate the regeneration of ganglion cells in the chicken retina. *Dev. Biol., 251*, 367–379.

Fischer, A. J., & Reh, T. A. (2003). Potential of Muller glia to become neurogenic retinal progenitor cells. *Glia, 43*, 70–76.

Fraichard, A., Chassande, O., Bilbaut, G., Dehay, C., Savatier, P., & Samarut, J. (1995). In vitro differentiation of embryonic stem cells into glial cells and functional neurons. *J. Cell Sci., 108*(Pt 10), 3181–3188.

Fuhrmann, S., Levine, E. M., & Reh, T. A. (2000). Extraocular mesenchyme patterns the optic vesicle during early eye development in the embryonic chick. *Development, 127*, 4599–4609.

Furukawa, T., Kozak, C. A., & Cepko, C. L. (1997). rax, a novel paired-type homeobox gene, shows expression in the anterior neural fold and developing retina. *Proc. Natl. Acad. Sci. U.S.A., 94*, 3088–3093.

Grindley, J. C., Davidson, D. R., & Hill, R. E. (1995). The role of Pax-6 in eye and nasal development. *Development, 121*, 1433–1442.

Halder, G., Callaerts, P., & Gehring, W. J. (1995). Induction of ectopic eyes by targeted expression of the eyeless gene in Drosophila. *Science, 267*, 1788–1792.

Haruta, M., Kosaka, M., Kanegae, Y., Saito, I., Inoue, T., Kageyama, R., et al. (2001). Induction of photoreceptor-specific phenotypes in adult mammalian iris tissue. *Nat. Neurosci., 4*, 1163–1164.

Hayes, S., Nelson, B. R., Buckingham, B., & Reh, T. A. (2007). Notch signaling regulates regeneration in the avian retina. *Dev. Biol., 312*, 300–311.

Hill, R. E., Favor, J., Hogan, B. L., Ton, C. C., Saunders, G. F., Hanson, I. M., et al. (1991). Mouse small eye results from mutations in a paired-like homeobox-containing gene. *Nature, 354*, 522–525.

Hirami, Y., Osakada, F., Takahashi, K., Okita, K., Yamanaka, S., Ikeda, H., et al. (2009). Generation of retinal cells from mouse and human induced pluripotent stem cells. *Neurosci. Lett., 458*, 126–131.

Hirano, M., Yamamoto, A., Yoshimura, N., Tokunaga, T., Motohashi, T., Ishizaki, K., et al. (2003). Generation of structures formed by lens and retinal cells differentiating from embryonic stem cells. *Dev. Dyn., 228*, 664–671.

Hollyfield, J. G. (1968). Differential addition of cells to the retina in Rana pipiens tadpoles. *Dev. Biol., 18*, 163–179.

Holt, C. E., Bertsch, T. W., Ellis, H. M., & Harris, W. A. (1988). Cellular determination in the Xenopus retina is independent of lineage and birth date. *Neuron, 1*, 15–26.

Ikeda, H., Osakada, F., Watanabe, K., Mizuseki, K., Haraguchi, T., Miyoshi, H., et al. (2005). Generation of Rx+/Pax6+ neural retinal precursors from embryonic stem cells. *Proc. Natl. Acad. Sci. U.S.A., 102*, 11331–11336.

Inoue, T., Kagawa, T., Fukushima, M., Shimizu, T., Yoshinaga, Y., Takada, S., et al. (2006). Activation of canonical Wnt pathway promotes proliferation of retinal stem cells derived from adult mouse ciliary margin. *Stem Cells, 24*, 95–104.

James, J., Das, A. V., Bhattacharya, S., Chacko, D. M., Zhao, X., & Ahmad, I. (2003). In vitro generation of early-born neurons from late retinal progenitors. *J. Neurosci., 23*, 8193–8203.

Jasoni, C. L., & Reh, T. A. (1996). Temporal and spatial pattern of MASH-1 expression in the developing rat retina demonstrates progenitor cell heterogeneity. *J. Comp. Neurol., 369*, 319–327.

Jasoni, C. L., Walker, M. B., Morris, M. D., & Reh, T. A. (1994). A chicken achaete-scute homolog (CASH-1) is expressed in a temporally and spatially discrete manner in the developing nervous system. *Development, 120*, 769–783.

Jean, D., Bernier, G., & Gruss, P. (1999). Six6 (Optx2) is a novel murine Six3-related homeobox gene that demarcates the presumptive pituitary/hypothalamic axis and the ventral optic stalk. *Mech. Dev., 84*, 31–40.

Karl, M. O., Hayes, S., Nelson, B. R., Tan, K., Buckingham, B., & Reh, T. A. (2008). Stimulation of neural regeneration in the mouse retina. *Proc. Natl. Acad. Sci. U.S.A., 105,* 19508–19513.

Kassen, S. C., Thummel, R., Campochiaro, L. A., Harding, M. J., Bennett, N. A., & Hyde, D. R. (2009). CNTF induces photoreceptor neuroprotection and Muller glial cell proliferation through two different signaling pathways in the adult zebrafish retina. *Exp. Eye Res., 88,* 1051–1064.

Kawasaki, H., Mizuseki, K., Nishikawa, S., Kaneko, S., Kuwana, Y., Nakanishi, S., et al. (2000). Induction of midbrain dopaminergic neurons from ES cells by stromal cell-derived inducing activity. *Neuron, 28,* 31–40.

Kobayashi, M., Toyama, R., Takeda, H., Dawid, I. B., & Kawakami, K. (1998). Overexpression of the forebrain-specific homeobox gene six3 induces rostral forebrain enlargement in zebrafish. *Development, 125,* 2973–2982.

Kubota, R., Hokoc, J. N., Moshiri, A., McGuire, C., & Reh, T. A. (2002). A comparative study of neurogenesis in the retinal ciliary marginal zone of homeothermic vertebrates. *Brain Res. Dev. Brain Res., 134,* 31–41.

Lamba, D. A., Gust, J., & Reh, T. A. (2009). Transplantation of human embryonic stem cell-derived photoreceptors restores some visual function in Crx-deficient mice. *Cell Stem Cell, 4,* 1–7.

Lamba, D. A., Karl, M. O., Ware, C. B., & Reh, T. A. (2006). Efficient generation of retinal progenitor cells from human embryonic stem cells. *Proc. Natl. Acad. Sci. U.S.A., 103,* 12769–12774.

Lamba, D. A., McUsic, A., Hirata, R. K., Wang, P. R., Russell, D., & Reh, T. A. (2010). Generation, purification and transplantation of photoreceptors derived from human induced pluripotent stem cells. *PLoS One, 5,* e8763.

Lee, H. S., Bong, Y. S., Moore, K. B., Soria, K., Moody, S. A., & Daar, I. O. (2006). Dishevelled mediates ephrinB1 signalling in the eye field through the planar cell polarity pathway. *Nat. Cell Biol., 8,* 55–63.

Lee, S. H., Lumelsky, N., Studer, L., Auerbach, J. M., & McKay, R. D. (2000). Efficient generation of midbrain and hindbrain neurons from mouse embryonic stem cells. *Nat. Biotechnol., 18,* 675–679.

Li, H., Tierney, C., Wen, L., Wu, J. Y., & Rao, Y. (1997). A single morphogenetic field gives rise to two retina primordia under the influence of the prechordal plate. *Development, 124,* 603–615.

Li, S., Mo, Z., Yang, X., Price, S. M., Shen, M. M., & Xiang, M. (2004). Foxn4 controls the genesis of amacrine and horizontal cells by retinal progenitors. *Neuron, 43,* 795–807.

Lillien, L., & Cepko, C. (1992). Control of proliferation in the retina: temporal changes in responsiveness to FGF and TGF alpha. *Development, 115,* 253–266.

Liu, C., Wang, Y., Smallwood, P. M., & Nathans, J. (2008). An essential role for Frizzled5 in neuronal survival in the parafascicular nucleus of the thalamus. *J. Neurosci., 28,* 5641–5653.

Livesey, F. J., & Cepko, C. L. (2001). Vertebrate neural cell-fate determination: lessons from the retina. *Nat. Rev. Neurosci., 2,* 109–118.

Loosli, F., Koster, R. W., Carl, M., Krone, A., & Wittbrodt, J. (1998). Six3, a medaka homologue of the Drosophila homeobox gene sine oculis is expressed in the anterior embryonic shield and the developing eye. *Mech. Dev., 74,* 159–164.

Loosli, F., Winkler, S., & Wittbrodt, J. (1999). Six3 overexpression initiates the formation of ectopic retina. *Genes Dev., 13,* 649–654.

Lopez-Rios, J., Gallardo, M. E., Rodriguez de Cordoba, S., & Bovolenta, P. (1999). Six9 (Optx2), a new member of the six gene family of transcription factors, is expressed at early stages of vertebrate ocular and pituitary development. *Mech. Dev., 83,* 155–159.

Lund, R. D., Ono, S. J., Keegan, D. J., & Lawrence, J. M. (2003). Retinal transplantation: progress and problems in clinical application. *J. Leukoc. Biol., 74,* 151–160.

Mack, A. F., & Fernald, R. D. (1993). Regulation of cell division and rod differentiation in the teleost retina. *Brain Res. Dev. Brain Res., 76,* 183–187.

MacLaren, R. E., Pearson, R. A., MacNeil, A., Douglas, R. H., Salt, T. E., Akimoto, M., et al. (2006). Retinal repair by transplantation of photoreceptor precursors. *Nature, 444,* 203–207.

Marquardt, T., Ashery-Padan, R., Andrejewski, N., Scardigli, R., Guillemot, F., & Gruss, P. (2001). Pax6 is required for the multipotent state of retinal progenitor cells. *Cell, 105,* 43–55.

Masse, K., Bhamra, S., Eason, R., Dale, N., & Jones, E. A. (2007). Purine-mediated signalling triggers eye development. *Nature, 449,* 1058–1062.

Mathers, P. H., Grinberg, A., Mahon, K. A., & Jamrich, M. (1997). The Rx homeobox gene is essential for vertebrate eye development. *Nature, 387,* 603–607.

Matsuo, I., Kuratani, S., Kimura, C., Takeda, N., & Aizawa, S. (1995). Mouse Otx2 functions in the formation and patterning of rostral head. *Genes Dev., 9,* 2646–2658.

Matsuo, T., Osumi-Yamashita, N., Noji, S., Ohuchi, H., Koyama, E., Myokai, F., et al. (1993). A mutation in the Pax-6 gene in rat small eye is associated with impaired migration of midbrain crest cells. *Nat. Genet., 3,* 299–304.

Maurus, D., Heligon, C., Burger-Schwarzler, A., Brandli, A. W., & Kuhl, M. (2005). Noncanonical Wnt-4 signaling and EAF2 are required for eye development in Xenopus laevis. *Embo J., 24,* 1181–1191.

Meno, C., Ito, Y., Saijoh, Y., Matsuda, Y., Tashiro, K., Kuhara, S., et al. (1997). Two closely-related left-right asymmetrically expressed genes, lefty-1 and lefty-2: their distinct expression domains, chromosomal linkage and direct neuralizing activity in Xenopus embryos. *Genes Cells, 2*, 513–524.

Meyer, J. S., Katz, M. L., Maruniak, J. A., & Kirk, M. D. (2004). Neural differentiation of mouse embryonic stem cells in vitro and after transplantation into eyes of mutant mice with rapid retinal degeneration. *Brain Res., 1014*, 131–144.

Meyer, J. S., Shearer, R. L., Capowski, E. E., Wright, L. S., Wallace, K. A., McMillan, E. L., et al. (2009). Modeling early retinal development with human embryonic and induced pluripotent stem cells. *Proc. Natl. Acad. Sci. U.S.A., 106*, 16698–16703.

Moore, K. B., Mood, K., Daar, I. O., & Moody, S. A. (2004). Morphogenetic movements underlying eye field formation require interactions between the FGF and ephrinB1 signaling pathways. *Dev. Cell, 6*, 55–67.

Moshiri, A., McGuire, C. R., & Reh, T. A. (2005). Sonic hedgehog regulates proliferation of the retinal ciliary marginal zone in posthatch chicks. *Dev. Dyn., 233*, 66–75.

Moshiri, A., & Reh, T. A. (2004). Persistent progenitors at the retinal margin of ptc+/− mice. *J. Neurosci., 24*, 229–237.

Nelson, B. R., & Reh, T. A. (2008). Relationship between Delta-like and proneural bHLH genes during chick retinal development. *Dev. Dyn., 237*, 1565–1580.

Okabe, S., Forsberg-Nilsson, K., Spiro, A. C., Segal, M., & McKay, R. D. (1996). Development of neuronal precursor cells and functional postmitotic neurons from embryonic stem cells in vitro. *Mech. Dev., 59*, 89–102.

Okada, T. S. (1980). Cellular metaplasia or transdifferentiation as a model for retinal cell differentiation. *Curr. Top Dev. Biol., 16*, 349–380.

Oliver, G., Mailhos, A., Wehr, R., Copeland, N. G., Jenkins, N. A., & Gruss, P. (1995). Six3, a murine homologue of the sine oculis gene, demarcates the most anterior border of the developing neural plate and is expressed during eye development. *Development, 121*, 4045–4055.

Ooto, S., Akagi, T., Kageyama, R., Akita, J., Mandai, M., Honda, Y., et al. (2004). Potential for neural regeneration after neurotoxic injury in the adult mammalian retina. *Proc. Natl. Acad. Sci. U.S.A., 101*, 13654–13659.

Osakada, F., Ikeda, H., Mandai, M., Wataya, T., Watanabe, K., Yoshimura, N., et al. (2008). Toward the generation of rod and cone photoreceptors from mouse, monkey and human embryonic stem cells. *Nat. Biotechnol., 26*, 215–224.

Osakada, F., Jin, Z. B., Hirami, Y., Ikeda, H., Danjyo, T., Watanabe, K., et al. (2009). In vitro differentiation of retinal cells from human pluripotent stem cells by small-molecule induction. *J. Cell Sci., 122*, 3169–3179.

Osakada, F., Ooto, S., Akagi, T., Mandai, M., Akaike, A., & Takahashi, M. (2007). Wnt signaling promotes regeneration in the retina of adult mammals. *J. Neurosci., 27*, 4210–4219.

Park, C. M., & Hollenberg, M. J. (1989). Basic fibroblast growth factor induces retinal regeneration in vivo. *Dev. Biol., 134*, 201–205.

Park, C. M., & Hollenberg, M. J. (1991). Induction of retinal regeneration in vivo by growth factors. *Dev. Biol., 148*, 322–333.

Park, C. M., & Hollenberg, M. J. (1993). Growth factor-induced retinal regeneration in vivo. *Int. Rev. Cytol., 146*, 49–74.

Pera, E. M., Wessely, O., Li, S. Y., & DeRobertis, E. M. (2001). Neural and head induction by insulin-like growth factor signals. *Dev. Cell, 1*, 655–665.

Perron, M., Kanekar, S., Vetter, M. L., & Harris, W. A. (1998). The genetic sequence of retinal development in the ciliary margin of the Xenopus eye. *Dev. Biol., 199*, 185–200.

Pittack, C., Grunwald, G. B., & Reh, T. A. (1997). Fibroblast growth factors are necessary for neural retina but not pigmented epithelium differentiation in chick embryos. *Development, 124*, 805–816.

Pittack, C., Jones, M., & Reh, T. A. (1991). Basic fibroblast growth factor induces retinal pigment epithelium to generate neural retina in vitro. *Development, 113*, 577–588.

Porter, F. D., Drago, J., Xu, Y., Cheema, S. S., Wassif, C., Huang, S. P., et al. (1997). Lhx2, a LIM homeobox gene, is required for eye, forebrain, and definitive erythrocyte development. *Development, 124*, 2935–2944.

Prada, C., Puga, J., Perez-Mendez, L., Lopez, R., & Ramirez, G. (1991). Spatial and temporal patterns of neurogenesis in the chick retina. *Eur. J. Neurosci., 3*, 559–569.

Qin, Z., Barthel, L. K., & Raymond, P. A. (2009). Genetic evidence for shared mechanisms of epimorphic regeneration in zebrafish. *Proc. Natl. Acad. Sci. U.S.A., 106*, 9310–9315.

Radtke, N. D., Aramant, R. B., Petry, H. M., Green, P. T., Pidwell, D. J., & Seiler, M. J. (2008). Vision improvement in retinal degeneration patients by implantation of retina together with retinal pigment epithelium. *Am. J. Ophthalmol., 146*, 172–182.

Raymond, P. A., Barthel, L. K., Bernardos, R. L., & Perkowski, J. J. (2006). Molecular characterization of retinal stem cells and their niches in adult zebrafish. *BMC Dev. Biol., 6*, 36.

447

Raymond, P. A., & Hitchcock, P. F. (1997). Retinal regeneration: common principles but a diversity of mechanisms. *Adv. Neurol., 72*, 171–184.

Reh, T. A. (1987). Cell-specific regulation of neuronal production in the larval frog retina. *J. Neurosci., 7*, 3317–3324.

Reh, T. A., & Cagan, R. L. (1994). Intrinsic and extrinsic signals in the developing vertebrate and fly eyes: viewing vertebrate and invertebrate eyes in the same light. *Perspect. Dev. Neurobiol., 2*, 183–190.

Reh, T. A., & Constantine-Paton, M. (1983). Qualitative and quantitative measures of plasticity during the normal development of the Rana pipiens retinotectal projection. *Brain Res., 312*, 187–200.

Reh, T. A., Nagy, T., & Gretton, H. (1987). Retinal pigmented epithelial cells induced to transdifferentiate to neurons by laminin. *Nature, 330*, 68–71.

Reyer, R. W. (1983). Availability time of tritium-labeled DNA precursors in newt eyes following intraperitoneal injection of 3H-thymidine. *J. Exp. Zool., 226*, 101–121.

Richard-Parpaillon, L., Heligon, C., Chesnel, F., Boujard, D., & Philpott, A. (2002). The IGF pathway regulates head formation by inhibiting Wnt signaling in Xenopus. *Dev. Biol., 244*, 407–417.

Sakaguchi, D. S., Janick, L. M., & Reh, T. A. (1997). Basic fibroblast growth factor (FGF-2) induced trans-differentiation of retinal pigment epithelium: generation of retinal neurons and glia. *Dev. Dyn., 209*, 387–398.

Sakaguchi, D. S., van Hoffelen, S. J., & Young, M. J. (2003). Differentiation and morphological integration of neural progenitor cells transplanted into the developing mammalian eye. *Ann. N.Y. Acad. Sci., 995*, 127–139.

Sakami, S., Hisatomi, O., Sakakibara, S., Liu, J., Reh, T. A., & Tokunaga, F. (2005). Downregulation of Otx2 in the dedifferentiated RPE cells of regenerating newt retina. *Brain Res. Dev. Brain Res., 155*, 49–59.

Sidman, R. L. (1961). Histogenesis of the mouse retina. Studies with [3H] thymidine. In *The Structure of the Eye* (pp. 487–506). New York: Academic Press.

Simeone, A., Acampora, D., Mallamaci, A., Stornaiuolo, A., d'Apice, M. R., Nigro, V., et al. (1993). A vertebrate gene related to orthodenticle contains a homeodomain of the bicoid class and demarcates anterior neuroectoderm in the gastrulating mouse embryo. *Embo J., 12*, 2735–2747.

Spence, J. R., Madhavan, M., Ewing, J. D., Jones, D. K., Lehman, B. M., & del Rio-Tsonis, K. (2004). The hedgehog pathway is a modulator of retina regeneration. *Development, 131*, 4607–4621.

Stone, L. S. (1950). Neural retina degeneration followed by regeneration from surviving retinal pigment cells in grafted adult salamander eyes. *Anat. Rec., 106*, 89–109.

Stone, L. S. (1957). Further experiments on lens regeneration from retina pigment cells in adult newt eyes. *J. Exp. Zool., 136*, 75–87.

Stone, L. S., & Steinitz, H. (1957). Regeneration of neural retina and lens from retina pigment cell grafts in adult newts. *J. Exp. Zool., 135*, 301–317.

Stroeva, O. G., & Mitashov, V. I. (1983). Retinal pigment epithelium: proliferation and differentiation during development and regeneration. *Int. Rev. Cytol., 83*, 221–293.

Sugie, Y., Yoshikawa, M., Ouji, Y., Saito, K., Moriya, K., Ishizaka, S., et al. (2005). Photoreceptor cells from mouse ES cells by co-culture with chick embryonic retina. *Biochem. Biophys. Res. Commun., 332*, 241–247.

Tabata, Y., Ouchi, Y., Kamiya, H., Manabe, T., Arai, K., & Watanabe, S. (2004). Specification of the retinal fate of mouse embryonic stem cells by ectopic expression of Rx/rax, a homeobox gene. *Mol. Cell Biol., 24*, 4513–4521.

Takeda, M., Takamiya, A., Jiao, J. W., Cho, K. S., Trevino, S. G., Matsuda, T., et al. (2008). alpha-Aminoadipate induces progenitor cell properties of Muller glia in adult mice. *Invest. Ophthalmol. Vis. Sci., 49*, 1142–1150.

Taylor, M., & Reh, T. A. (1990). Induction of differentiation of rat retinal, germinal, neuroepithelial cells by dbcAMP. *J. Neurobiol., 21*, 470–481.

Thomson, J. A., Itskovitz-Eldor, J., Shapiro, S. S., Waknitz, M. A., Swiergiel, J. J., Marshall, V. S., et al. (1998). Embryonic stem cell lines derived from human blastocysts. *Science, 282*, 1145–1147.

Thummel, R., Kassen, S. C., Enright, J. M., Nelson, C. M., Montgomery, J. E., & Hyde, D. R. (2008). Characterization of Muller glia and neuronal progenitors during adult zebrafish retinal regeneration. *Exp. Eye Res., 87*, 433–444.

Toy, J., & Sundin, O. H. (1999). Expression of the optx2 homeobox gene during mouse development. *Mech. Dev., 83*, 183–186.

Toy, J., Yang, J. M., Leppert, G. S., & Sundin, O. H. (1998). The optx2 homeobox gene is expressed in early precursors of the eye and activates retina-specific genes. *Proc. Natl. Acad. Sci. U.S.A., 95*, 10643–10648.

Tropepe, V., Coles, B. L., Chiasson, B. J., Horsford, D. J., Elia, A. J., McInnes, R. R., et al. (2000). Retinal stem cells in the adult mammalian eye. *Science, 287*, 2032–2036.

Turner, D. L., Snyder, E. Y., & Cepko, C. L. (1990). Lineage-independent determination of cell type in the embryonic mouse retina. *Neuron, 4*, 833–845.

Voronina, V. A., Kozhemyakina, E. A., O'Kernick, C. M., Kahn, N. D., Wenger, S. L., Linberg, J. V., et al. (2004). Mutations in the human RAX homeobox gene in a patient with anophthalmia and sclerocornea. *Hum. Mol. Genet., 13*, 315–322.

Wan, J., Zheng, H., Chen, Z. L., Xiao, H. L., Shen, Z. J., & Zhou, G. M. (2008). Preferential regeneration of photoreceptor from Muller glia after retinal degeneration in adult rat. *Vision Res., 48*, 223–234.

Wan, J., Zheng, H., Xiao, H. L., She, Z. J., & Zhou, G. M. (2007). Sonic hedgehog promotes stem-cell potential of Muller glia in the mammalian retina. *Biochem. Biophys. Res. Commun., 363*, 347–354.

Wetts, R., & Fraser, S. E. (1988). Multipotent precursors can give rise to all major cell types of the frog retina. *Science, 239*, 1142–1145.

Wu, J., O'Donnell, M., Gitler, A. D., & Klein, P. S. (2006). Kermit 2/XGIPC, an IGF1 receptor interacting protein, is required for IGF signaling in *Xenopus* eye development. *Development, 133*, 3651–3660.

Xu, Y., Baldassare, M., Fisher, P., Rathbun, G., Oltz, E. M., Yancopoulos, G. D., et al. (1993). LH-2: a LIM/homeodomain gene expressed in developing lymphocytes and neural cells. *Proc. Natl. Acad. Sci. U.S.A., 90*, 227–231.

Yoshizaki, T., Inaji, M., Kouike, H., Shimazaki, T., Sawamoto, K., Ando, K., et al. (2004). Isolation and transplantation of dopaminergic neurons generated from mouse embryonic stem cells. *Neurosci. Lett., 363*, 33–37.

Yurco, P., & Cameron, D. A. (2007). Cellular correlates of proneural and Notch-delta gene expression in the regenerating zebrafish retina. *Vis. Neurosci., 24*, 437–443.

Zhang, L., Mathers, P. H., & Jamrich, M. (2000). Function of Rx, but not Pax6, is essential for the formation of retinal progenitor cells in mice. *Genesis, 28*, 135–142.

Zhao, X., Das, A. V., Soto-Leon, F., & Ahmad, I. (2005). Growth factor-responsive progenitors in the postnatal mammalian retina. *Dev. Dyn., 232*, 349–358.

Zhao, X., Liu, J., & Ahmad, I. (2002). Differentiation of embryonic stem cells into retinal neurons. *Biochem. Biophys. Res. Commun., 297*, 177–184.

Zuber, M. E., Gestri, G., Viczian, A. S., Barsacchi, G., & Harris, W. A. (2003). Specification of the vertebrate eye by a network of eye field transcription factors. *Development, 130*, 5155–5167.

Zuber, M. E., & Harris, W. A. (2006). Formation of the eye field. In E. Sernagor, S. Eglen, W. Harris, & R. Wong (Eds.), *Retinal Development* (pp. 8–29). Cambridge: Cambridge University Press.

Zuber, M. E., Perron, M., Philpott, A., Bang, A., & Harris, W. A. (1999). Giant eyes in *Xenopus laevis* by overexpression of XOptx2. *Cell, 98*, 341–352.

CHAPTER 25

Somatic Cells: Growth and Expansion Potential of T Lymphocytes

Rita B. Effros
Department of Pathology and Laboratory Medicine and UCLA AIDS Institute, David Geffen School of Medicine at UCLA, Los Angeles, CA, USA

INTRODUCTION

All human somatic cells are restricted in their overall proliferative potential by an intrinsic programmed barrier to unlimited cell division (Hayflick and Moorhead, 1961). This stringent property, known as replicative (or cellular) senescence, is thought to serve as one of many safeguards to maintain cellular integrity necessitated by the extended longevity of humans. Indeed, a restriction in the number of cell divisions may serve as a protection against the potential for multiple mutations that are required for the development of a cancer cell from a cell that is normal, suggesting that replicative may have evolved as a tumor suppressor mechanism (Effros et al., 2005; Campisi and d'Adda, 2007).

Despite the importance of preventing uncontrolled growth, for some cell types, the replicative senescence cellular program can lead to deleterious consequences, particularly by old age. A prime example of this is the human immune system, which requires an extraordinary expansion of lymphocytes, due to the low frequency of cells that can respond to a particular foreign pathogen. Under most circumstances, the limited proliferative potential of T lymphocytes, the cells that are the main players in controlling infections and cancer, does not hamper primary or even secondary immune responses (Effros and Pawelec, 1997). However, by old age and/or during certain chronic viral infections and cancers, there is an accumulation of clones of T-cells that show signs of having reached their maximum replicative limit.

This chapter will discuss the nature and underlying mechanism of replicative senescence in human T lymphocytes, a specific facet of immune system activity that seems particularly well suited to regenerative medicine approaches. Indeed, one of the signature changes associated with aging is the decline in immune function. Immune system failure is believed to underlie the increased risk of morbidity and mortality from influenza and other infections. Ironically, vaccination, which is intended to offer protection from infection, fails to elicit adequate antibody responses in older persons, providing further evidence of the diminished immune function. There is also increasing documentation indicating that many of the pathologies and diseases associated with aging have an immune component, or, in some cases, even an immune-based etiology (Effros et al., 2003). In fact, a cluster of immune parameters (including high proportions of T-cells with characteristics of replicative senescence) that correlates with early all-cause mortality has been identified in longitudinal studies on persons

451

aged 80 and older. These and other studies suggest that improved health and quality of life may be possible if the immune system of the elderly is either prevented from "aging" or can be rejuvenated in some way. The immune exhaustion associated with chronic HIV disease is another situation that might also be amenable to similar therapeutic strategies. In the sections below, we will provide an overview of the immune system, a summary of the features of T-cell replicative senescence, and evidence demonstrating the presence and consequences of high proportions of senescent T-cells *in vivo*. Finally, approaches to reverse or retard this process, in order to regenerate a more functional immune system, will be discussed.

T LYMPHOCYTES: CRITICAL FOR INFECTION AND CANCER IMMUNITY

The immune system is a complex and highly integrated network of cells and lymphoid organs that functions to protect the body from foreign pathogens. Immunity is generated by two interacting components; namely, the innate and the adaptive immune systems. The innate immune response is capable of dealing with certain pathogens in a rapid, albeit somewhat non-specific, manner. By contrast, the adaptive immune response takes longer to develop, but has the advantage of exquisite specificity and long-term memory. Indeed, this anamnestic response is the basis for the efficacy of vaccines.

All the cellular components of both the innate and adaptive immune systems, including B lymphocytes, T lymphocytes, monocytes, and dendritic cells, are derived from primitive stem cells in the bone marrow. One of the noteworthy aspects of T and B lymphocytes, the main players in adaptive immunity, is the presence of antigen receptors on the surface of each cell that confer the ability to recognize a specific region of a particular pathogen. These antigen receptors are generated during the complex transition from hematopoietic stem cells to mature lymphocytes by an intricate process of cutting and splicing that leads to random joining of DNA segments from several different gene families (Janeway et al., 2001). The end result of this process is that each lymphocyte expresses a unique antigen receptor, and, if that lymphocyte becomes activated as a result of an encounter with the appropriate antigen, the identical receptor is expressed on all the resulting daughter cells.

The generation of antigen receptors by this stochastic process leads to an extremely large repertoire of antigen specificities, thereby enabling the immune system to have broad coverage over multiple and varied types of pathogens. However, precisely because of the huge spectrum of antigen specificities within each individual, the number of lymphocytes that can respond to any single pathogen is extremely small, leading to the requirement for massive cell division and clonal expansion of the few cells whose receptors recognize the invading pathogen, or, in the case of cancer, a tumor-specific antigen.

Whereas B cells and T-cells generate their antigen receptors by similar processes, they function in quite distinct ways when that antigen is encountered (Janeway et al., 2001). B cells produce soluble proteins called antibodies, which can neutralize or otherwise inactivate pathogens that are present within the blood. T-cells, on the other hand, only recognize pathogens that have already infected other cells. In the case of a viral infection, for example, the infected cells become decorated with components of the virus, indicating to the immune system that the cell is no longer normal and must be eliminated. Those cytotoxic T-cells whose receptors recognize the specific viral antigens on the surface of the infected cell become activated and then undergo massive cell division and migrate into the tissues, where they actually kill infected or otherwise abnormal cells, thereby controlling the infection. Once the antigen-specific T-cells complete their function, most of the expanded cell population dies by apoptosis, leaving only a few memory cells to handle possible future encounters with the same antigen. Thus, proliferation and the ability to undergo repeated rounds of clonal expansion is a critical feature of effective T-cell function.

STUDIES OF HUMAN T-LYMPHOCYTE GROWTH IN CELL CULTURE

Aging is associated with increased severity of infections and a steep rise in cancer incidence. Extensive research has traced the immune defects responsible for these changes to the T lymphocyte compartment (Miller, 1996; Swain et al., 2005; Effros et al., 2006). This chapter will focus on studies performed on the so-called cytotoxic T-cells (also known as CD8 T-cells), which are the immune cells responsible for killing cells that are perceived as "non-self," due to infection or to cancer.

As noted above, due to the random nature of DNA regions utilized in creating the T-cell antigen receptor, there is a huge repertoire of different T-cell specificities. This broad spectrum of specificities is necessary to counter the huge universe of potential pathogens, or neoantigens associated with a tumor. The corollary to this is that T-cells of any given specificity are low in frequency, requiring extensive and rapid clonal expansion in order to reach the numbers needed for an effective response to pathogens. Although T-cells can divide faster than any other vertebrate cell type, the extensive cell division is not without consequences. Indeed, some T-cells can actually reach the end stage of replicative senescence, particularly by old age, but also in younger people during certain chronic infections. The limited proliferative potential of somatic cells is the reason why most current approaches to regenerative medicine rely on stem cells, which have unlimited expansion potential. However, if it becomes possible to manipulate the process of replicative senescence in normal somatic cells to allow increased proliferation, this might greatly expand the field of regenerative medicine to include specific types of differentiated cells with known functions. In the case of T-cells, the added advantage would be the ability to focus on cells directed at a specific viral or tumor antigen. With these goals in mind, our laboratory has focused on analyzing the process of replicative senescence in cytotoxic T-cells, and on developing methods to retard or prevent this process.

The *in vitro* model system used for our studies was developed in an effort to mimic the *in vivo* biology of human cytotoxic (CD8) T-cells. Although the system is isolated from the normal *in vivo* environment, it has the unique advantage of allowing longitudinal analysis of the same population of T-cells over time, which would be technically impossible to do in humans *in vivo*. The basic protocol of our *in vitro* model is to isolate peripheral blood mononuclear cells from venous blood samples, and to stimulate the cells with either irradiated foreign (allogeneic) tumor cells or with antibodies to the T-cell receptor. In some studies, virus-specific CD8 T-cell cultures were established from blood samples of HIV-infected individuals (Dagarag et al., 2004). Irrespective of the mode of stimulation, after a period of 2–3 weeks, the vigorous cell proliferation subsided and the cells became quiescent. The cycle of stimulation-proliferation-quiescence is repeated multiple times until the culture reaches an irreversible final stage of quiescence that cannot be overcome by further stimulation or by the addition of growth factors (Perillo et al., 1989, 1993). This terminal state is known as replicative senescence. The overall finding from numerous different laboratories, using a variety of modes of stimulation, is that human T-cells are able to undergo a limited number of replications, after which they cease dividing (Perillo et al., 1989; Adibzadeh et al., 1996). It is important to note that this end stage of replicative senescence does not imply loss of viability. Indeed, with appropriate feeding, senescent cells remain viable and metabolically active for several months (Wang et al., 1994; Spaulding et al., 1999).

CHARACTERISTICS OF SENESCENT T LYMPHOCYTES

The state of irreversible growth arrest is the most easily discernible characteristic of replicative senescence. Thus, senescent cultures are initially identified by the inability of the cells to enter the cell cycle. However, the functional, genetic, and phenotypic alterations associated with senescence may be at least as important to the biology of cells as the inability to proliferate (Campisi, 1997).

453

For CD8 T lymphocytes, one of the major changes observed in cultures that have reached replicative senescence is resistance to apoptosis, a property they share with other types of senescent cells (Wang et al., 1994). Whereas CD8 T lymphocytes from early passage cultures undergo brisk apoptosis in response to a variety of stimuli (e.g. mild heat shock, antibodies to Fas or to the T-cell receptor), the descendants of those cells that have reached senescence show significantly reduced ability to undergo apoptosis and increased expression of the anti-apoptotic protein, Bcl2 (Spaulding et al., 1999). This change in the ability to initiate timely and efficient programmed cell death is highly relevant to effective immune function *in vivo*, since elimination of the massive numbers of activated virus-specific CD8 T-cells is an essential event in the resolution of an immune response (Effros and Pawelec, 1997).

Another notable characteristic of CD8 T-cell replicative senescence in cell culture is an alteration in the pattern of cytokine production (Effros et al., 2005). Cytokine secretion by T-cells is essential for cell-cell communication and efficient immune function. Our cell culture studies show that, as the cultures progress to senescence, there is an increasing concentration of two pro-inflammatory cytokines in the culture medium. Specifically, the levels of both TNF-alpha (TNFα) and IL-6 increase progressively as the cells reach senescence (Effros et al., 2005). These two cytokines are often associated with frailty in the elderly (Hubbard et al., 2009), as well as with increased maturation and activation of bone-resorbing osteoclasts (Arron and Choi, 2000). Moreover, TNFα serum levels in HIV-infected persons are closely linked to adverse disease outcomes (Haissman et al., 2009; Ross et al., 2009). A second important change in cytokine secretion is the anti-viral cytokine, IFN-gamma (IFNγ), which CD8 T-cells secrete in conjunction with their cytotoxic function. With progressive cell divisions in culture, HIV-specific CD8 T-cells show significantly reduced production and secretion of IFNγ, along with reduced lytic capacity and diminished production of perforin, a protein involved in cytotoxicity (Dagarag et al., 2003, 2004; Yang et al., 2005).

In comparison to early passage cells, senescent T-cells also show a significantly blunted upregulation of the hsp 70 protein in response to a mild, brief heat stress (Effros et al., 1994a). Finally, as cells age in culture, they show increased microsatellite instability, an indicator of reduced DNA mismatch repair capacity, which is capable of rectifying errors in DNA replication (Krichevsky et al., 2004). Thus, as T-cells progress to the end stage of replicative senescence in cell culture, they are altered in a variety of processes reflecting cellular integrity and defense.

Arguably, one of the most significant changes associated with T-cell replicative senescence in cell culture is the complete and irreversible loss of expression of the major signaling molecule, CD28 (Effros et al., 1994b; Vallejo et al., 1998). This so-called co-stimulatory molecule is an integral component of the immunological synapse, and is involved in a variety of T-cell functions, including activation, proliferation, stabilization of cytokine messenger RNA levels, and glucose metabolism (Shimizu et al., 1992; Holdorf et al., 2000; Sansom, 2000; Frauwirth et al., 2002). Importantly, the absence of CD28 expression is in marked contrast to the sustained expression of a variety of T-cell-specific cell-surface markers reflecting lineage, memory, and cell-cell adhesion.

TELOMERES AND TELOMERASE

Telomeres, the repeated DNA sequences that cap the ends of linear chromosomes, shorten with each cell division, due to a particular feature of DNA replication (Watson, 1972; Olovnikov, 1973). Telomere loss and a critically short telomere length (i.e. 5–7 kb) are intimately involved in the DNA damage signal involved in the cell cycle arrest associated with replicative senescence (Vaziri et al., 1993). Telomerase is a reverse transcriptase-like enzyme that functions to elongate telomeres (Blackburn, 2005). High levels of telomerase are present in the developing embryo, but, after birth, telomerase activity is retained only in stem cells and germ cells. Although most normal somatic cells lack telomerase, immune cells are able to upregulate telomerase activity in concert with activation (Hiyama et al., 1995; Bodnar et al., 1996; Weng et al., 1996). Therefore, it seemed somewhat paradoxical that T-cells, which produced high levels of telomerase activity in response to stimulation, should be restricted in their overall proliferative potential and undergo telomere loss in cell culture.

To carefully analyze telomerase dynamics in T-cells, long-term cultures were followed over time and tested for telomerase activity and telomere length at various points along the trajectory to senescence. Our studies showed that the overall loss of telomere sequences occurs at a rate of 50–100 bp/cell division, as had been shown for a variety of cell types (Harley et al., 1990; Vaziri et al., 1993). During the period following activation, telomerase activity is as high as that present in tumor cells and telomere length is actually maintained (Bodnar et al., 1996). Nonetheless, this high telomerase activity induced in response to the first and second encounters with antigen is not sustained during subsequent stimulations. In fact, by the fourth antigenic stimulation, CD8 T-cells show no detectable telomerase activity (Valenzuela and Effros, 2002). Interestingly, loss of telomerase activity parallels the loss of CD28 expression in CD8 T-lymphocyte cultures. Moreover, blocking CD28 stimulation results in marked inhibition of activation-induced telomerase activity. Thus, three key elements in the CD8 T-lymphocyte replicative senescence program — CD28 expression, telomere loss, and telomerase — are intimately connected, an observation that will become important in developing strategies to retard this process.

SENESCENT T LYMPHOCYTES ARE PRESENT *IN VIVO*

Telomere length, telomerase activity, and CD28 expression in PBMC or isolated populations of T lymphocytes are increasingly being used as biomarkers of disease, immune status, and overall health. The importance of telomerase in lymphocyte function is underscored by multiple immune abnormalities in persons with the autosomal dominant form of dyskeratosis congenita, a disease caused by a mutation in the gene encoding the RNA component of telomerase (Knudson et al., 2005). By contrast, enhanced telomerase activity and increased telomere length, due to another type of mutation in a telomerase gene, are associated with the extreme longevity of centenarians (Atzmon et al., 2010). In terms of diseases, telomere shortening has been documented in individuals with Down syndrome (Vaziri et al., 1993) and with Alzheimer's disease (Panossian et al., 2002; Jenkins et al., 2006). Accelerated telomere shortening also occurs in patients with coronary heart disease (Spyridopoulos et al., 2009), atherosclerosis (Samani et al., 2001), and premature myocardial infarction (Brouilette et al., 2003). In addition, insulin resistance and type 2 diabetes are associated with telomere loss (Demissie et al., 2006; Tentolouris et al., 2007). Finally, even psychological stress has been linked to decreased telomerase and telomere loss (Epel et al., 2004, 2006; Damjanovic et al., 2007). Overall, these reports suggest that telomere loss and/or reduced telomerase activity can serve as useful biomarkers or early warning systems of physiological or mental stress.

The permanent loss of CD28 expression in senescent T-cell cultures constitutes yet another type of biomarker that has been used to identify senescent T lymphocytes *in vivo*. Flow cytometry analysis of peripheral blood samples has clearly demonstrated that persons aged 70–90 have high proportions of CD8 T-cells that lack CD28 expression. Indeed, in some elderly persons, more than 50% of the CD8 T-cells within the total peripheral blood T-cell pool do not express the CD28 molecule (Effros et al., 1994b). Cells in this category show minimal proliferative activity and have shorter telomeres than CD28-expressing CD8 T-cells from the same donor (Effros et al., 2006). *Ex vivo* analysis of distinct populations of peripheral blood cells has shown that, compared with other T-cell subsets and with B cells, the $CD8^+CD28^-$ T-lymphocyte subset has both lower telomerase activity and the shortest telomere length (Lin et al., 2010). Moreover, it has been shown that the higher the percentage of $CD8^+CD28^-$ T lymphocytes, the shorter the telomere length of the total PBMC population PBMC (Lin et al., 2010). Thus, by a variety of criteria, the $CD8^+CD28^-$ T lymphocyte subset isolated from *ex vivo* PBMC samples resembles CD8 T lymphocytes that are driven to replicative senescence in cell culture.

It should be emphasized that the *in vivo* presence of cells with markers indicative of replicative senescence is not restricted to old age — the proportion of these cells, which is <1% at birth, increases progressively over the entire lifespan (Azuma et al., 1993; Effros et al., 1996; Boucher et al., 1998). Moreover, these putatively senescent CD8 T-cells are significantly greater in persons infected with viruses that establish chronic/latent infections, such as HIV-1 (Borthwick et al., 1994; Brinchmann et al., 1994; Jennings et al., 1994). Indeed, one study showed that the telomere length of senescent CD8 T lymphocytes in 40-year-olds who are HIV-positive is in the same range as total PBMC from uninfected 90-year-olds (Effros et al., 1996). Similarly, telomere shortening in antigen-specific CD8 T lymphocytes is seen during the latent/chronic phase of Epstein-Barr virus (EBV) infection (Hathcock et al., 1998; Maini et al., 1999), presumably due to the downregulation of telomerase activity associated with repeated antigen-driven proliferation (Valenzuela and Effros, 2002).

Senescent CD8 T-cells have also been reported in the context of certain forms of cancer. For example, in advanced renal carcinoma, the proportion of CD8 T-cells that express CD57 (a marker present on the majority of $CD8^+CD28^-$ T lymphocytes) has predictive value with respect to patient survival (Characiejus et al., 2002). Also, in patients with head and neck tumors, it has been shown that the $CD8^+CD28^-$ T-cell subset undergoes expansion during the period of tumor growth, and decreases after tumor resection, consistent with the notion that the increased antigenic burden may cause extensive proliferation in the tumor-reactive cells (Tsukishiro et al., 2003).

What is the driving force for the generation of senescent CD8 T-cells *in vivo*? The most likely cause of the excessive division of certain CD8 T-cells in the intact organism is chronic antigenic stimulation, which could be the result of long-term exposure to antigens associated with latent viral infections as well as certain tumor antigens. It has been suggested that latent infection with several herpes viruses, which are endemic and persist throughout life in infected individuals, is the main culprit (Pawelec et al., 2004). In older persons, the primary source of chronic CD8 T-lymphocyte stimulation seems to be CMV (Pawelec et al., 2009). Clinical data on bone marrow and organ transplant recipients indicate that, under conditions of immunosuppression, both CMV and other latent herpes viruses are often reactivated. Moreover, these patients show increased incidence of EBV lymphomas. In the elderly, many of whom are also immunocompromised, another herpes virus, varicella zoster, is often reactivated, manifesting itself as shingles. Reactivation rarely occurs in healthy individuals with normal immune systems, suggesting that maintaining latency requires active participation by the immune system, and that the constant and prolonged CD8 T-cell activity may drive certain virus-specific T-cells to senescence.

SENESCENT T LYMPHOCYTES AND HEALTH

The presence of senescent CD8 T-cells *in vivo* may have a variety of effects on the proper function of both the immune system as well as other organ systems. In terms of immune function, senescent CD8 T-cells undoubtedly influence the quality and composition of the memory T-cell pool. Due to the property of apoptosis resistance, once senescent CD8 T-cells are generated, they persist, leading to their progressive accumulation over time. Since homeostatic mechanisms are believed to independently regulate the memory and naïve T-cell pools (Freitas and Rocha, 2000), a high proportion of senescent T-cells will result in the reduced proportion of proliferation-competent, non-senescent memory T-cells. The fact that CD28-negative ($CD28^-$) T-cells are usually part of oligoclonal expansions (Posnett et al., 1994; Schwab et al., 1997) would presumably also lead to a reduction in the overall spectrum of antigenic specificities within the T-cell pool. The repertoire of antigenic specificities is, in fact, reduced in elderly persons who have high proportions of CD8 T-cells lacking CD28 (Ouyang et al., 2003).

A more direct effect of senescent CD8 T-cells is their putative suppressive activity on other cell types. For example, a population of CD8$^+$CD28$^-$ T-cells generated in the course of *in vitro* and *in vivo* immunizations has been shown to suppress immune reactivity by affecting the process of antigen presentation (Cortesini et al., 2001). In the context of organ transplantation, the suppression may actually work to the benefit of the patient, by suppressing immune-mediated organ rejection. Indeed, donor-specific CD8$^+$CD28$^-$ T-cells are detectable in the peripheral blood of those patients with stable function of heart, liver, and kidney transplants, whereas no such cells are found in patients undergoing acute rejection (Cortesini et al., 2001). By contrast, in other situations, the suppression of immune reactivity by CD8$^+$CD28$^-$ T-cells may be maladaptive. An illustration of the possible negative outcome of this suppression emerges from the repeated observed correlation between poor antibody responses to influenza vaccines in the elderly and the presence of high proportions of senescent CD8 T-cells (Goronzy et al., 2001; Saurwein-Teissl et al., 2002).

Additional effects of CD8 T-cells with a senescent phenotype have also been reported. CD8 T-cells that express a marker known as CD57 (the expression of which is associated with loss of CD28) exert suppressive influences on effector functions of HIV-specific cytotoxic T lymphocytes (CTLs) (Sadat-Sowti et al., 1994) and CD8$^+$CD28$^-$ T-cells also accumulate and mediate liver damage in hepatitis C infection (Kurokohchi et al., 2003). A novel cellular interaction between CD8$^+$CD28$^-$ T lymphocytes and endothelial cells has recently been suggested by *in vitro* experiments, which, if confirmed *in vivo*, would have major implications for HIV pathogenesis. It has been reported that primary human endothelial cells that are exposed to culture supernatants from CD28$^-$, but not CD28$^+$, T-cells show increased expression of a series of cell surface molecules that are specific markers of Kaposi's sarcoma (KS) (Alessandri et al., 2003). The endothelial cells also acquire proliferative and morphological features of KS cells. The effect is mediated by several soluble factors, including TNFα, which, as noted above, is significantly increased in cultures of senescent CD8 T-cells.

A variety of pathological conditions have been correlated with the presence of senescent CD8 T-cells. For example, a population of TNFα-producing CD8$^+$CD28$^-$ T-cells has been identified in patients with cervical cancer (Pilch et al., 2002). Expanded populations of CD8$^+$CD28$^-$ T-cells are present in anklylosing spondylitis patients, and, in fact, correlate with a more severe course of this autoimmune disease (Schirmer et al., 2002). Cells with the same phenotype accumulate in persons with coronary artery disease, suggesting some chronic antigenic exposure related to atherosclerosis (Jonasson et al., 2003). Finally, as noted above, there is a progressive expansion of CD8$^+$CD28$^-$ T-cells in patients with head and neck tumors. Interestingly, the proportion of these cells is reduced upon tumor resection (Tsukishiro et al., 2003). The common thread in many of these reported accumulations of CD8$^+$CD28$^-$ T-cells is chronic antigenic stimulation, be it by virus, alloantigen, autoantigen, or tumor-associated antigen.

The regulatory effects of senescent CD8 T-cells are not restricted to the immune system. For example, there is accumulating evidence indicating that bone biology is directly linked to immune system activity, and that chronic immune activation is associated with bone loss (Arron and Choi, 2000). The CD8 T-cell subset, in particular, has been implicated in both bone resorption activity (Buchinsky et al., 1996; John et al., 1996) and osteoporotic fractures in the elderly (Pietschmann et al., 2001). One of the central regulators of bone resorption is expressed on and secreted by activated T-cells. This molecule, known as "RANKL" (receptor activator of NFkB ligand) binds to RANK on osteoclasts, inducing these bone-resorbing cells to mature and become activated (Kong et al., 2000). Under normal circumstances, the bone-resorbing activity induced by RANKL is kept in check by IFNγ, a cytokine also produced by the activated T-cells (Takayanagi et al., 2000). However, senescent CD8 T lymphocytes show reduced ability to produce IFNγ (Dagarag et al., 2004). In addition to enhancing bone-resorbing activity, chronically activated T lymphocytes may also modulate bone mass by producing cytokines that inhibit the bone-forming activity of osteoblasts (Lorenzo, 2000). Thus, T-cell replicative senescence may disrupt bone homeostasis via multiple pathways.

APPROACHES TO RETARD REPLICATIVE SENESCENCE

As noted above, shorter telomeres and/or reduced telomerase activity of either total PBMC or various subsets of lymphocytes are increasingly being documented to correlate with a variety of diseases. Given the intimate relationship of telomerase with replicative senescence dynamics, and the emerging picture of pleiotropic negative effects exerted by senescent CD8 T-cells *in vivo*, many of the approaches to modulate the process are based on telomerase. Targeting telomerase is consistent with several recent studies showing that high telomerase activity is associated with several favorable health outcomes. In HIV disease, increased telomerase activity in the virus-specific CD8 T-lymphocyte subset is present in elite controllers of the infection (Lichterfeld et al., 2008). In addition, a genetic variation in telomerase, leading to increased activity of the enzyme and longer telomeres, has recently been reported for centenarians (Atzmon et al., 2010). Telomerase-based approaches to retard the process of replicative senescence have been successful in several *in vitro* studies that focused on CD8 T lymphocytes from HIV-infected persons. Gene transduction with the catalytic component of telomerase (hTERT) led to telomere length stabilization and reduced expression of the $p16^{INK4A}$ and $p21^{WAF1}$ cell cycle inhibitors, implicating both of these proteins in the senescence program (Dagarag et al., 2004). Indeed, the transduced cultures showed indefinite proliferation, with no signs of change in growth characteristics or karyotypic abnormalities. In terms of protective immune function, the "telomerized" HIV-specific CD8 T-cells were able to maintain the production of IFNγ for extended periods and showed significantly enhanced capacity to inhibit HIV replication.

Telomerase enhancement has also been achieved using non-genetic strategies, which would offer more practical approaches to therapeutic interventions in the elderly. Pharmacologic enhancement of telomerase has the important advantage over gene therapy approaches of allowing control over the dose and timing. We recently showed that short-term exposure to a small-molecule telomerase activator led to a significant enhancement of telomerase activity in T-cells from both healthy and HIV-infected persons. For HIV-specific CD8 T lymphocytes, the increased telomerase activity was accompanied by increased proliferation and, importantly, significant enhancement of a variety of anti-viral functions (Fauce et al., 2008).

Several additional categories of non-genetic telomerase-enhancing treatments show preliminary promise in cell culture studies. We have recently reported that inhibition of TNFα in cell cultures of CD8 T lymphocytes allows for increased telomerase activity, prolonged maintenance of CD28 expression, and enhanced proliferation (Parish et al., 2009). Estrogen or modified "designer" versions of the hormone may also have the desired effect. It is well established that estrogen is able to enhance telomerase activity in reproductive tissues. The complex formed when estrogen binds to its receptors migrates to the nucleus and functions as a transcription factor. In normal ovarian epithelial cells, this complex actually binds to the hTERT promoter region (Misiti et al., 2000). It has been known for some time that T-cells can bind to estrogen via specific estrogen receptors. Thus, we tested whether pre-incubation of T-cells to 17β-estradiol prior to activation might augment telomerase activity. Our preliminary data suggest that estrogen does, in fact, enhance T-cell telomerase activity (Effros et al., 2005). Thus, therapeutic approaches that are based on telomerase modulation would seem to be promising candidates for clinical interventions that are aimed at reversing or retarding the process of replicative senescence in T-cells. Interestingly, there is preliminary evidence that even lifestyle changes can affect T-cell telomerase activity (Ornish et al., 2008).

Several possible alternative strategies for regenerating/rejuvenating the immune system are also worth considering. It has been repeatedly observed that a large proportion of senescent CD8 T cells are directed at CMV (Pawelec et al., 2004, 2009). Therefore, childhood vaccination against this virus might offer a practical preventive approach for eliminating one of the major driving forces for generating senescent CD8 T-cells (Pawelec et al., 2004). Alternatively, actual physical removal of senescent cells from the circulation might stimulate both the expansion of more functional memory cells and the production of naïve T-cells by the thymus.

CONCLUDING REMARKS

Normal somatic cells have limited expansion potential, a feature that has a dramatic effect on certain cells within the immune system. Indeed, the presence of senescent T-cells *in vivo* is associated with a variety of maladaptive consequences during aging, HIV infection, and cancer. Moreover, those forms of immunotherapy for cancer that are dependent on continued expansion of functional anti-tumor CD8 T-cells will also be severely limited by the innately restricted expansion potential of immune cells. Therefore, manipulation of the process of replicative senescence in CD8 T lymphocytes constitutes a novel approach to regenerating functional immune cells. This strategy is relevant to a wide range of clinical scenarios and broadens the spectrum of approaches to regenerative medicine.

Acknowledgments

The research described in this chapter was made possible by the following sources of support: National Institutes of Health, University-wide AIDS Research Program, UC Discovery grant, Geron Corporation, and Telomerase Activator (TA) Therapeutics, Ltd.

References

Adibzadeh, M., Pohla, H., Rehbein, A., & Pawelec, G. (1996). The T cell in the ageing individual. *Mechs. Ageing and Dev., 91*, 145–154.

Alessandri, G., Fiorentini, S., Licenziati, S., Bonafede, M., Monini, P., Ensoli, B., et al. (2003). CD8(+)CD28(−) T lymphocytes from HIV-1-infected patients secrete factors that induce endothelial cell proliferation and acquisition of Kaposi's sarcoma cell features. *J. Interferon Cytokine Res., 23*(9), 523–531.

Arron, J. R., & Choi, Y. (2000). Bone versus immune system. *Nature, 408*, 535–536.

Atzmon, G., Cho, M., Cawthon, R. M., Budagov, T., Katz, M., Yang, X., et al. (2010). Genetic variation in human telomerase is associated with telomere length in Ashkenazi centenarians. *Proc. Natl. Acad. Sci. U.S.A., 107* (Suppl. 1), 1710–1717.

Azuma, M., Phillips, J. H., & Lanier, L. L. (1993). CD28− T lymphocytes: antigenic and functional properties. *J. Immunol., 150*, 1147–1159.

Blackburn, E. H. (2005). Telomeres and telomerase: their mechanisms of action and the effects of altering their functions. *FEBS Lett., 579*(4), 859–862.

Bodnar, A. G., Kim, N. W., Effros, R. B., & Chiu, C. P. (1996). Mechanism of telomerase induction during T cell activation. *Exp. Cell Res., 228*, 58–64.

Borthwick, N. J., Bofill, M., & Gombert, W. M. (1994). Lymphocyte activation in HIV-1 infection II. Functional defects of CD28− T cells. *AIDS, 8*, 431–441.

Boucher, N., Defeu-Duchesne, T., Vicaut, E., Farge, D., Effros, R. B., & Schachter, F. (1998). CD28 expression in T cell aging and human longevity. *Exp. Geronol., 33*, 267–282.

Brinchmann, J. E., Dobloug, J. H., Heger, B. H., Haaheim, L. L., Sannes, M., & Egeland, T. (1994). Expression of costimulatory molecule CD28 on T cells in human immunodeficiency virus type 1 infection: functional and clinical correlations. *J. Infect. Dis., 169*, 730–738.

Brouilette, S., Singh, R. K., Thompson, J. R., Goodall, A. H., & Samani, N. J. (2003). White cell telomere length and risk of premature myocardial infarction. *Arterioscler. Thromb. Vasc. Biol., 23*(5), 842–846.

Buchinsky, F. J., Ma, Y., Mann, G. N., Rucinski, B., Bryer, H. P., Romero, D. F., et al. (1996). T lymphocytes play a critical role in the development of cyclosporin A-induced osteopenia. *Endocrinology, 137*(6), 2278–2285.

Campisi, J., & d'Adda, F. (2007). Cellular senescence: when bad things happen to good cells. *Nat. Rev. Mol. Cell. Biol., 8*(9), 729–740.

Campisi, J. (1997). The biology of replicative senescence. *Eur. J. Cancer, 33*(5), 703−709.

Characiejus, D., Pasukoniene, V., Kazlauskaite, N., Valuckas, K. P., Petraitis, T., Mauricas, M., et al. (2002). Predictive value of CD8highCD57+ lymphocyte subset in interferon therapy of patients with renal cell carcinoma. *Anticancer Res., 22*(6B), 3679−3683.

Cortesini, R., LeMaoult, J., Ciubotariu, R., & Cortesini, N. S. (2001). CD8+CD28− T suppressor cells and the induction of antigen-specific, antigen-presenting cells-mediated suppression of Th reactivity. *Immunol. Rev., 182*, 201−206.

Dagarag, M. D., Evazyan, T., Rao, N., & Effros, R. B. (2004). Genetic manipulation of telomerase in HIV-specific CD8+ T cells: enhanced anti-viral functions accompany the increased proliferative potential and telomere length stabilization. *J. Immunol., 173*, 6303−6311.

Dagarag, M. D., Ng, H., Lubong, R., Effros, R. B., & Yang, O. O. (2003). Differential impairment of lytic and cytokine functions in senescent HIV-1-specific cytotoxic T lymphocytes. *J. Virol., 77*, 3077−3083.

Damjanovic, A. K., Yang, Y., Glaser, R., Kiecolt-Glaser, J. K., Nguyen, H., Laskowski, B., et al. (2007). Accelerated telomere erosion is associated with a declining immune function of caregivers of Alzheimer's disease patients. *J. Immunol., 179*(6), 4249−4254.

Demissie, S., Levy, D., Benjamin, E. J., Cupples, L. A., Gardner, J. P., Herbert, A., et al. (2006). Insulin resistance, oxidative stress, hypertension, and leukocyte telomere length in men from the Framingham Heart Study. *Aging Cell, 5*(4), 325−330.

Effros, R. B., Cai, Z., & Linton, P. J. (2003). CD8 T cells and aging. *Crit. Rev. Immunol., 23*(1−2), 45−64.

Effros, R. B., Allsopp, R., Chiu, C. P., Wang, L., Hirji, K., Harley, C. B., et al. (2006). Shortened telomeres in the expanded CD28−CD8+ subset in HIV disease implicate replicative senescence in HIV pathogenesis. *AIDS/Fast Track, 10*, F17−F22.

Effros, R. B., Zhu, X., & Walford, R. L. (1994a). Stress response of senescent T lymphocytes: reduced hsp70 is independent of the proliferative block. *J. Gerontol., 49*, B65−B70.

Effros, R. B., Boucher, N., Porter, V., Zhu, X., Spaulding, C., Walford, R. L., et al. (1994b). Decline in CD28+ T cells in centenarians and in long-term T cell cultures: a possible cause for both in vivo and in vitro immunosenescence. *Exp. Gerontol., 29*, 601−609.

Effros, R. B., Dagarag, M. D., Spaulding, C. C., & Man, J. (2005). The role of CD8 T cell replicative senescence in human aging. *Immunol. Rev., 205*, 147−157.

Effros, R. B., & Pawelec, G. (1997). Replicative senescence of T lymphocytes: does the Hayflick limit lead to immune exhaustion? *Immunol. Today, 18*, 450−454.

Epel, E. S., Blackburn, E. H., Lin, J., Dhabhar, F. S., Adler, N. E., Morrow, J. D., et al. (2004). Accelerated telomere shortening in response to life stress. *Proc. Natl. Acad. Sci. U.S.A., 101*(49), 17312−17315.

Epel, E. S., Lin, J., Wilhelm, F. H., Wolkowitz, O. M., Cawthon, R., Adler, N. E., et al. (2006). Cell aging in relation to stress arousal and cardiovascular disease risk factors. *Psychoneuroendocrinology, 31*(3), 277−287.

Fauce, S. R., Jamieson, B. D., Chin, A. C., Mitsuyasu, R. T., Parish, S. T., Ng, H. L., et al. (2008). Telomerase-based pharmacologic enhancement of anitviral function of human CD8+ T lymphocytes. *J. Immunol., 181*, 7400−7406.

Frauwirth, K. A., Riley, J. L., Harris, M. H., Parry, R. V., Rathmell, J. C., Plas, D. R., et al. (2002). The CD28 signaling pathway regulates glucose metabolism. *Immunity, 16*(6), 769−777.

Freitas, A. A., & Rocha, B. (2000). Population biology of lymphocytes: the flight for survival. *Annu. Rev. Immunol., 18*, 83−111.

Goronzy, J. J., Fulbright, J. W., Crowson, C. S., Poland, G. A., O'Fallon, W. M., & Weyand, C. M. (2001). Value of immunological markers in predicting responsiveness to influenza vaccination in elderly individuals. *J. Virol., 75*(24), 12182−12187.

Haissman, J. M., Vestergaard, L. S., Sembuche, S., Erikstrup, C., Mmbando, B., Mtullu, S., et al. (2009). Plasma cytokine levels in Tanzanian HIV-1-infected adults and the effect of antiretroviral treatment. *J. Acquir. Immune Defic. Syndr., 52*(4), 493−497.

Harley, C., Futcher, A. B., & Greider, C. (1990). Telomeres shorten during ageing of human fibroblasts. *Nature, 345*, 458−460.

Hathcock, K. S., Weng, N. P., Merica, R., Jenkins, M. K., & Hodes, R. (1998). Antigen-dependent regulation of telomerase activity in murine T cells. *J. Immunol., 160*(12), 5702−5706.

Hayflick, L., & Moorhead, P. (1961). The serial cultivation of human diploid cell strains. *Exp. Cell Res., 25*, 585−621.

Hiyama, K., Hirai, Y., Kyoizumi, S., Akiyama, M., Hiyama, E., Piatyszek, M. A., et al. (1995). Activation of telomerase in human lymphocytes and hematopoietic progenitor cells. *J. Immunol., 155*(8), 3711−3715.

Holdorf, A. D., Kanagawa, O., & Shaw, A. S. (2000). CD28 and T cell co-stimulation. *Rev. Immunogenet., 2*(2), 175−184.

Hubbard, R. E., O'Mahony, M. S., Savva, G. M., Calver, B. L., & Woodhouse, K. W. (2009). Inflammation and frailty measures in older people. *J. Cell Mol. Med.,* 13, 3103−3109.

Janeway, C. A., Jr., Travers, P., & Walpert, M. S. M. (2001). *The Immune System in Health and Disease* (5th ed.). New York: Garland Publishing, Inc.

Jenkins, E. C., Velinov, M. T., Ye, L., Gu, H., Li, S., Jenkins, E. C., Jr., et al. (2006). Telomere shortening in T lymphocytes of older individuals with Down syndrome and dementia. *Neurobiol. Aging,* 27(7), 941−945.

Jennings, C., Rich, K., Siegel, J. N., & Landay, A. (1994). A phenotypic study of CD8+ lymphocyte subsets in infants using three-color flow cytometry. *Clin. Immunol. Immunopath.,* 71, 8−13.

John, V., Hock, J. M., Short, L. L., Glasebrook, A. L., & Galvin, R. J. (1996). A role for CD8+ T lymphocytes in osteoclast differentiation in vitro. *Endocrinology,* 137(6), 2457−2463.

Jonasson, L., Tompa, A., & Wikby, A. (2003). Expansion of peripheral CD8+ T cells in patients with coronary artery disease: relation to cytomegalovirus infection. *J. Intern. Med.,* 254(5), 472−478.

Knudson, M., Kulkarni, S., Ballas, Z. K., Bessler, M., & Goldman, F. (2005). Association of immune abnormalities with telomere shortening in autosomal-dominant dyskeratosis congenita. *Blood,* 105(2), 682−688.

Kong, Y. Y., Boyle, W. J., & Penninger, J. M. (2000). Osteoprotegerin ligand: a regulator of immune responses and bone physiology. *Immunol. Today,* 21, 445−502.

Krichevsky, S., Pawelec, G., Gural, A., Effros, R. B., Globerson, A., Yehuda, D. B., et al. (2004). Age related micro-satellite instability in T cells from healthy individuals. *Exp. Gerontol.,* 39(4), 507−515.

Kurokohchi, K., Masaki, T., Arima, K., Miyauchi, Y., Funaki, T., Yoneyama, H., et al. (2003). CD28-negative CD8-positive cytotoxic T lymphocytes mediate hepatocellular damage in hepatitis C virus infection. *J. Clin. Immunol.,* 23(6), 518−527.

Lichterfeld, M., Mou, D., Cung, T. D., Williams, K. L., Waring, M. T., Huang, J., et al. (2008). Telomerase activity of HIV-1-specific CD8+ T cells: constitutive upregulation in controllers and selective increase by blockade of PD ligand 1 in progressors. *Blood,* 112(9), 3679−3687.

Lin, J., Epel, E., Cheon, J., Kroenke, C., Sinclair, E., Bigos, M., et al. (2010). Analyses and comparisons of telomerase activity and telomere length in human T and B cells: insights for epidemiology of telomere maintenance. *J. Immunol. Methods,* 352(1−2), 71−80.

Lorenzo, J. (2000). Interactions between immune and bone cells: new insights with many remaining questions. *J. Clin. Invest.,* 106(6), 749−752.

Maini, M. K., Soares, M. V., Zilch, C. F., Akbar, A. N., & Beverley, P. C. (1999). Virus-induced CD8+ T cell clonal expansion is associated with telomerase up-regulation and telomere length preservation: a mechanism for rescue from replicative senescence. *J. Immunol.,* 162(8), 4521−4526.

Miller, R. A. (1996). The aging immune system: primer and prospectus. *Science,* 273(5271), 70−74.

Misiti, S., Nanni, S., Fontemaggi, G., Cong, Y. S., Wen, J., Hirte, H. W., et al. (2000). Induction of hTERT expression and telomerase activity by estrogens in human ovary epithelium cells. *Mol. Cell Biol.,* 20(11), 3764−3771.

Olovnikov, A. M. (1973). A theory of marginotomy: the incomplete copying of template margin in enzymatic synthesis of polynucleotides and biological significance of the phenomenon. *J. Theoret. Biol.,* 41, 181−190.

Ornish, D., Lin, J., Daubenmier, J., Weidner, G., Epel, E., Kemp, C., et al. (2008). Increased telomerase activity and comprehensive lifestyle changes: a pilot study. *Lancet Oncol.,* 9(11), 1048−1057.

Ouyang, Q., Wagner, W. M., Wikby, A., Walter, S., Aubert, G., Dodi, A. I., et al. (2003). Large numbers of dysfunctional CD8+ T lymphocytes bearing receptors for a single dominant CMV epitope in the very old. *J. Clin. Immunol.,* 23(4), 247−257.

Panossian, L., Porter, V. R., Valenzuela, H. F., Masterman, D., Reback, E., Cummings, J., et al. (2002). Telomere shortening in T cells correlates with Alzheimer's disease status. *Neurobiol. Aging,* 24, 77−84.

Parish, S. T., Wu, J. E., & Effros, R. B. (2009). Modulation of T lymphocyte replicative senescence via TNF-α inhibition: role of caspase-3. *J. Immunol.,* 182(7), 4237−4243.

Pawelec, G., Akbar, A., Caruso, C., Effros, R., Grubeck-Loebenstein, B., & Wikby, A. (2004). Is immunosenescence infectious? *Trends Immunol.,* 25(8), 406−410.

Pawelec, G., Derhovanessian, E., Larbi, A., Strindhall, J., & Wikby, A. (2009). Cytomegalovirus and human immunosenescence. *Rev. Med. Virol.,* 19(1), 47−56.

Perillo, N. L., Naeim, F., Walford, R. L., & Effros, R. B. (1993). The in vitro senescence of human lymphocytes: failure to divide is not associated with a loss of cytolytic activity or memory T cell phenotype. *Mech. Ageing Dev.,* 67, 173−185.

Perillo, N. L., Walford, R. L., Newman, M. A., & Effros, R. B. (1989). Human T lymphocytes possess a limited in vitro lifespan. *Exp. Gerontol.,* 24, 177−187.

Pietschmann, P., Grisar, J., Thien, R., Willheim, M., Kerschan-Schindl, K., Preisinger, E., et al. (2001). Immune phenotype and intracellular cytokine production of peripheral blood mononuclear cells from postmenopausal patients with osteoporotic fractures. *Exp. Gerontol., 36*(10), 1749–1759.

Pilch, H., Hoehn, H., Schmidt, M., Steiner, E., Tanner, B., Seufert, R., et al. (2002). CD8+CD45RA+CD27−CD28− T-cell subset in PBL of cervical cancer patients representing CD8+ T-cells being able to recognize cervical cancer associated antigens provided by HPV 16 E7. *Zentralbl. Gynakol., 124*(8–9), 406–412.

Posnett, D. N., Sinha, R., Kabak, S., & Russo, C. (1994). Clonal populations of T cells in normal elderly humans: the T cell equivalent to "benign monoclonal gammopathy". *J. Exp. Med., 179*, 609–618.

Ross, A. C., Rizk, N., O'Riordan, M. A., Dogra, V., El Bejjani, D., Storer, N., et al. (2009). Relationship between inflammatory markers, endothelial activation markers, and carotid intima-media thickness in HIV-infected patients receiving antiretroviral therapy. *Clin. Infect. Dis., 49*(7), 1119–1127.

Sadat-Sowti, B., Parrot, A., Quint, L., Mayaud, C., Debre, P., & Autran, B. (1994). Alveolar CD8+CD57+ lymphocytes in human immunodeficiency virus infection produce an inhibitor of cytotoxic functions. *Am. J. Respir. Crit. Care Med., 149*(4 Pt 1), 972–980.

Samani, N. J., Boultby, R., Butler, R., Thompson, J. R., & Goodall, A. H. (2001). Telomere shortening in atherosclerosis. *Lancet, 358*(9280), 472–473.

Sansom, D. M. (2000). CD28, CTLA-4 and their ligands: who does what and to whom? *Immunology, 101*(2), 169–177.

Saurwein-Teissl, M., Lung, T. L., Marx, F., Gschosser, C., Asch, E., Blasko, I., et al. (2002). Lack of antibody production following immunization in old age: association with CD8(+)CD28(−) T cell clonal expansions and an imbalance in the production of Th1 and Th2 cytokines. *J. Immunol., 168*(11), 5893–5899.

Schirmer, M., Goldberger, C., Wurzner, R., Duftner, C., Pfeiffer, K. P., Clausen, J., et al. (2002). Circulating cytotoxic CD8(+) CD28(−) T cells in ankylosing spondylitis. *Arthritis Res., 4*(1), 71–76.

Schwab, R., Szabo, P., Manavalan, J. S., Weksler, M. E., Posnett, D. N., Pannetier, C., et al. (1997). Expanded CD4+ and CD8+ T cell clones in elderly humans. *J. Immunol., 158*(9), 4493–4499.

Shimizu, Y., van Seventer, G., Ennis, E., Newman, W., Horgan, K., & Shaw, S. (1992). Crosslinking of the T cell-specific accessory molecules CD7 and CD28 modulates T cell adhesion. *J. Exp. Med., 175*, 577–582.

Spaulding, C. S., Guo, W., & Effros, R. B. (1999). Resistance to apoptosis in human CD8+ T cells that reach replicative senescence after multiple rounds of antigen-specific proliferation. *Exp. Gerontol., 34*, 633–644.

Spyridopoulos, I., Hoffmann, J., Aicher, A., Brummendorf, T. H., Doerr, H. W., Zeiher, A. M., et al. (2009). Accelerated telomere shortening in leukocyte subpopulations of patients with coronary heart disease: role of cytomegalovirus seropositivity. *Circulation, 120*(14), 1364–1372.

Swain, S., Clise-Dwyer, K., & Haynes, L. (2005). Homeostasis and the age-associated defect of CD4 T cells. *Semin. Immunol., 17*(5), 370–377.

Takayanagi, H., Ogasawara, K., Hida, S., Chiba, T., Murata, S., Sato, K., et al. (2000). T-cell-mediated regulation of osteoclastogenesis by signalling cross-talk between RANKL and IFN-gamma. *Nature, 408*(6812), 600–605, (Comment in: Nature *408*(6812), 535–536 UI: 21003806).

Tentolouris, N., Nzietchueng, R., Cattan, V., Poitevin, G., Lacolley, P., Papazafiropoulou, A., et al. (2007). White blood cells telomere length is shorter in males with type 2 diabetes and microalbuminuria. *Diabetes Care, 30*(11), 2909–2915.

Tsukishiro, T., Donnenberg, A. D., & Whiteside, T. L. (2003). Rapid turnover of the CD8(+)CD28(−) T-cell subset of effector cells in the circulation of patients with head and neck cancer. *Cancer Immunol. Immunother., 52*(10), 599–607.

Valenzuela, H. F., & Effros, R. B. (2002). Divergent telomerase and CD28 expression patterns in human CD4 and CD8 T cells following repeated encounters with the same antigenic stimulus. *Clin. Immunol., 105*(2), 117–125.

Vallejo, A. N., Nestel, A. R., Schirmer, M., Weyand, C. M., & Goronzy, J. J. (1998). Aging-related deficiency of CD28 expression in CD4+ T cells is associated with the loss of gene-specific nuclear factor binding activity. *J. Biol. Chem., 273*(14), 8119–8129.

Vaziri, H., Schachter, F., Uchida, I., Wei, L., Zhu, X., Effros, R., et al. (1993). Loss of telomeric DNA during aging of normal and trisomy 21 human lymphocytes. *Am. J. Hum. Genet., 52*, 661–667.

Wang, E., Lee, M. J., & Pandey, S. (1994). Control of fibroblast senescence and activation of programmed cell death. *J. Cell Biochem., 54*, 432–439.

Watson, J. D. (1972). Origin of concatemeric T7 DNA. *Nature New Biol., 239*, 197–199.

Weng, N. P., Levine, B. L., June, C. H., & Hodes, R. J. (1996). Regulated expression of telomerase activity in human T lymphocyte development and activation. *J. Exp. Med., 183*(6), 2471–2479.

Yang, O. O., Lin, H., Dagarag, M., Ng, H. L., Effros, R. B., & Uittenbogaart, C. H. (2005). Decreased perforin and granzyme B expression in senescent HIV-1-specific cytotoxic T lymphocytes. *Virology, 332*(1), 16–19.

CHAPTER 26

Mechanical Determinants of Tissue Development

Jonathan A. Kluge[*], Gary G. Leisk[**], David L. Kaplan[***]
*McKay Orthopaedic Research Laboratory, University of Pennsylvania, Philadelphia, PA, USA
**Department of Mechanical Engineering, Tufts University, Medford, MA, USA
***Department of Biomedical Engineering, Tufts University, Medford, MA, USA

INTRODUCTION

In recent years, increasing evidence has suggested that mechanics plays a key role in fetal development and is an important determinant in functional homeostasis of human tissues. Simultaneously, newer technologies in the field of regenerative medicine have focused on providing directed cellular influences with mechanical signals in mind, in order to more closely recapitulate normal development and send the maturing tissue construct along a regenerative path. However, mimicking normal fetal development is challenging, considering the complex orchestration of many signaling events, which are varied in their type and spatial-temporal profiles. Furthermore, *in vivo* environmental conditions, adult tissue mechanical loading, and complex signaling critical to cell function and tissue development are difficult to quantify, let alone replicate in a controlled setting. By studying animal models during development and establishing *in vitro* research paradigms that probe relevant cellular responses, however, an improved understanding and ability to define the sequence of major developmental events and their mechanical underpinnings can be achieved. This insight can serve as a roadmap to guide the design of human tissues and organs. Moreover, emerging literature on the mechanical influences on progenitor cell commitments and the well-established literature on signal mechanotransduction pathways provide the fundamentals necessary to interpret tissue- or organ-scale observations. In the attempt to optimize tissue-engineered constructs *in vitro* or predict their performance *in vivo*, significant benefits are anticipated from the improved delineation of developmental landmarks and the required cellular machinery that responds to mechanical inputs.

In this chapter we describe the mechanical determinants for developing tissues for readers interested in several topics, including developmental biology, mechanotransduction, and the use of important signaling factors such as active mechanical loading or substrate mechanical properties in tissue engineering. Due to the breadth of overlapping research fields covered in this chapter, we offer an introduction to seminal work in order to appropriately guide the reader. We start with a brief discussion of the fundamentals, mainly by introducing the stimuli-responsive machinery and pathways. We then describe the role of mechanics in tissue-level development, with particular emphasis on related *in vitro* systems and animal models of development. Finally, at the conclusion of each subsection, we summarize the knowledge gained from studying these model systems towards regulating and enhancing tissue regeneration *in vitro*.

463

Principles of Regenerative Medicine. DOI: 10.1016/B978-0-12-381422-7.10026-4

FUNDAMENTALS: CELL MACHINERY AND SIGNALING IN MECHANOBIOLOGY

Cells contain a multi-component internal architecture that can both transduce and adapt to applied loads, providing versatile responses to external mechanical stimuli (Ingber, 2002). Figure 26.1 shows a schematic of this loading scenario at various levels, including full-tissue, the cell-ECM compartment, and intracellular responses. The adaptation of full tissues to applied loads, both functionally and structurally, suggests that cell-level responses may ultimately translate to new tissue formation or tissue remodeling (Mow and Huiskes, 2005). Such stimuli can spur additional tissue production (via proliferation and metabolism) or further specify a cell population to a terminal phenotype (via signaling or differentiation). As external stimuli are connected to cell responses, it becomes important to define mechanically sensitive cell components and how their stimulation can lead to downstream signaling responses. The following sections will review fundamental biochemical principles of mechanobiology; mainly the stimuli-responsive components of the cell and the overall process by which cells integrate mechanical signals to direct tissue-specific growth or cell fate.

Cell surface receptors

Cells within living tissues transduce mechanical forces by using a variety of mechanisms. Although the signaling processes of mechanotransduction are complex, involving many different molecules and pathways, they may be activated by similar mechanisms in response to a variety of incoming signals (Huang et al., 2004; Mammoto and Ingber, 2009). Both membrane-bound proteins, the first-responders to external stimuli, and the inner supportive

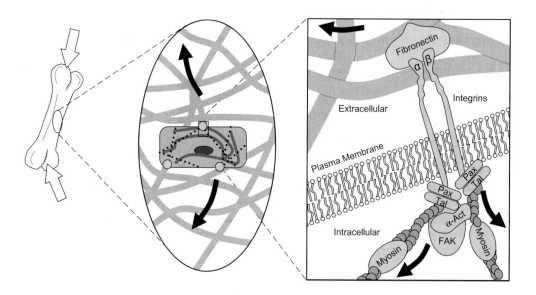

FIGURE 26.1

Diagrammatic view of mechanical force propagation (dark arrows) throughout tissue from macroscopic to nanoscopic levels. Human connective tissue, such as bone, is affected by repeated external loads (left). The underlying extracellular matrix (mostly collagen, represented by random fibrils) is subjected to these forces, which are then transferred to cells through insoluble ligands (fibronectin, circles at cell periphery). Surrounded by ECM, the cell contains its own substructure: lines, both continuous and dotted, represent the cytoskeletal microtubules and actin microfilaments, respectively (middle). The result of external forces is intracellular signaling: recruitment of bridging proteins, paxillin (Pax), α-actinin (α-Act), and talin (Tal), and signaling molecules, such as focal adhesion kinase (FAK), to the site of developing integrin-mediated focal adhesions. The result is a force balance that presumably affects actin-bound signaling proteins (actin helices with connecting myosin) (right).

cell structures (the cytoskeleton) are vehicles by which signals can be received and transduced from the host tissue towards the cell interior.

The cell makes contact with its surrounding ECM through "adhesions," a term used to describe an array of protein-mediated molecular links. The membrane portion of a cell ECM adhesion contains integrins, which are glycoprotein heterodimers of α and β subunits that bind to specific sequences on ECM molecules through a large extracellular domain (Geiger et al., 2001). Extracellularly, ligands can act as part of the ECM adhesion receptors such as fibronectin (α_5 and β_1), vitronectin (α_v and β_3), and various collagens (α_1 and β_1), which are all supplemented by membrane-bound non-integrin proteoglycan components such as syndecan-4 and CD44 (Geiger et al., 2001).

Intracellularly, these integrins interact with plaque proteins, which act as bridges that connect the integrins to the cytoskeleton or signaling molecules. Several intracellular multi-molecular proteins serve to link the actin non-muscle myosin portion of the cell cytoskeleton to membrane integrins; these linkers include α-actinin and talin, among others. Following the occupation of integrins by their ligands, the initial step in reinforcing adhesions involves the clustering of integrin molecules. Focal complexes, small focused or dot-like structures associated with the cell lamellipodia (thin, flat extensions at the cell periphery), are typically the first structures formed at cell/ECM junctions, and are either transient or evolve into more stable focal adhesions. The formation of more stable adhesion is thought to be generated internally by responses of the cytoskeleton to applied forces (Geiger et al., 2009). Intracellular activators (such as talin) are thought to interact with either the α or β subunit tail of integrins and induce their separation, thereby causing further conformational changes that open the binding site on the headpiece and allow the integrin to form this enhanced focal adhesion (Giancotti, 2003). Since the cytoplasmic segment undergoes conformational changes via intracellular linkers and the extracellular segment is controlled by ECM interactions, integrin molecules appear to have two functions: to regulate the extracellular binding activity from inside the cell (inside-out signaling) and to elicit intracellular changes through ECM binding (outside-in signaling) (Giancotti and Ruoslahti, 1999).

Mechanosensitive membranes and ion channels

The lipid bilayer is believed to be one of the major mechanotransduction components of a cell, and was likely one of the first mechanosensitive cellular components to arise through evolution as a protective mechanism against large concentration or swelling pressure gradients (Hamill and Martinac, 2001) and large shear forces (Hoger et al., 2002) encountered in multicellular organisms. When a lipid bilayer, such as the outer plasma membrane in eukaryotic cells, is deformed by a force, two changes may occur. Disturbance of the lateral force balance around a lipid bilayer may first lead to conformational changes of transmembrane proteins (such as integrins) by indirect association with the bilayer. A second response may stem from membrane perturbations that cause immediate changes in local curvatures, which segregates the lipid bilayer into alternating regions of outwardly-located flat sections and inwardly-located curved sections (Janmey and Weitz, 2004). It is believed that either change in membrane states can activate ion channels, which, in turn, respond with changes in their permeability, or to intracellular signaling via membrane-bound linker proteins (Hamill and Martinac, 2001; Kung, 2005). Whether the forces affect membrane proteins or lipids, and whether the magnitude of these forces can be correlated to channel properties, is not clearly understood (Janmey and Weitz, 2004).

The original study of channel gating (regulatory open-and-close mechanisms) in eukaryotic cells was with *Xenopus* oocytes, which could be activated for a latent response with patch-clamp techniques (McBride and Hamill, 1999). A long history of patch-clamp studies have illustrated not only the prevalence of these channels across many eukaryotic species but also their key influence on cell volume regulation and the possibility that tight seal formation could lead to

465

mechanosensitivity in isolated K^+ channels (Hamill and Martinac, 2001). Many of these experiments were designed in order to control membrane tension by suction pressure in a micropipette attached to a small region of the cell membrane. In these experiments, increased suction pressure to just below that which would cause the membrane to rupture was shown to increase pore dimensions and cause large calcium ion (Ca^{2+}) influx through channels of large conductance (Zhang et al., 2000). This "pressure relief valve" mechanism, in addition to its role in mechanosensation, can be seen as a cell's natural defense against large osmotic gradients. Similar mechanisms have been linked to Ca^{2+} channel activation in the stereocilia of hair cells in the inner ear and fluctuations in intracellular ion concentration of endothelial cells (Hamill and Martinac, 2001; Huang et al., 2004).

Although the existence of stretch-activated ion channels has been well documented for "specialized" cells (i.e. human cells associated with auditory and visual functions as examples) and to a lesser extent in non-specialized cells (Hua et al., 2010), the mechanism(s) behind their function are not clear, nor are their connections with an intact, functioning cytoskeleton (Hamill and Martinac, 2001). For example, it is still possible to recruit focal adhesion-related signaling factors and linking proteins to the cytoskeleton following adherent substrate stretch in the case where the cellular cytoplasm and apical cellular membrane are removed (Sawada and Sheetz, 2002). Similarly, in the case of intact cell membranes of endothelial cells, recent evidence suggests that actin stress fibers can activate these channels via external perturbation (Hayakawa et al., 2008). These and other results indicate that, in addition to ion channels, there are other mechanisms that enable cells to sense physical forces. These mechanisms, in conjunction with transmembrane proteins discussed earlier, are part of the underlying cell cytoskeletal structure.

Cytoskeleton

The cytoskeleton is not merely a randomly configured collection of molecules; instead, it is believed that each different cytoskeletal molecule is interconnected in a unique way to maintain mechanical signaling pathways. Furthermore, the mechanical behavior of the whole cell is driven by both the cytoskeletal elements found just below the surface of the plasma membrane and also the internal cytoskeletal network. The role of the cytoskeleton as a support and shape-retaining structure has long been recognized, but it is now known that it can also provide directed signals to intracellular elements, and is thus capable of inducing endogenous changes (Ingber, 2004).

Structural molecules that make up the cytoskeleton can be broken down into three major groups: actin-based microfilaments associated with the cell's cortical cytoskeleton (adhesion complexes), stiff hollow tubulin-based microtubules that radiate from an organizational center (centrosome), and thick intermediate filaments, such as lamin, vimentin, and keratin, that integrate with the cell's nucleus and attachment sites (desmosomes). Formation of larger and stronger cytoskeletal structures is possible when these molecules are supplemented by other proteins, such as actin-bound myosin chains, which combine to form stress fibers around the cell periphery. Together, these structural elements organize throughout the cell interior to form a complex network (Huang et al., 2004).

In order to account for the collective role of these cytoskeletal elements and their force-sensing capacity, a tensegrity model was proposed (Chen et al., 1997; Ingber, 2006). This model assumes that compressive-bearing struts (microtubules and ECM adhesions) are resisted by the pull of surrounding tensile elements (microfilaments, intermediate filaments), adapting the principles used in the design of a ship mast and riggings, as linear reinforcing elements that can be linked together to form tension-resistant scaffolding around a hull (cell membrane). Cytoskeletal elements can thereby exchange energy with their surroundings, permitting their shapes to fluctuate as they bend and twist in response to transverse loading. According to the tensegrity model, many cytoskeletal molecules in their natural state have a certain level of prestress or isometric tension, generated by the contractile function of actin and myosin sliding, osmotic forces, and/or forming new ECM adhesions. The balance of these internal stresses can be visually confirmed when the plating of cells on a flexible substrate does not lead to its distortion, or when the cutting of a cell leads to spontaneous retraction of its intracellular cytoskeleton (Mammoto and Ingber, 2010). A tensegrity model would also explain the unique ability of a cell to perform two specialized tasks: (1) the capacity to rapidly transduce signals over long intracellular distances and (2) the ability to connect external stimuli (about focal adhesions) to transcriptional changes that start in the nucleus.

Immediately after mechanical signals are sensed by surface integrins, the cytoskeleton will realign in the direction of the applied tensional stimulus (Mammoto and Ingber, 2010). The rate of force transduction is extremely rapid; while normal diffusion of small ions such as Ca^{2+} should pass through the cytoplasm at rates close to 1 μm per second, mechanical signals are estimated to transmit along stressed cytoskeletal filaments at rates close to 30 m per second (Wang et al., 2009). This also implies that rates of force transfer between membrane-bound receptors and intracellular cytoskeletal proteins can exceed the rate of transient actin/microtubule turnover (Wang et al., 2009).

Further promoting this view on signal transduction is the fact that actin and tubulin are dynamic polymers; their fundamental protein building blocks can both polymerize and depolymerize, depending on the conditions, changing the length of the filaments in the process. Rapid depolymerization releases the contents of the microtubule to the cytoplasm and permits it, or nearby microtubules, to start reconstruction elsewhere. These mechanisms allow a cell to constantly adjust its internal prestress, and thus alter the tightness with which the cytoskeletal lattice is held together.

Rapid cytoskeletal realignment and construction will affect the many cytoskeletally-linked enzymes and substrates that mediate protein synthesis, glycolysis, and signal transduction. It is believed that, if the cytoskeleton and its immobilized proteins distort without breaking following focal adhesion stimulation, then the attached molecules must similarly change shape. Altered biophysical properties may result in altered local thermodynamic properties or altered kinetic behavior, just as a spring would change its vibration frequency following distortion (Ingber, 2008). For instance, it has recently been shown that stretching of single talin molecules (a model integrin-cytoskeletal linker) causes unfolding and conformational changes to attached vinculin in isolation (del Rio et al., 2009). It has also been shown that forces as low as 3—5 pN are sufficient to unfold cryptic subdomains in fibronectin, which can then lead to fibronectin fibril formation (Geiger et al., 2001). Similarly, the altered cytoskeleton has been shown to influence protein synthesis by destabilizing cytoskeleton-associated mRNAs at the intersections of actin filaments, and through polymerization at vertices within highly triangulated microfilament networks (Bassell et al., 1994; Wang et al., 2009). Taken together, these findings suggest that the cytoskeleton not only facilitates rapid information transfer and shape adjustment in response to external stimuli, but is also directly responsible for initiating signaling cascades through its stretch-responsive nature.

The hard-wired nucleus

The second unique cell feature explained by the tensegrity model is that proximal stimuli can perturb distal cell features as part of a rich hierarchy of structural connections. The nucleus may be "hard-wired" to integrins, such that distortions of adhesion complexes result in synchronized realignment of structural elements to the nuclear envelope, via the underlying lamin network (Huang et al., 2004). Mitochondria, located distal to focal adhesions connected to cytoplasmic microtubules, have been shown to immediately release reactive oxygen species and activate various signaling molecules including nuclear factor-κB and vascular cell-adhesion molecule 1 in response to surface integrin stimulation (Hu et al., 2005). It has also been shown that ligand-coated beads can be used to pull focal adhesions of cultured cells at very high rates, and that the nucleoli will immediately deform and elongate in the direction of applied force. This same phenomenon was also observed in cultures with extracted membranes, but with intact intracellular components, suggesting that this signal was transduced directly through the cytoskeletal lattice, and not through a signaling cascade (Maniotis et al., 1997).

If such connections can penetrate from the outermost to innermost regions of the cell, then the tensed intermediate filaments that connect to the nucleus and its proximal network may be a route by which the signals are delivered (Ingber, 2008). Stress mapping of the cell cytoplasm has shown that stress concentrations appear >50 μm away from the point of origin, a pattern that is inhibited by actin network alterations and thus prestress disruption (Na et al., 2008). The effects of force propagation through this route are presumably extensive, and these proposed mechanisms have been delineated by recent experimentation, as detailed elsewhere (Wang et al., 2009).

Briefly, in response to direct channel distortion or tugging of linker cytoskeletal components, ion flux may occur through nuclear pores, similarly to the above-described cell membrane mechanism. Alternatively, nuclear binding structures called LINC (linker of nucleoskeleton and cytoskeleton) are likely participants in direct force transduction to the nucleus (Padmakumar et al., 2005). As part of LINC, the actin cytoskeleton binds to rod-like nuclear membrane proteins called nesprins, which themselves contain domains that mediate *sun* protein binding. Sun, located on the inner nuclear membrane, may bind to lamin A on the embedded nuclear scaffold or connect this series of linkages to the nuclear pores (Dahl et al., 2008). Inside the nucleus, laminins may bind to the nuclear isoform of the protein titin, and thus affect the supportive lattice inside the nucleus, or bind to other linkers known to affect chromatin modification, regulate transcriptional factors, or bind directly to dsDNA (Wang et al., 2009). Mutations in the lamin family of genes can cause a variety of human diseases, presumably since its disruption can lead to enhanced nuclear deformability and thus enhanced mechanosensitiviy (Lammerding et al., 2004).

Signaling molecules

One of the most highly studied signaling molecules implicated in mechanotransduction includes the phosphorylated kinase protein focal adhesion kinase (FAK, also known as protein tyrosine kinase 2); by following the response of FAK, for example, one can appreciate the complexity of one of many signaling pathways and the importance of focal adhesion in initiating cell responses. For several years, FAK has been associated with both the growth and the disassembly of integrin-based focal adhesion sites (Geiger et al., 2001). More recently, the relationship of FAK with various GTPase proteins (Rho, Rac, and Cdc42) and indirect association with integrins through bridging proteins have been elucidated, and place FAK at the forefront of several intracellular signaling pathways (Mitra et al., 2005).

To briefly illustrate these channels, a simple schematic is provided that links FAK to intracellular activity (Fig. 26.2). Formation of new integrin-mediated focal complexes or the transduction of forces through integrins may lead to activation of FAK signaling mechanisms, mainly recruitment of other focal adhesion proteins. One such mechanism, responsible for the assembly and disassembly of focal complexes, is the ability of FAK to control phosphorylation of the bridging protein α-actinin, which can cross-link and tether actin/myosin stress fibers (Mitra et al., 2005). A second mechanism, although not completely understood, may be the activation of the Rho effector diaphanous (mDia), which leads to the stabilization of the cytoskeleton (i.e. microtubules) at the leading edge of migrating cells (Geiger et al., 2001). A third mechanism is the activation and/or inhibition of the various GTPase proteins that lead to regulation of cytoskeletal extensions, such as stress fibers, lamellipodia, and filopodia. A final mechanism is the formation and disassembly of cell-cell (cadherin-based) connections, providing an added route for solute

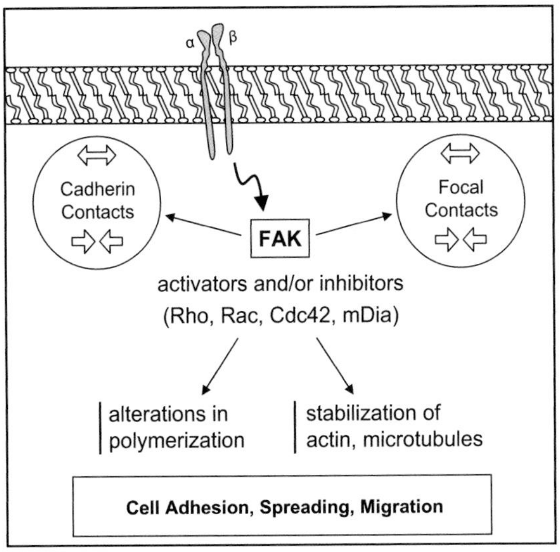

FIGURE 26.2

Cell signaling mediated by integrin responses at membrane interfaces, for example due to changes in external mechanical states, leads to intracellular cascades. FAK plays a central role in these responses. Mediation of cell interactions with the external environment are summarized, including responses such as cell adhesion, spreading, and movement based on changes in focal contacts related to cell-matrix adhesion and cadherins related to cell-cell mediated interactions. FAK functions to recruit other focal contact proteins or their regulators, leading to changes in the internal structure through polymerization and stabilization of cytoskeletal elements. All of the events illustrated occur in a complex symphony of orchestrated events to modulate internal and external changes in response to changes in external mechanical signaling. The figure is in part patterned after Figure 1 from Mitra et al., 2005.

exchange and signaling (Mitra et al., 2005). The importance of relationships between the cytoskeleton and molecular signaling pathways will gain further emphasis in the following discussions.

CELL-BASED DEVELOPMENT AND MECHANICAL FORCE BALANCING

In order to study cell responses to mechanical stimuli or passive cell interactions with their substrates, a variety of experimental systems have been developed to precisely control the mechanical "treatments" given to cells in isolation or to simulate normal cell-cell contacts. As a result, the past several years have seen an explosion of data linking mechanics to development across every stage (from embryonic to committed terminal phenotypes) and from cells encountering a variety of different loading scenarios. In this section, we describe observations on mechanotransduction specific to cells at various stages along the developmental path and the implications for tissue-level control of morphogenesis.

Cell studies of ESCs

The earliest-stage embryo, consisting of cells (embryonic stem cells (ESCs)) and thus lacking any defined extracellular matrix, is the earliest stage at which mechanical force balancing leads to cellular cues and pattern formation. Recent studies have shown that murine embryonic stem cells cultured *in vitro* are more sensitive to deformations and more easily spread on soft substrates than their differentiated counterparts (Chowdhury et al., 2010), where *oct3/4* gene downregulation was thought to reflect the long-term stretch-mediated differentiation of these cells. The more uncommitted ES cells were unable to generate the same level of myosin-II- and actin polymerization-dependent endogenous forces. Inhibition studies showed that both rho kinase (ROCK) and upstream signalers of FAK (but not Rac) were involved in these integrin-mediated pathways. Interestingly, ESCs also showed optimal responsiveness (i.e. spreading area) to 1 Hz loading frequency, as has been reported for many downstream progenitor and committed phenotypes (Kaunas et al., 2005; Hu and Wang, 2006).

The early-stage substrate or stretch sensitivity of ESCs may be conserved upon differentiation towards more committed phenotypes. Embryo-derived cardiomyocytes, although still immature precursors for cardiac tissues (and thus a popular source for generating beating cardiac tissue), will only maintain the synchronous contractility necessary for significant actomyosin striated force generation on substrates with a stiffness equal to or less than the developing myocardial microenvironment (Engler et al., 2008). Rates of formation and assembly of cytoskeletal proteins vimentin, filamin, and myosin in cardiomyocytes were likewise shown to be sensitive to substrate stiffness. However, ROCK inhibition can restore contractility on more rigid substrates (>10 kPa) through a reduction of stress fibers, which are atypical of cardiomyocytes found *in vivo* (Jacot et al., 2008). These studies have shown that individual cells from the earliest progenitor populations have phenotypic-stage-dependent responses to the stiffness and mechanical activity of their surroundings, but this does not describe how those effects lead to morphological shifts.

The formation of the lung buds from planar layers of epithelium illustrates how, from an albeit simplified perspective, force balancing can help guide the development of new tissue structures from early progenitors (Moore et al., 2005). It was observed that cells located in tight clusters on relatively high turnover regions of epithelium demonstrated more pronounced budding outgrowth, whereas alternating regions of higher ECM deposition and lower proliferation remained constant in size. As this pattern is repeated many times and propagates forward, expanded patterns can emerge. Furthermore, bud formation could be modulated through regulation of cytoskeletal tension (for example, by ROCK inhibitor using agent Y27632), leading to a loss of differential basement membrane thickness. This phenomenon and underlying mechanism are paralleled in the case of capillary sprout growth during angiogenesis (Ingber, 2002). Furthermore, since bud outgrowth is triggered before cell proliferation rate increases, one may hypothesize that active mechanical loading is not needed in order to derive new patterns; instead, local variations in passive substrate properties (elasticity, stiffness) may be the only stimuli necessary to trigger further specification. To gain insight into the mechanisms and timing of such developmental events and their mechanical determinants, various animal models have proven useful in determining these effects *in situ* as a means of expanding on the above-mentioned *in vitro* work.

Embryonic specification using animal modeling

There is evidence that, as early as gastrulation, the mechanical forces imparted by cells upon one another have transcriptional consequences that lead to germ layer formation (Mammoto and Ingber, 2010). Germ layer cells migrate and proliferate extensively during this phase, forming first a two-layer ectoderm/endoderm complex, followed by the formation of a third primary layer, the mesoderm. Each layer contains the first forms of ECM (laminin, fibronectin, type IV collagen), including the basement membrane separating epithelial cell layers, to act as a relatively rigid (albeit still capable of adaptive shape change) template from which spatial force gradients may arise (Ingber, 2006).

With the current inability to study further stages of human embryonic development *in situ*, most studies on the morphogenesis of the early embryo have taken place in a variety of non-terrestrial animal models, including *Drosophila melanogaster, Caenorhabditis elegans, Xenopus laevis,* and zebrafish (*Tub longfin*), owing to their ease of study. The three germ layers and their constituent cells have been shown in zebrafish to be intrinsically stiffer than suggested for the ESCs (Chowdhury et al., 2010) and all have different degrees of relative stiffness (Krieg et al., 2008). Several morphogens, including Nodal and transforming growth factor-β (TGF-β), were shown to regulate interlayer cell contractility by disrupting cell-cell (cadherin-based) and intracellular (cytoskeleton-based) tension (Krieg et al., 2008).

Drosophila models have been valuable for defining how cell-generated contractile forces and layer boundaries are maintained during compartmentalization. *Drosophila* embryos contain cells with myosin-based cytoskeletal features similar to human cells, which, when disrupted, can lead to cell stiffening and isolation, rather than the motility necessary for population mixing and new organotypic feature formation (Landsberg et al., 2009; Monier et al., 2010). Imparting traction forces on surrounding ECM or cells is facilitated by way of actin-rich filopodia (Hogan et al., 2004); *Drosophila* make use of these structures both to guide adjacent cells away from immobile planar orientations (Bertet et al., 2004) and to force organization that eventually gives rise to the paraxial somatic mesoderm (Zhou et al., 2009).

C. elegans has been useful for demonstrating the origins of tissue folding and bud formation (Sawyer et al., 2010), confirming some of the mechanisms described for the *in vitro* lung bud formation detailed earlier (Moore et al., 2005). The increased presence of myosin II causes constriction and apical cell shortening of folded cell layers, which is thought to be activated by the Wnt signaling cascade (Lee et al., 2006).

MSC mechanobiology

As a further developmental step during embryonic development, the formation of the mesenchyme gives rise to a variety of connective and skeletal tissues that undergo active or passive mechanical loading functions during post-natal development and adulthood. In adult life, these varied loading scenarios are routinely encountered when sequestered mesenchymal stem cell (MSC) populations are recruited from their bone marrow origins and coerced into a variety of niches with differing stiffness and levels of activity (bone vs. cartilage vs. neural) (Engler et al., 2006). The relative ease of isolation, lineage pluripotency, and clinical relevance of MSCs has generated great interest in their practical implementation for various regenerative medicine strategies (Pittenger et al., 1999). Accompanying this interest has been the curiosity in understanding how these pluripotent (yet partially committed) cell lines can be finely controlled to dictate both temporary states and terminal paths of differentiation.

For several years, it has been shown that control of MSC shape via *in vitro* control of substrate stiffness (Engler et al., 2004, 2006) or available contact surface area (Pittenger et al., 1999; McBeath et al., 2004; Kurpinski et al., 2006) can determine a stable path of differentiation. The design of such *in vitro* systems facilitating this work is described elsewhere in this book. Culture substrates for these studies are typically designed by some combination of polymer design (to control stiffness), surface topographical patterning using soft lithography techniques, and surface modification by coupling of specific ECM ligands. These modifications can control the organization, spacing, and specificity of cell binding.

Study of MSC responsiveness to varied substrate stiffness reveals that lineage potential may be equally if not more sensitive to passive mechanical cues than to growth factors or chemotactic factors (Engler et al., 2004, 2006). In the absence of directive chemical cues, the morphology of MSCs plated onto relatively rigid synthetic substrates and/or substrates with large available binding surface area appear flattened, and this represents a phenotype typical of differentiated osteoblasts (McBeath et al., 2004; Engler et al., 2006). Conversely, when plated on soft substrates in the presence of chemically inert (growth-promoting) media, MSCs assume a more spindle-like elongated phenotype, commensurate with differentiated neuronal cells (Engler et al., 2006). Thus, one guiding principle that has emerged by comparing the relative "softness" of various progenitors is that differentiation *in vitro* via substrate cues can only be induced by careful force balancing; furthermore, progressively increased substrate stiffness is required with each increased developmental step (from ~500 Pa in ESCs to ~12 kPa in skeletal muscle cells) (Chowdhury et al., 2010). This resultant substrate "guidance" has been termed cellular *durotaxis* (Engler et al., 2006), and is now considered a new principle for incorporation in material designs for regenerative medicine (Reilly and Engler, 2010). These new designs represent another step towards translating and incorporating lessons learned from the field of mechanobiology into regenerative medicine.

LATER ORGAN SPECIFICATION DURING DEVELOPMENT

During later stages of development, after gastrulation but before birth, specific tissues, organs, and physiological systems continue to be regulated, in part, by mechanical forces through many of the same mechanisms previously described in this text. Mounting evidence suggests that force balancing and active loading play a role in the development of every system, regardless of the mechanical environment after birth, including kidneys, lungs, and mammary tissues (Stokes et al., 2002; Cohen and Larson, 2006; Nelson et al., 2006; Adamo et al., 2009; Vasilyev et al., 2009). So far, however, the majority of research has centered on tissues that normally undergo significant routine mechanical loading during post-natal life, including those generating or subjected to blood flow and those responsible for skeletal movement of the human body. The development of these major systems will be reviewed here with emphasis on the animal models employed.

Cardiovascular development

The embryonic heart is the first functioning organ in the human body (Keller, 2007), and related cardiovascular organs are influenced by the mechanical environment provided by this pulsatile flow, either via shear forces (le Noble et al., 2004) or gradients in pressure leading to circumferential strain in blood vessels (Lucitti et al., 2006). The immediate formation of a primitive heart tube and subsequent chamber, valve, and vascular branching occurs in concert with these mechanical forces (Keller et al., 2007). Confocal microscopy has allowed researchers to look within the developing embryos over time to probe both spatial and temporal patterns resulting from these mechanical roots. One such study in zebrafish found that, rather than the peristaltic pumping movement common to many species, flow is facilitated by a physical valveless hydro-impedence model (Forouhar et al., 2006). In this model, constructive interference and thus "suction" forces were produced by overlapping incident and reflective pressure waves, presumably generated by the mechanical action of proximal tube myocytes. Likewise, when using confocal imaging to study GFP-labeled plexus endothelial cells in an embryonic mouse model, the authors found that by reducing blood viscosity (and thereby decreasing shear forces) they could limit vascular remodeling into branched phenotypes and trigger signaling cascades within resident endothelial cells (Lucitti et al., 2006).

In response to fluid-induced shear stress, monolayers of endothelial cells (ECs) have been shown to change morphology and become torpedo shaped, aligned in the fluid flow direction. It is thought that this morphology is protective, and can serve to minimize shear acting directly to the nucleus (Hazel and Pedley, 2000). From micropipette aspiration studies and atomic force microscopy studies, the nucleus of a sheared cell also appears stiffer, although the mechanism of this shift is unknown (Deguchi et al., 2005; Mathur et al., 2007). From this cell shape change, it would follow that the cytoskeletal backbones of ECs undergo major alterations, in which stress fibers reinforce the EC membrane. Because of such cytoskeletal responses, shear stresses acting on the luminal cell membrane of ECs *in vivo* are transmitted to the basal attachment sites (Satcher and Dewey, 1996). It is unclear whether focal complex enhancement is solely driven by basal side integrin activation, or whether further support is also provided by the translocation of inactive apical side integrins to the basal membrane following shear stress (Shyy and Chien, 1997).

In either case, the development of focal adhesions will lead to recruitment of cytoplasmic signaling molecules and mitogen-activated protein kinase (MAPK) signaling pathways. Focal adhesion sites, such as the cytoskeleton, align their shape parallel to the flow direction without changing their overall contact area (Helmke and Davies, 2002). Activated luminal cell surface mechanisms (stretch-activated Ca^{2+} or K ion channels) have been linked to EC shear strain response. Similarly, G-protein activation due to distortions of the plasma membrane from shear has also been documented (Helmke and Davies, 2002). These and the integrin-dependent mechanisms are part of either the inside-out or outside-in signaling routes that develop from an EC's complex response to shear.

In the engineering of cardiovascular tissue, it is believed that bioreactor design should involve laminar fluid flows that induce a uniform distribution of shear stress and laminar convective mass transfer. Rotating-wall bioreactors have been used to establish engineered cardiac tissues that are structurally and functionally superior to those grown in static or mixed flasks (Barron et al., 2003; Martin et al., 2004). Other bioreactors have been used that include strain actuation, mimicking the dynamic mechanical stimuli present *in vivo*. For example, it is thought that, since arteries experience axial strains through connective tissue, tubular scaffolds that represent a cardiovascular vessel should experience the same strain. In addition, circumferential strains can be provided by a pulsatile force through the tissue scaffold, mimicking pulsatile blood flow in actual arteries (McCulloch et al., 2004). We and others have used these effects to greatly improve the ECM deposition and upregulation of relevant ECM markers specific to mature vasculature (Gong and Niklason, 2008; Zhang et al., 2009). In turn, improved designs can lead to enhanced repair options through augmented structural stability and vessel patency *in vivo* (Lovett et al., 2007, 2009).

Musculoskeletal joint formation

The formation of musculoskeletal joints in the absence of proper mechanical cues has provided another interesting view on the careful force balance required during development. Early animal models of embryogenesis including chick and pig showed that paralysis of the developing limb resulted in a loss of proper bone or cartilage formation, as reviewed elsewhere (Estes et al., 2004). Mouse knockout models have been useful in studying defect-specific morphogenic comparisons with anatomical similarities to human systems. Joint development in growing limbs involves a condensation of early progenitors at the point of limb separation, otherwise known as the inter-zone. Cells in this specialized niche are characterized by a "flat," non-chondrogenic phenotype that aligns perpendicular to the long bone direction (Mitrovic, 1977). Accompanying this loss of chondrogenic morphology, these cells express fewer chondrogenic transcripts, and instead express new sets of genes affiliated with the Wnt signaling cascade (Hartmann and Tabin, 2001).

Joint "cavitation", or zonal severance and segregation, follows the initial joint specification, and results in the morphogenesis of the entire articular space, including cartilage layer separation and synovium. When knockout mouse models devoid of fully functioning muscle progenitors (stripped of their migratory capacity) were developed, cavitation was delayed and their inter-zone failed to form properly at all articulating surfaces (Cohen and Larson, 2006). This lack of proper segregation was due to a lack of muscle contractility (by lack of β-catenin expression), a cell-cell force deemed necessary to properly organize the intact chondrocyte precursor population. By removing functional muscle contractions, similar findings in chick embryos were reported long ago (Murray and Drachman, 1969), but until the development of specific knockout variants was poorly understood.

Recent "in vitro" studies have similarly implicated muscle cell contractility and in the normal morphogenesis of engineered cartilage tissue. By co-culturing chondrocytes with muscle cells, pro-chondrogenic biochemical signals were enhanced, leading to the promotion of cartilage matrix production (Cairns et al., 2010). Cells cultured cooperatively in identical media formulations maintain phenotypes of the independent (separate) cultures, in co-cultures muscle cells were observed to "corral" chondrocytes into a compacted colony. Increased chondroctye condensation thereby leads to a rounder chondrocyte phenotype and a synergistically increased production of glycosaminoglycans and collagen type II, both specific hallmarks of chondrogenesis (Cairns et al., 2009).

Like a secondary cell population, a hydrogel or other stiff three-dimensional matrix can be used to encapsulate and stabilize the synthetic phenotype of isolated chondrocytes. Using these systems, researchers have studied the effects of external mechanical loading on cartilage development (Mauck et al., 2000), or the cooperative effects with various growth factors in the case of progenitor MSCs (Mauck et al., 2006). The rapid or acute response of chondrocytes to mechanical stimulation was studied *in vitro* using a two-dimensional monolayer model, and revealed that substrate stretch induced membrane hyperpolarization within 20 minutes. *In vitro*, this study confirmed that tyrosine phosphorylation of both paxillin and FAK was induced within 1 minute of initiation of stretch, and led to the eventual signaling cascade inducing small conductance K channels (Millward-Sadler and Salter, 2004). In three dimensions, unconfined dynamic compression likewise upregulates glycosaminoglycan and collagen II expression and deposition over unloaded controls, leading to increased functional load-bearing tissue-engineered materials (Hung et al., 2004). The effects of compression on stem cells appear, however, to depend on the level of commitment; MSCs appear to differentiate towards chondrogenic lineage; however, ESCs from embryoid bodies instead downregulate these chondrogenic markers (Guilak et al., 2009). Future studies will undoubtedly focus on the precise discrepancies between the various progenitor cell lines (and their varying levels of "softness") for their ability to transduce mechanical signals and thus use mechanical conditioning as a directive cue towards functional tissue development.

CONCLUSIONS

Mechanical forces play a crucial role in tissue development, function, and repair *in vivo*. Thus, the design of smart cell-sensitive polymers and novel biomimetic bioreactors to impart complex mechanical forces to cells and tissues *in vitro*, and in general those designs that exploit knowledge of mechanotransduction, can offer important options to improve functional tissue engineering. These inputs must be considered within the context of the cells used in these systems, as well as the overall clinical question at hand, to generate functional tissues *in vitro* for utility *in vivo*. It is clear that the road ahead is challenging, yet promise rests in ongoing efforts to continually translate these findings to more macroscopic systems until full regeneration can be realized for the future of clinically relevant therapeutics.

References

Adamo, L., Naveiras, O., Wenzel, P. L., McKinney-Freeman, S., Mack, P. J., Gracia-Sancho, J., et al. (2009). Biomechanical forces promote embryonic haematopoiesis. *Nature, 459,* 1131–1135.

Barron, V., Lyons, E., Stenson-Cox, C., McHugh, P. E., & Pandit, A. (2003). Bioreactors for cardiovascular cell and tissue growth: a review. *Ann. Biomed. Eng., 31,* 1017–1030.

Bassell, G. J., Powers, C. M., Taneja, K. L., & Singer, R. H. (1994). Single mRNAs visualized by ultrastructural in situ hybridization are principally localized at actin filament intersections in fibroblasts. *J. Cell Biol., 126,* 863–876.

Bertet, C., Sulak, L., & Lecuit, T. (2004). Myosin-dependent junction remodelling controls planar cell intercalation and axis elongation. *Nature, 429,* 667–671.

Cairns, D. M., Lee, P. G., Uchimura, T., Seufert, C. R., Kwon, H., & Zeng, L. (2010). The role of muscle cells in regulating cartilage matrix production. *J. Orthop. Res., 28,* 529–536.

Chen, C. S., Mrksich, M., Huang, S., Whitesides, G. M., & Ingber, D. E. (1997). Geometric control of cell life and death. *Science, 276,* 1425–1428.

Chowdhury, F., Na, S., Li, D., Poh, Y. C., Tanaka, T. S., Wang, F., et al. (2010). Material properties of the cell dictate stress-induced spreading and differentiation in embryonic stem cells. *Nat. Mater., 9,* 82–88.

Cohen, J. C., & Larson, J. E. (2006). Cystic fibrosis transmembrane conductance regulator (CFTR) dependent cytoskeletal tension during lung organogenesis. *Dev. Dyn., 235,* 2736–2748.

Dahl, K. N., Ribeiro, A. J., & Lammerding, J. (2008). Nuclear shape, mechanics, and mechanotransduction. *Circ. Res., 102,* 1307–1318.

Deguchi, S., Maeda, K., Ohashi, T., & Sato, M. (2005). Flow-induced hardening of endothelial nucleus as an intracellular stress-bearing organelle. *J. Biomech., 38,* 1751–1759.

me
Tal

S1

It i
bo
ma
cel
pei
the
de

del Rio, A., Perez-Jimenez, R., Liu, R., Roca-Cusachs, P., Fernandez, J. M., & Sheetz, M. P. (2009). Stretching single talin rod molecules activates vinculin binding. *Science, 323,* 638–641.

Engler, A. J., Carag-Krieger, C., Johnson, C. P., Raab, M., Tang, H. Y., Speicher, D. W., et al. (2008). Embryonic cardiomyocytes beat best on a matrix with heart-like elasticity: scar-like rigidity inhibits beating. *J. Cell Sci., 121,* 3794–3802.

Engler, A. J., Griffin, M. A., Sen, S., Bonnemann, C. G., Sweeney, H. L., & Discher, D. E. (2004). Myotubes differentiate optimally on substrates with tissue-like stiffness: pathological implications for soft or stiff microenvironments. *J. Cell. Biol., 166,* 877–887.

Engler, A. J., Sen, S., Sweeney, H. L., & Discher, D. E. (2006). Matrix elasticity directs stem cell lineage specification. *Cell, 126,* 677–689.

Estes, B. T., Gimble, J. M., & Guilak, F. (2004). Mechanical signals as regulators of stem cell fate. *Curr. Top. Dev. Biol., 60,* 91–126.

Forouhar, A. S., Liebling, M., Hickerson, A., Nasiraei-Moghaddam, A., Tsai, H. J., Hove, J. R., et al. (2006). The embryonic vertebrate heart tube is a dynamic suction pump. *Science, 312,* 751–753.

Geiger, B., Bershadsky, A., Pankov, R., & Yamada, K. M. (2001). Transmembrane extracellular matrix-cytoskeleton crosstalk. *Nat. Rev. Mol. Cell Biol., 2,* 793–805.

Geiger, B., Spatz, J. P., & Bershadsky, A. D. (2009). Environmental sensing through focal adhesions. *Nat. Rev. Mol. Cell Biol., 10,* 21–33.

Giancotti, F. G. (2003). A structural view of integrin activation and signaling. *Dev. Cell, 4,* 149–151.

Giancotti, F. G., & Ruoslahti, E. (1999). Integrin signaling. *Science, 285,* 1028–1032.

Gong, Z., & Niklason, L. E. (2008). Small-diameter human vessel wall engineered from bone marrow-derived mesenchymal stem cells (hMSCs). *Faseb J., 22,* 1635–1648.

Guilak, F., Cohen, D. M., Estes, B. T., Gimble, J. M., Liedtke, W., & Chen, C. S. (2009). Control of stem cell fate by physical interactions with the extracellular matrix. *Cell Stem Cell, 5,* 17–26.

Hamill, O. P., & Martinac, B. (2001). Molecular basis of mechanotransduction in living cells. *Physiol. Rev., 81,* 685–740.

Hartmann, C., & Tabin, C. J. (2001). Wnt-14 plays a pivotal role in inducing synovial joint formation in the developing appendicular skeleton. *Cell, 104,* 341–351.

Hayakawa, K., Tatsumi, H., & Sokabe, M. (2008). Actin stress fibers transmit and focus force to activate mechanosensitive channels. *J. Cell. Sci., 121,* 496–503.

Hazel, A. L., & Pedley, T. J. (2000). Vascular endothelial cells minimize the total force on their nuclei. *Biophys. J., 78,* 47–54.

Helmke, B. P., & Davies, P. F. (2002). The cytoskeleton under external fluid mechanical forces: hemodynamic forces acting on the endothelium. *Ann. Biomed. Eng., 30,* 284–296.

Hogan, C., Serpente, N., Cogram, P., Hosking, C. R., Bialucha, C. U., Feller, S. M., et al. (2004). Rap1 regulates the formation of E-cadherin-based cell-cell contacts. *Mol. Cell Biol., 24,* 6690–6700.

Hoger, J. H., Ilyin, V. I., Forsyth, S., & Hoger, A. (2002). Shear stress regulates the endothelial Kir2.1 ion channel. *Proc. Natl. Acad. Sci. U.S.A., 99,* 7780–7785.

Hu, S., Chen, J., Butler, J. P., & Wang, N. (2005). Prestress mediates force propagation into the nucleus. *Biochem. Biophys. Res. Commun., 329,* 423–428.

Hu, S., & Wang, N. (2006). Control of stress propagation in the cytoplasm by prestress and loading frequency. *Mol. Cell. Biomech., 3,* 49–60.

Hua, S. Z., Gottlieb, P. A., Heo, J., & Sachs, F. (2010). A mechanosensitive ion channel regulating cell volume. *Am. J. Physiol. Cell Ph., 298,* C1424–C1430.

Huang, H., Kamm, R. D., & Lee, R. T. (2004). Cell mechanics and mechanotransduction: pathways, probes, and physiology. *Am. J. Physiol. Cell Ph., 287,* C1–C11.

Hung, C. T., Mauck, R. L., Wang, C. C., Lima, E. G., & Ateshian, G. A. (2004). A paradigm for functional tissue engineering of articular cartilage via applied physiologic deformational loading. *Ann. Biomed. Eng., 32,* 35–49.

Ingber, D. E. (2002). Mechanical signaling and the cellular response to extracellular matrix in angiogenesis and cardiovascular physiology. *Circ. Res., 91,* 877–887.

Ingber, D. E. (2004). The mechanochemical basis of cell and tissue regulation. *Mech. Chem. Biosyst., 1,* 53–68.

Ingber, D. E. (2006). Mechanical control of tissue morphogenesis during embryological development. *Int. J. Dev. Biol., 50,* 255–266.

Ingber, D. E. (2008). Tensegrity and mechanotransduction. *J. Bodyw. Mov. Ther., 12,* 198–200.

Jacot, J. G., McCulloch, A. D., & Omens, J. H. (2008). Substrate stiffness affects the functional maturation of neonatal rat ventricular myocytes. *Biophys. J., 95,* 3479–3487.

Griffith, D. L., Keck, P. C., Sampath, T. K., Rueger, D. C., & Carlson, W. D. (1996). Three-dimensional structure of recombinant human osteogenic protein-1: structural paradigm for the transforming growth factor-β superfamily. *Proc. Natl. Acad. Sci. U.S.A., 93,* 878−883.

Hangody, L., Feczko, P., Bartha, L., Bodo, G., & Kish, G. (2001). Mosaicplasty for the treatment of articular defects of the knee and ankle. *Clin. Orthop. Relat. Res., 391*(Suppl.), S328−S336, (review).

Hayashi, H., Abdollah, S., Qui, Y., Cai, J., Xu, Y. Y., Grinnell, B. W., et al. (1997). The MAD-related protein Smad 7 associates with the TGF-β receptor and functions as an antagonist of TGF-β signalling. *Cell, 89,* 1165−1173.

Heldin, C. H., Miyazono, K., & ten Dijke, P. (1997). TGF-β signaling from cell membrane to nucleus through Smad proteins. *Nature, 390,* 465−471.

Hemmati-Brivanlou, A., Kelly, O. G., & Melton, D. A. (1994). Follistatin an antagonist of activin is expressed in the Spemann organizer and displays direct neuralizing activity. *Cell, 77,* 283−295.

Hollinger, J., Mayer, M., Buck, D., Zegzula, H., Ron, E., Smith, J., et al. (1996). Poly ("-hydroxy acid") carrier for delivering recombinant human bone morphogenetic protein-2 for bone regeneration. *J. Contr. Release, 39,* 287−304.

Hubbell, J. A. (1995). Biomaterials in tissue engineering. *Biotechnology, 13,* 565−575.

Hunzinker, E. B., & Rosenberg, L. C. (1996). Repair of partial-thickness defects in articular cartilage: cell recruitment from the synovial membrane. *J. Bone Joint Surg., 78-A,* 721−733.

Johnson, R. L., & Tabin, C. J. (1997). Molecular models for vertebrate limb development. *Cell, 90,* 979−990.

Khouri, R. K., Koudsi, B., & Reddi, A. H. (1991). Tissue transformation into bone *in vivo. JAMA, 266,* 1953−1955.

Kim, W. S., Vacanti, J. P., Cima, L., Mooney, D., Upton, J., Puelacher, W. C., et al. (1994). Cartilage engineered in predetermined shapes employing cell transplantation on synthetic biodegradable polymers. *Plast. Reconstr. Surg., 94,* 233−237.

Kozarsky, K. F., & Wilson, J. M. (1993). Gene therapy: adenovirus vectors. *Curr. Opin. Genetic. Dev., 3,* 499−503.

Kuznetsov, S. A., Friedenstein, A. J., & Robey, P. G. (1997). Factors required for bone marrow stromal fibroblast colony formation *in vitro. Br. J. Haematol., 97*(3), 561−570.

Lacroix, P. (1945). Recent investigations on the growth of bone. *Nature, 156,* 576.

Langer, R., & Vacanti, J. P. (1993). Tissue engineering. *Science, 260,* 930−932.

Lemaire, P., & Gurdon, J. B. (1994). Vertebrate embryonic inductions. *Bio. Essays, 16*(9), 617−620.

Livnah, O., Stura, E. A., Johnson, D. L., Middleton, S. A., Mulcahy, L. S., Wrighton, N. D., et al. (1996). Functional mimicry of a protein hormone by a peptide agonist: the EPO receptor complex at 2.8 C. *Science, 273,* 464−471.

Luo, G., Hoffman, M., Bronckers, A. L. J., Sohuki, M., Bradley, A., & Karsenty, G. (1995). BMP-7 is an inducer of morphogens and is also required for eye development, and skeletal patterning. *Gene. Dev., 9,* 2808−2820.

Luyten, F., Cunningham, N. S., Ma, S., Muthukumaran, S., Hammonds, R. G., Nevins, W. B., et al. (1989). Purification and partial amino acid sequence of osteogenin, a protein initiating bone differentiation. *J. Biol. Chem., 265,* 13377−13380.

Luyten, F. P., Yu, Y. M., Yanagishita, M., Vukicevic, S., Hammonds, R. G., & Reddi, A. H. (1992). Natural bovine osteogenin and recombinant BMP-2B are equipotent in the maintenance of proteoglycans in bovine articular cartilage explant cultures. *J. Biol. Chem., 267,* 3685−3691.

Lyons, K. M., Hogan, B. L. M., & Robertson, E. J. (1995). Colocalization of BMP-2 and BMP-7 RNA suggest that these factors cooperatively act in tissue interactions during murine development. *Mech. Dev., 50,* 71−83.

Ma, S., Chen, G., & Reddi, A. H. (1990). Collaboration between collagenous matrix and osteogenin is required for bone induction. *Ann. N.Y. Acad. Sci., 580,* 524−525.

McPherson, J. M. (1992). The utility of collagen-based vehicles in delivery of growth factors for hard and soft tissue wound repair. *Clin. Mater., 9,* 225−234.

Melton, D. A. (1991). Pattern formation during animal development. *Science, 252,* 234−241.

Morsy, M. A., Mitani, K., Clemens, P., & Caskey, T. (1993). Progress toward human gene therapy. *JAMA, 270*(19), 2338−2345.

Mosbach, K., & Ramstrom, O. (1996). The emerging technique of molecular imprinting and its future impact on biotechnology. *Biotechnology, 14,* 163−170.

Mow, V. C., Ratcliffe, A., & Poole, A. R. (1992). Cartilage and diarthrodial joints as paradigms for hierarchical materials and structures. *Biomaterials, 13,* 67−97.

Mulligan, R. C. (1993). The basic science of gene therapy. *Science, 260,* 926−932.

Nakahara, H., Goldberg, V. M., & Caplan, A. I. (1991). Culture-expanded human periosteal-derived cells exhibit osteochondral potential *in vivo. J. Orthop. Res., 9*(4), 465−476.

Nishitoh, H., Ichijo, H., Kimura, M., Matsumoto, T., Makishima, F., Yamaguchi, A., et al. (1996). Identification of type I and type II serine/threonine kinase receptors for growth and differentiation factor-5. *J. Biol. Chem., 271,* 21345−21352.

Owen, M. E., & Friedenstein, A. J. (1988). Marrow derived osteogenic precursors. *Stromal stem cells: CIBA Foundation Symposium, 136,* 42−60.

Ozkaynak, E., Rueger, D. C., Drier, E. A., Corbett, C., Ridge, R. J., Sampath, T. K., et al. (1990). OP-1 cDNA encodes an osteogenic protein in the TGF-β family. *EMBO J., 9,* 2085−2093.

Paralkar, V. M., Nandedkar, A. K. N., Pointers, R. H., Kleinman, H. K., & Reddi, A. H. (1990). Interaction of osteogenin, a heparin binding bone morphogenetic protein, with type IV collagen. *J. Biol. Chem., 265,* 17281−17284.

Paralkar, V. M., Vukicevic, S., & Reddi, A. H. (1991). Transforming growth factor-β type I binds to collagen IV of basement membrane matrix: implications for development. *Dev. Biol., 143,* 303−308.

Paralkar, V. M., Weeks, B. S., Yu, Y. M., Kleinman, H. K., & Reddi, A. H. (1992). Recombinant human bone morphogenetic protein 2B stimulates PC12 cell differentiation: potentiation and binding to type IV collagen. *J. Cell Biol., 119,* 1721−1728.

Piccolo, S., Sasai, Y., Lu, B., & de Robertis, E. M. (1996). Dorsoventral patterning in *Xenopus*: inhibition of ventral signals by direct binding of chordin to BMP-4. *Cell, 86,* 589−598.

Pittenger, M. F., Mackay, A. M., Beck, S. C., et al. (1999). Multilineage potential of adult human mesenchymal stem cells. *Science, 284*(5411), 143−147.

Prockop, D. J. (1997). Marrow stromal cells and stem cells for non hematopoietic tissues. *Science, 276,* 71−74.

Rath, N. C., & Reddi, A. H. (1979). Collagenous bone matrix is a local mitogen. *Nature, 278,* 855−857.

Reddi, A. H., & Huggins, C. B. (1972). Biochemical sequences in the transformation of normal fibroblasts in adolescent rat. *Proc. Natl. Acad. Sci. U.S.A., 69,* 1601−1605.

Reddi, A. H., & Anderson, W. A. (1976). Collagenous bone matrix-induced endochondral ossification and hemopoiesis. *J. Cell Biol., 69,* 557−572.

Reddi, A. H. (1981). Cell biology and biochemistry of endochondral bone development. *Collagen Rel. Res., 1,* 209−226.

Reddi, A. H. (1984). Extracellular matrix and development. In K. A. Piez & A. H. Reddi (Eds.), *Extracellular Matrix Biochemistry* (pp. 375−412). New York: Elsevier.

Reddi, A. H. (1994). Bone and cartilage differentiation. *Curr. Opinion Gen. Dev., 4,* 937−944.

Reddi, A. H. (1997). Bone morphogenetic proteins: an unconventional approach to isolation of first mammalian morphogens. *Cytokine Growth Factor Rev., 8,* 11−20.

Reddi, A. H. (1998). Role of morphogenetic proteins in skeletal tissue engineering and regeneration. *Nat. Biotechnol., 16,* 247−252.

Ripamonti, U., Ma, S., & Reddi, A. H. (1992). The critical role of geometry of porus hydroxyapatite delivery system induction of bone by osteogenin, a bone morphogenetic protein. *Matrix, 12,* 202−212.

Ripamonti, U. (1996). Osteoinduction in porous hydroxyapatite implanted in heterotopic sites of different animal models. *Biomaterials, 17,* 31−35.

Ripamonti, U., van den Heever, B., Sampath, T. K., Tucker, M. M., Rueger, D. C., & Reddi, A. H. (1996). Complete regeneration of bone in the baboon by recombinant human osteogenic protein-1 (hOP-1, bone morphogenetic protein-7). *Growth Factors, 123,* 273−289.

Ruoslahti, E., & Pierschbacher, M. D. (1987). New perspectives in cell adhesion: RGD and integrins. *Science, 238,* 491−497.

Sampath, T. K., & Reddi, A. H. (1981). Dissociative extraction and reconstitution of bone matrix components involved in local bone differentiation. *Proc. Natl. Acad. Sci. U.S.A., 78,* 7599−7603.

Sampath, T. K., & Reddi, A. H. (1983). Homology of bone inductive proteins from human, monkey, bovine, and rat extracellular matrix. *Proc. Natl. Acad. Sci. U.S.A., 80,* 6591−6595.

Senn, N. (1989). On the healing of aseptic bone cavities by implantation of antiseptic decalcified bone. *Am. J. Med. Sci., 98,* 219−240.

Storm, E. E., Huynh, T. V., Copeland, N. G., Jenkins, N. A., Kingsley, D. M., & Lee, S.-J. (1994). Limb alterations in brachypodism mice due to mutations in a new member of TGF-β superfamily. *Nature, 368,* 639−642.

ten Dijke, P., Yamashita, H., Sampath, T. K., Reddi, A. H., Riddle, D., Heldin, C. H., & Miyazono, K. (1994). Identification of type I receptors for OP-1 and BMP-4. *J. Biol. Chem., 269,* 16986−16988.

Tsumaki, N., Tanaka, K., Krikawa-Hirasawa, E., Nakase, T., Kimura, T., Thomas, J. T., et al. (1999). Role of CDMP-1 in skeletal morphogenesis: promotion of mesenchymal cell recruitment and chondrocyte differentiation. *J. Cell. Biol., 144,* 161−173.

Urist, M. R. (1965). Bone: formation by autoinduction. *Science, 150,* 893−899.

Vukicevic, S., Luyten, F. P., Kleinman, H. K., & Reddi, A. H. (1990). Differentiation of canalicular cell processes in bone cells by basement membrane matrix components: regulation by discrete domains of laminin. *Cell, 63,* 437−445.

491

Vukicevic, S., Kopp, J. B., Luyten, F. P., & Sampath, K. (1996). Induction of nephrogenic mesenchyme by osteogenic protein-1 (bone morphogenetic protein-7). *Proc. Natl. Acad. Sci. U.S.A., 92,* 9021–9026.

Weintroub, S., & Reddi, A. H. (1988). Influence of irradiation on the osteoinductive potential of demineralized bone matrix. *Calcif. Tissue Int., 42,* 255–260.

Weintroub, S., Weiss, J. F., Catravas, G. N., & Reddi, A. H. (1990). Influence of whole body irradiation and local shielding on matrix-induced endochondral bone differentiation. *Calcif. Tissue Int., 46,* 38–45.

Weiss, R. E., & Reddi, A. H. (1980). Synthesis and localization of fibronectin during collagenous matrix mesenchymal cell interaction and differentiation of cartilage and bone *in vivo. Proc. Natl. Acad. Sci. U.S.A., 77,* 2074–2078.

Winnier, G., Blessing, M., Labosky, P. A., & Hogan, B. L. M. (1996). Bone morphogenetic protein-4 is required for mesoderm formation and patterning in the mouse. *Gene. Dev., 9,* 2105–2116.

Wozney, J. M., Rosen, V., Celeste, A. J., Mitsock, L. M., Whittiers, M., Kriz, W. R., et al. (1988). Novel regulators of bone formation: molecular clones and activities. *Science, 242,* 1528–1534.

Zhang, H., & Bradley, A. (1996). Mice deficient of BMP-2 are nonviable and have defects in amnion/chorion and cardiac development. *Development, 122,* 2977–2986.

Zimmerman, L. B., Jesus-Escobar, J. M., & Harland, R. M. (1996). The Spemann organizer signal Noggin binds and inactivates bone morphogenetic protein-4. *Cell, 86,* 599–606.

Zuk, P. A., Zhu, M., Mizuno, H., et al. (2001). Multilineage cells from human adipose tissue: implications for cell-based therapies. *Tissue Eng., 7*(2), 211–228.

Zvaifler, N. J., Marinova-Mutafchieva, L., Adams, G., Edwards, C. J., Moss, J., et al. (2000). Mesenchymal precursor cells in the blood of normal individuals. *Arthritis Res., 2*(6), 477–488.

CHAPTER 28

Physical Stress as a Factor in Tissue Growth and Remodeling

Joel D. Boerckel, Christopher V. Gemmiti, Yash M. Kolambkar, Blaise D. Porter, Robert E. Guldberg
Woodruff School of Mechanical Engineering, Georgia Institute of Technology

INTRODUCTION

The goal of tissue engineering and regenerative medicine is to restore or regenerate damaged and degenerate tissues, many of which have explicit mechanical functions or are regulated by mechanical factors. Indeed, physical stimuli are essential for proper morphogenesis, maintenance, and repair of numerous tissue types, including bone, cartilage, and blood vessels, all of which are primary targets for tissue engineers. Therefore, being able to mathematically describe and understand the role of physical stress and strain in tissue repair is essential to regenerating functional, mechanically competent tissues.

The role of physical stresses and strains in regulating tissue growth and remodeling has been of tremendous interest to investigators for well over 100 years. Although somewhat unfairly to his contemporary colleagues, Julius Wolff is often credited with the concept that tissue structure or form follows from its function (i.e. Wolff's law). At the time, Wolff's law was simply based on the apparent correspondence between anatomical observations of trabecular bone organization and estimations of principal stress directions due to functional loading conditions. The recognition that adaptation of tissue structure and composition is cell mediated was not made until later by other investigators. These early observations spawned the interdisciplinary field of mechanobiology, focused on identifying mechanisms by which mechanical signals are transduced into cellular activity, and emphasized the need to consider the effects of physical factors on tissue growth and remodeling as an important part of strategies for tissue regeneration.

Many different cell types from various tissues have been shown to be sensitive to mechanical stimuli in one form or another. The effects of physiological mechanical signals on cells and tissues can be beneficial, playing a central role in the maintenance of tissue structural integrity via remodeling processes. Alterations in mechanical signals can also contribute to the development of pathological conditions. For example, local sheer stresses play a key role in the development and localization of atherosclerotic lesions. Likewise, the progression of osteoarthritis is due to a vicious cycle of cartilage matrix degradation and increased local stresses. In bone, the mechanical environment also has important clinical implications in the development of osteoporosis, stress fractures, total joint implant loosening, and bone loss during space flight.

Therefore, approaching a tissue engineering problem from a mechanical perspective includes four steps: (1) quantitatively describe the native mechanical environment, (2) understand the

Principles of Regenerative Medicine. DOI: 10.1016/B978-0-12-381422-7.10028-8

role of mechanical factors in the tissue of interest, (3) manipulate mechanical conditions to enhance function or regeneration, and (4) quantitatively evaluate the degree of functional restoration of the engineered tissue. This chapter will present an overview of the tools and concepts required in each.

DESCRIBE THE MECHANICAL ENVIRONMENT

As mentioned, cells respond and adapt to their local mechanical environment, but, to understand this phenomenon, the mechanical environment must be usefully described. This requires an appropriate theoretical framework that can accurately model the salient features of the tissue of interest and describe the local mechanical stimuli experienced by a cell either *in situ* or in a tissue-engineered construct. The framework must therefore describe the properties and behavior of the tissue or construct, incorporate the loads (boundary conditions) that will be applied, and justify the assumptions made.

It is useful to view tissues as a structural hierarchy through which functional loads are transmitted down to the cellular level (Fig. 28.1). In bone, for example, applied joint and muscle forces result in stresses and strains within the mineralized tissue that can be defined at different scale levels, from the whole bone level down to submicron mineral crystals embedded within collagen molecules. At each hierarchical level, it is convenient to assume that everything below that level is a continuum (i.e. there is a finite mass density at every point within the material). This simplification allows material properties to be expressed at a given hierarchical level in terms of constitutive equations. As described in the next section, constitutive equations define the relationship between stresses and strains at each level. Cells sense and respond to local stresses or strains produced by forces transmitted from the macro level down through the complex structural hierarchy to the cellular level. Cell-mediated adaptational changes in tissue structure and composition subsequently alter the local stresses and strains resulting from functionally applied loads, thus providing a regulatory feedback mechanism. It is important to note that the sensitivity of the cellular response to mechanical stimuli can be altered by a variety of non-mechanical factors such as age, disease, as well as numerous biochemical factors.

Strain and stress definitions

STRAIN

Strain is a normalized measure of deformation. Consider the simple case of a thin rectangular piece of tissue being axially loaded by a force, as shown in Figure 28.2A. The axial force increases the length of the tissue, but at the same time decreases its width and thickness.

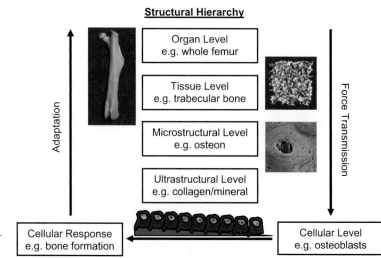

FIGURE 28.1
Force transmission through the structural hierarchy of bone to the cellular level resulting in cell-mediated adaptation of tissue structure and composition.

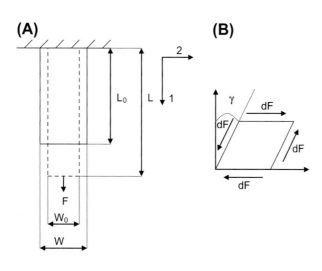

(A) **(B)**

FIGURE 28.2
(A) Axial and transverse strains associated with uniaxial tensile loading. (B) Sheer strain associated with torsional or sheer loading.

Engineering strain is defined as the change in a dimension of the tissue normalized by its original dimension, and is given in the axial direction by

$$\varepsilon_{11} = \frac{L - L_0}{L_0} \tag{1}$$

Another important deformation parameter is the Poisson's ratio v, which is defined as the ratio of lateral strain to axial strain, and is given in this case by

$$v = -\frac{\varepsilon_{22}}{\varepsilon_{11}} = -\frac{\frac{W-W_0}{W_0}}{\frac{L-L_0}{L_0}} \tag{2}$$

Poisson's ratio is a measure of the tendency for a material body to try to retain its total volume as it is deformed. When $v = 0.5$, the material is said to be incompressible (e.g. water), and does not undergo a volume change after deformation. The typical value of v for tissues is between 0.2 and 0.45. Thus, a tissue subjected to tensile deformation and strain would increase in volume slightly. In contrast to normal strains, sheer strains due to simple sheer forces, dF, or from torsional loading, for example, produce a change in shape but not volume, as shown in Figure 28.2B. Measurement of the angle of sheer deformation, γ_{12}, allows calculation of sheer strain, as given by

$$\varepsilon_{12} = \frac{\gamma_{12}}{2} \tag{3}$$

The complex deformations created by forces acting in multiple directions necessitate the generalization of deformation to three-dimensional space. Deformation in three-dimensional can be expressed by the deformation gradient tensor, **F**. Consider the body shown in Figure 28.3A undergoing a deformation from the reference state to a deformed configuration. If one follows the particles P_1 and P_2, they move from positions X_{P1} and X_{P2} to x_{P1} and x_{P2}, respectively. There will also be a similar one-to-one mapping of other particles in the reference and deformed configurations. Thus, the deformation of the body can be written as a function relating the current state (lower case) to the reference state (upper case):

$$\mathbf{x} = f(X) \tag{4}$$

In scalar form, this would involve three equations:

$$\begin{aligned} x_1 &= f_1(X_1, X_2, X_3), \\ x_2 &= f_2(X_1, X_2, X_3), \\ x_3 &= f_3(X_1, X_2, X_3), \end{aligned} \tag{5}$$

where 1, 2, and 3 correspond to the three directions of the coordinate system.

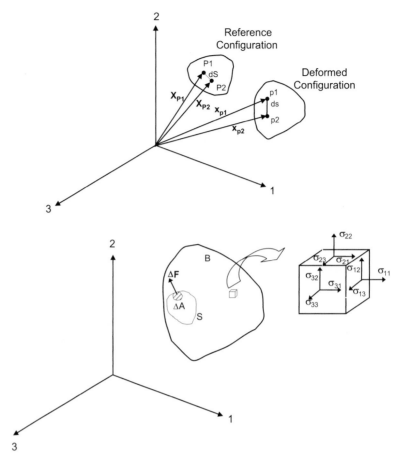

FIGURE 28.3
(A) Deformation of a three-dimensional body from a reference configuration to a deformed configuration. (B) Stress on a surface element, and the nine stress components defining the stress state at a point.

The displacement vector is given by

$$\mathbf{u} = \mathbf{x} - \mathbf{X} \tag{6}$$

The deformation gradient $\mathbf{F}$ is then defined as

$$\mathbf{F} = \frac{\partial \mathbf{x}}{\partial \mathbf{X}} \tag{7}$$

In matrix form, the deformation gradient can be written as

$$[F] = \begin{bmatrix} \frac{\partial x_1}{\partial X_1} & \frac{\partial x_1}{\partial X_2} & \frac{\partial x_1}{\partial X_3} \\ \frac{\partial x_2}{\partial X_1} & \frac{\partial x_2}{\partial X_2} & \frac{\partial x_2}{\partial X_3} \\ \frac{\partial x_3}{\partial X_1} & \frac{\partial x_3}{\partial X_2} & \frac{\partial x_3}{\partial X_3} \end{bmatrix}, \tag{8}$$

and is related to the gradient of displacement by the following expression, in which $\mathbf{I}$ is the identity tensor:

$$F = \frac{\partial u}{\partial X} + I \tag{9}$$

The engineering strains as defined above are appropriate to use when the strains in the material are small (typically less than 5%). However, the analysis of large deformations, as frequently observed for soft tissues under functional loading conditions, requires use of other strain measures. When the deformation is large, a useful measure of deformation is the Green (or Lagrangian) strain, $\mathbf{E}$, which is defined as:

$$\mathbf{E} = \frac{1}{2}[\mathbf{F}^T\mathbf{F} - \mathbf{I}] = \frac{1}{2}[\mathbf{D} + \mathbf{D}^T + \mathbf{D}^T\mathbf{D}] \tag{10}$$

where $D = \frac{\partial u}{\partial X}$ and the superscript T stands for the transpose of the matrix form of the second order tensor.

Consider the segment P_1P_2 of length dS that has deformed to p_1p_2 with length ds. In this case, the one-dimensional Green strain becomes:

$$E = \frac{1}{2}\left(\frac{ds^2 - dS^2}{dS^2}\right) \tag{11}$$

If the deformation under consideration is small, as is typically the case for bone and most structural engineering materials, the quadratic term, $\mathbf{D}^T\mathbf{D}$, in the Green strain can be neglected to give the infinitesimal (engineering) strain tensor, ε:

$$\varepsilon = \frac{1}{2}[\mathbf{D} + \mathbf{D}^T], \tag{12}$$

which, in one-dimension, gives us the familiar expression of engineering strain in a uniaxial test:

$$\varepsilon = \frac{L - L_0}{L_0} \tag{13}$$

To get a feel for the relative values of these strain measures, consider the following example of uniaxial elongation of our rectangular tissue having original length of 5 cm. In one case, the tissue is stretched to a final length of 5.05 cm (small strain), whereas in the second case it is elongated to 10 cm (large strain).

	Case I (L = 5.05 cm)	Case II (L = 10 cm)
Green strain $\left(E = \frac{1}{2}\frac{L^2 - L_0^2}{L_0^2}\right)$	0.01005	1.5
Engineering strain $\left(\varepsilon = \frac{L - L_0}{L_0}\right)$	0.01	1.0

Thus, we see that, for the small deformations, the different strain definitions give approximately the same value and engineering strain is reasonably accurate, whereas for large deformations the strain definitions yield very different values due to neglect of the higher order terms in the engineering strain definition.

STRESS

Stress is a measure of the intensity of internal force developed in a material upon application of an external force. Consider the force $\Delta\mathbf{F}$ acting on a small surface element of area ΔA in Figure 28.3B. This element lies on the surface S, which is part of the larger body B. As ΔA tends to zero, the ratio $\frac{\Delta F}{\Delta A}$ tends to a finite limit $\frac{dF}{dA}$, which is defined as the stress on the surface element.

Consider an infinitesimal cube in the body as shown in Figure 28.3B. Due to the external force applied on the body, internal forces are applied on the surface of the cube. Each internal force can be resolved into its three components and normalized by the area to give three stress components on each face. The volume of the cube can be continuously decreased such that the cube collapses to a point. The nine stress components define the second order stress tensor, and completely describe the stress state at this point. Using equilibrium conditions, we can show that the stress tensor is symmetric, that is $\sigma_{ij} = \sigma_{ji}$; thus the stress tensor has only six independent components. If a stress component acts in a direction perpendicular to the surface it acts on, it is referred to as a normal stress. On the other hand, if it is parallel to the surface, it is called a sheer stress. Thus, σ_{11}, σ_{22}, and σ_{33} are normal stresses, while σ_{12}, σ_{23}, and σ_{31} are sheer stresses. Normal stresses can change both the volume and shape of the body, while sheer stresses induce only shape change.

The first Piola-Kirchoff stress tensor, **P**, is defined as the force acting in the current configuration, $\Delta\mathbf{F}$, on an oriented area in the reference configuration, ΔA, in the limit as ΔA tends to zero. In a typical experiment, the force is constantly measured, but the cross-sectional area is not. Thus, the first Piola-Kirchoff stress is an easy quantity to compute as the undeformed cross-sectional area can be measured prior to loading. However, when considering the balance of forces in the deformed body at equilibrium, the deformed area Δa of the surface element is required for the stress definition. The Cauchy stress, or "true" stress, is thus defined as the limit of $\sigma = \frac{\Delta F}{\Delta \alpha}$ as Δa tends to zero. While the difference between Δa and ΔA is negligible for small deformations, the difference becomes significant for large deformations, and choice of stress definition is important. A third definition, the second Piola-Kirchoff stress tensor, **S**, is often used in constitutive modeling, and is defined as the limit of $S = \frac{\Delta F'}{\Delta A}$ as ΔA tends to infinity, where $\Delta\mathbf{F}'$ is the current force, $\Delta\mathbf{F}$, mapped onto the reference configuration, and is therefore defined completely in terms of the reference state. All three stress definitions can be related to one another by the deformation gradient (see Malvern, 1969 for details).

Constitutive relations

These definitions enable us to begin establishing a theoretical framework in which to describe the material behavior of the tissue of interest in response to mechanical forces. The material behavior will be expressed as a constitutive equation, a mathematical model that specifies the relationship between stress and strain. At this stage, assumptions must be made regarding the behavior of the material, which will dictate the framework selected. These assumptions will be tested by experimental validation of the model. The goal of constitutive modeling is to describe the important features of a tissue's material behavior in the simplest and most mathematically useful way possible. For example, the simplest constitutive equation is that for linearly elastic, homogeneous, isotropic materials, where there is a linear, reversible relationship between stress and strain and the material properties do not vary by position or direction. For such materials, we are limited to discussion of small deformations. This gives the familiar equation, $\sigma = E\varepsilon$, where **E** is the elastic modulus, and σ and ε are the engineering stress and strain, respectively. Many engineering materials (i.e. steel) can be modeled in this way; however, most biological materials are more complex.

Bone is, mechanically speaking, one of the simpler tissue types, and can be modeled as linear elastic, though it is highly inhomogeneous and anisotropic (the properties vary by both position and direction). This behavior can be described quite well using generalized Hooke's law, which can be written in indicial notation as:

$$\sigma_{ij} = C_{ijkl}\varepsilon_{kl}, \tag{14}$$

where C_{ijkl} is a fourth order tensor describing the material properties and contains 81 constants. However, due to symmetry arguments and thermodynamic constraints, the number of independent constants is reduced to 21. If the six stress and strain components are written in the form of a column matrix, the material tensor can be represented by a matrix called the stiffness matrix:

$$C = \begin{bmatrix} C_{11} & C_{12} & C_{13} & C_{14} & C_{15} & C_{16} \\ & C_{22} & C_{23} & C_{24} & C_{25} & C_{26} \\ & & C_{33} & C_{34} & C_{35} & C_{36} \\ & & & C_{44} & C_{45} & C_{46} \\ & & & & C_{55} & C_{56} \\ & & & & & C_{66} \end{bmatrix}, \tag{15}$$

where the other side of the diagonal is symmetric (i.e. $C_{ij} = C_{ji}$). The above stiffness matrix represents a fully anisotropic linear elastic material, for which 21 constants must be determined experimentally to fully characterize the material behavior. Fortunately, many tissues have some degree of material symmetry. For example, trabecular bone is frequently described using an orthotropic material model, which consists of three mutually orthogonal planes of symmetry that coincide with the chosen reference coordinate system. This reduces the number of independent constants to nine; they are related to the Young's moduli (E_1, E_2, E_3) and sheer moduli (G_1, G_2, G_3) and the Poisson's ratios (v_{12}, v_{23}, v_{31}) in the three planes. In the case of a fully isotropic material, which has infinite planes of symmetry, these equations reduce to the basic mechanics of materials expression of Hooke's law, with only two independent constants, E and v.

For most biological materials (e.g. cartilage, tendon, blood vessels, etc.) the constitutive models must describe an expanding array of features. Unlike most engineering materials, most tissues are non-linear, viscoelastic (i.e. time-dependent), inhomogeneous, and anisotropic, and experience large deformations under physiologic loads. It can be very difficult, indeed, to include every one of these features in a three-dimensional model. Fortunately, reasonable assumptions can be made to simplify the approach and yield useful information about the most important features of the tissue at hand. For example, Y. C. Fung made the observation that, in many soft tissues, after several "preconditioning" cycles, the loading and unloading curves reach a steady state and the final curves lose much of their rate-dependence (Fung, 1993). He termed this characteristic behavior "pseudo-elasticity." Thus, many tissues such as skin, tendon, and blood vessels can often be modeled as non-linear, pseudo-elastic, homogeneous, and anisotropic, with large deformations. Using these assumptions and applying the second law of thermodynamics (via the Clausius-Duhem entropy inequality), it can be shown that the second Piola-Kirchoff stress, $\mathbf{S}$, is directly related to the derivative of the strain potential, or strain energy function, W, with respect to the Green strain, $\mathbf{E}$:

$$\mathbf{S} = \rho_0 \frac{\partial W}{\partial \mathbf{E}}, \tag{16}$$

where ρ_0 is the mass density. Now, provided we know a proper strain energy function for a given material, we have a direct relationship between the stress and the strain. Selection of a proper functional form for W is an active area of research, and can account for anisotropy, incompressibility, and non-linearity (Humphrey et al., 1990; Fung, 1993). One commonly used model, proposed by Fung (1967, 1973), assumes that the exponential relationship observed in one-dimension can be extended to three-dimensions:

$$W = \exp[Q(\mathbf{E}) - 1], \tag{17}$$

where $Q(\mathbf{E})$ is a polynomial function of the strain components, whose functional form allows for different degrees of anisotropy.

Classically, constitutive models were phenomenological in nature and not derived from microstructure; however, many current investigators are developing multi-scale models that incorporate microstructural composition, fiber-matrix interactions, and fiber orientation to predict both elastic and plastic behavior (Natali et al., 2005; Hansen et al., 2009; Maceri et al., 2010). Such models feature increasing complexity, and closed-form solutions are often impossible, though modern computing power allows for solution of the incremental constitutive relations.

Many biological tissues exhibit other characteristics such as time-dependence (viscoelasticity) that are essential to their function *in vivo* and must be incorporated to accurately describe tissue properties. Articular cartilage, for example, exhibits multi-phasic behavior in which the matrix composition and interstitial fluid-matrix interactions are essential to conferring the lubricating and impact-absorbing properties of the tissue. See textbooks by Fung and Cowin for further reading on constitutive modeling of time-dependent tissues (Fung, 1965; Cowin and Doty, 2007).

Finally, validity of assumptions and predictive capabilities of constitutive models must be evaluated by experiment. Often, simple experiments (e.g. tension, compression, biaxial stretch) can be used to validate the models before application, but the capability of the models to predict essential behavior under near-physiologic conditions is critical.

Boundary value problems

Once the tissue properties and behavior are known, determining the actual stresses and strains experienced by cells within the extracellular matrix requires solution of the boundary value problems defined by the physical field equations, the constitutive behavior, and the boundary conditions. Boundary conditions are simply the loads and deformations applied further up in the structural hierarchy that induce local stresses and strains at the level of interest. For classical engineering materials, the boundary value problems and admissible assumptions have been identified, and most current research focuses on their solutions; however, in the field of biomechanics, the formulation of boundary value problems, including constitutive models and boundary conditions, remains an active area of research with great promise for enhancing our understanding of biological materials (Fung, 1993).

UNDERSTAND THE ROLE OF MECHANICAL STIMULI

With the necessary tools established and a theoretical framework chosen, we can begin to understand how mechanical stimuli affect organs, tissues, and cells. Mechanical loads are essential for proper morphogenesis (see Chapter 27), and maintenance of normal tissue structure and function in a wide variety of tissues. In many systems, physical stimuli may also be pathogenic, inducing damage or disease, depending on the type, magnitude, and/or frequency of the stimulus. Understanding these factors is therefore important to regenerating mechanosensitive tissues.

In many tissues, subcellular components to whole tissues are coupled to actively adapt the structure and properties according to the mechanical environment. The role of mechanical stimulation has been studied and described at many hierarchical levels, linking gross tissue remodeling to mechanotransduction, the cellular and subcellular responses that convert mechanical stimuli into chemical signals.

Tissue remodeling

We will first consider bone adaptation as an example of tissue-level remodeling to physical stresses. The mechanisms behind bone tissue remodeling were first described by Frost, who suggested that modeling and remodeling are mediated by basic multicellular units (BMUs), made up of osteoblasts, osteoclasts, and osteocytes (Frost, 1963; Mackie, 2003; Knothe Tate et al., 2004; Robling et al., 2006). Osteoblasts lay down bone matrix and osteoclasts degrade it within a highly regulated and intertwining milieu of chemical signals. Osteocytes, residing within bone matrix and communicating with other cells through the lacunocanalicular network, are thought to be the primary mechanosensors that transduce mechanical signals into chemical signals. These adaptations are stimulated by changes in the mechanical loading history.

Turner and others have studied numerous mechanical variables affecting bone adaptation (Lanyon and Rubin, 1984; Rubin and Lanyon, 1984, 1985; Turner et al., 1994, 1995) and have proposed three rules for load-induced adaptation (Turner, 1998). First, bone adapts to dynamic but not static strains. Experimental observations revealed that the strain stimulus, or the strain needed to induce adaptation, was proportional to both strain magnitude and frequency:

$$E = k_1 \varepsilon f, \tag{18}$$

where E is the strain stimulus, k is a proportionality constant, ε is the peak-to-peak strain magnitude, and f is the loading frequency (Turner et al., 1995). Second is the principle of diminishing returns; that is, as loading duration is increased, the bone formation response tends to level off. This effect was mathematically described by Carter and colleagues as:

$$S = k_2 \left[\sum_{j=1}^{n} N_j \sigma_j^m \right]^{\frac{1}{m}}, \tag{19}$$

where k_2 is a constant, N is the number of loading cycles per day, σ is the effective stress, and m is a constant weighting factor, which has been estimated at 3.5–4, based on published data (Carter et al., 1987; Turner, 1998). Finally, bone adaptation is error-driven such that bone cells accommodate to "normal" strain waveforms, but adapt to abnormal strain changes (Lanyon, 1992). This has been described mathematically as:

$$\frac{\partial M}{\partial t} = B\{\varphi - F\}, \tag{20}$$

where M is bone mass, t is time, φ is the local stress/strain state, and B and F are constants that describe the "normal" load state (Fyhrie and Schaffler, 1995). Thus, $\varphi - F$ represents the error function driving force for bone mass adaptation.

These rules apply to both mechanical stimulation of new bone formation and disuse-induced bone resorption. Astronauts, for example, experience significant reductions in bone mass when the local stress/strain state, φ, becomes less than the normal earth-bound state, F; because of reduced gravitational loads, the negative error function drives bone resorption. This tightly regulated system can also become pathogenic in osteoporosis, in which the communication between constituents of the BMU is disrupted, and more bone is resorbed than can be replaced, leading to a decrease in bone mass and skeletal fragility.

In blood vessels, hemodynamic forces play multiple important roles in the regulation of vascular cells (Riha et al., 2005). Pulsatile intramural pressures produce cyclic strain within vessel walls, and blood flows exert sheer stresses on the lumen walls. These two types of physical stimuli influence the phenotype and activity of smooth muscle cells and endothelial cells within the vasculature. A tremendous amount of recent research has been directed towards studying hemodynamic effects given the potential implications for prevention or treatment of atherosclerosis as well as vascular tissue engineering. Arteries are capable of remodeling their structure in response to changes in their mechanical environment. A chronic increase in systemic blood pressure induces an increase in vessel wall thickness and area, while reduced pressure leads to a decrease in vessel dimensions (Arner et al., 1984).

In cartilage, normal joint loading produces compressive, tensile, and sheer forces that deform the cells (chondrocytes) and induce interstitial fluid flows and streaming potentials throughout the matrix (Mow and Ratcliffe, 1997). These mechanical, chemical, and electric signals prominently influence the metabolism of the chondrocytes. As articular cartilage in adults is devoid of a blood supply, mechanical deformations are of critical importance to facilitate flow of nutrients and waste products into and out of the tissue. Mechanical deformations also serve to maintain the tissue's proper matrix composition, organization, and mechanical properties. It is generally accepted that static or constant compression/pressure results in loss and/or reduction of synthesis of proteoglycans and DNA in a nearly dose-dependent manner (Li et al., 2001). Dynamic compression has been shown to positively modulate proteoglycan synthesis and this stimulation is heavily influenced by both the frequency and amplitude of the compressive waveform (Li et al., 2001). Similarly, dynamic tissue sheer also has a pronounced effect on matrix components in a frequency- and amplitude-dependent manner (Jin et al., 2001).

Abnormal joint loads have been shown to induce changes in composition, structure, and mechanical properties of articular cartilage. Disuse studies, for example, that use casting or other means of immobilization have demonstrated a loss of matrix constituents such as proteoglycans and a reduction in tissue thickness and mechanical properties (Akeson et al.,

1987). Conversely, moderate exercise may have beneficial effects on maintaining healthy articular cartilage (Lane, 1996). Dynamic compression modulates biomarkers implicated in important disease states (e.g. osteoarthritis) such as cartilage oligomeric matrix protein (COMP) (Piscoya et al., 2005), but high impact loading or altered joint loading due to instability or injury is recognized as a significant risk factor for the development and progression of osteoarthritis (Buckwalter, 1995a; Lane, 1996; Piscoya et al., 2005). These studies suggest that there is a range of local stresses and strains that promote healthy tissue homeostasis, but loading conditions that are abnormally high or low can trigger catabolic responses and a loss of tissue function.

Mechanotransduction

So how are local mechanical signals transduced into cellular responses that affect tissue growth, repair, and remodeling? The process of mechanotransduction can be divided into four stages (Gooch et al., 1998), as shown in Figure 28.5. These are: (1) force transmission, (2) mechanotransduction, (3) signal propagation, and (4) cellular response. The first stage refers to the transmission of the force from the point it is applied to the cell surface. The second corresponds to the sensory action of the cells in sensing mechanical stimuli, and transducing them into a biochemical signal, which is propagated inside the cell in the third stage. Finally, the cell responds to the intracellular signal by modulating gene expression, completing the mechanotransduction process.

In the first stage of mechanotransduction, applied forces are converted into local stimuli that may be detected by cells. Transmitted forces can cause direct cellular deformation by deforming the surrounding extracellular matrix. Applied forces may also result in local fluid flow and/or hydrostatic pressures. For example, compression of articular cartilage generates hydrostatic pressure that can regulate chondrocyte metabolism. Dynamic compression of cartilage induces fluid flow through the matrix and exposes cells to local sheer stresses. The relative importance of these different types of local stimuli *in vivo* is not clear due to the difficulty of isolating each kind of mechanical stimulus. However, extensive research has been done to study the effects of various forms of mechanical stimuli on cells *in vitro*. These include tensile stretch, compression, hydrostatic pressure, and fluid-flow-induced sheer stress, applied either statically or dynamically. These studies have allowed investigators to identify potential mechanotransduction mechanisms.

The next stage of mechanotransduction occurs at the plasma membrane of the cell, and it is here that the cell detects the external signal and converts it into an intracellular signal. The plasma membrane contains numerous receptors and ion channels that can serve as sensors of the mechanical stimuli. The key structures in this interaction are the mechanosensitive (also known as stretch-activated) ion channels, integrin receptors, and other plasma membrane receptors.

Mechanosensitive ion channels (Sachs, 1991; Hamill and Martinac, 2001; Martinac, 2004) are thought to be important to many cell types including chondrocytes (Wright et al., 1996; Guilak and Hung, 2005), osteoblasts (Charras and Horton, 2002), endothelial cells (Davies, 1995), and cardiac myocytes (Hu and Sachs, 1997). Experiments involving direct perturbation of the chondrocyte membrane have implicated such ion channels in the increase in concentration of cytosolic calcium ion (Guilak et al., 1999), which is a second messenger and has well-known intracellular effects (Carafoli, 1987; Otey et al., 1993; Faber and Sah, 2003).

Integrins are heterodimeric transmembrane proteins that bind to extracellular matrix proteins and cluster together leading to the assembly of focal adhesions, at which the cell contacts the ECM. Focal adhesions intracellularly associate with α-actinin (Otey et al., 1993), talin (Critchley, 2004), tensin (Bockholt and Burridge, 1993), and other cytoskeletal binding proteins as well as signaling molecules such as focal adhesion kinase (Schaller et al., 1995). Due to their associations with both structural and signaling proteins, integrins are well placed to act as transducers of physical stimuli, and have been implicated as a link between the extracellular and intracellular environments for a variety of cell types that allow transmission of inside-out and outside-in signals capable of modulating cell behavior (Wright et al., 1997; Pelham and Wang, 1999; Jalali et al., 2001; Aikawa et al., 2002; Martinez-Lemus et al., 2003). Wright et al. reported that the transduction pathways involved in the hyperpolarization response of human articular chondrocytes *in vitro* after cyclical pressure-induced strain involve $\alpha_5\beta_1$ integrin, which they suggest to be an important chondrocyte mechanoreceptor (Wright et al., 1997). Externally applied forces would cause changes in the conformations of the ECM molecules that would affect their binding to integrins, and modify the force balance within focal adhesions. It is thought that increased tension within focal adhesions can trigger increased integrin clustering and FAK phosphorylation (Sieg et al., 1999; Katsumi et al., 2004), which initiates a signal cascade resulting in altered gene expression.

In addition to integrins, the plasma membrane is host to other receptors for specific ECM proteins such as collagen, aggrecan, and hyaluronic acid, which may also be able to sense extracellular forces due to their interactions with their ligands. G protein coupled receptors (GPCRs) may also act as mechanotransducers or be activated secondarily to other pathways, as the consequences of the G protein stimulation of the PLC-IP$_3$ pathway have been observed in mechanically stimulated cells (Davies, 1995; Reich et al., 1997).

Primary cilia — microtubule-based, flagella-like extensions of the membrane — have been identified recently as potent mechanosensors (Whitfield, 2008). First identified in the late 1800s, and thought to be a functionless vestige, the primary cilium has recently been implicated as a mechanism for mechanosensation in numerous cell types including kidney (Nauli et al., 2003), bone (Whitfield, 2003; Malone et al., 2007), and cartilage (Poole, 1997). Jacobs and colleagues identified primary cilia in both osteocytes and osteoblasts, and proposed that primary cilia sense lacunocanalicular fluid flow caused by bone loading (Malone et al., 2007). They demonstrated that primary cilia are required for osteocyte and osteoblast response to fluid sheer stress, inducing expression of genes (OPN, COX-2, OPG/RANKL) and production of second messengers (PGE$_2$) associated with bone remodeling. Primary cilia have also been shown to cause intracellular Ca^{2+} release by GPCR proteolysis of PKD1 to activate Runx2 and IP$_3$ production (Chauvet et al., 2004). Further research is required to fully understand the mechanisms through which primary cilia respond to mechanical stimuli.

The third stage of mechanotransduction is signal propagation, in which the signal generated at the plasma membrane in the second stage is propagated within the cell. This is usually carried out using the same machinery that the cell uses for responding to biochemical stimuli. Signal propagation is initiated by second messengers such as Ca^{2+}, cAMP, and mitogen-activated protein kinase (MAPK). Activated kinases subsequently phosphorylate transcription factors, leading to changes in gene expression.

Cytoplasmic calcium serves as a ubiquitous signal for regulation of important cellular processes such as cell growth, differentiation, protein synthesis, and even cell death. Numerous studies have found an increase in cytosolic Ca^{2+} concentration due to mechanical loading in a variety of cell types (Hung et al., 1997; Edlich et al., 2001; Sharma et al., 2002; Donahue et al., 2003). This may be due to the opening of mechanosensitive Ca^{2+} channels as discussed above or secondary to another mechanotransduction mechanism. The intracellular Ca^{2+} concentration can also be elevated by release of calcium from intracellular stores through the IP$_3$/DAG pathway (Berridge, 1987). This pathway can be triggered by GPCRs leading to the activation of the enzyme phospholipase C (PLC). PLC cleaves the phosphoinositide PIP$_2$ to generate two second messengers: diacylglycerol (DAG) and inositol trisphosphate (IP$_3$). After diffusing though the cytosol, IP$_3$ interacts with and opens Ca^{2+}

channels in the membrane of the endoplasmic reticulum, causing release of Ca^{2+} into the cytosol. One of the various cellular responses induced by a rise in cytosolic Ca^{2+} is recruitment of protein kinase C (PKC) to the plasma membrane, where it is activated by DAG. The activated kinase can phosphorylate various proteins, including transcription factors, leading to gene activation. Ca^{2+} is also known to bind to the small cytosolic protein calmodulin to form a complex that interacts with and modulates the activity of other enzymes and transcription factors. Ca^{2+} influx is known to activate certain K^+ channels, thus affecting membrane potential (Wright et al., 1992; Faber and Sah, 2003), and has been shown to be necessary for integrin-dependent tyrosine phosphorylation of focal adhesion-associated molecules (Alessandro et al., 1998).

The cyclic nucleotide cAMP is produced by adenylyl cyclases, which are in turn activated by GPCRs. Protein kinase A (PKA), which consists of two catalytic subunits and two regulatory subunits, is the most well-known cAMP effector. Binding of cAMP to the regulatory subunits releases the catalytic subunits, which are then free to phosphorylate substrates (Dumaz and Marais, 2005). cAMP, along with intracellular Ca^{2+}, has been implicated in the regulation of gene expression in response to static compression of cartilage explants (Valhmu et al., 1998; Fitzgerald et al., 2004). Boo et al. demonstrated that sheer stress stimulates phosphorylation of eNOS and thus nitric oxide (NO) production in bovine aortic endothelial cells in a PKA-dependent manner (Boo et al., 2002).

The recently identified β-catenin pathway has sparked great interest as a mechanotransduction mechanism. Intracellular β-catenin is normally controlled by binding to a "destruction complex," containing glycogen synthase kinase (GSK-3β) (Robinson et al., 2006). Under mechanical stimulation, cells produce small molecules known as Wnts that bind to the membrane receptor complex of LRP5/6 and Frizzled to phosphorylate GSK-3β, resulting in deactivation of the destruction complex (Mao et al., 2001; Staal et al., 2002). This allows stabilization of intracellular β-catenin, which translocates to the nucleus to initiate gene expression, and, in osteoblasts, for example, induces bone formation (Case et al., 2003; Norvell et al., 2004). Interestingly, mechanical stimulation of osteocytes has also been demonstrated to activate the β-catenin pathway independently of Wnt signaling through NO and phosphatydil inositol-3 kinase (PI3-K) (Santos et al., 2010). This pathway has potential to provide novel targets for intervention in bone remodeling pathologies and to manipulate the response of cells to mechanical stimuli.

The final stage of mechanotransduction is the altered response of the cell, which may include changes in matrix synthesis/degradation, proliferation, differentiation, apoptosis, cell alignment, and migration. The effectors of the mechanotransduction pathways are the various transcription factors, which are activated by the events discussed previously. Numerous studies on vascular cells have shown activation of transcription factors such as AP-1, CRE, and NF-κβ in response to cyclic strain (Kakisis et al., 2004). The activated transcription factors interact with the promoter and enhancer regions of various genes to mediate transcription. This results in an increase in expression of genes such as Cox-2, VEGF, TGF-β3, and eNOS (Kakisis et al., 2004), which orchestrate the cellular responses. Lee et al. demonstrated that vascular smooth muscle cells respond to mechanical strain by increasing specific proteoglycan synthesis and aggregation (Lee et al., 2001). It is known that mechanical loading of osteocytes results in anabolic responses such as the expression of c-fos, insulin-like growth factor-I (IGF-I), and osteocalcin (Mikuni-Takagaki, 1999). Elevations in Ca^{2+} activate a Ca^{2+}/calmodulin-dependent protein kinase, which causes increased c-fos expression, which is a pro-growth transcription factor. Calcineurin, a Ca^{2+}/calmodulin-activated phosphatase, dephosphorylates and activates the NF-AT family of transcription factors. Different NF-ATs, expressed in different cells including those of the heart, cartilage, and bone, serve as tissue-specific activators of cell growth and differentiation (Crabtree, 1999; Iqbal and Zaidi, 2005).

MECHANICALLY STIMULATE TO ENHANCE REGENERATION

Replacing tissues that serve a significant biomechanical function has proven exceptionally challenging (Butler et al., 2000). For mechanosensitive cells and tissues, it may be possible to manipulate the mechanical environment, either *in vivo* or in a bioreactor, to enhance the integration, degradation, or activity of a tissue-engineered construct.

Mechanical stimulation *in vivo*

Bone repair is acutely responsive to mechanics. In fracture healing, the local mechanical environment determines both the course and success of healing (Goodship and Kenwright, 1985). Though it was once held that complete immobilization was needed for successful fracture healing and that the resorptive effect of disuse was necessary to release calcium for callus mineralization (Baker, 1934), it is now known that limited physical activity can promote tissue repair and restoration of function (Buckwalter, 1995b).

Despite strong evidence that the local mechanical environment acutely influences bone healing, few studies to date have attempted to directly assess the effects of *in vivo* stresses on tissue-engineered constructs following implantation. Case et al. investigated the effects of controlled intermittent compressive deformation on cellular constructs using a hydraulic bone chamber device implanted into the distal femoral metaphyses of rabbits (Case et al., 2003) (Fig. 28.4A). Constructs receiving four weeks of daily mechanical loading at 0.5 Hz were found to have nine-fold more new bone formation compared to contralateral control constructs that did not receive loading. Similarly, Boerckel et al. presented a rat bone segmental defect model in which axially compliant fixation plates allow transfer of ambulatory loads to the defect, resulting in an increase in bone formation and mechanical properties (Boerckel et al., 2009) (Fig. 28.4B). These studies demonstrate the important role that the *in vivo* mechanical environment can play in the repair and integration of an implanted tissue-engineered construct.

505

Mechanical stimulation *in vitro*

Many tissues bear tremendous stress and strain over repeated loading cycles *in vivo* while maintaining normal function. To date, no engineered construct has been developed *in vitro* possessing the same biomechanical properties as its *in situ* counterpart. One approach to address this challenge is the use of physiologically inspired mechanical forces to transmit stimuli to developing constructs *in vitro*. Since these tissues normally experience a dynamic environment *in vivo*, the rationale is that the application of mechanical forces such as compression or sheer stress

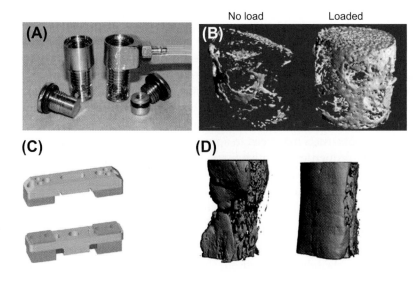

FIGURE 28.4

(A) Hydraulic bone chamber implant used to apply cyclic compressive loading to tissue-engineered constructs *in vivo*. Implanted constructs (B) receiving the mechanical stimulus (right) had nine-fold more new bone formation than no-load controls (left). (C) Stiff (top) and axially compliant (below) fixation plates. (D) Compliant fixation (right) resulted in enhanced bone formation over stiff fixation (left).

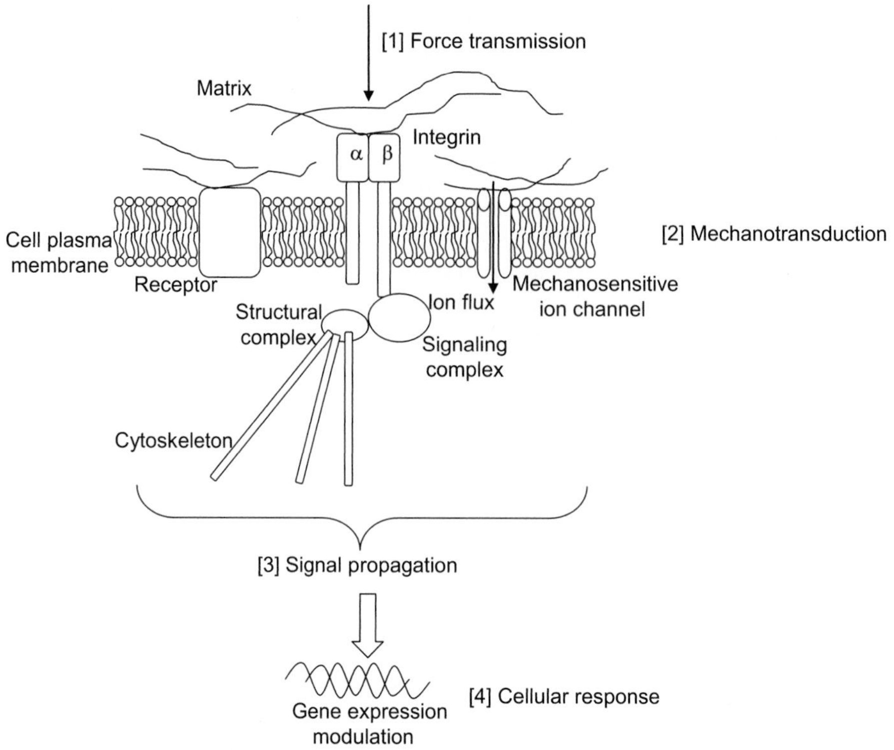

FIGURE 28.5
Schematic showing the four stages of mechanotransduction: [1] force transmission, [2] mechanotransduction, [3] signal propagation, and [4] cellular response. See text for details.

will stimulate the cells of the engineered construct to secrete and organize the proper matrix proteins required to reproduce the native tissue mechanical function.

Perhaps the tissues of the body most subjected to mechanical forces are those of musculo-skeletal and cardiovascular origin. Consequently, orthopedic and cardiovascular tissue-engineered constructs represent the bulk of the research in which mechanical forces have been applied to developing tissues *in vitro*. Cartilage, bone, tendon, ligament, blood vessels, heart valves, and muscle have been cultured *in vitro* under the influence of mechanical forces. The remainder of this section will discuss select examples from the orthopedic and cardiovascular fields that use the *in vivo* environment as inspiration to mechanically condition tissue-engineered constructs *in vitro*.

CARTILAGE BIOREACTORS

Cartilage bioreactors commonly apply compression and/or sheer forces to modulate construct matrix composition and mechanical properties. While many different tissue-engineering models exist for cartilage (e.g. alginate, agarose, pellet/micro-mass, scaffold, and scaffold-free culture), the mechanical properties necessary to withstand the complex and demanding *in vivo* mechanical environment have yet to be recapitulated. For clinical success, it has been suggested that tissue-engineered constructs may need to approximate the matrix composition, organization, and biomechanical properties of native tissue in order to promote construct integration and load-bearing capability *in vivo* (Hung et al., 2004). Bioreactor systems have produced encouraging results indicating that *in vitro* mechanical conditioning of tissue-engineered constructs is a promising approach to reproducing native tissue properties.

As one example, a novel dual-chambered, parallel-plate flow bioreactor system has been used to apply controlled sheer stresses to the surface of cartilaginous constructs grown *de novo* from primary bovine articular chondrocytes without the aid of a scaffold (Fig. 28.6).

The "parallel-plate" design refers to the top bioreactor surface and tissue-engineered construct face, which form two parallel walls separated by a defined distance that creates a flow channel. Fluid is flowed through the channel, resulting in a parabolic velocity profile. Consequently, a sheer stress is applied that is maximal at the upper wall and tissue surface; this is commonly referred to as Poiseuille flow (Fox and McDonald, 1992). One can estimate the wall sheer stress (τ_w) by the following equation:

$$\tau_w = \frac{6\mu Q}{bh^2} \tag{21}$$

where μ is the media viscosity, Q is the volumetric flow rate, b is the flow chamber width, and h is the fluid gap height.

Chondrocytes are seeded on to a semi-permeable membrane and, following a static pre-culture period, fluid-induced sheer stress is applied to the construct. The application of flow significantly increases type II collagen compared to static (no flow) controls, as well as both Young's modulus and ultimate strength (Gemmiti and Guldberg, 2006). This study suggests that flow-induced sheer stresses may be an effective functional tissue-engineering strategy for modulating matrix composition and mechanical properties *in vitro*.

BONE BIOREACTORS

Without a vascular blood supply *in vitro*, nutrient delivery to cells throughout three-dimensional tissue-engineered constructs grown in static culture must occur by simple diffusion alone. As a result, attempts to engineer bone greater than 1 mm in thickness usually result in a thin shell of viable tissue and cells localized at the periphery (Gersbach et al., 2004). It has been theorized that this effect is due to suboptimal mass transport conditions and a lack of mechanical stimulation in static culture. Therefore, tissue culture systems that provide dynamic media flow around or within tissue-engineered constructs have been designed to enhance nutrient and waste exchange *in vitro* (Bujia et al., 1995). In addition to enhancing mass transport, fluid flow applies sheer stresses to the cells within the scaffolds.

507

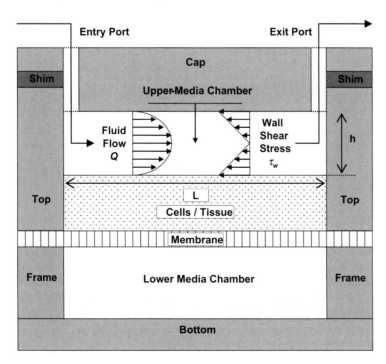

FIGURE 28.6
Dual-chambered parallel-plate bioreactor system that applies controlled sheer stresses to the surface of cartilaginous construct slabs.

The effects of flow-mediated sheer on cells have been studied in two-dimensional monolayer cultures. Continuous fluid flow applied to osteoblasts *in vitro* has been shown to alter bone-related gene expression and cellular phenotype (Ogata, 2000). Parallel plate flow experiments have shown that bone cells cultured in monolayer are highly responsive to flow-mediated sheer stresses. Sheer stresses in the range of 5–15 dynes/cm^2 affect osteoblast proliferation as well as production of nitric oxide (NO) and prostaglandin E$_2$ (PGE$_2$), suggesting that sheer stress is an important regulator of osteoblast function (McAllister et al., 2000). Pulsatile and oscillatory flow conditions applied to osteoblasts using *in vitro* parallel-plate flow chambers have also been shown to increase gene expression, intracellular calcium concentration, and the production of NO and PGE$_2$ in comparison to static controls (Klein-Nulend et al., 1997; Bakker et al., 2001). Furthermore, cell responsiveness has been reported to vary with fluid flow rate and frequency (Jacobs et al., 1998; Edlich et al., 2001). Proposed mechanisms for the stimulation of cells by fluid flow include increased mass transport, generation of streaming potentials, and application of sheer stresses to the cell membranes (McAllister and Frangos, 1999; Bakker et al., 2001). Although these studies were performed using two-dimensional cell culture systems for short-term experiments, they suggest that variable flow conditions may also have differential effects in three-dimensional tissue culture systems.

Such tissue culture systems may be useful to engineer thicker, more uniform bone graft substitutes for implantation or as test bed models that simulate aspects of the *in vivo* environment. While many different bioreactor systems have been developed, perfusion bioreactors in particular have shown significant increases in both cell viability and mineralized matrix formation on large three-dimensional constructs *in vitro*. In a recent study, micro-CT has been used to quantify mineralized matrix production within perfused and statically cultured marrow progenitor cells seeded on large polymer scaffolds (6.35 mm diameter, 9 mm thick) (Porter et al., 2005). Statically cultured constructs were found to have mineralized matrix localized only to the periphery of the constructs. In contrast, perfused constructs were found to have a several-fold increase in mineralized matrix production distributed throughout the constructs (Fig. 28.7).

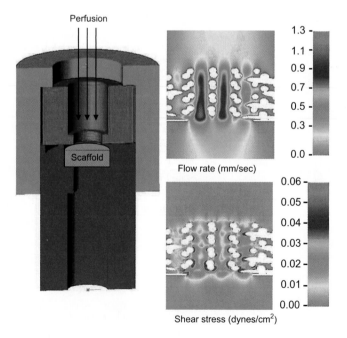

FIGURE 28.7
Perfusion bioreactor system (left) for production of mineralized constructs for bone defects. Computational fluid dynamics simulation of flow rate and sheer stresses within the three-dimensional scaffold porosity (right).

BLOOD VESSEL BIOREACTORS

Following the same rationale for mechanical conditioning of orthopedic engineered tissues, cardiovascular tissues can also be enhanced by *in vitro* mechanical stimulation. Cardiovascular tissues reside in a dynamic environment that can be mimicked *in vitro* using bioreactors and mechanical loading systems to deliver the physiologically inspired environmental cues.

In vivo, the pulsatile flow of blood imparts cyclic strains and sheer stresses to the vessel's constituents, which respond in a variety of ways to these mechanical signals. Endothelial cells are uniquely situated in the lumen and are directly in contact with the flowing blood, which causes a sheer stress to be applied to the cells. Consequently, these rapidly responding, mechanosensitive cells attain an elongated shape, aligning their long axis with the direction of flow. Sensing of the sheer via cell-surface receptors, ion channels, or integrins leads to secretion and/or activation of a number of signaling molecules, such as NO, endothelial nitric oxide synthase (eNOS), kinases, and transcription factors (Takahashi et al., 1997; Fisslthaler et al., 2000; Fisher et al., 2001). Perhaps most importantly, the fluid-induced sheer stress confers a protective effect on the vessel by decreasing the probability of atherosclerosis (Traub and Berk, 1998). Indeed, areas of irregular blood flow (i.e. velocity, direction, and sheer stress) have been implicated as sites of increased atherosclerosis (Papadaki et al., 1999). Sheer stress also modulates smooth muscle cells' production of signaling molecules (such as NO) (Papadaki et al., 1998a) and gene transcription levels of cell-surface receptors (Papadaki et al., 1998b).

Tissue-engineered blood vessels (TEBVs) aim to reproduce cellular and mechanical properties of the native vessel in order to be an effective replacement. However, similarly to other engineered tissues, those cultured in static conditions fall short of native tissue properties. The concept of mechanical stimulation of tissue-engineered blood vessels to enhance matrix organization and mechanical properties began in the mid-1990s, and has been an active topic of research since. Historically, three types of TEBVs have received the greatest attention for application of *in vitro* cyclic strains. First, collagen gel-derived TEBVs were exposed to cyclic circumferential strains by Nerem and Tranquillo and their colleagues. Cummings et al. demonstrated an increase in mechanical properties and altered cellular function in response to cyclic mechanical strain of smooth muscle cell-seeded collagen I matrices. Likewise, Seliktar et al. showed that 10% cyclic strain of cell-seeded collagen gels induced MMP-2-mediated remodeling, yielding improved mechanical properties (Seliktar et al., 2003). More recently, Gleason and colleagues presented a microstructurally motivated continuum mechanics model that combines numerous factors affecting the success of the tissue-engineered blood vessel approach, such as cell type, matrix composition, mechanical stimulus, and their interacting effects on TEBV growth, remodeling, and mechanics (Raykin et al., 2009). Such studies point both to the challenge of optimizing this approach and to the importance of biomechanics-informed rational experimental design.

Second, the effects of cyclic stretching on biodegradable polymeric scaffolds have also been investigated. Recently, Gong and Niklason reported that cyclic strain of cell-seeded constructs enhanced mesenchymal stem cell differentiation towards a smooth muscle cell phenotype and induced a more normal tissue composition (Gong and Niklason, 2008).

Finally, Auger and colleagues presented a method of self-assembly in which cells are cultured in two dimensions to excrete their own extracellular matrix, forming a tissue sheet, which is then rolled into a TEBV. When exposed to uniaxial stretch, the cells realigned along the axis of strain and improved the contractile capacity of the resulting TEBVs (Grenier et al., 2006).

Other experiments have demonstrated that exposing tissue-engineered vascular grafts to fluid-induced sheer stress has been shown to increase endothelial cell adherence (Ott and Ballermann, 1995) and proliferation (Imberti et al., 2002) and alter tissue morphology and mechanical properties (Niklason et al., 2001). Cyclic mechanical strains cause an increase in collagen (types I and III) transcription by smooth muscle cells (Leung et al., 1976); an increase in mechanical properties (strength and stiffness), attributed to an increase in remodeling enzymes such as matrix metalloproteinase-2 (Seliktar et al., 2001); and an increase in matrix and cellular organization (Seliktar et al., 2001; Imberti et al., 2002). Subjecting smooth-muscle-cell-impregnated constructs to dynamic mechanical stress not only causes ultrastructural and orientational changes in the cell phenotype and matrix, but can also induce cells to shift from a synthetic to a contractile state (Kanda and Matsuda, 1994). Similar constructs (smooth muscle cells seeded into polyglycolic acid meshes) exposed to pulsatile radial stresses of 165 beats per minute (analogous to fetal heart rates) and 5% radial strain produce constructs with burst pressures in excess of 2,000 mm Hg, increased collagen deposition, and desirable histological characteristics (Niklason et al., 1999).

While great strides have been made in the field of tissue-engineered vascular grafts, a completely successful graft still has yet to be identified. However, as the field continues to progress and learn more about the *in vivo* environment, those cues can be translated to more realistic conditioning techniques for *in vitro*-grown constructs. This mechanical stimulation is critical to remodeling the graft to possess proper mechanical properties as well as matrix composition and organization. The same can be said for cartilage and bone. Thus, mechanical conditioning in an *in vitro* setting has proven to be a powerful technique to increase the similarity of tissue-engineered constructs to the native tissues they aim to replace.

EVALUATE FUNCTIONAL RESTORATION

Assessment of functional regeneration is an essential benchmark for establishing a successful tissue-engineering strategy. Often, qualitative, indirect measures of regeneration are presented without evaluation of biomechanical integrity. For tissues and structures whose primary function is to bear physical loads, mechanical testing is an essential measure of repair. In cartilage regeneration, for example, many studies present only compositional and morphological assessments of healing, without direct measurement of mechanical function. While composition and morphology are important measures, the true indicator of regeneration is whether the regenerated tissue recapitulates normal tissue behavior, including both the monotonic elastic and viscoelastic properties. Though often experimentally difficult, establishment of standards for mechanical evaluation of engineered tissues will be a significant contribution. For tissue-engineering approaches to long bone defect healing, torsion testing is an experimentally facile and analytically simple method of determining the structural properties of the regenerated tissue, and allows for direct comparison with age-matched uninjured limbs.

Other functional considerations include assessment of the long-term consequences and temporal remodeling of tissue-engineered constructs. These include evaluation of scaffold degradation by-products, extent of remodeling to native architecture, and restoration of native material behavior.

CONCLUSIONS

In conclusion, this chapter has presented four steps for successfully approaching a tissue-engineering problem from the perspective of physical mechanics: (1) quantitatively describe the native mechanical environment, (2) understand the role of mechanical factors in the tissue of interest, (3) manipulate mechanical conditions to enhance function or regeneration, and (4) quantitatively evaluate the degree of functional biomechanical restoration of the engineered tissue. While all of these considerations are active areas of research, regenerating tissues that serve a significant biomechanical function continues to prove exceptionally challenging (Butler et al., 2000). It is clear that tissue regeneration strategies must take into consideration the complex and demanding *in vivo* mechanical environment into which tissue-engineered constructs are implanted. Fortunately, a wealth of knowledge is now available to tissue engineers about how local stresses and strains affect cell function within tissues. Integration of this knowledge into strategies for tissue replacement or regeneration will be key to achieving the goal of long-term functional restoration in patients.

References

Aikawa, R., et al. (2002). Integrins play a critical role in mechanical stress-induced p38 MAPK activation. *Hypertension, 39*(2), 233–238.

Akeson, W. H., et al. (1987). Effects of immobilization on joints. *Clin. Orthop. Relat. Res., 219*, 28–37.

Alessandro, R., et al. (1998). Endothelial cell spreading on type IV collagen and spreading-induced FAK phosphorylation is regulated by Ca2+ influx. *Biochem. Biophys. Res. Commun., 248*, 635–640.

Arner, A., Malmqvist, U., & Uvelius, B. (1984). Structural and mechanical adaptations in rat aorta in response to sustained changes in arterial pressure. *Acta Physiol. Scand., 122*(2), 119–126.

Baker, A. H. T. U. M. J. (1934). Non-union in fractures. *The Ulster Medical Journal, 3*(4), 277–283.

Bakker, A. D., et al. (2001). The production of nitric oxide and prostaglandin E(2) by primary bone cells is shear stress dependent. *J. Biomech., 34*(5), 671–677.

Berridge, M. (1987). Inositol trisphosphate and diacylglycerol: two interacting second messengers. *Annu. Rev. Biochem., 56*, 159–193.

Bockholt, S., & Burridge, K. (1993). Cell spreading on extracellular matrix proteins induces tyrosine phosphorylation of tensin. *J. Biol. Chem., 268*, 14565–14567.

Boerckel, J. D., et al. (2009). In vivo model for evaluating the effects of mechanical stimulation on tissue-engineered bone repair. *J. Biomech. Eng., 131*(8), 084502.

Boo, Y., et al. (2002). Shear stress stimulates phosphorylation of endothelial nitric-oxide synthase at Ser1179 by Akt-independent mechanisms: role of protein kinase A. *J. Biol. Chem., 277*(5), 3388–3396.

Buckwalter, J. A. (1995a). Activity vs. rest in the treatment of bone, soft tissue and joint injuries. *Iowa Orthop. J., 15*, 29–42.

Buckwalter, J. A. (1995b). Osteoarthritis and articular cartilage use, disuse, and abuse: experimental studies. *J. Rheumatol. Suppl., 43*, 13–15.

Bujia, J., et al. (1995). Engineering of cartilage tissue using bioresorbable polymer fleeces and perfusion culture. *Acta Otolaryngol., 115*(2), 307–310.

Butler, D. L., Goldstein, S. A., & Guilak, F. (2000). Functional tissue engineering: the role of biomechanics. *J. Biomech. Eng., 122*(6), 570–575.

Carafoli, E. (1987). Intracellular calcium homeostasis. *Annu. Rev. Biochem., 56*, 395–433.

Carter, D. R., Fyhrie, D. P., & Whalen, R. T. (1987). Trabecular bone density and loading history: regulation of connective tissue biology by mechanical energy. *J. Biomech., 20*(8), 785–794.

Case, N. D., et al. (2003). Bone formation on tissue-engineered cartilage constructs in vivo: effects of chondrocyte viability and mechanical loading. *Tissue Eng., 9*(4), 587–596.

Charras, G., & Horton, M. (2002). Single cell mechanotransduction and its modulation analyzed by atomic force microscope indentation. *Biophys. J., 82*, 2970–2981.

Chauvet, V., et al. (2004). Mechanical stimuli induce cleavage and nuclear translocation of the polycystin-1 C terminus. *J. Clin. Invest., 114*(10), 1433–1443.

Cowin, S. C., & Doty, S. B. (2007). *Tissue Mechanics*. New York: Springer. p. 682.

Crabtree, G. (1999). Generic signals and specific outcomes: signaling through Ca2+, calcineurin, and NF-AT. *Cell, 96*, 611–614.

Critchley, D. (2004). Cytoskeletal proteins talin and vinculin in integrin-mediated adhesion. *Biochem. Soc. Trans., 32*(5), 831–836.

Davies, P. (1995). Flow-mediated endothelial mechanotransduction. *Physiol. Rev., 75*(3), 519–560.

Donahue, S., Donahue, H., & Jacobs, C. R. (2003). Osteoblastic cells have refractory periods for fluid-flow-induced intracellular calcium oscillations for short bouts of flow and display multiple low-magnitude oscillations during long-term flow. *J. Biomech., 36*(1), 35–43.

Dumaz, N., & Marais, R. (2005). Integrating signals between cAMP and the RAS/RAF/MEK/ERK signalling pathways. *FEBS J., 272*(14), 3491–3504.

Edlich, M., et al. (2001). Oscillating fluid flow regulates cytosolic calcium concentration in bovine articular chondrocytes. *J. Biomech., 34*(1), 59–65.

Faber, E. S. L., & Sah, P. (2003). Calcium-activated potassium channels: multiple contributions to neuronal function. *The Neuroscientist, 9*(3), 181–194.

Fisher, A. B., et al. (2001). Endothelial cellular response to altered shear stress. *Am. J. Physiol. Lung Cell Mol. Physiol.,* *281*(3), L529–L533.

Fisslthaler, B., et al. (2000). Phosphorylation and activation of the endothelial nitric oxide synthase by fluid shear stress. *Acta Physiol. Scand., 168*(1), 81–88.

Fitzgerald, J., et al. (2004). Mechanical compression of cartilage explants induces multiple time-dependent gene expression patterns and involves intracellular calcium and cyclic AMP. *J. Biol. Chem., 279*(19), 19502–19511.

Fox, R., & McDonald, A. (1992). *Introduction to Fluid Mechanics* (4th ed.). New York: John Wiley & Sons, Inc.

Frost, H. M. (1963). *Bone Remodelling Dynamics.* Springfield: Thomas. p. 175.

Fung, Y. C. (1993). *Biomechanics: Mechanical Properties of Living Tissues* (2nd ed.). New York: Springer-Verlag. p. 568.

Fung, Y. C. (1973). Biorheology of soft tissues. *Biorheology, 10*(2), 139–155.

Fung, Y. C. (1967). Elasticity of soft tissues in simple elongation. *Am. J. Physiol., 213*(6), 1532–1544.

Fung, Y. C. (1965). *Foundations of Solid Mechanics.* Englewood Cliffs: Prentice-Hall. p. 525.

Fyhrie, D. P., & Schaffler, M. B. (1995). The adaptation of bone apparent density to applied load. *J. Biomech., 28*(2), 135–146.

Gemmiti, C. V., & Guldberg, R. E. (2006). Fluid flow increases type ii collagen deposition and tensile mechanical properties in bioreactor-grown tissue-engineered cartilage. *Tissue Eng., 12*(3), 469–479.

Gersbach, C. A., et al. (2004). Runx2/Cbfa1-genetically engineered skeletal myoblasts mineralize collagen scaffolds in vitro. *Biotechnol. Bioeng., 88*(3), 369–378.

Gong, Z., & Niklason, L. E. (2008). Small-diameter human vessel wall engineered from bone marrow-derived mesenchymal stem cells (hMSCs). *FASEB J., 22*(6), 1635–1648.

Gooch, K., et al. (1998). Mechanical forces and growth factors utilized in tissue engineering. In C. Patrick, A. Mikos, & L. McIntire (Eds.), *Frontiers in Tissue Engineering.* Amsterdam: Elsevier Science.

Goodship, A. E., & Kenwright, J. (1985). The influence of induced micromovement upon the healing of experimental tibial fractures. *J. Bone Joint Surg. Br., 67*(4), 650–655.

Grenier, G., et al. (2006). Mechanical loading modulates the differentiation state of vascular smooth muscle cells. *Tissue Eng., 12*(11), 3159–3170.

Guilak, F., & Hung, C. T. (2005). Physical regulation of cartilage metabolism. In V. C. Mow, & R. Huiskes (Eds.), *Basic Orthopaedic Biomechanics and Mechano-Biology* (3rd ed.) (pp. 404–409). Philadelphia, PA: Lippincott Williams and Wilkins.

Guilak, F., et al. (1999). Mechanically induced calcium waves in articular chondrocytes are inhibited by gadolinium and amiloride. *J. Ortho. Res., 17*(3), 421–429.

Hamill, O. P., & Martinac, B. (2001). Molecular basis of mechanotransduction in living cells. *Physiol. Rev., 81,* 685–740.

Hansen, L., Wan, W., & Gleason, R. L. (2009). Microstructurally motivated constitutive modeling of mouse arteries cultured under altered axial stretch. *J. Biomech. Eng., 131*(10), 101015.

Hu, H., & Sachs, F. (1997). Stretch-activated ion channels in the heart. *J. Mol. Cell Cardiol., 29,* 1511–1523.

Humphrey, J. D., Strumpf, R. K., & Yin, F. C. (1990). Determination of a constitutive relation for passive myocardium: I. A new functional form. *J. Biomech. Eng., 112*(3), 333–339.

Hung, C., et al. (1997). Intracellular calcium response of ACL and MCL ligament fibroblasts to fluid-induced shear stress. *Cell Signal., 9*(8), 587–594.

Hung, C. T., et al. (2004). A paradigm for functional tissue engineering of articular cartilage via applied physiologic deformational loading. *Ann. Biomed. Eng., 32*(1), 35–49.

Imberti, B., et al. (2002). The response of endothelial cells to fluid shear stress using a co-culture model of the arterial wall. *Endothelium, 9*(1), 11–23.

Iqbal, J., & Zaidi, M. (2005). Molecular regulation of mechanotransduction. *Biochem. Biophys. Res. Commun., 328,* 751–755.

Jacobs, C. R., et al. (1998). Differential effect of steady versus oscillating flow on bone cells. *J. Biomech., 31*(11), 969–976.

Jalali, S., et al. (2001). Integrin-mediated mechanotransduction requires its dynamic interaction with specific extracellular matrix (ECM) ligands. *Proc. Natl. Acad. Sci., 98*(3), 1042–1046.

Jin, M., et al. (2001). Tissue shear deformation stimulates proteoglycan and protein biosynthesis in bovine cartilage explants. *Arch. Biochem. Biophys., 395*(1), 41–48.

Kakisis, J., Liapis, C., & Sumpio, B. (2004). Effects of cyclic strain on vascular cells. *Endothelium, 11*(1), 17–28.

Kanda, K., & Matsuda, T. (1994). Mechanical stress-induced orientation and ultrastructural change of smooth muscle cells cultured in three-dimensional collagen lattices. *Cell Transplant., 3*(6), 481–492.

Katsumi, A., et al. (2004). Integrins in mechanotransduction. *J. Biol. Chem., 279*(13), 12001–12004.

Klein-Nulend, J., et al. (1997). Pulsating fluid flow stimulates prostaglandin release and inducible prostaglandin G/H synthase mRNA expression in primary mouse bone cells. *J. Bone Miner. Res., 12*(1), 45–51.

Knothe Tate, M. L., et al. (2004). The osteocyte. *Int. J. Biochem. Cell Biol., 36*(1), 1–8.

Lane, N. E. (1996). Physical activity at leisure and risk of osteoarthritis. *Ann. Rheum. Dis., 55*(9), 682–684.

Lanyon, L. E., & Rubin, C. T. (1984). Static vs dynamic loads as an influence on bone remodelling. *J. Biomech., 17*(12), 897–905.

Lanyon, L. E. (1992). The success and failure of the adaptive response to functional load-bearing in averting bone fracture. *Bone, 13*(Suppl. 2), S17–S21.

Lee, R., et al. (2001). Mechanical strain induces specific changes in the synthesis and organization of proteoglycans by vascular smooth muscle cells. *J. Biol. Chem., 276*(17), 13847–13851.

Leung, D. Y., Glagov, S., & Mathews, M. B. (1976). Cyclic stretching stimulates synthesis of matrix components by arterial smooth muscle cells in vitro. *Science, 191*(4226), 475–477.

Li, K. W., et al. (2001). Growth responses of cartilage to static and dynamic compression. *Clin. Orthop. Relat. Res., 391*(Suppl.), S34–S48.

Maceri, F., Marino, M., & Vairo, G. (2010). A unified multiscale mechanical model for soft collagenous tissues with regular fiber arrangement. *J. Biomech., 43*(2), 355–363.

Mackie, E. J. (2003). Osteoblasts: novel roles in orchestration of skeletal architecture. *Int. J. Biochem. Cell Biol., 35*(9), 1301–1305.

Malone, A. M., et al. (2007). Primary cilia mediate mechanosensing in bone cells by a calcium-independent mechanism. *Proc. Natl. Acad. Sci. U.S.A., 104*(33), 13325–13330.

Malvern, L. E. (1969). *Introduction to the Mechanics of a Continuous Medium*. Englewood Cliffs: Prentice-Hall. p. 713.

Mao, J., et al. (2001). Low-density lipoprotein receptor-related protein-5 binds to Axin and regulates the canonical Wnt signaling pathway. *Mol. Cell, 7*(4), 801–809.

Martinac, B. (2004). Mechanosensitive ion channels: molecules of mechanotransduction. *J. Cell Sci., 117*, 2449–2460.

Martinez-Lemus, L., et al. (2003). Integrins as unique receptors for vascular control. *J. Vasc. Res., 40*(3), 211–233.

McAllister, T. N., & Frangos, J. A. (1999). Steady and transient fluid shear stress stimulate NO release in osteoblasts through distinct biochemical pathways. *J. Bone Miner. Res., 14*(6), 930–936.

McAllister, T. N., Du, T., & Frangos, J. A. (2000). Fluid shear stress stimulates prostaglandin and nitric oxide release in bone marrow-derived preosteoclast-like cells. *Biochem. Biophys. Res. Commun., 270*(2), 643–648.

Mikuni-Takagaki, Y. (1999). Mechanical responses and signal transduction pathways in stretched osteocytes. *J. Bone Miner. Metab., 17*(1), 57–60.

Mow, V., & Ratcliffe, A. (1997). Structure and function of articular cartilage and meniscus. In V. Mow & W. Hayes (Eds.), *Basic Orthopaedic Biomechanics* (pp. 113–177). Philadelphia: Lippincott-Raven Publishers.

Natali, A. N., et al. (2005). Anisotropic elasto-damage constitutive model for the biomechanical analysis of tendons. *Med. Eng. Phys., 27*(3), 209–214.

Nauli, S. M., et al. (2003). Polycystins 1 and 2 mediate mechanosensation in the primary cilium of kidney cells. *Nat. Genet., 33*(2), 129–137.

Niklason, L. E., et al. (1999). Functional arteries grown in vitro. *Science, 284*(5413), 489–493.

Niklason, L. E., et al. (2001). Morphologic and mechanical characteristics of engineered bovine arteries. *J. Vasc. Surg., 33*(3), 628–638.

Norvell, S. M., et al. (2004). Fluid shear stress induces beta-catenin signaling in osteoblasts. *Calcif. Tissue Int., 75*(5), 396–404.

Ogata, T. (2000). Fluid flow-induced tyrosine phosphorylation and participation of growth factor signaling pathway in osteoblast-like cells. *J. Cell. Biochem., 76*(4), 529–538.

Otey, C., et al. (1993). Mapping of the alpha-actinin binding site within the beta 1 integrin cytoplasmic domain. *J. Biol. Chem., 268*(28), 21193–21197.

Ott, M. J., & Ballermann, B. J. (1995). Shear stress-conditioned, endothelial cell-seeded vascular grafts: improved cell adherence in response to in vitro shear stress. *Surgery, 117*(3), 334–339.

Papadaki, M., et al. (1998a). Nitric oxide production by cultured human aortic smooth muscle cells: stimulation by fluid flow. *Am. J. Physiol., 274*(2 Pt 2), H616–H626.

Papadaki, M., et al. (1998b). Differential regulation of protease activated receptor-1 and tissue plasminogen activator expression by shear stress in vascular smooth muscle cells. *Circ. Res., 83*(10), 1027–1034.

Papadaki, M., et al. (1999). Fluid shear stress as a regulator of gene expression in vascular cells: possible correlations with diabetic abnormalities. *Diabetes Res. Clin. Pract., 45*(2–3), 89–99.

Pelham, R. J., & Wang, Y. (1999). High resolution detection of mechanical forces exerted by locomoting fibroblasts on the substrate. *Mol. Biol. Cell., 10*(4), 935–945.

Piscoya, J. L., et al. (2005). The influence of mechanical compression on the induction of osteoarthritis-related biomarkers in articular cartilage explants. *Osteoarthritis Cartilage, 13*(12), 1092–1099.

Poole, C. A. (1997). Articular cartilage chondrons: form, function and failure. *J. Anat., 191*(Pt 1), 1–13.

Porter, B. D., et al. (2005). Perfusion significantly increases mineral production inside 3-D PCL composite scaffolds. In Proceedings of the American Society for Mechanical Engineering Summer Bioengineering Meeting. *Vail, CO.*

Raykin, J., Rachev, A. I., & Gleason, R. L., Jr. (2009). A phenomenological model for mechanically mediated growth, remodeling, damage, and plasticity of gel-derived tissue engineered blood vessels. *J. Biomech. Eng., 131*(10), 1010–1016.

Reich, K. M., et al. (1997). Activation of G proteins mediates flow-induced prostaglandin E2 production in osteoblasts. *Endocrinology, 138*(3), 1014–1018.

Riha, G. M., et al. (2005). Roles of hemodynamic forces in vascular cell differentiation. *Ann. Biomed. Eng., 33*(6), 772–779.

Robinson, J. A., et al. (2006). Wnt/beta-catenin signaling is a normal physiological response to mechanical loading in bone. *J. Biol. Chem., 281*(42), 31720–31728.

Robling, A. G., Bellido, T., & Turner, C. H. (2006). Mechanical stimulation in vivo reduces osteocyte expression of sclerostin. *J. Musculoskelet. Neuronal Interact., 6*(4), 354.

Rubin, C. T., & Lanyon, L. E. (1984). Regulation of bone formation by applied dynamic loads. *J. Bone Joint Surg. Am., 66*(3), 397–402.

Rubin, C. T., & Lanyon, L. E. (1985). Regulation of bone mass by mechanical strain magnitude. *Calcif. Tissue Int., 37*(4), 411–417.

Sachs, F. (1991). Mechanical transduction by membrane ion channels: a mini review. *Mol. Cell Biochem., 104*, 57–60.

Santos, A., et al. (2010). Early activation of the beta-catenin pathway in osteocytes is mediated by nitric oxide, phosphatidyl inositol-3 kinase/Akt, and focal adhesion kinase. *Biochem. Biophys. Res. Commun., 391*(1), 364–369.

Schaller, M., et al. (1995). Focal adhesion kinase and paxillin bind to peptides mimicking beta integrin cytoplasmic domains. *J. Cell. Biol., 130*, 1181–1187.

Seliktar, D., Nerem, R. M., & Galis, Z. S. (2003). Mechanical strain-stimulated remodeling of tissue-engineered blood vessel constructs. *Tissue Eng., 9*(4), 657–666.

Seliktar, D., Nerem, R. M., & Galis, Z. S. (2001). The role of matrix metalloproteinase-2 in the remodeling of cell-seeded vascular constructs subjected to cyclic strain. *Ann. Biomed. Eng., 29*(11), 923–934.

Sharma, R., et al. (2002). Intracellular calcium changes in rat aortic smooth muscle cells in response to fluid flow. *Ann. Biomed. Eng., 30*(3), 371–378.

Sieg, D., Hauck, C., & Schlaepfer, D. (1999). Required role of focal adhesion kinase (FAK) for integrin-stimulated cell migration. *J. Cell Sci., 112*(16), 2677–2691.

Staal, F. J., et al. (2002). Wnt signals are transmitted through N-terminally dephosphorylated beta-catenin. *EMBO Rep., 3*(1), 63–68.

Takahashi, M., et al. (1997). Mechanotransduction in endothelial cells: temporal signaling events in response to shear stress. *J. Vasc. Res., 34*(3), 212–219.

Traub, O., & Berk, B. C. (1998). Laminar shear stress: mechanisms by which endothelial cells transduce an atheroprotective force. *Arterioscler. Thromb. Vasc. Biol., 18*(5), 677–685.

Turner, C. H., et al. (1994). Mechanical loading thresholds for lamellar and woven bone formation. *J. Bone Miner. Res., 9*(1), 87–97.

Turner, C. H., Owan, I., & Takano, Y. (1995). Mechanotransduction in bone: role of strain rate. *Am. J. Physiol., 269*(3 Pt 1), E438–E442.

Turner, C. H. (1998). Three rules for bone adaptation to mechanical stimuli. *Bone, 23*(5), 399–407.

Valhmu, W., et al. (1998). Load-controlled compression of articular cartilage induces a transient stimulation of aggrecan gene expression. *Arch. Biochem. Biophys., 353*(1), 29–36.

Whitfield, J. F. (2003). Primary cilium – is it an osteocyte's strain-sensing flowmeter? *J. Cell Biochem., 89*(2), 233–237.

Whitfield, J. F. (2008). The solitary (primary) cilium – a mechanosensory toggle switch in bone and cartilage cells. *Cell Signal, 20*(6), 1019–1024.

Wright, M., et al. (1996). Effects of intermittent pressure-induced strain on the electrophysiology of cultured human chondrocytes: evidence for the presence of stretch-activated membrane ion channels. *Clin. Sci., 90* (1), 61–71.

Wright, M., et al. (1997). Hyperpolarisation of cultured human chondrocytes following cyclical pressure-induced strain: evidence of a role for alpha 5 beta 1 integrin as a chondrocyte mechanoreceptor. *J. Ortho. Res., 15*(5), 742–747.

Wright, M., Stockwell, R., & Nuki, G. (1992). Response of plasma membrane to applied hydrostatic pressure in chondrocytes and fibroblasts. *Connect. Tissue Res., 28*(1–2), 49–70.

CHAPTER 29

Intelligent Surfaces for Cell-Sheet Engineering

Takanori Iwata[*,**], **Masayuki Yamato**[*], **Teruo Okano**[*]
* Institute of Advanced Biomedical Engineering and Science
** Department of Oral and Maxillofacial Surgery, Tokyo Women's Medical University,
Kawada-cho, Shinjuku-ku, Tokyo, Japan

INTRODUCTION

Tissue engineering, a concept proposed by Vacanti and Langer (Langer and Vacanti, 1993), was widely accepted in the field of regenerative medicine to create several kinds of tissues by the combination of cells, scaffolds, and growth factors. However, the limitations of this concept have been exposed due to graft rejections induced by inflammation during degradation of scaffolds, necrosis, or low stability of transplanted cells (Yang et al., 2005). To overcome these problems, we have developed a temperature-responsive polymer called N-isopropylacryla-mide (PIPAAm) for the surface of cell culture dishes (Yamada et al., 1990). At temperatures lower than 32°C, the lower critical solution temperature (LCST), PIPAAm is fully hydrated with an extended-chain conformation; however, at temperatures higher than LCST, PIPAAm is extensively dehydrated and compact. Cells generally adhere to hydrophobic surfaces, but not to hydrophilic surfaces. Our laboratory used PIPAAm to develop temperature-responsive culture dishes by grafting PIPAAm onto tissue culture-grade polystyrene dishes by irradiation with an electron beam (Okano et al., 1993). The advantage of this temperature-responsive dish is the possibility of harvesting intact cells and proteins with low-temperature treatment. Just by reducing the temperature to less than 32°C, cells spontaneously detach from the surface of dishes. Compared to retrieving cells with enzymes such as trypsin or dispase, loss of cell viability and degradation of surface proteins are minimized. Recent reports have clearly demonstrated that various kinds of single cells can also be harvested from temperature-responsive dishes with high functionality as well as intact proteins (Nakajima et al., 2001; Collier et al., 2002; Ishii et al., 2009). In addition, the preserved subcellular matrix proteins of harvested cell sheets provide the adhesive properties for stacking. Therefore, cell sheets can be stabilized at the recipient sites without any glues or sutures. It is also possible to stack cell sheets by layering multiple cell sheets to create thick tissues.

In this chapter, we will introduce the intelligent surface of PIPAAm and the practical applications of cell-sheet engineering for regenerative medicine.

INTELLIGENT SURFACE OF N-ISOPROPYLACRYLAMIDE (PIPAAm)

We have developed an intelligent cell culture surface that responds to small temperature changes to release cells. A temperature-responsive polymer, poly(N-isopropylacrylamide) (PIPAAm), is grafted onto commercial cell culture dish surfaces using electron beam irradia-tion (Okano et al., 1993). PIPAAm is a nanoscale material, which is not toxic for cells. This surface is slightly hydrophobic under cell culture conditions at 37°C, but readily becomes

517

Principles of Regenerative Medicine. DOI: 10.1016/B978-0-12-381422-7.10029-X

hydrated and hydrophilic below its LCST, 32°C. Thus, the attachment and detachment of cells on the culture surface can be controlled by simple temperature change. Cells can adhere, spread, and proliferate similarly to those on ungrafted tissue culture grade polystyrene surfaces at 37°C, and cells detach from the surface by reducing temperature below LCST, allowing cell harvest from the culture surfaces without the use of proteolytic enzymes. The applications of this technology enable the retrieval of cells as a sheet, for example keratinocytes (Yamato et al., 2001), corneal epithelial cell sheets (Nishida et al., 2004a), oral mucosal epithelial cells (Hayashida et al., 2005; Ohki et al., 2006), etc. These epithelial cells were multi-layered and preserved intact proteins such as E-cadherin and laminin 5 (Yamato et al., 2001), which were destroyed in dispase treatment. Epithelial cell sheets are now utilized for regenerative medicine (Nishida et al., 2004b; Ohki et al., 2009) as well as for the investigation of intact multi-layered epithelial cell sheets. Recent studies revealed epithelial cell sheets can be fabricated with temperature-responsive culture inserts without feeder layers (Murakami et al., 2006a,b) to exclude xenogeneic factors for animal-free cell transplantation.

CELL-SHEET ENGINEERING

To fabricate thick tissues, cell sheets fabricated with the intelligent surface can be applied because cell sheets connect with each other in a short time (Fig. 29.1A). A study showed that bilayer cardiomyocyte sheets were completely coupled 46 ± 3 minutes (mean $\pm$ SEM) after initial layering (Haraguchi et al., 2006), suggesting multi-layered cell sheets can communicate and synchronize as functional tissues. Based on this study, multi-layered transplantation was performed (Shimizu et al., 2006a). When more than three cardiomyocyte sheets were layered and transplanted into the subcutaneous space, the appearance of fibrosis and disordered vasculature indicated the presence of fibrotic areas within the transplanted laminar structures. Although the rapid establishment of microvascular networks occurred within the engineered tissues, this formation of new vessels was not able to rescue tissues with increased thicknesses above 80 μm. Multiple-step transplantation at one- or two-day intervals established the rapid neovascularization of the engineered myocardial tissues with more than 1 mm thickness (Shimizu et al., 2006a); these results directed us to fabricate prevascularized cell sheets. Recent studies showed that the combination of different type of cells, an endothelial cell sheet sandwiched with other types of cell sheets, induced prevascularization *in vitro*, which may allow the graft to survive. Three-dimensional manipulation of fibroblast cell sheets and micro-patterned endothelial cells with the gelatin-coated stacking manipulator induced a microvascular-like network within a 5-day culture *in vitro* (Tsuda et al., 2007). Non-patterned endothelial cell sheets and other types of cell sheets with a fibrin gel manipulator also induced prevascular networks both *in vitro* (Asakawa et al., 2010) and *in vivo* (Sasagawa et al., 2010) (Fig. 29.2).

TRANSPLANTATION IN ANIMAL MODELS

From the beginning of this century, various kinds of cells have been extracted, cultured on the temperature-responsive dishes, and fabricated as cell sheets. Transplantation was performed and the effectiveness of cell sheets was evaluated in most of the studies (Fig. 29.1B).

Corneal regeneration

Limbal stem-cell deficiency resulting from ocular trauma or diseases causes corneal opacification and visual loss. To recruit limbal stem cells, a novel cell-sheet manipulation technology using temperature-responsive culture surfaces was developed (Nishida et al., 2004a). The results showed multi-layered corneal epithelial sheets were successfully fabricated and their characteristics were similar to those of native tissues, and corneal surface reconstruction in rabbits was highly successful. For patients who suffer from unilateral limbal stem deficiency, corneal epithelial cell sheets can be cultured from autologous limbal stem cells. When the objective is the repair of bilateral corneal stem cell deficiency, and when the reduction of risks

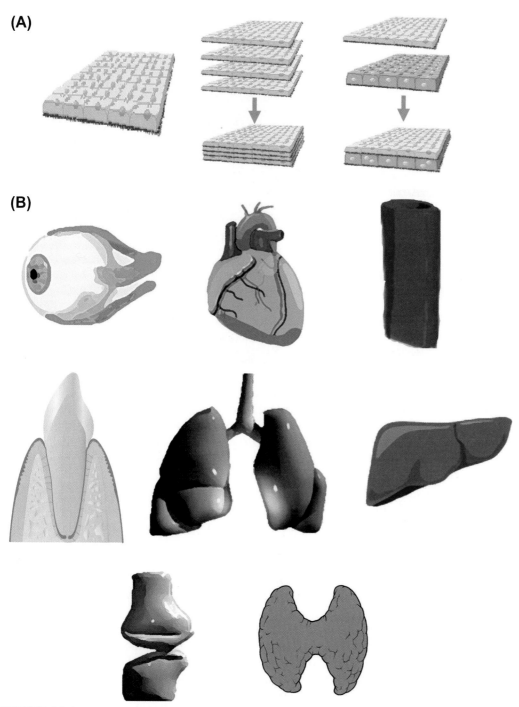

FIGURE 29.1

(A) Strategy of cell sheet engineering. Mono-layer of homogeneous cell sheet is suitable to reconstruct epithelial tissues (left panel). Multi-layered sheets of homogeneous cells are useful to create thick tissues (center panel). Multi-layered sheets with several types of cells might be effective to fabricate laminar structures such as liver and kidney (right panel). (B) Expansion of cell sheet engineering. Human clinical trials have been started in cornea, heart, and esophagus (top, left to right). Animal studies have revealed the efficacy of cell sheet engineering and clinical trials are expected in periodontal tissue (middle left), lung (middle center), liver (middle right), cartilage (low left), and thyroid (low right).

(A)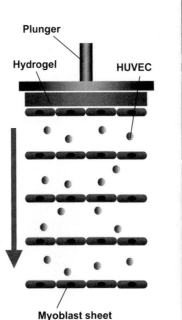

(B) **Four days after the co-culture**

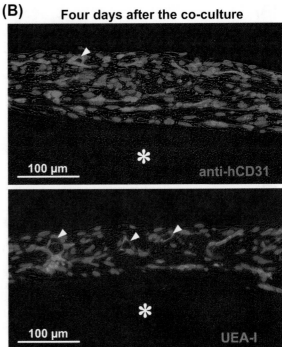

FIGURE 29.2
Prevascularization of five-layer myoblast sheets constructs in vitro. (A) Schematic illustration showing a five-layer myoblast sheet construct containing human umbilical vein endothelial cells (HUVECs) sandwiched between cell layers. (B) HUVECs in the five-layer myoblast sheet constructs were stained with anti-human CD31 antibody (the green color in the upper photograph) or UEA-I (the red color in the lower photograph). Nuclei were counterstained with Hoechst 33342 (the blue color). Four days after the co-culture, the HUVECs were networked through cell layers and formed capillary-like structures (the white arrowheads) in the sandwich constructs. Asterisk indicates the location of fibrin gel as supporting material. Reprinted from Sasagawa et al. (2010), Copyright © 2009, with permission from Elsevier.

while manipulating human eyes is desired, autologous oral mucosal epithelial cells are cultured to successfully obtain autologous oral mucosal epithelial cell sheets, which contain both cell-to-cell junctions and extra cellular matrix proteins, and which can be transplanted without the need for any carrier substrate or sutures. Therefore, oral mucosal epithelial sheets were examined as an alternative cell source to expand the opportunity for autologous transplantation (Hayashida et al., 2005). Autologous transplantation to rabbit corneal surfaces successfully reconstructed the corneal surface, with restoration of transparency. Four weeks after transplantation, epithelial stratification was similar to that in the corneal epithelium, although the keratin expression profile retained characteristics of the oral mucosal epithelium.

Cardiac regeneration

To enhance the function of cardiac tissue, neonatal rat cardiomyocyte sheets have been fabricated and their characteristics examined (Shimizu et al., 2002). When four sheets were layered, engineered constructs were macroscopically observed to pulse spontaneously. When they were transplanted into subcutaneous heart tissue-like structures, neovascularization within contractile tissues was observed. Long-term survival of pulsatile cardiac grafts was confirmed over up to 1 year (Shimizu et al., 2006b). The next study was performed to create thick tissue (Shimizu et al., 2006a). However, the thickness limit for layered cell sheets in subcutaneous tissue was ~80 µm (three layers). To overcome this limitation, repeated transplantation of triple-layer grafts was performed and multi-step transplantation created ~1 mm thick myocardium with a well-organized microvascular network. Other types of cell sheets were also examined in terms of improving cardiac function. Adipose-derived mesenchymal stem cells (Miyahara et al., 2006) and skeletal myoblasts (Hata et al., 2006; Kondoh et al., 2006; Hoashi et al., 2009) were transplanted as cell sheets and results showed the effectiveness in cardiac repairs. Recently, pre-vascularized cell sheets were created and neovascularization was observed in the multi-layered cell sheets as described above.

Cartilage regeneration

Chondrocyte sheets applied to cartilage regeneration have been prepared with the cell-sheet technique using temperature-responsive culture dishes. Layered chondrocyte sheets were able to maintain the phenotype of cartilage, and could be attached to the sites of cartilage damage, which acted as a barrier to prevent a loss of proteoglycan from these sites and to protect them from catabolic factors in the joint (Kaneshiro et al., 2006). Chondrocyte sheets with a consistent cartilaginous phenotype and adhesive properties were confirmed and may lead to a new strategy for cartilage regeneration (Mitani et al., 2009).

Esophageal regeneration

With the recent development of endoscopic submucosal dissection (ESD), large esophageal cancers can be removed with a single procedure, with few limits on the resectable range. However, after aggressive ESD, a major complication that arises is post-operative inflammation and stenosis that can considerably affect the patient's quality of life. Therefore, a novel treatment combining ESD and the endoscopic transplantation of tissue-engineered cell sheets created using autologous oral mucosal epithelial cells was examined in a canine model (Ohki et al., 2006). Results showed the effectiveness of a novel combined endoscopic approach for the potential treatment of esophageal cancers that can effectively enhance wound healing and possibly prevent post-operative esophageal stenosis.

Hepatocyte regeneration

Hepatic tissue sheets transplanted into the subcutaneous space resulted in efficient engraftment to the surrounding cells, with the formation of two-dimensional hepatic tissues that stably persisted for longer than 200 days (Ohashi et al., 2007). The engineered hepatic cell sheet also showed several characteristics of liver-specific functionality and bilayered sheets enhanced the effects more.

Fibroblast transplantation for sealing air leaks

In thoracic surgery, the development of post-operative air leaks is the most common cause of prolonged hospitalization. To seal the lung leakage, autologous fibroblast sheets have been put on the defects and showed to be effective in permanently sealing air leaks in a dynamic fashion (Kanzaki et al., 2007). Using almost the same procedures, pleural defects were also closed using fibroblast sheets (Kanzaki et al., 2008).

Mesothelial cells for prevention of post-operative adhesions

Post-operative adhesions often cause severe complications such as bowel obstruction and abdominopelvic pain. Mesothelial cell sheets have been examined to see whether they can prevent post-operative adhesions in a canine model (Asano et al., 2006). Mesothelial cells were harvested from tunica vaginalis (Asano et al., 2005) and cell sheets were fabricated on a fibrin gel. The results showed that mesothelial cell sheet is effective for preventing post-operative adhesion formation.

Periodontal regeneration

Periodontal regeneration has been performed with periodontal ligament cell sheets. First, the regeneration of periodontal ligament was observed both in canine and rat models (Akizuki et al., 2005; Hasegawa et al., 2005). Following these studies, culture condition was optimized and osteoinductive medium proved appropriate to regenerate thick cementum (Flores et al., 2008), and then complete regeneration was performed with the combination of cell sheets and β-tricalcium phosphate (Fig. 29.3) (Iwata et al., 2009).

Retinal pigment epithelial (RPE) cell regeneration

The retinal pigment epithelium (RPE) plays an important role in maintaining a healthy neural retina. RPE cell sheets have been fabricated and shown to be of monolayer structure, similar to the native RPE, with intact cell-to-cell junctions (Kubota et al., 2006). In a transplantation

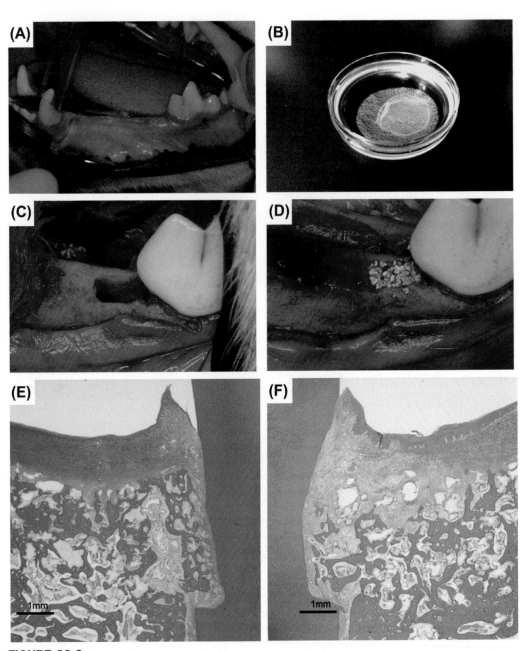

FIGURE 29.3
Periodontal regeneration by multi-layered periodontal ligament (PDL) cell sheets combined with β-tricalcium phosphate. All premolars were extracted 5 weeks before the transplantation. Spontaneous bone healing occurred (A) and canine periodontal cell sheets were detached and triple layered with a wet sheet of woven polyglycolic acid (PGA) (B). Three-wall infrabony defects (5 × 5 × 4 mm in depth, mesio-distal width, and bucco-lingual width, respectively) were created surgically on the mesial side of bilateral mandibular first molars (C). Root cementum was removed completely with curettes and conditioning with 19% EDTA was performed for 2 minutes to enhance the cell attachment. Three-layered PDL cell sheets supported by PGA sheets were trimmed to the size of defects, and applied to the exposed root surfaces in the experimental group, while only PGA sheets were applied in the control group. Infrabony defects were filled with porous β-tricalcium phosphate (D) in both groups. Results showed that functionally well-oriented periodontal fibers with newly formed bone were seen only in the experimental group (E). In contrast, limited bone formation with poor fibers was observed in the control group (F) (bar, 1 mm; Azan staining). Reprinted from Iwata et al. (2009), Copyright © 2009, with permission from Elsevier.

study, RPE cell sheets attached to the host tissues in the subretinal space more effectively than with the injection of isolated cell suspensions (Yaji et al., 2009).

Urothelial regeneration

Augmentation cystoplasty using gastrointestinal flaps may induce severe complications such as lithiasis, urinary tract infection, and electrolyte imbalance. The use of viable, contiguous urothelial cell sheets cultured *in vitro* should enable us to avoid these complications. Urothelial cell sheets have been created and their structures shown to be appropriate (Shiroyanagi et al., 2003). Urothelial cell sheets have been autografted onto dog demucosalized gastric flaps successfully, with no suture or fixation, generating a multi-layered urothelium *in vivo* (Shiroyanagi et al., 2004). The novel intact cell-sheet grafting method rapidly produced native-like epithelium *in vivo*.

Islet regeneration

To establish a novel approach for diabetes mellitus, pancreatic islet cell sheets have been fabricated and transplanted (Shimizu et al., 2009). Laminin-5 was coated on the temperature-responsive dishes to enhance the initial cell attachment and specific molecules, such as insulin and glucagon, were positive in the recipient site (Fig. 29.4).

Thyroid regeneration

The cells from rat thyroid have been spread on the temperature-responsive culture dishes, and cell sheets were created (Arauchi et al., 2009). Rats were exposed to total thyroidectomy as hypothyroidism models and received thyroid cell-sheet transplantation one week after total thyroidectomy. Transplantation of the thyroid cell sheets was able to restore thyroid function one week after the cell-sheet transplantation and the improvement was maintained.

CLINICAL SETTINGS

The three clinical trials listed below have already started in Japan.

Corneal reconstruction

The first clinical trial of cell-sheet engineering concerned corneal reconstruction using autologous mucosal epithelial cells and the results were published in 2004 (Nishida et al., 2004b). Oral mucosal tissue was harvested from four patients with bilateral total corneal stem cell deficiencies. Cells were then cultured for 2 weeks on mitomycin C-treated 3T3 feeder layer and transplanted directly to the denuded corneal surfaces without sutures. Results showed that complete re-epithelialization of the corneal surfaces occurred and the vision of all patients was restored. A clinical trial of the epithelial cell sheets for corneal reconstruction has also been carried out in France.

Endoscopic treatment of esophageal ulceration

Based on a canine model (Ohki et al., 2006), oral mucosal epithelial cell sheets are utilized for the endoscopic treatment of esophageal ulceration after endoscopic submucosal dissection. In a clinical study conducted by Tokyo Women's Medical University, favorable outcomes were observed in reconstruction of the esophageal epithelium, suppression of the inflammatory reaction, and prevention of esophageal stenosis (Ohki et al., 2009).

Improvement of left ventricular function in patients with dilated and ischemic cardiomyopathy

Appropriate therapies for severe heart diseases such as ischemic heart disease and dilated cardiomyopathy have long been sought; however, radical therapies have proved elusive. The regenerated cardiac patch consists of a cell sheet cultured from patient-derived cells

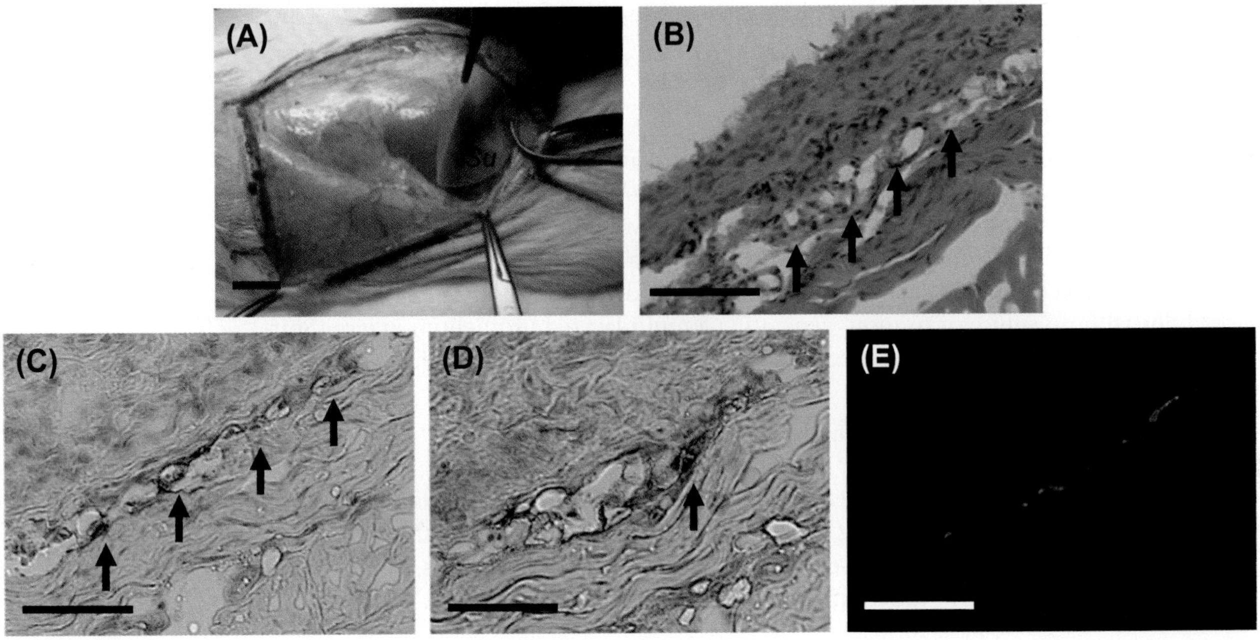

FIGURE 29.4

Transplantation of an engineered sheet of islet cells into the subcutaneous space. (A) A sheet of islet cells attached to a support membrane (Su) was transplanted into the subcutaneous space of the Lewis rat. Five minutes after transplantation, the islet cell sheet was found to be well attached to the surrounding tissue and the Su was removed. (B—D) Microscopic observation of the engineered monolayer sheet of islet cells (arrows) engrafted in the subcutaneous space after 7 days following the transplantation procedure. Transplanted tissues from Lewis rats were processed into 5 μm-thick sections, and either (B) stained for H&E or immunostained for insulin (C) and glucagon (D). (E) PKH26 red fluorescent cell membrane labeling for viable islet cells within the engineered islet sheet. Scale bar = 1 cm (A); 50 μm (B—E). Reprinted from Shimizu et al. (2009), Copyright © 2009, with permission from Elsevier.

(e.g. myoblasts collected from the patient's thigh) using temperature-responsive culture dishes. The patch is transplanted into the affected part of the heart, where the use of intact cell sheets allows transplanted cells to achieve a higher engraftment rate than by cell injection. The innovative technological advances embodied in regenerated cardiac patch therapy promise greater therapeutic benefits than could be obtained with conventional approaches. Osaka University is presently conducting a clinical trial of the patch for dilated cardiomyopathy. The first patient to receive a patch was successfully treated and discharged from the hospital without requiring a ventricular assistive device.

DISCUSSION AND COMMENTARY

Tissue engineering using biodegradable scaffolds and injection of cell suspension is limited and alternative strategies have recently been proposed. "Cell-sheet engineering" specifically refers to the application of temperature-responsive polymer to the surface of cell culture dishes to overcome the problems. Temperature-responsive dishes, which are now commercially available as UpCell™ Surface, enable the harvesting of cells without enzymes and permit the cell sheets to be readily manipulated, transferred, layered, or fabricated, because they adhere rapidly to other surfaces, cell sheets, and recipient sites. Cell-sheet engineering has already been tested in clinical settings such as corneal reconstruction (Nishida et al., 2004b), endoscopic treatment of esophageal ulceration (Ohki et al., 2009), and improvement of left ventricular function in patients with dilated and ischemic cardiomyopathy (Hoashi et al., 2009).

In this chapter, the principle of temperature-responsive dishes has been described. Knowledge of the nature of temperature-responsive dishes can assist in experiments. Strict control of temperature prevents unexpected detachment of cells. Cells should be confluent before transplantation, or cell sheets tend to be fragile. On the other hand, when cells are spread at low density, intact single cells with intelligent surfaces, such as microglia (Nakajima et al., 2001), osteoclasts (Ishii et al., 2009), and macrophages (Collier et al., 2002), can be harvested.

In addition, the applications of cell-sheet engineering for regenerative medicine have been mentioned. Various types of cells have been examined and most of them shown to have improved the functions of recipients, suggesting that cell-sheet engineering can be an alternative strategy for the therapy of tissue engineering (Yang et al., 2007). For the growing of cell-sheet engineering, peripheral devices have also been invented, such as a transporter of temperature-responsive dishes, which keeps the temperature at 36°C for >30 hours (Nozaki et al., 2008), and a cell-sheet transplantation device (Maeda et al., 2009).

The implementation of robotic systems that allow the safe mass production of sterile cell sheets automatically, as well as further collaboration between researchers and medical professionals, will make "cell-sheet engineering" the cutting edge solution for regenerative medicine (Elloumi-Hannachi et al., 2010).

References

Akizuki, T., Oda, S., Komaki, M., Tsuchioka, H., Kawakatsu, N., Kikuchi, A., et al. (2005). Application of periodontal ligament cell sheet for periodontal regeneration: a pilot study in beagle dogs. *J. Periodontal Res., 40*, 245–251.

Arauchi, A., Shimizu, T., Yamato, M., Obara, T., & Okano, T. (2009). Tissue-engineered thyroid cell sheet rescued hypothyroidism in rat models after receiving total thyroidectomy comparing with nontransplantation models. *Tissue Eng., Part A, 15*, 3943–3949.

Asakawa, N., Shimizu, T., Tsuda, Y., Sekiya, S., Sasagawa, T., Yamato, M., et al. (2010). Pre-vascularization of in vitro three-dimensional tissues created by cell sheet engineering. *Biomaterials, 31*(14), 3903–3909.

Asano, T., Takazawa, R., Yamato, M., Kageyama, Y., Kihara, K., & Okano, T. (2005). Novel and simple method for isolating autologous mesothelial cells from the tunica vaginalis. *BJU Int., 96*, 1409–1413.

Asano, T., Takazawa, R., Yamato, M., Takagi, R., Iimura, Y., Masuda, H., et al. (2006). Transplantation of an autologous mesothelial cell sheet prepared from tunica vaginalis prevents post-operative adhesions in a canine model. *Tissue Eng., 12*, 2629–2637.

Collier, T. O., Anderson, J. M., Kikuchi, A., & Okano, T. (2002). Adhesion behavior of monocytes, macrophages, and foreign body giant cells on poly (N-isopropylacrylamide) temperature-responsive surfaces. *J. Biomed. Mater. Res., 59*, 136–143.

Elloumi-Hannachi, I., Yamato, M., & Okano, T. (2010). Cell sheet engineering: a unique nanotechnology for scaffold-free tissue reconstruction with clinical applications in regenerative medicine. *J. Inter. Med., 267*, 54–70.

Flores, M. G., Yashiro, R., Washio, K., Yamato, M., Okano, T., & Ishikawa, I. (2008). Periodontal ligament cell sheet promotes periodontal regeneration in athymic rats. *J. Clin. Periodontol., 35*, 1066–1072.

Haraguchi, Y., Shimizu, T., Yamato, M., Kikuchi, A., & Okano, T. (2006). Electrical coupling of cardiomyocyte sheets occurs rapidly via functional gap junction formation. *Biomaterials, 27*, 4765–4774.

Hasegawa, M., Yamato, M., Kikuchi, A., Okano, T., & Ishikawa, I. (2005). Human periodontal ligament cell sheets can regenerate periodontal ligament tissue in an athymic rat model. *Tissue Eng., 11*, 469–478.

Hata, H., Matsumiya, G., Miyagawa, S., Kondoh, H., Kawaguchi, N., Matsuura, N., et al. (2006). Grafted skeletal myoblast sheets attenuate myocardial remodeling in pacing-induced canine heart failure model. *J. Thorac. Cardiovasc. Surg., 132*, 918–924.

Hayashida, Y., Nishida, K., Yamato, M., Watanabe, K., Maeda, N., Watanabe, H., et al. (2005). Ocular surface reconstruction using autologous rabbit oral mucosal epithelial sheets fabricated ex vivo on a temperature-responsive culture surface. *Invest. Ophthalmol. Vis. Sci., 46*, 1632–1639.

Hoashi, T., Matsumiya, G., Miyagawa, S., Ichikawa, H., Ueno, T., Ono, M., et al. (2009). Skeletal myoblast sheet transplantation improves the diastolic function of a pressure-overloaded right heart. *J. Thorac. Cardiovasc. Surg., 138*, 460–467.

Ishii, K. A., Fumoto, T., Iwai, K., Takeshita, S., Ito, M., Shimohata, N., et al. (2009). Coordination of PGC-1beta and iron uptake in mitochondrial biogenesis and osteoclast activation. *Nat. Med., 15*, 259–266.

Iwata, T., Yamato, M., Tsuchioka, H., Takagi, R., Mukobata, S., Washio, K., et al. (2009). Periodontal regeneration with multi-layered periodontal ligament-derived cell sheets in a canine model. *Biomaterials, 30*, 2716–2723.

Kaneshiro, N., Sato, M., Ishihara, M., Mitani, G., Sakai, H., & Mochida, J. (2006). Bioengineered chondrocyte sheets may be potentially useful for the treatment of partial thickness defects of articular cartilage. *Biochem. Biophys. Res. Commun., 349*, 723–731.

Kanzaki, M., Yamato, M., Yang, J., Sekine, H., Kohno, C., Takagi, R., et al. (2007). Dynamic sealing of lung air leaks by the transplantation of tissue engineered cell sheets. *Biomaterials, 28,* 4294–4302.

Kanzaki, M., Yamato, M., Yang, J., Sekine, H., Takagi, R., Isaka, T., et al. (2008). Functional closure of visceral pleural defects by autologous tissue engineered cell sheets. *Eur. J. Cardiothorac. Surg., 34,* 864–869.

Kondoh, H., Sawa, Y., Miyagawa, S., Sakakida-Kitagawa, S., Memon, I. A., Kawaguchi, N., et al. (2006). Longer preservation of cardiac performance by sheet-shaped myoblast implantation in dilated cardiomyopathic hamsters. *Cardiovasc. Res., 69,* 466–475.

Kubota, A., Nishida, K., Yamato, M., Yang, J., Kikuchi, A., Okano, T., et al. (2006). Transplantable retinal pigment epithelial cell sheets for tissue engineering. *Biomaterials, 27,* 3639–3644.

Langer, R., & Vacanti, J. P. (1993). Tissue engineering. *Science, 260,* 920–926.

Maeda, M., Yamato, M., Kanzaki, M., Iseki, H., & Okano, T. (2009). Thoracoscopic cell sheet transplantation with a novel device. *J. Tissue Eng. Regen. Med., 3,* 255–259.

Mitani, G., Sato, M., Lee, J. I., Kaneshiro, N., Ishihara, M., Ota, N., et al. (2009). The properties of bioengineered chondrocyte sheets for cartilage regeneration. *BMC Biotechnol., 9,* 17.

Miyahara, Y., Nagaya, N., Kataoka, M., Yanagawa, B., Tanaka, K., Hao, H., et al. (2006). Monolayered mesenchymal stem cells repair scarred myocardium after myocardial infarction. *Nat. Med., 12,* 459–465.

Murakami, D., Yamato, M., Nishida, K., Ohki, T., Takagi, R., Yang, J., et al. (2006a). Fabrication of transplantable human oral mucosal epithelial cell sheets using temperature-responsive culture inserts without feeder layer cells. *J. Artif. Organs, 9,* 185–191.

Murakami, D., Yamato, M., Nishida, K., Ohki, T., Takagi, R., Yang, J., et al. (2006b). The effect of micropores in the surface of temperature-responsive culture inserts on the fabrication of transplantable canine oral mucosal epithelial cell sheets. *Biomaterials, 27,* 5518–5523.

Nakajima, K., Honda, S., Nakamura, Y., Lopez-Redondo, F., Kohsaka, S., Yamato, M., et al. (2001). Intact microglia are cultured and non-invasively harvested without pathological activation using a novel cultured cell recovery method. *Biomaterials, 22,* 1213–1223.

Nishida, K., Yamato, M., Hayashida, Y., Watanabe, K., Maeda, N., Watanabe, H., et al. (2004a). Functional bioengineered corneal epithelial sheet grafts from corneal stem cells expanded ex vivo on a temperature-responsive cell culture surface. *Transplantation, 77,* 379–385.

Nishida, K., Yamato, M., Hayashida, Y., Watanabe, K., Yamamoto, K., Adachi, E., et al. (2004b). Corneal reconstruction with tissue-engineered cell sheets composed of autologous oral mucosal epithelium. *N. Engl. J. Med., 351,* 1187–1196.

Nozaki, T., Yamato, M., Inuma, T., Nishida, K., & Okano, T. (2008). Transportation of transplantable cell sheets fabricated with temperature-responsive culture surfaces for regenerative medicine. *J. Tissue Eng. Regen. Med., 2,* 190–195.

Ohashi, K., Yokoyama, T., Yamato, M., Kuge, H., Kanehiro, H., Tsutsumi, M., et al. (2007). Engineering functional two- and three-dimensional liver systems in vivo using hepatic tissue sheets. *Nat. Med., 13,* 880–885.

Ohki, T., Yamato, M., Murakami, D., Takagi, R., Yang, J., Namiki, H., et al. (2006). Treatment of oesophageal ulcerations using endoscopic transplantation of tissue-engineered autologous oral mucosal epithelial cell sheets in a canine model. *Gut, 55,* 1704–1710.

Ohki, T., Yamato, M., Ota, M., Murakami, D., Takagi, R., Kondo, M., et al. (2009). Endoscopic transplantation of human oral mucosal epithelial cell sheets – world's first case of regenerative medicine applied to endoscopic treatment. *Gastrointes. Endosc., 69,* AB253–AB254.

Okano, T., Yamada, N., Sakai, H., & Sakurai, Y. (1993). A novel recovery system for cultured cells using plasma-treated polystyrene dishes grafted with poly(N-isopropylacrylamide). *J. Biomed. Mater. Res., 27,* 1243–1251.

Sasagawa, T., Shimizu, T., Sekiya, S., Haraguchi, Y., Yamato, M., Sawa, Y., et al. (2010). Design of prevascularized three-dimensional cell-dense tissues using a cell sheet stacking manipulation technology. *Biomaterials, 31*(7), 1646–1654.

Shimizu, H., Ohashi, K., Utoh, R., Ise, K., Gotoh, M., Yamato, M., et al. (2009). Bioengineering of a functional sheet of islet cells for the treatment of diabetes mellitus. *Biomaterials, 30,* 5943–5949.

Shimizu, T., Sekine, H., Yang, J., Isoi, Y., Yamato, M., Kikuchi, A., et al. (2006a). Polysurgery of cell sheet grafts overcomes diffusion limits to produce thick, vascularized myocardial tissues. *FASEB J., 20,* 708–710.

Shimizu, T., Sekine, H., Isoi, Y., Yamato, M., Kikuchi, A., & Okano, T. (2006b). Long-term survival and growth of pulsatile myocardial tissue grafts engineered by the layering of cardiomyocyte sheets. *Tissue Eng., 12,* 499–507.

Shimizu, T., Yamato, M., Isoi, Y., Akutsu, T., Setomaru, T., Abe, K., et al. (2002). Fabrication of pulsatile cardiac tissue grafts using a novel 3-dimensional cell sheet manipulation technique and temperature-responsive cell culture surfaces. *Circ. Res., 90,* e40.

Shiroyanagi, Y., Yamato, M., Yamazaki, Y., Toma, H., & Okano, T. (2003). Transplantable urothelial cell sheets harvested noninvasively from temperature-responsive culture surfaces by reducing temperature. *Tissue Eng., 9,* 1005–1012.

Shiroyanagi, Y., Yamato, M., Yamazaki, Y., Toma, H., & Okano, T. (2004). Urothelium regeneration using viable cultured urothelial cell sheets grafted on demucosalized gastric flaps. *BJU Int.*, *93*, 1069–1075.

Tsuda, Y., Shimizu, T., Yamato, M., Kikuchi, A., Sasagawa, T., Sekiya, S., et al. (2007). Cellular control of tissue architectures using a three-dimensional tissue fabrication technique. *Biomaterials*, *28*, 4939–4946.

Yaji, N., Yamato, M., Yang, J., Okano, T., & Hori, S. (2009). Transplantation of tissue-engineered retinal pigment epithelial cell sheets in a rabbit model. *Biomaterials*, *30*, 797–803.

Yamada, N., Okano, T., Sakai, H., Karikusa, F., Sawasaki, Y., & Sakurai, Y. (1990). Thermo-responsive polymeric surfaces; control of attachment and detachment of cultured cells. *Die Makromolekulare Chemie, Rapid Communications*, *11*, 571–576.

Yamato, M., Utsumi, M., Kushida, A., Konno, C., Kikuchi, A., & Okano, T. (2001). Thermo-responsive culture dishes allow the intact harvest of multilayered keratinocyte sheets without dispase by reducing temperature. *Tissue Eng.*, *7*, 473–480.

Yang, J., Yamato, M., Kohno, C., Nishimoto, A., Sekine, H., Fukai, F., et al. (2005). Cell sheet engineering: recreating tissues without biodegradable scaffolds. *Biomaterials*, *26*, 6415–6422.

Yang, J., Yamato, M., Shimizu, T., Sekine, H., Ohashi, K., Kanzaki, M., et al. (2007). Reconstruction of functional tissues with cell sheet engineering. *Biomaterials*, *28*, 5033–5043.

CHAPTER 30

Applications of Nanotechnology for Regenerative Medicine

Benjamin S. Harrison, Sirinrath Sirivisoot
Wake Forest Institute for Regenerative Medicine, Wake Forest University, Medical Center BLVD, Winston-Salem, NC, USA

INTRODUCTION

While organ transplants have provided renewed life in individuals with failed organs, the reality is that the demand of organs far exceeds the available supply. The potential ability to build organs is therefore an attractive option to fill the deficit in the supply of organs available. In constructing regenerative therapies, there will be a need to develop new tools to aid in the engineering of neo-organs. From chemistry, physics, and biology disciplines has emerged the field of nanotechnology, which has the potential to provide the tools needed to accelerate the engineering of organs.

Nanotechnology is a bottom-up approach that focuses on assembling simple elements to form complex structures. According to the United States National Nanotechnology Initiative, nanotechnology is broadly defined as "the understanding and control of matter at dimensions of roughly 1 to 100 nanometers, where unique phenomena enable novel applications." Nanomaterials are those with at least one dimension in nanometer scale. Nanotechnology can be understood as a technology of design, fabrication, and applications of nanostructures and nanomaterials, as well as the fundamental understanding of its physical properties and phenomena (Cao, 2004). At the nanometer scale, where many biological processes operate, nanotechnology can provide the tools to probe and even direct these biological processes. Thus, nanotechnology could potentially repair damaged parts, cure diseases, and even actively monitor and respond to the needs of the body. The broad potential of nanotechnology is owed to the fact that cells and the extracellular matrix possess a multitude of nanodimensionality, which affects cell behaviors (e.g. adhesion, proliferation, differentiation). Cells, typically microns in diameter, are composed of numerous nanosized components all working together to create a highly organized, self-regulating machine. For example, the cell surface is composed of ion channels that regulate the coming and going of ions such as calcium and potassium in and out of the cell. Enzyme reactions, protein dynamics, and DNA all possess some aspect of nanodimensionality. These nanodimensional components control how cells produce the extracellular matrix (ECM) including its composition and architecture. The extracellular matrix that cells interact with also abounds with nanosized features that influence the behaviors of other cells and tissues. These nanosized features, such as fiber diameter and pores, in concert with the intrinsic properties of the matrix itself, control the mechanical strength, the adhesiveness of the cells to the matrix, cell proliferation, and the shape of the ECM.

Principles of Regenerative Medicine. DOI: 10.1016/B978-0-12-381422-7.10030-6

529

Nanomaterials allow for different functional components to be contained together in a single unit. For example, therapeutic, targeting, contrast, and/or bio-compatibilizing components can be combined. A description of the different components of a nanocarrier can be found in Table 30.1. These components can be added or removed to create the desired effect without necessarily compromising the overall function of the particle. This is inherently different from the high costs approach to drug development, where a small change in molecular structure can dramatically influence the pharmokinetics and even potency of the drug.

The ability to readily combine different components into a small physical space is not the only advantage of nanoscale materials. For example, quantum effects become more prominent at the nanoscale, which can result in high optical absorptivities, large photostabilities, or unusual magnetic properties that can be used to enhance cellular imaging (Zhang et al., 2002; Medintz et al., 2005). Besides imaging, these quantum effects can allow for novel methods of drug delivery-triggered light, electric, or magnetic fields (Yuan et al.). Therefore, there is great potential for using quantum dots in imaging, diagnostics, therapy, bioconjugation, and drug delivery for regenerative medicine.

Nanotechnology's impact on regenerative medicine will be through development of multi-functional tools to enhance effectiveness of implants, cell therapies, and tissue engineering. Since nanotechnology is at the interface of modern physical science and medicine, new and unconventional ideas will be developed, capable of bringing about major revolutions in science and medicine. Therapies developed using nanotechnology could someday minimize or eliminate the side-effects of drugs through targeted delivery and will provide real-time, and

TABLE 30.1	A typical Nanocarrier of Image Contrast and/or Therapeutic Agents is Composed of Six Components
Binder	All the different components are held together using a binder. The binder may be an inert piece of the nanocarrier; however, it also often serves another purpose. The binder may also be the image contrasting agents. For example, iron nanoparticles and quantum dots serve as the core for the attachment of the other components. Polymers such as polyglycolic acid may serve as the binder of the therapeutic but also the biocompatiblizing agent.
Biocompatibilization	This component makes the nanocarrier compatible with the biological environment. It does this by minimizing aggregation of the nanocarrier and increases the lifetime of the carrier by avoiding the defense mechanisms of the biological systems such as the reticuloendothelial system.
Imaging contrast	This component provides the means for imaging modalities to observe the nanocarrier. These contrasting agents may be observed using optical, magnetic, ultrasound, and scintillating methods.
Sensor	The sensor or trigger is used to alter the behavior of the nanocarrier once it has been deployed. For example, near-infrared light or electromagnetic radiation may be used to accelerate the release of a therapeutic or cause rapid localized heating as part of a therapy. Chemical sensors such as polymers that are pH or ion sensitive may also provide feedback to the nanocarrier in delivery of its payload.
Targeting	This component provides the means of driving the nanocarrier to its desired location. There are two types of targeting: passive and active. Passive targeting incorporates only non-specific targeting agents, which may be useful for determining microenvironment permeability or areas of increased angiogenesis. Active targeting uses ligands or antibodies that bind to specific receptors at the target site. Active targeting aids in obtaining higher concentrations of therapeutics and contrasting agents at the desired site. Also, multiple targeting agents can be bound to the nanocarrier, allowing lower binding affinity molecules to be used to increase binding probabilities.
Therapeutics	Bioactive agents such as drugs or DNA are typical payloads of the nanocarrier. Drugs that are incapable of penetrating cellular membranes or hydrophobic drugs that cannot be administered systemically by themselves can be contained within the nanocarrier awaiting release in a controlled manner. Other novel properties of nanoparticles have also shown promise as hyperthermic agents.

even non-invasive, monitoring of the disease and tissue repair. In this chapter we will examine the impact nanotechnology will have on regenerative medicine related to cellular therapies and biomaterial control, which play an important role for implant design and tissue engineering.

NANOTECHNOLOGY AS A MULTI-FUNCTIONAL TOOL FOR CELL-BASED THERAPIES

There is much excitement driving research into cell-based therapies to regenerate tissue function. However, many questions still remain as to how the cells behave once placed *in vivo*. One way to better understand how cells behave *in vivo* is through the use of nanoparticles as a multi-functional tool to improve monitoring or even potentially modify cell behavior. For example, with the enormous self-repair potential of stem cells, it is important to be able to locate, recruit, and signal these cells to begin the regeneration process.

Improving non-invasive monitoring methods is particularly desirable since current methods of evaluating cell treatments typically involve destructive or invasive techniques such as tissue biopsies. Traditional non-invasive methods such as MRI and PET, which rely heavily on contrast agents, lack the specificity or resident time to be a viable option for cell tracking. However, *in vitro* and *in vivo* visualization of nanoscale systems can be carried out using a variety of clinically relevant modalities such as fluorescence microscopy, single photon emission computed tomography (SPECT), positron emission tomography (PET), magnetic resonance imaging (MRI), four-photon microscopy, near-infrared surface-enhanced Raman scattering, X-ray fluorescence micro- and nano-probe imaging, coherent X-ray diffraction imaging, ultrasound, and radiotracing such as gamma scintigraphy (Hong et al., 2009). Nanoparticulate imaging probes include semiconductor quantum dots, magnetic and magnetofluorescent nanoparticles, gold nanoparticles, and nanoshells, among others. Nanoparticles that are novel intravascular or cellular probes are being developed for diagnostic (imaging) and therapeutic (drug and gene delivery) purposes (Fig. 30.1) (Heller et al., 2005). There are a growing number of nanomaterials being used to probe aspects of the tissue regeneration process, such as monitoring angiogenesis (Winter et al., 2003), apoptosis (Jung et al., 2004; Sosnovik et al., 2009), and tissue viability (Sosnovik et al., 2009). These nanoparticles can play a critical role in future regenerative medicine, especially in the areas of target-specific drug and gene delivery.

Quantum dots (QDs) are one class of nanomaterial that is receiving special attention. QD are inorganic nanocrystals that possess physical dimensions between 2 and 10 nanometers, composed of a core of a semiconductor material. Quantum dots are tunable in a broad spectrum of colors by varying particle size or composition. They also possess strong and narrow symmetrical emission spectra and usually have high photochemical stability. The emission wavelength is controlled by the size of the nanocrystal and can be tuned throughout the visible spectrum to the near-infrared region ($>670\,$nm). Early live cell experiments using fluorescent quantum dots sparked interest in using nanoparticles for immunocytochemical and immunohistochemical assays as well as for cell tracking (Akerman et al., 2002; Tokumasu and Dvorak, 2003; Sukhanova et al., 2004). A significant advantage of quantum dots is their increased photostability (typically 10–1,000 times more stable) compared to organic dyes. This allows quantum dots and the cells or proteins attached to them to be tracked over longer periods of time. Tumor cells labeled with QDs have been intravenously injected into mice and successfully followed using fluorescence microscopy (Gao et al., 2004; Voura et al., 2004). As passive imaging agents, quantum dots can be used for imaging microvascularity in animals since PEG-coated quantum dots injected into mice have shown good tissue perfusion and appear to be biocompatible (Ballou et al., 2004). Moreover, the capability to modify both surface chemistry and size of quantum dots allows for multiple targeting analyses within the same cell. Quantum dots can be functionalized with biomolecules, which interact with

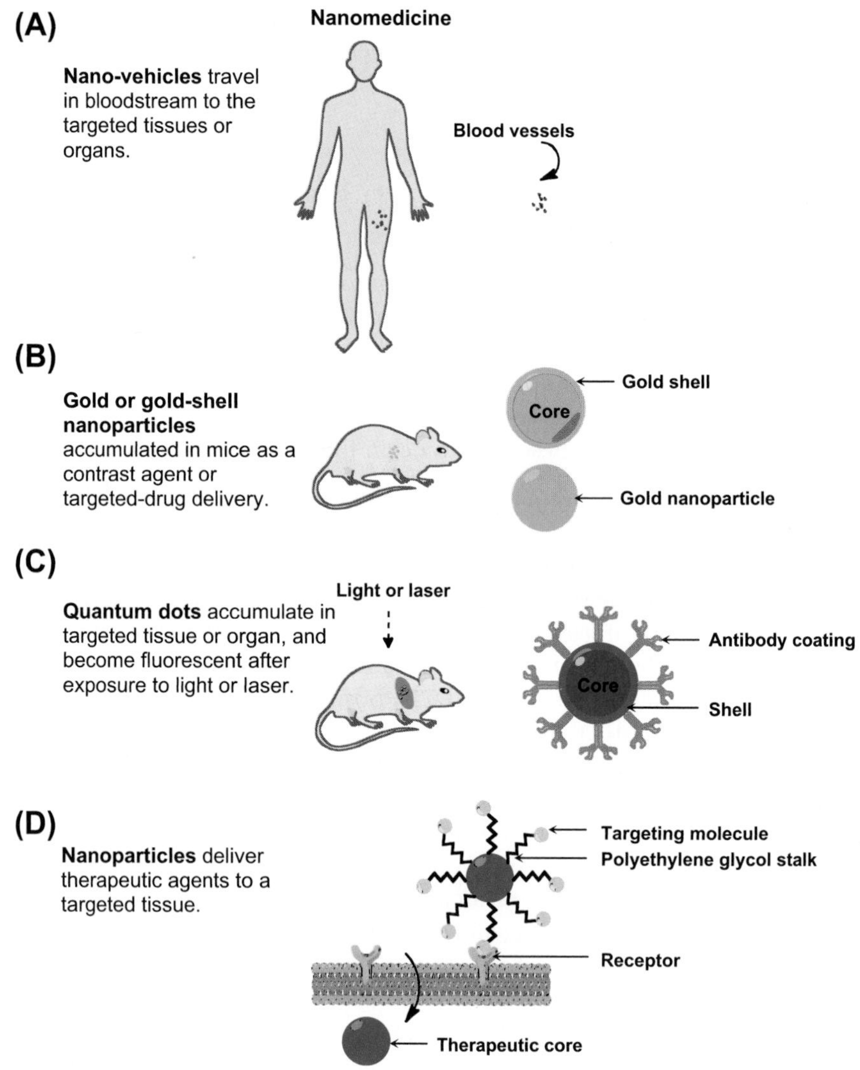

(A) Nanomedicine

Nano-vehicles travel in bloodstream to the targeted tissues or organs.

Blood vessels

(B)

Gold or gold-shell nanoparticles accumulated in mice as a contrast agent or targeted-drug delivery.

Gold shell

Core

Gold nanoparticle

(C)

Light or laser

Quantum dots accumulate in targeted tissue or organ, and become fluorescent after exposure to light or laser.

Antibody coating

Core

Shell

(D)

Nanoparticles deliver therapeutic agents to a targeted tissue.

Targeting molecule
Polyethylene glycol stalk

Receptor

Therapeutic core

FIGURE 30.1

Nanomedicine overview. (A) Nanostructured vehicles reach targeted tissues through a bloodstream. (B) Nanoshells or gold nanoparticles can be used as contrast agents for medical imaging and as vehicles for drug and gene delivery. (C) Quantum dot with a semiconductor nanocrystal core emits fluorescent light. (D) Polyethylene glycol-nanoparticles functionalized with targeting molecules deliver therapeutics via receptors and membranes. Modified from Kateb et al. (2007).

532

a biological entity through electrostatic or hydrogen bonding, to suit their targets. For example, when quantum dots are coated with trimethoxysilylpropyl urea and acetate groups, they have shown their ability to bind with the nuclear membrane (Bruchez et al., 1998). Wu et al. used quantum dots to image cell surface markers (Her2), cytoplasmic proteins (actin and microtubules), and nuclear antigens (Wu et al., 2003). CdSe-CdS core-shell nanocrystals linked covalently with biotin have been used as the secondary antibody, binding F-actin filaments in 3T3 mouse fibroblasts that were labeled with phalloidin-biotin and streptavidin (Bruchez et al., 1998). Quantum dots represent just one novel class of nanomaterials whose ability to aid in imaging cells could help develop better regenerative therapies.

Other nanoparticles are showing promise for optical cell tracking and imaging. For instance, nanosized tubes of carbon known as carbon nanotubes possess optical transitions in the near-infrared that can be used for tracking cells. However, unlike quantum dots, which are typically composed of heavy metals such as cadmium, carbon nanotubes are made of carbon, an abundant element in nature. Carbon nanotubes possess large aspect ratios with nanometer diameters and lengths ranging from submicrons to millimeters. These tubes can contain a single wall of carbon or multiple walls (typically 3–10) of carbon, commonly called single wall carbon nanotubes (SWNTs) or multi-wall carbon nanotubes (MWNTs), respectively. The versatile chemistry of carbon nanotubes in combination with their intrinsic optical properties can lead to a multifunctional nanoplatform for multimodality molecular imaging and therapy (Fig. 30.2) (Hong et al., 2009).

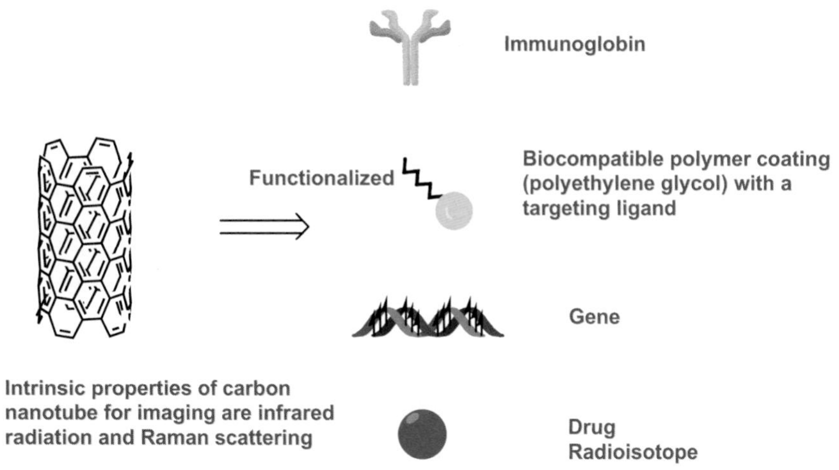

 Immunoglobin

 Functionalized

Biocompatible polymer coating
(polyethylene glycol) with a
targeting ligand

 Gene

**Intrinsic properties of carbon
nanotube for imaging are infrared
radiation and Raman scattering**

Drug
Radioisotope

FIGURE 30.2
Multifunctional carbon nanotube-based platform
functionalized with antibody, polymer coating, ligand,
drug, gene, and radioisotope for a multimodality
imaging and multiple therapeutic delivery.

The infrared spectrum between 900 and 1300 nm is an important optical window for biomedical applications because of the lower optical absorption (greater penetration depth of light) and small auto-fluorescent background. Like quantum dots, carbon nanotubes possess good photostability and can be imaged over long periods of time using Raman scattering and fluorescence microscopy. Single-wall carbon nanotubes dispersed in a Pluorinc surfactant can be readily imaged through fluorescence microscopy after being ingested by mouse peritoneal macrophage-like cells. The small size of the SWNT makes it possible for 70,000 nanotubes to be ingested, where they can remain stable for weeks inside 3T3 fibroblasts and murine myoblast stem cells (Cherukuri et al., 2004; Heller et al., 2005). Having such a high concentration of carbon nanotubes within a cell without distributing the cell behavior means such probes could be used for studying cell proliferation and stem cell differentiation, even through repeated cells. While such nanomaterials have yet to reach clinical applications, it does show the potential for non-invasive optical imaging.

Still, there is much to learn about how carbon nanotubes interact with cells. For example, it has been shown that SWNTs and double-walled carbon nanotubes can trigger immunological responses (Salvador-Morales et al., 2006). However, MWNTs reportedly do not result in proliferative or cytokine changes *in vitro*. Some studies have shown that the size and composition of carbon nanotubes must be carefully controlled to promote their biocompatibility and to prevent body immune reaction (Kateb et al., 2007). Brown et al. suggested that MWNTs enter the cell by either mechanically translocating through the lipid bilayer or by incomplete or frustrated phagocytosis, which occurs when MWNTs are too large for the cell to phagocytose (Brown et al., 2007). It has been suggested that if MWNTs penetrate into cell membranes they can promote reactive oxygen species (ROS) generation and cause cell death (Nel et al., 2006). Others have shown that, if MWNTs enter via incomplete phagocytosis, the cell can release digestive enzymes from the phagosome into extracellular regions and cause chronic inflammation (Zeidler-Erdely et al., 2006). It is unclear with certainty what properties of carbon nanotubes will have the most impact on biological systems whether it is their chemical structure, length and aspect ratio, surface area, degree of aggregation, extent of oxidation, surface topology, bound functional groups, and catalyst residues/produced impurities (Foldvari and Bagonluri, 2008). For instance, heparinizing CNT reduced or eliminated complement activation (Murugesan et al., 2006). Other reports imply that the accumulation of carbon nanotubes within cells depends on the functional groups and the functionalization degree (Wang et al., 2004; Lacerda et al., 2008; Schipper et al., 2008; Yang et al., 2008). Even though understanding how carbon nanotubes interact with cells is still in its infancy, these examples illustrate that nanomaterials have the potential for large impact in cell biology and by extension regenerative medicine.

Along with optical contrast agents, magnetic nanoparticles have been used to track cells and report on cell behavior. Many nanoparticle contrasting agents are based on superparamagnetic iron oxide nanoparticles and some have already been approved as clinical MRI contrast agents. When placed into a magnetic field, magnetic nanoparticles create perturbations of the external field that significantly reduce the spin-spin relaxation time (T2) of the nearby environment generating MR contrast. Typically, these probes consist of a magnetic iron oxide core surface functionalized with an agent, such as dextran or other polymers, to prevent aggregation and to enhance stability and solubility. Sizes of these particles can range from one nanometer to hundreds of nanometers in diameter. Magnetic iron oxide nanoparticles and their composites are emerging as novel contrast agents for MRI and are much more sensitive than conventional gadolinium-based contrast agents (Chemaly et al., 2005).

When used in conjunction with HIV-Tat and polyArginine peptides, these particles are readily taken up by many cell types (Dodd et al., 2001; Zhao et al., 2002). For example, super-paramagnetic iron oxide (SPIO)-labeled rat mesenchymal stem cells injected into rats could be imaged and tracked to the liver and kidneys (Bos et al., 2004). Another example of a composite nanoparticle is the triple-labeled (magnetic, fluorescent, and isotope) SPIO that can be readily internalized by hematopoietic stem and neural progenitor cells and not affect their potential for viability, proliferation, or differentiation (Lewin et al., 2000). A third example of functionalized nanoparticles for imaging includes using an antitransferrin receptor monoclonal antibody-functionalized nanoparticle to label oligodendrocyte progenitor cells by targeting the transferring receptors on the cells (Bulte et al., 1999). The oligodendrocyte progenitor cells, which as shown previously significantly myelinate a large area in the central nervous system (Duncan and Milward, 1995), were transplanted into the spinal cord of myelin-deficient rats. Since they were labeled with nanoparticles, they could be tracked easily using MRI and the extent of myelination could be determined.

Apoptosis is commonly detected by using the binding of annexin V to externalized phosphatidylserine. This binding event is the basis of optical and radiolabel methods for detecting apoptotic cells and can be bound to iron nanoparticles for sensing using MRI. It has been demonstrated that tumor-bearing mice injected with SPIO particles bearing apoptotic sensing proteins showed a sharp decrease in the $T2*$ weight image corresponding to the location of the tumor (Zhao et al., 2001). This demonstrated that nanomaterials can be used to create high-specificity MRI contrast agents for apoptotic cells. Such results are encouraging because they show that nanomaterials can be used not only for imaging the physical location of cells but also to provide information on the biological state of cells.

While MRI has revolutionized our way of visualization *in vivo*, allowing cells to be tracked non-invasively, it is difficult to quantify the MRI signals and provide real quantification of cell numbers. The difficulty arises because MRI contrasting agents that are based on paramagnetic gadolinium and iron metals are not directly detected by the scanner but are indirectly detected by their influence on surrounding water molecules. However, the use of perfluoronated nanoparticles has recently been shown to be a new way to provide quantitative numbers to MRI since the fluorine nuclei (^{19}F) can be directly detected (Morawski et al., 2004; Ahrens et al., 2005). Since endogenous fluorine is negligible in the body, ^{19}FMRI is capable of directly detecting fluorine against a dark background similarly to radiotracers and fluorescent dyes. While this has been demonstrated with dendritic cells, similar results should be obtainable using other cell types.

Besides imaging enhancements, nanotechnology can produce carriers for delivery of therapeutics for aiding the regeneration process. For example, biodegradable nanoparticles can deliver drugs, growth factors, and other bioactive agents to cells and tissue (Panyam and Labhasetwar, 2003). Nanomaterials can be used as immensely powerful tools for gene delivery in specific differentiation of stem cells. Gold nanoparticles (20 nm in diameter) conjugated with a DNA-poly-ethylenimine complex were patterned on a solid surface (glass) and used as nanoscaffolds for the delivery of DNA into hMSCs through reverse transfection (Uchimura et al., 2007). The development of safe and efficient gene delivery systems, which can lead to high levels of gene expression within stem cells, is a strong indicator for the effective implementation of regenerative therapies (Solanki et al., 2008). Nanodelivery vehicles possess three distinct advantages over conventional drug delivery methods. First, nanoparticles, due to their small size, are able to bypass biological barriers such as cell membranes and the blood brain barrier (BBB), allowing greater concentrations of therapeutics to be delivered. Second, nanocarriers can be functionalized with active targeting agents to allow selective delivery of bioactive agents. Third, drug delivery systems can incorporate nanotriggers for non-invasive delivery of therapeutic agents. These sensitive triggers can be activated using *in vivo* signals such as pH, ion concentration, and temperature or external sources such as near-infrared light, ultrasound, and magnetic fields.

Nanotechnology can provide powerful new tools for non-invasive tracking of cells in engineered tissues. As was also mentioned at the outset, the real benefits of nanotechnology are the multifunctional tools that it can bring. As nanotechnology progresses, new nanomaterials and techniques are being developed regarding cellular imaging and drug delivery that will better equip those practicing regenerative medicine to reach their goals. Cellular therapies for regenerative medicine would benefit from nanotechnology since tracking of implanted cells would provide the means to better evaluate the viability of engineered tissues and help in understanding the biodistribution and migration pathways of transplanted cells. Nanotechnology would also allow better and more intelligent control of the bioactive factors which can influence cellular therapies. The potential of nanotechnology for impacting regenerative medicine is great, creating the hope of individualized and targeted therapies.

NANOTECHNOLOGY AS A MULTI-FUNCTIONAL TOOL FOR BIOMATERIAL CONTROL

Biomaterials play an important role in regenerative medicine because they make up a large component of implants and tissue scaffolds. Biocompatible scaffolds can provide temporary structural support guiding cell growth, assist the transportation of essential nutrients, and facilitate the formation of functional tissues and organs. Increasing evidence shows that the nature of the biomaterial greatly affects long-term success of biomedical implants and short-term wound healing response. Substrate features such as the chemical composition and surface morphology affect the viability, adhesion, morphology, and motility of cells. Therefore, controlling the three-dimensional structure and surface composition of a biomaterial is important to promoting normal tissue growth or minimizing foreign body response.

To illustrate the importance of controlling the biomaterial surface, one can examine the use of implants to repair bone defects. Currently, there are several strategies for repairing large bone defects including using implants made of metal, plastic, ceramics, and or graphing of tissue. However, there are limitations to these biomaterials. Autographs can be expensive and difficult to handle, and may have physical limitations in their use. Allographs are also expensive and carry additional risks of an autoimmune response and disease transmission. Bone tissue engineering seeks to develop strategies to heal bone loss due to trauma or disease without the limitations and drawbacks of current clinical autografting and allografting treatments (Langer and Vacanti, 1993; Mistry and Mikos, 2005).

While metal and plastics mitigate many of the aforementioned risks, implants made from these materials, instead of integrating with bone, often form soft undesirable fibrous tissue. This is especially true with surfaces that are uniform and non-porous. This mechanical mismatch between tissue leads to wear of the implant that either aggravates or in some cases leads to cell death in nearby tissue, causing implant failure. However, inclusion of nanosized particles into implant materials, for example, has been shown to increase osteoblast adhesion (Kay et al., 2002). While this may be partially due to increased surface area, other factors may be involved, such as controlling protein adsorption. For instance, on carbon nanofiber surfaces, osteoblast adhesion was greater than other competitive cell types, possibly due to the fact that the aspect ratio and physical shape of these fibers mimic the crystalline hydroxyapatite structures of natural of nature bone (i.e. hydroxyapatite crystal dimensions from 50 to 100 nm in length and 1 to 10 nm in diameter) (Price et al., 2003). Sitharaman et al. demonstrated after 12 weeks that bone formation in defects (4 mm in diameter and 8 mm in depth) containing ultrashort-SWNT/poly(propylene fumarate) scaffolds had significantly higher (about 200% increase) bone volumes than poly(propylene fumarate) (PPF) scaffolds alone (Sitharaman et al., 2008). The histological sections of the ultrashort-SWNT/PPF implants showed increased collagen matrix production along with decreased foreign body giant cell density when compared to PPF scaffolds.

Taking advantage of the electroactive properties of carbon nanotubes, scaffolds could be formed that could be electrically conductive and thus stimulate cells contained on the scaffolds. For example, applying an alternating current to a nanocomposite of polylactic acid and multi-walled carbon nanotubes resulted in an increase in osteoblast proliferation by 46% and a greater than 300% increase in calcium production (Supronowicz et al., 2002). Also, upregulation of collagen I (a major component in organic bone formation), osteonectin, and osteocalcin was observed. Such results suggest that nanocomposites could accelerate the bone regeneration process.

For neuronal regeneration, carbon nanotube scaffolds could guide neurite growth into a specific neural bundle or network. Functionalized carbon nanotubes (f-CNT) have been able to guide neurite growth by providing a platform for the growth cone to grasp onto, instead of relying on, physiosorption alone (Hu et al., 2004). Zhang et al. suggested that positively charged carbon nanotubes are suitable to use as a template and patterned guide to grow an elaborate and controlled neuronal network (Zhang et al., 2005). The number of neurite growth cones, length of neurite outgrowths, and the degree of branching on positively charged polyethyleneimine f-CNT templates were significantly higher than on neutral or negatively charged CNT substrates (Hu et al., 2005). Lovat et al. demonstrated an increase in spontaneous post-synaptic currents in hippocampal neurons grown on a CNT substrate even when the neurons were randomly spaced apart from the substrate, suggesting that electric coupling had occurred between neurons and the CNT (Lovat et al., 2005). Such examples demonstrate that carbon nanotubes are potentially useful materials that can serve as both a supportive matrix and a conduit for delivering electrical signals.

Nanomaterials, like carbon nanotubes, are part of a growing new class of multifunctional biomaterial – smart biomaterials. Unlike passive structural biomaterials, smart biomaterials are designed to interact with their environment either by responding to changes in their surroundings or by stimulating or surpressing specific cellular behavior. They can change their shape, porosity, or hydrophilicity based on changes in temperature (Gan et al., 2005), pH (Bulmus et al., 2003), or external stimuli such as electric (Lahann et al., 2003) or magnetic (Jordan et al., 1999) fields. Such control of the biomaterial behavior through nanotechnology could create a major shift in the way biomaterials are used.

Examples of some techniques used for creating nanostructured surfaces for tissue engineering are shown in Table 30.2. The current paradigm to tissue regeneration is to isolate a patient's cells and then expand the cell population outside the body and finally place or seed the cells onto scaffold-like biomaterials before implantation. This method of engineered tissue using two different cell types has met with great success (Atala et al., 2006). Ideally, one would want to directly implant a biomaterial into the patient that would then selectively recruit the correct cell types. This approach would be especially important for engineering organs with very elaborate structures.

Another area where nanotechnology can impact the effectiveness of biomaterial surfaces is affecting stem cell differentiation within the engineered tissue. The unique properties of

TABLE 30.2 Examples of Tissue Scaffolds Created using Nanofabrication Techniques

Technique	Tissue scaffold prepared
Lithography	Nerve (Gabay et al., 2005)
Electrospinning	Heart (Zong et al., 2005), nerve (Yang et al., 2005), bone (Fujihara et al., 2005)
Self-assembly	Nerve (Ellis-Behnke et al., 2006)
Polymer demixing	Bone (Kim et al., 2005; Liao et al., 2004; Kikuchi et al., 2001; Du et al., 1999)
Solvent casting	Bladder (Pattison et al., 2005; Thapa et al., 2003a,b)
Salt leaching	Bladder (Pattison et al., 2005; Thapa et al., 2003a,b)

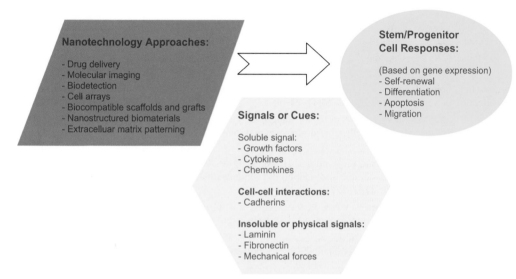

FIGURE 30.3
Regulation of stem cell fate corresponding to applications of nanotechnology and environmental signals *(modified from Solanki et al., 2008).*

537

nanomaterials and nanostructures can be particularly useful in controlling intrinsic stem cell signals and in dissecting the mechanisms underlying embryonic and adult stem cell behavior (Fig. 30.3) (Solanki et al., 2008). Currently, blends of expensive growth factors are used to guide the differentiation of stem cells. With the ability to control the surface morphology and chemistry at the nanoscale, nanobiomaterials may eliminate the need to culture different cell types for reassembly into an engineered tissue as they can recruit the body's own stem cells and differentiate them into the correct phenotype (Silva et al., 2004).

Biomaterials play an important role in regenerative medicine through their use in implants and tissue scaffolds. Nanotechnology is poised to provide the tools for rapidly increasing the pace of biomaterials development. Through the ability to control the nanostructure of a biomaterial, better understanding and control of cell behaviors will result, creating better regenerative therapies. The timeline of the impact of nanotechnology on biomaterial development as it relates to regenerative medicine will first be felt through better-performing, longer-lasting implants, and will eventually give way to smart biomaterials that can be implanted and direct the regenerative process at the cellular level.

CONCLUSION

As nanotechnology continues to grow, it will provide new and powerful tools that will revolutionize regenerative medicine. The most significant impact nanotechnology will have on regenerative medicine is that it will help in providing a detailed understanding and control of biology. Already the young field has demonstrated significant advances over traditional imaging, sensing, and structural technologies. Many of these advantages stem from the capability of nanomaterials to be multifunctional. These advances help in tackling one of most significant challenges faced in designing new biomedical technologies — targeting biological functions while at the same time avoiding non-specific effects. While there have been challenges for some time, nanotechnology provides us with the means to successfully negotiate these challenges and create new innovations in regenerative medicine.

References

Ahrens, E. T., Flores, R., Xu, H. Y., & Morel, P. A. (2005). In vivo imaging platform for tracking immunotherapeutic cells. *Nat. Biotechnol., 23*(8), 983–987.

Akerman, M. E., Chan, W. C. W., Laakkonen, P., Bhatia, S. N., & Ruoslahti, E. (2002). Nanocrystal targeting in vivo. *Proc. Natl. Acad. Sci. U.S.A., 99*(20), 12617–12621.

Atala, A., Bauer, S. B., Soker, S., Yoo, J. J., & Retik, A. B. (2006). Tissue-engineered autologous bladders for patients needing cystoplasty. *Lancet, 367*(9518), 1241–1246.

Ballou, B., Lagerholm, B. C., Ernst, L. A., Bruchez, M. P., & Waggoner, A. S. (2004). Noninvasive imaging of quantum dots in mice. *Biocon. Chem., 15*(1), 79–86.

Bos, C., Delmas, Y., Desmouliere, A., Solanilla, A., Hauger, O., Grosset, C., et al. (2004). In vivo MR imaging of intravascularly injected magnetically labeled mesenchymal stem cells in rat kidney and liver. *Radiology, 233*(3), 781–789.

Brown, D. M., Kinloch, I. A., Bangert, U., Windle, A. H., Walter, D. M., Walker, G. S., et al. (2007). An in vitro study of the potential of carbon nanotubes and nanofibres to induce inflammatory mediators and frustrated phagocytosis. *Carbon, 45*(9), 1743–1756.

Bruchez, M., Jr., Moronne, M., Gin, P., Weiss, S., & Alivisatos, A. P. (1998). Semiconductor nanocrystals as fluorescent biological labels. *Science (New York, NY), 281*(5385), 2013–2016.

Bulmus, V., Woodward, M., Lin, L., Murthy, N., Stayton, P., & Hoffman, A. (2003). A new pH-responsive and glutathione-reactive, endosomal membrane-disruptive polymeric carrier for intracellular delivery of biomolecular drugs. *J. Controlled Rel., 93*(2), 105–120.

Bulte, J. W., Zhang, S., van Gelderen, P., Herynek, V., Jordan, E. K., Duncan, I. D., et al. (1999). Neurotransplantation of magnetically labeled oligodendrocyte progenitors: magnetic resonance tracking of cell migration and myelination. *Proc. Natl. Acad. Sci. U.S.A., 96*(26), 15256–15261.

Cao, G. (2004). *Nanostructures & Nanomaterials: Synthesis, Properties & Applications.* London: Imperial College Press.

Chemaly, E. R., Yoneyama, R., Frangioni, J. V., & Hajjar, R. J. (2005). Tracking stem cells in the cardiovascular system. *Trends Cardiovasc. Med., 15*(8), 297–302.

Cherukuri, P., Bachilo, S. M., Litovsky, S. H., & Weisman, R. B. (2004). Near-infrared fluorescence microscopy of single-walled carbon nanotubes in phagocytic cells. *J. Am. Chem. Soc., 126*(48), 15638–15639.

Dodd, C. H., Hsu, H. C., Chu, W. J., Yang, P. G., Zhang, H. G., Mountz, J. D., et al. (2001). Normal T-cell response and in vivo magnetic resonance imaging of T-cells loaded with HIV transactivator-peptide-derived superparamagnetic nanoparticles. *J. Immunol. Methods, 256*(1–2), 89–105.

Du, C., Cui, F. Z., Zhu, X. D., & de Groot, K. (1999). Three-dimensional nano-HAp/collagen matrix loading with osteogenic cells in organ culture. *J. Bio. Mat. Res., 44*(4), 407–415.

Duncan, I. D., & Milward, E. A. (1995). Glial cell transplants: experimental therapies of myelin diseases. *Brain Path. (Zurich, Switzerland), 5*(3), 301–310.

Ellis-Behnke, R. G., Liang, Y. X., You, S. W., Tay, D. K. C., Zhang, S. G., So, K. F., et al. (2006). Nano neuro knitting: peptide nanofiber scaffold for brain repair and axon regeneration with functional return of vision. *Proc. Natl. Acad. Sci. U.S.A., 103*(13), 5054–5059.

Foldvari, M., & Bagonluri, M. (2008). Carbon nanotubes as functional excipients for nanomedicines: II. Drug delivery and biocompatibility issues. *Nanomedicine, 4*(3), 183–200.

Fujihara, K., Kotaki, M., & Ramakrishna, S. (2005). Guided bone regeneration membrane made of polycaprolactone/calcium carbonate composite nano-fibers. *Biomaterials, 26*(19), 4139–4147.

Gabay, T., Jakobs, E., Ben-Jacob, E., & Hanein, Y. (2005). Engineered self-organization of neural networks using carbon nanotube clusters. *Physica A, 350*(2–4), 611–621.

Gan, D. J., & Lyon, L. A. (2005). Tunable swelling kinetics in core-shell hydrogel nanoparticles. *J. Am. Chem. Soc., 123*(31), 7511–7517.

Gao, X. H., Cui, Y. Y., Levenson, R. M., Chung, L. W. K., & Nie, S. M. (2004). In vivo cancer targeting and imaging with semiconductor quantum dots. *Nat. Biotechnol., 22*(8), 969–976.

Heller, D. A., Baik, S., Eurell, T. E., & Strano, M. S. (2005). Single-walled carbon nanotube spectroscopy in live cells: towards long-term labels and optical sensors. *Advanced Mat., 17*(23), 2793.

Hong, H., Gao, T., & Cai, W. B. (2009). Molecular imaging with single-walled carbon nanotubes. *Nano Today, 4*(3), 252–261.

Hu, H., Ni, Y. C., Mandal, S. K., Montana, V., Zhao, N., Haddon, R. C., et al. (2005). Polyethyleneimine function-alized single-walled carbon nanotubes as a substrate for neuronal growth. *J. Phys. Chem. B, 109*(10), 4285–4289.

Hu, H., Ni, Y. C., Montana, V., Haddon, R. C., & Parpura, V. (2004). Chemically functionalized carbon nanotubes as substrates for neuronal growth. *Nano Letters, 4*(3), 507–511.

Jordan, A., Scholz, R., Wust, P., Fahling, H., & Felix, R. (1999). Magnetic fluid hyperthermia (MFH): cancer treat-ment with AC magnetic field induced excitation of biocompatible superparamagnetic nanoparticles. *J. Magn. Magn. Mat., 201*, 413–419.

Jung, H. I., Kettunen, M. I., Davletov, B., & Brindle, K. M. (2004). Detection of apoptosis using the C2A domain of synaptotagmin I. *Bioconj. Chem., 15*(5), 983–987.

Kateb, B., van Handel, M., Zhang, L., Bronikowski, M. J., Manohara, H., & Badie, B. (2007). Internalization of MWCNTs by microglia: possible application in immunotherapy of brain tumors. *NeuroImage, 37*(Suppl. 1), S9–S17.

Kay, S., Thapa, A., Haberstroh, K. M., & Webster, T. J. (2002). Nanostructured polymer/nanophase ceramic composites enhance osteoblast and chondrocyte adhesion. *Tissue Eng., 8*(5), 753–761.

Kikuchi, M., Itoh, S., Ichinose, S., Shinomiya, K., & Tanaka, J. (2001). Self-organization mechanism in a bone-like hydroxyapatite/collagen nanocomposite synthesized in vitro and its biological reaction in vivo. *Biomaterials, 22* (13), 1705–1711.

Kim, S. S., Park, M. S., Jeon, O., Choi, C. Y., & Kim, B. S. (2005). Poly(lactide-co-glycolide)/hydroxyapatite composite scaffolds for bone tissue engineering. *Biomaterials, 27*(8), 1399–1409.

Lacerda, L., Soundararajan, A., Singh, R., Pastorin, G., Al-Jamal, K. T., Turton, J., et al. (2008). Dynamic imaging of functionalized multi-walled carbon nanotube systemic circulation and urinary excretion. *Advanced Mat., 20*(2), 225.

Lahann, J., Mitragotri, S., Tran, T. N., Kaido, H., Sundaram, J., Choi, I. S., et al. (2003). A reversibly switching surface. *Science, 299*(5605), 371–374.

Langer, R., & Vacanti, J. P. (1993). Tissue engineering. *Science (New York, NY), 260*(5110), 920–926.

Lewin, M., Carlesso, N., Tung, C. H., Tang, X. W., Cory, D., Scadden, D. T., et al. (2000). Tat peptide-derivatized magnetic nanoparticles allow in vivo tracking and recovery of progenitor cells. *Nat. Biotechnol., 18*(4), 410–414.

Liao, S. S., Cui, F. Z., Zhang, W., & Feng, Q. L. (2004). Hierarchically biomimetic bone scaffold materials: Nano-HA/collagen/PLA composite. *J. Biomed. Mat. Res. Part B – Applied Biomat., 69B*(2), 158–165.

Lovat, V., Pantarotto, D., Lagostena, L., Cacciari, B., Grandolfo, M., Righi, M., et al. (2005). Carbon nanotube substrates boost neuronal electrical signaling. *Nano Letters, 5*(6), 1107–1110.

Medintz, I. L., Uyeda, H. T., Goldman, E. R., & Mattoussi, H. (2005). Quantum dot bioconjugates for imaging, labelling and sensing. *Nat. Mat., 4*(6), 435–446.

Mistry, A. S., & Mikos, A. G. (2005). Tissue engineering strategies for bone regeneration. *Adv. Biochem. Eng./Biotechnol., 94*, 1–22.

Morawski, A. M., Winter, P. M., Yu, X., Fuhrhop, R. W., Scott, M. J., Hockett, F., et al. (2004). Quantitative "magnetic resonance immunohistochemistry" with ligand-targeted F-19 nanoparticles. *Magn. Reson. Med., 52*(6), 1255–1262.

Murugesan, S., Park, T. J., Yang, H., Mousa, S., & Linhardt, R. J. (2006). Blood compatible carbon nanotubes–nano-based neoproteoglycans. *Langmuir, 22*(8), 3461–3463.

Nel, A., Xia, T., Madler, L., & Li, N. (2006). Toxic potential of materials at the nanolevel. *Science (New York, NY), 311* (5761), 622–627.

Panyam, J., & Labhasetwar, V. (2003). Biodegradable nanoparticles for drug and gene delivery to cells and tissue. *Adv. Drug Deliv. Rev., 55*(3), 329–347.

Pattison, M. A., Wurster, S., Webster, T. J., & Haberstroh, K. M. (2005). Three-dimensional, nano-structured PLGA scaffolds for bladder tissue replacement applications. *Biomaterials, 26*(15), 2491–2500.

Price, R. L., Waid, M. C., Haberstroh, K. M., & Webster, T. J. (2003). Selective bone cell adhesion on formulations containing carbon nanofibers. *Biomaterials, 24*(11), 1877–1887.

Salvador-Morales, C., Flahaut, E., Sim, E., Sloan, J., Green, M. L., & Sim, R. B. (2006). Complement activation and protein adsorption by carbon nanotubes. *Mol. Immunol., 43*(3), 193–201.

Schipper, M. L., Nakayama-Ratchford, N., Davis, C. R., Kam, N. W., Chu, P., Liu, Z., et al. (2008). A pilot toxicology study of single-walled carbon nanotubes in a small sample of mice. *Nat. Nanotechnol., 3*(4), 216–221.

Silva, G. A., Czeisler, C., Niece, K. L., Beniash, E., Harrington, D. A., Kessler, J. A., et al. (2004). Selective differ-entiation of neural progenitor cells by high-epitope density nanofibers. *Science, 303*(5662), 1352–1355.

539

Sitharaman, B., Shi, X., Walboomers, X. F., Liao, H., Cuijpers, V., Wilson, L. J., et al. (2008). In vivo biocompatibility of ultra-short single-walled carbon nanotube/biodegradable polymer nanocomposites for bone tissue engineering. *Bone, 43*(2), 362–370.

Solanki, A., Kim, J. D., & Lee, K. B. (2008). Nanotechnology for regenerative medicine: nanomaterials for stem cell imaging. *Nanomedicine (London, England), 3*(4), 567–578.

Sosnovik, D. E., Garanger, E., Aikawa, E., Nahrendorf, M., Figuiredo, J. L., Dai, G., et al. (2009). Molecular MRI of cardiomyocyte apoptosis with simultaneous delayed-enhancement MRI distinguishes apoptotic and necrotic myocytes in vivo: potential for midmyocardial salvage in acute ischemia. *Circulation, 2*(6), 460–467.

Sukhanova, A., Devy, M., Venteo, L., Kaplan, H., Artemyev, M., Oleinikov, V., et al. (2004). Biocompatible fluorescent nanocrystals for immunolabeling of membrane proteins and cells. *Analyt. Biochem., 324*(1), 60–67.

Supronowicz, P. R., Ajayan, P. M., Ullmann, K. R., Arulanandam, B. P., Metzger, D. W., & Bizios, R. (2002). Novel current-conducting composite substrates for exposing osteoblasts to alternating current stimulation. *J. Biomed. Mat. Res., 59*(3), 499–506.

Thapa, A., Webster, T. J., & Haberstroh, K. M. (2003a). Polymers with nano-dimensional surface features enhance bladder smooth muscle cell adhesion. *J. Biomed. Mat. Res., Part A, 67A*(4), 1374–1383.

Thapa, A., Miller, D. C., Webster, T. J., & Haberstroh, K. M. (2003b). Nano-structured polymers enhance bladder smooth muscle cell function. *Biomaterials, 24*(17), 2915–2926.

Tokumasu, F., & Dvorak, J. (2003). Development and application of quantum dots for immunocytochemistry of human erythrocytes. *J. Microsc. – Oxford, 211*, 256–261.

Uchimura, E., Yamada, S., Uebersax, L., Fujita, S., Miyake, M., & Miyake, J. (2007). Method for reverse transfection using gold colloid as a nano-scaffold. *J. Biosci. Bioeng., 103*(1), 101–103.

Voura, E. B., Jaiswal, J. K., Mattoussi, H., & Simon, S. M. (2004). Tracking metastatic tumor cell extravasation with quantum dot nanocrystals and fluorescence emission-scanning microscopy. *Nat. Med., 10*(9), 993–998.

Wang, H., Wang, J., Deng, X., Sun, H., Shi, Z., Gu, Z., et al. (2004). Biodistribution of carbon single-wall carbon nanotubes in mice. *J. Nanosci. Nanotechnol., 4*(8), 1019–1024.

Winter, P. M., Caruthers, S. D., Kassner, A., Harris, T. D., Chinen, L. K., Allen, J. S., et al. (2003). Molecular imaging of angiogenesis in nascent vx-2 rabbit tumors using a novel alpha(v)beta(3)-targeted nanoparticle and 1.5 tesla magnetic resonance imaging. *Cancer Res., 63*(18), 5838–5843.

Wu, X., Liu, H., Liu, J., Haley, K. N., Treadway, J. A., Larson, J. P., et al. (2003). Immunofluorescent labeling of cancer marker Her2 and other cellular targets with semiconductor quantum dots. *Nat. Biotechnol., 21*(1), 41–46.

Yang, F., Murugan, R., Wang, S., & Ramakrishna, S. (2005). Electrospinning of nano/micro scale poly(L-lactic acid) aligned fibers and their potential in neural tissue engineering. *Biomaterials, 26*(15), 2603–2610.

Yang, S. T., Fernando, K. A., Liu, J. H., Wang, J., Sun, H. F., Liu, Y., et al. (2008). Covalently PEGylated carbon nanotubes with stealth character in vivo. *Small (Weinheim an der Bergstrasse, Germany), 4*(7), 940–944.

Yuan, Q., Hein, S., & Misra, R.D.K. New generation of chitosan-encapsulated ZnO quantum dots loaded with drug: Synthesis, characterization and in vitro drug delivery response. *Acta Biomaterialia.*, In press, corrected proof.

Zeidler-Erdely, P. C., Calhoun, W. J., Ameredes, B. T., Clark, M. P., Deye, G. J., Baron, P., et al. (2006). In vitro cytotoxicity of Manville Code 100 glass fibers: effect of fiber length on human alveolar macrophages. *Particle Fibre Toxicol., 3*, 5.

Zhang, X., Prasad, S., Niyogi, S., Morgan, A., Ozkan, M., & Ozkan, C. S. (2005). Guided neurite growth on patterned carbon nanotubes. *Sensor. Actuat. B – Chem., 106*(2), 843–850.

Zhang, Y., Kohler, N., & Zhang, M. Q. (2002). Surface modification of superparamagnetic magnetite nanoparticles and their intracellular uptake. *Biomaterials, 23*(7), 1553–1561.

Zhao, M., Beauregard, D. A., Loizou, L., Davletov, B., & Brindle, K. M. (2001). Non-invasive detection of apoptosis using magnetic resonance imaging and a targeted contrast agent. *Nat. Med., 7*(11), 1241–1244.

Zhao, M., Kircher, M. F., Josephson, L., & Weissleder, R. (2002). Differential conjugation of tat peptide to superparamagnetic nanoparticles and its effect on cellular uptake. *Bioconj. Chem., 13*(4), 840–844.

Zong, X. H., Bien, H., Chung, C. Y., Yin, L. H., Fang, D. F., Hsiao, B. S., et al. (2005). Electrospun fine-textured scaffolds for heart tissue constructs. *Biomaterials, 26*(26), 5330–5338.

PART 3

Biomaterials for Regenerative Medicine

CHAPTER 31

Design Principles in Biomaterials and Scaffolds

Hyukjin Lee, Hyun Jung Chung, Tae Gwan Park
Department of Biological Sciences, Korea Advanced Institute of Science and Technology, Daejeon, Korea

INTRODUCTION

Tissue or organ transplantation is severely limited by the problems of donor shortage and immune rejection by patients. Tissue engineering strategies allow the transplantation of cells from a patient's own tissue to regenerate damaged tissue or organ without causing immune responses. For cell transplantation, extracted cells are often required to be cultivated on a large scale to attain a sufficient cell seeding density. During culture, the *in vitro* culture conditions play pivotal roles in proliferation and differentiation of the cells. Three-dimensional biomaterial scaffolds are firstly developed for the temporary substrate to grow cells in an organized fashion. Although direct injection or implantation of *in vitro* cultured cells is often performed, using a suspension of single cells is doubtful for the successful regeneration of impaired tissues. It is also well established that the three-dimensional organization of cells often related to cellular attachments affects the fate of cellular development. As a result, biodegradable and biocompatible polymers have been widely used to fabricate three-dimensional scaffolds for tissue engineering.

543

In the past, biomaterial scaffolds were mainly used for temporary prosthetic devices to fill the void spaces after tissue necrosis or surgery. However, current biomaterials aim to mimic the role of natural extracellular matrix (ECM), which can support cell adhesion, differentiation, and proliferation. ECM-mimicking biomaterial scaffolds should be designed considering the following requirements. First, suitable biomaterials must be selected for particular applications (Mikos and Langer, 1993b; Athanasiou and Agrawal, 1996; Lutolf and Hubbell, 2005). This is analogous to the effort to build up the target-specific biological scaffolds. Second, biomaterial scaffolds require a highly open porous structure with good interconnectivity, yet possessing sufficient mechanical strength for cellular in- or outgrowth (Cima and Langer, 1991). Third, the surface of fabricated scaffolds must be able to support cellular attachment, proliferation, and differentiation (Varkey and Uludag, 2004; Peattie and Prestwich, 2006; Vasita and Katti, 2006). Fourth, drug or cytokine releasing scaffolds are ideal for modulating tissue regeneration since cytokines such as growth factors and other small molecules have fundamental roles in growing functional living tissues (Niemann, 2005; Raghunath and Seifalian, 2005; Keilhoff and Wolf, 2006). Harmony of the above considerations is essential to fulfill the requirements of excellent biological scaffolds, thereby inducing synergic effects on successful tissue repair.

This chapter focuses on recent developments in fabricating biomimetic, ECM-like porous scaffolds useful for tissue engineering. Our experiences in designing novel biomaterials and innovating scaffold fabrication techniques are highlighted here along with the work of other leading researchers. Novel fabrication methods and designing strategies are elucidated; for example, generating the macroporous biodegradable scaffolds, the surface modification of biodegradable scaffolds to enhance cellular attachment and biological activity, and the incorporation of bioactive molecules within the scaffold systems. A number of excellent reviews are available of synthetic biomaterials for medical applications and tissue engineering (Peppas and Langer, 1994; Ratner, 1996; Uhrich, 1999; Sakiyama-Elbert and Hubbell, 2001).

SELECTION OF BIOMATERIALS

Natural biomaterials have been extensively used for tissue engineering since they have advantages over synthetic materials such as similarity with natural ECM. For example, alginate, chitosan, collagen and its derivatives, fibrin, heparin, and hyaluronic acid (HA) have been investigated for the fabrication of three-dimensional scaffolds (Rosso and Barbarisi, 2005). However, difficulty in adjusting the properties and their source-related immunogenicity remains a problem. In contrast, synthetic biomaterials composed of artificially synthesized polymers, although most reveal poor biocompatibility, can be designed with precise control of their physiochemical properties to give better performance when biomedically applied. Aliphatic polyesters and polyanhydrides are the most commonly used synthetic polymers for tissue engineering and drug delivery. By combining hydrophilic and hydrophobic segments within the structure of the polymers, a variety of synthetic biomaterials with the desired mechanical properties and degradation behaviors can be generated.

BIODEGRADABLE SYNTHETIC POLYMERS
Aliphatic polyesters

Aliphatic polyesters are synthetic biomaterials approved by the Food and Drug Administration (FDA) that have been widely used for biomedical applications such as surgical sutures and bone fixing screws. Poly(α-hydroxyl esters) such as poly(L-lactic acid) (PLLA), poly(lactic-co-glycolic acid) (PLGA), and polycaprolactone (PCL) can be synthesized by ring-opening polymerization of monomers, resulting in biodegradable polymers with hydrolytically cleavable bonds along the polymer backbone. When these synthetic polymers are implanted in the body, hydrolysis of the polymer backbone reduces the molecular weight of the polymer and their degraded products such as lactic and glycolic acids are metabolized in the body (Fig. 31.1). Based on their biocompatibility and safety record in humans, these polyesters have been used extensively for drug delivery and tissue engineering applications (Saltzman, 1999; Putman, 2001).

Aliphatic polyesters typically lack chemical functionality for modification with biological molecules. For the introduction of functional groups in the polymer backbone, Barrera and Langer (1993) reported the use of a novel monomer to incorporate amine groups into poly-lactic acid (PLA) polymers. Poly(lactic acid-co-lysine) was synthesized by the copolymerization of cyclic lactide and its analog containing the lysine. This novel amine-containing PLA showed similar biocompatibility while providing additional sites for further chemical modifications.

Polyanhydrides

Another class of degradable biopolymers is polyanhydrides. Unlike polyesters, which predominately show a bulk-erosion process, polyanhydrides exhibit a surface-erosion process that is particularly useful for sustained drug delivery systems. Leong et al. demonstrated the use of polyanhydrides based on sebacic acid (SA) and p-carboxyphenoxyproane (CPP) (Leong et al., 1985). By combining hydrophilic SA and hydrophobic CPP, the rate of surface erosion can be controlled from days to years. In addition, these polyanhydrides exhibit great biocompatibility and excellent *in vivo* performance for potential biomedical applications.

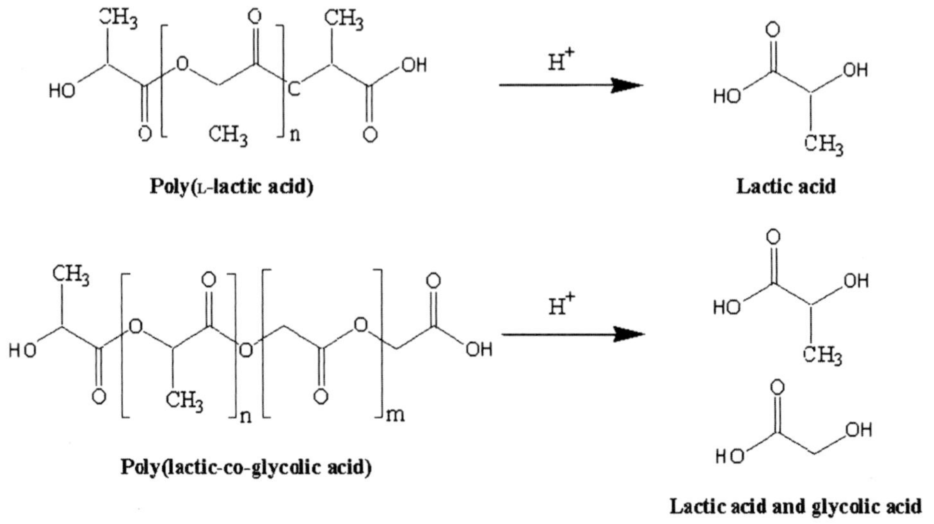

FIGURE 31.1

Structure of poly(L-lactic acid) and poly(lactic-co-glycolic acid) and their degradation products; acid hydrolysis of PLLA and PLGA to give lactic and glycolic acid.

DESIGN PRINCIPLES OF BIOLOGICAL SCAFFOLDS

Fabrication of macroporous biodegradable scaffolds

Along with the selection of materials, fabrication methods are also critical for designing biological scaffolds. For tissue regeneration, highly open-porous polymeric scaffolds are often required for high-density cell seeding and efficient nutrient and oxygen transport. Various methods to fabricate highly porous and biodegradable polymeric scaffolds have been reported and are listed in Table 31.1.

Briefly illustrating a few techniques, compressed polyglycolic acid (PGA) meshes made of non-woven PGA fibers have been widely used for soft tissue regeneration (Freed and Langer, 1993). Random coiling and heat treatment of PGA fibers can generate highly open porous and interconnected structures with a high surface to volume ratio. However, the mechanical strength of these meshes is insufficient for hard tissue regeneration (Mikos and Langer, 1993a).

To enhance the mechanical properties of compressed PGA meshes, Mooney and Langer (1996b) demonstrated that a mixed solution of PLLA and PLGA can be applied to the compressed PGA meshes. A mixture of PLLA and PLGA dissolved in organic solvent was

TABLE 31.1 List of Fabrication Methods for Preparation of Highly Porous Biodegradable Scaffolds

Fabrication methods	Materials	References
Compressed mesh of non-woven fibers	PGA, PLGA	Freed & Langer, 1993; Mikos & Langer, 1993a; Mooney & Langer, 1996b
Solvent casting/salt leaching	PLLA, PLGA	Mikos & Langer, 1993b; Mikos & Vacanti, 1994
CO_2 expansion	PLGA	Mooney & Langer, 1996a; Harris & Mooney, 1998
Emulsion freeze drying	PLGA	Whang & Nuber, 1995
Phase separation	PLLA, PLGA	Lo & Leong, 1996; Schugens & Teyssie, 1996; Nam & Park, 1999a
Three-dimensional imprinting	PLLA, PLGA	Park & Griffith, 1998

sprayed throughout the compressed PGA meshes. As the organic solvent evaporated, dried PLLA/PLGA strengthened the cross regions of the fibers and enhanced the mechanical properties of the compressed meshes. However, this method exhibited reduced surface to volume ratio of the meshes and difficulty in matching the degradation rate of the surface-coated and bulk materials.

In addition, the solvent casting/salt-leaching technique has been extensively exploited for fabricating scaffolds for tissue engineering (Mikos and Langer, 1993b; Mikos and Vacanti, 1994). PLGA dissolved in an organic solvent with salt particles is placed in a mold to produce a polymer/salt mixture, which is immersed in water to remove the salt particles and generate open-pore structures. The scaffolds prepared often show a dense surface layer and poor interconnectivity between the macropores, which reduces cell seeding into the scaffolds *in vitro* and causes non-uniform distribution of the seeded cells. This results in poor cell viability and tissue in-growth when implanted *in vivo*.

In order to resolve the problems of the salt-leaching techniques, Nam and Park (2000) utilized PLLA paste containing ammonium bicarbonate salt particles, which acts as a gas-foaming agent as well as a salt-leaching porogen to fabricate highly interconnected porous biodegradable scaffolds (Fig. 31.2). Sodium bicarbonate salt with acidic excipients has been widely used for effervescent gas evolving oral tablets. Ammonium bicarbonate salt produces gaseous ammonia and carbon dioxide upon contact with an acidic aqueous solution such as citric acid and/or incubation at elevated temperatures, and therefore could be incorporated into a biodegradable gel paste prepared by dissolving high-molecular-weight PLLA in an organic solvent. The resultant putty paste was easy to shape into different geometry and could be immersed in hot water solution and directly dried under vacuum to remove or leach out the salt particles while concurrently generating gaseous ammonia and carbon dioxide. This would provide highly interconnected pores within a solidifying polymer scaffold, resulting in an open-porous structure without any surface skin layer on either side of the scaffolds (Fig. 31.3).

Macroporous PLGA scaffolds with controlled degradation rates were fabricated using the gas-foaming/salt-leaching method and investigated (Yoon and Park, 2001). Unlike semi-crystalline PLLA, amorphous PLGA could form a gel-like paste in an organic solvent even at high concentration. PLGA was dissolved in an organic solvent such as chloroform, and then precipitated in a non-solvent, ethanol. Resulting precipitates exhibited a gel-like property such that the paste could be molded or hand-shaped in any desirable dimensions. In this study, instead of incubating the scaffolds in a hot water bath or vacuum oven, citric acid solution was used to control the porosity of scaffolds as well as the mechanical properties. Using citric acid, carbon dioxide and ammonia gases could be generated at room temperature, and the concentration of citric acid in the solution could be varied to control the porosity of scaffolds.

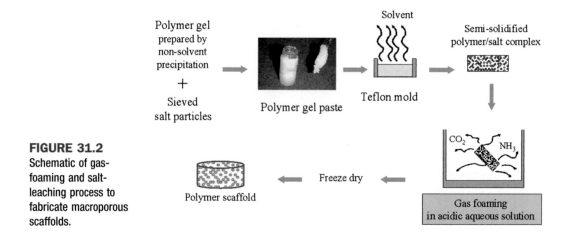

FIGURE 31.2
Schematic of gas-foaming and salt-leaching process to fabricate macroporous scaffolds.

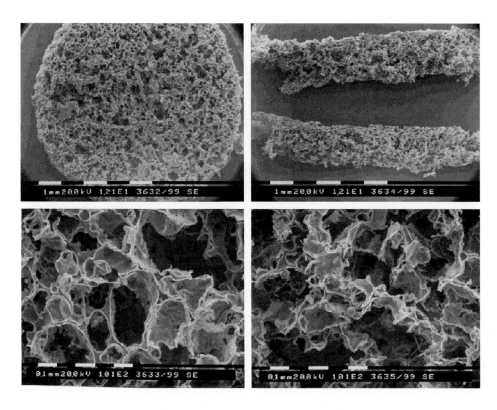

FIGURE 31.3
SEM images of macroporous scaffolds fabricated by gas-foaming and salt-leaching process. Uniform interconnectivity and high porosity are observed on both surface (left) and cross-section (right) of scaffolds.

Results showed that an increase in citric acid concentration produced scaffolds with higher porosity. In addition, degradation and swelling behaviors of PLGA scaffolds with different compositions were investigated. Macroporous scaffolds with various compositions of lactic and glycolic acid were incubated in phosphate buffered solution (pH 7.4) at 37°C. During the incubation period, significant swelling of the scaffolds was observed depending on the composition, and the change in dimension and morphology was caused by the accelerated degradation of PLGA scaffold, which could generate more water-adsorbing small-molecular-weight PLGA oligomers within the degrading scaffolds (Fig. 31.4).

As an alternative to salt-leaching and gas-forming fabrication, electrospinning has received much attention for fabricating polymeric ultrafine nanofibers to build three-dimensional tissue engineering scaffolds (Kim and Park, 2008; Yoo and Park, 2009). Nanofibrous biodegradable scaffolds would have definite advantages for cell attachment, proliferation, and differentiation because they resemble ECM structures. Previously, Kim and Park (2006) demonstrated ECM mimicking nanofiber mesh for tissue engineering applications. The amine-terminated PLGA dissolved in a mixture of DMF/THF solvent was ejected through a nozzle by

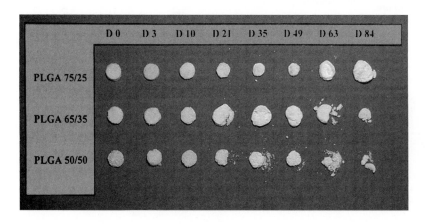

FIGURE 31.4
Photographs of different PLGA scaffolds after hydrolytic degradation in PBS at 37°C. With increasing composition of glycolic acid, rapid degradation and swelling of scaffolds are observed.

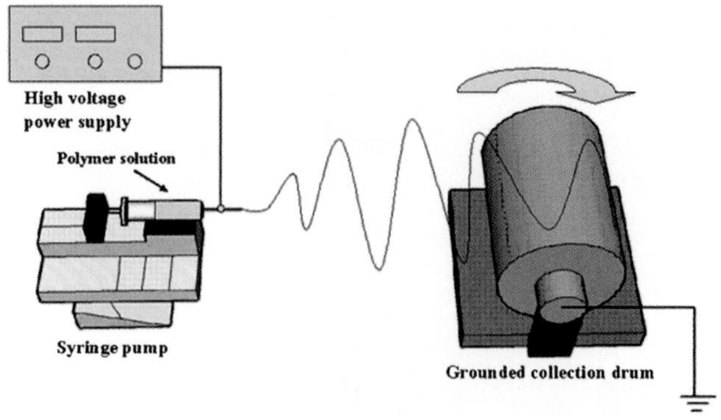

FIGURE 31.5
Schematic of electrospinning (left) and a SEM image of electrospun PLGA nanofiber (right).

an electrostatic force, resulting in the formation of non-woven fabrics. During electrospinning, the solvent evaporated and the charged polymer nanofibers were deposited on a grounded collector. The resultant structure was a three-dimensional, randomly oriented nanofiber network mesh with a highly macroporous architecture (Fig. 31.5). *In vitro* cell culture revealed that the resulting nanofiber ranging from 300 to 1,000 nm provided an excellent environment for cellular attachment, proliferation, and differentiation.

As an emerging non-invasive tissue engineering material, injectable solid scaffolds prepared from porous microspheres have attracted much attention. Chung and Park (2008) demonstrated the use of macroporous PLGA microcarriers for injectable chondrocyte delivery (Fig. 31.6). Macroporous microcarriers with a highly interconnected porous structure were fabricated using a gas foaming method during a double-emulsion and solvent-evaporation process. The size of microcarriers ranged from 170 to 500 μm with pores of ~30 μm. These macroporous microcarriers supported the three-dimensional growth of chondrocytes within the scaffolds in dynamic spinner culture conditions. Compared to two-dimensional mono-layer culture, cartilage phenotypes (type II collagen and aggrecan expression) were well maintained during cultivation and the cell-microcarrier constructs were readily injectable

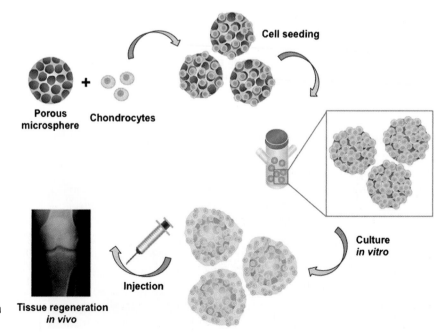

FIGURE 31.6

Use of injectable porous scaffold microspheres for cartilage tissue engineering. As an example, primary chondrocytes are seeded into the porous microspheres, cultured *in vitro*, and then injected into the cartilage defect site for tissue regeneration *in vivo*.

through a syringe needle for cell therapy. Recently, cellular aggregates were formed using porous microspheres with a size of ~50 μm and mesenchymal stem cells for adipose tissue regeneration (Chung and Park, 2010). The mesenchymal stem cell aggregates were cultured and differentiated *in vitro* to form adipose-like micro-tissues, which showed high regenerative potential for adipose tissue formation *in vivo*.

Surface immobilization of bioactive molecules on macroporous biodegradable scaffolds

The surface modification of scaffolds is essential since the microenvironment of the body cannot see the bulk property of biomaterials, but the surface of biomaterials. In the past, a major issue concerned with biomaterials was the biocompatibility of the materials upon injection or implantation *in vivo*. Only a few biomaterials are known to be free of causing acute inflammation. As a result, the surfaces of fouling devices were modified with non-protein-adsorbing materials such as polyethylene glycol (PEG) to hide the implants from the body.

Since many cell adhesive peptides in the ECM dictate cellular behaviors, the immobilization of various bioactive ligands on the surface of biomaterials was attempted for actively mimicking physiological conditions, thereby increasing cytocompatibility and biological functionality when the biomaterials are implanted in the body. A number of surface modification methods were developed, such as chemical oxidation and etching, plasma and corona discharge, radiation and UV grafting, partial hydrolysis, protein adsorption, and conjugation/immobilization of bioactive ligands (Rasmussen and Whitesides, 1977; Ramsey and Binkowski, 1984; Weisz and Schnaar, 1991; Gao and Langer, 1998; Nam and Park, 1999b; Otsuka and Kataoka, 2000; Chung and Park, 2007).

As an example, we demonstrated galactose-modified PLGA macroporous scaffolds for culturing hepatocytes *in vitro* (Park, 2002; Yoon and Park, 2002). When selecting bioactive molecules for immobilization, ligands for cell membrane receptors have a pivotal role since these ligands are associated with cellular signaling pathways and activities such as cell migration, proliferation, and differentiation. Moreover, cell-specific ligands help to initiate binding and attachment of cells on modified scaffolds. For instance, galactose is a specific ligand for asialoglycoprotein receptor in hepatocytes. Galactose-modified PLGA was prepared by conjugation of end aminated PLGA with lactobionic acid using dicyclohexyl carbodiimide/*N*-hydroxysuccinimide (DCC/NHS) coupling agents (Fig. 31.7). The galactosylated PLGA was then processed to form films and macroporous scaffolds to examine hepatocyte-specific cellular binding to the modified surface. Albumin secretion was quantified as well for validating cellular functionality. For cell-specific binding, galactose-modified films were also fabricated and the attachment of hepatocytes on films was observed. Results showed that hepatocytes were more selectively attached to the galactose-modified films compared to the non-specific glucose-modified films. Additionally, it was demonstrated that conjugation of galactose on the PLGA surface resulted in higher cell viability as compared to control PLGA films.

The idea of mimicking an *in vivo* system using short bioactive peptide sequences such as arginine–glycine–aspartic acid (RGD) has been applied for decades. Surface modification with RGD sequences has been widely used for enhancing cellular attachment and growth (Yoon and Park, 2004). Cell adhesive ligands such as RGD are abundant in collagen and play vital roles for cellular attachment via integrin-mediated binding to ECM. There are a number of excellent reviews demonstrating the effects of RGD in tissue engineering. For instance, Langer and coworkers published a comprehensive review of creating biomimetic microenvironments using adhesive peptides (Shakesheff and Langer, 1998). Continuing the mimicking of biological surfaces, selecting bioactive ligands is crucial for each application. For cartilage tissue engineering, a microenvironment with high water content, similar to native cartilage, is required. HA is a naturally occurring non-sulfated glycosaminoglycan (GAG) composed of *N*-acetyl-D-glucosamine and D-glucuronic acid that is a major constituent of ECM and

549

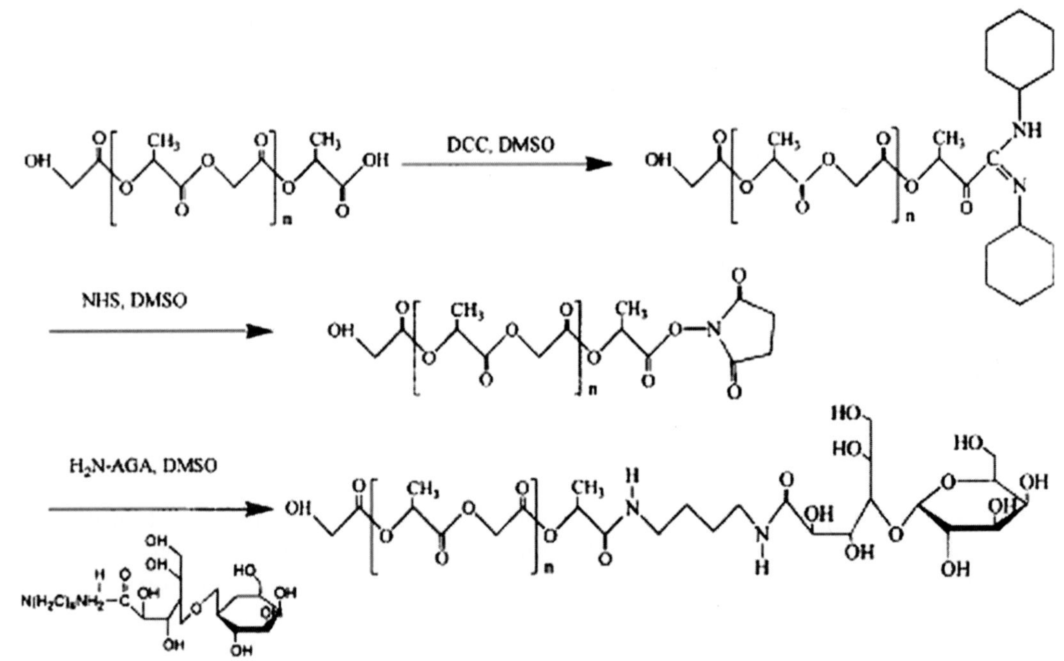

FIGURE 31.7
Synthesis of galactosylated PLGA.

abundantly expressed in cartilage. In addition, HA is known to have vital roles in various biological functions of chondrocytes such as regulating adhesion and motility, and mediating cell proliferation and differentiation (Larsen and Balazs, 1992). There are a number of publications on the effects of HA on proliferation and phenotypic expression of chondrocytes (Chow and Knudson, 1995; Lindenhayn and Sit, 1999).

From the reasons above, Yoo and Park fabricated HA-modified PLGA macroporous scaffold (Yoo and Park, 2005). As previously described, a macroporous structure of PLGA was obtained from the gas-foaming/salt-leaching process and the surface of these materials was chemically conjugated with HA. Amine end-capped PLGA was synthesized and mixed with PLGA to form biodegradable scaffolds. To expose the amine groups on the surface, fabricated scaffolds were purged into the HA solution with EDC/NHS coupling agents (Fig. 31.8). The resulting HA-coated PLGA macroporous scaffolds exhibited higher chondrocyte proliferation, probably via interaction of CD44 with HA, and initiated increased production of GAG, as compared to PLGA alone, while enhancing type II collagen and aggrecan gene expression.

Sustained release of bioactive molecules from macroporous scaffolds

In many tissue engineering applications using stem cells, specific cellular differentiation is often required to achieve the expression of desirable phenotypes and the secretion of functional proteins and carbohydrates. To satisfy the above requirements, the *in situ* local delivery of cytokines such as growth factors and molecular drugs within cell-seeded scaffolds has been pursued since the sustained release of bioactive molecules is known to stimulate cell proliferation, differentiation, and the secretion of desirable proteins. There have been multiple reports on local delivery of growth factors within the scaffold such as epidermal growth factor (Mooney and Langer, 1996c), transforming growth factor (TGF) (Behof and Jansen, 2002), vascular endothelial growth factor (VEGF) (Wissink and Feijen, 2000; Richardson and Mooney, 2001), basic fibroblast growth factor (b-FGF) (Royce and Marra, 2004), and bone morphogenic proteins (BMPs) (Lee and Battle, 1994; Whang and Healy, 2000). These scaffolds were able to stimulate embedded cells to express tissue-specific phenotypes in mRNA level and

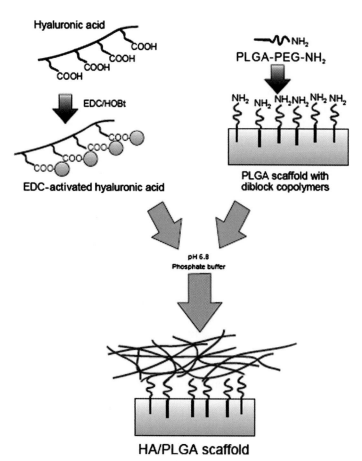

FIGURE 31.8
Schematic of surface modification of PLGA biodegradable scaffold with HA.

induce the production of functional ECM corresponding to the desired applications. In addition, the sustained release of plasmid DNA for transfecting neighboring cells was also investigated (Chun and Park, 2004, 2005).

One of the emerging fields of drug delivery is the local delivery of small drug molecules such as steroid analogs from biodegradable scaffolds in a sustained manner. Dexamethasone is a family of glucocortiocoids that exhibits various inhibitory effects on the inflammation process and the proliferation of smooth muscle cells (Reil and Gelabert, 1999; Hickey and Moussy, 2002). As well, dexamethasone is commonly used along with specific growth factors to induce stem cell differentiation toward osteoblasts or chondrocyte-like cells (Peter and Mikos, 1998). To investigate the effects of sustained release of dexamethasone, Yoon and Park (2003) fabricated dexamethasone-releasing macroporous scaffolds composed of PLGA. Hydrophobic dexamethasone was incorporated into the PLGA polymer solution and the macroporous scaffolds were fabricated by the gas-foaming/salt-leaching method. Due to bulk degradation of PLGA, dexamethasone was slowly released out in a zero order fashion without an initial burst effect. The bioactivity of released dexamethasone was established by culturing smooth muscle cells with/without dexamethasone-releasing scaffolds. The results showed a large decrease in smooth muscle cell proliferation with increase in the concentration of dexamethasone. The suppression of lymphocyte activation or anti-inflammation activity by dexamethasone released from the scaffolds was also validated with different concentrations of dexamethasone.

With continuing development of synthetic biomaterials for drug delivery systems, biodegradable scaffolds can also be utilized as a gene carrier for sustained release of plasmid DNA, oligodeoxyribonucleotides (ODN), and siRNA. By delivering growth factor and other cytokine-related genes, transfected cells can be genetically controlled and used in tissue repair. In addition, transfected cells can trigger neighboring cells to proliferate and differentiate to cells with specific phenotypes for specific tissue engineering applications. Conventional gene delivery carriers usually express highly positive charges such that the charge-charge interaction between negatively charged DNA molecules and the carriers can form a tight ionic complex. However, excess use of highly positive polymer species such as polyethyleneimine (PEI), poly (L-lysine) (PLL), and positively charged fatty acids can cause severe cytotoxicity and reduces the biocompatibility of gene carriers. Although a single injection of naked plasmid DNA can induce appreciable protein expression, enhancement of transfection efficiency and sustained release of nucleic acid drugs are highly demanded.

To achieve a high level of specific protein synthesis, sustained release of naked DNA is a promising approach to overcome the low transfection efficiency. Therefore, PLGA macro-porous scaffolds for sustained release of plasmid DNA were fabricated by the thermally induced phase separation method (TIPS) (Chun and Park, 2004). In this study, homogeneous polymer solution at elevated temperature was phase separated into polymer rich and polymer poor domains by lowering the solution temperature with subsequent lyophilization of solvent, generating a microcellular structure (Fig 31.9). In order to encapsulate plasmid DNA within scaffolds, PLGA was dissolved in 1,4-dioxane and mixed with plasmid DNA dissolved in deionized water followed by quenching in liquid nitrogen and solvent lyophilization. To control the release of encapsulated plasmid DNA, the effects of higher quenching temperature (annealing) and addition of PLGA-grafted PLL were subsequently examined. The resulting scaffolds with the encapsulated DNA could slowly release the DNA in an intact form for over 20 days. Furthermore, higher quenching temperature produced larger pore formation within the scaffolds, giving a rapid release of plasmid DNA while the addition of PLGA-grafted PLL lowered the release profiles. Lastly, the bioactivity of released plasmid DNA was confirmed by the high level of luciferase expression in cells.

As described earlier, biomimetic scaffolds have received much interest (Park, 2002; Yoo and Park, 2005). Since natural ECM plays pivotal roles in various biological events, functions of ECM component such as HA and heparin have been investigated. For tissue engineering, angiogenesis—the sprouting of microvessels from existing ones—is crucial for cell-scaffold implantation since a lack of blood supply results in poor delivery of oxygen and nutrient, causing necrosis of implanted cells. To enhance angiogenesis at implanted sites, angiogenic growth factors have been applied in various fashions (Wissink and Feijen, 2000; Richardson and Mooney, 2001). A common way of incorporating growth factors is mixing them with the polymer solution and casting them to form scaffolds or films. However, the use of organic solvent causes a critical problem in maintaining the bioactivity of growth factors. To amend this problem, Mok and Park (2008) demonstrated a novel protein solubilization technique in organic solvent using PEG. It was found that proteins and PEG could form stable nano-sized complexes in organic solvents by non-covalent interactions such as hydrogen bonding. Based on the PEG-assisted protein solubilization, a water-free protein microencapsulation within

FIGURE 31.9
Cross-sectional SEM images of PLGA scaffolds fabricated by TIPS methods quenching in liquid nitrogen (A) and annealing at −20°C (B). Note that increasing annealing temperature generates larger pores for rapid release of encapsulated plasmid DNA.

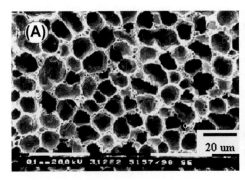
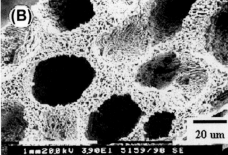

PLGA microspheres was performed. Bovine serum albumin (BSA) and recombinant human growth hormone (rhGH) were successfully encapsulated within the polymer matrix using a spray drying method and the structural and functional integrities of these proteins were verified.

Heparin is a negatively charged polysaccharide and widely used as an anticoagulation agent to enhance biocompatibility of implanted devices. In natural ECM, heparin plays a role as a reservoir for controlled secretion of growth factors since it has a high binding affinity with various growth factors such as VEGF, TGF-β, and b-FGF. Heparin stabilizes the released growth factors and concentrates them in the local targeted areas. Exploiting the unique biological functions of heparin, heparin-modified injectable PLGA microscaffolds were fabricated for the sustained release of b-FGF (Fig. 31.10). By synthesizing PLGA microspheres with free surface amine groups, carboxylic groups of heparin can be covalently conjugated on the surface of PLGA scaffolds. Soluble b-FGFs were readily bound to the heparin, resulting in high loading efficiency. At last, *in vitro* studies showed that the bound b-FGF was released in a sustained manner in bioactive form (Yoon and Park, 2006).

SUMMARY AND CONCLUSION

Design of biomaterials and scaffolds is a complex interdisciplinary subject. Biodegradable and erodible biomaterials serve as scaffolds and drug delivery devices for applications in regenerative medicine. Natural biomaterials have already been clinically used for many years by trial-and-error material selection, while synthetic biomaterials have also begun to be applied recently. The use of biomaterials requires the understanding of the differences in structure and properties between these implanted materials and their interaction with the host's tissues. *In vivo* tolerance of early biomaterials helped to initiate a rapid development of more complex biomimetic systems. Especially, the development of synthetic polymers has enabled the engineering of biomaterials with tailored properties and functions. For biomedical applications, scaffolds must be designed according to their specific purposes considering the complex functions and interactions of cells, cytokines, and the scaffold. Among the synthetic biomaterials, aliphatic polyesters have been widely utilized for many years and offer excellent design

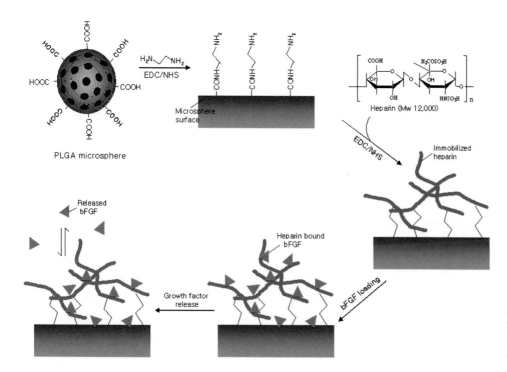

FIGURE 31.10
Schematic of heparin-immobilized porous PLGA microsphere for local delivery of angiogenic growth factors.

versatility and biocompatibility. The complicated requirements of a biomaterial allowed the development of more sophisticated designs of scaffolds such as highly macroporous scaffolds for facilitating nutrient and oxygen transfer; functionalization with specific biological ligands on the surface for promoting cell attachment, proliferation, and differentiation; and finally release of cytokines to manipulate the functions of encapsulated cells or hosts tissues. Further scientific and technological advances will envision the development of more ideal scaffolds that are specifically designed for each purpose in a wide range of applications in regenerative medicine.

References

Athanasiou, K. A., & Agrawal, C. M. (1996). Sterilization, toxicity, biocompatibility and clinical applications of polylactic acid/polyglycolic acid copolymers. *Biomaterials, 17,* 93–102.

Barrera, D., & Langer, R. (1993). Synthesis and RGD peptide modification of a new biodegradable copolymer: poly (lactic acid-co-lysine). *J. Am. Chem. Soc., 115,* 11010–11011.

Behof, J. W. M., & Jansen, J. A. (2002). Bone formation in transforming growth factor b-1-coated porous poly (propylene fumarate) scaffolds. *J. Biomed. Mater. Res., 60,* 241–251.

Chow, G., & Knudson, W. (1995). Increased expression of CD44 in bovine articular chondrocytes by catabolic cellular mediators. *J. Biol. Chem., 270,* 27734–27741.

Chun, K. W., & Park, T. G. (2004). Controlled release of plasmid DNA from biodegradable scaffolds fabricated by a thermally induced phase separation method. *J. Biomater. Polymer, 15,* 1341–1353.

Chun, K. W., & Park, T. G. (2005). Controlled release of plasmid DNA from photo-crosslinked pluronic hydrogels. *Biomaterials, 26,* 3319–3326.

Chung, H. J., & Park, T. G. (2007). Surface engineered and drug releasing pre-fabricated scaffolds for tissue engineering. *Adv. Drug Deliv. Res., 59,* 249–262.

Chung, H. J., & Park, T. G. (2008). Highly open porous biodegradable microcarriers: *in vitro* cultivation of chondrocytes for injectable delivery. *Tissue Engineering, 14,* 607–615.

Chung, H. J., & Park, T. G. (2010). Fabrication of adipose-derived mesenchymal stem cell aggregates using biodegradable porous microspheres for injectable adipose tissue regeneration. *J. Biomater. Sci. Polym.,* DOI: 10.1163/092050609X12580983495681.

Cima, L. G., & Langer, R. (1991). Tissue engineering by cell transplantation using biodegradable polymer substrates. *J. Biomech. Eng., 113,* 143–151.

Freed, L. E., & Langer, R. (1993). Neocartilage formation *in vitro* and *in vivo* using cells cultured on synthetic biodegradable polymers. *J. Biomed. Mater. Res., 27,* 11–23.

Gao, J., & Langer, R. (1998). Surface hydrolysis of poly(glycolic acid) meshes increases the seeding density of vascular smooth muscle cells. *J. Biomed. Mater. Res., 42,* 417–424.

Harris, L. D., & Mooney, D. J. (1998). Open pore biodegradable matrices formed with gas foaming. *J. Biomed. Mater. Res., 42,* 396–402.

Hickey, T., & Moussy, F. (2002). Evaluation of a dexamethasone/PLGA microsphere system designed to suppress the inflammatory tissue response to implantable medical devices. *J. Biomed. Mater. Res., 61,* 180–187.

Keilhoff, G., & Wolf, G. (2006). Transdifferentiation of mesenchymal stem cells into schwann cell like myelinating cells. *Eur. J. Cell Biol., 85,* 11–24.

Kim, T. K., & Park, T. G. (2006). Biomimicking extracellular matrix: cell adhesive RGD peptide modified electrospun poly(D, L-lactic-coglycolic acid) nanofiber mesh. *Tissue Eng., 12,* 221–233.

Kim, T. K., & Park, T. G. (2008). Macroporous and nanofibrous hyaluronic acid/collagen hybrid scaffold fabricated by concurrent electrospinning and deposition/leaching of salt particles. *Acta Biomaterialia, 4,* 1611–1619.

Larsen, N. E., & Balazs, E. A. (1992). Effect of hyaluronan on cartilage and chondrocyte cultures. *J. Orthop. Res., 10,* 23–32.

Lee, S. C., & Battle, M. A. (1994). Healing of large segmental defects in rat femurs is aided by rhBMP-2 in PLGA matrix. *J. Biomed. Mater. Res., 28,* 1149–1156.

Leong, K., Brott, B., & Langer, R. (1985). Bioerodible polyanhydrides as drug carrier matrixes. I: Characterization, degradation, and release characteristics. *J. Biomed. Mater. Res., 19,* 941–955.

Lindenhayn, K., & Sit, M. (1999). Retention of hyaluronic acid in alginate beads: aspects for *in vitro* cartilage engineering. *J. Biomed. Mater. Res., 44,* 149–155.

Lo, H., & Leong, K. W. (1996). Poly(l-lactic acid) foams with cell seeding and controlled release capacity. *J. Biomed. Mater. Res., 30,* 475–484.

Lutolf, M. P., & Hubbell, J. A. (2005). Synthetic biomaterials as instructive extracellular microenvironments for morphogenesis in tissue engineering. *Nat. Biotechnol., 23,* 47–55.

Mikos, A. G., & Langer, R. (1993a). Preparation of poly(glycolic acid) bonded fiber structures for cell attachment and transplantation. *J. Biomed. Mater. Res., 27*, 183–189.

Mikos, A. G., & Langer, R. (1993b). Laminated three-dimensional biodegradable foams for use in tissue engineering. *Biomaterials, 14*, 323–330.

Mikos, A. G., & Vacanti, J. P. (1994). Preparation and characterization of poly(l-lactic acid) foams. *Polymer, 35*, 1068–1077.

Mok, H. J., & Park, T. G. (2008). Water-free microencapsulation of proteins within PLGA microparticles by spray drying using PEG-assisted protein solubilization technique in organic solvent. *Eur. J. Pharm. Biopharm., 70*, 138–144.

Mooney, D. J., & Langer, R. (1996a). Novel approach to fabricate porous sponges of poly(d, l-lactic-coglycolic acid) without the use of organic solvents. *Biomaterials, 17*, 1417–1422.

Mooney, D. J., & Langer, R. (1996b). Stabilized polyglycolic acid fiber-based tubes for tissue engineering. *Biomaterials, 17*, 114–124.

Mooney, D. J., & Langer, R. (1996c). Localized delivery of epidermal growth factor improves the survival of transplanted hepatocytes. *Biotechnol. Bioeng., 50*, 422–429.

Nam, Y. S., & Park, T. G. (1999a). Porous biodegradable polymeric scaffolds prepared by thermally induced phase separation. *J. Biomed. Mater. Res., 47*, 9–17.

Nam, Y. S., & Park, T. G. (1999b). Adhesion behaviors of hepatocytes cultured onto biodegradable polymer surface modified by alkali hydrolysis process. *J. Biomater. Sci. Polymer, 10*, 1145–1158.

Nam, Y. S., & Park, T. G. (2000). A novel fabrication method for macroporous scaffolds using gas foaming salt as porogen additive. *J. Biomed. Mater. Res., 53*, 1–7.

Niemann, C. (2005). Controlling the stem cell niche: right time, right place, right strength. *BioEssays, 28*, 1–5.

Otsuka, H., & Kataoka, K. (2000). Surface characterization of functional polylactide through the coating with heterobifunctional poly(ethylene glycol)/polylacide block copolymers. *Biomacromolecules, 1*, 29–48.

Park, A., & Griffith, L. G. (1998). Integration of surface modification and 3D fabrication techniques to prepare patterned poly(l-lactide) substrates allowing regionally selective cell adhesion. *J. Biomater. Sci. Polymer, 9*, 89–110.

Park, T. G. (2002). Perfusion culture of hepatocytes within galactose-derivatized biodegradable poly(lactide-co-glycolide) scaffolds prepared by gas foaming of effervescent salts. *J. Biomed. Mater. Res., 59*, 127–135.

Peattie, R. A., & Prestwich, G. D. (2006). Dual growth factor induced angiogenesis *in vivo* using hyaluronan hydrogel implants. *Biomaterials, 27*, 1868–1875.

Peppas, N. A., & Langer, R. (1994). New challenges in biomaterials. *Science, 263*, 1715–1720.

Peter, S. J., & Mikos, A. G. (1998). Osteoblastic phenotype of rat marrow stromal cells cultured in the presence of dexamethasone, b-glycerolphosphate, and L-ascorbic acid. *J. Cell Biochem, 71*, 55–62.

Putnam, D. (2001). Polymer-based gene delivery with low cytotoxicity by a unique balance of side-chain termini. *Proc. Natl. Acad. Sci., U.S.A., 98*, 1200–1205.

Raghunath, J., & Seifalian, A. M. (2005). Advancing cartilage tissue engineering: the application of stem cell technology. *Curr. Opin. Biotechnol, 16*, 503–509.

Ramsey, W. S., & Binkowski, N. J. (1984). Surface treatments and cell attachment. *Vitro, 20*, 802–808.

Rasmussen, J. R., & Whitesides, G. M. (1977). Introduction, modification, and characterization of functional groups on the surface of low density polyethylene films. *J. Am. Chem. Soc., 99*, 4736–4745.

Ratner, B. D. (1996). *Biomaterials Science.* San Diego: Academic Press. pp. 11–35.

Reil, T. D., & Gelabert, H. A. (1999). Dexamethasone suppresses vascular smooth muscle cell proliferation. *J. Surg. Res., 85*, 109–114.

Richardson, T. P., & Mooney, D. J. (2001). Polymeric system for dual growth factor delivery. *Nat. Biotechnol., 19*, 1029–1034.

Rosso, F., & Barbarisi, A. (2005). Smart materials as scaffolds for tissue engineering. *J. Cell Physiol., 203*, 465–470.

Royce, S. M., & Marra, K. G. (2004). Incorporation of polymer microspheres within fibrin scaffolds for controlled delivery of FGF-1. *J. Biomater. Sci. Polymer, 15*, 1327–1336.

Sakiyama-Elbert, S. E., & Hubbell, J. A. (2001). Functional biomaterials: design of novel biomaterials. *Annu. Rev. Mater. Res., 31*, 183–201.

Saltzman, W. M. (1999). Delivering tissue regeneration. *Nat. Biotechnol., 17*, 534–535.

Schugens, C., & Teyssie, P. (1996). Poly-lactide macroporous biodegradable implants for cell transplantation. II. Preparation of polylactide foams by liquid–liquid phase separation. *J. Biomed. Mater. Res., 30*, 449–461.

Shakesheff, K., & Langer, R. (1998). Creating biomimetic micro-environment with synthetic polymer–peptide hybrid molecules. *J. Biomater. Sci. Polymer, 9*, 507–518.

Uhrich, K. E. (1999). Polymeric systems for controlled drug release. *Chem. Rev., 99*, 3181–3198.

Varkey, M., & Uludag, H. (2004). Growth factor delivery for bone tissue repair: an update. *Expert. Opin. Drug Deliv.,* *1,* 19–36.

Vasita, R., & Katti, D. S. (2006). Growth factor delivery systems for tissue engineering: a materials perspective. *Expert. Rev. Med. Dev., 1,* 29–47.

Weisz, O. A., & Schnaar, R. L. (1991). Hepatocyte adhesion to carbohydrate-derived surfaces II. Regulation of cytoskeletal organization and cell morphology. *J. Cell Biol., 115,* 495–504.

Whang, K., & Healy, K. E. (2000). A biodegradable polymer scaffold for delivery of osteotropic factors. *Biomaterials, 21,* 2535–2551.

Whang, K., & Nuber, G. A. (1995). Novel methods to fabricate bioabsorbable scaffolds. *Polymer, 36,* 837–842.

Wissink, M. J. B., & Feijen, J. (2000). Improved endothelialization of vascular grafts by local release of growth factor from heparinized collagen matrices. *J. Contr. Release, 64,* 103–114.

Yoon, J. J., & Park, T. G. (2001). Degradation behaviors of biodegradable macroporous scaffolds prepared by gas foaming of effervescent salts. *J. Biomed. Mater. Res., 55,* 401–408.

Yoon, J. J., & Park, T. G. (2002). Surface immobilization of galactose onto aliphatic biodegradable polymers for hepatocyte culture. *Biotech. Bioeng., 78,* 1–10.

Yoon, J. J., & Park, T. G. (2003). Dexamethasone releasing biodegradable polymer scaffolds fabricated by a gas foaming/salt leaching method. *Biomaterials, 24,* 2323–2329.

Yoon, J. J., & Park, T. G. (2004). Immobilization of cell adhesive RGD peptide onto the surface of highly porous biodegradable polymer scaffolds fabricated by gas foaming/salt leaching method. *Biomaterials, 25,* 5613–5620.

Yoon, J. J., & Park, T. G. (2006). Heparin-immobilized biodegradable scaffolds for local and sustained release of angiogenic growth factor. *J. Biomed. Mater. Res., Part A, 79,* 934–942.

Yoo, H. S., & Park, T. G. (2005). Hyaluronic acid modified biodegradable scaffolds for cartilage tissue engineering. *Biomaterials, 26,* 1925–1933.

Yoo, H. S., & Park, T. G. (2009). Surface-functionalized electrospun nanofibers for tissue engineering and drug delivery. *Adv. Drug Del. Rev., 61,* 1033–1042.

CHAPTER 32

Natural Origin Materials for Bone Tissue Engineering — Properties, Processing, and Performance

V.M. Correlo[*,†], J.M. Oliveira[*,†], J.F. Mano[*,†], N.M. Neves[*,†], R.L. Reis[*,†]

* 3B's Research Group — Biomaterials, Biodegradables and Biomimetics, University of Minho, Headquarters of the European Institute of Excellence on Tissue Engineering and Regenerative Medicine, Taipas, Guimarães, Portugal
[†] IBB — Institute for Biotechnology and Bioengineering, PT Associated Laboratory, Guimarães, Portugal

INTRODUCTION

Bone injuries, mainly resulting from an increasingly aged population, degenerative diseases, or traumatic injuries, compromise significantly the quality of life of humanity, resulting in an increasingly significant socio-economic problem. Current options to treat these injuries are unsatisfactory as they rely on the use of autografts, allografts, and an assortment of synthetic or biomimetic materials and devices. Each of these options has significant limitations, such as the need for an additional surgery, limited supply, inadequate size and shape, and morbidity associated with the donor site (Salgado et al., 2005; van Gaalen et al., 2007). All those limitations lead to the need for the development of innovative approaches to aid skeletal tissue repair and reconstruction. It is in this context that tissue engineering emerged as an alternative approach to repair and regenerate damaged human tissues, avoiding the need for a permanent prosthesis (Mistry and Mikos, 2005; Nesic et al., 2006; Chung and Burdick, 2008). Tissue engineering has the potential to address these clinical needs and new treatment concepts. The engineered substitute should structurally and morphologically resemble the native tissue and be able to perform similar biological functions, eliminating problems of donor site scarcity, immune rejection, and pathogen transfer.

Tissue engineering can be subdivided into different strategies; the most used strategy applied for the regeneration of hard tissue (such as bone) combines the use living cells, biologically active molecules, and a temporary three-dimensional (3D) porous scaffolds (Hutmacher et al., 2007). Since load bearing applications require porous structures with improved mechanical performance, we believe that solid porous structures can be more adequate for connective tissue engineering applications than matrices from hydrogels. Therefore, special attention will be given to processing techniques used in the preparation of foams and meshes,

Principles of Regenerative Medicine. DOI: 10.1016/B978-0-12-381422-7.10032-X

in vitro and *in vivo* performance, alone or in combination with cells, in the context of bone tissue engineering.

NATURAL-BASED POLYMERS

Natural polymers are widely spread in nature. Those polymers are formed during the growth cycles of many organisms, being obtained from renewable sources such as plants, animals, or microorganisms. A large variety of natural polymers are available with potential interest for the production of scaffolds due to, as natural components of living structures, their biological and chemical similarities to natural tissues. Moreover, natural polymers have the advantage of being prone or susceptible to enzymatic or hydrolytic degradation, which may indicate the great susceptibility of these materials to being metabolized by the physiological mechanisms (Gomes et al., 2007).

Starch

Starch is the predominant energy-storing compound in many plants. It can be found in storage organs such as roots and tubers in a granular form. Most of the granules are oval and vary in size from 1 to 110 µm depending on the starch source (Hoover, 2001). By far the largest source of starch is corn (maize) with other commonly used sources being wheat, potato, tapioca, and rice.

The structure and composition of native starches vary with the botanical sources, but all granules consist of two types of α-glucan polymers; that is, amylose and amylopectin (Hoover, 2001; Tester et al., 2004). Amylose, the minor constituent, is defined as a relative long and linear polymer consisting mainly of $\alpha(1\rightarrow4)$ linked D-glucopyranosyl units. Amylopectin, the major component, is a branched polysaccharide composed of hundreds of short $(1\rightarrow4)$-α-glucan chains, which are interlinked by $(1\rightarrow6)$-α-linkages (Buléon et al., 1998; Hoover, 2001; Tester et al., 2004). Figure 32.1 shows the typical structure of amylose and amylopectin macromolecules.

Starch contributes 50–70% of the energy in the human diet, providing a direct source of glucose, which is an essential substrate in brain and red blood cells for generating metabolic energy (Copeland et al., 2009). The human body can degrade starch by using specific enzymes including α-amylase present in saliva and also in the blood plasma. Starch degradation products are oligosaccharides that can be metabolized to produce energy. Other enzymes involved in starch degradation are β-amylase, α-glucosidases, and other debranching enzymes (Martins et al., 2008a).

The crystallinity of native starch granules can vary from about 15% for high-amylose starches to about 45–50% for waxy starches (Copeland et al., 2009). The processing of native starch to be used in varied applications requires both disrupting and melting the semicrystalline granular structure of native starches. The modification method is usually referred to as

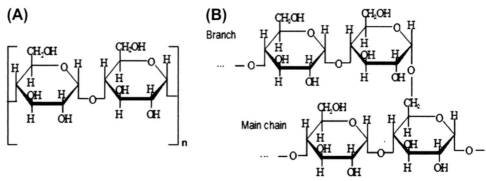

FIGURE 32.1
Chemical structure of amylose (A) and amylopectin with $\alpha(1\rightarrow6)$ branch point (B).

gelatinization. Gelatinization occurs during heating in the presence of a sufficient quantity of moisture. In those conditions the starch granules absorb water and swell, losing irreversibly their crystallinity and structural organization (Sousa et al., 2008).

Products from pure starch or from thermoplastic starch (starch with disrupted granular structure) are usually brittle and moisture sensitive, thus strongly limiting their potential fields of application. One possible way to overcome these limitations is to blend starch with other biodegradable polymers. Several polymeric systems have already been obtained by blending native maize starch with: (1) ethylene-vinyl alcohol (SEVA-C), (2) cellulose acetate (SCA), (3) polycaprolactone (SPCL), and (4) poly(lactic acid) (SPLA) (Reis and Cunha, 2001). These blends were originally proposed by Reis and co-workers (Reis and Cunha, 1995; Reis et al., 1996a, 1996b, 1997) as potential alternatives for various tissue applications including connective tissues. Starch-based blends and composites were shown to be non-cytotoxic and potentially biocompatible (Marques et al., 2002, 2005), and were proposed for several biomedical applications, including bone cements (Boesel et al., 2004), drug delivery systems (Balmayor et al., 2009), bone fixation devices (Sousa et al., 2000), and tissue engineering scaffolding (Gomes et al., 2008; Duarte et al., 2009; Salgado et al., 2009).

PROCESSING METHODS

Due to the thermoplastic behavior of the starch-based blends and composites, it is possible to produce 3D porous scaffolds using traditional melt-based technologies, such as compression molding combined with particulate leaching (Gomes et al., 2002) and injection molding (Gomes et al., 2001; Neves et al., 2005) or extrusion with blowing agents (Gomes et al., 2002; Salgado et al., 2004). This processing routine offers the unique advantage of avoiding the use of solvents, which sometimes are detrimental in the biomedical field.

Gomes et al. (2002) produced scaffolds from a blend of starch with cellulose acetate (SCA) by a method consisting of extrusion with different types and amounts of blowing agents. The porous structure of the samples results from the gases released by the thermal decomposition of the blowing agents (BA) during processing. Thus, by using different types and amounts of BA it was possible to obtain scaffolds with different sizes of porosity in the range of 50–500 µm. The same approach was used to produce scaffolds from a blend of corn starch/ethylene-vinyl alcohol (SEVA-C) (Salgado et al., 2004). The developed porous structures had 60% porosity with pore sizes between 200 and 900 µm with an acceptable degree of inter-connectivity. The limitation of this method lies in the difficulty on controlling tightly the pore size and its interconnectivity.

Gomes et al. (2002) used other melt-based technology, compression molding and particulate leaching, to produce scaffolds from the same starch-based blend (SCA). The obtained scaffolds have an open network of pores throughout the sample with sizes ranging from 10 to 500 µm and an average porosity of about 50%. The advantage of this technique is the possibility of tightly controlling the percentage of porosity and pore size simply by varying the amount and size of the leachable particles.

SPCL-(starch with ε-polycaprolactone, 30:70%) and SPLA-(starch with poly(lactic acid), 30:70%) based scaffolds were prepared by a fibre-bonding process using fibres obtained by melt-spinning (Gomes et al., 2008). The two types of scaffolds produced by this method exhibited a typical fibre-mesh structure, with a fibre diameter of approximately 180 µm for SPCL (Fig. 32.2) and 210 µm for SPLA, with highly interconnected pores and a porosity of approximately 75%. Both types of scaffolds exhibited enhanced mechanical performance in comparison with most scaffolds obtained using other biodegradable polymers aimed at tissue engineering applications. Moreover, using the fiber bonding method, different porosities of the fiber meshes scaffolds can be obtained using different amounts (by weight) of fibers (Gomes et al., 2006). The typical morphology of SPCL scaffolds obtained by fiber bonding is shown in Figure 32.2.

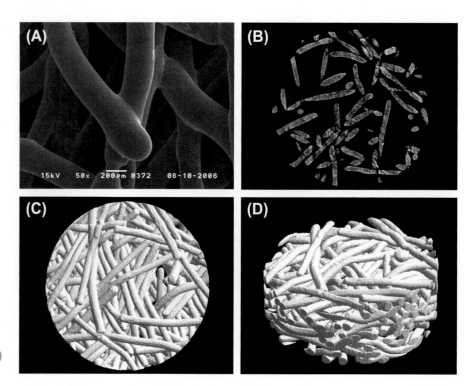

FIGURE 32.2
Morphology of a SPCL-based scaffold obtained by fibre bonding. (A) SEM micrograph; (B) 2D microCT image; and (C,D) respective 3D microCT images.

Duarte et al. (2009) proposed the use of supercritical fluid technology — supercritical immersion precipitation — as a clean and environmentally friendly approach to preparing porous scaffolds from starch and poly(l-lactic acid) blends and composites for tissue engineering applications. The obtained structures are highly porous and interconnected and their morphology can be considered as a bicontinuous structure composed of macropores ($\geq$75 μm) and micropores with sizes ranging from 10 to 20 μm, with the surfaces appearing very rough. Therefore, the impregnation with Bioglass did not affect the porosity or the interconnectivity of the starch-based scaffolds (Duarte et al., 2010a). The results obtained enabled the conclusion that the pressure was the parameter that most affected the porosity, interconnectivity, and pore size of the scaffolds produced by this technique (Duarte et al., 2009). Supercritical fluid technology can be also used to prepare drug-loaded starch-based porous scaffolds in a one-step and clean process (Duarte et al., 2010b). Scaffolds prepared by this method and loaded with dexamethasone showed a sustained release over 21 days with morphology comparable to the unloaded ones.

A novel hierarchical starch-based scaffold, obtained by the combination of rapid prototyping (RP) and electrospinning techniques, was developed with the objective of overcoming the high number of cells needed to attain sufficient adherent cells to the RP scaffolds (Martins et al., 2009a). These scaffolds were characterized by a 3D structure of parallel aligned rapid prototyped microfibers (average fiber diameter 300 μm) periodically intercalated by randomly distributed electrospun nanofibers (fiber diameters in the range of 400 nm to 1.4 μm). Those systems were design to improve the cell seeding efficiency of the systems obtained by RP.

STARCH IN BONE TISSUE ENGINEERING APPLICATIONS

Several studies reported in the literature have shown that starch-based scaffolds can be used on bone tissue engineering strategies by promoting the attachment, proliferation, and differentiation of bone marrow stromal cells (Gomes et al., 2003, 2006) and endothelial cells (Santos et al., 2007). It has been also demonstrated that the use of starch-based scaffolds, in conjunction with fluid flow bioreactor culture, minimizes diffusion constraints and provides mechanical stimulation to the marrow stromal cells, leading to enhancement of the differentiation towards the development of bone-like mineralized tissue (Gomes et al., 2006).

Nevertheless, for a bone cell-scaffold construct to be successful it is necessary to establish a viable and functional vascular network. With this objective, Santos et al. (2007) demonstrated that starch-based fiber mesh scaffolds are an excellent substrate for the growth of human endothelial cells (ECs) required for the vascularization process. These findings, coupled with those reported for bone marrow cells, suggest that starch-based scaffolds may have a high potential for use as a scaffold material to obtain vascularized bone tissue engineering applications.

Recently, it was shown that prevascular structures were induced by co-culturing outgrowth endothelial cells (OECs) with primary osteoblasts on SPCL scaffolds, which were achieved without additional supplementation of culture medium with angiogenic growth factors (Fuchs et al., 2009). Additionally, in cellular constructs consisting of OECs and primary osteoblasts on SPCL scaffolds implanted subcutaneously into a nude mouse model, OECs formed vascular structures closely associated with the scaffold material and embedded in a rich extracellular matrix produced by the primary osteoblasts. These results provide enhanced evidence of the great performance of those biomaterial structures in the context of bone tissue engineering.

Aiming at mimicking the conditions found *in vivo*, Martins et al. (2009b) studied the influence of both α-amylase and lipase on the degradation of SPCL fiber meshes as a function of immersion time and its effect on the osteogenic differentiation of rat bone marrow stromal cells. Results indicated that culture medium supplemented with enzymes enhanced cell proliferation after 16 days of culture and that lipase positively influenced osteoblastic differentiation of MSCs and promoted matrix mineralization. Furthermore, *in vivo* studies have also shown that different starch-based scaffolds (SCA, SEVA-C, and SEVAC/CaP) implanted into bone defects created on the distal femur integrated with host tissue at the defect site and surrounding marrow, indicating their good biocompatibility. Early connective tissue developing at the bone/scaffold interface could be characterized as an early form of bone tissue (Salgado et al., 2007).

Chitosan

Chitin is a homopolymer of β(1→4)-linked *N*-acetyl-D-glucosamine residues (Fig. 32.3A). Chitin is a natural polysaccharide and it is the principal structural component of the exoskeleton of invertebrates such as crustaceans, insects, and spiders, and can also be found in the cell walls of most fungi and many algae (Shi et al., 2006).

Chitosan is obtained from the alkaline deacetylation of the biopolymer chitin, the second most abundant polysaccharide in nature. Structurally, chitosan is a linear polysaccharide consisting of *N*-glucosamine (deacetylated unit) and *N*-acetyl glucosamine (acetylated unit) units linked by β(1→4) glycosidic bonds (Fig. 32.3B). The degree of deacetylation (DD) is the glucosamine/*N*-acetyl glucosamine ratio and usually can vary, depending on the source, from 30 to 95%. The degree of crystallinity of chitosan is mainly controlled by the degree of deacetylation being maximum for both chitin (i.e. 0% deacetylated) and fully deacetylated

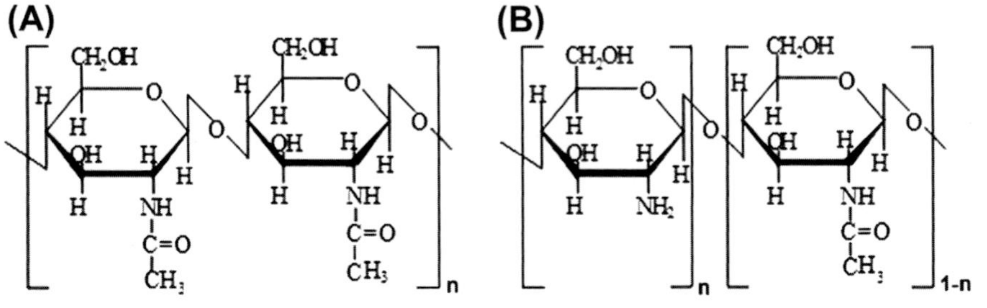

561

FIGURE 32.3
Chemical structure of chitin (A) and chitosan (B).

forms (100% chitosan), and minimum for intermediate degrees of deacetylation (Kim et al., 2008b).

Chitosan is degraded by means of hydrolysis and lysozyme has been shown to be the primary agent of its degradation, *in vivo* (Vårum et al., 1997). The degradation rate is inversely related to the percentage of crystallinity which is, as referred, controlled mainly by the degree of deacetylation. Highly deacetylated forms (e.g. 85%) exhibit the lowest degradation rates and may last several months *in vivo* (Yang et al., 2007). The degradation products are chitosan oligosaccharides of variable length.

Chitosan is insoluble in aqueous solutions above pH 7, but, in dilute acids, the free amino groups are protonated and the molecule becomes fully soluble below around pH 5. The pH-dependent solubility of chitosan provides a convenient mechanism for its processing under mild conditions. For example, viscous solutions can be extruded and gelled in high-pH solutions. Freeze-drying techniques also allow highly porous structures to be obtained by means of freezing a polymer solution (-20°C and -196°C), followed by the removal of solvent through lyophilization. Thus, chitosan can be easily processed into films and porous scaffolds (Madihally and Matthew, 1999). If we consider its biological properties, such as biocompatibility (VandeVord et al., 2002) as well as immunological, antibacterial, and wound-healing activity and processability, chitosan is one of the most appealing biomaterials for prospective applications in tissue engineering. In addition, much of the potential of chitosan as a biomaterial for tissue engineering can also be partially justified by its structural similarity to glycosaminoglycans (GAGs), as it possesses similar glucosamine residues to those of major components of the cartilage ECM (di Martino et al., 2005). Since GAGs' properties include many specific interactions with growth factors, receptors, and adhesion proteins, this suggests that the analogous structure in chitosan may also have related bioactivities (Suh and Matthew, 2000). Moreover, the cationic nature of chitosan also allows it to interact with anionic GAGs, proteoglycans, and other negatively charged species. This property can be of great interest since it may serve as a mechanism for retaining or accumulating these molecules within a scaffold during colonization or after an *in vivo* implantation (Madihally and Matthew, 1999).

PROCESSING METHODS

A very interesting property of chitosan is that it can be transformed into 3D highly porous structures with a high degree of interconnectivity using various technologies. For example, porous scaffolds can be produced by lyophilizing a frozen solution of chitosan powder dissolved in acetic acid (Madihally and Matthew, 1999; Nettles et al., 2002). The obtained scaffolds have porosities of $\sim 80\%$ and median pore diameters of $\sim 68\,\mu$m. The mean pore diameters can be controlled within the range of $1-250\,\mu$m by varying the freezing conditions (Madihally and Matthew, 1999). Figure 32.4 shows the typical morphology of chitosan scaffolds obtained by freeze-drying a 3 wt% chitosan solution. The main limitation of those structures is that the mechanical properties are very low, eventually compromising its use for connective tissue applications.

Abdel-Fattah et al. (2007) used chitosan with different degrees of deacetylation to produce microspheres and 3D porous matrices via a sintered microsphere technique. The median pore size and porosity level of the obtained scaffolds were $\sim 200\,\mu$m and $\sim 20\%$, respectively. The porosity of the scaffolds was considered to be too low and that was attributed to the large size of the microspheres. A similar particle aggregation approach was followed by Malafaya et al. (2008) to produce chitosan-based scaffolds intended to promote neo-vascularization. The microCT quantitative analyses of porosity, median pore size, and interconnectivity were $\sim 28\%$, $\sim 265\,\mu$m, and 95%, respectively. Moreover, it was concluded that pores of the scaffolds produced by applying this technique will allow for tissue ingrowths resembling those of trabecular bone. The main limitation of this technique is the level of porosity that may be considered insufficient.

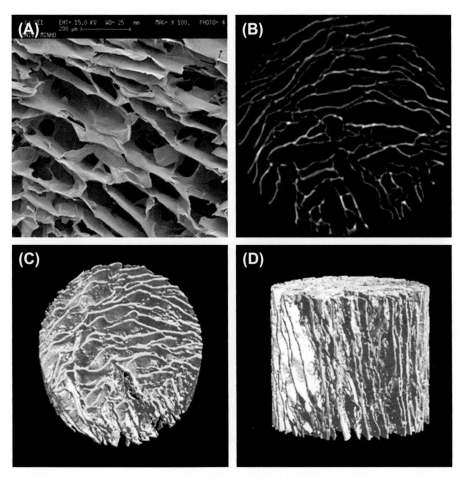

FIGURE 32.4
Morphology of a chitosan scaffold obtained by freeze-drying. (A) SEM micrograph; (B) microCT 2D image; and (C,D) respective microCT 3D images.

Another strategy to overcome the mechanical properties of chitosan is to incorporate chitosan, for example by free-drying, into a mechanically stronger scaffold (Prabaharan et al., 2007). This would allow provision of a chitosan environment for cells and tissues in a more robust porous structure.

Wet spinning is one of the most used methods to produce natural fibers, and was used to prepare chitosan fibers and 3D fiber meshes (Tuzlakoglu et al., 2004). The obtained scaffolds had an average pore size in the range of 100–500 µm, which is ideal for bone-related applications.

Until very recently, the methods to produce chitosan scaffolds were based on the use of solvents. Nevertheless, Correlo et al. (2008), produced chitosan/polyester 3D porous scaffolds suitable for supporting the adhesion, proliferation, and osteogenic differentiation of mouse MSCs (Costa-Pinto et al., 2008), using for the first time a melt-based processing technology. All the scaffolds were produced by melt-based compression molding followed by salt leaching. By using this technique it was possible to obtain scaffolds with distinct properties concerning porosity, pore size, interconnectivity, and mechanical performance by varying the porogen particle size and amount and by varying the ratio and type of aliphatic polyester used in the blend.

In a different concept, Martins et al. (2008b) reported the production of chitosan-based scaffolds with the ability to form a porous structure *in situ* due to the degradation promoted by specific enzymes present in the human body (α-amylase and lysozyme). The *in vitro* formation of pores, controlled by the location of the "sacrifice" phase (native starch), was evident, although pore formation is expected to occur more rapidly *in vivo*, due to the presence of other enzymes and cells.

CHITOSAN IN BONE TISSUE ENGINEERING APPLICATIONS

In vitro tests have shown that chitosan-based scaffolds support the adhesion and proliferation of osteoblasts (Fakhry et al., 2004; Tuzakoglu et al., 2004). Costa-Pinto et al. (2008) confirmed that chitosan/polyester-based scaffolds can support adhesion, viability/proliferation, and osteogenic differentiation of a mouse mesenchymal stem cell line (BMC-9) and therefore are promising scaffolds to be used in the bone tissue engineering field.

The inclusion of bioactive ceramics in the scaffold composition can confer osteoconductive and even osteoinductive properties on the final structure that will guide bone formation. *In vitro* studies have shown that the use of chitosan-based scaffolds prepared with biphasic calcium phosphate (BCP) for culture of MSCs and preosteoblasts increased bone tissue formation (Sendemir-Urkmez and Jamison, 2007). In a different study (Zhao et al., 2006), the incorporation of hydroxyapatite (HAp) into chitosan/gelatin composite scaffolds promoted *in vitro* initial adhesion and enhanced osteogenic differentiation of hMSC. Porous scaffolds from collagen-chitosan-HAp were characterized by possessing high histocompatibility and suitable material to be used as a bone substitute (Wang et al., 2008d). Nano-HAp/chitosan/carboxymethyl cellulose composites have shown promising properties in terms of being used as bone repair materials (Liuyan et al., 2008). *In vivo* studies have shown that a composite consisting of calcium phosphate cement (CPC), chitosan fibers, and gelatin displays the ability to form new bone more rapidly, with faster bioresorption as compared to that for pure CPC (Pan and Jiang, 2008).

Several reports (Kong et al., 2006; Tuzakoglu and Reis, 2007; Manjubala et al., 2008) on the literature have shown that the development of apatite coatings on polymeric scaffolds using biomimetic approaches enhances cell adhesion and proliferation, and is an interesting method to enhance biomaterials properties aimed at bone tissue engineering applications. Chitosan-based fiber mesh scaffolds with a bone-like apatite coating have been shown to posses an higher cell adhesion and proliferation as compared to the uncoated samples (Tuzakoglu et al., 2007). Mineralized chitosan scaffolds with HAp nanocrystals at their surface and within the pore channels induced the formation of extracellular matrix but did not significantly influence the growth of human osteoblasts (SaOs-2) (Manjubala et al., 2008).

The performance of tissue engineering constructs can be greatly enhanced through the incorporation of bioactive agents. Chitosan scaffolds have been modified with RGDS enhancing the attachment of rat osteosarcoma (ROS) cells (Ho et al., 2005). In a different study, chitosan/collagen scaffold loaded with adenoviral vector encoding human bone morphogenetic proteins (BMP7) were implanted into defects on both sides of the mandible (Zhang et al., 2007). The results have shown that the scaffold containing Ad-BMP7 exhibited the higher ALP activity, and the expression of osteopontin and bone sialoprotein were up-regulated. Moreover, in defects around the implant, the bone formation in Ad-BMP7 scaffolds was enhanced when compared with other scaffolds. In addition, *in vivo* analyses using a mouse implantation model have shown that, although there was a large migration of neutrophils into the implantation area, minimal signs of any inflammatory reaction in the chitosan scaffolds were observed. This study demonstrated that chitosan has a high degree of biocompatibility in this specific application (VandeVord et al., 2002). Recently, chitosan-based scaffolds implanted into rat muscle-pockets showed a mild inflammatory response commonly observed in the implantation of foreign bodies, *in vivo* (Malafaya et al., 2008). In addition, neo-vascularization of the implants created by new blood vessels formation was clear even after only 2 weeks of implantation. This process evolved significantly for a longer period of time, showing that the scaffolds' characteristics promoted the integration into the host tissue.

Polyhydroxyalkanoates

Poly([R]-hydroxyalkanoates) (PHAs) are a family of polyesters that can be produced by a large number of bacteria as carbon storage compounds under metabolic stress, such as limitation of an essential nutrient (Chen, 2005; Furrer et al., 2008). PHA chemical and physical properties depend on the monomeric composition, which is determined by the producing microorganism and its nutrition (Furrer et al., 2008). PHAs are composed of 3-, 4-, or rarely 5-hydroxy fatty acid monomers, which form linear polyesters. The general chemical structure is shown in Figure 32.5.

To date, more than 100 different monomers have been reported, but most of the studies are focused on poly 3-hydroxybutyrate (PHB), the simplest and the first to be discovered (Zinn et al., 2001). PHB is a semi-crystalline thermoplastic with a melting temperature of $\sim 177°C$ and a glass transition temperature of $\sim 4°C$. It can be processed by melting or solvent-based technologies (Chen, 2005). PHB is a relatively stiff and brittle material, which limits its application in the biomaterials field (Chen and Wu, 2005). To overcome the limitation in properties and the narrow processing window of PHB, it has been proposed to use copolymers of 3-hydroxybutyrate and 3-hydroxyvalerate (PHBV), poly 4-hydroxybutyrate (P4HB), and copolymers of 3-hydroxybutyrate and 3-hydroxyhexanoate (PHBHHx) (Nair and Laurencin, 2007).

In general, PHAs are biodegradable and possesses potential biocompatibility (Chen and Wu, 2005). PHAs can be degraded within 3—9 months by many microorganisms into carbon dioxide and water. Their primary breakdown products are 3-hydroxyacids, which are naturally found in the human body (R-3-hydroxybutyric acid is a normal constituent of blood). However, PHB has a rather slow degradation in the body due to its high crystallinity. Thus, medical studies are more focused on the copolymer PHBV, which is less crystalline and thus undergoes degradation at a much faster rate (Zinn et al., 2001).

PROCESSING METHODS

Electrospinning was successfully used to produce ultrafine mats from PHB, PHBV, and their 50/50 w/w blend to be used as bone scaffolds (Sombatmankhong et al., 2007). In a different study, electrospinning was used to fabricate both nanofibers with an average size in the range of 300—500 nm and nanofibrous membranes from PHBHHx (Cheng et al., 2008). An improved method, consisting of combining conventional electrospinning with a gas-jet, was developed to produce PHB-based scaffolds containing nanosized HAp and possessing an ECM-like topography (Guan et al., 2008).

PHBV and PHBV/HAp composite scaffolds can be fabricated using the emulsion freezing/freeze-drying method (Sultana and Wang, 2006). These scaffolds were characterized as being highly porous, with pore size ranging from several microns to around 300 μm, and with an interconnected porous structure. Moreover, the incorporation of HAp, besides promoting osteoconductivity, lowered the crystallinity of PHBV matrix and enhanced the mechanical properties of the composite scaffolds.

PHBHHx scaffolds were prepared by directional freezing and phase-separation (Lin et al., 2008). The scaffolds were characterized as possessing a uniaxial microtubular structure able of guiding cell growth and possessing anisotropic mechanical properties. Moreover, it was possible to adjust the structure and mechanical properties of the scaffolds by changing the PHBHHx concentration, solvent, and freezing temperature. In turn, Wang et al. (2008a)

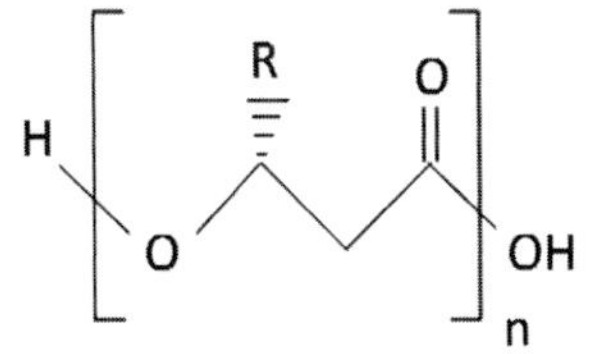

FIGURE 32.5
Chemical structure of poly ([R]-hydroxyalkanoates).

produced PHBHHx scaffolds using a solvent-lyophilizing method. The scaffolds had a non-directional porous structure, with an average pore size of 100 μm and a porosity of 90%.

PHB and PHBHHx scaffolds with or without addition of HAp can be produced by combining solvent casting with salt leaching (Wang et al., 2004, 2005). PHBV scaffolds were also prepared by combining freeze-drying and particulate leaching, aiming at producing a scaffold with high porosity and uniform pore sizes (Köse et al., 2003a). The obtained scaffolds were further treated with rf-oxygen plasma to modify their surface chemistry and hydrophilicity. The results have shown that the pore size, porosity, and pore morphology could be controlled by the polymer concentration, presence and size of leachable solutes, and surface modification. Sun et al. (2005) used a solvent-free technique consisting of compression molding, thermal processing, and the salt particulate leaching method to produce PHBV scaffolds. These scaffolds exhibited macroporous structure with interconnected open pores with size varying from 30 to 300 μm and a porosity of $80 \pm 1.2\%$. A similar approach was followed to prepare composite scaffolds of PHBV with bioactive wollastonite (Li and Chang, 2004).

POLYHYDROXYALKANOATES IN BONE TISSUE ENGINEERING APPLICATIONS

It has been shown that PHB scaffolds seeded with human maxillary osteoblasts can induce ectopic bone formation (Mai et al., 2006). Rat marrow osteoblasts were cultured on PHBV scaffolds and bone formation was investigated *in vitro* over a period of 60 days (Köse et al., 2003b). The results showed that osteoblasts could grow inside the scaffolds and lead to mineralization, making them suitable to be used in bone tissue engineering.

The *in vitro* biocompatibility of PHB, PHBHHx, and PLA scaffolds for growth of osteoblasts was investigated (Wang et al., 2004). It was found that PHBHHx had a better performance on attachment and proliferation of bone marrow cells than PLA scaffolds.

Recently, it was also demonstrated that 3-hydroxybutyrate (3HB), one of the degradation products of PHA, supported *in vitro* differentiation of murine osteoblast MC3T3-E1 in direct proportion to its concentration (Zhao et al., 2007). This study also revealed that 3HB administration can become an effective agent against osteoporosis.

The incorporation of HAp into PHA allows the production of bioactive and biodegradable composite scaffolds for bone tissue engineering applications. An *in vitro* study (Guan et al., 2008) has shown that PHB scaffolds containing nanosized HAp had positive effects on attachment, proliferation, and differentiation of BMSCs. Similar results were obtained by Wang et al. (2005), showing that the addition of HAp to PHB increased the mechanical properties and osteoblast responses including cell growth and alkaline phosphatase activity. Nevertheless, it was found that the addition of HAp to PHBHHx had an adverse effect on the biological performance of the composite scaffolds.

Collagen

Collagen (Fig. 32.6) is the most abundant structural protein in the connective tissue ECM and acts as the natural scaffold for cell attachment in the body. It gives mechanical stability,

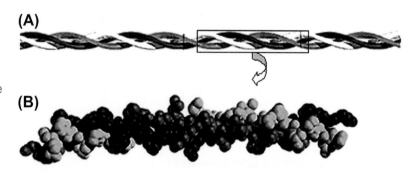

(A)

(B)

FIGURE 32.6
Overall collagen structure: van der Waals model of the helical structure of collagen (A) and three single chains intertwined into a triple-stranded helix (B). (Adapted from Fratzl, P. (2008). Collagen: Structure and Mechanics, an Introduction. In Collagen, Structure and Mechanics (Fratzl, P., ed.), pp. 1—12. Springer, New York).

strength, and toughness to a range of tissues. In special cases such as bone and dentin, the stiffness is improved by the inclusion of minerals (Fratzl, 2008).

The hallmark of a collagen is a molecule that consists of three polypeptide chains (α chains), each having a general amino acid motif of $(-Gly-X-Y)_n$, where the residues for X and Y are frequently the amino acids proline (Pro) and hydroxyproline (Hyp), with Gly-Prol-Hyp being the most common triplet found in collagen (Chau et al., 2007; Beckman et al., 2008). This repeating sequence allows the chains to form a right-handed triple-helical structure, with all glycine residues buried within the core of the protein and the residues X and Y exposed on the surface (Chau et al., 2007; Beckman et al., 2008). The individual triple helices are usually arranged in fibrils, which provide high tensile strength.

To date, more than 20 genetically distinct types of collagen have been identified. Collagens I—V are the five major types, with type I collagen being the most abundant form, and can be isolated from adult connective tissues, including skin, tendons, and bone (Beckman et al., 2008; Hulmes, 2008). Type II collagen can be found in cartilage, the developing cornea, and in the vitreous body of the eye (Wahl and Czernuszka, 2006). It can be isolated and purified from various animal species by enzyme treatment,with bovine collagen being the most commonly used type of collagen. Nevertheless, as an alternative to the possible complications (e.g. risk of bovine spongiform encephalopathy (BSE) transmission) that can arise from using collagen of bovine origin, collagen obtained from other sources such as porcine (Mimura et al., 2008) or, even safer, from marine species (Song et al., 2006) is often used in the production of scaffolds. Human recombinant collagen has also become available and can be used in scaffolds manufacturing (Wang et al., 2008b). Collagen possesses several advantages including biodegradability, low immunogenicity, and the ability to promote cellular attachment and growth, making it an attractive component of tissue engineering scaffolds (Lee et al., 2001). In the human body, collagen is degraded largely through the activity of the metalloprotease collagenase and some serine proteases (Gomes et al., 2007). By altering the degree to which it is crosslinked, it is possible to adapt the mechanical properties and degradation rate of collagen. Furthermore, the abundance of functional groups along its polypeptide backbone makes it highly receptive to the binding of genes, growth factors, and other biological molecules (Harley and Gibson, 2008).

PROCESSING METHODS

Collagen has been widely used to prepare scaffolds for bone and cartilage tissue engineering. The most used process to produce collagen scaffolds is freeze drying. 3D scaffolds consisting of mineralized type I collagen, with a composition that mimics the extracellular matrix of bone tissue, were generated by a freeze drying process, where the pore size could be controlled by temperature and by the freezing velocity (Gelinsky et al., 2008).

An improved method was developed by O'Brien and co-workers (2004) consisting of using a freeze drying process whereby a suspension of collagen and glycosaminoglycans (GAG) in acetic acid is cooled at a constant rate to a final temperature of freezing to produce collagen-GAG scaffolds with homogeneous pore structure. It was also found that, by varying the final freezing temperatures, a range of homogeneous collagen-GAG scaffolds with different mean pore sizes could be produced (O'Brien et al., 2005). In a different study, Chen et al. (2004, 2006) used a freeze-drying technique to produce PLGA-collagen hybrid meshes by forming collagen microsponges in the opening of PLGA knitted meshes. Recently, Kanungo et al. (2008) produced collagen-GAG scaffolds with varying mineral content via a triple co-precipitation method followed by freeze-drying. Although the scaffolds have been shown to possess pore size adequate for bone growth, the mechanical properties were lower than those for mineralized scaffolds made by other techniques, as well as cortical and cancellous bone. In a different study, a solvent casting/particulate leaching process was used to produce PLGA-collagen scaffolds (Lee et al., 2006). These scaffolds are characterized as being easy to fabricate and possessing a porous structure with a consistent interconnectivity throughout the entire scaffold.

Electrospinning has been also used to produce a 3D nanofibrous matrix of type I collagen aiming at mimic as close as possible the native extracellular matrix (ECM) (Sefcik et al., 2008).

To overcome some limitations of scaffolds related to cell migration and nutrients, diffusion constraints, collagen, and collagen/HAp scaffolds were produced, combining critical point drying and solid freeform technologies (Sachlos et al., 2003, 2006). These methods allow control of the pore size, biodegradation, and mechanical properties of the produced scaffolds and the incorporation of microchannels vasculature on their interior that can overcome the diffusion limitations of current foam scaffolds (Wahl and Gesnuska, 2006). *In vitro* and *in vivo* studies demonstrated that these scaffolds supported osteogenesis and chondrogenesis using HBMSCs and that the introduction of microchannels to scaffolds architecture enhanced chondrogenesis (Dawson et al., 2008). Aiming at generating scaffolds closely resembling the natural ECM components of bone, this processing method has been modified to produce collagen/HAp scaffolds (Sachlos et al., 2008).

Microfabrication is a promising technique for generating high-precision scaffolds. Chin et al. (2008) developed a microfabrication strategy based on gelling collagen-based components inside a microfluidic device, that allows well-controlled pore sizes inside the scaffold to be obtained. This approach has some disadvantages, including the need for dehydrating the scaffold before peeling it off and the delicate handling required for manipulating those thin hydrogel structures.

Liu et al. (2008) produced a gradient collagen/nano-HAp composite scaffold, with a Ca-rich side and a Ca-depleted side, aimed at applications in tissues with gradient properties, such as osteochondral bone.

The controlled release of signaling molecules, such as growth factors, from the scaffolds is critical for tissue repair, by providing cell guidance and development. PLGA-based microspheres encapsulating a model protein were imbedded in collagen and collagen/HAp scaffolds (Ungaro et al., 2006).

COLLAGEN IN BONE TISSUE ENGINEERING APPLICATIONS

Type I collagen is the major organic component of the ECM in bone and can play an important role in bone tissue engineering. Type I collagen (bovine) is the basis of several commercial products including Collapat II, Healos, Collagraft, and Biostite, among others (Wahl and Czernuszka, 2006). Several studies demonstrated that the use of type I collagen matrices can promote osteogenic differentiation and mineralization of marrow stromal cells and human adipose stem cells (Byrne et al., 2008; Kakudo et al., 2008; Sefcik et al., 2008). Another study demonstrated that a collagen scaffold (Gingistat®) is suitable for supporting distribution of MSCs and their commitment to form bone tissue (Donzelli et al., 2007). *In vitro* evaluation of the degradation time of the collagen sponge (degraded in 4–5 weeks) suggested that its use in *in vivo* experiments may be hindered by the complete dissolution of the scaffolds prior to the healing process and before bone formation is completed.

Tierney et al. (2009) examined the effects of varying collagen concentrations and crosslink densities on the biological, structural, and mechanical properties of collagen/GAG scaffolds for bone tissue engineering. The results indicated that doubling the collagen content to 1% and dehydrothermally crosslinking the scaffold at 150°C for 48 h enhances its mechanical and biological properties, making it more attractive for use in bone tissue engineering.

Aiming at mimicking the microstructure and composition of bone ECM, several studies (Bernhardt et al., 2008; Dawson et al., 2008; Pek et al., 2008) have been conducted to produce scaffolds based on type I collagen combined with HAp. The collagen-HAp composite scaffolds supported the osteogenic differentiation of human bone marrow-derived stromal cells (hBMSCs) both *in vitro* and *in vivo*. Additionally, extensive new osteoid formation of the implant was observed in the areas of vasculature, *in vivo*. Another study demonstrated that porous nanocomposite scaffolds containing collagen fibers and synthetic apatite nanocrystals successfully healed critical-sized defects in the femur of Wistar rats and on the tibia of Yorkshire-Landrace pigs (Pek et al., 2008).

Silk fibroin

Silks are natural fibrous proteins produced by a variety of species, including lepidoptera, scorpions, and spiders. Silks differ widely in composition, structure, and properties, depending on the specific source. The most extensively characterized silks are produced from the domesticated silkworm, *Bombyx mori*. These have been the main silk-like material used in biomedical applications, particularly as sutures (Altman et al., 2003; Wang et al., 2006).

Silkworm silk is composed of a filament core protein termed fibroin (the major component), and sericin, a water-soluble glue-like protein that bind the fibroin fibers together (MacIntosh et al., 2008). Silk fibroin is a hydrophobic glycoprotein and consists of heavy and light chain polypeptides of ~ 350 and ~ 25 kDa, respectively, linked by a disulfide bond (Tanaka et al., 1993; Kundu et al., 2008). The light chain, which is linked to the heavy chain, plays only a marginal role in the fiber. The larger heavy chain is glycine (Gly) rich and most of its amino acid composition consists of Gly, alanine (Ala), and serine (Ser) (Zhou et al., 2001).

In the solid state, silk fibroin from *B. mori* can assume two distinct crystalline structures, namely silk I (before spinning) and silk II (after spinning) (Yao et al., 2004). Structure determination of silk I is difficult because any attempt to study it causes its conversion from the silk I form to the silk II form. A structural model (Fig. 32.7) was proposed by Asakura and colleagues (Yao et al., 2004), and can be seen as a repeated β-turn type II structure. Regarding the silk II structure, although there are some reports suggesting the possession of some intrinsic structural disorder, diatoms are basically in accordance with Marsh's antiparallel β-sheet model based on a fiber diffraction study (Yao and Asakura, 2008).

The motivation for using fibroin in biomedical and tissue engineering applications derives from the unique mechanical properties of these fibers, the versatility in its processing, as well as its

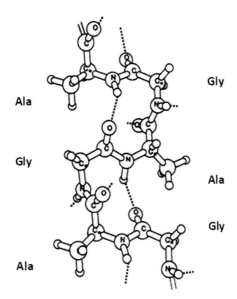

FIGURE 32.7
The conformation of a repeated β-turn type II — like molecules as a model for silk I. (Adapted from Zhou, C. Z., Confalonieri, F., Jacquet, M., Perasso, R., Li, Z. G. & Janin, J. (2001). Silk fibroin: Structural implications of a remarkable amino acid sequence. Proteins-Structure Function and Genetics 44, 119—122).

biocompatibility and low inflammatory response (Panilaitis et al., 2003; Meinel et al., 2005a; Wang et al., 2006, 2008c). Immunogenic reactions to silk sutures have been largely attributable to the sericin proteins (Panilaitis et al., 2003). Twenty-five to thirty percent of the silk cocoon is composed of sericins that can be removed by boiling the material in an alkaline solution to obtain purified silk fibroin (Vepari and Kaplan, 2007). While silk is an FDA-approved biomaterial defined by the US Pharmacopeia as non-degradable, fibroin is proteolytically degraded with predictable long-term degradation characteristics (Altman et al., 2003; Horan et al., 2005). The susceptibility to proteolytic hydrolysis of silk fibroin structures can be enhanced by using an all-aqueous process for their preparation (Kim et al., 2005). Recently, *in vivo* studies have also demonstrated that, besides the processing method, the processing variables (silk fibroin concentration and pore size) also affect the degradation rate (Wang et al., 2008c).

PROCESSING METHODS

It is well known that silk fibroin can be processed into various products with different morphologies by using different methods. Nazarov et al., (2004) used three fabrication techniques, freeze-drying, salt leaching, and gas foaming, to form porous 3D silk biomaterial matrices. Freeze-dried scaffolds, processed with 15% methanol or 15% 2-propanol, at −20 or −80°C, presented highly interconnected and porous structures with an average pore size in the range of $50 \pm 20\,\mu m$. Moreover, some of the scaffolds formed two distinct layers, an upper layer with flakelike pores and a bottom layer that was more condensed and compact. The salt-leached scaffolds used NaCl as a porogen and formed pores with a size in the range of $202 \pm 112\,\mu m$. Although the pores were larger in comparison to those generated by the freeze-drying method, the pore structure was not highly interconnected in contrast to those obtained by lyophilization. Scaffolds formed by gas-foaming (ammonium bicarbonate was used as the porogen) showed a highly interconnected open pore morphology with diameters in the range of $155 \pm 114\,\mu m$. Furthermore, the gas-foamed process did not leave a skin layer at the surface of the scaffold, leading to the conclusion that the scaffolds formed by gas foaming were the most promising of the matrices prepared. Aiming to avoid the use of organic solvents or harsh chemicals, Kim et al. (2005) reported the formation of 3D silk fibroin porous scaffolds prepared by an all-aqueous process. By adjusting both the concentration of silk fibroin in water and NaCl particle size, it was possible to control the morphological and functional properties of the scaffolds.

A salt-leaching method with sieved NaCl crystals has been used by Hofmann et al. (2007), aiming at engineering a 3D silk fibroin scaffold with separate domains of different pore diameters on a single scaffold. The produced scaffolds had a total porosity of ~95% but mixed pore sizes; i.e. on one side of the scaffold the pore diameter ranged between 112 and 224 μm, and those on the other ranged between 400 and 500 μm.

Due to the high surface area, fibers with nanoscale diameters can provide benefits in tissue engineering applications. Thus, electrospinning has been used in the fabrication of silk-based biomaterial scaffolds, and, to improve the processability of silk solutions, polyethylene oxide (PEO) can be blended with the aqueous solution of fibroin (Jin et al., 2004). Moreover, biomimetic alignment of fibers can be achieved using a cylindrical collector and controlled as a function of its speed during electrospinning of a silk fibroin solution blended with PEO (Meinel et al., 2009). However, nanofibrous scaffolds developed through the electrospinning technique might have structural limits for cell proliferation because the characteristic pore size is too small for the cells to grow inside. To overcome this limitation, nanofibrous silk fibroin scaffolds have been prepared via electrospinning followed by a salt-leaching method (Baek et al., 2008). The obtained scaffolds had uniformly distributed pores and high porosity (about 94%), but low pore interconnectivity, in a range of pore size from 58 to 930 μm.

Recently, direct ink writing was used to fabricate microperiodic scaffolds of regenerated silk fibroin (Ghosh et al., 2008). The method consisted of extruding an ink (fibroin solution) in a layer-by-layer fashion through a fine nozzle to produce a 3D array of silk fibroin fibers 5 μm in diameter — much finer than those produced by other rapid prototyping methods.

SILK FIBROIN IN BONE TISSUE ENGINEERING APPLICATIONS

Silk fibroin is an attractive biomaterial in which mechanical performance and biological interactions are major factors for success, including in its application for bone tissue engineering. Silk fibroin scaffolds were shown to be suitable substrates to engineer bone-like tissue *in vitro* (Meinel et al., 2004a,b). Human mesenchymal stem cells (hMSCs) cultured on porous and 3D silk scaffolds were differentiated into bone depositing cells *in vitro*, which resulted in the formation of a trabecular-like network of tissue-engineered bone structures (Meinel et al., 2005a). It was also demonstrated that the implantation of silk fibroin scaffolds cultured with human mesenchymal stem cells pre-differentiated along an osteoblastic lineage promoted bone formation in critical-sized cranial (Meinel et al., 2005b) and mid-femoral segmental (Meinel et al., 2006) defects.

Aiming at improving bone tissue engineering outcomes, BMP-2 was loaded in porous silk fibroin scaffolds (Karageorgiou et al., 2006). The results showed that BMP-2 induced hMSCs to undergo osteogenic differentiation when the seeded scaffolds were cultured in medium supplemented with osteogenic stimulants. Moreover, when implanted in critical-sized cranial defects in mice, scaffolds loaded with BMP-2 and seeded with hMSCs resulted in significant bone ingrowths. *In vitro* bone formation from hMSCs was significantly improved by incorporating functional factors, such as BMP-2 and a bone-like mineral HAp, into silk fibroin scaffolds (Li et al., 2006; Kim et al., 2008a).

Bone is a complex hierarchical structure having variable pore sizes. Aiming at mimicking the physiological tissue morphology, Hofmann et al. (2007) engineered different bone-like structures using scaffolds with small pores (112—224 mm in diameter) on one side and large pores (400—500 mm) on the other. MicroCT analysis revealed the pore structure of the newly formed tissue and results suggested that the structure of tissue-engineered bone was controlled by the underlying scaffold geometry.

One of the major challenges in bone repair and regeneration is vascularization. One of the current approaches to improving the vascularization of bone tissue-engineered constructs is pre-vascularization by including endothelial cells. Outgrowth endothelial cells (OECs) isolated and expanded from heterogeneous human peripheral blood cultures were investigated regarding their ability to serve as an autologous cell source for the endothelialization of 3D silk fibroin scaffolds (Fuchs et al., 2006). Results have shown both a close interaction of OECs with the scaffolds and the formation of microvessel-like structures induced by angiogenic stimuli involved in the processes of neo-vascularization. More recently, Fuchs et al. (2009) have shown that OECs in co-culture with human primary osteoblasts on silk fibroin scaffolds formed highly organized pre-vascular structures.

NATURAL-BASED CERAMICS

Ceramics can be defined as inorganic and non-metallic compounds. Silica-based bioactive glasses and calcium phosphate ceramics have long been known to promote or support bone formation. Currently, those materials are used in the clinic as bone defect-filling biomaterials. Herein, we will highlight the routes and the natural origin sources for obtaining these materials, and review their application in bone tissue engineering strategies.

Calcium phosphates

Calcium phosphates such as hydroxyapaptite (HAp), which is represented by the chemical formula $(Ca_{10}(PO_4)_6(OH)_2)$, possesses a structure that resembles the primary mineral component of bone hydroxyapatite. For some time now, bone substitute materials composed of HAp have been available for use in orthopaedics due to their good biocompatibility, bioactivity, high osteoconductive and/or osteo-inductive capacity, non-toxicity, non-inflammatory behavior, and non-immunogenic properties. The abundant calcium carbonate ($CaCO_3$), which is not as interesting as calcium phosphates from the regenerative application point of view, is the calcium precursor material for obtaining different ceramics. Consequently, there is a growing interest in finding new sources of this inorganic compound. Some examples of species (Olin et al., 1991; Elsinger and Leal, 1996; Rahimi et al., 1997; Fuchs et al., 2006; Heinemann et al., 2006; Lemos et al., 2006; Li et al., 2006; Jacob et al., 2008; Sethmann and Wörheide, 2008; Yaping and Yu, 2008; Kamenos et al., 2009; Born et al., 2010) possessing a calcium carbonate skeleton with potential for applications in regenerative medicine strategies are listed in Table 32.1.

Marine origin nano-organized ceramics provide an abundant source of novel bone replacement materials, and also are a source of inspiration for the development of novel biomimetic composites.

Natural corals are composed by an organic matrix with calcified nodes. The inorganic part is very interesting from the commercial point of view. Actually, coral skeletal carbonate partially converted to HAp material is already commercially available (Interpore Cross Inc.). Figure 32.8 shows the variety of red algae that the 3B's Research Group (Portugal) has been proposing to be used as a source of calcium carbonate aimed at finding application in bone tissue engineering strategies.

Shells mainly consist of calcium carbonate forming multilayered microstructures, and a small amount of organic component (1−5 wt%), mainly located within the inter-crystalline

TABLE 32.1 Organisms that possess a calcium carbonate structure with potential interest in the regenerative medicine field.

Organisms	Sources	Species	Potential application(s)	References
Marine	Corals	*Coralline officinallis*	Bone filler and scaffolds	146; 147; 148
		Lithothamnion glaciale	Bone filler or scaffolds	149
		Phymatholithon calcareum	Bone filler or scaffolds	149
		Isidella sp. (Bamboo corals)	Bone filler and scaffolds	150
	Sponges (Calcarea)	Calcareus sponge spicules from triactines of *Pericharax heteroraphis*	Precursor material for bioceramic coatings	151
		Verongula gigantea	Bone filler	150
	Mollusk shells	Nacre from *Haliotis* (abalone);*Mytilus galloprovincialis* and *Ostrea edulis* (oysters); and *Pinctada maxima* (bivalve)	Precursor material for bioceramic coatings	152; 153; 154; 155
	Fish (fish bones)	*Prionace glauca* (blue shark)	Bone filler and precursor material for bioceramic coatings	156

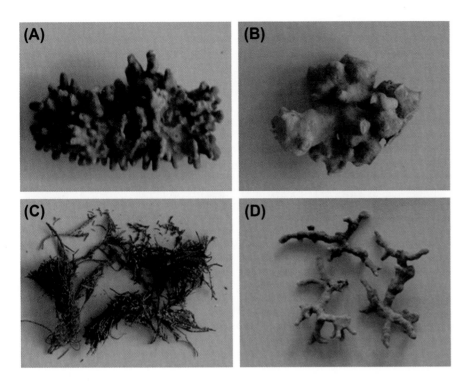

FIGURE 32.8
Photographs of different red algae used as sources of calcium carbonate. (A,B) *Lithothamnion glaciale*; (C) *Coralline officinallis*; (D) *Phymatholithon calcareum*.

boundaries. Despite this composition, and owing to the special composite microstructure, mollusk shells present an enhancement in toughness by three orders of magnitude with respect to non-biogenic calcium carbonate (Currey, 1977; Jackson et al., 1988). Some shells are composed by an inner nacreous layer consisting of aragonite in polygonal tablets (a calcium carbonate), 10–20 μm wide and 0.5 μm thick, between thin sheets of organic matrix. Nacre is a natural composite consisting of 95–98 wt% of inorganic and 2–5 wt% of organic matter. This matrix is formed by a protein-polysaccharide and limits the thickness of the crystals and is structurally important in the mechanical design of the shell (Addadi and Weiner, 1992; Levi-Kalisman et al., 2001).

As reported elsewhere (Born et al., 2010) sponges are also fascinating species because of its hierarchical organization, i.e. fibrous skeletons (Demospongiae) and mineralized spicules, which may contain amorphous silica (Demospongiae and Hexactinellida) or calcium carbonate (Calcarea).

PROCESSING METHODS

There are many sources of calcium carbonate, but coral skeletal carbonate (Fig. 32.9) has been attracting a great deal of attention as a precursor in the preparation of substitute materials for orthopaedics and dentistry. This is mainly due to its unique architecture, i.e. porosity, pore size and pore interconnectivity, which have been shown (Kuhne et al., 1994; Laine et al., 2008) to be key issues in bone tissue regeneration.

Besides its microstructure, other characteristics play a key role in the *in vivo* performance of these biomaterials. Microstructural composition and mechanical properties are key issues that also need to be considered. In this respect, it has been reported (Ben-Nissan, 2003) that marine-derived calcium carbonate skeletons are unsuitable for most applications due to its fast dissolution rate and poor stability. As previously mentioned, to circumvent these limitations, several authors (Ben-Nissan et al., 2004; Balázsi et al., 2007) have shown that it is possible to convert the hard calcium carbonate skeleton of mineralized algae into calcium phosphates. Using a hydrothermal method, it has already become possible to partially convert coralline

573

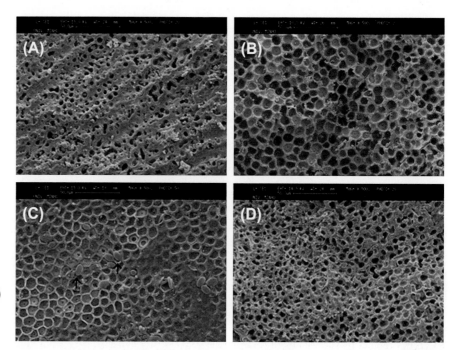

FIGURE 32.9
SEM micrographs of different red algae used as sources of calcium carbonate. (A,B) *Lithothamnion glaciale*; (C) *Coralline officinallis*; (D) *Phymatholithon calcareum*. Black arrows indicate the diatom organisms.

calcium carbonate to HAp. A major bottleneck has been the difficulty in converting the coral carbonate skeletons into calcium phosphates without destroying its native architecture. Oliveira et al. (2007) reported different routes to convert the calcium carbonate skeleton of *Coralline officinallis* red algae into calcium-phosphates. This work showed that, by performing a combined treatment (thermal and chemical), it is possible to obtain a calcium-phosphate material with HAp nanocrystallites, while the native microstructure of the red algae was preserved. First, red algae particulates (Fig. 32.10) free of the organic phase were obtained by heat treatment at 400°C for 3 h, in a furnace. This temperature was chosen since it has been reported (Sivakumar et al., 1996) that at higher temperatures carbonate phases may decompose.

At higher magnification (Fig. 32.11) it is possible to observe that by selecting the burning temperature and time it was possible to preserve the native red algae architecture. From Figure 32.11, we can distinguish the different microporous architectures of the several red algae. It is also possible to detect the existence of nano-sized channels responsible for the pores' interconnectivity (Fig. 32.11A,D).

In the same work, the conversion of the calcium carbonate skeleton into calcium-phosphates was achieved following the hydrothermal exchange strategy (Roy and Linnehan, 1974) (Eq. (1)):

$$10CaCO_3 + 6(NH_4)2HPO_4 + 2H_2O \rightarrow Ca_{10}(PO_4)_6(OH)_2 + 6(NH_4)2CO_3 + 4H_3CO_3 \quad (1)$$

This step forward seems very promising in terms of developing adequate algae-derived calcium-phosphate particulates (Fig. 32.12) to find applications as bone filler and for tissue engineering scaffolding.

Walsh et al. (2008) prepared coralline-derived HAp by developing a low-pressure hydro-thermal process. The synthesis method consisted of using ambient pressure at a low temperature of 100°C in a highly alkaline environment. Results have shown that the synthesized HAp maintained the unique microporous structure of the original algae. Therefore, in order to convert carbonate phases into HAp using the hydrothermal method, we should bear in mind: (1) to remove the organic matter from algae by burning or using chemical methods, (2) to avoid decompose carbonate phases, i.e. use relatively low temperatures because the carbonate

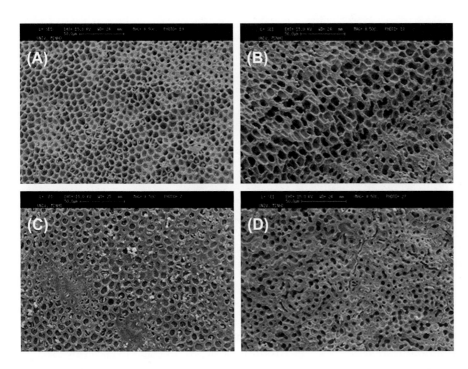

FIGURE 32.10
SEM micrographs of different red algae
after heat treatment at 400°C for 3 h. (A,B)
Lithothamnion glaciale; (C) *Coralline officinallis*;
(D) *Phymatholithon calcareum*.

phases readily decompose to CO_2 and Calcium oxide upon processing at high temperatures
and (3) to preserve the original algae morphology.

It was shown that nacre coatings or seashells may be transformed into apatite using mild
chemical methodologies (Vecchio et al., 2007; Guo and Zhou, 2008). Such strategies may be
interesting if it is intended to keep the internal hierarchical structure of the natural composite.
It is also clear that nacre, or nacre-based materials, may find applications in the biomedical
field, namely in the orthopedic or dental areas. Several strategies have been proposed to

575

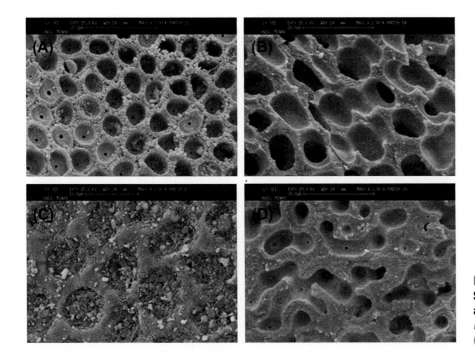

FIGURE 32.11
SEM micrographs of different red algae
after heat treatment at 400°C for 3 h. (A,B)
Lithothamnion glaciale; (C) *Coralline officinallis*;
(D) *Phymatholithon calcareum*.

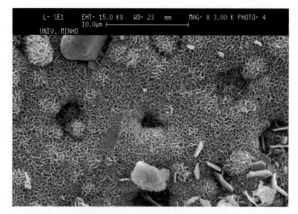

FIGURE 32.12

SEM micrograph of *Coralline officinallis* particulate after performing the heat treatment at 400°C for 3 h, followed by chemical treatment with $(NH_4)_2HPO_4$ for 28 days. It is possible to observe the HAp nanocrystallites in the coral surface.

synthetically produce materials with a similar structure to nacre, based on the concept of producing nanolaminates combining inorganic and organic layers or domains. As reviewed elsewhere (Luz and Mano, 2009), the methodologies may be organized into (1) covalent self-assembling or bottom-up approaches, (2) electrophoretic deposition, and (3) layer-by-layer or template inhibition. Osteoconductive nacre-based coatings may be produced by using, for example, bioactive nanoparticles in their construction. As an example, the layer-by-layer methodology was used to build up nanostructured hybrid coatings by sequential deposition of bioactive glass-ceramic nanoparticles, exhibiting a negatively charged surface, and chitosan, a positively charged macromolecule (Couto et al., 2009). Such biodegradable coatings promoted the precipitation of apatite *in vitro* and are believed to have potential to be used in a series of orthopaedic applications.

Natural HAp has also been extracted from biowaste such as bovine bone (Bakarat et al., 2009). Three different processes have been applied to obtain the natural HAp: (1) thermal decomposition, (2) subcritical water, and (3) alkaline hydrothermal processes.

CALCIUM PHOSPHATES IN BONE TISSUE ENGINEERING APPLICATIONS

As mentioned in the previous sections of this chapter, *in situ* tissue regeneration and tissue implantation require the use of temporary matrices for tissue growth. Coral-derived materials have been widely used in granules and blocks for bone filling and in bone tissue engineering scaffolding (Finn et al., 1980; Bay et al., 1993; Elsinger and Leal, 1996; Wolfe et al., 1999; Turhani et al., 2005a). For example, Turhani et al. (2005b) demonstrated that C GRAFT/ Algipore (The Clinician's Preference LLC, Golden, CO), a bone substitute obtained from *Coralline officinallis*, supported the proliferation and differentiation of human osteoblast-like cells, *in vitro*. The biochemical data confirmed the osteogenic potential of C GRAFT/Algipore material, showing that it might be a suitable candidate for use as scaffolds in tissue engineering strategies, *in vivo*. Similarly, Norman et al. (1994) demonstrated that coralline HAp supports the growth of osteoblast-like cells, *in vitro*. Many authors (Martin et al., 1989, 1993; Bay et al., 1993; Orr et al., 2001; Ewers, 2005) have been showing the excellent biological and mechanical performance of marine algae-derived bone-forming materials, *in vivo*. Despite those works, the bone-regenerative approach requires that these temporary matrices can provide a support for osteoblast or stem cell functions, and thus needs to prove to be osteogenic, *in vivo*. Okumura et al. (1991) investigated the osteogenic potential of coralline HAp materials. In their-work, coralline HAp alone (control) and HAp scaffolds seeded with rat marrow cells were implanted subcutaneously in the back of Fischer rats for time periods of up to 24 weeks. While control ceramics (without cells) showed no bone formation, the ceramics/ marrow cells constructs showed *de novo* bone formation throughout the pore regions, and proceeded centripetally towards the centre of the pores (bonding osteogenesis) over time. In turn, Cui et al. (2007) investigated the performance of using adipose-derived stem cells (ASCs) and coral scaffolds in repairing a cranial bone defect in a canine model, and followed up the outcome for up to 6 months. This work showed that the defects were either repaired with ASC-coral constructs or with coral alone. MicroCT and histological data revealed that the defect area was repaired by typical bone tissue when using the ASC-coral constructs, while only minimal bone formation with fibrous connection was observed when using the coral alone. Thus, successful repair of bone defects by combining coral-derived scaffolds and cells with an osteogenic potential is a feasible strategy for bone regeneration that has been successfully reported in various works (Hamilton et al., 1992; Gravel et al., 2006; Geiger et al., 2007; Mygind et al., 2007). More information on these subjects may be found elsewhere (Cancedda et al., 2007; Liao et al., 2008; Ganey et al., 2009).

Similarly to other nature-origin calcium phosphate materials, nacre itself also integrates well into bone tissue (Atlan et al., 1999; Berland et al., 2005) and has been shown to stimulate the differentiation of stem cells into the osteoblastic lineage (Rousseau et al., 2008; Zhu et al., 2008).

Silicates

Nanocomposites are formed by combining natural polymers and inorganic solids and show at least one nanosized dimension. Biohybrid materials possessing inorganic nanometer-sized solids exhibit improved structural and functional properties of great interest for different applications, including regenerative medicine. Silica sol-gel chemistry is quite common in materials science. In this section, we emphasize the relevance of silicates in biohybrid materials development, flexibility, and processability.

Sponges and diatom, besides being inspiring and having biomimetic potential, can be also a source of silica and natural polymers or precursors for the development of new biomaterials. As previously mentioned, sponges (e.g. Demospongiae and Hexactinellida) may contain amorphous silica spicules (Born et al., 2010). Siliceous spicules are rod-like glassy spikes consisting of an axial filament surrounded by several hundred concentric layers of hydrated silica. At a lower scale, these layers are made of densely-packed silica nanoparticles in the 50—100 nm range. In turn, diatom (unicellular algae with cell walls of shape resembling a Petri dish that are indicated by black arrows in Fig. 32.9C) are also predominantly composed of a biomineral derived from hydrated silica (SiO_2). Actually, they can accumulate silica at intracellular concentrations up to 250 times higher than that in the surrounding media. These are classified as centric and pennate diatoms and can be distinguished from each other on the basis of cellular symmetry; i.e. centric diatoms are radially symmetrical, whereas pennate diatoms are elongated and bilaterally symmetrical. Further details on the physiology of these organisms may be found elsewhere (Zurzolo and Bowels, 2001).

PROCESSING METHODS

A wide variety of natural-origin materials have been proposed as implants or fillers for tissue regeneration, with a special emphasis on nanocomposites for bone tissue engineering. These hybrid materials generally incorporate natural inorganic nanofillers that can produce a reinforcing effect in the biopolymer matrix, thereby resulting in improved mechanical properties. Few silicate materials have been combined with natural-origin polymers to produce hybrid scaffolds. Instead, silica-based nanocomposites processed as nanospheres by means of spray-drying methods have been developed to be used as drug carriers (Boissière et al., 2007). Others (Daniel-da-Silva et al., 2008) been shown to be possible developing nanosized silica/κ-carrageenan hydrogels with potential for application in tissue engineering and cell encapsulation.

Silicate fibers prepared by combining the electrospinning method and sol-gel process have been also produced as scaffolds for bone tissue engineering (Shinji et al., 2006). In all these works, amorphous silica was synthesized *in situ*. Nevertheless, the use of natural-origin hydrated silica obtained from sponges and diatom (e.g. siliceous spicules) for preparing these types of materials may be envisaged due to their bioactive potential (Madhumathi et al., 2009).

Silicate including bioglass, $CaSiO_3$, and Ca-Si-M (M = Mg, Zn, Ti, Zr) ceramics has attracted a great deal of interest for bone tissue repair and dental applications. It is known that silicate glass can be converted to HAp by soaking the substrates in a solution of K_2HPO_4 with a pH value of 9.0 at 37°C. Recently, Hayakawa et al. (2009) proposed a new strategy for self-assembling one-dimensional HAp nanorods into organized superstructures. The nanometer-scale rod array of HAp crystals was successfully prepared by soaking calcium-containing silicate glass substrates in Na_2HPO_4 aqueous solution at 80°C for various periods up to 14 days.

SILICATE IN BONE TISSUE ENGINEERING APPLICATIONS

Few silicate materials of natural origin have been used for bone repair/regeneration purposes. Sepiolite is a natural magnesium silicate of sedimentary origin, and is one of the few examples that have been studied until now. Herrera et al. (1995) developed a bone substitute by combining this inorganic compound with collagen. In this work, the resulting material was implanted in rat calvarial defects and did not induce any toxic effect or necrosis. In turn, Madhumathi et al. (2009) produced hybrid composite scaffolds of chitin containing nano-silica. This early-stage work revealed that such hybrid scaffolds were bioactive in simulated body fluid (SBF) and biocompatible when tested with an MG 63 cell line, *in vitro*.

Diopside ($CaMgSi_2O_6$) powders and dense ceramics are also promising bioactive materials for bone repair. Wu et al. (2010) fabricated diopside scaffolds using the polymer sponge template method and showed that this material possesses enhanced mechanical strength and mechanical stability, and decreased degradation rate compared to bioglass and $CaSiO_3$. *In vitro* studies have demonstrated that diopside scaffolds supported human osteoblastic-like cells proliferation and the expression of ALP activity, suggesting that these could be promising bioactive materials for bone tissue engineering.

A great example of scaffolds produced using sponges as a source of biomaterials was reported by Ehrlich et al., (2007). This work reported the successful development of 3D hybrid scaffolds consisting of silica and collagen obtained from *Chondrosia reniformis*. *In vitro* studies revealed that the silica/collagen hybrid materials supported the adhesion, proliferation, and osteogenic differentiation of human mesenchymal stem cells.

As a concluding note, it is worth stating that all these research works are inspiring examples in the search for natural-origin species as sources or precursors of inorganic and organic materials. These will allow new ways to be opened for the development of adequate scaffolds, shaping the future of bone tissue engineering strategies.

References

Abdel-Fattah, W. I., Jiang, T., El-Bassyouni, G. E. T., & Laurencin, C. T. (2007). Synthesis, characterization of chitosans and fabrication of sintered chitosan microsphere matrices for bone tissue engineering. *Acta Biomater.*, 3, 503−514.

Addadi, L., & Weiner, S. (1992). Control and design principles in biological mineralization. *Angew. Chem. Int. Engl.*, 31, 153−169.

Altman, G. H., Diaz, F., Jakuba, C., Calabro, T., Horan, R. L., Chen, J., et al. (2003). Silk-based biomaterials. *Biomaterials*, 24, 401−416.

Atlan, G., Delattre, O., Berland, S., LeFaou, A., Nabias, G., Cot, D., & Lopez, E. (1999). Interface between bone and nacre implants in sheep. *Biomaterials*, 20, 1017−1022.

578

Baek, H. S., Park, Y. H., Ki, C. S., Park, J.-C., & Rah, D. K. (2008). Enhanced chondrogenic responses of articular chondrocytes onto porous silk fibroin scaffolds treated with microwave-induced argon plasma. *Surface Coatings Technol., 202,* 5794–5797.

Balázsi, C., Wéber, F., Kövér, Z., Horváth, E., & Németh, C. (2007). Preparation of calcium-phosphate bioceramics from natural resources. *J. Eur. Ceramic Soc., 27,* 1601–1606.

Balmayor, E. R., Tuzlakoglu, K., Azevedo, H. S., & Reis, R. L. (2009). Preparation and characterization of starch-poly-ε-caprolactone microparticles incorporating bioactive agents for drug delivery and tissue engineering applications. *Acta Biomater., 5,* 1035–1045.

Barakat, N. A. M., Khil, M. S., Omran, A. M., Sheikh, F. A., & Kim, H. Y. (2009). Extraction of pure natural hydroxyapatite from the bovine bones bio waste by three different methods. *J. Mater. Process. Technol., 209,* 3408–3415.

Bay, B. K., Martin, R. B., Sharkey, N. A., & Chapman, M. W. (1993). Repair of large cortical defects with block coralline hydroxyapatite. *Bone, 14,* 225–230.

Beckman, M. J., Shields, K. J., & Diegelmannm, R. F. (2008). Collagen. In G. E. Wnek, & G. L. Bowlin (Eds.), *Encyclopedia of Biomaterials and Biomedical Engineering* (pp. 628–638). New York: Taylor and Francis.

Ben-Nissan, B. (2003). Natural bioceramics: from coral to bone and beyond. *Curr. Opin. Solid State Mater. Sci., 7,* 283–288.

Ben-Nissan, B., Milev, A., & Vago, R. (2004). Morphology of sol-gel derived nano-coated coralline hydroxyapatite. *Biomaterials, 25,* 4971–4975.

Berland, S., Delattre, O., Borzeix, S., Catonne, Y., & Lopez, E. (2005). Nacre/bone interface changes in durable nacre endosseous implants in sheep. *Biomaterials, 26,* 2767–2773.

Bernhardt, A., Lode, A., Boxberger, S., Pompe, W., & Gelinsky, M. (2008). Mineralized collagen — an artificial, extracellular bone matrix — improves osteogenic differentiation of bone marrow stromal cells. *J. Mater. Sci. Mater. Med., 19,* 269–275.

Boesel, L. F., Mano, J. F., & Reis, R. L. (2004). Optimization of the formulation and mechanical properties of starch based partially degradable bone cements. *J. Mater. Sci. Mater. Med., 15,* 73–83.

Boissière, M., Allouche, J., Chanéac, C., Brayner, R., Devoisselle, J.-M., Livage, J., & Coradin, T. (2007). Potentialities of silica/alginate nanoparticles as hybrid magnetic carriers. *Int. J. Pharm., 344,* 128–134.

Born, R., Ehrlich, H., Bazhenov, V., & Shapkin, N. P. (2010). Investigation of nanoorganized biomaterials of marine origin. *Arabian J. Chem., 3,* 27–32.

Boutinguiza, M., Lusquiños, F., Comesaña, R., Riveiro, A., Quintero, F., & Pou, J. (2007). Production of microscale particles from fish bone by gas flow assisted laser ablation. *Appl. Surface Sci., 254,* 1264–1267.

Buléon, A., Colonna, P., Planchot, V., & Ball, S. (1998). Starch granules: structure and biosynthesis. *Int. J. Biol. Macromol., 23,* 85–112.

Byrne, E. M., Farrell, E., McMahon, L. A., Haugh, M. G., O'Brien, F. J., Campbell, V. A., et al. (2008). Gene expression by marrow stromal cells in a porous collagen-glycosaminoglycan scaffold is affected by pore size and mechanical stimulation. *J. Mater. Sci. Mater. Med., 19,* 3455–3463.

Cancedda, R., Giannoni, P., & Mastrogiacomo, M. (2007). A tissue engineering approach to bone repair in large animal models and in clinical practice. *Biomaterials, 28,* 4240–4250.

Chau, D. Y. S., Collighan, R. J., & Griffin, M. (2007). Collagen: structure and modification for biomedical applications. In P. J. Pannone (Ed.), *Trends in Biomaterials Research* (pp. 143–190). New York: Nova Science Publishers.

Chen, G.-Q. (2005). Polyhydroxyalkanoates. In R. Smith (Ed.), *Biodegradable Polymers for Industrial Applications* (pp. 32–56). Cambridge: Woodhead Publishing.

Chen, G.-Q., & Wu, Q. (2005). The application of polyhydroxyalkanoates as tissue engineering materials. *Biomaterials, 26,* 6565–6578.

Chen, G., Liu, D., Maruyama, N., Ohgushi, H., Tanaka, J., & Tateishi, T. (2004). Cell adhesion of bone marrow cells, chondrocytes, ligament cells and synovial cells on a PLGA-collagen hybrid mesh. *Mater. Sci. Eng. C, 24,* 867–873.

Chen, G., Tanaka, J., & Tateishi, T. (2006). Osteochondral tissue engineering using a PLGA-collagen hybrid mesh. *Mater. Sci. Eng. C, 26,* 124–129.

Cheng, M.-L., Lin, C.-C., Su, H.-L., Chen, P.-Y., & Sun, Y.-M. (2008). Processing and characterization of electrospun poly(3-hydroxybutyrate-co-3-hydroxyhexanoate) nanofibrous membranes. *Polymer, 49,* 546–553.

Chin, C. D., Khanna, K., & Sia, S. K. (2008). A microfabricated porous collagen-based scaffold as prototype for skin substitutes. Biomed. *Microdev., 10,* 459–467.

Chung, C., & Burdick, J. A. (2008). Engineering cartilage tissue. *Adv. Drug Deliv. Rev., 60,* 243–262.

Copeland, L., Blazek, J., Salman, H., & Tang, M. C. (2009). Form and functionality of starch. *Food Hydrocolloids, 23,* 1527–1534.

Correlo, V. M., Boesel, L. F., Pinho, E., Costa-Pinto, A. R., Silva, M. L. A. D., Bhattacharya, M., et al. (2008). Melt-based compression-molded scaffolds from chitosan-polyester blends and composites: morphology and mechanical properties, *91*, 489–504.

Costa-Pinto, A. R., Salgado, A. J., Correlo, V. M., Sol, P., Bhattacharya, M., Charbord, P., et al. (2008). Adhesion, proliferation, and osteogenic differentiation of a mouse mesenchymal stem cell line (BMC9) seeded on novel melt-based chitosan/polyester 3D porous scaffolds. *Tissue Engin. A, 14*, 1049–1057.

Couto, D. S., Alves, N. M., & Mano, J. F. (2009). Multilayer coatings combining chitosan with bioactive glass nanoparticles. *J. Nanosci. Nanotechnol., 9*, 1741–1748.

Cui, L., Liu, B., Liu, G., Zhang, W., Cen, L., Sun, J., et al. (2007). Repair of cranial bone defects with adipose derived stem cells and coral scaffold in a canine model. *Biomaterials, 28*, 5477–5486.

Currey, J. D. (1977). Mechanical properties of mother of pearl in tension. *Proc. R. Soc. Lond B., 196*, 443–463.

Daniel-da-Silva, A. L., Pinto, F., Lopes-da-Silva, J. A., Trindade, T., Goodfellow, B. J., & Gil, A. M. (2008). Rheological behavior of thermoreversible κ-carrageenan/nanosilica gels. *J. Colloid Interface Sci., 320*, 575–581.

Dawson, J. I., Wahl, D. A., Lanham, S. A., Kanczler, J. M., Czernuszka, J. T., & Oreffo, R. O. C. (2008). Development of specific collagen scaffolds to support the osteogenic and chondrogenic differentiation of human bone marrow stromal cells. *Biomaterials, 29*, 3105–3116.

di Martino, A., Sittinger, M., & Risbud, M. V. (2005). Chitosan: a versatile biopolymer for orthopaedic tissue-engineering. *Biomaterials, 26*, 5983–5990.

Donzelli, E., Salvadè, A., Mimo, P., Viganò, M., Morrone, M., Papagna, R., Carini, F., et al. (2007). Mesenchymal stem cells cultured on a collagen scaffold: *in vitro* osteogenic differentiation. *Arch. Oral Biol., 52*, 64–73.

Duarte, A. R. C., Mano, J. F., & Reis, R. L. (2009). Preparation of starch-based scaffolds for tissue engineering by supercritical immersion precipitation. *J. Supercritical Fluids, 49*, 279–285.

Duarte, A. R. C., Caridade, S. G., Mano, J. F., & Reis, R. L. (2010a). Processing of novel bioactive polymeric matrixes for tissue engineering using supercritical fluid technology. *Mater. Sci. Eng. C, 29*, 2110–2115.

Duarte, A. R. C., Mano, J. F., & Reis, R. L. (2010b). Dexamethasone-loaded scaffolds prepared by supercritical-assisted phase inversion. *Acta Biomater., 5*, 2054–2062.

Ehrlich, H., Worch, H., Custódio, M. R., Lôbo-Hajdu, G., Hajdu, E., & Muricy, G. (2007). Sponges as natural composites: from biomimetic potential to development of new biomaterials. In *Porifera Research: Biodiversity, Innovation and Sustainability* (pp. 303–312). Rio de Janeiro: Série Livros 28.

Elsinger, E. C., & Leal, L. (1996). Coralline hydroxyapatite bone graft substitutes. *J. Foot Ankle Surg., 35*, 396–399.

Ewers, R. (2005). Maxilla sinus grafting with marine algae derived bone forming material: a clinical report of long-term results. *J. Oral Maxillofac. Surg., 63*, 1712–1723.

Fakhry, A., Schneider, G. B., Zaharias, R., & Senel, S. (2004). Chitosan supports the initial attachment and spreading of osteoblasts preferentially over fibroblasts. *Biomaterials, 25*, 2075–2079.

Finn, R. A., Bell, W. H., & Brammer, J. A. (1980). Interpositional "Grafting" with autogenous bone and coralline hydroxyapatite. *J. Maxillofac. Surg., 8*, 217–227.

Fratzl, P. (2008). Collagen: structure and mechanics, an introduction. In P. Fratzl (Ed.), *Collagen, Structure and Mechanics* (pp. 1–12). New York: Springer.

Fuchs, S., Motta, A., Migliaresi, C., & Kirkpatrick, C. J. (2006). Outgrowth endothelial cells isolated and expanded from human peripheral blood progenitor cells as a potential source of autologous cells for endothelialization of silk fibroin biomaterials. *Biomaterials, 27*, 5399–5408.

Fuchs, S., Ghanaati, S., Orth, C., Barbeck, M., Kolbe, M., Hofmann, A., et al. (2009). Contribution of outgrowth endothelial cells from human peripheral blood on *in vivo* vascularization of bone tissue-engineered constructs based on starch polycaprolactone scaffolds. *Biomaterials, 30*, 526–534.

Furrer, P., Zinn, M., & Panke, S. (2008). Polyhydroxyalkanoate and its potential for biomedical applications. In N. M. Neves, J. F. Mano, M. E. Gomes, A. P. Marques, H. S. Azevedo, & R. L. Reis (Eds.), *Natural-based Polymers for Biomedical Applications* (pp. 416–445). Cambridge: Woodhead Publishing.

Ganey, T., Hutton, W., & Meisel, H. J. (2009). Osteoconductive carriers for integrated bone repair. *SAS J., 3*, 108–112.

Geiger, F., Lorenz, H., Xu, W., Szalay, K., Kasten, P., Claes, L., et al. (2007). VEGF producing bone marrow stromal cells (BMSC) enhance vascularization and resorption of a natural coral bone substitute. *Bone, 41*, 516–522.

Gelinsky, M., Welzel, P. B., Simon, P., Bernhardt, A., & König, U. (2008). Porous three-dimensional scaffolds made of mineralized collagen: preparation and properties of a biomimetic nanocomposite material for tissue engineering of bone. *Chem. Eng. J, 137*, 84–96.

Ghosh, S., Parker, S. T., Wang, X. Y., Kaplan, D. L., & Lewis, J. A. (2008). Direct-write assembly of microperiodic silk fibroin scaffolds for tissue engineering applications. *Adv. Func. Mater., 18*, 1883–1889.

Gomes, M. E., Ribeiro, A. S., Malafaya, P. B., Reis, R. L., & Cunha, A. M. (2001). A new approach based on injection molding to produce biodegradable starch-based polymeric scaffolds: morphology, mechanical and degradation behaviour. *Biomaterials, 22,* 883–889.

Gomes, M. E., Godinho, J. S., Tchalamov, D., Cunha, A. M., & Reis, R. L. (2002). Alternative tissue engineering scaffolds based on starch: processing methodologies, morphology, degradation and mechanical properties. *Mater. Sci. Eng. C, 20,* 19–26.

Gomes, M. E., Sikavitsas, V. I., Behravesh, E., Reis, R. L., & Mikos, A. G. (2003). Effect of flow perfusion on the osteogenic differentiation of bone marrow stromal cells cultured on starch-based three-dimensional scaffolds. *J. Biomed. Mater. Res. A, 67A,* 87–95.

Gomes, M. E., Holtorf, H. L., Reis, R. L., & Mikos, A. G. (2006). Influence of the porosity of starch-based fiber mesh scaffolds on the proliferation and osteogenic differentiation of bone marrow stromal cells cultured in a flow perfusion bioreactor. *Tissue Eng., 12,* 801–809.

Gomes, M. E., Azevedo, H. S., Malafaya, P. B., Silva, S. S., Oliveira, J. M., Silva, G. A., et al. (2007). Natural polymers in tissue engineering applications. In C. Van Blitterswijk, A. Lindahl, P. Thomsen, D. Williams, J. Hubbell, & R. Cancedda (Eds.), *Textbook on Tissue Engineering* (pp. 145–192). Amsterdam: Elsevier.

Gomes, M. E., Azevedo, H. S., Moreira, A. R., Ella, V., Kellomaki, M., & Reis, R. L. (2008). Starch-poly (epsilon-caprolactone) and starch-poly(lactic acid) fibre-mesh scaffolds for bone tissue engineering applications: structure, mechanical properties and degradation behaviour. *J. Tissue Eng. Regen. Med., 2,* 243–252.

Gravel, M., Gross, T., Vago, R., & Tabrizian, M. (2006). Responses of mesenchymal stem cell to chitosan-coralline composites microstructured using coralline as gas forming agent. *Biomaterials, 27,* 1899–1906.

Guan, D., Chen, Z., Huang, C., & Lin, Y. (2008). Attachment, proliferation and differentiation of BMSCs on gas-jet/electrospun nHAP/PHB fibrous scaffolds. *Appl. Surface Sci., 255,* 324–327.

Guo, Y., & Zhou, Y. (2008). Transformation of nacre coatings into apatite coatings in phosphate buffer solution at low temperatures. *J. Biomed. Mater. Res. A, 86A,* 510–521.

Hamilton, H. E., Christianson, M. D., Williams, J. P., & Thomas, R. A. (1992). Evaluation of vascularization of coralline hydroxyapatite ocular implants by magnetic resonance imaging. *Clin. Imaging, 16,* 243–246.

Harley, B. A. C., & Gibson, L. J. (2008). *In vivo* and *in vitro* applications of collagen-GAG scaffolds. *Chem. Eng. J., 137,* 102–121.

Hayakawa, S., Li, Y., Tsuru, K., Osaka, A., Fujii, E., & Kawabata, K. (2009). Preparation of nanometer-scale rod array of hydroxyapatite crystal. *Acta Biomater., 5,* 2152–2160.

Heinemann, F., Treccani, L., & Fritz, M. (2006). Abalone nacre insoluble matrix induces growth of flat and oriented aragonite crystals. *Biochem. Biophys. Res. Commun., 344,* 45–49.

Herrera, J., Olmo, N., Turnay, J., Sicilia, A., Bascones, A., Gavilanes, J., et al. (1995). Implantation of sepiolite-collagen complexes in surgically created rat calvaria defects. *Biomaterials, 16,* 625–631.

Ho, M. H., Wang, D. M., Hsieh, H. J., Liu, H. C., Hsien, T. Y., Lai, J. Y., et al. (2005). Preparation and characterization of RGD-immobilized chitosan scaffolds. *Biomaterials, 26,* 3197–3206.

Hofmann, S., Hagenmüller, H., Koch, A. M., Müller, R., Vunjak-Novakovic, G., Kaplan, D. L., et al. (2007). Control of *in vitro* tissue-engineered bone-like structures using human mesenchymal stem cells and porous silk scaffolds. *Biomaterials, 28,* 1152–1162.

Hoover, R. (2001). Composition, molecular structure, and physicochemical properties of tuber and root starches: a review. *Carbohydrate Polymers, 45,* 253–267.

Horan, R. L., Antle, K., Collette, A. L., Wang, Y., Huang, J., Moreau, J. E., et al. (2005). *In vitro* degradation of silk fibroin. *Biomaterials, 26,* 3385–3393.

Hulmes, D. J. S. (2008). Collagen diversity, synthesis and assembly. In P. Fratzl (Ed.), *Collagen, Structure and Mechanics* (pp. 15–41). New York: Springer.

Hutmacher, D. W., Schantz, J. T., Lam, C. X., Tan, K. C., & Lim, T. C. (2007). State of the art and future directions of scaffold-based bone engineering from a biomaterials perspective. *J. Tissue Eng. Regen. Med., 1,* 245–260.

Jackson, A. P., Vincent, J. F. V., & Turner, R. M. (1988). The mechanical design of nacre. *Proc. R. Soc. Lond B, 234,* 415–440.

Jacob, D. E., Soldati, A. L., Wirth, R., Huth, J., Wehrmeister, U., & Hofmeister, W. (2008). Nanostructure, composition and mechanisms of bivalve shell growth. *Geochim. Cosmochim. Acta, 72,* 5401–5415.

Jin, H.-J., Chen, J., Karageorgiou, V., Altman, G. H., & Kaplan, D. L. (2004). Human bone marrow stromal cell responses on electrospun silk fibroin mats. *Biomaterials, 25,* 1039–1047.

Kakudo, N., Shimotsuma, A., Miyake, S., Kushida, S., & Kusumoto, K. (2008). Bone tissue engineering using human adipose-derived stem cells and honeycomb collagen scaffold. *J. Biomed. Mater. Res. A, 84A,* 191–197.

Kamenos, N. A., Cusack, M., Huthwelker, T., Lagarde, P., & Scheibling, R. E. (2009). Mg-lattice associations in red coralline algae. *Geochim. Cosmochim. Acta, 73,* 1901–1907.

581

Kanungo, B. P., Silva, E., Vliet, K. V., & Gibson, L. J. (2008). Characterization of mineralized collagen-glycosaminoglycan scaffolds for bone regeneration. *Acta Biomater., 4,* 490–503.

Karageorgiou, V., Tomkins, M., Fajardo, R., Meinel, L., Snyder, B., Wade, K., et al. (2006). Porous silk fibroin 3D scaffolds for delivery of bone morphogenetic protein-2 *in vitro* and *in vivo*. *J. Biomed. Mater. Res. A, 78A,* 324–334.

Kim, U.-J., Park, J., Joo Kim, H., Wada, M., & Kaplan, D. L. (2005). Three-dimensional aqueous-derived biomaterial scaffolds from silk fibroin. *Biomaterials, 26,* 2775–2785.

Kim, H. J., Kim, U. J., Kim, H. S., Li, C. M., Wada, M., Leisk, G. G., et al. (2008a). Bone tissue engineering with premineralized silk scaffolds. *Bone, 42,* 1226–1234.

Kim, I. Y., Seo, S. J., Moon, H. S., Yoo, M. K., Park, I. Y., Kim, B. C., et al. (2008b). Chitosan and its derivatives for tissue engineering applications. *Biotechnol. Adv, 26,* 1–21.

Kong, L. J., Gao, Y., Lu, G. Y., Gong, Y. D., Zhao, N. M., & Zhang, X. F. (2006). A study on the bioactivity of chitosan/nano-hydroxyapatite composite scaffolds for bone tissue engineering. *Eur. Polymer J, 42,* 3171–3179.

Köse, G. T., Kenar, H., HasIrcI, N., & HasIrcI, V. (2003a). Macroporous poly(3-hydroxybutyrate-co-3-hydroxyvalerate) matrices for bone tissue engineering. *Biomaterials, 24,* 1949–1958.

Köse, G. T., Korkusuz, F., Korkusuz, P., Purali, N., Özkul, A., & Hasirci, V. (2003b). Bone generation on PHBV matrices: an *in vitro* study. *Biomaterials, 24,* 4999–5007.

Kuhne, J.-H., Bart, R., Frisch, B., Hamme, C., Jansson, V., & Zimmer, M. (1994). Bone formation in coralline hydroxyapatite: effects of pore size studied in rabbits. Acta Orthop. *Scand, 65,* 246–252.

Kundu, S. C., Dash, B. C., Dash, R., & Kaplan, D. L. (2008). Natural protective glue protein, sericin bioengineered by silkworms: potential for biomedical and biotechnological applications. *Prog. Polymer Sci., 33,* 998–1012.

Laine, J., Labady, M., Albornoz, A., & Yunes, S. (2008). Porosities and pore sizes in coralline calcium carbonate. *Materials Characterization, 59,* 1522–1525.

Lee, C. H., Singla, A., & Lee, Y. (2001). Biomedical applications of collagen. *Int. J. Pharm, 221,* 1–22.

Lee, S. J., Lim, G. J., Lee, J.-W., Atala, A., & Yoo, J. J. (2006). *In vitro* evaluation of a poly(lactide-co-glycolide)-collagen composite scaffold for bone regeneration. *Biomaterials, 27,* 3466–3472.

Lemos, A. F., Rocha, J. H. G., Quaresma, S. S. F., Kannan, S., Oktar, F. N., Agathopoulos, S., et al. (2006). Hydroxyapatite nano-powders produced hydrothermally from nacreous material. *J. Eur. Ceramic Soc., 26,* 3639–3646.

Levi-Kalisman, Y., Falini, G., Addadi, L., & Weiner, S. (2001). Structure of the nacreous organic matrix of a bivalve mollusk shell examined in the hydrated state using cryo-TEM. *J. Struct. Biol., 135,* 8–17.

Li, C., Vepari, C., Jin, H.-J., Kim, H. J., & Kaplan, D. L. (2006). Electrospun silk-BMP-2 scaffolds for bone tissue engineering. *Biomaterials, 27,* 3115–3124.

Li, H., & Chang, J. (2004). Fabrication and characterization of bioactive wollastonite/PHBV composite scaffolds. *Biomaterials, 25,* 5473–5480.

Liao, S., Chan, C. K., & Ramakrishna, S. (2008). Stem cells and biomimetic materials strategies for tissue engineering. *Mater. Sci. Eng. C, 28,* 1189–1202.

Lin, F., Li, Y., Jin, J., Cai, Y., Wei, K., & Yao, J. (2008). Deposition behavior and properties of silk fibroin scaffolds soaked in simulated body fluid. *Mater. Chem. Phys., 111,* 92–97.

Liu, L., Kong, X. D., Cai, Y. R., & Yao, J. M. (2008). Degradation behavior and biocompatibility of nano-hydroxyapatite/silk fibroin composite scaffolds. *Acta Chim. Sin, 66,* 1919–1923.

Liuyun, J., Yubao, L., Li, Z., & Jianguo, L. (2008). Preparation and properties of a novel bone repair composite: nano-hydroxyapatite/chitosan/carboxymethyl cellulose. *J. Mater. Sci. Mater. Med., 19,* 981–987.

Luz, G. M., & Mano, J. F. (2009). Biomimetic design of materials and biomaterials inspired by the structure of nacre. *Phil. Transac. R. Soc. A, 367,* 1587–1605.

MacIntosh, A. C., Kearns, V. R., Crawford, A., & Hatton, P. V. (2008). Skeletal tissue engineering using silk biomaterials. *J. Tissue Eng. Regen. Med., 2,* 71–80.

Madhumathi, K., Sudheesh Kumar, P. T., Kavya, K. C., Furuike, T., Tamura, H., Nair, S. V., et al. (2009). Novel chitin/nanosilica composite scaffolds for bone tissue engineering applications. *Int. J. Biol. Macromol., 45,* 289–292.

Madihally, S. V., & Matthew, H. W. T. (1999). Porous chitosan scaffolds for tissue engineering. *Biomaterials, 20,* 1133–1142.

Mai, R., Hagedorn, M. G., Gelinsky, M., Werner, C., Turhani, D., Späth, H., et al. (2006). Ectopic bone formation in nude rats using human osteoblasts seeded poly(3)hydroxybutyrate embroidery and hydroxyapatite-collagen tapes constructs. *J. Cranio-Maxillofac. Surg., 34,* 101–109.

Malafaya, P. B., Santos, T. C., van Griensven, M., & Reis, R. L. (2008). Morphology, mechanical characterization and *in vivo* neo-vascularization of chitosan particle aggregated scaffolds architectures. *Biomaterials, 29,* 3914–3926.

Manjubala, I., Ponomarev, I., Wilke, I., & Jandt, K. D. (2008). Growth of osteoblast-like cells on biomimetic apatite-coated chitosan scaffolds. *J. Biomed. Mater. Res. B Appl. Biomater., 84B,* 7–16.

Marques, A. P., Reis, R. L., & Hunt, J. A. (2002). The biocompatibility of novel starch-based polymers and composites: *in vitro* studies. *Biomaterials, 23,* 1471–1478.

Marques, A. P., Reis, R. L., & Hunt, J. A. (2005). An *in vivo* study of the host response to starch-based polymers and composites subcutaneously implanted in rats. *Macromol. Biosci., 5,* 775–785.

Martin, R. B., Chapman, M. W., Holmes, B. E., Sartoris, D. J., Shors, E. C., Gordon, J. E., et al. (1989). Effects of bone ingrowth on the strength and non-invasive assessment of a coralline hydroxyapatite material. *Biomaterials, 10,* 481–488.

Martin, R. B., Chapman, M. W., Sharkey, N. A., Zissimos, S. L., Bay, B., & Shors, E. G. (1993). Bone ingrowth and mechanical properties of coralline hydroxyapatite 1 yr after implantation. *Biomaterials, 14,* 341–348.

Martins, A. M., Alves, C. M., Reis, R. L., Mikos, A. G., & Kasper, F. K. (2008a). Towards osteogenic differentiation of marrow stromal cells and *in vitro* production of mineralized extracellular matrix onto natural scaffolds. In Understanding., Controlling Protein, R. Bizios, & D. Puleo (Eds.), *Biological Interactions on Materials Surfaces* (pp. 263–281). San Diego: Springer.

Martins, A. M., Santos, M. I., Azevedo, H. S., Malafaya, P. B., & Reis, R. L. (2008b). Natural origin scaffolds with *in situ* pore forming capability for bone tissue engineering applications. *Acta Biomater., 4,* 1637–1645.

Martins, A., Chung, S., Pedro, A. J., Sousa, R. A., Marques, A. P., Reis, R. L., et al. (2009a). Hierarchical starch-based fibrous scaffold for bone tissue engineering applications. *J. Tissue Eng. Regen. Med., 3,* 37–42.

Martins, A. M., Pham, Q. P., Malafaya, P. B., Sousa, R. A., Gomes, M. E., Raphael, R. M., et al. (2009b). The role of lipase and alpha-amylase in the degradation of starch/poly(ε-caprolactone) fiber meshes and the osteogenic differentiation of cultured marrow stromal cells. *Tissue Engin. A, 15,* 295–305.

Meinel, L., Karageorgiou, V., Fajardo, R., Snyder, B., Shinde-Patil, V., Zichner, L., et al. (2004a). Bone tissue engineering using human mesenchymal stem cells: effects of scaffold material and medium flow. *Ann. Biomed. Eng., 32,* 112–122.

Meinel, L., Karageorgiou, V., Hofmann, S., Fajardo, R., Snyder, B., Li, C. M., et al. (2004b). Engineering bone-like tissue *in vitro* using human bone marrow stem cells and silk scaffolds. *J. Biomed. Mater. Res. A, 71A,* 25–34.

Meinel, L., Fajardo, R., Hofmann, S., Langer, R., Chen, J., Snyder, B., et al. (2005a). Silk implants for the healing of critical size bone defects. *Bone, 37,* 688–698.

Meinel, L., Hofmann, S., Karageorgiou, V., Kirker-Head, C., McCool, J., Gronowicz, G., et al. (2005b). The inflammatory responses to silk films *in vitro* and *in vivo*. *Biomaterials, 26,* 147–155.

Meinel, L., Betz, O., Fajardo, R., Hofmann, S., Nazarian, A., Cory, E., et al. (2006). Silk based biomaterials to heal critical sized femur defects. *Bone, 39,* 922–931.

Meinel, A. J., Kubow, K. E., Klotzsch, E., Garcia-Fuentes, M., Smith, M. L., Vogel, V., et al. (2009). Optimization strategies for electrospun silk fibroin tissue engineering scaffolds. *Biomaterials, 30,* 3058–3067.

Mimura, T., Imai, S., Kubo, M., Isoya, E., Ando, K., Okumura, N., et al. (2008). A novel exogenous concentration-gradient collagen scaffold augments full-thickness articular cartilage repair. *Osteoarthritis Cartilage, 16,* 1083–1091.

Mistry, A. S., & Mikos, A. G. (2005). Tissue engineering strategies for bone regeneration. In *Regenerative Medicine II: Clinical and Preclinical Applications, Vol 94* (pp. 1–22). New York: Springer.

Mygind, T., Stiehler, M., Baatrup, A., Li, H., Zou, X., Flyvbjerg, A., et al. (2007). Mesenchymal stem cell ingrowth and differentiation on coralline hydroxyapatite scaffolds. *Biomaterials, 28,* 1036–1047.

Nair, L. S., & Laurencin, C. T. (2007). Biodegradable polymers as biomaterials. *Prog. Polymer Sci., 32,* 762–798.

Nazarov, R., Jin, H. J., & Kaplan, D. L. (2004). Porous 3D scaffolds from regenerated silk fibroin. *Biomacromolecules, 5,* 718–726.

Nesic, D., Whiteside, R., Brittberg, M., Wendt, D., Martin, I., & Mainil-Varlet, P. (2006). Cartilage tissue engineering for degenerative joint disease. *Adv. Drug Deliv. Rev., 58,* 300–322.

Nettles, D. L., Elder, S. H., & Gilbert, J. A. (2002). Potential use of chitosan as a cell scaffold material for cartilage tissue engineering. *Tissue Eng., 8,* 1009–1016.

Neves, N. M., Kouyumdzhiev, A., & Reis, R. L. (2005). The morphology, mechanical properties and ageing behavior of porous injection molded starch-based blends for tissue engineering scaffolding. *Mater. Sci. Eng. C, 25,* 195–200.

Norman, M. E., Elgendy, H. M., Shors, E. C., El-Amin, S. F., & Laurencin, C. T. (1994). An in-vitro evaluation of coralline porous hydroxyapatite as a scaffold for osteoblast growth. *Clin. Mater., 17,* 85–91.

O'Brien, F. J., Harley, B. A., Yannas, I. V., & Gibson, L. (2004). Influence of freezing rate on pore structure in freeze-dried collagen-GAG scaffolds. *Biomaterials, 25,* 1077–1086.

O'Brien, F. J., Harley, B. A., Yannas, I. V., & Gibson, L. J. (2005). The effect of pore size on cell adhesion in collagen-GAG scaffolds. *Biomaterials, 26,* 433–441.

Okumura, M., Ohgushi, H., & Tamai, S. (1991). Bonding osteogenesis in coralline hydroxyapatite combined with bone marrow cells. *Biomaterials, 12,* 411—416.

Olin, P. S., Ettel, R. G., & Schaffer, E. M. (1991). Improved pontic/tissue relationships using porous coralline hydroxyapatite block. *J. Prosthet. Dent., 66,* 234—238.

Oliveira, J. M., Grech, J. M. R., Leonor, I. B., Mano, J. F., & Reis, R. L. (2007). Calcium-phosphate derived from mineralized algae for bone tissue engineering applications. *Mater. Lett., 61,* 3495—3499.

Orr, T. E., Villars, P. A., Mitchell, S. L., Hsu, H. P., & Spector, M. (2001). Compressive properties of cancellous bone defects in a rabbit model treated with particles of natural bone mineral and synthetic hydroxyapatite. *Biomaterials, 22,* 1953—1959.

Pan, Z. H., & Jiang, P. P. (2008). Assessment of the suitability of a new composite as a bone defect filler in a rabbit model. *J. Tissue Eng. Regen. Med., 2,* 347—353.

Panilaitis, B., Altman, G. H., Chen, J., Jin, H.-J., Karageorgiou, V., & Kaplan, D. L. (2003). Macrophage responses to silk. *Biomaterials, 24,* 3079—3085.

Pek, Y. S., Gao, S., Arshad, M. S. M., Leck, K.-J., & Ying, J. Y. (2008). Porous collagen-apatite nanocomposite foams as bone regeneration scaffolds. *Biomaterials, 29,* 4300—4305.

Prabaharan, M., Rodriguez-Perez, M. A., de Saja, J. A., & Mano, J. F. (2007). Preparation and characterization of poly (l-lactic acid)-chitosan hybrid scaffolds with drug release capability. *J. Biomed. Mater. Res. B Appl. Biomater., 81B,* 427—434.

Rahimi, F., Maurer, B. T., & Enzweiler, M. G. (1997). Coralline hydroxyapatite: a bone graft alternative in foot and ankle surgery. *J. Foot Ankle Surg., 36,* 192—203.

Reis, R. L., & Cunha, A. M. (1995). *12th European Conference on Biomaterials.* Oporto: Portugal.

Reis, R. L., & Cunha, A. M. (2001). Starch and starch based thermoplastics. In K. H. Jurgen Buschow, R. W. Cahn, M. C. Flemings, B. Ilschner, E. J. Kramer, & S. Mahajan (Eds.), *Encyclopedia of Materials Science and Technology* (pp. 8810—8816). Amsterdam: Elsevier Science.

Reis, R. L., Cunha, A. M., Allan, P. S., & Bevis, M. J. (1996a). Mechanical behavior of injection-molded starch-based polymers. *Polymers Adv. Technol., 7,* 784—790.

Reis, R. L., Mendes, S. C., Cunha, A. M., & Bevis, M. J. (1996b). *Cambridge Polymer Conference — Partnership in Polymers.* Cambridge: England.

Reis, R. L., Cunha, A. M., Allan, P. S., & Bevis, M. J. (1997). Structure development and control of injection-molded hydroxylapatite-reinforced starch/EVOH composites. *Adv. Polymer Technol., 16,* 263—277.

Rousseau, M., Boulzaguet, H., Biagianti, J., Duplat, D., Milet, C., Lopez, E., et al. (2008). Low molecular weight molecules of oyster nacre induce mineralization of the MC3T3-E1 cells. *Biomed. Mater. Res., 85A,* 487—497.

Roy, D. M., & Linnehan, S. K. (1974). Hydroxyapatite formed from coral skeletal carbonate by hydrothermal exchange. *Nature, 247,* 220—222.

Sachlos, E., Gotora, D., & Czernuszka, J. T. (2006). Collagen scaffolds reinforced with biomimetic composite nano-sized carbonate-substituted hydroxyapatite crystals and shaped by rapid prototyping to contain internal microchannels. *Tissue Eng., 12,* 2479—2487.

Sachlos, E., Reis, N., Ainsley, C., Derby, B., & Czernuszka, J. T. (2003). Novel collagen scaffolds with predefined internal morphology made by solid freeform fabrication. *Biomaterials, 24,* 1487—1497.

Sachlos, E., Wahl, D. A., Triffitt, J. T., & Czernuszka, J. T. (2008). The impact of critical point drying with liquid carbon dioxide on collagen-hydroxyapatite composite scaffolds. *Acta Biomater., 4,* 1322—1331.

Salgado, A. J., Coutinho, O. P., & Reis, R. L. (2004). Novel starch-based scaffolds for bone tissue engineering: cytotoxicity, cell culture, and protein expression. *Tissue Eng., 10,* 465—474.

Salgado, A. J., Gomes, M. E., Coutinho, O. P., & Reis, R. L. (2005). Bone and articular cartilage tissue engineering: the biological components. In R. L. Reis, & J. San Roman (Eds.), *Biodegradable Systems in Tissue Engineering and Regenerative Medicine* (pp. 457—478). Boca Raton: CRC Press.

Salgado, A. J., Coutinho, O. P., Reis, R. L., & Davies, J. E. (2007). *In vivo* response to starch-based scaffolds designed for bone tissue engineering applications. *J. Biomed. Mater. Res. A, 80A,* 983—989.

Salgado, A. J., Sousa, R. A., Fraga, J. S., Pego, J. M., Silva, B. A., Malva, J. O., et al. (2009). Effects of starch/poly-caprolactone-based blends for spinal cord injury regeneration in neurons/glial cells viability and proliferation. *J. Bioactive Compatible Polymers, 24,* 235—248.

Santos, M. I., Fuchs, S., Gomes, M. E., Unger, R. E., Reis, R. L., & Kirkpatrick, C. J. (2007). Response of micro- and macrovascular endothelial cells to starch-based fiber meshes for bone tissue engineering. *Biomaterials, 28,* 240—248.

Sefcik, L. S., Neal, R. A., Kaszuba, S. N., Parker, A. M., Katz, A. J., Ogle, R. C., et al. (2008). Collagen nanofibres are a biomimetic substrate for the serum-free osteogenic differentiation of human adipose stem cells. *J. Tissue Eng. Regen. Med., 2,* 210—220.

Sendemir-Urkmez, A., & Jamison, R. D. (2007). The addition of biphasic calcium phosphate to porous chitosan scaffolds enhances bone tissue development *in vitro*. *J. Biomed. Mater. Res. A, 81A*, 624–633.

Sethmann, I., & Wörheide, G. (2008). Structure and composition of calcareous sponge spicules: a review and comparison to structurally related biominerals. *Micron, 39*, 209–228.

Shi, C. M., Zhu, Y., Ran, X. Z., Wang, M., Su, Y. P., & Cheng, T. M. (2006). Therapeutic potential of chitosan and its derivatives in regenerative medicine. *J. Surg. Res., 133*, 185–192.

Shinji, S., Yusuke, Y., Tetsu, Y., & Koei, K. (2006). Prospective use of electrospun ultra-fine silicate fibers for bone tissue engineering. *Biotechnol. J., 1*, 958–962.

Sivakumar, M., Kumar, T. S. S., Shantha, K. L., & Rao, K. P. (1996). Development of hydroxyapatite derived from Indian coral. *Biomaterials, 17*, 1709–1714.

Sombatmankhong, K., Sanchavanakit, N., Pavasant, P., & Supaphol, P. (2007). Bone scaffolds from electrospun fiber mats of poly(3-hydroxybutyrate), poly(3-hydroxybutyrate-co-3-hydroxyvalerate) and their blend. *Polymer, 48*, 1419–1427.

Song, E., Yeon Kim, S., Chun, T., Byun, H.-J., & Lee, Y. M. (2006). Collagen scaffolds derived from a marine source and their biocompatibility. *Biomaterials, 27*, 2951–2961.

Sousa, R. A., Kalay, G., Reis, R. L., Cunha, A. M., & Bevis, M. J. (2000). Injection molding of a starch/EVOH blend aimed as an alternative biomaterial for temporary applications. *J. Appl. Polymer Sci., 77*, 1303–1315.

Sousa, R. A., Correlo, V. M., Chung, S., Neves, N. M., Mano, J. F., & Reis, R. L. (2008). Processing of thermoplastic natural-based polymers: an overview of starch based blends. In N. M. Neves, J. F. Mano, M. E. Gomes, A. P. Marques, H. S. Azevedo, & R. L. Reis (Eds.), *Handbook of Natural-based Polymers for Biomedical Applications* (pp. 85–105). Cambridge: CRC Press.

Suh, J. K. F., & Matthew, H. W. T. (2000). Application of chitosan-based polysaccharide biomaterials in cartilage tissue engineering: a review. *Biomaterials, 21*, 2589–2598.

Sultana, N., & Wang, M. (2006). *5th Asian-Australian Conference on Composite Materials (ACCM-5)*. Hong Kong: Peoples R China.

Sun, J., Wu, J., Li, H., & Chang, J. (2005). Macroporous poly(3-hydroxybutyrate-co-3-hydroxyvalerate) matrices for cartilage tissue engineering. *Eur. Polymer J., 41*, 2443–2449.

Tanaka, K., Mori, K., & Mizuno, S. (1993). Immunological identification of the major disulfide-linked light component of silk fibroin. *J. Biochem., 114*, 1–4.

Tester, R. F., Karkalas, J., & Qi, X. (2004). Starch – composition, fine structure and architecture. *J. Cereal Sci., 39*, 151–165.

Tierney, C. M., Haugh, M. G., Liedl, J., Mulcahy, F., Hayes, B., & O'Brien, F. J. (2009). The effects of collagen concentration and crosslink density on the biological, structural and mechanical properties of collagen-GAG scaffolds for bone tissue engineering. *J. Mechan. Behav. Biomed. Mater., 2*, 202–209.

Turhani, D., Watzinger, E., Weienböck, M., Cvikl, B., Thurnher, D., Wittwer, G., et al. (2005a). Analysis of cell-seeded 3-dimensional bone constructs manufactured *in vitro* with hydroxyapatite granules obtained from red algae. *J. Oral Maxillofac. Surg., 63*, 673–681.

Turhani, D., Cvikl, B., Watzinger, E., Weißenböck, M., Yerit, K., Thurnher, D., et al. (2005b). *In vitro* growth and differentiation of osteoblast-like cells on hydroxyapatite ceramic granule calcified from red algae. *J. Oral Maxillofac. Surg., 63*, 793–799.

Tuzlakoglu, K., & Reis, R. L. (2007). Formation of bone-like apatite layer on chitosan fiber mesh scaffolds by a biomimetic spraying process. *J. Mater. Sci. Mater. Med., 18*, 1279–1286.

Tuzlakoglu, K., Alves, C. M., Mano, J. F., & Reis, R. L. (2004). Production and characterization of chitosan fibers and 3-D fiber mesh scaffolds for tissue engineering applications. *Macromol. Biosci., 4*, 811–819.

Ungaro, F., Biondi, M., d'Angelo, I., Indolfi, L., Quaglia, F., Netti, P. A., et al. (2006). Microsphere-integrated collagen scaffolds for tissue engineering: effect of microsphere formulation and scaffold properties on protein release kinetics. *J. Controlled Release, 113*, 128–136.

van Gaalen, S., Kruyt, M., Meijer, G., Mistry, A., Mikos, A. G., van den Beucken, J., et al. (2007). Tissue engineering of bone. In C. van Blitterswijk, A. Lindahl, P. Thomsen, D. Williams, J. Hubbell, & R. Cancedda (Eds.), *Textbook on Tissue Engineering* (pp. 559–610). Amsterdam: Elsevier.

VandeVord, P. J., Matthew, H. W. T., DeSilva, S. P., Mayton, L., Wu, B., & Wooley, P. H. (2002). Evaluation of the biocompatibility of a chitosan scaffold in mice. *J. Biomed. Mater. Res., 59*, 585–590.

Vårum, K. M., Myhr, M. M., Hjerde, R. J. N., & Smidsrød, O. (1997). *In vitro* degradation rates of partially N-acetylated chitosans in human serum. *Carbohydrate Res., 299*, 99–101.

Vecchio, K. S., Zhang, X., Massie, J. B., Wang, M., & Kim, C. W. (2007). Conversion of bulk seashells to biocompatible hydroxyapatite for bone implants. *Acta Biomater., 3*, 910–918.

Vepari, C., & Kaplan, D. L. (2007). Silk as a biomaterial. *Prog. Polymer Sci., 32*, 991–1007.

585

Wahl, D. A., & Czernuszka, J. T. (2006). Collagen-hydroxyapatite composites for hard tissue repair. *Eur. Cells Mater.,* *11*, 43–56.

Wahl, D. A., Sachlos, E., Liu, C. Z., & Czernuszka, J. T. (2006). *20th Conference of the European-Society-for-Biomaterials, Nantes, France.*

Walsh, P. J., Buchanan, F. J., Dring, M., Maggs, C., Bell, S., & Walker, G. M. (2008). Low-pressure synthesis and characterization of hydroxyapatite derived from mineralize red algae. *Chem. Eng. J.,* *137*, 173–179.

Wang, Y., Kim, H.-J., Vunjak-Novakovic, G., & Kaplan, D. L. (2006). Stem cell-based tissue engineering with silk biomaterials. *Biomaterials,* *27*, 6064–6082.

Wang, Y., Bian, Y.-Z., Wu, Q., & Chen, G.-Q. (2008a). Evaluation of three-dimensional scaffolds prepared from poly (3-hydroxybutyrate-co-3-hydroxyhexanoate) for growth of allogeneic chondrocytes for cartilage repair in rabbits. *Biomaterials,* *29*, 2858–2868.

Wang, Y., Cui, F. Z., Hu, K., Zhu, X. D., & Fan, D. D. (2008b). Bone regeneration by using scaffold based on mineralized recombinant collagen. *J. Biomed. Mater. Res. B Appl. Biomater.,* *86B*, 29–35.

Wang, Y., Rudym, D. D., Walsh, A., Abrahamsen, L., Kim, H.-J., Kim, H. S., et al. (2008c). *In vivo* degradation of three-dimensional silk fibroin scaffolds. *Biomaterials,* *29*, 3415–3428.

Wang, Y., Zhang, L., Hu, M., Liu, H., Wen, W., Xiao, H., et al. (2008d). Synthesis and characterization of collagen-chitosan-hydroxyapatite artificial bone matrix. *J. Biomed. Mater. Res. A,* *86*, 244–252.

Wang, Y.-W., Wu, Q., & Chen, G.-Q. (2004). Attachment, proliferation and differentiation of osteoblasts on random biopolyester poly(3-hydroxybutyrate-co-3-hydroxyhexanoate) scaffolds. *Biomaterials,* *25*, 669–675.

Wang, Y.-W., Wu, Q., Chen, J., & Chen, G.-Q. (2005). Evaluation of three-dimensional scaffolds made of blends of hydroxyapatite and poly(3-hydroxybutyrate-co-3-hydroxyhexanoate) for bone reconstruction. *Biomaterials,* *26*, 899–904.

Wolfe, S. W., Pike, L., Slade, J. F., & Katz, L. D. (1999). Augmentation of distal radius fracture fixation with coralline hydroxyapatite bone graft substitute. *J. Hand Surg.,* *24*, 816–827.

Wu, C., Ramaswamy, Y., & Zreiqat, H. (2010). Porous diopside (CaMgSi2O6) scaffold: a promising bioactive material for bone tissue engineering. *Acta Biomater.,* *6*, 2237–2245.

Yang, Y. M., Hu, W., Wang, X. D., & Gu, X. S. (2007). The controlling biodegradation of chitosan fibers by N-acetylation *in vitro* and *in vivo*. *J. Mater. Sci. Mater. Med.,* *18*, 2117–2121.

Yao, J., & Asakura, T. (2008). Silks. In G. E. Wnek, & G. L. Bowlin (Eds.), *Encyclopedia of Biomaterials and Biomedical Engineering* (pp. 2442–2449). New York: Taylor and Francis.

Yao, J. M., Nakazawa, Y., & Asakura, T. (2004). Structures of Bombyx mori and Samia cynthia ricini silk fibroins studied with solid-state NMR. *Biomacromolecules,* *5*, 680–688.

Yaping, G., & Yu, Z. (2008). Transformation of nacre coatings into apatite coatings in phosphate buffer solution at low temperature. *J. Biomed. Mater. Res. A,* *86A*, 510–521.

Zhang, Y. F., Song, J. H., Shi, B., Wang, Y. N., Chen, X. H., Huang, C., et al. (2007). Combination of scaffold and adenovirus vectors expressing bone morphogenetic protein-7 for alveolar bone regeneration at dental implant defects. *Biomaterials,* *28*, 4635–4642.

Zhao, F., Grayson, W. L., Ma, T., Bunnell, B., & Lu, W. W. (2006). Effects of hydroxyapatite in 3-D chitosan-gelatin polymer network on human mesenchymal stem cell construct development. *Biomaterials,* *27*, 1859–1867.

Zhao, Y., Zou, B., Shi, Z., Wu, Q., & Chen, G.-Q. (2007). The effect of 3-hydroxybutyrate on the *in vitro* differentiation of murine osteoblast MC3T3-E1 and *in vivo* bone formation in ovariectomized rats. *Biomaterials,* *28*, 3063–3073.

Zhou, C. Z., Confalonieri, F., Jacquet, M., Perasso, R., Li, Z. G., & Janin, J. (2001). Silk fibroin: structural implications of a remarkable amino acid sequence. *Proteins Structure Func. Genet.,* *44*, 119–122.

Zhu, L. Q., Wang, H. M., Xu, J. H., Wei, D., Zhao, W. Q., Wang, X. X., et al. (2008). Effects of titanium implant surface coated with natural nacre on MC3T3E1 cell line *in vitro*. *Prog. Biochem. Biophys.,* *35*, 671–675.

Zinn, M., Witholt, B., & Egli, T. (2001). Occurrence, synthesis and medical application of bacterial polyhydroxyalkanoate. *Adv. Drug Deliv. Rev.,* *53*, 5–21.

Zurzolo, C., & Bowler, C. (2001). Exploring bioinorganic pattern formation in diatoms. A story of polarized trafficking. *Plant Physiol.,* *127*, 1339–1345.

CHAPTER 33

Synthetic Polymers

M.C. Hacker[*]**, A.G. Mikos**[†]
*Institute of Pharmacy, Pharmaceutical Technology, University of Leipzig, Leipzig, Germany
[†] Department of Bioengineering, Rice University, Houston, TX, USA

INTRODUCTION

Regenerative medicine is an emerging, interdisciplinary approach to repairing or replacing damaged or diseased tissues and organs. In order to reestablish tissue and organ function impaired by disease, trauma, or congenital abnormalities, regenerative medicine employs cellular therapies, tissue engineering strategies, and artificial or biohybrid organ devices. Typically, these techniques rely on combinations of cells, genes, morphogens, or other biological building blocks with bioengineered materials and technologies to address tissue or organ insufficiency.

Materials used in these approaches range from metals and ceramics, to natural and synthetic polymers, as well as micro- and nanocomposites thereof. When used in a three-dimensional context, these materials are processed into micro- and/or nanoporous cell carriers, typically known as scaffolds, of various structures and properties, a topic that is discussed elsewhere in this book. This chapter focuses exclusively on synthetic polymers used in regenerative medicine. Some synthetic derivatives of natural materials are briefly discussed where appropriate. Accompanying the various facets of regenerative medicine, a plethora of synthetic polymers with different compositions and physicochemical properties have already been developed and investigated; however, research is still ongoing. Synthetic materials play a key role in many applications of regenerative medicine, including implants, tissue engineering scaffolds, and orthopedic fixation devices. In a broader sense, sutures, drug delivery systems, non-viral gene delivery vectors, and sensors made from synthetic polymers are further examples.

This chapter provides a structural overview of these synthetic polymers and discusses their physicochemical characteristics, structure-property relationships, applications, and limitations. Synthetic polymers that are hydrolytically labile and erode (biodegradable polymers) as well as those that are bioinert and remain unchanged after implantation (non-degradable polymers) are considered. It is the authors' intention to provide a thorough overview of the synthetic material classes available. Some polymer classes are briefly mentioned and their chemical structures are provided; other more relevant materials are discussed in more detail. For most polymer classes and properties, reviews are referenced to present guidance for further reading.

Biomaterial history in general can be best organized into four eras: prehistory, the era of the surgeon hero (first-generation biomaterials), designed biomaterials and engineered devices (second-generation biomaterials), and the contemporary era leading into the new millennium (third-generation biomaterials) (Hench and Polak, 2002; Ratner, 2004). As far back as 600 AD, the use of dental implants made from materials such as sea shells or iron was reported. Also, there is evidence that sutures have been used for as long as 32,000 years to close large wounds.

The word "biomaterials," however, was first introduced within the last 50 years. Almost at the same time, aided by rapid advancements in industrial polymer development and synthesis, the exploration of synthetic polymers for biomedical applications began. The development of plastic contact lenses, utilizing primarily poly(methyl methacrylate), started around 1936, and the first data on implantation of nylon as a suture was reported in 1941. This development was accompanied by studies on the biocompatibility of the new materials. From the beginning, differences in foreign body reaction to the materials became apparent. Additives such as plasticizers, unpolymerized reactants, and degradation products were discussed as possible causes, leading to awareness of polymer quality for biomedical applications and biocompatibility testing.

At the end of World War II, a wide variety of durable high-performance metal, ceramic, and especially polymeric materials was available to inspiring surgeons to break new grounds in replacing diseased or damaged body parts. Materials including silicones, polyurethanes, Teflon, nylon, methacrylates, titanium, and stainless steel were available "off the shelf" for surgeons to apply to medical problems (Ratner, 2004). Primarily medical and dental practitioners, driven by the vision to replace lost organ or tissue functionality, made use of minimal regulatory constraints to develop and improvise replacements, bridges, conduits, and even organ systems based on such materials. Those pioneering approaches laid the foundation for novel procedures and engineered biomaterials. Such early implants made from industrial materials available "off the shelf" were often poorly biocompatible, in many cases due to insufficient purity. With a developing understanding of the immune system and foreign body reaction, a first generation of materials was developed during the 1960s and 1970s by engineers and scientists for use inside the human body. The primary goal of early biomaterial development was to achieve a suitable combination of physical properties to match those of the replaced tissue with a minimal toxic response in the host (Hench, 1980). Following this paradigm, more than 50 implanted devices made from 40 different materials were in clinical use in 1980. In the early 1980s, research began to shift from materials that exclusively exhibited a bioinert tissue response to materials that actively interacted with their environment. Another advance in this second generation was the development of biodegradable materials that exhibited controllable chemical breakdown into non-toxic degradation products, which were either metabolized or directly eliminated. Biodegradable synthetic polymers were designed to resolve the interface problem, since the foreign material is ultimately replaced by regenerating tissues and eventually the regeneration site is histologically indistinguishable from the host tissue. Resorbable polymers were routinely used clinically as sutures by 1984. Other applications in fracture fixation aids or drug-delivery devices emerged quickly. Despite considerable clinical success of bioinert, bioactive, and resorbable implants, there is still a high long-term prostheses failure rate and need for revision surgery (Ratner, 2004).

Improvements of first- and second-generation biomaterials have been limited for one main reason: unlike living tissue, artificial biomaterials cannot respond to changing physiological loads or biochemical stimuli. This limits the lifetime of artificial body parts. To overcome these limitations, a third generation of biomaterials is being developed that involves molecular tailoring of resorbable polymers for specific cellular responses. By immobilizing specific biomolecules, such as signaling molecules or cell-specific adhesion peptides or proteins, onto a material it is possible to mimic the extracellular matrix (ECM) environment and provide a cell-adhesive surface (Hench and Polak, 2002; Drotleff et al., 2004; Lutolf and Hubbell, 2005; Patterson et al., 2010). Biomimetic surfaces are promising tools to control cell adhesion, implant integration, cell differentiation, and tissue development. Synthetic polymer matrices can also be tailored to deliver drugs, signaling molecules, and genetic code and thus provide versatile technologies for regenerative medicine (Saltzman and Olbricht, 2002; Segura and Shea, 2002; Tabata, 2003). Constantly expanding knowledge of the basic biology of stem cell differentiation and the corresponding signaling pathways as well as tissue development provides the basis for molecular design of scaffolds. In tissue engineering attempts, which aim at regenerating lost or defective tissue by transplanting *in vitro*-engineered tissue constructs based on a patient's own cells, one no longer strives to closely match scaffold mechanical properties to those of the replaced tissue. It is rather considered important that the transplanted construct is engineered to be steadily remodeled *in vivo* to resemble the histological and mechanical properties of the surrounding tissue (Nerem, 2006). Due to this paradigm shift, mechanically labile hydrogels, especially injectable systems that can be used to directly encapsulate cells, have gained great importance as a basis for biomimetic cell carriers. Hydrogels are characterized by a high water content that allows encapsulated cells to survive and enables sufficient passive transport of nutrients, oxygen, and wastes. Hydrogel-forming materials typically offer functional groups for chemical modifications, and their degradation can be controlled by chemical composition and crosslinking content.

Following a brief overview on synthesis techniques, inert and biodegradable synthetic polymers representative of all three generations will be presented in the subsequent sections. Their structure, synthesis, physicochemical properties, and applications will be described.

POLYMER SYNTHESIS

Polymerization reactions for the synthesis of organic polymers are often categorized into chain-growth polymerizations and step-growth polymerizations depending on how the chemical process of chain formation proceeds. The synthesis of polymers with a carbon-carbon backbone, such as polyolefins and polyacrylates, typically follows a chain-growth mechanism (Reimschuessel, 1975). Chain-growth polymerizations involve the steps of chain initiation, chain propagation, and termination. Characteristics of this type of polymerization are that chain growth occurs only by addition of monomers to the active chain end, generally at a very high speed, and that only monomers and polymers are present during the reaction. Depending on the nature of the reactive centre of the propagation chains, chain-growth reactions are subdivided into radical, ionic (anionic or cationic), or transition-metal-mediated (coordinative, insertion) polymerizations. Suitable monomers contain unsaturated carbon-carbon bonds (double or triple) or are cyclic molecules with a sufficiently high ring strain. For the industrial synthesis of polyolefins, for example, free radical and transition-metal-mediated polymerizations are commonly employed. Unlike radical polymerization, transition-metal-coordinated mechanisms, for example with Ziegler-Natta catalysts, allow for control of polymer tacticity (Soga and Shiono, 1997). A milestone in radical chain-growth polymerization history was the development of controlled or living radical polymerization techniques that allow for precise control of polymer composition and architecture and yield polymeric products with low polydispersity (Braunecker and Matyjaszewski, 2007).

Polymers that contain heteroatoms in the main chain are typically synthesized by a step-growth mechanism. During step-growth, polymer molecular weight increases through the reaction of any two molecular species, i.e. monomers, oligomers, and polymer chains. In contrast to chain-growth, monomers disappear early on during the reaction and polymer molecular weight increases slowly over the course of the reaction, which can last up to days. Typical polymerization types that follow a step-growth mechanism are polycondensation reactions and polyaddition reactions. In condensation reactions, small molecules such as water, alcohols, and hydrochloric acid are eliminated during step-growth. Polyethylene terephthalate; polyamides, such as nylon; and poly(propylene fumarate) (Kasper et al., 2009) are examples of polymers that are synthesized by condensation reactions between carboxylic acid derivatives and diols or diamines (nylon). Most polyanhydrides are also synthesized by polycondensation reactions (Leong et al., 1987). Polyaddition reactions follow a similar mechanism as nucleophilic groups react with electrophilic moieties during polymer chain build-up. In contrast to condensation reactions, addition reactions do not produce any small molecules. During polyurethane synthesis, for example, diisocyanate monomers are reacted with diamines or dihydroxy-terminated molecules in the presence of catalysts under the formation of urethane and urea groups, respectively, to build up polymer chains (Król, 2007).

Ring-opening polymerizations (ROPs) also yield polymers with heteroatoms in the main chain and are used for the synthesis of polyamines; polyethers, such as poly(ethylene glycol)s; and most biodegradable polyesters including polylactides, polyglycolides, and copolymers (Albertsson and Varma, 2003). ROPs can follow chain-growth and step-growth kinetics and are executed in melts or solutions in the presence of catalysts and heat.

Driven by the advances in drug design through combinatorial approaches in small molecule chemistry, similar techniques have been adapted to polymerization chemistries (Goldberg et al., 2008). Through the systematic screening of libraries of polymeric materials that have similar chemistries but are synthesized from a series of different monomers and comonomers in various combinations, structure-property relations can be identified and polymer properties can be fine-tuned for specific applications. Polymer properties that are screened in such approaches include the materials' glass transition temperature, degradative properties, air-water contact angle, mechanical properties, cytocompatibility, and cell proliferation.

NON-DEGRADABLE SYNTHETIC POLYMERS

A common characteristic of most non-degradable synthetic polymers is their biological inertness (Hench and Polak, 2002). These materials were developed to reduce to a minimum the host response to the biomaterial. Non-degradable synthetic polymers provide the basis for a plethora of medical devices as diverse as suture materials, orthopedic implants, fracture fixation devices, and catheters and dialysis tubing. These materials are also applied as implantable carriers for the long-term delivery of drugs, e.g. contraceptive hormones. Despite their excellent biological inertness and well-adjustable mechanical properties, orthopedic implants made from non-degradable synthetic polymers and non-degradable bone cements ultimately fail at a high rate from problems at the interface arising from a lack of integration with the surrounding tissue, infections, or bone resorption caused by stress shielding (Bobyn et al., 1992; Jacobs et al., 1993).

Major groups of non-degradable synthetic polymers are highlighted in the following paragraphs.

Polymers with a -C—C- backbone
POLYETHYLENE AND DERIVATIVES
Poly(ethylene), poly(propylene), and poly(styrene)

Poly(ethylene) (PE) (Fig. 33.1A), poly(propylene) (PP) (Fig. 33.1B), and poly(styrene) (PS) (Fig. 33.1C) are ubiquitous industrial polymers and have been applied as biomaterials. All three thermoplastic polymers, which only consist of carbon, are synthesized by direct polymerization of their corresponding monomers. While PE can be synthesized by radical or ionic polymerization of ethylene, special organometallic catalysts are required to polymerize propylene to useful PP. PE and PP are classified into several different categories based on their density, branching, and molecular weight. These parameters significantly influence the crystallinity and mechanical properties of the polymers. PE has been used for the production of catheters. High-density PE, which is characterized by a low degree of branching and thus strong intermolecular forces and tensile strength, has been processed into highly durable hip prostheses. A three-dimensional fabric comprising PE fibers and coated with hydroxyapatite was used to regenerate hyaline cartilage in osteochondral defects in rabbit knees and showed successful biocompatibility (Hasegawa et al., 1999). The best-known application for PP is its use for syringe bodies. Copolymers of PE and vinyl acetate (poly(ethylene-*co*-vinyl acetate), PEVAc) (Fig. 33.1D) are widely used in non-degradable drug delivery devices (Langer, 1990). PEVAc is one of the most biocompatible implant materials (Langer et al., 1981) and has been

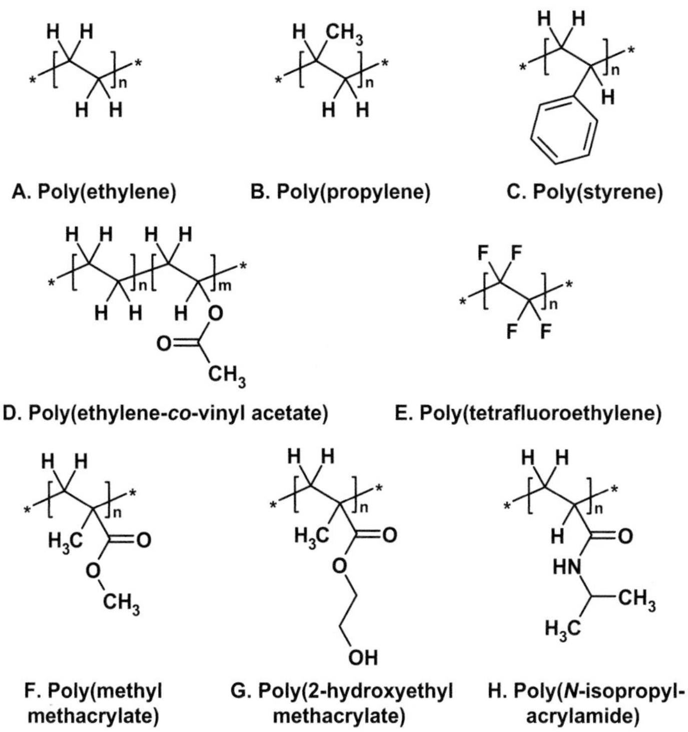

FIGURE 33.1

Chemical structures of non-degradable synthetic polymers (I).

approved by the FDA for use in implanted and topically applied devices. Ocusert and Progestasert are prominent examples of PEVAc-based drug delivery systems (Chandrasekaran et al., 1978).

PS is a hard and brittle polymer used for the fabrication of tissue culture flasks and dishes. By copolymerization with butadiene, copolymers with improved elasticity are synthesized that are used for the fabrication of catheters and medical devices for perfusion and dialysis.

Poly(tetrafluoroethylene)

Poly(tetrafluoroethylene) (PTFE) (Fig. 33.1E), well known as Teflon (DuPont), can be synthesized from liquid tetrafluoroethylene by radical polymerization and through fluorination of polyethylene. Among known polymers, PTFE has the lowest coefficient of friction, has excellent resistance to chemicals, and is hemocompatible. Porous PTFE fiber meshes (Goretex) have become a popular synthetic vascular graft material (Xue and Greisler, 2003).

POLY(METH)ACRYLATES AND POLYACRYLAMIDES

Poly(meth)acrylate hydrogels have found applications in medical devices, especially for ocular applications (e.g. contact lenses and intraocular lenses), as drug delivery systems, and as cell delivery systems (Langer and Peppas, 1981; Lloyd et al., 2001; Peppas et al., 2000). Three major types, poly(methyl methacrylate), poly(2-hydroxyethyl methacrylate), and poly(N-isopropyla-crylamide), are discussed in more detail.

A variety of (meth)acrylate and acrylamide monomers with different functional groups is available; thus, poly(meth)acrylates and polyacrylamides of different chemical compositions can be synthesized. Together with the free carboxylic acid moieties of (meth)acrylic acid, the presentation of different functional groups and charges along copolymer chains or within crosslinked hydrogels is possible. Using an imprinting technique, these moieties can be oriented in such a way that pouches are created that interact non-covalently with molecules, e.g. drugs or therapeutic peptides and proteins, by ionic interactions, hydrogen bonds, π-π interactions, and hydrophobic interactions (Mosbach and Ramstrom, 1996; Tunc et al., 2006). Besides intelligent hydrogels for controlled drug release, this technology has impact on microfluidic devices, biomimetic sensors, intelligent polymeric membranes (Ulbricht, 2006), and analyte-sensitive materials (Byrne et al., 2002).

Poly(methyl methacrylate)

Poly(methyl methacrylate) (PMMA) (Fig. 33.1F) is a non-degradable polyacrylate and is the most commonly applied non-metallic implant material in orthopedics. After being used as an essential ingredient in making dentures, PMMA was introduced to orthopedic surgery in the mid 1950s (Saha and Pal, 1984). PMMA tissue biocompatibility became further apparent when Plexiglas fragments were accidentally implanted in the eyes and other body tissues of World War II fighter pilots during aircraft crashes.

PMMA can be *in situ* polymerized and crosslinked from a slurry containing PMMA and methyl methacrylate monomers and is so used as a common bone grafting material, mainly in the fixation of orthopedic prosthetic materials for hips, knees, and shoulders (Kenny and Buggy, 2003). PMMA-based bone cements can be mixed with inorganic ceramics or bioactive glass to modulate curing kinetics and enforce mechanical properties. Antibiotics can be loaded within the cement to reduce the risk of prosthesis-related infection. Significant drawbacks of self-curing PMMA cements include that they are not degraded, that their high curing temperatures and toxic monomers can cause necrosis of the surrounding tissue, and that the cements show limited interactions with the surrounding bone (Hendriks et al., 2004). Therefore, development of alternative injectable bone cements is directed towards biodegradable materials with improved curing properties and osteoconductive interfaces (Yaszemski et al., 1996; Hendriks et al., 2004).

Due to its excellent bio- and hemocompatibility and ease of manipulation, PMMA is used in many medical devices, including blood pumps and dialyzers. Its optical properties make it a candidate material for implantable ocular lenses and hard contact lenses (Lloyd et al., 2001). PMMA also offers physical and coloring properties that are beneficial for denture fabrication (Hendriks et al., 2004).

Poly(2-hydroxyethyl methacrylate)

Poly(2-hydroxyethyl methacrylate) (PHEMA) (Fig. 33.1G) was the first hydrogel successfully employed for biological use (Wichterle and Lim, 1960). PHEMA has become the major component of most soft contact lenses and is also part of intraocular lenses (Lloyd et al., 2001). Due to their free hydroxyl groups, PHEMA gels contain relatively high amounts of water, facilitating the diffusion of solutes and oxygen. PHEMA has excellent biocompatibility, which initiated the development of a plethora of HEMA-containing copolymers. Hydrogels fabricated from PHEMA and copolymers have been intensively characterized for controlled drug delivery applications (Mack et al., 1987; Lu and Anseth, 1999) and employed for biomedical uses. PHEMA gels, which have limited mechanical properties, have been used in attempts to reconstruct female breasts and nasal cartilages, and as artificial corneas as well as wound dressings (Young et al., 1998).

In a subcutaneous rabbit model, porous PHEMA sponges promoted significant cellular ingrowth and neovascularization in combination with good cytocompatibility (Chirila et al., 1993). Recently, a mineralization technique has been demonstrated that exposes carboxylate groups on crosslinked PHEMA hydrogel scaffolds, promoting calcification (Song et al., 2003).

Poly(N-isopropylacrylamide)

Poly(N-isopropylacrylamide) (PNiPAAm) (Fig. 33.1H) has gained great significance for injectable applications in drug and cell delivery using minimally invasive techniques due to its unique physicochemical properties (Hoffman, 2002). PNiPAAm undergoes (lower critical) phase separation resulting in the formation of an opaque hydrogel in response to a temperature above 32°C, the material's lower critical solution temperature (LCST). This thermoresponsive behavior is the result of strong hydrogen bonds between the polymer and water molecules and the specific molecular orientations of these bonds due to the molecular structure of the polymer. The formation of hydrogen bonds between the polymer and the solvent lowers the free energy of the solution. Due to the hydrophobic N-isopropyl residues in PNiPAAm, the hydrogen bonds between water and the amide functionality require specific molecular orientation. Such ordered structures lead to negative entropy changes and positive contributions to the free energy. Since the enthalpic contribution to the free energy is temperature-dependant, the formation of strong but specifically oriented hydrogen bonds is no longer thermodynamically favored above a certain temperature. Consequently, PNiPAAm dissolves in water below the LCST. At and above the LCST, the polymer chains partially desolvate and undergo a coil-to-globule transition resulting in colloidal aggregation that may lead to gel formation or polymer precipitation (Schild and Tirrell, 1990; Schild, 1992). Hydrogels formed by linear PNiPAAm at 32°C are instable and collapse substantially as the temperature is increased above the LCST. The synthesis of crosslinked networks and copolymers, typically with hydrophilic building blocks, has resulted in materials that demonstrate reversible thermogelation and form hydrogels without significant syneresis at body temperature. Different PNiPAAm-containing copolymers for cell delivery have been synthesized with acrylic acid, poly(ethylene glycol), hyaluronic acid, and gelatin (Stile et al., 1999; Ohya et al., 2001; Hoffman, 2002; Morikawa and Matsuda, 2002). Detailed information is available for the *in vitro* and *in vivo* use of gelatin-PNiPAAm conjugates for the regeneration of articular cartilage (Ibusuki et al., 2003a,b).

POLYETHERS

Poly(ethylene glycol) (PEG) (Fig. 33.2A), often also called poly(ethylene oxide) (PEO), is a non-degradable polyether of the monomer ethylene glycol. Technically, PEG and PEO should not be used as synonyms, since PEO is synthesized from the monomer ethylene oxide and is typically terminated by only one hydroxyl group and an initiator fragment. Commonly, "PEG" is used to refer to the polymer with molecular weight less than 50,000 Da while "PEO" is used for higher molecular weights. PEG is water soluble and solutions of its high-molecular-weight form can be categorized as a hydrogel. PEG hydrogels for biomedical applications are typically composed of polymer chains that are crosslinked. These crosslinked networks frequently contain chemical bonds between the PEG chains and the crosslinkable moieties, which are prone to aqueous hydrolysis and are therefore characterized as biodegradable systems. The molecular weight of the PEG chains crosslinked in such hydrogels is below a threshold molecular weight to allow for complete resorption by renal elimination of the individual chains. Consequently, these systems are discussed with biodegradable polymers on page 610.

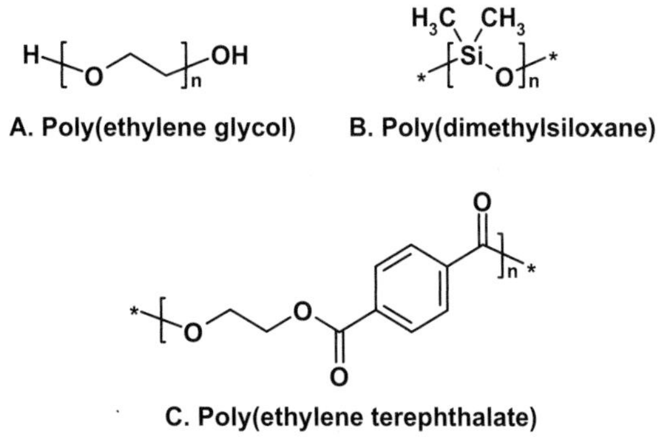

593

A. Poly(ethylene glycol)   B. Poly(dimethylsiloxane)

C. Poly(ethylene terephthalate)

FIGURE 33.2

Chemical structures of non-degradable synthetic polymers (II).

Favorable characteristics of PEG and PEO are their high hydrophilicity, bioinertness, and outstanding biocompatibility, which make them candidate biomaterials. PEG and PEO are frequently used as hydrophilic polymeric building blocks in copolymers with more hydrophobic degradable or non-degradable polymers for drug delivery (Jeong et al., 1997), gene delivery, tissue engineering scaffolds, medical devices, and implants. PEG has also been immobilized on polymeric biomaterial surfaces to make them resistant to protein absorption and cell adhesion. These effects are attributed to highly hydrated PEG chains on the polymer surfaces that exhibit steric repulsion based on an osmotic or entropic mechanism. Attempts to benefit from this phenomenon include the design of long-circulating nanoparticles or liposomes (Gref et al., 1997, 2000; Photos et al., 2003; Vonarbourg et al., 2006) and PEGylated enzymes or proteins with prolonged functional residence time *in vivo* compared to unmodified biomolecules (Roberts et al., 2002; Harris and Chess, 2003).

A variety of PEG-containing block copolymers for injectable drug delivery have been developed over the last decades (Ruel-Gariepy and Leroux, 2004). The most prominent class comprises triblock copolymers composed of two hydrophilic PEO blocks and one hydrophobic poly(propylene oxide) (PPO), also known as Pluronics or poloxamers. These materials are designed to show similar phase transition behavior to the thermogelling PNiPAAm-containing materials (see p. 593). Poloxamers have been intensively investigated for the delivery of drugs and proteins (Jeong et al., 2002). Since poloxamers are non-degradable, biodegradable structural analogues have been synthesized and are described within the next chapter on biodegradable synthetic polymers (see p. 603).

POLYSILOXANES

Polysiloxanes, or silicones, are a general category of polymers consisting of a silicon-oxygen backbone with organic groups, typically methyl groups, attached to the silicon atoms (Colas and Curtis, 2004). Certain organic side groups can be used to link two or more chains together. By varying the -Si-O- chain length, side groups, and crosslinking extent, silicones with properties ranging from liquids to hard plastics can be synthesized. Silicone synthesis typically involves the hydrolysis of chlorosilanes into linear or cyclic siloxane oligomers, which are then polymerized into polysiloxanes by polycondensation or polymerization, respectively. The most common polysiloxane is linear poly(dimethylsiloxane) (PDMS) (Fig. 33.2B).

Polysiloxanes, which are characterized by unique material properties combining biocompatibility and biodurability, have found widespread application in healthcare (Curtis and Colas, 2004). The materials' high biodurability is a result of other material properties such as hydrophobicity, low surface tension, and chemical and thermal stability. Silicone surfaces have been found to inhibit blood from clogging for many hours and have been therefore used for the fabrication of silicone-coated needles, syringes, and other blood-collecting instruments. Silicone materials have also been employed as heart valves and as components in kidney dialysis and blood-oxygenator and heart-bypass machines due to their hemocompatibility. Silicone elastomers have found application in numerous catheters, shunts, drains, and tubular implants, such as artificial urethra. Significant orthopedic applications of silicones are hand and foot joint implants. The most prominent application of silicones is their extensive use as cosmetic implants in aesthetic and reconstructive plastic surgery. Prosthetic silicone implants are available for the breast, scrotum, chin, nose, cheek, calf, and buttock. Different silicone materials, including slightly crosslinked silicone gels, are combined to achieve a natural feel. Controversy was aroused regarding the safety of popular silicone gel-filled breast implants in the early 1990s. These discussions initially involved increased risk for breast cancer, then progressed to autoimmune connective tissue disease and continued to evolve to the frequency of local or surgical complications such as rupture, infection, or capsular contracture. To date, no epidemiology study has indicated that the rate of breast cancer has significantly increased in women with silicone breast implants (Silverman et al., 1996). Similarly, studies on autoimmune or connective tissue disease have agreed on a lack of causal association between breast implants and these diseases (Sanchez-Guerrero et al., 1995; Lewin and Miller, 1997). A safety concern that has been controversially discussed recently involves the amount of platinum (part of catalysts used during silicone synthesis) that is released from silicone implants and accumulated in the host organism (Arepalli et al., 2002; Brook, 2006). Other mentioned complications, especially implant rupture, are persisting problems; in 1992, the FDA restricted the use of silicone gel-filled implants. Since that time, the implants may be used only under certain controlled conditions. The premarket approval, an application for marketing a device, has only been approved for two saline-filled breast implants and no silicone gel-filled implants by the FDA as of 2004 (US Food and Drug Administration, 2004).

Polysiloxane gels, combining the high oxygen permeability of silicone and the comfort and clinical performance of conventional polyacrylate hydrogels, enabled the fabrication of soft, gas permeable contact lenses for extended wear. In contrast to conventional hydrogels, silicone gels make the lens surface highly hydrophobic and less "wettable," which frequently results in discomfort and dryness during lens wear. Surface modifications of the silicones or the addition of conventional hydrogels are suitable strategies to compensate for the hydrophobicity.

Overall, polysiloxanes have displayed expanded medical application since the 1960s and today are one of the most thoroughly tested and important biomaterials.

OTHER NON-DEGRADABLE POLYMERS
Poly(ethylene terephthalate)

Poly(ethylene terephthalate) (PET) (Fig. 33.2C), a linear polyester synthesized by poly-condensation of terephthalic acid and ethylene glycol, is typically processed into fiber meshes. These meshes are applied as vascular grafts (Xue and Greisler, 2003) or used to reinforce prostheses.

Hydrolytically stable polyurethanes

Polyurethanes (PUs) are a heterogeneous class of polymers that consist of organic units joined by urethane links (Fig. 33.3). Generally, PUs can be synthesized from a bischloroformate and a diamine or by reacting a diisocyanate with a dihydroxy component. PUs used in biomedical applications typically have a segmented structure that results in useful physicochemical properties (Boretos and Pierce, 1967). Such segmented PUs or PU copolymers are elastomers composed of alternating polydispersed "soft" and "hard" segments. These two segments are thermodynamically incompatible and phase-segregate, resulting in discrete, crystalline domains of the associated "hard" segments surrounded by a continuous, amorphous phase of "soft" segments. The segregated domains are stabilized by interchain hydrogen bonds and are responsible for the materials' mechanical properties (Gunatillake et al., 2003). Segmented PUs are synthesized in a two-step process that provides control over polymer architecture (Fig. 33.3A). The first step involves the synthesis of an isocyanate-terminated prepolymer from a diisocyanate (D in Fig. 33.3) and a hydroxyl group-terminated polyether or polyester (P in Fig. 33.3). The prepolymer and excess diisocyanate are then reacted with a hydroxy or amine group-terminated chain extender (C in Fig. 33.3) to generate the final PU (Fig. 33.3A).

A chain extender terminated with hydroxy groups yields segmented polyurethanes, while a diamine extender yields polyurethane-urea (Fig. 33.3B). The "hard" segment of the PU copolymer is composed of the diisocyanate and the chain extender, while the "soft" segment contains the polymeric segment introduced during the first step. The extent of phase separation is dependent on molecular weights, chemistry, and relative percentages of the building blocks (Fromstein and Woodhouse, 2006).

After almost 50 years of use in biomedical applications, PUs remain one of the most popular groups of biomaterials for the fabrication of medical devices. Their popularity results from a wide range of versatility with regard to tailoring their physicochemical and mechanical

595

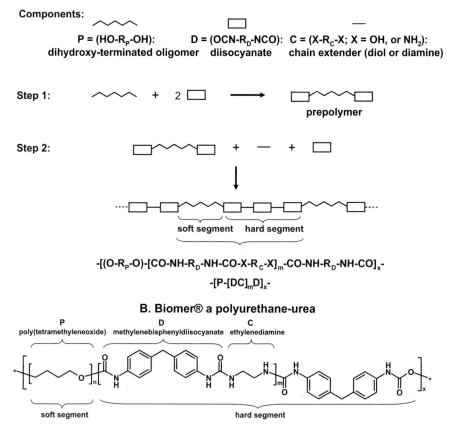

A. Polyurethane synthesis

Components:

P = (HO-R$_P$-OH):   D = (OCN-R$_D$-NCO):   C = (X-R$_C$-X; X = OH, or NH$_2$):
dihydroxy-terminated oligomer   diisocyanate   chain extender (diol or diamine)

Step 1:   ⌇⌇⌇ + 2 ▭ ⟶ ▭⌇⌇⌇⌇▭
prepolymer

Step 2:   ▭⌇⌇⌇▭ + — + ▭

soft segment   hard segment

-[(O-R$_P$-O)-[CO-NH-R$_D$-NH-CO-X-R$_C$-X]$_m$-CO-NH-R$_D$-NH-CO]$_x$-

-[P-[DC]$_m$D]$_x$-

B. Biomer® a polyurethane-urea

P
poly(tetramethyleneoxide)   D
methylenebisphenyldiisocyanate   C
ethylenediamine

soft segment   hard segment

FIGURE 33.3

General synthesis scheme (A) and an example structure (B) for polyurethanes.

properties, blood and tissue compatibility, and degradative properties by altering block copolymer composition. Biomedical PUs are used in numerous medical devices, such as breast implants, catheters, vascular and aortic grafts, pacemaker leads, artificial heart valves, and artificial hearts and have been found to perform well in a variety of *in vivo* applications. PUs often have better blood and tissue compatibilities in comparison to numerous other synthetic polymers. The efficient removal of impurities from the polymer synthesis, such as catalyst residues and low-molecular-weight oligomers, has been found to critically determine PU biocompatibility (Gogolewski, 1989).

Traditional PUs, such as Biomer (P: polytetramethylene oxide; D: methylene bisphenylene-diisocyante; C: ethylenediamine) (Fig. 33.2D), were materials of first choice. However, the assumption of polyetherurethane non-degradability had to be revised following well-documented failures of pacemaker leads and breast implant coatings containing PUs in the late 1980s. Although PUs can be designed to be stable against hydrolysis, these materials have been shown to degrade in the biological environment by mechanisms including oxidation and enzyme- and cell-mediated degradation (Howard, 2002; Santerre et al., 2005; Fromstein and Woodhouse, 2006). Oxidation of PUs can be initiated by peroxides, free radicals, and enzymes. Metal-catalyzed oxidation was found to be most frequently associated with pacemaker lead failure. Another important oxidation-driven problem with long-term PU implants is environmental stress cracking. It has also been found that PU surfaces become coated with a protein layer that enhances the adhesion of macrophages. The macrophages, activated by proteins of the complement family, release oxidative factors that accelerate degradation of the polymer (Stokes et al., 1995).

Chemical design criteria for biostable PUs have been identified. To increase the degree of interchain hydrogen bonding, on which biostability depends in part, low-molecular-weight oligomeric diols (P) are preferred as building blocks. To avoid oligomer hydrolysis, oligoethers are favored over oligoesters. Aromatic diisocyanates (D) have been found to yield more biostable PUs than aliphatic diisocyanates. The use of a diamine chain extender (C) instead of a dihydroxy-terminated one typically results in stronger polyurethane-urea, but polymer fabrication is often hampered due to solubility problems. The use of soft segment building blocks with high crystallinity, such as polycaprolactone, or silicone-based oligomers are also assumed to improve polymer biostability (Fromstein and Woodhouse, 2006).

PUs can be surface modified to reduce the risk of thrombosis or improve the interactions with cells and tissues. Different strategies, including adsorption, covalent grafting, and the use of self-assembled monolayers, have been applied to distribute proteins, such as fibronectin, or adhesion peptides, which contain the integrin-binding peptide motif RGD, across the PU surface (Lin et al., 1994; Fromstein and Woodhouse, 2006).

BIODEGRADABLE SYNTHETIC POLYMERS FOR REGENERATIVE MEDICINE

Biodegradable synthetic polymers offer a number of advantages over non-degradable materials for applications in regenerative medicine. Like all synthetic polymers, they can be synthesized at reproducible quality and purity and fabricated into various shapes with desired bulk and surface properties. Specific advantages include the ability to tailor mechanical properties and degradation kinetics to suit various applications. Clinical applications for biodegradable synthetic polymers are manifold and traditionally include resorbable sutures, drug delivery systems, and orthopedic fixation devices such as pins, rods, and screws (Behravesh et al., 1999). More recently, synthetic biodegradables were widely explored as artificial matrices for tissue engineering applications (Seal et al., 2001; Nguyen and West, 2002; Salgado et al., 2004). For such applications, the mechanical properties of the scaffolds, which are determined by the constitutive polymer, should functionally mimic the properties of the tissue to be regenerated. Ultimately, the polymeric support is designed to degrade while transplanted or invading cells proliferate, lay down extracellular matrix, and form coherent tissue that, in the ideal case, is functionally, histologically, and mechanically indistinguishable from the surrounding tissue. To engineer scaffolds suitable for different applications, a wide variety of biodegradable polymers is required ranging from pliable, elastic materials for soft tissue regeneration to stiff materials that can be used in load-bearing tissues such as bone. In addition to the mechanical properties, the degradation kinetics of polymer and ultimately scaffold also have to be tailored to suit various applications.

The major classes of synthetic, biodegradable polymers are briefly reviewed and their potential in regenerative medicine is discussed below.

POLYESTERS

Polyesters have been attractive for biomedical applications because of their ease of degradation by primarily non-enzymatic hydrolysis of ester linkages along the backbone. Additionally, degradation products can be resorbed through the metabolic pathways in most cases, and there is the potential to tailor the structure to alter degradation rates (Gunatillake and Adhikari, 2003).

A vast majority of biodegradable polymers studied belong to the polyester family (Middleton and Tipton, 2000). Polyester fibers, which also became popular with the textile industry, were used as resorbable sutures (Freed et al., 1994). Promising observations regarding biocompatibility of the materials lead to applications in drug delivery, orthopedic implants, and most recently tissue engineering scaffolds, particularly for orthopedic applications (Heller, 1984; Amecke et al., 1992; Hubbell, 1995; Behravesh et al., 1999; Webb et al., 2004).

Poly(α-hydroxy acids)

The family of polyesters can be subdivided according to the structure of the monomers. In poly (α-hydroxy acids), each monomer carries two functionalities, a carboxylic acid and a hydroxyl group, located at the carbon atom next to the carboxylic acid (α-position), that form ester bonds. Poly(α-hydroxy acids) are linear thermoplastic elastomers that are typically synthesized by ring-opening polymerization of cyclic dimers of the building blocks (Gupta and Kumar, 2007). Poly(lactic acid) (PLA) (Fig. 33.4A), poly(glycolic acid) (PGA) (Fig. 33.4B), and a range of their copolymers (poly(lactic-co-glycolic acid), PLGA) (Fig. 33.4C) are prominent representatives not only of biodegradable polyesters but of biodegradables in general. The cyclic dimers that are polymerized during PLA and PGA synthesis are called lactide and glycolide, respectively. Therefore, the polymers are often named polylactides or polyglycolides. For reasons of consistency with the general term poly(α-hydroxy acids), the terms poly(lactic acid) and poly(glycolic acid) will be used here. Poly(α-hydroxy acids) have a long history of use as synthetic biodegradable materials in a number of clinical applications. Initially, resorbable sutures were made from these materials (Cutright et al., 1971). Later, poly(α-hydroxy acids) were the basis for controlled release systems for drugs and proteins (Juni and Nakano, 1987; Brannon-Peppas, 1995; Jain, 2000) and orthopedic fixation devices. Langer and coworkers have pioneered the development of these polymers in the form of porous scaffolds for tissue engineering (Langer and Vacanti, 1993).

Due to the chiral nature of lactic acid, several forms of poly(lactid acid) exist: poly(L-lactid acid) (PLLA), for example, is synthesized from dilactid in the L form. The polymerization of racemic dilactide leads to poly(D,L-lactic acid) (PD,LLA), which is an amorphous polymer. PLLA, in contrast, is a semicrystalline polymer with a crystallinity of around 37%. PLLA is characterized by

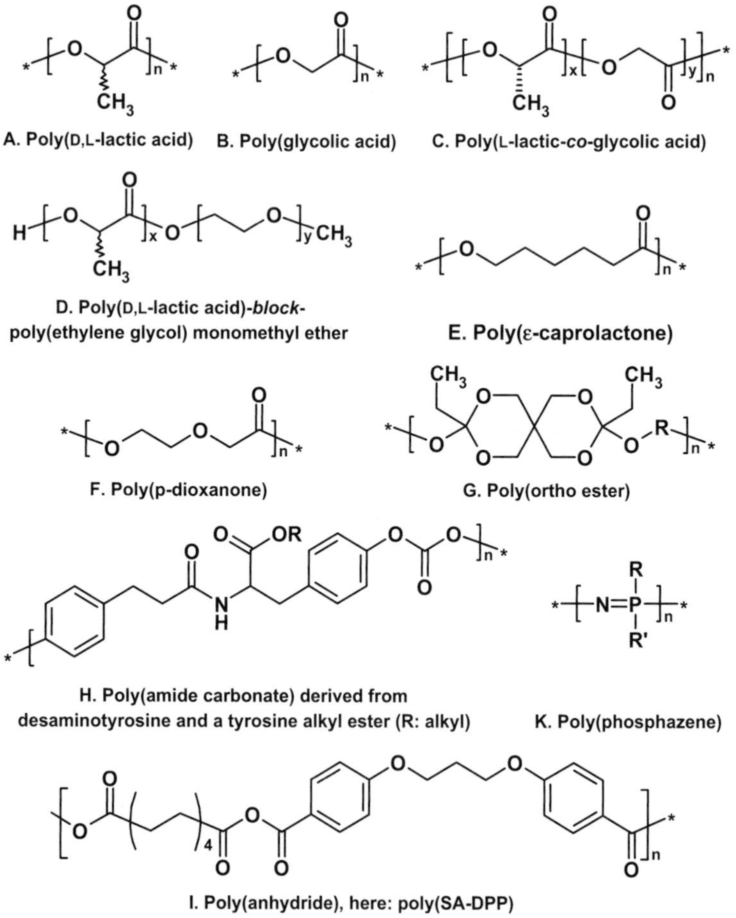

FIGURE 33.4

Chemical structures of biodegradable synthetic polymers.

a glass transition temperature between 50–80°C and a melting temperature between 173–178°C. Amorphous PD,LLA is typically used in drug delivery applications, while semi-crystalline PLLA is preferred in applications where high mechanical strength and toughness are required, e.g. for sutures and orthopedic devices. Poly(glycolic acid) is also a semicrystalline polymer with a higher crystallinity of 46–52%. Thermal characteristics of PGA are glass transition and melting temperatures of 36 and 225°C, respectively. Because of its high crystallinity, PGA, unlike PLA, is not soluble in most organic solvents; the exceptions are highly fluorinated and highly toxic organic solvents such as hexafluoroisopropanol. Consequently, common processing techniques for PGA include melt extrusion, injection, and compression molding.

PLA, PGA, and PLGA undergo homogeneous erosion via ester linkage hydrolysis into the degradation products lactic acid and glycolic acid, which are both natural metabolites that are fully metabolized and excreted as carbon dioxide and water. Degradation of poly(α-hydroxy acid)s was found to show typical characteristics of bulk erosion. Bulk erosion occurs when water penetrates the entire structure, and the device degrades simultaneously (Goepferich, 1996). During the initial stages of degradation almost no mass loss can be detected. Analysis of the average molecular weight of the polymer bulk over the same period, however, reveals a steady decrease in molecular weight. Once the polymer chains throughout the bulk are degraded below a certain threshold, the water-soluble degradation products are washed out and the system collapses accompanied by significant mass loss. Due to its well-accessible ester group, PGA degrades rapidly in aqueous media. PGA sutures typically lose their mechanical strength over a period of 2–4 weeks post-operatively (Reed and Gilding, 1981). In order to adapt these properties to a wider range of applications, copolymers with more hydrophobic PLA were synthesized and investigated. The two main series are those of PLLGA (Fig. 33.4C) and PDLLGA. It has been shown that the range of compositions from 25 to 70% glycolic acid (GA) for L-lactic acid (L-LA)/GA and from 0 to 70% GA for DL-LA/GA are amorphous (Miller et al., 1977; Gilding and Reed, 1979; Sawhney and Hubbell, 1990; Li, 1999; Middleton and Tipton, 2000; Gunatillake and Adhikari, 2003). For the PLLGA copolymers, the rate of hydrolysis was found to be slower at either extreme of the copolymer composition range. It is generally accepted that intermediate PLGA copolymers have a shorter half-life *in vivo* than either homopolymer. Besides polymer composition, the rate of degradation is affected by factors such as configurational structure, copolymer ratio, crystallinity, molecular weight, morphology, stresses, amount of residual monomer, bulk porosity, and site of implantation (Gunatillake and Adhikari, 2003).

Multiple *in vitro* and *in vivo* studies that were conducted on the biocompatibility of PLA, PLGA, and PGA have generally revealed satisfying results (Athanasiou et al., 1996). Consequently, PLA, PLGA copolymers, and PGA are among the few biodegradable polymers with FDA approval for human clinical use.

Concerns with poly(α-hydroxy esters) typically focus on the accumulation of acidic degradation products within the polymer bulk that can have detrimental effects on encapsulated drugs in delivery applications (Brunner et al., 1999; Lucke et al., 2002; Houchin and Topp, 2008) or can cause late non-infectious inflammatory responses when released in a sudden burst upon structure breakdown (Simon et al., 1997). This adverse reaction can occur weeks and months postoperatively and might need operative drainage. This is a major concern in orthopedic applications, where implants of considerable size would be required, which may result in release of degradation products with high local acid concentrations. Inflammatory response to poly(α-hydroxy acids) were found to be also triggered by the release of small particles during degradation that were phagocytosed by macrophages and multinucleated giant cells (Anderson and Shive, 1997; Xia and Triffitt, 2006). In general, implant size as well as surface properties appear to be critical factors with regard to biocompatibility. Fewer concerns seem to exist towards the application of poly(α-hydroxy acids) in soft tissues compared to hard tissue applications (Athanasiou et al., 1996).

Poly(α-hydroxy acids) were the materials of choice when one of the key concepts of tissue engineering, the *de novo* engineering of tissue by combining isolated cells and three-dimensional macroporous cell carriers *in vitro*, was first realized and developed (Langer and Vacanti, 1993; Freed et al., 1997; Mooney and Mikos, 1999). Polymers based on lactic and glycolic acid are still popular scaffold materials, especially for orthopedic applications such as bone, cartilage, and meniscus, as outlined in several reviews (Agrawal et al., 2000; Hutmacher, 2000; Seal et al., 2001). Limitations of this class of materials include insufficient mechanical properties with regard to load-bearing applications (Webb et al., 2004) and inflammatory or cytotoxic events due to the above-mentioned accumulation of acidic products during degradation.

In order to cover a broader range of mechanical and physicochemical properties, such as water absorption, polymer degradation, and polymer-drug interactions, block copolymers containing PLA and hydrophilic PEO or PEG were synthesized for drug delivery applications (Bouillot et al., 1998). Solid particulate systems from these block copolymers were found to be almost invisible to the immune system due to the hydrophilic PEG chains that swell on the surface (Gref et al., 1994; Bazile et al., 1995) (see pp. XXX) (Fig. 33.4D). The stealthiness of such surfaces is mainly caused by the suppression of protein adsorption, which also inhibits cell adhesion. Investigations of cell adhesion to PEG-PLA diblock copolymer surfaces revealed that cell adhesion can be controlled and cell differentiation can be modulated by the PEG content (Lieb et al., 2003). With the objective to specifically control cell-polymer interactions, PEG-PLA copolymers were further developed to allow for the covalent attachment of signaling molecules (Cannizzaro et al., 1998; Tessmar et al., 2003). Since these polymers were insoluble in water, they could be processed into macroporous scaffolds for tissue engineering applications (Hacker et al., 2003).

Polylactones

The most prominent and thoroughly investigated polylactone is poly(ε-caprolactone) (PCL) (Fig. 33.4E), an aliphatic, semicrystalline polyester with interestingly low glass transition temperature ($-60°C$) and melting temperature ($59-64°C$) (Middleton and Tipton, 2000). PCL is considered biocompatible (Matsuda et al., 2003). PCL is prepared by the ring-opening polymerization of the cyclic monomer ε-caprolactone, and is compatible with a range of other polymers. Catalysts, such as stannous octoate, are used to catalyze the polymerization, and low-molecular-weight alcohols can be used as initiator and to control the molecular weight of the polymer. ε-Caprolactone can be copolymerized with numerous other monomers. Copolymers with PLA and PEG are probably the most noteworthy and have been investigated extensively (Pitt et al., 1979, 1981; Cerrai et al., 1994; Petrova et al., 1998). PCL degrades at a much slower rate than PLA and is therefore most suitable for the development of long-term, implantable drug delivery systems. Aforementioned copolymers of caprolactone with dilactide were synthesized to accelerate degradation rates (Middleton and Tipton, 2000). Tubular, highly permeable poly(L-lactide-*co*-ε-caprolactone) guides were found to be suitable for regeneration and functional reinnervation of large gaps in injured nerves (Rodriguez et al., 1999). While this study focuses on tissue regeneration, the application of PCL in drug delivery devices is still far more common (Sinha et al., 2004). With increasing popularity of electrospinning, a lab-scale technique that allows for the fabrication of non-woven meshes composed of nano- and/or microfibers (Pham et al., 2006), PCL might find its way into cell-based therapies since slowly degrading polymers are preferred for this technique to ensure sufficient stability of the fibers (Yoshimoto et al., 2003).

Poly(p-dioxanone) (Fig. 33.4F), another polylactone, and its copolymers with lactide, glycolide, and/or trimethylene carbonate are synthesized by catalyzed ring-opening polymerization and have been used in a number of clinical applications ranging from suture materials to bone fixation devices (Wang et al., 1998; Yang et al., 2002).

600

Poly(diol citrates)

Poly(diol citrates) are a group of elastomeric polyester networks that are synthesized from citric acid and various low- and high-molecular-weight diols by polycondensation reaction without using exogenous catalysts (Yang et al., 2006). Poly(1,8-octanediol-*co*-citrate) (POC), one of the first poly(diol citrates), demonstrated mechanical properties, such as tensile strength, Young's modulus, and elongation at break that justify applications in ligament reconstruction and vascular engineering (Yang et al., 2004). Variations in chemical composition, especially diol chemistry, allowed for the synthesis of a variety of biodegradable elastomers covering a range of mechanical and degradative properties (Yang et al., 2006). With regard to vascular tissue engineering, POC showed good hemocompatibility and exhibited decreased platelet adhesion and clotting relative to poly(L-lactide-*co*-glycolide) and expanded poly(tetrafluoro ethylene) (Motlagh et al., 2007). Endothelial cell attachment and differentiation were supported without any modification of the surface. In order to improve the mechanical properties of the POC elastomer, unsaturated acrylate and fumarate diols were added during the condensation reaction and moieties for secondary crosslinking were introduced (Zhao and Ameer, 2009).

Polyorthoesters

Polyorthoesters (POEs) (Fig. 33.4G) were developed by the Alza Corporation and SRI International in 1970 in the search for a new biodegradable polymer for drug delivery applications (Heller et al., 2002). Since then, polymer synthesis has been improved over four generations. POEs are synthesized by condensation or addition reactions typically involving dialcohols and monomeric orthoesters or diketene acetals, respectively. The use of triethylene glycol as the diol component produced predominantly hydrophilic polymers, whereas hydrophobic materials could be obtained by using 1,10-decanediol. Orthoester is a functional group containing three alkoxy groups attached to one carbon atom. In POEs, two of the three alkoxy groups are typically part of a cyclic acetal (Fig. 33.4G).

POEs have been synthesized that degrade by surface erosion, which is characterized by a constant decrease of bulk mass while polymer molecular weight within the polymer bulk is preserved (Burkersroda et al., 2002). It is known that materials built from functional groups with short hydrolysis half-lives and low water diffusivity tend to be surface eroding. Polymers that exhibit surface erosion can be used to fabricate drug delivery systems that, at a high aspect to volume ratio (e.g. as for wafers), release loaded drugs at a constant rate.

The addition of lactide segments to the POE structure resulted in self-catalyzed erosion and allowed for tunable degradation times ranging from weeks to months (Ng et al., 1997). POEs provide the material platform for a variety of drug delivery applications including the treatment of post-surgical pain, osteoarthritis, and ophthalmic diseases as well as the delivery of proteins and DNA. Block copolymers of poly(ortho ester) and poly(ethylene glycol) have been prepared, and their use as drug delivery matrices or as colloidal structures for tumor targeting are being explored (Heller et al., 2002).

Initial biocompatibility studies revealed that POEs provoked little inflammation and were largely absorbed by 4 weeks. In contrast, poly(DL-lactic acid) (PLA) degraded slower and provoked a chronic inflammation with multinuclear giant cells, macrophages with engulfed material, and proliferating fibroblasts within the same model. Ossicles with bone marrow had formed in the implants of POE in combination with demineralized bone. In PLA/demineralized bone implants the bone formation was inhibited (Andriano et al., 1999; Solheim et al., 2000).

Polycarbonates

Polycarbonates have become interesting biomaterials due to their excellent mechanical strength and good processability. Since pure polycarbonates degrade extremely slowly under physiological conditions, polyiminocarbonates (Kohn and Langer, 1986) and tyrosine-based polycarbonates (Pulapura and Kohn, 1992) (Fig. 33.4H) have been engineered to yield biodegradable polymers of good mechanical strength (Engelberg and Kohn, 1991) for use in drug delivery and orthopedic applications. Degradation of most polycarbonates is controlled by the hydrolysis of the carbonate group, which yields two alcohols and carbon dioxide, thus alleviating the problem of acid bursting seen in polyesters (Gunatillake and Adhikari, 2003). Structural variation of the pendant side groups allows for the preparation of polymers with different mechanical properties, degradation rates, as well as cellular response. Polycarbonates that contain a pendant ethyl ester group have been shown to be osteoconductive and to possess mechanical properties sufficient for load bearing bone fixation. Long-term (48 week) *in vivo* degradation kinetics and host bone response to tyrosine-derived polycarbonates were investigated using a canine bone chamber model (Choueka et al., 1996). Histological sections revealed intimate contact between bone and the tested polycarbonates. It was concluded that, from a degradation-biocompatibility perspective, the tyrosine-derived polycarbonates appear to be comparable, if not superior, to PLA in this model.

AMINO ACID-DERIVED POLYMERS, POLY(AMINO ACIDS), AND PEPTIDES

Amino acids are an interesting building block for polymers due to the biocompatibility of the degradation products and the degradability of the amide or ester bonds by which amino acids are typically polymerized or integrated in copolymers. Early studies on pure poly(amino acids) revealed significant concerns with the materials' immunogenicity and mechanical properties (Bourke and Kohn, 2003). To improve those unfavorable properties, amino acids have been used as monomeric building blocks in polymers that have a backbone structure different from natural peptides. Based on polymer structure and chemistry, four major groups have been used to classify such "non-peptide amino-acid-based polymers," namely: (1) synthetic polymers with amino acid side chains, (2) copolymers of natural amino acids and non-amino acid monomers, (3) block copolymers containing peptide or poly(amino acid) blocks, and (4) pseudo-poly(amino acids).

As in tyrosine-derived polycarbonates (see pp. 601), L-tyrosine is the predominantly employed amino acid in the synthesis of tyrosine-derived polyarylates and polyesters. These copolymers exhibit excellent engineering properties, and polymer systems can be designed whose members show exceptional strength (polycarbonates), flexibility and elastomeric behavior (polyarylates), or water-solubility and self-assembly properties (copolymers with PEG). Poly (DTE carbonate) (DTE: desaminotyrosyl-tyrosine ethyl ester) (Fig. 33.4H, R: -CH_2CH_3) exhibits a high degree of tissue compatibility and is currently being evaluated for possible clinical uses by the US Federal Drug Administration (Bourke and Kohn, 2003).

A combinatorial library of degradable tyrosine-derived polyarylates has been synthesized by copolymerizing 14 different tyrosine-derived diphenols and eight different aliphatic diacids in all possible combinations, resulting in 112 distinct polymers (Brocchini, 2001). Significant differences were observed in the mechanical properties of the polymers and fibroblast proliferation assays on these materials. This illustrates that such combinatorial approaches provide a library of related polymers that encompasses a broad range of properties and permits the systematic study of material-dependent biological responses in order to choose a suitable material for a specific application.

Solid-phase peptide synthesis, pioneered by Merrifield, and genetic engineering allow for the automated and highly efficient synthesis of peptides of a predefined sequence. In contrast to synthetic poly(amino acids), which are traditionally composed of a single amino acid and were found to be highly immunogenic in most cases, synthetic peptides have become an important polymer class for biomedical applications. Specifically, peptides and petide-amphiphiles that undergo self-assembly-driven *in situ* gelation in response to temperature, pH, or chemical stimuli are of interest as these materials can be minimally invasively implanted starting from aqueous solutions (Stupp et al., 1997; Meyer and Chilkoti, 1999; Hartgerink et al., 2001).

Genetically engineered elastin-like polypeptides, which are composed of a pentapeptide repeat and undergo inverse temperature phase transition, have been used to encapsulate chondrocytes. The cell culture studies showed that cartilaginous tissue formation, characterized by the biosynthesis of sulfated glycosaminoglycans and collagen, was supported (Betre et al., 2002).

Self-assembled peptide-amphiphiles, which form hydrogels composed of nanofibers resembling the native ECM components, have been demonstrated to be cytocompatible in cell encapsulation studies (Beniash et al., 2005). Peptide nanostructures designed through self-assembly strategies and supramolecular chemistry have the potential to combine bioactivity with biocompatibility (Webber et al., 2010). In addition, such structures can be used to deliver proteins, nucleic acids, drugs, and cells.

Peptide-amphiphile nanofibers were shown to promote *in vitro* proliferation and osteogenic differentiation of marrow stromal cells (Hosseinkhani et al., 2006). Towards dental tissue engineering, dental stem cells were recently encapsulated in peptide-amphiphile hydrogels containing adhesion peptides and enzyme-cleavable sites. The cells proliferated and differentiated within the gels and remodelled the matrices (Galler et al., 2008).

POLYURETHANES

Polyurethanes (PUs) represent a major class of synthetic elastomers that have excellent mechanical properties and good biocompatibility. PUs have been evaluated for a variety of medical devices and implants, particularly for long-term implants.

Knowledge gained about the mechanisms of PU biodegradation in response to implant failures throughout the 1990s has been translated to form a new class of bioresorbable materials (Santerre et al., 2005). Recent research has utilized the flexible chemistry and diverse mechanical properties of PUs to design degradable polymers for a variety of regenerative applications. Segmented PUs with varied molecular structure have been synthesized to control rates of hydrolysis (Skarja and Woodhouse, 2001; Santerre et al., 2005). To obtain biodegradable, segmented PUs, significant changes were required to be made to the structural components historically used for their synthesis. Traditional aromatic diisocyanates (D; cf. Fig. 33.3) can yield toxic or carcinogenic degradation products when part of a degradable PU; therefore, linear diisocyanates, such as lysine-diisocyanate that yields the non-toxic degradation product lysine, are preferred. The soft segment, typically composed of an oligomeric diol (P; cf. Fig. 33.3), is typically the block of the PU used to modify the degradation rate. Biodegradable PUs have been synthesized with a variety of soft segments including PEO, degradable polyesters such as PLA, PGA, or PCL, and combinations thereof. Other strategies focus on the copolymers' hard segments. PUs were synthesized that contain enzyme-sensitive linkages introduced with the chain extender (C; cf. Fig. 33.3). For example, the use of a phenylalanine diester chain extender yielded a PU that showed susceptibility to enzyme-mediated degradation upon exposure to chymotrypsin and trypsin.

Saad et al. investigated cell and tissue interactions with a series of degradable polyesterurethanes. *In vivo* investigations showed that all test polymers exhibited favorable tissue compatibility and degraded significantly during the course of one year (Saad et al., 1997). Polyurethane-urea matrices were shown to allow vascularization and tissue infiltration *in vivo* (Ganta et al., 2003). The flexible chemistry and diverse mechanical properties of PU materials allowed researchers to design degradable polymers for the regeneration of tissues as varied as neurons, vasculature, smooth muscle, cartilage, and bone (Xue and Greisler, 2003; Zhang et al., 2003; Santerre et al., 2005).

BLOCK COPOLYMERS OF POLYESTERS OR POLYAMIDES WITH POLY (ETHYLENE GLYCOL)

Amphiphilic block copolymers of biodegradable polymers with poly(ethylene glycol) (PEG) have become popular materials for injectable drug delivery applications (Jeong et al., 2002).

Inspired by the thermoresponsive behavior observed for non-degradable A-B-A-type triblock copolymers composed of hydrophilic poly(ethylene oxide) (PEO) (block A) and hydrophobic poly(propylene oxide) (PPO) (block B), polymer development focused on synthesizing biodegradable analogs of these poloxamers (or Pluronics) that were water-soluble at ambient temperature and formed stable hydrogels at body temperature. Biodegradable block copolymers were synthesized by substituting the hydrophobic PPO block with a biodegradable polymer block, such as PLA or PCL (Jeong et al., 1997; Lee et al., 2001; Ruel-Gariepy and Leroux, 2004).

Biodegradable, physically crosslinkable block copolymers of inverse structure, that is, B-A-B triblock copolymers with two biodegradable hydrophobic polymer blocks (block B) and a hydrophilic PEO block, have also been investigated as protein delivery systems (Kissel et al., 2002).

POLYANHYDRIDES

Drug delivery technologies rely on engineered polymers that degrade in a well-controllable and adjustable fashion (Langer, 1990). Increasing understanding of erosion mechanisms led to a demand for synthetic polymers that contain a hydrolytically labile backbone while limiting water diffusion within the polymer bulk significantly to confine erosion to the polymer-water interface. Such surface-eroding polymers allow for the fabrication of drug delivery devices that erode at constant velocity at any time during erosion, thereby releasing incorporated drugs at constant rates (Gopferich and Tessmar, 2002). Polyanhydrides were engineered following this paradigm by selecting the anhydride linkage, one of the least hydrolytically stable chemical bonds available, to connect the building hydrophobic monomers.

Polyanhydrides (Fig. 33.4I) have been synthesized by various techniques, including melt condensation, ring opening polymerization, interfacial condensation, dehydrochlorination, and dehydrative coupling agents (Kumar et al., 2002). Solution polymerization traditionally yielded low-molecular-weight polymers. Different dicarboxylic acid monomers have been polymerized to yield polyanhydrides with various physicochemical properties. Examples are linear, aromatic, fatty acid-based dicarboxylic acid monomers, and fatty acid-terminated polyanhydrides. Polyanhydrides made from linear sebacic acid (SA) and aromatic 1,3-bis (p-carboxyphenoxy) propane (CPP) (Fig. 33.4I) have been engineered to deliver carmustine (BNCU), an anticancer drug, to sites in the brain following primary resection of a malignant glioma (Westphal et al., 2003). Poly(SA-CPP) hydrolyzes into non-toxic degradation products and the local chemotherapy with BCNU wafers was shown to be well tolerated and to offer a survival benefit to patients with newly diagnosed malignant glioma.

The chemical composition of a polyanhydride can be used to custom-design its degradation properties. While polyanhydrides from linear monomers, such as poly(SA), degrade within a few days, polymerized aromatic dicarboxylic acids, such as poly(1,6-bis(p-carboxyphenoxy) hexane), degrade much more slowly (up to a year) (Temenoff and Mikos, 2000). The structural versatility of polyanhydrides in combination with their unique degradation and erosion properties make them precious materials for numerous medical, biomedical, and pharmaceutical applications in which degradable polymers that allow for perfect erosion control are needed (Gopferich and Tessmar, 2002). With regard to tissue engineering applications, polyanhydrides are also interesting polymers due to their degradative properties and their good biocompatibility (Katti et al., 2002). The use of polyanhydrides in load-bearing orthopedic applications, however, is restricted due to limited mechanical properties. Poly(anhydrides-co-imides), which were developed in order to combine the good mechanical properties of polyimides with the degradative properties of polyanhydrides, were shown to meet compressive strengths comparable to human bone (Uhrich et al., 1995) and displayed good osteocompatibility (Ibim et al., 1998).

Photopolymerizable polyanhydrides have been synthesized with the objective to combine high strength, controlled degradation, and minimal invasive techniques for orthopedic applications and were shown to be osteocompatible (Anseth et al., 1999). Depending on the chemical composition, these materials reached compressive and tensile strengths similar to those of cancellous bone (Muggli et al., 1999).

An interesting strategy for the controlled release of bioactive substances has been recently explored with poly(anhydride-esters). Bioactive substances, such as anti-inflammatory drugs (Bryers et al., 2006) and antiseptics (Schmeltzer and Uhrich, 2006) have been used as monomers or co-monomers for polyanhydrides. Upon polymer degradation, the active substances were released from the polymer bulk in a controlled manner.

POLYPHOSPHAZENES

Polyphosphazenes (Fig. 33.4K), which are polymers containing a high-molecular-weight backbone of alternating phosphorus and nitrogen atoms with two organic side groups attached to each phosphorus atom, are a relatively new heterogenic class of biomaterials. Because different synthetic pathways allow for a tremendous variety of derivatives, phosphazene polymers exhibit a very diverse spectrum of chemical and physical properties. Due to this variety, these polymers are suitable for many biomedical applications ranging from templates for nerve regeneration and cardiovascular and dental uses to implantable and controlled-release devices (Andrianov and Payne, 1998; Langone et al., 1995; Schacht et al., 1996).

The best-studied and most important route to polyphosphazenes, whose synthesis is generally more involved than that for most petrochemical biomaterials but offers unique flexibility, is a macromolecular substitution route. A reactive polymeric intermediate, poly (dichlorophosphazene), is typically synthesized by a thermal ring opening cationic polymerization of hexachlorocyclotriphosphazene in bulk at 250°C that yields a polydisperse high-molecular-weight product. The intermediate is reacted with low-molecular-weight organic nucleophiles resulting in stable, substituted polyphosphazenes, which in this case are also termed poly(organo)phosphazenes. Depending on the substituent chemistry, the polyphosphazene is more or less susceptible to hydrolysis. Biodegradable hydrophobic polyphosphazenes have been synthesized using imidazolyl, ethylamino, oligopeptides, amino acid esters, and depsipeptide groups (dimers composed of an amino acid and a glycolic or lactic ester) as hydrolysis-sensitive side groups. Hydrolytic degradation products include free side group units, phosphate, and ammonia due to backbone degradation (Andrianov and Payne, 1998). Hydrogel-forming, hydrophilic polyphosphazenes can be synthesized through the introduction of small, hydrophilic side groups, such as glucosyl, glyceryl, or methylamino side groups. Ionic side groups yield polymers that form hydrogels upon ionic complexation with multivalent ions (Allcock and Kwon, 1989). Hydrophilic, water-soluble polyphosphazenes with amphiphilic side groups, such as poly(bis(methoxyethoxyethoxy)phosphazene) (Fig. 33.4K, R,R': -OCH$_2$CH$_2$OCH$_2$CH$_2$OCH$_3$), display an LCST (see p. 593) and are responsive to changes in temperature and ionic strength (Lee, 1999). Both hydrophilic and hydrophobic polyphosphazenes have demonstrated their potential as biocompatible materials for controlled protein delivery. Ionic polyphosphazenes have been explored as vaccine delivery systems and poly(di(carboxylatophenoxy)phosphazene) has demonstrated a remarkable adjuvant activity on the immunogenicity of inactivated influenza virions and commercial trivalent influenza vaccine in the soluble state (Andrianov and Payne, 1998).

Porous scaffolds from biodegradable polyphosphazenes have been shown to be good substrates for osteoblast-like cell attachment and growth with regard to skeletal tissue regeneration (Laurencin et al., 1996). Tubular polyphosphazene nerve guides were investigated in a rat sciatic nerve defect. After 45 days, a regenerated nerve fiber bundle was found bridging the nerve stumps in all cases (Langone et al., 1995).

BIODEGRADABLE CROSSLINKED POLYMER NETWORKS

The chemical crosslinking of individual, linear polymer chains results in networks of increased stability. This concept has been extensively explored for applications in regenerative medicine and most likely represents the concept of choice for modern biomaterial research, especially if polymer crosslinking can be conducted inside a tissue defect (Temenoff and Mikos, 2000). The crosslinking of hydrophobic polymers or monomers results in tough polymer networks that can be used for orthopedic fixation. Poly(methyl methacrylate) (PMMA) (Fig. 33.1F), the main component in injectable bone cements, is the most prominent example. Due to their hydrophobicity, the precursors are typically injected as a moldable liquid or paste free of additional solvents. *In situ* crosslinking can be initiated thermally or photo-chemically by UV-rich light. Both ways of initiation are also applicable to hydrophilic injectable systems that form highly swollen gels (hydrogels) as a result of precursor crosslinking. In contrast to hydrophobic networks, which scarcely swell in the presence of water, injectable hydrogels are characterized by a high water content and diffusivity, which allow for the direct encapsulation of cells and sufficient transport of oxygen, nutrients, and waste. Hydrophobic networks, however, often require the addition of a leachable porogen, such as salt particles, to facilitate cell migration and tissue ingrowth. Generally, injectable polymer systems have considerable advantages over prefabricated implants or tissue engineering scaffolds, which include the ability to fill irregularly shaped defects with minimal surgical intervention (Peter et al., 1998a).

A number of demanding requirements have to be fulfilled by synthetic materials for applications in regenerative medicine. In addition to physicochemical properties that fit the application site, the polymer and any adjuvant component that is required to formulate an *in situ* crosslinkable system have to be biocompatible. Ideally, the resulting network should also have the ability to support cell growth and proliferation early in the tissue regeneration process (Temenoff and Mikos, 2000).

The crosslinkable synthetic polymers that will be discussed in the following sections are reactive polyesters. The main chemical functionality involved in the chemical crosslinking mechanisms is the polarized, electron-poor double bond, such as in vinylsulfones and in esters of acrylic acid, methacrylic acid, and fumaric acid. Other chemically or thermally crosslinkable macromonomer functional groups are styryl, coumarin, and phenylazide and will not be discussed here (Hou et al., 2004).

Crosslinked polyesters

Fumarate-based polymers The development of fumarate-based polyesters for biomedical applications started around 20 years ago. Fumaric acid is a naturally occurring metabolite, which is found in the tri-carboxylate cycle (Krebs cycle), and is composed of a reactive double bond available for chemically crosslinking reactions. These characteristics make fumaric acid a candidate building block for crosslinkable polymers. The first and most comprehensively investigated fumarate-based copolymer is the biodegradable copolyester poly(propylene fumarate) (PPF) (Fig. 33.5A). PPF was first polymerized from fumaric acid and propylene oxide (Domb et al., 1990). Mikos and coworkers optimized the synthesis of PPF and broadly investigated tissue compatibility and applications of PPF both *in vitro* and *in vivo*. Synthesis progressed to polymerization of copolymeriz fumaryl chloride with 1,2-propanediol (propylene glycol) (Peter et al., 1999b) and now involves the transesterification of diethylfumarate with propylene glycol and subsequent polycondensation of the diester intermediate bis(2-hydroxypropyl) fumarate (PF) (Shung et al., 2003). A variety of methods to synthesize PPF have been explored, and each results in different polymer molecular weights and properties (Peter et al., 1997a). PPF has been developed as an alternative to PMMA bone cements. PPF can be injected as a viscous liquid and thermally crosslinked *in vivo*, eliminating the need for direct exposure of the defect site to light. Typically, PPF is crosslinked with either methyl methacrylate (MMA) or *N*-vinyl pyrrolidone (NVP) monomers and benzoyl peroxide

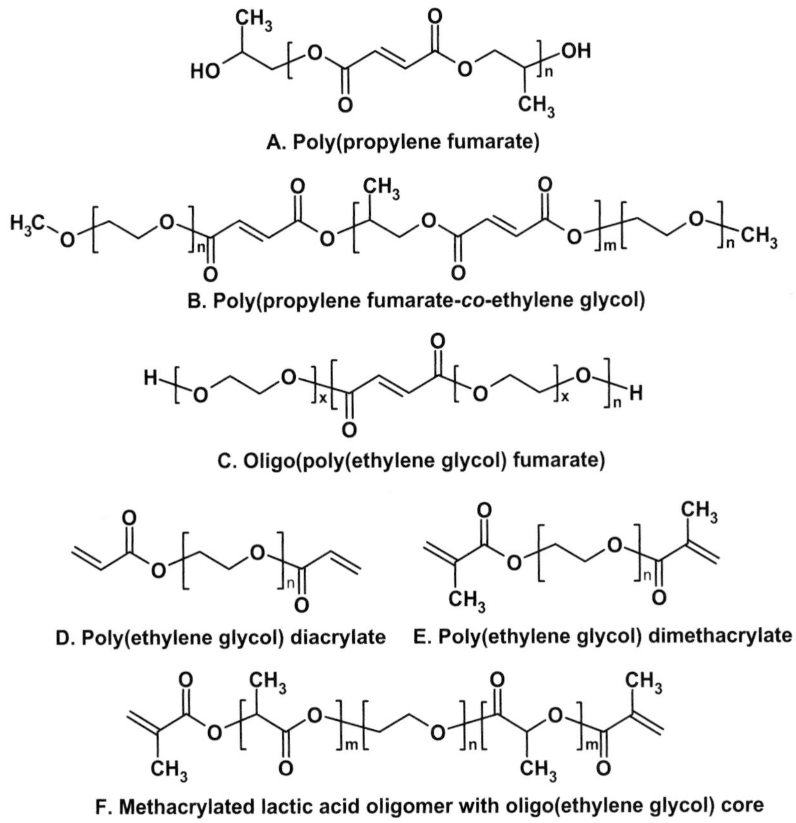

FIGURE 33.5

Chemical structures of synthetic polymers for the fabrication of crosslinked biodegradable networks.

A. Poly(propylene fumarate)

B. Poly(propylene fumarate-*co*-ethylene glycol)

C. Oligo(poly(ethylene glycol) fumarate)

D. Poly(ethylene glycol) diacrylate   E. Poly(ethylene glycol) dimethacrylate

F. Methacrylated lactic acid oligomer with oligo(ethylene glycol) core

as a radical initiator (Gresser et al., 1995; Frazier et al., 1997). Depending on the ratio of initiator, monomer, and PPF, the curing time can be controlled to between 1 and 121 minutes. Compared to PMMA, which is not resorbable and suffers from the fact that its high curing temperatures (94°C) can cause necrosis of the surrounding tissue, the curing temperature of PPF has been shown to never exceed 48°C (Peter et al., 1997b, 1999a). PPF can also be photo-crosslinked via the electron-poor double bonds along the backbone. Typical formulations include NVP, diethylfumarate, or PF-diacrylate (PF-DA) as co-monomers together with a photoinitiator, such as bis(2,4,6-trimethylbenzoyl) phenylphosphine oxide (He et al., 2001; Fisher et al., 2001, 2002a). The mechanical properties of PPF, which are dependent on composition, synthesis condition, and crosslinking density, are already promising. However, these materials are probably not sufficient for load-bearing applications, especially when used as macroporous scaffolds (Peter et al., 1998a; Fisher et al., 2002a; Timmer et al., 2003). One strategy to further strengthen PPF scaffolds includes the incorporation of nano-particulate fillers. Reinforced PPF composites have been synthesized using aluminum oxide-based ceramic nanoparticles and chemically modified single walled carbon nanotubes (SWNTs). For just 0.05 wt% loading with the latter, a 74% increase was recorded for the compressive modulus and a 69% increase for the flexural modulus as compared to plain PPF/PF-DA (Shi et al., 2005). The chemical integrations of alumoxane nanoparticles in crosslinked PPF/PF-DA networks resulted in a significantly increased flexural modulus (Horch et al., 2004). Both the PPF/alumoxane nanocomposites and the PPF/SWNT nanocomposites have been processed into macroporous tissue engineering scaffolds (Shi et al., 2007; Mistry et al., 2009) and showed good biocompatibility *in vitro* (Mistry et al., 2007; Shi et al., 2008) and *in vivo* (Sitharaman et al., 2008; Mistry et al., 2010). The ultra-short SWNT-reinforced porous biodegradable PPF scaffolds were implanted in rabbit femoral condyles and in subcutaneous pockets. By micro-computed tomography, histology, and histomorphometry at 4 and 12 weeks after implantation, favorable hard and soft tissue responses were detected. At 12 weeks, a three-fold greater bone tissue ingrowth was seen in defects containing the nanocomposite scaffolds, suggesting that the presence of ultra-short SWNT may render nanocomposite scaffolds bioactive, assisting osteogenesis (Sitharaman et al., 2008).

Micro-particulate ceramic materials, such as β-tricalcium phosphate (β-TCP), have also been employed as inorganic filler to improve mechanical properties of composite scaffolds and to improve the material's osteoconductivity (Peter et al., 2000). The composite scaffolds exhibit increased compressive strengths in the range of 2–30 MPa, and β-TCP reinforcement delayed scaffold disintegration significantly *in vivo* (Peter et al., 1998b). This subcutaneous rat implantation study also revealed a mild initial inflammatory response and formation of a fibrous capsule around the implant at 12 weeks. A deleterious long-term inflammatory response was not observed. Rabbit *in vivo* studies also revealed biocompatibility of photo-crosslinked PPF scaffolds in both soft and hard tissues (Fisher et al., 2002b).

PPF hydrolytically degrades along the ester bonds in its backbone. Degradation time was found to be dependent on polymer structure as well as other components, such as fillers. *In vitro* studies identified the time needed to reach 20% original mass, ranging from around 84 (PPF/β-TCP composite) to over 200 days (PPF/CaSO$_4$ composite) (Temenoff and Mikos, 2000).

In order to broaden the application spectrum for *in situ* crosslinkable PPF, block copolymers with hydrophilic poly(ethylene glycol) (PEG) of different compositions were synthesized. Poly (propylene fumarate-*co*-ethylene glycol) (P(PF-*co*-EG)) (Fig. 33.5B) was synthesized from PPF and PEG in a transesterification reaction catalyzed by antimony trioxide; propylene glycol was removed by condensation (Suggs et al., 1997). Behravesh et al. have modified the synthesis to yield well-defined ABA-type triblock copolymers from 2 moles monomethoxy-PEG and 1 mole PPF (Behravesh et al., 2002a). Generally, P(PF-*co*-EG) copolymers are hydrophilic polymers with specific properties including crystallinity and mechanical characteristics being dependent on the molecular weights of the individual blocks and the copolymer. As a result, platelet attachment to P(PF-*co*-EG) hydrogels was significantly reduced as compared to the PPF homopolymer, making these copolymers candidate materials when direct biomaterial-blood contact is inevitable, for example for vascular grafts (Suggs et al., 1999b). Most P(PF-*co*-EG) copolymers are amphiphiles and soluble in water, making them candidate materials for injectable applications. A-B-A-type copolymers were found to show thermoreversible properties, comparable to other PEG-containing triblock copolymers discussed above. The thermogelling properties of P(PF-*co*-EG) were dependent on the PEG molecular weight and salt concentration and the physical gelation temperature could be adjusted to values below body temperature (Behravesh et al., 2002a). In addition, the hydrophobic PPF block is highly unsaturated and available for additional chemical crosslinking, which could result in stiff crosslinked networks suitable for the production of prefabricated cell carriers. *In vitro* degradation studies of macroporous, crosslinked P(PF-*co*-EG) scaffolds revealed considerable mass loss and swelling over 12 weeks. In these studies, the degradation rate was mainly dependent on the content of the PEG-DA crosslinker and almost unaffected by construct porosity. Overall, the results indicated a bulk degradation mechanism of the macroporous constructs (Behravesh et al., 2002b). In a subcutaneous rat model, P(PF-*co*-EG) hydrogels demonstrated good initial biocompatibility followed by development and maturation of a fibrous capsule, which is very often seen for polymeric implants (Suggs et al., 1999a). Overall, the reported *in vitro* cytocompatibility and *in vivo* biocompatibility assays suggest that P(PF-*co*-EG) hydrogels have potential for use as injectable biomaterials. Fisher et al. have demonstrated the suitability of thermoresponsive P(PF-*co*-EG) hydrogels for chondrocyte delivery towards the regeneration of articular cartilage defects (Fisher et al., 2004).

Similarly to previously discussed, stealthy, PEG-containing biodegradables, PEG-content and hydrophilicity of crosslinked P(PF-*co*-EG) hydrogels are critical factors affecting cell adhesion (Tanahashi and Mikos, 2002). Low-adhesive hydrogels allow for a controlled surface or bulk modification with adhesion molecules to specifically enhance cell adhesion. P(PF-*co*-EG) hydrogels have been modified by covalent integration of agmatine (Tanahashi and Mikos, 2003) and the adhesion peptide GRGDS (Behravesh et al., 2003). Significantly increased numbers of smooth muscle cells and marrow stromal cells were found adhered as compared to the unmodified networks.

An exclusively hydrophilic fumarate-based macromer is oligo(poly(ethylene glycol) fumarate) (OPF) (Fig. 33.5C). OPF macromers have been synthesized from PEG and fumaryl chloride by a simple condensation reaction in the presence of triethylamine. OPF crosslinking, with or without the addition of a crosslinker such as PEG-DA, can be initiated photochemically (Jo et al., 2001) or thermally (Temenoff et al., 2002). In contrast to chemically crosslinked PPF and P(PF-*co*-EG), which both form rigid polymer networks with low water content, crosslinked OPF networks exhibit typical properties of hydrogels. Gel characteristics mainly depended on the molecular weight of PEG and reactant ratio (Jo et al., 2001). Crosslinked OPF hydrogels degrade hydrolytically along the ester bonds between fumaric acid and PEG, resulting in increased polymer swelling and decreased dry weight. The weight loss of OPF hydrogels was dependent on their crosslinking density (Shin et al., 2003c). Studies investigating the mechanical properties revealed that crosslinked OPF hydrogels made from low-molecular-weight PEG (1000 Da) swelled less, were stiffer, and elongated less before fracture when compared to hydrogels composed of longer PEG chains. OPF hydrogels can also be combined in layers to form biphasic gels, with each phase having different material properties (Temenoff et al., 2002). *In vitro* investigation of the cytotoxicity of each component of OPF hydrogel formulations and the resulting crosslinked network were conducted employing marrow stromal cells (MSCs). After 24 h, the MSCs maintained more than 75% viability for OPF concentrations below 25% (w/v). A high MW (3400 Da) PEG-DA crosslinker demonstrated significantly higher viability compared to lower MW (575 Da) PEG-DA. Leachable products from crosslinked OPF hydrogels were found to have minimal adverse effects on MSC viability (Shin et al., 2003a). The *in vivo* bone and soft tissue compatibility of OPF hydrogels was demonstrated using a rabbit model (Shin et al., 2003c). Based on these promising biocompatibility data, OPF-based hydrogels were investigated as injectable drug, DNA, and cell delivery devices. Crosslinked OPF hydrogels that encapsulate gelatin microparticles were developed as a means of simultaneously delivering two chondrogenic proteins, insulin-like growth factor-1 (IGF-1) and transforming growth factor-β1 (TGF-β1) (Holland et al., 2005b). Similar systems were implanted into osteochondral defects in the rabbit model. No evidence of prolonged inflammation was observed, and hyaline cartilage was found filling the chondral region of the defect at 14 weeks. The subchondral region was filled with bony tissue and completely integrated with the surrounding bone. The newly formed surface tissue stained positive for Safranin O and displayed promising chondrocyte organization (Holland et al., 2005a). Kasper et al. developed and characterized composites of OPF and cationized gelatin microspheres that release plasmid DNA in a sustained, controlled manner *in vivo* (Kasper et al., 2005). In order to control cell adhesion to the hydrophilic hydrogels, RGD adhesion peptide modified OPF hydrogels have been developed (Shin et al., 2002). OPF hydrogels have also been shown to be useful as injectable cell delivery vehicles for bone regeneration. MSCs were then directly combined with the OPF hydrogel precursors and encapsulated during thermal crosslinking. In the presence of osteogenic supplements, MSC differentiation in these hydrogels was apparent by day 21. At day 28, mineralized matrix could be seen throughout the hydrogels (Temenoff et al., 2004a). Hydrogel properties have been identified to affect osteogenic differentiation within these systems (Temenoff et al., 2004b). Recent studies focused on the combination of cell and growth factor delivery using injectable OPF formulations (Park et al., 2005).

Reactive cyclic acetal polymers Motivated by persistent concerns over adverse effects of acidic degradation products that are liberated from degrading bulks of polyesters, polylactones, and polyanhydrides *in vivo*, degradable cyclic acetal polymers have been developed (Falco et al., 2008). Synthetic polymers based upon acetals, cyclic acetals, and ketals can be designed to biodegrade hydrolytically and produce alcohols, carbonyls and aldehydes depending on the structure of the monomers. Relatively simple polymeric cyclic acetal networks can be fabricated by radical polymerization of the monomer 5-ethyl-5-(hydroxymethyl)-β,β-dimethyl-1, 3-dioxane-2-ethanol diacrylate (EHD). The resulting networks are hydrophobic and do not swell in water. Using the traditional salt leaching technique, macroporous biodegradable scaffolds can be fabricated that supported myoblast adhesion and proliferation and have been investigated for muscular tissue engineering (Falco et al., 2007). Networks with increased hydrophilicity have been designed by using hydrophilic PEG macromonomers (Kaihara et al., 2008) or through the incorporation of PEG in the cyclic acetal building blocks (Kaihara et al., 2007). Such biodegradable hydrogels can be formulated as an injectable system that can be crosslinked *in vivo* under cytocompatible conditions and have been loaded with osteogenic BMP-2 and investigated for the repair of craniofacial defects (Betz et al., 2008).

Current synthetic biomaterials for tissue engineering applications are sufficient, yet they are far from ideal. Biomaterials based upon polyesters and polyanhydrides possess distinctive properties and are used extensively in clinical practice. While synthetic biomaterials can be tailored to meet many tissue engineering and drug delivery needs, many are not biologically inert. In an effort to develop alternative materials, extensive research is being done to synthesize polymers that have more desirable degradation properties. Cyclic acetals are an increasingly versatile group of materials that can be utilized for both soft and hard tissue repair. Properties of cyclic acetal biomaterials have been controlled by varying fabrication parameters to create highly hydrophobic EH networks. These networks have been shown to support a viable osteoprogenitor and myoblast cell population. Alternatively, water-swellable EH-PEG hydrogels were able to sustain an encapsulated osteoprogenitor cell population for up to 7 days *in vitro* as well as deliver BMP-2 to bone *in vivo*. Finally, in an effort to create a more organized hydrogel structure, EHD and PEG were copolymerized to form PECA. PECA hydrogels have been shown to be a favorable material for both drug delivery and tissue engineering applications. Other groups of biomaterials are based upon polyacetals and polyketals, and have shown potential in drug delivery applications due to their pH-dependent degradation properties. The development of alternative synthetic polymers, such as those described here, is a critical step for the future success of many tissue engineering and drug delivery applications.

Polymers containing acrylate, methacrylate, or vinylsulfone functionalities Precursors for crosslinked biodegradable polyester networks that bear vinylsulfone, acrylate, or methacrylate functionalities include PEG-DA (Fig. 33.5D), PEG-dimethacrylate (Fig. 33.5E), PEG vinylsulfones, diacrylated PLA-PEG-PLA block copolymers, acrylic modified PVA, methacrylate-modified dextran, and acrylated chitosan (Hoffman, 2002; Nguyen and West, 2002; Hou et al., 2004). Since the last two are synthetic derivatives of natural macromolecules, they are not discussed further. Besides such hydrophilic, natural macromolecules, which are considered candidate building blocks based on their inherent biocompatibility, PEG is the most prominent synthetic component of crosslinked polymer networks due to its biocompatibility and inertness. As described above, PEG is hydrophilic and does not promote cell adhesion. To improve cell adhesion to crosslinked PEG hydrogels, adhesion peptides containing the tripeptide motif RGD have been incorporated (Hern and Hubbell, 1998; Burdick and Anseth, 2002; Gonzalez et al., 2004). Recent research on engineered hydrogels has been focused on mimicking the invasive characteristics of native extracellular matrices by including substrates for matrix metalloproteinases (MMPs) in addition to integrin-binding sites. PEG hydrogels crosslinked in part by MMP-sensitive linkers were made degradable and invasive for cells via cell-secreted MMPs (Lutolf et al., 2003a). Critical-sized defects in rat crania were completely infiltrated by cells and were remodeled into bony tissue within 5 weeks when the above-mentioned gels were loaded with recombinant human bone morphogenetic protein-2 and implanted in the defect site. As in natural extracellular matrices, which sequester a variety of cellular growth factors and act as a local depot for them, invading cells were presented with a mitogen that, in this case, specifically promoted bone regeneration (Lutolf et al., 2003b). The PEG-based hydrogels used in these studies were fabricated by a conjugate addition reaction between vinylsulfone-functionalized branched PEG and thiol-bearing peptides under almost physiological conditions.

In order to enhance the initial mechanical stability and biodegradability of crosslinked PEG-based hydrogels, oligomeric biodegradable lipophilic blocks, such as oligo(lactic acid) (Burdick et al., 2001) (Fig. 33.5F) and oligo(ε-caprolactone) (Davis et al., 2003), were included in the crosslinkable polymeric precursors. In a critical-size cranial defect model, porous crosslinked poly(ethylene glycol(2)-lactic acid(10)) scaffolds in combination with osteoinductive growth factors have shown potential as an *in situ* forming synthetic bone graft material (Burdick et al., 2003).

Photopolymerized (meth)acrylated biodegradable hydrogels have been used in a wide range of biomedical applications. As described above, limited interactions with proteins are characteristic for hydrophilic surfaces. Consequently, applications such as the use of crosslinked hydrogels as barriers applied after tissue injury to improve wound healing or as cell encapsulation materials that immunoisolate transplanted cells capitalize on this property (Cruise et al., 1999; Nguyen and West, 2002). Islets of Langerhans encapsulated in PEG-DA hydrogels and transplanted in order to develop a bioartificial endocrine pancreas are a prominent example of the latter applications. The hydrogels are permeable for nutrients, oxygen, and metabolic products, allowing the entrapped islets to survive and to secrete insulin that is released by diffusion. Hydrophilic tissue barriers from crosslinked polyesters, such as poly (ethylene glycol-*co*-lactic acid) diacrylate, have been used to prevent thrombosis and restenosis following vascular injury and post-operative adhesion formation following many abdominal and pelvic surgical procedures.

Crosslinked hydrophilic polyesters are also promising depots for local drug delivery because of their compatibility with hydrophilic, macromolecular drugs, such as proteins or oligonucleotides. The materials' good tissue and hemocompatibility even allow for intravascular applications (An and Hubbell, 2000). Drug release from crosslinked hydrogels generally can be well controlled by adjusting swelling, crosslink density, and polymer degradation (Peppas et al., 1999, 2000; Davis and Anseth, 2002).

Photopolymerized (meth)acrylated polymer networks have also been widely explored for injectable tissue engineering (Hoffman, 2002; Varghese and Elisseeff, 2006). Elisseeff and coworkers employed PEG-DA scaffolds for cartilage engineering by encapsulating chondrocytes, MSCs, and embryonic stem cells. In these studies, the crosslinked PEG-based hydrogels served as an efficient scaffold for anchorage-independent cells and promoted tissue formation. Photogelation, which offers good spatial and temporal control of hydrogel curing, has been used to control the spatial organization of different cell types within a three-dimensional system for osteochondral defect regeneration by sequentially polymerizing multiple cell/hydrogel layers. In an attempt to promote hydrogel-tissue integration, a tissue-initiated polymerization technique has been developed that utilizes *in situ*-generated tyrosyl radicals to initiate photogelation of an injectable macromer solution (Varghese and Elisseeff, 2006).

Traditionally, photopolymerization occurs by directly exposing materials to UV or visible light in accessible cavities or during invasive surgery. For PEG-dimethacrylate hydrogels, it has been shown that light, which penetrates tissue including skin, can cause a photopolymerization indirectly (transdermal photopolymerization). *In vivo* studies revealed that gels can be polymerized in 3 minutes with no harm to imbedded chondrocytes and subsequent cartilaginous tissue formation as indicated by increasing GAG and collagen contents (Elisseeff et al., 1999). In deep crevices, as they may be found in larger orthopedic defects, problems are expected to arise from limited light penetration and inconsistent photopolymerization. For those applications, thermally induced crosslinking techniques appear to be advantageous (Temenoff and Mikos, 2000).

Hydrogel forming macromonomers containing other functionalities The development of injectable hydrophilic macromonomers that can be cross-copolymerized to hydrogels under physiological conditions using cytocompatible chemistries has become a major focus in biomaterial development. The process started with the development of protocols to use photo- or heat-initiated free radical polymerization of hydrophilic, typically PEG-based, macromonomers in the presence of cells (Temenoff and Mikos, 2000; Shin et al., 2003a). Over the years, several alternative strategies have been explored employing specific addition reactions (Patterson et al., 2010), classical bioconjugation chemistry and "click" chemistry (Lutz and Börner, 2008; van Dijk et al., 2009), as well as enzymatic conjugation (Liu et al., 2009). Specific examples include the Michael-type addition between thiol groups of designed peptides and multi-arm PEG vinyl sulfone (Lutolf et al., 2003a), or the conjugation reaction between amine groups and succinimidyl esters that was utilized for the fabrication of transparent PEG-hydrogels for ocular applications from branched PEG-succinimidyl propionates and bi- or multi- functional PEG-amines (Brandl et al., 2007). Recently, a complex engineering approach was presented for the direct fabrication of biologically functionalized gels with ideal structures that can be photopatterned to generate specific microenvironments *in situ*, and all in the presence of cells (Deforest et al., 2009). In this approach an enzymatically degradable peptide macromer was reacted with a multiarm PEG-azide through a copper-free click chemistry that allowed for the direct encapsulation of cells. Subsequently, biological functionalities, e.g. adhesion peptides, were introduced within the gel by a thiol-ene photocoupling chemistry in real time and with micrometre-scale resolution.

APPLICATIONS OF SYNTHETIC POLYMERS

Synthetic polymers play a vital role in biomedical applications, including nano-, micro-, and macroscopic drug and gene delivery devices (Brannon-Peppas, 1995; Hubbell, 1998; Uhrich et al., 1999; Panyam and Labhasetwar, 2003), orthopedic fixation devices (Bostman and Pihla-jamaki, 2000), cosmetic and prosthetic implants (Behravesh et al., 1999), and as artificial matrices for tissue engineering applications (Seal et al., 2001). The interested reader may be directed to the referenced reviews that provide in-depth insight into current trends and technologies. Researchers have sought to develop and clinically explore third-generation biomaterials (Hench and Polak, 2002) that are designed to control protein adsorption, cell adhesion and differentiation, implant integration, and foreign body reaction, and to develop biomimetic synthetic materials (Shin et al., 2003b; Drotleff et al., 2004; Lutolf and Hubbell, 2005; Patterson et al., 2010).

CONCLUSION/SUMMARY

Synthetic biomaterials have progressed from testing "off-the-shelf" plastics not developed for biomedical purposes, to a field of synergistic research by engineers, scientists, and physicians dedicated to tailoring material properties for specific applications. Most recent trends shift the focus towards biology in order to first understand and then mimic physiological interactions and signaling.

Hydrogels, especially injectable systems, enjoy increasing attention due to the comfort of their application, their structural similarity to native extracellular matrix, and their good compatibility for direct cell encapsulation due to high water content. It is no longer believed in tissue engineering that the biomaterial itself has to provide mechanical properties comparable to the diseased tissue; the polymer rather has to promote defect-site remodeling and tissue regeneration *in vivo* in such a way that the regenerated tissue is histologically and functionally indistinguishable from the surrounding tissue. Hydrogels might be superior to hydrophobic polymers in that regard, as they can degrade, faster resolving the problem of non-functional fibrous tissue formation on the polymer-tissue interface. Also, hydrogel breakdown can be synchronized with cell proliferation and migration by using enzymatically cleavable crosslinker.

Besides providing tailored degradative properties, synthetic materials for regenerative medicine should allow for minimally invasive application techniques, integrate well with the surrounding tissue, and promote cell adhesion, migration, and finally differentiation. The development and thorough characterization of injectable biodegradables provides the foundation for injectable tissue regeneration. *In situ* gelation or polymerization concepts will still have to be developed and optimized with regard to cytocompatibility and stability of the resulting construct. The implementation of biomimetic design strategies will allow us to control and custom-design cell-biomaterial interactions in order to guide tissue formation from transplanted cells. Strategies based on gene delivery or gene-activating biomaterials also have great potential in regenerative medicine but the long-term safety of such therapies remains to be proven.

Overall, the advances that have been made in the field of biomaterial synthesis and design of physicochemical properties during the last 50 years in conjunction with the rapidly increasing knowledge of cell biology concerning adhesion, migration, differentiation, and signaling will reveal design concepts for improved injectable, biomimetic polymer-based formulations for tissue engineering applications.

References

Agrawal, C. M., Athanasiou, K. A., & Heckman, J. D. (2000). Biodegradable PLA-PGA polymers for tissue engineering in orthopaedics. *Mater. Sci. Forum, 1997*, 115–128.

Albertsson, A. C., & Varma, I. K. (2003). Recent developments in ring opening polymerization of lactones for biomedical applications. *Biomacromolecules, 4*, 1466–1486.

Allcock, H. R., & Kwon, S. (1989). Ionically cross-linkable polyphosphazene: poly[bis(carboxylatophenoxy)phosphazene] and its hydrogels and membranes. *Macromolecules, 22*, 75–79.

Amecke, B., Bendix, D., & Entenmann, G. (1992). Resorbable polyesters: composition, properties, applications. *Clin. Mater., 10*, 47–50.

An, Y., & Hubbell, J. A. (2000). Intraarterial protein delivery via intimally-adherent bilayer hydrogels. *J. Control. Release, 64*, 205–215.

Anderson, J. M., & Shive, M. S. (1997). Biodegradation and biocompatibility of PLA and PLGA microspheres. *Adv. Drug Deliv. Rev., 28*, 5–24.

Andriano, K. P., Tabata, Y., Ikada, Y., & Heller, J. (1999). In vitro and in vivo comparison of bulk and surface hydrolysis in absorbable polymer scaffolds for tissue engineering. *J. Biomed. Mater. Res., 48*, 602–612.

Andrianov, A. K., & Payne, L. G. (1998). Protein release from polyphosphazene matrices. *Adv. Drug Deliv. Rev., 31*, 185–196.

Anseth, K. S., Shastri, V. R., & Langer, R. (1999). Photopolymerizable degradable polyanhydrides with osteocompatibility. *Nat. Biotechnol., 17*, 156–159.

Arepalli, S. R., Bezabeh, S., & Brown, S. L. (2002). Allergic reaction to platinum in silicone breast implants. *J. Long Term Eff. Med. Implants, 12*, 299–306.

Athanasiou, K. A., Niederauer, G. G., & Agrawal, C. M. (1996). Sterilization, toxicity, biocompatibility and clinical applications of polylactic acid/polyglycolic acid copolymers. *Biomaterials, 17*, 93–102.

Bazile, D., Prud'homme, C., Bassoullet, M. T., Marlard, M., Spenlehauer, G., & Veillard, M. (1995). Stealth Me.PEG-PLA nanoparticles avoid uptake by the mononuclear phagocytes system. *J. Pharm. Sci., 84*, 493–498.

Behravesh, E., Yasko, A. W., Engel, P. S., & Mikos, A. G. (1999). Synthetic biodegradable polymers for orthopaedic applications. *Clin. Orthop. Relat. Res.*, S118–S129.

Behravesh, E., Shung, A. K., Jo, S., & Mikos, A. G. (2002a). Synthesis and characterization of triblock copolymers of methoxy poly(ethylene glycol) and poly(propylene fumarate). *Biomacromolecules, 3*, 153–158.

Behravesh, E., Timmer, M. D., Lemoine, J. J., Liebschner, M. A. K., & Mikos, A. G. (2002b). Evaluation of the in vitro degradation of macroporous hydrogels using gravimetry, confined compression testing, and microcomputed tomography. *Biomacromolecules, 3*, 1263–1270.

Behravesh, E., Zygourakis, K., & Mikos, A. G. (2003). Adhesion and migration of marrow-derived osteoblasts on injectable in situ crosslinkable poly(propylene fumarate-co-ethylene glycol)-based hydrogels with a covalently linked RGDS peptide. *J. Biomed. Mater. Res. A, 65*, 260–270.

Beniash, E., Hartgerink, J. D., Storrie, H., Stendahl, J. C., & Stupp, S. I. (2005). Self-assembling peptide amphiphile nanofiber matrices for cell entrapment. *Acta Biomater., 1*, 387–397.

Betre, H., Setton, L. A., Meyer, D. E., & Chilkoti, A. (2002). Characterization of a genetically engineered elastin-like polypeptide for cartilaginous tissue repair. *Biomacromolecules, 3*, 910–916.

Betz, M. W., Modi, P. C., Caccamese, J. F., Coletti, D. P., Sauk, J. J., & Fisher, J. P. (2008). Cyclic acetal hydrogel system for bone marrow stromal cell encapsulation and osteodifferentiation. *J. Biomed. Mater. Res. A, 86*, 662–670.

Bobyn, J. D., Mortimer, E. S., Glassman, A. H., Engh, C. A., Miller, J. E., & Brooks, C. E. (1992). Producing and avoiding stress shielding: laboratory and clinical observations of noncemented total hip arthroplasty. *Clin. Orthop. Relat. Res.*, 79–96.

Boretos, J. W., & Pierce, W. S. (1967). Segmented polyurethane: a new elastomer for biomedical applications. *Science, 158*, 1481–1482.

Bostman, O., & Pihlajamaki, H. (2000). Clinical biocompatibility of biodegradable orthopaedic implants for internal fixation: a review. *Biomaterials, 21*, 2615–2621.

Bouillot, P., Petit, A., & Dellacherie, E. (1998). Protein encapsulation in biodegradable amphiphilic microspheres. I. Polymer synthesis and characterization and microsphere elaboration. *J. Appl. Polym. Sci., 68*, 1695–1702.

Bourke, S. L., & Kohn, J. (2003). Polymers derived from the amino acid tyrosine: polycarbonates, polyarylates and copolymers with poly(ethylene glycol). *Adv. Drug Deliv. Rev., 55*, 447–466.

Brandl, F., Henke, M., Rothschenk, S., Gschwind, R., Breunig, M., Blunk, T., et al. (2007). Poly(ethylene glycol)-based hydrogels for intraocular applications. *Adv. Eng. Mater., 9*, 1141–1149.

Brannon-Peppas, L. (1995). Recent advances on the use of biodegradable microparticles and nanoparticles in controlled drug delivery. *Int. J. Pharm., 116*, 1–9.

Braunecker, W. A., & Matyjaszewski, K. (2007). Controlled/living radical polymerization: features, developments, and perspectives. *Progr. Polymer Sci., 32*, 93–146.

Brocchini, S. (2001). Combinatorial chemistry and biomedical polymer development. *Adv. Drug Deliv. Rev., 53*, 123–130.

Brook, M. A. (2006). Platinum in silicone breast implants. *Biomaterials, 27*, 3274–3286.

Brunner, A., Mader, K., & Gopferich, A. (1999). pH and osmotic pressure inside biodegradable microspheres during erosion. *Pharm. Res., 16*, 847–853.

Bryers, J. D., Jarvis, R. A., Lebo, J., Prudencio, A., Kyriakides, T. R., & Uhrich, K. (2006). Biodegradation of poly (anhydride-esters) into non-steroidal anti-inflammatory drugs and their effect on *Pseudomonas aeruginosa* biofilms in vitro and on the foreign-body response in vivo. *Biomaterials, 27*, 5039–5048.

Burdick, J. A., & Anseth, K. S. (2002). Photoencapsulation of osteoblasts in injectable RGD-modified PEG hydrogels for bone tissue engineering. *Biomaterials, 23*, 4315–4323.

Burdick, J. A., Philpott, L. M., & Anseth, K. S. (2001). Synthesis and characterization of tetrafunctional lactic acid oligomers: a potential in situ forming degradable orthopaedic biomaterial. *J. Polym. Sci. A, 39*, 683–692.

Burdick, J. A., Frankel, D., Dernell, W. S., & Anseth, K. S. (2003). An initial investigation of photocurable three-dimensional lactic acid-based scaffolds in a critical-sized cranial defect. *Biomaterials, 24*, 1613–1620.

Burkersroda, F. V., Schedl, L., & Gopferich, A. (2002). Why degradable polymers undergo surface erosion or bulk erosion. *Biomaterials, 23*, 4221–4231.

Byrne, M. E., Park, K., & Peppas, N. A. (2002). Molecular imprinting within hydrogels. *Adv. Drug Deliv. Rev., 54*, 149–161.

Cannizzaro, S. M., Padera, R. F., Langer, R., Rogers, R. A., Black, F. E., Davies, M. C., et al. (1998). A novel biotinylated degradable polymer for cell-interactive applications. *Biotechnol. Bioeng., 58*, 529–535.

Cerrai, P., Guerra, G. D., Lelli, L., Tricoli, M., Sbarbati del Guerra, R., Cascone, M. G., et al. (1994). Poly(ester-ether-ester) block copolymers as biomaterials. *J. Mater. Sci. Mater. Med., 5*, 33–39.

Chandrasekaran, S. K., Capozza, R., & Wong, P. S. L. (1978). Therapeutic systems and controlled drug delivery. *J. Membr. Sci., 3*, 271–286.

Chirila, T. V., Constable, I. J., Crawford, G. J., Vijayasekaran, S., Thompson, D. E., Chen, Y. C., et al. (1993). Poly (2-hydroxyethyl methacrylate) sponges as implant materials: in vivo and in vitro evaluation of cellular invasion. *Biomaterials, 14*, 26–38.

Choueka, J., Charvet, J. L., Koval, K. J., Alexander, H., James, K. S., Hooper, K. A., et al. (1996). Canine bone response to tyrosine-derived polycarbonates and poly(L-lactic acid). *J. Biomed. Mater. Res., 31*, 35–41.

Colas, A., & Curtis, J. (2004). Silicone biomaterials: history and chemistry. In B. D. Ratner, A. S. Hoffman, F. J. Schoen, & J. E. Lemons (Eds.), *Biomaterials Science. An Introduction to Materials in Medicine* (2nd Ed.) (pp. 80–86). San Diego, CA: Academic Press.

Cruise, G. M., Hegre, O. D., Lamberti, F. V., Hager, S. R., Hill, R., Scharp, D. S., et al. (1999). In vitro and in vivo performance of porcine islets encapsulated in interfacially photopolymerized poly(ethylene glycol) diacrylate membranes. *Cell Transplant., 8,* 293–306.

Curtis, J., & Colas, A. (2004). Medical applications of silicones. In B. D. Ratner, A. S. Hoffman, F. J. Schoen, & J. E. Lemons (Eds.), *Biomaterials Science. An Introduction to Materials in Medicine* (2nd Ed.) (pp 697–707). San Diego, CA: Academic Press.

Cutright, D. E., Beasley, I. I. I., & Perez, B. (1971). Histologic comparison of polylactic and polyglycolic acid sutures. *Oral Surg. Oral Med. Oral Pathol., 32,* 165–173.

Davis, K. A., & Anseth, K. S. (2002). Controlled release from crosslinked degradable networks. *Crit. Rev. Ther. Drug Carrier Syst., 19,* 385–423.

Davis, K. A., Burdick, J. A., & Anseth, K. S. (2003). Photoinitiated crosslinked degradable copolymer networks for tissue engineering applications. *Biomaterials, 24,* 2485–2495.

Deforest, C. A., Polizzotti, B. D., & Anseth, K. S. (2009). Sequential click reactions for synthesizing and patterning three-dimensional cell microenvironments. *Nat. Mater., 8,* 659–664.

Domb, A. J., Laurencin, C. T., Israeli, O., Gerhart, T. N., & Langer, R. (1990). Formation of propylene fumarate oligomers for use in bioerodible bone cement composites. *J. Polym. Sci. A, 28,* 973–985.

Drotleff, S., Lungwitz, U., Breunig, M., Dennis, A., Blunk, T., Tessmar, J., et al. (2004). Biomimetic polymers in pharmaceutical and biomedical sciences. *Eur. J. Pharm. Biopharm., 58,* 385–407.

Elisseeff, J., Anseth, K., Sims, D., McIntosh, W., Randolph, M., & Langer, R. (1999). Transdermal photo-polymerization for minimally invasive implantation. *Proc. Natl. Acad. Sci. U.S.A. 96,* 3104–3107.

Engelberg, I., & Kohn, J. (1991). Physico-mechanical properties of degradable polymers used in medical applications: a comparative study. *Biomaterials, 12,* 292–304.

Falco, E. E., Roth, J. S., & Fisher, J. P. (2007). EH Networks as a scaffold for skeletal muscle regeneration in abdominal wall hernia repair. *J. Surg. Res., 149,* 76–83.

Falco, E. E., Patel, M., & Fisher, J. P. (2008). Recent developments in cyclic acetal biomaterials for tissue engineering applications. *Pharm. Res., 25,* 2348–2356.

Fisher, J. P., Holland, T. A., Dean, D., Engel, P. S., & Mikos, A. G. (2001). Synthesis and properties of photocross-linked poly(propylene fumarate) scaffolds. *J. Biomater. Sci. Polym. Ed., 12,* 673–687.

Fisher, J. P., Dean, D., & Mikos, A. G. (2002a). Photocrosslinking characteristics and mechanical properties of diethyl fumarate/poly(propylene fumarate) biomaterials. *Biomaterials, 23,* 4333–4343.

Fisher, J. P., Vehof, J. W. M., Dean, D., van der Waerden, J. P., Holland, T. A., Mikos, A. G., et al. (2002b). Soft and hard tissue response to photocrosslinked poly(propylene fumarate) scaffolds in a rabbit model. *J. Biomed. Mater. Res., 59,* 547–556.

Fisher, J. P., Jo, S., Mikos, A. G., & Reddi, A. H. (2004). Thermoreversible hydrogel scaffolds for articular cartilage engineering. *J. Biomed. Mater. Res. A, 71,* 268–274.

Frazier, D. D., Lathi, V. K., Gerhart, T. N., & Hayes, W. C. (1997). Ex vivo degradation of a poly(propylene glycol-fumarate) biodegradable particulate composite bone cement. *J. Biomed. Mater. Res., 35,* 383–389.

Freed, L. E., Vunjak, N. G., Biron, R. J., Eagles, D. B., Lesnoy, D. C., Barlow, S. K., et al. (1994). Biodegradable polymer scaffolds for tissue engineering. *Biotechnology (NY), 12,* 689–693.

Freed, L. E., Langer, R., Martin, I., Pellis, N. R., & Vunjak-Novakovic, G. (1997). Tissue engineering of cartilage in space. *Proc. Natl. Acad. Sci. USA, 94,* 13885–13890.

Fromstein, J. D., & Woodhouse, K. A. (2006). Polyurethane biomaterials. In G. E. Wnek, & G. L. Bowlin (Eds.), *Encyclopedia of Biomaterials and Biomedical Engineering* (2nd. Ed.) (pp. 2304). New York: Marcel Dekker.

Galler, K. M., Cavender, A., Yuwono, V., Dong, H., Shi, S., Schmalz, G., et al. (2008). Self-assembling peptide amphiphile nanofibers as a scaffold for dental stem cells. *Tissue Eng. A, 14,* 2051–2058.

Ganta, S. R., Piesco, N. P., Long, P., Gassner, R., Motta, L. F., Papworth, G. D., et al. (2003). Vascularization and tissue infiltration of a biodegradable polyurethane matrix. *J. Biomed. Mater. Res. A, 64,* 242–248.

Gilding, D. K., & Reed, A. M. (1979). Biodegradable polymers for use in surgery — polyglycolic/poly(actic acid) homo- and copolymers: 1. *Polymer, 20,* 1459–1464.

Gogolewski, S. (1989). Selected topics in biomedical polyurethanes. A review. *Colloid Polym. Sci., 267,* 757–785.

Goldberg, M., Mahon, K., & Anderson, D. (2008). Combinatorial and rational approaches to polymer synthesis for medicine. *Adv. Drug Deliv. Rev., 60,* 971–978.

Gonzalez, A. L., Gobin, A. S., West, J. L., McIntire, L. V., & Smith, C. W. (2004). Integrin interactions with immobilized peptides in polyethylene glycol diacrylate hydrogels. *Tissue Eng., 10,* 1775–1786.

Göpferich, A. (1996). Polymer degradation and erosion. Mechanisms and applications. *Eur. J. Pharm. Biopharm., 42,* 1–11.

Göpferich, A., & Tessmar, J. (2002). Polyanhydride degradation and erosion. *Adv. Drug Deliv. Rev., 54,* 911–931.

Gref, R., Minamitake, Y., Peracchia, M. T., Trubetskoy, V., Torchilin, V., & Langer, R. (1994). Biodegradable long-circulating polymeric nanospheres. *Science, 263,* 1600–1603.

Gref, R., Minamitake, Y., Peracchia, M. T., Domb, A., Trubetskoy, V., Torchilin, V., et al. (1997). Poly(ethylene glycol)-coated nanospheres: potential carriers for intravenous drug administration. *Pharm. Biotechnol., 10,* 167–198.

Gref, R., Luck, M., Quellec, P., Marchand, M., Dellacherie, E., Harnisch, S., et al. (2000). 'Stealth' corona-core nanoparticles surface modified by polyethylene glycol (PEG): influences of the corona (PEG chain length and surface density) and of the core composition on phagocytic uptake and plasma protein adsorption. *Colloids Surf. B, 18,* 301–313.

Gresser, J. D., Hsu, S. H., Nagaoka, H., Lyons, C. M., Nieratko, D. P., Wise, D. L., et al. (1995). Analysis of a vinyl pyrrolidone/poly(propylene fumarate) resorbable bone cement. *J. Biomed. Mater. Res., 29,* 1241–1247.

Gunatillake, P. A., & Adhikari, R. (2003). Biodegradable synthetic polymers for tissue engineering. *Eur. Cell. Mater., 5,* 1–16.

Gunatillake, P. A., Martin, D. J., Meijs, G. F., McCarthy, S. J., & Adhikari, R. (2003). Designing biostable polyurethane elastomers for biomedical implants. *Aust. J. Chem., 56,* 545–557.

Gupta, A. P., & Kumar, V. (2007). New emerging trends in synthetic biodegradable polymers – Polylactide: A critique. *Eur. Polym. J., 43,* 4053–4074.

Hacker, M., Tessmar, J., Neubauer, M., Blaimer, A., Blunk, T., Gopferich, A., et al. (2003). Towards biomimetic scaffolds: anhydrous scaffold fabrication from biodegradable amine-reactive diblock copolymers. *Biomaterials, 24,* 4459–4473.

Harris, J. M., & Chess, R. B. (2003). Effect of pegylation on pharmaceuticals. *Nat. Rev. Drug Discov., 2,* 214–221.

Hartgerink, J. D., Beniash, E., & Stupp, S. I. (2001). Self-assembly and mineralization of peptide-amphiphile nanofibers. *Science, 294,* 1684–1688.

Hasegawa, M., Sudo, A., Shikinami, Y., & Uchida, A. (1999). Biological performance of a three-dimensional fabric as artificial cartilage in the repair of large osteochondral defects in rabbit. *Biomaterials, 20,* 1969–1975.

He, S., Timmer, M. D., Yaszemski, M. J., Yasko, A. W., Engel, P. S., & Mikos, A. G. (2001). Synthesis of biodegradable poly(propylene fumarate) networks with poly(propylene fumarate)-diacrylate macromers as crosslinking agents and characterization of their degradation products. *Polymer, 42,* 1251–1260.

Heller, J. (1984). Biodegradable polymers in controlled drug delivery. *Crit. Rev. Ther. Drug Carrier Syst., 1,* 39–90.

Heller, J., Barr, J., Ng, S. Y., Abdellauoi, K. S., & Gurny, R. (2002). Poly(ortho esters): synthesis, characterization, properties and uses. *Adv. Drug Deliv. Rev., 54,* 1015–1039.

Hench, L. L. (1980). Biomaterials. *Science, 208,* 826–831.

Hench, L. L., & Polak, J. M. (2002). Third-generation biomedical materials. *Science, 295,* 1014–1017.

Hendriks, J. G. E., van Horn, J. R., van der Mei, H. C., & Busscher, H. J. (2004). Backgrounds of antibiotic-loaded bone cement and prosthesis-related infection. *Biomaterials, 25,* 545–556.

Hern, D. L., & Hubbell, J. A. (1998). Incorporation of adhesion peptides into nonadhesive hydrogels useful for tissue resurfacing. *J. Biomed. Mater. Res., 39,* 266–276.

Hoffman, A. S. (2002). Hydrogels for biomedical applications. *Adv. Drug Deliv. Rev., 54,* 3–12.

Holland, T. A., Bodde, E. W. H., Baggett, L. S., Tabata, Y., Mikos, A. G., & Jansen, J. A. (2005a). Osteochondral repair in the rabbit model utilizing bilayered, degradable oligo(poly(ethylene glycol) fumarate) hydrogel scaffolds. *J. Biomed. Mater. Res. A, 75,* 156–167.

Holland, T. A., Tabata, Y., & Mikos, A. G. (2005b). Dual growth factor delivery from degradable oligo(poly(ethylene glycol) fumarate) hydrogel scaffolds for cartilage tissue engineering. *J. Control. Release, 101,* 111–125.

Horch, R. A., Shahid, N., Mistry, A. S., Timmer, M. D., Mikos, A. G., & Barron, A. R. (2004). Nanoreinforcement of poly(propylene fumarate)-based networks with surface modified alumoxane nanoparticles for bone tissue engineering. *Biomacromolecules, 5,* 1990–1998.

Hosseinkhani, H., Hosseinkhani, M., Tian, F., Kobayashi, H., & Tabata, Y. (2006). Osteogenic differentiation of mesenchymal stem cells in self-assembled peptide-amphiphile nanofibers. *Biomaterials, 27,* 4079–4086.

Hou, Q. P., de Bank, P. A., & Shakesheff, K. M. (2004). Injectable scaffolds for tissue regeneration. *J. Mater. Chem., 14,* 1915–1923.

Houchin, M. L., & Topp, E. M. (2008). Chemical degradation of peptides and proteins in PLGA: a review of reactions and mechanisms. *J. Pharm. Sci., 97,* 2395–2404.

Howard, G. T. (2002). Biodegradation of polyurethane: a review. *Int. Biodeterior. Biodegradation, 49,* 245–252.

Hubbell, J. A. (1995). Biomaterials in tissue engineering. *Biotechnology (NY), 13,* 565–576.

Hubbell, J. A. (1998). Synthetic biodegradable polymers for tissue engineering and drug delivery. *Curr. Opin. Solid State Mater. Sci., 3,* 246–251.

Hutmacher, D. W. (2000). Scaffolds in tissue engineering bone and cartilage. *Biomaterials, 21,* 2529–2543.

616

Ibim, S. E. M., Uhrich, K. E., Attawia, M., Shastri, V. R., El-Amin, S. F., Bronson, R., et al. (1998). Preliminary in vivo report on the osteocompatibility of poly(anhydride- co-imides) evaluated in a tibial model. *J. Biomed. Mater. Res., 43*, 374–379.

Ibusuki, S., Fujii, Y., Iwamoto, Y., & Matsuda, T. (2003a). Tissue-engineered cartilage using an injectable and in situ gelable thermoresponsive gelatin: fabrication and in vitro performance. *Tissue Eng., 9*, 371–384.

Ibusuki, S., Iwamoto, Y., & Matsuda, T. (2003b). System-engineered cartilage using poly(*N*-isopropylacrylamide)-grafted gelatin as in situ-formable scaffold: in vivo performance. *Tissue Eng., 9*, 1133–1142.

Jacobs, J. J., Sumner, D. R., & Galante, J. O. (1993). Mechanisms of bone loss associated with total hip replacement. *Orthop. Clin. North Am., 24*, 583–590.

Jain, R. A. (2000). The manufacturing techniques of various drug loaded biodegradable poly(lactide-co-glycolide) (PLGA) devices. *Biomaterials, 21*, 2475–2490.

Jeong, B., Bae, Y. H., Lee, D. S., & Kim, S. W. (1997). Biodegradable block copolymers as injectable drug-delivery systems. *Nature, 388*, 860–862.

Jeong, B., Kim, S. W., & Bae, Y. H. (2002). Thermosensitive sol-gel reversible hydrogels. *Adv. Drug Deliv. Rev., 54*, 37–51.

Jo, S., Shin, H., Shung, A. K., Fisher, J. P., & Mikos, A. G. (2001). Synthesis and characterization of oligo(poly (ethylene glycol) fumarate) macromer. *Macromolecules, 34*, 2839–2844.

Juni, K., & Nakano, M. (1987). Poly(hydroxy acids) in drug delivery. *Crit. Rev. Ther. Drug Carrier Syst., 3*, 209–232.

Kaihara, S., Matsumura, S., & Fisher, J. P. (2007). Synthesis and properties of poly [poly (ethylene glycol)-co-cyclic acetal] based hydrogels. *Macromolecules, 40*, 7625–7632.

Kaihara, S., Matsumura, S., & Fisher, J. P. (2008). Synthesis and characterization of cyclic acetal based degradable hydrogels. *Eur. J. Pharm. Biopharm., 68*, 67–73.

Kasper, F. K., Kushibiki, T., Kimura, Y., Mikos, A. G., & Tabata, Y. (2005). In vivo release of plasmid DNA from composites of oligo(poly(ethylene glycol)fumarate) and cationized gelatin microspheres. *J. Control. Release, 107*, 547–561.

Kasper, F. K., Tanahashi, K., Fisher, J. P., & Mikos, A. G. (2009). Synthesis of poly(propylene fumarate). *Nat. Protoc., 4*, 518–525.

Katti, D. S., Lakshmi, S., Langer, R., & Laurencin, C. T. (2002). Toxicity, biodegradation and elimination of poly-anhydrides. *Adv. Drug Deliv. Rev., 54*, 933–961.

Kenny, S. M., & Buggy, M. (2003). Bone cements and fillers: a review. *J. Mater. Sci. Mater. Med., 14*, 923–938.

Kissel, T., Li, Y., & Unger, F. (2002). ABA-triblock copolymers from biodegradable polyester A-blocks and hydro-philic poly(ethylene oxide) B-blocks as a candidate for in situ forming hydrogel delivery systems for proteins. *Adv. Drug Deliv. Rev., 54*, 99–134.

Kohn, J., & Langer, R. (1986). Poly(iminocarbonates) as potential biomaterials. *Biomaterials, 7*, 176–182.

Król, P. (2007). Synthesis methods, chemical structures and phase structures of linear polyurethanes. Properties and applications of linear polyurethanes in polyurethane elastomers, copolymers and ionomers. *Prog. Mater. Sci., 52*, 915–1015.

Kumar, N., Langer, R. S., & Domb, A. J. (2002). Polyanhydrides: an overview. *Adv. Drug Deliv. Rev., 54*, 889–910.

Langer, R. (1990). New methods of drug delivery. *Science, 249*, 1527–1533.

Langer, R. S., & Peppas, N. A. (1981). Present and future applications of biomaterials in controlled drug delivery systems. *Biomaterials, 2*, 201–214.

Langer, R., & Vacanti, J. P. (1993). Tissue engineering. *Science, 260*, 920–926.

Langer, R., Brem, H., & Tapper, D. (1981). Biocompatibility of polymeric delivery systems for macromolecules. *J. Biomed. Mater. Res., 15*, 267–277.

Langone, F., Lora, S., Veronese, F. M., Caliceti, P., Parnigotto, P. P., Valenti, F., et al. (1995). Peripheral nerve repair using a poly(organo)phosphazene tubular prosthesis. *Biomaterials, 16*, 347–353.

Laurencin, C. T., El-Amin, S. F., Ibim, S. E., Willoughby, D. A., Attawia, M., Allcock, H. R., et al. (1996). A highly porous three-dimensional polyphosphazene polymer matrix for skeletal tissue regeneration. *J. Biomed. Mater. Res., 30*, 133–138.

Lee, J. W., Hua, F. J., & Lee, D. S. (2001). Thermoreversible gelation of biodegradable poly(ε-caprolactone) and poly (ethylene glycol) multiblock copolymers in aqueous solutions. *J. Control. Release, 73*, 315–327.

Lee, S. B. (1999). A new class of biodegradable thermosensitive polymers. 2. hydrolytic properties and salt effect on the lower critical solution temperature of poly(organophosphazenes) with methoxypoly(ethylene glycol) and amino acid esters as side groups. *Macromolecules, 32*, 7820–7827.

Leong, K. W., Simonte, V., & Langer, R. (1987). Synthesis of polyanhydrides: melt-polycondensation, dehydro-chlorination, and dehydrative coupling. *Macromolecules, 20*, 705–712.

Lewin, S. L., & Miller, T. A. (1997). A review of epidemiologic studies analyzing the relationship between breast implants and connective tissue diseases. *Plast. Reconstr. Surg., 100,* 1309–1313.

Li, S. (1999). Hydrolytic degradation characteristics of aliphatic polyesters derived from lactic and glycolic acids. *J. Biomed. Mater. Res., 48,* 342–353.

Lieb, E., Tessmar, J., Hacker, M., Fischbach, C., Rose, D., Blunk, T., et al. (2003). Poly(D, L-lactic acid)-poly(ethylene glycol)-monomethyl ether diblock copolymers control adhesion and osteoblastic differentiation of marrow stromal cells. *Tissue Eng., 9,* 71–84.

Lin, H. B., Sun, W., Mosher, D. F., Garcia-Echeverria, C., Schaufelberger, K., Lelkes, P. I., & Cooper, S. L. (1994). Synthesis, surface, and cell-adhesion properties of polyurethanes containing covalently grafted RGD-peptides. *J. Biomed. Mater. Res., 28,* 329–342.

Liu, S. Q., Tay, R., Khan, M., Rachel Ee, P. L., Hedrick, J. L., & Yang, Y. Y. (2009). Synthetic hydrogels for controlled stem cell differentiation. *Soft Matter, 6,* 67–81.

Lloyd, A. W., Faragher, R. G. A., & Denyer, S. P. (2001). Ocular biomaterials and implants. *Biomaterials, 22,* 769–785.

Lu, S., & Anseth, K. S. (1999). Photopolymerization of multilaminated poly(HEMA) hydrogels for controlled release. *J. Control. Release, 57,* 291–300.

Lucke, A., Kiermaier, J., & Gopferich, A. (2002). Peptide acylation by poly(α-hydroxy esters). *Pharm. Res., 19,* 175–181.

Lutolf, M. P., & Hubbell, J. A. (2005). Synthetic biomaterials as instructive extracellular microenvironments for morphogenesis in tissue engineering. *Nat. Biotechnol., 23,* 47–55.

Lutolf, M. P., Lauer-Fields, J. L., Schmoekel, H. G., Metters, A. T., Weber, F. E., Fields, G. B., et al. (2003a). Synthetic matrix metalloproteinase-sensitive hydrogels for the conduction of tissue regeneration: engineering cell-invasion characteristics. *Proc. Natl. Acad. Sci. U.S.A. 100,* 5413–5418.

Lutolf, M. P., Weber, F. E., Schmoekel, H. G., Schense, J. C., Kohler, T., Muller, R., et al. (2003b). Repair of bone defects using synthetic mimetics of collagenous extracellular matrices. *Nat. Biotechnol., 21,* 513–518.

Lutz, J. F., & Börner, H. G. (2008). Modern trends in polymer bioconjugates design. *Progr. Polymer Sci., 33,* 1–39.

Mack, E. J., Okano, T., & Kim, S. W. (1987). Biomedical applications of poly(2-hydroxyethyl methacrylate) and its copolymers. In N. A. Peppas (Ed.), *Hydrogels in Medicine and Pharmacy, Vol. II* (pp. 65–93).. Boca Raton, FL, USA: CRC Press.

Matsuda, T., Nagase, J., Ghoda, A., Hirano, Y., Kidoaki, S., & Nakayama, Y. (2003). Phosphorylcholine-endcapped oligomer and block co-oligomer and surface biological reactivity. *Biomaterials, 24,* 4517–4527.

Meyer, D. E., & Chilkoti, A. (1999). Purification of recombinant proteins by fusion with thermally-responsive polypeptides. *Nat. Biotechnol., 17,* 1112–1115.

Middleton, J. C., & Tipton, A. J. (2000). Synthetic biodegradable polymers as orthopedic devices. *Biomaterials, 21,* 2335–2346.

Miller, R. A., Brady, J. M., & Cutright, D. E. (1977). Degradation rates of oral resorbable implants (polylactates and polyglycolates): rate modification with changes in PLA/PGA copolymer ratios. *J. Biomed. Mater. Res., 11,* 711–719.

Mistry, A. S., Mikos, A. G., & Jansen, J. A. (2007). Degradation and biocompatibility of a poly(propylene fumarate)-based/alumoxane nanocomposite for bone tissue engineering. *J. Biomed. Mater. Res. A, 83,* 940–953.

Mistry, A. S., Cheng, S. H., Yeh, T., Christenson, E., Jansen, J. A., & Mikos, A. G. (2009). Fabrication and in vitro degradation of porous fumarate-based polymer/alumoxane nanocomposite scaffolds for bone tissue engineering. *J. Biomed. Mater. Res. A, 89,* 68–79.

Mistry, A. S., Pham, Q. P., Schouten, C., Yeh, T., Christenson, E. M., Mikos, A. G., et al. (2010). In vivo bone biocompatibility and degradation of porous fumarate-based polymer/alumoxane nanocomposites for bone tissue engineering. *J. Biomed. Mater. Res. A, 92,* 451–462.

Mooney, D. J., & Mikos, A. G. (1999). Growing new organs. *Sci. Am., 280,* 60–65.

Morikawa, N., & Matsuda, T. (2002). Thermoresponsive artificial extracellular matrix: N-isopropylacrylamide-graft-copolymerized gelatin. *J. Biomater. Sci. Polym. Ed., 13,* 167–183.

Mosbach, K., & Ramstrom, O. (1996). The emerging technique of molecular imprinting and its future impact on biotechnology. *Nat. Biotechnol., 14,* 163–170.

Motlagh, D., Allen, J., Hoshi, R., Yang, J., Lui, K., & Ameer, G. (2007). Hemocompatibility evaluation of poly(diol citrate) in vitro for vascular tissue engineering. *J. Biomed. Mater. Res. A, 82,* 907–916.

Muggli, D. S., Burkoth, A. K., & Anseth, K. S. (1999). Crosslinked polyanhydrides for use in orthopedic applications: degradation behavior and mechanics. *J. Biomed. Mater. Res., 46,* 271–278.

Nerem, R. M. (2006). Tissue engineering: the hope, the hype, and the future. *Tissue Eng., 12,* 1143–1150.

Ng, S. Y., Vandamme, T., Taylor, M. S., & Heller, J. (1997). Synthesis and erosion studies of self-catalyzed poly(ortho ester)s. *Macromolecules, 30,* 770–772.

Nguyen, K. T., & West, J. L. (2002). Photopolymerizable hydrogels for tissue engineering applications. *Biomaterials, 23*, 4307–4314.

Ohya, S., Nakayama, Y., & Matsuda, T. (2001). Thermoresponsive artificial extracellular matrix for tissue engineering: hyaluronic acid bioconjugated with poly(*N*-isopropylacrylamide) grafts. *Biomacromolecules, 2*, 856–863.

Panyam, J., & Labhasetwar, V. (2003). Biodegradable nanoparticles for drug and gene delivery to cells and tissue. *Adv. Drug Deliv. Rev., 55*, 329–347.

Park, H., Temenoff, J. S., Holland, T. A., Tabata, Y., & Mikos, A. G. (2005). Delivery of TGF-β1 and chondrocytes via injectable, biodegradable hydrogels for cartilage tissue engineering applications. *Biomaterials, 26*, 7095–7103.

Patterson, J., Martino, M. M., & Hubbell, J. A. (2010). Biomimetic materials in tissue engineering. *Mater. Today, 13*, 14–22.

Peppas, N. A., Keys, K. B., Torres-Lugo, M., & Lowman, A. M. (1999). Poly(ethylene glycol)-containing hydrogels in drug delivery. *J. Control. Release, 62*, 81–87.

Peppas, N. A., Bures, P., Leobandung, W., & Ichikawa, H. (2000). Hydrogels in pharmaceutical formulations. *Eur. J. Pharm. Biopharm., 50*, 27–46.

Peter, S. J., Miller, M. J., Yaszemski, M. J., & Mikos, A. G. (1997a). Poly(propylene fumarate). In A. J. Domb, J. Kost, & D. M. Wiseman (Eds.), *Handbook of Biodegradable Polymers* (pp. 87–97).. Amsterdam: Harwood Academic.

Peter, S. J., Nolley, J. A., Widmer, M. S., Merwin, J. E., Yaszemski, M. J., Yasko, A. W., et al. (1997b). In vitro degradation of a poly(propylene fumarate)/β-tricalcium phosphate composite orthopaedic scaffold. *Tissue Eng., 3*, 207–215.

Peter, S. J., Miller, M. J., Yasko, A. W., Yaszemski, M. J., & Mikos, A. G. (1998a). Polymer concepts in tissue engineering. *J. Biomed. Mater. Res., 43*, 422–427.

Peter, S. J., Miller, S. T., Zhu, G., Yasko, A. W., & Mikos, A. G. (1998b). In vivo degradation of a poly(propylene fumarate)/β-tricalcium phosphate injectable composite scaffold. *J. Biomed. Mater. Res., 41*, 1–7.

Peter, S. J., Kim, P., Yasko, A. W., Yaszemski, M. J., & Mikos, A. G. (1999a). Crosslinking characteristics of an injectable poly(propylene fumarate)/β-tricalcium phosphate paste and mechanical properties of the crosslinked composite for use as a biodegradable bone cement. *J. Biomed. Mater. Res., 44*, 314–321.

Peter, S. J., Suggs, L. J., Yaszemski, M. J., Engel, P. S., & Mikos, A. G. (1999b). Synthesis of poly(propylene fumarate) by acylation of propylene glycol in the presence of a proton scavenger. *J. Biomater. Sci. Polym. Ed., 10*, 363–373.

Peter, S. J., Lu, L., Kim, D. J., & Mikos, A. G. (2000). Marrow stromal osteoblast function on a poly(propylene fumarate)/β-tricalcium phosphate biodegradable orthopaedic composite. *Biomaterials, 21*, 1207–1213.

Petrova, T., Manolova, N., Rashkov, I., Li, S., & Vert, M. (1998). Synthesis and characterization of poly(oxyethylene)-poly(caprolactone) multiblock copolymers. *Polym. Int., 45*, 419–426.

Pham, Q. P., Sharma, U., & Mikos, A. G. (2006). Electrospinning of polymeric nanofibers for tissue engineering applications: a review. *Tissue Eng., 12*, 1197–1211.

Photos, P. J., Bacakova, L., Discher, B., Bates, F. S., & Discher, D. E. (2003). Polymer vesicles in vivo: correlations with PEG molecular weight. *J. Control. Release, 90*, 323–334.

Pitt, C. G., Jeffcoat, A. R., Zweidinger, R. A., & Schindler, A. (1979). Sustained drug delivery systems. I. The permeability of poly(ε-caprolactone), poly(DL-lactic acid), and their copolymers. *J. Biomed. Mater. Res., 13*, 497–507.

Pitt, G. G., Gratzl, M. M., Kimmel, G. L., Surles, J., & Sohindler, A. (1981). Aliphatic polyesters II. The degradation of poly (DL-lactide), poly (ε-caprolactone), and their copolymers in vivo. *Biomaterials, 2*, 215–220.

Pulapura, S., & Kohn, J. (1992). Tyrosine-derived polycarbonates: backbone-modified 'pseudo'-poly(amino acids) designed for biomedical applications. *Biopolymers, 32*, 411–417.

Ratner, B. D. (2004). A history of biomaterials. In B. D. Ratner, A. S. Hoffman, F. J. Schoen, & J. E. Lemons (Eds.), *Biomaterials Science. An Introduction to Materials in Medicine* (2nd Ed.) (pp 10–19). San Diego, CA: Academic Press.

Reed, A. M., & Gilding, D. K. (1981). Biodegradable polymers for use in surgery — poly(glycolic)/poly(lactic acid) homo and copolymers: 2. In vitro degradation. *Polymer, 22*, 494–498.

Reimschuessel, H. K. (1975). General aspects in polymer synthesis. *Environ. Health Perspect., 11*, 9–20.

Roberts, M. J., Bentley, M. D., & Harris, J. M. (2002). Chemistry for peptide and protein PEGylation. *Adv. Drug Deliv. Rev., 54*, 459–476.

Rodriguez, F. J., Gomez, N., Perego, G., & Navarro, X. (1999). Highly permeable polylactide-caprolactone nerve guides enhance peripheral nerve regeneration through long gaps. *Biomaterials, 20*, 1489–1500.

Ruel-Gariepy, E., & Leroux, J. C. (2004). In situ-forming hydrogels — review of temperature-sensitive systems. *Eur. J. Pharm. Biopharm., 58*, 409–426.

Saad, B., Hirt, T. D., Welti, M., Uhlschmid, G. K., Neuenschwander, P., & Suter, U. W. (1997). Development of degradable polyesterurethanes for medical applications: in vitro and in vivo evaluations. *J. Biomed. Mater. Res., 36*, 65–74.

Saha, S., & Pal, S. (1984). Mechanical properties of bone cement: a review. *J. Biomed. Mater. Res., 18*, 435–462.

Salgado, A. J., Coutinho, O. P., & Reis, R. L. (2004). Bone tissue engineering: state of the art and future trends. *Macromol. Biosci., 4*, 743–765.

Saltzman, W. M., & Olbricht, W. L. (2002). Building drug delivery into tissue engineering. *Nat. Rev. Drug Discov., 1*, 177–186.

Sanchez-Guerrero, J., Colditz, G. A., Karlson, E. W., Hunter, D. J., Speizer, F. E., & Liang, M. H. (1995). Silicone breast implants and the risk of connective-tissue diseases and symptoms. *N. Engl. J. Med., 332*, 1666–1670.

Santerre, J. P., Woodhouse, K., Laroche, G., & Labow, R. S. (2005). Understanding the biodegradation of poly-urethanes: from classical implants to tissue engineering materials. *Biomaterials, 26*, 7457–7470.

Sawhney, A. S., & Hubbell, J. A. (1990). Rapidly degraded terpolymers of dl-lactide, glycolide, and epsilon-capro-lactone with increased hydrophilicity by copolymerization with polyethers. *J. Biomed. Mater. Res., 24*, 1397–1411.

Schacht, E., Vandorpe, J., Dejardin, S., Lemmouchi, Y., & Seymour, L. (1996). Biomedical applications of degradable polyphosphazenes. *Biotechnol. Bioeng., 52*, 102–108.

Schild, H. G. (1992). Poly(*N*-isopropylacrylamide): experiment, theory and application. *Progr. Polymer Sci., 17*, 163–249.

Schild, H. G., & Tirrell, D. A. (1990). Microcalorimetric detection of lower critical solution temperatures in aqueous polymer solutions. *J. Phys. Chem., 94*, 4352–4356.

Schmeltzer, R. C., & Uhrich, K. E. (2006). Synthesis and characterization of antiseptic-based poly(anhydride-esters). *Polymer Bull., 57*, 281–291.

Seal, B. L., Otero, T. C., & Panitch, A. (2001). Polymeric biomaterials for tissue and organ regeneration. *Mat. Sci. Eng. R, 34*, 147–230.

Segura, T., & Shea, L. D. (2002). Surface-tethered DNA complexes for enhanced gene delivery. *Bioconjug. Chem., 13*, 621–629.

Shi, X., Hudson, J. L., Spicer, P. P., Tour, J. M., Krishnamoorti, R., & Mikos, A. G. (2005). Rheological behaviour and mechanical characterization of injectable poly(propylene fumarate)/single-walled carbon nanotube composites for bone tissue engineering. *Nanotechnology, 16*, S531–S538.

Shi, X., Sitharaman, B., Pham, Q. P., Liang, F., Wu, K., Edward Billups, W., et al. (2007). Fabrication of porous ultra-short single-walled carbon nanotube nanocomposite scaffolds for bone tissue engineering. *Biomaterials, 28*, 4078–4090.

Shi, X., Sitharaman, B., Pham, Q. P., Spicer, P. P., Hudson, J. L., Wilson, L. J., et al. (2008). In vitro cytotoxicity of single-walled carbon nanotube/biodegradable polymer nanocomposites. *J. Biomed. Mater. Res. A, 86*, 813–823.

Shin, H., Jo, S., & Mikos, A. G. (2002). Modulation of marrow stromal osteoblast adhesion on biomimetic oligo [poly(ethylene glycol) fumarate] hydrogels modified with Arg-Gly-Asp peptides and a poly(ethyleneglycol) spacer. *J. Biomed. Mater. Res., 61*, 169–179.

Shin, H., Temenoff, J. S., & Mikos, A. G. (2003a). In vitro cytotoxicity of unsaturated oligo[poly(ethylene glycol) fumarate] macromers and their cross-linked hydrogels. *Biomacromolecules, 4*, 552–560.

Shin, H., Jo, S., & Mikos, A. G. (2003b). Biomimetic materials for tissue engineering. *Biomaterials, 24*, 4353–4364.

Shin, H., Quinten Ruhe, P., Mikos, A. G., & Jansen, J. A. (2003c). In vivo bone and soft tissue response to injectable, biodegradable oligo(poly(ethylene glycol) fumarate) hydrogels. *Biomaterials, 24*, 3201–3211.

Shung, A. K., Behravesh, E., Jo, S., & Mikos, A. G. (2003). Crosslinking characteristics of and cell adhesion to an injectable poly(propylene fumarate-co-ethylene glycol) hydrogel using a water-soluble crosslinking system. *Tissue Eng., 9*, 243–254.

Silverman, B. G., Brown, S. L., Bright, R. A., Kaczmarek, R. G., Arrowsmith-Lowe, J. B., & Kessler, D. A. (1996). Reported complications of silicone gel breast implants: an epidemiologic review. *Ann. Intern. Med., 124*, 744–756.

Simon, J. A., Ricci, J. L., & di Cesare, P. E. (1997). Bioresorbable fracture fixation in orthopedics: a comprehensive review. Part II. Clinical studies. *Am. J. Orthop., 26*, 754–762.

Sinha, V. R., Bansal, K., Kaushik, R., Kumria, R., & Trehan, A. (2004). Poly-[epsilon]-caprolactone microspheres and nanospheres: an overview. *Int. J. Pharm., 278*, 1–23.

Sitharaman, B., Shi, X., Walboomers, X. F., Liao, H., Cuijpers, V., Wilson, L. J., et al. (2008). In vivo biocompatibility of ultra-short single-walled carbon nanotube/biodegradable polymer nanocomposites for bone tissue engineering. *Bone, 43*, 362–370.

Skarja, G. A., & Woodhouse, K. A. (2001). In vitro degradation and erosion of degradable, segmented polyurethanes containing an amino acid-based chain extender. *J. Biomater. Sci. Polym. Ed., 12*, 851–873.

Soga, K., & Shiono, T. (1997). Ziegler-Natta catalysts for olefin polymerizations. *Progr. Polymer Sci., 22*, 1503–1546.

Solheim, E., Sudmann, B., Bang, G., & Sudmann, E. (2000). Biocompatibility and effect on osteogenesis of poly (ortho ester) compared to poly(DL-lactic acid). *J. Biomed. Mater. Res., 49,* 257–263.

Song, J., Saiz, E., & Bertozzi, C. R. (2003). A new approach to mineralization of biocompatible hydrogel scaffolds: an efficient process toward 3-dimensional bonelike composites. *J. Am. Chem. Soc., 125,* 1236–1243.

Stile, R. A., Burghardt, W. R., & Healy, K. E. (1999). Synthesis and characterization of injectable poly(*N*-isopropylacrylamide)-based hydrogels that support tissue formation in vitro. *Macromolecules, 32,* 7370–7379.

Stokes, K., Mcvenes, R., & Anderson, J. M. (1995). Polyurethane elastomer biostability. *J. Biomater. Appl., 9,* 321–354.

Stupp, S. I., LeBonheur, V., Walker, K., Li, L. S., Huggins, K. E., Keser, M., et al. (1997). Supramolecular materials: self-organized nanostructures. *Science, 276,* 384–389.

Suggs, L. J., Payne, R. G., Yaszemski, M. J., Alemany, L. B., & Mikos, A. G. (1997). Synthesis and characterization of a block copolymer consisting of poly(propylene fumarate) and poly(ethylene glycol). *Macromolecules, 30,* 4318–4323.

Suggs, L. J., Shive, M. S., Garcia, C. A., Anderson, J. M., & Mikos, A. G. (1999a). In vitro cytotoxicity and in vivo biocompatibility of poly(propylene fumarate-co-ethylene glycol) hydrogels. *J. Biomed. Mater. Res., 46,* 22–32.

Suggs, L. J., West, J. L., & Mikos, A. G. (1999b). Platelet adhesion on a bioresorbable poly(propylene fumarate-co-ethylene glycol) copolymer. *Biomaterials, 20,* 683–690.

Tabata, Y. (2003). Tissue regeneration based on growth factor release. *Tissue Eng., 9,* S5–15.

Tanahashi, K., & Mikos, A. G. (2002). Cell adhesion on poly(propylene fumarate-co-ethylene glycol) hydrogels. *J. Biomed. Mater. Res., 62,* 558–566.

Tanahashi, K., & Mikos, A. G. (2003). Protein adsorption and smooth muscle cell adhesion on biodegradable agmatine-modified poly(propylene fumarate-co-ethylene glycol) hydrogels. *J. Biomed. Mater. Res. A, 67,* 448–457.

Temenoff, J. S., & Mikos, A. G. (2000). Injectable biodegradable materials for orthopedic tissue engineering. *Biomaterials, 21,* 2405–2412.

Temenoff, J. S., Athanasiou, K. A., LeBaron, R. G., & Mikos, A. G. (2002). Effect of poly(ethylene glycol) molecular weight on tensile and swelling properties of oligo(poly(ethylene glycol) fumarate) hydrogels for cartilage tissue engineering. *J. Biomed. Mater. Res., 59,* 429–437.

Temenoff, J. S., Park, H., Jabbari, E., Conway, D. E., Sheffield, T. L., Ambrose, C. G., et al. (2004a). Thermally crosslinked oligo(poly(ethylene glycol) fumarate) hydrogels support osteogenic differentiation of encapsulated marrow stromal cells in vitro. *Biomacromolecules, 5,* 5–10.

Temenoff, J. S., Park, H., Jabbari, E., Sheffield, T. L., LeBaron, R. G., Ambrose, C. G., et al. (2004b). In vitro osteogenic differentiation of marrow stromal cells encapsulated in biodegradable hydrogels. *J. Biomed. Mater. Res. A, 70,* 235–244.

Tessmar, J., Mikos, A., & Gopferich, A. (2003). The use of poly(ethylene glycol)-block-poly(lactic acid) derived copolymers for the rapid creation of biomimetic surfaces. *Biomaterials, 24,* 4475–4486.

Timmer, M. D., Ambrose, C. G., & Mikos, A. G. (2003). Evaluation of thermal- and photo-crosslinked biodegradable poly(propylene fumarate)-based networks. *J. Biomed. Mater. Res. A, 66,* 811–818.

Tunc, Y., Hasirci, N., Yesilada, A., & Ulubayram, K. (2006). Comonomer effects on binding performances and morphology of acrylate-based imprinted polymers. *Polymer, 47,* 6931–6940.

US Food and Drug Administration (FDA) (2004). *FDA Breast Implant Consumer Handbook 2004.* Center for Devices and Radiological Health. Available online: http://www.fda.gov/cdrh/breastimplants/indexbip.html

Uhrich, K. E., Gupta, A., Thomas, T. T., Laurencin, C. T., & Langer, R. (1995). Synthesis and characterization of degradable poly(anhydride-co-imides). *Macromolecules, 28,* 2184–2193.

Uhrich, K. E., Cannizzaro, S. M., Langer, R. S., & Shakesheff, K. M. (1999). Polymeric systems for controlled drug release. *Chem. Rev., 99,* 3181–3198.

Ulbricht, M. (2006). Advanced functional polymer membranes. *Polymer, 47,* 2217–2262.

van Dijk, M., Rijkers, D. T. S., Liskamp, R. M. J., van Nostrum, C. F., & Hennink, W. E. (2009). Synthesis and applications of biomedical and pharmaceutical polymers via click chemistry methodologies. *Bioconjug. Chem., 20,* 2001–2016.

Varghese, S., & Elisseeff, J. (2006). Hydrogels for musculoskeletal tissue engineering. *Adv. Polym. Sci., 203,* 95–144.

Vonarbourg, A., Passirani, C., Saulnier, P., & Benoit, J. P. (2006). Parameters influencing the stealthiness of colloidal drug delivery systems. *Biomaterials, 27,* 4356–4373.

Wang, H., Dong, J. H., Qiu, K. Y., & Gu, Z. W. (1998). Synthesis of poly(1,4-dioxan-2-one-co-trimethylene carbonate) for application in drug delivery systems. *J. Polym. Sci. A, 36,* 1301–1307.

Webb, A. R., Yang, J., & Ameer, G. A. (2004). Biodegradable polyester elastomers in tissue engineering. *Expert Opin. Biol. Ther., 4,* 801–812.

Webber, M. J., Kessler, J. A., & Stupp, S. I. (2010). Emerging peptide nanomedicine to regenerate tissues and organs. *J. Intern. Med., 267,* 71−88.

Westphal, M., Hilt, D. C., Bortey, E., Delavault, P., Olivares, R., Warnke, P. C., et al. (2003). A phase 3 trial of local chemotherapy with biodegradable carmustine (BCNU) wafers (Gliadel wafers) in patients with primary malignant glioma. *Neuro-Oncology, 5,* 79−88.

Wichterle, O., & Lim, D. (1960). Hydrophilic gels for biological use. *Nature, 185,* 117−118.

Xia, Z., & Triffitt, J. T. (2006). A review on macrophage responses to biomaterials. *Biomed. Mater., 1,* R1−R9.

Xue, L., & Greisler, H. P. (2003). Biomaterials in the development and future of vascular grafts. *J. Vasc. Surg., 37,* 472−480.

Yang, J., Webb, A. R., & Ameer, G. A. (2004). Novel citric acid-based biodegradable elastomers for tissue engineering. *Adv. Mater., 16,* 511−516.

Yang, J., Webb, A. R., Pickerill, S. J., Hageman, G., & Ameer, G. A. (2006). Synthesis and evaluation of poly(diol citrate) biodegradable elastomers. *Biomaterials, 27,* 1889−1898.

Yang, K. K., Li, X. L., & Wang, Y. Z. (2002). Poly(p-dioxanone) and its copolymers. *J. Macromol. Sci. Poly. Rev., 42,* 373−398.

Yaszemski, M. J., Payne, R. G., Hayes, W. C., Langer, R., & Mikos, A. G. (1996). In vitro degradation of a poly (propylene fumarate)-based composite material. *Biomaterials, 17,* 2127−2130.

Yoshimoto, H., Shin, Y. M., Terai, H., & Vacanti, J. P. (2003). A biodegradable nanofiber scaffold by electrospinning and its potential for bone tissue engineering. *Biomaterials, 24,* 2077−2082.

Young, C. D., Wu, J. R., & Tsou, T. L. (1998). Fabrication and characteristics of polyHEMA artificial skin with improved tensile properties. *J. Membr. Sci., 146,* 83−93.

Zhang, J., Doll, B. A., Beckman, E. J., & Hollinger, J. O. (2003). A biodegradable polyurethane-ascorbic acid scaffold for bone tissue engineering. *J. Biomed. Mater. Res. A, 67,* 389−400.

Zhao, H., & Ameer, G. A. (2009). Modulating the mechanical properties of poly(diol citrates) via the incorporation of a second type of crosslink network. *J. Appl. Polym. Sci., 114,* 1464−1470.

CHAPTER 34

Biological Scaffolds for Regenerative Medicine

Alexander Huber, Stephen F. Badylak
McGowan Institute for Regenerative Medicine, University of Pittsburgh, Pittsburgh,
PA, USA

INTRODUCTION

The ultimate goal of regenerative medicine is the restoration and replacement of damaged or missing tissues to the structural and functional state that existed prior to injury or disease. Scaffold materials guide and facilitate the temporal and spatial organization of functional and site-specific neotissues through the process of cell attachment, migration, proliferation, and/or differentiation. A large variety of scaffold materials — composed of either synthetic polymers or biological materials — have been used in preclinical animal studies and in human clinical applications for the treatment of various tissue defects (Table 34.1). The known chemistry of synthetic scaffold materials, including their accessibility to controlled chemical modification and the fact that they can be manufactured to yield two- and three-dimensional scaffolds of almost any mechanical or structural specifications, has made the use of such materials commonplace as surgical meshes for many surgical applications. Intact extracellular matrices (ECMs) as well as their purified individual components have also been used as surgical materials and as inductive templates for the functional reconstruction of damaged or missing tissues. This chapter will evaluate biological scaffold materials in comparison to conventional synthetic scaffold materials, with a focus on intact acellular ECM scaffold materials.

623

BIOLOGICAL SCAFFOLD MATERIALS

All biological scaffold materials currently used in medical applications and regenerative medicine approaches are derived from naturally occurring materials produced by the resident cells of each tissue and organ; specifically, the extracellular matrix (ECM). The composition of ECM is tissue-specific, highly dynamic, and crucially important in organ and tissue development, homeostasis, and response to injury (Bissell et al., 1982).

Individual components of the ECM such as proteins, glycosaminoglycans, glycoproteins, and small molecules (Bosman and Stamenkovic, 2003; Badylak, 2004, 2005), or the intact matrix itself, can be harvested and prepared for use as scaffold materials. While individual ECM components, such as collagen and fibronectin, have been used to modify synthetic scaffold materials to promote their interaction with and integration into host tissues (Shin et al., 2003; Morra, 2006), they have also been used to create both naturally derived scaffold materials and combination products with synthetic materials as biohybrid devices (Stamm et al., 2004; Hong et al., 2008).

SCAFFOLD MATERIALS FROM INDIVIDUAL ECM COMPONENTS

Collagens, glycosaminoglycans, chitosans, and other components of the extracellular matrix have been used as implantable scaffold materials. Collagen — the most common and

Principles of Regenerative Medicine. DOI: 10.1016/B978-0-12-381422-7.10034-3

TABLE 34.1 Synthetic and Biologic Scaffold Materials in Regenerative Medicine

	Synthetic polymers		Biologic materials	
			Purified ECM components	**Intact tissue-derived ECM**
Example(s)	*Non-degradable:* Polypropylene (PP) Polyethylene (PE) Polytetrafluoroethylene (PTFE) Others	*Degradable:* Poly (ethylene glycol) (PEG) Poly-L-lactide (PLLA) Polycaprolactone (PCL) Poly(lactic-*co*-glycolic acid) (PLGA) Others	Collagen Chitosan Hyaluronan Alginate Agarose Others	Urinary bladder matrix (UBM) Subintestinal submucosa (SIS) Dermis Pericardium Fascia lata Cell culture-derived Others
Advantage(s)	Predictable mechanical and physical properties Well characterized biochemistry Accessibility to controlled chemical modifications and various processing techniques to meet site-specific requirements		Predictable mechanical, physical, and biologic properties Well characterized biochemistry Tailored towards multiple endpoints (i.e. biomechanics, biologics)	Heterogenic mechanical, physical, and biologic properties between and within preparations
Disadvantage(s)	Undesired immunogenic effect(s), often tailored towards single endpoint (i.e. biomechanics)		Limited accessibility to structural modification, which may also lead to undesirable physiological effects (non-degradable vs. degradable scaffolds)	Great mechanical and physical heterogenicity Limited accessibility to structural modification, which may also lead to undesirable physiological effects (non-degradable vs. degradable scaffolds)
Reference(s)	Puskas and Chen, 2004	Ma et al., 2007; Ma, 2008	Mano et al., 2007; Ma, 2008	Mano et al., 2007

abundant naturally occurring scaffold material — is a highly conserved protein that is ubiquitous among mammalian species, accounting for approximately 30% of all body proteins (van der Rest and Garrone, 1991). It can be extracted from various tissues such as tendons, ligaments, and other connective tissues, solubilized, and reconstituted into fibers of various geometries that can, in turn, be fashioned into a variety of shapes and sizes to mimic body structures such as heart valves, blood vessels, and skin (Glowacki and Mizuno, 2008).

Collagen provides considerable mechanical strength in its natural polymeric state. The necessary and required mechanical and physical properties of tissue engineered products for use in cardiovascular, orthopedic, and other body systems often depend upon chemical manipulation of collagen-based scaffolds to achieve the desired mechanical properties (Badylak, 2005). In addition to the structural properties of scaffold materials, collagen-containing implants also provide functional properties important in cellular attachment, proliferation, and differentiation, all of which contribute to a regenerative wound healing response (Cornelius et al., 1998; Maeshima et al., 2000; Brennan et al., 2006).

Bovine and porcine type I collagen is readily available and has been used in injectable form or as solid scaffolds in numerous clinical applications. Examples of collagen scaffold materials include Contigen® (C. R. Bard, Inc., Covington, GA), CosmoDerm®, Zyplast®, Zyderm® (INAMED Aesthetics/Allergan Inc., Santa Barbara, CA), CollaGUARD™/Collieva™ and CollaRx® (Innocoll Inc., Ashburn, VA), Condro-Gide (Geistlich Pharma AG, Wolhusen Switzerland), and MenaFlex™ (formerly Collagen Meniscal Implant (CMI®), ReGen Biologicals, Inc., Hackensack, NJ).

The lack of an adverse immune response to the use of xenogeneic collagen in implantable scaffold materials has been attributed to the common nature of amino acid sequences and surface epitopes between species (Boyd et al., 1991; Garrone et al., 1993; Beier et al., 1996). Allogeneic and xenogeneic collagen is generally recognized as "self" tissue when used as a biological scaffold material regardless of its species of origin, and it is subjected to the fundamental biological processes of degradation and integration into adjacent host tissues when left in its native ultrastructure. However, structural modifications designed to alter the rate of degradation and remodeling (e.g. crosslinking) may impair the desired healing and regenerative response.

Chitosans are the second most abundant biopolymer in nature and represent a family of biodegradable cationic polysaccharides consisting of glucosamine and randomly distributed *N*-acetylglucosamine (Domish et al., 2001). Chitosans are derived by the alkaline *N*-deacetylation of chitin isolated from fungal mycelium or invertebrate exoskeletons. Commercially available chitosans vary in their molecular weights and degrees of deacetylation depending on preparation procedures (Madihally and Matthew, 1999; Domish et al., 2001; Cao et al., 2005). They can readily be processed into numerous scaffold geometries (sponges, membranes, beads, etc.), while meeting the desired biomechanical (i.e. mechanical strength, elasticity) and biological (i.e. level of porosity, cell attachment and proliferation, rate of degradation) requirements at the site of implantation (Aiedeh et al., 1997; Madihally and Matthew, 1999; Chow and Khor, 2000; Domish et al., 2001; Ho et al., 2004; Cao et al., 2005; Freier et al., 2005; Geng et al., 2005). Chitosans are cationic in nature and carry the *N*-acetylglucosamine moieties found on glycosaminoglycans. These chemical similarities and their ability to interact directly with glycosaminoglycans and other negatively charged particles are assumed to play an important role in the processes of neotissue formation, including cell adhesion, migration, proliferation, and differentiation (Suzuki et al., 2008).

The *in vivo* implantation of various chitosan-based materials results in an acute to subacute inflammatory reaction (Nishimura et al., 1994; Peluso et al., 1994; Muzzarelli et al., 1988, 1989; Damour et al., 1994; Muzzarelli, 1997; Suh and Matthew, 2000; Khor and Lim, 2003; di Martino et al., 2005). While serving as a strong chemoattractant for neutrophils for the first week after implantation, neotissue formation takes place without the establishment of a classic foreign body response (Suh and Matthew, 2000). Granulation tissue accompanied by robust angiogenesis in response to chitosan implantation has been reported (Chen et al., 2006). Chitosan has been used as a conduit for guided peripheral nerve regeneration (Jenq and Coggeshall, 1987; Aebischer et al., 1990; Knoops et al., 1990; Kim et al., 1993; den Dunnen et al., 1995; Rodriguez et al., 1999; Wang et al., 2005) and as a scaffold for the treatment of experimentally induced skin wounds with good results (Ueno et al., 1999, 2001a, b; Chen et al., 2002; Tanabe et al., 2002; Mizuno et al., 2003). Cartilage repair (di Martino et al., 2005) and bone tissue engineering applications (Lee et al., 2002; Bumgardner et al., 2003a, b; Wang et al., 2004) have also been investigated.

Chitosans are currently being used in a variety of biomedical applications in humans including wound healing, wound/burn dressing, ophthalmology, and drug delivery (Dutta et al., 2004). Examples of chitosan scaffold materials currently being investigated for clinical applications in humans include BST-Gel® and BST-CarGel® (Bio Synthec Canada Inc., Canada) for cartilage repair and the HemCon® Patch (HemCon Medical Technologies Inc., Portland, OR) for wound dressing.

The glycosaminoglycan, hyaluronic acid (HA), has been extensively investigated as a natural scaffold material for tissue reconstruction in addition to the aforementioned materials. HA can be found in the ECMs of various different tissues, for example skin and cartilage (Entwhistle et al., 1995; Hodde et al., 1996), and non-animal-derived sources (Band, 1998) and has been used in injectable form in numerous clinical applications in humans. Examples of hyaluronic acid-based materials include JuveDerm® (INAMED Aesthetics/Allergan Inc., Santa Barbara, CA), Restylane®, and Perlane® (Medicis Aesthetics Inc., Scottdale, AZ).

Other less commonly used naturally derived ECM components for scaffold construction include hydroxyapatite (Yoshikawa et al., 2009), alginate (Umeda et al., 2009), and agarose (Gunja et al., 2009).

SCAFFOLD MATERIALS COMPOSED OF INTACT EXTRACELLULAR MATRICES

The individual ECM components mentioned above – in addition to synthetic polymers – have led to the development of a number of scaffold materials that may be useful in tissue regeneration. However, the use of a homogeneous scaffold material such as collagen or hyaluronic acid to promote the reconstruction of structurally and functionally complex and heterogeneous organs and tissues may be suboptimal. In contrast, intact ECM consists of all the structural and functional molecules secreted by the resident cells. Perhaps more importantly, this diverse collection of molecules is arranged in the unique three-dimensional architecture of the native tissue if processed appropriately. Scaffold materials composed of intact ECM have become useful in numerous clinical applications.

Intact ECM can be isolated from a large variety of different tissues, including heart valves, blood vessels, skin, nerves, skeletal muscle, tendons, ligaments, small intestine, urinary bladder, and liver. These biological scaffolds can be harvested from several different species including tissues from human, porcine, bovine, and equine (Badylak et al., 2009), or from cells grown *in vitro* (Datta et al., 2005) (Table 34.2). As a result, ECM scaffold materials harvested from different tissues have unique structural, functional, and molecular characteristics. For instance, ECM scaffolds composed of porcine small intestinal submucosa (SIS–ECM) consist of ~90% of collagen, the vast majority collagen type I, and minor amounts of the collagen types (Col) III, IV, V, and VI (Badylak et al., 1995). On the other hand, ECM scaffolds composed of porcine urinary bladder matrix (UBM–ECM), while featuring the same collagen types as SIS–ECM, contain greater amounts of Col III, as well as Col VII, an important component of the epithelial basement membrane (Brown et al., 2006). ECM isolated from different tissues also differs in the amount and distribution of glycosaminoglycans (GAGs), including heparin, heparin sulfate, chondroitin sulfates, and hyaluronic acid (Entwhistle et al., 1995; Hodde et al., 1996); adhesion molecules such as fibronectin and laminin (Schwarzbauer, 1999; Hodde et al., 2002; Brown et al., 2006); the proteoglycan decorin and the glycoproteins biglycan and entactin (Badylak et al., 2009); as well as various growth factors (Roberts et al., 1988; Kagami et al., 1998; Bonewald, 1999) including transforming growth factor-b (TGF-b) (Voytik-Harbin et al., 1997; McDevitt et al., 2003), basic fibroblast growth factor (b-FGF) (Voytik-Harbin et al., 1997; Hodde et al., 2005), and vascular endothelial growth factor (VEGF) (Hodde et al., 2001). ECM scaffold materials are specific in their protein make-up from location to location within various tissues; for example endocrine versus exocrine loci within the pancreas or the valvular versus mural loci within the heart. It is assumed that the preservation of the intact ECM composition as well as its intrinsic ultrastructure and three-dimensional architecture – in particular its collagen fiber architecture – are fundamentally important in processes such as cell recruitment, migration, proliferation, and differentiation during neotissue formation *in vivo* (Brown et al., 2006; Sellaro et al., 2007; Gong et al., 2008; Hosokawa et al., 2008).

The ECM manufacturing process is designed to remove any cellular material without adversely affecting the composition, mechanical integrity, and biological activity of the remaining ECM.

TABLE 34.2 Examples of Intact Scaffold Materials Currently Used in Various Clinical Applications in Humans

Product	Company	Material	Processing	Form
AlloDerm	Lifecell	Human skin	Natural	Dry sheet
AlloPatch®	Musculoskeletal Transplant Foundation	Human fascia lata	Natural	Dry sheet
Axis™ Dermis	Mentor	Human dermis	Natural	Dry sheet
Bard® Dermal Allograft	Bard	Cadaveric human dermis	Natural	Dry sheet
CuffPatch™	Arthrotek	Porcine small intestinal submucosa (SIS)	Crosslinked	Hydrated sheet
DurADAPT™	Pegasus Biologicals	Horse pericardium	Crosslinked	Dry sheet
Dura-Guard®	Synovis Surgical	Bovine pericardium	Crosslinked	Hydrated sheet
Durasis® SIS	Cook	Porcine small intestinal submucosa (SIS)	Natural	Dry sheet
Durepair®	TEI Biosciences	Fetal bovine skin	Natural	Dry sheet
FasLata®	Bard	Cadaveric fascia lata	Natural	Dry sheet
GraftJacket®	Wright Medical Tech	Human skin	Natural	Dry sheet
Oasis®	Healthpoint	Porcine small intestinal submucosa (SIS)	Natural	Dry sheet
OrthADAPT™	Pegasus Biologicals	Horse pericardium	Crosslinked	Dry sheet
Pelvicol®	Bard	Porcine dermis	Crosslinked	Hydrated sheet
Peri-Guard®	Synovis Surgical	Bovine pericardium	Crosslinked	Dry sheet
Permacol™	Tissue Science Laboratories	Porcine skin	Crosslinked	Hydrated sheet
PriMatrix™	TEI Biosciences	Fetal bovine skin	Natural	Dry sheet
Restore™	DePuy	Porcine small intestinal submucosa (SIS)	Natural	Dry sheet
Stratasis®	Cook	Porcine small intestinal submucosa (SIS)	Natural	Dry sheet
SurgiMend™	TEI Biosciences	Fetal bovine skin	Natural	Dry sheet
Surgisis®	Cook	Porcine small intestinal submucosa (SIS)	Natural	Dry sheet
Suspend™	Mentor	Human fascia lata	Natural	Dry sheet
TissueMend®	TEI Biosciences	Fetal bovine skin	Natural	Dry sheet
Vascu-Guard®	Synovis Surgical	Bovine pericardium	Crosslinked	Dry sheet
Veritas®	Synovis Surgical	Bovine pericardium	Crosslinked	Hydrated sheet
Xelma™	Molnlycke	ECM protein, PGA, water		gel
Xenform™	TEI Biosciences	Fetal bovine skin	Natural	Dry sheet
Zimmer Collagen Patch®	Tissue Science Laboratories	Porcine dermis	Crosslinked	Hydrated sheet

The process generally includes multiple mechanical and biochemical procedures, i.e. liberation of desired tissues from surrounding tissues, decellularization, disinfection, lyophilization and/or hydrolyzation, and terminal sterilization. Most commercially available, intact ECM scaffold materials are processed into a sheet prior to decellularization by methods that include trimming and spreading of the original tissue to facilitate the removal of cellular components and debris. Additionally, the decellularization of whole organs has also been performed successfully through the perfusion of the tissue's vascular network (Ott et al., 2008; Wainwright et al., 2009).

Commonly used methods of decellularization include a combination of physical and chemical treatments, e.g. sonication, agitation, freeze-thawing, and washes with various proteolytic detergents and solvents (reviewed in Gilbert et al., 2006). The decellularization process effectively removes xenogeneic and allogeneic cellular antigens that may be recognized as foreign by the host and result in an adverse inflammatory response or overt immune-mediated rejection (Ross et al., 1993; Erdag and Morgan, 2004; Gock et al., 2004). As described earlier, the molecules of the ECM are highly conserved between species and are well tolerated by xenogeneic recipients (Bernard et al., 1983a, b; Constantinou and Jimenez, 1991; Exposito et al., 1992). Residual amounts of DNA and certain immunogenic species-specific antigens, such as galactosyl-a-1,3-galactose (α-Gal epitope), have been shown to be present in ECM scaffolds, but fail to activate complement or bind IgM antibody, possibly due to the small amount and widely scattered distribution of the antigen (McPherson et al., 2000; Raeder et al., 2002; Daly et al., 2009; Gilbert et al., 2009).

Further processing may include lyophilization (freeze drying) or vacuum pressing prior to terminal sterilization to avoid leaching of soluble factors, for example VEGF and b-FGF, and extending the product's shelf life. The production of multilaminate forms of the ECM also improves the device's handling and allows the construction of three-dimensional constructs in the form of tubes (Badylak et al., 2005), cones (Nieponice et al., 2006), and multilaminate sheets (Dejardin et al., 2001; Freytes et al., 2004; Gilbert et al., 2008a). Lyophilized ECM scaffolds can be processed further to yield powders (Gilbert et al., 2005), liquids, or hydrogels (Brightman et al., 2000; Freytes et al., 2008a) for use as injectable scaffolds in minimally invasive surgeries (Lundy et al., 2003; Wood et al., 2005; Choi et al., 2009) or in combination with ECM in sheet conformation to produce sheet-powder hybrid scaffolds.

While each of the processing steps will change the overall composition and structure of the prepared ECM compared to those found *in vivo*, intact ECM preparations retain a multitude of structurally and functionally active proteins (Freytes et al., 2004, 2005, 2008b; Gilbert et al., 2008b).

The ability of an ECM harvested from one tissue to function as a biological scaffold material for the same or different tissue may vary and is dependent on its preparation and modification procedures. Intact acellular ECM scaffolds derived from human dermis (Wainwright, 1995; Isch et al., 2001; Clemons et al., 2003), porcine and human urinary bladder (UBM-ECM) (Duel et al., 1996; Atala, 1998; Dahms et al., 1998), porcine small intestinal submucosa (SIS-ECM) (Oeschlager et al., 2003; Wang et al., 2003; Badylak, 2004; Derwin et al., 2004; Musahl et al., 2004), porcine heart valves (Cohn et al., 1989; Hammermeister et al., 1993; Simon et al., 2003), and bovine dermis (Barber et al., 2006; Coons and Barber, 2006), among others, have all been used in human clinical applications (de Ugarte et al., 2004; Alpert et al., 2005; Dedecker et al., 2005; Helton et al., 2005; Jones et al., 2005a, b; Smith et al., 2005; Zalavras et al., 2006). Additional applications have been investigated in preclinical animal studies including the use of porcine SIS-ECM in the repair of the Achilles tendon (Hodde et al., 1997), the anterior cruciate ligament (Badylak et al., 1999a), abdominal wall (Badylak et al., 2001), reconstruction of the lower urinary tract (Kropp, 1995; Kropp et al., 1996, 1998), and the treatment of dermal wounds (Lindberg and Badylak, 2001). Similarly, UBM-ECM has been used in the repair of the esophagus (Badylak et al., 2005) and myocardium (cardiac muscle) (Badylak et al., 2005; Kochupura et al., 2005).

ECM scaffolds that are not chemically cross-linked are rapidly degraded *in vivo*. Typically, 50% of a non-crosslinked SIS-ECM scaffold is degraded within a period of 1 month post-implantation and the scaffold is usually completely degraded within a 3-month timeframe, as demonstrated in the repair of a urinary bladder defect or the Achilles tendon (Badylak et al., 1998; Record et al., 2001; Gilbert et al., 2007). ECM degradation leads to an initial decrease in overall strength during the early phase of *in vivo* remodeling, followed by an increase in strength due to the deposition of site-specific ECM and the formation of functional site-appropriate neotissue by infiltrating cells in response to their experienced mechanical stresses (Musahl et al., 2004; Badylak et al., 2001, 2005; Record et al., 2001; Liang et al., 2006). Soluble factors within ECM scaffold materials, that is, growth factors, and the release of biologically active cryptic peptides resulting from degradation of the ECM material

(Sarikaya et al., 2002; Li et al., 2004; Brennan et al., 2006; Reing et al., 2009), are thought to be directly involved in the processes of neotissue formation including angiogenesis, mononuclear cell infiltration, cell proliferation, cell migration, and cell differentiation (Voytik-Harbin et al., 1997; Badylak et al., 1999b, 2002; Badylak, 2002; Valentin et al., 2006). The release of soluble factors along with the rapid degradation of the ECM appear to be essential processes for constructive remodeling to occur. This fact is highlighted by an altered remodeling profile in clinical applications using scaffolds that have been chemically crosslinked using glutaralde-hyde, carbodiimide or hexamethylene-diisocyanate (Ratner, 2004), or non-chemical methods. While chemical crosslinking can increase the mechanical strength and reduce the rate of scaffold degradation of an ECM scaffold, this modification results in the formation of a chronic, pro-inflammatory, foreign body type of tissue response and a reduced level of constructive remodeling (Valentin et al., 2006; Badylak and Gilbert, 2008; Badylak et al., 2008). Similarly to the host response to synthetic polymer scaffolds, an adverse host response is heralded by a predominance of M1 macrophages, a high level of pro-inflammatory cytokines, and the formation of a Th-1 (cell mediated rejection) response (Mosser, 2003; Martinez et al., 2008). In contrast, intact, non-chemically modified ECM scaffold materials show a host immune response characteristic of accommodation and integration, as demonstrated by the increased presence of alternatively activated M2 macrophages resulting in low levels of pro-inflammatory cytokines and the establishment of a Th-2 type of response (Allman et al., 2001, 2002; Brown et al., 2009). The initial host response to intact ECM scaffolds appears to be critically important in determining subsequent processes including scaffold degradation, release of matricryptic peptides, cell recruitment, and angiogenesis, among others (Haviv et al., 2005; Sarikaya et al., 2002; Brennan et al., 2006; Beattie et al., 2009).

While several different intact ECM scaffold materials derived from a number of different tissue and animal sources have been used successfully in various clinical applications, current research is directed at a better understanding of the scaffold requirements with regard to their biochemical and structural composition and the molecular events that lead to constructive and functional tissue remodeling.

SUMMARY

Intact, acellular ECMs and their individual components have become a valuable source of scaffold materials for clinical application in humans. Contrary to conventional synthetic scaffold materials, whose main purpose is the structural support at the implantation site, ECM scaffolds participate in a dynamic remodeling process, acting as temporary inducers of site-specific constructive remodeling. Intact ECMs have demonstrated great potential in a large number of clinical applications. However, optimization with regard to the appropriate source tissue for ECM isolation, processing methods, scaffold design, area of clinical application, and a greater understanding of the biochemical events leading to the desired outcome are still required to achieve the full potential of these biological scaffold materials.

References

Aebischer, P., Guenard, V., & Valentini, R. F. (1990). The morphology of regenerating peripheral nerves is modulated by the surface microgeometry of polymeric guidance channels. *Brain Res., 531,* 211−218.

Aiedeh, K., Gianasi, E., Orienti, I., & Zecchi, V. (1997). Chitosan microcapsules as controlled release systems for insulin. *J. Microencapsulation, 14,* 567−576.

Allman, A. J., McPherson, T. B., Merrill, L. C., Badylak, S. F., & Metzger, D. W. (2002). The Th2-restricted immune response to xenogeneic small intestinal submucosa does not influence systemic protective immunity to viral and bacterial pathogens. *Tissue Eng., 8,* 53−62.

Allman, A. J., McPherson, T. B., Badylak, S. F., Merrill, L. C., Kallakury, B., Sheehan, C., et al. (2001). Xenogeneic extracellular matrix grafts elicit a TH2-restricted immune response. *Transplantation, 71,* 1631−1640.

Alpert, S. A., Cheng, E. Y., Kaplan, W. E., Snodgrass, W. T., Wilcox, D. T., & Kropp, B. P. (2005). Bladder neck fistula after the complete primary repair of exstrophy: a multi-institutional experience. *J. Urol., 174*(4 Pt 2), 1687−1689, discussion 1689−1690.

Atala, A. (1998). Tissue engineering in urologic surgery. *Urol. Clin. N. Am., 25,* 39–50.

Badylak, S. E. (2002). The extracellular matrix as a scaffold for tissue reconstruction. *Semin. Cell Dev. Biol., 13,* 377–383.

Badylak, S. F. (2004). Xenogeneic extracellular matrix as a scaffold for tissue reconstruction. *Transplant Immunol., 12,* 367–377.

Badylak, S. F. (2005). Regenerative medicine and developmental biology: the role of the extracellular matrix. *Anatom. Rec. New Anatomist, 287,* 36–41.

Badylak, S. F., & Gilbert, T. W. (2008). Immune response to biological scaffold materials. *Semin. Immunol., 20,* 109–116.

Badylak, S. F., Tullius, R., Kokini, K., Shelbourne, K. D., Klootwyk, T., Voytik, S. L., et al. (1995). The use of xenogeneic small intestinal submucosa as a biomaterial for Achilles tendon repair in a dog model. *J. Biomed. Mater. Res., 29,* 977–985.

Badylak, S. F., Kropp, B., McPherson, T., Liang, H., & Snyder, P. W. (1998). Small intestinal submucosa: a rapidly resorbed bioscaffold for augmentation cystoplasty in a dog model. *Tissue Eng., 4,* 379–387.

Badylak, S., Arnoczky, S., Plouhar, P., Haut, R., Mendenhall, V., Clarke, R., et al. (1999a). Naturally occurring extracellular matrix as a scaffold for musculoskeletal repair. *Clin. Orthop. Rel. Res., 367,* S333–S343.

Badylak, S., Liang, A., Record, R., Tullius, R., & Hodde, J. (1999b). Endothelial cell adherence to small intestinal submucosa: an acellular bioscaffold. *Biomaterials, 20,* 2257–2263.

Badylak, S., Kokini, M., Tullius, B., & Whitson, B. (2001). Strength over time of a resorbable bioscaffold for body wall repair in a dog model. *J. Surg. Res., 99,* 282–287.

Badylak, S., Kokini, K., Tullius, B., Simmons-Byrd, A., & Morff, R. (2002). Morphologic study of small intestinal submucosa as a body wall repair device. *J. Surg. Res., 103,* 190–202.

Badylak, S. F., Vorp, D. A., Spievack, A. R., Simmons-Byrd, A., Hanke, J., Freytes, D. O., et al. (2005). Esophageal reconstruction with ECM and muscle tissue in a dog model. *J. Surg. Res., 128,* 87–97.

Badylak, S. F., Valentin, J. E., Ravindra, A. K., McCabe, G. P., & Stewart-Akers, A. M. (2008). Macrophage phenotype as a determinant of biological scaffold remodeling. *Tissue Eng. A, 14,* 1835–1842.

Badylak, S. F., Freytes, D. O., & Gilbert, T. W. (2009). Extracellular matrix as a biological scaffold material: structure and function. *Acta Biomater., 5,* 1–13.

Band, P. A. (1998). Hyaluronan derivatives: chemistry and clinical applications. In T. C. Laurent (Ed.), *The Chemistry, Biology and Medical Applications of Hyaluronan and its Derivatives* (pp. 33–42). London: Portland Press.

Barber, F. A., Herbert, M. A., & Coons, D. A. (2006). Tendon augmentation grafts: biomechanical failure loads and failure patterns. *Arthroscopy, 22,* 534–538.

Beattie, A. J., Gilbert, T. W., Guyot, J. P., Yates, A. J., & Badylak, S. F. (2009). Chemoattraction of progenitor cells by remodeling extracellular matrix scaffolds. *Tissue Eng. A, 15,* 1119–1125.

Beier, F., Eerola, I., Vuorio, E., Luvalle, P., Reichenberger, E., Bertling, W., et al. (1996). Variability in the upstream promoter and intron sequences of the human, mouse, and chick type X collagen genes. *Matrix Biol., 15,* 415–422.

Bernard, M. P., Chu, M. L., Myers, J. C., Ramirez, F., Eikenberry, E. F., & Prockop, D. J. (1983a). Nucleotide sequences of complementary deoxyribonucleic acids for the pro alpha 1 chain of human type I procollagen. Statistical evaluation of structures that are conserved during evolution. *Biochemistry, 22,* 5213–5223.

Bernard, M. P., Myers, J. C., Chu, M. L., Ramirez, F., Eikenberry, E. F., & Prockop, D. J. (1983b). Structure of a cDNA for the pro alpha 2 chain of human type I procollagen. Comparison with chick cDNA for pro alpha 2(I) identifies structurally conserved features of the protein and the gene. *Biochemistry, 22,* 1139–1145.

Bissell, M. J., Hall, H. G., & Parry, G. (1982). How does the extracellular matrix direct gene expression? *J. Theoret. Biol., 99,* 31–68.

Bonewald, L. F. (1999). Regulation and regulatory activities of transforming growth factor beta. *Crit. Rev. Eukaryotic Gene Exp., 9,* 33–44.

Bosman, F. T., & Stamenkovic, I. (2003). Functional structure and composition of the extracellular matrix. *J. Pathol., 200,* 423–428.

Boyd, C. D., Christiano, A. M., Pierce, R. A., Stolle, C. A., & Deak, S. B. (1991). Mammalian tropoelastin: multiple domains of the protein define an evolutionary divergent amino acid sequence. *Matrix, 11,* 235–241.

Brennan, E. P., Reing, J., Chew, D., Myers-Irvin, J. M., Young, E. J., & Badylak, S. F. (2006). Antibacterial activity within degradation products of biological scaffolds composed of extracellular matrix. *Tissue Eng., 12,* 2949–2955.

Brightman, A. O., Rajwa, B. P., Sturgis, J. E., McCallister, M. E., Robinson, J. P., & Voytik-Harbin, S. L. (2000). Time-lapse confocal reflection microscopy of collagen fibrillogenesis and extracellular matrix assembly *in vitro. Biopolymers, 54,* 222–234.

Brown, B., Lindberg, K., Reing, J., Stolz, D. B., & Badylak, S. F. (2006). The basement membrane component of biological scaffolds derived from extracellular matrix. *Tissue Eng., 12,* 519–526.

Brown, B. N., Valentin, J. E., Stewart-Akers, A. M., McCabe, G. P., & Badylak, S. F. (2009). Macrophage phenotype and remodeling outcomes in response to biological scaffolds with and without a cellular component. *Biomaterials, 30*, 1482–1491.

Bumgardner, J. D., Wiser, R., Gerard, P. D., Bergin, P., Chestnutt, B., Marin, M., et al. (2003a). Chitosan: potential use as a bioactive coating for orthopaedic and craniofacial/dental implants. *J. Biomater. Sci. Polymer Ed., 14*, 423–438.

Bumgardner, J. D., Wiser, R., Elder, S. H., Jouett, R., Yang, Y., & Ong, J. L. (2003b). Contact angle, protein adsorption and osteoblast precursor cell attachment to chitosan coatings bonded to titanium. *J. Biomater. Sci. Polymer Ed., 14*, 1401–1409.

Cao, W., Cheng, M., Ao, Q., Gong, Y., Zhao, N., & Zhang, X. (2005). Physical, mechanical and degradation properties, and schwann cell affinity of cross-linked chitosan films. *J. Biomater. Sci. Polymer Ed., 16*, 791–807.

Chen, X. G., Wang, Z., Liu, W. S., & Park, H. J. (2002). The effect of carboxymethyl-chitosan on proliferation and collagen secretion of normal and keloid skin fibroblasts. *Biomaterials, 23*, 4609–4614.

Chen, Y. L., Chen, H. C., Lee, H. P., Chan, H. Y., & Hu, Y. C. (2006). Rational development of GAG-augmented chitosan membranes by fractional factorial design methodology. *Biomaterials, 27*, 2222–2232.

Choi, J. S., Yang, H. J., Kim, B. S., Kim, J. D., Kim, J. Y., Yoo, B., et al. (2009). Human extracellular matrix (ECM) powders for injectable cell delivery and adipose tissue engineering. *J. Controlled Release, 139*, 2–7.

Chow, K. S., & Khor, E. (2000). Novel fabrication of open-pore chitin matrixes. *Biomacromolecules, 1*, 61–67.

Clemons, J. L., Myers, D. L., Aguilar, V. C., & Arya, L. A. (2003). Vaginal paravaginal repair with an AlloDerm graft. *Am. J. Obstet. Gynecol., 189*, 1612–1618, discussion 1618–1619.

Cohn, L. H., Collins, J. J., Jr., DiSesa, V. J., Couper, G. S., Peigh, P. S., Kowalker, W., et al. (1989). Fifteen-year experience with 1678 Hancock porcine bioprosthetic heart valve replacements. *Ann. Surg., 210*, 435–442, discussion 442–443.

Constantinou, C. D., & Jimenez, S. A. (1991). Structure of cDNAs encoding the triple-helical domain of murine alpha 2 (VI) collagen chain and comparison to human and chick homologues. Use of polymerase chain reaction and partially degenerate oligonucleotide for generation of novel cDNA clones. *Matrix, 11*, 1–9.

Coons, D. A., & Barber, F. A. (2006). Tendon graft substitutes-rotator cuff patches. Sports Med. *Arthrosc. Rev., 14*, 185–190.

Cornelius, L. A., Nehring, L. C., Harding, E., Bolanowski, M., Welgus, H. G., Kobayashi, D. K., et al. (1998). Matrix metalloproteinases generate angiostatin: effects on neovascularization. *J. Immunol., 161*, 6845–6852.

Dahms, S. E., Piechota, H. J., Dahiya, R., Lue, T. F., & Tanagho, E. A. (1998). Composition and biomechanical properties of the bladder acellular matrix graft: comparative analysis in rat, pig and human. *Br. J. Urol., 82*, 411–419.

Daly, K. A., Stewart-Akers, A. M., Hara, H., Ezzelarab, M., Long, C., Cordero, K., et al. (2009). Effect of the alphaGal epitope on the response to small intestinal submucosa extracellular matrix in a nonhuman primate model. *Tissue Eng., 15*, 3877–3888.

Damour, O., Gueugniaud, P. Y., Berthin-Maghit, M., Rousselle, P., Berthod, F., Sahuc, F., et al. (1994). A dermal substrate made of collagen–GAG–chitosan for deep burn coverage: first clinical uses. *Clin. Mater., 15*, 273–276.

Datta, N., Holtorf, H. L., Sikavitsas, V. I., Jansen, J. A., & Mikos, A. G. (2005). Effect of bone extracellular matrix synthesized *in vitro* on the osteoblastic differentiation of marrow stromal cells. *Biomaterials, 26*, 971–977.

de Ugarte, D. A., Choi, E., Weitzbuch, H., Wulur, I., Caulkins, C., Wu, B., et al. (2004). Mucosal regeneration of a duodenal defect using small intestine submucosa. *Am. Surg., 70*, 49–51.

Dedecker, F., Grynberg, M., & Staerman, F. (2005). Small intestinal submucosa (SIS): perspectives en chirurgie urogenitale. *Prog. Urol., 15*, 405–410.

Dejardin, L. M., Arnoczky, S. P., Ewers, B. J., Haut, R. C., & Clarke, R. B. (2001). Tissue-engineered rotator cuff tendon using porcine small intestine submucosa – histologic and mechanical evaluation in dogs. *Am. Sports Med., 29*, 175–184.

den Dunnen, W. F., van der Lei, B., Robinson, P. H., Holwerda, A., Pennings, A. J., & Schakenraad, J. M. (1995). Biological performance of a degradable poly(lactic acid-epsilon-caprolactone) nerve guide: influence of tube dimensions. *J. Biomed. Mater. Res., 29*, 757–766.

Derwin, K., Androjna, C., Spencer, E., Safran, O., Bauer, T. W., Hunt, T., et al. (2004). Porcine small intestine submucosa as a flexor tendon graft. *Clin. Orthop. Rel. Res., 423*, 245–252.

di Martino, A., Sittinger, M., & Risbud, M. V. (2005). Chitosan: a versatile biopolymer for orthopaedic tissue-engineering. *Biomaterials, 26*, 5983–5990.

Dornish, M., Kaplan, D., & Skaugrud, O. (2001). Standards and guidelines for biopolymers in tissue-engineered medical products: ASTM alginate and chitosan standard guides. American Society for Testing and Materials. *Ann NY Acad. Sci., 944*, 388–397.

Duel, B. P., Hendren, W. H., Bauer, S. B., Mandell, J., Colodny, A., Peters, C. A., et al. (1996). Reconstructive options in genitourinary rhabdomyosarcoma. *J. Urol., 156*, 1798–1804.

Dutta, K. P., Dutta, J., & Tripathi, V. S. (2004). Chitin and chitosan: chemistry, properties and application. *J. Sci. Ind. Res., 63*, 20–31.

Entwistle, J., Zhang, S., Yang, B., Wong, C., Li, Q., Hall, C. L., et al. (1995). Characterization of the murine gene encoding the hyaluronan receptor RHAMM. *Gene, 163*, 233–238.

Erdag, G., & Morgan, J. R. (2004). Allogeneic versus xenogeneic immune reaction to bioengineered skin grafts. *Cell Transplantation, 13*, 701–712.

Exposito, J. Y., d'Alessio, M., Solursh, M., & Ramirez, F. (1992). Sea urchin collagen evolutionarily homologous to vertebrate pro-alpha 2(I) collagen. *J. Biol. Chem., 267*, 15559–15562.

Freier, T., Montenegro, R., Shan Koh, H., & Shoichet, M. S. (2005). Chitin-based tubes for tissue engineering in the nervous system. *Biomaterials, 26*, 4624–4632.

Freytes, D. O., Rundell, A. E., Vande Geest, J., Vorp, D. A., Webster, T. J., & Badylak, S. F. (2005). Analytically derived material properties of multilaminated extracellular matrix devices using the ball-burst test. *Biomaterials, 26*, 5518–5531.

Freytes, D. O., Badylak, S. F., Webster, T. J., Geddes, L. A., & Rundell, A. E. (2004). Biaxial strength of multilaminated extracellular matrix scaffolds. *Biomaterials, 25*, 2353–2361.

Freytes, D. O., Martin, J., Velankar, S. S., Lee, A. S., & Badylak, S. F. (2008a). Preparation and rheological characterization of a gel form of the porcine urinary bladder matrix. *Biomaterials, 29*, 1630–1637.

Freytes, D. O., Tullius, R. S., Valentin, J. E., Stewart-Akers, A. M., & Badylak, S. F. (2008b). Hydrated versus lyophilized forms of porcine extracellular matrix derived from the urinary bladder. *J. Biomed. Mater. Res. A, 87*, 862–872.

Garrone, R., Exposito, J. Y., Franc, J. M., Franc, S., Humbert-David, N., Qin, L., et al. (1993). Phylogenesis of the extracellular matrix. *C.R. Seances Soc. Biol. Fil., 187*, 114–123.

Geng, X., Kwon, O. H., & Jang, J. (2005). Electrospinning of chitosan dissolved in concentrated acetic acid solution. *Biomaterials, 26*, 5427–5432.

Gilbert, T. W., Stolz, D. B., Biancaniello, F., Simmons-Byrd, A., & Badylak, S. F. (2005). Production and characterization of ECM powder: implications for tissue engineering applications. *Biomaterials, 26*, 1431–1435.

Gilbert, T. W., Sellaro, T. L., & Badylak, S. F. (2006). Decellularization of tissues and organs. *Biomaterials, 27*, 3675–3683.

Gilbert, T. W., Stewart-Akers, A. M., Simmons-Byrd, A., & Badylak, S. F. (2007). Degradation and remodeling of small intestinal submucosa in canine Achilles tendon repair. *J. Bone Joint Surg., 89A*, 621–630.

Gilbert, T. W., Nieponice, A., Spievack, A. R., Holcomb, C. J., Gilbert, S., & Badylak, S. F. (2008a). Repair of the thoracic wall with an extracellular matrix scaffold in a canine model. *J. Surg. Res., 147*, 61–67.

Gilbert, T. W., Wognum, S., Joyce, E. M., Freytes, D. O., Sacks, M. S., & Badylak, S. F. (2008b). Collagen fiber alignment and biaxial mechanical behavior of porcine urinary bladder derived extracellular matrix. *Biomaterials, 29*, 4775–4782.

Gilbert, T. W., Freund, J. M., & Badylak, S. F. (2009). Quantification of DNA in biological scaffold materials. *J. Surg. Res., 152*, 135–139.

Glowacki, J., & Mizuno, S. (2008). Collagen scaffolds for tissue engineering. *Biopolymers, 89*, 338–344.

Gock, H., Murray-Segal, L., Salvaris, E., Cowan, P., & d'Apice, A. J. (2004). Allogeneic sensitization is more effective than xenogeneic sensitization in eliciting Gal-mediated skin graft rejection. *Transplantation, 77*, 751–753.

Gong, J., Sagiv, O., Cai, H., Tsang, S. H., & del Priore, L. V. (2008). Effects of extracellular matrix and neighboring cells on induction of human embryonic stem cells into retinal or retinal pigment epithelial progenitors. *Exp. Eye Res., 86*, 957–965.

Gunja, N. J., Huey, D. J., James, R. A., & Athanasiou, K. A. (2009). Effects of agarose mould compliance and surface roughness on self-assembled meniscus-shaped constructs. *J. Tissue Eng. Regen. Med., 3*, 521–530.

Hammermeister, K. E., Sethi, G. K., Henderson, W. G., Oprian, C., Kim, T., & Rahimtoola, S. (1993). A comparison of outcomes in men 11 years after heart-valve replacement with a mechanical valve or bioprosthesis. Veterans Affairs Cooperative Study on Valvular Heart Disease. *New Engl. J. Med., 328*, 1289–1296.

Haviv, F., Bradley, M. F., Kalvin, D. M., Schneider, A. J., Davidson, D. J., Majest, S. M., et al. (2005). Thrombospondin-1 mimetic peptide inhibitors of angiogenesis and tumor growth: design, synthesis, and optimization of pharmacokinetics and biological activities. *J. Med. Chem., 48*, 2838–2846.

Helton, W. S., Fisichella, P. M., Berger, R., Horgan, S., Espat, N. J., & Abcarian, H. (2005). Short-term outcomes with small intestinal submucosa for ventral abdominal hernia. *Arch. Surg., 140*, 549–560, discussion 560–562.

Ho, M. H., Kuo, P. Y., Hsieh, H. J., Hsien, T. Y., Hou, L. T., Lai, J. Y., et al. (2004). Preparation of porous scaffolds by using freeze-extraction and freeze-gelation methods. *Biomaterials, 25*, 129–138.

Hodde, J. P., Badylak, S. F., Brightman, A. O., & Voytik-Harbin, S. L. (1996). Glycosaminoglycan content of small intestinal submucosa: a bioscaffold for tissue replacement. *Tissue Eng., 2*, 209–217.

632

Hodde, J. P., Badylak, S. F., & Shelbourne, K. D. (1997). The effect of range of motion on remodeling of small intestinal submucosa (SIS) when used as an Achilles tendon repair material in the rabbit. *Tissue Eng., 3,* 27–37.

Hodde, J. P., Record, R. D., Liang, H. A., & Badylak, S. F. (2001). Vascular endothelial growth factor in porcine-derived extracellular matrix. *Endothelium J. Endoth. Cell Res., 8,* 11–24.

Hodde, J., Record, R., Tullius, R., & Badylak, S. (2002). Fibronectin peptides mediate HMEC adhesion to porcine-derived extracellular matrix. *Biomaterials, 23,* 1841–1848.

Hodde, J. P., Ernst, D. M., & Hiles, M. C. (2005). An investigation of the long-term bioactivity of endogenous growth factor in OASIS Wound Matrix. *J. Wound Care, 14,* 23–25.

Hong, H., Dong, G. N., Shi, W. J., Chen, S., Guo, C., & Hu, P. (2008). Fabrication of biomatrix/polymer hybrid scaffold for heart valve tissue engineering *in vitro. Asaio J., 54,* 627–632.

Hosokawa, T., Betsuyaku, T., Odajima, N., Suzuki, M., Mochitate, K., Nasuhara, Y., et al. (2008). Role of basement membrane in EMMPRIN/CD147 induction in rat tracheal epithelial cells. *Biochem. Biophys. Res. Commun., 368,* 426–432.

Isch, J. A., Engum, S. A., Ruble, C. A., Davis, M. M., & Grosfeld, J. L. (2001). Patch esophagoplasty using AlloDerm as a tissue scaffold. *J. Pediatr. Surg., 36,* 266–268.

Jenq, C. B., & Coggeshall, R. E. (1987). Permeable tubes increase the length of the gap that regenerating axons can span. *Brain Res., 408,* 239–242.

Jones, J. S., Rackley, R. R., Berglund, R., Abdelmalak, J. B., DeOrco, G., & Vasavada, S. P. (2005a). Porcine small intestinal submucosa as a percutaneous mid-urethral sling: 2-year results. *BJU Int., 96,* 103–106.

Jones, J. S., Vasavada, S. P., Abdelmalak, J. B., Liou, L., Ahmed, E. S., Zippe, C. D., et al. (2005b). Sling may hasten return of continence after radical prostatectomy. *Urology, 65,* 1163–1167.

Kagami, S., Kondo, S., Loster, K., Reutter, W., Urushihara, M., Kitamura, A., et al. (1998). Collagen type I modulates the platelet-derived growth factor (PDGF) regulation of the growth and expression of beta1 integrins by rat mesangial cells. *Biochem. Biophys. Res. Commun., 252,* 728–732.

Khor, E., & Lim, L. Y. (2003). Implantable applications of chitin and chitosan. *Biomaterials, 24,* 2339–2349.

Kim, D. H., Connolly, S. E., Zhao, S., Beuerman, R. W., Voorhies, R. M., & Kline, D. G. (1993). Comparison of macropore, semipermeable, and nonpermeable collagen conduits in nerve repair. *J. Reconstruct. Microsurg., 9,* 415–420.

Knoops, B., Hurtado, H., & van den Bosch de Aguilar, P. (1990). Rat sciatic nerve regeneration within an acrylic semipermeable tube and comparison with a silicone impermeable material. *J. Neuropathol. Exp. Neurol., 49,* 438–448.

Kochupura, P. V., Azeloglu, E. U., Kelly, D. J., Doronin, S. V., Badylak, S. F., Krukenkamp, I. B., et al. (2005). Tissue-engineered myocardial patch derived from extracellular matrix provides regional mechanical function. *Circulation, 112,* 1144–1149.

Kropp, B. P. (1995). Experimental assessment of small-intestinal submucosa as a bladder wall substitute – reply. *Urology, 46,* 400.

Kropp, B. P., Sawyer, B. D., Shannon, H. E., Rippy, M. K., Badylak, S. F., Adams, M. C., et al. (1996). Characterization of small intestinal submucosa regenerated canine detrusor: assessment of reinnervation, *in vitro* compliance and contractility. *J. Urol., 156,* 599–607.

Kropp, B. P., Ludlow, J. K., Spicer, D., Rippy, M. K., Badylak, S. F., Adams, M. C., et al. (1998). Rabbit urethral regeneration using small intestinal submucosa onlay grafts. *Urology, 52,* 138–142.

Lee, J. Y., Nam, S. H., Im, S. Y., Park, Y. J., Lee, Y. M., Seol, Y. J., et al. (2002). Enhanced bone formation by controlled growth factor delivery from chitosan-based biomaterials. *J. Controlled Release, 78,* 187–197.

Li, F., Li, W., Johnson, S. A., Ingram, D. A., Yoder, M. C., & Badylak, S. F. (2004). Low-molecular-weight peptides derived from extracellular matrix as chemoattractants for primary endothelial cells. *Endothelium J. Endoth. Cell Res., 11,* 199–206.

Liang, R., Woo, S. L. Y., Takakura, Y., Moon, D. K., Jia, F. Y., & Abramowitch, S. D. (2006). Long-term effects of porcine small intestine submucosa on the healing of medial collateral ligament: a functional tissue engineering study. *J. Orthop. Res., 24,* 811–819.

Lindberg, K., & Badylak, S. F. (2001). Porcine small intestinal submucosa (SIS): a bioscaffold supporting *in vitro* primary human epidermal cell differentiation and synthesis of basement membrane proteins. *Burns, 27,* 254–266.

Lundy, D. S., Casiano, R. R., McClinton, M. E., & Xue, J. W. (2003). Early results of transcutaneous injection laryngoplasty with micronized acellular dermis versus type-I thyroplasty for glottic incompetence dysphonia due to unilateral vocal fold paralysis. *J. Voice, 17,* 589–595.

Ma, P. X. (2008). Biomimetic materials for tissue engineering. *Adv. Drug Deliv. Rev., 60,* 184–198.

Ma, Z., Mao, Z., & Gao, C. (2007). Surface modification and property analysis of biomedical polymers used for tissue engineering. *Colloids Surfaces B Biointerfaces, 60,* 137–157.

Madihally, S. V., & Matthew, H. W. (1999). Porous chitosan scaffolds for tissue engineering. *Biomaterials, 20,* 1133–1142.

Maeshima, Y., Colorado, P. C., Torre, A., Holthaus, K. A., Grunkemeyer, J. A., Ericksen, M. B., et al. (2000). Distinct antitumor properties of a type IV collagen domain derived from basement membrane. *J. Biol. Chem., 275,* 21340–21348.

Mano, J. F., Silva, G. A., Azevedo, H. S., Malafaya, P. B., Sousa, R. A., Silva, S. S., et al. (2007). Natural origin biodegradable systems in tissue engineering and regenerative medicine: present status and some moving trends. *J. R. Soc. Interface, 4,* 999–1030.

Martinez, F. O., Sica, A., Mantovani, A., & Locati, M. (2008). Macrophage activation and polarization. *Front. Biosci., 13,* 453–461.

McDevitt, C. A., Wildey, G. M., & Cutrone, R. M. (2003). Transforming growth factor-beta1 in a sterilized tissue derived from the pig small intestine submucosa. *J. Biomed. Materials Res. A, 67,* 637–640.

McPherson, T. B., Liang, H., Record, R. D., & Badylak, S. F. (2000). Galalpha(1,3)Gal epitope in porcine small intestinal submucosa. *Tissue Eng., 6,* 233–239.

Mizuno, K., Yamamura, K., Yano, K., Osada, T., Saeki, S., Takimoto, N., et al. (2003). Effect of chitosan film containing basic fibroblast growth factor on wound healing in genetically diabetic mice. *J. Biomed. Mater. Res. A, 64,* 177–181.

Morra, M. (2006). Biochemical modification of titanium surfaces: peptides and ECM proteins. *Eur. Cell Mater., 12,* 1–15.

Mosser, D. M. (2003). The many faces of macrophage activation. *J. Leukocyte Biol., 73,* 209–212.

Musahl, V., Abramowitch, S. D., Gilbert, T. W., Tsuda, E., Wang, J. H., Badylak, S. F., et al. (2004). The use of porcine small intestinal submucosa to enhance the healing of the medial collateral ligament — a functional tissue engineering study in rabbits. *J. Orthop. Res., 22,* 214–220.

Muzzarelli, R., Baldassarre, V., Conti, F., Ferrara, P., Biagini, G., Gazzanelli, G., et al. (1988). Biological activity of chitosan: ultrastructural study. *Biomaterials, 9,* 247–252.

Muzzarelli, R., Biagini, G., Pugnaloni, A., Filippini, O., Baldassarre, V., Castaldini, C., et al. (1989). Reconstruction of parodontal tissue with chitosan. *Biomaterials, 10,* 598–603.

Muzzarelli, R. A. (1997). Human enzymatic activities related to the therapeutic administration of chitin derivatives. *Cell Molec. Life Sci., 53,* 131–140.

Nieponice, A., Gilbert, T. W., & Badylak, S. F. (2006). Reinforcement of esophageal anastomoses with an extracellular matrix scaffold in a canine model. *Ann. Thor. Surg., 82,* 2050–2058.

Nishimura, K., Nishimura, S., Nishi, N., Saiki, I., Tokura, S., & Azuma, I. (1984). Immunological activity of chitin and its derivatives. *Vaccine, 2,* 93–99.

Oelschlager, B. K., Barreca, M., Chang, L., & Pellegrini, C. A. (2003). The use of small intestine submucosa in the repair of paraesophageal hernias: initial observations of a new technique. *Am. J. Surg., 186,* 4–8.

Ott, H. C., Matthiesen, T. S., Goh, S. K., Black, L. D., Kren, S. M., Netoff, T. I., et al. (2008). Perfusion-decellularized matrix: using nature's platform to engineer a bioartificial heart. *Nature Med., 14,* 213–221.

Peluso, G., Petillo, O., Ranieri, M., Santin, M., Ambrosio, L., Calabro, D., et al. (1994). Chitosan-mediated stimulation of macrophage function. *Biomaterials, 15,* 1215–1220.

Puskas, J. E., & Chen, Y. H. (2004). Biomedical application of commercial polymers and novel polyisobutylene-based thermoplastic elastomers for soft tissue replacement. *Biomacromolecules, 5,* 1141–1154.

Raeder, R. H., Badylak, S. F., Sheehan, C., Kallakury, B., & Metzger, D. W. (2002). Natural anti-galactose alpha1,3 galactose antibodies delay, but do not prevent the acceptance of extracellular matrix xenografts. *Transplant Immunology, 10,* 15–24.

Ratner, B. D. (2004). *Biomaterials Science: An Introduction to Materials in Medicine* (2nd ed). Amsterdam: Academic Press.

Record, R. D., Hillegonds, D., Simmons, C., Tullius, R., Rickey, F. A., Elmore, D., et al. (2001). *In vivo* degradation of C-14-labeled small intestinal submucosa (SIS) when used for urinary bladder repair. *Biomaterials, 22,* 2653–2659.

Reing, J. E., Zhang, L., Myers-Irvin, J., Cordero, K. E., Freytes, D. O., Heber-Katz, E., et al. (2009). Degradation products of extracellular matrix affect cell migration and proliferation. *Tissue Eng. A, 15,* 605–614.

Roberts, R., Gallagher, J., Spooncer, E., Allen, T. D., Bloomfield, F., & Dexter, T. M. (1988). Heparan sulphate bound growth factors: a mechanism for stromal cell mediated haemopoiesis. *Nature, 332*(6162), 376–378.

Rodriguez, F. J., Gomez, N., Perego, G., & Navarro, X. (1999). Highly permeable polylactide-caprolactone nerve guides enhance peripheral nerve regeneration through long gaps. *Biomaterials, 20,* 1489–1500.

Ross, J. R., Kirk, A. D., Ibrahim, S. E., Howell, D. N., Baldwin, W. M., III, & Sanfilippo, F. P. (1993). Characterization of human anti-porcine "natural antibodies" recovered from ex vivo perfused hearts — predominance of IgM and IgG2. *Transplantation, 55,* 1144–1150.

Sarikaya, A., Record, R., Wu, C. C., Tullius, B., Badylak, S., & Ladisch, M. (2002). Antimicrobial activity associated with extracellular matrices. *Tissue Eng., 8*, 63–71.

Schwarzbauer, J. (1999). Basement membranes: putting up the barriers. *Curr. Biol., 9*, R242–R244.

Sellaro, T. L., Ravindra, A. K., Stolz, D. B., & Badylak, S. F. (2007). Maintenance of hepatic sinusoidal endothelial cell phenotype *in vitro* using organ-specific extracellular matrix scaffolds. *Tissue Eng., 13*, 2301–2310.

Shin, H., Jo, S., & Mikos, A. G. (2003). Biomimetic materials for tissue engineering. *Biomaterials, 24*, 4353–4364.

Simon, P., Kasimir, M. T., Seebacher, G., Weigel, G., Ullrich, R., Salzer-Muhar, U., et al. (2003). Early failure of the tissue engineered porcine heart valve SYNERGRAFT in pediatric patients. *Eur. J. Cardio-Thor. Surg., 23*, 1002–1006, discussion 1006.

Smith, A. M., Walsh, R. M., & Henderson, J. M. (2005). Novel bile duct repair for bleeding biliary anastomotic varices: case report and literature review. *J. Gastrointest. Surg., 9*, 832–836.

Stamm, C., Khosravi, A., Grabow, N., Schmohl, K., Treckmann, N., Drechsel, A., et al. (2004). Biomatrix/polymer composite material for heart valve tissue engineering. *Ann. Thor. Surg., 78*, 2084–2092.

Suh, J. K., & Matthew, H. W. (2000). Application of chitosan-based polysaccharide biomaterials in cartilage tissue engineering: a review. *Biomaterials, 21*, 2589–2598.

Suzuki, D., Takahashi, M., Abe, M., Sarukawa, J., Tamura, H., Tokura, S., et al. (2008). Comparison of various mixtures of beta-chitin and chitosan as a scaffold for three-dimensional culture of rabbit chondrocytes. *J. Mater. Sci. Mater. Med., 19*, 1307–1315.

Tanabe, T., Okitsu, N., Tachibana, A., & Yamauchi, K. (2002). Preparation and characterization of keratin-chitosan composite film. *Biomaterials, 23*, 817–825.

Ueno, H., Yamada, H., Tanaka, I., Kaba, N., Matsuura, M., Okumura, M., et al. (1999). Accelerating effects of chitosan for healing at early phase of experimental open wound in dogs. *Biomaterials, 20*, 1407–1414.

Ueno, H., Nakamura, F., Murakami, M., Okumura, M., Kadosawa, T., & Fujinag, T. (2001a). Evaluation effects of chitosan for the extracellular matrix production by fibroblasts and the growth factors production by macrophages. *Biomaterials, 22*, 2125–2130.

Ueno, H., Murakami, M., Okumura, M., Kadosawa, T., Uede, T., & Fujinaga, T. (2001b). Chitosan accelerates the production of osteopontin from polymorphonuclear leukocytes. *Biomaterials, 22*, 1667–1673.

Umeda, H., Kanemaru, S. I., Yamashita, M., Ohno, T., Suehiro, A., Tamura, Y., et al. (2009). In situ tissue engineering of canine skull with guided bone regeneration. *Acta Otolaryngol., 129*, 1509–1518.

Valentin, J. E., Badylak, J. S., McCabe, G. P., & Badylak, S. F. (2006). Extracellular matrix bioscaffolds for orthopaedic applications. A comparative histologic study. *J. Bone Joint Surg., 88A*, 2673–2686.

van der Rest, M., & Garrone, R. (1991). Collagen family of proteins. *FASEB J., 5*, 2814–2823.

Voytik-Harbin, S. L., Brightman, A. O., Kraine, M. R., Waisner, B., & Badylak, S. F. (1997). Identification of extractable growth factors from small intestinal submucosa. *J. Cell. Biochem., 67*, 478–491.

Wainwright, D. J. (1995). Use of an acellular allograft dermal matrix (AlloDerm) in the management of full-thickness burns. *Burns, 21*, 243–248.

Wainwright, J. M., Czajka, C. A., Patel, U. B., Freytes, D. O., Tobita, K., Gilbert, T. W., et al. (2009). Preparation of cardiac extracellular matrix from an intact porcine heart. *Tissue Eng. C Methods, 15*, 605–614.

Wang, J., de Boer, J., & de Groot, K. (2004). Preparation and characterization of electrodeposited calcium phosphate/chitosan coating on Ti6Al4V plates. *J. Dent. Res., 83*, 296–301.

Wang, X., Hu, W., Cao, Y., Yao, J., Wu, J., & Gu, X. (2005). Dog sciatic nerve regeneration across a 30-mm defect bridged by a chitosan/PGA artificial nerve graft. *Brain, 128*(Pt 8), 1897–1910.

Wang, Z. Q., Watanabe, Y., & Toki, A. (2003). Experimental assessment of small intestinal submucosa as a small bowel graft in a rat model. *J. Pediatr. Surg., 38*, 1596–1601.

Wood, J. D., Simmons-Byrd, A., Spievack, A. R., & Badylak, S. F. (2005). Use of a particulate extracellular matrix bioscaffold for treatment of acquired urinary incontinence in dogs. *J. Am. Vet. Med. Assoc., 226*, 1095–1097.

Yoshikawa, H., Tamai, N., Murase, T., & Myoui, A. (2009). Interconnected porous hydroxyapatite ceramics for bone tissue engineering. *J. R. Soc. Interface, 6*, S341–S348.

Zalavras, C. G., Gardocki, R., Huang, E., Stevanovic, M., Hedman, T., & Tibone, J. (2006). Reconstruction of large rotator cuff tendon defects with porcine small intestinal submucosa in an animal model. *J. Shoulder Elbow Surg., 15*, 224–231.

635

CHAPTER 35

Hydrogels in Regenerative Medicine

Justin M. Saul[*,†], **David F. Williams**[*,‡,§,¶,**]
*Wake Forest Institute for Regenerative Medicine, Wake Forest University Health Sciences, Winston-Salem, NC, USA
†Virginia Tech-Wake Forest University School of Biomedical Engineering and Sciences, Wake Forest University Health Sciences, Winston-Salem, NC, USA
‡Christiaan Barnard Department of Cardiothoracic Surgery, Cape Town, South Africa
§University of New South Wales, Graduate School of Biomoedical Engineering, Sydney, Australia
¶Tsinghua University, Beijing, China, Shanghai Jiao Tong University, China
**University of Liverpool, Liverpool, UK

INTRODUCTION: RELEVANCE OF HYDROGELS TO REGENERATIVE MEDICINE

Hydrogels are crosslinked polymeric networks containing hydrophilic groups that promote swelling due to interaction with water (Peppas, 1986). While hydrogels are heavily used in the field of regenerative medicine, their application to biomedical systems is not new. In fact, it has been suggested that they were truly the first polymer materials to be developed for use in man (Kopecek, 2007). They have been in use for clinical applications since the 1960s, initially for use in ocular applications including contact lenses and intraocular lenses due to their favorable oxygen permeability and lack of irritation leading to inflammation and foreign body response, which was observed with other plastics (Wichterle and Lim, 1960). Before the concept of tissue engineering and regenerative medicine had gained traction, hydrogels were used for cell encapsulation (Lim and Sun, 1980). They have also been utilized extensively in the clinic for wound healing applications due to their oxygen permeability, high water content, and ability to shield wounds from external agents. Perhaps the largest research focus and utility of hydrogels has been found in their use as controlled release systems. This combination of controlled release and cell encapsulation has led to increasing uses of hydrogels in regenerative medicine applications.

Hydrogels used in regenerative applications can be based on naturally or synthetically derived polymers. By most definitions, native tissues, particularly the extra-cellular matrix, are hydrogels and derivatives of these and other naturally based systems are in widespread use. Natural hydrogels are generally regarded as having favorable biodegradation products compared to some synthetic polymers as the monomeric degradation products are typically amino acids or saccharide units. In contrast, synthetics offer wide flexibility in terms of mechanical properties, water swelling, degradation rates, ionic charge, and other important parameters. Table 35.1 describes several prominent synthetic and natural hydrogels. These natural systems are derived from mammals, crustaceans, plants, and bacteria and are typically polypeptides or polysaccharides. Table 35.1, which highlights uses of hydrogels in regenerative

Principles of Regenerative Medicine. DOI: 10.1016/B978-0-12-381422-7.10035-5

TABLE 35.1 Hydrogels Used in Regenerative Medicine and Medical Technology Applications

Hydrogel Material [Abbreviation]	Description	Examples of applications	References
Poly(ethylene glycol) diacrylate [PEG]	Widely used, flexible synthetic polymer with low protein adsorption that is photo-crosslinkable	Neural Cartilage	Mahoney and Anseth, 2006 Bryant et al., 2004
Poly(2-hydroxyethyl methacrylate) [pHEMA]	Non-degradable polymer modifiable for degradation (Atzet et al., 2008)	Intraocular lenses Nerve guidance	Wichterle and Lim, 1960 Dalton et al., 2002; Flynn et al., 2003
Oligo-(polypropylene fumarate) [OPF]		Bone Cartilage Nerve	Guo et al., 2010 Park et al., 2009 Hausner et al., 2007; Dadsetan et al., 2009
Poly(N-isopropylacrylamide) [pNIPAAM]	Synthetic material with lower critical solution temperature (LCST) near physiological temperature	Corneal sheets	Lai et al., 2007
Collagen	Most prevalent protein in mammals characterized by proline-lysine-glycine repeating units and various isoforms	Bone Tendon	Hesse et al., 2010 Abousleiman et al., 2008
Fibrin	Protein involved in clot formation and platelet binding in blood coagulation cascade	Surgical glue Neural regeneration Bone Cartilage	Brennan, 1991 Kalbermatten et al., 2009 Lutolf et al., 2003; Arrighi et al., 2009 Ho et al., 2010
Keratin	Protein derived from intermediate filaments of eukaryotes	Neural regeneration Hemostasis	Sierpinski et al., 2008 Aboushwareb et al., 2009
Silk	Biopolymer derived from spiders, *Bombyx mori*, and other sources	Cartilage Bone	Aoki et al., 2003 Fini et al., 2005
Agarose	Thermoreversible linear polysaccharide derived from red algae	Cartilage Neural regeneration	Mauck et al., 2000; Ng et al., 2005; Bian et al., 2010 Dodla and Bellamkonda, 2006; Stokols and Tuszynski, 2006
Alginate	Second most abundant polysaccharide on earth; derived from seaweed and contains β-D-mannuoronate and β-L-guluronate sub-units (1,4 linkage)	Cell encapsulation Bone Ovary follicles	Lim and Sun, 1980 Alsberg et al., 2003 Jin et al., 2010
Chitin	β-(1,4)-N-acetyl glucosamine polysaccharide	Neural regeneration Cartilage	Freier et al., 2005 Hoemann et al., 2005
Chitosan	Deacetylated chitin	Wound dressings	Neuffer et al., 2004; Wedmore et al., 2006
Hyaluronic acid	β-(1,3) glucuronic acid and β (1,4)-N-acetylglucosamine	Cartilage Vocal cord Bone	Grigolo et al., 2001 Farran et al., 2010 Kim et al., 2007

Continued

TABLE 35.1 continued

Hydrogel Material [Abbreviation]	Description	Examples of applications	References
	glycosaminoglycan found in connective tissue	Wound healing	Ghosh et al., 2006
Methyl cellulose	Cellulose (polysaccharide) with hydroxyl groups substituted with methoxyl groups to disrupt cellulose crysallinity and provide solubility	Neural regeneration Nucleus pulposa	Stabenfeldt et al., 2006 Reza and Nicoll, 2010
Bacterial cellulose	Non-degradable cellulosic material produced by *Acetobacter xylinum*	Vascular graft Bone	Esguerra et al., 2010 Grande et al., 2009

medicine, primarily shows the use of single-component systems. However, combinatorial uses of hydrogels to obtain desirable properties of each component are widely investigated.

The diversity and flexibility of natural and synthetic hydrogels makes it impossible to consider every type of hydrogel. The goal of this chapter is to provide an overview of the basic theory of hydrogels, describe important uses and applications in regenerative medicine, and consider their continued importance in research and the clinic.

BACKGROUND AND THEORY

Classification

Several methods can be used to classify hydrogels including network structure or porosity, physical structure, source, and crosslink type (Peppas, 2004). In tissue engineering and regenerative medicine, the porosity of the scaffold materials is often of considerable importance. Hydrogels can be classified according to their *network structure* as macroporous (pores of ~10−200 μm), microporous (pores of ~1−10 μm), and non-porous (pores of < 1 μm). Clearly, if used for approaches in which cells must infiltrate the scaffold, macroporous scaffolds must be used. Non-porous scaffolds have low rates of diffusion, making cell viability a significant concern and preventing infiltration of cells not pre-encapsulated in the hydrogel.

Classification by *physical structure* is also fairly intuitive for regenerative applications. Hydrogels can be considered amorphous, semicrystalline, hydrogen bonded, or complexation products. They can be classified according to their *source*, that is, whether they are naturally derived or synthetic. Synthetic systems have a plethora of options and may be homopolymer or copolymer systems. Further, synthetic polymers can also be synthesized to form interpenetrating polymer networks in which the two polymer sheets are physically entangled. Naturally derived systems may be polysaccharide or polypeptide-based and derived from numerous sources.

More common in the polymer chemistry literature is classification according to the *ionic charge* of the hydrogel. The gels may be neutral, anionic, cationic, or ampholytic. The charge properties become crucial in considering their interaction with physiological environments and cells. Ionotropic gels are ionic polymers that contain a balancing multivalent counterion. One example of this is an alginate hydrogel. Alginate in its native form is anionic but is not a hydrogel. In the presence of calcium counterions it forms a physical hydrogel through ionic interactions.

Hydrogels can also be classified according to the type of *crosslink*. The two broadest categorizations are chemical or physical crosslinks. In general, a chemical crosslink is defined as a covalent interaction at a point of overlap or junction (see below). Physical crosslinks may be the physical entanglement of the polymer chains, interpenetrating polymer networks, and other secondary forces (e.g. hydrophobic interactions, hydrogen bonding, ionic interactions, electrostatic interactions) (Hoffman, 2002).

Theory

An in-depth description of the polymer science theory governing the formation and behavior of hydrogels is beyond the scope of this text. Below we briefly describe several of the key parameters that are of particular importance to scientists working with hydrogels in regenerative medicine applications. More complete descriptions of the theoretical polymer science basis for hydrogels are given in several other locations (Peppas, 1986; Lowman and Peppas, 1999). Nonetheless, the principles that govern their formation are highly relevant to regenerative medicine applications as these properties control (among other things) swelling, interaction with cells, and drug delivery kinetics.

Numerous mathematical models have been developed to estimate the properties of hydrogels and to predict processes related to controlled release (e.g. diffusion parameters and rates of release). Much of the discussion below is taken from review articles on the topic (Lowman and Peppas, 1999).

Three important parameters to consider in defining hydrogels are: (1) the *volume fraction* in the swollen state, (2) the *crosslink density*, and (3) the *porosity of the hydrogel*. These parameters can be described mathematically.

The volume fraction is simply the volume fraction occupied by the polymer and gives a sense of the amount of water in the gel. Mathematically, it has been defined as:

$$v_{2,s} = \frac{\text{Volume of polymer}}{\text{Volume of swollen gel}} = \frac{1}{Q}$$

Where Q is another useful parameter known as the volume degree of swelling. Although the polymer volume can be estimated from the density of the polymer, it is sometimes easier to determine the polymer weight. A second set of parameters based on the weights can therefore be defined akin to the volume fractions above:

$$m_{2,s} = \frac{\text{Mass of polymer}}{\text{Mass of swollen gel}} = \frac{1}{q}$$

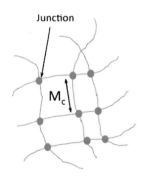

FIGURE 35.1
Simplified schematic of hydrogel structure. Junctions indicate points of crosslinks, with the distance between the junctions indicating the molecular weight of between crosslinks, M_c. Junctions may be physical entanglements or chemical or ionic crosslinks. Adapted from Peppas and Barr-Howell (1986), with permission.

The effective molecular weight between crosslinks can be inferred from the description in Figure 35.1. This figure shows a schematic of a hydrogel polymer network (Peppas and Barr-Howell, 1986). Areas of "overlap" in Figure 35.1 are known as junctions and indicate either physical or chemical crosslinks. The average molecular weight between crosslinks is known as the effective molecular weight between crosslinks, M_c. This value can be used to determine the crosslink density based on the known molecular weight of the polymer repeat units, M_o, as:

$$X = \frac{M_c}{M_o}$$

Several other parameters of interest are those of porosity, tortuosity, and the diffusion coefficients through the gel. Porosity is the volume of the gel that is not physically occupied by the polymer itself. This parameter is often of particular interest to tissue engineers working with macroporous or microporous gels as the porosity is an indication of the volume of the material available for cell infiltration or seeding. Within regenerative medicine, the role that cells play in remodeling a matrix is often important, so materials with higher porosity are viewed as favorable, potentially allowing a larger number of the cellular ("functional" units) of the system to be present. Tortuosity is the path that a molecule or cell must navigate in order to penetrate the gel. Both the porosity and tortuosity are important in determining the rate of diffusion within hydrogels. Diffusion is the main transport mechanism of nutrients and oxygen to cells within the gel, of metabolic waste products out of the gel, as well as the release of therapeutic agents from the gel or from cells within the gel.

UTILITY IN REGENERATIVE MEDICINE

Hydrogels currently have a wide range of applications from consumer products to electronics and biosensors. Within the biomedical field, the earliest applications of hydrogels were for ocular applications including intraocular lenses (Cavanagh et al., 1980), soft contact lenses (Wichertle and Lim, 1960; Dreifus and Wichertle, 1964), and corneal repair (Sendele et al., 1983). They have also been used as suture, as dental materials, in biosensors, and as coatings for catheters and defibrillators, and for wound care products.

The original use of hydrogels came as stand-alone biomaterials or as devices for controlled release applications. Later, they became the focus of cell-encapsulation approaches. The wedding of these three uses for hydrogels has propelled their use in the field of regenerative medicine as scientists use their properties to direct cell attachment, migration, and differentiation. The application of these individual roles of hydrogels and their combinatorial approaches in regenerative medicine are described in more detail below.

Hydrogels as biomaterials

Although the field of biomaterials is clearly moving from the use of inert materials that minimize host response to more bioactive and integrative materials, the original use of hydrogels came as stand-alone biomaterials for ophthalmic and blood-contacting applications. Indeed these applications remain important in the biomedical field. At the time of their discovery, one of the appealing aspects of hydrogels was their minimal foreign body response compared to other semi-crystalline polymers (Wichertle and Lim, 1960). This stems from the observation that hydrogels typically have low levels of protein adsorption. For example, poly (ethylene glycol) (PEG) is widely used on drug delivery vehicles due to its ability to provide long circulation times, likely through steric effects on complement proteins. Similarly, PEG-diacrylate gels are known to have low protein adsorption (DeLong et al., 2005b), as are various other synthetic-based hydrogels (Horbett, 1986) due to steric effects and possibly their hydrophilic surfaces. Low protein adsorption is generally associated with poor cell attachment. Further, synthetic hydrogels and many natural hydrogels (particularly polysaccharides such as alginate and agarose) lack peptidic sequences that would promote cell attachment via integrin binding.

While this lack of biological activity seems counter-intuitive in regenerative medicine, hydrogels are highly labile in terms of chemical and physical modifications that can be used to modulate cellular response in a more bioactive fashion (see below). Thus, while hydrogels remain highly significant as stand-alone biomaterials in minimizing host response, they are also useful for the direction of cell response to materials in a more bioactive fashion.

The method of hydrogel formation is particularly important for maintaining bioactivity of molecules and for minimizing any detrimental effects to cells associated with the gels. Typically, a monomer or non-crosslinked polymer is found in the solution (sol) phase. Upon application of some initiation conditions, the sol phase forms the hydrogel (gel) phase; that is, it undergoes the sol-gel transition. For hydrogels containing bioactive molecules or cells, it is clear that high temperatures and many monomers, solvents, and polymerization initiators cannot be used due to inactivation of bioactives or cytotoxicity. Therefore, approaches that achieve the sol-gel transition with minimal effect on bioactive molecules and cells have been developed and are described here and in other sections of this textbook.

Photopolymerization is commonly used to achieve gelation for regenerative medicine applications. Photopolymerization reactions for hydrogel formation have been performed for poly(vinyl alcohol) (Bader and Rochefort, 2008), polysaccharide-based materials, poly (2-hydroxyethylmethacrylate) (Bae et al., 2006; Ayhan and Ozkan, 2007; Faxalv et al., 2010), and modified PEG-collagen (Bayramoglu et al., 2010) among others. Poly(ethylene glycol) or PEG diacrylate gels are the most common gel systems used for the formation of photopolymerized hydrogels (Mann and West, 2002). The presence of pi bonds in the diacrylate terminal ends of PEG provides the chemical moieties for the reaction. Photoinitiators used for biomaterial applications have been classified as photolytic or hydrogen abstraction (Nguyen and West, 2002). Photolytic groups include free radical initiators including acetophenone derivatives widely used for hydrogel formation. One advantage of photopolymerizable hydrogels is their ability to be polymerized *in situ*. Materials that spontaneously undergo the sol-gel transition in response to light or physiological temperature can be maintained in solution phase until *in vivo* injection followed by gel formation at the desired location. Other more robust forms of achieving gelation have been described, such as the use of click chemistry (Malkoch et al., 2006; Crescenzi et al., 2007; Testa et al., 2009); however, the effects of this chemistry and the copper initiators on cells have not been established.

Hydrogels have been coupled with most modern approaches to biomaterial scaffold fabrication. Cell-hydrogel constructs with alginate or poly(ethylene oxide)-pluronic-poly(ethylene oxide) block polymer as the hydrogels have been used with BioPlotter and solid free-form fabrication systems (Cohen et al., 2006; Fedorovich et al., 2008). One drawback to hydrogel systems is a lack of mechanical integrity associated with certain applications.

Non-woven electrospun fibers (Ji et al., 2006) and woven hydrogel/cell mixtures (Moutos et al., 2007) have been reported that may allow for materials of greater mechanical integrity. More complex approaches to achieving spatial regulation of cells within hydrogel biomaterials include photopatterning and photolithographic techniques of cell adhesion and migration molecules to direct cell attachment and material response (see below) (Hahn et al., 2006; Bryant et al., 2007). The role of topographical cues in cell fate is becoming well established and hydrogels are useful for the creation of three-dimensional topographical cues such as nanopillars (Kim et al., 2006) to promote cell migration, and recently through the introduction of enzyme-assisted photolithography to achieve spatial functionalization of hydrogels that may provide insight into the topographical designs necessary for larger tissue constructs (Gu and Tang, 2010).

Hydrogels for controlled release

Over the past 30 years, numerous systems have been developed for delivery of therapeutic agents within the context of the medical device field with the goal of achieving long-term, zero-order release of therapeutic agents. These systems can generally be classified as swelling-controlled, diffusion-controlled, or chemically controlled. Examples include osmotic pumps (swelling-controlled), transdermal patches (diffusion-controlled reservoir matrix), and drug eluting stents (chemically controllable erodible system).

Within the field of regenerative medicine, it desirable to have release of therapeutic agents occur from a system that degrades with time so that secondary retrieval and removal of a device (e.g. osmotic pump) is not required. As such, those systems described above as medical devices are usually not employed within the context of regenerative medicine. Rather, implantable or injectable bulk hydrogels are a preferential alternative. The primary therapeutic agent of interest is the release of growth factors. However, the delivery of small molecules (e.g. antibiotics) and nucleic acids have also been widely employed, often in conjunction with cell therapies.

A drawback to this bulk hydrogel approach is that the complexity provided by a device (e.g. an osmotic pump) is lost, making zero-order release difficult to achieve over the time periods that can be achieved with medical devices. However, in principle, regenerative medicine approaches seek to stimulate a physiological response to a chemical cue such as growth factor release. Following the initial stimulus, the regenerated tissue should move toward normal physiological function, thereby obviating the need for long-term release of the chemical cue as is desirable in the traditional medical device field.

As a controlled release platform, several parameters can be used to modulate the release of the therapeutic agent. Because these systems are examples of diffusion-mediated release, the porosity and tortuosity of the system significantly affect the release of therapeutic agents. The ionic nature of many hydrogel systems also lends itself to the sustained release of counter-charged molecules. More generally, affinity between the hydrogel and therapeutic agent (whether through ionic charge, hydrophobic effects, or protein-protein interactions) can be used to modulate the rates of release.

The ability to modulate the release through external control is also of considerable importance. Examples of these types of systems include increasing binding between the gel and therapeutic, the use of environmental controls such as temperature, pH, or enzymes, or the application of external energy sources such as ultrasound, light, or electrical fields to promote or mitigate release. Such approaches allow for the tunable, responsive, and/or pulsatile release profiles.

The hydrophilic groups of hydrogels that lead to high water content and therefore low protein adsorption or cell binding also provide chemical flexibility for the covalent binding of molecules to the hydrogel backbone. This is described in more detail below for matrix-immobilized ligands. However, this approach can also be used to control the rate of release of therapeutic agents. One example of this approach is in making use of the heparin-binding domains found on many growth factors. By coupling heparin to the hydrogel via these reactive groups, it is possible to maintain the association of heparin-binding growth factors with the hydrogel for a longer timeframe than achieved with diffusion-mediated release (Sakiyama-Elbert and Hubbell, 2000).

A more sophisticated controlled release system applicable to hydrogels and other polymers is that of cell-demanded liberation and is akin to a pendant-chain system (Langer, 1990). The premise of this approach is that growth factors can be tethered to the backbone of the hydrogel (Lutolf et al., 2003). The backbone of the hydrogel or the pendant-chain tether itself can contain sequences cleavable by matrix-metalloproteases (MMPs). Diffusion-mediated release from the hydrogel or cleavage of the growth factor from the tether occur only as cells infiltrate the scaffold, hence the term cell-demanded liberation. In one version of this system, the 121 isoform of VEGF was engineered to contain the factor XIII substrate amino acid sequence, for binding to a fibrin hydrogel, and an $\alpha 2$ plasmin-inhibitor that is MMP-cleavable (Ehrbar et al., 2004). As cells infiltrate the scaffold, they produce MMPs, leading to cleavage of the VEGF from the hydrogel to promote vascularization of the construct in conjunction with cell infiltration.

The most well known and studied thermally responsive system is the *N*-isopropylacrylamide or pNIPAAM. pNIPAAM has a lower critical solution temperature (LCST) of approximately 32°C (for the linear form). Therefore, below 32°C the polymer is hydrophilic and promotes cell attachment and adhesion. Above 32°C, the polymer becomes hydrophobic, leading to detachment of cells. This approach has therefore been exploited for use in the creation of cell sheets for various tissue engineering applications including cornea (Nishida et al., 2004a, b; Hsiue et al., 2006; Lai et al., 2007), cardiac grafts (Shimizu et al., 2001, 2002a, b), urothelium (Shiroyanagi et al., 2003, 2004), and skin (Yamato et al., 2001), among others.

In addition to intrinsic physiological control mechanisms such as pH, enzymes, and temperature, release of therapeutic agents can also be controlled through extrinsic mechansisms or energy sources. In one example of these systems, pHEMA has been used as the drug reservoir for the release of small molecules (ciprofloxacin antibiotic, molecular weight ∼ 330 Da) and pHEMA/2-hydroxyethylacrylate PEG-dimethylacrylate has been used to release larger molecules (insulin, molecular weight ∼ 5.8 kDa). This system used a coating of methylene chains to prevent passive diffusion of drug during periods without the application of ultrasound and was compatible with ultrasonic energies that are clinically relevant (43 kHz and 1.1 MHz) (Kwok et al., 2001). Electrically responsive hydrogels are also an active area of research. Although an important area of research for drug delivery applications, the use of electrically responsive hydrogels has not found widespread use in regenerative medicine due to the need for application of electrical field and/or electrode implantation (Murdan, 2003). Nonetheless, extrinsically mediated systems and feedback-controlled systems provide added control and sophistication that may find utility, particularly in endocrine-related tissues.

In summary, hydrogel systems can provide a mechanism to achieve near zero-order release for finite time periods, which is advantageous as a secondary procedure for device removal is not required. Hydrogels can also be used to achieve on-demand or pulsatile release. Because they can serve as a physical matrix for cells or for cell encapsulation, they can also be considered controlled release systems by slowly releasing therapeutic agents produced by cells.

Cell association with hydrogels

CELL ENCAPSULATION

Hydrogels have been in use for nearly 30 years as a system to encapsulate cells (Lim and Moss, 1981). For example, endocrine disorders such as diabetes that result from autoimmunity against hormone-secreting cells are intensely studied for regenerative therapies. The ability to achieve a functional effect may be attainable not through whole organ replacement but through delivery of cells microencapsulated in hydrogels (Fig. 35.2), which can promote cell viability, controlled release, and protection from immune response to implanted cells. While the spatial distribution of cells is clearly important in the engineering of *de novo* functional tissues, the more general non-spatially controlled encapsulation of cells still provides the potential for a significant clinical impact.

The incorporation of cellular components provides the opportunity to achieve zero-order or physiologically responsive release over prolonged periods of time. Alginate microspheres are among the most studied hydrogel carriers for microencapsulation of cells. The mechanical properties of these gels can be modulated depending on the divalent cation used to achieve crosslinking. For example, the use of barium or strontium instead of calcium leads to more rigid gels. These systems also provide a barrier to immune response, providing the potential to use autologous stem cells, allogeneic differentiated cells, or even xenografted cells. They can also be coated with polycations such as poly-L-lysine or poly-ornithine for stabilization and to regulate solute release from the capsules (Orive et al., 2006). With these semi-permeable

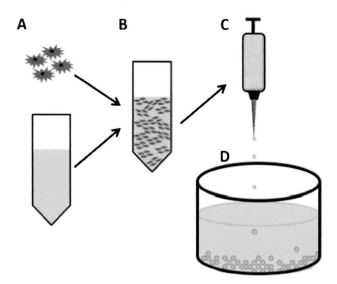

FIGURE 35.2
Schematic of the process of cell microencapsulation with sol-gel transition of hydrogels. (A) In this process, cultured or isolated cells are dissolved in a solution-phase hydrogel (e.g. alginate) and (B) mixed into a viscous cell suspension. (C) The solution-phase hydrogel/cell mixture is then extruded (e.g. by pressure) from a droplet microencapsulator (typically a syringe, possibly with some electric field applied to regulate droplet size). (D) The droplets are collected in a solution promoting gelation (e.g. calcium-containing solution for alginate). Regulation of drop size allows single or multiple cell encapsulation and mechanical properties can be regulated through the hydrogel precursor concentration or the gelling agent.

membrane coatings, the encapsulated cell/microbead system becomes akin to the traditional reservoir drug delivery systems but with the ability to respond to physiological stimuli. Problems with foreign body response and issues of achieving sufficient nutrient supply and waste removal remain challenges to long-term patency of these systems, but many of the approaches to achieve vascularization currently under investigation may provide more immediate benefit to these types of cell-encapsulation therapies for numerous hormone-related deficiencies.

Within the last decade, approaches have been developed to provide more control over methods to achieve encapsulation of cells. Two particularly important approaches include *in situ* gelation and photoencapsulation. In situ gelation systems may depend on pH, temperature, or light as the initiating species. Temperature and light are the primary mechanisms of *in situ* gelation initiators used in regenerative medicine with pH systems being used less frequently to avoid exposing cells to caustic environments. Particularly advantageous for regenerative medicine applications are reverse thermogelation compounds that are in solution at low temperatures but gel at higher temperatures, preferably in the physiological range. That is, compounds that undergo the sol-gel transition with increasing temperature. Important natural hydrogels that are reverse thermogelation compounds include collagen, carboxymethylcellulose, and certain combinations of hydrogels. Important synthetic hydrogels known to undergo *in situ* gelation include poly(N-isopropylacrlamide) and block co-polymer systems consisting of poly(ethylene glycol) in combination with Pluronic or PLGA (Jeong et al., 2002). The ability of hydrogels to undergo sol-gel transitions at increasing temperatures depends on the balance of intermolecular forces including hydrogen bonding and hydrophobic interactions as well as water content and crosslinking density. The applications of these systems are numerous, but include both hard and soft tissues (Chen et al., 2003; Jain et al., 2006).

Photo-encapsulation can be a subclass of *in situ* gelling materials. Photo-initiators are widely used in dental applications to cure resins. Photo-initiators have also been used for the formation of hydrogels for approximately 20 years (Sawhney et al., 1993) as a solution-phase material can be injected and then cured to the gel phase *in situ*. Various initiators have been

645

employed, but those that are active in the ultraviolet range are most widely reported in part because they are less likely to polymerize in response to ambient light and have higher reported crosslinking efficiencies (Bryant and Anseth, 2006). The use of polymerization initiators as well as the use of UV-light have been shown to have detrimental effects on cells (Fedorovich et al., 2009). However, alternate initiators such as those active in the infrared range may mitigate some of the UV effects, and the ability to achieve spatially controlled encapsulation of cells within hydrogels is clearly important for achieving structure-function relationships (Elisseeff et al., 2000; Williams et al., 2003).

SPATIAL PATTERNING

As described above, hydrogels can be utilized to achieve three-dimensional spatial patterning of cells. One important development in the use of hydrogels has been the advent of inkjet printing technology, first described by Boland's group in 2004 (Roth et al., 2004; Xu et al., 2005). Originally used for high-throughput screening of drug compounds (Lemmo et al., 1998) and then for the printing of nucleic acids to solid substrates (Goldman and Gonzalez, 2000), this approach has since garnered more attention for the spatial patterning of cells that may resemble native tissue. In the initial iterations of this work, ink from inkjet printer cartridges was replaced with cells suspended in a non-crosslinked polymer solution. Alginates in the absence of calcium is the most common gel system although there are reports using collagen (Xu et al., 2005), fibrin (Campbell et al., 2005; Cui and Boland, 2009), polyurethanes (Zhang et al., 2008), and polyacrylamide (Ilkhanizadeh et al., 2007). In each case, the cells are suspended in a non-crosslinked form of the polymer solution and printed into a solution containing the crosslinking agent. In the case of alginate this involves printing into a calcium solution whereas fibrinogen can be printed into a thrombin solution. Cells are printed in a two-dimensional pattern, and three-dimension structures are formed through a layer-by-layer approach. More recently, true three-dimension bioprinters have been developed (Nishiyama et al., 2009) and the ability to print from an array of cartridges is also under development. The role of gradients in directing migration and/or differentiation is an area of interest in many aspects of regenerative medicine. Inkjet printers provide a medium for the development of these types of gradients (Ilkhanizadeh et al., 2007; Cai et al., 2009; Miller et al., 2009). Current drawbacks to this approach are the lack of mechanical integrity and questions regarding the thermal effects on cells. Although the technology has not shown a noticeable effect on cell viability (Xu et al., 2006b), more subtle effects on gene regulation through the heat shock protein (HSP) family are yet to be fully characterized.

Dielectrophoresis is an approach useful for screening and may also be applicable to the formation of scaffold materials on scales relevant to organ engineering (Lin et al., 2006). This approach allows for the patterning of cells within hydrogel constructs via the application of uniform electrical field. One drawback to the approach is that relatively weak hydrogels are required to allow manipulation and patterning of the cells for the dielectrophoretic field. An approach has been described to overcome this, which treats hydrogels as a composite system wherein cells are patterned via electrical field within an agarose or PEG gel and surrounded by a bulk-phase material with the desired mechanical properties, thus providing an approach to scale up this technique to larger-sized constructs (Albrecht et al., 2007). It is unclear whether the application of electrical field or the application of high temperatures in inkjet printing will be less deleterious to cells, but the approach demonstrates the ability to achieve spatial patterning through other approaches.

The ability to spatially pattern cells on or within hydrogels is also important in considering approaches to achieve high-throughput screening of hydrogels to assess the role of hydrogel type, topographical signals, soluble chemical cues, immobilized chemical cues, mechanical properties, and other parameters on cells. Several approaches have been developed toward this end and will likely be important tools in providing more systematic study of the role of hydrogels (and other biomaterials) on cell phenotype and genotype. One approach to high-throughput analysis is the use of the dip pen lithographic technique with hydrogels (Baird et al., 2008). This technique allows for the rapid printing of gels with or without cells and is compatible with soluble and matrix-immobilized chemical cues. While this technique is likely limited to screening assays, the knowledge imparted is important for a rational approach to the design of larger-scale constructs.

HYDROGELS AS SUBSTRATES FOR PROMOTING CELL ATTACHMENT, GROWTH, AND DIFFERENTIATION

It is increasingly clear that both physical (e.g. mechanical properties and topographical cues) as well as chemical cues have a significant impact on cell functions such as attachment, proliferation, migration, and differentiation. As described above, one advantage of hydrogel systems is the ability to achieve three-dimensional constructs of materials and cells that begin to recapitulate the architecture of native tissue. Because their high water content and low protein adsorption typically lead to poor cell attachment (in the absence of native or added binding motifs), these systems also provide the ability to isolate specific effects such as a particular ligand or a range of mechanical properties. That is, they provide the means to conduct systematic evaluations of certain biological ligands, combinations of ligands, or the effects of specific mechanical properties on cell response. Hydrogels have many parameters than can be altered to modulate and study the environmental cues associated with the material. These alterable parameters can generally be divided into chemical and physical parameters. A summary of approaches to modulate these parameters is shown in Figure 35.3.

There are two basic methods to provide chemical cues to cells via hydrogels: soluble factor delivery and matrix-immobilized presentation. Parameters and methods that control soluble factor delivery are described on pages 642–644. The primary function of matrix-bound cues is to promote cell attachment and to provide signaling cues to direct cell behavior. The role of specific ligands is discussed elsewhere in this text, but Table 35.2 lists specific examples of matrix-immobilized molecules on hydrogels.

It is noteworthy that most of the gels shown in Table 35.2 are based on poly(ethylene glycol) or other gels to which cell attachment is poor. One approach to achieving better cell attachment and biological function is to mix hydrogels that have some inherent biological activity with those that do not to provide a type of composite material that has desirable properties (La Gatta et al., 2009). An approach that is slightly more complex but can provide greater control over the presentation of binding molecules is to incorporate cell binding motifs or other cell-guiding peptides via surface immobilization into gels that lack these components. It should be noted, however, that hydrogels based on natural materials that provide good cell attachment (e.g. collagen) can benefit from several of these techniques. For example, the use of gradient techniques should be more broadly applicable to all hydrogels in an effort to direct cell migration, and photolithographic approaches to pattern hydrogels are important for achieving spatial organization of multiple cell type tissue constructs. Lastly, the spatial distribution of specific cell types is an important consideration in attempts to create *de novo* complex tissues due to well-known physiological structure-function relationships. The ability to achieve cell-specific attachment in a spatially defined fashion is key to achieving this physiology. As such, spatial distribution of peptides or proteins that promote the selective (or preferably specific) attachment of certain cell types based on the principles highlighted in Table 35.2 is also of considerable importance (Hubbell et al., 1991; Mann and West, 2002).

Approaches to achieve surface immobilization include selective adsorption and covalent crosslinking. Selective adsorption is difficult to achieve for protein molecules on hydrogels due to their low protein adsorption. Certain hydrogels may allow adsorption through electrostatic interactions and larger particles (see PEI–DNA complexes in Table 35.2) allow more interactions with the material surface, thereby promoting adsorption through multiple weak interactions. Non-covalent but strong interactions such as avidin-biotin crosslinking has also been used as it is a straightforward process without reaction by-products. However, covalent crosslinking is clearly the most widely used approach to achieving matrix immobilization of bioactive molecules.

For covalent crosslinking, the molecule of interest may be reacted with a chemically labile side group on the hydrogel in the solution phase prior to gelation. Photo-crosslinking is now more widely used, particularly for cellular applications to avoid chemical reaction by products and to prevent loss of activity of the bioactive molecule during the gelation process. Most hydrogels are compatible to some degree with photo-crosslinking as they are somewhat translucent. Thicker constructs, however, may require the use of layered constructs to achieve desired coupling throughout the hydrogel. Photo-crosslinking is also useful as it is compatible with lithographic techniques that allow control over cell spatial distribution. References to representative papers on these approaches are provided in Table 35.2.

In addition to their compatibility with soluble and matrix-bound chemical cues, parameters of hydrogels can be altered to change the mechanical properties of the gel. Table 35.3 lists several of these parameters. Changes in these parameters clearly affect the mechanical properties of the gels, which in turn affect cell systems. For example, low moduli PEG gels have been shown to promote markers of early cardiomyocyte differentiation in an embryonal carcinoma cell

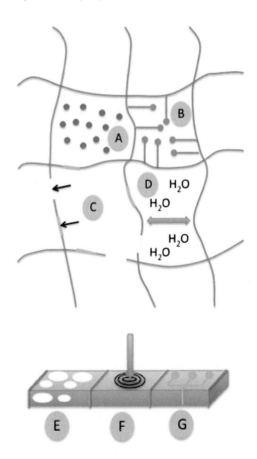

FIGURE 35.3

Schematic highlighting several key mechanisms and properties of hydrogels important in regenerative medicine applications. Additional details of these are provided in the text. Delivery of growth factors and other compounds may be provided via soluble factors (A) or matrix-bound cues (B). Delivery of soluble factors may also be controlled by the rate of degradation of the hydrogel (C) through hydrolysis of certain regions (arrows) or other specific enzymatic mechanisms. The water content and swelling (D) also affect soluble factor delivery, protein adsorption, and cell attachment. The porosity of the scaffold (E) and mechanical properties (F) such as compressive modulus are known to have significant effects on cellular infiltration. Lastly, hydrogels can be modified through numerous patterning techniques such as photolithography to provide micropatterned substrates to assess and regulate the microenvironment of cells.

TABLE 35.2 Examples and Applications of Matrix-immobilized Ligands with Hydrogel Systems

Application	Hydrogel component(s)	Ligand and immobilization method	Reference(s)
Cell attachment	Poly(ethylene glycol) (PEG)	Arginine-glycine-aspartic acid (RGD) modified with acryol groups for reaction with poly(ethylene diacrylate) hydrogel	Hern and Hubbell, 1998
Vascularization of three-dimensional constructs	Collagenase-degradable PEG	Vascular endothelial growth factor (VEGF) and arginine-glycine-aspartic acid-serine (RGDS) covalently coupled to PEG by attaching acry-moiety to ligand for reaction to PEG during polymerization	Leslie-Barbick et al., 2009
Improve cell adhesion to non-adhesive gels	PEG	Collagen-mimetic peptide co-polymerized into hydrogel	Lee et al., 2006
Delivery of plasmid DNA transgene growth factor expression	Hyaluronic acid/PEG-collagen	Polyethylenimine-DNA complexes coupled via neutravidin-biotin or non-specific adsorption	Segura et al., 2005
Delivery of plasmid DNA transgene growth factor expression	Fibrin	Polyethylenimine-DNA complexes adsorbed non-specifically with high affinity	Saul et al., 2007
Guided tissue regeneration with focus on bone	Oligo-poly(ethylene glycol)-fumarate	Osteopontin-derived peptide covalently coupled to hydrogel via PEG linker	Shin et al., 2004
Cell attachment to hydrogels	Proteolytically-degradable PEG	RGDS and other cell-adhesive peptides conjugated to PEG linker and photocoupled to hydrogel	Mann et al., 2001; Mann and West, 2002
Cell attachment for soft tissue constructs in trauma applications	Alginate	RGD covalently coupled to alginate backbone	Halberstadt et al., 2002
Osteoblast adhesion for bone tissue engineering	Poly(propylene fumarate-co-ethylene glycol)	RGDS covalently coupled to hydrogel via PEG linker	Behravesh et al., 2003
Directing cell attachment and migration	PEG	RGDS or basic fibroblast growth factor (bFGF) covalently immobilized to hydrogel in gradient fashion	Delong et al., 2005a, b
Cell adhesion and neurite outgrowth	Agarose	RGDS covalently immobilized to hydrogel via photo-cross-linking	Luo and Shoichet, 2004
Cell binding and vascularization of tissue constructs	PEG	RGDS and vascular endothelial growth factor covalently coupled to hydrogel via photo-cross-linker in pattern defined by photo-masking lithography	Moon et al., 2009

649

model (Kraehenbuehl et al., 2008). Proliferation of neural stem cells has been shown to be inversely proportional to the matrix modulus of alginate hydrogels (Banerjee et al., 2009). Matrix modulus in conjunction with mechanical stimulation has been shown to have effects on chondrocyte morphology in PEG gels (Bryant et al., 2004; Villanueva et al., 2009). Other parameters that affect the mechanical properties and the cellular response include the porosity of the hydrogel, which can be modulated through the use of porogens. Effects of the porosity have also been demonstrated to affect mesenchymal stem cell proliferation (Dadsetan et al.,

2008) and mineralization in osteogenic medium (Keskar et al., 2009). Thus, although chemical regulatory mechanisms are perhaps better understood, it is clear that the mechanical nature of the materials must be given equal consideration.

APPLICATIONS OF HYDROGELS IN REGENERATIVE MEDICINE

As described above, hydrogels are applicable for fundamental studies on the role of topographical patterning, microfluid flow, and high-throughput screening. They are also in use or under investigation as a repair or replacement strategy for virtually all tissues including both hard and soft tissues. It is not possible to describe every tissue and application, but the goal of this section is to highlight several areas in which hydrogels are used in a translational light.

Skin and wound healing

The FDA's searchable database (1979 to present) indicates that more than 510,000 applications have been granted for the use of hydrogels in treating wounds than any for other medical device application. This is in part due to the shear number of wound healing products as well as the overlap in mechanical property similarities between skin and most hydrogels. Beyond the mechanical properties, there are numerous advantages to the use of hydrogels in these systems. Namely, hydrogels for wound healing applications are:

- Permeable to oxygen to prevent necrosis of remaining and newly forming tissue.
- Able to retain moisture at the site of injury to promote healing.
- Compatible with therapeutic agents important to wound healing including antimicrobials, steroids, and growth factors.
- Able to stimulate cellular response and infiltration with minimal foreign body response and subsequent scarring.

An FDA-approved dressing that makes use of a number of concepts described earlier is Oxyzyme, and there are other similar market-approved products. This proprietary hydrogel contains glucose oxidase, which, in the presence of oxygen, leads to the formation of gluconic acid and hydrogen peroxide. The hydrogen peroxide decomposes to water and oxygen, thereby further improving the oxygenation of the tissue. This product also contains iodine to inhibit bacterial growth and contamination of the wound (Thorn et al., 2006; Queen et al., 2007). Preclinically, bilayered chitosan hydrogels have been used in a third degree porcine burn model (Boucard et al., 2007) and the application of matrix-immobilized fibronectin on hyaluronic acid has been shown to promote fibroblastic wound healing response (Ghosh et al., 2006). These studies indicate that the next generation of clinical hydrogel wound dressings will utilize current-generation research principles.

Musculoskeletal

Some of the earliest-envisioned applications for hydrogels in tissue engineering involved replacements for gel-like tissues such as articular cartilage (Corkhill et al., 1990; Oka et al., 1990; Noguchi et al., 1991). These tissues have high levels of glycosaminoglycans and therefore have a high water content, making them a type of natural hydrogel. Hydrogels encapsulating chondrogenic cells are a promising approach to cartilage repair. Examples of chondrocyte-encapsulating materials include alginate (Elisseeff et al., 2002), oligo(polyethylene glycol) fumarate (OPF) with soluble TGFβ1 delivered from gelatin microspheres (Park et al., 2005), and fibrin-PEG-poly(lactic acid) hydrogels, which are currently in large animal trials (Lind et al., 2008). Indeed, for articular cartilage, an extremely wide range of hydrogels has been used, in part because suitable treatments have not be found. An interesting approach that is applicable to cartilage repair is the use of adhesive hydrogels that contain chondroitin sulfate in combination with methacrylate and aldehydes to bridge the material with native tissue (Wang et al., 2007).

TABLE 35.3 Effect of Hydrogel Parameters on Mechanical Properties

Parameter	Effect	Relevant reference(s)
Water content	Inversely correlated with crosslink density. Decreased water content leads to increase in compressive modulus	Metters et al., 2000; Bryant and Anseth, 2002
Crosslink density	Inversely correlated with water content. Increased crosslink density leads to increase in compressive modulus	Metters et al., 2000; Bryant et al., 2004; Kuo and Ma, 2008; Villanueva et al., 2009
Degradation	Decrease in compressive modulus as gel degrades	Bryant and Anseth, 2002
Molecular weight	Increase in molecular weight for oligo (polyethylene glycol) leads to decrease in tensile modulus, increase in toughness, and a higher percent elongation	Temenoff et al., 2002
Charge	For anionic hydrogels, increasing surface charge leads to swelling and decrease in modulus	Beebe et al., 2000; Yew et al., 2007[*]; Shang et al., 2008
Electrical field	Hydrogel deswells with application of electrical field leading to increase in modulus	Yew et al., 2007[.*]; Shang et al., 2008

[*]Yew et al. (2007) is a theoretical model based on experimental results from Beebe et al. (2000).

Hydrogels do not possess the mechanical properties of native bone tissue and are not applied directly for bridging critically sized gaps without fixation. However, the controlled release capabilities of hydrogels have been widely exploited for the delivery of growth factors (and plasmid DNA encoding for growth factors) that promote robust bone formation. For cell-free hydrogels, this generally involves delivery of bone morphogenetic protein 2 (BMP-2) or 7 (BMP-7 also known as OP-1) as these compounds are FDA approved. Recent examples of this approach include delivery of BMP-2 via an *in situ* gelling hyaluronic acid/polyvinyl alcohol hydrogel (Bergman et al., 2009), and BMP-2 delivery from a gelatin hydrogel for a rabbit segmental bone defect model (Yamamoto et al., 2006) and regeneration of skull tissue in non-human primates (Takahashi et al., 2007). It might be argued that any material that delivers the potent mitogen BMP-2 will lead to successful bone regeneration. However, hydrogels or other materials that minimize ectopic bone formation meet an important criteria for bone regeneration systems. The controlled release of BMP-2 on timescales most beneficial for promoting regeneration is also important. Unfortunately, this timescale is not well defined, though it appears that there exists a threshold level of BMP-2 that can achieve bone regeneration.

Neural regeneration

Hydrogels are becoming more widely used with encapsulation of Schwann cells and neural progenitor cells to promote neural regeneration via cell-based trophic support. However, hydrogels have been and remain widely investigated as stand-alone materials for neural regeneration applications as well. This is because the regeneration of neuronal axons is different from many other tissues in that the cell body is typically located proximally to the site of injury, requiring migration of a part of the cell (i.e. the axon) to and then through the site of injury.

In the peripheral nervous system, hydrogels are under investigation as fillers for nerve conduits composed of natural materials such as collagen (e.g. Neuragen Nerve Guide) or synthetic materials (e.g. Silastic). Examples of hydrogels used as fillers include agarose, fibrin, and keratin, among others. The role of charge (Dillon et al., 1998) and mechanical properties

651

(Balgude et al., 2001) have been elucidated for peripheral nerve systems. Some natural materials appear to promote regeneration simply through the provision of the physical matrix (Sierpinski et al., 2008), which allows for Schwann cell infiltration and axonal extension through gels. Others have developed hydrogel scaffolds of higher mechanical integrity that are also promising for neural regeneration through the provision of physical guidance cues (Flynn et al., 2003; Bozkurt et al., 2007). Glial and neuronal axon migration through hydrogels has also been enhanced through delivery of soluble growth factors such as NGF from fibrin (Wood and Sakiyama-Elbert, 2008) as well as through the presentation of laminin or laminin-based peptides that are matrix-immobilized (Yu et al. 1999; Dodla and Bellamkonda, 2006). Hydrogels have also been created to provide topographical cues in three dimensions alone or in combination with matrix-bound ligands (Flynn et al., 2003; Yu and Shoichet, 2005), indicating the integration of chemical and mechanical properties in guiding neural regeneration.

These principles are also applicable to spinal cord injuries (SCIs). However, injectable gel systems are advantageous in the case of SCIs as complete transection of the spinal cord is rare and materials that can be injected in a less invasive fashion are desirable. Those hydrogels that can gel *in situ* may be particularly appealing as they are injectable yet provide additional mechanical integrity (Jain et al., 2006).

Liver

The mechanical properties of hydrogels could indicate that they would be best suited to the development of the visceral organs. Indeed, hydrogels are widely used both for fundamental studies at the cell level and in more translational applications toward the development of functional units. As indicated in the previous sections, hydrogels are used extensively as controlled release systems, often for the delivery of factors to promote angiogenesis. Due to diffusional limitations in large tissues, one approach has been the formation of smaller functional constructs. The simplest example of this is the microencapsulation approach described above.

More complex approaches are being developed toward the creation of functional tissues through provision of the correct structural properties of these tissues. In one example of this, hepatocytes encapsulated in thin collagen gel sheets have been shown to be vascularized and demonstrate functional markers (Zhao et al., 2009). In a more sophisticated approach, PEG hydrogels were functionalized to promote cell adhesion and processed into complex shapes reminiscent of native liver structure, demonstrating functionality (Liu Tsang et al., 2007). In each case, the hydrogels serve the role of promoting cell engraftment through encapsulation and also likely provide for improved diffusional profiles to allow for bioreactor perfusion (Liu Tsang et al., 2007) and vascularization (Zhao et al., 2009). These approaches are certainly applicable to other visceral organs. Recently, multi-layered hydrogels have been developed from natural hydrogels such as alginate and hyaluronic acid (Ladet et al., 2008) as well as synthetic materials (Kizilel et al., 2006). Such multi-layer systems provide the opportunity to juxtapose cells in structural orientations that mimic native tissue.

The ideal approach to the generation of complex tissues such as the liver and other visceral organs certainly remains in doubt, but it is clear that hydrogel-based technologies such as these of creating small functional units as well as other encapsulation and printing technologies provide potential solutions alone or in combination with other approaches.

Reproductive medicine

An area in which the mechanical properties of hydrogels has recently proven useful is in the tissue engineering of ovarian follicles. Females facing radiotherapy or chemotherapy are at increased risk of future infertility. The ability to grow and preserve immature ovarian follicles in culture has been suggested as an approach to minimize the risk of reintroducing cancer cells to patients as may occur with the use of ovarian tissue cryopreservation (Xu et al., 2006a). Alginate hydrogels have been used to provide three-dimensional context to immature ovarian follicles, allowing growth *in vitro* and also allowing for stable cryopreservation (Amorim et al., 2009). This approach has resulted in the live birth of fertilized embryos in a mouse model (Xu et al., 2006a). Early studies have utilized alginate due to its mechanical properties and compatibility with cell encapsulation as described above. Recently, alginate-fibrin hydrogels have shown increased numbers of meiotically competent oocytes (Shikanov et al., 2009), indicating that optimization of the hydrogel chemical and mechanical properties may lead to further improvements in this promising technique.

PROSPECTS AND CONCLUSIONS

Hydrogels are clearly an important component in the repertoire of biomaterial approaches to regenerative medicine with applications in nearly every organ system currently under exploration. The ability to modify hydrogels with bioactive molecules has proven fortuitous in the era of bioactive compound delivery as it allows for the integration of mechanical properties that mimic native tissue with bound and soluble chemical cues. In looking to the future of hydrogels in regenerative medicine applications, there are several opportunities for the development of these systems.

It should be clear from this chapter that there is a very large number of material parameters that can be altered for numerous applications. While many hydrogels allow a systematic evaluation of these parameters in isolation, the ability to compile a set of parameters that are optimal or near optimal for a particular application is not readily accomplished. The use of statistical analysis and parametric analysis, and the development of sophisticated engineering models in combination with some of the high-throughput screening methods described will be important in the design of regenerative medicine therapies involving hydrogels (Comisar et al., 2007).

The use and characterization of hydrogels for the delivery of therapeutic agents that behave differently from traditionally delivered molecules is also of increasing importance. Examples include the use of hydrogels for the delivery of nucleic acids or gaseous materials. The advent of the gene-activated matrix (GAM) has spurred methods to achieve delivery of DNA, RNA, antisense, siRNA, or other nucleotides from hydrogels. Efficient delivery of these molecules to their cellular site of action has been of interest to the gene delivery community, and recent approaches to promoting viral (Schek et al., 2004) or non-viral (Segura et al., 2005) delivery of nucleic acids from hydrogel scaffolds (Kasper et al., 2006) will likely take on increasing importance. The lack of oxygen and nutrient supply to tissue-engineered constructs is also well documented (Johnson et al., 2007). While delivery of factors promoting vascularization is widely studied, other alternatives such as the direct delivery of oxygen have been investigated more recently (Oh et al., 2009). The role of hydrogels in regulating release of these types of molecules will also be an important consideration.

Another challenge in regard to hydrogels is obtaining the mechanical integrity necessary for tissue engineering applications. As described previously, one advantage of hydrogel systems is that they provide a matrix comparable in modulus to native extracellular matrix for many soft tissues. However, the mechanical integrity of many commonly used biodegradable hydrogels diminishes with time and the starting properties may be insufficient to withstand mechanically challenging bioreactor systems that are increasingly used for perfusion as well as mechanical and electrical stimulation. Increasing the polymer concentration within the gels provides additional mechanical support but reduces nutrient and oxygen diffusion, thereby inhibiting cell viability with the material. One approach to providing mechanical stability while promoting cell infiltration into hydrogels is through the use of templating approaches. These approaches include the use of micellar structures (Texter, 2009) and sacrificial polymer beads

(Linnes et al., 2007), or other geometries (Flynn et al., 2003; Lam et al., 2009) that mimic native tissue architecture. This approach promotes cell infiltration or encapsulation throughout the scaffold and allows diffusion to occur in a fashion similar to macroporous hydrogels. This approach may provide the mechanical integrity necessary for bioreactors such as those used for ligament (Noth et al., 2005; Kahn et al., 2008), tendon (Saber et al., 2010), and muscle (Moon et al., 2008). To date, limited hydrogels have been used in these mechanically challenging systems (Pfister et al., 2006; Nicodemus and Bryant, 2008). However, modifications to natural polymers to allow bioreactor pre-conditional in conjunction with their advantageous cell adhesion and differentiation motifs may prove of great utility in regenerative medicine applications.

It is difficult to generalize the host response to hydrogels due to differences in the chemistry of the materials, leachates, processing techniques, and methods of sterilization. However, most experts would agree that the difference in the host response to hydrogels compared to traditional polymer systems was and remains a key point of interest for these materials. Monocyte and macrophage response to traditional polymers involves protein adsorption, monocyte/macrophage response, and (typically) fibrous encapsulation. This is generally considered an acceptable response for a material (Williams, 1987). However, hydrogels may be well positioned to exploit the body's native machinery to achieve a more desirable host response to the material and cells or tissues associated with it. For example, it has been suggested that certain biological hydrogels and porous hydrogels may promote an alternative macrophage phenotype that is anti-inflammatory and may even lead to recruitment of progenitor cell populations that can transdifferentiate for the promotion of tissue remodeling within and around the hydrogel (Lee et al., 2008; Piterina et al., 2009; Ratner and Atzet, 2009). While further studies are needed to elucidate the mechanisms, the methods described in this chapter provide the tools to harness and manipulate these processes.

References

Aboushwareb, T., Eberli, D., Ward, C., Broda, C., Holcomb, J., Atala, A., et al. (2009). A keratin biomaterial gel hemostat derived from human hair: evaluation in a rabbit model of lethal liver injury. *J. Biomed. Mater. Res. B Appl. Biomater., 90*, 45−54.

Abousleiman, R. I., Reyes, Y., McFetridge, P., & Sikavitsas, V. (2008). Tendon tissue engineering using cell-seeded umbilical veins cultured in a mechanical stimulator. *Tissue Eng. A, 15*, 787−795.

Albrecht, D. R., Underhill, G. H., Mendelson, A., & Bhatia, S. N. (2007). Multiphase electropatterning of cells and biomaterials. *Lab. Chip., 7*, 702−709.

Alsberg, E., Kong, H. J., Hirano, Y., Smith, M. K., Albeiruti, A., & Mooney, D. J. (2003). Regulating bone formation via controlled scaffold degradation. *J. Dent. Res., 82*, 903−908.

Amorim, C. A., van Langendonckt, A., David, A., Dolmans, M. M., & Donnez, J. (2009). Survival of human pre-antral follicles after cryopreservation of ovarian tissue, follicular isolation and *in vitro* culture in a calcium alginate matrix. *Hum. Reprod., 24*, 92−99.

Aoki, H., Tomita, N., Morita, Y., Hattori, K., Harada, Y., Sonobe, M., Wakitani, S., et al. (2003). Culture of chondrocytes in fibroin-hydrogel sponge. *Biomed. Mater. Eng., 13*, 309−316.

Arrighi, I., Mark, S., Alvisi, M., von Rechenberg, B., Hubbell, J. A., & Schense, J. C. (2009). Bone healing induced by local delivery of an engineered parathyroid hormone prodrug. *Biomaterials, 30*, 1763−1771.

Atzet, S., Curtin, S., Trinh, P., Bryant, S., & Ratner, B. (2008). Degradable poly(2-hydroxyethyl methacrylate)-co-polycaprolactone hydrogels for tissue engineering scaffolds. *Biomacromolecules, 9*, 3370−3377.

Ayhan, F., & Ozkan, S. (2007). Gentamicin release from photopolymerized PEG diacrylate and pHEMA hydrogel discs and their *in vitro* antimicrobial activities. *Drug Deliv., 14*, 433−439.

Bader, R. A., & Rochefort, W. E. (2008). Rheological characterization of photopolymerized poly(vinyl alcohol) hydrogels for potential use in nucleus pulposus replacement. *J. Biomed. Mater. Res. A, 86*, 494−501.

Bae, K. H., Yoon, J. J., & Park, T. G. (2006). Fabrication of hyaluronic acid hydrogel beads for cell encapsulation. *Biotechnol. Prog., 22*, 297−302.

Baird, I. S., Yau, A. Y., & Mann, B. K. (2008). Mammalian cell-seeded hydrogel microarrays printed via dip-pin technology. *Biotechniques, 44*, 249−256.

Balgude, A. P., Yu, X., Szymanski, A., & Bellamkonda, R. V. (2001). Agarose gel stiffness determines rate of DRG neurite extension in 3D cultures. *Biomaterials, 22,* 1077–1084.

Banerjee, A., Arha, M., Choudhary, S., Ashton, R. S., Bhatia, S. R., Schaffer, D. V., et al. (2009). The influence of hydrogel modulus on the proliferation and differentiation of encapsulated neural stem cells. *Biomaterials, 30,* 4695–4699.

Bayramoglu, G., Kayaman-Apohan, N., Akcakaya, H., Vezir Kahraman, M., Erdem Kuruca, S., & Gungor, A. (2010). Preparation of collagen modified photopolymers: a new type of biodegradable gel for cell growth. *J. Mater. Sci. Mater. Med., 21,* 761–775.

Beebe, D. J., Moore, J. S., Bauer, J. M., Yu, Q., Liu, R. H., Devadoss, C., et al. (2000). Functional hydrogel structures for autonomous flow control inside microfluidic channels. *Nature, 404,* 588–590.

Behravesh, E., Zygourakis, K., & Mikos, A. G. (2003). Adhesion and migration of marrow-derived osteoblasts on injectable *in situ* crosslinkable poly(propylene fumarate-co-ethylene glycol)-based hydrogels with a covalently linked RGDS peptide. *J. Biomed. Mater. Res. A, 65,* 260–270.

Bergman, K., Engstrand, T., Hilborn, J., Ossipov, D., Piskounova, S., & Bowden, T. (2009). Injectable cell-free template for bone-tissue formation. *J. Biomed. Mater. Res. A, 91,* 1111–1118.

Bian, L., Fong, J. V., Lima, E. G., Stoker, A. M., Ateshian, G. A., Cook, J. L., et al. (2010). Dynamic mechanical loading enhances functional properties of tissue-engineered cartilage using mature canine chondrocytes. *Tissue Eng. A, 16,* 1781–1790.

Boucard, N., Viton, C., Agay, D., Mari, E., Roger, T., Chancerelle, Y., et al. (2007). The use of physical hydrogels of chitosan for skin regeneration following third-degree burns. *Biomaterials, 28,* 3478–3488.

Bozkurt, A., Brook, G. A., Moellers, S., Lassner, F., Sellhaus, B., Weis, J., et al. (2007). *In vitro* assessment of axonal growth using dorsal root ganglia explants in a novel three-dimensional collagen matrix. *Tissue Eng., 13,* 2971–2979.

Brennan, M. (1991). Fibrin glue. *Blood Rev., 5,* 240–244.

Bryant, S. J., & Anseth, K. S. (2002). Hydrogel properties influence ECM production by chondrocytes photo-encapsulated in poly(ethylene glycol) hydrogels. *J. Biomed. Mater. Res., 59,* 63–72.

Bryant, S. J., & Anseth, K. S. (2006). Photopolymerization of hydrogel scaffolds. In J. Elisseeff, & P. X. Ma (Eds.), *Scaffolding in Tissue Engineering* (pp. 71–90). Boca Raton, FL: CRC Press.

Bryant, S. J., Anseth, K. S., Lee, D. A., & Bader, D. L. (2004). Crosslinking density influences the morphology of chondrocytes photoencapsulated in PEG hydrogels during the application of compressive strain. *J. Orthop. Res., 22,* 1143–1149.

Bryant, S. J., Cuy, J. L., Hauch, K. D., & Ratner, B. D. (2007). Photo-patterning of porous hydrogels for tissue engineering. *Biomaterials, 28,* 2978–2986.

Cai, K., Dong, H., Chen, C., Yang, L., Jandt, K. D., & Deng, L. (2009). Inkjet printing of laminin gradient to investigate endothelial cellular alignment. *Colloids Surf. B Biointerfaces, 72,* 230–235.

Campbell, P. G., Miller, E. D., Fisher, G. W., Walker, L. M., & Weiss, L. E. (2005). Engineered spatial patterns of FGF-2 immobilized on fibrin direct cell organization. *Biomaterials, 26,* 6762–6770.

Cavanagh, H. D., Bodner, B. I., & Wilson, L. A. (1980). Extended wear hydrogel lenses. Long-term effectiveness and costs. *Ophthalmology, 87,* 871–876.

Chen, F., Mao, T., Tao, K., Chen, S., Ding, G., & Gu, X. (2003). Injectable bone. *Br. J. Oral Maxillofac. Surg., 41,* 240–243.

Cohen, D. L., Malone, E., Lipson, H., & Bonassar, L. J. (2006). Direct freeform fabrication of seeded hydrogels in arbitrary geometries. *Tissue Eng., 12,* 1325–1335.

Comisar, W. A., Kazmers, N. H., Mooney, D. J., & Linderman, J. J. (2007). Engineering RGD nanopatterned hydrogels to control preosteoblast behavior: a combined computational and experimental approach. *Biomaterials, 28,* 4409–4417.

Corkhill, P. H., Trevett, A. S., & Tighe, B. J. (1990). The potential of hydrogels as synthetic articular cartilage. *Proc. Inst. Mech. Eng. H, 204,* 147–155.

Crescenzi, V., Cornelio, L., di Meo, C., Nardecchia, S., & Lamanna, R. (2007). Novel hydrogels via click chemistry: synthesis and potential biomedical applications. *Biomacromolecules, 8,* 1844–1850.

Cui, X., & Boland, T. (2009). Human microvasculature fabrication using thermal inkjet printing technology. *Biomaterials, 30,* 6221–6227.

Dadsetan, M., Hefferan, T. E., Szatkowski, J. P., Mishra, P. K., Macura, S. I., Lu, L., et al. (2008). Effect of hydrogel porosity on marrow stromal cell phenotypic expression. *Biomaterials, 29,* 2193–2202.

Dadsetan, M., Knight, A. M., Lu, L., Windebank, A. J., & Yaszemski, M. J. (2009). Stimulation of neurite outgrowth using positively charged hydrogels. *Biomaterials, 30,* 3874–3881.

Dalton, P. D., Flynn, L., & Shoichet, M. S. (2002). Manufacture of poly(2-hydroxyethyl methacrylate-co-methyl methacrylate) hydrogel tubes for use as nerve guidance channels. *Biomaterials, 23,* 3843–3851.

DeLong, S. A., Gobin, A. S., & West, J. L. (2005a). Covalent immobilization of RGDS on hydrogel surfaces to direct cell alignment and migration. *J. Control. Release, 109,* 139–148.

DeLong, S. A., Moon, J. J., & West, J. L. (2005b). Covalently immobilized gradients of bFGF on hydrogel scaffolds for directed cell migration. *Biomaterials, 26,* 3227–3234.

Dillon, G. P., Yu, X., Sridharan, A., Ranieri, J. P., & Bellamkonda, R. V. (1998). The influence of physical structure and charge on neurite extension in a 3D hydrogel scaffold. *J. Biomater. Sci. Polym., 9,* 1049–1069.

Dodla, M. C., & Bellamkonda, R. V. (2006). Anisotropic scaffolds facilitate enhanced neurite extension *in vitro. J. Biomed. Mater. Res. A, 78,* 213–221.

Dreifus, M., & Wichterle, O. (1964). Clinical experiences with hydrogel contact lenses. *Cesk. Oftalmol., 20,* 393–399.

Ehrbar, M., Djonov, V. G., Schnell, C., Tschanz, S. A., Martiny-Baron, G., Schenk, U., et al. (2004). Cell-demanded liberation of VEGF121 from fibrin implants induces local and controlled blood vessel growth. *Circ. Res., 94,* 1124–1132.

Elisseeff, J., McIntosh, W., Anseth, K., Riley, S., Ragan, P., & Langer, R. (2000). Photoencapsulation of chondrocytes in poly(ethylene oxide)-based semi-interpenetrating networks. *J. Biomed. Mater. Res., 51,* 164–171.

Elisseeff, J. H., Lee, A., Kleinman, H. K., & Yamada, Y. (2002). Biological response of chondrocytes to hydrogels. *Ann. N.Y. Acad. Sci., 961,* 118–122.

Esguerra, M., Fink, H., Laschke, M. W., Jeppsson, A., Delbro, D., Gatenholm, P., et al. (2010). Intravital fluorescent microscopic evaluation of bacterial cellulose as scaffold for vascular grafts. *J. Biomed. Mater. Res. A, 93,* 140–149.

Farran, A. J., Teller, S. S., Jha, A. K., Jiao, T., Hule, R. A., Clifton, R. J., et al. (2010). Effects of matrix composition, microstructure, and viscoelasticity on the behaviors of vocal fold fibroblasts cultured in three-dimensional hydrogel networks. *Tissue Eng. A, 16,* 1247–1261.

Faxalv, L., Ekblad, T., Liedberg, B., & Lindahl, T. L. (2010). Blood compatibility of photografted hydrogel coatings. *Acta Biomater., 6,* 2599–2608.

Fedorovich, N. E., de Wijn, J. R., Verbout, A. J., Alblas, J., & Dhert, W. J. (2008). Three-dimensional fiber deposition of cell-laden, viable, patterned constructs for bone tissue printing. *Tissue Eng. A, 14,* 127–133.

Fedorovich, N. E., Oudshoorn, M. H., van Geemen, D., Hennink, W. E., Alblas, J., & Dhert, W. J. (2009). The effect of photopolymerization on stem cells embedded in hydrogels. *Biomaterials, 30,* 344–353.

Fini, M., Motta, A., Torricelli, P., Giavaresi, G., Nicoli Aldini, N., Tschon, M., et al. (2005). The healing of confined critical size cancellous defects in the presence of silk fibroin hydrogel. *Biomaterials, 26,* 3527–3536.

Flynn, L., Dalton, P. D., & Shoichet, M. S. (2003). Fiber templating of poly(2-hydroxyethyl methacrylate) for neural tissue engineering. *Biomaterials, 24,* 4265–4272.

Freier, T., Montenegro, R., Shan Koh, H., & Shoichet, M. S. (2005). Chitin-based tubes for tissue engineering in the nervous system. *Biomaterials, 26,* 4624–4632.

Ghosh, K., Ren, X. D., Shu, X. Z., Prestwich, G. D., & Clark, R. A. (2006). Fibronectin functional domains coupled to hyaluronan stimulate adult human dermal fibroblast responses critical for wound healing. *Tissue Eng., 12,* 601–613.

Goldmann, T., & Gonzalez, J. S. (2000). DNA-printing: utilization of a standard inkjet printer for the transfer of nucleic acids to solid supports. *J. Biochem. Biophys. Methods, 42,* 105–110.

Grande, C. J., Torres, F. G., Gomez, C. M., & Bano, M. C. (2009). Nanocomposites of bacterial cellulose/hydroxyapatite for biomedical applications. *Acta Biomater., 5,* 1605–1615.

Grigolo, B., Roseti, L., Fiorini, M., Fini, M., Giavaresi, G., Aldini, N. N., et al. (2001). Transplantation of chondrocytes seeded on a hyaluronan derivative (hyaff-11) into cartilage defects in rabbits. *Biomaterials, 22,* 2417–2424.

Gu, Z., & Tang, Y. (2010). Enzyme-assisted photolithography for spatial functionalization of hydrogels. *Lab Chip., 10,* 1946–1951.

Guo, X., Park, H., Young, S., Kretlow, J. D., van den Beucken, J. J., Baggett, L. S., et al. (2010). Repair of osteochondral defects with biodegradable hydrogel composites encapsulating marrow mesenchymal stem cells in a rabbit model. *Acta Biomater., 6,* 39–47.

Hahn, M. S., Taite, L. J., Moon, J. J., Rowland, M. C., Ruffino, K. A., & West, J. L. (2006). Photolithographic patterning of polyethylene glycol hydrogels. *Biomaterials, 27,* 2519–2524.

Halberstadt, C., Austin, C., Rowley, J., Culberson, C., Loebsack, A., Wyatt, S., et al. (2002). A hydrogel material for plastic and reconstructive applications injected into the subcutaneous space of a sheep. *Tissue Eng., 8,* 309–319.

Hausner, T., Schmidhammer, R., Zandieh, S., Hopf, R., Schultz, A., Gogolewski, S., et al. (2007). Nerve regeneration using tubular scaffolds from biodegradable polyurethane. *Acta Neurochir., 100*(Suppl.), 69–72.

Hern, D. L., & Hubbell, J. A. (1998). Incorporation of adhesion peptides into nonadhesive hydrogels useful for tissue resurfacing. *J. Biomed. Mater. Res., 39,* 266–276.

656

Hesse, E., Hefferan, T. E., Tarara, J. E., Haasper, C., Meller, R., Krettek, C., et al. (2010). Collagen type I hydrogel allows migration, proliferation, and osteogenic differentiation of rat bone marrow stromal cells. *J. Biomed. Mater. Res. A, 94*, 442–449.

Ho, S. T., Cool, S. M., Hui, J. H., & Hutmacher, D. W. (2010). The influence of fibrin based hydrogels on the chondrogenic differentiation of human bone marrow stromal cells. *Biomaterials, 31*, 38–47.

Hoemann, C. D., Sun, J., Legare, A., McKee, M. D., & Buschmann, M. D. (2005). Tissue engineering of cartilage using an injectable and adhesive chitosan-based cell-delivery vehicle. *Osteoarthritis Cartilage, 13*, 318–329.

Hoffman, A. S. (2002). Hydrogels for biomedical applications. *Adv. Drug Deliv. Rev., 54*, 3–12.

Horbett, T. A. (1986). Protein adsorption to hydrogels. In N. A. Peppas (Ed.), *Hydrogels in Medicine and Pharmacy* (pp. 127–171). Boca Raton, FL: CRC Press.

Hsiue, G. H., Lai, J. Y., Chen, K. H., & Hsu, W. M. (2006). A novel strategy for corneal endothelial reconstruction with a bioengineered cell sheet. *Transplantation, 81*, 473–476.

Hubbell, J. A., Massia, S. P., Desai, N. P., & Drumheller, P. D. (1991). Endothelial cell-selective materials for tissue engineering in the vascular graft via a new receptor. *Biotechnology (NY), 9*, 568–572.

Ilkhanizadeh, S., Teixeira, A. I., & Hermanson, O. (2007). Inkjet printing of macromolecules on hydrogels to steer neural stem cell differentiation. *Biomaterials, 28*, 3936–3943.

Jain, A., Kim, Y. T., McKeon, R. J., & Bellamkonda, R. V. (2006). In situ gelling hydrogels for conformal repair of spinal cord defects, and local delivery of BDNF after spinal cord injury. *Biomaterials, 27*, 497–504.

Jeong, B., Kim, S. W., & Bae, Y. H. (2002). Thermosensitive sol–gel reversible hydrogels. *Adv. Drug Deliv. Rev., 54*, 37–51.

Ji, Y., Ghosh, K., Li, B., Sokolov, J. C., Clark, R. A., & Rafailovich, M. H. (2006). Dual-syringe reactive electrospinning of crosslinked hyaluronic acid hydrogel nanofibers for tissue engineering applications. *Macromol. Biosci., 6*, 811–817.

Jin, S. Y., Lei, L., Shikanov, A., Shea, L. D., & Woodruff, T. K. (2010). A novel two-step strategy for *in vitro* culture of early-stage ovarian follicles in the mouse. *Fertil Steril., 93*, 2633–2639.

Johnson, P. C., Mikos, A. G., Fisher, J. P., & Jansen, J. A. (2007). Strategic directions in tissue engineering. *Tissue Eng., 13*, 2827–2837.

Kahn, C. J., Vaquette, C., Rahouadj, R., & Wang, X. (2008). A novel bioreactor for ligament tissue engineering. *Biomed. Mater. Eng., 18*, 283–287.

Kalbermatten, D. F., Pettersson, J., Kingham, P. J., Pierer, G., Wiberg, M., & Terenghi, G. (2009). New fibrin conduit for peripheral nerve repair. *J. Reconstr. Microsurg., 25*, 27–33.

Kasper, F. K., Jerkins, E., Tanahashi, K., Barry, M. A., Tabata, Y., & Mikos, A. G. (2006). Characterization of DNA release from composites of oligo(poly(ethylene glycol) fumarate) and cationized gelatin microspheres *in vitro. J. Biomed. Mater. Res. A, 78*, 823–835.

Keskar, V., Marion, N. W., Mao, J. J., & Gemeinhart, R. A. (2009). *In vitro* evaluation of macroporous hydrogels to facilitate stem cell infiltration, growth, and mineralization. *Tissue Eng. A, 15*, 1695–1707.

Kim, D. H., Kim, P., Song, I., Cha, J. M., Lee, S. H., Kim, B., et al. (2006). Guided three-dimensional growth of functional cardiomyocytes on polyethylene glycol nanostructures. *Langmuir, 22*, 5419–5426.

Kim, J., Kim, I. S., Cho, T. H., Lee, K. B., Hwang, S. J., Tae, G., et al. (2007). Bone regeneration using hyaluronic acid-based hydrogel with bone morphogenic protein-2 and human mesenchymal stem cells. *Biomaterials, 28*, 1830–1837.

Kizilel, S., Sawardecker, E., Teymour, F., & Perez-Luna, V. H. (2006). Sequential formation of covalently bonded hydrogel multilayers through surface initiated photopolymerization. *Biomaterials, 27*, 1209–1215.

Kopecek, J. (2007). Hydrogel biomaterials: a smart future? *Biomaterials, 28*, 5185–5192.

Kraehenbuehl, T. P., Zammaretti, P., van der Vlies, A. J., Schoenmakers, R. G., Lutolf, M. P., Jaconi, M. E., et al. (2008). Three-dimensional extracellular matrix-directed cardioprogenitor differentiation: systematic modulation of a synthetic cell-responsive PEG-hydrogel. *Biomaterials, 29*, 2757–2766.

Kuo, C. K., & Ma, P. X. (2008). Maintaining dimensions and mechanical properties of ionically crosslinked alginate hydrogel scaffolds *in vitro. J. Biomed. Mater. Res. A, 84*, 899–907.

Kwok, C. S., Mourad, P. D., Crum, L. A., & Ratner, B. D. (2001). Self-assembled molecular structures as ultrasonically responsive barrier membranes for pulsatile drug delivery. *J. Biomed. Mater. Res., 57*, 151–164.

La Gatta, A., Schiraldi, C., Esposito, A., d'Agostino, A., & de Rosa, A. (2009). Novel poly(HEMA-co-METAC)/alginate semi-interpenetrating hydrogels for biomedical applications: synthesis and characterization. *J. Biomed. Mater. Res. A, 90*, 292–302.

Ladet, S., David, L., & Domard, A. (2008). Multi-membrane hydrogels. *Nature, 452*, 76–79.

Lai, J. Y., Chen, K. H., & Hsiue, G. H. (2007). Tissue-engineered human corneal endothelial cell sheet transplantation in a rabbit model using functional biomaterials. *Transplantation, 84*, 1222–1232.

Lam, M. T., Huang, Y. C., Birla, R. K., & Takayama, S. (2009). Microfeature guided skeletal muscle tissue engineering for highly organized 3-dimensional free-standing constructs. *Biomaterials, 30,* 1150–1155.

Langer, R. (1990). New methods of drug delivery. *Science, 249,* 1527–1533.

Lee, H. J., Lee, J. S., Chansakul, T., Yu, C., Elisseeff, J. H., & Yu, S. M. (2006). Collagen mimetic peptide-conjugated photopolymerizable PEG hydrogel. *Biomaterials, 27,* 5268–5276.

Lee, S. J., van Dyke, M., Atala, A., & Yoo, J. J. (2008). Host cell mobilization for *in situ* tissue regeneration. *Rejuvenation Res., 11,* 747–756.

Lemmo, A. V., Rose, D. J., & Tisone, T. C. (1998). Inkjet dispensing technology: applications in drug discovery. *Curr. Opin. Biotechnol., 9,* 615–617.

Leslie-Barbick, J. E., Moon, J. J., & West, J. L. (2009). Covalently-immobilized vascular endothelial growth factor promotes endothelial cell tubulogenesis in poly(ethylene glycol) diacrylate hydrogels. *J. Biomater. Sci. Polym., 20,* 1763–1779.

Lim, F., & Moss, R. D. (1981). Microencapsulation of living cells and tissues. *J. Pharm. Sci., 70,* 351–354.

Lim, F., & Sun, A. M. (1980). Microencapsulated islets as bioartificial endocrine pancreas. *Science, 210,* 908–910.

Lin, R. Z., Ho, C. T., Liu, C. H., & Chang, H. Y. (2006). Dielectrophoresis based-cell patterning for tissue engineering. *Biotechnol. J., 1,* 949–957.

Lind, M., Larsen, A., Clausen, C., Osther, K., & Everland, H. (2008). Cartilage repair with chondrocytes in fibrin hydrogel and MPEG polylactide scaffold: an *in vivo* study in goats. *Knee Surg. Sports Traumatol. Arthrosc., 16,* 690–698.

Linnes, M. P., Ratner, B. D., & Giachelli, C. M. (2007). A fibrinogen-based precision microporous scaffold for tissue engineering. *Biomaterials, 28,* 5298–5306.

Liu Tsang, V., Chen, A. A., Cho, L. M., Jadin, K. D., Sah, R. L., DeLong, S., et al. (2007). Fabrication of 3D hepatic tissues by additive photopatterning of cellular hydrogels. *Faseb. J., 21,* 790–801.

Lowman, A. M., & Peppas, N. A. (1999). Hydrogels. In E. Mathiowitz (Ed.), *Encyclopedia of Controlled Drug Delivery* (pp. 397–418). New York: Wiley.

Luo, Y., & Shoichet, M. S. (2004). Light-activated immobilization of biomolecules to agarose hydrogels for controlled cellular response. *Biomacromolecules, 5,* 2315–2323.

Lutolf, M. P., Lauer-Fields, J. L., Schmoekel, H. G., Metters, A. T., Weber, F. E., Fields, G. B., et al. (2003). Synthetic matrix metalloproteinase-sensitive hydrogels for the conduction of tissue regeneration: engineering cell-invasion characteristics. *Proc. Natl. Acad. Sci. U.S.A., 100,* 5413–5418.

Mahoney, M. J., & Anseth, K. S. (2006). Three-dimensional growth and function of neural tissue in degradable polyethylene glycol hydrogels. *Biomaterials, 27,* 2265–2274.

Malkoch, M., Vestberg, R., Gupta, N., Mespouille, L., Dubois, P., Mason, A. F., et al. (2006). Synthesis of well-defined hydrogel networks using click chemistry. *Chem. Commun. (Camb.),* 2774–2776.

Mann, B. K., & West, J. L. (2002). Cell adhesion peptides alter smooth muscle cell adhesion, proliferation, migration, and matrix protein synthesis on modified surfaces and in polymer scaffolds. *J. Biomed. Mater. Res., 60,* 86–93.

Mann, B. K., Gobin, A. S., Tsai, A. T., Schmedlen, R. H., & West, J. L. (2001). Smooth muscle cell growth in photopolymerized hydrogels with cell adhesive and proteolytically degradable domains: synthetic ECM analogs for tissue engineering. *Biomaterials, 22,* 3045–3051.

Mauck, R. L., Soltz, M. A., Wang, C. C., Wong, D. D., Chao, P. H., Valhmu, W. B., et al. (2000). Functional tissue engineering of articular cartilage through dynamic loading of chondrocyte-seeded agarose gels. *J. Biomech. Eng., 122,* 252–260.

Metters, A. T., Anseth, K. S., & Bowman, C. N. (2000). Fundamental studies of a novel, biodegradable PEG-b-PLA hydrogel. *Polymer., 41,* 3993–4004.

Miller, E. D., Phillippi, J. A., Fisher, G. W., Campbell, P. G., Walker, L. M., & Weiss, L. E. (2009). Inkjet printing of growth factor concentration gradients and combinatorial arrays immobilized on biologically relevant substrates. *Comb. Chem. High Throughput Screen, 12,* 604–618.

Moon du, G., Christ, G., Stitzel, J. D., Atala, A., & Yoo, J. J. (2008). Cyclic mechanical preconditioning improves engineered muscle contraction. *Tissue Eng. A, 14,* 473–482.

Moon, J. J., Hahn, M. S., Kim, I., Nsiah, B. A., & West, J. L. (2009). Micropatterning of poly(ethylene glycol) diacrylate hydrogels with biomolecules to regulate and guide endothelial morphogenesis. *Tissue Eng. A, 15,* 579–585.

Moutos, F. T., Freed, L. E., & Guilak, F. (2007). A biomimetic three-dimensional woven composite scaffold for functional tissue engineering of cartilage. *Nat. Mater., 6,* 162–167.

Murdan, S. (2003). Electro-responsive drug delivery from hydrogels. *J. Control. Rel., 92,* 1–17.

Neuffer, M. C., McDivitt, J., Rose, D., King, K., Cloonan, C. C., & Vayer, J. S. (2004). Hemostatic dressings for the first responder: a review. *Mil. Med., 169,* 716–720.

Ng, K. W., Wang, C. C., Mauck, R. L., Kelly, T. A., Chahine, N. O., Costa, K. D., et al. (2005). A layered agarose approach to fabricate depth-dependent inhomogeneity in chondrocyte-seeded constructs. *J. Orthop. Res., 23,* 134–141.

Nguyen, K. T., & West, J. L. (2002). Photopolymerizable hydrogels for tissue engineering applications. *Biomaterials, 23,* 4307–4314.

Nicodemus, G. D., & Bryant, S. J. (2008). The role of hydrogel structure and dynamic loading on chondrocyte gene expression and matrix formation. *J. Biomech., 41,* 1528–1536.

Nishida, K., Yamato, M., Hayashida, Y., Watanabe, K., Maeda, N., Watanabe, H., et al. (2004a). Functional bioengineered corneal epithelial sheet grafts from corneal stem cells expanded ex vivo on a temperature-responsive cell culture surface. *Transplantation, 77,* 379–385.

Nishida, K., Yamato, M., Hayashida, Y., Watanabe, K., Yamamoto, K., Adachi, E., et al. (2004b). Corneal reconstruction with tissue-engineered cell sheets composed of autologous oral mucosal epithelium. *N. Engl. J. Med., 351,* 1187–1196.

Nishiyama, Y., Nakamura, M., Henmi, C., Yamaguchi, K., Mochizuki, S., Nakagawa, H., et al. (2009). Development of a three-dimensional bioprinter: construction of cell supporting structures using hydrogel and state-of-the-art inkjet technology. *J. Biomech. Eng., 131,* 035001-1–035001-6.

Noguchi, T., Yamamuro, T., Oka, M., Kumar, P., Kotoura, Y., Hyon, S., et al. (1991). Poly(vinyl alcohol) hydrogel as an artificial articular cartilage: evaluation of biocompatibility. *J. Appl. Biomater., 2,* 101–107.

Noth, U., Schupp, K., Heymer, A., Kall, S., Jakob, F., Schutze, N., et al. (2005). Anterior cruciate ligament constructs fabricated from human mesenchymal stem cells in a collagen type I hydrogel. *Cytotherapy, 7,* 447–455.

Oh, S. H., Ward, C. L., Atala, A., Yoo, J. J., & Harrison, B. S. (2009). Oxygen generating scaffolds for enhancing engineered tissue survival. *Biomaterials, 30,* 757–762.

Oka, M., Noguchi, T., Kumar, P., Ikeuchi, K., Yamamuro, T., Hyon, S. H., et al. (1990). Development of an artificial articular cartilage. *Clin. Mater., 6,* 361–381.

Orive, G., Tam, S. K., Pedraz, J. L., & Halle, J. P. (2006). Biocompatibility of alginate-poly-l-lysine microcapsules for cell therapy. *Biomaterials, 27,* 3691–3700.

Park, H., Temenoff, J. S., Holland, T. A., Tabata, Y., & Mikos, A. G. (2005). Delivery of TGF-beta1 and chondrocytes via injectable, biodegradable hydrogels for cartilage tissue engineering applications. *Biomaterials, 26,* 7095–7103.

Park, H., Temenoff, J. S., Tabata, Y., Caplan, A. I., Raphael, R. M., Jansen, J. A., et al. (2009). Effect of dual growth factor delivery on chondrogenic differentiation of rabbit marrow mesenchymal stem cells encapsulated in injectable hydrogel composites. *J. Biomed. Mater. Res. A, 88,* 889–897.

Peppas, N. A. (1986). *Hydrogels in Medicine and Pharmacy.* Boca Raton, FL: CRC Press.

Peppas, N. A. (2004). Hydrogels. In B. D. Ratner, A. S. Hoffman, F. J. Schoen, & J. E. Lemons (Eds.), *Biomaterials Science: An Introduction to Materials in Medicine* (pp. 100–106). San Diego, CA: Elsevier Academic Press.

Peppas, N. A., & Barr-Howell, B. D. (1986). Characterization of the crosslinked structure of hydrogels. In N. A. Peppas (Ed.), *Hydrogels in Medicine and Pharmacy* (pp. 27–56). Boca Raton, FL: CRC Press.

Pfister, B. J., Iwata, A., Taylor, A. G., Wolf, J. A., Meaney, D. F., & Smith, D. H. (2006). Development of transplantable nervous tissue constructs comprised of stretch-grown axons. *J. Neurosci. Methods, 153,* 95–103.

Piterina, A. V., Cloonan, A. J., Meaney, C. L., Davis, L. M., Callanan, A., Walsh, M. T., et al. (2009). ECM-based materials in cardiovascular applications: inherent healing potential and augmentation of native regenerative processes. *Int. J. Mol. Sci., 10,* 4375–4417.

Queen, D., Coutts, P., Fierheller, M., & Sibbald, R. G. (2007). The use of a novel oxygenating hydrogel dressing in the treatment of different chronic wounds. *Adv. Skin Wound Care, 20,* 200–206.

Ratner, B. D., & Atzet, S. (2009). Hydrogels for healing. In R. Barbucci (Ed.), *Hydrogels: Biological Properties and Applications* (pp. 43–51). Milan: Springer.

Reza, A. T., & Nicoll, S. B. (2010). Characterization of novel photocrosslinked carboxymethylcellulose hydrogels for encapsulation of nucleus pulposus cells. *Acta Biomater., 6,* 179–186.

Roth, E. A., Xu, T., Das, M., Gregory, C., Hickman, J. J., & Boland, T. (2004). Inkjet printing for high-throughput cell patterning. *Biomaterials, 25,* 3707–3715.

Saber, S., Zhang, A. Y., Ki, S. H., Lindsey, D. P., Smith, R. L., Riboh, J., et al. (2010). Flexor tendon tissue engineering: bioreactor cyclic strain increases construct strength. *Tissue Eng. A, 16,* 2085–2090.

Sakiyama-Elbert, S. E., & Hubbell, J. A. (2000). Development of fibrin derivatives for controlled release of heparin-binding growth factors. *J. Control. Rel., 65,* 389–402.

Saul, J. M., Linnes, M. P., Ratner, B. D., Giachelli, C. M., & Pun, S. H. (2007). Delivery of non-viral gene carriers from sphere-templated fibrin scaffolds for sustained transgene expression. *Biomaterials, 28,* 4705–4716.

659

Sawhney, A. S., Pathak, C. P., & Hubbell, J. A. (1993). Interfacial photopolymerization of poly(ethylene glycol)-based hydrogels upon alginate-poly(l-lysine) microcapsules for enhanced biocompatibility. *Biomaterials, 14,* 1008—1016.

Schek, R. M., Hollister, S. J., & Krebsbach, P. H. (2004). Delivery and protection of adenoviruses using biocompatible hydrogels for localized gene therapy. *Mol. Ther., 9,* 130—138.

Segura, T., Chung, P. H., & Shea, L. D. (2005). DNA delivery from hyaluronic acid-collagen hydrogels via a substrate-mediated approach. *Biomaterials, 26,* 1575—1584.

Sendele, D. D., Abelson, M. B., Kenyon, K. R., & Hanninen, L. A. (1983). Intracorneal lens implantation. *Arch. Ophthalmol., 101,* 940—944.

Shang, J., Shao, Z., & Chen, X. (2008). Electrical behavior of a natural polyelectrolyte hydrogel: chitosan/carboxymethylcellulose hydrogel. *Biomacromolecules, 9,* 1208—1213.

Shikanov, A., Xu, M., Woodruff, T. K., & Shea, L. D. (2009). Interpenetrating fibrin-alginate matrices for *in vitro* ovarian follicle development. *Biomaterials, 30,* 5476—5485.

Shimizu, T., Yamato, M., Akutsu, T., Shibata, T., Isoi, Y., Kikuchi, A., et al. (2002a). Electrically communicating three-dimensional cardiac tissue mimic fabricated by layered cultured cardiomyocyte sheets. *J. Biomed. Mater. Res., 60,* 110—117.

Shimizu, T., Yamato, M., Isoi, Y., Akutsu, T., Setomaru, T., Abe, K., et al. (2002b). Fabrication of pulsatile cardiac tissue grafts using a novel 3-dimensional cell sheet manipulation technique and temperature-responsive cell culture surfaces. *Circ. Res., 90,* e40.

Shimizu, T., Yamato, M., Kikuchi, A., & Okano, T. (2001). Two-dimensional manipulation of cardiac myocyte sheets utilizing temperature-responsive culture dishes augments the pulsatile amplitude. *Tissue Eng., 7,* 141—151.

Shin, H., Zygourakis, K., Farach-Carson, M. C., Yaszemski, M. J., & Mikos, A. G. (2004). Attachment, proliferation, and migration of marrow stromal osteoblasts cultured on biomimetic hydrogels modified with an osteopontin-derived peptide. *Biomaterials, 25,* 895—906.

Shiroyanagi, Y., Yamato, M., Yamazaki, Y., Toma, H., & Okano, T. (2003). Transplantable urothelial cell sheets harvested noninvasively from temperature-responsive culture surfaces by reducing temperature. *Tissue Eng., 9,* 1005—1012.

Shiroyanagi, Y., Yamato, M., Yamazaki, Y., Toma, H., & Okano, T. (2004). Urothelium regeneration using viable cultured urothelial cell sheets grafted on demucosalized gastric flaps. *BJU Int., 93,* 1069—1075.

Sierpinski, P., Garrett, J., Ma, J., Apel, P., Klorig, D., Smith, T., et al. (2008). The use of keratin biomaterials derived from human hair for the promotion of rapid regeneration of peripheral nerves. *Biomaterials, 29,* 118—128.

Stabenfeldt, S. E., Garcia, A. J., & LaPlaca, M. C. (2006). Thermoreversible laminin-functionalized hydrogel for neural tissue engineering. *J. Biomed. Mater. Res. A, 77,* 718—725.

Stokols, S., & Tuszynski, M. H. (2006). Freeze-dried agarose scaffolds with uniaxial channels stimulate and guide linear axonal growth following spinal cord injury. *Biomaterials, 27,* 443—451.

Takahashi, Y., Yamamoto, M., Yamada, K., Kawakami, O., & Tabata, Y. (2007). Skull bone regeneration in nonhuman primates by controlled release of bone morphogenetic protein-2 from a biodegradable hydrogel. *Tissue Eng., 13,* 293—300.

Temenoff, J. S., Athanasiou, K. A., LeBaron, R. G., & Mikos, A. G. (2002). Effect of poly(ethylene glycol) molecular weight on tensile and swelling properties of oligo(poly(ethylene glycol) fumarate) hydrogels for cartilage tissue engineering. *J. Biomed. Mater. Res., 59,* 429—437.

Testa, G., di Meo, C., Nardecchia, S., Capitani, D., Mannina, L., Lamanna, R., et al. (2009). Influence of dialkyne structure on the properties of new click-gels based on hyaluronic acid. *Int. J. Pharm., 378,* 86—92.

Texter, J. (2009). Templating hydrogels. *Colloid Polym. Sci., 287,* 313—321.

Thorn, R. M., Greenman, J., & Austin, A. (2006). An *in vitro* study of antimicrobial activity and efficacy of iodine-generating hydrogel dressings. *J. Wound Care, 15,* 305—310.

Villanueva, I., Klement, B. J., von Deutsch, D., & Bryant, S. J. (2009). Cross-linking density alters early metabolic activities in chondrocytes encapsulated in poly(ethylene glycol) hydrogels and cultured in the rotating wall vessel. *Biotechnol. Bioeng., 102,* 1242—1250.

Wang, D. A., Varghese, S., Sharma, B., Strehin, I., Fermanian, S., Gorham, J., et al. (2007). Multifunctional chondroitin sulphate for cartilage tissue-biomaterial integration. *Nat. Mater., 6,* 385—392.

Wedmore, I., McManus, J. G., Pusateri, A. E., & Holcomb, J. B. (2006). A special report on the chitosan-based hemostatic dressing: experience in current combat operations. *J. Trauma, 60,* 655—658.

Wichterle, O., & Lim, D. (1960). Hydrophilic gels for biological use. *Nature, 185,* 117—118.

Williams, C. G., Kim, T. K., Taboas, A., Malik, A., Manson, P., & Elisseeff, J. (2003). *In vitro* chondrogenesis of bone marrow-derived mesenchymal stem cells in a photopolymerizing hydrogel. *Tissue Eng., 9,* 679—688.

Williams, D. F. (1987). Definitions in biomaterials: proceedings of a consensus conference of the European Society for Biomaterials. In D. F. Williams (Ed.), *European Society for Biomaterials* (p. 67). Chester: Elsevier.

Wood, M. D., & Sakiyama-Elbert, S. E. (2008). Release rate controls biological activity of nerve growth factor released from fibrin matrices containing affinity-based delivery systems. *J. Biomed. Mater. Res. A, 84*, 300–312.

Xu, M., Kreeger, P. K., Shea, L. D., & Woodruff, T. K. (2006a). Tissue-engineered follicles produce live, fertile offspring. *Tissue Eng., 12*, 2739–2746.

Xu, T., Jin, J., Gregory, C., Hickman, J. J., & Boland, T. (2005). Inkjet printing of viable mammalian cells. *Biomaterials, 26*, 93–99.

Xu, T., Gregory, C. A., Molnar, P., Cui, X., Jalota, S., Bhaduri, S. B., et al. (2006b). Viability and electrophysiology of neural cell structures generated by the inkjet printing method. *Biomaterials, 27*, 3580–3588.

Yamamoto, M., Takahashi, Y., & Tabata, Y. (2006). Enhanced bone regeneration at a segmental bone defect by controlled release of bone morphogenetic protein-2 from a biodegradable hydrogel. *Tissue Eng., 12*, 1305–1311.

Yamato, M., Utsumi, M., Kushida, A., Konno, C., Kikuchi, A., & Okano, T. (2001). Thermo-responsive culture dishes allow the intact harvest of multilayered keratinocyte sheets without dispase by reducing temperature. *Tissue Eng., 7*, 473–480.

Yew, Y. K., Ng, T. Y., Li, H., & Lam, K. Y. (2007). Analysis of pH and electrically controlled swelling of hydrogel-based micro-sensors/actuators. *Biomed. Microdev., 9*, 487–499.

Yu, T. T., & Shoichet, M. S. (2005). Guided cell adhesion and outgrowth in peptide-modified channels for neural tissue engineering. *Biomaterials, 26*, 1507–1514.

Yu, X., Dillon, G. P., & Bellamkonda, R. B. (1999). A laminin and nerve growth factor-laden three-dimensional scaffold for enhanced neurite extension. *Tissue Eng., 5*, 291–304.

Zhang, C., Wen, X., Vyavahare, N. R., & Boland, T. (2008). Synthesis and characterization of biodegradable elastomeric polyurethane scaffolds fabricated by the inkjet technique. *Biomaterials, 29*, 3781–3791.

Zhao, Y., Xu, Y., Zhang, B., Wu, X., Xu, F., Liang, W., et al. (2009). *In vivo* generation of thick, vascularized hepatic tissue from collagen hydrogel-based hepatic units. *Tissue Eng. C Methods, 16*, 653–659.

CHAPTER 36

Surface Modification of Biomaterials

Andrés J. García
Woodruff School of Mechanical Engineering, Petit Institute for Bioengineering and
Bioscience, Georgia Institute of Technology, Atlanta, GA, USA

INTRODUCTION
Biomaterial interfaces in regenerative medicine

Biomaterials, either synthetic (e.g. polymers, metals, ceramics) or natural (e.g. proteins, polysaccharides), play central roles in tissue engineering and regenerative medicine applications by providing (1) three-dimensional scaffolds to support cellular activities, (2) matrices for delivery of therapeutic agents (e.g. drugs, proteins, DNA, siRNA), and (3) functional device components (e.g. mechanical supports, sensing/stimulating elements, non-thrombogenic surfaces, diffusional barriers). The bulk properties of the biomaterial are critical determinants of the biological performance of the material (Ratner et al., 2004). For example, the mechanical properties of a vascular substitute, including elastic modulus, ultimate tensile stress, and compliance, dictate the ability of this tissue construct to support the applied mechanical loads associated with blood flow. On the other hand, the biological response to a biomaterial is governed by the material surface properties, primarily surface chemistry and structure. Protein adsorption/activation and cell adhesion, events that regulate host responses to materials, occur at the biomaterial-tissue interface, and the physicochemical properties of the material surface modulate these biological events (Anderson, 2001; Anderson et al., 2008). For instance, the chemical properties of the surface of a vascular substitute control blood compatibility (i.e. protein adsorption, platelet adhesion, thrombogenicity, patency). Hence, modification of biomaterial surfaces represents a promising route to engineer biofunctionality at the material-tissue interface in order to modulate biological responses without altering material bulk properties.

Overview of surface modification strategies

Numerous surface modification approaches have been developed for all classes of materials to modulate biological responses and improve device performance. Applications include reduction of protein adsorption and thrombogenicity; control of cell adhesion, growth, and differentiation; modulation of fibrous encapsulation and osseointegration; improved wear and/or corrosion resistance; and potentiation of electrical conductivity (Ratner et al., 2004). Surface modifications fall into two general categories: (1) physicochemical modifications involving alterations to the atoms, compounds, or molecules on the surface, and (2) surface coatings consisting of a different material from the underlying support. Physicochemical modifications include chemical reactions (e.g. oxidation, reduction, silanization, acetylation), etching, and mechanical roughening/polishing and patterning (Fig. 36.1). Overcoating

663

Principles of Regenerative Medicine. DOI: 10.1016/B978-0-12-381422-7.10036-7

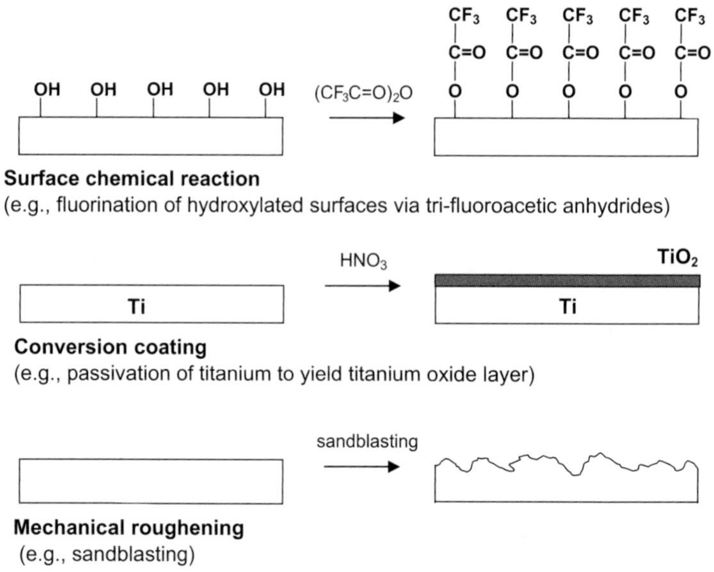

FIGURE 36.1
Schematic representations of common physicochemical surface modifications of biomaterials.

alterations comprise grafting (including tethering of biomolecules), non-covalent and covalent coatings, and thin film deposition (Fig. 36.2).

While the specific requirements of the surface modification approach vary with application, several characteristics are generally desirable. Thin surface modifications are preferred for most applications since thicker coatings often negatively influence the mechanical and functional properties of the material. Ideally, the surface modification should be confined to the outermost molecular layer ($\sim$ 10–15 Å), but, in practice, thicker layers (10–100 nm) are used to ensure uniformity, durability, and functionality. Stability of the modified surface is a critical requirement for adequate biological performance. Surface stability not only refers to mechanical durability (i.e. resistance to cracking, delamination, debonding) but also chemical stability, especially in aggressive, chemically active environments such as biological milieu. Several types of surface rearrangements, such as translation of surface atoms or molecules in response to environmental factors and mobility of bulk molecules to the surface and vice versa, readily occur in polymers and ceramics following exposure to biological fluids.

Given the uniquely reactive nature and mobility/rearrangement of surfaces, as well as the tendency of surfaces to readily contaminate, rigorous analyses of surface treatments are essential to surface modification strategies. Surface analyses technologies generally focus on characterizing topography, chemistry/composition, and surface energy (Woodruff and Delchar, 1994) (Table 36.1). Important considerations for these surface analysis technologies include operational principles (impact of high energy particles/X-rays under ultrahigh vacuum, adsorption or emission spectroscopies), depth of analysis, sensitivity, and resolution. For most applications, several analysis techniques must be used to obtain a complete description of the surface.

PHYSICOCHEMICAL SURFACE MODIFICATIONS

Physicochemical modifications involve alterations to the atoms, compounds, or molecules on the material surface (Fig. 36.1).

Chemical modifications

Countless chemical reactions, including UV/laser irradiation and etching reactions to clean, alter, or crosslink surface groups, have been developed to modify biomaterial surfaces

(Ratner and Hoffman, 2004). Non-specific reactions yield a distribution of chemically distinct groups at the surface, and the resulting surface is complex and difficult to characterize due to the presence of different chemical species in various concentrations. Nevertheless, non-specific chemical reactions are widely used in biomaterials processing. Examples of non-specific reactions include radio-frequency glow discharge in different plasmas (e.g. oxygen, nitrogen, argon), corona discharge in air, oxidation of metals, and acid-base treatments of polymers. In contrast, specific chemical reactions target particular chemical moieties on the surface to convert them into another functional group with few side (unwanted) reactions. Acetylation, fluorination of hydroxylated surfaces via tri-fluoroacetic anhydrides, silanization of hydroxylated surfaces, and incorporation of glycidyl groups into polysiloxanes are examples of specific chemical reactions. In addition, various chemical methods exist to tether biomacromolecules onto available anchoring groups on surfaces, as described on pages 201-218.

Reaction of metal surfaces to produce an oxide-rich layer that conveys corrosion resistance, passivation, and improved wear and adhesive properties (also referred to as conversion coatings) are common surface modifications in metallic biomaterials. For example, nitric acid treatment of titanium and titanium alloys to generate titanium oxide layers is regularly performed on titanium-based medical devices, and the excellent biocompatibility properties of titanium are attributed to this oxide layer (Albrektsson et al., 1983). Implantation of ions

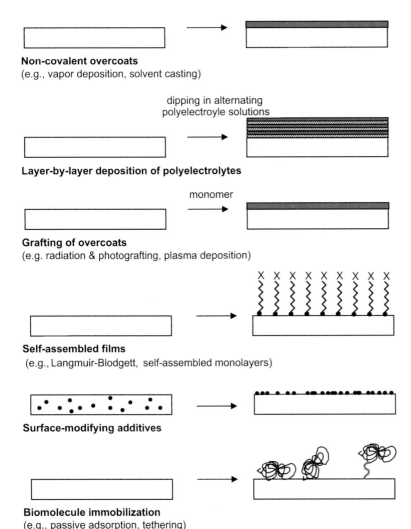

Non-covalent overcoats
(e.g., vapor deposition, solvent casting)

dipping in alternating
polyelectroyle solutions

Layer-by-layer deposition of polyelectrolytes

monomer

Grafting of overcoats
(e.g. radiation & photografting, plasma deposition)

X X X X X X X X X

Self-assembled films
(e.g., Langmuir-Blodgett, self-assembled monolayers)

Surface-modifying additives

Biomolecule immobilization
(e.g., passive adsorption, tethering)

FIGURE 36.2
Schematic representations of common overcoating technologies for surface modification.

TABLE 36.1 Common Surface Analysis Techniques

	Principle	Operation	Spatial resolution	Info. depth	Sensitivity	Texture	Chemical composition info.			
							Elements	Compounds	Isotopes	Additional
Contact angle	Liquid wetting of surfaces	Air Liquid	NA	3–20 Å	NA	Indirect				Surface energy
AFM	Records interatomic forces between tip and sample	Air Aqueous	Atomic	NA	Single atom	Yes	No	No	No	
SEM	Secondary electron emission caused by electron bombardment is imaged	Vacuum	40 Å	5–10 Å	High	Yes	No	No	No	Crystallinity
EDXA	X-ray emission caused by electron bombardment	Vacuum	40 Å	1 μm	10^{-7} g/cm^2	No	Z > 5	No	No	
AES	Auger electron emission caused by electron bombardment	Vacuum	100 Å	15–50 Å	10^{-10} g/cm^2 0.1 atom %	No	Z > 3	Chemical shift	No	
XPS	X-rays cause emission of photoelectrons with characteristic energies	Vacuum	10 μm	10–150 Å	10^{-10} g/cm^2 0.1 atom %	No	Z > 3	Chemical shift (excellent)	No	
SIMS	Ion bombardment causes secondary ion emission	Vacuum	3–10 μm	10 Å	10^{-13} g/cm^2	No	All	Yes	Yes	
FTIR	Molecular vibrations resulting from adsorption of IR radiation	Air Aqueous (ATR)	10 μm	<1 μm	1 mole %	No	Indirect	Vibration frequency	No	Monolayer orientation

ATR=attenuated total reflectance

into surfaces via a beam of accelerated ions has been applied to modify the surface properties of mostly metals and ceramics. For example, ion beam implantation of nitrogen into titanium and boron and carbon into stainless steel improves wear resistance and fatigue life, respectively (Sioshansi, 1987). In addition, recent evidence suggests that ion beam implantation of silicone and silver can also enhance the blood compatibility and infection resistance of silicone rubber catheters (Bambauer et al., 2004).

Topographical modifications

The size and shape of topographical features on a surface influence cellular and host responses to the material. For example, surface macro- and micro-texture alters cell adhesion, spreading, and alignment (Curtis and Wilkinson, 1998; Flemming et al., 1999) and can regulate cell phenotypic activities, including neurite extension and osteoblastic differentiation (Boyan et al., 1996; Jansen and von Recum, 2004). Moreover, surface topography can have significant *in vivo* effects. For instance, implant porosity modulates bone and soft tissue ingrowth (Yamamoto et al., 2000; Pilliar, 2005), and surface texture alters epithelial downgrowth responses to percutaneous devices and inflammatory reactions and fibrous encapsulation to materials implanted subcutaneously (Chehroudi et al., 1989; Brauker et al., 1995; Chehroudi and Brunette, 2002). While specific surface texture parameters that elicit particular biological responses have been identified in several cases, the mechanisms generating these behaviors remain poorly understood.

Methods for generating surface texture can be grouped into approaches for engineering either roughness or topography (Fig. 36.3). Surface *roughness* indicates a random or complex pattern of features of varying amplitude and spacing, typically on a scale smaller than a cell (10–20 μm). On the other hand, surface *topography* refers to patterns of well-defined, controlled features on the surface. Surface roughness has been traditionally modified via sandblasting, plasma spraying, and mechanical polishing, and it is the non-specific nature of these processes that renders surfaces with random or complex topographies. Ion beam and electric arc (for conductive materials) texturing approaches have also been applied to modulate surface roughness. For generating controlled topographies, micro- and nano-machining techniques have been exploited using silicon, glass, and polymers as substrate materials (Flemming et al., 1999). Photolithography combined with reactive plasma and ion etching has been extensively applied to generate surfaces with well-defined topographies. This technique allows the preparation of machined silicon and polymeric substrates and silicon templates that can then be used as molds to transfer features to polymers via solvent casting or injection molding. Similarly, LIGA (in German, "lithographie, galvanoformung, abformung" (lithography, electroplating, and molding)), electron beam, and laser machining have been used to manufacture defined topographical features on various materials. More recently, hot embossing imprint lithography has been applied for low cost and rapid fabrication of micro- and nano-scale features on biomedically relevant polymers (Charest et al., 2004).

OVERCOATING TECHNOLOGIES

Coating strategies rely on the deposition of a surface layer consisting of a different composition from the underlying base material (Fig. 36.2). These surface modification approaches include non-covalent and covalent coatings (Ratner and Hoffman, 2004).

667

surface roughness   surface topography

FIGURE 36.3
Surface roughness and topography.

Non-covalent coatings

Major advantages of non-covalent coatings include simple application and the ability to coat a variety of different base materials. Examples of common non-covalent coating methods are solvent-casting, and vapor deposition of metals, parylene, and carbons. In the Langmuir-Blodgett deposition method, one or more highly ordered layers of surfactant molecules (e.g. phospholipids, amphiphiles) are placed at the surface of the base material via assembly at the air-water interface and compression of the surfactant molecules. Langmuir-Blodgett films exhibit high order and uniformity and provide flexibility in incorporating a wide range of chemistries. The stability of these films can be improved by crosslinking or internally polymerizing the surfactant molecules following film formation. Another surface modification strategy that takes advantage of intermolecular interactions is the deposition of multilayer polyelectrolytes (e.g. poly(styrenesulfonate)/poly(allylamine), hyaluronic acid/chitosan). In this simple layer-by-layer method, a charged surface is sequentially dipped into alternating aqueous solutions of polyelectrolytes of opposite charge in order to deposit multilayers of a polyelectrolyte complex. Another elegant strategy for surface modification is the use of surface-modifying additives. These molecules are blended in the bulk material during fabrication but will spontaneously rise to and concentrate at the surface due to the driving force to minimize interfacial energy.

Covalent coatings

Covalent coating methodologies rely on direct tethering of overcoats onto the base material to improve film stability and adherence. Radiation grafting, both with ionizing radiation and high-energy electron beams, and photografting have been extensively pursued to modify polymer substrates in order to introduce chemically reactable groups into inert hydrophobic polymers and polymerize overcoats onto the base support (Ratner and Hoffman, 2004). In principle, the radiation breaks chemical bonds in the base material into free radicals and other reactive species, which are then exposed to a monomer. The monomer reacts with the reactive species at the surface and propagates as a free radical chain reaction into a surface grafted polymer. These strategies allow for generation of a wide range of surface chemistries, and unique graft co-polymers can be synthesized by combining different monomers. Plasma deposition (also referred to as glow discharge deposition) via radio frequency or microwave has also been extensively applied to biomaterial surface modification (Hoffman, 1988). In particular, radio frequency glow discharge (RFGD) plasma deposition has received considerable attention because it can generate continuous (relatively free of pin holes and voids) conformal coatings that can be applied to many different types of supports (metals, ceramics, polymers) with complex geometries. In addition, these films exhibit good adherence to the substrate and can be engineered to present different functionalities, although the resulting chemistry is complex and ill-defined. In contrast to these relatively low-energy/low-temperature plasmas, high-energy/high-temperature plasmas have also been used to apply inorganic surface modifications onto inorganic substrates. For example, calcium phosphate ceramic particles, such as hydroxyapatite, have been deposited via flame spraying onto titanium and cobalt chrome orthopedic implants to improve osseointegration (Gruner, 2005).

Coatings consisting of self-assembled monolayers (SAMs) have gained significant attention as robust surface modification agents (Ulman, 1991; Mrksich and Whitesides, 1995). These films spontaneously assemble; form highly ordered, well-defined surfaces with excellent chemical stability; and provide a wide range of available surface functionalities. The basic structure of molecules that form SAMs is an anchoring "head" group, organic chain backbone, and functional "tail" group. Common SAM systems are alkanethiols on coinage metals (gold, silver), n-akyl silanes on hydroxylated supports (glass, silica), and phosphoric acid or phosphate groups on titanium or tantalum surfaces. Assembly of these organic chains into highly ordered structures is driven by the strong adsorption of the anchoring "head" group of the monolayer constituent to the surface and van der Waals interactions of the backbone chains.

TABLE 36.2 Biomedical and Biotechnological Applications of Immobilized Biomolecules

Biomolecule	Applications
Heparin	Blood-compatible surfaces; growth factor immobilization
Fibronectin, collagen, RGD peptides	Cell adhesion and function in biosensors; arrays, devices, and tissue-engineered constructs
Antibodies	Biosensors; bioseparations; anti-cancer treatments
DNA plasmids, antisense oligonucleotides, siRNA	Gene therapy for a multitude of diseases; DNA probes
Growth factor proteins and peptides	Anti-cancer treatments; treatments for auto-immune and inflammatory conditions; enhanced wound repair
Enzymes	Biosensors; bioreactors; anti-cancer treatments; antithrombotic surfaces
Drugs and antibiotics	Antithrombotic agents; anti-cancer treatments; anti-hyperplasia treatments; anti-infection/inflammation treatments
Polysaccharides	Non-fouling supports for biosensors and bioseparations

The order and stability of the SAMs are strongly influenced by the length of the backbone chain, and, in the case of alkanethiols, molecules with backbones between 9 and 24 methylene groups assemble well on gold. Importantly, the terminal functional group is presented at the surface-solution interface and controls the physicochemical properties of the SAM.

BIOLOGICAL MODIFICATION OF SURFACES

Biomolecules (e.g. cell receptor ligands, enzymes, antibodies, pharmacological agents, lipids, nucleic acids) have been immobilized onto and within biomaterial supports for numerous therapeutic, diagnostic, and bioprocess applications. Table 36.2 lists several examples of biological modifications to surfaces for biomedical and biotechnological applications. The rationale for these hybrid materials integrating synthetic and biological components is to convey biofunctionality and hence engineer materials that elicit desired biological responses or have attributes associated with biosystems. One of the earliest examples of this strategy is the immobilization of heparin onto polymer surfaces to improve blood compatibility. More recently, drug eluting stents (stents coated with a polymeric layer loaded with anti-hyperplasia drugs) have been developed to reduce restenosis and improve patency. Another example of a widely used biological modification strategy is the immobilization of adhesive ligands, either adsorbed proteins (e.g. fibronectin, laminin) or tethered synthetic oligopeptides (e.g. RGD), on synthetic and natural supports to promote cell adhesion and function in various tissue engineering and regenerative medicine applications (Lutolf and Hubbell, 2005). Important considerations in the modification of surfaces with biological molecules include the density, distribution (uniform vs. clustered), and activity (e.g. orientation, conformation, accessibility) of the immobilized biomolecule.

Three major methods are used to immobilize biomolecules onto biomaterial surfaces: physical adsorption, physical entrapment, and covalent immobilization (Fig. 36.4) (Hoffman and Hubbell, 2004). Passive physisorption of biomacromolecules (i.e. proteins, polysaccharides, nucleic acids) is a simple yet efficient method to render surfaces biologically active. Everyday applications include coating of synthetic materials with extracellular matrix proteins, such as fibronectin and collagen, to improve cell adhesion. As discussed in Chapter 56, protein adsorption is a complex, dynamic energy-driven process involving hydrophobic interactions,

669

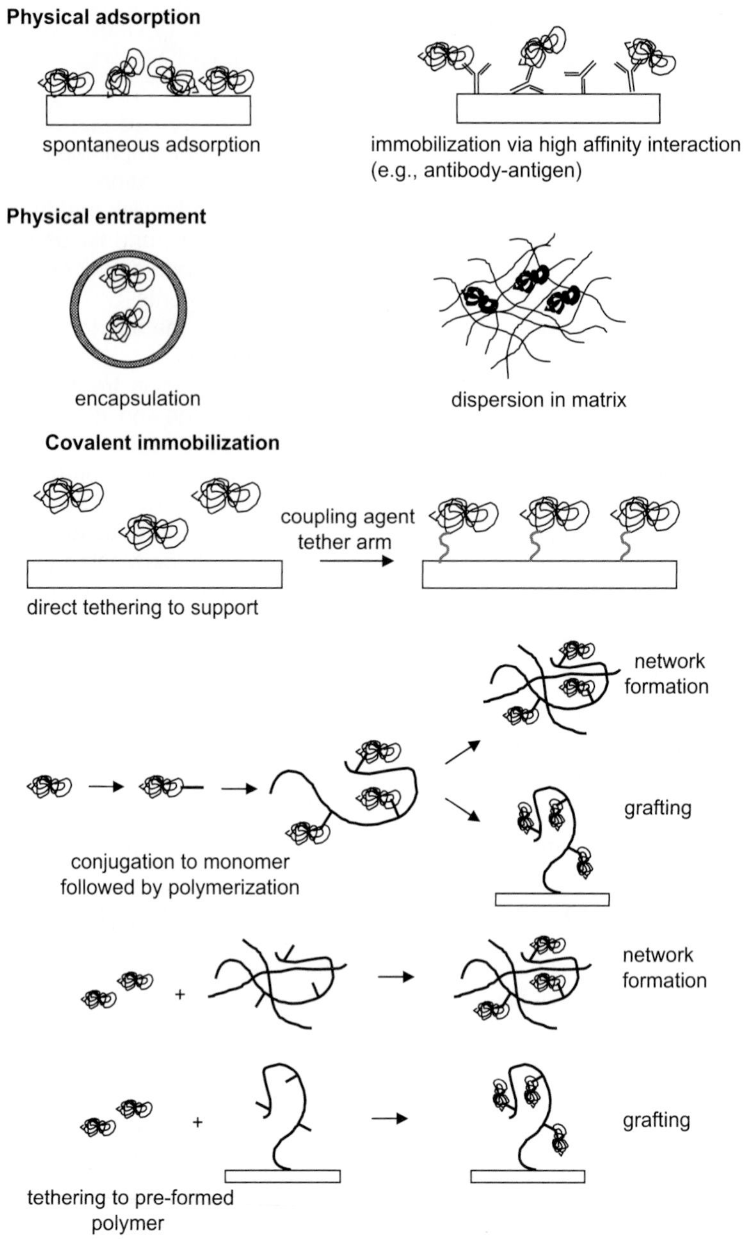

Physical adsorption

spontaneous adsorption

immobilization via high affinity interaction (e.g., antibody-antigen)

Physical entrapment

encapsulation

dispersion in matrix

Covalent immobilization

coupling agent tether arm

direct tethering to support

network formation

grafting

conjugation to monomer followed by polymerization

network formation

grafting

tethering to pre-formed polymer

FIGURE 36.4
Schematic diagram of methods for immobilizing biomolecules onto and within biomaterials.

electrostatic interactions, hydrogen bonding, and van der Waals forces. Protein parameters such as primary structure, size, and structural stability as well as surface properties including surface energy and chemistry influence the biological activity of the adsorbed biomacromolecules. It is important to point out that these biologically modified surfaces can undergo further modifications, such as displacement of adsorbed proteins and cell-mediated deposition and remodeling of matrix components, in the biological milieu. As an approach to improve the stability of these modified surfaces, the biological molecules can be crosslinked following adsorption. Finally, the use of high-affinity interactions, for example avidin-biotin and antibody-antigen, represents a special case of these physical immobilization methods that is particularly important in diagnostics and bioprocessing.

Physical entrapment methods rely on diffusive barriers or matrix systems to control the transport or availability of the biomolecule. For example, entrapment of enzymes within sol-gels with nano-scale porosites and drug or protein therapeutics within encapsulation matrices provide technologies for enhanced stability, separation or recovery of the biological agent, and regulated delivery kinetics. The encapsulation systems can be engineered to permanently isolate the biomolecule or degrade in non-specific (e.g. hydrolysis) or specific (e.g. enzymatic degradation) fashions for controlled release kinetics.

An extensive and diverse group of strategies has been developed to covalently immobilize or tether biomolecules to soluble or solid supports (Fig. 36.4) (Weetall, 1976; Hoffman and Hubbell, 2004). Soluble polymers functionalized with biomolecules can then be polymerized into a network or grafted onto a solid support. These strategies rely on coupling reactions between groups in the biomolecule (-NH_2, -COOH, -SH) and the biomaterial support, and often involve crosslinkers or coupling agents such as CNBr, carbodiimides, and N-hydroxy-sulfosuccinimide. Furthermore, recently developed chemoselective reactions, such as "click" chemistry, provide unparalleled control over the tethering and presentation of biomolecules. In many instances, the biomolecule is covalently immobilized via a spacer arm (e.g. poly-ethylene glycol) that provides increased steric freedom and activity. Additionally, the tether arm can be designed to be hydrolytically or enzymatically labile in order to allow for release of the tethered biomolecule. As expected, the properties of the underlying biomaterial support play central roles in the tethering efficiency and resulting biological activity of the immobilized biomolecule. In some cases, the surface needs to be modified via the techniques described above to introduce reactive groups for the subsequent immobilization step. For example, inert surfaces can be modified by overcoating with a polymeric adlayer that then presents anchoring groups suitable for immobilization of biomolecules. For many biomedical and biotechno-logical applications, it is desirable to tether biomolecules within a protein adsorption-resistant (non-fouling) background in order to eliminate effects associated with non-specific protein adsorption. This is particularly important in biomaterials and regenerative medicine appli-cations in which inflammatory responses to non-specifically adsorbed proteins limit biolog-ical performance. Poly(ethelyne glycol) (PEG) (-$[CH_2CH_2O]_n$) groups have proven to have the most protein-resistant functionality and remain the standard (Hoffman, 1999). A strong correlation exists between PEG chain density and length and resistance to protein adsorption, and consequently cell adhesion. Other hydrophilic polymers, such as poly(2-hydroxyethyl methacrylate), polyacrylamide, and phosphoryl choline polymers, also resist protein adsorption. In addition, mannitol, oligomaltose, and taurine groups have emerged as prom-ising moieties to prevent protein adsorption. Critical considerations for these approaches include the ability to apply these coatings to existing medical-grade materials and the long-term stability of these coatings in terms of non-fouling character and whether protein adsorption can be reduced below threshold levels for suppression of non-specific biological responses. Emerging strategies combining passive and active approaches (e.g. biotherapeutics that modulate blood clotting or inflammation, self-cleaning materials) provide promising approaches to effectively reduce protein adsorption while enabling controlled presentation of bioactive molecules.

SURFACE CHEMICAL PATTERNING

While the surface chemical and biological modification strategies described above were presented in the context of a uniform surface, many of these technologies can be used to generate surfaces that present chemical or biological functionalities in distinct geometrical patterns. Important applications of patterned surfaces include protein and oligonucleotide arrays, biosensors, and cell-based arrays (Hubbell, 2004). In many instances, these patterned substrates contain spatially-defined domains presenting biomolecules surrounded by a non-fouling background. Photolithography and other techniques relying on exposure through masked patterns or direct surface exposure (e.g. laser or electron beam) in combination with chemical reaction or grafting are often used to generate chemically patterned surfaces. Recently, "soft" lithography methods such as microcontact printing and microfluidic fluid exposure have been applied to produce micropatterned substrates in high throughput and low cost, and without the need of a cleanroom environment (Whitesides et al., 2001).

CONCLUSION AND FUTURE PROSPECTS

Surface modifications of biomaterials represent promising routes to engineer biofunctionality at the material-tissue interface in order to modulate biological responses without altering material bulk properties. Countless technologies have been developed to create (1) physico-chemical modifications involving alterations to the chemical groups on the surface and (2) coatings consisting of a different material from the underlying support, including immobilized biomolecules. These approaches hold tremendous promise to enhance biomaterial performance in regenerative medicine. Future structure-function analyses on the effects of specific surface chemistries, topographies, and biological modifications on *in vivo* responses in particular healing and regenerative environments will further advance the understanding of host responses to implanted devices. These insights will result in the identification of surface modifications that synergize with biological elements (e.g. cells, growth and differentiation factors) to enhance tissue repair and regeneration. It is anticipated that technical breakthroughs in synthetic chemistry, biofunctionalization, micro- and nano-fabrication, and surface characterization will lead to the engineering of advanced, bioactive materials. In particular, complex patterns of bioligand presentation, such as clusters, gradients, temporal exposure, and multiple ligands, are expected to provide unparalleled control over cellular activities and healing responses.

References

Albrektsson, T., Branemark, P. I., Hansson, H. A., Kasemo, B., Larsson, K., Lundstorm, I., et al. (1983). The interface zone of inorganic implants in vivo: titanium implants in bone. *Ann. Biomed. Eng., 11*, 1–27.

Anderson, J. M. (2001). Biological responses to materials. *Annu. Rev. Mater. Res., 31*, 81–110.

Anderson, J. M., Rodriguez, A., & Chang, D. T. (2008). Foreign body reaction to biomaterials. *Semin. Immunol., 20*, 86–100.

Bambauer, R., Latza, R., Bambauer, S., & Tobin, E. (2004). Large bore catheters with surface treatments versus untreated catheters for vascular access in hemodialysis. *Artif. Organs, 28*, 604–610.

Boyan, B. D., Hummert, T. W., Dean, D. D., & Schwartz, Z. (1996). Role of material surfaces in regulating bone and cartilage cell response. *Biomaterials, 17*, 137–146.

Brauker, J. H., Carr-Brendel, V. E., Martinson, L. A., Crudele, J., Johnston, W. D., & Johnson, R. C. (1995). Neovascularization of synthetic membranes directed by membrane microarchitecture. *J. Biomed. Mater. Res., 29*, 1517–1524.

Charest, J. L., Bryant, L. E., Garcia, A. J., & King, W. P. (2004). Hot embossing for micropatterned cell substrates. *Biomaterials, 25*, 4767–4775.

Chehroudi, B., & Brunette, D. M. (2002). Subcutaneous microfabricated surfaces inhibit epithelial recession and promote long-term survival of percutaneous implants. *Biomaterials, 23*, 229–237.

Chehroudi, B., Gould, T. R., & Brunette, D. M. (1989). Effects of a grooved titanium-coated implant surface on epithelial cell behavior in vitro and in vivo. *J. Biomed. Mater. Res., 23*, 1067–1085.

Curtis, A. S., & Wilkinson, C. D. (1998). Reactions of cells to topography. *J. Biomater. Sci. Polym. Ed., 9*, 1313–1329.

Flemming, R. G., Murphy, C. J., Abrams, G. A., Goodman, S. L., & Nealey, P. F. (1999). Effects of synthetic micro- and nano-structured surfaces on cell behavior. *Biomaterials, 20*, 573–588.

Gruner, H. (2005). Thermal spray coating on titanium. In D. M. Brunette, P. Tengvall, M. Textor, & P. Thomsen (Eds.), *Titanium in Medicine* (pp. 375–416). Berlin: Springer-Verlag.

Hoffman, A. S. (1988). Biomedical applications of plasma gas discharge processes. *J. Appl. Polymer Sci. Appl. Polymer Symp., 42*, 251–267.

Hoffman, A. S. (1999). Non-fouling surface technologies. *J. Biomater. Sci. Polym. Ed., 10*, 1011–1014.

Hoffman, A. S., & Hubbell, J. A. (2004). Surface-immobilized biomolecules. In B. D. Ratner, A. S. Hoffman, F. J. Schoen, & J. E. Lemons (Eds.), *Biomaterials Science: An Introduction to Materials in Medicine* (pp. 225–233). San Diego: Academic Press.

Hubbell, J. A. (2004). Biomaterials science and high-throughput screening. *Nat. Biotechnol., 22*, 828–829.

Jansen, J. A., & von Recum, A. F. (2004). Textured and porous materials. In B. D. Ratner, A. S. Hoffman, F. J. Schoen, & J. E. Lemons (Eds.), *Biomaterials Science: An Introduction to Materials in Medicine* (pp. 218−225). San Diego: Academic Press.

Lutolf, M. P., & Hubbell, J. A. (2005). Synthetic biomaterials as instructive extracellular microenvironments for morphogenesis in tissue engineering. *Nat. Biotechnol., 23*, 47−55.

Mrksich, M., & Whitesides, G. M. (1995). Patterning self-assembled monolayers using microcontact printing: a new technology for biosensors? *Trends Biotechnol., 13*, 228−235.

Pilliar, R. M. (2005). Cementless implant fixation − toward improved reliability. *Orthop. Clin. North Am., 36*, 113−119.

Ratner, B. D., & Hoffman, A. S. (2004). Physicochemical surface modification of materials used in medicine. In B. D. Ratner, A. S. Hoffman, F. J. Schoen, & J. E. Lemons (Eds.), *Biomaterials Science: An Introduction to Materials in Medicine* (pp. 201−218). San Diego: Elsevier/Academic Press.

Ratner, B. D., Hoffman, A. S., Schoen, F. J., & Lemons, J. E. (2004). *Biomaterials Science: An Introduction to Materials in Medicine.* San Diego: Elsevier/Academic Press.

Sioshansi, P. (1987). Surface modification of industrial components by ion implantation. *Mater. Sci. Eng., 90*, 373−383.

Ulman, A. (1991). *An Introduction to Ultrathin Organic Films: From Langmuir−Blodgett to Self-Assembly.* San Diego: Academic Press.

Weetall, H. H. (1976). Covalent coupling methods for inorganic support materials. *Meth. Enzymol., 44*, 134−148.

Whitesides, G. M., Ostuni, E., Takayama, S., Jiang, X., & Ingber, D. E. (2001). Soft lithography in biology and biochemistry. *Annu. Rev. Biomed. Eng., 3*, 335−373.

Woodruff, D. P., & Delchar, T. A. (1994). *Modern Techniques of Surface Science.* Cambridge: Cambridge University Press.

Yamamoto, M., Tabata, Y., Kawasaki, H., & Ikada, Y. (2000). Promotion of fibrovascular tissue ingrowth into porous sponges by basic fibroblast growth factor. *J. Mater. Sci. Mater. Med., 11*, 213−218.

CHAPTER 37

Histogenesis in Three-dimensional Scaffolds

Melissa K. McHale, Nicole M. Bergmann, Jennifer L. West
Department of Bioengineering, Rice University, Houston, TX, USA

THE NEED FOR REPLACEMENT TISSUES

Replacement of diseased tissues and organs is one of the biggest problems facing the medical industry. According to the US Scientific Registry of Transplant Recipients, there were over 97,000 patients on the National Organ Registry at the end of 2007 (OPTN/SRTR, 2008). Of the patients awaiting life-saving transplants, approximately 10% died before they could receive donor organs. In addition, tissue disease and organ failure lead to an estimated 8 million surgical procedures annually in the USA (Angelos et al., 2003). From these statistics, one fact is clear — the need for replacement organs far outweighs the supply, and the discrepancy is only expected to worsen as populations rise and life expectancies increase.

Synthetic materials and tissue grafting have been employed throughout history in an effort to address the need for tissue substitutes, though each is marred by seemingly insurmountable challenges. In the case of synthetic replacements, the primary mode of failure comes through attack via the host's foreign body response. Disease transmission, in the case of xenografts or allografts, and donor site morbidity from autografts leave these tissues underutilized, though the primary constraint, as with whole organs, is a limited supply. Thus, tissue engineering, with its goal of recapitulating biological structure and function, has emerged as the most viable solution to the lack of suitable replacement tissues.

The idea that tissue function can be restored is as old as the medical profession. Historically, one of the most popular sites of implant was the nose, which, other than the hand, was the body part most likely to be injured in battle (Wallace, 1978). As a consequence, many cultures were performing successful nasal operations thousands of years ago. A pioneering technique for nasal tissue regeneration was developed in 1596 when a flap of skin from the arm, complete with underlying vessels, was reconnected to the face. In many cases, these tissues were able to reconnect with the local blood supply and successfully regenerate (Ang, 2005). By the nineteenth century, numerous surgeons had successfully transplanted skin between individuals (Hauben, 1985; Ang, 2005) and with World War II came the advent of large-scale tissue replacement using synthetic materials. Faced with high numbers of battlefield casualties, surgeons began experimenting with artificial materials for tissue replacement and augmentation in efforts to reduce battlefield mortality (DeBakey, 1946). Although most of these attempts were unsuccessful for various reasons, the idea of replacing lost tissue with synthetic materials endured and continues to be utilized today.

Through the successes and challenges of previous replacement strategies, the medical community has gained a heightened appreciation of the complexity of biological tissues. It is now recognized that each tissue is a carefully organized aggregate of cells and other factors

Principles of Regenerative Medicine. DOI: 10.1016/B978-0-12-381422-7.10037-9

specialized to perform a particular function (Wnek and Bowlin, 2008), and that the complexity increases as tissues come together to form organs. While still relying to a large degree on artificial materials and devices, the focus of tissue replacement research is moving toward regenerative medicine and total biological solutions. Along those lines, this chapter will focus on efforts to promote histogenesis in three-dimensional (3D) scaffold matrices.

TISSUE COMPONENTS

Biological tissues are composed of three components: cells, the extracellular matrix (ECM), and the signaling systems that are encoded by genes in the nuclei of the cells and then activated through cues from the ECM or other cells (Shin et al., 2003) (Fig. 37.1). Together, the three components interact in balance to form tissues and organs, and mimicking these interactions is the focus of tissue engineering.

From an engineering design standpoint, the tissues of the human body are infinitely complex, and the dynamic processes that constitute organs are not confined to a single dimension. At the most basic level, cells reside in a complex ECM environment of proteins, carbohydrates, and other structural molecules. Cells respond to changes in their environment (e.g. injury or disease) by secreting proteases, growth factors, and other signaling molecules, which in turn lead to intra- and intercellular communications dictating the initiation of processes from proliferation and protein synthesis, to migration and apoptosis.

The ECM can be broadly defined as a natural scaffold that supports tissues and organs, but it should not be viewed as merely providing strength and physical reinforcement. This matrix, which is composed of both a fibrillar and an amorphous component, is now known to be intricately involved in the events that lead to tissue formation (Reid and Zern, 1993). These two broad components interact with cells via cell surface receptors and various membrane proteins to influence processes such as cell orientation, growth and differentiation, and the secretion of bioactive molecules (Reid and Zem, 1993). Cellular responses to changes in the tissue environment may result from direct interaction with the matrix — as when physical stresses are transmitted through fibrillar ECM components to cell focal adhesion complexes within the cell — or they may arise as a consequence of the release of signaling molecules that are stored in the ECM reservoir. Understanding the complexity of tissue components and their roles in natural processes, such as embryogenesis and wound healing, will help guide tissue engineering methodologies and lead to more functional replacement options.

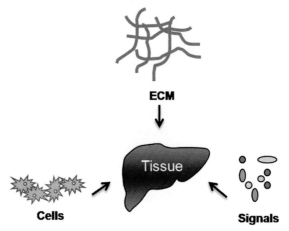

FIGURE 37.1
The basic components of mammalian tissue.

REGENERATION OF DISEASED TISSUES

Regeneration is defined as the synthesis of physiological tissue with the objective to restore lost biological function. This is in contrast with the notion of tissue repair, which is simply the closure of a wound to allow for the return of homeostatic function at the site of injury (Yannas, 2005a). Repair generally results in the contraction of tissue and formation of a scar, which contains undifferentiated epithelial tissue with little or no physiological function. While many tissue engineering strategies have been influenced by the principles at work in the wound healing process (especially with regard to cell-cell and cell-matrix interactions), the formation of scar tissue is regarded as an adverse event that directly inhibits the regeneration of new tissues (Yannas, 2005b).

The adult human, unlike an embryo, has very limited capacity for self-regeneration. While the precise reason for this inability is not understood, research demonstrates that the removal of ECM, as in disease or injury, disrupts normal tissue regeneration and directly leads to the formation of scar tissue (Yannas, 2005a). This observation signifies the importance of a structural support to direct and guide cell-mediated regeneration. Indeed, studies have shown that for large tissue defects the implantation of cells alone is ineffective at initiating new tissue formation (Yannas, 2005b). Depending on the size and location of tissue damage, cells must work in conjunction with the ECM to initiate repair or regeneration. Again, it is valuable to consider the natural process of wound healing when attempting to delineate these interactions.

Tissues are composed of three interconnected layers — the epithelium, basal lamina, and stroma. The epithelium is composed entirely of cells with little or no matrix component. In contrast, the basal lamina contains only ECM, while the stroma is composed of cells, ECM, and blood vessels (Wetzels et al., 1991) (Fig. 37.2). Both the epithelia and basal lamina can regenerate without inducing scar formation, but the stroma is usually non-regenerative, and injury to this layer directly leads to the repair state (Yannas, 2004). When tissue injury occurs, the inflammatory response is swift and quickly results in recruitment of fibroblasts and myofibroblasts to the wound bed (Yannas, 2004). The destruction of stromal tissue signals myofibroblasts to secrete collagen fibrils in an attempt to stabilize the wound. Unlike the random fibril orientation achieved by fibroblasts in normal tissue formation, myofibroblasts deposit collagen in a highly organized fashion along stress planes (Yannas, 2005a). The stress forces generated by these contractile cells are generally in one dimension and both the myofibroblasts and matrix fibers align parallel to the force (Ng et al., 2005). Deposition of ECM under stress results in contraction of the wound bed, bringing healthy tissue closer together and thus reducing the size of the defect to be healed. Unfortunately, the highly aligned scar fibrils inhibit subsequent infiltration by additional cell types, thereby limiting the extent to which the scar can be remodeled into functional, regenerated tissue.

677

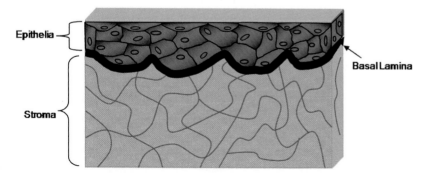

Epithelia

Basal Lamina

Stroma

FIGURE 37.2
The three layers of tissue. The epithelia is the outermost layer of tissue and is composed mainly of cells. The basal lamina is a thin ECM membrane that separates the epithelia from the stroma. The stroma is mainly composed of ECM fibers and is generally unable to spontaneously regenerate.

With the goal of limiting wound contraction and subsequent scar tissue formation, much research has focused on implanting exogenous materials into the defect site to replace the lost matrix component. As an example, Yannas and colleagues placed a collagen scaffold into a full-thickness skin wound and were able to block wound contraction and greatly reduce scar formation while inducing partial skin regeneration (Ehrlich and Hunt, 1968). The collagen construct was shown to greatly reduce the inflammatory response and, consequently, the number of myofibroblasts recruited to the wound bed. In control studies, inhibition of contraction and subsequent tissue regeneration were not evident in defects treated with cytokines, cell suspensions, or in artificial scaffolds known to induce a high inflammatory response (Yannas, 2005b). In addition, inhibition of contraction by steroids or other chemicals (Hedrick and Daniels, 2003) is not sufficient for regeneration, demonstrating the complex influence of scaffold materials in the healing process.

Though the need for a scaffold material has been validated, a complete tissue replacement strategy is composed of more than a single component. Successful histogenesis requires the presence of cells capable of producing new tissue matrix, and must include the biochemical factors necessary for the cells to live and function. Furthermore, several important design parameters related to the scaffold itself have become evident. In addition to being biocompatible, scaffold materials should have the appropriate chemical structure and micro-architecture to support cell viability and tissue formation. As tissues mature, precisely tuned scaffold degradation will allow transfer of mechanical properties to the newly formed matrix and materials permissive to vascular ingrowth will encourage incorporation with surrounding tissues, providing a blood supply for nourishment.

DESIGN PARAMETERS FOR HISTOGENESIS

Tissue engineers have recognized certain components critical to regeneration design strategies. Methodologies generally start by including the three components of tissue: cells, matrix, and signaling factors. From here, constituents are specialized based on specific tissue needs. In general, researchers seek to develop a scaffold material that can support viability of the appropriate cell type while acting as a temporary substitute for the ECM. Over time this surrogate matrix will ideally be replaced by functional replacement tissue.

Cell sources

Because scaffold materials alone are not sufficient to stimulate regeneration of all tissue defects, cell sourcing is an important consideration in the histogenesis design strategy. Host cells have been shown to infiltrate acellular matrices in some instances, but generally this infiltration distance is limited to a few microns. By implanting cells into the biomaterial scaffold, it is believed that the time for tissue infiltration by the host will be minimized and the cells can secrete bioactive factors *in vitro* that will encourage autologous ECM formation. Thus, a combination of cells and scaffold serves as the most common implant construct.

A variety of cell types have been investigated for regeneration applications including differentiated cells, adult-derived stem cells, and embryonic stem cells (Hedrick and Daniels, 2003). Mature cells have historically come from three sources: autologous, allogeneic, or xenogeniec, though xenogeniec cell transplantations have mostly been abandoned due to concerns over immune rejection and cross-species disease transmission. Allogeneic cells are harvested from healthy adult donor organs and then expanded *in vitro*. Scaffolds with allogeneic cells are subject to immune rejection but these cells have been successful in skin regeneration for burn patients (Horch et al., 2005). Autologous cells biopsied from a patient, expanded *in vitro* and then seeded onto a tissue scaffold prior to re-implantation into the same individual, are generally viewed as the ideal replacement in terms of compatibility. However, in many patients the extent of tissue damage or disease makes it difficult to obtain a sufficient supply of healthy cells (Hedrick and Daniels, 2003).

Recently the field has seen an increase in the number of progenitor cells employed for regeneration applications. These stem cells have varying degrees of potential depending on their source. In general, adult-derived cells are considered multipotent and have the capacity to generate a limited number of cell types, while pluripotent, embryonic cells can be differentiated into cells of all three germ layers (Brignier and Gewirtz, 2010). As an example, murine embryonic stem cells have been differentiated *in vitro* into various tissue cells including neurons, myocytes, and chondrocytes (Vats et al., 2002). This differentiation was initiated *in vitro* through the use of media containing cytokines and growth factors specific to the target cell lineage. Osteoblast differentiation has been accomplished in a similar manner and also by incubating embryonic stem cells in preconditioned media transferred from a culture of mature bone cells (Buttery et al., 2001). In a recent study, Ameer and colleagues obtained circulating adult progenitors from blood and guided the cells to create a functional endothelium layer on their engineered vascular construct (Allen et al., 2010). This work demonstrates the potential of engineering functional, autologous vessel substitutes.

Porosity

In most types of synthetic materials, successful histogenesis requires a porous microstructure. Scaffold porosity, pore size, and the overall pore structure all have important effects upon tissue formation and infiltration into biomaterial constructs. Interconnecting pores facilitate the loading of cells into scaffold materials while the increased internal surface area provides sites for attachment and spreading. By allowing diffusion of metabolites, oxygen, and growth factors into and out of the material, an open network structure permits cell viability and proliferation, and allows room for the deposition of cell-secreted proteins. *In vivo*, a porous structure serves to encourage infiltration of host cells, and, importantly, the extension of vascular processes, which offer nourishment to developing tissue.

Methods of creating pores or increasing the porosity of biomaterials include gas foaming (Montjovent et al., 2005), salt leaching (Gross et al., 2004), freeze drying (Moscato et al., 2006), electrospinning (He et al., 2005), and the recent inclusion of biodegradable hydrogel microparticles (Kim et al., 2009). As an example, a microporous polyurethaneurea has been developed using a gas foaming method. In these studies, sodium bicarbonate is added to a mixture of a polyurethaneurea copolymer in dimethylformamide. The polymer is cast in a mold and the solvent evaporated. The cast polymer is next placed into an acidic solution to react with the salt. Both the removal of the salt and the bubbles formed during this foaming process contribute to the porosity of the resulting scaffold (Jun and West, 2005b). The final material is homogenous in structure with surface pore morphology and pore volume fraction similar to those seen throughout the bulk (Fig. 37.3). In addition, incorporation of the laminin-derived peptide, YIGSR, permits cell adhesion to the otherwise non-adhesive polymer matrix without changing the overall porosity of the scaffold (Fig. 37.4).

It is important to consider how variations in pore architecture affect cell-material interactions. Experimental studies have led to the determination of a characteristic interaction parameter,

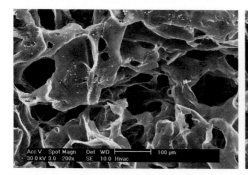

FIGURE 37.3
Microporous polyurethaneurea copolymer scaffold. The scaffold formed by gas foaming exhibits a similar porous morphology on the surface (left) and in the bulk (right) (Jun and West, 2005a).

FIGURE 37.4
Microporous polyurethaneurea scaffold modified for cell adhesion. The original scaffold porosity (left) was unaltered by addition of the cell-adhesive peptide YIGSR (right) (Jun and West, 2005a).

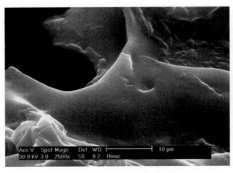
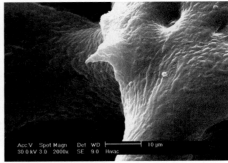

Φ_c, which relates the number of cells bound, N_c, to the available surface area, A, of the scaffold according to equation (1) (Yannas, 2005a):

$$\Phi_c = \frac{N_c}{A} \qquad (1)$$

By careful analysis, it can be shown that two matrices with identical chemistries but differing pore diameters have vastly different abilities to provide for cellular attachment. For example, scaffolds containing pores of 300 µm have a much lower internal surface area than matrices with pore diameters of less than 50 µm and, as such, bind a much lower number of cells in the matrix. Because of the low cellular infiltration, tissue formation in these constructs is slow and in certain cases inhibited due to lack of interactions between cells (Yannas, 2005a). From this calculation, it can be assumed that there is a maximum pore diameter that will allow cells sufficient area to attach and divide, while at the same time maintaining the critical intra-cellular communication that leads to histogenesis. This parameter will be influenced by scaffold material properties (e.g. hydrophobicity) and limited by the size of the cell in each application. In general, research supports that pores larger than 10 µm (the average cell diameter) will permit cell infiltration (Agrawal and Ray, 2001).

In terms of successful histogenesis, both pore size and the degree of porosity are tissue- and scaffold-specific. For example, *in vivo* osteogenesis occurs in biomaterials of high porosity (>70%) with average pore sizes >300 µm (Karageorgiou and Kaplan, 2005). However, in skin regeneration, successful scaffolds need only to exhibit pore sizes of 20–125 µm (Yannas et al., 1989). This discrepancy can be explained by the low vascular requirements of the skin and its convenient juxtaposition to an ample supply of atmospheric oxygen. There is, however, an upper limit in porosity and pore size set by constraints associated with mechanical properties. An increase in the void volume results in a reduction in mechanical strength of the scaffold, which can be detrimental in applications where regenerated tissues must support significant mechanical loads (e.g. long bones, heart valves, and articular cartilage) (Yannas, 2004). The extent to which the porosity of a scaffold can be increased while still allowing it to meet tissue mechanical requirements is dependent on many factors, including the intrinsic makeup of the biomaterial and the processing conditions used in fabrication (Karageorgiou and Kaplan, 2005).

As histogenesis progresses and gives way to organogenesis, the impact of scaffold pore structure on material degradation and tissue vacularization become apparent. The size and distribution of pores within a scaffold greatly influence the manner and rate of *in vivo* degradation (Lu et al., 2000), which can impact tissue formation and construct mechanical integrity. In materials susceptible to hydrolytic cleavage, for example, the access of water molecules to the interior of a scaffold is limited by porosity. Similar parallels exist for matrices subject to enzymatic degradation, which rely on interaction with cell-secreted molecules for dissolution. A final, important consideration is the influence of pore structure on the establishment of a blood supply in newly developing tissue. In early stages of histogenesis, nutrients, metabolites, and other factors essential to cell survival pass freely through scaffold pores. As these pores fill with new tissue, however, a functioning vasculature is necessary. New, "designer" tissue-engineered scaffolds are composed of precisely controlled porous architectures that support and guide vessel ingrowth during tissue development (Hollister, 2005). The concepts of degradation and microvascularization are discussed in more detail in the following sections.

Degradation

Though non-degradable biomaterials have had success in many medical devices, complications, due primarily to chronic foreign body responses, have yet to be overcome. For this reason, the ideal regeneration construct is one that that can eventually be replaced by native tissues. Furthermore, the degradation rate of the construct is intrinsic to the success of the implant. This means that the material dissolution should complement tissue synthesis in order to ensure suitable mechanical stability during the process of histogenesis. The necessary scaffold residence time is tissue specific and must account for the time required for cells to adequately populate the scaffold and deposit a stable ECM. If a biomaterial degrades before sufficient ECM deposition has occurred, cells will lose important physiochemical factors for tissue regeneration and repair is likely to occur, resulting in scar formation. However, if the scaffold residence time is too long, ECM deposition and cell proliferation will be suppressed. An important balance must be established for successful regeneration.

In most currently employed scaffold materials, degradation in the *in vivo* aqueous environment occurs via hydrolysis of chemical bonds in the base material. Chemical functionalities, molecular weight, and degree of crosslinking determine the degradation characteristics. For example, higher-molecular-weight materials tend to degrade more slowly over time, as do materials with a higher hydrophobicity and crystallinity. Using a combination of these factors, predictable degradation profiles can be developed to match expected tissue formation rates. However, the consequences of material dissolution must be considered. As mentioned above, scaffolds that undergo bulk erosion can become rapidly unstable due to formation of large pores with low mechanical stability (Lu et al., 2000). Additionally, the degradation products of some scaffolds can be toxic not only to cells of the surrounding tissue, but to vital organs of the lymphatic system. For example, the degradation of the frequently studied polylactic acid and polyglycolic acid scaffolds results in a marked pH drop in the local vicinity due to the release of acidic degradation products (Martin et al., 1996; Lu et al., 2000). The pH decrease can be detrimental to cells and organs and over time can lead to an inflammatory response with possible capsule formation and tissue necrosis (Sung et al., 2004).

As an alternative to hydrolytic degradation, many investigators are developing smart materials that can be dynamically remodeled during histogenesis via cell-mediated processes. These scaffolds are designed to mimic the degradation of natural ECM proteins, which are subject to matrix metalloproteinases (MMPs) and serine proteases that are either secreted or activated by most cell types. Since proteolysis-induced degradation is required for cell migration and invasion, researchers have had success in introducing synthetic hydrogels that are sensitive to cell proteases. Hydrogels containing amino acid sequences that can be degraded by plasmin (Halstenberg et al., 2002), MMPs (Kim et al., 2005), or both of these protease families (West and Hubbell, 1999; Mann et al., 2001a; Raeber et al., 2005), all exhibit sustained degradation upon cellular infiltration.

In our own lab we have fabricated MMP-degradable hydrogels that become fluorescent when degraded by cell proteases (Lee et al., 2005). These PEG-based hydrogels are synthesized with MMP-degradable segments in the polymer backbone that are labeled with fluorescent, self-quenching tags. Thus, intact substrates show no fluorescence, but, upon degradation by cell proteases, quantifiable fluorescence is emitted. Cells seeded in these fluorogenic substrates are able to cleave the degradable hydrogel matrix as visualized by a marked increase in fluorescence in the areas immediately around the cell (Fig. 37.5). In addition, cell migration trails could be seen in the hydrogels. It is believed these materials will contribute to the understanding of cell migration and cell-mediated scaffold degradation.

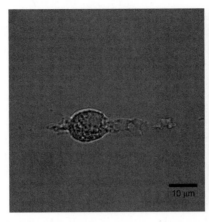
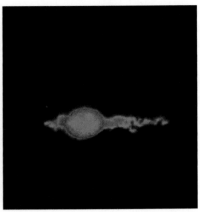

FIGURE 37.5
Fibroblast encapsulated within a fluorogenic substrate. DIC image (left) and fluorescent image (right) showing green fluorescence generated by material degradation around cell (red).

Biomolecular factors

In many cases, seeding cells inside a porous scaffold is not sufficient to induce tissue regeneration because the material does not contain the chemical cues necessary to promote cellular remodeling events. Thus, researchers have attempted to actively modify biomaterials at the molecular level by incorporating cell-specific biomolecules. One strategy is to encapsulate molecules such as peptides or proteins into biomaterial carriers so that these molecules can be released from the material to trigger or modulate new tissue formation (Shin et al., 2003). Another approach involves physically or chemically modifying scaffolds with specific cell-binding peptides to increase cellular interaction with the substrate. Cell-binding peptides are short amino acid sequences derived from much longer native ECM proteins that have been identified as able to incur specific, predictable interactions with cell receptors. Essentially, the peptides function to mimic the signaling dynamic between the ECM and cells, and, because many synthetic scaffold materials are not inherently adhesive to cells, introduction of such sequences can be critical to encouraging cell retention and subsequent tissue formation (Mann et al., 1999). The most well studied cell adhesion peptide, arginine-glycine-aspartic acid-serine (RGDS), has been widely used to encourage cells of various types to interact with otherwise non-adhesive synthetic matrices (Hern and Hubbell, 1998) (Fig. 37.6). Other amino acid sequences have been found to promote adhesion by specific cell phenotypes including endothelial cells (Gobin and West, 2003b; Heilshorn et al., 2003; Jun and West 2005a,b), smooth muscle cells (Gobin and West, 2003b), neural cells (Adams et al., 2005), and osteoblasts (Benoit and Anseth, 2005).

Various growth factors have also been employed in efforts to enhance the process of histogenesis. Because they play key roles in tissue differentiation and repair, molecules such as epidermal growth factor (EGF), platelet-derived growth factor (PDGF), and transforming growth factor-β1 (TGF-β1) are popular for use in these applications. Recently, the release of PDGF from a polyurethane scaffold has been shown to encourage healing of wounds in a rat

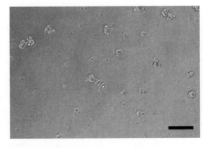
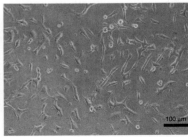

FIGURE 37.6
Peptide modification promotes cell adhesion. The non-adhesive nature of PEG hydrogels (left) can be significantly altered by inclusion of an RGDS peptide to promote cell attachment (right). Images courtesy of Christy Franco, Rice University.

model (Li et al., 2009), while Griffith and colleagues used tethered EGF to promote osteogenic differentiation in efforts to improve connective tissue regeneration (Marcantonio et al., 2009). TGF-β1 has many roles in histogenesis (Letterio and Roberts, 1998) and has received particular attention for aiding the differentiation of stem cells and the development of new vasculature *in vivo* (Bohnsack and Hirschi, 2004). In other work, basic fibroblast growth factor (bFGF) and nerve growth factor (NGF) were immobilized in fibrin scaffolds to facilitate cellular recruitment and differentiation (Sakiyama-Elbert and Hubbell, 2000). Biomolecules have also been covalently coupled to PEG-based materials (Gobin and West, 2003a; DeLong et al., 2005b; Leslie-Barbick et al., 2009; Moon et al., 2009) with vascular endothelial growth factor (VEGF) showing potential to drive endothelial cell tubulogenesis. Further, Delong and West formed gradients of bFGF and observed cellular alignment and migration directly influenced by growth factor presentation (DeLong et al., 2005b) (Fig. 37.7).

Importance of microvasculature

One of the biggest limitations to histogenesis in 3D scaffolds is the lack of a functional vascular system. The most successful engineered materials to date have been for tissues, such as skin, that are thin enough to be supported by diffusion from the host vascular, and cartilage, which is relatively avascular and as such contains cells that are tolerant of anoxic conditions. All other tissues require some form of vascular system to permit long-term survival of cells within the material. There are two primary strategies for establishing a blood supply in implanted 3D scaffolds. Vessels can either grow into the construct from host tissue, or they can be preformed *in vitro* and interconnect with host vasculature upon implantation.

The premise behind the first strategy is to encourage vessels to enter an avascular construct from the host tissue by stimulating angiogenesis. To encourage this process, scaffolds can be fabricated with precisely designed pore structures or surface chemistries that support ingrowth (Hollister, 2005). In some cases pro-angiogenic factors are immobilized on or released from implants in order to encourage ingrowth of vessels from host tissue. Recent work from our laboratory showed extensive infiltration of functional vessels into a VEGF-laden, PEG-based hydrogel that had been implanted in the mouse cornea (Moon et al., 2010). The limitation to these strategies is the time required for vessels to extend into the entirety of the engineered construct. At an extension rate of approximately 5 μm/h (Laschke et al., 2009), new vessels will not reach the center of large scaffolds until several days after implantation, leaving any cells at these locations without sufficient supplies of oxygen and nutrients.

A second option is to create preformed vascular networks *in vitro* that are capable of anastomosing with host vascular upon implantation. This process of connecting two independent

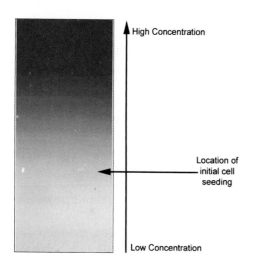

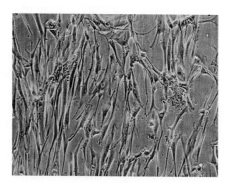

FIGURE 37.7
A gradient of bFGF immobilized in a hydrogel scaffold. Cells seeded on hydrogels containing a bFGF gradient aligned and migrated along the axis of growth factor immobilization (DeLong et al., 2005b).

vascular networks is called inosculation and is the mechanism primarily responsible for the successes in plastic surgery and skin transplantations. To generate microvascular networks *in vitro*, researchers seed scaffolds with cells known to participate in vasculogenesis including endothelial cells, stem cells, and pericytes. With appropriate biochemical and/or physical stimulation, these cells will self-assemble into capillary-like structures. As an example, successful vascular networks have been formed by endothelial cells in fibrin scaffolds (Montano et al., 2010) and by adult and cord blood-derived progenitor cells in Matrigel (Melero-Martin et al., 2008). In each study, the preformed vessels were functional upon implantation *in vivo* and, in the case of the endothelial cells, immature capillaries were further stabilized by host mural cells. Other work along these lines suggests that providing relevant physiomechanical stimulation *in vitro* will aid in developing functional prevascularized networks in engineered constructs (Burg et al., 2000; Sudo et al., 2009). A final approach exploits the body's own ability to form blood vessels. A variety of scaffold materials have been implanted in highly vascularized anatomical sites, where they are incubated for up to 3 weeks as host vessels infiltrate. Once the vessel network is established, the scaffold is explanted, loaded with cells, and finally implanted into the site targeted for regeneration (Laschke et al., 2008).

The need for adequate blood flow in 3D scaffolds is readily apparent and of concern to all researchers interested in regeneration strategies. Advances of late are quite promising, though much work remains to be done. Inosculation of preformed microvascular networks with host vascular, for instance, provides functional transport of engineered materials much more rapidly than some other methods, but is still too slow for very sensitive tissue applications. A combination of this approach with pro-angiogenic strategies may encourage connection in a shorter time period, thus leading to better long-term regeneration of target tissue.

SYNTHETIC MATERIALS FOR HISTOGENESIS OF NEW ORGANS

Biomaterials investigated as scaffolds for histogenesis include natural polymers such as collagen and fibrin as well as a range of synthetic substrates. While natural matrices have certain advantages in that their chemical composition is generally amenable to cell growth, batch to batch variations in substrate quality and performance can make their use in clinical regeneration applications problematic. As such, the control and flexibility of synthetic materials make them attractive alternatives. As alluded to previously, control is an important element in tailoring the scaffold's material properties for appropriate cell-material interactions. The following sections highlight two popular classes of synthetic materials: hydrolytically degradable polymers and hydrogels.

Hydrolytically degradable polymers

The most widely used polymers for cellular scaffold materials are polylactic acid (PLA), polyglycolic acid (PGA), or a combination of these two polymers (PLGA). PLA, PGA, and PLGA are aliphatic esters that possess good biocompatibility (Li, 1999) and have been used as drug delivery materials to administer biomolecules during tissue regeneration (Brannon-Peppas and Vert, 2000; Whang et al., 2000). These polymers are also among the few synthetic polymers approved by the US Food and Drug Administration (FDA) for certain human clinical applications. PGA is extremely hydrophilic in nature and, consequently, will lose its mechanical strength within 2–4 weeks of implantation (Reed and Gilding, 1981). PLA, however, contains an additional methyl group and as a result is more hydrophobic. Degradation of PLA scaffolds can take from months to years (Pitt et al., 1981; Brannon-Peppas and Vert, 2000). In addition, the degradation rates of these polymers can be tailored by using copolymer blends (PLGA), which give distinct degradation profiles (Ma, 2004; Brannon-Peppas and Vert, 2000). However, these scaffolds undergo acid-catalyzed hydrolysis and bulk erosion, which have the potential to result in structural instability and interruption of the regeneration process (Moran, 1998).

Polyanhydrides have been synthesized for a number of biomedical applications including tissue engineering and drug delivery (Burkoth and Anseth, 2000). Polyanhydride scaffolds exhibit excellent biocompatibility and contain a large aliphatic component that possesses an ester group that makes the material subject to surface erosion (Davis et al., 2003). This deliberate surface erosion is mechanistically different from bulk hydrolysis and can be exploited to synthesize biomaterial scaffolds that have very predictable degradation profiles. In addition, the erosion of only the surface of the material allows anhydrides to maintain structural integrity in support of histogenesis. Because they exhibit mechanical properties similar to bone and are ideal scaffolds for tissue infiltration, anhydrides have been widely employed as scaffolds for *in vivo* bone regeneration (Anseth et al., 1999; Muggli et al., 1999; Burkoth and Anseth, 2000). Polyanhydride networks can also be combined with other polymers to change their degradation and structural characteristics. Jiang and Zhu (1999) showed that anhydride polymers could be polymerized in the presence of PEG to form a crosslinked network with both hydrophobic and hydrophilic components. The hydrophilic PEG chains increase uptake of water to in turn drive the hydrolysis of the ester bond in the hydrophobic anhydride. As such, the degradation properties can be tailored by altering the amount of PEG in the scaffold material.

Hydrogels

Hydrogels, which contain up to 90% water, are another widely studied class of materials for tissue regeneration. These materials are appealing because the polymer properties are controllable and reproducible (Peppas, 2004) and the large water uptake promotes excellent biocompatibility. In many cases hydrogel mechanical properties resemble those of native tissue and can be systematically controlled for specific applications. In addition, several hydrogel monomers contain vinyl chemical moieties, which are conducive to various free-radical-initiated polymerizations schemes that can be employed to generate solid substrate materials. Photointiation, for example, allows for polymers to be formed using specific wavelengths of light. Using this method, many researchers have had success forming complex, 3D structures of varying stiffnesses. For example, polyacrylamide hydrogels have been shown to induce regeneration of soft tissue in facial defects (von Buelow et al., 2005), and 2-hydroxyethylmethacrylate has had good success as a fibrillar support for nerve regeneration (Flynn et al., 2003).

Among the most studied hydrogel materials is crosslinked PEG, which has been approved by the FDA for use in certain medical applications (Drury and Mooney, 2003). As with other hydrogels, the hydrophilic nature of PEG discourages cell and protein adhesion and therefore results in a low instance of immunorejection by the host. By changing the monomer chain length, adding biological molecules, or utilizing copolymers, researchers have generated a wide array of PEG hydrogel formulations suitable for many different tissue engineering applications. To render an otherwise blank slate amenable to histogenesis, various biomemetic peptides and growth factors have been incorporated into the PEG hydrogel matrix (Mann et al., 2001b; Gonzalez et al., 2004; DeLong et al., 2005a,b; Leslie-Barbick et al., 2009). These modifications have been successful in achieving selective cell adhesion and promoting the accumulation of secreted tissue matrix. As mentioned previously, similar methods have been employed to encourage vasculogenesis and angiogenesis in PEG hydrogel materials. In addition, these hydrogel materials show promise as small-diameter vascular grafts (Hahn et al., 2007). Incorporation of peptides subject to proteolytic cleavage in the backbone of the PEG polymer chain renders the scaffolds subject to cell-mediated remodeling, giving these materials an additional advantage as histogenesis conduits.

685

FUTURE DIRECTIONS IN 3D SCAFFOLDS: 3D MICROFABRICATION

New advances in the biomaterial field are providing tissue engineers with the means to generate complex and highly specialized 3D scaffolds. One of the earliest examples of such architecture was developed by Griffith and colleagues for hepatocyte culture and liver regeneration. Using a rapid printing technique, microporous PLGA scaffolds were fabricated by directing solvent streams onto polymer granules in a precisely controlled manner (Kim et al., 1998). The hepatocytes seeded upon these constructs exhibited increased metabolic rates that more closely mimicked cells *in vivo*. In other work, 3D, microporous PLGA foams were prepared by drilling with dies of a specific size. The dimensions of these cylindrical scaffolds were reproducible with millimeter precision, and, when placed *in vivo*, the materials supported bone regeneration in non-healing defect models (Karp et al., 2003, 2004). Porous scaffolds have also been micro-patterned for vascular tissue engineering applications (Sarkar et al., 2006).

Several researchers have used photopolymerization techniques to mold and pattern hydrogel scaffolds for better control of cell-substrate interactions (Bryant et al., 2007). Peppas reports micropatterning of PEG hydrogels using UV polymerization to generate many different substrate morphologies on the order of 100 µm (Peppas and Ward, 2004). Liu and Bhatia also photopatterned PEG hydrogels using a layer-by-layer method to generate a 3D scaffold for hepatocytes (Tsang et al., 2007).

Laser-based patterning of hydrogels is a relatively new technique for generating complex 3D microenvironments inside hydrogel materials and natural constructs. Liu et al. used a laser ablation technique to form lines, holes, and interconnected grids in collagen matrices (Liu et al., 2005), while growth factors and peptides were patterned by Roy and colleagues using laser-based stereolithography (Mapili et al., 2005) inside a PEG hydrogel. Biomolecules have also been patterned inside agarose hydrogels (Luo and Shoichet, 2004). In this case, RGDS peptides were patterned in cylindrical shapes within the hydrogel material. After 3 days,

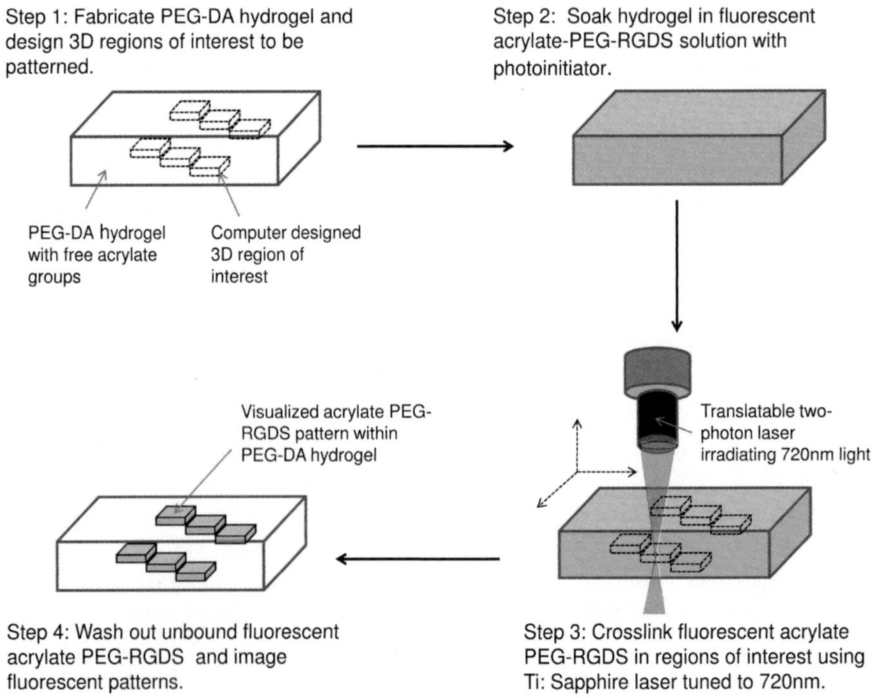

FIGURE 37.8

Laser scanning lithography patterning of PEG hydrogels. Precisely defined patterned areas are generated in 3D hydrogels by using a confocal microscope laser to crosslink photosensitive materials. Schematic courtesy of Joseph C. Hoffman, Rice University.

Step 1: Fabricate PEG-DA hydrogel and design 3D regions of interest to be patterned.

PEG-DA hydrogel with free acrylate groups

Computer designed 3D region of interest

Step 2: Soak hydrogel in fluorescent acrylate-PEG-RGDS solution with photoinitiator.

Translatable two-photon laser irradiating 720nm light

Step 3: Crosslink fluorescent acrylate PEG-RGDS in regions of interest using Ti: Sapphire laser tuned to 720nm.

Visualized acrylate PEG-RGDS pattern within PEG-DA hydrogel

Step 4: Wash out unbound fluorescent acrylate PEG-RGDS and image fluorescent patterns.

 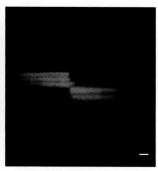

FIGURE 37.9
LSL pattern of fluorescently labeled RGDS in a PEG hydrogel. The fluorescent peptide (red) is visible in the bulk hydrogel (black) after patterning. Scale bar is 10 µm. Image courtesy of Joseph C. Hoffman, Rice University.

neuronal cells seeded on the surface of the materials were shown to have migrated into the hydrogel in only the selectively patterned areas. Additional studies of cell migration in hydrogel materials were conducted with micro-patterned PEG-based materials functionalized with several different bioactive moieties (Hahn et al., 2005; Lee et al., 2008; Moon et al., 2009). In the process of laser scanning lithography (LSL), photosensitive peptides or proteins are covalently incorporated into 3D hydrogels with the precision of a confocal microscope laser (Fig. 37.8). The technique is capable of generating features from 1 µm to 1 mm, and can be extended to include multiple bioactive moieties in a single substrate. The image in Figure 37.9 illustrates the 3D nature of a patterned ligand. These precisely fabricated regeneration matrices provide great opportunities for controlled tissue growth.

CONCLUSIONS

687

The need for replacement tissues and organs is driving tissue engineers to develop materials and strategies capable of generating biologically functional substitutes. The study of natural processes, such as wound healing, have provided insights into the complex mechanisms of tissue regeneration and have allowed researchers to prioritize design parameters for 3D scaffolds. At the same time, advances in biomaterial synthesis and modification, as well as a better understanding of the signaling molecules important in tissue synthesis, are providing a wealth of tools for regeneration strategies. In a systematic approach to histogenesis, Nettles et al. developed a method of neural network analysis in which a self-organizing map delineates the relationships between scaffold parameters, such as crosslink density, and tissue outcomes (Nettles et al., 2010). The investigators employed this tool with the goal of optimizing and accelerating the design of a cartilage tissue substitute. Tools like these will help to focus the work of tissue engineers going forward. The last decade has seen good success in developing substitutes for skin and cartilage. Recent advances in scaffold microvascularization techniques will aid in progressing the field to larger, more complex target tissues.

References

Adams, D. N., Kao, E. Y. C., Hypolite, C. L., Distefano, M. D., Hu, W. S., & Letourneau, P. C. (2005). Growth cones turn and migrate up an immobilized gradient of the laminin IKVAV peptide. *J. Neurobiol., 62*, 134–147.

Agrawal, C. M., & Ray, B. R. (2001). Biodegradable polymeric scaffolds for musculoskeletal tissue engineering. *J. Biomed. Mater. Res., 55*, 141–150.

Allen, J. B., Khan, S., Lapidos, K. A., & Ameer, G. A. (2010). Toward engineering a human neoendothelium with circulating progenitor cells. *Stem Cells, 28*, 318–328.

Ang, G. C. (2005). History of skin transplantation. *Clin. Dermatol., 23*, 320.

Angelos, P., Lafreniere, R., Murphy, T. F., & Rosen, W. (2003). Ethical issues in surgical treatment and research. *Curr. Probl. Surg., 40*, 353–448.

Anseth, K. S., Shastri, V. R., & Langer, R. (1999). Photopolymerizable degradable polyanhydrides with osteo-compatibility. *Nature Biotechnol., 17,* 156–159.

Benoit, D. S. W., & Anseth, K. S. (2005). The effect on osteoblast function of colocalized RGD and PHSRN epitopes on PEG surfaces. *Biomaterials, 26,* 5209–5220.

Bohnsack, B. L., & Hirschi, K. K. (2004). Red light, green light: signals that control endothelial cell proliferation during embryonic vascular development. *Cell Cycle, 3,* 1506–1511.

Brannon-Peppas, L., & Vert, M. (2000). Polylactic and polyglycolic acids as drug delivery carriers. In D. Wise, L. Brannon-Peppas, A. M. Klibanov, R. Langer, A. G. Mikos, & N. A. Peppas et al. (Eds.), *Handbook of Pharmaceutical Controlled Release Technology* (pp. 99–130). New York: CRC Press.

Brignier, A. C., & Gewirtz, A. M. (2010). Embryonic and adult stem cell therapy. *J. Allergy Clin. Immunol., 125,* S336–S344.

Bryant, S. J., Cuy, J. L., Hauch, K. D., & Ratner, B. D. (2007). Photo-patterning of porous hydrogels for tissue engineering. *Biomaterials, 28,* 2978–2986.

Burg, K. J., Holder, W. D., Jr., Culberson, C. R., Beiler, R. J., Greene, K. G., Loebsack, A. B., et al. (2000). Comparative study of seeding methods for three-dimensional polymeric scaffolds. *J. Biomed. Mater. Res., 52,* 576.

Burkoth, A. K., & Anseth, K. S. (2000). A review of photocrosslinked polyanhydrides: in situ forming degradable networks. *Biomaterials, 21,* 2395–2404.

Buttery, L. D. K., Bourne, S., Xynos, J. D., Wood, H., Hughes, F. J., Hughes, S. P. F., et al. (2001). Differentiation of osteoblasts and *in vitro* bone formation from murine embryonic stem cells. *Tissue Eng., 7,* 89–99.

Davis, K. A., Burdick, J. A., & Anseth, K. S. (2003). Photoinitiated crosslinked degradable copolymer networks for tissue engineering applications. *Biomaterials, 24,* 2485–2495.

DeBakey, M. S., & Simeone, F. A. (1946). Battle injuries of the arteries. *Am. J. Surg, 123,* 534.

DeLong, S. A., Gobin, A. S., & West, J. L. (2005a). Covalent immobilization of RGDS on hydrogel surfaces to direct cell alignment and migration. *J. Control. Release, 109,* 139.

DeLong, S. A., Moon, J. J., & West, J. L. (2005b). Covalently immobilized gradients of bFGF on hydrogel scaffolds for directed cell migration. *Biomaterials, 26,* 3227–3234.

Drury, J. L., & Mooney, D. J. (2003). Hydrogels for tissue engineering: scaffold design variables and applications. *Biomaterials, 24,* 4337–4351.

Ehrlich, H. P., & Hunt, T. K. (1968). Effects of cortisone and vitamin A on wound healing. *Ann. Surg., 167,* 324.

Flynn, L., Dalton, P. D., & Shoichet, M. S. (2003). Fiber templating of poly(2-hydroxyethyl methacrylate) for neural tissue engineering. *Biomaterials, 24,* 4265–4272.

Gobin, A. S., & West, J. L. (2003a). Effects of epidermal growth factor on fibroblast migration through biomimetic hydrogels. *Biotechnol. Prog., 19,* 1781–1785.

Gobin, A. S., & West, J. L. (2003b). Val-ala-pro-gly, an elastin-derived non-integrin ligand: smooth muscle cell adhesion and specificity. *J. Biomed. Mater. Res. A, 67A,* 255–259.

Gonzalez, A. L., Gobin, A. S., West, J. L., McIntire, L. V., & Smith, C. W. (2004). Integrin interactions with immobilized peptides in polyethylene glycol diacrylate hydrogels. *Tissue Eng., 10,* 1775–1786.

Gross, K. A., & Rodriguez-Lorenzo, L. M. (2004). Biodegradable composite scaffolds with an interconnected spherical network for bone tissue engineering. *Biomaterials, 25,* 4955–4962.

Hahn, M. S., Miller, J. S., & West, J. L. (2005). Laser scanning lithography for surface micropatterning on hydrogels. *Adv. Mater., 17,* 2939.

Hahn, M. S., McHale, M. K., Wang, E., Schmedlen, R. H., & West, J. L. (2007). Physiologic pulsatile flow bioreactor conditioning of poly(ethylene glycol)-based tissue engineered vascular grafts. *Ann. Biomed. Eng., 35,* 190–200.

Halstenberg, S., Panitch, A., Rizzi, S., Hall, H., & Hubbell, J. A. (2002). Biologically engineered protein-graft-poly (ethylene glycol) hydrogels: a cell adhesive and plasm in-degradable biosynthetic material for tissue repair. *Biomacromolecules, 3,* 710–723.

Hauben, D. J. (1985). The history of free skin transplant operations. *Acta Chir. Plast., 27,* 66.

He, W., Ma, Z. W., Yong, T., Teo, W. E., & Ramakrishna, S. (2005). Fabrication of collagen-coated biodegradable polymer nanofiber mesh and its potential for endothelial cells growth. *Biomaterials, 26,* 7606–7615.

Hedrick, M. H., & Daniels, E. J. (2003). The use of adult stem cells in regenerative medicine. *Clinics In. Plastic Surgery, 30,* 499–505.

Heilshorn, S. C., DiZio, K. A., Welsh, E. R., & Tirrell, D. A. (2003). Endothelial cell adhesion to the fibronectin CS5 domain in artificial extracellular matrix proteins. *Biomaterials, 24,* 4245–4252.

Hern, D. L., & Hubbell, J. A. (1998). Incorporation of adhesion peptides into nonadhesive hydrogels useful for tissue resurfacing. *J. Biomed. Mater. Res., 39,* 266–276.

Hollister, S. J. (2005). Porous scaffold design for tissue engineering. *Nat. Mater., 4,* 518–524.

688

Horch, R. E., Kopp, J., Kneser, U., Beier, J., & Bach, A. D. (2005). Tissue engineering of cultured skin substitutes. *J. Cell. Mol. Med., 9*, 592–608.

Jiang, H. L., & Zhu, K. J. (1999). Preparation, characterization and degradation characteristics of polyanhydrides containing poly(ethylene glycol). *Polymer Int., 48*, 47–52.

Jun, H.-W., & West, J. L. (2005a). Endothelialization of microporous YIGSR/PEG-modified polyurethaneurea. *Tissue Eng., 11*, 1133–1140.

Jun, H.-W., & West, J. L. (2005b). Modification of polyurethaneurea with PEG and YIGSR peptide to enhance endothelialization without platelet adhesion. *J. Biomed. Mater. Res., 72B*, 131–139.

Karageorgiou, V., & Kaplan, D. (2005). Porosity of 3D biomaterial scaffolds and osteogenesis. *Biomaterials, 26*, 5474.

Karp, J. M., Rzeszutek, K., Shoichet, M. S., & Davies, J. E. (2003). Fabrication of precise cylindrical three-dimensional tissue engineering scaffolds for *in vitro* and *in vivo* bone engineering applications. *J. Craniofac. Surg., 14*, 317–323.

Karp, J. M., Sarraf, F., Shoichet, M. S., & Davies, J. E. (2004). Fibrin-filled scaffolds for bone-tissue engineering: an *in vivo* study. *J. Biomed. Mater. Res. A, 71A*, 162–171.

Kim, J., Yaszemski, M. J., & Lu, L. (2009). Three-dimensional porous biodegradable polymeric scaffolds fabricated with biodegradable hydrogel porogens. *Tissue Eng. C Methods, 15*, 583–594.

Kim, S. S., Utsunomiya, H., Koski, J. A., Wu, B. M., Cima, M. J., Sohn, J., et al. (1998). Survival and function of hepatocytes on a novel three-dimensional synthetic biodegradable polymer scaffold with an intrinsic network of channels. *Ann. Surg., 228*, 8–13.

Kim, S., Chung, E. H., Gilbert, M., & Healy, K. E. (2005). Synthetic MMP-13 degradable ECMs based on poly(*N*-isopropylacrylamide-co-acrylic acid) semi-interpenetrating polymer networks. I. Degradation and cell migration. *J. Biomed. Mater. Res. A, 75A*, 73–88.

Laschke, M. W., Rucker, M., Jensen, G., Carvalho, C., Mulhaupt, R., Gellrich, N. C., et al. (2008). Improvement of vascularization of PLGA scaffolds by inosculation of in situ-preformed functional blood vessels with the host microvasculature. *Ann. Surg., 248*, 939–948.

Laschke, M. W., Vollmar, B., & Menger, M. D. (2009). Inosculation: connecting the life-sustaining pipelines. *Tissue Eng. B Rev., 15*, 455–465.

Lee, S. H., Miller, J. S., Moon, J. J., & West, J. L. (2005). Proteolytically degradable hydrogels with a fluorogenic substrate for studies of cellular proteolytic activity and migration. *Biotechnol. Prog., 21*, 1736–1741.

Lee, S. H., Moon, J. J., & West, J. L. (2008). Three-dimensional micropatterning of bioactive hydrogels via two-photon laser scanning photolithography for guided 3D cell migration. *Biomaterials, 29*, 2962–2968.

Leslie-Barbick, J. E., Moon, J. J., & West, J. L. (2009). Covalently-immobilized vascular endothelial growth factor promotes endothelial cell tubulogenesis in poly(ethylene glycol) diacrylate hydrogels. *J. Biomater. Sci. Polym. Ed., 20*, 1763–1779.

Letterio, J. J., & Roberts, A. B. (1998). Regulation of immune responses by TGF-beta. *Annu. Rev. Immunol., 16*, 137–161.

Li, B., Davidson, J. M., & Guelcher, S. A. (2009). The effect of the local delivery of platelet-derived growth factor from reactive two-component polyurethane scaffolds on the healing in rat skin excisional wounds. *Biomaterials, 30*, 3486–3494.

Li, S. M. (1999). Hydrolytic degradation characteristics of aliphatic polyesters derived from lactic and glycolic acids. *J. Biomed. Mater. Res., 48*, 342–353.

Liu, Y. M., Sun, S., Singha, S., Cho, M. R., & Gordon, R. J. (2005). 3D femtosecond laser patterning of collagen for directed cell attachment. *Biomaterials, 26*, 4597–4605.

Lu, L., Peter, S. J., Lyman, M. D., Lai, H. L., Leite, S. M., Tamada, J. A., et al. (2000). *In vitro* and *in vivo* degradation of porous poly(dl-lactic-co-glycolic acid) foams. *Biomaterials, 21*, 1837–1845.

Luo, Y., & Shoichet, M. S. (2004). A photolabile hydrogel for guided three-dimensional cell growth and migration. *Nature Mater., 3*, 249–253.

Ma, P. X. (2004). Scaffolds for tissue fabrication. *Mater. Today, 7*, 30–40.

Mann, B. K., Tsai, A. T., Scott-Burden, T., & West, J. L. (1999). Modification of surfaces with cell adhesion peptides alters extracellular matrix deposition. *Biomaterials, 20*, 2281–2286.

Mann, B. K., Gobin, A. S., Tsai, A. T., Schmedlen, R. H., & West, J. L. (2001a). Smooth muscle cell growth in photopolymerized hydrogels with cell adhesive and proteolytically degradable domains: synthetic ECM analogs for tissue engineering. *Biomaterials, 22*, 3045.

Mann, B. K., Schmedlen, R. H., & West, J. L. (2001b). Tethered-TGF-[beta] increases extracellular matrix production of vascular smooth muscle cells. *Biomaterials, 22*, 439–444.

Mapili, G., Lu, Y., Chen, S. C., & Roy, K. (2005). Laser-layered microfabrication of spatially patterned functionalized tissue-engineering scaffolds. *J. Biomed. Mater. Res. B Appl. Biomater., 75B*, 414–424.

Marcantonio, N. A., Boehm, C. A., Rozic, R. J., Au, A., Wells, A., Muschler, G. F., et al. (2009). The influence of tethered epidermal growth factor on connective tissue progenitor colony formation. *Biomaterials, 30,* 4629–4638.

Martin, C., Winet, H., & Bao, J. Y. (1996). Acidity near eroding polylactide-polyglycolide *in vitro* and *in vivo* in rabbit tibial bone chambers. *Biomaterials, 17,* 2373.

Melero-Martin, J. M., de Obaldia, M. E., Kang, S. Y., Khan, Z. A., Yuan, L., Oettgen, P., & Bischoff, J. (2008). Engineering robust and functional vascular networks *in vivo* with human adult and cord blood-derived progenitor cells. *Circ. Res., 103,* 194–202.

Montano, I., Schiestl, C., Schneider, J., Pontiggia, L., Luginbuhl, J., Biedermann, T., et al. (2010). Formation of human capillaries *in vitro*: the engineering of prevascularized matrices. *Tissue Eng. A, 16,* 269–282.

Montjovent, M. O., Mathieu, L., Hinz, B., Applegate, L. L., Bourban, P. E., Zambelli, P. Y., et al. (2005). Biocompatibility of bioresorbable poly(l-lactic acid) composite scaffolds obtained by supercritical gas foaming with human fetal bone cells. *Tissue Eng., 11,* 1640–1649.

Moon, J. J., Hahn, M. S., Kim, I., Nsiah, B. A., & West, J. L. (2009). Micropatterning of poly(ethylene glycol) diacrylate hydrogels with biomolecules to regulate and guide endothelial morphogenesis. *Tissue Eng. A, 15,* 579–585.

Moon, J. J., Saik, J. E., Poche, R. A., Leslie-Barbick, J. E., Lee, S. H., Smith, A. A., Dickinson, M. E., & West, J. L. (2010). Biomimetic hydrogels with pro-angiogenic properties. *Biomaterials, 31,* 3840–3847.

Moran, J. M. B. (1998). Fabrication and characterization of PLA/PGA composites for cartilage tissue engineering. *Tissue Eng., 4,* S498.

Moscato, S., Cascone, M. G., Lazzeri, L., Danti, S., Mattii, L., Dolfi, A., & Bernardini, N. (2006). Morphological features of ovine embryonic lung fibroblasts cultured on different bioactive scaffolds. *J. Biomed. Mater. Res. A, 76A,* 214–221.

Muggli, D. S., Burkoth, A. K., & Anseth, K. S. (1999). Crosslinked polyanhydrides for use in orthopedic applications: degradation behavior and mechanics. *J. Biomed. Mater. Res., 46,* 271–278.

Nettles, D. L., Haider, M. A., Chilkoti, A., & Setton, L. A. (2010). Neural network analysis identifies scaffold properties necessary for *in vitro* chondrogenesis in elastin-like polypeptide biopolymer scaffolds. *Tissue Eng. A, 16,* 11–20.

Ng, C. P., Hinz, B., & Swartz, M. A. (2005). Interstitial fluid flow induces myofibroblast differentiation and collagen alignment *in vitro*. *J. Cell Sci., 118,* 4731–4739.

OPTN/SRTR. (2008). *2008 Annual Report of the US Organ Procurement and Transplantation Network and Scientific Registry of Transplant Recipients: Transplant Data 1998–2007.* Rockville, MD and Richmond, VA: HHS.

Peppas, N. A. (2004). Devices based on intelligent biopolymers for oral protein delivery. *Int. J. Pharm., 277,* 11–17.

Peppas, N. A., & Ward, J. H. (2004). Biomimetic materials and micropatterned structures using iniferters. *Adv. Drug Deliv. Rev., 56,* 1587–1597.

Pitt, C. G., Gratzl, M. M., Kimmel, G. L., Surles, J., & Schindler, A. (1981). Aliphatic polyesters. 2. The degradation of poly(dl-lactide), poly(epsilon-caprolactone), and their copolymers *in vivo*. *Biomaterials, 2,* 215–220.

Raeber, G. P., Lutolf, M. P., & Hubbell, J. A. (2005). Molecularly engineered PEG hydrogels: a novel model system for proteolytically mediated cell migration. *Biophys. J., 89,* 1374–1388.

Reed, A. M., & Gilding, D. K. (1981). Biodegradable polymers for use in surgery – poly(glycolic)/poly(lactic acid) homo and copolymers: 2. In vitro degradation. *Polymer, 22,* 494–498.

Reid, L. M., & Zern, M. A. (1993). *Extracellular Matrix Chemistry and Biology.* New York: Marcel Dekker.

Sakiyama-Elbert, S. E., & Hubbell, J. A. (2000). Development of fibrin derivatives for controlled release of heparin-binding growth factors. *J. Control. Rel., 65,* 389–402.

Sarkar, S., Lee, G. Y., Wong, J. Y., & Desai, T. A. (2006). Development and characterization of a porous micro-patterned scaffold for vascular tissue engineering applications. *Biomaterials, 27,* 4775–4782.

Shin, H., Jo, S., & Mikos, A. G. (2003). Biomimetic materials for tissue engineering. *Biomaterials, 24,* 4353–4364.

Sudo, R., Chung, S., Zervantonakis, I. K., Vickerman, V., Toshimitsu, Y., Griffith, L. G., et al. (2009). Transport-mediated angiogenesis in 3D epithelial coculture. *FASEB J., 23,* 2155–2164.

Sung, H. J., Meredith, C., Johnson, C., & Galis, Z. S. (2004). The effect of scaffold degradation rate on three-dimensional cell growth and angiogenesis. *Biomaterials, 25,* 5735–5742.

Tsang, V. L., Chen, A. A., Cho, L. M., Jadin, K. D., Sah, R. L., DeLong, S., et al. (2007). Fabrication of 3D hepatic tissues by additive photopatterning of cellular hydrogels. *FASEB J., 21,* 790–801.

Vats, A., Tolley, N. S., Polak, J. M., & Buttery, L. D. K. (2002). Stem cells: sources and applications. *Clin. Otolaryngol., 27,* 227–232.

von Buelow, S., von Heimburg, D., & Pallua, N. (2005). Efficacy and safety of polyacrylamide hydrogel for facial soft-tissue augmentation. *Plast. Reconstr. Surg., 116,* 1137–1146.

Wallace, A. F. (1978). The early development of pedicle flaps. *J. R. Soc. Med., 71,* 834.

West, J. L., & Hubbell, J. A. (1999). Polymeric biomaterials with degradation sites for proteases involved in cell migration. *Macromolecules, 32*, 241–244.

Wetzels, R. H. W., Robben, H. C. M., Leigh, I. M., Schaafsma, H. E., Vooijs, G. P., & Ramaekers, F. C. S. (1991). Distribution patterns of type-Vii collagen in normal and malignant human tissues. *Am. J. Pathol., 139*, 451–459.

Whang, K., Goldstick, T. K., & Healy, K. E. (2000). A biodegradable polymer scaffold for delivery of osteotropic factors. *Biomaterials, 21*, 2545–2551.

Wnek, G. E., & Bowlin, G. L. (2008). *Encyclopedia of Biomaterials and Biomedical Engineering*. New York: Informa Healthcare USA.

Yannas, I. V. (2004). Synthesis of tissues and organs. *Chembiochem., 5*, 26–39.

Yannas, I. V. (2005a). Facts and theories of induced organ regeneration. In *Regenerative Medicine I: Theories, Models and Methods* (pp. 1–38). Berlin: Springer.

Yannas, I. V. (2005b). Similarities and differences between induced organ regeneration in adults and early foetal regeneration. *J. R. Soc. Interface, 2*, 403–417.

Yannas, I. V., Lee, E., Orgill, D. P., Skrabut, E. M., & Murphy, G. F. (1989). Synthesis and characterization of a model extracellular matrix that induces partial regeneration of adult mammalian skin. *Proc. Natl. Acad. Sci. U.S.A., 86*, 933.

CHAPTER 38

Biocompatibility and Bioresponse to Biomaterials

James M. Anderson
Pathology, Macromolecular Science, and Biomedical Engineering, Case Western Reserve University, Cleveland, OH, USA

INTRODUCTION

Biocompatibility is generally defined as the ability of a biomaterial or medical device to perform with an appropriate host response in a specific application. Bioresponse or biocompatibility assessment (i.e. evaluation of biological responses) is considered to be a measure of the magnitude and duration of the adverse alterations in homeostatic mechanisms that determine the host response. From a practical point of view, the evaluation of biological responses to a medical device is carried out to determine that the medical device performs as intended and presents no significant harm to the patient. The goal of bioresponse evaluation is to predict whether a biomaterial or medical device presents potential harm to the patient. In regenerative medicine, biomaterials are utilized in a wide variety of ways ranging from carriers of genetic material to tissue-engineered implants that may contain autologous, allogeneic, or xenogeneic genetic materials, cells, and scaffold materials. Scaffolds may be composed of synthetic or modified-natural materials. A tissue-engineered implant is a biological-biomaterial combination in which some component of tissue has been combined with a biomaterial to create a device for the restoration or modification of tissue or organ function. Thus, tissue-engineered devices having a biological component(s) require an expanded perspective and understanding of biocompatibility and biological response evaluation. The purpose of this chapter is to provide an overview of this expanded perspective. It must be understood that each unique tissue-engineered device requires a unique set of experiments to determine its biological responses and biocompatibility.

This chapter presents an overview of host responses that must be considered in determining the biocompatibility of tissue-engineered devices that utilize biomaterials. The three major responses that must be considered for biocompatibility assessment are: (1) inflammation, (2) wound healing, and (3) immunological reactions or immunity. For the purposes of biological response evaluation, the immunological reactions or immunity are considered to be immunotoxicity.

Pathologists use the terminology of inflammation and immunity to describe adverse tissue reactions whereas immunologists commonly refer to inflammation as innate immunity and activation of the immune system as being acquired immunity. Tissue/material interactions are a series of responses that are initiated by the implantation procedure, as well as by the presence of the biomaterial, medical device, or tissue-engineered device. In this chapter, we divide the

693

Principles of Regenerative Medicine. DOI: 10.1016/B978-0-12-381422-7.10038-0

series of tissue/material responses into inflammation (innate immunity) and wound healing, and immunotoxicity. Following implantation, early, transient tissue/material responses include injury (implantation), blood-materials interactions, provisional matrix formation, and the temporal sequence of inflammation and wound healing including acute inflammation, chronic inflammation, granulation tissue development, foreign body reaction, and ultimately fibrosis/fibrous capsule (scar) development. Immunotoxicity is any adverse effect on the function or structure of the immune system or other systems as a result of an immune system dysfunction. Two significant failure mechanisms of tissue-engineered devices are fibrosis/fibrous capsule (scar) development surrounding and infiltrating the tissue-engineered device, or the initiation of acquired or cellular immunity by the biological component of the tissue-engineered device. It must also be considered that the biological component and the biomaterial component in a tissue-engineered device may act in concert or synergistically to facilitate either of these failure mechanisms.

INFLAMMATION (INNATE IMMUNITY) AND WOUND HEALING

The process of implantation of a biomaterial or tissue-engineered device results in injury to tissues or organs (Anderson, 1988, 1993, 2001; Gallin and Synderman, 1999; Anderson et al., 2008; Kumar et al., 2010). It is this injury and the subsequent perturbation of homeostatic mechanisms that lead to the inflammatory responses, foreign body reaction, and wound healing. The response to injury is dependent on multiple factors that include the extent of injury, loss of basement membrane structures, blood-material interactions, provisional matrix formation, extent or degree of cellular necrosis, and extent of the inflammatory response. The organ or tissue undergoing implantation may play a significant role in the response. These events, in turn, may affect the extent or degree of granulation tissue formation, foreign body reaction, and fibrosis or fibrous capsule (scar) development (Broughton et al., 2006). These events are summarized in Table 38.1. These host reactions for biocompatible biomaterials are considered to be normal. It is noteworthy that these host reactions are also tissue-dependent, organ-dependent, and species-dependent. These dependencies thus provide perspectives on the biological response evaluation and the ultimate determination of biocompatibility. It is important to recognize that these reactions occur or are initiated early — that is, within 2—3 weeks of the time of implantation — and undergo resolution rather quickly, leading to fibrosis or fibrous capsule formation.

Blood-material interactions and initiation of the inflammatory response

Blood-material interactions and the inflammatory response are intimately linked, and, in fact, early responses to injury involve mainly blood and the vasculature (Anderson, 1988, 1993, 2001; Gallin and Synderman, 1999; Anderson et al., 2008; Kumar et al., 2010). Regardless of the tissue into which a biomaterial is implanted, the initial inflammatory response is activated by injury to vascularized connective tissue. Because blood and its components are involved in the initial inflammatory responses, thrombus and/or blood clot also form. Thrombus formation involves activation of the extrinsic and intrinsic coagulation systems, the

TABLE 38.1 Sequence of Host Reactions

Injury
Blood-material interactions
Provisional matrix formation
Acute inflammation
Chronic inflammation
Granulation tissue
Foreign body reaction
Fibrosis/fibrous capsule development

complement system, the fibrinolytic system, the kinin-generating system, and platelets. Thrombus or blood clot formation on the surface of a biomaterial is related to the well-known Vroman effect of protein adsorption. From a wound healing perspective, blood protein deposition on a biomaterial surface is described as provisional matrix formation.

Although injury initiates the inflammatory response, released chemicals from plasma, cells, and injured tissue mediate the response (Salthouse, 1976; Weisman et al., 1980; Gallin and Synderman, 1999; Kumar et al., 2010). Important classes of chemical mediators of inflammation are presented in Table 38.2. Several important points must be noted in order to understand the inflammatory response and how it relates to biomaterials. First, although chemical mediators are classified on a structural or functional basis, different mediator systems interact and provide a system of checks and balances regarding their respective activities and functions. Second, chemical mediators are quickly inactivated or destroyed, suggesting that their action is predominantly local (i.e. at the implant site). Third, generally acid, lyosomal proteases and oxygen-derived free radicals produce the most significant damage or injury. These chemical mediators are also important in the degradation of biomaterials. Phagolysosomes in macrophages can have acidity as low as a pH of 4 and direct microelectrode studies of this acid environment have determined pH levels as low as 3.5. Moreover, only several hours are necessary to achieve these acid levels following adhesion of macrophages (Silver et al., 1988; Jankowski et al., 2002; Haas, 2007).

The predominant cell type present in the inflammatory response varies with the age of the injury. In general, neutrophils, commonly called polymorphonuclear leukocytes or polys, predominate during the first several days following injury and then are replaced by monocytes as the predominant cell type. Three factors account for this change in cell type: (1) Neutrophils are short-lived and disintegrate and disappear after 24—48 h; neutrophil emigration is of short duration because chemotactic factors for neutrophil migration are activated early in the inflammatory response. (2) Following emigration from the vasculature, monocytes differentiate into macrophages, and these cells are very long-lived (up to months). (3) Monocyte emigration may continue for days to weeks, depending on the injury and implanted biomaterial, and chemotactic factors for monocytes are activated over longer periods of time.

TABLE 38.2 Important Chemical Mediators of Inflammation Derived from Plasma, Cells, or Injured Tissue

Mediators	Examples
Vasoactive agents	Histamine, serotonin, adenosine, endothelial derived relaxing factor (EDRF), prostacyclin, endothelin, thromboxane a_2
Plasma proteases	
Kinin system	Bradykinin, kallikrein
Complement system	C3a, C5a, C3b, C5b—C9
Coagulation/fibrinolytic system	Fibrin degradation products, activated Hageman factor (FXIIA), tissue plasminogen activator (tPA)
Leukotrienes	Leukotriene B_4 (LTB$_4$), hydroxyeicosatetranoic acid (HETE)
Lysosomal proteases	Collagenase, elastase
Oxygen-derived free radicals	H_2O_2, superoxide anion, nitric oxide
Platelet activating factors	Cell membrane lipids
Cytokines	Interleukin-1 (IL-1), TNF
Growth factors	PDGF, fibroblast growth factor (FGF), transforming growth factor (TGF-α or TGF-β), epithelial growth factor (EGF)

Provisional matrix formation

Injury to vascularized tissue in the implantation procedure leads to immediate development of the provisional matrix at the implant site. This provisional matrix consists of fibrin, produced by activation of the coagulative and thrombosis systems, and inflammatory products released by the complement system, activated platelets, inflammatory cells, and endothelial cells (Clark et al., 1982; Tang and Eaton, 1993; Tang, 1998; Gorbet and Sefton, 2004). These events occur early, within minutes to hours following implantation of a medical device. Components within or released from the provisional matrix, that is, fibrin network (thrombosis or clot), initiate the resolution, reorganization, and repair processes such as inflammatory cell and fibroblast recruitment. Platelets, activated during the fibrin network formation, release platelet factor 4, platelet-derived growth factor (PDGF), and transforming growth factor β (TGF-β), which contribute to fibroblast recruitment (Wahl et al., 1989; Riches, 1998). Monocytes and lymphocytes, upon activation, generate additional chemotactic factors including LTB$_4$, PDGF, and TGF-β to recruit fibroblasts.

The provisional matrix is composed of adhesive molecules such as fibronectin and thrombospondin bound to fibrin as well as platelet granule components released during platelet aggregation. Platelet granule components include thrombospondin, released from the platelet α-granule, and cytokines including TGF-α, TGF-β, PDGF, platelet factor 4, and platelet-derived endothelial cell growth factor. The provisional matrix is stabilized by the crosslinking of fibrin by factor XIIIa.

The provisional matrix appears to provide both structural and biochemical components to the process of wound healing. The complex three-dimensional structure of the fibrin network with attached adhesive proteins provides a substrate for cell adhesion and migration. The presence of mitogens, chemoattractants, cytokines, and growth factors within the provisional matrix provides for a rich milieu of activating and inhibiting substances for various cellular proliferative and synthetic processes.

The provisional matrix may be viewed as a naturally derived, biodegradable, sustained release system in which mitogens, chemoattractants, cytokines, and growth factors are released to control subsequent wound healing processes (Dvorak et al., 1987; Ignotz et al., 1987; Muller et al., 1987; Wahl et al., 1987; Madri et al., 1988; Sporn and Roberts, 1988; Broadley et al., 1989). In spite of the rapid increase in our knowledge of the provisional matrix and its capabilities, our knowledge of the control of the formation of the provisional matrix and its effect on subsequent wound healing events is poor.

Temporal sequence of inflammation and wound healing

Inflammation is generally defined as the reaction of vascularized living tissue to local injury. Inflammation serves to contain, neutralize, dilute, or wall off the injurious agent or process. In addition, it sets into motion a series of events that may heal and reconstitute the implant site through replacement of the injured tissue by regeneration of native parenchymal cells, formation of fibroblastic scar tissue, or a combination of these two processes (Gallin and Synderman, 1999; Kumar et al., 2010).

The sequence of events following implantation of a biomaterial is illustrated in Figure 38.1. The size, shape, and chemical and physical properties of the biomaterial and the physical dimensions and properties of the prosthesis or device may be responsible for variations in the intensity and time duration of the inflammatory and wound healing processes. Thus, intensity and/or time duration of inflammatory reaction may characterize the biocompatibility of a biomaterial or device.

Classically, the biocompatibility of an implanted material has been described in terms of the morphological appearance of the inflammatory reaction to the material; however, the inflammatory response is a series of complex reactions involving various types of cells, the

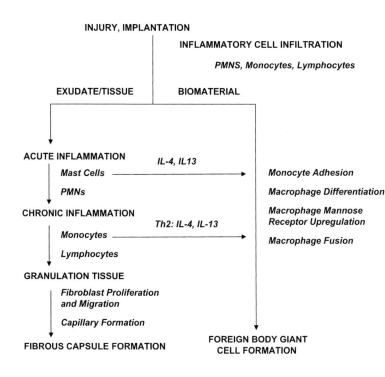

INJURY, IMPLANTATION

INFLAMMATORY CELL INFILTRATION

PMNS, Monocytes, Lymphocytes

EXUDATE/TISSUE BIOMATERIAL

ACUTE INFLAMMATION   *IL-4, IL13*

Mast Cells ──────────────────▶ *Monocyte Adhesion*

PMNs   *Macrophage Differentiation*

CHRONIC INFLAMMATION   *Macrophage Mannose Receptor Upregulation*
                    Th2: IL-4, IL-13

Monocytes ──────────────────▶ *Macrophage Fusion*

Lymphocytes

GRANULATION TISSUE

Fibroblast Proliferation and Migration

Capillary Formation

FIBROUS CAPSULE FORMATION   **FOREIGN BODY GIANT CELL FORMATION**

FIGURE 38.1
Sequence of events involved in inflammatory and wound healing responses leading to FBGC formation. This shows the potential importance of mast cells in the acute inflammatory phase and Th2 lymphocytes in the transient chronic inflammatory phase with the production of IL-4 and IL-13, which can induce monocyte/macrophage fusion to form FBGCs.

densities, activities, and functions of which are controlled by various endogenous and auto-coid mediators. The simplistic view of the acute inflammatory response progressing to the chronic inflammatory response may be misleading with respect to biocompatibility studies and the inflammatory response to implants. *In vivo* studies using the cage implant system show that monocytes and macrophages are present in highest concentrations when neutrophils are also at their highest concentrations; that is, the acute inflammatory response (Marchant et al., 1983; Spilizewski et al., 1985). Neutrophils have short lifetimes — hours to days — and disappear from the exudates more rapidly than do macrophages, which have lifetimes of days to weeks to months. Eventually macrophages become the predominant cell type in the exudates, resulting in a chronic inflammatory response. Monocytes rapidly differentiate into macrophages, the cells principally responsible for normal wound healing in the foreign body reaction. Classically, the development of granulation tissue has been considered to be part of chronic inflammation, but, because of unique tissue-material interactions, it is preferable to differentiate the foreign body reaction — with its varying degree of granulation tissue development, including macrophages, fibroblasts, and capillary formation — from chronic inflammation.

Acute inflammation

Acute inflammation is of relatively short duration, lasting from minutes to days, depending on the extent of injury. The main characteristics of acute inflammation are the exudation of fluid and plasma proteins (edema) and the emigration of leukocytes (predominantly neutrophils). Neutrophils and other motile white cells emigrate or move from the blood vessels to the perivascular tissues and the injury (implant) site (Henson and Johnston, 1987; Malech and Gallin, 1987; Ganz, 1988).

The accumulation of leukocytes, in particular neutrophils and monocytes, is the most important feature of the inflammatory reaction. Leukocytes accumulate through a series of processes including margination, adhesion, emigration, phagocytosis, and extracellular release of leukocyte products (Jutila, 1990). Increased leukocytic adhesion in inflammation involves specific interactions between complementary "adhesion molecules" present on the leukocyte and endothelial surfaces (Cotran and Pober, 1990; Pober and Cotran, 1990). The surface expression of these adhesion molecules is modulated by inflammatory agents; mechanisms of interaction include stimulation of leukocyte adhesion molecules (C5a, LTB$_4$), stimulation of endothelial adhesion molecules (IL-1), or both effects of tumor necrosis factor-α (TNF-α). Integrins comprise a family of transmembrane glycoproteins that modulate cell-matrix and cell-cell relationships by acting as receptors to extracellular protein ligands and also as direct adhesion molecules (Hynes, 1992). An important group of integrins (adhesion molecules) on leukocytes include the CD11/CD18 family of adhesion molecules. Inflammatory mediators (i.e. cytokines) stimulate a rapid increase in these adhesion molecules on the leukocyte surface as well as increased leukocyte adhesion to endothelium. Leukocyte-endothelial cell interactions are also controlled by endothelial-leukocyte adhesion molecules (ELAMs, E-selectins) or intracellular adhesion molecules (ICAM-1, ICAM-2), and vascular cell adhesion molecules (VCAMs) on endothelial cells (Butcher, 1991).

Inflammatory cell emigration is controlled in part by chemotaxis, which is the unidirectional migration of cells along a chemical gradient. A wide variety of exogenous and endogenous substances have been identified as chemotactic agents (Henson, 1971, 1980; Weisman et al., 1980; Henson and Johnston, 1987; Malech and Gallin, 1987; Ganz, 1988; Weiss, 1989; Cotran and Pober, 1990; Jutila, 1990; Paty et al., 1990; Pober and Cotran, 1990; Butcher, 1991; Hynes, 1992). Important to the emigration or movement of leukocytes is the presence of specific receptors for chemotactic agents on the cell membranes of leukocytes. These and other receptors may also play a role in the activation of leukocytes. Following localization of leukocytes at the injury (implant) site, phagocytosis and the release of enzymes occur following activation of neutrophils and macrophages. The major role of the neutrophils in acute inflammation is to phagocytose microorganisms and foreign materials. Phagocytosis is seen as a three-step process in which the injurious agent undergoes recognition and neutrophil attachment, engulfment, and killing or degradation. With regard to biomaterials, engulfment and degradation may or may not occur depending on the properties of the biomaterial.

Although biomaterials are not generally phagocytosed by neutrophils or macrophages because of the size disparity (i.e. the surface of the biomaterial is greater than the size of the cell), certain events in phagocytosis may occur. The process of recognition and attachment is expedited when the injurious agent is coated by naturally occurring serum factors called opsonins. The two major opsonins are IgG and the complement-activated fragment, C3b. Both of these plasma-derived proteins are known to adsorb to biomaterials, and neutrophils and macrophages have corresponding cell membrane receptors for these opsonization proteins. These receptors may also play a role in the activation of the attached neutrophil or macrophage. Because of the size disparity between the biomaterial surface and the attached cell, "frustrated phagocytosis" may occur (Henson, 1971, 1980). This process does not involve engulfment of the biomaterial but does cause the extracellular release of leukocyte products in an attempt to degrade the biomaterial. Neutrophils adherent to complement-coated and immunoglobulin-coated non-phagocytosable surfaces may release enzymes by direct extrusion or exocytosis from the cell (Henson, 1971, 1980). The amount of enzyme released during this process depends on the size of the polymer particle, with larger particles inducing greater amounts of enzyme release. This suggests that the specific mode of cell activation in the inflammatory response in tissue is dependent upon the size of the implant and that a material in a phagocytosable form (e.g. powder or particulate) may provoke a degree of inflammatory response different from that of the same material in a non-phagocytosable form (e.g. film).

Tissue-engineered constructs containing biomaterial scaffolds alone, or with cells and/or chemokines, growth factors, or other biological components, are thus subjected to an aggressive microenvironment that may quickly compromise the intended function of the construct (Babensee et al., 1998).

Chronic inflammation

Chronic inflammation is less uniform histologically than is acute inflammation. In general, chronic inflammation is characterized by the presence of monocytes and lymphocytes with the early proliferation of blood vessels and connective tissue (Williams and Williams, 1983; Johnston, 1988; Gallin and Synderman, 1999; Browder et al., 2000; Kumar et al., 2010). It must be noted that many factors modify the course and histological appearance of chronic inflammation.

Persistent inflammatory stimuli lead to chronic inflammation. Although the chemical and physical properties of the biomaterial may lead to chronic inflammation, motion in the implant site by the biomaterial may also produce chronic inflammation. The chronic inflammatory response to biomaterials is confined to the implant site. Inflammation with the presence of mononuclear cells, including lymphocytes and plasma cells, is given the designation chronic inflammation, whereas the foreign body reaction with granulation tissue development is considered the normal wound healing response to implanted biomaterials (i.e. the normal foreign body reaction). Chronic inflammation with biocompatible materials is usually of very short duration (i.e. a few days).

Lymphocytes and plasma cells are involved principally in immune reactions and are key mediators of antibody production and delayed hypersensitivity responses. Their roles in non-immunological injuries and inflammation are largely unknown (Brodbeck et al., 2005; MacEwan et al., 2005; Revell, 2008). Little is known regarding immune responses and cell-mediated immunity to synthetic biomaterials. The role of macrophages must be considered in the possible development of immune responses to synthetic biomaterials. Macrophages and dendritic cells process and present the antigen to immunocompetent cells and thus are key mediators in the development of immune reactions.

The macrophage is probably the most important cell in chronic inflammation because of the great number of biologically active products its produces (Johnston, 1988). Important classes of products produced and secreted by macrophages include neutral proteases, chemotactic factors, arachidonic acid metabolites, reactive oxygen metabolites, complement components, coagulation factors, growth-promoting factors, and cytokines (Anderson and Jones, 2007; Jones et al., 2007, 2008).

Growth factors such as PDGF, FGF, TGF-β, TGF-α/EGF, and IL-1 or TNF are important to the growth of fibroblasts and blood vessels and the regeneration of epithelial cells. Growth factors, released by activated cells, stimulate production of a wide variety of cells; initiate cell migration, differentiation, and tissue remodeling; and may be involved in various stages of wound healing (Mustoe et al., 1987; Wahl et al., 1989; Fong et al., 1990; Sporn and Roberts, 1990; Golden et al., 1991; Kovacs, 1991). It is clear that there is a lack of information regarding interaction and synergy among various cytokines and growth factors and their abilities to exhibit chemotactic, mitogenic, and angiogenic properties.

Granulation tissue

Within 1 day following implantation of a biomaterial (i.e. injury), the healing response is initiated by the action of monocytes and macrophages, followed by proliferation of fibroblasts and vascular endothelial cells at the implant site, leading to the formation of granulation tissue, the hallmark of healing inflammation. Granulation tissue derives its name from the pink, soft granular appearance on the surface of healing wounds, and its characteristic histological features include the proliferation of new small blood vessels and fibroblasts. Depending on the extent of injury, granulation tissue may be seen as early as 3–5 days following implantation of a biomaterial.

The new small blood vessels are formed by budding or sprouting of pre-existing vessels in a process known as neovascularization or angiogenesis (Ziats et al., 1985; Thompson et al.,

699

1988; Maciag, 1990; Browder et al., 2000; Nguyen and d'Amore, 2001). This process involves proliferation, maturation, and organization of endothelial cells into capillary tubes. Fibroblasts also proliferate in developing granulation tissue and are active in synthesizing collagen and proteoglycans. In the early stages of granulation tissue development, proteoglycans predominate; later, however, collagen — especially type I collagen — predominates and forms the fibrous capsule. Some fibroblasts in developing granulation tissue may have features of smooth muscle cells. These cells are called myofibroblasts and are considered to be responsible for the wound contraction seen during the development of granulation tissue.

Macrophage interactions

The inflammatory and immune systems overlap considerably through the activity and phenotypic expression of macrophages that are derived from blood-borne monocytes. Monocytes and macrophages belong to the mononuclear phagocytic system (MPS) (Table 38.3). Cells in the MPS may be considered as resident macrophages in the respective tissues that take on specialized functions that are dependent on their tissue environment. From this perspective, the host defense system may be seen as blood-borne or circulating inflammatory and immune cells as well as mononuclear phagocytic cells that reside in specific tissues with specialized functions. In the inflammatory and immune responses, the macrophage plays a pivotal role in both the induction and effector phases of these responses.

Two factors that play a role in monocyte/macrophage adhesion and activation and foreign body giant cell (FBGC) formation are the surface chemistry of the substrate onto which the cells adhere and the protein adsorption that occurs before cell adhesion. These two factors have been hypothesized to play significant roles in the inflammatory and wound healing responses to biomaterials and medical devices *in vivo*.

Macrophage interactions with biomaterials are initiated when blood-borne monocytes in the early, transient responses migrate to the implant site and adhere to the blood protein adsorbed biomaterial through monocyte-integrin interactions. Following adhesion, adherent monocytes differentiate into macrophages that may then fuse to form FBGCs. Figure 38.2 demonstrates the progression from circulating blood monocyte to tissue macrophage to FBGC development that is most commonly observed. Because of the progression of monocytes to macrophages to FBGCs (Fig. 38.2), the following discussion of macrophage interactions also includes perspectives on how macrophages are formed (i.e. monocyte adhesion) and what happens to macrophages on biomaterial surfaces (i.e. FBGC formation) (McNally and Anderson, 1994, 1995).

TABLE 38.3 The Mononuclear Phagocytic System	
Tissues	**Cells**
Implant sites	Inflammatory macrophages, FBGCs
Liver	Kupffer cells
Lung	Alveolar macrophages
Connective tissue	Histiocytes
Bone marrow	Macrophages
Spleen and lymph nodes	Fixed and free macrophages
Serous cavities	Pleural and peritoneal macrophages
Nervous system	Microglial cells
Bone	Osteoclasts
Skin	Langerhans' cells, dendritic cells
Lymphoid tissue	Dendritic cells

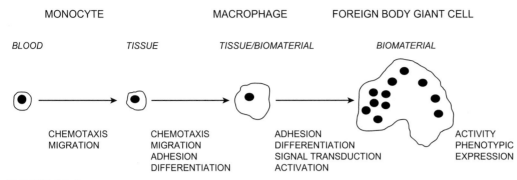

MONOCYTE   MACROPHAGE   FOREIGN BODY GIANT CELL

BLOOD   TISSUE   TISSUE/BIOMATERIAL   BIOMATERIAL

CHEMOTAXIS   CHEMOTAXIS   ADHESION   ACTIVITY
MIGRATION   MIGRATION   DIFFERENTIATION   PHENOTYPIC
               ADHESION   SIGNAL TRANSDUCTION   EXPRESSION
               DIFFERENTIATION   ACTIVATION

FIGURE 38.2

***In vivo* transition from blood-borne monocyte to biomaterial adherent monocyte/macrophage to FBGC at the tissue/biomaterial interface.** Little is known regarding the indicated biological responses that are considered to play important roles in the transition to FBGC development.

Material surface property-dependent blood protein adsorption occurs immediately upon surgical implantation of a biomaterial and it is the protein-modified biomaterial that inflammatory cells subsequently encounter. Monocytes express receptors for various blood components, but they recognize naturally occurring foreign surfaces by receptors for opsonins such as fragments of complement component C3. Complement activation by biomaterials has been well documented (Nilsson et al., 2007). Exposure to blood during biomaterial implantation may permit extensive opsonization with the labile fragment C3b and the rapid conversion of C3b to its hemolytically inactive but nevertheless opsonic and more stable form, C3bi. C3b is bound by the CD35 receptor, but C3bi is recognized by distinct receptors, CD11b/CD18 and CD11c/CD18 on monocytes (McNally and Anderson, 1994). Fibrinogen, a major plasma protein that adsorbs to biomaterials, is another ligand for these receptors that together with CD11a/CD18 constitutes a subfamily of integrins that is restricted to leukocytes (McNally and Anderson, 1994, 1995). Studies with monoclonal antibodies to their common β_2 subunit (CD 18) and distinct α chains have implicated CD11b/CD18 and CD11c/CD18 in monocyte/macrophage responses. Other potential adhesion-mediating proteins that adsorb to biomaterials include IgG, which may interact with monocytes via various receptors and fibronectin, for which monocytes also express multiple types of receptors (Jenney and Anderson, 2000; McNally and Anderson, 2002; McNally et al., 2007, 2008).

FBGC formation and interactions

The foreign body reaction is composed of FBGCs and the components of granulation tissue, which consist of macrophages, fibroblasts, and capillaries in varying amounts, depending upon the form and topography of the implanted material (Brodbeck and Anderson, 2009). Relatively flat and smooth surfaces, such as those found on breast prostheses, have a foreign body reaction that is composed of a layer of macrophages one to two cells in thickness. Relatively rough surfaces, such as those found on the outer surfaces of expanded poly (tetrafluroethylene) (ePTFE) vascular prostheses or poly(methyl methacrylate) (PMMA) bone cement, have a foreign body reaction composed of several layers of macrophages and FBGCs at the surface. Fabric materials generally have a surface response composed of macrophages and FBGCs with varying degrees of granulation tissue subjacent to the surface response.

As previously discussed, the form and topography of the surface of the biomaterial determines the composition of the foreign body reaction. With biocompatible materials, the composition of the foreign body reaction in the implant site may be controlled by the surface properties of the biomaterial, the form of the implant, and the relationship between the surface area of the biomaterial and the volume of the implant. For example, high surface-to-volume implants such as fabrics or porous materials will have higher ratios of macrophages and FBGCs in the implant site than will smooth-surface implants, which will have fibrosis as a significant component of the implant site.

The foreign body reaction consisting mainly of macrophages and/or FBGCs may persist at the tissue-implant interface for the lifetime of the implant (Chambers and Spector; 1982; Rae, 1986; Anderson, 1988, 1993; Greisler, 1988). Generally, fibrosis (i.e. fibrous encapsulation) surrounds the biomaterial or implant with its interfacial foreign body reaction, isolating the implant and foreign body reaction from the local tissue environment. Early in the inflammatory and wound healing response, the macrophages are activated upon adherence to the material surface (Jones et al., 2007). Although it is generally considered that the chemical and physical properties of the biomaterial are responsible for macrophage activation, the nature of the subsequent events regarding the activity of macrophages at the surface is not clear. Tissue macrophages, derived from circulating blood monocytes, may coalesce to form multinucleated FBGCs. FBGCs containing large numbers of nuclei are typically present on the surface of biomaterials. Although these FBGCs may persist for the lifetime of the implant, it is not known whether they remain activated, releasing their lysosomal constituents, or become quiescent. FBGCs have been implicated in the biodegradation of polymeric medical devices (Zhao et al., 1990, 1991; Wiggins et al., 2001).

Figure 38.1 demonstrates the sequence of events involved in inflammation and wound healing when medical devices are implanted. In general, the neutrophil (PMN) predominant acute inflammatory response and the lymphocyte/monocyte predominant chronic inflammatory response resolve quickly (i.e. within 2 weeks), depending on the type and location of the implant. Studies utilizing IL-4 demonstrate the role for Th2 helper lymphocyte and mast cells in the development of the foreign body reaction at the tissue/material interface (Zdolsek et al., 2007). Th2 helper lymphocytes have been described as "anti-inflammatory" based on their cytokine profile, of which IL-4 is a significant component. Th2 helper lymphocytes also produce IL-13 which has a similar effect to IL-4 on FBGC formation. In this regard, it is noteworthy that anti-IL-4 antibody does not inhibit IL-13-induced FBGC formation, nor does anti-IL-13 antibody inhibit IL-4-induced FBGC formation. In IL-4 and IL-13 FBGC culture systems, the macrophage mannose receptor (MMR) has been identified as critical to the fusion of macrophages in the formation of FBGC (McNally et al., 1996; DeFife et al., 1997). FBGC formation can be prevented by competitive inhibitors of MMR activity (i.e. α-mannan) or inhibitors of glycoprotein processing that restrict MMR surface expression.

Regarding the effect of lymphocytes on the foreign body reaction, recent studies have demonstrated that interactions with lymphocytes enhance adherent macrophage and FBGC production of pro-inflammatory cytokines. Interactions through indirect (paracrine) signaling showed a significant pro-inflammatory effect in enhancing adherent macrophage/FBGC at early time points, whereas interactions via direct (juxtacrine) mechanisms dominated at later time points. Furthermore, lymphocytes prefer interactions with adherent macrophages and FBGCs, as opposed to biomaterial surfaces, resulting in lymphocyte activation (Chang et al., 2008, 2009a, 2009b). In vivo studies utilizing clinically synthetic biomaterials have demonstrated that there is a quantitative increase in T-cells following secondary biomaterial exposure, but the T-cell subset distribution does not change, indicating non-specific recruitment of T-cells rather than an adaptive immune response. Studies in T-cell-deficient mice have shown no change in the foreign body giant cell formation. In vitro studies with clinical synthetic biomaterials showed no expression of the activation markers CD69 and CD25 and lymphocyte proliferation was not identified (Rodriguez et al., 2008, 2009a,b; Rodriguez and Anderson, 2010). Results from these in vivo and in vitro studies do not suggest an adaptive immune response with clinically relevant biomaterials as T-cell markers, CD25 and CD69, were not upregulated and T-cell cytokines, IL-2 and interferon-γ, not present.

FIBROSIS AND FIBROUS ENCAPSULATION

The end-stage healing response to biomaterials is generally fibrosis or fibrous encapsulation. However, tissue-engineered devices may be exceptions to this general statement (e.g. porous materials inoculated with parenchymal cells or porous materials implanted into bone).

Repair of implant sites involves two distinct processes: regeneration, which is the replacement of injury tissue by parenchymal cells of the same type, or replacement by connective tissue that constitutes the fibrous capsule. These processes are generally controlled by either (1) the proliferative capacity of the cells in the tissue receiving the implant and the extent of injury as it relates to the destruction or (2) persistence of the tissue framework of the implant site. The regenerative capacity of cells permits classification into three groups: labile, stable (or expanding), and permanent (or static) cells. Labile cells continue to proliferate throughout life, stable cells retain this capacity but do not normally replicate, and permanent cells cannot reproduce themselves after birth. Perfect repair with restitution of normal structure theoretically occurs only in tissue consisting of stable and labile cells, whereas all injuries to tissues composed of permanent cells may give rise to fibrosis and fibrous capsule formation with very little restitution of the normal tissue or organ structure. Tissues composed of permanent cells (e.g. nerve cells, skeletal muscle cells, and cardiac muscle cells) most commonly undergo an organization of the inflammatory exudates, leading to fibrosis. Tissues composed of stable cells (e.g. parenchymal cells of the liver, kidney, and pancreas), mesenchymal cells (e.g. fibroblasts, smooth muscle cells, osteoblasts, and chondroblasts), and vascular endothelial and labile cells (e.g. epithelial cells and lymphoid and hematopoietic cells) may also follow this pathway to fibrosis or may undergo resolution of the inflammatory exudates, leading to restitution of the normal tissue structure. The condition of the underlying framework or supporting stroma of the parenchymal cells following an injury plays an important role in the restoration of normal tissue structure. Retention of the framework may lead to restitution of the normal tissue structure, whereas destruction of the framework most commonly leads to fibrosis. It is important to consider the species-dependent nature of the regenerative capacity of cells. For example, cells from the same organ or tissue but from different species may exhibit different regenerative capacities and/or connective tissue repair.

The extent of provisional matrix formation is an important factor as it is related to wound healing by first or second intention. First intention (primary union) wound healing occurs when there is minimal to no space between the tissue and device whereas second intention (secondary union) wound healing occurs when a large space, providing for extensive provisional matrix formation, is present. Obviously, inappropriate or inadequate preparation of the implant site leading to extensive provisional matrix formation may predispose the implant to failure through mechanisms related to fibrous capsule formation.

The inflammatory (innate) and immune (adaptive) responses have common components. It is possible to have inflammatory responses only with no adaptive immune response. In this situation, both humoral and cellular components that are shared by both types of responses may only participate in the inflammatory response. Table 38.4 indicates the common components of the inflammatory (innate) and immune (adaptive) responses. Macrophages and dendritic cells are known as professional antigen-presenting cells responsible for the initiation of the adaptive immune response.

Many investigators have considered macrophages and dendritic cells as being distinctly different antigen-presenting cells (APCs). Hume has summarized evidence that dendritic cells are part of the mononuclear phagocyte system and are derived from the same common

TABLE 38.4	Common Components of the Inflammatory (Innate) and Immune (Adaptive) Responses

Components
    Complement cascade components
    Immunoglobulins
Cellular components
    Macrophages
    NK (natural killer) cells
    Dendritic cells
    Cells with dual phagocytic and antigen-presenting capabilities

macrophage precursor, they are responsive to the same growth factors, express the same surface markers, and have no unique adaptation for antigen presentation that is not shared by other macrophages (Hume, 2008).

IMMUNOTOXICITY (ACQUIRED IMMUNITY)

The acquired or adaptive immune system acts to protect the host from foreign agents or materials and is usually initiated through specific recognition mechanisms and the ability of humoral and cellular components to recognize the foreign agent or material as being "non-self" (Coligan et al., 1992; Burleson et al., 1995; Smialowicz and Holsapple, 1996; Janeway and Travers, 1997; Rose, 1997; Sefton et al., 2008). Generally, the adaptive immune system may be considered as having two components: humoral or cellular. Humoral components include antibodies, complement components, cytokines, chemokines, growth factors, and other soluble mediators. These components are synthesized by cells of the immune response and, in turn, function to regulate the activity of these same cells and provide for communication between different cells in the cellular component of the adaptive immune response. Cells of the immune system arise from stem cells in the bone marrow (B lymphocytes) or the thymus (T lymphocytes) and differ from each other in morphology, function, and the expression of cell-surface antigens. They share the common features of maintaining cell-surface receptors that assist in the recognition and/or elimination of foreign materials. Regarding tissue-engineered devices, the adaptive immune response may recognize the biological components, modifications of the biological components, or degradation products of the biological components, commonly known as antigens, and initiate immune response through humoral or cellular mechanisms.

Components of the humoral immune system play important roles in the inflammatory responses to foreign materials. Antibodies and complement components C3b and C3bi adhere to foreign materials, act as opsonins, and facilitate phagocytosis of the foreign materials by neutrophils and macrophages that have cell-surface receptors for C3b. Complement component C5a is a chemotactic agent for neutrophils, monocytes, and other inflammatory cells and facilitates the immigration of these cells to the implant site. The complement system is composed of classic and alternative pathways that eventuate in a common pathway to produce the membrane attack complex (MAC), which is capable of lysing microbial agents. The complement system (i.e. complement cascade) is closely controlled by protein inhibitors in the host cell membrane that may prevent damage to host cells. This inhibitory mechanism may not function when non-host cells are used in tissue-engineered devices.

T (thymus-derived) lymphocytes are significant cells in the cell-mediated adaptive immune response and their cell-adhesion molecules play a significant role in lymphocyte migration, activation, and effector function. The specific interaction of cell membrane adhesion molecules, sometimes also called ligands or antigens, with antigen-presenting cells (APCs) produces, specific types of lymphocytes with specific functions. Table 38.5 indicates cell types

TABLE 38.5 Cell Types and Function in the Adaptive Immune System

Cell type	Motor function
Macrophages (APC)	Process and present antigen to immunocompetent T-cells
	Phagocytosis
	Activated by cytokines (i.e. IFN-γ) from other immune cells
T-cells	Interact with APCs and are activated through two required cell membrane interactions
	Facilitate target cell apoptosis
	Participate in transplant rejection (type IV hypersensitivity)
B-cells	Form plasma cells that secrete immunoglobulins (IgG, IgA, and IgE)
	Participate in antigen-antibody complex mediated tissue damage (type III hypersensitivity)
Dendritic cells (APC)	Process and present antigen to immunocompetent T-cells
	Utilize Fc receptors for IgG to trap antigen-antibody complexes
NK cells (non-T, non-B lymphocytes)	Innate ability to lyse tumor, virus infected, and other cells without previous sensitization
	Mediates T- and B-cell function by secretion of IFN-γ

and function in the adaptive immune response. Obviously, the functions of these cells are more numerous than those indicated in Table 38.5 but the major function of these cells is provided to indicate similarities and differences in the interaction and responsiveness of these cells. Effector T-cells (Table 38.6) are produced when their antigen-specific receptors and either the CD4 or the CD8 co-receptors bind to peptide-MHC (major histocompatibility complex) complexes. A second, co-stimulatory, signal is also required and this is provided by the interaction of the CD28 receptor on the T-cell and the B7.1 and B7.2 glycoproteins of the immunoglobulin superfamily present on APCs. B lymphocytes bind soluble antigens through their cell-surface immunoglobulin and thus can function as professional APCs by internalizing the soluble antigens and presenting peptide fragments of these antigens as MHC: peptide complexes. Once activated, T-cells can synthesize the T-cell growth factor IL-2 and its receptor. Thus, activated T-cells secrete and respond to IL-2 to promote T-cell growth in an autocrine fashion.

Cytokines are the messenger molecules of the immune system. Most cytokines have a wide spectrum of effects, reacting with many different cell types, and some are produced by several

TABLE 38.6 Effector T lymphocytes in Adaptive Immunity

Th1 helper cells	CD41
	Pro-inflammatory
	Activation of macrophages
	Produces IL-2, interferon-γ (IFN-γ), IL-3, TNF-α, GM-CSF, macrophage chemotactic factor (MCF), migration inhibitor factor (MIF)
	Induces IgG2a
Th2 helper cells	CD41
	Anti-inflammatory
	Activation of B-cells to make antibodies
	Produces IL-4, IL-5, IL-6, IL-10, IL-3, GM-CSF, and IL-13
	Induces IgG1
Cytotoxic T-cells (CTL)	CD81
	Induces apoptosis of target cells
	Produces IFN-γ, TNF-β, and TNF-α
	Releases cytotoxic proteins

TABLE 38.7 Selected Cytokines and their Effects

Cytokine	Effect
IL-1, TNF-α, INF-γ, IL-6	Mediate natural immunity
IL-1, TNF-α, IL-6	Initiate non-specific inflammatory responses
IL-2, IL-4, IL-5, IL-12, IL-15 and TGF-β	Regulate lymphocyte growth, activation, and differentiation
IL-2 and IL-4	Promote lymphocyte growth and differentiation
IL-10 and TGF-β	Downregulate immune responses
IL-1, INF-γ, TNF-α, and MIF	Activate inflammatory cells
IL-8	Produced by activated macrophages and endothelial cells
	Chemoattractant for neutrophils
MCP-1, MIP-α, and RANTES	Chemoattractant for monocytes and lymphocytes
GM-CSF and G-CSF	Stimulate hematopoiesis
IL-4 and IL-13	Promote macrophage fusion and foreign body giant cell formation

different cell types. Table 38.7 presents common categories of cytokines and lists some of their general properties. It should be noted that, while cytokines can be subdivided into functional groups, many cytokines such as IL-1, TNF-α, and IFN-γ are pleotropic in their effects and regulate, mediate, and activate numerous responses by various cells.

Immunotoxicity is any adverse effect on the function or structure of the immune system or other systems as a result of an immune system dysfunction (Langone, 1998). Adverse or immunotoxic effects occur when humoral or cellular immunity needed by the host to defend itself against infections or neoplastic disease (immunosuppression) or unnecessary tissue damage (chronic inflammation, hypersensitivity, or autoimmunity) is compromised. Potential immunological effects and responses that may be associated with one or more of these effects are presented in Table 38.8. Hypersensitivity responses are classified on the basis of the immunological mechanism that mediates the response. There are four types: type I (anaphylactic), type II (cytotoxic), type III (immune complex), and type IV (cell-mediated delayed hypersensitivity). Hypersensitivity is considered to be increased reactivity to an antigen to which a human or animal has been previously exposed, with an adverse rather than a protective effect. Hypersensitivity is a synonym for allergy. Type I (anaphylactic) reactions and type IV (cell-mediated delayed hypersensitivity) reactions are the most common (Nebeker et al., 2006). Types II and III reactions are relatively rare and are less likely to occur with medical devices and biomaterials; however, with tissue-engineered devices containing potential antigens (i.e. proteins), extracellular matrix (ECM) components, and/or cells, types II and III reactions must be considered in biological response evaluations.

TABLE 38.8 Potential Immunological Effects and Responses

Effects	Responses
Hypersensitivity	Histopathological changes
Type I — anaphylactic	Humoral responses
Type II — cytotoxic	Host resistance
Type III — immune complex	Clinical symptoms
Type IV — cell-mediated (delayed)	Cellular responses
Chronic inflammation	T-cells
Immunosuppression	NK cells
Immunostimulation	Macrophages
Autoimmunity	Granulocytes

Type I (anaphylactic) hypersensitivity reactions are mediated by IgE antibodies, which are cytotropic and affect the immediate release of basoactive amines and other mediators from basophils and mast cells followed by recruitment of other inflammatory cells. Type IV cell-mediated (delayed) hypersensitivity responses involve sensitized T lymphocytes that release cytokines and other mediators that lead to cellular and tissue injury. Type IV hypersensitivity (cell-mediated) reactions are initiated by specifically sensitized T lymphocytes. This reaction includes the classic delayed-type hypersensitivity reaction initiated by CD4+ T-cells and direct cell cytotoxicity mediated by CD8+ T-cells. The less common type II (cytotoxic) hypersensitivity involves the formation and binding of IgG and/or IgM to antigens on target cell surfaces that facilitate phagocytosis of the target cell or lysis of the target cell by activated complement components. Type II hypersensitivity (cytotoxic) is mediated by antibodies directed toward antigens present on the surface of cells or other tissue components. Three different antibody-dependent mechanisms may be involved in this type of reaction: complement-dependent reactions, antibody-dependent reactions, cell-mediated cytotoxicity, or antibody-mediated cellular dysfunction. Type III immune complex hypersensitivity is present when circulating antigen-antibody complexes activate complement whose components are chemotactic for neutrophils that release enzymes and other toxic moieties and mediators leading to cellular and tissue injury.

Immunological reactions that occur with organ transplant rejection also offer insight into potential immune responses to tissue-engineered devices. Mechanisms involved in organ transplant rejection include T-cell-mediated reactions by direct and indirect pathways and antibody-mediated reactions. Immune responses may be avoided or diminished by using autologous or isogeneic cells in cell/polymer scaffold constructs. The use of allogeneic or xenogenic cells incorporated into the device requires prevention of immune rejection by immune suppression of the host, induction of tolerance in the host, or immunomodulation of the tissue-engineered construct. The development of tissue-engineered constructs by immunoisolation using polymer membranes and the use of non-host cells have been compromised by immune responses. In this concept, a polymer membrane is used to encapsulate non-host cells or tissues, thus separating them from the host immune system. However, antigens shed by encapsulated cells were released from the device and initiated immune responses (Brauker, 1992; Brauker et al., 1995; Babensee et al., 1998).

Although exceptionally minimal and superficial in its presentation, the previously discussed humoral and cell-mediated immune responses demonstrate the possibility that any known tissue-engineered construct may undergo immunological tissue injury. To date, our understanding of immune mechanisms and their interactions with tissue-engineered constructs is markedly limited. One of the obvious problems is that preliminary studies are generally carried out with non-human tissues and immune reactions result when tissue-engineered constructs from one species are used in testing the device in another species. Ideally, tissue-engineered constructs would be prepared from cells and tissues of a given species and subsequently tested in that species. While this approach does not guarantee that immune responses will not be present, the probability of immune responses in this type of situation is markedly decreased.

The following examples provide perspective to these issues. They further demonstrate the detailed and in-depth approach that must be taken to appropriately and adequately evaluate tissue-engineered constructs or devices and their potential adverse responses.

The inflammatory response considered to be immunotoxic is persistent chronic inflammation. With biomaterials, controlled release systems, and tissue-engineered devices, potential antigens capable of stimulating the immune response may be present and these agents may facilitate a chronic inflammatory response that is of extended duration (weeks, months). Regarding immunotoxicity, it is this persistent chronic inflammation that is of concern as immune granuloma formation and other serious immunological reactions such as autoimmune disease may occur. Thus, in biological response evaluation, it is important to discriminate between the short-lived chronic inflammation that is a component of the normal inflammatory and healing responses versus long-term, persistent chronic inflammation that may indicate an adverse immunological response.

Immunosuppression may occur when antibody and T-cell responses (adaptive immune response) are inhibited. Potentially significant consequences of this type of response are frequent, and serious infections result from reduced host defense. Edelman and colleagues have demonstrated that incorporating endothelial cells into three-dimensional collage matrices has a downregulating effect on the humeral and cellular immune response elicited by the endothelial cells (Methe et al., 2008). The strong MHC dominant immune response that occurs when endothelial cells are the primary component of an implant can be significantly reduced when the endothelial cells are embedded in the three-dimensional collagen matrix. The endothelial cells, while retaining many of the characteristics of quiescent endothelial cells, evoke no significant humeral or cellular immune response in immuno-competent animals and additionally reduce the memory response to previous free endothelial cell implants. These studies are significant and they demonstrate the influence of spatial matrix formation as well as matrix composition on endothelial cell immunophenotype. Thus, modulation of the matrix structure may be helpful in designing vascular conduits for tissue-engineered devices.

Utilizing ECM scaffolds prepared under different conditions, Badylak and colleagues have determined the participation of different macrophage phenotypes in the degradation and remodeling of the extracellular matrix scaffolds, demonstrating that the properties of the matrix can control the innate and, possibly, the acquired immune responses to ECM scaffolds (Badylak et al., 2008; Brown et al., 2009; Valentin et al., 2009).

Immunostimulation may occur when unintended or inappropriate antigen-specific or non-specific activation of the immune system is present. From a biomaterial and controlled release system perspective, antibody and/or cellular immune responses to a foreign protein may lead to unintended immunogenicity. Enhancement of the immune response to an antigen by a biomaterial with which it is mixed *ex vivo* or *in situ* may lead to adjuvancy, which is a form of immunostimulation. This effect must be considered when biodegradable controlled release systems are designed and developed for use as vaccines (Jones, 2008a,b,c).

Patients implanted with polyurethanes used for left ventricular assist devices experience B-cell hyperreactivity and allosensitization (Schuster et al., 2002). There is evidence that T lymphocytes can be activated in response to biomaterials. T lymphocytes cultured in the presence of polyurethane particles from the flexible diaphragms of left ventricular assist devices (LVADs) resulted in intracellular calcium flux, CD40 ligand expression, and nuclear translocation of nuclear factor of activated T-cells (NFAT). NFAT translocation was reduced by a calcineurin inhibitor and CD40 ligand expression was reduced by both a calcineurin inhibitor and CD25 blockade indicating interleukin-2 (IL-2)-dependent activation pathways (Schuster et al., 2001, 2002). T lymphocytes in response to polyurethane particles exhibited classic activation indicators; that is, calcium flux, translocation of transcription factors, upregulation of activation cell surface markers, and proliferation.

Differences in human leukocyte antigen (HLA) gene inheritance can result in major histocompatibility complex (MHC) diversity. MHC loci are among the most genetically variable loci in humans. The MHC class II proteins (DP, DQ, DR) are found on APC. Diversity in MHCII proteins results in individual variability in antigen presentation and, in turn, immune responses. Because of this diversity, individuals mount immune responses to different epitopes of pathogens. LVAD recipients who are predisposed to develop B-cell hyperreactivity have HLA-DR3 expression, indicating that lymphocyte responses to biomaterials are variable and dependent on the individual's genetic profile (Itescu et al., 2003). It is possible that only individuals with certain MHCII receptors can interact with biomaterials in a mechanism that results in a lymphocyte response.

Autoimmunity is the immune response to the body's own constituents, which are considered in this response to be autoantigens. An autoimmune response, indicated by the presence of autoantibodies or T lymphocytes that are reactive with host tissue or cellular antigens may, but not necessarily, result in autoimmune disease with chronic, debilitating, and sometimes life-threatening tissue and organ injury.

Representative tests for the evaluation of immune responses are given in Table 38.9. Table 38.9 is not all-inclusive and other tests may be applicable. The examples presented in Table 38.9 are only representative of the large number of tests that are currently available (Coligan et al., 1992; Burleson et al., 1995; Smialowicz and Holsapple, 1996; Rose et al., 1997). Table 38.9 is informative but incomplete as in the future direct and indirect markers of immune response may be validated and their predictive value documented, thus providing new tests for immunotoxicity. Direct measures of immune system activity by functional assays are the most important types of tests for immunotoxicity. Functional assays are generally more important than tests for soluble mediators, which are more important than phenotyping. Signs of illness may be important in *in vivo* experiments but symptoms may also have a significant role in studies of immune function in clinical trials and postmarket studies.

As with any type of test for biological response evaluation, immunotoxicity tests should be valid and have been shown to provide accurate, reproducible results that are indicative of the effect being studied and are useful in a statistical analysis. This implies that appropriate control groups are also included in the study design.

Immunogenicity involving a specific immune response to a biomaterial is an important consideration as it may lead to serious adverse effects. For example, a foreign, non-human, protein may induce IgE antibodies that cause an anaphylactic (type I) hypersensitivity reaction. An example of this type of response is latex protein found in latex gloves. Low-molecular-weight compounds such as chemical accelerators used in the manufacture of latex gloves may also induce a T-cell-mediated (type IV) reaction resulting in contact dermatitis. Tests for type I (e.g. antigen-specific IgE) and type IV (e.g. guinea pig) maximization tests, hypersensitivity should be considered for materials with the potential to cause these allergic reactions. In

TABLE 38.9 Representative Tests for the Evaluation of Immune Responses

Functional assays	Soluble mediators
Skin testing	Antibodies
Immunoassays (e.g. ELISA)	Complement
Lymphocyte proliferation	Immune complexes
Plaque-forming cells	Cytokine patterns (T-cell subsets)
Local lymph node assay	Cytokines (IL-1, IL-1ra, TNF-α, IL-6, TGF-β, IL-4, IL-13)
Mixed lymphocyte reaction	Chemokines
Tumor cytotoxicity	Basoactive amines
Antigen presentation	**Signs of illness**
Phagocytosis	
Degranulation	Allergy
Resistance to bacteria, viruses, and tumors	Skin rash
Phenotyping	Urticaria
	Edema
Cell-surface markers	Lymphadenopathy
MHC markers	

addition to hypersensitivity reactions, a device may elicit autoimmune responses (i.e. antibodies or T-cells) that react with the body's own constituents. An autoimmune response may lead to the pathological consequences of an autoimmune disease. For example, a foreign protein may induce IgG or IgM antibodies that cross-react with a human protein and cause tissue damage by activating the complement system. In a similar fashion, a biomaterial or controlled release system that has a gel or oil constituent may act as an adjuvant leading to the induction of an autoimmune response. Even if an autoimmune response (autoantibodies and/ or autoreactive T lymphocytes) is suggested in preclinical testing, it is difficult to obtain convincing evidence that a biomaterial or controlled release system causes autoimmune disease in animals. Therefore, routine testing for induction of autoimmune disease in animal models is not recommended.

Babensee and co-workers have tested the hypothesis that the biomaterial component of a medical device, by promoting an inflammatory response, can recruit APCs (e.g. macrophages and dendritic cells) and induce their activation, thus acting as an adjuvant in the immune response to foreign antigens originating from the histological component of the device (Babensee et al., 2002; Matzell and Babensee, 2004; Babensee, 2008). Utilizing polystyrene and polylactic-glycolic acid microparticles and polylactic-glycolic scaffolds together with their model antigen, ovalbumin, in a mouse model for 18 weeks, Babensee et al. demonstrated that a persistent humoral immune response that was Th2 helper T-cell dependent, as determined by the IgG1, was present. These findings indicated that activation of CD41 T-cells and the proliferation and isotype switching of B-cells had occurred. A Th1 immune response characterized by the presence of IgG2a was not identified. Moreover, the humoral immune responses for all three types of microparticles were similar, indicating that the production of antigen-specific antibodies was not material chemistry-dependent in this model. Babensee suggests that the presence of the biomaterial functions as an adjuvant for initiation and promotion of the immune response and augments the phagocytosis of the antigen with expression of MHC class II and co-stimulatory molecules on APCs with the presentation of antigen to CD41 T-cells.

Babensee and co-workers have identified differential levels of dendritic cell maturation on different biomaterials used in combination products (Babensee and Paranjpe, 2005; Bennewitz and Babensee, 2005; Yoshida and Babensee, 2004, 2006; Yang and Jones, 2009). The effect of biomaterials on dendritic cell maturation, and the associated adjuvant effect, is a novel biocompatibility selection and design criteria for biomaterials to be used in combination products in which immune consequences are potential complications or outcomes.

Badylak and colleagues have carried out extensive studies on the utilization of xenogeneic ECM as a scaffold for tissue reconstruction (Allman et al., 2002; Badylak, 2004; Badylak and Gilbert, 2008). Use of the small intestinal submucosa (SIS) ECM in animals has indicated a restricted Th2-type immune response. The presence of natural antibodies to the terminal galactose-α1,3-galactose (α-gal) epitope is considered to be a major barrier to xenotransplantation in humans. Cell membranes of all animals except humans express this epitope and naturally occurring antibodies mediate hyperacute or delayed rejection of transplanted organs through complement fixation or antibody dependence cell-mediated cytotoxicity. While ECM derived from porcine tissues, SIS, contains small amounts of the gal epitope, it appears that the quantity or distribution of this epitope and/or the subtype of immunoglobulin response to the epitope is such that complement activation does not occur (McPherson et al., 2000). In addition, the resorbable characteristics of this non-chemically crosslinked ECM scaffold demonstrate constructive tissue remodeling and deposition of new matrix whereas chemically crosslinked ECM leads to active inflammation and eventually scar formation.

The role of Th1 and Th2 lymphocytes in cell-mediated immune responses to xenografts has been examined. Activation of the Th1 pathway leads to macrophage activation, stimulation of complement fixing antibody isotypes, and differentiation of CD8+ cells to a cytotoxic type phenotype that is associated with both allogeneic and xenogeneic transplant rejection. The Th2 lymphocyte response does not activate macrophages and leads to production of non-complement fixing antibody isotypes and usually is associated with transplant acceptance.

The use of appropriate animal models is an important consideration in the safety evaluation of controlled release systems that may contain potential immunoreactive materials (Greenwald and Diamond, 1988; Cohen and Miller, 1994; Rose, 1997). A recently published study involving the *in vivo* evaluation of recombinant human growth hormone in poly(lactic-co-glycolic acid) (PLGA) microspheres demonstrates the appropriate use of various animal models to evaluate biological responses and the potential for immunotoxicity. Utilizing biodegradable PLGA microspheres containing recombinant human growth hormone (rhGH), Cleland et al. used rhesus monkeys, transgenic mice expression rhGH, and normal control (Balb/C) mice in their *in vivo* studies (Cleland et al., 1997). Rhesus monkeys were utilized for serum assays in the pharmacokinetic study of rhGH release as well as tissue responses to the injected microcapsule formulation. Placebo injection sites were also utilized and a comparison of the injection sites from rhGH PLGA microspheres and placebo PLGA microspheres demonstrated a normal inflammatory and wound healing response with a normal focal foreign body reaction. To further examine the tissue response, transgenic mice were utilized to assess the immunogenicity of the rhGH PLGA formulation. Transgenic mice expressing a heterologous protein have been previously used for assessing the immunogenicity of sequence or structural mutant proteins (Stewart et al., 1989; Stewart, 1993). With the transgenic animals, no detectable antibody response to rhGH was found. In contrast, the Balb/C control mice had a rapid onset of high titer antibody response to the rhGH PLGA formulation. This study points out the appropriate utilization of animal models to not only evaluate biological responses but also one type of immunotoxicity (immunogenicity) of controlled release systems.

The focus in tissue engineering traditionally has been on modulating the fate of transplanted-host-cell populations that directly participate in the reconstruction of tissues. However, new materials for tissue engineering are now being considered that give greater control over the inflammatory and immune responses (Jeong et al., 2006). Biomimetic strategies based on viruses and bacteria are now being utilized for the development of immune evasive biomaterials (Novak et al., 2009). Materials are being investigated that can promote tolerance to specific antigens and cells by directly signaling antigen-presenting cells (APCs), such as dendritic cells, or by releasing growth factors or cytokines that promote tolerance. On the other hand, materials might promote a destructive immune response by directly providing immunity-promoting signals or by releasing insoluble factors. This approach could be used to combat infections and cancer (Chan and Mooney, 2008).

CONCLUSION

Tissue-engineered devices are biological-biomaterial combinations in which some component of tissue has been combined with a biomaterial to create a device for the restoration or modification of tissue or organ function. The biocompatibility and bioresponse require the ultimate achievement of four significant goals if these devices are to function adequately and appropriately in the host environment. These goals are: (1) restoration of the target tissue with its appropriate function and cellular phenotypic expression, (2) inhibition of the macrophage and FBGC foreign body response that may degrade or adversely modify device function, (3) inhibition of scar and fibrous capsule formation that may be deleterious to the function of the device, and (4) inhibition of immune responses that may inhibit the proposed function of the device and ultimately lead to the destruction of the tissue component of the tissue-engineered device. This chapter has presented a brief and limited overview of mechanisms and biological responses that determine biocompatibility: inflammation, wound healing, and immunotoxity. Given the unique nature of the combination of tissue component and biomaterial in tissue-engineered devices, coupled with the species differences in biological responses, a significant future challenge in the development of tissue-engineered devices is the construction and utilization of a unique set of tests that will ensure that the four goals indicated above are achieved for the lifetime of the device in its *in vivo* environment in humans.

References

Allman, A. J., McPherson, T. B., Merrill, L. C., Badylak, S. F., & Metzger, D. W. (2002). The Th2-restricted immune response to xenogeneic small intestinal submucosa does not influence systemic protective immunity to viral and bacterial pathogens. *Tiss. Eng., 8*, 53–62.

Anderson, J. M. (1988). Inflammatory response to implants. *ASAIO, 11*, 101–107.

Anderson, J. M. (1993). Mechanisms of inflammation and infection with implanted devices. *Cardiovasc. Pathol., 2*, 199S–208S.

Anderson, J. M. (2001). Biological responses to materials. *Ann. Rev. Mater. Res., 31*, 81–110.

Anderson, J. M., & Jones, J. A. (2007). Phenotypic dichotomies in the foreign body reaction. *Biomaterials, 28*, 5114–5120.

Anderson, J. M., Rodriguez, A., & Chang, D. T. (2008). Foreign body reaction to biomaterials. *Semin. Immunol., 20*, 86–100.

Babensee, J. E. (2008). Interaction of dendritic cells with biomaterials. *Semin. Immunol., 20*, 101–108.

Babensee, J. E., & Paranjpe, A. (2005). Differential levels of dendritic cell maturation on different biomaterials used in combination products. *J. Biomed. Mater. Res., 74A*, 503–510.

Babensee, J. E., Anderson, J. M., McIntire, L. V., & Mikos, A. G. (1998). Host response to tissue-engineered devices. *Adv. Drug Del. Rev., 33*, 111–139.

Babensee, J. E., Stein, M. M., & Moore, L. K. (2002). Interconnections between inflammatory and immune responses in tissue engineering. *Ann. N.Y. Acad. Sci., 961*, 360–363.

Badylak, S. F. (2004). Xenogeneic extracellular matrix as a scaffold for tissue reconstruction. *Transpl. Immunol., 12*, 367–377.

Badylak, S. F., & Gilbert, T. W. (2008). Immune response to biological scaffold materials. *Semin. Immunol., 20*, 109–116.

Badylak, S. F., Valentin, J. E., Ravindra, A. K., McCabe, G. P., & Stewart-Akers, A. M. (2008). Macrophage phenotype as a determinant of biological scaffold remodeling. *Tissue Eng., 14*, 1835–1842.

Bennewitz, N. L., & Babensee, J. E. (2005). The effect of the physical form of poly(lactic- *co-* glycolic acid) carriers on the humoral immune response to co-delivered antigen. *Biomaterials, 26*, 2991–2999.

Brauker, J. (1992). Neovascularization of immuno-isolation membranes: the effect of membrane architecture and encapsulated tissue. In *First International Congress of Cell Transplant Society, A138* (pp 163). Immunoisolation and Bioartificiality.

Brauker, J. H., Carr-Brendel, V. E., Martinson, L. A., Crudele, J., Johnston, W. D., & Johnson, R. C. (1995). Neovascularization of synthetic membranes directed by membrane microarchitecture. *J. Biomed. Mater. Res., 29*, 1517–1524.

Broadley, K. N., Aquino, A. M., Woodward, S. C., Buckley-Sturrock, A., Sato, Y., Rifkin, D. B., et al. (1989). Monospecific antibodies implicate basic fibroblast growth factor in normal wound repair. *Lab. Invest., 61*, 571–575.

Brodbeck, W. G., & Anderson, J. M. (2009). Giant cell formation and function. *Curr. Opin. Hematol, 16*, 53–57.

Brodbeck, W. G., MacEwan, M., Colton, E., Meyerson, H., & Anderson, J. M. (2005). Lymphocytes and the foreign body response: lymphocyte enhancement of macrophage adhesion and fusion. *J. Biomed. Mater. Res., 74A*, 222–229.

Broughton, G., Janis, J. E., & Attinger, C. E. (2006). The basic science of wound healing. *Plast. Reconstr. Surg., 117* (Suppl.), 12S–34S.

Browder, T., Folkman, J., & Pirie-Shepherd, S. (2000). The hemostatic system as a regulator of angiogenesis. *J. Biol. Chem., 275*, 1521–1524.

Brown, B. N., Valentin, J. E., Stewart-Akers, A. M., McCabe, G. P., & Badylak, S. F. (2009). Macrophage phenotype and remodeling outcomes in response to biological scaffolds with and without a cellular component. *Biomaterials, 30*, 1482–1491.

Burleson, G. R., Dean, J. H., & Munson, A. E. (Eds.). (1995). *Methods in Immunotoxicology*. New York: Wiley-Liss.

Butcher, E. C. (1991). Leukocyte-endothelial cell recognition: three (or more) steps to specificity and diversity. *Cell, 67*, 1033–1036.

Chambers, T. J., & Spector, W. G. (1982). Inflammatory giant cells. *Immunobiology, 161,* 283–289.

Chan, G., & Mooney, D. J. (2008). New materials for tissue engineering: towards greater control over the biological response. *Trends Biotechnol., 26,* 382–392.

Chang, D. T., Jones, J. A., Meyerson, H., Colton, E., Kwon, I. K., Matsuda, T., et al. (2008). Lymphocyte/macrophage interactions: biomaterial surface-dependent cytokine, chemokine, and matrix protein production. *J. Biomed. Mater. Res., 87A,* 676–687.

Chang, D. T., Saidel, G. M., & Anderson, J. M. (2009a). Dynamic systems model for lymphocyte interactions with macrophages at biomaterial surfaces. *Cell. Molec. Bioeng., 2,* 573–590.

Chang, D. T., Colton, E., & Anderson, J. M. (2009b). Paracrine and juxtacrine lymphocyte enhancement of adherent macrophage and foreign body giant cell activation. *J. Biomed. Mater. Res., 89A,* 490–498.

Chang, D. T., Colton, E., Matsuda, T., & Anderson, J. M. (2009c). Lymphocyte adhesion and interactions with biomaterial adherent macrophages and foreign body giant cells. *J. Biomed. Mater. Res., 91A,* 1210–1220.

Clark, R. A., Lanigan, J. M., DellePelle, P., Manseau, E., Dvorak, H. F., & Colvin, R. B. (1982). Fibronectin and fibrin provide a provisional matrix for epidermal cell migration during wound reepithelialization. *J. Invest. Dermatol., 79,* 264–269.

Cleland, J. L., Duenas, E., Daugherty, A., Marian, M., Yang, J., Wilson, M., et al. (1997). Recombinant human growth hormone poly(lactic-co-glycolic acid) (PLGA) microspheres provide a long lasting effect. *J. Contr. Rel., 49,* 193–205.

Cohen, I. R., & Miller, A. (Eds.). (1994). *Autoimmune Disease Models: A Guidebook.* New York: Academic Press.

Coligan, J. E., Kruisbeek, A. M., Magulies, D. H., Shevach, E. M., & Strober, R. (Eds.). (1992). *Current Protocols in Immunology.* New York: Greene Publishing Associates and Wiley Interscience.

Cotran, R. S., & Pober, J. S. (1990). Cytokine-endothelial interactions in inflammation, immunity, and vascular injury. *J. Am. Soc. Nephrol., 1,* 225–235.

DeFife, K. M., McNally, A. K., Colton, E., & Anderson, J. M. (1997). Interleukin-13 induces human monocyte/macrophage fusion and macrophage mannose receptor expression. *J. Immunol., 158,* 319–328.

Dvorak, H. F., Harvey, V. S., Estrella, P., Brown, L. F., McDonagh, J., & Dvorak, A. M. (1987). Fibrin containing gels induce angiogenesis. Implications for tumor stroma generation and wound healing. *Lab. Invest., 57,* 673–686.

Fong, Y., Moldawer, L. L., Shires, G. T., & Lowry, S. F. (1990). The biological characteristics of cytokines and their implication in surgical injury. *Surg. Gynecol. Obstet., 170,* 363–378.

Gallin, J. I., & Synderman, R. (Eds.). (1999). *Inflammation: Basic Principles and Clinical Correlates.* New York: Raven Press.

Ganz, T. (1988). Neutrophil receptors. In "Neutrophils and Host Defense" (R.I. Lehrer, moderator). *Ann. Intern. Med, 109,* 127–142.

Golden, M. A., Au, Y. P., Kirkman, T. R., Wilcox, J. N., Raines, E. W., Ross, R., et al. (1991). Platelet-derived growth factor activity and RNA expression in healing vascular grafts in baboons. *J. Clin. Invest., 87,* 406–414.

Gorbet, M. B., & Sefton, M. V. (2004). Biomaterial-associated thrombosis: roles of coagulation factors, complement, platelets and leukocytes. *Biomaterials, 25,* 5681–5703.

Greenwald, R. A., & Diamond, H. S. (Eds.). (1988). *Handbook of Animal Models for the Rheumatic Diseases, Vol. I.* Boca Raton, FL: CRC Press.

Greisler, H. (1988). Macrophage-biomaterial interactions with bioresorbable vascular prostheses. *Trans. Am. Soc. Artif. Intern. Organs, 34,* 1051–1057.

Haas, A. (2007). The phagosome: compartment with a license to kill. *Traffic, 8,* 311–330.

Henson, P. M. (1971). The immunologic release of constituents from neutrophil leukocytes. II. Mechanisms of release during phagocytosis, and adherence to nonphagocytosable surfaces. *J. Immunol., 197,* 1547–1557.

Henson, P. M. (1980). Mechanisms of exocytosis in phagocytic inflammatory cells. *Am. J. Pathol., 101,* 494–511.

Henson, P. M., & Johnston, R. B., Jr. (1987). Tissue injury in inflammation: oxidants, proteinases, and cationic proteins. *J. Clin. Invest., 79,* 669–674.

Hume, D. A. (2008). Macrophages as APC and the dendritic cell myth. *J. Immunol., 181,* 5829–5835.

Hynes, R. O. (1992). Integrins: versatility, modulation, and signaling in cell adhesion. *Cell, 69,* 11–25.

Ignotz, R., Endo, T., & Massague, J. (1987). Regulation of fibronectin and type I collagen mRNA levels by transforming growth factor-beta. *J. Biol. Chem., 262,* 6443–6446.

Itescu, S., Schuster, M., Burke, E., Ankersmit, J., Kocher, A., Deng, M., John, R., & Lietz, K. (2003). Immunobiological consequences of assist devices. *Cardiol. Clin., 21,* 119–133, ix–x.

Janeway, C. A., Jr., & Travers, P. (1997). *Immunobiology: The Immune System in Health and Disease* (3rd ed.). New York: Current Biology — Garland.

Jankowski, A., Scott, C. C., & Grinstein, S. (2002). Determinants of the phagosomal pH in neutrophils. *J. Biol. Chem., 277,* 6059–6066.

713

Jenney, C. R., & Anderson, J. M. (2000). Adsorbed IgG: a potent adhesive susbstrate for human macrophages. *J. Biomed. Mater. Res., 50,* 281–290.

Jeong, H. J., Lee, S. A., Moon, P. D., Na, H. J., Park, R. K., Um, J. Y., et al. (2006). Alginic acid has anti-anaphylactic effects and inhibits inflammatory cytokine expression via suppression of nuclear factor-kappaB activation. *Clin. Exp. Allergy, 36,* 785–794.

Johnston, R. B., Jr. (1988). Monocytes and macrophages. *N. Engl. J. Med., 318,* 747–752.

Jones, J. A., Chang, D. T., Meyerson, H., Colton, E., Kwon, I. K., Matsuda, T., & Anderson, J. M. (2007). Proteomic analysis and quantification of cytokines and chemokines from biomaterial surface-adherent macrophages and foreign body giant cells. *J. Biomed. Mater. Res., 83A,* 585–596.

Jones, J. A., McNally, A. K., Chang, D. T., Qin, L. A., Meyerson, H., Colton, E., et al. (2008). Matrix metalloproteinases and their inhibitors in the foreign body reaction on biomaterials. *J. Biomed. Mater. Res., 84A,* 158–166.

Jones, K. S. (2008a). Assays on the influence of biomaterial on allogeneic rejection in tissue engineering. *Tissue Engineering: Part B, 14,* 407–417.

Jones, K. S. (2008b). Effects of biomaterial-induced inflammation on fibrosis and rejection. *Semin. Immunol., 20,* 130–136.

Jones, K. S. (2008c). Biomaterials as vaccine adjuvants. *Biotechnol. Prog., 24,* 807–814.

Jutila, M. A. (1990). Leukocyte traffic to sites of inflammation. *APMIS, 100,* 191–201.

Kovacs, E. J. (1991). Fibrogenic cytokines: the role of immune mediators in the development of scar tissue. *Immunol. Today, 12,* 17–23.

Kumar, V., Abbas, A. K., Fauston, N., & Aster, J. C. (Eds.). (2010). *Robbins and Cotran Pathologic Basis of Disease* (8th ed.) (pp. 43–110). Saunders/Elsevier.

Langone, J. J. (1998). Immunotoxicity Testing Guidance, *Draft Document.* Office of Science and Technology, Center for Devices and Radiological Health, Food and Drug Administration.

MacEwan, M. R., Brodbeck, W. G., Matsuda, T., & Anderson, J. M. (2005). *Student Research Award In the Undergraduate Degree Candidate category, 30th Annual Meeting of the Society for Biomaterials, Memphis, Tennessee, April 27–30, 2005. Monocyte/lymphocyte interactions and the foreign body response: in vitro effects of biomaterial surface chemistry.* In: *J. Biomed. Mater. Res., 74,* 285–293.

Maciag, T. (1990). Molecular and cellular mechanisms of angiogenesis. In V. T. DeVita, S. Hellman, & S. Rosenberg (Eds.), *Important Advances in Oncology* (pp. 85–103). Philadelphia: Lippincott.

Madri, J. A., Pratt, B. M., & Tucker, A. M. (1988). Phenotypic modulation of endothelial cells by transforming growth factor-beta depends upon the composition and organization of the extracellular matrix. *J. Cell Biol., 106,* 1375–1384.

Malech, H. L., & Gallin, J. I. (1987). Current concepts: immunology. Neutrophils in human diseases. *N. Engl. J. Med., 317,* 687–694.

Marchant, R., Hiltner, A., Hamlin, C., Rabinovitch, A., Slobodkin, R., & Anderson, J. M. (1983). *In vivo* biocompatibility studies: I. The cage implant system and a biodegradable hydrogel. *J. Biomed. Mater. Res., 17,* 301–325.

Matzell, M. M., & Babensee, J. E. (2004). Humoral immune responses to model antigen co-delivered with biomaterials used in tissue engineering. *Biomaterials, 25,* 295–304.

McNally, A. K., & Anderson, J. M. (1994). Complement C3 participation in monocyte adhesion to different surfaces. *Proc. Natl. Acad. Sci. U.S.A., 91,* 10119–10123.

McNally, A. K., & Anderson, J. M. (1995). Interleukin-4 induces foreign body giant cells from human monocytes/macrophages. Differential lymphokine regulation of macrophage fusion leads to morphological variants of multinucleated giant cells. *Am. J. Pathol., 147,* 1487–1499.

McNally, A. K., & Anderson, J. M. (2002). Beta1 and beta2 integrins mediate adhesion during macrophage fusion and multinucleated foreign body giant cell formation. *Am. J. Pathol., 160,* 621–630.

McNally, A. K., DeFife, K. M., & Anderson, J. M. (1996). Interleukin-4-induced macrophage fusion is prevented by inhibitors of mannose receptor activity. *Am. J. Pathol., 149,* 975–985.

McNally, A. K., MacEwan, S. R., & Anderson, J. M. (2007). Alpha subunit partners to Beta1 and Beta2 integrins during IL-4-induced foreign body giant cell formation. *J. Biomed. Mater. Res., 82A,* 568–574.

McNally, A. K., Jones, J. A., MacEwan, S. R., Colton, E., & Anderson, J. M. (2008). Vitronectin is a critical protein adhesion substrate for IL-4-induced foreign body giant cell formation. *J. Biomed. Mater. Res., 86A,* 535–543.

McPherson, T. B., Liang, H., Record, R. D., & Badylak, S. F. (2000). Galα(1,3)Gal epitope in porcine small intestinal submucosa. *Tiss. Eng., 6,* 233–239.

Methe, H., Hess, S., & Edelman, E. R. (2008). The effect of three-dimensional matrix-embedding of endothelial cells on the humoral and cellular immune response. *Semin. Immunol., 20,* 117–122.

Muller, G., Behrens, J., Nussbaumer, U., Böhlen, P., & Birchmeier, W. (1987). Inhibitor action of transforming growth factor beta on endothelial cells. *Proc. Natl. Acad. Sci. U.S.A., 84,* 5600–5604.

Mustoe, T. A., Pierce, G. F., Thomason, A., Gramats, P., Sporn, M. B., & Deuel, T. F. (1987). Accelerated healing of incisional wounds in rats induced by transforming growth factor. *Science, 237,* 1333–1336.

Nebeker, J. R., Virmani, R., Bennett, C. L., Hoffman, J. M., Samore, M. H., Alvarez, J., et al. (2006). Hypersensitivity cases associated with drug-eluting coronary stents. *J. Am. Coll. Cardiol., 47,* 175–181.

Nguyen, L. L., & d'Amore, P. A. (2001). Cellular interactions in vascular growth and differentiation. *Int. Rev. Cytol., 204,* 1–48.

Nilsson, B., Ekdahl, K. N., Mollnes, T. E., & Lambris, J. D. (2007). The role of complement in biomaterial-induced inflammation. *Mol. Immunol., 44,* 82–94.

Novak, M. T., Bryers, J. D., & Reichert, W. M. (2009). Biomimetic strategies based on viruses and bacteria for the development of immune evasive biomaterials. *Biomaterials, 30,* 1989–2005.

Paty, P. B., Graeff, R. W., Mathes, S. J., & Hunt, T. K. (1990). Superoxide production by wound neutrophils: evidence for increased activity of the NADPH oxidase. *Arch. Surg., 125,* 65–69.

Pober, J. S., & Cotran, R. S. (1990). The role of endothelial cells in inflammation. *Transplantation, 50,* 537–544.

Rae, T. (1986). The macrophage response to implant materials. *Crit. Rev. Biocompatibility, 2,* 97–126.

Revell, P. A. (2008). The combined role of wear particles, macrophages, and lymphocytes in the loosening of total joint prostheses. *J. R. Soc. Interface, 5,* 1263–1278.

Riches, D. W. F. (1988). Macrophage involvement in wound repair, remodeling, and fibrosis. In R. A. P. Clark, & P. M. Henson (Eds.), *The Molecular and Cellular Biology of Wound Repair* (pp. 213–239). New York: Plenum Press.

Rodriguez, A., & Anderson, J. M. (2010). Evaluation of clinical biomaterial surface effects on T lymphocyte activation. *J. Biomed. Mater. Res., 92A,* 214–220.

Rodriguez, A., Voskerician, G., Meyerson, H., MacEwan, S. R., & Anderson, J. M. (2008). T cell subset distributions following primary and secondary implantation at subcutaneous biomaterial implant sites. *J. Biomed. Mater. Res., 85A,* 556–565.

Rodriguez, A., Meyerson, H., & Anderson, J. M. (2009a). Quantitative *in vivo* cytokine analysis at synthetic biomaterial implant sites. *J. Biomed. Mater. Res., 89A,* 152–159.

Rodriguez, A., MacEwan, S. R., Meyerson, H., Kirk, J. T., & Anderson, J. M. (2009b). The foreign body reaction in T-cell-deficient mice. *J. Biomed. Mater. Res., 90A,* 106–113.

Rose, N. R. (1997). In M. S. Lefell, A. D. Donnenberg, & N. R. Rose (Eds.), *Immunologic Diagnosis of Autoimmune Disease* (pp. 111–123). Boca Raton, FL: CRC Press.

Rose, N. R., de Mecario, E. C., Folds, J. D., Lane, H. C., & Nakamura, R. M. (1997). *Manual of Clinical Laboratory Immunology.* Washington, DC: ASM Press.

Salthouse, T. N. (1976). Cellular enzyme activity at the polymer-tissue interface: a review. *J. Biomed. Mater. Res., 10,* 197–229.

Schuster, M., Kocher, A., Lietz, K., Ankersmit, J., John, R., Edwards, N., et al. (2001). Induction of CD40 ligand expression in human T cells by biomaterials derived from left ventricular assist device surfaces. *Transplant. Proc., 33,* 1960–1961.

Schuster, M., Kocher, A., John, R., Hoffman, M., Ankersmit, J., Lietz, K., et al. (2002). B-cell activation and allosensitization after left ventricular assist device implantation is due to T-cell activation and CD40 ligand expression. *Hum. Immunol., 63,* 211–220.

Sefton, M. V., Babensee, J. E., & Woodhouse, K. A. (2008). Innate and adaptive immune responses in tissue engineering. *Semin. Immunol., 20,* 83–85.

Silver, I. A., Murrills, R., & Etherington, D. J. (1988). Microelectrode studies on the acid environment beneath adherent macrophages and osteoclasts. *Exp. Cell Res., 175,* 266–276.

Smialowicz, R. J., & Holsapple, M. P. (1996). *Experimental Immunotoxicology.* Boca Raton, FL: CRC Press.

Spilizewski, K. L., Marchant, R. E., Hamlin, C. R., Anderson, M., Tice, T. R., Dappert, T. O., et al. (1985). The effect of hydrocortisone acetate loaded poly(dl-lactide) films on the inflammatory response. *J. Contr. Rel., 2,* 197–203.

Sporn, M. B., & Roberts, A. B. (1988). Peptide growth factors are multifunctional. *Nature, 332*(6161), 217–219.

Sporn, M. B., & Roberts, A. B. (1990). *Peptide Growth Factors and Their Receptors I.* New York: Springer.

Stewart, T. A. (1993). Models of human endocrine disorders in transgenic rodents. *Trends Endocrinol. Metab., 4,* 136–141.

Stewart, T. A., Hollinghead, P. G., Pitts, S. L., Chang, R., Martin, L. E., & Oakley, H. (1989). Transgenic mice as a model to test the immunogenicity of proteins altered by site-specific mutagenesis. *Mol. Biol. Med., 6,* 275–281.

Tang, L. (1998). Mechanism of pro-inflammatory fibrinogen-biomaterial interactions. *J. Biomater. Sci. Polym. Ed., 9,* 1257–1266.

Tang, L., & Eaton, J. W. (1993). Fibrin(ogen) mediates acute inflammatory responses to biomaterials. *J. Exp. Med., 178,* 2147–2156.

715

Thompson, J. A., Anderson, K. D., DiPetro, J. M., Zweibel, J. A., Zmaetta, M., Anderson, W. F., & Maciag, T. (1988). Site-directed neovessel formation *in vivo*. *Science, 241*, 1349—1352.

Valentin, J. E., Stewart-Akers, A. M., Gilbert, T. W., & Badylak, S. F. (2009). Macrophage participation in the degradation and remodeling of extracellular matrix scaffolds. *Tissue Eng., 15*, 1687—1694.

Wahl, S. M., Hunt, D. A., Wakefield, L. M., Roberts, A. B., & Sporn, M. B. (1987). Transforming growth factor-beta (TGF-β) induces monocyte chemotaxis and growth factor production. *Proc. Natl. Acad. Sci. U.S.A., 84*, 5788—5792.

Wahl, S. M., Wong, H., & McCartney-Francis, N. (1989). Role of growth factors in inflammation and repair. *J. Cell Biochem., 40*, 193—199.

Weisman, G., Smolen, J. E., & Krochak, H. M. (1980). Release of inflammatory mediators from stimulated neutrophils. *N. Engl. J. Med, 303*, 27—34.

Weiss, S. J. (1989). Tissue destruction by neutrophils. *N. Engl. J. Med., 320*, 365—376.

Wiggins, M. J., Wilkoff, B., Anderson, J. M., & Hiltner, A. (2001). Biodegradation of polyether polyurethane inner insulation in bipolar pacemaker leads. *J. Biomed. Mater. Res. (Appl. Biomater.), 58*, 302—307.

Williams, G. T., & Williams, W. J. (1983). Granulomatous inflammation — a review. *J. Clin. Pathol., 36*, 723—733.

Yang, D., & Jones, K. S. (2009). Effect of alginate on innate immune activation of macrophages. *J. Biomed. Mater. Res., 90A*, 411—418.

Yoshida, M., & Babensee, J. E. (2004). Poly(lactic-co-glycolic acid) enhances maturation of human monocyte-derived dendritic cells. *J. Biomed. Mater. Res., 71*, 45—54.

Yoshida, M., & Babensee, J. E. (2006). Differential effects of agarose and poly(lactic-co-glycolic acid) on dendritic cell maturation. *J. Biomed. Mater. Res., 79*, 393—408.

Zdolsek, J., Eaton, J. W., & Tang, L. (2007). Histamine release and fibrinogen adsorption mediate acute inflammatory responses to biomaterial implants in humans. *J. Translational Med., 5*, 31—37.

Zhao, Q., Agger, M. P., Fitzpatrick, M., Anderson, J. M., Hiltner, A., Stokes, K., et al. (1990). Cellular interactions with biomaterials: *in vivo* cracking of pre-stressed Pellethane 2363—80A. *J. Biomed. Mater. Res., 24*, 621—637.

Zhao, Q., Topham, N., Anderson, J. M., Hiltner, A., Lodoen, G., & Payet, C. R. (1991). Foreign-body giant cells and polyurethane biostability: *in vivo* correlation of cell adhesion and surface cracking. *J. Biomed. Mater. Res., 25*, 177—183.

Ziats, N. P., Miller, K. M., & Anderson, J. M. (1985). *In vitro* and *in vivo* interactions of cells with biomaterials. *Biomaterials, 9*, 5—13.

CHAPTER 39

Designing Tunable Artificial Matrices for Stem Cell Culture

Elizabeth F. Irwin[*]**, Jacob F. Pollock**[*]**, David V. Schaffer**[†,‡]**, Kevin E. Healy**[*,§]
[*] Department of Bioengineering
[†] Department of Chemical Engineering
[‡] The Helen Wills Neuroscience Institute
[§] Department of Materials Science and Engineering, University of California at Berkeley, Berkeley, CA, USA

INTRODUCTION

Embryonic stem (ES) cells, induced pluripotent stem (iPS) cells, and adult stem cells can generate a myriad of different cells types in the body and thus have enormous potential for use in regenerative medicine. *In vivo*, the fate of these stem cells, that is, to remain undifferentiated or to differentiate into a particular cell type, is determined in large part by their local microenvironment, where the regulatory signaling mechanisms include cell-cell interactions; cell-matrix interactions; growth factors; cytokines; and the physiochemical nature of the environment, including the oxygen tension, osmolarity, and pH. The local microenvironment, however, is highly dynamic, not only as organismal development progresses, but also within the fluctuating nature of adult tissue.

Adult stem cells grow in niches and are often maintained in a dormant, multipotent state where they retain the ability to either self-renew or divide. These cells receive signals through a diverse population of neighboring differentiated cell types, which secrete growth factors and cytokines and organize the extracellular matrix (Fuchs et al., 2004). Niche cells thereby provide an environment that isolates stem cells from differentiation stimuli, apoptotic stimuli, and excessive stem cell proliferation that could lead to cancer (Moore and Lemischka, 2006). However, with tissue injury and other processes associated with tissue turnover, the surrounding microenvironment actively signals to these adult stem cells to begin to proliferate, self-renew, and/or differentiate to form new tissues.

The fate of the inner cell mass, the *in vivo* precursors of ES cells, is likewise determined by a complex sequence of signaling from the local environment, which in this case provides a more dynamic set of chemical and mechanical signals that orchestrate tissue formation and differentiation. During the earliest stages of this process, inner cell mass constituents interact with the matrix as it guides fundamental processes of development including migration in the early embryo, and the modulation of growth and differentiation programs of cells (Zagris, 2001). Subsequently, as groups of cells form tissues, they experience not only morphogen patterns but also tension, compression, and shear forces, and these mechanical forces can regulate the expression of various genes (Brouzes and Farge, 2004).

717

Principles of Regenerative Medicine. DOI: 10.1016/B978-0-12-381422-7.10039-2

In order to grow stem cells and tissues *in vitro*, it is necessary to understand and attempt to reproduce the complex microenvironment presented to these cells *in vivo*; however, current culture technologies are not sufficient to mimic this dynamic and intricate natural environment. This chapter focuses on progress in the design and characterization of artificial matrices that attempt to recapitulate microenvironmental cues for *in vitro* stem cell culture and differentiation.

THE EXTRACELLULAR MATRIX
Physical properties of the extracellular matrix

The natural extracellular matrix (ECM) provides a network of chemically and physically associated proteins and polysaccharides that allow cells to attach, migrate, and proliferate and also presents biochemical and physical signals affecting cell fate (Roskelley et al., 1995). A schematic of these interactions is shown in Figure 39.1. In addition, the ECM is not a static entity but instead provides a very dynamic environment whose components are locally secreted and restructured by cells. For example, as they move through the matrix, cells deposit new proteins as well as locally cleave proteins by releasing metalloproteinases (Streuli, 1999). In addition, the remodeling of the ECM is even more rapid in developing tissues, a process particularly relevant to embryonic and fetal stem cells (Zagris, 2001).

Collagen scaffolds, which have been extensively studied, exemplify the basic architecture of the ECM. In tissues such as bone, cartilage, and tendons, collagen is arranged in fibrils to provide tensile strength. In contrast, in epithelial tissue, collagen forms a network of fibers as a basement membrane (Bosman and Stamenkovic, 2003), an open structure that allows rapid diffusion of nutrients, metabolites, and hormones between the blood and constituent cells. In addition to these two structures, there is huge variability in collagen structure from tissue to tissue, as their complex architectures are composed of more than 28 genetically distinct collagen molecules (Martin et al., 1985; Gordon and Hahn, 2010).

In addition to the collagen network in the ECM, there exists long chain glycoaminoglycans (GAGs) and adhesive proteins, including fibronectin, tenascin, and laminin. GAGs are highly hydrated and provide some compressive strength to the network. Adhesive proteins present an immense number of physically immobilized and non-immobilized signals to the cells. Fibronectin, for example, is an important protein in guiding cell attachment and migration

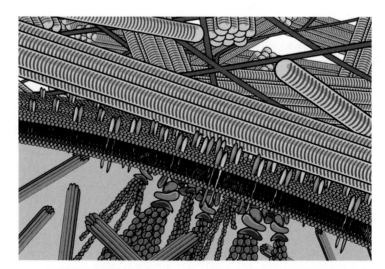

FIGURE 39.1
Schematic of the mechanical interaction between a cell and the surrounding ECM. Integrin receptors engage binding sites on structural ECM proteins, bridging the cytoskeleton of the cell with the surrounding matrix. Integrin binding at the surface can influence structural rearrangements in both the cytoplasm and the nucleus.

during embryonic development, where its absence leads to defects in mesodermal, neural tube, and vascular development. Similarly, laminin has been shown to affect cell migration and differentiation in numerous systems (Kubota et al., 1988).

Cell adhesion and mechanotransduction

Cell adhesion to the ECM is crucial for both development and tissue maintenance (de Arcangelis and Georges-Labouesse, 2000). Cell adhesion events mediate cell spreading, migration, neurite extension, muscle-cell contraction, cell-cycle progression, and differentiation (Giancotti and Ruoslahti, 1999). Cells adhere to distinct adhesion domains on the ECM (Ruoslahti and Pierschbacher, 1987) through cell-surface receptors, primarily from the integrin family (Hynes, 2002). Upon engagement, these receptors provide chemical and mechanical signals to the cell that can lead to altered gene expression and in some cases cell fate, including apoptosis, migration, differentiation, and proliferation.

Integrins are a family of approximately 25 membrane-spanning heterodimeric proteins containing ligand-binding regions on the outer-membrane region and microfilaments-docking domains within the ectodomain. They are composed of α ($\sim$120 kD) and the β subunits ($\sim$180 kD), and each combination has a different binding affinity and signaling properties. Most integrins are expressed on a wide variety of cell types, and most cells express several integrin receptors.

Integrins signal both from the outside-in (binding of the integrin with the ECM induces intracellular signaling events) and the inside-out (the binding activity and expression of integrins is regulated by the cell) (Giancotti and Ruoslahti, 1999). Following activation (via a binding event on the cytoplasmic domain), the focal adhesion complexes (FACs) binds actin-associated proteins such as talin, vinculin, zyxin, and paxillin and provides a direct physical link to the cytoskeleton, which also links to the nuclear scaffold. Integrin binding at the surface can therefore influence structural rearrangements in both the cytoplasm and the nucleus (Geiger et al., 2001). This is particularly relevant since mechanical signals can potentially travel faster than signals that are mediated via diffusion either across or through the cell.

Once bound to the ECM, integrins enable cells to sense the physical and mechanical properties of the matrix. As a result, changes in physical or mechanical properties of the matrix can activate signaling pathways including MAPK and JNK that direct cell-cycle progression and differentiation, and this conversion of physical signals into biochemical signals is termed mechanotransduction.

Mechanics of the natural ECM

In vivo, tissues including bone, arteries, and brain naturally have distinct moduli (Black and Hastings, 1998), where the modulus of each is primarily defined by the properties of the ECM. Accordingly, *ex situ* measurements of natural tissues demonstrate the wide range of stiffnesses of different tissues in the body. Engler et al. (2006) observed that the osteoid matrix that surrounds osteoblasts in culture had a Young's modulus of $\sim$27 $\pm$ 10 kPa, much stiffer than other tissues in the body. In addition, a few years prior, Engler et al. (2004) sectioned arteries *ex situ* and measured an intermediate stiffness with a measured Young's modulus of 5–8 kPa. Finally, Elkin et al. (2007) and Lu et al. (2006) made measurements of native brain tissue, which had a much lower modulus of $\sim$500 Pa. The different moduli of these tissues *in vivo* indicate there may be a significant role of the stiffness of the matrix in cell fate and cell behavior.

DEVELOPING ARTIFICIAL MATRICES WITH TUNABLE MODULI FOR STEM CELL CULTURE

The physical and chemical properties of the ECM play a key role in stem cell fate; therefore, the field of tissue engineering has the difficult task of creating artificial matrices that impart the desired signals to the cells in direct contact with those matrices. In addition, since the natural microenvironment is dynamic in nature, it may be necessary to create tunable systems that the user is able to modify, for example to direct progressive processes such as cell fate specification or tissue organization. This section focuses on designing matrices for *in vitro* stem cell culture, both for maintaining stem cell self-renewal and differentiating stem cells into a variety of cell lineages.

Physical structure of the matrix

The physical structure, or microarchitecture, of an artificial matrix must provide appropriate physical signals, present or allow access to biochemical cues, and permit nutrient and waste exchange. Synthetic matrices should mimic some aspects of the natural properties of the collagen scaffold and adjacent proteins of the ECM, which constitutes a highly hydrated and fibrous network that supports cell attachment, migration, and other functions.

One approach to mimicking the physical structure of the ECM is the creation of matrices of nanofibers prepared with electrospun polymers. In 2003, Yoshimoto et al. grew mesenchymal stem cells (MSCs) on scaffolds created by electrospinning poly(ε-caprolactone) (PCL). They demonstrated increased osteogenic differentiation on the nanofiber matrices compared to standard tissue culture surfaces. In 2006, Nur-E-Kamal et al. (2006) cultured mouse ES (mES) cells on a synthetic polyamide matrix whose three-dimensional (3D) nanofibrillar organization resembled the ECM/basement membrane and found that this surface greatly enhanced proliferation and self-renewal compared to propagation on tissue culture surfaces without nanofibers. This important work indicated that mimicking the nanofiber structure of the ECM can yield enhanced cell behavior, and subsequent work (discussed on p. 9723) has built upon these efforts to incorporate material designs that allow tuning of a wide range of mechanical properties.

Another approach to the design of artificial matrices for stem cells is to employ a hydrogel to mimic the physical properties of the natural ECM. Hydrogels emulate the high water content and porous nature of most natural soft tissues. In addition, scaffolds have been designed to enable cells to proteolytically cleave certain domains of the network as they move through it, allowing for the creation of pores (West and Hubbell, 1999; Schense et al., 2000; Kim et al., 2005; Raeber et al., 2005; Levesque and Shoichet, 2007). Hydrogels can thus provide the diverse physical properties of an artificial matrix, while also providing a system that can be chemically and mechanically tuned for a desired application.

Choice of material

For the desired microarchitecture, a variety of both natural and synthetic polymer chemistries can be employed to create the nanofiber or hydrogel systems discussed above that offer different material characteristics.

NATURAL MATERIALS

Naturally occurring polymer components from the ECM can be isolated and employed as artificial microenvironments for stem cell culture. For decades, natural materials — including alginate (Barralet et al., 2005), chitosan (Azab et al., 2006), hyaluronic acid (Masters et al., 2004), collagen (Butcher and Nerem, 2004), laminin, fibronectin, and fibrin (Eyrich et al., 2007) — have been used as matrices for a variety of primary cells and cell lines. As they are part of the natural ECM, these proteoglycan and protein molecules contain binding sequences to engage cell surface receptors and allow for cell attachment and migration. However, disadvantages of using natural, isolated materials include lot-to-lot variability in the signals the matrix presents to the cells, the potential transfer of immunogens to cells, and the potential for viral or bacterial contamination. Therefore, the primary value of most work using natural materials has been to elucidate the roles of natural ECM molecule(s) on cell fate.

720

In 2005, Battista et al. employed collagen, fibronectin (FN), and laminin (LM) matrices for the culture of mES cell-derived embryoid bodies (EBs) in an attempt to direct stem cell behavior. They showed that the composition of the matrix plays an important role in EB development, where high collagen concentrations inhibited EB differentiation, FN constructs stimulated endothelial cell differentiation and vascularization, and LM constructs increased the percentage of cells that differentiated into beating cardiomyocytes. This work indicates that there is key regulatory "information," i.e. ligands, and possibly physical cues within these ECM proteins that regulate mES cell fate decisions.

In addition, several groups have utilized surface arrays for the high-throughput analysis of how different combinations of natural ECM molecules can impact stem cell fate. Flaim et al. (2005) designed a platform to study the effects of 32 different combinations of collagen I, collagen III, collagen IV, laminin, and fibronectin on the differentiation of mouse ES cells towards an early hepatic fate. They identified ECM combinations that impacted both hepatocyte function and ES cell differentiation. Soen et al. (2006) printed mixtures of ECM components and signaling factors on a glass surface to generate an array of immobilized "molecular microenvironments" and found the composition of the microenvironment affected the degree of differentiation of primary human neural precursor cells. These studies provide fundamental information that increases our understanding of the role of matrix composition on stem cell behavior, which can be harnessed to design synthetic hydrogels that can offer more precise control over matrix properties and signals presented to cells.

SYNTHETIC MATERIALS

In contrast to natural materials, synthetic polymer hydrogels offer improved control, repeatability, safety, and scalability. However, it can be challenging to functionalize synthetic materials with the highly complex bioactivities of natural materials. Synthetic materials used commonly for tissue engineering of all cell types include poly(glycolic acid), poly(lactic acid), and their copolymers; polyethylene glycol (PEG) (Sawhney et al., 1993); polyvinyl (PVA) (Martens and Anseth, 2000); polyNIPAAm (Stile et al., 2004); polyacrylamides; and poly-acrylates. Several reviews describe the use of these synthetic polymer matrices for the growth of anchorage-dependent, differentiated cells (Lutolf and Hubbell, 2005; Lin and Anseth, 2009; Tibbitt and Anseth, 2009), and these chemistries likewise have potential for use in scaffolds for stem cell culture. When selecting polymer chemistry for a particular application, design parameters include the toxicity of the material to the cells, hydrophilicity, swelling behavior, degradation properties, interactions the polymer chains have with neighboring cells, biofunctionalization (discussed on p. 722), and crosslinking properties (discussed on p. 723).

Hydrogel matrices have been used to support the potency and differentiation of stem cells. Biodegradable polymer scaffolds were employed for the differentiation of hES cells into 3D structures with characteristics of developing neural tissues, cartilage, or liver (Levenberg et al., 2003). These synthetic matrices were superior to their natural counterparts in scalability, repeatability, and control over design parameters. In addition, advanced screening methods have been employed to identify hydrogel surfaces for the self-renewal of pluripotent stem cells (Yang et al., 2009; Derda et al., 2010; Hook et al., 2010). However, there is still a great deal of work to be done to evaluate various hydrogels and their influence on stem cell behavior, particularly engineering them with the biochemical and mechanical signals inherent in natural matrix.

SYNTHETIC MATERIALS WITH BIOACTIVE LIGANDS

Synthetic hydrogels can be modified with bioactive ligands to allow cells to attach, proliferate, and/or differentiate upon otherwise inert surfaces as shown schematically in Figure 39.2. Extensive work has been performed to identify binding sequences in natural ECM molecules

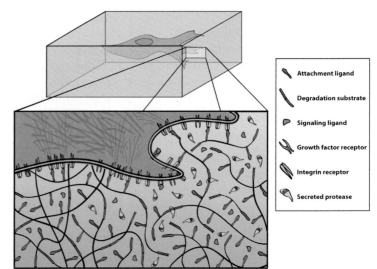

FIGURE 39.2

Schematic of a cell embedded in a 3D synthetic hydrogel. Integrin receptors bind to pendant cell-binding domains and growth factor receptors bind to soluble ligands. Cell-secreted proteases enzymatically cleave substrates incorporated into the polymer network, locally degrading it.

(Derda et al., 2007) and then generate short peptides or small recombinant proteins encompassing these sequences for incorporation into artificial matrices (Orner et al., 2004; Derda et al., 2007). The resulting bioactive materials can support cell receptor-ligand adhesion, which enables cells to sense and respond to the stiffness of the matrix. However, many key parameters must be tuned to control cell and stem cell behavior, including the ligand identity, presentation, and density (Nowakowski et al., 2004; Shin et al., 2004; Yim and Leong, 2005; Beckstead et al., 2006).

Peptides are typically conjugated to hydrogels using either primary amines or sulfhydryl groups on the peptides themselves. The method of conjugation, as well as spacer-arm length, can be varied to modulate the steric accessibility of the peptide sequence to the cell. In addition, it can be difficult to generate synthetic analogues of the complex motifs that natural ECM proteins present. In some cases, the natural conformation of the binding site on the protein can be more closely approximated by cyclizing short peptide sequences that are otherwise linear (Schense et al., 2000). The most common molecules used to mediate cell attachment to synthetic matrices are short peptide sequences (usually between 6 and 15 amino acids) containing the consensus sequence arginine-glycine-aspartate (RGD), which is present in several ECM proteins. Many studies have tethered this peptide to hydrogels and demonstrated its ability to bind anchorage-dependent cells through a subset of RGD-binding integrins, such as $\alpha_v\beta_3$ (Massia and Hubbell, 1991; Hubbell, 1995). In another approach, Meng et al. (2010) physisorbed short peptide sequences to cell culture plates to bind specific integrins identified on hES cells to create a completely synthetic defined cell culture system for medium-term self-renewal of hES cells.

Using synthetic matrices functionalized with bioactive ligands, Saha et al. (2008) demonstrated both long-term self-renewal and multipotent differentiation of neural stem cells on interpenetrating networks of polyacrylamide and polyethylene glycol functionalized with a 15-mer RGD sequence isolated from bone rat sialoprotein (bsp-RGD15). In another example, Li et al. (2006) were able to maintain hESC pluripotency employing synthetic semi-interpenetrating polymer networks (sIPNs) functionalized with the same peptide. In addition, Hwang et al. (2006) differentiated hES cells into mesenchymal stem cells and then evaluated their chondrogenic capacity upon encapsulation in poly(ethylene glycol)-diacrylate (PEGDA) hydrogels functionalized with an RGD peptide. This combination yielded neocartilage within 3 weeks of culture. In another study, Ferreira et al. (2007) formed a 3D matrix from the natural polymer dextran, then functionalized the material with RGD peptides and microencapsulated VEGF. They were able to increase the fraction of cells displaying a vascular marker by 20-fold compared to spontaneously differentiated EBs and propose that this hydrogel enables the derivation of vascular cells in large quantities.

CREATING MATRICES WITH TUNABLE MODULI

Mechanical design parameters for artificial matrices include elasticity, compressibility, visco-elastic behavior, and tensile strength. Controlling the mechanical properties of a material at the cellular level can help elicit a desired cell response, and, in addition, the bulk mechanical properties of the matrix must be controlled such that the matrix is able to withstand loads that may be involved in downstream applications.

The mechanical properties of hydrogels can be varied and controlled via chemical synthesis and processing. Hydrogels are composed of long, hydrophilic polymer chains either physically entangled or chemically crosslinked to form a network, and their mechanical properties can be chemically altered by controlling crosslinking density (entanglements or chemical crosslinks). For the AAm gels described on page 725, the input crosslinker (bisacrylamide) concentration of the AAm gels was varied, and a linear relationship between input crosslinker density and gel modulus was found. Based on prior work (see p. 725) (Engler et al., 2006; Saha et al., 2007, 2008; Boonen et al., 2009), tuning the crosslink density of hydrogels may aid in designing systems to support stem cell self-renewal or differentiation, depending on the desired application. An increasing number of studies have illustrated a role of stiffness in regulating stem cell function in two dimensions, and initial evidence to date indicates that the mechanical properties of a material are likely to also influence stem cell behavior in 3D (Banerjee et al., 2009).

Although a direct correlation with matrix stiffness and behavior of hES or iPS cells has yet to be demonstrated, Li et al. (2006) proposed that the soft mechanical properties of their hydrogels improve the self-renewal of hES cells on their defined, synthetic hydrogels. In this work, pNIPAAm hydrogels functionalized with bsp-RGD15 and with a complex shear modulus of ~50–100 Pa (depending on the frequency of the measurement) and were able to maintain pluripotency in the short-term. Future studies are very likely to focus on analyzing the effects of stiffness and other mechanical properties on the self-renewal, lineage commitment, and differentiation of numerous cell types, providing additional key design parameters to control cell function for downstream applications.

723

CHARACTERIZATION OF MATRIX MECHANICS

The mechanical properties of synthetic and natural matrices are typically characterized by either atomic force microscopy or rheology, and each is addressed in further detail.

Atomic force microscopy

The mechanical stiffness of two-dimensional (2D) hydrogels can be characterized by force mode atomic force microscopy (AFM). AFMs have been used widely as microindenters to probe the physical properties of the materials (Burnham and Colton, 1989; Tao et al., 1992; Rotsch et al., 1997; Domke and Radmacher, 1998; Dimitriades et al., 2002; Irwin et al., 2008). In force mode, the AFM tip is indented into the surface, and the deflection of the cantilever is measured as shown in Figure 39.3. To reduce strain at the point of contact, and ensure uniform curvature at the point of contact, a bead can be attached to the cantiliver tip as shown.

The AFM collects data by reflecting a laser off a cantilever with a known spring constant. The laser is reflected into a photodiode (detector) as the tip (on the end of a cantilever) is indented into the surface and bends in response to the force between the tip and the sample. A constant force is maintained on the sample by the tip by using a feedback loop with piezoelectric translators that adjust the z-axis of the stage.

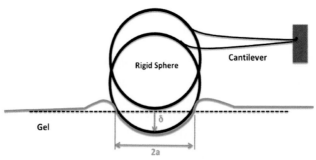

FIGURE 39.3
Schematic of the indentation of a gel sample with a rigid sphere using an AFM. Adapted from Dimitriadis et al. (2002).

The elastic response of the underlying material is analyzed by applying Hertzian mechanical models to the slope of the force-displacement curves to estimate the elastic modulus and other material properties or structural parameters (Dimitriades et al., 2002). The indentation curves are then analyzed with a Hertzian mechanics model with the following relationship:

$$F = (2/\Pi)[E/\left(1 - v^2\right)]\delta^2 \quad \tan(\alpha)$$ (1)

where F is the applied force, E is the elastic modulus, v is the Poisson ratio of the material, δ is the indentation depth, and α is the angle of the indenting cone. This model assumes infinite depth of the sample and therefore the indentation of the tip into the material must be less than 10% of the film thickness or the stiffness of the underlying substrate may be sensed by the AFM tip.

Rheology

The mechanical properties of engineered 3D hydrogel systems, like natural biological tissues, are viscoelastic and are typically characterized using rheological techniques. The mechanical characteristics of such materials are intermediate between an ideal solid and an ideal liquid and are dependent on loading rate and history. Rheometry measures the flow and deformation behavior of materials under stress, for example by using rotational parallel-plate devices.

Oscillatory strain-controlled parallel-plate rheometers apply a sinusoidal shear strain to a hydrogel and measure the resulting stress (torque) response. The ratio of the amplitudes and phase difference of the stress and strain waves provide the storage (elastic), G', and loss (viscous), G'', moduli. The phase angle, $\delta = $ arc tan (G''/G'), indicates the degree to which a material is like an elastic solid or a viscous liquid, while the complex modulus, $G^* = |$sqrt $[(G')2 + (G'')2]|$, indicates the overall resistance to shear deformation. These properties are measured over a range of frequencies to determine their dependence on loading rate. Strain sweeps are performed to define the linear viscoelastic regime and yield point of the material.

Rheology is particularly applicable for the analysis of environmentally responsive and *in situ*-forming hydrogels for tissue engineering and cell biology. Kinetic changes in mechanical properties can be measured as the materials transition from liquid to solid. Liquids are indicated by low, frequency-dependent elastic moduli and high phase angles while solids have high, frequency-independent elastic moduli and low phase angles.

ROLE OF MATRIX MECHANICS IN STEM CELL BEHAVIOR

The biochemistry, physical architecture, and modulus of the microenvironment are all important parameters in influencing cell behavior. Cells are traditionally cultured on tissue culture polystyrene (TCPS) and glass, which have Young's moduli of $\sim 10^8$ and $\sim 10^{10}$ Pa, respectively: values that are orders of magnitude higher than the moduli of most natural tissues, ∼ 102−105 Pa. Given this mismatch in mechanical properties, and given that it has been demonstrated in multiple anchorage-dependent cell types that a material's modulus impacts cell morphology, cytoskeletal formation, and gene expression, this key parameter must be considered in the design of cell culture systems.

Pelham and Wang developed a system to evaluate the effect of material stiffness on cell behavior (Pelham and Wang, 1998). This system was composed of 2D, variable moduli polyacrylamide (pAAm) gels functionalized with collagen to allow for cell attachment, and has been since employed by multiple research laboratories to demonstrate modulus dependent behavior of a variety of different anchorage-dependent cell types. These gels, which greatly contrast with the current tissue culture polystyrene used for standard cell culture that is orders of magnitude more rigid (∼ GPa), provided moduli that more accurately matched those of native tissue. In 1998, Pelham and Wang first demonstrated that fibroblast and epithelial cell behavior were regulated by the mechanical properties of the underlying synthetic matrix on which the cells were cultured. They found that both focal adhesion and cytoskeletal formation depended on the stiffness of the underlying pAAm gels. In 2000, Thomas and Dimilla. cultured human SNB-19 glioblastoma cells on poly(methylphenyl)siloxane (PDMS) films of variable moduli and showed that the average projected cell area decreases by over 60% with a two-orders-of-magnitude increase in compliance. Lo et al. (2000) cultured fibroblasts on pAAm gels with a spatial gradient in modulus and demonstrated that the fibroblasts preferentially migrated to the stiffer areas of the gel, a process termed durotaxis, indicating the cells were able to sample the stiffness of the underlying substrate. In addition, by applying mechanical strain to the substrate with a microneedle, they demonstrated that cell movement is also guided by strain in the substrate. In 2004, Engler et al. employed the same pAAm gel system and demonstrated that matrix stiffness affected the cell spreading, actin cytoskeletal formation, and focal adhesion organization of smooth muscle cells (SMCs), where stiffer gels cause an increase in all three.

Several groups have been able to show that, in addition to varying a number of properties of differentiated cells, the stiffness of the matrix can regulate the lineage commitment processes of adult stem cells. In 2006, the lab of Engler et al. used the pAAm gel system to test the effect of matrix stiffness of the differentiation of adult stem cells. A variation in stiffness alone was able to control lineage commitment, where softer matrices resulted in neurogenic commitment, intermediate stiffnesses yielded myogenic commitment, and finally the stiffest matrices resulted in osteogenic commitment. This work suggests that mesenchymal stem cells differentiate according to the stiffness of the environment in which they were cultured. Reflecting this observation, Saha et al. (2008) demonstrated that adult neural stem cell differentiation also depended on the stiffness of the underlying matrix. In this work, an interpenetrating polymer network (IPN) of AAm and PEG functionalized with bspRGD(15) was employed to demonstrate that softer gels (100−500 Pa) greatly favored differentiation into neurons, whereas harder gels (1,000−10,000 Pa) promoted glial cultures. Recently, Boonen et al. cultured muscle progenitor cells (MPCs) onto pAAm gels of varying stiffness and found that proliferation and differentiation were influenced by elasticity (Boonen et al., 2009). An intermediate stiffness of 21 kPa was optimal for the proliferation of MPCs, where only gels with elasticities greater than 3 kPa led to maturation with cross-striations and contractions. Collectively, these studies demonstrate that the stiffness of the matrix is a crucial parameter in designing matrices for stem cells; stiffness collaborates with soluble cues to direct lineage commitment of the cells.

CONCLUSIONS AND FUTURE DIRECTIONS

In engineering matrices for stem cell culture, it is evident that ligand identity and presentation, as well as material architecture and mechanical properties, are key design parameters in controlling stem cell fate. Although there have been significant advances in the design and synthesis of artificial ECMs, there is still a great need for more sophisticated scaffolds that play an active role in guiding tissue regeneration and functional adaptation of the newly formed tissue. In particular, while it is recognized as a signal to the cells, the modulus of the material is still not often varied and optimized for the particular application. Future work is likely to increasingly tap into this and other opportunities that materials offer to afford greater control over cell fate and function and thereby enhance the potential of numerous downstream applications for stem cell engineering.

References

Azab, A. K., Orkin, B., Doviner, V., Nissan, A., Klein, M., Srebnik, M., et al. (2006). Crosslinked chitosan implants as potential degradable devices for brachytherapy: in vitro and in vivo analysis. *J. Control. Release, 111*, 281–289.

Banerjee, A., Arha, M., Choudhary, S., Ashton, R. S., Bhatia, S. R., Schaffer, D. V., et al. (2009). The influence of hydrogel modulus on the proliferation and differentiation of encapsulated neural stem cells. *Biomaterials, 30*, 4695–4699.

Barralet, J. E., Wang, L., Lawson, M., Triffitt, J. T., Cooper, P. R., & Shelton, R. M. (2005). Comparison of bone marrow cell growth on 2D and 3D alginate hydrogels. *J. Mater. Sci. Mater. Med., 16*, 515–519.

Battista, S., Guarnieri, D., Borselli, C., Zeppetelli, S., Borzacchiello, A., Mayol, L., et al. (2005). The effect of matrix composition of 3D constructs on embryonic stem cell differentiation. *Biomaterials, 26*, 6194–6207.

Beckstead, B. L., Santosa, D. M., & Giachelli, C. M. (2006). Mimicking cell–cell interactions at the biomaterial-cell interface for control of stem cell differentiation. *J. Biomed. Mater. Res., 79A*, 94–103.

Black, J., & Hastings, G. (1998). *Handbook of Biomaterial Properties.* New York: Chapman & Hall.

Boonen, K. J. M., Rosaria-Chak, K. Y., Baaijens, F. P. T., van der Schaft, D. W. J., & Post, M. J. (2009). Essential environmental cues from the satellite cell niche: optimizing proliferation and differentiation. *Am. J. Physiol. Cell Physiol., 296*, C1338–C1345.

Bosman, F. T., & Stamenkovic, I. (2003). Functional structure and composition of the extracellular matrix. *J. Pathol., 200*, 423–428.

Brouzes, E., & Farge, E. (2004). Interplay of mechanical deformation and patterned gene expression in developing embryos. *Curr. Opin. Genet. Dev., 14*, 367–374.

Burnham, N. A., & Colton, R. J. (1989). Measuring the nanomechanical properties and surface forces of materials using an atomic force microscope. *J. Vacuum Sci. Technol. A, 7*, 2906–2913.

Butcher, J. T., & Nerem, R. M. (2004). Porcine aortic valve interstitial cells in three-dimensional culture: comparison of phenotype with aortic smooth muscle cells. *J. Heart Valve Dis., 13*, 478–485.

de Arcangelis, A., & Georges-Labouesse, E. (2000). Integrin and ECM functions – roles in vertebrate development. *Trends Genet., 16*, 389–395.

Derda, R., Li, L. Y., Orner, B. P., Lewis, R. L., Thomson, J. A., & Kiessling, L. L. (2007). Defined substrates for human embryonic stem cell growth identified from surface arrays. *Acs Chem. Biol., 2*, 347–355.

Derda, R., Musah, S., Orner, B. P., Klim, J. R., Li, L. Y., & Kiessling, L. L. (2010). High-throughput discovery of synthetic surfaces that support proliferation of pluripotent cells. *J. Am. Chem. Soc., 132*, 1289–1295.

Dimitriadis, E. K., Horkay, F., Kachar, B., & Chadwick, R. S. (2002). Measurement of the elastic modulus of polymer gels using the atomic force microscope. *Abstr. Papers Am. Chem. Soc., 224*, U417–U417.

Domke, J., & Radmacher, M. (1998). Measuring the elastic properties of thin polymer films with the atomic force microscope. *Langmuir, 14*, 3320–3325.

Elkin, B. S., Azeloglu, E. U., Costa, K. D., & Morrison, B. (2007). Mechanical heterogeneity of the rat hippocampus measured by atomic force microscope indentation. *J. Neurotrauma, 24*, 812–822.

Engler, A., Bacakova, L., Newman, C., Hategan, A., Griffin, M., & Discher, D. (2004). Substrate compliance versus ligand density in cell on gel responses. *Biophysical J., 86*, 617–628.

Engler, A. J., Sen, S., Sweeney, H. L., & Discher, D. E. (2006). Matrix elasticity directs stem cell lineage specification. *Cell, 126*, 677–689.

Eyrich, D., Wiese, H., Mailer, G., Skodacek, D., Appel, B., Sarhan, H., et al. (2007). In vitro and in vivo cartilage engineering using a combination of chondrocyte-seeded long-term stable fibrin gels and polycaprolactone-based polyurethane scaffolds. *Tissue Engin., 13*, 2207–2218.

Ferreira, L. S., Gerecht, S., Fuller, J., Shieh, H. F., Vunjak-Novakovic, G., & Langer, R. (2007). Bioactive hydrogel scaffolds for controllable vascular differentiation of human embryonic stem cells. *Biomaterials, 28*(17), 2706–2717.

Flaim, C. J., Chien, S., & Bhatia, S. N. (2005). An extracellular matrix microarray for probing cellular differentiation. *Nature Methods, 2*, 119–125.

Fuchs, E., Tumbar, T., & Guasch, G. (2004). Socializing with the neighbors: stem cells and their niche. *Cell, 116,* 769–778.

Geiger, B., Bershadsky, A., Pankov, R., & Yamada, K. M. (2001). Transmembrane extracellular matrix-cytoskeleton crosstalk. *Nat. Rev. Mol. Cell Biol., 2,* 793–805.

Giancotti, F. G., & Ruoslahti, E. (1999). Transduction – integrin signaling. *Science, 285,* 1028–1032.

Gordon, M. K., & Hahn, R. A. (2010). Collagens. *Cell Tissue Res., 339*(1), 247–257.

Hook, A. L., Anderson, D. G., Langer, R., Williams, P., Davies, M. C., & Alexander, M. R. (2010). High throughput methods applied in biomaterial development and discovery. *Biomaterials, 31,* 187–198.

Hubbell, J. A. (1995). Biomaterials in tissue engineering. *Bio-Technology, 13,* 565–576.

Hwang, N. S., Varghese, S., Zhang, Z., & Elisseeff, J. (2006). Chondrogenic differentiation of human embryonic stem cell-derived cells in arginine-glycine-aspartate modified hydrogels. *Tissue Eng., 12,* 2695–2706.

Hynes, R. O. (2002). Integrins: bidirectional, allosteric signaling machines. *Cell, 110,* 673–687.

Irwin, E. F., Saha, K., Rosenbluth, M., Gamble, L. J., Castner, D. G., & Healy, K. E. (2008). Modulus-dependent macrophage adhesion and behavior. *J. Biomater. Sci. Polymer Ed., 19,* 1363–1382.

Kim, S., Chung, E. H., Gilbert, M., & Healy, K. E. (2005). Synthetic MMP-13 degradable ECMs based on poly(*N*-isopropylacrylamide-co-acrylic acid) semi-interpenetrating polymer networks. I. Degradation and cell migration. *J. Biomed. Mater. Res. Part A, 75A,* 73–88.

Kubota, Y., Kleinman, H. K., Martin, G. R., & Lawley, T. J. (1988). Role of laminin and basement-membrane in the morphological-differentiation of human-endothelial cells into capillary-like structures. *J. Cell Biol., 107,* 1589–1598.

Levenberg, S., Huang, N. F., Lavik, E., Rogers, A. B., Itskovitz-Eldor, J., & Langer, R. (2003). Differentiation of human embryonic stem cells on three-dimensional polymer scaffolds. *Proc. Natl. Acad. Sci. U.S.A., 100,* 12741–12746.

Levesque, S. G., & Shoichet, M. S. (2007). Synthesis of enzyme-degradable, peptide-cross-linked dextran hydrogels. *Bioconjugate Chem., 18,* 874–885.

Li, Y. J., Chung, E. H., Rodriguez, R. T., Firpo, M. T., & Healy, K. E. (2006). Hydrogels as artificial matrices for human embryonic stem cell self-renewal. *J. Biomed. Mater. Res., 79A,* 1–5.

Lin, C. C., & Anseth, K. S. (2009). PEG hydrogels for the controlled release of biomolecules in regenerative medicine. *Pharm. Res., 26,* 631–643.

Lo, C. M., Wang, H. B., Dembo, M., & Wang, Y. L. (2000). Cell movement is guided by the rigidity of the substrate. *Biophys. J., 79,* 144–152.

Lu, Y. B., Franze, K., Seifert, G., Steinhauser, C., Kirchhoff, F., Wolburg, H., et al. (2006). Viscoelastic properties of individual glial cells and neurons in the CNS. *Proc. Natl. Acad. Sci. U.S.A., 103,* 17759–17764.

Lutolf, M. P., & Hubbell, J. A. (2005). Synthetic biomaterials as instructive extracellular microenvironments for morphogenesis in tissue engineering. *Nature Biotechnol., 23,* 47–55.

Martens, P., & Anseth, K. S. (2000). Characterization of hydrogels formed from acrylate modified poly(vinyl alcohol) macromers. *Polymer, 41,* 7715–7722.

Martin, G. R., Timpl, R., Muller, P. K., & Kuhn, K. (1985). The genetically distinct collagens. *Trends Biochem. Sci., 10* (7), 285–287.

Massia, S. P., & Hubbell, J. A. (1991). An Rgd spacing of 440nm is sufficient for integrin alpha-v-beta-3-mediated fibroblast spreading and 140nm for focal contact and stress fiber formation. *J. Cell Biol., 114,* 1089–1100.

Masters, K. S., Shah, D. N., Walker, G., Leinwand, L. A., & Anseth, K. S. (2004). Designing scaffolds for valvular interstitial cells: cell adhesion and function on naturally derived materials. *J. Biomed. Mater. Res., 71A,* 172–180.

Meng, Y., Eshghi, S., Li, Y. J., Schmidt, R., Schaffer, D. V., & Healy, K. E. (2010). Characterization of integrin engagement during defined human embryonic stem cell culture. *Faseb J., 24,* 1056–1065.

Moore, K. A., & Lemischka, I. R. (2006). Stem cells and their niches. *Science, 311,* 1880–1885.

Nowakowski, G. S., Dooner, M. S., Valinski, H. M., Mihaliak, A. M., Quesenberry, P. J., & Becker, P. S. (2004). A specific heptapeptide from a phage display peptide library homes to bone marrow and binds to primitive hematopoietic stem cells. *Stem Cells, 22,* 1030–1038.

Nur-E-Kamal, A., Ahmed, I., Kamal, J., Schindler, M., & Meiners, S. (2006). Three-dimensional nanofibrillar surfaces promote self-renewal in mouse embryonic stem cells. *Stem Cells, 24,* 426–433.

Orner, B. P., Derda, R., Lewis, R. L., Thomson, J. A., & Kiessling, L. L. (2004). Arrays for the combinatorial exploration of cell adhesion. *J. Am. Chem. Soc., 126,* 10808–10809.

Pelham, R. J., & Wang, Y. L. (1998). Cell locomotion and focal adhesions are regulated by substrate flexibility. *Proc. Natl. Acad. Sci. U.S.A., 95,* 12070.

Raeber, G. P., Lutolf, M. P., & Hubbell, J. A. (2005). Molecularly engineered PEG hydrogels: a novel model system for proteolytically mediated cell migration. *Biophys. J., 89,* 1374–1388.

Roskelley, C. D., Srebrow, A., & Bissell, M. J. (1995). A hierarchy of Ecm-mediated signaling regulates tissue-specific gene-expression. *Curr. Opin. Cell Biol., 7,* 736–747.

Rotsch, C., Braet, F., Wisse, E., & Radmacher, M. (1997). AFM imaging and elasticity measurements on living rat liver macrophages. *Cell Biol. Int., 21*, 685–696.

Ruoslahti, E., & Pierschbacher, M. D. (1987). New perspectives in cell-adhesion – Rgd and integrins. *Science, 238*, 491–497.

Saha, K., Irwin, E. F., Kozhukh, J., Schaffer, D. V., & Healy, K. E. (2007). Biomimetic interfacial interpenetrating polymer networks control neural stem cell behavior. *J. Biomed. Mater. Res., 81A*, 240–249.

Saha, K., Keung, A. J., Irwin, E. F., Li, Y., Little, L., Schaffer, D. V., et al. (2008). Substrate modulus directs neural stem cell behavior. *Biophys. J., 95*, 4426–4438.

Sawhney, A. S., Pathak, C. P., & Hubbell, J. A. (1993). Bioerodible hydrogels based on photopolymerized poly (ethylene glycol)-co-poly(alpha-hydroxy acid) diacrylate macromers. *Macromolecules, 26*, 581–587.

Schense, J. C., Bloch, J., Aebischer, P., & Hubbell, J. A. (2000). Enzymatic incorporation of bioactive peptides into fibrin matrices enhances neurite extension. *Nature Biotechnol., 18*, 415–419.

Shin, H., Zygourakis, K., Farach-Carson, M. C., Yaszemski, M. J., & Mikos, A. G. (2004). Modulation of differentiation and mineralization of marrow stromal cells cultured on biomimetic hydrogels modified with Arg-Gly-Asp containing peptides. *J. Biomed. Mater. Res., 69A*, 535–543.

Soen, Y., Mori, A., Palmer, T. D., & Brown, P. O. (2006). Exploring the regulation of human neural precursor cell differentiation using arrays of signaling microenvironments. *Molecular Systems Biology, 2*, 37.

Stile, R. A., Chung, E., Burghardt, W. R., & Healy, K. E. (2004). Poly(*N*-isopropylacrylamide)-based semi-interpenetrating polymer networks for tissue engineering applications. Effects of linear poly(acrylic acid) chains on rheology. *J. Biomater. Sci. Polymer, 15*, 865–878.

Streuli, C. (1999). Extracellular matrix remodelling and cellular differentiation. *Curr. Opin. Cell Biol., 11*, 634–640.

Tao, N. J., Lindsay, S. M., & Lees, S. (1992). Measuring the microelastic properties of biological material. *Biophys. J., 63*, 1165–1169.

Thomas, T. W., & DiMilla, P. A. (2000). Spreading and motility of human glioblastoma cells on sheets of silicone rubber depend on substratum compliance. *Med. Biol. Eng. Comput., 38*, 360–370.

Tibbitt, M. W., & Anseth, K. S. (2009). Hydrogels as extracellular matrix mimics for 3D cell culture. *Biotechnol. Bioeng., 103*, 655–663.

West, J. L., & Hubbell, J. A. (1999). Polymeric biomaterials with degradation sites for proteases involved in cell migration. *Macromolecules, 32*, 241–244.

Yang, F., Mei, Y., Langer, R., & Anderson, D. G. (2009). High throughput optimization of stem cell microenvironments. *Combin. Chem. High Throughput Screening, 12*, 554–561.

Yim, E. K. F., & Leong, K. W. (2005). Proliferation and differentiation of human embryonic germ cell derivatives in bioactive polymeric fibrous scaffold. *J. Biomater. Sci. Polymer, 16*, 1193–1217.

Yoshimoto, H., Shin, Y. M., Terai, H., & Vacanti, J. P. (2003). A biodegradable nanofiber scaffold by electrospinning and its potential for bone tissue engineering. *Biomaterials, 24*, 2077–2082.

Zagris, N. (2001). Extracellular matrix in development of the early embryo. *Micron, 32*, 427–438.

PART **4**

Therapeutic Applications

SECTION A

Cell Therapy

CHAPTER 40

Biomineralization and Bone Regeneration

Jiang Hu, Xiaohua Liu, Peter X. Ma
Department of Biologic and Materials Sciences, University of Michigan, Ann Arbor, MI, USA

INTRODUCTION

Biomineralization is the process by which mineral crystals are deposited in the matrix of living organisms. This process gives rise to inorganic-based skeletal structures such as bone during development, which is a complex and dynamic organ with both structural and metabolic functions. However, ectopic biomineralization often causes severe diseases, such as calcification of vascular tissues, which leads to atherosclerotic lesions (Rumberger et al., 1995). This chapter will focus on orthotopic bone formation and bone regeneration. Bone defects, caused by tumor or trauma, are a major health problem. There is an enormous clinical need to develop safe and effective modalities to stimulate bone regeneration. Tissue engineering offers a promising new approach in facilitating bone formation by recapitulating the natural process of bone development/healing using engineering techniques. This chapter will briefly describe the biological processes of bone development and fracture repair, summarizing the current applications of stem cells and growth/differentiation factors involved in bone regeneration, and then focus on the principles of design and fabrication of scaffolds.

DEVELOPMENT AND FRACTURE HEALING OF BONE
Development of bone

Bone formation proceeds in two different ways: endochondral ossification, which is a complex, multistep process requiring the sequential formation and degradation of cartilaginous templates for the developing bones, and intramembranous ossification, which occurs through the direct differentiation of precursor cells into osteoblasts (Karaplis, 2002).

Limb development involves a complex series of events that first define embryological zones for future endochondral bone development and subsequently induce cartilage and bone of precisely defined structures. These processes are regulated by a variety of signals including soluble growth/differentiation factors and cell-cell and cell-extracelluar matrix (ECM) interactions, all of which are orchestrated by an underlying genetic program. At the cellular level, the development of bone involves restrictions in lineage potential of multipotent mesenchymal precursor cells by controlling the cellular transcriptional program. This process can be broadly divided into two phases: an initial commitment phase, at which cells that will eventually form bone are committed in defined time and space, and the subsequent differentiation phase, when the necessary cellular phenotypes are induced to construct bone tissues. The flat bones of the skull form through an intramembranous process. Although the precursor cells in the skull are derived from the neural crest, these cells are regulated by many of the same signaling molecules found in the limb development.

733

Fracture healing of bone

Like embryological development of skeleton, fracture repair involves multiple factors and the establishment of a specific morphogenetic field to drive the differentiation of precursor cells, and, in some ways, can be considered a recapitulation of bone development (Gerstenfeld et al., 2003). After a fracture happens, the initial inflammatory response recruits activated macrophages and polymorphonuclear neutrophils (PMNs) to the damaged sites. Under the control of multiple factors secreted by macrophages, an initial hematoma is formed. Then, granulation tissue fibroblasts proliferate to form a blastema. Osteoprogenitors, migrating from periosteum, surrounding soft tissues, and the bone marrow space at the damaged sites, differentiate into chondrocytes and osteoblasts and form bone tissues. This process is induced and controlled by multiple soluble growth/differentiation factors. Among these, fibroblast growth factors (FGFs), insulin-like growth factors (IGFs), and platelet-derived growth factors (PDGFs), which are distributed in the soft callus early in the fracture healing, act as mitogenic factors to promote precursor cells proliferation, while other differentiation factors such as bone morphogenetic proteins (BMPs) are more responsible for differentiation of chondrocytes and osteoblasts present later in the healing tissues.

PRINCIPLES OF BONE TISSUE ENGINEERING

For bone regeneration therapy to be successful, sufficient mesenchymal precursor cells must be either recruited or implanted directly to the damaged sites, and these cells must be given the appropriate signals to grow and differentiate in a controlled manner. Current clinically applicable therapies for bone defect repair include bone grafts and allogenic bone matrix implantation. Bone grafts, containing viable bone cells and osteoprogenitors, as well as growth/differentiation factors, are considered to be the "gold standard." However, bone regeneration after bone grafting is quite variable, probably because of differences in the quality of the bone graft (Parikh, 2002). In addition, severe morbidity may occur at donor sites. Allogenic bone matrix provides a bone-like ECM and a crude source of growth/differentiation factors. These inductive factors may attract appropriate oseteoprogenitors to the regeneration site and stimulate their differentiation into osteoblast cells. However, osteoinductive activity of allogenic bone matrix is commonly inconsistent, primarily because it contains variable and often low levels of growth/differentiation factors, which are partially inactivated during processing (Iwata et al., 2002). There is also a potential risk of disease transmission if the matrix is not appropriately processed. In contrast, tissue engineering affords a new way for bone regeneration, having the advantage of combining the use of precisely engineered scaffolds, the appropriate osteoprogenitor cells, and related growth/differentiation factors (Liu and Ma, 2004). If a damaged tissue to be repaired has high activity in terms of regeneration, new tissue can form in a biodegradable scaffold directly by precursor cells infiltrating from the surrounding tissues. However, non-union or delayed union fracture sites are often too large or inflamed and associated with significant scarring that may limit the migration of osteogenic precursors. Also, some bone damage sites are related to low concentrations of growth/differentiation factors. Additional components such as mesenchymal stem cells (MSCs) and BMPs are required in these cases.

STEM CELLS IN BONE TISSUE ENGINEERING

Stem cells are defined as cellular populations with two critical properties: self-renewal to produce daughter stem cells with identical potentialities and the ability to differentiate along one or more lineages (Wagers and Weissman, 2004). Potential sources of stem cells for bone tissue engineering include embryonic stem cells (ESCs), induced pluripotent stem cells (iPSCs), and adult MSCs.

Embryonic stem cells

ESCs offer a potentially unlimited supply of cells that may be driven down specific lineages, giving rise to all cell types in the body (Thomson et al., 1998). ESCs can be driven to differentiate into osteoblast cells *in vitro*. In one method, osteogenic cells are derived from three-dimensional (3D) cell spheroids called embryoid bodies (EBs) (Bielby et al., 2004). EBs can be formed through suspension or hanging drop methods from single cell suspension. Since EBs mimic the structure of the developing embryo and recapitulate many of the stages involved during its differentiation, they create suitable conditions to drive ESCs to differentiate into precursor cells of all three germ layers. Then, EBs are dispersed and committed cells are further cultured in monolayer to be induced to osteogenic cells under the presence of exogenous factors such as dexamethasone (DEX), L-ascorbic acid (AA), and sodium-β-glycerophosphate (βgP). DEX has been demonstrated to stimulate osteogenic differentiation for precursor cells derived from multiple tissues, AA is used to promote collagen secretion and deposition, and βgP is used to mineralize the deposited matrix. Besides chemical cues, ECM also plays an important role in directing ESCs differentiation. We found nano-fibrous matrices, mimicking the architecture of natural collagenous matrices, promoted osteogenic differentiation of mouse ESCs (Smith et al., 2009) and human ESCs (unpublished data). Alternatively, undifferentiated ESCs or dispersed EBs can be directly seeded into 3D scaffolds and driven to multiple tissues (Levenberg et al., 2003) for later implantation. However, one of the major challenges for the use of hESCs in the repair of the defective tissues is the development of efficient strategies to fully direct cell differentiation into specific lineages, since a heterogeneous population of cells differentiated from hESCs may cause teratoma formation or inferior tissue organization. Recently, methods that can generate a more homogeneous cell population have been developed (Barberi et al., 2005; Hwang et al., 2006; Barberi et al., 2007; Brown et al., 2009). These methods have shown that a human embryonic stem cells-derived mesenchymal stem cells (hESCs-MSCs) population can be further induced along a chondrogenic (Hwang et al., 2006) or osteogenic (Brown et al., 2009) route.

Induced pluripotent stem cells

iPSCs are somatic cells reprogrammed to exhibit pluripotent properties. Mouse skin fibroblasts are first reprogrammed to iPSCs by overexpression of a set of four key transcription factors (Takahashi and Yamanaka, 2006). Later, adult human cells were reprogrammed (Takahashi et al., 2007; Yu et al., 2007). The technology to generate iPSCs is rapidly evolving, with small molecules employed to replace some transcription factors (Huangfu et al., 2008; Shi et al., 2008). Recently, multiple cells types have been derived from human iPSCs. However, the application of iPSCs to bone tissue engineering remains to be explored.

Mesenchymal stem cells

MSCs are an ideal stem cell source for cell therapies because of their easy purification, amplification, and multipotency, and low immunogenicity. MSCs were first identified in 1966 by Friedenstein and co-workers, who isolated bone/cartilage-forming progenitor cells from rat bone marrow cells with fibroblast-like morphology (Friedenstein et al., 1966). Although MSCs have been isolated from a number of tissues, including the fetal blood, umbilical cord blood (Lee et al., 2004), liver, adipose, and bone marrow (Campagnoli et al., 2001), the most studied and accessible source of MSCs is the bone marrow. Within the bone marrow, MSCs are estimated to comprise 0.001—0.1% of the total population of nucleated cells, which can be selected from other nucleated cells by their adherence property to plastic flasks in culture and can be expanded extensively for multiple passage numbers *in vitro* without loss of phenotype. Unlike hemopoietic stem cells (HSCs), which can be defined by specific surface markers, MSCs only express a number of non-specific surface markers. MSCs express neither hemopoietic (CD34, CD45, CD14) nor endothelial cell marker (CD31), but a large number of adhesion molecules (CD44, CD29, CD90) and stromal cell markers (SH-2, SH-3, SH-4), and some cytokine receptors. These MSC markers can be collectively used to identify isolated MSCs in culture. Some enrichment strategies are also developed based on the selection of cells positive for STRO-1 (Simmons and Torokstorb, 1991) and SH-2 (Barry et al., 1999) markers. MSCs can be driven down along mesenchymal cellular pathways, including osteogenic, chondrogenic, and adipogenic lineages, when placed in appropriate *in vitro* or *in vivo* environments (Pittenger et al., 1999) or on 3D scaffolds (Hu et al., 2009). Osteogenic differentiation is stimulated under the supplement of DEX, AA, and βgP. Under these culture conditions, MSCs upregulate alkaline phosphatase, osteocalcin, and osteopontin expressions, and also calcium deposition within the ECM. For bone regeneration *in vivo*, bone-marrow-derived MSCs have been demonstrated to facilitate bone repair when implanted locally, commonly on an artificial matrix, such as hydroxyapatite (HAP) scaffold (Kasten et al., 2005) in craniotomy and long-bone defects. In addition to multipotency, the low immunogenicity property of MSCs make the cells applicable for allogenic implantation (Barry et al., 2005). Another clinically applicable MSC source is white adipose. Like bone marrow, adipose tissue is mesodermally derived with a stromal part containing microvascular endothelial cells, smooth muscle cells, and MSCs. These cells can be enzymatically isolated from adipose tissue and separated from the buoyant adipocytes by centrifugation. A more homogeneous population can be selected and expanded under culture conditions favorable for MSC growth (Zuk et al., 2002). This population, called adipose tissue-derived stem cells (ADSCs), shares many of the characteristics of its counterpart in bone marrow, including extensive proliferative potential and multipotency (de Ugarte et al., 2003). ADSCs can be obtained in large numbers at high frequency from white adipose tissue with minimal morbidity, representing another potential clinically useful source of MSCs for bone tissue engineering.

GROWTH/DIFFERENTIATION FACTORS IN BONE TISSUE ENGINEERING

Many growth/differentiation factors are used in bone tissue engineering. Among these, BMPs have the unique ability to stimulate the differentiation of mesenchymal precursor cells to chondrocytes and osteoblasts, and induce formation of new bone at both ectopic and orthotopic sites.

Bone morphogenetic proteins

It was observed that demineralized bone matrix (DBM) is able to induce ectopic bone formation in subcutaneous and intramuscular pockets in rodents (Urist, 1965). Isolation of the bone-inducing substance revealed certain proteins termed BMPs or osteogenetic proteins (OPs) (Wozney et al., 1988). BMPs belong to the transforming growth factor-β (TGFβ) superfamily, which consists of a group of related peptide growth factors. More than 40 related members of this family have been identified, including 15 BMPs (de Caestecker, 2004). They are further divided into subfamilies according to their amino acid sequence similarities. BMPs consist of dimers that are interconnected by seven disulfide bonds. This dimerization is a prerequisite for bone induction. BMPs are active both as homodimers that consist of two identical chains and as heterodimers consisting of two different chains (Granjeiro et al., 2005). Compared to other known growth factors, BMP-2 (Boyne et al., 2005) and BMP-7 (Vaccaro et al., 2005) have the most robust osteoinductive activity as observed in both preclinical animal studies and in human trials.

Regional growth/differentiation factors' delivery

A simple method for bone regeneration is to supply growth factors such as BMPs to the site of defect for cell proliferation and differentiation in a controllable manner. Bone tissue regeneration is sometimes induced by use of growth/differentiation factors in soluble form, but the amount applied is much higher than that under normal physiological conditions, commonly at milligram level, which may cause adverse effects. Drug delivery systems are currently under development that allow for the controlled release of proteins, either encapsulated in poly (D,L-lactic acid-co-glycolic acid) (PLGA) microspheres (Weber et al., 2002) or incorporated into collagen carriers (Murata et al., 2000). We have recently developed technologies to immobilize nanospheres encapsulating growths factors onto the porous nano-fibrous 3D scaffolds (Wei et al., 2006, 2007). Single or multiple growth factors can be released in a temporally and spatially controlled manner while maintaining the architecture of scaffolds. The release kinetics of each factor can be individually controlled using a specific nanosphere formulation.

Regional gene therapy

Regional gene therapy offers another approach to delivery growth/differentiation factors to the healing sites. Transfected cells express growth/differentiation factors for a sustained period. Viral vectors and non-viral vectors are presently being investigated as potential gene delivery vehicles to enhance bone repair. In addition, MSCs themselves can be used as gene transfer carriers. Not only a source of BMPs after transfection, the cells also directly respond to BMPs and participate in bone formation after implantation, which may be important at some damage sites, where the supply of endogenous osteogenic precursors is limited. MSCs transfected with adenoviruses encoding BMPs have been shown to stimulate bone regeneration in several experimental models (Wang et al., 2003). Although recombinant adenovirus can be produced in high titers, and can easily infect both dividing and non-dividing cells at high efficiency (Spector et al., 2000), the immune response to the adenoviral proteins is a major obstacle to the adaptation of this approach to treat non-lethal diseases such as bone defect in humans. In contrast, non-viral vectors are easier to produce and have better chemical stability. However, the *in vivo* transfection efficiency of current available non-viral vectors such as liposome and poly (ethylenimine) (PEI) is low (Lollo et al., 2000). New vectors and delivery methods that increase transfection efficiency (Woodrow et al., 2009) are being developed in this field.

Combination of growth/differentiation factors

At any time during bone development or fracture healing, multiple growth/differentiation factors are functioning in a coordinated manner. Therefore, combinations of bioactive factors might synergistically stimulate bone regeneration. Angiogenic factors and BMPs can act synergistically. To examine possible interactions between BMPs and angiogenic signals in bone regeneration, Peng and co-workers used muscle-derived stem cell (MDSC) lines genetically modified to express BMP-4 or vascular endothelial growth factor (VEGF) (Peng et al., 2002). VEGF by itself had no effect on the osteogenic activity of MDSC. However, it acted synergistically with BMP-4 to increase recruitment of mesenchymal precursor cells and to enhance cell survival, thus stimulating bone formation in a calvarial defect.

SCAFFOLDS FOR BONE TISSUE ENGINEERING
Scaffolding design criteria for bone tissue engineering

In bone tissue engineering, the scaffold plays a critical role in supporting cell adhesion, migration, proliferation, differentiation, and mineralized bone tissue formation (Ma, 2003, 2004; Liu and Ma, 2004). Scaffolds for bone regeneration should meet certain criteria to serve these functions (Liu and Ma, 2004; Ma, 2004, 2008; Smith and Ma, 2004; Smith et al., 2008). First of all, the scaffold should have a controlled porous architecture to allow for cell growth, tissue regeneration, and vascularization. High interconnectivity between pores is desirable for uniform cell seeding and distribution, and the diffusion of nutrients to and metabolites away from the cell/scaffold constructs. The scaffold should have adequate mechanical stability to provide a suitable environment for new bone tissue formation. The scaffold degradation rate must be tuned to match the rate of new bone tissue formation. Furthermore, the scaffold should be osteoconductive to enhance osteoblast attachment, migration, and differentiated function. A variety of processing technologies have been developed to fabricate porous 3D polymeric scaffolds for bone regeneration. These techniques include solvent casting/particulate leaching (Mikos et al., 1994; Thomson et al., 1995), gas foaming (Mooney et al., 1996; Hile et al., 2000), emulsion freeze-drying (Whang et al., 1995), electrospinning (Li et al., 2002;

Matthews et al., 2002), rapid prototyping (Giordano et al., 1996; Sun et al., 2004), and thermally induced phase separation (Zhang and Ma, 1999a; Ma and Zhang, 2001). Several review papers have addressed the scaffolding fabrication methods, their advantages, and disadvantages (Hutmacher, 2000; Chaikof et al., 2002; Liu and Ma, 2004). This chapter does not intend to be exhaustive in detailing various processing techniques. Instead, it will focus on illustrating how to achieve the above scaffolding design goals through certain engineering methods. Important issues for scaffolding design, such as porosity, interconnectivity, mechanical strength, morphology, and surface properties, will be emphasized using examples from our group and others.

Porous scaffolds with high interconnectivity

Porosity and interconnectivity between pores are important scaffold parameters. Porous scaffolds with high interconnectivity are desirable for uniform cell seeding and distribution. Solvent casting/particulate leaching is a simple and the most commonly used method to fabricate porous scaffolds for bone tissue engineering (Mikos et al., 1994). The method involves casting a mixture of polymer solution and porogen in a mold, drying the mixture, and subsequently leaching the porogen with water to obtain a porous structure. Usually, water-soluble particulates such as NaCl are used as the porogen materials. This method is simple to operate, and the pore size and porosity of the scaffold can be adequately controlled by the particle size of the added salt and the salt/polymer ratio. However, the limited interpore connectivity is not desirable for uniform cell seeding and tissue growth. A new technique has been developed to fabricate scaffolds with spherical pore shape and well-controlled interpore connectivity by using paraffin spheres as pore-generating materials (Ma and Choi, 2001). The created new scaffold has a homogeneous foam skeleton and high porosity (Fig. 40.1). The control of porosity and the pore size can be achieved by changing the concentration of the polymer solution, the number of the casting steps, and the size of the paraffin spheres. The degree of interconnectivity is finely tuned by the heat treatment time to bond paraffin spheres, which is critical to uniform cell seeding, tissue ingrowth, and regeneration.

Composite scaffolds for bone tissue engineering

Although poly(α-hydroxy acids), such as poly(L-lactic acid) (PLLA), poly(glycolic acid) (PGA), and PLGA, have been widely used to fabricate scaffolds for bone tissue engineering, the disadvantages of these materials are the weak mechanical properties and insufficient osteo-conductivity. On the other hand, HAP, bioglass, and calcium phosphate have been

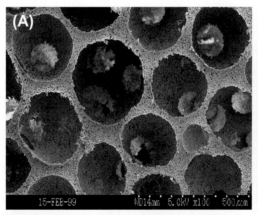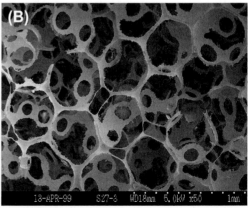

FIGURE 40.1
SEM micrographs of poly(α-hydroxy acids) scaffolds. (A) PLLA foams prepared with paraffin spheres with a size range of 250—420 μm (×250). (B) PLGA foams prepared with paraffin spheres with a size range of 420—500 μm (×50). From Ma and Choi, 2001; copyright 2001 by Mary Ann Liebert, Inc. Reprinted with permission.

demonstrated to have good osteoconductivity and bone-bonding ability. They also have been shown to enhance mineralized new bone formation when implanted into bone defects (Hench, 1998; Suchanek and Yoshimura, 1998). However, the application of ceramics alone in bone tissue engineering is limited because of their fragility and low degradability in biological environment. Polymer/ceramic composite scaffolds may enhance both mechanical properties and osteoconductivity. Highly porous poly(α-hydroxy acids)/HAP scaffolds have been created through a thermally induced phase separation technique (Zhang and Ma, 1999a; Ma et al., 2001). These composite scaffolds showed significant improvement in compressive modulus and compressive yield strength over pure polymer scaffolds. Compared to pure polymer scaffolds, in which cell ingrowth and tissue matrix formation were limited to the periphery of the scaffold, the composite scaffolds supported uniform cell seeding, cell ingrowth, and tissue formation throughout the scaffold (Fig. 40.2). Further examination revealed that polymer/ HAP scaffolds had a higher osteoblast survival rate, more uniform cell distribution and growth, enhanced bone-specific gene expression, and improved new tissue formation (Ma et al., 2001). Another strategy is to prepare bone-like apatite-coated composite scaffold by immersing polymeric scaffolds in a simulated body fluid (SBF) (Zhang and Ma, 1999b). In this approach, pre-fabricated polymeric scaffolds are incubated in SBF at 37°C to allow the *in situ* apatite formation on the inner pore wall surface of the 3D scaffold. After incubation, large amounts of apatite particles are formed uniformly on the scaffold pore walls (Fig. 40.3). The apatite particles formed using this method are similar to the apatite of natural bone based on EDS, FTIR, and XRD analyses (Zhang and Ma, 1999b). It has been observed that the growth of apatite crystals was affected greatly by the polymer materials, porous structure, and ionic concentration of SBF, as well as the pH value (Zhang and Ma, 2004). Biomimetic deposition of bone-like apatite is not only of direct interest for the development of a composite scaffold but also for assessing the calcification function of existing scaffolds (Wei and Ma, 2004).

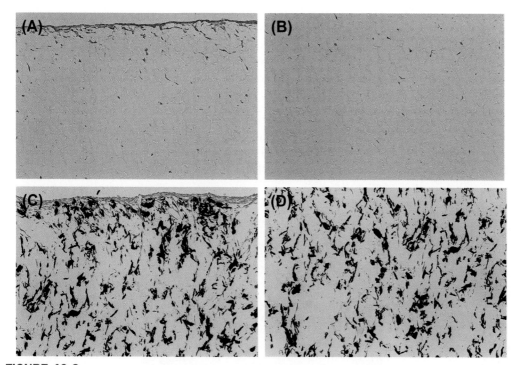

FIGURE 40.2

Osteoblastic cell distribution in highly porous PLLA and PLLA/HAP composite scaffolds 1 week after cell seeding (von Kossa's silver nitrate staining; original magnification ×100). (A) The surface area of an osteoblast-PLLA construct. (B) The center of an osteoblast-PLLA construct. (C) The surface area of an osteoblast-PLLA/HAP construct. (D) The center of an osteoblast-PLLA/HAP construct. From Ma et al., (2001); copyright 2001 by John Wiley & Sons, Inc. Reprinted with permission.

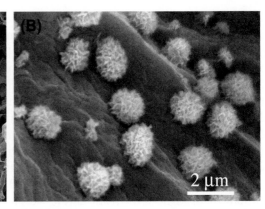

FIGURE 40.3

SEM micrographs of a PLLA scaffold incubated in SBF for 30 days. Original magnifications: (A) ×100; (B) ×10,000. From Zhang and Ma, (1999b); copyright 1999 by John Wiley & Sons, Inc. Reprinted with permission.

Nano-fibrous scaffolds for bone tissue engineering

It is well known that the ECM environment plays an integral role in regulating cell behavior with respect to morphology, cytoskeletal structure, and functionality (Aumailley and Gayraud, 1998; Rosso et al., 2004). Thus, it is often beneficial that the scaffold replicates the cells' natural ECM environment (Ma, 2008). Collagen is the main ECM component of bone, and its nano-fibrous architecture has long been known to play a role in cell adhesion, growth, and differentiated function in tissue cultures (Grinnell, 1982; Strom and Michalopoulos, 1982). To mimic the nano-fibrous architecture of collagen, a novel liquid-liquid phase separation technique has been developed to fabricate nano-fibrous PLLA (NF-PLLA) matrices (Ma and Zhang, 1999). The synthetic NF-PLLA matrix is composed of interconnected fibrous network with a fiber diameter ranging from 50 to 500 nm, which is in the same range as that of collagen matrix (Fig. 40.4). When combined with porogen-leaching techniques, 3D macroporous architectures can be built in the nano-fibrous matrices (Zhang and Ma, 2000; Chen and Ma, 2004; Wei et al., 2006). These synthetic analogs of natural ECM combine the advantages of the synthetic biodegradable polymers and the nano-scale architecture similar to the natural ECM. A new phase separation technique has been developed to create 3D nano-fibrous gelatine (NF-gelatin) scaffolds (Liu et al., 2009). Compared to commercial gelatine foam (Gelfoam), the NF-gelatin scaffold showed great dimensional stability to support bone tissue regeneration.

Synthetic NF polymer scaffolds have been found to selectively enhance protein adsorption and therefore osteoblastic cell adhesion (Woo et al., 2003). They also promote the osteogenic differention of primary mouse osteoblastic cells (Woo et al., 2007), human amniotic fluid-

FIGURE 40.4

SEM micrographs of a PLLA fibrous matrix prepared from 2.5% (wt/v) PLLA/THF solution at a gelation temperature of 8°C. From Ma and Zhang, (1999); copyright 1999 by John Wiley & Sons, Inc. Reprinted with permission.

derived stem cells (Sun et al., 2010), mouse ESCs (Smith et al., 2009), and human ESCs (Smith et al., 2010). Multiple signaling pathways may be related to the NF matrix effects. In one study, it was found that bone sialoprotein (BSP) gene expression level was greatly enhanced when osteoblastic cells were cultured on NF matrices, and the effect was correlated with the downregulation of the small GTPase RhoA activities (Hu et al., 2008).

Surface modification of nano-fibrous scaffolds

Surface properties as well as scaffolding architecture are important for a desirable scaffold in tissue engineering (Boyan et al., 1996; Liu et al., 2005a, b). The interactions of cells with the scaffolding materials take place on the material surface; therefore, the nature of the surface can directly affect cellular response, ultimately influencing the rate and quality of new tissue formation. Although a variety of synthetic biodegradable polymers have been used as tissue engineering scaffolding materials, they often lack of biological recognition on the material surface. Surface modification methods have been developed to promote cell-material interactions (Neff et al., 1998; Mann et al., 1999; Lenza et al., 2002). However, most of the surface modification methods thus far are applicable to 2D films or very thin 3D constructs. A novel surface modification method based on electrostatic layer-by-layer self-assembly technique has been recently introduced to modify true 3D scaffolding (especially nano-fibrous 3D scaffolding) surface (Liu et al., 2005a). As mentioned above, NF-PLLA scaffolds fabricated by thermally induced phase separation technique mimic the physical structure of natural collagen matrix. To mimic the chemical composition of collagen matrix, gelatin (derived from collagen by hydrolysis) is incorporated onto the surface of NF-PLLA scaffolds by the electrostatic self-assembly technique. The NF-PLLA scaffolds are first activated in an aqueous poly(diallyldimethylammonium chloride) (PDAC) solution to obtain stable positively charged surfaces. After washing the scaffolds with water, the scaffolds are dipped into gelatin solution for a designated time and then washed with water. The scaffolds are again exposed to PDAC solution. Following the same washing procedure, the scaffolds are dipped into gelatin solution and rinsed with water again. The growth of PDAC/gelatin bilayers is accomplished by repeating the same cycle. Multilayers of gelatin are deposited on the NF-PLLA surfaces after crosslinking and drying. The amount of gelatin on the surface is controlled by the number of assembled polyelectrolyte bilayers, and increases linearly with the bilayer number after the first two bilayers. The wettability of the scaffold is controlled by varying the nature of the outmost layer. The surface-modified NF-PLLA scaffolds mimick both the chemical composition and architecture of collagen matrix, and have been demonstrated to significantly improve cell adhesion and proliferation (Fig. 40.5).

741

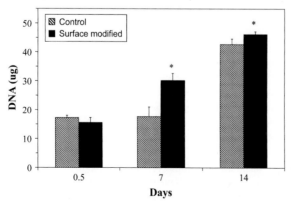

FIGURE 40.5

The proliferation of osteoblasts cultured on control NF-PLLA scaffolds and surface-modified NF-PLLA scaffolds (four bilayers of PDAC/gelatin). 2×10^6 cells were seeded on each scaffold (*$p < 0.05$ between surface-modified and control groups). From Liu et al., (2005a); copyright 2005 by American Scientific Publishers. Reprinted with permission.

CONCLUSIONS

Bone development and fracture healing are complex processes controlled by multiple factors. Bone tissue engineering offers a promising new approach for bone regeneration by mimicking these natural processes, and combining the stem cells and growth/differentiation factors together with supportive scaffolds in a controlled manner. Stem cells offer an ideal source for generating bone-forming cells and are especially desired for therapies to treat large defects and damaged sites with limited osteoprogenitor cells. Growth/differentiation factors can be used to stimulate bone regeneration by drug delivery or gene therapy approaches, and it is proposed that combinations of appropriate factors may have synergistic effects. Scaffolds play important roles in bone tissue engineering. Many characteristic parameters (e.g. porosity, interconnectivity, mechanical strength, morphology, and surface properties) should be carefully considered for the design and fabrication of scaffolds to meet the needs of a specific tissue engineering application. Mimicking the natural bone matrix structure and composition represents a new biomimetic scaffold design approach. As scientists learn more about cellular interactions with materials and growth/differentiation factors, it is likely that scaffolds will be designed to controllably manipulate stem or osteoblastic cell function to enable the development of more advanced bone regeneration therapies.

References

Aumailley, M., & Gayraud, B. (1998). Structure and biological activity of the extracellular matrix. *J. Molec. Med., 76*, 253–265.

Barberi, T., Willis, L. M., Socci, N. D., & Studer, L. (2005). Derivation of multipotent mesenchymal precursors from human embryonic stem cells. *Plos. Med., 2*, 554–560.

Barberi, T., Bradbury, M., Dincer, Z., Panagiotakos, G., Socci, N. D., & Studer, L. (2007). Derivation of engraftable skeletal myoblasts from human embryonic stem cells. *Nat. Med., 13*, 642–648.

Barry, F. P., Boynton, R. E., Haynesworth, S., Murphy, J. M., & Zaia, J. (1999). The monoclonal antibody SH-2, raised against human mesenchymal stem cells, recognizes an epitope on endoglin (CD105). *Biochem. Biophys. Res. Commun., 265*, 134–139.

Barry, F. P., Murphy, J. M., English, K., & Mahon, B. P. (2005). Immunogenicity of adult mesenchymal stem cells: lessons from the fetal allograft. *Stem Cells Devel., 14*, 252–265.

Bielby, R. C., Boccaccini, A. R., Polak, J. M., & Buttery, L. D. K. (2004). In vitro differentiation and in vivo mineralization of osteogenic cells derived from human embryonic stem cells. *Tissue Eng., 10*, 1518–1525.

Boyan, B. D., Hummert, T. W., Dean, D. D., & Schwartz, Z. (1996). Role of material surfaces in regulating bone and cartilage cell response. *Biomaterials, 17*, 137–146.

Boyne, P. J., Lilly, L. C., Marx, R. E., Moy, P. K., Nevins, M., Spagnoli, D. B., et al. (2005). De novo bone induction by recombinant human bone morphogenetic protein-2 (rhBMP-2) in maxillary sinus floor augmentation. *J. Oral Maxillofac. Surg., 63*, 1693–1707.

Brown, S. E., Tong, W., & Krebsbach, P. H. (2009). The derivation of mesenchymal stem cells from human embryonic stem cells. *Cells Tissues Organs, 189*, 256–260.

Campagnoli, C., Roberts, I. A. G., Kumar, S., Bennett, P. R., Bellantuono, I., & Fisk, N. M. (2001). Identification of mesenchymal stem/progenitor cells in human first-trimester fetal blood, liver, and bone marrow. *Blood, 98*, 2396–2402.

Chaikof, E. L., Matthew, H., Kohn, J., Mikos, A. G., Prestwich, G. D., & Yip, C. M. (2002). Biomaterials and scaffolds in reparative medicine. *Ann. NY Acad. Sci., 961*, 96–105.

Chen, V. J., & Ma, P. X. (2004). Nano-fibrous poly(l-lactic acid) scaffolds with interconnected spherical macropores. *Biomaterials, 25*, 2065–2073.

de Caestecker, M. (2004). The transforming growth factor-beta superfamily of receptors. *Cytokine Growth Fact. Rev., 15*, 1–11.

de Ugarte, D. A., Morizono, K., Elbarbary, A., Alfonso, Z., Zuk, P. A., Zhu, M., et al. (2003). Comparison of multi-lineage cells from human adipose tissue and bone marrow. *Cells Tissues Organs, 174*, 101–109.

Friedenstein, A. J., Piatetzky-Shapiro, I. I., & Petrakova, K. V. (1966). Osteogenesis in transplants of bone marrow cells. *J. Embryol. Exp. Morphol., 16*, 381.

Gerstenfeld, L. C., Cullinane, D. M., Barnes, G. L., Graves, D. T., & Einhorn, T. A. (2003). Fracture healing as a post-natal developmental process: molecular, spatial, and temporal aspects of its regulation. *J. Cell. Biochem., 88*, 873–884.

Giordano, R. A., Wu, B. M., Borland, S. W., Cima, L. G., Sachs, E. M., & Cima, M. J. (1996). Mechanical properties of dense polylactic acid structures fabricated by three dimensional printing. *J. Biomater. Sci. Polymer Ed.*, 8, 63–75.

Granjeiro, J. M., Oliveira, R. C., Bustos-Valenzuela, J. C., Sogayar, M. C., & Taga, R. (2005). Bone morphogenetic proteins: from structure to clinical use. *Brazilian J. Med. Biol. Res.*, 38, 1463–1473.

Grinnell, F. (1982). Cell-collagen interactions – overview. *Meth. Enzymol.*, 82, 499–503.

Hench, L. L. (1998). Bioceramics. *J. Am. Ceramic Soc.*, 81, 1705–1728.

Hile, D. D., Amirpour, M. L., Akgerman, A., & Pishko, M. V. (2000). Active growth factor delivery from poly (d, l-lactide-co-glycolide) foams prepared in supercritical CO_2. *J. Control. Release*, 66, 177–185.

Hu, J., Liu, X. H., & Ma, P. X. (2008). Induction of osteoblast differentiation phenotype on poly(l-lactic acid) nanofibrous matrix. *Biomaterials*, 29, 3815–3821.

Hu, J., Feng, K., Liu, X. H., & Ma, P. X. (2009). Chondrogenic and osteogenic differentiations of human bone marrow-derived mesenchymal stem cells on a nanofibrous scaffold with designed pore network. *Biomaterials*, 30, 5061–5067.

Huangfu, D., Maehr, R., Guo, W., Eijkelenboom, A., Snitow, M., Chen, A. E., et al. (2008). Induction of pluripotent stem cells by defined factors is greatly improved by small-molecule compounds. *Nat. Biotechnol.*, 26, 795–797.

Hutmacher, D. W. (2000). Scaffolds in tissue engineering bone and cartilage. *Biomaterials*, 21, 2529–2543.

Hwang, N. S., Varghese, S., Zhang, Z., & Elisseeff, J. (2006). Chondrogenic differentiation of human embryonic stem cell-derived cells in arginine-glycine-aspartate modified hydrogels. *Tissue Eng.*, 12, 2695–2706.

Iwata, H., Sakano, S., Itoh, T., & Bauer, T. W. (2002). Demineralized bone matrix and native bone morphogenetic protein in orthopaedic surgery. *Clin. Orthop. Rel. Res.*, 395, 99–109.

Karaplis, A. C. (2002). Embryonic development of bone and the molecular regulation of intramembranous and endochondral bone formation. In J. P. Bilezikian, L. G. Raisz, & G. A. Rodan (Eds.), *Principles of Bone Biology, Vol. 1* (pp. 33–58). San Diego: Academic Press.

Kasten, P., Vogel, J., Luginbuhl, R., Niemeyer, P., Tonak, M., Lorenz, H., et al. (2005). Ectopic bone formation associated with mesenchymal stem cells in a resorbable calcium deficient hydroxyapatite carrier. *Biomaterials*, 26, 5879–5889.

Lee, O. K., Kuo, T. K., Chen, W. M., Lee, K. D., Hsieh, S. L., & Chen, T. H. (2004). Isolation of multipotent mesenchymal stem cells from umbilical cord blood. *Blood*, 103, 1669–1675.

Lenza, R. F. S., Vasconcelos, W. L., Jones, J. R., & Hench, L. L. (2002). Surface-modified 3D scaffolds for tissue engineering. *J. Mater. Sci. Mater. Med.*, 13, 837–842.

Levenberg, S., Huang, N. F., Lavik, E., Rogers, A. B., Itskovitz-Eldor, J., & Langer, R. (2003). Differentiation of human embryonic stem cells on three-dimensional polymer scaffolds. *Proc. Natl. Acad. Sci. U.S.A.*, 100, 12741–12746.

Li, W. J., Laurencin, C. T., Caterson, E. J., Tuan, R. S., & Ko, F. K. (2002). Electrospun nanofibrous structure: a novel scaffold for tissue engineering. *J. Biomed. Mater. Res.*, 60, 613–621.

Liu, X., & Ma, P. X. (2004). Polymeric scaffolds for bone tissue engineering. *Ann. Biomed. Eng.*, 32, 477–486.

Liu, X. H., Smith, L. A., Wei, G., Won, Y. J., & Ma, P. X. (2005a). Surface engineering of nano-fibrous poly(l-lactic acid) scaffolds via self-assembly technique for bone tissue engineering. *J. Biomed. Nanotechnol.*, 1, 54–60.

Liu, X. H., Won, Y. J., & Ma, P. X. (2005b). Surface modification of interconnected porous scaffolds. *J. Biomed. Mater. Res.*, 74A, 84–91.

Liu, X., Smith, L. A., Hu, J., & Ma, P. X. (2009). Biomimetic nanofibrous gelatin/apatite composite scaffolds for bone tissue engineering. *Biomaterials*, 30, 2252–2258.

Lollo, C. P., Banaszczyk, M. G., & Chiou, H. C. (2000). Obstacles and advances in non-viral gene delivery. *Curr. Opin. Molec. Ther.*, 2, 136–142.

Ma, P. X. (2003). Tissue engineering. In J. I. Kroschwitz (Ed.), *Encyclopedia of Polymer Science and Technology*. Hoboken: John Wiley & Sons.

Ma, P. X. (2004). Scaffolds for tissue fabrication. *Mater. Today*, 7, 30–40.

Ma, P. X. (2008). Biomimetic materials for tissue engineering. *Adv. Drug Deliv. Rev.*, 60, 184–198.

Ma, P. X., & Choi, J. W. (2001). Biodegradable polymer scaffolds with well-defined interconnected spherical pore network. *Tissue Eng.*, 7, 23–33.

Ma, P. X., & Zhang, R. Y. (1999). Synthetic nano-scale fibrous extracellular matrix. *J. Biomed. Mater. Res.*, 46, 60–72.

Ma, P. X., & Zhang, R. Y. (2001). Microtubular architecture of biodegradable polymer scaffolds. *J. Biomed. Mater. Res.*, 56, 469–477.

Ma, P. X., Zhang, R. Y., Xiao, G. Z., & Franceschi, R. (2001). Engineering new bone tissue in vitro on highly porous poly(alpha-hydroxyl acids)/hydroxyapatite composite scaffolds. *J. Biomed. Mater. Res.*, 54, 284–293.

Mann, B. K., Tsai, A. T., Scott-Burden, T., & West, J. L. (1999). Modification of surfaces with cell adhesion peptides alters extracellular matrix deposition. *Biomaterials*, 20, 2281–2286.

Matthews, J. A., Wnek, G. E., Simpson, D. G., & Bowlin, G. L. (2002). Electrospinning of collagen nanofibers. *Biomacromolecules, 3*, 232–238.

Mikos, A. G., Thorsen, A. J., Czerwonka, L. A., Bao, Y., Langer, R., Winslow, D. N., et al. (1994). Preparation and characterization of poly(l-lactic acid) foams. *Polymer, 35*, 1068–1077.

Mooney, D. J., Baldwin, D. F., Suh, N. P., Vacanti, L. P., & Langer, R. (1996). Novel approach to fabricate porous sponges of poly (d, l-lactic-co-glycolic acid) without the use of organic solvents. *Biomaterials, 17*, 1417–1422.

Murata, M., Maki, F., Sato, D., Shibata, T., & Arisue, M. (2000). Bone augmentation by onlay implant using recombinant human BMP-2 and collagen on adult rat skull without periosteum. *Clin. Oral Implants Res., 11*, 289–295.

Neff, J. A., Caldwell, K. D., & Tresco, P. A. (1998). A novel method for surface modification to promote cell attachment to hydrophobic substrates. *J. Biomed. Mater. Res., 40A*, 511–519.

Parikh, S. N. (2002). Bone graft substitutes in modern orthopedics. *Orthopedics, 25*, 1301–1309.

Peng, H. R., Wright, V., Usas, A., Gearhart, B., Shen, H. C., Cummins, J., & Huard, J. (2002). Synergistic enhancement of bone formation and healing by stem cell-expressed VEGF and bone morphogenetic protein-4. *J. Clin. Invest., 110*, 751–759.

Pittenger, M. F., Mackay, A. M., Beck, S. C., Jaiswal, R. K., Douglas, R., Mosca, J. D., et al. (1999). Multilineage potential of adult human mesenchymal stem cells. *Science, 284*(5411), 143–147.

Rosso, F., Giordano, A., Barbarisi, M., & Barbarisi, A. (2004). From cell-ECM interactions to tissue engineering. *J. Cell. Physiol., 199*, 174–180.

Rumberger, J. A., Simons, D. B., Fitzpatrick, L. A., Sheedy, P. F., & Schwartz, R. S. (1995). Coronary-artery calcium area by electron-beam computed-tomography and coronary atherosclerotic plaque area – a histopathologic correlative study. *Circulation, 92*, 2157–2162.

Shi, Y., Do, J. T., Desponts, C., Hahm, H. S., Scholer, H. R., & Ding, S. (2008). A combined chemical and genetic approach for the generation of induced pluripotent stem cells. *Cell Stem Cell, 2*, 525–528.

Simmons, P. J., & Torokstorb, B. (1991). Identification of stromal cell precursors in human bone-marrow by a novel monoclonal-antibody, Stro-1. *Blood, 78*, 55–62.

Smith, L. A., & Ma, P. X. (2004). Nano-fibrous scaffolds for tissue engineering. *Colloid Surface B, 39*, 125–131.

Smith, L. A., Liu, X. H., & Ma, P. X. (2008). Tissue engineering with nano-fibrous scaffolds. *Soft Matter, 4*, 2144–2149.

Smith, L. A., Liu, X. H., Hu, J., & Ma, P. X. (2009). The influence of three-dimensional nanofibrous scaffolds on the osteogenic differentiation of embryonic stem cells. *Biomaterials, 30*, 2516–2522.

Smith, L. A., Liu, X., Hu, J., & Ma, P. X. (2010). The enhancement of human embryonic stem cell osteogenic differentiation with nano-fibrous scaffolding. *Biomaterials, 31*(21), 5526–5535.

Spector, J. A., Mehrara, B. J., Luchs, J. S., Greenwald, J. A., Fagenholz, P. J., Saadeh, P. B., et al. (2000). Expression of adenovirally delivered gene products in healing osseous tissues. *Ann. Plast. Surg., 44*, 522–528.

Strom, S. C., & Michalopoulos, G. (1982). Collagen as a substrate for cell-growth and differentiation. *Meth. Enzymol., 82*, 544–555.

Suchanek, W., & Yoshimura, M. (1998). Processing and properties of hydroxyapatite-based biomaterials for use as hard tissue replacement implants. *J. Mater. Res., 13*, 94–117.

Sun, H. L., Feng, K., Hu, J., Soker, S., Atala, A., & Ma, P. X. (2010). Osteogenic differentiation of human amniotic fluid-derived stem cells induced by bone morphogenetic protein-7 and enhanced by nanofibrous scaffolds. *Biomaterials, 31*, 1133–1139.

Sun, W., Darling, A., Starly, B., & Nam, J. (2004). Computer-aided tissue engineering: overview, scope and challenges. *Biotechnol. Appl. Biochem., 39*, 29–47.

Takahashi, K., & Yamanaka, S. (2006). Induction of pluripotent stem cells from mouse embryonic and adult fibroblast cultures by defined factors. (see comment). *Cell, 126*, 663–676.

Takahashi, K., Tanabe, K., Ohnuki, M., Narita, M., Ichisaka, T., Tomoda, K., et al. (2007). Induction of pluripotent stem cells from adult human fibroblasts by defined factors. *Cell, 131*, 861–872.

Thomson, J. A., Itskovitz-Eldor, J., Shapiro, S. S., Waknitz, M. A., Swiergiel, J. J., Marshall, V. S., et al. (1998). Embryonic stem cell lines derived from human blastocysts. *Science, 282*(5391), 1145–1147.

Thomson, R. C., Yaszemski, M. J., Powers, J. M., & Mikos, A. G. (1995). Fabrication of biodegradable polymer scaffolds to engineer trabecular bone. *J. Biomater. Sci. Polymer Ed., 7*, 23–38.

Urist, M. R. (1965). Bone – formation by autoinduction. *Science, 150*(3698), 893–899.

Vaccaro, A. R., Patel, T., Fischgrund, J., Anderson, D. G., Truumees, E., Herkowitz, H., et al. (2005). A 2-year follow-up pilot study evaluating the safety and efficacy of op-1 putty (rhbmp-7) as an adjunct to iliac crest autograft in posterolateral lumbar fusions. *Eur. Spine J., 14*, 623–629.

Wagers, A. J., & Weissman, I. L. (2004). Plasticity of adult stem cells. *Cell, 116*, 639–648.

744

Wang, J. C., Kanim, L. E. A., Yoo, S., Campbell, P. A., Berk, A. J., & Lieberman, J. R. (2003). Effect of regional gene therapy with bone morphogenetic protein-2-producing bone marrow cells on spinal fusion in rats. *J. Bone Joint Surg., 85A*, 905—911.

Weber, F. E., Eyrich, G., Gratz, K. W., Maly, F. E., & Sailer, H. F. (2002). Slow and continuous application of human recombinant bone morphogenetic protein via biodegradable poly(lactide-co-glycolide) foamspheres. *Int. J. Oral Maxillofac Surg., 31*, 60—65.

Wei, G. B., & Ma, P. X. (2004). Structure and properties of nano-hydroxyapatite/polymer composite scaffolds for bone tissue engineering. *Biomaterials, 25*, 4749—4757.

Wei, G. B., Jin, Q. M., Giannobile, W. V., & Ma, P. X. (2006). Nano-fibrous scaffold for controlled delivery of recombinant human PDGF-BB. *J. Control. Release, 112*, 103—110.

Wei, G. B., Jin, Q. M., Giannobile, W. V., & Ma, P. X. (2007). The enhancement of osteogenesis by nano-fibrous scaffolds incorporating rhBMP-7 nanospheres. *Biomaterials, 28*, 2087—2096.

Whang, K., Thomas, C. H., Healy, K. E., & Nuber, G. (1995). A novel method to fabricate bioabsorbable scaffolds. *Polymer, 36*, 837—842.

Woo, K. M., Chen, V. J., & Ma, P. X. (2003). Nano-fibrous scaffolding architecture selectively enhances protein adsorption contributing to cell attachment. *J. Biomed. Mater. Res., 67A*, 531—537.

Woo, K. M., Jun, J. H., Chen, V. J., Seo, J. Y., Baek, J. H., Ryoo, H. M., et al. (2007). Nano-fibrous scaffolding promotes osteoblast differentiation and biomineralization. *Biomaterials, 28*, 335—343.

Woodrow, K. A., Cu, Y., Booth, C. J., Saucier-Sawyer, J. K., Wood, M. J., & Saltzman, W. M. (2009). Intravaginal gene silencing using biodegradable polymer nanoparticles densely loaded with small-interfering RNA. *Nature Mater., 8*, 526—533.

Wozney, J. M., Rosen, V., Celeste, A. J., Mitsock, L. M., Whitters, M. J., Kriz, R. W., et al. (1988). Novel regulators of bone-formation — molecular clones and activities. *Science, 242*(4885), 1528—1534.

Yu, J., Vodyanik, M. A., Smuga-Otto, K., Antosiewicz-Bourget, J., Frane, J. L., Tian, S., et al. (2007). Induced pluripotent stem cell lines derived from human somatic cells. *Science, 318*(5858), 1917—1920.

Zhang, R. Y., & Ma, P. X. (1999a). Poly(alpha-hydroxyl acids) hydroxyapatite porous composites for bone-tissue engineering. I. Preparation and morphology. *J. Biomed. Mater. Res., 44*, 446—455.

Zhang, R. Y., & Ma, P. X. (1999b). Porous poly(l-lactic acid)/apatite composites created by biomimetic process. *J. Biomed. Mater. Res., 45*, 285—293.

Zhang, R. Y., & Ma, P. X. (2000). Synthetic nano-fibrillar extracellular matrices with predesigned macroporous architectures. *J. Biomed. Mater. Res., 52*, 430—438.

Zhang, R. Y., & Ma, P. X. (2004). Biomimetic polymer/apatite composite scaffolds for mineralized tissue engineering. *Macromol. Biosci., 4*, 100—111.

Zuk, P. A., Zhu, M., Ashjian, P., de Ugarte, D. A., Huang, J. I., Mizuno, H., et al. (2002). Human adipose tissue is a source of multipotent stem cells. *Molec. Biol. Cell, 13*, 4279—4295.

CHAPTER 41

Cell Therapy for Blood Substitutes

Shi-Jiang Lu[*,†], **Qiang Feng**[*,†], **Feng Li**[*,†], **Erin A. Kimbrel**[*,†], **Robert Lanza**[*,‡]

* Stem Cell & Regenerative Medicine International, Worcester, MA, USA
† Department of Applied Bioscience, Cha University, Seoul, Korea
‡ Advanced Cell Technology, Inc., Worcester, MA, USA

INTRODUCTION

For decades, supplies of transfusable blood components have failed to keep pace with increasing clinical demand. This critical shortfall has prompted efforts to develop safe and effective blood substitutes that can be produced from non-immunoreactive sources and in limitless quantities. This chapter summarizes recent efforts to develop alternative sources for red blood cells (RBCs) and platelets, two of the blood's most critical life-savings elements.

RBCs, the oxygen-carrying component of the blood, are transfused in over half of all anemic patients admitted to intensive care units in the USA (Corwin et al., 1995, 2004; Littenberg et al., 1995) and it is estimated that nearly 5 million patients receive approximately 14 million units of RBCs per year in the USA alone (Whitaker and Henry, 2005). Limitations in the supply of RBCs can have potentially life-threatening consequences for patients, especially for those who have rare or unusual blood types with massive blood loss due to trauma or other emergency situations. Unfortunately, the supply of transfusable RBCs, especially "universal" donor type (O)Rh-negative, is often insufficient, particularly in the battlefield environment and/or following major natural disasters due to the lack of blood type information and the limited time required for life-saving transfusion. Moreover, the low prevalence of (O)Rh-negative blood type in the general population (<8% in Western countries and <0.3% in Asia) further intensifies the consequences of blood shortages for emergency situations where blood typing may not be possible.

Platelets, anucleate discoid-shaped cell fragments released from megakaryocytes (MKs), are essential to hemostasis, the biological process that stops blood loss. Platelets adhere to damaged blood vessels and trigger a series of biochemical changes that stimulate clot formation and vascular repair. In cases of thrombocytopenia (platelet counts are less than $150 \times 10^3/\mu l$), the increased risk of bleeding can have life-threatening consequences (Patel et al., 2005). In the USA, approximately 1.5 million platelet transfusions are performed annually to protect patients, including those treated with chemotherapy or stem cell trans-plantations, from the risk of thrombocytopenia and its related dangers (Kaushansky, 2008). Unfortunately, becomes refractory to the therapy approximately one out of every three patients who require repeated platelet transfusions (Slichter et al., 2005; Hod and Schwartz, 2008) and, while both immunological and non-immunological complications may be to blame, HLA alloimmunization is the primary cause of refractoriness (Hod and Schwartz, 2008). Over the past few decades, a steady increase in demand for platelets in combination with their limited shelf-life has presented a constant challenge for blood centers and donor-dependent

Principles of Regenerative Medicine. DOI: 10.1016/B978-0-12-381422-7.10041-0

programs. As with RBCs, there is a vast and continuous need for functional, transfusable platelets, especially in times of emergency. Pluripotent stem cells may be able to serve as an alternative source for producing transfusable RBCs and platelets. Here, we will review recent progress in developing these life-saving blood substitutes, including some of our own efforts in unlocking the potential use of human embryonic stem cells and induced pluripotent stem cells in these endeavors.

RED BLOOD CELLS

Erythropoiesis

Erythropoiesis is the highly regulated, multistep process by which the body generates mature red blood cells, or erythrocytes. During mammalian development, erythropoiesis consists of two major waves: (1) primitive erythropoiesis, which is initiated in the yolk sac with the generation of large *nucleated* erythroblasts, and (2) definitive erythropoiesis, which arises from the fetal liver with the development of smaller *enucleated* erythrocytes (Ney, 2006). Definitive erythropoiesis in fetal liver features the production of enucleated RBCs that quickly become dominant in the circulation. Hemoglobin switching from embryonic types ($\zeta_2\varepsilon_2$) to fetal types ($\alpha_2\gamma_2$) also occurs at the initiation of definitive erythropoiesis (Kovach et al., 1967; Brotherton et al., 1979; Stamatoyannopoulos, 2005). However, recent reports show that yolk sac-derived primitive erythroblasts can also enucleate in the circulation of a mouse embryo and persist throughout gestation (Kingsley et al., 2004; Fraser et al., 2007).

In adults, all blood cell types including lymphocytes, myeloid cells, and RBCs are derived from hematopoieitc stem cells (HSCs) residing in the bone marrow. The initial differentiation of a multipotential HSC into a common myeloid progenitor (CMP) determines its capacity to further differentiate into granulocytes, erythrocytes, megakaryocytes, and macrophages but not lymphoid cells. As the CMP continues to differentiate, it undergoes significant expansion and will eventually commit to one particular lineage. Erythroid lineage commitment leads to the appearance of the pronormoblast (also called the proerythroblast or rubriblast). The pronormoblast will then pass through early, intermediate, and late normoblast (erythroblast) stages, prior to expelling its nucleus and becoming a reticulocyte. Upon exiting the bone marrow, reticulocytes enter the blood circulation and become fully mature RBCs. The various stages of erythropoiesis can be distinguished by characteristic morphological features in the cell cytoplasm and nucleus, which become evident after Wright-Giemsa staining. Additionally, using an *in vitro* colony-forming assay, CMP progenitors can be identified by their ability to form a characteristic colony-forming unit, called the CFU-GEMM, while early erythroid progenitors give rise to burst-forming units-erythroid (BFU-E), and late erythroid progenitors give rise to colony-forming units-erythroid (CFU-E) in this assay (Ney, 2006).

"Universal" blood generated by modifying RBC surface antigens

While Karl Lansteiner's Nobel Prize-winning discovery of ABO blood groups occurred over a century ago, there are now 30 known human blood group systems and the complicated issues surrounding blood type incompatibility continue to frustrate clinicians and scientists. To circumvent these issues, researchers have been trying to develop universal blood for decades, primarily through chemical modifications on the surface of RBCs. Among these efforts, Goldstein and co-workers demonstrated that group B erythrocytes can be enzymatically converted to group O, and that the converted cells survived normally when transfused in to A, B, and O individuals (Goldstein et al., 1982). Since this groundbreaking discovery, there have been coordinated efforts to identify both clinically and economically viable alternative enzymes capable of catalyzing blood group conversions. Recently, new classes of bacterial exoglycosidases have been discovered that can enzymatically perform group O conversions with faster kinetics (Liu et al., 2007; Olsson and Clausen, 2008), yet the likelihood of this technology playing a major role in blood transfusion practice is unclear. Currently, there are no reports showing the enzymatic conversion of other important blood typing groups such as the Rh factor. With the momentum of the rapidly advancing stem cell field in recent years, the concept of (O)Rh-negative RBCs derived from pluripotent stem cells clearly offers an attractive option for the future of blood transfusions.

RBCs generated from adult stem cells *in vitro*

Erythrocytes have been derived from a variety of primary stem cell sources including umbilical cord blood (CB), peripheral blood (PB), and bone marrow (BM). CD34+ cells from CB, PB, and BM have been isolated and differentiated into erythrocytes with 95% purity after a little over a week of culture using EPO, stem cell factor (SCF), and interleukin-3 (IL-3) (Giarratana et al., 2005; Leberbauer et al., 2005; Miharada et al., 2006). Co-culturing with mouse MS-5 stromal cells line or human mesenchymal stem cells facilitates enucleation, the hallmark of mature RBCs. The stem cell-derived erythrocytes have similar properties to normal RBCs, including membrane deformation capacity, intrinsic enzymatic activity, and balanced adult/fetal forms of hemoglobin that can bind and release oxygen. Additionally, these erythrocytes have been found to survive *in vivo* in NOD/SCID mice, being detectable 3 days after transplantation. Stem cells from PB and BM have limited expansion capacity (29,000- and 16,500-fold, respectively) compared to those from CB (140,000-fold). Yet, cells from adult PB and BM are often easier to obtain and display mature forms of hemoglobin, whereas CB sources are more difficult to obtain and cells derived from them only express fetal globins. Despite their potential utility, however, these primary cells still represent donor-limited sources of blood substitutes.

RBCs generated from human embryonic stem cells

Human embryonic stem cells (hESCs) represent an alternative stem cell source for generating blood components, one whose capacity for expansion far exceeds that of BM, PB, or even CB. Hematopoietic precursors as well as more mature progeny representing erythroid, granulocyte, macrophage, megakaryocytic, and lymphoid lineages have all been identified in differentiating hESC culture systems (Kaufman et al., 2001; Lu et al., 2004; Zhan et al., 2004; Qiu et al., 2005; Vodyanik et al., 2005; Wang et al., 2005; Chang et al., 2006). Therefore, many groups have focused their efforts on trying to steer the *in vitro* differentiation of entire hESC cultures into specific blood cell types, such as RBCs or megakaryocytes/platelets (Gaur et al., 2006; Lu et al., 2008a; Takayama et al., 2008).

The controlled differentiation of hESCs into erythrocytes has primarily been achieved by either embryoid body (EB) formation or by co-culturing with stromal cells followed by isolation of CD34+ cells and further expansion/differentiation. While somewhat different in approach, both systems have encountered the same obstacles in generating fully mature adult RBCs. For example, Chang et al. generated erythroid cells from hESCs by isolating and expanding non-adherent cells of day-14 EBs for an additional 15 to 56 days (Chang et al., 2006). The resulting cells coexpressed high levels of embryonic ε- and fetal γ-globins but little or no adult β-globin. In addition, the cells had not enucleated. Using a stroma co-culture method, erythroid cells could be generated by culturing hESCs with FHB-hTERT human fetal liver stromal cells for 14−35 days, isolating CD34+ cells and further differentiating them in a four-step culture system. In steps 1 and 2, cocktails of cytokines were used to promote the proliferation and maturation of erythroid precursors. In steps 3 and 4, erythroid cells were transferred onto mouse BM stromal cells (MS5) to facilitate terminal maturation (Olivier et al., 2006). While these erythrocytes were generated on a relatively large scale (0.5 to 5×10^7 cells), the resulting cells had similar problems to those generated by the EB method; they mainly expressed embryonic ε− and fetal γ-globin isoforms, with only a trace amount of adult β-globin being detected. Despite these reports, other studies suggest that specific types of stroma can, in fact, facilitate the expression of adult β-globin in developing erythrocytes. Using immunostaining with globin chain-specific monoclonal antibodies, Ma et al. showed that almost 100% of hESC-derived erythrocytes expressed the adult β-globin chain after co-culture with murine fetal liver-derived stromal cells (mFLCs) (Ma et al., 2008). Yet, the majority of the cells still had not enucleated, a problem that does not seem to have an easily explainable mechanism or simple solution. For example, a study published in 2005 showed that co-culturing CD34+ CB, PB, and BM cells on MS5 stroma produced RBCs with up to 100% enucleation (Giarratana et al., 2005), while a year later Miharada et al. reported that up to 77% of cord blood-derived erythrocytes could achieve enucleation without the use of any stromal cells (Miharada et al., 2006).

The mechanism(s) by which stromal cells may facilitate erythrocyte enucleation are largely unknown, and therefore it remains to be determined whether or not stroma co-culture will be absolutely required for enucleation of hESC-derived erythrocytes. Stroma may secrete important soluble factors, provide critical cell-cell contact, and/or engulf nuclei and other organelles. Some studies suggest that the process of enucleation involves a critical asymmetric cell division (Chasis et al., 1989) and that Rac GTPases and their effector, mDia2, help extrude the pyncnotic nucleus during this process (Ji et al., 2008). Other studies show that, *in vivo*, contact with macrophages is important and the erythroblast-macrophage-protein, Emp, plays a critical role in enucleation (Hanspal and Hanspal, 1994; Hanspal et al., 1998; Soni et al., 2006). Clearly, further studies will be required in order to determine how to improve the efficiency of both enucleation and globin switching *in vitro*. While co-culture with stroma may help circumvent problems with these processes, production of clinical-grade RBC substitutes will demand stroma-free culture conditions.

Not only must *in vitro*-generated RBC substitutes be fully matured and enucleated to use in the clinic, but they must also be capable of large-scale production. Related to this issue, we recently developed a strategy that efficiently and reproducibly generates functional hemangioblasts (the common precursor cell to all hematopoietic and endothelial cell lineages) using a serum-free culture system and have been able to do so with high purity (>95%) and on a relatively large scale (Lu et al., 2007, 2008b). One of the characteristics of the primitive hemangioblast cell is its highly efficient generation of large BFU-E (Fig. 41.1(A,B)), CFU-GEMM (Fig. 41.1(C)) and CFU-E (Fig. 41.1(D)) colonies when cultured in methylcellulose-based medium. This prompted us to investigate whether or not hemangioblasts could be used as an intermediate cell source to generate large, clinically relevant quantities of blood components, such as erythrocytes and platelets. The overall strategy, as depicted in Figure 41.2, employs a basic three-step approach to proceed from hESCs to the final blood cell products.

Using the hemangioblast system, we have generated functional RBCs (blood types A, B, O, and both RhD+ and RhD−) on a large scale from multiple hESC lines (Lu et al., 2008a). Three critical elements allowed us to do this: (1) the efficient generation of hemangioblasts without disruption of their colony-forming environment, (2) expansion of hemangioblasts to erythroblasts in a high cell density (Fig. 41.3(A,B)), and (3) culture in semi-solid methylcellulose-based media to provide optimal conditions for maximum expansion and high erythroid purity

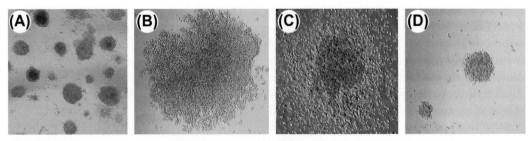

FIGURE 41.1
Human ESC-derived hemangioblasts showed tremendous CFU capability when cultured in CFC-medium. (A,B) early BFU-E (×4, ×10). (C) Large CFU-GEMM (×10). (D) Late CFU-E (×10).

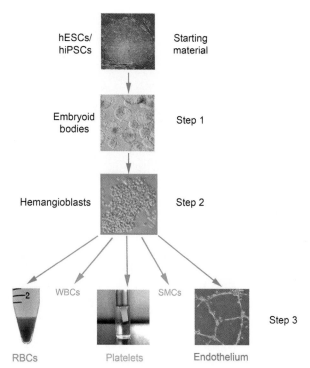

FIGURE 41.2
Graphic illustration of hemangioblastic derivatives such as RBCs and platelets from pluripotent stem cells *in vitro*.
A dramatic expansion in cell number between steps 1 and 2 makes this system amenable to large-scale production. In addition
to RBCs and platelets, derivatives such as white blood cells (WBCs), smooth muscle cells (SMCs), and endothelium may also be
generated using this *in vitro* differentiation system.

(Fig. 41.3(A)). We generated approximately 10^{10} to 10^{11} erythroid cells per six-well plate of
hESCs (Lu et al., 2008a), which is over a thousand-fold more efficient than previously reported
methods (Olivier et al., 2006). Oxygen equilibrium curves of erythroid cells from days 19 to 21
of differentiation were comparable to normal transfusible RBCs and responded to changes in
pH and 2,3-diphosphoglyerate. During the course of our studies, we found that extended
in vitro culture facilitated further maturation of these erythroid cells, inducing a progressive
decrease in size, increased expression of the erythrocyte cell surface marker, glycophorin A
(CD235a), as well as chromatin and nuclear condensation. When the extended culture was
performed on OP9 stromal cells, it resulted in the extrusion of the pycnotic nucleus in up to
65% of cells and the generation of enucleated erythrocytes with a diameter of approximately
6–8 µm (Fig. 41.4). At this stage, the erythrocyte population was nearly 100% positive for
glycophorin A (Fig. 41.3(C)), had a very high content of hemoglobin (Fig. 41.3(D)), and
expressed ABO antigen (Fig. 41.3(E)). Although the cells were found to express fetal and
embryonic globin chains, globin chain-specific PCR and immunofluorescence showed that,
after extended culture, expression of adult β-globin increased from 0 to 15% (Fig. 41.3(F)).
Overall, these results show that it is feasible to differentiate and mature hESCs into functional
oxygen-carrying erythrocytes on a large scale. The identification of an hESC line with an O(−)
genotype would permit the production of ABO- and RhD-compatible (and pathogen-free)
"universal donor" RBCs. While considerable effort is still needed to bring hESC-derived RBCs
to clinical trials, these efforts certainly provide a promising lead.

Can RBCs be generated from human induced pluripotent stem cells?

The successful reprogramming of somatic cells into a pluripotent state has been achieved by
ectopic expression of various combinations of transcription factors such as Oct4, Sox2, Klf4,
c-Myc, LIN28, and Nanog cells (Takahashi et al., 2007; Yu et al., 2007). The derivation of

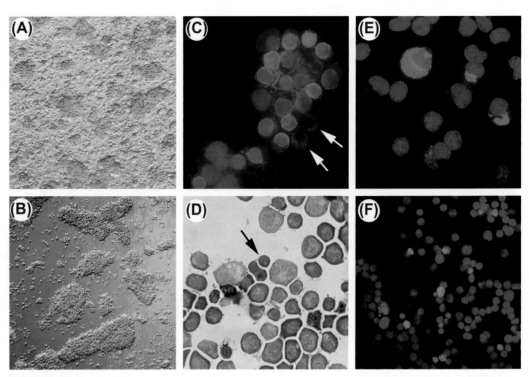

FIGURE 41.3
(A,B) Morphology of a typical high-density expansion/differentiation of immature hESC-derived erythrocytes cultured with medium containing methylcellulose (×10). (C) Maturing hESC-derived erythrocytes stain positive for erythrocyte marker CD235a, with a small fraction of cells showing enucleation (arrows). (D) Giemsa/benzidine double staining of maturing hESC-derived erythrocytes showing high content of intracellular hemoglobin (brown). Arrow indicates a fully enucleated red blood cell (×100). (E) Maturing hESC-derived erythrocytes express A type surface antigen (red, ×100 magnification). (F) Some hESC-derived erythrocytes express beta chain hemoglobin (green, ×40) after elongated culture *in vitro*. Cellular nuclei in all fluorescence images were stained with DAPI (blue).

these induced pluripotent stem cells (iPSCs) is less controversial than that of hESCs and, thus, they open up an exciting new route to obtain pluripotent stem cells. Moreover, the fact that iPSCs can be produced in a patient-specific manner will eliminate the issue of immuno-rejection in cell, tissue, or organ replacement therapies in the future. We and others have demonstrated that human iPSCs can successfully differentiate into erythrocytes that possess classic morphology, express glycophorin A, and have abundant hemoglobin content (Choi et al., 2009; Lengerke et al., 2009; Ye et al., 2009; Feng et al., 2010). However, our recent study has revealed some intrinsic molecular and cellular abnormalities in the iPSC derivatives such as increased apoptosis, limited CFU capability, and limited expansion (Feng et al., 2010). The exact cause(s) of these abnormalities is unclear at this time but may be due to alterations caused by the modified genome of virally reprogrammed cells. More research will be needed to determine whether iPSCs generated with viral-free methods have the same abnormalities as the virally reprogrammed ones or whether they offer a better alternative for the generation of patient-specific blood replacement cells.

Where do we go from here?

The manufacture of safe and effective red blood cell substitutes will help alleviate many of the risks, complications, and hardships associated with donor-dependent RBC sources. As summarized here, significant progress has been made towards this end by manipulating the differentiation potential of hESCs and iPSCs and driving them towards erythrocyte development. *In vitro* differentiation systems that can be scaled up for mass production of RBC substitutes have already been developed and the hemangioblast methodology described in

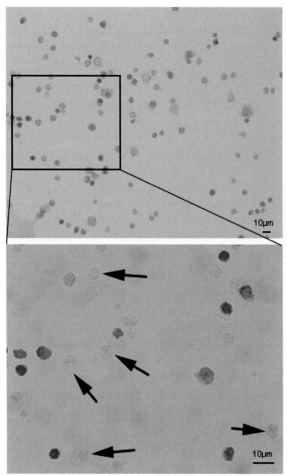

FIGURE 41.4

Enucleation of *in vitro*-cultured erythrocytes derived from hESCs. (top) Low-magnification image of Giemsa-stained enucleated erythrocytes (×20). (bottom) High-magnification (×100) image of enucleated erythrocytes (black arrows).

this chapter represents one such possibility. Despite many exciting advances with *in vitro* culture systems, problems associated with the final stages of erythrocyte maturation, namely enucleation and globin switching, will still need to be fully resolved before hESC/iPSC-derived RBCs can be produced in a stroma-free manner, scaled up for mass production, and brought to the clinic.

MEGAKARYOCYTES AND PLATELETS
Megakaryopoiesis

Due to their role in clot formation and blood vessel repair, platelets are essential for cessation of bleeding, and maintaining an abundant supply of them is vitally important. Megakaryocytes, the large multinucleate precursors to platelets, provide a constant, renewable source of platelets to the blood system and are themselves produced through a process called megakaryopoiesis. As previously mentioned, hematopoietic stem cells differentiate into CMPs, which then undergo significant proliferation and expansion as they further mature and differentiate. Once exclusively committed to the megakaryocyte lineage, however, they only retain limited expansion capacity. On a cellular level, megakaryocyte maturation involves many changes such as an increase in expression of the cell surface markers, GPIIb/IIIa (also known as CD41 or αIIb/βIII integrin receptor) and GPIb/GPIX/GPV receptors, and an increase in cytoplasmic and nuclear mass, with cells expanding to 50–100 μm in diameter. The increase in nuclear mass encompasses several rounds of endomitosis, that is, chromosome replication without cytokinesis (Lordier et al., 2008), which results in nuclear polyploidization and cells with up to 128N (Tomer, 2004).

In vivo, the drastic increase in cytoplasmic mass that occurs during maturation facilitates production of several thousand platelets per mature megakaryocyte (Long, 1998) as it allows the accumulation of several components crucial for platelet biogenesis, such as α granules, dense bodies, and platelet-associated proteins (e.g. vWF, PF4) (Schmitt et al., 2001; Deutsch and Tomer, 2006). In addition, the intracellular membrane system (demarcation membrane system or DMS) undergoes various structural and functional changes that are necessary for platelet formation. Recently, it has been reported that SDF/CXCR4 signaling encourages maturing megakaryocytes to migrate close to blood vessels (Lane et al., 2000; Kowalska et al., 1999; Majka et al., 2000; Avecilla et al., 2004) and this potentially helps explain why morphologically distinct mature megakaryocytes are often observed in close proximity to the bone marrow sinusoidal vascular cavity surrounded by endothelial cells.

On a molecular level, thrombopoietin (TPO) is the primary physiological regulator of megakaryocyte and platelet generation (Kaushansky, 2008); however, numerous other cytokines, growth factors, and small molecules have been found to be important as well. Cytokines including IL3, IL6, IL9, IL11, BMP4, FL, and SCF have been reported to synergistically stimulate the proliferation of megakaryocyte progenitors (Gordon and Hoffman, 1992; Deutsch and Tomer, 2006), while TGF-β1 and PF4 have been found to inhibit their development and maturation (Eslin et al., 2004). Downstream of cytokine signaling pathways, various transcription factors help orchestrate the progressive lineage commitment and maturation of developing megakaryocytes. For example, the GATA1/FOG1 complex synergistically controls the expression of megakaryocyte-specific genes including integrin αIIb (CD41a) (Gaines et al., 2000; Wang et al., 2002). Targeted disruption of either GATA1 or FOG1 can lead to severe anemia and embryonic lethality (Fujiwara et al., 1996), whereas disruption of GATA1 in a megakaryocyte lineage-restricted manner results in significant thrombocytopenia and differentiation defects in megakaryocytes (Shivdasani et al., 1997). Evidence from *in vitro* studies and mouse genetic models has identified other transcription factors such as GATA2, Fli1, NFE2, and RUNX1 as being critical regulators of megakaryopoiesis as well (reviewed by Pang et al., 2005).

Biogenesis of platelets

While not mutually exclusive, two models have been proposed to describe the assembly and release of platelets from megakaryocytes into the blood stream, a process known as thrombopoiesis (reviewed by Kosaki, 2005; Patel et al., 2005). In the global fragmentation model, cellular processes termed proplatelets undergo mutual detachment to produce small functional platelets within the megakaryocyte cytoplasm, which then explodes and releases the cell's entire content of platelets simultaneously. This theory, conceived in the 1970s, is supported by electron microscopy of mature megakaryocytes, as well as *in vivo* and *ex vivo* observations of platelet release. More recent experimental evidence lends support to the proplatelet model of platelet biogenesis. In this model, proplatelets actually extend from the mature megakaryocyte cell body, traverse the vascular endothelium, and enter the bone marrow sinusoids, where the shear force of the blood stream facilitates release of nascent platelets from proplatelet ends. On a molecular level, thrombopoiesis is thought to be a highly coordinated process, with sophisticated reorganization of membrane and microtubules and precise distributions of granules and organelles (Junt et al., 2007). It also appears as though localized apoptosis may play important roles in proplatelet formation and platelet release (Clarke et al., 2003). Despite these recent advances in our understanding of platelet biogenesis, mechanistic details underlying membrane reorganization, initiation of proplatelets, transportation of platelet organelles and secretary granules, and control of platelet size remain to be elucidated.

Production of megakaryocytes and platelets from adult stem cells

Various sources of adult HSCs have been used for studying megakaryocyte/ platelet differentiation. Choi and colleagues were the first to report that functional platelets could be produced from CD34+ HSC isolated from peripheral blood (Choi et al., 1995). During 10–12 days of culture, aplastic canine serum was used to promote *in vitro* megakaryocyte lineage commitment/maturation and was then replaced by human serum to enhance platelet production from the megakaryocytes. Platelets generated from this system demonstrated aggregation capacity when stimulated with either adenosine diphosphate (ADP) or thrombin, the physiological agonists for normal blood platelets. Subsequent studies have shown that HSCs from PB, BM, and CB are also all capable of producing megakaryocytes and functional platelets (Norol et al., 1998; Bruno et al., 2003; Ungerer et al., 2004).

Matsunaga et al. used an *in vitro* culture system to demonstrate the feasibility of producing functional platelets for clinical use (Matsunaga et al., 2006). Using this protocol, the calculated yield of platelets from 5×10^6 CD34+ cells was $1.26–1.68 \times 10^{11}$, which is equivalent to approximately three units of donor-derived platelets. However, these calculations were based on extrapolating data from experiments using just 500 starting CD34+ HSCs. A 3D culture system described by Sullenbarger et al. provides additional evidence that it is possible to produce platelets in scalable quantities (Sullenbarger et al., 2009). In this system, about 20 platelet-like particles were produced from each input CD34+ cell. Both of the above protocols consisted of three distinct culture steps, the first of which amplifies hematopoietic progenitors from cord blood CD34+ HSCs over the course of approximately 14 days. The second step, lasting ∼5–7 days, encourages megakaryocyte differentiation while the third step alters culture conditions to enhance platelet biogenesis. Other protocols have described two-step culture methods, with the first step consisting of hematopoietic progenitor amplification and the second step lumping megakaryocyte/platelet production together; however, these systems appear to be less efficient than the three-step methods of Matsunaga and Sullenbarger.

Megakaryocytes and platelets generated from human embryonic stem cells

Following initial successes with murine ES cells (Eto et al., 2002; Fujimoto et al., 2003; Nishikii et al., 2008), megakaryocyte differentiation from hESCs has been achieved in co-culture systems with animal stromal cells (Gaur et al., 2006; Takayama et al., 2008). Using hESCs co-cultured with either OP9 or C3H10T1/2 stroma, Takayama et al. reported production of 4.8×10^6 functional platelets from 10^5 hESCs (Takayama et al., 2008). After initial induction of hematopoietic differentiation, single cells from hESC-derived structures called ES sacs were replated onto stroma to allow megakaryocyte differentiation and platelet production during 24 days of culture.

As mentioned in the first part of this chapter, our group has produced an *in vitro* hemangioblast differentiation system that enables production of functional erythrocytes without stromal cells (Lu et al., 2008a). We have adopted this system for *in vitro* production of megakaryocytes and have observed significantly improved efficiency as compared to other methods, obtaining approximately 100 CD41+ megakaryocytes from one input hESC (Li et al., 2009). *In vitro*-derived megakaryocytes from this system are shown in Figure 41.5. Upon further maturation, the hESC-derived megakaryocytes gave rise to platelet-like particles which, when treated with thrombin, demonstrated comparable functionalities (e.g. spreading and clot formation/retraction) to normal human blood platelets (Li et al., 2009). Yet again, as observed with RBCs, stromal co-culture enhances megakaryocyte maturation and efficiency of

FIGURE 41.5

Megakaryocytes generated from hESCs *in vitro*. Polyploid nuclei are evident in Giemsa staining (left panel) and represent a key morphological feature of mature megakaryocytes. Immunofluorescent staining (right panel) shows CD41 cell surface marker expression (green), DAPI (blue)-positive nuclei, and granular von Willebrand factor (vWF) in the cytoplasm (red) of mature megakaryocytes (merged image).

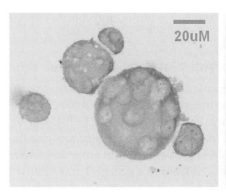
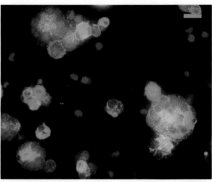

20uM

Green CD41
Blue DAPI
Red vWF

platelet production. Collectively, these results indicate that it is feasible to induce the *in vitro* differentiation and maturation of hESCs into functional platelets and this is an important step in generating large-scale and donorless supplies of platelets for clinical use.

Improving the efficiency for *in vitro* platelet production

The studies described above provide an important proof of principle for the *in vitro* manufacturing of functional platelets from HSCs or hESCs, yet the efficiency of platelet production will need to be significantly improved in order to achieve clinically relevant yields. A potential work flow diagram for mass *in vitro* platelet production is depicted in Figure 41.6. Considering the optimal *in vivo* capacity of megakaryocyte development from HSCs and platelet production from megakaryocytes, both the initial hematopoietic amplification stage and downstream platelet biogenesis could stand to be optimized. Strategies to increase the efficiency of megakaryocyte/platelet production include the development of novel culture systems that mimic the *in vivo* bone marrow microenvironment as well as the optimization of media formulations and concentrations of cytokines, small molecule mimetics, and nutrients. For instance, TPO receptor agonists such as YM477 and AMG531 have been found to enhance megakaryopoiesis and thrombopoiesis (Broudy and Lin, 2004; Fukushima-Shintani et al., 2008). In addition, physiological parameters such as pH, media viscosity, and oxygen levels

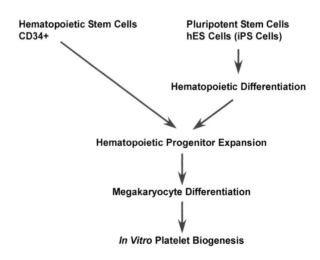

FIGURE 41.6

***In vitro* biogenesis of platelets from different stem cell sources.** The expansion at the hematopoietic progenitor stage and the efficiency of platelet biogenesis are the key stages for large-scale *in vitro* production of platelets.

may all be optimized for increased platelet biogenesis. Lastly, *in vivo* observations that helped formulate the proplatelet model of platelet biogenesis suggest that shear force could play an important role in platelet release (Patel et al., 2005; Junt et al., 2007). Adaptation of such mechanical force in culture systems may also significantly promote proplatelet growth and platelet release.

PERSPECTIVES

Limitations in the supply of RBCs and platelets can have potentially life-threatening consequences for transfusion-dependent patients, particularly those with unusual/rare blood types, and those who develop alloimmunization. Although cord blood, bone marrow, and peripheral blood have been investigated as sources of progenitors for the generation of large-scale transfusable RBCs and platelets, (Giarratana et al., 2005; Leberbauer et al., 2005; Matsunaga et al., 2006; Miharada et al., 2006; Sullenbarger et al., 2009), it is clear that, even after expansion and differentiation, these progenitors represent donor-limited sources of RBCs and platelets. hESCs and iPSCs, especially virus-free human iPSCs (Kaji et al., 2009; Kim et al., 2009; Woltjen et al., 2009; Yu et al., 2009), represent a new source of stem cells that can be propagated and expanded *in vitro* indefinitely, providing a potentially inexhaustible and donorless source of RBCs and platelets for human therapy. For hESCs, the ability to create banks of cell lines with matched or reduced incompatibility could potentially decrease or eliminate the need for immunosuppressive drugs and/or immunomodulatory protocols (e.g. O negative lines for the generation of universal RBCs). Inasmuch as iPSCs could potentially be produced from a patient's own cells, they carry enormous potential as an alternative source of stem cells for treating human diseases related to many different organ systems, and could eliminate tissue incompatibility issues altogether.

References

Avecilla, S. T., Hattori, K., Heissig, B., Tejada, R., Liao, F., Shido, K., Jin, D. K., et al. (2004). Chemokine-mediated interaction of hematopoietic progenitors with the bone marrow vascular niche is required for thrombopoiesis. *Nat. Med., 10*, 64–71.

Brotherton, T. W., Chui, D. H., Gauldie, J., & Patterson, M. (1979). Hemoglobin ontogeny during normal mouse fetal development. *Proc. Natl. Acad. Sci. U.S.A., 76*, 2853–2857.

Broudy, V. C., & Lin, N. L. (2004). AMG531 stimulates megakaryopoiesis *in vitro* by binding to Mpl. *Cytokine, 25*, 52–60.

Bruno, S., Gunetti, M., Gammaitoni, L., Dane, A., Cavalloni, G., Sanavio, F., et al. (2003). *In vitro* and *in vivo* megakaryocyte differentiation of fresh and ex-vivo expanded cord blood cells: rapid and transient megakaryocyte reconstitution. *Haematologica, 88*, 379–387.

Chang, K. H., Nelson, A. M., Cao, H., Wang, L., Nakamoto, B., Ware, C. B., et al. (2006). Definitive-like erythroid cells derived from human embryonic stem cells coexpress high levels of embryonic and fetal globins with little or no adult globin. *Blood, 108*, 1515–1523.

Chasis, J. A., Prenant, M., Leung, A., & Mohandas, N. (1989). Membrane assembly and remodeling during reticulocyte maturation. *Blood, 74*, 1112–1120.

Choi, E. S., Nichol, J. L., Hokom, M. M., Hornkohl, A. C., & Hunt, P. (1995). Platelets generated *in vitro* from proplatelet-displaying human megakaryocytes are functional. *Blood, 85*, 402–413.

Choi, K., Yu, J., Smuga-Otto, K., Salvagiotto, G., Rehrauer, W., Vodyanik, M. A., et al. (2009). Hematopoietic and endothelial differentiation of human induced pluripotent stem cells. *Stem Cells, 27*, 559–567.

Clarke, M. C., Savill, J., Jones, D. B., Noble, B. S., & Brown, S. B. (2003). Compartmentalized megakaryocyte death generates functional platelets committed to caspase-independent death. *J. Cell Biol., 160*, 577–587.

Corwin, H. L., Parsonnet, K. C., & Gettinger, A. (1995). RBC transfusion in the ICU. Is there a reason? *Chest, 108*, 767–771.

Corwin, H. L., Gettinger, A., Pearl, R. G., Fink, M. P., Levy, M. M., Abraham, E., et al. (2004). The CRIT Study: anemia and blood transfusion in the critically ill – current clinical practice in the United States. *Crit. Care Med., 32*, 39–52.

Deutsch, V. R., & Tomer, A. (2006). Megakaryocyte development and platelet production. *Br. J. Haematol., 134*, 453–466.

Eslin, D. E., Zhang, C., Samuels, K. J., Rauova, L., Zhai, L., Niewiarowski, S., et al. (2004). Transgenic mice studies demonstrate a role for platelet factor 4 in thrombosis: dissociation between anticoagulant and antithrombotic effect of heparin. *Blood, 104,* 3173–3180.

Eto, K., Murphy, R., Kerrigan, S. W., Bertoni, A., Stuhlmann, H., Nakano, T., et al. (2002). Megakaryocytes derived from embryonic stem cells implicate CalDAG-GEFI in integrin signaling. *Proc. Natl. Acad. Sci. U.S.A., 99,* 12819–12824.

Feng, Q., Lu, S.-J., Klimanskaya, I., Gomes, I., Kim, D., Chung, Y., et al. (2010). Hemangioblastic derivatives from human induced pluripotent stem cells exhibit limited expansion and early senescence. *Stem Cells, 28,* 704–712.

Fraser, S. T., Isern, J., & Baron, M. H. (2007). Maturation and enucleation of primitive erythroblasts during mouse embryogenesis is accompanied by changes in cell-surface antigen expression. *Blood, 109,* 343–352.

Fujimoto, T. T., Kohata, S., Suzuki, H., Miyazaki, H., & Fujimura, K. (2003). Production of functional platelets by differentiated embryonic stem (ES) cells *in vitro. Blood, 102,* 4044–4051.

Fujiwara, Y., Browne, C. P., Cunniff, K., Goff, S. C., & Orkin, S. H. (1996). Arrested development of embryonic red cell precursors in mouse embryos lacking transcription factor GATA-1. *Proc. Natl. Acad. Sci. U.S.A., 93,* 12355–12358.

Fukushima-Shintani, M., Suzuki, K., Iwatsuki, Y., Abe, M., Sugasawa, K., Hirayama, F., et al. (2008). AKR-501 (YM477) in combination with thrombopoietin enhances human megakaryocytopoiesis. *Exp. Hematol., 36,* 1337–1342.

Gaines, P., Geiger, J. N., Knudsen, G., Seshasayee, D., & Wojchowski, D. M. (2000). GATA-1- and FOG-dependent activation of megakaryocytic alpha IIB gene expression. *J. Biol. Chem., 275,* 34114–34121.

Gaur, M., Kamata, T., Wang, S., Moran, B., Shattil, S. J., & Leavitt, A. D. (2006). Megakaryocytes derived from human embryonic stem cells: a genetically tractable system to study megakaryocytopoiesis and integrin function. *J. Thromb. Haemost, 4,* 436–442.

Giarratana, M. C., Kobari, L., Lapillonne, H., Chalmers, D., Kiger, L., Cynober, T., et al. (2005). Ex vivo generation of fully mature human red blood cells from hematopoietic stem cells. *Nat. Biotechnol., 23,* 69–74.

Goldstein, J., Siviglia, G., Hurst, R., Lenny, L., & Reich, L. (1982). Group B erythrocytes enzymatically converted to group O survive normally in A, B, and O individuals. *Science, 215,* 168–170.

Gordon, M. S., & Hoffman, R. (1992). Growth factors affecting human thrombocytopoiesis: potential agents for the treatment of thrombocytopenia. *Blood, 80,* 302–307.

Hanspal, M., & Hanspal, J. S. (1994). The association of erythroblasts with macrophages promotes erythroid proliferation and maturation: a 30-kD heparin-binding protein is involved in this contact. *Blood, 84,* 3494–3504.

Hanspal, M., Smockova, Y., & Uong, Q. (1998). Molecular identification and functional characterization of a novel protein that mediates the attachment of erythroblasts to macrophages. *Blood, 92,* 2940–2950.

Hod, E., & Schwartz, J. (2008). Platelet transfusion refractoriness. *Br. J. Haematol., 142,* 348–360.

Ji, P., Jayapal, S. R., & Lodish, H. F. (2008). Enucleation of cultured mouse fetal erythroblasts requires Rac GTPases and mDia2. *Nat. Cell Biol., 10,* 314–321.

Junt, T., Schulze, H., Chen, Z., Massberg, S., Goerge, T., Krueger, A., et al. (2007). Dynamic visualization of thrombopoiesis within bone marrow. *Science, 317,* 1767–1770.

Kaji, K., Norrby, K., Paca, A., Mileikovsky, M., Mohseni, P., & Woltjen, K. (2009). Virus-free induction of pluripotency and subsequent excision of reprogramming factors. *Nature, 458,* 771–775.

Kaufman, D. S., Hanson, E. T., Lewis, R. L., Auerbach, R., & Thomson, J. A. (2001). Hematopoietic colony-forming cells derived from human embryonic stem cells. *Proc. Natl. Acad. Sci. U.S.A., 98,* 10716–10721.

Kaushansky, K. (2008). Historical review: megakaryopoiesis and thrombopoiesis. *Blood, 111,* 981–986.

Kim, D., Kim, C. H., Moon, J. I., Chung, Y. G., Chang, M. Y., Han, B. S., Ko, S., et al. (2009). Generation of human induced pluripotent stem cells by direct delivery of reprogramming proteins. *Cell Stem Cell, 4,* 472–476.

Kingsley, P. D., Malik, J., Fantauzzo, K. A., & Palis, J. (2004). Yolk sac-derived primitive erythroblasts enucleate during mammalian embryogenesis. *Blood, 104,* 19–25.

Kosaki, G. (2005). *In vivo* platelet production from mature megakaryocytes: does platelet release occur via proplatelets? *Int. J. Hematol., 81,* 208–219.

Kovach, J. S., Marks, P. A., Russell, E. S., & Epler, H. (1967). Erythroid cell development in fetal mice: ultrastructural characteristics and hemoglobin synthesis. *J. Mol. Biol., 25,* 131–142.

Kowalska, M. A., Ratajczak, J., Hoxie, J., Brass, L. F., Gewirtz, A., Poncz, M., et al. (1999). Megakaryocyte precursors, megakaryocytes and platelets express the HIV co-receptor CXCR4 on their surface: determination of response to stromal-derived factor-1 by megakaryocytes and platelets. *Br. J. Haematol., 104,* 220–229.

Lane, W. J., Dias, S., Hattori, K., Heissig, B., Choy, M., Rabbany, S. Y., et al. (2000). Stromal-derived factor 1-induced megakaryocyte migration and platelet production is dependent on matrix metalloproteinases. *Blood, 96,* 4152–4159.

Leberbauer, C., Boulme, F., Unfried, G., Huber, J., Beug, H., & Mullner, E. W. (2005). Different steroids co-regulate long-term expansion versus terminal differentiation in primary human erythroid progenitors. *Blood, 105*, 85–94.

Lengerke, C., Grauer, M., Niebuhr, N. I., Riedt, T., Kanz, L., Park, I. H., et al. (2009). Hematopoietic development from human induced pluripotent stem cells. *Ann. NY Acad. Sci., 1176*, 219–227.

Li, F., Lu, S-J., Feng, Q., & Lanza, R. (2009). Large scale generation of functional megakaryocytes from human embryonic stem cells (hESCs) under stromal-free conditions. *Blood, 114*, 2540 (ab).

Littenberg, B., Corwin, H., Gettinger, A., Leichter, J., & Aubuchon, J. (1995). A practice guideline and decision aid for blood transfusion. *Immunohematology, 11*, 88–94.

Liu, Q. P., Sulzenbacher, G., Yuan, H., Bennett, E. P., Pietz, G., Saunders, K., et al. (2007). Bacterial glycosidases for the production of universal red blood cells. *Nat. Biotechnol., 25*, 454–464.

Long, M. W. (1998). Megakaryocyte differentiation events. *Semin. Hematol., 35*, 192–199.

Lordier, L., Jalil, A., Aurade, F., Larbret, F., Larghero, J., Debili, N., et al. (2008). Megakaryocyte endomitosis is a failure of late cytokinesis related to defects in the contractile ring and Rho/Rock signaling. *Blood, 112*, 3164–3174.

Lu, S.-J., Li, F., Vida, L., & Honig, G. R. (2004). CD34+CD38- hematopoietic precursors derived from human embryonic stem cells exhibit an embryonic gene expression pattern. *Blood, 103*, 4134–4141.

Lu, S. J., Feng, Q., Caballero, S., Chen, Y., Moore, M. A., Grant, M. B., et al. (2007). Generation of functional hemangioblasts from human embryonic stem cells. *Nat. Methods, 4*, 501–509.

Lu, S. J., Feng, Q., Park, J. S., Vida, L., Lee, B. S., Strausbauch, M., et al. (2008a). Biologic properties and enucleation of red blood cells from human embryonic stem cells. *Blood, 112*, 4475–4484.

Lu, S. J., Luo, C., Holton, K., Feng, Q., Ivanova, Y., & Lanza, R. (2008b). Robust generation of hemangioblastic progenitors from human embryonic stem cells. *Regen. Med., 3*, 693–704.

Ma, F., Ebihara, Y., Umeda, K., Sakai, H., Hanada, S., Zhang, H., et al. (2008). Generation of functional erythrocytes from human embryonic stem cell-derived definitive hematopoiesis. *Proc. Natl. Acad. Sci. U.S.A., 105*, 13087–13092.

Majka, M., Janowska-Wieczorek, A., Ratajczak, J., Kowalska, M. A., Vilaire, G., Pan, Z. K., et al. (2000). Stromal-derived factor 1 and thrombopoietin regulate distinct aspects of human megakaryopoiesis. *Blood, 96*, 4142–4151.

Matsunaga, T., Tanaka, I., Kobune, M., Kawano, Y., Tanaka, M., Kuribayashi, K., et al. (2006). Ex vivo large-scale generation of human platelets from cord blood CD34+ cells. *Stem Cells, 24*, 2877–2887.

Miharada, K., Hiroyama, T., Sudo, K., Nagasawa, T., & Nakamura, Y. (2006). Efficient enucleation of erythroblasts differentiated *in vitro* from hematopoietic stem and progenitor cells. *Nat. Biotechnol., 24*, 1255–1256.

Ney, P. A. (2006). Gene expression during terminal erythroid differentiation. *Curr. Opin. Hematol., 13*, 203–208.

Nishikii, H., Eto, K., Tamura, N., Hattori, K., Heissig, B., Kanaji, T., et al. (2008). Metalloproteinase regulation improves *in vitro* generation of efficacious platelets from mouse embryonic stem cells. *J. Exp. Med., 205*, 1917–1927.

Norol, F., Vitrat, N., Cramer, E., Guichard, J., Burstein, S. A., Vainchenker, W., et al. (1998). Effects of cytokines on platelet production from blood and marrow CD34+ cells. *Blood, 91*, 830–843.

Olivier, E. N., Qiu, C., Velho, M., Hirsch, R. E., & Bouhassira, E. E. (2006). Large-scale production of embryonic red blood cells from human embryonic stem cells. *Exp. Hematol., 34*, 1635–1642.

Olsson, M. L., & Clausen, H. (2008). Modifying the red cell surface: towards an ABO-universal blood supply. *Br. J. Haematol., 140*, 3–12.

Pang, L., Weiss, M. J., & Poncz, M. (2005). Megakaryocyte biology and related disorders. *J. Clin. Invest., 115*, 3332–3338.

Patel, S. R., Hartwig, J. H., & Italiano, J. E., Jr. (2005). The biogenesis of platelets from megakaryocyte proplatelets. *J. Clin. Invest., 115*, 3348–3354.

Qiu, C., Hanson, E., Olivier, E., Inada, M., Kaufman, D. S., Gupta, S., et al. (2005). Differentiation of human embryonic stem cells into hematopoietic cells by coculture with human fetal liver cells recapitulates the globin switch that occurs early in development. *Exp. Hematol., 33*, 1450–1458.

Schmitt, A., Guichard, J., Masse, J. M., Debili, N., & Cramer, E. M. (2001). Of mice and men: comparison of the ultrastructure of megakaryocytes and platelets. *Exp. Hematol., 29*, 1295–1302.

Shivdasani, R. A., Fujiwara, Y., McDevitt, M. A., & Orkin, S. H. (1997). A lineage-selective knockout establishes the critical role of transcription factor GATA-1 in megakaryocyte growth and platelet development. *EMBO J., 16*, 3965–3973.

Slichter, S. J., Davis, K., Enright, H., Braine, H., Gernsheimer, T., Kao, K. J., et al. (2005). Factors affecting post-transfusion platelet increments, platelet refractoriness, and platelet transfusion intervals in thrombocytopenic patients. *Blood, 105*, 4106–4114.

Soni, S., Bala, S., Gwynn, B., Sahr, K. E., Peters, L. L., & Hanspal, M. (2006). Absence of erythroblast macrophage protein (Emp) leads to failure of erythroblast nuclear extrusion. *J. Biol. Chem., 281*, 20181–20189.

Stamatoyannopoulos, G. (2005). Control of globin gene expression during development and erythroid differentiation. *Exp. Hematol., 33*, 259–271.

Sullenbarger, B., Bahng, J. H., Gruner, R., Kotov, N., & Lasky, L. C. (2009). Prolonged continuous *in vitro* human platelet production using three-dimensional scaffolds. *Exp. Hematol., 37*, 101–110.

Takahashi, K., Tanabe, K., Ohnuki, M., Narita, M., Ichisaka, T., Tomoda, K., et al. (2007). Induction of pluripotent stem cells from adult human fibroblasts by defined factors. *Cell, 131*, 861–872.

Takayama, N., Nishikii, H., Usui, J., Tsukui, H., Sawaguchi, A., Hiroyama, T., et al. (2008). Generation of functional platelets from human embryonic stem cells *in vitro* via ES-sacs, VEGF-promoted structures that concentrate hematopoietic progenitors. *Blood, 111*, 5298–5306.

Tomer, A. (2004). Human marrow megakaryocyte differentiation: multiparameter correlative analysis identifies von Willebrand factor as a sensitive and distinctive marker for early (2N and 4N) megakaryocytes. *Blood, 104*, 2722–2727.

Ungerer, M., Peluso, M., Gillitzer, A., Massberg, S., Heinzmann, U., Schulz, C., Munch, G., et al. (2004). Generation of functional culture-derived platelets from CD34+ progenitor cells to study transgenes in the platelet environment. *Circ. Res., 95*, e36–e44.

Vodyanik, M. A., Bork, J. A., Thomson, J. A., & Slukvin, I. I. (2005). Human embryonic stem cell-derived CD34+ cells: efficient production in the coculture with OP9 stromal cells and analysis of lymphohematopoietic potential. *Blood, 105*, 617–626.

Wang, L., Menendez, P., Shojaei, F., Li, L., Mazurier, F., Dick, J. E., et al. (2005). Generation of hematopoietic repopulating cells from human embryonic stem cells independent of ectopic HOXB4 expression. *J. Exp. Med., 201*, 1603–1614.

Wang, X., Crispino, J. D., Letting, D. L., Nakazawa, M., Poncz, M., & Blobel, G. A. (2002). Control of megakaryocyte-specific gene expression by GATA-1 and FOG-1: role of Ets transcription factors. *EMBO J., 21*, 5225–5234.

Whitaker, B. I., & Henry, R. (2005). *2005 Nationwide Blood Collection and Utilization Survey Report*. Washington, DC: National Blood Data Resource Center, US Department of Health and Human Services.

Woltjen, K., Michael, I. P., Mohseni, P., Desai, R., Mileikovsky, M., Hamalainen, R., et al. (2009). piggyBac transposition reprograms fibroblasts to induced pluripotent stem cells. *Nature, 458*, 766–770.

Ye, Z., Zhan, H., Mali, P., Dowey, S., Williams, D. M., Jang, Y. Y., et al. (2009). Human induced pluripotent stem cells from blood cells of healthy donors and patients with acquired blood disorders. *Blood, 114*, 5473–5480.

Yu, J., Vodyanik, M. A., Smuga-Otto, K., Antosiewicz-Bourget, J., Frane, J. L., Tian, S., et al. (2007). Induced pluripotent stem cell lines derived from human somatic cells. *Science, 318*, 1917–1920.

Yu, J., Hu, J., Smuga-Otto, K., Tian, S., Stewart, R., Slukvin, I. I., et al. (2009). Human induced pluripotent stem cells free of vector and transgene sequences. *Science, 324*, 797–801.

Zhan, X., Dravid, G., Ye, Z., Hammond, H., Shamblott, M., Gearhart, J., et al. (2004). Functional antigen-presenting leucocytes derived from human embryonic stem cells *in vitro*. *Lancet, 364*, 163–171.

CHAPTER 42

Articular Cartilage

Lily Jeng[*,†]**, Francois Ng kee Kwong**[**]**, Myron Spector**[*,‡]

* Tissue Engineering, VA Boston Healthcare System, Boston, MA, USA
† Department of Biological Engineering, Massachusetts Institute of Technology, Cambridge, MA, USA
‡ Department of Orthopaedic Surgery, Brigham and Women's Hospital, Harvard Medical School, Boston, MA, USA
** Histopathology Department, Cambridge University Hospitals NHS Foundation Trust, Cambridge, UK

INTRODUCTION

The treatment of lesions in articular cartilage has benefited greatly from regenerative medicine strategies. While the complete regeneration of articular cartilage has not yet been achieved, such treatments have been shown to provide meaningful symptomatic relief at least for a time. In some cases, such treatments have enabled patients to return to an active lifestyle, even returning to professional athletics. In other cases with extensive cartilage degeneration, these treatments may delay the need for total joint arthroplasty. The promising results of the many new biomaterials, cell sources, and regulatory molecules being investigated in pre-clinical animal models will likely result in the production of more robust reparative tissue in articular cartilage defects that better meets the functional demands placed on joints.

TYPES OF ARTICULAR CARTILAGE DEFECTS THAT PRESENT IN THE CLINIC

Cartilage defects are a common source of pain and/or loss of function in patients presenting to the orthopedic clinic. While any joint can be affected, the joint most commonly affected is by far the knee. A chondral lesion was found in 63% of a large series of over 31,000 arthroscopic procedures performed in patients with a symptomatic knee (Curl et al., 1997). Articular cartilage damage is often associated with meniscal and anterior cruciate ligament injuries (Shelbourne et al., 2003).

These defects can be divided according to their etiology or morphology. Focal injuries typically occur as a result of a sporting injury and hence tend to affect the younger population. Focal defects can be further subdivided into chondral or osteochondral lesions, depending on the depth of the defect. Chondral lesions, also known as partial thickness lesions, lie entirely within the cartilage and do not penetrate into the subchondral bone. In the adult, defects of this nature do not regenerate because of the lack of cells that could participate in the repair process. Osteochondral defects penetrate through the vascularized subchondral bone and some spontaneous repair occurs as mesenchymal chondroprogenitor cells invade the lesion and form cartilage. However, full-thickness defect repair is only transient and the novel tissue formed does not have the functional properties of native hyaline cartilage (Shapiro et al., 1993).

On the other hand, degenerative chondral changes typically occur in the older population as a result of arthritic changes. They often involve a large area of the affected joint, but start off as a focal lesion initially.

761

Principles of Regenerative Medicine. DOI: 10.1016/B978-0-12-381422-7.10042-2

RATIONALE FOR CELL THERAPY

Articular cartilage has a limited capacity for self-regeneration after injury. This was recognized as early as in 1743 by Hunter who stated that cartilage "once destroyed is not repaired." This is because none of the normal inflammatory and reparative processes of the body are available to repair the tissue. This itself is a result of its isolation from the systemic regulation, and lack of blood vessels and nerve supply (Mankin, 1982).

Furthermore, chondrocytes that are surrounded by an extracellular matrix cannot freely migrate to the site of injury from an intact healthy site, unlike most tissues (Buckwalter and Mankin, 1998), and there is no provisional fibrin clot filling the defect into which cells can migrate. Full-thickness defects induce mesenchymal chondroprogenitor cells to differentiate into repair tissue, but this is predominantly fibrous in nature and degenerates with time.

The two major problems that need to be addressed in repair of articular cartilage are the filling of the defect void with a tissue that has the same mechanical properties as articular cartilage and the promotion of successful integration between the repair tissue and the native articular cartilage and calcified cartilage. Even a small defect caused by mechanical damage will fail to heal and degenerate over time, progressing to osteoarthritis (OA).

Conventional surgical techniques of cartilage repair are partially successful in alleviating symptoms, but fail to regenerate tissue anywhere similar in nature to native articular cartilage. There was no promising solution to this problem until Brittberg et al. introduced a cell-based therapy in which culture-expanded chondrocytes were transplanted into defects, raising the expectations of a breakthrough in repairing damaged articular cartilage (Brittberg et al., 1994). While there are other cartilage repair methods being researched, this chapter will focus on cell therapies that could be conjoined with tissue engineering strategies.

762

CURRENT CELL THERAPIES AVAILABLE IN THE CLINIC

These techniques are generally aimed at delivering chondrogenic cells to the cartilage defect, in the form of tissues containing precursor cells (e.g. the periosteum or perichondrium); chondrocytes isolated from a biopsy of healthy cartilage and expanded in number *in vitro*; or marrow-derived chondroprogenitor cells provided with access to the defect through small holes perforating the subchondral bone plate (viz., microfracture).

Periosteal transplantation

Rubak initially described this technique in a rabbit model of cartilage defect (Rubak, 1982). He used a periosteal flap to cover the defects. The defects were repaired and filled after 4 weeks with a hyaline-like cartilage whereas the empty control defect showed fibrocartilage-like repair tissue. The first clinical study was published by Niedermann et al., who reported successful results in all of their four initially treated patients (Niedermann et al., 1985).

Perichondrial transplantation

Autologous perichondrium has also been employed for cartilage repair (Homminga et al., 1989, 1990, 1991). Perichondrium, taken from the cartilaginous covering of the rib, is placed into the chondral defect of the affected joint. The first clinical study of this approach was performed by Homminga et al. (1990). A major shortcoming of perichondrial grafting is the limited availability of large grafts. Graft size is limited to the rib size, so that several rib perichondrial grafts have to be harvested to fill a large defect. Additionally, endochondral ossification and delamination of the cartilage from the subchondral bone plate are potentially significant limitations to the long-term efficacy of this repair.

Autologous chondrocyte implantation (ACI)

Since first published in 1994 (Brittberg et al., 1994), techniques of cell isolation, expansion in culture, and implantation have remained essentially the same. Cartilage (150–300 mg) is harvested arthroscopically from a minimally load-bearing area of the upper aspect or the medial condyle of the affected knee. The biopsy is then transported to a laboratory facility using a transport media. Chondrocytes are isolated using standard techniques. After a certain period of cell expansion (11–21 days (Peterson et al., 2000), depending upon the growth kinetics) a certain number of cells (e.g. minimally 12 million for Genzyme's Carticel procedure) are provided in a serum-free and gentamycin-free transport medium.

Using a medial or lateral parapatellar incision, the defect is debrided to the level of normal-appearing surrounding cartilage. The integrity of the tidemark needs to be maintained in order to avoid infiltration of undifferentiated mesenchymal stem cells (MSCs), which could contribute to the formation of fibrocartilagenous repair tissue (Brittberg, 1999). A periosteal flap is harvested from the anterior aspect of the proximal tibia or distal femur, formed to the shape of the lesion, and sutured to the rim of the defect. The chondrocyte suspension is subsequently injected under the periosteal flap and the border of periosteal cover sealed using fibrin glue. Postoperative rehabilitation protocols generally involve continuous passive motion and limited weight bearing for an extended time. Cooperation of the patient in this respect is essential for a favorable outcome and is difficult to control. This contributes to difficulty in evaluating outcome data.

There are several second-generation variations of this technique. In one variation, an off-the-shelf porcine type I/III collagen membrane has been used in place of the periosteal membrane (Bartlett et al., 2005). Its outer surface is smooth, facilitating a low-friction surface. Its inner surface, which is porous due to large gaps among collagen fibers, can accommodate the seeding of cultured chondrocytes, in a procedure referred to as "matrix-assisted autologous chondrocyte implantation" (MACI). In other work, a hyaluronan-based sponge-like scaffold seeded with autologous chondrocytes has been used to fill a cartilage defect site (Manfredini et al., 2007). Both studies found comparable short-term outcomes among the various methods.

Microfracture

An array of marrow-stimulation techniques, including abrasion arthroplasty, drilling, and microfracture, have been used to treat cartilage defects. Microfracture, a technique commonly used in the clinic, involves the creation of small holes in the underlying bone to allow blood, bone marrow, and marrow-derived MSCs into the cartilage defect site (Rodrigo et al., 1994). More recently, a collagen membrane has been applied over microfracture-treated defects in order to contain the blood clot and marrow-derived MSCs, in the procedure referred to as "autologous matrix-induced chondrogenesis" (AMIC).

CELL THERAPIES

An optimal cell source should have the following characteristics: no immunorejection, no tumorigenicity, immediate availability, availability in pertinent quantities, controlled cellular proliferation rate, predictable, and consistent chondrogenic potential as well as controlled integration into the surrounding tissues.

Autologous versus allogeneic

An autologous source of stem cells is most desirable as cells are collected from each patient, thereby eliminating complications associated with immune rejection of allogeneic tissue. Even with an autologous system, challenges exist in ensuring a safe and reproducible product. Genzyme established a quality assurance program based on US FDA Good Manufacturing Practice regulations, which was reviewed in Mayhew et al. (1998).

Limitations of the autologous approach in obtaining stem cells and the desire to obtain "marketable products" that could benefit as many patients as possible have provided incentives for the development of generic cell lines, which can be taken off the shelf as, and when, needed for patient treatment. These universal cells would have the following advantages: (1) availability through the development of large cell banks, (2) consistency and efficacy because only cells with desirable characteristics and controlled critical parameters are selected and amplified, and (3) sterility and assurance of compatibility through extensive safety testing.

Until recently, it was difficult to envision utilization of allogeneic generic cells in orthopedics as it was believed that their transplantation would require immunosuppressive drugs to reduce associated risks of rejection. However, cultured MSCs have been recently shown to exhibit a poorly immunogenic phenotype (Tse et al., 2003). *In vivo*, a single intravenous administration of MSCs led to a modest, but significant, prolongation of skin graft survival (Bartholomew et al., 2002). These data have greatly enhanced the therapeutic appeal of MSCs because they raised the possibility of creating universal cell lines. Indeed, allogeneic adult stem cells are already being investigated in patients with meniscal injuries, in an FDA-approved clinical trial (http://www.osiristx.com/).

Intra-operative versus culture expanded

Intra-operative cell-based therapies have the advantage of being less time consuming and less costly than *ex vivo* therapies. *Ex vivo* therapies also have the disadvantage of involving an additional harvesting step. The advantage of an *ex vivo* technique is that the surgeon can select specific cells (i.e. bone marrow cells or stem cells) and the cellular delivery vehicle for specific clinical problems. It is also safer than an *in vivo* strategy when working with viruses for gene therapy because no viral particles or DNA complexes are injected directly into the body. In addition, *ex vivo* strategies have a high efficiency of cell transduction.

Chondrocytes

METHODS FOR INTRA-OPERATIVE CELL THERAPY

The minced cartilage technique is a recent development, and not yet in use in the clinic (McCormick et al., 2008). For this procedure, autologous cartilage tissue is minced into small pieces, placed on a scaffold carrier, and applied to the defect site. By mincing the tissue and spreading it evenly on the scaffold, a larger surface area compared to ACI can be created and implanted for cartilage repair in the same surgical session. Preclinical studies suggest promising results; Lu et al. found that chondral defects in goats treated with cartilage fragments on a resorbable scaffold revealed hyaline-like tissue after 6 months (Lu et al., 2006), and Frisbie et al. observed superior results with minced cartilage-treated defects in horses compared to ACI-treated defects at 12 months (Frisbie et al., 2009).

CULTURE EXPANSION

Cell source

Chondrocytes comprise the single cellular component of adult hyaline cartilage and are considered to be terminally differentiated, thus being highly specialized. In articular cartilage, their main function is to maintain the cartilage matrix, synthesizing types II, IX, and XI collagen; the large aggregating proteoglycan, aggrecan (which consists of glycosaminoglycans (GAG) attached to a protein core); the smaller proteoglycans, biglycan and decorin; and specific and non-specific matrix proteins that are expressed at defined stages during growth and development. This makes them a suitable cell type for a cell-based treatment of chondral defects. Chondrocytes, in addition to cells from other musculoskeletal tissues, have also been shown to synthesize lubricin (Schumacher et al., 1994). Lubricin is an important glycoprotein in the joint, having been found to act as boundary lubrication at the surfaces of articular cartilage (Schmidt et al., 2007) and to provide chondroprotective effects by dissipating strain energy resulting from loading (Jay et al., 2007). However, lubricin has not yet been extensively studied in tissue-engineered constructs.

In addition to articular chondrocytes, chondrocytes can be harvested from several other locations, including auricular, meniscal, nasoseptal, and costal (Panossian et al., 2001; van Osch et al., 2004; Isogai et al., 2006; Chung et al., 2008). Several *in vitro* and *in vivo* studies have compared chondrocytes from different sources. One study compared human articular and auricular chondrocytes in alginate hydrogel constructs and found that gene expression levels for types I and II collagen and aggrecan were comparable between the two (Malicev et al., 2009). Our lab examined goat articular, auricular, and meniscal chondrocytes in type I and II collagen scaffolds and noted increased histogenesis, with no elastin, in the auricular constructs and the presence of lubricin in the articular and auricular constructs (Zhang and Spector, 2009). Similarly, Xu et al. found that constructs made with auricular chondrocytes resulted in greater histogenesis compared to articular and costal chondrocytes (Xu et al., 2004). They also showed that the mechanical properties of the new cartilage matrix in the auricular chondrocyte constructs were superior to those found in the other two. In another study, pig articular, auricular, and costal chondrocytes were observed to produce neocartilaginous tissue that formed mechanically functional bonds with surrounding cartilage disks (Johnson et al., 2004). These findings suggest that chondrocytes from multiple sources may have potential for articular cartilage repair.

Culture methods

Chondrocytes can be isolated from cartilage by digesting the extracellular matrix. The cells are allowed to attach to a flat surface as a monolayer and cultured using standard cell culture techniques. Several soluble regulators have been shown to aid in the induction and development of cartilage formation, including fibroblast growth factor (FGF)-2, insulin-like growth factor (IGF)-1, and transforming growth factor (TGF)-β, and these molecules are often included in the culture medium. Freshly isolated articular chondrocytes exhibit their specific phenotype in culture for at least several days to weeks. However, chondrocytes are available in very limited quantities, comprising only about 2% of the tissue by volume, and they do not readily proliferate *in vitro*. Cells from a younger population have been found to undergo 0.3 doublings per day, using a standardized and validated approach for culturing cells for later implantation (Mayhew et al., 1998). Even lower proliferation rates are obtained in older patients and arthritic cartilage (Peterson et al., 2000). Another report demonstrates the rapid replicative senescence of articular chondrocytes (Martin and Buckwalter, 2001).

While the steps involved in the isolation and expansion of articular chondrocytes for autologous chondrocyte implantation (ACI) are quite similar among various commercial and academic laboratories, there may be important differences. One such difference is the use of the patient's own serum for culturing the cells, as described originally by Brittberg et al. (1994). One commercial enterprise, Genzyme Biosurgery (Cambridge, Massachusetts, USA), uses approved and validated fetal bovine serum (FBS), instead of the patient's serum, in the culture media. Another potentially important difference is that Genzyme needs to freeze and store the isolated cells in order to allow for verification of adequate insurance coverage prior to the implantation procedure. A recent study has indicated that this freeze-thaw cycle may adversely affect the outcome of the procedure (Perka et al., 2000). Cryopreserved chondrocytes seeded into polymer scaffolds yielded an 85% repair of an osteochondral defect in rabbits, whereas 100% of the defects treated with non-cryopreserved cells were filled.

One of the indirect benefits of culturing cells *ex vivo* is that it offers a model system in which to make discoveries and study the effects of various factors that might not otherwise be explored *in vivo*. For example, our lab noted substantial chondrocyte-mediated contraction of cell-seeded scaffolds cultured *in vitro*, an observation that may not have been found in an *in vivo* system, and investigation into this phenomenon revealed that chondrocytes expressed

765

α-smooth muscle actin, an isoform responsible for contraction of smooth muscle cells and myofibroblasts (Lee et al., 2000a). *Ex vivo* culture also allows for investigation of the effects of substrate mechanical properties on chondrogenesis (Lee et al., 2001; Bryant et al., 2004; Engler et al., 2006). In one such study, the degree of crosslinking and time in culture of chondrocyte-seeded scaffolds were varied, and a modest correlation between equilibrium modulus and GAG density of the constructs was noted, with a linear coefficient of determination of 0.6 (Pfeiffer et al., 2008). *In vitro* studies involving the culturing of cells may serve as a basis for future tissue-engineered constructs.

One of the major disadvantages of culturing articular chondrocytes in monolayer is that they tend to dedifferentiate. Once chondrocytes are deprived of their three-dimensional environment, their phenotype switches to a more fibroblastic cell form, expressing types I and III collagen, instead of cartilage-specific type II collagen (Goldring et al., 1986, 1988; Saadeh et al., 1999). Also, our own studies demonstrated in a canine model that the harvesting of articular cartilage predisposes the other cartilage in the same joint to changes associated with early OA (Lee et al., 2000b). While the lesion itself in a knee joint may serve to induce such osteoarthritic changes in the joint, the additional surgical procedure of harvesting cartilage may exacerbate the condition. There is, then, a compelling need for an alternative cell source for a cell-based cartilage repair procedure.

Mesenchymal stem cells

MSCs isolated from the bone marrow and other sources can provide an alternative and abundant supply of cells for cartilage repair procedures. Adult marrow stromal cells are being investigated for the treatment of defects in connective tissues using cell and gene therapy and tissue-engineering approaches (see for reviews Caplan, 1991; Prockop, 1997). Differentiation of such cells can be obtained *in vitro* by changing the culture conditions after their expansion or *in vivo* as a consequence of the new "physiological" microenvironment in the transplant area.

WHOLE MARROW IMPLANTS
Safety of whole marrow injected/implanted in human subjects

Whole autologous and allogeneic bone marrow has been injected and implanted into human subjects for decades to treat myriad medical problems with no adverse events associated with the MSC subpopulation present. Of note, for example, is the procedure in which up to 1 liter of whole bone marrow is routinely infused into the bone marrow transplant patient. This infusion contains a small but significant proportion of MSCs and does not seem to have any significant side-effects.

Marrow-stimulation techniques, such as microfracture, have been used for decades to introduce marrow-derived MSCs into the joint. While these procedures have not yielded lasting symptomatic relief, they demonstrate that the presence of endogenous bone marrow-derived MSCs in the joint does not lead to adverse clinical sequelae.

Since the early days of bone grafting, autogenous marrow has been known to be of value in improving the osteogenic response (Salama et al., 1973). Whole autogenous marrow has been implanted in various sites in the body with no untoward clinical findings. In more recent years, bone marrow and bone marrow fractions including the stromal cell population have been injected percutaneously to treat non-unions in human subjects (Connolly, 1998). There have been no adverse events reported.

Efficacy of Whole Marrow for Cartilage Repair in Clinical Trials

Clinical trials studying the use of whole marrow for cartilage repair are currently ongoing. Initial clinical results from one such study, by BioSyntech Canada Inc. (Laval, Quebec, Canada), found that treatment of microfracture-treated defects with chitosan-glycerol phosphate mixed with autologous whole blood resulted in statistically significant improvement of repair tissue compared to microfracture treatment alone at the 12-month follow-up period in 22 patients. Final results for all 80 patients enrolled in the study are expected later in 2010.

METHODS FOR CULTURE EXPANSION
Cell sources

Human MSCs are commonly obtained from bone marrow, but they have also been reported to be present in adipose tissue, synovium, periosteum, dermis, muscles, and peripheral blood (de Bari et al., 2001, 2003; Young et al., 2001). They have the potential to differentiate along different lineages including those forming bone, cartilage, fat, muscle, and nerve. Several studies have compared the chondrogenic potential of MSCs from these different sources, with varying results. One study found that synovium-derived MSCs had greater *in vitro* chondrogenic potential than those from bone marrow, periosteum, adipose, and muscle (Yoshimura et al., 2007), and another study found that synovium- and marrow-derived MSCs had greater *in vivo* chondrogenic potential than those from adipose tissue and muscle (Koga et al., 2008). However, the fact that there are many more safety and efficacy studies of marrow-derived MSCs compared to MSCs from other sources and the ease of procurement mean that bone marrow remains the popular choice.

Culture procedures

MSCs represent a minor fraction of the total nucleated cell population in the marrow. They can be plated and enriched using standard cell culture techniques. Frequently, the whole marrow sample is subjected to fractionation on a density gradient solution such as Ficoll, after which the cells are plated at densities ranging from 1×10^4 cells/cm^2 to 0.4×10^6 cells/cm^2 (Pittenger et al., 1999; Lodie et al., 2002; McBride et al., 2003). Cells are generally cultured in basal medium such as Dulbecco's modified Eagle's medium (low glucose) in the presence of 10% FBS (Pittenger et al., 1999). MSCs in culture have a fibroblastic morphology and adhere to the tissue culture substrate. Primary cultures are usually maintained for 12–16 days, during which time the non-adherent hematopoietic cell fraction is depleted. Optimal expansion of MSCs from marrow requires the pre-selection of FBS. As MSCs are expanded in large-scale culture for human applications, it will be important to identify defined growth media, without or with reduced FBS, to ensure more reproducible culture techniques and enhanced safety.

SAFETY OF MSCS IN CLINICAL TRIALS

Early clinical tests using bone marrow-derived MSCs in humans as a means of treating a variety of conditions have reported no adverse side-effects. For example, Osiris Therapeutics' double-blind safety trials examined a range of doses (0.5, 1.6, and 5 million cells/kg) of allogeneic human MSCs injected intravenously into 53 patients with myocardial infarction (Hare et al., 2009). The results showed that "adverse event rates were similar between the hMSC-treated (5.3 per patient) and placebo-treated (7.0 per patient) groups." Another clinical pilot study investigated the safety of injecting 1 million cells/kg of autologous MSCs via lumbar puncture to treat 30 patients with spinal cord injury (Pal et al., 2009). No serious adverse effects were reported at follow-up — 3 patients at 3 years, 10 patients at 2 years, and 10 patients at 1 year. One clinical study administered two applications of autologous MSCs into seven traumatic brain injury patients, 0.01–1 billion cells applied directly to the injured area during the primary operation followed by 0.1–10 billion additional cells infused intravenously, and no toxicity was noted during the 6-month follow-up period (Zhang et al., 2008).

EFFICACY OF MSCS FOR CARTILAGE REPAIR IN CLINICAL STUDIES

That MSCs may yield results comparable to autologous chondrocytes was supported by *in vitro* studies (Kavalkovich et al., 2002) that have demonstrated that MSC cultures undergoing chondrogenesis synthesize GAG at levels significantly higher than explant cultures or primary chondrocyte cultures. This strategic approach follows the supposition that regeneration can be facilitated by the recapitulation of certain phases of embryonic development, and that these stem cells will allow for such, whereas fully differentiated cells will not.

Numerous pre-clinical animal studies (Wakitani et al., 1994; Im et al., 2001; Zhou et al., 2004; Oshima et al., 2005; Yanai et al., 2005) have demonstrated that the reparative tissue formed in articular cartilage defects following MSC implantation shows marked improvement compared to controls, often appearing hyaline-like in nature and expressing type II collagen. However, there are only a few early clinical studies examining the efficacy of MSCs for cartilage repair applications.

In one study (Wakitani et al., 2002), 12 patients with osteoarthritis undergoing high tibial osteotomies were treated with MSCs. These cells in bone marrow aspirates were expanded *ex vivo*, embedded in collagen gel, transplanted into articular cartilage defects, and covered with autologous periosteum. Forty-two weeks after transplantation, the arthroscopic and histological grading score was better in the experimental group than in the cell-free control group, but no significant difference in clinical improvement was noted between the two groups.

The same group also examined healing of cartilage defects in the patellae of a 26-year-old female and a 44-year-old male (Wakitani et al., 2004b). Clinical symptoms (pain and walking ability) were improved in both patients during the 4- to 5-year follow-up period. Arthroscopic results 1–2 years after transplantation indicated that the defects were filled with fibrocartilage.

In another study (Wakitani et al., 2007), nine full-thickness cartilage defects in the patello-femoral joint of three patients were treated with autologous culture-expanded MSCs. "Six months after transplantation, the patients' clinical symptoms had improved and the improvements have been maintained over the follow-up periods (17–27 months). Histology of the first patient 12 months after the transplantation revealed that the defect had been repaired with the fibrocartilaginous tissue. Magnetic resonance imaging of the second patient 1 year after transplantation revealed complete coverage of the defect, but we were unable to determine whether or not the material that covered the defects was hyaline cartilage."

One study investigated the use of autologous MSCs in a 20 × 30-mm full-thickness cartilage defect of an athlete. Similarly to the other studies, MSCs were expanded *in vitro* for 4 weeks, embedded in a collagen gel, transplanted into the defect, and covered with periosteum. "Seven months after surgery, arthroscopy revealed the defect to be covered with smooth tissues. Histologically, the defect was filled with a hyaline-like type of cartilage tissue which stained positively with Safranin-O. One year after surgery, the clinical symptoms had improved significantly. The patient had reattained his previous activity level and experienced neither pain nor other complications."

Characterization of phenotype

IDENTIFICATION AND THERAPEUTIC USE OF THE ADHERENT CELL POPULATION FROM BONE MARROW

Numerous studies have investigated characteristics of the stromal cell population of marrow that includes the MSC (Barry and Murphy, 2004). Many of these studies have characterized the MSC on the basis of selected surface proteins (Barry et al., 1999, 2001; Reyes and Verfaillie, 2001; Young et al., 2001; Gronthos et al., 2003). Related studies have attempted to isolate more purified subpopulations of MSCs using cell sorting for selected surface markers, including CD105(+), CD45(−), GlyA(−) (Reyes and Verfaillie, 2001); endoglin (Majumdar et al., 2003); and Stro-1 (Gronthos and Simmons, 1995).

For the purpose of a cartilage repair therapeutic agent, there are no data that indicate that any specific subpopulation of MSCs would be safer and more efficacious than the entire adherent cell population.

Embryonic stem cells

Embryonic stem cells (ESCs) are less differentiated than MSCs, with the potential to generate all cell types in the body, and have the almost unlimited capacity to self-renew in culture. However, the use of these cells for cartilage applications is not as common as other cell types. Early research in this area has demonstrated the chondrogenic potential of ES cells (Kramer et al., 2006). The optimal culture conditions needed to differentiate ES cells down the chondrogenic lineage have not been identified. This area is actively being researched; for example, one study showed that TGF-β1, insulin, ascorbic acid, and BMP-2 — many of the same regulators used in the culture of chondrocytes and MSCs — induced chondrogenic differentiation of ES cells, as evidenced by the expression of collagen type IIB and aggrecan (zur Nieden et al., 2005). Other studies have found that co-culture of ES cells with articular chondrocytes enhances the chondrocytic commitment of the ES cells (Hwang et al., 2008; Bigdeli et al., 2009).

A few *in vivo* studies have shown that ES cells (Wakitani et al., 2004a; Dattena et al., 2009) and ES cell-derived chondrogenic cells (Fecek et al., 2008) implanted into animals resulted in improved healing and the formation of cartilaginous tissue. However, Wakitani et al. found teratoma formation in the knee joints of mice with severe combined immunodeficiency following transplantation of ES cells into the joints (Wakitani et al., 2003), highlighting important safety concerns regarding the use of ES cells. It also raises the question of whether ES cells should be pre-differentiated *in vitro* prior to transplantation. Clearly, much more basic research is needed into ES cells before their efficacy for cartilage tissue engineering can be evaluated.

CELL-SCAFFOLD IMPLANTS

Owing to the now well-established phenomenon of dedifferentiation of chondrocytes in monolayer culture, there has been increasing interest in three-dimensional systems of culture and delivery of cells to the chondral defect. These systems can provide an environment for growth more similar to native tissue and hence contribute to the phenotypic stability of the chondrocytes. Ideally, the material should be biodegradable and degrade at the same rate as new tissue is formed. A scaffold also provides an increased surface area for cell attachment. Choosing the right scaffold in cartilage repair requires consideration of a number of factors. The scaffold should:

1. Support cartilage-specific matrix production (collagen type II and aggrecan). Our previous studies showed that there is a considerable difference in performance among scaffolds, even if only changing the collagen type, pore size, or method of cross-linking.
2. Provide enough mechanical support for early mobilization of the treated joint.
3. Allow for migration of cells to achieve bonding to the adjacent host tissue.

The physical presentation of scaffolds can vary, such as porous sponge-like scaffolds or gel-like hydrogel scaffolds. In a comparison of several matrix materials (synthetic materials such as polylactic acid and naturally occurring polymers such as collagen gel and porous collagen), Grande et al. showed a marked variability of the chondrocyte response (Grande et al., 1997). Bioabsorbable polymers such as polyglycolic acid (PGA) enhanced proteoglycan synthesis, whereas collagen matrices stimulated synthesis of collagen.

Sponge-like scaffolds

Sponge-like scaffolds have been studied extensively for cartilage tissue engineering. Oliveira et al. compared two different synthetic scaffolds, a starch-polycaprolactone fiber mesh scaffold and a polyglycolic acid scaffold (Oliveira et al., 2007). The scaffolds were seeded with bovine chondrocytes and cultured for 6 weeks. The starch-polycaprolactone constructs revealed cells with chondrocyte morphology distributed throughout the scaffolds and the presence of extracellular matrix that stained positively for types I and II collagen and glycosaminoglycans. The polyglycolic acid constructs revealed some similar features, but a necrotic center was noted, suggesting that the starch-polycaprolactone scaffold may be more biocompatible than the polyglycolic acid one.

Breinan et al. examined the use of porous collagen sponge scaffolds on the healing of articular cartilage defects in a canine model previously developed for ACI (Breinan et al., 2000). In the articular surface of the trochlear grooves of 12 adult mongrel dogs, two 4-mm diameter defects were made to the depth of the tidemark. Four dogs were assigned to each treatment group: (1) micro-fracture treatment, (2) micro-fracture with a type II collagen scaffold placed in the defect, and (3) a type II collagen scaffold seeded with cultured autologous chondrocytes. After 15 weeks, the defects were studied histologically. Data quantified on histological cross sections included area or linear percentages of specific tissue types filling the defect, integration of reparative tissue with the calcified and the adjacent cartilage, and integrity of the subchondral plate. Total defect filling averaged 56–86%, with the greatest amount found in the dogs in the micro-fracture group implanted with a type II collagen matrix. The profiles of tissue types for the dogs in each treatment group were similar: the tissue filling the defect was predominantly fibrocartilage, with the balance being fibrous tissue. There were no significant differences in the percentages of the various tissue types among the three groups. There was a significant correlation between the degree to which the calcified cartilage layer and subchondral bone were disrupted and the amount of tissue filling the defect. Moreover, when it formed, hyaline cartilage most frequently occurred superficial to intact calcified cartilage.

Injectable hydrogel matrices

A novel biomaterial that has been receiving increasing attention in the past few years is the injectable hydrogel. Injectable gels have the potential of minimizing the need for invasive surgery and can instead be administered using intra-articular injections. These hydrogels are generally designed to be liquids at room temperature, which form into solid scaffolds *in situ* through triggers such as enzymatic crosslinking or thermosensitivity. This provides the advantage of forming matrices that conform to irregularly shaped defects. Cells can also be easily incorporated and encapsulated within the scaffold. These three-dimensional polymeric networks have high water content and allow for the transport of nutrients and waste through the matrix.

Various materials, including collagen (Laude et al., 2000), chitosan and hyaluronic acid (Tan et al., 2009), and polyethylene glycol and chondroitin sulfate (Varghese et al., 2008), have been investigated for injectable hydrogel formulations. Many of the hydrogels have also undergone characterization, such as gelation time, microstructure, mechanical properties, and degradation rate. Recent pre-clinical trials have also begun to report on the performance of injectable gels in animal models such as rabbits and goats. A rabbit study (Funayama et al., 2008) in which full-thickness defects were injected with chondrocytes in type II collagen gel found good bonding of the scaffold with the surrounding tissue and synthesis of type II collagen through the 24-week study. In another study (Nettles et al., 2008), osteochondral defects were treated with crosslinkable elastin-like polypeptide (ELP) gel or left as untreated controls. Three months after treatment, ELP-filled defects showed better integration than controls and supported cell infiltration and matrix formation. However, by 6 months, untreated defects showed greater tissue filling, better integration, and more proteoglycan-rich matrix, suggesting that more work is needed on injectable gels for long-term benefits.

GENE THERAPY

Protein-based therapies, such as the use of TGF-β in the culture medium, are expensive to administer effectively. Many of these molecules have a short half-life and would require frequent administration, and targeted delivery can be difficult, leading to the use of high dosages that could have serious unintentional side-effects for other tissues. Advances in molecular biology have allowed for the use of the gene precursors of these proteins for gene therapy (Shuler et al., 2000; Brower-Toland et al., 2001; Cucchiarini et al., 2005; Madry et al., 2005; Guo et al., 2006; Capito and Spector, 2007; Tong and Yao, 2007), enabling the body to produce sustained, regulated levels of proteins in a local area.

Naked DNA is generally not taken up by cells, necessitating the use of gene delivery agents, including viral and non-viral vectors. One group (Kaul et al., 2006) transfected chondrocytes using a non-viral, lipid-mediated agent carrying the FGF-2 gene and implanted the genetically engineered cells into osteochondral defects. They found enhanced articular cartilage repair after 14 weeks without adverse effects. Another group studied the effects of chondrocytes transduced using an adenovirus vector encoding for IGF-1 on healing in equine cartilage defects (Goodrich et al., 2007). They found an increased amount of hyaline-like tissue filling the defects compared to controls filled with thin and irregular fibrous tissue. In another study, adeno-associated virus-TGF-β1-transduced MSCs implanted into osteochondral defects of athymic rats resulted in improved cartilage repair at 12 weeks (Pagnotto et al., 2007). Since articular cartilage is a naturally avascular tissue, there has also been emerging interest in genetically modifying to overexpress angiogenesis inhibitors, such as Flt-1 (Kubo et al., 2009) and endostatin (Sun et al., 2009), in order to regulate angiogenesis during the repair process. These promising findings warrant further investigation of genetically modified cells for cartilage repair.

CURRENT CLINICAL OUTCOMES

By far, the most commonly used cell therapy for cartilage repair is ACI, first reported by Brittberg et al. (1994). This was a case series of 23 patients treated in Sweden for symptomatic cartilage defects. Thirteen patients had femoral condylar defects, ranging in size from 1.6 to 6.5 cm^2, due to trauma or osteochondritis dissecans. Seven patients had patellar defects. Ten patients had previously been treated with shaving and debridement of unstable cartilage. Cartilage was harvested arthroscopically from a minimally load-bearing area of the upper aspect or the medial condyle of the affected knee. Chondrocytes were isolated and culture expanded in a cell culture laboratory. In a second procedure, following a medial or lateral parapatellar incision, the defect was debrided and a periosteal flap was harvested and sutured to the rim of the defect. Finally, the chondrocyte suspension was injected under the periosteal flap.

Follow-up of the patients was over 16–66 months, with a mean of 39 months. Initially, the transplants eliminated knee locking and reduced pain and swelling in all patients. After 3 months, a repeat arthroscopy showed that the transplants were level with the surrounding tissue and spongy when probed, with visible borders. A repeat arthroscopic examination showed that in many instances the transplants had the same macroscopic appearance as they had earlier but were firmer when probed and similar in appearance to the surrounding cartilage. Two years after transplantation, 14 of the 16 patients with femoral condylar transplants had good-to-excellent results. Two patients required a second operation, because of severe central wear in the transplants, with locking and pain. A mean of 36 months after transplantation, the results were excellent or good in two of the seven patients with patellar transplants, fair in three, and poor in two; two patients required a second operation because of severe chondromalacia. Biopsies showed that 11 of the 15 femoral transplants and one of the seven patellar transplants had the appearance of "hyaline-like" cartilage. These results and the fact that a commercial service for culturing autologous chondrocytes was established led to a dramatic increase in the use of this cell-based therapy for cartilage repair.

More recently, a cohort study examined the clinical results of ACI treatment of chondral lesions in the patellofemoral joint (Pascual-Garrido et al., 2009). Sixty-two patients, average age of

31.8 years (ranging from 15.8 to 49.4 years), presented with a mean defect size of 4.2 cm^2. The average follow-up period was 4 years (range 2–7 years). Mean improvements were noted from pre- to postoperative scores using established outcome scales, including the Lysholm, International Knee Documentation Committee, Knee Injury and Osteoarthritis Outcome Score (KOOS), Short Form (SF)-12 Physical, Cincinnati, and Tegner. Only 7.7% of the cases were deemed clinical failures.

Another multicenter cohort study evaluated the clinical success of ACI in 154 patients with prior failed cartilage treatments, namely marrow stimulation or debridement (Zaslav et al., 2009). Forty-eight months after ACI treatment, 76% of the cases were considered successes and 24% were considered failures. Mean improvements in the pre- to postoperative scores were observed using established outcome scales, including the KOOS, Modified Cincinnati Overall Knee Score, and SF-36 Overall Physical Health Score. The results of this study suggest that ACI can provide clinically meaningful improvement of moderate to large chondral lesions that have been previously treated.

There have also been a number of clinical trials comparing ACI with the conventional methods of cartilage repair. One such study compared ACI with microfracture at a 5-year follow-up interval in 80 patients with a cartilage defect on the femoral condyle (Knutsen et al., 2007). Similarly to the other studies, clinical outcomes were evaluated using established scales, including the Lysholm, SF-36, International Cartilage Repair Society, and Tegner. Radiographs were examined using the Kellgren and Lawrence systems. At the follow-up period, 77% of the cases were clinical successes, and 23% of both the ACI and microfracture cases were failures. Radiographs indicated that one-third of the patients exhibited early signs of osteoarthritis. Overall, there was no statistically significant difference between the two treatment groups in the study.

In a non-randomized cohort study, 80 patients, average age of 29.8 years, with cartilage defects in the femoral condyles or trochlea, were treated with second-generation ACI using a hyluronan-based scaffold or microfracture (Kon et al., 2009). Five years post-operation, clinical results indicated statistically significant improvement compared to pre-operative results using the International Knee Documentation Committee score. Comparing the scores for the two groups, the ACI-treated group resulted in better improvement compared to the microfracture-treated group. Sport activity resumption was also noted, and it was observed to be similar for the two groups 2 years after treatment, to have remained stable for the ACI group at 5 years, and to have worsened for the microfracture group at 5 years.

Biological (including tissue engineering) therapies for the treatment of cartilage defects have progressed significantly and are becoming important modalities of treatment in orthopedic surgery. However, for all these therapies long-term outcome is unknown, and there is a lack of controlled studies comparing the different treatment options.

CONCLUSION

Regenerating cartilage tissue *in vivo* is likely to remain challenging over the next few years. However, cell-based therapies have already shown great promise in being better at regenerating the damaged tissue than conventional surgical techniques. These techniques can be further improved upon by investigating the role of scaffolds in trials in repairing cartilage defects. Alternative cell sources, such as stem cells derived from bone marrow, may also provide an improvement in the quality of tissue regenerated.

Acknowledgments

This work was supported by the US Department of Veterans Affairs.

References

Barry, F. P., & Murphy, J. M. (2004). Mesenchymal stem cells: clinical applications and biological characterization. *Int. J. Biochem. Cell Biol., 36,* 568–584.

Barry, F. P., Boynton, R. E., Haynesworth, S., Murphy, J. M., & Zaia, J. (1999). The monoclonal antibody SH-2, raised against human mesenchymal stem cells, recognizes an epitope on endoglin (CD105). *Biochem. Biophys. Res. Commun., 265,* 134–139.

Barry, F., Boynton, R., Murphy, M., & Zaia, J. (2001). The SH-3 and SH-4 antibodies recognize distinct epitopes on CD73 from human mesenchymal stem cells. *Biochem. Biophys. Res. Commun., 289,* 519–524.

Bartholomew, A., Sturgeon, C., Siatskas, M., Ferrer, K., McIntosh, K., Patil, S., et al. (2002). Mesenchymal stem cells suppress lymphocyte proliferation *in vitro* and prolong skin graft survival *in vivo. Exp. Hematol., 30,* 42–48.

Bartlett, W., Skinner, J. A., Gooding, C. R., Carrington, R. W. J., Flanagan, A. M., Briggs, T. W. R., et al. (2005). Autologous chondrocyte implantation versus matrix-induced autologous chondrocyte implantation for osteochondral defects of the knee. *J. Bone Joint Surg., 87B,* 640–645.

Bigdeli, N., Karlsson, C., Strehl, R., Concaro, S., Hyllner, J., & Lindahl, A. (2009). Coculture of human embryonic stem cells and human articular chondrocytes results in significantly altered phenotype and improved chondrogenic differentiation. *Stem Cells, 27,* 1812–1821.

Breinan, H. A., Martin, S. D., Hsu, H. P., & Spector, M. (2000). Healing of canine articular cartilage defects treated with microfracture, a type-II collagen matrix, or cultured autologous chondrocytes. *J. Orthop. Res., 18,* 781–789.

Brittberg, M. (1999). Autologous chondrocyte transplantation. *Clin. Orthop. Rel. Res.,* S147–S155.

Brittberg, M., Lindahl, A., Nilsson, A., Ohlsson, C., Isaksson, O., & Peterson, L. (1994). Treatment of deep cartilage defects in the knee with autologous chondrocyte transplantation. *New Engl. J. Med., 331,* 889–895.

Brower-Toland, B. D., Saxer, R. A., Goodrich, L. R., Mi, Z. B., Robbins, P. D., Evans, C. H., et al. (2001). Direct adenovirus-mediated insulin-like growth factor I gene transfer enhances transplant chondrocyte function. *Human Gene Ther., 12,* 117–129.

Bryant, S. J., Bender, R. J., Durand, K. L., & Anseth, K. S. (2004). Encapsulating chondrocytes in degrading PEG hydrogels with high modulus: engineering gel structural changes to facilitate cartilaginous tissue production. *Biotechnol. Bioeng., 86.* 747–755.

Buckwalter, J. A., & Mankin, H. J. (1998). Articular cartilage: tissue design and chondrocyte-matrix interactions. *Instructional Course Lectures, Vol. 47,* 477–486, American Academy of Orthopaedic Surgeons, Rosemont.

Capito, R. M., & Spector, M. (2007). Collagen scaffolds for nonviral IGF-1 gene delivery in articular cartilage tissue engineering. *Gene Ther., 14,* 721–732.

Caplan, A. I. (1991). Mesenchymal stem-cells. *J. Orthop. Res., 9,* 641–650.

Chung, C., Erickson, I. E., Mauck, R. L., & Burdick, J. A. (2008). Differential behavior of auricular and articular chondrocytes in hyaluronic acid hydrogels. *Tissue Eng. Part A, 14,* 1121–1131.

Connolly, J. F. (1998). Clinical use of marrow osteoprogenitor cells to stimulate osteogenesis. *Clin. Orthop. Rel. Res.,* S257–S266.

Cucchiarini, M., Madry, H., Ma, C. Y., Thurn, T., Zurakowski, D., Menger, M. D., et al. (2005). Improved tissue repair in articular cartilage defects *in vivo* by rAAV-mediated overexpression of human fibroblast growth factor 2. *Molec. Ther., 12,* 229–238.

Curl, W. W., Krome, J., Gordon, E. S., Rushing, J., Smith, B. P., & Poehling, G. G. (1997). Cartilage injuries: a review of 31,516 knee arthroscopies. *Arthroscopy, 13,* 456–460.

Dattena, M., Pilichi, S., Rocca, S., Mara, L., Casu, S., Masala, G., et al. (2009). Sheep embryonic stem-like cells transplanted in full-thickness cartilage defects. *J. Tissue Eng. Regen. Med., 3,* 175–187.

de Bari, C., Dell'Accio, F., & Luyten, F. P. (2001). Human periosteum-derived cells maintain phenotypic stability and chondrogenic potential throughout expansion regardless of donor age. *Arthritis Rheum., 44,* 85–95.

de Bari, C., Dell'Accio, F., Vandenabeele, F., Vermeesch, J. R., Raymackcrs, J. M., & Luyten, F. P. (2003). Skeletal muscle repair by adult human mesenchymal stem cells from synovial membrane. *J. Cell Biol., 160,* 909–918.

Engler, A. J., Sen, S., Sweeney, H. L., & Discher, D. E. (2006). Matrix elasticity directs stem cell lineage specification. *Cell, 126,* 677–689.

Fecek, C., Yao, D. G., Kacorri, A., Vasquez, A., Iqbal, S., Sheikh, H., et al. (2008). Chondrogenic derivatives of embryonic stem cells seeded into 3D polycaprolactone scaffolds generated cartilage tissue *in vivo. Tissue Eng. Part A, 14,* 1403–1413.

Frisbie, D. D., Lu, Y., Kawcak, C. E., DiCarlo, E. F., Binette, F., & McIlwraith, C. W. (2009). In vivo evaluation of autologous cartilage fragment-loaded scaffolds implanted into equine articular defects and compared with autologous chondrocyte implantation. *Am. J. Sports Med., 37,* 71S–80S.

Funayama, A., Niki, Y., Matsumoto, H., Maeno, S., Yatabe, T., Morioka, H., Yanagimoto, S., et al. (2008). Repair of full-thickness articular cartilage defects using injectable type II collagen gel embedded with cultured chondrocytes in a rabbit model. *J. Orthopaed. Sci., 13,* 225–232.

Goldring, M. B., Sandell, L. J., Stephenson, M. L., & Krane, S. M. (1986). Immune interferon suppresses levels of procollagen messenger-RNA and type-ii collagen-synthesis in cultured human articular and costal chondrocytes. *J. Biol. Chem., 261,* 9049–9056.

Goldring, M. B., Birkhead, J., Sandell, L. J., Kimura, T., & Krane, S. M. (1988). Interleukin-1 suppresses expression of cartilage-specific type-ii and type-ix collagens and increases type-i and type-iii collagens in human chondrocytes. *J. Clin. Invest., 82,* 2026–2037.

Goodrich, L. R., Hidaka, C., Robbins, P. D., Evans, C. H., & Nixon, A. J. (2007). Genetic modification of chondrocytes with insulin-like growth factor-1 enhances cartilage healing in an equine model. *J. Bone Joint Surg., 89B,* 672–685.

Grande, D. A., Halberstadt, C., Naughton, G., Schwartz, R., & Manji, R. (1997). Evaluation of matrix scaffolds for tissue engineering of articular cartilage grafts. *J. Biomed. Mater. Res., 34,* 211–220.

Gronthos, S., & Simmons, P. J. (1995). The growth-factor requirements of Stro-1-positive human bone-marrow stromal precursors under serum-deprived conditions in-vitro. *Blood, 85,* 929–940.

Gronthos, S., Zannettino, A. C. W., Hay, S. J., Shi, S. T., Graves, S. E., Kortesidis, A., et al. (2003). Molecular and cellular characterisation of highly purified stromal stem cells derived from human bone marrow. *J. Cell Sci., 116,* 1827–1835.

Guo, X. D., Zheng, Q. X., Yang, S. H., Shao, Z. W., Yuan, Q., Pan, Z. Q., et al. (2006). Repair of full-thickness articular cartilage defects by cultured mesenchymal stem cells transfected with the transforming growth factor beta(1) gene. *Biomed. Mater., 1,* 206–215.

Hare, J. M., Traverse, J. H., Henry, T. D., Dib, N., Strumpf, R. K., Schulman, S. P., et al. (2009). A randomized, double-blind, placebo-controlled, dose-escalation study of intravenous adult human mesenchymal stem cells (prochymal) after acute myocardial infarction. *J. Am. Coll. Cardiol., 54,* 2277–2286.

Homminga, G. N., Vanderlinden, T. J., Terwindtrouwenhorst, E. A. W., & Drukker, J. (1989). Repair of articular defects by perichondrial graft — experiments in the rabbit. *Acta Orthop. Scand., 60,* 326–329.

Homminga, G. N., Bulstra, S., Bouwmeester, P. S. M., & Vanderlinden, A. J. (1990). Perichondrial grafting for cartilage lesions of the knee. *J. Bone Joint Surg., 72,* 1003–1007.

Homminga, G. N., Bulstra, S. K., Kuijer, R., & Vanderlinden, A. J. (1991). Repair of sheep articular-cartilage defects with a rabbit costal perichondrial graft. *Acta Orthop. Scand., 62,* 415–418.

Hwang, N. S., Varghese, S., & Elisseeff, J. (2008). Derivation of chondrogenically-committed cells from human embryonic cells for cartilage tissue regeneration. *Plos One, 3,* e2498.

Im, G. I., Kim, D. Y., Shin, J. H., Hyun, C. W., & Cho, W. H. (2001). Repair of cartilage defect in the rabbit with cultured mesenchymal stem cells from bone marrow. *J. Bone Joint Surg., 83B,* 289–294.

Isogai, N., Kusuhara, H., Ikada, Y., Ohtani, H., Jacquet, R., Hillyer, J., Lowder, E., et al. (2006). Comparison of different chondrocytes for use in tissue engineering of cartilage model structures. *Tissue Eng., 12,* 691–703.

Jay, G. D., Torres, J. R., Warman, M. L., Laderer, M. C., & Breuer, K. S. (2007). The role of lubricin in the mechanical behavior of synovial fluid. *Proc. Natl. Acad. Sci. U.S.A., 104,* 6194–6199.

Johnson, T. S., Xu, J. W., Zaporojan, V. V., Mesa, J. M., Weinand, C., Randolph, M. A., et al. (2004). Integrative repair of cartilage with articular and nonarticular chondrocytes. *Tissue Eng., 10,* 1308–1315.

Kaul, G., Cucchiarini, M., Arntzen, D., Zurakowski, D., Menger, M. D., Kohn, D., et al. (2006). Local stimulation of articular cartilage repair by transplantation of encapsulated chondrocytes overexpressing human fibroblast growth factor 2 (FGF-2) in vivo. *J. Gene Med., 8,* 100–111.

Kavalkovich, K. W., Boynton, R. E., Murphy, J. M., & Barry, F. (2002). Chondrogenic differentiation of human mesenchymal stem cells within an alginate layer culture system. *In Vitro Cell. Devel. Biol. Animal, 38,* 457–466.

Knutsen, G., Drogset, J. O., Engebretsen, L., Grontvedt, T., Isaksen, V., Ludvigsen, T. C., et al. (2007). A randomized trial comparing autologous chondrocyte implantation with microfracture. *J. Bone Joint Surg., 89A,* 2105–2112.

Koga, H., Muneta, T., Nagase, T., Nimura, A., Ju, Y. J., Mochizuki, T., et al. (2008). Comparison of mesenchymal tissues-derived stem cells for *in vivo* chondrogenesis: suitable conditions for cell therapy of cartilage defects in rabbit. *Cell Tissue Res., 333,* 207–215.

Kon, E., Gobbi, A., Filardo, G., Delcogliano, M., Zaffagnini, S., & Marcacci, M. (2009). Arthroscopic second-generation autologous chondrocyte implantation compared with microfracture for chondral lesions of the knee: prospective non-randomized study at 5 years. *Am. J. Sports Med., 37,* 33–41.

Kramer, J., Bohrnsen, F., Schlenke, P., & Rohwedel, J. (2006). Stem cell-derived chondrocytes for regenerative medicine. *Transplantation Proc., 38,* 762–765.

Kubo, S., Cooper, G. M., Matsumoto, T., Phillippi, J. A., Corsi, K. A., Usas, A., et al. (2009). Blocking vascular endothelial growth factor with soluble Flt-1 improves the chondrogenic potential of mouse skeletal muscle-derived stem cells. *Arthritis Rheum., 60,* 155–165.

Laude, D., Odlum, K., Rudnicki, S., & Bachrach, N. (2000). A novel injectable collagen matrix: *in vitro* characterization and *in vivo* evaluation. *J. Biomech. Eng. Trans. ASME, 122,* 231–235.

Lee, C. R., Breinan, H. A., Nehrer, S., & Spector, M. (2000a). Articular cartilage chondrocytes in type I and type II collagen-GAG matrices exhibit contractile behavior *in vitro*. *Tissue Eng.*, 6, 555–565.

Lee, C. R., Grodzinsky, A. J., Hsu, H. P., Martin, S. D., & Spector, M. (2000b). Effects of harvest and selected cartilage repair procedures on the physical and biochemical properties of articular cartilage in the canine knee. *J. Orthop. Res.*, 18, 790–799.

Lee, C. R., Grodzinsky, A. J., & Spector, M. (2001). The effects of cross-linking of collagen-glycosaminoglycan scaffolds on compressive stiffness, chondrocyte-mediated contraction, proliferation and biosynthesis. *Biomaterials*, 22, 3145–3154.

Lodie, T. A., Blickarz, C. E., Devarakonda, T. J., He, C. F., Dash, A. B., Clarke, J., et al. (2002). Systematic analysis of reportedly distinct populations of multipotent bone marrow-derived stem cells reveals a lack of distinction. *Tissue Eng.*, 8, 739–751.

Lu, Y. L., Dhanaraj, S., Wang, Z. W., Bradley, D. M., Bowman, S. M., Cole, B. J., et al. (2006). Minced cartilage without cell culture serves as an effective intraoperative cell source for cartilage repair. *J. Orthop. Res.*, 24, 1261–1270.

Madry, H., Kaul, G., Cucchiarini, M., Stein, U., Zurakowski, D., Remberger, K., et al. (2005). Enhanced repair of articular cartilage defects *in vivo* by transplanted chondrocytes overexpressing insulin-like growth factor I (IGF-I). *Gene Ther.*, 12, 1171–1179.

Majumdar, M. K., Keane-Moore, M., Buyaner, D., Hardy, W. B., Moorman, M. A., McIntosh, K. R., & Mosca, J. D. (2003). Characterization and functionality of cell surface molecules on human mesenchymal stem cells. *J. Biomed. Sci.*, 10, 228–241.

Malicev, E., Kregar-Velikonja, N., Barlic, A., Alibegovic, A., & Drobnic, M. (2009). Comparison of articular and auricular cartilage as a cell source for the autologous chondrocyte implantation. *J. Orthop. Res.*, 27, 943–948.

Manfredini, M., Zerinati, F., Gildone, A., & Faccini, R. (2007). Autologous chondrocyte implantation: a comparison between an open periosteal-covered and an arthroscopic matrix-guided technique. *Acta Orthop. Belg.*, 73, 207–218.

Mankin, H. J. (1982). The response of articular-cartilage to mechanical injury. *J. Bone Joint Surg.*, 64, 460–466.

Martin, J. A., & Buckwalter, J. A. (2001). Roles of articular cartilage aging and chondrocyte senescence in the pathogenesis of osteoarthritis. *Iowa Orthop. J.*, 21, 1–7.

Mayhew, T. A., Williams, G. R., Senica, M. A., Kuniholm, G., & du Moulin, G. C. (1998). Validation of a quality assurance program for autologous cultured chondrocyte implantation. *Tissue Eng.*, 4, 325–334.

McBride, C., Gaupp, D., & Phinney, D. G. (2003). Quantifying levels of transplanted murine and human mesenchymal stem cells *in vivo* by real-time PCR. *Cytotherapy*, 5, 7–18.

McCormick, F., Yanke, A., Provencher, M. T., & Cole, B. J. (2008). Minced articular cartilage-basic science, surgical technique, and clinical application. *Sports Med. Arthroscopy Rev.*, 16, 217–220.

Nettles, D. L., Kitaoka, K., Hanson, N. A., Flahiff, C. M., Mata, B. A., Hsu, E. W., Chilkoti, A., et al. (2008). In situ crosslinking elastin-like polypeptide gels for application to articular cartilage repair in a goat osteochondral defect model. *Tissue Eng. Part A*, 14, 1133–1140.

Niedermann, B., Boe, S., Lauritzen, J., & Rubak, J. M. (1985). Glued periosteal grafts in the knee. *Acta Orthop. Scand.*, 56, 457–460.

Oliveira, J. T., Crawford, A., Mundy, J. M., Moreira, A. R., Gomes, M. E., Hatton, P. V., et al. (2007). A cartilage tissue engineering approach combining starch-polycaprolactone fibre mesh scaffolds with bovine articular chondrocytes. *J. Mater. Sci. Mater. Med.*, 18, 295–302.

Oshima, Y., Watanabe, N., Matsuda, K., Takai, S., Kawata, M., & Kubo, T. (2005). Behavior of transplanted bone marrow-derived GFP mesenchymal cells in osteochondral defect as a simulation of autologous transplantation. *J. Histochem. Cytochem.*, 53, 207–216.

Pagnotto, M. R., Wang, Z., Karpie, J. C., Ferretti, M., Xiao, X., & Chu, C. R. (2007). Adeno-associated viral gene transfer of transforming growth factor-beta 1 to human mesenchymal stem cells improves cartilage repair. *Gene Ther.*, 14, 804–813.

Pal, R., Venkataramana, N. K., Jaan, M., Bansal, A., Balaraju, S., Chandra, R., Dixit, A., et al. (2009). Ex vivo-expanded autologous bone marrow-derived mesenchymal stromal cells in human spinal cord injury/paraplegia: a pilot clinical study. *Cytotherapy*, 11, 897–911.

Panossian, A., Ashiku, S., Kirchhoff, C. H., Randolph, M. A., & Yaremchuk, M. J. (2001). Effects of cell concentration and growth period on articular and ear chondrocyte transplants for tissue engineering. *Plast. Reconstr. Surg.*, 108, 392–402.

Pascual-Garrido, C., Slabaugh, M. A., L'Heureux, D. R., Friel, N. A., & Cole, B. J. (2009). Recommendations and treatment outcomes for patellofemoral articular cartilage defects with autologous chondrocyte implantation prospective evaluation at average 4-year follow-up. *Am. J. Sports Med.*, 37, 33S–41S.

Perka, C., Sittinger, M., Schultz, O., Spitzer, R. S., Schlenzka, D., & Burmester, G. R. (2000). Tissue engineered cartilage repair using cryopreserved and noncryopreserved chondrocytes. *Clin. Orthop. Rel. Res.*, 245–254.

Peterson, L., Minas, T., Brittberg, M., Nilsson, A., Sjogren-Jansson, E., & Lindahl, A. (2000). Two- to 9-year outcome after autologous chondrocyte transplantation of the knee. *Clin. Orthop. Rel. Res.*, 212–234.

Pfeiffer, E., Vickers, S. M., Frank, E., Grodzinsky, A. J., & Spector, M. (2008). The effects of glycosaminoglycan content on the compressive modulus of cartilage engineered in type II collagen scaffolds. *Osteoarthritis Cartilage, 16*, 1237–1244.

Pittenger, M. F., Mackay, A. M., Beck, S. C., Jaiswal, R. K., Douglas, R., Mosca, J. D., et al. (1999). Multilineage potential of adult human mesenchymal stem cells. *Science, 284*, 143–147.

Prockop, D. J. (1997). Marrow stromal cells as stem cells for nonhematopoietic tissues. *Science, 276*, 71–74.

Reyes, M., & Verfaillie, C. M. (2001). Characterization of multipotent adult progenitor cells, a subpopulation of mesenchymal stem cells. *Ann. NY Acad. Sci., 938*, 231–233; discussion 233–235.

Rodrigo, J. J., Steadman, J. R., & Silliman, H. A. (1994). Improvement of full-thickness chondral defect healing in the human knee after debridement and microfracture using continuous passive motion. *Am. J. Knee Surg., 7*, 109–116.

Rubak, J. M. (1982). Reconstruction of articular-cartilage defects with free periosteal grafts – an experimental-study. *Acta Orthop. Scand., 53*, 175–180.

Saadeh, P. B., Brent, B., Mehrara, B. J., Steinbrech, D. S., Ting, V., Gittes, G. K., & Longaker, M. T. (1999). Human cartilage engineering: chondrocyte extraction, proliferation, and characterization for construct development. *Ann. Plast. Surg., 42*, 509–513.

Salama, R., Burwell, R. D., & Dickson, I. R. (1973). Recombined grafts of bone and marrow. The beneficial effect upon osteogenesis of impregnating xenograft (heterograft) bone with autologous red marrow. *J. Bone Joint Surg., 55B*, 402–417.

Schmidt, T. A., Gastelum, N. S., Nguyen, Q. T., Schumacher, B. L., & Sah, R. L. (2007). Boundary lubrication of articular cartilage – role of synovial fluid constituents. *Arthritis Rheum., 56*, 882–891.

Schumacher, B. L., Block, J. A., Schmid, T. M., Aydelotte, M. B., & Kuettner, K. E. (1994). A novel proteoglycan synthesized and secreted by chondrocytes22 of the superficial zone of articular cartilage. *Arch. Biochem. Biophys., 311*, 144–152.

Shapiro, F., Koide, S., & Glimcher, M. J. (1993). Cell origin and differentiation in the repair of full-thickness defects of articular-cartilage. *J. Bone Joint Surg., 75A*, 532–553.

Shelbourne, K. D., Jari, S., & Gray, T. (2003). Outcome of untreated traumatic articular cartilage defects of the knee – a natural history study. *J. Bone Joint Surg., 85A*, 8–16.

Shuler, F. D., Georgescu, H. I., Niyibizi, C., Studer, R. K., Mi, Z. B., Johnstone, B., et al. (2000). Increased matrix synthesis following adenoviral transfer of a transforming growth factor beta(1) gene into articular chondrocytes. *J. Orthop. Res., 18*, 585–592.

Sun, X.-D., Jeng, L., Bolliet, C., Olsen, B. R., & Spector, M. (2009). Non-viral endostatin plasmid transfection of mesenchymal stem cells via collagen scaffolds. *Biomaterials, 30*, 1222–1231.

Tan, H. P., Chu, C. R., Payne, K. A., & Marra, K. G. (2009). Injectable in situ forming biodegradable chitosan-hyaluronic acid based hydrogels for cartilage tissue engineering. *Biomaterials, 30*, 2499–2506.

Tong, J. C., & Yao, S. L. (2007). Novel scaffold containing transforming growth factor-beta 1 DNA for cartilage tissue engineering. *J. Bioactive Compatible Polymers, 22*, 232–244.

Tse, W. T., Pendleton, J. D., Beyer, W. M., Egalka, M. C., & Guinan, E. C. (2003). Suppression of allogeneic T-cell proliferation by human marrow stromal cells: implications in transplantation. *Transplantation, 75*, 389–397.

van Osch, G., Mandl, E. W., Jahr, H., Koevoet, W., Nolst-Trenite, G., & Verhaar, J. A. (2004). Considerations on the use of ear chondrocytes as donor chondrocytes for cartilage tissue engineering. *Biorheology, 41*, 411–421.

Varghese, S., Hwang, N. S., Canver, A. C., Theprungsirikul, P., Lin, D. W., & Elisseeff, J. (2008). Chondroitin sulfate based niches for chondrogenic differentiation of mesenchymal stem cells. *Matrix Biol., 27*, 12–21.

Wakitani, S., Goto, T., Pineda, S. J., Young, R. G., Mansour, J. M., Caplan, A. I., et al. (1994). Mesenchymal cell-based repair of large, full-thickness defects of articular-cartilage. *J. Bone Joint Surg., 76A*, 579–592.

Wakitani, S., Imoto, K., Yamamoto, T., Saito, M., Murata, N., & Yoneda, M. (2002). Human autologous culture expanded bone marrow mesenchymal cell transplantation for repair of cartilage defects in osteoarthritic knees. *Osteoarthr. Cartilage, 10*, 199–206.

Wakitani, S., Takaoka, K., Hattori, T., Miyazawa, N., Iwanaga, T., Takeda, S., et al. (2003). Embryonic stem cells injected into the mouse knee joint form teratomas and subsequently destroy the joint. *Rheumatology, 42*, 162–165.

Wakitani, S., Aoki, H., Harada, Y., Sonobe, M., Morita, Y., Mu, Y., Tomita, N., et al. (2004a). Embryonic stem cells form articular cartilage, not teratomas, in osteochondral defects of rat joints. *Cell Transplantation, 13*, 331–336.

Wakitani, S., Mitsuoka, T., Nakamura, N., Toritsuka, Y., Nakamura, Y., & Horibe, S. (2004b). Autologous bone marrow stromal cell transplantation for repair of full-thickness articular cartilage defects in human patellae: two case reports. *Cell Transplantation, 13*, 595–600.

Wakitani, S., Nawata, M., Tensho, K., Okabe, T., Machida, H., & Ohgushi, H. (2007). Repair of articular cartilage defects in the patello-femoral joint with autologous bone marrow mesenchymal cell transplantation: three case reports involving nine defects in five knees. *J. Tissue Eng. Regen. Med., 1*, 74–79.

Xu, J. W., Zaporojan, V., Peretti, G. M., Roses, R. E., Morse, K. B., Roy, A. K., et al. (2004). Injectable tissue-engineered cartilage with different chondrocyte sources. *Plast. Reconstr. Surg., 113*, 1361–1371.

Yanai, T., Ishii, T., Chang, F., & Ochiai, N. (2005). Repair of large full-thickness articular cartilage defects in the rabbit – the effects of joint distraction and autologous bone-marrow-derived mesenchymal cell transplantation. *J. Bone Joint Surg., 87B*, 721–729.

Yoshimura, H., Muneta, T., Nimura, A., Yokoyama, A., Koga, H., & Sekiya, I. (2007). Comparison of rat mesenchymal stem cells derived from bone marrow, synovium, periosteum, adipose tissue, and muscle. *Cell Tissue Res., 327*, 449–462.

Young, H. E., Steele, T. A., Bray, R. A., Hudson, J., Floyd, J. A., Hawkins, K., et al. (2001). Human reserve pluripotent mesenchymal stem cells are present in the connective tissues of skeletal muscle and dermis derived from fetal, adult, and geriatric donors. *Anatomical Rec., 264*, 51–62.

Zaslav, K., Cole, B., Brewster, R., DeBerardino, T., Farr, J., Fowler, P., et al. (2009). A prospective study of autologous chondrocyte implantation in patients with failed prior treatment for articular cartilage defect of the knee results of the study of the treatment of articular repair (STAR) clinical trial. *Am. J. Sports Med., 37*, 42–55.

Zhang, L., & Spector, M. (2009). Comparison of three types of chondrocytes in collagen scaffolds for cartilage tissue engineering. *Biomed. Mater., 4*, 045012.

Zhang, Z. X., Guan, L. X., Zhang, K., Zhang, Q., & Dai, L. J. (2008). A combined procedure to deliver autologous mesenchymal stromal cells to patients with traumatic brain injury. *Cytotherapy, 10*, 134–139.

Zhou, G. D., Wang, X. Y., Miao, C. L., Liu, T. Y., Zhu, L., Liu, D. L., et al. (2004). Repairing porcine knee joint osteochondral defects at non-weight bearing area by autologous BMSC. *Zhonghua Yi Xue Za Zhi, 84*, 925–931.

zur Nieden, N.I., Kempka, G., Rancourt, D. E., & Ahr, H. J. (2005). Induction of chondro-, osteo- and adipogenesis in embryonic stem cells by bone morphogenetic protein-2: Effect of cofactors on differentiating lineages. *BMC Dev. Biol., 5*, 1.

CHAPTER 43

Myoblast Transplantation in Skeletal Muscles

Daniel Skuk, Jacques P. Tremblay
Research Unit on Human Genetics, CHUL Research Center, Quebec, QC, Canada

INTRODUCTION

Transplantation of myogenic cells is an approach under study to treat diseases of the skeletal muscle. Myoblasts were the first cells to be proposed for this therapeutic approach (Partridge et al., 1978). The term "myoblast" defines the proliferating mononuclear progenitor cells that fuse with analogous cells to form myotubes, the sincytial stage that precedes myofiber formation. This applies to every case in which this phenomenon occurs; that is during skeletal muscle histogenesis *in utero*, in post-natal myofiber regeneration, and *in vitro* culture of skeletal muscle cells. The starting point of this therapeutic approach can be traced to 1978, when Partridge, Grounds, and Sloper proposed that "in subjects suffering from inherited recessive myopathies, muscle function might be restored if normal myoblasts could be made to fuse with defective muscle fibers" (Partridge et al., 1978). Duchenne muscular dystrophy (DMD) has being the main target among these myopathies. This is due to its relative frequency (a prevalence of 50 cases per million in the male population) and severity: progressive skeletal-muscle degeneration in the limbs and trunk during childhood and adolescence, leading to motion loss, respiratory insufficiency, and ultimately death by respiratory or cardiac complications.

The saga of myoblast transplantation in clinical myology is an example of the importance of a preclinical basis to design new therapeutic protocols. It was after a few animal experiments in the 1980s that several groups undertook clinical trials of myoblast transplantation in the 1990s (Skuk, 2004). Lacking sufficient preclinical background to plan the strategies of cell implantation and even the strategies to control acute rejection, these clinical trials reported scarce and very modest results at the molecular level. In fact, these trials were conducted in the hope that some myoblasts injected in a few sites of a skeletal muscle would be able to diffuse throughout the muscle and spontaneously fuse with so many myofibers that a therapeutic effect would be obtained. The subsequent research demonstrated that this hope was unrealistic, and the lesson to take from this experience is that researchers need to know the behavior of the grafted cells in appropriate experimental conditions to design clinical protocols without improvisation.

This chapter aims to address the current knowledge that could be useful for future clinical applications of myoblast transplantation in the treatment of skeletal muscle diseases. For this reason, priority will be given to observations made in humans and non-human primates. The chapter is organized to address three main questions: (1) why myoblasts meet the properties needed for the aim of the treatment, (2) how myoblasts can be properly delivered to the target tissues, and (3) how we can ensure the long-term survival of the graft.

Principles of Regenerative Medicine. DOI: 10.1016/B978-0-12-381422-7.10043-4

MYOBLASTS AS CANDIDATE CELLS FOR TRANSPLANTATION INTO SKELETAL MUSCLES

In cell transplantation, the implanted elements could be either differentiated cells or precursor cells with the ability to differentiate into the former. In the skeletal muscle, the differentiated elements of the parenchyma are useless for transplantation: myofibers are very long syncytia that cannot be properly manipulated or implanted. The possibility of cell transplantation is offered by the specific stem cell of the skeletal muscle, the "satellite cell." A main function of satellite cells is to repair myofiber damage. They are normally quiescent in the periphery of myofibers until an injury produces focal or total necrosis of a nearby myofiber. This necrosis triggers a process that involves the removal of myofiber debris by phagocytic cells and the activation of dormant satellite cells, which differentiate into proliferating myoblasts (Betz et al., 1966). The adjective "adult" is sometimes applied to these myoblasts (Yablonka-Reuveni and Nameroff, 1990) to distinguish them from the embryonic or fetal myoblasts that form skeletal muscles during histogenesis. Satellite cells can be isolated from skeletal muscle biopsies by standard cell-culture techniques and can be expanded as myoblasts *in vitro*, maintaining their capacity to fuse into myotubes that will differentiate into myofibers (Konigsberg, 1960). This ease regarding obtaining the cells and proliferating them *in vitro* facilitated the use of myoblasts for strategies of myogenic-cell transplantation.

Potentially useful properties of grafted myoblasts

Lipton and Schultz reported for the first time in 1979 the two main properties of exogenous myoblasts implanted into skeletal muscles, that is, the fusion of the grafted cells with the recipient's myofibers and the formation of new small myofibers by the fusion of the grafted cells among themselves (Lipton and Schultz, 1979). The first property allows gene complementation, that is, myofibers in which exogenous myogenic cells fused will express proteins coded by the exogenous and endogenous nuclei (Watt et al., 1982), and will thus be referred to as "mosaic" or "hybrid" (Kikuchi et al., 1980). Through gene complementation, grafted myoblasts can act as vehicles of therapeutic genes, for example by introducing normal genomes in the genetically abnormal myofiber of a patient with a recessive genetic myopathy. The second property opens the door to the possibility of forming new myofibers in patients in which the skeletal-muscle parenchyma has been lost. A third property of myoblast transplantation was reported later: the possibility of giving rise to donor-derived satellite cells.

GENE COMPLEMENTATION

The first experimental demonstration of this principle was reported by Partridge et al. in 1989. They transplanted normal mouse myoblasts in *mdx* mice (which, as DMD patients, have a myopathy caused by the deficiency of the protein named *dystrophin*) and observed later dystrophin expression in several myofibers. The same observation was repeated soon by other researchers (Karpati et al., 1989; Kinoshita et al., 1994) and is now routine in research (Skuk and Tremblay, 2003; Skuk, 2004). Other proteins restored by normal myoblast transplantation in mouse models of muscular dystrophies were *merosin* in *dy/dy* mice (model of merosin-deficient congenital muscle dystrophy) (Vilquin et al., 1996) and *dysferlin* in SJL mice (model of limb-girdle muscular dystrophy 2B) (Leriche-Guerin et al., 2002). In humans, occasional observations of improved dystrophin expression following normal-myoblast allotransplantation in DMD patients were reported during the clinical trials conducted in the 1990s (Huard et al., 1992; Karpati et al., 1993; Tremblay et al., 1993; Mendell et al., 1995). However, these results were not conclusive and the majority of patients gave negative results. It was not until more recent clinical trials, based on preclinical studies in non-human primates, that it was shown that donor-derived dystrophin can be expressed systematically in the muscles of DMD patients implanted with normal myoblasts (Fig. 43.1) (Skuk et al., 2004, 2006b, 2007a).

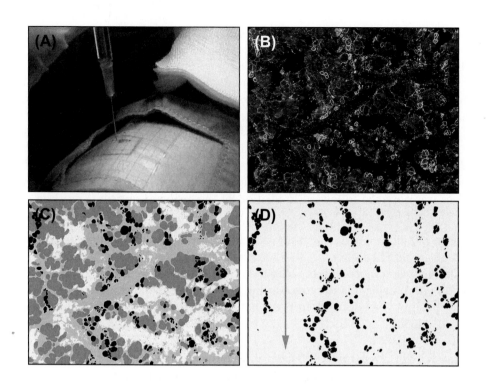

FIGURE 43.1

Allotransplantation of normal adult myoblasts in 1 cm^3 of the tibialis anterior of a patient with DMD, in a phase 1A clinical trial conducted by the authors. (A) Several parallel intramuscular cell injections are being done with a 100-μl Hamilton syringe. The cells are delivered homogeneously during the needle withdrawal, and the density of cell injections is controlled with the help of a sterile transparent dressing with a grid. (B) The whole cross-section of a biopsy done 1 month later in a cell-grafted site is shown stained for fluorescent immunodetection of dystrophin. (C) A schematic representation illustrates the distribution of the dystrophin-positive myofiber profiles: in black, dystrophin-positive myofibers; in dark gray, muscle tissue; in light gray, connective tissue; the white spaces inside the biopsy correspond to fat tissue. (D) The dystrophin-positive myofiber profiles are distributed roughly in parallel axes corresponding with the original trajectories of the injections (indicated by the arrow), although some dispersion is observed. Scale bar = 1 mm.

A critical factor complicating the technique of myogenic-cell implantation is that the intracellular proteins encoded by a single nucleus in a myofiber do not diffuse across the syncytium. On the contrary, they remain localized near the nucleus of origin, in a region named the "nuclear domain" (Pavlath et al., 1989). This restriction is produced by the limited diffusion of both the mRNA (Ralston and Hall, 1992) and protein (Hall and Ralston, 1989). Consequently, donor-derived proteins are expressed only in the myofiber segments where fusion of the donor's cells was produced. The size of the nuclear domain depends on the capacity of the protein to diffuse or to remain anchored to stationary cellular components (Hall and Ralston, 1989). As an example, single injections of β-galactosidase-labeled normal myoblasts into *mdx* mice produced dystrophin expression throughout roughly 500 μm of the myofiber, in contrast with 1500 μm for β-galactosidase (Kinoshita et al., 1998). The longer domain of β-galactosidase should be attributed to the solubility of this enzyme, spreading more than dystrophin, which remains attached to the cytoskeleton.

FORMATION OF NEW MYOFIBERS

In DMD and other myopathies, the progressive worsening of muscle weakness is produced by the steady and irreversible loss of myofibers. An ideal treatment for the advanced stages of these diseases may include not only molecular correction but also restoration of muscle mass.

The potential of myoblast transplantation to restore skeletal muscle mass and strength was reported in mice following acute severe muscle destruction (Almaddine et al., 1994; Wernig et al., 1995, 2000; Irintchev et al., 1997). These experimental conditions, however, are not similar to those of a degenerative myopathy, and how myogenic-cell transplantation might form new functional tissue in skeletal muscles that have degenerated to fibrosis and/or fat substitution remains insufficiently studied. A study in mice suggested that it could be possible to create myotubes within the adipose tissue (Satoh et al., 1992). Otherwise, neo-muscles were formed ectopically after subcutaneous implantation of myoblasts in mice, in spite of the absence of a previous endomysial support (Irintchev et al., 1998). In *mdx* mice, formation of new myofibers through the fusion of the implanted myoblasts among themselves was observed following irradiation of the recipient muscle (Kinoshita et al., 1996b).

Progressing from these few observations to a clinically functional procedure remains a challenge, among other factors because these results were obtained in mice, which have intrinsically greater muscle regeneration capacity than primates (Borisov et al., 1999). A clinical observation encouraging this research was the presence of small dystrophin-positive myofibers, putatively neo-formed, in DMD patients transplanted with normal myoblasts (Skuk et al., 2006b).

FORMATION OF DONOR-DERIVED SKELETAL MUSCLE PROGENITOR CELLS

Transplants of mouse and human myoblasts into mouse muscles showed that some grafted myoblasts remained as mononuclear cells able to participate later in muscle regeneration and also to proliferate and form myotubes *in vitro* (Yao and Kurachi, 1993; Gross and Morgan, 1999; Ehrhardt et al., 2007; Xu et al., 2010). Morphological studies showed specifically that donor-derived satellite cells were produced by the transplantation of mouse myoblasts (Irintchev et al., 1997, 1998; Heslop et al., 2001) and human myoblasts (Brimah et al., 2004; Negroni et al., 2009; Xu et al., 2010) in mouse muscles.

Some observations suggest that this phenomenon could also occur in humans. Donor-derived mononuclear cells were detected in the muscles of DMD patients that received myoblast allotransplantations (Gussoni et al., 1997; Skuk et al., 2006b), and some of the donor-derived nuclei were in locations susceptible to correspond to satellite cells (Skuk et al., 2006b). This may imply that the potential therapeutic effect of myoblast transplantation is not limited to the early fusion of the implanted cells, but should also ensure a permanent source of normal satellite cells to participate in muscle hypertrophy and regeneration. Moreover, this could mean that the percentage of myofibers expressing donor-derived dystrophin may increase over time, if a similar process as that described in *mdx* mice regarding the expansion of clusters of myofibers expressing revertant dystrophin is produced (Yokota et al., 2006).

CELL IMPLANTATION

Once appropriate cells for transplantation are produced, the following step is to deliver them to the target tissue in such a way that they will reach a therapeutic objective. Two main routes have been explored for cell transplantation in myology: direct intramuscular injection and intravascular infusion.

To date, intravascular delivery has seemed to be efficient only using a specific type of cells called "mesoangioblasts," and in animal models of particular myopathies (Sampaolesi et al., 2003, 2006). Other myogenic cells so far require direct intramuscular injection. This is notably the case for adult myoblasts, but also for other cell types more recently reported, sometimes considered as stem cells, derived or not from the skeletal muscle (Deasy et al., 2001; Bacou et al., 2004; Zheng et al., 2007; Negroni et al., 2009).

Density of cell injections

The main constraint of the intramuscular route for myoblast transplantation is that the injected cells contribute to muscle regeneration essentially around the injection trajectories, and mostly with the myofibers damaged during the injection. This is clearly observed in non-human primates, in which each myoblast injection leads a strip of hybrid myofiber profiles in histological muscle cross-sections (Figs 43.2 and 43.3) (Skuk et al., 2000, 2002; Quenneville et al., 2007b). An almost similar pattern, although sometimes less defined, was observed following injections of normal myoblasts in DMD patients (Fig. 43.1) (Skuk et al., 2004, 2006b).

Since protein expression is limited to nuclear domains, which are very short in the case of dystrophin, cell injections must be placed very close to each other and must reach the whole muscle to obtain a significant homogeneous expression of donor-derived proteins throughout a skeletal muscle. Consequently, the extent of expression of a donor-derived protein in

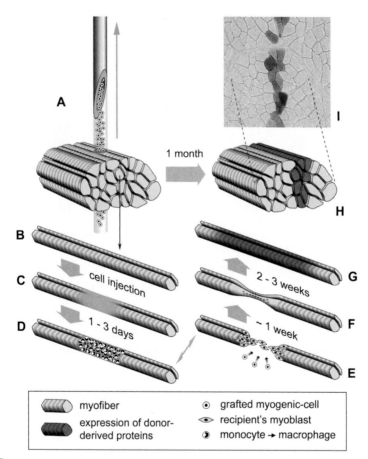

783

FIGURE 43.2
Representation of the mechanism allowing the incorporation of the grafted myoblasts in the recipient's myofibers.
In monkeys, the grafted cells are labeled with a gene coding for β-galactosidase and the result of the graft after 1 month is evaluated through the β-galactosidase expression in the myofibers (I, dark staining). (A) A cell injection traversing a muscle fascicle and delivering the cells homogeneously during the needle withdrawal. (B) to (G) illustrate the process of grafted-cell uptake in two myofibers isolated from this fascicle. Myofibers are damaged by the injection needle (B) and suffer segmental necrosis (C). The necrotic region is invaded by circulating monocytes (D), which become macrophages with two main functions: phagocytosis of the necrotic debris and release of factors helping myofiber regeneration. Myofiber regeneration is done by the activation of the recipient satellite cells, which proliferate as myoblasts that fuse together (E). The regenerative process recruits grafted myoblasts (E). The nuclei of the grafted cells that participated in myoblast fusion are integrated in the myotubes that fill the gap lead by segmental myofiber necrosis (F). Later, these donor-derived nuclei allow the expression of donor-derived proteins throughout a restricted length of the myofiber (D). This process leads to restricted regions of donor-protein expression in the fascicle (H), which are observed as bands of β-galactosidase expression in cross-sections of the cell-grafted muscle (I).

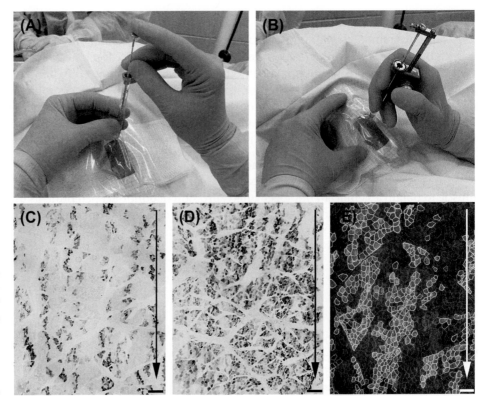

FIGURE 43.3

Myoblast transplantation in non-human primates. Myoblasts are implanted through parallel close intramuscular injections using either a Hamilton syringe operated manually (A) or a repeating dispenser with a Hamilton syringe (B). As in humans, the density of cell injections is controlled with a sterile transparent dressing with a 5-mm grid. After 1 month, the fusion of the grafted cells with the recipient's myofibers is analyzed in cross-sections of muscle biopsies, using histological techniques to detect the transgenic proteins that labeled the grafted cells (C—E). The images show cross-sections of monkey muscles grafted with transgenic myoblasts labeled either with β-galactosidase (C, D) or a micro-dystrophin coupled to a peptide tag (E). Myofibers expressing donor-derived proteins are respectively detected by histochemical detection of β-galactosidase (C, D, dark staining) or fluorescent immunohistological detection of the peptide tag (E). The distribution of the myofibers expressing donor-derived proteins reproduces the pattern of the original cell injection trajectories (indicated by the arrows). The density of β-galactosidase-positive myofibers is higher in (D) than in (C), because the density of cell injections was higher: 25 per cm^2 in (C) and 100 per cm^2 in (D). Scale bars: C,D = 500 μm; E = 200 μm.

a skeletal muscle will depend on the density of the matrix of cell injections, that is on the number of injections per surface of the matrix (Fig. 43.3) (Skuk et al., 2002). The highest percentages of dystrophin-positive myofibers following normal-myoblast transplants in DMD patients were obtained when at least 100 cell injections per cm^2 were done (Skuk et al., 2004, 2006b). This form of cell implantation was denominated "high-density injections" (Skuk, 2004) to make a difference with the earlier clinical trials of myoblast transplantation, which performed few injections away from each other.

Technical approaches for intramuscular transplantation

In a clinical context, a high-density injections protocol with manually operated syringes (Figs 43.1 and 43.3) is adequate only for small volumes of muscle (Skuk et al., 2004, 2006b). Done in that way, the method is slow and needs much concentration to permanently ensure the depth of the intramuscular cell delivery, becoming excessively time-consuming and technically exigent for large volumes of muscle. A first attempt to partially alleviate this problem was to adapt for cell injection some laboratory dispensers for repetitive delivery of small volumes of liquid (Skuk et al., 2006a). A mono-syringe dispenser became our routine instrument for myoblast transplantations in monkeys (Fig. 43.3) (Skuk et al., 2006a; Quenneville et al., 2007b), having been used for myoblast transplants in a DMD patient (Skuk et al., 2007a) and adopted by other research teams (Cesar et al., 2008). However, its clinical use is limited to relatively small muscles, and the precision needed to deliver the cells through a thick skin remains a challenge. Consequently, we determined on the necessity of developing specific instruments for the percutaneous intramuscular injection of cells in a clinical setting. A first semi-manual device was created to deliver very small quantities of cell suspension, homogeneously throughout the intramuscular trajectory of several needles at the same time, avoiding wasting in skin and hypodermis (Richard et al., 2010).

Potential risks of the cell injection procedure

A protocol of high-density injections may involve risks that need to be avoided or controlled. These risks could be local and systemic and, according to the experience in non-human primates, are limited to the first post-transplantation days.

Locally, a monkey's biceps brachium is swollen the first day post-transplantation but reaches its pre-transplantion diameter after 5 days (Skuk et al., 2000). This implies risks of a compartment syndrome in muscles enclosed in a rigid osteofascial space. The monkey's biceps brachium tolerates this treatment well, but for muscles such as the tibialis anterior it may be necessary to proceed cautiously in humans. The first limited tests in a DMD patient were well-tolerated (Skuk et al., 2007a).

Systemically, extensive muscle damage releases intracellular metabolites such as myoglobin and potassium. This implies risks of cardiac arrhythmia in the case of severe hyperkalemia and acute renal failure if myoglobinuria is produced. Both phenomena were not observed following high-density cell injections in the biceps brachii of monkeys (Skuk et al., 2000). Indeed, maintaining muscle damage in each transplant session under the potentially hazardous limits might prevent this problem. As an example, high-density injections throughout two biceps brachii in monkeys increased serum creatine-kinase to 2000 U/l (Skuk et al., 2000) while the risk of acute renal failure is considered to be at 16,000 U/l (Ward, 1988). This problem was not observed in a first test in humans, but the volumes of muscle injected were proportionally smaller than those used in monkeys (Skuk et al., 2007a).

Trying to improve the efficiency of myoblast injections

Lower density of cell injections is desirable but, to reach this objective, the volume of muscle expressing a therapeutic protein (e.g. dystrophin) following each single cell injection must be increased. This could be achieved by: (1) forcing the implanted cells to fuse with other myofibers than those around the injection trajectory, and/or (2) increasing the nuclear domain of the therapeutic protein. This last possibility has rarely been investigated: only one study in *mdx* mice reported a three-fold increase of the nuclear domain of dystrophin after transplantation of myoblasts overexpressing dystrophin (Kinoshita et al., 1998).

Two factors may explain why the implanted cells fuse mostly with the myofiber regions reached by the injections: (1) the implanted cells lack the capacity to move through the tissue and/or (2) myofiber regeneration is mandatory to attract the transplanted cells to fuse but this occurs only in the myofiber regions damaged by the injection.

Addressing the first factor, various studies wanted promotion of the intramuscular diffusion of the grafted myoblasts by inducing in them the secretion of enzymes that degrade the extracellular matrix (Caron et al., 1999; El-Fahime et al., 2002; Lafreniere et al., 2006, 2009; Mills et al., 2007a,b). Several of these studies improved the migration capacity of myoblasts under experimental conditions *in vitro* and *in vivo* in mice, and even enhanced the transplantation success. However, the first tests in monkeys were disappointing and suggested that an increasing migration capacity in the implanted myoblasts does not augment per se the success of their transplantation (Lafreniere et al., 2009). In fact, evidence in monkeys indicates that the grafted myoblasts have an intrinsic capacity to migrate several millimeters into the muscle, but that they do this essentially to fuse with regenerating myofibers in a mechanically damaged tissue (Skuk, 2004).

Addressing the second factor, other experiments increased the number of regenerating myofibers to favor the uptake of the grafted myoblasts. Local injection of myotoxins such as phospholipases derived from snake venoms (Kinoshita et al., 1994; Vilquin et al., 1995) and local anesthetics (Cantini et al., 1994; Pin and Merrifield, 1997) were used efficiently in mice for this purpose, and snake venom phospholipases are now routinely employed for myogenic-cell transplantation in mice. Intense muscular exercise produced extensive myofiber necrosis and almost doubled the success of myoblast transplantation in dystrophic mice (Bouchentouf et al., 2008). On the other hand, inhibiting the proliferation of the recipient's satellite cells favors the participation of the grafted myoblasts in myofiber regeneration. This is achieved in mice by exposing the recipient muscles to high doses of ionizing radiation prior to cell transplantation (Morgan et al., 1990; Alameddine et al., 1994; Kinoshita et al., 1994; Vilquin et al., 1995; Wernig et al., 2000). Cryoinjury of recipient muscles necroses myofibers and satellite cells and also has been used in mice as a pretreatment to favor the implanted cells (Wernig et al., 1995; Irintchev et al., 1997; Brimah et al., 2004).

So far, only the co-injection of myoblasts and myotoxic phospholipases has improved the success of myoblast transplantation in non-human primates. However, this was observed only when myoblasts and myotoxins were highly concentrated in small volumes of muscle (Skuk et al., 1999, 2000). Moreover, with the exception of local anesthetics or muscle exercise, it seems difficult to expect that the other procedures would be accepted for human use. Even then, increasing muscle damage would reduce the volume of muscle to be treated in a session, considering the risks of a rhabdomyolysis-like phenomenon.

CELL-GRAFT SURVIVAL

Once a good delivery of the appropriate cells is obtained, their survival in the recipient must be ensured. The post-transplantation survival of myogenic-cells should be analyzed at two periods: initial and long-term.

Early survival

There is a consensus that an important mortality occurs among the grafted myoblasts rapidly after their intramuscular implantation, that is, mainly within the first 3 days post-transplantation. This was deduced by the progressive loss of different donor-cell markers after myoblast transplantation in mice (Beauchamp et al., 1994; Huard et al., 1994; Guerette et al., 1997b; Qu et al., 1998; Beauchamp et al., 1999; Skuk et al., 2003). In addition, morphological evidence of apoptosis and necrosis was found among the grafted myoblasts early after implantation (Skuk et al., 2003).

This early cell death does not prevent the success of myoblast transplantation. Indeed, not all cells die and in some experiments the proliferation of the surviving cells compensated totally (Skuk et al., 2003) or partially (Beauchamp et al., 1999) the cell death in mice. The process of death and proliferation of the grafted cells during the first post-transplantation days is not well understood, and the studies approaching this topic show contradictions, probably because of methodological differences (Skuk et al., 2003). First experiments in mice blamed cells of the acute inflammatory reaction for killing the implanted myoblasts (Guerette et al., 1997a,b), but later studies found no evidence of neutrophil, macrophage, or natural killer cell responsibility in this death (Sammels et al., 2004). It was also postulated that the survival of the graft could be due to a special small subpopulation of cells that specifically avoid the early cell death and proliferate greatly (Beauchamp et al., 1999; Cousins et al., 2004). Other mechanisms proposed as responsible for causing apoptosis among the grafted myoblasts were hypoxia (Bouchentouf et al., 2008) and anoikis (Bouchentouf et al., 2007).

A rapid and well-determined mechanism of grafted-cell death detected in monkeys is the ischemic necrosis of the central part of the pockets of implanted myoblasts (Fig. 43.4) (Skuk et al., 2007b). The implanted cells form avascular accumulations and their survival depends on the oxygen and nutrients diffusing from the surrounding tissue. Since this diffusion is limited, only a peripheral cell layer of $\sim$100–200 µm survives, the amount of central necrosis depending on the size of the grafted-cell pocket.

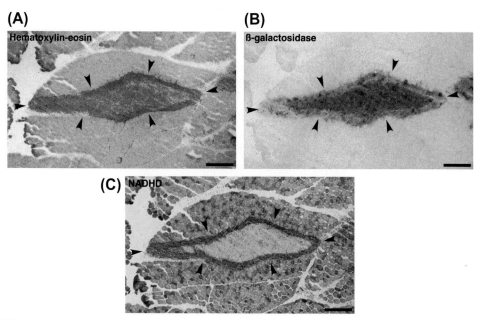

(A) Hematoxylin-eosin

(B) β-galactosidase

(C) NADHD

FIGURE 43.4
Ischemic necrosis in a pocket of intramuscularly implanted myoblasts in a monkey, 1 day post-transplantation. The same region is shown through serial cross-sections stained with hematoxylin-eosin (A), β-galactosidase histochemical detection (B), and histochemical detection of the oxidative enzyme nicotinamide adenine dinucleotide reduced diaphorase (NADHD) (C). Arrowheads circumscribe the pocket of implanted cells. (A) Shows a dense accumulation of mononuclear cells into the muscle and (B) confirms that these are the β-galactosidase-positive grafted myoblasts. Two regions are clearly delimited in the pocket of implanted cells in (C): a peripheral ring strongly NADHD-positive (living cells with oxidative activity) and a core almost devoid of NADHD reaction (necrosed cells without oxidative activity). Scale bars = 500 µm.

787

Long-term survival

The principal challenge of the long-term survival of myoblast grafts is acute rejection in inadequately immunosuppressed allotransplantations (Skuk et al., 2000; Kinoshita et al., 1996a). Acute rejection in the context of myoblast transplants was extensively studied in mice since the first description of lymphocyte infiltration and grafted myoblast disappearance soon after allogeneic transplantation (Jones, 1979). Subsequent studies identified CD8+ and CD4+ lymphocytes in these infiltrates (Guerette et al., 1995a; Irintchev et al., 1995; Wernig and Irintchev, 1995) and expression of IL-2 receptors, Th-1 cytokine, and granzyme B (Guerette et al., 1995b, 1996). Insufficiently immunosuppressed monkeys also exhibited CD8+ infiltration following myoblast allotransplantation, with lymphocyte invasion of myofibers expressing donor proteins (Kinoshita et al., 1996a; Skuk et al., 1999, 2002).

Ensuring cell survival in the recipient

Early cell death among grafted myoblasts is presently misunderstood and so far has not been efficiently prevented. However, since this death does not devastate the population of implanted myoblasts, which seems well restored by the proliferation of the surviving cells, the only potential benefit of inhibiting it would be, in theory, a reduction in the number of cells to be injected. Conversely, the massive ischemic central necrosis can be prevented or reduced by ensuring the formation of accumulations of grafted myoblasts in which most cells are within 100–200 µm of the surrounding tissue (Skuk et al., 2007b).

Acute rejection precludes the success of myoblast allotransplantation. This is controlled by pharmacological immunosuppression, but a careful selection of the drugs is required because some of them kill and/or inhibit the differentiation of the grafted myoblasts (for reviews, see Skuk and Tremblay, 2003; Skuk, 2004). The best results of myoblast allotransplantation in mice were reported using tacrolimus (Kinoshita et al., 1994) and for this reason it became the immunosuppressant of choice for myoblast allotransplantation in monkeys (Kinoshita et al., 1995, 1996a; Skuk et al., 1999, 2000, 2002, 2006a) and consequently in humans (Skuk et al., 2004, 2006b, 2007a). The first clinical trial of myoblast transplantation with tacrolimus immunosuppression involved a follow-up of only 1 month (Skuk et al., 2004, 2006b), but a patient in which tacrolimus was continued during 18 months showed preservation of donor-derived dystrophin during that time: 27.5% of myofibers expressed donor-derived dystrophin 1 month post-transplantation and 34.5% 18 months post-transplantation (Skuk et al., 2007a).

Given that pharmacological immunosuppression has severe secondary effects, one of the main objectives in clinical transplantation is to create immune tolerance, that is, long-term specific unresponsiveness to grafts with preservation of immune reactions against other foreign antigens. In mice, a transient immunosuppression can be sufficient to develop immune tolerance to myoblast allotransplants in some strains (Pavlath et al., 1994) or to delay acute rejection for months in others (Hall and Ralston, 1989; Pavlath et al., 1994). However, the first tests of immunosuppression withdrawal in monkeys caused rapid acute rejection of the hybrid myofibers (Skuk et al., 2000), confirming that immune tolerance is more easily obtained in mice than in monkeys or humans (Hale et al., 2005). Specific protocols to develop immune tolerance for myoblast transplantations were tested in mice. Central tolerance via mixed chimerism (Camirand et al., 2004; Stephan et al., 2006) was more successful than peripheral tolerance combining donor-specific transfusion and administration of anti-CD154 antibodies (Camirand et al., 2002). However, central tolerance in the context of myogenic-cell allotransplantation may not include neoantigens appearing in the hybrid myofibers and would also need peripheral-tolerance mechanisms to avoid acute rejection (Camirand et al., 2008).

Another approach to avoid immunosuppression in this context is the autotransplantation of myoblasts genetically corrected *ex vivo* (Quenneville and Tremblay, 2006). For DMD, genetic correction could be done either by introducing a functional dystrophin gene in the patient's myoblasts (depending on the capacity of the vector the transduced gene may encode for the full-length protein or for a truncated dystrophin) or through inducing the translation of a truncated functional dystrophin following targeted alternative RNA splicing. Tests in mice with different viral vectors or transfection approaches to introduce the therapeutic transgene in myoblasts were considered promising for a future development of this approach (Floyd et al., 1998; Moisset et al., 1998; Bujold et al., 2002; Quenneville et al., 2004, 2007a,b). The feasibility of this strategy was also tested in macaques (Quenneville et al., 2007b). Some of these tactics, however, leave the possibility of a rejection due to incompatibility of minor antigens, essentially neoantigens from the product of the transgene incorporated in the cells (Ohtsuka et al., 1998).

CONCLUSIONS

To reach a therapeutic objective, cell transplantation needs basically three conditions: a cell that meets the properties required for the treatment, a method of cell implantation that ensures that the treatment could be clinically relevant, and conditions to ensure the graft survival throughout the life of the recipient. So far, adult myoblasts are the only myogenic cells that have passed a clinical test of cell transplantation in skeletal muscles, restoring dystrophin in variable amounts of myofibers in DMD patients. Intramuscular implantation through a protocol of high-density injections is the only method that has so far produced relevant percentages of hybrid myofibers in non-human primates and humans. Finally, a tacrolimus-based immunosuppression is the only method that has been successfully tested in non-human primates and humans to ensure the survival of the hybrid myofibers produced by the myoblast allografts.

Some important challenges remain in order to render this technique applicable in the clinics with any expected benefit for patients. A main challenge is to further enhance the success of myoblast transplantation while, at the same time, reducing the density of cell injections. Another challenge is a main problem in the global context of transplantation, that is, to minimize as much as possible the toxicity of the methods needed to control graft rejection. This could be achieved by refining immunosuppressive protocols or by developing efficient approaches of immune tolerance. Cell transplantation also offers the alternative to manipulate the cells before transplantation to avoid immunosuppression, opening the possibility of autotransplantation of genetically corrected cells. Finally, the capacity to restore the functional parenchyma in skeletal muscles that have degenerated to fibrosis and/or fat substitution remains unsolved and barely studied.

References

Alameddine, H. S., Louboutin, J. P., Dehaupas, M., Sebille, A., & Fardeau, M. (1994). Functional recovery induced by satellite cell grafts in irreversibly injured muscles. *Cell Transplant., 3*, 3—14.

Bacou, F., el Andalousi, R. B., Daussin, P. A., Micallef, J. P., Levin, J. M., Chammas, M., et al. (2004). Transplantation of adipose tissue-derived stromal cells increases mass and functional capacity of damaged skeletal muscle. *Cell Transplant., 13*, 103—111.

Beauchamp, J. R., Morgan, J. E., Pagel, C. N., & Partridge, T. A. (1994). Quantitative studies of efficacy of myoblast transplantation. *Muscle Nerve*, S261.

Beauchamp, J. R., Morgan, J. E., Pagel, C. N., & Partridge, T. A. (1999). Dynamics of myoblast transplantation reveal a discrete minority of precursors with stem cell-like properties as the myogenic source. *J. Cell Biol., 144*, 1113—1122.

Betz, E. H., Firket, H., & Reznik, M. (1966). Some aspects of muscle regeneration. *Int. Rev. Cytol., 19*, 203—227.

Borisov, A. B. (1999). Regeneration of skeletal and cardiac muscle in mammals: do nonprimate models resemble human pathology? *Wound Repair Regen., 7*, 26—35.

Bouchentouf, M., Benabdallah, B. F., Rousseau, J., Schwartz, L. M., & Tremblay, J. P. (2007). Induction of anoikis following myoblast transplantation into SCID mouse muscles requires the Bit1 and FADD pathways. *Am. J. Transplant., 7*, 1491—1505.

Bouchentouf, M., Benabdallah, B. F., Bigey, P., Yau, T. M., Scherman, D., & Tremblay, J. P. (2008). Vascular endothelial growth factor reduced hypoxia-induced death of human myoblasts and improved their engraftment in mouse muscles. *Gene Ther., 15*, 404—414.

Brimah, K., Ehrhardt, J., Mouly, V., Butler-Browne, G. S., Partridge, T. A., & Morgan, J. E. (2004). Human muscle precursor cell regeneration in the mouse host is enhanced by growth factors. *Hum. Gene Ther., 15*, 1109—1124.

Bujold, M., Caron, N., Camiran, G., Mukherjee, S., Allen, P. D., Tremblay, J. P., et al. (2002). Autotransplantation in mdx mice of mdx myoblasts genetically corrected by an HSV-1 amplicon vector. *Cell Transplant., 11*, 759—767.

Camirand, G., Caron, N. J., Turgeon, N. A., Rossini, A. A., & Tremblay, J. P. (2002). Treatment with anti-CD154 antibody and donor-specific transfusion prevents acute rejection of myoblast transplantation. *Transplantation, 73*, 453—461.

Camirand, G., Rousseau, J., Ducharme, M. E., Rothstein, D. M., & Tremblay, J. P. (2004). Novel Duchenne muscular dystrophy treatment through myoblast transplantation tolerance with anti-CD45RB, anti-CD154 and mixed chimerism. *Am. J. Transplant., 4*, 1255—1265.

Camirand, G., Stephan, L., Rousseau, J., Sackett, M. K., Caron, N. J., Mills, P., et al. (2008). Central tolerance to myogenic cell transplants does not include muscle neoantigens. *Transplantation, 85*, 1791—1801.

Cantini, M., Massimino, M. L., Catani, C., Rizzuto, R., Brini, M., & Carraro, U. (1994). Gene transfer into satellite cell from regenerating muscle: bupivacaine allows beta-Gal transfection and expression *in vitro* and *in vivo*. *In Vitro Cell. Dev. Biol. Anim., 30A*, 131—133.

Caron, N. J., Asselin, I., Morel, G., & Tremblay, J. P. (1999). Increased myogenic potential and fusion of matrilysin-expressing myoblasts transplanted in mice. *Cell Transplant., 8*, 465—476.

Cesar, M., Roussanne-Domergue, S., Coulet, B., Gay, S., Micallef, J. P., Chammas, M., et al. (2008). Transplantation of adult myoblasts or adipose tissue precursor cells by high-density injection failed to improve reinnervated skeletal muscles. *Muscle Nerve, 37*, 219—230.

Cousins, J. C., Woodward, K. J., Gross, J. G., Partridge, T. A., & Morgan, J. E. (2004). Regeneration of skeletal muscle from transplanted immortalized myoblasts is oligoclonal. *J. Cell Sci., 117*, 3259—3269.

Deasy, B. M., Jankowski, R. J., & Huard, J. (2001). Muscle-derived stem cells: characterization and potential for cell-mediated therapy. *Blood Cells Mol. Dis., 27*, 924–933.

Ehrhardt, J., Brimah, K., Adkin, C., Partridge, T., & Morgan, J. (2007). Human muscle precursor cells give rise to functional satellite cells *in vivo. Neuromuscul. Disord., 17*, 631–638.

El Fahime, E., Mills, P., Lafreniere, J. F., Torrente, Y., & Tremblay, J. P. (2002). The urokinase plasminogen activator: an interesting way to improve myoblast migration following their transplantation. *Exp. Cell Res., 280*, 169–178.

Floyd, S. S., Jr., Clemens, P. R., Ontell, M. R., Kochanek, S., Day, C. S., Yang, J., et al. (1998). Ex vivo gene transfer using adenovirus-mediated full-length dystrophin delivery to dystrophic muscles. *Gene Ther., 5*, 19–30.

Gross, J. G., & Morgan, J. E. (1999). Muscle precursor cells injected into irradiated mdx mouse muscle persist after serial injury. *Muscle Nerve, 22*, 174–185.

Guerette, B., Asselin, I., Vilquin, J. T., Roy, R., & Tremblay, J. P. (1995a). Lymphocyte infiltration following allo- and xenomyoblast transplantation in mdx mice. *Muscle Nerve, 18*, 39–51.

Guerette, B., Roy, R., Tremblay, M., Asselin, I., Kinoshita, I., Puymirat, J., & Tremblay, J. P. (1995b). Increased granzyme B mRNA after alloincompatible myoblast transplantation. *Transplantation, 60*, 1011–1016.

Guerette, B., Tremblay, G., Vilquin, J. T., Asselin, I., Gingras, M., Roy, R., et al. (1996). Increased interferon-gamma mRNA expression following alloincompatible myoblast transplantation is inhibited by FK506. *Muscle Nerve, 19*, 829–835.

Guerette, B., Asselin, I., Skuk, D., Entman, M., & Tremblay, J. P. (1997a). Control of inflammatory damage by anti-LFA-1: increase success of myoblast transplantation. *Cell Transplant., 6*, 101–107.

Guerette, B., Skuk, D., Celestin, F., Huard, C., Tardif, F., Asselin, I., et al. (1997b). Prevention by anti-LFA-1 of acute myoblast death following transplantation. *J. Immunol., 159*, 2522–2531.

Gussoni, E., Blau, H. M., & Kunkel, L. M. (1997). The fate of individual myoblasts after transplantation into muscles of DMD patients. *Nat. Med., 3*, 970–977.

Hale, D. A., Dhanireddy, K., Bruno, D., & Kirk, A. D. (2005). Induction of transplantation tolerance in non-human primate preclinical models. *Philos. Trans. R. Soc. Lond. B Biol. Sci., 360*, 1723–1737.

Hall, Z. W., & Ralston, E. (1989). Nuclear domains in muscle cells. *Cell, 59*, 771–772.

Heslop, L., Beauchamp, J. R., Tajbakhsh, S., Buckingham, M. E., Partridge, T. A., & Zammit, P. S. (2001). Transplanted primary neonatal myoblasts can give rise to functional satellite cells as identified using the Myf5 (nlacZl+) mouse. *Gene Ther., 8*, 778–783.

Huard, J., Bouchard, J. P., Roy, R., Malouin, F., Dansereau, G., Labrecque, C., et al. (1992). Human myoblast transplantation: preliminary results of 4 cases. *Muscle Nerve, 15*, 550–560.

Huard, J., Acsadi, G., Jani, A., Massie, B., & Karpati, G. (1994). Gene transfer into skeletal muscles by isogenic myoblasts. *Hum. Gene Ther., 5*, 949–958.

Irintchev, A., Zweyer, M., & Wernig, A. (1995). Cellular and molecular reactions in mouse muscles after myoblast implantation. *J. Neurocytol., 24*, 319–331.

Irintchev, A., Langer, M., Zweyer, M., Theisen, R., & Wernig, A. (1997). Functional improvement of damaged adult mouse muscle by implantation of primary myoblasts. *J. Physiol. (Lond.), 500*, 775–785.

Irintchev, A., Rosenblatt, J. D., Cullen, M. J., Zweyer, M., & Wernig, A. (1998). Ectopic skeletal muscles derived from myoblasts implanted under the skin. *J. Cell Sci., 111*, 3287–3297.

Jones, P. H. (1979). Implantation of cultured regenerate muscle cells into adult rat muscle. *Exp. Neurol., 66*, 602–610.

Karpati, G., Pouliot, Y., Zubrzycka-Gaarn, E., Carpenter, S., Ray, P. N., Worton, R. G., et al. (1989). Dystrophin is expressed in mdx skeletal muscle fibers after normal myoblast implantation. *Am. J. Pathol., 135*, 27–32.

Karpati, G., Ajdukovic, D., Arnold, D., Gledhill, R. B., Guttman, R., Holland, P., et al. (1993). Myoblast transfer in Duchenne muscular dystrophy. *Ann. Neurol., 34*, 8–17.

Kikuchi, T., Doerr, L., & Ashmore, C. R. (1980). A possible mechanism of phenotypic expression of normal and dystrophic genomes on succinic dehydrogenase activity and fiber size within a single myofiber of muscle transplants. *J. Neurol. Sci., 45*, 273–286.

Kinoshita, I., Vilquin, J. T., Guerette, B., Asselin, I., Roy, R., & Tremblay, J. P. (1994). Very efficient myoblast allotransplantation in mice under FK506 immunosuppression. *Muscle Nerve, 17*, 1407–1415.

Kinoshita, I., Vilquin, J. T., Gravel, C., Roy, R., & Tremblay, J. P. (1995). Myoblast allotransplantation in primates. *Muscle Nerve, 18*, 1217–1218.

Kinoshita, I., Roy, R., Dugre, F. J., Gravel, C., Roy, B., Goulet, M., et al. (1996a). Myoblast transplantation in monkeys: control of immune response by FK506. *J. Neuropathol. Exp. Neurol., 55*, 687–697.

Kinoshita, I., Vilquin, J. T., & Tremblay, J. P. (1996b). Mechanism of increasing dystrophin-positive myofibers by myoblast transplantation: study using mdx/beta-galactosidase transgenic mice. *Acta Neuropathol., 91*, 489–493.

Kinoshita, I., Vilquin, J. T., Asselin, I., Chamberlain, J., & Tremblay, J. P. (1998). Transplantation of myoblasts from a transgenic mouse overexpressing dystrophin produced only a relatively small increase of dystrophin-positive membrane. *Muscle Nerve, 21*, 91–103.

Konigsberg, I. R. (1960). The differentiation of cross-striated myofibrils in short term cell culture. *Exp. Cell Res., 21*, 414–420.

Lafreniere, J. F., Mills, P., Bouchentouf, M., & Tremblay, J. P. (2006). Interleukin-4 improves the migration of human myogenic precursor cells *in vitro* and *in vivo*. *Exp. Cell Res., 312*, 1127–1141.

Lafreniere, J. F., Caron, M. C., Skuk, D., Goulet, M., Cheikh, A. R., & Tremblay, J. P. (2009). Growth factor co-injection improves the migration potential of monkey myogenic precursors without affecting cell transplantation success. *Cell Transplant., 18*, 719–730.

Leriche-Guerin, K., Anderson, L. V., Wrogemann, K., Roy, B., Goulet, M., & Tremblay, J. P. (2002). Dysferlin expression after normal myoblast transplantation in SCID and in SJL mice. *Neuromuscul. Disord., 12*, 167–173.

Lipton, B. H., & Schultz, E. (1979). Developmental fate of skeletal muscle satellite cells. *Science, 205*, 1292–1294.

Mendell, J. R., Kissel, J. T., Amato, A. A., King, W., Signore, L., Prior, T. W., et al. (1995). Myoblast transfer in the treatment of Duchenne's muscular dystrophy. *N. Engl. J. Med., 333*, 832–838.

Mills, P., Dominique, J. C., Lafreniere, J. F., Bouchentouf, M., & Tremblay, J. P. (2007a). A synthetic mechano growth factor E peptide enhances myogenic precursor cell transplantation success. *Am. J. Transplant., 7*, 2247–2259.

Mills, P., Lafreniere, J. F., Benabdallah, B. F., El Fahime el, M., & Tremblay, J. P. (2007b). A new pro-migratory activity on human myogenic precursor cells for a synthetic peptide within the E domain of the mechano growth factor. *Exp. Cell Res., 313*, 527–537.

Moisset, P. A., Skuk, D., Asselin, I., Goulet, M., Roy, B., Karpati, G., et al. (1998). Successful transplantation of genetically corrected DMD myoblasts following ex vivo transduction with the dystrophin minigene. *Biochem. Biophys. Res. Commun., 247*, 94–99.

Morgan, J. E., Hoffman, E. P., & Partridge, T. A. (1990). Normal myogenic cells from newborn mice restore normal histology to degenerating muscles of the mdx mouse. *J. Cell Biol., 111*, 2437–2449.

Negroni, E., Riederer, I., Chaouch, S., Belicchi, M., Razini, P., di Santo, J., et al. (2009). *In vivo* myogenic potential of human CD133(+) muscle-derived stem cells: a quantitative study. *Mol. Ther., 17*, 1234–1240.

Ohtsuka, Y., Udaka, K., Yamashiro, Y., Yagita, H., & Okumura, K. (1998). Dystrophin acts as a transplantation rejection antigen in dystrophin-deficient mice: implication for gene therapy. *J. Immunol., 160*, 4635–4640.

Partridge, T. A., Grounds, M., & Sloper, J. C. (1978). Evidence of fusion between host and donor myoblasts in skeletal muscle grafts. *Nature, 273*, 306–308.

Partridge, T. A., Morgan, J. E., Coulton, G. R., Hoffman, E. P., & Kunkel, L. M. (1989). Conversion of mdx myofibres from dystrophin-negative to -positive by injection of normal myoblasts. *Nature, 337*, 176–179.

Pavlath, G. K., Rich, K., Webster, S. G., & Blau, H. M. (1989). Localization of muscle gene products in nuclear domains. *Nature, 337*, 570–573.

Pavlath, G. K., Rando, T. A., & Blau, H. M. (1994). Transient immunosuppressive treatment leads to long-term retention of allogeneic myoblasts in hybrid myofibers. *J. Cell Biol., 127*, 1923–1932.

Pin, C. L., & Merrifield, P. A. (1997). Developmental potential of rat L6 myoblasts *in vivo* following injection into regenerating muscles. *Dev. Biol., 188*, 147–166.

Qu, Z., Balkir, L., van Deutekom, J. C., Robbins, P. D., Pruchnic, R., & Huard, J. (1998). Development of approaches to improve cell survival in myoblast transfer therapy. *J. Cell Biol., 142*, 1257–1267.

Quenneville, S. P., & Tremblay, J. P. (2006). Ex vivo modification of cells to induce a muscle-based expression. *Curr. Gene Ther., 6*, 625–632.

Quenneville, S. P., Chapdelaine, P., Rousseau, J., Beaulieu, J., Caron, N. J., Skuk, D., et al. (2004). Nucleofection of muscle-derived stem cells and myoblasts with phiC31 integrase: stable expression of a full-length-dystrophin fusion gene by human myoblasts. *Mol. Ther., 10*, 679–687.

Quenneville, S. P., Chapdelaine, P., Rousseau, J., & Tremblay, J. P. (2007a). Dystrophin expression in host muscle following transplantation of muscle precursor cells modified with the phiC31 integrase. *Gene Ther., 14*, 514–522.

Quenneville, S. P., Chapdelaine, P., Skuk, D., Paradis, M., Goulet, M., Rousseau, J., et al. (2007b). Autologous transplantation of muscle precursor cells modified with a lentivirus for muscular dystrophy: human cells and primate models. *Mol. Ther., 15*, 431–438.

Ralston, E., & Hall, Z. W. (1992). Restricted distribution of mRNA produced from a single nucleus in hybrid myotubes. *J. Cell Biol., 119*, 1063–1068.

Richard, P. L., Gosselin, C., Laliberte, T., Paradis, M., Goulet, M., Tremblay, J. P., & Skuk, D. (2010). A first semi-manual device for clinical intramuscular repetitive cell injections. *Cell Transplant, 19*, 67–78.

Sammels, L. M., Bosio, E., Fragall, C. T., Grounds, M. D., van Rooijen, N., & Beilharz, M. W. (2004). Innate inflammatory cells are not responsible for early death of donor myoblasts after myoblast transfer therapy. *Transplantation, 77*, 1790–1797.

Sampaolesi, M., Torrente, Y., Innocenzi, A., Tonlorenzi, R., d'Antona, G., Pellegrino, M. A., et al. (2003). Cell therapy of alpha-sarcoglycan null dystrophic mice through intra-arterial delivery of mesoangioblasts. *Science, 301*, 487–492.

Sampaolesi, M., Blot, S., d'Antona, G., Granger, N., Tonlorenzi, R., Innocenzi, A., et al. (2006). Mesoangioblast stem cells ameliorate muscle function in dystrophic dogs. *Nature, 444*, 574–579.

Satoh, A., Labrecque, C., & Tremblay, J. P. (1992). Myotubes can be formed within implanted adipose tissue. *Transplant. Proc., 24*, 3017–3019.

Skuk, D. (2004). Myoblast transplantation for inherited myopathies: a clinical approach. *Expert Opin. Biol. Ther., 4*, 1871–1885.

Skuk, D., & Tremblay, J. P. (2003). Myoblast transplantation: the current status of a potential therapeutic tool for myopathies. *J. Muscle Res. Cell Motil., 24*, 285–300.

Skuk, D., Roy, B., Goulet, M., & Tremblay, J. P. (1999). Successful myoblast transplantation in primates depends on appropriate cell delivery and induction of regeneration in the host muscle. *Exp. Neurol., 155*, 22–30.

Skuk, D., Goulet, M., Roy, B., & Tremblay, J. P. (2000). Myoblast transplantation in whole muscle of non-human primates. *J. Neuropathol. Exp. Neurol., 59*, 197–206.

Skuk, D., Goulet, M., Roy, B., & Tremblay, J. P. (2002). Efficacy of myoblast transplantation in non-human primates following simple intramuscular cell injections: toward defining strategies applicable to humans. *Exp. Neurol., 175*, 112–126.

Skuk, D., Caron, N. J., Goulet, M., Roy, B., & Tremblay, J. P. (2003). Resetting the problem of cell death following muscle-derived cell transplantation: detection, dynamics and mechanisms. *J. Neuropathol. Exp. Neurol., 62*, 951–967.

Skuk, D., Roy, B., Goulet, M., Chapdelaine, P., Bouchard, J. P., Roy, R., et al. (2004). Dystrophin expression in myofibers of Duchenne muscular dystrophy patients following intramuscular injections of normal myogenic cells. *Mol. Ther., 9*, 475–482.

Skuk, D., Goulet, M., & Tremblay, J. P. (2006a). Use of repeating dispensers to increase the efficiency of the intramuscular myogenic cell injection procedure. *Cell Transplant., 15*, 659–663.

Skuk, D., Goulet, M., Roy, B., Chapdelaine, P., Bouchard, J. P., Roy, R., et al. (2006b). Dystrophin expression in muscles of Duchenne muscular dystrophy patients after high-density injections of normal myogenic cells. *J. Neuropathol. Exp. Neurol., 65*, 371–386.

Skuk, D., Goulet, M., Roy, B., Piette, V., Cote, C. H., Chapdelaine, P., et al. (2007a). First test of a 'high-density injection' protocol for myogenic cell transplantation throughout large volumes of muscles in a Duchenne muscular dystrophy patient: eighteen months follow-up. *Neuromuscul. Disord., 17*, 38–46.

Skuk, D., Paradis, M., Goulet, M., & Tremblay, J. P. (2007b). Ischemic central necrosis in pockets of transplanted myoblasts in non-human primates: implications for cell-transplantation strategies. *Transplantation, 84*, 1307–1315.

Stephan, L., Pichavant, C., Bouchentouf, M., Mills, P., Camirand, G., Tagmouti, S., et al. (2006). Induction of tolerance across fully mismatched barriers by a nonmyeloablative treatment excluding antibodies or irradiation use. *Cell Transplant., 15*, 835–846.

Tremblay, J. P., Malouin, F., Roy, R., Huard, J., Bouchard, J. P., Satoh, A., et al. (1993). Results of a triple blind clinical study of myoblast transplantations without immunosuppressive treatment in young boys with Duchenne muscular dystrophy. *Cell Transplant., 2*, 99–112.

Vilquin, J. T., Asselin, I., Guerette, B., Kinoshita, I., Roy, R., & Tremblay, J. P. (1995). Successful myoblast allotransplantation in mdx mice using rapamycin. *Transplantation, 59*, 422–426.

Vilquin, J. T., Kinoshita, I., Roy, B., Goulet, M., Engvall, E., Tome, F., et al. (1996). Partial laminin alpha2 chain restoration in alpha2 chain-deficient dy/dy mouse by primary muscle cell culture transplantation. *J. Cell Biol., 133*, 185–197.

Ward, M. M. (1988). Factors predictive of acute renal failure in rhabdomyolysis. *Arch. Intern. Med., 148*, 1553–1557.

Watt, D. J., Lambert, K., Morgan, J. E., Partridge, T. A., & Sloper, J. C. (1982). Incorporation of donor muscle precursor cells into an area of muscle regeneration in the host mouse. *J. Neurol. Sci., 57*, 319–331.

Wernig, A., & Irintchev, A. (1995). "Bystander" damage of host muscle caused by implantation of MHC-compatible myogenic cells. *J. Neurol. Sci., 130*, 190–196.

Wernig, A., Irintchev, A., & Lange, G. (1995). Functional effects of myoblast implantation into histoincompatible mice with or without immunosuppression. *J. Physiol. (Lond.), 484*, 493–504.

Wernig, A., Zweyer, M., & Irintchev, A. (2000). Function of skeletal muscle tissue formed after myoblast transplantation into irradiated mouse muscles. *J. Physiol. (Lond.), 522*, 333–345.

Xu, X., Xu, Z., Xu, Y., & Cui, G. (2010). Effects of mesenchymal stem cell transplantation on extracellular matrix after myocardial infarction in rats. *Coron. Artery Dis., 16,* 245—255.

Yablonka-Reuveni, Z., & Nameroff, M. (1990). Temporal differences in desmin expression between myoblasts from embryonic and adult chicken skeletal muscle. *Differentiation, 45,* 21—28.

Yao, S. N., & Kurachi, K. (1993). Implanted myoblasts not only fuse with myofibers but also survive as muscle precursor cells. *J. Cell Sci., 105,* 957—963.

Yokota, T., Lu, Q. L., Morgan, J. E., Davies, K. E., Fisher, R., Takeda, S., & Partridge, T. A. (2006). Expansion of revertant fibers in dystrophic mdx muscles reflects activity of muscle precursor cells and serves as an index of muscle regeneration. *J. Cell Sci., 119,* 2679—2687.

Zheng, B., Cao, B., Crisan, M., Sun, B., Li, G., Logar, A., et al. (2007). Prospective identification of myogenic endothelial cells in human skeletal muscle. *Nat. Biotechnol., 25,* 1025—1034.

CHAPTER 44

Clinical Islet Transplantation

Juliet A. Emamaullee, Michael McCall, A.M. James Shapiro
Department of Surgery, University of Alberta, Edmonton, Alberta, Canada

INTRODUCTION
Background

Diabetes is a disease that results from impaired glucose metabolism. Approximately 90% of diabetes is caused by a defect in insulin production and/or utilization (type 2 diabetes mellitus (T2DM)), while the more severe form, type 1 diabetes mellitus (T1DM), is caused by a complete loss of the insulin-producing β-cells within the islets of Langerhans of the pancreas. Diabetes currently affects more than 200 million patients worldwide and is projected to afflict at least 5% of the global adult population by the year 2025 (King et al., 1998). As the incidence of diabetes increases, the cost of treating these patients has skyrocketed, consuming between 7 and 13% of heathcare expenditure in developed countries (World Health Organization, 2002). Since the discovery of insulin in 1921, diabetes has become a treatable condition, and the life expectancy of patients with diabetes has been greatly improved. However, even with diligent blood glucose monitoring and insulin administration, the metabolic abnormalities associated with diabetes can lead to many chronic secondary complications, including nephropathy, retinopathy, peripheral neuropathy, coronary ischemia, stroke, amputation, erectile dysfunction, and gastroparesis (National Diabetes Data Group, 1995). In the USA, patients with diabetes represent 8% of those who are legally blind, 30% of all patients on dialysis due to end-stage renal disease, and 20% of all patients receiving kidney transplants (National Diabetes Data Group, 1995). The Diabetes Control and Complications Trial (DCCT) was conducted to determine whether intensive blood glucose regulation by frequent insulin injection or pump could prevent these long-term complications in patients with diabetes (Diabetes Control and Complications Trial Reasearch Group, 1990, 1993; Keen, 1994). Results from the DCCT and the subsequent Epidemiology of Diabetes Interventions and Complications study (EDIC) have clearly demonstrated that this approach improved but did not normalize glycosylated hemoglobin levels (HbA1C) and significantly protected against cardiovascular disease, nephropathy, neuropathy, and retinopathy (Diabetes Control and Complications Trial Reasearch Group, 1990; Keen, 1994; Nathan et al., 2003, 2005). However, the consequence of improved glycemic control was a three-fold increased risk of serious hypoglycemic reactions leading to recurrent seizures and coma (Keen, 1994; Diabetes Control and Complications Trial Reasearch Group, 1995). Recent improvements in the size and sensitivity of insulin pumps have increased their utility, but the creation of implantable devices has been more challenging. Also, while insulin pump therapy can improve HbA1C levels compared to multiple daily injections of insulin, pumps may malfunction and thus still necessitate frequent blood glucose monitoring by the user (Owen, 2006; National Library of Medicine, 2010). Recent developments in closed loop pump devices have prompted a number of ongoing clinical trials (National Library of Medicine, (2010).

Principles of Regenerative Medicine. DOI: 10.1016/B978-0-12-381422-7.10044-6

While advances in the formulation, half-life, and administration of insulin have markedly improved the quality of life and long-term survival of patients with diabetes, it has long been recognized that the restoration of an adequate islet mass would provide the maximum benefit to diabetic patients, leading to a true physiological correction of the diabetic state. In the early 1960s, great advances were made in the field of renal transplantation due to improved immunosuppressive therapies (azathioprine and corticosteroids), which prompted the first attempts in whole pancreas transplantation (Merrill et al., 1963; Murray et al., 1963). First introduced by Kelly and Lillehei in 1966, early attempts were associated with high mortality rates and poor graft survival, with less than 3% graft function at 1 year post-transplant (Kelly et al., 1967). The risk profile and long-term outcomes in whole pancreas transplantation have been greatly improved by recent improvements in surgical technique, including portal venous and enteric endocrine drainage and steroid-free maintenance immunosuppression (Newell et al., 1996, Kendall et al., 1997). To date more than 30,000 pancreas transplants have been performed worldwide for end-stage renal disease (simultaneous kidney and pancreas or pancreas after kidney transplantation) or less frequently for severe hypoglycemic unawareness (pancreas transplant alone). Data collected in the International Pancreas Transplant Registry (IPTR) have shown that only 50% of patients who have undergone pancreas-alone transplantation remain insulin independent at 5 years, despite recent improvements in surgical technique and immunosuppression (Larsen, 2004; Gruessner et al., 2005, 2008). Also, less than 30% of the approximately 6,000 cadaveric pancreata donated each year are transplanted due to strict donor criteria and requirements for short cold ischemic time (Larsen, 2004; US Scientific Registry of Transplant Recipients and the Organ Procurement and Transplantation Network, 2005). The surgical risks and requirement for lifelong immunosuppression have reserved pancreas-alone transplantation only for those diabetic patients with the most severe and life-threatening disease, despite strong evidence that the procedure can prolong life, reverse established nephropathy, and improve quality of life. Since the major surgical complications in whole pancreas transplantation are related to the exocrine function of the pancreas, which is not necessary to restore euglycemia in diabetic patients, it has long been recognized that β-cell replacement could be achieved with implantation of isolated pancreatic islets. Since this approach involves transplantation of a cellular graft that would be implanted using minimally invasive techniques, it would avoid the risks associated with major surgery, resulting in a more widely available treatment for patients with diabetes.

History of islet transplantation

The concept of islet transplantation actually preceded the discovery of insulin in 1921 by nearly 30 years (Fig. 44.1). In 1893, physicians in Bristol attempted to treat a young boy suffering from diabetic ketoacidosis by transplanting fragments of a freshly slaughtered sheep's pancreas (Williams, 1894). While the graft ultimately failed in the absence of immunosuppression, the patient's health did temporarily improve, which suggested that cells within the pancreas could restore euglycemia. After the discovery of insulin, it was thought that exogenous insulin replacement would be an effective treatment for patients with T1DM, and therefore islet transplantation was not actively pursued. However, as insulin therapy transformed T1DM from an acute health crisis to a chronic disease, it became apparent that insulin injections could not prevent the onset of debilitating and life-threatening secondary complications. As the first series of whole pancreas transplants in the late 1960s were associated with poor morbidity and mortality, isolated islet transplantation gained a renewed interest (Sutherland et al., 2001). The first successful islet isolations and subsequent transplantation into chemically induced diabetic rodents were pioneered by Dr. Paul Lacy at Washington University in St. Louis, which immediately sparked interest in the implementation of clinical trials (Lacy and Kostianovsky, 1967; Ballinger and Lacy, 1972; Reckard et al., 1973). While euglycemia was routinely obtained in animal models of islet transplantation, clinical islet transplantation struggled to find success for most of the 1970s and 1980s. During this time,

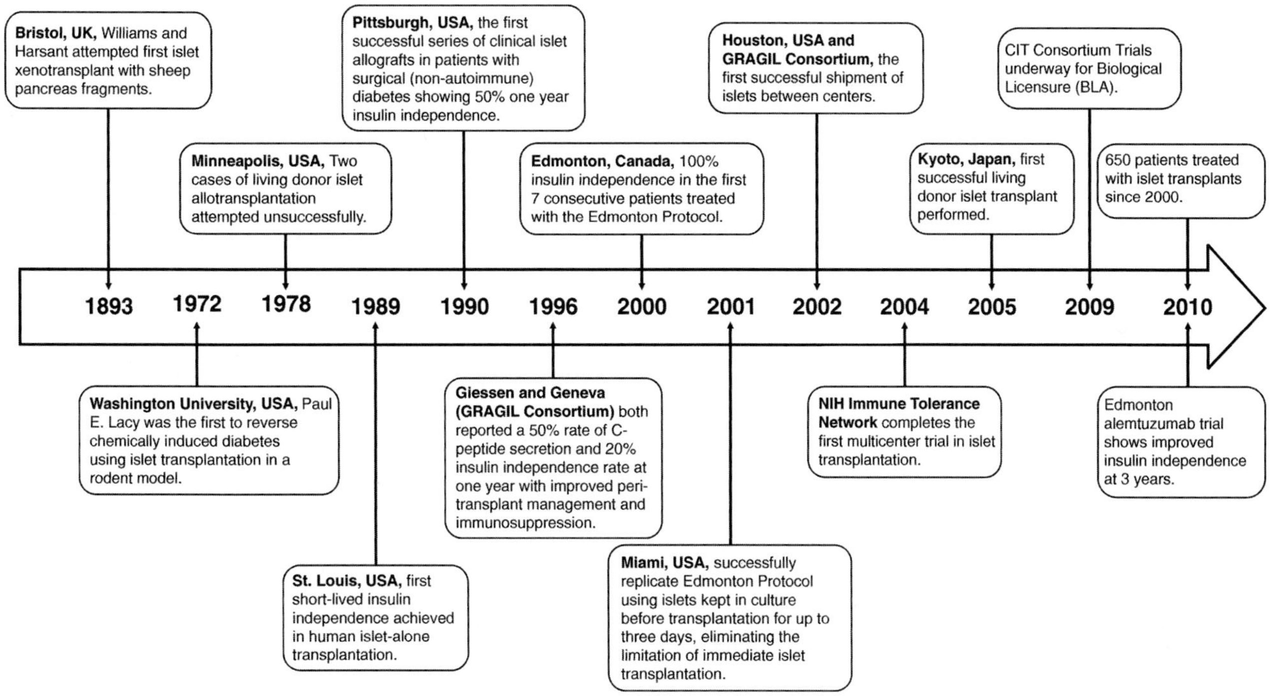

FIGURE 44.1
Timeline of notable advances in the history of islet transplantation.

unpurified islets were infused into the portal vein, leading to many serious complications including portal vein thrombosis, portal hypertension and disseminated intravascular coagulation (Walsh et al., 1982). While working in Lacy's group, Dr. Camillo Ricordi developed the "automated method" for high-yield islet isolation in 1989 (Ricordi et al., 1989). This represented a major turning point in the field and led to the report that Lacy's group had achieved short-lived insulin independence in a patient with T1DM who had received an islet graft following a previous kidney transplant (Scharp et al., 1990). The following year, the group lead by Ricordi at the University of Pittsburgh reported the first series of clinical islet allografts that demonstrated improved insulin independence rates of 50% at 1 year, in subjects who underwent cluster islet-liver transplants for abdominal malignancies in the setting of surgical-induced (non-autoimmune) diabetes (Ricordi et al., 1989; Tzakis et al., 1990). Although this represented a major advance in the field of islet transplantation, these results could not be reproduced in patients with T1DM, the key patient population in need of β-cell replacement (Ricordi et al., 1992). In the late 1990s, the European GRAGIL consortium reported the first modestly successful insulin independence rates of 20% at 1 year in patients with T1DM, which could be attributed to improved peri-transplant management and immunosuppressive drug regimens (Benhamou et al., 2001). Since the results from Pittsburgh and the GRAGIL consortium were obtained in patients who had previously received a kidney transplant, there was no additional risk in terms of immunosuppression to the patients after receiving an islet graft (Ricordi et al., 1992; Benhamou et al., 2001). An international registry held in Giessen, Germany, has maintained a comprehensive record of previous clinical attempts at islet transplantation globally, and, of the total world experience of over 450 attempts at clinical islet transplantation prior to 2000, less than 8% of subjects achieved insulin independence (Brendel et al., 2001). After three decades of research, the 1-year insulin independence rates in clinical islet transplantation were still too low to justify the risks associated with portal infusion and lifelong immuno-suppression in the majority of patients with T1DM (Seechi et al., 1991; Gross et al., 1998; Hering, 1999; Benhamou et al., 2001; Brendel et al., 2001).

The Edmonton Protocol

Shapiro and colleagues at the University of Alberta developed a new protocol in 1999 that was designed for patients with "brittle diabetes" who experienced extreme difficulty in managing their blood glucose levels ("glucose lability") and/or severe hypoglycemic unawareness (Shapiro et al., 2000). The so-called "Edmonton Protocol" was unique compared to previous attempts in clinical islet transplantation in its high targeted islet mass, with a mean of approximately 13,000 islet equivalents (IE)/kg recipient body weight, often derived from two (or occasionally more) fresh islet preparations, and in its immunosuppression strategy, with emphasized avoidance of corticosteroids and use of potent immunosuppression with combined sirolimus, tacrolimus, and anti-CD25 antibody to protect against rejection and recurrent autoimmunity (Shapiro et al., 2000). This approach lead to dramatic improvements in islet allograft survival, with all of the first seven patients achieving sustained independence from insulin (Shapiro et al., 2000). More than 125 consecutive patients have received islet transplants at the University of Alberta since 1999, and the 1-year insulin independence rate remains steady at approximately 80% after completed transplants (<13,000 IE/kg). The results obtained at the University of Alberta have been replicated at other centers as part of an international multicenter trial through the Immune Tolerance Network, but each center's success has varied greatly depending on its previous experience and skill in islet isolation and immunosuppressive management (Shapiro et al., 2003). The Miami group has demonstrated that islets can be cultured for up to 3 days pre-transplant or shipped and transplanted at a remote facility (Houston) with similar success as freshly isolated islets when transplanted using Edmonton-like immunosuppression (Goss et al., 2002, 2004). The GRAGIL Network (a Swiss-French consortium) has also demonstrated the benefits of centralized islet processing facilities that can service a broader network of centers throughout Europe (Benhamou et al., 2001; Kempf et al., 2005).

Based upon the success of the Edmonton group, islet transplantation has been funded in Alberta, Canada as an accepted clinical standard of care since 2001. Progress in this area has been slower in the USA, but large registration trials are currently moving forward to secure a Biological License and therefore reimbursement, which will make a significant difference to the availability of islets for transplantation in that country. The recent success of clinical islet transplantation has encouraged many centers around the world to implement a program, and since 2000 more than 650 patients have been transplanted using recent variants of the Edmonton Protocol in almost 50 centers worldwide (International Islet Transplant Registry, 2005).

Despite this success, the current requirement for lifelong immunosuppression in islet-alone transplantation has restricted its availability to patients with T1DM and severe hypoglycemia or glycemic lability. The benefit of islet transplantation in patients with T2DM has not been determined, since many of these patients are overweight and/or insulin resistant and thus would require a large islet mass to meet their metabolic demands. Many patients require two or occasionally three islet implant procedures in order to achieve insulin independence, although insulin independence following single donor infusion has been steadily improving at our center and others (Hering et al., 2004, 2005). While C-peptide secretion (>0.5 ng/ml) has been maintained in 88% of islet graft recipients beyond 3 years in Edmonton, initial data on the long-term insulin-independence rates demonstrated that only 50% of recipients remained off insulin at 3 years, with less than 10% off insulin at 5 years post-transplant (Ryan et al., 2005). The causes of this discrepancy between insulin independence and maintenance of C-peptide status are likely complex. While rejection (acute or chronic) and recurrent auto-immunity may be responsible for graft loss, it is probable that other, non-immune-mediated, damage occurs, such as chronic toxicity from sirolimus/tacrolimus and failure of islet regeneration or transdifferentiation due to the antiproliferative effects of sirolimus. Perhaps the most important component of decaying graft function over time is the concept of islet "burn-out" from constant metabolic stimulation, since only a marginal mass of islets actually engraft in most subjects. In clinical islet transplantation thus far, the risks of malignancy, post-transplant lymphoma, and life-threatening sepsis have been minimal, but fears of these complications limit a broader application in patients with less severe forms of diabetes including children. Moreover, a number of immunosuppression-related side-effects have been encountered, including dyslipidemia, mouth ulceration, peripheral edema, fatigue, and ovarian cysts and menstrual irregularities in female subjects, which can be dose- or drug-limiting in some patients (Ryan et al., 2002). Thus, while dramatic improvements in outcomes following islet transplantation have been observed, extensive refinements in clinical protocols are needed both to improve safety and to enhance success with single donor islet infusions.

CLINICAL ISLET TRANSPLANTATION
Patient assessment and selection

Clinical islet transplantation is associated with a number of risks, including procedural complications such as bleeding or portal vein thrombosis, and those associated with lifelong immunosuppression, that is, infection or malignancy. For these reasons, patients selected as islet recipients must have severe, life-threatening diabetic complications that justify the risks of transplantation. Two T1DM patient populations have been identified as suitable candidates for islet transplantation: those individuals that experience frequent, severe, and recurrent hypoglycemic unawareness, and those patients with highly unstable blood glucose control despite an optimized insulin regimen (glycemic lability).

When patients are evaluated for islet transplantation, their metabolic status and diabetes-related secondary complications should be carefully characterized so that those patients who would receive the greatest benefit despite the requirement for life-long immunosuppression are selected. First and foremost, islet transplantation is reserved for patients with C-peptide negative (<0.3 ng/ml) T1DM. Recipients with elevated BMI (>30 kg/m^2) or those >90 kg are generally excluded, as their metabolic demand may not be met by the transplanted islet mass. As mentioned previously, the current indications for islet-alone transplantation include severe hypoglycemic unawareness and/or glycemic lability. To assess these symptoms, Ryan et al. developed an objective scoring system to measure the severity of hypoglycemia (the HYPO score) and the Lability Index (LI), which is based upon the changes in blood glucose over time (Ryan et al., 2004b). Current selection criteria for islet-alone transplantation include a HYPO score > 1047 (90th percentile), LI > 433 mmol/L^2/h·week^{-1} (90th percentile), or a composite with the HYPO score > 423 (75th percentile) and LI > 329 (75th percentile) (Ryan et al., 2005). Since patients with poor diabetes compliance or an inadequate baseline insulin regimen are likely to benefit from improved design of their insulin dosing regimens, patients selected for transplant should have a plasma HbA1C < 10%. In an effort to reduce the risk of serious procedural and immunosuppressive drug-related complications, the patient's cardiac and renal function should be carefully assessed. Selected recipients should have adequate cardiac function including blood pressure < 160/100 mmHg, no evidence of myocardial infarction in the 6 months prior to assessment, no angiographic evidence of non-correctable coronary artery disease, and left ventricular ejection fraction (LVEF) > 30% as measured by echocardiogram. To eliminate patients who are better candidates for simultaneous kidney-pancreas transplantation or those who may experience adverse renal function as a result of tacrolimus or sirolimus therapy, selected recipients should have no evidence of macroscopic proteinuria (<300 mg/24 h) and a calculated glomerular filtration rate (GFR) > 80 (> 70 in females) mL/min/1.73 m^2. Proliferative retinopathy should be stabilized prior to transplantation, as acute correction of glycemic control may lead to accelerated retinopathy. Finally, to reduce the risk of antibody-mediated graft rejection, potential recipients should be screened for panel reactive antibody assays (PRA) and determined to be < 20%.

Although several locations have been tested as potential implantation sites for islet grafts, the high level of graft function and ease of delivery associated with infusion into the portal circulation of the liver have led to this being the transplantation site of choice in clinical protocols (Kemp et al., 1973). There are two accepted approaches for implanting purified islets into the liver by way of the portal vein. While surgical laparotomy and cannulation of the portal vein were most often used in the early islet transplant programs, current protocols routinely employ the percutaneous transhepatic approach to implant donor islets in cadaveric islet transplantation (Fig. 44.2A) (Ryan et al., 2005). Compared to surgical laparotomy, this procedure is minimally invasive and thus can be performed using local anesthesia, combined with opiate analgesia and hypnotics given as pre-medication. Access to the portal vein is achieved by percutaneous transhepatic approach using a combination of ultrasound and fluoroscopy to guide the radiologist. A branch of the right portal vein is cannulated, and a catheter is positioned proximal to the confluence of the portal vein, which is confirmed with a portal venogram (Owen et al., 2003). The risk of portal vein thrombosis is reduced by inclusion of unfractionated heparin (70 units/kg) in the islet preparation. Islets are then infused, aseptically, into the main portal vein under gravity, with regular monitoring of portal venous pressure (by an indirect pressure transducer) before, during, and after the infusion. The risk of bleeding after percutaneous portal access has now been eliminated by plugging of the catheter tract with thrombostatis paste (Avitene™) at the time of catheter withdrawal. This allows therapeutic heparin to be given (targeting a PTT of 60–80 seconds) within hours of islet implantation, further reducing the risk of portal venous thrombosis and potentially reducing

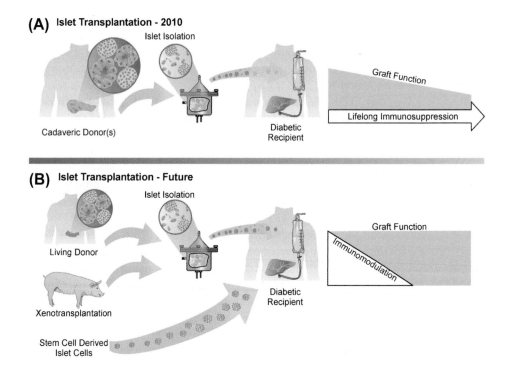

FIGURE 44.2

The islet transplant procedure — present and future. Islet transplantation, in its current form (A), has provided insulin independence in most diabetic patients at 1 year post-transplant, but this procedure is currently limited by the availability of suitable cadaveric donors and the requirement for lifelong immunosuppression. In the future (B), islet transplantation could be made available to a broader range of diabetic patients through the usage of alternative tissue sources, such as living donors, xenogeneic donors, or stem cell-derived β-cells. Also, as novel immunomodulatory therapies are identified, tolerance induction strategies can be developed that will prolong graft function and allow for the reduction or complete withdrawal of immunosuppressive drug therapy.

the instant blood-mediated inflammatory reaction (IBMIR). An ultrasound examination should be performed at 1 day and 1 week post-transplant to rule out intraperitoneal hemorrhage and to confirm that the portal vein is patent and has normal flow.

A large hemangioma present on the right side of the liver may preclude safe access to the portal system. Potentially, a left-sided percutaneous approach may be considered, or alternatively a surgical laparotomy and cannulation of a mesenteric venous tributary of the portal system should be considered. In this situation, complete surgical control is in place to prevent uncontrolled bleeding. Another advantage includes the potential for use of a dual lumen catheter for cannulation of a mesenteric vein (i.e. dual lumen 9Fr Broviac line), which allows for continuous monitoring of portal pressure during islet infusion. Still, this surgical approach should only be considered when the percutaneous transhepatic approach cannot be utilized, as it does present several major disadvantages, including the requirement for a surgical incision, formation of adhesions, and the risk of wound infection and wound herniation, which may be exacerbated when the drug sirolimus is used post-transplant, as this drug interferes with wound healing.

Risks to the recipient

SURGICAL COMPLICATIONS

There are two potentially serious procedural complications in islet transplantation: bleeding from the catheter tract created by the percutaneous transhepatic approach, and portal vein thrombosis, particularly when large volumes of tissue are infused. Adverse bleeding events were noted early in the development of several clinical islet transplant programs (including our own), but these have been completely avoided in the past 100 consecutive procedures with the routine use of effective methods to seal and ablated the transhepatic portal catheter tract on egress when the catheter is withdrawn. We currently advocate injection of Avitene® paste (1 g Avitene powder mixed with 3 ml of radiological contrast media and 3 ml of saline – approximately 0.5–1.0 ml of this paste is injected into the liver tract for a length of at least 5 cm) (Villiger et al., 2005). The use of purified islet allograft preparations has not resulted in main portal vein thrombosis in the Edmonton Program, but thrombosis of a right or left branch, or peripheral segmental vein, has been encountered in approximately 5% of patients. Other rarely observed procedural side-effects have included fine needle gallbladder puncture, arteriovenous fistulae (which may require selective embolization), and steatosis in the hepatic parenchyma, which generally does not present any clinical complications or require intervention (Bhargava et al., 2004).

IMMUNOSUPPRESSIVE THERAPY AND COMPLICATIONS

Islet transplantation for T1DM represents a unique challenge in immunosuppression, as both alloimmunity and islet-specific autoimmunity must be effectively controlled to preserve graft function. An additional important consideration is that many of the immunosuppressive agents used in solid organ transplantation since the 1960s, particularly corticosteroids, are known to be toxic to islets. Previously in the Edmonton Protocol, the induction agent daclizumab (anti-CD25 (IL-2R) antibody) was administered intravenously immediately prior to transplantation and again at 2 weeks post-transplant (1 mg/kg). Maintenance immunosuppression was achieved using sirolimus with a low dose of tacrolimus. This regimen, described initially at the University of Alberta, has been successfully replicated at other centers as part of a multicenter Islet Transport Network (ITN) trial (Shapiro et al., 2003, 2005b).

Currently our approach is to use T-depletional induction therapy with thymoglobulin (cumulative total dose of 6 mg/kg i.v. over 3 days) for first transplants, and basiliximab (Simulect™) 20 mg i.v. on days 0 and 4 for subsequent transplants (daclizumab has recently been taken off the market since its patent expired). We, and others, have found that the more standard post-transplant combination of tacrolimus (level 6–10 ng/ml) and mycophenolate mofetil (up to 2 g per day in divided dose as tolerated) is much better tolerated than sirolimus, and the 3- and 5-year outcome data suggest much more graft durability in terms of sustained insulin independence when T-depletion is combined with tacrolimus/mycophenolate mofetil. More recently we have explored the role of alemtuzumab induction (30 mg i.v. before first or subsequent islet infusion), combined with tacrolimus and mycophenolate maintenance, and again have found durable islet graft function at 3 years without falloff in insulin independence, which has been very encouraging. Again this approach has been well tolerated, but we have seen two cases of opportunistic infection (nocardia and aseptic meningitis), suggesting the potential risk of over-immunosuppression with such an approach, and clearly dosing of maintenance immunosuppression requires further optimization. As a result, we have essentially stopped using sirolimus as maintenance immunosuppression, as we have found it to be extremely poorly tolerated at high dose in this patient population.

We are participating with Emory University presently in a two-center trial of costimulation blockade with belatacept in clinical islet translatation, as part of the National Institutes of Health Clinical Islet Transplantation Consortium trials. Thus far the therapy has been well tolerated and early outcomes look promising when mycophenolate mofetil monotherapy is used in maintenance together with ongoing belatacept. We do not regard this as a tolerogenic protocol, and do have some reservations about the potency of mycophenolate in the longer term, but further data is pending. Certainly such an approach will open up the opportunity for patients with underlying limited renal reserve to undergo islet transplantation — and these patients are currently excluded as tacrolimus maintenance would clearly be detrimental to renal function.

In addition to the Edmonton Protocol immunosuppression described above, alternative regimens have been reported. The Minnesota Group, led by Dr. Bernhard Hering, has utilized antithymocyte globulin and etanercept (anti-tumor necrosis factor-α antibody) induction with a combination of sirolimus and mycophenolate mofetil $\pm$ low dose tacrolimus for maintenance, or hOKT3γ1(Ala-Ala) (humanized anti-CD3 antibody) and sirolimus induction with sirolimus and reduced-dose tacrolimus for maintenance (Hering et al., 2004, 2005). In some instances, alternative immunosuppressive agents have been used because of drug intolerance or other side-effects. Islet patients often possess mild preexisting renal impairment as a result of longstanding diabetes, and this renal dysfunction may be exacerbated with calcineurin inhibitor therapy, even at the low doses involved in the Edmonton Protocol. The drug sirolimus may also have nephrotoxic side-effects, which may be compounded when used in combination with a calcineurin inhibitor drug (Kaplan et al., 2004; Senior et al., 2005). For these reasons, renal status must be monitored diligently in all patients following islet transplantation. In addition to its recognized nephrotoxicity, tacrolimus is associated with gastrointestinal side-effects that may lead to episodic diarrhea. Neurotoxicity may be seen with tacrolimus but is often avoided in low-dose regimens (Gruessner et al., 1996). Sirolimus is associated with neutropenia and mouth ulceration, but these side-effects can be reduced with lower target trough levels and tablet formulations. In the context of islet transplantation, sirolimus has been linked to a number of side-effects including dyslipidemia, small bowel ulceration, peripheral edema, and the development of ovarian cysts or menstrual cycle irregularities in female recipients (Molinari et al., 2005; Ryan et al., 2005).

While chronically immunosuppressed patients are at risk for developing all types of malignancy, squamous epithelial cancers most commonly occur and are most readily treatable. The lifetime risk of lymphoma is estimated to be 1—2% in transplant recipients, but this risk is likely to be reduced in islet recipients, as these patients are generally not treated with glucocorticoids or OKT3.

FUTURE CHALLENGES
Overcoming tissue shortage

In its current form, islet transplantation is reserved for patients with the most severe forms of diabetes, which in reality constitute a small fraction of all patients with T1DM. Even with the relatively small patient population selected for islet transplantation, the waitlist time for patients in Edmonton, which has access to organs from a large geographic region, ranges from 6 months to 2 years depending on blood group. As islet transplantation becomes more suitable for a broader range of diabetic patients and as the incidence of diabetes increases, there will be an even more severe shortage of islet tissue for transplantation. Presently, clinical islet programs rely on the scarce supply of pancreas organs derived exclusively from heart-beating, brain-dead cadavers. Compared to organs procured for whole pancreas transplantation, which must fall within very strict donor criteria, organs obtained for islet transplantation tend to be more "marginal" and come from older, less stable donors. Furthermore, the pancreas is particularly susceptible to toxicity from the circulating products of severe brain injury, hemodynamic instability, and inotropic support in a brain-dead organ donor. The quality of the pancreas is further degraded by cold ischemic injury during transportation, which inevitably results in islet damage and loss. Contreras et al. demonstrated a marked reduction in islet recovery and in islet viability in experimental islet transplantation using tissue derived following brain death compared to healthy rodent donors, highlighting this issue, and recently his group has confirmed these findings using human islets (Contreras et al., 2003). Similarly, a strong relationship between islet recovery and donor stability has been demonstrated (Lakey et al., 1996). Once the pancreas is in the isolation lab, the extensive processing and purification steps during processing result in further islet destruction and loss, often resulting in at best 60% recovery of the estimated 10^7 IE/pancreas (Tsujimura et al., 2004). As a result, nearly all islet recipients require islets derived from two cadaveric donors. Thus, a rapidly growing area of islet transplant research involves the development of improved cadaveric or alternative islet tissue sources for transplantation.

LIVING DONOR ISLET TRANSPLANTATION

One approach to alleviating islet tissue demand would be to make use of living donors for islet transplantation. Living donor programs in kidney, liver, and lung transplantation have moved forward successfully at most leading transplant centers worldwide, in an attempt to meet the growing demand for donor organs and to improve clinical outcomes. Given the rapid, global acceptance of cadaveric islet transplantation over the past 5 years, it is likely that living donor islet transplantation will soon be offered to patients listed in cadaveric islet transplant programs. Despite remarkable progress in clinical islet transplantation since 1999, islet supply and functional viability remain to be significant challenges when islets are derived from cadaveric organ donors, even at the most experienced centers (Contreras et al., 2003). In the living donor setting, the distal half-pancreas could be procured under "ideal" circumstances, without exposure of the pancreas to hemodynamic instability or inotropic drugs, and the pancreas would be processed immediately without prolonged cold ischemia. Thus, the potency of islets derived from a living donor source is assumed to be far superior to cadaveric tissue. Living donor islet transplantation represents a unique opportunity to overcome donor organ shortage and procure the islet tissue under perfect conditions, with closer HLA matching between donor and recipient. Furthermore, the living donor islet transplant setting will provide a unique opportunity to develop protocols for pre-transplant recipient conditioning for donor-specific tolerance induction.

While cadaveric islet transplantation has been an active area of clinical research involving more than 1000 patients in the past 30 years, only three cases of living donor islet allo-transplantation have been reported (Sutherland et al., 1980; Matsumoto et al., 2005). The first two clinical attempts at living donor islet allo-transplantation were carried out in 1978 by Sutherland and colleagues at the University of Minnesota (Sutherland et al., 1980). While neither recipient achieved sustained islet function, these pioneering efforts were truly remarkable given the early stage of clinical islet transplant development at the time. The immunosuppression available was primitive by current standards (azathioprine and high-dose steroids), and the islets were isolated using suboptimal conditions, prior to the development of the Ricordi chamber and the sophisticated purification schemes currently used in clinical islet transplantation. The dramatic improvement in clinical outcomes obtained in cadaveric islet transplantation since 2000 has renewed interest in the development of living donor islet transplantation. The first living donor islet transplantation case attempted since the introduction of the Edmonton Protocol was carried out at the University of Kyoto in early 2005, as a collaboration between the Japanese and Edmonton programs (Matsumoto et al., 2005). The recipient, a 27-year-old female, developed C-peptide negative, unstable diabetes following chronic pancreatitis as a child. Her 56-year-old mother was approved to be the donor, and islets were purified from the distal pancreas (47% as measured pre-operatively by CT volumetry) obtained during an open laparotomy. There were no surgical complications in either donor or recipient. The unpurified islet mass (408,114 IE (8,200 IE/kg) in a volume of 9.5 ml after tissue digestion) was transplanted into the portal vein using the percutaneous approach under full systemic heparinization. Edmonton Protocol-style immunosuppression was started pre-transplant using sirolimus and low-dose tacrolimus (started 7 days pre-transplant), anti-IL2R antibody (given 4 days pre-transplant and on the day of transplant), and anti-TNFα blockade induction (infliximab; given 1 day pre-transplant). Insulin therapy in the recipient was discontinued at 22 days post-transplant, and this patient continues to be insulin independent with excellent glycemic control and a normal HbA1C more than 1 year post-transplant (Matsumoto et al., 2006). The donor has presented no evidence of glucose intolerance and has maintained normal HbA1C values since the procedure.

While no definitive conclusions can be drawn from this single successful case of living donor islet allo-transplantation, results from living donor islet auto-transplantation suggest that the insulin independence may be achieved routinely with significantly less IE/kg recipient body weight than has been required for cadaveric allografts thus far. It is widely accepted that over 70% of patients will remain insulin free following islet autotransplantation if an islet mass exceeding 300,000 IE ($\geq$2,500 IE/kg) is transplanted, compared to the 13,000 IE/kg that is often required to achieve insulin independence with cadaveric islet preparations (Gruessner et al., 2004). Recent reports from the Minnesota group have shown that clinical islet autografts have a significantly lower rate of metabolic decay over time, even with a smaller islet implantation mass (Sutherland et al., 2008). Despite the potential risks for a living donor in terms of surgically induced diabetes and surgical complications, the demand for islet tissue and relative ease of implementation of living donor protocols into established islet transplant programs are likely to move this approach forward rapidly.

XENOTRANSPLANTATION

Living donor islet transplantation may circumvent the wait for suitable donor tissue in some diabetic patients, but the risks to the donor and the possibility of insufficient islet yield to obtain insulin dependence remain significant concerns. Identification of a renewable xenogeneic source of islets would avoid the requirement for human islet donors altogether and could provide enough tissue to transplant diabetic patients as often as required. Pigs are particularly attractive as a xenogeneic islet donors since they are widely available, produce insulin that is functional in humans, and could be selected for certain donor characteristics. Of all types of experimental xenotransplantation, islet transplantation is probably the closest to clinical application. Over the past decade, a number of small clinical trials in islet transplantation using porcine islets have been reported, but few have resulted in reduced insulin requirements and no patients have achieved prolonged insulin independence (Groth et al., 1994; Elliott et al., 2000; Valdes-Gonzalez et al., 2005; Hering and Walawalkar, 2009). Despite these setbacks, islet xenotransplantation using porcine tissue has remained an active area of research, and progress has been made over the past several years in experimental islet xenotransplantation using pre-clinical non-human primate models (Cardona et al., 2006; Hering et al., 2006; Rood et al., 2006; van der Windt et al., 2009). The generation of α1,3-galactosyltransferase-deficient pigs has provided a source of islet tissue lacking the major xenoantigens causing hyperacute rejection in pig-to-human xenotransplantation (Phelps et al., 2003). Still, it remains to be determined whether the transmission of endogenous retroviruses or other zoonotic infections from pig to human can be completely avoided in xenotransplantation, even with the establishment of highly monitored "clean" pig colonies (Fishman and Patience, 2004).

While significant advances have been made in the area of islet xenotransplantation, it is unclear whether enough data have been generated to justify the move toward large-scale clinical trials. However, there are reports that clinical trials are ongoing in centers in China and Russia (Rood and Cooper, 2006).

STEM CELL TRANSPLANTATION

Unlike solid organ transplantation, which requires a complex vascularized tissue structure to restore function in a recipient, islet transplantation could be achieved through the development of a renewable source of stem cell-derived β-cells. Substantial research efforts have been made in identifying suitable islet precursor cells that could be differentiated into an unlimited source of insulin-producing β-cells, but difficulties in producing physiologically regulated insulin secretion and control of proliferation have made progress in this area difficult to achieve (reviewed in Bonner-Weir and Weir, 2005; Otonkoski et al., 2005).

The quest for a renewable source of insulin-producing cells has led researchers to consider many possible origins for these cells. The pancreas itself contains progenitor cells capable of β-cell repopulation in the event of injury (Xu et al., 2003; Seaberg et al., 2004). Given the proper environment and transcription factors, these cells can be directly re-programmed into cells that closely resemble β-cells (Zhou et al., 2008). Some exciting data has been reported using genetically modified human fetal hepatocytes, but data in large animal models is lacking (Zalzman et al., 2003, 2005).

Others have explored using hematopoietic stem cells as precursors to insulin-producing cells. This includes attempts to utilize bone marrow-derived cells in addition to umbilical cord blood (UCB). This was initially a very exciting area of study since UCB is easily obtained and would avoid some of the ethical implications associated with the use of stem cells. Unfortunately, the early animal studies did not show any conclusive evidence of endogenous β-cell replenishment after hematopoietic stem cell injection (Ianus et al., 2003; Kodama et al., 2003; Suri et al., 2006). Even so, clinical studies have been conducted in type 1 diabetic patients. Haller et al. utilized stored autologous UCB infusions in newly diagnosed type 1 diabetics, showing reduced insulin requirements and lower HbA_{1c} (Haller et al., 2008). A further study employing hematopoietic stem cells resulted in the majority of the 23 newly diagnosed type 1 diabetics who received these cells achieving insulin independence and elevated c-peptide levels (Couri et al., 2009).

Embryonic stem cells (ESCs), due to their pluripotency and ability to self-renew, have received an enormous amount of research attention in recent years. Since 2000, researchers have attempted to find the optimal set of conditions and signals to differentiate them into an insulin-producing cellular population. A number of the early attempts were conducted using rodent ESCs, and, while promising initially, they were limited by cell homogeneity, immaturity, low number of insulin-positive cells, and a lack of glucose sensitivity (Soria et al., 2000; Assady et al., 2001; Lumelsky et al., 2001; Hori et al., 2002). It was not until 2004 that an effective differentiation strategy was discovered, paving the way for differentiation of human ESCs into cells that contained both insulin and C-peptide (Kubo et al., 2004; d'Amour et al., 2006). Further refinement of the strategy allowed these cells to become glucose-sensitive, showing the ability to ameliorate diabetes in a rodent model (Kroon et al., 2008). Unfortunately, a number of issues need to be addressed regarding ESCs, including their tendency to form teratomas, but altogether they could potentially overcome the need to rely on organ donation as a source of islet tissue.

The challenge of reproducing the highly differentiated neuroendocrine β-cell phenotype has proven significant, and more investigation in this area is required before stem cell-derived islets will see broad clinical application. Even as progress is made in this area, political and ethical issues may ultimately prevent the timely application of this technology in human subjects.

OPTIMAL TRANSPLANTATION SITE

There continues to be a significant amount of debate revolving around the optimal islet transplantation site. While the liver has become the implantation site of choice, receiving more than 90% of clinical islet grafts, it may not provide the best chance for long-term islet survival.

The islet isolation process subjects islets to significant ischemic and physical injury, rendering them susceptible to post-transplantation stresses. Islets require ready access to oxygen and glucose and benefit from close proximity to a good vascular supply since their revascularization is not immediate and their capacity for diffusion is limited. As an endocrine tissue, islets require a means to sample representative glucose levels and be able to deliver insulin through a relevant route to its target tissues. Ideally, a transplanted islet should reside in a site with minimal immunological attack and low levels of post-transplantation β-cell apoptosis, such as that induced by the IBMIR. From a surgical standpoint, it would be advantageous to have a transplant site that afforded minimal procedural complications and allowed one to monitor islets after implantation.

The portal vein/liver site has become the standard site in the majority of islet transplants. An early rodent study showed this site to be superior with respect to the number of autologous islets required to reverse hyperglycemia (Kemp et al., 1973). However, further studies showed that there is an eventual loss of islet function even in the absence of allo- or autoimmune attack (Alejandro et al., 1986). Being a vascular site, intraportal islets are additionally subjected to IBMIR, leading to significant β-cell apoptosis and islet loss. Finally, while islet infusion is relatively straightforward, there are possible complications including bleeding and thrombosis. While this site has indeed allowed islet transplantation to reach amazing clinical success, there are clear reasons why a search continues for alternative sites.

Although the kidney subcapsular space has become the site of choice for many researchers employing a mouse islet transplant model, it has never shown promise in clinical practice. This is likely for multiple reasons including the poor blood supply and relative lack of oxygen in this space coupled with the difficulty in surgical access. The pancreas, while a tempting site theoretically with its high oxygen content and proximity to endogenous islet location, is relatively invasive to access and may potentiate the autoimmune attack of transplanted islets through the priming of local lymph nodes. While the latter has not yet been proven, the usefulness of this site is nullified in the case of islet autotransplantation. The formation of an omental pouch, created surgically using omentum and the parietal peritoneum, has shown efficacy in both rat (Kin et al., 2003) and dog (Ao et al., 1992) models of diabetes. Although necessitating a higher number of islets to reverse diabetes (as compared to the renal subcapsular site), the omentum is a very vascular site and is a possible location for the implantation of islet encapsulation devices. A related structure in mice, the epididymal fat pat, has been used successfully to transplant embryonic endocrine progenitor cells (Wszola et al., 2009). In addition, a pouch could be created laparoscopically, minimizing the morbidity of surgery. Further research needs to be completed to determine the long-term survival of islets at this potentially useful site.

Researchers have recently shown that islets can be transplanted into the gastric submucosal space (GSMS) (Echeverri et al., 2009; Wszola et al., 2009). This site has many potential benefits including avoidance of IBMIR, a rich oxygen supply, and a high oxygen tension. In a pre-clinical animal study (Echeverri et al., 2009), it was shown that diabetic pigs receiving islets endoscopically into the GSMS faired better than pigs receiving intraportal islets. Pigs in the former group showed less early islet loss and received less insulin in order to maintain normoglycemia. This has become an exciting possibility for an extra-portal site of islet graft deposition; further research should shed light onto the potential for long-term graft survival at this site.

In the end, the question becomes "what is the optimal site for islet transplantation?" This is a difficult question to answer, as there are a number of criteria that need evaluation in order to determine the site's effectiveness, and it is unlikely that any one site will fulfill all of them (reviewed by Merani et al., 2008). The physiology (oxygen content and vascularity), endocrine function, immunological appropriateness, and surgical and technical aspects all need to be considered and studied before a definitive answer can be given. For the time being, the portal vein will remain the choice for clinical islet transplantation; however, there is significant clinical potential for one of the previously mentioned alternatives to supplant it in the not-so-distant future.

Improving engraftment post-transplant

In clinical islet transplantation, islets derived from multiple donors are often required to achieve insulin independence, which suggests that a significant portion of the transplanted islets must fail to engraft and become functional. It has been estimated that up to 70% of the transplanted β-cell mass may be destroyed in the early post-transplant period (Davalli et al., 1995; Biarnes et al., 2002; Ryan et al., 2005). Since this profound loss has been observed in both immunodeficient and syngeneic islet transplantation models, islet survival is likely regulated by non-immune-mediated stimuli. Following isolation, the islet microvasculature is completely disrupted, and, upon implantation into the portal circulation, hypoxia persists while the islets revascularize, which can take up to 2 weeks (Dionne et al., 1993; Carlsson et al., 2001, 2002; Giuliani et al., 2005). During this engraftment period, the islets are continuously exposed to immunosuppressive drugs including tacrolimus and sirolimus, which are known to adversely impact β-cell survival and function (Hyder et al., 2005). These negative effects are likely compounded by the proximity of the transplanted islets and high concentrations of these drugs in the hepatoportal circulation, further degrading β-cell mass over time (Desai et al., 2003; Shapiro et al., 2005a).

Another process that may influence islet engraftment and survival in the early post-transplant period has been termed the "instant blood-mediated inflammatory reaction" (IBMIR). Islets have been shown to naturally express tissue factor, a protein that acts as a receptor and cofactor for factor VII, an important mediator of the coagulation cascade (Moberg et al., 2002). Isolated human islets release tissue factor along with glucagon and insulin, which ultimately leads to platelet activation and binding at the surface of the islets. This causes the formation of a fibrin capsule around the islet and disruption of the islet morphology (Bennet et al., 1999; Moberg et al., 2002; Ozmen et al., 2002). Most of this process has been characterized using an *in vitro* tubing loop model, so the true impact of this process in the clinical setting has yet to be fully characterized. However, examination of serum in patients undergoing islet transplantation has shown that a statistically significant increase in the serum concentration of thrombin/anti-thrombin complexes is present almost immediately following portal infusion, with peak levels occurring at 15 minutes, even when there was no clinical evidence of portal hypertension or intraportal thrombosis (Moberg et al., 2002). Given that platelet activation is one of the primary contributing factors in the generation of an inflammatory response, IBMIR is probably one of the important early processes in islet transplantation that elicits an immune response (Rabinovitch and Suarez-Pinzon, 1998; Moberg et al., 2002).

Many studies targeted at enhancing islet survival during the early post-transplant period have been published, and a variety of different strategies have been tested. Some groups have aimed to enhance revascularization with vascular endothelial growth factor (VEGF), but these studies

807

have not yet demonstrated that this approach significantly improves islet graft survival (Narang et al., 2004). Anti-coagulation strategies using injection of activated protein C or inhibition of thrombin have been studied as a means to inhibit IBMIR, but these interventions have shown only a modest benefit in a series of *in vivo* studies in animal models (Ozmen et al., 2002; Contreras et al., 2004; Goto et al., 2008). Clinical studies designed to prevent IBMIR are currently under investigation and should provide more insight into this area.

Since the processes described above involve both extracellular (i.e. IBMIR) and intracellular (i.e. hypoxia) stimuli leading to β-cell death, another approach to preserve β-cell mass in the early post-transplant period has been to directly inhibit the apoptotic triggers that ultimately lead to loss of islet mass post-transplant. A variety of strategies have been explored in the experimental setting, and, while promising data have been generated *in vitro*, demonstration of *in vivo* benefit to islet graft survival has been more elusive (Dupraz et al., 1999, 2000; Cottett et al., 2001, 2002; Cattan et al., 2003; Klein et al., 2004). Many studies have described inhibition of a variety of apoptosis-associated proteins, including cFLIP (cellular FLICE-inhibitory protein; prevents caspase-8 activation), A20 (inhibits NF-kB activation), Bcl-2, and Bcl-XL (mitochondria-associated anti-apoptotic proteins) (Dupraz et al., 1999, 2000; Grey et al., 1999, 2003; Cottett et al., 2001, 2002; Klein et al., 2004). A20 has shown promise, as its overexpression reduced the islet mass required in syngeneic islet transplantation in mice (Grey et al., 1999, 2003). Recently, investigations using XIAP (X-linked inhibitor of apoptosis protein), which inhibits the downstream effector caspases that function in the final common pathway of apoptosis, have demonstrated promise in both human and rodent models of engraftment and in promoting murine islet allograft survival (Emamaullee et al., 2005a, 2005b; Plesner et al., 2005). However, this area of research is currently limited by its requirement for genetic manipulation of islet tissue pre-transplant, which has proven to be quite variable and difficult to achieve in human islets. Also, these genetic alterations are most often regulated with viral vectors, which represent a highly controversial reagent for clinical use, especially in immunosuppressed transplant recipients.

Recently, our group has investigated the use of small molecule peptidyl pan caspase inhibitors to promote β-cell survival during the post-transplant engraftment period. These data demonstrate that euglycemia can be achieved in >90% of transplant recipients after an 80–90% reduction in islet implant mass using mouse or human islets in a non-allogeneic transplant model (Emamaullee et al., 2007, 2008). We are currently investigating this approach in pre-clinical large animal models and anticipate starting trials within our clinical islet program in the coming years.

Improved immunomodulation: towards donor-specific tolerance

One unique component of islet transplantation in patients with T1DM is the possibility of recurrent autoimmunity, which may elevate the demand for immunosuppression. Indeed, it has been well established using a rodent model of T1DM, the non-obese diabetic (NOD) mouse, that control of recurrent autoimmune reactivity to β-cells is one of the most difficult obstacles to overcome in islet transplantation (reviewed by Rossini et al., 2001; Pearson et al., 2003). Although it has been quite challenging to study recurrent autoimmunity in clinical patients, some evidence exists to suggest that levels of autoantibodies to GAD (glutamic acid decarboxylase) and IA-2 increase following islet transplantation, although the direct impact of this phenomenon on graft survival is not yet clear (Jaeger et al., 2000; Bosi et al., 2001). If recurrent autoimmunity does alter immunosuppressive drug functional thresholds, this presents yet another problem in the context of islet transplantation, as many of the drugs are directly β-cell toxic. In fact, up to 15% of non-diabetic patients who receive solid organ grafts can develop post-transplant diabetes as a result of calcineurin inhibitor therapy (i.e. tacrolimus) or steroids (i.e. prednisone) (Jindal et al., 1997; Djamali et al., 2003). Most patients that are candidates for islet transplantation have had disregulated diabetes for many years, and as such their renal status may be somewhat impaired (Shapiro et al., 2000). This leads to an increased susceptibility to the deleterious renal side-effects of these immunosuppressive drugs, and thus limits the extent to which the dose can be increased to preserve graft function (Ryan et al., 2004a). It is therefore likely that immunosuppressive drugs either contribute to β-cell loss over time directly via toxicity, or indirectly by incomplete protection against recurrent autoimmunity and/or alloreactivity.

Direct control of recurrent autoimmunity may enhance long-term graft function in islet transplantation. Attempts have been made to control autoimmunity at the time of diabetes onset, using various immunosuppressive agents such as azathioprine, prednisone, cyclosporin A, or anti-thymocytic globulin, but no significant benefit was observed (Elliott et al., 1981; Eisenbarth et al., 1985; Silverstein et al., 1988; Bougneres et al., 1990). Recent clinical studies using a modified anti-CD3 (hOKT3γ1(Ala-Ala) in patients with new onset T1DM have demonstrated that this treatment significantly improved C-peptide responses in these patients, which persisted for up to 2 years following treatment (Herold et al., 2005). Incorporation of this induction agent into clinical islet transplant protocols has suggested that it may enhance insulin independence rates following single donor infusion, which may be related to its ability to curtail β-cell autoimmunity in these patients (Hering et al., 2004). Continued development of therapies targeted at regulation of autoimmunity will allow further refinement of immunosuppression protocols for islet transplantation in the future.

In all types of transplantation, the ultimate goal is to develop therapeutic protocols that involve a brief period of treatment only during the initial post-transplant period, followed by the complete withdrawal of all immunosuppressive drugs. This phenomenon has been termed "operational tolerance," since it may involve a passive ignorance of the graft or a more active T-cell tolerance to the graft antigens. In experimental transplantation, the difference in these two types of response is quite important and can be measured using re-transplantation of donor-type or third-party tissue, with tolerance resulting in acceptance of the donor-type graft and rejection of the third-party graft. In the clinical setting, however, the distinction may not be so critical, as both ignorance and tolerance would allow for reduction or withdrawal of immunosuppressive therapies. The most widely studied pathway to tolerance induction involves the inhibition of T-cell costimulation following T-cell receptor ligation. During an immune response, a T-cell must receive "signal 2" through interactions between its surface molecule CD28 and CD80 or CD86 on the antigen-presenting cell to become fully activated. In order to disrupt this interaction, the extracellular portion of CTLA-4, which has a higher affinity for CD80/CD86 than CD28, has been artificially fused with human Fcγ to produce the soluble molecule CTLA4-Ig, designed for therapeutic purposes. CTLA4-Ig has been recognized for its potent immunoregulatory activity in murine models of T1DM, where treatment of young NOD mice dramatically reduced the incidence of T1DM (Lenschow et al., 1995). Our lab and others have demonstrated that CTLA4-Ig treatment in allogeneic islet transplantation can prolong graft survival but does not induce tolerance (Levisetti et al., 1997; Kirk et al., 1997; Benhamou, 2002; Casey et al., 2002). A new high-affinity version of CTLA4-Ig called belatacept or LEA29Y has been developed for clinical use and has shown considerable promise in promoting allograft survival in non-human primates and in clinical renal transplantation (Adams et al., 2002, 2005; Vincenti et al., 2005; Emamaullee et al., 2009). These studies have generated considerable excitement for this approach, and clinical trials using belatacept in clinical islet transplantation are underway at our center, in conjunction with Emory University.

A second costimulatory pathway that has been examined in transplantation involves the interaction between CD40 on antigen-presenting cells and CD40L (CD154) on T-cells, leading to T-cell activation. This interaction also promotes B-cell differentiation and the activation of antigen-presenting cells including macrophages and dendritic cells. Blockade of this pathway using anti-CD154 therapies demonstrated considerable promise in promoting tolerance induction in primate models early on, but further testing of the potent anti-CD154 blocking antibody (Hu5C8) has been halted due to unexpected thromboembolic complications in clinical trials (Kirk et al., 1997; Kenyon et al., 1999; Kirk et al., 1999; Kawai et al., 2000). Recent development of therapeutic antibodies targeting the CD40 molecule appear to avoid this negative side-effect and should prove to be important in future clinical tolerance induction protocols in islet transplantation (Adams et al., 2005).

SUMMARY AND CONCLUSIONS

β-cell replacement through islet transplantation presents the best opportunity to treat T1DM and prevent the long-term serious complications associated with this disease. The concept of islet transplantation is not new, but investigators struggled to achieve success in establishing insulin independence until the introduction of the Edmonton Protocol in 2000. This has provided hope for many patients with diabetes, but islet transplantation, in its current form, is reserved only for those patients with the most severe disease. Data from our program and others have continued to demonstrate that 80% of recipients may attain and maintain insulin independence at 1 year post-transplant, and improvements in induction and maintenance immunosuppression regimens have greatly improved long-term insulin independence rates as compared to the initial series, where most patients resumed insulin injections, albeit with a much lower insulin requirement than before receiving an islet graft. Importantly, patients that do exhibit partial islet function avoid both glycemic lability and hypoglycemic unawareness, which greatly improves the quality of life for many patients.

However, even with improved single donor success rates, the current requirement for islets derived from one to two or more cadaveric donors severely limits the current availability of this procedure. There are multiple opportunities for intervention throughout the entire process, from pancreas procurement, shipment, and islet processing, through to strategies for enhanced islet survival after implantation. Priority areas for clinical trials currently include expansion of living donor protocols, interventions to impede the IBMIR process, and the use of non-dia-betogenic and more "islet-friendly" immunosuppressive and tolerance-induction strategies to effectively control both auto- and alloimmunity. Strategies targeted at preserving β-cell mass throughout the process will have a substantial and immediate impact on islet transplantation by reducing the amount of islet tissue necessary to reverse diabetes. Once some of these obstacles are overcome, islet transplantation will become available to a broader population of patients with T1DM (Fig. 44.2B), especially those early in the progression of their disease who will benefit most as the development of serious chronic secondary complications could be avoided.

References

Adams, A. B., Shirasugi, N., Durham, M. M., Strobert, E., Anderson, D., Rees, P., et al. (2002). Calcineurin inhibitor-free CD28 blockade-based protocol protects allogeneic islets in non-human primates. *Diabetes, 51*, 265–270.

Adams, A. B., Shirasugi, N., Jones, T. R., Durham, M. M., Strobert, E. A., Cowan, S., et al. (2005). Development of a chimeric anti-CD40 monoclonal antibody that synergizes with LEA29Y to prolong islet allograft survival. *J. Immunol., 174*, 542–550.

Alejandro, R., Cutfield, R. G., Shienvold, F. L., Polonsky, K. S., Noel, J., Olson, L., et al. (1986). Natural history of intrahepatic canine islet cell autografts. *J. Clin. Invest., 78*. 1339–1348.

Ao, Z., Matayoshi, K., Yakimets, W. J., Katyal, D., Rajotte, R. V., & Warnock, G. L. (1992). Development of an omental pouch site for islet transplantation. *Transplant. Proc., 24*, 2789.

Assady, S., Maor, G., Amit, M., Itskovitz-Eldor, J., Skorecki, K. L., & Tzukerman, M. (2001). Insulin production by human embryonic stem cells. *Diabetes, 50*, 1691–1697.

Ballinger, W. F., & Lacy, P. E. (1972). Transplantation of intact pancreatic islets in rats. *Surgery, 72*, 175–186.

Benhamou, P. Y. (2002). Immunomodulation with CTLA4-Ig in islet transplantation. *Transplantation, 73*, S40–S42.

Benhamou, P. Y., Oberholzer, J., Toso, C., Kessler, L., Penfornis, A., Bayle, F., et al. (2001). Human islet transplantation network for the treatment of Type I diabetes: first data from the Swiss-French GRAGIL consortium (1999–2000). Groupe de Recherche Rhin Rhjne Alpes Geneve pour la transplantation d'Ilots de Langerhans. *Diabetologia, 44*, 859–864.

Bennet, W., Sundberg, B., Groth, C. G., Brendel, M. D., Brandhorst, D., Brandhorst, H., et al. (1999). Incompatibility between human blood and isolated islets of Langerhans: a finding with implications for clinical intraportal islet transplantation? *Diabetes, 48,* 1907–1914.

Bhargava, R., Senior, P. A., Ackerman, T. E., Ryan, E. A., Paty, B. W., Lakey, J. R., et al. (2004). Prevalence of hepatic steatosis after islet transplantation and its relation to graft function. *Diabetes, 53,* 1311–1317.

Biarnes, M., Montolio, M., Nacher, V., Raurell, M., Soler, J., & Montanya, E. (2002). Beta-cell death and mass in syngeneically transplanted islets exposed to short- and long-term hyperglycemia. *Diabetes, 51,* 66–72.

Bonner-Weir, S., & Weir, G. C. (2005). New sources of pancreatic beta-cells. *Nat. Biotechnol., 23,* 857–861.

Bosi, E., Braghi, S., Maffi, P., Scirpoli, M., Bertuzzi, F., Pozza, G., et al. (2001). Autoantibody response to islet transplantation in type 1 diabetes. *Diabetes, 50,* 2464–2471.

Bougneres, P. F., Landais, P., Boisson, C., Carel, J. C., Frament, N., Boitard, C., et al. (1990). Limited duration of remission of insulin dependency in children with recent overt type I diabetes treated with low-dose cyclosporin. *Diabetes, 39,* 1264–1272.

Brendel M, H. B., Schulz A, Bretzel R. (2001). *International Islet Transplant Registry Report.* University of Giessen, Germany. p. 1.

Cardona, K., Korbutt, G. S., Milas, Z., Lyon, J., Cano, J., Jiang, W., et al. (2006). Long-term survival of neonatal porcine islets in non-human primates by targeting costimulation pathways. *Nat. Med., 12,* 304–306.

Carlsson, P. O., Palm, F., Andersson, A., & Liss, P. (2001). Markedly decreased oxygen tension in transplanted rat pancreatic islets irrespective of the implantation site. *Diabetes, 50,* 489–495.

Carlsson, P. O., Palm, F., & Mattsson, G. (2002). Low revascularization of experimentally transplanted human pancreatic islets. *J. Clin. Endocrinol. Metab., 87,* 5418–5423.

Casey, J. J., Lakey, J. R., Ryan, E. A., Paty, B. W., Owen, R., O'Kelly, K., et al. (2002). Portal venous pressure changes after sequential clinical islet transplantation. *Transplantation, 74,* 913–915.

Cattan, P., Rottembourg, D., Cottet, S., Tardivel, I., Dupraz, P., Thorens, B., et al. (2003). Destruction of conditional insulinoma cell lines in NOD mice: a role for autoimmunity. *Diabetologia, 46,* 504–510.

Contreras, J. L., Eckstein, C., Smyth, C. A., Sellers, M. T., Vilatoba, M., Bilbao, G., et al. (2003). Brain death significantly reduces isolated pancreatic islet yields and functionality *in vitro* and *in vivo* after transplantation in rats. *Diabetes, 52,* 2935–2942.

Contreras, J. L., Eckstein, C., Smyth, C. A., Bilbao, G., Vilatoba, M., Ringland, S. E., et al. (2004). Activated protein C preserves functional islet mass after intraportal transplantation: a novel link between endothelial cell activation, thrombosis, inflammation, and islet cell death. *Diabetes, 53,* 2804–2814.

Cottet, S., Dupraz, P., Hamburger, F., Dolci, W., Jaquet, M., & Thorens, B. (2001). SOCS-1 protein prevents Janus Kinase/STAT-dependent inhibition of beta cell insulin gene transcription and secretion in response to interferon-gamma. *J. Biol. Chem., 276,* 25862–25870.

Cottet, S., Dupraz, P., Hamburger, F., Dolci, W., Jaquet, M., & Thorens, B. (2002). cFLIP protein prevents tumor necrosis factor-alpha-mediated induction of caspase-8-dependent apoptosis in insulin-secreting betaTc-Tet cells. *Diabetes, 51,* 1805–1814.

Couri, C. E., Oliveira, M. C., Stracieri, A. B., Moraes, D. A., Pieroni, F., Barros, G. M., et al. (2009). C-peptide levels and insulin independence following autologous nonmyeloablative hematopoietic stem cell transplantation in newly diagnosed type 1 diabetes mellitus. *JAMA, 301,* 1573–1579.

d'Amour, K. A., Bang, A. G., Eliazer, S., Kelly, O. G., Agulnick, A. D., Smart, N. G., et al. (2006). Production of pancreatic hormone-expressing endocrine cells from human embryonic stem cells. *Nat. Biotechnol., 24,* 1392–1401.

Davalli, A. M., Ogawa, Y., Ricordi, C., Scharp, D. W., Bonner-Weir, S., & Weir, G. C. (1995). A selective decrease in the beta cell mass of human islets transplanted into diabetic nude mice. *Transplantation, 59,* 817–820.

Desai, N. M., Goss, J. A., Deng, S., Wolf, B. A., Markmann, E., Palanjian, M., et al. (2003). Elevated portal vein drug levels of sirolimus and tacrolimus in islet transplant recipients: local immunosuppression or islet toxicity? *Transplantation, 76,* 1623–1625.

Diabetes Control and Complications Trial Research Group. (1990). Diabetes Control and Complications Trial (DCCT). Update. DCCT Research Group. *Diabetes Care, 13,* 427–433.

Diabetes Control and Complications Trial Research Group. (1993). The effect of intensive treatment of diabetes on the development and progression of long-term complications in insulin-dependent diabetes mellitus. *N. Engl. J. Med., 329,* 977–986.

Diabetes Control and Complications Trial Research Group. (1995). Adverse events and their association with treatment regimens in the diabetes control and complications trial. *Diabetes Care, 18,* 1415.

Dionne, K. E., Colton, C. K., & Yarmush, M. L. (1993). Effect of hypoxia on insulin secretion by isolated rat and canine islets of Langerhans. *Diabetes, 42,* 12–21.

Djamali, A., Premasathian, N., & Pirsch, J. D. (2003). Outcomes in kidney transplantation. *Semin. Nephrol., 23,* 306–316.

811

Dupraz, P., Rinsch, C., Pralong, W. F., Rolland, E., Zufferey, R., Trono, D., et al. (1999). Lentivirus-mediated Bcl-2 expression in betaTC-tet cells improves resistance to hypoxia and cytokine-induced apoptosis while preserving *in vitro* and *in vivo* control of insulin secretion. *Gene Ther., 6*, 1160–1169.

Dupraz, P., Cottet, S., Hamburger, F., Dolci, W., Felley-Bosco, E., & Thorens, B. (2000). Dominant negative MyD88 proteins inhibit interleukin-1beta /interferon-gamma -mediated induction of nuclear factor kappa B-dependent nitrite production and apoptosis in beta cells. *J. Biol. Chem., 275*, 37672–37678.

Echeverri, G. J., McGrath, K., Bottino, R., Hara, H., Dons, E. M., van der Windt, D. J., et al. (2009). Endoscopic gastric submucosal transplantation of islets (ENDO-STI): technique and initial results in diabetic pigs. *Am. J. Transplant., 9*, 2485–2496.

Eisenbarth, G. S., Srikanta, S., Jackson, R., Rabinowe, S., Dolinar, R., Aoki, T., et al. (1985). Anti-thymocyte globulin and prednisone immunotherapy of recent onset type 1 diabetes mellitus. *Diabetes Res., 2*, 271–276.

Elliott, R. B., Crossley, J. R., Berryman, C. C., & James, A. G. (1981). Partial preservation of pancreatic beta-cell function in children with diabetes. *Lancet, 2*, 631–632.

Elliott, R. B., Escobar, L., Garkavenko, O., Croxson, M. C., Schroeder, B. A., McGregor, M., et al. (2000). No evidence of infection with porcine endogenous retrovirus in recipients of encapsulated porcine islet xenografts. *Cell Transplant., 9*, 895–901.

Emamaullee, J. A., Liston, P., Korneluk, R. G., Shapiro, A. M. J., & Elliott, J. (2005a). XIAP overexpression in islet beta-cells enhances engraftment and minimizes hypoxia-reperfusion injury. *Am. J. Transplant., 5*, 1297–1305.

Emamaullee, J. A., Rajotte, R. V., Lakey, J. R. T., Liston, P., Korneluk, R. G., Shapiro, A. M. J., et al. (2005b). XIAP overexpression in human islets prevents early post-transplant apoptosis and reduces the islet mass needed to treat diabetes. *Diabetes, 54*, 2541–2548.

Emamaullee, J. A., Stanton, L., Schur, C., & Shapiro, A. M. (2007). Caspase inhibitor therapy enhances marginal mass islet graft survival and preserves long-term function in islet transplantation. *Diabetes, 56*, 1289–1298.

Emamaullee, J. A., Davis, J., Pawlick, R., Toso, C., Merani, S., Cai, S. X., et al. (2008). The caspase selective inhibitor EP1013 augments human islet graft function and longevity in marginal mass islet transplantation in mice. *Diabetes, 57*, 1556–1566.

Emamaullee, J., Toso, C., Merani, S., & Shapiro, A. M. (2009). Costimulatory blockade with belatacept in clinical and experimental transplantation – a review. *Expert Opin. Biol. Ther., 9*, 789–796.

Fishman, J. A., & Patience, C. (2004). Xenotransplantation: infectious risk revisited. *Am. J. Transplant., 4*, 1383–1390.

Giuliani, M., Moritz, W., Bodmer, E., Dindo, D., Kugelmeier, P., Lehmann, R., et al. (2005). Central necrosis in isolated hypoxic human pancreatic islets: evidence for postisolation ischemia. *Cell Transplant., 14*, 67–76.

Goss, J. A., Schock, A. P., Brunicardi, F. C., Goodpastor, S. E., Garber, A. J., Soltes, G., et al. (2002). Achievement of insulin independence in three consecutive type-1 diabetic patients via pancreatic islet transplantation using islets isolated at a remote islet isolation center. *Transplantation, 74*, 1761–1766.

Goss, J. A., Goodpastor, S. E., Brunicardi, F. C., Barth, M. H., Soltes, G. D., Garber, A. J., et al. (2004). Development of a human pancreatic islet-transplant program through a collaborative relationship with a remote islet-isolation center. *Transplantation, 77*, 462–466.

Goto, M., Tjernberg, J., Dufrane, D., Elgue, G., Brandhorst, D., Ekdahl, K. N., et al. (2008). Dissecting the instant blood-mediated inflammatory reaction in islet xenotransplantation. *Xenotransplantation, 15*, 225–234.

Grey, S. T., Arvelo, M. B., Hasenkamp, W., Bach, F. H., & Ferran, C. (1999). A20 inhibits cytokine-induced apoptosis and nuclear factor kappaB-dependent gene activation in islets. *J. Exp. Med., 190*, 1135–1146.

Grey, S. T., Longo, C., Shukri, T., Patel, V. I., Csizmadia, E., Daniel, S., et al. (2003). Genetic engineering of a suboptimal islet graft with A20 preserves beta cell mass and function. *J. Immunol., 170*, 6250–6256.

Gross, C. R., Limwattananon, C., & Matthees, B. J. (1998). Quality of life after pancreas transplantation: a review. *Clin. Transplant., 12*, 351–361.

Groth, C. G., Korsgren, O., Tibell, A., Tollemar, J., Moller, E., Bolinder, J., et al. (1994). Transplantation of porcine fetal pancreas to diabetic patients. *Lancet, 344*, 1402–1404.

Gruessner, A. C., & Sutherland, D. E. (2005). Pancreas transplant outcomes for United States (US) and non-US cases as reported to the United Network for Organ Sharing (UNOS) and the International Pancreas Transplant Registry (IPTR) as of June 2004. *Clin. Transplant., 19*, 433–455.

Gruessner, A. C., & Sutherland, D. E. (2008). Pancreas transplant outcomes for United States (US) cases as reported to the United Network for Organ Sharing (UNOS) and the International Pancreas Transplant Registry (IPTR). *Clin. Transpl.*, 45–56.

Gruessner, R. W., Burke, G. W., Stratta, R., Sollinger, H., Benedetti, E., Marsh, C., et al. (1996). A multicenter analysis of the first experience with FK506 for induction and rescue therapy after pancreas transplantation. *Transplantation, 61*, 261–273.

Gruessner, R. W., Sutherland, D. E., Dunn, D. L., Najarian, J. S., Jie, T., Hering, B. J., et al. (2004). Transplant options for patients undergoing total pancreatectomy for chronic pancreatitis. *J. Am. Coll. Surg., 198*, 559–567, discussion 568–569.

Haller, M. J., Viener, H. L., Wasserfall, C., Brusko, T., Atkinson, M. A., & Schatz, D. A. (2008). Autologous umbilical cord blood infusion for type 1 diabetes. *Exp. Hematol., 36,* 710–715.

Hering, B, R. C (1999). Islet transplantation for patients with Type 1 diabetes: results, research priorities, and reasons for optimism. *Graft, 2,* 12.

Hering, B. J., & Walawalkar, N. (2009). Pig-to-non-human primate islet xenotransplantation. *Transpl. Immunol., 21,* 81–86.

Hering, B. J., Kandaswamy, R., Harmon, J. V., Ansite, J. D., Clemmings, S. M., Sakai, T., et al. (2004). Transplantation of cultured islets from two-layer preserved pancreases in type 1 diabetes with anti-CD3 antibody. *Am. J. Transplant., 4,* 390–401.

Hering, B. J., Kandaswamy, R., Ansite, J. D., Eckman, P. M., Nakano, M., Sawada, T., et al. (2005). Single-donor, marginal-dose islet transplantation in patients with type 1 diabetes. *JAMA, 293,* 830–835.

Hering, B. J., Wijkstrom, M., Graham, M. L., Hardstedt, M., Aasheim, T. C., Jie, T., et al. (2006). Prolonged diabetes reversal after intraportal xenotransplantation of wild-type porcine islets in immunosuppressed non-human primates. *Nat. Med., 12,* 301–303.

Herold, K. C., Gitelman, S. E., Masharani, U., Hagopian, W., Bisikirska, B., Donaldson, D., et al. (2005). A single course of anti-CD3 monoclonal antibody hOKT3γ1(Ala-Ala) results in improvement in C-peptide responses and clinical parameters for at least 2 years after onset of type 1 diabetes. *Diabetes, 54,* 1763–1769.

Hori, Y., Rulifson, I. C., Tsai, B. C., Heit, J. J., Cahoy, J. D., & Kim, S. K. (2002). Growth inhibitors promote differentiation of insulin-producing tissue from embryonic stem cells. *Proc. Natl. Acad. Sci. U.S.A., 99,* 16105–16110.

Hyder, A., Laue, C., & Schrezenmeir, J. (2005). Effect of the immunosuppressive regime of Edmonton protocol on the long-term in vitro insulin secretion from islets of two different species and age categories. *Toxicol. In Vitro, 19,* 541–546.

Ianus, A., Holz, G. G., Theise, N. D., & Hussain, M. A. (2003). *In vivo* derivation of glucose-competent pancreatic endocrine cells from bone marrow without evidence of cell fusion. *J. Clin. Invest., 111,* 843–850.

International Islet Transplant Registry (2005)

Jaeger, C., Brendel, M. D., Eckhard, M., & Bretzel, R. G. (2000). Islet autoantibodies as potential markers for disease recurrence in clinical islet transplantation. *Exp. Clin. Endocrinol. Diabetes, 108,* 328–333.

Jindal, R. M., Sidner, R. A., & Milgrom, M. L. (1997). Post-transplant diabetes mellitus. The role of immunosuppression. *Drug Saf., 16,* 242–257.

Kaplan, B., Schold, J., Srinivas, T., Womer, K., Foley, D. P., Patton, P., et al. (2004). Effect of sirolimus withdrawal in patients with deteriorating renal function. *Am. J. Transplant., 4,* 1709–1712.

Kawai, T., Andrews, D., Colvin, R. B., Sachs, D. H., & Cosimi, A. B. (2000). Thromboembolic complications after treatment with monoclonal antibody against CD40 ligand. *Nat. Med., 6,* 114.

Keen, H. (1994). The Diabetes Control and Complications Trial (DCCT). *Health Trends, 26,* 41–43.

Kelly, W. D., Lillehei, R. C., Merkel, F. K., Idezuki, Y., & Goetz, F. C. (1967). Allotransplantation of the pancreas and duodenum along with the kidney in diabetic nephropathy. *Surgery, 61,* 827–837.

Kemp, C. B., Knight, M. J., Scharp, D. W., Ballinger, W. F., & Lacy, P. E. (1973). Effect of transplantation site on the results of pancreatic islet isografts in diabetic rats. *Diabetologia, 9,* 486–491.

Kempf, M. C., Andres, A., Morel, P., Benhamou, P. Y., Bayle, F., Kessler, L., et al. (2005). Logistics and transplant coordination activity in the GRAGIL Swiss-French multicenter network of islet transplantation. *Transplantation, 79,* 1200–1205.

Kendall, D. M., Rooney, D. P., Smets, Y. F., Salazar Bolding, L., & Robertson, R. P. (1997). Pancreas transplantation restores epinephrine response and symptom recognition during hypoglycemia in patients with long-standing type I diabetes and autonomic neuropathy. *Diabetes, 46,* 249–257.

Kenyon, N. S., Chatzipetrou, M., Masetti, M., Ranuncoli, A., Oliveira, M., Wagner, J. L., et al. (1999). Long-term survival and function of intrahepatic islet allografts in rhesus monkeys treated with humanized anti-CD154. *Proc. Natl. Acad. Sci. U.S.A., 96,* 8132–8137.

Kin, T., Korbutt, G. S., & Rajotte, R. V. (2003). Survival and metabolic function of syngeneic rat islet grafts transplanted in the omental pouch. *Am. J. Transplant., 3,* 281–285.

King, H., Aubert, R. E., & Herman, W. H. (1998). Global burden of diabetes, 1995–2025: prevalence, numerical estimates, and projections. *Diabetes Care, 21,* 1414–1431.

Kirk, A. D., Harlan, D. M., Armstrong, N. N., Davis, T. A., Dong, Y., Gray, G. S., et al. (1997). CTLA4-Ig and anti-CD40 ligand prevent renal allograft rejection in primates. *Proc. Natl. Acad. Sci. U.S.A., 94,* 8789–8794.

Kirk, A. D., Burkly, L. C., Batty, D. S., Baumgartner, R. E., Berning, J. D., Buchanan, K., et al. (1999). Treatment with humanized monoclonal antibody against CD154 prevents acute renal allograft rejection in non-human primates. *Nat. Med., 5,* 686–693.

Klein, D., Ribeiro, M. M., Mendoza, V., Jayaraman, S., Kenyon, N. S., Pileggi, A., et al. (2004). Delivery of Bcl-XL or its BH4 domain by protein transduction inhibits apoptosis in human islets. *Biochem. Biophys. Res. Commun.,* *323*, 473–478.

Kodama, S., Kuhtreiber, W., Fujimura, S., Dale, E. A., & Faustman, D. L. (2003). Islet regeneration during the reversal of autoimmune diabetes in NOD mice. *Science, 302,* 1223–1227.

Kroon, E., Martinson, L. A., Kadoya, K., Bang, A. G., Kelly, O. G., Eliazer, S., et al. (2008). Pancreatic endoderm derived from human embryonic stem cells generates glucose-responsive insulin-secreting cells *in vivo. Nat. Biotechnol., 26,* 443–452.

Kubo, A., Shinozaki, K., Shannon, J. M., Kouskoff, V., Kennedy, M., Woo, S., et al. (2004). Development of definitive endoderm from embryonic stem cells in culture. *Development, 131,* 1651–1662.

Lacy, P. E., & Kostianovsky, M. (1967). Method for the isolation of intact islets of Langerhans from the rat pancreas. *Diabetes, 16,* 35–39.

Lakey, J. R., Warnock, G. L., Rajotte, R. V., Suarez-Alamazor, M. E., Ao, Z., Shapiro, A. M., et al. (1996). Variables in organ donors that affect the recovery of human islets of Langerhans. *Transplantation, 61,* 1047–1053.

Larsen, J. L. (2004). Pancreas transplantation: indications and consequences. *Endocr. Rev., 25,* 919–946.

Lenschow, D. J., Ho, S. C., Sattar, H., Rhee, L., Gray, G., Nabavi, N., et al. (1995). Differential effects of anti-B7–1 and anti-B7–2 monoclonal antibody treatment on the development of diabetes in the nonobese diabetic mouse. *J. Exp. Med., 181,* 1145–1155.

Levisetti, M. G., Padrid, P. A., Szot, G. L., Mittal, N., Meehan, S. M., Wardrip, C. L., et al. (1997). Immunosuppressive effects of human CTLA4Ig in a non-human primate model of allogeneic pancreatic islet transplantation. *J. Immunol., 159,* 5187–5191.

Lumelsky, N., Blondel, O., Laeng, P., Velasco, I., Ravin, R., & McKay, R. (2001). Differentiation of embryonic stem cells to insulin-secreting structures similar to pancreatic islets. *Science, 292,* 1389–1394.

Matsumoto, S., Okitsu, T., Iwanaga, Y., Noguchi, H., Nagata, H., Yonekawa, Y., et al. (2005). Insulin independence after living-donor distal pancreatectomy and islet allotransplantation. *Lancet, 365,* 1642–1644.

Matsumoto, S., Okitsu, T., Iwanaga, Y., Noguchi, H., Nagata, H., Yonekawa, Y., et al. (2006). Follow-up study of the first successful living donor islet transplantation. *Transplantation, 82,* 1629–1633.

Merani, S., Toso, C., Emamaullee, J., & Shapiro, A. M. (2008). Optimal implantation site for pancreatic islet transplantation. *Br. J. Surg., 95,* 1449–1461.

Merrill, J. P., Murray, J. E., Takacs, F. J., Hager, E. B., Wilson, R. E., & Dammin, G. J. (1963). Successful transplantation of kidney from a human cadaver. *JAMA, 185,* 347–353.

Moberg, L., Johansson, H., Lukinius, A., Berne, C., Foss, A., Kallen, R., et al. (2002). Production of tissue factor by pancreatic islet cells as a trigger of detrimental thrombotic reactions in clinical islet transplantation. *Lancet, 360,* 2039–2045.

Molinari, M., Al-Saif, F., Ryan, E. A., Lakey, J. R., Senior, P. A., Paty, B. W., et al. (2005). Sirolimus-induced ulceration of the small bowel in islet transplant recipients: report of two cases. *Am. J. Transplant., 5,* 2799–2804.

Murray, J. E., Merrill, J. P., Harrison, J. H., Wilson, R. E., & Dammin, G. J. (1963). Prolonged survival of human-kidney homografts by immunosuppressive drug therapy. *N. Engl. J. Med., 268,* 1315–1323.

Narang, A. S., Cheng, K., Henry, J., Zhang, C., Sabek, O., Fraga, D., et al. (2004). Vascular endothelial growth factor gene delivery for revascularization in transplanted human islets. *Pharm. Res., 21,* 15–25.

Nathan, D. M., Lachin, J., Cleary, P., Orchard, T., Brillon, D. J., Backlund, J. Y., et al. (2003). Intensive diabetes therapy and carotid intima-media thickness in type 1 diabetes mellitus. *N. Engl. J. Med., 348,* 2294–2303.

Nathan, D. M., Cleary, P. A., Backlund, J. Y., Genuth, S. M., Lachin, J. M., Orchard, T. J., et al. (2005). Intensive diabetes treatment and cardiovascular disease in patients with type 1 diabetes. *N. Engl. J. Med., 353,* 2643–2653.

National Diabetes Data Group (US). (1995). *National Institute of Diabetes and Digestive and Kidney Diseases (US) and National Institutes of Health (US)* (2nd ed). In *Diabetes in America.* Bethesda, MD: NIH publication no. 95–1468, National Institutes of Health, National Institute of Diabetes and Digestive and Kidney Diseases.

National Library of Medicine (2010). www.clinicaltrials.gov

Newell, K. A., Bruce, D. S., Cronin, D. C., Woodle, E. S., Millis, J. M., Piper, J. B., et al. (1996). Comparison of pancreas transplantation with portal venous and enteric exocrine drainage to the standard technique utilizing bladder drainage of exocrine secretions. *Transplantation, 62,* 1353–1356.

Otonkoski, T., Gao, R., & Lundin, K. (2005). Stem cells in the treatment of diabetes. *Ann. Med., 37,* 513–520.

Owen, S. (2006). Pediatric pumps: barriers and breakthroughs. *Diabetes Educ., 32,* 29S–38S.

Owen, R. J., Ryan, E. A., O'Kelly, K., Lakey, J. R., McCarthy, M. C., Paty, B. W., et al. (2003). Percutaneous transhepatic pancreatic islet cell transplantation in type 1 diabetes mellitus: radiologic aspects. *Radiology, 229,* 165–170.

Ozmen, L., Ekdahl, K. N., Elgue, G., Larsson, R., Korsgren, O., & Nilsson, B. (2002). Inhibition of thrombin abrogates the instant blood-mediated inflammatory reaction triggered by isolated human islets: possible application of the thrombin inhibitor melagatran in clinical islet transplantation. *Diabetes, 51*, 1779−1784.

Pearson, T., Markees, T. G., Serreze, D. V., Pierce, M. A., Wicker, L. S., Peterson, L. B., et al. (2003). Islet cell autoimmunity and transplantation tolerance: two distinct mechanisms? *Ann. NY Acad. Sci., 1005*, 148−156.

Phelps, C. J., Koike, C., Vaught, T. D., Boone, J., Wells, K. D., Chen, S. H., et al. (2003). Production of alpha 1,3-galactosyltransferase-deficient pigs. *Science, 299*, 411−414.

Plesner, A., Liston, P., Tan, R., Korneluk, R. G., & Verchere, C. B. (2005). The X-linked inhibitor of apoptosis protein enhances survival of murine islet allografts. *Diabetes, 54*, 2533−2540.

Rabinovitch, A., & Suarez-Pinzon, W. L. (1998). Cytokines and their roles in pancreatic islet beta-cell destruction and insulin-dependent diabetes mellitus. *Biochem. Pharmacol., 55*, 1139−1149.

Reckard, C. R., Ziegler, M. M., & Barker, C. F. (1973). Physiological and immunological consequences of transplanting isolated pancreatic islets. *Surgery, 74*, 91−99.

Ricordi, C., Lacy, P. E., & Scharp, D. W. (1989). Automated islet isolation from human pancreas. *Diabetes, 38* (Suppl. 1), 140−142.

Ricordi, C., Tzakis, A. G., Carroll, P. B., Zeng, Y. J., Rilo, H. L., Alejandro, R., et al. (1992). Human islet isolation and allotransplantation in 22 consecutive cases. *Transplantation, 53*, 407−414.

Rood, P. P., & Cooper, D. K. (2006). Islet xenotransplantation: are we really ready for clinical trials? *Am. J. Transplant., 6*, 1269−1274.

Rood, P. P., Buhler, L. H., Bottino, R., Trucco, M., & Cooper, D. K. (2006). Pig-to-non-human primate islet xenotransplantation: a review of current problems. *Cell Transplant., 15*, 89−104.

Rossini, A. A., Mordes, J. P., Greiner, D. L., & Stoff, J. S. (2001). Islet cell transplantation tolerance. *Transplantation, 72*, S43−S46.

Ryan, E. A., Lakey, J. R., Paty, B. W., Imes, S., Korbutt, G. S., Kneteman, N. M., et al. (2002). Successful islet transplantation: continued insulin reserve provides long-term glycemic control. *Diabetes, 51*, 2148−2157.

Ryan, E. A., Paty, B. W., Senior, P. A., & Shapiro, A. M. (2004a). Risks and side-effects of islet transplantation. *Curr. Diab. Rep., 4*, 304−309.

Ryan, E. A., Shandro, T., Green, K., Paty, B. W., Senior, P. A., Bigam, D., et al. (2004b). Assessment of the severity of hypoglycemia and glycemic lability in type 1 diabetic subjects undergoing islet transplantation. *Diabetes, 53*, 955−962.

Ryan, E. A., Paty, B. W., Senior, P. A., Bigam, D., Alfadhli, E., Kneteman, N. M., et al. (2005). Five-year follow-up after clinical islet transplantation. *Diabetes, 54*, 2060−2069.

Scharp, D. W., Lacy, P. E., Santiago, J. V., McCullough, C. S., Weide, L. G., Falqui, L., et al. (1990). Insulin independence after islet transplantation into type I diabetic patient. *Diabetes, 39*, 515−518.

Seaberg, R. M., Smukler, S. R., Kieffer, T. J., Enikolopov, G., Asghar, Z., Wheeler, M. B., et al. (2004). Clonal identification of multipotent precursors from adult mouse pancreas that generate neural and pancreatic lineages. *Nat. Biotechnol., 22*, 1115−1124.

Secchi, A., di Carlo, V., Martinenghi, S., la Rocca, E., Caldara, R., Spotti, D., et al. (1991). Effect of pancreas transplantation on life expectancy, kidney function and quality of life in uraemic type 1 (insulin-dependent) diabetic patients. *Diabetologia, 34*(Suppl. 1), S141−S144.

Senior, P. A., Paty, B. W., Cockfield, S. M., Ryan, E. A., & Shapiro, A. M. (2005). Proteinuria developing after clinical islet transplantation resolves with sirolimus withdrawal and increased tacrolimus dosing. *Am. J. Transplant., 5*, 2318−2323.

Shapiro, A. M., Lakey, J. R., Ryan, E. A., Korbutt, G. S., Toth, E., Warnock, G. L., et al. (2000). Islet transplantation in seven patients with type 1 diabetes mellitus using a glucocorticoid-free immunosuppressive regimen. *N. Engl. J. Med., 343*, 230−238.

Shapiro, A. M., Ricordi, C., & Hering, B. (2003). Edmonton's islet success has indeed been replicated elsewhere. *Lancet, 362*, 1242.

Shapiro, A. M., Gallant, H. L., Hao, E. G., Lakey, J. R., McCready, T., Rajotte, R. V., et al. (2005a). The portal immunosuppressive storm: relevance to islet transplantation? *Ther. Drug Monit., 27*, 35−37.

Shapiro, A. M., Lakey, J. R., Paty, B. W., Senior, P. A., Bigam, D. L., & Ryan, E. A. (2005b). Strategic opportunities in clinical islet transplantation. *Transplantation, 79*, 1304−1307.

Silverstein, J., Maclaren, N., Riley, W., Spillar, R., Radjenovic, D., & Johnson, S. (1988). Immunosuppression with azathioprine and prednisone in recent-onset insulin-dependent diabetes mellitus. *N. Engl. J. Med., 319*, 599−604.

Soria, B., Roche, E., Berna, G., Leon-Quinto, T., Reig, J. A., & Martin, F. (2000). Insulin-secreting cells derived from embryonic stem cells normalize glycemia in streptozotocin-induced diabetic mice. *Diabetes, 49*, 157−162.

Suri, A., Calderon, B., Esparza, T. J., Frederick, K., Bittner, P., & Unanue, E. R. (2006). Immunological reversal of autoimmune diabetes without hematopoietic replacement of beta cells. *Science, 311*, 1778−1780.

Sutherland, D. E., Matas, A. J., Goetz, F. C., & Najarian, J. S. (1980). Transplantation of dispersed pancreatic islet tissue in humans: autografts and allografts. *Diabetes, 29*(Suppl. 1), 31−44.

Sutherland, D. E., Gruessner, R. W., Dunn, D. L., Matas, A. J., Humar, A., Kandaswamy, R., et al. (2001). Lessons learned from more than 1,000 pancreas transplants at a single institution. *Ann. Surg., 233,* 463−501.

Sutherland, D. E., Gruessner, A. C., Carlson, A. M., Blondet, J. J., Balamurugan, A. N., Reigstad, K. F., et al. (2008). Islet autotransplant outcomes after total pancreatectomy: a contrast to islet allograft outcomes. *Transplantation, 86,* 1799−1802.

Tsujimura, T., Kuroda, Y., Avila, J. G., Kin, T., Oberholzer, J., Shapiro, A. M., et al. (2004). Influence of pancreas preservation on human islet isolation outcomes: impact of the two-layer method. *Transplantation, 78,* 96−100.

Tzakis, A. G., Ricordi, C., Alejandro, R., Zeng, Y., Fung, J. J., Todo, S., et al. (1990). Pancreatic islet transplantation after upper abdominal exenteration and liver replacement. *Lancet, 336,* 402−405.

US Scientific Registry of Transplant Recipients and the Organ Procurement and Transplantation Network (2005). *Statistical Data Reported by the US Scientific Registry of Transplant Recipients and the Organ Procurement and Transplantation Network,* Vol. 2005.

Valdes-Gonzalez, R. A., Dorantes, L. M., Garibay, G. N., Bracho-Blanchet, E., Mendez, A. J., Davila-Perez, R., et al. (2005). Xenotransplantation of porcine neonatal islets of Langerhans and Sertoli cells: a 4-year study. *Eur. J. Endocrinol., 153,* 419−427.

van der Windt, D. J., Bottino, R., Casu, A., Campanile, N., Smetanka, C., He, J., et al. (2009). Long-term controlled normoglycemia in diabetic non-human primates after transplantation with hCD46 transgenic porcine islets. *Am. J. Transplant., 9,* 2716−2726.

Villiger, P., Ryan, E. A., Owen, R., O'Kelly, K., Oberholzer, J., Saif, F. A., et al. (2005). Prevention of bleeding after islet transplantation: lessons learned from a multivariate analysis of 132 cases at a single institution. *Am. J. Transplant., 5,* 2992−2998.

Vincenti, F., Larsen, C., Durrbach, A., Wekerle, T., Nashan, B., Blancho, G., et al. (2005). Costimulation blockade with belatacept in renal transplantation. *N. Engl. J. Med., 353,* 770−781.

Walsh, T. J., Eggleston, J. C., & Cameron, J. L. (1982). Portal hypertension, hepatic infarction, and liver failure complicating pancreatic islet autotransplantation. *Surgery, 91,* 485−487.

Williams, P. (1894). Notes on diabetes treated with extract and by grafts of sheep's pancreas. *BMJ, 2,* 1303−1304.

World Health Organization (2002). Diabetes: the cost of diabetes. Fact Sheet No. 236. http://www.who.int/mediacentre/factsheets/fs236/en/print.html

Wszola, M., Berman, A., Fabisiak, M., Domagala, P., Zmudzka, M., Kieszek, R., et al. (2009). TransEndoscopic Gastric SubMucosa Islet Transplantation (eGSM-ITx) in pigs with streptozotocine induced diabetes − technical aspects of the procedure − preliminary report. *Ann. Transplant., 14,* 45−50.

Xu, Z. L., Mizuguchi, H., Mayumi, T., & Hayakawa, T. (2003). Regulated gene expression from adenovirus vectors: a systematic comparison of various inducible systems. *Gene, 309,* 145−151.

Zalzman, M., Gupta, S., Giri, R. K., Berkovich, I., Sappal, B. S., Karnieli, O., et al. (2003). Reversal of hyperglycemia in mice by using human expandable insulin-producing cells differentiated from fetal liver progenitor cells. *Proc. Natl. Acad. Sci. U.S.A., 100,* 7253−7258.

Zalzman, M., Anker-Kitai, L., & Efrat, S. (2005). Differentiation of human liver-derived, insulin-producing cells toward the beta-cell phenotype. *Diabetes, 54,* 2568−2575.

Zhou, Q., Brown, J., Kanarek, A., Rajagopal, J., & Melton, D. A. (2008). *In vivo* reprogramming of adult pancreatic exocrine cells to beta-cells. *Nature, 455,* 627−632.

Tissue Therapy

CHAPTER 45

Fetal Tissues

Ryan P. Dorin[*], **Chester J. Koh**[*,†]
[*] University of Southern California (USC) Institute of Urology, Keck School of Medicine, USC, Los Angeles
[†] Division of Pediatric Urology and the Developmental Biology, Regenerative Medicine, and Surgery Program, Childrens Hospital Los Angeles, Los Angeles, California, USA

INTRODUCTION

The field of regenerative medicine aims to replace damaged, diseased, or malformed tissue through the development of biological substitutes that can restore and maintain normal function. In following the principles of cell biology, transplantation, materials science, and engineering, many current strategies for regenerative medicine depend upon a sample of autologous cells from the diseased organ of the host. Usually, a small piece of donor tissue is dissociated into individual cells, and either implanted directly into the host or expanded in culture, attached to a support matrix, and then reimplanted into the host after expansion (Oberpenning et al., 1999). The use of autologous cells prevents immunologic rejection, and thus the deleterious side effects of immunosuppressive medications can be avoided. Ideally, both structural and functional tissue replacement will occur with minimal complications.

However, for many patients with extensive end-stage organ failure, a tissue biopsy may not yield enough healthy cells for expansion and transplantation. In other instances, primary autologous human cells cannot be expanded from a particular organ, such as the pancreas. In these situations, stem cells present an alternative source of cells from which the desired tissue can be derived. Combining the regenerative medicine techniques discovered over the past few decades with this potentially abundant source of versatile cells could lead to a novel source of replacement organs for transplantation.

Embryonic stem cells exhibit two remarkable properties: the ability to proliferate in an undifferentiated but pluripotent state (self-renew), and the ability to differentiate into many specialized cell types (Brivanlou et al., 2003). They can be isolated by immunosurgery from the inner cell mass of the embryo during the blastocyst stage (5 days post-fertilization), and are usually grown on feeder layers consisting of mouse embryonic fibroblasts or human feeder cells (Richards et al., 2002). More recent reports have shown that these cells can be grown without the use of a feeder layer (Amit et al., 2003), thus avoiding the exposure of these human cells to mouse viruses and proteins. These cells have demonstrated longevity in culture by maintaining their undifferentiated state for at least 80 passages when grown using current published protocols (Thomson et al., 1998; Reubinoff et al., 2000).

Human embryonic stem cells have been shown to differentiate into cells from all three embryonic germ layers *in vitro*. Skin and neurons have been formed, indicating ectodermal differentiation (Schuldiner et al., 2000, 2001; Reubinoff et al., 2001; Zhang et al., 2001). Blood, cardiac cells, cartilage, endothelial cells, and muscle have been formed, indicating mesodermal differentiation (Kaufman et al., 2001; Kehat et al., 2001; Levenberg et al., 2002). Also, pancreatic cells have been formed, indicating endodermal differentiation (Assady et al., 2001).

Principles of Regenerative Medicine. DOI: 10.1016/B978-0-12-381422-7.10045-8

In addition, as further evidence of their pluripotency, embryonic stem cells can form embryoid bodies, which are cell aggregations that contain all three embryonic germ layers while in culture, and can form teratomas *in vivo* (Itskovitz-Eldor et al., 2000).

However, the harvesting of human embryonic stem cells requires the destruction of human embryos, which has raised significant ethical and political concerns in the USA. As stated in the National Institutes of Health (NIH) testimony before Congress in April 2003, only 11 stem cell lines were currently available, which had deleterious effects on the progress of stem cell research in the USA (Kennedy, 2003). Recent policy changes have begun to loosen the restrictions on embryonic stem cell research in the USA. However, many of the approved cell lines were grown in the presence of mouse cells (feeder cells), which can supply many needed growth factors but also expose the human cells to potential contamination from mouse viruses or proteins. This may render the current cell lines unsuitable for human therapeutic purposes.

These barriers to progress in embryonic stem cell research have spawned the search for alternative sources of stem cells, and the use of fetal tissues such as umbilical cord blood as a source of stem cells may overcome some of the political and ethical controversies surrounding embryonic stem cells. Additionally, engineering tissue from fetal cells could allow for ready repair of birth defects detected *in utero*. Ideally, following prenatal detection of a particular defect, autologous fetal stem cells could be harvested from the amnion or fetus, expanded and engineered in tissue culture, then used for surgical repair in the pre- or neonatal period. Tissues engineered from this source would be far superior to synthetic or allograft materials, in that they would be non-immunogenic and physiologically similar to the native fetal organ or tissue. Engineering fetal tissues in this manner was the subject of the first reported experiments in the use of fetal tissues for structural and functional replacement, which were conducted in large animal models (Fauza et al., 1998).

STEM CELLS DERIVED FROM FETAL TISSUES

Fetal stem cells are not a new concept and in fact they have been in clinical use over the past 20 years, though not consistently in the field of tissue engineering. These cells display many properties that make them superior to adult cells for use in regenerative medicine applications, including greater plasticity in differentiation potential, faster growth in culture, and increased survival at low oxygen tension. Fetal cells have also been observed to produce high levels of angiogenic and trophic factors, resulting in improved growth *in vivo* and facilitating regeneration of surrounding host tissues (Turner and Fauza, 2009).

The three most reliable sources to date of abundant fetal stem cells are the placenta, amniotic fluid, and umbilical cord blood. These sources are also attractive in that their stem cells are obtained in a minimally invasive manner from the fetus. Samples from amniotic fluid and the placenta are retrieved using the common prenatal diagnostic procedures of amniocentesis and chorionic villous sampling, whose fetal complication rates are estimated at 0.5% (Evans and Wapner, 2005). Umbilical cord blood is easily obtained at the time of birth and can be easily preserved in cord blood banks. Indeed, there are currently over 400,000 units of cord blood banked worldwide (Kurtzberg, 2009). Other fetal progenitor cell sources that have been investigated include bone marrow (Michejda, 2004), neural tissue (Lindvall and Bjorklund, 2004), liver (Dabeva and Shafritz, 2003), kidney (Al-Awqati and Oliver, 2002; Rollini et al., 2004), and lung (in't Anker et al., 2003).

Amniotic fluid is unique among these sources in that it contains a wide array of cell types, owing to its constantly changing composition throughout gestation. Transport of fluid across fetal skin, respiratory secretions, fetal urination, and fetal swallowing and gastrointestinal (GI) excretions all contribute to the makeup of amniotic fluid. This results in the presence of fetal skin, respiratory, urinary tract, and GI tract cells in the aminotic fluid mileu, as well as embryonic cells from all three germ layers — endoderm, mesoderm, and ectoderm. Other fetal cell types may also be present in certain pathologic states, such as nerve cells in the case of a fetus with a neural tube defect. The presence of pluripotent stem cells in amniotic fluid was suggested in a study by Sancho and colleagues in 1993, who demonstrated differentiation of aminotic fluid cells to myoblasts using viral transfection of a gene regulating myogenesis (Sancho et al., 1993). The ability of these cells, known as mesenchymal stem cells (MSCs), to differentiate into cells of all three germ cell layers has been demonstrated much more recently, making them well suited to tissue engineering applications (Tsai et al., 2006; de Coppi et al., 2007; Holden, 2007).

IMMUNOLOGICAL CONSIDERATIONS

Fetal cells have long been known to exist in a microchimeric state in females during pregnancy, and it appears that microchimerism persists until decades later (O'Donoghue et al., 2004). Since fetal MSCs do not express human leukocyte antigen (HLA) class II antigens and may not express HLA class I antigens as well, this may help to explain this phenomenon. Since these early fetal stem cells appear to have a pre-immune status, this may be ideal for allogenic transplantation/mismatch situations, as both undifferentiated and differentiated fetal MSCs do not elicit alloreactive lymphocyte proliferation (Gotherstrom et al., 2004).

ETHICAL CONSIDERATIONS: FETUS AND OOCYTES

The ethics of transplantation of human fetal tissue continues to be a topic of heated debate. Concerned religious and political groups have lobbied against important funding for research in the use of fetal tissues, thus restricting progress in the field. Although the use of tissue from non-viable fetuses would potentially be less controversial, this tissue is usually unsuitable because of associated pathology such as chromosomal anomalies (Abouna, 2003). The use of fetal tissue sources that otherwise would be discarded, such as umbilical cord blood and amniotic fluid, may provide a more feasible way of circumventing ethical opposition to the field.

A similar debate on the ethics of donated tissue has arisen in the area of therapeutic cloning and oocyte donation, and perhaps lessons can be learned from the resulting discussions (Magnus and Cho, 2005). Some areas for discussion include ethical oversight of collaborations between scientists working in countries with different standards, the protection of tissue donors, and the avoidance of unrealistic expectations arising from the research.

REGENERATIVE MEDICINE APPLICATIONS OF FETAL TISSUES

Stem cells derived from fetal tissues have been utilized in regenerative medicine applications for many organ systems, and some recent investigations are highlighted below:

Neural tissue

Neural tissue regeneration is a complex biological phenomenon for which many laboratories have investigated time and resources in the quest for novel treatments for diseases such as Parkinson's and Alzheimer's. In the case of peripheral nerve injuries, regeneration may be spontaneous if the injury is small. Larger injuries, however, must be surgically treated, typically with nerve grafts harvested from elsewhere in the body (Schmidt and Leach, 2003). Furthermore, regeneration of the central nervous system after injury or disease remains an elusive goal of regenerative medicine. This has encouraged research into stem cell sources for regeneration of neural tissue. One early investigation demonstrated that multipotent neural stem cells, once implanted in a murine brain, are capable of differentiation into multiple neural cell types appropriate to their site of implantation. The authors concluded that such cells could be used for repair of central nervous system defects (Snyder et al., 1992). A later study reported that the use of murine neural stem cells engrafted onto a polyglocolic acid scaffold and implanted into injured rat spinal cords resulted in significant improvement in ambulation, suggesting that these stem cells could be used for the treatment of spinal cord injury (Teng et al., 2002).

As discussed above, fetal stem cells may prove to be an abundant source of stem cells for this type of therapy, and have been investigated for neuroregenerative applications, especially in the case of Parkinson's disease. Since 1990, there have been over a dozen open label trials on the transplantation of fetal mesencephalic tissue into the substantia nigra of human Parkinson's patients, with significant long-term clinical improvement in many cases, often resulting in discontinuation of medication (Lindvall and Bjorklund, 2004). Neural tube defects (NTDs), and specifically spina bifida, represent another common and often devastating neurologic disorder that may be amenable to treatment with fetal tissues. Typical treatment for open NTDs is surgical closure in the neonatal period, with significant persistence of neurologic deficits throughout the lifetime of the patient. A recent study in an ovine model reported that delivery of murine neural fetal stem cells to the spinal cord defect, in addition to surgical closure, resulted in engraftment of the stem cells in an undifferentiated state to the sites of injury, with production of neurotrophic factors that could encourage regeneration of the spinal cord (Fauza et al., 2008). Future research into this technology will likely focus on improving consistency of outcomes and the manufacturing of a large, ideally suited supply of stem cells for implantation.

Since fetal brain tissue is unlikely to be a practical and ethically feasible large-scale source of these cells, investigations into more readily available sources, such as umbilical cord blood, have been undertaken. Potential uses of umbilical cord blood for the treatment of hypoxia-induced encephalopathy have been elucidated in preliminary studies, which have shown that cord blood mononuclear cells and mesenchymal stem cells may have a therapeutic potential through multiple mechanisms acting locally in the central nervous system, and possibly in peripheral organs, when implanted in hypoxic animal models (Pimentel-Coelho et al., 2010). These findings may lead to novel future therapies for those with neurodegenerative disorders and neurological injuries.

Heart

Current experimental efforts in the cardiac regenerative medicine field have focused on cellular cardiomyoplasty, myocardial tissue engineering, and myocardial regeneration as alternative approaches to whole organ transplantation (Krupnick et al., 2004). As it is presently understood, the mature human heart is not known to contain any stem cells; thus, regenerative medicine strategies from fetal tissue may hold the key for novel forms of therapy for patients with end-stage heart failure.

Cardiomyoplasty, which involves direct injection of stem cells into the myocardium, has been investigated in several different laboratories. One investigation demonstrated the efficacy of mesenchymal stem cells derived from amniotic fluid and directly injected into a rat model of myocardial injury. When compared with mesenchymal stem cells derived from adult bone marrow, those derived from amniotic fluid demonstrated greater differentiation into endothelial cells and cardiomyocytes (Iop et al., 2008). Another recent investigation reported significant improvement in cardiac function in post-myocardial infarction rats following injection of hematopoietic precursor cells derived from human umbilical cord blood (Schlechta et al., 2010). This study was also encouraging in that the precursor cells were effective following *in vitro* expansion and demonstrated successful homing to the site of injury following peripheral injection, suggesting these cells can be expanded to large quantities in the laboratory and thus provide a practical and abundant source of cells for tissue repair.

With regards to fetal heart tissue, several preliminary studies have been reported. Embryonic cardiomyocytes were shown to have the ability to remodel the abdominal aorta into a spontaneous pulsating apparatus and to function as a vascular pump (Okamura et al., 2002).

Another research area of interest in cardiac research is valvular interstitial cells. Most valvular interstitial cells in normal valves are quiescent with a fibroblast-like phenotype. However, valvular interstitial cells in developing, diseased, adapting, and engineered valves are adjusted to a dynamic environment through the activation of these cells and secretion of proteolytic enzymes. They appear to be mediating extracellular matrix remodeling ("developing/remodeling/activated" phenotype), which is then followed by a normalization of phenotype (Rabkin-Aikawa et al., 2004). The presence of interstitial cells in the core of engineered heart valves is important to physiological valvular function, as is the presence of endothelial cells on the outer surface to prevent thrombosis when in contact with blood. Successful myocardial tissue engineering of heart valves containing these two cell types has been reported using human amniotic fluid stem cells. The investigators expanded these cells in culture, and sorted them into two phenotypes based on membrane protein expression. One phenotype was coated as an inner layer on a biodegradable polymer heart valve scaffold and conditioned in a bioreactor, and then the other phenotype was coated on the construct as an outer layer. Following *in vitro* maturation, the engineered valve was examined with microscopy and immunohistochemistry, which demonstrated interstitial cells in the inner layer, and endothelial cell surfaces. *In vitro* testing of the valves demonstrated satisfactory opening and closing at half of physiological conditions. The authors concluded that amniotic fluid-derived stem cells could likely be used to engineer functioning heart valves for ready repair of congenital heart malformations in the neonatal period using these methods (Schmidt et al., 2007). These technologies, though very promising, await further testing in human clinical trials.

Lung

For lung tissue engineering, pulmonary cell replacement therapeutics is being studied for treating respiratory diseases such as cystic fibrosis, severe chronic obstructive pulmonary disease (COPD), and idiopathic pulmonary fibrosis. Fetal stem cells have been utilized in a variety of techniques aimed at regeneration and repair, including cell injection therapy, tissue engineering, and modulation of the inflammatory response to injury.

The successful differentiation of mouse embryonic stem cells into respiratory cell lineages and pulmonary structures has previously been demonstrated (Denham et al., 2006), while other studies have shown that these cells can play a role in the regeneration of damaged lung tissue (Weiss et al., 2008). The use of more abundant stem cell sources, such as umbilical cord blood, for these purposes has also been investigated. Sueblinvong et al. reported the culturing of cord blood in specialized airway growth medium followed by systemic injection into mice (Sueblinvong et al., 2008). On subsequent biopsy, these cells were found in the mouse airway, and expressed phenotypic characteristics of human airway epithelium. The authors concluded that mesenchymal stem cells derived from umbilical cord blood are a viable source of cells for airway remodeling.

Engineering lung tissue for transplantation is an ambitious endeavor due to the complex architecture of lung parenchyma. However, several investigators have utilized fetal tissue to successfully create such structures *in vitro*. One report, using a rat model, demonstrated implantation of fetal lung tissue fragments into adult lungs following partial resection (Kenzaki et al., 2006). The grafts showed maturation into neonatal lung tissue with integration into the adult rat circulatory and respiratory system. These results suggested the feasibility of successful lung regeneration utilizing fetal tissues. Fetal stem cells, specifically amniotic MSCs, have also been used to engineer cartilaginous grafts for repair of tracheal defects in a large animal model (Kunisaki et al., 2006). The authors reported that the grafts became lined with respiratory epithelium *in vivo*, and that the animals were able to breathe spontaneously post-operatively. Further investigation into these techniques could result in the production of an ideal graft for repair of congenital tracheal defects.

Another interesting area of research has focused on the immunomodulatory properties of MSCs. These cells have demonstrated secretion of growth factors and cytokines that promote tissue repair and remodeling, as well as suppression of allogeneic T-cell proliferation. These properties have led to investigations on the effects of these cells on tissue healing following injury (Sueblinvong and Weiss, 2009). One study in a mouse model, where these cells were injected systemically following intratracheal bleomycin administration, resulted in decreased lung fibrosis and inflammation (Ortiz et al., 2003). Interestingly, secretion of interleukin-1 (IL-1) receptor antagonist by the MSCs may have accounted for a significant portion of the effect, as only minimal engraftment as lung epithelial cells was observed (Ortiz et al., 2007). These observations have led to further studies into the potential therapeutic effects of mesenchymal stem cells in inflammatory diseases of the lung and other organs.

Liver

The scarcity of donor livers, as well as the risk of complications associated with liver transplantation, has created the need for alternative methods of liver regeneration. Fetal tissues may provide viable alternative methods of hepatic tissue replacement.

The transplantation of mature hepatocytes into diseased livers may provide regenerative capacity, but adult hepatocytes are limited by their ability to grow in culture and by concerns of immunogenicity. Fetal liver cells, however, have been readily grown in culture and display a more versatile differentiation potential. With the development of improved techniques, cell transplantation technology, instead of whole organ transplantation, could present a more efficient and feasible treatment for many liver failure patients. Indeed, studies have shown that replacement of only 10% of normal liver mass may be sufficient for the correction of many of the congenital enzymatic defects that adversely affect pediatric liver failure patients (Asonuma et al., 1992). Furthermore, there have been reports describing the successful injection of cadaveric hepatocytes for the treatment of glycogen storage disease (Muraca et al., 2002) and Crigler-Najjar syndrome (Fox et al., 1998). Progenitor cells from fetal liver cells have been isolated and transplanted, where up to 10% of a normal liver was repopulated (Dabeva and Shafritz, 2003; Rollini et al., 2004). However, further studies are necessary to determine the regenerative and functional capabilities of these cells in the liver as well as in other mesenchymal tissues.

Skin

Skin tissue engineering was one of the early organ systems to which regenerative medicine techniques were applied, often in situations when autologous skin grafting is insufficient or not available. As a result, engineered dermal tissue could be the key to providing sufficient healthy donor skin for engraftment for patients with large burn surface areas. Additionally, the ability of fetal cells to stimulate skin regeneration has also been investigated.

The majority of current research in skin tissue engineering focuses on the synthesis of complex three-dimensional (3D) polymer scaffolds containing functional biomolecules to which cells are introduced, leading to scaffold/skin constructs for regeneration. Hohlfeld and associates developed fetal skin constructs to improve healing of severe burns (Hohlfeld et al., 2005). Their simple techniques provided complete treatment without traditional skin grafting, showing that fetal skin cells might have great potential to treat burns and eventually acute and chronic wounds of other types. Sun et al. showed that co-culture with fibroblasts enables keratinocytes and endothelial cells to proliferate without serum, and that keratinocytes and endothelial cells appear to self-organize according to the native epidermal-dermal structure given the symmetry-breaking field of an air-liquid interface (Sun et al., 2005). Kaviani et al. consistently isolated subpopulations of fetal mesenchymal cells from human amniotic fluid and rapidly expanded them *in vitro* (Kaviani et al., 2003). These human mesenchymal amniocytes attach firmly to both polyglycolic acid polymer and acellular human dermis, and thus it was hypothesized that amniotic fluid may be a valuable and practical cell source for fetal tissue engineering.

Another potential source for skin tissue engineering is human umbilical cord-derived stem cells. Preliminary reports have described the use of cord blood cells combined with a fibrin/platelet glue for treatments of human skin wounds, with significant subjective improvement (Valbonesi et al., 2004). Importantly, no evidence of a host immune response was seen. Several other studies have demonstrated successful growth of many of the complex components of human skin from umbilical cord blood, including small blood vessels (Wu et al., 2004). Further studies into the use of this and other types of fetal tissue, both for engineering of skin replacements and for acceleration of wound healing, are ongoing.

Muscle

Disorders affecting muscle structure and function, such as muscular dystrophy, afflict many patients with devastating consequences. The ability to replace poorly functioning muscle tissue with engineered healthy muscle would therefore be an important breakthrough in regenerative medicine. Human umbilical cord blood (UCB) has been regarded as a possible cell source for muscle tissue engineering because of its hematopoietic and non-hematopoietic (mesenchymal) potential. Gang and associates demonstrated that UCB-derived mesenchymal stem cells possess the potential of skeletal myogenic differentiation and that these cells could be a suitable source for skeletal muscle repair and muscle tissue engineering (Gang et al., 2004).

A later report described the differentiation of human fetal mesenchymal stem cells (hfMSCs) into skeletal muscle fibers by co-culturing the cells with the compound galectin-1 (Chan et al., 2006). Two-thirds of the stem cells differentiated into a skeletal muscle phenotype utilizing this method. When adult-derived mesenchymal stem cells were subjected to the same medium, differentiation did not occur. Additionally, when the fetal mesenchymal cell-derived muscle cells were transplanted into a murine model of muscular dystrophy, a four-fold increase in muscle fiber formation was noted when compared with unstimulated mesenchymal stem cells. This same group also reported on the intravascular injection of human fetal mesenchymal stem cells into dystrophic mouse fetuses (Chan et al., 2007). Widespread long-term engraftment (19 weeks) in multiple organs was seen, with a predilection for muscle compared with non-muscle tissues as evidence of myogenic differentiation of hfMSCs in skeletal and myocardial muscle.

Kopenen et al. investigated the use of umbilical cord blood-derived stem cells for regeneration of muscle tissue damaged by ischemic insult (Koponen et al., 2007). Mesenchymal and CD 133+ stem cells derived from cord blood were genetically enhanced with vascular endothelial growth factor gene transduction, then injected directly into the hindlimbs of mice that had suffered surgically induced ischemia to those limbs. Histologic analysis of the limbs at 3 weeks demonstrated increased regeneration in the stem cell-injected mice. In summary, these studies have demonstrated the potential for muscle regeneration utilizing pluripotent fetal stem cells.

Bone

Human fetal cells have been envisioned for use in bone tissue engineering and in the regeneration of adult skeletal tissue (Montjovent et al., 2004). To construct bioengineered bone structures, vascularized bone grafts theoretically have many advantages over non-vascularized cadaveric and autologous free grafts. However, the availability of these grafts can be extremely limited, resulting in the need for other sources of bone tissue. A recent report demonstrated the engineering of bone grafts from amniotic MSCs and electrospun nanofibers in a rabbit model (Steigman et al., 2009). Amniotic fluid-derived MSCs were harvested, labeled, and seeded onto biodegradable electrospun nanofiber scaffolds. The constructs were maintained in an osteogenic medium for 4–7 months, then used to repair full-thickness sternal defects spanning two or three intercostal spaces. Two months after repair, radiographs demonstrated complete closure of the defects, and CT scans demonstrated increased density of the grafts, suggesting *in vivo* growth and remodeling. The authors concluded that amniotic fluid-derived stem cells may be a viable source of tissue for chest wall repair.

The repair of chest wall defects using fetal tissue was also investigated in a separate study using fetal hyaline cartilage constructs. Fuchs et al. showed that, *in vivo*, engineered implants retained their hyaline characteristics for up to 10 weeks after implantation, but then remodeled into fibrocartilage by 12 weeks postoperatively (Fuchs et al., 2003). Mononuclear inflammatory infiltrates surrounding residual polymer fibers were noted in all of the specimens but they were most prominent in the acellular controls. As a result, the authors concluded that fetal bone tissue engineering may have utility in the treatment of severe congenital chest wall defects at birth.

Another topic of interest has been the use of fetal stem cells to treat congenital defects of bone development such as osteogenesis imperfecta. Human MSCs obtained via cardiocentesis from the blood of first-trimester fetuses were transplanted into the peritoneum of fetal osteogenesis imperfecta mice while *in utero* (Guillot et al., 2008). At time of delivery, the treated mice demonstrated a two-thirds reduction in long bone fractures, with fewer fractures per mouse and an increased proportion of mice having no fractures at all. Transplantation was also associated with increased bone strength, thickness, and length. More transplanted MSCs were found in bone tissues compared with other organs, especially at areas of active bone formation and fracture healing. In addition, the transplanted cells expressed osteoblast lineage genes and produced the bone structural protein osteopontin. These findings have significant implications for the successful future treatments of this debilitating and currently incurable disorder.

826

Blood

Hematological malignancies, such as leukemia and lymphoma, afflict millions of adults and children every year, and are often deadly. Bone marrow transplantation is a well established treatment for these diseases, achieving cure by replacement of diseased blood progenitor cells with healthy stem cells. Unfortunately, partly owing to a shortage of registered donors, many patients in need of marrow transplantation are able to find a matched unrelated donor. This has sparked a keen interest in other sources of hematopoietic stem cells, including fetal bone marrow (Michejda, 2004) and umbilical cord blood (UCB) stem cells. UCB offers several significant advantages over bone marrow as a source for treating these patients, including less strict HLA-matching criteria and prompt availability of units for transplantation. This results in significantly less treatment delay, thus reducing disease-related morbidity and the need for other potentially toxic treatments (Brunstein and Weisdorf, 2009).

Human UCB stem cells have been utilized extensively in the treatment of pediatric patients with hematological malignancies. Treatment of children (instead of adults) has been the predominant avenue of interest owing largely to the smaller quantities of UCB cells needed. Multiple studies comparing UCB transplant with adult unrelated bone marrow transplant have shown similar success rates (Barker et al., 2001; Eapen et al., 2007). Additionally, rates of graft versus host disease are generally lower with UCB transplantation when compared with adult bone marrow (Rocha et al., 2000). This may be attributed to the relative immunologic naivete of neonatal cells. These findings have led to many pediatric oncological centers utilizing UCB over adult bone marrow as their preferred source of unrelated hematopoietic stem cells for transplantation.

Recently, interest has increased in the use of UCB stem cells for the treatment of adults with hematological malignancies. Early results were less favourable, but this may have been related to many centers offering UCB transplantation as a last resort after a prolonged search for

HLA-matched bone marrow. Several contemporary series directly comparing UCB cell transplant with adult bone marrow transplant have demonstrated similar survival rates, with some studies showing superior outcomes in patients receiving UCB grafts (Takahashi et al., 2004; Kumar et al., 2008). Evolving transplantation techniques, such as the double UCB platform and reduced intensity conditioning, are extending the utility of UCB transplantation to older patients (Brunstein and Weisdorf, 2009). These promising results, combined with the relative abundance of UCB, make the use of this fetal tissue an exciting future direction for cancer treatment.

Pancreas

Fetal pancreatic tissue has been suggested as a possible cell source for islet replacement therapy in type 1 diabetes mellitus. While this tissue usually consists of a small amount of beta-cells, a raft of immature and/or progenitor cells may have the potential to proliferate and differentiate into functional insulin-producing cells. Suen et al. showed that both the expansion and differentiation of fetal islet-like cell clusters could be enhanced in the presence of appropriate growth factors and microenvironments (Suen et al., 2005). Their data indicated that *in vivo* exendin-4 treatment may have enhanced the growth and differentiation of fetal mice islet-like cell clusters, and thus promoted the functional maturation of the graft after transplantation. Zhang et al. showed that monoclonal side population (SP) progenitors were isolated from human fetal pancreas, which may be a novel method of purifying pancreatic progenitor cells from human tissues (Zhang et al., 2005b). Zhang and colleagues also isolated nestin-positive cells isolated from human fetal pancreas and discovered that these cells possess the characteristics of pancreatic progenitor cells since they have highly proliferative potential and the capability of differentiation into insulin-producing cells *in vitro* (Zhang et al., 2005a). Interestingly, the nestin-positive pancreatic progenitor cells shared many of the same phenotypic markers as bone marrow-derived MSCs.

More accessible fetal tissue sources for the engineering of insulin-producing cells are also under investigation. Wu et al. reported the isolation of primitive stromal cells from umbilical cord Wharton's jelly, followed by induction of differentiation of these cells into insulin-producing cells (Wu et al., 2009). When compared with similarly induced bone marrow-derived mesenchymal stem cells, these cells showed higher proliferation rates and increased secretion of insulin, leading the authors to conclude that umbilical cord tissue is a superior source for the engineering of pancreatic tissue. Denner et al. recently reported the successful use of umbilical cord blood stem cells to engineer insulin and C-peptide-producing cells by expansion and differentiation in culture utilizing a protocol designed to differentiate murine embryonic stem cells toward a pancreatic phenotype (Denner et al., 2007). However, these encouraging developments are only the first steps in producing viable clinical treatments, and clinical trials are likely several years in to the future.

Kidney

End-stage renal disease is common, progressive, and a major cause of morbidity worldwide. The ability to slow or reverse the loss of glomerular function is a major goal of contemporary medicine, as the affected population grows with the increasing incidence of diabetes, hypertension, and other associated illnesses. One common cause of renal insufficiency, acute ischemic renal disease, has been the subject of much recent research. Although the mechanism of injury and recovery is increasingly being delineated, as it involves the migration of stem cells from the bone marrow and the growth and differentiation of renal progenitor cells for repair (Duffield and Bonventre, 2005; Lin et al., 2005), therapeutic options other than supportive measures are still lacking. As a result, there has been much recent interest in augmenting the recovery from renal injury utilizing stem cells. Many laboratories have investigated the ability of bone marrow stem cell therapy to stimulate renal repair. Lin et al., using an animal model, determined that peripheral injection of bone marrow cells resulted in detection of these cells in the kidney at 5 days post-injury (Lin, 2008). However, the fraction of cells involved in renal regeneration remained very low, and no improvement in blood urea nitrogen was noted. The authors concluded that strategies to improve transmigration of injected stem cells and to speed their conversion to renal progenitors must be developed in order for stem cell injection therapy to prove clinically useful in ischemic renal injury. Another way in which stem cells may be therapeutic in kidney injury is via a paracrine effect. Togel and colleagues have demonstrated renal protection by mesenchymal stem cells in a rat model of ischemic renal injury (Lange et al., 2005; Togel et al., 2005). Microscopic analysis demonstrated that the bulk of injected stem cells did not reach the recovering renal tubules, but were localized to glomerular capillaries, and exerted their effects via inhibition of pro-inflammatory cytokines and promotion of anti-inflammatory cytokines. A later study by the same group showed that the injected mesenchymal stem cells decreased renal apoptosis in mice afflicted with ischemic renal injury, likely via local production of vascular endothelial growth factor, human growth factor, and insulin-like growth factor-1 (Togel et al., 2007).

Fetal stem cells may also possess similar capabilities for renal regeneration, and have been investigated for the engineering of *de novo* kidney tissue. One report described the development of a model for an artificial glomerulus utilizing human umbilical cord blood endothelial progenitor cells (Vu et al., 2009). Following expansion in culture, the UCB cells were used to endothelialize the inner surface of a hemofilter. The finished construct demonstrated good adhesion of the cells, and the cells expressed many phenotypic qualities of endothelial cells. These promising results imply that UCB stem cells could be successfully utilized in the construction of artificial kidneys in the future.

Bladder

The bladder serves as a reservoir for the storage of urine and it maintains a low intraluminal pressure as it fills under normal conditions. Bladder reconstruction has been attempted with both natural materials and synthetic polymers. For bladder regeneration, Ram-Liebig et al. investigated the optimum conditions for the proliferation of urothelial cells, in order to obtain confluent coverage of large surfaces of biocompatible membranes, and for their terminal differentiation (Ram-Liebig et al., 2004). They concluded that the mitogenic effects of the extracellular matrix content of biological membranes and fibroblastic inductive factors were synergistic with each other, and may be able to compensate for a low fetal calf serum concentration and the absence of other additives. They found that lowering the fetal calf serum concentration to 1% in the culture medium inhibited the proliferation of urothelial cells and permitted their terminal differentiation.

Several congenital and acquired diseases of the bladder may need, due to lack or destruction of functional tissue, mechanically stable biomaterials as cell carriers for the engineering of these tissues. Collagen scaffolds have some advantageous characteristics for tissue engineering purposes because of their capacity to induce tissue regeneration and their biocompatibility. Recently, Danielsson and colleagues evaluated cell growth by WST-1 proliferation assay and showed improved growth of bladder cells when modified collagen scaffolds were used (Danielsson et al., 2006). The cell penetration assessed by histology showed similar results on both modified and native scaffolds. *In vivo* studies in athymic mice showed the presence of the fluorescent-labeled transplanted smooth muscle cells in the cell-scaffold constructs until day 3. Thereafter, angiogenesis was noted and infiltration of mouse fibroblasts and polymorphonuclear cells was observed. Nyirady and colleagues characterized the developmental changes to the normal bladder by examining the *in vitro* contractile properties of the fetal sheep detrusor smooth muscle bladder at different gestational ages (Nyirady et al., 2005). They found that fetal development between 65 and 140 days in the sheep was associated with increased contractile activation, which correlated with an increase of muscle development in the earlier stages (65—110 days). In later stages (110—140 days), muscle development appeared to be complete but functional innervation of the tissue was still noted.

CONCLUSIONS

The use of fetal tissue for regenerative medicine purposes has been investigated for essentially every organ system, and some applications, especially in the area of hematopoietic cells, have been in clinical use for several years. There are currently unresolved ethical and moral issues regarding the use of some fetal tissues, but the increasing use and comparative ubiquity of umbilical cord blood- and amniotic fluid-derived stem cells may make such ethical debates largely historical. Future refinements in the harvest, storage, differentiation, and transplantation of these tissues, in conjunction with regenerative medicine techniques already in use, may revolutionize the treatment of many currently incurable diseases with the use of regenerative medicine techniques.

References

Abouna, G. M. (2003). Ethical issues in organ and tissue transplantation. *Exp. Clin. Transplant., 1*, 125—138.

Al-Awqati, Q., & Oliver, J. A. (2002). Stem cells in the kidney. *Kidney Int., 61*, 387—395.

Amit, M., Margulets, V., Segev, H., Shariki, K., Laevsky, I., Coleman, R., et al. (2003). Human feeder layers for human embryonic stem cells. *Biol. Reprod., 68*, 2150—2156.

Asonuma, K., Gilbert, J. C., Stein, J. E., Takeda, T., & Vacanti, J. P. (1992). Quantitation of transplanted hepatic mass necessary to cure the gunn rat model of hyperbilirubinemia. *J. Pediatr. Surg., 27*, 298—301.

Assady, S., Maor, G., Amit, M., Itskovitz-Eldor, J., Skorecki, K. L., & Tzukerman, M. (2001). Insulin production by human embryonic stem cells. *Diabetes, 50*, 1691—1697.

Barker, J. N., Davies, S. M., DeFor, T., Ramsay, N. K., Weisdorf, D. J., & Wagner, J. E. (2001). Survival after transplantation of unrelated donor umbilical cord blood is comparable to that of human leukocyte antigen-matched unrelated donor bone marrow: results of a matched-pair analysis. *Blood, 97*, 2957—2961.

Brivanlou, A. H., Gage, F. H., Jaenisch, R., Jessell, T., Melton, D., & Rossant, J. (2003). Stem cells. Setting standards for human embryonic stem cells (comment). *Science, 300*(5621), 913—916.

Brunstein, C. G., & Weisdorf, D. J. (2009). Future of cord blood for oncology uses. *Bone Marrow Transpl., 44*, 699—707.

Chan, J., O'Donoghue, K., Gavina, M., Torrente, Y., Kennea, N., & Mehmet, H. (2006). Galectin-1 induces skeletal muscle differentiation in human fetal mesenchymal stem cells and increases muscle regeneration. *Stem Cells, 24*, 187—191.

Chan, J., Waddington, S. N., O'Donoghue, K., Kurata, H., Guillot, P. V., Gotherstrom, C., et al. (2007). Widespread distribution and muscle differentiation of human fetal mesenchymal stem cells after intrauterine transplantation in dystrophic mdx mouse. *Stem Cells, 25*, 875—884.

Dabeva, M. D., & Shafritz, D. A. (2003). Hepatic stem cells and liver repopulation. *Semin. Liver Dis., 23*, 349—362.

Danielsson, C., Ruault, S., Basset-Dardare, A., & Frey, P. (2006). Modified collagen fleece, a scaffold for transplantation of human bladder smooth muscle cells. *Biomaterials, 27*, 1054—1060.

de Coppi, P., Bartsch, G., Jr., Siddiqui, M. M., Xu, T., Santos, C. C., Perin, L., et al. (2007). Isolation of amniotic stem cell lines with potential for therapy. *Nat. Biotechnol., 25*, 100—106.

Denham, M., Cole, T. J., & Mollard, R. (2006). Embryonic stem cells form glandular structures and express surfactant protein-C following culture with dissociated fetal respiratory tissue. *Am. J. Physiol. Lung Cell Mol. Physiol., 290*, L1210—L1215.

Denner, L. F., Bodenburg, Y., Zhao, J. G., Howe, M., Cappo, J., Tilton, R. G., et al. (2007). Directed engineering of umbilical cord blood stem cells to produce C-peptide and insulin. *Cell Prolif., 40*, 367—380.

Duffield, J. S., & Bonventre, J. V. (2005). Kidney tubular epithelium is restored without replacement with bone marrow-derived cells during repair after ischemic injury. *Kidney Int., 68*, 1956—1961.

Eapen, M., Rubinstein, P., Zhang, M. J., Stevens, C., Kurtzberg, J., Scaradavou, A., et al. (2007). Outcomes of transplantation of unrelated donor umbilical cord blood and bone marrow in children with acute leukaemia: a comparison study. *Lancet, 369*, 1947—1954.

Evans, M. I., & Wapner, R. J. (2005). Invasive prenatal diagnostic procedures. *Semin Perinatol., 29*, 215—218.

Fauza, D. O., Fishman, S. J., Mehegan, K., & Atala, A. (1998). Videofetoscopically assisted fetal tissue engineering: bladder augmentation. *J. Pediatr. Surg., 33*, 7—12.

Fauza, D. O., Jennings, R. W., Teng, Y. D., & Snyder, E. Y. (2008). Neural stem cell delivery to the spinal cord in an ovine model of fetal surgery for spina bifida. *Surgery, 144,* 367–373.

Fox, I. J., Chowdhury, J. R., Kaufmann, S. S., Goetzen, T. C., Chowdhury, N. R., Warkentin, P. I., et al. (1998). Treatment of the Crigler–Najjar syndrome type I with hepatocyte transplantation. *N. Engl. J. Med., 338,* 1422–1426.

Fuchs, J. R., Terada, S., Hannouche, D., Ochoa, E. R., Vacanti, J. P., & Fauza, D. O. (2003). Fetal tissue engineering: chest wall reconstruction. *J. Pediatr. Surg., 38,* 1188–1193.

Gang, E. J., Jeong, J. A., Hong, S. H., Hwang, S. H., Kim, S. W., Yang, I. H., et al. (2004). Skeletal myogenic differentiation of mesenchymal stem cells isolated from human umbilical cord blood. *Stem Cells, 22,* 617–624.

Gotherstrom, C., Ringden, O., Tammik, C., Zetterberg, E., Westgren, M., & le Blanc, K. (2004). Immunologic properties of human fetal mesenchymal stem cells. *Am. J. Obstet. Gynecol., 190,* 239–245.

Guillot, P. V., Abass, O., Bassett, J. H., Shefelbine, S. J., Bou-Gharios, G., Chan, J., et al. (2008). Intrauterine transplantation of human fetal mesenchymal stem cells from first-trimester blood repairs bone and reduces fractures in osteogenesis imperfecta mice. *Blood, 111,* 1717–1725.

Hohlfeld, J., de Buys Roessingh, A., Hirt-Burri, N., Chaubert, P., Gerber, S., Scaletta, C., et al. (2005). Tissue engineered fetal skin constructs for paediatric burns. *Lancet, 366*(9488), 840–842.

Holden, C. (2007). Stem cells. Versatile stem cells without the ethical baggage? *Science, 315,* 170.

in 't Anker, P. S., Noort, W. A., Scherjon, S. A., Kleijburg-van der Keur, C., Kruisselbrink, A. B., van Bezooijen, R. L., et al. (2003). Mesenchymal stem cells in human second-trimester bone marrow, liver, lung, and spleen exhibit a similar immunophenotype but a heterogeneous multilineage differentiation potential. *Haematologica, 88,* 845–852.

Iop, L., Chiavegato, A., Callegari, A., Bollini, S., Piccoli, M., Pozzobon, M., et al. (2008). Different cardiovascular potential of adult- and fetal-type mesenchymal stem cells in a rat model of heart cryoinjury. *Cell Transplant., 17,* 679–694.

Itskovitz-Eldor, J., Schuldiner, M., Karsenti, D., Eden, A., Yanuka, O., Amit, M., et al. (2000). Differentiation of human embryonic stem cells into embryoid bodies compromising the three embryonic germ layers. *Mol. Med., 6*(2), 88–95.

Kaufman, D. S., Hanson, E. T., Lewis, R. L., Auerbach, R., & Thomson, J. A. (2001). Hematopoietic colony-forming cells derived from human embryonic stem cells. *Proc. Natl. Acad. Sci. U.S.A., 98,* 10716–10721.

Kaviani, A., Guleserian, K., Perry, T. E., Jennings, R. W., Ziegler, M. M., & Fauza, D. O. (2003). Fetal tissue engineering from amniotic fluid. *J. Am. Coll. Surg., 196,* 592–597.

Kehat, I., Kenyagin-Karsenti, D., Snir, M., Segev, H., Amit, M., Gepstein, A., et al. (2001). Human embryonic stem cells can differentiate into myocytes with structural and functional properties of cardiomyocytes (comment). *J. Clin. Invest., 108,* 407–414.

Kennedy, D. (2003). Stem cells: still here, still waiting (comment). *Science, 300*(5621), 865.

Kenzaki, K., Sakiyama, S., Kondo, K., Yoshida, M., Kawakami, Y., Takehisa, M., et al. (2006). Lung regeneration: implantation of fetal rat lung fragments into adult rat lung parenchyma. *J. Thorac. Cardiovasc. Surg., 131,* 1148–1153.

Koponen, J. K., Kekarainen, T., E Heinonen, S., Laitinen, A., Nystedt, J., Laine, J., et al. (2007). Umbilical cord blood-derived progenitor cells enhance muscle regeneration in mouse hindlimb ischemia model. *Mol. Ther., 15,* 2172–2177.

Krupnick, A. S., Kreisel, D., Riha, M., Balsara, K. R., & Rosengard, B. R. (2004). Myocardial tissue engineering and regeneration as a therapeutic alternative to transplantation. *Curr. Top. Microbiol. Immunol., 280,* 139–164.

Kumar, P., Defor, T. E., Brunstein, C., Barker, J. N., Wagner, J. E., Weisdorf, D. J., et al. (2008). Allogeneic hematopoietic stem cell transplantation in adult acute lymphocytic leukemia: impact of donor source on survival. *Biol. Blood Marrow Transplant., 14,* 1394–1400.

Kunisaki, S. M., Freedman, D. A., & Fauza, D. O. (2006). Fetal tracheal reconstruction with cartilaginous grafts engineered from mesenchymal amniocytes. *J. Pediatr. Surg., 41,* 675–682.

Kurtzberg, J. (2009). Update on umbilical cord blood transplantation. *Curr. Opin. Pediatr., 21,* 22–29.

Lange, C., Togel, F., Ittrich, H., Clayton, F., Nolte-Ernsting, C., Zander, A. R., et al. (2005). Administered mesenchymal stem cells enhance recovery from ischemia/reperfusion-induced acute renal failure in rats. *Kidney Int., 68,* 1613–1617.

Levenberg, S., Golub, J. S., Amit, M., Itskovitz-Eldor, J., & Langer, R. (2002). Endothelial cells derived from human embryonic stem cells. *Proc. Natl. Acad. Sci. U.S.A., 99,* 4391–4396.

Lin, F. (2008). Renal repair: role of bone marrow stem cells. *Pediatr. Nephrol., 23,* 851–861.

Lin, F., Moran, A., & Igarashi, P. (2005). Intrarenal cells, not bone marrow-derived cells, are the major source for regeneration in postischemic kidney. *J. Clin. Invest., 115,* 1756–1764.

Lindvall, O., & Bjorklund, A. (2004). Cell therapy in Parkinson's disease. *NeuroRx, 1,* 382–393.

Magnus, D., & Cho, M. K. (2005). Ethics. Issues in oocyte donation for stem cell research. *Science, 308*(5729), 1747−1748.

Michejda, M. (2004). Which stem cells should be used for transplantation? *Fetal Diagn. Ther., 19,* 2−8.

Montjovent, M. O., Burri, N., Mark, S., Federici, E., Scaletta, C., Zambelli, P. Y., et al. (2004). Fetal bone cells for tissue engineering. *Bone, 35,* 1323−1333.

Muraca, M., Gerunda, G., Neri, D., Vilei, M. T., Granato, A., Feltracco, P., et al. (2002). Hepatocyte transplantation as a treatment for glycogen storage disease type Ia. *Lancet, 359,* 317−318.

Nyirady, P., Thiruchelvam, N., Godley, M. L., David, A., Cuckwo, P. M., & Fry, C. H. (2005). Contractile properties of the developing fetal sheep bladder. *Neurourol. Urodyn., 24,* 276−281.

O'Donoghue, K., Chan, J., de la Fuente, J., Kennea, N., Sandison, A., Anderson, J. R., et al. (2004). Microchimerism in female bone marrow and bone decades after fetal mesenchymal stem-cell trafficking in pregnancy (see comment). *Lancet, 364*(9429), 179−182.

Oberpenning, F., Meng, J., Yoo, J. J., & Atala, A. (1999). De novo reconstitution of a functional mammalian urinary bladder by tissue engineering (see comment). *Nat. Biotechnol., 17,* 149−155.

Okamura, S., Suzuki, A., Johkura, K., Ogiwara, N., Harigaya, M., Yokouchi, T., et al. (2002). Formation of the biopulsatile vascular pump by cardiomyocyte transplants circumvallating the abdominal aorta. *Tissue Eng., 8,* 201−211.

Ortiz, L. A., Gambelli, F., McBride, C., Gaupp, D., Baddoo, M., Kaminski, N., et al. (2003). Mesenchymal stem cell engraftment in lung is enhanced in response to bleomycin exposure and ameliorates its fibrotic effects. *Proc. Natl. Acad. Sci. U.S.A., 100,* 8407−8411.

Ortiz, L. A., Dutreil, M., Fattman, C., Pandey, A. C., Torres, G., Go, K., et al. (2007). Interleukin 1 receptor antagonist mediates the anti-inflammatory and antifibrotic effect of mesenchymal stem cells during lung injury. *Proc. Natl. Acad. Sci. U.S.A., 104,* 11002−11007.

Pimentel-Coelho, P. M., & Mendez-Otero, R. (2010). Cell therapy for neonatal hypoxic-ischemic encephalopathy. *Stem Cells Dev., 19*(3), 299−310.

Rabkin-Aikawa, E., Farber, M., Aikawa, M., & Schoen, F. J. (2004). Dynamic and reversible changes of interstitial cell phenotype during remodeling of cardiac valves. *J. Heart Valve Dis., 13,* 841−847.

Ram-Liebig, G., Meye, A., Hakenberg, O. W., Haase, M., Baretton, G., & Wirth, M. P. (2004). Induction of proliferation and differentiation of cultured urothelial cells on acellular biomaterials. *BJU Int., 94,* 922−927.

Reubinoff, B. E., Pera, M. F., Fong, C. Y., Trounson, A., & Bongso, A. (2000). Embryonic stem cell lines from human blastocysts: somatic differentiation *in vitro* (comment). *Nat. Biotechnol., 18,* 399−404; erratum: *18*(5), 559.

Reubinoff, B. E., Itsykson, P., Turetsky, T., Pera, M. F., Reinhartz, E., Itzik, A., et al. (2001). Neural progenitors from human embryonic stem cells (comment). *Nat. Biotechnol., 19,* 1134−1140.

Richards, M., Fong, C. Y., Chan, W. K., Wong, P. C., & Bongso, A. (2002). Human feeders support prolonged undifferentiated growth of human inner cell masses and embryonic stem cells (comment). *Nat. Biotechnol., 20,* 933−936.

Rocha, V., Wagner, J. E., Jr., Sobocinski, K. A., Klein, J. P., Zhang, M. J., Horowitz, M. M., et al. (2000). Graft-versus-host disease in children who have received a cord-blood or bone marrow transplant from an HLA-identical sibling. Eurocord and International Bone Marrow Transplant Registry Working Committee on Alternative Donor and Stem Cell Sources. *N. Engl. J. Med., 342,* 1846−1854.

Rollini, P., Kaiser, S., Faes-van't Hull, E., Kapp, U., & Leyvraz, S. (2004). Long-term expansion of transplantable human fetal liver hematopoietic stem cells. *Blood, 103,* 1166−1170.

Sancho, S., Mongini, T., Tanji, K., Tapscott, S. J., Walker, W. F., Weintraub, H., et al. (1993). Analysis of dystrophin expression after activation of myogenesis in amniocytes, chorionic-villus cells, and fibroblasts. A new method for diagnosing Duchenne's muscular dystrophy. *N. Engl. J. Med., 329,* 915−920.

Schlechta, B., Wiedemann, D., Kittinger, C., Jandrositz, A., Bonaros, N. E., Huber, J. C., et al. (2010). *Ex-vivo* expanded umbilical cord blood stem cells retain capacity for myocardial regeneration. *Circ. J., 74,* 188−194.

Schmidt, C. E., & Leach, J. B. (2003). Neural tissue engineering: strategies for repair and regeneration. *Annu. Rev. Biomed. Eng., 5,* 293−347.

Schmidt, D., Achermann, J., Odermatt, B., Breymann, C., Mol, A., Genoni, M., et al. (2007). Prenatally fabricated autologous human living heart valves based on amniotic fluid derived progenitor cells as a single cell source. *Circulation, 116,* 164−170.

Schuldiner, M., Yanuka, O., Itzkovitz-Eldor, J., Melton, D. A., & Benvenisty, N. (2000). Effects of eight growth factors on the differentiation of cells derived from human embryonic stem cells. *Proc. Natl. Acad. Sci. U.S.A., 97,* 11307−11312.

Schuldiner, M., Eiges, R., Eden, A., Yanuka, O., Itskovitz-Eldor, J., Goldstein, R. S., et al. (2001). Induced neuronal differentiation of human embryonic stem cells. *Brain Res., 913,* 201−205.

Snyder, E. Y., Deitcher, D. L., Walsh, C., Arnold-Aldea, S., Hartwieg, E. A., & Cepko, C. L. (1992). Multipotent neural cell lines can engraft and participate in development of mouse cerebellum. *Cell, 68*, 33–51.

Steigman, S. A., Ahmed, A., Shanti, R. M., Tuan, R. S., Valim, C., & Fauza, D. O. (2009). Sternal repair with bone grafts engineered from amniotic mesenchymal stem cells. *J. Pediatr. Surg., 44*, 1120–1126.

Sueblinvong, V., & Weiss, D. J. (2009). Cell therapy approaches for lung diseases: current status. *Curr. Opin. Pharmacol., 9*, 268–273.

Sueblinvong, V., Loi, R., Eisenhauer, P. L., Bernstein, I. M., Suratt, B. T., Spees, J. L., et al. (2008). Derivation of lung epithelium from human cord blood-derived mesenchymal stem cells. *Am. J. Respir. Crit. Care Med., 177*, 701–711.

Suen, P. M., Li, K., Chan, J. C., & Leung, P. S. (2005). *In vivo* treatment with glucagon-like peptide 1 promotes the graft function of fetal islet-like cell clusters in transplanted mice. *Int. J. Biochem. Cell Biol., 38*, 951–960.

Sun, T., Mai, S., Norton, D., Haycock, J. W., Ryan, A. J., & MacNeil, S. (2005). Self-organization of skin cells in three-dimensional electrospun polystyrene scaffolds. *Tissue Eng., 11*(7–8), 1023–1033.

Takahashi, S., Iseki, T., Ooi, J., Tomonari, A., Takasugi, K., Shimohakamada, Y., et al. (2004). Single-institute comparative analysis of unrelated bone marrow transplantation and cord blood transplantation for adult patients with hematologic malignancies. *Blood, 104*, 3813–3820.

Teng, Y. D., Lavik, E. B., Qu, X., Park, K., Ourednik, J., Zurakowski, D., et al. (2002). Functional recovery following traumatic spinal cord injury mediated by a unique polymer scaffold seeded with neural stem cells. *Proc. Natl. Acad. Sci. U.S.A., 99*, 3024–3029.

Thomson, J. A., Itskovitz-Eldor, J., Shapiro, S. S., Waknitz, M. A., Swiergiel, J. J., Marshall, V. S., et al. (1998). Embryonic stem cell lines derived from human blastocysts (comment). *Science, 282*(5391), 1145–1147; erratum: *282*(5395), 1827.

Togel, F., Hu, Z., Weiss, K., Isaac, J., Lange, C., & Westenfelder, C. (2005). Administered mesenchymal stem cells protect against ischemic acute renal failure through differentiation-independent mechanisms. *Am. J. Physiol. Renal Physiol., 289*, F31–F42.

Togel, F., Weiss, K., Yang, Y., Hu, Z., Zhang, P., & Westenfelder, C. (2007). Vasculotropic, paracrine actions of infused mesenchymal stem cells are important to the recovery from acute kidney injury. *Am. J. Physiol. Renal Physiol., 292*, F1626–F1635.

Tsai, M. S., Hwang, S. M., Tsai, Y. L., Cheng, F. C., Lee, J. L., & Chang, Y. J. (2006). Clonal amniotic fluid-derived stem cells express characteristics of both mesenchymal and neural stem cells. *Biol. Reprod., 74*, 545–551.

Turner, C. G. B., & Fauza, D. O. (2009). Fetal tissue engineering. *Clin. Perinatol., 36*, 473–488.

Valbonesi, M., Giannini, G., Migliori, F., Dalla Costa, R., & Dejana, A. M. (2004). Cord blood (CB) stem cells for wound repair. Preliminary report of 2 cases. *Transfus. Apher. Sci., 30*, 153–156.

Vu, D. M., Masuda, H., Yokoyama, T. A., Fujimura, S., Kobori, M., Ito, R., et al. (2009). CD133+ endothelial progenitor cells as a potential cell source for a bioartificial glomerulus. *Tissue Eng. Part A, 15*, 3173–3182.

Weiss, D. J., Kolls, J. K., Ortiz, L. A., Panoskaltsis-Mortari, A., & Prockop, D. J. (2008). Stem cells and cell therapies in lung biology and lung diseases. *Proc. Am. Thorac. Soc., 5*, 637–667.

Wu, L. F., Wang, N. N., Liu, Y. S., & Wei, X. (2009). Differentiation of Wharton's jelly primitive stromal cells into insulin-producing cells in comparison with bone marrow mesenchymal stem cells. *Tissue Eng. Part A, 15*, 2865–2873.

Wu, X., Rabkin-Aikawa, E., Guleserian, K. J., Perry, T. E., Masuda, Y., & Sutherland, F. W. H. (2004). Tissue-engineered microvessels on three-dimensional biodegradable scaffolds using human endothelial progenitor cells. *Am. J. Physiol. Heart Circ. Physiol., 287*, 480–487.

Zhang, S. C., Wernig, M., Duncan, I. D., Brustle, O., & Thomson, J. A. (2001). *In vitro* differentiation of transplantable neural precursors from human embryonic stem cells (comment). *Nat. Biotechnol., 19*, 1129–1133.

Zhang, L., Hong, T. P., Hu, J., Liu, Y. N., Wu, Y. H., & Li, L. S. (2005a). Nestin-positive progenitor cells isolated from human fetal pancreas have phenotypic markers identical to mesenchymal stem cells. *World J. Gastroenterol., 11*, 2906–2911.

Zhang, L., Hu, J., Hong, T. P., Liu, Y. N., Wu, Y. H., & Li, L. S. (2005b). Monoclonal side population progenitors isolated from human fetal pancreas. *Biochem. Biophys. Res. Comm., 333*, 603–608.

832

CHAPTER 46

Engineering of Large Diameter Vessels

Masood A. Machingal, Saami K. Yazdani, George J. Christ
Wake Forest Institute for Regenerative Medicine, Winston-Salem, NC, USA

PREVALENCE OF CARDIOVASCULAR DISEASES AND NEED FOR TISSUE-ENGINEERED BLOOD VESSELS

Cardiovascular diseases are the leading cause of death in the USA (CDC, 2007). Total cost for cardiovascular diseases and stroke is estimated at $400 billion per year (American Heart Association, 2006) and is expected to increase. Abnormal vascular function contributes to coronary artery disease, stroke, peripheral arterial disease, renal insufficiency, and diabetic neuropathy. In 2003 alone, nearly 500,000 coronary artery bypass graft surgeries, and over 100,000 lower extremity bypass procedures were performed (www.americanheart.org; Birkmeyer et al., 2002). Important risk factors for vascular disease include older age, hypertension, hyperlipidemia, smoking, diabetes, and chronic renal insufficiency (Collins et al., 2003). Population trends are unfavorable with respect to vascular disease, as the US population is aging, diabetes is reaching epidemic proportions, and chronic renal disease, especially end-stage renal disease (ESRD), is now epidemic (McClellan, 1994; Gilbertson et al., 2005). With respect to ESRD, there is a significant unmet medical need for autologous dialysis vascular access grafts. Such grafts are clearly of relatively large diameter when mature (>6 mm), and thus represent an important target for the relatively large diameter tissue-engineered blood vessels (TEBVs) that are the subject of this report. Among cardiovascular problems, atherosclerosis is the most common cause of premature mortality, with more than two of every five Americans dying of cardiovascular disease (American Heart Association, 2004). Coronary artery atherosclerosis is estimated to cause over a million myocardial infarctions annually in the USA alone (National Center for Health Statistics, 1994).

CLINICALLY UNMET NEED FOR IMPROVED DIALYSIS VASCULAR ACCESS

The 21st annual report on the ESRD program in USA (US Renal Data System, 2009) summarizes the clinical challenges and outcomes on ESRD patients. In fact, USRDS predicts that the number of ESRD patients will increase drastically; to 800,000 in 10 years (Fig. 46.1) resulting in higher health care expenditure for ESRD. Presently only a mature native radial artery to cephalic vein fistula achieves the ideal access route of blood circulation for hemodialysis. A close alternative is another site of native arteriovenous fistula (AVF) within the upper extremity, for example, an upper arm brachial artery to cephalic or basilic vein fistulas. Figure 46.2 demonstrates the types of hemodialysis approaches currently used. Regardless, only 33% of hemodialysis patients in the USA achieve dialysis via a native AVF, while the majority requires a prosthetic polytetrafluoroethylene (PTFE)

Principles of Regenerative Medicine. DOI: 10.1016/B978-0-12-381422-7.10046-X

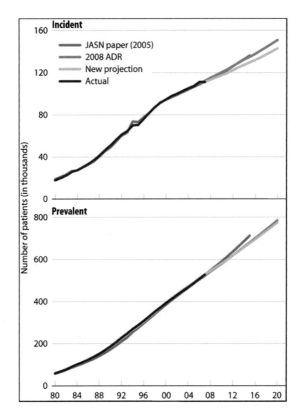

FIGURE 46.1

Projected counts of incident & prevalent ESRD patients through 2020. *(Sources: U.S. Renal Data System 2009* Annual Data Report: Atlas of End-stage renal disease in the United States, U.S. Renal Data System 2008 Annual Data Report (2008 ADR) and *J Am Soc Nephrol* **16**, 3736–41 (JASN paper 2005).

artery to vein bypass grafts (AVBG, 41%) or chronic indwelling central venous catheters (McClellan, 1994; Kohler and Kirkman, 1999; National Kidney Foundation, 2000, 2001; Hsu et al., 2004). A detailed discussion of the limitations of PTFE is well beyond the scope of this report. Suffice it to say, that stenosis is the most common problem, and moreover, the presence of PTFE creates a foreign body response (Kohler and Kirkman, 1999; Huber et al., 2003, 2004). In addition, endothelialization occurs only within the first 1−2 cm at anastomoses, and furthermore, prosthetic materials are prone to infection. In fact, chronic cannulation with needles inserted through the skin and left in place for hours during dialysis predisposes to frequent graft infection (National Kidney Foundation, 2002; Basaran et al., 2003; Huber et al., 2004; Neville et al., 2004). Failure of the lumen surface to heal in PTFE grafts may also predispose to hematogenous seeding. Finally, as PTFE does not regenerate, the graft wall deteriorates over time from chronic puncture, predisposing to pseudoaneurysm formation, skin breakdown, cannulation site bleeding, and graft infection. Certainly, the unique hemodynamic profile of the AVF environment poses significant challenges for vascular replacement.

For all of the aforementioned reasons, creating an autologous blood vessel of the appropriate geometry for AVBG directly addresses many of the limitations of the PTFE grafts currently used for dialysis access. Certainly, a cellularized vessel wall with luminal endothelial coverage is likely to be more resistant to thrombosis and infection. Furthermore, the cellularized wall of a mature, bioengineered vessel should allow healing at puncture sites, and hence prevent vessel wall deterioration and provide resistance to infection superior to PTFE. Moreover, the engineered blood vessel is likely to have a compliance profile better matched to the outflow vein than PTFE, which in turn should reduce the extent of outflow venous stenosis (Kohler and

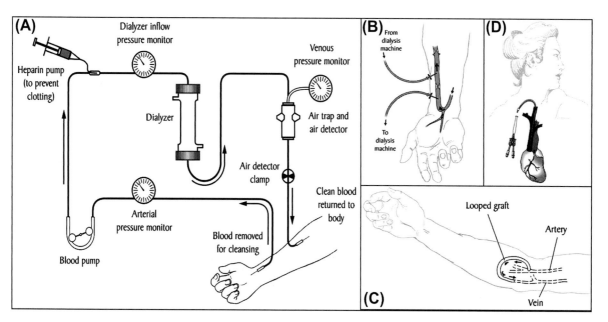

FIGURE 46.2

Schematic diagram of hemodialysis process and types of vascular access approaches. (A) Schematic diagram of hemodialysis. Blood removed for dialysis travels via blood pump, heparin pump that prevents clotting, dialyzer, pressure monitor, venous pressure monitor, and air trap detectors and returns to body as clean blood. (B) Arteriovenous fistula created with radial artery and cephalic vein being connected to dialysis machine using two needles inserted in the fistula. Flow of blood is indicated by indicated by the arrows in fistula and fistula needle access. (C) Arteriovenous graft created by connecting artery and vein is via a looped graft. Flow of blood is indicated by Dialysis access is achieved by placing the needles on the graft. Blood flow is indicated by arrows. (D) Temporary hemodialysis access achieved by a venous catheter inserted through the skin near the collar bone. The catheter is connected to the large vein from the heart. *(Source: National Kidney and Urological Disease Information Clearing House. http://kidney.niddk.nih.gov/kudiseases/pubs/hemodialysis/index.htm)*

Kirkman, 1999). All of these properties are prerequisites for the next generation of dialysis vascular access, and are summarized in Table 46.1.

In this regard, it is important to point out that relatively little attention has been paid to the importance of the medial smooth muscle cells (SMC) layer to TEBV function, or to the possibility that the presence of SMC in the vessel wall may promote accelerated maturation of TEBV (both *in vitro* and *in vivo*). Both of these beneficial properties of smooth muscle have important implications for the further development and clinical translation of TEBV. When viewed from this perspective, the creation of TEBV for dialysis vascular access provides an extraordinary opportunity to further examine the role of the SMC in TEBV. To this end, the objective of this report is to address how the presence of the SMC can help to meet the physiological characteristics/demands of the bioengineered vessels that would be required for such clinical success, and furthermore, to outline one currently envisioned strategy for achieving this goal. We also provide an introduction to the concept of regenerative pharmacology and its importance to the further development of TEBV.

TABLE 46.1 Properties of ideal vascular access graft for dialysis

Anti-thrombogenic
Anti-inflammatory
Resistant to injury and intimal hyperplasia
Mechanical and structural similarities with native vessels
Ability to remodel and maintain structural integrity over long time
Suitable geometry to achieve high blood flow without turbulence
Ability to withstand multiple needlestick punctures over large period of time
Ability to respond to pharmacological/physiological stimulus *in vivo*
Available to patient in a clinically relevant time period
Affordable and commercially feasible manufacturing technology

VASCULAR PHYSIOLOGY RELEVANT TO TISSUE-ENGINEERED BLOOD VESSELS

Blood is carried from the heart to the capillaries by the arteries, and then returned via the venous circulation. The magnitude of the cardiovascular problems described above has certainly served to focus most tissue-engineered blood vessel research on the arterial side of the vascular tree, which will also remain the subject of this report. In that regard, the arterial vascular tree can be subdivided into three general types of arteries, based both on their location in the vascular tree, and on the functions that they serve. As blood is moved away from the heart, it moves from large, elastic arteries that have strictly function (e.g. aorta) to more medium-sized muscular arteries that have a distributive function, and eventually to small muscular arteries and arterioles, which provide the majority of the resistive function. The lumen-to-wall ratio decreases as one moves down the vascular tree away from the heart, and so does the ratio of the elastic component to the smooth muscle component (Boulpaep, 2003). Regardless of the considerable differences in function, the vessel wall in all three types of arteries possesses three distinct layers (tunics) which are the intima, media, and adventitia (Fig. 46.3). The innermost layer encountered when traversing the vessel wall from the luminal side is the tunica intima, which is in direct contact with moving blood. The intima is covered by the endothelium, which in turn, resides on a thick basement membrane referred to as the internal elastic lamina. The endothelium provides the antithrombogenic surface that ensures continuous laminar blood flow. The middle layer in the vessel wall is the tunica media. The media is composed of smooth muscle cells embedded in a matrix of collagen, elastin, and proteoglycans, the ratio and composition of which varies along the vascular tree (see below).

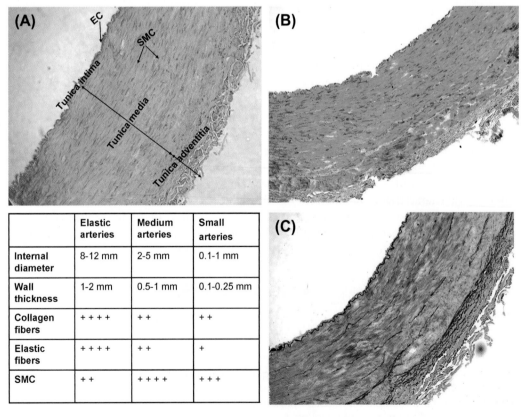

	Elastic arteries	Medium arteries	Small arteries
Internal diameter	8-12 mm	2-5 mm	0.1-1 mm
Wall thickness	1-2 mm	0.5-1 mm	0.1-0.25 mm
Collagen fibers	+ + + +	+ +	+ +
Elastic fibers	+ + + +	+ +	+
SMC	+ +	+ + + +	+ + +

FIGURE 46.3

Structure and composition of arterial wall. Representative H&E (A), Masson's Trichrome (B), and Verhoeff—Van Gieson elastin (C) staining of ovine carotid artery illustrating the major components of vessel wall. *(Adapted from Boulpaep, 2003.)*

The media resides between the internal elastic lamina and the tunica externa (i.e. adventitia). The adventitia represents the outermost portion of the vessel wall, and is primarily comprised of loose connective tissue, fibroblasts, and small nerve fibers. Of note, nerve fibers rarely penetrate the adventitial-medial smooth muscle cell border.

The physiological characteristics of each vessel depends on its location in the vascular tree. Of note, there is no native vessel that mimics the physiological characteristics of the proposed dialysis vascular access graft (i.e. AVF). While arteriovenous anastomoses are quite common in the circulation (e.g. for rapid shunting of blood in the skin for heat exchange), categorizing the behavior of the AVF as proposed herein is somewhat unique. In fact, arteriovenous anastomoses occur naturally between small muscular arteries and venules to bypass the capillary network and enable rapid shunting of blood. The proposed bioengineered AVF that is described here would be a much larger vessel (>6 mm), and therefore has some unique characteristics. Thus, the ideal AVF must possess some hybrid characteristics, for example, the compliance of large elastic arteries and perhaps the tone of large to medium sized muscular arteries. The main goal of these bioengineered vessels is to maintain a non-thrombogenic and non-proliferative surface, while retaining the ability to adapt and remodel to external stimuli, and moreover, be able to heal in response to repetitive puncture wounds (i.e. 3×/week). Clearly, to incorporate all of these features will require the presence of both smooth muscle cells and endothelial cells. A brief review of the phenotypic and functional characteristics of these two vascular wall cell types most directly pertinent to TEBV is provided below.

Endothelial cells

There are many excellent reviews on endothelial cells and the reader is referred to a few of these for more details (Cines et al., 1998; Michiels, 2003; Aird, 2006). Endothelial cells (EC) line the entire vascular tree and provide a functional barrier between blood and the vascular wall cells and tissue parenchyma. Perhaps more importantly, they serve as a biologically active lining of the blood vessels and play a critical role in the control of vascular tone. Regulation of vascular tone is accomplished via a variety of endothelium-derived vasoactive substances. Some important endothelium-derived vascorelaxants include nitric oxide (NO), prostacyclin (PGI_2) and endothelium-derived hyperpolarizing factor (EDHF). The endothelium also provides an important source of constrictor substances, such as endothelin-1, superoxide anions/radicals, angiotensin II, thromboxane A_2, and endoperoxides. These are synthesized and released in response to a wide variety of environmental and mechanical stimuli. In addition to the regulation of vascular tone, the endothelium is also responsible for the maintenance of vessel wall permeability (i.e. regulating the flow of nutrients, biological molecules), as well as the balance between coagulation and fibrinolysis, the composition of the subendothelial matrix, the adhesion and extravasation of leukocytes and the mediation of inflammatory processes in the vascular wall. Prevention of thrombotic events is accomplished by maintaining a healthy monolayer of endothelial cells that retain the ability to secrete antithrombotic agents such as NO, PGI_2, tissue plasminogen activator (tPA), and thrombomodulin. All of these endothelial cell functions are controlled via membrane bound proteins, junctional proteins, and a variety of cell surface receptors, and they are critical to circulatory homeostasis, and thus, to normal organ function.

Smooth muscle cells

Vascular myocytes are interposed between the variable autonomic innervation on one side of the vessel (adventitial or abluminal side), and the endothelium on the other. This anatomical arrangement has important mechanistic implications for coordinated vessel function, as the size of the medial smooth muscle cell layer varies from a single cell in the terminal arteriole, to numerous relatively concentric layers of muscle such as those that encircle the large elastic and muscular arteries. Nonetheless, the role of myocytes in most vessels is similar; that is, it maintains vessel tone at some partial level of contractility, with the ability to become further constricted, or relaxed, as the physiological necessities of the vessels dictate. More importantly, contraction and cell seeding conditions and bioreactor TEBV preconditioning protocols, to selection of the appropriate animal model for implantation, needs to be thoroughly evaluated. Each of these steps has a critical impact on the TEBV process, which will likely vary with each TEBV for each indication. Certainly, with respect to the best "recipe" for TEBV, the devil is in the details. Below we provide some basic concepts, features and requirements for each step in the process.

Scaffolds

Various synthetic and naturally derived biomaterials have been used in constructing vascular grafts (reviewed by Ravi et al., 2009) but none have proven entirely satisfactory. The goal is always the same; that is, to develop a reproducible, biocompatible scaffold similar to that characteristic of native vasculature. With respect to the synthetic constructs, polymers and electrospun scaffolds are both very attractive options due to the control one has over composition, architecture, and the reproducibility of the manufacturing process. The current generations of polymers are mostly biodegradable, and include polylactic acid (PLA), poly-glycolic acid (PGA), polyhydroxyalkanoate (PHA), and polydioxanone (PDS). These polymers can be used singly or in combination, to optimize the desired mechanical performance and biocompatibility of the graft. Similarly to polymers, electrospinning techniques can take advantage of a variety of materials to create scaffolds. Electrospinning involves creation of an electromagnetic field by using a high-voltage source. Exposure to a high voltage causes poly-mers in volatile solvents to elongate and splay into small fibers, and be drawn/sprayed onto a grounded surface (i.e. a mandrel), where they can be spun into tubular structures. By controlling the characteristics of individual fiber formation during the electrospinning process, as well as the rotational speed of the mandrel (Stitzel et al., 2006), structural characteristics such as porosity and geometry can be precisely controlled. Thus, from a commercial perspective, synthetic scaffolds are very attractive for the clinical translation of TEBV.

However, from a biological perspective, decellularized vessels (i.e. natural scaffolds), possess a biochemical composition, ultrastructural architecture, and biomechanics that are similar to native vessels. Not surprisingly, decellularized, collagen-based, vascular scaffolds derived from porcine blood vessels have been successfully used for TEBV *in vivo* (Kaushal et al., 2001). Similar approaches have been used in a variety of clinical applications for developing tissue-engineered vascular patches (Cho et al., 2005), heart valves (Lichtenberg et al., 2006), and bladders (Gabouev et al., 2003). To summarize, while synthetic scaffolds will undoubtedly provide an important source of "off the shelf" scaffold material for clinical TEBV, the natural scaffold still provides the ultimate "gold" standard with respect to the biological requirements and characteristics of native vessels required to guide the development of the TEBV *in vivo*. The TEBV strategy outlined below utilizes the decellularized scaffold.

Step 1: Removal of cells from mature arteries produces a collagen-based scaffold that is amenable for seeding and growth of vascular cells. Prior work has established a working protocol for preparation of scaffolds from animal arteries using a multi-step decellularization process. Details of the procedure can be found in previous literature that shows the overall concept (Kaushal et al., 2001; Amiel et al., 2006; Yazdani et al., 2009). Decellularized scaffolds preserve their extracellular matrix architecture, including internal and external elastin layers and several layers of collagen. Moreover, the decellularization process removes all cellular components, maintaining only the collagen and elastin components. The quantity and distribution of collagen and elastin in a vascular scaffold is vital when considering scaffold material in developing TEBV.

The mechanical characteristics of vascular grafts play a significant influence in the long-term patency of the implant. In fact, compliance mismatch is thought to be one of the most important factors predisposing prosthetic vascular grafts to intimal hyperplasia, thrombosis and occlusion. If the TEBV is stiff, then flow disturbances and tissue vibration may predispose to hyperplasia. Conversely, a TEBV that is too compliant may result in the formation of an

838

840

FIGL
Vascu
In the
contra
respo
dilatio
press
the 5(
deple
contra
over t
phenc
impor
functi

aneurysm. As such, we have rigorously analyzed the biomechanical characteristics of the decellularized scaffolds. To measure compliance, decellularized vascular scaffolds were immersed in a water bath, cannulated at one end, and pressurized, while recording the diameter change using a digital camera. Burst strength testing and stress-strain measurements demonstrate that the decellularization process does not disturb the mechanical integrity to the extent that failure might occur *in vivo* (Yazdani et al., 2009).

Cell source

Step 2: There are numerous potential cell sources available for cellularizing the synthetic or naturally derived scaffolds. The strategy that we are currently pursuing is to isolate progenitor cells from circulating blood, and expand them to obtain the EC and SMC that are required for TEBV. The overall concept is to utilize cell-selective markers to isolate and expand the progenitor cells, prior to differentiation and further proliferation for seeding purposes. This process is well characterized with respect to differentiation of endothelial cells from endothelial progenitor cells, but further research is required in order to obtain similar procedures for derivation of smooth muscle cells from circulating muscle progenitor cells. The latter work is ongoing in our group.

Cell seeding and preconditioning

Steps 3 and 4: The final steps in creating TEBV involve the development of a bioreactor system for cell seeding and preconditioning; that is to expose TEBV to the *in vivo* conditions they will face upon implantation. Seeding TEBV consists of depositing cells (EC and/or SMC) onto a three-dimensional scaffold to achieve a confluent monolayer of EC at the inner surface and/or SMC on the outside. A variety of approaches have been attempted in seeding both the endothelium and smooth muscle cells, and recently published studies have demonstrated highly evolved bioreactor systems to produce and monitor the mechanical forces required for cell seeding and/or preconditioning (Thompson et al., 2002; Barron et al., 2003; Mironov et al., 2003; McCulloch et al., 2004; Narita et al., 2004; Williams and Wick, 2004; Portner et al., 2005; Soletti et al., 2006).

The theory behind the use of bioreactors for TEBV derives from studies which have demonstrated that mechanical stress accelerated cell and tissue growth and phenotypic differentiation (Braddon et al., 2002; Nerem, 2003; Jeong et al., 2005; Kurpinski et al., 2006). In this regard, a properly designed bioreactor system provides physiologically relevant stress in a 3D tissue, accelerating tissue maturation and development functional properties. While we are unaware of any published studies which document that bioreactor preconditioning *per se* is capable of producing a relatively mature and fully functional vessel *in vitro*, this certainly seems an area worthy of further investigation. It corresponds to intuition that implantation of a more mature functional TEBV would accelerate tissue formation and maturation *in vivo*; thereby providing a quicker restoration of function, and presumably, promoting more widespread clinical applications. Regardless of the precise operational concept, a bioreactor system for development of TEBV should be capable of the following functions:

- Permitting static and/or dynamic seeding.
- Providing and monitoring physiological flow rate and pressure.
- Capable of dynamic data display and recording (archival).
- Providing physiological axial and circumferential stress.
- Providing an external bath.
- Maintaining desired concentration of gases and nutrients in the culture medium.
- Maintaining temperature and sterility.
- Be easily portable and accessible for transportation and use in surgical procedure.

Obviously, the optimal preconditioning protocol(s) required to seed and mature TEBV are still being developed. However, Fig. 46.6 shows the general features of a bioreactor system, while Fig. 46.7 shows some preliminary results with SMC seeding on decellularized scaffolds.

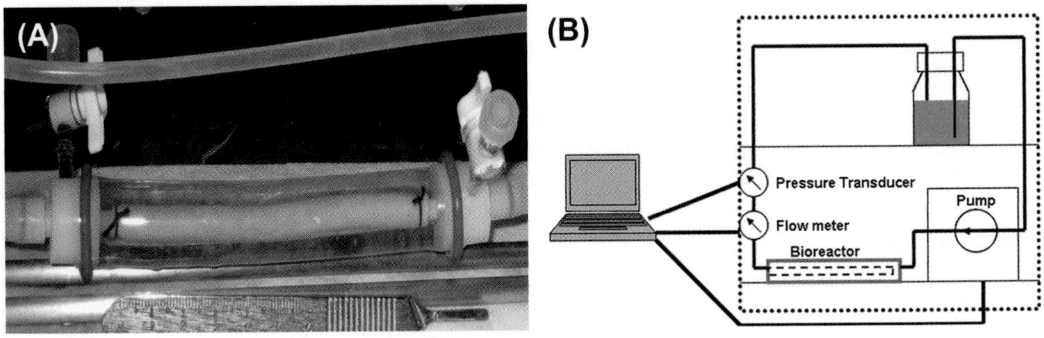

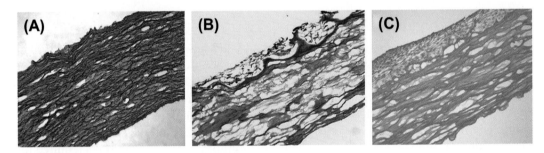

FIGURE 46.6
Bioreactor system. (A) Bioreactor flow system containing the scaffold seeded with EC in the luminal side and SMC with abluminal side. Bioreactor provides an external media bath, optical access, a bypass system, control over flow and pressure conditions, and the ability to maintain sterility. (B) Schematic diagram of bioreactor flow diagram set up and computer controls.

FIGURE 46.7
Cell seeding of decellularized constructs. H&E staining of decellularized carotid artery (A), statically seeded SMC on decellularized constructs after 48 h (B) and after long-term (3-4 weeks) bioreactor preconditioning (C).

We continue to investigate and optimize the impact of various bioreactor protocols on the efficiency of cell seeding and the phenotypic differentiation of ECs and SMCs. Major parameters of interest include rotational speed of scaffold during seeding, optimal cell seeding density and time course of cell seeding protocol, and duration of bioreactor preconditioning period (i.e. days or weeks). Clearly, further development and refinement of the bioreactor system is required, but unequivocally, such development holds intrinsic scientific value, and moreover, will likely be required to ensure the widespread clinical application of TEBV.

CURRENT STATUS OF LARGE DIAMETER TISSUE-ENGINEERED VASCULAR GRAFTS

Since the seminal work of Weinberg and Bell (1986), the complexities associated with clinical translation of the TEBV technology have become quite apparent. As is clear from the aforementioned discussion, development of TEBV is a complex process that varies widely depending on the scaffolding materials, source and type of cells used, methods for bioreactor preconditioning *in vitro*, and finally, the *in vivo* animal models chosen for "proof of concept" studies. In this regard, many excellent reviews have been devoted to TEBV development for both small and large diameter grafts (Ratcliffe, 2000; Tiwari et al., 2001; Rabkin and Schoen, 2002; Teebken and Haverich, 2002; Sales et al., 2005; Vara et al., 2005; Isenberg et al., 2006; Bordenave et al., 2008; Aper et al., 2009). A summary of the main findings of many of the studies conducted are listed in Table 46.2. Below we briefly review a few efforts devoted to large diameter TEBV. Note that, for the purposes of this report, we consider small diameter grafts to be <4 mm.

Study	Scaffold / approach	Diameter (mm)	Cell type	Seeding / culture time	Conditioning	Animal model	Outcome	Comments
Leyh et al. 2006	Acellular ovine pulmonary artery	n/a	EC (ovine carotid artery)	(4 hours(static) + 12 hours (0.1 RPM)) × 3	N/A	Ovine (pulmonary artery)	100% patency at 6 months (n = 5), increase in diameter was observed	N/A
Dahl et al. 2007	PGA	3	SMC (porcine carotid)	30 minute static seeding	peristaltic pump (strain 1.5% @ 2.8 Hz) 7-8 weeks	N/A	N/A	EC and SMC were seeded with success
L'Heureux et al. 2007	autologous cell sheet approach	4.8	FB, EC	6-9 months; mean 7.5;	N/A	human (arterio venous shunt)	N/A	grafts were mechanically stable
McAllister et al. 2009	autologous cell sheet approach	4.8	FB, EC	6-9 months; mean 7.5;	N/A	human (arterio venous shunt)	78% patency after 1 month (n = 9); 60% patency after 3 months (n = 8)	N/A
Mirensky et al. 2009	PLLA mesh with e-caprolactone and L-lactide	0.7	human SMC and EC	6 days	N/A	infrarenal aortic interposition in mice upto 1 year	graft remained patent till 1 year	N/A
Hibino et al. 2010	PGA and e-caprolactone or L-lactide	12-24	BMC	2-4 hours	N/A	human (mean age 5.5 years -as extra-cardiac cavo-pulmonary conduit)	Mean follow up of 5.8 years: No graft failure due to aneurysm, graft rupture, infection or calcification; 24% of patients had stenosis (n = 25)	N/A
Gauvin et al. 2010	cell sheet approach	4.5	SMC, dermal & saphenous vein FB	28-42 days	N/A	N/A	N/A	Developed single step fabrication method to produce graft with SMC media and FB adventitia with improved mechanical properties
Wang et al. 2010	PGA	4	human ADSC	7 days	75 beat/min, 5% radial distention for 8 weeks	N/A	N/A	Adipose derived stem cells were differentiated into SMC pathway for development of TEBV with improved mechanical strength
Zhao et al. 2010	Acellular ovine carotid	3-4	autologous sheep bone marrow MSC	1 week	N/A	sheep carotid artery interposition 2,5 months	N/A	remained patent and non thrombogenic until five months

Shin'oka et al. (2005) have developed a TEBV from a PGA/PLLA scaffold seeded with autologous bone marrow cells on the luminal surface to treat pediatric patients with congenital heart defects. The preclinical studies revealed that seeded TEBV remain patent for up to 6 months with no sign of stenosis or dilation (Watanabe et al., 2001; Opitz et al., 2004; Hibino et al., 2010). Moreover, when retrieved, the endothelium of the vessel stained positive for functional endothelial-specific surface marker (Factor VIII). In a ground breaking clinical study, the peripheral pulmonary arteries of 23 patients were replaced with large-diameter autologous seeded biodegradable scaffolds (PGA/PLLA + autologous bone marrow cells) (12–24 mm diameter). Long-term follow-up of these seminal clinical studies (mean 5.8 years) has shown no complications; such as thrombosis, stenosis or obstruction associated with the implants. Importantly, these results demonstrate the potential of TEBV to remodel, grow, and remain patent in a growing patient Hibino et al. (2010).

Opitz et al. (2004) investigated the development of a tissue-engineered graft for aortic replacement (15 mm). The challenges of a bioengineered aorta clearly present a significant departure from TEBV investigations conducted elsewhere in the vascular system. The scaffold for these studies was constructed from poly-4-hydroxybutyrate (P-4-HB, Tepha Inc., Cambridge, MA), endothelialized and impregnated with SMCs within a bioreactor system. Dynamic preconditioning of the scaffold for 2 weeks resulted in a TEBV with a rupture force of approximately 80% of the native ovine aorta, the target replacement arterial segment. *In vivo* experiments of the TEBV demonstrated that the implanted grafts remained patent for up to 3 months, followed by significant dilation and thrombus formation of the graft likely due to insufficient elastic fiber synthesis.

Preclinical studies by L'Heureux et al. (1998) used SMCs, fibroblasts, and EC to engineer a polymer-free blood vessel that had improved mechanical properties and performed reasonably well *in vivo* (three out of six implanted grafts remained patent after 7 days). More recently, L'Heureux et al. (2006) have implanted autologous TEBV extracted from fibroblasts for up to 8 months in rats, canine, and primate models (3–6 mm in diameter). These grafts demonstrated tissue integration, suitable mechanical properties, and formation of vasa vasorum. L'Heuruex and colleagues have also pioneered the use of their proprietary, autologous, cell sheet based method for developing large diameter vascular access grafts in human clinical trials on patients with ESRD (L'Heureux et al., 2007; McAllister et al., 2009). For these studies, a sheet of extracellular matrix was derived from human fibroblasts and wrapped around a mandrel. Upon fusion of this multilayered tubular construct, the structure was removed from the mandrel, and the inner layers were devitalized and seeded with endothelial cells, while the outer layers were preserved as a living adventitia-like structure. The average time needed to produce these grafts was ~7.5 months, indicating that this technology may not be applicable to more urgent clinical needs. Results from this safety study were recently reported from 10 ESRD patients after a minimum of 6 months of follow-up (McAllister et al., 2009).

PROGRESS TOWARDS MORE NATIVE-LIKE TISSUE-ENGINEERED BLOOD VESSELS: REGENERATIVE PHARMACOLOGY

Initial studies in the field were understandably more focused on developing TEBV with improved patency and mechanical stability. Not surprisingly then, to date, a majority of the TEBV studies have focused on histology, molecular biology and biomechanics as the major endpoints of investigation. However, clearly the ultimate goal of vascular tissue engineering is to develop grafts that can mimic, as much as possible, the physiology/pharmacology of native blood vessels, and the term "regenerative pharmacology" has been coined to describe the methods for both calibrating progress toward such a biomimetic, as well as the pharmacological methods that will be used to influence/guide the process (Andersson and Christ, 2007) (Fig. 46.8).

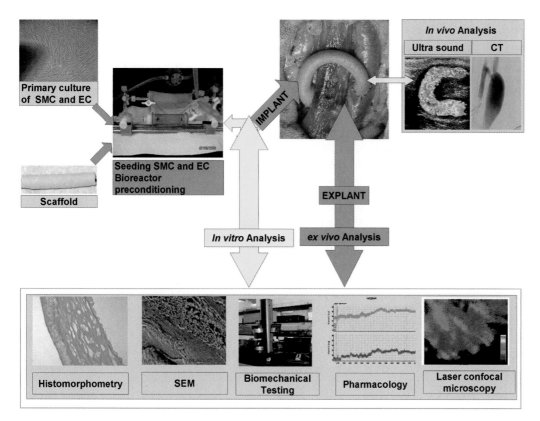

FIGURE 46.8
Overview of interdisciplinary approaches for developing and evaluating TEBV.

As discussed before (see p. 838), blood vessels maintain tone at some partial level of contractility, enabling further contraction or relaxation in a coordinated manner based on the physiological necessities/environment of the vessel of interest. Hence the ideal TEBV is expected to respond to physiologically relevant circulating neurotransmitters, hormones and metabolic factors released from surrounding tissues in a fashion similar to native vessels. Consistent with this supposition, there has been more recent interest in evaluating the physiological/pharmacological characteristics of these grafts *in vitro*; both prior to implantation *in vivo* and following retrieval. These data are summarized in Table 46.3. Not surprisingly, the physiological properties and pharmacological profiles of TEBV grafts can vary as widely as the cells sources, scaffolding systems, bioreactor preconditioning protocols and *in vivo* models used for creating them. Some of this variability is related to the presence of mitigating factors that obscure direct comparisons. For example, the studies highlighted in Table 46.3 reflect different incubation times *in vitro* as well as different durations of implantation *in vivo*. Furthermore, in some cases the magnitude of the observed vasomotor responses was differentially quantified. While such circumstances prohibit sweeping generalizations at this point, it is obvious from a review of Table 46.3 that it is possible to recapitulate many physiologically relevant aspects of receptor-mediated signal transduction in TEBV, even if the absolute magnitude of the contractile responses observed in TEBV are generally much lower than that observed for their corresponding native vessels. Moreover, there may be several reasons for TEBV being physiologically/pharmacologically numb to certain stimuli and these include, among others, the high series resistance of some scaffolding materials to contraction, the low number of SMC in comparison to the resistance of the scaffold/graft material, the state of differentiation of SMC, lack of syncytial communication between SMC, absence of receptors or uncoupling of these receptors with intracellular signal transduction events (i.e. calcium mobilization) or lack of organized collagen/elastin architecture. All of aforementioned

TABLE 46.3 Regenerative pharmacology of engineered vessels

Authors and year	Scaffolding approach	Test system	Contractile Agonists tested	Relaxant Agonists tested
Niklason et al. 1999, Dahl et al. 2007	PGA	*In vitro* ex *vivo*	0.1 mM Prostaglandin F_{2a}	N/A
Swartz et al. 2005	Fibrin gel		118 mM KCl 1 µM Phenylephrine 300 mM U-46619	0.1 µM Sodium nitroprusside
L'Heureux et al. 2001	Self assembled cell sheets	*In vitro*	100 µM Histamine 1 µM Bradykinin 3 mM ATP	0.1 µM Sodium nitroprusside 10 µM Forskolin
Laflamme et al. 2005 & 2006	Self assembled cell sheets	*In vitro*	50 mM KCl 3 mM ATP 0.1 µM Endothelin -1,2 and 3 0.1 µM IRL-1620	N/A
Kaushal et al. 2001	Decellularized artery scaffold	*In vitro* ex *vivo*	80 mM KCl 10 µM Phenylephrine 10 µM 5-hydroxytryptamine	100 µM Sodium nitroprusside 100 µM Acetylecholine 1 mM Calcium ionophore A23187
Amiel et al. 2006	Decellularized artery scaffold	ex *vivo*	N/A	1 mM Sodium nitroprusside 1 mM Calcium ionophore A23187
Yazdani et al. 2009	Decellularized artery scaffold	*In vitro*	60 mM KCl	N/A
Neff et al. 2010	Decellularized artery scaffold	ex *vivo*	60 mM KCl 10 µM Phenlyephrine 10 uM 5-hydroxytryptamine	N/A

considerations will need to be incorporated into the rationale used to design the next generation of TEBV prototypes. Thus, while it is too early yet to know whether a completely native-like TEBV is truly possible, the initial results seem quite encouraging.

References

Aird, W. C. (2006). Endothelial cell heterogeneity and atherosclerosis. *Curr. Atheroscler. Rep., 8,* 69–75.

American Heart Association. (2004). *Heart disease and stroke statistics—2004 update.* Dallas, Tex: American Heart Association; 2003.

Amiel, G. E., Komura, M., Shapira, O., Yoo, J. J., Yazdani, S., Berry, J., et al. (2006). Engineering of blood vessels from acellular collagen matrices coated with human endothelial cells. *Tissue Eng., 12,* 2355–2365.

Andersson, K. E., & Christ, G. J. (2007). Regenerative pharmacology: the future is now. *Mol. Interv., 7,* 79–86.

Aper, T., Haverich, A., & Teebken, O. (2009). New developments in tissue engineering of vascular prosthetic grafts. *Vasa, 38,* 99–122.

Badylak, S. F., Lantz, G. C., Coffey, A., & Geddes, L. A. (1989). Small intestinal submucosa as a large diameter vascular graft in the dog. *J. Surg. Res., 47,* 74–80.

Bank, A. J., Wilson, R. F., Kubo, S. H., Holte, J. E., Dresing, T. J., & Wang, H. (1995). Direct effects of smooth muscle relaxation and contraction on *in vivo* human brachial artery elastic properties. *Circ. Res., 77,* 1008–1016.

Barra, J. G., Armentano, R. L., Levenson, J., Fischer, E. I., Pichel, R. H., & Simon, A. (1993). Assessment of smooth muscle contribution to descending thoracic aortic elastic mechanics in conscious dogs. *Circ. Res., 73,* 1040–1050.

Barron, V., Lyons, E., Stenson-Cox, C., McHugh, P. E., & Pandit, A. (2003). Bioreactors for cardiovascular cell and tissue growth: a review. *Ann. Biomed. Eng., 31,* 1017–1030.

Basaran, O., Karakayali, H., Emiroglu, R., Belli, S., & Haberal, M. (2003). Complications and long-term follow-up of 4416 vascular access procedures. *Transplant. Proc., 35,* 2578–2579.

Birkmeyer, J. D., Siewers, A. E., Finlayson, E. V., Stukel, T. A., Lucas, F. L., Batista, I., et al. (2002). Hospital volume and surgical mortality in the United States. *N. Engl. J. Med., 346,* 1128–1137.

Bordenave, L., Menu, P., & Baquey, C. (2008). Developments towards tissue-engineered, small-diameter arterial substitutes. *Exp. Rev. Med. Devices, 5,* 337–347.

Borschel, G. H., Huang, Y. C., Calve, S., Arruda, E. M., Lynch, J. B., Dow, D. E., et al. (2005). Tissue engineering of recellularized small-diameter vascular grafts. *Tissue Eng., 11,* 778–786.

Boulpaep, E. L. (2003). Arteries and veins. In Medical Physiology., E. L. Boulpaep, & W. F. Boron (Eds.) (pp. 447–462). NY: Saunders, NY.

Braddon, L. G., Karoyli, D., Harrison, D. G., & Nerem, R. M. (2002). Maintenance of a functional endothelial cell monolayer on a fibroblast/polymer substrate under physiologically relevant shear stress conditions. *Tissue Eng.*, *8*, 695–708.

Brink, P. (2000). Gap junction voltage dependence. A clear picture emerges. *J. Gen. Physiol.*, *116*, 11–12.

Campbell, J. H., Efendy, J. L., & Campbell, G. R. (1999). Novel vascular graft grown within recipient's own peritoneal cavity. *Circ. Res.*, *85*, 1173–1178.

CDC. (2007). Prevalence of self-reported cardiovascular disease among persons aged > or =35 years with diabetes – United States, 1997-2005. *MMWR Morb. Mortal. Wkly. Rep.*, *56*, 1129–1132.

Cho, S. W., Park, H. J., Ryu, J. H., Kim, S. H., Kim, Y. H., Choi, C. Y., et al. (2005). Vascular patches tissue-engineered with autologous bone marrow-derived cells and decellularized tissue matrices. *Biomaterials*, *26*, 1915–1924.

Christ, G., & Wingard, C. (2005). Calcium sensitization as a pharmacological target in vascular smooth-muscle regulation. *Curr. Opin. Invest. Drugs*, *6*, 920–933.

Christ, G. J., Spray, D. C., el-Sabban, M., Moore, L. K., & Brink, P. R. (1996). Gap junctions in vascular tissues. Evaluating the role of intercellular communication in the modulation of vasomotor tone. *Circ. Res.*, *79*, 631–646.

Christ, G. J., Wang, H. Z., Venkateswarlu, K., Zhao, W., & Day, N. S. (1999). Ion channels and gap junctions: their role in erectile physiology, dysfunction, and future therapy. *Mol. Urol.*, *3*, 61–73.

Cines, D. B., Pollak, E. S., Buck, C. A., Loscalzo, J., Zimmerman, G. A., McEver, R. P., et al. (1998). Endothelial cells in physiology and in the pathophysiology of vascular disorders. *Blood*, *91*, 3527–3561.

Collins, A. J., Kasiske, B., Herzog, C., Chen, S. C., Everson, S., Constantini, E., et al. (2003). Excerpts from the United States Renal Data System 2003 Annual Data Report: Atlas of End-stage Renal Disease in the United States. *Am. J. Kidney Dis.*, *42*, A5–7, S1–S230.

Collins, A. J., Foley, R. N., Herzog, C., Chavers, B. M., Gilbertson, D., Ishani, A., et al. (2010). Excerpts from the US Renal Data System 2009 Annual Data Report. *Am. J. Kidney Dis.*, *55*, S1–420, A6-7.

Dahl, S. L., Rhim, C., Song, Y. C., & Niklason, L. E. (2007). Mechanical properties and compositions of tissue engineered and native arteries. *Ann. Biomed. Eng.*, *35*, 348–355.

del Valle-Rodriguez, A., Calderon, E., Ruiz, M., Ordonez, A., Lopez-Barneo, J., & Urena, J. (2006). Metabotropic Ca^{2+} channel-induced Ca($^{2+}$) release and ATP-dependent facilitation of arterial myocyte contraction. *Proc. Natl. Acad. Sci. U.S.A*, *103*, 4316–4321.

Gabouev, A. I., Schultheiss, D., Mertsching, H., Koppe, M., Schlote, N., Wefer, J., et al. (2003). *In vitro* construction of urinary bladder wall using porcine primary cells reseeded on acellularized bladder matrix and small intestinal submucosa. *Int. J. Artif. Organs*, *26*, 935–942.

Gauvin, R., Ahsan, T., Larouche, D., Levesque, P., Dube, J., Auger, F. A., et al. (2010). A novel single-step self-assembly approach for the fabrication of tissue-engineered vascular constructs. *Tissue Eng. Part A.*, *16*, 1737–1747.

Gilbertson, D. T., Liu, J., Xue, J. L., Louis, T. A., Solid, C. A., Ebben, J. P., et al. (2005). Projecting the number of patients with end-stage renal disease in the United States to the year 2015. *J. Am. Soc. Nephrol.*, *16*, 3736–3741.

Haddock, R. E., & Hill, C. E. (2005). Rhythmicity in arterial smooth muscle. *J. Physiol.*, *566*, 645–656.

Haefliger, J. A., Nicod, P., & Meda, P. (2004). Contribution of connexins to the function of the vascular wall. *Cardiovasc. Res.*, *62*, 345–356.

Hibino, N., McGillicuddy, E., Matsumura, G., Ichihara, Y., Naito, Y., Breuer, C., et al. (2010). Late-term results of tissue-engineered vascular grafts in humans. *J. Thorac. Cardiovasc. Surg.*, *139*, 431–436, e1–2.

Hoerstrup, S. P., Zund, G., Sodian, R., Schnell, A. M., Grunenfelder, J., & Turina, M. I. (2001). Tissue engineering of small caliber vascular grafts. *Eur. J. Cardiothorac. Surg.*, *20*, 164–169.

Hoerstrup, S. P., Cummings Mrcs, I., Lachat, M., Schoen, F. J., Jenni, R., Leschka, S., et al. (2006). Functional growth in tissue-engineered living, vascular grafts: follow-up at 100 weeks in a large animal model. *Circulation*, *114*, I159–I166.

Hsu, C. Y., Vittinghoff, E., Lin, F., & Shlipak, M. G. (2004). The incidence of end-stage renal disease is increasing faster than the prevalence of chronic renal insufficiency. *Ann. Intern. Med.*, *141*, 95–101.

Huber, T. S., Carter, J. W., Carter, R. L., & Seeger, J. M. (2003). Patency of autogenous and polytetrafluoroethylene upper extremity arteriovenous hemodialysis accesses: a systematic review. *J. Vasc. Surg.*, *38*, 1005–1011.

Huber, T. S., Hirneise, C. M., Lee, W. A., Flynn, T. C., & Seeger, J. M. (2004). Outcome after autogenous brachial-axillary translocated superficial femoropopliteal vein hemodialysis access. *J. Vasc. Surg.*, *40*, 311–318.

Huynh, T., Abraham, G., Murray, J., Brockbank, K., Hagen, P. O., & Sullivan, S. (1999). Remodeling of an acellular collagen graft into a physiologically responsive neovessel. *Nat. Biotechnol.*, *17*, 1083–1086.

Isenberg, B. C., Williams, C., & Tranquillo, R. T. (2006). Small-diameter artificial arteries engineered. *in vitro. Circ. Res., 98*, 25–35.

Jarajapu, Y. P., & Knot, H. J. (2005). Relative contribution of Rho kinase and protein kinase C to myogenic tone in rat cerebral arteries in hypertension. *Am. J. Physiol. Heart Circ. Physiol., 289*, H1917–H1922.

Jeong, S. I., Kwon, J. H., Lim, J. I., Cho, S. W., Jung, Y., Sung, W. J., et al. (2005). Mechano-active tissue engineering of vascular smooth muscle using pulsatile perfusion bioreactors and elastic PLCL scaffolds. *Biomaterials, 26*, 1405–1411.

Kaushal, S., Amiel, G. E., Guleserian, K. J., Shapira, O. M., Perry, T., Sutherland, F. W., et al. (2001). Functional small-diameter neovessels created using endothelial progenitor cells expanded *ex vivo. Nat. Med., 7*, 1035–1040.

Kohler, T. R., & Kirkman, T. R. (1999). Dialysis access failure: a sheep model of rapid stenosis. *J. Vasc. Surg., 30*, 744–751.

Kuecherer, H. F., Just, A., & Kirchheim, H. (2000). Evaluation of aortic compliance in humans. *Am. J. Physiol. Heart Circ. Physiol., 278*, H1411–H1413.

Kurpinski, K., Chu, J., Hashi, C., & Li, S. (2006). Anisotropic mechanosensing by mesenchymal stem cells. *Proc. Natl. Acad. Sci. U.S.A., 103*, 16095–16100.

Laflamme, K., Roberge, C. J., Labonte, J., Pouliot, S., d'Orleans-Juste, P., Auger, F. A., et al. (2005). Tissue-engineered human vascular media with a functional endothelin system. *Circulation, 111*, 459–464.

Laflamme, K., Roberge, C. J., Pouliot, S., d'Orleans-Juste, P., Auger, F. A., & Germain, L. (2006). Tissue-engineered human vascular media produced in vitro by the self-assembly approach present functional properties similar to those of their native blood vessels. *Tissue Eng., 12*, 2275–2281.

Lagaud, G., Davies, K. P., Venkateswarlu, K., & Christ, G. J. (2002a). The physiology, pathophysiology and therapeutic potential of gap junctions in smooth muscle. *Curr. Drug Targets, 3*, 427–440.

Lagaud, G., Karicheti, V., Knot, H. J., Christ, G. J., & Laher, I. (2002b). Inhibitors of gap junctions attenuate myogenic tone in cerebral arteries. *Am. J. Physiol. Heart Circ. Physiol., 283*, H2177–H2186.

Leyh, R. G., Wilhelmi, M., Rebe, P., Ciboutari, S., Haverich, A., & Mertsching, H. (2006). Tissue engineering of viable pulmonary arteries for surgical correction of congenital heart defects. *Ann. Thorac. Surg., 81*, 1466–1470, discussion 1470–1.

L'Heureux, N., Paquet, S., Labbe, R., Germain, L., & Auger, F. A. (1998). A completely biological tissue-engineered human blood vessel. *FASEB J., 12*, 47–56.

L'Heureux, N., Stoclet, J. C., Auger, F. A., Lagaud, G. J., Germain, L., & Andriantsitohaina, R. (2001). A human tissue-engineered vascular media: a new model for pharmacological studies of contractile responses. *FASEB J., 15*, 515–524.

L'Heureux, N., Dusserre, N., Konig, G., Victor, B., Keire, P., Wight, T. N., et al. (2006). Human tissue-engineered blood vessels for adult arterial revascularization. *Nat. Med., 12*, 361–365.

L'Heureux, N., Dusserre, N., Marini, A., Garrido, S., de la Fuente, L., & McAllister, T. (2007). Technology insight: the evolution of tissue-engineered vascular grafts – from research to clinical practice. *Nat. Clin. Pract. Cardiovasc. Med., 4*, 389–395.

Lichtenberg, A., Tudorache, I., Cebotari, S., Suprunov, M., Tudorache, G., Goerler, H., et al. (2006). Preclinical testing of tissue-engineered heart valves re-endothelialized under simulated physiological conditions. *Circulation, 114*, I559–I565.

McAllister, T. N., Maruszewski, M., Garrido, S. A., Wystrychowski, W., Dusserre, N., Marini, A., et al. (2009). Effectiveness of haemodialysis access with an autologous tissue-engineered vascular graft: a multicentre cohort study. *Lancet, 373*, 1440–1446.

McClellan, W. M. (1994). Epidemic end-stage renal disease in the United States. *Artif. Organs, 18*, 413–415.

McCulloch, A. D., Harris, A. B., Sarraf, C. E., & Eastwood, M. (2004). New multi-cue bioreactor for tissue engineering of tubular cardiovascular samples under physiological conditions. *Tissue Eng., 10*, 565–573.

Michiels, C. (2003). Endothelial cell functions. *J. Cell Physiol., 196*, 430–443.

Mirensky, T. L., Nelson, G. N., Brennan, M. P., Roh, J. D., Hibino, N., Yi, T., et al. (2009). Tissue-engineered arterial grafts: long-term results after implantation in a small animal model. *J. Pediatr. Surg, 44*, 1127–1132, discussion 1132–3.

Mironov, V., Kasyanov, V., McAllister, K., Oliver, S., Sistino, J., & Markwald, R. (2003). Perfusion bioreactor for vascular tissue engineering with capacities for longitudinal stretch. *J. Craniofac. Surg., 14*, 340–347.

Moosmang, S., Schulla, V., Welling, A., Feil, R., Feil, S., Wegener, J. W., et al. (2003). Dominant role of smooth muscle L-type calcium channel Cav1.2 for blood pressure regulation. *EMBO J., 22*, 6027–6034.

Narita, Y., Hata, K., Kagami, H., Usui, A., Ueda, M., & Ueda, Y. (2004). Novel pulse duplicating bioreactor system for tissue-engineered vascular construct. *Tissue Eng., 10*, 1224–1233.

National Center for Health Statistics Data Line. (1994). *Public Health Rep., 109*, 713–714.

National Kidney Foundation. (2000). Clinical practice guidelines for nutrition in chronic renal failure. K/DOQI, National Kidney Foundation. *Am. J. Kidney Dis., 35*, S1−140.

National Kidney Foundation. (2001). III. NKF-K/DOQI Clinical practice guidelines for vascular access: update 2000. *Am. J. Kidney Dis., 37*, S137−S181.

National Kidney Foundation. (2002). K/DOQI clinical practice guidelines for chronic kidney disease: evaluation, classification, and stratification. *Am. J. Kidney Dis., 39*, S1−266.

Neff, L. P., Tillman, B. S., Yazadani, S., Machingal, M. A., Yoo, J., Soker, S., et al. (2010). Vascular smooth muscle enhances functionality of tissue engineered blood vessels in. *vivo. J. Vasc. Surg., 51*, 4S.

Nerem, R. M. (2003). Role of mechanics in vascular tissue engineering. *Biorheology, 40*, 281−287.

Neville, R. F., Abularrage, C. J., White, P. W., & Sidawy, A. N. (2004). Venous hypertension associated with arteriovenous hemodialysis access. *Semin. Vasc. Surg., 17*, 50−56.

Niklason, L. E., Gao, J., Abbott, W. M., Hirschi, K. K., Houser, S., Marini, R., et al. (1999). Functional arteries grown in vitro. *Science, 284*, 489−493.

Niklason, L. E., Abbott, W., Gao, J., Klagges, B., Hirschi, K. K., Ulubayram, K., et al. (2001). Morphologic and mechanical characteristics of engineered bovine arteries. *J. Vasc. Surg., 33*, 628−638.

Opitz, F., Schenke-Layland, K., Richter, W., Martin, D. P., Degenkolbe, I., Wahlers, T., et al. (2004). Tissue engineering of ovine aortic blood vessel substitutes using applied shear stress and enzymatically derived vascular smooth muscle cells. *Ann. Biomed. Eng., 32*, 212−222.

Portner, R., Nagel-Heyer, S., Goepfert, C., Adamietz, P., & Meenen, N. M. (2005). Bioreactor design for tissue engineering. *J. Biosci. Bioeng., 100*, 235−245.

Rabkin, E., & Schoen, F. J. (2002). Cardiovascular tissue engineering. *Cardiovasc. Pathol., 11*, 305−317.

Ratcliffe, A. (2000). Tissue engineering of vascular grafts. *Matrix Biol., 19*, 353−357.

Ravi, S., Qu, Z., & Chaikof, E. L. (2009). Polymeric materials for tissue engineering of arterial substitutes. *Vascular, 17* (Suppl 1), S45−S54.

Safar, M. E., Blacher, J., Mourad, J. J., & London, G. M. (2000). Stiffness of carotid artery wall material and blood pressure in humans: application to antihypertensive therapy and stroke prevention. *Stroke, 31*, 782−790.

Sales, K. M., Salacinski, H. J., Alobaid, N., Mikhail, M., Balakrishnan, V., & Seifalian, A. M. (2005). Advancing vascular tissue engineering: the role of stem cell technology. *Trends Biotechnol., 23*, 461−467.

Shin'oka, T., Matsumura, G., Hibino, N., Naito, Y., Watanabe, M., Konuma, T., et al. (2005). Midterm clinical result of tissue-engineered vascular autografts seeded with autologous bone marrow cells. *J. Thorac. Cardiovasc. Surg., 129*, 1330−1338.

Shum-Tim, D., Stock, U., Hrkach, J., Shinoka, T., Lien, J., Moses, M. A., et al. (1999). Tissue engineering of autologous aorta using a new biodegradable polymer. *Ann. Thorac. Surg., 68*, 2298−2304, discussion 2305.

Soletti, L., Nieponice, A., Guan, J., Stankus, J. J., Wagner, W. R., & Vorp, D. A. (2006). A seeding device for tissue-engineered tubular structures. *Biomaterials, 27*, 4863−4870.

Stitzel, J., Liu, J., Lee, S. J., Komura, M., Berry, J., Soker, S., et al. (2006). Controlled fabrication of a biological vascular substitute. *Biomaterials, 27*, 1088−1094.

Swartz, D. D., Russell, J. A., & Andreadis, S. T. (2005). Engineering of fibrin-based functional and implantable small-diameter blood vessels. *Am. J. Physiol. Heart Circ. Physiol., 288*, H1451−H1460.

Teebken, O. E., & Haverich, A. (2002). Tissue engineering of small diameter vascular grafts. *Eur. J. Vasc. Endovasc. Surg., 23*, 475−485.

Teebken, O. E., Bader, A., Steinhoff, G., & Haverich, A. (2000). Tissue engineering of vascular grafts: human cell seeding of decellularised porcine matrix. *Eur. J. Vasc. Endovasc. Surg., 19*, 381−386.

Teebken, O. E., Pichlmaier, A. M., & Haverich, A. (2001). Cell seeded decellularised allogeneic matrix grafts and biodegradable polydioxanone-prostheses compared with arterial autografts in a porcine model. *Eur. J. Vasc. Endovasc. Surg., 22*, 139−145.

Thompson, C. A., Colon-Hernandez, P., Pomerantseva, I., MacNeil, B. D., Nasseri, B., Vacanti, J. P., et al. (2002). A novel pulsatile, laminar flow bioreactor for the development of tissue-engineered vascular structures. *Tissue Eng., 8*, 1083−1088.

Thom, T., Haase, N., Rosamond, W., Howard, V. J., Rumsfeld, J., Manolio, T., et al. (2006). Heart disease and stroke statistics − 2006 update: a report from the American Heart Association Statistics Committee and Stroke Statistics Subcommittee. *Circulation, 113*, e85−151.

Tiwari, A., Salacinski, H. J., Hamilton, G., & Seifalian, A. M. (2001). Tissue engineering of vascular bypass grafts: role of endothelial cell extraction. *Eur. J. Vasc. Endovasc. Surg., 21*, 193−201.

Vara, D. S., Salacinski, H. J., Kannan, R. Y., Bordenave, L., Hamilton, G., & Seifalian, A. M. (2005). Cardiovascular tissue engineering: state of the art. *Pathol. Biol. (Paris), 53*, 599−612.

Wang, H. Z., Day, N., Valcic, M., Hsieh, K., Serels, S., Brink, P. R., et al. (2001). Intercellular communication in cultured human vascular smooth muscle cells. *Am. J. Physiol. Cell Physiol., 281*, C75–C88.

Wang, C., Cen, L., Yin, S., Liu, Q., Liu, W., Cao, Y., et al. (2010). A small diameter elastic blood vessel wall prepared under pulsatile conditions from polyglycolic acid mesh and smooth muscle cells differentiated from adipose-derived stem cells. *Biomaterials, 31*, 621–630.

Watanabe, H., Takahashi, M., Hayashi, J., Sugawara, M., Miyamura, H., & Eguchi, S. (2001). Long-term results of valve-retaining homograft and xenograft patch for transannular reconstruction of the right ventricular outflow tract in tetralogy of Fallot. *Kyobu Geka, 54*, 618–623.

Weinberg, C. B., & Bell, E. (1986). A blood vessel model constructed from collagen and cultured vascular cells. *Science, 231*, 397–400.

Williams, C., & Wick, T. M. (2004). Perfusion bioreactor for small diameter tissue-engineered arteries. *Tissue Eng., 10*, 930–941.

Williams, C., & Wick, T. M. (2005). Endothelial cell-smooth muscle cell co-culture in a perfusion bioreactor system. *Ann. Biomed. Eng., 33*, 920–928.

Xu, J., Ge, H., Zhou, X., Yang, D., Guo, T., He, J., et al. (2005). Tissue-engineered vessel strengthens quickly under physiological deformation: application of a new perfusion bioreactor with machine vision. *J. Vasc. Res., 42*, 503–508.

Yang, J., Motlagh, D., Webb, A. R., & Ameer, G. A. (2005). Novel biphasic elastomeric scaffold for small-diameter blood vessel tissue engineering. *Tissue Eng., 11*, 1876–1886.

Yazdani, S. K., Watts, B., Machingal, M., Jarajapu, Y. P., Van Dyke, M. E., & Christ, G. J. (2009). Smooth muscle cell seeding of decellularized scaffolds: the importance of bioreactor preconditioning to development of a more native architecture for tissue-engineered blood vessels. *Tissue Eng. Part A, 15*, 827–840.

Zhao, Y., Zhang, S., Zhou, J., Wang, J., Zhen, M., Liu, Y., et al. (2010). The development of a tissue-engineered artery using decellularized scaffold and autologous ovine mesenchymal stem cells. *Biomaterials, 31*, 296–307.

CHAPTER 47

Engineering of Small-Diameter Vessels

Brett C. Isenberg[*], **Chrysanthi Williams**[†], **Robert T. Tranquillo**[‡]
[*] Department of Biomedical Engineering, Boston University, Boston, MA
[†] Bose Corporation, ElectroForce Systems Group, Eden Prairie, MN
[‡] Department of Biomedical Engineering, University of Minnesota, Minneapolis, MN, USA

INTRODUCTION

Cardiovascular disease

Since 1918, cardiovascular disease has been the number one killer in the USA, claiming over 830,000 persons in 2006 alone, with coronary heart disease being the single largest killer of American males and females (American Heart Association, 2010). Coronary heart disease is the result of a progressive narrowing of the lumen of the coronary artery, which supplies blood to the heart wall, due to the accumulation of cholesterol-lipid-calcium plaques on the inner walls of the vessel that ultimately restrict adequate blood supply from reaching the heart wall. Atherosclerosis is a specific form of arteriosclerosis (hardening of the arteries) and is a multi-factorial disease that is influenced by diet, cigarette smoking, diabetes, high blood pressure, and exercise (Burke et al., 1997). Several theories have been formulated to explain the localized nature of atherosclerosis. Fluid mechanical theories predict that atherogenesis occurs in areas that have a relatively complex geometry, a fairly large Reynolds number, and a lower than average wall shear stress throughout the pulsatile cycle. Mechanical views blame sites of high stress concentration such as bifurcations, constrictions, increased radius of curvature, saddle shape, areas surrounding a small side branch, and bending of the wall (Fung, 1996).

The disease begins with the focal eccentric accumulation of lipid in the intima with intracellular lipid visible mainly in macrophages and smooth muscle cells (SMCs) over time. This leads to the formation of a fatty streak, which is composed of SMCs, matrix fibers, and lipids. At a later stage, the fatty streak becomes the preatheroma, which contains multiple extracellular lipid pools, as well as collagen and elastin fibers accumulated beneath the endothelium. The subendothelial zone may subsequently become more organized to form the fibrous cap, which resembles the normal media layer in structure and thickness, and does not contain any macrophages or lipids. As the lipid pools coalesce into lipid cores, the intima becomes disorganized, and this lesion type is termed an atheroma. As the disease develops, the lesion becomes stratified due to the increasing amount of fibrous tissue in deep and superficial layers, and the localization of lipid cores between the fibrous regions, forming fibroatheromas (Glagov et al., 1995). However, as the lesion enlarges, the artery also enlarges by an outward bulging of the wall beneath the growing plaque to compensate for the narrowing that has occurred. Lumen stenosis becomes evident when the plaque takes up approximately 40% or more of the lumen area (Bassiouny et al., 1997).

Principles of Regenerative Medicine. DOI: 10.1016/B978-0-12-381422-7.10047-1

Intimal thickening or hyperplasia could also be a response to vascular intimal injury. When the vascular wall is injured, SMCs proliferate in and migrate from the media to the intima and synthesize extracellular matrix (ECM) proteins. SMCs undergo dedifferentiation, lose their ability to contract, gain the capacity to divide, and increase ECM synthesis (Assoian and Marcantonio, 1996). SMCs residing in the intima lose their thick myosin-containing filaments and greatly increase the amount of organelles involved in protein synthesis, such as rough endoplasmic reticulum and Golgi apparatus. The migratory and proliferative activity of SMCs is regulated by both growth promoters, such as platelet-derived growth factor, basic fibroblast growth factor, and interleukin 1, and inhibitors, such as heparan sulfate, nitric oxide, and transforming growth factor-β. Intimal SMCs may return to a non-proliferative state when either the overlying endothelial layer is re-established or the abnormal chronic endothelial stimulation ceases.

Atheromas in advanced disease almost always undergo patchy or massive calcification, and atherosclerotic lesions cause clinical disease by one of the following mechanisms: slow narrowing of the intima that results in ischemia of the tissues perfused by the involved vessels; sudden occlusion of the lumen by superimposed thrombosis or hemorrhage into an atheroma; thrombosis followed by embolism; or weakening of the wall of a vessel, causing an aneurysm or rupture (Schoen, 1994).

Several approaches are taken to treat atherosclerotic cardiovascular disease of small-caliber arteries (<6 mm), and the most common ones are briefly described next.

Clinical therapies

Balloon angioplasty or percutaneous transluminal coronary angioplasty is a procedure used to dilate narrowed arteries. A catheter is inserted with a deflated balloon at its tip into the narrowed vessel; the balloon is inflated, compressing the plaque and enlarging the inner diameter of the artery; and then the balloon is deflated and the catheter removed. About 70–90% of these procedures also involve the placement of a stent, which is a wire mesh tube that is initially collapsed to a small diameter, placed over a balloon catheter, and moved into the area of the blockage. When the balloon is inflated, the stent expands, locks in place, and holds the artery open. Concerns with stents include injury to the vessel wall during insertion, and acute thrombosis or intimal hyperplasia as consequences of injury (Didisheim and Watson, 1996). Drug-eluting stents that slowly release an anti-mitotic drug from the wall of the stent in order to prevent restenosis have become extremely popular due to their ability to dramatically reduce the amount of neointimal thickening observed following a stenting procedure compared with bare metal stents (Guyton, 2006; Kivela and Hartikainen, 2006). While initially hailed as a major advance in medicine, their long-term advantage and mortality in multivessel coronary artery disease remain unclear (Daemen and Serruys, 2007a,b). Unfortunately, as these drugs prevent reocclusion by suppressing the proliferation of SMCs, they also inhibit endothelial cell (EC) proliferation and thus greatly inhibit re-endothelialization of the vessel, requiring patients to be placed on long-term anti-coagulant therapy and increasing the risk for thrombus formation (Valgimigli et al., 2009).

Coronary artery bypass graft operation, first performed in 1964, is an invasive procedure that is done to reroute, or "bypass," blood around occluded arteries and improve the supply of blood and oxygen to the heart. Grafts commonly used include the great saphenous vein from the leg, internal thoracic artery (often referred to as the internal mammary artery) from the chest, radial artery from the arms, and sometimes arteries from the stomach. Although 70–82% of the saphenous vein substitutes remain patent in 5 years (Lytle et al., 1985) and 61% after 10 years (Goldman et al., 2004), stenosis due to intimal hyperplasia, which is a flow-restricting lesion, and limited availability are important limitations. Internal thoracic and radial arteries are often used in the coronary circulation and are preferred for key artery branches because they tend to remain open longer but also have limited availability (Cameron et al., 1996; Conte, 1998). Although bypass surgery is a common procedure, it carries some serious risks, such as heart attack, stroke, or even death.

Autologous vessels remain the gold standard for bypass grafts; however, many patients do not have a vessel suitable for use due to vascular disease, amputation, or previous harvest. Significant work has been performed towards use of synthetic materials (e.g. Dacron, Teflon) to serve as small-diameter conduits, which account for a vast majority of the demand. Unfortunately, the use of synthetic grafts for such applications has generally proved inadequate largely due to acute thrombogenicity of the graft, anastomotic intimal hyperplasia, aneurysm formation, infection, and progression of atherosclerotic disease (Conte, 1998). Considerable effort has been devoted to improving the performance of synthetic grafts, primarily by modifying their surface characteristics and/or coating their lumens with ECs in order to prevent thrombus formation. While some of these approaches have been successful in the short-term, complications such as thrombosis, infection, and graft failure ultimately arise with time.

Tissue engineering

The shortcomings of the therapies described above have led an increasing number of researchers to pursue the development of a living blood vessel substitute that closely mimics the native artery. Among the desired properties of a blood vessel substitute are adequate mechanical strength, controllable adaptation to changing hemodynamics, compliant elasticity, long-term fatigue strength and durability, low thrombogenicity, biocompatibility, suturability, easy handling, and low cost with essentially zero tolerance for failure (Mitchell and Niklason, 2003). Although some tissue engineers aim at reproducing the arterial wall architecture and function, most are striving toward restoring function with an arterial replacement possessing a composition and architecture that confers the required functional properties, not necessarily a complete replica of the native artery.

Small-diameter tissue-engineered grafts are typically composed of cells and a scaffold. The scaffold serves as a temporary template that provides the required geometry and has such properties to allow the cells to remodel the scaffold and deposit their own ECM proteins (Fig 47.1). The choice of cells and scaffold is paramount, and the optimal combination is unclear as of yet. However, the *in vitro* development of vascular grafts is not feasible without the appropriate environmental cues. The following sections address three major components in the design and fabrication of a tissue-engineered small-diameter vascular graft: cells, scaffolds, and environmental stimuli.

CELL SOURCING CONSIDERATIONS

Cell sourcing is a critical component of a tissue-engineered graft. In order to select the appropriate source, several questions must be answered. What cell type, species, passage, differentiation stage, and donor age should be chosen? How many cell types and which should be used? In what spatial and temporal fashion should the different cell types be introduced to the scaffold? The cell types residing in native blood vessels, that is, ECs, SMCs, and fibroblasts, or their progenitors, are the obvious candidates as cell types.

Endothelial cells

Endothelial cells (ECs) form a monolayer that lines the entire vascular system and have a remarkable capacity to adjust their number and arrangement in order to suit local requirements. ECs play an important role in tissue homeostasis, fibrinolysis and coagulation (thrombogenicity), regulation of vascular tone, growth regulation of other cell types, regulation of nutrient transport across the vessel wall, translation of mechanical signals from the blood flow to the other regions of the vessel wall, and blood cell activation and migration during physiological and pathological processes (Risau, 1995; Shireman and Pearce, 1996; Aird, 2006; Liebner et al., 2006).

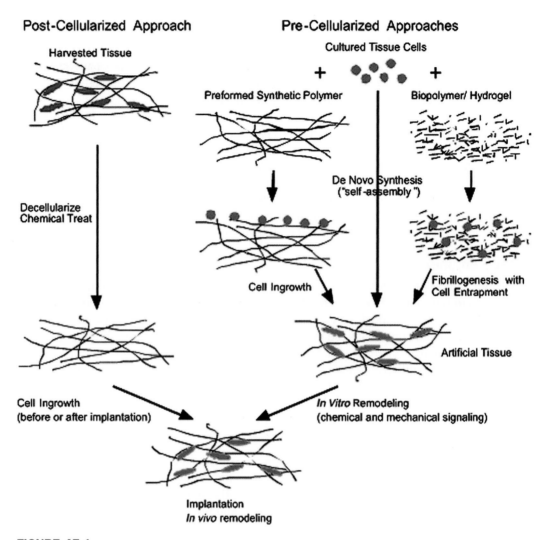

FIGURE 47.1

The general approach to tissue engineering. A scaffold is combined with tissue cells, which subsequently remodel the scaffold by synthesizing ECM to form an engineered tissue.

Controlling thrombogenicity is of paramount importance in vascular repair; thus, providing a functional endothelium must be central to the successful development of a vascular graft. Thrombogenicity is an imprecisely defined vascular property, but it implies the qualitative and quantitative assessment of platelet and fibrin deposition on the vascular luminal surface (endothelium). A vessel that is thrombogenic may be so for a variety of reasons, many or most of which are likely related to endothelial dysfunction. Normal quiescent endothelium exhibits limited or absent expression of secreted and cell-associated procoagulant proteins, including platelet adhesogens (e.g. P-selectin) and activators of thrombin generation (tissue factor). Membrane phospholipid asymmetry is maintained in healthy endothelium in order to prevent exposure of the highly pro-thrombotic aminophospholipids that support the assembly of coagulation enzymatic complexes. Conversely, the loss of normal anti-thrombotic or anti-fibrinolytic mechanisms, or the loss or inhibition of mechanisms that prevent platelet adhesion, may also induce thrombogenicity. In tissue-engineered vessels, the physical detachment of ECs resulting in exposure of the procoagulant sub-endothelial surface may be at least as important as EC activation as a mechanism for thrombogenicity.

Due to the critical role ECs play in determining the thrombogenicity of a vascular graft, EC sourcing dictates to a large extent the patency of a graft (Heyligers et al., 2005). There is a high demand for ECs that can be isolated from the patient requiring bypass surgery to eliminate the need for long-term anti-coagulation therapy and graft rejection. The main EC sources currently explored are umbilical vein ECs (l'Heureux et al., 1998; McKee et al., 2003), venous ECs, mesothelial cells as an alternative to ECs, and progenitor ECs. Venous ECs have been successfully isolated from a single saphenous vein biopsy and used to engineer vascular grafts (Grenier et al., 2003). However, EC isolation from short (2.5-cm-long) vein segments without any SMC contamination has proven difficult, and vein biopsies are invasive procedures.

Blood outgrowth ECs have been collected from human peripheral blood through the outgrowth of a marrow-derived, transplantable circulating, putative endothelial progenitor (Lin et al., 2000, 2002; Sieminski et al., 2005). These cells have a cobblestone morphology, take up aLDL, have Weibel-Palade bodies, express multiple endothelial markers (CD34, VE-cadherin, CD31, P1H12, $\alpha_v\beta_3$ integrin, β_1 integrin, thrombomodulin, von Willebrand Factor, flk-1), have a quiescent phenotype, and hold promise as an alternative EC source. Yoder et al. argue that not all blood outgrowth ECs are true endothelial progenitors and, thus, for cardiovascular therapeutic applications, it is crucial to distinguish between cells that form monocyte-like cells with low proliferative potential that fail to form perfusable microvessels *in vivo* (and are actually of hematopoietic lineage) and a rare, clonally distinct population of cells, known as endothelial colony-forming cells, that are highly proliferative and form perfusable vessels *in vivo* (Yoder et al., 2007).

Mesothelial cells line serosal cavities and most internal organs in the body and share many characteristics and functions with ECs (Herrick and Mutsaers, 2004). Although this EC source is promising, more work is ongoing to fully characterize the phenotype of these cells and their potential of providing an alternative source of EC-like cells (Campbell et al., 1999).

857

Smooth muscle cells

Vascular SMCs perform many functions, including vasoconstriction and dilation in response to normal or pharmacological stimuli; synthesis of various types of collagen, elastin, and proteoglycans; elaboration of growth factors and cytokines; and migration and proliferation. SMCs are capable of expressing a range of phenotypes or alterations in character (Chamley-Campbell and Campbell, 1981; Thyberg et al., 1990; Owens, 1995). Modulations in cell phenotype may occur as a result of cell-cell interactions, alterations of ECM, or in response to other signals such as hormones (Stegemann and Nerem, 2003). At one end of this phenotypic spectrum are SMCs in contractile state with 80–90% of the cytoplasmic volume occupied with contractile apparatus (Tagami et al., 1986). Organelles such as rough endoplasmic reticulum, Golgi, and free ribosomes are few in number and located in the perinuclear region. SMCs in the contractile phenotypic state exhibit reduced proliferation and matrix production. At the other end of the spectrum is the synthetic state, which is seen in development, repair, and pathological conditions. SMCs with synthetic phenotype proliferate and actively produce ECM proteins, and their cytoplasm contains few filament bundles but large amounts of rough endoplasmic reticulum, Golgi, and free ribosomes (Ross, 1971). Aortic SMCs in synthetic state synthesize four-fold the amount of collagen and five-fold the amount of sulfated glycosaminoglycan, and twice as much non-collagen protein as compared to contractile SMCs under the same conditions and are not related to cell proliferation (Campbell, 1985). Contractile SMCs switch to a more synthetic phenotype *in vitro* (Thyberg et al., 1985), and this change may be irreversible depending on the culture conditions (Chamley-Campbell and Campbell, 1981; Stadler et al., 1989). Vascular SMCs have been successfully used to engineer small-diameter vascular grafts (see sections below), although a non-invasive method of harvesting these cells from patients is impossible.

Fibroblasts

Fibroblasts and SMCs share several functions such as collagen, elastin, and proteoglycan synthesis and contractile behavior. In the normal adult, some of these functions are specifically performed by the fibroblast (e.g. collagen synthesis) or the SMC (e.g. contractility), but during development or pathological conditions this can change. SMCs secrete collagen during development and during the formation of an atheromatous plaque, whereas contractility may be exerted by fibroblasts during wound contraction (Desmouliere and Gabbiani, 1995). Therefore, fibroblasts (termed myofibroblasts) can acquire SMC-like features during wound contraction and disease and express SMC-specific markers in particular situations. Fibroblasts can be easily harvested from patients through a simple skin biopsy (Normand and Karasek, 1995), making these cells an attractive alternative to SMCs for vascular graft tissue engineering if they prove to provide sufficient function.

Stem cells

Stem cells have become an increasingly popular cell source for vascular tissue engineering applications due to their unique regenerative properties and potential to persist in allogenic environments. Riha et al. have written a thorough review of the use of stem cells in vascular tissue engineering (Riha et al., 2005) and full discussion of this important topic is beyond the scope of this work. Below is a brief discussion of a few uses of stem cells in vascular tissue engineering applications utilizing bone marrow- and adipose-derived mesenchymal stem cells.

An attractive EC source, and the subject of much research, has been circulating EC progenitor cells (EPCs) (Matsumura et al., 2003; Cho et al., 2004; Allen et al., 2010). EPCs are found in post-natal bone marrow, have high proliferative capacity, and can differentiate into mature ECs (Hristov et al., 2003). In contrast to mature ECs, which have low proliferative capacity and limited ability to replace damaged ECs, EPCs are active participants in the repair of denuded endothelium. EPCs have been successfully used to line engineered vascular grafts that remained patent post-implantation in animal models (Kaushal et al., 2001; Matsumura et al., 2003; Matsuda, 2004; Cho et al., 2005). The main drawback of using EPCs is low yield (Matsuda, 2004), which becomes a critical factor when multiple grafts are needed for bypass surgery.

Bone marrow-derived mesenchymal stem cells have been used as an alternative source of SMCs. Progenitor cells were either seeded into poly(lactic-*co*-glycolic acid) (PLGA) scaffolds and implanted in the peritoneal cavity of athymic mice (Cho et al., 2004) or exposed to cyclic strain (Hamilton et al., 2004; Gong and Niklason, 2008); in both cases, cells expressed markers of SMC differentiation. Lim et al. used bone marrow-derived cells to seed poly(lactide-co-epsilon-caprolactone) scaffolds that were subsequently implanted into dogs for 8 weeks and observed the formation of SMC and EC layers, significant collagen deposition, and native levels of nitric acid synthase expression (Lim et al., 2008). Simper et al. isolated circulating smooth muscle progenitor cells from human blood and showed evidence of SMC outgrowth through the expression of smooth muscle α-actin, myosin heavy chain, and calponin (Simper et al., 2002).

Adipose-derived mesenchymal stem cells (ASCs) are yet another attractive potential source for stem cells capable of differentiating into both ECs and SMCs due to their abundance and relative ease of isolation (DiMuzio and Tulenko, 2007). Harris et al. isolated ACSs from a number of patients undergoing vascular surgery and differentiated them *in vitro* using a medium containing angiotensin II, sphingosylphosphorylcholine, or transforming growth factor-β 1 for up to 3 weeks (Harris et al., 2009). These cells were seeded onto decellularized saphenous veins where they attached, proliferated, and demonstrated upregulation of SMC differentiation markers calponin, caldesmon, and myosin heavy chain. These results indicate that ASCs could be a readily available source of autologous cells for vascular tissue engineering.

SCAFFOLDS FOR SMALL-DIAMETER TISSUE-ENGINEERED VESSELS

The design of an appropriate scaffold in fabricating a cellularized tubular construct is crucial as it is the scaffold that provides the initial spatial templating and structural support for the developing tissue and it must support cell adhesion, proliferation, and matrix synthesis. Currently, there are three broad classes of scaffold being investigated: (1) synthetic polymer scaffolds, (2) hydrogels or biopolymer scaffolds, and (3) decellularized tissues. Each has associated advantages and disadvantages that are discussed below (Fig. 47.1).

Synthetic scaffolds

Much of the work in the engineering of vascular grafts has focused on the use of biodegradable synthetic polymer scaffolds. The main advantage of these scaffolds is that they can provide the initial strength necessary for implantation while being biodegradable and can be readily processed into tubes. The obvious disadvantage is that synthetic biomaterials may elicit an immune response. In addition, direct cellularization is often difficult because the conditions required for polymer synthesis/scaffold formation are typically cytotoxic (there are some exceptions, such as PEG-based scaffolds (Seliktar et al., 2004; DeLong et al., 2005)). Semi-crystalline polymers such as polyglycolic (PGA) and poly-L-lactic acid (PLLA) degrade by bulk hydrolysis. Degradation occurs first in the amorphous domains, which are more accessible to water, and crystallinity gradually increases, resulting in a highly crystalline material that is much more resistant to hydrolysis than the starting material. The increase in crystallinity is believed to occur due to an increased mobility of the partially degraded polymer chains, which enables a realignment of the polymer chains into a more ordered crystalline state (Anderson, 1995). PLGA copolymers degrade via non-specific hydrolytic scission of their ester bonds to reform the monomers lactic acid and glycolic acid. Other factors such as pH, heat, and carboxypeptidases may also contribute to the degradation process. *In vivo* and *in vitro* experiments with PLGA copolymers have studied their degradation and biocompatibility (Lu et al., 2000). PLLA is thought to degrade via simple hydrolysis, whereas PGA may also be subjected to enzyme-mediated hydrolysis. PGA absorption takes 6–17 weeks and the tensile strength falls to about 10% after 3 weeks, depending on the molecular weight, while the more hydrophobic PLLA can require several years to degrade and lose significant mechanical strength. Only mild inflammatory responses have been caused by PLGA polymers and in some instances phagocytic and giant cells have been observed. There is also concern that the local acidity following degradation can induce cell dedifferentiation (Niklason et al., 1999), as well as residual polymer fragments leading to stress concentrations in the wall of engineered vessels, which significantly reduce their ultimate tensile strength (Dahl et al., 2007). PLGA polymers have been approved by the FDA for suture materials, bone plates and screws, and cardiovascular woven meshes. Thus, these polymers have been used in various tissue engineering applications.

Biodegradable polymeric scaffolds have been used successfully in vascular tissue engineering. Niklason et al. seeded PGA scaffolds with bovine aortic SMCs in a bioreactor system for 8 weeks under pulsatile conditions and subsequently applied bovine aortic ECs for 3 days with continuous flow (Niklason et al., 1999). The resulting endothelium stained positive for von Willebrand factor and platelet endothelial cell adhesion molecule (PECAM), and the SMCs expressed smooth muscle α-actin and calponin. The grafts showed high SMC density and collagen production, had burst pressure of over 2000 mmHg, and contracted in response to serotonin, endothelin-1, and prostaglandin $F_{2\alpha}$. Pulsatility increased wall thickness, collagen production, and suture retention strength. Implantation of scaffolds seeded with autologous ECs and SMCs and cultured *in vitro* under pulsatile flow into the right saphenous artery of miniature swine resulted in patent grafts after 4 weeks although decreased flow was observed. Limitations of this approach were low elastin production compared to that of native vessels and the presence of de-differentiated SMCs around residual polymer fragments.

Hoerstrup et al. coated PGA meshes with a thin layer of poly-4-hydroxybutyrate, seeded them with ovine myofibroblasts under static conditions for 4 days, seeded them subsequently with ovine ECs, and cultured them in a pulse duplicator bioreactor for up to 28 days (Hoerstrup et al., 2001). DNA and collagen content increased continuously for 21 days but small amounts of matrix were produced that resulted in low mechanical strength. In a separate study with human cells, culture conditions were optimized with ascorbic acid and basic fibroblast growth factor for increased collagen production (Hoerstrup et al., 2000).

Shin'oka et al. implanted a polycaprolactone-polylactic acid copolymer reinforced with woven PGA that was cultured with autologous cells for 10 days *in vitro* in the right pulmonary artery of a 4-year-old girl in Japan (Shin'oka et al., 2001). Seven months later there was no evidence of graft occlusion or aneurysm formation. Although the demands for a blood vessel substitute of the pulmonary circulation are not as high as those for the systemic circulation, the successful implantation of a completely tissue-engineered graft in humans is still a very exciting accomplishment.

Recently, electrospinning of degradable polymers to form scaffolds with precise fibrillar architectures has become an increasingly popular method of scaffold fabrication (see extensive reviews by Kumbar et al. (2008), Ifkovits et al. (2009), and Sell et al. (2009) for more information). The ability to produce synthetic matrices with known fibril diameter and orientation at the nanoscale gives researchers an additional degree of control over matrix properties, enabling the fine tuning of scaffolds for optimum cell growth and tissue formation. As with most synthetic biomaterial applications, cellularization of the scaffolds can be challenging; however, electrospraying of cells concurrently with electrospinning of biodegradable polymers has demonstrated that it is possible to obtain direct cellularization in electrospinning applications (Stankus et al., 2007). To date, most electrospinning-based vascular tissue engineering applications have focused on searching for these optimum combinations of synthetic or biopolymeric materials and processing conditions that yield scaffolds with acceptable material properties and biocompatibility. These materials include the commonly used biodegradable scaffolding materials such as PLLA and polycaprolactone (Dong et al., 2008; Chen et al., 2009; He et al., 2009; Nottelet et al., 2009; Chung et al., 2010), polyurethanes (Grasl et al., 2009; Nieponice et al., 2010; Soletti et al., 2010; Theron et al., 2010), silk (discussed in the next section) (Zhang et al., 2008, 2009; McClure et al., 2009), as well as collagen/elastin mixed with various synethetic polymers (Lee et al., 2007; Smith et al., 2008; Thomas et al., 2009; Tillman et al., 2009; McClure et al., 2010). Of particular interest are tubular, bilayered scaffolds fabricated from a class of biodegrable poly(ester-urethane)urea (PEUU) elastomers (Soletti et al., 2010). These scaffolds are composed of a highly porous inner layer of PEUU formed by thermally induced phase separation and a fibrous, electronspun circumferentially aligned outer layer of PEUU. The resulting scaffolds have displayed material properties that are remarkably similar to native vessels. *In vivo* assessment of these constructs seeded with muscle-derived stem cells for 8 weeks in the abdominal aorta of rats demonstrated that bilayered constructs were effective at preventing aneurysm formation as compared with single-layer constructs without the electrospun layer (Nieponice et al., 2010).

Biological scaffolds

Biopolymers, typically a reconstituted type I collagen gel or fibrin gel, are formed with and compacted by tissue cells, where an appropriately applied mechanical constraint to the compaction yields circumferential alignment of fibrils and cells characteristic of the arterial media (l'Heureux et al., 1993; Barocas et al., 1998; Seliktar et al., 2000). It is this last feature that is most attractive about a biopolymer-based tissue-engineered artery. This follows from two axioms: (1) that native artery function, particularly mechanical function, depends on structure (particularly alignment of the SMCs and collagen fibers in the medial layer) as much as it depends on composition, and (2) that the tissue-engineered artery should serve as a functional remodeling template, so that, while providing function during the remodeling, the artificial tissue also provides a template for the alignment of the growing tissue (Tranquillo, 2002). Cells entrapped in a biopolymer gel exert traction on the network of fibrils. When this contraction occurs around a non-adhesive mandrel, fibrils and cells become circumferentially aligned (Barocas et al., 1998).

Collagen gels have been previously used to engineer small-diameter grafts (Weinberg and Bell, 1986; Seliktar et al., 2000) but possessed insufficient mechanical strength for arterial replacements. Attempts at improving the mechanical strength of collagen gels have been moderately successful (Tranquillo et al., 1996; Girton et al., 1999; Seliktar et al., 2000). Huynh et al. used submucosal collagen isolated from porcine small intestine coated with type I bovine collagen and treated the inner surface with heparin-benzalkonium to prevent coagulation (Huynh et al., 1999). The collagen layers were cross-linked to increase the mechanical strength, and the grafts remained patent and thrombi-free up to 13 weeks when implanted in rabbits. However, the response of the human cardiovascular system to animal collagens remains unknown.

In contrast to collagen-based constructs, fibrin gels are used as provisional scaffolds that are expected to be completely remodeled by the entrapped cells. Fibrin, which can be readily obtained from plasma (Gilbert et al., 2001), is ideally suited for this role as it is the temporary structural protein used in the native wound healing process in which an initial fibrin scaffold is slowly replaced by more permanent matrix proteins (e.g. collagen). Cells entrapped in fibrin gel are able to proliferate and deposit collagen (Fig. 47.2) and elastic fibers to a greater extent compared to cells entrapped in collagen gel (Grassl et al., 2002; Long and Tranquillo, 2003), resulting in stronger and stiffer tissues (Grassl et al., 2003). SMCs in fibrin produce around three to five times more collagen than SMCs in collagen, depending on the concentration of an inhibitor used to control fibrinolysis (Grassl et al., 2002). Collagen fibrils produced by the cells adopt the alignment of the contracted fibrin fibrils, that is, when "media-equivalents" are fabricated such that the SMC-induced fibrin contraction results in circumferential alignment, the collagen subsequently produced by the SMC is also circumferentially aligned (Grassl et al., 2003).

Ross and Tranquillo (2003) performed a study characterizing the tissue growth and development process that occurs *in vitro* with this system (Fig. 47.3). Following fibrin gel contraction during week 1, peak rates of SMC proliferation, collagen production, and tropoelastin production occur between weeks 1–4. Organized, cross-linked collagen and elastic fibers replace the degrading fibrin over weeks 3–5 and are manifested as increased mechanical strength. The peak rate of SMC proliferation (weeks 1–2) precedes that for maximum collagen production (weeks 2–4), which is consistent with the 3-week time point of maximum expression of collagen type I and III from qRT-PCR. Insoluble elastin quantification reveals that the majority of elastic fibers are produced by week 4, which is also consistent with the qRT-PCR data showing a dramatic downregulation of tropoelastin expression by week 4, indicating elastogenesis occurs during the early stages of tissue growth and development. There is a strong

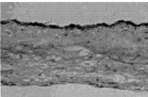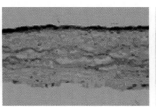

FIGURE 47.2
Remodeling of fibrin disks revealed by sections stained with Masson's trichrome stain: fibrin is pink-red, collagen is green, cell nuclei are purple. Top surface of disk is up and adherent surface of disk is down. Thickness (vertical dimension) is ~150 μm.

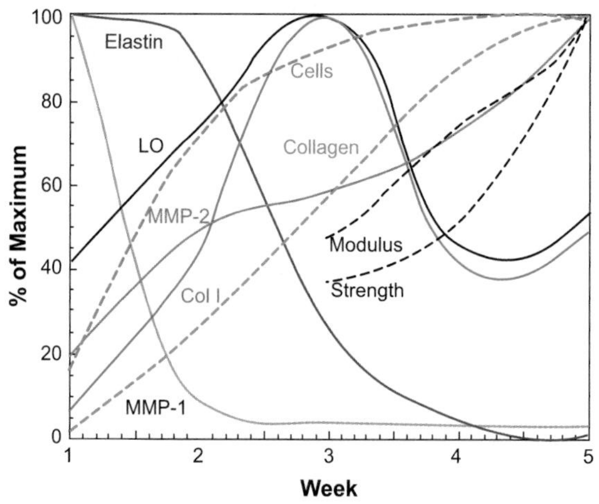

FIGURE 47.3

Time-course of fibrin remodeling into tissue by neonatal SMCs. Gene expression (curves with labels on left), SMC and collagen content, and mechanical properties are plotted together as a percentage of peak value during the 5-week incubation period.

upregulation of lysyl oxidase expression during weeks 1—3 with a peak in expression at week 3, correlating with the phases of collagen and tropoelastin production. Mechanical strength doubles over weeks 4—5 when production of collagen and elastic fibers and expression of lysyl oxidase are subsiding. This may be due in part to the more organized collagen fibrils evident from the histological sections in weeks 3—5.

Fibrin-based grafts were successfully implanted into ovine jugular veins for 15 weeks (Swartz et al., 2005; Yao et al., 2005). Explanted grafts were populated with a monolayer of ECs, circumferentially aligned SMCs, and significant amounts of collagen and elastin fibers. Although these pre-implantation grafts possessed significant amounts of residual fibrin and were not sufficiently strong for the arterial circulation, this *in vivo* study holds great promise for fibrin gel-based vascular grafts.

Silk has recently emerged as a potential scaffold for vascular tissue engineering. While it is not a biopolymer in the same sense as collagen and fibrin (i.e. it is not an *in vivo* structural protein), it is a natural fibrillar protein that possesses unique material properties (e.g. high tensile strength, biocompatibility, biodegradability) that make it an intriguing choice for use in tissue engineering applications (Sell et al., 2009). Zhang et al. demonstrated that reconstituted silk fibroin from silkworm cocoons could be electrospun into tubular scaffolds that supported human aortic EC and SMC attachment, proliferation, and matrix synthesis (Zhang et al., 2008). In a follow-up study, the researchers placed these silk-based constructs into a pulsatile bioreactor and showed a significant increase in SMC proliferation, matrix production, and EC alignment; however, pre-coating of the scaffolds with Matrigel prior to seeding was necessary to ensure SMC and EC attachment under flow (Zhang et al., 2009). While these results are promising, considerable work remains to be done, particularly in regards to the mechanical properties of these constructs. While silk is extremely strong, it is not very compliant, which can potentially cause problems *in vivo* where compliance mismatch can cause anastomotic intimal hyperplasia, a major source of small-diameter graft failure (Conte, 1998).

Decellularized tissue scaffolds

Decellularized tissues have the advantage of being entirely composed of natural ECM, giving them numerous potential advantages in terms of mechanical properties and biocompatibility (Schmidt and Baier, 2000). Decellularized vascular tissue has been studied extensively, showing benefits from grossly retaining the structure and composition of a native vessel following decellularization; however, decellularization can adversely impact the tissue, resulting in reduced ultimate tensile strength and compliance (Dahl et al., 2003; Williams et al., 2009a). Decellularized vessels have been used as scaffolds, with (Hodde et al., 2002; Amiel et al., 2006) or without recellularization (Hiles et al., 1995; Martin et al., 2005), prior to implantation. Although decellularized grafts may have a shorter route to the clinic by avoiding graft rejection and immune response due to the absence of cells, recellularization often improves patency. Indeed, 4-week patency rates of recellularized rat arteries with ECs was 89% compared to 29% for acellular controls (Borschel et al., 2005).

Others have sought to enhance graft function by improving EC recruitment and retention on implanted xenogenic decellurized grafts by applying various surface treatments and have demonstrated that fully formed EC layers significantly reduce thrombus formation and improve long-term patency (Kerdjoudj et al., 2008; Zhou et al., 2009). In order to avoid rejection associated with implanting non-autologous tissue, Yang et al. have investigated the possibility of implanting dogs with decellularized allogenic blood vessels seeded with autologous ECs and SMCs (Yang et al., 2009). Vessels seeded with the autologous cells showed 100% patency at 6 months compared with 50% patency for unseeded vessels. Interestingly, decellularized porcine carotid arteries were implanted into goats for up to 12 months with 19 of the 20 vessels remaining patent and repopulated by host SMCs and ECs (Kim et al., 2007). These two seemingly contradictory results serve to highlight that the issues of autologous/ allogenic/xenogenic transplants of decellularized tissues are not well understood, and there are clearly species-specific differences in response to the use of decellularized vessels, making it difficult to draw any broader conclusions about the utility of these approaches for vascular tissue engineering.

While most studies involving decellularized tissue have focused on using vascular tissue, a number of researchers have also explored the idea of using alternative tissue sources to engineer vascular grafts. In a hybrid approach, decellularized porcine small intestinal submucosa was seeded with human umbilical vein ECs for 2 weeks to allow the deposition of a basement membrane and subsequently removed (Woods et al., 2004). ECs were then reseeded and showed enhanced organization of adherens cell junctions, increased metabolic activity, downregulation of pro-inflammatory prostaglandin PGI_2, and decreased adhesion of resting or activated human platelets compared to a non-conditioned graft. Vessels that have been stripped of all components except elastin using an alkaline extraction process and further treated with penta-galloyl glucose (to prevent rapid degeneration *in vivo*) have shown promise as vascular grafts (Chuang et al., 2009). These processed vessels demonstrated burst pressures in excess of 800 mmHg and EC/fibroblast adhesion *in vitro*, while *in vivo* they supported cell infiltration and matrix deposition, and were resistant to elastin degradation. Decellularized ureters have also been examined as potential small-diameter vascular grafts due to their ease of harvest and appropriate size (Narita et al., 2008). Reseeded decellularized ureters showed good cell infiltration, possessed mechanical compliance similar to that of native carotid arteries, and remained patent as carotid artery replacements in dogs for 24 weeks.

SCAFFOLD-FREE APPROACHES
Cell sheets

A unique approach was taken by l'Heureux et al. (1998), who created a tissue-engineered blood vessel made exclusively of cultured human cells and their matrix without any starting scaffold (Fig. 47.4). Human vascular SMCs were cultured on tissue culture plastic to produce a cohesive cellular sheet (media layer) that was wrapped around a mandrel, and a similar sheet of human skin fibroblasts was subsequently wrapped around the media to create the adventitia. After a minimum of 8 weeks maturation, the mandrel was removed and the lumen was seeded with human umbilical vein ECs. The overall culture period was 3 months, excluding cell expansion. The resulting graft had a burst strength of over 2,000 mmHg and good

863

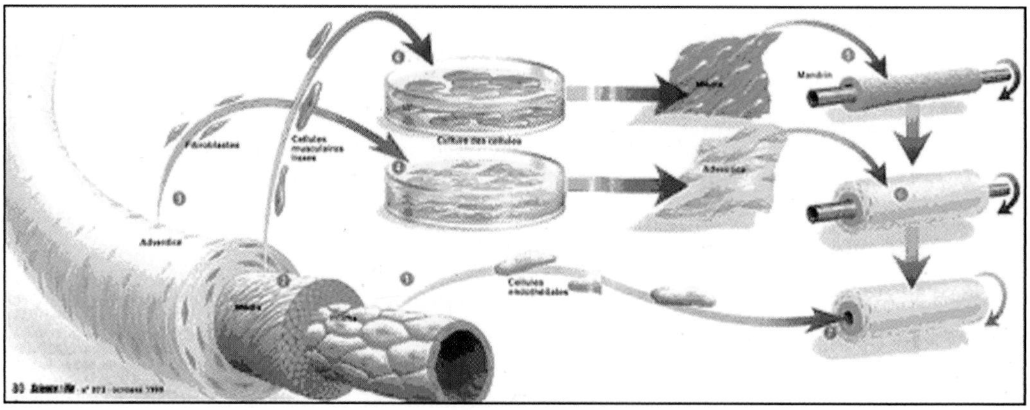

FIGURE 47.4
Cell sheet assembly method for the development of three-layered vascular grafts.

handling and suturability when implanted in a canine model. The media layer was also tested with pharmacological stimuli and showed contractile/relaxation responses (l'Heureux et al., 2001). In another study where human blood vessels were engineered with a similar approach, abdominal interpositional graft patency was 85% in nude rats with time points up to 225 days (l'Heureux et al., 2006). Lack of significant elastogenesis remains an issue with this approach; however, when such grafts were made using SMCs that had been transduced with an isoform of the proteoglycan versican, tropoelastin content, elastin cross-linking, burst strength, collagen bundle thickness, and elasticity all increased as compared to grafts that had not been transduced (Keire et al., 2009).

More recently, grafts created by Cytograft using this technology have entered human clinical safety trials as arterio-venous shunts in patients with end-stage renal disease (l'Heureux et al., 2007b; Konig et al., 2009; McAllister et al., 2009). The completely autologous tissue-engineered grafts that were prepared for this study demonstrated burst pressures of well over 3000 mmHg, which compares favorably to that of the internal thoracic artery. Grafts were implanted into nine patients, with seven grafts remaining patent after 1 month, five remaining patent for at least 6 months, and one remaining patent for at least 20 months. Although the cell sheet-based tissue engineering technique has been very successful, the overall development time is long (6—9 months), and scale-up capabilities to meet patient demand do not appear straightforward without regulatory changes (l'Heureux et al., 2007a). Cell sheet engineering using thermoresponsive substrates (Yang et al., 2007), particularly patterned substrates (Isenberg et al., 2008; Williams et al., 2009b), may help in this regard by dramatically speeding up the cell sheet production process by growing large numbers of thinner sheets (i.e. shorter time in culture) in parallel that can be rapidly stacked and rolled into tubular constructs (Kubo et al., 2007).

In vivo construct fabrication

Harvesting cells and expanding them in culture oftentimes result in a phenotypic change and may prevent robust tissue formation *in vitro*. To overcome these limitations, researchers have attempted recruiting cells *in vivo* and using the body as the bioreactor. Campbell et al. developed grafts in the recipient's peritoneal cavity by inserting silastic tubing as a mandrel in the peritoneal cavity of rats or rabbits. The intima of the grafts became populated with mesothelial cells and the media with myofibroblasts within 2 weeks (Campbell et al., 1999). Although this innovative method of graft development needs to be further evaluated in long-term *in vivo* experiments to establish patency, it bypasses complicated cell and scaffold sourcing-related issues. In a different study, six different kinds of polymeric rods were inserted subcutaneously into rabbits for up to 3 months (Nakayama et al., 2004). The tissue formed on these biotubes was mostly collagen-rich ECM and fibroblasts, and endothelialization of the resulting grafts was not addressed. Another approach is the implantation of a cell-free scaffold to recruit cells from the host. Hyaluronan-based grafts implanted in rat abdominal aortas were populated with ECs and SMCs within 21 days and contained collagen and elastic fibers (Lepidi et al., 2006).

SMOOTH MUSCLE CELL-ENDOTHELIAL CELL CO-CULTURES

ECs used to be regarded solely as a lining of all blood vessels that forms a barrier to blood. However, it is now known that ECs are involved in numerous cell signaling pathways and communicate with SMCs via heterocellular junctions and other mediators (Davies, 1986). Intermittent fenestrations in the internal elastic lamina are 0.5–1.5 μm in large vessels and 0.1–0.45 μm in capillaries, thereby allowing direct contact between the two cell types (Saunders and d'Amore, 1992). EC-SMC co-culture experiments have revealed an EC effect on SMC proliferation (Campbell and Campbell, 1986; Fillinger et al., 1993; Axel et al., 1996; Yoshida et al., 1996; Waybill et al., 1997; Vouyouka et al., 2004), migration (Casscells, 1992), phenotype (Brown et al., 2005), and ECM production (see below). ECs are believed to recruit SMC-pericytes to newly formed blood vessels during vasculogenesis (Hungerford and Little, 1999). One study suggests that a small percentage of primary ECs may give rise to SMCs via transdifferentiation *in vitro* (Frid et al., 2002). However, the origin of these SMC-like cells *in vivo* remains unclear and bone marrow-derived progenitor cells may be implicated. Although further work is required to establish whether transdifferentiation of ECs into SMCs is possible, these two cell types clearly interact significantly *in vivo*.

Three-dimensional co-culture studies of ECs and SMCs in collagen gels show that SMCs affect not only EC proliferation but also EC alignment and elongation (Ziegler et al., 1995; Imberti et al., 2002). ECs seeded onto SMC-contracted collagen gel were growth-inhibited relative to tissue culture plastic, and proliferation was further reduced in the presence of physiological shear flow for 24 h (Ziegler et al., 1995). In a co-culture study where ECs were seeded directly on SMCs, a lower (and decreasing over time) attachment efficiency of ECs was observed when they were seeded on proliferating compared to quiescent SMCs (Lavender et al., 2005). Therefore, cell signaling between ECs and SMCs is a two-way communication.

ECs also affect SMC matrix deposition, but these effects appear less studied. Culturing rabbit aortic SMC monolayers in conditioned medium from confluent bovine aortic ECs stimulates glycosaminoglycans synthesis by the SMCs up to 120% within 24 h (Merrilees et al., 1990) and keeps the SMC in a differentiated, contractile phenotype. In a different study with pig and rat aortic ECs and SMCs, hyaluronic acid and sulfated glycosaminoglycans production increased in co-culture compared to separate cultures of the two cell types (Merrilees and Scott, 1981). In contrast, collagen synthesis and collagen type I expression decrease in EC-SMC *in vitro* co-culture models (Powell et al., 1997). ECs have also been shown to increase gene expression of vascular endothelial growth factor, platelet-derived growth factors AA and BB, and transforming growth factor β, and decrease expression of basic fibroblast growth factor when co-cultured with SMCs (Heydarkhan-Hagvall et al., 2003).

When ECs and SMCs were seeded into a tubular polyglycolic acid scaffold and co-cultured in a perfusion bioreactor, 15-day co-culture increased cell proliferation, decreased collagen and proteoglycan deposition, and led to higher SMC expression of contractile proteins compared to 2-day co-cultures (Williams and Wick, 2005). Lavender et al. investigated the effects of various substrates and medium formulations on their ability to successfully co-culture SMCs and ECs under steady, laminar flow conditions (Lavender et al., 2005). They found that medium conditions that yielded a quiescent SMC population allowed for the direct culture of a distinct, confluent, and adherent EC monolayer on top of the SMC layer for up to 10 days. Furthermore, under these conditions, ECs increased their rate of Ac-LDL uptake, and platelet-EC adhesion molecule (PECAM) expression in EC borders was decreased. It has also been shown that, when human umbilical cord ECs are co-cultured with human umbilical cord

SMCs under disturbed flow conditions in a vertical-step flow chamber, neutrophil, peripheral blood lymphocyte, and monocyte adhesion to ECs and transmigration through the EC monolayers are significantly increased compared to controls in the absence of SMCs (Chen et al., 2006).

BIOREACTOR CONDITIONING

Many studies have elucidated the importance of bioreactors for improving cell seeding, extracellular matrix production, tissue architecture, and differentiation compared to static culture techniques (Kim et al., 1998; Niklason et al., 1999; Burg et al., 2000; Carrier et al., 2002; Davisson et al., 2002; Pei et al., 2002; Baguneid et al., 2004; McFetridge et al., 2004; Opitz et al., 2004b; Williams and Wick, 2004; Engbers-Buijtenhuijs et al., 2006). Seliktar et al. developed a dynamic mechanical conditioning bioreactor, in which up to four constructs can be mounted on a distensible silicone tube (Seliktar et al., 2000). The intraluminal pressure was regulated to produce a 10% cyclic change in the outer diameter of each silicone tube, and it was shown that mechanical preconditioning of SMC-seeded collagen gels for up to 8 days improved their mechanical strength and histological organization (Seliktar et al., 2000). In two follow-up studies, they also demonstrated that such conditioning upregulated matrix metalloproteinase-2 activity (Seliktar et al., 2001) and altered collagen and elastin gene expression (Seliktar et al., 2003). In a similar study, Isenberg and Tranquillo examined the long-term (5 weeks) effects of cyclic distension on SMC-compacted, ribose cross-linked collagen gels (Isenberg and Tranquillo, 2003). The cyclically loaded constructs were found to be 1.7-fold stronger and 2.2-fold stiffer than their static counterparts, while loading parameters such as strain, stretch time, and relaxation time all influenced construct mechanical properties. In addition, deposition of significant amounts of insoluble elastin was detected, which was a surprising finding given that the adult rat aortic SMCs used in this study typically lack significant elastogenic potential in static culture (McMahon et al., 1985; Johnson et al., 1995). More recently, the same lab has demonstrated that incrementally increasing the cyclic strain amplitude of fibrin-based constructs from 5 to 15% over a 3-week period dramatically increased stiffness and strength over static and constant 10% strain amplitude (Syedain et al., 2008), indicating that cells undergoing a constant cyclic loading had adapted to the mechanical stimulation while those experiencing a periodic increase in loading remained active in tissue growth and remodeling.

Perhaps the most impressive demonstration of the effect of bioreactor conditioning on synthetic polymer-based graft development has been the work of Niklason et al. discussed above (Niklason et al., 1999); however, a number of similar studies using other synthetic polymer scaffolds have also shown that mechanical stimulation can improve matrix synthesis and mechanical properties in these systems. Rabbit SMCs seeded into PLCL scaffolds and exposed to 5% radial distension and 25 mmHg pressure at 1 Hz exhibited higher proliferation and collagen deposition, significant cell alignment in the radial direction, and upregulation of smooth muscle α-actin compared to static controls (Jeong et al., 2005). Stekelenburg et al. have shown that PGA/poly-4-hydroxybutyrate/fibrin hybrid scaffolds seeded with myofibroblasts cyclically conditioned for 28 days increased burst strength and collagen content (Stekelenburg et al., 2009), while Solan et al. have described similar results for SMCs on PGA scaffolds at various levels of pulse frequency (Solan et al., 2009).

Bioreactors have also been used to optimize cell seeding, especially for ECs. Baguneid et al. developed a bioreactor in which SMC-seeded scaffolds were exposed to pulsatile shear stress (9.32 dynes/cm^2 average, 32.1 dynes/cm^2 maximum, at 120 mmHg systolic pressure) for 7 days prior to EC seeding (Baguneid et al., 2004). Endothelialized grafts were then exposed to 1 h of physiological shear stress, during which time EC retention decreased from 100% to approximately 75% (Baguneid et al., 2004). These results show that a more gradual increase of shear stress over a longer period of time is needed to maximize EC retention. Niklason et al.

gradually increased flow rate in the lumen of endothelialized grafts from 1.98 to 6 ml/min or 0.1 to 0.3 dynes/cm^2 (Niklason et al., 1999). Clearly, mechanical stimulation of ECs was minimal, and ECs were not exposed to a physiological-like environment prior to implantation, yet the ECs stained positively for vWF and PECAM (Niklason et al., 1999). Finally, basement membrane formation has been shown to have a significant effect on EC adhesion and retention (Baguneid et al., 2004). Fibrin-based grafts fabricated from rat aortic SMC or human dermal fibroblasts were seeded with rat aortic ECs (achieving ~99% EC surface coverage 2 days post-seeding) and placed in a pulsatile flow bioreactor (Isenberg et al., 2006a). The constructs were exposed to steady and pulsatile flow, and EC elongation and alignment were observed in the flow direction, but only when the flow was in the laminar regime (Re < 2100). EC surface coverage remained high (> 95%) in the presence of pulsatile flow up to (at least) 10 dynes/cm^2 for 48 h, indicating that ECs were highly adherent to the grafts, which were found to have fibronectin and laminin localized at the lumenal surface prior to EC seeding. Both static and flow-conditioned media-equivalents expressed von Willebrand factor, a marker of properly functioning ECs, and no acute thrombus formation when exposed to whole blood, suggesting that ECs exposed to flow in the bioreactor were in a normal, non-activated phenotype.

CONCLUSION AND OUTLOOK

Significant advances have been made towards the development of a small-diameter vascular graft, although the challenges remain substantial. Development of vascular substitutes is time-consuming and in most cases one graft is produced at a time. This approach raises the issue of just how efficient and cost-effective the process can be, and also how reproducibility can be ensured (Ratcliffe and Niklason, 2002). A functional small-diameter vascular graft possesses appropriate mechanical properties, including physiological compliance and viscoelasticity and, critically, adequate burst strength, without any propensity for permanent creep that leads to aneurysm. It also possesses physiological transport properties, such as appropriate permeability to plasma and proteins. Finally, it exhibits physiological properties, such as vasoconstriction/relaxation responses, insofar as these responses indicate a physiological SMC phenotype, and, critically, a non-thrombogenic endothelium. From a practical standpoint, suturability and simplicity of handling are necessary, and, from a commercial standpoint, it must be fabricated in a process that scales well with quantity and be a product that can be shipped and stored.

Meeting all criteria simultaneously remains a challenge. For example, high burst strength is often associated with compliance mismatch (l'Heureux et al., 1998), which can lead to intimal hyperplasia at the suture line. Conversely, collagen-based constructs that possess physiological compliance have lacked high burst strength (Girton et al., 2000). Fibrin-based constructs yield higher burst strengths and physiological compliance (Isenberg et al., 2006b), although there is no accepted standard for what constitutes a minimum burst pressure at implantation. Notably, no approach has yet resulted in all the key features of the media layer, namely circumferential alignment of SMCs, collagen fibers, and elastin lamellae. In fact, mature (i.e. crosslinked) elastin fibers have only been reported in the self-assembly approach, and in association with fibroblasts, not SMCs (l'Heureux et al., 1998). The developmental downregulation of elasto-genesis in SMCs creates a major hurdle (McMahon et al., 1985; Johnson et al., 1995). Indeed, elastic recoil is critical to abolish permanent creep and is conferred by elastin lamellae in the large elastic arteries (Opitz et al., 2004a; Patel et al., 2006), whereas lamellae are less prominent in smaller diameter muscular arteries, which are the targets of vascular tissue engineering. It remains to be seen whether other ECM can confer both elasticity and physiological compliance in the absence of elastin lamellae.

This question is related to a broader challenge for the field of tissue engineering: the need for a predictive basis for the optimal combination of cell source/scaffold/stimulation/bioreactor. This will hinge on a more complete understanding of how the cell integrates the various signals at the cellular and molecular level. This understanding will translate into biophysical models that relate cell cycle regulation and the production and assembly of ECM components in response to these integrated signals, and ultimately into multi-scale mechanical models that relate the evolving ECM at the molecular level to macroscopic mechanical and functional properties. There are recent continuum mechanical models of vascular growth and remodeling that are aimed in this direction (Taber, 2001; Humphrey and Rajagopal, 2003; Gleason and Humphrey, 2004; Niklason et al., 2010). Furthermore, technologies such as organ printing (Mironov et al., 2009; Norotte et al., 2009) and nano/microfabrication (Isenberg et al., 2008; Mironov et al., 2008; Williams et al., 2009b; Zorlutuna et al., 2009) are emerging with the promise of building tissues from the bottom up by precisely organizing and orienting cells and ECM in defined configurations over multiple length scales (see review by Mironov et al. (2008) for more information). Ultimately, the growth and remodeling that occur following implantation in response to signals, which the tissue engineer has little or no control over, will determine the success of tissue-engineered vessels. There is scant information about how growth and remodeling depend on the properties at implantation.

Furthermore, there is no imminent solution to the extreme immunogenicity of non-autologous ECs. Even if a construct could be pre-fabricated from non-autologous SMCs, fibroblasts, or other tissue cells types, it would still take many days to weeks to isolate and expand the patient's ECs to the numbers required for seeding a construct of useful length, for example with circulating EC progenitor cells (Hristov et al., 2003; Matsumura et al., 2003; Cho et al., 2005) or blood outgrowth endothelial cells (Lin et al., 2000, 2002), both of which possess high proliferative capacity and can differentiate into mature ECs. The associated time lag, however, might limit the applicability of vascular grafts fabricated with these cell sources to patients with anticipated repeat procedures. The optimal sources for SMCs and ECs remain to be determined, but economic and regulatory considerations would favor pre-fabrication of small-diameter vascular grafts from non-autologous, genetically unmodified cells.

References

Aird, W. C. (2006). Endothelial cell heterogeneity and atherosclerosis. *Curr. Atheroscler. Rep., 8,* 69—75.

Allen, J. B., Khan, S., Lapidos, K. A., & Ameer, G. A. (2010). Toward engineering a human neoendothelium with circulating progenitor cells. *Stem Cells, 28,* 318—328.

American Heart Association. (2010). Heart Disease and Stroke Statistics — 2010 Update. http://www.americanheart.org/statistics

Amiel, G. E., Komura, M., Shapira, O., Yoo, J. J., Yazdani, S., Berry, J., et al. (2006). Engineering of blood vessels from acellular collagen matrices coated with human endothelial cells. *Tissue Eng., 12,* 2355—2365.

Anderson, J. M. (1995). Perspectives on the *in vivo* responses of biodegradable polymers. In J. Hollinger (Ed.), *Biomedical Applications of Synthetic Biodegradable Polymers* (pp. 223—234). Boca Raton: CRC Press.

Assoian, R. K., & Marcantonio, E. E. (1996). The extracellular matrix as a cell cycle control element in atherosclerosis and restenosis. *J. Clin. Invest., 98,* 2436—2439.

Axel, D. I., Brehm, B. R., Wolburg-Buchholz, K., Betz, E. L., Koveker, G., & Karsch, K. R. (1996). Induction of cell-rich and lipid-rich plaques in a transfilter coculture system with human vascular cells. *J. Vasc. Res., 33,* 327—339.

Baguneid, M., Murray, D., Salacinski, H. J., Fuller, B., Hamilton, G., Walker, M., et al. (2004). Shear-stress preconditioning and tissue-engineering-based paradigms for generating arterial substitutes. *Biotechnol. Appl. Biochem., 39*(Pt 2), 151—157.

Barocas, V. H., Girton, T. S., & Tranquillo, R. T. (1998). Engineered alignment in media-equivalents: magnetic prealignment and mandrel compaction. *J. Biomech. Eng., 120,* 660—666.

Bassiouny, H., Sakaguchi, Y., & Glagov, S. (1997). Non-invasive imaging of atherosclerosis. In M. Mercuri, D. D. McPherson, H. Bassiouny, & S. Glagov (Eds.), *Carotid Atherosclerosis: From Induction to Complication* (pp. 3—26). Kluwer Academic Publishers.

Borschel, G. H., Huang, Y. C., Calve, S., Arruda, E. M., Lynch, J. B., Dow, D. E., et al. (2005). Tissue engineering of recellularized small-diameter vascular grafts. *Tissue Eng., 11*(5—6), 778—786.

Brown, D. J., Rzucidlo, E. M., Merenick, B. L., Wagner, R. J., Martin, K. A., & Powell, R. J. (2005). Endothelial cell activation of the smooth muscle cell phosphoinositide 3-kinase/Akt pathway promotes differentiation. *J. Vasc. Surg., 41,* 509—516.

Burg, K. J., Holder, W. D., Jr., Culberson, C. R., Beiler, R. J., Greene, K. G., Loebsack, A. B., et al. (2000). Comparative study of seeding methods for three-dimensional polymeric scaffolds. *J. Biomed. Mater. Res., 52*, 576.

Burke, A. P., Farb, A., Malcom, G. T., Liang, Y. H., Smialek, J., & Virmani, R. (1997). Coronary risk factors and plaque morphology in men with coronary disease who died suddenly. *N. Engl. J. Med., 336*, 1276–1282.

Cameron, A., Davis, K. B., Green, G., & Schaff, H. V. (1996). Coronary bypass surgery with internal-thoracic-artery grafts — effects on survival over a 15-year period. *N. Engl. J. Med., 334*, 216–219.

Campbell, G. (1985). *CJH: Phenotypic Modulation of Smooth Muscle Cells in Primary Culture. Vacular Smooth Muscle in Culture.* Boca Raton, FL: CRC Press.

Campbell, J. H., & Campbell, G. R. (1986). Endothelial cell influences on vascular smooth muscle phenotype. *Annu. Rev. Physiol., 48*, 295–306.

Campbell, J. H., Efendy, J. L., & Campbell, G. R. (1999). Novel vascular graft grown within recipient's own peritoneal cavity. *Circ. Res., 85*, 1173–1178.

Carrier, R. L., Rupnick, M., Langer, R., Schoen, F. J., Freed, L. E., & Vunjak-Novakovic, G. (2002). Perfusion improves tissue architecture of engineered cardiac muscle. *Tissue Eng., 8*, 175–188.

Casscells, W. (1992). Migration of smooth muscle and endothelial cells. Critical events in restenosis. *Circulation, 86*, 723–729.

Chamley-Campbell, J. H., & Campbell, G. R. (1981). What controls smooth muscle phenotype? *Atherosclerosis, 40* (3–4), 347–357.

Chen, C. N., Chang, S. F., Lee, P. L., Chang, K., Chen, L. J., Usami, S., et al. (2006). Neutrophils, lymphocytes, and monocytes exhibit diverse behaviors in transendothelial and subendothelial migrations under coculture with smooth muscle cells in disturbed flow. *Blood, 107*, 1933–1942.

Chen, F., Su, Y., Mo, X., He, C., Wang, H., & Ikada, Y. (2009). Biocompatibility, alignment degree and mechanical properties of an electrospun chitosan-P(LLA-CL) fibrous scaffold. *J. Biomater. Sci., 20*, 2117–2128.

Cho, S. W., Kim, I. K., Lim, S. H., Kim, D. I., Kang, S. W., Kim, S. H., et al. (2004). Smooth muscle-like tissues engineered with bone marrow stromal cells. *Biomaterials, 25*, 2979–2986.

Cho, S. W., Lim, S. H., Kim, I. K., Hong, Y. S., Kim, S. S., Yoo, K. J., et al. (2005). Small-diameter blood vessels engineered with bone marrow-derived cells. *Ann. Surg., 241*, 506–515.

Chuang, T. H., Stabler, C., Simionescu, A., & Simionescu, D. T. (2009). Polyphenol-stabilized tubular elastin scaffolds for tissue engineered vascular grafts. *Tissue Eng., 15*, 2837–2851.

Chung, S., Ingle, N. P., Montero, G. A., Kim, S. H., & King, M. W. (2010). Bioresorbable elastomeric vascular tissue engineering scaffolds via melt spinning and electrospinning. *Acta Biomater., 6*, 1958–1967.

Conte, M. S. (1998). The ideal small arterial substitute: a search for the Holy Grail? *Faseb J., 12*, 43–45.

Daemen, J., & Serruys, P. W. (2007a). Drug-eluting stent update 2007: part I. A survey of current and future generation drug-eluting stents: meaningful advances or more of the same? *Circulation, 116*, 316–328.

Daemen, J., & Serruys, P. W. (2007b). Drug-eluting stent update 2007: part II: Unsettled issues. *Circulation, 116*, 961–968.

Dahl, S. L., Koh, J., Prabhakar, V., & Niklason, L. E. (2003). Decellularized native and engineered arterial scaffolds for transplantation. *Cell Transplant., 12*, 659–666.

Dahl, S. L., Rhim, C., Song, Y. C., & Niklason, L. E. (2007). Mechanical properties and compositions of tissue engineered and native arteries. *Ann. Biomed. Eng., 35*, 348–355.

Davies, P. F. (1986). Vascular cell interactions with special reference to the pathogenesis of atherosclerosis. *Lab. Invest., 55*, 5–24.

Davisson, T., Sah, R. L., & Ratcliffe, A. (2002). Perfusion increases cell content and matrix synthesis in chondrocyte three-dimensional cultures. *Tissue Eng., 8*, 807–816.

DeLong, S. A., Moon, J. J., & West, J. L. (2005). Covalently immobilized gradients of bFGF on hydrogel scaffolds for directed cell migration. *Biomaterials, 26*, 3227–3234.

Desmouliere, A., & Gabbiani, G. (1995). Smooth muscle cell and fibroblast biological and functional features: similarities and differences. In S. M. Schwartz, & R. P. Mecham (Eds.), *The Vascular Smooth Muscle Cell: Molecular and Biological Responses to the Extracellular Matrix* (pp. 329–359). New York: Academic Press.

Didisheim, P., & Watson, J. T. (1996). Cardiovascular applications. In B. D. Ratner, A. S. Hoffman, F. J. Schoen, & J. E. Lemons (Eds.), *Biomaterials Science: An Introduction to Materials in Medicine* (pp. 283–296). New York: Academic Press.

DiMuzio, P., & Tulenko, T. (2007). Tissue engineering applications to vascular bypass graft development: the use of adipose-derived stem cells. *J. Vasc. Surg., 45*(Suppl A), A99–A103.

Dong, Y., Yong, T., Liao, S., Chan, C. K., & Ramakrishna, S. (2008). Long-term viability of coronary artery smooth muscle cells on poly(L-lactide-co-epsilon-caprolactone) nanofibrous scaffold indicates its potential for blood vessel tissue engineering. *J. R. Soc. Interface/R. Soc., 5*, 1109–1118.

Engbers-Buijtenhuijs, P., Buttafoco, L., Poot, A. A., Dijkstra, P. J., de Vos, R. A., Sterk, L. M., et al. (2006). Biological characterization of vascular grafts cultured in a bioreactor. *Biomaterials, 27,* 2390−2397.

Fillinger, M. F., O'Connor, S. E., Wagner, R. J., & Cronenwett, J. L. (1993). The effect of endothelial cell coculture on smooth muscle cell proliferation. *J. Vasc. Surg., 17,* 1058−1067, discussion 1067−1068.

Frid, M. G., Kale, V. A., & Stenmark, K. R. (2002). Mature vascular endothelium can give rise to smooth muscle cells via endothelial-mesenchymal transdifferentiation: *in vitro* analysis. *Circ. Res., 90,* 1189−1196.

Fung, Y. C. (1996). *Biomechanics Circulation.* New York: Springer.

Gilbert, A., Evtushenko, M., & Nair, H. (2001). Purification of fibrinogen and virus removal using preparative electrophoresis. *Ann. NY Acad. Sci., 936,* 625−629.

Girton, T. S., Oegema, T. R., & Tranquillo, R. T. (1999). Exploiting glycation to stiffen and strengthen tissue equivalents for tissue engineering. *J. Biomed. Mater. Res., 46,* 87−92.

Girton, T. S., Oegema, T. R., Grassl, E. D., Isenberg, B. C., & Tranquillo, R. T. (2000). Mechanisms of stiffening and strengthening in media-equivalents fabricated using glycation. *J. Biomech. Eng., 122,* 216−223.

Glagov, S., Bassiouny, H. S., Giddens, D. P., & Zarins, C. K. (1995). Pathobiology of plaque modeling and complication. *Surg. Clin. North Am., 75,* 545−556.

Gleason, R. L., & Humphrey, J. D. (2004). A mixture model of arterial growth and remodeling in hypertension: altered muscle tone and tissue turnover. *J. Vasc. Res., 41,* 352−363.

Goldman, S., Zadina, K., Moritz, T., Ovitt, T., Sethi, G., Copeland, J. G., et al. (2004). Long-term patency of saphenous vein and left internal mammary artery grafts after coronary artery bypass surgery: results from a Department of Veterans Affairs Cooperative Study. *J. Am. Coll. Cardiol., 44,* 2149−2156.

Gong, Z., & Niklason, L. E. (2008). Small-diameter human vessel wall engineered from bone marrow-derived mesenchymal stem cells (hMSCs). *Faseb J., 22,* 1635−1648.

Grasl, C., Bergmeister, H., Stoiber, M., Schima, H., & Weigel, G. (2009). Electrospun polyurethane vascular grafts: *in vitro* mechanical behavior and endothelial adhesion molecule expression. *J. Biomed. Mater. Res., 93A,* 716−723.

Grassl, E. D., Oegema, T. R., & Tranquillo, R. T. (2002). Fibrin as an alternative biopolymer to type-I collagen for the fabrication of a media equivalent. *J. Biomed. Mater. Res., 60,* 607−612.

Grassl, E. D., Oegema, T. R., & Tranquillo, R. T. (2003). A fibrin-based arterial media equivalent. *J. Biomed. Mater. Res., 66A,* 550−561.

Grenier, G., Remy-Zolghadri, M., Guignard, R., Bergeron, F., Labbe, R., Auger, F. A., et al. (2003). Isolation and culture of the three vascular cell types from a small vein biopsy sample. *In Vitro Cell Dev. Biol. Anim., 39*(3−4), 131−139.

Guyton, R. A. (2006). Coronary artery bypass is superior to drug-eluting stents in multivessel coronary artery disease. *Ann. Thorac. Surg., 81,* 1949−1957.

Hamilton, D. W., Maul, T. M., & Vorp, D. A. (2004). Characterization of the response of bone marrow-derived progenitor cells to cyclic strain: implications for vascular tissue-engineering applications. *Tissue Eng., 10*(3−4), 361−369.

Harris, L. J., Abdollahi, H., Zhang, P., McIlhenny, S., Tulenko, T. N., & Dimuzio, P. J. (2009). Differentiation of adult stem cells into smooth muscle for vascular tissue engineering. *J. Surg. Res.,* in press.

He, W., Ma, Z., Teo, W. E., Dong, Y. X., Robless, P. A., Lim, T. C., et al. (2009). Tubular nanofiber scaffolds for tissue engineered small-diameter vascular grafts. *J. Biomed. Mater. Res., 90,* 205−216.

Herrick, S. E., & Mutsaers, S. E. (2004). Mesothelial progenitor cells and their potential in tissue engineering. *Int. J. Biochem. Cell Biol., 36,* 621−642.

Heydarkhan-Hagvall, S., Helenius, G., Johansson, B. R., Li, J. Y., Mattsson, E., & Risberg, B. (2003). Co-culture of endothelial cells and smooth muscle cells affects gene expression of angiogenic factors. *J. Cell Biochem., 89,* 1250−1259.

Heyligers, J. M., Arts, C. H., Verhagen, H. J., de Groot, P. G., & Moll, F. L. (2005). Improving small-diameter vascular grafts: from the application of an endothelial cell lining to the construction of a tissue-engineered blood vessel. *Ann. Vasc. Surg., 19,* 448−456.

Hiles, M. C., Badylak, S. F., Lantz, G. C., Kokini, K., Geddes, L. A., & Morff, R. J. (1995). Mechanical properties of xenogeneic small-intestinal submucosa when used as an aortic graft in the dog. *J. Biomed. Mater. Res., 29,* 883−891.

Hodde, J. P., Record, R. D., Tullius, R. S., & Badylak, S. F. (2002). Retention of endothelial cell adherence to porcine-derived extracellular matrix after disinfection and sterilization. *Tissue Eng., 8,* 225−234.

Hoerstrup, S. P., Zund, G., Schnell, A. M., Kolb, S. A., Visjager, J. F., Schoeberlein, A., et al. (2000). Optimized growth conditions for tissue engineering of human cardiovascular structures. *Int. J. Artif. Organs., 23,* 817−823.

Hoerstrup, S. P., Zund, G., Sodian, R., Schnell, A. M., Grunenfelder, J., & Turina, M. I. (2001). Tissue engineering of small caliber vascular grafts. *Eur. J. Cardiothorac. Surg., 20,* 164−169.

Hristov, M., Erl, W., & Weber, P. C. (2003). Endothelial progenitor cells: mobilization, differentiation, and homing. *Arterioscler. Thromb. Vasc. Biol.*, 23, 1185–1189.

Humphrey, J. D., & Rajagopal, K. R. (2003). A constrained mixture model for arterial adaptations to a sustained step change in blood flow. *Biomech. Model Mechanobiol.*, 2, 109–126.

Hungerford, J. E., & Little, C. D. (1999). Developmental biology of the vascular smooth muscle cell: building a multilayered vessel wall. *J. Vasc. Res.*, 36, 2–27.

Huynh, T., Abraham, G., Murray, J., Brockbank, K., Hagen, P. O., & Sullivan, S. (1999). Remodeling of an acellular collagen graft into a physiologically responsive neovessel. *Nat. Biotechnol.*, 17, 1083–1086.

Ifkovits, J. L., Sundararaghavan, H. G., & Burdick, J. A. (2009). Electrospinning fibrous polymer scaffolds for tissue engineering and cell culture. *J. Vis. Exp.*, doi:10.3971/1589.

Imberti, B., Seliktar, D., Nerem, R. M., & Remuzzi, A. (2002). The response of endothelial cells to fluid shear stress using a co-culture model of the arterial wall. *Endothelium*, 9, 11–23.

Isenberg, B. C., & Tranquillo, R. T. (2003). Long-term cyclic distention enhances the mechanical properties of collagen-based media-equivalents. *Ann. Biomed. Eng.*, 31, 937–949.

Isenberg, B. C., Williams, C., & Tranquillo, R. T. (2006a). Endothelialization and flow conditioning of fibrin-based media-equivalents. *Ann. Biomed. Eng.*, 34, 971–985.

Isenberg, B. C., Williams, C., & Tranquillo, R. T. (2006b). Small-diameter artificial arteries engineered *in vitro*. *Circ. Res.*, 98, 25–35.

Isenberg, B. C., Tsuda, Y., Williams, C., Shimizu, T., Yamato, M., Okano, T., et al. (2008). A thermoresponsive, microtextured substrate for cell sheet engineering with defined structural organization. *Biomaterials*, 29, 2565–2572.

Jeong, S. I., Kwon, J. H., Lim, J. I., Cho, S. W., Jung, Y., Sung, W. J., et al. (2005). Mechano-active tissue engineering of vascular smooth muscle using pulsatile perfusion bioreactors and elastic PLCL scaffolds. *Biomaterials*, 26, 1405–1411.

Johnson, D. J., Robson, P., Hew, Y., & Keeley, F. W. (1995). Decreased elastin synthesis in normal development and in long-term aortic organ and cell cultures is related to rapid and selective destabilization of mRNA for elastin. *Circ. Res.*, 77, 1107–1113.

Kaushal, S., Amiel, G. E., Guleserian, K. J., Shapira, O. M., Perry, T., Sutherland, F. W., et al. (2001). Functional small-diameter neovessels created using endothelial progenitor cells expanded *ex vivo*. *Nat. Med.*, 7, 1035–1040.

Keire, P. A., l'Heureux, N., Vernon, R. B., Merrilees, M. J., Starcher, B., Okon, E., et al. (2009). Expression of Versican Isoform V3 in the absence of ascorbate improves elastogenesis in engineered vascular constructs. *Tissue Eng*, 16(2), 501–512.

Kerdjoudj, H., Berthelemy, N., Rinckenbach, S., Kearney-Schwartz, A., Montagne, K., Schaaf, P., et al. (2008). Small vessel replacement by human umbilical arteries with polyelectrolyte film-treated arteries: *in vivo* behavior. *J. Am. Coll. Cardiol.*, 52, 1589–1597.

Kim, B. S., Putnam, A. J., Kulik, T. J., & Mooney, D. J. (1998). Optimizing seeding and culture methods to engineer smooth muscle tissue on biodegradable polymer matrices. *Biotechnol. Bioeng.*, 57, 46–54.

Kim, W. S., Seo, J. W., Rho, J. R., & Kim, W. G. (2007). Histopathologic changes of acellularized xenogenic carotid vascular grafts implanted in a pig-to-goat model. *Int. J. Artif. Organs.*, 30, 44–52.

Kivela, A., & Hartikainen, J. (2006). Restenosis related to percutaneous coronary intervention has been solved? *Ann. Med.*, 38, 173–187.

Konig, G., McAllister, T. N., Dusserre, N., Garrido, S. A., Iyican, C., Marini, A., et al. (2009). Mechanical properties of completely autologous human tissue engineered blood vessels compared to human saphenous vein and mammary artery. *Biomaterials*, 30, 1542–1550.

Kubo, H., Shimizu, T., Yamato, M., Fujimoto, T., & Okano, T. (2007). Creation of myocardial tubes using cardio-myocyte sheets and an in vitro cell sheet-wrapping device. *Biomaterials*, 28, 3508–3516.

Kumbar, S. G., James, R., Nukavarapu, S. P., & Laurencin, C. T. (2008). Electrospun nanofiber scaffolds: engineering soft tissues. *Biomed. Mat.*, 3, 034002.

l'Heureux, N., Germain, L., Labbe, R., & Auger, F. A. (1993). *In vitro* construction of a human blood vessel from cultured vascular cells: a morphologic study. *J. Vasc. Surg.*, 17, 499–509.

l'Heureux, N., Paquet, S., Labbe, R., Germain, L., & Auger, F. A. (1998). A completely biological tissue-engineered human blood vessel. *Faseb J.*, 12, 47–56.

l'Heureux, N., Stoclet, J. C., Auger, F. A., Lagaud, G. J., Germain, L., & Andriantsitohaina, R. (2001). A human tissue-engineered vascular media: a new model for pharmacological studies of contractile responses. *Faseb J.*, 15, 515–524.

l'Heureux, N., Dusserre, N., Konig, G., Victor, B., Keire, P., Wight, T. N., et al. (2006). Human tissue-engineered blood vessels for adult arterial revascularization. *Nat. Med.*, 12, 361–365.

l'Heureux, N., Dusserre, N., Marini, A., Garrido, S., de la Fuente, L., & McAllister, T. (2007a). Technology insight: the evolution of tissue-engineered vascular grafts — from research to clinical practice. *Nature Clin. Pract., 4*, 389–395.

l'Heureux, N., McAllister, T. N., & de la Fuente, L. M. (2007b). Tissue-engineered blood vessel for adult arterial revascularization. *N. Engl. J. Med., 357*, 1451–1453.

Lavender, M. D., Pang, Z., Wallace, C. S., Niklason, L. E., & Truskey, G. A. (2005). A system for the direct co-culture of endothelium on smooth muscle cells. *Biomaterials, 26*, 4642–4653.

Lee, S. J., Yoo, J. J., Lim, G. J., Atala, A., & Stitzel, J. (2007). *In vitro* evaluation of electrospun nanofiber scaffolds for vascular graft application. *J. Biomed. Mater. Res., 83*, 999–1008.

Lepidi, S., Grego, F., Vindigni, V., Zavan, B., Tonello, C., Deriu, G. P., et al. (2006). Hyaluronan biodegradable scaffold for small-caliber artery grafting: preliminary results in an animal model. *Eur. J. Vasc. Endovasc. Surg., 32*(4), 411–417

Liebner, S., Cavallaro, U., & Dejana, E. (2006). The multiple languages of endothelial cell-to-cell communication. *Arterioscler. Thromb. Vasc. Biol., 26*, 1431–1438.

Lim, S. H., Cho, S. W., Park, J. C., Jeon, O., Lim, J. M., Kim, S. S., et al. (2008). Tissue-engineered blood vessels with endothelial nitric oxide synthase activity. *J. Biomed. Mater. Res., 85*, 537–546.

Lin, Y., Weisdorf, D. J., Solovey, A., & Hebbel, R. P. (2000). Origins of circulating endothelial cells and endothelial outgrowth from blood. *J. Clin. Invest., 105*, 71–77.

Lin, Y., Chang, L., Solovey, A., Healey, J. F., Lollar, P., & Hebbel, R. P. (2002). Use of blood outgrowth endothelial cells for gene therapy for hemophilia A. *Blood, 99*, 457–462.

Long, J. L., & Tranquillo, R. T. (2003). Elastic fiber production in cardiovascular tissue-equivalents. *Matrix Biol., 22*, 339–350.

Lu, L., Peter, S. J., Lyman, M. D., Lai, H. L., Leite, S. M., Tamada, J. A., et al. (2000). *In vitro* and *in vivo* degradation of porous poly(DL-lactic-co-glycolic acid) foams. *Biomaterials, 21*, 1837–1845.

Lytle, B. W., Loop, F. D., Cosgrove, D. M., Ratliff, N. B., Easley, K., & Taylor, P. C. (1985). Long-term (5 to 12 years) serial studies of internal mammary artery and saphenous vein coronary bypass grafts. *J. Thorac. Cardiovasc. Surg., 89*, 248–258.

Martin, N. D., Schaner, P. J., Tulenko, T. N., Shapiro, I. M., Dimatteo, C. A., Williams, T. K., et al. (2005). *In vivo* behavior of decellularized vein allograft. *J. Surg. Res., 129*, 17–23.

Matsuda, T. (2004). Recent progress of vascular graft engineering in Japan. *Artif. Organs, 28*, 64–71.

Matsumura, G., Miyagawa-Tomita, S., Shin'oka, T., Ikada, Y., & Kurosawa, H. (2003). First evidence that bone marrow cells contribute to the construction of tissue-engineered vascular autografts *in vivo*. *Circulation, 108*, 1729–1734.

McAllister, T. N., Maruszewski, M., Garrido, S. A., Wystrychowski, W., Dusserre, N., Marini, A., et al. (2009). Effectiveness of haemodialysis access with an autologous tissue-engineered vascular graft: a multicentre cohort study. *Lancet, 373*(9673), 1440–1446.

McClure, M. J., Sell, S. A., Ayres, C. E., Simpson, D. G., & Bowlin, G. L. (2009). Electrospinning-aligned and random polydioxanone-polycaprolactone-silk fibroin-blended scaffolds: geometry for a vascular matrix. *Biomed. Mater., 4*, 55010.

McClure, M., Sell, S., Simpson, D., Walpoth, B., & Bowlin, G. (2010). A three-layered electrospun matrix to mimic native arterial architecture using polycaprolactone, elastin, and collagen: A preliminary study. *Acta Biomater., 6*(7), 2422–2433.

McFetridge, P. S., Bodamyali, T., Horrocks, M., & Chaudhuri, J. B. (2004). Endothelial and smooth muscle cell seeding onto processed *ex vivo* arterial scaffolds using 3D vascular bioreactors. *Asaio J., 50*, 591–600.

McKee, J. A., Banik, S. S., Boyer, M. J., Hamad, N. M., Lawson, J. H., Niklason, L. E., et al. (2003). Human arteries engineered *in vitro*. *EMBO Rep., 4*, 633–638.

McMahon, M. P., Faris, B., Wolfe, B. L., Brown, K. E., Pratt, C. A., Toselli, P., et al. (1985). Aging effects on the elastin composition in the extracellular matrix of cultured rat aortic smooth muscle cells. *In Vitro Cell. Dev. Biol., 21*, 674–680.

Merrilees, M. J., & Scott, L. (1981). Interaction of aortic endothelial and smooth muscle cells in culture. Effect on glycosaminoglycan levels. *Atherosclerosis, 39*, 147–161.

Merrilees, M. J., Campbell, J. H., Spanidis, E., & Campbell, G. R. (1990). Glycosaminoglycan synthesis by smooth muscle cells of differing phenotype and their response to endothelial cell conditioned medium. *Atherosclerosis, 81*, 245–254.

Mironov, V., Kasyanov, V., & Markwald, R. R. (2008). Nanotechnology in vascular tissue engineering: from nano-scaffolding towards rapid vessel biofabrication. *Trends Biotechnol., 26*, 338–344.

Mironov, V., Visconti, R. P., Kasyanov, V., Forgacs, G., Drake, C. J., & Markwald, R. R. (2009). Organ printing: tissue spheroids as building blocks. *Biomaterials, 30*, 2164–2174.

Mitchell, S. L., & Niklason, L. E. (2003). Requirements for growing tissue-engineered vascular grafts. *Cardiovasc. Pathol., 12*, 59–64.

Nakayama, Y., Ishibashi-Ueda, H., & Takamizawa, K. (2004). *In vivo* tissue-engineered small-caliber arterial graft prosthesis consisting of autologous tissue (biotube). *Cell Transplant., 13*, 439–449.

Narita, Y., Kagami, H., Matsunuma, H., Murase, Y., Ueda, M., & Ueda, Y. (2008). Decellularized ureter for tissue-engineered small-caliber vascular graft. *J. Artif. Organs., 11*, 91–99.

Nieponice, A., Soletti, L., Guan, J., Hong, Y., Gharaibeh, B., Maul, T. M., et al. (2010). *In vivo* assessment of a tissue-engineered vascular graft combining a biodegradable elastomeric scaffold and muscle-derived stem cells in a rat model. *Tissue Eng. Part A, 16*(4), 1215–1223.

Niklason, L. E., Gao, J., Abbott, W. M., Hirschi, K. K., Houser, S., Marini, R., et al. (1999). Functional arteries grown *in vitro*. *Science, 284*(5413), 489–493.

Niklason, L. E., Yeh, A. T., Calle, E. A., Bai, Y., Valentin, A., & Humphrey, J. D. (2010). Regenerative medicine special feature: enabling tools for engineering collagenous tissues integrating bioreactors, intravital imaging, and biomechanical modeling. *Proc. Natl. Acad. Sci. U.S.A., 107*, 3335–3339.

Normand, J., & Karasek, M. A. (1995). A method for the isolation and serial propagation of keratinocytes, endothelial cells, and fibroblasts from a single punch biopsy of human skin. *In Vitro Cell Dev. Biol. Anim., 31*, 447–455.

Norotte, C., Marga, F. S., Niklason, L. E., & Forgacs, G. (2009). Scaffold-free vascular tissue engineering using bioprinting. *Biomaterials, 30*, 5910–5917.

Nottelet, B., Pektok, E., Mandracchia, D., Tille, J. C., Walpoth, B., Gurny, R., et al. (2009). Factorial design optimization and *in vivo* feasibility of poly(epsilon-caprolactone)-micro- and nanofiber-based small diameter vascular grafts. *J. Biomed. Mater. Res. A, 89*, 865–875.

Opitz, F., Schenke-Layland, K., Cohnert, T. U., Starcher, B., Halbhuber, K. J., Martin, D. P., et al. (2004a). Tissue engineering of aortic tissue: dire consequence of suboptimal elastic fiber synthesis *in vivo*. *Cardiovasc. Res., 63*, 719–730.

Opitz, F., Schenke-Layland, K., Richter, W., Martin, D. P., Degenkolbe, I., Wahlers, T., et al. (2004b). Tissue engineering of ovine aortic blood vessel substitutes using applied shear stress and enzymatically derived vascular smooth muscle cells. *Ann. Biomed. Eng., 32*, 212–222.

Owens, G. K. (1995). Regulation of differentiation of vascular smooth muscle cells. *Physiol. Rev., 75*, 487–517.

Patel, A., Fine, B., Sandig, M., & Mequanint, K. (2006). Elastin biosynthesis: the missing link in tissue-engineered blood vessels. *Cardiovasc. Res., 71*, 40–49.

Pei, M., Solchaga, L. A., Seidel, J., Zeng, L., Vunjak-Novakovic, G., Caplan, A. I., et al. (2002). Bioreactors mediate the effectiveness of tissue engineering scaffolds. *Faseb J., 16*, 1691–1694.

Powell, R. J., Hydowski, J., Frank, O., Bhargava, J., & Sumpio, B. E. (1997). Endothelial cell effect on smooth muscle cell collagen synthesis. *J. Surg. Res., 69*, 113–118.

Ratcliffe, A., & Niklason, L. E. (2002). Bioreactors and bioprocessing for tissue engineering. *Ann. NY Acad. Sci., 961*, 210–215.

Riha, G. M., Lin, P. H., Lumsden, A. B., Yao, Q., & Chen, C. (2005). Review: application of stem cells for vascular tissue engineering. *Tissue Eng., 11*(9–10), 1535–1552.

Risau, W. (1995). Differentiation of endothelium. *Faseb J., 9*, 926–933.

Ross, J. J., & Tranquillo, R. T. (2003). ECM gene expression correlates with *in vitro* tissue growth and development in fibrin gel remodeled by neonatal smooth muscle cells. *Matrix Biol., 22*, 477–490.

Ross, R. (1971). The smooth muscle cell. II. Growth of smooth muscle in culture and formation of elastic fibers. *J. Cell Biol., 50*, 172–186.

Saunders, K. B., & d'Amore, P. A. (1992). An *in vitro* model for cell-cell interactions. *In Vitro Cell. Dev. Biol., 28A*(7–8), 521–528.

Schmidt, C. E., & Baier, J. M. (2000). Acellular vascular tissues: natural biomaterials for tissue repair and tissue engineering. *Biomaterials, 21*, 2215–2231.

Schoen, F. J. (1994). Blood vessels. Robbins Pathologic Basis of Disease. In R. S. Cotran, V. Kumar, & S. L. Robbins (Eds.), *Robbins Pathologic Basis of Disease* (pp. 507–551). New York: W.B. Saunders.

Seliktar, D., Black, R. A., Vito, R. P., & Nerem, R. M. (2000). Dynamic mechanical conditioning of collagen-gel blood vessel constructs induces remodeling *in vitro*. *Ann. Biomed. Eng., 28*, 351–362.

Seliktar, D., Nerem, R. M., & Galis, Z. S. (2001). The role of matrix metalloproteinase-2 in the remodeling of cell-seeded vascular constructs subjected to cyclic strain. *Ann. Biomed. Eng., 29*, 923–934.

Seliktar, D., Nerem, R. M., & Galis, Z. S. (2003). Mechanical strain-stimulated remodeling of tissue-engineered blood vessel constructs. *Tissue Eng., 9*, 657–666.

Seliktar, D., Zisch, A. H., Lutolf, M. P., Wrana, J. L., & Hubbell, J. A. (2004). MMP-2 sensitive, VEGF-bearing bioactive hydrogels for promotion of vascular healing. *J. Biomed. Mater. Res. A, 68*, 704–716.

Sell, S. A., McClure, M. J., Garg, K., Wolfe, P. S., & Bowlin, G. L. (2009). Electrospinning of collagen/biopolymers for regenerative medicine and cardiovascular tissue engineering. *Adv. Drug Deliv. Rev., 61*, 1007–1019.

Shin'oka, T., Imai, Y., & Ikada, Y. (2001). Transplantation of a tissue-engineered pulmonary artery. *N. Engl. J. Med.*, *344*, 532–533.

Shireman, P. K., & Pearce, W. H. (1996). Endothelial cell function: biologic and physiologic functions in health and disease. *AJR Am. J. Roentgenol.*, *166*, 7–13.

Sieminski, A. L., Hebbel, R. P., & Gooch, K. J. (2005). Improved microvascular network *in vitro* by human blood outgrowth endothelial cells relative to vessel-derived endothelial cells. *Tissue Eng.*, *11*, 1332–1345.

Simper, D., Stalboerger, P. G., Panetta, C. J., Wang, S., & Caplice, N. M. (2002). Smooth muscle progenitor cells in human blood. *Circulation*, *106*, 1199–1204.

Smith, M. J., McClure, M. J., Sell, S. A., Barnes, C. P., Walpoth, B. H., Simpson, D. G., et al. (2008). Suture-reinforced electrospun polydioxanone–elastin small-diameter tubes for use in vascular tissue engineering: a feasibility study. *Acta Biomater.*, *4*, 58–66.

Solan, A., Dahl, S. L., & Niklason, L. E. (2009). Effects of mechanical stretch on collagen and cross-linking in engineered blood vessels. *Cell Transplant.*, *18*, 915–921.

Soletti, L., Hong, Y., Guan, J., Stankus, J. J., El-Kurdi, M. S., Wagner, W. R., et al. (2010). A bilayered elastomeric scaffold for tissue engineering of small diameter vascular grafts. *Acta Biomater.*, *6*, 110–122.

Stadler, E., Campbell, J. H., & Campbell, G. R. (1989). Do cultured vascular smooth muscle cells resemble those of the artery wall? If not, why not? *J. Cardiovasc. Pharmacol.*, *14*(Suppl. 6), S1–S8.

Stankus, J. J., Soletti, L., Fujimoto, K., Hong, Y., Vorp, D. A., & Wagner, W. R. (2007). Fabrication of cell microintegrated blood vessel constructs through electrohydrodynamic atomization. *Biomaterials*, *28*, 2738–2746.

Stegemann, J. P., & Nerem, R. M. (2003). Altered response of vascular smooth muscle cells to exogenous biochemical stimulation in two- and three-dimensional culture. *Exp. Cell Res.*, *283*, 146–155.

Stekelenburg, M., Rutten, M. C., Snoeckx, L. H., & Baaijens, F. P. (2009). Dynamic straining combined with fibrin gel cell seeding improves strength of tissue-engineered small-diameter vascular grafts. *Tissue Eng.*, *15*, 1081–1089.

Swartz, D. D., Russell, J. A., & Andreadis, S. T. (2005). Engineering of fibrin-based functional and implantable small-diameter blood vessels. *Am. J. Physiol. Heart Circ. Physiol.*, *288*, H1451–H1460.

Syedain, Z. H., Weinberg, J. S., & Tranquillo, R. T. (2008). Cyclic distension of fibrin-based tissue constructs: evidence of adaptation during growth of engineered connective tissue. *Proc. Natl. Acad. Sci. U.S.A.*, *105*, 6537–6542.

Taber, L. A. (2001). Biomechanics of cardiovascular development. *Annu. Rev. Biomed. Eng.*, *3*, 1–25.

Tagami, M., Nara, Y., Kubota, A., Sunaga, T., Maezawa, H., Fujino, H., et al. (1986). Morphological and functional differentiation of cultured vascular smooth-muscle cells. *Cell Tissue Res.*, *245*, 261–266.

Theron, J. P., Knoetze, J. H., Sanderson, R. D., Hunter, R., Mequanint, K., Franz, T., et al. (2010). Modification, crosslinking and reactive electrospinning of a thermoplastic medical polyurethane for vascular graft applications. *Acta Biomater.*, *6*(7), 2434–2447.

Thomas, V., Zhang, X., & Vohra, Y. K. (2009). A biomimetic tubular scaffold with spatially designed nanofibers of protein/PDS bio-blends. *Biotechnol. Bioeng.*, *104*, 1025–1033.

Thyberg, J., Nilsson, J., Palmberg, L., & Sjolund, M. (1985). Adult human arterial smooth muscle cells in primary culture. Modulation from contractile to synthetic phenotype. *Cell. Tissue Res.*, *239*, 69–74.

Thyberg, J., Hedin, U., Sjolund, M., Palmberg, L., & Bottger, B. A. (1990). Regulation of differentiated properties and proliferation of arterial smooth muscle cells. *Arteriosclerosis*, *10*, 966–990.

Tillman, B. W., Yazdani, S. K., Lee, S. J., Geary, R. L., Atala, A., & Yoo, J. J. (2009). The *in vivo* stability of electrospun polycaprolactone-collagen scaffolds in vascular reconstruction. *Biomaterials*, *30*, 583–588.

Tranquillo, R. T. (2002). The tissue-engineered small-diameter artery. *Ann. NY Acad. Sci.*, *961*, 251–254.

Tranquillo, R. T., Girton, T. S., Bromberek, B. A., Triebes, T. G., & Mooradian, D. L. (1996). Magnetically oriented tissue-equivalent tubes: application to a circumferentially orientated media-equivalent. *Biomaterials*, *17*, 349–537.

Valgimigli, M., Airoldi, F., & Zimarino, M. (2009). Stent choice in primary percutaneous coronary intervention: drug-eluting stents or bare metal stents? *J. Cardiovasc. Med.*, *10*(Suppl. 1), S17–S26.

Vouyouka, A. G., Jiang, Y., & Basson, M. D. (2004). Pressure alters endothelial effects upon vascular smooth muscle cells by decreasing smooth muscle cell proliferation and increasing smooth muscle cell apoptosis. *Surgery*, *136*, 282–290.

Waybill, P. N., Chinchilli, V. M., & Ballermann, B. J. (1997). Smooth muscle cell proliferation in response to co-culture with venous and arterial endothelial cells. *J. Vasc. Interv. Radiol.*, *8*, 375–381.

Weinberg, C. B., & Bell, E. (1986). A blood vessel model constructed from collagen and cultured vascular cells. *Science*, *231*(4736), 397–400.

Williams, C., & Wick, T. M. (2004). Perfusion bioreactor for small diameter tissue-engineered arteries. *Tissue Eng.*, *10*, 930–941.

874

Williams, C., & Wick, T. M. (2005). Endothelial cell-smooth muscle cell co-culture in a perfusion bioreactor system. *Ann. Biomed. Eng., 33*, 920–928.

Williams, C., Liao, J., Joyce, E. M., Wang, B., Leach, J. B., Sacks, M. S., et al. (2009a). Altered structural and mechanical properties in decellularized rabbit carotid arteries. *Acta biomaterialia, 5*, 993–1005.

Williams, C., Tsuda, Y., Isenberg, B. C., Yamato, M., Shimizu, T., Okano, T., et al. (2009b). Aligned cell sheets grown on thermo-responsive substrates with microcontact printed protein patterns. *Adv. Mater., 21*, 2161.

Woods, A. M., Rodenberg, E. J., Hiles, M. C., & Pavalko, F. M. (2004). Improved biocompatibility of small intestinal submucosa (SIS) following conditioning by human endothelial cells. *Biomaterials, 25*, 515–525.

Yang, D., Guo, T., Nie, C., & Morris, S. F. (2009). Tissue-engineered blood vessel graft produced by self-derived cells and allogenic acellular matrix: a functional performance and histologic study. *Ann. Plast. Surg., 62*, 297–303.

Yang, J., Yamato, M., Shimizu, T., Sekine, H., Ohashi, K., Kanzaki, M., et al. (2007). Reconstruction of functional tissues with cell sheet engineering. *Biomaterials, 28*, 5033–5043.

Yao, L., Swartz, D. D., Gugino, S. F., Russell, J. A., & Andreadis, S. T. (2005). Fibrin-based tissue-engineered blood vessels: differential effects of biomaterial and culture parameters on mechanical strength and vascular reactivity. *Tissue Eng., 11*, 991–1003.

Yoder, M. C., Mead, L. E., Prater, D., Krier, T. R., Mroueh, K. N., Li, F., et al. (2007). Redefining endothelial progenitor cells via clonal analysis and hematopoietic stem/progenitor cell principals. *Blood, 109*, 1801–1809.

Yoshida, H., Nakamura, M., Makita, S., & Hiramori, K. (1996). Paracrine effect of human vascular endothelial cells on human vascular smooth muscle cell proliferation: transmembrane co-culture method. *Heart Vessels, 11*, 229–233.

Zhang, X., Baughman, C. B., & Kaplan, D. L. (2008). *In vitro* evaluation of electrospun silk fibroin scaffolds for vascular cell growth. *Biomaterials, 29*, 2217–2227.

Zhang, X., Wang, X., Keshav, V., Wang, X., Johanas, J. T., Leisk, G. G., et al. (2009). Dynamic culture conditions to generate silk-based tissue-engineered vascular grafts. *Biomaterials, 30*, 3213–3223.

Zhou, M., Liu, Z., Wei, Z., Liu, C., Qiao, T., Ran, F., et al. (2009). Development and validation of small-diameter vascular tissue from a decellularized scaffold coated with heparin and vascular endothelial growth factor. *Artif. Organs, 33*, 230–239.

Ziegler, T., Alexander, R. W., & Nerem, R. M. (1995). An endothelial cell–smooth muscle cell co-culture model for use in the investigation of flow effects on vascular biology. *Ann. Biomed. Eng., 23*, 216–225.

Zorlutuna, P., Elsheikh, A., & Hasirci, V. (2009). Nanopatterning of collagen scaffolds improve the mechanical properties of tissue engineered vascular grafts. *Biomacromolecules, 10*, 814–821.

875

CHAPTER 48

Cardiac Tissue

Milica Radisic, Michael V. Sefton
Institute of Biomaterials and Biomedical Engineering, Department of Chemical Engineering and Applied Chemistry, University of Toronto, Ontario, Canada

INTRODUCTION: FROM TISSUES TO ORGANS: KEY GOALS AND ISSUES

Nearly 8 million people in the USA have suffered from myocardial infarction, with 800,000 new cases occurring each year (American Heart Association, 2004). Myocardial infarction results in the substantial death of cardiomyocytes in the infarct zone followed by pathological remodeling of the heart. The remodeling process involves cardiac dilation, wall thinning, and severe deterioration of contractile function leading to congestive heart failure in more than 500,000 patients in the USA each year (American Heart Association, 2004). Conventional therapies are limited by the substantial inability of myocardium to regenerate after injury (Soonpaa and Field, 1998) and the shortage of organs available for transplantation. This chapter will focus on describing cell- and tissue-based therapies that have been considered as novel treatment options (Reinlib and Field, 2000).

Regardless of the approach to regenerative medicine or the scope of the application (a vascular graft, a pediatric valve, or an entire heart), there are three overlapping therapeutic goals — the three R's:

- Make tissue and organ **replacement** safer, more effective, and more widely available.
- **Repair** tissues and organs without having to replace them.
- Enable tissues and organs to **regenerate** so that repair and regeneration become one and the same.
  Furthermore, the problems of reaching these goals can be summarized (Table 48.1) in three categories (here largely in the context of tissue engineering) (Sefton, 2002; Sefton et al., 2005):
  - **Cell number.** What is the source of cells to be used and how will large numbers be generated? How will they be supplied with nutrients and oxygen (and have wastes removed) within a device of reasonable volume?
  - **Cell function.** How will the scaffold, extracellular matrix, and diffusible factors interact to generate the desired cell phenotype? How will the engineered tissue/organ integrate with the host to ensure a functional outcome?
  - **Cell durability.** What will happen over the long-term as remodeling and/or the host immune/inflammatory system responds to the new tissue?

In order to replace, repair, or regenerate cardiovascular tissue, these central issues of regenerative medicine will need to be addressed. Some of these issues (Table 48.1) reflect the fundamental nature of how an organ is different from a tissue: the large size and three-dimensional (3D) structure and the presence of multiple cell types that work in unison. Beyond these largely scientific challenges, there are the no less critical, practical questions of

877

TABLE 48.1 Critical Issues Associated with Tissue Engineering a Heart

	Objective	Critical issues
Cell number	~ 300 g of cells (3×10^{11} cells) ~ 200 mL O_2/h[a]	Cell source/purity Vascularization
Function	Cellular phenotype (multiple cell types) Coordinated muscle contraction Pump blood Connect to circulation	Microenvironment (soluble and insoluble factors) Pacemaker and electrical conduction Valves and conduits Biomechanical elasticity and strength Non-thrombogenicity
Durability	Fatigue resistance Hypoxia and disease tolerance Host tolerance	Biocompatibility Remodeling Innate/adaptive immune response

Manufacturing and quality control
Ethical, legal, and social issues
Imaging and non-invasive diagnostics
Regulatory and public policy issues

Reproduced with permission from Sefton et al., 2005.
[a]Based on moderate activity (Burton, 1972).

manufacturing, sterilization, storage, and distribution, and the regulatory and public policy issues that will need to be addressed before such therapies can be made available to the patients who are expected to benefit. Furthermore, we will also need new imaging or other non-invasive strategies to monitor the success (or not) of these therapies: that is, to enable the translation into clinical practice.

CELL AND GENE THERAPY
Cell therapy

Treatment options for heart failure and myocardial infarction are limited by the inability of adult cardiomyocytes to proliferate and regenerate injured myocardium. Cell injection has thus emerged as an alternative treatment option. In animal models, injection of fetal or neonatal cardiomyocytes improved left ventricular function and ventricle thickness, thus attenuating pathological ventricular remodeling (Reinecke et al., 1999; Muller-Ehmsen et al., 2002a,b). Differentiated cardiomyocytes are indeed an ideal cell source for injection or tissue engineering, since they contain a developed contractile apparatus and can integrate through gap junctions and intercalated discs with the host cardiomyocytes. However, large numbers of clinically relevant autologous cardiomyocytes are unavailable.

In searching for an appropriate cell source (Table 48.2), regeneration of infarcted myocardium has been attempted in animal models by transplantation of skeletal myoblasts (Dorfman et al., 1998), as well as cardiomyocytes derived from embryonic stem cells (Klug et al., 1996) and bone marrow-derived mesenchymal stem cells (Toma et al., 2002). For a review of cell therapy approaches see Laflamme and Murry (2005).

The obvious advantage of skeletal myoblasts is that they can be harvested from the patient and expanded *in vitro*. However, mature skeletal myoblasts do not express gap junctional proteins; thus, they are incapable of functionally integrating with the host myocardium. This was the most likely reason for the occurrence of arrhythmias in 4 out of 10 patients in a Phase I clinical trial of autologous skeletal myoblast transplantation (Menasche et al., 2003). For further information on myoblast clinical trails see Laflamme and Murry (2005).

TABLE 48.2 Cell Sources for Cardiac Tissue Engineering, and some of their Advantages and Disadvantages

Cell sources	Advantages	Disadvantages
Adult cardiac cells	Target cell source	Little proliferative or developmental potential, limited resource
Fetal cardiac cells	Some proliferative potential, appropriate developmental potential; demonstrated efficacy	Limited resource; ethical considerations
Endothelial progenitor cells	Some proliferative potential; may elicit *in vivo* healing through indirect mechanisms	Appropriate developmental potential yet to be demonstrated; may not be appropriate for larger tissue replacement or *in vitro* tissue engineering
Adult bone marrow-derived cells	Significant *in vitro* proliferative potential; some demonstration of efficacy	Appropriate developmental potential to be demonstrated; safety tolerance after *in vitro* culture to be determined
Embryonic stem cells iPS cells	Significant *in vitro* proliferative potential; demonstration of efficacy; appropriate developmental potential; sustainable resource	*In vitro* culture may introduce genetic changes; safety tolerance after *in vitro* culture and differentiation to be determined

Reproduced with permission from Sefton et al., 2005

Hematopoietic stem (HS) cells from bone marrow were tested for their ability to contribute to the regeneration of infarcted myocardium. Although Anversa and colleagues (Orlic et al., 2001) reported that HS cells injected into the peri-infarct zone in mice with acute myocardial infarction (MI) gave rise to cardiomyocytes regenerating ~68% of the infarct, these results could not be reproduced by other groups (Balsam et al., 2004; Murry et al., 2004). Instead the studies suggest that HS cells differentiate into blood cells (Murry et al., 2004; Nygren et al., 2004), and occasionally fuse with host cardiomyocytes. The discrepancy may lie in the different techniques used.

Bone marrow mesenchymal stem cells (MSCs) have also been considered as a cell source for myocardial repair. When injected directly into the hearts of mice (Toma et al., 2002) and pigs (Shake et al., 2002) post-infarction, the cells attenuated pathological ventricle remodeling and expressed cardiac markers. Contribution of cell fusion to these events remains to be determined. Bone marrow mononuclear cells (consisting of both HS cells and MSCs) were evaluated in clinical trials (for a review see Dimmeler et al., 2005). In general, the initial clinical studies indicate that bone marrow transplantation is safe and contributed to the increase in ejection fraction (Chen et al., 2004; Wollert et al., 2004) although the mechanism of the effect is unclear. The main advantage of bone marrow as a cell source is that it can be harvested from the patient; however, the frequency of stem cells is generally low (<0.1%).

The main outcome of the first-generation clinical trials, using mostly bone marrow cells or peripheral mononuclear cells, is a small but significant improvement in functional properties (e.g. 3% improvement in the ejection fraction (Wollert and Drexler, 2005b; Lipinski et al., 2007; Martin-Rendon et al., 2008)). Further work is required to determine the optimal timing of cell transfer, dose, and delivery methods in order to achieve appreciable effects on the functional properties of the heart (Wollert and Drexler, 2005a).

Recent emerging work suggests that the heart may contain resident progenitor cells. This is an exciting possibility, as resident progenitors may be an ideal source of autologous cardiomyocytes. However, it appears that there is more than one heart cell subpopulation that fits the description of a cardiac progenitor. C-kit+ cells isolated from adult rat hearts and expanded under limited dilution gave rise to cardiomyocytes, smooth muscle cells, and endothelial cells when injected into ischemic myocardium (Beltrami et al., 2003). Oh et al. (2003) reported Sca-1 as a marker of resident cardiac progenitors, and expression of cardiac markers upon treatment with 5-azacytidine. LIM homeodomain islet 1 transcription factor (isl1+) was also identified as a marker of resident cardiac progenitor cells (Laugwitz et al., 2005). The isl1+ cells from mouse hearts were propagated in culture and they differentiated into functional cardiac myocytes when in contact with terminally differentiated cardiomyocytes. It remains to be determined whether the progenitors, regardless of their marker, can be isolated from adult human biopsies and whether sufficient numbers of cardiomyocytes ($>10^8$ cells/patient) can be generated *in vitro*.

Embryonic stem cells (ESCs) have enormous proliferative potential, and in combination with nuclear transfer can generate autologous cells. However, the main technical concern in utilization of ESCs is that the presence of a single undifferentiated cell *in vivo* can potentially yield teratomas (Laflamme and Murry, 2005). Highly pure populations of cardiomyocytes (~99.6%) can be generated using a neomyocin-resistant transgene driven by a cardiac marker promoter (Klug et al., 1996; Zandstra et al., 2003). Upon injection into hearts, the ES cell-derived cardiomyocytes formed stable intracardiac grafts (Klug et al., 1996) and improved contractile function (Etzion et al., 2001). Electromechanical integration of the cardiomyocytes derived from human ES cells with the host myocardium has also been reported (Kehat et al., 2001).

In 2006, Yamanaka reported the generation of pluripotent stem cells from mouse fibroblasts. These cells, termed induced pluripotent stem (iPS) cells, were first generated from mouse embryonic and adult fibroblasts by retroviral transduction of four transcription factors, namely Oct3/4 (also known as Pou5f1), Sox2, Klf4, and c-Myc, and then by selection for Fbx15 expression (Takahashi and Yamanaka, 2006). These cells were similar to ESCs in morphology, proliferation, and teratoma formation, but different in gene expression, DNA methylation patterns, and the ability to produce adult chimeras (Okita et al., 2007). Importantly, these cells have significant proliferative capacity and capability to give rise to bona fide cardiomyocytes (Zhang et al., 2009). iPS cells hold a great potential as they give rise to patient-specific cells while avoiding the ethical issues surrounding ESCs (Takahashi et al., 2007). It remains to be determined whether cardiac patches can be engineered using iPS cells as a source of cardiomyocytes.

Besides focusing on restoration of contractile function through injection of myogenic cells, regeneration of infarcted myocardium has also been attempted through injection of endothelial cell progenitors (Kocher et al., 2001). The regeneration is based on the improvements in infarct neovasculature that lead to improved perfusion and ultimately improved left ventricular (LV) function.

In most of the cases described above, the cells were suspended in an appropriate liquid (saline or culture medium) followed by intramyocardial or coronary injection. The main challenges associated with this procedure are poor survival of the injected cells (Muller-Ehmsen et al., 2002b) and washout from the injection site (Reffelmann and Kloner, 2003). According to some estimates, 90% of the cells delivered through a needle leak out of the injection site (Muller-Ehmsen et al., 2002a, 2002b). In addition, a significant number of cells (~90%) die within days after injection (Zhang et al., 2001; Muller-Ehmsen et al., 2002b). Thus, developing improved delivery and localization methods (e.g. hydrogels) and effective anti-death strategies could significantly improve effectiveness of cell injection procedures.

Gene therapy

Gene therapy approaches are based on either delivering exogenous genes capable of expressing therapeutic proteins, or on the blockade of genes involved in the pathological process. The genes can be delivered using non-viral vectors (such as naked plasmids, liposome formulation, and synthetic peptides) or recombinant viruses. Replication-defective recombinant viruses are significantly more effective in gene transfer to myocardium compared to the non-viral vectors, which are limited by high degradation rate and low genomic integration (Melo et al., 2004a). However, viruses sometimes lead to immune reaction, and there is a small risk that they may become proliferative.

In an early work aimed at converting the non-contractile scar tissue into tissue capable of contraction, Murry and colleagues (1996) used adenovirus to transfer MyoD, a myogenic determination gene, into granulation tissue of rat myocardium post-infarction. *In vitro*, gene transfer converted fibroblasts into skeletal muscle cells. Similar results (i.e. expression of MyoD , myogenin, and embryonic isoform of myosin heavy chain (MHC)) were observed *in vivo* after transfection with high doses of virus (10^{10} pfu). Restoration of contractile function has also been attempted by normalization of β-adregenic receptor signalling. In rabbits, intracoronary delivery of β2-adregenic receptor gene led to improvements in left ventricular and hemodynamic function (Maurice et al., 1999). Using a similar approach, β-adregenic receptor signalling was rescued in ventricular myocytes from patients with heart failure. Calcium signalling was another target for gene therapy aimed at restoration of contractile function (see review in Hajjar et al., 2000). Intracoronary delivery of SERCA2a genes in a rat model of heart failure improved long-term survival, restored systolic and diastolic function, and improved Ca^{2+} ATP-ase activity (del Monte et al., 2001). Antisense inhibition of phospholamban was shown to improve contractility of cardiomyocytes from end-stage heart failure patients (del Monte et al., 2002).

Gene therapies for acute myocardial infarction were limited by the available delivery techniques. In general, the time it takes for transcription and translation is too long for a successful intervention in acute MI (Melo et al., 2004b). However, individuals at risk may benefit from preventive strategies that protect from ischemia/reperfusion injury. In that respect, overexpression of antioxidant enzyme systems (HO-1), heat shock proteins, and survival genes (Bcl-2 Akt) was demonstrated to be beneficial in small animal models (Melo et al., 2004b).

A novel gene therapy approach was reported for treatment of acute myocardial infarction and chronic ischemia. Intramyocardial injection of naked DNA encoding human sonic hedgehog preserved LV function, enhanced neovascularization, and reduced fibrosis and cardiac apoptosis. Sonic hedgehog is a morphogen and a crucial regulator of organ development during embryogenesis; thus, transient reconstruction of embryonic signalling had a beneficial effect on tissue repair and neovascularization (Kusano et al., 2005).

Gene therapy was also utilized to treat ischemia in patients with coronary artery disease who were not eligible for standard treatment options such as percutaneous angioplasty or surgical vascularization. A number of pre-clinical and clinical trials focused on overexpression of VEGF, FGF, and hepatocyte growth factor in an attempt to improve collateral blood vessel formation (Melo et al., 2004b). Although functional improvements were reported in large animals, phase II and III clinical trials failed to conclusively prove efficacy and the long-term therapeutic effect (Yla-Herttuala et al., 2004; Markkanen et al., 2005). Although the safety record was excellent in all of the trials, the following reasons were considered as possible causes for disappointing results in efficacy: a wrong dose, a less-than-optimal route of administration, an inefficient delivery system, an insufficient duration of the treatment, selection of an appropriate animal model in pre-clinical trials, as well as selection of an appropriate patient group.

In recent years, considerable excitement was generated regarding the ability of gene therapy to treat cardiac arrhythmias via transfection of genes targeting specific ion channels (Cho and Marban, 2010; Gepstein, 2010).

While all of the above-mentioned limitations are technical in nature, targeting a single gene as most commonly used in gene therapy may have conceptual limitations as well. Most pathological process are complex and involve expression or downregulation of multiple genes. In many instances, this genetic complexity is not well understood and thus it is difficult to predict *a priori* what the ultimate effect of overexpression or blockade of a single gene will be. In this respect, combination of gene and cell therapy may be a preferred approach in the treatment of heart diseases.

One of the major limitations of cell therapy approaches is low cell survival. Thus, transfecting the injected cells with agents that enhance angiogenesis or cell survival may benefit the cell injection procedure. Once in the appropriate location, the cells may contribute to the contractile function and adjust appropriately to the complex physiological stimuli of the local milieu. Li and colleagues demonstrated that injection of $VEGF_{165}$ transfected cardiomyocytes into cryoinjured rat myocardium sustained VEGF expression and increased capillary density in the border zone as well as regional blood flow within the scar (Yau et al., 2001). Most other studies also focused on the injection of cardiomyocytes expressing growth factors (for a review see Fazel et al., 2005) consistently reported that a combination of cell and gene therapy results in improved angiogenesis and functional properties in comparison with cell therapy alone.

SCAFFOLD-BASED APPROACHES

While small infarcts may be treated with cell therapy, larger areas of damaged tissue will require excision and replacement with a cardiac patch. The time post-infarction is critical in the success of any regeneration strategy. Upon myocardial infarction, a vigorous inflammatory response is elicited and dead cells are removed by marrow-derived macrophages. Over the subsequent weeks to months, fibroblasts and endothelial cells proliferate, forming granulation tissue and ultimately dense collagenous scar. Formation of scar tissue severely reduces the contractile function of the myocardium and leads to ventricle wall thinning and dilatation, remodeling, and ultimately heart failure. The best regeneration strategy thus depends on the time post-infarction; that is, new and old infarcts most likely cannot be treated using the same approach.

Cell injection strategies will work best if applied shortly after MI. Application of cells and growth factors within hours and days after MI has the potential of directing the wound repair process so that the minimum amount of scar tissue is formed, the contractile function is maintained in the border zone, and pathological remodeling is attenuated. Tissue engineering strategies will work in the acute phase as well, but may be more necessary after scar has formed. Then larger areas of heart must be replaced or augmented and this is potentially where a scaffold-based approach may be most useful.

The scaffold approaches can be divided into: (1) hydrogel approaches where cells are either encapsulated and cultivated *in vitro* or injected directly into MI without pre-culture, and (2) porous and fibrous 3D scaffold approaches where scaffolds are seeded with cells and in most cases cultivated *in vitro* prior to utilization as cardiac patches. Natural extracellular matrix (ECM) may also serve as a scaffold. In addition, repair of the heart with biomaterials alone or constructs made by cell self-assembly has also been performed.

Cell-free cardiac patches

Patients with large transmural akynetic scars often benefit from the Dor procedure (endo-ventricular circular patch plasty) (Dor et al., 1989; di Donato et al., 1997). In some cases, however, the success of this procedure is temporary, thus motivating the need for viable tissue patches. In this procedure the scar tissue is excised and the ventricle is closed using a circular Dacron (polyethylene terephtalate) patch lined with endocardium.

Another strategy to address pathological remodeling and prevent heart failure is a CorCap cardiac support device. CorCap is an implant-grade polyethylene terepthalate mesh that is wrapped around the heart ventricle to prevent further dilatation and support contractile function. In clinical trials, it was demonstrated that it resulted in improved quality of life, as well as improved heart size and shape (Starling and Jessup, 2004). Recent studies suggest it was safe to use and effective in reducing the heart size in the setting of dilated cardiomyopathy (Bredin and Franco-Cereceda, 2010). Clinical trials are underway to evaluate its safety and effectiveness when placed at the time of the restrictive mitral annuloplasty (Rubino et al., 2009).

In vitro cultivation of cell-based cardiac patches

DECELLULARIZED NATIVE EXTRACELLULAR MATRIX (ECM)-BASED SCAFFOLDS

In a pioneering study, Taylor and colleagues utilized the ECM of the native rat heart as a scaffold for cardiac tissue engineering (Ott et al., 2008). This approach enabled them to preserve the underlying geometry and create an ideal natural template for tissue engineering of the heart (Fig. 48.1). The authors decellularized adult (12-weeks-old) cadaveric Fisher rat hearts by coronary perfusion with detergents (Fig. 48.1A). In addition to ECM, the vasculature was also preserved and it was perfusable (Fig. 48.1B). The structure of the ventricles, atria, and heart valves were all preserved. Cardiomyocytes were then isolated from the neonatal rats and reseeded onto the structure. Vascular perfusion with the oxygenated media was provided via the peristaltic pump. In a sub-group of samples, rat aortic endothelial cells were injected into the aorta, in order to recellularize the vasculature. Macroscopic contractions were observed by day 4 of cultivation, while pump function of about 2.4 mmHg was generated at day 8 under electrical stimulation. Clearly the performance was only a small fraction (~2%) of the native heart but this was nonetheless a milestone in cardiac tissue development.

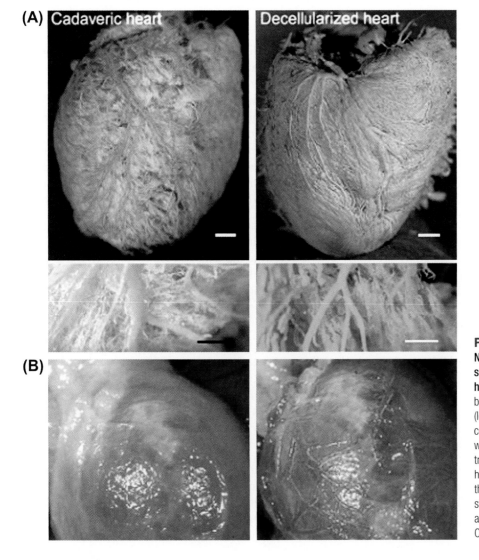

FIGURE 48.1
Native heart ECM can serve as a suitable substrate for tissue engineering of a whole heart. (A) Low-magnification (upper row; scale bar, 1,000 μm) and high-magnification views (lower row; scale bar, 250 mm) of coronary corrosion casts of cadaveric and decellularized whole adult rat hearts. (B) Upon heterotropic transplantation of the decellularized whole rat heart before (left), blood can be seen flowing through the preserved decellularized vascular structures shortly after unclamping of the host aorta (right). Reproduced with permission from Ott et al., 2008.

SELF-ASSEMBLY

In cardiac tissue engineering approaches, most studies suggest that some type of scaffold, an inductive 3D matrix, is necessary to support assembly of cardiac tissue *in vitro*. An important scaffold-free approach includes stacking of confluent monolayers of cardiomyocytes (Shimizu et al., 2002). Although cardiac patches obtained in this way generate high active force, engineering patches more than two or three cell layers thick remains a problem. Shimizu and colleagues also described the polysurgery approach, whereby vascularized cardiac grafts can be created by sequential layering of cell sheets in multiple surgeries spaced at the 1- to 3-day intervals (Shimizu et al., 2006). Although this approach demonstrates that thick tissues (~ 1 cm) can in principle be created from cell sheets, the approach will be difficult to implement in the clinical setting.

Contractile organoids, 24 mm long and 100 μm thick, were fabricated by self-organization (Baar et al., 2005). Cardiomyocytes were cultivated on a PDMS surface coated with laminin. As laminin degraded, the confluent monolayer detached from the periphery of the substrate, moving towards the center and wrapping around a string placed in the center of the plate until a cylindrical contractile organoid was formed.

Murry and colleagues recently managed to obtain a cardiac patch based on human ESC-derived cardiomyocytes by self-assembly of isolated cells in orbitally mixed dishes (Stevens et al., 2009b), essentially creating cell aggregates that could be deployed as a patch.

HYDROGELS

The most important example of hydrogel-based cardiac tissue engineering includes the work of Eschenhagen, Zimmerman, and colleagues. Cardiomyocytes were cast in growth factor-supplemented collagen gels and cultivated in the presence of cyclic mechanical stretch (Eschenhagen et al., 1997; Fink et al., 2000; Zimmermann et al., 2000a, 2002a, 2002b). The main advantage of the hydrogel approach is the higher active force generated by such cardiac tissues, compared to the force generated by tissues on porous or fibrous 3D scaffolds. In addition, collagen and laminin are the main components of the myocardial extracellular matrix; thus, they are supportive of cardiomyocyte attachment and elongation. However, the main challenge remains tailoring the shape and dimensions of such tissues. One interesting approach to address this issue is the use of extruded collagen type I tubes (Yost et al., 2004).

A technique that can potentially combine the advantages of the hydrogel approach with ease in tailoring tissue shape and size is inkjet printing. Cardiac constructs based on feline cardiomyocytes were created by printing cell solution onto alginate and using calcium as a cross-linking agent (Xu et al., 2009). This approach may be particularly useful for co-culture as it enables precise control over cell location in the tissue construct.

POROUS SCAFFOLDS

Three-dimensional cardiac tissue constructs were successfully cultivated in dishes using a variety of scaffolds amongst which collagen sponges were the most common. In the pioneering approach of Li and colleagues, fetal rat ventricular cardiac myocytes were expanded after isolation, inoculated into collagen sponges, and cultivated in static dishes for up to 4 weeks (Li et al., 1999). The cells proliferated with time in culture and expressed multiple sarcomeres. Adult human ventricular cells were used in a similar system, although they exhibited no proliferation (Li et al., 2000)

Fetal cardiac cells were also cultivated on porous alginate scaffolds in static 96-well plates. After 4 days in culture the cells formed spontaneously beating aggregates in the scaffold pores (Leor et al., 2000). Cell seeding densities of the order of 10^8 cells/cm^3 were achieved in the alginate scaffolds using centrifugal forces during seeding (Dar et al., 2002). Neonatal rat cardiomyocytes formed spontaneously contracting constructs when inoculated in collagen sponges (tissue fleece) within 36 h after seeding (Kofidis et al., 2003a) and maintained their activity for up to 12 weeks. The contractile force increased upon addition of Ca^{2+} and epinephrine.

FIBROUS SCAFFOLDS

In a classical tissue engineering approach, fibrous polyglycolic acid (PGA) (Fig. 48.2A) scaffolds were combined with neonatal rat cardiomyocytes and cultivated in spinner flasks and rotating vessels (Carrier et al., 1999). The scaffold was 97% porous and consisted of non-woven PGA fibers 14 μm in diameter. This material has advantages from a clinical standpoint since it is found in biodegradable sutures.

Neonatal rat or embryonic chick ventricular myocytes were seeded onto PGA scaffolds by placing a dilute cell suspension in the spinner flasks and mixing for three days (50 rpm) (Carrier et al., 1999). Mixing in the spinner flasks (0, 50, or 90 rpm) had a significant effect on the construct metabolism and cellularity. Constructs cultivated in well-mixed flasks had significantly higher cellularity index and metabolic activity compared to the constructs cultivated in the static flasks. After 1 week of culture, constructs seeded with neonatal heart cells contained a peripheral tissue-like region (50–70 μm thick) in which cells stained positive for tropomyosin and organized in multiple layers in a 3D configuration (Bursac et al., 1999) (Fig. 48.2A,B). Electrophysiological studies conducted using a linear array of extracellular electrodes showed that the peripheral layer of the constructs exhibited relatively homogeneous electrical properties and sustained macroscopically continuous impulse propagation on a centimeter-size scale (Bursac et al., 1999). Constructs based on the cardiomyocytes enriched by preplating exhibited lower excitation threshold (ET), higher conduction velocity, higher maximum capture rate (MCR), and higher maximum and average amplitude of contraction.

Laminar flow conditions in rotating bioreactors further improved the PGA-based constructs. The cells in the peripheral layer expressed tropomyosin and had spatial distribution of connexin-43 comparable to the neonatal rat ventricle. The expression levels of cardiac proteins connexin-43, creatin kinase-MM, and sarcomeric myosin heavy chain were lower in rotating bioreactor-cultivated constructs compared to the neonatal rat ventricle but higher than in the spinner flask-cultivated constructs (Papadaki et al., 2001). It is important to note that in both spinner flasks and rotating bioreactors the center of the constructs was mostly acellular due to the oxygen diffusional limitations.

Electrospun scaffolds (Fig. 48.2C) have gained significant attention as they enable control over structure at sub-micron levels as well as control over mechanical properties, both of which are important for cell attachment and contractile function. Entcheva and colleagues (Zong et al., 2005) used electrospinning to fabricate oriented biodegradable non-woven poly(lactide) (PLA) scaffolds. Neonatal rat cardiomyocytes cultivated on oriented PLA matrices had remarkably well-developed contractile apparatus (Fig. 48.2D) and exhibited electrical activity.

COMBINATION APPROACHES

To combine the benefits of the presence of naturally occurring extracellular matrix (laminin) and the stability of porous scaffolds, neonatal rat cardiomyocytes were inoculated into collagen sponges or synthetic poly(glycerol sebacate) scaffolds (PGS) using Matrigel (Radisic et al., 2006). The main advantage of a collagen sponge is that it supports cell attachment and differentiation. However, the scaffold tends to swell when placed in culture medium; thus, creation of a parallel channel array resembling a capillary network is difficult. For that purpose a biodegradable elastomer (Wang et al., 2002) with high degree of flexibility was used (Fig. 48.2I,J).

Freed and colleagues have reported mechanical stimulation of hybrid cardiac grafts based on knitted hyularonic acid-based fabric and fibrin (Boublik et al., 2005) (Fig. 48.2G,H). The grafts

886

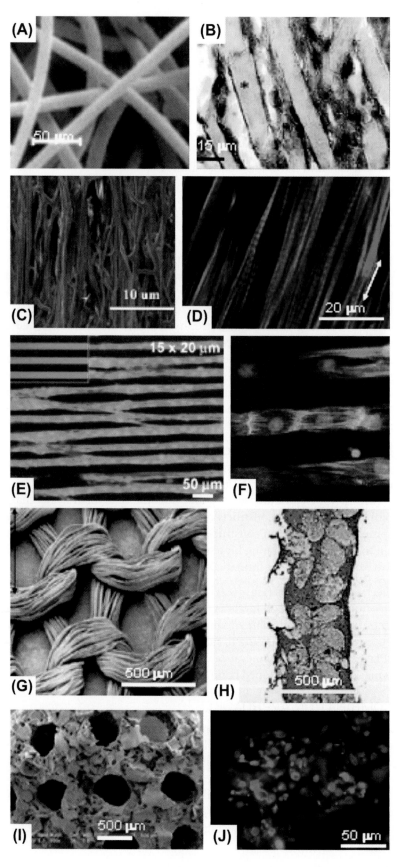

FIGURE 48.2

Representative scaffolds used in cardiac tissue engineering. (A) Scanning electron micrograph of a non-woven fibrous PGA scaffold used in a classical approach by Freed and colleagues. (B) Immunohistochemical staining for tropomyosin in constructs based on surface-hydrolyzed PGA seeded with neonatal rat cardiomyocytes and cultivated in rotating vessels for 1 week (with permission from Papadaki et al., 2001). (C) Scanning electron micrograph of a fibrous PLA scaffold obtained by electrospinning followed by uniaxial stretching. (D) Neonatal rat cardiomyocytes cultured on oriented PLA scaffolds exhibited well-developed contractile apparatus (actin-green) (with permission from Zong et al., 2005). (E) Thin PLGA films patterned with laminin using microcontact printing (inset; 15 μm laminin lanes spaced 20 μm apart) and seeded with neonatal rat cardiomyocytes (actin filaments — red; nuclei — blue). (F) Immunohistochemical staining illustrates elements of intercalated disks (N-cadherin — yellow; actin filaments — red) (with permission from McDevitt et al., 2002). (G) Scanning electron micrograph of the knitted Hylonect fabric; arrow indicates the direction of cyclic stretch applied during culture. (H) Cross-section of a construct sampled 2 h after cell seeding, showing the multifilament yarn (arrow) and immunohistochemical staining for cardiac troponin. (I) Neonatal rat cardiomyocytes were inoculated into the scaffold using fibrin (with permission from Boublik et al., 2005). (I) Parallel channel array bored in the PGS scaffolds using CO_2 laser/scanning engraving system. (J) Neonatal rat heart cells seeded onto channeled PGS scaffolds using Matrigel™ and cultivated in perfusion with 5.4vol% perfluorocarbon emulsion-supplemented culture medium (vimentin-stained fibroblasts — red; troponin I-stained cardiomyocytes — green; nuclei — blue) (with permission from Radisic et al., 2006).

exhibited mechanical properties comparable to those of native neonatal rat hearts. In a subcutaneous rat implantation model the constructs exhibited the presence of cardiomyocytes and blood vessel ingrowth after 3 weeks.

THIN FILMS AND MICROFABRICATION APPROACHES

A significant step towards a clinically useful cardiac patch was the cultivation of ES cell-derived cardiomyocytes on thin polyurethane films. Cells exhibited cardiac markers (actinin) and were capable of synchronous macroscopic contractions (Alperin et al., 2005). The orientation and cell phenotype could further be improved by microcontact printing of extracellular matrix components (e.g. laminin) as demonstrated for neonatal rat cardiomyocytes cultivated on thin polyurethane and PLA films (McDevitt et al., 2002, 2003) (Fig. 48.2E,F).

We have used microfluidic patterning of hyaluronic acid on glass substrates to create thin (10–15 μm diameter) several-millimeter-long cardiac organoids that exhibited spontaneous contractions and stained positive for troponin I, a cardiac marker (Khademhosseini et al., 2006).

In a recent study, Domian et al. identified distinct transcriptional signatures, including the expression of unique subsets of miRNAs, specific for the first and second heart fields in mouse embryos as well as embryonic stem cells. The mammalian heart is composed of a diversified set of muscle and non-muscle cells that are differentiated from the progenitor cells in either first or second heart fields, or a combination of the two. Using the fluorescence profiles of progenitor cells governed by the expression of Isl1-dependent enhancer of the Mef2c gene or cardiac-specific Nkx2.5 enhancer, the authors were able to isolate cells that had the maximum potential for differentiation into cardiomyocytes. These cells were able to align on micro-patterned surfaces and were subsequently used to engineer beating muscular thin films *in vitro* that could be paced by field stimulation at 0.5 and 1 Hz (Domian et al., 2009).

In another study, Feinberg et al. seeded a layer of neonatal rat ventricular cardiomyocytes on a polydimethylsiloxane membrane that could be detached from a thermo-sensitive poly (isopropylacrylamide) layer at room temperature. Called "muscular thin films," these cell-covered sheets could be designed to perform tasks such as gripping, pumping, walking, and swimming by careful tailoring of the tissue architecture, thin-film shape, and electrical-pacing protocol (Feinberg et al., 2007) (Fig. 48.3A).

Badie et. al. investigated yet another method to replicate the microstructure of heart tissue *in vitro*. The two-step method first involves imaging the heart using diffusion tensor magnetic resonance imaging (DTMRI). From the 3D reconstructed image, a specific 2D plane is chosen and the cardiac fiber directions on this plane are converted into soft-lithography photomasks, and later into fibronectin-coated polydimethylsiloxane sheets. Fibronectin patterns consisted of a matrix of $190 \, \mu m^2$ subregions, each composed of parallel lines 11–20 μm wide, spaced 2–8.5 μm apart and angled to match local DTMRI-measured fiber directions. By adjusting fibronectin line widths and spacing, cell elongation, gap junctional membrane distribution, and local cellular disarray were altered without affecting the cell direction. This approach enabled the systematic studies of intramural structure-function relationships in both healthy and structurally remodeled hearts (Badie and Bursac, 2009; Badie et al., 2009).

Scaffold structure can be used to effectively guide orientation of cardiomyocytes and yield anisotropic structure similar to the native myocardium even in the absence of specific physical cues such as electrical or mechanical stimulation. Freed and colleagues created an accordion-like scaffold (Fig. 48.3B) using laser boring of 250 μm-thick poly(glycerol sebacate) layer (Engelmayr et al., 2008). The accordion-like honeycomb was designed by overlapping two 200 by 200 μm squares at the angle of $45°$. The pore walls and struts were ~50 μm thick. The scaffolds were pre-treated with cardiac fibroblasts followed by seeding of enriched cardiomyocytes. During pre-treatment, rotating culture was used, while static culture was used upon cardiomyocyte seeding. At the end of cultivation, the authors obtained contractile cardiac

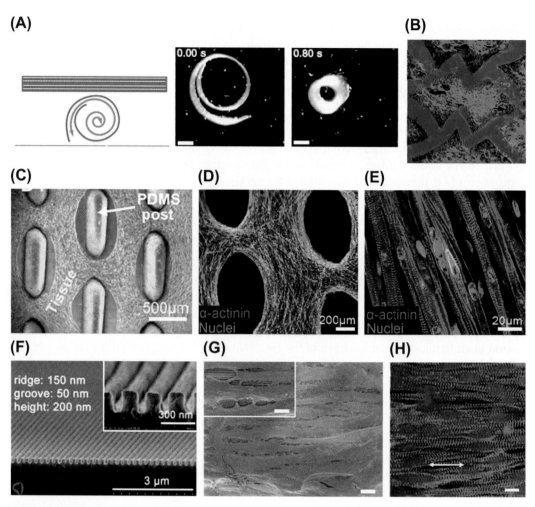

FIGURE 48.3

Scaffold microfabrication enables tissue engineering of anisotropic constructs. (A) Muscular thin films. The coiled strip made of thin layer of PDMS had anisotropic myocardium (on the concave surface) aligned along the rectangle length created by micropatterning of fibronectin lanes. As the cardiomyocytes contract, the coil moves from an uncoiled to coiled state. Scale bar = 1 mm (with permission from Feinberg et al., 2007). (B) An accordion-like scaffold was created by laser-boring the honeycomb macropores in PGS scaffolds. The macropores enabled cell elongation along the long axis of the pore and yielded non-isotropic mechanical properties (with permission from Engelmayr et al., 2008). (C—E) High-aspect-ratio PDMS posts guide gel compaction and assembly of anisotropic muscle tissue based on C2C12 myoblasts and a mixture of fibrinogen and collagen type I (with permission from Bian and Bursac, 2009; Bian et al., 2009). (F,G) Sub-micrometer features on micromolded poly (ethylene glycol) hydrogels guide cardiomyocyte elongation and orientation along the grooves. Scale bar in G = 5 μm (with permission from Kim et al., 2010). (H) The cell monolayer exhibits a remarkably well-developed contractile apparatus. Staining for sarcomeric α-actinin. Scale bar = 10 μm (with permission from Kim et al., 2010).

grafts with mechanical properties closely resembling those of the native rat right ventricle. In addition, the cells in the pores were aligned along the preferred direction.

Bursac and colleagues developed a cell/fibrin hydrogel micromolding approach where poly-dimethylsiloxane (PDMS) molds containing an array of elongated posts were used to fabricate relatively large neonatal rat skeletal muscle tissue networks (Fig. 48.3C—E). As the cells compacted the hydrogel, the presence of high-aspect-ratio posts forced them to elongate and align, thus imparting a high degree of anisotropy to the cells and the tissue. This approach is currently being extended to cultivation of cardiac patches based on mouse ESC-derived progenitor cells (Bian and Bursac, 2009; Bian et al., 2009).

Interestingly, a high degree of anisotropy, correlating with the high propagation velocities in the longitudinal direction (~35 cm/s), was achieved by cultivation of neonatal rat cardiomyocytes on micromolded poly(ethylene glycol) hydrogels with sub-micrometer features, specifically alternating 800 nm by 800 nm groves and ridges (Fig. 48.3F—H). The sub-micrometer features forced the cells to align focal adhesions along the grove/ridge direction and the cytoskeleton followed (Kim et al., 2010).

TISSUE AND ORGAN FUNCTION

Successful implantation of engineered tissues requires both maintenance of cellular pheno-type and the functional integration of the construct within the host tissue. As progress is made from the state of the art described above to the final goal, it will be necessary to ensure not only that engineered cardiac cells and tissues contract in unison with the surrounding native myocardium to produce the desired force but also that the graft is electrically integrated with the host, to prevent arrhythmogenesis.

Underlying such integration and the implicit control of the construct phenotype is the creation of the arborized networks (vessels, lymphatics, and nerves) needed to sustain large and complex tissue structures. Then there are the issues associated with blood compatibility, tissue remodeling, and more generally the immune and inflammatory responses to the new tissue or cells. Using autologous cells is an approach that is immunologically preferable, but it likely precludes the "off-the-shelf" concept behind much of the attraction of tissue engineering.

Mechanical elasticity and strength development

A critical feature of a heart is its mechanical characteristics. Simply speaking, the heart must pump blood at a mean pressure of roughly 100 mmHg. Hence, heart muscle must stretch in response to capillary filling pressure and eject a volume of blood that varies with demand. The latter requires a uniform and well-coordinated contraction that generates the required power. The mechanical fatigue limitations of a heart that must beat 3×10^8 times over 10 years must be compared with the flexural fatigue life of synthetic elastomeric materials, which is typically much lower.

It will be a significant challenge to replicate the complex architecture of the myocardium and its non-linear viscoelastic properties in both resting and activated states (Fung, 1993). While some constructs exhibit a significant burst strength and some groups are very advanced in the use of the tools of biomechanics to advance vascular graft (Nerem, 2003) or heart valve development, this area has received less attention than it deserves (Butler et al., 2000).

Tissue architecture and electrical conduction

The complexity of the electrical conduction pathways in the heart is just starting to gain attention in the tissue engineering literature. The cells need to form the appropriate intercel-lular connections and matrix arrangements to enable the directed beating of contracting cells to generate the forces required to pump blood (Akins, 2000). The proper formation of the intercalated disks between myocytes is also critical in enabling electrical pulses to be trans-mitted in the correct direction at normal speeds and in allowing suitable force transmission. The heart also contains specialized cells that participate in the electrical conduction routes found throughout the heart. These specialized cells are crucial to the coordination of the heart's contractile effort, and including them in the proper places in a regenerated substitute may be critical. There are clear differences between the rhythmic twitching of cultured cardiac cells en masse and the organized, efficient, regulated beating of the heart; only the latter will generate the force required to pump blood at systolic pressure levels. It is not difficult to envision the problems yet to be faced. Given the variety of electrical conduction-related diseases in a normal myocardium, there is good reason to suspect that simple mimicry of heart muscle may fall short of the goal.

Thrombogenicity and endothelialization

The need for blood compatibility is another crucial characteristic of cardiovascular constructs (McGuigan and Sefton, 2007). All biomaterials lack the desired non-thrombogenicity and most extracellular matrices initiate thrombosis; hence, endothelialization of the construct is another critical issue. Endothelial cells (ECs) have a reversible plasticity (Augustin-Voss et al., 1991; Lipton et al., 1991; Risau, 1995) and they can become activated (proliferative or adhesive to leukocytes) upon exposure to inflammatory cytokines (e.g. IL1, TNF) or to growth factors such as VEGF. Flow and the associated shear stress, normally in the range of 5–20 dyn/cm^2, elongate and align cells in the direction of flow (Eskin et al., 1984; Ives et al., 1986) and modify gene expression (McCormick et al., 2001) as well as many other functions including markers of anti-thrombogenicity.

ECs provide a hemocompatible surface by production of molecules that modulate platelet aggregation (e.g. prostacyclin), coagulation (thrombomodulin (Marcum et al., 1984; Esmon, 2000)) and fibrinolysis (Shen, 1998) (e.g. tissue plasminogen activator). They can be transformed into a pro-thrombotic surface, for example by the action of thrombin or through exposure to some biomaterials (Li et al., 1992; Cenni et al., 1993; 2000; Lu and Sipehia, 2001). Blood compatibility has been a key issue in the development of vascular grafts. Recent clinical success (Meinhart et al., 2001) has renewed enthusiasm for seeding grafts with endothelial cells. In some protocols, many of the pre-seeded cells are lost on implantation due to insufficient adhesion (Williams, 1995) and thus the protection from thrombosis provided by the cells is limited due to the incomplete cell coverage. The potential to exploit the presence of circulating endothelial cell progenitors has only begun to be explored (Rafii, 2000).

It is also worth noting the effects of the endothelium on the neighbouring tissue and the corresponding effects on EC phenotype. With vascular smooth endothelial cells (VSMCs), this bidirectional cross-talk is thought to be a critical regulator of vascular homeostasis (Korff et al., 2001): secretion and expression of molecules such as nitric oxide (Palmer et al., 1988), prostacyclin (Moncada, 1982), and endothelin (Mawji and Marsden, 2003) act on VSMC to regulate vessel tone. Meanwhile, VSMC inhibits EC endothelin 1 (ET-1) production to increase EC NO and eNOS expression (di Luozzo et al., 2000). Many other relevant systems (e.g. MMP secretion and matrix remodeling) are also affected by the interactions between ECs and other cell types. It is also worth noting that VEGF inhibits pericyte coverage under conditions of PDGF-mediated angiogenenis (Greenberg et al., 2008), complicating vascularization strategies built around VEGF delivery (see below).

Vascularization

The intrinsic nature of large cell-based constructs and the corresponding difficulty of supplying cells deep within the construct with nutrients is yet another problem. Diffusion is fine for 100 μm or so and low cell densities can extend this limit, but at the cost of making constructs too large to be useful. Thin or essentially 2D (e.g. a tube) constructs are feasible without an internal blood/nutrient supply. However, it is hard to combine cells at tissue densities $>10^8$ cells/cm^3 into large tissues without some sort of prevascularization or its alternative. Thus, a capillary network (and a lymphatic network) needs to be "engineered" as part of the creation of a larger structure.

In a cell-free approach, vascularization and improvement of LV function following MI were achieved by sustained release of bFGF incorporated into gelatin microspheres (Sakakibara et al., 2003), aFGF from ethylene vinyl acetate copolymer (Sellke and Simons, 1999), and bFGF from heparin-alginate beads (Harada et al., 1994). Mooney and colleagues have incorporated an endothelial cell mitogen (VEGF) into 3D porous poly(lactide-co-glycolide, PLG) scaffolds during fabrication (Sheridan et al., 2000) to promote scaffold vascularization. Sustained delivery of bioactive VEGF translated into a significant increase in blood vessel ingrowth in mice and the vessels appeared to integrate with the host vasculature. We are using microencapsulated $VEGF_{165}$ secreting cells (prepared by transfection of L929 cells) as a means of exploring this strategy, at least for microcapsules (Vallbacka et al., 2001). Of course, VEGF is but one angiogenic factor (Ahrendt et al., 1998) and issues associated with the functional maturity of the vessels and the need for multiple factors (e.g. Cheng and Sefton, 2009) may limit this strategy. In a third approach, Vacanti et al. micromachined a hierarchical branched network mimicking the vascular system in 2D. Silicon and Pyrex surfaces were etched with branching channels ranging from 500 μm to 10 μm in diameter (Kaihara et al., 2000) that were then seeded with rat hepatocytes and microvascular endothelial cells.

Recently, we covalently immobilized VEGF165 and angiopoietin 1 to porous collagen scaffolds in order to enhance scaffold vascularization *in vitro* and *in vivo* (Chiu and Radisic, 2010). Such covalent immobilization offers the advantages of prolonged signaling and lower total amount of growth factors required and offers the possibility of generating capillary-like structures in the tissue-engineered scaffolds *in vitro* (Fig. 48.4).

A prevascularized skeletal muscle was created (Levenberg et al., 2005) by co-culturing skeletal muscle cells with embryonic stem cell-derived endothelial cells and fibroblasts. It appeared that up to 40% of the engineered blood vessels "connected" to the host vasculature upon implantation, at least in this small animal model. Finally we note that we adapted endothelial seeding in a modular approach to create scalable and vascularized tissue constructs (Fig. 48.5B) (McGuigan and Sefton, 2006). Endothelial cells were seeded onto sub-micrometer-sized collagen gel cylindrical modules that contained a second cell (e.g. HepG2, smooth muscle cells, or, most relevant here, cardiomyocytes) (Leung and Sefton, 2010). With a view to creating uniform, scalable, and vascularized constructs, these modules were packed into a larger tube, formed into a sheet, or implanted directly, with interconnected channels lined with endothelial cells resulting from the random assembly of the modules. These channels connected with the host vasculature *in vivo* (Chamberlain et al., 2010), creating a perfuseable chimeric vasculature containing both host and donor cells and with host smooth muscle cell involvement. Embedded cardiomyocytes formed "contractile" structures near the periphery of modules although the density of such structures was relatively low (Leung and Sefton, 2010). Remodeling occurred *in vivo* (after peri-infarct injection or use as a patch) resulting in a well-distributed microvasculature (after 2 or 3 weeks in syngeneic animals) but the distribution of cardiac structures was again relatively low.

Host response and biocompatibility

Questions related to the immune and inflammatory response to tissue constructs are starting to draw attention. The host response to a tissue-engineered construct is manifested by the innate and adaptive immune systems, involving both plasma (e.g. complement) and cellular components (e.g. macrophages, T-cells, etc.) that are directed against engineered cells and grafts or the materials used in tissue constructs. This potent immune response is most often mediated by MHC mismatches between donor and host tissue in allogeneic transplantations. This response can also be manifested in situations where autologous cells or tissues are engineered to express therapeutic but foreign factors or if these autologous cells are placed in tissue constructs that themselves negatively impact immune consequences (Mikos et al., 1998).

Immunosuppressants have enabled the successful transplantation of kidneys, hearts, and other organs. With the advent of tissue engineering, new configurations of tissues and organs (often with an added biomaterial component) are being developed and our understanding of the immune and inflammatory response to these new therapies is being shown to be inadequate. Some xenogeneic cell transplants (mice to rat) survive in situations of cardiac repair despite the species differences (Saito et al., 2002) although this may be specific to the animal model or to cardiac repair. The longevity of a transplant is also dependent on the ability of somatic cells to

892

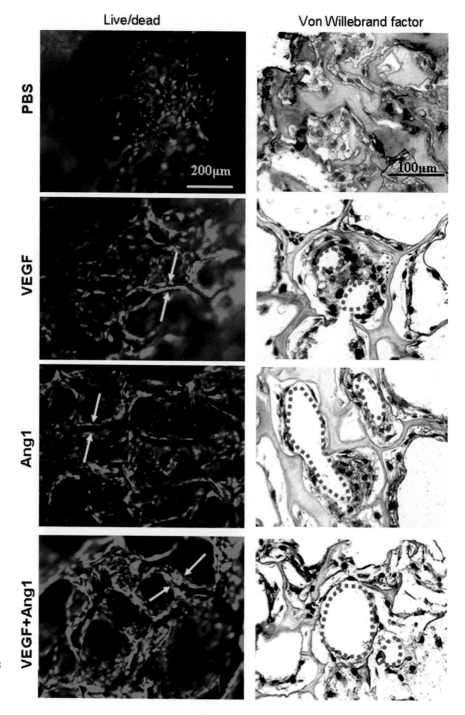

FIGURE 48.4
Co-immobilization of VEGF₁₆₅ and
Angiopoientin-1 enables creation of
capillary-like structure in scaffolds for
cardiac tissue engineering. Co-immobilized
growth factors were superior in comparison to
single growth factors. H5V endothelial cells
cultivated for 7 days on porous collagen
scaffolds. Live cells stain green and dead cells
stain red. Arrows indicate tube-like structures.
Reproduced with permission from Chiu and
Radisic, 2010.

withstand and respond to the stresses of implantation, rejection, and other injuries (Halloran and Melk, 2001). The classic "foreign body reaction" to biomaterials is well known, but the details of the molecular signals (complement regulatory proteins, MMPs) that accompany this phenomenon (in the context of biomaterials) are only beginning to be defined.

A variety of approaches have been undertaken or are in development to generate or to improve upon graft acceptance (Rossini et al., 1999). These approaches include methods to block the innate immune response such as by use of drugs or transferred genes to block NFκB signaling pathways, for example. Other methods to block the innate response include the use of

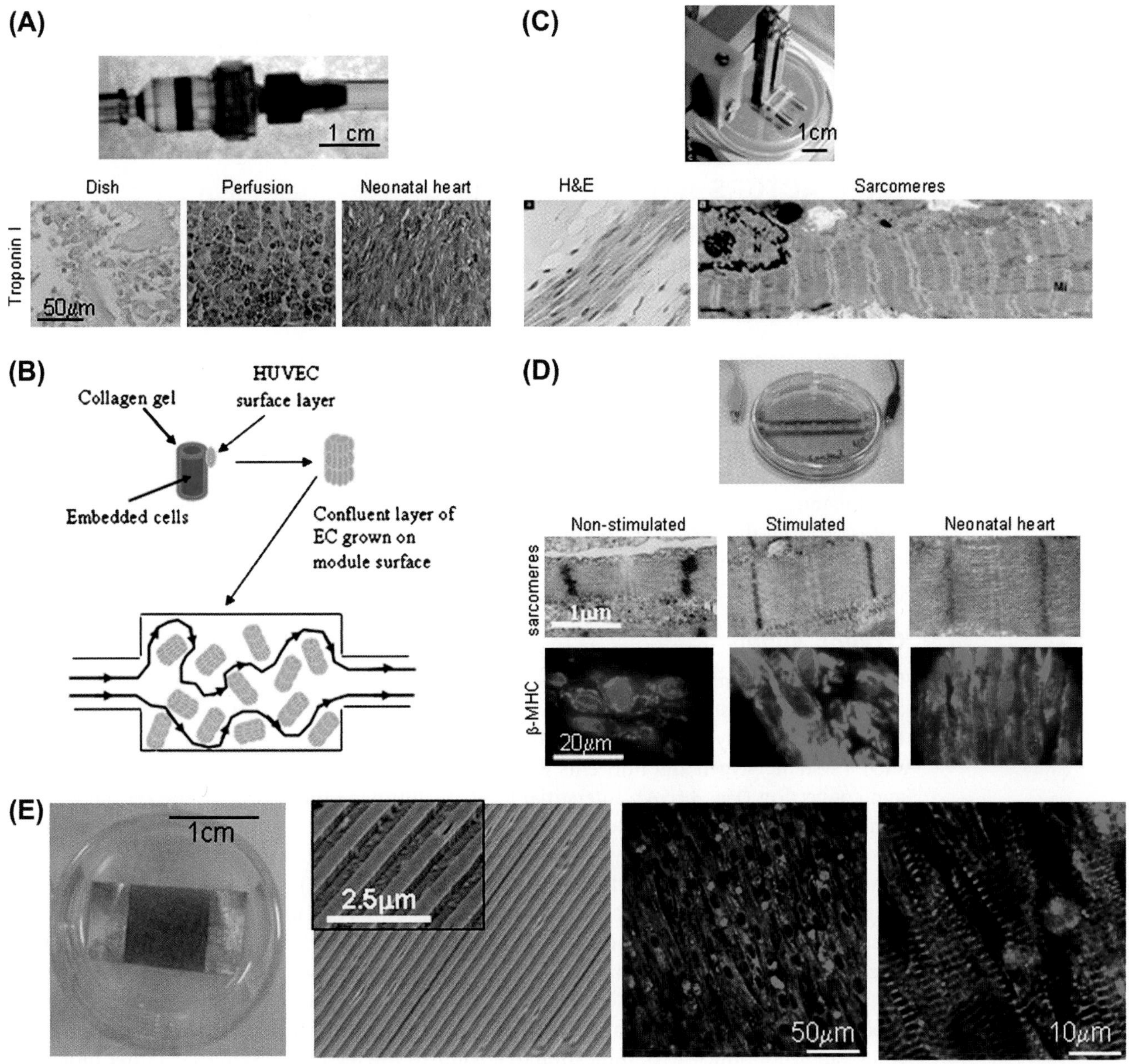

FIGURE 48.5

Cardiac tissue engineering culture systems focus on achieving adequate oxygen supply for highly metabolically active cells (A,B) and providing appropriate physical cues that lead to differentiated phenotype (C,D). (A) Direct culture medium perfusion of constructs based on neonatal rat cardiomyocytes inoculated into collagen sponges using Matrigel. Medium perfusion resulted in uniform cell distribution and maintenance of cell viability. Immunohistochemical staining illustrated cross-sectional distribution of cells expressing cardiac Troponin I (with permission from Radisic et al., 2004b). (B) Modular tissue engineering approach using sub-millimeter-sized endothelial cell seeded collagen modules assembled into a larger tube or construct (with permission from McGuigan and Sefton, 2006). (C) Zimmermann and Eschenhagen designed a bioreactor that provides cyclic mechanical stretch to engineered heart tissue based on neonatal rat cardiomyocytes and collagen gel. Mechanical stimulation yielded elongated cardiomyocytes with remarkably well-developed contractile apparatus (with permission from Zimmermann et al., 2002a). (D) Cardiac-like electrical field stimulation was applied to collagen sponges inoculated with suspension of neonatal rat cardiomyocytes in Matrigel™, resulting in differentiated phenotype and improved tissue assembly (with permission from Radisic et al., 2004a). (E) Alternating grooves and ridges were introduced by hot-embossing of polystyrene and electrical field stimulation cues were incorporated by electrodeposition of gold electrodes. Upon cultivation of neonatal rat cardiomyocyte they elongate along the topographical cues and exhibit a remarkably well-developed contractile apparatus. Immunostaining for sarcomeric α-actinin (with permission from Heidi Au et al., 2009).

antibodies to IL-1 or TNF or the use of anti-adhesion and anti-elastase antibodies. We must better understand the mechanism of the host response itself so that we can design better biomaterials, select or engineer more suitable cells or devise better strategies for controlling both innate and adaptive immune responses, and enable a functional integration of the new tissue with the host.

BIOREACTORS AND CONDITIONING

Major efforts in the development of bioreactors for tissue engineering of myocardium focus on: (1) providing sufficient oxygen supply for the highly metabolically active cardiomyocytes, and (2) providing appropriate physical stimuli necessary to reproduce complex structure at various length scales (subcellular to tissue).

The most common culture vessels utilized for tissue engineering of the myocardium include static or mixed dishes, static or mixed flasks, and rotating vessels. These bioreactors offer three distinct flow conditions (static, turbulent, and laminar) and therefore differ significantly in the rate of oxygen supply to the surface of the tissue construct. Oxygen transport is a key factor for myocardial tissue engineering due to the high cell density, very limited cell proliferation, and low tolerance of cardiac myocytes for hypoxia. In all configurations oxygen is supplied only by diffusion from the surface to the interior of the tissue construct, yielding a ~100 μm-thick surface layer of compact tissue capable of electrical signal propagation and an acellular interior (Radisic et al., 2005).

Oxygen supply

In an attempt to enhance mass transport within cultured constructs, a perfusion bioreactor that provides interstitial medium flow through the cultured construct at velocities similar to those found in native myocardium (~400−500 μm/s) was developed (Radisic et al., 2004b). In such a system, oxygen and nutrients were supplied to the construct interior by both diffusion and convection (Fig. 48.5A). Interstitial flow of culture medium through the central 5 mm-diameter by 1.5 mm-thick region resulted in physiological density of viable and differentiated, aerobically metabolizing cells. In response to electrical stimulation, perfused constructs contracted synchronously, had lower excitation threshold (ET), and recovered their baseline function levels of ET and MCR following treatment with a gap junctional blocker; dish-grown constructs exhibited arrhythmic contractile patterns and failed to recover their baseline MCR levels. These studies suggested that the immediate establishment and maintenance of interstitial medium flow markedly enhanced the control of oxygen supply to the cells and thereby enabled engineering of compact constructs. However, most cells in perfused constructs were round and mononucleated, indicating that some of the regulatory signals — either molecular or physical — were not present in the culture environment.

In another approach, a separate compartment for medium flow was created by perfusing channelled scaffolds in a configuration resembling the capillary network *in vivo*. Neonatal rat heart cells were inoculated into the pores of an elastic, highly porous scaffold (PGS) with a parallel channel array and perfused with a synthetic oxygen carrier (Oxygent™ in culture medium, perfluorocarbon (PFC) emulsion) (Radisic et al., 2006). Constructs cultivated with PFC emulsion had significantly higher DNA content, significantly lower excitation threshold, and higher relative presence of cardiac markers troponin I and connexin-43 (Western blot) compared to the culture medium alone. Cells were present throughout the construct volume. In this configuration, the presence of PFC emulsion further enhanced the oxygen supply to the cells by improving both axial (convective term) and radial (effective diffusivity) transport properties (Radisic et al., 2005).

Kofidis et al. supplied pulsatile flow to cardiomyocytes encapsulated in fibrin glue around a rat artery *in vitro* (Kofidis et al., 2003b). Dvir et al. designed a novel perfusion bioreactor that employs a distributing mesh upstream from the construct to provide homogeneous fluid flow and maximum exposure to perfusing medium (Dvir et al., 2006). This convective supply of oxygen led to increased cell viability in alginate scaffolds seeded with physiologically relevant cells (Dvir et al., 2006). In addition, pulsatile culture medium flow resulted in physiological cardiac hypertrophy via stimulation of the Erk pathway (Dvir et al., 2006, 2007). We combined mechanical stimulation and perfusion in a single system by utilizing a normally closed pinch valve at the outlet from the perfusion chamber. The valve was set to open at the frequency of 1 Hz. The build-up of the culture medium during the closed period resulted in tissue compression followed by relaxation at valve opening (Brown et al., 2008).

Differentiation

MECHANICAL STIMULATION

One significant approach to cardiac tissue engineering, established by Eschenhagen, Zimmerman, and colleagues (Eschenhagen et al., 1997; Fink et al., 2000; Zimmermann et al., 2000a, 2002b) involves the cultivation of neonatal rat heart cells in collagen gel or Matrigel, in the presence of growth factors. The cultured tissues are subjected to sustained mechanical strain. Under these conditions, cardiomyocytes and non-myocytes form 3D cardiac organoids, consisting of a well-organized and highly differentiated cardiac muscle syncytium, that exhibit contractile and electrophysiological properties of working myocardium. The first implantation experiments in healthy rats showed survival, strong vascularization, and signs of terminal differentiation of cardiac tissue grafts (Zimmermann et al., 2002b).

In the state-of-the-art approach by Eschenhagen and colleagues, neonatal rat cardiac cells were suspended in the collagen/Matrigel mix and cast into circular molds (Zimmermann et al., 2002c). After 7 days of static culture, the strips of cardiac tissue were placed around two rods of a custom-made mechanical stretcher and subjected to either unidirectional or cyclic stretch (Fig. 48.5C). Histology and immunohistochemistry revealed the formation of intensively interconnected, longitudinally oriented cardiac muscle bundles with morphological features resembling adult rather than immature native tissue. Primitive capillary structures were also detected. Cardiomyocytes exhibited well-developed ultrastructural features: sarcomeres arranged in myofibrils, with well-developed Z, I, A, H, and M bands, specialized cell-cell junctions, T tubules, as well as well-developed basement membrane. Contractile properties were similar to those measured for native tissue, with a high ratio of twitch to resting tension and strong β-adrenegenic response. Action potentials characteristic of rat ventricular myocytes were recorded.

ELECTRICAL STIMULATION

In native heart, mechanical stretch is induced by electrical signals. Contraction of the cardiac muscle is driven by the waves of electrical excitation (generated by pacing cells) that spread rapidly along the membranes of adjoining cardiac myocytes and trigger release of calcium, which in turn stimulates contraction of the myofibrils. Electro-mechanical coupling of the myocytes is crucial for their synchronous response to electrical pacing signals, resulting in contractile function and pumping of blood (Severs, 2000).

Cardiac constructs prepared by seeding collagen sponges with neonatal rat ventricular cells were electrically stimulated using suprathreshold square biphasic pulses (2 ms duration, 1 Hz, 5 V) (Radisic et al., 2004a). The stimulation was initiated after 1–5 days of scaffold seeding (3-day period was optimal) and applied for up to 8 days. Over only 8 days *in vitro*, electrical field stimulation induced cell alignment and coupling, increased the amplitude of synchronous construct contractions by a factor of seven, and resulted in a remarkable level of ultrastructural organization. Development of conductive and contractile properties of cardiac constructs was concurrent, with strong dependence on the initiation and duration of electrical stimulation. Aligned myofibers expressing cardiac markers were present in stimulated samples and neonatal heart (Fig. 48.5D). Stimulated samples had sarcomeres with clearly visible M, and Z lines, and H, I, and A bands. In most cells, Z lines were aligned, and the intercalated discs were positioned between two Z lines. Mitochondria (between myofibrils) and abundant glycogen were detected. In contrast, non-stimulated constructs had poorly developed cardiac-specific organelles and poor organization of ultrastructural features.

Subsequently, we applied biphasic electrical stimulation in order to further mimic conditions in the heart and demonstrated that electrical stimulation enhances the assembly of cardiac organoids based on multiple cell types, including fibroblasts, endothelial cells, and cardiomyocytes (Chiu et al., 2008). We have also developed cardiac microchips that combine electrical field stimulation and topographical cues. Specifically, mircrometer-sized groves and ridges were created by hot-embossing of polystyrene and placed between gold electrodes on a single chip (Fig. 48.5E). Simultaneous application of topographical cues and electrical field stimulation resulted in a remarkable level of cardiomyocyte alignment, elongation, and assembly of contractile apparatus (Heidi Au et al., 2009).

Hence, the *in vitro* application of a single but key *in vivo* factor progressively enhanced the functional tissue assembly and improved the properties of engineered myocardium at the cellular, ultrastructural, and tissue levels.

IN VIVO STUDIES

In situ cardiac tissue engineering via injection of cells in hydrogels

Over the past 5 years, hydrogels have gained much attention as vehicles for delivery of reparative cells into the myocardium, due to their injectability and ability to control cross-linking chemistry. General requirements for a hydrogel to be used in myocardial regeneration are: (1) biocompatible, (2) biodegradable, (3) injectable, so that it can be applied with a syringe in a minimally invasive manner, and (4) mechanically stable enough to withstand the beating environment of the heart. In addition, a biomaterial that can promote the attachment and survival of cells, and localize them at the infarction site, would address these current limitations of poor cell retention and survival.

Early studies relied on cell injection using natural hydrogels such as Matrigel (Balsam et al., 2004; Kofidis et al., 2005) or fibrin (Christman et al., 2004a,b; Ryu et al., 2005), reporting structural stabilization, reduced infarct size, and improved vascularization upon injection of undifferentiated ESCs (Balsam et al., 2004; Kofidis et al., 2005) or bone marrow cells (Christman et al., 2004a,b; Ryu et al., 2005). Alginate alone was demonstrated to reduce pathological remodeling and improve function (Landa et al., 2008), initiating commercialization efforts of this hydrogel. A synthetic, self-assembling peptide hydrogel (AcN-RARADA-DARARADADA-CNH) was also used, forming a nano-fibrous structure upon injection into the myocardium that promoted recruitment of endogenous ECs and supported survival of injected cardiomyocytes (CMs) (Davis et al., 2005). Insulin-like growth factor-1 (IGF) bound to the self-assembling peptide was demonstrated to improve grafting and survival of CMs injected into MI (Davis et al., 2006). Laflamme and Murry demonstrated that targeting of multiple pathways related to cell survival by encapsulating a number of biomolecules in Matrigel significantly increased the survival and grafting of the human ESC-derived CMs injected into infarcted rat hearts (Laflamme et al., 2007).

Zhang et. al. studied the effect of injecting CMs in a mixture of collagen type I and Matrigel (Zhang et al., 2006), the material used by Zimmerman et al. to create engineered heart tissue (Zimmermann et al., 2006), in MI-induced rats. An additional problem with the use of an ECM protein in this setting may be the immune response exhibited by rats to mouse protein (i.e. Matrigel is a basement membrane derived from mouse sarcoma). Positive connexin-43 staining was found in the cells in the biomaterial, and the biomaterial was also seen to improve the thickness and the function of the heart. The main drawback is that the material takes 1 h to gel, which could allow for significant cell loss, although no cell retention studies were conducted in these experiments.

A thermo-responsive chitosan-based gel was prepared and injected into the infarcted hearts of rats with and without mouse ESCs, resulting in cell retention and 4-week graft size being significantly higher than PBS + ESC control. In addition, heart function (measured through echocardiography), wall thickness, and micro-vessel density were all higher in chitosan-alone and chitosan + ESC groups than PBS + ESC control, with chitosan + ESC showing the greatest improvement 4 weeks after injection (Lu et al., 2009).

We have modified chitosan with the peptide QHREDGS derived from angiopoietin-1, the peptide sequence implicated in the survival response of muscle cells cultivated in the presence of this growth factor (Dallabrida et al., 2005). The chitosan was rendered photocrosslinkable by modification with azidiobenzoic acid (Az-chitosan) (Yeo et al., 2007). Neonatal rat heart cells cultivated on crosslinked films of Az-chitosan-QHREDGS attached, elongated, and remained viable while they exhibited lower attachment levels and decrease in viability when cultivated on the chitosan substrates modified with the scrambled peptide sequence (Rask et al., 2010). Interestingly, cells on Az-chitosan-QHREDGS were capable of resisting taxol-induced apoptosis, while those on Az-chitosan-RGDS were not (Rask et al., 2010).

Recent studies collectively indicate that an injection of hydrogel alone, without the reparative cells, may also attenuate pathological remodeling upon MI (Landa et al., 2008; Dobner et al., 2009; Fujimoto et al., 2009; Leor et al., 2009). For example, injection of alginate or collagen alone improved LV function and reduced cardiac remodeling post-infarction (Dai et al., 2005; Landa et al., 2008). It is suspected that by changing the ventricular geometry and mechanics, hydrogels reduce the elevated local wall stresses that have been implicated in pathological remodeling (Wall et al., 2006). Finite element modeling of wall stresses indicated that, upon injection of the material of elastic modulus 10–20 kPa in the infarct, the relationship between ejection fraction and the stroke volume/end-diastolic volume was improved. In addition, injections of the material in the border zone decreased end-systolic fiber stress proportionally to the volume and the stiffness of the injected material.

Two alginate biopolymers were modified to assess the therapeutic potential in rat MI models. Alginate modified with 0.025% v/v polypyrrole, a conductive polymer, injected into the infarct zone showed improved arteriogenesis at 5 weeks post-treatment and significantly enhanced infiltration of myofibroblasts into the infarct area when compared to saline-and alginate-only controls (Mihardja et al., 2008). Also, RGD conjugated alginate, and alginate alone, injected into the infarct zone showed improved LV function and increased arteriole density 5 weeks post-injection when compared to BSA in PBS control (Yu et al., 2009). Results from both studies again show the potential for non-cell-based therapies to treat chronic heart failure. In addition, many of the above-mentioned studies used a control group with just the acellular biomaterial and found that the material was able to produce some of the beneficial effects, but not all of those achieved with the cellular treatment.

We believe that properly tuning mechanical properties of a hydrogel and providing bioactive molecules may offer new cell-free treatment options for MI. The death of CMs by necrosis and apoptosis peaks at 6 h upon acute MI (Anversa et al., 1998). However, the persistent and progressive loss of CMs in neighboring areas of the infarct continues up to 60 days after the onset of MI. During this process, up to 35% of cells at the borders of subacute and old infarcts may become apoptotic (Yaoita et al., 2000), in comparison to only 1% in the remote regions of myocardium (Olivetti et al., 1996). Studies in rats and dogs demonstrated that CM loss by apoptosis persists for 1–4 months upon MI, correlating with the progressive worsening of the pump function. Thus, developing hydrogels that specifically prevent apoptosis of the heart cells (e.g. QHREDGS peptide modified chitosan) may result in new treatment options in the future, where hydrogel injected alone in the border zone, without the reparative cells, would act to both mechanically stabilize the ventricle and prevent further apoptosis of cardiomyocytes.

For example, it was shown that EC-induced CM protection post-infarction occurs through PDGF-BB signaling. Thus, binding PDGF-BB to the self-assembling nanofibers of RAD16-II (a peptide consisting of alternating RAD domains, AcN-RARADADARARADADA-CNH$_2$) hydrogel was evaluated as a potential therapeutic option. Sustained, targeted release of this signaling molecule to host myocardium was observed up to 14 days after injection. Injection of nanofibers with PDGF-BB at the site of infarct in rats decreased CM death and preserved systolic function post-MI, and showed (separately) a decrease in infarct size after ischemia/reperfusion (Hsieh et al., 2006).

The relative contribution of cells versus the injected biomaterial to the attenuation of pathological remodeling also needs to be assessed and the mechanism by which various cells induce functional improvements needs to be elucidated. While with the injection of contractile cardiomyocytes the expectation is that the cells will functionally couple to the host myocardium and contribute to contractile function, the same is not possible for non-cardiomyocytes. The exact mechanism by which non-myocytes impart the improvement in function and attenuation of pathological remodeling is still under debate but some researchers suggest that the transplantation of healthy cells results in the release of growth factors and other molecular signals, that is, the paracrine effect. These help with angiogenesis, cell survival, and recruitment of progenitors. One possible drawback of the biomaterial use is that the scaffold or the hydrogel may also take up space that would prevent a high tissue density until the material degrades.

Implantation of cardiac patches

While significant progress has been made in constructing *in vitro* cultivation systems and biomaterial scaffolds, fewer studies have focused on implantation of cell-based cardiac patches onto viable or injured myocardium (Fig. 48.6). In a pioneering study, Li and colleagues (1999) implanted a construct based on neonatal rat cardiomyocytes and collagen sponges onto the surface of the cryoinjured myocardium of Lewis rats (Fig. 48.6). The grafts were implanted 3 weeks post-infarction. After 5 weeks *in vivo*, the cells survived supported by the blood vessel ingrowth and integrated with the surrounding tissue. However, the graft did not improve left ventricular function.

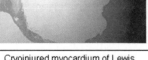 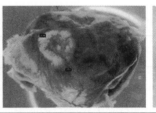 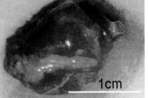

Cryoinjured myocardium of Lewis rats	Coronary artery occlusion in Sprague-Dawley rat myocardium	Uninjured heart of Fisher 344 rats
Constructs implanted after 3 weeks	Constructs implanted after 1 week	
Fetal (Lewis) rat cardiomyocytes in collagen sponge Cultivated under static condition for 7 days	Fetal (Sprague-Dawley) rat cardyomyocytes in porous aglinate scaffolds Cultivated under static conditions for 4 days	Neonatal (Fisher 344) rat cardyomyocytes in collagen type I gel Cultivation with mechanical stimulation for 12 days
After 5 weeks *in vivo* no significant difference compared to the controls	After 9 weeks *in vivo* attenuation of LV function and maintenance of contractile function in comparison to controls without construct	After 14 weeks *in vivo* the implant vascularized and improved the level of maturation Immunosupression was required
Li et al., 1999	Leor et al., 2000	Zimmermann et al., 2002a

FIGURE 48.6
Representative early studies investigating the effect of implantation of the cardiomyocyte-based constructs on the function of injured or viable hearts.

Attenuation of pathological remodeling (i.e prevention of ventricle dilatation and maintenance of contractile function) was observed in a study by Leor et al. (2000), where cardiac constructs based on neonatal rat cardiomyocytes and porous alginate scaffolds were implanted onto myocardium of Sprague-Dawley rats that underwent permanent main coronary artery occlusion (Fig. 48.3). The grafts were implanted 7 days after MI. After 9 weeks of implantation, the grafts demonstrated integration with host myocardium at the anchorage sites as well as inflammatory infiltrates and presence of fibrous collagen.

Zimmerman et al. (2002b) placed cardiac tissue rings cultivated in the presence of mechanical stimulation onto uninjured hearts of Fisher 344 rats for 14 days (Fig. 48.6). They noticed that, although both cells and collagen were isolated from Fisher rats, immunosuppression was required for maintenance of heart tissue upon implantation. In the absence of immunosuppression, even in the syngeneic approach, cardiac constructs completely degraded after only 2 weeks *in vivo*. It is unknown what exactly caused the response; it is possible that it was the remainder of serum or chick extract. Regardless, the finding has significant implications in the potential implantation of cardiac patches in clinical settings.

In order to decrease the potential immunogenicity of their engineered tissue, Zimmerman and colleagues discarded all xenogenic components from their culture (Naito et al., 2006). This included cultivating the EHTs in serum-free and Matrigel-free conditions. Mixed heart cell populations rather than cardiomyocyte-rich populations were utilized, and the culture medium was supplemented with triiodothyronine and insulin (Naito et al., 2006). Other studies have also established the need for non-immunogenic media. Schwarzkopf et al. used autospecies sera, in this case rat, for culturing of rat cardiomyocytes (Schwarzkopf et al., 2006). The metabolic activity of the cells was significantly higher than for the cells cultivated in conventional culture medium with fetal bovine serum.

Zimmermann et al. demonstrated integration and electrical coupling of a complex multi-loop graft to native myocardium in rats with LAD ligation (Fig. 48.7A−D). Functional improvement was demonstrated not to be merely a result of scar stabilization or paracrine effects (Fig. 48.7C,D) (Zimmermann et al., 2006). Functional integration of cardiac cell sheets to the heat-injured myocardium was also demonstrated (Furuta et al., 2006).

In addition to engineering the patches of myocardium, Zimmermann and colleagues designed the first biological assist device (Yildirim et al., 2007). The authors mechanically stimulated a hollow-spherical construct consisting of collagen I and neonatal rat cardiomyocytes until a beating pouch-like structure was created. The pouch was then placed over uninjured rat hearts in such a manner that the right and left ventricles were covered. Fourteen days after implantation, the pouch covered the epicardial surface of the heart and exhibited blood vessel ingrowth.

Badylak et al. implanted ECM derived from porcine urinary bladder into surgically created 2-cm^2 defects in the left ventricular free wall of dogs. Eight weeks following the implantation, the ECM patches showed higher regional systolic contraction compared to the control group, in which a material (Dacron) currently used for myocardial defects was paced. Histological analysis suggested that cardiomyocytes accounted for about 30% of the remodeled tissue in the ECM scaffolds (Badylak et al., 2006). In a recent study, it was found that this improvement in the heart function can be attributed to an increase in the myocyte content in the ECM patches between weeks 2 and 8. The relationship between the myocyte content and the extent of mechanical function was observed to be linear. There was also some evidence (decrease in cardiomyocyte diameter and increase in the overall area occupied by cardiomyocytes over time) that suggested a possibility of cardiomyocyte proliferation in the patches (Kelly et al., 2009).

Limitations related to the source of autologous cardiomyocytes motivated the studies that utilized non-myocyte-based patches for MI repair. Smooth muscle cells seeded into poly (ε-caprolactone-co-L-lactide) sponge reinforced with poly-L-actide fabric were used in

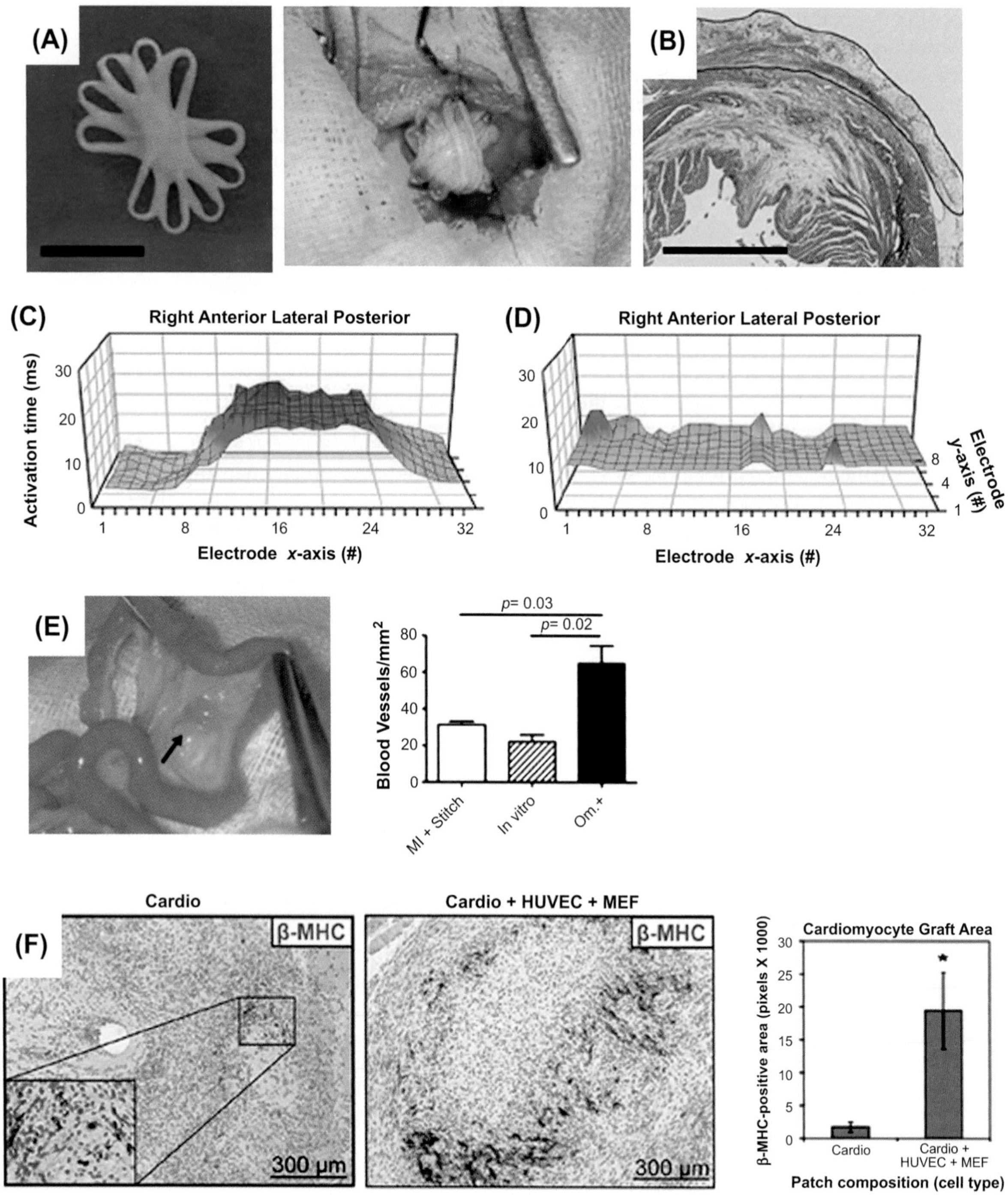

FIGURE 48.7

In vivo **integration of the engineered myocardium.** (A) A multi-looped, mechanically stimulated cardiac construct based on neonatal rat cardiomyocytes and collagen hydrogel was used to repair myocardial infarction in the rat heart. Scale bar = 10 mm. The patch integrated well with the surrounding myocardium as demonstrated by (B) histology (scale bar = 5 mm) and (C,D) the multielectrode recordings of impulse propagation. (C) In the sham-operated animal (with MI alone), significant prolongation of the epicardial activation time is noted, (D) while the hearts repaired using engineered heart tissue do not exhibit any prolongation (with permission from Zimmermann et al., 2006). (E) Prevascularization of cardiac constructs based on the alginate scaffolds and neonatal rat cardiomyocytes results in significantly higher vascularization upon implantation in the infarcted rat heart (with permission from Dvir et al., 2009). (F) Cardiac patches containing cardiomyocytes, human umbilical vein, and endothelial cells (Cardio + HUVEC + MEF) exhibit higher survival of cardiomyocytes (b-myosin heavy chain positive) upon implantation in the gluteus superficialis muscle of nude rats for 1 week compared to those containing cardiomyocytes alone (Cardio). The constructs were generated by cell self-assembly in orbitally mixed dishes (with permission from Stevens et al., 2009a).

a modified endoventricular circular patch plasty procedure (Dor procedure). Cell-seeded grafts resulted in improved left ventricular function (as assessed by echocardiography) compared to cell-free controls (Matsubayashi et al., 2003). A patch made of dermal fibroblasts seeded onto knitted Vicryl mesh (Dermagraft) was used in an attempt to increase angiogenesis upon MI. When placed over the infarcted regions on the hearts of SCID mice, the grafts improved microvessel density within the damaged myocardium (Kellar et al., 2001).

There appears to be a consensus regarding the requirement for multiple cell types, specifically fibroblasts and endothelial cells in addition to cardiomyocytes, for successful cardiac tissue graft survival and vascularization *in vivo*. In one approach, omentum was used to prevascularize cardiac patches based on neonatal rat cardiomyocytes and alginate scaffolds modified with angiogenic factors (Fig. 48.7E). Following excision and implantation into the infarcted rat myocardium, the vascularized cardiac patch showed structural and electrical integration into host myocardium and attenuated pathological remodeling of the ventricle significantly better than the *in vitro*-cultivated patch alone (Dvir et al., 2009).

In another strategy, a simultaneous tri-culture scaffold-free approach was used to generate the beating cardiac patches based on human ESC. Upon implantation into the hindlimb muscle of nude rats, these patches, composed only of enriched cardiomyocytes, did not survive to form significant grafts (Fig. 48.7F). However, patches containing endothelial cells (either human umbilical vein or hESC-derived endothelial cells) and fibroblasts in addition to cardiomyocytes persisted in the (non-infarcted) rat heart and resulted in 10-fold larger cell grafts compared with cardiomyocyte-only patches (Fig. 48.7F). The preformed human microvessels also anastomosed with the rat host coronary circulation and delivered blood to the grafts (Stevens et al., 2009a).

These studies demonstrated the feasibility of cardiac patch implantation, but further studies are necessary to estimate the effect of culture conditions and scaffold type on the *in vivo* outcome. Although significant progress has been made in the area of biomaterials and bioreactors, it is currently unknown which cultivation conditions and what biomaterial will best preserve contractile function and prevent pathological remodeling upon implantation. Thus, studies that investigate this in a systematic fashion and correlate *in vitro* parameters (e.g. force of contraction) to *in vivo* outcomes (e.g. fractional shortening) are required. The host response to the patch and the nature of the immunological situation further complicate these studies.

SUMMARY

Overall, the field of cardiac tissue engineering is very much in its infancy. Although the results to date are exceedingly encouraging, much remains to be done in order to develop clinically relevant approaches, let alone move towards a whole heart. Not surprisingly, an NIH task force (National Institutes of Health, 1999) has emphasized development of heart components such as a cardiac patch or a valve before "graduating" to whole heart engineering.

Since the *in vivo* studies conducted thus far used different cell sources, biomaterials, animal models, delivery times post-infarction, and experimental time frames, a direct comparison between the methods cannot be achieved. While all reported studies have shown some form of improvement, complete myocardial regeneration has not been achieved. Perhaps a valid question to be answered in the future is: What is the required level of myocardial regeneration in terms of survival and attenuation of symptoms? Complete regeneration is an ambitious goal that may not be required. Future studies must also increase their time frames, to better assess the long-term effects of these treatments.

However, significant progress has been made since the LIFE initiative embarked on the creation of the artificial heart in 1999. Functional viable cardiac patches have been engineered based on neonatal rat cardiomyocytes and more recently based on ES cell-derived cardiomyocytes.

Various biomaterials have been tested for this purpose and *in vitro* culture systems have been developed that enhance cardiac construct differentiation (mechanical and electrical stimulation) as well as improve cardiomyocyte survival at high density (medium perfusion). The discovery of induced pluripotent stem cells offers the possibility to engineer an autologous cardiac patch of clinically relevant size. While the completely artificial heart will remain a dream, the near future may bring a clinically relevant autologous cardiac patch as evidenced by the rapid progress in engineering of cardiac patches based on stem cell-derived cardiomyocytes. The work on recellularization of decellularized hearts may represent the first step towards "the heart in a box" envisioned more than a decade ago.

References

Ahrendt, G., Chickering, D. E., & Ranieri, J. P. (1998). Angiogenic growth factors: a reveiw for tissue engineering. *Tissue Eng., 4*, 117–130.

Akins, R. E. (2000). Prospects for use of cell implantation, gene therapy and tissue engineering in the treatment of myocardial disease and congenital heart defects. In K. Sames (Ed.), *Medizinische Regeneration und Tissue Engineering* (pp. 1–16). Landsberg, Germany: EcoMed.

Alperin, C., Zandstra, P. W., & Woodhouse, K. A. (2005). Polyurethane films seeded with embryonic stem cell-derived cardiomyocytes for use in cardiac tissue engineering applications. *Biomaterials, 26*, 7377–7386.

American Heart Association (2004). *Heart Disease and Stroke Statistics – 2004* http://www.americanheart.org/downloadable/heart/1079736729696HDSStats2004updateREV3-19-04.pdf.

Anversa, P., Cheng, W., Liu, Y., Leri, A., Redaelli, G., & Kajstura, J. (1998). Apoptosis and myocardial infarction. *Basic Res. Cardiol., 93*(Suppl 3), 8–12.

Augustin-Voss, H. G., Johnson, R. C., & Pauli, B. U. (1991). Modulation of endothelial cell surface glycoconjugate expression by organ-derived biomatrices. *Exp. Cell Res., 192*, 346–351.

Baar, K., Birla, R., Boluyt, M. O., Borschel, G. H., Arruda, E. M., & Dennis, R. G. (2005). Self-organization of rat cardiac cells into contractile 3-D cardiac tissue. *Faseb J., 19*, 275–277.

Badie, N., & Bursac, N. (2009). Novel micropatterned cardiac cell cultures with realistic ventricular microstructure. *Biophys. J., 96*, 3873–3885.

Badie, N., Satterwhite, L., & Bursac, N. (2009). A method to replicate the microstructure of heart tissue *in vitro* using DTMRI-based cell micropatterning. *Ann. Biomed. Eng., 37*, 2510–2521.

Badylak, S. F., Kochupura, P. V., Cohen, I. S., Doronin, S. V., Saltman, A. E., Gilbert, T. W., et al. (2006). The use of extracellular matrix as an inductive scaffold for the partial replacement of functional myocardium. *Cell Transplant., 15*(Suppl 1), S29–S40.

Balsam, L. B., Wagers, A. J., Christensen, J. L., Kofidis, T., Weissman, I. L., & Robbins, R. C. (2004). Hematopoietic stem cells adopt mature hematopoietic fates in ischemic myocardium. *Nature, 428*, 668–673.

Beltrami, A. P., Barlucchi, L., Torella, D., Baker, M., Limana, F., Chimenti, S., et al. (2003). Adult cardiac stem cells are multipotent and support myocardial regeneration. *Cell, 114*, 763–776.

Bian, W., & Bursac, N. (2009). Engineered skeletal muscle tissue networks with controllable architecture. *Biomaterials, 30*, 1401–1412.

Bian, W., Liau, B., Badie, N., & Bursac, N. (2009). Mesoscopic hydrogel molding to control the 3D geometry of bioartificial muscle tissues. *Nat. Protoc., 4*, 1522–1534.

Boublik, J., Park, H., Radisic, M., Tognana, E., Chen, F., Pei, M., et al. (2005). Mechanical properties and remodeling of hybrid cardiac constructs made from heart cells, fibrin, and biodegradable, elastomeric knitted fabric. *Tissue Eng., 11*, 1122–1132.

Bredin, F., & Franco-Cereceda, A. (2010). Midterm results of passive containment surgery using the acorn Cor Cap cardiac support device in dilated cardiomyopathy. *J. Card. Surg., 25*, 107–112.

Brown, M. A., Iyer, R. K., & Radisic, M. (2008). Pulsatile perfusion bioreactor for cardiac tissue engineering. *Biotechnol. Prog., 24*, 907–920.

Bursac, N., Papadaki, M., Cohen, R. J., Schoen, F. J., Eisenberg, S. R., Carrier, R., et al. (1999). Cardiac muscle tissue engineering: toward an *in vitro* model for electrophysiological studies. *Am. J. Physiol., 277*, H433–H444.

Burton, A. C. (1972). *Biophysical Basis of the Circulation*. Chicago, IL: Yearbook Medical Publishers.

Butler, D. L., Goldstein, S. A., & Guilak, F. (2000). Functional tissue engineering: the role of biomechanics. *J. Biomech. Eng., 122*, 570–575.

Carrier, R. L., Papadaki, M., Rupnick, M., Schoen, F. J., Bursac, N., Langer, R., et al. (1999). Cardiac tissue engineering: cell seeding, cultivation parameters and tissue construct characterization. *Biotechnol. Bioeng., 64*, 580–589.

Cenni, E., Ciapetti, G., Cavedagna, D., di Leo, A., & Pizzoferrato, A. (1993). Production of prostacyclin and fibrinolysis modulators by endothelial cells cultured in the presence of polyethylene terephthalate. *J. Biomed. Mater. Res., 27*, 1161–1164.

Cenni, E., Granchi, D., Ciapetti, G., Savarino, L., Corradini, A., & di Leo, A. (2000). Cytokine expression *in vitro* by cultured human endothelial cells in contact with polyethylene terephthalate coated with pyrolytic carbon and collagen. *J. Biomed. Mater. Res., 50*, 483–489.

Chamberlain, M. D., Gupta, R., & Sefton, M. V. (2010). Chimeric vessel tissue engineering driven by endothelialized modules. *Tissue Eng.* (submitted).

Chen, S. L., Fang, W. W., Ye, F., Liu, Y. H., Qian, J., Shan, S. J., et al. (2004). Effect on left ventricular function of intracoronary transplantation of autologous bone marrow mesenchymal stem cell in patients with acute myocardial infarction. *Am. J. Cardiol., 94*, 92–95.

Cheng, D., & Sefton, M. V. (2009). Dual delivery of placental growth factor and vascular endothelial growth factor from poly(hydroxyethyl methacrylate-co-methyl methacrylate) microcapsules containing doubly transfected luciferase-expressing L929 cells. *Tissue Eng. Part A, 15*, 1929–1939.

Chiu, L. L., & Radisic, M. (2010). Scaffolds with covalently immobilized VEGF and Angiopoietin-1 for vascularization of engineered tissues. *Biomaterials, 31*(2), 226–241.

Chiu, L. L., Iyer, R. K., King, J. P., & Radisic, M. (2008). Biphasic electrical field stimulation aids in tissue engineering of multicell-type cardiac organoids. *Tissue Eng. Part A.* Epub ahead of print.

Cho, H. C., & Marban, E. (2010). Biological therapies for cardiac arrhythmias: can genes and cells replace drugs and devices? *Circ. Res., 106*, 674–685.

Christman, K. L., Fok, H. H., Sievers, R. E., Fang, Q., & Lee, R. J. (2004a). Fibrin glue alone and skeletal myoblasts in a fibrin scaffold preserve cardiac function after myocardial infarction. *Tissue Eng., 10*, 403–409.

Christman, K. L., Vardanian, A. J., Fang, Q., Sievers, R. E., Fok, H. H., & Lee, R. J. (2004b). Injectable fibrin scaffold improves cell transplant survival, reduces infarct expansion, and induces neovasculature formation in ischemic myocardium. *J. Am. Coll. Cardiol., 44*, 654–660.

Dai, W., Wold, L. E., Dow, J. S., & Kloner, R. A. (2005). Thickening of the infarcted wall by collagen injection improves left ventricular function in rats: a novel approach to preserve cardiac function after myocardial infarction. *J. Am. Coll. Cardiol., 46*, 714–719.

Dallabrida, S. M., Ismail, N., Oberle, J. R., Himes, B. E., & Rupnick, M. A. (2005). Angiopoietin-1 promotes cardiac and skeletal myocyte survival through integrins. *Circ. Res., 96*, e8–e24.

Dar, A., Shachar, M., Leor, J., & Cohen, S. (2002). Cardiac tissue engineering optimization of cardiac cell seeding and distribution in 3D porous alginate scaffolds. *Biotechnol. Bioeng., 80*, 305–312.

Davis, M. E., Motion, J. P., Narmoneva, D. A., Takahashi, T., Hakuno, D., Kamm, R. D., et al. (2005). Injectable self-assembling peptide nanofibers create intramyocardial microenvironments for endothelial cells. *Circulation, 111*, 442–450.

Davis, M. E., Hsieh, P. C., Takahashi, T., Song, Q., Zhang, S., Kamm, R. D., et al. (2006). Local myocardial insulin-like growth factor 1 (IGF-1) delivery with biotinylated peptide nanofibers improves cell therapy for myocardial infarction. *Proc. Natl. Acad. Sci. U.S.A, 103*, 8155–8160.

del Monte, F., Williams, E., Lebeche, D., Schmidt, U., Rosenzweig, A., Gwathmey, J. K., et al. (2001). Improvement in survival and cardiac metabolism after gene transfer of sarcoplasmic reticulum Ca^{2+}-ATPase in a rat model of heart failure. *Circulation, 104*, 1424–1429.

del Monte, F., Harding, S. E., Dec, G. W., Gwathmey, J. K., & Hajjar, R. J. (2002). Targeting phospholamban by gene transfer in human heart failure. *Circulation, 105*, 904–907.

di Donato, M., Sabatier, M., Dor, V., Toso, A., Maioli, M., & Fantini, F. (1997). Akinetic versus dyskinetic post-infarction scar: relation to surgical outcome in patients undergoing endoventricular circular patch plasty repair. *J. Am. Coll. Cardiol., 29*, 1569–1575.

di Luozzo, G., Bhargava, J., & Powell, R. J. (2000). Vascular smooth muscle cell effect on endothelial cell endothelin-1 production. *J. Vasc. Surg., 31*, 781–789.

Dimmeler, S., Zeiher, A. M., & Schneider, M. D. (2005). Unchain my heart: the scientific foundations of cardiac repair. *J. Clin. Invest., 115*, 572–583.

Dobner, S., Bezuidenhout, D., Govender, P., Zilla, P., & Davies, N. (2009). A synthetic non-degradable polyethylene glycol hydrogel retards adverse post-infarct left ventricular remodeling. *J. Card. Fail., 15*, 629–636.

Domian, I. J., Chiravuri, M., van der Meer, P., Feinberg, A. W., Shi, X., Shao, Y., et al. (2009). Generation of functional ventricular heart muscle from mouse ventricular progenitor cells. *Science, 326*, 426–429.

Dor, V., Saab, M., Coste, P., Kornaszewska, M., & Montiglio, F. (1989). Left ventricular aneurysm: a new surgical approach. *Thorac. Cardiovasc. Surg., 37*, 11–19.

Dorfman, J., Duong, M., Zibaitis, A., Pelletier, M. P., Shum-Tim, D., Li, C., et al. (1998). Myocardial tissue engineering with autologous myoblast implantation. *J. Thorac. Cardiovasc. Surg., 116*, 744–751.

Dvir, T., Benishti, N., Shachar, M., & Cohen, S. (2006). A novel perfusion bioreactor providing a homogenous milieu for tissue regeneration. *Tissue Eng., 12*, 2843–2852.

Dvir, T., Levy, O., Shachar, M., Granot, Y., & Cohen, S. (2007). Activation of the ERK1/2 cascade via pulsatile interstitial fluid flow promotes cardiac tissue assembly. *Tissue Eng., 13*, 2185–2193.

Dvir, T., Kedem, A., Ruvinov, E., Levy, O., Freeman, I., Landa, N., et al. (2009). Prevascularization of cardiac patch on the omentum improves its therapeutic outcome. *Proc. Natl. Acad. Sci. U.S.A, 106*, 14990–14995.

Engelmayr, G. C., Jr., Cheng, M., Bettinger, C. J., Borenstein, J. T., Langer, R., & Freed, L. E. (2008). Accordion-like honeycombs for tissue engineering of cardiac anisotropy. *Nat. Mater., 7*, 1003–1010.

Eschenhagen, T., Fink, C., Remmers, U., Scholz, H., Wattchow, J., Woil, J., et al. (1997). Three-dimensional reconstitution of embryonic cardiomyocytes in a collagen matrix: a new heart model system. *FASEB J., 11*, 683–694.

Eskin, S. G., Ives, C. L., McIntire, L. V., & Navarro, L. T. (1984). Response of cultured endothelial cells to steady flow. *Microvasc. Res., 28*, 87–94.

Esmon, C. T. (2000). Regulation of blood coagulation. *Biochim. Biophys. Acta, 1477*, 349–360.

Etzion, S., Battler, A., Barbash, I. M., Cagnano, E., Zarin, P., Granot, Y., et al. (2001). Influence of embryonic cardiomyocyte transplantation on the progression of heart failure in a rat model of extensive myocardial infarction. *J. Mol. Cell. Cardiol., 33*, 1321–1330.

Fazel, S., Tang, G. H., Angoulvant, D., Cimini, M., Weisel, R. D., Li, R. K., et al. (2005). Current status of cellular therapy for ischemic heart disease. *Ann. Thorac. Surg., 79*, S2238–S2247.

Feinberg, A. W., Feigel, A., Shevkoplyas, S. S., Sheehy, S., Whitesides, G. M., & Parker, K. K. (2007). Muscular thin films for building actuators and powering devices. *Science, 317*, 1366–1370.

Fink, C., Ergun, S., Kralisch, D., Remmers, U., Weil, J., & Eschenhagen, T. (2000). Chronic stretch of engineered heart tissue induces hypertrophy and functional improvement. *FASEB J., 14*, 669–679.

Fujimoto, K. L., Ma, Z., Nelson, D. M., Hashizume, R., Guan, J., Tobita, K., et al. (2009). Synthesis, characterization and therapeutic efficacy of a biodegradable, thermoresponsive hydrogel designed for application in chronic infarcted myocardium. *Biomaterials, 30*, 4357–4368.

Fung, Y. C. (1993). *Biomechanics: Mechanical Properties of Living Tissues.* New York: Springer.

Furuta, A., Miyoshi, S., Itabashi, Y., Shimizu, T., Kira, S., Hayakawa, K., et al. (2006). Pulsatile cardiac tissue grafts using a novel three-dimensional cell sheet manipulation technique functionally integrates with the host heart, *in vivo. Circ. Res., 98*, 705–712.

Gepstein, L. (2010). Cell and gene therapy strategies for the treatment of postmyocardial infarction ventricular arrhythmias. *Ann. NY Acad. Sci., 1188*, 32–38.

Greenberg, J. I., Shields, D. J., Barillas, S. G., Acevedo, L. M., Murphy, E., Huang, J., et al. (2008). A role for VEGF as a negative regulator of pericyte function and vessel maturation. *Nature, 456*, 809–813.

Hajjar, R. J., del Monte, F., Matsui, T., & Rosenzweig, A. (2000). Prospects for gene therapy for heart failure. *Circ. Res., 86*, 616–621.

Halloran, P. F., & Melk, A. (2001). Renal senescence, cellular senescence, and their relevance to nephrology and transplantation. *Adv. Nephrol. Necker Hosp., 31*, 273–283.

Harada, K., Grossman, W., Friedman, M., Edelman, E. R., Prasad, P. V., Keighley, C. S., et al. (1994). Basic fibroblast growth factor improves myocardial function in chronically ischemic porcine hearts. *J. Clin. Invest., 94*, 623–630.

Heidi Au, H. T., Cui, B., Chu, Z. E., Veres, T., & Radisic, M. (2009). Cell culture chips for simultaneous application of topographical and electrical cues enhance phenotype of cardiomyocytes. *Lab. Chip., 9*, 564–575.

Hsieh, P. C. H., Davis, M. E., Gannon, J., MacGillivray, C., & Lee, R. T. (2006). Controlled delivery of PDGF-BB for myocardial protection using injectable self-assembling peptide nanofibers. *J. Clin. Invest., 116*, 237–248.

Ives, C. L., Eskin, S. G., & McIntire, L. V. (1986). Mechanical effects on endothelial cell morphology: *in vitro* assessment. *Vitro Cell. Dev. Biol., 22*, 500–507.

Kaihara, S., Borenstein, J., Koka, R., Lalan, S., Ochoa, E. R., Ravens, M., et al. (2000). Silicon micromachining to tissue engineer branched vascular channels for liver fabrication. *Tissue Eng., 6*, 105–117.

Kehat, I., Kenyagin-Karsenti, D., Snir, M., Segev, H., Amit, M., Gepstein, A., et al. (2001). Human embryonic stem cells can differentiate into myocytes with structural and functional properties of cardiomyocytes. *J. Clin. Invest., 108*, 407–414.

Kellar, R. S., Landeen, L. K., Shepherd, B. R., Naughton, G. K., Ratcliffe, A., & Williams, S. K. (2001). Scaffold-based three-dimensional human fibroblast culture provides a structural matrix that supports angiogenesis in infarcted heart tissue. *Circulation, 104*, 2063–2068.

Kelly, D. J., Rosen, A. B., Schuldt, A. J., Kochupura, P. V., Doronin, S. V., Potapova, I. A., et al. (2009). Increased myocyte content and mechanical function within a tissue-engineered myocardial patch following implantation. *Tissue Eng. Part A, 15*, 2189–2201.

Khademhosseini, A., Eng, G., Yeh, J., Kucharczyk, P. A., Langer, R., Vunjak-Novakovic, G., et al. (2006). Microfluidic patterning for fabrication of contractile cardiac organoids. *Biomed. Microdev., 12*, 2077–2091.

Kim, D. H., Lipke, E. A., Kim, P., Cheong, R., Thompson, S., Delannoy, M., et al. (2010). Nanoscale cues regulate the structure and function of macroscopic cardiac tissue constructs. *Proc. Natl. Acad. Sci. U.S.A, 107*, 565–570.

Klug, M. G., Soonpaa, M. H., Koh, G. Y., & Field, L. J. (1996). Genetically selected cardiomyocytes from differentiating embronic stem cells form stable intracardiac grafts. *J. Clin. Invest., 98*, 216–224.

Kocher, A. A., Schuster, M. D., Szabolcs, M. J., Takuma, S., Burkhoff, D., Wang, J., et al. (2001). Neovascularization of ischemic myocardium by human bone-marrow-derived angioblasts prevents cardiomyocyte apoptosis, reduces remodeling and improves cardiac function. *Nat. Med., 7*, 430–436.

Kofidis, T., Akhyari, P., Wachsmann, B., Mueller-Stahl, K., Boublik, J., Ruhparwar, A., et al. (2003a). Clinically established hemostatic scaffold (tissue fleece) as biomatrix in tissue- and organ-engineering research. *Tissue Eng., 9*, 517–523.

Kofidis, T., Lenz, A., Boublik, J., Akhyari, P., Wachsmann, B., Mueller-Stahl, K., et al. (2003b). Pulsatile perfusion and cardiomyocyte viability in a solid three-dimensional matrix. *Biomaterials, 24*, 5009–5014.

Kofidis, T., Lebl, D. R., Martinez, E. C., Hoyt, G., Tanaka, M., & Robbins, R. C. (2005). Novel injectable bioartificial tissue facilitates targeted, less invasive, large-scale tissue restoration on the beating heart after myocardial injury. *Circulation, 112*, I173–I177.

Korff, T., Kimmina, S., Martiny-Baron, G., & Augustin, H. G. (2001). Blood vessel maturation in a 3-dimensional spheroidal coculture model: direct contact with smooth muscle cells regulates endothelial cell quiescence and abrogates VEGF responsiveness. *Faseb J., 15*, 447–457.

Kusano, K. F., Pola, R., Murayama, T., Curry, C., Kawamoto, A., Iwakura, A., et al. (2005). Sonic hedgehog myocardial gene therapy: tissue repair through transient reconstitution of embryonic signaling. *Nat. Med., 11*, 1197–1204.

Laflamme, M. A., & Murry, C. E. (2005). Regenerating the heart. *Nat. Biotechnol., 23*, 845–856.

Laflamme, M. A., Chen, K. Y., Naumova, A. V., Muskheli, V., Fugate, J. A., Dupras, S. K., et al. (2007). Cardiomyocytes derived from human embryonic stem cells in pro-survival factors enhance function of infarcted rat hearts. *Nat. Biotechnol., 25*, 1015–1024.

Landa, N., Miller, L., Feinberg, M. S., Holbova, R., Shachar, M., Freeman, I., et al. (2008). Effect of injectable alginate implant on cardiac remodeling and function after recent and old infarcts in rat. *Circulation, 117*, 1388–1396.

Laugwitz, K. L., Moretti, A., Lam, J., Gruber, P., Chen, Y., Woodard, S., et al. (2005). Postnatal isl1+ cardioblasts enter fully differentiated cardiomyocyte lineages. *Nature, 433*, 647–653.

Leor, J., Aboulafia-Etzion, S., Dar, A., Shapiro, L., Barbash, I. M., Battler, A., et al. (2000). Bioengineered cardiac grafts: a new approach to repair the infarcted myocardium? *Circulation, 102*, III56–III61.

Leor, J., Tuvia, S., Guetta, V., Manczur, F., Castel, D., Willenz, U., et al. (2009). Intracoronary injection of *in situ* forming alginate hydrogel reverses left ventricular remodeling after myocardial infarction in Swine. *J. Am. Coll. Cardiol., 54*, 1014–1023.

Leung, B. M., & Sefton, M. V. (2010). A modular approach to cardiac tissue engineering. *Tissue Eng, 16*(10), 3207–3218.

Levenberg, S., Rouwkema, J., Macdonald, M., Garfein, E. S., Kohane, D. S., Darland, D. C., et al. (2005). Engineering vascularized skeletal muscle tissue. *Nat. Biotechnol., 23*, 879–884.

Li, J. M., Menconi, M. J., Wheeler, H. B., Rohrer, M. J., Klassen, V. A., Ansell, J. E., et al. (1992). Precoating expanded polytetrafluoroethylene grafts alters production of endothelial cell-derived thrombomodulators. *J. Vasc. Surg., 15*, 1010–1017.

Li, R.-K., Jia, Z. Q., Weisel, R. D., Mickle, D. A. G., Choi, A., & Yau, T. M. (1999). Survival and function of bioengineered cardiac grafts. *Circulation, 100*, II63–II69.

Li, R. K., Yau, T. M., Weisel, R. D., Mickle, D. A., Sakai, T., Choi, A., et al. (2000). Construction of a bioengineered cardiac graft. *J. Thorac. Cardiovasc. Surg., 119*, 368–375.

Lipinski, M. J., Biondi-Zoccai, G. G., Abbate, A., Khianey, R., Sheiban, I., Bartunek, J., et al. (2007). Impact of intracoronary cell therapy on left ventricular function in the setting of acute myocardial infarction: a collaborative systematic review and meta-analysis of controlled clinical trials. *J. Am. Coll. Cardiol., 50*, 1761–1767.

Lipton, B. H., Bensch, K. G., & Karasek, M. A. (1991). Microvessel endothelial cell transdifferentiation: phenotypic characterization. *Differentiation, 46*, 117–133.

Lu, A., & Sipehia, R. (2001). Antithrombotic and fibrinolytic system of human endothelial cells seeded on PTFE: the effects of surface modification of PTFE by ammonia plasma treatment and ECM protein coatings. *Biomaterials, 22*, 1439–1446.

Lu, W.-N., Lü, S.-H., Wang, H.-B., Li, D.-X., Duan, C.-M., Liu, Z.-Q., et al. (2009). Functional improvement of infarcted heart by co-injection of embryonic stem cells with temperature-responsive chitosan hydrogel. *Tissue Eng. Part A, 15*, 1437–1447.

905

Marcum, J. A., McKenney, J. B., & Rosenberg, R. D. (1984). Acceleration of thrombin-antithrombin complex formation in rat hindquarters via heparinlike molecules bound to the endothelium. *J. Clin. Invest.*, *74*, 341–350.

Markkanen, J. E., Rissanen, T. T., Kivela, A., & Yla-Herttuala, S. (2005). Growth factor-induced therapeutic angiogenesis and arteriogenesis in the heart – gene therapy. *Cardiovasc. Res.*, *65*, 656–664.

Martin-Rendon, E., Brunskill, S. J., Hyde, C. J., Stanworth, S. J., Mathur, A., & Watt, S. M. (2008). Autologous bone marrow stem cells to treat acute myocardial infarction: a systematic review. *Eur. Heart J.*, *29*, 1807–1818.

Matsubayashi, K., Fedak, P. W., Mickle, D. A., Weisel, R. D., Ozawa, T., & Li, R. K. (2003). Improved left ventricular aneurysm repair with bioengineered vascular smooth muscle grafts. *Circulation*, *108*(Suppl 1), II219–I1225.

Maurice, J. P., Hata, J. A., Shah, A. S., White, D. C., McDonald, P. H., Dolber, P. C., et al. (1999). Enhancement of cardiac function after adenoviral-mediated *in vivo* intracoronary beta2-adrenergic receptor gene delivery. *J. Clin. Invest.*, *104*, 21–29.

Mawji, I. A., & Marsden, P. A. (2003). Perturbations in paracrine control of the circulation: role of the endothelial-derived vasomediators, endothelin-1 and nitric oxide. *Microsc. Res. Tech.*, *60*, 46–58.

McCormick, S. M., Eskin, S. G., McIntire, L. V., Teng, C. L., Lu, C. M., Russell, C. G., et al. (2001). DNA microarray reveals changes in gene expression of shear stressed human umbilical vein endothelial cells. *Proc. Natl. Acad. Sci. U.S.A*, *98*, 8955–8960.

McDevitt, T. C., Angello, J. C., Whitney, M. L., Reinecke, H., Hauschka, S. D., Murry, C. E., et al. (2002). *In vitro* generation of differentiated cardiac myofibers on micropatterned laminin surfaces. *J. Biomed. Mater. Res.*, *60*, 472–479.

McDevitt, T. C., Woodhouse, K. A., Hauschka, S. D., Murry, C. E., & Stayton, P. S. (2003). Spatially organized layers of cardiomyocytes on biodegradable polyurethane films for myocardial repair. *J. Biomed. Mater. Res. A*, *66*, 586–595.

McGuigan, A. P., & Sefton, M. V. (2006). Vascularized organoid engineered by modular assembly enables blood perfusion. *PNAS*, *103*, 11461–11466.

McGuigan, A. P., & Sefton, M. V. (2007). The influence of biomaterials on endothelial cell thrombogenicity. *Biomaterials*, *28*, 2547–2571.

Meinhart, J. G., Deutsch, M., Fischlein, T., Howanietz, N., Froschl, A., & Zilla, P. (2001). Clinical autologous *in vitro* endothelialization of 153 infrainguinal ePTFE grafts. *Ann. Thorac. Surg.*, *71*, S327–S331.

Melo, L. G., Pachori, A. S., Kong, D., Gnecchi, M., Wang, K., Pratt, R. E., et al. (2004a). Gene and cell-based therapies for heart disease. *Faseb J.*, *18*, 648–663.

Melo, L. G., Pachori, A. S., Kong, D., Gnecchi, M., Wang, K., Pratt, R. E., et al. (2004b). Molecular and cell-based therapies for protection, rescue, and repair of ischemic myocardium: reasons for cautious optimism. *Circulation*, *109*, 2386–2393.

Menasche, P., Hagege, A. A., Vilquin, J. T., Desnos, M., Abergel, E., Pouzet, B., et al. (2003). Autologous skeletal myoblast transplantation for severe postinfarction left ventricular dysfunction. *J. Am. Coll. Cardiol.*, *41*, 1078–1083.

Mihardja, S. S., Sievers, R. E., & Lee, R. J. (2008). The effect of polypyrrole on arteriogenesis in an acute rat infarct model. *Biomaterials*, *29*, 4205–4210.

Mikos, A. G., McIntire, L. V., Anderson, J. M., & Babensee, J. E. (1998). Host response to tissue engineered devices. *Adv. Drug Deliv. Rev.*, *33*, 111–139.

Moncada, S. (1982). Eighth Gaddum Memorial Lecture. University of London Institute of Education, December 1980. Biological importance of prostacyclin. *Br. J. Pharmacol.*, *76*, 3–31.

Muller-Ehmsen, J., Peterson, K. L., Kedes, L., Whittaker, P., Dow, J. S., Long, T. I., et al. (2002a). Rebuilding a damaged heart: long-term survival of transplanted neonatal rat cardiomyocytes after myocardial infarction and effect on cardiac function. *Circulation*, *105*, 1720–1726.

Muller-Ehmsen, J., Whittaker, P., Kloner, R. A., Dow, J. S., Sakoda, T., Long, T. I., et al. (2002b). Survival and development of neonatal rat cardiomyocytes transplanted into adult myocardium. *J. Molec. Cell. Cardiol.*, *34*, 107–116.

Murry, C. E., Kay, M. A., Bartosek, T., Hauschka, S. D., & Schwartz, S. M. (1996). Muscle differentiation during repair of myocardial necrosis in rats via gene transfer with MyoD. *J. Clin. Invest.*, *98*, 2209–2217.

Murry, C. E., Soonpaa, M. H., Reinecke, H., Nakajima, H., Nakajima, H. O., Rubart, M., et al. (2004). Haematopoietic stem cells do not transdifferentiate into cardiac myocytes in myocardial infarcts. *Nature*, *428*, 664–668.

Naito, H., Melnychenko, I., Didie, M., Schneiderbanger, K., Schubert, P., Rosenkranz, S., et al. (2006). Optimizing engineered heart tissue for therapeutic applications as surrogate heart muscle. *Circulation*, *114*, I72–I78.

National Institutes of Health. (1999). Working Group on Tissuegenesis and Organogenesis for Heart, Lung and Blood Applications. http://www.nhbli.nih.gov/meetings/workshops/tissueg1.htm

Nerem, R. M. (2003). Role of mechanics in vascular tissue engineering. *Biorheology, 40*, 281–287.

Nygren, J. M., Jovinge, S., Breitbach, M., Sawen, P., Roll, W., Hescheler, J., et al. (2004). Bone marrow-derived hematopoietic cells generate cardiomyocytes at a low frequency through cell fusion, but not trans-differentiation. *Nat. Med., 10*, 494–501.

Oh, H., Bradfute, S. B., Gallardo, T. D., Nakamura, T., Gaussin, V., Mishina, Y., et al. (2003). Cardiac progenitor cells from adult myocardium: homing, differentiation, and fusion after infarction. *Proc. Natl. Acad. Sci. U.S.A, 100*, 12313–12318.

Okita, K., Ichisaka, T., & Yamanaka, S. (2007). Generation of germline-competent induced pluripotent stem cells. *Nature, 448*, 313–317.

Olivetti, G., Quaini, F., Sala, R., Lagrasta, C., Corradi, D., Bonacina, E., et al. (1996). Acute myocardial infarction in humans is associated with activation of programmed myocyte cell death in the surviving portion of the heart. *J. Mol. Cell. Cardiol., 28*, 2005–2016.

Orlic, D., Kajstura, J., Chimenti, S., Jakoniuk, I., Anderson, S. M., Li, B., et al. (2001). Bone marrow cells regenerate infarcted myocardium. *Nature, 410*, 701–705.

Ott, H. C., Matthiesen, T. S., Goh, S. K., Black, L. D., Kren, S. M., Netoff, T. I., et al. (2008). Perfusion-decellularized matrix: using nature's platform to engineer a bioartificial heart. *Nat. Med., 14*, 213–221.

Palmer, R. M., Ashton, D. S., & Moncada, S. (1988). Vascular endothelial cells synthesize nitric oxide from l-arginine. *Nature, 333*, 664–666.

Papadaki, M., Bursac, N., Langer, R., Merok, J., Vunjak-Novakovic, G., & Freed, L. E. (2001). Tissue engineering of functional cardiac muscle: molecular, structural and electrophysiological studies. *Am. J. Physiol.: Heart Circ. Physiol., 280*, H168–H178.

Radisic, M., Park, H., Shing, H., Consi, T., Schoen, F. J., Langer, R., et al. (2004a). Functional assembly of engineered myocardium by electrical stimulation of cardiac myocytes cultured on scaffolds. *Proc. Natl. Acad. Sci. U.S.A, 101*, 18129–18134.

Radisic, M., Yang, L., Boublik, J., Cohen, R. J., Langer, R., Freed, L. E., et al. (2004b). Medium perfusion enables engineering of compact and contractile cardiac tissue. *Am. J. Physiol.: Heart Circ. Physiol., 286*, H507–H516.

Radisic, M., Deen, W., Langer, R., & Vunjak-Novakovic, G. (2005). Mathematical model of oxygen distribution in engineered cardiac tissue with parallel channel array perfused with culture medium containing oxygen carriers. *Am. J. Physiol.: Heart Circ. Physiol., 288*, H1278–H1289.

Radisic, M., Park, H., Chen, F., Salazar-Lazzaro, J. E., Wang, Y., Dennis, R., et al. (2006). Biomimetic approach to cardiac tissue engineering: oxygen carriers and channeled scaffolds. *Tissue Eng., 12*, 2077–2091.

Rafii, S. (2000). Circulating endothelial precursors: mystery, reality, and promise. *J. Clin. Invest., 105*, 17–19.

Rask, F., Dallabrida, S. M., Ismail, N. S., Amoozgar, Z., Yeo, Y., & Radisic, M. (2010). Photocrosslinkable chitosan modified with angiopoietin-1 peptide, QHREDGS, promotes survival of neonatal rat heart cells. *J. Biomed. Mater. Res. Part A, 95*(1), 105–117.

Reffelmann, T., & Kloner, R. A. (2003). Cellular cardiomyoplasty — cardiomyocytes, skeletal myoblasts, or stem cells for regenerating myocardium and treatment of heart failure? *Cardiovasc. Res., 58*, 358–368.

Reinecke, H., Zhang, M., Bartosek, T., & Murry, C. E. (1999). Survival, integration, and differentiation of cardiomyocyte grafts: a study in normal and injured rat hearts. *Circulation, 100*, 193–202.

Reinlib, L., & Field, L. (2000). Cell transplantation as future therapy for cardiovascular disease? A workshop of the National Heart, Lung, and Blood Institute. *Circulation, 101*, e182–e187.

Risau, W. (1995). Differentiation of endothelium. *Faseb J., 9*, 926–933.

Rossini, A. A., Greiner, D. L., & Mordes, J. P. (1999). Induction of immunologic tolerance for transplantation. *Physiol. Rev., 79*, 99–141.

Rubino, A. S., Onorati, F., Santarpino, G., Pasceri, E., Santarpia, G., Cristodoro, L., et al. (2009). Neurohormonal and echocardiographic results after CorCap and mitral annuloplasty for dilated cardiomyopathy. *Ann. Thorac. Surg., 88*, 719–725.

Ryu, J. H., Kim, I. K., Cho, S. W., Cho, M. C., Hwang, K. K., Piao, H., et al. (2005). Implantation of bone marrow mononuclear cells using injectable fibrin matrix enhances neovascularization in infarcted myocardium. *Biomaterials, 26*, 319–326.

Saito, T., Kuang, J. Q., Bittira, B., Al-Khaldi, A., & Chiu, R. C. (2002). Xenotransplant cardiac chimera: immune tolerance of adult stem cells. *Ann. Thorac. Surg., 74*, 19–24, discussion 24.

Sakakibara, Y., Tambara, K., Sakaguchi, G., Lu, F., Yamamoto, M., Nishimura, K., et al. (2003). Toward surgical angiogenesis using slow-released basic fibroblast growth factor. *Eur. J. Cardiothorac. Surg., 24*, 105–111, discussion 112.

Schwarzkopf, R., Shachar, M., Dvir, T., Dayan, Y., Holbova, R., Leor, J., et al. (2006). Autospecies and post-myocardial infarction sera enhance the viability, proliferation, and maturation of 3D cardiac cell culture. *Tissue Eng., 12*, 3467–3475.

Sefton, M. V. (2002). Functional considerations in tissue-engineering whole organs. *Ann. NY Acad. Sci.*, *961*, 198–200.

Sefton, M. V., Zandstra, P., Bauwens, C. L., & Stanford, W. L. (2005). In P. J. del Nido, & S. J. Swanson (Eds.), *Sabiston and Spencer, Surgery of the Chest* (7th ed.). *Tissue regeneration*, Vol. I (pp. 817–831). New York: Saunders.

Sellke, F. W., & Simons, M. (1999). Angiogenesis in cardiovascular disease: current status and therapeutic potential. *Drugs*, *58*, 391–396.

Severs, N. J. (2000). The cardiac muscle cell. *Bioessays*, *22*, 188–199.

Shake, J. G., Gruber, P. J., Baumgartner, W. A., Senechal, G., Meyers, J., Redmond, J. M., et al. (2002). Mesenchymal stem cell implantation in a swine myocardial infarct model: engraftment and functional effects. *Ann. Thorac. Surg.*, *73*, 1919–1925, discussion 1926.

Shen, G. X. (1998). Vascular cell-derived fibrinolytic regulators and atherothrombotic vascular disorders (review). *Int. J. Mol. Med.*, *1*, 399–408.

Sheridan, M. H., Shea, L. D., Peters, M. C., & Mooney, D. J. (2000). Bioabsorbable polymer scaffolds for tissue engineering capable of sustained growth factor delivery. *J. Control Rel.*, *64*, 91–102.

Shimizu, T., Yamato, M., Isoi, Y., Akutsu, T., Setomaru, T., Abe, K., et al. (2002). Fabrication of pulsatile cardiac tissue grafts using a novel 3-dimensional cell sheet manipulation technique and temperature-responsive cell culture surfaces. *Circ. Res.*, *90*, e40–e48.

Shimizu, T., Sekine, H., Yang, J., Isoi, Y., Yamato, M., Kikuchi, A., et al. (2006). Polysurgery of cell sheet grafts overcomes diffusion limits to produce thick, vascularized myocardial tissues. *Faseb J.*, *20*, 708–710.

Soonpaa, M. H., & Field, L. J. (1998). Survey of studies examining mammalian cardiomyocyte DNA synthesis. *Circ. Res.*, *83*, 15–26.

Starling, R. C., & Jessup, M. (2004). Worldwide clinical experience with the CorCap Cardiac Support Device. *J. Card. Fail.*, *10*, S225–S233.

Stevens, K. R., Kreutziger, K. L., Dupras, S. K., Korte, F. S., Regnier, M., Muskheli, V., et al. (2009a). Physiological function and transplantation of scaffold-free and vascularized human cardiac muscle tissue. *Proc. Natl. Acad. Sci. U.S.A*, *106*, 16568–16573.

Stevens, K. R., Pabon, L., Muskheli, V., & Murry, C. E. (2009b). Scaffold-free human cardiac tissue patch created from embryonic stem cells. *Tissue Eng. Part A*, *15*, 1211–1222.

Takahashi, K., & Yamanaka, S. (2006). Induction of pluripotent stem cells from mouse embryonic and adult fibroblast cultures by defined factors. *Cell*, *126*, 663–676.

Takahashi, K., Tanabe, K., Ohnuki, M., Narita, M., Ichisaka, T., Tomoda, K., et al. (2007). Induction of pluripotent stem cells from adult human fibroblasts by defined factors. *Cell*, *131*, 861–872.

Toma, C., Pittenger, M. F., Cahill, K. S., Byrne, B. J., & Kessler, P. D. (2002). Human mesenchymal stem cells differentiate to a cardiomyocyte phenotype in the adult murine heart. *Circulation*, *105*, 93–98.

Vallbacka, J. J., Nobrega, J. N., & Sefton, M. V. (2001). Tissue engineering as a platform for controlled release of therapeutic agents: implantation of microencapsulated dopamine producing cells in the brains of rats. *J. Control Rel.*, *72*, 93–100.

Wall, S. T., Walker, J. C., Healy, K. E., Ratcliffe, M. B., & Guccione, J. M. (2006). Theoretical impact of the injection of material into the myocardium: a finite element model simulation. *Circulation*, *114*, 2627–2635.

Wang, Y., Ameer, G. A., Sheppard, B. J., & Langer, R. (2002). A tough biodegradable elastomer. *Nature Biotechnol.*, *20*, 602–606.

Williams, S. K. (1995). Endothelial cell transplantation. *Cell. Transplant.*, *4*, 401–410.

Wollert, K. C., & Drexler, H. (2005a). Cell therapy for the treatment of coronary heart disease: a critical appraisal. *Nat. Rev. Cardiol.*, *7*, 204–215.

Wollert, K. C., & Drexler, H. (2005b). Clinical applications of stem cells for the heart. *Circ. Res.*, *96*, 151–163.

Wollert, K. C., Meyer, G. P., Lotz, J., Ringes-Lichtenberg, S., Lippolt, P., Breidenbach, C., et al. (2004). Intracoronary autologous bone-marrow cell transfer after myocardial infarction: the BOOST randomized controlled clinical trial. *Lancet*, *364*, 141–148.

Xu, T., Baicu, C., Aho, M., Zile, M., & Boland, T. (2009). Fabrication and characterization of bio-engineered cardiac pseudo tissues. *Biofabrication*, *1*, 1–6.

Yaoita, H., Ogawa, K., Maehara, K., & Maruyama, Y. (2000). Apoptosis in relevant clinical situations: contribution of apoptosis in myocardial infarction. *Cardiovasc. Res.*, *45*, 630–641.

Yau, T. M., Fung, K., Weisel, R. D., Fujii, T., Mickle, D. A., & Li, R. K. (2001). Enhanced myocardial angiogenesis by gene transfer with transplanted cells. *Circulation*, *104*, I218–I222.

Yeo, Y., Geng, W., Ito, T., Kohane, D. S., Burdick, J. A., & Radisic, M. (2007). Photocrosslinkable hydrogel for myocyte cell culture and injection. *J. Biomed. Mater. Res. B Appl. Biomater.*, *81*, 312–322.

Yildirim, Y., Naito, H., Didie, M., Karikkineth, B. C., Biermann, D., Eschenhagen, T., et al. (2007). Development of a biological ventricular assist device: preliminary data from a small animal model. *Circulation*, *116*, I16–I123.

Yla-Herttuala, S., Markkanen, J. E., & Rissanen, T. T. (2004). Gene therapy for ischemic cardiovascular diseases: some lessons learned from the first clinical trials. *Trends Cardiovasc. Med., 14*, 295–300.

Yost, M. J., Baicu, C. F., Stonerock, C. E., Goodwin, R. L., Price, R. L., Davis, J. M., et al. (2004). A novel tubular scaffold for cardiovascular tissue engineering. *Tissue Eng., 10*, 273–284.

Yu, J., Gu, Y., Du, K. T., Mihardja, S., Sievers, R. E., & Lee, R. J. (2009). The effect of injected RGD modified alginate on angiogenesis and left ventricular function in a chronic rat infarct model. *Biomaterials, 30*, 751–756.

Zandstra, P. W., Bauwens, C., Yin, T., Liu, Q., Schiller, H., Zweigerdt, R., et al. (2003). Scalable production of embryonic stem cell-derived cardiomyocytes. *Tissue Eng., 9*, 767–778.

Zhang, J., Wilson, G. F., Soerens, A. G., Koonce, C. H., Yu, J., Palecek, S. P., et al. (2009). Functional cardiomyocytes derived from human induced pluripotent stem cells. *Circ. Res., 104*, e30–e41.

Zhang, M., Methot, D., Poppa, V., Fujio, Y., Walsh, K., & Murry, C. E. (2001). Cardiomyocyte grafting for cardiac repair: graft cell death and anti-death strategies. *J. Mol. Cell Cardiol., 33*, 907–921.

Zhang, P., Zhang, H., Wang, H., Wei, Y., & Hu, S. (2006). Artificial matrix helps neonatal cardiomyocytes restore injured myocardium in rats. *Artif. Organs, 30*, 86–93.

Zimmermann, W. H., Fink, C., Kralish, D., Remmers, U., Weil, J., & Eschenhagen, T. (2000). Three-dimensional engineered heart tissue from neonatal rat cardiac myocytes. *Biotechnol. Bioeng., 68*, 106–114.

Zimmermann, W. H., Didie, M., Wasmeier, G. H., Nixdorff, U., Hess, A., Melnychenko, I., et al. (2002a). Cardiac grafting of engineered heart tissue in syngenic rats. *Circulation, 106*, 1151–1157.

Zimmermann, W. H., Schneiderbanger, K., Schubert, P., Didie, M., Munzel, F., Heubach, J. F., et al. (2002b). Tissue engineering of a differentiated cardiac muscle construct. *Circ. Res., 90*, 223–230.

Zimmermann, W. H., Melnychenko, I., Wasmeier, G., Didie, M., Naito, H., Nixdorff, U., et al. (2006). Engineered heart tissue grafts improve systolic and diastolic function in infarcted rat hearts. *Nat Med., 12*, 452–458.

Zong, X., Bien, H., Chung, C. Y., Yin, L., Fang, D., Hsiao, B. S., et al. (2005). Electrospun fine-textured scaffolds for heart tissue constructs. *Biomaterials, 26*, 5330–5338.

CHAPTER 49

Regenerative Medicine in the Cornea

May Griffith[*,†], **Per Fagerholm**[*], **Neil Lagali**[*], **Malcolm A. Latorre**[‡], **Joanne Hackett**[*], **Heather Sheardown**[§]

* Department of Clinical and Experimental Medicine, Division of Cell Biology,
Linköping University, Linköping, Sweden
‡ Department of Biomedical Engineering, Linköping University, Linköping, Sweden
† University of Ottawa Eye Institute, Ottawa, Ontario, Canada
§ Department of Chemical Engineering, McMaster University, Hamilton, Ontario, Canada

INTRODUCTION: THE NEED FOR REGENERATIVE MEDICINE IN THE CORNEA

The cornea is the transparent window to the eye that transmits and focuses light into the eye for vision. The average human cornea is about 500 µm thick centrally and about 750 µm thick peripherally (Jonas and Holbach, 2005). It consists of three cellular layers: an outer epithelial layer; middle stroma comprising a hydrated extracellular matrix (ECM) with fibroblast-like cells (keratocytes); and an innermost monolayer of endothelial cells. The cornea is highly innervated. Being avascular, it is unique in its dependence upon its sensory nerves and their interactions with corneal cells for maintenance of tissue integrity and wound healing (Lambiase et al., 1999).

The corneal epithelium, which forms the main protective barrier, consists of stratified, non-keratinizing epithelial cells, with a total thickness of approximately 50 µm. The basal layer of the epithelium proliferates to replace the superficial cells lost at the anterior surface (Ren and Wilson, 1996), which are subsequently replenished by a population of corneal stem cells that reside within the corneal/scleral limbus (Nishida, 2005). Several anti-inflammatory and anti-microbial factors are secreted by the epithelium as an insoluble mucous layer that aids in maintaining the tear film (Sack et al., 2001). The corneal stroma, which makes up about 90% of the corneal thickness, consists mainly of type I collagen (13.6%), 0.9% glycosaminoglycans, and 80% water, making the stroma resemble a hydrogel. This "hydrogel" contains over 300 highly ordered lamellae of primarily type I collagen interspersed with stromal cells or keratocytes, and gives the cornea both its strength and transparency. Lastly, the single-cell-thick posterior endothelial layer is essential for the maintenance of stromal hydration and hence corneal transparency. It contains sodium/potassium ATPase pumps that circulate aqueous humor between the anterior chamber and stroma (Nishida, 2005).

Any irreversible damage or failure of corneal and/or limbal cells, and/or nerve damage due to trauma or infection can lead to loss of transparency and hence vision loss or blindness. According to the World Health Organization, diseases of the cornea are a major cause of vision loss, second only to cataracts as the leading cause of blindness (Whitcher et al., 2001). Corneal ulceration and ocular trauma are estimated to result in between 1.5 and 2 million new cases of blindness worldwide on an annual basis; corneal scarring resulting from

911

Principles of Regenerative Medicine. DOI: 10.1016/B978-0-12-381422-7.10049-5

measles is a leading cause of blindness in children. Corneal blindness is estimated to affect more than 10 million individuals worldwide (estimates from the Vision Share Consortium of Eye Banks, USA).

The most successful and widely accepted treatment worldwide for corneal blindness is full-thickness replacement of the damaged organ with an allograft, known as penetrating keratoplasty (PKP). While PKP is generally successful in the short-term, a 15% rejection rate leading to failure of 10% of grafts within 2 years has been reported in Sweden (Claesson et al., 2002). Graft failure rates are even greater in high-risk transplantation, autoimmune disease, alkali burns, and recurrent grafts (Williams et al., 2006). Moreover, graft survival 10–15 years following PKP is only about 55% (Williams et al., 2006). Severe cornea damage caused by conditions including alkali burns, severe dry eye, immunological disorders, stem cell deficiency, vascularization, or ocular diseases such as Stevens-Johnson syndrome (SJS), ocular citracial pemphigoid, and neurotropic scars secondary to herpes zoster ophthalmicus often result in the eye not being able to support corneal transplants (Trinkause-Randall, 2000; Khan et al., 2001). In these cases, reported success rates are much lower; in cases of repeated graft rejection, for example, the success rate of future transplantation drops to near zero (Khan et al., 2001). Additionally, as for all solid organ transplants, donor-derived infection is a serious complication and a leading concern in eye and tissue banking (O'Day, 1989; Remeijer et al., 2001; Hassan et al., 2008). More importantly, there is a severe shortage of donor tissue worldwide, which means millions of patients worldwide remain untreated.

While there have been many efforts to develop corneal substitutes to alleviate both the shortage and drawbacks of human donor tissues, allograft surgery involving full thickness corneal replacement by penetrating keratoplasty (PKP) in particular has remained the gold standard for a century. Lamellar keratoplasty (LKP) is an alternative surgical procedure requiring removal of only the damaged or diseased epithelium and stroma, leaving the endothelium intact, in cases where only the more superficial layers are damaged. Non-penetration of the aqueous humor reduces the rate of rejection and post-operative complications such as leakage, thereby improving long-term graft stability (Funnell et al., 2006; Ardjomand et al., 2007).

As is the case with many other organs, worldwide demand for donor corneas exceeds the supply to graft all patients on waiting lists, even in developed nations in Europe (Muraine et al., 2002). The case is worse in developing countries. Wait times are also expected to further increase with the aging population, as older corneas are less suitable for transplantation and these patients are more likely to require transplants. A further decrease in the availability of acceptable donor tissue is expected with the increasing incidence of infectious diseases, including HIV and hepatitis, as well as the growing popularity of laser *in situ* keratomileusis (LASIK) for correcting refractive errors. These surgically treated corneas are unacceptable for use as donor tissue. These issues are compounded in third world countries, where instances of corneal blindness are rising, yet the skills and resources to perform transplant surgeries are limited (Chirila, 2001). An additional serious disadvantage of cornea allograft transplantation is the possibility for transmission of infection. Person-to-person transmission of the rabies virus (Houff et al., 1979) and at least one case of Creutzfeldt-Jakob disease (Duffy et al., 1974) have been reported. Hepatitis B and C and HIV can be isolated from tears and there is concern about their possible transmission. Another concern is that transmission of as-yet-unknown pathogens could also occur. Within the last few years, there have been significant developments in both biomaterials and stem cell-based methods and combinations of both, to replace part or the full thickness of damaged or diseased corneas. In addition, *in situ* methods for stabilizing and restoring the cornea as alternatives to surgery are rapidly gaining acceptance, for example corneal crosslinking (Wollensak et al., 2003). A selection of these exciting new developments along with the artificial corneas or keratoprostheses (KPros) that have been tested or are in clinical use are discussed in this chapter.

KERATOPROSTHESES

Conventional keratoprostheses

Keratoprostheses (KPro) (commonly referred to as artificial corneas) are usually completely synthetic constructs designed to replace the central portion of an opaque cornea. As several recent, comprehensive reviews are available on keratoprostheses (Myung et al., 2008; Princz et al., 2009), we discuss only representative KPros that are currently used in clinical trials and focus more on the KPros that have been designed with biological components that render the devices more cell interactive.

Most KPro designs are based upon the "core and skirt" concept, with a transparent central optic surrounded by a porous, flexible skirt that enables cellular integration of the host tissue through fibroblast in-growth for anchorage. Although improving and well-retained, kerato-prostheses in general suffer from serious complications including retroprosthetic membrane formation, calcification, infection, glaucoma, and retinal detachment, so that their use is limited to cases where human donor grafting has failed repeatedly or is contraindicated.

The osteo-odonto keratoprosthesis (OOKP), developed by Strampelli (1963), consists of autologous tissue derived from tooth and bone that surrounds a central poly(methyl meth-acrylate) PMMA optic. Before implantation, the osteodental skirt is pre-implanted into the buccal mucosa to allow colonization of fibroblasts to support integration when implanted ocularly. This KPro has been one of the most successful as it has a low extrusion rate, due to the excellent integration of the skirt material (mostly hydroxyapatite) with the host tissue (Mehta et al., 2005). Complications associated with this KPro include retroprosthetic membrane formation, glaucoma, and decentration of the central optic, due to absorption of the osteo-dental skirt (Falcinelli et al., 2005). A number of synthetic osteodental analogues, such as aluminium oxide, hydroxyapatite ceramic and glass ceramic, and hydroxyapatite-coated carbon mesh, have now been developed, with varying degrees of success (Sandeman et al., 2009; Viitala et al., 2009).

A very promising KPro with a PMMA optic is the Boston KPro (previously the Dohlman-Doane KPro), of which there have been several iterations. There are two versions, with the most common one being the single collar button. This design consists of a front plate that is 5.5—7 mm in diameter, connected by a 3.5 mm stem piece to a 7 mm back plate. Despite potential complications such as glaucoma, this KPro has gained usage for patients who have had multiple graft rejections for a number of different conditions, including those with chemical burns, congenital glaucoma, and herpetic keratitis (Khan et al., 2007; Chew et al., 2009). The construction and choice of materials have been improved (Ament et al., 2009). The added use of a contact lens has proven successful and such a lens could eventually be constructed as a drug reservoir (Ciolino et al., 2009). The introduction of continuous topical Vancomycin therapy has reduced the incidence of bacterial endophthalmitis considerably (Durand and Dohlman, 2009). More recently, the glaucoma complication has been addressed by evaluation of cyklophotokoagulation (Rivier et al., 2009) and improvement of the shunting system that has been adapted for these implanted eyes (Dohlman et al., 2010).

The AlphaCor KPro is another prosthesis that has now undergone several iterations and multicentre trials. It is approved for use in a number of countries as an alternative to donor corneal tissue in patients contraindicated for conventional grafting (Hicks et al., 2003a). Implantation involves a two-stage procedure in which the device is placed within an intra-stromal pocket closed by suturing a conjunctival flap over the anterior surface of the cornea. After a period of 12 weeks, the device optic is exposed by removing the conjunctival flap. Complications included stromal melting, retroprosthetic membrane formation, optic damage, and poor biointegration (Hicks et al., 2003a,b, 2006). A comprehensive and unique program of data collection has allowed for ongoing review of complications and risk and protective factors. In early studies, active ocular simplex virus was found to be a contraindication (Hicks et al., 2002) but recent results, however, suggest that, with appropriate therapies, herpes simplex virus (HSV) does not exclude patients from AlphaCor treatment. Additionally, in approximately 20% of clinical trial cases, deposits either on the surface or within the hydrogel optic resulted in diminished vision; these are thought to be related to smoking, or adsorption of certain combinations of medications to the exposed hydrogel leading to calcium deposition (Hicks et al., 2004). However, studies indicate that elimination of the implicated medications effectively prevented calcium deposit formation in more recently implanted devices (Legeais et al., 1997).

Other core skirt KPros based on various materials including a porous semitransparent poly tetrafluoroethylene (PTFE) skirt and a central optic of poly vinyl pyrrolidone (PVP)-coated silicone rubber (poly(dimethyl siloxane) or PDMS) (Legeais et al., 1997; Legeais and Renard, 1998), or poly(butyl methacrylate), hexaethyleneglycolmethacrylate with a dimethacrylate crosslinker (Bruining et al., 2002) have been proposed.

Regenerative medicine applied to keratoprosthesis development

Over the years, KPro researchers have come to believe that it is important for the epithelium to cover the device in order to maintain the tear film, and prevent infection and extrusion of the implant (George and Pitt, 2002; Sweeney et al., 2003). However, since the majority of the materials utilized for traditional KPros are non-cell-adhesive, such as PMMA, poly(vinyl alcohol) (PVA), and PHEMA, improvements have been made to modify the ability of corneal cells to adhere, and migrate over the surface. Naturally occurring extracellular matrix proteins such as collagen, fibronectin, laminin, and other cell adhesive peptides such as IKVAV, YIGSR, and RGD have been grafted onto the KPros (Kobayashi and Ikada, 1991; Merrett et al., 2001; Aucoin et al., 2002; Jacob et al., 2005; Wallace et al., 2005), although other factors including pore size and surface topography (Johnson et al., 2000) can also impact device epithelialization.

More recent *in vitro* work suggests that corneal epithelial cell growth and adhesion were significantly enhanced by tethering of laminin or fibronectin adhesion promoting peptide (FAP) via flexible polyethylene glycol (PEG) chains, more so than by tethering of fibronectin or simple coating of the surface with matrix proteins (Jacob et al., 2005; Wallace et al., 2005). In several other studies, modification with fibronectin-based RGD(S) (Legeais and Renard, 1998; Bruining et al., 2002; George and Pitt, 2002), laminin-based YIGSR (Merrett et al., 2001; Aucoin et al., 2002), and a novel collagen-based peptide Gly—Pro—Nleu (Johnson et al., 2000) has been observed to improve epithelial cell adhesion to various surfaces *in vitro*. Surface modification with combinations of peptides, including the cell adhesion peptides RGDS and YIGSR as well as synergistic counterparts PHSRN and PDSGR, demonstrated that corneal epithelial cell adhesion is greatly improved on surfaces with the cell adhesion peptides and at least one of the counterparts (Aucoin et al., 2002).

Another strategy to improve epithelialization is through the use of growth factors. In particular, epidermal growth factor (EGF) is a potent stimulator of corneal epithelial cell proliferation and migration and is active in the wound healing process. The covalent binding of EGF to PDMS substrates via a PEG tether has been shown to significantly improve cell coverage of the polymer *in vitro* (Klenkler et al., 2005). This is likely correlated to the significantly greater production of various ECM proteins required for cell adhesion. Interestingly, modification with growth factor/ECM peptide combinations did not lead to significant increases in epithelialization *in vitro* despite expected amounts of peptide and growth factor on the surface (Fig. 49.1). Clearly, the interactions between the growth factor-modified polymer and the cells are complex and require further study but have significant potential to alter epithelialization of KPRO materials. Underlying surface modifications also appear to play a role in the extent of cell coverage as well as the density of the EGF on the surface and the presence of EGF in the cell culture medium. In contrast to stimulatory effects, epithelial cell attachment to certain parts of the keratoprosthesis must be inhibited to prevent epithelial downgrowth and retroprosthetic

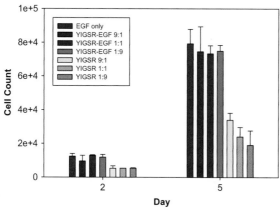

FIGURE 49.1

Effect of surfaces modified with combinations of cell adhesion peptides and epidermal growth factor (EGF) on corneal epithelial cell growth on silicone surfaces. Addition of cell adhesion peptides to the surfaces did not enhance cell numbers at early times relative to surfaces modified with EGF only. However, slightly lower relative cell numbers were observed on the peptide-only modified surfaces. By 5 days, there were clear differences between the EGF-containing and non-EGF-containing surfaces.

membrane formation. Transforming growth factor β (TGFβ) was investigated due to its previously demonstrated ability to inhibit epithelial growth and promote stromal keratocyte proliferation, and hence could potentially be useful for modification of the stromal implant surface. However, the results observed on TGFβ-modified PDMS surfaces *in vitro* were opposite to those expected; keratocyte adhesion was inhibited and epithelial cell growth enhanced by the surface treatment, indicating the complex nature of growth factor-cell interactions (Merrett et al., 2003). Grafting of PEG to PMMA implants, which typically exhibit high protein deposition and cell adhesion associated with retroprosthetic membrane formation, was investigated (Kim et al., 2001). The modification resulted in decreased keratocyte and inflammatory cell adhesion on the polymer surface *in vitro* and in rabbit experiments.

Permeability to oxygen and nutrients, also a key parameter for survival of cells adjacent to a polymeric implant, is the basis for the development of novel materials. In one study, interpenetrating networks of PDMS and hydrogels were found to have glucose permeability levels similar to those of the native cornea (Liu and Sheardown, 2005) and to support corneal epithelial cell adhesion (unpublished data). As well, these materials have been shown to be capable of drug release for periods of 2 weeks or more, which may ultimately be used to stimulate cell interactions (Fig. 49.2).

Novel perfluoropolyether-based materials with both oxygen and nutrient permeability have shown good success in corneal onlay applications, where a corneal onlay, which is a thin lenticule, is placed on top of the corneal stroma underneath the epithelium, or in a pocket, for refractive purposes. To enhance epithelial overgrowth, a 5–10 nm layer of collagen I was covalently immobilized on the anterior surface of each lenticule as a potential substrate material for a keratoprosthesis.

Jacob et al. (2005) coupled cell adhesion peptides and various cytokines to polymethacrylic acid-co-2-hydroxyethyl methacrylate (PHEMA/MAA). The bioactive factors examined included fibronectin, laminin, substance P, IGF-1 (insulin-like growth factor 1), and RGD. They compared the effects of these factors on corneal epithelial cell adhesion and growth rate and adhesion when the bioactive factors were directly coated on the surfaces, or if they were tethered through PEG spacers. They showed that the spacer molecules provided the correct microenvironment for the epithelial cells by exposure of the bioactive motifs, in order to allow the cells to reach confluence, compared to little or no epithelial growth on the surfaces that were only coated with the bioactive factors (Jacob et al., 2005). Of notable importance is that

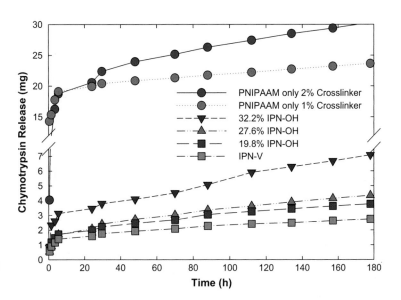

FIGURE 49.2
Cumulative release of a model protein (chymotrypsin) from silicone-hydrogel interpenetrating networks (IPNs). Compared to the PNIPAAM-only controls, the silicone PNIPAAM IPNs showed prolonged release of lower quantities of protein with a smaller burst. Furthermore, release could be controlled by altering the amount of hydrogel in the silicone matrix.

the peptides and factors are exposed to biodegradation, and, therefore, effort must be taken to ensure long-term attachment of the epithelial cells on the surface is maintained.

Myung et al. (2008) reported on a surface-patterned keratoprosthesis comprising a double network of PEG and poly(acrylic acid) (PAA). A recent version comprised a photolithographically patterned device comprising a PEG/PAA central core and a poly (hydroxyethyl acrylate (PHEA) micro-perforated skirt. Coupling of collagen type I to the hydrogel allowed for epithelial coverage in wound healing models both *in vitro* and *in vivo* in rabbits (Myung et al., 2009). The latest iteration, comprising a single-piece keratoprosthesis fabricated using a two-step polymerization process, is under investigation. First a core-skirt construct is fabricated by photolithographic polymerization of PEG. This is then followed by sequential polymerization and crosslinking of acrylic acid within the bulk of the PEG form (Myung et al., 2008).

FULLY CELL-BASED REGENERATIVE THERAPIES
Self-assembled cell-based constructs

Several groups have been developing corneal equivalents using completely natural materials as potentially implantable replacements. The model developed by the Laboratoire d'Organogenese Experimentale (LOEX) (Germain et al., 1999) uses a self-assembly approach whereby stromal cells are provided with the nutrients and appropriate factors such as ascorbic acid to induce production of sheets of collagen and other ECM macromolecules (Gaudreault et al., 2003). These sheets are stacked together and subsequently seeded with epithelial cells; the endothelial cell layer was not included in initial reconstructions although more recent work has focused on the optimization of the culture conditions for endothelial cells for the inclusion of this layer in the construct (Gagnon et al., 2005). Previous work with tissue-engineered blood vessels demonstrated that high tensile strength could be achieved by this method (Auger et al., 2002), suggesting that this might eventually be achieved in the corneal models as well. More recently, Carrier et al. (2008) reported on a model comprising a stroma that consisted of a combination of human corneal and dermal fibroblasts. According to the authors, the combination of the corneal and dermal fibroblasts was more conducive to the formation of a well-differentiated epithelium that showed higher re-epithelialization rates than just corneal fibroblasts alone. This model reproduced the microanatomy of the native human cornea. More importantly, this model was able to reproduce a mechanistically accurate wound healing process and is therefore useful as a tool for studying wound healing, or screening bioactive factors that could modulate wound healing, or as a pre-screen prior to animal testing.

Using a similar approach, Guo et al. (2007) fabricated and then characterized the ECM macromolecules deposited by primary human corneal fibroblasts in such self-assembled corneal substitutes. The average culture took 4 weeks to produce a multi-layered construct about 36 μm thick. These constructs were highly cellular and are morphologically similar to the stroma of mammalian corneas, with multiple, parallel layers of cells and small fibrillar ECM arrays. On average, the collagen fibrils were between 27 and 51 nm, with a mean of 38.1 ± 7.4 nm, compared to the 31 ± 0.8 nm reported in adult human corneas (Meek and Leonard, 1993).

Direct injection of stem cells

In mutant mice that lack lumican proteoglycan, the corneas have an opacity that resembles that of a scarred cornea due to a disrupted stromal organization. Du et al. (2009) isolated stem cells from the adult human corneal stroma and injected these into the corneal stroma of the lumican-deficient mice. In wild-type control animals, the injected human stem cells simply remained within the cornea without fusing with host cells or eliciting an immune T-cell response. Within the pathological corneas, however, the injected human stromal stem cells elaborated human corneal-specific extracellular matrix, including the proteoglycans lumican and keratocan. These accumulated in the treated corneas, restoring stromal thickness and collagen fibril defects in these pathological corneas. This resulted in both restoration of corneal thickness and transparency in these mutant mice to resemble healthy corneas in the wild-type animals. These promising results suggest that direct cell-based therapy could become an effective approach to treatment of human corneal blindness in the future.

BIOMATERIALS-ENHANCED CELL-BASED REGENERATION
Biomaterial scaffolds with cells

In many cases, only one corneal layer may be damaged. In general, the outermost epithelial layer is exposed to the environment, and may be prone to injury such as chemical burns or dry eye syndrome. The stem and progenitor cells that are normally responsible for affecting the repair may also be decimated and, hence, there have been various attempts to re-populate the cornea. Corneal stem cells from the surrounding limbus either from the undamaged contralateral eye (autograft) or from allogeneic sources can be obtained as explants. The explants are most frequently seeded on prepared human amniotic membranes (Nakamura et al., 2006) or fibrin substrates, including autologous fibrin (Rama et al., 2001; Han et al., 2002) and outgrowing cells are allowed to form sheets that are then transplanted onto the damaged eye. Other substrates tested as potential delivery vehicles include cross-linked recombinantly produced human collagen substrates (Dravida et al., 2008) and silk fibroin (Chirila et al., 2008) that support proliferation and differentiation of corneal epithelia from progenitor cells. A vitrified collagen membrane developed by McIntosh Ambrose et al. (2009) that achieved a tensile strength of 6.8 ± 1.5 MPa when hydrated (and 28.6 ± 7.0 MPa when dry) can be used for separate delivery of primary and progenitor cells from all three corneal layers.

In some patients, where both corneal surfaces are depleted of stem cells, for example 12 patients with Stevens-Johnson syndrome, chemical and thermal injury, pseudo-ocular cicatricial pemphigoid, and idiopathic ocular surface disorder, successful autologous reconstruction of the corneal surface by transdifferentiation of oral mucosal epithelium has been performed (Inatomi et al., 2006). The oral mucosal cells were cultured on human amniotic membranes and transplanted onto 15 eyes in 12 patients, with successful, stable outcomes. Kinoshita and co-workers (Inatomi et al., 2006) further suggest that the use of transdifferentiated, autologous epithelial precursor cells may be safer for ocular resurfacing than with allogeneic grafts, in particular for younger patients with the most severe ocular surface disorders. However, it should be noted that all transplanted eyes had some peripheral corneal neovascularization.

917

Multi-layered constructs of animal corneas have been described by several groups. The first description of a functional *in vitro* human corneal equivalent based on human cell lines that expressed biochemical markers and showed physiological function was reported by Griffith et al. (1999). This construct, comprising all three cellular layers of the cornea, consisted of immortalized human corneal cells within and on either side of a collagen-chondroitin sulphate C hydrogel. The construct was able to osmoregulate, and also respond to chemical stimuli by changes in gene expression and transparency. However, it was designed for *in vitro* toxicology as immortalized cells were used, and the scaffold itself was mechanically very weak. The more recent use of decellularized corneal stromas, for example from bovine corneas (Ponce Marquez et al., 2009) for seeding stromal cells, potentially allows for a much stronger substrate for reconstruction of a multi-layered corneal equivalent.

Cell-free biomimetic scaffolds as regeneration templates

While cell growth in two dimensions has been shown on the surfaces of many synthetic polymers, ingrowth or encapsulation (three-dimensional growth) of living cells has only been demonstrated in a few, fully synthetic polymers, particularly polyethylene oxide, polypropylene oxide, and poly(*N*-isopropylacrylamide) (PNiPAAm) (Lee and Mooney, 2001; Hoffman, 2002). In contrast, many natural biopolymer hydrogels, such as those based on alginate, fibrinogen-fibrin, chitosan, agarose, albumin, collagens, and their derivatives, are widely used to encapsulate living cells. Hydrogels of collagen I, the dominant biopolymer in the human cornea, are particularly attractive as matrix replacement-type scaffolds, partly because of their strength at relatively low concentrations, resulting from the virtually rigid rod properties of the collagen type I triple helix (Amis et al., 1985). In addition, collagen brings the cell attachment motif arginine–glycine–glutamic acid (RGD) (Pierschbacher and Ruoslahti, 1987). However, both the biodegradation resistance of collagen I and the strength of hydrogels in general at low concentrations (10wt/vol.%) need to be enhanced by chemical crosslinking (Hoffman, 2002).

A novel NiPAAm-based polymer [poly(*N*-isopropylacrylamide-co-acrylic acid-co-acryloxysuccinimide] or its YIGSR-modified analog (co-polymers abbreviated to terpolymer (TERP) and TERP5, respectively), was co-polymerized with type I bovine atelocollagen to give optically clear hydrogels that could be molded to the curvature and dimensions of a cornea (Li et al., 2003). Collagen-TERP5 hydrogels were sutured into one cornea of each of a series of mini-pigs as lamellar grafts, with contralateral untreated corneas and pig cornea allografts as controls. This study reported for the first time the regrowth of corneal, epithelial, and stromal cells into the implant to reconstitute corneal tissue as well as the restoration of tear film mucin, and regeneration of corneal nerves with concomitant recovery of touch sensitivity by 6 weeks post-operation. Allograft controls had no innervation or sensitivity at this time. Previous studies of restoration of touch sensitivity have indicated that only minimal function is detected even 10 years after partial-thickness lenticule transplantation from a human donor cornea (Kaminski et al., 2002). Using multifunctional dendrimers instead of TERP as collagen crosslinkers, Duan and Sheardown (2005, 2006) were not only able to show improved mechanical strength, but the presence of additional functional groups also allowed these gels to be modified with large and tunable amounts of biologically relevant functional groups. The maximum achievable YIGSR concentration of 3.1×10^{-2} mg/mg collagen is significantly greater than that obtained previously using the NIPAAM-based crosslinking agent at 1.6×10^{-6} mg/mg collagen (Li et al., 2005).

Liu et al. (2009) recently showed that biologically interactive corneal substitutes could also be fabricated from interpenetrating polymeric networks of collagen and a synthetic phosphorylcholine (lipid). In this case, one network comprised collagen (either porcine or recombinant human) crosslinked with 1-ethyl-3-(3-dimethyl aminopropyl) carbodiimide and *N*-hydroxysuccinimide. The other network comprised poly(ethylene glycol) diacrylate crosslinked

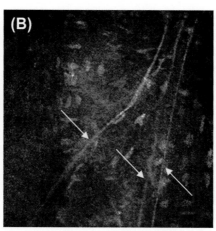
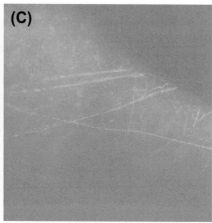

FIGURE 49.3

(A) Rabbit cornea at 9 months after implantation with a hydrogel comprising interpenetrating networks of recombinant human collagen and phosporylcholine (RHC-MPC). (B) Regenerated corneal nerves are present (arrows) and can be visualized non-invasively by *in vivo* confocal microscopy. (C) The nerves are also seen by immunohistochemical localization with *Tuj-1* antibody for nerve fibers, in a flat mount of the cornea.

2-methacryloyloxyethyl phosphorylcholine (MPC). The resulting hydrogels showed an overall increase in mechanical strength beyond that of either original component and enhanced stability against enzymatic digestion (by collagenase) or UV degradation. More importantly, these constructs retained the full biointeractive and cell-friendly properties of collagen in promoting corneal cell and nerve in-growth and regeneration in both normal animal models (despite MPC's known anti-adhesive properties) and alkali-burnt corneas. These hydrogels had refractive indices, white light transmission, and backscatter comparable or superior to those of the human cornea. Glucose and albumin permeability were also comparable to those of human corneas. The porcine collagen could be substituted with recombinant human collagen, resulting in a fully synthetic implant that is free from the potential risks of disease transmission (e.g. prions) present in animal source materials. Recent full-thickness collagen-MPC implants into guinea pig corneas showed for the first time, by electrophysiology, that the subtypes of corneal sensory nerves were regenerated within the implants by 8 months post-operative, that is, the nerves were functional (McLaughlin et al., 2010). This was in addition to the reconstitution of corneal tissue within the implant by in-growth of cells from endogenous progenitors. Similarly, recombinant human collagen-MPC implants in rabbits also showed extensive nerve regeneration (Fig. 49.3).

To date, only EDC and NHS crosslinked recombinant human collagen corneal substitutes have been tested in humans in a phase I clinical study in Sweden as lamellar grafts in 10 patients by the authors. 24-month post-operative results show the regeneration of epithelium and in-growth of stromal cells, anchoring the implants. Most significantly, as in animal studies that used healthy specimens, we show the presence of regenerating nerves within these pathological human corneas (Fagerholm et al., 2010). Two-year clinical results show implants have been stably retained without adverse reactions or need for long-term immunosuppression (Fig. 49.4) and therefore are suitable as temporary grafts or patches. However, longer-term monitoring and more extensive testing are needed to determine whether or not they will be useful as substitutes for donor tissue. In addition, further modifications, such as the use of interpenetrating networks, are likely needed to address the needs of a wider range of clinical indications.

Bioactive collagen-based corneal substitutes can also incorporate micro- or nanoparticles that would release a drug to possibly treat existing conditions, thereby extending their functionality to a wider number of clinical indications. For example, the incorporation of a porous silica dioxide nanoparticle-encapsulated anti-viral drug, Acyclovir, within a collagen-MPC hydrogel, was able to sustain drug release over 10 days to suppress viral activity *in vitro* (Bareiss et al., 2010). In the

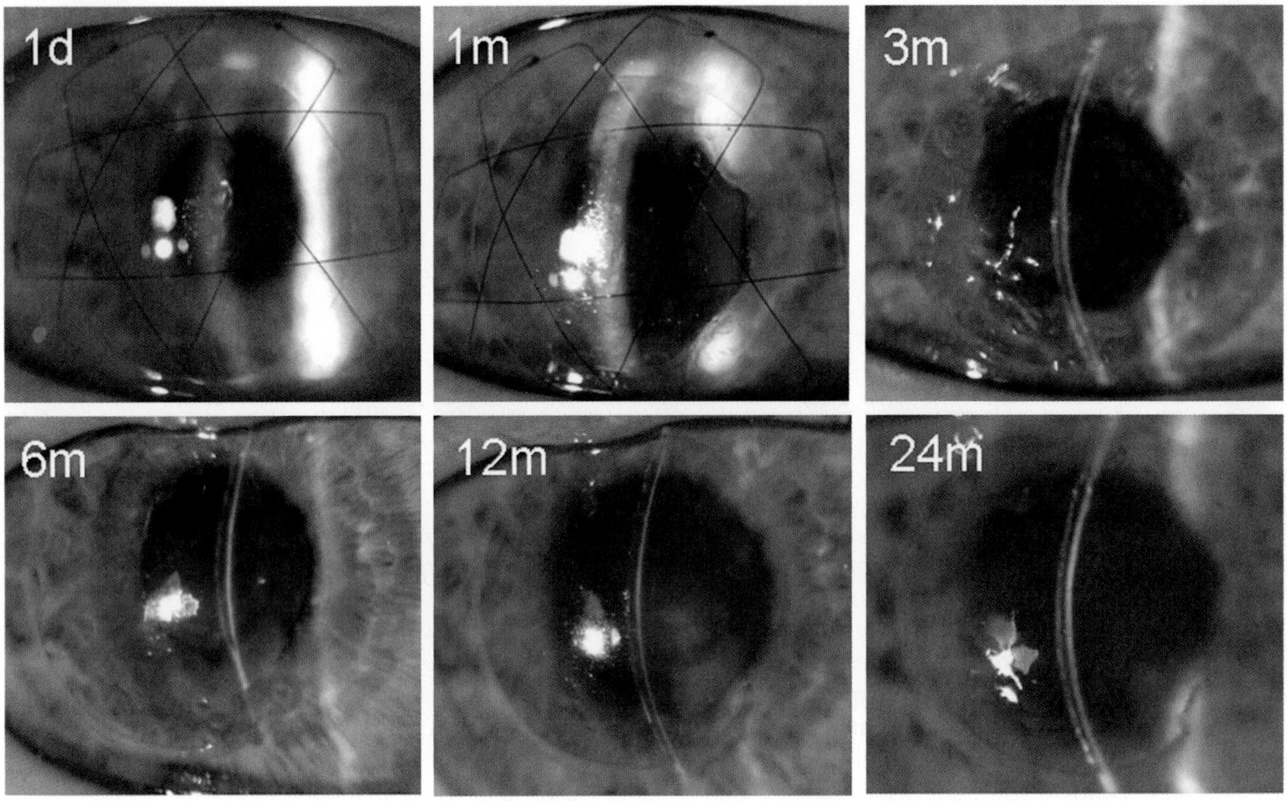

FIGURE 49.4
Slit lamp images of the cornea of a keratoconus patient who was implanted with a biosynthetic cardbodiimide crosslinked, recombinant human collagen corneal substitute at 1 day post-operation, and then at 1, 3, 6, 12, and 24 months after surgery.

future, such composite corneal constructs might be useful for prevention of viral reactivation and re-infection in high-risk transplants such as of herpetic corneas during transplantation surgery.

IN SITU REPAIR AND ENHANCEMENT OF WEAK CORNEAS

Keratoconus is a pathological condition that causes thinning and concomitant weakening of the cornea, causing it to bulge outwards as the disease progresses. Normally, treatment for keratoconus involves corneal transplantation, but a new method has been developed that approaches the problem from a materials viewpoint. This treatment utilizes a photosensitizer (riboflavin) and a light source (UVA) to crosslink the cornea *in situ* and has been applied in clinical studies (Wollensak et al., 2003; Caporossi et al., 2006; Seiler and Hafezi, 2006).

The crosslinking treatment was developed by Wollensak et al. (2003) and involved abrading the corneal epithelium to allow a riboflavin photosensitizer to diffuse throughout the stroma. This was followed by alternating bathing the eye with the photosensitizer and UVA exposure for a total of 30 minutes. The UVA illuminating power level used in clinical work was set to 3 mW/cm^2 at the surface of the eye and allows the stroma to be crosslinked to a depth of about 300 μm (Spoerl et al., 2007). Spoerl et al. (2007) postulate that the crosslinking was due to the photochemical reaction of collagen caused by the production of oxygen radicals by riboflavin and UVA light, inducing a change at the end of an amine group (lysine) (Raiskup et al., 2009). After the treatment, the new reactive groups can form new covalent bonds. In the first study, 23 eyes were treated by Wollensak et al. (2003). Of these, all cases of advancing keratoconus progress stopped in the follow-up period, ranging from 3 months to 4 years. Seventy percent of these eyes showed an average regression of 2.01 diopters, and 65% of the eyes indicated a slight improvement in visual acuity.

Caporossi et al. (2006) reported similar results with their clinical study of 10 eyes, where a decrease in keratoconus readings of 2.10 diopters and an improvement in visual acuity of 3.6 lines were achieved. In a 1-year clinical study where 163 corneas in 127 patients were UV crosslinked, 149 (91.4%) eyes of 114 patients had a clear cornea, while permanent, clinically significant haze developed in 14 eyes (8.6%) of 13 patients (Raiskup et al., 2009). Similar complications were observed in a second study of another 117 eyes of another 99 patients (Koller et al., 2009). The results suggest that this method, while generally effective, does carry a risk of haze development, particularly in cases of advanced keratoconus where the cornea is significantly thinned and protrudes more. However, very recent results suggest that the thinned corneas would be pre-operatively swelled using hypoosmolar riboflavin solutions prior to crosslinking, which would allow patients with thin corneas to receive treatment (Hafezi et al., 2009).

CONCLUSIONS AND FUTURE PERSPECTIVE

There have been significant developments in regenerative medicine-based approaches to corneal repair and regeneration. These include biomaterials and stem cell-based methods and combinations of both, to replace part or the full thickness of damaged or diseased corneas. These different approaches may soon be able to supplement the supply of post-mortem human corneas harvested for transplantation, thereby meeting the demand for donor corneas. In addition, the development of new *in situ* reinforcement methods holds a promise of applying a materials approach to repairing pathological corneal tissue without transplantation, while the direct injection of stem cells into pathological corneas suggests a purely cell-based therapeutic approach may also be viable.

References

Ament, J. D., Spur-Michaud, S. J., Dohlman, C. H., & Gipson, I. K. (2009). The Boston Keratoprosthesis: comparing corneal epithelial cell compatibility with titanium and PMMA. *Cornea, 28,* 808−1131.

Amis, E., Carriere, C., Ferry, J., & Veis, A. (1985). Effect of pH on collagen flexibility determined from dilute solution viscoelastic measurements. *Int. J. Biol. Macromol., 7,* 130−134.

Ardjomand, N., Hau, S., McAlister, J. C., Bunce, C., Galaretta, D., Tuft, S. J., et al. (2007). Quality of vision and graft thickness in deep anterior lamellar and penetrating corneal allografts. *Am. J. Ophthalmol., 143,* 228−235.

Aucoin, L., Griffith, C. M., Pleizier, G., Deslandes, Y., & Sheardown, H. (2002). Interactions of corneal epithelial cells and surfaces modified with cell adhesion peptide combinations. *J. Biomater. Sci. Polym. Ed., 13,* 447−462.

Auger, F. A., Remy-Zolghadri, M., Grenier, G., & Germain, L. (2002). A truly new approach for tissue engineering: the LOEX self-assembly technique. *Ernst Schering Res. Found. Workshop,* 73−88.

Bareiss, B., Ghorbani, M., Li, F., Blake, J. A., Scaiano, J. C., Zhang, J., et al. (2010). Controlled release of acyclovir through bioengineered corneal implants with silica nanoparticle carriers. *Open Tissue Eng. Regen. Med. J, 3,* 10−17.

Bruining, M. J., Pijpers, A. P., Kingshott, P., & Koole, L. H. (2002). Studies on new polymeric biomaterials with tunable hydrophilicity, and their possible utility in corneal repair surgery. *Biomaterials, 23,* 1213−1219.

Caporossi, A., Baiocchi, S., Mazzotta, C., Traversi, C., & Caporossi, T. (2006). Parasurgical therapy for keratoconus by riboflavin-ultraviolet type A rays induced crosslinking of corneal collagen: preliminary refractive results in an Italian study. *J. Cataract Refract. Surg., 32,* 837−845.

Carrier, P., Deschambeault, A., Talbot, M., Giasson, C. J., Auger, F. A., Guerin, S. L., et al. (2008). Characterization of wound reepithelialization using a new human tissue-engineered corneal wound healing model. *Invest. Ophthalmol. Vis. Sci., 49,* 1376−1385.

Chew, H. F., Ayres, B. D., Hammersmith, K. M., Rapuano, C. J., Laibson, P. R., Myers, J. S., et al. (2009). Boston keratoprosthesis outcomes and complications. *Cornea, 28,* 989−996.

Chirila, T. V. (2001). An overview of the development of artificial corneas with porous skirts and the use of PHEMA for such an application. *Biomaterials, 22,* 3311−3317.

Chirila, T. V., Barnard, Z., Zainuddina, Harkin, D. G., Schwab, I. R., & Hirst, L. W. (2008). *Bombyx mori* silk fibroin membranes as potential substrata for epithelial constructs used in the management of ocular surface disorders. *Tissue Eng. Part A, 14,* 1203−1211.

Ciolino, J. B., Hoare, T. R., Iwata, N. G., Behlau, I., Dohlman, C. H., Langer, R., et al. (2009). A drug-eluting contact lens. *Invest. Ophthalmol. Vis. Sci., 50,* 3346−3352.

Claesson, M., Armitage, W. J., Fagerholm, P., & Stenevi, U. (2002). Visual outcome in corneal grafts: a preliminary analysis of the Swedish Corneal Transplant Register. *Br. J. Ophthalmol., 86,* 174–180.

Dohlman, C. H., Grosskreutz, C. L., Chen, T. C., Pasquale, L. R., Rubin, P. A., Kim, E. C., et al. (2010). Shunts to divert aqueous humor to distant epithelialized cavities after keratoprosthesis surgery. *J. Glaucoma, 19,* 111–115.

Dravida, S., Gaddipati, S., Griffith, M., Merrett, K., Lakshmi, S., Sangwan, V. S., et al. (2008). A biomimetic scaffold for culturing limbal stem cells: promising alternative for clinical transplantation. *J. Tissue Eng. Regen. Med., 2,* 263–271.

Du, Y., Carlson, E. C., Funderburgh, M. L., Birk, D. E., Pearlman, E., Guo, N., et al. (2009). Stem cell therapy restores transparency to defective murine corneas. *Stem Cells, 27,* 1635–1642.

Duan, X., & Sheardown, H. (2005). Crosslinking of collagen with dendrimers. *J. Biomed. Mater. Res. A, 75,* 510–518.

Duan, X., & Sheardown, H. (2006). Dendrimer crosslinked collagen as a corneal tissue engineering scaffold: mechanical properties and corneal epithelial cell interactions. *Biomaterials, 27,* 4608–4617.

Duffy, P., Wolf, J., Collins, G., DeVoe, A. G., Streeten, B., & Cowen, D. (1974). Letter: possible person-to-person transmission of Creutzfeldt–Jakob disease. *N. Engl. J. Med., 290,* 692–693.

Durand, M. L., & Dohlman, C. H. (2009). Successful prevention of bacterial endophthalmitis in eyes with the Boston keratoprosthesis. *Cornea, 28,* 896–901.

Fagerholm, P., Lagali, N. S., Merrett, K., Jackson, W. B., Munger, R., Liu, Y., et al. (2010). A biosynthetic alternative to human donor tissue for inducing corneal regeneration: 24-month follow-up of a Phase 1 clinical study. *Sci. Transl. Med, 2,* 46ra61.

Falcinelli, G., Falsini, B., Taloni, M., Colliardo, P., & Falcinelli, G. (2005). Modified osteo-odonto-keratoprosthesis for treatment of corneal blindness: long-term anatomical and functional outcomes in 181 cases. *Arch. Ophthalmol., 123,* 1319–1329.

Funnell, C. L., Ball, J., & Noble, B. A. (2006). Comparative cohort study of the outcomes of deep lamellar keratoplasty and penetrating keratoplasty for keratoconus. *Eye (Lond.), 20,* 527–532.

Gagnon, N., Auger, F., & Germain, L. (2005). Porcine corneal endothelial cell culture improvement: effect of initial seeding density and presence of a feeder layer. *Invest. Ophthalmol. Vis. Sci., 46,* E-Abstract 5006.

Gaudreault, M., Carrier, P., Larouche, K., Leclerc, S., Giasson, M., Germain, L., et al. (2003). Influence of sp1/sp3 expression on corneal epithelial cells proliferation and differentiation properties in reconstructed tissues. *Invest. Ophthalmol. Vis. Sci., 44,* 1447–1457.

George, A., & Pitt, W. G. (2002). Comparison of corneal epithelial cellular growth on synthetic cornea materials. *Biomaterials, 23,* 1369–1373.

Germain, L., Auger, F. A., Grandbois, E., Guignard, R., Giasson, M., Boisjoly, H., et al. (1999). Reconstructed human cornea produced *in vitro* by tissue engineering. *Pathobiology, 67,* 140–147.

Griffith, M., Osborne, R., Munger, R., Xiong, X., Doillon, C. J., Laycock, N. L., et al. (1999). Functional human corneal equivalents constructed from cell lines. *Science, 286,* 2169–2172.

Guo, X., Hutcheon, A. E., Melotti, S. A., Zieske, J. D., Trinkaus-Randall, V., & Ruberti, J. W. (2007). Morphologic characterization of organized extracellular matrix deposition by ascorbic acid-stimulated human corneal fibroblasts. *Invest. Ophthalmol. Vis. Sci., 48,* 4050–4060.

Hafezi, F., Mrochen, M., Iseli, H. P., & Seiler, T. (2009). Collagen crosslinking with ultraviolet-A and hypoosmolar riboflavin solution in thin corneas. *J. Cataract Refract. Surg., 35,* 621–624.

Han, B., Schwab, I. R., Madsen, T. K., & Isseroff, R. R. (2002). A fibrin-based bioengineered ocular surface with human corneal epithelial stem cells. *Cornea, 21,* 505–510.

Hassan, S. S., Wilhelmus, K. R., Dahl, P., Davis, G. C., Roberts, R. T., Ross, K. W., et al. (2008). Infectious disease risk factors of corneal graft donors. *Arch. Ophthalmol., 126,* 235–239.

Hicks, C. R., Crawford, G. J., Tan, D. T., Snibson, G. R., Sutton, G. L., Gondhowiardjo, T. D., et al. (2002). Outcomes of implantation of an artificial cornea, AlphaCor: effects of prior ocular herpes simplex infection. *Cornea, 21,* 685–690.

Hicks, C. R., Crawford, G. J., Lou, X., Tan, D. T., Snibson, G. R., Sutton, G., et al. (2003a). Corneal replacement using a synthetic hydrogel cornea, AlphaCor: device, preliminary outcomes and complications. *Eye (Lond.) 17,* 385–392.

Hicks, C. R., Crawford, G. J., Tan, D. T., Snibson, G. R., Sutton, G. L., Downie, N., et al. (2003b). AlphaCor cases: comparative outcomes. *Cornea, 22,* 583–590.

Hicks, C. R., Chirila, T. V., Werner, L., Crawford, G. J., Apple, D. J., & Constable, I. J. (2004). Deposits in artificial corneas: risk factors and prevention. *Clin. Exp. Ophthalmol., 32,* 185–191.

Hicks, C. R., Crawford, G. J., Dart, J. K., Grabner, G., Holland, E. J., Stulting, R. D., et al. (2006). AlphaCor: clinical outcomes. *Cornea, 25,* 1034–1042.

Hoffman, A. S. (2002). Hydrogels for biomedical applications. *Adv. Drug Deliv. Rev., 54,* 3–12.

922

Houff, S. A., Burton, R. C., Wilson, R. W., Henson, T. E., London, W. T., Baer, G. M., et al. (1979). Human- to-human transmission of rabies virus by corneal transplant. *N. Engl. J. Med., 300,* 603—604.

Inatomi, T., Nakamura, T., Kojyo, M., Koizumi, N., Sotozono, C., & Kinoshita, S. (2006). Ocular surface reconstruction with combination of cultivated autologous oral mucosal epithelial transplantation and penetrating keratoplasty. *Am. J. Ophthalmol., 142,* 757—764.

Jacob, J. T., Rochefort, J. R., Bi, J., & Gebhardt, B. M. (2005). Corneal epithelial cell growth over tethered-protein/peptide surface-modified hydrogels. *J. Biomed. Mater. Res. B Appl. Biomater., 72,* 198—205.

Johnson, G., Jenkins, M., McLean, K. M., Griesser, H. J., Kwak, J., Goodman, M., et al. (2000). Peptoid-containing collagen mimetics with cell binding activity. *J. Biomed. Mater. Res., 51,* 612—624.

Jonas, J. B., & Holbach, L. (2005). Central corneal thickness and thickness of the lamina cribrosa in human eyes. *Invest. Ophthalmol. Vis. Sci., 46,* 1275—1279.

Kaminski, S. L., Biowski, R., Lukas, J. R., Koyuncu, D., & Grabner, G. (2002). Corneal sensitivity 10 years after epikeratoplasty. *J. Refract. Surg., 18,* 731—736.

Khan, B., Dudenhoefer, E. J., & Dohlman, C. H. (2001). Keratoprosthesis: an update. *Curr. Opin. Ophthalmol., 12,* 282—287.

Khan, B. F., Harissi-Dagher, M., Pavan-Langston, D., Aquavella, J. V., & Dohlman, C. H. (2007). The Boston keratoprosthesis in herpetic keratitis. *Arch. Ophthalmol., 125,* 745—749.

Kim, M. K., Park, I. S., Park, H. D., Wee, W. R., Lee, J. H., Park, K. D., et al. (2001). Effect of poly(ethylene glycol) graft polymerization of poly(methyl methacrylate) on cell adhesion. *In vitro* and *in vivo* study. *J. Cataract Refract. Surg., 27,* 766—774.

Klenkler, B. J., Griffith, M., Becerril, C., West-Mays, J. A., & Sheardown, H. (2005). EGF-grafted PDMS surfaces in artificial cornea applications. *Biomaterials, 26,* 7286—7296.

Kobayashi, H., & Ikada, Y. (1991). Corneal cell adhesion and proliferation on hydrogel sheets bound with cell-adhesive proteins. *Curr. Eye Res., 10,* 899—908.

Koller, T., Mrochen, M., & Seiler, T. (2009). Complication and failure rates after corneal crosslinking. *J. Cataract Refract. Surg., 35,* 1358—1362.

Lambiase, A., Rama, P., Aloe, L., & Bonini, S. (1999). Management of neurotrophic keratopathy. *Curr. Opin. Ophthalmol., 10,* 270—276.

Lee, K. Y., & Mooney, D. J. (2001). Hydrogels for tissue engineering. *Chem. Rev., 101,* 1869—1879.

Legeais, J. M., & Renard, G. (1998). A second generation of artificial cornea (Biokpro II). *Biomaterials, 19,* 1517—1522.

Legeais, J. M., Drubaix, I., Briat, B., Renard, G., & Pouliquen, Y. (1997). 2nd generation bio-integrated keratoprosthesis. Implantation in animals. *J. Fr. Ophtalmol., 20,* 42—48.

Li, F., Carlsson, D., Lohmann, C., Suuronen, E., Vascotto, S., Kobuch, K., et al. (2003). Cellular and nerve regeneration within a biosynthetic extracellular matrix for corneal transplantation. *Proc. Natl. Acad. Sci. U.S.A., 100,* 15346—15351.

Li, F., Griffith, M., Li, Z., Tanodekaew, S., Sheardown, H., Hakim, M., et al. (2005). Recruitment of multiple cell lines by collagen-synthetic copolymer matrices in corneal regeneration. *Biomaterials, 26,* 3093—3104.

Liu, L., & Sheardown, H. (2005). Glucose permeable poly (dimethyl siloxane) poly (N-isopropyl acrylamide) interpenetrating networks as ophthalmic biomaterials. *Biomaterials, 26,* 233—244.

Liu, W., Deng, C., McLaughlin, C. R., Fagerholm, P., Lagali, N. S., Heyne, B., et al. (2009). Collagen-phosphorylcholine interpenetrating network hydrogels as corneal substitutes. *Biomaterials, 30,* 1551—1559.

McIntosh Ambrose, W., Salahuddin, A., So, S., Ng, S., Ponce Marquez, S., Takezawa, T., et al. (2009). Collagen Vitrigel membranes for the *in vitro* reconstruction of separate corneal epithelial, stromal, and endothelial cell layers. *J. Biomed. Mater. Res. B Appl. Biomater., 90,* 818—831.

McLaughlin, C. R., Acosta, M. C., Luna, C., Liu, W., Belmonte, C., Griffith, M., et al. (2010). Regeneration of functional nerves within full thickness collagen-phosphorylcholine corneal substitute implants in guinea pigs. *Biomaterials, 31,* 2770—2778.

Meek, K. M., & Leonard, D. W. (1993). Ultrastructure of the corneal stroma: a comparative study. *Biophys. J., 64,* 273—280.

Mehta, J. S., Futter, C. E., Sandeman, S. R., Faragher, R. G., Hing, K. A., Tanner, K. E., et al. (2005). Hydroxyapatite promotes superior keratocyte adhesion and proliferation in comparison with current keratoprosthesis skirt materials. *Br. J. Ophthalmol., 89,* 1356—1362.

Merrett, K., Griffith, C. M., Deslandes, Y., Pleizier, G., & Sheardown, H. (2001). Adhesion of corneal epithelial cells to cell adhesion peptide modified pHEMA surfaces. *J. Biomater. Sci. Polym. Ed., 12,* 647—671.

Merrett, K., Griffith, C. M., Deslandes, Y., Pleizier, G., Dube, M. A., & Sheardown, H. (2003). Interactions of corneal cells with transforming growth factor beta 2- modified poly dimethyl siloxane surfaces. *J. Biomed. Mater. Res. A, 67,* 981—993.

Muraine, M., Toubeau, D., Menguy, E., & Brasseur, G. (2002). Analysing the various obstacles to cornea postmortem procurement. *Br. J. Ophthalmol., 86,* 864–868.

Myung, D., Duhamel, P. E., Cochran, J. R., Noolandi, J., Ta, C. N., & Frank, C. W. (2008). Development of hydrogel-based keratoprostheses: a materials perspective. *Biotechnol. Prog., 24,* 735–741.

Myung, D., Farooqui, N., Zheng, L. L., Koh, W., Gupta, S., Bakri, A., et al. (2009). Bioactive interpenetrating polymer network hydrogels that support corneal epithelial wound healing. *J. Biomed. Mater. Res. A, 90,* 70–81.

Nakamura, T., Inatomi, T., Sotozono, C., Ang, L. P., Koizumi, N., Yokoi, N., et al. (2006). Transplantation of autologous serum-derived cultivated corneal epithelial equivalents for the treatment of severe ocular surface disease. *Ophthalmology, 113,* 1765–1772.

Nishida, T. (2005). Fundamentals of cornea and external disease. In J. H. Krachmer, M. J. Mannis, & E. J. Holland (Eds.), *Cornea* (Vol.1) (2nd ed.) (pp. 3–26). Mosby, PA: Elsevier.

O'Day, D. M. (1989). Diseases potentially transmitted through corneal transplantation. *Ophthalmology, 96,* 1133–1137, discussion 1137–1138.

Pierschbacher, M. D., & Ruoslahti, E. (1987). Influence of stereochemistry of the sequence Arg-Gly-Asp-Xaa on binding specificity in cell adhesion. *J. Biol. Chem., 262,* 17294–17298.

Ponce Marquez, S., Martinez, V. S., McIntosh Ambrose, W., Wang, J., Gantxegui, N. G., Schein, O., et al. (2009). Decellularization of bovine corneas for tissue engineering applications. *Acta Biomater., 5,* 1839–1847.

Princz, M. A., Griffith, M., & Sheardown, H. (2009). Corneal tissue engineering versus synthetic artificial corneas. In T. V. Chirila (Ed.), *Biomaterials and Regenerative Medicine in Ophthalmology* (pp. 134–149). Cambridge, UK: CRC Press (Woodhead Publishing).

Raiskup, F., Hoyer, A., & Spoerl, E. (2009). Permanent corneal haze after riboflavin-UVA-induced crosslinking in keratoconus. *J. Refract. Surg., 25,* S824–S828.

Rama, P., Bonini, S., Lambiase, A., Golisano, O., Paterna, P., de Luca, M., et al. (2001). Autologous fibrin-cultured limbal stem cells permanently restore the corneal surface of patients with total limbal stem cell deficiency. *Transplantation, 72,* 1478–1485.

Remeijer, L., Maertzdorf, J., Doornenbal, P., Verjans, G. M., & Osterhaus, A. D. (2001). Herpes simplex virus 1 transmission through corneal transplantation. *Lancet, 357,* 442.

Ren, H., & Wilson, G. (1996). Apoptosis in the corneal epithelium. *Invest. Ophthalmol. Vis. Sci., 37,* 1017–1025.

Rivier, D., Paula, J., Kim, E., Dohlman, C. H., & Grosskreutz, C. (2009). Glaucoma and keratoprostesis surgery: role of adjunctive cyclophotokoagulation. *J. Glaucoma, 18,* 321–324.

Sack, R. A., Nunes, I., Beaton, A., & Morris, C. (2001). Host-defense mechanism of the ocular surfaces. *Biosci. Rep., 21,* 463–480.

Sandeman, S. R., Jeffery, H., Howell, C. A., Smith, M., Mikhalovsky, S. V., & Lloyd, A. W. (2009). The *in vitro* corneal biocompatibility of hydroxyapatite-coated carbon mesh. *Biomaterials, 30,* 3143–3149.

Seiler, T., & Hafezi, F. (2006). Corneal crosslinking-induced stromal demarcation line. *Cornea, 25,* 1057–1059.

Spoerl, E., Mrochen, M., Sliney, D., Trokel, S., & Seiler, T. (2007). Safety of UVA-riboflavin crosslinking of the cornea. *Cornea, 26,* 385–389.

Strampelli, B. (1963). Osteo-odontokeratoprosthesis. *Ann. Ottalmol. Clin. Ocul., 89,* 1039–1044.

Sweeney, D. F., Xie, R. Z., Evans, M. D., Vannas, A., Tout, S. D., Griesser, H. J., et al. (2003). A comparison of biological coatings for the promotion of corneal epithelialization of synthetic surface *in vivo*. *Invest. Ophthalmol. Vis. Sci., 44,* 3301–3309.

Trinkaus-Randall, V. (2000). Cornea. In R. Lanza, R. Langer, & J. Vacanti (Eds.), *Principles of Tissue Engineering* (pp. 471–491). San Diego: Academic Press.

Viitala, R., Franklin, V., Green, D., Liu, C., Lloyd, A., & Tighe, B. (2009). Towards a synthetic osteo-odonto-keratoprosthesis. *Acta Biomater., 5,* 438–452.

Wallace, C., Jacob, J. T., Stoltz, A., Bi, J., & Bundy, K. (2005). Corneal epithelial adhesion strength to tethered-protein/peptide modified hydrogel surfaces. *J. Biomed. Mater. Res. A, 72,* 19–24.

Whitcher, J. P., Srinivasan, M., & Upadhyay, M. P. (2001). Corneal blindness: a global perspective. *Bull. World Health Organ., 79,* 214–221.

Williams, K. A., Esterman, A. J., Bartlett, C., Holland, H., Hornsby, N. B., & Coster, D. J. (2006). How effective is penetrating corneal transplantation? Factors influencing long-term outcome in multivariate analysis. *Transplantation, 81,* 896–901.

Wollensak, G., Spoerl, E., & Seiler, T. (2003). Riboflavin/ultraviolet-a-induced collagen crosslinking for the treatment of keratoconus. *Am. J. Ophthalmol., 135,* 620–627.

924

CHAPTER 50

Alimentary Tract

Richard M. Day
Centre for Gastroenterology & Nutrition, Division of Medicine, University College London, London, UK

INTRODUCTION

The alimentary tract is a hollow organ starting at the mouth and terminating at the anus. It conducts a number of highly complex and diverse functions that are regulated by distinct cellular and functional differences along its length, which allow it to perform its primary function of providing the body with nutrients, water, and electrolytes. To achieve this, food must be propelled along at a rate that will allow efficient digestion and absorption to take place, whilst also enabling waste products to be excreted in a controlled manner. In addition to this, an important symbiotic relationship exists between bacterial species that colonize the alimentary tract and the host (Qin et al., 2010). Therefore, the surface of the alimentary tract leds to provide an important barrier against unwanted entry of organisms and toxins. When the barrier function is breached, the gut also functions as an immune organ to protect the host.

925

Due its complexity, dysfunction of the alimentary tract may result from a variety of congenital and acquired conditions. This chapter will discuss the current knowledge regarding tissue engineering of different components of the alimentary tract, highlighting successful strategies as well as failures and some of the obstacles that have yet to be overcome in this rapidly developing field.

ESOPHAGUS

The esophagus is a muscular tube approximately 25 cm long in adults. It functions primarily as a conduit that connects the pharynx to the stomach, providing coordinated peristaltic contractions in response to swallowing that propel food into the stomach. The esophageal mucosa is lined by stratified, squamous, non-keratinized epithelium. The submucosa contains muscle, nerve, blood vessels, lymphatics, and mucosal glands. The well-developed muscularis has two layers consisting of an outer longitudinal layer and an inner circular layer. Both layers are striated muscle in the upper portion and smooth muscle in the lower third, being continuous with the muscle layers of the stomach. The myenteric plexus exists between the muscle layers. The esophagus has no serosa and its vascular supply is less extensive compared with the intra-abdominal portions of the gut. Sphincters at the upper and lower ends of the esophagus ensure food is transferred appropriately between it and the pharynx or stomach. The upper esophageal sphincter, found in the upper 3–4 cm of the esophagus, and lower esophageal sphincter, located 2–5 cm above the gastroesophageal junction, remain tonically and strongly constricted to prevent air entering the esophagus during respiration between swallowing, and to prevent reflux of stomach contents into the esophagus between peristaltic waves, respectively. Regenerative medicine techniques are being explored for a number of conditions affecting the esophagus. Gastroesophageal reflux disease is one of the most

Principles of Regenerative Medicine. DOI: 10.1016/B978-0-12-381422-7.10050-1

common disorders affecting the gastrointestinal tract resulting from lower esophageal sphincter incompetence. Medical therapy is generally safe and effective in the majority of patients, but for patients where this fails anti-reflux surgery or endoscopic procedures that involve the injection of bulking materials have been used with limited success to narrow the lumen of the lower esophagus. More recently, the feasibility of using muscle precursor cells to restore gastroesophageal function in a model of gastroesophageal reflux disease has been explored (Fascetti-Leon et al., 2007). Muscle precursor cells isolated from expanded satellite cells derived from skeletal muscle fibers were injected into the gastroesophageal junction following cryoinjury. Histology showed an increase in myofibers at the site of injection that had fused into newly formed or pre-existing myofibers. Future studies will need to demonstrate that cells injected in this manner can contribute to a functional improvement of damaged esophageal sphincter, but the feasibility of using this approach offers a promising new therapy for this common condition.

Esophageal reconstruction is a requirement for congenital esophageal atresia, burns, malignancy, or severe benign disease. Surgical techniques currently available include stretching, circular myotomy, and interposition of stomach or colon, but these approaches are frequently associated with complications including stricture, leakage, elongation, and gastroesophageal reflux. Thus, an artificial esophageal construct has been sought for many years. To be effective, an esophageal construct must be implantable without rejection, be biocompatible to support appropriate tissue growth, and retain biomechanical characteristics of native esophageal tissue, that is be soft and elastomeric, whilst maintaining a tubular structure when implanted *in vivo*.

Attempts to tissue engineer replacement esophageal tissue have included both patch and circumferential implantation of constructs composed of synthetic as well as natural scaffold materials. To date, a full length of tissue-engineered esophagus has not been produced but a number of incremental advances towards this goal have been achieved.

926

Early attempts explored the use of a non-degradable prosthetic tube. Fukushima and colleagues used a Dacron tube as a substitute for the esophagus in a canine model (Fukushima et al., 1983). Lengths of Dacron tube measuring 5—7 cm in length were placed into the esophagus of dogs. Nearly half of the dogs survived over a year, with some remaining alive for 6 years. The tubes provided a substrate for the formation of a thin layer of squamous epithelium and submucosa near the anastomotic site, but this did not extend into the central portions of the tube where fibrous scarring without muscle or mucous glands was observed.

Surgical reconstruction techniques of the esophagus have generally been favored to date following resection for structures and malignancies. This has involved the transfer of small segments or patches of skin and other tissues on a vascular pedicle with fairly good results (Jurkiewicz, 1984; Harii et al., 1985; Kakegawa et al., 1987). Building on such findings, tissue-engineered sheets of autologous oral mucosal epithelial cells have been successfully transplanted by endoscopy in a canine model (Ohki et al., 2006). The transplanted sheets adhered to the underlying esophageal muscle layers created by endoscopic submucosal dissection and enhanced wound healing without post-operative stenosis. Because the interaction between the epithelium and mesenchymal cells is thought to reduce fibrosis and scarring that can cause stenosis, the authors of this study suggested this approach may offer a novel therapy to reduce scarring and prevent painful constriction that can be associated with endoscopic submucosal dissection for the removal of large esophageal cancers.

The use of scaffold materials consisting of meshes or sheets of collagen and silicone has resulted in moderate success in pre-clinical models, as well as providing useful information for subsequent developments (Shinhar et al., 1998; Yamamoto et al., 1999; Badylak et al., 2000, 2005; Lynen Jansen et al., 2004). A variety of acellular scaffolds consisting of extracellular matrix components have been explored because of their innate ability to promote cell attachment, growth, and cell-cell signaling between different tissue components being advantageous over synthetic scaffold materials. Decellularized esophageal tissue has been produced via repeated detergent-enzymatic treatment resulting in a scaffold with biocompatibility suitable for the growth of esophageal epithelial cells (Marzaro et al., 2006; Ozeki et al., 2006). Based on these pre-clinical findings it could be envisaged that human donor esophageal tissue might one day be used in a similar manner to that described for the successful tissue engineering of human airway tissue (Macchiarini et al., 2008).

To date, scaffold materials derived from small intestinal submucosa (SIS) have been investigated most widely to tissue engineer replacement esophageal tissue. SIS consists of extracellular matrix material harvested from porcine small intestine and has been used extensively in tissue-engineering experiments. Originally described by Matsumoto and colleagues in 1966 for use in large vein replacement in dogs, it has since been used as an effective scaffold for the regeneration of numerous tissues (Matsumoto et al., 1966; Badylak et al., 1989; Kropp et al., 1995; Dalla Vecchia et al., 1999; de Ugarte et al., 2003). It has been successfully applied to regenerative medicine applications in humans, including repair of hernias, diaphragms, and tympanic membranes, and for large wound coverage (Puccio et al., 2005; Spiegel and Kessler, 2005; Grethel et al., 2006; Smith and Campbell, 2006).

The success with using SIS as a scaffold to promote tissue regeneration appears to relate to the retention of collagen (types I, II, and V), growth factors (transforming growth factor, fibroblast growth factor 2, vascular endothelial growth factor), glycosaminoglycans (hyaluronic acid, chondroitin sulphate, heparin sulphate), proteoglycans, and glycoproteins (fibronectin) during the fabrication process (Hodde et al., 1996; Voytik-Harbin et al., 1997). The resulting scaffold has a composition closely resembling native tissue, making it ideally suited for the attachment and growth of new tissue.

Despite the apparent ideal compositional properties of SIS for a scaffold material, the extent of circumferential replacement of esophageal tissue appears to have an impact on the outcome of attempts to tissue engineer esophagus, with patches producing better results compared with tubular segments. Lopes and colleagues successfully used SIS patches to repair defects to the anterior wall of cervical or abdominal esophagus in rats without signs of stenosis over a 150-day time period (Lopes et al., 2006). Likewise, Badylak and colleagues used SIS patches (or urinary bladder submucosa) to repair esophageal defects created in dogs without clinical signs of esophageal dysfunction (Badylak et al., 2000). However, the latter study reported signs of stenosis in dogs receiving complete circumferential segmental grafts of SIS (Badylak et al., 2000). Doede and colleagues reported similar findings, with severe stenosis occurring when relatively short tubular lengths (4 cm) of SIS were used in alloplastic esophageal replacement in piglets (Doede et al., 2009). The esophageal lumen was collapsed except during the passage of saliva and ingesta. Thus, although scaffolds consisting of only extracellular matrix have shown the capacity to promote cell growth *in vitro* and tissue regeneration of patch defects *in vivo*, replacement of circumferential defects without stricture formation remains difficult to achieve. Epithelial-mesenchymal cell signaling is likely to play a key role in facilitating reconstruction of the esophageal construct after implantation. A similar effect has been shown in bladder reconstruction, where the presence of urothelium led to infiltration of fibroblasts into acellular matrices and apparent transdifferentiation into a smooth muscle phenotype (Master et al., 2003). Signaling from the mesenchymal cell population appears to be equally important in promoting growth of overlying epithelium (Rheinwald and Green, 1975). Moreover, the presence of epithelial-mesenchymal signaling may also prevent stricture formation in an esophageal construct, a problem frequently encountered with many of the scaffolds tested to date (Badylak et al., 2000; Nakase et al., 2008). Similar signaling properties have been demonstrated in bladder reconstruction where acellular collagen scaffolds seeded with urothelium and smooth muscle cells prevented tissue contraction (Yoo et al., 1998). Likewise, the interaction of muscle with the ablumenal surface of esophageal scaffolds at the time of implantation of partially circumferential grafts appears to have accounted for the reduced stricture formation observed in a canine model of esophageal reconstruction described by Badylak and colleagues (2005). It can be concluded from these observations that careful consideration of the order in which cells are added to the tissue-engineered construct will improve the likelihood of achieving a successful outcome.

In addition to SIS, gastric acellular matrix has been used as scaffold by Urita and colleagues to regenerate esophagus in a rat model (Urita et al., 2007). Grafts of gastric acellular matrix were used to patch defects in the abdominal esophagus and animals were sacrificed at time-points between 1 week and 18 months. Although regeneration of the muscle layer or lamina muscularis did not occur, there was no evidence of stenosis or dilatation at the graft site. The matrix obtained in this study was from whole stomachs, but the authors suggest gastric acellular matrix may provide an autologous source of naturally derived extracellular matrix scaffold in a clinical setting because the portion of stomach sacrificed to obtain the matrix is minimal. It remains to be seen whether this approach is feasible in a larger animal model, but the use of autologous acellular matrix scaffolds does avoid the concerns related to the use of xenogenic scaffold materials such as porcine-derived SIS. In addition to the risk of transmitting viral pathogens and prions, cultural and religious beliefs may also need to be considered when using acellular matrix scaffolds derived from certain species. Recently, extracellular matrix scaffold has been generated from ovine forestomach tissue to avoid these issues (Lun et al., 2010).

Although the esophagus can be considered as one of the less complex structures in the alimentary tract, there are several significant hurdles yet to be overcome before tissue engineering and clinical replacement of full-length esophageal segments become a clinical reality in humans. Unlike patch grafts, replacement of longer lengths of tissue will be unable to rely on adjacent esophagus to cover the surface area of larger scaffolds via guided tissue regeneration. Improved methods for isolating and expanding the different esophageal cell populations will therefore be a prerequisite for successful tissue engineering of larger constructs. Kofler and colleagues recently identified subsets of ovine esophageal epithelial cells that may help achieve this (Kofler et al., 2010). PCK-26-positive esophagus epithelial cells demonstrated high proliferative capacity and uniform coverage on collagen scaffolds, which the authors suggest could play an important role in successfully tissue engineering esophagus.

As with attempts to tissue engineer other large organs, the choice of scaffold materials available for esophageal tissue engineering is not matched by the number options available for inducing a vascular supply to retain tissue viability. The native esophagus has a poor vascular supply, which makes regenerating a complete length of tissue-engineered esophagus in the mediastinum highly unlikely to be successful. If neo-esophagus was formed in a heterotopic locale it would be difficult to maintain the vascular supply during transfer to the mediastinum. Without a sufficient vascular supply tissue, in-growth will be limited by nutrient diffusion. Although the presence of angiogenic growth factors in SIS has been reported, these are unlikely to provide stimulus for sufficient neovascularization (Hodde et al., 2001).

Failure to regenerate a muscle layer of reasonable thickness may not prove problematical for short or non-circumferential grafts, but for longer lengths of esophagus the presence of an innervated functional muscle layer will be essential to transport food bolus. A retrospective study investigating the temporal appearance and spatial distribution of nervous tissue in a canine model of esophageal reconstruction using porcine urinary bladder submucosa showed the presence of nerve tissue within sites of the remodeling scaffold (Agrawal et al., 2009). Although the study was unable to demonstrate whether the nervous tissue was functional or to distinguish between the various subsets of neurons, it opens up the possibility of using similar models to identify mechanisms that promote innervation that will facilitate the tissue engineering of functional tissue. Peristalsis of food is also dependent on the correct orientation of muscle fibers in the wall of the alimentary tract. To address this, promising results have been obtained with orientating smooth muscle tissue on unidirectional scaffolds for tissue-engineered esophagus in rats (Saxena et al., 2009). Orientated strands of smooth muscle mimicking the configurations found in the native organ were engineered when cells were seeded onto unidirectional scaffolds. These were assembled with esophageal epithelium to create a hybrid approach.

SMALL INTESTINE

In adults, the small intestine measures approximately 6 m in length from the duodenojejunal flexure to the ileocaecal valve. This consists of the jejunum (upper two-fifths) and ileum (lower three-fifths), but there is no definite point of transition. The small intestine is designed primarily for absorption of nutrients from the lumen. To facilitate this, the absorptive surface area of the intestinal mucosa contains a number of specialized features (folds of Kerckring, villi, microvilli) that increase the absorptive surface about 600-fold, resulting in a total surface area measuring about 250 m^2 — approximately the same surface area as a tennis court. The mucosa of the intestine is lined with epithelial cells and consists of the lamina propria, containing vascular and reticular stroma, large aggregates of lymphoid tissue called Peyer's patches, and a strip of smooth muscle called the muscularis mucosae. Intestinal stem cells reside at the base of epithelial invaginations called crypts in the mucosa and provide all four lineages of epithelial cells that line the intestine (Booth and Potten, 2000). Epithelial cells migrate out of the crypts, differentiating and maturing towards the lumen of the bowel where they become senescent over the course of a few days and are subsequently shed. Despite knowledge of their presence, few stem cell markers exist for intestinal epithelial stem cells. Because of this, identifying and isolating pure populations of intestinal epithelial stem cells remains difficult. Studies have shown that Musashi-1 may be a marker of intestinal stem cells (Kayahara et al., 2003; Potten et al., 2003). More recently, a Sox9(EGFP) mouse model has been used to enrich multipotent intestinal epithelial stem cells (Gracz et al., 2010). Using a culture system that mimics the native intestinal epithelial stem cell niche, these cells are capable of generating "organoids" that contain all four epithelial cell types of the small intestinal epithelium. Furthermore, the Sox9(EGFP) multipotent intestinal epithelial stem cells express CD24, which may facilitate their enrichment by FACS using widely available antibodies. Additional studies are needed to address whether these cells will be capable of regenerating intestinal tissue constructs.

Beneath the mucosa the small intestine contains other important tissue layers that contribute to its function. The submucosa consists of fibrous connective tissue that supplies blood and lymphatic vessels to the mucosa. The muscularis propria consists of an inner layer of circular muscle and an outer longitudinal muscle layer. The muscularis propria is covered by the adventitia, a layer of loose connective tissue, and the serosa, a mesothelial lining of peritoneum.

The small intestine is an essential component of the alimentary tract and cannot be replaced by transposing another part of the gut. Intestinal ischemia and bowel resection for tumors and inflammatory bowel disease can result in short bowel syndrome, when more than 75% of the small intestine is lost. Short bowel syndrome is often associated with intestinal failure and the requirement of life-long nutritional support (total parenteral nutrition), which is frequently accompanied by severe complications, such as liver failure, line sepsis, and poor long-term survival rates. The length of residual intestine is critical for these patients; thus, techniques for increasing absorptive surface area have been sought for many years. Surgical options for increasing the absorptive surface or slowing the transit time to enhance absorption have been reported but these approaches require longer residual intestinal segments and most have only limited long-term clinical success (Bianchi, 1999; Thompson, 1999; Weber, 1999; Javid et al., 2005). Small bowel transplantation is a viable option for some patients but this procedure has limitations including the availability of donor tissue, the need for long-term immunosuppression, graft versus host disease, and potential post-transplant lymphoproliferative disorder

(Botha and Horslen, 2006). The amount of small bowel required for successful nutritional rehabilitation is dependent on factors including the patient's age, the amount of small bowel present, the presence or absence of the ileocecal valve, and the amount of large bowel present. Therefore, small bowel elongation of just a few centimetres could allow many patients to become independent of total parenteral nutrition. This possibility has led to the concept of creating tissue-engineered neointestine as a therapeutic tool being an attractive option. Several different approaches, using either guided tissue regeneration or tissue engineering, have been taken towards regenerating intestine that have used combinations of a various synthetic and natural scaffold materials, different cell types, and surgical procedures.

Early attempts to patch bowel defects using the serosal surface of another piece of intestine resulted in it being covered with regenerated mucosa (Kobold and Than, 1963; Binnington et al., 1973). This paved the way for other researchers to investigate the use of increasingly elaborate biomaterials as scaffolds for the in-growth of neointestine via guided tissue regeneration. Harmon and colleagues used Dacron as a patch to repair defects in the ileum of rabbits (Harmon et al., 1979). Other non-degradable materials assessed included the placement of a polytetrafluoroethylene (PTFE) tube in continuity with the small bowel, resulting in some in-growth of mucosa onto the scaffold (Watson et al., 1980). Although such studies on the use of prosthetic materials as a patch for repair of bowel defects were initially thought to be unsuccessful because the materials used were non-resorbable, the use of non-resorbable materials for studying intestinal morphogenesis and regeneration continues to be of interest (Jwo et al., 2008). Despite this, the use of resorbable scaffold biomaterials for intestinal tissue engineering has become the predominant approach.

Chen and Badylak used SIS to patch partial defects in the small bowel wall of a canine model (Chen and Badylak, 2001). The majority of the dogs survived up to the time of elective necropsy (between 2 weeks and 1 year) with no evidence of intestinal dysfunction. The implanted SIS scaffold material was completely resorbed by 3 months and the resulting neointestine, created by guided tissue regeneration, was similar in appearance to normal small bowel. Histological evaluation showed the presence of mucosa, varying amounts of smooth muscle, sheets of collagen, and an outer serosal layer. However, in the same study attempts to use a tubular configuration of SIS were unsuccessful. The tubes either leaked or became obstructed and this occurred primarily because the SIS was unable to maintain lumenal patency when exposed to the moist luminal contents. Similar limitations were reported more recently by Pahari and colleagues, who also used guided tissue regeneration to create a segment of new intestine in rats using acellular dermal matrix (AlloDerm) rolled into tubes (Pahari et al., 2006). Animals that received the graft in continuity with the small intestine developed peritonitis whereas animals that received the graft as a blind-ended pouch to a defunctionalized jejunal limb survived up to 6 months after surgery and displayed a fully regenerated mucosa.

In an attempt to maintain an open lumen in the tissue-engineered intestine, Hori and colleagues reported that scaffolds composed of sheets of acellular collagen sponge wrapped on a temporary silicone stent and covered with omentum guided tissue regeneration of almost all layers of the gastrointestinal tract in a canine model, but only a thin muscularis mucosae was present and the muscularis propria was absent (Hori et al., 2001b). The same group also explored the addition of mesenchymal stem cells seeded onto a collagen scaffold, which it was hypothesized might differentiate "site-specifically" into muscle cells and regenerate the muscle layer (Hori et al., 2001b). Intestinal regeneration occurred but muscle regeneration in an organized manner was not observed. Wang and colleagues used a rat model to evaluate the feasibility of regenerating tubular intestine using sheets of rat-derived SIS wrapped around a silicone stent (Wang et al., 2005). The tubular graft was interposed in the middle of a Thiry-Vella loop (a defunctionalized segment of ileum that is brought out as a double ileostomy) in Lewis rats. The silicone stent was left in place for 3 weeks to maintain lumenal patency during tissue regeneration. At 4 weeks, an epithelial layer had begun to form and this completely covered the lumenal surface by 12 weeks. The neomucosa had a typical morphology containing goblet cells, Paneth cells, enterocytes, and enteroendocrine cells. Although the regenerated bowel contained bundles of smooth muscle-like cells, especially near the sites of anastomosis, the quantity and organization of the muscle layer differed from that found in native small intestine, being predominantly circular muscle with no longitudinal muscle. The use of a Thiry-Vella loop in the model created by Wang may have facilitated mucosal development in the neointestine by protecting it from alimentary transit and creating an isolated environment that avoids the food stream and digestive enzymes. An alternative model using dysfunctioned bowel has been used by Jwo and colleagues that involved grafting a 3 cm silicone tube into the bowel after Roux-en-Y bypass surgery (Jwo et al., 2008). Using a similar rat experimental model with porcine-derived SIS as a xenograft scaffold, Ansaloni and colleagues reported the presence of both circular and longitudinal muscle layers, together with innervation of the neointestine with myoenteric and submucosal plexi into a 3 cm tubular graft (Ansaloni et al., 2006). Lee and colleagues observed only minimal intestinal regeneration in a rat model used for evaluating SIS scaffolds. From this they concluded that SIS scaffolds alone were not sufficient to regenerate small intestine and suggested the use of appropriate progenitor cells is probably necessary to facilitate the regeneration of small intestine (Lee et al., 2008a).

The value of combining intestinal tissue with a polymer scaffold for intestinal tissue engineering was recognized by the Boston group in the late 1980s (Vacanti et al., 1988). Since then, this group has reported on a number of important studies investigating the development and refinement of intestinal tissue engineering techniques. Key to much of the success of their work was the prior demonstration by Tait and colleagues that intestinal tissue could be separated by enzymatic digestion to produce organoid units (Tait et al., 1994). These clusters of cells contained all the elements of the intestinal mucosa including stem cells and mesenchyme, which could be used to regenerate intestinal neomucosa expressing digestive enzyme activities and glucose transport capacity similar to that of age-matched native intestinal mucosa. When organoid units were subcutaneously grafted, they displayed different epithelial populations consistent with epithelial transit amplifying and stem cell populations (Slorach et al., 1999).

The Boston group demonstrated that transplantation of organoid units onto biodegradable polymer scaffolds followed by implantation into the omentum of syngeneic adult animals resulted in the formation of neointestinal cysts attached to a vascular pedicle with mucosa facing a lumen that contained mucoid material (Choi and Vacanti, 1997). The mucosa of the neointestine created with this technique showed morphological similarities to native intestine, including the formation of a primitive crypt-villus axis lined with columnar epithelial cells and goblet cells, and a polarized epithelium with brush border enzyme sucrase expressed at the apical surface and laminin at the basolateral surface. The same study also showed the neomucosa exhibited similar transepithelial resistance to native intestine (Choi et al., 1998). A major step forward in intestinal tissue engineering was the investigation of the functionality of the neointestine. Initially the cyst-like structures were anastomosed to native jejunum in adult rats to provide continuity with the native intestinal tract (Kaihara et al., 1999). This resulted in a more developed neomucosa, with significant increases in villus number, villus height, and surface length of the cyst compared with non-anastomosed cysts. The authors postulated that anastomosis may have facilitated neomucosal growth in the cysts by drainage of lumenal contents, or via stimulatory factors present in the lumenal contents of the native intestine in continuity with the neomucosa. The anastomosed neointestine was also shown to express Na^+-dependent glucose transporter SGLT1 (Tavakkolizadeh et al., 2003) and a mucosal immune system with intraepithelial and lamina propria immunocytes similar to that of native jejunum (Perez et al., 2002). The native small intestine has a great adaptive and compensatory capacity in response to massive small bowel resection, which is considered to be controlled by humoral factors. The mucosa of the neointestine was also shown to possess this adaptive capacity following massive small bowel resection, resulting in a significant regenerative

931

stimulus for the morphogenesis and differentiation of the tissue-engineered intestine (Kim et al., 1999). An improvement of intestinal function, capable of facilitating patient recovery after massive small bowel resection, was putatively demonstrated when cysts containing neointestine were anastomosed to native small bowel at the time of an 85% enterectomy in rats (Grikscheit et al., 2004). The study showed that animals with tissue-engineered intestine returned to their preoperative weight more rapidly compared with animals undergoing small bowel resection alone. These findings are of significance since they are the first to suggest that tissue-engineered intestine may provide a therapeutic intervention for the management of patients with short bowel syndrome. Whilst it is tempting to speculate that the observed effects were due to neointestine restoring absorptive function after small bowel resection, the mechanism underlying the beneficial effects remains uncertain (Warner, 2004). It has been postulated that the amount of intestine replaced by the anastomosed neointestine (approximately 4 cm) was far shorter than the amount resected, probably equating to approximately 10% of the original length, and is unlikely to have added sufficient mucosal surface area to account for the increase in post-operative weight observed. Furthermore, the improved nutrition may have resulted from the tissue-engineered intestine slowing intestinal transit, leading to increased absorption and weight gain, a principle that could be achieved with simpler remedial surgical procedures. Moreover, a significant drawback with this approach is the need for large amounts of donor tissue to harvest a sufficient number of organoid units to seed each scaffold that will generate a comparatively short length of neointestine that is likely to offer limited therapeutic value (Warner, 2004). A solution might exist with the use of yet-unexplored alternative sources of intestinal epithelial stem cells, for example bone marrow-derived cells and pluripotent stem cells circulating in the peripheral blood (Zhao et al., 2003; Rizvi et al., 2006). These issues need to be resolved before translation into humans can occur.

932

Another important aspect of intestinal tissue engineering is the ability for the neointestine to repair, regenerate, and remodel. The latter is particularly important when considering the use of engineered intestinal tissue for children, in whom the length of the intestine increases significantly during development. Epithelial cells are responsive to an expanding array of mitogens, including epidermal growth factor, hepatocyte growth factor, fibroblast growth factor, neurotensin, growth hormone, transforming growth factor, interleukin-11, glucagon-like peptide-2 (GLP-2), and glutamine (Walters, 2004). To date, the trophic effects of only GLP-2 have been evaluated on neointestinal growth (Ramsanahie et al., 2003). GLP-2 is an endogenous regulatory peptide with potent trophic effects on intestinal mucosal growth and an ability to modulate the expression of Na^+-glucose cotransporter 1 (SGLT1). Adult rats with neointestinal implants that received subcutaneous injections of a GLP-2 analog twice daily for 10 days had enhanced mucosal growth and increased expression of SGLT1 compared with control rats. These findings indicate the neointestine is capable of responding to external regulator signals that could be used to further expand the surface of the neointestine.

Despite a number of studies reporting the creation of tissue-engineered constructs in preclinical models that resemble native small intestine occurring over the past two decades, the clinical impact of these studies in humans has been negligible. One reason for this is the lack of suitable models for observing intestinal tissue regeneration and improved intestinal function on a scale that can be feasibly translated into humans. The list of suitable models capable of achieving this is limited to a few species. Recently, this problem has partly been addressed by investigating intestinal tissue engineering in a large animal model, using autologous tissue, designed to emulate the conditions required for human therapy (Sala et al., 2009). In this study, scaffolds were seeded with organoid units isolated from the jejunum of 6-week-old piglets and implanted into the omentum of the same animal. However, this study provides only limited information on issues related to the scaling-up of a technique for use in humans since the neointestine was not anastomosed to the native intestine and the scaffolds used were a similar size to those used in previous small animal models.

Establishment of a functional mucosal barrier is an essential element of intestinal tissue engineering for which scalability needs to be considered. Although transepithelial resistance of the mucosa created in neointestinal cysts is similar to that of native intestine (Choi et al., 1998), the creation of larger intestinal constructs will require rapid coverage of the scaffold surface to ensure the barrier function is established. This process might be accelerated by the inclusion of materials in the scaffold that promote epithelial cell spreading. Yoshida and colleagues recently investigated the effect of transplanting organoid units onto denuded colonic mucosa of syngeneic recipient rats (Yoshida et al., 2009). The addition of 50 ng/ml basic fibroblast growth factor (bFGF) facilitated neomucosal growth and improved restoration of intestinal epithelial cell coverage over the denuded mucosa compared with the control group. Other approaches might involve the inclusion of inorganic materials into hybrid scaffolds, such as bioactive glass, which has been shown to increase epithelial cell migration via bFGF in an indirect manner (Moosvi and Day, 2009). As well as stimulating regeneration of the mucosa, delivery of bFGF may also provide a strategy for regenerating the muscularis propria. Local delivery of bFGF from scaffolds, via either incorporation into the collagen coating of scaffolds or encapsulation into microspheres, was also shown to increase the engraftment and density of seeded smooth muscle cells and blood vessel formation after 28 days' implantation in the omentum of rats (Lee et al., 2008c).

Rapid vascular in-growth into the tissue-engineered intestine will be essential to maintain the viability and engraftment of cells seeded on the scaffold. Gardner-Thorpe and colleagues observed that tissue-engineered intestine exhibited lower levels of bFGFG and VEGF and a fixed capillary density compared with native juvenile bowel (Gardner-Thorpe et al., 2003). This led the same group to evaluate a polymeric microsphere system to deliver encapsulated VEGF and stimulate angiogenesis in the maturing neointestine (Rocha et al., 2008). Capillary density in the muscular and connective tissue layers was significantly increased in the presence of microspheres containing VEGF, as was the size and weight of the constructs. Interestingly, the rate of epithelial cell proliferation was also increased in constructs implanted with VEGF-releasing microspheres, possibly related to the improved vascularization of the construct providing greater nutritional support to the rapidly proliferating epithelium. The need for neovascularization is not restricted to engineering tissues of the alimentary tract and a number of different approaches are being used to tackle this problem (Day et al., 2004). It remains to be seen whether any of these approaches will provide a sufficient stimulus to promote arteriogenesis required for sufficient vascularization of larger tissue constructs. Furthermore, a functional lymphatic system in the neointestine is also essential to establish normal nutrient absorption, fluid homeostasis, and immunological functions. Lymphangiogenesis is reported to occur in the neointestine created by the organoid unit-cyst model in rats (Duxbury et al., 2004). Although angiogenesis has been demonstrated in intestinal tissue engineering using small animal models, it is not certain whether the provision of thin-walled endothelium-lined structures will be sufficient to support the functionality of a larger tissue construct. Therefore, techniques to promote the formation of medium sized blood vessels via arteriogenesis are likely to be required to facilitate complete integration of large-scale intestinal constructs with a functional capacity. The small bowel has an extensive vascular system fed by arcades of arteries in the mesentery derived from the superior mesenteric artery. Translation of the existing tissue-engineering models to a scale suitable for implantation into humans will require the formation of a similar vascular system consisting of medium-sized blood vessels to maintain viability of a larger tissue construct as well as enable absorption of fluid and dissolved nutrient material from the intestine into the portal blood, which will require a vascular system similar to that found in native intestine. One approach to enable immediate perfusion of the tissue-engineered construct might involve utilizing the existing vascular system in decelluarized tissue. Preservation of the vascular structure in decellularized porcine small bowel has been used to engineer tissue with an innate vascularization (Mertsching et al., 2009). The decellularized scaffold was repopulated by endothelial cells and exhibited patent vessels after arterial and venous microanastomosis. This approach would also take advantage of the beneficial properties of SIS that could potentially be retained.

Improved methods for seeding and maturing larger tissue-engineered intestinal constructs will be needed to ensure the limited cells available are delivered efficiently and in a uniform manner to the tissue construct. These obstacles may be overcome with the development of bioreactor systems that will assist with the long-term culture and bioengineering of tissues by providing an *in vitro* environment that is similar to normal physiological conditions. Kim and colleagues have designed a perfusion bioreactor for intestinal tissue engineering (Kim et al., 2007). Techniques used to tissue engineer vascular grafts might also provide solutions that can be translated to intestinal tissue engineering. For example, centrifugal casting onto decellularized laser-porated natural scaffolds has been reported to enable the rapid fabrication of tubular tissue in a bioreactor-free manner (Kasyanov et al., 2009).

The type of scaffold material chosen for tissue engineering is an important consideration. It must allow seeded cells to rapidly engraft and proliferate whilst enabling tissue perfusion of nutrients and remodeling to ensure complete integration with the host. Compared with natural extracellular matrix-derived scaffolds, biodegradable synthetic polymer scaffolds provide more control over intrinsic properties, such as scaffold architecture, degradation rates, and mechanical properties. Chen and colleagues evaluated different polyester scaffolds with respect to their mucosal engraftment rates, mucosal morphology, compliance, and structural properties (Chen et al., 2006). Engraftment was affected by variations in the polymer constructs, processing techniques, and material properties of the scaffolds. The maximum surface area of the scaffolds covered by neomucosa after 4 weeks' implantation was 36%, indicating that further refinements of the scaffolds used for intestinal tissue engineering will be required to improve efficiency in the process.

The topography of scaffolds used for small bowel tissue engineering may influence the properties of cells grown on their surface. The function of the geometry of the crypt-villus micro-environment in regulating intestinal cell proliferation and differentiation has been explored by Wang and colleagues (2009). Caco-2 cells migrating over microwell structure showed increased metabolic activity and lower levels of differentiation compared with cells cultured on flat surfaces, suggesting that that the structure of crypts may play a role in retaining a proliferative phenotype. Likewise, scaffold architecture is a parameter that can be used to enhance the mass transfer of nutrients to ensure the viability of tissue is maintained. Lee and colleagues fabricated scaffolds with a high surface area to volume ratio using 3D printing technology (Lee et al., 2008b). The growth of smooth muscle cells *in vitro* was found to be influenced by the geometry of the scaffold. Scaffolds with small villi (0.5 mm) had increased cell density compared with scaffolds containing large villi (1 mm) after 14 days of culture.

Instilling peristaltic activity to the tissue-engineered intestine to establish gut motility will require both correctly orientated smooth muscle cell regeneration and re-innervation. In addition to regulating peristalsis, the nerve system in the intestine also controls villi activity and modulation of secretions from gut epithelial cells. Due to their cellular composition, intestinal organoid units seeded into polymer scaffolds to create cyst structures containing mucosa cannot regenerate the organized thick layers of circular and longitudinal smooth muscle required for peristalsis, so alternative approaches to regenerate this critical component of the intestine will need to be identified. Gut endocrine cells play a key role in regulating gastrointestinal activity by releasing serotonin, secretin, cholecystokinin, gastrin, and enteroglucagon and will be an essential component of the tissue-engineered intestine. Nakase and colleagues investigated the regeneration of endocrine cells and the nerve system in a canine patch model of tissue-engineered small intestine using a collagen sponge scaffold loaded with autologous gastric smooth muscle cells (Nakase et al., 2007). At 24 weeks after implantation of the scaffolds into the middle of an isolated ileal loop, the location and number of endocrine cells stained positive for chromogranin A were almost identical to native mucosa. Nerve fibers were present in the regenerated smooth muscle layer and villi, but the myenteric plexus of Auerbach and the submucosal plexus of Meissner were not visible. The density of smooth muscle cells implanted into the scaffolds did not affect the thickness of the regenerated smooth muscle layer, which remained approximately half that of the native smooth muscle layer, indicating that other cues will be necessary to increase its thickness. The authors suggested that the thickness of the muscle layer might be limited by the blood supply available to the regenerating tissue, which might be increased by the delivery of angiogenic factors from the scaffold. Grikscheit and colleagues have also reported that ganglion cells were distributed in the locality of the Auerbach and Meissner's plexi in tissue-engineered small intestine (Grikscheit et al., 2004).

Small bowel tissue engineering remains in the early stages of clinical development and has yet to provide a clear demonstration of improvement in nutrient absorption that will be of value in a clinical therapeutic setting in humans. Whilst none of the models used to date have unequivocally demonstrated functional neointestine with peristaltic activity, they do indicate that it is feasible to engineer tubular segmental replacement of small bowel that incorporates innervated smooth muscle layers. Refinement of existing techniques should yield further improvements in the tissue-engineered intestinal construct.

STOMACH, COLON, AND ANUS

The stomach, colon, and anus are not vital to life, but their losses are associated with significant morbidities.

The size and shape of the stomach vary depending on its contents. The stomach wall contains outer longitudinal and inner circular layers of smooth muscle, with an innermost layer of oblique muscle fibers. These layers facilitate important functions including storage of ingested food in the stomach until it can be accommodated in the lower portion of the alimentary tract, mixing of the food to form chyme, and regulation of food transit into the small intestine at an optimal rate for digestion and absorption in the small intestine. Stomach emptying is controlled by the gastric food volume and the release of the hormone gastrin, as well as feedback signals from the duodenum.

Insufficient stomach mass, which may arise from gastrectomy or congenital microgastria, is associated with increased patient morbidity and therefore a number of surgical reconstructive strategies have been proposed, including jejunal interposition and pouch formation.

The possibility of tissue engineering stomach tissue to patch a partial gastrectomy has been explored in a canine model using a two-part sheet composed of an outer layer of collagen sponge and a temporary inner silicone sheet to protect the collagen from degradation by the acidic stomach juices and to provide mechanical support (Hori et al., 2001a). The silicone sheet was removed endoscopically 4 weeks after placement. Evidence of stomach regeneration was observed at 4 weeks and complete coverage of the scaffold had occurred by 16 weeks, confirmed by the presence of mucosa and a thin muscular layer. Acid production capacity was present in the regenerated stomach wall but the contractile response to acetylcholine was poor (Hori et al., 2002a). To overcome technical difficulties for suturing and endoscopic removal of the silicone sheet in this model, the same group created a tissue-engineered sheet without silicone that had sufficient strength to allow suturing and resist anastomotic dehiscence (Araki et al., 2009). The silicone sheet was replaced by a biodegradable co-polymer of poly(D,L-lactide) and epsilon-caprolactone (PDLCL) on the mucosal side of the collagen scaffold, both of which were completely absorbed at 16 weeks implantation. Although regeneration of the stomach mucosa was observed, the replacement of the silicone sheet with PDLCL did not provide sufficient mechanical strength to prevent significant shrinkage of the scaffold.

The Boston group evaluated the feasibility of creating new stomach tissue with the same technique described for regenerating small intestine (Grikscheit et al., 2003a). Stomach organoid units were harvested from neonatal and adult rats. The organoid units were seeded onto the same type of polymer scaffold tubes previously used by the group to form constructs that were implanted into the omentum of adult syngeneic rats. At four weeks the construct was anastomosed to the small intestine. Histology of the tissue-engineered stomach tissue was similar to native stomach, with gastric pits, squamous epithelium, and positive staining for α-actin smooth muscle in the muscularis and gastrin indicating the presence of a well-developed gastric epithelium. The same group have subsequently reported the creation of tissue-engineered stomach constructs in an autologous large animal model (Sala et al., 2009).

The colon is important for water and sodium resorption and as a storage pouch for waste products. Patients who undergo total colectomy are at risk of significant morbidities (Papa et al., 1997). The surgical creation of an ileal pouch to create a reservoir provides only a limited solution and patients may still suffer from inflammation of the pouch (pouchitis), malabsorption, diarrhea, cramping, abdominal pain, and fever (Meagher et al., 1998).

The Boston group have applied the technique of organoid unit transplantation to create tissue-engineered colon (Grikscheit et al., 2003b). Organoid units were harvested from the sigmoid colon of neonatal Lewis rats, adult rats, and tissue-engineered colon itself, seeded onto a polymer scaffold, and implanted into the omentum of syngeneic adult Lewis rats. Tissue-engineered colon was successfully generated by each of the tissue sources used and the resulting neocolon architecture was similar to that of native tissue. The muscularis propria stained positively for smooth muscle actin and acetylcholinesterase was detected in the lamina propria in a linear distribution with presence of ganglion cells. *In vitro* Ussing chamber studies indicated appropriate colonic transport parameters and barrier function. When anastomosed to the native bowel, there was gross evidence of fluid absorption by the tissue-engineered colon.

Controlled storage and timely disposal of feces relies largely on the appropriate function of sphincter muscles that constrict the anal canal and maintain fecal continence. Fecal incontinence is a common disease, particularly in aging societies where it has a huge impact on quality of life and incurs colossal health costs. Conservative estimates indicate approximately 2% of community-dwelling adults suffer from regular fecal incontinence (Perry et al., 2002; Nelson, 2004). This figure increases to 50% in the institutionalized and geriatric population (Nelson, 2004). The commonest causes are obstetric or iatrogenic trauma, congenital anal malformation, and neuropathic degeneration (Chatoor et al., 2007; Dudding et al., 2008). In women, obstetric injury is the commonest cause, and in a large proportion of these patients there is a recognized third- or fourth-degree tear, a complication that is found in about 2% of deliveries (Eskander and Shet, 2009). Conservative treatments for fecal incontinence are ineffective in patients with any more than mild symptoms and surgical interventions produce poor long-term benefit, with frequent complications (Tan et al., 2007).

Cell therapy technology has been extensively investigated for urethral sphincter deficiency (Huard et al., 2002; Kwon et al., 2006; Lecoeur et al., 2007; Mitterberger et al., 2007). Despite holding considerable promise, cell therapy for incontinence affecting the alimentary tract remains relatively unexplored in humans (Frudinger et al., 2010). Whilst the technical feasibility of injecting autologous myoblasts for treating fecal incontinence in humans has been demonstrated, the study by Frudinger and colleagues was unable to demonstrate integration of cells into the damaged sphincter or improvement of functional integrity (Frudinger et al., 2010).

A fibrin-based bioengineered *in vitro* model of the internal anal sphincter that demonstrates physiological functionality has been described that is likely to be of value in studying complex physiological mechanisms underlying sphincter malfunction (Hecker et al., 2005; Somara et al., 2009). The bioengineered sphincter has been surgically implanted into the subcutaneous tissues of syngeneic mice and responds to the local delivery of basic fibroblast growth factor, resulting in improved muscle viability, vascularization, and survival of the graft

(Hashish et al., 2010). These promising findings suggest further refinements to this technique might provide a viable future option for human therapy of fecal incontinence.

Regenerative medicine may also offer solutions to conditions where existing medical and surgical procedures have failed. A condition where this affects the alimentary tract is perianal fistulas that result from a connection between the anal canal and the perianal skin surface, creating an abnormal passageway for the discharge of pus, blood, and in some cases feces. The condition has an incidence in the range of 1.2–2.8 cases per 10,000 and is a cause of significant morbidity. The goals of fistula treatment are eradication of perineal sepsis and fistula closure, whilst posing a minimal risk of causing sphincter muscle damage.

One of the difficulties in treating perianal fistulas is the avoidance of abscess formation due to healing of the skin before closure of the tract. To address this, collagen anal fistula plugs have been devised for the treatment of fistulas. Although early studies reported good healing rates, with little or no risk to continence, long-term follow-up has revealed variable and disappointing success rates (24–78%) (Adamina et al., 2010). Reports of the plugs failing due to dislodgment from the tracts indicate this approach may not provide an ideal scaffold material to promote guided tissue regeneration and closure of the tract (Adamina et al., 2010). A possible solution to this problem is the use of scaffold materials that provide both optimal conditions for rapid cell infiltration when implanted into tissue cavities and mechanical strength to maintain an open scaffold structure (Blaker et al., 2008).

CONCLUSION

The alimentary tract is a complex organ that is essential for maintaining physiological homeostasis. Tissue engineering and regenerative medicine for hollow visceral organs have been proposed as a way of replacing damaged or diseased tissue and have recently been demonstrated in humans with bladder (Atala et al., 2006) and tracheal (Macchiarini et al., 2008) tissue. Although these "first-in-man" studies have rightly attracted much attention, there are significant obstacles to be overcome until a similar approach becomes routine in the comparatively complex structures of the alimentary tract.

The past few decades have delivered a series of important tissue engineering studies that have utilized the innate ability of the alimentary tract to regenerate. Further studies are needed to demonstrate these approaches are transferable and of clinical value to humans. Fundamental challenges, such as scalability, have yet to be resolved; this will be necessary to allow the translation of promising results obtained in small animal models into pre-clinical models applicable to humans. This will require refinement of the scaffolds to be used and the ability to seed limited quantities of cells available in an efficient manner onto the scaffold. These are not insurmountable problems and the prospect of tissue engineering being applied to alimentary tract in humans is likely to occur in the near future.

References

Adamina, M., Hoch, J. S., & Burnstein, M. J. (2010). To plug or not to plug: a cost-effectiveness analysis for complex anal fistula. *Surgery, 147*, 72–78.

Agrawal, V., Brown, B. N., Beattie, A. J., Gilbert, T. W., & Badylak, S. F. (2009). Evidence of innervation following extracellular matrix scaffold-mediated remodeling of muscular tissues. *J. Tissue Eng. Regen. Med., 3*, 590–600.

Ansaloni, L., Bonasoni, P., Cambrini, P., Catena, F., de Cataldis, A., Gagliardi, S., et al. (2006). Experimental evaluation of Surgisis as scaffold for neointestine regeneration in a rat model. *Transplant Proc., 38*, 1844–1848.

Araki, M., Tao, H., Sato, T., Nakajima, N., Hyon, S. H., Nagayasu, T., et al. (2009). Development of a new tissue-engineered sheet for reconstruction of the stomach. *Artif. Organs, 33*(10), 818–826.

Atala, A., Bauer, S. B., Soker, S., Yoo, J. J., & Retik, A. B. (2006). Tissue-engineered autologous bladders for patients needing cystoplasty. *Lancet, 367*(9518), 1241–1246.

Badylak, S. F., Lantz, G. C., Coffey, A., & Geddes, L. A. (1989). Small intestinal submucosa as a large diameter vascular graft in the dog. *J. Surg. Res., 47*, 74–80.

Badylak, S., Meurling, S., Chen, M., Spievack, A., & Simmons-Byrd, A. (2000). Resorbable bioscaffold for esophageal repair in a dog model. *J. Pediatr. Surg., 35,* 1097–1103.

Badylak, S. F., Vorp, D. A., Spievack, A. R., Simmons-Byrd, A., Hanke, J., Freytes, D. O., et al. (2005). Esophageal reconstruction with ECM and muscle tissue in a dog model. *J. Surg. Res., 128,* 87–97.

Bianchi, A. (1999). Experience with longitudinal intestinal lengthening and tailoring. *Eur. J. Pediatr. Surg., 9,* 256–259.

Binnington, H. B., Siegel, B. A., Kissane, J. M., & Ternberg, J. L. (1973). A technique to increase jejunal mucosa surface area. *J. Pediatr. Surg., 8,* 765–769.

Blaker, J. J., Pratten, J., Ready, D., Knowles, J. C., Forbes, A., & Day, R. M. (2008). Assessment of antimicrobial microspheres as a prospective novel treatment targeted towards the repair of perianal fistulae. *Aliment. Pharmacol. Ther., 28,* 614–622.

Booth, C., & Potten, C. S. (2000). Gut instincts: thoughts on intestinal epithelial stem cells. *J. Clin. Invest., 105,* 1493–1499.

Botha, J. F., & Horslen, S. P. (2006). Small bowel transplantation: literature review 2003–2005. *Pediatr. Transplant., 10,* 7–16.

Chatoor, D. R., Taylor, S. J., Cohen, C. R., & Emmanuel, A. V. (2007). Faecal incontinence. *Br. J. Surg., 94,* 134–144.

Chen, M. K., & Badylak, S. F. (2001). Small bowel tissue engineering using small intestinal submucosa as a scaffold. *J. Surg. Res., 99,* 352–358.

Chen, D. C., Avansino, J. R., Agopian, V. G., Hoagland, V. D., Woolman, J. D., Pan, S., et al. (2006). Comparison of polyester scaffolds for bioengineered intestinal mucosa. *Cells Tissues Organs, 184,* 154–165.

Choi, R. S., & Vacanti, J. P. (1997). Preliminary studies of tissue-engineered intestine using isolated epithelial organoid units on tubular synthetic biodegradable scaffolds. *Transplant. Proc., 29,* 848–851.

Choi, R. S., Riegler, M., Pothoulakis, C., Kim, B. S., Mooney, D., Vacanti, M., et al. (1998). Studies of brush border enzymes, basement membrane components, and electrophysiology of tissue-engineered neointestine. *J. Pediatr. Surg., 33,* 991–996.

Dalla Vecchia, L., Engum, S., Kogon, B., Jensen, E., Davis, M., & Grosfeld, J. (1999). Evaluation of small intestine submucosa and acellular dermis as diaphragmatic prostheses. *J. Pediatr. Surg., 34,* 167–171.

Day, R. M., Boccaccini, A. R., Shurey, S., Roether, J. A., Forbes, A., Hench, L. L., et al. (2004). Assessment of polyglycolic acid mesh and bioactive glass for soft-tissue engineering scaffolds. *Biomaterials, 25,* 5857–5866.

de Ugarte, D. A., Puapong, D., Roostaeian, J., Gillis, N., Fonkalsrud, E. W., Atkinson, J. B., et al. (2003). Surgisis patch tracheoplasty in a rodent model for tracheal stenosis. *J. Surg. Res., 112,* 65–69.

Doede, T., Bondartschuk, M., Joerck, C., Schulze, E., & Goernig, M. (2009). Unsuccessful alloplastic esophageal replacement with porcine small intestinal submucosa. *Artif. Organs, 33,* 328–333.

Dudding, T. C., Vaizey, C. J., & Kamm, M. A. (2008). Obstetric anal sphincter injury: incidence, risk factors, and management. *Ann. Surg., 247,* 224–237.

Duxbury, M. S., Grikscheit, T. C., Gardner-Thorpe, J., Rocha, F. G., Ito, H., Perez, A., et al. (2004). Lymphangiogenesis in tissue-engineered small intestine. *Transplantation, 77,* 1162–1166.

Eskandar, O., & Shet, D. (2009). Risk factors for 3rd and 4th degree perineal tear. *J. Obstet. Gynaecol., 29,* 119–122.

Fascetti-Leon, F., Malerba, A., Boldrin, L., Leone, E., Betalli, P., Pasut, A., et al. (2007). Murine muscle precursor cells survived and integrated in a cryoinjured gastroesophageal junction. *J. Surg. Res., 143,* 253–259.

Frudinger, A., Kolle, D., Schwaiger, W., Pfeifer, J., Paede, J., & Halligan, S. (2010). Muscle derived cell injection to treat anal incontinence due to obstetric trauma: pilot study with one-year follow-up. *Gut, 59,* 56–61.

Fukushima, M., Kako, N., Chiba, K., Kawaguchi, T., Kimura, Y., Sato, M., et al. (1983). Seven year follow-up study after the replacement of the esophagus with an artificial esophagus in the dog. *Surgery, 93,* 70–77.

Gardner-Thorpe, J., Grikscheit, T. C., Ito, H., Perez, A., Ashley, S. W., Vacanti, J. P., et al. (2003). Angiogenesis in tissue-engineered small intestine. *Tissue Eng., 9,* 1255–1261.

Gracz, A. D., Ramalingam, S., & Magness, S. T. (2010). Sox9-expression marks a subset of CD24-expressing small intestine epithelial stem cells that form organoids *in vitro. Am. J. Physiol. Gastrointest. Liver Physiol., 298,* G590–G600.

Grethel, E. J., Cortes, R. A., Wagner, A. J., Clifton, M. S., Lee, H., Farmer, D. L., Harrison, M. R., Keller, R. L., & Nobuhara, K. K. (2006). Prosthetic patches for congenital diaphragmatic hernia repair: Surgisis vs. Gore-Tex. *J. Pediatr. Surg., 41,* 29–33.

Grikscheit, T., Srinivasan, A., & Vacanti, J. P. (2003a). Tissue-engineered stomach: a preliminary report of a versatile *in vivo* model with therapeutic potential. *J. Pediatr. Surg., 38,* 1305–1309.

Grikscheit, T. C., Ochoa, E. R., Ramsanahie, A., Alsberg, E., Mooney, D., Whang, E. E., et al. (2003b). Tissue-engineered large intestine resembles native colon with appropriate *in vitro* physiology and architecture. *Ann. Surg., 238,* 35–41.

Grikscheit, T. C., Siddique, A., Ochoa, E. R., Srinivasan, A., Alsberg, E., Hodin, R. A., et al. (2004). Tissue-engineered small intestine improves recovery after massive small bowel resection. *Ann. Surg.*, *240*, 748–754.

Harii, K., Ebihara, S., Ono, I., Saito, H., Terui, S., & Takato, T. (1985). Pharyngoesophageal reconstruction using a fabricated forearm free flap. *Plast. Reconstr. Surg.*, *75*, 463–476.

Harmon, J. W., Wright, J. A., Noel, J., & Cogan, M. (1979). Fate of Dacron protheses in the small bowel of rabbits. *Surg. Forum*, *30*, 365–366.

Hashish, M., Raghavan, S., Somara, S., Gilmont, R. R., Miyasaka, E., Bitar, K. N., et al. (2010). Surgical implantation of a bioengineered internal anal sphincter. *J. Pediatr. Surg.*, *45*, 52–58.

Hecker, L., Baar, K., Dennis, R. G., & Bitar, K. N. (2005). Development of a three-dimensional physiological model of the internal anal sphincter bioengineered *in vitro* from isolated smooth muscle cells. *Am. J. Physiol. Gastrointest. Liver Physiol.*, *289*, G188–G196.

Hodde, J. P., Badylak, S. F., Brightman, A. O., & Voytik-Harbin, S. L. (1996). Glycosaminoglycan content of small intestinal submucosa: a bioscaffold for tissue replacement. *Tissue Eng.*, *2*, 209–217.

Hodde, J. P., Record, R. D., Liang, H. A., & Badylak, S. F. (2001). Vascular endothelial growth factor in porcine-derived extracellular matrix. *Endothelium*, *8*, 11–24.

Hori, Y., Nakamura, T., Matsumoto, K., Kurokawa, Y., Satomi, S., & Shimizu, Y. (2001a). Experimental study of in situ tissue engineering of the stomach by an acellular collagen sponge scaffold graft. *ASAIO J.*, *47*, 206–210.

Hori, Y., Nakamura, T., Matsumoto, K., Kurokawa, Y., Satomi, S., & Shimizu, Y. (2001b). Tissue engineering of the small intestine by acellular collagen sponge scaffold grafting. *Int. J. Artif. Organs*, *24*, 50–54.

Hori, Y., Nakamura, T., Kimura, D., Kaino, K., Kurokawa, Y., Satomi, S., et al. (2002a). Functional analysis of the tissue-engineered stomach wall. *Artif. Organs*, *26*, 868–872.

Hori, Y., Nakamura, T., Kimura, D., Kaino, K., Kurokawa, Y., Satomi, S., et al. (2002b). Experimental study on tissue engineering of the small intestine by mesenchymal stem cell seeding. *J. Surg. Res.*, *102*, 156–160.

Huard, J., Yokoyama, T., Pruchnic, R., Qu, Z., Li, Y., Lee, J. Y., et al. (2002). Muscle-derived cell-mediated *ex vivo* gene therapy for urological dysfunction. *Gene Ther.*, *9*, 1617–1626.

Javid, P. J., Kim, H. B., Duggan, C. P., & Jaksic, T. (2005). Serial transverse enteroplasty is associated with successful short-term outcomes in infants with short bowel syndrome. *J. Pediatr. Surg.*, *40*, 1019–1023.

Jurkiewicz, M. J. (1984). Reconstructive surgery of the cervical esophagus. *J. Thorac. Cardiovasc. Surg.*, *88*, 893–897.

Jwo, S. C., Chiu, J. H., Ng, K. K., & Chen, H. Y. (2008). Intestinal regeneration by a novel surgical procedure. *Br. J. Surg.*, *95*, 657–663.

Kaihara, S., Kim, S. S., Benvenuto, M., Choi, R., Kim, B. S., Mooney, D., et al. (1999). Successful anastomosis between tissue-engineered intestine and native small bowel. *Transplantation*, *67*, 241–245.

Kakegawa, T., Machi, J., Yamana, H., Fujita, H., & Tai, Y. (1987). A new technique for esophageal reconstruction by combined skin and muscle flaps after failure in primary colonic interposition. *Surg. Gynecol. Obstet.*, *164*, 576–578.

Kasyanov, V. A., Hodde, J., Hiles, M. C., Eisenberg, C., Eisenberg, L., de Castro, L. E., et al. (2009). Rapid biofabrication of tubular tissue constructs by centrifugal casting in a decellularized natural scaffold with laser-machined micropores. *J. Mater. Sci. Mater. Med.*, *20*, 329–337.

Kayahara, T., Sawada, M., Takaishi, S., Fukui, H., Seno, H., Fukuzawa, H., et al. (2003). Candidate markers for stem and early progenitor cells, Musashi-1 and Hes1, are expressed in crypt base columnar cells of mouse small intestine. *FEBS Lett.*, *535*, 131–135.

Kim, S. S., Kaihara, S., Benvenuto, M. S., Choi, R. S., Kim, B. S., Mooney, D. J., et al. (1999). Regenerative signals for intestinal epithelial organoid units transplanted on biodegradable polymer scaffolds for tissue engineering of small intestine. *Transplantation*, *67*, 227–233.

Kim, S. S., Penkala, R., & Abrahimi, P. (2007). A perfusion bioreactor for intestinal tissue engineering. *J. Surg. Res.*, *142*, 327–331.

Kobold, E. E., & Thal, A. P. (1963). A simple method for management of experimental wounds of the duodenum. *Surg. Gynecol. Obstet.*, *116*, 340–344.

Kofler, K., Ainoedhofer, H., Hollwarth, M. E., & Saxena, A. K. (2010). Fluorescence-activated cell sorting of PCK-26 antigen-positive cells enables selection of ovine esophageal epithelial cells with improved viability on scaffolds for esophagus tissue engineering. *Pediatr. Surg. Int.*, *26*, 97–104.

Kropp, B. P., Badylak, S., & Thor, K. B. (1995). Regenerative bladder augmentation: a review of the initial preclinical studies with porcine small intestinal submucosa. *Adv. Exp. Med. Biol.*, *385*, 229–235.

Kwon, D., Kim, Y., Pruchnic, R., Jankowski, R., Usiene, I., de Miguel, F., et al. (2006). Periurethral cellular injection: comparison of muscle-derived progenitor cells and fibroblasts with regard to efficacy and tissue contractility in an animal model of stress urinary incontinence. *Urology*, *68*, 449–454.

Lecoeur, C., Swieb, S., Zini, L., Riviere, C., Combrisson, H., Gherardi, R., et al. (2007). Intraurethral transfer of satellite cells by myofiber implants results in the formation of innervated myotubes exerting tonic contractions. *J. Urol.*, *178*, 332–337.

Lee, M., Chang, P. C., & Dunn, J. C. (2008a). Evaluation of small intestinal submucosa as scaffolds for intestinal tissue engineering. *J. Surg. Res., 147,* 168–171.

Lee, M., Wu, B. M., & Dunn, J. C. (2008b). Effect of scaffold architecture and pore size on smooth muscle cell growth. *J. Biomed. Mater. Res., 87A,* 1010–1016.

Lee, M., Wu, B. M., Stelzner, M., Reichardt, H. M., & Dunn, J. C. (2008c). Intestinal smooth muscle cell maintenance by basic fibroblast growth factor. *Tissue Eng. Part A, 14,* 1395–1402.

Lopes, M. F., Cabrita, A., Ilharco, J., Pessa, P., & Patricio, J. (2006). Grafts of porcine intestinal submucosa for repair of cervical and abdominal esophageal defects in the rat. *J. Invest. Surg., 19,* 105–111.

Lun, S., Irvine, S. M., Johnson, K. D., Fisher, N. J., Floden, E. W., Negron, L., et al. (2010). A functional extracellular matrix biomaterial derived from ovine forestomach. *Biomaterials, 31,* 4517–4529.

Lynen Jansen, P., Klinge, U., Anurov, M., Titkova, S., Mertens, P. R., & Jansen, M. (2004). Surgical mesh as a scaffold for tissue regeneration in the esophagus. *Eur. Surg. Res., 36,* 104–111.

Macchiarini, P., Jungebluth, P., Go, T., Asnaghi, M. A., Rees, L. E., Cogan, T. A., et al. (2008). Clinical transplantation of a tissue-engineered airway. *Lancet, 372,* 2023–2030.

Marzaro, M., Vigolo, S., Oselladore, B., Conconi, M. T., Ribatti, D., Giuliani, S., et al. (2006). *In vitro* and *in vivo* proposal of an artificial esophagus. *J. Biomed. Mater. Res., 77A,* 795–801.

Master, V. A., Wei, G., Liu, W., & Baskin, L. S. (2003). Urothlelium facilitates the recruitment and trans-differentiation of fibroblasts into smooth muscle in acellular matrix. *J. Urol., 170,* 1628–1632.

Matsumoto, T., Holmes, R. H., Burdick, C. O., Heisterkamp, C. A., & O'Connell, T. J. (1966). The fate of inverted segment of small bowel used for the replacement of major veins. *Surgery, 60,* 739–743.

Meagher, A. P., Farouk, R., Dozois, R. R., Kelly, K. A., & Pemberton, J. H. (1998). J Ileal pouch–anal anastomosis for chronic ulcerative colitis: complications and long-term outcome in 1310 patients. *Br. J. Surg., 85,* 800–803.

Mertsching, H., Schanz, J., Steger, V., Schandar, M., Schenk, M., Hansmann, J., et al. (2009). Generation and transplantation of an autologous vascularized bioartificial human tissue. *Transplantation, 88,* 203–210.

Mitterberger, M., Pinggera, G. M., Marksteiner, R., Margreiter, E., Plattner, R., Klima, G., et al. (2007). Functional and histological changes after myoblast injections in the porcine rhabdosphincter. *Eur. Urol., 52,* 1736–1743.

Moosvi, S. R., & Day, R. M. (2009). Bioactive glass modulation of intestinal epithelial cell restitution. *Acta Biomater., 5,* 76–83.

Nakase, Y., Nakamura, T., Kin, S., Nakashima, S., Yoshikawa, T., Kuriu, Y., et al. (2007). Endocrine cell and nerve regeneration in autologous in situ tissue-engineered small intestine. *J. Surg. Res., 137,* 61–68.

Nakase, Y., Nakamura, T., Kin, S., Nakashima, S., Yoshikawa, T., Kuriu, Y., et al. (2008). Intrathoracic esophageal replacement by in situ tissue-engineered esophagus. *J. Thorac. Cardiovasc. Surg., 136,* 850–859.

Nelson, R. L. (2004). Epidemiology of fecal incontinence. *Gastroenterology, 126*(Suppl. 1), S3–S7.

Ohki, T., Yamato, M., Murakami, D., Takagi, R., Yang, J., Namiki, H., et al. (2006). Treatment of oesophageal ulcerations using endoscopic transplantation of tissue-engineered autologous oral mucosal epithelial cell sheets in a canine model. *Gut, 55,* 1704–1710.

Ozeki, M., Narita, Y., Kagami, H., Ohmiya, N., Itoh, A., Hirooka, Y., et al. (2006). Evaluation of decellularized esophagus as a scaffold for cultured esophageal epithelial cells. *J. Biomed. Mater. Res., 79A,* 771–778.

Pahari, M. P., Raman, A., Bloomenthal, A., Costa, M. A., Bradley, S. P., Banner, B., et al. (2006). A novel approach for intestinal elongation using acellular dermal matrix: an experimental study in rats. *Transplant. Proc., 38,* 1849–1850.

Papa, M. Z., Karni, T., Koller, M., et al. (1997). Avoiding diarrhea after subtotal colectomy with primary anastomosis in the treatment of colon cancer. *J. Am. Coll. Surg., 184,* 269–272.

Perez, A., Grikscheit, T. C., Blumberg, R. S., Ashley, S. W., Vacanti, J. P., & Whang, E. E. (2002). Tissue-engineered small intestine: ontogeny of the immune system. *Transplantation, 74,* 619–623.

Perry, S., Shaw, C., McGrother, C., et al. (2002). Prevalence of faecal incontinence in adults aged 40 years or more living in the community. *Gut, 50,* 480–484.

Potten, C. S., Booth, C., Tudor, G. L., Booth, D., Brady, G., Hurley, P., et al. (2003). Identification of a putative intestinal stem cell and early lineage marker: musashi-1. *Differentiation, 71,* 28–41.

Puccio, F., Solazzo, M., & Marciano, P. (2005). Comparison of three different mesh materials in tension-free inguinal hernia repair: prolene versus Vypro versus Surgisis. *Int. Surg., 90,* S21–S23.

Qin, J., Li, R., Raes, J., Arumugam, M., Burgdorf, K. S., Manichanh, C., et al. (2010). A human gut microbial gene catalogue established by metagenomic sequencing. *Nature, 464,* 59–65.

Ramsanahie, A., Duxbury, M. S., Grikscheit, T. C., Perez, A., Rhoads, D. B., Gardner-Thorpe, J., et al. (2003). Effect of GLP-2 on mucosal morphology and SGLT1 expression in tissue-engineered neointestine. *Am. J. Physiol. Gastrointest. Liver Physiol., 235,* G1345–G1352.

Rheinwald, J. G., & Green, H. (1975). Serial cultivation of strains of human epidermal keratinocytes: the formation of keratinizing colonies from single cells. *Cell, 6,* 331–343.

Rizvi, A. Z., Swain, J. R., Davies, P. S., Bailey, A. S., Decker, A. D., Willenbring, H., Grompe, M., et al. (2006). Bone marrow-derived cells fuse with normal and transformed intestinal stem cells. *Proc. Natl. Acad. Sci. U.S.A., 103,* 6321—6325.

Rocha, F. G., Sundback, C. A., Krebs, N. J., Leach, J. K., Mooney, D. J., Ashley, S. W., et al. (2008). The effect of sustained delivery of vascular endothelial growth factor on angiogenesis in tissue-engineered intestine. *Biomaterials, 29,* 2884—2890.

Sala, F. G., Kunisaki, S. M., Ochoa, E. R., Vacanti, J., & Grikscheit, T. C. (2009). Tissue-engineered small intestine and stomach form from autologous tissue in a preclinical large animal model. *J. Surg. Res., 156,* 205—212.

Saxena, A. K., Kofler, K., Ainodhofer, H., & Hollwarth, M. (2009). Esophagus tissue engineering: hybrid approach with esophageal epithelium and unidirectional smooth muscle tissue component generation *in vitro*. *J. Gastrointest. Surg., 13,* 1037—1043.

Shinhar, D., Finaly, R., Niska, A., & Mares, A. J. (1998). The use of collagen-coated vicryl mesh for reconstruction of the canine cervical esophagus. *Pediatr. Surg. Int., 13,* 84—87.

Slorach, E. M., Campbell, F. C., & Dorin, J. R. (1999). A mouse model of intestinal stem cell function and regeneration. *J. Cell Sci., 112*(Pt 18), 3029—3038.

Smith, M. D., & Campbell, R. M. (2006). Use of a biodegradable patch for reconstruction of large thoracic cage defects in growing children. *J. Pediatr. Surg., 41,* 46—49.

Somara, S., Gilmont, R. R., Dennis, R. G., & Bitar, K. N. (2009). Bioengineered internal anal sphincter derived from isolated human internal anal sphincter smooth muscle cells. *Gastroenterology, 137,* 53—61.

Spiegel, J. H., & Kessler, J. L. (2005). Tympanic membrane perforation repair with acellular porcine submucosa. *Otol. Neurotol., 26,* 563—566.

Tait, I. S., Flint, N., Campbell, F. C., & Evans, G. S. (1994). Generation of neomucosa *in vivo* by transplantation of dissociated postnatal small intestinal epithelium. *Differentiation, 56,* 91—100.

Tan, J. J., Chan, M., & Tjandra, J. J. (2007). Evolving therapy for fecal incontinence. *Dis. Colon Rectum, 50,* 1950—1967.

Tavakkolizadeh, A., Berger, U. V., Stephen, A. E., Kim, B. S., Mooney, D., Hediger, M. A., et al. (2003). Tissue-engineered neomucosa: morphology, enterocyte dynamics, and SGLT1 expression topography. *Transplantation, 75,* 181—185.

Thompson, J. S. (1999). Surgical approach to the short-bowel syndrome: procedures to slow intestinal transit. *Eur. J. Pediatr. Surg., 9,* 263—266.

Urita, Y., Komuro, H., Chen, G. P., Shinya, M., Kaneko, S., Kaneko, M., et al. (2007). Regeneration of the esophagus using gastric acellular matrix: an experimental study in a rat model. *Pediatr. Surg. Int., 23,* 21—26.

Vacanti, J. P., Morse, M. A., Saltzman, W. M., Domb, A. J., Perez-Atayde, A., & Langer, R. (1988). Selective cell transplantation using bioabsorbable artificial polymers as matrices. *J. Pediatr. Surg., 23,* 3—9.

Voytik-Harbin, S. L., Brightman, A. O., Kraine, M. R., Waisner, B., & Badylak, S. F. (1997). Identification of extractable growth factors from small intestinal submucosa. *J. Cell Biochem., 67,* 478—491.

Walters, J. R. (2004). Cell and molecular biology of the small intestine: new insights into differentiation, growth and repair. *Curr. Opin. Gastroenterol., 20,* 70—76.

Wang, L., Murthy, S. K., Fowle, W. H., Barabino, G. A., & Carrier, R. L. (2009). Influence of micro-well biomimetic topography on intestinal epithelial Caco-2 cell phenotype. *Biomaterials, 30,* 6825—6834.

Wang, Z. Q., Yusuhiro, W., Noda, T., Yoshida, A., Ovama, T., & Toki, A. (2005). Morphologic evaluation of regenerated small bowel by small intestinal submucosa. *J. Pediatr. Surg., 40,* 1898—1902.

Warner, B. W. (2004). Tissue engineered small intestine: a viable clinical option? *Ann. Surg., 240,* 755—756.

Watson, L. C., Friedman, H. I., Griffin, D. G., Norton, L. W., & Mellick, P. W. (1980). Small bowel neomucosa. *J. Pediatr. Surg., 28,* 280—291.

Weber, T. R. (1999). Isoperistaltic bowel lengthening for short bowel syndrome in children. *Am. J. Surg., 178,* 600—604.

Yamamoto, Y., Nakamura, T., Shimizu, Y., Takimoto, Y., Matsumoto, K., Kiyotani, T., et al. (1999). Experimental replacement of the thoracic esophagus with a bioabsorbable collagen sponge scaffold supported by a silicone stent in dogs. *ASAIO J., 45,* 311—316.

Yoo, J. J., Meng, J., Oberpenning, F., & Atala, A. (1998). Bladder augmentation using allogenic bladder submucosa seeded with cells. *Urology, 51,* 221—225.

Yoshida, A., Noda, T., Tani, M., Oyama, T., Watanabe, Y., Kiyomoto, H., et al. (2009). The role of basic fibroblast growth factor to enhance fetal intestinal mucosal cell regeneration *in vivo*. *Pediatr. Surg. Int., 25,* 691—695.

Zhao, Y., Glesne, D., & Huberman, E. (2003). A human peripheral blood monocyte-derived subset acts as pluripotent stem cells. *Proc. Natl. Acad. Sci. U.S.A., 100,* 2426—2431.

CHAPTER 51

Extracorporeal Renal Replacement

Kimberly A. Johnston[*], **H. David Humes**[*,†]
[*] Innovative Biotherapies, Ann Arbor, MI, USA
[†] Department of Internal Medicine, University of Michigan, Ann Arbor, MI, USA

INTRODUCTION

The kidney is unique in that it is the first organ for which long-term *ex vivo* substitutive therapy has been made available and lifesaving. Renal failure prior to the era of hemodialysis and transplantation resulted in certain death, and this outcome of renal failure is still common outside the industrialized world. In the USA, 526,343 patients were listed as having end-stage renal disease (ESRD) by the 2007 US Renal Data System (USRDS) database, of whom 308,910 were receiving maintenance dialysis (US Renal Data System, 2004). The prevalence of ESRD in the USA is rising at approximately 8% per year (Neilson et al., 1997; US Renal Data System, 2004). The financial cost of dialysis is immense, estimated at $54,900 per hemodialysis patient per year and $46,121 per peritoneal dialysis patient per year. In contrast, transplant patients cost an average of $17,227 per patient per year (US Renal Data System, 2004). The higher cost of maintenance dialysis when compared with transplantation does not translate into better results; annual mortality for patients listed for transplant and awaiting a kidney is 6.3%, compared with only 3.8% for patients listed for transplant who did receive a kidney (US Renal Data System, 2004). While organ transplantation provides the best prognosis for survival, demand vastly outweighs the availability of donated organs.

Current dialysis therapies include hemodialysis (HD), hemofiltration, and peritoneal dialysis (PD). Dialysis provides clearance of small molecules by diffusive flow across a semipermeable membrane and control of volume status by bulk flow of water and solutes through that membrane. These short-term effects are sufficient to abrogate the lethal acidosis, volume overload, and uremic syndromes that accompany renal failure but do not protect the patient from the increased mortality associated with dialysis-treated renal failure in either the acute or chronic form. These methodologies all address water and electrolyte balance — functional replacement of the kidney. However, they fail to provide for the lost endocrine function. Thus, the metabolic, endocrine, and immune roles of the functioning kidney are candidate mechanisms for the difference in survival noted above. The dialytic clearance of glutathione, a key tripeptide in free radical scavenging and protection against oxidant stress; the negative nitrogen balance and energy loss in the clearance of peptides and amino acids in dialysate; loss of oxidative deamination and gluconeogenesis in the tubule cell; and loss of cytokine and hormone metabolic activity by the kidney each impose substantial stress upon the dialyzed patient and as such are appropriate targets for improved renal replacement therapy.

943

Principles of Regenerative Medicine. DOI: 10.1016/B978-0-12-381422-7.10051-3

REQUIREMENTS OF A RENAL REPLACEMENT DEVICE

Filtration is accomplished by the glomerulus, a tuft of capillaries supported by a basement membrane and specialized epithelial cells called podocytes. The renal proximal tubule, a hollow tube of cells surrounded by capillaries, receives the filtrate from the glomerulus and accomplishes the bulk of reclamation of salt, water, glucose, small proteins, amino acids, glutathione, and other substances. The tubule also performs metabolic functions, including excretion of acid as ammonia and hydroxylation of 25-hydroxy-vitamin-D_3 among others. Intermittent hemodialysis is thought to replace the filtration function of the glomerulus and advances in hemodialysis and hemofiltration have focused on emulation of glomerular physiology. Recent attention has been drawn to duplicating the function of the proximal tubule. The transport of solutes and water is accomplished by ATP-driven electrolyte transporters in the luminal cell membrane. Reabsorption of small proteins and peptides in the filtrate stream is accomplished by membrane-bound proteases and specific amino acid transport proteins within the luminal membrane of the tubule cell. These amino acids are either used for protein and peptide synthesis in the tubule cell or transported into the capillaries for transport to and use by the body. The diversity and specificity of the functions of the proximal tubule cell argue against the development of an electromechanical or polymeric substitute, and so a number of years ago our research group turned its attention to the isolation and culture of renal proximal tubule cells, which research has culminated in the hollow-fiber bioreactor discussed below.

RENAL PROXIMAL TUBULE CELL SOURCING AND FABRICATION OF A BIOREACTOR

Critical to providing organ function replacement through cell therapy is the isolation and growth *in vitro* of specific cells from adult tissue. These cells must have stem cell-like characteristics, with a high capacity for self-renewal and the ability to differentiate under defined conditions into specialized cells to develop the correct structure and functional components of a physiological organ system. Methodology to isolate and grow renal proximal tubule progenitor cells from adult pig kidneys has been reported (Humes and Cielinski, 1992; Humes et al., 1996). These studies were promoted by clinical and experimental observations suggesting that progenitor cells of renal proximal tubules must exist, as tubule cells have the ability to regenerate after severe nephrotoxic or ischemic injury. Porcine cells were utilized, as the pig has been considered the best source of organs for both xenotransplantation and cell therapy devices due to its anatomic and physiological similarities to human tissue and the relative ease of breeding large numbers of pigs in closed herds. However, reports of the ability of porcine endogenous retroviruses (PERVs) to infect human cells in co-culture *in vitro* have raised concerns about the potential, but currently unquantifiable, risk of transmission of viral elements from porcine tissue to humans in xenotransplantation or cell therapy devices (le Tissier et al., 1997; Paradis et al., 1999). Accordingly, Humes and colleagues fabricated bioreactors containing human renal epithelial cells isolated from kidneys donated for transplantation but found unsuitable for such purpose because of anatomic or fibrotic defects. These cells performed well in studies to assess viability, durability, and physiological performance (Humes et al., 2002).

The renal assist device (RAD) is a bioreactor containing proximal tubule cells grown in confluent monolayers along the inner surface of the hollow fibers of a conventional hemofiltration cartridge. Within this multifiber unit, proximal tubule cells not only maintain transport properties but also differentiated metabolic and endocrine functions. The non-biodegradability and the pore size of the hollow fibers allow the membranes to act as both scaffolds for the cells and as an immunoprotective barrier. Completed experiments have successfully scaled up to a clinically applicable device with the use of commercially available high-flux hemofiltration hollow fiber cartridges.

Experiments have tested the transport and metabolic functions of cells grown intraluminally within these cartridges with membrane surface areas of 97 cm^2 to 0.4 m^2 (Humes et al., 1999b). Starting with a hemofilter cartridge, the intraluminal surface of the hollow fibers was coated with pronectin L, a synthetic protein with multiple cell attachment sites found in the extracellular protein laminin. Renal proximal tubule progenitor cells were then seeded at a density of 10^5 cells/ml into the intracapillary space. The seeded cartridge was connected to the bioreactor perfusion system, in which the extracapillary space was filled with culture media and the intracapillary space perfused with similar media. After 7—14 days, light microscopy revealed a confluent monolayer of tubule cells grown on the inner surface of the fibers, and electron microscopy identified differentiated epithelial characteristics, including microvilli, endocytic vesicles, and extensive basolateral infoldings. The cells retain vectorial fluid transport properties as a result of Na,K-ATPase; differentiated active transport properties, including active electrolyte, bicarbonate, and glucose transport; differentiated metabolic activities, including intraluminal glutathione breakdown and ammonia production; and the important endocrinological conversion of 25-OH-vitD$_3$ to 1,25-(OH)$_2$-vitD$_3$.

ULTRAFILTRATION MEMBRANE DEVELOPMENT

In parallel with progress in stem cell work, there has been considerable interest in applying novel technology to membrane engineering. The filtration barrier in the kidney has been extensively studied. This barrier is widely considered to be trilaminate, with an endothelial cell layer, a basement membrane, and an elaborate epithelial layer bearing a specialized cell-cell junction called the glomerular slit diaphragm. Unfortunately, the cell considered responsible for the permselectivity barrier in the kidney, the glomerular podocyte, is a terminally differentiated cell with limited regenerative capacity. Damaged cells are not replaced by expansion of neighboring podocytes. Similarly, primary cultures of podocytes do not assume a differentiated phenotype in the laboratory dish, nor do they easily divide and expand in number. Despite progress with a conditionally transformed cell line derived from mouse podocytes, there seems little immediate prospect of a cell-based bioartificial glomerulus. Without a purely biological ultrafiltration unit on the horizon, advances in HD membrane development have been aimed at attempting to better reproduce the physiological process of glomerular ultrafiltration using synthetic membranes.

Membrane materials are diverse, ranging from regenerated cellulose filters to metals and ceramics to modern-day polymers. Critical to the choice of materials for current dialysis membranes, biocompatibility with blood is a major concern, as well as cost and manufacturing. Synthetic polymers have become the dominant membrane materials used due a combination of these factors. The most common polymers in the manufacturing of synthetic membranes are non-degradable polymers: cellulose derivatives; nitrates; polyesters, polysulfones; polyacrilonitrile derivatives; polyamides; polyimides; polyolefins such as polyethylene, polypropylene or polyvinylchloride; and fluorinated polymers such as polytetrafluoroethylene, and polyvinylidinefluoride.

Dialysis membrane clearance is based on concentration differences rather than convective separation of small solutes and low-molecular-weight proteins from large serum proteins and blood elements. In an attempt to recapitulate glomerular ultrafiltration and removal of "middle molecules," synthetic membranes with larger pore sizes and high water permeability have been developed. These so-called "high-flux" membranes are prepared with hydrophobic base materials, including polyacrylnitrile, polysulfone, polyethersulfone, or polymide, with various hydrophilic components. Recent membrane development has focused on increasing pore size while sharpening the molecular weight cutoff of high-flux membranes to maximize removal of low-molecular-weight proteins. Removal of a distinct class of uremic toxins, such as β2-microglobulin, factor D, leptin, and adrenomedullin, while minimizing the loss of albumin, could improve treatment outcomes of patients with ESRD. This idea has spurned the creation of superflux or protein-leaking membranes. These membranes provide greater clearances for low-molecular-weight proteins and small protein-bound solutes, such as homocysteine and advanced glycation end products, but with significantly higher loss of albumin than high-flux dialysis membranes. The overall benefits for patients on chronic HD still require more extensive evaluation.

Micromechanical systems (MEMS) are increasingly used to develop novel membrane technology. Current polymer membranes for renal replacement are performance limited. In general, such membranes are fairly thick or employ a multilayer scaffold for mechanical support, and they have a distribution of pore sizes rather than a regular array of uniform pores. Pores in conventional polymer membranes tend to either be roughly cylindrical, have a round orifice terminating a larger channel, or have a structure resembling an open cell sponge. These structures provide less than optimal geometries for membrane filtration for two reasons. First, a wide dispersion in pore sizes within a membrane leads to imperfect retention of molecules larger than the mean pore size of the membrane. Reducing the mean pore size of the membrane may partially solve this problem; however, it has the undesired effect of reducing the hydraulic permeability of the membrane. Second, the round shape of conventional pores dictates a dependence of hydraulic permeability on pore radius. In contrast, a pore that is slit-shaped allows steric hindrance to solute passage dictated by the smallest critical dimension of the pore, while increasing hydraulic permeability based on the long dimension of the pore. Consequently, it might be predicted that filtration structures with parallel slit-shaped pores will have superior performance when compared to structures with round pores. The glomerular filtration barrier also imposes an electrostatic restriction on solute passage. This function has been variously attributed to the proteins within the slit diaphragm, the glomerular basement membrane, and the glycocalyx of the glomerular endothelial cell. In regard to artificial membranes, a double thick electrical layer related to the nanometer-scale pore size itself contributes to the rejection of charged solutes.

Novel silicon nanopore membranes with 10–100 nm × 45 μm slit pores, approximating the glomerular slit diaphragm, have been prototyped by an innovative process based on MEMS technology. Silicon chips bearing 1 × 1 mm arrays of approximately 10^4 slit pores were fabricated via sacrificial layer techniques. The pore structure is defined by deposition and patterning of a polysilicon film on the silicon wafer. The critical submicron pore dimension is defined by the thickness of a sacrificial SiO_2 layer, which can be grown with unprecedented control to within ±1 nm.

Preliminary data on the transport properties of MEMS membranes are encouraging. Measured hydraulic permeabilities correlated well with theoretical predictions for flow-through slit-shaped pipes, also known as Hele-Shaw flows. The observed albumin sieving coefficient data provide encouragement that protein permselectivity is also feasible with this technology. Recent laboratory data have validated the possibilities of these membranes as scaffolding for a renal tubule cell bioreactor (Ward et al., 2001).

TRANSPORT AND METABOLIC CHARACTERISTICS OF HOLLOW-FIBER BIOREACTORS

As initial experiments using the single hollow-fiber model were promising, the design was scaled up to use commercially available polysulfone hollow-fiber dialysis cartridges from the manufacturers of the single hollow fibers. Single hollow-fiber measurements of transport and metabolic activity were repeated with 97 cm^2 and 0.4 m^2 surface area cartridges. Further exploration of the metabolic and transport characteristics of the cultured proximal tubule cells was assessed. The transport of glucose, bicarbonate, and glutathione excretion was measured in the absence and presence of a known inhibitor of an enzyme essential for the reabsorption. In each case, there was evidence of active transport and specific inhibition (Humes et al., 1999b).

The synthesis and secretion of ammonia into the tubule is essential for renal excretion of an acid load, as it buffers secreted protons. Proximal tubule cells are able to upregulate their ammoniagenesis in response to a decline in pH, and the proximal tubule cells in the bioreactor demonstrated a stepwise increase in ammonia production with changes in pH (Humes et al., 1999b). The experiments detailed above were performed in our laboratory with porcine tubule cells that demonstrated similar results in culture, attachment, and activity with human proximal tubule cells from cadaveric organs. The final selection of cell type for use in a renal tubule device not only rests on supply and safety of cells, but also depends on the ability of cells to participate in the homeostasis of the host. The above data suggest that our laboratory has successfully isolated and cultured renal proximal tubule cells, established stable confluent monolayers within hollow-fiber bioreactors, and scaled the initial construct to a level approximating the number of proximal tubule cells in a single kidney.

PRECLINICAL CHARACTERIZATION OF THE RENAL TUBULE ASSIST DEVICE/BIOARTIFICIAL KIDNEY

In keeping with its role as a metabolically active replacement for the renal proximal tubule, an extracorporeal circuit was devised that recapitulated nephron anatomy. The bioartificial kidney setup consists of a filtration device (a conventional hemofilter) followed in series by the renal tubule assist device (RAD) unit. Specifically, blood is pumped out of a patient using a peristaltic pump. The blood then enters the fibers of a hemofilter, where ultrafiltrate is formed and delivered into the fibers of the tubule lumens within the RAD downstream to the hemofilter. Processed ultrafiltrate exiting the RAD is collected and discarded as "urine." The filtered blood exiting the hemofilter enters the RAD through the extracapillary space port and disperses among the fibers of the device. Upon exiting the RAD, the processed blood travels through a third pump and is delivered back to the patient. Heparin is delivered continuously into the blood before entering the RAD to diminish clotting within the device. The RAD is oriented horizontally and maintained at 37°C throughout its operation to ensure optimal functionality of the cells. The tubule unit is able to maintain viability because metabolic substrates and low-molecular-weight growth factors are delivered to the tubule cells from the ultrafiltration unit and the blood in the extracapillary space. Furthermore, immunoprotection of the cells grown within the hollow fiber is achieved because of the impenetrability of immunoglobulins and immunologically competent cells through the hollow fibers. Rejection of the cells, therefore, does not occur.

This extracorporeal circuit containing the RAD was initially tested on uremic dogs with bilateral nephrectomies (Humes et al., 1999a). The animals were treated with either a RAD or a sham control cartridge daily for either 7 or 9 h for three successive days or for 24 h continuously. The RADs maintained viability and functionality throughout the study period. Fluid and small solutes, including blood urea nitrogen (BUN), creatinine (Cr), and electrolytes, were adequately controlled in both groups, but potassium and BUN levels were more easily controlled by RAD treatment. Furthermore, active reabsorption of K^+, $HCO3^-$, and glucose and excretion of ammonia were accomplished only in RAD treatments. Glutathione reclamation from UF exceeded 50% in the RAD. Finally, uremic animals receiving cell therapy attained normal 1,25-$(OH)_2$-vitD$_3$ levels, whereas sham treatment resulted in a further decline from the already low plasma levels. Thus, these experiments clearly showed that the combination of a synthetic hemofiltration cartridge and a RAD in an extracorporeal circuit successfully replaced filtration, transport, and metabolic and endocrinological functions of the kidney in acutely uremic dogs.

CLINICAL EXPERIENCE WITH A HUMAN RENAL TUBULE ASSIST DEVICE

Encouraging preclinical data led to FDA approval for an investigational new drug and phase I/II clinical trials. The first human clinical study of the bioartificial kidney containing human cells was carried out in 10 ICU patients with AKI receiving CVVH (Humes et al., 2004). This study demonstrated that the RAD can be used safely for up to 24 h. Cardiovascular stability was maintained, and increased native renal function, as determined by elevated urine outputs, temporally correlated with RAD treatment. All patients were critically ill with acute kidney injury (AKI) and multiple organ failure (MOF), with predicted hospital mortality rates between 80 and 95%, Six of the 10 treated patients survived past 30 days, with mortality reduced to 40%. The human renal tubule cells contained in the RAD demonstrated differentiated metabolic and endocrinological activity in this *ex vivo* situation, including glutathione degradation and endocrinological conversion of 25-OH-vitD$_3$ to 1,25-(OH)$_2$-vitD$_3$. Plasma cytokine levels suggest that RAD therapy produces dynamic and individualized responses in patients depending on their unique pathophysiological conditions. For the subset of patients who had excessive proinflammatory levels, RAD treatment resulted in significant declines in granulocyte-colony stimulating factor (G-CSF), IL-6, IL-10, and especially IL-6/IL-10 ratio, suggesting a greater decline in IL-6 relative to IL-10 levels and a less proinflammatory state. These favorable phase I/II trial results led to a randomized, controlled, open-label phase II trial conducted at 12 clinical sites in the US (Tumlin et al., 2008). Fifty-eight patients with ARF requiring CVVH in the ICU were randomized (2:1) to receive CVVH + RAD ($n = 40$) or CVVH alone ($n = 18$). Despite the critical nature and life-threatening illnesses of the patients enrolled in this study, the addition of the RAD to CVVH resulted in a substantial clinical impact on survival compared with the conventional CVVH-treatment group. RAD treatment for up to 72 h promoted a statistically significant survival advantage over 180 days of follow-up in ICU patients with AKI and demonstrated an acceptable safety profile. Cox proportional hazards models suggested that the risk of death was approximately 50% of that observed in the CRRT-alone group. A follow-up phase IIb study to evaluate a commercial manufacturing process was not completed due to difficulties with the manufacturing process and clinical study design. This approach will be further evaluated when an improved scale-up manufacturing process is established.

CELL THERAPY OF ACUTE RENAL FAILURE DUE TO SEPSIS

After a series of experiments demonstrating bioactivity, longevity, and systemic activity of the proximal tubule cells in a large animal model, a series of experiments was designed to examine the impact of cell therapy on the course of sepsis complicated by renal failure (Fissell et al., 2002a, 2003). After two initial studies supported a systemic effect and hemodynamic benefit from cell therapy in large animal models of sepsis, our laboratory pursued further evidence that cell therapy with renal proximal tubule cells alters the physiological response to sepsis (Humes et al., 2003). A porcine model of septic shock was developed from the previous work (Natanson et al., 1989; Natanson, 1990; Dinarello, 1991). Purpose-bred pigs were anesthetized and administered an intraperitoneal dose of bacteria, causing shock and renal failure. An hour later continuous venovenous hemofiltration (CVVH) was initiated with either cell or sham RAD. Urine output and mean arterial pressure declined within the first few hours after insult. Cell-treated animals survived 9.0 ± 10.83 h versus 5.1 ± 10.4 h ($P \leq 0.005$) for sham-treated animals. Serum cytokines were similar between the two groups, with the striking exception of interleukin (IL)-6 and interferon (IFN)-γ. Treatment with the cell RAD resulted in significantly lower plasma levels of both IL-6 ($P \leq 0.04$) and INF-γ ($P \leq 0.02$) throughout the experimental time course compared to sham RAD exposure. This controlled trial of cell therapy of renal failure in a realistic animal model of sepsis has several findings not immediately expected from *a priori* assumptions regarding renal function. Heretofore, although renal failure has been strongly associated with poor outcome in hospitalized patients, and chronic renal failure is associated with specific defects in humoral and cellular immunity, a direct immunomodulatory effect of the kidney had not been accepted. In this trial, clear differences in survival and clear differences in a serum cytokine associated with mortality in sepsis were found between groups: The increased mortality in renal failure appears to be not attributable to inadequate solute clearance, but may arise from other bioactivity of the kidney.

BIOARTIFICIAL KIDNEY IN END-STAGE RENAL DISEASE

A bioartificial kidney for long-term use in ESRD, similar to short-term use in acute renal failure (ARF), would integrate tubular cell therapy and the filtration function of a hemofilter. As noted above, ESRD patients on conventional renal replacement therapy are at high risk for cardio-vascular and infectious diseases. A recent clinical trial failed to show survival benefit from increased doses of hemodialysis above what is now standard care (Eknoyan et al., 2002), suggesting that there are important metabolic derangements not adequately treated with conventional dialytic treatment. Data from the survival of renal transplant recipients, which far exceed those from the survival of age-, sex-, and risk-matched controls awaiting transplant, also suggest that there is some metabolic function provided by the kidney that transcends this organ's filtration function. Patients with ESRD display elevated levels of C-reactive protein (CRP), an emerging clinical marker, and pro-inflammatory cytokines, including IL-1, IL-6, and tumor necrosis factor (TNF) (Bologa et al., 1998; Kimmel et al., 1998; Zimmermann et al., 1999). All these parameters are associated with enhanced mortality in ESRD patients. Specifically, IL-6 has been identified as a single predictive factor closely correlated with mortality in hemodialysis patients (Bologa et al., 1998). Although all ESRD patients could conceivably benefit from a bioartificial kidney, patients in the inflammatory stage who display elevated levels of certain markers of chronic inflammation (most notably IL-6 and CRP) would likely benefit most and will be the target population for clinical study in the near future.

For the ESRD patient population, however, there are obvious limitations in using an extra-corporeal RAD connected to a hemofiltration circuit. Ideally, a bioartificial kidney suitable for long-term use in ESRD patients would be capable of performing continuously, like the native kidney, to reduce risks from fluctuations in volume status, electrolytes, and solute concen-trations and to maintain acid-base and uremic toxin regulation. Such treatment requires the design and manufacture of a compact implantable or wearable dialysis apparatus and the development of miniaturized renal tubule cell devices with long service lifetimes. The ideal design of the next-generation RAD would be like that of an implantable device such as the pacemaker.

Attempts have been made to develop wearable dialysis systems to improve the portability of renal replacement therapies. Gura et al. (2005) have published research into a light-weight, wearable, continuous ambulatory ultrafiltration device consisting of a hollow fiber hemofilter, a battery-operated pulsatile pump, and two micropumps to control heparin administration and ultrafiltration. This device regenerates dialysate with activated carbon, immobilized urease, zirconium hydroxide, and zirconium phosphate, similar to the once commercially available REDY dialysis system. Ronco and Fecondini (2007) have described a wearable continuous PD system consisting of a double lumen dialysate line with a peritoneal catheter, a miniaturized rotary pump, a circuit for dialysate regeneration, and a handheld computer as a remote control. These systems still rely on inconvenient dialysate and expensive dialysis regeneration devices and/or dialyzers, but they promise to improve the convenience of dialysis.

In contrast to wearable dialysis systems, a hybrid bioartificial kidney integrates tubule cell and filtration functions. The first bioartificial kidney, consisting of a passive hemofilter and an active renal tubule cell bioreactor, has consistently demonstrated excellent safety and effec-tiveness in animal studies and FDA-approved human clinical trials, as described above (Ward et al., 2001; Humes et al., 2003, 2004; Tumlin et al., 2008). A major drawback of the current version of the bioartificial kidney is its large size, owing to the requisite extracorporeal circuit with peristaltic pumps to provide driving pressure for hemofiltration.

A new smaller and more durable RAD is currently being developed by Humes and colleagues. In collaboration, Fissell and colleagues are developing a nanopore membrane to replace the filtration function of the glomerulus without the hemofilters and mechanical pumps of existing dialysis machines. A filtration device based on nanopore membrane technology would be implantable (Fissell et al., 2002b, 2006; Magistrelli, 2004). Further refinement of the RAD would be encouraging for ESRD patients because, in principle, such a tissue-engineered device could be free of dialysate or replacement fluid while providing functions of healthy kidney that are not offered by current dialysis strategies. The combination of cell therapy and solute clearance could be a viable renal replacement therapy, conferring dialysis independence to the patient.

IMMUNOMODULATORY EFFECT OF THE RENAL TUBULE ASSIST DEVICE

As described earlier, RAD treatment altered systemic circulating cytokine levels in animal and human experiments. In endotoxin-challenged and gram-negative peritonitis uremic dog models, plasma levels of IL-10 were significantly higher in RAD-treated animals (Fissell et al., 2002a, 2003). The role of IL-10 in regulating immune response continues to be elucidated, but data suggest that IL-10 levels influence outcome in endotoxin shock and gram-negative sepsis. Several reports have demonstrated that administration of recombinant IL-10 is protective against gram-negative septic shock in murine sepsis models (Walley et al., 1996; Matsumoto et al., 1998). Another study in a similar model demonstrated that administration of antibodies to IL-10 was associated with higher mortality (Marchant et al., 1994). The mechanism underlying the link between proximal tubule function and IL-10 levels remains to be detailed, but preliminary data suggest that renal production of IL-6 induces liver production of IL-10 (Kielar et al., 2002). In gram-negative septic pigs without nephrectomy, RAD treatment significantly reduced plasma circulating levels of IL-6 and INF-γ (Humes et al., 2003). The difference in IL-6 concentrations is especially noteworthy, since the plasma elevations of this proinflammatory cytokine have been directly correlated to outcome in patients with SIRS (Pinksy et al., 1993). The lower concentration of plasma INF-γ may be important due to its central role in the inflammatory response. INF-γ stimulates B-cell antibody production, enhances polymorpholeukocyte phagocytosis, and activates monocytes and macrophages to release proinflammatory cytokines (Bone, 1991; Redmond et al., 1991; Joyce et al., 1994). Excessive rates of INF-γ production by NK cells have correlated with progression to lethal endotoxin shock in mice (Emoto et al., 2002). Further support for an immunomodulatory role of renal tubule cells has been suggested in the phase I/II clinical trial of the RAD containing human renal tubule cells (Humes et al., 2004). The patients treated in this study had a wide spectrum of plasma cytokine levels. The subset of patients who presented with very high plasma cytokine levels and who were treated for an adequate period showed that RAD treatment resulted in significant reductions in G-CSF, IL-6, and IL-10 levels. The greater relative reduction in IL-6/IL-10 ratio suggests renal tubule cell therapy may rebalance the excessive proinflammatory response with the concurrent anti-inflammatory response. These results are consistent with an immunomodulatory role for the RAD in patients with acute tubular necrosis and multiorgan failure. To further evaluate the RAD's influence on local inflammation in tissue and distant organ dysfunction, especially in the lungs, a recent study compared bronchoalveolar lavage (BAL) fluid from cell-RAD-treated and non-cell, sham-treated groups in a pig model with septic shock with AKI (Humes et al., 2007). The levels of total protein in BAL were significantly higher in sham control animals than in the RAD group (143 ± 111 compared to 78 ± 110 µg/ml, respectively; $P > 0.05$). Proinflammatory cytokines, including IL-6 and IL-8, were markedly elevated in the control group. These results demonstrate an important role for renal epithelial cells in ameliorating multiorgan injury in sepsis by influencing microvascular injury and the local proinflammatory response. A more promising direction to improve outcome of AKI is to better understand and interrupt the pathophysiological processes that are activated in AKI, resulting in distant multi-organ dysfunction and eventually death. AKI results in a profound inflammatory response state resulting in microvascular dysfunction in distant organs (Okusa, 2002; Simmons et al., 2004). Leukocyte activation plays a central role in these acute inflammatory states. Disruption of the activation process of circulating leukocytes may limit microvascular damage and multi-organ dysfunction (Maroszynska and Fiedor, 2000). The RAD appears to influence systemic leukocyte activation and the balance of inflammatory cytokines and may alter the proinflammatory state of AKI and, ultimately, improve morbidity and mortality. Our group has recently developed a novel synthetic membrane embedded in an extracorporeal device to bind and inhibit circulating leukocytes. This "selective cytopheretic inhibitory device" (SCD) mimics immunomodulation and duplicates RAD efficacy. The SCD improved septic shock survival times in preclinical animal models and improved the survival outcome of ICU patients with multi-organ failure in a small exploratory, randomized, double-blinded, multicenter trial (Ding et al., 2008; Humes et al., 2008).

THE BIOARTIFICIAL RENAL EPITHELIAL CELL SYSTEM

A next-generation RAD, the bioartificial renal epithelial cell system (BRECS), was designed by Humes and colleagues to achieve the support of 10-fold more cells in less than one-third of the volume of previous RAD designs, along with the capacity to cryopreserve the full unit in order to facilitate distribution. The polysulfone hollow fibers used as the scaffold in previous RAD designs were limited in cell attachment surface area and were prone to fracture during freeze-thaw. The cell-seeding scaffold for the BRECS, niobium-coated carbon disks, were selected based on their biologically inert, non-biodegradable, and favorable thermo-mechanical properties. Growth of adequate cell numbers to achieve a therapeutic impact was allowed due to disks' high surface area. The other BRECS components, a polycarbonate housing, gasket, nuts, bolts, and access ports, were all carefully selected and thoroughly tested to withstand cryogenic temperatures, while still maintaining an uncompromised, sterile internal BRECS environment.

In laboratory studies, the BRECS was able to be cryopreserved for long-term storage at $-140°C$, transitioned to $-80°C$ for short-term storage, thawed, and maintained at $37°C$ for clinical application, accompanied by a loss of no greater than 10% of the cell dose (Buffington et al., 2009). These data demonstrate that the BRECS is the first single device that can serve as a culture vessel to maintain cells, reach cryopreservation temperatures as a full unit, and, lastly, be reconstituted to provide cell therapy. Having this storage capacity makes both emergent and acute use feasible.

CONCLUSION

Despite all the advances in renal replacement therapies, a portable, continuous, dialysate-free artificial kidney remains the holy grail of renal tissue engineering. The enabling platform technologies discussed in this review advance this goal from a dream to the laboratory bench and even to the bedside. Future research in renal tissue engineering will need to focus on reproducing mechanisms of whole-body homeostasis. A high priority must be given to sensing and regulating extracellular fluid volume, even if only at the crude level of having the patient weigh him- or herself daily and adjust ultrafiltration and reabsorption by the bioartificial kidney. Chemical-field effect transistors (ChemFETs) offer the possibility of measuring electrolyte levels in a protein-free ultrafiltrate and reading out the potassium level to the patient, who could then alter diet or treat him- or herself with potassium-absorbing resins.

The critical building blocks of an autonomous bioartificial kidney are advancing rapidly with revolutionary clinical trials currently underway at multiple medical centers. The technology with which to adapt these advances to a more autonomous, dialysate-free system is under development. In addition, progress has been made in the field of cryopreservation and thus the ability to manufacture, store, and distribute bioartificial organs is advancing. The next decade, like the previous, will likely see quantum advances in renal tissue engineering.

References

Bologa, R. M., Levine, D. M., Parker, T. S., Cheigh, J. S., Serur, D., Stenzel, K. H., et al. (1998). Interleukin-6 predicts hypoalbuminemia, hypocholesterolemia, and mortality in hemodialysis patients. *Am. J. Kidney Dis., 32,* 107–114.

Bone, R. C. (1991). The pathogenesis of sepsis. *Ann. Intern. Med., 115,* 457–469.

Buffington, D. A., Hageman, G., Wang, M., Ding, F., Song, J., Jung, J., et al. (2009). (Abstract). Design of a compact cryopreservable bioartificial renal cell system. *J. Am. Soc. Nephrol., 20,* 27A.

Dinarello, C. A. (1991). The proinflammatory cytokines interleukin-1 and tumor necrosis factor and the treatment of the septic shock syndrome. *J. Infect. Dis., 163,* 1177–1184.

Ding, F., Song, J. H., Lou, L., Rojas, A., Reoma, J. L., Cook, K. E., et al. (2008). A novel selective cytopheretic inhibitory device (SCD) inhibits circulating leukocyte activation and ameliorates multiorgan dysfunction in a porcine model of septic shock. (Abstract.). *J. Am. Soc. Nephrol., 19,* 458A.

Eknoyan, G., Beck, G. J., Cheung, A. K., Daugirdas, J. T., Greene, T., Kusek, J. W., et al., for the Hemodialysis (HEMO) Study Group. (2002). Effect of dialysis dose and membrane flux in maintenance hemodialysis. *N. Engl. J. Med., 347,* 2010–2019.

Emoto, M., Miyamoto, M., Yoshizawa, I., Emoto, Y., Schaible, U. E., Kita, E., et al. (2002). Critical role of NK cells rather than Vα14+NKT cells in lipopolysaccharide-induced lethal shock in mice. *J. Immunol., 169,* 1426–1432.

Fissell, W. H., Dyke, D. B., Buffington, D. A., Weitzel, W. F., Westover, A. J., MacKay, S. M., et al. (2002a). Bioartificial kidney alters cytokine response and hemodynamics in endotoxin-challenged uremic animals. *Blood Purif., 20,* 55–60.

Fissell, W. H., Humes, H. D., Roy, S., & Fleischman, A. (2002b). Initial characterization of a nanoengineered ultrafiltration membrane. *J. Am. Soc. Nephrol., 13,* 602A.

Fissell, W. H., Lou, L., Abrishami, S., Buffington, D. A., & Humes, H. D. (2003). Bioartificial kidney ameliorates gram-negative bacteria-induced septic shock in uremic animals. *J. Am. Soc. Nephrol., 14,* 454–461.

Fissell, W. H., Manley, S., Westover, A., Humes, H. D., Fleischman, A. J., & Roy, S. (2006). Differentiated growth of human renal tubule cells on thin-film and nanostructured materials. *ASAIO J., 52,* 221–227.

Gura, V., Beizai, M., Ezon, C., & Polaschegg, H. D. (2005). Continuous renal replacement therapy for end-stage renal disease. The wearable artificial kidney (WAK). *Contrib. Nephrol., 149,* 325.

Humes, H. D., & Cieslinski, D. A. (1992). Interaction between growth factors and retinoic acid in the induction of kidney tubulogenesis in tissue culture. *Exp. Cell Res., 201,* 8–15.

Humes, H. D., Krauss, J. C., Cieslinski, D. A., & Funke, A. J. (1996). Tubulogenesis from isolated single cells of adult mammalian kidney: clonal analysis with a recombinant retrovirus. *Am. J. Physiol., 271*(1 Pt 2), F42–F49.

Humes, H. D., Buffington, D. A., MacKay, S. M., Funke, A. J., & Weitzel, W. F. (1999a). Replacement of renal function in uremic animals with a tissue-engineered kidney. *Nat. Biotechnol., 17,* 451–455.

Humes, H. D., MacKay, S. M., Funke, A. J., & Buffington, D. A. (1999b). Tissue engineering of a bioartificial renal tubule assist device: *in vitro* transport and metabolic characteristics. *Kidney Int., 55,* 2502–2514.

Humes, H. D., Fissell, W. H., Weitzel, W. F., Buffington, D. A., Westover, A. J., MacKay, S. M., et al. (2002). Metabolic replacement of kidney function in uremic animals with a bioartificial kidney containing human cells. *Am. J. Kidney Dis., 39,* 1078–1087.

Humes, H. D., Buffington, D. A., Lou, L., Abrishami, S., Wang, M., Xia, J., et al. (2003). Cell therapy with a tissue-engineered reduces the multiple-organ consequences of septic shock. *Crit. Care Med., 31,* 2421–2428.

Humes, H. D., Weitzel, W. F., Bartlett, R. H., Swaniker, F. C., Paganini, E. P., Luderer, J. R., et al. (2004). Initial clinical results of the bioartificial kidney containing human cells in ICU patients with acute renal failure. *Kidney Int., 66,* 1578–1588.

Humes, H. D., Buffington, D. A., Lou, L., Wang, M., & Abrishami, S. (2007). Renal cell therapy ameliorates pulmonary abnormalities in a large animal model of septic shock and acute renal injury. *J. Am. Soc. Nephrol., 18,* A382.

Humes, H. D., Dillon, J., Tolwani, A., Cremisi, H., Wali, R., Murray, P., et al. (2008). A novel selective cytopheretic inhibitory device (SCD) improves mortality in ICU patients with acute kidney injury (AKI) and multiorgan failure (MOF) in a phase II clinical study. *J. Am. Soc. Nephrol., 19,* 458A.

Joyce, D. A., Gibbons, D. P., Green, P., Steer, J. H., Feldmann, M., & Brennan, F. M. (1994). Two inhibitors of pro-inflammatory cytokine release, interleukin-10 and interleukin-4, have contrasting effects on release of soluble p75 tumor necrosis factor receptor by cultured monocytes. *Eur. J. Immunol., 24,* 2699–2705.

Kielar, M., Jeyarajah, D. R., & Lu, C. Y. (2002). The regulation of ischemic acute renal failure by extrarenal organs. *Curr. Opin. Nephrol. Hypertens., 11,* 451–457.

Kimmel, P. L., Phillips, T. M., Simmens, S. J., Peterson, R. A., Weihs, K. L., Alleyne, S., et al. (1998). Immunologic function and survival in hemodialysis patients. *Kidney Int., 54,* 236–244.

le Tissier, P., Stoye, J. P., Takeuchi, Y., Patience, C., & Weiss, R. A. (1997). Two sets of human-tropic pig retrovirus. *Nature, 389,* 681–682.

Magistrelli, J. M. (2004). Investigating Fluid Flow Through Silicon Nanoporous Membranes. Master's thesis. Cleveland: Case Western Reserve University.

Marchant, A., Bruyns, C., Vandenabeele, P., Ducarne, M., Gerard, C., Delvaux, A., et al. (1994). Interleukin-10 controls interferon-γ and tumor necrosis factor production during experimental endotoxemia. *Eur. J. Immunol., 24,* 1167–1171.

Maroszynska, I., & Fiedor, P. (2000). Leukocytes and endothelium interaction as rate limiting step in the inflammatory response and a key factor in the ischemia-reperfusion injury. *Ann. Transplant., 5,* 5–11.

Matsumoto, T., Tateda, K., Miyazaki, S., Furuya, N., Ohno, A., Ishii, Y., et al. (1998). Effect of interleukin-10 on gut-derived sepsis caused by *Pseudomonas aeruginosa* in mice. *Antimicrob. Agents Chemother., 42,* 2853–2857.

Natanson, C. (1990). Studies using a canine model to investigate the cardiovascular abnormality of and potential therapies for septic shock. *Clin. Res., 38,* 206–214.

Natanson, C., Danner, R. L., Elin, R. J., Hosseini, J. M., Peart, K. W., Banks, S. M., et al. (1989). Role of endotoxemia in cardiovascular dysfunction and mortality: *Escherichia coli* and *Staphylococcus aureus* challenges in a canine model of human septic shock. *J. Clin. Invest., 83,* 243–251.

Neilson, E. G., Hull, A. R., Wish, J., Neylan, J. F., Sherman, D., & Suki, W. N. (1997). The Ad Hoc Committee Report on estimating the future workforce and training requirements for nephrology. *J. Am. Soc. Nephrol., 8*(5 Suppl. 9), S1–S4.

Okusa, M. D. (2002). The inflammatory cascade in acute ischemic renal failure. *Nephron, 90,* 133–138.

Paradis, K., Langford, G., Long, Z., Heneine, W., Sandstrom, P., Switzer, W. M., et al., The XEN 111 Study Group, Otto, E. (1999). Search for cross-species transmission of porcine endogenous retrovirus in patients treated with living pig tissue. *Science, 285,* 1236–1241.

Pinsky, M. R., Vincent, J. L., Deviere, J., Alegre, M., Kahn, R. J., & Dupont, E. (1993). Serum cytokine levels in human septic shock. *Chest, 103,* 565–576.

Redmond, H. P., Chavin, K. D., Bromberg, J. S., & Daly, J. M. (1991). Inhibition of macrophase-activating cytokines is beneficial in the acute septic response. *Ann. Surg., 214,* 502–508.

Ronco, C., & Fecondini, L. (2007). The Vicenza wearable artificial kidney for peritoneal dialysis (ViWAK PD). *Blood Purif., 25,* 383–388.

Simmons, E. M., Himmelfarb, J., Sezer, M. T., Chertow, G. M., Mehta, R. L., Paganini, E. P., et al., for the PICARD study group. (2004). Plasma cytokine levels predict mortality in patients with acute renal failure. *Kidney Int., 65,* 1357–1365.

Tumlin, J., Wali, R., Williams, W., Murray, P., Tolwani, A. J., Vinnikova, A. K., et al. (2008). Efficacy and safety of renal tubule cell therapy for acute renal failure. *J. Am. Soc. Nephrol., 19*(5), 1034–1040.

US Renal Data System. (2004). USRDS 2002 Annual Data Report, Atlas of End-Stage Renal Disease in the United States. Bethesda, MD: National Institutes of Health, National Institute of Diabetes and Digestive and Kidney Diseases.

Walley, K., Lukacs, N., Standiford, T., Streiter, R., & Kunkel, S. (1996). Balance of inflammatory cytokines related to severity and mortality of murine sepsis. *Infect. Immunol., 64,* 4733–4738.

Ward, R. A., Leypoldt, J. K., Clark, W. R., Ronco, C., Mishkin, G. J., & Paganini, E. P. (2001). What clinically important advances in understanding and improving dialyzer function have occurred recently? *Semin. Dial., 14,* 160–174.

Zimmermann, J., Herrlinger, S., Pruy, A., Metzger, T., & Wanner, C. (1999). Inflammation enhances cardiovascular risk and mortality in hemodialysis patients. *Kidney Int., 55,* 648–658.

CHAPTER 52

Tissue Engineering of the Reproductive System

Stefano Da Sacco, Laura Perin, Roger E. De Filippo
Department of Urology, Children's Hospital Los Angeles, University of Southern California Keck School of Medicine, Los Angeles, CA, USA

INTRODUCTION

Human organ replacement is limited by a donor shortage, and problems of tissue compatibility and rejection. Tissue engineering combines the principles and methods of the life sciences with those of engineering to elucidate a fundamental understanding of structure–function relationships of normal and diseased tissues, in order to facilitate the development of materials and methods to repair damaged or diseased tissues, and to create entire tissue replacements.

A large number of materials, including naturally derived and synthetic polymers, have been utilized to facilitate prostheses for the genitourinary system.

Natural biomaterials such as amniotic membranes, bladder accellular matrices, and collagen constructs have been investigated in recent years in *in vitro* and *in vivo* assays (Roth and Kropp, 2009). The advantages of natural scaffolds have been evaluated in many studies. The presence of bioactive molecules (cytokines, vascular endothelial growth factor, basic fibroblast growth factor) was shown to bring a real advantage for tissue growth and vascularization when compared to synthetic scaffolds, as shown by Azzarello et al. (2007). However, these molecules can also give rise to inflammatory and rejection processes. Hyaluronic acid has been proposed by Cartwright et al. (2006) to overcome, or at least mitigate, the logistic response. In addition to the immunoresponse issue, the architecture of natural scaffolds may greatly change based on the tissue source and this has been hypothesized as a reason for the inconsistent results in animal models (Roth and Kropp, 2009).

In the last few decades, research into a suitable scaffold has been a priority for the tissue engineering field. Porous, absorbable matrices made of natural or synthetic polymers are currently being investigated as scaffolds for genitourinary tissue transplantation. These biodegradable polymers include poly(glycolic-acid) (PGA), polylactide (PLA), poly(glycolide-co-lactide) (PGLA), poly(caprolactone) (PCL), poly(glycolide-co-ε-caprolactone), collagen, alginate, hyaluronate, and laminin. These scaffolds require proper biocompatibility, degradability, mechanical stability, high surface area/volume ratio, and proper pore size. A highly porous scaffold is desirable to allow large numbers of cells to seed or migrate throughout the material and the pore size affects both tissue ingrowth and the internal surface area available for cell attachment (Mikos et al., 1994).

Nanotechnology techniques have been successfully applied by Mondalek et al. (2008) on porcine small intestinal mucosa to obtain a more homogeneous structure. In addition, other researchers are investigating the use of nanotechnologies for improved engraftment on synthetic tissue grafts (McManus et al., 2007; Harrington et al., 2008).

Principles of Regenerative Medicine. DOI: 10.1016/B978-0-12-381422-7.10052-5

3. Maintenance of urothelial cells:
   a. Replace medium with fresh warm (37°C) medium every 3 days, depending on cell density.
   b. Trypsinize cells when 80–90% confluent.

4. Subculture of corporal smooth muscle cells:
   a. Remove medium and add 1ml of phosphate-buffered saline–ethylenediaminetetraacetic acid (PBS–EDTA) (0.5 M) to each well or 10 ml to each 10-cm culture plate. Observe the cells under a phase contrast microscope.
   b. When cell–cell junctions are separated for the majority of the cells, remove PBS–EDTA and add 300 μl of trypsin–EDTA to each well or 5 ml to each 10-cm culture plate.
   c. Periodically agitate the plates. When 80–90% of the cells are detached, add 30 μl of soybean trypsin inhibitor (Gibco/BRL, 294 mg) to 20 ml of PBS to each well, or 700 μl to each 10-cm plate. Add 0.5 ml of medium to each well or 3 ml to each 10-cm plate.
   d. Aspirate and centrifuge the cell suspension at 1000 rpm for 4 min, and remove the supernatant.
   e. Resuspend cells and count the number of viable cells by means of trypan blue exclusion.
   f. Aliquot the desired number of cells on the plate and place the cells in the incubator.

Protocol I.B: bladder smooth muscle cell culture

1. Materials and medium:
   a. Tissue source: bladder tissue.
   b. Medium: DMEM, 10% fetal bovine serum (FBS), and antibiotic (penicillin (100 U/ml), streptomycin (100 μg/ml), and amphotericin B (0.25 μg/ml)).

2. Tissue harvest:
   a. Obtain fresh bladder tissue specimen.
   b. Use sharp tenotomy scissors to cut muscle tissue into small fragments (2–3 mm).
   c. Space muscle fragments evenly on a cell culture plate (100 mm).
   d. Allow muscle fragments to dry and adhere to the plate (5–10 min).
   e. Add 15 ml of DMEM and incubate for 5 days.
   f. Change medium on the sixth day and remove non-adherent tissue fragments.
   g. When small islands of cell colony are formed, remove the tissue fragments and change the medium.
   h. When sufficient cells are grown, trypsinize, count, and plate them onto new plates.

3. Maintenance of bladder smooth muscle cells:
   a. Feed cells every 3 days, depending on the cell density.
   b. Trypsinize cells when 80–90% confluent.

4. Subculture of bladder smooth muscle cells:
   a. Remove medium and add 10 ml of PBS–EDTA (0.5 M) for 4 min. Confirm the separation of cell junction under a phase contrast microscope.
   b. Remove PBS–EDTA and add 5 ml of trypsin–EDTA.
   c. Add 5 ml of medium when 80–90% of the cells lift under microscope.
   d. Aspirate the cell suspension into a 15-ml test tube.
   e. Aliquot the desired number of cells on the plate and makeup the volume of medium to a total of 10 ml.
   f. Place the cells in the incubator.

Cells were seeded onto non-woven meshes of PGA (Cilento et al., 1995). Partial urethrectomies were performed in rabbits and a segment of the polymer mesh was interposed to form the neourethra in each animal. Retrograde urethrograms showed no evidence of stricture formation. Histological examination of the neourethras demonstrated complete re-epithelialization of the polymer mesh implanted sites by day 14, and re-epithelialization continued for the entire duration of the study. Polymer fiber degradation was evident 14 days after implantation.

Scriven has reported the successful application of human urothelial cells for the formation of multi-layered epithelium when seeded onto de-epithelialized bladder stroma (Scriven et al., 1997).

More recently, innovative works in the field have proved the capacity to harvest and expand bladder-washed urothelial cells (Maurer et al., 2005). In addition, progenitor cells isolated by bladder wash were cultured and expanded and were able to give rise to urothelium and smooth muscle cells (Zhang et al., 2007).

The recent advances in knowledge of tissue sources and culturing conditions for the isolation and expansion of urothelial cells are promising for future human regenerative applications. However, Scriven showed how some cell surface markers, such as cytokeratins, were not expressed in cells seeded onto the de-epithelialized bladder stroma, maybe due to an absence of mechanical stress and/or incomplete maturation (Scriven et al., 1997).

The work of Scriven underlined and confirmed the absolute importance of the right engineered scaffold and culture conditions to enhance cell growth and maturation within the matrices. A suitable scaffold must present low immunogenicity, high vascularization capacity, and versatility.

A variety of synthetic grafts composed of silicone, Teflon, or polyvinyl have been proposed for urethral reconstruction. Erosion, dislodgment, fistula, stenosis, extravasation, and calcification have been associated with synthetic grafts (Guzman, 1999; Vozzi et al., 2002). Tissues from other sources have been used, such as genital and extragenital skin flaps or grafts, bladder mucosal grafts or grafts from buccal regions, tunica vaginalis, and peritoneum (Humby, 1941; Ehrlich et al., 1989; Dessanti et al., 1992). Various acellular biomaterials have been used experimentally (in animal models) for urethral tissue regeneration, including PGA, and acellular collagen-based matrices from small intestine, omentum, and bladder (Bazeed et al., 1983; Atala et al., 1992; Kropp et al., 1998; Chen et al., 1999; Sievert et al., 2000). Some of these biomaterials have also been seeded with autologous cells for urethral reconstruction. Acellular collagen matrices derived from bladder submucosa have been used experimentally and clinically. In animal studies, segments of the urethra were resected and replaced with acellular matrix grafts in an onlay fashion. The animals were able to void through the neourethras (Chen et al., 1999). These results were confirmed clinically in a series of patients with hypospadias and urethral stricture disease (Atala, 1999c; El-Kassaby et al., 2003).

In 2007, Ribeiro-Filho et al. proved in seven patients the feasibility of organ-specific acellular tissue, as demonstrated in animals by Sievert et al. (2000). In addition, following Weiser's studies on small intestinal submucosa (Weiser et al., 2003), clinical trials have been performed. Outcomes have been contradictory in different studies on adults while better results were obtained in young patients (Hauser et al., 2006; Atala, 2007; Fiala et al., 2007; Palminteri et al., 2007).

Unfortunately, the above techniques are not applicable for tubularized urethral repairs (Atala, 2007). The collagen matrices are able to replace urethral segments when used in an onlay fashion. However, if a tubularized repair is needed, the collagen matrices need to be seeded with autologous cells (De Filippo et al., 2002a, b).

Silicone, Teflon, and polyvinyl have been associated with side-effects (Hakky, 1976, 1977; Anwar et al., 1984). Biodegradable substitutes such as a polyglactin fiber mesh tube coated with poly(hydroxybutyric acid) and hyaluronan benzyl ester have been used experimentally. Complete regeneration of the urethral epithelium and the adjacent connective tissue was achieved as a consequence of the fact that the scaffolds guided urothelial and connective tissue regeneration (Italiano et al., 1991). Free grafts of tubularized peritoneum were used as urethral tissue substitutes experimentally in rabbits. Organized multilayered graft epithelialization occurred; however, fistulae formed in two of the animals (Shaul et al., 1996). Later, porcine small intestine submucosa (SIS) was used for urethral repair in a rabbit model to determine whether this material can evoke urethral regeneration. The SIS onlay grafts were shown to promote regeneration of the normal rabbit epithelium supported by a vascularized collagen and smooth muscle backing (Kropp et al., 1998).

More recently, Nuininga et al. partially resected a 0.5–1-cm segment of the native urethra in 24 rabbits and a novel molecularly defined collagen-based biocompatible and biodegradable matrix graft was sewn into place and compared with SIS. They did not notice any differences between the two biomatrices and the major advantage is that the new biometrics proposed can be modulated in different ways such as variation in the porous matrix structures, incorporation of growth factors, and binding of glycosaminoglycans (Nuininga et al., 2003).

A naturally derived acellular collagen-based tissue substitute was developed from donor porcine bladder (see Protocol II) and was able, upon *in vivo* implantation, to form bladder tissue similar to the native bladder (Yoo et al., 1998b).

Protocol II: acellular collagen matrix preparation

1. Isolate the submucosa from the muscular and serosal layers by means of microdissection techniques.

2. Treat tissue with distilled water in a magnetic stirring flask set at moderate speed for 24–48 h at 4°C.

3. Remove distilled water and treat with Triton X-100 (0.5%) and ammonium hydroxide (0.05%) in fresh distilled water for 72 h in a stirring flask at 4°C.

4. Wash with distilled water in a stirring flask for 24–48 h at 4°C. After this washing step, take a small piece of tissue for histological analysis to confirm any cellular remnants. Tissue matrix is usually decellularized at this time.

5. After confirmation of decellularization, wash with distilled water in a stirring flask for 24–48 h at 4°C. Tissue retaining cellular components should undergo an additional cycle of treatment. Repeat Steps 4 and 5, and perform another histological analysis.

6. After the washing cycle with distilled water, rinse with PBS overnight.

7. Freeze-dry the tissue sample overnight.

8. Pack the samples and sterilize in ethylene oxide.

9. Store until used. When ready to use equilibrate the tissue in PBS.

The use of decellularized matrix alone was evaluated in rabbits by Fu et al. (2007) in an animal model of urethral defects. The acellular matrix, when compared with epidermal cell seeded acellular matrix, performed poorly and was shown to be less suitable for urethral reconstruction.

More recently, Dorin et al. (2008) evaluated urethral replacement with the use of unseeded matrices transplanted in rabbits. Their work showed a normal engraftment of resident urothelial cells for defects up to 0.5 cm with vascularization of the transplanted matrix.

However, greater defects (up to 3 cm) showed normal cellular regeneration only at the matrix edges while the center exhibited collagen deposition and fibrosis after 2 weeks from the transplantation, proving that the size of the damage is a limiting factor in the use of acellular matrices for the regeneration of extensive urethral damage.

Penis

The indication for extended phalloplastic procedures results from severe congenital malformation, penile tissue loss from malignancies, trauma or other diseases, and gender dysphoria. Owing to the shortage of autologous penile tissue, multiple staged surgeries using non-genital tissues and silicone prostheses have been the mainstay in phallic reconstruction. However, graft failure and prothesis-related complications remain a problem. Replacement of penile tissue with alternative materials is challenging due to the unique anatomical architecture of the corporal body and autologous tissue availability is still poor.

Non-genital tissue sources have been used over the years; however, complications such as infection, graft failure, and donor site morbidity have posed continuing problems (Goodwin and Scott, 1952; Puckett and Montie, 1978; Chang and Hwang, 1984; Gilbert et al., 1988; Horton and Dean, 1990; Sharaby et al., 1995). The ability to engineer penile tissue composed of autologous cells would be beneficial.

ANATOMY

The anatomy of the penis is complex and comprises primarily three separate cylinders. The two paired cylinders called the corpora cavernosa make up the majority of the bulk and the erectile functioning of the penis. Each of these cylinders is encased in a very tough thick sheath called the tunica albuginea. The third cylinder of the penis is called the corpus spongiosum, and it contains the urethra. The tissue around this erectile body is much thinner, and the cylinder actually sits in a groove created by the other two cylinders. As this structure approaches the end of the penis, it becomes swollen and is known as the glans, or the head of the penis. As this layer gets closer to the body, it expands to form the bulb. Covering all three of these cylinders is a thick tough membrane called Buck's fascia. Finally, a final layer covers this area called Colles' fascia, or the superficial layer. This is actually continuous with the abdominal wall and makes this whole supporting structure of the penis very tough, allowing it to take quite a bit of force and trauma.

The shaft is covered by nearly hairless skin. Under the skin lies the dense connective tissue of penile fascia. The tunica albuginea encircles all three corpora, divides the corpora proximally, but is incomplete distally.

The corpora cavernosa are paired columns of erectile tissue located dorsally. Each column consists of a network of large venous sinuses separated by dense connective tissue septae, the trabeculae. The corpus spongiosum has a similar arrangement to the corpora cavernosa except it contains the penile urethra. It has the same arterial and venous relationship as the corpora cavernosa.

CORPUS CAVERNOSUM RECONSTRUCTION

Although consisting of only two important functional cell types (i.e. smooth muscle and endothelial cells), tissue engineering of autologous penile tissue remains a challenge. Our initial effort was focused on the formation of corporal tissue, since corpus cavernosum is one of the major tissue components of the phallus. Human corporal smooth muscle cells were isolated, grown, and expanded in culture (see Protocol I.A). The cells were seeded on bio-degradable PGA polymers for implantation. Multilayers of corporal smooth muscle cells were identified grossly and histologically. This study provided the evidence that cultured human corporal smooth muscle cells could be used in conjunction with biodegradable polymers to create cavernosal smooth muscle tissue *in vivo*.

Protocol I.A: corpus cavernosal smooth muscle cell culture

1. Materials and medium:
    a. Tissue source: human corpus cavernosum.
    b. Medium: DMEM, 10% FBS, and antibiotic (penicillin (100 U/ml), streptomycin (100 μg/ml), amphotericin B (0.25 μg/ml)).

2. Tissue harvest:
    a. Obtain fresh cavernosal tissue specimen.
    b. Use sharp tenotomy scissors to cut muscle tissue into small fragments (2–3 mm).
    c. Space muscle fragments evenly onto a cell culture plate (100 mm).
    d. Allow muscle fragments to dry and adhere to the plate (5–10 min).
    e. Add 15 ml of DMEM and incubate for 5 days.
    f. Change medium on the sixth day and remove non-adherent tissue fragments.
    g. When small islands of cell colony are formed, remove the tissue fragments and change the medium.
    h. When sufficient cells are grown, trypsinize, count, and plate the cells onto new plates.

3. Maintenance of corporal smooth muscle cells:
    a. Feed cells every 3 days, depending on the cell density.
    b. Trypsinize cells when 80–90% confluent.

4. Subculture of corporal smooth muscle cells:
    a. Remove medium and add 10 ml of PBS–EDTA (0.5 M) over 4 min. Confirm the separation of cell junction under a phase contrast microscope.
    b. Remove PBS–EDTA and add 5 ml of trypsin–EDTA.
    c. Add 5 ml of medium when 80–90% of the cells lift under the microscope.
    d. Aspirate the cell suspension into a 15-ml test tube.
    e. Centrifuge the cells at 1000 rpm for 4 min and remove the supernatant.
    f. Resuspend cells and use trypan blue exclusion to count viable cells.
    g. Aliquot the desired number of cells in the plate and makeup the volume of medium to a total of 10 ml.
    h. Place the cells in the incubator.

In a subsequent study, we investigated the possibility of developing corporal tissue by combining smooth muscle and endothelial cells. Normal human cavernosal smooth muscle cells and ECV 304 human endothelial cells were seeded on biodegradable polymer for implantation (Park et al., 1999). ECV 304 endothelial cells were used in the study to allow the investigator to distinguish the implanted cells from the host endothelial cells. The retrieved structures showed formation of distinct tissue structures, consisting of organized smooth muscle tissue adjacent to endothelial cells. The presence of vascular structures was evident. Each cell type was confirmed by means of various assessment methods. This study showed that human corporal muscle and endothelial cells seeded on biodegradable polymer scaffolds are able to form vascularized cavernosal tissue when implanted *in vivo*.

We developed a naturally derived collagen matrix, which is structurally similar to the native corporal architecture (Falke et al., 2003). Acellular collagen matrices, derived from rabbit corpora, were obtained by means of a cell lysis technique (see Protocol II for urethral acellular matrix preparation). Human corpus cavernosal muscle and endothelial cells were grown and expanded in culture (Protocol I.B). We have used human capillary cells, isolated from newborn foreskin via *Ulex europaeus* I (UEA-I)-coated Dynabeads (Jackson et al., 1990; Kraling and Bischoff, 1998). Primary human cavernosal smooth muscle and endothelial cells were seeded in a stepwise fashion. Cavernosal smooth muscle cells were initially seeded on the collagen matrices at a concentration of 30×10^6 cells/ml. Endothelial cells were then seeded at a concentration of 3×10^6 cells/ml. Cell matrices seeded with corporal cells were implanted *in vivo*. The implanted cell matrices showed neovascularity into the sinusoidal spaces by 1 week after implantation. Increased organization of smooth muscle and endothelial cells lining the sinusoidal walls was observed at 2 weeks and continued with time. The matrices were covered with the appropriate cell architecture 4 weeks after implantation (Atala, 1999b). This study demonstrates that human cavernosal smooth muscle and endothelial cells seeded on three-dimensional (3D) acellular collagen matrices derived from donor corpora are able to form a well-vascularized corporal architecture *in vivo*.

Protocol I.B: human endothelial cell culture from foreskin

1. Materials and media:
   a. Medium A (for primary culture and first passage after UEA-I bead selection): 38.5 ml of endothelial basal medium (EBM) 131, 10 ml of 20% FBS, 0.5 ml (2 mM) L-glutamine, 0.5 ml of PFS (antibiotic–antimycotic), 0.5 ml (0.5 mM) dibutyryl-cyclic adenosine-39,59-cyclic monophosphate (AMP), and 50 µl (1 µg/ml) hydrocortisone.
   b. Medium B (for passage 2 and all following passages): endothelial basal medium 131, $1\times$ GPS, 10% FBS, and 2 µg/ml basic fibroblast growth factor (25 µg/ml stock solution).
   c. Gelatin coating (1% Difco Bacto Gelatin in PBS): dissolve gelatin in PBS; autoclave to sterilize, and filter to remove particles.

2. Processing foreskin:
   a. Prepare foreskin collecting medium: 450 ml of DMEM, 25 ml of PBS (5%), 20 ml of antibiotic–antimycotic (400 U/ml penicillin, 400 µg/ml streptomycin, 1 µg/ml fungizone), 5 ml of L-glutamine (2 mM), and 1 ml of gentamicin sulfate (100 µg/ml).
   b. Place the collecting medium with the foreskin in a culture plate.
   c. Rinse 2–3 times with the collecting medium.
   d. Add 30 ml of collecting medium to a new 50-ml Falcon tube. Add an additional 2 ml of antibiotic–antimycotic.
   e. Separate the skin and subcutaneous tissue with a sterile scalpel blade and transfer the segments into the collecting medium in a 50-ml tube.
   f. Agitate the segments in the collecting medium at room temperature for at least 4–5 h to kill bacteria and spores that reside on the skin.

3. Isolation of endothelial cells:
   a. Prepare digestion solution: 7.5 ml of 1:250 trypsin, 2.7 ml of 0.5 M EDTA, pH 8.0, and 40 ml of Hanks' balanced salt solution (HBSS).
   b. Prepare 10 µl HBSS without Ca^{2+} and Mg^{2+}: 40 g of NaCl, 2 g of KCl, 240 mg of Na_2HPO_4, 300 mg of KH_2PO_4, 1750 mg of $NaHCO_3$, 5 g of glucose, and 100 mg of phenol red.
   c. Prepare wash solution (HBBS with $1 \times Ca^{2+}$ and Mg^{2+}): 50 ml of 10 µl HBSS, 92.7 mg of $CaCl_2$ $2H_2O$ (1.26 mM final), 100 mg of $MgSO_4$ $7H_2O$ (0.8 mM final), 25 ml of FBS (5% final), and 5 ml of PSF (antibiotic–antimycotic).
   d. Coat a Petri dish (100 mm) for each one or two foreskins with 8.0 ml of 1% gelatin–PBS. Remove excessive gelatin before plating.
   e. Autoclave a Teflon homogenizer (2.5-cm diameter) and gauze.
   f. Remove the collecting medium from the foreskin segments.
   g. Transfer the tissue segments into a sterile culture plate (100 mm).
   h. Cut the foreskin segments into 4-mm^2 fragments with a sterile scalpel blade.
   i. Transfer the tissue fragments to a sterile 50-ml Falcon tube and add 6.0 ml of digestion solution for 1–2 foreskins. Agitate vigorously at 37°C for 10 min.

j. Allow the skin fragments to sediment by gravitational force and aspirate the digestion medium. Wash once with 20 ml of wash solution, swirl vigorously, and remove the wash solution.

k. Add 10 ml of fresh wash solution and squeeze the fragments with the homogenizer.

l. Filter through 8–10 layers of sterile gauze into a 50-ml Falcon tube (mesh filter).

m. Repeat Steps k and l, and collect the expelled cells into the same Falcon tube.

n. Centrifuge cells at 1000 rpm for 10 min at room temperature.

o. Aspirate the supernatant and plate the cells with 10 ml of EBM 131 (culture medium A) in a gelatin-coated culture dish (100 mm). Place the cells in an incubator overnight with 5% CO_2.

p. Wash the cells vigorously three or four times with PBS. Feed the cells with 10 ml of culture medium A.

q. Change the medium every 2 days. The primary culture will be subconfluent after 7–8 days. They will be ready for the UEA-I isolation procedure at this point.

4. UEA-I selection of endothelial cells:

a. Coating of Dynabeads with UEA: mix together 250 µl of Dynabeads (4×10^8 beads/ml, M-450, tosylactivated, 50 µg of unconjugated UEA-I (Vector, L-1060), and 225 µl of 0.5 M boric acid, pH 9.5. The bead/lectin ratio should be 2.0×10^6 beads per microgram of lectin. The Dynabeads to boric acid ratio with lectin should be 1:1.

b. Reconstitute the UEA-I with 1 ml of sterile PBS-0.1 mM $CaCl_2$ to 2 mg/ml and store at 4°C (UEA-I is quite stable); 50 µg = 25 µl.

c. Mix Dynabeads, lectin, and boric acid in a sterile 2.0-ml screw-cap tube and agitate on a rotor at room temperature overnight.

d. Pipette the bead–lectin mixture (in 10 ml of HBSS) into a 15-ml Falcon tube. Wash with 10 ml of HBSS (plus Ca^{2+}/Mg^{2+}, 1% bovine serum albumin (BSA)) on the rotator for 15 min at room temperature.

e. Place the tube in a magnetic particle concentrator (MPC) (MPC-1, Dynal) and wait 1 min for the beads to be collected onto the magnet. Aspirate the supernatant with a Pasteur pipette. Take the tube out of the MPC, rinse three times at room temperature for 15 min, and once overnight at 4°C.

f. Resuspend the beads in 250 µl of HBSS (plus Ca^{2+}/Mg^{2+}, 5% FBS, $1 \times$ PBS) and store at 4°C in a sterile 2.0-ml screw-cap tube. The beads will be stable for several months.

5. Purification of endothelial cells from primary cultures:

a. Trypsinize subconfluent cell cultures (7–8 days) with $1 \times$ trypsin–EDTA.

b. Centrifuge the trypsinize cells at 208 g (1000 rpm) for 10 min.

c. Resuspend the cell pellet from one 100-mm Petri dish in 190 µl of HBSS buffer. Pipette up and down several times with a 200 µl pipetman to break up the cell clusters. Transfer the cell suspension into a sterile 2-ml screw-cap tube and add 5 µl UEA-I-coated Dynabeads.

d. Incubate cells and the beads for 3–5 min. Hold the tube in your hand and roll it gently between your palms to keep the beads in suspension. Endothelial cells and beads will form visible tiny clusters.

e. Transfer the cell–bead mixture to a 15-ml Falcon tube. Add 5 ml of HBSS buffer and pipette the cells several times up and down with the buffer. Place the Falcon tube into the MPC and collect the beads onto the magnet for about 1 min. Aspirate the wash solution with a Pasteur pipette while the tube is in the MCP. Take the tube out of the MCP. Repeat this wash four times with 5 ml HBSS wash buffer.

f. Resuspend the cells in 6 ml of EBM 131 growth medium A and place 3 ml onto each gelatin-coated 60-mm Petri dish. This passage is designated as passage 1. Let the cells grow to confluence at 37°C and 5% CO_2. Change the medium every 3–4 days or twice a week.

g. When endothelial cells become confluent, trypsinize, and split the cells 1:3—1:4. From now on (passage 2 and all the following passages), endothelial cells are cultured in growth medium B.

h. The endothelial cells should be fed every 2—3 days and split every 5—7 days (at least once a week).

The viability of engineered scaffolds was proved *in vivo* in rabbits in which the corpus cavernosum was replaced with an acellular collagen matrix. The animals showed normal erectile function and the ability to mate after 1 month; compared to the control animals, which were implanted with acellular matrix itself and were unable to reach erection (Chen et al., 2005).

PENILE PROTHESIS FOR RECONSTRUCTION

Early attempts at penile reconstruction involved the use of rib cartilage as a stiffener but this method was discouraged due to the unsatisfactory functional and cosmetic results (Frumpkin, 1944; Goodwin and Scott, 1952; Small, 1976; Bretan, 1989). However, biocompatibility has been a problem in some patients (Kardar and Pettersson, 1995; Nukui et al., 1997). Of the tissue existing in the human body, cartilage would serve as an ideal prothesis for penile reconstruction, owing to its biomechanical properties (Yoo et al., 2000).

Initial studies performed in our laboratory showed that chondrocytes suspended in biocompatible polymers form cartilage structures when implanted *in vivo* (Atala et al., 1993). A feasibility study of engineering natural penile prothesis made with cartilage was attempted. Chondrocytes, harvested from bovine articular cartilage tissue, were grown and seeded onto preformed cylindrical PGA polymer rods for implantation *in vivo* (Yoo et al., 1998a). Chondrocytes were seeded onto preformed cylindrical PGA polymer rods at a concentration of 50×10^6 chondrocytes/cm^3. The cell-polymers were implanted *in vivo*.

The retrieved implants formed milky white rod-shaped cartilaginous structures, maintaining their preimplantation size and shape. Biomechanical properties of the engineered cartilage rods, including compression, tension, and bending, showed that the cartilage tissues were readily elastic and could withstand high degrees of pressure. These results indicate that the engineered cartilage rods possessed the mechanical properties required to maintain penile rigidity. Histomorphological analyses confirmed the presence of mature and well-formed cartilage in all the cell-seeded implants.

In a subsequent study using an autologous system, the feasibility of applying the engineered cartilage rods *in situ* was investigated (Yoo et al., 1999). Autologous cartilages harvested from rabbit ear were dissected into small fragments (2×2 mm^2). The technique describe in Protocol II.A was used to harvest chondrocytes under sterile conditions (Atala et al., 1993, 1994). The chondrocytes were expanded until sufficient cell quantities were available. The cells were trypsinized, collected, washed, and counted for seeding onto performed poly (L-lactic acid)-coated PGA polymer rods at a concentration of 50×10^6 chondrocytes/cm^3. The chondrocyte—polymer scaffolds were implanted in the corporal spaces of rabbits. The implants were retrieved and analyzed grossly and histologically 1, 2, 3, and 6 months after surgery.

Gross examination showed the presence of well-formed milky white cartilage structures within the corpora at 1 month. There was no evidence of erosion or infection in any of the implant sites. Histological analysis demonstrated the presence of mature and well-formed chondrocytes in the retrieved implants. Autologous chondrocytes seeded on preformed biodegradable polymer structures are able to form cartilage structures within the corpus cavernosum. Subsequent studies were performed to assess long-term functionality of the cartilage. Animals were able to cope and reproduce as reported by Atala (2006). The technology appears to be useful for the creation of autologous penile prostheses.

Testes

In males, androgens, in particular testosterone, are known to have many important physiological actions, including effects on muscle, bone, central nervous system, prostate, bone marrow, and sexual function. Testicular dysfunction and hypogonadal disorders evolve from different pathophysiological conditions such as Klinefelter's syndrome, bilateral mump orchitis, toxic damage from alcohol or chemotherapy, and orchiectomy (Griffen and Willson, 1998). Patients with such conditions require lifelong androgen replacement therapy to maintain physiological levels of serum testosterone. Such therapy may increase muscle strength, stabilize bone density, improve osteoporosis, and restore secondary sexual characteristics, including libido and erectile function (Bhasin and Bremner, 1997).

ANATOMY

The testes are two glandular organs, which secrete the semen, suspended in the scrotum by the spermatic cords. In mammals, the testes are located outside the body due to the fact that spermatogenesis in mammals is more efficient at a temperature somewhat less than the core body temperature (37°C for humans). When the temperature needs to be lowered, the cremasteric muscle relaxes and the testicles are lowered away form the warm body and are able to cool.

Under a tough fibrous shell, the tunica albuginea, the testis contains very fine coiled tubes called seminiferous tubules. The tubes are lined with a layer of cells that, from puberty into old age, produce sperm cells.

From the cellular point of view the human testis is a complex organ comprising germ cells and a variety of somatic cells such as Sertoli, Leydig, endothelial, fibroblast, macrophage, and peritubular myoid cells.

Testicles are components of both the reproductive system (being gonads) and the endocrine system (being endocrine glands). The testis has two functions: spermatogenesis, which occurs in the seminiferous tubules, and secretion of steroid hormones (androgens) by Leydig cells in the interstitial tissue.

TRANSPLANTATION OF TESTES

Anorchia, acquired or congenital, is a condition that requires the use of testicular prostheses and testosterone supplementation. Use of prostheses, even if not essential for the life of a patient, has been shown to improve personal body image and satisfaction. Major problems occurring after transplantation of prostheses are infection and inflammation, and long-term safety concerns have been rising in the last few years (Raya-Rivera et al., 2008). In addition, prostheses are unable to restore testosterone production. Pharmacological treatments are available but side-effects and pharmacokinetics are unsatisfactory (Raya-Rivera et al., 2008). For these reasons, scientists are investigating the possibility of tissue-engineered testicular replacements.

The first authenticated gonadal transplantation is attributed to an eighteenth century Scottish anatomist, John Hunter, who grafted chicken testes to the body cavity of birth male and female hosts. Full details of this work have not survived, and is difficult to evaluate its outcome. Berhold was the first to report on a successful testicular transplant, since he used autografts and avoided the risk of rejection (Berhold, 1849). A century later, interest in testicular transplant increased as a result of the misapprehension that somatic aging is caused by withdrawal of sex hormones. Lydston published a series of testicular transplantation experiments performed in his patients (Lydston, 1916).

Voronoff in 1923 was the first to use chimpanzee and baboon organs for treating patients (Brinster and Zimmermann, 1994). This approach was taken by other surgeons, but none of them used microsurgery to join blood vessels of the graft to the host's circulation, resulting in ischemic necrosis preceded by organ rejection. Later, successful testicular transplantation was achieved when the ischemia time was reduced to less than an hour by using vascular anastomosis in dogs (Attaran et al., 1966; Lee et al., 1971; Gittes et al., 1972). The first convincing human testicular transplant was published by Silber (1978), who grafted a patient with a testis from the patient's genetically identical twin brother. However, with time the stringent requirements for success have precluded a surge in demand for this operation. Moreover, carefully conducted grafting trials failed to confirm the former claim; the new synthetic sex steroids were shown not to affect the lifespan of experimental animals (Parkes, 1966). Nevertheless, testicular transplantation may still be regarded as having clinical potential, for example in carriers of genetic disease, who can receive normal germ cells from donors.

TRANSPLANTATION OF TESTICULAR TISSUE

The problems arising from the size of the testis and its fibrous capsule led some transplanters to use sliced or minced organs. Kearns (1941), who reimplanted testicular tissue subcutaneously in a victim of accidental castration, reported the most plausible case (Keams, 1941). According to this report, testosterone was being produced by the autograft, but without the normal architecture of the seminiferous epithelium, it hard to understand how germ cell transfer could have restored spermatogenesis. Therefore, efforts to develop tissue grafting for the purpose of improving testosterone levels in hypogonadal men are more likely to succeed than are attempts at restoring fertility. The former goal appears to be simple, requiring the transfer of interstitial cells (Leydig cells), which are readily isolated from the donor testes by means of collagenase. Interstitial cells grafted in castrated rodents resulted in partial restoration of body weight, and testosterone levels above those of controls (Fox et al., 1973; Boyle et al., 1975; Tai and Sun, 1993). A number of vehicles and several implantation sites for interstitial cells have been tried, but none fully replaced testicular androgen production. However, in 2009, Sun reported the successful restoration of androgen production in prepubertal rats undergoing Leydig cell transplantation. Serum testosterone levels were reported as normal 12 weeks post-transplantation with a high rate of survival and functionality of transplanted cells (Sun et al., 2009).

TESTOSTERONE DELIVERY SYSTEMS

The main goal of androgen replacement therapy is to maintain physiological levels of serum testosterone and also its metabolites, dihydrotestosterone and estradiol. Hypogonadal states secondary to hypothalamic–pituitary disorders, gonadal abnormalities, and defects in androgen action or secretion may benefit from androgen replacement.

Implants, consisting of pellets implanted subcutaneously, were the first testosterone replacement therapies to be used. Implants grant a constant release of unmodified testosterone over time but require surgery; pain and the occasional extrusion of the pellets affect their compliance (Nieschlag et al., 2004).

Androgen replacement modalities include oral administration of testosterone tablets or capsules (Franchimont et al., 1978; Snyder and Lawrence, 1980; Sokol et al., 1982; Canteril, 1984; Fujioka et al., 1986; Stuenkel et al., 1991; Chang, 1993; Bennett, 1998; Ferrini and Barrett-Connor, 1998; McClellan and Goa, 1998; Wilson et al., 1998). When taken orally, testosterone preparations are largely rendered metabolically inactive during the "first pass" through the liver. This metabolic inactivation requires large oral doses of testosterone (200 ng/day) to reach normal serum levels and may be toxic to the liver, leading to hepatitis, hepatoma, or hepatocarcinoma (Gooren, 1994; Bagatelle and Bremner, 1996). Use of oral testosterone derivatives is associated with hepatotoxicity (Nieschlag et al., 2004).

Parenteral depot preparation includes testosterone enanthate and testosterone cypionate, given intramuscularly, with slow-release, oil-based injection vehicles every 10–21 days, but fluctuation in testosterone levels may produce significant swings in mood, libido, and sexual function (Sokol et al., 1982). Products with short- and long-acting esters have been combined without any improvement in duration. New formulations of testosterone undecanoate and testosterone buciclate have been shown to be effective in the long-term (up to 16 weeks) but the long duration of the effect make them less suitable for older men if any side-effect occur (Nieschlag et al., 2004).

Transdermal testosterone therapy includes both scrotal and non-scrotal patches. Testoderm and androderm are multilayered skin patches that deliver measured doses of testosterone across the scrotal skin, acting as a result of the 5α-reductase activity present within this site. When used in non-scrotal skin, the patch has to be applied twice daily, reducing the frequency of administration. However, despite these advantages, the transdermal systems have been associated with adverse effects, such as transient erythema, pruritis, induration, burning, rash, and skin necrosis (Hogan and Maibach, 1990; Bennett, 1998; McClellan and Goa, 1998). Transdermal gels have been commercialized in recent years and have maintained serum testosterone within normal levels. However, dihydrotestosterone serum levels have been shown to rise after application. In addition, precautions in skin contact, to avoid transfer of testosterone to others, and being unable to bathe or shower for at least 6 h after the application reduce compliance with these products.

Buccal delivery tablets have been developed in recent years following preliminary studies to assess their feasibility. Advantages of this delivery system are: (1) it avoids the first pass metabolism, (2) constant serum testosterone levels are reached within 24 h without super-normal peaks, and (3) the medication was generally well tolerated, with mild to moderate skin irritation on the application site as the most common, though rare, adverse effect (Nieschlag et al., 2004).

Long-term exogenous testosterone therapy has been associated with several complications, such as fluid and nitrogen retention, erythropoiesis, hypertension, and bone-density changes. In addition, fluctuating serum testosterone levels may occur, and frequent treatments may be required. Due to these problems, alternate treatment modalities, involving more physiological and longer-acting systems for androgen delivery, have been pursued.

CELL ENCAPSULATION FOR TESTOSTERONE THERAPY

Cell transplantation has long been proposed as a treatment for several diseases involving hormone or protein deficiencies. Cell rejection by the host immune system, however, has limited the use of this strategy. Encapsulation of living cells in a protective, biocompatible, and semi-permeable polymeric membrane has been proven to be an effective method of immunoprotection of the desired cells, regardless of the type of recipient (allograft, xenograft) (Chang, 1998). Most of the implantation work using microencapsulated cells as delivery vehicles employs two polymers: sodium alginate and poly(L-lysine) (PLL) (Lim and Sun, 1980). Alginate microcapsules have been used for various applications (Chang, 1998; Joki et al., 2001), particularly for the encapsulation of the pancreatic islet cells/or insulin delivery (Lim and Sun, 1980; Wang et al., 1997), and recombinant cells have served for the delivery of therapeutic gene products (Tai and Sun, 1993).

The Leydig cells of the testes are the major source of testosterone in men (95%). Implantation of heterologous Leydig cells has been proposed as a method for chronic testosterone replacement. However, these approaches were limited by tissue and cell failure to produce long-term testosterone and dissemination of the implanted cells. Therefore, encapsulation of Leydig cells might be useful for testosterone replacement therapy. Such a system might be able to stimulate the normal diurnal pattern of testosterone release by the testes, thereby avoiding side-effects such as those associated with chemically modified testosterone administration. Leydig cell transplantation may be also beneficial not only for testosterone replacement but also for the secretion of other associated hormones and growth factors such as melanocytes, β-andorpilin, prostaglandins, insulin-like growth factor 1 (IGF-1), and interleukins (Verhoeven, 1992).

An alternative approach was published in 2003 by Atala and his team, showing the efficacy of Leydig cell encapsulation within alginate/poly-L-lysine/alginate microspheres. The cells were able to release testosterone *in vitro* and *in vivo* when transplanted into castrated rats (Machluf et al., 2003).

METHODS FOR ENCAPSULATION

Microencapsulation is currently the optimal immunoisolation technique. Different approaches and polymers are being used for encapsulating cells and tissue for therapeutic applications. The technique of microencapsulation used by our laboratory utilizes two polymers: highly purified calcium-alginate (Pronova, Norway) and low-molecular-weight (23.6 kDa, Sigma) PLL. This procedure is described as follows.

Protocol: cell encapsulation

1. Isolated cells are suspended in sodium alginate (1.2%) (60% glucuronic acid content) in 0.9% saline for 5 min.

2. The cell–alginate suspension is extruded through a 22-gauge airjet-needle into a calcium chloride–4-(2-hydroxyethyl)-1-piperazineethanesulfonic acid ($CaCl_2$–HEPES) solution (1.5%).

3. The beads are stirred for 20 min in the $CaCl_2$–HEPES solution.

4. Gelled droplets are transferred to ecno-colums (Bio-Rad) and decanted.

5. The columns are filled with 15 ml of PLL solution in 0.9% saline, sealed, and rotated gently for 12 min.

6. The PLL solution is decanted from the columns and washed three times with HEPES solution.

7. A 0.125% alginate solution is added, and the mixture is rotated for 10 min. Then the alginate solution is decanted and the supernatant is washed three times with HEPES prior to culturing.

FEMALE

Vagina

A variety of pathological and congenital disorders affects the vagina and requires extensive surgical intervention. Vaginal reconstruction is an uncommon and a challenging procedure that varies considerably by specialty, with plastic surgeons and gynecologists generally recommending skin graft/dilation procedures and pediatric urologists recommending bowel vaginoplasty (Rajimwale et al., 2004). Various procedures have been used in the past for vaginal reconstruction and different tissue sources have been employed for reconstructive surgery. Traditionally, the reconstructions have been performed with non-urological tissues or synthetic prostheses, due to the paucity of available vaginal tissue, such as gastrointestinal segments (Leong and Ong, 1972; Hendren and Atala, 1994), skin (Draper and Stark, 1956), peritoneum (Hutschenreiter et al., 1978), omentum (Goldstein et al., 1967), pericardium (Kambic et al., 1992), and dura (Kelami, 1971). However, the use of non-vaginal tissue for surgical reconstruction is not ideal in terms of normal vaginal function (De Filippo et al., 2003). Tissue engineering may offer a solution for challenging cases when a shortage of local tissue exists. While tissue engineering has been applied to many tissue–organ reconstructions, there is a paucity of information regarding the engineering of female reproductive and genital tissues.

This section summarizes the known and recently developed tissue engineering applications for total vaginal reconstruction.

ANATOMY

The vagina (Kelami, 1971) is a muscular, highly expandable, tubular cavity that connects the vulva on the outside to the cervix of the uterus on the inside. The vagina consists of an internal mucous lining and a muscular coat separated by a layer of erectile tissue. It does not have any glands and is kept moist by the lubrication provided by the cervical and uterine glands.

The vagina is an extremely elastic canal connected at the upper part to the cervix of the uterus. The vaginal mucous membrane continues with the lining of the uterus and is covered by a stratified squamous epithelial layer. A loose tissue constitutes the submucosal layer and contains blood vessels, nerve fibers, and lymphatic ducts (Gray, 1918).

The muscular coat is formed by two distinct layers of muscle fibers: an external longitudinal layer and a weaker internal circular one. Decussating fasciculi interconnect the two muscular layers. An additional layer of muscular cells, called bulbospongiosus (bulbocavernosus in old references), is present in the lower part of the vagina and is formed by a band of striped muscular fibers. Connective tissue, containing a large plexus of blood vessels, forms the outermost layer and connects the vagina with the rectum, bladder, and other pelvic structures.

The erectile tissue is composed of loose connective tissue containing a plexus of veins and muscular fibers from the internal circular layer.

The thickness of the vagina's lining is directly connected with the fluctuations of various hormones released by the ovaries. Ridges, known as vaginal rugae, allow the expansion of the vaginal cavity during coitus and pregnancy.

VAGINAL TISSUE ENGINEERING

Clinically related studies have already demonstrated encouraging results with regard to the applicability of tissue engineering in genitourinary reconstruction (Atala, 1999a). In this study expanded cells of muscle and epithelial cells seeded onto PGA scaffolds were co-cultured for 24–48 h and implanted subcutaneously into athymic mice. The cells were able to survive and replicate *in vivo* for prolonged periods. By the sixth week of implantation the constructs were shown to organize into a distinguishable layer of both the vaginal epithelial and smooth muscle cell types. Penetrating native vasculature was also observed. Further analysis of the tissue-engineered vaginal constructs has been shown to produce contractile forces similar to those seen with native vaginal tissue when simulated with a series of electrical impulses.

Protocol: methods of cell culture

1. Materials and medium:
   a. Tissue source: vaginal tissue from New Zealand White rabbits.
   b. Medium:
      i. Smooth muscle cells – DMEM supplemented with 10% FBS.
      ii. Epithelial cells–keratinocytes – keratinocyte serum-free medium (K-SFM) supplemented with bovine pituitary extract and EGF.

2. Tissue harvest and cell culture:
   a. Obtain vaginal tissue.
   b. Wash the specimen several times with PBS containing EDTA.

3. Smooth muscle:
   a. Mechanically microdissect the muscle from the seromuscular layer with sterile instruments.

b. Individually place small portions of the dissected samples onto culture dishes, and allow them to dry and adhere to the surface.

c. Incubate the pieces with medium at 37°C in 5% CO_2 undisturbed until a sufficient colony of progenitor cells grows from the tissue islets.

d. Remove the tissue explants by gentle suction when sufficient a amount of cells is established.

4. Epithelial cells:

a. Digest the vaginal specimen with collagenase type IV by immersing it in the enzymatic solution and shake vigorously for 30 min at 37°C.

b. Centrifuge the cell-fluid suspension at low revolutions for 5 min.

c. Resuspend the supernatant in K-SFM and distribute onto culture dishes.

5. Cell expansion:

a. Remove the culture medium and wash the cells with PBS—EDTA.

b. Incubate the cells with a 0.05% trypsin—EDTA solution (0.5 g trypsin and 0.2 g EDTA per 1.0 liter of stock solution) and monitor under the microscope until cell separation is observed.

c. With a pipette gently transfer the cell—trypsin solution in to a 50-ml Falcon tube with serum containing medium to inactivate the trypsin.

d. Centrifuge the cells at 1,500 rpm for 5 min.

e. Resuspend the cell pellet into a predetermined volume of fresh medium and partition equally among several more culture dishes for expansion.

6. Cell maintenance:

a. Replace the medium with fresh warm (37°C) medium every 24—48 h.

Epithelial and smooth muscle cells can be subsequently seeded into polymer scaffolds and implanted into nude mice. De Filippo et al. reported the successful application of biodegradable cell-seeded scaffold for the reconstruction of rabit vaginal tissue (De Filippo et al., 2008).

Uterus

Tissue engineering is a relatively new and rapidly expanding field of biological research. It is also a clinically applicable discipline that aims to provide a repository of alternative tissue sources when reconstructive surgery is necessary (Skalak and Fox, 1988). Congenital malformations of the uterus may have profound implications clinically. With developing aspects of tissue engineering it may be possible to solve this kind of problem in the future (Fig. 52.2).

971

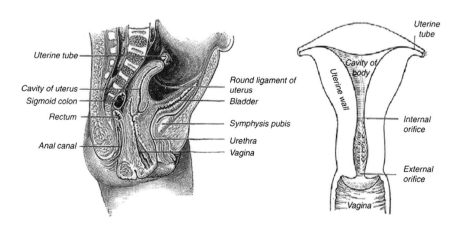

FIGURE 52.2
Anatomy of the uterus.

ANATOMY

The uterus is a pear-shaped cavity situated in the pelvic cavity between the bladder and the rectum. The upper part (fundus) opens into the fallopian tubes, one for each side, while the vagina delimits its lower area, called the cervix.

The uterus weighs from 30 to 40 g and is about 7.5 cm long, 5 cm wide in its upper part, with a thickness of about 2.5 cm (Gray, 1918).

The uterus is formed by three layers: an external layer, the perimetrium; a muscular coat, the myometrium; and an internal coat, the endometrium.

The outermost layer, the perimetrium, is formed of serous coat. It derives from the peritoneum and surrounds the fundus and the intestinal surface of the uterus but does not cover the cervix.

The myometrium is formed by a muscular coat of layered muscular fibers, intermixed with nerves, blood vessels, areolar tissue, and lymphatic vessels. The myometrium itself is divided into three layers: an external, a middle, and an internal layer; the external and middle layers are constituted of muscular coat proper while the internal myometrium layer is mostly formed of hypertrophied muscularis mucosae. The internal layer, the endometrium, consists of connective tissue and columnar epithelium.

Innervation of the uterus is primarily vasomotor with little parasympathetic input. Two arteries, the uterine (from the hypogastric) and the ovarian (from the abdominal aorta), carry the blood to the uterus. They present a tortuous course and frequent anastomoses. Veins, characterized by their large size, correspond to the arteries. During pregnancy arteries and veins convey blood to and from the intervillous space of the placenta.

UTERINE TISSUE RECONSTRUCTION

The first study of tissue engineering of human uterine smooth muscle cells was reported in 2003 (Atala, 2004). In this study, primary cell lines were initiated from human myomerium obtained at the time of term cesarean delivery. Cells were seeded onto a polyglactin-910 (Vicryl) mesh and maintained in culture. This system provides a 3D myocyte culture where cells are attached to each other instead of to a culture dish and grown under controlled conditions. Similar experiments have been reported for urinary bladder (Vozzi et al., 2002) and vascular smooth muscle cells (Dessanti et al., 1992).

In addition to this, double-mesh experiments were performed to build thicker sections of tissue. The mechanical strength of the bridging myocytes was determined by hanging the two-mesh complexes in the muscle bath, with one mesh fixed and the other attached to the force transducer. The constructs were able to maintain a maximum force of 5 g/cm^2. The bridging myocytes were also tested for contractile activity by hanging a two-mesh complex in the muscle bath and applying 2–3 g of force. Addition of oxytocin (100 nM) to the bath produced small, irregular contractions, which remained stable for 25 min. Addition of 140 mM KCl to a final concentration of about 50 mM resulted in the loss of contractile behavior. Although no repetitive pattern reminiscent of human labor was observed, these observations represent the first example of a group of cultured human uterine myocytes exhibiting coordinated contraction.

Protocol: uterine cell culture

1. Materials and medium:
   a. Tissue source: human myometrium.
   b. Medium: DMEM supplemented with 10% FBS.

2. Tissue harvest:
   a. Obtain human myocytes from the upper margin of the uterine incision.

b. Mince the collected tissue.

c. Perform double digestion at 37°C for 45 min each.
   i. Prepare and perform the first digestion containing collagenase-dispase (1.5 mg/ml), trypsin inhibitor (1 mg/ml), and BSA (2 mg/ml) in calcium-free Hank's solution.
   ii. Prepare and perform the second digestion containing collagenase (1 mg/ml), trypsin inhibitor (0.3 mg/ml), and bovine serum albumin (2 mg/ml) in the same Hanks' solution.

d. Centrifuge the cell-digestion solution mix at low revolutions for 5 min, wash with PBS, and resuspend in culture medium.

e. Culture the cells onto culture flasks in an atmosphere of 95% O_2 and 5% CO_2 at 37°C.

3. Cell expansion:
   a. Follow the protocol for vaginal cell culture expansion.

4. Cell maintenance:
   a. Replace the medium with fresh warm (37°C) medium every 2–3 days.

In a subsequent study the possibility of engineering functional uterine tissue using autologous cells was investigated (Kim et al., 1999). Autologous rabbit uterine smooth muscle and epithelial cells were harvested, then grown and expanded in culture. These cells were seeded onto uterine-shaped, biodegradable polymer scaffolds, which were then used for subtotal uterine tissue replacement in the corresponding autologous animals. Upon retrieval 6 months after implantation, histological, immunocytochemical, and Western blot analyses confirmed the presence of normal uterine tissue components. Biomechanical analyses and organ bath studies showed that the functional characteristics of these tissues were similar to those of normal uterine tissue.

Many other works were published reporting successful generation of endometrial tissue. Lü et al. reported the generation of endometrial tissue by seeding endometrial and stromal cells on a collagen-GAG scaffold. The engineered endometrium was able to sustain a co-cultured mouse embryo development up to gastrulation and advanced stages (Lü et al., 2009). However, the presence of myometrium should be considered essential for the correct functionality of the uterine tissue. In 2009, Lü et al. reported the successful re-creation of murine-engineered uterine tissue comprising an epithelial, a stromal, and a muscular layer. They showed the formation of three-layered tissue, which was able to increase the development rate and quality of murine embryos when compared with controls.

Ovaries

ANATOMY

The ovaries are oval-shaped gonads, homologous to the testes in the male (Gray, 1918). They are located in the lateral wall of the pelvis, one on either side of the uterus, to which they are attached by a broad bundle of ligaments. Ovaries are directly connected to the uterus by the fallopian tubes. A layer of columnar cells, known as the germinal epithelium of Waldeyer, covers their surface. Vesicular ovarian follicles are imbedded within stromal tissue and blood vessels. In particular, blood is supplied by arteries departing from the abdominal aorta. Veins are parallel to the arteries and form a complex network known as the pampiniform plexus.

Vesicular ovarian follicles are formed since birth, but their development and maturation only occur between puberty and menopause. Prior to sexual development the ovaries are small and their follicles are imbedded in a thick cortical layer. During puberty, under the influence of different hormonal signals, the ovaries grow in size. Furthermore, vascularization increases to fully supply blood to the now functional reproductive organs and the follicles are developed in greater numbers (Fig. 52.3).

973

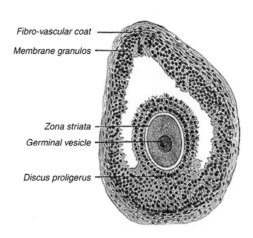

Fibro-vascular coat
Membrane granulos
Zona striata
Germinal vesicle
Discus proligerus

FIGURE 52.3
Anatomy of the ovary.

Many of the follicles never reach full development, while some gradually approach the ovary surface and burst, releasing the ovum and the liquid content that is carried by cilia movements to the fallopian tube.

After the discharge of the ovum the lining of the follicle is thrown into folds, and vascular processes grow inward from the surrounding tissue. In this way the space is filled up and the corpus luteum is formed.

The arteries of the ovaries anastomose freely in the mesosalpinx, which traverse the mesovarium and enter the hilum of the ovary.

IN VITRO CULTURE OF OVARIAN FOLLICLES

The fundamental role of the ovary is to produce oocytes capable of fertilization and subsequent development into viable offspring (Wang et al., 2003). A number of pathological conditions such as polycystic ovarian syndrome (PCOS), premature ovarian failure, and definitive sterility (post-oncotherapy) may affect ovarian function and severely compromise the reproductive potential of the ovaries.

For the preservation of fertility in women or young girls, cryopreservation of ovaries has been proposed; however, there is a critical limitation in obtaining a sufficient supply of meiotically competent oocytes (Cortvrindt et al., 1996). Furthermore, engraftment is often impaired by post-grafting ischemia—reperfusion—induced damage (Amorim et al., 2009). In order to overcome these limitations numerous studies have been performed in the last two decades, aiming to find the right conditions and approaches for the isolation, culture, and *in vitro* maturation of ovarian follicles.

In the mid 1990s Eppig and O'Brien (1996) showed that an *in vitro* matured rodent oocyte had the potential to give birth to a newborn when implanted. The culture system was then improved to give rise to multiple offspring (O'Brien et al., 2003).

In the last few years, *in vitro* culture methods involving tissue-engineered matrices have been developed to study the maturation of ovarian follicles (Pangas et al., 2003). Unlike the two-dimensional culture systems supporting the production of immature mouse follicles or granulose cell—oocyte complexes where the granulose cells attach to the culture substrate and migrate away from the oocyte (Spears et al., 1994; Cortvrindt et al., 1996; Rowley et al., 1999; Smitz and Cortvrind, 2002; Kreeger et al., 2006), this research study has developed a 3D culture system for mouse granulose—oocyte complexes, which maintains cell—cell connections and provides an environment that supports follicle development (Wang et al., 2003).

Protocol: follicle isolation and culture

1. Materials and medium:
   a. Tissue source: C57BL/6 × CBA F_1 mouse.
   b. Medium: αMEM supplemented with 3 mg/ml BSA, 5 μg/ml insulin, 5 μg/ml transferrin, and 5 μg/ml selenium.

2. Tissue harvest and culture:
   a. Obtain two-layered (100–130 μm) and multilayered secondary follicles (150–180 μm) using insulin gauge needles in L-15 medium, while maintaining them at 37°C and pH 7.
   b. Encapsulate the follicles into alginate or alginate-ECM matrices.
      i. Suspend droplets (2–3 μl) of alginate or alginate-ECM solution on a polypropylene mesh (0.1 mm opening).
      ii. Pipette a single follicle into each droplet in a minimal amount of medium.
   c. After all the droplets are filled, immerse the mesh in sterile 50 mM $CaCl_2$ for 2 min.
   d. Rinse the mesh in L-15 medium.
   e. Plate individual beads in 96-well plates in 100 μl of culture medium.
   f. Culture the follicles at 37°C in 5% CO_2 for 8 days.
   g. Change half of the media volume every 2 days.

However, although culture of the whole ovarian follicle works well with rodents where the follicles develop in the first days after birth, primates and other animals have prolonged follicular development, increasing the difficulty of maintaining viable follicles in culture (Smitz et al., 2010). In addition, enzymatic dissociation of primate follicles is harder, whereas the technique is well suited for rodents (Wandji et al., 1996). Recently it was shown that human primordial follicles can be grown within cortical pieces and develop to the multilaminar stage in as much as 6 days (Telfer et al., 2008). Further culture up to 10 days seems to give rise to follicles capable of further differentiation to the antral stage.

However, follicles are thought to develop *in vivo* for up to 8 months from the primordial to the pre-ovulatory stage. The short time required to develop human ovarian follicles *in vitro* makes this technique a viable option for possible treatment, but it is essential to assay their safety and viability and improve the culturing conditions for normal development (Smitz et al., 2010). In particular, while the ovarian cortex, after removal of the stromal cells, is increasing its growth rate to the multilaminar stage, inhibition of further follicle development must be controlled for in order to allow further development to the antral stage (Telfer et al., 2008).

Based on the experience in the culture of mouse follicles, the use of 3D extracellular matrices for culture of human ovarian follicles was investigated (Abir et al., 2001), showing an increased size of follicles cultured in a 3D collagen system. Alginate matrices have been used to successfully grow secondary human follicles (Smitz et al., 2010). In particular, alginate matrices have the capacity to maintain the 3D structure of the follicles without compromising the physiological expansion of the oocyte cellular proliferation and antrum formation. In addition, actin organization, cell–cell interactions and permeability to growth factors are key factors in the successful application of alginate systems (Smitz et al., 2010). Furthermore, in 2009, Almorim et al. (2009) were able to culture previously cryopreserved human follicles. Culture was performed in a 3D system of alginate beads with a reported survival rate of about 90%. However, despite the successful outcome in the use of alginate 3D systems for human follicle culture and development, further studies are necessary to assess morphological and functional variations in the cultured follicles, as well as safety and viability.

References

Abir, R., Fisch, B., Nitke, S., Okon, E., Raz, A., & BenRafael, Z. (2001). Morphological study of fully and partially isolated early human follicles. *Fertil. Steril., 75*, 141–146.

Amorim, C. A., Van Langendonckt, A., David, A., Dolmans, M. M., & Donnez, J. (2009). Survival of human pre-antral follicles after cryopreservation of ovarian tissue, follicular isolation and *in vitro* culture in a calcium alginate matrix. *Hum. Reprod., 24*, 92–99.

Anwar, H., Dave, B., & Seebode, J. J. (1984). Replacement of partially resected canine urethra by polytetrafluoro-ethylene. *Urology, 24*, 583–586.

Atala, A. (1997). Tissue engineering in the genitourinary system. In A. Atala, & D. Mooney (Eds.), *Tissue Engineering* (p. 149). Boston: Birkhauser Press.

Atala, A. (1999a). Engineering tissues and organs. *Curr. Opin. Urol., 9*, 517–526.

Atala, A. (1999b). Future perspectives in reconstructive surgery using tissue engineering. *Urol. Clin. N. Am., 26*, 157–165.

Atala, A. (1999c). Tissue engineering applications for erectile dysfunction. *Int J. Impot. Res., 11*(Suppl. 1), S41–S47.

Atala, A. (2004). Tissue engineering and regenerative medicine: concepts for clinical application. *Rejuvenation Res., 7*, 15–31.

Atala, A. (2006). Recent developments in tissue engineering and regenerative medicine. *Curr. Opin. Pediatr., 18*, 167–171.

Atala, A. (2007). Engineering tissues, organs and cells. *J. Tissue Eng. Regen. Med., 1*, 83–96 (review).

Atala, A., Vacanti, J. P., Peters, C. A., Mandell, J., Retik, A. B., & Freeman, M. R. (1992). Formation of urothelial structures *in vivo* from dissociated cells attached to biodegradable polymer scaffolds *in vitro*. *J. Urol., 148*, 658–662.

Atala, A., Cima, L. G., Kim, W., Paige, K. T., Vacanti, J. P., Retik, A. B., et al. (1993). Injectable alginate seeded with chondrocytes as a potential treatment for vesicoureteral reflux. *J. Urol., 150*, 745–747.

Atala, A., Kim, W., Paige, K. T., Vacanti, C. A., & Retik, A. B. (1994). Endoscopic treatment of vesicoureteral reflux with a chondrocyte-alginate suspension. *J. Urol., 152*, 641–643.

Attaran, S. E., Hodges, C. V., Crary, L. S., Jr., Vangalder, G. C., Lawson, R. K., & Ellis, L. R. (1966). Homotransplants of the testis. *J. Urol., 95*, 387–389.

Azzarello, J., Ihnat, M. A., Kropp, B. P., Warnke, L. A., & Lin, H. K. (2007). Assessment of angiogenic properties of biomaterials using the chicken embryo chorioallantoic membrane assay. *Biomed. Mater., 2*, 55–61.

Bagatelle, C., & Bremner, W. (1996). Drug therapy: androgen in men, use and abuses. *N. Engl. J. Med., 334*, 707–714.

Bazeed, M. A., Thuroff, J. W., & Schmidt, R. A. (1983). New treatment for urethral strictures. *Urology, 21*, 53–57.

Bennett, N. J. (1996). A burn-like lesion caused by a testosterone transdermal system. *Burns, 24*, 478–480.

Berhold, A. (1849). Transplantation der hoden. *Arch. Anat. Physiol. Wiss. Med., 16*, 42.

Bhasin, S., & Bremner, W. J. (1997). Clinical review 85: Emerging issues in androgen replacement therapy. *J. Clin. Endocrinol. Metab., 82*, 3.

Bolland, F., & Southgate, J. (2008). Bio-engineering urothelial cells for bladder tissue transplant. *Expert Opin. Biol. Ther., 8*, 1039–1049.

Boyle, P. F., Fox, M., & Slater, D. (1975). Transplantation of interstitial cells of the testis: effect of implant site, graft mass and ischaemia. *Br. J. Urol., 47*, 891–898.

Bretan, P. N., Jr. (1989). History of the prosthetic treatment of impotence. *Urol. Clin. N. Am., 16*, 1.

Brinster, R. L., & Zimmermann, J. W. (1994). Spermatogenesis following male germ-cell transplantation. *Proc. Natl. Acad. Sci. USA, 91*, 11298.

Canteril, J. (1984). Which testosterone therapy? *Clin. Endocrinol., 21*, 97.

Cartwright, L. M., Shou, Z., Yeger, H., & Farhat, W. A. (2006). Porcine bladder acellular matrix porosity: impact of hyaluronic acid and lyophilization. *J. Biomed. Mater. Res. A, 77*, 180–184.

Chang, T. M. (1993). Bioencapsulation in biotechnology. *Biomater. Artif. Cell Immobil. Biotechnol., 21*, 291–297.

Chang, T. M. (1998). Pharmaceutical and therapeutic applications of artificial cells including microencapsulation. *Eur. J. Pharm. Biopharm., 45*, 3–8.

Chang, T. S., & Hwang, W. Y. (1984). Forearm flap in one-stage reconstruction of the penis. *Plast. Reconstr. Surg., 74*, 251.

Chen, F., Yoo, J. J., & Atala, A. (1999). Acellular collagen matrix as a possible off the shelf biomaterial for urethral repair. *Urology, 54*, 407–410.

Chen, K.L., Yoo, J.J., & Atala, A. (2005). Total penile corpora cavernosa replacement using tissue engineering techniques. In *Regenerate Proceedings*. April 2005, Atlanta.

Cilento, B. G., Freeman, M. R., & Schneck, F. X. (1994). Phenotypic and cytogenetic characterization of human bladder urothelia expanded *in vitro. J. Urol., 152*, 665–670.

Cilento, B., Retik, A., & Atala, A. (1995). Uretheral reconstruction using a polymer mesh. *J. Urol., 153*, 371A.

Cortvrindt, R., Smitz, J., & Van Steirteghem, A. C. (1996). *In-vitro* maturation, fertilization and embryo development of immature oocytes from early preantral follicles from prepuberal mice in a simplified culture system. *Hum. Reprod., 11*, 2656–2666.

De Filippo, R. E., Pohl, H. G., Yoo, J. J., & Atala, A. (2002a). Total penile urethral replacement with autologous cell-seeded collagen matrices. *J. Urol., 168*, 1789 (abstract).

De Filippo, R. E., Yoo, J. J., & Atala, A. (2002b). Urethral replacement using cell seeded tubularized collagen matrices. *J. Urol., 168*, 1789.

De Filippo, R. E., Yoo, J. J., & Atala, A. (2003). Engineering of vaginal tissue *in vivo. Tissue Eng., 9*, 301–306.

De Filippo, R. E., Bishop, C. E., Filho, L. F., Yoo, J. J., & Atala, A. (2008). Tissue engineering a complete vaginal replacement from a small biopsy of autologous tissue. *Transplantation, 86*, 208–214.

Dessanti, A., Rigamonti, W., Merulla, V., Falchetti, D., & Caccia, G. (1992). Autologous buccal mucosa graft for hypospadias repair: an initial report. *J. Urol., 147*, 1081.

Dorin, R. P., Pohl, H. G., De Filippo, R. E., Yoo, J. J., & Atala, A. (2008). Tubularized urethral replacement with unseeded matrices: what is the maximum distance for normal tissue regeneration? *World J. Urol., 26*, 323–326.

Draper, J. W., & Stark, R. B. (1956). End results in the replacement of mucous membrane of the urinary bladder with thick-split grafts of skin. *Surgery, 39*, 434–440.

Ehrlich, R., Reda, E., & Kyle, M. (1989). Complications of bladder mucosal graft. *J. Urol., 142*, 626–627.

El-Kassaby, A. W., Retik, A. B., Yoo, J. J., & Atala, A. (2003). Urethral stricture repair with an "off the shelf" collagen matrix. *J. Urol., 169*, 170.

Eppig, J. J., & O'Brien, M. J. (1996). Development *in-vitro* of mouse oocytes from primordial follicles. *Biol. Reprod., 54*, 197–207.

Falke, G., Yoo, J. J., Kwon, T. G., Moreland, R., & Atala, A. (2003). Formation of corporal tissue architecture *in vivo* using human cavernosal muscle and endothelial cells seeded on collagen matrices. *Tissue Eng., 9*, 871–879.

Ferrini, R. L., & Barrett-Connor, E. (1998). Sex hormones and age: a cross-sectional study of testosterone and estradiol and their bioavailable fractions in community-dwelling men. *Am. J. Epidemiol., 147*, 750.

Fiala, R., Vidlar, A., Vrtal, R., Belej, K., & Student, V. (2007). Porcine small intestinal submucosa graft for repair of anterior urethral strictures. *Eur. Urol., 51*, 1702–1708, discussion 1708.

Fox, M., Boyle, P. F., & Hammonds, J. C. (1973). Transplantation of interstitial cells of the testis. *Br. J. Urol., 45*, 696–701.

Franchimont, P., Kicovic, P. M., Mattei, A., & Roulier, R. (1978). Effects of oral testosterone undecanoate in hypogonadal male patients. *Clin. Endocrinol. (Oxf.), 9*, 313.

Frumpkin, A. (1944). Reconstruction of male genitalia. *Am. Rev. Sov. Med., 2*, 14.

Fu, Q., Deng, C. L., Liu, W., & Cao, Y. L. (2007). Urethral replacement using epidermal cell seeded tubular acellular bladder collagen matrix. *BJU Int., 99*, 1162–1165.

Fujioka, M., Shinohara, Y., Baba, S., Irie, M., & Inoue, K. (1986). Pharmacokinetic properties of testosterone propionate in normal men. *J. Clin. Endocrinol. Metab., 63*, 1361.

Gilbert, D. A., Williams, M. W., Horton, C. E., Terzis, J. K., Winslow, B. H., Gilbert, D. M., et al. (1988). Phallic reinnervation via the pudendal nerve. *J. Urol., 140*, 295–298.

Gittes, R. F., Altwein, J. E., Yen, S. S., & Lee, S. (1972). Testicular transplantation in the rat: long-term gonadotropin and testosterone radioimmunoassays. *Surgery, 72*, 187.

Goldstein, M. B., Dearden, L. C., & Gualtieri, V. (1967). Regeneration of subtotally cystectomized bladder patched with omentum: an experimental study in rabbits. *J. Urol., 97*, 664–668.

Goodwin, W. E., & Scott, W. W. (1952). Phalloplasty. *J. Urol., 68*, 903.

Gooren, L. J. (1994). A ten-year safety study of the oral androgen testosterone undecanoate. *J. Androl., 15*, 212.

Gray, H. (1918). *Anatomy of the Human Body.* Philadelphia: Lea & Febiger.

Griffen, J., & Willson, J. (1998). Disorders of the testes and the male reproductive tract. In J. D. Willson, et al. (Eds.), *Williams Textbook of Endocrinology* (p. 819). Philadelphia, PA: Saunders.

Guhe, C., & Follmann, W. (1994). Growth and characterization of porcine urinary bladder epithelial cells *in vitro. Am. J. Physiol., 266*, F298–F308.

Guzman, L. (1999). Neourethra with rectum, posterior sagittal approach. In R. M. Ehrlich, & G. J. Alter (Eds.), *Reconstructive and Plastic Surgery of the External Genitalia: Adult and Pediatric* (p. 101). Philadelphia, PA: Saunders.

Hakky, S. I. (1976). Urethral replacement by dacron mesh. *Lancet, ii*, 1192.

Hakky, S. I. (1977). The use of fine double siliconized dacron in urethral replacement. *Br. J. Urol., 49*, 167.

Harrington, D. A., Sharma, A. K., Erickson, B. A., & Cheng, E. Y. (2008). Bladder tissue engineering through nanotechnology. *World J. Urol., 26,* 315–322.

Hauser, S., Bastian, P. J., Fechner, G., & Muller, S. C. (2006). Small intestine submucosa in urethral stricture repair in a consecutive series. *Urology, 68,* 263–266.

Hendren, W. H., & Atala, A. (1994). Use of bowel for vaginal reconstruction. *J. Urol., 152,* 752–755, discussion 756–757.

Hendren, W. H., & Reda, E. F. (1986). Bladder mucosa graft for construction of male urethra. *J. Pediatr. Surg., 21,* 189–192.

Hogan, D. J., & Maibach, H. I. (1990). Adverse dermatologic reactions to transdermal drug delivery systems. *J. Am. Acad. Dermatol., 22,* 811.

Horton, C. E., & Dean, J. A. (1990). Reconstruction of traumatically acquired defects of the phallus. *World J. Surg., 14,* 757–762.

Humby, G. (1941). A one-stage operation for hypospadia. *Br. J. Surg., 29,* 84.

Hutschenreiter, G., Rumpelt, H. J., Klippel, K. F., & Hohenfellner, R. (1978). The free peritoneal transplant as substitute for the urinary bladder wall. *Invest. Urol., 15,* 375.

Italiano, G., Abatangelo, G., Jr., Calabrò, A., Abatangelo, G., Sr., Zanoni, R., O'Regan, M., et al. (1997). Reconstructive surgery of the urethra: a pilot study in the rabbit on the use of hyaluronan benzyl ester (Hyaff-11) biodegradable grafts. *Urol. Res., 25,* 137.

Jackson, C. J., Garbett, P. K., Nissen, B., & Schrieber, L. (1990). Binding of human endothelium to Ulex europaeus I-coated Dynabeads: application to the isolation of microvascular endothelium. *J. Cell Sci., 96,* 257.

Joki, T., Machluf, M., Atala, A., Zhu, J., Seyfried, N. T., Dunn, I. F., et al. (2001). Continuous release of endostatin from microencapsulated engineered cells for tumor therapy. *Nat. Biotechnol., 19,* 35.

Kambic, H., Kay, R., Chen, J. F., Matsushita, M., Harasaki, H., & Zilber, S. (1992). Biodegradable pericardial implants for bladder augmentation: a 2.5-year study in dogs. *J. Urol., 148,* 539–543.

Kardar, A., & Pettersson, B. A. (1995). Penile gangrene: a complication of penile prosthesis. *Scand. J. Urol. Nephrol., 29,* 355.

Kearns, W. (1941). Successful autoplastic graft following accidental castration. *Ann. Surg., 114,* 886–890.

Kelami, A. (1971). Lyophilized human dura as a bladder wall substitute: experimental and clinical results. *J. Urol., 105,* 518–522.

Kim, B. S., Nikolovski, J., Bonadio, J., Smiley, E., & Mooney, D. J. (1999). Engineered smooth muscle tissues: regulating cell phenotye with the scaffold. *Exp. Cell Res., 251,* 318.

Kraling, B. M., & Bischoff, J. (1998). A simplified method for growth of human microvascular endothelial cells results in decreased senescence and continued responsiveness to cytokines and growth factors. *In Vitro Cell Dev. Biol. Anim., 34,* 308–315.

Kreeger, P. K., Deck, J. W., Woodruff, T. K., & Shea, L. D (2006). The *in vitro* regulation of ovarian follicle development using alginate-extracellular matrix gels. *Biomaterials, 27,* 714–723.

Kreft, M. E., Hudoklin, S., & Sterle, M. (2005). Establishment and characterization of primary and subsequent subcultures of normal mouse urothelial cells. *Folia Biol. (Praha), 51,* 126–132.

Kropp, B. P., Ludlow, J. K., Spicer, D., Rippy, M. K., Badylak, S. F., Adams, M. C., et al. (1998). Rabbit urethral regeneration using small intestinal submucosa onlay grafts. *Urology, 52,* 138–142.

Kurzrock, E. A., Lieu, D. K., de Graffenried, L. A., & Isseroff, R. R. (2005). Rat urothelium: improved techniques for serial cultivation, expansion, freezing and reconstitution onto acellular matrix. *J Urol., 173,* 281–285.

Lee, S., Tung, K. S., & Orloff, M. J. (1971). Testicular transplantation in the rat. *Transplant. Proc., 3,* 586.

Leong, C. H., & Ong, G. B. (1972). Gastrocystoplasty in dogs. *Aust. NZ J. Surg., 41,* 272–279.

Liebert, M., Wedemeyer, G., Abruzzo, L. V., Kunkel, S. L., Hammerberg, C., Cooper, K. D., et al. (1991). Stimulated urothelial cells produce cytokines and express an activated cell surface antigenic phenotype. *Semin. Urol., 9,* 124–130.

Liebert, M., Hubbel, A., Chung, M., Wedemeyer, G., Lomax, M. I., Hegeman, A., et al. (1997). Expression of mal is associated with urothelial differentiation *in vitro*: identification by differential display reverse-transcriptase polymerase chain reaction. *Differentiation, 61,* 177–185.

Lim, F., & Sun, A. M. (1980). Microencapsulated islets as bioartificial endocrine pancreas. *Science, 210,* 908.

Lü, S. H., Wang, H. B., Liu, H., Wang, H. P., Lin, Q. X., Li, D. X., et al. (2009). Reconstruction of engineered uterine tissues containing smooth muscle layer in collagen/matrigel scaffold *in vitro. Tissue Eng. Part A., 15,* 1611–1618.

Lydston, G. (1916). Sex gland implantation. Additional cases and conclusions to date. *JAMA, 66,* 1540.

Machluf, M., Orsola, A., Boorjian, S., Kershen, R., & Atala, A. (2003). Microencapsulation of Leydig cells: a system for testosterone supplementation. *Endocrinology, 144,* 4975.

Maurer, S., Feil, G., & Stenzl, A. (2005). *In vitro* stratified urothelium and its relevance in reconstructive urology. *Urology, 44,* 738–742.

McClellan, K. J., & Goa, K. L. (1998). Transdermal testosterone. *Drugs, 55,* 253.

McManus, M., Boland, E., Sell, S., Bowen, W., Koo, H., Simpson, D., et al. (2007). Electrospun nanofibre fibrinogen for urinary tract tissue reconstruction. *Biomed. Mater., 2,* 257–262.

Mikos, A. G., Lyman, M. D., Freed, L. E., & Langer, R. (1994). Wetting of poly(L-lactic acid) and poly(DL-lactic-co-glycolic acid) foams for tissue culture. *Biomaterials, 15,* 55.

Mondalek, F. G., Lawrence, B. J., Kropp, B. P., Grady, B. P., Fung, K. M., Madihally, S. V., et al. (2008). The incorporation of poly(lactic-co-glycolic) acid nanoparticles into porcine small intestinal submucosa biomaterials. *Biomaterials, 29,* 1159–1166.

Nieschlag, E., Behre, H. M., Bouchard, P., Corrales, J. J., Jones, T. H., Stalla, G. K., et al. (2004). Testosterone replacement therapy: current trends and future directions. *Hum. Reprod., 10,* 409–419.

Nuininga, J. E., van Moerkerk, H., Hanssen, A., Hulsbergen, C. A., Oosterwijk-Wakka, J., Oosterwijk, E., et al. (2003). Rabbit urethra replacement with a defined biomatrix or small intestinal submucosa. *Eur. Urol., 44,* 266.

Nukui, F., Okamoto, S., Nagata, M., Kurokawa, J., & Fukui, J. (1997). Complications and reimplantation of penile implants. *Int. J. Urol., 4,* 52–54.

O'Brien, M. J., Pendola, J. K., & Eppig, J. J. (2003). A revised protocol for *in-vitro* development of mouse oocytes from primordial follicles dramatically improves their developmental competence. *Biol. Reprod., 68,* 1682–1686.

Ozcan, M., & Kahveci, R. (1987). One-stage repair of distal and midpenile hypospadia by modified Hodgson III technique. *Eur. J. Plast. Surg., 10,* 159.

Palminteri, E., Berdondini, E., Colombo, F., & Austoni, E. (2007). Small intestinal submucosa (SIS) graft urethroplasty: short-term results. *Eur. Urol., 51,* 1695–1701, discussion 1701.

Pangas, S. A., Saudye, H., Shea, L. D., & Woodruff, T. K. (2003). Novel approach for the three-dimensional culture of granulosa cell–oocyte complexes. *Tissue Eng., 9,* 1013–1021.

Park, H. J., Yoo, J. J., Kershen, R. T., Moreland, R., & Atala, A. (1999). Reconstitution of human corporal smooth muscle and endothelial cells *in vivo. J. Urol., 162,* 1106–1109.

Parkes, A. S. (1966). The rise of reproductive endocrinology. *J. Endocrinol., 34,* 1926–1940.

Puckett, C. L., & Montie, J. E. (1978). Construction of male genitalia in the transsexual, using a tubed groin flap for the penis and a hydraulic inflation device. *Plast. Reconstr. Surg., 61,* 523.

Puthenveettil, J. A., Burger, M. S., & Reznikoff, C. A. (1999). Replicative senescence in human uroepithelial cells. *Adv. Exp. Med. Biol., 462,* 83–91.

Rajimwale, A., Furness, P. D., III, Brant, W. O., & Koyle, M. A. (2004). Vaginal construction using sigmoid colon in children and young adults. *BJU Int., 94,* 115.

Raya-Rivera, A. M., Baez, C., Atala, A., & Yoo, J. J. (2008). Tissue engineered testicular prostheses with prolonged testosterone release. *World J. Urol., 26,* 335–351.

Ribeiro-Filho, L. A., Mitrw, A. I., & Sarkis, A. S. (2007). Cadaveric organ-specific acellular matrix for urethral reconstruction in humans. *J. Urol. Suppl., 177,* 12.

Roth, C. C., & Kropp, B. P. (2009). Recent advances in urologic tissue engineering. *Curr. Urol. Rep., 10,* 112–119.

Rowley, J. A., Madlambayan, G., & Mooney, D. J. (1999). Alginate hydrogels as synthetic extracellular matrix materials. *Biomaterials, 20,* 45.

Scriven, S. D., Booth, C., Thomas, D. F., Trejdosiewicz, L. K., & Southgate, J. (1997). Reconstitution of human urothelium from monolayer cultures. *J. Urol., 158,* 1147–1152.

Sharaby, J. S., Benet, A. E., & Melman, A. (1995). Penile revascularization. *Urol. Clin. N. Am., 22,* 821.

Shaul, D. B., Xie, H. W., Diaz, J. F., Mahnovski, V., & Hardy, B. E. (1996). Use of tubularized peritoneal free grafts as urethral substitutes in the rabbit. *J. Pediatr. Surg., 31,* 225–228.

Sievert, K. D., Bakircioglu, M. E., Nunes, L., Tu, R., Dahiya, R., & Tanagho, E. A. (2000). Homologous acellular matrix graft for urethral reconstruction in the rabbit: histological and functional evaluation. *J. Urol., 163,* 1958–1965.

Silber, S. J. (1978). Transplantation of a human testis for anorchia. *Fertil. Steril., 30,* 181.

Skalak, R., & Fox, C. (1988). *Tissue Engineering.* New York: Alan R. Liss.

Small, M. P. (1976). Small-Carrion penile prosthesis. A new implant for management of impotence. *Mayo Clin. Proc., 51,* 336.

Smitz, J. E., & Cortvrind, R. G. (2002). The earliest stages of folliculogenesis *in vitro. Reproduction, 123,* 185–202.

Smitz, J., Dolmans, M. M., Donnez, J., Fortune, J. E., Hovatta, O., Jewgenow, K., et al. (2010). Current achievements and future research directions in ovarian tissue culture, *in vitro* follicle development and transplantation: implications for fertility preservation. *Hum. Reprod, 16,* 395–414, *Update.*

Snyder, P. J., & Lawrence, D. A. (1980). Treatment of male hypogonadism with testosterone enanthate. *J. Clin. Endocrinol. Metab., 51,* 1335.

Sokol, R. Z., Palacios, A., Campfield, L. A., Saul, C., & Swerdloff, R. S. (1982). Comparison of the kinetics of injectable testosterone in eugonadal and hypogonadal men. *Fertil. Steril., 37,* 425.

Spears, N., Boldand, N. I., Murray, A. A., & Gosden, R. G. (1994). Mouse oocytes derived from *in vitro* grown primary ovarian follicles are fertile. *Hum. Reprod., 9,* 527–532.

Stuenkel, C. A., Dudley, R. E., & Yen, S. S. (1991). Sublingual administration of testosterone-hydroxypropyl-beta-cyclodextrin inclusion complex simulates episodic androgen release in hypogonadal men. *J. Clin. Endocrinol. Metab., 72,* 1054.

Sun, J., Xi, Y. B., Zhang, Z. D., Shen, P., Li, H. Y., Yin, M. Z., et al. (2009). Leydig cell transplantation restores androgen production in surgically castrated prepubertal rats. Asian J. *Andrology, 11,* 405–409.

Tai, I. T., & Sun, A. M. (1993). Microencapsulation of recombinant cells: a new delivery system for gene therapy. *FASEB J., 7,* 1061.

Telfer, E. E., McLaughlin, M., Ding, C., & Thong, K. J. (2008). A two-step, serum-free culture system supports development of human oocytes from primordial follicles in the presence of activin. *Hum. Reprod., 23,* 1151–1158.

Tobin, M., Freeman, M., & Atala, A. (1994). Maturation response of normal human urothelial cells in culture is dependent on extracellular matrix and serum additives. *Surg. Forum, 45,* 786.

Truschel, S. T., Ruiz, W. G., Shulman, T., Pilewski, J., Sun, T. T., Zeidel, M. L., et al. (1999). Primary uroepithelial cultures. A model system to analyze umbrella cell barrier function. *J. Biol. Chem., 274,* 15020.

Verhoeven, G. (1992). Local control system within the testis. In B. Tindall (Ed.), *Bailliere's Clinical Endocrinology and Metabolism* (p. 313). London: Bailliere Tindal.

Vozzi, G., Flaim, C., Ahluwalia, A., & Bhatia, S. (2002). Microfabricated PLGA scaffolds: a comparative study for application to tissue engineering. *Mat. Sci. Eng., 20,* 43.

Wandji, S.-A., Eppig, J. J., & Fortune, J. E. (1996). FSH and growth factors affect the growth and endocrine function *in-vitro* of granulosa cells of bovine preantral follicles. *Theriogenology, 45,* 817–832.

Wang, T., Lacík, I., Brissová, M., Anilkumar, A. V., Prokop, A., Hunkeler, D., et al. (1997). An encapsulation system for the immunoisolation of pancreatic islets. *Nat. Biotechnol., 15,* 358.

Wang, T., Koh, C. J., Yoo, J. J., & Atala, A. (2003). Creation of an engineered uterus for surgical reconstruction. Presented at *Proceedings of the American Academy of Pediatrics. Section on Urology.* New Orleans, LA: (pp.4–7).

Weiser, A. C., Franco, I., Herz, D. B., Silver, R. I., & Reda, E. F. (2003). Single layered small intestinal submucosa in the repair of severe chordee and complicated hypospadias. *J. Urol., 170,* 1593–1595, discussion 1595.

Wilson, D. E., Meikle, A. W., Boike, S. C., Fairless, A. J., Etheredge, R. C., & Jorkasky, D. K. (1998). Bioequivalence assessment of a single 5 mg/day testosterone transdermal system versus two 2.5 mg/day systems in hypogonadal men. *J. Clin. Pharmacol., 38,* 54.

Xu, Y., Qiao, Y., Sa, Y., Zhang, H., Zhang, X., Zhang, J., et al. (2002). An experimental study of colonic mucosal graft for urethral reconstruction. *Chin. Med. J. (Engl.), 115,* 1163–1165.

Yoo, J. J., Lee, I., & Atala, A. (1998a). Cartilage rods as a potential material for penile reconstruction. *J. Urol., 160,* 1164.

Yoo, J. J., Meng, J., Oberpenning, F., & Atala, A. (1998b). Bladder augmentation using allogenic bladder submucosa seeded with cells. *Urology, 51,* 221–225.

Yoo, J. J., Park, H. J., Lee, I., & Atala, A. (1999). Autologous engineered cartilage rods for penile reconstruction. *J. Urol., 162,* 1119.

Yoo, J. J., Park, H. J., & Atala, A. (2000). Tissue-engineering applications for phallic reconstruction. *World J. Urol., 18,* 62.

Zhang, Y. Y., Ludwikowski, B., Hurst, R., & Frey, P. (2001). Expansion and long-term culture of differentiated normal rat urothelial cells *in vitro*. *In Vitro Cell Dev. Biol. Anim., 37,* 419–429.

Zhang, Y. Y., McNeill, E., Soker, S., Yoo, J. J., & Atala, A. (2007). A novel cell source for urologic tissue reconstruction. *J. Urol. Suppl., 177,* 238 (abstract 710).

CHAPTER 53

Cartilage Tissue Engineering

Qiongyu Guo, Jennifer H. Elisseeff
Department of Biomedical Engineering, Johns Hopkins University, Baltimore, MD, USA

CARTILAGE AND CARTILAGE REPAIR

Cartilage is a connective tissue that functions to provide form, strength, and support. There are three types of cartilage distinguished by their molecular components in the extracellular matrix (ECM), their anatomical location, and their function. Hyaline (articular) cartilage has a white glassy appearance and is found primarily in joints. Its ECM is mainly composed of water, proteoglycans, and type II collagen. Hyaline cartilage functions to provide stable movement with minimal friction. It demonstrates an excellent ability to provide resistance to compression and cushion the impact caused by physical load during movement (Temenoff and Mikos, 2000). Elastic cartilage is distinguished by the presence of elastin in the ECM. Elastic cartilage provides a structural function, represented by the support it provides in the airtube and external ear. Lastly, fibrocartilage has a higher proportion of type I collagen in the matrix. Fibrocartilage is found at the end of tendons and ligaments in apposition to bone, providing tensile strength and countering compression and shear forces (Benjamin and Ralphs, 2004).

All of the three types of cartilage feature a sparse cellularity, limited blood supply, and lack of neural innervations. Due to their intrinsically poor reparative capabilities, once defects, even very small ones, are initiated in cartilage, the degradation process is progressive (Hinziker, 2009; van Osch et al., 2009). One of the irreversible consequences of the destruction of articular cartilage is arthritis, a leading cause of disability. Osteoarthritis (OA), the most common type of arthritis, is widespread globally in 60–70% of people older than 65 years of age (Sarzi-Puttini et al., 2005; Dillon et al., 2006; Xie et al., 2007). Over 21 million people are suffering from this disease in the USA, and 10% of cases are estimated to be caused by previous trauma to the weight-bearing joints, which is classified as post-traumatic arthritis (PTA) (Furman et al., 2006). PTA develops not only in old people, but also in young people suffering the results of previous trauma. The disease causes significant pain, disability, and morbidities, strongly affecting an individual's capacity to live a full and active life.

Various surgical treatment options are available for focal cartilage repair. The microfracture technique is frequently used in patients with previously untreated cartilage defects due to its low cost and minimally invasive procedure (Mithoefer et al., 2006). This technique employs subchondral drilling to initiate cartilage regeneration by inducing bleeding, including mesenchymal progenitor cells from the bone marrow, into the lesion site. After this procedure, the repair tissue appears to be a cartilage-like substitute but is mainly composed of fibrocartilage, which shows inferior quality and duration as compared to the native hyaline cartilage. Osteochondral autografting or mosaicplasty is a technique of autotransplantation in which osteochondral plugs are harvested from non-weight-bearing or low-weight-bearing regions of the joint and implanted into defects that have been prepared and sized. Survival of

Principles of Regenerative Medicine. DOI: 10.1016/B978-0-12-381422-7.10053-7

hyaline cartilage has been reported in 85% of patients, with a 91% good to excellent clinical outcome reported by patients followed for 3–6 years (Hangody et al., 1994, 2001). However, the cartilage autografts suffer from many problems including limited donor tissue availability, donor site injury, scarring, and pain. In 1994, an innovative therapeutic option was proposed by Brittberg et al. using a cell-based therapy, called autologous chondrocyte implantation (ACI), for localized cartilage injuries (Brittberg et al., 1994; Brittberg, 2008). This technique allows a cell suspension of *in vitro* expanded autologous articular chondrocytes to synthesize new cartilaginous matrix under a surgically closed periosteal flap in the defect site. Although the clinical outcomes of the standard ACI methods are encouraging, numerous potential disadvantages are associated with this technique, including donor site morbidity, risk of leakage of transplanted chondrocytes, complexity of the surgical procedure (Marcacci et al., 2002), uneven distribution of the cell suspension in the transplanted site (Sohn et al., 2002), periosteal hypertrophy (Haddo et al., 2004), and dedifferentiation of the chondrocyte phenotype during *in vitro* monolayer culture (Benya and Shaffer, 1982; Kimura et al., 1984).

TISSUE ENGINEERING FOR CARTILAGE REPAIR

In order to overcome the treatment obstacles of the available surgical options for cartilage repair, the reconstruction of cartilage using tissue engineering has attracted tremendous attention. Tissue engineering is a multidisciplinary field that applies the principles of engineering, life sciences, cell and molecular biology to the development of biological substitutes that restore, maintain, and improve tissue function (Mooney and Mikos, 1999). Three general components are involved in tissue engineering: (1) reparative cells that can form a functional matrix, (2) an appropriate scaffold for transplantation and support, and (3) bioreactive molecules, such as cytokines and growth factors, that will support and choreograph formation of the desired tissue (Sharma and Elisseeff, 2004). These three components may be used individually or in combination to regenerate organs or tissues. In addition, environmental factors, including mechanical stimulation and shear forces, play important roles in the reconstruction of engineered tissues by creating biological cues that exert an effect on cells (Concaro et al., 2009).

Cell type

Different cell sources are available to provide reparative tissue including differentiated cells, mesenchymal stem cells (MSCs), and embryonic progenitor cells. Chondrocytes and MSCs are the two most investigated cell sources for cartilage tissue engineering. Chondrocytes are readily available as they can be isolated from human cartilage and cultured *ex vivo*. For over a decade, chondrocytes have been expanded *ex vivo* for clinical applications as an FDA approved therapy (Brittberg et al., 1994). However, one of the major limitations of chondrocytes is a tendency to rapidly dedifferentiate in monolayer culture (Darling et al., 2004; Darling and Athanasiou, 2005). Tissue culture material and scaffold type can influence chondrocyte phenotype. A flat shape of chondrocytes is associated with their expansion in fibrocartilage phenotype, while a round shape suggests being associated with a cell synthesis mode. Three-dimensional (3D) culture scaffolds can help preserve the round-shape chondrocyte phenotype and promote chondrogenesis by producing increased type II collagen and decreased type I collagen compared to chondrocytes in monolayer culture (Benya and Shaffer, 1982; Freed et al., 1993).

MSCs represent a viable alternative to chondrocytes as a cell source for cartilage tissue engineering (Pittenger et al., 1999). MSCs have the advantage of being able to be expanded *in vitro* in an undifferentiated state, while retaining the ability to differentiate after exposure to suitable stimuli (Song et al., 2004). To create distinct tissue types, specific control over the induction and maintenance of stem cell differentiation is imperative. Like chondrocytes, a 3D culture environment for cartilage engineering with MSCs is also superior to monolayer culture. Winter et al. compared chondrogeneic gene expression and morphology from MSCs derived from bone marrow and adipose tissue (Winter et al., 2003). The study demonstrated similar partial differentiation in monolayer culture; however, bone marrow-derived MSCs improved chondrogenesis in 3D culture. This study's results combined with 3D culture results and established MSC isolation techniques resulted in the majority of research using bone marrow-derived MSCs.

Bioscaffold in cartilage repair

Tissue engineering scaffolds are designed to provide a 3D environment to support and direct cellular processes in their migration, proliferation, and differentiation toward functional tissue. The selection of bioscaffolds for cartilage engineering requires excellent mechanical properties to support cellular functions, biocompatibility, capability of waste and nutrient transport, and sufficient structural integrity for joint reconstruction. Both natural and synthetic materials have been applied as cartilage tissue engineering scaffolds in a variety of forms, including fibrous structures, porous sponges, woven or non-woven meshes, and hydrogels.

NATURAL SCAFFOLDS

Collagen

Collagen is the primary structural protein found in both bone and cartilage (Eyre, 2002; Eyre et al., 2006). As such, collagen-based scaffolds are theoretically capable of supporting chondrocyte attachment and function. They are also biocompatible and biodegradable. Collagen scaffolds have been used in a wide variety of forms such as gels, membranes, and sponges into which cells and/or bioactive factors may be introduced (Pieper et al., 2002; Frenkel and Di Cesare, 2004). Pieper et al. utilized a cross-linked porous type II collagen sponge to support the proliferation and differentiation of chondrocytes under cell culture condition up to 14 days (Pieper et al., 2002). Yokoyama et al. cultured MSCs in a collagen gel matrix in a chondrogeneic medium supplemented with bone morphogenetic protein-2 (BMP-2), transforming growth factor-β3 (TGF-β3), and dexamethasone (Yokoyama et al., 2005). The constructs were characterized by a downregulation of type I collagen, and upregulation of type II collagen and the cartilage-related proteoglycans aggrecan, biglycan, and decorin. The maximum size of cartilaginous tissue produced was 7 mm in diameter and 0.5 mm in thickness, still too small for partial-thickness cartilage repair. The cell-based studies indicate some of the disadvantages of collagen scaffolds. Collagen gels allow for uniform mixing of cells and matrix, and for extensive molding and shaping of tissue, but tend to be fragile until new matrix is laid down. Solid collagen scaffolds such as membranes or sponges exhibit greater initial mechanical strength, but at the cost of less flexibility in shaping and a greater risk of non-uniform cell seeding. Collagen remains a useful scaffold with which to study 3D cell culture, but the disadvantages noted above weigh against its use in clinical applications.

Hyaluronic acid

Hyaluronic acid (HA) is a polysaccharide that is naturally found both in the ECM of articular cartilage and in synovial fluid. It is composed of alternating residues of *N*-acetyl-D-glucosamine and D-glucoronic acid. As with collagen, interest focused on HA as a potential scaffold for cartilage engineering due to its intimate association with chondrocytes *in vivo*. Intra-articular HA injection has been used to treat symptoms of osteoarthritis with very large world markets and sales. The HA has been shown to have a stimulatory effect on chondrocyte production of type II collagen and proteoglycan (Akmal et al., 2005). A novel use of HA was reported in which HA was modified by methacrylation to form a photo-cross-linkable polymer, which was then used to encapsulate chondrocytes for *in vitro* and *in vivo* culture (Nettles et al., 2004). Chondrocytes encapsulated within this matrix retained their phenotype and generated type II collagen.

Alginates

Alginates are polysaccharides derived from seaweed. They comprise a family of linear mannuronate/guluronate copolymers that differ in their specific sequences and overall compositions (Rowley et al., 1999). When exposed to a divalent cation (usually calcium for sake of biocompatibility), the linear alginate polymers ionically cross-link to form a porous hydrogel. This allows the uniform seeding of chondrocytes and bioactive factors within the alginate hydrogel, as well as their release, if desired, by exposure to a cation chelating agent such as EDTA. Alginates may also be covalently modified in order to enhance properties such as cell adhesion (Sultzbaugh and Speaker, 1996; Rowley et al., 1999; Alsberg et al., 2001) or to fix bioactive factors in place (Suzuki et al., 2000; Gerard et al., 2005; Ma et al., 2005). The clinical translation of alginates as *in vivo* scaffold for cartilage repair may be limited by the potential calcification of the constructs (Ma et al., 2005). On the other hand, compared to monolayer cell culture of chondrocytes, alginate matrices provide a convenient means to help preserve or re-establish characteristic chondrocyte phenotype and matrix production during *in vitro* expansion (Diduch et al., 2000; Homicz et al., 2003; Chia et al., 2005; Hsieh-Bonassera et al., 2009).

Chitosan

Chitosan is a polysaccharide, this time derived from chitin (found in arthropod exoskeletons), that has been partially or fully deacetylated. It is composed of linear chains of β-linked D-glucosamine residues. Chitosan has been studied both as a scaffold and as a controlled delivery system for bioactive factors (Lee et al., 2004; Hoemann et al., 2005b). There is interest in chitosan as a cell delivery vehicle as it demonstrates good biocompatibility, and some formulations exhibit the property of temperature-dependent gelation, in that they are liquid at room temperature but gel when exposed to physiological temperatures (Chenite et al., 2000). In addition, the degree of deacetylation of chitosan directly influences the degradation rate of the constructs as well as the inflammatory response. A lower degree of deacetylation was associated with an increased degradation rate and host inflammatory response. Thermosetting chitosan constructs injected subcutaneously into nude mice supported chondrocyte growth and matrix production, although the constructs were mechanically inferior to native cartilage (Hoemann et al., 2005b). Chitosan constructs were also injected into osteochondral defects created in rabbit knees. Retention of the constructs in the defects was observed at 1 week despite full mobility and weight-bearing.

Composite scaffolds using chitosan combined with alginate and/or hyaluronic acid have also been investigated. Li et al. cultured HTB-94 chondrocytes in interconnected 3D porous chitosan—alginate scaffolds and found a promoted cell proliferation and enhanced phenotype expression of chondrocytes in these scaffolds compared to chitosan-only scaffolds (Li and Zhang, 2005). Yamane et al. observed higher cell adhesivity, proliferation, and aggrecan synthesis in HA-coated chitosan hybrid polymer fiber sheet than that in chitosan fiber sheet (Yamane et al., 2005). Recently, Tan et al. developed injectable *in situ* forming composite hydrogels consisting of chitosan and HA for cartilage tissue engineering (Tan et al., 2009). Hsu et al. evaluated a chitosan—alginate—hyaluronate scaffold modified with a protein containing an arginine-glycine-aspartic acid (RGD)-modified adhesion peptide motif (Hsu et al., 2004). It was noted that glycosaminoglycan and collagen synthesis was greater in the chitosan—alginate—HA—RGD scaffold as compared to chitosan—alginate and chitosan—alginate—HA scaffolds.

SYNTHETIC SCAFFOLDS

Bioscaffolds derived from natural materials, compared to synthetic scaffolds, potentially allow for better regulation of cell adhesion and matrix production of the resident cells (Hubbell, 2003). However, biological scaffolds have a greater risk of contamination or immune reaction than synthetic scaffolds. In addition, biological materials are notoriously difficult to generate in large quantities with acceptable consistency, and often exhibit poor mechanical characteristics (Frenkel and De Cesare, 2004). Synthetic materials generally avoid these problems. These materials are created *de novo* and provide precise control over the structural properties, mechanical properties, and rates of resorption with a great deal of batch-to-batch consistency (Drury and Mooney, 2003).

The most common synthetic polymers in use at this time are polyglycolic acid (PGA), polylactic acid (PLA), polyethylene oxide (PEO), and various derivatives and copolymers based on these entities (Frenkel and De Cesare, 2004). These biodegradable polymers have a long history of medical usage, and are able to be fabricated and processed in a variety of ways (Sharma and Elisseeff, 2004). These materials provide the scaffolds for the adherence, growth, differentiation, and matrix production of chondrocytes or MSCs (Lu et al., 2001; Riley et al., 2001; Lynn et al., 2004; Klein et al., 2005). In general, these materials exhibit many properties ideal for the production of engineered tissue: a high surface area to volume ratio if processed correctly, sufficient porosity to allow for nutrient and waste diffusion, the potential for surface modification, and the ability to control their degradation rate via selection and modification of their chemical composition (Muschler et al., 2004). In particular, the ability to specifically control the rate of degradation is important. First, the scaffold must provide sufficient mechanical strength when first implanted, but should ultimately degrade to allow for replacement by growing tissue. If degradation is too rapid, then there is a risk of cell loss, scaffold failure, and inflammation of surrounding tissue due to rapid release of acidic breakdown products (Lu et al., 2001). Conversely, an overly slow rate of scaffold degradation would likely impede tissue incorporation.

Synthetic scaffolds have been processed in a variety of configurations, from preformed fibers, meshes, and membranes, to photopolymerized injectable gels. Preformed solid scaffolds are seeded *in vitro* by incubation in a cell suspension. These scaffolds may be applied to large, shallow, or open defects. Li et al. manufactured an electrospun nanofibrous polylactic-co-glycolic acid (PLGA)/poly-ε-caprolactone (PCL) amalgam to better mimic the natural extracellular architecture of cartilage (Li et al., 2002). Their recent study found that a nanofibrous scaffold was more favorable to promote cell expansion and matrix deposition over microfibrous scaffolds for cartilage tissue engineering (Li et al., 2006). The solid synthetic scaffolds have the potential to be modified with natural materials to improve biological characteristics. Enhanced cell attachment and proliferation has been achieved in various composite scaffolds combining synthetic polymers with natural materials, including PLA sponge incorporated with cell-seeded alginate (Caterson et al., 2001), PLGA or poly-L-lactic acid (PLLA) sponge filled with collagen microsponge (Chen et al., 2004; Hsu et al., 2006), macroporous PLGA scaffold conjugated with HA on the porous surface (Yoo et al., 2005). Generally, natural materials have difficulty making mechanically strong engineered cartilage with thickness comparable to the partial-thickness and full-thickness articular cartilage defects. Chen et al. successfully prepared a unique composite web with adjustable thickness from 0.2 mm to 8 mm featuring web-like collagen microsponges formed in a mechanically strong knitted PLGA mesh (Chen et al., 2003).

Nevertheless, obtaining suitable cell densities and uniform cell seeding continues to be a challenge (Lu et al., 2001). A considerable amount of recent work by our group and others has focused on liquid polymer solutions that are polymerized or cross-linked *in situ* after incorporation of cells and bioactive factors. Such solutions allow a uniform incorporation of cells throughout the scaffold and development of minimally invasive application techniques (Sims et al., 1996; Elisseeff et al., 1999a, b; Xu et al., 2004). Finally, these *in situ* polymerizable solutions offer the possibility of precise control of the final shape and composition of the scaffold. For example, there has been considerable work by our group and others studying bilayered constructs in which one layer contains MSCs and the other contains chondrocytes, in order to approximate the cell—cell interactions that would occur in native tissue. Recapitulating the zonal architecture of native cartilage has also been investigated using sequentially photopolymerized hydrogel layers to generate an engineered tissue that more closely approximates normal cartilage (Nettles et al., 2004; Elisseeff et al., 1999a, b; Mercier et al., 2004; Alhadlaq and Mao, 2005).

Novel polymers based on self-assembling synthetic peptides have been studied as a potential scaffold with internal microstructure closely resembling ECM for cartilage tissue engineering. These peptides spontaneously form hydrogels in response to changes in their environment, such as alterations in pH or ionic strength (Kisiday et al., 2002). The nanofiber structure in these hydrogels is approximately three orders of magnitude smaller than that of most polymer fibers, and more closely approximates the structure of native ECM. These materials have the potential for extensive modification by incorporation of peptide domains that influence cell adhesion, differentiation, and proliferation (Holmes, 2002). 3D culture of chondrocytes in peptide hydrogels results in maintenance of chondrocyte phenotype and secretion of cartilage-specific matrix, with increased proliferation and improved mechanical characteristics as compared to chondrocytes cultured in agarose (Kisiday et al., 2002, 2004).

Biological factors

Biological factors are commonly applied to guide cellular differentiation, migration, adhesion, and gene expression (Bottaro et al., 2002; Sekiya et al., 2002). These factors include soluble biochemical signals, transfection of gene vectors, and cell—cell interactions. Soluble signaling molecules have been used to instruct cells to proliferate, differentiate, and generate cartilage matrix during cartilage tissue reconstruction. The signaling molecules of growth factors have been investigated intensively, especially TGF-β superfamily, several BMPs, insulin-like growth factor (IGF)-1, fibroblast growth factors (FGFs), and epidermal growth factor (EGF). Wang et al. found TGF-β3 combined with dexamethasone to be essential for MSC chondrogenesis in 3D silk scaffold and yielded cellular spacing and type II collagen distribution similar to that of native articular cartilage tissue (Wang et al., 2005). Identifying the correct factors as well as the timing and amount of their release plays a large role in the efficacy of tissue differentiation. Byers et al. reported enhanced biomechanical and biochemical maturation of tissue-engineered cartilage constructs through transient exposure to TGF-β3 under serum-free conditions (Byers et al., 2008). BMP-2 causes chondrogeneic differentiation in early embryonic distal digit formation, but causes cell death in a later phase (Zou and Niswander, 1996; Caplan, 2003). IGF-1 has been shown to a potent inducer for cartilage matrix generation by enhancing the deposition of collagen and proteoglycan of chondrocytes (Jenniskens et al., 2006). In addition, integrins represent molecules that adhere to cell surface receptors and influence cell morphology, migration, and signal transmission (Bottaro et al., 2002). These molecules may be integrated with scaffolds and theoretically combine with growth factors to have a synergistic effect on intercellular signaling to regulate cellular migration, proliferation, and differentiation. It should be emphasized at this time that many signals, signaling pathways, and the rationale behind physiological design remain to be elucidated. Furthermore, the effect of the signaling molecule may depend on the effector cell's location within tissue; that is, the effect on a cell at the tissue edge may be different compared to that on a cell within the tissue center.

Many biological factors have limited half-lives, leading to administration difficulties to achieve therapeutic concentrations at sites of cartilage damage. Gene therapy techniques are being developed to deliver therapeutic genes encoding necessary gene products to cells at the site of cartilage injury to synthesize biological factors of interest for sustained local expression (Steinert et al., 2008). In contrast to measuring and monitoring growth factor administration, gene transfer provides a local and sustained supply of bioactive proteins. Gene therapy has encountered obstacles with delivery methods; however, upon development of a reliable delivery technique, genetic engineering will likely interface with tissue engineering (Nussenbaum et al., 2004).

Bioreactors

A fundamental characteristic of many musculoskeletal tissues is their responsiveness to mechanical stimuli (Hung et al., 2004). Articular cartilage is subject to complex forces through its range of motion, including shear, compression, and hydrostatic pressure, and it is reasonable to expect that those forces affect the growth and functioning of chondrocytes. In addition, one of the key observations regarding *in vitro* cartilage growth is the need for high cell densities within the scaffold, on the order of 20 to 100×10^6 cells/ml (Lu et al., 2001). The high cell density coupled with the need to maintain the cells in a 3D construct lead to potential problems with nutrient and waste transport, particularly as the constructs get larger and more matrix is deposited. Static culture which relies on passive diffusion of nutrients and waste may be inadequate to serve the needs of metabolically active tissue.

In order to enhance the biochemical and mechanical properties of engineered cartilage tissues, bioreactors have been developed to provide adequate mass transfer and mechanical stimulation. On the other hand, 3D culture of chondrocytes in bioreactors creates an isolated *in vitro* system to measure the effect of mass flow and dynamic mechanical loading on cartilage formation (Demarteau et al., 2003). Various bioreactor systems have been applied, including rotating-wall vessel, direct perfusion bioreactor, compression bioreactor, and spinner flask (Concaro et al., 2009). Pei et al. cultured bovine chondrocytes in a variety of preformed scaffolds in static conditions and in a rotating bioreactor system (Pei et al., 2002). Constructs cultivated in the bioreactor system demonstrated more uniform cell seeding, greater cell numbers, and enhanced chondrogenesis when compared with their static counterparts. Raimondi et al. utilized a novel forced-perfusion bioreactor system in order to expose the inner portions of their chondrocyte constructs to bulk fluid flow and hydrodynamic stresses, as opposed to the rotating bioreactors which expose the surface only to fluid stresses and convective mass transport (Raimondi et al., 2002). The constructs cultivated in the bioreactor demonstrated greater cell proliferation and better structural integrity than in static conditions. Vunjak-Novakovic et al. demonstrated that dynamic laminar flow patterns on chondrocytes grown on a PGA scaffold resulted in higher fractions of collagen and glycosaminoglycan as well as improved mechanical and electromechanical properties when compared to chondrocytes grown in static culture or turbulent flow conditions (Vunjak-Novakovic et al., 1999). Kisiday et al. used an alternating day mechanostimulation on chondrocytes to increase proteoglycan accumulation (Kisiday et al., 2004). Waldman et al. compared shear with compression stimulation and demonstrated a greater effect on ECM molecule synthesis with shear forces (Waldman et al., 2003). Increased ECM translates to an increased load-bearing capacity and stiffness by the cartilage construct (Vunjak-Novakovic et al., 1999; Mauck et al., 2003; Waldman et al., 2003). Nevertheless, high shear conditions promote apoptosis in chondrocytes resulting in matrix degradation (Healy et al., 2005). Hung et al. demonstrated that mechanical stimulation affected gene expression and biochemical and mechanical properties of bovine articular chondrocytes cultured in agarose. It was shown that the effect was proportional to the frequency of stimulation (which was varied between 0.005 and 1 Hz) and was synergistic with TGF-1 and IGF-1 (Hung et al., 2004). Elder et al. demonstrated enhanced chondrogenesis of chick embryo mesenchymal cells in agarose culture when subjected to cyclic loading at a frequency between 0.15 and 0.33 Hz (Elder et al., 2001). Prior groups had noted a similar stimulatory effect of moderate dynamic stress applied to cultured chondrocytes, within a range of 0.1–1 Hz. Static compression tended to have an inhibitory effect, and dynamic compression frequencies outside the range of 0.1–1 Hz tended to be inhibitory or have no effect (Buschmann et al., 1995; Lee and Bader, 1997; Mauck et al., 2000).

Combining mechanical stimulation of chondrocytes with growth factor administration enhances matrix synthesis to a greater degree than either variable alone. Mauck et al. demonstrated that dynamic deformational loading combined with TGF-1 or IGF-1 increased ECM production in a 3D scaffold (Mauck et al., 2003). The synergy may be a result of

exists between the *in vivo* studies and *in vitro* testing and optimization for the clinical translation of engineered cartilage. The *in vitro* methods should be standardized to provide clear results to develop successful clinical applications for cartilage tissue engineering (Song et al., 2004). There are several important basic questions that remain to be answered. What are the optimal types, amounts, and timing of the growth factor milieu? Will small molecular drugs work effectively for cartilage tissue engineering since many biological factors are complex and exhibit delivery problems? Is transdifferentiation possible — does stem cell plasticity exist? *In vitro* models provide the isolated environment necessary to clearly define genetic programming and signaling pathways that are involved in chondrogenesis. As basic research determines these steps it will allow cartilage tissue engineering to translate to the clinical realm (Caplan and Bruder, 2001; Tuan et al., 2003).

Other areas of intense scrutiny include exploring ways to implement complex tissue elements such as spatial organization and vasculature in engineered tissues. The current work on osteochondral tissue generation has been discussed above. However, as the effects of 3D organization of tissues become more fully appreciated, there will be a need to regenerate those structures for the purposes of controlled laboratory study and clinical application. With the use of novel polymers that gel under controlled conditions there is the potential for fine control of the shape of engineered tissue, as well as of 3D spatial arrangement of heterogeneous cell populations within the scaffold. This has already been demonstrated using photolithographic methods to control hydrogel configuration and cellular organization (Liu and Bhatia, 2002). A clinical application was demonstrated by Naumann et al., in which computer-aided design techniques were used to fashion an HA scaffold seeded with chondrocytes into the shape of a human ear. The construct demonstrated an acceptable shape, as well as evidence of cartilage production *in vitro*, but mechanical properties were not tested (Naumann et al., 2003). Another clinical application of computer-assisted arthroplasty was shown by Sidler et al., in which a bony defect was made in the talus of a human cadaver ankle joint. The defect was then analyzed by computed tomography, and an implant was fashioned using a computer-aided manufacturing device (Sidler et al., 2005).

Of course, there are many more questions and challenges that remain before the promise of tissue engineering is fully realized. The contributions of scientists in fields as diverse as cell/molecular biology, materials science, chemistry, and mathematics will be required in order to answer these questions.

References

Akmal, M., Singh, A., Anand, A., Kesani, A., Aslam, N., Goodship, A., et al. (2005). The effects of hyaluronic acid on articular chondrocytes. *J. Bone Joint Surg., 87B,* 1143–1149.

Alhadlaq, A., & Mao, J. J. (2005). Tissue-engineered osteochondral constructs in the shape of an articular condyle. *J. Bone Joint Surg., 87A,* 936–944.

Alhadlaq, A., Elisseeff, J. H., Hong, L., Williams, C. G., Caplan, A. I., Sharma, B., et al. (2004). Adult stem cell driven genesis of human-shaped articular condyle. *Ann. Biomed. Eng., 32,* 911–923.

Alsberg, E., Anderson, K. W., Albeiruti, A., Franceschi, R. T., & Mooney, D. J. (2001). Cell-interactive alginate hydrogels for bone tissue engineering. *J. Dent. Res., 80,* 2025–2029.

Benjamin, M., & Ralphs, J. R. (2004). Biology of fibrocartilage cells. *Int. Rev. Cytol., 233,* 1–45.

Benya, P. D., & Shaffer, J. D. (1982). Dedifferentiated chondrocytes reexpress the differentiated collagen phenotype when cultured in agarose gels. *Cell, 30,* 215–224.

Bottaro, D. P., Liebmann-Vinson, A., & Heidaran, M. A. (2002). Molecular signaling in bioengineered tissue microenvironments. *Ann. NY Acad. Sci., 961,* 143–153.

Brittberg, M. (2008). Autologous chondrocyte implantation — technique and long-term follow-up. *Injury, 39* (Suppl. 1), S40–S49.

Brittberg, M., Lindahl, A., Nilsson, A., Ohlsson, C., Isaksson, O., & Peterson, L. (1994). Treatment of deep cartilage defects in the knee with autologous chondrocyte transplantation. *N. Engl. J. Med., 331,* 889–895.

Buschmann, M. D., Gluzband, Y. A., Grodzinsky, A. J., & Hunziker, E. B. (1995). Mechanical compression modulates matrix biosynthesis in chondrocyte/agarose culture. *J. Cell Sci., 108,* 1497–1508.

Byers, B. A., Mauck, R. L., Chiang, I. E., & Tuan, R. S. (2008). Transient exposure to transforming growth factor beta 3 under serum-free conditions enhances the biomechanical and biochemical maturation of tissue-engineered cartilage. *Tissue Eng., 14A,* 1821–1834.

Caplan, A. I. (2003). Embryonic development and the principles of tissue engineering. *Novartis Found. Symp., 249,* 17–25, discussion 25–33, 170–174, 239–141.

Caplan, A. I., & Bruder, S. P. (2001). Mesenchymal stem cells: building blocks for molecular medicine in the 21st century. *Trends Mol. Med., 7,* 259–264.

Caterson, E. J., Nesti, L. J., Li, W. J., Danielson, K. G., Albert, T. J., Vaccaro, A. R., et al. (2001). Three-dimensional cartilage formation by bone marrow-derived cells seeded in polylactide/alginate amalgam. *J. Biomed. Mater. Res., 57,* 394–403.

Chen, G., Sato, T., Ushida, T., Hirochika, R., Shirasaki, Y., Ochiai, N., et al. (2003). The use of a novel PLGA fiber/collagen composite web as a scaffold for engineering of articular cartilage tissue with adjustable thickness. *J. Biomed. Mater. Res., 67A,* 1170–1180.

Chen, G., Sato, T., Ushida, T., Ochiai, N., & Tateishi, T. (2004). Tissue engineering of cartilage using a hybrid scaffold of synthetic polymer and collagen. *Tissue Eng., 10,* 323–330.

Chenite, A., Chaput, C., Wang, D., Combes, C., Buschmann, M. D., Hoemann, C. D., et al. (2000). Novel injectable neutral solutions of chitosan form biodegradable gels *in situ. Biomaterials, 21,* 2155–2161.

Chia, S. H., Homicz, M. R., Schumacher, B. L., Thonar, E. J., Masuda, K., Sah, R. L., et al. (2005). Characterization of human nasal septal chondrocytes cultured in alginate. *J. Am. Coll. Surg., 200,* 691–704.

Concaro, S., Gustavson, F., & Gatenholm, P. (2009). Bioreactors for tissue engineering of cartilage. *Adv. Biochem. Eng. Biotechnol., 112,* 125–143.

Darling, E. M., & Athanasiou, K. A. (2003). Biomechanical strategies for articular cartilage regeneration. *Ann. Biomed. Eng., 31,* 1114–1124.

Darling, E. M., & Athanasiou, K. A. (2005). Rapid phenotypic changes in passaged articular chondrocyte subpopulations. *J. Orthop. Res., 23,* 425–432.

Darling, E. M., Hu, J. C., & Athanasiou, K. A. (2004). Zonal and topographical differences in articular cartilage gene expression. *J. Orthop. Res., 22,* 1182–1187.

Demarteau, O., Jakob, M., Schafer, D., Heberer, M., & Martin, I. (2003). Development and validation of a bioreactor for physical stimulation of engineered cartilage. *Biorheology, 40,* 331–336.

Diduch, D. R., Jordan, L. C., Mierisch, C. M., & Balian, G. (2000). Marrow stromal cells embedded in alginate for repair of osteochondral defects. *Arthroscopy, 16,* 571–577.

Dillon, C. F., Rasch, E. K., Gu, Q. P., & Hirsch, R. (2006). Prevalence of knee osteoarthritis in the United States: arthritis data from the Third National Health and Nutrition Examination Survey 1991–94. *J. Rheumatol., 33,* 2271–2279.

Drury, J. L., & Mooney, D. J. (2003). Hydrogels for tissue engineering: scaffold design variables and applications. *Biomaterials, 24,* 4337–4351.

Elder, S. H., Goldstein, S. A., Kimura, J. H., Soslowsky, L. J., & Spengler, D. M. (2001). Chondrocyte differentiation is modulated by frequency and duration of cyclic compressive loading. *Ann. Biomed. Eng., 29,* 476–482.

Elisseeff, J., Anseth, K., Sims, D., McIntosh, W., Randolph, M., & Langer, R. (1999a). Transdermal photopolymerization for minimally invasive implantation. *Proc. Natl. Acad. Sci. USA, 96,* 3104–3107.

Elisseeff, J., Anseth, K., Sims, D., McIntosh, W., Randolph, M., Yaremchuk, M., et al. (1999b). Transdermal photopolymerization of poly(ethylene oxide)-based injectable hydrogels for tissue-engineered cartilage. *Plast. Reconstr. Surg., 104,* 1014–1022.

Eyre, D. (2002). Collagen of articular cartilage. *Arthritis Res., 4,* 30–35.

Eyre, D. R., Weis, M. A., & Wu, J. J. (2006). Articular cartilage collagen: an irreplaceable framework? *Eur. Cell. Mater., 12,* 57–63.

Freed, L. E., Marquis, J. C., Nohria, A., Emmanual, J., Mikos, A. G., & Langer, R. (1993). Neocartilage formation *in vitro* and *in vivo* using cells cultured on synthetic biodegradable polymers. *J. Biomed. Mater. Res., 27,* 11–23.

Frenkel, S. R., & Di Cesare, P. E. (2004). Scaffolds for articular cartilage repair. *Ann. Biomed. Eng., 32,* 26–34.

Furman, B. D., Olson, S. A., & Guilak, F. (2006). The development of posttraumatic arthritis after articular fracture. *J. Orthop. Trauma, 20,* 719–725.

Gerard, C., Catuogno, C., Amargier-Huin, C., Grossin, L., Hubert, P., Gillet, P., et al. (2005). The effect of alginate, hyaluronate and hyaluronate derivatives biomaterials on synthesis of non-articular chondrocyte extracellular matrix. *J. Mater. Sci. Mater. Med., 16,* 541–551.

Gilbert, S. J., Singhrao, S. K., Khan, I. M., Gonzalez, L. G., Thomson, B. M., Burdon, D., et al. (2009). Enhanced tissue integration during cartilage repair *in vitro* can be achieved by inhibiting chondrocyte death at the wound edge. *Tissue Eng. Part A, 15,* 1739–1749.

Grande, D. A., Breitbart, A. S., Mason, J., Paulino, C., Laser, J., & Schwartz, R. E. (1999). Cartilage tissue engineering: current limitations and solutions. *Clin. Orthop. Rel. Res.*, S176–S185.

Haddo, O., Mahroof, S., Higgs, D., David, L., Pringle, J., Bayliss, M., et al. (2004). The use of chondrogide membrane in autologous chondrocyte implantation. *Knee, 11*, 51–55.

Hangody, L., Kish, G., Karpati, Z., Udvarhelyi, I., Szigeti, I., & Bely, M. (1998). Mosaicplasty for the treatment of articular cartilage defects: application in clinical practice. *Orthopedics, 21*, 751–756.

Hangody, L., Feczko, P., Bartha, L., Bodo, G., & Kish, G. (2001). Mosaicplasty for the treatment of articular defects of the knee and ankle. *Clin. Orthop. Rel. Res.*, S328–S336.

Healy, Z. R., Lee, N. H., Gao, X., Goldring, M. B., Talalay, P., Kensler, T. W., et al. (2005). Divergent responses of chondrocytes and endothelial cells to shear stress: cross-talk among COX-2, the phase 2 response, and apoptosis. *Proc. Natl. Acad. Sci. USA, 102*, 14010–14015.

Hoemann, C. D., Hurtig, M., Rossomacha, E., Sun, J., Chevrier, A., Shive, M. S., et al. (2005a). Chitosan-glycerol phosphate/blood implants improve hyaline cartilage repair in ovine microfracture defects. *J. Bone Joint Surg. Am., 87*, 2671–2686.

Hoemann, C. D., Sun, J., Legare, A., McKee, M. D., & Buschmann, M. D. (2005b). Tissue engineering of cartilage using an injectable and adhesive chitosan-based cell-delivery vehicle. *Osteoarthritis Cartilage, 13*, 318–329.

Hoemann, C. D., Sun, J., McKee, M. D., Chevrier, A., Rossomacha, E., Rivard, G. E., et al. (2007). Chitosan-glycerol phosphate/blood implants elicit hyaline cartilage repair integrated with porous subchondral bone in micro-drilled rabbit defects. *Osteoarthritis Cartilage, 15*, 78–89.

Holmes, T. C. (2002). Novel peptide-based biomaterial scaffolds for tissue engineering. *Trends Biotechnol., 20*, 16–21.

Homicz, M. R., Chia, S. H., Schumacher, B. L., Masuda, K., Thonar, E. J., Sah, R. L., et al. (2003). Human septal chondrocyte redifferentiation in alginate, polyglycolic acid scaffold, and monolayer culture. *Laryngoscope, 113*, 25–32.

Hsieh-Bonassera, N. D., Wu, I., Lin, J. K., Schumacher, B. L., Chen, A. C., Masuda, K., et al. (2009). Expansion and redifferentiation of chondrocytes from osteoarthritic cartilage: cells for human cartilage tissue engineering. *Tissue Eng. Part A, 15*, 3513–3523.

Hsu, S. H., Whu, S. W., Hsieh, S. C., Tsai, C. L., Chen, D. C., & Tan, T. S. (2004). Evaluation of chitosan-alginate-hyaluronate complexes modified by an RGD-containing protein as tissue-engineering scaffolds for cartilage regeneration. *Artif. Organs, 28*, 693–703.

Hsu, S. H., Chang, S. H., Yen, H. J., Whu, S. W., Tsai, C. L., & Chen, D. C. (2006). Evaluation of biodegradable polyesters modified by type II collagen and Arg-Gly-Asp as tissue engineering scaffolding materials for cartilage regeneration. *Artif. Organs, 30*, 42–55.

Hubbell, J. A. (2003). Materials as morphogenetic guides in tissue engineering. *Curr. Opin. Biotechnol., 14*, 551–558.

Hung, C. T., Mauck, R. L., Wang, C. C., Lima, E. G., & Ateshian, G. A. (2004). A paradigm for functional tissue engineering of articular cartilage via applied physiologic deformational loading. *Ann. Biomed. Eng., 32*, 35–49.

Hunziker, E. B. (2009). The elusive path to cartilage regeneration. *Adv. Mater., 21*, 3419–3424.

Iwasa, J., Engebretsen, L., Shima, Y., & Ochi, M. (2009). Clinical application of scaffolds for cartilage tissue engineering. *Knee Surg. Sports Traumatol. Arthrosc., 17*, 561–577.

Jenniskens, Y. M., Koevoet, W., de Bart, A. C., Weinans, H., Jahr, H., Verhaar, J. A., et al. (2006). Biochemical and functional modulation of the cartilage collagen network by IGF1, TGFbeta2 and FGF2. *Osteoarthritis Cartilage, 14*, 1136–1146.

Khan, I. M., Gilbert, S. J., Singhrao, S. K., Duance, V. C., & Archer, C. W. (2008). Cartilage integration: evaluation of the reasons for failure of integration during cartilage repair. A review. *Eur. Cell Mater., 16*, 26–39.

Kimura, T., Yasui, N., Ohsawa, S., & Ono, K. (1984). Chondrocytes embedded in collagen gels maintain cartilage phenotype during long-term cultures. *Clin. Orthop. Rel. Res., 186*, 231–239.

Kisiday, J., Jin, M., Kurz, B., Hung, H., Semino, C., Zhang, S., et al. (2002). Self-assembling peptide hydrogel fosters chondrocyte extracellular matrix production and cell division: implications for cartilage tissue repair. *Proc. Natl. Acad. Sci. USA, 99*, 9996–10001.

Kisiday, J. D., Jin, M., DiMicco, M. A., Kurz, B., & Grodzinsky, A. J. (2004). Effects of dynamic compressive loading on chondrocyte biosynthesis in self-assembling peptide scaffolds. *J. Biomech., 37*, 595–604.

Klein, A. M., Graham, V. L., Gulleth, Y., & Lafreniere, D. (2005). Polyglycolic acid/poly-L-lactic acid copolymer use in laryngotracheal reconstruction: a rabbit model. *Laryngoscope, 115*, 583–587.

Kramer, J., Bohrnsen, F., Lindner, U., Behrens, P., Schlenke, P., & Rohwedel, J. (2006). *In vivo* matrix-guided human mesenchymal stem cells. *Cell. Mol. Life Sci., 63*, 616–626.

Lee, D. A., & Bader, D. L. (1997). Compressive strains at physiological frequencies influence the metabolism of chondrocytes seeded in agarose. *J. Orthop. Res., 15*, 181–188.

Lee, J. E., Kim, S. E., Kwon, I. C., Ahn, H. J., Cho, H., Lee, S. H., et al. (2004). Effects of a chitosan scaffold containing TGF-beta1 encapsulated chitosan microspheres on *in vitro* chondrocyte culture. *Artif. Organs, 28*, 829–839.

Li, W. J., Laurencin, C. T., Caterson, E. J., Tuan, R. S., & Ko, F. K. (2002). Electrospun nanofibrous structure: a novel scaffold for tissue engineering. *J. Biomed. Mater. Res., 60*, 613–621.

Li, W. J., Jiang, Y. J., & Tuan, R. S. (2006). Chondrocyte phenotype in engineered fibrous matrix is regulated by fiber size. *Tissue Eng., 12*, 1775–1785.

Li, Z., & Zhang, M. (2005). Chitosan-alginate as scaffolding material for cartilage tissue engineering. *J. Biomed. Mater. Res., 75A*, 485–493.

Liu, V. A., & Bhatia, S. N. (2002). Three-dimensional photopatterning of hydrogels containing living cells. *Biomed. Microdev., 4*, 257–266.

Lu, L., Zhu, X., Valenzuela, R. G., Currier, B. L., & Yaszemski, M. J. (2001). Biodegradable polymer scaffolds for cartilage tissue engineering. *Clin. Orthop. Rel. Res., 11*(Suppl.), S251–S270.

Lu, Y., Dhanaraj, S., Wang, Z., Bradley, D. M., Bowman, S. M., Cole, B. J., et al. (2006). Minced cartilage without cell culture serves as an effective intraoperative cell source for cartilage repair. *J. Orthop. Res., 24*, 1261–1270.

Lynn, A. K., Brooks, R. A., Bonfield, W., & Rushton, N. (2004). Repair of defects in articular joints. Prospects for material-based solutions in tissue engineering. *J. Bone Joint Surg., 86B*, 1093–1099.

Ma, H. L., Chen, T. H., Low-Tone Ho, L., & Hung, S. C. (2005). Neocartilage from human mesenchymal stem cells in alginate: implied timing of transplantation. *J. Biomed. Mater. Res., 74A*, 439–446.

Marcacci, M., Zaffagnini, S., Kon, E., Visani, A., Iacono, F., & Loreti, I. (2002). Arthroscopic autologous chondrocyte transplantation: technical note. *Knee Surg. Sports Traumatol. Arthrosc., 10*, 154–159.

Marcacci, M., Berruto, M., Brocchetta, D., Delcogliano, A., Ghinelli, D., Gobbi, A., et al. (2005). Articular cartilage engineering with Hyalograft C: 3-year clinical results. *Clin. Orthop. Rel. Res., 435*, 96–105.

Mauck, R. L., Soltz, M. A., Wang, C. C., Wong, D. D., Chao, P. H., Valhmu, W. B., et al. (2000). Functional tissue engineering of articular cartilage through dynamic loading of chondrocyte-seeded agarose gels. *J. Biomech. Eng., 122*, 252–260.

Mauck, R. L., Nicoll, S. B., Seyhan, S. L., Ateshian, G. A., & Hung, C. T. (2003). Synergistic action of growth factors and dynamic loading for articular cartilage tissue engineering. *Tissue Eng., 9*, 597–611.

Mercier, N. R., Costantino, H. R., Tracy, M. A., & Bonassar, L. J. (2004). A novel injectable approach for cartilage formation *in vivo* using PLG microspheres. *Ann. Biomed. Eng., 32*, 418–429.

Mithoefer, K., Williams, R. J., 3rd, Warren, R. F., Potter, H. G., Spock, C. R., Jones, E. C., et al. (2006). Chondral resurfacing of articular cartilage defects in the knee with the microfracture technique. Surgical technique. *J. Bone Joint Surg., 88A*(Suppl. 1 Pt 2), 294–304.

Mizuno, S., Tateishi, T., Ushida, T., & Glowacki, J. (2002). Hydrostatic fluid pressure enhances matrix synthesis and accumulation by bovine chondrocytes in three-dimensional culture. *J. Cell Physiol., 193*, 319–327.

Mooney, D. J., & Mikos, A. G. (1999). Growing new organs. *Sci. Am., 280*, 60–65.

Moroni, L., & Elisseeff, J. H. (2008). Biomaterials engineered for integration. *Mater. Today, 11*, 44–51.

Muschler, G. F., Nakamoto, C., & Griffith, L. G. (2004). Engineering principles of clinical cell-based tissue engineering. *J. Bone Joint Surg., 86A*, 1541–1558.

Naumann, A., Aigner, J., Staudenmaier, R., Seemann, M., Bruening, R., Englmeier, K. H., et al. (2003). Clinical aspects and strategy for biomaterial engineering of an auricle based on three-dimensional stereolithography. *Eur. Arch. Otorhinolaryngol., 260*, 568–575.

Nettles, D. L., Vail, T. P., Morgan, M. T., Grinstaff, M. W., & Setton, L. A. (2004). Photocrosslinkable hyaluronan as a scaffold for articular cartilage repair. *Ann. Biomed. Eng., 32*, 391–397.

Nussenbaum, B., Teknos, T. N., & Chepeha, D. B. (2004). Tissue engineering: the current status of this futuristic modality in head neck reconstruction. *Curr. Opin. Otolaryngol. Head Neck Surg., 12*, 311–315.

Obradovic, B., Martin, I., Padera, R. F., Treppo, S., Freed, L. E., & Vunjak-Novakovic, G. (2001). Integration of engineered cartilage. *J. Orthop. Res., 19*, 1089–1097.

Pei, M., Solchaga, L. A., Seidel, J., Zeng, L., Vunjak-Novakovic, G., Caplan, A. I., et al. (2002). Bioreactors mediate the effectiveness of tissue engineering scaffolds. *FASEB J., 16*, 1691–1694.

Pieper, J. S., van der Kraan, P. M., Hafmans, T., Kamp, J., Buma, P., van Susante, J. L., et al. (2002). Crosslinked type II collagen matrices: preparation, characterization, and potential for cartilage engineering. *Biomaterials, 23*, 3183–3192.

Pittenger, M. F., Mackay, A. M., Beck, S. C., Jaiswal, R. K., Douglas, R., Mosca, J. D., et al. (1999). Multilineage potential of adult human mesenchymal stem cells. *Science, 284*, 143–147.

Raimondi, M. T., Boschetti, F., Falcone, L., Fiore, G. B., Remuzzi, A., Marinoni, E., et al. (2002). Mechanobiology of engineered cartilage cultured under a quantified fluid-dynamic environment. *Biomech. Model Mechanobiol., 1*, 69–82.

Riley, S. L., Dutt, S., De La Torre, R., Chen, A. C., Sah, R. L., & Ratcliffe, A. (2001). Formulation of PEG-based hydrogels affects tissue-engineered cartilage construct characteristics. *J. Mater. Sci. Mater. Med., 12*, 983–990.

Rowley, J. A., Madlambayan, G., & Mooney, D. J. (1999). Alginate hydrogels as synthetic extracellular matrix materials. *Biomaterials, 20,* 45–53.

Sarzi-Puttini, P., Cimmino, M. A., Scarpa, R., Caporali, R., Parazzini, F., Zaninelli, A., et al. (2005). Osteoarthritis: an overview of the disease and its treatment strategies. *Semin. Arthritis Rheum., 35,* 1–10.

Sekiya, I., Vuoristo, J. T., Larson, B. L., & Prockop, D. J. (2002). *In vitro* cartilage formation by human adult stem cells from bone marrow stroma defines the sequence of cellular and molecular events during chondrogenesis. *Proc. Natl. Acad. Sci. USA, 99,* 4397–4402.

Sharma, B., & Elisseeff, J. H. (2004). Engineering structurally organized cartilage and bone tissues. *Ann. Biomed. Eng., 32,* 148–159.

Shieh, A. C., & Athanasiou, K. A. (2003). Principles of cell mechanics for cartilage tissue engineering. *Ann. Biomed. Eng., 31,* 1–11.

Sidler, R., Kostler, W., Bardyn, T., Styner, M. A., Sudkamp, N., Nolte, L., et al. (2005). Computer-assisted ankle joint arthroplasty using bio-engineered autografts. *Med. Image Comput. Assist. Interv., 8,* 474–481.

Sims, C. D., Butler, P. E., Casanova, R., Lee, B. T., Randolph, M. A., Lee, W. P., et al. (1996). Injectable cartilage using polyethylene oxide polymer substrates. *Plast. Reconstr. Surg., 98,* 843–850.

Sohn, D. H., Lottman, L. M., Lum, L. Y., Kim, S. G., Pedowitz, R. A., Coutts, R. D., et al. (2002). Effect of gravity on localization of chondrocytes implanted in cartilage defects. *Clin. Orthop. Rel. Res., 394,* 254–262.

Song, L., Baksh, D., & Tuan, R. S. (2004). Mesenchymal stem cell-based cartilage tissue engineering: cells, scaffold and biology. *Cytotherapy, 6,* 596–601.

Steinert, A. F., Noth, U., & Tuan, R. S. (2008). Concepts in gene therapy for cartilage repair. *Injury, 39*(Suppl. 1), S97–S113.

Sultzbaugh, K. J., & Speaker, T. J. (1996). A method to attach lectins to the surface of spermine alginate micro-capsules based on the avidin biotin interaction. *J. Microencapsul., 13,* 363–376.

Suzuki, Y., Tanihara, M., Suzuki, K., Saitou, A., Sufan, W., & Nishimura, Y. (2000). Alginate hydrogel linked with synthetic oligopeptide derived from BMP-2 allows ectopic osteoinduction *in vivo. J. Biomed. Mater. Res., 50,* 405–409.

Tan, H., Chu, C. R., Payne, K. A., & Marra, K. G. (2009). Injectable *in situ* forming biodegradable chitosan-hyaluronic acid based hydrogels for cartilage tissue engineering. *Biomaterials, 30,* 2499–2506.

Temenoff, J. S., & Mikos, A. G. (2000). Review: Tissue engineering for regeneration of articular cartilage. *Biomaterials, 21,* 431–440.

Tuan, R. S., Boland, G., & Tuli, R. (2003). Adult mesenchymal stem cells and cell-based tissue engineering. *Arthritis Res. Ther., 5,* 32–45.

van de Breevaart Bravenboer, J., In der Maur, C. D., Bos, P. K., Feenstra, L., Verhaar, J. A., Weinans, H., et al. (2004). Improved cartilage integration and interfacial strength after enzymatic treatment in a cartilage transplantation model. *Arthritis Res. Ther., 6,* R469–R476.

van Osch, G. J., Brittberg, M., Dennis, J. E., Bastiaansen-Jenniskens, Y. M., Erben, R. G., Konttinen, Y. T., et al. (2009). Cartilage repair: past and future — lessons for regenerative medicine. *J. Cell Mol. Med., 13,* 792–810.

Vunjak-Novakovic, G., Martin, I., Obradovic, B., Treppo, S., Grodzinsky, A. J., Langer, R., et al. (1999). Bioreactor cultivation conditions modulate the composition and mechanical properties of tissue-engineered cartilage. *J. Orthop. Res., 17,* 130–138.

Vunjak-Novakovic, G., Obradovic, B., Martin, I., & Freed, L. E. (2002). Bioreactor studies of native and tissue engineered cartilage. *Biorheology, 39,* 259–268.

Waldman, S. D., Spiteri, C. G., Grynpas, M. D., Pilliar, R. M., Hong, J., & Kandel, R. A. (2003). Effect of bio-mechanical conditioning on cartilaginous tissue formation *in vitro. J. Bone Joint Surg., 85A*(Suppl. 2), 101–105.

Wang, D. A., Varghese, S., Sharma, B., Strehin, I., Fermanian, S., Gorham, J., et al. (2007). Multifunctional chon-droitin sulphate for cartilage tissue–biomaterial integration. *Nat. Mater., 6,* 385–392.

Wang, Y., Kim, U. J., Blasioli, D. J., Kim, H. J., & Kaplan, D. L. (2005). *In vitro* cartilage tissue engineering with 3D porous aqueous-derived silk scaffolds and mesenchymal stem cells. *Biomaterials, 26,* 7082–7094.

Westreich, R., Kaufman, M., Gannon, P., & Lawson, W. (2004). Validating the subcutaneous model of injectable autologous cartilage using a fibrin glue scaffold. *Laryngoscope, 114,* 2154–2160.

Winter, A., Breit, S., Parsch, D., Benz, K., Steck, E., Hauner, H., et al. (2003). Cartilage-like gene expression in differentiated human stem cell spheroids: a comparison of bone marrow-derived and adipose tissue-derived stromal cells. *Arthritis Rheum., 48,* 418–429.

Xie, F., Thumboo, J., Fong, K. Y., Lo, N. N., Yeo, S. J., Yang, K. Y., et al. (2007). Direct and indirect costs of oste-oarthritis in Singapore: a comparative study among multiethnic Asian patients with osteoarthritis. *J. Rheumatol., 34,* 165–171.

Xu, J. W., Zaporojan, V., Peretti, G. M., Roses, R. E., Morse, K. B., Roy, A. K., et al. (2004). Injectable tissue-engineered cartilage with different chondrocyte sources. *Plast. Reconstr. Surg., 113,* 1361–1371.

Yamane, S., Iwasaki, N., Majima, T., Funakoshi, T., Masuko, T., Harada, K., et al. (2005). Feasibility of chitosan-based hyaluronic acid hybrid biomaterial for a novel scaffold in cartilage tissue engineering. *Biomaterials, 26,* 611–619.

Yokoyama, A., Sekiya, I., Miyazaki, K., Ichinose, S., Hata, Y., & Muneta, T. (2005). *In vitro* cartilage formation of composites of synovium-derived mesenchymal stem cells with collagen gel. *Cell Tissue Res., 322,* 289–298.

Yoo, H. S., Lee, E. A., Yoon, J. J., & Park, T. G. (2005). Hyaluronic acid modified biodegradable scaffolds for cartilage tissue engineering. *Biomaterials, 26,* 1925–1933.

Zou, H., & Niswander, L. (1996). Requirement for BMP signaling in interdigital apoptosis and scale formation. *Science, 272,* 738–741.

CHAPTER 54

Functional Tissue Engineering of Ligament and Tendon Injuries

Savio L-Y. Woo, Alejandro J. Almarza, Sinan Karaoglu, Rui Liang, Matthew B. Fisher
Musculoskeletal Research Center, Department of Bioengineering, University of Pittsburgh, Pittsburgh, PA, USA

INTRODUCTION

Tendons and ligaments are soft connective tissues composed of closely packed, parallel collagen fiber bundles which connect bone to muscle and bone to bone, respectively. These unique tissues serve essential roles in the musculoskeletal system by transferring tensile loads to guide motion and stabilize diarthrodial joints. Injuries to tendons, such as the patellar tendon (PT) of the knee, or ligaments, such as the collateral and cruciate ligaments of the knee, upset the balance between mobility and stability of this joint. These injuries are often manifested in abnormal knee kinematics and damage to other tissues in and around the joint such as meniscus and articular cartilage, which may lead to morbidity, pain, and osteoarthritis. With the high incidence of ligament and tendon injuries in sports- and work-related activities, improvements on healing and repair of these tissues are of great interest (Beaty, 1999).

Interestingly, there is a dramatic variability in the propensity for ligaments to heal within the same knee joint, namely the medial collateral ligament (MCL) and anterior cruciate ligament (ACL). Clinical and laboratory studies have shown that injuries to the MCL generally heal sufficiently well such that non-surgical management has become the treatment of choice (Frank et al., 1983; Indelicato, 1983; Jokl et al., 1984; Woo et al., 1987; Kannus, 1988; Scheffler et al., 2001). While most structural properties of the femur–MCL–tibia complex (FMTC) are restored within weeks, the mechanical properties of the healed MCL (i.e. the stress–strain curve) remain very different from those of the normal MCL, as are the altered histomorphological appearance (e.g. uniform distribution of small collagen fibrils) and biochemical composition (e.g. elevated type III and V collagens) (Adachi and Hayashi, 1986; Birk et al., 1990; Weiss et al., 1991; Frank et al., 1992, 1997; Hart et al., 1992, 2000; Marchant et al., 1996; Nakamura et al., 2000; Niyibizi et al., 2000; Birk, 2001).

For the ACL, it is well known that a midsubstance tear will not heal and the success of non-surgical management is limited. Thus, surgical reconstruction of the ACL using autografts harvested from the PT or hamstring tendons is recommended. Issues affecting patient outcome from the use of bone–PT–bone (BPTB) autografts include a persistent palpable defect in the tendon, anterior knee pain, arthrofibrosis, changes to the remaining PT, and PT adhesion to adjacent tissues (i.e. the fat pad) (Coupens et al., 1992; Svensson et al., 2005). The problems associated with hamstring tendon autografts include slower healing due to the

Principles of Regenerative Medicine. DOI: 10.1016/B978-0-12-381422-7.10054-9

development of a soft tissue to bone interphase, less long-term stability of the knee (Freedman et al., 2003), significant hamstring muscle weakness (Marder, 1991; Aune et al., 2001), as well as the increased prevalence of bone tunnel enlargement after reconstruction (Nebelung et al., 1998; Clatworthy et al., 1999; Jansson et al., 1999; Feller and Webster, 2003). Hence, functional tissue engineering (FTE) efforts are aiming to improve the suboptimal properties of the healing MCL, as well as the issues related to ACL graft harvest and healing following reconstruction. With the knowledge gained, it is hoped that the same principles could be applied to aid the repair of other ligaments and tendons (Huang et al., 1993; Badylak et al., 1995; Hildebrand and Frank, 1998; Woo et al., 1999; Nakamura et al., 2000; Spindler et al., 2002; Shimomura et al., 2003).

Thus, FTE offers many attractive approaches to enhance ligament and tendon healing. The goal is not only to restore the normal histomorphological appearance, biochemistry, and mechanical properties of the healing ligament, but most importantly to restore its normal joint function. In this chapter, we will review the properties of normal and healing ligaments and tendons, and discuss the current FTE methods, which include the use of growth factors, gene delivery, stem cell therapy, and the use of scaffolding as well as external mechanical stimuli, aimed at enhancing tendon and ligament healing. To conclude, new technologies and research avenues that have the potential to enhance treatment strategies for ligament and tendon injuries are suggested.

NORMAL LIGAMENTS AND TENDONS
Biology

Ligaments and tendons consist of collagen, proteoglycans, elastin, glycolipids, water (65–70% of the total weight), and cells. Both tissues are hypocellular with <5% of the total volume occupied by cells (Woo et al., 2000; Lo et al., 2002; Hildebrand et al., 2004) and are relatively hypovascular (Manske, 1988; Bray et al., 1996; Lo et al., 2002). The cells in these tissues, fibroblasts and tenocytes, are arranged in rows along the fibers of the extracellular matrix. Both cell types produce both the fibrillar and the non-fibrillar components of the extracellular matrix and may also reabsorb collagen fibers (Birk and Trelstad, 1984; Maffulli, 1999).

Roughly 70–80% of the dry weight of normal tendon or ligament is composed of type I collagen. Histologically, collagen displays a crimp pattern, which refers to a regular, wavy pattern of the matrix when viewed in unloaded conditions (Amiel et al., 1984; Woo et al., 2000; Thornton et al., 2002; Hildebrand et al., 2004). In both ligaments and tendons, there is a bimodal distribution of collagen fibril diameters. One group of fibrils measures between 40 and 75 nm in diameter, the other is between 100 and 150 nm (Dyer and Enna, 1976; Eyden and Tzaphlidou, 2001; Goh et al., 2003). It has been proposed that a bimodal diameter distribution endows tendons and ligaments with better functional properties. The incorporation of a high fraction of small diameter fibrils would ensure a better interfibrillar binding by virtue of their higher surface/volume ratio, whereas the strength requirements would be satisfied by the inclusion of large diameter fibrils. A bimodal distribution would also improve fibril packing, with the smaller fibrils wedging themselves into the spaces left among the larger ones (Ottani et al., 2001).

There are many other collagen types, including III, V, X, XI, and XII, which exist only in minor amounts in ligaments and tendons. The significance of some of these minor collagen types has recently been elucidated. For example, type V collagen is believed to exist in association with type I collagen and serves as a regulator of collagen fibril diameter (Linsenmayer et al., 1993; Birk and Mayne, 1997), whereas type III collagen is needed for wound healing (Liu et al., 1995). Our research has further identified that type XII collagen provides lubrication between collagen fibers (Niyibizi et al., 1995). Lastly, collagen types IX, X, and XI have been identified to exist with type II collagen at the fibrocartilaginous zone of the ligament—bone and tendon–bone interface (Niyibizi et al., 1996; Sagarriga Visconti et al., 1996; Fukuta et al., 1998). It is hypothesized that these collagens exist in this zone to minimize the stress concentrations when loads are transmitted from soft tissue to bone (Cooper and Misol, 1970; Matyas et al., 1995).

The ground substance constituents of tendons or ligaments make up only a small percentage of the total dry tissue weight but are nevertheless quite significant because of their ability to imbibe water. The water and proteoglycans provide lubrication and spacing which are crucial to the gliding function of fibers in the tissue matrix. Elastin, which is present in ligaments and tendons in a few percent by weight, allows the tissue to return to its prestretched length following physiological loading, but the detailed significance has yet to be elucidated. Collectively, these constituents serve to maintain fiber orientation and separation for optimal load distributions.

Although ligaments and tendons are morphologically similar to each other, there are some biochemical differences. When compared with tendons, ligaments are more metabolically active. They have more cellular nuclei, a higher DNA content, and greater amounts of reducible cross-links between collagen fibers (Amiel et al., 1984). Ligaments are composed primarily of water (65–70% of wet weight), cells and collagen (70–80% of dry weight). The most abundant collagen is type I collagen (90% dry weight); type III collagen (8% dry weight) and type V collagen (12% dry weight) are other major components (Linsenmayer et al., 1993; Birk and Mayne, 1997). Collagen types II, IX, X, XI, and XII have also been found to be present (Niyibizi et al., 1996; Sagarriga Visconti et al., 1996; Fukuta et al., 1998; Woo et al., 2006). Tendons, on the other hand, generally contain less water (55% of wet weight), and slightly more type I collagen (85% of dry weight), along with much smaller amounts of other collagens, such as collagens type III, V, XII, and XIV (Goh et al., 2003).

Biomechanics

999

The major function of ligaments and tendons include maintaining the proper anatomical alignment of the skeleton and guiding joint motions. They accomplish this by transmitting forces along their longitudinal axis; hence their biomechanical properties are measured in uniaxial tension. They demonstrate non-linear behavior, which is governed by the recruitment of collagen. This allows ligaments to maintain normal joint laxity in response to low loads, and also to stiffen dramatically in response to high loads, preventing excessive joint displacements. Ligaments and tendons also exhibit time- and history-dependent viscoelastic behavior which could be attributed to the complex interactions of tissue constituents such as collagen, proteoglycans, water, and ground substance (Fung et al., 1972; Woo et al., 1981b). The viscoelastic properties of ligaments and tendons are important and clinically relevant. For instance, during regular activities such as walking and jogging, tissues have the ability to soften over time. This phenomenon reduces the susceptibility to damage related to fatigue (Frank and Jackson, 1997).

However, following injury, ligaments and tendons generally fail to recover their normal mechanical and viscoelastic behaviors. Thus, abnormal joint kinematics result which can directly lead to excessive forces in surrounding tissues (e.g. articular cartilage). This can lead to further injury to other structures through either traumatic mechanisms or degeneration (i.e. osteoarthritis). As the ultimate goal of functional tissue engineering is to restore the function of ligaments and tendons, and thereby the function of the injured joint, it is necessary to understand their normal mechanical behavior and contributions to joint function. Testing methodologies for this purpose include: (1) functional testing, which involves determining the contribution of the ligament or tendon to joint kinematics (i.e. the *in situ* forces in response to external loading conditions), and (2) tensile testing, which provides an assessment of the structural properties of the bone–ligament–bone complex or muscle–tendon–bone complex and mechanical properties of the tissue substance.

TENSILE TESTING

Tendons are generally long and can be tested in their isolated state using sinusoidal-shape or frozen grips to limit slippage. Isolated ligaments, on the other hand, are short in length, making it difficult to clamp them independently. Hence, a tensile test is generally conducted on the entire bone−ligament−bone complex (e.g. femur−MCL−tibia complex or FMTC) with the tissue insertion sites left anatomically intact. With cross-sectional area (CSA) measurements and the utilization of tissue markers to measure tissue strain, the structural properties of the bone−
ligament−bone complex as well as mechanical properties of the ligament substance can be measured from a load to failure test (Woo et al., 1983; Lee and Woo, 1988).

Structural properties (Fig. 54.1) of the bone−ligament−bone complex (i.e. a load−elongation curve) are generally described by four parameters including stiffness (slope of the linear portion of the load−elongation curve), ultimate load (maximum load at which the complex fails), ultimate elongation (elongation corresponding to with the maximum load), and energy absorbed at failure (area under the curve to the maximum load). These data reflect the behavior of the entire bone−ligament−bone complex which includes tissue size, orientation of collagen fibers to applied loads, and the contribution of the bony insertions (Woo et al., 1983).

Measuring the mechanical properties (Fig. 54.2) of the ligament substance (i.e. a stress−strain curve), on the other hand, requires knowledge of the CSA of the ligament, commonly measured using a laser micrometer system (Lee and Woo, 1988), and tissue strain, commonly measured using video techniques to track two or more reflective markers placed on the tissue midsubstance (Scheffler et al., 2001). Stress in the tissue is obtained by dividing load by the cross-sectional area and strain is obtained by computing change in marker distance during the test relative to their original distance. Parameters describing the mechanical properties of the ligaments and tendons (Fig. 54.2) include tangent modulus (slope of the linear portion of the stress−strain curve), tensile strength (stress at failure), ultimate strain (strain corresponding to the tensile strength), and strain energy density (area under the stress−strain curve until failure). These data represent the quality of the tissue, irrespective of tissue size.

The viscoelastic properties of ligaments and tendons include stress relaxation (decrease in stress over time in response to a constant elongation) and creep (increase in elongation over time in response to a constant load). In addition, they also display a phenomenon called "hysteresis" in response to cyclic loading (Fig. 54.3). This results from a loss of internal energy causing the loading and unloading paths to be different. The area of hysteresis reduces as the tissue undergoes several cycles of loading and unloading and the tissue is said to be "preconditioned," a state desired for a tissue prior to mechanical testing. Non-linear viscoelastic models such as the Quasi-linear Viscoelastic Theory:

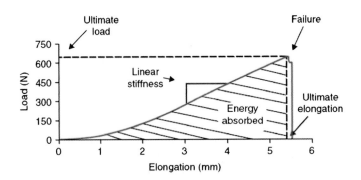

FIGURE 54.1
A typical load−elongation curve representing the structural properties of the femur−anterior medial bundle−tibia complex of the human anterior cruciate ligament.

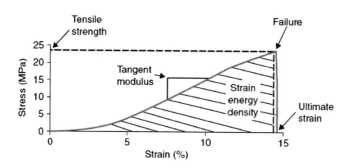

FIGURE 54.2

A typical stress—strain curve representing the mechanical properties of the anterior medial bundle of the human anterior cruciate ligament.

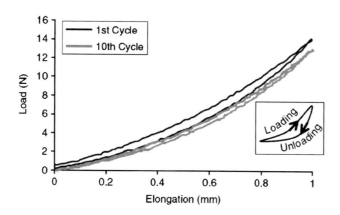

FIGURE 54.3

Hysteresis loops (1st and 10th cycles) obtained from cyclic loading of the femur—anterior medial bundle—tibia complex of the human anterior cruciate ligament in uniaxial tension. The area of hysteresis, i.e. the area between the loading and unloading curves, decreases with repetitive cycling, demonstrating the phenomenon of "preconditioning."

$$\sigma(t) = \int_{-\infty}^{t} G(t - \tau)\frac{\partial\sigma^e(\varepsilon)}{\partial\varepsilon}\frac{\partial\varepsilon}{\partial\tau}\partial\tau$$

and Single Integral Finite Strain Theory:

$$T = -pI + C_0\{[1 + \mu I(t)]B(t) - \mu B^2(t)\} - C_0(1 - \gamma)^*$$
$$\int_{0}^{t} G(t - s)\{[1 + \mu I(s)]B(t) - \mu F(t)C(s)F^T(s)\}ds$$

have been utilized to model these behaviors in ligaments and tendons (Fung et al., 1972; Woo et al., 1981b; Johnson et al., 1994).

The basic testing methodologies described in this section have been utilized for decades to examine soft tissues and much work has been done to define the appropriate testing procedures such as specimen orientation (Woo et al., 1991), handling, storage, and hydration (Woo et al., 1986a). They have led to important findings regarding the physiological changes associated with growth and development (Woo et al., 1986b, 1991), the adaptation of ligaments and tendons to mobility (Woo et al., 1981a, c 1982, 1987a), and the effects of injury and treatment (Woo et al., 1987b; Weiss et al., 1991).

CONTRIBUTION TO JOINT FUNCTION

Joint motion is governed by the direction and magnitude of externally applied loads, ligament forces, contact between joint surfaces, and muscle activity. For the knee, motions include a combination of translations: proximal-distal, medial-lateral, and anterior-posterior, and rotations: internal-external, flexion-extension and varus-valgus. In total, these translations and rotations describe motion in 6 degrees of freedom (DOF).

While evaluating joint function, it is important to note that constraining DOF of the knee can have a significant impact on the results obtained (Inoue et al., 1987; Livesay et al., 1997). When knee motion was allowed in all directions, sectioning the MCL only resulted in small increases in valgus laxity (21%), suggesting that the ACL plays a significant role as a joint restraint to this knee motion. However, when anterior-posterior translation and internal-external rotation were constrained, valgus laxity increased significantly (171%) following sectioning of the MCL. For this reason, it is important to have a testing device which allows for unconstrained knee motion.

For more than a decade, a robotic/universal force-moment sensor (UFS) testing system, which was developed by our research center, has been used to study knee kinematics as well as to directly measure the *in situ* forces in the knee ligaments in response to external loading conditions (Fig. 54.4) (Fujie et al., 1993, 1995; Rudy et al., 1996). This methodology has been utilized to study many variables of ACL reconstruction. Most recently the limitations of single bundle reconstructions to restore rotatory stability along with the potential advantages of an anatomical reconstruction were demonstrated (Woo et al., 2002; Yagi et al., 2002). In addition, knee function following an isolated MCL injury in a goat model has been studied (Scheffler et al., 2001; Abramowitch et al., 2003a).

HEALING OF LIGAMENTS AND TENDONS

The events of healing of ligaments and tendons can be roughly divided into four overlapping phases: hemorrhage, inflammation, repair (proliferation), and remodeling. Following injury, the hemorrhagic and inflammatory phases occur over the first several days. Minutes after the ligament injury, blood collects and forms a platelet-rich fibrin clot at the injury site. The hemorrhage phase of the injury forms a lattice for many following cellular events. Triggered by cytokines released within the clot, polymononuclear leukocytes and lymphocytes appear

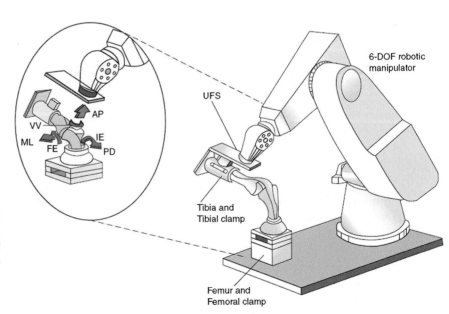

FIGURE 54.4
Schematic drawing illustrating the six degrees of freedom of motion of the human knee joint, indicating anterior-posterior (AP), medial-lateral (ML), and proximal-distal (PD) translation as well as flexion-extension (FE), and internal-external (IE), and varus-valgus (VV) rotation. *(Permission obtained from Allen et al., 1999.)*

within several hours. These cells respond to autocrine and paracrine signals to expand the inflammatory response and recruit other types of cells to the wound (Frank et al., 1994).

The reparative phase takes place over the first couple of weeks to months following the injury. During this phase, fibroblasts recruited to the injury site start forming healing tissue. Growth factors, including transforming growth factor-β (TGF-β) and platelet-derived growth factor (PDGF) isoforms, are likely to be involved in modulating the healing environment in favor of effectively repairing the damaged ligament substance (Murphy et al., 1994). Meanwhile, increased neovascularization brings in circulating cells and nutrients to further enhance the healing process. The blood clot quickly turns into newly formed healing tissue which is composed of an aggregation of cells surrounded by a matrix whose histomorphological appearance and biochemical composition are different from those of the uninjured ligament. It is notably characterized by a homogeneous distribution of smaller diameter collagen fibrils, which is in stark contrast to the bimodal distribution of the normal ligament (Frank et al., 1992, 1997; Hart et al., 1992, 2000; Nakamura et al., 2000). Biochemically, it contains increased amounts of proteoglycans, a higher ratio of type V to type I collagen, and a decrease in the number of mature collagen cross-links.

The proliferative phase gives way to the remodeling phase. It occurs from months to years after the injury and is characterized by decreasing cellularity, decreasing levels of collagen type III, and realigning of the matrix to respond better to the forces applied to the tissues. On the other hand, the diameters of collagen fibrils have been found to remain small and levels of collagen type V have been found to remain elevated for years after injury (Adachi and Hayashi, 1986; Birk et al., 1990; Frank et al., 1992, 1997; Hart et al., 1992, 2000; Marchant et al., 1996; Nakamura et al., 2000; Niyibizi et al., 2000; Birk, 2001). Interestingly, the type V collagen has been shown to play a central role in the regulation of the lateral growth of collagen fibrils. The elevated type V collagen could be involved in the lack of large collagen fibrils which in turn are associated with the inferior mechanical properties of healing tissue (Parry et al., 1978; Doillon et al., 1985).

1003

Medial collateral ligament of the knee

The healing process of the medial collateral ligament follows this general wound healing pathway and has been well studied. Thus, it serves as a good model for the histological, biochemical, and biomechanical events. The process of ligament healing is also greatly impacted by the selection of treatment (Clayton et al., 1968; Tipton et al., 1970; O'Donoghue et al., 1971; Woo et al., 1987b, 1990). Laboratory and clinical studies have shown mobilization is superior to immobilization (Woo et al., 1987b; Inoue et al., 1990; Weiss et al., 1991). Interestingly, non-operative repairs have an equivalent healing outcome to surgical repairs. A severe "mop-end" injury model in the rabbit, developed in our research center, that causes a midsubstance tear and damage at the insertion sites (Weiss et al., 1991), along with non-operative treatment with mobilization, was used to compare to surgical repair with mobilization. After 52 weeks of healing, there were no significant differences in varus-valgus rotation of the knee, *in situ* force of the MCL, or tensile properties between the repaired and non-repaired MCL (Weiss et al., 1991). Based on these and other studies, clinical management has shifted from surgical repair with immobilization to non-operative management with early controlled range-of-motion exercises as soon as pain subsides (Reider et al., 1994; Indelicato, 1995).

While the MCL heals with non-operative treatment and the stiffness of the healing FMTC begins to approach normal levels by 52 weeks after injury, the cross-sectional area of the healed tissue increases with time, measuring as much as two and a half times its normal size (Inoue et al., 1990). The mechanical properties of the healing MCL midsubstance remain consistently inferior to those of the normal ligament and do not change with time. Thus, the healing process involves making a larger quantity of lower quality ligamentous tissue. Moreover, the rate of healing between the ligament midsubstance and the insertion sites is asynchronous with the insertion sites demonstrating a lower stiffness and strength resulting from injury as well as a lack of stress during the healing process.

There has been evidence that indicates that activity level influences the rate of healing (Abramowitch et al., 2003a). The goat model is a more clinically relevant model to study ligament healing than the rabbit due to its more robust activity level. In addition, its large size and the previously published success of ACL reconstructions using this animal (Ng et al., 1996) make it attractive to study more complex multiple-ligament injuries. Generally, these models display similar trends. However, it was noted that the stiffness and ultimate load of the healing goat FMTC are closer to control values at earlier time periods when compared to data from the rabbit model, suggesting that activity level may influence the healing response.

Anterior cruciate ligament reconstruction

With the ultimate goal of ACL reconstruction being to restore knee function, the success of these procedures is dependent on a number of surgical, biomechanical, and biological factors. The most popular choice is autografts from the PT, i.e. BPTB, or hamstring tendons, i.e. semitendinosous plus gracilis tendons. Allografts, including the Achilles tendon, BPTB, and hamstring tendons, have seen limited use except in revision surgery or for multiple ligamentous injuries. The BPTB graft is generally considered the "gold standard" for ACL reconstruction because it facilitates better fixation and bone-to-bone healing inside the bone tunnels (Jones, 1970; Lambert, 1983; Noyes et al., 1984; Kurosaka et al., 1987; Aglietti et al., 1992; Cooper et al., 1993). However, the major drawback is that the open defect of the donor site remains visible and is not completely healed for months (Coupens et al., 1992; Rubinstein et al., 1994; Cerullo et al., 1995; Nixon et al., 1995). This contributes to a higher incidence of complications including donor site morbidity, patella baja, arthrofibrosis, adhesion to the fat pad, and patellofemoral pain (Paulos et al., 1987; Tibone and Antich, 1988; Sachs et al., 1989; Shelbourne et al., 1991; Breitfuss et al., 1996; Kartus et al., 1999).

Efforts have been made to examine the healing PT after harvest, using animal models, after removal of the central third. Studies have found a deterioration of PT structural properties with a concomitant increase in the CSA of the PT tissue (Kamps et al., 1994; Linder et al., 1994; Beynnon et al., 1995; Awad et al., 2003; Tohyama et al., 2003). Specifically in the rabbit model, the ultimate load of the entire BPTB complex decreased by 38% (Beynnon et al., 1995), while a CSA increase of 83–108% was observed at 12 weeks post-harvesting (Awad et al., 2003; Tohyama et al., 2003). For the central healing tissue, its tangent modulus and ultimate tensile strength were measured to be only 15% and 18% of controls, respectively, after 26 weeks (Awad et al., 2003). The mechanical properties of the remaining PT tissues also deteriorated compared to sham controls after 24 weeks (Tohyama et al., 2003).

Following implantation, the autograft becomes inflamed and necrotic leading to a decrease in graft stiffness and strength (Tohyama and Yasuda, 2000). The graft undergoes revascularization and repopulation with fibroblasts followed by a remodeling period with restructuring of collagen fibers and proteoglycans (Arnoczky et al., 1982). Further more, bone-to-bone healing and tendon-to-bone healing within the femoral and tibial tunnels revealed that there was complete incorporation of the bone block by 6 weeks, but incomplete incorporation at the tendon–bone interface (Papageorgiou et al., 2001). For the latter, the failure mode of the femur–graft–tibia complex (FGTC) consistently occurred as a pull-out from the tibial tunnel. Over time, however, experimental animal studies show that the FGTC gradually shows improvement (Ballock et al., 1989; Butler et al., 1989; Gerich et al., 1996), but its structural properties failed to be restored to levels of the intact femur–ACL–tibia complex (FATC) even after 12 months (Clancy et al., 1981; Ballock et al., 1989; Butler et al., 1989). It is agreed that accelerating graft incorporation and healing may lead to an earlier return to sports and normal activities, and therefore has become a goal of tissue engineering efforts, as discussed later in this chapter.

In addition to the graft selection, other important surgical decisions include tunnel placement, graft tension, and fixation. There has been a substantial amount of research which is focused on the impact of these variables at time-zero and following various periods of healing in both human and animal models (Yagi et al., 2002; Abramowitch et al., 2003b; Loh et al., 2003). All these parameters can lead to various degrees of graft tunnel motion which may affect graft integration and graft healing. Ultimately, these factors impact the post-operative rehabilitation and the return to normal activities and sports (Tsuda et al., 2002). Thus, it is important to take into account the large changes in properties of the graft after implantation.

Combined ligamentous injuries

Combined ACL/MCL injuries occur frequently and the best methods for treatment are still controversial. Some surgeons elect to surgically reconstruct the ACL without addressing the MCL, while others advocate reconstruction of the ACL with repair of the MCL. Regardless of the treatment modality, clinical and basic science studies continue to show that the outcome of this injury is worse than for an isolated MCL injury.

Our research center has elucidated the effects of ACL deficiency on the healing of the injured MCL using canine and goat models (Woo et al., 1990; Ohno et al., 1995; Yamaji et al., 1996), and thus ACL reconstruction has been suggested. Further more, repairing the MCL in combination with ACL reconstruction resulted in reduced valgus laxity and improved the structural properties of the FMTC in early stages, but the long-term effect became minimal (Ohno et al., 1995; Yamaji et al., 1996). These data further suggest that only reconstruction of the ACL is necessary for successful healing of the MCL after a combined ACL/MCL injury.

Recently, a larger animal model, i.e. the goat knee, was used in our research center to examine the function of the knee and quality of the healing MCL after a combined ACL/MCL injury treated with ACL reconstruction (Abramowitch et al., 2003c). These results confirmed that valgus rotation was twice that for an isolated MCL injury. Moreover, the structural properties of the FMTC and tangent modulus of the MCL substance are all substantially lower than for the isolated MCL injury (Scheffler et al., 2001; Abramowitch et al., 2003c). These results demonstrate a clear need to enhance ligament healing after such a severe knee injury, requiring improved treatment strategies.

1005

APPLICATION OF FUNCTIONAL TISSUE ENGINEERING

Functional tissue engineering emphasizes the importance of biomechanical considerations in the design and development of cell- and matrix-based implants for soft and hard tissue repair. Musculoskeletal tissues, especially ligaments and tendons, are accustomed to being mechanically challenged, therefore tissue-engineered constructs used to replace these tissues after injury or disease must meet these requirements. By combining the fields of molecular biology, biochemistry, and biomechanics, novel therapeutic approaches (e.g. growth factors, gene transfer/gene therapy, cell therapy, and biological scaffolds) offer the possibilities for improvement of the treatment of ligament and tendon injuries. The following will be a brief review of the currently available approaches to enhance ligament and tendon healing.

Growth factors

Growth factors can induce wide-ranging effects on cell function including migration, proliferation, and protein synthesis. The application of exogenous growth factors is based on the premise that they can promote ligament regeneration that will lead to a biologically and biomechanically superior healed ligament substance. Many studies, *in vitro* and *in vivo*, have tried to define the role of growth factors in ligament and tendon healing and to determine appropriate strategies for the use of growth factors for these structures (Steenfos, 1994; Duffy et al., 1995; Panossian et al., 1997; Sciore et al., 1998).

IN VITRO STUDIES

Cell culture or tissue explant methodologies involving the addition of exogenous growth factors have been the major study designs. Measured responses include cell proliferation, synthesis of extracellular matrix proteins such as collagen, proteoglycans, tissue remodeling enzymes, and cell migration or chemotaxis. In our research center, the effects of eight different growth factors on the MCL and ACL fibroblast culture were determined for proliferation and extracellular matrix production (Ohno et al., 1995; Deie et al., 1997; Marui et al., 1997; Scherping et al., 1997).

In terms of cell proliferation, PDGF-BB, epidermal growth factor (EGF), and basic fibroblast growth factor (bFGF) have been found to have a significant effect on cell proliferation and caused greater proliferation in MCL fibroblasts than in ACL fibroblasts (Scherping et al., 1997). We found that the proliferation of MCL and ACL fibroblasts from skeletally immature rabbits increased by 7.6 times in response to EGF and 5.6 times in response to bFGF (Ohno et al., 1995). The same study in skeletally mature rabbits showed that IGF and bFGF also had significant effects on fibroblast proliferation in both cell types, but the difference was less pronounced (Scherping et al., 1997). The biological role of TGF-β1 in ligament healing has also been recently addressed. Studies in our research center on fibroblast proliferation in skeletally immature rabbits demonstrated that TGF-β1 stimulated proliferation of MCL fibroblasts 1.3–1.4 times greater than in ACL fibroblasts (Ohno et al., 1995). Subsequent studies on skeletally mature rabbits showed little effect of TGF-β1 on cell proliferation for either fibroblast type. Comparison of these results suggests that age has a significant effect on the ability of growth factors to stimulate fibroblast proliferation (Scherping et al., 1997). The effect of TGF-β1 on canine ACL fibroblast proliferation was shown to be dose dependent because smaller doses acted synergistically with PDGF whereas higher concentrations inhibited the stimulatory effect of PDGF (Desrosiers et al., 1995). These findings show the complex interactions of growth factors to enhance proliferation of fibroblasts.

In terms of *in vitro* protein synthesis in MCL and ACL fibroblasts, collagen synthesis increased 160% over controls in both MCL and ACL fibroblasts treated with TGF-β1, and the majority of this increase was for type I collagen (Marui et al., 1997). The relative increase in protein production was similar for both cell types, but the absolute increase in protein synthesis was twice as much for MCL fibroblasts as for ACL fibroblasts. These data suggest that TGF-β1 may improve ligament healing by increasing matrix synthesis during the proliferative and remodeling phases (Marui et al., 1997). Similar results have also been found by other investigators (Desrosiers et al., 1995). These studies illustrate the ability of TGF-β1 to increase the production of ECM by fibroblasts *in vitro*.

In vitro models, however, are limited in the extent that they cannot reproduce the complex interplay of signals affected by growth factors in the intricate process of ligament or tendon healing. Differences in the effects of different growth factors on cell proliferation and matrix synthesis suggest that wound healing depends on a highly integrated biochemical network of cell signaling events with intrinsic stimulatory and inhibitory feedback loops. Thus, in addition to providing a better physiological model, *in vivo* studies are critical to defining the interaction of biology and biomechanics and the degree to which healed ligament or tendon substance restores the structural and mechanical properties of the native tissue.

IN VIVO STUDIES

In vitro studies showed that EGF and PDGF-BB have the greatest effect on ligament fibroblast proliferation, whereas TGF-β1 superiorly promotes extracellular matrix production. These growth factors were then applied *in vivo* at different dosages, in isolation and in combination, for an MCL injury in the rabbit model. It was found that a higher dose of PDGF-BB improved the structural properties of the femur–MCL–tibia complex compared to a lower dose of PDGF-BB, demonstrating that the effects of PDGF-BB were dose dependent (Woo et al., 1998).

However, the mechanical properties of the ligament substance remained unchanged from untreated controls, demonstrating that the improved structural properties resulted from a larger quantity of tissue instead of tissue with improved quality. In contrast, the combination of EGF or PDGF-BB plus TGF-β1 did not lead to additional improvements in MCL healing compared with PDGF-BB alone.

Other investigators have also found that higher doses of PDGF demonstrated a plateau effect in improving the structural properties of the healed ligament. Furthermore, administration of PDGF for more than 24 h after the injury markedly decreased the efficacy of growth factors in improving MCL healing (Batten et al., 1996). Additionally, administration of TGF-β2 to the healing rabbit MCL improved the mechanical stiffness, but not the load at failure, of the ligament scar (Spindler et al., 2003). These data show that *in vivo* application of growth factors is indeed more complex than *in vitro* studies.

One possible approach to improve the *in vivo* application might be to combine growth factors with gene transfer technology. Bone morphogenetic protein-2 (AdBMP-2) delivered to the bone–tendon interface using a gene transfer technique has been shown to improve the integration of semitendinosous tendon grafts in rabbits (Martinek et al., 2002). The stiffness (29.0 ± 7.1 N/mm vs. 16.7 ± 8.3 N/mm) and the ultimate load (108.8 ± 50.8 N vs. 45.0 ± 18.0 N) were also significantly increased in the specimens with AdBMP-2 compared to untreated controls. Hence, the enhancement of tendon to bone healing is a promising approach to accelerate the patient's return to activity.

Based on these studies, an optimal therapy of introducing growth factors to injury sites is still an open question. While promising, timing of application, mode of delivery, and dosage remain as major hurdles that need to be crossed before success can be achieved *in vivo*.

Gene therapy

Gene therapy is a potential approach to improve ligament and tendon healing. Foreign nucleic acids, gene transfer, can be introduced into cells to alter protein synthesis or induce the expression of therapeutic proteins. Modern gene therapy relies on mammalian viruses and cationic liposomes as delivery vectors, and both have been developed to deliver genes into host tissue via the direct and indirect methods. Direct gene transfer involves *in vivo* injection of the delivery vector into the host tissue. Indirect gene transfer involves *in vitro* transduction of host cells with the desired gene, and subsequent replantation of these cultured cells *in vivo*. Studies have shown that PT fibroblasts can be transduced with the LacZ marker gene both directly using an adenovirus liposomal vector, and indirectly using a retrovirus, with the expression of the transferred genes persisting for 6 weeks following the application (Gerich et al., 1996). PT fibroblasts staining positively for β-galactosidase were subsequently found to migrate and incorporate into the tendon tissue following injection. In our research center, we sought to determine whether genes could be transduced into MCL and ACL fibroblasts and if ligament injury affected gene transfer and expression (Hildebrand et al., 1999). When both the direct and indirect methods were employed using adenovirus and BAG retrovirus, respectively, it was found that both techniques resulted in expression of the LacZ marker gene by fibroblasts from intact as well as injured ligaments. Gene expression lasted longer (6 weeks) with the direct method than with the indirect technique (3 weeks). Fibroblasts from injured ligaments showed transduction both in the wound site and in the ligament substance. There was no difference in the duration of gene expression by fibroblasts from intact and injured ligaments, suggesting that injury does not affect gene transfer or expression (Hildebrand et al., 1999).

Newer techniques for gene transfer have recently been reported. Gene transfer using liposomal vectors may reduce the adverse immune responses seen when using viral vectors. Antisense gene therapy involving blocking the transcription or translation of specific genes which may be excessively expressed within healing tissue has been proposed. By the binding of antisense

1007

oligodeoxynucleotides (ODNs) to target DNA, investigators have performed direct transfer of an HVJ—liposome complex containing a labeled ODN for the protein decorin, which has been shown to inhibit type I collagen fibril formation in prior *in vitro* studies (Nakamura et al., 2000). Histological analysis was performed 24 h after direct injection into rabbit MCL specimens which were 2 weeks post-injury. Quantitative analysis revealed a transduction efficiency of 62% at 1 day and 23% at 7 days. Significant suppression of decorin mRNA expression was seen at both 2 days (42.7%) and 2 weeks (60.3%) (Nakamura et al., 2000).

In our research center, we have evaluated the efficacy of utilizing ODNs to regulate the over-production of collagens III and V (Shimomura et al., 2003; Woo et al., 2004). Normal human patellar tendon fibroblasts (HPTFs) were transfected with antisense collagen III or V ODNs by mixing with Lipofectamine. The uptake of the ODNs was detected as early as 1 h and as late as 3 days after delivery. The relative expression of collagen V mRNA was reduced to $67.8 \pm 5.1\%$ of missense levels. Also, preliminary RT-PCR results showed that the inhibitory effects of the collagen III antisense ODNs were most dominant at day 1 as the type III collagen mRNA level was $38.9 \pm 19.6\%$ of missense controls. At days 3 and 7, differences could not be observed. These results suggested that antisense gene therapy can indeed be a potential functional tissue engineering approach to enhance the quality of ligaments and tendons.

Despite these promising results, several obstacles currently impede the practical implementation of gene transfer as a biological intervention in ligament healing. The immune reaction against these antigens decreases the expression of the introduced gene (Tripathy et al., 1996). In addition, retroviral infection of fibroblasts often leads to shut-off of the promoter region, which adversely affects expression of the incorporated gene (Krall et al., 1994). Thus, delivery of the ODNs to the appropriate target and reproducibility of the results remain great challenges.

Newer strategies in the evolving field of gene transfer include the search for more effective and less immunogenic vectors, modification of promoters to ensure gene expression after incorporation, and temporary and self-limiting gene expression regulation tailored to the changing environment of the healing ligament. As the complex steps involved in gene expression and regulation are further elucidated, the potential therapeutic efficacy of gene transfer is likely to enjoy practical application.

Cell therapy

Cell therapy is another potential strategy to enhance ligament and tendon healing. Studies have focused on the application of mesenchymal stem cells (MSCs), bone marrow-derived cells (BMDCs), and synovial tissue-derived fibroblasts into the healing site Young et al., 1998; Watanabe et al., 2002). BMDCs have been shown to play an important role in wound healing (Badiavas et al., 2003; Galiano et al., 2004; Mathews et al., 2004) and can be obtained in high numbers with relative ease (Awad et al., 2003; Juncosa-Melvin et al., 2005). In one study, autologous marrow-derived progenitor cells were seeded on a collagen gel, and subsequently contracted onto a pretensioned suture (Young et al., 1998). The resulting tissue prosthesis was then implanted into the rabbit Achilles tendon gap defect. Significantly greater structural and mechanical properties were seen after the implantation. The treated tissues had a significantly larger cross-sectional area, and their collagen fibers appeared to be better aligned than those in the matched controls.

Recently, new methods geared toward PT healing have tried to fill the central third PT defect with collagen gels filled with BMDCs, in which BMDCs were expanded *in vitro*. These MSC—collagen composites were implanted into full-thickness, full-length, central defects created in the PTs of rabbits. The healing PTs treated with MSC—collagen gels were one-fourth of the maximum stress of the normal central portion of the PT. The modulus and maximum stress of the repair tissues grafted with MSC—collagen gels increased at significantly faster rates than did natural repairs over time. Thus, overall improved mechanical properties were seen when compared to non-treated defects (Awad et al., 1999, 2003; Butler et al., 2005; Juncosa-Melvin et al., 2005). This particular cell therapy is attractive because the use of autogenous cells would minimize the immune response at the injury site.

Alternatively, fibroblasts, myoblasts, and bone marrow cells have also been transplanted into injured ligaments following the induction of marker genes or stimulation by growth factors *in vitro*. These results show great potential *in vivo* (Day et al., 1997; Caplan et al., 1998; Hildebrand et al., 1999; Watanabe et al., 2002). However, issues remain on cell therapy. MSCs from bone marrow are relatively few in number and their numbers decrease further after transplantation. Thus, it is essential to develop *in vitro* techniques to expand MSCs without altering their differentiation potential. Further more, scaffolds may help to prevent the loss of these cells and may further enhance the effect already provided by this strategy.

Scaffolding

An ideal scaffold should provide a suitable mechanical and biological environment for cells to migrate into, guide the formation of the newly synthesized extracellular matrix, and then slowly degrade such that the new matrix begins to bear the mechanical loads. Both synthetic and naturally occurring biological scaffolds are commonly utilized in FTE of ligaments and tendons (Aragona et al., 1983; Dunn et al., 1992; Badylak et al., 1999; Bourke et al., 2004). The major advantage of using synthetic polymers as scaffolds is their ease of fabrication and reproducibility. A structure can be created that mimics the structure of a ligament or tendon, and appropriate proteins (e.g. growth factors) can be incorporated into the scaffold during manufacturing. However, performance of current synthetic grafts *in vivo* have been disappointing (Bellincampi et al., 1998; Guidoin et al., 2000); thus, current work has focused on the use of synthetic scaffolds seeded with fibroblasts (Cao et al., 1994; Lin et al., 1999). Alternatively, a number of naturally occurring biological scaffolds, including bovine pericardium (Integra Life Science), human dermal collagen (Alloderm), and porcine small intestinal submucosa (SIS), have been used. The SIS has shown most promising results in enhancing healing of both ligaments and tendons (Badylak et al., 1995, 1999; Dejardin et al., 1999, 2001; Musahl et al., 2004; Liang et al., 2006), as it possesses a structural hierarchy that is naturally arranged, and it is mostly composed of collagen type I. Further, 40% of the SIS degrades within 1 month *in vivo* (Record et al., 2001) and its by-products have been shown to be chemo-attractants for cells (including BMDCs) (Badylak et al., 2001; Li et al., 2004; Zantop et al., 2005). Moreover, it contains many bioactive agents (growth factors, fibronectin, and so on) (Voytik-Harbin et al., 1997; McPherson et al., 2000; Hodde et al., 2002) and causes a limited inflammatory reaction (Allman et al., 2002).

Based on these positive findings, our research center conducted multidisciplinary studies to determine the effect of SIS treatment on MCL in the short and long-term (12 weeks and 26 weeks). The mechanistic hypothesis is that SIS would act as a guidance of neoligament tissue formation, limiting the cross-sectional growth of the healing tissue, thereby increasing the mechanical demand. The tissue would respond by decreasing the production of collagen type V, which in turn would lead to an increase in collagen fibril diameters and improved collagen orientation in the healing ligament. Finally, these changes would result in an improvement in the mechanical properties of SIS-treated MCLs (Fig. 54.5).

It was observed that the histomorphological appearance and biochemical composition of the healing ligament were indeed closer to the normal ligament when compared to the non-treated ligament. Biochemical analysis revealed that the collagen type V/I ratio decreased with SIS treatment, while TEM (transmission electron microscopy) showed a heterogeneous distribution of large collagen fibril diameters (Fig. 54.6), and ultimately mechanical properties of the healing MCL were improved. Most importantly, the effects of SIS treatment persisted even up to 26 weeks (Liang et al., 2006). In addition, the CSA in the SIS-treated group decreased by 28% and the tangent modulus increased by 33% compared to the non-treated

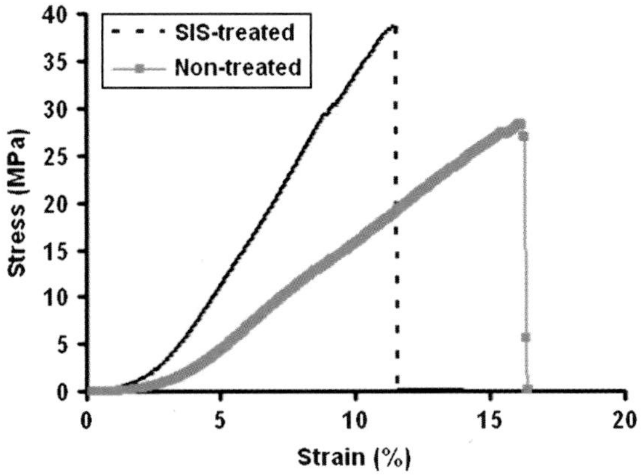

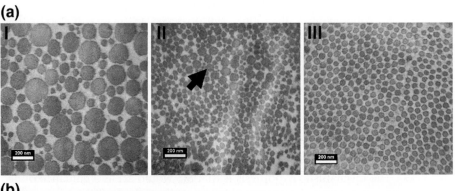

FIGURE 54.5

Typical stress—strain curves for SIS-treated and non-treated groups at 12 weeks post-injury. *(Permission obtained from Liang et al., 2006.)*

group. Also, the stress at failure was 49% higher than in the non-treated group. These findings demonstrate that a layer of SIS can act as a guide for neoligament formation by inducing better organization and limiting cross-sectional growth of the healing ligament. This, in turn, requires the mechanical properties of the SIS-treated ligament to improve through the formation of larger collagen fibrils resulting from a lower collagen type V/I ratio.

We have extended the use of SIS on PT healing after the central third was harvested for ACL reconstruction (Fig. 54.7). Since SIS has a preferred collagen alignment (Sacks and Gloeckner 1999), it has the potential for contact guidance and for promoting cells to produce a newly more aligned deposited matrix (Brunette, 1986; Clark et al., 1990; Chen et al., 1998;

(a)

I II III

(b)

I II

FIGURE 54.6

Transmission electron micrographs (×70,000) of collagen fibrils in (a) sham-operated MCL (I), SIS-treated MCL (II), and non-treated MCL (III) at 26 weeks post-injury. The arrow indicates the appearance of large fibrils between cells in the SIS-treated MCL. (b) The TEM appearance of both large and small fibrils (heterogeneity) in the pericellular area in the SIS-treated MCL (I) and non-treated MCL (II). The arrow indicates the large fibrils surrounding a cell process. F indicates a fibroblast.
(Permission obtained from Liang et al., 2006.)

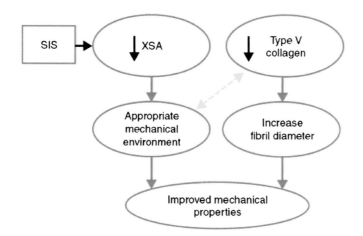

FIGURE 54.7
Schematic of possible mechanism of action of SIS on MCL healing response.

Walboomers et al., 1999), and as a result, a concomitant set of improved mechanical and viscoelastic properties. Additionally, the chemoattractant degradation products (Li et al., 2004) and bioactive agents of SIS could enhance the rate of healing. Finally, the SIS scaffold could also form a barrier between the healing PT and the underlying fat pad, limiting adhesion formation to permit motion between them to take place. The maintenance of stress and motion would help the homeostasis of the remaining PT (Woo et al., 1981a, 1982). Together, these effects will limit problems associated with poor healing, such as excessive hypertrophy of the remaining PT, and limiting the deterioration of its mechanical properties.

In a preliminary study, the effects of SIS on healing of a central third defect (3 mm width) of the rabbit PT were investigated. By 12 weeks, the SIS-treated group contained a large number of spindle-shaped cells with more organized collagen matrix, while the non-treated group had a sparse distribution of cells with only patches of collagen. After the healing PT tissue was dissected, the CSA was 61% greater in the SIS-treated group compared to those in the non-treated group (5.0 ± 2.0 mm^2 vs. 3.1 ± 1.2 mm^2, respectively). SIS treatment also showed higher stiffness (33.9 ± 14.0 N/mm vs. 24.3 ± 14.9 N/mm, or 38%) and ultimate load (67.7 ± 25.8 N vs. 43.8 ± 27.4 N, or 58%) compared to non-treatment. This study demonstrated that SIS treatment shows the potential to increase the quantity of healing PT tissue and structural properties of the healing central BPTB complex after a surgically created central third PT defect. Thus, the results of morphology, histology, and structural properties are very encouraging for further investigation of this application.

The success of this approach had prompted us to combine this SIS-bioscaffold with the SIS in a hydrogel form to heal a surgically transected ACL following primary repair using a goat model. It is well known that primary repair of ACL alone does not mount a good healing response after injury. With the new treatment, we have found ACL healing with continuous neotissue at 12 weeks. Its CSA and shape were similar to those of the sham-operated ACL. Morphologically, its collagen fibers were aligned with spindle-shaped fibroblasts. Functionally, the SIS-treated ACL could help to reduce the anterior−posterior knee instability while its *in situ* forces were similar to the control ACL. Further more, the structural properties of the FATC when tested in uniaxial tension showed that tensile stiffness could reach approximately 50% of the normal FATC, which was comparable to the results found for ACL reconstruction. It is also important to note that we were using the SIS from genetically modified αGal-deficient pigs. With the use of alpha αGal(−) SIS, it would significantly reduce the potential immunological reaction in humans when porcine SIS is used in clinical applications.

Recent studies have also shown the feasibility of enhancing ACL graft integration following reconstruction using a triphasic scaffold (Spalazzi et al., 2006). Fibroblast and osteoblasts were seeded on a section of the scaffold that mimicked the native environment of the ligament insertion and bone, respectively. It was found that the specific cell types populated and thrived in their respective phase and also each migrated into the middle phase, while each cell type expressed the appropriate type of genes for its particular matrix. Collectively these approaches demonstrate that scaffolds have many potential applications for improving the treatment of injured ligaments and tendons.

Mechanical factors

Progress in FTE has also included the elucidation of the importance of mechanical stimuli on cells and on tissue development and remodeling (Huang et al., 1993; Banes et al., 1995; Eastwood et al., 1998; Hsieh et al., 2000; Altman et al., 2002; Wang et al., 2003b). Cyclic stretching of cells from ligaments and tendons *in vitro* has been shown to cause increases in collagen synthesis (Desrosiers et al., 1995; Hsieh et al., 2000) and changes in intra-cellular processes (i.e. different regulation of metabolic and inflammatory genes and calcium signaling) (Banes et al., 1995, 1999; Archambault et al., 2002; Ralphs et al., 2002; Wang et al., 2003b; Yang et al., 2004).

Interestingly, fibroblast alignment can be a result of the external mechanical environment. When fibroblasts are grown in microgrooved silicone surfaces instead of smooth culture surfaces, they become elongated and aligned within the microgrooves through contact guidance. Most importantly, the extracellular matrix that the cells produce is also aligned along the microgroove direction (Fig. 54.8) (Wang et al., 2003a).

Cells also align along the direction of the maximum principal strain in collagen gels (Eastwood et al., 1998). Similarly, *in vitro* studies have shown that multidimensional mechanical strains applied to BMDCs embedded in a collagen gel upregulated the gene expression of collagen types I and III and tenascin-C, which are typically expressed in fibroblasts (Altman et al., 2002).

Recently, our research center developed a uni-axial stretching system to study the effects of the mechanical stimuli on cells seeded on the bioscaffold (SIS). In order to first understand the effect of elongating the SIS on the alignment of cells, the scaffold was stretched for 24 h to 15% of its original length and was then seeded with cells. After 5 days, collagen fibers of the scaffold were more aligned compared to non-stretched controls. In addition, the cells seeded on the scaffold demonstrated a preferred alignment along the direction of stretch. Preliminary results have also been obtained for continuous cyclic stretching (15% at 1 Hz for 4 h per day for

1012

FIGURE 54.8
Randomly aligned cells cultured on a smooth dish (upper left). Aligned cells cultured on a dish etched with microgroove (upper right). Randomly aligned matrix produced by cells cultured on a smooth dish (lower left). Aligned matrix produced by cells cultured on a dish etched with microgrooves (lower right). *(Permission obtained from Wang et al., 2003a.)*

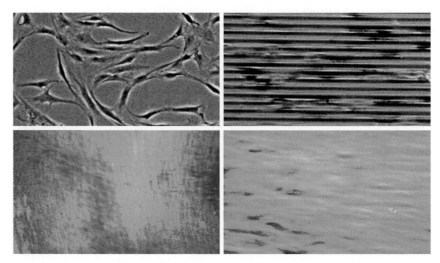

5 days) of SIS. Again, it was observed that the collagen fiber organization of the SIS improved to a more aligned state when the SIS was both seeded and stretched, and the cells were also aligned along the stretching direction. Gene expression analysis is underway to determine the differences in matrix protein expression between the cyclic group and the constant elongated group. Since tendon and ligament fibroblasts are aligned with collagen fibrils *in vivo*, it is hoped that mechanical stimuli can align cells within the scaffold and produce a better organized collagen matrix that may further enhance the healing response when implanted *in vivo*.

SUMMARY AND FUTURE DIRECTIONS

In this review, the biomechanical and biological problems facing healing and repair of ligament and tendon injuries were discussed. There have already been tremendous improvements to clinical treatment paradigms based on studies that have established a fundamental understanding of healing following ligament or tendon injury and the benefits of controlled mobilization. Nevertheless, many issues still remain.

For ligaments and tendons that display a good healing potential after injury, the major challenges are recovering their normal ultrastructural appearance, biochemical composition, and mechanical properties. Specifically, increasing the fibril diameters of healing tissues by limiting the production of type V collagen and decorin as well as improving the alignment of healing tissue by guiding the organization of newly produced matrix are important steps to be taken. By manipulating the healing response at the molecular and cellular level and guiding tissue formation, FTE approaches may offer the potential to restore the properties of healing tissue to normal levels.

We are particularly interested in bioscaffolds, such as the porcine SIS. When applied to a healing ligament or tendon *in vivo*, it serves as a substrate that provides contact guidance for cells to form more aligned collagen fibers with a concomitant improvement in mechanical and viscoelastic properties when compared to non-treated controls. Further, the chemoattractant degradation products and bioactive agents of SIS could enhance the rate of healing (Li et al., 2004), allowing better maintenance of stress and motion-dependent homeostasis. More excitingly, the SIS can be modified *in vitro* by seeding BMDCs on the scaffold and applying cyclic stretching in order to increase its alignment. Hence, when applied *in vivo*, the tissue-engineered scaffold could serve to accelerate the initiation of the healing process by improving the production and orientation of collagen that ultimately will help to make a better neo-ligament or tendon.

On the other hand, for ligaments and tendons that do not heal following injury and require surgical reconstruction using replacement grafts (e.g. ACL reconstruction), the major challenge is to promote a remodeling response such that the graft maintains sufficient stiffness and strength to provide functional stability of the joint. Most importantly, enhancing the rate of integration of tendon–bone interfaces during early graft incorporation may permit an earlier and more aggressive post-operative rehabilitation (Chen et al., 2002). These complex issues may require a combination of approaches including gene and cell therapies as well as biological scaffolds. Indeed, treatment of grafts with adenoviral-BMP-2-vector (AdBMP-2) has shown some potential (Martinek et al., 2002) in both canine and rabbit models. Additionally, other biological tissues such as periosteum have been used to enhance the interface between tendon and bone with some success (Chen et al., 2002). All these results suggest an exciting potential for clinical application.

Indeed, FTE has generated many exciting developments. For example, there is a class of biodegradable metallic scaffolds, namely porous magnesium or magnesium oxide, that have the advantage of initial stiffness to provide the needed stability for the ligament to heal while performing its function. The degradation rate of these "smart" scaffolds could also be controlled as they are replaced by the neotissue. Further more, protein coating of these

1013

biodegradable metallic scaffolds could also be done for better tissue integration and for control release of growth factors and cytokines to sustain tissue healing as well as to guide tissue regeneration.

To translate the knowledge gained about a particular gene, protein, or cell to a clinical application will require experts from many disciplines to work in a seamless fashion. One of the roles of biomedical engineers within this framework would be to help link interactions of the functions of molecules to cells, cells to tissues, tissues to organs, and organs to the body. When biologists, biomedical engineers, clinicians, and experts from other disciplines work together this will result in better therapies, leading to the injured ligaments and tendons healing with properties closer to those of normal ligaments and tendons. Efforts of such a team-based approach on the new developments of FTE will bring a bright future to the outcome of healing of ligament and tendon injuries.

ACKNOWLEDGMENTS

Financial support was provided by the National Institutes of Health grants AR41820 and AR39683.

References

Abramowitch, S. D., Papageorgiou, C. D., Debski, R. E., Clineff, T. D., & Woo, S. L-Y. (2003a). A biomechanical and histological evaluation of the structure and function of the healing medial collateral ligament in a goat model. *Knee Surg. Sports Traumatol. Arthrosc., 11*, 155–162.

Abramowitch, S. D., Papageorgiou, C. D., Withrow, J. D., Gilbert, T. W., & Woo, S. L-Y. (2003b). The effect of initial graft tension on the biomechanical properties of a healing ACL replacement graft: a study in goats. *J. Orthop. Res., 21*, 708–715.

Abramowitch, S. D., Yagi, M., Tsuda, E., & Woo, S. L-Y (2003c). The healing medial collateral ligament following a combined anterior cruciate and medial collateral ligament injury – a biomechanical study in a goat model. *J. Orthop. Res., 21*, 30–1124.

Adachi, E., & Hayashi, T. (1986). *In vitro* formation of hybrid fibrils of type V collagen and type I collagen. Limited growth of type I collagen into thick fibrils by type V collagen. *Connect. Tissue Res., 14*, 257–266.

Aglietti, P., Buzzi, R., D'Andria, S., & Zaccherotti, G. (1992). Arthroscopic anterior cruciate ligament reconstruction with patellar tendon. *Arthroscopy, 8*, 510–516.

Allen, C. R., Livesay, G. A., Wong, E. K., & Woo, S. L-Y (1999). Injury and reconstruction of the anterior cruciate ligament and knee osteoarthritis. *Osteoarthritis Cartilage, 7*, 110–121.

Allman, A. J., McPherson, T. B., Merrill, L. C., Badylak, S. F., & Metzger, D. W. (2002). The Th2-restricted immune response to xenogeneic small intestinal submucosa does not influence systemic protective immunity to viral and bacterial pathogens. *Tissue Eng., 8*, 53–62.

Altman, G. H., Horan, R. L., Martin, I., Farhadi, J., Stark, P. R., Volloch, V., et al. (2002). Cell differentiation by mechanical stress. *FASEB J., 16*, 270–272.

Amiel, D., Frank, C., Harwood, F., Fronek, J., & Akeson, W. (1984). Tendons and ligaments: a morphological and biochemical comparison. *J. Orthop. Res., 1*, 257–265.

Aragona, J., Parsons, J. R., Alexander, H., & Weiss, A. B. (1983). Medial collateral ligament replacement with a partially absorbable tissue scaffold. *Am. J. Sports Med., 11*, 228–233.

Archambault, J., Tsuzaki, M., Herzog, W., & Banes, A. J. (2002). Stretch and interleukin-1beta induce matrix metalloproteinases in rabbit tendon cells *in vitro*. *J. Orthop. Res., 20*, 36–39.

Arnoczky, S. P., Tarvin, G. B., & Marshall, J. L. (1982). Anterior cruciate ligament replacement using patellar tendon. An evaluation of graft revascularization in the dog. *J. Bone Joint Surg. Am., 64*, 217–224.

Aune, A. K., Holm, I., Risberg, M. A., Jensen, H. K., & Steen, H. (2001). Four-strand hamstring tendon autograft compared with patellar tendon-bone autograft for anterior cruciate ligament reconstruction. A randomized study with two-year follow-up. *Am. J. Sports Med., 29*, 722–728.

Awad, H. A., Butler, D. L., Boivin, G. P., Smith, F. N., Malaviya, P., Huibregtse, B., et al. (1999). Autologous mesenchymal stem cell-mediated repair of tendon. *Tissue Eng., 5*, 267–277.

Awad, H. A., Boivin, G. P., Dressler, M. R., Smith, F. N., Young, R. G., & Butler, D. L. (2003). Repair of patellar tendon injuries using a cell-collagen composite. *J. Orthop. Res., 21*, 420–431.

Badiavas, E. V., Abedi, M., Butmarc, J., Falanga, V., & Quesenberry, P. (2003). Participation of bone marrow derived cells in cutaneous wound healing. *J. Cell Physiol., 196,* 245–250.

Badylak, S. F., Tullius, R., Kokini, K., Shelbourne, K. D., Klootwyk, T., Voytik, S. L., et al. (1995). The use of xenogeneic small intestinal submucosa as a biomaterial for Achilles tendon repair in a dog model. *J. Biomed. Mater. Res., 29,* 977–985.

Badylak, S., Arnoczky, S., Plouhar, P., Haut, R., Mendenhall, V., Clarke, R., et al. (1999). Naturally occurring extracellular matrix as a scaffold for musculoskeletal repair. *Clin. Orthop. Rel. Res.* S333–S343.

Badylak, S. F., Park, K., Peppas, N., McCabe, G., & Yoder, M. (2001). Marrow-derived cells populate scaffolds composed of xenogeneic extracellular matrix. *Exp. Hematol., 29,* 1310–1318.

Ballock, R. T., Woo, S. L-Y., Lyon, R. M., Hollis, J. M., & Akeson, W. H. (1989). Use of patellar tendon autograft for anterior cruciate ligament reconstruction in the rabbit: a long-term histologic and biomechanical study. *J. Orthop. Res., 7,* 474–485.

Banes, A. J., Tsuzaki, M., Hu, P., Brigman, B., Brown, T., Almekinders, L., et al. (1995). PDGF-BB, IGF-I and mechanical load stimulate DNA synthesis in avian tendon fibroblasts. *in vitro. J. Biomech., 28,* 1505–1513.

Banes, A. J., Weinhold, P., Yang, X., Tsuzaki, M., Bynum, D., Bottlang, M., et al. (1999). Gap junctions regulate responses of tendon cells ex vivo to mechanical loading. *Clin. Orthop. Rel. Res.* S356–S370.

Batten, M. L., Hansen, J. C., & Dahners, L. E. (1996). Influence of dosage and timing of application of platelet-derived growth factor on early healing of the rat medial collateral ligament. *J. Orthop. Res., 14,* 736–741.

Beaty, J. (1999). Knee and leg: soft tissue trauma. In E. A. Arendt (Ed.), *OKU Orthopaedic Knowledge Update.* Rosemont, IL: American Academy of Orthopaedic Surgeons. pp. xix, 442.

Bellincampi, L. D., Closkey, R. F., Prasad, R., Zawadsky, J. P., & Dunn, M. G. (1998). Viability of fibroblast-seeded ligament analogs after autogenous implantation. *J. Orthop. Res., 16,* 414–420.

Beynnon, B. D., Proffer, D., Drez, D. J., Jr., Stankewich, C. J., & Johnson, R. J. (1995). Biomechanical assessment of the healing response of the rabbit patellar tendon after removal of its central third. *Am. J. Sports Med., 23,* 452–457.

Birk, D. E. (2001). Type V collagen: heterotypic type I/V collagen interactions in the regulation of fibril assembly. *Micron, 32,* 223–237.

Birk, D. E., & Mayne, R. (1997). Localization of collagen types I, III and V during tendon development. Changes in collagen types I and III are correlated with changes in fibril diameter. *Eur. J. Cell Biol., 72,* 352–361.

Birk, D. E., & Trelstad, R. L. (1984). Extracellular compartments in matrix morphogenesis: collagen fibril, bundle, and lamellar formation by corneal fibroblasts. *J. Cell Biol., 99,* 2024–2033.

Birk, D. E., Fitch, J. M., Babiarz, J. P., Doane, K. J., & Linsenmayer, T. F. (1990). Collagen fibrillogenesis *in vitro*: interaction of types I and V collagen regulates fibril diameter. *J. Cell Sci., 95,* 649–657.

Bourke, S. L., Kohn, J., & Dunn, M. G. (2004). Preliminary development of a novel resorbable synthetic polymer fiber scaffold for anterior cruciate ligament reconstruction. *Tissue Eng., 10,* 43–52.

Bray, R. C., Rangayyan, R. M., & Frank, C. B. (1996). Normal and healing ligament vascularity: a quantitative histological assessment in the adult rabbit medial collateral ligament. *J. Anat., 188,* 87–95.

Breitfuss, H., Frohlich, R., Povacz, P., Resch, H., & Wicker, A. (1996). The tendon defect after anterior cruciate ligament reconstruction using the midthird patellar tendon – a problem for the patellofemoral joint? *Knee Surg. Sports Traumatol. Arthrosc., 3,* 194–198.

Brunette, D. M. (1986). Fibroblasts on micromachined substrata orient hierarchically to grooves of different dimensions. *Exp. Cell Res., 164,* 11–26.

Butler, D. L., Grood, E. S., Noyes, F. R., Olmstead, M. L., Hohn, R. B., Arnoczky, S. P., et al. (1989). Mechanical properties of primate vascularized vs. nonvascularized patellar tendon grafts; changes over time. *J. Orthop. Res., 7,* 68–79.

Butler, D. L., Juncosa-Melvin, N., Shearn, J., Galloway, M., Boivin, G., & Gooch, C. (2005). Evaluation of an MSC-based tissue engineered construct to improve patellar tendon repair. *Summer Bioengineering Conference, Vail, CO.*

Cao, Y., Vacanti, J. P., Ma, X., Paige, K. T., Upton, J., Chowanski, Z., et al. (1994). Generation of neo-tendon using synthetic polymers seeded with tenocytes. *Transplant. Proc., 26,* 3390–3392.

Caplan, A. I., Reuben, D., & Haynesworth, S. E. (1998). Cell-based tissue engineering therapies: the influence of whole body physiology. *Adv. Drug Deliv. Rev., 33,* 3–14.

Cerullo, G., Puddu, G., Gianni, E., Damiani, A., & Pigozzi, F. (1995). Anterior cruciate ligament patellar tendon reconstruction: it is probably better to leave the tendon defect open! *Knee Surg. Sports Traumatol. Arthrosc., 3,* 14–17.

Chen, C. S., Mrksich, M., Huang, S., Whitesides, G. M., & Ingber, D. E. (1998). Micropatterned surfaces for control of cell shape, position, and function. *Biotechnol. Prog., 14,* 356–363.

Chen, C. H., Chen, W. J., & Shih, C. H. (2002). Enveloping of periosteum on the hamstring tendon graft in anterior cruciate ligament reconstruction. *Arthroscopy, 18,* 27E.

Clancy, W. G., Jr., Narechania, R. G., Rosenberg, T. D., Gmeiner, J. G., Wisnefske, D. D., & Lange, T. A. (1981). Anterior and posterior cruciate ligament reconstruction in rhesus monkeys. *J. Bone Joint Surg., 63A*, 1270–1284.

Clark, P., Connolly, P., Curtis, A. S., Dow, J. A., & Wilkinson, C. D. (1990). Topographical control of cell behaviour: II. Multiple grooved substrata. *Development, 108*, 635–644.

Clatworthy, M. G., Annear, P., Bulow, J. U., & Bartlett, R. J. (1999). Tunnel widening in anterior cruciate ligament reconstruction: a prospective evaluation of hamstring and patella tendon grafts. *Knee Surg. Sports Traumatol. Arthrosc., 7*, 138–145.

Clayton, M. L., Miles, J. S., & Abdulla, M. (1968). Experimental investigations of ligamentous healing. *Clin. Orthop. Rel. Res., 61*, 146–153.

Cooper, D. E., Deng, X. H., Burstein, A. L., & Warren, R. F. (1993). The strength of the central third patellar tendon graft. A biomechanical study. *Am. J. Sports Med., 21*, 818–823, discussion 823–824.

Cooper, R. R., & Misol, S. (1970). Tendon and ligament insertion. A light and electron microscopic study. *J. Bone Joint Surg., 52A*, 1–20.

Coupens, S. D., Yates, C. K., Sheldon, C., & Ward, C. (1992). Magnetic resonance imaging evaluation of the patellar tendon after use of its central one-third for anterior cruciate ligament reconstruction. *Am. J. Sports Med., 20*, 325–332.

Day, C. S., Kasemkijwattana, C., Menetrey, J., Floyd, S. S., Jr., Booth, D., Moreland, M. S., et al. (1997). Myoblast-mediated gene transfer to the joint. *J. Orthop. Res., 15*, 894–903.

Deie, M., Marui, T., Allen, C. R., Hildebrand, K. A., Georgescu, H. I., Niyibizi, C., et al. (1997). The effects of age on rabbit MCL fibroblast matrix synthesis in response to TGF-beta 1 or EGF. *Mech. Ageing Dev., 97*, 121–130.

Dejardin, L. M., Arnoczky, S. P., & Clarke, R. B. (1999). Use of small intestinal submucosal implants for regeneration of large fascial defects: an experimental study in dogs. *J. Biomed. Mater. Res., 46*, 203–211.

Dejardin, L. M., Arnoczky, S. P., Ewers, B. J., Haut, R. C., & Clarke, R. B. (2001). Tissue-engineered rotator cuff tendon using porcine small intestine submucosa. Histologic and mechanical evaluation in dogs. *Am. J. Sports Med., 29*, 175–184.

Desrosiers, E. A., Methot, S., Yahia, L., & Rivard, C. H. (1995). [Responses of ligamentous fibroblasts to mechanical stimulation]. *Ann. Chir., 49*, 768–774.

Doillon, C. J., Dunn, M. G., Bender, E., & Silver, F. H. (1985). Collagen fiber formation in repair tissue: development of strength and toughness. *Coll. Rel. Res., 5*, 481–492.

Duffy, F. J., Jr., Seiler, J. G., Gelberman, R. H., & Hergrueter, C. A. (1995). Growth factors and canine flexor tendon healing: initial studies in uninjured and repair models. *J Hand Surg., 20A*, 645–649.

Dunn, M. G., Tria, A. J., Kato, Y. P., Bechler, J. R., Ochner, R. S., Zawadsky, J. P., et al. (1992). Anterior cruciate ligament reconstruction using a composite collagenous prosthesis. A biomechanical and histologic study in rabbits. *Am. J. Sports Med., 20*, 507–515.

Dyer, R. F., & Enna, C. D. (1976). Ultrastructural features of adult human tendon. *Cell Tissue Res., 168*, 247–259.

Eastwood, M., Mudera, V. C., McGrouther, D. A., & Brown, R. A. (1998). Effect of precise mechanical loading on fibroblast populated collagen lattices: morphological changes. *Cell Motil. Cytoskeleton, 40*, 13–21.

Eyden, B., & Tzaphlidou, M. (2001). Structural variations of collagen in normal and pathological tissues: role of electron microscopy. *Micron, 32*, 287–300.

Feller, J. A., & Webster, K. E. (2003). A randomized comparison of patellar tendon and hamstring tendon anterior cruciate ligament reconstruction. *Am. J. Sports Med., 31*, 564–573.

Frank, C. B., & Jackson, D. W. (1997). The science of reconstruction of the anterior cruciate ligament. *J. Bone Joint Surg., 79A*, 1556–1576.

Frank, C., Woo, S. L-Y., Amiel, D., Harwood, F., Gomez, M., & Akeson, W. (1983). Medial collateral ligament healing. A multidisciplinary assessment in rabbits. *Am. J. Sports Med., 11*, 379–389.

Frank, C., McDonald, D., Bray, D., Bray, R., Rangayyan, R., Chimich, D., et al. (1992). Collagen fibril diameters in the healing adult rabbit medial collateral ligament. *Connect. Tiss. Res., 27*, 251–263.

Frank, C. B., Bray, R. C., Hart, D. A., Shirve, N. G., Loitz, B. J., Matyas, J. R., et al. (1994). Soft tissue healing. In F. H. Fu, C. D. Harner, & G. Vince Kelly (Eds.), *Knee Surgery* (pp. 189–229). Baltimore, MD: Williams & Wilkins.

Frank, C., McDonald, D., & Shrive, N. (1997). Collagen fibril diameters in the rabbit medial collateral ligament scar: a longer term assessment. *Connect. Tissue Res., 36*, 261–269.

Freedman, K. B., D'Amato, M. J., Nedeff, D. D., Kaz, A., & Bach, B. R., Jr. (2003). Arthroscopic anterior cruciate ligament reconstruction: a metaanalysis comparing patellar tendon and hamstring tendon autografts. *Am. J. Sports Med., 31*, 2–11.

Fujie, H., Mabuchi, K., Woo, S. L., Livesay, G. A., Arai, S., & Tsukamoto, Y. (1993). The use of robotics technology to study human joint kinematics: a new methodology. *J. Biomech. Eng., 115*, 211–217.

Fujie, H., Livesay, G. A., Woo, S. L., Kashiwaguchi, S., & Blomstrom, G. (1995). The use of a universal force-moment sensor to determine in-situ forces in ligaments: a new methodology. *J. Biomech. Eng., 117*, 1–7.

Fukuta, S., Oyama, M., Kavalkovich, K., Fu, F. H., & Niyibizi, C. (1998). Identification of types II, IX and X collagens at the insertion site of the bovine achilles tendon. *Matrix Biol., 17*, 65–73.

Fung, Y. C., Perrone, N., Anliker, M., University of California San Diego. and United States. Office of Naval Research. (1972). *Biomechanics, Its Foundations and Objectives*. Englewood Cliffs, NJ: Prentice-Hall.

Galiano, R. D., Tepper, O. M., Pelo, C. R., Bhatt, K. A., Callaghan, M., Bastidas, N., et al. (2004). Topical vascular endothelial growth factor accelerates diabetic wound healing through increased angiogenesis and by mobilizing and recruiting bone marrow-derived cells. *Am. J. Pathol., 164*, 1935–1947.

Gerich, T. G., Kang, R., Fu, F. H., Robbins, P. D., & Evans, C. H. (1996). Gene transfer to the rabbit patellar tendon: potential for genetic enhancement of tendon and ligament healing. *Gene Ther., 3*, 1089–1093.

Goh, J. C., Ouyang, H. W., Teoh, S. H., Chan, C. K., & Lee, E. H. (2003). Tissue-engineering approach to the repair and regeneration of tendons and ligaments. *Tissue Eng., 9*(Suppl. 1), S31–S44.

Guidoin, M. F., Marois, Y., Bejui, J., Poddevin, N., King, M. W., & Guidoin, R. (2000). Analysis of retrieved polymer fiber based replacements for the ACL. *Biomaterials, 21*, 2461–2474.

Hart, D. A., Nakamura, N., Marchuk, L., Hiraoka, H., Boorman, R., Kaneda, Y., et al. (2000). Complexity of determining cause and effect *in vivo* after antisense gene therapy. *Clin. Orthop., 379*, S242–S251.

Hart, R. A., Woo, S. L-Y., & Newton, P. O. (1992). Ultrastructural morphometry of anterior cruciate and medial collateral ligaments: an experimental study in rabbits. *J. Orthop. Res., 10*, 96–103.

Hildebrand, K. A., & Frank, C. B. (1998). Scar formation and ligament healing. *Can. J. Surg., 41*, 425–429.

Hildebrand, K. A., Deie, M., Allen, C. R., Smith, D. W., Georgescu, H. I., Evans, C. H., et al. (1999). Early expression of marker genes in the rabbit medial collateral and anterior cruciate ligaments: the use of different viral vectors and the effects of injury. *J. Orthop. Res., 17*, 37–42.

Hildebrand, K. A., Frank, C. B., & Hart, D. A. (2004). Gene intervention in ligament and tendon: current status, challenges, future directions. *Gene Ther., 11*, 368–378.

Hodde, J., Record, R., Tullius, R., & Badylak, S. (2002). Fibronectin peptides mediate HMEC adhesion to porcine-derived extracellular matrix. *Biomaterials, 23*, 1841–1848.

Hsieh, A. H., Tsai, C. M., Ma, Q. J., Lin, T., Banes, A. J., Villarreal, F. J., et al. (2000). Time-dependent increases in type-III collagen gene expression in medical collateral ligament fibroblasts under cyclic strains. *J. Orthop. Res., 18*, 220–227.

Huang, D., Chang, T. R., Aggarwal, A., Lee, R. C., & Ehrlich, H. P. (1993). Mechanisms and dynamics of mechanical strengthening in ligament-equivalent fibroblast-populated collagen matrices. *Ann. Biomed. Eng., 21*, 289–305.

Indelicato, P. A. (1983). Non-operative treatment of complete tears of the medial collateral ligament of the knee. *J. Bone Joint Surg., 65A*, 323–329.

Indelicato, P. (1995). Isolated medial collateral ligament injuries in the knee. *J. Am. Acad. Orthop. Surg., 3*, 9–14.

Inoue, M., McGurk-Burleson, E., Hollis, J. M., & Woo, S. L-Y. (1987). Treatment of the medial collateral ligament injury. I: The importance of anterior cruciate ligament on the varus-valgus knee laxity. *Am. J. Sports Med., 15*, 15–21.

Inoue, M., Woo, S. L-Y., Gomez, M. A., Amiel, D., Ohland, K. J., & Kitabayashi, L. R. (1990). Effects of surgical treatment and immobilization on the healing of the medial collateral ligament: a long-term multidisciplinary study. *Connect. Tissue Res., 25*, 13–26.

Jansson, K. A., Harilainen, A., Sandelin, J., Karjalainen, P. T., Aronen, H. J., & Tallroth, K. (1999). Bone tunnel enlargement after anterior cruciate ligament reconstruction with the hamstring autograft and endobutton fixation technique. A clinical, radiographic and magnetic resonance imaging study with 2 years follow-up. *Knee Surg. Sports Traumatol. Arthrosc., 7*, 290–295.

Johnson, G. A., Tramaglini, D. M., Levine, R. E., Ohno, K., Choi, N. Y., & Woo, S. L-Y. (1994). Tensile and visco-elastic properties of human patellar tendon. *J. Orthop. Res., 12*, 796–803.

Jokl, P., Kaplan, N., Stovell, P., & Keggi, K. (1984). Non-operative treatment of severe injuries to the medial and anterior cruciate ligaments of the knee. *J. Bone Joint Surg., 66A*, 741–744.

Jones, K. G. (1970). Reconstruction of the anterior cruciate ligament using the central one-third of the patellar ligament. *J. Bone Joint Surg., 52A*, 838–839.

Juncosa-Melvin, N., Boivin, G. P., Galloway, M. T., Gooch, C., West, J. R., Sklenka, A. M., et al. (2005). Effects of cell-to-collagen ratio in mesenchymal stem cell-seeded implants on tendon repair biomechanics and histology. *Tissue Eng., 11*, 448–457.

Kamps, B. S., Linder, L. H., DeCamp, C. E., & Haut, R. C. (1994). The influence of immobilization versus exercise on scar formation in the rabbit patellar tendon after excision of the central third. *Am. J. Sports Med., 22*, 803–811.

Kannus, P. (1988). Long-term results of conservatively treated medial collateral ligament injuries of the knee joint. *Clin. Orthop. Rel. Res., 103*–112.

Kartus, J., Magnusson, L., Stener, S., Brandsson, S., Eriksson, B. I., & Karlsson, J. (1999). Complications following arthroscopic anterior cruciate ligament reconstruction. A 2–5-year follow-up of 604 patients with special emphasis on anterior knee pain. *Knee Surg. Sports Traumatol. Arthrosc., 7*, 2–8.

Krall, W. J., Challita, P. M., Perlmutter, L. S., Skelton, D. C., & Kohn, D. B. (1994). Cells expressing human glucocerebrosidase from a retroviral vector repopulate macrophages and central nervous system microglia after murine bone marrow transplantation. *Blood, 83*, 2737–2748.

Kurosaka, M., Yoshiya, S., & Andrish, J. T. (1987). A biomechanical comparison of different surgical techniques of graft fixation in anterior cruciate ligament reconstruction. *Am. J. Sports Med., 15*, 225–229.

Lambert, K. L. (1983). Vascularized patellar tendon graft with rigid internal fixation for anterior cruciate ligament insufficiency. *Clin. Orthop. Rel. Res.,* 85–89.

Lee, T. Q., & Woo, S. L-Y. (1988). A new method for determining cross-sectional shape and area of soft tissues. *J. Biomech. Eng., 110*, 110–114.

Li, F., Li, W., Johnson, S., Ingram, D., Yoder, M., & Badylak, S. (2004). Low-molecular-weight peptides derived from extracellular matrix as chemoattractants for primary endothelial cells. *Endothelium, 11*, 199–206.

Liang, R., Woo, S. L-Y., Takakura, Y., Moon, D. K., Jia, F. Y., & Abramowitch, S. (2006). Long-term effects of porcine small intestine submucosa on the healing of medial collateral ligament: a functional tissue engineering study. *J. Orthop. Res., 24*, 811–819.

Lin, V. S., Lee, M. C., O'Neal, S., McKean, J., & Sung, K. L. (1999). Ligament tissue engineering using synthetic biodegradable fiber scaffolds. *Tissue Eng., 5*, 443–452.

Linder, L. H., Sukin, D. L., Burks, R. T., & Haut, R. C. (1994). Biomechanical and histological properties of the canine patellar tendon after removal of its medial third. *Am. J. Sports Med., 22*, 136–142.

Linsenmayer, T. F., Gibney, E., Igoe, F., Gordon, M. K., Fitch, J. M., Fessler, L. I., et al. (1993). Type V collagen: molecular structure and fibrillar organization of the chicken alpha 1(V) NH$_2$-terminal domain, a putative regulator of corneal fibrillogenesis. *J. Cell Biol., 121*, 1181–1189.

Liu, S. H., Yang, R. S., al-Shaikh, R., & Lane, J. M. (1995). Collagen in tendon, ligament, and bone healing. A current review. *Clin. Orthop. Rel. Res.,* 265–278.

Livesay, G. A., Rudy, T. W., Woo, S. L-Y., Runco, T. J., Sakane, M., Li, G., et al. (1997). Evaluation of the effect of joint constraints on the *in situ* force distribution in the anterior cruciate ligament. *J. Orthop. Res., 15*, 278–284.

Lo, I. K., Ou, Y., Rattner, J. P., Hart, D. A., Marchuk, L. L., Frank, C. B., et al. (2002). The cellular networks of normal ovine medial collateral and anterior cruciate ligaments are not accurately recapitulated in scar tissue. *J. Anat., 200*, 283–296.

Loh, J. C., Fukuda, Y., Tsuda, E., Steadman, R. J., Fu, F. H., & Woo, S. L-Y. (2003). Knee stability and graft function following anterior cruciate ligament reconstruction: comparison between 11 o'clock and 10 o'clock femoral tunnel placement. 2002 Richard O'Connor Award paper. *Arthroscopy, 19*, 297–304.

Maffulli, N. (1999). Rupture of the Achilles tendon. *J. Bone Joint Surg., 81A*, 1019–1036.

Manske, P. R. (1988). Flexor tendon healing. *J. Hand Surg., 13B*, 237–245.

Marchant, J. K., Hahn, R. A., Linsenmayer, T. F., & Birk, D. E. (1996). Reduction of type V collagen using a dominant-negative strategy alters the regulation of fibrillogenesis and results in the loss of corneal-specific fibril morphology. *J. Cell Biol., 135*, 1415–1426.

Marder, R. A. (1991). Arthroscopic-assisted reconstruction of the anterior cruciate ligament. *West. J. Med., 155*, 172.

Martinek, V., Latterman, C., Usas, A., Abramowitch, S., Woo, S. L-Y., Fu, F. H., et al. (2002). Enhancement of tendon-bone integration of anterior cruciate ligament grafts with bone morphogenetic protein-2 gene transfer: a histological and biomechanical study. *J. Bone Joint Surg., 84A*, 1123–1131.

Marui, T., Niyibizi, C., Georgescu, H. I., Cao, M., Kavalkovich, K. W., Levine, R. E., et al. (1997). Effect of growth factors on matrix synthesis by ligament fibroblasts. *J. Orthop. Res., 15*, 18–23.

Mathews, V., Hanson, P. T., Ford, E., Fujita, J., Polonsky, K. S., & Graubert, T. A. (2004). Recruitment of bone marrow-derived endothelial cells to sites of pancreatic beta-cell injury. *Diabetes, 53*, 91–98.

Matyas, J. R., Anton, M. G., Shrive, N. G., & Frank, C. B. (1995). Stress governs tissue phenotype at the femoral insertion of the rabbit MCL. *J. Biomech., 28*, 147–157.

McPherson, T. B., Liang, H., Record, R. D., & Badylak, S. F. (2000). Galalpha(1,3)Gal epitope in porcine small intestinal submucosa. *Tissue Eng., 6*, 233–239.

Murphy, P. G., Loitz, B. J., Frank, C. B., & Hart, D. A. (1994). Influence of exogenous growth factors on the synthesis and secretion of collagen types I and III by explants of normal and healing rabbit ligaments. *Biochem. Cell Biol., 72*, 403–409.

Musahl, V., Abramowitch, S. D., Gilbert, T. W., Tsuda, E., Wang, J. H., Badylak, S. F., et al. (2004). The use of porcine small intestinal submucosa to enhance the healing of the medial collateral ligament — a functional tissue engineering study in rabbits. *J. Orthop. Res., 22*, 214–220.

Nakamura, N., Hart, D. A., Boorman, R. S., Kaneda, Y., Shrive, N. G., Marchuk, L. L., et al. (2000). Decorin antisense gene therapy improves functional healing of early rabbit ligament scar with enhanced collagen fibrillogenesis *in vivo. J. Orthop. Res., 18*, 517–523.

Nebelung, W., Becker, R., Merkel, M., & Ropke, M. (1998). Bone tunnel enlargement after anterior cruciate ligament reconstruction with semitendinosus tendon using Endobutton fixation on the femoral side. *Arthroscopy, 14,* 810–815.

Ng, G. Y., Oakes, B. W., Deacon, O. W., McLean, I. D., & Eyre, D. R. (1996). Long-term study of the biochemistry and biomechanics of anterior cruciate ligament-patellar tendon autografts in goats. *J. Orthop. Res., 14,* 851–856.

Nixon, R. G., SeGall, G. K., Sax, S. L., Cain, T. E., & Tullos, H. S. (1995). Reconstitution of the patellar tendon donor site after graft harvest. *Clin. Orthop. Rel. Res.* 162–171.

Niyibizi, C., Visconti, C. S., Kavalkovich, K., & Woo, S. L-Y. (1995). Collagens in an adult bovine medial collateral ligament: immunofluorescence localization by confocal microscopy reveals that type XIV collagen predominates at the ligament–bone junction. *Matrix Biol., 14,* 743–751.

Niyibizi, C., Sagarrigo Visconti, C., Gibson, G., & Kavalkovich, K. (1996). Identification and immunolocalization of type X collagen at the ligament–bone interface. *Biochem. Biophys. Res. Commun., 222,* 584–589.

Niyibizi, C., Kavalkovich, K., Yamaji, T., & Woo, S. L. (2000). Type V collagen is increased during rabbit medial collateral ligament healing. *Knee Surg. Sports Traumatol. Arthrosc., 8,* 281–285.

Noyes, F. R., Butler, D. L., Grood, E. S., Zernicke, R. F., & Hefzy, M. S. (1984). Biomechanical analysis of human ligament grafts used in knee-ligament repairs and reconstructions. *J. Bone Joint Surg., 66A,* 344–352.

O'Donoghue, D. H., Frank, G. R., Jeter, G. L., Johnson, W., Zeiders, J. W., & Kenyon, R. (1971). Repair and reconstruction of the anterior cruciate ligament in dogs. Factors influencing long-term results. *J. Bone Joint Surg., 53A,* 710–718.

Ohno, K., Pomaybo, A. S., Schmidt, C. C., Levine, R. E., Ohland, K. J., & Woo, S. L-Y. (1995). Healing of the medial collateral ligament after a combined medial collateral and anterior cruciate ligament injury and reconstruction of the anterior cruciate ligament: comparison of repair and nonrepair of medial collateral ligament tears in rabbits. *J. Orthop. Res., 13,* 442–449.

Ottani, V., Raspanti, M., & Ruggeri, A. (2001). Collagen structure and functional implications. *Micron, 32,* 251–260.

Panossian, V., Liu, S. H., Lane, J. M., & Finerman, G. A. (1997). Fibroblast growth factor and epidermal growth factor receptors in ligament healing. *Clin. Orthop. Rel. Res.* 173–180.

Papageorgiou, C. D., Ma, C. B., Abramowitch, S. D., Clineff, T. D., & Woo, S. L-Y. (2001). A multidisciplinary study of the healing of an intraarticular anterior cruciate ligament graft in a goat model. *Am. J. Sports Med., 29,* 620–626.

Parry, D., Barnes, G., & Craig, A. (1978). A comparision of the size distribution of collagen fibrils in connective tissues as a function of age and a possible relation between fibril size distribution and mechanical properties. *Proc. R. Soc. Lond. (Biol.), 203,* 305–321.

Paulos, L. E., Rosenberg, T. D., Drawbert, J., Manning, J., & Abbott, P. (1987). Infrapatellar contracture syndrome. An unrecognized cause of knee stiffness with patella entrapment and patella infera. *Am. J. Sports Med., 15,* 331–341.

Ralphs, J. R., Waggett, A. D., & Benjamin, M. (2002). Actin stress fibres and cell–cell adhesion molecules in tendons: organization *in vivo* and response to mechanical loading of tendon cells *in vitro*. *Matrix Biol., 21,* 67–74.

Record, R. D., Hillegonds, D., Simmons, C., Tullius, R., Rickey, F. A., Elmore, D., et al. (2001). *In vivo* degradation of ^{14}C-labeled small intestinal submucosa (SIS) when used for urinary bladder repair. *Biomaterials, 22,* 2653–2659.

Reider, B., Sathy, M. R., Talkington, J., Blyznak, N., & Kollias, S. (1994). Treatment of isolated medial collateral ligament injuries in athletes with early functional rehabilitation. A five-year follow-up study. *Am. J. Sports Med., 22,* 470–477.

Rubinstein, R. A., Jr., Shelbourne, K. D., VanMeter, C. D., McCarroll, J. C., & Rettig, A. C. (1994). Isolated autogenous bone-patellar tendon-bone graft site morbidity. *Am. J. Sports Med., 22,* 324–327.

Rudy, T. W., Livesay, G. A., Woo, S. L-Y., & Fu, F. H. (1996). A combined robotic/universal force sensor approach to determine *in situ* forces of knee ligaments. *J. Biomech., 29,* 1357–1360.

Sachs, R. A., Daniel, D. M., Stone, M. L., & Garfein, R. F. (1989). Patellofemoral problems after anterior cruciate ligament reconstruction. *Am. J. Sports Med., 17,* 760–765.

Sacks, M. S., & Gloeckner, D. C. (1999). Quantification of the fiber architecture and biaxial mechanical behavior of porcine intestinal submucosa. *J. Biomed. Mater. Res., 46,* 1–10.

Sagarriga Visconti, C., Kavalkovich, K., Wu, J., & Niyibizi, C. (1996). Biochemical analysis of collagens at the ligament–bone interface reveals presence of cartilage-specific collagens. *Arch. Biochem. Biophys., 328,* 135–142.

Scheffler, S. U., Clineff, T. D., Papageorgiou, C. D., Debski, R. E., Benjamin, C., & Woo, S. L-Y. (2001). Structure and function of the healing medial collateral ligament in a goat model. *Ann. Biomed. Eng., 29,* 173–180.

Scherping, S. C., Jr., Schmidt, C. C., Georgescu, H. I., Kwoh, C. K., Evans, C. H., & Woo, S. L-Y. (1997). Effect of growth factors on the proliferation of ligament fibroblasts from skeletally mature rabbits. *Connect. Tissue Res., 36,* 1–8.

Sciore, P., Boykiw, R., & Hart, D. A. (1998). Semiquantitative reverse transcription-polymerase chain reaction analysis of mRNA for growth factors and growth factor receptors from normal and healing rabbit medial collateral ligament tissue. *J. Orthop. Res., 16,* 429–437.

Shelbourne, K. D., Wilckens, J. H., Mollabashy, A., & DeCarlo, M. (1991). Arthrofibrosis in acute anterior cruciate ligament reconstruction. The effect of timing of reconstruction and rehabilitation. *Am. J. Sports Med., 19,* 332–336.

Shimomura, T., Jia, F., Niyibizi, C., & Woo, S. L-Y. (2003). Antisense oligonucleotides reduce synthesis of procollagen alpha1 (V) chain in human patellar tendon fibroblasts: potential application in healing ligaments and tendons. *Connect. Tissue Res., 44,* 167–172.

Spalazzi, J. P., Doty, S. B., Levine, W. N., & Lu, H. H. (2006). *Co-culture of Human Fibroblasts and Osteoblasts on a Tri-phasic Scaffold for Interface Tissue Engineering.* Chicago: Orthopedic Research Society.

Spindler, K. P., Dawson, J. M., Stahlman, G. C., Davidson, J. M., & Nanney, L. B. (2002). Collagen expression and biomechanical response to human recombinant transforming growth factor beta (rhTGF-beta2) in the healing rabbit MCL. *J. Orthop. Res., 20,* 318–324.

Spindler, K. P., Murray, M. M., Detwiler, K. B., Tarter, J. T., Dawson, J. M., Nanney, L. B., et al. (2003). The biomechanical response to doses of TGF-beta 2 in the healing rabbit medial collateral ligament. *J. Orthop. Res., 21,* 245–249.

Steenfos, H. H. (1994). Growth factors and wound healing. *Scand. J. Plast. Reconstr. Surg. Hand Surg., 28,* 95–105.

Svensson, M., Kartus, J., Christensen, L. R., Movin, T., Papadogiannakis, N., & Karlsson, J. (2005). A long-term serial histological evaluation of the patellar tendon in humans after harvesting its central third. *Knee Surg. Sports Traumatol. Arthrosc., 13,* 398–404.

Thornton, G. M., Boorman, R. S., Shrive, N. G., & Frank, C. B. (2002). Medial collateral ligament autografts have increased creep response for at least two years and early immobilization makes this worse. *J. Orthop. Res., 20,* 346–352.

Tibone, J. E., & Antich, T. J. (1988). A biomechanical analysis of anterior cruciate ligament reconstruction with the patellar tendon. A two year followup. *Am. J. Sports Med., 16,* 332–335.

Tipton, C. M., James, S. L., Mergner, W., & Tcheng, T. K. (1970). Influence of exercise on strength of medial collateral knee ligaments of dogs. *Am. J. Physiol., 218,* 894–902.

Tohyama, H., & Yasuda, K. (2000). The effects of stress enhancement on the extracellular matrix and fibroblasts in the patellar tendon. *J. Biomech., 33,* 559–565.

Tohyama, H., Yasuda, K., Kitamura, Y., Yamamoto, E., & Hayashi, K. (2003). The changes in mechanical properties of regenerated and residual tissues in the patellar tendon after removal of its central portion. *Clin. Biomech. (Bristol, Avon), 18,* 765–772.

Tripathy, S. K., Black, H. B., Goldwasser, E., & Leiden, J. M. (1996). Immune responses to transgene-encoded proteins limit the stability of gene expression after injection of replication-defective adenovirus vectors. *Nat. Med., 2,* 545–550.

Tsuda, E., Fukuda, Y., Loh, J. C., Debski, R. E., Fu, F. H., & Woo, S. L-Y. (2002). The effect of soft-tissue graft fixation in anterior cruciate ligament reconstruction on graft-tunnel motion under anterior tibial loading. *Arthroscopy, 18,* 960–967.

Voytik-Harbin, S. L., Brightman, A. O., Kraine, M. R., Waisner, B., & Badylak, S. F. (1997). Identification of extractable growth factors from small intestinal submucosa. *J. Cell Biochem., 67,* 478–491.

Walboomers, X. F., Croes, H. J., Ginsel, L. A., & Jansen, J. A. (1999). Contact guidance of rat fibroblasts on various implant materials. *J. Biomed. Mater. Res., 47,* 204–212.

Wang, J. H., Jia, F., Gilbert, T. W., & Woo, S. L-Y. (2003a). Cell orientation determines the alignment of cell-produced collagenous matrix. *J. Biomech., 36,* 97–102.

Wang, J. H., Jia, F., Yang, G., Yang, S., Campbell, B. H., Stone, D., et al. (2003b). Cyclic mechanical stretching of human tendon fibroblasts increases the production of prostaglandin E2 and levels of cyclooxygenase expression: a novel *in vitro* model study. *Connect. Tissue Res., 44,* 128–133.

Watanabe, N., Woo, S. L-Y., Papageorgiou, C., Celechovsky, C., & Takai, S. (2002). Fate of donor bone marrow cells in medial collateral ligament after simulated autologous transplantation. *Microsc. Res. Tech., 58,* 39–44.

Weiss, J. A., Woo, S. L-Y., Ohland, K. J., Horibe, S., & Newton, P. O. (1991). Evaluation of a new injury model to study medial collateral ligament healing: primary repair versus nonoperative treatment. *J. Orthop. Res., 9,* 516–528.

Woo, S. L-Y., Gelberman, R. H., Cobb, N. G., Amiel, D., Lothringer, K., & Akeson, W. H. (1981a). The importance of controlled passive mobilization on flexor tendon healing. A biomechanical study. *Acta Orthop. Scand., 52,* 615–622.

Woo, S. L-Y., Gomez, M. A., & Akeson, W. H. (1981b). The time and history-dependent viscoelastic properties of the canine medical collateral ligament. *J. Biomech. Eng., 103,* 293–298.

Woo, S. L-Y., Gomez, M. A., Amiel, D., Ritter, M. A., Gelberman, R. H., & Akeson, W. H. (1981c). The effects of exercise on the biomechanical and biochemical properties of swine digital flexor tendons. *J. Biomech. Eng., 103,* 51–56.

Woo, S. L-Y., Gomez, M. A., Woo, Y. K., & Akeson, W. H. (1982). Mechanical properties of tendons and ligaments. II. The relationships of immobilization and exercise on tissue remodeling. *Biorheology, 19*, 397–408.

Woo, S. L-Y., Gomez, M. A., Seguchi, Y., Endo, C. M., & Akeson, W. H. (1983). Measurement of mechanical properties of ligament substance from a bone-ligament-bone preparation. *J. Orthop. Res., 1*, 22–29.

Woo, S. L-Y., Orlando, C. A., Camp, J. F., & Akeson, W. H. (1986a). Effects of postmortem storage by freezing on ligament tensile behavior. *J. Biomech., 19*, 399–404.

Woo, S. L-Y., Orlando, C. A., Gomez, M. A., Frank, C. B., & Akeson, W. H. (1986b). Tensile properties of the medial collateral ligament as a function of age. *J. Orthop. Res., 4*, 133–141.

Woo, S. L-Y., Gomez, M. A., Sites, T. J., Newton, P. O., Orlando, C. A., & Akeson, W. H. (1987a). The biomechanical and morphological changes in the medial collateral ligament of the rabbit after immobilization and remobilization. *J. Bone Joint Surg., 69A*, 1200–1211.

Woo, S. L-Y., Inoue, M., McGurk-Burleson, E., & Gomez, M. A. (1987b). Treatment of the medial collateral ligament injury. II: Structure and function of canine knees in response to differing treatment regimens. *Am. J. Sports Med., 15*, 22–29.

Woo, S. L-Y., Young, E. P., Ohland, K. J., Marcin, J. P., Horibe, S., & Lin, H. C. (1990). The effects of transection of the anterior cruciate ligament on healing of the medial collateral ligament. A biomechanical study of the knee in dogs. *J. Bone Joint Surg., 72A*, 382–392.

Woo, S. L-Y., Hollis, J. M., Adams, D. J., Lyon, R. M., & Takai, S. (1991). Tensile properties of the human femur-anterior cruciate ligament-tibia complex. The effects of specimen age and orientation. *Am. J. Sports Med., 19*, 217–225.

Woo, S. L-Y., Smith, D. W., Hildebrand, K. A., Zeminski, J. A., & Johnson, L. A. (1998). Engineering the healing of the rabbit medial collateral ligament. *Med. Biol. Eng. Comput., 36*, 359–364.

Woo, S. L-Y., Hildebrand, K., Watanabe, N., Fenwick, J. A., Papageorgiou, C. D., & Wang, J. H. (1999). Tissue engineering of ligament and tendon healing. *Clin. Orthop. Rel. Res.*, S312–S323.

Woo, S. L-Y., An, K.-N., Frank, C., Livesay, G., Ma, C., & Zeminski, J. (2000). Anatomy, biology and biomechanics of tendon and ligaments. In T. A. Einhorn, S. R. Simon, & American Academy of Orthopaedic Surgeons. (Eds.), *Orthopaedic Basic Science: Biology and Biomechanics of the Musculoskeletal System* (pp. 581–616). Rosemont, IL: American Academy of Orthopaedic Surgeons.

Woo, S. L-Y., Kanamori, A., Zeminski, J., Yagi, M., Papageorgiou, C., & Fu, F. H. (2002). The effectiveness of reconstruction of the anterior cruciate ligament with hamstrings and patellar tendon. A cadaveric study comparing anterior tibial and rotational loads. *J. Bone Joint Surg., 84A*, 907–914.

Woo, S. L-Y., Jia, F., Zou, L., & Gabriel, M. T. (2004). Functional tissue engineering for ligament healing: potential of antisense gene therapy. *Ann. Biomed. Eng., 32*, 342–351.

Woo, S. L-Y., Abramowitch, S. D., Kilger, R., & Liang, R. (2006). Biomechanics of knee ligaments: injury, healing, and repair. *J. Biomech., 39*, 1–20.

Yagi, M., Wong, E. K., Kanamori, A., Debski, R. E., Fu, F. H., & Woo, S. L-Y. (2002). Biomechanical analysis of an anatomic anterior cruciate ligament reconstruction. *Am. J. Sports Med., 30*, 660–666.

Yamaji, T., Levine, R. E., Woo, S. L-Y., Niyibizi, C., Kavalkovich, K. W., & Weaver-Green, C. M. (1996). Medial collateral ligament healing one year after a concurrent medial collateral ligament and anterior cruciate ligament injury: an interdisciplinary study in rabbits. *J. Orthop. Res., 14*, 223–227.

Yang, G., Crawford, R. C., & Wang, J. H. (2004). Proliferation and collagen production of human patellar tendon fibroblasts in response to cyclic uniaxial stretching in serum-free conditions. *J. Biomech., 37*, 1543–1550.

Young, R. G., Butler, D. L., Weber, W., Caplan, A. I., Gordon, S. L-Y., & Fink, D. J. (1998). Use of mesenchymal stem cells in a collagen matrix for Achilles tendon repair. *J. Orthop. Res., 16*, 406–413.

Zantop, T., Gilbert, T. W., Yoder, M., & Badylak, S. (2005). Extracellular matrix scaffolds attract bone marrow-derived cells in a mouse model of Achilles tendon reconstruction. *J. Orthop. Res., 24*, 1299–1309.

1021

CHAPTER 55

Central Nervous System

Brian G. Ballios[*], **M. Douglas Baumann**[†], **Michael J. Cooke**[†], **Molly S. Shoichet**[*,†,‡]
[*] Institute of Medical Science
[†] Department of Chemical Engineering and Applied Chemistry
[‡] Institute of Biomaterials and Biomedical Engineering, University of Toronto, Toronto, Ontario, Canada

INTRODUCTION

The human central nervous system (CNS) is a network of more than 100 billion individual nerve cells that control our actions, sense our surroundings, and define who we are. The CNS is comprised of the brain, spinal cord, and retina. The functional units of the CNS are neurons which are unique in their ability to transmit and store information, yet are also vulnerable to injury. Most mature neurons are post-mitotic cells incapable of cell division, so their destruction can leave a patient with a severe functional deficit. The CNS has several unique anatomical and physiological characteristics: it is enclosed in bony structures (skull and vertebrae) making it difficult to access; it has a special vascular system which serves as a barrier for drug delivery; and it has limited capacity for repair. Injury to the CNS can take many forms, but the endpoint of neuronal injury or loss is always the same: permanent functional impairment. The anatomical location of the lesion and the limited capacity for repair are two factors of significance in CNS trauma. Injury of several cubic centimeters of parenchyma may be clinically silent; that is, unapparent to both patient and clinician (if in the frontal cortex), severely disabling (if in the spinal cord), or fatal (if in the brainstem) (Kumar et al., 2005).

1023

Injuries to the CNS include stroke, traumatic brain and spinal cord injury, and retinal degeneration. These conditions affect millions worldwide, and cover the entire age spectrum. CNS injury in the pediatric population includes predominantly traumatic or congenital defects, while in an adult/senior population the injuries are traumatic or degenerative. Stroke is the third leading cause of death in the western world, and approximately 3.5 million people live with chronic brain damage from stroke in the USA alone (Lo et al., 2003). The incidence of traumatic spinal cord injury is 10,000 per year in North America (Tator, 1996) while the prevalence is approximated at 755 per one million people (Wyndaele and Wyndaele, 2006). Retinal degenerative conditions, such as age-related macular degeneration which is the leading cause of irreversible blindness in developed nations, account for an estimated 200,000 new cases annually in the USA. For all of these conditions acute and chronic populations pose different treatment challenges.

Despite the complexity of the CNS anatomy, physiology and pathobiology, progress in tissue engineering in recent years has led to new approaches for treatment and new hope. This chapter focuses on tissue engineering strategies, including those based on drug delivery, cell delivery, and physical constructs, all of which are aimed at achieving functional recovery after CNS injury. First, the commonalities in the response to injury are reviewed in different areas of

Principles of Regenerative Medicine. DOI: 10.1016/B978-0-12-381422-7.10055-0

the CNS. Then various therapeutic strategies under investigation for the treatment of CNS injury are introduced, namely: drug therapy, cell therapy, and physical constructs. The second half of the chapter covers the application of tissue therapy to three case studies in CNS pathology, namely: stroke, spinal cord injury, and retinal degeneration. Common to these conditions are the currently irreversible functional impairment and overarching psychological, quality of life, and financial costs. The case studies will also illustrate similarities and highlight important differences in applying therapies to various CNS locales.

WOUND RESPONSE IN THE CNS

The CNS is considered to be immunoprivileged, which impacts the wound response. An immune-privileged state prevents uncontrolled inflammation of vital organs in the event of injury. This is achieved through a number of mechanisms including a specialized membrane known as the blood–brain barrier (BBB) that prevents entry of circulating immune cells. The BBB controls the influx of molecules into the brain. There are three main components: (1) endothelial cells joined by tight junctions, which limit the paracellular flux of hydrophilic molecules, (2) astrocytic end-feet, which tightly ensheath the vessels walls and help to maintain the tight junctions, and (3) pericytes, which are important for structural integrity and formation of the tight junctions. The BBB permits diffusion of small lipophilic molecules such as O_2 and CO_2 along their concentration gradients. In contrast, the transport of large molecules is controlled. Nutrients (e.g. glucose) cross the BBB via transporters and endocytosis mediates the uptake of larger molecules (e.g. insulin) (Ballabh et al., 2004). Although the inflammation mediated by microglia within the CNS aids the healing process, the relatively muted response allows dead and dying cells to continue to secrete factors which contribute to the progression of injury. In stroke, dying cells trigger an immune response leading to inflammatory cell activation and infiltration. Reperfusion with oxygenated blood results in the production of reactive oxygen species (ROS), which stimulates the secretion of inflammatory cytokines from ischemic cells. Activated inflammatory cells release metalloproteinases (MMPs) which leads to the disruption of the BBB and subsequent secondary injury by allowing serum and blood to enter the brain (Wang et al., 2007). Similarly to stroke, the immune response following spinal cord injury (SCI) potentiates the injury. The inflammatory response is initiated by peripherally derived immune cells, such as macrophages, and activated glial cells, such as microglia, that migrate into the injury. Macrophages and microglia release cytokines, tumor necrosis factor, and interleukins 1, 6, and 10 (IL-1, -6, -10). The released cytokines induce the expression of additional cytokines, chemokines, nitric oxide, and reactive oxygen, and subsequently cause cell death (Keane et al., 2006).

One of the similarities between the brain, spinal cord, and retina is the gliotic response to injury. Under normal conditions glial cells help to protect the tissue from damage. During embryogenesis microglia are highly activated and remove dead cells and debris that naturally occur during the development of the brain. In the eye, the principal glial cells are the Muller cells, which support neuronal activity and integrity of the blood–retinal barrier (BRB) as well as protect retinal ganglion cells from glutamate and nitric oxide neurotoxicity (Garcia et al., 2002; Limb et al., 2010). Injury results in gliosis, the proliferation of glial cells, as a mechanism to minimize damage and protect uninjured tissue (Dyer and Cepko, 2000; Fisher and Reh, 2001; Duggan et al., 2009). Microglia become activated following stroke and migrate to the lesion site where they release a wide range of soluble factors that include cytotoxins, neurotrophins, and immunomodulatory factors, and clear cellular debris by phagocytosis. It is clear that reactive astrocytes are important in minimizing tissue damage after injury in other locations in the CNS, as their ablation after SCI exacerbates the severity of the injury (Faulkner et al., 2004). Although glial cells have a beneficial effect by minimizing further tissue damage, they also form a glial scar. The glial scar contains inhibitory factors such as chondrotin sulfate proteoglycan and myelin-associated glycoproteins (Fawcett and Asher, 1999), which prevent axonal growth.

Host regeneration is dependent upon the existence of endogenous cells with the capacity to proliferate and differentiate. It was originally thought that repair of the CNS by endogenous mechanisms was unlikely as neurogenesis had ceased prior to adulthood. It has now been demonstrated that neurogenesis does in fact occur in adulthood. The two main regions that are known to contribute to neurogenesis are the subventricular zone (SVZ) of the lateral ventricles and the subgranular zone (SGZ) of the hippocampus (Weiss et al., 1996; Palmer et al., 1997). Neural stem cells have also been found in the lumbar/sacral segment of the spinal cord. These quiescent cells are found in the ependymal layer of the subependymal zone of the central canal (Weiss et al., 1996). Stem cells have also been identified in the eye. Retinal stem cells are found in the pigmented ciliary epithelium, have been shown to proliferate *in vitro*, and are capable of differentiating into retinal-specific cell types (Tropepe et al., 2000). Following injury to the spinal cord, spinal cord ependymal-derived stem/progenitor cells are stimulated to proliferate and cells positive for neural precursor markers (e.g. Sox2 and nestin) were detected in the lesioned parenchyma (Foret et al., 2010). No reports have shown functional recovery by endogenous stem cells. Following stroke endogenous stem cells have been stimulated with erythropoietin and epidermal growth factor to proliferate and then migrate to the area of damage, where they promote tissue repair (Kolb et al., 2007). Although endogenous stem cells are stimulated by injury, this is not sufficient to induce recovery after stroke (Leker, 2009). Although stem cells have been isolated from the adult mammalian eye, there are currently no examples of their promoting functional recovery after injury.

In summary, the endogenous repair mechanisms in CNS wound response, such as proliferation of quiescent stem cell populations, are insufficient to accomplish significant functional repair on their own. The largely similar wound response across the CNS is reflected in common therapeutic approaches to the different regions such that drug therapy, cell therapy, and physical constructs are used in each of the brain, spinal cord, and retina. Differences in wound response and anatomy, however, necessitate tailoring a therapy to the tissue target and pathological state.

1025

THERAPEUTIC STRATEGIES IN THE CNS

Broadly, the therapeutic strategies in the CNS can be classified as drug therapy, cell therapy, or physical construct. This section gives a broad overview of these approaches. Further detail is given in the case studies of CNS pathology.

Drug therapy

The use of drug delivery as a reparative and regenerative strategy after CNS injury has been investigated in animal models and clinically. Molecules have been delivered systemically and injected locally, the latter to the epidural space, intrathecal space (or into the ventricles), and directly into the cord. Intraspinal injection can cause tissue damage and has not been widely pursued. Systemic delivery is only suitable for molecules that can cross the BBB or blood--spinal cord barrier (BSCB) but the high doses required can result in undesirable side-effects. Notwithstanding this limitation, the majority of drugs delivered to the CNS have been through systemic means, although local delivery of molecules that do not readily cross the BSCB have been pursued for applications in SCI. Strategies for local delivery to the brain are discussed on pages 1029–1031 for treatment of stroke and to the spinal cord on pages 1032–1035 after SCI. The goals of drug delivery in the CNS are to limit degeneration (neuroprotection) and promote regeneration (neuroregeneration).

Neuroprotection is strictly defined as a long-lasting increase in functional ability but the term is also applied to increased axonal sparing and reduced lesion volume. Much attention has been paid to the steroid methylprednisolone and GM-1 ganglioside, although both showed only limited success in clinical trials (Ramer et al., 2005). Methylprednisolone is valued as an anti-inflammatory (Bracken et al., 1998) while GM-1 ganglioside is purported to inhibit apoptosis and promote axonal sprouting (Geisler et al., 2001). Other molecules targeting the host inflammatory response include a tetracycline derivative, minocycline (Stirling et al., 2005), and the cytokines IL-10 (Bethea et al., 1999) and erythropoietin (Eid and Brines, 2002). Each of these molecules probably acts on a variety of cellular targets, such as reduced dieback of certain neuronal populations or oligodendrocytes and inhibition of microglial activation (Eid and Brines, 2002; Ramer et al., 2005). Growth factors best known for their regenerative potential can also be neuroprotective. Nerve growth factor (NGF), brain-derived neurotrophic growth factor (BDNF), neurotrophin-3 (NT-3), ciliary neurotrophic factor (CNTF), and glial cell-line-derived neurotrophic factor (GDNF) enhance the survival of either or both of sensory and motor neurons (Sendtner et al., 1990; Matheson et al., 1997; Nakahara et al., 1996; Lee et al., 1999; Teng et al., 1999).

Neuroregeneration comprises elongation of existing axons, sprouting and growth of new axons from neural cell soma, remyelination, plasticity among surviving connections, and functional recovery. Current therapies are focused on axonal growth as a precursor to functional recovery and are divided into two classes by their mode of action. Some, like the growth factors, act on regeneration, associated genes (RAGs) such as growth-associated protein-43 (GAP-43) and activating transcription factor-3 (ATF-3; Hannila and Filbin, 2008) to stimulate axonal sprouting and growth. Outgrowth of motor and/or sensory neurons has been reported for NGF (Ramer et al., 2000), BDNF (Braun et al., 1996), NT-3 (Ramer et al., 2000), GDNF (Matheson et al., 1997), CNTF (Oyesiku et al., 1996), the fibroblast growth factors (FGF-1 and FGF-2) (Romero et al., 2001), and others. The conditions for controlled regeneration are not yet clear, with side-effects (Christensen et al., 1997; Romero et al., 2001) and conflicting conclusions (Nakahara et al., 1996; Oudega and Hagg, 1996) having been reported. RAG stimulating agents are predicted to act in synergy with one another (Schwab, 2004) and with a second class of neuroregenerative molecules which target growth inhibitory glycoproteins and proteoglycans. These in turn activate the Rho-ROCK pathway. For example, myelin-associated glycoprotein (MAG) and NogoA are native to the CNS and bind to the Nogo (NgR) and $p75^{NTR}$ cell surface receptors, thereby triggering Rho-ROCK and growth inhibition (Hannila and Filbin, 2008). Blocking this interaction with anti-NogoA or anti-NgR allows neurite outgrowth to resume (Oertle et al., 2003). Rho activation (and growth inhibition) can be inhibited by increased intracellular concentrations of the nucleotide cAMP (Hannila and Filbin, 2008). Delivery of cAMP or Rolipram, an inhibitor of cAMP degradation, has also been shown to enhance neuroregeneration *in vivo* (Nikulina et al., 2004; Pearse et al., 2004).

Cell therapy

Cell therapy relies on transplanted or host cells either producing the desired therapeutic molecule over a prolonged time to promote endogenous repair or replacing lost tissue that will integrate with the host tissue. There are essentially two methods of cell therapy: (1) cell transplantation, where the cells are derived from autogeneic, allogeneic, or xenogeneic sources, and (2) cell stimulation, where the endogenous cell population is stimulated to promote repair. Success in experimentation with cell transplantation in other disorders such as diabetes has stimulated investigations for its use in neurodegenerative diseases and disorders in clinical settings. These include conditions such as chronic pain (Buchser et al., 1996), Parkinson's disease (PD) (Freed et al., 1990 Lindvall et al., 1990), amyotrophic lateral sclerosis (ALS) (Aebischer et al., 1990), and spinal cord injury (Kigerl and Popovich, 2006), among others.

Despite some clinical success with PD, chronic pain, and ALS, cell-loaded membranes are not used clinically due to inconsistent therapeutic results. One of the major challenges with the transplantation of allogeneic or xenogeneic cells in cellular replacement therapy is cell survival due to the host's immune response (Jones et al., 2004). Limited availability of allogeneic donor cells is also an issue: even committed neural precursors isolated from fetal tissue do not proliferate to the quantities necessary for clinical utility. In addition, there are legislative provisions and ethical concerns surrounding the use of embryonic/fetal tissue. An alternative strategy involves the directed differentiation of stem cells, which have the unlimited ability for growth in culture. Two types of allogeneic stem cell sources have emerged with the potential for therapeutic application in the CNS: embryonic stem cells and adult neural stem cells. In addition, induction of pluripotent cells from adult somatic cells might allow for the generation of patient-specific tissue grafts derived from the patient's own cells.

Embryonic stem (ES) cells are pluripotent cells isolated from the inner cell mass of pre-implantation blastocysts. The inner mass cells proliferate and differentiate to become the variety of cells that make up the three germ layers of developing humans and they have the potential to become any cell in the body. This propagation potential can be recapitulated in culture and differentiation of these cells can be directed in a lineage-specific manner to yield terminally differentiated cells (Richards et al., 2002). Safety is an important concern in the clinical application of ES cells, as their proliferative ability may lead to tumor and mass formation (Amariglio et al., 2009). Pre-differentiation of these cells is one means to diminish the implied risk of transplanting these pluripotent cells directly (Sharp and Keirstead, 2007). However, in the absence of directed differentiation towards a particular cellular fate, impure populations of differentiated cells can arise (Keirstead et al., 2005). Significantly, in early 2009 the US FDA approved the beginning of clinical trials for human embryonic stem (hES) cells (Alper, 2009). This study will inject a suspension containing hESC-derived oligodendrocyte progenitor cells directly into the lesion site. These cells have demonstrated remyelinating and nerve growth properties (Keirstead et al., 2005). This exciting development represents an important milestone in CNS tissue therapy as the first FDA-approved clinical trial for an hESC product. These trials were delayed in 2009 as a result of the identification of microscopic cysts in the regenerating injury site in animal models. These cysts are reported to be non-proliferative and confined to the injury sites and did not lead to any adverse consequences to the animals. The trials were subsequently allowed to proceed and the first patient was enrolled in October 2010.

1027

Neural stem-progenitor cells (NSPCs) reside in distinct regions of both fetal and adult CNS, namely the subventricular zone of the lateral ventricles (Morshead et al., 1994; Chiasson et al., 1999), subgranular zone of the dentate gyrus of the hippocampus (Cameron et al., 1993; Kuhn et al., 1996; Palmer et al., 1997), and ependymal layer of the spinal cord central canal (Weiss et al., 1996; Horner et al., 2000). These cells undergo proliferation and maturation throughout adult life (Bambakidis et al., 2003). NSPCs can be isolated from embryonic or adult CNS and expanded in culture. These cells have been shown to survive, differentiate into all classes of cells in the CNS, integrate, repair tissue, and improve neurological function after transplantation into animal models of CNS pathology or injury (Gage et al., 1995; Akiyama et al., 2001; Ben-Hur et al., 2003), including multiple sclerosis (Pluchino et al., 2003) and traumatic spinal cord injury (Parr et al., 2007). Stimulation of endogenous stem cells is a relatively new approach that is being investigated in the laboratory. The idea of stimulating endogenous stem cells was demonstrated by Derek van der Kooy's laboratory in spinal tissue (Martens et al., 2002). Here, specific factors (EGF, FGF-2, heparin) were delivered to the ventricle to locally stimulate neural precursor cells. The intrathecal drug delivery strategy described on page 1033 may be an appropriate methodology to locally stimulate a specific stem cell population in the context of spinal disease.

Recently, the derivation of induced pluripotent stem (iPS) cells (Okita et al., 2007) from dermal fibroblasts, among other tissues such as liver and stomach (Aoi et al., 2008), has made it possible to start with adult terminally differentiated cells and give them embryonic stem cell-like properties. Autogeneic cells negate the detrimental immune response seen in allogeneic grafts that results in improperly HLA-matched samples. The exploration of this approach may eventually lead to the development of patient-specific neural grafts grown from the patient's own cells. In addition, Freda Miller's laboratory has shown the promise of transplanted cells derived from skin precursors (Toma et al., 2001; Fernandes et al., 2004) for the treatment of neurological disorders (Fernandes et al., 2006), including spinal cord injury (Biernaskie et al., 2007).

Considering the practical aspects of stem cell replacement therapy, and the desire to transplant a pure cellular population, investigators are considering how this might be achieved while minimizing patient risks identified above. Purification of cell populations can be achieved using cell-sorting devices with appropriate fluorescent labels or reporters to isolate cells with a uniform therapeutic potential. However, the insertion of a fluorescent marker or an antibiotic resistance gene for survival selection often involves the potential for aberrant genomic manipulation (Keirstead et al., 2005). Although a useful tool for fundamental research, any potential for genomic manipulation which increases tumorigenic risk is undesirable for clinical cell replacement therapy (CRT) as per current US FDA guidelines (Halme and Kessler, 2006; Fink, 2009). Development of clinical CRT will require refinement of the methods of stem cell isolation and lineage directed differentiation in culture to encourage the production of pure populations, side-stepping the complexity of sorting impure populations.

The relatively poor survival of transplanted cells (Le Belle and Svendsen, 2002) suggests that a combination of cellular therapeutic and drug/biomolecular delivery could enhance survival and integration of the transplanted cell population. A recent report from the laboratory of Michael Fehlings has shown that combining delivery of neural stem/progenitor cells with chondroitinase ABC and growth factors enhances cellular integration and functional recovery in a rat model of chronic spinal cord injury (Karimi-Abdolrezaee et al., 2010). Chondroitinase ABC optimizes the host environment for cell engraftment by removing inhibitory chondroitin sulfate proteoglycans (CSPGs) in the glial scar, while the growth factors EGF, FGF-2, and PDGF-AA enhance stem cell integration and differentiation to oligodendrocytes. These systems might benefit from an engineered composite vehicle that could deliver both the cells and drugs to the therapeutic target in the CNS.

Physical constructs

Physical constructs for tissue therapy in the CNS take the form of scaffolds to support cell growth and, in the spinal cord, tubular scaffolds to assist nerve regeneration. The scaffold may alternatively function solely as a drug delivery system to the surrounding tissue (Lee et al., 2006; Vaisman et al., 2010) or serve a dual role with cells contained within the scaffold (Ma et al., 2007). Injectable scaffolds are valued because they conform to the available tissue space and are thought to promote tissue integration. Polysaccharide-based cell delivery/infiltration scaffolds have been developed for use in the brain (Tate et al., 2001; Stabenfeldt et al., 2006), retina (Ballios et al., 2010), and spinal cord (Jain et al., 2006). Implantable scaffolds offer unique control over pore (Wong et al., 2008) or fiber architecture (Nisbet et al., 2009) that may lead to enhanced tissue engraftment. Implantation of a porous poly(N-2-hydroxypropyl methacrylamide) hydrogel resulted in functional recovery in acute and chronic models of SCI in the rat (Woerly et al., 2001, 2004).

The design of nerve guidance channels for spinal cord repair is meant to mimic the peripheral nerve tissue by providing a permissive environment for regeneration. Considerable effort has been expended in the optimization of tube fabrication methods, including polymer extrusion, casting in molds, immersion of a polymer solution-coated mandrel in a non-solvent, and centrifugal casting of phase-separated polymerization mixtures. These techniques vary in their ability to produce tubes with uniform properties, with centrifugal casting remaining the most versatile option and molding being the most simple (see Tsai et al., 2004; Katayama et al., 2006, and references therein). One of the most popular classes of materials is hydrogels such as

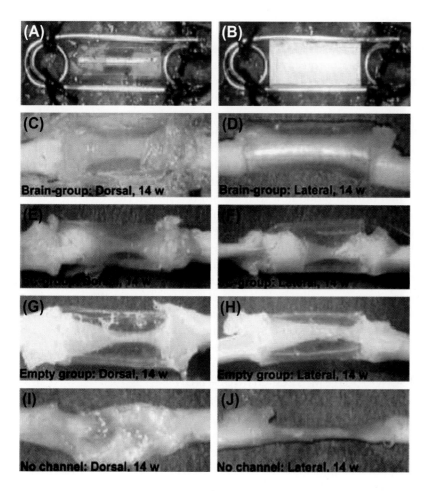

FIGURE 55.1
Surgical procedure for extramedullary chitosan channel implantation after spinal cord transection **(A,B).** At 14 weeks, gross appearance of the implanted channel (C—H) or spinal cord transection alone (I,J), from the dorsal and lateral aspects. In each panel, rostral is on the left. (A) Dorsal view of the transected spinal cord stumps placed within the transparent chitosan channel. The gap between the stumps is approximately 3.5 mm. (B) The spinal fusion with wire for spinal stabilization and the expanded polytetrafluoroethylene membrane placed on the dorsal aspect of the channel for prevention of scar invasion. (C—H) There is a tissue bridge inside the channels in the brain (C,D), spinal cord (E,F), and empty-channel groups (G,H). The brain group has the thickest bridge (C,D). In the no-channel group, there is pale connective tissue between the stumps (I,J). *(Reprinted with permission from Nomura et al., 2008b.)*

poly(2-hydroxyethyl methacrylate-*co*-methyl methacrylate) (P(HEMA/MMA)) (Dalton et al., 2002). P(HEMA/MMA) in particular is highly biocompatible and has shown great promise in nerve guidance tubes in the CNS and peripheral nervous system (PNS) (Tsai et al., 2004; Katayama et al., 2006). P(HEMA/MMA) is non-degradable and investigators have turned to materials such as poly(L-lactide) (Oudega et al., 2001), poly(hydroxybutyrate) (Novikov et al., 2002), chitosan (Kim et al., 2008; Nomura et al., 2008a), and collagen (Paino and Bunge, 1991), which are degradable *in vivo*. Figure 55.1 illustrates the tissue bridge which forms when a nerve guidance channel is implanted in a rat transection model of SCI. In this case the nerve guidance channel was seeded with brain or spinal cord-derived NSPCs (Nomura et al., 2008b).

CASE STUDIES IN TISSUE THERAPY IN THE CNS
Stroke

Cerebrovascular disease is the third leading cause of death in the USA. In terms of morbidity and mortality it is one of the most common neurological disorders and is defined as an abnormality of the brain caused by a pathological process involving its blood supply (Hossmann, 2006). Stroke is the clinical designation used to apply to cerebrovascular diseases, particularly when they are seen acutely. Pathophysiologically, strokes result from two processes: hypoxia/ischemia/infarction resulting from impairment of blood supply, or hemorrhage resulting from CNS blood vessel rupture. This case study will focus on hypoxic/ischemic stroke in which the brain is starved of the vital oxygen and nutrients required for proper function.

PHARMACOLOGICAL THERAPY

Pharmacological treatment of stroke can be broadly classified into two categories, neuro-protective and thrombolytic. Neuroprotective treatments focus on prevention of further damage and preservation of as much tissue as possible. Despite many promising animal studies, clinical use of neuroprotective strategies has been unsuccessful and has been conducted using calcium antagonists, glutamate antagonists, AMPA antagonists, opioid antagonists, GABA agonists, free radical scavengers, anti-inflammatory agents, and membrane stabilizers (Labiche and Grotta, 2004). Although pharmacological therapies have not yet shown neuroprotection, their application for thrombolysis has shown more clinical promise. The goal of thrombolysis treatment is to remove the blood clot (that results in ischemia) so that normal blood flow to the brain is restored. However, there is an increased risk of hemorrhagic complications with this treatment (Murray et al., 2010). Thrombolysis treatment promotes the conversion of the proenzyme plasminogen in the blood stream into an active enzyme plasmin, a protease that degrades blood plasma proteins associated with blood clotting, most notably, fibrin. Currently, the only FDA approved treatment for stroke is tissue plasminogen activator (tPA). Tissue plasminogen activator activates plasminogen by catalyzing its conversion into plasmin. tPA has under-gone clinical trials where it is intravenously administered safely (Koennecke et al., 2001) and has been shown to improve the outcome in ischemic stroke when administered 4.5 h following stroke (Stemer and Lyden, 2010).

DRUGS FOR ENDOGENOUS STEM CELL STIMULATION

Following stroke, neural stem cells within the SVZ are known to elevate levels of receptors for growth factors. For example, the GDNF receptor (Kobayashi et al., 2006) and EGF receptor (Alagappan et al., 2009) are increased. This suggests that growth factors could be administered to promote the stimulation of endogenous repair. It has been demonstrated in animal models that administration of endothelial growth factor (EGF) and erythropoietin (EPO) can stim-ulate endogenous stem cells to replace damaged tissue after middle cerebral artery occlusion (MCAO) (Kolb et al., 2007). A study was conducted where EGF was replaced with beta-human chorionic gonadotropin (hCG) and a significant decrease in lesion volume was seen in addition to a functional improvement (Belayev et al., 2009). A clinical study has demonstrated that the sequential administration of beta-hCG and EPO 24–48 h after stroke is safe (Cramer et al., 2010). Another growth factor that has promise for stimulation of stem cells is granulocyte-colony stimulating factor (G-CSF). Cells in the SVZ express the receptor for G-CSF and upon subcutaneous administration of this factor cells were stimulated to proliferate, resulting in an improved functional recovery (Sprigg et al., 2006). An alternative strategy to stimulating the endogenous brain stem cells is to stimulate circulating stem cells, which will home to the site of injury. G-CSF has been administered intravenously to rats subjected to MCAO and has shown a decrease in infarct volume (Schabitz et al., 2003). A clinical study has shown that subcutaneous administration of G-CSF is a safe and feasible method to mobilize bone marrow stem cells (Sprigg et al., 2006).

CELL TRANSPLANTATION

Due to the limited endogenous regeneration of the brain, investigations have been carried out to transplant exogenous cells. NSPCs have been administered intravenously to mice subjected to MCAO. The cells migrate to the injured side of the brain but, of all NSPCs injected, only 0.09% were found in the brain yet many were detected in numerous other organs (Bacigaluppi et al., 2009). Localized delivery to restrict the transplanted cells to the site of injury will result in a greater number of cells remaining at the injury site. Due to the varying size of lesions, an injectable hydrogel that fills the irregular void is desirable. To administer cells locally they can be encapsulated into a hydrogel delivery vehicle. Such delivery vehicles have been investigated for injection into the tissue defect after traumatic brain injury (Tate et al., 2001) (Fig. 55.2).

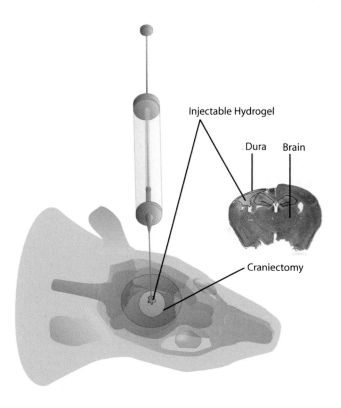

FIGURE 55.2

Injectable delivery vehicle for cell delivery to the injured brain. This conceptual drawing shows injection of the space-filling vehicle into the cystic cavity formed after traumatic or stroke-induced brain damage. A craniectomy is performed to expose the injury site and a hydrogel can be injected to fill the cavity. *(Reprinted with permission from Reichert, 2008.)*

Currently, the survival of transplanted cells is limited to 1–10% of the injected population (Kallur et al., 2006; Hicks et al., 2009). To improve cell survival, cells can be co-transplanted in a scaffold. Natural scaffolds such as Matrigel have been used to deliver neural progenitor cells (NPCs) and compared to cells delivered in artificial cerebrospinal fluid (aCSF). Matrigel increased the number of cells at the transplantation site (Jin et al., 2010). Matrigel is an undefined mixture of extracellular matrix proteins and growth factors derived from mouse sarcoma and thus could hinder clinical use. To elucidate whether individual components of Matrigel can improve cell survival, cells have been transplanted in collagen I (Lu et al., 2007) and a mixture of collagen I and laminin or fibronectin (Tate et al., 2009), both of which were found to improve cell survival. Scaffolds with defined chemical compositions are more easily transferred to the clinic, which often leads investigators to focus on synthetic materials. One such material is poly (D,L-lactic acid-co-glycolic acid) (PLGA). Cells were attached to the surface of microparticles containing encapsulated NGF so that the particles provided physical support and controlled release of a growth factor to improve cell survival following transplantation (Mahoney and Saltzman, 2001). When stem cells are delivered to the site of injury they must be differentiated into the correct cell types. Transplantation of undifferentiated and pre-differentiated cells has been compared in a rat model. Pre-differentiation of hNPCs can produce cultures that are 60% neurons but transplantation of these pre-differentiated hNPCs into the intact hippocampus decreased the survival rate compared to undifferentiated NPCs. However, the total number of mature neurons in the transplantation site is increased in comparison to undifferentiated cells (Le Belle et al., 2004).

A combination therapy using growth factors and stem cells has been proposed to improve the efficacy of cell transplantation. Combined treatment of EPO and mesenchymal stem cells (MSCs) results in increased proliferation of cells along the lateral ventricle wall and greater functional recovery than EPO or MSC treatment alone (Esneault et al., 2008).

It has been proposed that rehabilitation can contribute to regeneration following stem cell transplantation. To study the effects of rehabilitation, animals are housed in "enriched environments" where they are allowed access to a variety of objects to stimulate exploratory behavior in addition to free access to individual running wheels (Hicks et al., 2007). In combination with NSPC transplantation, enriched housing has been found to enhance functional recovery, enhance proliferation of endogenous progenitor cells, and increase cell migration to the injury site (Hicks et al., 2007). In contrast, it has also been demonstrated that when rats were transplanted with neural precursor cells derived from hES cells, enriched housing had no significant effect on functional recovery (Hicks et al., 2009). The different effects observed with "enriched housing" require further studies into the type of enrichment the animals are receiving in order to draw more equivocal conclusions on its therapeutic utility.

Experimental evidence has provided hope for the treatment of stroke; however, this promise has not yet been translated into the clinic. Thrombosis treatment has had the best clinical success, yet the short window for treatment limits the utility. Experimentally, stem cell treatment appears to be a viable method to repair damaged tissue. Further studies are needed to determine how to control cell survival and cell differentiation, and to investigate the clinical safety of stem cell delivery to the injured brain.

Traumatic spinal cord injury

Traumatic spinal cord injury (SCI) is described as resulting from compression or transection (laceration) of the cord. Compression injuries comprise 70% of clinical cases and partial or complete transection 30% (Norenberg et al., 2004). Compression of the spinal cord may be caused by dislocation of a vertebra and pinching of the cord between the anterior and posterior faces of the vertebral foramen in adjacent vertebrae. In spinal cord trauma, the level of cord injury determines the extent of neurological manifestations and the prognosis. Damage to thoracic and lower vertebrae can lead to paraplegia and cervical injury may result in quadriplegia. Spinal cord injuries above the fourth cervical vertebra can also lead to respiratory compromise. Segmental damage to the ascending/descending white matter tracts accounts for the principal clinical deficits.

The initial primary injury causes uncontrolled necrotic cell death, inflammation, ischemia, and hypoxia. These processes persist for weeks and drive neurons and oligodendrocytes to apoptosis, increasing the volume of injured tissue and forming a spindle-shaped cystic cavity (Hausmann, 2003). This secondary biochemical cascade is also evident after transection, although a cavity does not form. In both cases the injured tissue is isolated from the environment by reactive astrocytes through formation of a glial scar (Hausmann, 2003). Tissue therapy after SCI is classified as neuroprotective, reducing the biochemical damage of secondary injury; or neuroregenerative, restoring lost function through regrowth and reinnervation of axons.

COMPRESSION INJURY

Strategies to treat compression injury have focused on drug delivery for neuroprotection and, more recently, neuroregeneration, using the molecules discussed on page 1026. Systemic delivery is often not feasible due to the filtering action of the BBB, which prevents many molecules from reaching the target tissue. Localized release circumvents the BBB and is the preferred preclinical delivery strategy, an approach that was applied in recent trials of anti-NogoA (Novartis) and Cethrin® (Alseres). These were the first clinical uses of local release after acute SCI. Drugs are most often delivered to the epidural space, intrathecal space, or directly into the cord, with these regions shown schematically in Figure 55.3.

Epidural delivery is the least invasive strategy and was used to deliver Cethrin® from a fibrin sealant deposited onto the exposed dura mater (McKerracher and Higuchi, 2006). Fibrin,

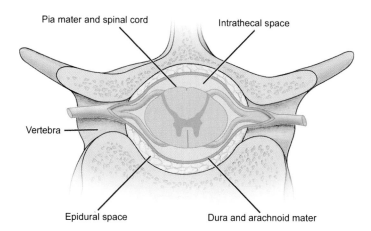

FIGURE 55.3

Gross anatomy of the spinal cord. Neural tissue is protected by the vertebrae, dura, arachnoid, and pia mater. Cerebrospinal fluid flows through the intrathecal space. Drugs are most often delivered to the epidural and intrathecal spaces. *(Copyright (2005) by Michael Corrin, reproduced with permission of the artist.)*

hydrogel (Paavola et al., 1998), or liposomal (Paavola et al., 2000) delivery strategies are preferred because the matrix localizes the drug load at a desired location and can prolong drug release. Release is typically sustained for up to a day by drug diffusion from the matrix. The disadvantage of epidural delivery is that the therapeutic molecule must cross the dura mater, arachnoid mater, and fluid-filled intrathecal space before reaching the damaged spinal cord, and thus there is very limited tissue penetration.

Intrathecal injection bypasses the two meninges and is the location of choice for many drug delivery strategies. Continuous intrathecal infusion is most often achieved from an indwelling catheter and pump system. The pump may be osmotic or battery powered, and implanted or external. A subcutaneously implanted pump is shown in Figure 55.4. Implanted osmotic mini-pumps have been used in many preclinical *in vivo* studies up to 28 days (Namiki et al., 2000; Oertle et al., 2003; Weinmann et al., 2006; Gonzenbach et al., 2008; Onose et al., 2009). Anti-NogoA was delivered by intrathecal infusion in animal studies and to certain patients in the Novartis trial (ClinicalTrials.gov, 2009). Continuous infusion provides excellent dose control but prolonged catheterization carries the risk of infection and catheter tips have been shown to compress and scar the spinal cord (Jones and Tuszynski, 2001). Alternatives include the bolus injection of a sustained release formulation such as drug-loaded polymer particles, which are distributed throughout the intrathecal space (Lagarce et al., 2005); and hydrogels which gel and remain localized at the site of injection. Work within Molly Shoichet's laboratory has focused on the development of such a hydrogel-based drug delivery system, shown in Figure 55.5. Early research showed that injected collagen remained localized and was safe as determined by behavioral, histological, and immunohistochemical techniques for both uninjured and SCI animals (Jimenez Hamann et al., 2003). Subsequent work with a blend of hyaluronic acid and methylcellulose (HAMC) resulted in a device demonstrating *in vitro* erythropoietin release for 1 day and localization of the hydrogel *in vivo* for between 4 and 7 days (Kang et al., 2009). Drug loaded poly(lactide-*co*-glycolide) nanoparticles were included in a second formulation of HAMC to both sustain drug release and stabilize the hydrogel, resulting in HAMC composites that were stable for more than 50 days and sustained drug release for 28 days *in vitro* (Baumann et al., 2009). The safety of this material was supported by a study which demonstrated that SCI rats which received composite HAMC showed no difference in motor function, cavity volume, area of inflammation, or astrogliosis relative to controls (Baumann et al., 2010).

1033

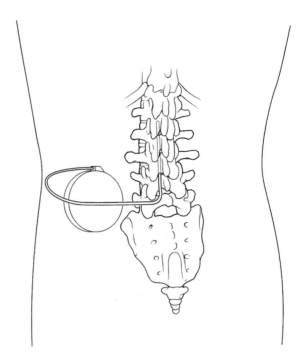

FIGURE 55.4
Infusions pumps may be implanted under the skin in a patient's abdomen and an intraspinal catheter tunneled
subcutaneously into the epidural or intrathecal space. Models used in animal studies are often powered by osmotic
pressure, those in clinical use by batteries or pressurized propellant. *(Copyright (2006) by Michael Corrin, reproduced with
permission of the artist.)*

1034

FIGURE 55.5
When injected into the
intrathecal space a hydrogel-
based drug delivery system can
localize and modulate drug
release at the site of injury. This
route is preferred over epidural
delivery when the diffusive barrier
presented by the dura mater is
significant, as is the case when
delivering large-molecular-weight
therapeutics. *(Copyright (2005) by
Michael Corrin, reproduced with
permission of the artist.)*

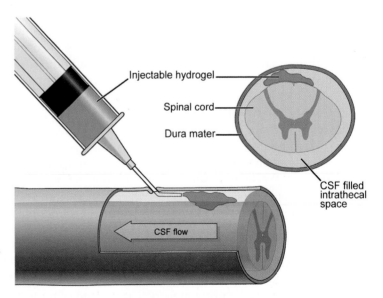

Intraspinal delivery affords the greatest possible degree of localization but this route is typically
not pursued for drug delivery because of the damage caused by the needle track. There is
significant interest in intraspinal cell delivery and investigators have used many of the cell types
introduced on page 1027. In acute injury models, cells are injected into the tissue as
a suspension in buffer (Parr et al., 2007; Mothe et al., 2008) or polymer matrix (Itosaka et al.,
2009) near the site of injury. Cell scaffolds have also been injected or implanted into the cystic
cavity in a chronic model of SCI (Nomura et al., 2008a; Hejel et al., 2010).

TRANSECTION INJURY

Treatment strategies for transection injury utilize physical constructs to provide a permissive environment for regeneration. When the cord is completely severed the stumps can be placed within nerve guidance channels in a strategy analogous to those pursued in peripheral nerve repair, where there has been clinical success. Tube dimensions, wall thickness, porosity, and mechanical strength are all important physical factors that influence the suitability of tubes for CNS repair applications. Matching the mechanical properties, and specifically modulus, of the tube to the tissue has been shown to be important to avoid necrotic tissue at the interface (Millesi et al., 1995; Dalton et al., 2001). Tubes are fabricated from the materials described on pages 1028–1029 to meet these criteria and increasingly, to function as platforms for drug and/or cell delivery. BDNF has been released from PLGA foams *in vivo* (Patist et al., 2004) and the model drug alkaline phosphatase has been released from chitosan tube/PLGA microparticle composites *in vitro* (Kim et al., 2008). In this system the drug-loaded particles were coated onto the tube surface, providing a mechanism to deliver multiple molecules at different rates according to the particle formulation used. Concerning cell delivery, Schwann cells (Montgomery and Robson, 1990; Oudega et al., 2001; Novikov et al., 2002), astrocytes (Montgomery and Robson, 1990), and neural stem/progenitor cells (Nomura et al., 2008b; Zahir et al., 2008) have been included in the inner lumen of tubes, either adherent to the inner surface or suspended in a hydrogel within the tube. Cell-adhesive hydrogels such as dilute collagen (Midha et al., 2001) and laminin-functionalized agarose (Bellamkonda et al., 1995) have shown promise in this application. Hydrogels alone are also used when the cord is partially or fully transected to fill the tissue defect and promote bridging of the gap (Woerly et al., 2001; Horn et al., 2007; Hejcl et al., 2008).

Retinal degeneration

Retinal disease leads to permanent loss of visual function for which there is no definitive treatment. Vision is a major quality-of-life issue for individuals and while diseases of sight do not attract as much attention as those which are life-threatening, polls conducted in the past by the Gallup Organization revealed blindness was the second most feared disease by Americans, after cancer.

Age-related macular degeneration (AMD) is the most common cause of irreversible vision loss in the USA (Kaufman, 2009), predominantly affecting seniors over 65. Diabetic retinopathy is the primary cause of blindness in middle-aged working adults (Klein, 2007). Retinitis pigmentosa (RP) shows a pediatric onset and is the leading cause of inherited retinal degeneration-associated blindness (Shintani et al., 2009). Irreversible photoreceptor death or loss is common to all of these pathologies. The search for cures is made more urgent by the knowledge that, as populations in the developed world age, the incidence of blindness due to retinal degeneration is expected to rise (Lee et al., 1998; Congdon et al., 2003).

Current therapies are focused on pharmacotherapy. These therapies show promise in limiting the pathophysiological advancement of the disease but are not a restorative approach to vision loss. To understand pharmacological therapies for retinal degeneration it is important to understand the pathogenesis of a typical degenerative condition such as AMD. It is commonplace to describe AMD as "dry" (atrophic) or "wet" (exudative). Dry-AMD is identified by diffuse, discrete deposits in Bruch's membrane (drusen) and atrophy of the retinal pigment epithelium (RPE). Choroidal neovascular membranes originating from the underlying choriocapillaris are the hallmark of wet AMD. These vessels are leaky, and exudate can form macular scars that directly destroy both photoreceptors and the RPE. The RPE is essential for homeostatic photoreceptor outer segment phagocytosis, and disruption of the RPE is associated with photoreceptor death (Arden et al., 2005; Rakoczy et al., 2006; Ding et al., 2009). There have been recent advances in the treatment of wet AMD with anti-vascular endothelial growth factor (anti-VEGF) therapies (Rosenfeld et al., 2006; Menon and Walters,

1035

2009). Examples of currently available drugs include ranibizumab (Lucentis®), bevacizumab (Avastin®) (Pedersen et al., 2009; Carneiro et al., 2010; Gupta et al., 2010), and pegaptanib (Macugen®). Other experimental treatments in AMD and diabetic retinopathy focus on eliminating oxidative damage using bioactive molecules such as advanced glycosylation end-product inhibitors and antioxidants (Comer and Ciulla, 2005). Gene therapies might also present a method for targeting defects that result from inherited retinal disorders such as Leber's congenital amaurosis (Bainbridge et al., 2008), retinitis pigmentosa (Tschernutter et al., 2005), and ocular albinism (Surace et al., 2005).

Cell transplantation is an alternate strategy which holds promise for restoration of vision. The most relevant clinical studies currently being conducted in patients with retinal degeneration are retinal sheet transplants. This procedure relies on an immature retinal sheet taken from fetal tissue to extend processes and integrate with the degenerating recipient retina. The rationale is that spared inner cells of the retina only require synaptic connections with functional photoreceptors to complete the visual circuit. Various types of tissue have been allografted: fetal RPE cells in patients with AMD (Algveve et al., 1996, 1999), and neural retinal cells in patients with RP (Das et al., 1996). These studies have shown some subjective improvement in function (Kaplan et al., 1997; Radtke et al., 1999; Humayun et al., 2000; Berger et al., 2003). Fetal sheet transplantation has thus far shown limited potential for repair but is one of few therapeutic options for those with retinal degeneration.

Preclinical research on cell transplantation is directed toward the transplantation of cell suspensions. Immature neurons are capable of migrating and differentiating during neural development, and many studies have focused on the integration of these brain-derived neural progenitors transplanted into the retina. However, transplantation to the adult retina showed limited donor cell integration (Young et al., 2000; Sakaguchi et al., 2005). It was assumed that the adult retina lacks the extrinsic cues that are present during development to aid donor cell integration (West et al., 2009). The inability of neural precursors to differentiate into photo-receptors suggested that a more appropriate cell source for transplantation might be neural retinal progenitor cells (RPCs). These cells are native to the retinal microenvironment and therefore may more readily generate photoreceptors (Klassen et al., 2004 a, b). Morphological differentiation and integration have been observed with transplanted RPCs into the retina of adult pigs (Warfvinge et al., 2005). Recently, MacLaren and colleagues demonstrated the integration of functional photoreceptors into the retina of adult mice, but only when the donor cells were post-mitotic photoreceptor precursors (MacLaren et al., 2006). These transplants showed functional synaptic activity and the ability to recover light response when transplanted into the $rho^{-/-}$ mouse, a model of selective photoreceptor loss. This study suggests that the ontogenetic stage of the transplanted cell is likely crucial for successful integration into the outer nuclear layer of the retina, and association with other photoreceptors (Fig. 55.6).

Recent studies suggest stem cell transplantation shows promise for reconstituting damaged cell populations in the retina (Klassen et al., 2004a; Enzmann et al., 2009). Embryonic stem cells, adult stem cells and induced pluripotent stem (iPS) cells are the three main classes of stem cells. They represent sources for therapeutic cell populations for retinal repair. Each has their own advantages and hurdles that must be cleared before they can be utilized for therapy (Daley and Scadden, 2008). While ES and iPS cells can be differentiated into cells of any of the three germ layers of the adult organism, they are inefficient in differentiation along a particular lineage and their uncontrolled proliferative ability is a safety concern in clinical application (Amariglio et al., 2009). The absence of a method to reliably direct these cells along a single lineage while avoiding the risk of tumor formation has restricted their use in humans (Belmonte et al., 2009; Clarke and van der Kooy, 2009). To date, various protocols have been developed for enriching the portion of ESCs which differentiate along a retinal lineage (Banin et al., 2006; Lamba et al., 2006, 2009; Osakada et al., 2008). These cells show the ability to integrate into models of retinal degeneration (Lamba et al., 2006, 2010) and restore some visual function in models of

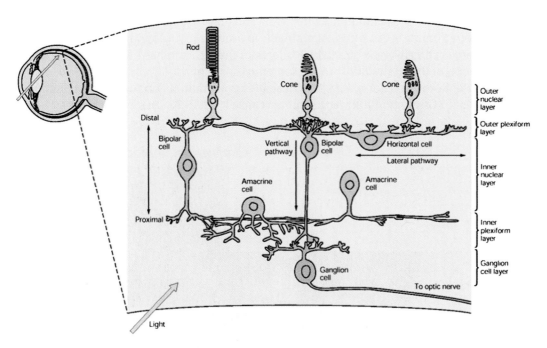

FIGURE 55.6

The human retina possesses a laminar structure. Photoreceptors (rods and cones) lie in the outer nuclear layer, interneurons (bipolar, horizontal, and amacrine cells) in the inner nuclear layer, and ganglion cells in the ganglion cell layer. Photoreceptors, bipolar cells, and horizontal cells make connections with each other in the outer plexiform layer. Bipolar, amacrine, and ganglion cells make connections in the inner plexiform layer. Information flows from outer lamina to inner lamina, as well as laterally via horizontal and amacrine cells. *(Reprinted with permission from Kandel et al., 2000.)*

photoreceptor dystrophism (Lamba et al., 2009). But the absence of a method to purify these populations means a mixed population of differentiated cells is transplanted (Sharp and Keirstead, 2007). Concerning iPS cells, the current methods of somatic cell reprogramming are still slow and inefficient (Belmonte et al., 2009). Also, without going through a definitive retinal stem cell type, these methods cannot be certain to exclude non-retinal cell types. Multipotent adult stem cells are not considered as tumorigenic as ES/iPS cells but lack the potency to generate all cell types. Amplification of adult stem cell populations *in vitro* to yield clinically useful numbers is the major challenge, and will require an understanding of the *in vivo* stem cell niche (Scadden, 2006; Morrison and Spralding, 2008). Van der Kooy and colleagues reported isolation of a stem cell in the adult mouse eye in 2000 that was multipotential for all retinal cell types (Tropepe et al., 2000). Four years later, the isolation of human retinal stem cells (RSCs) was reported (Coles et al., 2004) from donor eyes ranging from early post-natal to seventh decade of life. These cells were able to survive, migrate, and integrate into the developing neural retina. A major limitation of using these cells is that photoreceptors make up only a small proportion (~10%) of differentiated cells *in vitro*. Stable transfection of genes known to be important in determining photoreceptor fate has greatly increased the number of photo-receptors obtained under differentiation conditions, as well as shown integration of functional photoreceptors following transplantation into developing retina (Inoue et al., 2010).

Used as a transplant model, the developing mouse host eye is permissive to donor cell inte-gration but the adult retina presents unique challenges for integration. Cell death, leakage, and migration from the transplantation site occur when cells are delivered as a suspension in saline (Klassen et al., 2004a). To overcome these barriers in the adult, interdisciplinary studies including the combination of regular tissue culture with tissue engineering are being pursued. These have included the delivery of retinal progenitors, isolated during development, on solid biomaterial scaffolds (Tomita et al., 2005; Hogg et al., 2006; Tao et al., 2007; Neeley et al., 2008; Redenti et al., 2009). These reports are an important advancement but the scaffolds employed do not match the modulus of the retina and may lack the flexibility required for sub-retinal delivery (Tomita et al., 2005). A minimally invasive, injectable and *in situ* biodegradable hydrogel-based cell delivery vehicle which addresses this concern has been developed for transplantation of the adult RSCs into the subretinal space of adult mice (Ballios et al., 2010). The hydrogel allowed for normal RSC survival and proliferation, and for greater, continuous integration of RSCs into the RPE versus saline vehicle (Fig. 55.7). This cell delivery strategy

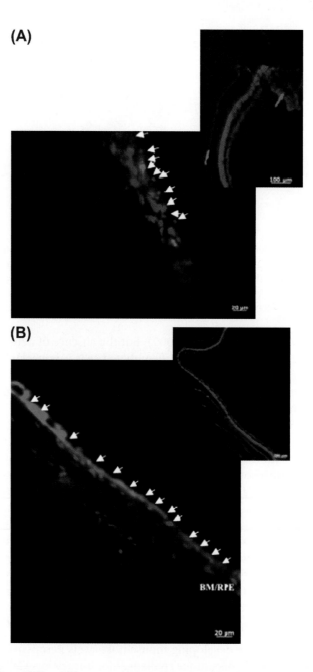

FIGURE 55.7
Subretinal transplantation of GFP + RSC progeny in the vehicle, a physical blend of hyaluronan (HA) and methylcellulose (MC) — HAMC — assayed at 4 weeks post-transplantation. (A) Transplantation of GFP+ RSCs in saline shows non-contiguous cellular integration and localized cellular aggregates (inset upper right) atop Bruch's membrane (BM), suggestive of aggregation pre- or post-transplantation. (B) Transplantation of GFP+ RSCs in HAMC shows contiguous areas of RPE integration over large areas of retina (inset upper right), suggesting HAMC maintains cellular distribution during injection and prevents aggregation pre- or post-transplantation. Arrowheads indicate location of individual nuclei of transplanted cells (Hoechst nuclear stain, blue). *(Reprinted with permission from Ballios et al., 2010.)*

may be useful for the treatment of advanced maculopathy, where large areas of RPE are destroyed (Hogg et al., 2006). (These advances in solid and injectable scaffolds are a new dimension to the treatment of retinal disease. Translation to the clinic will depend on improved visual function which will result from greater cell survival and host tissue integration.

CONCLUSIONS AND OUTLOOK

Tissue therapy in the treatment of CNS disorders and injury is at the forefront of techniques which address the root cause of functional deficits. Important advance have been made in many disorders including stroke, spinal cord injury, and retinal degeneration. The combination of drug delivery, cell delivery, and physical constructs is a promising approach to create a permissive environment populated with cells primed for regeneration. For this reason, future efforts will continue to see a merging of combination strategies of cells and biological factors with biomaterials-based scaffolds important for cell survival and integration — two key limitations to cell transplantation today. Optimization of these strategies is still required. Cell delivery requires determining the optimum population for transplantation that produces significant integration with existing neuronal circuitry and functional recovery. Continued innovation is required in the field of scaffold design to provide the optimal microenvironment for cell survival, differentiation, and integration. The biochemical and mechanical properties of the scaffold in addition to cell—cell interactions are all active areas of research in biomimetic scaffold design.

For endogenous stem cell stimulation, the challenges associated with cell transplantation are negated; however, there are other challenges. Advances in drug delivery technology will allow temporal and spatial control of released factors, yet determining which factors to release and when remains to be done. Moreover, for local release, new delivery strategies will be required that are minimally invasive, ensure diffusion through CNS tissue and thus encourage local stimulation of stem cells. Since many cells have common receptors, uniquely stimulating the endogenous stem cells may be difficult. Of course, stimulating stem cells to promote repair is the ultimate challenge. The complexity of the CNS and its numerous protective barriers has made the intricacies of its function difficult to understand and repair. Notwithstanding these challenges, the promise of regenerative medicine in the CNS can build on the successes in other tissues, to advance science to clinical application in the future.

References

Aebischer, P., Guenard, V., & Valentini, R. F. (1990). The morphology of regenerating peripheral nerves is modulated by the surface microgeometry of polymeric guidance channels. *Brain Res., 531,* 211—218.

Akiyama, Y., Honmou, O., Kato, T., Uede, T., Hashi, K., & Kocsis, J. D. (2001). Transplantation of clonal neural precursor cells derived from adult human brain establishes functional peripheral myelin in the rat spinal cord. *Exp. Neurol., 167,* 27—39.

Alagappan, D., Lazzarino, D. A., Felling, R. J., Balan, M., Kotenko, S. V., & Levison, S. W. (2009). Brain injury expands the numbers of neural stem cells and progenitors in the SVZ by enhancing their responsiveness to EGF. *ASN Neuro., 1.*

Algvere, P. V., Berglin, L., Gouras, P., & Sheng, Y. (1996). Human fetal RPE transplants in age related macular degeneration (ARMD). *Invest. Ophthalmol. Vis. Sci., 37,* S96.

Algvere, P. V., Gouras, P., & Kopp, E. D. (1999). Long-term outcome of RPE allografts in non-immunosuppressed patients with AMD. *Eur. J. Ophthalmol., 9,* 217—230.

Alper, J. (2009). Geron gets green light for human trial of ES cell-derived product. *Nat. Biotechnol., 27,* 213—214.

Amariglio, N., Hirshberg, A., Scheithauer, B. W., et al. (2009). Donor-derived brain tumor following neural stem cell transplantation in an ataxia telangiectasia patient. *PLoS Med., 6,* e1000029.

Aoi, T., Yae, K., Nakagawa, M., et al. (2008). Generation of pluripotent stem cells from adult mouse liver and stomach cells. *Science, 321,* 699—702.

Arden, G. B., Sidman, R. L., Arap, W., & Schlingemann, R. O. (2005). Spare the rod and spoil the eye. *Br. J. Ophthalmol., 89,* 764—769.

Bacigaluppi, M., Pluchino, S., Peruzzotti Jametti, L., et al. (2009). Delayed post-ischaemic neuroprotection following systemic neural stem cell transplantation involves multiple mechanisms. *Brain, 132,* 2239–2251.

Bainbridge, J. W. B., Smith, A. J., Barker, S. S., et al. (2008). Effect of gene therapy on visual function in Leber's congenital amaurosis. *N. Engl. J. Med., 358,* 2231–2239.

Ballabh, P., Braun, A., & Nedergaard, M. (2004). The blood–brain barrier: an overview: structure, regulation, and clinical implications. *Neurobiol. Dis., 16,* 1–13.

Ballios, B. G., Cooke, M. J., van der Kooy, D., & Shoichet, M. S. (2010). A hydrogel-based stem cell delivery system to treat retinal degenerative diseases. *Biomaterials, 31,* 2555–2564.

Bambakidis, N. C., Wang, R.-Z., Franic, L., & Miller, R. H. (2003). Sonic hedgehog-induced neural precursor proliferation after adult rodent spinal cord injury. *J. Neurosurg., 99,* 70–75.

Banin, E., Obolensky, A., Idelson, M., et al. (2006). Retinal incorporation and differentiation of neural precursors derived from human embryonic stem cells. *Stem Cells, 24,* 246–257.

Baumann, M. D., Kang, C. E., Stanwick, J. C., et al. (2009). An injectable drug delivery platform for sustained combination therapy. *J. Control. Rel., 138,* 205–213.

Baumann, M. D., Kang, C. E., Tator, C. H., & Schoichet, M. S. (2010). Intrathecal drug delivery of a polymeric nanocomposite hydrogel after spinal cord injury. *Biomaterials, 31,* 7631–7639.

Belayev, L., Khoutorova, L., Zhao, K. L., Davidoff, A. W., Moore, A. F., & Cramer, S. C. (2009). A novel neurotrophic therapeutic strategy for experimental stroke. *Brain Res., 1280,* 117–123.

Bellamkonda, R., Ranieri, J. P., & Aebischer, P. (1995). Laminin oligopeptide derivatized agarose gels allow three-dimensional neurite extension. *in vitro. J. Neurosci. Res., 41,* 501–509.

Belmonte, J. C. I., Ellis, J., Hochedlinger, K., & Yamanaka, S. (2009). Induced pluripotent stem cells and reprogramming: seeing the science through the hype. *Nat. Rev. Genet., 10,* 878–883.

Ben-Hur, T., Einstein, O., Mizrachi-Kol, R., et al. (2003). Transplanted multipotential neural precursor cells migrate into the inflamed white matter in response to experimental autoimmune encephalomyelitis. *Glia, 41,* 73–80.

Berger, A. S., Tezel, T. H., Del, P. L. V., & Kaplan, H. J. (2003). Photoreceptor transplantation in retinitis pigmentosa: short-term follow-up. *Ophthalmology, 110,* 383–391.

Bethea, J. R., Nagashima, H., Acosta, M. C., et al. (1999). Systemically administered interleukin-10 reduces tumor necrosis factor-alpha production and significantly improves functional recovery following traumatic spinal cord injury in rats. *J. Neurotrauma, 16,* 851–863.

Biernaskie, J., Sparling, J. S., Liu, J., et al. (2007). Skin-derived precursors generate myelinating Schwann cells that promote remyelination and functional recovery after contusion spinal cord injury. *J. Neurosci., 27,* 9545–9559.

Bracken, M. B., Shepard, M. J., Holford, T. R., et al. (1998). Methylprednisolone or tirilazad mesylate administration after acute spinal cord injury: 1-year follow up. Results of the third National Acute Spinal Cord Injury randomized controlled trial. *J. Neurosurg., 89,* 699–706.

Braun, S., Croizat, B., Lagrange, M. C., Warter, J. M., & Poindron, P. (1996). Neurotrophins increase motoneurons' ability to innervate skeletal muscle fibers in rat spinal cord–human muscle cocultures. *J. Neurol. Sci., 136,* 17–23.

Buchser, E., Goddard, M., Heyd, B., et al. (1996). Immunoisolated xenogenic chromaffin cell therapy for chronic pain. Initial clinical experience. *Anesthesiology, 85,* 1005–1012, discussion 29A–30A.

Cameron, H. A., Woolley, C. S., McEwen, B. S., & Gould, E. (1993). Differentiation of newly born neurons and glia in the dentate gyrus of the adult rat. *Neuroscience, 56,* 337–344.

Carneiro, C. M., Falco, M. S., Brando, E. M., & Falco-Reis, F. M. (2010). Intravitreal bevacizumab for neovascular age-related macular degeneration with or without prior treatment with photodynamic therapy: one-year results. *Retina, 30,* 85–92.

Chiasson, B. J., Tropepe, V., Morshead, C. M., & van der kooy, D. (1999). Adult mammalian forebrain ependymal and subependymal cells demonstrate proliferative potential, but only subependymal cells have neural stem cell characteristics. *J. Neurosci., 19,* 4462–4471.

Christensen, M. D., & Hulsebosch, C. E. (1997). Spinal cord injury and anti-NGF treatment results in changes in CGRP density and distribution in the dorsal horn in the rat. *Exp. Neurol., 147,* 463–475.

Clarke, L., & van der Kooy, D. (2009). A safer stem cell: inducing pluripotency. *Nat. Med., 15,* 1001–1002.

ClinicalTrials.gov. Acute safety, tolerability, feasibility and pharmacokinetics of intrathecally administered ATI355 in patients with acute SCI. 2009

Coles, B. L. K., Angenieux, B., Inoue, T., et al. (2004). Facile isolation and the characterization of human retinal stem cells. *Proc. Natl. Acad. Sci. USA, 101,* 15772–15777.

Comer, G. M., & Ciulla, T. A. (2005). Current and future pharmacological intervention for diabetic retinopathy. *Expert Opin. Emerg. Drugs, 10,* 441–455.

Congdon, N. G., Friedman, D. S., & Lietman, T. (2003). Important causes of visual impairment in the world today. *JAMA, 290,* 2057–2060.

Cramer, S. C., Fitzpatrick, C., Warren, M., et al. (2010). The Beta-hCG+Erythropoietin in Acute Stroke (BETAS) Study. A 3-center, single-dose, open-label, noncontrolled, phase IIa safety trial. *Stroke, 41*, 745—750.

Daley, G. Q., & Scadden, D. T. (2008). Prospects for stem cell-based therapy. *Cell, 132*, 544—548.

Dalton, P. D., & Shoichet, M. S. (2001). Creating porous tubes by centrifugal forces for soft tissue application. *Biomaterials, 22*, 2661—2669.

Dalton, P. D., Flynn, L., & Shoichet, M. S. (2002). Manufacture of poly(2-hydroxyethyl methacrylate-co-methyl methacrylate) hydrogel tubes for use as nerve guidance channels. *Biomaterials, 23*, 3843—3851.

Das, T. P., Del Cerro, M., Lazar, E. S., et al. (1996). Transplantation of neural retina in patients with retinitis pigmentosa. *Invest. Ophthalmol. Vis. Sci., 37*, S96.

Ding, X., Patel, M., & Chan, C-C. (2009). Molecular pathology of age-related macular degeneration. *Prog. Retin. Eye Res., 28*, 1—18.

Duggan, P. S., Siegel, A. W., Blass, D. M., et al. (2009). Unintended changes in cognition, mood, and behavior arising from cell-based interventions for neurological conditions: ethical challenges. *Am. J. Bioethics, 9*, 31—36.

Dyer, M. A., & Cepko, C. L. (2000). Control of Muller glial cell proliferation and activation following retinal injury. *Nat. Neurosci., 3*, 873—880.

Eid, T., & Brines, M. (2002). Recombinant human erythropoietin for neuroprotection: what is the evidence? *Clin. Breast Cancer, 3*(Suppl. 3), S109—S115.

Enzmann, V., Yolcu, E., Kaplan, H. J., & Ildstad, S. T. (2009). Stem cells as tools in regenerative therapy for retinal degeneration. *Arch. Ophthalmol., 127*, 563—571.

Esneault, E., Pacary, E., Eddi, D., et al. (2008). Combined therapeutic strategy using erythropoietin and mesenchymal stem cells potentiates neurogenesis after transient focal cerebral ischemia in rats. *J. Cereb. Blood Flow Metab., 28*, 1552—1563.

Faulkner, J. R., Herrmann, J. E., Woo, M. J., Tansey, K. E., Doan, N. B., & Sofroniew, M. V. (2004). Reactive astrocytes protect tissue and preserve function after spinal cord injury. *J. Neurosci., 24*, 2143—2155.

Fawcett, J. W., & Asher, R. A. (1999). The glial scar and central nervous system repair. *Brain Res. Bull., 49*, 377—391.

Fernandes, K. J., McKenzie, I. A., Mill, P., et al. (2004). A dermal niche for multipotent adult skin-derived precursor cells. *Nat. Cell Biol., 6*, 1082—1093.

Fernandes, K. J., Kobayashi, N. R., Gallagher, C. J., et al. (2006). Analysis of the neurogenic potential of multipotent skin-derived precursors. *Exp. Neurol., 201*, 32—48.

Fink, J. D. W. (2009). FDA regulation of stem cell-based products. *Science, 324*, 1662—1663.

Fischer, A. J., & Reh, T. A. (2001). Muller glia are a potential source of neural regeneration in the postnatal chicken retina. *Nat. Neurosci., 4*, 247—252.

Foret, A., Quertainmont, R., Botman, O., et al. (2010). Stem cells in the adult rat spinal cord: plasticity after injury and treadmill training exercise. *J. Neurochem., 112*, 762—772.

Freed, C. R., Breeze, R. E., Rosenberg, N. L., et al. (1990). Transplantation of human fetal dopamine cells for Parkinson's disease. Results at 1 year. *Arch. Neurol., 47*, 505—512.

Gage, F. H., Coates, P. W., Palmer, T. D., et al. (1995). Survival and differentiation of adult neuronal progenitor cells transplanted to the adult brain. *Proc. Natl. Acad. Sci. USA, 92*, 11879—11883.

Garcia, M., Forster, V., Hicks, D., & Vecino, E. (2002). Effects of Muller glia on cell survival and neuritogenesis in adult porcine retina *in vitro*. *Invest. Ophthalmol. Vis. Sci., 43*, 3735—3743.

Geisler, F. H., Coleman, W. P., Grieco, G., & Poonian, D. (2001). The Sygen multicenter acute spinal cord injury study. *Spine, 26*, S87—S98.

Gonzenbach, R. R., & Schwab, M. E. (2008). Disinhibition of neurite growth to repair the injured adult CNS: focusing on Nogo. *Cell Mol. Life Sci., 65*, 161—176.

Gupta, B., Elagouz, M., & Sivaprasad, S. (2010). Intravitreal bevacizumab for choroidal neovascularization secondary to causes other than age-related macular degeneration. *Eye, 24*, 203—213.

Halme, D. G., & Kessler, D. A. (2006). FDA regulation of stem-cell-based therapies. *N. Engl. J. Med., 355*, 1730—1735.

Hannila, S. S., & Filbin, M. T. (2008). The role of cyclic AMP signaling in promoting axonal regeneration after spinal cord injury. *Exp. Neurol., 209*, 321—332.

Hausmann, O. N. (2003). Post-traumatic inflammation following spinal cord injury. *Spinal Cord, 41*, 369—378.

Hejcl, A., Lesny, P., Pradny, M., et al. (2008). Biocompatible hydrogels in spinal cord injury repair. *Physiol. Res., 57*, S121—S132.

Hejcl, A., Sedy, J., Kapcalova, M., et al. (2010). HPMA-RGD hydrogels seeded with mesenchymal stem cells improve functional outcome in chronic spinal cord injury. *Stem Cells Dev., 19*, 1535—1546.

Hicks, A. U., Hewlett, K., Windle, V., et al. (2007). Enriched environment enhances transplanted subventricular zone stem cell migration and functional recovery after stroke. *Neuroscience, 146*, 31—40.

Hicks, A. U., Lappalainen, R. S., Narkilahti, S., et al. (2009). Transplantation of human embryonic stem cell-derived neural precursor cells and enriched environment after cortical stroke in rats: cell survival and functional recovery. *Eur. J. Neurosci., 29*, 562–574.

Hogg, R. E., & Chakravarthy, U. (2006). Visual function and dysfunction in early and late age-related maculopathy. *Prog. Retin. Eye Res., 25*, 249–276.

Horn, E. M., Beaumont, M., Shu, X. Z., et al. (2007). Influence of cross-linked hyaluronic acid hydrogels on neurite outgrowth and recovery from spinal cord injury. *J. Neurosurg. Spine, 6*, 133–140.

Horner, P. J., Power, A. E., Kempermann, G., et al. (2000). Proliferation and differentiation of progenitor cells throughout the intact adult rat spinal cord. *J. Neurosci., 20*, 2218–2228.

Hossmann, K. A. (2006). Pathophysiology and therapy of experimental stroke. *Cell. Mol. Neurobiol., 26*, 1057–1083.

Humayun, M. S., De, J. E., Jr., Del, C. M., et al. (2000). Human neural retinal transplantation. *Invest. Ophthalmol. Vis. Sci., 41*, 3100–3106.

Inoue, T., Coles, B. L., Dorval, K., et al. (2010). Maximizing functional photoreceptor differentiation from adult human retinal stem cells. *Stem Cells, 28*, 489–500.

Itosaka, H., Kuroda, S., Shichinohe, H., et al. (2009). Fibrin matrix provides a suitable scaffold for bone marrow stromal cells transplanted into injured spinal cord: a novel material for CNS tissue engineering. *Neuropathology, 29*, 248–257.

Jain, A., Kim, Y. T., McKeon, R. J., & Bellamkonda, R. V. (2006). *In situ* gelling hydrogels for conformal repair of spinal cord defects, and local delivery of BDNF after spinal cord injury. *Biomaterials, 27*, 497–504.

Jimenez Hamann, M. C., Tsai, E. C., Tator, C. H., & Shoichet, M. S. (2003). Novel intrathecal delivery system for treatment of spinal cord injury. *Exp. Neurol., 182*, 300–309.

Jin, K., Mao, X., Xie, L., et al. (2010). Transplantation of human neural precursor cells in Matrigel scaffolding improves outcome from focal cerebral ischemia after delayed postischemic treatment in rats. *J. Cereb. Blood Flow Metab., 30*, 534–544.

Jones, K. S., Sefton, M. V., & Gorczynski, R. M. (2004). *In vivo* recognition by the host adaptive immune system of microencapsulated xenogeneic cells. *Transplantation, 78*, 1454–1462.

Jones, L. L., & Tuszynski, M. H. (2001). Chronic intrathecal infusions after spinal cord injury cause scarring and compression. *Microsc. Res. Tech., 54*, 317–324.

Kallur, T., Darsalia, V., Lindvall, O., & Kokaia, Z. (2006). Human fetal cortical and striatal neural stem cells generate region-specific neurons *in vitro* and differentiate extensively to neurons after intrastriatal transplantation in neonatal rats. *J. Neurosci. Res., 84*, 1630–1644.

Kandel, E. R., Schwartz, J. H. & Jessel, T. M. (2000). *Principles of Neural Science.* New York: McGraw-Hill. pp. xxvi, 515.

Kang, C. E., Poon, P. C., Tator, C. H., & Shoichet, M. S. (2009). A new paradigm for local and sustained release of therapeutic molecules to the injured spinal cord for neuroprotection and tissue repair. *Tissue Eng. Part A, 15*, 595–604.

Kaplan, H. J., Tezel, T. H., Berger, A. S., Wolf, M. L., & Del, P. L. V. (1997). Human photoreceptor transplantation in retinitis pigmentosa: a safety study. *Arch. Ophthalmol., 115*, 1168–1172.

Karimi-Abdolrezaee, S., Eftekharpour, E., Wang, J., Schut, D., & Fehlings, M. G. (2010). Synergistic effects of transplanted adult neural stem/progenitor cells, chondroitinase, and growth factors promote functional repair and plasticity of the chronically injured spinal cord. *J. Neurosci., 30*, 1657–1676.

Katayama, Y., Montenegro, R., Freier, T., Midha, R., Belkas, J. S., & Shoichet, M. S. (2006). Coil-reinforced hydrogel tubes promote nerve regeneration equivalent to that of nerve autografts. *Biomaterials, 27*, 505–518.

Kaufman, S. R. (2009). Developments in age-related macular degeneration: diagnosis and treatment. *Geriatrics, 64*, 16–19.

Keane, R. W., Davis, A. R., & Dietrich, W. D. (2006). Inflammatory and apoptotic signaling after spinal cord injury. *J. Neurotrauma, 23*, 335–344.

Keirstead, H. S., Nistor, G., Bernal, G., et al. (2005). Human embryonic stem cell-derived oligodendrocyte progenitor cell transplants remyelinate and restore locomotion after spinal cord injury. *J. Neurosci., 25*, 4694–4705.

Kigerl, K., & Popovich, P. (2006). Drug evaluation: ProCord – a potential cell-based therapy for spinal cord injury. *IDrugs, 9*, 354–360.

Kim, H., Tator, C. H., & Shoichet, M. S. (2008). Design of protein-releasing chitosan channels. *Biotechnol. Prog., 24*, 932–937.

Klassen, H., Sakaguchi, D. S., & Young, M. J. (2004a). Stem cells and retinal repair. *Prog. Retinal Eye Res., 23*, 149–181.

Klassen, H. J., Ng, T. F., Kurimoto, Y., et al. (2004b). Multipotent retinal progenitors express developmental markers, differentiate into retinal neurons, and preserve light-mediated behavior. *Invest. Ophthalmol. Vis. Sci., 45*, 4167–4173.

Klein, B. E. K. (2007). Overview of epidemiologic studies of diabetic retinopathy. *Ophthalm. Epidemiol.*, *14*, 179−183.

Kobayashi, T., Ahlenius, H., Thored, P., Kobayashi, R., Kokaia, Z., & Lindvall, O. (2006). Intracerebral infusion of glial cell line-derived neurotrophic factor promotes striatal neurogenesis after stroke in adult rats. *Stroke*, *37*, 2361−2367.

Koennecke, H. C., Nohr, R., Leistner, S., & Marx, P. (2001). Intravenous tPA for ischemic stroke team performance over time, safety, and efficacy in a single-center, 2-year experience. *Stroke*, *32*, 1074−1078.

Kolb, B., Morshead, C., Gonzalez, C., et al. (2007). Growth factor-stimulated generation of new cortical tissue and functional recovery after stroke damage to the motor cortex of rats. *J. Cereb. Blood Flow Metab.*, *27*, 983−997.

Kuhn, H. G., Dickinson-Anson, H., & Gage, F. H. (1996). Neurogenesis in the dentate gyrus of the adult rat: age-related decrease of neuronal progenitor proliferation. *J. Neurosci.*, *16*, 2027−2033.

Kumar, V., Abbas, A. K., Fausto, N., Robbins, S. L., & Cotran, R. S. (2005). *Robbins and Cotran Pathologic Basis of Disease*. Philadelphia: Elsevier Saunders. pp. xv, 1525.

Labiche, L. A., & Grotta, J. C. (2004). Clinical trials for cytoprotection in stroke. *NeuroRx*, *1*, 46−70.

Lagarce, F., Faisant, N., Desfontis, J. C., et al. (2005). Baclofen-loaded microspheres in gel suspensions for intra-thecal drug delivery: *in vitro* and *in vivo* evaluation. *Eur. J. Pharm. Biopharm.*, *61*, 171−180.

Lamba, D. A., Karl, M. O., Ware, C. B., & Reh, T. A. (2006). Efficient generation of retinal progenitor cells from human embryonic stem cells. *Proc. Natl. Acad. Sci. USA*, *103*, 12769−12774.

Lamba, D. A., Gust, J., & Reh, T. A. (2009). Transplantation of human embryonic stem cell-derived photoreceptors restores some visual function in Crx-deficient mice. *Cell. Stem Cell*, *4*, 73−79.

Lamba, D. A., McUsic, A., Hirata, R. K., Wang, P. R., Russell, D., & Reh, T. A. (2010). Generation, purification and transplantation of photoreceptors derived from human induced pluripotent stem cells. *PLoS ONE*, *5*, e8763.

Le Belle, J. E., & Svendsen, C. N. (2002). Stem cells for neurodegenerative disorders: where can we go from here? *BioDrug*, *16*, 389−401.

Le Belle, J. E., Caldwell, M. A., & Svendsen, C. N. (2004). Improving the survival of human CNS precursor-derived neurons after transplantation. *J. Neurosci. Res.*, *76*, 174−183.

Lee, J., Jallo, G. I., Penno, M. B., et al. (2006). Intracranial drug-delivery scaffolds: biocompatibility evaluation of sucrose acetate isobutyrate gels. *Toxicol. Appl. Pharmacol.*, *215*, 64−70.

Lee, P., Wang, C. C., & Adamis, A. P. (1998). Ocular neovascularization: an epidemiologic review. *Surv. Ophthalmol.*, *43*, 245−269.

Lee, T. T., Green, B. A., Dietrich, W. D., & Yezierski, R. P. (1999). Neuroprotective effects of basic fibroblast growth factor following spinal cord contusion injury in the rat. *J. Neurotrauma*, *16*, 347−356.

Leker, R. R. (2009). Fate and manipulations of endogenous neural stem cells following brain ischemia. *Expert Opin. Biol. Ther.*, *9*, 1117−1125.

Limb, A. G., & Jayaram, H. (2010). Regulatory and pathogenic roles of Müller glial cells in retinal neovascular processes and their potential for retinal regeneration. *Exp. Approach. Diab. Retinopathy*, *20*, 98−108.

Lindvall, O., Brundin, P., Widner, H., et al. (1990). Grafts of fetal dopamine neurons survive and improve motor function in Parkinson's disease. *Science*, *247*, 574−577.

Lo, E. H., Dalkara, T., & Moskowitz, M. A. (2003). Mechanisms, challenges and opportunities in stroke. *Nat. Rev. Neurosci.*, *4*, 399−415.

Lu, D., Mahmood, A., Qu, C., Hong, X., Kaplan, D., & Chopp, M. (2007). Collagen scaffolds populated with human marrow stromal cells reduce lesion volume and improve functional outcome after traumatic brain injury. *Neurosurgery*, *61*, 596−602, discussion 602−603.

Ma, J., Tian, W. M., Hou, S. P., Xu, Q. Y., Spector, M., & Cui, F. Z. (2007). An experimental test of stroke recovery by implanting a hyaluronic acid hydrogel carrying a Nogo receptor antibody in a rat model. *Biomed. Mater.*, *2*, 233−240.

MacLaren, R. E., Pearson, R. A., MacNeil, A., et al. (2006). Retinal repair by transplantation of photoreceptor precursors. *Nature*, *444*, 203−207.

Mahoney, M. J., & Saltzman, W. M. (2001). Transplantation of brain cells assembled around a programmable synthetic microenvironment. *Nat. Biotechnol.*, *19*, 934−939.

Martens, D. J., Seaberg, R. M., & Van Der Kooy, D. (2002). *In vivo* infusions of exogenous growth factors into the fourth ventricle of the adult mouse brain increase the proliferation of neural progenitors around the fourth ventricle and the central canal of the spinal cord. *Eur. J. Neurosci.*, *16*, 1045−1057.

Matheson, C. R., Carnahan, J., Urich, J. L., Bocangel, D., Zhang, T. J., & Yan, Q. (1997). Glial cell line-derived neurotrophic factor (GDNF) is a neurotrophic factor for sensory neurons: comparison with the effects of the neurotrophins. *J. Neurobiol.*, *32*, 22−32.

McKerracher, L., & Higuchi, H. (2006). Targeting Rho to stimulate repair after spinal cord injury. *J. Neurotrauma*, *23*, 309−317.

Menon, G., & Walters, G. (2009). New paradigms in the treatment of wet AMD: the impact of anti-VEGF therapy. *Eye*, *23*(Suppl. 1), 1−7.

Midha, R., Shoichet, M. S., Dalton, P. D., et al. (2001). Tissue engineered alternatives to nerve transplantation for repair of peripheral nervous system injuries. *Transplant. Proc., 33*, 612–615.

Millesi, H., Zoch, G., & Reihsner, R. (1995). Mechanical properties of peripheral nerves. *Clin. Orthop. Rel. Res.*, 76–83.

Montgomery, C. T., & Robson, J. A. (1990). New method of transplanting purified glial cells into the brain. *J. Neurosci. Meth., 32*, 135–141.

Morrison, S. J., & Spradling, A. C. (2008). Stem cells and niches: mechanisms that promote stem cell maintenance throughout life. *Cell, 132*, 598–611.

Morshead, C. M., Reynolds, B. A., Craig, C. G., et al. (1994). Neural stem cells in the adult mammalian forebrain: a relatively quiescent subpopulation of subependymal cells. *Neuron, 13*, 1071–1082.

Mothe, A. J., Kulbatski, I., Parr, A., Mohareb, M., & Tator, C. H. (2008). Adult spinal cord stem/progenitor cells transplanted as neurospheres preferentially differentiate into oligodendrocytes in the adult rat spinal cord. *Cell Transplant., 17*, 735–751.

Murray, V., Norrving, B., Sandercock, P. A., Terent, A., Wardlaw, J. M., & Wester, P. (2010). The molecular basis of thrombolysis and its clinical application in stroke. *J. Intern. Med., 267*, 191–208.

Nakahara, Y., Gage, F. H., & Tuszynski, M. H. (1996). Grafts of fibroblasts genetically modified to secrete NGF, BDNF, NT-3, or basic FGF elicit differential responses in the adult spinal cord. *Cell Transplant., 5*, 191–204.

Namiki, J., Kojima, A., & Tator, C. H. (2000). Effect of brain-derived neurotrophic factor, nerve growth factor, and neurotrophin-3 on functional recovery and regeneration after spinal cord injury in adult rats. *J. Neurotrauma, 17*, 1219–1231.

Neeley, W. L., Redenti, S., Klassen, H., et al. (2008). A microfabricated scaffold for retinal progenitor cell grafting. *Biomaterials, 29*, 418–426.

Nikulina, E., Tidwell, J. L., Dai, H. N., Bregman, B. S., & Filbin, M. T. (2004). The phosphodiesterase inhibitor rolipram delivered after a spinal cord lesion promotes axonal regeneration and functional recovery. *Proc. Natl. Acad. Sci. USA, 101*, 8786–8790.

Nisbet, D. R., Rodda, A. E., Horne, M. K., Forsythe, J. S., & Finkelstein, D. I. (2009). Neurite infiltration and cellular response to electrospun polycaprolactone scaffolds implanted into the brain. *Biomaterials, 30*, 4573–4580.

Nomura, H., Baladie, B., Katayama, Y., Morshead, C. M., Shoichet, M. S., & Tator, C. H. (2008a). Delayed implantation of intramedullary chitosan channels containing nerve grafts promotes extensive axonal regeneration after spinal cord injury. *Neurosurgery, 63*, 127–141, discussion 141–143.

Nomura, H., Zahir, T., Kim, H., et al. (2008b). Extramedullary chitosan channels promote survival of transplanted neural stem and progenitor cells and create a tissue bridge after complete spinal cord transection. *Tissue Eng., 14A*, 649–665.

Norenberg, M. D., Smith, J., & Marcillo, A. (2004). The pathology of human spinal cord injury: defining the problems. *J. Neurotrauma., 21*, 429–440.

Novikov, L. N., Novikova, L. N., Mosahebi, A., Wiberg, M., Terenghi, G., & Kellerth, J. O. (2002). A novel biodegradable implant for neuronal rescue and regeneration after spinal cord injury. *Biomaterials, 23*, 3369–3376.

Oertle, T., van der Haar, M. E., Bandtlow, C. E., et al. (2003). Nogo-A inhibits neurite outgrowth and cell spreading with three discrete regions. *J. Neurosci., 23*, 5393–5406.

Okita, K., Ichisaka, T., & Yamanaka, S. (2007). Generation of germline-competent induced pluripotent stem cells. *Nature, 448*, 313–317.

Onose, G., Anghelescu, A., Muresanu, D. F., et al. (2009). A review of published reports on neuroprotection in spinal cord injury. *Spinal Cord, 47*, 716–726.

Osakada, F., Ikeda, H., Mandai, M., et al. (2008). Toward the generation of rod and cone photoreceptors from mouse, monkey and human embryonic stem cells. *Nat. Biotechnol., 26*, 215–224.

Oudega, M., & Hagg, T. (1996). Nerve growth factor promotes regeneration of sensory axons into adult rat spinal cord. *Exp. Neurol., 140*, 218–229.

Oudega, M., Gautier, S. E., Chapon, P., et al. (2001). Axonal regeneration into Schwann cell grafts within resorbable poly(alpha-hydroxyacid) guidance channels in the adult rat spinal cord. *Biomaterials, 22*, 1125–1136.

Oyesiku, N. M., & Wigston, D. J. (1996). Ciliary neurotrophic factor stimulates neurite outgrowth from spinal cord neurons. *J. Comp. Neurol., 364*, 68–77.

Paavola, A., Tarkkila, P., Xu, M., Wahlstrom, T., Yliruusi, J., & Rosenberg, P. (1998). Controlled release gel of ibuprofen and lidocaine in epidural use – analgesia and systemic absorption in pigs. *Pharm. Res., 15*, 482–487.

Paavola, A., Kilpelainen, I., Yliruusi, J., & Rosenberg, P. (2000). Controlled release injectable liposomal gel of ibuprofen for epidural analgesia. *Int. J. Pharm., 199*, 85–93.

Paino, C. L., & Bunge, M. B. (1991). Induction of axon growth into Schwann cell implants grafted into lesioned adult rat spinal cord. *Exp. Neurol., 114*, 254–257.

Palmer, T. D., Takahashi, J., & Gage, F. H. (1997). The adult rat hippocampus contains primordial neural stem cells. *Mol. Cell. Neurosci., 8*, 389–404.

ISOTROPIC NERVE GRAFTS FOR REGENERATION

This section on isotropic grafts has been divided into four subsections, to discuss the four components influencing peripheral nerve regeneration.

Natural and synthetic scaffolds for nerve repair

A scaffold can consist of two components. The first is a tubular structure serving as a "guidance channel" and the second consists of scaffold elements that are inside the tubular structure. In general, scaffolds for nerve repair should support axonal proliferation, have low antigenicity, support vascularization, be porous for oxygen diffusion, and avoid long-term compression. The scaffold can be made from natural or synthetic materials.

NATURAL MATERIALS AS SCAFFOLDS

Isotropic natural materials used as scaffolds include veins, skeletal muscle fibers, and collagen. Although these materials support nerve regeneration, they do not provide any direction to the axons. Autologous vein grafts have been shown to provide a good environment for axonal regeneration in short nerve gaps (Wang et al., 1993; Ferrari et al., 1999). However, use of vein grafts for long nerve gaps has been less successful, because of collapse of veins due to their thin walls, and constriction due to the surrounding scar tissue (Chiu and Strauch, 1990). In order to prevent vein grafts from collapsing and improve their performance, intraluminal space fillers such as autologous Schwann cells (SCs), collagen, and muscle fibers have been used. Collagen-filled vein grafts were found to promote better axonal growth than empty vein grafts for a 15-mm nerve gap in rabbits (Choi et al., 2005a). Similarly, SC-seeded venous grafts supported axonal growth and performed better than unseeded grafts to repair 40-mm nerve gaps (Zhang et al., 2002) and 60-mm nerve gaps in rabbits (Strauch et al., 2001). The principal drawback of this approach is that it requires the availability of the relevant amount of live autologous SCs (up to 8 million cells/ml) that are difficult to obtain. Muscle–vein combined grafts, in which the muscle fibers are inserted in veins, were used in 10-mm-long nerve defects in rats, and found to promote axonal regeneration comparable to that of syngenic nerve grafts (Geuna et al., 2004). Although the muscle–vein grafts were able to promote nerve regeneration in 55-mm-long nerve defects in rabbits, they were not comparable to nerve autografts (Geuna et al., 2004). Autologous muscle–vein combined grafts have been used clinically in humans to bridge nerve gaps ranging from 5 to 60 mm. The results were scored as "poor," "satisfactory," "good" and "very good," based on recovery of sensory and motor functions. Of the 21 lesions repaired (in 20 patients), 10 were lesions of the sensory nerves and 11 were mixed nerve lesions. All lesions in the sensory nerves, except one greater than 30 mm, showed "good" to "very good" recovery. All lesions in the mixed nerves showed "satisfactory" to "good" recovery of motor and sensory functions (Battiston et al., 2000).

Although autogenous/natural materials have shown encouraging results when used for nerve repair, they still have certain drawbacks. In the case of autogenous grafts, the drawbacks include the need for a second surgery, the loss of function at the donor site, and neuropathic pain at the donor site. Allografts have problems related to preservation and immunorejection. In order to avoid these problems, grafts made of artificial/synthetic materials have been used.

SYNTHETIC SCAFFOLDS FOR NERVE REPAIR

Unlike natural scaffolds, synthetic scaffolds are advantageous because they can be tailored in terms of their mechanical, chemical, and structural properties to augment nerve regeneration. Among the artificial materials, synthetic tubular NGCs have shown the most promising results so far (Fig. 56.2). Some of the commonly used synthetic scaffolds are given in Table 56.1. The use of NGCs reduces tension at the suture line, protects the regenerating axons from the infiltrating scar tissue, and directs the sprouting axons toward their distal targets. The luminal space of NGCs can be filled with growth-promoting matrix, growth factors, and/or appropriate

FIGURE 56.2

A schematic of a synthetic nerve guidance channel (NGC). The NGC, sutured to the nerve ends, is filled with hydrogel, filaments, cells, neurotrophic factors, and ECM proteins. For an isotropic graft, there would be no filaments and the other components would be distributed uniformly. For an anisotropic graft, there may be filaments, and the other components would be aligned longitudinally or in increasing concentration from proximal to distal nerve end.

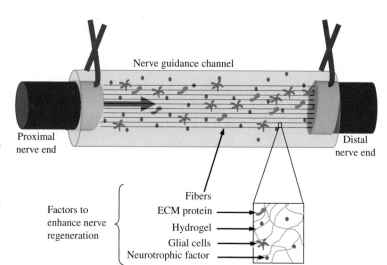

cells. In some cases of nerve repair, NGCs have been used to intentionally leave a small gap between the injured nerve ends, to allow accumulation of cytokines, growth factors, and cells (Dahlin and Lundborg, 2001). The NGCs can be used as an excellent experimental tool, to precisely control the distance between the nerve stumps, test the fluid and tissue entering the channel, and vary the properties of the channel. Although NGCs prevent regenerating nerve fibers from wandering, they do not direct axonal growth microscopically. Hence, for the purposes of this chapter, NGCs have been considered as isotropic scaffolds.

Nerve regeneration in silicone NGCs has been studied in detail (Williams et al., 1983). Within a few hours of nerve repair using an NGC, the tube fills with serum exuded by the cut blood vessels in the nerve ends. This fluid contains neurotrophic factors, as well as several cytokines and inflammatory cells such as macrophages. The macrophages help remove the myelin and axonal debris formed due to injury. The fluid also contains the clot-forming protein, fibrin. Within days, the fibrin coalesces and forms a longitudinally oriented fibrin cable bridging the two nerve ends. Without the formation of the fibrin cable, axonal regeneration cannot occur, thus making the fibrin cable formation a very critical step. The fibrin cable is then invaded by cells migrating from the proximal and distal nerve stumps, including fibroblasts, macrophages, SCs, and endothelial cells (which form capillaries and larger vessels). Axons from the proximal end grow into the fibrin matrix and are engulfed in the cytoplasm of SCs. Some of these axons then reach the distal nerve end and get myelinated. In inert silicone tubes of 10 mm or shorter, these processes occur spontaneously. However, it is generally accepted that impermeable, inert NGCs such as silicone do not support regeneration across defects larger than 10 mm, without the presence of exogenous growth factors. The regeneration process can be improved by various approaches like changing properties of the tube (permeability, porosity, texture, and electric charge characteristics), addition of matrices, neurotrophic factors, ECM molecules, and cells (Valentini and Aebischer, 1997). These strategies augment nerve regeneration by affecting the sequence of events that lead to the bridging of the nerve gap.

Based on the porosity and/or degradability of the material used, NGCs can be classified as impermeable, semi-permeable, and resorbable (Table 56.2). Silicone tube is an example of an impermeable NGC since it does not permit movement of molecules across the tube walls. Porosity affects the movement of soluble factors, oxygen, and waste products, into and out of the NGCs, which is vital for nerve regeneration. Examples of semi-permeable tubes are polysulphone (PS) and polyacrylonitrile/polyvinylchloride (PAN/PVC). Nerves regenerated in semi-permeable tubes featured more myelinated axons and less connective tissue (Uzman and Villegas, 1983; Aebischer et al., 1989a). PAN/PVC channels with a molecular weight cutoff of

TABLE 56.1 Classification of nerve grafts: examples and references

1. Isotropic grafts:
Have uniform distribution of one
or more of the four components
A: Scaffolds:

Natural materials	Veins (Wang et al., 1993; Ferrari et al., 1999), muscle fibers (Geuna et al., 2004)
Synthetic materials: NGCs	PLA (Cai et al., 2005), PLLA, PGA, PAN/PVC (Uzman and Villegas, 1983)
Gels	Agarose (Yu and Bellakonda, 2003), alginate (Suzuki et al., 1999)
B: Neurotrophic factors	NGF (Levi-Montalcini, 1987; Thoenen et al., 1987), BDNF (Sendtner et al., 1992), IGF (Glazner et al., 1993), FGF (Gospodarowicz et al., 1987)
C: ECM proteins	Laminin (Yu and Bellamkonda, 2003), fibronectin (Chen et al., 2000), collagen (Choi et al., 2005b)
D: Support cells	SCs (Guenard et al., 1992), fibroblasts (Nakahara et al., 1996), stem cells (Ansselin et al., 1997; Choi et al., 2005a)

2. Anisotropic grafts:
Have directional distribution of one
or more of the four components
A: Scaffolds:

Aligned filaments	Collagen (Suzuki et al., 1999; Yoshi et al., 2003), PLLA (Ngo et al., 2003)
Magnetically aligned gels	Fibrin (Dahlin and Lundborg, 2001), collagen (Ceballos et al., 1999; Dubey et al., 2001)
B: Neurotrophic factors	NGF (Cao and Shoichet, 2003; Kapur and Schoichet, 2004), BDNF (Cao and Schoichet, 2003), CNTF, FGF
C: ECM proteins	Laminin (Saneinejad and Schoichet, 1998; Kam et al., 2001), fibronectin, collagen
D: Support cells	SCs (Hadlock et al., 2000; Rutkowski et al., 2004), fibroblasts, stem cells

3. Autologous nerve grafts:
Have all the four components: scaffolds, (Gospodarowicz et al., 1987; Nichols et al., 2004)
neurotrophic factors, ECM proteins, and cells

4. Nerve allografts:
Acellular grafts, but are structurally similar to (Evans et al., 1999; Midha et al., 2001)
autologous nerve grafts

TABLE 56.2 Classification of nerve guidance conduits/channels (NGCs) based on porosity and degradability

Porosity	Degradability	Example and reference
Impermeable	Non-degradable	Silicone (Lundborg et al., 1982)
Semipermeable	Non-degradable	PS (Yu and Bellamkonda, 2003), PAN/PVC (Uzman and Villegas, 1983; Aebischer et al., 1989a)
Resorbable	Degradable	PLA (Cai et al., 2005), PGA

50 kilodaltons support regeneration even in the absence of a distal nerve stump (Aebischer et al., 1989a). Examples of bioresorbable tubes are polylactic acid (PLA), polyglycolic acid (PGA), poly(L-lactide-co-glycolide) (PLGA), poly(lactide-co-caprolactone) (PLC), and poly-(3-hydroxybutyrate) (PHB). The use of bioresorbable tubes negates the need for a second surgery to remove the implant and prevents long-term compression of the nerve. However, it is critical that the degradation of the tube does not allow fibroblasts to invade the lumen space before regeneration occurs as this may prevent axons from regenerating.

INCLUSION OF HYDROGELS AS SCAFFOLDS

NGCs can be filled with gels to support axonal elongation. Here we briefly describe some of the isotropic gels used for nerve regeneration.

Agarose gels

Agarose is a polysaccharide derived from red agar and is widely used in gel electrophoresis and gel chromatography. SeaPrep® agarose hydrogel has been shown to support neurite extension from a variety of neurons in a non-immunogenic manner (Bellamkonda et al., 1995; Labrador, 1995; Dillon et al., 2000). Agarose gels also allow molecules to be covalently linked to the gels through functional groups on their polysaccharide chains. For example, laminin protein or fragments of laminin can be covalently coupled to SeaPrep agarose gels to enhance their ability to support neurite extension (Yu et al., 1999). Although agarose gels support neurite growth on their own, coupling of molecules, such as laminin, significantly enhances the gels' ability to promote neurite extension.

Collagen gels

Collagen gels and filaments have been used to promote PNS regeneration (scaffolds with collagen filaments will be discussed later in the anisotropic scaffolds, see pp. 1054–1056). Collagen gel can be used to fill the intraluminal space of a vein graft to prevent it from collapsing and improve its nerve repair efficiency. In collagen-filled vein grafts, the number and diameter of myelinated axons were significantly increased compared to vein grafts without collagen gel (Choi et al., 2005a). Nerve repair with silicone tubes can be significantly improved by filling them with collagen gel. Collagen tubes filled with collagen gel have promoted more rapid nerve sprouting and better morphology, than saline-filled collagen tubes (Satou et al., 1986). In some cases, collagen gels have hindered regeneration (Valentini et al., 1987). This negative effect, presumably due to gel remnants blocking diffusion and axonal elongation, might be overcome by reducing the concentration of the collagen gel (Labrador et al., 1998).

Hyaluronic acid, an ECM component, is associated with decreased scarring and improved fibrin matrix formation. It is hypothesized that during the fibrin matrix phase of regeneration, hyaluronic acid organizes the extracellular matrix into a hydrated open lattice, thereby facilitating migration of the regenerated axons (Seckel et al., 1995). Hyaluronan-based tubular conduits, used for peripheral nerve regeneration, resulted in more myelinated axons and higher nerve conduction velocities than silicone tubes filled with saline (Wang et al., 1998), with little cytotoxicity (Jansen et al., 2004) upon degradation.

Other gels used *in vivo* to promote nerve regeneration include Matrigel, alginate gels, fibrin gels, and heparin sulfate gels (Madison et al., 1988; Suzuki et al., 1999; Dubey et al., 2001).

Neurotrophic factors for nerve regeneration

Neurotrophic factors are produced in the target organs and by Schwann cells in response to injury. These neurotrophic factors help maintain the target organ–nerve synapse. A nerve injury usually results in disruption of communication between the target organs and the neuronal cell body, and leads to Wallerian degeneration (breakdown of myelin sheath and axons). Due to cytokines released during Wallerian degeneration, SCs are activated and produce neurotrophins such as nerve growth factor (NGF) and brain-derived neurotrophic factor (BDNF). Although many other trophic factors, including insulin-like growth factor (IGF), fibroblast growth factor (FGF), and ciliary neurotrophic factor (CNTF), have been shown to be involved in the promotion of nerve regeneration (Gospodarowicz et al., 1987; Glazner et al., 1993), it is believed that they are released from SCs following mechanical damage to the cells.

NGF is produced in the target organs of sensory and sympathetic nerves in the PNS, and has been shown to stimulate and promote the survival of sensory ganglia and nerves, including spinal sensory nerves and sciatic nerves (Levi-Montalcini, 1987; Thoenen et al., 1987). BDNF is expressed in very low levels in intact adult peripheral nerves, but is upregulated following injury. BDNF is effective in promoting the survival and outgrowth of not only sensory and sympathetic nerves, but also motor nerves (Lee et al., 2003).

Neurotrophic factors are likely an important part of future clinical therapies for peripheral nerve injuries/diseases. In diseases in which the functions of SCs are severely suppressed (multiple sclerosis, for example) or when acellular grafts (containing no viable SCs) are used, application of neurotrophic factors could be highly effective in facilitating nerve regeneration. Various studies have utilized the functions of NGF to promote nerve regeneration. Hubble and Sakiyama-Elbert have developed a fibrin matrix that immobilizes heparin molecules by electrostatic interactions, which in turn immobilizes heparin binding growth factors. The fibrin matrix, when implanted *in vivo*, releases the bound growth factor due to fibrin degradation. This system was used to deliver NGF (Lee et al., 2003) for peripheral nerve regeneration *in vivo*, and basic fibroblast growth factor (bFGF (Sakiyama-Elbert and Hubbell, 2000) for neurite extension from chick dorsal root ganglia (DRG) *in vitro*. Fibrin—heparin—NGF matrix was observed to promote nerve regeneration comparable to syngenic nerve grafts over a 13-mm nerve gap in rats. Fibrin matrix that released bFGF enhanced neurite extension from DRG by 100% compared to unmodified fibrin matrix.

1053

Extracellular matrix molecules for nerve regeneration

Insoluble ECM molecules, such as laminin, fibronectin, and certain forms of collagen, promote axonal extension and, therefore, are excellent candidates for incorporation into the lumen of NGCs. Agarose gels cross-linked with laminin showed enhanced neurite extension from chick DRG *in vitro* (Yu et al., 1999). Agarose gels cross-linked with laminin and soluble NGF showed nerve regeneration comparable to autografts over a 10-mm gap in rats (Yu and Bellamkonda, 2003). However, axonal extension in the laminin gels depends on concentrations of laminin gels. High concentrations of laminin hinder regeneration (Labrador et al., 1998). Matrigel, a gel containing collagen type IV, laminin and glycosaminoglycans, supports some degree of regeneration over a long nerve gap in adult rats, when introduced into the lumen of NGCs (Madison et al., 1988). Similarly, a gel mixture containing laminin, collagen, and fibronectin, significantly improved nerve regeneration compared to saline-filled silicone channels (Chen et al., 2000).

Seeding neuronal support cells for nerve regeneration

In the PNS, SCs are support cells that wrap around the axons. SCs form a multilamellar sheath of myelin, a phospholipid-containing substance, around axons that serves as an insulator and increases nerve conduction velocity. An individual SC may ensheath several unmyelinated axons, but only one myelinated axon, within its cytoplasm.

In NGCs used for nerve regeneration, formation of fibrin cable, migration of SCs, and longitudinal arrangement of SCs (known as bands of Büngner) are necessary processes for axonal regeneration. For nerve gaps less than a critical length (10 mm) these processes occur spontaneously, leading to axonal regeneration. However, for nerve gaps greater than 10 mm, spontaneous nerve regeneration does not occur, due to failure of formation of a fibrin cable and bands of Büngner (Lundborg, 1988). SCs of uninjured nerves are quiescent. Following nerve axotomy, the SCs become "reactive" and produce a number of neurotrophic factors, including NGF, BDNF, and CNTF (Thonas et al., 1998). They also synthesize and secrete ECM molecules such as laminin, which is known to modulate neurite outgrowth and express a variety of other cell adhesion molecules. All these components have been suggested to play roles in supporting neuronal survival and axonal regeneration.

Using SCs in NGCs bypasses the fibrin cable formation step, accelerates the formation of bands of Büngner, and introduces a persistent source of neurotrophic factors, leading to more efficient nerve repair. This could decrease the time required by the axons to reconnect to the target organ, as well as increase the distance over which regeneration occurs. SCs isolated from the peripheral nerve of a patient, and expanded *in vitro*, could be used to treat the patient's nerve injuries. Addition of SCs has been shown to significantly improve the performance of various scaffolds, such as empty NGCs, collagen gels, venous nerve grafts and muscle grafts, as compared to control scaffolds without SCs (Ansselin et al., 1997; Strauch et al., 2001; Keilhoff et al., 2005). The ability of SC-seeded NGCs to promote regeneration was found to be dependent on the SCs seeding density, and immunocompatibilty between donors and host (Guenard et al., 1992). For syngenic SCs, it was observed that increasing the seeding density improves the nerve regeneration outcome. Heterologous SCs elicited a strong immune reaction, impeding the nerve regeneration. Performance of SC-seeded NGCs was further improved by designing longitudinally aligned channels in the tube to resemble acellular nerve grafts. This will be discussed in more detail in the section on anisotropic scaffolds (see pp. 1054–1056).

As an alternative to SCs, other cells could be used as such or genetically modified, to produce desired levels of neurotrophic factors, or to express specific ECM molecules. Fibroblasts, genetically modified to produce NGF, BDNF, neurotrophin-3 (NT-3), and bFGF, showed promising results in central nervous system (CNS) regeneration (Nakahara et al., 1996). Olfactory ensheathing cells have been shown to promote regeneration of cut nerves in the adult rat spinal cord (Li et al., 2003). Although these are examples of CNS regeneration, the genetically modified cells can be used for PNS regeneration also. Addition of bone marrow stromal cells to NGCs has shown improved regeneration over empty NGCs (Choi et al., 2005b). Similarly, pluripotent stem cells derived from hair follicles have shown improvements in rats (Amoh et al., 2005). However, the difficulties of isolating and culturing these cells from the patient prior to surgery could limit this approach for some surgical procedures.

ANISOTROPIC NERVE GRAFTS

The four essential elements of nerve grafts — scaffolds, neurotrophic factors, ECM molecules, and cells — can be presented in an aligned fashion, so as to orient the regenerating axons towards their distal targets. Several anisotropic scaffolds have been fabricated to affect neuronal behavior. Neuronal as well as glial cells respond to the underlying topographical cues in a particular manner. Distinct cellular processes including migration, polarization, and gene expression have been shown to be affected by the isotropy of tissue engineered scaffolds (Chew et al., 2008). While the exact mechanisms of how topographical cues affect cell behavior are yet to be elucidated, they have been hypothesized to affect protein attachment, orient cell cytoskeleton, and affect downstream gene regulation (Gugutkov et al., 2009; Hoffman-Kim et al., 2010). In this section, studies involving nerve grafts that provide directional guidance are discussed.

Aligned anisotropic scaffolds

We hypothesize that the superior performance of autologous nerve grafts is due to its cellular components, and its longitudinally aligned structure. The longitudinally aligned structure of the degenerating nerves in the autografts provides contact guidance and direction to the regenerating nerves. In an attempt to mimic autografts, longitudinally patterned or oriented

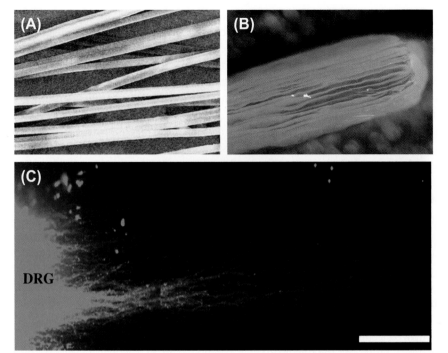

FIGURE 56.3
Nanofilaments for contact guidance-mediated growth. (A) SEM image of PAN-MA nanofilaments, with diameter of 400—700 nm. (B) Three-dimensional nanofilament-based scaffold, along with agarose gel, embedded in a polysulfone NGC can be used to direct neurite growth *in vitro*, or axonal growth *in vivo*. (C) Chick DRG extending neurites along horizontally oriented PAN-MA filaments, *in vitro* (scale bar = 1 mm). *(Images courtesy of Dr Young-tae Kim, Georgia Institute of Technology.)*

1055

gels and filaments to guide and accelerate the regenerating axons have been designed (Fig. 56.3). It has been shown that a poly(acrylonitrile-*co*-methylacrylate) (PAN-MA) nano-filament-based scaffold alone can facilitate regeneration across a 17-mm nerve gap in rats (Kim et al., 2008). Many other combinations of materials have been used, such as collagen filaments embedded in collagen tubes (Yoshii et al., 2003), laminin-coated collagen fibers in collagen tubes (Matsumoto et al., 2000), laminin-fibronectin double-coated collagen fibers in collagen tubes (Tong et al., 1994), poly(L-lactide) (PLLA) filaments in silicone tubes and PLA tubes (Ngo et al., 2003; Cai et al., 2005), and polyglycolic acid (PGA) fibers in chitosan tubes (Wang et al., 2005). All have been found to significantly improve regeneration compared to saline-filled tubes. In addition to synthetic filaments, magnetically aligned fibrin and collagen type I gels have been used to provide directional guidance to neurites *in vitro* (Dubey et al., 1999, 2001) and axons *in vivo* (Ceballos et al., 1999).

Neurotrophic factors

Neurotrophic factors have been delivered mostly in an isotropic manner *in vivo*. However, *in vitro* studies have suggested that gradients of neurotrophic factors can direct growth cones towards the source of neurotrophic factor (Gundersen and Barrett, 1979). Insoluble and soluble gradients of NGF, NT-3, and BDNF have been shown to direct the growth of neurites from PC12 cells toward increasing concentrations of neurotrophic factors (Cao and Shoichet, 2003; Kapur and Shoichet, 2004). Anisotropic scaffolds created by gradients of laminin coupled to agarose along with sustained delivery of NGF from lipid microtubes matched autograft perfomance in terms of anxonal diameter distribution (Dodla and Bellamkonda, 2006, 2008). Therefore, anisotropic scaffolds having gradients of neurotrophic factors, along with other components, might be an important tool for PNS regeneration.

Extracellular matrix molecules

Gels containing ECM molecules, such as laminin, collagen, fibronectin, and glycosamino-glycans, have been widely used to make isotropic scaffolds for nerve regeneration (discussed on pages 1049—1054). ECM molecules promote axonal growth by the mechanism of differential adhesion, wherein axons preferentially grow on substrates of ECM molecules due to the presence of specific cell surface receptors. *In vitro* experiments with spatial patterns of whole ECM molecules (Kam et al., 2001) or their peptide derivates (Saneinejad and Shoichet, 1998), have been used to direct the growth of neurites, as well as enhance neurite extension. *In vivo*, ECM protein-coated fibers have been used to enhance nerve regeneration, where the fibers provide the contact guidance for regenerating axons and the ECM protein provides the adhesive substrate (Tong et al., 1994; Matsumoto et al., 2000). *In vitro* experiments have demonstrated that gradients of ECM proteins could orient and enhance neurite outgrowth toward increasing concentrations of ECM molecules (Saneinejad and Shoichet, 1998; Kam et al., 2001; Dertinger et al., 2002; Adams et al., 2005).

Cell-seeded longitudinally aligned NGCs

Neuronal-growth supporting cells can be incorporated with longitudinally aligned filaments and gels in NGCs to enhance nerve regeneration. Since support cells synthesize ECM proteins and neurotrophic factors, aligned cells often result in directionally aligned ECM. Biodegradable conduits of a copolymer of lactic and glycolic acids (PLGA) with longitudinally aligned channels have been used for nerve regeneration (Hadlock et al., 2000). The channels, with the lumen coated with laminin, and seeded with SCs, showed regeneration comparable to nerve autografts over a 7-mm nerve gap in rats. PLA tubes, with a micropatterned inner lumen, and seeded with SCs, showed better nerve regeneration compared to unpatterned tubes with SCs (Rutkowski et al., 2004). However, the disadvantages of cell-seeded NGCs include the need for prolonged isolation and cell culture to prepare cells for implantation, high cell yield, and high cellular morbidity.

NATURAL NERVE GRAFTS

A common source of nerve grafts is the sural nerve, which is easy to obtain, has the appropriate diameter for most grafting needs, and is relatively indispensable. Other graft sources include the anterior branch of the medial ante-brachial cutaneous nerve, the lateral femoral cutaneous nerve, and the superficial radial sensory nerve (Sunderland, 1991). However, a motor nerve has a preference for a motor pathway (i.e. motor nerve graft) and shows inferior regeneration if a sensory nerve graft, such as sural nerve, is used. Similarly, a mixed nerve shows superior regeneration with either a mixed nerve or a motor nerve graft as compared to a sensory nerve graft (Nichols et al., 2004). Therefore, clinical outcomes might be improved by using alternatives to sensory nerve grafts in the reconstruction of a mixed nerve. However, there are relatively few expendable motor/mixed nerves in the human body that could be used as graft materials. Therefore, a more feasible alternative would be to use nerve allografts or biosynthetic graft materials. Cadavers are an abundant source of graft materials and avoid the complications of harvesting autografts. However, cadaveric nerve allografts require maintenance and can be used only with immunosuppressive therapy. Withdrawal of the immunosuppressant leads to profound loss of axons in the allografts. The axonal loss is most profound in mixed nerve allografts as compared to motor nerve allografts, followed by sensory nerve allografts (Midha et al., 2001).

Allografts, cold-preserved and/or freeze-thawed to prevent immune-rejection by the host body, perform better than fresh allografts in terms of axon density, fiber diameter, and nerve conduction velocity (Evans et al., 1999). Using natural materials (nerve grafts) for regeneration is ideal. However, it has been shown that if autografts or allografts are preserved for too long, their ability to support regeneration supporting is compromised (Gulati, 1996). Also, the pretreated allografts do not perform as well as autografts (Evans et al., 1999).

Although nerve autografts are used as "gold standard," the lack of functional recovery even with autografts remains an important clinical problem. Techniques utilized to improve the performance of the nerve autografts include treatments to either remove the inhibitory molecules, such as chondroitin sulfate proteoglycans (CSPGs), or provide factors for axonal growth, such as NT-4/5 or BDNF. CSPG molecules have a core protein structure with glycosaminoglycan (GAG) side-chains composed of chondroitin sulfate. Due to their large size and negative charge, the GAGs of CSPGs are thought to hinder neurite access to growth-promoting matrix molecules and also repel the axons, thereby inhibiting their growth (Properzi et al., 2003). It has been shown that CSPGs are upregulated almost seven-fold in the distal segment of peripheral nerve following transection nerve injury (Zuo et al., 1998). The upregulated CSPGs contribute significantly to inhibition of neurite sprouting and, consequently, growth into the distal nerve segment. Treatment of the injury site with chondroitinase ABC, which digests away the inhibitory CSPGs, increased the neurite ingrowth into the distal nerve segment several-fold, as compared to controls without any chondroitinase ABC treatment (Zuo et al., 2002). Treatment with chondroitinase ABC, however, did not improve neurite ingrowth in a crush injury model, suggesting that CSPGs are not upregulated in a crush injury.

Syngenic nerve grafts treated with chondroitinase ABC, heprinase I, heparinase III, or keratanase enzymes have shown significantly improved axonal ingrowth from the proximal nerve end into the nerve graft, as compared to untreated controls (Groves et al., 2005). Autografts treated with a combination of all these four enzymes showed the most significant neurite growth into the graft. However, the combination treatment was not significantly different from the arithmetic sum of the individual treatments. This suggests that molecules such as heparan sulfate proteoglycan and keratan sulfate proteoglycan also contribute to inhibition of neurite growth apart from CSPGs, and the pathways/mechanisms of inhibition for each of these molecules might be independent of each other. In another study, exogenous supply of BDNF and NT-4/5 was found to increase the number of axons regenerating into the nerve graft as compared to untreated nerve grafts (English et al., 2005). These techniques, used in clinical applications, could lead to better results with autografts/allografts.

Acellular grafts have been developed to address the problem of host response to allografts as well as the shortage of autografts. Cadaveric nerves are decellularized while keeping the ECM structure intact for growth of nerves. Acellular cadaveric nerves from rats were able to bridge 10-mm nerve injury. In the same study, acelluar grafts treated with bNFG and VGEF showed increased axonal diameter and neurovascularization at 1 month (Kim et al., 2004). Other studies have also shown that incorporating VGEF and NGF in acellular grafts promoted Schwann cell migration, increased the number of axons, and improved neovacularization compared to controls (Sondell et al., 1999). In longer gaps, 40 mm in Rhesus monkeys, acellular allogenic grafts seeded with autologous Schwann cells enhanced nerve density and fiber counts (Hu et al., 2007). Functionality of acellular grafts is dependent on the structural integrity of the graft after it has been processed for removal of cells and immunogenic components. Laboratories have developed chemical processing techniques that focus on preserving the integrity of the ECM structure. Studies with optimized acellular grafts created by Schmidt's group suggest that axon density after 84 days was dependent on graft structure and content (Hudson et al., 2004). In another study, processed acellular grafts held their laminin structure and promoted better functional results compared to nerve conduits in bridging nerve gaps (Whitlock et al., 2009).

ELECTRICAL STIMULATION

A major challenge to a successfully regeneration is to avoid muscle atrophy due to the length of time it requires to bridge the nerve gap and innervate the target muscle. To this end, we have discussed several strategies to promote nerve regeneration throughout the chapter. Electrical stimulation is another technique that has been used to promote nerve regeneration. Previous work has shown that electrical stimulation to the soleus nerve of rabbits after a crush injury promoted twitch force, tetanic tension, and muscle action potential in soleus muscle, indicating enhanced axonal growth (Nix and Hopf, 1983). To elucidate how electrical stimulation accelerated nerve growth, other groups evaluated its effects on production of growth factors as well as other cellular responses. Stimulating motor neurons at 20 Hz for 1 h accelerated sprouting of axons after nerve injury (Al-Majed et al., 2000a,b; Brushart et al., 2002). After electrical stimulation, BDNF and trkB receptor expression was upregulated in motor neurons (Al-Majed et al., 2000a). Similar observations were observed by another group where they observed increased neurotrophin production after electrical stimulation following nerve repair using a nerve allograft (English et al., 2007). Thus, future strategies can use electrical stimulation in conjugation with other scaffolds discussed throughout this chapter to accelerate successful nerve repair following a peripheral nerve injury.

ANIMAL MODELS

Traditionally, nerve regeneration studies have involved the use of various animal models, such as mice, rats, swine, canine, sheep, and non-human primates. Rat or mouse models are used initially to determine the efficacy of the various treatments. If the results are encouraging, they are followed by experiments with larger animal models. For PNS regeneration studies, the most commonly studied nerve models are the sciatic nerve and its branches, the tibial and the peroneal nerves. Other models include the cavernosal nerve and the facial nerve. The most common nerve injury model is the single-anastomosis model where the injury and repair are done on one sciatic nerve and the contralateral sciatic nerve is used as a control. This model is useful when the nerve gap is not more than 20 mm. The second version is the cross (double) anastomosis model, where both contralateral sciatic nerves are transected; the proximal end of the right sciatic nerve is then sutured to one end of an implanted tube and the distal end of the left sciatic nerve is inserted into the other end of the tube (Lundborg et al., 1982). This model allows the study of gaps in excess of 25 mm. Although very convenient, the rodent models suffer from the serious drawback that they present only short nerve gaps for regeneration studies. In order for a regeneration technique to be successfully applied in clinical trials, the nerve gap model has to be more than 40 mm in length. To create a long nerve gap model, rabbits (Geuna et al., 2004), cats (Suzuki et al., 1999), dogs (Matsumoto et al., 2000), sheep (Lawson and Glasby, 1998), and non-human primates (Ahmed et al., 1999) have been used. The large animal models are an important intermediary step before clinical application of experimental therapeutic approaches.

CONCLUSION

In spite of significant advances in research into the development of synthetic NGCs, nerve autografts are still considered the first-choice strategy for nerve repair, especially in the case of long nerve gaps. However, performance of autografts themselves has been unsatisfactory. Using autografts generally results in a good recovery of sensory functions but negligible return of motor functions. Hence, there will be continued interest in ideas to further enhance the performance of autografts by various treatments, such as chondroitinase ABC, NT-4/5, and BDNF. However, shortage of autografts and allografts is a big hindrance for their usage. This shortage can be overcome only by developing synthetic alternatives to autografts. Modulating the spatiotemporal distribution of the four components of grafts germane to regeneration can potentially improve the potential outcomes with these grafts. Ongoing rapid advances in cell biology, cell culture techniques, genetic engineering, and biomaterials research, are likely to provide new tools to improve regeneration using NGCs, and the day an engineered construct performs as well as autografts may be near (AxoGen Inc., 2006).

References

Adams, D. N., Kao, E. Y-C., Hypolite, C. L., Distefano, M. D., Hu, W-S., & Letourneau, P. C. (2005). Growth cones turn and migrate up an immobilized gradient of the laminin IKVAV peptide. *J. Neurobiol., 62*, 134–147.

Aebischer, P., Guenard, V., & Brace, S. (1989a). Peripheral nerve regeneration through blind-ended semipermeable guidance channels: effect of molecular weight cutoff. *J. Neurosci., 9*, 3590–3595.

Ahmed, Z., Brown, R. A., Ljungberg, C., Wiberg, M., & Terenghi, G. (1999). Nerve growth factor enhances peripheral nerve regeneration in non-human primates. *Scand. J. Plast. Reconstr. Surg. Hand Surg., 33*, 393–401.

Albert, E. (1885). Einige Operationen an Nerven. *Wien Med., 26*, 1285.

Al-Majed, A. A., Brushart, T. M., & Gordon, T. (2000a). Electrical stimulation accelerates and increases expression of BDNF and trkB mRNA in regenerating rat femoral motoneurons. *Eur. J. Neurosci., 12*, 4381–4390.

Al-Majed, A. A., Neumann, C. M., Brushart, T. M., & Gordon, T. (2000b). Brief electrical stimulation promotes the speed and accuracy of motor axonal regeneration. *J. Neurosci., 20*, 2602–2608.

Amoh, Y., Li, L., Campillo, R., Kawahara, K., Katsuoka, K., Penman, S., & Hoffman, R. M. (2005). Implanted hair follicle stem cells form Schwann cells that support repair of severed peripheral nerves. *Proc. Natl. Acad. Sci. USA, 102*, 17734–17738.

Ansselin, A. D., Fink, T., & Davey, D. F. (1997). Peripheral nerve regeneration through nerve guides seeded with adult Schwann cells. *Neuropathol. Appl. Neurobiol., 23*, 387–398.

AxoGen Inc. (2006). http://www.axogeninc.com

Battiston, B., Tos, P., Cushway, T. R., & Geuna, S. (2000). Nerve repair by means of vein filled with muscle grafts I. Clinical results. *Microsurgery, 20*, 32–36.

Bellamkonda, R. V., Ranieri, J. P., Bouche, N., & Aebischer, P. (1995). Hydrogel-based three-dimensional matrix for neural cells. *J. Biomed. Mater. Res., 29*, 663–671.

Brushart, T. M., Hoffman, P. N., Royall, R. M., Murinson, B. B., Witzel, C., & Gordon, T. (2002). Electrical stimulation promotes motoneuron regeneration without increasing its speed or conditioning the neuron. *J. Neurosci., 22*, 6631–6638.

Cai, J., Peng, X., Nelson, K. D., Eberhart, R., & Smith, G. M. (2005). Permeable guidance channels containing microfilament scaffolds enhance axon growth and maturation. *J. Biomed. Mater. Res. A, 75*, 374–386.

Cao, X., & Shoichet, M. S. (2003). Investigating the synergistic effect of combined neurotrophic factor concentration gradients to guide axonal growth. *Neuroscience, 122*, 381–389.

Ceballos, D., Navarro, X., Dubey, N., Wendelschafer-Crabb, G., Kennedy, W. R., & Tranquillo, R. T. (1999). Magnetically aligned collagen gel filling a collagen nerve guide improves peripheral nerve regeneration. *Exp. Neurol., 158*, 290–300.

Chen, Y. S., Hsieh, C. L., Tsai, C. C., Chen, T. H., Cheng, W. C., Hu, C. L., et al. (2000). Peripheral nerve regeneration using silicone rubber chambers filled with collagen, laminin and fibronectin. *Biomaterials, 21*, 1541–1547.

Chew, S. Y., Mi, R., Hoke, A., & Leong, K. W. (2008). The effect of the alignment of electrospun fibrous scaffolds on Schwann cell maturation. *Biomaterials, 29*, 653–661.

Chiu, D. T., & Strauch, B. (1990). A prospective clinical evaluation of autogenous vein grafts used as a nerve conduit for distal sensory nerve defects of 3 cm or less. *Plast. Reconstr. Surg., 86*, 928–934.

Choi, B. H., Zhu, S. J., Kim, S. H., Huh, J. H., Lee, S. H., & Jung, J. H. (2005a). Nerve repair using a vein graft filled with collagen gel. *J. Reconstr. Microsurg., 21*, 267–272.

Choi, B. H., Zhu, S. J., Kim, B. Y., Huh, J. Y., Lee, S. H., & Jung, J. H. (2005b). Transplantation of cultured bone marrow stromal cells to improve peripheral nerve regeneration. *Int. J. Oral Maxillofac. Surg., 34*, 537–542.

Dahlin, L. B., & Lundborg, G. (2001). Use of tubes in peripheral nerve repair. *Neurosurg. Clin. N. Am., 12*, 341–352.

Dertinger, S. K., Jiang, X., Li, Z., Murthy, V. N., & Whitesides, G. M. (2002). Gradients of substrate-bound laminin orient axonal specification of neurons. *Proc. Natl. Acad. Sci. USA, 99*, 12542–12547.

Dillon, G. P., Yu, X., & Bellamkonda, R. V. (2000). The polarity and magnitude of ambient charge influences three-dimensional neurite extension from DRGs. *J. Biomed. Mater. Res., 51*, 510–519.

Dodla, M. C., & Bellamkonda, R. V. (2006). Anisotropic scaffolds facilitate enhanced neurite extension *in vitro*. *J. Biomed. Mater. Res. A, 78*, 213–221.

Dodla, M. C., & Bellamkonda, R. V. (2008). Differences between the effect of anisotropic and isotropic laminin and nerve growth factor presenting scaffolds on nerve regeneration across long peripheral nerve gaps. *Biomaterials, 29*, 33–46.

Dubey, N., Letourneau, P. C., & Tranquillo, R. T. (1999). Guided neurite elongation and Schwann cell invasion into magnetically aligned collagen in simulated peripheral nerve regeneration. *Exp. Neurol., 158*, 338–350.

Dubey, N., Letourneau, P. C., & Tranquillo, R. T. (2001). Neuronal contact guidance in magnetically aligned fibrin gels: effect of variation in gel mechano-structural properties. *Biomaterials, 22*, 1065–1075.

English, A. W., Meador, W., & Carrasco, D. I. (2005). Neurotrophin-4/5 is required for the early growth of regenerating axons in peripheral nerves. *Eur. J. Neurosci., 21*, 2624–2634.

English, A. W., Schwartz, G., Meador, W., Sabatier, M. J., & Mulligan, A. (2007). Electrical stimulation promotes peripheral axon regeneration by enhanced neuronal neurotrophin signaling. *Dev. Neurobiol., 67*, 158–172.

Evans, G. R. (2001). Peripheral nerve injury: a review and approach to tissue engineered constructs. *Anat. Rec., 263,* 396—404.

Evans, P. J., MacKinnon, S. E., Midha, R., Wade, J. A., Hunter, D. A., Nakao, Y., et al. (1999). Regeneration across cold preserved peripheral nerve allografts. *Microsurgery, 19,* 115—127.

Ferrari, F., De Castro Rodrigues, A., Malvezzi, C. K., Dal Pai Silava, M., & Padvoni, C. R. (1999). Inside-out vs. standard vein graft to repair a sensory nerve in rats. *Anat. Rec., 256,* 227—232.

Geuna, S., Tos, P., Battiston, B., & Giacobini-Robecchi, M. G. (2004). Bridging peripheral nerve defects with muscle—vein combined guides. *Neurol. Res., 26,* 139—144.

Glazner, G. W., Lupien, S., Miller, J. A., & Ishii, D. N. (1993). Insulin-like growth factor II increases the rate of sciatic nerve regeneration in rats. *Neuroscience, 54,* 791—797.

Gluck, T. (1880). Ueber Neuroplastik auf dem Wege der Transplantation. *Arch. Klin. Chir., 25,* 606—616.

Gospodarowicz, D., Ferrara, N., Schweigerer, L., & Neufeld, G. (1987). Structural characterization and biological functions of fibroblast growth factor. *Endocrinol. Rev., 8,* 95—114.

Groves, M. L., McKeon, R., Werner, E., Nagarsheth, M., Meador, W., & English, A. W. (2005). Axon regeneration in peripheral nerves is enhanced by proteoglycan degradation. *Exp. Neurol., 195,* 278—292.

Guenard, V., Kleitman, N., Morrissey, T. K., Bunge, R. P., & Aebischer, P. (1992). Syngeneic Schwann cells derived from adult nerves seeded in semipermeable guidance channels enhance peripheral nerve regeneration. *J. Neurosci., 12,* 3310—3320.

Gugutkov, D., Gonzalez-Garcia, C., Rodriguez Hernandez, J. C., Altankov, G., & Salmeron-Sanchez, M. (2009). Biological activity of the substrate-induced fibronectin network: insight into the third dimension through electrospun fibers. *Langmuir, 25,* 10893—10900.

Gulati, A. K. (1996). Peripheral nerve regeneration through short- and long-term degenerated nerve transplants. *Brain Res., 742,* 265—270.

Gundersen, R. W., & Barrett, J. N. (1979). Neuronal chemotaxis: chick dorsal-root axons turn toward high concentrations of nerve growth factor. *Science, 206,* 1079—1080.

Hadlock, T., Sundback, C., Hunter, D., Cheney, M., & Vacanti, J. P. (2000). A polymer foam conduit seeded with Schwann cells promotes guided peripheral nerve regeneration. *Tissue Eng., 6,* 119—127.

Hoffman-Kim, D., Mitchel, J. A., & Bellamkonda, R. V. (2010). Topography, cell response, and nerve regeneration. *Annu. Rev. Biomed. Eng., 12,* 203—231.

Hu, J., Zhu, Q. T., Liu, X. L., Xu, Y. B., & Zhu, J. K. (2007). Repair of extended peripheral nerve lesions in Rhesus monkeys using acellular allogenic nerve grafts implanted with autologous mesenchymal stem cells. *Exp. Neurol., 204,* 658—666.

Hudson, T. W., Zawko, S., Deister, C., Lundy, S., Hu, C. Y., Lee, K., et al. (2004). Optimized acellular nerve graft is immunologically tolerated and supports regeneration. *Tissue Eng., 10,* 1641—1651.

Jansen, K., van der Werff, J. F., van Wachem, P. B., Nicolai, J. P., de Leij, L. F., & van Luyn, M. J. (2004). A hyaluronan-based nerve guide: *in vitro* cytotoxicity, subcutaneous tissue reactions, and degradation in the rat. *Biomaterials, 25,* 483—489.

Kam, L., Shain, W., Turner, J. N., & Bizios, R. (2001). Axonal outgrowth of hippocampal neurons on micro-scale networks of polylysine-conjugated laminin. *Biomaterials, 22,* 1049—1054.

Kapur, T. A., & Shoichet, M. S. (2004). Immobilized concentration gradients of nerve growth factor guide neurite outgrowth. *J. Biomed. Mater. Res. A, 68,* 235—243.

Keilhoff, G., Pratsch, F., Wolf, G., & Fansa, H. (2005). Bridging extra large defects of peripheral nerves: possibilities and limitations of alternative biological grafts from acellular muscle and Schwann cells. *Tissue Eng., 11,* 1004—1014.

Kim, B. S., Yoo, J. J., & Atala, A. (2004). Peripheral nerve regeneration using acellular nerve grafts. *J. Biomed. Mater. Res. A, 68,* 201—209.

Kim, Y. T., Haftel, V. K., Kumar, S., & Bellamkonda, R. V. (2008). The role of aligned polymer fiber-based constructs in the bridging of long peripheral nerve gaps. *Biomaterials, 29,* 3117—3127.

Kirk, E. G., & Lewis, D. (1915). Fascial tubulization in the repair of nerve defects. *JAMA, 65,* 486—492.

Kline, D. G., Kim, D., Midha, R., Harsh, C., & Tiel, R. (1998). Management and results of sciatic nerve injuries: a 24-year experience. *J. Neurosurg., 89,* 13—23.

Labrador, R. O., Buti, M., & Navarro, X. (1995). Peripheral nerve repair: role of agarose matrix density on functional recovery. *Neuroreport, 6,* 2022—2026.

Labrador, R. O., Buti, M., & Navarro, X. (1998). Influence of collagen and laminin gels concentration on nerve regeneration after resection and tube repair. *Exp. Neurol., 149,* 243—252.

Lawson, G. M., & Glasby, M. A. (1998). Peripheral nerve reconstruction using freeze—thawed muscle grafts: a comparison with group fascicular nerve grafts in a large animal model. *J.R. Coll. Surg. Edinb., 43,* 295—302.

Lee, A. C., Yu, V. M., Lowe, J. B., III, Brenner, M. J., Hunter, D. A., Mackinnon, S. E., et al. (2003). Controlled release of nerve growth factor enhances sciatic nerve regeneration. *Exp. Neurol., 184*, 295–303.

Levi-Montalcini, R. (1987). The nerve growth factor 35 years later. *Science, 237*, 1154–1162.

Li, Y., Decherchi, P., & Raisman, G. (2003). Transplantation of olfactory ensheathing cells into spinal cord lesions restores breathing and climbing. *J. Neurosci., 23*, 727–731.

Lundborg, G. (1988). *Nerve Injury and Repair.* New York: Longman Group UK.

Lundborg, G., Dahlin, L. B., Danielsen, N., Gelberman, R. H., Longo, F. M., Powell, H., C., et al. (1982). Nerve regeneration in silicone chambers: influence of gap length and of distal stump components. *Exp. Neurol., 76*, 361–375.

Madison, R. D., Da Silva, C. F., & Dikkes, P. (1988). Entubulation repair with protein additives increases the maximum nerve gap distance successfully bridged with tubular prostheses. *Brain Res., 447*, 325–334.

Matsumoto, K., Ohnishi, K., Kiyotani, T., Sekine, T., Ueda, H., Nakamura, T., et al. (2000). Peripheral nerve regeneration across an 80-mm gap bridged by a polyglycolic acid (PGA)-collagen tube filled with laminin-coated collagen fibers: a histological and electrophysiological evaluation of regenerated nerves. *Brain Res., 868*, 315–328.

Midha, R., Nag, S., Munro, C. A., & Ang, L. C. (2001). Differential response of sensory and motor axons in nerve allografts after withdrawal of immunosuppressive therapy. *J. Neurosurg., 94*, 102–110.

Millesi, H., Meissl, G., & Berger, A. (1972). The interfascicular nerve-grafting of the median and ulnar nerves. *J. Bone Joint Surg., 54A*, 7727–7750.

Nakahara, Y., Gage, F. H., & Tuszynski, M. H. (1996). Grafts of fibroblasts genetically modified to secrete NGF, BDNF, NT-3, or basic FGF elicit differential responses in the adult spinal cord. *Cell Transplant., 5*, 191–204.

Ngo, T. T., Waggoner, P. J., Romero, A. A., Nelson, K. D., Eberhart, R. C., & Smith, G. M. (2003). Poly(L-lactide) microfilaments enhance peripheral nerve regeneration across extended nerve lesions. *J. Neurosci. Res., 72*, 227–238.

Nichols, C. M., Brenner, M. J., Fox, I. K., Tung, T. H., Hunter, D. A., Rickman, S. R., et al. (2004). Effects of motor versus sensory nerve grafts on peripheral nerve regeneration. *Exp. Neurol., 190*, 347–355.

Nix, W. A., & Hopf, H. C. (1983). Electrical stimulation of regenerating nerve and its effect on motor recovery. *Brain Res., 272*, 21–25.

Payr, E. (1900). Beitrage zur Technik der Blutgefass und Nervennaht nebst Mittheilungen uber die Vervendung eines resorbibaren Metalles in der Chirurgie. *Arch. Klin. Chir., 62*, 67.

Properzi, F., Asher, R. A., & Fawcett, J. W. (2003). Chondroitin sulphate proteoglycans in the central nervous system: changes and synthesis after injury. *Biochem. Soc. Trans., 31*, 335–336.

Rutkowski, G. E., Miller, C. A., Jeftinija, S., & Mallapragada, S. K. (2004). Synergistic effects of micropatterned biodegradable conduits and Schwann cells on sciatic nerve regeneration. *J. Neural Eng., 1*, 151–157.

Sakiyama-Elbert, S. E., & Hubbell, J. A. (2000). Development of fibrin derivatives for controlled release of heparin-binding growth factors. *J. Control. Release, 65*, 389–402.

Saneinejad, S., & Shoichet, M. S. (1998). Patterned glass surfaces direct cell adhesion and process outgrowth of primary neurons of the central nervous system. *J. Biomed. Mater. Res., 42*, 13–19.

Satou, T., Nishida, S., Hiruma, S., Tanji, K., Takahashi, M., Fujita, S., et al. (1986). A morphological study on the effects of collagen gel matrix on regeneration of severed rat sciatic nerve in silicone tubes. *Acta Pathol. Jpn., 36*, 199–208.

Seckel, B. R., Jones, D., Hekimian, K. J., Wang, K. K., Chakalis, D. P., & Costas, P. D. (1995). Hyaluronic acid through a new injectable nerve guide delivery system enhances peripheral nerve regeneration in the rat. *J. Neurosci. Res., 40*, 318–324.

Sendtner, M., Holtmann, B., Kolbeck, R., Thoenen, H., & Barde, Y. A. (1992). Brain-derived neurotrophic factor prevents the death of motoneurons in newborn rats after nerve section. *Nature, 360*, 757–759.

Sondell, M., Lundborg, G., & Kanje, M. (1999). Vascular endothelial growth factor stimulates Schwann cell invasion and neovascularization of acellular nerve grafts. *Brain Res., 846*, 219–228.

Strauch, B., Rodriguez, D. M., Diaz, J., Yu, H. L., Kaplan, G., & Weinstein, D. E. (2001). Autologous Schwann cells drive regeneration through a 6-cm autogenous venous nerve conduit. *J. Reconstr. Microsurg., 17*, 589–595.

Sunderland, S. (1991). *Nerve Injuries and their Repair: A Critical Appraisal.* New York: Churchill Livingstone.

Suzuki, Y., Tanihara, M., Ohnishi, K., Suzuki, K., Endo, K., & Nishimura, Y. (1999). Cat peripheral nerve regeneration across 50-mm gap repaired with a novel nerve guide composed of freeze-dried alginate gel. *Neurosci. Lett., 259*, 75–78.

Terzis, J., Faibisoff, B., & Williams, B. (1975). The nerve gap: suture under tension vs. graft. *Plast. Reconstr. Surg., 56*, 166–170.

Thoenen, H., Barde, Y. A., Davies, A. M., & Johnson, J. E. (1987). Neurotrophic factors and neuronal death. *Ciba Found Symp., 126*, 82–95.

Thonas, P. K., Okajima, S., & Terzis, J. K. (1998). Utrastructure and cellular biology of nerve regeneration. *J. Reconstr. Microsurg., 14*, 423–436.

Tong, X. J., Hirai, K., Shimada, H., Mizutani, Y., Izumi, T., Toda, N., et al. (1994). Sciatic nerve regeneration navigated by laminin-fibronectin double coated biodegradable collagen grafts in rats. *Brain Res., 663*, 155–162.

Uzman, B. G., & Villegas, G. M. (1983). Mouse sciatic nerve regeneration through semi-permeable tubes: a quantitative model. *J. Neurosci., 9*, 325–338.

Valentini, R. F., & Aebischer, P. (1997). Strategies for the engineering of peripheral nervous tissue regeneration. In R. P. Lanza, & W. L. Chick (Eds.), *Principles of Tissue Engineering* (pp. 671–684). Austin, TX: R.G. Landes Company.

Valentini, R. F., Aebischer, P., Winn, S. R., & Galletti, P. M. (1987). Collagen- and laminin-containing gels impede peripheral nerve regeneration through semipermeable nerve guidance channels. *Exp. Neurol., 98*, 350–356.

Wang, K. K., Costas, P. D., Bryan, D. J., Jones, D. S., & Seckel, B. R. (1993). Inside-out vein graft promotes improved nerve regeneration in rats. *J. Reconstr. Microsurg., 14*, 608–618.

Wang, K. K., Nemeth, I. R., Seckel, B. R., Chakalis-Haley, D. P., Swann, D. A., Kuo, J. W., et al. (1998). Hyaluronic acid enhances peripheral nerve regeneration *in vivo. Microsurgery, 18*, 270–275.

Wang, X., Hu, W., Cao, Y., Yao, J., Wu, J., & Gu, X. (2005). Dog sciatic nerve regeneration across a 30-mm defect bridged by a chitosan/PGA artificial nerve graft. *Brain, 128*, 1897–1910.

Weiss, P., & Taylor, A. C. (1946). Guides for nerve regeneration across nerve gaps. *J. Neurosurg., 3*, 275–282.

Whitlock, E. L., Tuffaha, S. H., Luciano, J. P., Yan, Y., Hunter, D. A., Magill, C. K., et al. (2009). Processed allografts and type I collagen conduits for repair of peripheral nerve gaps. *Muscle Nerve, 39*, 787–799.

Williams, L. R., Longo, F. M., Powell, H. C., Lundborg, G., & Varon, S. (1983). Spatial–temporal progress of peripheral nerve regeneration within a silicone chamber: parameters for a bioassay. *J. Comp. Neurol., 218*, 460–470.

Yoshii, S., Oka, M., Shima, M., Taniguchi, A., & Akagi, M. (2003). Bridging a 30-mm nerve defect using collagen filaments. *J. Biomed. Mater. Res. A, 67*, 467–474.

Yu, X., & Bellamkonda, R. V. (2003). Tissue-engineered scaffolds are effective alternatives to autografts for bridging peripheral nerve gaps. *Tissue Eng., 9*, 421–430.

Yu, X., Dillon, G. P., & Bellamkonda, R. B. (1999). A laminin and nerve growth factor-laden three-dimensional scaffold for enhanced neurite extension. *Tissue Eng., 5*, 291–304.

Zhang, F., Blain, B., Beck, J., Zhang, J., Chen, Z., Chen, Z. W., et al. (2002). Autogenous venous graft with one-stage prepared Schwann cells as a conduit for repair of long segmental nerve defects. *J. Reconstr. Microsurg., 18*, 295–300.

Zuo, J., Hernandez, Y. J., & Muir, D. (1998). Chondroitin sulfate proteoglycan with neurite-inhibiting activity is upregulated following peripheral nerve injury. *J. Neurobiol., 34*, 41–54.

Zuo, J., Neubauer, D., Graham, J., Krekoski, C. A., Ferguson, T. A., & Muir, D. (2002). Regeneration of axons after nerve transection repair is enhanced by degradation of chondroitin sulfate proteoglycan. *Exp. Neurol., 176*, 221–228.

1062

CHAPTER 57

Tissue Engineering of Skin

Fiona Wood
Burns Service of Western Australia, Burn Injury Research Unit, University of Western
Australia, McComb Research Foundation, Western Australia

INTRODUCTION

Each one of us is a self-organizing mass of multiple cell types. From fertilization of the embryo our tissue structures develop until an adult morphology is achieved. At that point our capacity for self-organization is directed to maintaining that morphology in the face of the insults of our daily life and the processes of aging. When a given insult overwhelms our capacity to repair by regeneration the result is scar repair (Ferguson and O'Kane, 2004).

We know that tissues retain a variable ability to heal by regeneration (Martinez-Hernandez, 1988). With respect to the skin, in all but trivial injuries the capacity for regeneration is swamped, triggering the cellular mechanisms which result in scar formation. Burn injury is notorious for aggressive scars, compromising the individual functionally, cosmetically, and psychologically (Rockwell et al., 1989; Rumsey et al., 2003).

It has been well described that to heal a wound we need a source of cells capable of differentiation into the given tissue type and an extracellular matrix capable of supporting the cell migration, proliferation, and differentiation (Bannasch et al., 2003). We and others have spent considerable resources researching these areas to improve the speed and quality of wound healing to reduce the scar (Deitch et al., 1983; Wood, 2003). The question we ask is: "How can we harness the technology of tissue engineering (TE) of skin to provide a controlled repair to restore the original morphology?"

TE can be defined as the application of engineering principles to biological systems (Johnson et al., 2007). The healing of skin has been the subject of writings back to ancient times with attempts to stimulate healing, protect the surface while healing, and even replace the skin surface (Roupé et al., 2010). So our recent explorations into using TE principles in skin repair are based on a significant history. With the increasing knowledge and understanding of the skin structure and function along with the developing TE techniques can we provide a regenerative repair avoiding scar?

We know that the skin provides the barrier to the external environment as a dynamic, complex three-dimensional structure made up of cells from all embryological layers. Integral to its functions are the vessels and nerves within the cell/matrix construct. Furthermore, skin is specialized over different body sites, adapting to local functional needs demonstrated by the macroscopic differences seen between areas, e.g. the eyelid compared with the palm of hand. The varying capacity to respond to injury is seen with the greater scar potential of sternal and deltoid areas (Mustoe et al., 2002). So as we go forward in developing tissue engineering solutions to skin's it is timely to take stock of the skin functions and interactions. How will the skin changes over the different body sites influence the TE needs? Collaboration with

1063

specialists in bioinformatics will be essential in understanding the implication of changes in genes, genetic expression, and the phenotypic expression we need to guide to in the tissue constructs (Smiley et al., 2005).

In the therapeutic use of TE the understanding of the etiology of the skin defect and pathophysiology of the patient will identify what needs to be replicated, rebuilt, and replaced. Implementation into clinical practice hinges on the TE being problem-driven, providing a practical, timely, cost-effective solution to the clinical problem (Shakespeare and Shakespeare, 2002). TE of skin has developed to this point in response to clinical need of, for example, skin repair in major burns (Wood et al., 2006a, b), chronic ulcers (Llames et al., 2004), and giant nevi (Whang et al., 2005). It is clear that the current "gold standard" of skin grafting will always leave a scar (Mosier and Gibran, 2009). The development of TE gives the opportunity to tailor the repair to the defect with the understanding that one solution will not fit all. TE offers innovative solutions providing a spectrum of clinical solutions from strategies to facilitate wound healing *in situ* to multilayered skin constructs including multiple cell types.

The development of TE of skin is intimately linked to the vision of facilitating scar-free healing (Atiyeh and Costaglioa, 2007). It has broad implications post-trauma, -surgery, and -fibrosing pathologies where the common outcome is a functional compromise due to contracture and loss of normal tissue architecture. In the developed world it is estimated that 100 million patients acquire scars each year, some of which cause considerable problems, as a result of 55 million elective operations and 25 million operations after trauma. Within this number it is estimated that there are in the region of 11 million keloid scars and four million burn scars, 70% of which occur in children (Bayat, 2003).

We are living in a time where science and technology are advancing at an exponential rate. Harnessing the power of that science and technology into clinical practice presents an everincreasing challenge. We are all aware of the latest breakthrough holding promise to improve the quality of life from the health to the environmental sectors. However, it is worth noting that the exponential growth is possible in controlled systems where experiments can be designed to investigate a single variable; here lies the greatest challenge, for example, in burn injury research. The design of clinical trials is dogged by the complexities of assessing the extent of injury, the individual's response, and the availability of validated outcome measures.

Clinical practice is a fusion of experience and knowledge with the development of medical subspecialization directed to a targeted problem-solving approach and has facilitated great advances in clinical care. However, this should not be at the expense of a broad general knowledge gaining insight into potential links and facilitating cross-fertilization. It is essential to link the tissue engineer with the clinical specialist to ensure the opportunities, risks, and benefits of TE skin are understood to facilitate appropriate clinical trials. By collaboration between disciplines there are real opportunities for improvements in clinical care translating to improved outcomes for patients.

The road from the bench to the bedside is a long one and navigates the areas of regulation, commercialization, reimbursement, and clinical trial design, to name a few. Also, translational research itself is an area in need of research and audit. The investigation of drivers and barriers to the implementation of TE skin solutions is an area of increasing effort. It is vital to learn from history and understand the current situation to continue developing innovative TE solutions but also innovative solutions within health systems, to action timely translation into clinical practice.

This chapter will:

- explore the functions of the skin and injury responses we need to understand in order to harness the TE technology effectively
- identify the needs which could be supplied by TE strategies
- discuss currently available technologies

- demonstrate TE skin solutions in clinical practice
- highlight the areas of need for further development.

DEVELOPMENT, ANATOMY, AND FUNCTION OF SKIN

Skin is commonly described as a multilayered physical barrier composed chiefly of the surface cellular epidermis and relatively acellular dermis. Exploring the development, anatomy, and skin functions demonstrates its complexity and will guide our TE endeavors.

> "There is no magician's mantle to compare with the skin in its diverse roles of waterproof, overcoat, sunshade, suit of armour and refrigerator. Sensitive to the touch of a feather, to temperature and pain, withstanding the wear and tear of three score years and ten and executing its own running repairs."
>
> **(Lockhart et al., 1965)**

During development the initial covering of the embryo is the periderm, which is thought to have a barrier function expressing tight junctions and interacts with the amniotic fluid having surface microvilli. It is interesting to note that the periderm cells express keratins also associated with migratory epidermal cells seen in wound healing. During the first 6 weeks the epidermis becomes two-layered, with the outer periderm and the inner developing epidermis. During this time there is no dermis but a subepidermal cellular layer with deposition of basement membrane type IV collagen and laminin by 5 weeks. By 8 weeks there is evidence of vascular development.

During the embryonic fetal transition stage at 9–10 weeks several changes are seen:

- epidermal cells express keratins 1 and 10, associated with differentiation
- maturation of the basement membrane with the development of cell adhesion and expression of integrins α6 and β4
- rapid deposition of dermal matrix
- migration of melanocytes from the neural crest
- Langerhans' cells are detected originating in the fetal thymus and bone marrow
- Merkel cells are seen initially in the epidermis and subsequently in the dermis.

1065

The early fetal period from 11 to 14 weeks sees the development of the hair follicles within the skin. The skin adnexal structures continue to develop and mature into the mid-fetal period at 15–20 weeks. As the fetus grows in the late period, 20–40 weeks, acceleration of stratum corneum can be identified in specific regions, including the palms, soles, face, and scalp. There is a close association between keratinization and hair follicle development with the stratum corneum developing initially in the perifollicular regions. The mature stratum corneum is a structure which develops in the late fetal period; a combination of the terminally differentiated keratinocytes forming a cornified envelope and lipid extrusion from the abundant lamellar bodies of the granular layer keratinocytes.

Embryologically the neural tissue and the epidermis are derived from ectoderm. By the end of the fourth week of embryonic development the neural ectoderm has separated from the surface ectoderm forming the neural tube. The nerve endings will never, under normal conditions, be exposed to the external environment. The skin forms the interface and has developed as the interactive responsive surface. In the following weeks mesoblastic cells from the neural crest migrate into the skin as melanoblasts and the early nerve fibers develop as the vasculature migrates into the mesodermal elements destined to become dermis. A close association is seen in development with the skin, developing as a tactile organ providing the feedback information from the surface to the developing CNS. The understanding of neural plasticity of the CNS and the peripheral nerve field underpins the self-organization principle. The co-development is put forward as an explanation of the co-evolution of the human CNS and the skin as an adaptive dynamic interface.

In the fully developed skin there are cells from all three embryological layers in a complex framework of extracellular matrix (ECM). There are functions common to all skin areas, but also there has been adaptation to the specific functions of given body sites which are seen even at the early fetal stages.

The mature epidermis is composed primarily of keratinocytes arising from a layer of basal cells situated on the basement membrane. As the keratinocytes differentiate they form a stratified squamous epithelium. As the cells undergo terminal differentiation they lose their nuclei and form a highly cross-linked protein-based layer of keratin. The basal cells are in intimate contact with terminal dendrites: synapse-like structures have been described between nerve endings and keratinocytes. The melanocytes are situated in the basal layer with the melanosomes being transferred to the differentiating keratinocytes giving the color of the skin due to pigment load. The cells linking to the immune system, Langerhans' giant cells and dendritic cells, are also present in the epidermis. The epidermis is specialized to the body sites, most noticeably with the thickened cornified layer of the sole and palm.

The dermis is attached to the epidermis at the dermal–epidermal junction by the basement membrane morphologically arranged as the rete pegs, which are exaggerated at glaberous skin areas. The dermis is mainly connective tissue, predominantly collagen, with elastin seen in the superficial papillary dermis. The fibroblast is the cell that produces the ECM, which is specialized over differing body sites with areas such as the groin and axilla being more elastic than the thicker dermis of the back. Cells of hematopoietic origin such as lymphoctyes and macrophages migrate into the dermis and are involved in surveillance. The neural and vascular networks maintain the skin and facilitate the functions of the dynamic interactive skin interface (Hoath and Maibach, 2003).

The investigation of wound healing in parallel with an understanding of skin development and functions gives us the opportunity to further develop innovative methods of TE. The skin has developed specifically in relation to the multiple functions it performs. As an active organ it is responsive to changes in the external and internal environment, pivotal in maintaining the body's homeostasis. Our knowledge of skin functions is still growing and includes:

- semi-permeable barrier, overcoat, suit of armor
- thermoregulation, refrigerator
- antibacterial, waterproof
- UV protection, sunshade
- sensory receptor, sensitive to the touch of a feather, to temperature, and to pain
- self-regenerating, withstanding the wear and tear of a lifetime
- capable of rapid repair, executing its own running repairs
- immune modulation
- psychological interaction
- vitamin production.

The loss of skin integrity can result in severe morbidity and even mortality. The body needs a barrier to the atmosphere to maintain homeostasis. The production of the stratum corneum can be mimicked in tissue culture by exposing a sheet of keratinocytes in culture to an air–liquid interface (Kalyanaraman and Boyce, 2007), but it requires maturation *in situ* to fully develop the "smart material" of the anucleated cell bodies and the extracellular lipoproteins moisturized by vitamin E-producing sebum.

The keratinocytes produce surface proteins which are antimicrobial in the first line of defense against the colonizing bacteria on the skin surface (Bardan et al., 2004). The expression of these proteins changes as the keratinocyte is stressed, as in wounding or culturing, such that protection from microbial invasion is highlighted.

The multiple and specific sensory inputs to the skin are pivotal in the regulation of the body's temperature, immune responses, and the psychological responses via neural and neuro-endocrine control systems. Recent animal studies have highlighted the sensory role of the skin in normal development, with touch being associated with growth potential of the internal solid organs (Hoath and Maibach, 2003).

The skin is also profoundly influenced by the pathophysiology of the individual, with cutaneous changes in anatomy and physiology being linked to many disease processes.

It is with this expanding body of knowledge with respect to the skin that we engage the fields of TE to provide solutions for skin defects to maintain function and avoid scarring. To do so we also need a working knowledge of the skin response to injury and tissue loss; the processes of wound healing.

Briefly, wound healing in skin is a complex series of cascading events which has been described in three overlapping stages from the initial inflammation to tissue formation and subsequent tissue remodeling (Greenhalgh, 2005; Singer and Clark, 1999).

The initial response is clot formation to achieve hemostasis. The activation of platelets releases the contents of their alpha granules, resulting in the activation of the clotting cascade and the release of adhesive proteins forming the matrix of the clot, e.g. fibrin and chemotactic factors and growth factors into the wounded area. The coagulation pathway links to the activation of the complement pathway's facilitation of the recruitment of neutrophils needed to facilitate the inflammatory response with the removal of cellular debris and microorganisms.

The in-growth of new vessels as granulation tissue is initiated and the keratinocytes at the wound edge mobilize to commence re-epithelialization. Macrophages migrate to the wound, releasing multiple protein growth factors as the wound response progresses to the repair phase. Both hemopoietic and mesenchymal stem cells from the circulation are attracted into the wound with fibroblasts producing ECM, some as myofibroblasts associated with wound contraction.

There is an increased interest in the interactions between the keratinocytes and underlying fibroblasts as the matrix is remodeled and the new basement membrane develops. As the keratinocyte is exposed to collagen it secretes collagenase and as the basement membrane integrity is restored the cells revert to their normal phenotype in the situation where healing is achieved without scarring.

In situations of more extensive tissue damage the fibroblast is of a scar phenotype with the production of disordered collagen. The interactions between the ECM and the fibroblasts respond to the changes as healing progresses from tissue repair to remodeling. The initial migratory fibroblast transitions to the profibrotic phenotype producing ECM proteins. In the remodeling phase the cell number within the dermis or scar falls as the cells undergo apoptosis.

A knowledge of the wound-healing progression over time allows the TE skin solution to be clinically integrated into the process and may be directed at a number of target strategies: the control of cells in growth, the genetic manipulation of cells to express a given phenotype, seeding of the retained dermis with cells from the dermal—epidermal junction, removal of the full thickness of the area of compromised skin, and replacement with a TE construct.

THE POTENTIAL PREREQUISITE REQUIREMENTS FOR TE SKIN SOLUTIONS

The skin matures from the softness of the newborn to the skin in old age with the loss of elasticity (Fuchs, 2007). We believe that the young heal rapidly but scar aggressively, in contrast to the elderly, who heal more slowly and scar less. Regeneration of the skin without functional

1067

or aesthetic deficit, rather than enhanced repair, remains the ultimate goal of wound healing therapies (Boyce, 2001). However, the degree of scarring and the quality of the repair are highly dependent on the time taken to heal, with faster healing correlating to improved outcome (Deitch et al., 1983). The availability of the TE skin for timely use is a key factor balancing the issues of allograft to autograft, biological to non-biological solutions. The differences in wound healing responses with age and etiology of the defect will have implications when harvesting donor tissue for TE and will potentially influence the choice of TE technique (Horch et al., 2005).

The skin surface is continually replaced under normal conditions and the morphology is retained over the years. However, the capacity to regenerate and self-organize becomes overwhelmed in all but trivial injuries such that the repair forms a scar which all too often is debilitating, both physically and psychologically. The traditional approach to reduce the time to healing of a skin defect has been to skin graft. Full-thickness skin grafts (FTG) will give the best scar result but appropriate donor sites to match with recipient sites can lead to skin mismatch, as seen when a FTG from the groin is used to release a contracture on the palm of the hand. The donor site availability for split-thickness skin grafting (SSG) may be limited in size when large body surface areas are compromised as in burn injury or giant nevi. The area of cover of a given donor site can be increased by meshing or expanding as in the Meek technique (Munster and Smith-Meek, 1994). The expansion of the SSG is associated with small areas of the wound healing by secondary intention and a poorer scar outcome. It is the desire to eliminate, or at least reduce the scar with reduction of donor site morbidity, that has driven TE of skin over the last three decades.

The essential factors to achieve healing are:

- a source of cell's capable of differentiating into the tissue
- an extracellular matrix capable of supporting the cells.

With an increased understanding of the skin physiology and interactions with the internal and external environment we need to also consider:

- the three-dimensional spacial information of the area under repair
- feedback from the surface to facilitate self-organization.

The ideal needs for a TE skin replacement continue to be debated and are related to the clinical indication for use. In our group we have been guided by the following basic requirements (Martin, 1997):

- rapidly available
- autologous
- site-matched
- reliable wound adherence
- minimal donor site morbidity
- clinically manageable
- improved quality of scar
- affordable.

For TE to be successful it is clear that an in-depth working knowledge of the biology of the tissue is essential. There are a number of cell types within the skin, each with specific and often interrelated functions; it is fundamental to the success of TE to have an understanding of the essential information on how the cells relate to the other cells and the ECM of the skin.

The ECM scaffold is integral to tissue integrity, we know that the physical shape and chemical composition of the environment of a cell influence its phenotype. The field of bioinformatics may well assist in refining design in the future (Powell et al., 2010). In the design of the TE solution the understanding of this relationship will lead to increased clinical success.

From the engineering perspective there are technologies which will facilitate innovative clinical solutions such as the advanced modeling and fabrication for scaffold manufacture. The development of bioreactors to maintain viability and expand cell numbers associated with scale up in an appropriately regulated laboratory is essential for cell-culture-based techniques (Kalyanaraman and Boyce, 2007).

It is also clear from the experiences of the last three decades that many of the proposed TE skin solutions are disruptive technologies. It is essential that the TE skin should not only be designed for a clinical problem, be reproducible and reliable, but also be linked to an education and training program such that it realizes its full clinical potential.

CURRENT TE SKIN TECHNOLOGIES

Engineering principles have been applied to skin for many years with the development of medical devices to harvest SSG accurately and mesh the skin to allow expansion. The practice of tissue expansion is a well-established surgical tool for the development of skin by subcutaneous insertion of inflatable devices *in vivo*, which can be serially enlarged with the resulting development of the skin as demonstrated by cell proliferation (Argenta, 1984). The tissue-expanded skin has all layers and the complete characteristics of the donor site including retention of functional innervation (Wood and McMahon, 1989). There is an increasing interest in the concept of tissue expansion *in vitro* with a full-thickness skin biopsy put under tension in a bioreactor system to maintain its viability and facilitate cell replication resulting in tissue growth. However, these solutions are limited by time, area, and in some cases donor site and scar outcome. There remains the need to provide rapid large surface area cover with the ultimate goal of healing by regeneration, not scar repair.

The initial approach to TE skin was to separate the layers and consider the epidermis and dermis as separate problems.

The work by Green in the 1970s was focused on the culture of keratinocytes into cell sheets suitable for grafting (Rheinwald and Green, 1975). The solution to the large surface area burn was to harvest cells from a non-injured donor site and undertake laboratory-based tissue expansion. The resulting sheets of cells could be used to close the wound as would a traditional SSG. However, there were problems with the time taken to culture in the laboratory, fragility, adherence to the wound surface, and durability over time as only the epidermis was replaced, in addition to the cost (Wood et al., 2006a, b). In trying to solve some of these problems there has been development in the area of subconfluent cell transfer on a number of cell culture surfaces (Chester et al., 2004) in addition to the delivery of cells in suspension as an aerosol (Currie et al., 2003; Fredriksson et al., 2008). The subconfluent cells have a more reliable adherence and are available in a shorter timeframe from 3 weeks for sheets to 5 days for subconfluent cultures (Hernon et al., 2006; Johnen et al., 2008). The process of harvesting cells from the dermal–epidermal junction by enzymatic and physical dissociation has been used for immediate delivery of a non-cultured cell population to the wound (Wood, 2008). The cells are a mixed cell population in the same ratio as seen in the normal skin construct as there has been no selection of cell populations seen when culturing. The maintenance of the melanocytes allows the development of appropriate pigmentation (Navarro et al., 2001). The cells adhere, migrate, and proliferate across the wound surface and then differentiate and self-organize into a mature epidermis. The scar outcome is intimately linked to the underlying wound bed, which will be discussed in a later section.

The development of a suitable dermal scaffold was also the focus of the TE field (Klama-Baryła et al., 2008). Topographical features are known to influence cell behavior through a phenomenon known as "contact guidance," and alteration in the size of the surface detail can elicit different cell responses (Freytes et al., 2009). Running parallel to the epidermal cell culture was work by Yannas and Burke on dermal replacement, which culminated in the

1069

commercially available product Integra (Yannas et al., 1975). The concept of tissue-guided regeneration within an architectural framework is in current clinical practice. The underpinning research on the composition and construction demonstrated the importance of considering both aspects; a combination of bovine collagen coated with GAG but with a pore size of less than 60 μm or greater than 100 μm resulting in disordered granulation tissue, with the optimal pore size resulting in the migrating cells expressing a reticular dermal fibroblast phenotype. The main drawback with Integra is that is addresses only the dermal aspect, with the outer layer on silicone acting as a pseudo-epidermis for the period on vascularization, usually 3 weeks before a second surgical procedure is needed to repair the epidermis (Heitland et al., 2004). The epidermal repair is with a thin SSG which may be meshed to cover a larger area than the donor site with epidermal cells to speed the time to healing and reduce the mesh scar pattern (Navarro et al., 2000). The two-stage problem has been addressed by trying to reduce the time to vascularization or by the development of constructs that can be used with an SSG at the same procedure; Apligraft, Matriderm, and Pelnec are dermal templates marketed to provide appropriate topography and matrix properties to promote cell migration into the wound, improve healing, and reduce scarring as a one-stage procedure with SSG (Bannasch et al., 2008). An alternative is to use our knowledge of healing as we see cells migrating from areas of the skin adnexal structures in the dermis to form the new epidermal layer; introduced cells harvested from the dermal–epidermal junction seeded into the Integra will migrate and organize into a new epidermis with an established dermal–epidermal junction within 3 weeks (Wood et al., 2007). The use of Integra is a clear demonstration of the potential of tissue-guided regeneration with the expression of cell phenotype guided by the morphology and chemistry of the matrix (Orgill et al., 1999).

It is well established that cells change their phenotype in response to changes in their environment (Smiley et al., 2006; Takahashi and Yamanaka, 2006). The development of suitable technologies to generate an optimal environment for wound healing is key to enhancing cell response to tissue injury, reducing the time to heal and improving the outcome. Our knowledge of the ECM–cell interactions is increasing with the recognition of the cell signaling by nanoscale structures on cell surfaces with roles in attachment and cell migration (Black et al., 2005). Developments in nanotechnology have opened up possibilities in TE to improve on scaffold design, but relatively little is known about how changes in topography at the nanoscale affect cell behavior (Fleming et al., 1999). The scaffolds can be manufactured to address specific skin functions: to protect the injury from loss of fluid and proteins, enable the removal of exudates, inhibiting exogenous microorganism invasion, and improve the aesthetic appearance of the wound site (Curtis and Wilkinson, 2001). Current scaffolds are generally matrices of synthetic and/or natural polymers, fabricated by various techniques including (Wilkinson et al., 2002) solvent-casting, gas foaming, electrospinning, phase separation, freeze drying, melt molding, and solid free-form fabrication (Smith and Ma, 2004). Key to their performance is reproducibility, with control over pore size and the distribution of pores, removal of residual toxic organic solvents, and the control of the inflammatory and immune responses due to polymer degradation and the associated by-products (Chin et al., 2009).

Our group has explored anodic aluminum oxide (AAO) as a potential scaffold or template in TE (Parkinson et al., 2009). The self-organized oxide growth under controlled conditions generates a densely packed hexagonal array of uniform-size nanopores aligned perpendicular to the surface of the AAO film. The size of the pores can be nanoengineered by manipulating the anodization time and voltage, the anodizing electrolyte, and/or the time of post-chemical etching. Aluminum oxide is well known for its biocompatibility in the human body; it is inert, stable, and non-reactive, making it suitable for TE applications. The engineering of surfaces to manipulate healing is a rapidly expanding area, with the use of interactive dressing systems in widespread clinical use. With the realization of the impact of surface topography and chemistry on cell expression, and the developing nanoengineering techniques such as electrospinning and electrospraying, there is increasing interest in smart surface technology skin healing. Biocompatible polymeric self-assembling nanofiber constructs have the advantage of a large surface area, which can be linked to bioactive compounds. The release of the bioactive compounds can be controlled by intrinsic factors such as, in a hydrogel, release kinetics or extrinsic release triggers (Yamato, 2003). Recent exciting advances have been made in the area of nanocubes, nanocages, and nanorods, as primary candidates to be studied for the photo-therapeutic release of bioactive agents (Nath and Hyun, 2004). Clinically, the results of single cytokine applications have been disappointing and study of the natural healing processes is a result of complex cascade interactions over time (Eppley et al., 2003). It should be hardly surprising that we cannot achieve with a single cytokine administration what is the result of cell–ECM interactions. The aim with the advancing technology is to mimic the structure and function of the ECM, with the ability to adapt over time to the changing environment of the healing wound (Liu et al., 2004).

Bringing together the scaffold and cellular components has been successful with the development of multilayered constructs seeded with multiple cell types (Boyce, 2001; Sheridan et al., 2001). Clinical series have been presented demonstrating a soft supple skin but with the persistent problem of poor color match (Boyce et al., 2006). The understanding of the interdependence of the cells of the dermal–epidermal junction may well be key in the development of skin constructs with the appropriate melanocyte function (Mizoguchi et al., 2004). The main drawback in the clinical use of complex laboratory-based constructs is the time taken in the laboratory (Meana et al., 1998). However, the potential to use such technology in timed reconstructive surgery as opposed to acute trauma is beneficial with the ability to tailor the skin construct to the planned defect. The three-dimensional distribution of cells within the wound has been addressed by the innovative use of the "ink jet printing" technology, with the cells "printed," controlled by the shape and depth of the defect. The cell type within the system can be changed with the fibroblasts being laid down prior to the keratinocytes (Lamme et al., 2000). ECM and other proteins can also be introduced into the system, such as hair-based keratins, chemical-processing substrates from biological origins to develop innovative scaffold solutions to enhance cell performance within the constructs.

The time to availability of a TE technique is a key driver to the clinical utilization (Carsin et al., 2000). The use of allograft materials allows an "off the shelf" approach, whereas autograft material may take time in the expansion phase. The use of scaffolds alone can provide an advanced wound management system to facilitate healing, and can replace a tissue defect guiding repair as it is replaced or provide a permanent solution. The use of cells alone can also provide a surface epithelium which can modulate the underlying healing, form a mosaic of cells of intrinsic and extrinsic origin guiding repair, or provide a permanent surface. The combination of the two elements can provide an advanced skin repair solution, but as the differentiation of the construct is more advanced the time taken in the processes is prolonged. Consideration of the time taken has led to the investigation of construct being used in the immature form differentiating *in situ*. It is clear that TE provides a range of innovative solutions which are useful in a range of wounding/skin replacement areas. The effective use of the technology hinges on the clinician understanding the wound preparation and the aim of the repair. The wound assessment drives the initial clinical decisions directing management in terms of resuscitation, tissue salvage, infection control, and then planning the repair. The understanding of the range of TE solutions is essential in appropriate clinical implementation.

TE SKIN SOLUTIONS IN CLINICAL PRACTICE

The following clinical case of an extensive 65% total body surface area (TBSA) burn injury associated with multiple fractures and pneumothorax is used to demonstrate the decision-making and options available in current practice. The initial stabilization and resuscitation is life-saving, along with attention to infection control using Acticoat, a nanocrystalline silver dressing. Once stable, surgical debridement is planned to excise the areas of skin which cannot be salvaged. Although the timing of debridement is vigorously debated with respect to controlling the ongoing inflammation and improving outcome potential, a key element in the decision-making is what is available to cover the debrided wounds, either as a temporary measure or as a permanent replacement. A full-thickness wound requires dermal and epidermal repair for the optimal outcome. In large-surface-area wounds the standard SSG is not possible in one procedure due to restricted donor sites; the SSG can be meshed to achieve healing in a larger area but the healing of the interstices by secondary intention often leaves an unsightly meshed pattern scar. Cadaver allograft has been widely used but provides only a temporary solution and will require serial cover as the donor sites heal. A combination of allograft dermis and sheet cultured epithelial autograft (CEA) has been reported to result in a composite repair with retention of the dermal elements. The CEA sheets may take 21 days to culture, or more timely availability can be achieved using preconfluent cultures on carriers or delivered as suspension. The dermis can be replaced now with a number of off-the-shelf products such as Matriderm, Pelnec, or Apligraft. The TE technique most widely reported to date is Integra. The use of composite TE skin could be considered to augment the second stage of the repair with Integra, as it takes time in laboratory preparation.

In this case Integra was used to reconstruct the dermis in the areas of full-thickness excision, as seen in Figure 57.1, where the Integra is held in place with a combination of staples and dressings to facilitate take. In areas where dermis could be salvaged cells were harvested from the dermal—epidermal junction of uninjured skin using a ReCell kit with Biobrane as a dressing. The main drawback with Integra is the period of vascularization of 3 weeks prior to proceeding to repair of the epidermis. The epidermis was repaired using a thin meshed SSG 1 to 3 in combination with a non-cultured cell suspension from the ReCell kit to reduce the meshed pattern scar. The healing is seen in Figure 57.2 with the mash pattern fading well in the upper compared to the lower abdomen. At the time of scar maturity in Figure 57.3, the scar situation demonstrates the current issues faced by our patients:

- contour deflect due to removed subcutaneous fat layer
- persistent mesh pattern in the lower scar
- mismatch of pigmentation
- contracture bands distorting the anatomy
- the repair is a scar.

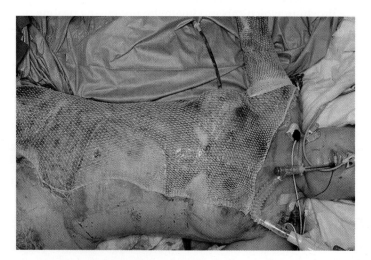

FIGURE 57.1
Post-surgical debridement of full-thickness burn of 65% TBSA and replacement with Integra dermal scaffold held in place with Fastinet, area of partial-thickness burn on the abdomen treated with autologous cells under Biobrane, and non-injured skin donor site on the left flank.

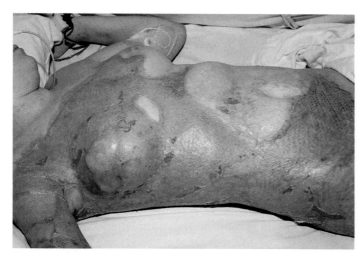

FIGURE 57.2
Post second procedure to repair the epidermal layer using a combination of meshed split-thickness graft with a non-cultured autologous cell spray.

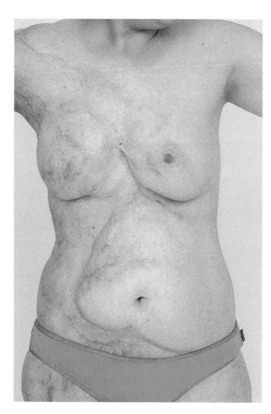

FIGURE 57.3
Two years post-injury, prior to planned reconstructive surgery.

There has clearly been progress with the development of TE techniques intimately linked with the advancing survival and quality of scar in patients with major burn injuries. However, there is a clear need to continue to develop TE with the aim of total three-dimensional soft tissue and skin replacement.

THE FUTURE

It is the vision of scarless healing which has led to the exploration of regeneration and the interplay between genes (Goessler et al., 2006), cells, and tissues. Pleuripotential stem cells are present within each individual; the drivers of the cells down a regenerative path are as yet unknown but an exciting area of research with promise for the future (Madlambayan and Rogers, 2006). The introduction of allograft stem cells such as mesenchymal stem cells may provide an alternative source of cell regeneration (Wu et al., 2007). Understanding that every intervention from the time of injury influences the scar worn for life has driven research in a multitude of directions. TE of skin is an exciting area which already has made a significant clinical impact (Metcalfe and Ferguson, 2007). However, we are far from the routine provision of technologies and strategies to provide site-matched fully functional skin (MacNeil, 2007). Great progress has been made in the areas of dermal templates and cell-based therapies and bringing the two elements together in skin constructs. The clinical implementation of TE skin solutions, along with the use of interactive surface dressing systems, has improved outcomes.

The task is far from completed:

- skin adnexal structures are elusive
- timely availability of TE skin remains problematic

and more work is needed in:

- harnessing the explosion in smart material technology
- understanding the drivers to tissue-guided regeneration
- understanding the concept of self-organization and the bioinformatics behind morphology
- investigating the impact of neural plasticity and its role in scarless healing
- understanding the barriers to clinical translation
- developing regulatory pathways for novel solutions to ensure safety but timely availability.

At the high-tech end of the spectrum we may consider bringing together laser surface imaging linked to fabrication to build bioreactors the shape of the defect with the "smart" cytokine-loaded scaffold materials tailored to the correct three-dimensional shape. Cells of the appropriate body site could then be introduced into the scaffold by cell printing methods in the individualized bioreactor and flow of tissue culture medium used to induce small-vessel formation. At the time of transplant application of external techniques such as infrared could control the release of biologically active molecules from the "smart" scaffold surface to ensure, for instance, reinnervation and restoration of function with the capacity to integrate into the body. An alternative to the individualized bioreactor could be the wound itself prepared by "smart" surface sand seeded with cells directly. With the understanding of the drivers to self-organization they could be used to enhance *in situ* tissue-guided regeneration.

CONCLUSION

Over the last three decades the concept of TE has been an area of active research, investigating innovative solutions. Combinations of three-dimensonal engineered scaffolds with cells to produce tissue over time provide the fourth dimension of skin repair techniques.

The fundamental elements required for primary healing are:

- a source of cells capable of differentiation into the lost tissue
- an architectural framework for cells to migrate into and express the appropriate phenotype
- three-dimensional spatial information of the damaged tissue and the relationship to the surrounding viable non-injured tissue interface
- a feedback mechanism to guide self-organization.

In July 2005 the *Medical Journal of Australia* published a vision of clinical care in a number of disciplines in 50 years' time.

"Assessment is key in understanding the extent of injury.

Debridement is focused on tissue salvage.

Reconstruction balances repair with regeneration.

Investigation of multimodality, multiscale characterisation, including confocal microscopy and synchrotron technology will quantify assessment.

Debridement using autolytic inflammatory control techniques with image guided physical methods will ensure the vital tissue frameworks are retained.

Tissue guided regeneration afforded by self-assembly nano-particles will provide the framework to guide cells to express the appropriate phenotype in reconstruction.

To solve the clinical problem a multi-disciplinary scientific approach is needed to ensure the quality of the scar is worth the pain of survival."

It is of note that already many of the technologies highlighted are available and in need of research to move along the innovation pathway to ensure safe implementation into healthcare systems. Progress requires collaboration at all stages from basic science, clinical trial design to health economics, driven by improved clinical outcomes. Translation of new technologies into health systems requires the rigor of a research framework to identify and measure the impact of innovation in communication and education. Close working relationships between basic research and clinical service delivery are essential.

If the solution to the problem is scarless healing by a regenerative repair process and the aim is to improve the outcome from injury by restoration of function, then the future must blend the long-term vision with incremental short-term improvements.

There has been great progress in TE of skin to date, and it is an exciting area which offers tangible clinical solutions with an enormous potential for further improvement. The challenge we face now is to capitalize on that tradition and link with the opportunities afforded by the unprecedented growth in science and technology, to ensure the quality of the scar outcome is worth the pain of survival.

References

Argenta, L. C. (1984). Controlled tissue expansion in reconstructive surgery. *Br. J. Plast. Surg., 37*, 520–526.

Atiyeh, B. S., & Costagliola, M. (2007). Cultured epithelial autograft (CEA) in burn treatment: three decades later. *Burns, 33*, 405–413.

Bannasch, H., Fohn, M., Unterberg, T., Bach, A. D., Weyand, B., & Stark, G. B. (2003). Skin tissue engineering. *Clin. Plast. Surg, 30*, 573–579.

Bannasch, H., Unterberg, T., Fohn, M., Weyand, B., Horch, R. E., & Stark, G. B. (2008). Cultured keratinocytes in fibrin with decellularised dermis close porcine full-thickness wounds in a single step. *Burns, 34*, 1015–1020.

Bardan, A., Nizet, V., & Gallo, R. L. (2004). Antimicrobial peptides and the skin. *Expert Opin. Biol. Ther., 4*, 543–549.

Bayat, A. (2003). Skin scarring. *BMJ, 326*, 88–92.

Black, A. F., Bouez, C., Perrier, E., Schlotmann, K., Chapuis, F., & Damour, O. (2005). Optimization and characterization of an engineered human skin equivalent. *Tissue Eng., 11*, 723–733.

Boyce, S. T. (2001). Design principles for composition and performance of cultured skin substitutes. *Burns, 27*, 523–533.

Boyce, S. T., Kagan, R. J., Greenhalgh, D. G., Warner, P., Yakuboff, K. P., & Palmieri, T. (2006). Cultured skin substitutes reduce requirements for harvesting of skin autograft for closure of excised, full-thickness burns. *J. Trauma., 60*, 821–829.

Carsin, H., Ainaud, P., Le Bever, H., Rives, J., Lakhel, A., & Stephanazzi, J. (2000). Cultured epithelial autografts in extensive burn coverage of severely traumatized patients: a five year single-center experience with 30 patients. *Burns, 26,* 379–387.

Chester, D. L., Balderson, D. S., & Papini, R. P. (2004). A review of keratinocyte delivery to the wound bed. *J. Burn Care Rehabil., 25,* 266–275.

Chin, S. F., Iyer, K. S., Saunders, M., St. Pierre, T. S., Buckley, C., Paskevicius, M., & Raston, C. L. (2009). Encapsulation and sustained release of curcumin using superparamagnetic silica reservoirs. *Chem. Eur. J., 15,* 5661–5669.

Currie, L. J., Martin, R., Sharpe, J. R., & James, S. E. (2003). A comparison of keratinocyte cell sprays with and without fibrin glue. *Burns, 29,* 677–685.

Curtis, A., & Wilkinson, C. (2001). Nanotechniques and approaches in biotechnology. *Trends Biotechnol., 19,* 97–101.

Deitch, E. A., Wheelahan, T. M., Rose, M. P., Clothier, J., & Cotter, J. (1983). Hypertrophic burn scars: analysis of variables. *J. Trauma., 23,* 895–898.

Eppley, B. L., Woodell, J. E., & Higgins, J. (2003). Platelet quantification and growth factor analysis from platelet-rich plasma: implications for wound healing. *Plast. Reconstr. Surg., 114,* 1502–1508.

Ferguson, M. W., & O'Kane, S. (2004). Scar-free healing: from embryonic mechanisms to adult therapeutic intervention. *Philos. Trans. R. Soc. Lond. B Biol. Sci., 359,* 839–850.

Fleming, R. G., Murphy, C. J., Abrams, G. A., Goodman, S. L., & Nealey, P. F. (1999). Effects of synthetic micro- and nano-structured surfaces on cell behaviour. *Biomaterials, 20,* 573–588.

Fredriksson, C., Kratz, G., & Huss, F. (2008). Transplantation of cultured human keratinocytes in single cell suspension: a comparative *in vitro* study of different application techniques. *Burns, 34,* 212–219.

Freytes, D. O., Wan, L. Q., & Vunjak-Novakovic, G. (2009). Geometry and force control of cell function. *J. Cell Biochem., 108,* 1047–1058.

Fuchs, E. (2007). Scratching the surface of skin development. *Nature, 445,* 834–842.

Goessler, U. R., Riedel, K., Hormann, K., & Riedel, F. (2006). 'Perspectives of gene therapy in stem cell tissue engineering.' *Cells Tissues Organs., 183,* 169–179.

Greenhalgh, D. G. (2005). Models of wound care. *J. Burn Care Rehabil., 26,* 293–395.

Heitland, A., Piatkowski, A., Noah, E. M., & Pallua, N. (2004). Update on the use of collagen/glycosaminoglycate skin substitute – six years of experiences with artificial skin in 15 German burn centers. *Burns, 30,* 471–475.

Hernon, C. A., Dawson, R. A., Freedlander, E., Short, R., Haddow, D. B., Brotherston, M., et al. (2006). Clinical experience using cultured epithelial autografts leads to an alternative methodology for transferring skin cells from the laboratory to the patient. *Regen. Med., 1,* 809–821.

Hoath, S. B., & Maibach, H. I. (2003). *Neonatal Skin: Structure and Function.* New York: Marcel Dekker.

Horch, R. E., Kopp, J., Kneser, U., Beier, J., & Bach, A. D. (2005). Tissue engineering of cultured skin substitutes. *J. Cell Mol. Med., 9,* 592–608.

Johnen, C., Steffen, I., Beichelt, D., Brautigam, K., Witascheck, T., Toman, N., et al. (2008). Culture of subconfluent human fibroblasts and keratinocytes using biodegradable transfer membranes. *Burns, 34,* 655–663.

Johnson, P. C., Mikos, C. G., Fisher, J. P., & Jansen, J. A. (2007). Strategic directions in tissue engineering. *Tissue Eng., 13,* 2827–2837.

Kalyanaraman, B., & Boyce, S. (2007). Assessment of an automated bioreactor to propagate and harvest keratinocytes for fabrication of engineered skin substitutes. *Tissue Eng., 13,* 983–993.

Klama-Baryła, A., Glik, J., Kawecki, M., Nowak, M., & Sieroń, A. L. (2008). Skin substitutes – the application of tissue engineering in burn treatment Part 1. *J. Orthop. Trauma Sur. Rel. Res., 3,* 96–103.

Lamme, E. N., Van Leeuwen, R. T., Brandsma, K., Van Marle, J., & Middelkoop, E. (2000). Higher numbers of autologous fibroblasts in an artificial dermal substitute improve tissue regeneration and modulate scar tissue formation. *J. Pathol., 190,* 595–603.

Liu, H., Mao, J., Yao, K., Yang, G., Cui, L., & Cao, Y. (2004). A study on a chitosan-gelatin-hyaluronic acid scaffold as artificial skin *in vitro* and its tissue engineering applications. *J. Biomater. Sci. Polym. Ed., 15,* 25–40.

Llames, S. G., Del Rio, M., Larcher, F., Garcia, E., Garcia, M., Escamez, M. J., et al. (2004). Human plasma as a dermal scaffold for the generation of a completely autologous bioengineered skin. *Transplantation, 77,* 350–355.

Lockhart, R. D., Hamilton, G. F., & Fyfe, F. W. (1965). *Anatomy of the Human Body.* Philadelphia: J.B. Lippincott.

MacNeil, S. (2007). Progress and opportunities for tissue-engineered skin. *Nature, 445,* 874–880.

Madlambayan, G., & Rogers, I. (2006). Umbilical cord-derived stem cells for tissue therapy: current and future uses. *Regen. Med., 1,* 777–778.

Martin, P. (1997). Wound healing — aiming for perfect skin regeneration. *Science, 276,* 75–81.

Martinez-Hernandez, A. (1988). Repair, regeneration, and fibrosis. In E. Rubin, & J. L. Farber (Eds.), *Pathology* (pp. 66–95). Philadelphia: J.B. Lippincott.

Meana, A., Iglesias, J., Del Rio, M., Larcher, F., Madrigal, B., Fresno, M. F., et al. (1998). Large surface of cultured human epithelium obtained on a dermal matrix based on live fibroblast-containing fibrin gels. *Burns, 24,* 621–630.

Metcalfe, A. D., & Ferguson, M. W. (2007). Tissue engineering of replacement skin: the crossroads of biomaterials, wound healing, embryonic development, stem cells and regeneration. *J. R. Soc. Interface, 4,* 413–437.

Mizoguchi, M., Suga, Y., Sanmano, B., Ikeda, S., & Ogawa, H. (2004). Organotypic culture and surface plantation using umbilical cord epithelial cells: morphogenesis and expression of differentiation markers mimicking cutaneous epidermis. *J. Dermatol. Sci., 35,* 199–220.

Mosier, M. J., & Gibran, N. S. (2009). Surgical excision of the burn wound. *Clin. Plastic Surg., 36,* 617–625.

Munster, A. M., & Smith-Meek, M. (1994). The effect of early surgical intervention on mortality and cost effectiveness in burn care 1978–1991. *Burns, 20,* 61–64.

Mustoe, T. A., Cooter, R. D., Gold, M. H., Hobbs, F. D., Ramelet, A. A., Shakespeare, P. G., et al. (2002). International clinical recommendations on scar management. *Plast. Reconstr. Surg., 110,* 560–571.

Nath, N., & Hyun, J. (2004). Surface engineering strategies for control of protein and cell interactions. *Surface Sci., 570,* 98–110.

Navarro, F. A., Stoner, M. L., Parks, C. S., Huertas, J. C., Lee, H. B., Wood, F. M., et al. (2000). Sprayed keratinocyte suspensions accelerate epidernal coverage in porcine microwound model. *J. Burn Care Rehabil., 21,* 513–518.

Navarro, F. A., Stoner, M. L., Lee, H. B., Park, C. S., Wood, F. M., & Orgill, D. P. (2001). Melanocyte repopulation in full-thickness wounds using a cell spray apparatus. *J. Burn Care Rehabil., 22,* 41–46.

Orgill, D. P., Straus, F. H., II, & Lee, R, C (1999). The use of collagen-GAG membranes in reconstructive surgery. *Ann. N. Y. Acad. Sci., 888,* 233–248.

Parkinson, L. G., Giles, N. L., Adcroft, K. F., Fear, M. W., Wood, F. M., & Poinern, G. E. (2009). The potential of nanoporous anodic aluminium oxide membranes to influence skin wound repair. *Tissue Eng. Part A, 15,* 3753–3756.

Powell, H. M., McFarland, K. L., Butler, D. L., Supp, D. M., & Boyce, S. T. (2010). Uniaxial strain regulates morphogenesis, gene expression, and tissue strength in engineered skin. *Tissue Eng. Part A, 16,* 1083–1092.

Rheinwald, J. G., & Green, H. (1975). Serial cultivation of strains of human epidermal keratinocytes: the formation of keratinizing colonies from single cells. *Cell, 6,* 331–343.

Rockwell, W. B., Cohen, I. K., & Ehrlich, H. P. (1989). Keloids and hypertrophic scars: a comprehensive review. *Plast. Reconstr. Surg., 84,* 827–837.

Roupé, K. M., Nybo, M., Sjöbring, U., Alberius, P., Schmidtchen, A., & Sørensen, O. E. (2010). Injury is a major inducer of epidermal innate immune responses during wound healing. *J. Invest. Dermatol., 130,* 910–916.

Rumsey, N., Clarke, A., & White, P. (2003). Exploring the psychosocial concerns of outpatients with disfiguring conditions. *J. Wound Care, 12,* 247–252.

Shakespeare, P., & Shakespeare, V. (2002). Survey: use of skin substitute materials in UK burn treatment centres. *Burns, 28,* 295–297.

Sheridan, R. L., Morgan, J. R., Cusick, J. L., Petras, L. M., Lydon, M. M., & Tompkins, R. G. (2001). Initial experience with a composite autologous skin substitute. *Burns, 27,* 421–424.

Singer, A. J., & Clark, R. A. (1999). Cutaneous wound healing. *N. Engl. J. Med., 341,* 738–746.

Smiley, A. K., Klingenberg, J. M., Aronow, B. J., Boyce, S. T., Kitzmiller, W. J., & Supp, D. M. (2005). Microarray analysis of gene expression in cultured skin substitutes compared with native human skin. *J. Invest. Dermatol., 125,* 1286–1301.

Smiley, A. K., Klingenberg, J. M., Boyce, S. T., & Supp, D. M. (2006). Keratin expression in cultured skin substitutes suggests that the hyperproliferative phenotype observed *in vitro* is normalized after grafting. *Burns, 32,* 135–138.

Smith, L. A., & Ma, P. X. (2004). Nano-fibrous scaffolds for tissue engineering. *Colloid Surfaces B: Biointerfaces, 39,* 125–131.

Takahashi, K., & Yamanaka, S. (2006). Induction of pluripotent stem cells from mouse embryonic an adult fibroblast cultures by defined factors. *Cell, 126,* 663–676.

Whang, K., Kim, M., Song, W., & Cho, S. (2005). Comparative treatment of giant congenital melanocytic nevi with curettage or Er:YAG laser ablation alone versus with cultured epithelial autografts. *Dermatol. Surg., 31,* 1660–1667.

Wilkinson, C. D. W., Riehle, M., Wood, M., Gallagher, J. O., & Curtis, A. S. G. (2002). The use of materials patterned on a nano- and micro-metric scale in cellular engineering. *Mater. Sci. Eng. C, 19,* 263–269.

Wood, F. M. (2003). Clinical potential of cellular autologous epithelial suspension. *Wounds, 15,* 16–22.

Wood, F. (2008). Alternative delivery of keratinocytes for epidermal replacement. In D. P. Orgill, C. Blanco, & M. Joseph (Eds.), *Biomaterials for Treating Skin Loss.* Cambridge: Woodhead Publishing.

Wood, F. M., & McMahon, S. B. (1989). The response of the peripheral nerve field to controlled soft tissue expansion. *Br. J. Plast. Surg., 42*, 682–686.

Wood, F. M., Kolybaba, M. L., & Allen, P. (2006a). The use of cultured epithelial autograft in the treatment of major burn injuries: a critical review of the literature. *Burns, 32*, 395–401.

Wood, F. M., Kolybaba, M. L., & Allen, P. (2006b). The use of cultured epithelial autograft in the treatment of major burn wounds: eleven years of clinical experience. *Burns, 32*, 538–544.

Wood, F. M., Stoner, M. L., Fowler, B. V., & Fear, M. W. (2007). The use of a non-cultured autologous cell suspension and Integra dermal regeneration template to repair full-thickness skin wounds in a porcine model: a one-step process. *Burns, 33*, 693–700.

Wu, Y., Wang, J., Scott, P. G., & Tredget, E. E. (2007). Bone marrow-derived stem cells in wound healing: a review. *Wound Repair Regen., 1*(Suppl. 15), S18–S26.

Yamato, M. (2003). Cell sheet engineering: from temperature-responsive culture surfaces to clinics. *Eur. Cell Mater., 6*, 420–428.

Yannas, I. V., Burke, J. F., Huang, C., & Gordon, P. L. (1975). Correlation of *in vivo* collagen degradation rate with *in vitro* measurements. *J. Biomed. Mater. Res., 9*, 623–628.

CHAPTER 58

Regenerative Medicine of the Respiratory Tract

Martin A. Birchall[*], **Sam Janes**[†], **Paolo Macchiarini**[‡,§]
* University College London, Centre for Stem Cells and Regenerative Medicine and UCL Ear Institute, Royal National Throat Nose and Ear Hospital, London
† Centre for Respiratory Research, Rayne Building, University College London, London, UK
‡ Cardiothoracic Surgery, Hospital Careggi, Florence, Italy
§ University College London, London, UK

INTRODUCTION

Airway and breathing are the two most immediate functions of mammalian life. The respiratory tract represents the interface between the body and the atmosphere around us, and thus represents a first line of defense against inhaled challenges such as potential allergens, pollutants, and cigarette smoke. Sequentially, it is responsible for filtering, humidifying air (nose), speech, and protecting the respiratory tract from inhalation of ingested materials and saliva (larynx), conducting air and mucus (tracheobronchial tree), and gas exchange (alveoli). Disorders affecting these functions can have devastating effects on quality and duration of life, and our ability to manage some of the most serious of these with conventional treatments is severely limited.

Regenerative medicine has considerable potential to bridge this gap between what present therapy can achieve and the restoration of normal existence. The most striking example of this was the restoration of normal function (and life) to a 31-year-old mother whose tracheobronchial tree had been ravaged by tuberculosis. In 2008, a stem-cell-based, tissue-engineered airway implant was prepared and transplanted with immediate normalization of lung function. However, there is a long way to go before this success may be translated into a raft of treatments for persons with severe disease at all levels of the respiratory tract. We review progress at each of these levels (Fig. 58.1).

While a crucial conduit for life, the trachea is a relatively simple, hollow organ, without intrinsic motility and complex neuromuscular actions, unlike the larynx or esophagus for example. Consequently, the trachea is, in some ways, the ideal organ to engineer and, thus, a perfect place to start to explore the therapeutic potential of regenerative medicine approaches to the airways. Approaches pursued to date center on the regeneration of a tube-like cartilage graft combining isolated cells and scaffolds with predetermined shapes and mechanical properties. It has rapidly become clear, however, that success depends on cell type used, culture conditions, and the scaffold's chemical and mechanical properties.

Understanding of stem cell biology of the lower respiratory tract has been driven by the unmet clinical needs of fibrotic and malignant disease. These requirements need a different approach to the more mechanical techniques applied to large airway reconstruction. The larynx is

Principles of Regenerative Medicine. DOI: 10.1016/B978-0-12-381422-7.10058-6

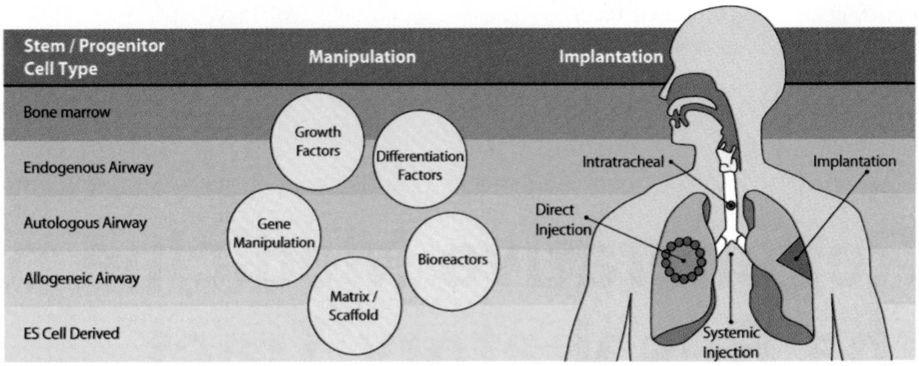

FIGURE 58.1
Schematic representation of the respiratory tract showing potential interventions by regenerative medicine techiniques.

a highly complex structure requiring functional neuromuscular units, joints, and a highly specialized mucosa for speech. All of these different demands are being addressed by various groups.

To ensure consistency in this rapidly evolving field, all terms used in this review are based upon the glossary defined by Mason and Dunnill (2008).

THE NOSE AND SINUSES

Both disease and therapy in rhinology (that part of medicine dealing with the nose and sinuses) can result in structural and functional defects. Reconstructive surgery has often required autologous cartilage grafting in order to correct skeletal defects. Traumato and disease of the facial musculoskeletal system result in considerable morbidity, which has also been addressed previously by means of synthetic prosthetic implant devices. However, all of these strategies are failing to deal with the increasing demand for treatments for this spectrum of clinical problems. There are many reasons for this, including the shortage of suitable graft material, problems in harvesting sufficient material to fill large defects adequately, implant biocompatibility and implant failure. Alternative approaches are needed therefore to address this growing problem (Griffiths and Naughton, 2002).

Olfaction and olfactory ensheathing cells

There are many potential therapeutic use of cells engineered to support regeneration. Of special interest to rhinologists are olfactory ensheathing cells. These are a class of glial cells which support regeneration of olfactory receptor neurons in the olfactory epithelium. When transplanted to other regions of the brain they may support regeneration of other classes of neurons (Au and Roskams, 2003). They also support axon regeneration and have been successfully employed in rat models of spinal cord regeneration (DeLucia et al., 2003). Periosteal stem cells and muscle stem cells have been engineered to express bone morphogenetic proteins (BMPs), sonic hedgehog homologs (Shhs), or vascular endothelial growth factor (VEGF) and then used in animal models of bone repair (Peng et al., 2002; Grande et al., 2003). Expression of the growth factors improved the quality of repair. The mammalian olfactory system has the distinction of being the only part of the central nervous system undergoing continuous renewal. Olfactory receptor neurons have half-lives of a few weeks and are renewed by division of stem cells in the olfactory epithelium which give rise both to neurons and to supporting cells (Jang et al., 2003; Murray et al., 2003). Their number is self-regulated by secretion of inhibitory factors including growth differentiation factor 11 (GDF-11; Wu et al., 2003) and fibroblast growth factors (FGFs) and BMPs (Calof et al., 2002) from the olfactory neurons. Newly generated receptor neurons extend axons through the cribriform plate to accurately reinnervate particular olfactory glomeruli specific for each odor receptor type (c. 400 in humans, c. 900 in rodents). Their regeneration is promoted by growth factors secreted by olfactory ensheathing cells (OECs) located beneath the olfactory epithelium (Au and Roskams, 2003). OEC are a form of glia, and there is much interest in their possible therapeutic use to stimulate axon regeneration in other regions of the brain (Au and Roskams, 2003; DeLucia et al., 2003; Li et al., 2003; Lipson et al., 2003). Not only are olfactory receptor neurons continuously renewed, but so too are their target neurons in the olfactory bulb. These form from neural stem cells in the subventricular zones of the lateral ventricles (Doetsch et al., 1999), and then migrate as a chain of neural precursor cells (Lois and Alvarez-Buylla, 1994; Lois et al., 1996) to form neurons in the olfactory bulb.

Nasal cartilage

Although articular cartilage has a low capacity for repair, mesenchymal stem cells with the capacity to form cartilage, bone, or adipose tissue can be extracted from it (Tallheden et al., 2003). Autologous transplantation of amplified stem cells to repair damaged cartilage is reported to have good outcomes (Lindahl et al., 2003; Peterson et al., 2003) which may in future be enhanced by engineering them to express growth factors (Grande et al., 2003), and by transplanting them embedded in a matrix rather than in suspension (Ochi et al., 2002; Kuriwaka et al., 2003). Human nasal septum is a good source of chondrogenic cells with the potential to be used to engineer transplants for restorative surgery in otolaryngology (Lavezzi et al., 2002). Furthermore, for reasons which remain unclear, cranial and bronchial cartilage has the capacity to heal, a phenomenon absent, but highly desirable, in adult articular cartilage. Current treatment methods have focused on autologous cell transplantation (mesenchymal stem cells (MSCs) or chondrocytes) with or without supporting scaffolds. Autologous chondrocyte transplantation was first used in humans in 1987 and showed stable long-term results in patients with single condyle lesions in particular. It was felt that a better understanding of the repair mechanism induced by cultured chondrocytes and the regulatory mechanisms controlling differentiation would be of benefit in developing new cartilage treatments (Lindahl et al., 2003). Brittberg et al. (2003) introduced a new cell technology, in which cultured chondrocyte were transplanted into defects, raising the expectation of repairing damaged articular cartilage and highlighting the importance of the microenvironment. Concerns over the maintenance of the chondrocyte phenotype in monolayer culture *in situ*, the leaking of chondrocytes from the primary site, and uneven distribution in the three-dimensional (3D) space were raised and have led to alternative methods of cartilage regeneration being investigated.

In 2008, MSC-derived chondrocytes were used successfully to seed the external surface decellularized human donor trachea which formed the world's first stem-cell-based organ transplant in a 30-year-old woman with end-stage tracheobronchial stenosis (Macchiarini et al., 2008). This important first step in the clinical application of stem cells for airway reconstruction shows clearly what should be possible for the generation of nasal cartilage, for example in patients with rhinectomy, large septal defects or, theoretically, for cosmetic purposes.

Differentiation of embryonic stem (ES) cells has been directed towards specific phenotypes, including chondrocytes, by modification of culture conditions, *in vitro*, previously. Co-culture systems have been shown to drive and promote the differentiation of cells towards the mature phenotype, highlighting the potential influence of the microenvironment. Chondrogenic differentiation has now been demonstrated *in vitro* and *in vivo* with human ES cells (Vats et al., 2006). A deeper understanding of the mechanisms of stem cell activation may generate technologies for controlled regeneration of particular tissues, using their intrinsic stem cell populations. The demonstration that bone marrow MSCs is can be activated to differentiate into specific cell types and to home to particular tissues offers the possibility of stimulating tissue regeneration by intravenous infusion of appropriate cell populations. Cultured stem cells may be surgically grafted into sites where regeneration is required. One can only speculate on the speed with which these predictions can become clinical reality.

Nasal epithelium

Nasal epithelial tissue was used as a source of respiratory epithelial cells to seed a human tracheal graft (Macchiarini et al., 2008). Although cultured bronchial cells were ultimately employed, the nasal cells proliferated well for several weeks, suggesting the existence of progenitor or "stem-cell-like" cells in this tissue. Such cells are known to exist in the trachea and bronchi (Snyder et al., 2009), and if their presence in the nose can be confirmed, then the potential exists to engineer single cells of whole sheets of functional epithelium for therapeutic purposes. It is likely that developing treatments that either engineer or regenerate nasal mucosa will build on developments in the lower respiratory tree, as described below.

Tissue-engineered scaffolds for rhinological application

Despite the relative successes in the use of biomaterials in otorhinolaryngology (ORL), there remain significant challenges where the desired parameters cannot be met using current materials. Other areas of the field can be said to be still in their infancy. For example, considering ossicular prostheses, while biocompatibility parameters are being established, little attention has been addressed to optimizing the acoustic properties of the material used. Similarly in the field of tissue engineering, optimal materials and parameters are still to be established. Significant improvements in implant performance are associated with a corresponding development in materials, but for each new application of a biomaterial, testing to ensure safety and efficacy must be done.

The most difficult and costly development for a new medical implant involves materials that have no history as biomaterials, because of the costly and lengthy process of of the preclinical process required for clinical trial accreditation by the FDA and equivalent organizations elsewhere. The development of new implants for use in rhinology requires proactive engagement by clinicians with bioengineers in order to highlight potential clinical applications. Areas of ORL which have been problematical for biomaterials include repair of stenotic trachea, replacement of the larynx, total tympanic membrane replacement, accelerated mucosal healing after sinus surgery, and reducing fibrous tissue formation after middle ear surgery. Biodegradable implants that promote tissue regeneration would obviate the concern about long-term implant failure due to mechanical mismatch at the implant–tissue interface and show great promise for some of the above applications. For permanent implants, seeding with cells, such as those described above, is increasingly recognized as essential if excessive scarring and stenosis are not to occur.

THE LARYNX

Of all the parts of the airway, the larynx represents the greatest challenge for regenerative medicine. Its main function is as a multilevel sphincter to prevent aspiration of food, drink, and saliva into the lower airway, while still permitting respiration. In humans, the larynx lies low in the neck, compatible with erect stance, and permitting speech. Although all vertebrates phonate to a greater or lesser extent, not all use the larynx as the primary sound source. However, in humans, the importance of verbal communication is such that the human vocal cords have unique biomechanical properties and the nerves and muscles contain more fine motor control fibers than any other part of the body, apart from the external ocular muscles alone. The unmet clinical needs in laryngology that regenerative medicine might be designed to address reflect this range of functions and complexity.

Framework replacement

Partial laryngeal reconstruction is required for stenosis in children and adults and after resection of carcinomas in adults. Conventional means of restoring subglottic lumen include endoscopic techniques, which need repeating, large segmental resections, or the insertion of rib cartilage grafts (Smith et al., 2010). In children, the latter are small enough to survive, but in adults, insertion may need to be staged, for example by implantation in sternomastoid muscle (Nouraei et al., 2008). In all of these cases there is a donor scar, which in girls can present a significant cosmetic challenge, and an "off-the-shelf" alternative material would be desirable. Sato and colleagues used collagen sponge soaked in autologous blood in rabbits and then in patients requiring resection of the anterior laryngeal wall for carcinoma. They obtained reasonable early functional results, but scarring was considerable, and there was a lack of good anatomical conformity (Nakamura et al., 2009).

Vocal cords

There is no present effective treatment for scarred or destroyed vocal cords. The standing waves in the unique biomechanical environment of the superficial lamina propria have eluded replication by current reconstructive techniques. However, this is a target ripe for biomaterial design. Thus, Thibeault and colleagues have combined immortalized laryngeal fibroblast cells with a hyaluronic acid-based biomaterial, and this could be delivered by injection, in theory (Thibeault et al., 2009). As well as potentially permitting the restoration of a normal mucosal wave and thus normal speech, these materials may present an ideal means of medializing the vocal cords in patients with unilateral vocal cord paresis (current injectables are reabsorbed or have the potential to cause undesirable foreign body reactions). Decellularization protocols, as used to generate scaffolds elsewhere, are not effective for vocal cords, or at least the speech-generating part, as the most superficial layers are destroyed during decellularization (Birchall, unpublished observations). Thus, another means of regeneration of larger defects is required. Another approach would be to design polymers engineered to have similar biomechanical properties to the superficial lamina propria contained within either a purely synthetic or, more likely, composite cellular and synthetic "bag" (Sahiner et al., 2008).

Total laryngeal replacement

Total laryngeal replacement is a tall order, but has been regarded as the Holy Grail by laryngologists since the early 1950s. Laryngeal allotransplantation has been a qualified success in a handful of patients, of whom only one is reported at the time of writing, but has the significant drawback of requiring immunosuppression (Strome et al., 2001). If only one neuromuscular–joint unit can be preserved in a larynx otherwise destroyed by tumor or disease, then hypothetically the remainder could be reconstructed using synthetic or biological scaffolds with seeded cells, stem cell or otherwise. Proof-of-principle was shown by the operation of subtotal laryngectomy (Pearson, 1981), which has now largely been abandoned in developed countries (though persisting in some developing countries, for good practical reasons) due to the lack of conformity of relocated autologous tissue compared to that likely to be possible with tissue-engineered replacements with appropriate anatomical and bio-mechanical properties.

Technologies to regenerate whole muscles are now described (Yazdani et al., 2009). Again, either synthetic or biological scaffolds may form the basis for myoblast progenitor cell proliferation and differentiation. The use of dynamic bioreactor "preconditioning" causes fiber-type differentiation into the appropriate MyHC subtypes (slow versus various forms of fast fibers) and, crucially, orientation of myotubes. In this way, muscles generating up to 70% of native muscle power have been described (Moon et al., 2008; Yazdani et al., 2009).

For muscles to be sustained and appropriately stimulated it is necessary to achieve reinner-vation. Nerves and vessels tend to grow back into tissue-engineered structures together, which is not surprising since some important neurotrophic factors are also angiogenic (Emanueli et al., 2002). Further, the application of Schwann cells engineered from mesenchymal stem cells (Wakao et al., 2010), or of olfactory ensheathing cells (Li et al., 2003; Lipson et al., 2003) would also be reasonable approaches to explore, as both have shown promise in improving speed and accuracy of nerve regeneration. Nonetheless, since direct repair of cut laryngeal nerves usually results in non-functional "mass movement," or synkinesis (Crumley, 2000), there are serious doubts about whether appropriate muscle movement would be generated by such strategies. Considerable work has to be done to explore this question in particular. However, if such work does not demonstrate the feasibility of regenerative approaches to functional cranial nerve regeneration, although they are likely to provide necessary neurotrophic stimuli to sustain muscle bulk and tone, then an alternative could be provided by electrical stimulation. Thus, groups in the USA have been able to restore laryngeal opening by using modified cardiac pacemaker devices (Zealear et al., 2009). These remain bulky, crude, and subject to displacement and corrosion, but such technology is advancing rapidly (Muller, unpublished observations) and may partner with nanotechnology to produce externally programmable, biointegrated stimulators which will ultimately solve this problem.

THE TRACHEA

The trachea serves as a conduit for ventilation and to clear tracheal and bronchial secretions. It is a large-bore, thin-walled, and cartilage-reinforced tube that connects the larynx with the main bronchi. The trachea is nearly, but not quite, cylindrical, being flattened posteriorly; it measures approximately 11 cm in length, and its diameter, from side to side, is from 2 to 2.5 cm, larger in men. Human trachea is composed of C-shaped incomplete cartilaginous rings (Minnich and Mathisen, 2007) that occupy the anterior and lateral walls and prevent collapse during inspiration. The tube is completed by fibrous tissue and smooth muscle. The lumen is lined by ciliated pseudostratified columnar epithelium (Minnich and Mathisen, 2007). Tracheal epithelium provides a physical and immunological barrier to the external environment and regulates functions including fluid and ion transport and mucociliary clearance (Lopez-Vidriero, 1989). To date, the only curative treatment of any tracheal lesions involving less than 50% of the total length of the trachea has been achieved by segmental resection followed by tracheal mobilization and primary end-to-end anastomosis in adults and/or slide tracheoplasty in children with long-segment tracheal stenosis (Elliott et al., 2008). However, few long-term data are available, and, particularly if stenting is required, multiple interventions are needed with high individual and health service costs. For more extended lesions, surgery is unsuitable, because resection would not leave enough length of the native airway for primary reconstruction. Patients are, unfortunately, treated with endoluminal therapies that are palliative at best (Elliott et al., 2008). Attempts at a surgical cure have met with limited experimental and clinical success, and no satisfactory tracheal substitute is yet available. The use of prosthetic materials, stents, tissue flaps, autografts, or a combination of these methods has been reported, but complications are legion: migration, dislodgment, material degradation/failure, bacterial chronic infection, obstruction because of granulated tissue, stenosis, necrosis, anastomosis failure, erosion of major blood vessels, need for life-long immunosuppression, lack of appropriate donor sources, lack of adequate vascularization and epithelium. Many of these proposed solutions also involve complex multistage procedures (Macchiarini, 1998; Grillo, 2002; Doss et al., 2007; Elliott et al., 2008).

Tracheal scaffolds

At the present state of our development of biodegradable synthetic scaffolds, with all their potential for design modifications and an off-the-shelf solution, there remain strong arguments for continuing to apply biological matrices (Doss et al., 2007). They maintain the natural extracellular matrix composition, do not release toxic biodegradable products or induce inflammation, and play an active part in the regulation of cell behavior affecting cell proliferation, migration, differentiation, and thus tissue regeneration (Doss et al., 2007; Macchiarini et al., 2008). Therefore, much recent work on tissue-engineered tracheae has focused on the application of decellularized matrices. A tissue-engineered fibromuscular patch was used successfully to repair a tracheal defect in a 58-year-old man who had undergone a right complete carinal pneumonectomy (Baiguera et al., 2010). The tissue-engineered airway patch was constructed by seeding autologous muscle cells and fibroblasts on a collagen network, obtained from a decellularized porcine proximal jejunal segment, and subsequent incubation of the construct in a bioreactor for 3 weeks. Postoperative endoscopies showed that the graft was airtight, surfaced with autologous ciliated respiratory epithelium, and neovascularized (Walles et al., 2005).

Johnson et al. (1982) devised a simple and fast detergent-enzymatic method for processing biological scaffolds. This permitted removal of all important cellular components, especially major histocompatibility complex antigens, without significantly altering the matrix structure or the basement membrane complex. We used a tracheal matrix decellularized in a manner based on these principles and reseeded it with endothelial progenitor cells to allow revascularization, plus bone marrow progenitor cells, autologous costal chondrocytes, and respiratory epithelial cells. We demonstrated that by using this approach, it was possible to obtain all functioning tracheal cellular elements: proliferation of chondrocytes, production of a cartilaginous matrix, production of muscular tissue by differentiated muscular cells, and the presence of ciliated epithelial cells. Moreover, this result suggested that long (10–15 cm) vascularized scaffolds could indeed be engineered and would be able to support all relevant cell types (Johnson et al., 1982).

The same decellularization method was used to obtain porcine decellularized tracheal matrices able to support *in vitro* adhesion and proliferation of auricular chondrocytes and tracheal epithelial cells, in a manner superior to that observed with formalin-, thimerosal-, or acetone-treated matrices described by others. Moreover, constructs, implanted as allograft or as xenograft, are easily revascularized and do not induce immunological responses in any models tested (Bader et al., 2010). Recently, decellularized porcine tracheae, recellularized with mesenchymal stem-cell-derived chondrocytes and bronchial epithelial cells, provided evidence that only when both cell types reseed the tracheal matrix will construct implantation be successful (Bader et al., 2010).

Recellularized tracheal scaffolds

Prerequisites for the ideal tracheal graft include (Macchiarini, 1998; Elliott et al., 2008) lateral rigidity, longitudinal flexibility, that they should be airtight, tissue integrated, biocompatible, non-immunogenic, non-toxic, resistant to bacterial colonization, free from the need for immunosuppression, permanent, easy to implant, and provide a platform of ciliated respiratory epithelium. We recently successfully applied a tissue-engineered tracheal construct clinically. A decellularized human donor trachea, colonized by the recipient's epithelial cells and chondrogenic MSCs, was implanted in a female patient with end-stage airway disease (Macchiarini et al., 2008). The detergent-enzymatic method employed used cycles of distilled water, sodium deoxycholate solution, and DNAase-I, which preserves the biomechanical profile of the trachea (Doss et al., 2007). Mesenchymal stem cells were used to recellularize the tracheal graft. We had previously devised methods for culturing airway epithelial cells and applied these to nasal and bronchial cells in culture. In practice, the cells of nasal origin proliferated too quickly, resulting in apoptosis before decellularization was completed. Thus, bronchial cells were used to seed the graft.

Clinical outcomes of stem-cell-based tissue-engineered tracheal graft

Although 3D computed tomography reconstruction showed a significant improvement in the airway appearance from a near-total collapse (preoperatively) to wide patency (at 6 months

1085

post-operatively), at 8 months, a ventral collapse of the most proximal 1 cm of the graft was observed. Hypotheses for why this might have occurred include external compression from the aortic arch superiorly (the graft was pulled against the arch due to the effects of previous surgery), local ischemia, the migration of the stem-cell-derived chondrocytes into the endo-luminal surface of the graft, and the wider lumen created to accommodate the carina at the proximal anastomosis compared to the narrow bronchial end (wound tension). Although the patient remained asymptomatic at this time, a temporary endoluminal stent was placed as a precaution and removed 6 months later since when there has been no recurrence. We are exploring the effects of hypothetical interventions, including the application of biomaterial coating, to try and prevent such proximal stenosis in future cases. This patient is working full-time with normal lung function at 2 years' follow-up.

Future of regenerative medicine of the trachea

This clinical experience shows that a cellular, tissue-engineered airway with appropriate mechanical properties allows normal functioning, and is free from the risks of rejection. However, it also highlights the scientific, clinical, and commercial bottlenecks standing in the way of full integration of this regenerative medicine technology into routine clinical care. Required advances in tracheal tissue engineering include optimization of the process of human donor tracheal graft recovery, transport, biobanking and decellularization, ensuring a reproducible supply of epithelial cells and chondrocytes, clinical grade, automated bio-reactors, and appropriate commercial models for varying healthcare systems. Critically, a physiologically relevant strategy for inducing truly functional angiogenesis (more specifi-cally, development of arterioles) after tracheal implantation will permit the application of larger grafts and similar grafts with greater security. All of these factors require attention if we are to successfully scale up the existing tracheal implantation technique from one that can be used for individual patients to one that can be rolled out to a large population in a simple, reproducible, and routine technique that is both commercially effective and socially acceptable.

SMALL AIRWAYS AND LUNG PARENCHYMA

Air space destruction and enlargement of the distal airways are characteristic in emphysema. Chronic obstructive pulmonary disease (COPD), caused largely by smoking in the west, is the fourth highest cause of death in the USA and approximately 25% of COPD patients display the clinical and radiological signs of emphysema. Current therapy for the end-stage disease is lung volume reduction surgery to try and improve the patient's air flow characteristics, followed by transplantation. The shortage of organ donation limits transplantation to a tiny fraction of the sufferers and new imaginative routes to organ regeneration are desperately needed.

Replacing the smaller airways and lung parenchyma is an extreme challenge. The basic numerous small tubular structures are not replaceable and hence the growth of endogenous or exogenous cells on scaffold/matrix is the only likely way that improvements in gas exchange will take place. Implants would ultimately have to achieve a highly organized 3D structure that must connect to existing small airways and blood vessels to allow adequate gas exchange. These implants would also need to take on other physiological requirements of the lung including compressibility and stretch.

Attempts to regenerate the distal lung are not new. Indeed, Sorokin demonstrated that fetal lungs could differentiate *in vitro* and produce almost all the components of the adult lung in the 1960s (Sorokin, 1961). More recently several groups have investigated the use of growth factors that encourage the expansion of endogenous cells within the lung. For example, the delivery of retinoic acid has led to the reversal of anatomical and physiological signs of emphysema in a rat model (Massaro and Massaro, 1997). Keratinocyte growth factor has led to an amelioration of damage in a fibrosis injury model (Sugahara et al., 1998). The progressive understanding of those signaling pathways that control lung stem cell fate will hopefully lead to other candidates that may increase repair by stimulating endogenous cells.

The true goal for tissue regeneration, however, has to be to form pulmonary alveolar structures with the combined use of scaffold, dissociated lung cells, and regulatory factors.

Perhaps the most successful example of tissue engineering to date has come from the use of Gelfoam® (Pfizer, New York) sponges. These act as a scaffold and matrix for cell culture experiments and a number of studies have used this technique for studies of fetal lung cell growth and differentiation *in vitro*. These experiments have demonstrated the formation of a polarized epithelial cell layer with characteristics of alveolar-like structures (Xu et al., 1996, 1999; Mourgeon et al., 1999; Nakamura et al., 2000). Within these 3D cultures plated cells are able to differentiate and form surfactant-producing cells (Douglas et al., 1976) and extracellular matrix molecules (Xu et al., 1996, 1999; Mourgeon et al., 1999; Nakamura et al., 2000). Importantly, studies examining how endothelial cells grow in Gelfoam have also shown the development of small vessels (Centra et al., 1992). Gelfoam is especially interesting for tissue engineers as it is already used clinically as a hemostatic and hence has FDA approval.

A very intriguing study by Andrade and colleagues again used Gelfoam supplemented with fetal rat lung cells (presuming them to have progenitor function) (Andrade et al., 2007). This was injected into normal rat lung parenchyma. The investigators found that the sponge did not induce inflammation and that, as hoped, the sponge formed alveolar-like structures. Within the sponge lung cells survived for at least 35 days and combined immunohistochemistry and bromodeoxyuridine (BrdU) labeling demonstrated epithelial and endothelial cells that were proliferating. Interestingly, the cells at the border with the surrounding lung underwent the most differentiation with evidence of neovascularization. The sponge was found to have degraded after several months, in theory allowing a temporary scaffold to be inserted into the lung to encourage new lung growth. It was, however, interesting that the authors noted that it was conceivable that the majority of cells within the Gelfoam were actually derived from the recipient rather than the exogenous cells placed with the Gelfoam, and that the implanted fetal cells may act more as manufacturer of growth factors encouraging local cell expansion. This is an important hypothesis and needs further investigation as the expansion of a local pool of endogenous stem or progenitors into a new scaffold has significant implications for the field and the formation of "autografts."

A more local treatment effect was achieved by Shigemura and co-workers using bone marrow-derived cells (Shigemura et al., 2006). This team performed lung reduction surgery on a rat model of elastase-induced emphysema. They subsequently implanted hepatocyte growth factor (HGF) secreting adipose tissue-derived stromal cells (ASCs) with a sealant material (polyglycolic acid felt sheet; PGAF) as a scaffold. The rats underwent pulmonary resection followed by a sealant material sheet of PGAF with or without the addition of ASCs being placed over the cut surface. The team found that both alveolar and vascular regeneration were significantly accelerated after 1 week, with improved gas exchange and exercise tolerance (Shigemura et al., 2006).

A final route of lung regeneration is simply to systemically deliver cells that can engraft in the lung and take on lung cell phenotypes. This has recently been intensively investigated and early studies showed both hematopoietic and mesenchymal bone marrow cells could engraft in the lung and take on an epithelial phenotype. More recent studies, however, have found this difficult to reliably reproduce and now many groups are looking at delivering gene therapy using bone marrow-derived cells (Loebinger and Janes, 2007). While this is exciting in the possible amelioration of lung diseases, lung regeneration still seems unproven using this technique.

Hence a combination approach of scaffold and cell-based therapy seems most likely to succeed. Considerable hurdles are yet to be overcome and the factors that encourage cell differentiation, expansion, and their interaction with scaffolds are still poorly understood.

References

Andrade, C. F., Wong, A. P., Waddell, T. K., Keshavjee, S., & Liu, M. (2007). Cell-based tissue engineering for lung regeneration. *Am. J. Physiol. Lung Cell. Mol. Physiol., 292*, L510–L518.

Au, E., & Roskams, A. J. (2003). Olfactory ensheathing cells of the lamina propria *in vivo* and *in vitro*. *Glia, 41*, 224–236.

Bader, A., & Macchiarini, P. (2010). Moving towards *in situ* tracheal regeneration: the bionic tissue engineered transplantation approach. *J. Cell Mol. Med., 14*, 1877–1889.

Baiguera S., Damasceno K. L., & Macchiarini P. (2010). Detergent-enzymatic method for bioengineering human airways. In: C. Lee, & K. Uygun (Eds.). *Organ Perfusion and Culture Methodology. Methods in Bioengineering.* Boston: Artech House.

Brittberg, M., Peterson, L., Sjogren-Jansson, E., Tallheden, T., & Lindahl, A. (2003). Articular cartilage engineering with autologous chondrocyte transplantation. A review of recent developments. *J. Bone Joint Surg., 85A* (Suppl. 3), 109–115.

Calof, A. L., Bonnin, A., Crocker, C., Kawauchi, S., Murray, R. C., Shou, J., et al. (2002). Progenitor cells of the olfactory receptor neuron lineage. *Microsc. Res. Tech., 58*, 176–188.

Centra, M., Ratych, R. E., Cao, G. L., Li, J., Williams, E., Taylor, R. M., et al. (1992). Culture of bovine pulmonary artery endothelial cells on Gelfoam blocks. *FASEB J., 6*, 3117–3121.

Crumley, R. L. (2000). Laryngeal synkinesis revisited. *Ann. Otol. Rhinol. Laryngol., 109*, 365–371.

DeLucia, T. A., Conners, J. J., Brown, T. J., Cronin, C. M., Khan, T., & Jones, K. J. (2003). Use of a cell line to investigate olfactory ensheathing cell-enhanced axonal regeneration. *Anat. Rec., 271B*, 61–70.

Doetsch, F., Caille, I., Lim, D. A., Garcia-Verdugo, J. M., & Alvarez-Buylla, A. (1999). Subventricular zone astrocytes are neural stem cells in the adult mammalian brain. *Cell, 97*, 703–716.

Doss, A. E., Dunn, S. S., Kucera, K. A., et al. (2007). Tracheal replacements: Part 2. *ASAIO J., 53*, 631.

Douglas, W. H., & Teel, R. W. (1976). An organotypic *in vitro* model system for studying pulmonary surfactant production by type II alveolar pneumonocytes. *Am. Rev. Respir. Dis., 113*, 17–23.

Elliott, M., Hartley, B. E., Wallis, C., & Roebuck, D. (2008). Slide tracheoplasty. *Curr. Opin. Otolaryngol. Head Neck Surg., 16*, 75.

Emanueli, C., Salis, M. B., Pinna, A., Graiani, G., Manni, L., & Madeddu, P. (2002). Nerve growth factor promotes angiogenesis and arteriogenesis in ischemic hindlimbs. *Circulation, 106*, 2257–2262.

Grande, D. A., Mason, J., Light, E., & Dines, D. (2003). Stem cells as platforms for delivery of genes to enhance cartilage repair. *J. Bone Joint Surg., 85A*(Suppl. 2), 111–116.

Griffiths, L. G., & Naughton, G. (2002). Tissue engineering – current challenges and expanding opportunities. *Science, 295*, 1009.

Grillo, H. C. (2002). Tracheal replacement: a critical review. *Ann. Thorac. Surg., 73*, 1995–2004.

Jang, W., Youngentob, S. L., & Schwob, J. E. (2003). Globose basal cells are required for reconstitution of olfactory epithelium after methyl bromide lesion. *J. Comp. Neurol., 460*, 123–140.

Johnson, P. C., Duhamel, R. C., Meezan, E., & Brendel, K. (1982). Preparation of cell-free extracellular matrix from human peripheral nerve. *Muscle Nerve, 5*, 335–344.

Kuriwaka, M., Ochi, M., Uchio, Y., Maniwa, S., Adachi, N., Mori, R., et al. (2003). Optimum combination of monolayer and three-dimensional cultures for cartilage-like tissue engineering. *Tissue Eng., 9*, 41–49.

Lavezzi, A., Mantovani, M., della Berta, L. G., & Matturri, L. (2002). Cell kinetics of human nasal septal chondrocytes *in vitro*: importance for cartilage grafting in otolaryngology. *J. Otolaryngol., 31*, 366–370.

Li, Y., Decherchi, P., & Raisman, G. (2003). Transplantation of olfactory ensheathing cells into spinal cord lesions restores breathing and climbing. *J. Neurosci., 23*, 727–731.

Lindahl, A., Brittberg, M., & Peterson, L. (2003). Cartilage repair with chondrocytes: clinical and cellular aspects. *Novartis Found. Symp., 249*, 175–186, discussion, 186–189, 234–238, 239–241.

Lipson, A. C., Widenfalk, J., Lindqvist, E., Ebendal, T., & Olson, L. (2003). Neurotrophic properties of olfactory ensheathing glia. *Exp. Neurol., 180*, 167–171.

Loebinger, M. R., & Janes, S. M. (2007). Stem cells for lung disease. *Chest, 132*, 279–285.

Lois, C., & Alvarez-Buylla, A. (1994). Long-distance neuronal migration in the adult mammalian brain. *Science, 264*, 1145–1148.

Lois, C., Garcia-Verdugo, J. M., & Alvarez-Buylla, A. (1996). Chain migration of neuronal precursors. *Science, 271*, 978–981.

Lopez-Vidriero, M. T. (1989). Mucus as a natural barrier. *Respiration, 55*, 28.

Macchiarini, P. (1998). Tracheal transplantation: beyond the replacement of a simple conduit. *Eur. J. Cardiothorac. Surg., 14*, 621.

Macchiarini, P., Jungebluth, P., Go, T., Asnaghi, M. A., Rees, L. E., Cogan, T. A., et al. (2008). Clinical transplantation of a tissue-engineered airway. *Lancet, 372,* 2023–2230.

Mason, C., & Dunnill, P. (2008). Regenerative medicine glossary. *Regen. Med., 4*(Suppl. 4), S1–88.

Massaro, G. D., & Massaro, D. (1997). Retinoic acid treatment abrogates elastase-induced pulmonary emphysema in rats. *Nat. Med., 3,* 675–677.

Minnich, D. J., & Mathisen, D. J. (2007). Anatomy of the trachea, carina, and bronchi. *Thorac. Surg. Clin., 17,* 571.

Moon, du. G., Christ, G., Stitzel, J. D., Atala, A., & Yoo, J. J. (2008). Cyclic mechanical preconditioning improves engineered muscle contraction. *Tissue Eng., 14A,* 473–482.

Mourgeon, E., Xu, J., Tanswell, A. K., Liu, M., & Post, M. (1999). Mechanical strain-induced posttranscriptional regulation of fibronectin production in fetal lung cells. *Am. J. Physiol., 277,* L142–L149.

Murray, R. C., Navi, D., Fesenko, J., Lander, A. D., & Calof, A. L. (2003). Widespread defects in the primary olfactory pathway caused by loss of Mash1 function. *J. Neurosci., 23,* 1769–1780.

Nakamura, T., Liu, M., Mourgeon, E., Slutsky, A., & Post, M. (2000). Mechanical strain and dexamethasone selectively increase surfactant protein C and tropoelastin gene expression. *Am. J. Physiol. Lung Cell. Mol. Physiol., 278,* L974–L980.

Nakamura, T., Sato, T., Araki, M., Ichihara, S., Nakada, A., Yoshitani, M., et al. (2009). *In situ* tissue engineering for tracheal reconstruction using a luminar remodeling type of artificial trachea. *J. Thorac. Cardiovasc. Surg., 138,* 811–819.

Nouraei, S. A., Nouraei, S. M., Sandison, A., Howard, D. J., & Sandhu, G. S. (2008). The prefabricated sternohyoid myocartilagenous flap: a reconstructive option for treating recalcitrant adult laryngotracheal stenosis. *Laryngoscope, 118,* 687–691.

Ochi, M., Uchio, Y., Kawasaki, K., Wakitani, S., & Iwasa, J. (2002). Transplantation of cartilage-like tissue made by tissue engineering in the treatment of cartilage defects of the knee. *J. Bone Joint Surg., 84B,* 571–578.

Pearson, B. W. (1981). Subtotal laryngectomy. *Laryngoscope, 91,* 1904–1912.

Peng, H., Wright, V., Usas, A., Gearhart, B., Shen, H. C., Cummins, J., et al. (2002). Synergistic enhancement of bone formation and healing by stem cell-expressed VEGF and bone morphogenetic protein-4. *J. Clin. Invest., 110,* 751–759.

Peterson, L., Minas, T., Brittberg, M., & Lindahl, A. (2003). Treatment of osteochondritis dissecans of the knee with autologous chondrocyte transplantation: results at two to ten years. *J. Bone Joint Surg., 85A*(Suppl. 2), 17–24.

Sahiner, N., Jha, A. K., Nguyen, D., & Jia, X. (2008). Fabrication and characterization of cross-linkable hydrogel particles based on hyaluronic acid: potential application in vocal fold regeneration. *J. Biomater. Sci. Polym., 19,* 223–243.

Shigemura, N., Okumura, M., Mizuno, S., Imanishi, Y., Matsuyama, A., Shiono, H., et al. (2006). Lung tissue engineering technique with adipose stromal cells improves surgical outcome for pulmonary emphysema. *Am. J. Respir. Crit. Care Med., 174,* 199–1205.

Smith, L. P., Zur, K. B., & Jacobs, I. N. (2010). Single- vs double-stage laryngotracheal reconstruction. *Arch. Otolaryngol. Head Neck Surg., 136,* 60–65.

Snyder, J. C., Teisanu, R. M., & Stripp, B. R. (2009). Endogenous lung stem cells and contribution to disease. *J. Pathol., 217,* 254–264.

Sorokin, S. (1961). A study of the development in organ cultures of mammalian lungs. *Dev. Biol., 3,* 60–83.

Strome, M., Stein, J., Esclamado, R., Hicks, D., Lorenz, R. R., Braun, W., et al. (2001). Laryngeal transplantation and 40-month follow-up. *N. Engl. J. Med., 344,* 676–679.

Sugahara, K., Iyama, K., Kuroda, M. J., & Sano, K. (1998). Double intratracheal instillation of keratinocyte growth factor prevents bleomycin-induced lung fibrosis in rats. *J. Pathol., 186,* 90–98.

Tallheden, T., Dennis, J. E., Lennon, D. P., Sjogren-Jansson, E., Caplan, A. I., & Lindahl, A. (2003). Phenotypic plasticity of human articular chondrocytes. *J. Bone Joint Surg., 85A*(Suppl. 2), 93–100.

Thibeault, S. L., Klemuk, S. A., Smith, M. E., Leugers, C., & Prestwich, G. (2009). *In vivo* comparison of biomimetic approaches for tissue regeneration of the scarred vocal fold. *Tissue Eng., 15A,* 1481–1487.

Vats, A., Bielby, R., Tolley, N. S., Dickinson, S., Boccaccini, A., Hollander, A., et al. (2006). Chondrogenic differentiation of human embryonic stem cells: the effect of the micro-environment. *Tissue Eng., 12,* 1687–1697.

Wakao, S., Hayashi, T., Kitada, M., Kohama, M., Matsue, D., Teramoto, N., et al. (2010). Long-term observation of auto-cell transplantation in non-human primate reveals safety and efficiency of bone marrow stromal cell-derived Schwann cells in peripheral nerve regeneration. *Exp. Neurol., 223,* 537–547.

Walles, T., Biancosino, C., Zardo, P., Macchiarini, P., Gottlieb, J., & Mertsching, H. (2005). Tissue remodeling in a bioartificial fibromuscular patch following transplantation in a human. *Transplantation, 80,* 284–285.

Wu, H. H., Ivkovic, S., Murray, R. C., Jaramillo, S., Lyons, K. M., Johnson, J. E., et al. (2003). Autoregulation of neurogenesis by GDF11. *Neuron, 37,* 197–207.

Xu, J., Liu, M., Liu, J., Caniggia, I., & Post, M. (1996). Mechanical strain induces constitutive and regulated secretion of glycosaminoglycans and proteoglycans in fetal lung cells. *J. Cell Sci.*, *109*, 1605–1613.

Xu, J., Liu, M., & Post, M. (1999). Differential regulation of extracellular matrix molecules by mechanical strain of fetal lung cells. *Am. J. Physiol.*, *276*, L728–L735.

Yazdani, S. K., Watts, B., Machingal, M., Jarajapu, Y. P., Van Dyke, M. E., & Christ, G. J. (2009). Smooth muscle cell seeding of decellularized scaffolds: the importance of bioreactor preconditioning to development of a more native architecture for tissue-engineered blood vessels. *Tissue Eng.*, *15A*, 827–840.

Zealear, D. L., Kunibe, I., Nomura, K., Billante, C., Singh, V., Huang, S., et al. (2009). Rehabilitation of bilaterally paralyzed canine larynx with implantable stimulator. *Laryngoscope*, *119*, 1737–1744.

(a)

Distal

Phalanges

Middle

Proximal

bone-cartilage

bone-ligament

bone-tendon

tendon-muscle

(b)

thumb. The IP joints are hinge joints located between
limited to flexion and extension.

To engineer digit tissues that restore normal function, a
necessary components, has to be constructed. One of t
a complex tissue is the need to generate each tissue co
function in a coordinated manner. Most importantly, a
mechanically dissimilar materials such as bone and car
tendon, or tendon and muscle, is an additional challer

Tissue engineering strategy: cell sourcing

Stem cells have been proposed as an alternative cell so
tissues. While autologous somatic cells are preferable, th
and require multiple tissue biopsies in order to obtain
engineering. For this reason, stem cells have been prop
various tissue regeneration applications. Human embry
proliferate in the undifferentiated state (self-renewal) ar
many specialized cell types (Brivanlou et al., 2003). Ho
controversies surrounding their use have hindered resea
through therapeutic cloning (also called somatic cell nu
proposed as an alternative method to bypass these ethi
taken from an adult somatic cell and placed into an en
ulated to begin dividing. Once the cell begins dividing,
method employs nuclear transfer techniques to generate
tissues, such as skeletal muscle, cardiac muscle, and kid

Adult stem cells are also being considered as a new cell
limitations have not been fully evaluated, stem cells from
cord, amniotic fluid, and bone marrow have been used
(Bartsch et al., 2005; Deasy et al., 2005; Moise, 2005; Tl
using adult stem cells to generate multiple tissue pheno
have been encouraging, and suggest feasible approaches i
anatomical structures consisting of multiple cell lineage
These cells are capable of self-renewal and can be readily e
losing their self-renewal capacity. They can differentiate
lineages such as osteoblasts, chondrocytes, tenocytes, fib

CHAPTER 59

The Digit: Engineering of Phalanges and Small Joints

Makoto Komura[*], **Jaehyun Kim**[†], **Anthony Atala**[†], **James J. Yoo**[†], **Sang Jin Lee**[†]
[*] Department of Pediatric Surgery, The University of Tokyo Hospital, Tokyo, Japan
[†] Wake Forest Institute for Regenerative Medicine, Wake Forest University Health
Sciences, Medical Center Boulevard, Winston-Salem, NC, USA

INTRODUCTION

The function of the digits is critically important in human beings. The delicate and complex
anatomy of the digit is reflected in its sophisticated mechanical and sensory functions.
However, because of its frequent use, it is susceptible to disorders such as traumatic injuries
and arthritic joint destruction. Importantly, any type of digital dysfunction is detrimental
to a patient's daily activities, but the complete loss of one or more digits through amputation is particularly devastating. Digital amputation is a common traumatic injury, with
an estimated 61,000 patients seen in emergency departments in the USA every year
(Dubernard et al., 1999), and this leads to substantial socioeconomic costs. In addition,
loss of the integrity of large portions of the digits due to such injury often results in
significant and extensive functional disability that affects the quality of life and work ability
of the patient.

For amputations where the removed digit can be located and is in a clean and fresh state,
immediate reimplantation is considered the treatment of choice. This involves microsurgical
anastomosis of nerves, blood vessels, bone, and tendon. Functional recovery is achievable if
the blood supply to the distal portion of the reimplanted digit is restored within a few hours
and sensory/motor nerves are connected to the stump. The success rate for immediate digital
reimplantation is reported to be up to 90% at specialized centers (Patradul et al., 1998).
Unfortunately, however, patients who are eligible for immediate digit reimplantation make up
only a small subset of the total cases.

For patients who are missing whole or partial digits in the hand as a result of traumatic injuries
or congenital absence, one of the surgical options for finger reconstruction includes the
transplantation of a vascularized autogenous joint from the foot (Mathes et al., 1980; Wray
et al., 1981; Tsay et al., 1982). This technique has been well established since the first report by
Buncke et al. (1966). However, this complex multiple-stage treatment is associated with
possible impairment in foot stability from the sacrifice of a toe, as well as only partial functional recovery and poor cosmetic outcomes. For these reasons, many patients facing loss of an
injured digit choose amputation as the final treatment.

Principles of Regenerative Medicine. DOI: 10.1016/B978-0-12-381422-7.10059-8

Prosthetic joint reconstr
unlike large limb prosth
prostheses are primarily
use of artificial prostheti
biological materials, infe
for pediatric or adolesce
tissues with time.

Despite the rapid techn
reconstruction are sever
a prolonged treatment p
a limited restoration of
tions. However, recent ti
using the concept of cel
intervention. Restoring s
already been successful
Yoo et al., 1999; Amiel
2002; De Filippo et al.,
Atala et al., 2006). In tu
strategies has provided a
to engineer physiologica
ligament, tendon, and n
appendages such as finge
of missing limbs. This c
tissues, provides an over
for phalanges and small
reconstruction.

PRINCIPLES OF R
AND SMALL JOIN
Reconstruction of (

Current treatments to re
materials or the utilizatic
remodel and integrate w
autograft material is a pr
limited donor bone sup
resorption during healin
response due to genetic (

With advances in bioma
considered the pathway
treatments. Tissue engine
scaffolds, cells, and signa
cells can theoretically rep
in vivo.

The digit is a complex stru
significant tissue that mu
provides support and the
The bone in a digit is a p
three phalanges, and the
fingers are the metacarpo
interphalangeal (DIP) jo

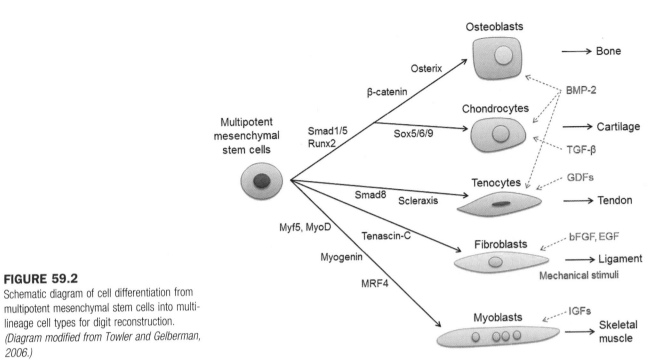

FIGURE 59.2
Schematic diagram of cell differentiation from multipotent mesenchymal stem cells into multilineage cell types for digit reconstruction. *(Diagram modified from Towler and Gelberman, 2006.)*

Because digits are complex composite tissues, the use of stem cells may be necessary in order to engineer these body parts.

Tissue engineering strategy: scaffold design and fabrication

The current advances in tissue engineering would not have been possible without the innovative design and fabrication of several generations of scaffolds that act as a temporary support for cells and facilitate the regeneration of tissues. Scaffolds provide three-dimensional (3D) architecture for regenerating tissues by supplying cell anchorage sites, mechanical stability, and structural guidance. *In vivo*, they provide an interface that allows a construct to respond to physiological and biological changes and guide the remodeling of the extracellular matrix (ECM) to allow integration with the surrounding native tissue. As the majority of mammalian cell types are anchorage dependent, scaffolds also provide a cell-adhesion substrate that can deliver cells to targeted regions of the body (Kim and Mooney, 1998) with high loading efficiency and the proper shape and dimensions. Furthermore, bioactive signals such as cell-adhesion peptides and growth factors can be loaded along with cells to help regulate cellular function and tissue maturation. To fulfill these roles, an enormous number of biomaterials have been studied experimentally and clinically (Kim and Mooney, 1998; Isogai et al., 1999; Amiel et al., 2001; Vacanti et al., 2001; Pariente et al., 2002; El-Kassaby et al., 2003; Falke et al., 2003; Derwin et al., 2004; Kim et al., 2004; Atala et al., 2006; Murray et al., 2006).

While initial success in developing single tissue types with these strategies has been reported, there is an increasing demand for methods that will allow the formation of more complex structures that contain multiple tissue types that have coordinated function. This is especially important in the engineering of digit tissues, in which bone, cartilage, ligament, tendon, and muscle must be conjoined in order to transmit forces across the entire length of the digit. Therefore, development of a system for engineering these complex multiple tissues is critical for the functional reconstruction of digit tissues.

Engineering of multiple tissues

Complex structures such as the digit are composed of two or more dissimilar tissues, each with different mechanical and biological properties. Therefore, reconstruction of these structures requires the engineering of not only hard tissue but also soft tissues such as cartilage, ligament, and tendon, which connect bone to bone and muscle to bone, respectively. Each of these tissues is composed of a distinct cellular population, and all of the tissues must operate in unison to facilitate physiological function and maintain tissue homeostasis. Thus, it is not surprising that the transition between various tissue types is characterized by a high level of heterogeneous structural organization that is crucial for joint function. This interface contains a gradient of mechanical properties and has a number of functions, from mediating load transfer between two distinct types of tissue to sustaining the heterotypic cellular communications required for interface function and homeostasis (Benjamin et al., 1986; Blevins et al., 1997; Lu and Jiang, 2006). In light of this complexity, functional tissue engineering must incorporate strategic biomimicry to facilitate the formation of the tissue-to-tissue interface and enable seamless graft integration.

BONE—CARTILAGE

The bone—cartilage construct has been extensively studied in orthopedics (Martin et al., 2007). A tissue engineering approach aimed at mimicking the physiological properties and structure of two different tissues (cartilage and bone) using a cell-scaffold construct would be ideal for repair and regeneration of limbs and digits. When single (monophasic) scaffolds are used (Vacanti and Upton, 1994; Temenoff and Mikos, 2000), the natural environment is not adequately replicated and new tissue formation is incomplete. Therefore, there have been several attempts to develop biphasic composite constructs for bone—cartilage engineering by combining two distinct scaffold materials. In one such study, Hung's team developed anatomically shaped osteochondral constructs by casting a layer of chondrocyte-seeded agarose gel on top of devitalized trabecular bone (Hung et al., 2003). While the accumulation of a cartilage-like matrix was observed in both phases of the composite constructs after 42 days in static culture, the peak compressive material properties only reached one-eighth of those of native tissue over the same period of time. Freed's team formed bilayered constructs by suturing engineered cartilage grown for 6 weeks on non-woven polyglycolide (PGA) scaffolds to a subchondral support made of Collagraft® sponge (Schaefer et al., 2002). These constructs were implanted into osteochondral defects in rabbit knees and allowed to remodel over a period of 6 months. Histological analysis revealed good integration with host bone, but poor integration with host cartilage using this technique.

Similarly, Caplan's team assembled a bilayered scaffold using a hyaluronan sponge (HYAFF®-11) for cartilage regeneration and a calcium phosphate ceramic sponge for bone regeneration (Gao et al., 2001). Each layer was seeded individually with mesenchymal stem cells (MSCs) before the two sponges were bonded together using fibrin glue. After 6 weeks of *in vivo* culture in syngeneic rats, the group reported finding bone ingrowth in the lower layer and immature fibrocartilage in the upper layer. Ratcliffe's group used a 3D printing system (TheriForm™) to form a composite scaffold consisting of a poly(lactide-*co*-glycolide) (PLGA)/polylactide (PLA) mix in the upper cartilage region, and a PLGA/tricalcium phosphate (TCP) mix in the lower bone region (Sherwood et al., 2002). Results from this study showed that these scaffolds possessed robust initial mechanical properties in the bone region, while still supporting the production of cartilage-like matrix proteins from seeded chondrocytes in the upper cartilage region.

In other studies, Freed's team has used multilayered concentric constructs as model systems to examine the integrative repair of engineered bone—cartilage composites (Tognana et al., 2005a). In these studies, engineered constructs were made using bovine calf chondrocytes seeded on hyaluronan benzyl ester non-woven mesh, which were then press-fitted into adjacent tissue rings made of articular cartilage, devitalized bone, or vital bone. The constructs were then cultured in rotating bioreactors for up to 8 weeks, at which point their structure, biomechanical properties, and composition were assessed. The type of adjacent tissue was found to significantly affect construct adhesion, modulus, proteoglycan content,

1095

and collagen content, with the strongest integrative repair observed for constructs cultured adjacent to bone, rather than cartilage (Tognana et al., 2005a). When implanted *in vivo*, these constructs showed similar behavior, although the properties of the constructs (e.g. moduli, proteoglycan content) were higher *in vivo* than *in vitro*. Certain histological characteristics, such chondrocyte alignment and bone remodeling, were observed only *in vivo* (Tognana et al., 2005b). Such composite systems provide novel model systems to test the potential integrative properties of engineered constructs with native tissues *in vitro* or *in vivo*.

Biphasic composite scaffolds manufactured by image-based design (IBD) and solid free-form (SFF) fabrication have been used to simultaneously generate bone and cartilage in discrete regions and provide for the development of a stable interface between cartilage and subchondral bone (Schek et al., 2004). PLLA/hydroxyapatite composite scaffolds were seeded with fibroblasts transduced with an adenovirus expressing bone morphogenetic protein-7 (BMP-7) in the ceramic phase. In addition, fully differentiated chondrocytes were seeded onto the polymeric phase. After subcutaneous implantation into mice, it was found that the biphasic scaffolds promoted the simultaneous growth of bone, cartilage, and a mineralized interface tissue. Within the ceramic phase, the pockets of tissue generated included blood vessels, marrow stroma, and adipose tissue. Overall, it appears that the use of biphasic scaffolds combined with gene and cell therapy is a promising approach to regenerate osteochondral defects.

BONE—LIGAMENT

The transition from ligament to bone tissue represents a significant challenge for functional interface engineering. Initial attempts to improve ligament to bone grafting focused on augmenting the surgical graft with a material that would encourage bone tissue growth (Chen et al., 2003; Kyung et al., 2003; Mutsuzaki et al., 2004; Tien et al., 2004; Huangfu and Zhao, 2007). Although these methods have improved osteointegration between the ligament graft and the bone tunnel, these efforts do not result in the regeneration of the fibrocartilage interface. Thus, a systematic and controlled approach which uses a biomimetic stratified scaffold to direct the growth of this multitissue interface may overcome these shortcomings, as it can be designed to recapitulate the inherent complexity of the ligament to bone interface. Laurencine and his team reported a multiphased design of a synthetic anterior cruciate ligament (ACL) graft fabricated from 3D braiding of PLGA fibers, with a ligament proper as well as two bony regions (Cooper et al., 2005). *In vitro* (Lu et al., 2005) and *in vivo* (Cooper et al., 2007) evaluation demonstrated scaffold biocompatibility, healing, and adequate mechanical strength in a rabbit model. Recently, Richmond and co-workers developed a multi-region, porous knitted silk ACL graft which was evaluated in a goat model with promising results (Altman et al., 2008). Using a cell-based approach, Larkin's team reported that it is possible to form bone—ligament—bone constructs by introducing engineered bone segments to ligament monolayers (Ma et al., 2009). These promising results demonstrate that biomimetic stratified scaffold design coupled with spatial control over the distribution of interface relevant cell populations can lead to the formation of cell type- and phase-specific matrix heterogeneity *in vitro* and *in vivo*, with a fibrocartilage-like interface formed in culture. These observations not only validate the feasibility of using this stratified scaffold for promoting biological fixation of ACL grafts to bone, but also highlight the potential for continuous multitissue regeneration on a single scaffold system.

In summary, current strategies in ligament-to-bone interface tissue engineering tackle the difficult problem of soft tissue-to-bone integration by pre-engineering the soft tissue-to-bone interface *ex vivo* through stratified scaffold design for multitissue regeneration. They also focus on the less challenging task of bone-to-bone integration *in vivo*. Moreover, functional and integrative ligament repair may be achieved by coupling both cell-based and scaffold-based approaches.

BONE—TENDON

Tendon and ligaments are responsible for linking the musculoskeletal system; tendons transfer loads between muscles and bones, while ligaments bind bone to bone to restrict their motion and maintain joint stability (Martin et al., 1998). The tissue comprising the interface of tendon and bone varies considerably depending on the particular attachment point and the type of joint. Thus, a major consideration in developing constructs to engineer joint tissue is the development of an optimal scaffolding system that will allow the creation of the zonal, fibrocartilaginous tissue that bridges tendon and bone (Benjamin et al., 1986; Blevins et al., 1997). Therefore, the biomimetic scaffold design and multilineage cell culture methods previously discussed for the ligament-to-bone interface are also applicable for the regeneration of tendon-to-bone insertions, such as those seen in the rotator cuff. However, while tendon-to-bone and ligament-to-bone insertions are physiologically and biochemically similar, the tissue engineering strategy applied is expected to differ, as the two interfaces do vary in terms of loading environment, mineral distribution, and surgical repair methods that can be used to repair damage. In addition, the subsequent healing response of these two different interfaces is different, and this must be considered when developing an engineering strategy.

Several groups have evaluated the feasibility of integrating tendon grafts with bone or biomaterials through the formation of an anatomical insertion site. By surgically reattaching the Achilles tendon to bone, Hurwitz's team reported that cellular reorganization occurred at the reattachment site, along with the formation of non-mineralized and mineralized fibro-cartilage-like regions (Fujioka et al., 1998). In addition, Chao's team was able to successfully promote supraspinatus tendon integration with a metallic implant using a bone marrow-infused bone graft (Inoue et al., 2002). These promising results indicate that the tendon-to-bone interface may be regenerated and underscore the need for functional grafting solutions that can promote biological fixation.

MUSCLE—TENDON

Engineering of muscle—tendon interfaces has not been extensively studied. However, the muscle—tendon junction (MTJ) is an important interface in the musculoskeletal system because it acts as the mechanical bridge between skeletal muscle and bone. It is likely that strategies for engineering the muscle—tendon junction will arise from existing efforts in both muscle tissue engineering and tendon tissue engineering. A main issue in designing muscle—tendon constructs is that engineered muscle must be efficiently interfaced with tendon to provide proper load transfer to bone, which may prove difficult given the task of replicating the natural membrane interdigitation and folding at the MTJ.

As the tendon joins the muscle to bone, the MTJ, which connects muscle to tendon, acts as a bridge to distribute mechanical loads (Yang and Temenoff, 2009). One tissue engineering approach includes the culturing of myoblasts in collagen gels *in vitro* to form contractile muscle constructs with fibrils that terminate in a manner similar to the native MTJ (Swasdison and Mayne, 1991, 1992). Adopting a cell-based approach, Larkin and colleagues evaluated a novel self-organizing system for *in vitro* myotendinous junction formation by co-culturing skeletal muscle constructs with engineered tendon constructs (Larkin et al., 2006). The aforementioned studies demonstrate the promise of the biodegradable scaffold system for tendon-to-bone interface tissue engineering as well as the potential of harnessing cellular interaction for engineering both tendon-to-bone and muscle-to-tendon interface and, ultimately, functional and integrative tendon repair.

Stimulation of angiogenesis

As engineering of a viable tissue is the prime goal of cell-based technology, obtaining adequate vascular supply after implantation is critical to long-term cell survival and function of cell-constructs, not only at the margin but also at the center of the tissue grafts (Folkman and

1097

Hochberg, 1973). In fact, the growth of a new microvascular system remains one of the major limitations to the successful introduction of tissue engineering products to clinical practice (Nomi et al., 2002). Accordingly, the focus of much research in tissue engineering has grown to include a better understanding of angiogenesis. Numerous efforts have been made to overcome this limitation and attempts to enhance angiogenesis within the host tissue have been pursued using several approaches. These include the delivery of growth factors and cytokines that play central regulatory roles in the process of angiogenesis, which is thought to induce ingrowth of capillaries and blood vessels into an engineered implant, thus diminishing hypoxia-related cell damage. The delivery of such angiogenic factors has been achieved either by incorporating the desired factors into the scaffold material to be used or by genetic modification of the cells to be used in the engineering process, which forces the cells to express factors such as vascular endothelial growth factor (VEGF). VEGF is one of the most potent angiogenic factors (Schuch et al., 2002; De Coppi et al., 2005). A recent study by Mooney's group evaluated controlled release of VEGF by incorporating VEGF directly into PLGA scaffolds or by incorporating VEGF encapsulated in PLGA microspheres into scaffolds (Ennett et al., 2006). VEGF incorporated into scaffolds resulted in rapid release of the cytokine, whereas the pre-encapsulated group showed a delayed release. In addition, both systems showed negligible release of VEGF into the systemic circulation, yet their use led to enhanced local angiogenesis *in vivo* for up to 21 days. These studies demonstrated the delivery of VEGF in a controlled and localized fashion *in vivo*. This angiogenic factor delivery system was applied to bone regeneration and its potent ability to enhance angiogenesis within implanted scaffolds was followed by enhanced bone regeneration. This outlines a novel approach for engineering tissues in hypovascular environments (Kaigler et al., 2006). Another mechanism of VEGF delivery was tested by Soker's team, in which human vascular endothelial cells (ECs) and skeletal myoblasts transfected with adenovirus encoding the gene for VEGF were injected subcutaneously in athymic mice (De Coppi et al., 2005). The transfected cells formed a vascularized muscle tissue mass, while the non-transfected cells resulted in less angiogenesis and led to the growth of a significantly smaller tissue mass. This study demonstrates that the use of cells producing biological factors can be another powerful tool in tissue engineering. Another approach from Langer's group has developed a method of formation and stabilization of endothelial vessel networks *in vitro* in engineered skeletal muscle tissue (Levenberg et al., 2005). They used a 3D multiculture system consisting of myoblasts, embryonic fibroblasts, and EC co-seeded on highly porous, biodegradable polymer scaffolds. These results showed that prevascularization of the implants improved angiogenesis and cell survival within the scaffolds. Moreover, they emphasized that co-cultures with EC and muscle cells may also be important for inducing differentiation of engineered tissues.

CURRENT APPROACHES FOR RECONSTRUCTING DIGIT TISSUES: PHALANGES AND SMALL JOINTS

Formation of phalanges and small joints by tissue engineering

The goal of engineering phalanges and small joints is to achieve functionally and aesthetically normal tissues that allow restoration of adequate dexterity for daily activities. To permit this capability, construction of composite constructs composed of the proper cellular and tissue components is critical. However, due to the complex structure of the phalange and its joints, which are composed of various cell types and ECMs, it is difficult to engineer such structures to function in a synchronous, integrated fashion. Toward this goal, initial efforts were focused on separately engineering each individual tissue in the phalange, such as bone, cartilage, ligament, and tendon (Amiel et al., 2001; Kim et al., 2002, 2004; Lanza et al., 2002; Yiou et al., 2003; Derwin et al., 2004; Lee et al., 2006; Wood et al., 2006). However, later groups attempted to engineer an entire phalanx at once, and they achieved the formation of a tenocapsule surrounding subchondral bone and mature cartilage (Isogai et al., 1999). This was achieved by seeding bovine periosteum on biodegradable PLGA that was sutured with sheets of PGA and seeded with chondrocytes and tenocytes. Subsequently, these constructs, which were config-ured to the shape of human distal and middle phalanges, were implanted into the subcuta-neous space of athymic mice and followed for up to 60 weeks. The retrieved phalangeal constructs showed the formation of bone, cartilage and tendon tissues, which were confirmed by PCR, histology and immunohistochemistry. Moreover, the implanted phalanges main-tained the shape of human phalanges (Landis et al., 2005). It was later demonstrated through *in situ* hybridization that the initial bovine genotype was maintained over 20 weeks of implantation in the host mice (Chubinskaya et al., 2004). This experiment demonstrated that phalangeal tissues can be engineered *ex situ* using a biodegradable polymeric scaffold and composite cell system. Vacanti and his team reported the case of patient who had a partial avulsion of the distal phalanx of the thumb. They replaced this phalanx with a tissue-engineered distal phalanx. The procedure resulted in the restoration of a stable and bio-mechanically sound thumb of normal length, without pain and complications that are often associated with harvesting a bone graft (Vacanti et al., 2001).

Kim's team also demonstrated that porous PLGA polymer was an effective synthetic bio-degradable scaffold for the tissue engineering of a hand phalanx (Sedrakyan et al., 2006). The shape and size of a small human distal phalanx was fabricated using a solvent cast/leaching particulate method. Fetal calf periosteum was sutured around the PLGA copolymer, leaving the distal end uncovered. Articular chondrocytes were injected into the distal end of the implant, which was then surgically placed into the dorsal subcutaneous space of an athymic mouse. After 8 and 16 weeks, growth of bone and cartilage was demonstrated by histological evalu-ation, which revealed the homogeneous presence of these tissues with tissue differentiation and matrix production.

We have designed scaffolds that could deliver different cell types to target sites and would allow joint movement (Komura et al., 2008). PGA polymer was chosen as a scaffold for the cartilage and bone components of the phalanx, and acellular collagen matrix derived from bladder submucosa was chosen as a scaffold material for muscle, as it allows movement. The PGA polymer was composed of two tubular pieces connected with non-absorbable sutures, which served as a hinge between the two bony scaffolds. Multiple thin strips of acellular collagen matrix were attached to the distal ends of the bony scaffolds. This configuration was designed to allow for contraction, relaxation, and movement of the composite-engineered digit-like struc-ture. In the initial experiments, the PGA polymer scaffolds were seeded with bovine chon-drocytes/osteoblasts, and bovine skeletal muscle cells were seeded onto the collagen matrix. The cell-seeded composites were incubated *in vitro* for 2 weeks. Histologically, abundant muscle cells were present throughout the collagen matrix and the PGA polymers seeded with chondrocytes showed the presence of whitish ECM. Western blot analysis confirmed tissue identity and confirmed expression of muscle-specific genes such as actin, desmin, and tropomyosin in the muscle component, and expression of collagen type I in the chondrocyte-seeded PGA.

Subsequently, the feasibility of engineering skeletal muscle and cartilage/bone composite structures was studied *in vivo*. Composite scaffolds consisting of synthetic PGA polymers and naturally derived collagen matrices obtained from the bladder submucosa were constructed. Bovine chondrocytes and osteoblasts were seeded onto the PGA polymer matrices and skeletal muscle cells onto the collagen matrices. The scaffolds containing both cell types were analyzed *in vitro* for cell viability and tissue formation. The engineered digits were then implanted subcutaneously in athymic mice and followed for up to 6 months. The cells seeded on the composite digit constructs readily attached to their designated region of the scaffold and remained viable. Grossly, the implanted scaffolds formed muscle and cartilage or bone tissues adjacent to each other. Each tissue type was confirmed histologically and immunohisto-chemically using cell specific antibodies. Biomechanical studies showed that the cartilage tissue was elastic and could withstand high degrees of pressure, which suggested the ability to preserve structural integrity. Physiological organ bath studies of the retrieved muscle tissues showed adequate contractility in response to electric field stimulation. These findings show that different tissue types can be engineered simultaneously using a composite scaffold system. The tissues retained their respective phenotypic and functional characteristics independently of the other. This study demonstrates that the engineering of functional digit tissues may be feasible.

Recently, we have also developed a continuous, integrated, dual scaffolding system that has regional variations in mechanical properties that mimic the trends seen in native MTJs (Ladd et al., 2009). Two different polymer solutions, polycaprolactone (PCL)/collagen and poly (L-lactide) (PLLA)/collagen blends were simultaneously electrospun onto opposite ends of a cylindrical mandrel to create a scaffold for muscle–tendon construct. The dual scaffold is biocompatible, displays a nanofiber architecture with fiber diameters ranging from 450 to 550 nm, and exhibits regional variations in mechanical properties with moduli ranging from 4.5 to 26.0 MPa. Moreover, video strain analysis revealed the differences in strain profile during scaffold testing, with the muscle side and tendon side displaying average maximum strains of 24.3% and 2.6%, respectively. *In vitro* experiments showed that the scaffold accommodates C2C12 myoblasts and allows myotube formation. Thus, we believe these scaffolds are attractive candidates for use in the formation of MTJs.

To determine the feasibility of replacing a missing digit segment, *in situ* implantation of digit segments was performed in a rabbit model. Autologous muscle cells and chondrocytes were grown, expanded, and seeded on the composite scaffolds to generate constructs consisting of interconnecting bony segments, muscle, and tendon. Digit segments containing interconnected joints were excised and replaced with the engineered digit segments of the same length and caliber. The forelimb with the engineered digit was placed in a cast for approximately 4 weeks in order to protect the wound and enhance tissue maturation *in situ*. We showed that the engineered digit segments were able to form cartilage, muscle, and tendon upon retrieval. Scaffolds without cells failed to form tissue structures (Komura et al., 2008).

SUMMARY AND FUTURE DIRECTIONS

Regeneration of homogeneous tissues using tissue engineering techniques has been successful in some cases. However, there is an increasing need for the development of composite tissue systems that function in a coordinated manner so that complex structures, such as digits and limbs, can be reconstructed, and the design of such systems presents a unique challenge because bone, articular cartilage, ligament, tendon, and skeletal muscle tissues must be regenerated simultaneously and function in a coordinated manner. Three types of tissues (bone, cartilage, and tendon) were recently integrated in a single construct for the reconstruction of phalanx (Isogai et al., 1999). However, the ultimate success of a reconstruction process depends on its ability to be used clinically. Although individual or partial replacement of phalanges has been somewhat successful in a few studies, achieving a fully functional engineered digit requires further investigation. The development of methods to integrate all tissue components, including the vascular and neural network, into a compact compartment of the distal limb would accelerate the progress of this research. Thus, establishment of multiple cell sources and development of an intelligent composite scaffolding system that would allow for the enhanced formation of individual tissue types in a controlled manner are critical. However, recent progress made towards the engineering of digit tissues suggests that achieving functional phalangeal tissues may have an expanded role in the clinic.

References

Altman, G. H., Horan, R. L., Weitzel, P., & Richmond, J. C. (2008). The use of long-term bioresorbable scaffolds for anterior cruciate ligament repair. *J. Am. Acad. Orthop. Surg.*, 16, 177–187.

Amiel, G. E., Yoo, J. J., Kim, B. S., & Atala, A. (2001). Tissue-engineered stents created from chondrocytes. *J. Urol.*, 165, 2091–2095.

Atala, A., Bauer, S. B., Soker, S., Yoo, J. J., & Retik, A. B. (2006). Tissue-engineered autologous bladders for patients needing cystoplasty. *Lancet, 367,* 1241–1246.

Bartsch, G., Yoo, J. J., De Coppi, P., Siddiqui, M. M., Schuch, G., Pohl, H. G., et al. (2005). Propagation, expansion, and multilineage differentiation of human somatic stem cells from dermal progenitors. *Stem Cells Dev., 14,* 337–348.

Benjamin, M., Evans, E. J., & Copp, L. (1986). The histology of tendon attachments to bone in man. *J. Anat., 149,* 89–100.

Blevins, F. T., Djurasovic, M., Flatow, E. L., & Vogel, K. G. (1997). Biology of the rotator cuff tendon. *Orthop. Clin. North Am., 28,* 1–16.

Brivanlou, A. H., Gage, F. H., Jaenisch, R., Jessell, T., Melton, D., & Rossant, J. (2003). Stem cells. Setting standards for human embryonic stem cells. *Science, 300,* 913–916.

Buncke, H. J., Buncke, C. M., & Schulz, W. P. (1966). Immediate Nicoladoni procedure in the Rhesus monkey, or hallux-to-hand transplantation, utilizing microminiature vasuclar anatomoses. *Br. J. Plast. Surg., 19,* 332–337.

Chen, C. H., Chen, W. J., Shih, C. H., Yang, C. Y., Liu, S. J., & Lin, P. Y. (2003). Enveloping the tendon graft with periosteum to enhance tendon-bone healing in a bone tunnel: a biomechanical and histological study in rabbits. *Arthroscopy, 19,* 290–296.

Chubinskaya, S., Jacquet, R., Isogai, N., Asamura, S., & Landis, W. J. (2004). Characterization of the cellular origin of a tissue-engineered human phalanx model by *in situ* hybridization. *Tissue Eng., 10,* 1204–1213.

Cooper, J. A., Lu, H. H., Ko, F. K., Freeman, J. W., & Laurencin, C. T. (2005). Fiber-based tissue-engineered scaffold for ligament replacement: design considerations and *in vitro* evaluation. *Biomaterials, 26,* 1523–1532.

Cooper, J. A., Jr., Sahota, J. S., Gorum, W. J., 2nd, Carter, J., Doty, S. B., & Laurencin, C. T. (2007). Biomimetic tissue-engineered anterior cruciate ligament replacement. *Proc. Natl. Acad. Sci. USA, 104,* 3049–3054.

De Coppi, P., Delo, D., Farrugia, L., Udompanyanan, K., Yoo, J. J., Nomi, M., et al. (2005). Angiogenic gene-modified muscle cells for enhancement of tissue formation. *Tissue Eng., 11,* 1034–1044.

De Filippo, R. E., Yoo, J. J., & Atala, A. (2003). Engineering of vaginal tissue *in vivo*. *Tissue Eng., 9,* 301–306.

Deasy, B. M., Gharaibeh, B. M., Pollett, J. B., Jones, M. M., Lucas, M. A., Kanda, Y., et al. (2005). Long-term self-renewal of postnatal muscle-derived stem cells. *Mol. Biol. Cell, 16,* 3323–3333.

Derwin, K., Androjna, C., Spencer, E., Safran, O., Bauer, T. W., Hunt, T., et al. (2004). Porcine small intestine submucosa as a flexor tendon graft. *Clin. Orthop. Rel. Res., 423,* 245–252.

Dubernard, J. M., Owen, E., Herzberg, G., Lanzetta, M., Martin, X., Kapila, H., et al. (1999). Human hand allograft: report on first 6 months. *Lancet, 353,* 1315–1320.

El-Kassaby, A. W., Retik, A. B., Yoo, J. J., & Atala, A. (2003). Urethral stricture repair with an off-the-shelf collagen matrix. *J. Urol., 169,* 170–173, discussion 173.

Ennett, A. B., Kaigler, D., & Mooney, D. J. (2006). Temporally regulated delivery of VEGF *in vitro* and *in vivo*. *J. Biomed. Mater. Res. A, 79,* 176–184.

Falke, G., Yoo, J. J., Kwon, T. G., Moreland, R., & Atala, A. (2003). Formation of corporal tissue architecture *in vivo* using human cavernosal muscle and endothelial cells seeded on collagen matrices. *Tissue Eng., 9,* 871–879.

Folkman, J., & Hochberg, M. (1973). Self-regulation of growth in three dimensions. *J. Exp. Med., 138,* 745–753.

Fujioka, H., Thakur, R., Wang, G. J., Mizuno, K., Balian, G., & Hurwitz, S. R. (1998). Comparison of surgically attached and non-attached repair of the rat Achilles tendon–bone interface. Cellular organization and type X collagen expression. *Connect. Tissue Res., 37,* 205–218.

Gao, J., Dennis, J. E., Solchaga, L. A., Awadallah, A. S., Goldberg, V. M., & Caplan, A. I. (2001). Tissue-engineered fabrication of an osteochondral composite graft using rat bone marrow-derived mesenchymal stem cells. *Tissue Eng., 7,* 363–371.

Huangfu, X., & Zhao, J. (2007). Tendon-bone healing enhancement using injectable tricalcium phosphate in a dog anterior cruciate ligament reconstruction model. *Arthroscopy, 23,* 455–462.

Hung, C. T., Lima, E. G., Mauck, R. L., Takai, E., LeRoux, M. A., Lu, H. H., et al. (2003). Anatomically shaped osteochondral constructs for articular cartilage repair. *J. Biomech., 36,* 1853–1864.

Inoue, N., Ikeda, K., Aro, H. T., Frassica, F. J., Sim, F. H., & Chao, E. Y. (2002). Biologic tendon fixation to metallic implant augmented with autogenous cancellous bone graft and bone marrow in a canine model. *J. Orthop. Res., 20,* 957–966.

Isogai, N., Landis, W., Kim, T. H., Gerstenfeld, L. C., Upton, J., & Vacanti, J. P. (1999). Formation of phalanges and small joints by tissue-engineering. *J. Bone Joint Surg., 81,* 306–316.

Isogai, N., Asamura, S., Higashi, T., Ikada, Y., Morita, S., Hillyer, J., et al. (2004). Tissue engineering of an auricular cartilage model utilizing cultured chondrocyte-poly(L-lactide-epsilon-caprolactone) scaffolds. *Tissue Eng., 10,* 673–687.

Kaigler, D., Wang, Z., Horger, K., Mooney, D. J., & Krebsbach, P. H. (2006). VEGF scaffolds enhance angiogenesis and bone regeneration in irradiated osseous defects. *J. Bone Miner. Res., 21,* 735–744.

Kim, B. S., & Mooney, D. J. (1998). Development of biocompatible synthetic extracellular matrices for tissue engineering. *Trends Biotechnol, 16,* 224–230.

Kim, B. S., Yoo, J. J., & Atala, A. (2002). Engineering of human cartilage rods: potential application for penile prostheses. *J. Urol., 168,* 1794–1797.

Kim, B. S., Yoo, J. J., & Atala, A. (2004). Peripheral nerve regeneration using acellular nerve grafts. *J. Biomed. Mater. Res. A, 68,* 201–209.

Komura, M., Nomi, M., Soker, S., Yoo, J. J., & Atala, A. (2008). Functional reconstruction of-engineered digits. *Tissue Eng., 14A,* 855.

Kwon, T. G., Yoo, J. J., & Atala, A. (2002). Autologous penile corpora cavernosa replacement using tissue engineering techniques. *J. Urol., 168,* 1754–1758.

Kyung, H. S., Kim, S. Y., Oh, C. W., & Kim, S. J. (2003). Tendon-to-bone tunnel healing in a rabbit model: the effect of periosteum augmentation at the tendon-to-bone interface. *Knee Surg. Sports Traumatol. Arthrosc., 11,* 9–15.

Ladd, M. R., Aboushwareb, T. A., Lee, S. J., Atala, A., & Yoo, J. J. (2009). Integrated scaffolding system for the engineering of muscle–tendon junction. *J. Am. Coll. Surg., 209,* S62.

Landis, W. J., Jacquet, R., Hillyer, J., Lowder, E., Yanke, A., Siperko, L., et al. (2005). Design and assessment of a tissue-engineered model of human phalanges and a small joint. *Orthod. Craniofac. Res., 8,* 303–312.

Lanza, R. P., Chung, H. Y., Yoo, J. J., Wettstein, P. J., Blackwell, C., Borson, N., et al. (2002). Generation of histo-compatible tissues using nuclear transplantation. *Nat. Biotech., 20,* 689–696.

Larkin, L. M., Calve, S., Kostrominova, T. Y., & Arruda, E. M. (2006). Structure and functional evaluation of tendon–skeletal muscle constructs engineered *in vitro. Tissue Eng., 12,* 3149–3158.

Lee, S. J., Lim, G. J., Lee, J. W., Atala, A., & Yoo, J. J. (2006). *In vitro* evaluation of a poly(lactide-co-glycolide)–collagen composite scaffold for bone regeneration. *Biomaterials, 27,* 3466–3472.

Levenberg, S., Rouwkema, J., Macdonald, M., Garfein, E. S., Kohane, D. S., Darland, D. C., et al. (2005). Engineering vascularized skeletal muscle tissue. *Nat. Biotech., 23,* 879–884.

Lu, H. H., & Jiang, J. (2006). Interface tissue engineering and the formulation of multiple-tissue systems. *Adv. Biochem. Eng. Biotechnol., 102,* 91–111.

Lu, H. H., Cooper, J. A., Jr., Manuel, S., Freeman, J. W., Attawia, M. A., Ko, F. K., et al. (2005). Anterior cruciate ligament regeneration using braided biodegradable scaffolds: *in vitro* optimization studies. *Biomaterials, 26,* 4805–4816.

Ma, J., Goble, K., Smietana, M., Kostrominova, T., Larkin, L., & Arruda, E. M. (2009). Morphological and functional characteristics of three-dimensional engineered bone–ligament–bone constructs following implantation. *J. Biomech. Eng., 131,* 101–117.

Martin, I., Miot, S., Barbero, A., Jakob, M., & Wendt, D. (2007). Osteochondral tissue engineering. *J. Biomech., 40,* 750–765.

Martin, R. B., Burr, D. B., & Sharkey, N. A. (Eds.) (1998). *Skeletal Tissue Mechanics.* New York: Springer.

Mathes, S. J., Buchannan, R., & Weeks, P. M. (1980). Microvascular joint transplantation with epiphyseal growth. *J. Hand Surg, 5,* 586–589.

Moise, K. J., Jr. (2005). Umbilical cord stem cells. *Obstet. Gynecol., 106,* 1393–1407.

Murray, M. M., Forsythe, B., Chen, F., Lee, S. J., Yoo, J. J., Atala, A., et al. (2006). The effect of thrombin on ACL fibroblast interactions with collagen hydrogels. *J. Orthop. Res., 24,* 508–515.

Mutsuzaki, H., Sakane, M., Nakajima, H., Ito, A., Hattori, S., Miyanaga, Y., et al. (2004). Calcium-phosphate-hybridized tendon directly promotes regeneration of tendon–bone insertion. *J. Biomed. Mater. Res. A, 70,* 319–327.

Nomi, M., Atala, A., Coppi, P. D., & Soker, S. (2002). Principals of neovascularization for tissue engineering. *Mol. Aspects Med., 23,* 463–483.

Oberpenning, F., Meng, J., Yoo, J. J., & Atala, A. (1999). *De novo* reconstitution of a functional mammalian urinary bladder by tissue engineering. *Nat. Biotech., 17,* 149–155.

Pariente, J. L., Kim, B. S., & Atala, A. (2002). *In vitro* biocompatibility evaluation of naturally derived and synthetic biomaterials using normal human bladder smooth muscle cells. *J. Urol., 167,* 1867–1871.

Patradul, A., Ngarmukos, C., & Parkpian, V. (1998). Distal digital replantations and revascularizations. 237 digits in 192 patients. *J. Hand Surg., 23,* 578–582.

Pereira, B. P., Kour, A. K., Leow, E. L., & Pho, R. W. (1996). Benefits and use of digital prostheses. *J. Hand Surg., 21,* 222–228.

Schaefer, D., Martin, I., Jundt, G., Seidel, J., Heberer, M., Grodzinsky, A., et al. (2002). Tissue-engineered composites for the repair of large osteochondral defects. *Arthritis Rheum., 46,* 2524–2534.

Schek, R. M., Taboas, J. M., Segvich, S. J., Hollister, S. J., & Krebsbach, P. H. (2004). Engineered osteochondral grafts using biphasic composite solid free-form fabricated scaffolds. *Tissue Eng., 10,* 1376–1385.

1102

Schuch, G., Machluf, M., Bartsch, G., Jr., Nomi, M., Richard, H., Atala, A., et al. (2002). *In vivo* administration of vascular endothelial growth factor (VEGF) and its antagonist, soluble neuropilin-1, predicts a role of VEGF in the progression of acute myeloid leukemia *in vivo*. *Blood, 100*, 4622–4628.

Sedrakyan, S., Zhou, Z., Perin, L., Leach, K., Mooney, D., & Kim, T. H. (2006). Tissue enginering of a small hand phalanx with a porously casted polylactic acid-polyglycolic acid copolymer. *Tissue Eng., 12*, 2675–2683.

Sherwood, J. K., Riley, S. L., Palazzolo, R., Brown, S. C., Monkhouse, D. C., Coates, M., et al. (2002). A three-dimensional osteochondral composite scaffold for articular cartilage repair. *Biomaterials, 23*, 4739–4751.

Swasdison, S., & Mayne, R. (1991). *In vitro* attachment of skeletal muscle fibers to a collagen gel duplicates the structure of the myotendinous junction. *Exp. Cell Res., 193*, 227–231.

Swasdison, S., & Mayne, R. (1992). Formation of highly organized skeletal muscle fibers *in vitro*. Comparison with muscle development *in vivo*. *J. Cell Sci., 102*, 643–652.

Temenoff, J. S., & Mikos, A. G. (2000). Review: Tissue engineering for regeneration of articular cartilage. *Biomaterials, 21*, 431–440.

Tholpady, S. S., Llull, R., Ogle, R. C., Rubin, J. P., Futrell, J. W., & Katz, A. J. (2006). Adipose tissue: stem cells and beyond. *Clin. Plast. Surg., 33*, 55–62.

Tien, Y. C., Chih, T. T., Lin, J. H., Ju, C. P., & Lin, S. D. (2004). Augmentation of tendon–bone healing by the use of calcium-phosphate cement. *J. Bone Joint Surg., 86B*, 1072–1076.

Tognana, E., Chen, F., Padera, R. F., Leddy, H. A., Christensen, S. E., Guilak, F., et al. (2005a). Adjacent tissues (cartilage, bone) affect the functional integration of engineered calf cartilage *in vitro*. *Osteoarthritis Cartilage, 13*, 129–138.

Tognana, E., Padera, R. F., Chen, F., Vunjak-Novakovic, G., & Freed, L. E. (2005b). Development and remodeling of engineered cartilage–explant composites *in vitro* and *in vivo*. *Osteoarthritis Cartilage, 13*, 896–905.

Towler, D. A., & Gelberman, R. H. (2006). The alchemy of tendon repair: a primer for the (S)mad scientist. *J. Clin. Invest., 116*, 863–866.

Tsay, T. M., Jupitor, J. B., Kutz, J. E., & Kleinert, H. E. (1982). Vascularized autogenous whole joint transfer in the hand – a clinical study. *J. Hand Surg., 7*, 335–342.

Vacanti, C. A., & Upton, J. (1994). Tissue-engineered morphogenesis of cartilage and bone by means of cell transplantation using synthetic biodegradable polymer matrices. *Clin. Plast. Surg., 21*, 445–462.

Vacanti, C. A., Bonassar, L. J., Vacanti, M. P., & Shufflebarger, J. (2001). Replacement of an avulsed phalanx with tissue-engineered bone. *N. Engl. J. Med., 344*, 1511–1514.

Wood, F. M., Kolybaba, M. L., & Allen, P. (2006). The use of cultured epithelial autograft in the treatment of major burn wounds: eleven years of clinical experience. *Burns, 32*, 538–544.

Wray, R. C., Jr., Mathes, S. M., Young, V. L., & Weeks, P. M. (1981). Free vascularized whole-joint transplants with ununited epiphyses. *Plast. Reconstr. Surg., 67*, 519–525.

Yang, P. J., & Temenoff, J. S. (2009). Engineering orthopedic tissue interfaces. *Tissue Eng. B Rev., 15*, 127–141.

Yiou, R., Yoo, J. J., & Atala, A. (2003). Restoration of functional motor units in a rat model of sphincter injury by muscle precursor cell autografts. *Transplantation, 76*, 1053–1060.

Yoo, J. J., Park, H. J., Lee, I., & Atala, A. (1999). Autologous engineered cartilage rods for penile reconstruction. *J. Urol., 162*, 1119–1121.

1103

CHAPTER 60

Intracorporeal Kidney Support

Jae Hyun Bae[*,†]**, Tamer Aboushwareb**[*,‡]**, Anthony Atala**[*,‡]**, James J. Yoo**[*,§]

*Wake Forest Institute for Regenerative Medicine, Wake Forest University School of Medicine, Winston-Salem, NC, USA
†Department of Urology, Korea University Medical Center, Seoul, South Korea
‡Department of Urology, Wake Forest University Health Sciences, Winston-Salem, NC, USA
§Joint Institute for Regenerative Medicine, Kyungpook National University Hospital, Daegu, South Korea

INTRODUCTION

End-stage renal failure is a devastating disease which involves multiple organs in affected individuals. Although currently available treatment modalities, including dialysis and transplantation, can prolong survival for many patients, problems such as donor shortage, complications and graft failure remain a continued concern (Chazan et al., 1991; Cohen et al., 1994; Feldman et al., 1996; Amiel et al., 2000a). Numerous investigative efforts have been commenced in order to improve, restore, or replace renal function. Cell-based approaches for kidney tissue regeneration have been proposed recently as an alternative method. In this chapter we describe various cell-based approaches to achieve functional intracorporeal kidney support.

The kidney is considered one of the more challenging organs to reconstruct, due to its complex structure and function. Normal renal function includes synthesis of 1,25 vitamin D3, erythropoietin, glutathione, and free radical scavenging enzymes. The kidney also participates in the catabolism of low-molecular-weight proteins and in the production and regulation of cytokines (Maack, 1992; Frank et al., 1993; Stadnyk, 1994). Because these functions cannot be replaced with dialysis therapy, long-term consequences, such as anemia and malnutrition, are prevalent in these patients.

The limitations of current therapies for renal functional augmentation have led investigators to pursue alternative therapeutic modalities. The concept of cell transplantation using tissue engineering techniques has been proposed as a method to improve, restore, or replace renal function (Atala, 1997, 1999; Amiel and Atala, 1999; Humes et al., 1999). The emergence of tissue engineering and regenerative medicine strategies has presented alternative possibilities for the management of pathological renal conditions (Fig. 60.1). Augmentation of either isolated or total renal function with kidney cell expansion may be a feasible solution. We have followed an approach which involves the development of intracorporeal support systems for renal functional replacement.

Principles of Regenerative Medicine. DOI: 10.1016/B978-0-12-381422-7.10060-4

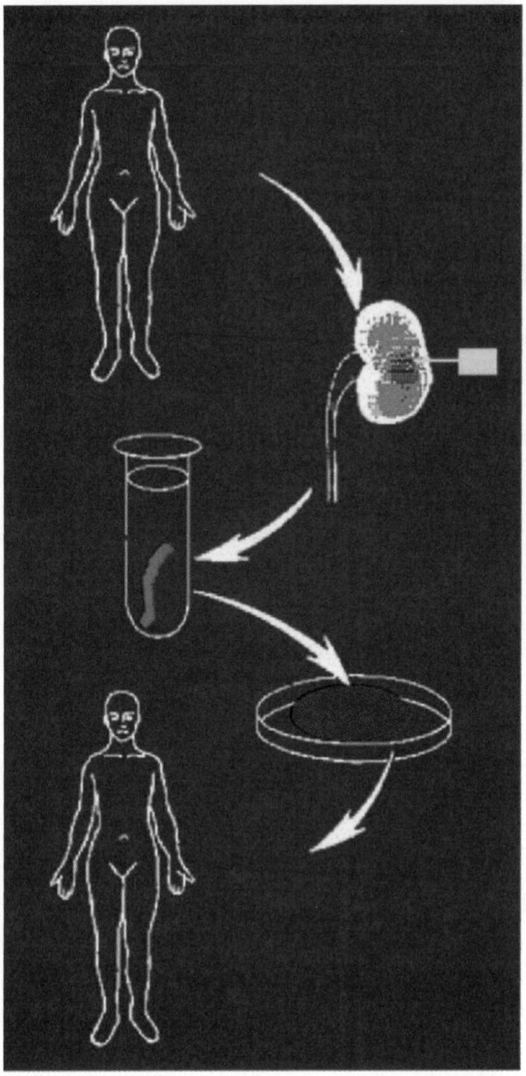

FIGURE 60.1

A strategy for engineering of renal tissue. A patient with end-stage renal failure undergoes a percutaneous biopsy. Renal cells are grown, expanded in culture, and seeded onto a 3D support system, which is then implanted into the same patient.

BASIC PRINCIPLES OF KIDNEY TISSUE REGENERATION

Components required to achieve partial or total kidney function are renal cells, three-dimensional (3D) scaffolds, and an *in vivo* environment. The challenge associated with renal cell culture is due to the unique structural and cellular heterogeneity present within the kidney. The system of nephrons and collecting ducts is composed of multiple functionally and phenotypically distinct segments. For this reason, appropriate conditions need to be provided for the long-term survival, differentiation, and growth of the cells. Extensive research has been performed in order to determine optimal growth conditions for renal cell enrichment (Milici et al., 1985; Carley et al., 1988; Horikoshi et al., 1988; Humes and Cieslinski, 1992; Schena, 1998). Isolation of particular cell types that produce specific factors, such as erythropoietin (EPO), may be a feasible approach for selective cell therapies. However, total renal function would not be achieved if specific cell types were separately isolated. To reconstitute kidney tissue that would deliver complete renal function, cells composing the functional nephron units may be preferable. Based on the literature and our experience, we were able to obtain optimal growth conditions for a stable cell culture system for kidney tissue reconstitution.

Renal cells grown in culture are able to maintain their cellular characteristics (Lanza et al., 2002). When primary renal cells are placed in a collagen-based 3D culture system, they are able to reconstitute into renal structures.

Recent efforts in the area of kidney tissue regeneration were focused toward the development of a reliable cell source. Multipotent or progenitor cells have been proposed as a promising source due to their potential to differentiate into several cell lineages. Bone marrow-derived human mesechymal stem cells have been shown to exhibit plasticity and differentiation potential into several different cell types (Prockop, 1997). These cells have been shown to participate in kidney development when they are placed in a rat embryonic niche that allows for continued exposure to a repertoire of nephrogenic signals (Yokoo et al., 2005). Another potential cell source is circulating stem cells derived from bone marrow, which are also known to participate at the site of kidney regeneration, and appear to differentiate into renal cell types, such as tubular and glomerular epithelial cells, podocytes, mesangial cells, and interstitial cells after renal injury (Ito et al., 2001; Poulsom et al., 2001; Gupta et al., 2002; Iwano et al., 2002; Kale et al., 2003; Lin et al., 2003; Rookmaaker et al., 2003). However, although bone marrow cells were found to contribute to regeneration of damaged glomerular endothelial cells, the major cell source of kidney regeneration was found to originate from intrarenal cells in an ischemic renal injury model (Ikarashi et al., 2005; Lin et al., 2005).

Although isolated renal cells are able to retain their phenotypic and functional characteristics in culture, transplantation of these cells *in vivo* may not result in structural remodeling. In addition, cell or tissue components may not be implanted in large volumes due to limited diffusion of oxygen and nutrients through a large tissue mass (Folkman and Hochberg, 1973). Thus, a cell-support matrix is necessary to allow diffusion of nutrients across the entire implant. A variety of synthetic and naturally derived materials has been examined in order to determine the ideal support structures for the regeneration of urological tissue (Tachibana et al., 1985; Atala et al., 1995, 2006; Oberpenning et al., 1999; El-Kassaby et al., 2003). Biodegradable synthetic materials, such as poly-lactic and glycolic acid polymers, have been used to provide structural support for cells. Synthetic materials can be easily fabricated and configured in a controlled manner. Naturally derived materials, such as collagen, laminin and fibronectin, are biocompatible and provide a similar extracellular matrix environment to normal tissue. For this reason, collagen-based scaffolds have been preferred in many tissue applications (Hubbell et al., 1991; Wald et al., 1993; Freed et al., 1994; Mooney et al., 1996).

1107

CREATION OF RENAL STRUCTURES *IN VIVO*

The kidney is responsible not only for urine excretion but for several other important metabolic functions. Our initial study involved investigating the feasibility of achieving renal cell growth, expansion and *in vivo* reconstitution using tissue engineering techniques (Atala et al., 1995). New Zealand White rabbits underwent nephrectomy and renal artery perfusion with a non-oxide solution which promoted iron particle entrapment in the glomeruli. Homogenization of the renal cortex and fractionation in 83 and 210 μm sieves with subsequent magnetic extraction yielded three separate purified suspensions of distal tubules, glomeruli, and proximal tubules. The cells were plated separately *in vitro* and seeded onto biodegradable polyglycolic acid scaffolds. Polymer scaffolds were implanted subcutaneously into host athymic mice. This included implants of proximal tubular cells, glomeruli, distal tubular cells, and a mixture of all three cell types. Polymers alone served as controls. Animals were sacrificed at 1 week, 2 weeks, and 1 month after implantation and the retrieved scaffolds were analyzed. An acute inflammatory phase and a chronic foreign body reaction were seen, accompanied by vascular ingrowth by 7 days after implantation. Histological examination demonstrated progressive formation and organization of the nephron segments within the polymer fibers with time. Renal cell proliferation in the cell–polymer scaffolds was detected by *in vivo* labeling of replicating cells with the thymidine analog bromodeoxyuridine (BrdU). BrdU incorporation into renal cell DNA was identified immunocytochemically with monoclonal anti-BrdU antibodies. These results demonstrated that renal specific cells can be successfully harvested, and that they can survive in culture and attach to artificial biodegradable polymers. The renal cell–polymer scaffolds can be implanted into host animals where the cells replicate and organize into nephron segments, as the polymer, which acts as a cell delivery vehicle, undergoes biodegradation.

The initial experiments demonstrated that implanted cell–polymer scaffolds gave rise to renal tubular structures. However, it was unclear whether the tubular structures reconstituted *de novo* from dispersed renal elements, or they merely represented fragments of donor tubules which survived intact. Further investigation was conducted in order to examine the tubular reconstitution process (Fung et al., 1996). Mouse renal cells were harvested, grown, and expanded in culture. Subsequently, single isolated cells were seeded on biodegradable polymers and implanted into syngeneic hosts. Renal epithelial cells were observed to reconstitute into tubular structures *in vivo*. Sequential analyses of the retrieved implants over time demonstrated that renal epithelial cells first organized into a cord-like structure with a solid center. Subsequent canalization into a hollow tube could be seen by 2 weeks. Histological examination with nephron segment-specific lactins showed successful reconstitution of proximal tubules, distal tubules, loops of Henle, collecting tubules, and collecting ducts. These results showed that single suspended cells are capable of reconstituting into tubular structures, with homogeneous cell types within each tubule.

REGENERATION OF FUNCTIONAL RENAL TISSUE *IN VIVO*

The kidneys are critical to body homeostasis because of their excretory, regulatory, and endocrine functions. The excretory function is initiated by filtration of blood at the glomerulus, and the regulatory function is provided by the tubular segments. Although our prior studies demonstrated that renal cells seeded on biodegradable polymer scaffolds are able to reconstitute into renal structures *in vivo*, complete renal function could not be achieved due to the type and structural configuration of polymers used. In our subsequent study we explored the feasibility of creating a functional artificial renal unit, wherein urine production could be achieved (Yoo et al., 1996). Mouse renal cells were harvested and expanded in culture. The cells were seeded onto a tubular device constructed from polycarbonate (4 μm pore size), connected at one end with a silastic catheter which terminated into a reservoir. The device was implanted in the subcutaneous space of athymic mice. Animals were sacrificed at 1, 2, 3, 4, and 8 weeks after implantation and the retrieved specimens were examined histologically and immunocytochemically. Fluid was collected from inside the implant, and uric acid and creatinine levels were determined.

Histological examination of the implanted device revealed extensive vascularization, formation of glomeruli, and highly organized tubule-like structures. Immunocytochemical staining with anti-osteopontin antibody, which is secreted by proximal and distal tubular cells and the cells of the thin ascending loop of Henle, stained the tubular sections. Immunohistochemical staining for alkaline phosphatase stained proximal tubule-like structures. Uniform staining for fibronectin in the extracellular matrix of newly formed tubes was observed. The fluid collected from the reservoir was yellow and contained 66 mg/dl uric acid (as compared to 2 mg/dl in plasma), suggesting that these tubules are capable of unidirectional secretion and concentration of uric acid. The creatinine assay performed on the collected fluid showed an 8.2-fold increase in concentration, as compared to serum. These results demonstrated that single cells form multicellular structures and become organized into functional renal units that are able to excrete high levels of solutes through a urine-like fluid.

To determine whether renal tissue could be formed using an alternative cell source, nuclear transplantation was performed to generate histocompatible tissues. The feasibility of engineering syngeneic renal tissues *in vivo* using cloned cells was investigated (Lanza et al., 2002). In this study nuclear material from bovine dermal fibroblasts was transferred into unfertilized enucleated donor bovine eggs. Renal cells from the cloned embryos were harvested, expanded *in vitro*, and seeded onto 3D renal devices (Fig. 60.2(a)). The devices were implanted into the back of the same steer from which the cells were cloned, and were retrieved 12 weeks later. Functioning renal units were created from cells cloned from bovine fibroblasts. Urine production and viability were demonstrated after transplantation back into the nuclear donor animal despite expression of a different mtDNA haplotype (Fig. 60.2(b)). Chemical analysis suggested unidirectional secretion and concentration of urea nitrogen and creatinine. The devices revealed formation of organized glomeruli and tubular structures (Fig. 60.2(c)). Immunohistochemical and reverse transcription–polymerase chain reaction (RT-PCR) analysis confirmed the expression of renal mRNA and proteins, whereas delayed-type hypersensitivity testing and *in vitro* proliferative assays showed that there was no rejection response to the cloned cells. This study indicates that the cloned renal cells are able to form and organize into functional tissue structures, which are genetically the same as the host. Generating immune-compatible cells using therapeutic cloning techniques is feasible and may be useful for the engineering of renal tissues for autologous applications.

In our previous study, we showed that renal cells seeded on synthetic renal devices with a collecting system are able to form functional renal structures with urine-like fluid excretion. However, a naturally derived tissue matrix with existing 3D kidney architecture would be preferable, in that it would allow for transplantation of a large number of cells for the creation of greater renal tissue volumes. We developed an acellular collagen-based kidney matrix, which is identical to the native renal architecture. In a subsequent study we investigated whether the collagen-based matrices could accommodate large volumes of renal cells which could proliferate and form kidney structures *in vivo* (Amiel et al., 2000b).

Acellular collagen matrices, derived from porcine kidneys, were obtained through a multiple step decellularization process. Serial evaluation of the matrix for cellular remnants was performed using histochemistry, scanning electron microscopy (SEM), and RT-PCR. Mice renal cells were harvested, grown, and seeded on 80 collagen matrices at a concentration of 30×10^6 cells/ml. Forty cell-matrix constructs grown *in vitro* were analyzed 3 days, and 1, 2, 4, and 6 weeks after seeding. The remaining 40 cell-matrices were implanted in the subcutaneous space of 20 athymic mice. The animals were sacrificed 3 days, and 1, 2, 4, 8, and 24 weeks after implantation for analyses. Gross, SEM, histochemical, immunocytochemical, and biochemical analyses were performed.

SEM and histological examination confirmed the acellularity of the processed matrix. RT-PCR performed on the kidney matrices demonstrated the absence of any RNA residues. Renal cells

1109

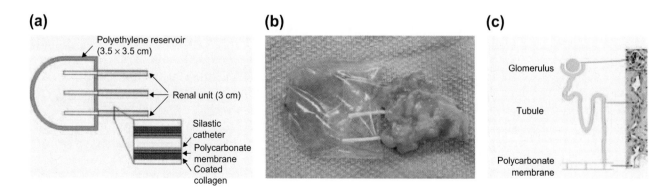

(a)

Polyethylene reservoir (3.5 × 3.5 cm)

Renal unit (3 cm)

Silastic catheter

Polycarbonate membrane

Coated collagen

(b)

(c)

Glomerulus

Tubule

Polycarbonate membrane

FIGURE 60.2

Formation of functional renal tissue *in vivo*. (a) Renal device. (b) Tissue-engineered renal unit shows the accumulation of urine-like fluid. (c) There was a clear unidirectional continuity between the mature glomeruli, their tubules, and the polycarbonate membrane.

seeded on the matrix adhered to the inner surface and proliferated to confluency 7 days after seeding, as demonstrated by SEM. Histochemical and immunocytochemical analyses performed using H&E, periodic acid Schiff, alkaline phosphatase, anti-osteopontin and anti-CD-31 identified stromal, endothelial, and tubular epithelial cell phenotypes within the matrix. Renal tubular and glomeruli-like structures were observed 8 weeks after implantation. 3-(4,5-Dimethylthiazol-2-yl)-2,5-diphenyltetrazolium (MTT) assays and radioactive thymidine incorporation assays performed 6 weeks after cell seeding demonstrated a cell population increase of 116% and 92%, respectively, as compared to the 2-week time points. This study demonstrates that renal cells are able to adhere and proliferate on the collagen-based kidney matrices. The renal cells reconstitute renal tubular and glomeruli-like structures. The collagen-based kidney matrix system seeded with renal cells may be useful in the future for augmenting renal function.

Our prior studies demonstrated that culture-expanded primary renal cells seeded on artificial renal devices with a collecting system are able to form functional renal structures with urine-like fluid excretion. However, creation of renal structures without the use of an artificial device system would be preferable. In addition, implantation procedures are invasive and may result in unnecessary complications. In a subsequent study we investigated the feasibility of creating 3D renal structures for *in situ* implantation within the native kidney tissue. Primary renal cells from 4-week-old mice were grown and expanded in culture. Culture-expanded renal cells were labeled with fluorescent markers and injected into mouse kidneys in a collagen gel scaffold for *in vivo* formation of renal tissues. Collagen injection without cells and sham-operated animals served as controls. *In vitro* reconstituted renal structures and *in vivo* implanted cells were retrieved and analyzed.

The implanted renal cells formed tubular and glomerular structures within the kidney tissue, as confirmed by the fluorescent markers. There was no evidence of renal tissue formation in the control and the sham-operated groups. These results demonstrate that single renal cells are able to reconstitute into organized kidney structures when placed in a collagen-based scaffolding system. The implanted renal cells are able to self-assemble into tubular and glomerular structures within the kidney tissue (Fig. 60.3). These findings suggest that this system may be the preferred approach to engineer functional kidney tissues for the treatment of end-stage renal disease.

Although the concept of renal cell therapy has been demonstrated in several studies in which implanted, culture-expanded cells show the formation of renal structures, the efficiency of the process of structural reconstitution could not be assessed upon implantation *in vivo*. Reconstitution of renal structure during the culture expansion stage followed by implantation was

(a) **(b)**

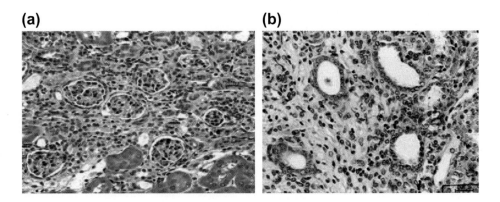

FIGURE 60.3
Reconstitution of kidney structures. The implanted renal cells self-assembled into (a) glomerular and (b) tubular structures within the kidney tissue.

proposed to provide a more controlled assessment of renal tissue *in vivo* (Joraku et al., 2005). A 3D collagen-based culture system was developed to facilitate the formation of 3D renal structures *in vitro*. After 1 week of growth, individual renal cells began to form renal structures resembling tubules and glomeruli. Histologically, these structures show phenotypic resemblance to native kidney structures. The reconstituted tubules stained positively for Tamm-Horsfall protein, which is expressed in the thick ascending limb of Henle's loop and distal convoluted tubules. This study shows that renal structures can be reconstituted in a 3D culture system, which may be used for renal cell therapy applications.

While many studies have been conducted to achieve improvements in total function, investigations have also been directed toward augmenting selective kidney functions. In one study, cultured renal cells were examined to determine whether EPO-producing cells derived from kidney tissue could be used for the treatment of renal failure-induced anemia (Aboushwareb et al., 2008). Renal cells from 7–10-day-old mice were culture expanded. The cells were characterized for EPO expression at each subculture stage. The levels of EPO expression were analyzed from renal cells incubated under normoxic and hypoxic conditions. The cultured renal cells expressed EPO at each subculture stage tested. This study indicates that EPO-producing cells may be used as a potential treatment option for anemia caused by chronic renal failure.

There has also been considerable progress in the development of stem cell-based therapies for renal failure. For example, the existence of non-tubular cells in adult mouse kidney that express stem cell antigen-1 (Sca-1) has been reported. This population of small cells includes a CD45-negative fraction that lacks hematopoietic stem cell lineage markers and resides in the renal interstitial space. In addition, these cells are enriched for β1-integrin, are cytokeratin negative, and show minimal expression of surface markers that typically are found on bone marrow-derived mesenchymal stem cells. Clonally derived lines can be differentiated into myogenic, osteogenic, adipogenic, and neural lineages. These renal Sca-1 cells were injected directly into the renal parenchyma of C57BL/6 wild-type mice, shortly after ischemic–reperfusion injury. After 1 month the injected cells had adopted a tubular phenotype and populated the renal tubule after ischemic injury. These adult kidney-derived cells may potentially contribute to kidney repair and may be important in the development of future regenerative medicine strategies (Dekel et al., 2003).

Another study demonstrated that parietal epithelial cells (PECs) in the Bowman's capsule exhibit coexpression of the stem cell markers CD24 and CD133 as well as expression of the stem cell-specific transcription factors Oct-4 and BmI-1. Lineage specific markers are absent in this population. This population, which was purified from cultured encapsulated glomeruli, revealed self-renewal potential and a high cloning efficiency. Under appropriate culture conditions, individual clones of $CD24^+CD133^+$ PEC could be induced to generate mature, functional tubular cells with phenotypic features of proximal and distal tubules, osteogenic cells, adipocytes, and neuronal cells. These cells were injected into severe combined immune-deficient (SCID) mice following rhabdomyolysis-induced acute renal failure. This treatment resulted in the regeneration of the tubular structures of different portions of the nephron, and it significantly ameliorated morphological and functional kidney damage. This study demonstrated the existence of resident multipotent progenitor cells in adult kidney and suggests a possible therapeutic use for these cells (Sagrinati et al., 2006).

CONCLUSION

Renal transplantation remains as the gold standard of treatment for end-stage renal failure. The increasing demand for donor organs has ignited tremendous efforts to seek alternative treatment modalities. Although it has been demonstrated that renal cells are able to reconstitute into functional kidney tissues *in vivo*, numerous challenges need to be addressed before these technologies can be moved to the clinic. Some of these challenges include the generation of a large tissue mass that would augment systemic renal function and become fully integrated into the host's vascular and excretory systems, and the development of a reliable renal failure model system for testing cell-based technologies. Research progress in the regeneration of kidney tissues has been somewhat successful in augmenting tissue function, but clinical application of this technology is still distant.

References

Aboushwareb, T., Egydio, F., Straker, L., Gyabaah, K., Atala, A., & Yoo, J. J. (2008). Erythropoietin producing cells for potential cell therapy. *World J. Urol., 26*, 295–300.

Amiel, G. E., & Atala, A. (1999). Current and future modalities for functional renal replacement. *Urol. Clin. North Am., 26*, 235–246, xi.

Amiel, G. E., Yoo, J. J., & Atala, A. (2000a). Renal therapy using tissue-engineered constructs and gene delivery. *World J. Urol., 18*, 71–79.

Amiel, G. E., Yoo, J. J., & Atala, A. (2000b). Renal tissue engineering using a collagen-based kidney matrix. *Tissue Eng., 6*(Suppl.), 685.

Atala, A. (1997). *Tissue Engineering in the Genitourinary System*. Boston: Birkhauser Press.

Atala, A. (1999). Future perspectives in reconstructive surgery using tissue engineering. *Urol. Clin. North Am., 26*, 157–165, ix–x.

Atala, A., Schlussel, R. N., & Retik, A. B. (1995). Renal cell growth *in vivo* after attachment to biodegradable polymer scaffolds. *J. Urol., 153*, 4.

Atala, A., Bauer, S. B., Soker, S., Yoo, J. J., & Retik, A. B. (2006). Tissue-engineered autologous bladders for patients needing cystoplasty. *Lancet, 367*, 1241–1246.

Carley, W. W., Milici, A. J., & Madri, J. A. (1988). Extracellular matrix specificity for the differentiation of capillary endothelial cells. *Exp. Cell Res., 178*, 426–434.

Chazan, J. A., Libbey, N. P., London, M. R., Pono, L., & Abuelo, J. G. (1991). The clinical spectrum of renal osteodystrophy in 57 chronic hemodialysis patients: a correlation between biochemical parameters and bone pathology findings. *Clin. Nephrol., 35*, 78–85.

Cohen, J., Hopkin, J., & Kurtz, J. (1994). *Infectious Complications After Renal Transplantation*. Philadelphia: W.B. Saunders.

Dekel, B., Burakova, T., Arditti, F. D., Reich-Zeliger, S., Milstein, O., Aviel-Ronen, S., et al. (2003). Human and porcine early kidney precursors as a new source for transplantation. *Nat. Med., 9*, 53–60.

El-Kassaby, A. W., Retik, A. B., Yoo, J. J., & Atala, A. (2003). Urethral stricture repair with an off-the-shelf collagen matrix. *J. Urol., 169*, 170–173, discussion 173.

Feldman, H. I., Kobrin, S., & Wasserstein, A. (1996). Hemodialysis vascular access morbidity. *J. Am. Soc. Nephrol., 7*, 523–535.

Folkman, J., & Hochberg, M. (1973). Self-regulation of growth in three dimensions. *J. Exp. Med., 138*, 745–753.

Frank, J., Engler-Blum, G., Rodemann, H. P., & Muller, G. A. (1993). Human renal tubular cells as a cytokine source: PDGF-B, GM-CSF and IL-6 mRNA expression *in vitro*. *Exp. Nephrol., 1*, 26–35.

Freed, L. E., Vunjak-Novakovic, G., Biron, R. J., Eagles, D. B., Lesnoy, D. C., Barlow, S. K., et al. (1994). Biodegradable polymer scaffolds for tissue engineering. *Biotechnology (NY), 12*, 689–693.

Fung, L. C. T., Elenius, K., Freeman, M., Donovan, M. J., & Atala, A. (1996). Reconstitution of poor EGFr-poor renal epithelial cells into tubular structures on biodegradable polymer scaffold. *Pediatrics, 98*(Suppl.), S631.

Gupta, S., Verfaillie, C., Chmielewski, D., Kim, Y., & Rosenberg, M. E. (2002). A role for extrarenal cells in the regeneration following acute renal failure. *Kidney Int., 62*, 1285–1290.

Horikoshi, S., Koide, H., & Shirai, T. (1988). Monoclonal antibodies against laminin A chain and B chain in the human and mouse kidneys. *Lab. Invest., 58*, 532–538.

Hubbell, J. A., Massia, S. P., Desai, N. P., & Drumheller, P. D. (1991). Endothelial cell-selective materials for tissue engineering in the vascular graft via a new receptor. *Biotechnology (NY), 9*, 568–572.

Humes, H. D., & Cieslinski, D. A. (1992). Interaction between growth factors and retinoic acid in the induction of kidney tubulogenesis in tissue culture. *Exp. Cell Res., 201*, 8–15.

Humes, H. D., Buffington, D. A., MacKay, S. M., Funke, A. J., & Weitzel, W. F. (1999). Replacement of renal function in uremic animals with a tissue-engineered kidney. *Nat. Biotechnol., 17*, 451–455.

Ikarashi, K., Li, B., Suwa, M., Kawamura, K., Morioka, T., Yao, J., et al. (2005). Bone marrow cells contribute to regeneration of damaged glomerular endothelial cells. *Kidney Int., 67,* 1925–1933.

Ito, T., Suzuki, A., Imai, E., Okabe, M., & Hori, M. (2001). Bone marrow is a reservoir of repopulating mesangial cells during glomerular remodeling. *J. Am. Soc. Nephrol., 12,* 2625–2635.

Iwano, M., Plieth, D., Danoff, T. M., Xue, C., Okada, H., & Neilson, E. G. (2002). Evidence that fibroblasts derive from epithelium during tissue fibrosis. *J. Clin. Invest., 110,* 341–350.

Joraku, A., Sullivan, C. A., Yoo, J. J., & Atala, A. (2005). Tissue engineering of functional salivary gland tissue. *Laryngoscope, 115,* 244–248.

Kale, S., Karihaloo, A., Clark, P. R., Kashgarian, M., Krause, D. S., & Cantley, L. G. (2003). Bone marrow stem cells contribute to repair of the ischemically injured renal tubule. *J. Clin. Invest., 112,* 42–49.

Lanza, R. P., Chung, H. Y., Yoo, J. J., Wettstein, P. J., Blackwell, C., Borson, N., et al. (2002). Generation of histocompatible tissues using nuclear transplantation. *Nat. Biotechnol., 20,* 689–696.

Lin, F., Moran, A., & Igarashi, P. (2005). Intrarenal cells, not bone marrow-derived cells, are the major source for regeneration in postischemic kidney. *J. Clin. Invest., 115,* 1756–1764.

Lin, F., Cordes, K., Li, L., Hood, L., Couser, W. G., Shankland, S. J., et al. (2003). Hematopoietic stem cells contribute to the regeneration of renal tubules after renal ischemia-reperfusion injury in mice. *J. Am. Soc. Nephrol., 14,* 1188–1199.

Maack, T. (1992). *Renal Handling of Proteins and Polypeptides.* New York: Oxford University Press.

Milici, A. J., Furie, M. B., & Carley, W. W. (1985). The formation of fenestrations and channels by capillary endothelium *in vitro. Proc. Natl. Acad. Sci. USA, 82,* 6181–6185.

Mooney, D. J., Mazzoni, C. L., Breuer, C., McNamara, K., Hern, D., Vacanti, J. P., et al. (1996). Stabilized polyglycolic acid fibre-based tubes for tissue engineering. *Biomaterials, 17,* 115–124.

Oberpenning, F., Meng, J., Yoo, J. J., & Atala, A. (1999). *De novo* reconstitution of a functional mammalian urinary bladder by tissue engineering. *Nat. Biotechnol., 17,* 149–155.

Poulsom, R., Forbes, S. J., Hodivala-Dilke, K., Ryan, E., Wyles, S., Navaratnarasah, S., et al. (2001). Bone marrow contributes to renal parenchymal turnover and regeneration. *J. Pathol., 195,* 229–235.

Prockop, D. J. (1997). Marrow stromal cells as stem cells for non-hematopoietic tissues. *Science, 276,* 71–74.

Rookmaaker, M. B., Smits, A. M., Tolboom, H., Van't Wout, K., Martens, A. C., Goldschmeding, R., et al. (2003). Bone-marrow-derived cells contribute to glomerular endothelial repair in experimental glomerulonephritis. *Am. J. Pathol., 163,* 553–562.

Sagrinati, C., Netti, G. S., Mazzinghi, B., Lazzeri, E., Liotta, F., Frosali, F., et al. (2006). Isolation and characterization of multipotent progenitor cells from the Bowman's capsule of adult human kidneys. *J. Am. Soc. Nephrol., 17,* 2443–2456.

Schena, F. P. (1998). Role of growth factors in acute renal failure. *Kidney Int. Suppl., 66,* S11–S15.

Stadnyk, A. W. (1994). Cytokine production by epithelial cells. *FASEB J., 8,* 1041–1047.

Tachibana, M., Nagamatsu, G. R., & Addonizio, J. C. (1985). Ureteral replacement using collagen sponge tube grafts. *J. Urol., 133,* 866–869.

Wald, H. L., Sarakinos, G., Lyman, M. D., Mikos, A. G., Vacanti, J. P., & Langer, R. (1993). Cell seeding in porous transplantation devices. *Biomaterials, 14,* 270–278.

Yokoo, T., Ohashi, T., Shen, J. S., Sakurai, K., Miyazaki, Y., Utsunomiya, Y., et al. (2005). Human mesenchymal stem cells in rodent whole-embryo culture are reprogrammed to contribute to kidney tissues. *Proc. Natl. Acad. Sci. USA, 102,* 3296–3300.

Yoo, J. J., Ashkar, S., & Atala, A. (1996). Creation of functional kidney structures with excretion of kidney-like fluid *in vivo. Pediatrics, 98S,* 605.

PART 5

Regulation and Ethics

CHAPTER 61

Ethical Considerations[*]

Ronald M. Green
Ethics Institute, Dartmouth College, Hanover, NH, USA

INTRODUCTION

Since the first development of human embryonic stem (hES) cells in 1998 (Shamblott et al., 1998; Thomson et al., 1998) there has been a consensus in the scientific community that pluripotent cells hold great promise for developing new treatments for a variety of serious and currently untreatable disease conditions (National Research Council, 2001; Office of Science Policy, 2001). However, because one of the leading forms of stem cell research involves the manipulation and destruction of human embryos, the field has also been a focus of ethical controversy and opposition. In the course of these debates, several challenging ethical questions have been raised. Scientists, clinicians, or patients using hES cells or therapies must formulate their answers to these questions. Society, too, must address them to determine the extent to which hES cell research may require oversight and regulation. This chapter presents these questions, and critically examines some of the leading answers that have been proposed to them on all sides of the controversy.

1117

IS IT NECESSARY TO USE HUMAN EMBRYOS?

In late 2007, teams led by Yamanaka in Japan and Thomson in the USA announced success in the use of gene transfer technology to produce induced pluripotent stem (iPS) cell lines (Takahashi et al., 2007; Yu et al., 2007). Some believe that this development ends debate about the appropriateness of using hES cell lines (Krauthammer, 2007). If we can directly manipulate somatic cells and perhaps even produce patient-specific (autologous) stem cells, why should we engage in the ethically controversial activity of manipulating and destroying human embryos? However, there are scientific and ethical considerations that challenge this view.

Scientifically, there is a question of whether iPS cell lines will prove suitable for use in human transplant and cell regeneration therapies. Current iPS cell lines exhibit high rates of tumorigenicity in mice, possibly a result of the use of retroviral vectors to carry pluripotency-inducing transcription factors, including the cancer-related factor c-Myc (Hyun et al., 2007). Active research is underway on replacement of c-Myc and the direct upregulation of the relevant genes (Okita et al., 2008; Yu et al., 2009). This may eventually overcome the problem, but until this is accomplished, iPS cell lines cannot ethically be used for human transplant purposes. However, they may be of value for the creation of model lines of disease-related cells.

Recent reports further suggest that the genomic reprogramming process in iPS cells may be less complete than that which takes place in the fertilized egg, causing the resulting cells to exhibit abnormal expansion and early senescence (Feng et al., 2010). These problems, too, may

---

[*]This chapter is a substantial revision and updating of the chapter "Ethical Considerations" that appeared in Lanza et al. (2009).

Principles of Regenerative Medicine. DOI: 10.1016/B978-0-12-381422-7.10061-6

eventually be overcome, but doing so will require a better understanding of the reprogramming process in embryonic cells. For this reason alone, the use of human embryos is likely to continue to be an important feature of stem cell research (Gurdon and Melton, 2008). At this time, there exist many useful hES cell lines created from donated human embryos. There are currently hundreds of thousands of frozen embryos remaining from infertility procedures that will likely be destroyed and that could be used for research (Weiss, 2003). If stem cell research involving embryos were halted, this vast resource would go to waste. Many people feel that it is unwise to foreclose any of the available paths to developing pluripotent cells for regenerative medicine research. In the words of the National Research Council, "The application of stem cell research to therapies for human disease will require much more knowledge about the biological properties of all types of stem cells" (National Research Council, 2001).

Also questionable is the widespread assumption that inducing pluripotency in somatic cells raises no ethical questions. It is true that protecting every body cell from gene manipulation would seem absurd. But is this so if the genetic manipulation of a cell renders it capable of eventual development into a human being? The work of Nagy and others shows that ES cells, when inserted into tetraploid embryos, are able to develop the placental material needed for further development (Nagy et al., 1990, 1993; Magill and Neaves, 2009). This possibility raises complex ethical questions. Is potentiality morally relevant if it is accompanied by such intensive technical interventions? *In vitro* fertilization (IVF) or cloned embryos also require intensive technical interventions. Those who favor iPS cell research but oppose the use of IVF or cloned embryos as sources of pluripotent stem cells will have to examine the consistency of their views. If it is acknowledged that the potentiality for full human development, however assisted, confers moral status on an entity, then iPS research may not be the ethical panacea that hES cell research opponents believe it to be.

Finally, it might be asked whether, even within the framework of hES cell research, we must destroy embryos. In 2005, the Bush administration's President's Council on Bioethics issued a White Paper encouraging research in alternate methods of hES cell derivation, including the use of arrested or developmentally non-viable embryos (President's Council on Bioethics, 2005). Several of these proposals, such as the deliberate genetic manipulation of embryos to prevent their normal development, are not free of ethical controversy (Green, 2007). A method involving single-cell blastomere biopsy has also been developed and implemented that could obviate the need to destroy the embryo in order to develop an hES cell line (Klimanskaya et al., 2006). This approach raises questions of risk to the manipulated embryos. It can only be clearly justified in the context of preimplantation genetic diagnosis, where cells are routinely removed for the diagnostic procedure. Hence, there is no easy route around the use of human embryos in pluripotent stem cell research.

IS IT MORALLY PERMISSIBLE TO DESTROY A HUMAN EMBRYO?

Human embryonic stem cell lines are usually made by chemically and physically disaggregating an early, blastocyst stage embryo, and removing its inner cell mass. At this stage, the embryo is composed of approximately 200 cells, including an outer layer of differentiated placental material, and the undifferentiated (pluripotent) cells of the inner cell mass. The embryo dies as a result of this procedure. New methods that permit the development of pluripotent cell lines without destroying an embryo have not yet replaced this standard method for developing hES cells, and most existing hES cell lines were created this way. Hence the question remains: may we intentionally kill a developing human being to expand scientific knowledge, and potentially provide medical benefits?

At one end of the spectrum are those who believe that, in moral terms, human life begins at conception when a new, self-developing genome comes into being. For those holding this view, the early embryo is not morally different from a child or an adult human being. It cannot be used in research that is not to its benefit, and it cannot be used without its consent

(Doerflinger, 1999; Linacre Centre for Healthcare Ethics, 2001; Pope John Paul II, 2003). Furthermore, proxy consent by parents in such cases is inadmissible, since it is an accepted rule of pediatric research that parents may not volunteer a child for risky studies that are not potentially beneficial to the child and that endanger the child's life. Many Roman Catholics, evangelical Protestants, and some Orthodox Jews take the position that life (morally) begins at conception, and they oppose hES cell research.

At the other end of the spectrum are those who believe that the embryo is not yet fully a human being in a moral sense. They hold a "developmental" or "gradualist" view of life's beginning (Ford, 1988; McCormick, 1991; Warren, 1997; Green, 2001a; Shannon, 2001). They do not deny that the early embryo is alive, and has the biological potential to become a person. Nevertheless, they believe that other features are needed for the full and equal protection we normally accord children and adults, and that these features only develop gradually across the full term of gestation. These features include such things as bodily form, and the ability to feel or think. Since it lacks organs, the early embryo cannot have these features or abilities. They also note that the very early embryo lacks human individuality, since it can still undergo twinning at this early stage, and two separate embryos with distinct genomes can fuse to become a single individual (Strain et al., 1998). Some dismiss this argument, maintaining that the possibility of twinning or fusion does not reduce the genetic uniqueness or moral status of the earliest developing cells (Guenin, 2008), but it is hard to see how the claim that "a person begins at conception" can easily withstand these biological facts. Finally, the very high mortality rate of such embryos (most never implant) reduces the force of the argument from potentiality (Hardy et al., 2001; Norwitz et al., 2001). Those who hold this developmental view do not agree on the classes of research that warrant the destruction of embryos, but most support some form of hES cell research. Their reasoning is that, although the early embryo merits some respect as a nascent form of human life, the lives and health of children and adults outweigh whatever claim it possesses (Lebacqz, 2001).

Each individual faced with involvement in hES cell research must arrive at his or her own answer to this first question. Legislators and others must also wrestle with these issues. Because American law (and the laws of most other nations) does not regard the early embryo as a person meriting the legal protections afforded to children and adults, it is hard to see how one can justify legal or regulatory prohibitions on privately financed hES cell research or clinical applications. Such prohibitions would interfere with individual liberty on grounds inconsistent with the reduced status of the embryo shown elsewhere in the law. However, because publicly funded research rests on more narrowly political considerations, including the way that a majority chooses to spend public funds, it may be expected that public support for hES cell research will depend on how a majority of citizens answer the question of the embryo's status. In the USA, the Republican administration of President George W. Bush prohibited in federally funded stem cell research the use of all but a small number of pre-existing hES cell lines. On March 9, 2009, President Barack Obama, a Democrat, issued an executive order reversing this stance and permitting the National Institutes of Health (NIH) to fund research on new hES cell lines derived from embryos remaining from infertility procedures.

MAY ONE BENEFIT FROM *OTHERS*' DESTRUCTION OF EMBRYOS?

If, as some maintain, the human embryo is a morally protectable entity that cannot be intentionally destroyed even for others' benefit, can one justify the use of a cell line produced from its destruction? That is, can downstream researchers, clinicians, or patients who value the embryo this much justify their *use* of stem cell lines that others have derived? This raises the more basic question of whether we can ever benefit from deeds with which we disagree morally, or regard as morally wrong (Kaveny, 2000; Green, 2002). It is also the question of when a connection with wrongdoing becomes complicity with it (Birnbacher, 2008).

1119

Why is it morally wrong to benefit from others' misconduct? One answer is that by doing so, we may encourage similar deeds in the future. This is most apparent in cases where our conduct directly instigates wrongdoing, such as when we receive stolen goods (Birnbacher, 2008). However, it seems less objectionable to benefit from others' wrongdoing when their deeds are independently undertaken and not in any way prompted or encouraged by us. For example, few would object to using the organs from a young victim of a gang killing to save the life of another dying child. The use of organs benefits one person and in no way encourages teen violence. Can similar logic apply to stem cell research using spare embryos remaining from infertility procedures?

It helps to remember that most embryos used to produce hES cell lines are left over from infertility procedures. Couples using IVF routinely create more embryos than can safely be implanted. There are hundreds of thousands of these embryos in cryogenic freezers in the USA and around the world (Weiss, 2003). Despite strenuous efforts, including some supported by the US government, few frozen embryos are adopted (Stolberg, 2001b; Office of Population Affairs, 2010). Most of these supernumerary embryos will be destroyed. In 1996, British law mandated the destruction of 3,600 such embryos (Ibrahim, 1996). This will continue regardless of whether some embryos are diverted to hES cell research. Does using hES cell lines made from these embryos encourage either the creation or destruction of such embryos?

Although a downstream researcher, clinician, or patient may abhor the deeds that led to the existence of an hES cell line, including the creation and destruction of excess human embryos in infertility medicine, nothing that a recipient of an hES cell line chooses to do is likely to alter, prevent, or discourage the continuing creation or destruction of human embryos, or make the existing lines go away. Those who use such embryos may also believe that if they refuse to use an hES cell line, they would forgo great therapeutic benefit. People in this position will struggle with the question of whether it is worthwhile to uphold a moral ideal when doing so has no practical effect, and when it threatens harm to others. However, anyone who believes that hES cell research is unlikely to be of therapeutic value will not experience a moral quandary in choosing to forgo use of hES cells (Birnbacher, 2008).

Religious views on the question of whether one may ever benefit from others' wrongdoing are diverse. The Roman Catholic moral tradition, despite its staunch opposition to complicity with wrongdoing, presents different answers, including some that permit one to derive benefit in particular cases (Miller, 1967; Ashley and O'Rourke, 1989; Green, 2002). This suggests that some researchers, clinicians, or patients who morally oppose the destruction of human embryos on religious or other grounds might ethically use hES cell lines derived from embryos otherwise slated for destruction. One critical question for those facing such a moral quandary, even when they agree that no encouragement to wrongdoing is involved, is whether they can personally tolerate dissonance between their competing moral commitments (Birnbacher, 2008).

It is noteworthy that in an August 2001 address to the nation, President George W. Bush adopted a version of the position that allows one to benefit from acts one morally opposes. Stating his belief that it is morally wrong to kill a human embryo for others' benefit, the president nevertheless permitted the use of existing stem cell lines on the grounds that the deaths of the embryos had already occurred (Bush, 2001). Similarly, Germany permits the importing and use of hES cell lines created before January 1, 2002, the date on which the Bundestag passed a law governing such matters (this date was subsequently moved to May 2007). These initiatives reflect the belief that it does not encourage further destruction of embryos to permit the use of previously generated cell lines. President Bush did not go so far as to permit the use of lines that could in the future be derived from embryos certainly slated for destruction because their progenitors choose never to use them. However, in July 2009, the NIH did just this. Implementing an Executive Order issued in March by President Obama, the NIH authorized funding for the use, but not the derivation of new hES cell lines from embryos remaining from infertility procedures (National Institutes of Health, 2009). Opponents of hES cell research have criticized this derivation-versus-use distinction as morally problematic. Apart from the moral issues, if the NIH wishes to fund hES cell research, this distinction was legally necessary because of the Dickey-Wicker amendment, which prohibits federal funding for research in which a human embryo is destroyed (Omnibus Appropriations Act, 2009). However, a ruling in August 2010 by Chief Judge Royce C. Lamberth of Federal District Court for the District of Columbia has thrown open to question whether federally funded hES cell research can proceed unless Dickey-wicker itself is changed or removed.

In summary, the downstream use of stem cells derived from embryos remains a source of moral controversy and disagreement. Researchers and clinicians who oppose embryo destruction will have to examine their own conscience in the light of the above considerations to determine how much they wish to associate themselves with the use of hES cell lines created in ways to which they morally object.

MAY WE CREATE AN EMBRYO IN ORDER TO DESTROY IT?

A fourth question takes us into even more controversial territory. Is it ever morally permissible to create an embryo deliberately to produce a stem cell line? This was done in the summer of 2002 at the Jones Institute in Norfolk, VA (Kolata, 2001). Those in favor of this research defend it on two grounds. First, they say that in the future, if we seek to develop stem cell lines with special properties, and perhaps closer genetic matches to tissue recipients, it may be necessary to produce stem cell lines to order using donor sperm and eggs. Second, they argue that it is ethically better to use an hES cell line created from embryos that have been produced for just this purpose, with the full and informed consent of their donor progenitors, than to use cell lines from embryos originally created for a different, reproductive purpose.

Those who believe the early embryo is our moral equal oppose the deliberate creation of embryos for research or clinical use. They are joined by some who do not share this view of the embryo's status, but who believe that it is morally repellant to deliberately create a potential human being only to destroy it (Annas et al., 1996). They argue that this research opens the way to the "instrumentalization" of all human life and the use of children or adult human beings as commodities. Some ask whether such research does not violate the Kantian principle that we should never use others as "a means only" (Green, 2001b).

On the other side of this debate are those who believe that the lesser moral status of the early embryo permits its creation and destruction for lifesaving research and therapies (National Institutes of Health, 1994; Davis, 1995). The proponents of this research direction ask why it is morally permissible to create supernumerary embryos in IVF procedures to help couples have children, but morally wrong to do the same thing to save a child's life. They are not persuaded by the reply that the status of the embryo is affected by its progenitors' intent, and that it is therefore permissible to create excess embryos for a "good" (reproductive) purpose, but not for a "bad" (research) purpose. They point out that the embryo is the same entity, and its status should not depend on its progenitors' intentions. We do not ordinarily believe that a child's rights are dependent on its parents' intent or degree of concern (Parens, 2001). They conclude that it is not parental intent that warrants the creation of excess of embryos in such cases, but the embryo's lesser moral status and the likelihood of significant human benefit from its use. These same considerations, they believe, justify deliberately creating embryos for stem cell research.

MAY WE CLONE HUMAN EMBRYOS?

This question arises in connection with a specific stem cell technology known as "human therapeutic cloning." It involves the deliberate creation of an embryo by somatic cell nuclear transfer technology (cloning) to produce an immunologically compatible (isogenic) hES cell line (Lanza et al., 2000).

1121

Immune rejection could occur if the embryo used to prepare a line of hES cells for transplant does not share the same genome as the recipient. This would be the case whether the cell line was created from a spare embryo, or from one made to order. Therapeutic cloning offers a way around this problem. In the case of a diabetic child, the mother could donate an egg whose nucleus would then be removed. A cell would be taken from the child's body, and its nucleus inserted into the egg cytoplasm. With stimulation, the reconstructed cell would divide, just like a fertilized egg. If the resulting embryo were transferred back to a womb, it could go on to birth and become a new individual — a clone of the child. But in therapeutic cloning, the blastocyst would be dissected and an hES cell line prepared. Growth factors could be administered to induce the cells to become replacement pancreatic cells for the child. Because these cells contain the child's own DNA, and even the same maternal mitochondrial DNA, they would not be subject to rejection. Research has shown that the small amount of alien RNA from the mitochondria of an egg other than the mother's would probably not provoke an immune response (Lanza et al., 2002).

Although this is a very promising technology, it raises a host of novel ethical questions. One is whether the embryonic organism produced in this way should be regarded as a "human embryo" in the accepted sense of that term (Nature Editorial, 2001). Those who believe that "life begins at conception" tend to answer this question affirmatively, even though cloned "embryos" are not the result of sexual fertilization (Doerflinger, 2003). They believe that it is no more permissible to create and destroy a cloned embryo than to do so with one produced by sexual fertilization. They base their view on the biological similarities between cloned and sexually produced embryos, and on the argument that both have the potential to become a human being. However, the very high mortality rate of cloned embryos suggests significant biological differences from sexually produced embryos. Furthermore, if the embryo's status rests on its potential, in an era of cloning, some degree of potentiality attaches to all bodily cells, which no one would argue should be withheld from use in biomedical research or therapy.

Therapeutic cloning implies the availability of human oocytes and raises the special question of ovulation induction. This is an invasive medical procedure, with both known and undetermined risks (Paulson, 1996; Rossing et al., 1996). Not only must egg donors be informed of these risks, but steps also must be taken to preserve the voluntary nature of their consent. This includes avoiding financial incentives that create an "undue influence" on donors' judgment. It also includes preventing them from being pressured into producing excess eggs or embryos for research in return for discounts on infertility services (Cohen, 2001).

Fears about coercion and the exploitation of poorer women or women of color have led some to oppose paid egg donation for research (Dresser, 2001; National Research Council, 2005). Several states, including California and Massachusetts, have passed laws prohibiting this practice. Nevertheless, experience has shown that women will not donate eggs for either reproductive or research purposes without adequate compensation (Klitzman and Sauer, 2009). In the face of these problems, New York State recently reversed the legal trend and approved payment for research egg donation.

Those who defend it point out that payment to research subjects involved in risky research is a common practice. They also ask why it is permissible to pay reproductive but not research egg donors (Crockin, 2010). The Ethics Committee of the American Society for Reproductive Medicine has supported payment for research egg donors. It conceptualizes this in terms of appropriate compensation for a donor's time, inconvenience, and discomfort, and not for the eggs themselves (Ethics Committee of the American Society for Reproductive Medicine, 2007).

Finally, there is a moral question specific to cloning itself. The more scientists are able to perfect therapeutic cloning, the more likely it is that they will sharpen the skills needed to accomplish reproductive cloning, which aims at the birth of a cloned child. There is a broad consensus in the scientific and bioethics communities that, for the foreseeable future, the state of cloning technology poses serious health risks to any child born as a result of it (Jaenisch and Wilmut, 2001). There are also serious, unresolved questions about the psychological welfare of such children (National Bioethics Advisory Commission, 1997). Finally, there is the possibility that embryos created for therapeutic cloning research might be diverted to reproductive cloning attempts. All these concerns raise the question: do we really want to develop cloning technology for the production of isogenic stem cells, if doing so hastens the advent of reproductive cloning (Weiss, 1999)?

In 2001, and again in 2003, the US House of Representatives answered "no" to this question, and passed a bill introduced by Rep. James Weldon that banned both reproductive and therapeutic cloning. Similar bills have been introduced in the Senate (Stolberg, 2001a, 2003). Although the Senate initiatives have stalled for some time, this may someday change (*New York Times*, 2003). If a Senate bill passes, therapeutic cloning research and therapies will be outlawed in the USA. Similar prohibitions are either in effect or being considered for passage in continental Europe (*BBC News*, 2006). This would leave only a relatively small number of countries, including Great Britain, Israel, and China, in which such research would be allowed (Israel Academy of Sciences and Humanities, 2001).

Those who oppose these prohibitions believe that therapeutic and reproductive cloning research can be decoupled. They observe that strict regulations and governmental oversight, of the sort provided in Great Britain by the Human Fertilisation and Embryology Authority, make it unlikely that embryos produced for therapeutic cloning will be diverted to reproductive purposes. They also point out that several researchers or groups with minimal qualifications in cloning research have announced their intent to clone a child, or have even tried to do so. Such attempts are likely to continue, regardless of whether therapeutic cloning research is banned. As a result, a ban on therapeutic cloning will not protect children, and will only have the negative effect of interrupting beneficial stem cell research.

WHAT ETHICAL GUIDELINES SHOULD APPLY TO THE USE OF HUMAN EMBRYONIC STEM CELLS OR INDUCED PLURIPOTENT STEM CELLS IN THE LABORATORY?

Mention of the need to prevent the diversion of cloned embryos to reproductive purposes raises the larger question of what guidelines should apply to the conduct of hES cell and iPS research in the laboratory. In August 2000, the US NIH released a series of guidelines for hES cell research that never went into effect, because they were largely pre-empted by President Bush's decision to limit hES cell research to existing cell lines (National Institutes of Health, 2000). Guidelines have also been developed by the Chief Medical Officer's Expert Group in Great Britain (Chief Medical Officer's Expert Group, 2000; Holland, 2001), by private ethics boards at the Geron Corporation (Geron Advisory Board, 1999) and Advanced Cell Technology (Green et al., 2002) in the USA, by committees of the National Research Council (Committee on Guidelines for Human Embryonic Stem Cell Research, National Research Council, 2005; National Research Council and Institute of Medicine of the National Academies, 2007, 2008), and by the International Society for Stem Cell Research (ISSCR, 2007). These various recommendations share several features.

SCRO review

First, there is agreement that, in addition to the usual scientific and review of research proposals by an Institutional Review Board (IRB), it is appropriate that there be another layer of review provided by a Stem Cell Research Oversight (SCRO) committee. SCROs are needed to review protocols involving special issues such as research that generates chimeric animals using pluripotent or multipotent human stem cells.

extensive manipulations, such as genetic alterations. The same is true of autologous versus allogenic use, and use for homologous versus non-homologous functions (Recommendations 8 and 9). Good manufacturing practice (GMP) procedures must be applied to manipulated products and for those destined for allogeneic use. Donors of cells should be screened for infectious diseases and the donor should give written informed consent that covers the likely storage, future manipulations, analyses, and uses of their cells, their commercial potential, and possible risks to the donor's privacy, including exposure of genetic information (Recommendation 3).

2. Preclinical studies are meant to provide evidence of product safety and proof-of-principle of therapeutic effect. This normally requires sufficient studies in animal models, including larger animals where structural tissue needs to be tested in a load-bearing model (Recommendations 11 and 14). Researchers must develop and implement a clear plan to assess the risks of tumorigenicity for any cell product (Recommendation 18). Cell cultures and animal models should be used to test the interaction of cells with drugs to which recipients will be exposed, including the immunosuppressants planned for recipients (Recommendation 19).

3. Clinical trials of stem-cell research must conform to internationally accepted principles governing the protection of human subjects, including regulatory oversight, peer review by an expert panel independent of the investigators and sponsors, fair subject selection, informed consent, and patient monitoring. In addition, there are special issues raised by stem cell research. Because of their undetermined risks stem cell-based therapies, as a rule, must aim at being clinically competitive or superior to existing therapies. Where there are already efficacious therapies for a disease condition, the stem cell-based intervention must be of low risk and offer a potential advantage (such as better functional outcome; single procedure versus life-long drug therapy). Greater risks are permissible where there is no efficacious therapy and where the disease condition is severely disabling and life threatening (Recommendation 25). Patients need to be informed that cell-derived products have never been tested before in humans, that researchers do not know whether they will work as hoped, and that, unlike many pharmacological products or even many implantable devices, stem cells may stay in the body and generate adverse effects for the lifetime of the patient. To respect their values, and because some subjects may have moral objections to the use of embryo-derived cells, subjects should be informed about the source of the cells (Recommendation 28). In order to determine the consequences of cellular implantation, and with consideration of cultural and familial sensitivities, subjects should be asked for autopsy in the event of death (Recommendation 31).

4. Speaking for the ISSCR, the task force condemned the widely prevalent practice of marketing unproven stem cell interventions. Yet it recognized a difference between "the commercial purveyance of unproven stem cell interventions and legitimate attempts at medical innovation outside the context of a formal clinical trial" (ISSCR, 2008). It would permit the latter in "exceptional circumstances" and only for "seriously ill patients who lack good medical alternatives." Such unproven stem-cell interventions require a written plan explaining, among other things, the intervention's scientific rationale, clinical justification, a description of procedures, and plans for follow-up, data collection, identification of adverse effects, and assessment of efficacy. An independent panel of experts should review this written plan, and the clinician's institution should be informed and supportive. The patient should be informed of the unproven nature of the intervention, there should be an action plan for handling adverse events, and provision should be made for insurance coverage or treatment of complications arising from the procedure (Recommendation 34). Clinicians performing such interventions should commit themselves to communication of outcomes to the scientific community and moving to a formal clinical trial following experience with at most a few patients (Recommendation 34).

1126

5. Considerations of social justice apply to all research involving human subjects but receive special importance in view of public involvement in this emergent research area. Among the recommendations of the task force are public engagement in the policy making of governmental agencies (Recommendation 37), fair allocation of benefits and risks, and the need to develop alternative models of intellectual property, licensing, product development, and public funding to promote fair and broad access to the new diagnostics and therapies (Recommendation 38). One justice consideration especially pertinent to stem cell research is the establishment of stem cell collections with genetically diverse sources of cell lines (Recommendation 38). The task force concludes its recommendations with an aspirational goal that companies, subject to their financial capability, should offer affordable therapeutic interventions to persons living in resource-poor countries who would not otherwise have access to these therapies. Universities and other institutions are asked to incorporate this requirement in their intellectual property licenses (Recommendation 39).

CONCLUSION

Fully answering all of the questions this chapter has raised would require an ethical treatise. However, in my view, the moral claims of the very early embryo do not outweigh those of children and adults that can be helped by hES cell and therapeutic cloning technologies. While iPS cell research is promising, reliance on it should not, in my view, replace the use of hES cell lines produced through proven techniques, and that are currently available. I also do not believe that therapeutic cloning research will lead to reproductive cloning, which we should take firm steps to forbid. As we move forward to clinical translational research, care should be taken that there is proper oversight for these novel and potentially harmful therapies. Some may disagree with these conclusions. Continuing dialogue about these questions, and clearer scientific research results, will bring us closer to a national consensus on these issues.

References

Annas, G., Caplan, A., & Elias, S. (1996). The politics of human-embryo research — avoiding ethical gridlock. *N. Engl. J. Med.*, 334, 1329—1332.

Ashley, B. M., & O'Rourke, K. D. (1989). *Health Care Ethics: A Theological Analysis* (3rd ed.). St Louis, MO: Catholic Health Care Association of the United States.

BBC News (2006). MEPs vote against stem cell research, 10 April. Available at: http://news.bbc.co.uk/1/hi/health/2932421.stm

Birnbacher, D. (2008). Embryonic stem cell research and the argument from complicity. *Reprod. BioMed Online, 18* (Suppl. 1), 12—16, Available at: www.rbmonline.com/Article/3605

Bonnicksen, A. (2009). *Chimeras, Hybrids, and Interspecies Research: Politics and Policymaking.* Georgetown: Georgetown University Press.

Bush, G. W. (2001). President's statement on funding stem cell research. *NY Times,* 10 August (late edn — final), A16.

Chief Medical Officer's Expert Group (2000). Stem Cell Research: Medical Progress with Responsibility. Available at: http://www. dh.gov.uk/en/Publicationsandstatistics/Publications/PublicationsPolic yAndGuidance/DH_4065084

Cohen, C. B. (2001). Leaps and boundaries: expanding oversight of human stem cell research. In S. Holland et al. (Eds.), *The Human Embryonic Stem Cell Debate* (pp. 209—222). Cambridge, MA: MIT Press.

Crockin, S. L. (2010). A legal defense for compensating research egg donors. *Cell Stem Cell, 6,* 99—102.

Davis, D. S. (1995). Embryos created for research purposes. *Kennedy Inst. Ethics J., 5,* 343—354.

Doerflinger, R. (1999). The ethics of funding embryonic stem cell research: a Catholic viewpoint. *Kennedy Inst. Ethics J., 9,* 137—150.

Doerflinger, R.M. (2003). Testimony of Richard M. Doerflinger on behalf of the Committee for Pro-Life Activities. National Conference of Catholic Bishops, Testimony of Richard Doerflinger, January 14, 2003 before the Health and Human Development Committee of the Delaware House of Representatives concerning Senate Bill No. 55. Cloning Prohibition and Research Protection Act. Available at: http:// www. cloninginformation.org/congressional_testimony/doerflinger_de.htm

Dresser, R. (2001). Letter to the editor. *JAMA, 285,* 1439.

Ethics Committee of The American Society for Reproductive Medicine (ASRM). (2007). Financial compensation of oocyte donors. *Fertil. Steril., 88,* 305—309.

Feng, Q., Lu, S., Klimanskaya, I., Gomes, I., Kim, D., Chung, Y., et al. (2010). Hemangioblastic derivatives from human induced pluripotent stem cells exhibit limited expansion and early senescence. *Stem Cells, 28,* 704—712.

Ford, N. M. (1988). *When Did I Begin?* Cambridge: Cambridge University Press.

Geron Advisory Board (1999). Research with human embryonic stem cells: ethical considerations. *Hastings Cent. Rep., 29,* 31—36.

Green, R. M. (2001a). *The Human Embryo Research Debates.* New York: Oxford University Press.

Green, R. M. (2001b). What does it mean to use someone as "a means only"? rereading Kant. *Kennedy Inst. Ethics J., 11,* 249—263.

Green, R. M. (2002). Benefiting from "evil": an incipient moral problem in human stem cell research. *Bioethics, 16,* 544—556.

Green, R. M. (2007). Policy forum: can we develop ethically universal embryonic stem lines? *Nature Rev. Genet., 8,* 480—485.

Green, R. M., Olsen DeVries, K., Bernstein, J., Goodman., K. W., Kaufmann, R. W., Kiessling, A. A., et al. (2002). Overseeing research on therapeutic cloning: a private ethics board responds to its critics. *Hastings Cent. Rep., 32,* 27—33.

Guenin, L. M. (2008). Ethical considerations. In A. Atala et al. (Eds.), *Principles of Regenerative Medicine* (1st ed.). (pp. 1334—1344). Burlington, MA: Academic Press.

Gurdon, J. B., & Melton, D. A. (2008). Nuclear reprogramming in cells. *Science, 322,* 1811—1822.

Hardy, K., Spanos, S., Becker, D., Iannelli, P., Winston., R. M. L., & Stark, J. (2001). From cell death to embryo arrest: mathematical models of human preimplantation embryo development. *Proc. Natl. Acad. Sci. USA, 98,* 1655—1660.

Holland, S. (2001). Beyond the embryo: a feminist appraisal of human embryonic stem cell research. In S. Holland et al. (Eds.), *The Human Embryonic Stem Cell Debate* (pp. 73—86). Cambridge, MA: MIT Press.

Hyder, N. (2009). Geron issues statement on halted stem cell trial. *Bionews.* 524 (6 September). Available at: http://www.bionews.org.uk/page_47650.asp

Hyun, I., Hochedlinger, K., Jaenisch, R., & Yamanaka, S. (2007). New advances in iPSC research do not obviate the need for human embryonic stem cells. *Cell Stem Cell, 1,* 367—368.

Ibrahim, Y. M. (1996). Ethical furor erupts in Britain: should embryos be destroyed? *NY Times,* 1 August (late edn), A1.

Israel Academy of Sciences and Humanities (2001). *Report of the Bioethics Advisory Committee of the Israel Academy of Sciences and Humanities on the Use of Embryonic Stem Cells for Therapeutic Research.* Jerusalem: Israel Academy of Sciences and Humanities.

ISSCR (International Society for Stem Cell Research) (2007). *Guidelines for the Conduct of Human Embryonic Stem Cell Research.* Available at: http://www.isscr.org/guidelines/index.htm

ISSCR (International Society for Stem Cell Research) (2008). *Guidelines for the Clinical Translation of Stem Cells,* December 3, 2008. Available at: http://www.isscr.org/clinical_trans/pdfs/ISSCRGLClinicalTrans.pdf

Jaenisch., R., & Wilmut, I. (2001). Don't clone humans! *Science, 291,* 2552.

Karpowicz, P., Cohen, C. B., & van der Kooy, D. (2005). Developing human—nonhuman chimeras in stem cell research: ethical issues and boundaries. *Kennedy Inst. Ethics J., 15,* 107—134.

Kaveny, M. C. (2000). Appropriation of evil: cooperation's minor image. *Theol. Stud., 61,* 280—313.

Klimanskaya, I., Chung, Y., Becker, S., Lu, S. J., & Lanza, R. (2006). Human embryonic stem cell lines derived from single blastomeres. *Nature, 444,* 481—485.

Klitzman, R., & Sauer, M. V. (2009). Payment of egg donors in stem cell research in the USA. *Reprod. Biomed. Online, 18,* 603—608.

Kolata, G. (2001). Researchers say embryos in labs are not available. *NY Times,* August 26 (late edn — final), A1.

Krauthammer, C. (2007). Stem cell vindication. *Wash. Post,* November 30 A23.

Lanza, R. P., Caplan, A. L., Silver, L. M., Cibelli, J. B., West, M. D., & Green, R. M. (2000). The ethical validity of using nuclear transfer in human transplantation. *JAMA, 284,* 3175—3179.

Lanza, R. P., Chung, H. Y., Yoo, J. J., Wettstein, P. J., Blackwell, C., Borson, N., et al. (2002). Generation of histo-compatible tissues using nuclear transplantation. *Nat. Biotechnol., 20,* 689—696.

Lanza., R., Gearhart, J., Hogan, B., & Melton, D. (2009). *Essentials of Stem Cell Biology* (2nd ed.). San Diego: Academic Press.

Lebacqz, K. (2001). On the elusive nature of respect. In S. Holland et al. (Eds.), *The Human Embryonic Stem Cell Debate* (pp. 149—162). Cambridge, MA: MIT Press.

Linacre Centre for Healthcare Ethics (2001). A Theologian's Brief on the Place of the Human Embryo within the Christian Tradition. Available at: http://www.lifeissues.net/writers/mis/mis_02christiantradition1. html#b1

Magill, G., & Neaves, W. B. (2009). Ontological and ethical implications of direct nuclear reprogramming. *Kennedy Inst. Ethics J., 19*, 23–32.

McCormick, R. A. (1991). Who or what is the preembryo? *Kennedy Inst. Ethics J., 1*, 1–15.

Miller, L. G. (1967). Scandal. In *New Catholic Encyclopedia, Vol. XII* (pp. 1112–1113). New York: McGraw-Hill.

Nagy, A., Gocza, E., Diaz, E. M., Prideaux, V. R., Ivanyi, E., Markkula, M., et al. (1990). Embryonic stem cells alone are able to support fetal development in the mouse. *Development, 110*, 815–821.

Nagy, A., Rossant, J., Nagy, R., Abramow-Newerly, W., & Roder, J. C. (1993). Derivation of completely cell culture-derived mice from early-passage embryonic stem cells. *Proc. Natl. Acad. Sci. USA, 90*, 8424–8428.

National Bioethics Advisory Commission (1997). *Cloning Human Beings*. Rockville, MD: National Bioethics Advisory Commission.

National Institutes of Health (1994). *Report of the Human Embryo Research Panel*. Bethesda, MD: National Institutes of Health.

National Institutes of Health (2000). *Guidelines for Research Using Human Pluripotent Stem Cells* (Effective August 25, 2000. 65 FR 5 1976) (Corrected November 21, 2000. 65 FR 69951). Bethesda, MD: National Institutes of Health.

National Institutes of Health (2009). Guidelines on Human Stem Cell Research. Available at: http://stemcells.nih. gov/policy/2009guidelines.htm

National Research Council (2001). *Stem Cells and the Future of Regenerative Medicine*. Washington, DC: National Academy Press.

National Research Council (2005). *Guidelines for Human Embryonic Stem Cell Research*. Washington, DC: National Academies Press.

National Research Council and Institute of Medicine of the National Academies (2007). *2007 Amendments to the National Academies' Guidelines for Human Embryonic Stem Cell Research*. Washington, DC: National Academies Press.

National Research Council and Institute of Medicine of the National Academies (2008). *2008 Amendments to the National Academies' Guidelines for Human Embryonic Stem Cell Research*. Washington, DC: National Academies Press.

Nature Editorial (2001). The meaning of life (editorial). *Nature, 412*, 255.

New York Times (2003). Cloning countdown (editorial). *NY Times*, March 1 (late edition – final), A18.

Norwitz, E. R., Shust, D. J., & Fisher, S. J. (2001). Implantation and the survival of early pregnancy. *N. Engl. J. Med., 345*, 1400–1408.

Okita, K., Nakagawa, M., Hyenjong, H., Ichisaka, T., & Yamanaka, S. (2008). Generation of mouse induced pluripotent stem cells without viral vectors. *Science, 322*, 949–953.

Office of Population Affairs, US Department of Health and Human Services (2010). Embryo Adoption. Available at: http://www.hhs.gov/opa/embryoadoption/index.html

Office of Science Policy (2001). *Stem Cells: Scientific Progress and Future Research Directions*. Bethesda, MD: National Institutes of Health.

Omnibus Appropriations Act (Dickey-Wicker Amendment) (2009). Public Law No. 111–8. Section 509 (a) (b). Available at: http://frwebgate.access.gpo.gov/cgi-bin/getdoc.cgi?dbname=111_cong_public_lawsanddocid=f: publ008.111

O'Rahilly, R., & Müller, F. (1992). *Human Embryology and Teratology*. New York: Wiley-Liss.

Parens, E. (2001). On the ethics and politics of embryonic stem cell research. In S. Holland et al. (Eds.), *The Human Embryonic Stem Cell Debate* (pp. 37–50). Cambridge, MA: MIT Press.

Paulson, R. J. (1996). Fertility drugs and ovarian epithelial cancer: is there a link? *J. Assist. Reprod. Genet., 13*, 751–756.

Pope John Paul II (2003). Address of John Paul II to the Members of the Pontifical Academy of Sciences, November 10, 2003. Available at: http://www.Vatican.va/holy_father/john_paul_ii/speeches/2003/November/documents/ hf_pii_spe~20011110 – academy-sciences_en.html

President's Council on Bioethics (PCBE). (2005). *White Paper: Alternative Sources of Human Pluripotent Stem Cells*. Washington, DC: President's Council on Bioethics. Available at: http://www.bioethics.gov/reports/

Rossing, M. A., Daling, J. R., Weiss, N. S., Moore, D. E., & Self, S. E. (1996). Ovarian tumors in a cohort of infertile women. *N. Engl. J. Med., 331*, 771–776.

Shamblott., M. J., Axelman, J., Wang, S., Bugg, E. M., Littlefield., J. W., Donovan, P. J., et al. (1998). Derivation of pluripotent stem cells from cultured human primordial germ cells. *Proc. Natl. Acad. Sci. USA, 95*, 13726–13731.

Shannon, T. (2001). From the micro to the macro. In S. Holland et al. (Eds.), *The Human Embryonic Stem Cell Debate* (pp. 177–184). Cambridge, MA: MIT Press.

Stolberg, S. G. (2001a). House backs ban on human cloning for any objective. *NY Times*, August 1, (late edn – final), A1.

Stolberg, S. G. (2001b). Some see new route to adoption in clinics full of frozen embryos. *NY Times*, February 25, A1.

Stolberg, S. G. (2003). House votes to ban all human cloning. *NY Times*, February 28, (late edn – final), A22.

Strain, L., Dean, J. C. S., Hamilton, M. P. R., & Bonthron, D. T. (1998). A true hermaphrodite chimera resulting from embryo amalgamation after *in vitro* fertilization. *N. Engl. J. Med.*, *338*, 166–169.

Sugarman, J., & Siegel, A. (2008). How to determine whether existing human embryonic stem cell lines can be used ethically. *Cell Stem Cell*, *3*, 238–239.

Takahashi, K., Tanabe, K., Ohnuki, M., Narita, M., Ichisaka, T., Tomoda, K., et al. (2007). Induction of pluripotent stem cells from adult human fibroblasts by defined factors. *Cell*, *131*, 861–872.

Thomson, J. A., Itskovitz-Eldor, J., Shapiro, S. S., Waknitz, M. A., Swiergiel, J. J., Marshall, V. S., et al. (1998). Embryonic stem cell lines derived from human blastocysts. *Science*, *282*, 1145–1147.

Warren, M. A. (1997). *Moral Status: Obligations to Persons and Other Living Things*. New York: Oxford University Press.

Weiss, R. (1999). Stem cell discovery grows into a debate. *Wash. Post*, October 9, A1, A8–A9.

Weiss, R. (2003). 400,000 human embryos frozen in US. Number at fertility clinics is far greater than previous estimates, survey finds. *Wash. Post*, May 8, A10.

Yu, J., Vodyanik, M. A., Smuga-Otto, K., Antosiewicz-Bourget, J., Frane, J. L., Tian, S., et al. (2007). Induced pluripotent stem cell lines derived from human somatic cells. *Science*, *318*, 1917–1920.

Yu, J., Hu, K., Smuga-Otto, K., Tian, S., Stewart, R., Slukvin, I. I., et al. (2009). Human induced pluripotent stem cells free of vector and transgene sequences. *Science*, *324*, 797–801.

CHAPTER 62

US Stem Cell Research Policy

Josephine Johnston
The Hastings Center, Garrison, NY, USA

INTRODUCTION

Since James A. Thomson and colleagues reported the isolation of pluripotent stem cells from human embryos in 1998 (Thomson et al., 1998), stem cell research has seldom been out of US headlines. Because isolating embryonic stem cells involves destroying embryos, some groups and individuals have opposed some or all of the research on moral grounds. This opposition has been translated into a number of policies and laws, including at a federal level. From 2001 until 2009, much embryonic stem cell research was ineligible for federal funding. In 2009, federal funding rules were relaxed somewhat, although research still may not use cells taken from embryos created for research purposes, including by cloning. Despite these restrictions, embryonic stem cell research has progressed in the USA using monies supplied by individual donors, charitable organizations, and states. After briefly discussing ethical and policy issues in adult and fetal stem cell research, this chapter will survey policy issues in embryonic stem cell research. I begin with a brief history of federal funding policies, which are compared to regulation strategies adopted in other nations active in the research, before considering oversight, donor, and commercialization issues that arise as the research moves forward.

SOURCES OF STEM CELLS

Stem cells are special kinds of cells that can regenerate themselves and make new, more specialized cells. For the purposes of this discussion, stem cells can be divided into three kinds based on the source of the cells: adult stem cells, fetal stem cells, and embryonic stem cells. In terms of ethics, politics, policy, and law, much depends on the source of the cells.

Adult stem cells are derived from the cells of adults and children. Although obtaining adult stem cells is not without its ethical issues, they are similar to those raised in other human subject research, the most important of which is the requirement for free and informed consent (President's Council on Bioethics (PCB), 2004). Because the ability of competent adults to consent to research enjoys wide acceptance, adult stem cell research has not been a major focus of ethical or political debate. It does, however, enter public consciousness as a less controversial alternative to embryonic stem cell research. For example, Catherine Verfaillie and colleagues announced in 2002 that they had isolated multipotent adult progenitor cells from bone marrow (Schwartz et al., 2002). Then, in 2007, scientists in Japan and the USA announced the creation of pluripotent stem cells from adult skin cells (Takahashi et al., 2007; Yu et al., 2007). These induced plurpipotent stem (iPS) cells are believed to be functionally very similar or identical to embryonic stem cells (for the purposes of this

Principles of Regenerative Medicine. DOI: 10.1016/B978-0-12-381422-7.10062-8

discussion, iPS cells are grouped with adult stem cells on account of their source). Opponents of embryonic stem cell research cite research like Takahashi's, Yu's and Verfaillie's as evidence that research using embryonic cells is unnecessary (Orr, 2002; US Conference of Catholic Bishops, 2009). However, adult stem cell scientists, including those using iPS cells, insist on the importance of pursuing research on both kinds of cell (Verfaillie et al., 2002; Hyun et al., 2007).

Pluripotent stem cells have also been extracted from the primordial reproductive tissue of aborted fetuses (Shamblott et al., 1998). Any source of stem cells that relies on women undergoing elective terminations is likely to be controversial in the USA simply because of the ongoing debate over the morality and legality of abortion. Nevertheless, researchers have used tissue from aborted fetuses since as early as the 1930s and federal funds are currently available for this kind of stem cell research.

It is unclear, however, whether pre-existing federal regulations and legislation apply to fetal stem cell research. Fetal research has been regulated in the USA since allegations of experiments on *in* and *ex utero* fetuses emerged in the early 1970s (National Bioethics Advisory Committee, 1999), around the same time as the Supreme Court's famous decision in the abortion case of Roe v. Wade (1973). In response to the allegations, human subject regulations were promulgated in 1975 requiring extra protections where federally funded research involves pregnant women, fetuses, and human *in vitro* fertilization. Among other things, these regulations forbid researchers from having any involvement in the woman's decision to terminate her pregnancy or from offering her financial inducements (45 CFR §46.204(h−i)). Although it has always been clear that these regulations applied to *in utero* fetal research, there has been some uncertainly about whether the regulations apply also to research that, like stem cell research, uses cadaveric fetal tissue (National Bioethics Advisory Committee, 1999). In March 2002, the Office for Human Research Protections issued guidance for research involving fetal stem cells in which it states that the research will only be subject to federal human subject protections where it involves "a living individual," which would seem to exclude most fetal stem cell research (Office of Human Research Protections, 2001).[1]

Likewise, the 1993 federal legislation on fetal tissue transplantation research likely does not apply to stem cell research using fetal cells unless the research also involves transplanting the cells into humans. Similar to the restrictions imposed by the federal human subject regulations, this legislation stipulates that no alteration in the timing, method, or procedures used to terminate the pregnancy be made solely for the purposes of obtaining the tissue (498A of the Public Health Service Act(42 USC 289g−l)). Even though both the federal regulations and the fetal tissue transplantation legislation likely do not apply to most fetal stem cell research, practices similar to those required by these laws will likely be conditions of Institutional Review Board (IRB) approval of much fetal stem cell research.

Various states also have laws affecting fetal stem cell research, which will apply to all researchers in those states regardless of their funding source. For example, six states ban research involving aborted fetuses[2] and some states ban paying for fetal remains.[3]

Despite fetal stem cell research's intimate connection with the controversial practice of induced abortion, this kind of stem cell research has seldom been the subject of public debate. Instead, debate has consistently focused on research that uses stem cells extracted from 4−7-day-old human embryos. Because any one cell in the very early human embryo can develop into a whole fetus, it is thought that embryonic stem cells will be able to repair damaged or diseased parts of the human body. The therapeutic potential of embryonic stem

[1] See also 45 CFR §46.206.
[2] Arizona, Indiana, Ohio, Oklahoma, South Dakota, and North Dakota.
[3] See, for example, Tennessee and Arkansas, as well as Section 10(a) and (b) of the Uniform Anatomical Gift Act.

cells, therefore, is thought to be enormous, but so is the moral peril. Extracting stem cells generally necessitates destroying the embryo.[4] For this reason, the research has been vigorously opposed by many individuals and groups, including (but not limited to) those who consider the early embryo to be a person or, if not a full person, an entity of sufficient moral significance that it should not be created for, or destroyed in, research. This opposition to destroying embryos in stem cell research has been reflected in the US policy on the federal funding of embryonic stem cell research.

EMBRYONIC STEM CELL RESEARCH: US LAW AND POLICY

At the time of writing, US regulation of embryonic stem cell research is limited to the federal funding policy contained in President Obama's March 9, 2009 Executive Order (Executive Order 13505, 2009), the National Institutes of Health Guidelines on Human Stem Cell Research (National Institutes of Health, 2009), and any applicable state laws (Andrews, 2004). The 2009 Executive Order and NIH (NIH) Guidelines establish the criteria for embryonic stem cell research that uses federal funds — they do not apply to research conducted using non-federal monies, such as might be supplied by an individual donor, a company, a charitable organization, or a state government. Nevertheless, because much basic biomedical research is traditionally funded by the federal government, federal funding policies have had a significant impact on embryonic stem cell research in the USA.

President Obama's Executive Order of March 2009 does little more than revoke the policy announced in 2001 by President George W. Bush, state that the NIH may fund research using embryonic stem cells, and request that NIH issue new guidance for such research. The particulars of the new federal funding policy are to be found in NIH's Guidelines on Human Stem Cell Research, announced 4 months later. Under these Guidelines, federal funding may be used to study embryonic stem cells derived from embryos "that were created using *in vitro* fertilization for reproductive purposes and were no longer needed for this purpose" (sometimes referred to as spare, surplus, or left-over embryos) and that were donated to research by the individuals who sought reproductive treatment. Donors must give their voluntary written consent for the embryos to be used in research, and documentation must be available to show, among other things, that the donors received no cash or in kind payment in exchange for making the donation and that there was a clear separation between the decision to create the embryos for reproductive purposes and the decision to donate them to research. The guidelines also establish a registry of embryonic stem cell lines that are eligible for use in federally funded research and describe the procedure for establishing eligibility for funding where cells are not already on the NIH registry. Federal funding is not available for research on cells taken from embryos that were created for research purposes, including by cloning (National Institutes of Health, 2009).

President Obama's 2009 Executive Order undid an extremely controversial 2001 policy made by President George W. Bush under which federal funds could only be used to study embry-onic stem cells that were already in existence as of 9p.m. eastern standard time on August 9, 2001. In addition, the cells must have been extracted from "spare" embryos that were donated with the informed consent of the donors, to whom no financial inducements were offered (The White House, 2001). To facilitate research on existing embryonic stem cells that met these criteria, the NIH set up a stem cell registry shortly after the policy was announced, listing embryonic stem cell lines eligible for use in federally funded research and available for shipping. In his 2001 address, the President stated that more than 60 embryonic stem cell lines

1133

---

[4] However, a method of extraction has been proposed where one cell is extracted by biopsy, leaving an embryo that could theoretically implant and develop just as a biopsied embryo does following preimplantation genetic diagnosis (Chung et al., 2006).

met his criteria, but for a variety of reasons only 22 lines were consistently listed in the registry as available for shipping.

The 2001 policy announced by President Bush superseded a policy announced by President Clinton a year earlier, in which he permitted the use of federal funds to study embryonic stem cells subject to a number of conditions, including that the cells had been extracted without using federal monies (PCB, 2004). That is, under President Clinton's policy, and under policies of Presidents Bush and Obama since, federal money could be used to study embryonic stem cells, but not to extract them from embryos (PCB, 2004). In opposing President Clinton's policy, some members of Congress complained that although it did not pay for embryo destruction, it nevertheless encouraged research that required the destruction of human embryos (PCB, 2004). The same critique could be applied today.

In formulating their respective policies, presidents Clinton, Bush, and Obama were bound by a longstanding policy against federal funding of any research involving the destruction of human embryos (PCB, 2004). Since 1996, this policy has appeared as a rider to the annual appropriation acts for the Departments of Labor, Education, and Health and Human Services. Since 1999, NIH has interpreted this policy, known as the Dickey Amendment, to prohibit them from funding derivation of embryonic stem cells, but not research on the cells once extracted. This interpretation was challenged in 2010, when a judge issued an interim injunction on federal funding of the research, which was later overturned pending resolution of the full case (the case is still pending at time of writing).

In his remarks on the 2001 policy, President Bush explicitly referred both to scientists' beliefs in the enormous therapeutic potential of embryonic stem cell research and to his own belief in the value of embryonic life. Of embryonic stem cell research, he noted: "At its core, this issue forces us to confront fundamental questions about the beginnings of life and the ends of science. It lies at a difficult moral intersection, juxtaposing the need to protect life in all its phases with the prospect of saving and improving life in all its stages." He called himself "a strong supporter of science and technology" but also noted that he believes "that human life is a sacred gift from our Creator" (Bush, 2001).

If President Bush intended his policy as a compromise between the value of scientific research and the value of embryonic life, it was one that left many scientists, disease groups, and others unsatisfied. Substantial criticism was directed at the 2001 policy. Those who oppose any research in which human embryos are destroyed argued that federal funding restrictions needed to be supplemented by a nationwide ban on creating embryos for use in research, including by cloning (National Right to Life Coalition, 2001; PCB, 2004). A more vocal opposition called the 2001 policy overly restrictive. Their arguments included that the President over-valued embryonic life (that it is not more important than embryonic stem cell research) (PCB, 2004), that the policy was arbitrary and inconsistent (PCB, 2004), that the cell lines in the NIH registry were of poor quality and inappropriate for long-term use (Dawson et al., 2003), and that the policy harmed American science by encouraging scientists to focus on research for which federal funding is available or to move to other countries to conduct their research (PCB, 2004). Although considerably less controversial, President Obama's Executive Order and the NIH Guidelines that followed have also been criticized for being too permissive by those opposed to embryonic stem cell research and too restrictive by those who believe federal funding should be available for research on stem cells extracted from embryos created for research purposes, including by cloning (O'Reilly, 2009).

Overall, however, stem cell policy under President Obama has so far received little political attention in contrast to the 2001 policy, and stem cell research generally during the Bush presidency, which became focal points for political debate. For example, both the research and federal policies were debated by the 2004 presidential candidates Senator John Kerry (the Democratic nominee) and President Bush (the Republican nominee) (Commission on Presidential Debates, 2004). Stem cell research was also a major issue at the Democratic national convention in 2004, where Ronald Regan Jr, son of former Republican President Ronald Regan, gave a speech calling for greater government support of embryonic stem cell research. In so far as they were motivated by disagreement with President Bush's 2001 policy, actions at a state level to fund research not eligible for federal funding (described below) can also be interpreted as political. To some extent, then, embryonic stem cell research became a partisan political issue during the first decade of the twenty-first century.

Other political activity during this time was, however, bipartisan, including two 2004 letters — one signed by over 200 congressional representatives and the other signed by nearly 60 senators — asking President Bush to relax federal funding restrictions on embryonic stem cell research. Similar bipartisan support was expressed for the proposed Stem Cell Research Enhancement Act, which sought to increase federal support of the research (American Association for the Advancement of Science, 2004). Republican politicians who supported changes in federal policy included Senator Orrin Hatch, who is in the unusual position of mixing opposition to abortion with support of embryonic stem cell research provided it uses only spare embryos (Hatch, 2003). During the 2008 Presidential campaign, then Senator Obama made it clear that, if elected, he would repeal President Bush's stem cell policy. His 2009 Executive Order was, therefore, in keeping with his campaign promise. The NIH guidelines that followed garnered bipartisan support, likely because they resulted in more research being funded but maintained some significant restrictions.

INTERNATIONAL COMPARISONS

Biomedical research is an international affair and embryonic stem cell research is no exception. Among those nations in which embryonic stem cell research is active, governments employ a variety of regulation strategies, most of which subject the research to some restrictions and oversight while still allowing it to move forward. No single method of regulation has prevailed internationally, although some patterns have emerged. For instance, many nations use national legislation to regulate stem cell research, often requiring oversight by a national stem cell research committee or licensing authority. Substantively, bans on creating embryos by cloning are common, although not universal, as are bans on creating embryos by fertilization except as part of fertility treatment, which means that research in many, although certainly not all, nations is limited to spare embryos. The federally funded/non-federally funded split currently employed in the USA is highly unusual by international standards.

In the UK, comprehensive legislation governing all research and medical use of human gametes and human embryos has existed since 1990 (Human Fertilisation and Embryology Act 1990 (UK)). A major feature of the legislation is that it institutes few bans, but requires that all collection and use of embryos and gametes be licensed and overseen by an independent body, called the Human Fertilisation and Embryology Authority (HFEA). The existence of this legislation and the HFEA before human embryonic stem cells were first isolated meant that the research already had a regulatory system into which it could immediately be slotted, obviating the need for significant new legislative action by the government in response to the research (although regulations were promulgated in 1991 to add three new purposes for which research on embryos is permitted, including increasing knowledge about serious disease) (Human Fertilisation and Embryology (Research Purposes) Regulations 1991). Some scholars have urged the USA to adopt a similar regulatory system, under which very few activities are banned outright but instead require a license and are subject to oversight (Parens and Knowles, 2003).

In neighboring Canada, national funding guidelines similar to the policy formulated by President Clinton were released in 2002 by the Canadian Institutes of Health Research (CIHR). Significant terms of these guidelines include that all research using CIHR funds is subject to national oversight by the Stem Cell Oversight Committee, that all embryonic stem cell lines generated using CIHR funding must be recorded in a national electronic registry and made available to other Canadian academic researchers at cost, and that stem cells can only be extracted from spare embryos (embryos cannot be created for research purposes, including by cloning).

The CIHR guidelines were followed in 2004 by the Assisted Human Reproduction Act, which, like the legislation in the UK, regulates much more than just embryonic stem cell research. Several provisions of the Act, however, directly impact embryonic stem cell research. In particular, the Act prohibits creating a cloned embryo and creating an embryo for "any purpose other than creating a human being or improving or providing instruction in assisted reproduction procedures." Therefore, the Act limits embryonic stem cell research in Canada to embryos originally created in the course of fertility treatment, although it does not specify that such embryos be surplus to the fertility needs of the donors (i.e. the embryos need not be "spare") (Johnston, 2006).

This distinction between stem cell research involving embryos created for use in research (including by cloning) and research involving embryos originally created for use in fertility treatment also appears in Australian legislation. Australia's Research Involving Human Embryos Act 2002 limits research to "excess ART embryos," which are defined as embryos created for use in assisted reproductive technology treatment that are now excess to the needs of the woman or couple for whom they were created (Research Involving Human Embryos Act, 2002 (Commonwealth of Australia)). Similar conditions are attached to research in other nations, including France, Denmark, Finland, Greece, the Netherlands, and Japan.

South Korea, a country active in stem cell research, is in the unusual position of permitting the creation of cloned embryos for research into rare or incurable diseases, but prohibiting the creation of embryos for research by *in vitro* fertilization. Under a 2004 law, revised in 2008, *in vitro* fertilization may only be used to create embryos for reproductive purposes, which can later be donated to research if not used. Cloned embryos may only be created for use in research into the treatment of rare or incurable diseases (Bioethics and Safety Act (South Korea), 2008). In Israel, another country active in stem cell research, no legislation regulates the field, although there is a law against implanting a cloned embryo (Prohibition of Genetic Intervention (Human Cloning and Genetic Manipulation of Reproductive Cells) Law, 5759-1999 (Israel)). Therefore, derivation of stem cells from embryos created for research use, including by cloning, is allowed (although to date there are no reports of Israeli scientists creating cloned embryos) (Walters, 2004). Under Singapore's 2004 legislation, research is permitted on spare embryos and embryos created for research, including by cloning (Human Cloning and Other Prohibited Practices Act (Singapore), 2004). In this international climate, current US policy is rather unusual because only federal funding rules and some state-level restrictions apply to US embryonic stem cell research (Knowles, 2004).

STATE AND PRIVATE FUNDING IN THE USA

In response to the 2001 limitations placed on the use of federal funds in embryonic stem cell research, private funders stepped up their support and advocates lobbied state governments to provide funds for the research. As a result of private funding, Harvard University's Douglas Melton and colleagues announced in 2004 that they had derived 17 new embryonic stem cells lines with support provided by the Juvenile Diabetes Research Foundation, the Howard Hughes Medical Institute, and Harvard University (Cowan et al., 2004). According to a 2005 special report in *Scientific American*, about $200 million of private money is spent on US stem cell research annually (Beardsley, 2005).

States have also moved to fund stem cell research. In 2004, New Jersey announced that it would begin funding local stem cell researchers through grants and by creating the Stem Cell Institute of New Jersey. Funding is also available in other states, including New York, Connecticut, Illinois, Maryland, Massachusetts, and Wisconsin. Thus far, however, the largest state initiative has been in California.

In November 2004, voters in California supported a proposition to allocate $3 billion over 10 years to embryonic stem cell research. The initiative, known as Proposition 71, authorized the state of California to sell $3 billion in general obligation bonds to provide funding for stem cell research and research facilities in California. Under the proposition, the funds are to be distributed as grants and loans to California-based institutions by the newly created Californian Institute for Regenerative Medicine, which is also required to establish regulatory standards for the research (Attorney General of California, 2004). Priority for funding is to go to research that would not be eligible for federal funding. Critics of the initiative called it fiscally irresponsible given the state's economy and other health and research needs. They also argued that the institute as structured lacked accountability (Attorney General of California, 2004). Nevertheless, 59% of voters approved the measure.

POLICY AS EMBRYONIC STEM CELL RESEARCH MOVES FORWARD

As embryonic stem cell research moves forward in the USA, even with limits on federal support, various ethical and policy questions arise. For example, should researchers create embryos in the laboratory by fertilization or cloning or should they only use spare embryos? Either way, they will need to interact with fertility clinics or with egg, sperm, or embryo donors, raising questions about how those interactions should be conducted. Should researchers pay fertility clinics for procuring gametes and embryos for stem cell research? How should gamete and embryo donors be approached to donate, precisely whose consent should be required, and should the donors be compensated? Once researchers extract cell lines, are they obliged to make those lines available to other researchers? Ought researchers to patent new cell lines or new stem cell-related processes? If they do obtain patents, what practices should they follow in licensing those lines or processes? Should researchers be allowed to mix human cells and animal cells in the creation of chimeras or hybrids?

GUIDELINES FROM THE NATIONAL ACADEMIES

Many of these issues were taken up by a panel convened in 2004 by the National Academies, a private, non-profit organization whose mission is to advise the nation on issues in science, engineering, and medicine. In general, the National Academies' reports and recommendations are very influential and their *Guidelines for Human Embryonic Stem Cell Research* (Committee on Guidelines for Human Embryonic Stem Cell Research, 2005) received significant media attention when they were released in April 2005. Although these guidelines do not carry the force of law they are very influential.

The guidelines are not the first document to speak to the conduct of embryonic stem cell research in the USA but at the time they were formulated it was not clear whether any of the previous guidance applied to most contemporary embryonic stem cell research because it was formulated either only for federally funded research or under a previous administration (and so could be considered out of date), or because it was simply too restrictive to meaningfully guide institutions and researchers that had made the decision to move forward with the research (Johnston, 2005). The guidelines were, therefore, received as filling a policy vacuum.

The committee that drafted the guidelines was asked to consider the use and derivation of stem cells from embryos originally created during fertility treatment, embryos created using donated eggs and sperm, and cloned embryos. It therefore did not engage in the debate over whether it is morally permissible to destroy human embryos in research. It also did not consider whether there is a moral difference between research that uses spare embryos and research that uses embryos created specifically for research purposes, including by cloning.

This debate is worth understanding, since the distinction appears in US and international policy. Briefly, creating embryos for research use is often opposed on the grounds that it is wrong to create human life for the purposes of destroying it (PCB, 2002). Using cloning techniques to generate embryos for use in research is often opposed on the same ground and on the ground that "cloning for research" could lead to "cloning to produce children," which is opposed by many stem cell scientists and policy makers (PCB, 2002). Research use of spare embryos, on the other hand, has received more support on the ground that the embryos would be destroyed anyway. Whether the intention of the original creator is sufficient reason for permitting research on spare embryos but not on embryos created for research use, has been questioned (Parens, 2001; Baylis et al., 2003).

Even if this moral issue can be resolved, spare embryos may not be satisfactory as the sole source of embryos for stem cell research because there may not be sufficient numbers available to meet demand. A 2003 survey of US fertility clinics reports that of the more than 400,000 embryos in frozen storage in the USA, only 2.8% have been donated to research (Hoffman et al., 2003). Other researchers have argued that these frozen embryos will not be genetically diverse enough for therapeutic purposes (Faden et al., 2003). Whether for these reasons or others, the National Academies committee proceeded on the basis that researchers could use spare embryos and embryos created for research use, including by cloning.

Overall, the National Academies committee recommended banning very little scientific activity. Instead, it recommended institutional review of protocols, oversight of the involvement of egg, sperm, and embryo donors, the establishment of stem cell banks, and the documentation of research activity.

Two recommendations were particularly significant. First, the committee recommended that much embryonic stem cell research be subject to a mixture of local and national oversight. Local oversight would occur at each institution engaged in embryonic stem cell research, which would establish an Embryonic Stem Cell Research Oversight (ESCRO) committee to oversee all issues related to the derivation and use of embryonic stem cells, review all proposals for scientific merit, maintain records of research that takes place at the institution, including registries of new cell lines, and educate investigators. Local IRBs would provide additional oversight. Even though much embryonic stem cell research will not, strictly speaking, need to go before an IRB,[5] the committee recommended that the procurement of egg, sperm, and embryos always be reviewed by an IRB, regardless of the applicability of federal regulations, and that IRBs never waive the requirement for informed consent from a person donating cells, eggs, sperm, or embryos to research, even where the federal human subject research regulations provide for such a waiver.

The guidelines also called for the establishment of a national oversight body to consider issues of practice and policy on an ongoing basis. Such a body would be similar to the UK's HFEA and Canada's Stem Cell Oversight Committee, although these national bodies also conduct some of the local review that the guidelines suggested be carried out by ESCRO committees and, to some extent, IRBs.[6]

The other significant set of recommendations addressed the involvement of egg, sperm, and embryo donors. In line with much guidance, law and regulation around the world, the guidelines recommended requiring consent from donors for research use. They went beyond previous US guidance, however, by extending this requirement to the women and men who contributed egg and sperm to an embryo but were not the intended parents of any resultant child. These donors are currently asked to consent to reproductive use of their gametes, but are not generally asked to consent also to research use. Accepting this requirement, the committee

---

[5] OHRP told the committee that merely asking couples whether they wish to donate their surplus embryos to research does not render them "human subjects of research," if no data on them is being gathered and if there is no substantive interaction with them other than gaining their consent.

[6] The UK Authority is also charged with overseeing assisted human reproduction, which is an area largely free from national regulation or oversight in the USA and is not addressed in these guidelines.

noted, might rule out the use of some embryos already created for fertility purposes that are now in frozen storage.

On the issue of compensating egg, sperm, and embryo donors, the guidelines noted the arguments in favor of compensation: paying egg and sperm donors is routine in the US fertility context and many Americans participating in other kinds of research are offered financial inducements to secure their participation. They acknowledged that arguments for compensating egg donors are particularly strong: "the invasiveness and risks of the procedure suggest that financial remuneration is most deserved, but at the same time there is a greater likelihood of enticing potential donors to do something that poses some risk to themselves."

Ultimately, the National Academies committee followed previous US guidelines and guidelines and laws from many other nations. It recommended that egg donors be reimbursed only for "direct expenses," and that no payment whatsoever be offered to sperm or embryo donors. They did allow, however, for reimbursement of fertility clinics for costs, including professional services, associated with obtaining consent and collecting eggs, sperm, or embryos.

In addition to cash payment, the guidelines recommended against compensation in kind. Donors are not to receive any benefit from their donation, including "personal medical benefit" (excepting autologous transplantation, where the donor receives stem cells derived from his or her donation). This rule would prevent a kind of egg or embryo sharing arrangement whereby women or couples receive cheaper or free fertility treatment in exchange for donating a portion of their eggs or embryos to stem cell research. Similar arrangements exist in the fertility context, where women or couples receive a discount if they donate some of their eggs to others undergoing fertility treatment.[7] This arrangement would help make fertility treatment available at a lower cost, but it would also more quickly exhaust the woman's or couple's supply of eggs or embryos, thereby possibly reducing their chances of achieving pregnancy (Nisker and White, 2005).

1139

In 2008 and 2010, the National Academy of Sciences' (NAS) Human Embryonic Stem Cell Research Advisory Committee revisited the issue of donor compensation. In both years, the Committee discussed the arguments for and against compensating gamete donors but retained its prohibition on compensating embryo donors. The guidelines were amended to make explicit that "actual lost wages" qualify as direct expenses for which gamete donors may be compensated, but that additional payments should not be permitted (Human Embryonic Stem Cell Research Advisory Committee, 2008).

COMPENSATING EGG DONORS: THE ARGUMENTS

After the NAS guidelines were issued, a controversy over the procuring of eggs in South Korea for use in cloning research brought additional attention to egg donation. In early 2006, it emerged that in research led by Woo-Suk Hwang of Seoul National University some egg donors were members of the research team, some egg donors were paid to donate, and researchers accompanied some egg donors as they underwent the extraction procedure (Johnston, 2006).

These facts about the donation process raised concerns about whether the women who donated eggs to Hwang's research did so completely voluntarily. Voluntariness is a core commitment of modern research ethics (World Medical Association, 2000), which generally translates into requirements that no one be pressured to participate in research and that each participant be able to withdraw from the research at any time without endangering ongoing medical care or the care-giving relationship with the researchers (World Medical Association,

[7] See, for example, the HFEA 2003 at Appendix A: Guidance for egg sharing arrangements.

2000). Researchers usually avoid enrolling family members and employees because they might reasonably feel significant pressure to participate. The commitment to voluntary participation, and specifically the derived right to withdraw from the research at any time, could also be under threat if researchers physically accompany volunteers through procedures as Hwang and his colleagues apparently did.

Voluntariness is also the major factor motivating bans on compensating egg donors, the argument being that the need for money could compel participation, especially, though not exclusively, among the poor. Nevertheless, as the National Academies committee already noted, many other research participants in the USA receive compensation in exchange for their involvement in research (Committee on Guidelines for Human Embryonic Stem Cell Research, 2005). Such payment, particularly where modest, is said to be not only a necessary incentive but also fair treatment of research subjects (after all, the researchers and their staff will be paid for the time and resources they contribute toward the research).

The concern about voluntariness is likely heightened because egg donation is time-consuming, and painful, and involves some risk. Egg donors are injected with drugs over weeks so that they super-ovulate (produce many eggs). The eggs are then removed from the woman either by inserting a hollow needle through her vagina or by laparoscopic surgery. Risks of the stimulation and egg-collection process include hot flashes, headaches, sleeplessness, mood alteration, ovarian hyper-stimulation syndrome, nausea, vomiting, pain, bleeding, and infection. There is even a controversy over a possible danger of ovarian cancer from the medications (Gurmankin, 2001). Could payment encourage women to donate eggs even though donation might be painful and pose a risk to their health? The answer is yes. However, whether the risk and pain are unacceptable is another issue. In theory at least, if an IRB approves research involving egg donation it has decided that the risk to donors is reasonable in relation to the importance of the knowledge that may reasonably be expected to result (45 CFR §46.111(a)). Compensation is not supposed to be offered to research participants in order to seduce them to take an unacceptable risk.

But voluntariness may not be the only concern about paying egg donors in stem cell research. There is also some opposition to paying anyone for providing bodily materials (rather than solely for their time and effort), for example as expressed in a federal law prohibiting payment for organ donation (although that law expressly does not apply to blood, sperm, or human eggs) (Uniform Anatomical Gift Act, 1987). The stance against sale of bodily materials is well defended in scholarly circles. Bioethicist Thomas Murray argued over 20 years ago that all donations of body parts, whether for research or for clinical treatment, should be gifts and not sales (Murray, 1987). Others, however, including Law Professor Lori Andrews, writing around the same time, countered that individuals "have the autonomy to treat their own (body) parts as property," particularly those parts of the body that they can regenerate (Andrews, 1986). The debate about compensating egg, sperm, and embryo donors continues today. Despite recommending against compensation, the National Academies' Guidelines left the issue open for revision.

COMMERCIALIZATION AND ACCESS

Another strong argument against paid donation in general is that it can add to the costs of conducting research and generating eventual treatments. But this argument works best if it is also applied to the scientists, institutions, and companies involved in the research and in any eventual treatment. Indeed, a commitment to scientific progress and widely available treatments in stem cell research might entail a commitment by all those involved to, for example, bank and widely distribute new cell lines, participate in international collaboration and adopt patenting and licensing practices designed to facilitate access (including possibly not patenting some discoveries at all) (Department of Health and Human Services, 1999). These concerns about secrecy, access, and commercialization in stem cell research mirror a larger debate in biomedical research in general (Krimsky, 2004; Johnston and Wasunna, 2007).

In terms of access, particular emphasis is often put on patenting and licensing practices, which can have an enormous impact on progress made by other researchers as well as on the availability of eventual treatments (Heller and Eisenberg, 1998). Indeed, a number of patents already attach to embryonic stem cell research, including the only patent to be issued in the world claiming a purified preparation of primate (including human) embryonic stem cells and a method for isolating them (US Patent 5,843,780.). This patent is currently the subject of a legal challenge.

Patenting and licensing issues were to some extent anticipated in California's Proposition 71, which included a provision requiring the establishment of standards in all grants and loans allowing the state to financially benefit from licenses, patents, and royalties resulting from the research activities funded under the measure (Attorney General of California, 2004). Regulations implementing these standards require that a portion of profits be returned to the state (17 California Code of Regulations, 2006).

CONCLUSION

As this brief introduction shows, embryonic stem cell research has not just generated significant scientific activity, but also led to the development of numerous policies. In the USA, as well as in some other nations, these policies have responded to the controversial nature of research involving the destruction of human embryos. The US federal funding policy of 2001, which limited the use of federal funds in embryonic stem cell research, generated a lot of criticism. In response to the 2001 policy's limits, funds for embryonic stem cell research were provided by private donors, charitable organizations, and states. In 2009, President Obama repealed the 2001 policy. Nevertheless, federal funding, while now more widely available, is still limited to research on cells derived from spare embryos. As research on embryonic stem cells taken from a variety of sources moves forward in the USA, some of it without federal funding, a range of policies are developing that respond to a number of important issues, including local and national oversight, the role of donors in the research, and the consequences of commercial interests.

ACKNOWLEDGMENTS

Parts of this chapter are based on two of the author's previous publications: "Paying egg donors: exploring the arguments," Hastings Center Report 2006; 36(1): 28–31 and "Stem cell protocols: the NAS guidelines are a useful start," Hastings Center Report 2005; 35(5): 16–17.

References

American Association for the Advancement of Science (AAAS) (2004). AAAS Policy Brief: Stem Cell Research. August 26, 2004. Available at: www.aaas.org/spp/cstc/briefs/stemcells/index.shtml

Andrews, L. B. (1986). My body, my property. *Hastings Cent. Rep.*, *16*(5), 28–38.

Andrews, L. B. (2004). Legislators as lobbyists: proposed state regulation of embryonic stem cell research, therapeutic cloning and reproductive cloning. In *President's Council on Bioethics (2004). Monitoring Stem Cell Research.* Washington, DC: President's Council on Bioethics.

Attorney General of California (2004). State of California: Stem Cell Research. Funding. Bonds. Initiative Constitutional Amendment and Statute (Proposition 71): Official Text and Summary. Available at: www.sos.ca.gov/elections/bp_nov04/prop_71_entire.pdf

Baylis, F. B., Beagan, B., Johnston, J., & Ram, N. (2003). Cryopreserved human embryos in Canada and their availability for research. *J. Obstet. Gynaecol. Can., 25*, 1026–1031.

Beardsley, S. (2005). A world of approaches to stem cells. *Financ. Times Sci. Am. (Spec. Rep.)*, June 27, A20–A21.

Bioethics and Safety Act No. 9100 (Republic of Korea) (2008). English translation. Available at: http://www.mbbnet.umn.edu/scmap/KoreanBioethics.pdf

Bush, President G. W. (2001). Remarks by President George W. Bush on stem cell research. In *President's Council on Bioethics (2004). Monitoring Stem Cell Research.* Washington, DC: President's Council on Bioethics, August 9, 2001.

Canadian Institutes of Health Research (March 12, 2002). *Human Pluripotent Stem Cell Research: guidelines for CIHR-funded research.* Canadian Institutes of Health Research, March 12, 2002. (Now superseded by: Canadian Institutes of Health Research. June 7, 2005). Available at: www.cihr-irsc.gc.ca/e/28216.html

Chung, Y., Klimanskaya, I., Becker, S., Marh, J., Lu, S. J., Johnson, J., et al. (2006). Embryonic and extraembryonic stem cell lines derived from single mouse blastomeres. *Nature, 439,* 216—219.

Commission on Presidential Debates (2004). The Second Bush—Kerry Presidential Debate (debate transcript). October 8, 2004. Available at: www.debates.org/pages/trans2004c.html

Committee on Guidelines for Human Embryonic Stem Cell Research, National Research Council (2005). *Guidelines for Human Embryonic Stem Cell Research.* Washington, DC: National Academies Press.

Cowan, C. A., Klimanskaya, I., McMahon, J., Atienza, J., Witmyer, J., Zucker, J. P., et al. (2004). Derivation of embryonic stem-cell lines from human blastocysts. *N. Engl. J. Med., 350,* 1353—1356.

Dawson, L., Bateman-House, A. S., Mueller Agnew, D., Bok, H., Brock, D. W., Chakravarti, A., et al. (2003). Safety issues in cell-based intervention trials. *Fertil. Steril., 80,* 1077—1085.

Department of Health and Human Services (DHHS) (1999). Principles and guidelines for recipients of NIH research grants and contracts on obtaining and disseminating biomedical research resources: final notice. *Fed. Reg., 64*(246), 72090—72096.

Executive Order 13505 (2009). Removing barriers to responsible scientific research involving human stem cells. *Fed. Reg., 74,* 10667.

Faden, R. R., Dawson, L., Bateman-House, A. S., Agnew, D. M., Bok, H., Brock, D. W., et al. (2003). Public stem cell banks: considerations of justice in stem cell research and therapy. *Hastings Cent. Rep., 33,* 13—27.

Gurmankin, A. D. (2001). Risk information provided to prospective oocyte donors in a preliminary phone call. *Am. J. Bioeth., 1*(4), 3—13.

Hatch, Senator Orrin, G. (2003). Promoting Ethical Regenerative Medicine Research and Prohibiting Immoral Human Reproductive Cloning (statement before the Senate Judiciary Committee Hearing on promoting ethical regenerative medicine research and prohibiting immoral human reproductive cloning). March 19, 2003. Available at: http://hatch.senate.gov/index.cfm? FuseAction=PressReleases.Detail&PressRelease_id=726

Heller, M. A., & Eisenberg, R. S. (1998). Can patents deter innovation? The anticommons in biomedical research. *Science, 280,* 698—701.

Hoffman, D. I., Zellman, G. L., Fair, C. C., Mayer, J. F., Zeitz, J. G., Gibbons, W. E., et al. (2003). Society for Assisted Reproduction Technology (SART) and RAND. Cryopreserved embryos in the United States and their availability for research. *Fertil. Steril., 79,* 1063—1069.

Human Cloning and Other Prohibited Practices Act 2004, Chapter 131B (Singapore).

Human Embryonic Stem Cell Research Advisory Committee (2008). *2008 Amendments to the National Academies' Guidelines for Human Embryonic Stem Cell Research.* Washington, DC: National Academies Press.

Human Fertilisation and Embryology Act 1990 (UK).

Human Fertilisation and Embryology (Research Purposes) Regulations 1991.

Hyun, I., Hochedlinger, K., Jaenisch, R., & Yamanaka, S. (2007). New advances in iPS cell research do not obviate the need for human embryonic stem cells. *Cell Stem Cell, 1*(4), 367—368.

Johnston, J. (2005). Stem cell protocols: the NAS guidelines are a useful start. *Hastings Cent. Rep., 35*(5), 16—17.

Johnston, J. (2006). Is research in Canada limited to "spare" embryos? *Health Law Rev., 14*(3), 3—13.

Johnston, J., & Wasunna, A. (2007). Patents, biomedical research, and treatments: examining concerns, canvassing solutions. *Hastings Cent. Rep., 37*(1), S1—S36 (Special Report).

Knowles, L. P. (2004). A regulatory patchwork — human ES cell research oversight. *Nat. Biotechnol., 22*(2), 157—163.

Krimsky, S. (2004). *Science in the Private Interest: Has the Lure of Profits Corrupted the Virtue of Biomedical Research?* Lanham, MD: Rowman & Littlefield.

Murray, T. H. (1987). Gifts of the body and the needs of strangers. *Hastings Cent. Rep., 17*(2), 30—38.

National Bioethics Advisory Commission (NBAC) (1999). *Ethical Issues in Human Stem Cell Research, Vol. 1.* Rockland, MD: National Bioethics Advisory Commission.

National Institutes of Health (2009). National Institutes of Health Guidelines for Human Stem Cell Research. *Fed. Reg., 74,* 32170—32175.

National Right to Life Coalition (NRLC) (2001). President Bush's Statement: NRLC Press Release. August 10, 2001. Available at: www.nrlc.org/press_releases_new/stemcelldno081010.htm

Nisker, J., & White, A. (2005). The CMA code of ethics and the donation of fresh embryos for stem cell research. *Can. Med. Assoc. J., 173,* 621—622.

Office of Human Research Protections (OHRP) (March 19, 2001). Guidance for Investigators and Institutional Review Boards Regarding Research Involving Human Embryonic Stem Cells, Germ Cells and Stem Cell-Derived Test Articles. Available at: www.hhs.gov/ohrp/humansubjects/guidance/stemcell.pdf

O'Reilly, K. B. (2009) NIH policy loosening stem cell research restrictions disappoints both sides in debate. *American Medical News*. May 4, 2009. Available at: http://www.ama-assn.org/amednews/2009/05/04/prl20504.htm.

Orr, R. D. (2002). The moral status of the embryonal stem cell: inherent or imputed? *Am. J. Bioeth.*, 2, 57–59.

Parens, E. (2001). On the ethics and politics of embryonic stem cell research. In S. Holland, K. LeBacqz, & L. Zoloth (Eds.), *The Human Embryonic Stem Cell Debate: Science, Ethics, and Public Policy*. Cambridge, MA: MIT Press.

Parens, E., & Knowles, L. P. (2003). Reprogenetics and public policy: reflections and recommendations. *Hastings Cent. Rep.*, 33(4), S1–S24.

President's Council on Bioethics (2002). *Human Cloning and Human Dignity: An Ethical Inquiry*. Washington, DC: President's Council on Bioethics.

President's Council on Bioethics (2004). *Monitoring Stem Cell Research*. Washington, DC: President's Council on Bioethics.

Research Involving Human Embryos Act 2002 (Commonwealth of Australia).

Roe v. Wade 410 US 113 (1973).

Schwartz, R. E., Reyes, M., Koodie, L., Jiang, Y., Blackstad, M., Lund, T., et al. (2002). Multipotent adult progenitor cells from bone marrow differentiate into functional hepatocyte-like cells. *J. Clin. Invest.*, 109, 1291–1302.

Shamblott, M. J., Axelman, J., Wang, S., Bugg, E. M., Littlefield, J. W., Donovan, P. J., et al. (1998). Derivation of pluripotent stem cells from cultured human primordial germ cells. *Proc. Natl. Acad. Sci. USA*, 95, 13726–13731.

Takahashi, K., Tanabe, K., Ohnuki, M., Narita, M., Ichisaka, T., Tomoda, K., et al. (2007). Induction of pluripotent stem cells from adult human fibroblasts by defined factors. *Cell*, 131, 861–867.

Thomson, J. A., Itskovitz-Eldor, J., Shapiro, S. S., Waknitz, M. A., Swiergiel, J. J., Marshall, V. S., et al. (1998). Embryonic stem cell lines derived from human blastocysts. *Science*, 282, 1145–1147.

Uniform Anatomical Gift Act, 1987.

US Conference of Catholic Bishops (2009). Scientific Experts Agree: Embryonic Stem Cells are Unnecessary for Medical Progress. Available at: http://www.usccb.org/prolife/issues/bioethic/fact401.shtml

US Patent 5,843,780.

Verfaillie, C. M., Pera, M. F., & Lansdrop, P. M. (2002). Stem cells: hype and reality. *Hematology (Am. Soc. Hematol. Educ. Program)*, 369–391.

Walters, L. (2004). Human embryonic stem cell research: an intercultural perspective. *Kennedy Inst. Ethics J.*, 14(1), 3–38.

World Medical Association (WMA) (2000). Declaration of Helsinki: ethical principles for medical research involving human subjects (reprinted). *JAMA*, 284, 3043–3045.

Yu, J., Vodyanik, M. A., Smuga-Otto, K., Antosiewicz-Bourget, J., Frane, J. L., Tian, S., et al. (2007). Induced pluripotent stem cell lines derived from human somatic cells. *Science*, 318, 1917–1920.

17 California Code of Regulations (2006) Section 100308. Revenue Sharing.

45 CFR §46.204(h–i).

45 CFR §46.111(a).

45 CFR §46.206.

498A of the Public Health Service Act (42 USC 289g–l).

CHAPTER 63

Overview of the FDA Regulatory Process

Kevin J. Whittlesey[*], **Mark H. Lee**[†], **Jiyoung M. Dang**[‡], **Maegen Colehour**[‡], **Judith Arcidiacono**[†], **Ellen Lazarus**[†], **David S. Kaplan**[‡], **Donald Fink**[†], **Charles N. Durfor**[‡], **Ashok Batra**[§], **Stephen L. Hilbert**[‖], **Deborah Lavoie Grayeski**[¶], **Richard McFarland**[†], **Celia Witten**[†]

[*] Office of the Commissioner, FDA, Silver Spring, MD, USA
[†] Center for Biologics Evaluation and Research, FDA, Rockville, MD, USA
[‡] Center for Devices and Radiological Health, FDA, Silver Spring, MD, USA
[§] SUNY-Syracuse, Syracuse, NY, USA; US Biotechnology & Pharma Consulting Group, Potomac, MD, USA
[‖] Children's Mercy Hospital, Kansas City, MO, USA
[¶] M Squared Associates, Inc., Alexandria, VA, USA

INTRODUCTION AND CHAPTER OVERVIEW

1145

The field of regenerative medicine encompasses a breathtaking array of interdisciplinary scientific approaches that address a broad spectrum of clinical needs. Recent advances in scientific knowledge related to cell biology, gene transfer therapy, biomaterials, immunology, and engineering principles applicable to biological systems place the regenerative medicine community in a position to address a number of challenging and critical health needs. These include treatment of disease conditions resulting from pancreas, liver, and kidney failure; structural cardiac valve repair; skin and wound repair; and orthopedic applications. Scientific challenges confronting this field include expanding the knowledge base in each discipline as well as developing an interdisciplinary approach for identifying and resolving key questions. The Food and Drug Administration's (FDA's) regulatory review process mirrors the scientific challenges with regard to development of review paradigms that cross scientific disciplines.

This chapter will provide a brief historical review of FDA and its organizational structure as well as discuss topics pertaining to the regulation of regenerative medicine products including possible regulatory pathways for combination products and relevant jurisdictional issues. Sources of information concerning FDA regulatory policies important to regenerative medicine product developers will also be discussed. It is essential for individuals, institutions, and companies, collectively referred to in FDA regulations as Sponsors (the term Sponsor for drugs and biologics is defined at 21 Code of Federal Regulations (CFR) 312.3(b), while Sponsor is similarly defined at 812.3(n) for devices), responsible for the clinical trials of regenerative medicine products to be aware of FDA regulatory policies and how to obtain the necessary information. Suggestions will also be provided as to how to effectively engage FDA during the development of a novel regenerative medicine product.

BRIEF LEGISLATIVE HISTORY OF FDA

Medical products regulated by FDA include human and animal drugs, medical devices, and biological products, such as vaccines, cellular and gene therapies, and blood products.

Principles of Regenerative Medicine. DOI: 10.1016/B978-0-12-381422-7.10063-X

Among the therapeutic agents of biological origin regulated by FDA are cellular therapies, including products derived in whole or part from human tissue and xenotransplants. In addition to medical products for human use, FDA regulates food other than meat and poultry, radiation-emitting products for consumer, medical, and occupational use, cosmetics, medical products for human use, and animal feed. FDA's role in medical product regulation extends throughout the entirety of the product life cycle. Depending on the product category, this may mean oversight, including review and inspection, of clinical trials, of the premarket product approval/clearance process, of manufacturing controls, controls over labeling, and registration and listing requirements. FDA also continues its oversight once a product is marketed in a variety of ways, including inspections, mandatory and voluntary post-approval (e.g. Phase IV) studies, and surveillance of adverse events reported to FDA.

FDA laws and regulations have developed over time, prompted partly in response to serious medical adverse events or by other public health and safety concerns. Early regulation of biological products was prompted in part by the death of 13 children in 1901 following administration of diphtheria antitoxin prepared from a source contaminated with tetanus. In response, Congress passed the Biologics Control Act in 1902. This act provided for regulation of viruses, serums, toxins, and analogous products; required licensing of manufacturing establishments and manufacturers; and provided the government with inspectional authority. The Act focused on requiring control of manufacturing processes for producing biological products, reflecting the extent to which the starting source material and the manufacturing process defined the final product.

In 1906, Congress passed the Federal Food and Drugs Act proposed in part in reaction to the meat-packing industry conditions described in Upton Sinclair's book *The Jungle*. While the primary focus of the Act was on food safety, the law also required that drugs be provided in accordance with standards of strength, quality, and purity unless otherwise specified in the label. Premarket review of new drugs was not required until the passage of the 1938 Food, Drug, and Cosmetic Act (FD&C Act), which repealed the earlier 1906 Federal Food and Drugs Act. In 1937, the sulfa drug Elixir Sulfanilamide, previously available only in tablet or powder form to treat streptococcal infections, was marketed as a liquid using diethylene glycol, an analog of antifreeze, as a formulating solvent. This change in formulation, made without the requirement for premarket review, resulted in over 100 deaths, many of whom were children, prompting the passage of the 1938 FD&C Act. The 1938 Act also put medical devices and cosmetics under FDA authority and authorized factory inspections.

The Public Health Service Act (PHS Act), passed in 1944, incorporated the 1902 Biologics Control Act and is the present legal basis for licensing of biological products. Because most biological products also meet the definition of "drugs" under the FD&C Act, they are also subject to regulation under that Act.

The requirement for premarket demonstration of efficacy and the authority for FDA oversight of clinical trials were provided by the Kefauver-Harris amendments to the FD&C Act in 1962. These amendments were prompted in part by the tragic adverse events resulting from use of thalidomide as a non-addictive prescription sedative. This drug, not approved as a sedative in the USA, when taken by pregnant women during the first trimester, resulted in thousands of birth defects for children born outside this country.

The Medical Device Amendments to the FD&C Act were passed in 1976, following reports of safety issues with respect to the Dalkon Shield intrauterine device. The Medical Devices Amendments included risk-based requirements for premarket notification or approval of medical devices. Prior to 1976, FDA authority was limited to taking action against marketed devices found to be unsafe or ineffective.

LAWS, REGULATIONS, AND GUIDANCE

The previous section summarized the history of laws that form the underpinning of FDA medical product regulation. This section provides a brief description of how laws are made and implemented, the procedures for promulgating regulations, and a description of how FDA develops and uses guidance documents.

Laws are created as an outcome of legislative activity conducted in the USA Senate and House of Representatives resulting in passage of a bill. Once Congress passes a bill, it becomes law if signed by the President. If the President vetoes the bill, it becomes law if two-thirds of the Senate and House of Representatives vote in favor of the bill. A federal law also is denoted as a public law and may contain a name, such as the FD&C or PHS Acts. These laws are then incorporated into the US Code (USC) which is updated every 6 years, with supplements published regularly to incorporate changes to statutes between updates. Drugs, biologics, and device laws can be found in the US Code at:

- Drugs and Devices: Title 21 Chapter 9
- Biologics: Title 42 Chapter 6A.

When laws are passed, government agencies, such as FDA, often implement them by promulgating regulations. Sometimes, an agency may elect to promulgate regulations on its own whereas other laws may explicitly require an agency to issue regulation. The process for making regulations must be performed in accordance to the Administrative Procedures Act (Title 5, USC, Chapter 5). This Act generally requires agencies, such as FDA, to provide public notice and opportunity for comment as part of the rule-making process.

FDA regulations are contained in the Code of Federal Regulations (CFR). Regulations for drugs, biologics, devices, and tissues, along with related regulations, may be found in various parts of Title 21 of the CFR. The following is a list of key regulatory provisions:

- Drugs: 21 CFR Parts 200–299, 300–369
- Biologics: 21 CFR Parts 600–680
- Devices: 21 CFR Parts 800–898
- Human Cells, Tissues, and Cellular and Tissue-based Products: 21 CFR Parts1270/1271
- Recalls: 21 CFR Part 7
- Informed Consent/Institutional Review Boards: 21 CFR Parts 50/56
- Financial Disclosure by Clinical Investigators: 21 CFR Part 54
- Good Laboratory Practice for Nonclinical Laboratory Studies: 21 CFR Part 58
- Good Guidance Practices: 21 CFR Part 10

Guidance documents are non-binding publications that describe FDA's interpretation of policy pertaining to a regulatory issue or set of issues related to:

- the design, production, labeling, promotion, manufacturing, and testing of regulated products
- the processing, content, and evaluation or approval of submissions
- inspection and enforcement policies.

Guidance documents, which are developed in accordance with Good Guidance Practices found at 21 CFR §10.115, are intended to clarify FDA's current thinking related to regulatory issues and procedures. Unlike regulations and laws, guidance documents are not enforceable. Therefore, Sponsors may elect to choose alternate approaches that still comply with existing laws and regulatory requirements. In most cases, guidance documents are issued in draft for public comment before implementation. In cases where prior public participation is not feasible or appropriate, FDA may issue a guidance document for immediate implementation without first seeking public comment. Many of the guidance documents referred to in this chapter, although available to the public, may be in draft form. The publication of draft guidance documents reflects FDA's efforts to convey up-to-date information to those involved in the developing field of regenerative medicine.

When considering development of a guidance document, FDA may freely discuss related issues with the public. In fact, FDA may hold a public meeting, advisory committee meeting, or workshop to obtain input on scientific issues. Finally, after receiving public input, FDA will evaluate submitted comments and finalize the document. Guidance documents are a useful way for FDA to communicate current thinking to the public. Within the arena of regenerative medicine, it is of value to be aware of both product-specific and cross-cutting guidance documents. Some of the more pertinent guidance documents to this field, such as those related to preclinical testing, manufacturing practices, and clinical trial design, are discussed in this chapter. In addition to guidance documents, FDA may refer to guidelines published by the International Conference on Harmonization (ICH). ICH is an international effort to harmonize regulatory requirements. ICH guidelines, similar to FDA guidance documents, are non-binding.

FDA ORGANIZATION AND JURISDICTIONAL ISSUES

Scientific development of regenerative medicine products involves extensive testing and planning prior to initiation of clinical trials. It can be helpful for individuals and organizations involved in product development to engage in early dialog with the appropriate FDA review unit in order to receive and consider FDA's comments on the design of the preclinical and clinical development plan. This section describes FDA's organizational structure and provides basic information regarding jurisdictional decisions made to determine the appropriate regulatory pathway for a broad range of products.

FDA consists of seven centers and the Office of the Commissioner. Three of the centers are responsible for regulating medical products for humans. The Center for Biologics Evaluation and Research (CBER) has jurisdiction over a variety of biological products, including blood and blood products, vaccines and allergenic products, and cellular, tissue, and gene therapies, as well as some related devices. The Center for Devices and Radiological Health (CDRH) has jurisdiction over diagnostic and therapeutic medical devices, administration of the Mammography Quality Standards Act (MQSA) program, and ensuring safety of radiation-emitting products. The Center for Drug Evaluation and Research (CDER) has jurisdiction over a variety of drug products, including small molecule drugs, and well-characterized biotechnology-derived drug products that include monoclonal antibodies and cytokines.

For many medical use products it is clear which center within FDA shall have primary jurisdiction for the premarket review. For other products, including some technologically novel products under development, determining which center has jurisdiction for review may be unclear. Important starting points for determining product jurisdiction are the formal regulatory definitions of biological products, drugs, devices, and combination products, as well as contacts with the agency. The formal definitions are as follows:

- Biological Product (42 USC 262(i)): A virus, therapeutic serum, toxin, antitoxin, vaccine, blood, blood component or derivative, allergenic product, or analogous product, or arsphenamine or derivative of arsphenamine (or any other trivalent organic arsenic compound), applicable to the prevention, treatment, or cure of a disease or condition of human beings.
- Drug (21 USC 321(g)(1)): (A) articles recognized in the official US Pharmacopeia, official Homeopathic Pharmacopeia of the USA, or official National Formulary, or any supplement to any of them; and (B) articles intended for use in the diagnosis, cure, mitigation, treatment, or prevention of disease in man or other animals; and (C) articles (other than food) intended to affect the structure or any function of the body of man or other animals; and (D) articles intended for use as a component of any articles specified in clause (A), (B), or (C).

- Device (21 USC 321(h)): An instrument, apparatus, implement, machine, contrivance, implant, *in vitro* reagent, or other similar or related article, including any component, part, or accessory, which is: (1) recognized in the official National Formulary, or the US Pharmacopeia, or any supplement to them, (2) intended for use in the diagnosis of disease or other conditions, or in the cure, mitigation, treatment, or prevention of disease, in man or other animals, or (3) intended to affect the structure or any function of the body of man or other animals, and which does not achieve its primary intended purposes through chemical action within or on the body of man or other animals and which is not dependent upon being metabolized for the achievement of its primary intended purposes.
- Combination Product (21 CFR 3.2(e)): (1) a product composed of two or more regulated components, that is, drug/device, biologic/device, drug/biologic, or drug/device/biologic, that are physically, chemically, or otherwise combined or mixed and produced as a single entity, (2) two or more separate products packaged together in a single package or as a unit and composed of drug and device products, device and biological products, or biological and drug products, (3) a drug, device, or biological product packaged separately that according to its investigational plan or proposed labeling is intended for use only with an approved individually specified drug, device, or biological product where both are required to achieve the intended use, indication, or effect and where upon approval of the proposed product the labeling of the approved product would need to be changed, for example, to reflect a change in intended use, dosage form, strength, route of administration, or significant change in dose, or (4) any investigational drug, device, or biological product packaged separately that according to its proposed labeling is for use only with another individually specified investigational drug, device, or biological product where both are required to achieve the intended use, indication, or effect.

FDA's Office of Combination Products (OCP), located in the Office of the Commissioner, has broad administrative overview responsibilities covering the regulatory life cycle of drug–device, drug–biologic, and device–biologic combination products. When jurisdiction is uncertain, Sponsors may contact OCP for assignment of primary regulatory review responsibility for combination and other medical products. Jurisdictional determinations are made following a formal submission process called a Request for Designation (RFD). The appropriate FDA center jurisdiction is determined by considering the primary mode of action of the product.

APPROVAL MECHANISMS AND CLINICAL STUDIES

There are several premarket approval pathways for medical products, depending on whether the product is a drug, biological product, or device. Approval pathways, explained in more detail below, include the Biologics License Application (BLA) for biologics and New Drug Application (NDA) for drugs. The Premarket Approval Application (PMA), Humanitarian Device Exemption (HDE) and 510(k) clearance mechanism are various regulatory pathways used for medical devices. Clarification on the type of application needed for a particular regenerative medicine product may be helpful to the Sponsor early in development, to enable the Sponsor to discuss the data needed for a marketing application during the planning stage.

A BLA is an application for licensure under the PHS Act; the approval standards set forth in the statute are demonstration that the product is safe, pure, and potent. Further information concerning the licensure of biological products is provided in "Guidance for Industry: Providing Clinical Evidence of Effectiveness for Human Drugs and Biologic Products" (US Food and Drug Administration, Center for Biologics Evaluation and Research, 1998b). A PMA is an application for approval for most Class III medical devices; the Sponsor must show reasonable assurance of safety and effectiveness (US Food and Drug Administration, Center for Devices and Radiological Health, 2002a). Under medical device regulation a product can also gain approval as an HDE, which is not a full marketing approval but requires demonstration of safety and probable benefit (US Food and Drug Administration, Center for Devices and Radiological Health, 2003). To qualify for this type of application, a Sponsor would need to first receive a designation from FDA Office of Orphan Products Development that the device is a Humanitarian Use Device (HUD), intended for treatment or diagnosis of a disease or condition that affects or is manifested in fewer than 4,000 individuals per year in the USA. The 510(k) clearance process applies to products that are "substantially equivalent" to a Class I or II (or in a few cases, a Class III) device already on the market.

Many, but not all, combination products are approved or cleared under one marketing application. For example, depending on the specific facts, including the primary mode of action of the product, a combination biological device could be licensed under the biologics authorities or approved under the medical device authorities. Following approval of a marketing application there are also post-marketing requirements such as reporting (US Food and Drug Administration, Center for Devices and Radiological Health, 2002b[1,2]). In addition, modifications to the product or labeling may require prior approval. FDA has published regulations and guidance documents that address submission and approval processes for modifications to marketed products (US Food and Drug Administration, Center for Devices and Radiological Health, 2002b, 2005a[3-5]). Compliance with manufacturing requirements is also an ongoing Sponsor obligation. FDA has issued a draft guidance document entitled "Draft Guidance for Industry and FDA: Current Good Manufacturing Practice (cGMPs) for Combination Products" which provides direction on applicable manufacturing requirements for combination products (US Food and Drug Administration, 2004). Due to the relatively new nature of regenerative medicine and its developmental status, post-approval topics will not be further discussed in this chapter.

In circumstances when clinical investigation is needed to evaluate the safety and efficacy of an investigational product prior to marketing approval, an Investigational New Drug (IND) application is required for drugs and biologics, and an Investigational Device Exemption (IDE) is generally required for devices (US Food and Drug Administration, Center for Biologics Evaluation and Research, 2005, 2006c; US Food and Drug Administration, Center for Devices and Radiological Health, 2010a). For both types of applications, the Sponsor needs to submit a description of the product and manufacturing process, preclinical studies, a clinical protocol, information on any other prior investigations such as human clinical studies, and a rationale for the study design. An Institutional Review Board (IRB) review and informed consent are also required. FDA has 30 days to review the application to determine if the study may proceed. The contents are specifically laid out in FDA regulations for each type of application. Requirements for the content of an IND can be found at 21 CFR 312.23 and for an IDE at 21 CFR 812.20.

For some products, there may be applicable guidance with respect to developing the manufacturing or the preclinical data to support the study. For example, the "Guidance for FDA Reviewers and Sponsors: Content and Review of Chemistry, Manufacturing, and Control (CMC) Information for Human Somatic Cell Therapy Investigational New Drug Applications (INDs)" discussed in the following section provides information on characterization and manufacturing of a cellular product to be submitted in an IND (US Food and Drug Administration, Center for Biologics Evaluation and Research, 2008a). Applicable regulations and

---

[1] Reporting for Biological Products: 21 CFR 314.80 and 21 CFR 314.81, 21 CFR 600.14, 21 CFR 600.80, 21 CFR 600. 81, 21 CFR 601.28, 21 CFR 601.70, and 21 CFR 601.93.

[2] Postmarketing Reports for Applications for FDA Approval to Market a New Drug (NDA): 21 CFR 314.80 and 314.81.

[3] PMA supplements: 21 CFR 814.39. (2007).

[4] Supplements and Changes to an Approved NDA: 21 CFR 314.70, 314.71, 314.72. (2007).

[5] BLA 21 CFR 814.39, 21 CFR 314.70, 21 CFR 314.71, 21 CFR 314.72.

guidance should be further consulted for information on adverse event reporting, labeling, study conduct and monitoring, and other topics related to requirements for conducting an IND (US Food and Drug Administration, Center for Biologics Evaluation and Research, 2005, 2006c). For information on general clinical study design and conduct issues, FDA has many guidance documents that may be helpful (US Food and Drug Administration, 2001b; US Food and Drug Administration, Center for Biologics Evaluation and Research, 2001). For some indications there may be guidance documents that apply across technologies, such as the "Guidance for Industry: Chronic Cutaneous Ulcer and Burn Wounds — Developing Products for Treatment" (US Food and Drug Administration, 2006). In addition, guidance documents not directly applicable for a specific product, indication, or technology may be worth consulting, as the documents may provide some insights into general clinical issues such as assessment parameters that may be of value.

MEETINGS WITH INDUSTRY, PROFESSIONAL GROUPS, AND SPONSORS

Although the terminology and procedures may vary, all three FDA Centers performing medical product review encourage meetings with Sponsors to address questions prior to a regulatory submission and at specific developmental milestones. When requesting a formal or informal meeting with FDA it is helpful to provide background information as well as specific discussion questions. Further information about formal meetings, such as what to include in a meeting request, and what type of information to include in the information package submitted prior to the meeting, is provided in "Guidance for Industry: Formal Meetings with Sponsors and Applicants for PDUFA Products" (US Food and Drug Administration, Center for Biologics Evaluation and Research, 2000). Early stage device meetings are addressed in "Early Collaboration Meetings under FDA Modernization Act (FDAMA); Final Guidance for Industry and CDRH Staff" (US Food and Drug Administration, Center for Biologics Evaluation and Research, 2001a).

FDA also interacts with organizations representing a group of interested parties (e.g. International Society for Cellular Therapy, American Association for Blood Banks, and Pharmaceutical Research and Manufacturers of America), which provides an opportunity to discuss topics of interest to FDA and the organization. These interactions can be very valuable for FDA and stakeholders as they are a way to better understand general issues of concern, as opposed to product-specific discussions with individual firms. In addition to such interactions and meetings with individual sponsors, FDA has various advisory committees that review available data and information, and make recommendations related to a variety of issues, many of which are pertinent to the field of regenerative medicine. Advisory committees will be discussed further in the "Advisory Committee Meetings" section.

REGULATIONS AND GUIDANCE OF SPECIAL INTEREST FOR REGENERATIVE MEDICINE

The topics discussed thus far have been of general applicability for medical product regulation: marketing pathways, clinical trial regulation, meetings, guidance development, and related topics. This section will review a few topics of particular interest to the scientific community engaged in development of regenerative medicine products: FDA regulations on human tissue products, product characterization for cellular products, FDA policy and guidance on xenotransplantation, and gene therapy.

Regulation of human cells and tissues intended for transplantation

An understanding of the regulations applicable to cells and tissues is important for developers of regenerative medicine products since human cells or tissues comprise the whole, or are a key component, of many products.

In 1997, noting the fragmented approach to regulation of human cell and tissue-based products, FDA issued the "Proposed Approach to the Regulation of Cellular and Tissue-Based Products" (US Food and Drug Administration, 1997). This document proposed a tiered risk-based approach to regulation of these products. According to the proposed approach, products posing less risk would be subject to the rules designed to minimize communicable disease risks, and additional regulatory requirements would be imposed on those products posing additional risk. The proposed approach to regulation of human tissues was implemented in three parts, collectively referred to as the "tissue rules": Establishment Registration and Listing, Donor Eligibility, and Good Tissue Practices (GTP). The complete set of rules went into effect on May 25, 2005, and are codified in 21 CFR Part 1271. The tissue rules derive from the statutory authority of section 361 of the PHS Act, which addresses control of spread of communicable diseases. Because the tissue rules apply to all human cellular and tissue-based products, it is important for Sponsors of regenerative medicine products to be aware of these rules, as well as the specific additional requirements for biologics or devices that may apply depending on the particular regulatory pathway for their products.

With some exceptions that are noted in the tissue rules, human cells or tissue intended for implantation, transplantation, infusion, or transfer into a human recipient are regulated as a human cell, tissue, and cellular and tissue-based product (HCT/P). Examples of HCT/Ps are: musculoskeletal tissue, skin, ocular tissue, human heart valves, dura mater, reproductive tissue, and hematopoietic stem/progenitor cells. Tissues specifically excluded are: vascularized organs, minimally manipulated bone marrow, blood products, xenografts, secreted or extracted products such as human milk and collagen, ancillary products, and *in vitro* diagnostic products.

The tissue rules require that tissue establishments do the following:

- register and list their HCT/Ps with FDA (21 CFR 1271 Subparts A and B)
- evaluate donors through screening and testing, to reduce risk of transmission of infectious diseases through tissue transplantation (21 CFR 1271 Subpart C)
- follow Current Good Tissue Practices to prevent the spread of communicable disease (21 CFR 1271 Subpart D).

Additional requirements for reporting, labeling, inspections, importation, and enforcement are described in 21 CFR 1271 Subparts E and F; these provisions apply only to HCT/Ps regulated solely under section 361 of the PHS Act, and therefore would not apply to most regenerative medicine products.

The Establishment Registration rule defines the circumstances under which a product would be subject to the tissue rules only (21 CFR 1270.10), and when there would be additional regulatory oversight such that a BLA, PMA, or other marketing application would be required (21 CFR 1271.20). Products that meet the following conditions are regulated by FDA solely under the tissue rules: (1) the tissue is not more than minimally manipulated, (2) the tissue is intended for homologous use, (3) the tissue is not combined with a drug or device (with certain exceptions), and (4) the tissue does not have a systemic effect and is not dependent upon the metabolic activity of living cells for its primary function (except for autologous use or allogeneic use in a first or second degree blood relative, or reproductive use). If all four of these conditions are not met, a marketing application is required. The Tissue Reference Group (TRG) handles various inquires from stakeholders concerning application of the tissue rules including generating recommendations for consideration for CBER, CDRH, and OCP regarding whether specific HCT/Ps meet the criteria specified in 21 CFR 1271.10 for regulation solely under Section 361 of the Public Health Service Act. Additional information and documents regarding these rules, as well as electronic forms for registration and listing, can be found on FDA's website (US Food and Drug Administration, Center for Biologics Evaluation and Research, 2006d).

A joint FDA–CDC workshop held in 2007 on the Processing of Orthopedic, Cardiovascular and Skin Allografts is of relevance to the regenerative medicine field. The workshop discussed pertinent information regarding current clinical practices and expectations for the graft, effect of processing on graft management, and adverse event assessment/management and challenges associated with current microbiological methods used for pre- and post-processing cultures and with disinfection/sterilization of tissues (US Food and Drug Administration, Center for Biologics Evaluation and Research, 2007b). Some of the manufacturing topics discussed at the workshop are addressed in draft guidance for industry entitled: "Draft Guidance for Industry: Current Good Tissue Practice (CGTP) and Additional Requirements for Manufacturers of HCT/Ps" (US Food and Drug Administration, Center for Biologics Evaluation and Research, 2009).

FDA has issued "Guidance for Industry: Eligibility Determination for Donors of Human Cells, Tissues, and Cellular and Tissue-Based Products" to assist establishments making donor eligibility determinations with complying with the Donor Eligibility rule (21 CFR 1271 Subpart C) (US Food and Drug Administration, Center for Biologics Evaluation and Research, 2007a). This guidance also incorporates and finalizes the content of "Guidance for Industry, Preventive Measures to Reduce the Possible Risk of Transmission of Creutzfeldt–Jakob Disease (CJD) and Variant Creutzfeldt–Jakob Disease (vCJD) by Human Cells, Tissues, and Cellular and Tissue-Based Products (HCT/Ps)."

Human cellular therapies

Many products in development for tissue repair or replacement are composed of cells or cells combined with a scaffold. The cell or tissue source and manufacturing process may vary greatly for different products. Despite the diversity in products, there are regulatory considerations that apply to all cellular preparations being developed as investigational regenerative medicine products intended for early phase clinical studies. Among these considerations are three that will be discussed briefly: control of the source material, demonstrated control of the manufacturing process, and characterizations of the cellular product that results from the manufacturing process.

The cell source will vary for different products and may be autologous or allogeneic, undifferentiated stem/progenitor cells, or terminally differentiated cells. Assuring the safety of source cellular materials used during manufacture of an investigational regenerative medicine product begins by determining the eligibility of the donors selected to provide the source material through screening and testing. This screening and testing is part of the tissue rules described earlier in this chapter. Autologous products are not required to comply with the donor screening and testing requirements in the tissue rules. However, if autologous tissue either is positive for specific pathogens or has not been screened or tested, it is recommended that manufacturers document if tissue culture methods could propagate or spread viruses or other adventitious agents to persons other than the recipient (US Food and Drug Administration, Center for Biologics Evaluation and Research, 2008a). Donor eligibility determination is required for all allogeneic donors of cells and tissues.

In addition to the possibility of infectious disease transmission, there are other aspects of the cell source that may raise substantial concerns. In addition to screening and testing donors for communicable disease agents, according to the "ICH Guidance on Quality of Biotechnological/Biological Products: Derivation and Characterization of Cell Substrates Used for Production of Biotechnological/Biological Products," FDA has suggested that Sponsors consider the importance of evaluating donor medical history information and the relevance of conducting specified molecular genetic testing as part of an overall comprehensive assessment program to establish the suitability of a specific cellular preparation for use in the manufacture of a regenerative medicine somatic cellular product (US Food and Drug Administration, Center for Biologics Evaluation and Research, 1998c). The rationale for and feasibility of

1153

collecting additional information about molecular genetic testing were discussed in a public meeting of FDA Biological response modifiers Advisory Committee (now known as Cellular, Tissue and Gene Therapies Advisory Committee (CTGTAC)) convened July 13–14, 2000 on the topic of "Human Stem Cells as Cellular Replacement Therapies for Neurological Disorders" (US Food and Drug Administration, Center for Biologics Evaluation and Research, 2006a). A description of the physiological source of the cellular material, including tissue of origin and phenotype such as hematopoietic, neuronal, fetal, or embryonic, conveys important information about the cells and their critical attributes.

Control of the manufacturing process helps provide assurance of the consistent, reproducible production of the cellular component. Often, manufacturing will involve a multistep process that must be performed using aseptic techniques to prevent introduction of microbial contamination (US Food and Drug Administration, 2004b). Many types of reagents may be used to manufacture the cellular component of a product, including those that promote cellular replication, induce differentiation, and those used to select targeted cell populations, specifically, serum, culture medium, peptides, cytokines, and monoclonal antibodies. It is essential that reagents be properly qualified (US Food and Drug Administration, Center for Biologics Evaluation and Research, 1993, 1997, 2008a, b). Demonstration of manufacturing control is evidenced by strict adherence to standard operating procedures and quality control assessment of manufacturing intermediates as well as testing of the final cellular preparation.

Due to inherent biological complexity it is unlikely that a unique biomarker or other single analytical test will be sufficient to permit full characterization of a cellular product. Accordingly, as recommended in "Guidance for Industry: Guidance for Human Somatic Cell Therapy and Gene Therapy," FDA asks Sponsors to provide documentation that their testing paradigm developed for the final cell product encompasses a multiparametric approach that may involve biological, biochemical/biophysical, and/or functional characterization (US Food and Drug Administration, Center for Biologics Evaluation and Research, 1998a, 2008a, b). Tests developed to demonstrate identity of the cell product (physical and chemical characteristics, identify the product as being what is designated on the label), purity (freedom from contaminants including residual reagents and unintended cell populations), and potency/biological activity (the specific ability of the cells, as indicated by appropriate laboratory tests, to effect a given result) should be conceived to determine the degree to which the characteristics of the manufactured cell preparation conform to desired and specified criteria (US Food and Drug Administration, Center for Biologics Evaluation and Research, 1998a, 2008a, b). This process can be challenging for a number of reasons. For example, the mechanism of action associated with a cell product may be incompletely understood and this constrains the ability to develop a specific potency assay. Direct assessment of potency for a cellular preparation may not be possible due to a lack of appropriate *in vitro* or *in vivo* assay systems. On February 9–10, 2006, FDA CTGTAC discussed this challenging topic and obtained input on alternative approaches for performing potency assessments of cellular therapy products (US Food and Drug Administration, Center for Biologics Evaluation and Research, 2006a).

In summary, assuring the safety of cell products that in and of themselves constitute a regenerative medicine product or that constitute a component of a product requires demonstrated control over each facet of the manufacturing process. This assurance begins with acquisition of the source material and is carried forward through manufacturing and characterization of the final cellular preparation using specified analytical tests based in large measure on the intrinsic biological properties of the cell product.

Xenotransplantation

The success of allogeneic organ transplantation has increased the demand for human cells, tissues, and organs. Scientific advances in the areas of immunology and molecular biology coupled with the growing worldwide shortage of transplantable organs have led to increased interest in xenotransplantation. In addition to the potential use of xenotransplantation to address the shortage of human organs for transplantation, there are increasing efforts to utilize other xenotransplantation products in the treatment of disease. An example of this is the use of encapsulated porcine pancreatic islet cells for the treatment of type 1 diabetes. Along with the promise of xenotransplantation are a number of challenges, including the potential risk of transmission of infectious agents from source animals to patients, and the spread of any zoonotic disease to the general public (Horvath-Arcidiacono et al., 2010). In addition, the potential exists for recombination or reassortment of source animal infectious agents, such as viruses, with non-pathogenic or endogenous human infectious agents, to form new pathogenic entities. These considerations demonstrate the need to proceed with caution in this area.

The US PHS Agencies including FDA, National Institutes of Health (NIH), Centers for Disease Control and Prevention (CDC), and Health Resources and Services Administration (HRSA) have worked together to address the risk of infectious disease transmission, publishing the "PHS Guideline on Infectious Disease Issues in Xenotransplantation" (US Public Health Service, 2001). This Guideline discusses xenotransplantation protocols, animal source, clinical issues, and public health issues. Following publication of the PHS Guideline, FDA published a Guidance document entitled "Guidance for Industry: Source Animal, Product, Preclinical and Clinical Issues Concerning the Use of Xenotransplantation Products in Humans" to build on the concepts in the PHS Guideline, and provide more specific advice regarding xenotransplantation product development and production, and xenotransplantation clinical trials (US Food and Drug Administration, 2003).

Xenotransplantation is defined in PHS Guideline and FDA Guidance as any procedure that involves the transplantation, implantation, or infusion into a human recipient of either live cells, tissues, or organs from a non-human animal source or human body fluids, cells, tissues, or organs that have had *ex vivo* contact with live non-human animal cells, tissues, or organs (US Public Health Service, 2001; US Food and Drug Administration, Center for Biologics Evaluation and Research, 2003).

Examples of xenotransplantation products provided in FDA Guidance are:

- transplantation of xenogeneic hearts, kidneys, or pancreatic tissue to treat organ failure, implantation of neural cells to ameliorate neurological degenerative diseases
- administration of human cells previously cultured *ex vivo* with live non-human animal antigen-presenting or feeder cells
- extracorporeal perfusion of a patient's blood or blood component through an intact animal organ or isolated cells contained in a device to treat liver failure.

Medical products that do not contain living cells are not considered to be xenotransplantation products by this definition. Therefore, products that include some common animal-derived components such as collagen, small intestinal submucosa (SIS), and heart valves do not automatically fall under this category. When a product does meet the definition, xenotransplantation guidelines are applied as appropriate to the specific product. FDA encourages any potential sponsor of a xenotransplantation product to familiarize themselves with available documents that can be found on FDA's website (US Food and Drug Administration, Center for Biologics Evaluation and Research, 2006e).

Gene therapy

FDA regulates human gene therapy products as biological products. The field of gene therapy holds great promise for treating a wide array of illnesses, from genetically inherited diseases such as cystic fibrosis or hemophilia, to heart disease, wound healing, AIDS, graft versus host disease, and cancer. The use of gene therapy in the area of tissue repair and tissue engineering is also being investigated.

1155

There are a number of safety issues associated with gene therapy, some of which are unique to this area. Safety issues specific to gene therapy trials include generation of replication competent virus, vector as well as transgene-associated immune responses, toxicity associated with transgene expression, and inadvertent germline transmission of vector. Two examples of gene therapy-specific risks are instructive: (1) high doses of adenovirus vector particles have been shown to induce toxicity under certain circumstances, and resulted in the death to a study subject in 1999, and (2) genomic integration of retroviral vectors has been shown to result in genotoxicity. In the latter case, five children developed leukemia and one died as a direct result of altered gene expression after vector integration. Detailed recommendations from FDA regarding what type of information to submit in an early phase study of gene therapy products are available in FDA "Guidance for FDA Reviewers and Sponsors: Content and Review of Chemistry, Manufacturing, and Control (CMC) Information for Human Gene Therapy Investigational New Drug Applications (INDs)" (US Food and Drug Administration, Center for Biologics Evaluation and Research, 2008b). This guidance covers product manufacturing and characterization information (including components and procedures), product testing (including microbiological testing, identity, purity, potency, and other testing), final release testing criteria, and product stability, giving specifics in these areas that are pertinent to gene therapy. Suggested preclinical testing includes tests designed to describe localization, and persistence of gene expression. For vectors intended for direct *in vivo* administration, demonstrating the extent of dissemination and gonadal distribution is suggested.

Gene therapies may differ from conventional drugs in that vector and transgene expression may persist for the lifetime of the subject. In these cases, there is a risk of delayed adverse effects. Indeed, the previously mentioned leukemias in a clinical study of gene therapy for the treatment of X-linked severe combined immunodeficiency (SCID) did not occur until approximately 3 years after exposure to the retroviral vector. These events highlight the need for assessment of long-term risk in research subjects. FDA has discussed these issues, noting that the assessment of risk is based on the persistence of vector sequences, the potential for integration into the host genome, and transgene-specific effects. FDA has published "Guidance for Industry: Gene Therapy Clinical Trials — Observing Subjects for Delayed Adverse Events" which addresses the duration and types of observations to be performed based on the patient population and the risks presented by the gene therapy product (US Food and Drug Administration, Center for Biologics Evaluation and Research, 2006b).

Although regulatory authority for gene therapy trials rests with FDA, NIH serves an important complementary role. In addition to funding a number of gene therapy research studies, NIH provides an important forum for open public deliberation on the scientific, ethical, and legal issues raised by recombinant DNA technologies and its basic and clinical research applications. This is accomplished through the Recombinant DNA Advisory Committee (RAC), an expert advisory committee to the NIH Director (US National Institutes of Health, 2010). Clinical studies discussed in this forum include studies funded by NIH, as well as industry-funded studies conducted at clinical sites receiving NIH funding for recombinant DNA research.

Cell—scaffold combination products

Cell—scaffold combination products often face unique product development challenges because of their inherent complexity. These products often combine metabolically active cells and extracellular matrix or other scaffold components into complex three-dimensional structures, making the manufacturing, characterization, and study of these products a challenge. Such a complex product that is derived from chemically or physically combining multiple entities cannot be defined solely by the characteristics of the individual components alone. Other factors such as product assembly and the resulting cell—scaffold interactions play critical roles in determining final product characteristics. Furthermore, these products are commonly designed to remodel *in vitro* during processing in bioreactors and/or *in vivo* post-transplantation during clinical use, thereby precluding complete functionality testing by product release. Packaging, shipping, and shelf-life for these dynamic cell—scaffold products are also non-trivial considerations.

As with other products, product safety and efficacy of cell—scaffold products need to be supported by a combination of appropriate *in vitro* and *in vivo* preclinical testing. FDA draws upon its extensive experience in regulating mammalian cell products and other tissue-derived products in evaluating product safety and efficacy. Many of the important tests needed for the individual components, such as sterility, mycoplasma, pyrogenicity/endotoxin, scaffold characteristics, cellular viability, identity, and purity, are applicable for the combined product as well. Demonstration of product potency and/or performance is also necessary and may require the development of new scientific techniques and assays (Lee et al., 2010). A critical consideration for developers of cell—scaffold combination products is determining which tests need to be conducted on individual components prior to assembly and which are most relevant after product assembly. As each product is often unique with its own set of potential safety issues for a specific clinical indication, preclinical evaluation of these products in the appropriate animal models provides important data to establish controls for product development and manufacture. In addition, it is strongly recommended that Sponsors engage FDA early in the product development process to obtain feedback on key considerations.

For many innovative products, such as cell—scaffold combinations used for regenerative medicine, the final product and instructions for use can be expected to undergo iterative modifications over time. Consequently, refinement of the product by the Sponsor and review of product modifications by FDA will be an ongoing process. It is critical for the Sponsor to have a good understanding of the product and key scientific and/or clinical issues that can affect safety and efficacy of the product, including the establishment of appropriate manufacturing controls to ensure product quality and consistency. When changes in composition or manufacturing of the cell and/or scaffold component of the combination product are made, it is essential that the Sponsor fully evaluate the impact of such changes to the final product's quality and function.

Case study: cell—scaffold wound healing skin constructs

Some of the earliest successes in developing cell—scaffold combination products for medical use produced skin-like constructs for wound management. These products are worth a special mention in this chapter because there are a number of FDA approved products currently on the market. These products are often composed of keratinocyte and/or fibroblast cells combined with a scaffold (e.g. animal collagen, glycosaminoglycans, or gauze). Cell—scaffold wound dressings for the skin that function primarily as physical wound coverings have generally been reviewed by CDRH.

Aspects of product manufacturing that are common to these products are cell, tissue, and scaffold material sourcing, product processing, in-process and final product tests, and quality control procedures to ensure lot-to-lot safety, effectiveness, and consistency. Considerations regarding product shipment and shelf-life are non-trivial in the manufacture and distribution of these skin constructs that contain living cells. Preclinical evaluation of these products, including study in animal models for wound healing, provide important data to establish controls in product development and manufacture.

Apligraf and Orcel are two examples of commercially available bilayered co-culture constructs composed of allogeneic neonatal keratinocyte and fibroblast cells on bovine collagen scaffolds. Apligraf is indicated for treatment of venous insufficiency and diabetic foot ulcers (US Food and Drug Administration, Center for Devices and Radiological Health, 2000a) and

Orcel is approved for treating split thickness donor site wounds on burn patients (US Food and Drug Administration, Center for Devices and Radiological Health, 2001). An HDE for Orcel use in a patient with recessive dystrophic epidermolysis bullosa (DEB) as an adjunct in covering wounds and donor sites after the surgical release of hand contracture and deformities was approved in 2001 (US Food and Drug Administration, Center for Devices and Radiological Health, 2002b). Lastly, Epicel consists of autologous keratinocytes (ranging from two to eight cell layers in thickness) layered on petrolatum gauze. Notably, Epicel is defined as a xeno-transplantation product because, as part of the manufacturing process, the keratinocytes are co-cultivated with a feeder layer of proliferation-arrested, murine 3T3 fibroblasts. The HDE for Epicel was approved in 2007 for use in patients who have deep dermal or full thickness burns comprising a total body surface area of greater than or equal to 30% (US Food and Drug Administration, Center for Devices and Radiological Health, 2007).

Information about the clinical performance of each product is available in the published literature and product labeling (Green and Rheinwald, 1975; Cazalet et al., 1988; Boyce et al., 1989; Hansbrough et al., 1992; Haeseker et al., 1993; Baird et al., 1998; Carsin et al., 2000; US Food and Drug Administration, Center for Devices and Radiological Health, 2001, 2005b; Currie et al., 2002; Lee et al., 2010). Summaries of Safety and Effectiveness Data (SSEDs) that provide information about the clinical studies, as well as other studies supporting approval, for each of these approved products are available on FDA's website using a searchable database (using the search term "product code MGR") (US Food and Drug Administration, Center for Devices and Radiological Health, 2010b). Information on approved HDEs and associated Summary of Safety and Probable Benefit are also available on FDA's website (US Food and Drug Administration, Center for Devices and Radiological Health, 2010c).

1158

PRECLINICAL DEVELOPMENT PLAN

For device and drug clinical trials, the goal of preclinical development studies is to establish a scientific rationale for the clinical investigation and to demonstrate an acceptable safety profile. Traditional pharmacology/toxicology safety studies are important to identify potential toxicity in target organs and tissues, and to obtain information on effective safe starting doses in humans as well as establishing a safety profile for dose escalation and/or clinical monitoring. For cellular therapy, gene therapy, and cell—scaffold combination products there are frequently additional product-specific safety questions that might need to be addressed prior to initiation of a clinical trial. For example, what is their potential to undergo unanticipated undesired changes in their characteristics, such as malignant trans-formation? For cell—scaffold constructs, are there safety issues associated with the implan-tation procedure or potential construct failure? Animal models have limitations that are confounded by anatomical as well as physiological differences between the animals and humans. For some products, *in vitro* analyses may play a critical role in product character-ization and evaluation for safety, efficacy, and potency. Non-destructive product testing and rapid characterization tests may also aid in product development. The goals, challenges, and methods of *in vitro* analyses of cell—scaffold products were discussed at a recent FDA-NIST co-sponsored workshop (McCright et al., 2009). Many of these products, because of their novelty, do not have an established paradigm for preclinical evaluation and Sponsors are therefore encouraged to discuss their development plan with FDA early in the development process.

In several specific clinical applications FDA has either had public discussion at an Advisory Committee meeting or published Guidance documents, and in either case it is valuable for Sponsors to be aware of these discussions or publications. For example, in March 2005 the CTGTAC discussed the manufacturing, preclinical, and clinical issues in the development of cellular therapies for repair and regeneration of articular joint surfaces (US Food and Drug

Administration, Center for Biologics Evaluation and Research, 1998c). In the preclinical session of this meeting the committee focused on addressing specific issues raised by FDA with respect to the range of the subset of products (device, cellular, and tissue-engineered (TE) products) that have been proposed for the repair and regeneration of articular cartilage. However, the issues can be readily generalized to a broader range of regenerative medicine products. The major issues discussed were:

- the limitations and capabilities of available animal models for predicting safety and clinical activity
- pivotal animal toxicology studies designed to support a clinical trial of a cellular cartilage repair product
- additional safety concerns for allogeneic cellular products (versus autologous products) that should be addressed in an *in vivo* study prior to clinical trials.

The committee arrived at a consensus that animal studies were needed to evaluate potential products prior to human administration; however, it also was evident that there was no single animal model that is adequate to test all of the hypotheses involved in development of these complex products. For example, the committee acknowledged small animals might be useful in assessing novel biomaterials or mechanisms of action in early stages of product development. Unfortunately, due primarily to anatomical and biomechanical considerations, large animals are needed to assess the potential clinical activity, therapeutic durability, and safety of a final product prior to initiation of clinical trials ("pivotal" toxicology studies). The committee also discussed the necessary study design features for pivotal toxicology studies and reached a general agreement that the design of such studies should incorporate:

- assessment of the mechanism of action of the product by either interim sacrifice of subgroups of animals or by non-lethal methods of assessment such as imaging or arthroscopy (particularly if this use in animals can provide data to support use of these modalities in clinical trials by allowing comparison between in-life and necropsy assessments)
- adequate duration prior to terminal sacrifice to demonstrate integration of the product with the native tissue
- histopathological examination of the site of implantation and surrounding regions, draining lymph nodes and remote organs with gross pathology at time of necropsy and histopathology of other organs as guided by data from preliminary studies.

The committee discussed the potential need to use analogous animal cells in lieu of human cells in testing the final product construct and the limitations posed by this approach in terms of extrapolating the data to human trials. In recognition of this concern the committee highlighted the need to understand the physiology of the animal cells relative to their human analogs. The committee also discussed the potential for additional safety concerns posed by use of allogeneic cells in these products.

CLINICAL DEVELOPMENT PLAN

The goal of the clinical development program is to establish product safety and efficacy. In the field of regenerative medicine, variability in the product, as well as the patient, poses unique challenges in clinical trial design and conduct. An additional challenge is the need, with many of these products, to observe their integration into the host over a prolonged period. Specific feedback regarding the adequacy of certain proposed studies and predictors of clinical benefit can be provided to Sponsors through the use of meetings with FDA at various development time points, and the use of a Special Protocol Assessment (for products regulated as a biologic) or an Agreement Meeting (for products regulated as a device) prior to initiation of their Phase III studies (US Food and Drug Administration, 2001a, 2002).

FDA's standards development program

Since its inception, the development and use of standards have been critical to the mission of FDA. The use of standards in FDA medical product regulation began with the 1906 Federal Food and Drugs Act. Drugs, defined in accordance with the standards of strength, quality, and purity in the US Pharmacopeia and the National Formulary, could not be sold in any other condition unless the specific variations from the applicable standards were plainly stated on the label (Federal Food and Drugs Act, 1906). In current times, Federal government agencies, including FDA, are encouraged, when practical, to use voluntary consensus standards, whether domestic or international, when performing regulatory activities in lieu of government-unique standards which are developed by the government for its own uses. Standard-setting activities include the development of performance characteristics, testing methodology, manufacturing practices, product standards, scientific protocols, compliance criteria, ingredient specifications, labeling, or other technical or policy criteria.

As with guidance document development, in which Good Guidance Practices describe FDA's procedures for developing and using guidance documents, there are specific regulations that describe FDA participation in outside standard-setting activities. Regulations governing this participation can be found in 21 CFR 10.95. Additionally, FDA's Staff Manual Guide (SMG) 9100.1 establishes Agency-wide policies and procedures related to standards management activities to assure a unified approach to standards within FDA. Constructive FDA participation in organizations responsible for developing standards applicable to the products regulated by the agency is considered essential.

The FDAMA of 1997 provides for the recognition of national and international standards in medical device reviews for IDEs, HDEs, PMAs, and 510(k)s (Marlowe and Phillips, 1998). A "recognized consensus standard" is a consensus standard that FDA has evaluated and recognized, in full or in part, for use in satisfying a regulatory requirement and that FDA has published in a Federal Register notice. A "consensus standard" is a standard developed by a private sector standards body using an open and transparent consensus process. Conformance with recognized consensus standards is strictly voluntary for a medical device manufacturer, who may choose either to conform to applicable recognized standards or to address relevant issues in another manner. Standards may also be used in support of non-device applications when appropriate and not in conflict with FDA regulation or Guidance. Lists of recognized standards, Guidance Documents, and Standard Operating Policies and Procedures (SOPP) can be found on FDA's website.

CDRH and CBER work actively with standards development organizations such as the American Society for Testing and Materials International (ASTM International). ASTM Committee F04 on Medical and Surgical Materials and Devices Division IV is actively engaged (e.g. attends meetings and workshops, participates in formal voting) in development of standards for tissue engineered medical products (TEMPs). F04 Division IV consists of six subcommittees: (1) Classification and Terminology, (2) Biomaterials and Biomolecules, (3) Cells and Tissue Engineered Constructs, (4) Assessment, (5) Adventitious Agent Safety, and (6) Cell Signaling. Currently, the ASTM TEMPs group has developed more than 25 published standards, including standard guides and test methods, and has approximately 30 draft standards under preparation. The first of these standards were developed for substrates, biomaterials, natural materials such as collagen, alginate, and chitosan, terminology, cells and cell processing, bone morphogenetic protein, assessment of adventitious agents, and test methods for characterizing biomaterials. These standards are reviewed on a regular basis by the appropriate ASTM subcommittee to ensure that the standards reflect the current scientific knowledge and FDA regulatory practices. Some examples of approved standards with which the TEMPs group was involved include standards for the classification of skin substitutes, the characterization and testing of biomaterial scaffolds, the quantification of cell viability within biomaterial scaffolds, and the *in vivo* repair of articular cartilage.

Another standards development organization with which FDA is involved is the International Standards Organization (ISO), a non-governmental international organization that develops consensus standards in collaboration with both the private and public sector. Standards for regenerative medicine/tissue engineering products are developed in TC 150 subcommittee (SC) 7, Tissue Engineered Medical Devices, and in TC 194, Biological Evaluation of Medical Devices. TC 194 SC 01 is responsible for Tissue Product Safety, and within SC 01 are four Working Groups (WGs): WG 01 Risk Assessment, Terminology, and Global Aspects, WG 02 Sourcing Controls, Collection, and Handling, WG 03 Elimination and/or Activation of Viruses and Transmissible Spongiform Encephalitis (TSE) Agents, and WG 04 TSE Elimination. ISO standards of particular interest to regenerative medicine products are those in the 10993 series (10993–1 through 20) (standards are available through ISO's website: http://www.iso.org).

FDA is actively engaged in standards development. Participation in standards development activities benefits the regenerative medicine community in that these activities can facilitate the development and maintenance of guidance for industry, address issues not addressed in FDA guidance documents, facilitate product design, and lead to international harmonization. Benefit to the regenerative medicine community also helps, in turn, the public. Thus, FDA plays a critical role in providing support for standards development activities.

ADVISORY COMMITTEE MEETINGS

As mentioned in Section "Meetings with Industry, Professional Groups, and Sponsors" above, because of the diversity of innovative technology evaluated by FDA review staff, FDA makes use of expert scientific Advisory Committees or Panels (for medical devices) to complement its internal review process. These advisors provide outside advice to contribute to scientific regulatory decision making. Outside experts can be asked to review data, or make recommendations about study designs across a product or clinical area; outside advisors can also be helpful at earlier stages of product development. Expertise on the advisory committee often includes scientific, statistical, and clinical experts, as well as consumer representation, patient advocates, and industry participation. Most meetings are public and there is an opportunity for public participation in the form of public comment.

There are 30 Advisory Committees and 18 Advisory Panels for medical devices (as well as a number of subcommittees and one Department of Health and Human Services (DHHS) Committee administered by CBER). The areas of responsibility for the panels and committees are divided along product lines. The Advisory Committee for cellular, tissue, and gene therapy products, as mentioned earlier, is the CTGTAC. This committee has discussed a number of areas in recent years that are of potential interest to product developers in the regenerative medicine area, including:

- Hematopoietic stem cells for hematopoietic reconstitution (February 2003)
- Allogeneic islet cell therapy for diabetes (October 2003)
- Somatic cell cardiac therapies (March 2004)
- Cellular products for joint surface repair (March 2005)
- Potency measures for cell, tissue and gene therapies (February 2006)
- Cellular therapies derived from human embryonic stem cells — considerations for pre-clinical safety testing and patient monitoring (April 2008)
- Animal models for porcine xenotransplantation products intended to treat type 1 diabetes or acute liver failure (May 2009)
- Clinical issues related to FDA draft guidance "Preparation of IDEs and INDs for Products Intended to Repair or Replace Knee Cartilage" (May 2009).

The presentations for each topic, as well as a transcript of the discussion, are available on FDA's website referenced at the end of this chapter (US Food and Drug Administration, Center for Biologics Evaluation and Research, 2006a).

As mentioned previously, the Medical Devices Advisory Committee consists of 18 panels that cover the medical specialty areas. Panel meetings are held regularly to discuss specific products. A few examples of past panel meetings are:

- Ear, Nose and Throat Devices Panel regarding an implantable hearing system (December 2009)
- Neurological Devices Panel regarding a spinal surgery sealant (May 2009)
- Orthopaedic and Rehabilitation Devices Panel regarding a bone putty (March 2009)
- Opthalmic Devices Panel regarding a visual prosthetic device (March 2009)
- General and Plastic Surgery Panel regarding dermal fillers (November 2008).

A complete list of upcoming Medical Devices Advisory Committee panel meetings and searchable archive of past meetings with agendas and materials can be found on FDA's website (US Food and Drug Administration, 2010).

FDA RESEARCH AND CRITICAL PATH SCIENCE

FDA recognizes the complexity of the scientific issues related to regenerative medicine products. FDA research laboratories play an important role in helping ensure that the agency stays abreast of the rapid change and development affecting the entire field as well as addressing specific regulatory science questions. In 1992, CBER researchers began systematic efforts to uncover mechanisms that control the behavior of cells subjected to various environmental stimuli, particularly those encountered during normal wound healing, regeneration, and prenatal development. Those studies led to the discovery of several growth factors and feedback mechanisms that help control these pathways (Hoang et al., 1996; Lin et al., 1997; Wang et al., 1997a, b; Thomas et al., 2006, 2009; Lenas et al., 2009a, b). Other research addressed the interactions between hematopoietic and mesenchymal cell lineages both *in vitro* and *in vivo* (Bauer et al., 1998; Ramadevi Raghunandan et al., 2008). The success of those efforts led to the recruitment of additional investigators in these areas. In 2004, FDA introduced the Critical Path Initiative to identify and support research priorities that are expected to advance innovation in medical products. The Critical Path Opportunities List "presents specific opportunities that, if implemented, can help speed the development and approval of medical products" (US Department of Health and Human Services, Food and Drug Administartion, 2006), and is available on FDA's website. A number of the research topics on the Critical Path Opportunities List have applications to regenerative medicine, such as developing characterization tools for cell therapy and tissue engineering, biomarkers for cardiovascular disease, and advanced imaging technologies. FDA labs are actively engaged in these and other research questions that will facilitate the advancement of the field of regenerative medicine.

A priority of the Critical Path Initiative is updating and modernizing techniques, to ensure that the agency and the research community have the tools necessary to bring safe and effective products to market. Efforts include promoting collaborations spanning multiple centers and regulatory jurisdictions across FDA as well as among FDA and other relevant organizations (e.g. other agencies, academic organizations, regulated industry). Developing research collaborations and the requisite infrastructure to support those and other efforts will facilitate review of combination products, which are often seen in regenerative medicine.

A recent example of collaborative work across FDA labs is a multi-investigator project at CBER using a battery of state-of-the-art analytic techniques chosen to provide complementary data on cell state and seeking to develop new biomarkers for cell therapies. The project aims to characterize mesenchymal stromal cells from a number of perspectives, including genetic stability, proteomic and phosphoproteomic analysis, microRNA analysis, mRNA profiling by microarray and quantitative polymerase chain reaction (PCR) analysis, chromatin immunoprecipitation, and examination of the potential for cells to mature and contribute to the formation of organs and tissues. Furthermore, the study will look for molecular differences between cells from early versus late passage numbers. Importantly, the same cells that go through this panel of tests will also be implanted in a mouse model of hindlimb ischemia, allowing for correlation of product characterization data with the *in vivo* outcome with regard to localization, differentiation, and functionality. This FDA research project may yield information that will be useful for product characterization, in-process testing, lot release criteria, developing comparability and stability protocols, and predicting cell fate and function after receiving a cell therapy.

One of the major issues associated with the clinical use of cellular therapies is predicting what happens to the cells after injection. Another FDA Critical Path research study at CBER will advance cell therapy by helping to develop methods for *in vivo* tracking and imaging of neural stem cells (NSCs) after transplantation. NSCs from adult, fetal, and embryonic sources have been proposed as treatments for degenerative conditions such as Parkinson's disease, and for repair of tissues damaged by stroke and spinal cord injury. Magnetic resonance imaging and single photon emission computed tomography are being used to qualitatively and quantitatively determine cellular location and persistence of engraftment. The goal of this project is to develop methods for evaluating biomarkers that may be predictive of NSC function.

FDA's research projects often involve collaborations with other federal and academic partners to employ new technologies to help address regulatory science questions. For example, a collaboration between FDA and the National Institute of Standards and Technology is using automated microscopy to characterize the differentiation of mesenchymal stem cells (MSCs). The goal is to improve the safety of MSC products by developing robust assays that can be used for in-process and lot release testing.

FDA critical path research also helps address some of the challenges faced in both product development and product evaluation. For example, following the observation of unexpected toxicity of adenoviral vector gene therapy in a clinical trial, CBER research provided insight into how adenovirus vectors cause toxicity and developed an animal model for gene therapy in the context of pre-existing liver disease (Smith et al., 2004). CBER researchers/regulators also worked with a consortium from industry and academia to develop reference material for adenoviral vector particles (Simek et al., 2002). FDA research is ongoing to understand the nature of toxicity of systemically delivered adenovirus and mechanisms for vector clearance to improve the safety of gene therapy trials (Smith et al., 2008).

Some FDA labs are engaged in research projects related to tissue. CDRH scientists are studying the effects of mechanical and electrical stimulation on cardiac cell cultures and how the parameters modulate cellular physiology. A CBER/CDRH collaboration is examining the relationship between encapsulation of chondrocytes in a scaffold material, with and without mechanical stress, on the status of several signaling pathways.

Additional FDA research projects are studying medical devices with implications for the regulation of regenerative medicine products. A CDRH research effort is examining the effects of contact with certain types of materials on cardiac cells, which has implications for biocompatibility regulations for regenerative medicine products with structural and other device components (Ravenscroft-Chang et al., 2010). An additional research project associated with cardiovascular disease is a collaboration between CDRH and CDER using cardiac myocyte cultures to study toxicological effects of certain drugs on cardiac tissue. A collaborative project between CDRH and other government researchers is examining the ability of wavelength-specific light to promote nerve regeneration in laboratory (Anders et al., 2008) and animal (Byrnes et al., 2005) models.

In summary, FDA research labs, with support through the Critical Path Initiative, provide an important source of in-house expertise in regenerative medicine and other cutting-edge technologies and research areas. Although by no means exhaustive, the examples provided above demonstrate the diverse range of topics under investigation at FDA. Especially in

1163

consideration of the rapid change and development of the regenerative medicine field, Critical Path research efforts ensure that FDA stays abreast of the current innovations. FDA's research programs provide an important source of the latest science to inform the regulatory process and bring safe regenerative medicine products to market.

OTHER COORDINATION EFFORTS

Due to the highly interdisciplinary nature of regenerative medicine, FDA recognizes the need to build collaborative efforts to successfully review these products in this area. As such, FDA is a partner in the Multi-Agency Tissue Engineering Science (MATES) Interagency Working Group. This partnership spanning more than a dozen federal agencies is designed to provide a forum to facilitate communication and coordination across the government regarding activities in tissue engineering and regenerative medicine. The full strategic plan can be found at the MATES website (Multi-Agency Tissue Engineering Science (MATES) Interagency Working Group, 2007).

In another example of interagency interaction, collaboration, and coordination, FDA has established Memoranda of Understanding (MOU) agreements with two different NIH institutes that involve FDA scientific review staff and NIH extramural research program officers. The MOUs co-signed by the National Institute of Neurological Disorders and Stroke (NINDS) and the National Heart, Lung and Blood Institute (NHLBI) incorporate safeguards to protect from disclosure shared, non-public information such as trade secrets and confidential commercial information, identities of study participants and other personal information, privileged and/ or pre-decisional agency information, research proposals, progress reports, and/or unpublished data or information protected for national security reasons. Under the MOU agreements, participants are able to hold unfettered discussions and exchange information that enables the respective agencies to maintain currency with respect to ongoing scientific activities that could impact regenerative medicine from both the laboratory and clinical research perspectives. Interagency MOU interactions help identify gaps in knowledge related to the state of available scientific information and familiarity with FDA regulatory expectations. This, in turn, contributes to identification of promising basic research with the potential for clinical translation.

There are also efforts to promote collaboration within the agency. For example, the FDA Commissioner's Fellowship Program (CFP) is facilitating collaboration related to the regulation of regenerative medicine across FDA's product jurisdictions. The CFP was established in 2008 to attract and retain new talent to the agency. Within its annual cohort of 50 fellows, a regenerative medicine fellowship program has been established. Regenerative medicine fellows work across both CBER and CDRH, to facilitate cross-agency collaboration and conduct research projects related to the regulation of regenerative medicine products.

Interaction between FDA and the scientific and regulated communities is an important area of collaboration. Workshops are one such example, such as the previously mentioned FDA-NIST co-sponsored session regarding cell–scaffold products. Workshops provide valuable opportunities for the agency to receive input from outside scientific experts and other stakeholders. An additional common example of this type of activity is liaison meetings, wherein FDA directly engages in dialog with professional societies (e.g. International Society for Stem Cell Research, American Association of Blood Banks) regarding scientific or regulatory issues related to a certain research area. All of these activities ensure that FDA receives continuous input on the latest scientific discoveries to inform the regulation of safe products.

CONCLUSION

The field of regenerative medicine is an exciting field with scientific advances leading to the promise of future therapies for current unmet medical needs for patients. The FDA regulatory approach to medical products evaluation includes an ongoing assessment of how the science of those products informs regulatory policy.

FDA looks to continue ongoing dialog with the scientific community and product Sponsors to continue to develop science-based regulatory review policies that are robust and predictable in order to meet the needs of the challenging array of products that are on the horizon.

References

Anders, J. J., Romanczyk, T. B., Ilev, I. K., Moges, H., Longo, L., Wu, X., et al. (2008). Light supports neurite outgrowth of human neural progenitor cells *in vitro*: the role of P2Y receptors. *IEEE J. Select. Topics Quantum Electron., 14*(1), 118–125.

Baird, L. G., Christenson, L., David, J., Du Moulin, G., Gentile, F. T., Omstead, D. R., et al. (1998). Voluntary guidance for the development of tissue-engineered products. *Tissue Eng., 4*, 239–266.

Bauer, S. R., Ruiz-Hidalgo, M. J., Rudikoff, E. K., Goldstein, J., & Laborda, J. (1998). Modulated expression of the epidermal growth factor-like homeotic protein dlk influences stromal-cell-pre-B cell interactions, stromal cell adipogenesis, and pre-B-cell interleukin-7 requirements. *Mol. Cell. Biol., 18*, 5247–5255.

Boyce, S. T., Cooper, M. L., Foreman, T. J., & Hansbrough, J. F. (1989). Burn wound closure with cultured autologous keratinocytes and fibroblasts attached to a collagen–glycosaminoglycan substrate. *JAMA, 262*, 2125–2130.

Byrnes, K. R., Waynant, R. W., Ilev, I. K., Wu, X., Barna, L., Smith, K., et al. (2005). Light promotes regeneration and functional recovery and alters the immune response after spinal cord injury. *Lasers Surg. Med., 36*, 171–185.

Carsin, H., Ainaud, P., Le Bever, H., Rives, J., Lakhel, A., Stephanazzi, J., et al. (2000). Cultured epithelial autografts in extensive burn coverage of severely traumatized pateients: a five year single-center experience with 30 patients. *Burns, 26*, 379–387.

Cazalet, C., Cherruau, B., Jaffray, P., Marien, M., Schlotterer, M., Toulon, A., et al. (1988). Preliminary clinical studies of a biological skin equivalent in burned patients. *Burns, 14*, 326–330.

Currie, L., Jones, I., & Martin, R. (2002). A guide to biological skin substitutes. *Br. J. Plast. Surg., 44*, 185–193.

Federal Food and Drugs Act, 1906, 34 Stat. 768, *repealed by* Food, Drug, and Cosmetic Act of 1938, 21 US C. Sec. 329(a).

Green, H., & Rheinwald, J. G. (1975). Serial cultivation of strains of human epidermal keratinocytes: the formation of keratinizing colonies from single cells. *Cell, 6*, 331–343.

Haeseker, B., Koch, R., & Teepe, R. G. (1993). Randomized trial comparing cryopreserved cultured epidermal allografts with tulle-gras in the treatment of split-thickness skin graft donor sites. *J. Trauma, 35*, 850–854.

Hansbrough, J. F., Dore, C., & Hansbrough, W. B. (1992). Clinical trials of a living dermal tissue replacement placed beneath meshed, split-thickness skin grafts on excised burn wounds. *J. Burn Care Rehabil., 13*, 519–529.

Hoang, B., Moos, M., Jr., Vukicevic, S., & Luyten, F. P. (1996). Primary structure and tissue distribution of Frzb, a novel protein related to Drosophila frizzled, suggest a role in human skeletal morphogenesis. *J. Biol. Chem., 271*, 26131–26137.

Horvath-Arcidiacono, J. Y. A, Evdokimov, E., Lee, M. H., Jones, J., Rudenko, L., Schneider, B., et al. (2010). Regulation of xenogeneic porcine pancreatic islets. *Xenotransplantation, 17*, 329–337.

Lee, M. H., Arcidiacono, J. A., Bilek, A. H., Wille, J. J., Hamill, C. A., Wonnacott, K. M., et al. (2010). Considerations for tissue-engineered and regenerative medicine product development prior to clinical trials in the USA. *Tissue Eng., 16B*, 41–54.

Lenas, P., Moos, M., Jr., & Luyten, F. P. (2009a). Developmental engineering: a new paradigm for design and manufacture of cell based products. Part I: From three-dimensional cell growth to biomimetics of *in vivo* development. *Tissue Eng., 15B*, 381–394.

Lenas, P., Moos, M., Jr., & Luyten, F. P. (2009b). Developmental engineering: a new paradigm for the design and manufacture of cell based products. Part II: From genes to networks: tissue engineering from the viewpoint of systems biology and network science. *Tissue Eng., 15B*, 395–422.

Lin, K., Wang, S., Julius, M. A., Kitajewski, J., Moos, M., Jr., & Luyten, F. P. (1997). The cysteine-rich frizzled domain of Frzb-1 is required and sufficient for the modulation of Wnt signaling. *Proc. Natl. Acad. Sci. USA, 94*, 11196–11200.

Marlowe, D. E., & Phillips, P. J. (1998). FDA recognition of consensus standards in the premarket notification program. In *Biomedical Instrumentation and Technology* (pp. 301–304). Philadelphia: Hanley & Belfus.

McCright, B., Dang, J. M., Hursh, D. A., Kaplan, D. S., Ballica, R., Benton, K., et al. (2009). Synopsis of the Food and Drug Administration – National Institute of Standards and Technology co-sponsored "*In Vitro* Analyses of Cell/Scaffold Products" Workshop. *Tissue Eng., 15A*, 455–460.

Multi-Agency Tissue Engineering Science (MATES) Interagency Working Group (2007). *Advancing Tissue Science and Engineering: A Foundation for the Future. A Multi-Agency Strategic Plan.* June 2007. Available at: http://www. tissueengineering.gov/advancing_tissue_science_&_engineering.pdf

Ramadevi Raghunandan, R., Ruiz-Hidalgo, M., Ettinger, R., Rudikoff, E., Riggins, P. S., Farnsworth, R., et al. (2008). *Dlk1* influences differentiation and function of B-lymphocytes. *Stem Cells Dev., 17*, 495−507.

Ravenscroft-Chang, M. S., Stohlman, J. M., Molnar, P., Natarajan, A., Canavan, H. E., Teliska, M., et al. (2010). Altered calcium dynamics in cardiac cells grown on silane-modified surfaces. *Biomaterials, 31*, 602−607.

Simek, S., Byrnes, A., & Bauer, S. (2002). FDA perspectives on the use of the adenovirus reference material. *Bioprocessing, 1*, 40−42.

Smith, J. S., Tian, J., Lozier, J. N., & Byrnes, A. P. (2004). Severe pulmonary pathology after intravenous administration of vectors in cirrhotic rats. *Mol. Ther., 9*, 932−941.

Smith, J. S., Xu, Z., Tian, J., Stevenson, S. C., & Byrnes, A. P. (2008). Interaction of systemically delivered adenovirus vector with Kupfer cells in mouse liver. *Hum. Gene Ther., 19*, 547−554.

Thomas, J. T., Prakash, D., & Moos, M., Jr. (2006). CDMP-1/GDF5 has specific processing requirements that restrict its action to joint surfaces. *J. Biol. Chem., 281*, 26725−26733.

Thomas, J. T., Canelos, P., Luyten, F. P., & Moos, M., Jr. (2009). XSMOC-1 Inhibits BMP signaling downstream of receptor binding and is essential for xenopus neurulation. *J. Biol. Chem., 284*, 18994−19005.

US Department of Health and Human Services, Food and Drug Administration (2006). Critical Path Opportunities List (March 2006). Available at: http://www.fda.gov/downloads/ScienceResearch/SpecialTopics/ CriticalPathInitiative/CriticalPathOpportunitiesReports/UCM077258.pdf

US Food and Drug Administration (1997). Proposed Approach to the Regulation of Cellular and Tissue-Based Products (February 28, 1997). Available at: http://www.fda.gov/downloads/BiologicsBloodVaccines/ GuidanceComplianceRegulatoryInformation/Guidances/Tissue/UCM062601.pdf

US Food and Drug Administration (2001a). Early Collaboration Meetings under the FDA Modernization Act (FDAMA), Final Guidance for Industry and for CDRH Staff (February 2001). Available at: http://www.fda.gov/ MedicalDevices/DeviceRegulationandGuidance/GuidanceDocuments/ucm073604.htm

US Food and Drug Administration (2001b). Regulatory Information: Guidances: Guidance for Industry: Acceptance of Foreign Clinical Studies (March 2001). Available at: http://www.fda.gov/RegulatoryInformation/Guidances/ ucm124932.htm

US Food and Drug Administration (2002). Guidance for Industry: Special Protocol Assessment (May 2002). Available at: http://www.fda.gov/downloads/Drugs/GuidanceComplianceRegulatoryInformation/Guidances/ UCM080571.pdf

US Food and Drug Administration (2004a). Draft Guidance for Industry and FDA: Current Good Manufacturing Practice (cGMPs) for Combination Products (September 2004). Available at: http://www.fda.gov/ RegulatoryInformation/Guidances/ucm126198.htm

US Food and Drug Administration (2004b). Guidance for Industry: Sterile Drug Products Produced by Aseptic Processing − Current Good Manufacturing Practice (September 2004). Available at: http://www.fda.gov/ downloads/Drugs/GuidanceComplianceRegulatoryInformation/Guidances/UCM070342.pdf

US Food and Drug Administration (2006). Guidance for Industry: Chronic Cutaneous Ulcer and Burn Wounds − Developing Products for Treatment (June 2006). Available at: http://www.fda.gov/downloads/Drugs/ GuidanceComplianceRegulatoryInformation/Guidances/UCM071324.pdf

US Food and Drug Administration (2010). Advisory Committee Calendar. Available at: http://www.fda.gov/ AdvisoryCommittees/Calendar/default.htm

US Food and Drug Administration, Center for Biologics Evaluation and Research (1993). Points to Consider in the Characterization of Cell Lines Used to Produce Biologicals (July 1993). Available at: http://www.fda.gov/ downloads/BiologicsBloodVaccines/GuidanceComplianceRegulatoryInformation/ OtherRecommendationsforManufacturers/UCM062745.pdf

US Food and Drug Administration, Center for Biologics Evaluation and Research (1997). Points to Consider in the Manufacture and Testing of Monoclonal Antibody Products for Human Use (February 1997). Available at: http://www.fda.gov/downloads/BiologicsBloodVaccines/GuidanceComplianceRegulatoryInformation/ OtherRecommendationsforManufacturers/UCM153182.pdf

US Food and Drug Administration, Center for Biologics Evaluation and Research (1998a). Guidance for Industry: Guidance for Human Somatic Cell Therapy and Gene Therapy (March 1998). Available at: http://www.fda.gov/ BiologicsBloodVaccines/GuidanceComplianceRegulatoryInformation/Guidances/CellularandGeneTherapy/ ucm072987.htm

US Food and Drug Administration, Center for Biologics Evaluation and Research (1998b). Guidance for Industry: Providing Clinical Evidence of Effectiveness for Human Drugs and Biologic Products (May 1998). Available at: http://www.fda.gov/downloads/Drugs/GuidanceComplianceRegulatoryInformation/Guidances/UCM078749. pdf

US Food and Drug Administration, Center for Biologics Evaluation and Research (1998c). ICH Guidance on Quality of Biotechnological/Biological Products: Derivation and Characterization of Cell Substrates Used for Production of Biotechnological/Biological Products (September 1998). Available at: http://www.fda.gov/downloads/RegulatoryInformation/Guidances/UCM129103.pdf

US Food and Drug Administration, Center for Biologics Evaluation and Research (2000). Guidance for Industry, Formal Meetings with Sponsors and Applicants for PDUFA Products (February 2000). Available at: http://www.fda.gov/downloads/Drugs/GuidanceComplianceRegulatoryInformation/Guidances/UCM079744.pdf

US Food and Drug Administration, Center for Biologics Evaluation and Research (2001). Science and Research Special Topics: Running Clinical Trials, Information Sheets, and Notices on Good Clinical Practice in FDA-Regulated Clinical (April 2001). Available at: http://www.fda.gov/ScienceResearch/SpecialTopics/RunningClinicalTrials/GuidancesInformationSheetsandNotices/default.htm

US Food and Drug Administration, Center for Biologics Evaluation and Research (2003). Guidance for Industry: Source Animal, Product, Preclinical, and Clinical Issues Concerning the Use of Xenotransplantation Products in Humans (April 2003). Available at: http://www.fda.gov/BiologicsBloodVaccines/GuidanceComplianceRegulatoryInformation/Guidances/Xenotransplantation/ucm074354.htm

US Food and Drug Administration, Center for Biologics Evaluation and Research (2005). Information on Submitting and Investigational New Drug Application for a Biological Product (June 2005). Available at: http://www.fda.gov/BiologicsBloodVaccines/DevelopmentApprovalProcess/InvestigationalNewDrugINDorDeviceExemptionIDEProcess/ucm094309.htm

US Food and Drug Administration, Center for Biologics Evaluation and Research (2006a). Cellular, Tissue and Gene Therapies Advisory Committee (Formerly Biological Response Modifiers Advisory Committee) (March 2006). Available at: http://www.fda.gov/AdvisoryCommittees/CommitteesMeetingMaterials/BloodVaccinesandOtherBiologics/CellularTissueandGeneTherapiesAdvisoryCommittee/default.htm

US Food and Drug Administration, Center for Biologics Evaluation and Research (2006b). Guidance for Industry: Gene Therapy Clinical Trials — Observing Subjects for Delayed Adverse Events (November 2006). Available at: http://www.fda.gov/BiologicsBloodVaccines/GuidanceComplianceRegulatoryInformation/Guidances/CellularandGeneTherapy/ucm072957.htm

US Food and Drug Administration, Center for Biologics Evaluation and Research (2006c). Guidances for Submission of INDs (January 2006). Available at: http://www.fda.gov/cber/ind/indpubs.htm

US Food and Drug Administration, Center for Biologics Evaluation and Research (2006d). Tissue & Tissue Products (April 2006). Available at: http://www.fda.gov/BiologicsBloodVaccines/TissueTissueProducts/default.htm

US Food and Drug Administration, Center for Biologics Evaluation and Research (2006e). Xenotransplantation Action Plan, FDA Approach to the Regulation of Xenotransplantation (April 2006). Available at: http://www.fda.gov/BiologicsBloodVaccines/Xenotransplantation/default.htm

US Food and Drug Administration, Center for Biologics Evaluation and Research (2007a). Guidance for Industry: Eligibility Determination for Donors of Human Cells, Tissues, and Cellular and Tissue-Based Products (August 2007). Available at: http://www.fda.gov/BiologicsBloodVaccines/GuidanceComplianceRegulatoryInformation/Guidances/Tissue/ucm073964.htm

US Food and Drug Administration, Center for Biologics Evaluation and Research, Center for Devices and Radiological Health (2007b). The Centers for Disease Control and Prevention: Processing of Orthopedic, Cardiovascular and Skin Allografts Workshop (October 2007). Available at: http://www.fda.gov/downloads/BiologicsBloodVaccines/NewsEvents/WorkshopsMeetingsConferences/TranscriptsMinutes/UCM054425.pdf

US Food and Drug Administration, Center for Biologics Evaluation and Research (2008a). Guidance for FDA Reviewers and Sponsors: Content and Review of Chemistry, Manufacturing, and Control (CMC) Information for Human Somatic Cell Therapy Investigational New Drug Applications (INDs) (April 2008). Available at: http://www.fda.gov/BiologicsBloodVaccines/GuidanceComplianceRegulatoryInformation/Guidances/Xenotransplantation/ucm074131.htm

US Food and Drug Administration, Center for Biologics Evaluation and Research (2008b). Guidance for FDA Reviewers and Sponsors: Content and Review of Chemistry, Manufacturing, and Control (CMC) Information for Human Gene Therapy Investigational New Drug Applications (INDs) (April 2008). Available at: http://www.fda.gov/BiologicsBloodVaccines/GuidanceComplianceRegulatoryInformation/Guidances/CellularandGeneTherapy/ucm072587.htm

US Food and Drug Administration, Center for Biologics Evaluation and Research (2009). Draft Guidance for Industry: Current Good Tissue Practice (CGTP) and Additional Requirements for Manufacturers of Human Cells, Tissues, and Cellular and Tissue-Based Products (HCT/Ps) (January 2009). Available at: http://www.fda.gov/BiologicsBloodVaccines/GuidanceComplianceRegulatoryInformation/Guidances/Tissue/ucm062693.htm

US Food and Drug Administration, Center for Devices and Radiological Health (2000a). Apligraf® (Grafskin) (June 2000). Available at: http://www.accessdata.fda.gov/scripts/cdrh/cfdocs/cfTopic/pma/pma.cfm?num=p950032s016

US Food and Drug Administration, Center for Devices and Radiological Health (2000b). Composite Cultured Skin (CCS) (February 2000). Available at: http://www.accessdata.fda.gov/scripts/cdrh/cfdocs/cftopic/pma/pma.cfm?num=H990013

US Food and Drug Administration, Center for Devices and Radiological Health (2001). OrCel Product Label (September 2001). Available at: http://www.accessdata.fda.gov/cdrh_docs/pdf/P010016c.pdf

US Food and Drug Administration, Center for Devices and Radiological Health (2002a). Device Advice: Premarket Approval (PMA) (November 2002). Available at: http://www.fda.gov/MedicalDevices/DeviceRegulationandGuidance/HowtoMarketYourDevice/PremarketSubmissions/PremarketApprovalPMA/default.htm

US Food and Drug Administration, Center for Devices and Radiological Health (2002b). PMA Postapproval Requirements: General Requirements (November 2002). Available at: http://www.fda.gov/MedicalDevices/DeviceRegulationandGuidance/HowtoMarketYourDevice/PremarketSubmissions/PremarketApprovalPMA/ucm050422.htm

US Food and Drug Administration, Center for Devices and Radiological Health (2003). Designating Humanitarian Use Devices (HUDs) (November 2003). Available at: http://www.fda.gov/MedicalDevices/DeviceRegulationandGuidance/GuidanceDocuments/ucm071473.htm

US Food and Drug Administration, Center for Devices and Radiological Health (2005a). Device Advice: Premarket Notification (510(k)) (February 2005). Available at: http://www.fda.gov/MedicalDevices/DeviceRegulationandGuidance/HowtoMarketYourDevice/PremarketSubmissions/PremarketNotification510k/default.htm

US Food and Drug Administration, Center for Devices and Radiological Health (2005b). CDRH Databases (December 2005). Available at: http://www.fda.gov/MedicalDevices/DeviceRegulationandGuidance/Databases/default.htm

US Food and Drug Administration, Center for Devices and Radiological Health (2007). Epicel® (cultured epidermal autografts) (October 2007). http://www.accessdata.fda.gov/scripts/cdrh/cfdocs/cftopic/pma/pma.cfm?num=H990002

US Food and Drug Administration, Center for Devices and Radiological Health (2010a). Device Advice: Investigational Device Exemption (IDE). Available at: http://www.fda.gov/MedicalDevices/DeviceRegulationandGuidance/HowtoMarketYourDevice/InvestigationalDeviceExemptionIDE/default.htm

US Food and Drug Administration, Center for Devices and Radiological Health (2010b). Device Approvals and Clearances. Available at: http://www.fda.gov/MedicalDevices/ProductsandMedicalProcedures/DeviceApprovalsandClearances/default.htm

US Food and Drug Administration, Center for Devices and Radiological Health (2010c). Device Approvals and Clearances: HDE Approvals. Available at: http://www.fda.gov/MedicalDevices/ProductsandMedicalProcedures/DeviceApprovalsandClearances/HDEApprovals/default.htm

US National Institutes of Health, Office of Biotechnology Activities (2010). Recombinant DNA and Gene Transfer (September 2010). Available at: http://oba.od.nih.gov/rdna_rac/rac_about.html

US Public Health Service (2001). PHS Guideline on Infectious Disease Issues in Xenotransplantation (January 2001). Available at: http://www.fda.gov/downloads/BiologicsBloodVaccines/GuidanceComplianceRegulatoryInformation/Guidances/Xenotransplantation/UCM092858.pdf

Wang, S., Krinks, M., & Moos, M., Jr. (1997a). Frzb-1, an antagonist of Wnt-1 and Wnt-8, does not block signaling by Wnts -3A, -5A, or -13.11. *Biochem. Biophys. Res. Commun., 236*, 502–504.

Wang, S., Krinks, M., Lin, K., Luyten, F. P., & Moos, M., Jr. (1997b). Frzb, a secretable protein expressed in the Spemann organizer, binds and inhibits Wnt-8. *Cell, 88*, 757–766.